PRIMATES IN PERSPECTIVE

PRIMATES IN PERSPECTIVE

Edited by

Christina J. Campbell
Pomona College

Agustín Fuentes
University of Notre Dame

Katherine C. MacKinnon
Saint Louis University

Melissa Panger
United States Environmental Protection Agency

Simon K. Bearder
Oxford Brookes University

New York   Oxford
OXFORD UNIVERSITY PRESS
2007

Oxford University Press, Inc., publishes works that further Oxford University's
objective of excellence in research, scholarship, and education.

Oxford   New York
Auckland   Cape Town   Dar es Salaam   Hong Kong   Karachi
Kuala Lumpur   Madrid   Melbourne   Mexico City   Nairobi
New Delhi   Shanghai   Taipei   Toronto

With offices in
Argentina   Austria   Brazil   Chile   Czech Republic   France   Greece
Guatemala   Hungary   Italy   Japan   Poland   Portugal   Singapore
South Korea   Switzerland   Thailand   Turkey   Ukraine   Vietnam

Copyright © 2007 by Oxford University Press, Inc.

Published by Oxford University Press, Inc.
198 Madison Avenue, New York, New York 10016

http://www.oup.com

Library of Congress Cataloging-in-Publication Data

Primates in perspective/edited by Christina J. Campbell . . . [et al.].
   p. cm.
   Includes bibliographical references.
   ISBN-13: 978-0-19-517133-4 (pbk.: alk. paper)
   ISBN-10: 0-19-517133-0 (pbk.: alk. paper)
   ISBN-13: 978-0-19-517134-1 (cloth: alk. paper)
   ISBN-10: 0-19-517134-9 (cloth: alk. paper)
   1. Primates. I. Campbell, Christina J.

QL737.P9P679 2006   2005049893
599.8—dc22

Printing number: 9 8 7 6 5 4 3 2 1

Printed in the United States of America
on acid-free paper

CONTENTS

PREFACE

Welcome to the first edited volume in nearly twenty years to provide a comprehensive overview of contemporary primate research. With significant improvements in methodology, dynamic shifts in theoretical perspectives, and increases in the number and types of primate studies conducted, primatology has continued to mature as a discipline over the past two decades. We thought the time had come for an up-to-date, inclusive volume that reflected these changes.

OUR APPROACH

We bring together here a broad range of primatologists who provide thorough reviews of their areas of expertise. We hope that the contributed chapters are accessible and useful to university students (undergraduate and graduate) and primate researchers alike, and that the chapters will stimulate new ideas and areas of research for the future. All chapters are written and edited in a manner that ensures the book is not only a valuable teaching asset, but also a vital reference tool for professionals. The 1987 edited volume *Primate Societies* (University of Chicago Press) set the standard for a primate overview and reference text. But since an updated edition of that text has not been published, nor has anything comparable appeared in the primate literature, we felt it important to create an inclusive volume that reflects the taxonomic, methodological, and theoretical changes that have occurred in primatology since the late 1980s.

We designed this volume so that it can be used by a wide range of people interested in the nonhuman primates. In particular, we feel the broad coverage of topical areas (taxonomy, methods, ecology, behavior, etc.) will be well-suited for general undergraduate primatology courses and graduate seminars. By providing extensive coverage of the primates via well-structured and inclusive overviews of taxonomy, methods, and theoretical issues, we remove the need for students and professors to deal with multiple books and/or photocopied readers. This text will also be of use as a reference source for professional primatologists, as it is an up-to-date, inclusive overview. The chapters will be useful for all primatologists looking for current and older literature pertaining to almost any area of interest within the discipline.

EXPLANATION OF THE ORGANIZATION

The book is divided into six main sections that reflect the major themes of primate research today; collectively they contain 44 chapters by 59 authors. We begin with the background section, which serves largely as a prologue to the study of primatology and to the evolutionary history of the primate order. We hope the chapters in this section will provide students with a comprehensive overview and introduction to the field (e.g., in the first week of an upper division primatology course).

Next, and with that background in mind, we present the taxonomy chapters, each of which focuses on the major taxonomic groups within the order Primates. The individual chapters include an overview of natural history, life history parameters, and the like, and do not revolve heavily around any theoretical framework. We hope students will find these chapters to be a valuable and complete introduction to the diversity of the living nonhuman primates. For our professional audience, this section provides up-to-date information and literature on a wide variety of taxa. For all readers, this portion serves as an important foundation for the rest of the book.

In the third section we include several chapters on methods. In order to gain a full appreciation for the study of primates, students must have some understanding of how research is carried out in the field. In addition, the final two chapters in this section will be of use to professional primatologists interested in learning more about recent advances in available technologies.

We then present chapters on reproduction. As a theoretical topic, primate reproduction is a vital component to a modern comprehensive-review book. The chapters included here introduce the student to many aspects of primate reproduction, and the up-to-date overviews and bibliographies provide a valuable reference source for professionals in a rapidly growing field of research.

The ecological topics of socioecology, diet, seed dispersal, predation, locomotion, and conservation are found in our next section. All are important aspects of the interface between primate populations and the habitats they occupy. Students and established researchers alike will find in-depth explanations and important theoretical arguments within these chapters.

Finally we present the reader with a section that covers varying aspects of primate social behavior and intelligence. The vast majority of nonhuman primates live in stable social groups throughout their lives, and exhibit complex behavioral relationships with other members of the group. Among mammals, primates are characterized by displaying an extended period of growth and development, during which virtually all social behaviors are learned. Thus, this section highlights the complex dynamics that characterize the social aspects of being a primate.

We feel the progression from a general introduction and background, to taxonomic overviews, then to methodological issues, and finally to topics in ecology and behavior, gives the book sound pedagogical structure and a user-friendly format in which to find information for classroom and/or professional use.

FEATURES

Throughout the volume we have ensured that there is a synthesis of modern data that will inform all readers of the "state of affairs" of contemporary nonhuman primate research. We have worked hard to ensure that this book is especially strong in behavioral ecology and theory. With specific taxonomic and behavioral chapters by leading researchers in the field, this volume provides unmatched detailed coverage of a multitude of topics. Because such incredible diversity is found among primates, the cursory taxonomic summaries found in most introductory textbooks can be of only limited value to students. Therefore, for students, these chapters provide a valuable and complete introduction to the primates, and an important foundation for the rest of the book. For professionals, this section provides up-to-date information and literature citations for a wide variety of taxa. No recently published primate volume available provides this type of overview.

We also include here varied theoretical positions on several topics in primatology (most notably in the reproduction, ecology, and social behavior and intelligence sections). We especially want student readers to be aware of the diversity of theoretical opinions that currently pervade our field. Dominant paradigms in modern primatology are well represented, as are less-entrenched positions that are nonetheless supported by existing data. Such a range of theoretical orientations is generally not found in primate texts, and we hope classroom discussion and critical evaluation of the topics will be facilitated by their inclusion here.

ACKNOWLEDGMENTS

When three of us first discussed this venture at the opening reception of the 71st Annual Meeting of the American Association of Physical Anthropologists in 2002, we knew it would end up involving an extremely large number of people. Our first round of acknowledgments and "thank yous" are graciously extended to all the authors who agreed to join us in this endeavor. The current volume would not have resulted in the astounding review of primate behavior and ecology without each and every one of them and the enormous effort they put into the excellent chapters that follow. In addition to their contributions, we greatly appreciate the authors' professional responses to our editorial suggestions and patience with the inevitable delays that accompany the production of such a large project.

Three people at Oxford University Press deserve special acknowledgment. Executive Editor Jan Beatty was extremely supportive from the beginning, offering advice, encouragement, and the occasional prodding in order to keep the volume on schedule. Associate Editor Talia Krohn's and Production Editor Barbara Mathieu's efforts in the actual publishing and organizational aspects of the project also helped immensely, especially in keeping track of missing files, figures, and tables. This book would certainly not be so well organized without the hard work and dedication they all contributed.

The following external reviewers provided invaluable comments and suggestions on the original proposal and earlier versions of the manuscript, and we extend our extreme appreciation for their time and interest: Deborah E. Blom (University of Vermont), Claud A. Bramblett (University of Texas, Austin), W. Scott McGraw (The Ohio State University), Laura Newell (University of Washington), Deborah Overdorff (University of Texas, Austin), Peter Rodman (University of California, Davis), Teryl Schessler (California State University, Fullerton), Russell H. Tuttle (The University of Chicago), Eric A. Worch (University of Michigan, Flint), and Patricia Wright (State University of New York, Stony Brook).

Meegan Anderson and Noelle Easterday spent many hours formatting each of the chapters with the financial support of the Department of Anthropology and the Dean of Arts and Letters at the University of Notre Dame. We thank them for making our lives easier.

We would be remiss if we did not respectfully acknowledge and extend appreciation to Barbara Smuts, Dorothy Cheney, Robert Seyfarth, Richard Wrangham and Thomas Struhsaker, the editors of *Primate Societies* (1987, University of Chicago Press). Their volume inspired us in part to undertake this project, and we sincerely hope that our book will be as influential to the primatological community as theirs has been.

Various funding agencies have contributed to the work presented in this volume. These include (but are not limited to) the National Science Foundation, the National Institutes of Health, the LSB Leakey Foundation, National Geographic, the Wenner Gren Foundation, the Margot Marsh Biodiversity Fund, the American Primatological Society, Primate Conservation International, the Smithsonian Institution, and anthropology, psychology, zoology, and biology departments at many different universities. We thank all of these agencies (and any others that are not mentioned) for their continued support of this important area of research. Many of the contributing authors wanted to add their own acknowledgments, but due to space limitations, they were unable to be included. We would like to take this opportunity to thank all those individuals who helped the authors produce the chapters that follow.

In addition to those who have our united thanks, there are many who have contributed to this volume through their support and encouragement of each of us individually. C. J. C. acknowledges Phyllis Dolhinow's mentorship and guidance. In addition, she thanks members of the Department of Anthropology at Pomona College (especially Gail Orozco), the Pomona College students who inspired her to work on this venture, and Mark Jenike for his ongoing encouragement. She also offers her undying gratitude to Tom Wake for his love and patience throughout this process.

A. F. thanks Devi Snively and Audrey, the wonderdog, for inspiration and assistance. He also thanks Phyllis

Dolhinow for her mentoring and guidance and all the students in his primate behavior classes at the University of California, Berkeley, Central Washington University, and the University of Notre Dame for keeping him on his toes and always wanting a better book.

K. C. M. thanks Phyllis Dolhinow, Linda Fedigan, and Bob Sussman for being exemplary mentors, true friends, and inspiring role models. She also thanks her fellow editors on this volume for patience, good humor, and intellectual robustness during the entire editorial process. She acknowledges her anthropology and primatology students at Saint Louis University, who continually inspire, amuse, and stimulate discussion in the classroom. Finally, she expresses extreme gratitude to Matt Wyczalkowski, as well as Caper, Basil, and Darby, for holding down the fort, yet again.

M. P. thanks Dian Fossey, Phyllis Dolhinow, Linda Wolfe, and Shirley McGreal for their years of inspiration and friendship. She also thanks Bernard Wood, the George Washington University Department of Anthropology, and the Center for the Advanced Study of Human Paleobiology for their support while working on this book. She is very grateful for the experiences she has had and the primates (human and nonhuman) she has been lucky enough to meet through her primatological journey. And a very special "thank you" to Norman Birchfield and Carter Panger Birchfield for their patience, support, and love.

S. K. B. thanks K. A. I. Nekaris, Kate Hill, Chris McDonaugh, and Iris Geens at Oxford Brookes University; the students and staff on the MSc in Primate Conservation; and research students at Brookes. He also acknowledges Bob Martin, Alan Dixson, Tom Butynski, and Gerald Doyle for their inspiration over the years. Last, but definitely not least, he expresses his deepest thanks to his wife Catherine.

Finally, we would all like to thank the nonhuman primates that continue to intrigue and inspire us!

Christina J. Campbell
Agustín Fuentes
Katherine C. MacKinnon
Melissa Panger
Simon K. Bearder

Contributors

Kate Arnold
School of Psychology
University of St. Andrews

Filippo Aureli
School of Biological and Earth Sciences
Liverpool John Moores University

Thad Q. Bartlett
Department of Anthropology
University of Texas at San Antonio

Simon K. Bearder (General Editor)
Department of Anthropology
Oxford Brookes University

Irwin S. Bernstein
Department of Psychology
University of Georgia

Gregory E. Blomquist
Department of Anthropology
University of Illinois

Debra Bolter
Department of Anthropology
University of California, Santa Cruz

Christine A. Caldwell
School of Psychology
University of Exeter

Christina J. Campbell (General Editor)
Department of Biology
Pomona College

Colin A. Chapman
Department of Zoology
University of Florida

Anthony Di Fiore
Department of Anthropology
New York University

Leslie J. Digby
Department of Biological
   Anthropology and Anatomy
Duke University

Phyllis Dolhinow
Department of Anthropology
University of California, Berkeley

Karin L. Enstam
Department of Anthropology
Sonoma State University

Peter J. Fashing
Department of Science and Conservation
Pittsburgh Zoo and Aquarium

Linda Marie Fedigan
Department of Anthropology
University of Calgary

Eduardo Fernandez-Duque
Zoological Society of San Diego
Center for Reproduction of
   Endangered Species

Stephen F. Ferrari
Department of Genetics
Universidade Federal do Pará

Agustín Fuentes (General Editor)
Department of Anthropology
University of Notre Dame

Pascal Gagneux
Conservation and Research for
   Endangered Species (CRES)
San Diego Zoological Society and
Department of Cellular and
   Molecular Medicine
University of California, San Diego

Paul A. Garber
Department of Anthropology
University of Illinois

Lisa Gould
Department of Anthropology
University of Victoria

Harold Gouzoules
Department of Psychology
Emory University

Sarah Gouzoules
Department of Anthropology
Emory University

Sharon Gursky
Department of Anthropology
Texas A&M University

Walter Hartwig
Department of Basic Sciences
Touro University College of
   Osteopathic Medicine

Michael A. Huffman
Primate Research Institute
Kyoto University

Lynne A. Isbell
Department of Anthropology
University of California, Davis

Katharine M. Jack
Department of Anthropology
Tulane University

Clifford J. Jolly
Department of Anthropology
New York University

Sonya M. Kahlenberg
Department of Anthropology
Harvard University

R. Craig Kirkpatrick
TRAFFIC East Asia

Cheryl D. Knott
Department of Anthropology
Harvard University

Joanna E. Lambert
Department of Anthropology
University of Wisconsin–Madison

Bill L. Lasley
Institute of Toxicology and
   Environmental Health
University of California, Davis

Steven R. Leigh
Department of Anthropology
University of Illinois

Katherine C. MacKinnon
   (General Editor)
Department of Sociology and Criminal
   Justice and Center for International
   Studies
Saint Louis University

Joseph H. Manson
Department of Anthropology
University of California, Los Angeles

Lynne E. Miller
Department of Anthropology
Mira Costa College

Anna Nekaris
School of Social Sciences and Law
Oxford Brookes University

Marilyn A. Norconk
Department of Anthropology
Kent State University

Deborah Overdorff
Department of Anthropology
University of Texas–Austin

Melissa Panger (General Editor)
George Washington University and
U.S. Environmental Protection Agency*

Joyce Parga
Department of Anthropology
University of Texas–Austin

Mary S. M. Pavelka
Department of Anthropology
University of Calgary

Elsworth Ray
Department of Anthropology
University of California, Berkeley

Martha M. Robbins
Department of Primatology
Max Planck Institute for Evolutionary
    Anthropology

Sabrina E. Russo
Center for Tropical Forest Science–
    Arnold Arboretum Asia Program
Harvard University

Wendy Saltzman
Department of Biology
University of California, Riverside

Michelle Sauther
Department of Anthropology
University of Colorado

Anne Savage
Disney's Animal Kingdom

Karen B. Strier
Department of Anthropology
University of Wisconsin–Madison

Rebecca Stumpf
Department of Anthropology
University of Illinois at
    Urbana–Champaign

Robert W. Sussman
Primate Biology Program and
    Department of Anthropology
Washington University

Bernard Thierry
Equipe d'Ethologie des Primates
Centre d'Ecologie, Physiologie, et
    Ethologie

Adrian Treves
Department of Forest Ecology and
    Management
University of Wisconsin

Andrew Whiten
Scottish Primate Research Group and
Centre for Social Learning and
    Cognitive Evolution
University of St. Andrews

Linda D. Wolfe
Department of Anthropology
East Carolina University

Adrienne Zihlman
Department of Anthropology
University of California, Santa Cruz

* The views presented here are the views of Melissa Panger and do not necessarily represent the views of the U.S. Environmental Protection Agency.

Introduction

Phyllis Dolhinow

Welcome to the field of primatology. It is now in the neighborhood of 50 years since I felt a special rush of excitement provoked by spotting a group of honest-to-god, real live monkeys running along rooftops next to a train station in India. I can recall it today as vividly as then. Happily, those infrangible feelings of intense engagement and fascination have not faltered; they continue to this day. I hazard a guess that you could consider this an addiction, and if you are very fortunate, you will acquire it.

Primatology has grown exponentially in the last 50 years, thanks to continued study of the nonhuman primates, prosimians, monkeys, and apes, in many different environments from rain forests to laboratories. Studies of different species, as well as groups within an individual species living in varied habitats under dissimilar conditions, reveal important variations in behavior among and within species. Even at the level of the individual primate, we are experiencing a new flow of information thanks to a fairly recent but steady infusion of techniques of investigation from other disciplines. Far-reaching vistas are opening into the hidden aspects of behavior: the neurological and physiological processes going on inside the animal out of range of human vision.

Kinship, the degree of relatedness among the members of groups, is an extremely important component for unraveling and analyzing social relations. Knowing how individuals are related biologically enables us to measure and evaluate the importance of family, lineages, friends, and strangers. For example, consider the question of paternity, knowing which male is the biological father of another animal. Fatherhood is a vital concern among humans, and nowadays it is relatively easy to resolve questions concerning human paternity. When investigators of nonhuman primates want to identify the father of a specific animal, the answer does not have to be puzzled out by observing many months of adult reproductive behavior. We can test data in the form of biological materials from the field using deoxyribonucleic acid (DNA) analysis.

Modern investigative techniques make it possible to uncover links in the relationship between biology and behavior that were previously invisible. We have windows into the internal mechanisms of male and female reproduction and are able to investigate critical components of the biochemistry of stress, loss, success, or failure. We can monitor physiological responses as an animal experiences changing levels of social power or control, as well as during the more mundane events of daily life. Most importantly, using the appropriate techniques allows biological investigations to occur in the field as well as in captivity. Today, we measure the unmeasurables of yesterday. Just imagine what will be available in the future.

Understandably, the newcomer to primatology might be intimidated by the masses of information already acquired and the technologies of investigation that are in place. Many of the methods and products of our research at the macroscopic as well as microscopic levels are carefully summarized in the following chapters. The wealth of information and techniques might induce a rather anxious state of epistemological aphasia, but, do not fret: that should not last long, and it does offer a remarkable variety of niches for investigation.

Why are we so fascinated by our cousins, the nonhuman primates? Likely in part because when we watch them, every one of us is apt to recognize something of ourselves in what we see. We share and express many very similar, if not at times the same, emotions. We often readily grasp the significance of their actions and interactions with each other. This, in itself, is intriguing. We empathize when they care for their rambunctious youngsters, play, fight, threaten, struggle to survive danger, or compete for a few desirable things or places. Remember, the recognition of these similarities simultaneously presents us with major enticements to project ourselves onto them, to attribute to them our feelings, our attitudes, our values, and our concerns when they may not be accurate. Actions, events, or interactions that bother, please, or alarm us may not have the same effects on them.

Consider one of a multitude of possible conundrums that come about when we try to make sense of what we see happening or not happening—yes, not happening. In a very tense situation, one of the highest-ranking, normally hypercontrolling members of a group of langur monkeys remains nonreactive, impassive. He just sits calmly and does not even appear to be aware of the mayhem around him. One thing is certain, he will not tell us what is happening, why he is not reacting in any overt manner to skirmishes he normally would resolve with a low-level threat or glare. What do we do? Keep watching. We watch long enough to become familiar with what the nonresponse might mean in the context of his life and experiences. We will be that much closer to solving this puzzle.

Projecting ourselves into other animals is precisely what we must avoid if we are to maintain a scientific, objective perspective and avoid anthropomorphism. Should we become aware that we cannot be entirely objective, we at least must be conscious of our potential biases. If we are not, we will make some basic assumptions about the nature and meaning of their behavior that will come back to haunt us and potentially undermine the validity of our observations and analyses. Although the nonhuman primates are intriguing and engrossing, we must be aware and resist the temptation to push the limits of credulity to recreate them in our own image. It is too easy to invest them with human concerns that are created by our unique abilities to reason and communicate linguistically. They are like us in many ways, some primates more so than others, but they are not us. Being cautious in no way diminishes the immense complexity of their lives. Enjoy them as they are; marvel at and investigate their lives and their place in nature.

Studies of primate evolution and subsequent taxonomic designations provide a necessary understanding of the degrees of relationship among primate species. They offer another tool for the analysis of general patterns of behavior far beyond those observed for, say, a single species. Having learned the taxonomic relationships of the primates, we can adjust to occasional changes in the assignment of units of taxonomic nomenclature. These usually, but not always, occur at the genus or species level. We all become accustomed to these unpredictable changes that accompany our growing understanding of the evolutionary relationship among the primates.

For the majority of primates, life takes place in a social setting. Social groups, ranging greatly in size, composition, and permanency, are the setting for most, if not all, of a primate's life. Every member, young to old, male and female, serves as a part of this vital unit within which maturation and development occur, affecting the life cycle of every individual. For the most part, learning is a social, experiential activity. Recognizing and documenting variability in the context of rearing is critical to understanding the details of development and the outcomes of experience throughout the remainder of life. Variances in behavioral development drive evolution by forming the individual. To identify and evaluate the multitudes of effective developmental factors requires longitudinal study of individuals of many species. Observations of animals of different ages over shorter periods of time fill out our appreciation of primate variability. They also identify significant components of development for research.

Who would anticipate that an adult female langur monkey could, and most likely would, change her caregiving style, not once, but several times during her reproductive years? Only the study of her offspring as they matured could demonstrate the impact of a female's attentions or inattentions on the next generation. There are so many variables affecting individuals as well as groups, populations, and species throughout individual lifetimes that it takes many years of research to record the range of these differences. Imagine undertaking to chronicle the prodigious range of phenomena displayed by all primates. Any general statement will provoke a slew of contrary examples, making generalization an untidy and vulnerable enterprise. However, take comfort: the chapters which follow provide excellent and detailed examples and summaries of much of the variability and many of the complexities of primate life.

How does one go about researching primates? First comes the planning, and that means deciding what it is that we want to know, as described in the Methods section. It will be most productive if we formulate our questions as hypotheses. We already have a great armamentarium of theoretical paradigms, models, and methods of observation and data analysis that are prefigured on what we have learned from past studies. These will serve to guide future investigations, including ways of conceptualizing which tools will be relevant to answering questions. Hypothesis-driven research is invaluable, but there have been and remain other forms of research. We can call the latter "natural history," or what we will. It is the first stage in the study of a species or a new primate group. It provides basic needed information on the demographic, social, and ecological dimensions of life. This basic first-round work is less ideologically intoxicating but necessary, and it requires appallingly hard work, sometimes for long periods of time in unromantic conditions. When research continues in locations away from the initial observation site, it is possible to identify and measure variability in diverse aspects of life. This is well illustrated in the following chapters on specific groups of primate taxa.

Not all data are created equal. The repertoire, a set of words or labels used to record what is seen and how the continuous flow of action and interaction is divided into measurable units, reflects the researcher's evaluation of what is important. Labels are given to the pieces of what happens that we wish to record and later to analyze. It is the researcher who establishes the accuracy or validity of data categories. The meaning of words and labels reflects the assumptions that have been made, a priori, about behavior. It is said that words are mightier than swords. Words can do just as much damage if they are misleading.

To describe animals "cheating" or "deceiving" is actually an "as-if" observation. Animal A acts as if it intended its actions to deceive or cheat animal B. However, if deception or cheating is supported with detailed descriptions of many such interactions, the interpretation is strengthened. Providing detailed observations underpinning the terms leaves it to the reader to agree or disagree with the application of such descriptive labels. Interpretation is a necessary element in understanding the behavior of an animal that cannot talk; and when such interpretations are fully explained and illustrated, they provide a firm platform for comparison and analysis. Ideally, a majority of the elements of behavior patterns that characterize one species ought to be comparable to those of other species. If they are not, then the value of comparative studies will be minimal. Of course, many species may display some gestures or vocalizations not shared by others, but a good percent of all actions and interactions are common to all of them.

The words we choose to describe and thereby understand the nonhuman primates are of particular concern when we seek to understand the root causes of their actions. Reconstructing the history and causes of primate behavior and social systems is both worthwhile and exceedingly challenging. As suggested in later chapters of this book, it requires new perspectives and creative ways of conceptualizing behavior systems. Existing social networks or ecological systems of behavior have a host of varied biotic and social causes and functions. Untangling and identifying the elements of these multiplex behaviors demands careful delineation and explicit characterization of the levels or points of view of inquiry. The layered, interwoven nature of social life invites study.

As quoted in Chapter 37, "Functional labels should not substitute for a full analysis of the causes of behavior." Using labels that refer to the presumed function of a behavior is likely to beget quagmires of misunderstanding unless these labels are appropriately applied. It is not sufficient to state that specific behavior Y is caused by X to do Z because "it makes sense," even if it does make apparent sense. To reason backward from effects, assumed outcomes, to probable cause creates seductive sequences; but the reasoning is going in the wrong direction. As stated above, it definitely is not sufficient to assert that an animal behaves "as if" anything. A case in point is the infrequent death of infants by aggression. Certainly, it occurs, in some locations far more often than in others. The data set remains sparse, especially when we recognize that the majority of deaths either were not soon after birth or were not witnessed and only assumed to have occurred. To incorporate these occurrences into a presumed evolved male reproductive strategy before sufficient evidence is acquired to prove or falsify the hypothesis begs the question. Alternative hypotheses are seldom considered, and the functional label of *infanticide* remains in constant use.

When it comes to analyses of living primates' behavior, two investigators may have drastically different ways of viewing and interpreting the same data set. This happens frequently and depends on a significant number of factors, such as training, experiences, and intellectual predispositions. More to the point, we call these *biases*, and we all have them. It is important to remember that even before data are analyzed there are many experiences and theoretical paradigms of the observer that are responsible for each researcher's initial recognition and collection of units of information that become the data of a study. Given our array of theoretical paraphernalia, what is the neophyte to do?

As in many disciplines, theory can easily outrun evidence when hypotheses are accepted before they have been investigated fully. Whereas theorizing and building models are fun and can be done from an armchair at home, they are the icing on the cake. Collecting evidence is also enjoyable, but it is labor-intensive and time-consuming. A hypothesis or, better yet, a set of alternative hypotheses generated by an initial study should be tested, ideally, on a body of data gathered from new groups that did not generate the hypotheses originally. With the constantly increasing amount of information from new and ongoing studies, it is possible to reanalyze correlations identified and accepted earlier. There are several excellent examples of such reanalysis in the chapters that follow, and they do not always uphold the earlier perceived truths.

Not all primatologists work in the field because not all research questions are best answered in the field. Consider briefly the appropriate role of investigations in captivity, which, of necessity, depends on the questions being asked. What do we want to see or to measure, and where are we most likely to achieve our objectives. Each research location, laboratory, colony, and field has advantages and disadvantages for investigating different questions. The behavior systems characteristic of social animals are fractally complex, with any small section of the whole likely to turn out as complicated as the broader system was considered to be. We have ample evidence of the tremendous variability that exists within the geographic range for many species in the field. In terms of variability, but in far different ways, life in captivity can be characterized by a world of different conditions from a tiny, bare individual cage with minimal or no communication with a conspecific to large, outdoor/indoor enclosures with natural vegetation housing an intact social group. Because there can be atypical environments in the field as well as under captive conditions, it is critically important to choose the research location carefully.

Observation conditions in the field often interrupt continuous observation of individuals or groups. Forests did not evolve with the convenience of the observer in mind, nor does the wandering of the animals make it any easier to follow them. The result is we cannot and do not witness as much as we need to if we are to characterize the animals' behavior fully, and the devil all too often does reside in the undetected details.

With appropriate planning, good baseline information on the subject species, and a considerable amount of time, it is possible to witness animals without interruption from birth to death in captivity. Also, providing our own lives and work schedules permit, we can watch generations unfold. Of course, we can also watch generations in the field; and at some sites for a few species, we have long-term information. Even in field locations where teams of workers have produced decades of data, limited observation conditions and the mobility of the animals themselves seldom allow monitoring of all the tiny details we want to know. There is no doubt that visibility is far more problematic when following an arboreal species than one spending appreciable time on the ground. It is definitely easier on the neck to follow the latter.

No matter where primates live or how easy or difficult it is to study them, our aim is to generate a real feeling and appreciation for the organism. We have a head start because they are so easy to identify with and intriguing to watch. What is it like for a young capuchin monkey making its way in a complex social world? How does this differ from the life experiences and challenges facing an immature chimpanzee or langur? These and all the other nonhuman primates are engaging. Each, in its own way, offers both a challenge and an obligation to those who study them to express and make real the myriad dimensions of their lives.

We have devoted many years to discovering all we can about the lives of our subjects in the diverse habitats of nature. We sample the range of living conditions, measure ecological parameters, and record everything considered important in their existence. Ironically and indisputably, it is humanity that has placed the lives of these extraordinary animals in danger. The most important achievement of our research is that it enables us to do the most we can to slow a potentially catastrophic decline in biotic diversity and decrease in the quality of life for the survivors. Some of the important steps we need to take to put primate conservation into effect are discussed in this book.

The following chapters offer an excellent selection of what we know about the nonhuman primates, their evolution, present great diversity, and what must be done in the future to assure their continued survival and prosperity. A prodigious amount of work is summarized, emphasizing not only how many humans have devoted decades of work but also how many disciplines have contributed directly to our understanding of primate life. There is a challenge for everyone, regardless of training or interests. We can seek the most microscopic and embedded data or perhaps search for antipodal symmetries. We have just begun and, as they say, the best is yet to come.

PART ONE

Background

1

A Brief History of Primate Field Studies

Robert W. Sussman

We might say that scientific interest in the natural behavior of primates began with an argument at the 1860 annual meeting of the British Association for the Advancement of Science. It was at this meeting that Thomas Henry Huxley defended Charles Darwin's *On the Origin of Species* from Bishop Wilberforce (Montagu 1959, Millar 1972; from an account written by Sir Charles Lyell, January 3, 1861). After a long-winded criticism of the theory of natural selection, the bishop turned to Huxley and begged to know "Was it through his grandfather or his grandmother that he claimed descent from a monkey?" However, Huxley had been preparing for such an attack and whispered to a friend "The Lord has delivered him into mine hands." He then answered, "If the question is put to me would I rather have a miserable ape for a grandfather or a man highly endowed by nature and possessing great means and influence for the mere purpose of introducing ridicule into a grave scientific discussion —I unhesitatingly affirm my preference for the ape."

Soon after, in 1863, Huxley's *Man's Place in Nature* was published. In this book, he put human and nonhuman primates into Darwin's story and compiled what could be thought of as the first "textbook" in biological anthropology. The book was organized into three sections: one was on comparative anatomy, illustrating the anatomical likeness of humans to other animals, and another included a summation of the human fossil remains known at the time; however, in the first section of the book, Huxley wrote the first systematic review of what was known about primates in their natural habitat. As might be expected, there was only anecdotal information available.

Immediately after Darwin, in the late nineteenth and early twentieth centuries, only a few scientists were interested in venturing into the forests to see what wild-living primates (and not those living in cages or already anatomical specimens) could tell us about human existence. The earliest of these adventurers were interested mainly in chimpanzees and gorillas. In the 1890s, Richard L. Garner (1848–1920), a zoologist and collector interested in the "speech" of monkeys and apes, went to Gabon, West Africa, to collect great apes and to observe them in the wild. Because it was commonly thought that gorillas were bold and violent, Garner (1896) armed himself and built a cage to sit in while waiting for the animals to come by. This he did much of the day and night for 112 successive days. A few

individual apes approached and quickly wandered off. Needless to say, little was learned about gorillas or chimpanzees on this venture. Indeed, we have only recently begun to learn about the natural behavior of lowland gorillas (*Gorilla gorilla gorilla*). Garner (1892) also was the first to use a phonograph to study primate vocalizations. Around the same time, Sir Arthur Keith (1866–1955) observed the naturalistic behavior of Asian primates in Thailand, particularly the gibbon (*Hylobates lar*). Mainly an anatomist, he related his observations on behavior and ecology to his field-based anatomical dissections (Sheeran 1997).

The first American woman to observe and describe the behavior of great apes in the field was Mary Hastings Bradley (188?–1976), a Chicago fiction writer and socialite. Her husband, Herbert Bradley, was a big-game hunter, who was asked to accompany Carl Akeley to the Virunga volcanoes in East Africa to collect gorilla (*G. g. beringei*) specimens for the American Museum of Natural History. She wanted to see and write about Africa, and she and her 5-year-old daughter Alice became an integral part of the expedition. (Mary Hastings Bradley's biography, along with her daughter's, is a story in itself, but alas for another time and place.) Akeley believed that the rumors of the gorilla's ferocity were greatly exaggerated and wanted not only to collect animals but also to observe and photograph them. Bradley (1922:131–132) was able to follow a group of gorillas in an afternoon:

> I had seen six gorillas, one of them, at least, and probably two, the demon male, and five gorillas had certainly seen us. And we had not been attacked on sight. Not one had beat his breast or roared or tried to ambush us! . . . We had never heard of any of the native(s) . . . being raided by gorillas. . . . Altogether Mr. Akeley's belief in the essential character of the gorilla was justified. He was simply the big monkey, the man ape, powerful beyond all words, dangerous when attacked, but not a bit the hellish demon or the malignant arch fiend!

Akeley returned to the volcanoes in 1926, this time to study the mountain gorilla and not to shoot it. However, he died at the beginning of the expedition and was buried in Albert National Park, Congo, which he had been instrumental in establishing the year before (Schaller 1964).

In 1929, Robert Mearns Yerkes (1876–1956) and his wife Ada Yerkes (1874–1963) published *The Great Apes*, a compilation of the current knowledge about primates. However, as in Huxley's volume written over 60 years earlier, there was no systematic field research to report in this volume, and concerning knowledge of primates to that date, they wrote that we know "next to nothing, with certainty, concerning their instincts, habits, other individual modes of behavior, mental life, and social relations" (p. 582).

In the same year, Robert Yerkes was funded to begin a great ape breeding facility in Orange Park, Florida, as part of the Yale Laboratories of Primate Biology. (This later became the Yerkes National Primate Research Center of Emory University, Atlanta, Georgia.) He sponsored two expeditions to Africa to study the behavior of apes in their native habitat, to aid in his captive breeding program and to collect some of these animals for his laboratories. In 1929, Harold C. Bingham set off to the Congo to conduct research on gorillas (*G. g. gorilla*), and a year later, Henry Nissen traveled to French Guinea, West Africa, to collect and study chimpanzees (*Pan troglodytes verus*). Yerkes, Bingham, and Nissen were psychologists; and their research on great apes was motivated mainly by an interest in the study of the evolution of intelligence. Yerkes had both scientific and political influence. He was a leader in the eugenics movement in the United States during the decades preceding World War II and a major proponent of the use of intelligence testing to restrict immigration into the United States and to support and rationalize racial prejudice. In 1923, he wrote a glowing forward to Carl Brigham's racist tome *A Study of American Intelligence*.

Nissen's expedition was fairly successful. He was able to observe chimpanzees on 49 of 64 days in the field, dispelling a number of widely believed but fictitious ideas about them. His research was published as a monograph, the earliest on primate field behavior (Nissen 1931). Bingham had little more success in his study of gorillas than did Garner before him. Unlike Akeley and Bradley, he believed anecdotal accounts of the gorilla's ferocity and thought it necessary to track the animals with a large entourage of guides, gun bearers, and porters. Bingham carried a rifle and his wife, a pistol. After about 2 weeks in the field, his party inadvertently surprised a group of feeding gorillas and was threatened by a large male. Bingham shot the animal and returned home about a month later, learning very little about the natural behavior of gorillas (Bingham 1932).

At about the same time that Nissen and Bingham were attempting to study great apes in western and central Africa, Sir Solly Zuckerman (1904–1993), a medical doctor, anatomist, endocrinologist, and chief scientific advisor to the British government for many years, was studying a captive colony of baboons in Regent's Park Zoo, London. In 1930, he returned to his native South Africa to add some information on natural populations of these creatures. However, he spent only 9 days observing chacma baboons (*Papio ursinus*) in the field. His monograph *The Social Life of Monkeys and Apes*, published in 1932, focused on the importance of the social behavior of primates and emphasized the fact that higher primates lived in permanent social groups. Zuckerman theorized that sex was the major reason for group living in primates; his theory is no longer accepted, and the factors underlying social organization in primates are still being debated.

There was, in fact, one earlier field study on baboons. This was by Eugene Marais (1871–1936), a South African journalist, lawyer, poet, and natural historian. Sometime around 1907, shortly after the Boer War, Marais retreated to a remote area of the Transvaal, Waterberg. Here, he and a young colleague studied a troop of habituated Chacma baboons (*P. ursinus*) for 3 years essentially from his backyard. It is interesting to note that Marais was worried during his observations because he "had no libraries and no means of checking what work had already been accomplished in this field" (Marais 1968:63). Little did he know that there had been no previous studies of this sort. However, until his unfinished monograph *The Soul of the Ape* appeared in 1968, long after his death, the primate field work of Marais remained relatively unknown (though Zuckerman made brief reference to his work).

Thus, although some field studies had been done during this early period, the individuals involved in the actual field work were not interested in dedicating their lives to the study of primates in remote habitats and none contributed to field primatology in any permanent way. However, C. Raymond Carpenter (1905–1975) was an exception. After earning his Ph.D. in psychology (on the sexual behavior of birds), Carpenter was granted a postdoctoral fellowship to work with Robert Yerkes at Yale. Yerkes and Frank M. Chapman, an ornithologist, convinced Carpenter to study monkeys at Chapman's field site, Barro Colorado, Panama. This began the first long-term involvement of a scientist in field primatology or, as Carpenter called it, "the naturalistic behavior of nonhuman primates" (Teleki 1981). Carpenter spent a number of months studying howlers (*Alouatta palliata*) on Barro Colorado and making some observations on spider monkeys (*Ateles geoffroyi*) in western Panama between 1931 and 1935 (Carpenter 1934, 1935). He focused mainly on social organization, social interactions, and group censuses.

In 1937, Carpenter was invited to collect the behavioral data on a multidisciplinary expedition to study gibbons (*H. lar*) in Thailand. The "Asian expedition" was sponsored by Harvard, Columbia, and Johns Hopkins Universities and included two physical anthropologists/anatomists who were to become very important to primatology, Adolph Schultz and Sherwood Washburn. Carpenter spent 4 months in the field studying white-handed gibbons, and this led to his monograph on this species, now considered a classic in the field (Carpenter 1940). In 1938, Carpenter was instrumental in exporting 450–500 rhesus monkeys (*Macaca mulatta*) from India and releasing them on Cayo Santiago, an island off the coast of Puerto Rico. This began the first long-term research on semicaptive primate populations. The colony is now part of the Caribbean Primate Research Center and is still an active research colony.

Because of World War II, field research essentially came to a halt between the late 1930s and early 1950s. In fact, before 1960, there was very little primate field research, though it is obvious that the natural behavior of primates was beginning to interest scientists. Between 1960 and 1965, a number of important scientific meetings and books appeared and the discipline of field primatology became solidly established. It has grown steadily since that time. Space will not allow me to detail the stages of this development and growth of the field. I thus simply will outline what I consider to be major events and influences.

Field research on primates during the 1950s stemmed mainly from four different areas. *(1)* Yellow fever had spread in many tropical areas during World War II, and nonhuman primates as well as humans were affected with the disease. Some studies of free-ranging primates were carried out in relation to the epidemiology of this disease, such as that done by the Virus Research Institute in Uganda (e.g., Haddow 1952). *(2)* Following Carpenter's research at the Smithsonian Institution field site of Barro Colorado, zoologist Charles Southwick and his professor Nicholas Collias recensused the howler monkeys (*A. palliata*) (Collias and Southwick 1952), and another zoologist, Stuart Altmann, studied the social behavior of the same species (Altmann 1959). Both Southwick and Altmann became major contributors to primate field studies during the 1960s and are still active in the field. *(3)* A few animal behaviorists and mammalogists teaching and residing in tropical countries did studies of primates as part of more general research on local fauna. For example, Niel Bolwig of Witwatersrand University studied baboons (*P. ursinus*) in South Africa and great ape (*Pan troglodytes schweinfurthii* and *G. g. beringei*) nests in East Africa (Bolwig 1959a,b), and Angus Booth of University College, Ghana, did sophisticated studies of synecology of West African primate communities in the early 1950s (e.g., Booth 1956). Booth surely would have had a major influence on primate field research if he had not died tragically in 1959 at the age of 30. *(4)* Under the leadership of Kinji Imanishi, in 1948, a number of Japanese animal behaviorists began to study the indigenous Japanese macaque (*Macaca fuscata*) (e.g., Imanishi 1953, Kawamura 1958). These were the first long-term studies of identified individuals of known kinship in natural primate populations and the beginning of what has turned out to be one of the most active centers of primate research. This group of researchers founded the first journal in primatology, *Primates*, which was first published in English in 1959.

Thus, studies of primates in their natural habitats began again in the 1950s, but field primatology did not really take off until the next decade. It appears that two developments in the 1950s were very important to the enormous growth of this subdiscipline in the 1960s. The first was the exchange of ideas and concepts between population biologists and anthropologists, especially concerning human evolution. This is exemplified by the Cold Spring Harbor Symposium of Quantitative Biology in June 1950. This meeting was attended by 129 of the most influential biologists and anthropologists in the world. The proceedings were published in 1951 and contained a number of papers calling for a more detailed look at primate behavior in interpreting nonhuman and human primate fossils. In 1953, a paper was published in *American Anthropologist*, coauthored by biologist George Bartholomew and anthropologist Joseph Birdsell (who had attended the 1950 symposium), entitled "Ecology and the Protohominids." This paper developed a method for reconstructing the behavior of early hominids by extrapolating from the behavior of extant primates and other mammals. The second development that stimulated primate field research was the discovery, in the early 1950s, that the Piltdown Man was a fake, accompanied by the concomitant acceptance of the australopithecines as true ancestors of humans. Piltdown Man was a skull discovered in 1912 and long thought to be the missing link between humans and apes, fitting the belief at that time that the earliest humans would have large brains and be more apelike in the rest of their anatomy. In 1953, it was discovered to be the spurious amalgam of a fairly recent human skull and the jaw of a female orangutan (Millar 1972). The discovery of this hoax ended any idea that there was a major evolutionary "gap" between human and nonhuman primates and emphasized the continuity between ourselves and our ancestors (Sussman 2000).

These developments stimulated biologists, and especially biological anthropologists, to think about the behavior and ecology of early humans and to see living primates as potential windows into the study of the evolution of human behavior. For example, the two preeminent australopithecine hunters, Raymond Dart and Louis Leakey, enlisted Rosalie Osborn, a former secretary, and Jill Donisthorpe, a journalist, to conduct a study of gorillas (*G. g. beringei*) at the Virunga volcanoes of Albert National Park in the mid-1950s (Schaller 1964). This study did not yield a great deal of information, but it was followed up by University of Wisconsin zoologist George Schaller's study of the Virunga gorillas in 1959, the first detailed study of a great ape (Schaller 1963). Soon after this, in 1960, Leakey sponsored and Robert Hinde, of Cambridge University, supervised Jane Goodall's study of the chimpanzees (*Pan t. schweinfurthii*) at Gombe.

By the mid-1950s and early 1960s, many conferences and books focused on the relationship between primate behavior and human evolution. For example, in 1953 a symposium was presented at the annual meeting of the American Association for the Advancement of Science entitled *The Non-Human Primates and Human Evolution*. The proceedings of this meeting (Gavan 1955) were dedicated to Earnest Hooton (1887–1954), an anthropologist who had been calling for primate field studies since his classic book *Man's Poor Relations* in 1942. He delivered a paper at this conference, shortly before his death, entitled "The Importance of Primate Studies in Anthropology." Other volumes to appear were *The Evolution of Man* (Tax 1960), *Social Life of Early Man* (Washburn 1961), *Ideas on Human Evolution* (Howells 1962), *Classification and*

Human Evolution (Washburn 1963), and *African Ecology and Human Evolution* (Howell and Bourliere 1963).

Except for the volume edited by Howells, each of these books developed out of international conferences. One of the prime movers for this interchange was the anthropologist Sherwood Washburn (1911–2000), a functional anatomist who had been a student of Hooton. Washburn was also a member of the Asian expedition with Carpenter and, as a graduate student, had spent a semester at Oxford, where he worked in Zuckerman's lab. He also participated in the 1950 Cold Spring Harbor symposium. Papers by Washburn, stressing the need for primate field research, appeared in each of these volumes, including now classic papers by Washburn and his student Irven DeVore on baboon (*Papio anubis*) and early human ecology and social behavior.

These works stimulated students to study primates in their natural habitats; by the early 1960s a number of conferences on free-ranging primates were held, and related books began to appear. The first two of these volumes to appear were based on international conferences held in 1962 in New York (Buettner-Janusch 1962) and London (Napier and Barnicot 1963). Other early collections were edited by Southwick (1963), DeVore (1965), Jay (1968), and Altmann (1967), the latter three based on meetings held between 1962 and 1965. The international journal *Folia Primatologica* began in 1963. The major figures in primatology at that time contributed to these books. Washburn was again a major catalyst for many of these meetings, and his influence on primate field biology cannot be overemphasized. In fact, the first eight dissertations in primatology after 1960 and 15 of the first 19 were stimulated by Washburn. By 1980, Washburn and his students had produced more than half of active anthropological primatologists (Gilmore 1981), and his influence continues to the present.

In a recent academic genealogy, mainly of American field primatologists, Washburn's lineage is the largest, including 222 (41.4%) of the 536 listed researchers (Kelley et al. 2005). Using this same database, we found that field primatology has been steadily growing since the mid-1950s. The number of field primatologists who graduated with doctoral degrees (mainly in the United States) from 1955 to 1964 was eight. This grew to 46 degrees granted from 1965 to 1974, 83 from 1975 to 1984, and 89 from 1985 to 1994. In the past 10 years (1995–2004), 151 doctoral degrees were granted in primate field research. Even more impressive is the fact that interest in field primatology has been steadily growing in many countries, especially in the tropical countries where primates occur naturally (e.g., Kinzey 1997).

Research in field primatology during the late 1950s and early 1960s was mainly descriptive natural history, with few comparative and quantitative, or problem-oriented, studies. However, by the 1970s and 1980s, field primatology moved into a problem-oriented phase (see Sussman 1979, Smuts et al. 1987). Altmann (1965) and the late K. R. L. Hall (1962, 1965) were the first to collect quantitative data on free-ranging primate populations. Hall urged the use of quanti-

tative methods, and at his suggestion, two of his colleagues in the Psychology Department at Bristol University, John Crook and Pelham Aldrich-Blake (1968), were the first to use the now common "scan sampling" method in a primate ecology field study. In 1974, Jeanne Altmann described various methods of collecting quantitative data on free-ranging primates. This is now one of the most referenced papers by field primatologists.

Problem-oriented studies focused mainly on determining relationships between behavior and morphology, on the one hand, and ecology and social structure, on the other, were an important component of field primatology during the 1970s and 1980s. More recently, many field primatologists have been formulating and testing theories developed out of classical sociobiology, such as those related to kin selection, reciprocal altruism, dominance and reproductive success, and the relationship of sexual selection to social organization and infanticide. In many cases, the theories and studies are elegant and elaborate, but the data are meager.

The philosopher of science F. S. C. Northrop (1965) suggests that any healthy scientific discipline goes through three stages during its development: the first stage involves analysis of the problem, the second is a descriptive natural history phase, and the third is the stage of postulationally prescribed theory, in which fundamental theories are tested. Although there is a movement in field primatology toward this final stage of enquiry, even today many primate species have not been studied in detail, and the most studied species normally are known only from a few localities (Sussman 2003a,b; Strier 2003).

The range of variation in the behavior and ecology of most free-ranging primates is still unknown, and basic natural history remains a necessary component of the subdiscipline. As stated by Northrup (1965:37–38):

> In fact, if one proceeds immediately to the deductively formulated type of scientific theory which is appropriate to the third stage of inquiry, before one has passed through the natural history type of science with its inductive Baconian method appropriate to the second stage, the result inevitably is immature, half-baked, dogmatic, and for the most part worthless theory.

With eugenics, social Darwinism, and the Piltdown Man being conventional wisdom during the first half of the twentieth century, these cautions still must be born in mind when developing theories in biology, the social sciences, and primate ecology and social behavior. Given the conservation status of most primate populations, let us hope that there is still enough time to learn more about these fascinating animals and to collect the data needed to properly test our theories.

REFERENCES

Altmann, J. (1974). Observational study of behavior: sampling methods. *Behavior* 49:227–267.

Altmann, S. A. (1959). Field observations on the howler monkey society. *J. Mammal.* 40:317–330.

Altmann, S. A. (1965). Sociobiology of rhesus monkeys. II. Stochastics of social communication. *J. Theor. Biol.* 8:490–522.

Altmann, S. A. (ed.) (1967). *Social Communication Among Primates*. University of Chicago, Chicago.

Bartholomew, G. A., Jr., and Birdsell, J. B. (1953). Ecology and the protohominids. *Am. Anthropol.* 55:481–498.

Bingham, H. C. (1932). Gorillas in a native habitat. Washington, DC: *Carnegie Inst. Publ.* 426:1–66.

Bolwig, N. (1959a). A study of the behaviour of the chacma baboon. *Behaviour* 14:136–163.

Bolwig, N. (1959b). A study of the nests built by mountain gorilla and chimpanzee. *S. Afr. J. Sci.* 55:286–291.

Booth, A. H. (1956). The distribution of primates in the Gold Coast. *J. W. Afr. Sci. Assoc.* 2:122–133.

Bradley, M. H. (1922). *On the Gorilla Trail*. D. Appleton, London.

Brigham, C. C. (1923). *A Study of American Intelligence*. Princeton University Press, Princeton, NJ.

Buettner-Janusch, J. (ed.) (1962). The relatives of man: modern studies on the relation of the evolution of nonhuman primates to human evolution. *Ann. N.Y. Acad. Sci.* 102:108–514.

Carpenter, C. R. (1934). A field study of the behavior and social relations of howling monkeys. *Comp. Psychol. Monogr.* 10:1–168.

Carpenter, C. R. (1935). Behavior of red spider monkeys in Panama. *J. Mammal.* 16:171–180.

Carpenter, C. R. (1940). A field study in Siam of the behavior and social relations of the gibbon (*Hylobates lar*). *Comp. Psychol. Monogr.* 16:1–212.

Collias, N., and Southwick, C. (1952). A field study of population density and social organization in howling monkeys. *Proc. Am. Phil. Soc.* 96:143–156.

Crook, J. H., and Aldrich-Blake, P. (1968). Ecological and behavioral contrasts between sympatric ground dwelling primates in Ethiopia. *Folia Primatol.* 8:192–227.

Darwin, C. (1859). *On the Origin of Species*. John Murray, London.

DeVore, I. (ed.) (1965). *Primate Behavior: Field Studies of Monkeys and Apes*. Holt, Rinehart and Winston, New York.

Garner, R. L. (1892). *The Speech of Monkeys*. Charles L. Webster, New York.

Garner, R. L. (1896). *Gorillas and Chimpanzees*. Osgood, McIlvaine, London.

Gavan, J. A. (ed.) (1955). *The Nonhuman Primates and Human Evolution*. Wayne State University, Detroit.

Gilmore, H. A. (1981) From Radcliffe-Brown to sociobiology: some aspects of the rise of primatology within physical anthropology. *Am. J. Phys. Anthropol.* 56:387–392.

Haddow, A. J. (1952). Field and laboratory studies on an African monkey, *Cercopithecus ascanius schmidti* Matschie. *Proc. Zool. Soc. Lond.* 122:297–394.

Hall, K. R. L. (1962). Numerical data, maintenance activities, and locomotion of the wild chacma baboon. *Proc. Zool. Soc. Lond.* 139:181–220.

Hall, K. R. L. (1965). Experiment and quantification in the study of baboon behavior in its natural habitat. In: Vagtborg, H. (ed.), *The Baboon in Medical Research*. University of Texas, San Antonio. pp. 43–61.

Hooton, H. (1942). *Man's Poor Relations*. Doubleday, New York.

Howell, F. C., and Bourliere, F. (eds.) (1963). *African Ecology and Human Evolution*. Aldine, Chicago.

Howells, W. W. (ed.) (1962). *Ideas on Human Evolution: Selected Essays 1949–1961*. Harvard University, Cambridge, MA.

Huxley, T. H. ([1863] 1959). *Man's Place in Nature*. University of Michigan, Ann Arbor.

Imanishi, K. (1953). Social behavior in Japanese monkeys, *Macaca fuscata. Psychologia* 1:47–54.

Jay, P. C. (ed.) (1968). *Primates: Studies in Adaptation and Variability*. Holt, Rinehart and Winston, New York.

Kawamura, S. (1958). The process of sub-culture propagation among Japanese macaques. *Primates* 2:43–60.

Kelley, E., Sussman, R. W., and Kelley, B. (2005). An academic genealogy on the history of American field primatologists. *Am. J. Phys. Anthropol.* 40:127–128.

Kinzey, W. G. (1997). New World primate studies. In: Spencer, F. (ed.), *History of Physical Anthropology: An Encyclopedia*. Garland, New York. pp. 743–748.

Marais, E. (1968). *The Soul of the Ape*. Atheneum, New York.

Millar, R. (1972). *The Piltdown Men*. St. Martin's Press, New York.

Montagu, A. (1959). Introduction to the Ann Arbor Paperback Edition. In: *Man's Place in Nature*. University of Michigan Press, Ann Arbor.

Napier, J. R., and Barnicot, N. A. (eds.) (1963). *The Primates*, Symp. Lond. Zool. Soc., No. 10.

Nissen, H. W. (1931). A field study of the chimpanzee. *Comp. Psychol. Monogr.* 8:1–122.

Northrup, F. S. C. (1965). *The Logic of the Sciences and Humanities*. Meridian, Cleveland, OH.

Schaller, G. B. (1963). *The Mountain Gorilla: Ecology and Behavior*. University of Chicago, Chicago.

Schaller, G. B. (1964). *The Year of the Gorilla*. University of Chicago, Chicago.

Sheeran, L. K. (1997). Asian apes. New World primate studies. In: Spencer, F. (ed.), *History of Physical Anthropology: An Encyclopedia*. Garland, New York. pp. 112–121.

Smuts, B., Cheney, D., Seyfarth, R., Wrangham, R., and Struhsaker, T. (eds.) (1987). *Primate Societies*. University of Chicago, Chicago.

Southwick, C. H. (ed.) (1963). *Primate Social Behavior*. Van Nostrand Reinhold, New York.

Strier, K. B. (2003). *Primate Behavioral Ecology*, 2nd ed. Allyn and Bacon, Boston.

Sussman, R. W. (ed.) (1979). *Primate Ecology: Problem-Oriented Field Studies*. Wiley, New York.

Sussman, R. W. (2000). Piltdown Man: the father of American field primatology. In: Strum, S., and Fedigan, L. (eds.), *Primate Encounters: Models of Science, Gender, and Society*. University of Chicago, Chicago. pp. 85–103.

Sussman, R. W. (2003a). *Primate Ecology and Social Structure. Lorises, Lemurs, and Tarsiers*, vol. 1. Pearson Custom, Boston.

Sussman, R. W. (2003b). *Primate Ecology and Social Structure. New World Monkeys*, vol. 2. Pearson Custom, Boston.

Tax, S. (ed.) (1960). *The Evolution of Man*. University of Chicago, Chicago.

Teleki, G. (1981). C. Raymond Carpenter, 1905–1975. *Am. J. Phys. Anthropol.* 56:383–386.

Washburn, S. L. (ed.) (1961). *Social Life of Early Man*. Aldine, Chicago.

Washburn, S. L. (ed.) (1963). *Classification and Human Evolution*. Aldine, Chicago.

Yerkes, R. M., and Yerkes, A. W. (1929). *The Great Apes*. Yale University, New Haven, CT.

Zuckerman, S. (1932). *The Social Life of Monkeys and Apes*. Routledge and Kegan Paul, London.

2

Primate Evolution

Walter Hartwig

INTRODUCTION

Primates comprise an order of Mammalia. Although no single anatomical trait unites them, their "place" in the natural world is *ordered* in the taxonomic sense of the word. Primates are, with few exceptions, generalist foragers with relatively large brains and sophisticated social systems. They interface with ecology cognitively as much as biologically and are not the primary or usual target of predators. This "exemption" from the adaptive spiral of predators and their prey is just one of numerous interpretive themes that emerge from a study of primate evolution.

Because the fossil record is incomplete, large gaps of inference in primate evolution are spanned either by stretching deductive reasoning beyond its capacity or by accepting inductive scenarios that can never be tested. Scholars must attend the data themselves, approach all interpretations critically, and look beyond the constraints of the historical moment. Do not be afraid to challenge traditional interpretations, but also be aware that answers to major and very interesting questions—e.g., How did primates originate? and How did the earliest New World monkeys get to South America?—will always be beyond the reckoning of the fossil record.

Accounts of primate evolution should be constructed from multiple lines of evidence, but the end result must be able to include the fossil record. That is, the fossil record is an essential fabric of our understanding of primate evolution, however fragmentary or undersampled it may be. With this in mind, it is equally important to regard aspects of the fossil record for what they are, not what we hope they are. We tend to promote the current "earliest evidence" of a lineage as its ancestral population, only to revise when the next earliest fossil is found. Having studied the interpretive follies of their predecessors, students of the next generation have the opportunity to be more objective about rendering the fossil record into evolutionary accounts rooted in other databases, molecular in particular.

Perhaps the most important, and most obvious, context of the primate fossil record is how it relates to the living species. Primate fossils represent the tangible record of the entire radiation through time and space. Living primates, on the other hand, constitute a negligible amount of the total evolutionary diversity of the order. They are precious, indeed, and distinct among the thousands of extinct species by virtue of still being. So, while living primates must have a fossil record, that does not mean that the fossils we find are closely related to living primates. This distinction is key to a meaningful understanding of both.

PRIMATE ORIGINS

How would we know the earliest primate if we held it in our hands? It would be a small mammal, given that all early mammals were small. It would have generalized gross anatomy, given that living primates today are the least specialized of all the mammalian radiations. Also, it would be 50–90 million years old, depending on which extrapolation we support. However, these are just attributes of what really interests us—the origin itself. We can never know the actual story, but historically theories have emphasized the physical characteristics that make primates different from other mammals: grasping thumbs and toes, stereoscopic vision. They have also assumed that only one aspect of early primate life, such as an arboreal habitat or foraging for insects by grabbing at them, could be the root of primate origins. We should think pluralistically as much as possible however.

Several recent publications explore primate origins in terms of the fossil record, genetic data, and molecular clocks of evolution (Hamrick 2001, Rasmussen 2002a,b, Sargis 2002, Tavaré et al. 2002, Martin 2003). The fossil record poorly represents tropical latitudes (where primates probably originated), especially in the time ranges believed to encompass the major early eutherian mammal radiations. Rather, provocative concepts of when primates arose and to what they are most closely related draw upon analysis of genes and models of how long it takes genetic distances to "evolve."

The Molecular Data

With enough lines of genetic evidence, such as mitochondrial deoxyribonucleic acid (DNA), nuclear DNA, amino acid sequences, and chromosomes, and with hundreds of separate vertebrate species included, broad pictures of genetic relatedness emerge. With a certain leap of faith, the genetic "distance" between two major types of mammal can be calibrated to real time based on confidence that the fossil record for the origin of the two types of mammal can be trusted (Foote et al. 1999, Murphy et al. 2001). This is sometimes referred to as the "molecular clock." Assuming that regions of the genetic code mutate at a constant rate, the "amount" of genetic difference between species can be measured as a unit of time; likewise, assuming that the fossil record accurately reflects some specific divergences between taxa, the combination of the genetic difference between those taxa and the presumed divergence date calibrates a "clock" for that analysis. While the relative genetic relatedness of orders of mammals is widely accepted, the accuracy of molecular clocks is widely disputed (Graur and Martin 2004).

Several different molecular studies suggest that primates may have originated 40 million years prior to the earliest known fossil primates (Arnason et al. 1998, Madsen et al. 2001, Murphy et al. 2001, Eizirik et al. 2001). The disconnect between the fossil record and the morphological record may be due to the lag between genotypic "roots" and phenotypic "branches," in the sense that genotypic markers of "primate-ness" may go much deeper in time than do recognizable primate morphological traits (Tavaré et al. 2002). Origin dates based on measurements of genetic similarity are likely to be the gold standard for the near future. The objective now is to narrow the ranges through replicating the analyses and expanding the samples.

Molecular studies also ally primates to other lineages, most frequently to tree shrews (Scandentia) and flying lemurs (Dermoptera) and less frequently to rodents, rabbits, and bats (Springer et al. 2003). The consistency with which these taxa fall close together in genetic comparisons suggests that they ultimately have a common ancestor.

The Comparative Data

The many theoretical constructs of primate origins have been based on how living primates may represent the current form of the ancestral adaptation or niche. A foundation of evolutionary biology that radiations are based on competitive exclusion requires that we identify what it is about primates, anatomically and/or behaviorally, that distinguishes them from other mammals. Circling around this definition will be species that converge on primate-ness and vice versa, thus presenting some amount of parallel evolution that we are willing to tolerate in our theory of primate origins.

Because most primate species depend on trees for survival, the first dominant theory of primate origins focused on adaptations for an arboreal habitat. The arboreal hypothesis argued that hallmark primate traits, such as stereoscopic vision and nails instead of claws, were basic adaptations to living in a three-dimensional and vertically stratified environment (see Rasmussen 2002a). This basic concept is logical but fails to explain the existence of other species which live in trees without these traits (e.g., squirrels, marsupials) or species with these traits which do not live in trees (e.g., stereoscopic vision in terrestrial carnivores).

Primate origin theories rooted in comparative anatomy were advanced by Cartmill (1974, 1992), who argued that a basic adaptation for visually preying upon small animals probably led the earliest primates to emerge from populations of other small insectivorous mammals. This would explain grasping hands and stereoscopic vision since no other mammals hunted in quite this same way. Lemelin (1999) and Hamrick (2001) studied proportional length of the extremities of several related nonprimate species and concluded that the unique proportions in primates conferred a selective advantage for exploiting the original fine branch/small food item–grasping niche. Sargis (2004) concluded that the grasping anatomy of some tree shrews reflects a model for the earliest primates, to which they are probably closely related.

A more inclusive ecological theory involving the contemporaneous rise of flowering plants, or angiosperms, was developed by Sussman (1991), who argued that this primary food source revolution would have opened numerous niches to animals capable of reaching the flowering parts, with or without stereoscopic vision. Fruit bats may represent the evolution of a mammal that reached the target differently (through flight, not climbing) but evolved similar visual specializations. The visual predation/manual grasping theory and the angiosperm theory are good examples of arguments that tried to minimize the amount of parallel examples in evolution but could not eliminate them.

Rasmussen (1990, 2002a) saw the tropical marsupial *Caluromys* as convergent on prosimian primates in several biological and behavioral traits. Using it as an analogy for early primates, he argued for tightly linked coevolution among insects, flowering plants, and the mammals that preyed upon them. Given a proliferation of a small-branch milieu, the earliest primates may have radiated into a niche that incorporated aspects of both the visual predation/manual grasping argument and the angiosperm theory.

Thinking critically about primate origins is a good exercise in logical reasoning and the tenets of evolutionary biology. The nearest relatives of living primates must be considered in this process, for if the superordinal group known as Archonta (tree shrews, primates, flying lemurs) is legitimate, then many anatomical aspects of arboreality were in place at the time primates diverged from the other groups. The arguments are not resolvable, of course, but analyzing them helps to expose assumptions we make about evolutionary history and how much convergence or parallelism we are willing to tolerate.

The Paleontological Data

The search for early primates in the fossil record has focused on Tertiary sites in the Northern Hemisphere, given that they are more abundant, have yielded primate fossils historically, and fall within the time range once believed to be the source age of primate origins. Although no definitive fossils of modern eutherian mammal orders are older than the Cretaceous–Tertiary boundary (65 million years ago), future investigations likely will focus on the tropical latitudes and earlier time ranges, given the signals of the genetic data.

In general, the boundary of what to call a primate in the early fossil record has included a taxonomic group called Plesiadapiformes (see Szalay and Delson 1979, Fleagle 1999). These species radiated widely in the Paleocene of North America and thus were in the "right place at the right time" for early theories of primate origins. The species lacked definitive primate traits, such as a postorbital bar of bone, forward-facing orbits, and generalized dentition; but some aspects of their auditory canal anatomy and speculations about their grasping abilities kept them under consideration as primate ancestors. With the shift to studying molecules and to the tropics (Martin 1993) coupled with earlier and earlier evidence of definitive anthropoids in the fossil record, Plesiadapiformes were sidelined as formal members of the primate line (e.g., Covert 2002, Hartwig 2002). Recent discoveries (Bloch and Boyer 2002) have expanded our knowledge of this radiation, however, and argue once again for a nearest-relative relationship to primates (Ravosa and Savakova 2004, Sargis 2004), as summarized by Kay (2004:840):

> . . . signs now point to the possibility that some or all Plesiadapiformes may be more closely related to primates than to any other living order of mammals. That makes them stem primates, even though they lacked the shared-derived features that are hallmarks of crown primates—"euprimates."

Fossils recognizable as primates of modern aspect (*euprimates*) are found in diverse assemblages beginning in the Eocene epoch, dated to 45–55 million years ago. These are the adapids and omomyids of the Northern Hemisphere (Gunnell and Rose 2002, Rasmussen 2002a), the early primates from northern Africa (Rasmussen 2002b, Seiffert et al. 2003), and the early anthropoids *Eosimias* and *Phenacopithecus* from China (Beard 2002, Dagosto 2002, Beard and Wang 2004). Hints of the earliest representatives of the euprimates may be *Teilhardina asiatica* from deposits in China estimated to be 55 million years old (Ni et al. 2004). Elucidating the story of primate origins from the fossil record is really about collecting as much data (fossils) as possible in order to understand the deep past. A reasonable current working consensus about the search for early primate fossils can be found in the concluding statement of Tavaré et al. (2002:728):

> Direct reading of the known fossil record suggests that primates originated during the Palaeocene in the northern continents and subsequently migrated southwards. An alternative interpretation is that primates originated earlier in the poorly documented southern continents and expanded northwards when climatic conditions permitted.

Summary

A good general principle of consensus may be to return to what it means to be a primate. Because many primate species are omnivorous, primate origins may be predicated on a dietary approach more generalized than just angiosperm or insect predation. The earliest primates may have been as frugivorous as they were insectivorous, and a large part of their divergence from other mammals may have been due to foraging on fruits where no other mammal could go. Thus, the best explanation for primate origins may be one in which both habit (visual predation, manual grasping) and habitat (the slight outer branches of the forest canopy and undergrowth) drove the evolution of characteristic primate features.

Much of our understanding of primate origins reflects our sense of how they radiated afterward. Primates have diversified into a great variety of habits and habitats due to the potential of their persistent generality, both anatomically and behaviorally, compared to other mammal species that are more specialized into the predator–prey pyramid. Behavioral novelty without loss of versatility was a central theme in primate origins, even if we cannot pin the sequence of adaptations to a particular scheme of phylogeny or discovery in the distant fossil record.

PRIMATE ADAPTIVE RADIATIONS

Fossil prosimians, monkeys, and apes are the only direct records of primate ancestry, even if they are often fragmentary and difficult to interpret. Finding fossils cumulatively adds to our understanding of primate evolution. However, our desire to interpret new fossils into an existing sense of primate evolution often leads to fruitless debate since the old interpretive framework was based on a different sense of the fossil record. The intractable limitations of the fossil record mean that some localities that have produced numerous fossils exert a tremendous influence over our interpretation of primate evolution, even though we understand that no single locality can sample the complete diversity of species living at any one time.

The Paleogene (35–65 Million Years Ago)

The oldest known fossil primates date to the 40–50 million year time range. They were small-bodied, both nocturnal and diurnal, and in evidence in North America, Europe, North Africa, and Asia. A broad picture of two types of primate at that time can be argued from the similarities among the fossils. New discoveries continue to challenge this broad picture, and further argue for Asia as a major area of early primate diversification (Table 2.1).

Historically speaking, some of the earliest discoveries of fossil primates were in exposures of the Eocene age (approximately 40 million years ago) in the northern latitudes (Europe and North America). These have come to be known as the

Table 2.1   A Geological Timeline of Key Fossils and Major Trends in Primate Evolution (Dates for Fossils Are Generalized)

GEOLOGICAL EPOCH (MILLIONS OF YEARS AGO, MYA)	NORTH AMERICA	EUROPE	ASIA	AFRICA	SOUTH AMERICA
Pleistocene (2.0–present)		Old World monkeys diminish, hominins arrive	Old World monkeys dominate tropical biomes of insular SE Asia; largest hominoid, *Gigantopithecus*, found in southern China and Vietnam	Old World monkeys and hominins proliferate on the continent, giant lemuroids radiate in Madagascar, hominoids limited to tropical belt	Poor fossil record except for large subfossil platyrrhines in Brazil and unusual species on Caribbean islands
Pliocene (5.0–2.0)		Old World monkeys proliferate, hominoids disappear	Poor fossil record, hominoid diversity probably declined	Old World monkeys greatly diversify, no obvious fossil pongids, earliest hominins (*Sahelanthropus*, *Orrorin*) at the Miocene–Pliocene boundary	No known primate fossils
Miocene (23.0–5.0)		*Dryopithecus* (10 Mya), *Ouranopithecus* (9 Mya) hominoids flourish in the middle and late Miocene, including the earliest known definite brachiator	*Sivapithecus* (9 Mya) hominoids flourish in the middle and late Miocene, including a probable direct ancestor of living orangutans; no known Old World monkeys	*Proconsul* (20 Mya), *Morotopithecus* (17 Mya), *Victoriapithecus* (16 Mya), *Kenyapithecus* (12 Mya), and *Nacholapithecus* (15 Mya), earliest Old World monkey; diverse radiation of large and small hominoids until late Miocene	*Homunculus* (16 Mya), *Stirtonia* (13 Mya), *Cebupithecia* (12 Mya), and *Neosaimiri* (12 Mya), diverse radiation in southern South America in early Miocene; numerous more modern-looking taxa from single middle Miocene site in Colombia
Oligocene (35.0–23.0)	Last surviving primate in North America	Remaining adapid and omomyid primates go extinct	Fragmentary anthropoid fossils in Myanmar	*Aegyptopithecus*, *Catopithecus*, and numerous other early anthropoids found in North Africa	*Branisella* (27 Mya), Bolivia; earliest known fossil New World monkeys
Eocene (55.0–35.0)	*Notharctus* and *Omomys*, two different types of prosimian-like primate, adapid, and omomyid, flourish	*Adapis* and *Teilhardina*, two different types of prosimian-like primate, adapid, and omomyid, flourish	*Teilhardina* (55 Mya), recently discovered earliest omomyid, in China; *Eosimias* (40 Mya, China) and *Pondaungia* (35 Mya, Burma) evidence early anthropoids	*Parapithecus* and *Algeripithecus* (36–40 Mya), early anthropoids; *Saharagalago* (40 Mya), earliest Lorisidae, all from northern Africa	
Paleocene (65.0–55.0)	Plesiadapiform fossils but no good candidates for earliest primate ancestor		South and SE Asia may be the most likely places to recover early primates	Localities are sparse, molecular data suggest that primates are not endemic African mammals	

Adapidae and Omomyidae (Gebo 2002, Gunnell and Rose 2002). The adapids bear a striking resemblance to extant Lemuridae, which are isolated on Madagascar. The omomyids are slightly smaller on average and bear a striking resemblance to the living nocturnal prosimians, especially *Tarsius*.

Despite being found only in northern latitudes, adapids and omomyids also have been linked to the origins of anthropoid primates (Fleagle and Kay 1994). Although this radiation of extinct primates is historically significant, recent discoveries of ancient anthropoids have prompted new theories of the time, place, and ancestry of their origins.

In the last 20 years, numerous primate fossils dating to the middle or late part of the Eocene (40–45 million years ago) have been found in North Africa and Asia (see Covert 2002). The record from North Africa is sparse everywhere except at the productive El Fayum locality, where Elwyn Simons and colleagues have been collecting fossils for several decades. The earliest deposits there record fragmentary remains of species believed to be related to later Old World adapoids. From smaller localities in Morocco and Mongolia, dental remains resembling primitive primates may date to the 50 million year range, which would make them among the most ancient fossils attributed to primates. The most recent contender for earliest fossil primate is a species of *Teilhardina*, an Old World omomyid, found in deposits in China believed to be 55 million years old (Ni et al. 2004).

Because living prosimians are perceived to be more like primitive primates than are the monkeys and apes, the possible phylogenetic relationship of these early fossil primates to living lemurs, lorises, and tarsiers is often debated (see Fleagle 1999, Covert 2002). The lorisiform fossils (*Karanisia* and *Saharagalago*) recently discovered from El Fayum deposits believed to be 40 million years old (Seiffert et al. 2003) and the equally ancient tarsier-like fossils in China (Gunnell and Rose 2002) are among the earliest primate remains Africa and Asia, and they suggest that the prosimian primates as we know them today had already diverged from one another 40 million years ago.

Origin and Evolution of Anthropoid Primates

Anthropoids differ from prosimians in having a relatively larger braincase, fewer teeth, and complete bony closure behind the eyes. Many aspects of their anatomy and behavior can be explained in terms of an emphasis on cognition, sociality, and learned behavior. The fossil record of the near tropics in the 40–45 million year time range has grown dramatically in the last 10 years, but with it have come controversial early anthropoid fossils that, predictably, are raising as many questions as answers.

Provocative fossil evidence of small anthropoid primates from Asia is not new. *Amphipithecus* and *Pondaungia*, known from jaw fragments discovered decades ago, represent two such genera from probable Eocene deposits in Myanmar (see Beard 2002). A majority of researchers accept them as early anthropoids, but new fossils of these and other taxa from the same area continue to fuel the debate (Ciochon and Gunnell 2002, Takai et al. 2001, Shigehara et al. 2002, Marivaux et al. 2003).

Over the last 10 years, a major survey effort in China has yielded fragmentary remains of what may be the oldest definitive anthropoid, *Eosimias* (Beard 2002, Beard and Wang 2004). This taxon is smaller than any known anthropoid (<200 g) and comes from deposits believed to be about 40 million years old. Jaw fragments and postcrania attributed to the genus show similarities to both adapids and later anthropoids. Given that it is difficult to define a shared and derived anatomical trait in early anthropoids, support for calling *Eosimias* an anthropoid depends largely on which aspects of its anatomy are emphasized. The same holds for all claims of early anthropoids. The most recent discoveries in this part of Asia include a new genus, *Phenacopithecus*, which, like many early anthropoid fossil species designations, is based only on isolated teeth (Beard and Wang 2004).

The Egyptian locality of El Fayum has dominated our understanding of early anthropoid evolution because of the numerous genera found there under exacting stratigraphic control for several decades (Simons and Rasmussen 1994). Historically, the fossils could be grouped into a primitive branch (the Parapithecidae, with three premolars) and an advanced branch (the Propliopithecidae, with two premolars), which reflected the roots of living New World monkeys and

living catarrhines, respectively. More recent cranial and dental discoveries, however, such as *Arsinoea* and *Proteopithecus* from lower in the stratigraphy at 36 million years ago, have shown that earlier anthropoids do not fall so hopefully into the ancestries of living monkeys.

Perhaps the most tantalizing early anthropoid fossil recovered at El Fayum is *Catopithecus*, which demonstrates a combination of primitive (unfused mandible) and advanced (full postorbital closure, two premolars) features that upset traditional divisions among early anthropoids (fused mandibles), platyrrhines (retention of three premolars), and catarrhines (two premolars). In this case, much of the cranial anatomy of *Catopithecus* is known; and in general aspect, it resembles expectations of an early catarrhine (Rasmussen 2002b). Its mosaic anatomy and well-dated age of 36 million years push again for a critical sense of how old we think primate lineages really are and how we believe anatomy evolves. As more fossils are uncovered at El Fayum, the picture of early primate evolution changes accordingly.

The Neogene (2–35 Million Years Ago) and Quaternary (0–2 Million Years Ago)

South America

How primates got to South America is one of the great questions of primate evolution. Assuming that anthropoids originated in the Old World, no scenario involving plate tectonics, rafting, vicariance biogeography, or island hopping is well supported (Hartwig 1994). The implausibility of a Tertiary migration of monkeys to the New World indirectly supports a very early origin of anthropoids, at a time when the southern continents were much closer together. However, given that primates and rodents are the only major Old World lineages to appear in South America, at approximately the same time according to the known fossil record, perhaps a miraculous transport via floating mats of vegetation is the only explanation, however fantastic it seems today.

The fossil record of New World monkeys is sparse but growing steadily (Rosenberger 2002). Fossils are known from southern South America, Bolivia, Chile, Colombia, and Brazil. The original source deposits in Colombia (Takai et al. 2003), Bolivia (Takai et al. 2000), and Argentina (Tejedor 2002, 2003) continue to be surveyed productively. New frontiers of discovery include western Amazonia, the Caatinga of eastern Brazil (Cartelle and Hartwig 1996, Hartwig and Cartelle 1996), and the Caribbean islands (MacPhee and Horovitz 2002, 2004). The record is relatively immature in the sense that each new discovery tempts the naming of a new ancient species, but it is developed enough to portray a broad outline of how New World monkeys radiated into their present distributions.

The earliest South American monkeys known so far have been found in Bolivia and southern South America, date to approximately 26 million years ago, and do not resemble any living forms (Takai and Anaya 1996, Fleagle and Tejedor 2002). Numerous small and fragmentary remains from Argentina and Chile dated to 25–15 million years ago constitute several species with some basic affinities to pitheciines and suggestive affinities to squirrel monkeys (*Saimiri*) and owl monkeys (*Aotus*). However, in general, the anatomy that is known is too derived to be ancestral to existing lineages. Rather, it seems more likely that these species are ancient offshoots of an as yet undiscovered ancestral platyrrhine lineage.

By contrast, the fossil New World monkeys from the Miocene of Colombia bear distinct similarities to living species. Dating from 16–12 million years ago, the fauna from La Venta, Colombia, includes more than a dozen different platyrrhine taxa (Kay et al. 1997). Potential ancestors of the lineages of *Alouatta* (*Stirtonia*), *Saimiri* (*Neosaimiri*), *Aotus* (*Aotus dindensis*, known mostly from dental and lower jaw remains), and *Pithecia* (*Cebupithecia*, known from a partial skeleton) have been nominated (Hartwig and Meldrum 2002).

The very productive Miocene locality of La Venta has been a boon to our awareness of extinct New World monkeys, but it also may distort our perspective of platyrrhine evolutionary history by virtue of being the only locality representing a major chunk of time and space. To the same extent that El Fayum stands for most of the early anthropoid evidence from Africa, our sense of how diverse New World monkeys were over the last 15 million years is largely the function of discovery within a single locality. Undeniably, the pattern of New World monkey evolution is conservative, but we must remember that vast areas of their current distribution are not fossiliferous and that large time gaps (the last 10 million years, for example) in their known fossil record persist.

The Quaternary fossil record holds a special interest because it documents biodiversity during the time of humans on the landscape (the last 2 million years). This time period also includes the well-documented global "ice ages," gradual but radical shifts in prevailing climatic conditions. Quaternary New World monkeys have been found on several Caribbean islands and in arid caves in eastern Brazil (MacPhee and Horovitz 2002). They are most interesting because they document anatomies that cannot be accommodated easily by existing phylogenies. Yet, they are the most "recent" examples of past biodiversity, so their uniqueness gives us reason to be cautious in using the known fossil record as a template of the past.

Two taxa from the Caribbean islands, *Xenothrix* and *Paralouatta*, continue to vex the experts because their cranial and dental anatomy can be interpreted as support for synapomorphic or convergent relationship to living species. The debate about their phylogenetic affinity (Rosenberger 1977, 1992, 2002; Horovitz and MacPhee 1999, MacPhee and Horovitz 2002, 2004) is an excellent capsule of how primate comparative anatomy can be analyzed competently toward opposing conclusions.

Another member of the Quaternary platyrrhine class, *Protopithecus* from Toca da Boa Vista in Brazil, merits a deeper look for the sake of its nearly complete preservation

(Hartwig and Cartelle 1996, MacPhee and Horovitz 2002). We often lament having to reconstruct a phylogeny from bits and pieces of individuals, but when an entire skeleton lies before us, we sometimes are equally frustrated. In this case, the largest known platyrrhine (20 kg, as estimated from the skeleton) presents a body very similar to the suspensory spider monkeys (*Ateles*), a cranium that is distinctly similar to *Alouatta*, and a face that is like neither. Does it belong on the atelin clade or the alouattin clade? Our temptation is always to link fossil forms as closely to living forms as possible, but even with this complete skeleton, it is not clear. The enduring lesson of *Protopithecus* may be that parallel evolution in closely related lineages is the rule, not the exception that we have tended to think it is.

Africa

From Oligocene deposits at El Fayum come large collections of definitive early catarrhine primates, such as *Aegyptopithecus* (see Fleagle 1999, Rasmussen 2002b). They are a good study in what a primitive Old World catarrhine might look like, before the time Cercopithecoidea and Hominoidea separated from one another. This glimpse is brief, however, and limited only to this location. After that, a major void in the fossil record leaves us guessing about the actual split of Cercopithecoidea and Hominoidea, both temporally and geographically.

Despite the fact that living cercopithecoids are relatively successful compared to other primate groups, their fossil record is limited to a few early specimens from the Miocene of Africa (Benefit and McCrossin 2002) and many specimens from the recent past (Pliocene and Pleistocene) in Africa, Europe, and Asia (Jablonski 2002). The fossil record of Cercopithecoidea also tracks a general trend in primate evolution in which hominoid taxonomic diversity dwindled over the last 10 million years while cercopithecoid diversity expanded, perhaps as a result of more temperate climates replacing more tropical ones over large areas of the Old World.

The fossil record of hominoid primates definitively reaches back to a landmark fossil from the early Miocene of the African Rift Valley, *Proconsul* (Harrison 2002). Like *Aegyptopithecus*, *Proconsul* "fits" expectations of an early member of a radiation, in this case the radiation of hominoids. It lacks the apparent dental specializations of the later cercopithecoids, and its skeletal configuration is primitive enough to have given rise to the unusual limb proportions of the living hominoids. The combination of its generalized anatomy and its early discovery in history makes it a strong reference taxon for the wide variety of other early through middle Miocene East African hominoid fossils.

Recent discoveries in Uganda that date to approximately the same time period as *Proconsul* are shedding new light on our sense of the basal hominoids. *Morotopithecus* (MacLatchy 2004) looks less like a base hominoid and more like a derived one, although this statement is qualified by the fact that the entire taxonomic radiation is based on preservation of primitive anatomical traits. The discovery of *Morotopithecus*, much like that of the early catarrhine *Catopithecus*, forces us to reconsider both when hominoids emerged and how much our sense of hominoid evolution has been influenced by thinking of *Proconsul* as its root.

Hominoids dominate the Miocene primate fossil record in East Africa, as exemplified by other "index" taxa such as *Kenyapithecus* (Ward and Duren 2002). Very well preserved specimens of new taxa, such as *Nacholapithecus* (Ishida et al. 2004), ensure that any synthesis of early African hominoid evolution is subject to quick revision. However, even as the diversity of African hominoid fossils continues to grow in the flush period of 20–12 million years ago, the signs of general shrinking of forest cover, proliferation of semiterrestrial monkeys, and decline of hominoid diversity at the close of the Miocene (12–5 million years ago) are unmistakable.

Eurasia

It is somewhat artificial to divide the fossil records of Africa and the rest of the Old World. Hominoids and cercopithecoids radiated widely throughout Europe and Asia shortly after they appear to have emerged in Africa. Their fates there seem to have been similar in the sense that cercopithecoids eventually succeeded and hominoids are now relegated to the still dwindling tropical forest cover in peninsular and insular southeast Asia. The Eurasian record does include one Miocene radiation all its own—the enigmatic pliopithecoids.

Fossils that are hard to interpret nonetheless make excellent study samples for evolutionary biology and systematics. They are hard to interpret, presumably, because they conflict with what either the theoretical basis or the data on hand, or both, lead us to expect. One of the best examples of such a radiation in the primate fossil record is the group of hominoid-like pliopithecoids from the Miocene of Europe and mainland Asia (Begun 2002a).

Pliopithecoids are united to each other almost by the resignation that they lack definitive traits that put them into the ancestry of the "mainstream" hominoids. They were catarrhines that may have radiated out of Africa before the root populations of later hominoids. They clearly found opportunities in the rich Miocene environments of Eurasia, then suffered the same fate as did hominoids when those environments shriveled. At times, pliopithecoids have been considered primitive hominoids. However, as is so common with the fossil record, eventually their presumed descendents were found in equally old deposits and their evolutionary position returned to enigmatic.

Hominoids spread widely throughout the southern latitudes of Europe, Asia, and Southeast Asia beginning about 15 million years ago (Begun 2002b, Kelley 2002). Eurasian hominoids such as *Dryopithecus*, *Ouranopithecus*, and *Sivapithecus* have been promoted as ancestral to every combination of pongids and hominids, suggesting at the very least migrations back and forth to Africa instead of just from it (Kordos and Begun 2002). After a zenith in the middle

Miocene, however, hominoid diversity dwindled ste. until the fossil record virtually disappears at the end of Miocene (Agusti et al. 2003). The living pongids and som Pleistocene fossils are the only evidence of the once predominant radiation.

Recent fossil discoveries of *Dryopithecus* and *Oreopithecus* have demonstrated that ape-like brachiation may date back to the earliest descendants of *Proconsul*, particularly in the form of a very long-armed skeleton of *Dryopithecus* from Europe (Begun 2002b). The locomotor repertoire of Miocene hominoids influences the probable scenarios for the ancestry of our own locomotion, which has long been a central debate in human evolution (Nakatsukasa 2004).

While the Miocene hominoid fossil record is rich in some ways, it is poor in others. Hominoids clearly radiated throughout Asia and into Southeast Asia and neighboring islands, but the critical period between the end of hominoid diversity and the emergence of human-like primates, probably 8–5 million years ago, is poorly sampled.

PHYLOGENIES AND LINEAGE RECONSTRUCTIONS

This book is concerned with the behavior and natural history of living primates. Grounding the different species and families in an evolutionary background helps us to make sense of the similarities and differences across taxa. The relatedness of species groups back through time is their *phylogeny*, a "family tree" in loose terms. Advances in molecular biology have enabled us to resolve primate family trees in a way that comparative anatomy and the fossil record never could.

Evolution of Living Prosimians

Phylogeny

The prosimian primates include the lemuroids, lorisids, galagids, and *Tarsius*. *Tarsius* has a dry rhinarium, as do anthropoids. This trait and other comparisons have led taxonomists to distinguish the Strepsirhini (lemuroids, lorisids, and galagids) from the Haplorhini (*Tarsius* and the anthropoids). Thus, phylogenetically, *Tarsius* may be more closely related to anthropoids than to the strepsirhine primates (Fig. 2.1). The prosimian association used here is based more on a composite sense of their behavioral ecology and niche position relative to anthropoids and other mammals.

The separation of anthropoids, *Tarsius*, and the strepsirhines probably followed shortly after the emergence of primates as a distinct form of early mammal. The fossil record now supports a separation of the lemuroid lineage from the lorisid and galagid lineage as early as 40 million years ago, and fossil tarsiers date from approximately the same time (see above). Molecular estimates push the dates even earlier (Tavaré et al. 2002, Yoder and Yang 2004). With this as a benchmark, it appears that the three prosimian lineages have deeply exclusive evolutionary pathways.

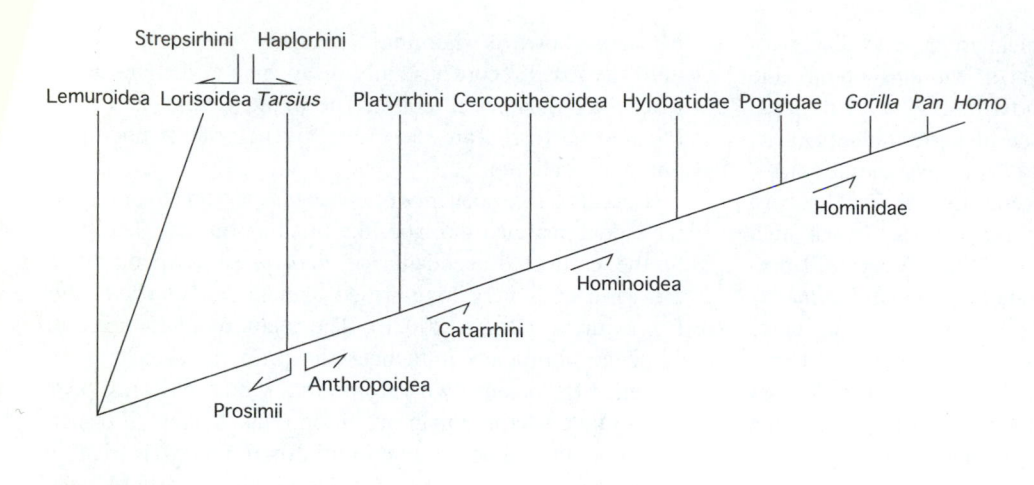

Figure 2.1 A macrophylogeny of living primates. This diagram applies higher taxonomic category terms to a standard cladogram design in order to express conventional interpretations of the relative time at which complementary categories arose or diverged from one another. Implicit in the diagram is the belief that major taxonomic groups of primates (e.g., anthropoids or catarrhines) experienced sequential dichotomous divisions (into platyrrhines and catarrhines as well as cercopithecoids and hominoids, respectively).

Adaptive Radiations

Because they are found only on the island of Madagascar, Lemuroidea is a natural target of evolutionary investigation. They include examples of the smallest living primate, *Microcebus* (30–60 g), and analogs of woodpeckers, rabbits, pandas, raccoons, monkeys, and apes among the ecological profiles of the five major family groups: Cheirogaleidae, Lemuridae, Lepilemuridae, Indriidae, and Daubentoniidae (Fleagle and Reed 2004). In the recent past, they radiated into a spectacular variety of large-animal niches, until climatic change and human arrival drove them to extinction (Godfrey and Jungers 2002).

It is tempting to portray the Malagasy strepsirhines as "living fossils" or as the best examples of primitive primates, given their skeletal anatomy, reliance on olfaction, and relative brain size to body size compared to anthropoids. It is more likely that Lemuroidea became functionally isolated on a large island early in its evolutionary history and differentiated according to the currents of natural selection that have prevailed (see Goodman and Benstead 2003). In the absence of strong competition from other mammals, lemuroid populations have "succeeded" as much as could be predicted (and as much as have the mainland African prosimians or anthropoids in terms of taxonomic diversity). They have paralleled the evolutionary pathway of anthropoids in terms of occupying similar niches on Madagascar, but as a group they have remained below the anthropoid "grade" of neural complexity in doing so.

The strepsirhines of mainland Africa and Asia are the nocturnal Lorisidae and Galagidae. They are found in tropical forest settings, small in body size, and committed to arboreal habitats. Lorisids are found in Africa and Asia, but galagids are found only in Africa. Unlike the Malagasy strepsirhines, they have evolved in direct association with other primate and mammalian competitors. This may explain why they remain nocturnal, solitary, small-bodied, and limited to tropical forests.

Tarsiers (*Tarsius*) are limited geographically to Southeast Asia but play a pivotal role in our understanding of primate evolution and systematics (Wright et al. 2003). They are in

many respects a typical prosimian primate—nocturnal, small-bodied, insectivorous, arboreal, and tropical (Gursky 2002). However, genetically and anatomically, they share special similarities to anthropoids, such as a haplorhine (dry skin) rhinarium, often designated as "advances." For example, although they are nocturnal, their adaptation derives from a retinal fovea characteristic of diurnal primates rather than the reflective eye layer (tapetum lucidum) found in strepsirhine primates. Like most primates, *Tarsius* is highly endangered because it depends on tropical forest for its habitat.

Evolution of New World Monkeys

Phylogeny

Taxonomy and phylogeny weave together well in the Platyrrhini compared to other major primate groups. The phylogenetic integrity of the four major clades is consistent with how the animals have been classified based on morphology, molecules, and ecology. The monophyly of the Atelinae (*Alouatta, Ateles, Lagothrix, Brachyteles*), Pitheciinae (*Pithecia, Chiropotes, Cacajao*), Cebinae (*Cebus, Saimiri*), and Callitrichinae (*Callithrix, Cebuella, Saguinus, Leontopithecus, Callimico*) is supported by numerous genetic and morphological studies (Schneider et al. 2001). The "problem" taxa in phylogenetic reconstructions, both morphological and molecular, consistently have been *Callicebus* and *Aotus*. In this volume, *Callicebus* is included in the Pitheciinae and *Aotus* is placed in its own subfamily (Aotinae).

Adaptive Radiations

The tight fit of the genera within their subfamilies and the relatively deep genetic distance between subfamilies suggest that modern platyrrhines are the result of a rapid burst of evolutionary differentiation followed by sustained gradual change. As with the Malagasy strepsirhines, it seems likely that once the ancestral population of platyrrhines occupied the isolated land mass of South America, they radiated quickly and widely into distinct basic niches. From there, the lineages have evolved subspecialties within those super-niches, preserving a close genetic relationship (see Norconk et al. 1996).

All platyrrhines depend on trees. They have evolved to exploit different nutrient zones within the tropical forest as well as different vertical habitats within the same zones. Some genera, such as *Cebus* and *Alouatta*, have dispersed widely but for different reasons. *Alouatta* forages on low-quality foliage and so can range into more marginal forested habitats. *Cebus* is the most social and "cognitive" of the New World monkeys and may range widely because of its very flexible and fluid resource exploitation strategies. Other genera, such as *Brachyteles*, found only in the narrow Atlantic Coastal Forest, are helpless victims of specific habitats that are diminishing around them. Platyrrhines are relatively small anthropoids, ranging in body size from the smallest of all anthropoids (*Cebuella*, the pygmy marmoset, at slightly over 100 g) to the 7–10 kg atelines.

The atelines are the large-bodied New World monkeys that occupy the uppermost canopies of the forest and exploit ripe fruits and leaves at the periphery. These conditions have led to their unique limb proportions for suspensory posture and brachiating locomotion—much like some of the hominoids. The pitheciines have radiated into a hard object–feeding niche that Rosenberger (1992) refers to as "sclerocarp foraging"—working through the tough outer coat of certain fruits and seeds to get to the nutritious pulp on the inside. The callitrichines have radiated into a lower canopy habitat zone better suited to small body size, vertical clinging and leaping, and, in the case of *Callithrix*, specialized foraging on nectar and tree gum. The cebines are omnivorous, highly social, and significantly more encephalized than are other New World monkeys (Hartwig 1996).

Evolution of Old World Monkeys

Phylogeny

The two major clades of Cercopithecoidea, the Cercopithecinae and the Colobinae, are each monophyletic as determined both morphologically and molecularly (Hoelzer et al. 2004). In this case, the molecular data have revealed unexpected close relationships between species groups that do not share obvious morphological similarities. For example, the West African mangabeys were thought to be a monophyletic assemblage of two species groups, *Cercocebus* and *Lophocebus*. Molecular studies and revisited morphological studies confirm, however, that the *Cercocebus* group is more closely related to the West African drills and mandrills (*Mandrillus*), long thought to be near relatives of the papionins (baboons). Likewise, the *Lophocebus* group is more closely related to the papionins than it is to the other *Cercocebus* group (see Fleagle and McGraw 1999, 2002). The unfolding of these relationships is one of the best examples of why genetic reconstructions of phylogeny should be the bases upon which morphological studies construct and test hypotheses.

Adaptive Radiations

The cercopithecoids have exploited the more generalis, habitat zones of both the tropical forest belts and the outlying savannas of Africa and Asia, as reflected in their classic quadrupedal skeletal profile that enables them to move effectively in both terrestrial and arboreal habitats. They are arguably the most successful primate radiation in terms of geographic distribution, numbers of species, and population sizes (see Whitehead and Jolly 2000).

At some point shortly after the divergence of catarrhines, the cercopithecoids diverged along two basic adaptive pathways. The cercopithecine Old World monkeys, exemplified by *Papio* and *Macaca*, stayed the course of the "generalized primate" in terms of anatomy and resource exploitation. They are the most widely dispersed of all primates. The colobines, by contrast, derived into a superniche based on more concentrated exploitation of local leafy food materials. As exemplified by *Colobus* and *Presbytis*, they are more committed to arboreal habitats, in general. They have speciated widely in the lush Southeast Asian and insular Asian tropical forests.

Evolution of Hominoids

The living pongids and hylobatids are instantly recognizable: chimpanzees, bonobos, gorillas, orangutans, and gibbons. However, these complex primates are just a fraction of the hominoid diversity that once stretched throughout the Old World tropics. This evolutionary context is key to understanding why they are so habitat-dependent and adaptively "overextended." In each case, the degree of specialization necessary to survive as a large-bodied, arboreal, highly cognitive, mostly frugivorous primate has cost them the anatomical variation and rate of reproductive turnover necessary to survive the loss of their immediate habitat.

Phylogeny

The basic assumption that hylobatids were the first branch of the hominoid family tree, *Pongo* the second, then some form of African pongid and human trichotomy has dominated depictions of hominoid phylogeny for several decades (Goodman et al. 2002). Molecular studies confirm that the Asian hominoids are less closely related to humans than are the African hominoids. Within the African pongid and hominid clade, the preponderance of data suggest that *Homo* and *Pan* are more closely related to each other than either clade is to *Gorilla*. This implies that morphological complexes thought to be synapomorphic between the African pongids, such as the knuckle-walking complex, either evolved in parallel between them or was present in the hominid ancestor and then reversed completely (e.g., Begun 2004). Indeed, the important lesson of advances in molecular genetics is that anatomical traits probably converged, evolved in parallel, and reversed much more often than we ever suspected.

Adaptive Radiations

Within the bounds of the general ape superniche, the hylobatids (gibbons and siamangs) have derived ultra-specializations for a brachiating lifestyle in the dense tropical forests of Southeast Asia. They effectively have lost the grasping ability of their hands in order to swing aggressively from their elongated upper limbs. They are the most speciose of the hominoids, with two major clades—*Hylobates* and *Symphalangus*—derived along subtle differences in dietary emphasis on ripe fruit and leaves.

Pongo is the other Asian hominoid and is restricted to the islands of Borneo and Sumatra, undoubtedly due to the aggravating effects of human encroachment. *Pongo* is virtually omnivorous, climbs quadrumanously, and is perhaps the most vertically perambulatory hominoid (Knott 1999). Its radiation would dominate the Asian tropical scene were it not for the faster-reproducing and less specialized cercopithecoids that radiated along with it.

The vestigial hominoids of Africa (excluding humans) are *Pan paniscus*, *P. troglodytes*, and *Gorilla*. As cognitive and complex as they are behaviorally, both *Pan* and *Gorilla* are on the rapid path to extinction. They are what remains after millennia of competition with African cercopithecoids, saddled the whole time with a slow reproductive rate and a morphological specialization for a life spent in or at least among the trees.

Gorilla probably represents the surviving hominoid that specialized into a mostly herbivorous, deep forest, reclusively arboreal niche (Taylor and Goldsmith 2003). *Pan* probably represents the surviving hominoid that stayed the course of moderate body size, multivory, and anatomical specialization to life in the trees. Out of this picture emerged humans, of course, probably as the surviving hominoid that exploited the patchy ecotones on the ever-expanding fringe of the forest, eventually specializing for a novel form of bipedal locomotion. As of now, the options were more productive for the human lineage and significantly more restrictive for *Pan* and *Gorilla*.

A SYNTHESIS

Through the fossil record, comparative anatomy, and molecular biology, we can reconstruct major aspects of how primates evolved and how living lineages relate to one another. The dates and places ascribed to major events in primate evolution continue to change as we collect more data and refine our analyses, but the broad patterns remain. Plurality rules the primate story. Primates exploit plural habitats, omnivorous diets, and fluid social group dynamics. A highly specialized primate species seems even more odd as we learn more about it, and the variation we have observed in other species expands by the same measure of further study.

In general, primates are a radiation of mammals in which growth of structures related to the brain has taken precedence over growth of structures related to the body.

Comparatively speaking, other lineages of mammals have "spent" their genetic variation on more specialized biological adaptations—complex anatomies for preying upon their food and/or to avoid becoming prey. For primates the evolutionary path has stayed a general course in which neural development, rather than somatic, has characterized the emergence of new taxa (see Bush and Allman 2004). In the broadest possible context, the later a lineage appears, the more encephalized it and its descendents appear to be.

Primates have evolved to exploit rich but ephemeral food resources in well-protected tropical forest microhabitats, such as terminal branches and emergent canopies. From this core superniche, primates radiated throughout the tropical zones of Africa, Eurasia, and the Americas over the last 50 million years. Within each lineage, more specialized species groups radiated into narrower niches as determined by the balance of opportunity and competition. Madagascar offered more opportunity and less competition to lemuroid strepsirhines, for example, than did continental Africa to lorisid and galagid strepsirhines. The generalized species groups, such as *Macaca* and *Papio*, have survived major biome shifts in forest cover and thrived in the balance. Hominoids, dominant for millions of years in a stable forest milieu, enjoyed almost runaway selection for cognition, learning, and extended life histories—all luxuries of a certain independence of the predator–prey cycle. In the face of a constrained milieu and unable to undo their life history pattern, hominoids evolved as expected: some lineages went extinct, some linger in the fragile remains of the original niche, and one abandoned its biological capacity completely in favor of its cognitive potential.

REFERENCES

Agusti, J., de Siria, A. S., and Garces, M. (2003). Explaining the end of the hominoid experiment in Europe. *J. Hum. Evol.* 45:145–153.

Arnason, U., Gullberg, A., and Janke, A. (1998). Molecular timing of primate divergences as estimated by two nonprimate calibration points. *J. Mol. Evol.* 47:718–727.

Beard, K. C. (2002). Basal anthropoids. In: Hartwig, W. C. (ed.), *The Primate Fossil Record*. Cambridge University Press, Cambridge. pp. 133–150.

Beard, K. C., and Wang, J. (2004). The eosimiid primates (Anthropoidea) of the Heti Formation, Yuanqu Basin, Shanxi and Henan Provinces, People's Republic of China. *J. Hum. Evol.* 46:401–432.

Begun, D. R. (2002a). Pliopithecoidea. In: Hartwig, W. C. (ed.), *The Primate Fossil Record*. Cambridge University Press, Cambridge. pp. 221–240.

Begun, D. R. (2002b). European hominoids. In: Hartwig, W. C. (ed.), *The Primate Fossil Record*. Cambridge University Press, Cambridge. pp. 339–368.

Begun, D. R. (2004). Knuckle-walking and the origin of human bipedalism. In: Meldrum, D. J., and Hilton, C. E. (eds.), *From Biped to Strider: The Emergence of Modern Human Walking, Running and Resource Transport*. Kluwer, New York. pp. 9–33.

Benefit, B. R., and McCrossin, M. L. (2002). The Victoriapithecidae, Cercopithecoidea. In: Hartwig, W. C. (ed.), *The Primate Fossil Record*. Cambridge University Press, Cambridge. pp. 241–253.

Bloch, J. I., and Boyer, D. M. (2002). Grasping primate origins. *Science* 298:1606–1610.

Bush, E. C., and Allman, J. M. (2004). The scaling of frontal cortex in primates and carnivores. *Proc. Natl. Acad. Sci. USA* 101:3962–3966.

Cartelle, C., and Hartwig, W. C. (1996). A new extinct primate among the Pleistocene megafauna of Bahia, Brazil. *Proc. Natl. Acad. Sci. USA* 93:6405–6409.

Cartmill, M. (1974). Rethinking primate origins. *Science* 184:436–443.

Cartmill, M. (1992). New views on primate origins. *Evol. Anthropol.* 1:105–111.

Ciochon, R. L., and Gunnell, G. F. (2002). Chronology of primate discoveries in Myanmar: influences on the anthropoid origins debate. *Ybk. Phys. Anthropol.* 45:2–35.

Covert, H. H. (2002). The earliest fossil primates and the evolution of prosimians: introduction. In: Hartwig, W. C. (ed.), *The Primate Fossil Record*. Cambridge University Press, Cambridge. pp. 13–20.

Dagosto, M. (2002). The origin and diversification of anthropoid primates. In: Hartwig, W. C. (ed.), *The Primate Fossil Record*. Cambridge University Press, Cambridge. pp. 125–132.

Eizirik, E., Murphy, W. J., and O'Brien, S. J. (2001). Molecular dating and biogeography of the early placental mammal radiation. *J. Hered.* 92:212–219.

Fleagle, J. G. (1999). *Primate Adaptation and Evolution*, 2nd ed. Academic Press, San Diego.

Fleagle, J. G., and Kay, R. F. (eds.) (1994). *Anthropoid Origins*. Plenum Press, New York.

Fleagle, J. G., and McGraw, W. S. (1999). Skeletal and dental morphology supports diphyletic origin of baboons and mandrills. *Proc. Natl. Acad. Sci. USA* 96:1157–1161.

Fleagle, J. G., and McGraw, W. S. (2002). Skeletal and dental morphology of African papionins: unmasking a cryptic clade. *J. Hum. Evol.* 42:267–292.

Fleagle, J. G., and Reed, K. E. (2004). The evolution of primate ecology: patterns of geography and phylogeny. In: Anapol, F., German, R. Z., and Jablonski, N. (eds.), *Shaping Primate Evolution: Form, Function and Behavior*. Cambridge University Press, New York. pp. 353–367.

Fleagle, J. G., and Tejedor, M. F. (2002). Early platyrrhines of southern South America. In: Hartwig, W. C. (ed.), *The Primate Fossil Record*. Cambridge University Press, Cambridge. pp. 161–174.

Foote, M., Hunter, J. P., Janis, C. M., and Sepkoski, J. J., Jr. (1999). Evolutionary and preservational constraints on origins of biologic groups: divergence times of eutherian mammals. *Science* 283:1310–1314.

Gebo, D. L. (2002). Adapiformes: phylogeny and adaptation. In: Hartwig, W. C. (ed.), *The Primate Fossil Record*. Cambridge University Press, Cambridge. pp. 21–44.

Godfrey, L. R., and Jungers, W. L. (2002). Quaternary fossil lemurs. In: Hartwig, W. C. (ed.), *The Primate Fossil Record*. Cambridge University Press, Cambridge. pp. 97–121.

Goodman, M., McConkey, E. H., and Page, S. L. (2002). Reconstructing human evolution in the age of genomic exploration. In: Harcourt, C. S., and Sherwood, B. R. (eds.), *New Perspectives in Primate Evolution and Behavior*. Westbu_ Publishing, Otley, UK. pp. 47–70.

Goodman, S. M., and Benstead, J. P. (eds.) (2003). *The Natural History of Madagascar*. University of Chicago Press, Chicago.

Graur, D., and Martin, W. (2004). Reading the entrails of chickens: molecular timescales of evolution and the illusion of precision. *Trends in Genetics*. 20:80–86.

Gunnell, G. F., and Rose, K. D. (2002). Tarsiiformes: evolutionary history and adaptation. In: Hartwig, W. C. (ed.), *The Primate Fossil Record*. Cambridge University Press, Cambridge. pp. 45–82.

Gursky, S. (2002). The behavioral ecology of the spectral tarsier, *Tarsius spectrum*. *Evol. Anthropol.* 11:226–234.

Hamrick, M. A. (2001). Primate origins: evolutionary change in digital ray patterning and segmentation. *J. Hum. Evol.* 40:339–351.

Harrison, T. (2002). Late Oligocene to middle Miocene catarrhines from Afro-Arabia. In: Hartwig, W. C. (ed.), *The Primate Fossil Record*. Cambridge University Press, Cambridge. pp. 311–338.

Hartwig, W. C. (1994). Patterns, puzzles and perspectives on platyrrhine origins. In: Corruccini, R. S., and Ciochon, R. L. (eds.), *Integrative Paths to the Past: Paleoanthropological Essays in Honor of F. Clark Howell*. Prentice-Hall, Englewood Cliffs, NJ. pp. 69–94.

Hartwig, W. C. (1996). Perinatal life history traits in New World monkeys. *Am. J. Primatol.* 40:99–130.

Hartwig, W. C. (ed.) (2002). *The Primate Fossil Record*. Cambridge University Press, Cambridge.

Hartwig, W. C., and Cartelle, C. (1996). A complete skeleton of the giant South American primate *Protopithecus*. *Nature* 381:307–311.

Hartwig, W. C., and Meldrum, D. J. (2002). Miocene platyrrhines of the northern neotropics. In: Hartwig, W. C. (ed.), *The Primate Fossil Record*. Cambridge University Press, Cambridge. pp. 175–188.

Hoelzer, G. A., Morales, J. C., and Melnick, D. J. (2004). Dispersal and population genetics of primate species. In: Chapais, B., and Berman, C. M. (eds.), *Kinship and Behavior in Primates*. Oxford University Press, Oxford. pp. 109–131.

Horovitz, I., and MacPhee, R. D. E. (1999). The Quaternary Cuban platyrrhine *Paralouatta varonai* and the origin of Antillean monkeys. *J. Hum. Evol.* 36:33–68.

Ishida, H., Kunimatsu, Y., Takano, T., Nakano, Y., and Nakatsukasa, M. (2004). *Nacholapithecus* skeleton from the middle Miocene of Kenya. *J. Hum. Evol.* 46:69–103.

Jablonski, N. G. (2002). Fossil Old World monkeys: the late Neogene radiation. In: Hartwig, W. C. (ed.), *The Primate Fossil Record*. Cambridge University Press, Cambridge. pp. 255–299.

Kay, R. F. (2003). Review of the primate fossil record. *Am. J. Hum. Biol.* 15:839–840.

Kay, R. F., Madden, R. H., Cifelli, R. L., and Flynn, J. J. (eds.) (1997). *Vertebrate Paleontology in the Neotropics*. Smithsonian Institution Press, Washington DC.

Kelley, J. (2002). The hominoid radiation in Asia. In: Hartwig, W. C. (ed.), *The Primate Fossil Record*. Cambridge University Press, Cambridge. pp. 369–384.

Knott, C. J. (1999). Orangutan behavior and ecology. In: Dolhinow, P., and Fuentes, A. (eds.), *The Nonhuman Primates*. Mayfield Press, Mountain View, CA. pp. 50–57.

Kordos, L., and Begun, D. R. (2002). Rudabanya: a late Miocene subtropical swamp deposit with evidence of the origin of the African apes and humans. *Evol. Anthropol.* 11:45–57.

Lemelin, P. (1999). Morphological correlates of substrate use in didelphid marsupials: implications for primate origins. *J. Zool.* 247:165–175.

MacLatchy, L. (2004). The oldest ape. *Evol. Anthropol.* 13:90–103.

MacPhee, R. D. E., and Horovitz, I. (2002). Extinct Quaternary platyrrhines of the Greater Antilles and Brazil. In: Hartwig, W. C. (ed.), *The Primate Fossil Record*. Cambridge University Press, Cambridge. pp. 189–199.

MacPhee, R. D. E., and Horovitz, I. (2004). New craniodental remains of the Quaternary Jamaican monkey *Xenothrix mcgregori* (Xenotrichini, Callicebinae, Pitheciidae), with a reconsideration of the *Aotus* hypothesis. *Am. Mus. Novit.* 3434:1–51.

Madsen, O., Scally, M., Douady, C. J., Kao, D. J., DeBry, R. W., Adkins, R., Amrine, H. M., Stanhope, M. J., de Jong, W. W., and Springer, M. S. (2001). Parallel adaptive radiations in two major clades of placental mammals. *Nature* 409:610–614.

Marivaux, L., Chaimanee, Y., Ducrocq, S., Marandat, B., Sudre, J., Soe, A. N., Tun S. T., Htoon, W., and Jaeger, J. J. (2003). The anthropoid status of a primate from the late middle Eocene Pondaung Formation (central Myanmar): tarsal evidence. *Proc. Natl. Acad. Sci. USA* 100:13173–13178.

Martin, R. D. (1993). Primate origins: plugging the gaps. *Nature* 363:223–234.

Martin, R. D. (2003). Combing the primate record. *Nature* 422:388–391.

Murphy, W. J., Eizirik, E., Johnson, W. E., Zhang, Y. P., Ryder, O. A., and O'Brien, S. J. (2001). Molecular phylogenetics and the origins of placental mammals. *Nature* 409:614–618.

Nakasukasa, M. (2004). Acquisition of bipedalism: the Miocene hominoid record and modern analogues for bipedal protohominids. *J. Anal.* 204:385–402.

Ni, X., Wang, Y., Hu, Y., and Li, C. (2004). A euprimate skull from the early Eocene of China. *Nature* 427:65–68.

Norconk, M. A., Rosenberger, A. L., and Garber, P. A. (1996). *Adaptive Radiations of Neotropical Primates*. Plenum Press, New York.

Rasmussen, D. T. (1990). Primate origins: lessons from a neotropical marsupial. *Am. J. Primatol.* 22:263–278.

Rasmussen, D. T. (2002a). The origin of primates. In: Hartwig, W. C. (ed.), *The Primate Fossil Record*. Cambridge University Press, Cambridge. pp. 5–10.

Rasmussen, D. T. (2002b). Early catarrhines of the African Eocene and Oligocene. In: Hartwig, W. C. (ed.), *The Primate Fossil Record*. Cambridge University Press, Cambridge. pp. 203–220.

Ravosa, M. J., and Savakova, D. G. (2004). Euprimate origins: the eyes have it. *J. Hum. Evol.* 46:357–364.

Rosenberger, A. L. (1977). *Xenothrix* and ceboid phylogeny. *J. Hum. Evol.* 6:461–481.

Rosenberger, A. L. (1992). Evolution of feeding niches in New World monkeys. *Am. J. Phys. Anthropol.* 88:525–562.

Rosenberger, A. L. (2002). Platyrrhine paleontology and systematics: the paradigm shifts. In: Hartwig, W. C. (ed.), *The Primate Fossil Record*. Cambridge University Press, Cambridge. pp. 151–160.

Sargis, E. J. (2002). Primate origins nailed. *Science* 298:1564–1565.

Sargis, E. J. (2004). New views on tree shrews: the role of tupaiids in primate supraordinal relationships. *Evol. Anthropol.* 13:56–66.

Schneider, H., Canavez, F. C., Sampaio, I., Moreira, F. A. M., Tagliaro, C. H., and Seuanez, H. N. (2001). Can molecular data place each neotropical monkey in its own branch? *Chromosoma* 109:515–523.

Seiffert, E. R., Simons, E. L., and Attia, Y. (2003). Fossil evidence for an ancient divergence of lorises and galagos. *Nature* 422:421–424.

Shigehara, N., Takai, M., Kay, R. F., Aung, A. K., Soe, A. N., Tun, S. T., Tsubamoto, T., and Thein, T. (2002). The upper dentition and face of *Pondaungia cotteri* from central Myanmar. *J. Hum. Evol.* 43:143–166.

Simons, E. L., and Rasmussen, D. T. (1994). A whole new world of ancestors: Eocene anthropoideans from Africa. *Evol. Anthropol.* 3:128–139.

Springer, M. S., Murphy, W. J., Eizirik, E., and O'Brien, S. J. (2003). Placental mammal diversification and the Cretaceous–Tertiary boundary. *Proc. Natl. Acad. Sci.* 100:1056–1061.

Sussman, R. (1991). Primate origins and the evolution of angiosperms, *Am. J. Primatol.* 23:209–223.

Szalay, F. S., and Delson, E. (1979). *Evolutionary History of the Primates*. Academic Press, New York.

Takai, M., and Anaya, F. (1996). New specimens of the oldest fossil platyrrhine, *Branisella boliviana*, from Salla, Bolivia. *Am. J. Phys. Anthropol.* 99:301–317.

Takai, M., Anaya, F., Shigehara, N., and Setoguchi, T. (2000). New fossil materials of the earliest New World monkey, *Branisella boliviana*, and the problem of platyrrhine origins. *Am. J. Phys. Anthropol.* 111:263–282.

Takai, M., Setoguchi, T., and Shigehara, N. (2003). New aotine fossils from the middle Miocene of La Venta, Colombia. *Am. J. Phys. Anthropol.* 35 (suppl.):205.

Takai, M., Shigehara, N., Aung, A. K., Tun, S. T., Soe, A. N., Tsubamoto, T., and Thein T. (2001). A new anthropoid from the latest middle Eocene of Pondaung, central Myanmar. *J. Hum. Evol.* 40:393–409.

Tavaré, S., Marshall, C. R., Will, O., Soligo, C., and Martin, R. D. (2002). Using the fossil record to estimate the age of the last common ancestor of extant primates. *Nature* 416:726–729.

Taylor, A. B., and Goldsmith, M. A. (2003). *Gorilla Biology: A Multidisciplinary Perspective*. Cambridge University Press, New York.

Tejedor, M. F. (2002). Primate canines from the early Miocene Pinturas Formation, southern Argentina. *J. Hum. Evol.* 43:127–141.

Tejedor, M. F. (2003). New fossil primate from Chile. *J. Hum. Evol.* 44:515–520.

Ward, S. C., and Duren, D. L. (2002). Middle and late Miocene African hominoids. In: Hartwig, W. C. (ed.), *The Primate Fossil Record*. Cambridge University Press, Cambridge. pp. 385–397.

Whitehead, P. F., and Jolly, C. J. (eds.) (2000). *Old World Monkeys*. Cambridge University Press, Cambridge.

Wright, P. C., Simons, E. L., and Gursky, S. (eds.) (2003). *Tarsiers: Past, Present and Future*. Rutgers University Press, New Brunswick, NJ.

Yoder, A. D., and Yang, Z. H. (2004). Divergence dates for Malagasy lemurs estimated from multiple gene loci: geological and evolutionary context. *Mol. Ecol.* 13:757–773.

PART TWO

The Primates

3

The Lorisiform Primates of Asia and Mainland Africa
Diversity Shrouded in Darkness
Anna Nekaris and Simon K. Bearder

INTRODUCTION

The primates known as galagos (or bushbabies), pottos (angwantibos and pottos), and lorises could easily vie for the position of "least known of all the primates." Despite the fact that the suborder Lorisiformes contains some of the most specialized primates, with a minimum of 34 species now recognized, some irresistible urge seems to possess the authors of textbooks to summarize what is known of this group in a hasty postscript to a chapter on their close cousins, the lemurs. One reason for this is that, unlike most lemurs, different taxa of lorisiforms can look very similar to each other (*cryptic* species), and for a long time they were misclassified as a few species and assumed to have little variation in genetics, behavior, and ecology. It is now known that superficial similarities are partly the result of extensive convergence due to the demands of nocturnal and arboreal niches and partly because members of each species recognize each other by more subtle visual, vocal, and olfactory signals. In this chapter, we intend to show that the nocturnal strepsirhines of Asia and Africa are a diverse group of primates and represent an untapped resource for the aspiring field biologist. One-sentence synopses, steeped in the literature of the 1960s, branding this enigmatic group as no more than acrobatic leapers and slow creepers (e.g., MacDonald 2001) seem to have hindered interest in their study in the wild, even though early biologists recognized great variability within this group (e.g., Gray 1863). The lorisiforms display a multitude of social systems, life histories, and locomotor strategies, a diversity evident despite the fact that only a handful of species have been studied in detail.

The lorisiform primates are widely dispersed in Africa (excluding Madagascar), southern Asia, and Southeast Asia. The relatively few long-term field studies that have been published on the galagos, pottos, and lorises are summarized in Table 3.1. Detailed behavior and ecological data are available for only 16 species, fewer than half of those currently recognized. In some cases where researchers have set out to study behavior, their projects were confounded by the discovery of too many new species (e.g., Honess 1996, Ambrose 1999). Instead, these studies have led to extensive useful descriptions of the presence/absence of species across a large geographical range, with morphological data gathered from trapping regimes (Oates and Jewell 1967; Honess 1996; Ambrose 1999; Perkin 2000, 2001a,b, 2002; Perkin et al. in press). Furthermore, despite advances in radio tracking, only nine species have been studied with this technology (Table 3.1) and only two studies have been able to take advantage of recent advances in molecular ecology (Pullen 2000, Pimley 2002). Clearly, an enormous avenue for research exists within this group.

Even what might appear to be the most fundamental questions regarding the evolutionary relationships among this group are far from resolved (Rasmussen and Nekaris 1998). For example, no consensus has yet been reached as to whether the pottos and lorises form a monophyletic clade to the exclusion of the galagos or if they form one of the most spectacular examples of parallel evolution among primates (Yoder et al. 2001b, Roos et al. 2004). Recent fossil discoveries have added new vigor to debates regarding the origins of the lorisiforms. Some authors contend that they may be among the most ancient of the living primates, with origins extending back to the Eocene (Seiffert et al. 2003, Martin 2003). Others propose an Asian origin for the Malagasy strepsirhines, with the deepest evolutionary relationships existing between the lemurs and lorises (Martin 2000, 2003; Tavare et al. 2002); yet another contrary view is that the lemurs are most closely related to the galagos (Charles-Dominique and Martin 1970, Roos et al. 2004). New molecular data have opened up questions about the genetic relationships between species. Mitochondrial deoxyribonucleic acid (DNA) analysis, for example, indicates that the galagos are not a single group of close relatives that have undergone recent speciation but can be divided into four deeply rooted clades which diverged over 30 million years ago (Bayes 1998). Further details of evolutionary relationships among galagos have been explored using comparisons of red blood cell enzymes (Masters et al. 1994,

Table 3.1   Taxonomy and Conservation Status

LATIN NAME	COMMON NAME	HABITAT	DISTRIBUTION	POPULATION DENSITY[1]	IUCN RED LIST STATUS[2]
Galaginae					
Galagoides demidovii	Demidoff's dwarf	Understory/forest edge	Bioko, Cameroon, Gabon, Ivory Coast, Nigeria, Uganda	0.16/hr and 50–80/km²	Not listed
G. thomasi	Thomas's dwarf	Forest/mid- to high canopy	Bioko, Cameroon, Gabon, Ivory Coast, Nigeria, Uganda	0.46–2.0/hr and 50–80/km²	Not listed
G. orinus	Mountain dwarf	Submontane–montane forest/ mid- to high canopy	Tanzania	0.4/hr, 2.7–5.4/hr	Endangered
G. zanzibaricus (*udzungwensis*)	Zanzibar lesser	Secondary forest/mid- to high canopy	Tanzania	12.0/hr	Endangered
G. rondoensis	Rondo dwarf	Cloud coastal forest/understory	Tanzania	3–6/hr, 3–10/hr	Endangered
G. sp. nov. 3	Ukinga or Rungwe dwarf	Montane forest	Tanzania		Critically endangered
G. cocos	Kenya coastal	Coastal forest/middle story	Kenya, Tanzania	170–180/km²	Not listed
G. granti	Mozambique lesser	Coastal forest/middle story	Tanzania		Data deficient, unknown trend
G. nyasae	Malawi lesser	Woodland	Malawi		Not listed
G. sp. nov. 1	Kalwe lesser	Forest/middle story	Malawi		Not listed
G. sp. nov. 2	Mt. Thyolo lesser	Montane forest	Malawi		Not listed
Galago senegalensis	Senegal lesser	Miombo, acacia woodland to forest/all strata	Senegal to Kenya	0.03–0.67/hr	Not listed
G. gallarum	Somali lesser	Acacia woodland, thicket/all strata	Kenya	1.0/hr	Low risk, trend unknown
G. moholi	Southern lesser	Acacia woodland/all strata	Botswana, Malawi, Namibia, South Africa, Tanzania		Not listed
G. matschiei	Spectacled	Forest/all strata	Uganda		Low risk, trend unknown
Euoticus elegantulus	Southern needle-clawed	Forest/mid- to high canopy	Cameroon, Gabon	15–20/km²	Low risk, trend unknown
E. pallidus	Northern needle-clawed	Forest/mid- to high canopy	Bioko, Cameroon	0.25/hr	Low risk, trend unknown
Sciurocheirus alleni	Allen's squirrel	Forest, forest edge/mid- to understory	Bioko, Cameroon	15/km²	Not listed
S. gabonensis	Gabon squirrel	Forest/mid- to understory	Cameroon, Gabon	15–20/km²	Low risk, trend unknown
Sciurocheirus sp. nov.	Makande squirrel	Forest/mid- to understory	Gabon		Not listed
Otolemur garnettii	Garnett's (small-eared) greater	Forest, farmland plantation/ mid- to high canopy	Kenya, Tanzania		Not listed
O. crassicaudatus	Thick-tailed greater	Woodland and forest edge/ mid- to high canopy	Malawi, South Africa, Tanzania, Zimbabwe		Not listed
O. monteiri	Silver greater	Woodland/unknown	Kenya		Not listed
Otolemur sp. nov.	Mwera (pygmy) greater	Woodland, farmland, plantation/mid- to high canopy	Tanzania		Not listed
Perodicticinae					
Perodicticus potto potto	Western potto	Secondary colonizing or flooded primary forest	Guinea, Guinea Bissau, Nigeria	?	Data deficient/not listed
P. p. edwardsi	Milne-Edwards or central potto	Swamp, lowland, mid-altitude montane rain forest	Nigeria, Zaire, Central African Republic	8–10/km², 4.7/km²	Data deficient/not listed
P. p. juju	S. Nigerian potto	Forest edge	Guinea Coast of Nigeria	?	Data deficient/not listed
P. p. faustus		Riverine forest	Congo Basin	?	Data deficient/not listed
P. p. ibeanus	Bosman's or eastern potto	Semimoist deciduous forest	Zaire, Burundi, Rwanda	0.04–0.26/hr and 1.8–17.7/km²	Data deficient/not listed
Arctocebus aureus	Golden angwantibo	Tree fall zones, forest edge, understory	Gabon	2/km²	Low risk, trends unknown
A. calabarensis	Calabar angwantibo	Tree fall zones, forest edge, understory	Cameroon, Gabon, Congo	0.7/km²	Low risk, trends unknown

Table 3.1   (cont'd)

LATIN NAME	COMMON NAME	HABITAT	DISTRIBUTION	POPULATION DENSITY[1]	IUCN RED LIST STATUS[2]
Lorisinae					
Loris lydekkerianus lydekkerianus	Mysore slender loris	Dry forest, acacia scrub jungle	South India	0.13–3.6/km² or 28/km²	Near threatened
L. l. malabaricus	Malabar slender loris	Rain forests, coastal forests	South India	?	Near threatened
L. l. nordicus	Northern Ceylon gray slender loris	Low-country dry zone, scrub forest, grassland	Sri Lanka	0.33–50/km²	Endangered
L. l. grandis	Highland Ceylon slender loris	Montane forest mixed with patana grassland	Sri Lanka	0.11–3.3/km²	Endangered
L. tardigradus tardigradus	Western Ceylon red slender loris	Lowland rain forest, intermonsoon forest	Sri Lanka	0.86–13/km²	Endangered
L. t. nycticeboides	Horton Plains slender loris	Montane rain and mist forests	Sri Lanka	0.08–0.16/km²	Critically endangered
Nycticebus bengalensis	Bengal slow loris	Bamboo forest mixed with hardwood trees, farmbush, mangrove swamps	Burma, Cambodia, China, India, Laos, Thailand, Vietnam	?	Data deficient, unknown trend
N. coucang	Greater slow loris	Tropical rain forest with continuous canopy	Sumatra, peninsular Malaysia, Thailand	?	Not listed
N. javanicus	Javan slow loris	Unknown	Indonesian Java	?	Data deficient, unknown trend
N. menagensis	Bornean slow loris	Unknown	Brunei, Indonesia, Malaysia	?	Not listed
N. pygmaeus	Pygmy slow loris	Bamboo forest mixed with hardwood trees, forest edge, dense scrub	Cambodia, China, Laos, Vietnam	8 seen during several night walks	Vulnerable, trend decreasing

[1] Because different survey methods were employed, some population densities are per kilometer squared, some are per kilometer, and some are a rate of animal encounters per hour.
[2] IUCN, World Conservation Union.

Masters and Brothers 2002) and highly repeated DNA sequences (Crovella et al. 1994, DelPero et al. 2000), which led to a new appreciation of the age and extent of their divergence. Finally, the Asian lorises have long been regarded as essential to understanding questions regarding the evolution of primate characteristics due to their having the greatest degree of orbital convergence of all primates (Cartmill 1972, Ross 1996). Only recently has the ecological significance of their visual adaptations been tested in the field (Nekaris in press, Bearder et al. in press).

With such a potential for discovery within this group, we hope that this summary of what is now known about lorisiform behavior and ecology will stimulate a new era of research. Our chapter contains more gaps than it provides answers, but these indicate a new direction for research on the strepsirhine primates in light of long-term field studies and surveys. Such research is urgently needed to help ensure their future protection.

TAXONOMY AND DISTRIBUTION

The Lorisiforms

Lorisiformes along with Lemuriformes of Madagascar comprise the infraorder Strepsirhini (Martin 1990). Although sometimes also classified as Prosimii along with the tarsiers, the strepsirhine primates are linked by a number of unique morphological traits, making them a monophyletic group to the exclusion of tarsiers, monkeys, and apes (Haplorhini). These traits include a moist nose, unfused mandibular symphasis and frontal bone, reduced upper incisors, a sloping talofibular facet (groove between the ankle and one of the lower limb bones), and a single grooming claw on the second digit of each hind foot. Living strepsirhines are further united by the possession of a *toothcomb*—a forward-pointing dental structure comprised of the lower incisors and canines used for both dietary and grooming purposes (Fleagle 1999).

Nocturnal primate taxonomy in general has gone through intense revision in the last few years (e.g., Nietsch and Kopp 1998, Yoder et al. 2001a, Pastorini et al. 2003). However, while the revision of lemur and tarsier taxonomy and the subsequent re-evaluation of their conservation status seem to have been accepted readily by the scientific community, the discovery of enormous taxonomic diversity among lorisiform primates has met with the same skepticism as acknowledgment of the diverse behavior within this group. We emphasize this point because even recent textbooks have updated their lemur and tarsier taxonomy to the exclusion of the galagos, pottos, and lorises (e.g., Dunbar and Barrett 2000, Falk 2000, MacDonald 2001, Boyd and Silk 2003).

Members of the infraorder Lorisiformes are currently conservatively classified as one superfamily (Lorisoidea) and one family (Lorisidae), comprising three distinctive subfamilies (Galaginae, Perodicticinae, and Lorisinae) (Rasmussen and Nekaris 1998, Grubb et al. 2003) (Table 3.2). The galagos and pottos are restricted to Africa and range in size from 55 to 2,000 g. The lorises are found in Asia and range in size from 85 to 1,850 g (Table 3.3).

The Galagines

Before 1979, the accepted taxonomy of the Galaginae was monogeneric (*Galago*) and contained only six species (Petter and Petter-Rousseaux 1979). This classification is still reported in a number of prominent texts, despite several published revisions expanding the number of species to 11 (Olson 1979, Nash et al. 1989), 17 (Bearder et al. 1995), and 25 (Bearder et al. 2003). This extraordinary diversity of galagos makes them comparable to the guenon group (see Chapter 15) in having one of the widest distributions and abundance of species found in Africa. For the galagos we adopt the most recent classification by Grubb et al. (2003; but see also Groves 2001) and divide the taxa into five genera (Table 3.2). Bioacoustic studies, using Paterson's (1985) mate recognition concept of species, have been at the forefront of this taxonomic revision (Masters 1988, 1991, 1998; Honess 1996; Bearder 1999; Ambrose 2003). This classification is also supported by behavioral studies (e.g., Harcourt and Bearder 1989); genetic research (Bayes 1998, Roos 2003); examination of hand, foot, and sexual organ morphology; and comparisons of hair scale structure (Dixson 1989, 1995, 1998; Anderson 1998, 2000, 2001; Anderson et al. 2000). In this chapter, we have carefully gone through previous studies, updating the taxonomy for each of them in the tables and text. This is extremely important as these animals are indeed distinct species but aspects of life history and morphology of even the best studied taxa are often reported, even in the more recent literature, under the wrong name (e.g., *Galago moholi*, which was formerly known as *G. senegalensis*).

Galagos are distributed across the whole of Africa south of the Sahara, with the exception of southern regions of South Africa. They occupy a very wide range of habitats, from near-desert conditions in Somalia and northern Kenya through subtropical savannahs, woodlands, riverine, and montane forests to dense tropical rain forests. Up to four species can occur in sympatry with each other, as well as with up to two pottos. One country, Tanzania, currently boasts 13 species of galagos within its borders and one species; *Otolemur garnettii* has been found in association with any of 14 other galago species in different parts of its geographic range. With huge tracts of rain forest yet to be surveyed for these nocturnal primates and considering their secretive habits and relatively cryptic characteristics, it will not be astonishing if further research adds to the complexity of this emerging picture.

Perodicticines and Lorisines

Similar diversity is now being uncovered in the Perodicticinae in Africa and the Lorisinae in Asia, each of which was once thought to comprise two monospecific genera, one gracile and the other robust (Yoder et al. 2001a). Despite being less vocal than the galagos, vocalizations have again yielded important taxonomic information, as have differences in behavior, morphology, facial markings, and genetic data (Coultas 2002, Nekaris and Jayewardene 2003, Roos 2003). It is almost without doubt that most of the currently recognized subspecies within these two subfamilies will be elevated to species level; for this reason, we report data regarding the pottos and lorises at the subspecific level.

In Africa, the gracile forms are now recognized as two species, the golden angwantibo (*Arctocebus aureus*) and the Calabar angwantibo (*A. calabarensis*), both confined to the rain forests of central Africa. The taxonomy of the robust forms, the pottos, is being reevaluated but currently consists of one species with five subspecies. Following Kingdon (1997), these are *Perodicticus potto potto* in West Africa, *P. p. juju* in Nigeria, *P. p. edwardsi* in Cameroon and Gabon, *P. p. faustus* in the Congo Basin, and *P. p. ibeanus* in the Eastern Democratic Republic of Congo, Uganda, and western Kenya. Schwartz (1996) recognizes another genus of potto, *Pseudopotto martini*, differing from *Perodicticus* in having relatively long upper first premolars, a reduced third molar, and a relatively longer tail. Other researchers doubt this taxon, suggesting the differences fall within the range of variation of *Perodicticus* (Sarmiento 1998); targeted searches in the wild have failed to yield any evidence of this putative genus (Pimley 2002).

Groves (2001) recognizes two gracile lorisines in south Asia, although we follow the subspecific classification of Osman Hill (1953). The red slender loris, the smallest of the lorisines, resides only in the lowland (*Loris tardigradus tardigradus*) and montane (*L. t. nycticeboides*) rain forests of Sri Lanka. Two subspecies of the larger gray slender loris also are found on this island: the northern Ceylonese slender loris (*L. lydekkerianus nordicus*) and the highland slender loris (*L. l. grandis*) (Nekaris and Jayewardene 2004). Southern India harbors an additional two slender loris taxa (Roonwal and Mohnot 1977, Schulze and Meier 1995a). The Malabar slender loris (*L. l. malabaricus*) is distributed in the wet southwest, including the Western Ghats, whereas the largest of all slender loris taxa, the Mysore slender loris (*L. l. lydekkerianus*), is distributed in the dry scrub forests of the southeast, including the Eastern Ghats.

The taxonomy of the robust Asian form, *Nycticebus*, is currently undergoing extensive revision, with some authorities recognizing three species based on morphology (e.g., Groves 1998, 2001) and others recognizing five species based on genetic analyses (Roos 2003). Here, we follow the taxonomy suggested by Roos (2003). The Bengal or northern slow loris (*Nycticebus bengalensis*) has the largest geographic range, including Burma, Cambodia, southern

Table 3.2   Long-Term Behavioral Studies of Lorisoid Primates; Data in Other Tables Are Drawn from These Studies Unless Otherwise Specified

TAXA	COUNTRY	STUDY SITE(S)	STUDY LENGTH (MONTHS)	RADIO TRACKING	REFERENCES
Galaginae					
Galagoides demidovii	Gabon	Makokou	42	Yes	Charles-Dominique 1972, 1977a
G. thomasi	Gabon	Makokou	42	–	Charles-Dominique 1977a
G. cocos	Kenya	Gedi Ruins National Monument	20	Yes	Harcourt and Nash 1986a,b; Nash 1986, 1993
G. cocos	Kenya	Diani	22	Yes	Harcourt 1984, 1986a; Harcourt and Nash 1986a,b; Harcourt and Bearder 1989
G. rondoensis	Tanzania	Litipo, Rondo, Ziwani	20	No	Honess 1996, Honess and Bearder 1996
G. zanzibaricus (*udzungwensis*)	Tanzania	Matundu	20	No	Honess 1996, Honess and Bearder 1996
G. granti	Tanzania	Rondo, Mtopwa	20	No	Honess 1996, Honess and Bearder 1996
G. orinus	Tanzania	Amani	20	No	Honess 1996, Honess and Bearder 1996
Galago moholi	S. Africa	Mosdene	24, 12, 11	Yes	Bearder 1969, 1987; Harcourt 1980; Bearder and Martin 1980a,b; Bearder and Doyle 1974; Charles-Dominique and Bearder 1979; Crompton 1980; Harcourt and Bearder 1989
G. moholi	S. Africa	Nylsvley Nature Reserve	18	Yes	Pullen 2000, Pullen et al. 2000
Euoticus elegantulus	Gabon	Makokou	42	No	Charles-Dominique 1977a
Sciurocheirus gabonensis	Gabon	Makokou	42	Yes	Charles-Dominique 1977a,b
S. alleni cameronensis	Cameroon	WWF Mt. Kupe Forest Reserve, Bakossiland[1]	22	Yes	Pimley 2002, Pimley et al. 2002, in press
Otolemur garnettii	Kenya	Gedi Ruins National Monument	20	Yes	Nash 1986, Nash and Harcourt 1986
O. garnettii	Kenya	Diani	22	Yes	Harcourt 1984, Nash and Harcourt 1986
O. crassicaudatus	S. Africa, Zimbabwe, Kwazula	Transval, Umtali, Mtunzini	15	Yes	Bearder 1974, Bearder and Doyle 1974b
O. crassicaudatus umbrosus	S. Africa	Soutsanberg Range	16, 11	No	Clark 1978a,b, 1985; Crompton 1980, 1983, 1984
O. crassicaudatus	S. Africa	Louis Trichard	12	No	Harcourt 1980, 1986b
Perodicticinae					
Perodicticus potto edwardsi	Cameroon	WWF Mt. Kupe Forest Reserve, Bakossiland	22	Yes	Pimley 2002, Pimley et al. 2002
P. p. edwardsi	Gabon	Makokou	42	Yes	Charles-Dominique 1974a,b, 1977a
Arctocebus aureus	Gabon	Makokou	42	No	Charles-Dominique 1977a
Lorisinae					
Loris lydekkerianus lydekkerianus	India	Ayyalur Interface Forestry Division	11–21	No	Nekaris 2001, 2002, 2003a,b, 2004; Nekaris and Rasmussen 2003; Rhadakrishna 2001; Rhadakrishna and Singh 2002, 2004
L. l. nordicus	Sri Lanka	Several sites in the north	5	No	Nekaris and Jayewardene 2003, Nekaris 2003b
L. tardigradus tardigradus	Sri Lanka	Masmullah Forest Reserve; Bangamukande Estate	8	No	Nekaris 2003b, Nekaris et al. in press, Nekaris and Jayewardene 2003
Nycticebus coucang coucang	Malaysia	Manjung District, Perak	27.5	Yes	Wiens 1995, 2002; Wiens and Zitzmann 2003a,b
N. c. coucang	Malaysia	Pasoh Forest Reserve, Sungai Tekam Forestry Concession	16	Limited	Barrett 1984
N. pygmaeus	Vietnam	Cuc Phuong National Park	24	Yes	Streicher 2003, Streicher 2004

[1] WWF, World Wildlife Fund.

Table 3.3   Physical Characteristics: Body Weights of Wild Caught Individuals, Unless Otherwise Noted

TAXA	ADULT MALE (G)			ADULT FEMALE (G)			BOTH SEXES (G)		
	AVERAGE	RANGE	N	AVERAGE	RANGE	N	AVERAGE	RANGE	N
Galaginae									
Galagoides demidovii	60	52–72	17	55	45–68	16	57	45–72	33
G. thomasi	82	74–88	6	75	59–85	6	78	59–88	12
G. orinus							89.6	74–98	3
G. zanzibaricus							149		23
G. rondoensis	69.2	60–73	7	66.5		3	69.1		7
G. cocos	150	130–183	35	137	118–155	38			
G. granti							134		5
Galago moholi	360[1]	±72	9	266[1]	±47	10	202[1]		1
G. moholi	186.1	±16.3	20	162.8	±16.3	20	200	177–250	
G. matschiei							196–225	210	
Euoticus elegantulus							300	270–360	39
E. pallidus								182–210	
Sciurocheirus alleni cameronensis	280.5	265–307	3	258	246–355	10	288	258–319	4
S. alleni (Bioko)	429		1	446	395–502	5	443	395–502	6
S. gabonensis							260	188–340	17
Otolemur garnettii	690–1,060	846	14	805	604–985	11	842	604–1,060	25
O. crassicaudatus	1,510		9	1,258		8			
Perodicticinae									
Perodicticus potto potto							600		
P. p. edwardsi (Gabon)							1,100	850–1,600	33
P. p. edwardsi (Cameroon)	1,502	938–1,795	8	1,572	1,407–1,858	4	1,524	938–1,858	12
P. p. ibeanus	920		1	861	847–875	3			
Arctocebus aureus							210	150–270	30
A. calabarensis	318	315–320	2	298	270–325	9			
Lorisinae									
Loris lydekkerianus lydekkerianus	294.4	267–322	4	260	227–292	6	273	227–322	10
L. l. malabaricus[1]	222.14	180–275	6	189	168–210	2			
L. l. nordicus		228–285	4		238–287	5			
L. l. grandis	204.1		1	238.1		1			
Loris tardigradus tardigradus	162	153–172	2	118	103–148	3	137	103–172	5
L. t. nycticeboides	140		1	190		1	165	140–190	2
Nycticebus bengalensis	1,134		1	1,400		1		1,588–1,605	
N. coucang	737	±111	8	637	±61	11			
N. menagensis								265–300	3
N. pygmaeus	418 ± 98	367–578	70	422 ± 88	360–543	97			

[1] Captive.

China, northeast India, Laos, northern Thailand, and Vietnam. The greater slow loris (*N. coucang coucang*) is found in Indonesia, Malaysia, and Thailand. The Javan slow loris (*N. javanicus*) occurs only in Java. The Bornean slow loris (*N. menagensis*) is found in Brunei, Indonesia, and Malaysia. Finally, the pygmy or lesser slow loris (*N. pygmaeus*) is found in Cambodia, China, Laos, and Vietnam. Most authorities do not recognize *N. intermedius* but class it together with *N. pygmaeus*. A recent study by Streicher (2003, 2004) revealed that the characteristics that distinguished *N. intermedius* were in fact seasonal coat and body weight changes of *N. pygmaeus* (Fig. 3.1).

PHYLOGENETIC RELATIONSHIPS

The evolutionary history of the lorisiform primates has been the subject of a comprehensive review (Rasmussen and Nekaris 1998) and will be only briefly recapped here. Until recently, most evidence for lorisiform origins pointed back to the Miocene of East Africa. Intense debate characterizes the subfamilial designation of the three best-known early Miocene forms, *Mioeuoticus*, *Progalago*, and *Komba*. Both cranial and postcranial features have allied these genera with either lorisines or galagines (Le Gros Clark 1956, Walker 1969, Gebo 1986, McCrossin 1992). Other authors have

Figure 3.1 Photographs illustrating different genera within the Lorisidae: (A) *Loris tardigradus tardigradus* (K. A. I. Nekaris); (B) *Nycticebus pygmaeus* (U. Streicher); (C) *Galagoides rondoensis* (A. W. Perkin); (D) *Arctocebus calabarensis* (C. Wild);

(F)

(E)

(G)

(H)

Figure 3.1 (cont'd) (E) *Sciurocheirus* sp. (L. Ambrose); (F) *Periodicticus potto edwardsi* (E. R. Pimley); (G) *Otolemur crassicaudatus* (S. Bearder); (H) *Galago moholi* (S. Bearder).

suggested that basal lorisiforms may have demonstrated a combination of lorisine cranial characteristics and galagine postcranial adaptations (Rasmussen and Nekaris 1998). Two newly discovered fossils may resolve this debate, making lorisiform origins even earlier than previously thought. *Karanisia* and *Saharagalago*, based on analysis of dental characteristics, are putative early lorises and galagos, respectively, from late Eocene sites at the Fayum Depression in Egypt (Seiffert et al. 2003). True, unrefuted lorisines (*Nycticeboides simpsoni*) and galagines (*Galago howelli* and *G. sadimensis*) occur in the fossil record of the late Miocene of Pakistan and early Plio-Pleistocene of Ethiopia and Kenya.

Some authors have attempted to resolve the phylogenetic relationships of the lorisiforms with molecular and morphological evidence (Bayes 1998, Masters and Brothers 2002,

Roos et al. 2004). Although the standard practice is to consider the Lorisinae a monophyletic group to the exclusion of the Galaginae, most molecular and morphological studies cannot resolve the position of the African pottos (Rasmussen and Nekaris 1998). Behavioral evidence allies Asian slender lorises more closely with galagos than with pottos (Bearder et al. 2002, Pimley 2002). It is not implausible that the galagos, pottos, and lorises share a common ancestor and form three monophyletic groups (Yoder et al. 2001a). Rasmussen and Nekaris (1998) and Nekaris and Rasmussen (2003) suggest that the cause for the divergence of these groups may have been a deviation in foraging strategies, with the galagos specializing on evasive prey, resulting in an emphasis on hearing and leaping, and the pottos and lorises concentrating on toxic prey, with a subsequent reliance on olfaction and a reduced basal metabolic rate, coinciding with slow

Table 3.4   Diet and Activity Budgets of Wild Lorisiform Primates Based on Long-Term Studies

TAXON	Diet					Activity Budget				
	ANIMAL PREY (%)	FRUIT (%)	GUM (%)	NECTAR (%)	OTHER (%)	REST (%)	TRAVEL (%)	FORAGE (%)	SOCIAL (%)	OTHER (%)
Galaginae										
Galagoides demidovii/thomasi[1]	70	19	10						25	
G. cocos	70	30	0	0						
G. moholi	52	0	48	0		4.5	25	63.9	5.9–18	0.6
Euoticus elegantulus	20	5	75	0					24	
Sciurocheirus gabonensis	25	73	0	2					14	
S. alleni cameronensis	55	55	0	0					0.6–30.5	
Otolemur garnettii	50	50	0	0		9.4	52.3	21	14.5	2.8
O. crassicaudatus	5	33	62	0					20	
Perodicticinae										
Perodicticus potto edwardsi	40	50			10				4	
P. p. edwardsi	11	67	22						0.2–44	
Arctocebus aureus	87	13							3	
Lorisinae										
Loris lydekkerianus lydekkerianus	96	1.2	2.8			36.4	35.6	26	23	2
L. l. nordicus	95	5				20.8	75.5 (includes forage)		49.5	3.7
L. tardigradus tardigradus	100.5					19.1	80 (includes forage)		43.7	0.9
Nycticebus coucang coucang	2.5	22.5	43.3	31.7		5.4	70.6	21	3	
N. c. coucang	29	71								
N. pygmaeus	33	–	63		4					

[1] During his study, Charles-Dominique did not recognize *G. thomasi* as a distinct species and, thus, all data were "lumped."

locomotion and life history. Further studies will surely elucidate these evolutionary relationships.

ECOLOGY AND BEHAVIOR

Diet

Only seven studies have focused in detail on the diet of lorisiforms, with one of these (Charles-Dominique 1977a) gaining most of its data from the analysis of stomach contents (Table 3.4). A number of brief studies provide us with preliminary knowledge of particular food preferences (e.g., Happold and Happold 1992, Tan and Drake 2001, A. W. Perkin, personal communication; A. B. Rylands, R. A. Mittermeier, and B. R. Konstant, unpublished report). Both direct observations and stomach content analysis have their limitations, but what is clear is that dietary choice among the lorisiforms is varied, including gum-eating specialists, highly frugivorous taxa, and some that are among the most faunivorous of all the primates.

Galagos are extremely varied in their diet, but all species appear to consume at least some gum; the ability to consume and digest gum may be a fundamental adaptation of this group (Bearder and Martin 1980a; Harcourt 1980, 1984; Nash 1989; Nash and Whitten 1998). Apart from this, the smaller-bodied taxa (e.g., *Galagoides demidovii*, *G.*

thomasi, *G. rondoensis*) rely more on insects, medium-sized taxa (e.g., *Galago moholi*, *Euoticus elegantulus*) add more exudates to their dietary repertoire, and the largest of the galagos (e.g., *Sciurocheirus* sp., *Otolemur* sp.) increase their intake of fruit. Non-toxic orthopterans and beetles comprise the invertebrate portion of galago diets (Bearder and Doyle 1974b, Harcourt and Nash 1986a), and the fruits eaten by galagines are in general sweet and soft (Charles-Dominique 1977a). Preliminary observations of a yet unnamed taxon in south-eastern Tanzania indicate yet another feeding behavior, that of consuming floral nectar, suggesting an important role by this primate in pollination, not unlike that of the Malagasy *Eulemur mongoz* or the greater slow loris *N. coucang* (see below) (A. W. Perkin, personal communication; A. B. Rylands, R. A. Mittermeier, and B. R. Konstant, unpublished report).

Galagos have adapted to their varied diet through a variety of morphological and behavioral adaptations. All galagos are capable of localizing animal prey with the help of their particularly large and independently mobile ears and frequently use this sense to detect prey items that are out of sight. They also search for insects visually and find sources of gum using their keen sense of smell (Bearder 1969, Charles-Dominique 1977a, Hladik 1979, Pariente 1979). As with all strepsirhines, insects are grabbed in the hands in a stereotyped fashion involving control of the whole hand as the individual fingers cannot be moved independently (Martin 1990). The toothcomb plays an important role in scraping gum from trees, and gum can be cleaned from between the teeth using a serrated cartilaginous sublingual, a second type of tongue located underneath the main tongue that is notched at the tip like a saw. Gum is processed in an elongated cecum containing microorganisms capable of digesting the complex polymerized sugars. *Euoticus* spp., which eat mainly gum, have additional specializations in the form of enlarged canines and premolars for exposing sources of gum and *keeled* (pointed) nails, allowing the animals to cling to large tree trunks and reach exudates that would otherwise be inaccessible (Osman Hill 1953, Charles-Dominique 1977a, Ambrose 1999). Galagos living in seasonal environments in South Africa may rely almost completely on carbohydrate-rich gum in the cold winters and reduce their activity accordingly (Bearder and Martin 1980a). Squirrel galagos (*Sciurocheirus* spp.), which usually feed on fallen fruits, are reported to eat rapidly and even swallow fruits whole, allowing them to fill their stomachs within minutes and retreat to areas safer from potential predators (Charles-Dominique 1977a).

Only limited observations are available of potto feeding behavior (Jewell and Oates 1969, Charles-Dominique 1977a, Oates 1984, Pimley 2002). In general terms, pottos (*Perodicticus* spp.) are mainly frugivorous but supplement their diet with a considerable amount of gums and animal prey, including ants, slow-moving arthropods, birds, and bats. Pottos are possessed with somewhat more powerful jaws than galagos and are able to consume fruits and stationary animal prey, in particular caterpillars and noxious beetles. Dietary conditioning is exhibited by all the African strepsirhines, whereby a young animal learns to eat by snatching food from its parent and examining novel food items with a curious, head-cocking movement (Bearder 1969, Charles-Dominique 1977a). This developmental behavior may be particularly important for the angwantibos (*Arctocebus* spp.), which process irritant prey in a specific manner (e.g., removing the hairs from caterpillars) before they can be consumed without discomfort.

In a detailed study of sympatric galagos (*Galagoides demidovii, Euoticus elegantulus, Sciurocheirus gabonensis*) and pottos (*P. p. edwardsi, A. aureus*) in Gabon, Charles-Dominique (1974a, 1977a) revealed classic dietary partitioning between nocturnal primate species that ensured they avoided competition. Species that spent most time in the forest canopy concentrated mainly on insects (*Galagoides*), gums (*Euoticus*), or fruits (*Perodicticus*). Species that preferred the undergrowth subsisted on caterpillars (*Arctocebus*) or fallen fruits (*Sciurocheirus*). Some years later, it was discovered that the dwarf galagos in Gabon were in fact two different species that live together throughout the tropical forests of central Africa (*Galagoides demidovii* and *G. thomasi*) (Wickings et al. 1998). Both these species prefer insects, but not surprisingly, one moves mainly in the canopy (*G. thomasi*) and the other is restricted to the

undergrowth (*G. demidovii*), where it consumes fast-moving insects in contrast to the noxious forms eaten by angwantibos (*A. aureus*). Similar separations occur between sympatric species in other parts of Africa. For example, in the Rondo Forest of southeastern Tanzania, Garnett's galagos (*O. garnettii*) forage in the canopy, Grant's galagos (*Galagoides granti*) use the middle story, and Rondo dwarf galagos (*G. rondoensis*) remain approximately 1 m above the ground and feed almost exclusively on insects and grubs from the leaf litter (Honess 1996).

Detailed observations have been made on the diet of three slender loris taxa. *L. l. lydekkerianus* was the focus of a long-term study (Nekaris and Rasmussen 2003), whereas *L. l. nordicus* and *L. t. tardigradus* were the subjects of short-term studies (Petter and Hladik 1970, Nekaris 2002, Nekaris and Jayewardene 2003). These studies concur that slender lorises are among the most faunivorous of primates (very like tarsiers, see Chapter 5). They specialize on prey of small size classes and are highly tolerant of toxic prey such as ants and darkling beetles (Tenibrionidae). Prey items are consistently eaten head first, followed by the animal lapping at the innards. Those insects which emit irritant sprays are removed individually from the colony, taken several meters away, and consumed while the loris slobbers, closes its eyes tightly, and shakes its head, all combined to produce what can be aptly termed a "disgust face." Although gum comprised a portion of the diet of *L. l. lydekkerianus*, it was not seen to be consumed by other taxa. Consumption of plant material was minimal to nonexistent. Vertebrates, particularly geckos and lizards, were consumed by all three taxa but comprised a large portion of the diet of *L. t. tardigradus* (Nekaris and Rasmussen 2003, Nekaris and Jayewardene 2003).

According to Barrett (1984), the greater slow loris (*N. coucang coucang*) predominantly eats fruit, supplemented by insects. A more detailed study of this species in Malaysia was conducted via direct observation and fecal analysis (Wiens 2002, Wiens and Zitzmann 2003a). This population consumed mainly nectar, gum, and sap, with fruit and arthropods comprising only a small portion of the diet. Nectar from the flowers of the Bertram palm (*Eugeissona tristis*) comprised more observations than any other dietary item, with animals spending up to 30 min feeding from these nectaries (Wiens 2002, Wiens and Zitzmann 2003a). Preliminary results on pygmy lorises (*N. pygmaeus*) suggested that they too rely on nectar (particularly *Saraca dives*) and gum, visit the same sites often, and leave noticeable gouges in the tree trunks (Tan and Drake 2001, Streicher 2004). As for galagos, gum is probably an important component of the diet during cold Vietnamese winters (Streicher 2004). Consumption of insects, including ants and moths, is relatively common; and processing of these prey items mirrors that of slender lorises (Streicher 2004). *N. pygmaeus* and *N. bengalensis*, sympatric in many parts of their range, are known from preliminary observations to share feeding sites; nothing is known about how they partition their niches (Duckworth 1994).

ACTIVITY PATTERNS AND LOCOMOTION

Activity

Very few activity budgets for the lorisiforms have been reported in the literature, but those that have are summarized in Table 3.4. What is clear is that all the lorisiforms are nocturnal in their activity patterns, with no diurnal or cathemeral species. Animals are not precluded from being active in daylight, however, and may do so in order to change position for thermoregulatory purposes, to eat during periods of intense food scarcity, and to avoid predators (Bearder et al. in press). What is becoming clearer is that, at least in more open habitats, activity patterns change with the amount of light available. *G. moholi*, for example, increased its behavior and range of travel patterns during the light moon and during periods of twilight, while *L. l. lydekkeri-anus* maintained activity regardless of moon phase (Bearder et al. 2002). Galago species living in closed forest, on the other hand, do not appear to be influenced by changes in the level of moonlight (Nash 1986). It is clear that further studies of nocturnal primates must take account of the importance of moonlight.

Locomotor Behavior

As the feature that is most often used to characterize this infraorder of strepsirhines, locomotion is possibly the best-studied aspect of their behavior (Table 3.5), forming the basis for entire field studies (Crompton 1980, 1983, 1984) and for numerous captive studies (e.g., Dykyj 1980, Glassman and Wells 1984, Oxnard et al. 1990, Ishida et al. 1992, Demes et al. 1998). A complex suite of morphological traits linked to locomotion differentiates the galagos from the pottos and lorises (Charles-Dominique and Bearder 1979). All taxa of galagos have long tails and elongated tarsal bones and are characterized by intermembral indices <100, whereas pottos and lorises, to varying degrees, have reduced or lost their tails and have intermembral indices close to 100 (Martin 1990). As a result, galagos can cross gaps by hopping and leaping, while lorises and pottos do this by stretching. Similarly, galagos usually evade predators by swift locomotion, whereas lorises and pottos have developed a suite of morphological characteristics that allow them to remain still for prolonged periods and to provide camouflage and protection if attacked (Charles-Dominique 1977a, Bearder 1987, Nekaris 2001). For example, both pottos and lorises exhibit features that allow for prolonged grip with no fatigue, such as shortened second digits on the hands and feet, highly mobile ankles and wrists, and retia mirabilia of the proximal limb vessels (Rasmussen and Nekaris 1998). Retia mirabilia are arteries and veins in the arms and legs that subdivide extensively to form networks of intertwining vessels which act as storage units, allowing blood to flow freely. Thus, the exchange of oxygen and waste materials in the muscles continues even though there is no bodily movement.

Perodicticus also possesses a scapular shield, a structure produced by a combination of raised apophyseal cervical spines, some of which protrude above the skin in the form of tubercles, which are covered by thick skin and bristles of sensory hair, which also extend to a wider nuchal region. This structure is used to provide defense against predators and possibly other pottos (Charles-Dominique 1977a). Slow lorises have developed an even more elaborate defense mechanism, that of being toxic. Before biting prey items or predators, slow lorises combine a secretion from brachial sebaceous glands with their saliva in order to produce a numbing poison, which can send humans into anaphylactic shock (Alterman 1995, Fry and Fry 2003). It is rumored that they also use this solution to cover their parked infants, although this is yet to be verified by field data.

Although vertical clinging and leaping is considered the quintessential galago locomotor mode, it is used by most galagos only to negotiate gaps between trees. Only a few taxa, such as *Sciurocheirus* spp. and *Galagoides rondoensis*, use it as their stereotypic mode of locomotion (Charles-Dominique 1977a, Honess 1996, Perkin 2002, Pimley 2002). In fact, *Otolemur* spp. rarely uses this mode of locomotion, although capable of leaping and bipedal hopping (Crompton 1983, Harcourt and Nash 1986a). These larger galagos are surprisingly monkey-like in their locomotion and regularly move quadrupedally through the trees on relatively broad and horizontal supports. Many of the smaller galagos (*Galagoides*) maneuver through the networks of tiny branches by quadrupedal running, climbing, and agile jumping (Charles-Dominique 1972, Ambrose 1999). Several taxa can cross the ground by walking or running (e.g., *Otolemur crassicaudatus*) or bipedal hopping (e.g., *O. garnettii*, *G. moholi*, and *G. moholi*), whereas others are strictly arboreal (e.g., *Galagoides* spp. and *Euoticus* spp.).

The absence of active leaping and the use of *cantilevering* (bridging or extending the body) to move across arboreal gaps are the key features that distinguish perodicticine and lorisine locomotion from that of the galagines (Sellers 1996). Both pottos (*Arctocebus* spp., *Perodicticus* spp.) and lorises (*Loris* spp., *Nycticebus* spp.) use their long bodies and flexible limbs to stretch across gaps in the canopy and, based on their body weights, require a certain gauge of branch to sustain their weight during the crossing (Charles-Dominique 1974b, Nekaris 2001). Unlike galagos, which can charge through the trees changing directions, the "slow-ness" of loris and potto locomotion comes from testing branches and having to back up and move position in the canopy to find a suitable crossing point (Charles-Dominique 1977a, Nekaris 2001). However, this progression need not be slow-paced. Captive studies have shown that the slender loris (*L. l. malabaricus*) is capable of a locomotor mode called the "race walk" (Demes et al. 1998). Wild *L. tardigradus* and *N. pygmaeus* regularly quadrupedally run and even negotiate gaps with mini-leaps, rearing up on their hindlegs and hurtling their bodies over gaps of several inches (Duckworth 1994, Nekaris and Stevens 2005). Even

Table 3.5   Locomotion and Habitat Use

TAXON	CHARACTERISTIC LOCOMOTION	USE OF STRATA/SUPPORTS WHEN ACTIVE	SLEEPING SITES	SLEEPING ASSOCIATIONS	HABITAT TYPE OF STUDY SITE
Galagoides demidovii	Fast-moving: mainly quadrupedal	0–5 m in dense secondary undergrowth, <1 cm fine branches and liane curtains, occur on roadside	Spherical leaf nest or dense vegetation, few sites	2–10 female w/offspring; male often sleeps alone	Primary equatorial rain forest
G. cocos	Fast-moving: mainly hopping and quadrupedal running	Ground to canopy (0–13 m), prefers undergrowth	Tree hollows, few sites	Male sleeps with one or 2 female w/offspring	Lowland dry forest on coral rag
G. rondoensis	Vertical clinging and leaping from thin stems	Low-diameter perches (<3 m), small vertical supports (3.0 cm)	Flat and leafy nests in high trees (5 m)	At least 3	Lowland evergreen and semievergreen forest
G. orinus	Quadrupedal running and walking	Canopy dwellers (20 m)	Nests of leaves and twigs set in lianes	At least 1–3	Natural montane evergreen forest
G. granti	Agile jumping, quadrupedal climbing and walking	Mid-strata (5–7.5 m), 8.0 cm diameter vertical supports	Tree holes	4–5 individuals	Lowland and coastal forest
G. zanzibaricus (*udzungwensis*)	Quadrupedal walking and running	Upper strata (10 m): thick secondary growth and vine tangles, small horizontal perches	Tree hollows or secondary growth tangles	Male & female pair and offspring	Natural lowland evergreen forest
Galago moholi	Active leaping, bipedal hopping	Ground to upper canopy, prefer lower strata (0–4 m), use small vertical supports	Flat leaf nest, tree hollow or branch fork in a thorn tree	1–8; males never together but with 2 or more female w/offspring	Acacia woodland savanna
G. moholi	Leaping and bipedal hopping	Ground to mid-canopy (1–4 m), can cross open ground	Tangled vegetation and tree holes at 1–2 m, human-made bee hives	At least 1–3	Savanna *Acacia* thorn scrub, *Cynometra* thicket, and open woodland
Euoticus elegantulus	Running, leaping, and climbing	Canopy (5–35 m), use large-caliber branches and vines	Branch fork in dense shelter of foliage	1–7	Primary equatorial rain forest
E. pallidus	Quadrupedal running and leaping	Prefer upper strata (4–12 m), use large horizontal supports	Branch fork in dense shelter of foliage	At least 1–4	Primary equatorial rain forest
Sciurocheirus gabonensis	Active leaping	Undergrowth (1–2 m), prefer vertical supports			Primary equatorial rain forest
Sciurocheirus sp. nov. (Makande Allen's Galago)	Vertical clinging and leaping	Lower–mid-canopy (0–5 m), vertical substrates <10 cm		1–4	Primary equatorial rain forest
S. alleni cameronensis	Vertical clinging and leaping	Ground to mid-canopy (0–5 m), 2–5 cm vertical supports	Tree holes and woody lianes at 1–4 m	At least 2–3	Secondary forest and farm bush
Otolemur garnettii	Quadrupedal running, leaping, and bipedal hopping	Mid- to upper canopy 50% of the time, horizontal substrates >5 cm	Tangled vegetation, hollows rarely, many sites	1–4, male and female w/offspring	Coastal forest
O. crassicaudatus	Monkey-like quadrupedal walking and running, some leaping and hopping	Low strata of canopy	Tangled vegetation or flat leaf nest, few sites	1–4, male and female w/offspring	Riverine forest
Perodicticinae					
Perodicticus potto edwardsi (Cameroon)	Slow climbing	Canopy at 6–10 m on 2–5 cm oblique branches	Leafy part of canopy at 10–30 m	1–3 individuals with male & female pairs	Farm bush, disturbed and secondary forest
P. p. edwardsi (Gabon)	Slow climbing	Canopy at 5–30 m, level branches and lianes of 1–15 cm diameter	Dense tangles or clumps with branches or forks	1–3, usually male sleeps alone, female w/offspring	Farm bush, primary and secondary forest
Arctocebus calabarensis	Slow climbing	Small branches, twigs, climbers	Dense vegetation	1–2, adults sleep alone, female w/offspring	Forest edges and tree fall zone
A. aureus	Slow climbing	0–5 m in undergrowth, <5 cm branches and lianes, use ground often	Dense vegetation	1–2, adults sleep alone, female w/offspring	Forest edges and tree fall zone
Lorisinae					
Loris lydekkerianus lydekkerianus	Quadrupedal climbing, walking, and cantilevering	Understory (<5 m), prefer oblique branches (1–5 cm), cross open ground and roads	Vine tangles, dense branches, few sites	1–7, female w/offspring and 1 or more male	Acacia scrub forest
L. l. nordicus	Quadrupedal climbing, walking, and cantilevering	Dense understory (<5 m), oblique branches (1–5 cm), cross open ground and roads	Vine tangles, dense branches, few sites	At least 1–6, female w/offspring and 1 or more male	Acacia scrub forest
L. tardigradus tardigradus	Quadrupedal climbing and running	Understory to canopy (0–15 m), prefer horizontal and vertical branches, and rely on vines	Vine tangles, dense branches, few sites	At least 1–4, female w/offspring and 1 male	Monsoon rain forest
Nycticebus coucang coucang	Slow climbing	Trunks, branches, and lianes <10 cm, mid- to upper canopy, will use understory in disturbed forest	Trees, palms, shrubs, lianes at 1.8–35 m; many sites	1–3, female w/offspring, sometimes 1 male	Primary forest, logged over forest, padang savanna
N. pygmaeus	Quadrupedal climbing and running	"Steady trails through vegetation;" come to ground if substrate is discontinuous	Dense scrub, or fairly exposed and high terminal branches	?	Forested limestone hills, plantation forest and scrub

Table 3.6   Range Size and Range Use Patterns for Wild Populations

TAXA	AVERAGE HOME RANGE (HA)	ADULT MALE (HA)	ADULT FEMALE (HA)	MALE OVERLAP	FEMALE OVERLAP	MALE & FEMALE OVERLAP	METHOD	INFERRED SOCIAL ORGANIZATION
Perodicticinae								
Perodicticus potto edwardsi (Pimley 2002)	28.4	30.6	31.5	P: 29%	P: 25%	P: 47%	Kernel	Semidispersed unimale/unifemale
P. p. edwardsi (Charles-Dominique 1977a)		17.8	7.5	A	Limited	P	mcp	Dispersed unimale, multiple female
Lorisinae								
Loris lydekkerianus lydekkerianus	2.5	3.6	1.59	P: 20%	P: 57%	P: 14%	mcp	Semidispersed multimale
L. l. nordicus				P	P	P		?
L. t. tardigradus				P	No data	P		?
Nycticebus coucang coucang (Wiens and Zitzmann 2003b)	2	0.8	2.1	No data	No data	P: 80.6%	mcp	Semidispersed unimale/unifemale
N. c. coucang (Wiens and Zitzmann 2003b)	6.4	7.35	4.8	No data	No data	P: 97.8%	mcp	Semidispersed unimale/unifemale
N. c. coucang (Wiens and Zitzmann 2003a)	18.1	22	10.4	No data	No data	P: 94.6%	mcp	Semidispersed unimale/unifemale
N. c. coucang (Barret 1984)			4.19	Rarely	P	P	mcp	?
N. pygmaeus	3.1							?
Galaginae								
Galagoides demidovii/thomasi		0.5–2.7	0.6–1.4	P	P	P	mcp	Dispersed multimale
G. cocos		2.2	1.8	P (slight)	P	P	mcp	Spatial monogamy
G. moholi		9.5–22.9	4.4–11.7	P	P	P	mcp	Dispersed multimale
Sciurocheirus gabonensis		30–50	8–16	A	P	P	mcp	Dispersed harem
S. alleni cameronensis		2.84	1.97	A	P: 58%	P: 31%	Kernel	Dispersed multimale
Otolemur garnettii		17	12	P: different age classes	P: different age classes	P: extensive overlap	mcp	Dispersed multimale

A, overlap absent; mcp, minimum convex polygon; P, overlap present.

P. p. edwardsi has been described to have a mini-leap, when it simply cannot negotiate a gap with any amount of stretch (Charles-Dominique 1977a).

Locomotion has been implicated as a factor affecting the ranging behavior of the lorisiforms, with galagos able to cross a larger home range and return to dispersed sleeping sites with greater regularity than the pottos and lorises (Charles-Dominique 1977a, 1977b; Oates 1984). Table 3.6 shows that, in fact, home ranges of similar-sized galagos and lorises are of comparable area. Despite initial suggestions that both pottos and lorises move as little as 10 m per night, studies conducted with all-night follows (Nekaris 2003a, Bearder et al. in press) and with radiotracking (Wiens and Zitzmann 2003b, Pimley et al. in press) have shown much more extensive ranging. For example, pottos (*P. p. edwardsi*) may move up to 6 km in one night, gray slender lorises (*L. lydekkerianus*) move several hundred meters, red slender lorises (*L. tardigradus*) travel up to 1 km per night, and greater slow lorises (*N. c. coucang*) travel up to 400 m per hour.

Habitat Use

As already noted when illustrating dietary partitioning, the lorisiforms also show a wide preference for use of both substrates and strata in the forest (e.g., Crompton 1983, Honess 1996, Ambrose 1999, Nekaris 2001, Pimley 2002, Nekaris et al. in press). Substrate size selection is almost always related to the body weight of the animal, with smaller animals moving on smaller-gauged twigs, branches, and lianas and larger animals negotiating sturdier supports with greater girth. An exception is made by *Euoticus*, which makes more use of large vertical supports (Charles-Dominique 1977a). A number of species (e.g., *S. gabonensis*, *Galagoides demidovii*, *A. aureus*, *L. lydekkerianus*) thrive in the undergrowth and in tree fall zones, whereas others (*G. orinus*, *N. coucang*) prefer the canopy. This ecological division is what allows the African lorisiforms in particular to occur in sympatry in many places throughout their range (Charles-Dominique 1977a) and may influence the distribution of sympatric Asian lorises (Duckworth 1994).

SOCIAL ORGANIZATION

Because of the difficulty of nocturnal observation, especially of taxa living in dense tropical rain forest, direct observations of social behavior may be limited (Sterling et al. 2000). Observations of associations between conspecifics, especially at sleeping sites, contribute to our knowledge of the social lives of lorisiforms; but for the most part, indirect observation has been more fruitful. Therefore, studies of communication and patterns of home range overlap have provided the bulk of our knowledge of social behavior, supplemented by a few studies using radio tracking.

Olfactory Communication

One of the most understudied areas of lorisiform social behavior is that of olfactory or chemical communication aided by an acute sense of smell and Jacobson's organ in the roof of the mouth, which senses liquid chemicals transferred from the moist nose (Schilling 1979, Martin 1990). Nocturnal lorisiforms communicate both with a number of specialized scent glands as well as with urine, which has also been shown to play an important role in enhancing an animal's grip during locomotion (Welker 1973, Harcourt 1981). The visual systems of nocturnal lorisiforms are highly sensitive and are supplemented by olfactory communication (Bearder et al. in press). The main advantage of olfactory communication via scent gland and urine marking in general is that it conveys information that is indirect and deferred in time, with a result that individuals do not have to come together in order to communicate. Although its prevalence has never been questioned, the difficulty of studying olfactory behavior has led to few systematic studies. Captive studies of pygmy lorises (Fisher et al. 2003a,b), Senegal galagos (Nash 1993), and thick-tailed galagos (Clark 1978a,b, 1982a,b) have shown the ability of nocturnal primates not only to differentiate the state of sexual receptivity of conspecifics using scent but also to recognize specific individuals of different age and sex classes. In fact, Clark (1985) suggested that the ability for fine olfactory differentiation contributed to increased gregariousness among *O. crassicaudatus*. In the only systematic study of olfactory behavior in free-ranging nocturnal lorisiformes, Charles-Dominique (1974b, 1977b) showed that, rather than using scent as trails, the sympatric taxa he studied scent-marked in specific areas, with clear signals serving for sexual attraction and avoidance.

Vocal Communication

More easily studied than olfaction, vocalizations have been invaluable for understanding the social behavior of galagos and, to a lesser degree, of lorises (Bearder et al. 1995, 2002; Honess 1996; Zimmermann et al. 1988; Zimmermann 1990, 1995a; Anderson et al. 2000; Coultas 2002). Since animals can always remain silent, their calls invariably reflect circumstances where they benefit in some way and, therefore, provide a strong clue to important aspects of their ecology and social behavior. For example, calls are given when it is advantageous to attract and maintain contact with companions, increase distance between rivals, warn kin of the presence of dangers, and warn potential predators that they have been detected. In the case of galagos in particular, the safety provided by living in trees at night and the ability to escape rapidly if detected means that they can communicate

effectively by sound even when they appear to be alone. They have a rich vocal repertoire of 8–25 structurally distinct calls, including sounds that are *discrete* (relatively invariable) and others that are *graded* (continuously changing from one form to another). Added to this, galagos are able to mix different calls into rapidly changing sequences that can sometimes last for over 30 min at a time. Calls are used during short-range social interactions, with some variation between animals of different age and sex; but each species also has some calls that are loud and used when mobbing predators, attracting partners, or repelling rivals. Fortunately for researchers, every species has one particular loud call that is common to both sexes and used to advertise their presence to companions and rivals. Since this call helps to bring mates together, it is invariably species-specific, remaining more or less constant across the entire geographical range of each species, thereby providing a convenient diagnostic tool for identifying new species (Courtenay and Bearder 1989; Masters 1991; Zimmermann 1995a,b; Anderson et al. 2000).

The less agile pottos and lorises as a group are not so obviously vocal, but unlike galagos, some of their calls include sounds in the ultrasonic range that remain inaudible to humans without a bat detector (Zimmerman 1985, Schulze and Meier 1995b, Nekaris and Jayewardene 2004). Still, some species, such as three slender loris taxa (*L. t. tardigradus*, *L. l. lydekkerianus*, and *L. l. nordicus*), are known to call throughout the night. Although the calls of Mysore slender lorises were not bioacoustically analyzed, they have several functions, including spacing, aggression, affiliation, and dawn assembly (Nekaris 2000, Bearder et al. 2002). At least six loud whistles with different functions have been identified for both *L. t. tardigradus* and *L. l. nordicus* (Coultas 2002). The latter species in captivity clearly uses one of these whistles for territorial spacing (Schulze and Meier 1995b), and one of these calls also has this function in the wild (Nekaris and Jayewardene 2003). Further studies of vocal repertoires within these species should prove to be rewarding.

Social Behavior

Nocturnal primates in general are typically described as solitary, despite extensive efforts by individuals studying them to dispel the use of this term (Charles-Dominique 1978, Bearder 1987). Sterling (1993) recommended that three components be used to aid in emphasizing the diversity of nocturnal primate social organization. The first of these, the social system, relates to social behavior and relationships within a group. Many of the lorisiforms engage in considerable amounts of social behavior. Table 3.4 compares the percentage of the active period that lorisiforms were seen together or in close proximity. This percentage does not include time spent communicating by scent or vocal communication, as described above. Although these figures also include mothers with their dependent offspring (e.g., *A. aureus*), a number of authors have pointed out that many adult nocturnal primates spend time together outside the breeding season, foraging and feeding (e.g., lesser galagos, Bearder and Martin 1980b, thick-tailed galagos, Clark 1985, Rhadakrishna and Singh 2002, Mysore slender lorises, Nekaris 2003a). Some taxa spend up to 50% of their time in social proximity with adult conspecifics. Variability also exists in choice of companions. Among *Galago moholi*, for example, females were the most common social partners (Bearder and Doyle 1974b), whereas in *L. l. lydekkerianus*, females formed positive affiliations only with multiple adult males (Nekaris 2002, 2003a; Rhadakrishna and Singh 2004). When compared with diurnal primates, the figures for social interactions among nocturnal primates fall well within the range of diurnal monkeys and apes (see Chapter 39). This is excluding the fact that most nocturnal lorisiforms sleep in close proximity (e.g., *Perodicticus* and *Nycticebus*) or in gregarious groups (most galagos and slender lorises) (Table 3.5), where social cohesion behaviors such as grooming and huddling take place.

Ranging

Determination of home range overlap via radio tracking or extended observation in open environments further elucidates the varied social relationships of the lorisiforms and defines the second descriptor recommended by Sterling (1993), that of the spacing system. Building on pioneering work by Bearder (1987), Müller and Thalmann (2000) have constructed a framework by which home range overlap, or spacing system, can be used to illustrate the diversity among nocturnal mammal social organization. In this framework, grouping systems can be cohesive and gregarious, dispersed yet social, or solitary, meaning no social contacts are made outside the mating system (Müller and Thalmann 2000, Sterling et al. 2000). Adult sex composition mirrors that seen among diurnal primate social organizations, with single male and female units, single male and multiple female groups, single female with multiple male groups, and multiple male and female groups. Nine long-term studies have been conducted, which have determined the size and degree of overlap of the home ranges of lorisiform primates (Table 3.6). In the case of rain forest primates, where observation by any other means might prove impossible, radio tracking has become invaluable to infer social organization based on spacing patterns.

Table 3.6 summarizes the inferred social organizations of those lorisiforms studied to date; social organization of these primates has also been the topic of two reviews (Bearder 1987, Müller and Thalmann 2000). Most galagos appear to exhibit a dispersed multimale system, whereby males have larger home ranges than females and females form matrilocal clusters of related females that may sleep together. These related females tend to be aggressive toward those from other groups, whereas males may be aggressive toward one another (Bearder and Doyle 1974a, Charles-Dominique 1974a, Bearder and Martin 1980b). These males may be of different types, relating to age and status. For example, smaller resident males may be tolerated by the larger territorial males, others may be constantly on the move ("floaters"), and finally some males remain solitary during the process of dispersing from their natal groups (Charles-Dominique 1972, Bearder 1987). A one male, multiple female system may be present in *S. gabonensis*, where males are exclusively associated with small groups of females and have nothing but extremely aggressive contact with other males (Charles-Dominique 1974a, 1977a,b). Another exception is found in *Galagoides cocos*, which may form one male/one female or one male/two or three female associations, although variability between study sites shows some convergence with the general multi-male social organization (Harcourt and Nash 1986b).

Two different systems have been shown for *P. p. edwardsi*, the only potto for which home range data are available. Charles-Dominique (1977a) studied this species in a restricted forest environment where no matriarchies were present, with female home ranges isolated from one another. However, males may overlap their ranges with more than one female but tend to avoid one another, probably using scent. Not enough data are available from this study in order to classify the social organization (Müller and Thalmann 2000). A more recent study of the same subspecies of potto found that males and females shared their home ranges to the exclusion of other male/female pairs. These same pairs also slept together or very near one another on most occasions, suggesting a single male/single female spacing system (Pimley 2002).

Mysore slender lorises (*L. l. lydekkerianus*) exhibited limited range overlap between females, who were aggressive at territorial boundaries. Male ranges were much larger than those of females. One or more adult males shared sleeping sites with females; males were aggressive only to males from other sleeping groups. The spacing indicates a single male/single female and single male/multiple female system but is also combined with promiscuous mating, suggesting a multimale/multifemale social organization (Nekaris 2003a). Greater slow lorises (*N. c. coucang*) appear to exhibit a single male/single female social organization, with the most common groupings being an adult male and female pair and their dependent offspring. This assessment corresponds with low testes volume for this taxon (Wiens and Zitzmann 2003b). Nevertheless, a polygynous mating system may exist (see below) (Elliot and Elliot 1967).

The final aspect recommended by Sterling (1993) as necessary to understand nocturnal primate social complexity is knowledge of the mating system, that is, which animal actually mates and produces offspring with another. The study of molecular ecology for the understanding of lemur mating systems has recently taken off (e.g., Fietz et al. 2000; Radespiel et al. 2001, 2002). Due to difficulties in gaining permits, only two such studies are available for the lorisiforms. A recent elegant study (Pullen 2000, Pullen et al. 2000) showed that, despite their spatial advantages and despite fathering a majority of offspring in the study population, "alpha" lesser galagos, *Galago moholi*, were not always the fathers of infants. Furthermore, not all twins were fathered by the same individuals (Pullen et al. 2000). These results are in line with both the testicular and copulatory evidence for this species, which suggests polygynandry. Pimley's (2002) molecular data for *P. p. edwardsi* at Mt. Kupe showed that offspring of mothers were not fathered by the male with which they were spatially paired. These data were in contrast to testicular volume data, which implied monogamy (see below), suggesting that the social system differed from the mating system (Pimley 2002).

REPRODUCTIVE STRATEGIES AND LIFE HISTORY

The life history strategies of the lorisiforms have been the focus of a number of captive studies (e.g., Manley 1966, 1967; Ehrlich and Musicant 1977; Doyle 1979; Izard and Rasmussen 1985; Rasmussen 1986; Rasmussen and Izard 1988; Ehrlich and Macbride 1989; Nash 1993; Weisenseel et al. 1998; Fitch-Snyder and Ehrlich 2003), yielding much of the information summarized in Table 3.7. A number of recent field studies, however, have supplemented the captive data, enhancing our knowledge of lorisiform life history parameters, mating behavior, mating systems, and infant care (Gursky and Nekaris 2003). Recent reviews have summarized in detail aspects of the development patterns of nocturnal primates (Nash 1993) and the reproductive biology of the African lorisiforms (Bearder et al. 2003) and the slender lorises (Nekaris 2003b).

A number of reproductive parameters characterize the galagos, pottos, and lorises. All taxa, with few exceptions, give birth to either singletons or twins, with twin births being known from more than half the taxa studied at present. A number of taxa have two litters per year. Little is known about survivorship ratios of lorisiforms in the wild, but when it is mentioned, it is not uncommon for only one infant out of a potential four to reach sexual maturity. *Infant parking* is common among the lorisiforms. In general, the practice is for the mother to leave her infants on a branch or in a tree hole while she goes off to forage. The only variation seems to be whether the infant is parked throughout the night (most pottos and lorises) or carried with the mother for short distances and cached in multiple sites throughout the night (most galagos). Variation across taxa also exists in whether or not infants cling to the fur while carried or are

Table 3.7 Reproductive and Life History Parameters

TAXA	INFANTS/YEAR[1]	LITTER SIZE[1]	PARKING[1]	INFANT CARRIAGE[1]	GESTATION (DAYS)	WEANING (DAYS)	WEIGHT AT BIRTH (G)	AGE AT SEXUAL MATURITY (MONTHS)	DISPERSING SEX[1]	BREEDING SEASON?
Galaginae										
Galagoides demidovii	1–2	1	Yes	Mouth	111–114	40–50	5–10	8–10	Male	—
G. thomasi	—	—	Yes	Mouth	111–114	—	5–12	—	—	—
G. cocos	2–4	1–2	Yes	Mouth	120	49	16.5	—	Male	—
G. rondoensis	2	1?	Yes	Mouth	—	—	—	—	—	—
G. zanzibaricus (udzungwensis)	2	1	Yes	Mouth	—	—	—	—	—	—
G. granti	—	—	Yes	Mouth	—	—	—	—	—	—
G. orinus	—	—	Yes	Mouth	—	—	—	—	—	—
Sciurocheirus gabonensis	1–2	1	Yes	Mouth	133	—	24	8–10	—	—
S. alleni cameronensis	1–2	1	Yes	Mouth	—	—	—	—	Male	Yes
Galago moholi	1–2	2	Yes	Mouth	120–126	—	11–12	8.5	Male	Jan–Feb/Oct–Nov
G. senegalensis	1–2	1	Yes	Mouth	141 ± 2	70–98	19 ± 2.6	12–18	Male	Feb–Mar/June–July
G. matschiei	—	—	Yes	Mouth	—	—	—	—	—	—
Euoticus elegantulus	1	1	No	Fur	—	—	—	—	—	—
E. pallidus	1	1	No	Fur	—	—	—	10	—	—
Otolemur garnettii	1	1	Yes	Mouth/fur	126–138	140	—	12–18	Male	Possibly Oct/Nov
O. crassicaudatus	2–3	1	Rare	Mouth/fur	136	70–134	—	18–24	Male	Possibly Oct/Nov
Perodicticinae										
Perodicticus potto edwardsi	1/(2)	1	Rare	Fur	197 (193–205)	120–180	52, 30–42	6	Male	No, Aug–Jan high rate
Arctocebus aureus	1	1	Yes	Fur	131–136	100–130[1]	24–30[1]	9–10	—	No
A. calabarensis	1–2	1	Yes	Fur	130	115	35	—	—	?, common Jan–Apr
Lorisinae										
Loris lydekkerianus lydekkerianus	1–4	1–2[1]	Yes	Fur	164 (160–166)[1]				Male or female	No[1]
L. l. malabaricus	1–4	1–2	Yes	Fur	166–169	120–150		11	—	No
L. l. nordicus	1–4	1–2[1]	Yes	Fur					—	No[1]
L. tardigradus tardigradus	1–4	1–2[1]	Yes	Fur	167–175				—	No[1]
L. t. nycticeboides		2[1]	?		174[1]				—	—
Nycticebus coucang coucang	1	1	Yes	Fur	165–175	85–180[1]	43.5	16–21	?	—
N. pygmaeus	1–4	1/2	Yes	Fur	[1]				—	—

[1] Data from wild animals or animals recently caught from the wild; all other data are from captive animals. —, no data available.

transported in the mother's mouth. Contrasting rates of life history among the galagos, pottos, and lorises are considered by some authors to be related to other locomotor and ecological differences among the three subfamilies (see Rasmussen and Nekaris 1998 for a review). The pottos and lorises are noted for having among the longest life history of any primates of their body size, including long gestation lengths followed by low birth weights and long periods of lactation, in contrast to galagos, which fall more in line with other primates of their body size (Martin 1990).

Interestingly, another feature uniting the lorisiforms is the absence of a single observation of infanticide in the wild. Although adults may kill infants under captive conditions, this has been shown to be due to stress or poor management rather than infanticide as an evolutionary strategy (Nekaris 2003a). In fact, male slender lorises regularly play with infants outside their sleeping groups. Males dispersing to a new area also show this behavior, even though it is highly unlikely that they are the fathers of infants. High reproductive output among twin-bearing lorisiforms with much opportunity for males to sire offspring suggests that infanticide has not played an important role in this infraorder (Manley 1966, Nekaris 2002, Bearder et al. 2003).

Dixson (1995, 1998) has pointed out that a number of features of the genital morphology and the copulatory behavior of nocturnal lorisiforms may provide evidence that the spacing system does not necessarily coincide with the mating system. For example, larger testes size or increase of testes size during a breeding season should be linked with a multiple male, multiple female mating system (*polygynandry*). The elaborate penile morphology of most lorisiforms might also serve to enhance female receptivity or genital lock or to break up copulatory plugs left by other males and might also provide a clue to the mating systems of these primates.

Information on dispersal is limited for most taxa. Many galagos appear to be matrilineal in their social organization, with males dispersing at sexual maturity and females either sharing a range with their mother or moving into a neighboring range (Bearder 1987). In Mysore slender lorises (*L. l. lydekkerianus*) and greater slow lorises (*N. c. coucang*), both males and females have been seen to disperse (Wiens 2002, Nekaris 2003b, Rhadakrishna and Singh 2004).

CONSERVATION STATUS

A cursory examination of Table 3.1 is enough to emphasize that very little is known about the conservation status of most nocturnal lorisiforms. Despite a number of surveys conducted for galagos (e.g., Honess 1996; Honess and Bearder 1996; Weisenseel et al. 1998; Butynski et al. 1998; Ambrose 1999; Ambrose and Perkin 2000; Perkin 2001a,b, 2002), pottos (Oates and Jewell 1967), and lorises (e.g., Barrett 1981; Duckworth 1994; Nekaris 1997; Singh et al. 1999, 2000; Fitch-Snyder and Vu 2002; Nekaris and Jayewardene 2004), a large proportion of species have been

described as "data-deficient." Where systematic studies have been conducted, they have almost always resulted in worrying conservation rankings (e.g., Nekaris 2003c). For example, one Sri Lankan loris (*L. t. nycticeboides*) and an unnamed species of galago (*Galagoides* sp. nov. 3) from Tanzania are considered critically endangered, and have been included on the recent list of the world's top 25 most endangered primates (Nekaris and Jayewardene 2004, Rylands personal communication). Although at this stage ranked as vulnerable, systematic surveys of pygmy lorises (*N. pygmaeus*) where virtually none have been seen suggest that they are more seriously threatened than the high availability in markets would suggest (Nekaris and Schulze 2004).

The paucity of studies on these African and Asian primates may lead them to be ignored at a time when they are facing severe human-induced threats. The bushmeat trade in Africa and the pet and biomedical trades in Asia are having detrimental effects on lorisiform populations (Ratajszczak 1998, Schulze and Groves 2004, Nekaris and Schulze 2004). Habitat loss in both Africa and Asia as a result of human population pressures also poses a severe threat to these species, which often go unconsidered in habitat development and planning (Erdelen 1988, Butynski 1996/97, Ratajszczak 1998). In Africa, human population growth rates are still increasing at 2.9% per annum (Butynski 1996/97). In Africa and Asia, clearing of the land for agriculture and deforestation for logging are the chief causes of forest loss (Mill 1995). Nocturnal prosimians may be at the greatest risk as they are asleep during the times of mass forest clearance, whereas other primates have the chance to flee. Sleeping nocturnal primates may be more easily burned alive or chopped down with the trees, collected, and sent to animal markets (Ratajszczak 1998, Schulze and Groves 2004). The tendency for lorises to cling to trees as they are cut, rather than fleeing, makes them an easy target for removal for the pet trade. Thus, whereas other animals can escape capture, lorises can be completely drained from areas of deforestation (Fitch-Snyder and Vu 2002, Streicher 2004). Furthermore, logging and human disturbance have been shown to adversely affect lorisiform density (Weisenseel et al. 1993, Nekaris and Jayewardene 2004). It is inappropriate to assume that healthy diurnal primate populations signify a healthy nocturnal primate population as the substrate and sleeping site requirements of these two groups of primates differ and surveys have often shown an inverse relationship in the presence of strepsirhine and haplorhine primates (e.g., Singh et al. 1999, 2000; Perkin 2001a; Nekaris and Jayewardene 2004).

Luckily, an increasing number of sanctuaries and reintroduction programs are being developed for Asian lorises, where the trade for pets and medicines is especially dire (Sanfey 2003, Nekaris and Schulze 2004, Streicher 2004). These sanctuaries operate in the face of stiff opposition from those who consider that priority should be given to "the more important" primates, such as gibbons and orangutans. Prosimians (strepsirhines and tarsiers) are in the unfortunate position of being relatively ignored by other conservation action groups because they are primates and ignored by primatologists because they are not anthropoids. With advances in the understanding of species-level biology and the uncovering of more and more species, there is a genuine chance that species can be lost or assigned the status of critically endangered before they are even named (Bearder 1999). Future studies of individual species, equivalent to those conducted for day-living primates, will ensure that this genetically diverse and interesting group is no longer excluded from conservation initiatives.

REFERENCES

Alterman, L. (1995). Toxins and toothcombs: potential allospecific chemical defenses in *Nycticebus* and *Perodicticus*. In: Alterman, L., Doyle, G. A., and Izard, M. K. (eds.), *Creatures of the Dark: The Nocturnal Prosimians*. Plenum Press, New York. pp. 413–424.

Ambrose, L. (1999). Species diversity in West and central African galagos (Primates, Galagonidae): the use of acoustic analysis [PhD thesis]. Oxford Brookes University, Oxford.

Ambrose, L. (2003). Three acoustic forms of Allen's galagos (Primates; Galagonidae) in the central African region. *Primates* 44:25–39.

Ambrose, L., and Perkin, A. W. (2000). A survey of nocturnal prosimians at Moca on Bioko Island, Equatorial Guinea. *Afr. Primates* 4:4–10.

Anderson, M. (1998). Comparative morphology and speciation in galagos. *Folia Primatol.* 69(suppl. 1):325–331.

Anderson, M. J. (2000). Penile morphology and classification of bush babies (subfamily Galagoninae). *Int. J. Primatol.* 21:815–836.

Anderson, M. J. (2001). The use of hair morphology in the classification of galagos (Primates, subfamily Galagonidae). *Primates* 42:113–121.

Anderson, M. J., Ambrose, L., Bearder, S. K., Dixson, A. F., and Pullen, S. (2000). Intraspecific variation in the vocalizations and hand pad morphology of southern lesser bush babies (*Galago moholi*): a comparison with *G. senegalensis*. *Int. J. Primatol.* 21:537–555.

Barrett, E. (1981). The present distribution and status of the slow loris in peninsular Malaysia. *Malays. Appl. Biol.* 10:205–211.

Barrett, E. (1984). The ecology of some nocturnal, arboreal mammals in the rainforests of peninsular Malaysia [PhD thesis]. Cambridge University, Cambridge.

Bayes, M. K. (1998). A molecular phylogenetic study of the galagos, strepsirhine primates and archontan mammals [PhD thesis]. Oxford Brookes University, Oxford.

Bearder, S. K. (1969). Territorial and intergroup behaviour of the lesser bushbaby, *Galago senegalensis moholi* (A Smith), in semi-natural conditions and in the field [MSc diss.]. University of Witwatersrand, Johannesburg.

Bearder, S. K. (1974). Aspects of the ecology and behaviour of the thick-tailed bushbaby *Galago crassicaudatus* [PhD thesis]. University of Witwatersrand, Johannesburg.

Bearder, S. K. (1987). Lorises, bushbabies and tarsiers: diverse societies in solitary foragers. In: Smuts, B. B., Cheney, D. L., Seyfarth, R. M., Wrangham, R. W., and Struhsaker, T. T. (eds.), *Primate Societies*. University of Chicago Press, Chicago. pp. 11–24.

Bearder, S. K. (1999). Physical and social diversity among nocturnal primates: a new view based on long term research. *Primates* 40:267–282.

Bearder, S. K., Ambrose, L., Harcourt, C., Honess, P., Perkin, A., Pimley, E., Pullen, S., and Svoboda, N. (2003). Species-typical patterns of infant contact, sleeping site use and social cohesion among nocturnal primates in Africa. *Folia Primatol.* 74:337–354.

Bearder, S. K., and Doyle, G. A. (1974a). Field and laboratory studies of social organisation in bushbabies (*Galago senegalensis*). *J. Hum. Evol.* 3:37–50.

Bearder, S. K., and Doyle, G. A. (1974b). Ecology of bushbabies *Galago senegalensis* and *Galago crassicaudatus*, with some notes on their behaviour in the field. In: Martin, R. D., Doyle, G. A., and Walker, A. C. (eds.), *Prosimian Biology*. Duckworth Press, London. pp. 109–130.

Bearder, S. K., Honess, P. E., and Ambrose, L. (1995). Species diversity among galagos with special reference to mate recognition. In: Alterman, L., Doyle, G., and Izard, M. K. (eds.), *Creatures of the Dark: The Nocturnal Prosimians*. Plenum Press, New York. pp. 331–352.

Bearder, S. K., and Martin, R. D. (1980a). Acacia gum and its use by bushbabies, *Galago senegalensis* (Primates: Lorisidae). *Int. J. Primatol.* 1:103–128.

Bearder, S. K., and Martin, R. D. (1980b). The social organisation of a nocturnal primate revealed by radio tracking. In: Amlaner, C. J., and MacDonald, D. W. (eds.), *A Handbook on Biotelemetry and Radio Tracking*. Pergamon Press, London. pp. 633–648.

Bearder, S. K., Nekaris, K. A. I., and Buzzell, C. A. (2002). Dangers of the night: are some primates afraid of the dark? In: Miller, L. E. (ed.), *Eat or Be Eaten: Predator Sensitive Foraging in Primates*. Cambridge University Press, Cambridge. pp. 21–43.

Bearder, S. K., Nekaris, K. A. I., and Curtis, D. J. (in press). A re-evaluation of the role of vision in the activity and communication of nocturnal primates. *Folia Primatol.*

Boyd, R., and Silk, J. B. (2003). *How Humans Evolved*. W. W. Norton, New York.

Butynski, T. M. (1996/97). African primate conservation—the species and the IUCN/SSC primate specialist group network. *Primate Conserv.* 17:87–100.

Butynski, T. M., Ehardt, C. L., and Struhsaker, T. T. (1998). Notes on two dwarf galagos (*Galagoides udzungwensis* and *Galagoides orinus*) in the Udzungwa Mountains, Tanzania. *Primate Conserv.* 18:69–75.

Cartmill, M. (1972). Arboreal adaptations and the origin of the order Primates. In: Tuttle, R. (ed.), *The Functional and Evolutionary Biology of the Primates*. Aldine Atherton Press, Chicago. pp. 97–122.

Charles-Dominique, P. (1972). Ecologie et vie sociale de *Galago demidovii* (Fisher 1808, Prosimii). *Z. Tierpsychol. Suppl.* 9:7–41.

Charles-Dominique, P. (1974a). Vie sociale de *Perodicticus potto* (Primates Lorisides). Étude de terrain en forêt equatorial de l'ouest africain au Gabon. *Mammalia* 38:355–379.

Charles-Dominique, P. (1974b). Ecology and feeding behaviour of five sympatric lorisids in Gabon. In: Martin, R. D., Doyle, G. A., and Walker, A. C. (eds.), *Prosimian Biology*. Duckworth Press, London. pp. 131–150.

Charles-Dominique, P. (1977a). *Ecology and Behaviour of Nocturnal Primates*. Duckworth Press, London.

Charles-Dominique, P. (1977b). Urine marking and territoriality in *Galago alleni* (Waterhouse, 1837-Lorisoidea, Primates)—a field study by radio-telemetry. *Z. Tierpsychol.* 43:113–138.

Charles-Dominique, P. (1978). Solitary and gregarious prosimians: evolution of social structures in primates. In: Chivers, D. J., and Joysey, K. A. (eds.), *Recent Advances in Primatology*, vol. 3. Academic Press, New York. pp. 139–149.

Charles-Dominique, P., and Bearder, S. K. (1979). Field studies of lorisid behavior: methodological aspects. In: Doyle, G. A., and Martin, R. D. (eds.), *The Study of Prosimian Behavior*. Academic Press, London. pp. 567–629.

Charles-Dominique, P., and Martin, R. D. (1970). Evolution of lorises and lemurs. *Nature* 227:257–260.

Clark, A. B. (1978a). Olfactory communication, *Galago crassicaudatus*, and the social life of prosimians. In: Chivers, D. J., and Joysey, K. A. (eds.), *Recent Advances in Primatology*. Evolution, vol. 3. Academic Press, New York. pp. 109–117.

Clark, A. B. (1978b). Sex ratio and local resource competition in a prosimian primate. *Science* 201:163–165.

Clark, A. B. (1982a). Scent marks as social signals in *Galago crassicaudatus* I. Sex and reproductive status as factors in signals and responses. *J. Chem. Ecol.* 8:1133–1151.

Clark, A. B. (1982b). Scent marks as social signals in *Galago crassicaudatus*. II. Discrimination between individuals by scent. *J. Chem. Ecol.* 8:1153–1165.

Clark, A. B. (1985). Sociality in a nocturnal "solitary" prosimian: *Galago crassicaudatus*. *Int. J. Primatol.* 6:581–600.

Coultas, D. S. (2002). Bioacoustic analysis of the loud call of two species of slender loris (*Loris tardigradus* and *L. lydekkerianus nordicus*) from Sri Lanka [MSc thesis]. Oxford Brookes University, Oxford.

Courtenay, D. O., and Bearder, S. K. (1989). The taxonomic status and distribution of bushbabies in Malawi with emphasis on the significance of vocalisations. *Int. J. Primatol.* 10:17–34.

Crompton, R. H. (1980). A leap in the dark: locomotor behaviour and ecology in *Galago senegalensis* and *Galago crassicaudatus* [PhD thesis]. Harvard University, Cambridge, MA.

Crompton, R. H. (1983). Age differences in locomotion in two subtropical Galaginae. *Primates* 24:241–259.

Crompton, R. H. (1984). Foraging, habitat structure, and locomotion in two species of *Galago*. In: Rodman, P. S., and Cant, J. G. H. (eds.), *Adaptations for Foraging in Non-Human Primates*. Columbia University Press, New York. pp. 73–111.

Crovella, S. J. C., Masters, J. C., and Rumpler, Y. (1994). Highly repeated DNA sequences as phylogenetic markers among Galaginae. *Am. J. Primatol.* 32:177–185.

DelPero, M., Masters, J. C., Zuccon, D., Cervella, P., Crovella, S., and Ardito, G. (2000). Mitochondrial sequences as indicators of generic classification in bush babies. *Int. J. Primatol.* 21:889–904.

Demes, B., Fleagle, J. G., and Lemelin, P. (1998). Myological correlates of prosimian leaping. *J. Hum. Evol.* 34:385–399.

Dixson, A. F. (1989). Effects of sexual selection upon the genitalia and copulatory behaviour in male primates. *Int. J. Primatol.* 10:47–55.

Dixson, A. F. (1995). Sexual selection and the evolution of copulatory behavior in nocturnal prosimians. In: Alterman, L., Doyle, G. A., and Izard, M. K. (eds.), *Creatures of the Dark: The Nocturnal Prosimians*. Plenum Press, New York. pp. 93–118.

Dixson, A. F. (1998). *Primate Sexuality: Comparative Studies of the Prosimians, Monkeys, Apes and Human Beings*. Oxford University Press, Oxford.

Doyle, G. A. (1979). Development of behaviour in prosimians with special reference to the lesser bushbaby, *Galago senegalensis moholi*. In: Doyle, G. A., and Martin, R. D. (eds.), *The Study of Prosimian Behaviour*. Academic Press, London. pp. 157–189.

Duckworth, J. W. (1994). Field sighting of the pygmy loris (*Nycticebus pygmaeus*) in Laos. *Folia Primatol.* 63:99–101.

Dunbar, R., and Barrett, L. (2000). *Cousins: Our Primate Relatives*. Bookman-Huntingdon, London.

Dykyj, D. (1980). Locomotion of the slow loris in a designed substrate context. *Am. J. Phys. Anthropol.* 52:577–586.

Ehrlich, A., and Macbride, L. (1989). Mother–infant interactions in captive slow lorises (*Nycticebus coucang*). *Am. J. Primatol.* 19:217–228.

Ehrlich, A., and Musicant, A. (1977). Social and individual behaviors in captive slow lorises (*Nycticebus coucang*). *Behaviour* 60:195–220.

Elliot, O., and Elliot, M. (1967). Field notes on the slow loris in Malaya. *J. Mammal.* 48:497–498.

Erdelen, W. (1988). Forest ecosystems and nature conservation in Sri Lanka. *Biol. Conserv.* 43:115–135.

Falk, D. (2000). *Primate Diversity*. Norton, New York.

Fietz, J., Zischler, H., Schwieg, C., Tomiuk, J., Dausmann, K. H., and Ganzhorn, J. U. (2000). High rates of extra-pair young in the pair-living fat-tailed dwarf lemur, *Cheirogaleus medius*. *Behav. Ecol. Sociobiol.* 49:8–17.

Fisher, H. S., Swaisgood, R. R., and Fitch-Snyder, H. (2003a). Odor familiarity and female preferences for males in a threatened primate, the pygmy loris *Nycticebus pygmaeus*: applications for genetic management of small populations. *Naturwissenschaften* 90:509–512.

Fisher, H. S., Swaisgood, R. R., and Fitch-Snyder, H. (2003b). Countermarking by male pygmy lorises (*Nycticebus pygmaeus*): do females use odor cues to select mates with high competitive ability? *Behav. Ecol. Sociobiol.* 53:123–130.

Fitch-Snyder, H., and Ehrlich, A. (2003). Mother–infant interactions in slow lorises (*Nycticebus bengalensis*) and pygmy lorises (*Nycticebus pygmaeus*). *Folia Primatol.* 74:259–271.

Fitch-Snyder, H., and Vu, N. T. (2002). A preliminary survey of lorises (*Nycticebus* spp.) in northern Vietnam. *Asian Primates* 8:1–3.

Fleagle, J. G. (1999). *Primate Adaptation and Evolution*. Academic Press, San Diego.

Fry, B. G., and Fry, A. (2003). The loris: a venomous primate. *Fauna* 4:8–11.

Gebo, D. L. (1986). Miocene lorisids: the foot evidence. *Folia Primatol.* 47:217–225.

Glassman, D. M., and Wells, J. P. (1984). Positional and activity behavior in a captive slow loris: a quantitative assessment. *Am. J. Primatol.* 7:121–132.

Gray, J. E. (1863). Revision of the species of lemuroid animals, with the description of some new species. *Proc. Zool. Soc. Lond.* 1863:129–152.

Groves, C. P. (1998). Systematics of tarsiers and lorises. *Primates* 39:13–27.

Groves, C. P. (2001). *Primate Taxonomy*. Smithsonian Institute Press, Washington DC.

Grubb, P., Butynski, T. M., Oates, J. F., Bearder, S. K., Disotell, T. R., Groves, C., and Struhsaker, T. (2003). An assessment of the diversity of African primates. *Int. J. Primatol.* 24:1301–1357.

Gursky, S., and Nekaris, K. A. I. (2003). An introduction to mating, birthing and rearing systems of nocturnal prosimians. *Folia Primatol.* 74:272–284.

Happold, D., and Happold, M. (1992). Termites as food for the thick-tailed bushbaby (*Otolemur crassicaudatus*) in Malawi. *Folia Primatol*. 58:118–120.

Harcourt, C. S. (1980). Behavioural adaptations of South African galagos [MSc diss.]. University of Witwatersrand, Johannesburg.

Harcourt, C. S. (1981). An examination of the function of urine washing in *Galago senegalensis*. *Z. Tierpsychol*. 55:119–128.

Harcourt, C. S. (1984). The behaviour and ecology of galagos in Kenyan coastal forest [PhD thesis]. University of Cambridge, Cambridge.

Harcourt, C. (1986a). *Galago zanzibaricus*: birth seasonality, litter size and perinatal behaviour of females. *J. Zool*. 210:451–457.

Harcourt, C. S. (1986b). Seasonal variation in the diet of South African galagos. *Int. J. Primatol*. 7:491–506.

Harcourt, C. S., and Bearder, S. K. (1989). A comparison of *Galago moholi* in South Africa with *Galago zanzibaricus* in Kenya. *Int. J. Primatol*. 10:35–45.

Harcourt, C. S., and Nash, L. T. (1986a). Species differences in substrate use and diet between sympatric galagos in two Kenyan coastal forests. *Primates* 27:41–52.

Harcourt, C. S., and Nash, L. T. (1986b). Social organization of galagos in Kenyan coastal forests, I. *Galago zanzibaricus*. *Am. J. Primatol*. 10:339–356.

Hladik, C. M. (1979). Diet and ecology of prosimians. In: Doyle, G. A., and Martin, R. D. (eds.), *The Study of Prosimian Behavior*. Academic Press, London. pp. 307–357.

Honess, P. E. (1996). Speciation among galagos (Primates, Galagidae) in Tanzanian forests [PhD thesis]. Oxford Brookes University, Oxford.

Honess, P. E., and Bearder, S. K. (1996). Descriptions of the dwarf galago species of Tanzania. *Afr. Primates* 2:75–79.

Ishida, H., Hirasaki, E., and Matano, S. (1992). Locomotion of the slow loris between discontinuous substrates. In: Matano, S., Tuttle, R. H., and Ishida, H. (eds.), *Topics in Primatology. Evolutionary Biology, Reproductive Endocrinology and and Virology*, vol. 3. University of Tokyo Press, Tokyo. pp. 139–152.

Izard, M. K., and Rasmussen, D. T. (1985). Reproduction in the slender loris (*Loris tardigradus malabaricus*). *Am. J. Primatol*. 8:153–165.

Jewell, P. A., and Oates, J. F. (1969). Ecological observations of the lorisoid primates of African lowland forest. *Zool. Afr*. 4:231–248.

Kingdon, J. (1997). *The Kingdon Field Guide to African Mammals*. Academic Press, London.

Le Gros Clark, W. (1956). *British Museum (Natural History). Fossil Mammals of Africa, No. 9: A Miocene Lemuroid Skull from East Africa*. British Museum, London. p. 6.

MacDonald, D. (2001). *The New Encyclopedia of Mammals*. Oxford University Press, Oxford.

Manley, G. H. (1966). Reproduction in lorisoid primates. *Symp. Zool. Soc. Lond*. 15:493–509.

Manley, G. H. (1967). Gestation periods in the Lorisidae. *Int. Zool. Ybk*. 7:80–81.

Martin, R. D. (1990). *Primate Origins and Evolution: A Phylogenetic Reconstruction*. Chapman and Hall, London.

Martin, R. D. (2000). Origins, diversity and relationship of lemurs. *Int. J. Primatol*. 21:1021–1049.

Martin, R. D. (2003). Combing the fossil record. *Nature* 422:388–391.

Masters, J. C. (1988). Speciation in the greater galagos (Prosimii: Galaginae): a review and synthesis. *Biol. J. Linn. Soc*. 34:149–174.

Masters, J. C. (1991). Loud calls of *Galago crassicaudatus* and *G. garnettii* and their relation to habitat structure. *Primates* 32:153–167.

Masters, J. C. (1998). Speciation in the lesser galagos. *Folia Primatol*. 69(suppl. 1):357–370.

Masters, J. C., and Brothers, D. J. (2002). Lack of congruence between morphological and molecular data in reconstructing the phylogeny of the Galagonidae. *Am. J. Phys. Anthropol*. 117:79–93.

Masters, J. C., Rayner, H., Ludewig, H., Zimmermann, E., Molez-Verriere, F., Vincent, F., and Nash, L. T. (1994). Phylogenetic relationships among the Galaginae as indicated by erythrocytic allozymes. *Primates* 35:177–190.

McCrossin, M. L. (1992). New species of bushbaby from the middle Miocene of Maboko Island, Kenya. *Am. J. Phys. Anthropol*. 89:215–233.

Mill, R. R. (1995). Regional overview: Indian subcontinent. In: *Centres of Plant Diversity: A Guide to Strategy for Their Conservation. Asia, Australia and the Pacific*, vol. 2. World Wildlife Fund for Nature and IUCN Press, Cambridge. pp. 62–135.

Müller, A. E., and Thalmann, U. (2000). Origin and evolution of primate social organisation: a reconstruction. *Biol. Rev. Camb. Philos. Soc*. 75:405–435.

Nash, L. T. (1986). Influence of moonlight level on traveling and calling patterns in two sympatric species of *Galago* in Kenya. In: Taub, D. M., and King, F. A. (eds.), *Current Perspectives in Primate Social Dynamics*. Van Nostrand Reinhold, New York. pp. 357–367.

Nash, L. T. (1989). Galagos and gummivory. *Hum. Evol*. 4:199–206.

Nash, L. T. (1993). Juveniles in nongregarious primates. In: Pereira, M. E., and Fairbanks, L. A. (eds.), *Juvenile Primates: Life History, Development, and Behavior*. Oxford University Press, Oxford. pp. 119–137.

Nash, L. T., Bearder, S. K., and Olson, T. (1989). Synopsis of galago species characteristics. *Int. J. Primatol*. 10:57–80.

Nash, L. T., and Harcourt, C. H. (1986). Social organization of galagos in Kenyan coastal forest II: *Galago garnettii*. *Am. J. Primatol*. 10:357–369.

Nash, L. T., and Whitten, P. L. (1998). Preliminary observations on the role of *Acacia* gum chemistry in *Acacia* utilization by *Galago senegalensis* in Kenya. *Am. J. Primatol*. 17:27–39.

Nekaris, K. A. I. (1997). A preliminary survey of the slender loris (*Loris tardigradus*) in south India. *Am. J. Phys. Anthropol. Suppl*. 24:176–177.

Nekaris, K. A. I. (2000). The socioecology of the Mysore slender loris (*Loris tardigradus lydekkerianus*) in Dindigul, Tamil Nadu, south India [PhD thesis]. Washington University, St. Louis.

Nekaris, K. A. I. (2001). Activity budget and positional behavior of the Mysore slender loris (*Loris tardigradus lydekkarianus*): implications for "slow climbing" locomotion. *Folia Primatol*. 72:228–241.

Nekaris, K. A. I. (2002). Slender in the night. *Nat. History* 2:54–59.

Nekaris, K. A. I. (2003a). Observations on mating, birthing and parental care in three taxa of slender loris in India and Sri Lanka (*Loris tardigradus* and *Loris lydekkerianus*). *Folia Primatol*. 74(suppl.):312–336.

Nekaris, K. A. I. (2003b). Spacing system of the Mysore slender loris (*Loris lydekkerianus lydekkerianus*). *Am. J. Phys. Anthropol*. 121:86–96.

Nekaris, K. A. I. (2003c). Rediscovery of the slender loris in Horton Plains National Park, Sri Lanka. *Asian Primates* 8:1–7.

Nekaris, K. A. I. (in press). Visual predation in the slender loris. *J. Hum. Evol.*

Nekaris, K. A. I., and Jayewardene, J. (2003). Pilot study and conservation status of the slender loris (*Loris tardigradus* and *Loris lydekkerianus*) in Sri Lanka. *Primate Conserv.* 19:83–90.

Nekaris, K. A. I., and Jayewardene, J. (2004). Distribution of slender lorises in four ecological zones in Sri Lanka. *J. Zool.* 262:1–12.

Nekaris, K. A. I., Liyanage, W. K. D. D., and Gamage, S. (in press). Relationship between forest structure and floristic composition and population density of the southwestern Ceylon slender loris (*Loris tardigradus tardigradus*) in Masmullah Forest, Sri Lanka. *Mammalia*.

Nekaris, K. A. I., and Rasmussen, D. T. (2003). Diet of the slender loris. *Int. J. Primatol.* 24:33–46.

Nekaris, K. A. I., and Schulze, H. (2004). Historical and recent developments of human–loris interactions in south and southeast Asia. Invited lecture for Primate Society of Great Britain winter meeting, London. *Primate Eye* 84:17–18.

Nekaris, K. A. I., and Stevens, N. J. (2005). All lorises are not slow: rapid arboreal locomotion in the newly recognised red slender loris (*Loris tardigradus tardigradus*) of southwestern Sri Lanka. *Am. J. Phys. Anthropol. Suppl.* 40:156.

Nietsch, A. A., and Kopp, M. L. (1998). Role of vocalization in species differentiation of Sulawesi tarsiers. *Folia Primatol.* 69(suppl. 1):371–378.

Oates, J. F. (1984). The niche of the potto, *Perodicticus potto. Int. J. Primatol.* 5:51–61.

Oates, J., and Jewell, P. A. (1967). Westerly extent of the range of three African lorisoid primates. *Nature* 215:778–779.

Olson, T. R. (1979). Studies on aspects of the morphology and systematics of the genus *Otolemur* (Coquerel, 1859) (Primates: Galagidae) [PhD thesis]. University of London, London.

Osman Hill, W. C. (1953). *Primates. Comparative Anatomy and Taxonomy. I. Strepsirhini.* Edinburgh University Press, Edinburgh.

Oxnard, C. E., Crompton, R. H., and Lieberman, S. S. (1990). *Animal Lifestyles and Anatomies.* University of Washington Press, Seattle.

Pariente, G. (1979). The role of vision in prosimian behaviour. In: Doyle, G. A., and Martin, R. D. (eds.), *The Study of Prosimian Behaviour.* Academic Press, London. pp. 411–459.

Pastorini, J., Thalmann, U., and Martin, R. D. (2003). A molecular approach to comparative phylogeography of extant Malagasy lemurs. *Proc. Natl. Acad. Sci. USA* 100:5879–5884.

Paterson, H. E. H. (1985). The recognition concept of species. In: Vbra, E. S. (ed.), *Species and Speciation.* Transvaal Museum, Pretoria. pp. 21–29.

Perkin, A. W. (2000). Bushbabies of Tanzania: an update. *Miombo* 20:4.

Perkin, A. W. (2001a). A field study on the conservation status and species diversity of galagos in the West Kilombero Scarp Forest Reserve. In: Doody, K. Z., Howell, K. M., and Fanning, E. (eds.), *West Kilombero Scarp Forest Reserve—Zoological Report.* Matumizi Endeleyu Mazingira, Iringa, Tanzania. pp. 149–159.

Perkin, A. W. (2001b). The taxonomic status and distribution of bushbabies (galagos) in the Uluguru Mountains, Tanzania. *Miombo* 23:5–7.

Perkin, A. W. (2002). The Rondo galago *Galagoides rondoensis* (Honess & Bearder, 1996): a primate conservation priority. *Primate Eye* 77:14–15.

Perkin, A. W., Bearder, S. K., Butynski, T., Bytebier, B., and Agwanda, B. (in press). The Taita Mountain dwarf galago *Galagoides* sp.: a new primate for Kenya. *J. East Afr. Nat. Hist.*

Petter, J. J., and Hladik, C. M. (1970). Observations sur le domaine vital et la densité de population de *Loris tardigradus* dans les forêts de Ceylan. *Mammalia* 34:394–409.

Petter, J., and Petter-Rousseaux, A. (1979). Classification of the prosimians. In: Doyle, G. A., and Martin, R. D. (eds.), *The Study of Prosimian Behaviour.* Academic Press, London. pp. 1–44.

Pimley, E. R. (2002). The behavioural ecology and genetics of the potto (*Perodicticus potto edwardsi*) and Allen's bushbaby (*Galago alleni cameronensis*) [PhD thesis]. University of Cambridge, Cambridge.

Pimley, E. R., Bearder, S. K., and Dixson, A. F. (2002). Patterns of ranging and social interactions in pottos (*Perodicticus potto edwardsi*) in Cameroon [abstract]. XIX Congress of the International Primatological Society, Beijing, PR China.

Pimley, E. R., Bearder, S. K., and Dixson, A. F. (in press). Examining the social organization of the Milne-Edward's potto *Perodicticus potto edwardsi. Am. J. Primatol.*

Pullen, S. L. (2000). Behavioural and genetic studies of the mating system in a nocturnal primate: the lesser galago (*Galago moholi*) [PhD thesis]. University of Cambridge, Cambridge.

Pullen, S. L., Bearder, S. K., and Dixson, A. F. (2000). Preliminary observations on sexual behavior and the mating system in free-ranging lesser galagos (*Galago moholi*). *Am. J. Primatol.* 51:79–88.

Radespiel, U., Dal Secco, V., Drogemuller, C., Braune, P., Labes, E., and Zimmermann, E. (2002). Sexual selection, multiple mating and paternity in grey mouse lemurs, *Microcebus murinus. Anim. Behav.* 63:259–268.

Radespiel, U., Funk, S. M., Zimmermann, E., and Bruford, M. W. (2001). Isolation and characterization of microsatellite loci in the grey mouse lemur (*Microcebus murinus*) and their amplification in the family Cheirogaleidae. *Mol. Ecol. Note* 1:16–18.

Rasmussen, D. T. (1986). Life history and behavior of slow lorises and slender lorises [PhD thesis]. Duke University, Durham, NC.

Rasmussen, D. T., and Izard, M. K. (1988). Scaling of growth and life-history traits relative to body size, brain size and metabolic rate in lorises and galagos (Lorisidae, Primates). *Am. J. Phys. Anthropol.* 75:357–367.

Rasmussen, D. T., and Nekaris, K. A. I. (1998). Evolutionary history of the lorisiform primates. *Folia Primatol.* 69(suppl. 1):250–285.

Ratajszczak, R. (1998). Taxonomy, distribution and status of the lesser slow loris *Nycticebus pygmaeus* and their implications for captive management. *Folia Primatol.* 69(suppl. 1):171–174.

Rhadakrishna, S. (2001). The social behavior of the Mysore slender loris (*Loris tardigradus lydekkerianus*) [PhD thesis]. University of Mysore, Manasagangotri.

Rhadakrishna, S., and Singh, M. (2002). Social behaviour of the slender loris (*Loris tardigradus lydekkerianus*). *Folia Primatol.* 73:181–196.

Rhadakrishna, S., and Singh, M. (2004). Reproductive biology of the slender loris (*Loris lydekkerianus lydekkerianus*). *Folia Primatol.* 75:1–13.

Roonwal, M. L., and Mohnot, S. M. (1977). *Primates of South Asia: Ecology, Sociobiology and Behavior*. Harvard University Press, Cambridge, MA.

Roos, C. (2003). Molekulare Phylogenie der Halbaffen, Schlankaffen, und Gibbons [diss.]. University of Munich, Munich.

Roos, C., Schmitz, J., and Zischler, H. (2004). Primate jumping genes elucidate strepsirrhine phylogeny. *Proc. Natl. Acad. Sci. USA* 101:10650–10654.

Ross, C. (1996). Adaptive explanation for the origins of the Anthropoidea (Primates). *Am. J. Primatol.* 40:205–230.

Sanfey, P. (2003). Study of re-released semi-free ranging slow and pygmy lorises in Thailand [MSc diss.]. Oxford Brookes University, Oxford.

Sarmiento, E. (1998). The validity of *Pseudopotto martini*. *Afr. Primates* 3:44–45.

Schilling, A. (1979). Olfactory communication in prosimians. In: Doyle, G. A., and Martin, R. D. (eds.), *The Study of Prosimian Behaviour*. Academic Press, London. pp. 461–542.

Schulze, H., and Groves, C. P. (2004). Asian lorises: taxonomic problems caused by illegal trade. In: Nadler, T., Streicher, U., and Thang Long, H. (eds.), *Conservation of Primates in Vietnam*. Frankfurt Zoological Society, Frankfurt. pp. 33–36.

Schulze, H., and Meier, B. (1995a). The subspecies of *Loris tardigradus* and their conservation status: a review. In: Alterman, L., Doyle, G. A., and Izard, M. K. (eds.), *Creatures of the Dark: The Nocturnal Prosimians*. Plenum Press, New York. pp. 193–209.

Schulze, H., and Meier, B. (1995b). Behaviour of captive *Loris tardigradus nordicus*: a qualitative description including some information about morphological bases of behavior. In: Alterman, L., Doyle, G. A., and Izard, M. K. (eds.), *Creatures of the Dark: The Nocturnal Prosimians*. Plenum Press, New York. pp. 221–250.

Schwartz, J. H. (1996). *Pseudopotto martini*: a new genus and species of extant lorisiform primate. *Anthropol. Papers Am. Mus. Nat. Hist.* 78:1–14.

Seiffert, E. F., Simon, E. L., and Attia, Y. (2003). Fossil evidence for an ancient divergence of lorises and galagos. *Nature* 422:421–424.

Sellers, W. (1996). A biomechanical investigation into the absence of leaping in the locomotor repertoire of the slender loris (*Loris tardigradus*). *Folia Primatol.* 67:1–14.

Singh, M., Kumar, M. A., Kumara, H. N., and Mohnot, S. M. (2000). Distribution and conservation of slender lorises in southern Andhra Pradesh, south India. *Int. J. Primatol.* 21:721–730.

Singh, M., Lindburg, D. G., Udhayan, A., Kumar, M. A., and Kumara, H. N. (1999). Status survey of the slender loris in Dindigul, Tamil Nadu, India. *Oryx* 33:31–37.

Sterling, E. J. (1993). Patterns of range use and social organization in aye-ayes (*Daubentonia madagascariensis*) on Nosy Mangabe. In: Kappeler, P. M., and Ganzhorn, J. U. (eds.), *Lemur Social Systems and Their Ecological Basis*. Plenum Press, New York. pp. 1–10.

Sterling, E. J., Nguyen, N., and Fashing, P. (2000). Spatial patterning in nocturnal prosimians: a review of methods and relevance to studies of sociality. *Am. J. Primatol.* 51:3–19.

Streicher, U. (2003). Saisonale Veränderungen in Fellzeichnung und Fellfärbung beim Zwergplumplori *Nycticebus pygmaeus* und irhe taxonomische Bedeutung. *Zool. Garten N.F.* 73:368–373.

Streicher, U. (2004). Aspects of the ecology and conservation of the pygmy loris *Nycticebus pygmaeus* in Vietnam [inaugural diss.]. Ludwig-Maximilians University, Munich.

Tan, C. L., and Drake, J. H. (2001). Evidence of tree gouging and exudate eating in pygmy slow lorises (*Nycticebus pygmaeus*). *Folia Primatol.* 72:37–39.

Tavare, S., Marshall, C. R., Will, O., Soligo, C., and Martin, R. D. (2002). Using the fossil record to estimate the age of the last common ancestor of extant primates. *Nature* 416:726–729.

Walker, A. C. (1969). The locomotion of the lorises, with special reference to the potto. *E. Afr. Wildlife J.* 7:1–5.

Weisenseel, K., Chapman, C. A., and Chapman, L. J. (1993). Nocturnal primates of Kibale Forest: effects of selective logging on prosimian densities. *Primates* 34:445–450.

Weisenseel, K. A., Izard, M. K., Nash, L. T., Ange, R. L., and Poorman-Allen, P. (1998). A comparison of reproduction in two species of *Nycticebus*. *Folia Primatol.* 69(suppl. 1):321–324.

Welker, C. (1973). Ethologic significance of the urine washing by *Galago crassicaudatus* E. Geoffroy, 1812 (Lorisiformes: Galagidae). *Folia Primatol.* 20:429–452.

Wickings, E. J., Ambrose, L., and Bearder, S. K. (1998). Sympatric populations of *Galagoides demidoff* and *G. thomasi* in the Haut Ogooué region of Gabon. *Folia Primatol.* 69(suppl. 1):389–393.

Wiens, F. (1995). Verhaltensbeobachtungen am plumplori, *Nycticebus coucang* (Primates: Lorisidae) im Freiland [diplomarbeit]. Universität Frankfurt am Main, Frankfurt am Main.

Wiens, F. (2002). Behavior and ecology of wild slow lorises (*Nycticebus coucang*): social organisation, infant care system and diet [PhD thesis]. Bayreuth University, Bayreuth.

Wiens, F., and Zitzmann, A. (2003a). Social dependence of infant slow lorises to learn diet. *Int. J. Primatol.* 24:1007–1021.

Wiens, F., and Zitzmann, A. (2003b). Social structure of the solitary slow loris *Nycticebus coucang* (Lorisidae). *J. Zool.* 261:35–46.

Yoder, A. D., Irwin, J. D., and Payseur, B. A. (2001a). Failure of the ILD to determine data combinability for slow loris phylogeny. *Syst. Biol.* 50:408–424.

Yoder, A. D., Rasoloarison, R. M., Goodman, S. M., Irwin, J. A., Atsalis, S., Ravosa, M., and Ganzhorn, J. U. (2001b). Remarkable species diversity in Malagasy mouse lemurs (Primates, *Microcebus*). *Proc. Natl. Acad. Sci. USA* 97:11325–11330.

Zimmerman, E. (1985). Vocalisations and associated behaviours in adult slow loris (*Nycticebus coucang*). *Folia Primatol.* 44:52–64.

Zimmermann, E. (1990). Differentiation of vocalisations in bushbabies (Galaginae, Prosimii, Primates) and the significance for assessing phylogenetic relationships. *Z. Zool. Syst. Evol.* 28:217–239.

Zimmermann, E. (1995a). Loud calls in nocturnal prosimians: structure, evolution and ontogeny. In: Zimmermann, E., Newman, J. D., and Jürgens, U. (eds.), *Current Topics in Primate Vocal Communication*. Plenum Press, New York. pp. 47–72.

Zimmermann, E. (1995b). Acoustic communication in nocturnal prosimians. In: Alterman, L., Doyle, G. A., and Izard, M. K. (eds.), *Creatures of the Dark: The Nocturnal Prosimians*. Plenum Press, New York. pp. 311–330.

Zimmermann, E., Bearder, S. K., Doyle, G. A., and Anderson, A. B. (1988). Variations in vocal patterns of Senegal and South African lesser bushbabies and their implications for taxonomic relationships. *Folia Primatol.* 51:87–105.

4

Lemuriformes

Lisa Gould and Michelle Sauther

INTRODUCTION

Behavioral and ecological research on the Malagasy primates began in the late 1950s, when Petter (1962) surveyed Madagascar's fauna and published preliminary information on several lemur species at a variety of sites around the island (Fig. 4.1). In the 1960s and 1970s, a number of primate biologists undertook the first in-depth studies of *Lemur catta* and *Propithecus verreauxi* at the Berenty site in the far south of the island (Jolly 1966) and comparative studies of *P. verreauxi* in the northwest and south (Richard 1973, 1974), *L. catta* and *Eulemur fulvus rufus* in the southwest (Sussman 1972, 1974), and *Indri* in the eastern rain forest (Pollock 1975, 1977). Some of the nocturnal lemurs were also studied for the first time in the 1970s by

Martin (1972a), who focused on *Microcebus*, while Charles-Dominique and Hladik (1971) documented early information on *Lepilemur*. On the nearby Comoro Islands, the only place outside of Madagascar where lemurs are found, Tattersall (1976, 1977b) conducted research on *Eulemur mongoz* on Moheli and Anjouan Islands as well as on the one subspecies of brown lemur not found on Madagascar, *E. f. mayottensis*, the Mayotte brown lemur, on the island of the same name.

In the 1970s, the political situation in Madagascar precluded most lemur research; but in the 1980s, many Malagasy,

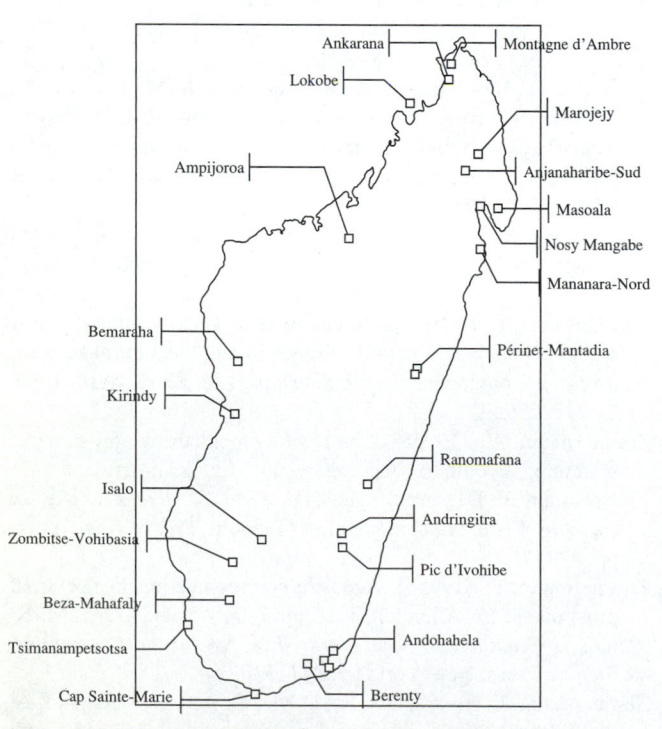

Figure 4.1 National parks and reserves in Madagascar. Based on a map from Madagascar: The Bradt Travel Guide, 7th ed., 2002. Reprinted with kind permission of the editor, H. Bradt.

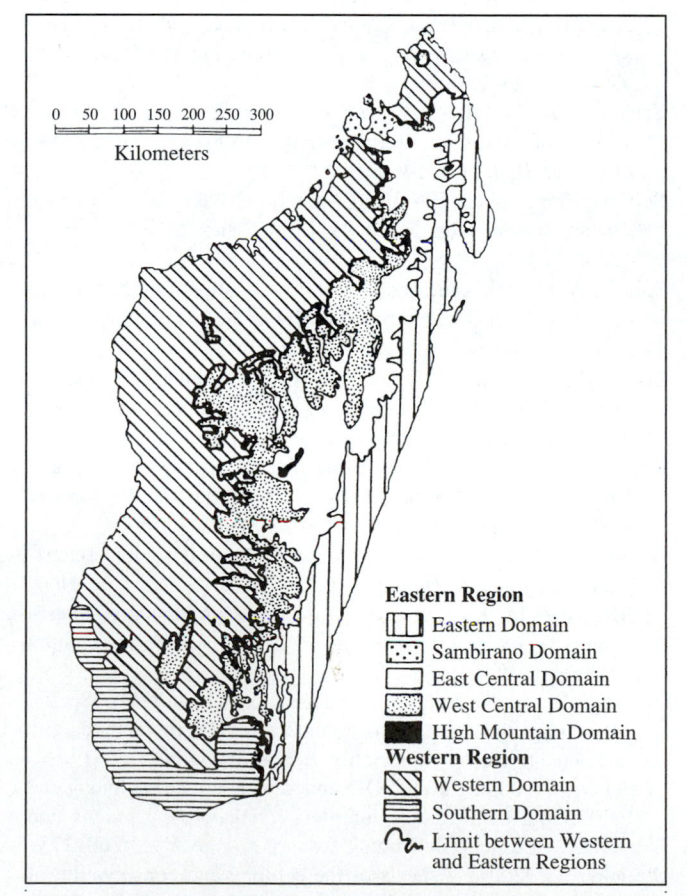

Figure 4.2 Map of vegetation zones in Madagascar. From Sussman (1999), based on Humber's (1955) original.

North American, European, and Japanese researchers began a new phase of field studies on most of the extant lemur species. Lemur research has grown remarkably in the past two and a half decades as topics such as geographic distribution; behavior and ecology of previously unknown species; correlations between climate, diet, and dominance patterns; long-term demographic and life history studies; hormonal correlates of behavior; health studies; and genetics have been and are being investigated.

In this chapter, we will first present information on the origins of the Malagasy primates, taxonomy and classification of the living lemurs, lemur morphology, and current explanations relating to the evolution of behavioral traits which distinguish lemurs from other primate groups. We will then give an overview of the ecology of the extant lemurs and end the chapter with a note concerning the extinct subfossil lemurs.

ORIGIN OF THE LEMURS

The landmass which is now Madagascar split from mainland Africa approximately 165 million years ago (mya) and drifted southward to its present position, 400 kms from mainland East Africa around 121 mya (Yoder et al. 2003). At this time, Madagascar was part of a larger landmass which included India. About 88 mya, the landmass split again, and India drifted northward, eventually colliding with Asia (Yoder et al. 2003). Madagascar has thus been isolated from any other landmass for 88 million years. How then did lemurs end up on Madagascar, when primates did not exist at the time of the landmass separation? Lemur ancestors arose in Africa either during the Eocene epoch (55–37 mya) or even earlier as Martin (2000) notes that deoxyribonucleic acid (DNA) sequencing suggests they may have begun to colonize Madagascar as early as 80 mya. The most accepted explanation as to how they got to Madagascar is via overwater dispersal, or "rafting," on large clumps of floating vegetation (Martin 1972b, 2000; Mittermeier et al. 1994; Yoder et al. 2003). It has been suggested that the ancestral lemurs survived this long journey because they may have had the ability to go into a state of torpor for lengths of time, which would have offset the problem of food shortages (Martin 1972b, 2000; Kappeler 2000). Tattersall (2004) argues that some form of "island hopping" may have occurred by means of paleogeographic "stepping stones" in the form of seabed exposures in the Mozambique channel at different times during the Tertiary and that these small islands would have reduced the distances between points of land that mammals crossing over to Madagascar would have had to travel. Whether today's lemurs arose from just one or several separate waves of migration (Yoder et al. 2003, Tattersall 2004), once on Madagascar these ancestral species underwent a large adaptive radiation over many millions of years, resulting in the living lemurs of today and the extinct (subfossil) lemurs.

CLASSIFICATION OF THE LEMURS

There are five taxonomic families of living Malagasy primates: the Lemuridae, Indriidae, Cheirogaleidae, Lepilemuridae (sometimes classified as Megaladapidae), and Daubentoniidae. Within these families we find 14 genera of extant lemurs, which represent 25% of all extant primate genera in the world (Fleagle 1999) and 43 living species. There are also six genera of very recently extinct lemurs (Burney 1997). Table 4.1 lists current extant lemur and extinct sub-fossil taxonomy.

Morphology

Like other strepsirhines, lemurs are morphologically characterized by a number of primitive features of the skull, including a postorbital bar rather than postorbital closure, a primitive mammalian nasal region, reduced upper incisors, and a toothcomb, made up of the lower incisors and canines

Table 4.1   Taxonomy of the Malagasy Primates

Extant lemurs

Family Lemuridae

Lemur catta (ring-tailed lemur)

Eulemur fulvus (**7 subspecies**)
  E. f. fulvus (common brown lemur), *E. f. rufus* (red-fronted brown lemur), *E. f. sanfordi* (Sanford's brown lemur), *E. f. albifrons* (white-fronted brown lemur), *E. f. collaris* (collared brown lemur), *E. f. albocollaris* (white-collared brown lemur), *E. f. mayottensis* (Mayotte brown lemur, on Mayotte Island in the Comoros)

Eulemur macaco (**2 subspecies**)
  E. m. macaco (black lemur), *E. m. flavifrons* (Sclater's black lemur)

Eulemur coronatus (crowned lemur)

Eulemur rubriventer (red-bellied lemur)

Eulemur mongoz (mongoose lemur)

Hapalemur griseus (**3 subspecies**)
  H. g. griseus (eastern lesser bamboo lemur), *H. g. occidentalis* (western lesser bamboo lemur), *H. g. alaotrensis* (Lac Alaotra bamboo lemur)

Hapalemur aureus (golden bamboo lemur)

Hapalemur simus (greater bamboo lemur)

Varecia variegata (**2 subspecies**)
  V. v. variegata (black and white ruffed lemur), *V. v. rubra* (red ruffed lemur)

Family Indriidae

Propithecus verreauxi (**4 subspecies**)
  P. v. verreauxi (Verreaux's sifaka), *P. v. deckeni* (Decken's sifaka), *P. v. coronatus* (crowned sifaka), *P. v. coquereli* (Coquerel's sifaka)

Propithecus diadema (**4 subspecies**)
  P. d. diadema (diademed sifaka), *P. d. edwardsi* (Milne-Edwards sifaka), *P. d. candidus* (silky sifaka), *P. d. perreri* (Perrier's sifaka)

Propithecus tattersalli (golden-crowned sifaka)

Indri indri (indri)

Avahi laniger (eastern woolly lemur)

Avahi occidentalis (western woolly lemur)

Avahi unicolor (unicolor avahi)

Table 4.1   (cont'd)

Family Lepilemuridae (also sometimes considered Megaladapidae)

Lepilemur mustelinus (weasel sportive lemur)

Lepilemur microdon (small-toothed sportive lemur)

Lepilemur leucopus (white-footed sportive lemur)

Lepilemur ruficaudatus (red-tailed sportive lemur)

Lepilemur edwardsi (Milne-Edwards sportive lemur)

Lepilemur dorsalis (gray-backed sportive lemur)

Lepilemur septentrionalis (northern sportive lemur)

Family Cheirogalidae

Microcebus murinus (gray mouse lemur)

Microcebus rufus (eastern rufous mouse lemur)

Microcebus myoxinus (pygmy mouse lemur)

Microcebus ravelobensis (golden-brown mouse lemur)

Microcebus tavaratra (northern rufous mouse lemur)

Microcebus sambiranensis (Sambirano mouse lemur)

Microcebus berthae (Berthe's mouse lemur)

Microcebus griseorufus (gray-brown mouse lemur)

Allocebus trichotis (hairy-eared dwarf lemur)

Cheirogaleus major (greater dwarf lemur)

Cheirogaleus crossleyi (Crossley's greater dwarf lemur)

Cheirogaleus ravus (large iron-gray dwarf lemur)

Cheirogaleus minisculus (lesser iron-gray dwarf lemur)

Cheirogaleus sibreei (Sibree's dwarf Lemur)

Cheirogaleus medius (fat-tailed dwarf lemur)

Cheirogaleus adapicaudatus (southern dwarf lemur)

Phaner furcifer (4 subspecies)
P. f. furcifer (eastern fork-marked lemur), *P. f. pallescens* (pale fork-marked lemur), *P. f. parienti* (Pariente's fork-marked lemur), *P. f. electromontis* (Amber Mountain fork-marked lemur), *Mirza coquereli* (Coquerel's dwarf lemur)

Family Daubentoniidae

Daubentonia madagascariensis (aye-aye)

Subfossil lemurs

Subfossil Lemuridae

Pachylemur insignis

Pachylemur jullyi

Subfossil Daubentoniidae

Daubentonia robusta

Subfossil Megaladapidae

Megaladapis grandidieri

Megaladapis madagascariensis

Megaladapis edwardsi

Family Paleopropithecidae (all subfossils)

Mesopropithecus globiceps

Mesopropithecus pithecoides

Mesopropithecus dolichobrachion

Babakotia radofilai

Paleopropithecus ingens

Paleopropithecus maximus

Archaeoindris fontoynontii

Source: Adapted from Tattersall (1982), Mittermeier et al. (1994), Groves (2000), Rasoloarison et al. (2000), Jungers et al. (2002).

(with the exception of the aye-aye, see below). They also possess a moist rhinarium, or "wet nose," a primitive mammalian trait which aids in their very keen sense of smell (Fleagle 1999). Scent marking is used by all species in various forms and contexts: scent glands are situated on heads and palms in male *Eulemur*, under wrist spurs in male *L. catta*, on chest glands in *Propithecus*, and in the anogenital area in all species, with the scent glands situated under the tail. Scent is used commonly to denote the presence of a group or individual and extensively during the mating season.

All lemurs possess a grooming claw on the second toe of each foot (Fleagle 1999).

The dental formula is variable. In both the cheirogaleids and the lemurids, the dental formula is 2.1.3.3. Lepilemurids lack permanent upper incisors, so their upper dental formula is 0.1.3.3. and the lower, 2.1.3.3. The indriids have quite different dentition from the above lemurs. Their dental formula is the same as that found in the Old World monkeys, apes, and hominids, 2.1.2.3; and their toothcomb includes the incisors only, not the canine as in the above taxa. The oddest dentition of all lemurs, and probably of all primates, is that of the aye-aye (*Daubentonia madagascariensis*), the only living member of the Daubentoniidae family. Its dental formula is 1.0.1.3, and the middle two incisors grow constantly, like those found in rodents and lagomorphs (rabbits and hares) (Fleagle 1999).

Mean weights of species captured in the wild can be found in Table 4.2.

Nocturnal Lemur Postcranial Morphology and Locomotion

Most nocturnal lemurs have relatively short, pointed snouts and large, moveable ears. The mouse lemurs (*Microcebus*) are branch runners. Their arms and legs are short relative to their trunk, and their tail is as long as their body. Dwarf lemurs (*Cheirogaleus*) have a shorter tail and legs that are longer than the arms. Sportive lemurs (*Lepilemur*) have an enlarged cecum, to help digest the cellulose in their mostly folivorous diet (Fleagle 1999). Sportive lemurs are not branch runners; rather, they travel by vertical leaping (Ganzhorn 1993). The woolly lemur (*Avahi*) is the only nocturnal indriid. It is much smaller than its close relatives, the diurnal sifakas (*Propithecus*) and indri (*Indri*); but its mode of locomotion is the same: vertical clinging and leaping. Using vertical clinging and leaping, the animal begins the leap with its back toward its destination tree, takes a large leap, twists in midair, and lands facing forward (Richard 1985). The legs of both *Lepilemur* and *Avahi* are much longer than their arms, a necessary adaptation for this type of locomotion (Fleagle 1999). The aye-aye, an extremely unusual-looking primate, is covered in black, shaggy hair and has enormous bat-like ears, a large bushy tail, and an extremely elongated third digit on its hands, which it uses in extractive foraging for grubs, insects under bark, and egg yolks (Erickson 1991, 1994; Fleagle 1999).

Table 4.2   Mean Body Weights of Wild-Caught Diurnal Lemur species

SPECIES	MEAN WEIGHT (AND/OR WEIGHT RANGE)	SOURCE
Diurnal lemurs		
Lemur catta	2.2 kg	Sussman 1991
Eulemur fulvus ssp.	1.8–2.6 kg	Glander et al. 1992, Mittermeier et al. 1994, Freed 1996, Terranova and Coffman 1997, Vasey 2000
E. mongoz	1.5 kg	Terranova and Coffman 1997
E. macaco	2.4 kg	Mittermeier et al. 1994
E. rubriventer	2.0 kg	Glander et al. 1992
E. coronatus	1.77 kg	Terranova and Coffman 1997
Hapalemur griseus griseus H. g. occidentalis	700–1,000 g	Mittermeier et al. 1994, Tan 1998
H. g. alaotrensis	1.2 kg	Mutschler 2002
H. aureus	1.5–1.6 kg	Glander et al. 1992, Tan 1998
H. simus	2.4 kg	Meier et al. 1987, Tan 1998
Varecia variegata	3–4.5 kg	Tattersall 1982, Morland 1993, Terranova and Coffman 1997, Britt et al. 2001, Vasey 2002
Propithecus verreauxi verreauxi	2.8 kg	Richard et al. 2002
P. diadema edwardsi	5–6.5 kg	Glander et al. 1992, Wright 1995
P. d. diadema	5–6 kg	Powzyk 1997
P. tattersalli	3.5 kg	Meyers and Wright 1993
Indri indri	6.5–6.9 kg	Powzyk 1997, Britt 2000
Nocturnal lemurs		
Cheirogaleus medius	Body weight changes seasonally (range = 75–200 g)	Hladik et al. 1980
C. major	Body weight changes seasonally (mean = 400 g)	Martin 1984
C. crossleyi	Body weight changes seasonally (mean = 400 g)	Petter et al. 1977
Allocebus trichotis	75–80 g	Meier and Albignac 1991
Mirza coquereli	300 g	Tattersall 1982
Microcebus murinus	50–90 g (mean = 62.3)	Martin 1973, Rasoloarison et al. 2000
M. rufus	50 g	Harcourt 1987
M. myoxinus	49 g	Rasoloarison et al. 2000
M. ravelobensis	71.7 g	Rasoloarison et al. 2000
M. tavaratra	61.1 g	Rasoloarison et al. 2000
M. sambirianensis	44.1 g	Rasoloarison et al. 2000
M. berthae	30.6 g	Rasoloarison et al. 2000
M. griseorufus	62.6 g	Rasoloarison et al. 2000
Phaner furcifer	360–500 g	Petter et al. 1977
Avahi laniger	900–1,200 g and 600–700 g	Razanahoera-Rakotomalala 1981, Petter et al. 1977
A. occidentalis	700–900 g	Razanahoera-Rakotomalala 1981
Daubentonia madagascariensis	3 kg	Tattersall 1982
Lepilemur mustelinus	1 kg	Jenkins 1987
L. dorsalis	500 g	Tattersall 1982
L. septentrionalis	700–800 g	Tattersall 1982
L. edwardsi	600–900 g	Tattersall 1982
L. leucopus	550 g	Petter et al. 1977, Petter and Petter-Rousseaux 1979
L. ruficaudatus	600–900 g	Petter et al. 1977, Petter and Petter-Rousseaux 1979
L. microdon	1 kg	Petter et al. 1977, Petter and Petter-Rousseaux 1979

Because there is very little sexual dimorphism in terms of body weight, "mean weight" is presented here as the actual mean in kilograms when data are combined for both males and females.

Postcranial Morphology and Locomotion of the Diurnal Lemurs

Lemur, *Eulemur*, and *Varecia*, are quadrupedal walkers and runners; but they also leap from branch to branch. *Varecia* also uses suspensory postures for feeding (Fleagle 1999). *L. catta*, the only species within the genus *Lemur*, spends about 30% of its time on the ground (Jolly 1966, Sussman 1974). Unlike other lemurs, the fleshy pads of its hands and feet extend upward to the wrist and beyond the heel.

The three *Hapalemur* species have shorter faces than the other lemurids. Their arms are short and legs are proportionally long (Jungers 1979), an adaptation to their primary mode of locomotion, vertical clinging and leaping, although the three species also move quadrupedally along branches when feeding (Fleagle 1999).

The longest leg in proportion to arm length is found in the indriids and is an adaptation to the vertical clinging and leaping mode of locomotion: the indriids are extraordinary leapers. Some sifaka (*Propithecus* spp.) occasionally come to the ground, particularly Verreaux's sifaka, which lives in dry forests. Because their legs are so long, they must move along the ground by hopping bipedally. Indriids also possess very long fingers and toes, which aid them when adopting suspensory feeding postures.

EVOLUTION OF UNIQUE BEHAVIORAL TRAITS IN MALAGASY PRIMATES, INCLUDING FEMALE DOMINANCE

Not only do lemurs exhibit aspects of morphology which differ from anthropoids but some unique behavioral traits have evolved in this group of primates which are not found in other strepsirhines or the haplorhines. These traits include female dominance in the majority of species, targeted female–female aggression, lack of notable sexual dimorphism, strict seasonal breeding (in all but two species, the aye-aye and the Lac Alaotra bamboo lemur), high infant mortality, and cathemerality (exhibiting both day and night activity). Why do we find such a combination of characteristics in the lemurs? A number of explanations have been offered and are explained in detail in Wright (1999) and Sussman (1999). These hypotheses are briefly presented below.

The *energy conservation hypothesis* involves a synthesis of explanations presented by Jolly (1984), Richard (1987), Young et al. (1990), Wright (1993), Sauther (1993), and Pereira (1993a,b). This hypothesis proposes that the combination of extreme and largely unpredictable climatic seasonality in Madagascar and high pre- and postreproductive costs have resulted in the evolution of female dominance. More specifically, strong food resource seasonality and climatic factors have resulted in energetic stress with respect to reproductive females. In group-living lemurs, all females in a social group are pregnant and lactating at the same time, which leads to strong female–female feeding competition during both gestation and lactation periods. Furthermore, in both group- and non-group-living lemur species, females give birth to altricial, quickly growing infants for which they must lactate. Female lemurs may have responded to such reproductive stress through the evolution of female priority of access to food resources, which can help them offset energy demands experienced in this situation. In addition to seasonal reproduction which is largely tied to availability of good weaning foods for growing infants, the relatively lengthy winter season in parts of Madagascar is proposed to have led to the evolution of seasonal energy storage in some species, strategies for temperature regulation, modulation of metabolic and growth rates and activity levels, and timing of aggressive behaviors.

Wright (1999) argues that not all of the unique behavioral traits found in lemurs are strictly related to the conservation of energy and suggests that some are more tailored to maximizing the extraction of scarce resources. She proposes that low basal metabolic rate, small group size, torpor, sperm competition, and seasonal breeding are adaptations related to energy conservation and that others, such as female dominance, weaning synchrony, fibrous diets, territoriality, and cathemerality (found in some species), have evolved as strategies to maximize the use of scarce resources resulting from seasonal resource shortages. Thus, Wright (1999) suggests that it may be appropriate to consider the energy conservation hypothesis as more of an "energy frugality" hypothesis.

The *evolutionary disequilibrium hypothesis* (van Schaik and Kappeler 1996) suggests that recent extinctions, particularly of large diurnal predators such as raptors, have allowed many lemur species to switch from a nocturnal activity pattern to diurnality (and cathemerality). Because of such extinctions, adaptation to a diurnal activity pattern is suggested to have occurred rapidly and recently: between 1,000 and 500 years ago. Also, the social systems of today's diurnal lemurs (species living in relatively small mixed-sex groups) may be an outgrowth of an ancestral nocturnal, monogamous condition; and pair-living animals may have been sufficiently tolerant, once diurnal, to form larger groups. van Schaik and Kappeler (1996) propose that cathemeral activity may be an ancestral and stable activity pattern among lemurs, or, conversely, may have evolved relatively recently as an occasional habit of nocturnal animals. With respect to the evolution of female dominance, van Schaik and Kappeler suggest that in group-living lemurs today female dominance may be a relic of pair-living in ancestral times since female priority of access to resources seems to be the case in pair-living species that do not exhibit sexual dimorphism and where male polygyny, and male–male competition, does not occur. They argue that the expansion of female feeding priority to overall female dominance in *group*-living lemurs (emphasis ours) suggests that female needs in ancestral monogamous species were greater in lemurs than in other primate taxa.

OVERVIEW OF THE ECOLOGY, SOCIAL ORGANIZATION, AND SOME ASPECTS OF BEHAVIOR OF THE EXTANT LEMURS

Mean group size and home range size of extant lemur species (Fig. 4.3) can be found in Table 4.3, and mean weights of wild-caught animals are presented in Table 4.2.

Lemuridae (*Lemur, Eulemur, Hapalemur, Varecia*)

Lemur catta (Ring-Tailed Lemur)

L. catta is one of the two lemur species that has been studied over the longest period of time, beginning with Jolly's (1966) pioneering work. Ring-tailed lemurs have been studied primarily at three sites in the south and

(A)

(B)

(C)

(D)

Figure 4.3 (A) *Lemur catta* (photo by L. Gould). (B) *Eulemur coronatus* (photo by B. Z. Freed). (C) *Hapalemur simus* (photo by D. Haring). (D) *Propithecus verreauxi verreauxi* (photo by M. L. Sauther).

(E)

(F)

(G)

(H)

Figure 4.3 *(cont'd)* (E) *Propithecus diadema edwardsi* (photo by S. Arrigo-Nelson). (F) *Varecia variegata variegata* (photo by S. Arrigo-Nelson). (G) *Lepilemur leucopus* (photo by L. Gould). (H) *Microcebus ravelobensis* (photo by U. Radespiel).

Table 4.3 Group Size, Home Range Size, and Habitat of Several Lemur Species at Sites in Madagascar

SPECIES	MEAN GROUP SIZE (OR RANGE OF GROUP SIZE)	MEAN HOME RANGE SIZE (RANGE)	HABITAT AND WHERE STUDIED	STUDIED BY OR CITED BY
Diurnal lemurs				
Lemur catta	11.5 and 16 (at two sites), range 3–27	6–35 depending on local habitat	South and southwestern riverine, xerophytic, spiny, and limestone forest: Beza Mahafaly Special Reseve and Berenty Reserve	Budnitz and Dainis 1975; Sussman 1977, 1991; Jolly et al. 2002; Koyama et al. 2002; Gould et al. 2003
Eulemur fulvus rufus (eastern)	6.8	100 ha	Southeastern submontane rain forest: Ranomafana National Park	Overdorff 1993a, Overdorff et al. 2003
E. f. rufus (western)	9.4	1–9 ha	Western dry deciduous forest: Mangoky River, Kirindy Forest	Sussman 1974, Kappeler and Erkert 2003
E. f. fulvus	12	~7 ha (west), >20 (east)	Northwestern dry deciduous forest, eastern rain forest: Ampijoroa (west), Andasibe (east)	Harrington 1975, Ganzhorn 1988, Mittermeier et al. 1994
E. f. sanfordi	Range 5–9	5–9 ha	Northern dry deciduous forest: Mt. d'Ambre	Freed 1996
E. f. albifrons	Range 7–11	13 ha	Northeastern rain forest: Masoala Peninsula	Vasey 2000
E. f. collaris	–	–	Southeastern rain forest and littoral forest: St. Luce	Mittermeier et al. 1994, Donati and Borgognini-Tarli 2002a
E. f. albocollaris	–	–	Restricted range, southeastern rain forest remnants	Tattersall 1982, Mittermeier et al. 1994
E. rubriventer	2–4	19 ha	Southeastern submontane rain forest: Ranomafana National Park	Overdorff 1993a
E. coronatus	Range 5–9	6.5–15.5 ha	Northern dry deciduous forest: Ankarana Reserve, Montagne d'Ambre National Park	Wilson et al. 1989, Freed 1996
E. macaco	10, range 5–14	3.5–7 ha	Northwestern dry forest: Sambirano, Nosy Be	Colquhoun 1993, Andrews and Birkenshaw 1998
E. mongoz	Range 3–8	2.8 ha (0.5–1.0)	Northwestern dry forest: Comoran Islands of Anjouan and Mohéli Humid forest: Ampijoroa Reserve	Tattersall 1977b, Harrington 1978, Curtis and Zaramody 1997
Hapalemur griseus griseus	Range 2–9	6–10 ha/14–20 ha	Eastern rain forest, southeastern submontane forest: Andasibe National Park, Ranomafana National Park	Wright 1986, Tan 1998, Grassi 2001
H. g. occidentalis	Range 1–4	26 ha	Isolated forest regions: western Madagascar, Manongarivo Reserve	Petter and Peyriéras 1970b, Tattersall 1982, Raxworthy and Rakotondraparany 1988
H. g. alaotrensis	Range 3–9	–	Reed beds: Lac Alaotra, east-central Madagascar	Mutschler 2002
H. aureus	Range 2–6	26–80 ha	Restricted range—found only at two sites in southeastern submontane rain forest: Ranomafana and Andringitra National Parks	Wright et al. 1987, Meier and Rumpler 1987, Mittermeier et al. 1994, Tan 1998
H. simus	Range 4–12	62 ha	Southeastern submontane rain forest: Ranomafana region, spotted in Andringitra National Park	Meier and Rumpler 1987, Wright et al. 1987, Tan 1998
Varecia variegata variegata	Range 2–6 and 8–16 (at Nosy Mangabe)	30–150 ha (depends on site and habitat disturbance)	Lowland, mid-altitude, and higher-altitude rain forests in northern, eastern, and east-central Madagascar: Nosy Mangabe, Ranomafana National Park, Manombo, Betampona	Morland 1991, Balko 1998, White 1989, Ratsimbazafy 2002, Britt et al. 2001
V. v. rubra	2–6 (Rigamonti)	25–58 ha	Northeastern rain forest: Masoala Peninsula	Rigamonti 1993, Vasey 2002
Propithecus verreauxi	2–14, mean = 6	3–10 ha	South, southwestern, western dry deciduous forest, spiny forest: Beza Mahafaly Reserve, Kirindy Reserve	Richard et al. 2002
P. v. coquereli	Range 3–10	?	Northwestern mixed-deciduous and evergreen forests, and brush and scrub forest: Ankarafatsika Reserve	Petter 1962, Albignac 1981b
P. v. coronatus	?	?	Northwestern Madagascar	Petter et al. 1977, Mittermeier et al. 1994

Table 4.3 (*cont'd*)

SPECIES	MEAN GROUP SIZE (OR RANGE OF GROUP SIZE)	MEAN HOME RANGE SIZE (RANGE)	HABITAT AND WHERE STUDIED	STUDIED BY OR CITED BY
P. v. deckeni	?	?	Western deciduous (fragments): Tsingy de Bemaraha Reserve	Mittermeier et al. 1994
P. diadema edwardsi	Range 2–9, mean = 5.3	400 ha	Southeastern submontane rain forest: Ranomafana National Park	Wright 1995, Pochron and Wright 2003
P. d. diadema	Range 3–8, mean = 4.83	33–42 ha	Eastern rain forest: Mantadia National Park	Powzyk 1997, Powzyk and Mowry 2003
P. d. perrieri	Range 2–6	30 ha	Northern dry forest: Analamera Reserve, Ankarana Reserve	Meyers and Ratsirarson 1989, Hawkins et al. 1990
P. d. candidus	Range 3–7, mean = 4.3	?	Northeastern humid forest: Marojejy Reserve	Safford et al. (unpub. report cited in Mittermeier et al. 1994)
P. tattersallli	Range 3–10, mean = 5	9–12 ha	Northeastern dry deciduous forest and semi-evergreen forest patches: Daraina region	Meyers 1993, Meyers and Wright 1993, Mittermeier et al. 1994, Vargas et al. 2002.
Indri indri	2 + offspring	34–40 ha	Eastern rain forest: Mantadia National Park	Powzyk 1997, Powzyk and Mowry 2003
Nocturnal lemurs				
Cheirogaleus medius	2 + offspring (monogamous)	4 ha	Dry deciduous forest: south and southwestern Madagascar	Hladik et al. 1980; Müller 1998, 1999; Fietz 1999
C. major	Solitary	?	Eastern lowland rain forest	Petter et al. 1977, Tattersall 1982
C. ravus	?	?	Eastern rain forest	Groves 2000
C. crossleyi	?	?	Northeastern Madagascar plateau	Groves 2000
C. adapicaudatus	?	?	Southern spiny forest	Groves 2000
C. minisculus	?	?	Central Madagascar: Ambositra	Groves 2000
C. sibreei	?	?		
Allocebus trichotis	Sleeps in groups of 2–6	?	Northern tropical evergreen forest	Meier and Albignac 1991
Mirza coquereli	Sleeps in groups up to 6 at some sites	~4 ha	Western coastal forests	Tattersall 1982
Microcebus murinus	Sleeps in groups of 1–15	0.22–3.2 ha (males), 0.24–1.8 ha (females)	Dry deciduous forest, spiny forest, littoral forest: western and southern Madagascar	Martin 1972a, 1973; Pagès-Feuillade 1989; Radespiel et al. 1998; Ramanamanjato and Ganzhorn 2001
M. rufus	?	?	Northwest and eastern rain forest	Tattersall 1982, Atsalis 2000
M. myoxinus	?	?	Dry forests: southern and western Madagascar	Petter et al. 1971, Rasoloarison et al. 2000
M. ravelobensis	Sleeps in groups	0.44–0.79 ha	Northwestern Madagascar	Petter 1962, Zimmerman et al. 1998
M. tavaratra	?	?	Northern dry deciduous forest: Ankanrana	Rasoloarison et al. 2000
M. berthae	Solitary/dispersed, do not sleep in groups	Male home ranges larger than female, size not known	Western dry deciduous forest: Kirindy, Andranomena, Analabe	Schmid and Kappeler 1994, Rasoloarison et al. 2000, Schwab 2000
M. griseorufus	?	?	Southestern dry deciduous forest, spiny forest: Beza Mahafaly	Rasoloarison et al. 2000, Rasoazanabary personal comm.
Phaner furcifer	Dispersed pairs (male–female pairs sleep together in nests)	3.8–4 ha	Humid, dry, and secondary forest: western and northern Madagascar	Petter and Petter-Rousseaux 1979, Schulke and Kappeler 2003
Avahi laniger	Male–female pairs and offspring	1–2 ha	Rain forest, coastal forest: eastern Madagascar	Albignac 1981a, Ganzhorn et al. 1985, Harcourt 1988
A. occidentalis	Male–female pairs and offspring	?	Dry deciduous forest: northwestern Madagascar	Petter et al. 1977, Tattersall 1977a
A. unicolor	Male–female pairs and offspring	?	Dry deciduous forest: northwestern Madagascar, Sambirano region	Thalmann and Geissmann 2000
Daubentonia madagascariensis	Solitary	mean = 35.6 (females), 170.3 (males)	Eastern, western, and northern Madagascar; primary rain forest, deciduous forest, secondary growth, cultivation, and dry scrub forest: Nosy Mangabe	Tattersall 1982, Sterling 1993, Sterling and Richard 1995
Lepilemur mustelinus	Solitary	1.5	Eastern rain forest	Ratsiraron and Rupler 1988

Table 4.3 *(cont'd)*

SPECIES	MEAN GROUP SIZE (OR RANGE OF GROUP SIZE)	MEAN HOME RANGE SIZE (RANGE)	HABITAT AND WHERE STUDIED	STUDIED BY OR CITED BY
L. dorsalis	Solitary	?	Humid forest, northwestern Madagascar: Nosy Be	Petter et al. 1977, Tattersall 1982
L. septentrionalis	Solitary	1 ha	Dry deciduous forest and humid forest: northern Madagascar	Tattersall 1982, Ratsirarson et al. 1987, Hawkins et al. 1990
L. edwardsi	Dispersed pairs (male–female pairs sleep together in tree holes)	1 ha	Dry deciduous forests: western Madagascar	Petter and Petter-Rousseaux 1979; Albignac 1981a,b; Warren 1994
L. leucopus	Solitary/dispersed (male–female pairs or mother–daughter pairs sleep together)	0.18–0.3 ha	Spiny and gallery forests: southern Madagascar, Beza Mahafaly Reserve	Petter and Petter-Rousseaux 1979, Sussman and Richard 1986
L. ruficaudatus	Solitary or male–female pairs	0.8 ha	Dry forest: western Madagascar	Petter and Petter-Rousseaux 1979
L. microdon	?	?	Eastern rain forest	Petter et al. 1977

southwest of Madagascar: Berenty, Beza Mahafaly, and Antserananomby.

While this species is found primarily in riverine, *xerophytic* (drought-adapted), and scrub forests in south and southwestern Madagascar (Jolly 1966, Sussman 1977), it is also found in spiny forests and low-lying limestone forests; and one population has even been found at Andringitra National Park in the central southeast, living above the tree line at an elevation of 2,500 m (Goodman and Langrand 1996).

L. catta has been defined as a very flexible "edge" species, able to withstand relatively extreme temperatures and to recover from serious droughts (Sussman 1977, Gould et al. 1999, Sauther et al. 1999). It is the most terrestrial species of lemur, spending up to 30% of the time on the ground (Jolly 1966, Sussman 1974).

Social Organization.   *L. catta* lives in multimale/multifemale groups (Jolly 1966, Sussman 1977). Group fission commonly occurs when groups reach a critical size, and they split along matrilines (Sussman 1991, Koyama 1991, Hood and Jolly 1995, Jolly et al. 2002, Gould et al. 2003). This species is female-philopatric, and males disperse at 3–4 years of age (Sussman 1992).

Diet and Feeding Ecology.   Ring-tailed lemurs have been described as "opportunistic omnivores" (Sauther et al. 1999), feeding on fruit (particularly tamarind), leaves and stems, flowers, some insects, and soil from both the ground and termite mounds (Jolly 1966, Sussman 1977, Sauther 1992, Sauther et al. 1999). Food resources are extremely seasonal, and the regions where *L. catta* are found often experience severe droughts (Jolly 1966, Sussman 1977, Sauther 1992, Gould et al. 1999). As a result, ring-tailed lemurs are very flexible and can switch their primary food resources to follow ecological unpredictability. For example, *L. catta* groups will expand their home ranges into those of other groups when particular seasonal resources are unavailable

in their own home range (Budnitz 1978, Jolly et al. 1993, Sussman 1991, Sauther and Sussman 1993).

Reproduction.   Reproductive synchrony is marked in this species and strongly tied to the specific nature of seasonal food resources (Jolly 1966; Sauther 1992, 1998). Average gestation length is 141 days (Sauther 1991). Although females normally give birth to a single infant, twinning in the wild has occasionally been reported (Koyama et al. 2002, Jolly et al. 2002, Bauer, personal communication). Infants are born near the end of the dry season in September and October and weaned at 4–5 months, during the rainy season, when weaning foods are available (Jolly 1966, Sussman 1977, Gould 1990, Sauther 1992). Alloparental care is common (Gould 1992).

Sauther (1994) found that pregnant females feed primarily on fruit and flowers but, when lactating, switch to low-cost, predictable, high-protein plant foods.

Infant mortality differs at the two different sites where it has been documented. Koyama et al. (2002) report 32%–37% at Berenty, where many groups are water-provisioned; but at Beza, where no provisioning occurs, a range of 52%–80% has been reported, depending on rainfall (Gould et al. 1999, 2003).

Eulemur

Within the *Eulemur* genus, we find five species and many subspecies. *E. fulvus* (the brown lemur) is the most geographically widespread, and as a result, there are seven subspecies (Table 4.1). All subspecies except *E. f. mayottensis*, the Mayotte brown lemur which lives on the island of Mayotte in the neighboring Comoros, inhabit continuous forest throughout Madagascar (Freed 1999).

Sexual Dichromatism.   One interesting morphological feature of *Eulemur* is that most species and subspecies are sexually dichromatic, making it easy to distinguish males from females, even in relatively high canopy. The extent of dichromatism ranges from completely different pelage color in *E. macaco macaco* and *E. m. flavifrons* (males black, females russet brown) to different-colored ventrums, beards, heads, and face markings in *E. fulvus spp.*, *E. rubriventer*, and *E. mongoz*. *E. f. fulvus* is the only lemur in this group that exhibits no sexual dichromatism (Harrington 1975, Mittermeier et al. 1994).

Reproduction. Gestation in *Eulemur* in the wild is reported to be between 120 and 126 days, and infants in most species are born between September and November (Colquhoun 1993, Mittermeier et al. 1994, Sussman 1999). Females normally give birth to just one infant, which, like most diurnal lemurs, can cling immediately. Interbirth intervals are 1 year, though Overdorff et al. (1999) found that in an *E. f. rufus* population which had been studied for 7 years, the mean interbirth interval between surviving offspring was 2.1 years. In *E. mongoz*, which is often monogamous, adult males frequently carry the infant after it is 2 weeks old (Curtis and Zaramody 1997).

Cathemerality. Cathemerality has been observed in all *Eulemur* species. Most species exhibit year-round cathemeral activity but with some seasonal variation (Overdorff 1988; Rasmussen 1999; Donati et al. 1999, 2001; Donati and Borgognini Tarli 2002b; Andrews and Birkenshaw 1998; Freed 1996; Overdorff and Rasmussen 1995; Kappeler and Erkert 2003). For example, *E. mongoz* exhibits greater nocturnal activity during the cooler, dry seasons, which may correlate with thermoregulation during long cool nights (Curtis et al. 1999). Low nocturnal light during the wet season compromises nighttime activity in *E. mongoz* (Rasmussen 1999), and Colquhoun (1998), Donati et al. (2001, 2002b), and Kappeler and Erkert (2003) stress that nocturnal activity in *E. m. macaco*, *E. f. collaris*, and *E. f. rufus* is strongly dependent on phases of the moon and available light. Rasmussen (1999) also suggests that in seasonally dry forests cathemeral activity may function as an antipredator strategy during times when canopy cover is thin.

Rasmussen (1999) divides cathemerality into three types: seasonal differences in day and night activity, found in *E. mongoz*; seasonal shift from diurnal to 24 hr activity, found in *E. f. fulvus* (Rasmussen 1999) and *E. f. rufus* (Donati et al. 1999); and year-round 24 hr activity, found in all *Eulemur* species that have been studied in rain forest habitats (Andrews and Birkenshaw 1998, Freed 1996, Overdorff 1988, Overdorff and Rasmussen 1995) as well as in some dry forest habitats (Kappeler and Erkert 2003).

Van Schaik and Kappeler (1996) propose that cathemerality may have occurred in formerly nocturnal taxa, due to an "evolutionary disequilibrium" related to human activities causing the subsequent extinction of both the aforementioned large-bodied lemurs as well as large raptors. These authors suggest that the extinctions of large raptors allowed for greater diurnal activity in the relatively small-bodied *Eulemur* species. Colquhoun (1993) suggests that cathemeral activity may be an ancestral trait for the entire *Eulemur* genus. Kappeler and

Erkert (2003) argue that cathemerality evolved from nocturnal ancestors, perhaps relatively recently, and may have occurred by adding some diurnal activity to a largely nocturnal baseline. Because cathemeral primates are primarily restricted to Madagascar, Kappeler and Erkhart suggest that the unusual aspects of Madagascar's ecology, outlined in Richard and Dewar (1991) and Wright (1999), have allowed for such a transition in activity pattern.

Eulemur fulvus (Brown Lemur, Seven Subspecies: Common Brown Lemur, Red-Fronted Brown Lemur, White-Fronted Brown Lemur, Sanford's Lemur, Collared Lemur, White-Collared Brown Lemur, Mayotte Brown Lemur)

Social Organization. Social organization of all subspecies is mixed-sex groups (Sussman 1974, 1999; Overdorff 1992, 1993a, 1996, 1998; Gerson 2001; Harrington 1975; Mittermeier et al. 1994; Vasey 2000). In *E. f. rufus* (rufous brown lemur), however, groups sometimes fission during periods of food scarcity (Overdorff 1998); and individuals of this subspecies also form strong affiliative dyadic relationships, primarily between males and females but also between other sex/age combinations (Overdorff 1998, Gerson 2001). Overdorff (1998) notes that dyads occurred more often in feeding contexts during the mating season and during periods of food scarcity and may be related to the unclear dominance hierarchies found in rufous brown lemurs, the distribution and density of food patches in the habitat, and vulnerability to predators.

Diet and Feeding Ecology. Diet in the wild has been noted in *E. f. fulvus* in a western dry forest (Harrington 1975); *E. f. rufus* in a southwestern dry forest (Sussman 1972, 1974); *E. f. rufus* at Ranomafana, a rain forest site (Overdorff 1992, 1993b, 1996); *E. f. sanfordi* in a northern dry forest at Montagne d'Ambre (Freed 1996, 1999); *E. f. collaris* in a southeastern littoral (coastal) forest (Donati and Borgognini Tarli 2002a); *E. f. albifrons* in a northern montane rain forest (Vasey 2000); and three populations in three different areas of the southeastern rain forests (Johnson and Overdorff 2002). The diet of all subspecies is described as highly frugivorous, but leaves, buds, flowers, invertebrates, and nectar are also consumed. In most areas where *E. fulvus* occurs, there can be marked seasonal fluctuation in amount and type of food resources available. During times of fruit scarcity (usually the dry season), animals include more leaves, flowers, and figs in their diets; and *E. f. rufus* groups have been observed to move well out of their home ranges during these periods to seek alternative resources or fission into smaller groups (Overdorff 1993a,b, 1996; Johnson and Overdorff 2002; Overdorff et al. 2003). Both common brown lemurs and rufous brown lemurs ingest a significant amount of tannins and alkaloids from unripe fruit and mature leaves in their diet, and Ganzhorn (1988) and Vasey (2000) suggest that tolerance of secondary compounds combined with ecological flexibility in *E. fulvus* spp. may explain the wide geographic range of this species. Vasey also notes that because of such flexibility, *E. fulvus* is able to avoid overt competition with sympatric lemur species.

Sympatry between *E. fulvus* and other lemur taxa has been documented in a number of geographic areas. Sympatry and polyspecific associations between *E. f. sanfordi* and *E. coronatus* are discussed below in the section on crowned lemurs, and information on sympatry and niche separation in *E. f. rufus* and *E. rubriventer* is presented in the section on *E. rubriventer*.

E. coronatus and E. f. sanfordi: Sympatry and Polyspecific Associations

Sanford's brown lemurs (*E. f. sanfordi*) and crowned lemurs (*E. coronatus*) are sympatric throughout the same region at the northern tip of Madagascar (Wilson et al. 1989; Freed 1996, 1999). Freed studied these two species at Montagne d'Ambre National Park in northern Madagascar, while Wilson's group focused on the two species in the unusual habitat of Ankarana, which is composed of dry forest growing on and around limestone karst pinnacles (*tsingy*) as well as xerophytic scrub and semideciduous dry forest.

Both species inhabit forests which vary in elevation, climate, structure, and disturbance. One difference, however, is that Sanford's lemurs are restricted to closed, continuous-canopy forest and share highly overlapping home ranges (Freed 1996). Both live in small, multimale/multifemale groups (Freed 1996, Wilson et al. 1989), but group cohesion and spacing differ by species. In the dry season, crowned lemur groups are less cohesive than those of Sanford's lemurs, and crowned lemurs sometimes divide into small foraging subgroups during the day (Freed 1996).

Both species are highly frugivorous; however, proportions of fruit and flowers differ between them, and both occasionally feed on leaves and insects.

At Ankarana, Wilson et al. (1989) found that the two species often fed together but did not travel in mixed-species groups. Conversely, Freed (1996) observed frequent polyspecific associations, the first report of such in sympatric lemurs. The two species tolerated the presence of each other well, and when interspecific agonism occurred (in 20%–25% of encounters), they were initiated by the Sanford's lemur group in response to feeding competition. Polyspecific associations varied according to season and were most frequent during the wet season. Freed suggests that both species benefit from one another's familiarity with food resources in different forest levels but not in relation to enhanced predator protection since there were few predators in the area and actual predation on these lemurs was rare.

E. mongoz (Mongoose Lemur)

E. mongoz occurs in the subhumid, seasonal forests of northwestern Madagascar as well as on two of the Comoro Islands: Anjouan and Moheli, where they were likely introduced by humans (Tattersall 1982, Mittermeier et al. 1994).

Social Organization.   Social organization of *E. mongoz* is variable as it has been observed in both pair-bonded (monogamous) family groups and larger mixed-sex groups (Harrington 1978, Tattersall 1977a, Curtis and Zaramody 1997). Offspring of both sexes disperse and establish their own social groups. Females leave the natal family unit at 27–30 months and males, at 31–42 months (Curtis and Zaramody 1997, 1998).

Diet and Feeding Ecology.   The mongoose lemur diet can be categorized as highly nectivorous during the dry season and frugivorous/folivorous during the wet season (Sussman and Tattersall 1976, Curtis and Zaramody 1997).

Reproduction.   Infants are born in October–November. At 3 weeks, infants began to explore the environment. Adult males frequently carry the infant between weeks 2 and 5, and at 9 weeks, infants begin to move and feed independently. Females give birth annually (Tattersall 1976, Curtis and Zaramody 1997).

E. m. macaco (Black Lemur) and E. m. flavifrons (Sclater's Black Lemur)

E. macaco exhibits marked sexual dichromatism: males are black with black ear tufts, and females are golden/reddish/rust brown with off-white ventrum and white ear tufts (Tattersall 1982, Mittermeier et al. 1994).

E. m. flavifrons and *E. m. macaco* × *E. m. flavifrons* hybrids are restricted to dry northwestern forests, just south of the geographic range of the black lemur. Sclater's black lemur differs from the black lemur in that it lacks tufted ears, but more strikingly, its eye color ranges from turquoise blue to gray, as opposed to the amber brown eyes of *E. m. macaco* (Koenders et al. 1985, Mittermeier et al. 1994). Hybrids exhibit either duller blue eyes and no beard or light brown eyes and a less prominent beard and ear tufts compared to the black lemur (Rabarivola et al. 1991).

Social Organization.   Social organization is multimale/multifemale. At Ambato Massif in the northwest, Colquhoun (1993) found that larger groups often fissioned into smaller sub-groups.

Diet and Feeding Ecology.   Marked wet and dry seasons occur in this area, and seasonal variation was noted with respect to dietary patterns. Fruit was the dominant food item during the rainy season, supplemented by mushrooms and millipedes. During the dry season, flowers, nectar, seed pods, and some leaves were eaten (Colquhoun 1993). Andrews and Birkenshaw (1998) found differences in daytime and nighttime feeding, with more variation in fruit species and leaves consumed in the day and more nectar consumed at night.

Cathemerality.   Colquhoun (1993) and Andrews and Birkenshaw (1998) noted year-round cathemeral activity in black lemurs, with nocturnal activity following phases of the moon. Cathemeral activity was seen more in the cooler, dry season and, as with *E. mongoz*, may be related to thermoregulation, allowing these lemurs to avoid cold stress by being physically active during cool nights in the dry season.

Reproduction. Females usually give birth annually to a single infant, in September or October, after a gestation period of 125–126 days (Colquhoun 1993, Mittermeier et al. 1994).

E. rubriventer (Red-Bellied Lemur)

One of the few pair-bonded lemurs, *E. rubriventer* has been closely studied at Ranomafana National Park by Overdorff (1992, 1993a,b, 1996).

Social Organization and Group Size. The red-bellied lemur lives in monogamous pairs with offspring and maintains exclusive use of its home range, actively defending the boundaries.

Diet, Feeding Ecology, and Sympatry with E. f. rufus. *E. rubriventer* is a highly frugivorous primate and includes some leaves and nectar in the diet (Overdorff 1992). At the Ranomafana site, *E. rubriventer* and *E. f. rufus* are sympatric. Even though the composition of their diets is similar, Overdorff (1992, 1993b) notes that *E. f. rufus* ate more unripe fruits, mature leaves, and insects than did *E. rubriventer*; and she suggests that *E. f. rufus* may have a higher tolerance than *E. rubriventer* for secondary compounds, which may also help with niche separation. The two species also used flowers in different ways: *E. rubriventer* licked flower nectar and *E. f. rufus* consumed the entire flower.

Both of these species may serve as pollinators for some of the plant species that they use, but Overdorff (1992) notes that *E. rubriventer* may be a more efficient pollinator since it does not destroy the reproductive parts of the flower. Overdorff (1996) suggests that the two sympatric species may avoid direct competition during periods of scarce resources by differing both their activity patterns and habitat use, and subtle and consistent differences in diet throughout the seasons allow these two species to coexist.

Reproduction. Females give birth to one infant annually, in September or October. As in some pair-bonded anthropoid species, male red-bellied lemurs help with infant care, often holding or carrying the infant. Males have been noted to carry infants up to 100 days (Overdorff 1993a, Mittermeier et al. 1994).

Hapalemur (Bamboo Lemurs)

All species of *Hapalemur* are highly unusual because they specialize on bamboo, a dietary focus not found in other primates. There are three species of *Hapalemur* and two subspecies (see Table 4.1). *H. aureus*, an extremely rare lemur so far found only in very small populations in two southeastern national parks (see Table 4.3), was discovered only in 1986 (Meier and Rumpler 1987, Wright et al. 1987).

Social Organization. *H. griseus griseus* is reported to have flexible social organization. Grassi (2001) found monogamous pairs as well as polygynous and multimale/multifemale social groups at her field site. Alaotran gentle lemurs (*H. g. alaotrensis*) also live in varying kinds of group: monogamous pairs, groups with two breeding females, and some groups with three adult males, though there is only one breeding male per group (Mutschler et al. 2000, Mutschler 2002). Mutschler found that both sexes disperse. Females leave their natal group as subadults, and males make their first migration as adults.

Tan (1998) notes that *H. aureus* live in monogamous pairs, and the one group of *H. simus* which Tan studied was multimale/multifemale, with three adult males, two adult females, and offspring.

Diet and Feeding Ecology. As their common names suggest, all *Hapalemur* species are bamboo specialists, and all three species ingest the cyanide found in the giant bamboo without harm, a remarkable dietary adaptation (Glander et al. 1989, Tan 1998). More than 85% of *Hapalemur* diets are made up of bamboo and grasses (Tan 1998, Mutschler 2002). Tan found that the three species are able to coexist sympatrically in the Ranomafana National Park habitat because each specializes on different parts of the bamboo plant. They also feed on several other plant species, fungus, and, at times, soil (Tan 1998, Grassi 2001). *H. griseus* and *H. aureus* both consume some fruit, and Grassi notes that both new and mature leaves were eaten by *H. griseus* at the higher-elevation Vato site in Ranomafana Park. She suggests that increased dietary diversity by female *H. g. griseus* during reproductive periods helps offset high metabolic needs.

Reproduction. *H. griseus* and *H. aureus* have similar gestation lengths of 137–140 and 138 days, respectively, while the gestation period of the larger *H. simus* is somewhat longer, at 149 days (Tan 2001, Grassi 2001).

The Alaotran gentle lemur does not have as strict and discrete a mating season as that found in most other lemur species. Mating season begins in September and ends in February (Mutschler 2002). Mutschler suggests that a year-round, consistent resource base is a key factor in the absence of strict breeding seasonality.

H. aureus mothers have been noted to nest their infants in thick foliage during the first 2 weeks of life (Tan 2001). Tan also found that *H. griseus* and *H. aureus* females both park and orally transport infants, but *H. simus* females carry their newborns.

Grassi (2001) reports high infant mortality (67%) in *H. g. griseus* at the Vato site, Ranomafana. Surviving infants were fully weaned by 5 months. There is a 1-year interbirth interval in *Hapalemur*.

Varecia (Ruffed Lemur)

The genus *Varecia* contains two subspecies: the black and white ruffed lemur (*V. variegata variegata*) and the red ruffed lemur (*V. v. rubra*), the latter having a very restricted range in the northern Masoala Peninsula. In all areas where these lemurs have been studied, populations have experienced occasional and sometimes devastating cyclones (Balko 1998; Ratsimbazafy 2001, 2002).

Social Organization. Both monogamy and multimale/multifemale mating systems have been reported (Morland 1991,

White 1989, Balko 1998, Britt 2000). Rigamonti's (1993) two study groups of red ruffed lemurs fissioned into subgroups of two or three animals. During the cool wet season, they lived in these small subgroups for several weeks at a time, and groups were cohesive in the transitional dry months. Ratsimbazafy (2002) found that after a severe cyclone black and white ruffed lemurs at Manombo on the southeastern coast foraged singly rather than as a group as 95% of larger trees in the area stopped fruiting.

Diet and Feeding Ecology.   *Varecia* is highly frugivorous, with fruit making up 75%–95% of the diet, and the remainder is comprised of nectar, flowers, and some leaves (Morland 1991, Britt 2000, Vasey 2000). During times of low fruit availability, *Varecia* will consume large amounts of young leaves (Balko 1998). After the above-mentioned cyclone at Manombo, ruffed lemurs at this site relied on fruit from nonendemic, invasive plant species, as well as fungus (Ratsimbazafy 2002). *Varecia* has been observed to come to the ground and ingest soil at particular times of the year (Morland 1991, White 1989, Britt 2000). Such geophagy may serve to neutralize secondary compounds in the diet as well as provide a source of minerals (Ganzhorn 1988, Britt 2000). Britt suggests that even though *Varecia* are marked frugivores, their ability to use other food items may be an important adaptation for dealing with low or absent fruit productivity during times of environmental stress because these lemurs live in areas of Madagascar where cyclones are common and important food trees can be destroyed.

Ruffed lemurs are considered important seed dispersers and pollinators in the eastern Madagascar rain forests (Balko 1998, Britt 2000), and Britt (2000) stresses that, as such, it is of utmost importance to develop conservation strategies that will aid in the survival of ruffed lemurs.

Reproduction.   Reproduction and infant care in *Varecia* differ from other diurnal lemurs. *Varecia* is the only diurnal prosimian in which females possess two sets of mammary glands and regularly exhibit multiple (two to four infants) births (Morland 1990, Mittermeier et al. 1994). Infants do not cling to the mother as do other diurnal lemur infants; rather, the mother transports infants one at a time by mouth and parks them in nests or in trees (Petter et al. 1977, Tattersall 1982). Morland (1990) noted frequent allo-parental care consisting of guarding infants at nest sites and allonursing. Nests are constructed by pregnant females a few weeks prior to parturition 10–20 m above ground (Morland 1990, Balko 1998). Furthermore, ruffed lemur infants develop more quickly than do other diurnal lemur offspring, and in captivity they have been noted to weigh up to 70% of adult body weight by 4 months of age (Pereira et al. 1987). In the wild, infants grow rapidly (Balko 1998) and appear to reach nearly adult size at about 6 months of age (Morland 1990).

Ruffed lemurs commonly experience cyclones in their geographic range. In Ratsimbazafy's (2001, 2002) study, females ceased reproduction for 3 years after a cyclone

destroyed most of their resource base. He suggests that plasticity in diet, small group size, solitary foraging, and reproductive cessation following a severe natural disaster are important reasons why *Varecia* groups can persist in such a highly disturbed habitat. Ratsimbazafy (2002) points out the link between environmental variability and female fertility in this species.

Gestation length is estimated at around 102 days in the wild, and infants are born in September and October (Morland 1990, Mittermeier et al. 1994).

Indriidae (*Propithecus* and *Indri*)

Propithecus (Sifakas)

Propithecus species are extraordinary vertical clingers and leapers. With three species and eight subspecies (Table 4.1), sifakas are found in many habitats and many regions of Madagascar, although some, like *P. tattersalli*, are found only in very restricted ranges and several are very rare, with populations threatened by habitat destruction.

The remainder of this section will focus on the four most-studied *Propithecus* species: *P. verreauxi verreauxi*, *P. tattersalli*, *P. diadema edwardsi*, and *P. d. diadema*.

P. v. verreauxi (Verreaux's Sifaka)

Social Organization.   Verreaux's sifaka is found in small multimale/multifemale groups; however, this species also fissions into small foraging parties at times. The social organization of Verreaux's sifaka has been sometimes been referred to as "neighborhoods" because of the fluidity of groups, the fact that males make temporary visits to adjacent groups, and the frequency of adult male intergroup transfer (Jolly 1966, Richard 1978, Richard et al. 1993).

Diet and Feeding Ecology.   Verreaux's sifaka lives in the dry west, south, and southwest of Madagascar and experiences dramatic shifts in seasonal resource distribution between the wet and dry seasons. Sifakas (and all indriids) are considered folivores; however, during the wet season, 60%–70% of their diet is made up of fruit and flowers, with young leaves accounting for 20%. During the dry season, mature leaves make up 70% of the diet, with fruit and flowers contributing only 20%. Bark makes up the remainder of the diet in both seasons (Richard 1978, Sussman 1999).

Reproduction.   Gestation in this species is 150–160 days (Petter-Rousseaux 1964). The infant is carried ventrally at first, then dorsally; and infants will continue to ride on the mother until 6–7 months (Jolly 1966). At the Beza Mahafaly site in southwestern Madagascar, Richard et al. (2002) found that more than half of the females in this population did not reproduce for the first time until they were 6 years old. Such a delay in reproduction for such a small primate is considered by Richard et al. (2002) to be "bet hedging," i.e. a slowing down of female reproductive life history where first births are later than expected and females reproduce into old age, also later than expected. This unusual reproductive strategy is suggested to be an evolutionary response to the climatic unpredictability in this species' geographic region, such as extremely varied annual rainfall patterns and frequent droughts (Richard et al. 2002). Infant mortality is high, averaging 52% in the first year of life (Richard et al. 2002). High mortality may be related to a combination of starvation after weaning in particularly dry years, hypothermia in the cold and dry season, disease, and predation. Adult males mate at 3–4 years of age (Richard et al. 2002).

P. tattersalli (Golden-Crowned Sifaka)

The golden-crowned sifaka has recently been reported to exist slightly outside of the original restricted area in the Daraina region of northeastern Madagascar, with an effective population size of these rare lemurs estimated at 2,520–3,960 individuals (Vargas et al. 2002).

Social Organization. The golden-crowned sifaka lives in small, multimale/multifemale groups (Meyers and Wright 1993).

Diet and Feeding Ecology. Seasonal variation in food resources occurs in *P. tattersalli*'s geographic range. Meyers and Wright (1993) note that the diet consists of immature and mature leaves (22% and 17%, respectively), 37% unripe fruit and seeds, 9% fruit pulp, and 13% flowers. These items peaked in availability in the wet season, but seeds, available year-round, formed the staple food item. Bark is also sometimes eaten during the dry season (Mittermeier et al. 1994).

Reproduction. In *P. tattersalli*, mating season occurs in late January and infants are born in late July. Weaning occurs at 5–6 months and is timed to coincide with peak immature leaf availability. As in many other lemur species, late lactation/weaning occurs in the early wet season so that infants have access to abundant weaning foods (Meyers and Wright 1993).

P. d. edwardsi (Milne-Edwards Sifaka)

A population of Milne-Edwards sifaka has been studied continuously since 1986 by Wright and her students and colleagues at Ranomafana National Park in southeastern Madagascar (see, e.g., Wright et al. 1987, Meyers and Wright 1993, Wright 1995, Hemingway 1996, Wright et al. 1997, Erhardt and Overdorff 1998, Overdorff et al. 2003, Pochron and Wright 2003, Pochron et al. 2004). Consequently, much is known about this species of rain forest sifaka.

Social Organization. Pochron and Wright (2003) and Pochron et al. (2004) report variable social organization in this species. Multimale/multifemale groups, unimale polygynous, polyandrous, and male–female pair groups have been observed. Pochron and Wright (2003) suggest that since females sometimes mate with males outside of their groups, such flexibility may reduce pressure for males to join groups with several females and result in the variability seen in group composition in this species.

Pochron and Wright (2003) suggest that Madagascar's harsh and unpredictable environment may have resulted in *P. d. edwardsi* living and foraging in small groups, which would reduce feeding competition yet help somewhat with predator protection (vs. living/foraging solitarily).

Diet and Feeding Ecology. Ripe fruit and seeds make up the majority (55%) of *P. d. edwardsi*'s diet, supplemented by vine leaves (15%), flowers (3%), and immature leaves (26%) (Meyers and Wright 1993). As in most lemur habitats, seasonal variation in resource availability is found in *P. d. edwardsi*'s habitat, with more immature leaves available during the wet season. Fruit production can vary annually (Meyers and Wright 1993).

Reproduction. Average gestation length is 179 days, 1 month longer than in the smaller *P. verreauxi* (Wright 1995). Infants are primarily independent by 7 months of age and fully weaned by 1 year. Most females begin reproducing at 4 years of age. However, Pochron et al. (2004) have found that only 24% of all females survive to the age of 4 years.

Average interbirth interval is 1.5 years, and average infant mortality is 50%.

P. d. diadema (Diademed Sifaka)

Powzyk (1997) studied sympatric *P. d. diadema* and *Indri* at Mantadia National Park in the eastern rain forest. She notes that in parts of their former distribution, diademed sifaka populations have disappeared due to overhunting or habitat destruction.

Social Organization. Diademed sifakas live in multimale/multifemale groups. Females choose mates within their group but have also been observed mating with novel males from other groups (Powzyk 1997).

Diet and Feeding Ecology. Diademed sifakas are primarily folivorous. Powzyk (1997) and Powzyk and Mowry (2003) note that 42% of their diet consists of immature leaves. They supplement their diet with fruits and flowers. Feeding differences between *P. d. diadema* and sympatric *Indri* at Mantadia are presented in the section on *Indri* below.

Reproduction. The average birth rate over 3 years was 0.50/year, and infant mortality over this period was 50% (Powzyk 1997).

Indri indri (Indri)

There is only one species of *Indri*, and the common name is also indri. In addition to its large size (see Table 4.2), *Indri* can be distinguished from most other lemurs by its rudimentary tail (Pollock 1975). It is also known because of its loud, wailing morning calls, which can be heard up to 3 km away (Pollock 1975, Mittermeier et al. 1994).

Indri has been characterized as the largest extant lemur species; however, Glander and Powzyk (1995) and Powzyk (1997) found that both *Indri* and *P. d. diadema* were similar in body weight, and Powzyk suggests that both species be considered the largest two extant lemurs. In both species, females weighed slightly more than males.

Geographic Range/Habitat. Indri inhabit the eastern rain forests, from near, but not in, the Masoala Peninsula in the northeast to east central Madagascar (Petter et al. 1977, Tattersall 1982).

Social Organization.   *Indri* is one of the few monogamous lemur species, living in pairs or small groups consisting of a pair and offspring (Pollock 1975, Powzyk 1997, Britt et al. 2001). The mated pair use morning calls to announce both their location to other pairs and their mated status (Powzyk 1997).

Diet and Feeding Ecology.   Like other indriids, *Indri*'s diet consists of leaves, flowers, fruit, bark, and seeds (Pollock 1975, Powzyk 1997, Britt et al. 2001, Powzyk and Mowry 2003). Powzyk found that leaves made up 71% of the diet at Mantadia. Britt et al. note that *Indri* at Betampona, farther north, fed on more mature leaves than those at Mantadia and that in the winter season they increased consumption of bark and fruit. Pollock (1975) notes that *Indri* regularly came to the ground to ingest earth; however, Powzyk and Mowry (2003) note that sympatric diademed sifaka at Mantadia engaged in geophagy twice as often as did *Indri*. Powzyk (1997) suggests that *Indri*'s specialization for plant fiber has allowed these two large-bodied diurnal lemur species to coexist in over 90% of their range.

Reproduction.   A single infant is born in May, after a gestation period of 120–150 days (Pollock 1975, Mittermeier et al. 1994). Infants are carried on the ventrum until 4 months, then carried dorsally until 8 months. Infants are weaned between 8 and 12 months (Pollock 1975). Powzyk (1997) calculated *Indri* average birth rates as 0.33/year and infant mortality as 0.67 over a 3-year period.

Nocturnal Lemurs

During the past 10 years, the nocturnal lemurs of Madagascar have been the focus of a number of behavioral and phylogenetic studies that have greatly expanded what is known of their socioecology and biology. Exciting new data on their behavioral ecology indicate that nocturnal prosimians live in complex societies and exhibit high interspecific diversity in lemur social and mating systems. Indeed, it has been suggested that nocturnal lemur social systems contain three types: gregarious, for animals living in cohesive groups; dispersed, for solitary foragers with social networks; and solitary, for completely solitary animals (Müller and Thalmann 2000).

There are three families that contain only nocturnal lemuriformes: Cheirogaleidae, Megaladapidae, and Daubentoniidae. The primarily diurnal Indriidae contains two nocturnal species: *Avahi laniger* and *A. occidentalis*.

Cheirogaleidae

Members of the family Cheirogaleidae are small, quadrupedal lemurs that sleep in nests of leaves or in tree holes during the day. The genus *Cheirogaleus* has recently undergone a number of taxonomic changes, and seven species are now recognized (Groves 2000). The scientific and common names for these species are found in Table 4.1. *Allocebus trichotis*, the hairy-eared dwarf lemur, was originally thought extinct but was rediscovered in 1989 (Meier and Albignac 1991). The genus *Mirza* is comprised of one species, *Mirza coquereli*, Coquerel's dwarf lemur. Fork-marked lemurs include *Phaner furcifer* and a number of newly described subspecies (see Groves and Tattersall 1991).

Microcebus (Mouse Lemurs)

Recent phylogenetic analyses of mitochondrial deoxyribonucleic acid (mtDNA) sequence data and newly collected mouse lemur specimens have also resulted in designations of several new species of *Microcebus* (Schmid and Kappeler 1994, Zimmerman et al. 1998, Rasoloarison et al. 2000, Yoder et al. 2000, Pastorini et al. 2001). There are now eight species of mouse lemur recognized, and these are listed in Table 4.1.

Cheirogaleus medius Group: C. medius and C. adipicaudatus (Dwarf Lemurs)

Social Organization.   While home ranges of *Cheirogaleus medius* and *C. adipicaudatus* may overlap, same-sexed individuals are intolerant of one another (Hladik et al. 1980). *C. medius* is monogamous and lives in dispersed family groups (Müller 1998, 1999; Fietz 1999). This species deals with seasonal variation in food resources by entering torpor. During torpor, nesting size is variable, from one to as many as five individuals sharing a nest in a hollow tree trunk. Males, however, emerge from torpor sooner than do females, and this may be a form of paternal investment as males patrol their home range and by doing so may maintain access to its resources for their family group (Müller 1999). Males dramatically lose weight during this time (Müller 1999).

Diet and Feeding Ecology.   *C. medius* focuses on a variety of high-quality foods, including fruits, nectar, vertebrates, and insects (Hladik 1979, Hladik et al. 1980, Wright and Martin 1995). Of particular note is their ability to store substantial fat in their tails (during which the volume of the tail triples), which is used during the torpid state in seasons of low food abundance (Hladik et al. 1980, Wright and Martin 1995).

Reproduction.   Fat-tailed dwarf lemurs are seasonal breeders. The gestation period is 61–64 days (Petter 1978, Hladik et al. 1980), and a female normally produces twins, although this can vary from one to four infants (Foerg 1982). In captivity, these lemurs become sexually mature in their first year of life (Foerg 1982); however, in the wild, sexual maturity may not occur until 2 years of age (Müller 1999).

Cheirogaleus major Group: C. major and C. crossleyi

Social Organization.   Little is known regarding the ranging or social behavior of the *C. major* group. They are essentially solitary and may nest together in groups of two (Petter et al. 1977).

Diet and Feeding Ecology. Like the *C. medius* group, these lemurs feed on young leaves, fruit, nectar, pollen, and insects (Petter et al. 1977) and can tolerate a medium level of tannins in their diet (Ganzhorn 1988). They also enter a period of torpor during the dry season and store fat in the tail to accommodate this period.

Reproduction. Gestation length is 70 days with two or three infants born in January (Petter-Rousseaux 1964, Petter et al. 1977). They are carried by the mother in her mouth as they are unable to cling at birth (Petter-Rousseaux 1964). Lactation lasts only 1.5 months, and infants develop quickly, being able to follow their mothers within a month and to eat fruit at about 25 days of age (Petter-Rousseaux 1964).

Allocebus trichotis (Hairy-Eared Dwarf Lemur)

Social Organization. As *Al. trichotis* has yet to be systematically studied in the wild, little information is available. It does sleep in tree holes in groups of two to six (Meier and Albignac 1991).

Diet and Feeding Ecology. Observations in captivity indicate that *Allocebus* feeds on fruit, honey, and locusts; and its long tongue is suggestive of nectar feeding (Meier and Albignac 1991). Seasonal body fat storage occurs but over the entire body, not just in the tail (Meier and Albignac 1991).

Reproduction. Little is known about its reproduction, but infants may be born in January or February (Meier and Albignac 1991).

Mirza coquereli (Coquerel's Dwarf Lemur)

Social Organization. Adult males are heavier than adult females, and this increases most dramatically prior to and during the mating season (Kappeler 1997). Female home ranges are 4 ha, remain stable over time, and overlap considerably, with little evidence of actively defended territories (Kappeler 1997). Male home ranges increase during the mating season and overlap with other home ranges only at that time (Kappeler 1997). Genetic data indicate that females are organized into matrilines, most females show philopatry, and dispersed multimale/multifemale is the social organization of this species (Kappeler et al. 2002).

Diet and Feeding Ecology. Coquerel's dwarf lemur has an eclectic diet that includes fruit, flowers, buds, gums, insects and insect secretions, spiders, frogs, chameleons, and small birds (Pagès 1980, Andrianarivo 1981).

Reproduction. Mating occurs in October, followed by a 3-month gestation (Petter-Rousseaux 1980). Infants develop quickly and can leave their nests after 1 month (Pagès 1980). Females may become reproductive within their first year (Kappeler 1997).

Microcebus

Considerable advances in the study of the behavior and ecology of *Microcebus* have revealed great flexibility in this genus. Many of the newer species have yet to be completely described, including *M. myoxinus*, *M. tavaratra*, *M. sambiranensis*, and *M. griseorufus*.

M. murinus (Gray Mouse Lemur)

Social Organization. The gray mouse lemur appears to exhibit a multimale/multifemale system within a dispersed social network (Fietz 1999, Radespiel 2000). While commonly observed foraging alone at night, during the nonmating season, this lemur sleeps in groups of up to 15 individuals in nests made of leaves or in tree hollows (Martin 1972a, 1973; Radespiel et al. 1998). Females will sleep with the same female partners, and these individuals often share home ranges; however, different female groups use nearly exclusive home ranges (Radespiel 2000). Males often sleep alone (Radespiel et al. 1998, Radespiel 2000), but during the mating season it is common to find mixed-sex groups in these nests, with a single male nesting with as many as seven females (Martin 1973). Home ranges overlap substantially (Barre et al. 1988, Fietz 1999, Radespiel 2000), and male home ranges are larger than those of females (Table 4.3) (Pagès-Feuillade 1989). Males prefer nests near those preferred by females (Rasoazanabary 2004). Preferred nests may have superior thermoregulation and protection from predators, and female nests are better insulated, suggesting that this may be a contested resource between the sexes (Radespiel et al. 1998, Schmid 1998). Genetic data indicate male-biased natal dispersal in this species (Radespiel et al. 2003b).

Diet and Feeding Ecology. The gray mouse lemur stores fat in its tail and may enter torpor, but time spent in torpor varies by sex; also, while females become inactive during periods of low food availability, males are more active (Rasoazanabary 2004). This species is omnivorous, but fruit and invertebrates are a major component of the diet. Other foods include flowers, nectar, leaves, sap and gum, homopteran larvae secretions, and small invertebrates (Martin 1972a, 1973; Petter 1978; Hladik 1979; Barre et al. 1988; Corbin and Schmid 1995). Insect prey is often caught on the ground (Martin 1972a, 1973).

Reproduction. Mating occurs in September, with a gestation of 59–62 days (Martin 1972a, Radespiel 2000). Normally, twins are born and are parked in tree holes and/or carried until 3 weeks of age. Infants develop quickly and exhibit adult behaviors by 2 months of age (Petter-Rousseaux 1964, 1980; Martin 1972a). In captivity, females first give birth at 18 months (Petter-Rousseaux 1964).

M. rufus (Eastern Rufous Mouse Lemur)

Social Organization. The brown mouse lemur remains understudied, and little is known of its social organization. Like other *Microcebus* species, it sleeps in tree holes or nests but may also use old birds' nests (Martin 1973). Mark–recapture data suggest overlapping home ranges and a multimale/multifemale social organization (Atsalis 2000).

Diet and Feeding Ecology. With a diet similar to that of *M. murinus*, this species consumes fruits, insects, and flowers

(Martin 1972a, Harcourt 1987) as well as, more rarely, young leaves (Ganzhorn 1988). *M. rufus* stores some fat in its tail and may enter torpor depending on the habitat (Atsalis 1998).

Reproduction. No data are available on its reproduction in the wild. This species is also difficult to maintain in captivity, but data from a breeding colony of wild-caught *M. rufus* indicate an estrous cycle of 59 days, 2.5 cycles per season, seasonal reproduction with a seasonal change in testicular size, and a gestation length of 56.5 days, with litter size ranging from one to three offspring. Mating behavior varied among pairs but copulation appeared to be limited to a single day per estrus (Wrogemann and Zimmermann 2001).

M. berthae (Berthe's Mouse Lemur)

Social Organization. Current data indicate that this newly discovered species is nongregarious and forages solitarily. It does not form sleeping associations but instead sleeps alone in a tangle of lianas rather than in self-constructed nests or tree holes. It has been suggested that such a sleeping pattern may occur as a result of both competition for nest sites from other nocturnal sympatric animals, including lemurs, as well as an antipredator strategy in this smallest of the living primates (Schwab 2000). Male home ranges appear to be larger than female home ranges at least during the mating season (Schwab 2000). Indirect data (e.g., changes in testicle size, presence of sperm plugs) suggest this species has a multimale mating system that includes promiscuous mating and sperm competition (Schwab 2000).

Diet and Feeding Ecology. In-depth studies of their feeding ecology have yet to be carried out, but males and females forage separately (Schwab 2000).

Reproduction. Female cycles are not synchronized during the mating season. Males are heavier than females during the mating period, but females are heavier than males during the nonreproductive season (Schwab 2000).

M. ravelobensis (Golden-Brown Mouse Lemur)

Social Organization. The golden-brown mouse lemur lives in a dispersed multimale/multifemale society with promiscuous mating (Radespiel et al. 2003a). Conspecifics interact frequently. Sleeping groups can contain only females or both females and males and are maintained over time even though sleeping sites may change (Radespiel et al. 2003a, Weidt et al. 2004). It is suggested that thermoregulation may explain such groupings and that sleeping groups are the basic social unit in brown mouse lemur society (Weidt et al. 2004).

Diet and Feeding Ecology. The diet is omnivorous and similar to that of *M. murinus* (Reimann 2002, Radespiel et al. 2003a). Individuals forage alone, remain active despite changes in environmental conditions, and do not appear to alter fat storage in their tails across different seasons and photoperiods (Randrianambinina et al. 2003). Daily torpor occurs in this species (Radespiel et al. 2003a).

Reproduction. There is a distinct mating season in this species, although females' estrus does not appear to be strongly synchronized (Schmelting et al. 2000, Randrianambinina et al. 2003).

Phaner furcifer (Fork-Marked Lemur, Four Subspecies: Eastern Fork-Marked Lemur, Pale Fork-Marked Lemur, Pariente's Fork-Marked Lemur, Amber Mountain Fork-Marked Lemur)

Social Organization. The highly vocal *Ph. furcifer* (a mean of 30 loud calls an hour emitted by males has been counted in a radius of about 200 m) can be found in holes in baobab trees, old *Mirza coquerli* nests, and leaf nests (Petter et al. 1971, 1975; Schulke and Kappeler 2003). Meetings between neighboring family groups occur where home ranges intersect, during which females may interact affiliatively with females of other groups (Schulke and Kappeler 2003). Males often interact agonistically with neighboring males and females during such encounters (Schulke and Kappeler 2003). Male–female pairs can maintain vocal contact throughout the night and may nest together during the day (Charles-Dominique and Petter 1980, Schulke and Kappeler 2003). *Phaner*'s social organization may be described as "dispersed pairs" because although there is pair stability for as many as 3 years and their territories overlap nearly completely, actual interaction between male and female pairs is very low (Schulke and Kappeler 2003).

Diet and Feeding Ecology. The fork-marked lemurs' primary food is gum, particularly from *Terminalia* trees; but they also consume insects, sap, buds, flowers, and insect exudates (Charles-Dominique and Petter 1980). Gum feeding correlates with this species' highly specialized toothcomb, which is used to create holes to access tree gum and sap (Charles-Dominique and Petter 1980).

Reproduction. Mating is in June (Charles-Dominique and Petter 1980), with a single infant born in November or December that is first carried and then rides on the mother's back (Petter et al. 1971, 1975; Charles-Dominique and Petter 1980).

Avahi Group (Woolly Lemurs)

Social Organization. Short-term studies indicate that the woolly lemur is monogamous. An adult male, female, and offspring make up the group; usually, it is encountered in pairs or trios, but as many as five individuals can be together (Pollock 1975, Petter and Charles-Dominique 1979, Albignac 1981a, Ganzhorn et al. 1985, Harcourt 1988, Thalmann 2001). Individuals forage alone but may meet throughout the night to groom and rest together (Harcourt 1988, Razanahoera-Rakotomalala 1981). Group members sleep together in dense foliage (Albignac 1981a).

Diet and Feeding Ecology. Although primarily active at night, woolly lemurs have also been observed feeding during the day (Ganzhorn et al. 1985). *Avahi* feeds primarily on leaves, an unexpected diet given its relatively small body size. Such folivory may explain its high level of resting during the evening (Albignac 1981a, Razanahoera-Rakotomalala 1981, Ganzhorn et al. 1985, Ganzhorn 1988, Harcourt 1988, Thalmann 2001). Males and females forage together and feed in the same trees (Thalmann 2001).

Reproduction.   Woolly lemurs give birth to a single infant in August or September. Infants initially cling to the mother's ventrum and then later ride on her back (Martin 1972b, Petter et al. 1977, Ganzhorn et al. 1985, Harcourt 1988).

Daubentonia madagascariensis (Aye-Aye)

Originally believed to be extinct, the highly specialized aye-aye was rediscovered in 1957 (Petter and Petter-Rousseaux 1959).

Social Organization.   The aye-aye builds its nest in the fork of trees and normally forages alone, but it can be found near other individuals (Petter et al. 1977; Iwano and Iwakawa 1988; Sterling 1992, 1993). Studies at Nosy Mangabe indicate that females have exclusive ranges and rarely interact with one another or do so aggressively (Sterling and Richard 1995). Males have large, overlapping ranges and interact both aggressively and affiliatively with one another (Sterling and Richard 1995). Male and female ranges overlap, and most interactions appear to be affiliative, with individuals communicating through vocalizations and scent marking (Sterling and Richard 1995).

Diet and Feeding Ecology.   The aye-aye exhibits a number of specializations, including continuously growing rodent-like incisors and a long and thin third digit that allows it to forage for wood-boring larvae and to feed on hard seeds of the genus *Canarium* (Sterling et al. 1994). It also focuses on other high-quality foods that include fruit, especially coconuts; adult insects; fungus; and nectar (Petter et al. 1977, Iwano and Iwakawa 1988, Sterling 1993, Sterling et al. 1994). This species is able to inhabit a wide variety of habitats, from rain forest to cultivated areas (especially coconut groves) (Tattersall 1982).

Reproduction.   Females in estrus give loud calls that attract males and will mate with some, but not all, attracted males (Sterling and Richard 1995). Aye-ayes are reported to give birth only every 2–3 years (Petter and Peyriéras 1970a,b; Petter et al. 1977). Births appear not to be seasonal, and infants may be weaned at 7 months (Petter and Peyriéras 1970a,b; Sterling 1993).

Lepilemur (Sportive Lemurs)

Lepilemur leucopus (White-Footed Sportive Lemur)

Social Organization.   Found primarily in the Didierea bush and southern dry forests of Madagascar, males and females may sleep in separate tree holes or bundled lianas; but in some studies, they are also found sleeping in pairs (Russell 1977, 1980). Females may share ranges with young offspring and

perhaps even with adult daughters (Charles-Dominique and Hladik 1971).

Diet and Feeding Ecology.   Highly folivorous, the white-footed sportive lemur focuses on low-quality leaves or flowers of the Didereaceae species *Alluadia procera* and *Alluadia ascendens* (Charles-Dominique and Hladik 1971, Hladik and Charles-Dominique 1974). While ingestion of feces (*coecotrophy*) has been reported in some studies (Hladik and Charles-Dominique 1974), it appears to be absent in others (Russell 1977, 1980).

Reproduction.   Mating in this species occurs in May–July, gestation is 4.5 months, with singleton births in September–November (Petter et al. 1977). Individuals are sexually mature at 18 months (Richard 1984).

Lepilemur edwardsi (Milne-Edwards Sportive Lemur)

Social Organization.   Found within the dry deciduous forests of western Madagascar, male and female *L. edwardsi* commonly sleep together in tree holes or near one another in separate holes (Albignac 1981b, Petter et al. 1977, Warren 1994, Rasoloharijaona et al. 2003). Two to three individuals may forage together and regularly engage in grooming bouts (Warren 1994). Current studies indicate this species may be characterized by dispersed monogamy, with each pair defending its home range by branch shaking and vocal displays (Rasoloharijaona 1998, Zimmermann 1998, Rasoloharijaona et al. 2003). Fidelity of these pairs may last as long as 4 years (Altrichter 2001).

Diet and Feeding Ecology.   As in other lepilemurs that have been studied in the wild, the Milne-Edwards sportive lemur forages solitarily and its diet is primarily leaves (Thalmann 2001), which are selected for their protein value and low alkaloid content (Ganzhorn 1988, 1993). Fruit, flowers, and fleshy seeds are also eaten but at much lower levels (Razanahoera-Rakotomalala 1981, Albignac 1981b, Ganzhorn 1988, Thalmann 2001).

Reproduction.   Females give birth at the end of September to a single infant (Rasoloharijaona et al. 2003). Infants are left in a tree hole or within dense foliage while the mother forages (Rasoloharijaona et al. 2003).

Subfossil Lemurs

About 17, or nearly one-third, of the known lemur species became extinct in the late Holocene due to anthropogenic effects (overhunting, habitat destruction) and aridification (Burney 1997, Dewar 1997). These extinct species are referred to as "subfossil" lemurs since most became extinct relatively recently, that is, within the first 1,000 years of the 2,000 years of human habitation on Madagascar (Simons et al. 1995, Fleagle 1999). All extinct species were larger than living lemurs, and this is the most striking difference between extant and extinct species. Some extinct lemurs are

Figure 4.4 Some subfossil lemur species, with extant *Indri*, one of the two largest living lemurs, shown for size comparison. Drawing by Stephen Nash, reprinted with kind permission of the artist.

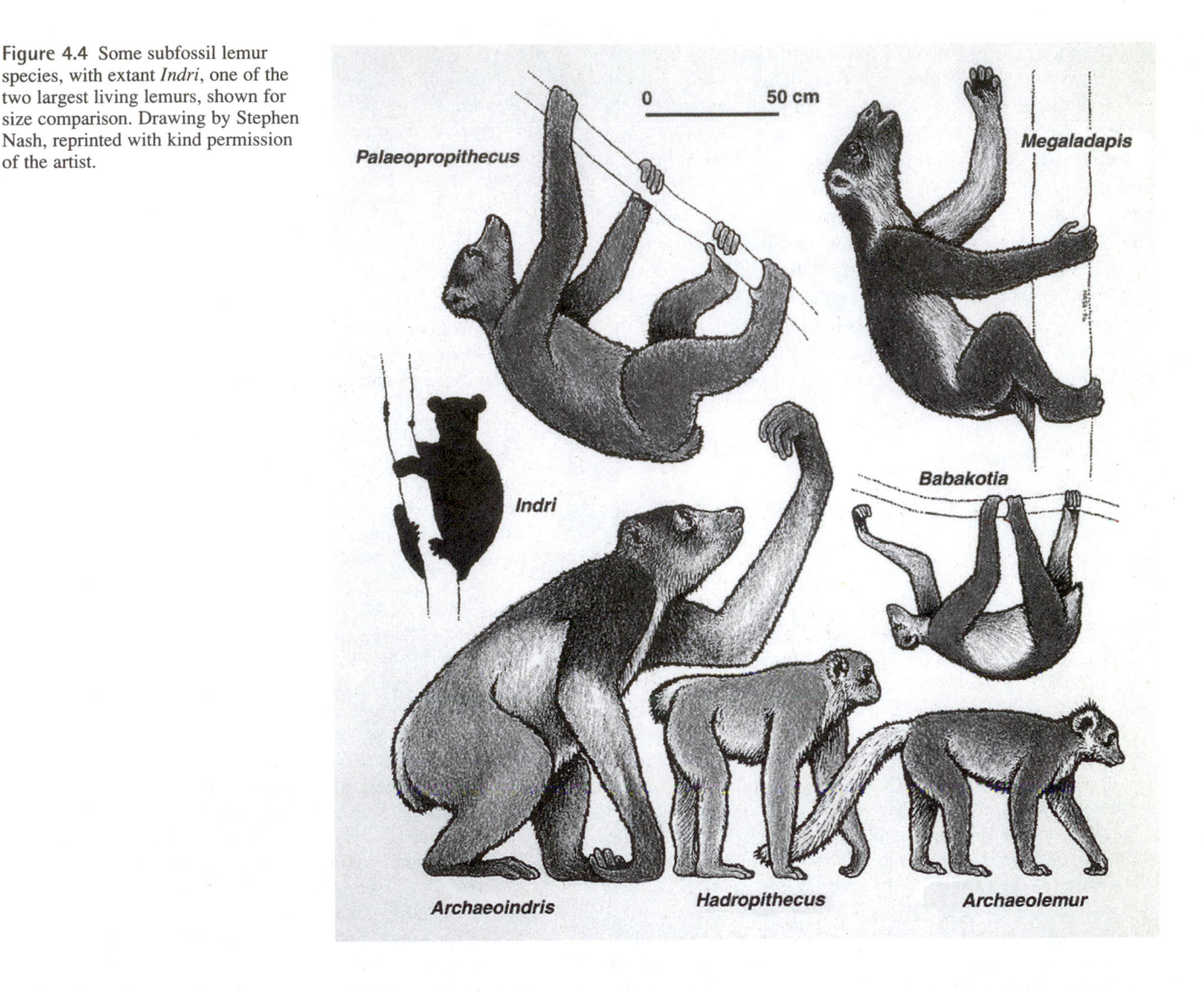

considered to be as large as the largest living anthropoids today (Fig. 4.4).

Godfrey et al. (1993) found that lack of sexual dimorphism is a morphological trend in all lemurs. Even in the largest and most terrestrial of these subfossil lemurs, there is little evidence for sexual dimorphism.

Walker (1967), Gingerich and Martin (1981), and Martin (1990) suggest, based on orbit size measurements and comparisons, that all subfossil lemur species were diurnal, although Jungers et al. (2002) note that the actual ancestral condition for all lemurs was nocturnality. Thus, the subfossil species became diurnal as a later adaptation during the speciation and adaptive radiation which occurred on Madagascar after the first lemur ancestors arrived.

Diet in subfossil lemurs has been inferred based on dental morphology and molar microwear analysis (Jungers et al. 2002, Rafferty et al. 2002, Godfrey et al. 2004). The very large *Megaladapis* species were likely arboreal leaf browsers (Godfrey et al. 1997, 2004). The large *Paleopropithecus* species (extinct relatives of the extant

Propithecus genus) were probably folivorous seed predators which supplemented their diet with a variety of fruits, similar to *Propithecus* today (Godfrey et al. 1997, 2004). The diet of the partially terrestrial *Hadropithecus* was misinterpreted for a number of years as being gramnivorous, rather like the gelada baboon (Jolly 1970, Mittermeier et al. 1994, Jungers et al. 2002); however, recent molar microwear analysis strongly suggests that the diet of this extinct lemur consisted of hard objects such as seeds but not grass seeds, rhizomes, or tubers (Rafferty et al. 2002, Godfrey et al. 2004). *Archaeolemur* species were likely hard-object feeders and may have been omnivorous with a seasonally diverse diet (Godfrey et al. 1997, Jungers et al. 2002, Rafferty et al. 2002). The dental morphology of *Pachylemur* suggests frugivory (Godfrey et al. 1997, 2004).

Despite the large body size of subfossil species, some exhibit skeletal characteristics that indicate some degree of arboreality; and Jungers et al. (2002) note that the large terrestrial species would have been awkard and slow-moving on the ground. *Paleopropithecus* would have used a suspensory posture. Jungers et al. suggest that the subfossil species were not vertical clingers and leapers (as is seen in extant indriids and *Lepilemur*) as they all had relatively short and robust hindlimbs.

While we will unfortunately never see these large and unusual extinct lemurs, research and conservation efforts are occurring in many areas of Madagascar today on extant species and new protected areas are being designated. Hopefully, these conservation programs, along with the continued work and training of both Malagasy and foreign primatologists, will inform us further as to the behavior and ecology of the lemurs and help protect these beautiful and fascinating primates through the twenty-first century and beyond.

REFERENCES

Albignac, R. (1981a). Variabilité dans l'organisation territoriale et l'écologie de Avahi laniger (Lémurien nocturne de Madagascar). *C. R. Acad Sci. III* 292:331–334.

Albignac, R. (1981b). Lemurine social and territorial organization in a northwestern Malagasy forest (restricted area of Ampijoroa). In: Chiarelli, A. B., and Corruccini, R. S. (eds.), *Primate Behavior and Sociobiology*. Springer-Verlag, Berlin. pp. 25–29.

Altrichter, H. (2001). Playback zur Erkennung und Diskrimination von Informationskategorien in den Sozialrufen von Edward-Wieselmakis (*Lepilemur edwardsi*) in NW-Madagascar [diplomarbeit]. Universität Osnabrück, Osnabrück.

Andrews, J. R., and Birkinshaw, C. R. (1998). A comparison between the daytime and night-time diet, activity and feeding height of the black lemur *Eulemur macaco* (Primates: Lemuridae), in Lokobe Forest, Madagascar. *Folia Primatol.* 69(suppl. 1):175–182.

Andrianarivo, A. J. (1981). Étude comparée de l'organisation sociale chez *Microcebus coquereli* [diss.]. University of Madagascar, Antananarivo.

Atsalis, S. (1998). Seasonal fattening and changes in activity levels in the brown mouse lemur (*Microcebus rufus*) in Ranomafana National Park, Madagascar. *Am. J. Primatol.* 45:165.

Atsalis, A. (2000). Spatial distribution and population composition of the brown mouse lemur (*Microcebus rufus*) in Ranomafana National Park, Madagascar, and its implications for social organization. *Am. J. Primatol.* 51:61–78.

Balko, E. A. (1998). A behaviorally plastic response to forest composition and logging disturbance by *Varecia variegata variegata* in Ranomafana National Park, Madagascar [PhD diss.]. State University of New York, Syracuse.

Barre, V., Lebac, A., Petter, J. J., and Albignac, R. (1988). Étude du Microcèbe par radiotracking dans la forêt de l'Ankarafantsika. In: Rakotovao, L., Barre, V., and Sayer, J. (eds.), *L'Equilibre des Ecosystèmes Forestiers a Madagascar: Actes d'un Séminaire International*. IUCN, Gland, Switzerland. pp. 61–71.

Britt, A. (2000). Diet and feeding behaviour of the black-and-white ruffed lemur (*Varecia variegata variegata*) in the Betampona Reserve, eastern Madagascar. *Folia Primatol.* 71:133–141.

Britt, A., Randriamandatonirina, N. J., Glasscock, K. D., and Iambana, B. R. (2001). Diet and feeding behaviour of *Indri indri* in a low-altitude rain forest. *Folia Primatol.* 73:225–239.

Budnitz, N. (1978). Feeding behavior of *Lemur catta* in different habitats. In: Bateson, P. P. G., and Klopfer, P. H. (eds.), *Perspectives in Ethology. Social Behavior*, vol. 3. Plenum Press, New York. pp. 85–108.

Budnitz, N., and Dainis, K. (1975). *Lemur catta*: ecology and behavior. In: Tattersall, I., and Sussman, R. W. (eds.), *Lemur Biology*, Plenum Press, New York. pp. 219–236.

Burney, D. A. (1997). Theories and facts regarding Holocene environmental change before and after human colonization. In: Goodman, S. M., and Patterson, B. D. (eds.), *Natural Change and Human Impact in Madagascar*. Smithsonian Institution Press, Washington DC. pp. 75–89.

Charles-Dominique, P., and Hladik, C. M. (1971). Le *Lepilemur* du sud de Madagascar: écologie, alimentation et vie sociale. *Terre Vie* 25:3–66.

Charles-Dominique, P., and Petter, J.-J. (1980). Ecology and social life of *Phaner furcifer*. In: Charles-Dominique, P., Cooper, H. M., Hladik, A., Hladik, C. M., Pages, E., Pariente, G. F., Petter-Rousseaux, A., Petter, J.-J., and Schilling, A. (eds.), *Nocturnal Malagasy Primates: Ecology, Physiology and Behavior*. Academic Press, New York. pp. 75–95.

Colquhoun, I. C. (1993). The socioecology of *Eulemur macaco*. In: Kappeler, P. M., and Ganzhorn, J. U. (eds.), *Lemur Social Systems and Their Ecological Basis*. Plenum Press, New York. pp. 11–23.

Colquhoun, I. C. (1998). Cathemeral behavior of *Eulemur macaco macaco* at Ambato Massif, Madagascar. *Folia Primatol.* 69(suppl. 1):22–34.

Corbin, G. D., and Schmid, J. (1995). Insect secretions determine habitat use patterns by a female lesser mouse lemur (*Microcebus murinus*). *Am. J. Primatol.* 37:317–324.

Curtis, D. J., and Zaramody, A. (1997). Monogamy and mate monopolization by females in *Eulemur mongoz*. *Primate Rep.* 48–2:16–17.

Curtis, D. J., and Zaramody, A. (1998). Group size, home range use, and seasonal variation in the ecology of *Eulemur mongoz*. *Int. J. Primatol.* 19:811–835.

Curtis, D. J., Zaramody, A., and Martin, R. D. (1999). Cathemerality in the mongoose lemur, *Eulemur mongoz*. *Am. J. Primatol.* 47:279–298.

Dewar, R. E. (1997). Were people responsible for the extinction of Madagacar's subfossils, and how will we ever know? In: Goodman, S. M., and Patterson, B. D. (eds.), *Natural Change and Human Impact in Madagascar*. Smithsonian Institution Press, Washington DC. pp. 364–377.

Donati, G., and Borgognini Tarli, S. M. (2002a). Feeding ecology of the collared brown lemur, *Eulemur fulvus collaris*, in the Sainte Luce Littoral Forest. *Folia Primatol.* 73:315.

Donati, G., and Borgognini Tarli, S. M. (2002b). The role of abiotic factors in influencing cathemeral activity of collared brown lemurs (*E. fulvus collaris*) in the Sainte Luce Littoral Forest. *Folia Primatol.* 73:305–306.

Donati, G., Lunardini, A., and Kappeler, P. M. (1999). Cathemeral activity of red-fronted brown lemurs (*Eulemur fulvus rufus*) in the Kirindy Forest. In: Rakotosamimanana, B., Rasamimanana, H., Ganzhorn, J. U., and Goodman, S. M. (eds.), *New Directions in Lemur Studies*. Kluwer/Plenum, New York. pp. 119–137.

Donati, G., Lunardini, A., Kappeler, P. M., and Borgognini Tarli, S. M. (2001). Nocturnal activity in the cathemeral red-fronted lemur (*Eulemur fulvus rufus*) with observations during a lunar eclipse. *Am. J. Primatol.* 53:69–78.

Erhardt, E. M., and Overdorff, D. J. (1998). Infanticide in *Propithecus diadema edwardsi*: an evaluation of the sexual selection hypothesis. *Into. J. Primatol.* 19:73–81.

Erickson, C. J. (1991). Percussive foraging in the aye-aye, *Daubentonia madagascariensis. Anim. Behav.* 41:793–801.

Erickson, C. J. (1994). Tap-scanning and extractive foraging in aye-ayes, *Daubentonia madagascariensis. Folia Primatol.* 62:125–135.

Fietz, J. (1999). Mating system of *Microcebus murinus. Am. J. Primatol.* 48:127–133.

Fleagle, J. G. (1999). *Primate Adaptation and Evolution*, 2nd ed. Academic Press, London.

Foerg, R. (1982). Reproduction in *Cheirogaleus medius. Folia Primatol.* 39:49–62.

Freed, B. Z. (1996). Co-occurrence among crowned lemurs (*Eulemur coronatus*) and Sanford's lemurs (*Eulemur fulvus sanfordi*) of Madagascar [PhD diss.]. Washington University, St. Louis.

Freed, B. Z. (1999). An introduction to the ecology of daylight-active lemurs. In: Dolhinow, P., and Fuentes, P. A. (eds.), *The Non-Human Primates*. Mayfield Press, Mountain View, CA. pp. 123–132.

Ganzhorn, J. U. (1988). Food partitioning among Malagasy primates. *Oecologica* 75:436–450.

Ganzhorn, J. U. (1993). Flexibility and constraints of *Lepilemur* ecology. In: Kappeler, P. M., and Ganzhorn, J. U. (eds.), *Lemur Social Systems and Their Ecological Basis*. Plenum Press, New York. pp. 153–166.

Ganzhorn, J. U., Abraham, J. P., and Razanahoera-Rakotomalala, M. (1985). Some aspects of the natural history and food selection of *Avahi laniger. Primates* 26:452–463.

Gerson, J. S. (2001). Social relationships in wild red-fronted brown lemurs (*Eulemur fulvus rufus*) [PhD diss.]. Duke University, Durham, NC.

Gingerich, P. D., and Martin, R. D. (1981). Cranial morphology and adaptations in Eocene Adapidae II: the Cambridge skull of *Adapis parisiensis. Am. J. Phys. Anthropol.* 56:235–257.

Glander, K. E., Powzyk, J. A. (1998). Morphometrics of wild *Indri indri* and *Propithecus diadema diadema. Folia Primatol.* 69(suppl. 1):399.

Glander, K. E., Wright, P. C., Daniels, P. S., and Merenlender, A. M. (1992). Morphometrics and testicle size of rain forest lemur species from southeastern Madagascar. *J. Hum. Evol.* 22:1–17.

Glander, K. E., Wright, P. C., Seigler, D. S., and Randriansolo, B. (1989). Consumption of cyanogenic bamboo by a newly discovered species of bamboo lemur. *Am. J. Primatol.* 19:199–124.

Godfrey, L. R., Jungers, W. L., Reed, K. E., Simons, E. L., and Chatrath, P. S. (1997). Subfossil lemurs: inferences about past and present primate communities. In: Goodman, S. M., and Patterson, B. D. (eds.), *Natural Change and Human Impact in Madagascar*. Smithsonian Institution Press, Washington DC. pp. 218–256.

Godfrey, L. R., Lyon, S. K., and Sutherland, M. R. (1993). Sexual dimorphism in large-bodied primates: the case of the subfossil lemurs. *Am. J. Phys. Anthropol.* 90:315–334.

Godfrey, L. R., Semprebon, G. M., Jungers, W. L., Sutherland, M. R., Simons, E. L., and Solounias, N. (2004). Dental use wear in extinct lemurs: evidence of diet and niche differentiation. *J. Hum. Evol.* 47:145–169.

Goodman, S. M., and Langrand, O. (1996). A high mountain population of the ringtailed lemur *Lemur catta* on the Andringitra Massif, Madagascar. *Oryx* 30:259–268.

Gould, L. (1990). The social development of free-ranging infant *Lemur catta* at Berenty Reserve, Madagascar. *Int. J. Primatol.* 11:297–317.

Gould, L. (1992). Alloparental care in free-ranging *Lemur catta* at Berenty Reserve, Madagascar. *Folia Primatol.* 58:72–83.

Gould, L., Sussman, R. W., and Sauther, M. L. (1999). Natural disasters and primate populations: the effects of a two-year drought on a naturally occurring population of ringtailed lemurs in southwestern Madagascar. *Int. J. Primatol.* 20:69–84.

Gould, L., Sussman, R. W., and Sauther, M. L. (2003). Demographic and life-history patterns in a population of ring-tailed lemurs (*Lemur catta*) at Beza Mahafaly Reserve, Madagascar: a 15-year perspective. *Am. J. Phys. Anthropol.* 120:182–194.

Grassi, C. (2001). The behavioral ecology of *Hapalemur griseus griseus*: the influence of microhabitat and population density on this small-bodied prosimian folivore [PhD diss.]. University of Texas, Austin.

Groves, C. P. (2000). The genus *Cheirogaleus*: unrecognized biodiversity in dwarf lemurs. *Int. J. Primatol.* 21:943–962.

Groves, C. P., and Tattersall, I. (1991). Geographical variation in the fork-marked lemur, *Phaner furcifer* (Mammalia, Primates). *Folia Primatol.* 56:39–49.

Harcourt, C. (1988). *Avahi laniger*: a study in inactivity. *Primate Eye* 35:9.

Harcourt, C. S. (1987). Brief trap/retrap study of the brown mouse lemur (*Microcebus rufus*). *Folia Primatol.* 49:209–211.

Harrington, J. E. (1975). Field observations of social behavior of *Lemur fulvus fulvus* (E. Geoffroy 1812). In: Tattersall, I., and Sussman, R. W. (eds.), *Lemur Biology*. Plenum Press, New York. pp. 259–279.

Harrington, J. E. (1978). Diurnal behavior of *Lemur mongoz* at Ampijoroa, Madagascar. *Folia Primatol.* 29:291–302.

Hawkins, A. F. A., Chapman, P., Ganzhorn, J. U., Bloxam, Q. M., Barlow, S. C., and Tonge, S. J. (1990). Vertebrate conservation in Ankarana Special Reserve, northern Madagascar. *Biol. Conserv.* 54:83–110.

Hemingway, C. A. (1996). Morphology and phenology of seeds and whole fruit eaten by Milre-Edwards sifaka. *Propithecus diadema edwardsi* in Ranomafana National Park, Madagascar. *Int. J. Primatol.* 17:637–659.

Hladik, C. M. (1979). Diet and ecology of prosimians. In: Doyle, G. A., and Martin, R. D. (eds.), *The Study of Prosimian Behavior*. Academic Press, New York. pp. 307–357.

Hladik, C. M., and Charles-Dominique, P. (1974). The behavior and ecology of the sportive lemur (*Lepilemur mustelinus*) in relation to its dietary peculiarities. In: Martin, R. D., Doyle, G. A., and Walker, A. C. (eds.), *Prosimian Biology*. Duckworth Press, London. pp. 23–37.

Hladik, C. M., Charles-Dominique, P., and Petter, J. J. (1980). Feeding strategies of five nocturnal prosimians in the dry forest of the west coast of Madagascar. In: Charles-Dominique, P., Cooper, H. M., Hladik, A., Hladik, C. M., Pages, E., Pariente, G. F., Petter-Rousseaux, A., Petter, J. J., and Schilling, A. (eds.), *Nocturnal Malagasy Primates: Ecology, Physiology and Behavior*. Academic Press, New York. pp. 41–73.

Hood, L. C., and Jolly, A. (1995). Troop fission in female *Lemur catta* at Berenty Reserve, Madagascar. *Int. J. Primatol.* 16:997–1016.

Iwano, T., and Iwakawa, C. (1988). Feeding behaviour of the aye-aye (*Daubentonia madagascariensis*) on nuts of Ramy (*Canarium madagascariensis*). *Folia Primatol.* 50:136–142.

Jenkins, P. D. (1987). Catalogue of primates in the British Museum (Natural History) and elsewhere in the British Isles Part IV: Suborder strepsirrhini, including the subfossil Madagascar lemurs and family Tarsiidae. London: British Museum (Natural History). px. 189.

Johnson, S., and Overdorff, D. J. (2002). Scarce season diet and keystone resources in three brown lemur (*Eulemur fulvus* spp.) populations. *Am. J. Primatol.* 56:65–66.

Jolly, A. (1966). *Lemur Behavior*. University of Chicago Press, Chicago.

Jolly, A. (1984). The puzzle of female feeding priority. In: Small, M. (ed.), *Female Primates: Studies by Women Primatologists*. Alan R. Liss, New York. pp. 197–215.

Jolly, A., Dobson, A., Rasamimanana, H. M., Walker, J., Solberg, M., and Perel, V. (2002). Demography of *Lemur catta* at Berenty Reserve, Madagascar: effects of troop size, habitat and rainfall. *Int. J. Primatol.* 23:327–353.

Jolly, A., Rasamimanana, H. R., Kinnaird, M. F., O'Brien, T. G., Crowley, H. M., Harcourt, C. S., Gardner, S., and Davidson, J. (1993). Territoriality in *Lemur catta* groups during the birth season at Berenty, Madagascar. In: Kappeler, P. M., and Ganzhorn, J. U. (eds.), *Lemur Social Systems and Their Ecological Basis*. Plenum Press, New York. pp. 85–109.

Jolly, C. J. (1970). *Hadropithecus*: A lemuroid small-object feeder. *Man* 5:619–626.

Jungers, W. L. (1979). Locomotion, limb proportions and skeletal allometry in lemurs and lorises. *Folia Primatol.* 32:8–28.

Jungers, W. L., Godfrey, L. R., Simons, E. L., Wunderlich, R. E., Richmond, B. G., and Chatrath, P. S. (2002). Ecomorphology and behavior of giant extinct lemurs from Madagascar. In: Plavcan, J. M., Kay, R. F., Jungers, W. L., and van Schaik, C. P. (eds.), *Reconstructing Behavior in the Primate Fossil Record*. Kluwer, New York. pp. 371–411.

Kappeler, P. M. (1997). Intrasexual selection in *Mirza coquereli*: evidence for scramble competition polygyny in a solitary primate. *Behav. Ecol. Sociobiol.* 45:115–127.

Kappeler, P. M. (2000). Lemur origins: rafting by groups of hibernators? *Folia Primatol.* 71:422–425.

Kappeler, P. M., and Erkert, H. G. (2003). On the move around the clock: correlates and determinants of cathemeral activity in wild redfronted lemurs (*Eulemur fulvus rufus*). *Behav. Ecol. Sociobiol.* 54:359–369.

Kappeler, P. M., Wimmer, B., Zinner, D., and Diethard, T. (2002). The hidden matrilineal structure of a solitary lemur: implications for primate social evolution. *Proc. R. Soc. Lond. B.* 269:1755–1763.

Koenders, L., Rumpler, Y., and Ratsirarson, J. (1985). *Lemur macaco flavifrons*: a rediscovered subspecies of primate. *Folia Primatol.* 44:210–215.

Koyama, N. (1991). Troop division and inter-troop relationships of ring-tailed lemurs (*Lemur catta*) at Berenty, Madagascar. In: Ehara, A., Kimura, T., and Iwamoto, M. (eds.), *Primatology Today*. Elsevier, Amsterdam. pp. 173–176.

Koyama, N., Nakamichi, M., Ichino, S., and Takahata, Y. (2002). Population and social dynamics changes in ring-tailed lemur troops at Berenty, Madagascar, between 1989–1999. *Primates* 43:291–314.

Martin, R. D. (1972a). A preliminary field study of the lesser mouse lemur (*Microcebus murinus* J. F. Miller 1777). *Z. Tierpsychol.* 9(suppl.):43–89.

Martin, R. D. (1972b). Adaptive radiation and behaviour of the Malagasy lemurs. *Phil. Trans. R. Soc. Lond. B.* 264:295–352.

Martin, R. D. (1973). A review of the behaviour and ecology of the lesser mouse lemur (*Microcebus murinus* J. F. Miller 1777). In: Michael, R. P., and Crook, J. H. (eds.), *Comparative Ecology and Behaviour of Primates*. Academic Press, London. pp. 1–68.

Martin, R. D. (1984). Dwarf and mouse lemurs. In: MacDonald, D. (ed.), *The Encyclopaedia of Mammals: 1*. George Allen and Unwin, London. p. 331.

Martin, R. D. (1990). *Primate Origins and Evolution*, Princeton University Press, Princeton, NJ.

Martin, R. D. (2000). Origins, diversity, and relationships of lemurs. *Int. J. Primatol.* 21:1021–1049.

Meier, B., and Albignac, R. (1991). Rediscovery of *Allocebus trichotis* Günther 1875 (Primates) in north east Madagascar. *Folia Primatol.* 56:57–63.

Meier, B., Albignac, R., Peyrieras, A., Rumpler, Y., and Wright, P. C. (1987). A new species of *Hapalemur* (Primates) from southeast Madagascar. *Folia Primatol.* 48:211–215.

Meier, B., and Rumpler, Y., (1987). Preliminary survey of *Hapalemur simus* and of a new species of *Hapalemur* in eastern Betsileo, Madagascar. *Primate Conserv.* 8:40–43.

Meyers, D. M. (1993). The effects of resource seasonality on the behavior and reproduction of the golden-crowned sifaka (*Propithecus tattersalli*, Simons, 1988) in three Malagasy forests [PhD diss.]. Duke University, Durham, NC.

Meyers, D. M., and Ratsirarson, J. (1989). Distribution and conservation of two endangered sifakas in northern Madagascar. *Primate Conserv.* 10:82–87.

Meyers, D. M., and Wright, P.C. (1993). Resource tracking: food availability and *Propithecus* seasonal reproduction. In: Kappeler, P. M., and Ganzhorn, J. U. (eds.), *Lemur Social Systems and Their Ecological Basis*. Plenum Press, New York. pp. 179–192.

Mittermeier, R. A., Tattersall, I., Konstant, W. R., Meyers, D. M., and Mast, R. B. (1994). *Lemurs of Madagascar. Conservation International Tropical Field Guide Series*. Conservation International, Washington DC.

Morland, H. S. (1990). Parental behavior and infant development in ruffed lemurs (*Varecia variegata*) in a northeast Madagascar rain forest. *Am. J. Primatol.* 20:253–265.

Morland, H. S. (1991). Preliminary report on the social organization of ruffed lemurs (*Varecia variegata variegata*) in a northeast Madagascar rain forest. *Folia Primatol.* 56:157–161.

Morland, H. S. (1993). Seasonal behavioral variation and its relationship to thermoregulation in ruffed lemurs (*Varecia variegata variegata*). In: Kappeler, P. M., and Ganzhorn, J. U. (eds.), *Lemur Social Systems and their Ecological Basis*. Plenum Press, New York. pp. 193–203.

Müller, A. E. (1998). A preliminary report on the social organisation of *Cheirogaleus medius* (Cheirogaleidae; Primates) in north-west Madagascar. *Folia Primatol.* 69:160–166.

Müller, A. E. (1999). Aspects of social life in the fat-tailed dwarf lemur (*Cheirogaleus medius*): inferences from body weights and trapping data. *Am. J. Primatol.* 49:265–280.

Müller, A., and Thalmann, U. (2000). Origin and evolution of primate social organization: a reconstruction. *Biol. Rev.* 75:405–435.

Mutschler, T. (2002). Alaotran gentle lemur: some aspects of its behavioral ecology. *Evol. Anth. Suppl* 1:101–104.

Mutschler, T., Nievergelt, C. M., and Feistner, A. T. C. (2000). Social organization of the Alaotran gentle lemur (*Hapalemur griseus alaotrensis*). *Am J. Primatol.* 50:9–24.

Overdorff, D. J. (1988). Preliminary report on the activity cycle and diet of the red-bellied lemur (*Eulemur rubriventer*) in Madagascar. *Am. J. Primatol.* 16:143–153.

Overdorff, D. J. (1992). Differential patterns of flower feeding by *Eulemur fulvus rufus* and *Eulemur rubriventer* in Madagascar. *Am. J. Primatol.* 28:191–203.

Overdorff, D. J. (1993a). Ecological and reproductive correlates to range use in red-bellied lemurs (*Eulemur rubriventer*) and rufous lemurs (*Eulemur fulvus rufus*). In: Kappeler, P. M., and Ganzhorn, J. U. (eds.), *Lemur Social Systems and Their Ecological Basis*. Plenum Press, New York. pp. 167–178.

Overdorff, D. J. (1993b). Similarities, differences, and seasonal patterns in the diets of *Eulemur rubriventer* and *Eulemur fulvus rufus* in the Ranomafana National Park, Madagascar. *Int. J. Primatol.* 14:721–753.

Overdorff, D. J. (1996). Ecological correlates to activity and habitat use of two prosimian primates: *Eulemur rubriventer* and *Eulemur fulvus rufus* in Madagascar. *Am. J. Primatol.* 40:327–342.

Overdorff, D. J. (1998). Are *Eulemur* species pair-bonded? Social organization and mating strategies in *Eulemur fulvus rufus* from 1988–1995 in southeast Madagascar. *Am. J. Phys. Anthropol.* 105:153–166.

Overdorff, D. J., Erhart, E. M., and Mutschler, T. (2003). Fission–fusion in *Eulemur fulvus rufus* in southeastern Madagascar. *Am. J. Primatol.* 60(suppl. 1):42–43.

Overdorff, D. J., Merenlender, A. M., Talata, P., Telo, A., and Forward, Z. A. (1999). Life history of *Eulemur fulvus rufus* from 1988–1995 in southeastern Madagascar. *Am. J. Phys. Anthropol.* 108:295–310.

Overdorff, D. J., and Rasmussen, M. A. (1995). Determinants of nighttime activity in "diurnal" lemurid primates. In: Alterman, L., Doyle, G. A., and Izard, M. K. (eds.), *Creatures of the Dark: The Nocturnal Prosimians*. Plenum Press, New York. pp. 61–74.

Pagès, E. (1980). Ethoecology of *Microcebus coquereli* during the dry season. In: Charles-Dominique, P., Cooper, H. M., Hladik, A., Hladik, C. M., Pagès, E., Pariente, C. F., Petter-Rousseaux, A., Peter, J. J., and Schilling, A. (eds.), *Nocturnal Malagasy Primates: Ecology, Physiology and Behavior*. Academic Press, New York. pp. 17–116.

Pagès-Feuillade, E. (1989). Modalités de l'occupation de l'espace et interindividuelles chez un prosimien nocturne malagache (*Microcebus murinus*). *Folia Primatol.* 50:204–220.

Pastorini, J., Martin, R. D., Ehresmann, P., Zimmermann, E., and Forstner, M. R. J. (2001). Molecular phylogeny of the lemur family Cheirogaleidae (primates) based on mitochondrial DNA sequences. *Mol. Phylogenet. Evol.* 19:45–56.

Pereira, M. E. (1993a). Agonistic interactions, dominance relations and ontogenetic trajectories in ringtailed lemurs. In: Pereira, M. E., and Fairbanks, L. A. (eds.), *Juvenile Primates: Life History, Development and Behavior*. Oxford University Press, New York. pp. 285–305.

Pereira, M. E. (1993b). Seasonal adjustment of growth rate and adult body weight in ringtailed lemurs. In: Kappeler, P. M., and Ganzhorn, J. U. (eds.), *Lemur Social Systems and Their Ecological Basis*. Plenum Press, New York. pp. 205–221.

Pereira, M. E., Klepper, A., and Simons, E. L. (1987). Tactics of care for young infants by forest-living ruffed lemurs (*Varecia variegata variegata*): ground nests, parking, and biparental guarding. *Am. J. Primatol.* 13:129–144.

Petter, J. J. (1962). Recherches dur l'ecologie et l'ethologie des lemuriens malagashes. Mémoires Museum National Histoire Naturelle, Paris, 27:1–146.

Petter, J. J. (1978). Ecological and physiological adaptations of five sympatric nocturnal lemurs to seasonal variation in food production. In: Chivers, D. J., and Herbert, J. (eds.), *Recent Advances in Primatology. 1: Behaviour*. Academic Press, London. pp. 211–223.

Petter, J. J., Albignac, R., and Rumpler, Y. (1977). Lemurine mammals (Primates, Prosimians). *Faune Madagascar*, 44:1–513.

Petter, J. J., and Charles-Dominique, P. (1979). Vocal communication in prosimians. In: Doyle, G. A., and Martin, R. D. (eds.), *The Study of Prosimian Behavior*. Academic Press, New York. pp. 247–305.

Petter, J. J., and Petter-Rousseaux, A. (1959). Contribution to the study of the aye-aye. *Naturaliste Malagache* 11:153–164.

Petter, J. J., and Petter-Rousseaux, A. (1979). Classification of the prosimians. In: Doyle, G. A., and Martin, R. D. (eds.), *The Study of Prosimian Behavior*. Academic Press, London. pp. 1–44.

Petter, J. J., and Peyriéras, A. (1970a). Nouvelle contribution a l'étude d'un lémurien malagache, le aye-aye (*Daubentonia madagascariensis* E. Geoffroy). *Mammalia* 34:167–193.

Petter, J. J., and Peyriéras, A. (1970b). Observation éco-étholgiques sur les lemuriens malagaches du genre Hapalemur. *Terre Vie* 24:356–382.

Petter, J. J., Schilling, A., and Pariente, G. (1971). Observations éco-éthologiques sur deux lémuriens malagaches nocturnes: *Phaner furcifer* et *Microcebus coquereli*. *Terre Vie* 25:287–327.

Petter, J. J., Schilling, A., and Pariente, G. (1975). Observations on behavior and ecology of *Phaner furcifer*. In: Tattersall, I., and Sussman, R. W. (eds.), *Lemur Biology*. Plenum Press, New York. pp. 209–218.

Petter-Rousseaux, A. (1964). Reproductive physiology and behavior of the Lemuroidea. In: Buettner-Janusch, J. (ed.), *Evolutionary and Genetic Biology of Primates*, vol. 2. Academic Press, New York. pp. 91–132.

Petter-Rousseaux, A. (1980). Seasonal activity rhythms, reproduction, and body weight variations in five sympatric nocturnal prosimians, in simulated light and climatic conditions. In: Charles-Dominique, P., Cooper, H. M., Hladik, A., Hladik, C. M., Pages, E., Pariente, G. F., Petter-Rousseaux, A., Petter, J. J., and Schilling, A. (eds.), *Nocturnal Malagasy Primates: Ecology, Physiology and Behavior*. Academic Press, New York. pp. 137–152.

Pochron, S. T., Tucker, W. T., and Wright, P. C. (2004). Demography, life history and social structure in *Propithecus diadema edwardsi* from 1986 to 2000 in Ranomafana National Park, Madagascar. *Am. J. Phys. Anthropol.* 125:61–72.

Pochron, S. T., and Wright, P. C. (2003). Variability in adult group compositions of a prosimian primate. *Beh. Ecol. Sociobiol.* 54:285–293.

Pollock, J. I. (1975). Field observations on *Indri indri*: a preliminary report. In: Tattersall, I., and Sussman, R. W. (eds.), *Lemur Biology*. Plenum Press, New York. pp. 287–311.

Pollock, J. I. (1977). The ecology and socioecology of feeding in *Indri indri*. In: Clutton-Brock, T. (ed.), *Primate Ecology:*

Studies of Feeding and Ranging Behavior in Lemurs, Monkeys and Apes. Academic Press, London. pp. 37–69.

Powzyk, J. A. (1997). The socioecology of two sympatric indriids: *Propithecus diadema diadema* and *Indri indri*, a comparison of feeding strategies and their possible repercussions on species-specific behavior [PhD diss.]. Duke University, Durham, NC.

Powzyk, J. A., and Mowry, C. B. (2003). Dietary and feeding differences between sympatric *Propithecus diadema diadema* and *Indri indri*. *Int. J. Primatol.* 24:1143–1162.

Rabarivola, C., Meyers, D., and Rumpler, Y. (1991). Distribution and morphological characteristics of intermediate forms between the black lemur (*Eulemur macaco macaco*) and the Sclater's lemur (*E. m. flavifrons*). *Primates* 32:269–273.

Radespiel, U. (2000). Sociality in the gray mouse lemur (*Microcebus murinus*) in northwestern Madagascar. *Am. J. Primatol.* 51:21–40.

Radespiel, U., Cepok, S., Zimmermann, E., and Zietemann, V. (1998). Sex-specific usage patterns of sleeping-sites in grey mouse lemurs (*Microcebus murinus*) in northwestern Madagascar. *Am. J. Primatol.* 46:77–84.

Radespiel, U., Ehresmann, P., and Zimmerman, E. (2003a). Species-specific usage of sleeping sites in two sympatric mouse lemur species (*Microcebus murinus* and *M. ravelobensis*) in northwestern Madagascar. *Am. J. Primatol.* 59:139–151.

Radespiel, U., Lutermann, H., Schmelting, B., Bruford, M. W., and Zimmerman, E. (2003b). Patterns and dynamics of sex-biased dispersal in a nocturnal primate, the grey mouse lemur, *Microcebus murinus*. *Anim. Behav.* 65:709–719.

Rafferty, K. L., Teaford, M. F., and Jungers, W. L. (2002). Molar microwear of subfossil lemurs: Improving the resolution of dietary inferences. *J. Hum. Evol.* 43:645–657.

Ramanamanjato, J. B., and Ganzhorn, J. U. (2001). Effects of forest fragmentation, introduced *Rattus rattus* and the role of exotic tree plantations and secondary vegetation for the conservation of an endemic rodent and a small lemur in littoral forests of southeastern Madagascar. *Anim. Conserv.* 4:175–183.

Randrianambinina, B., Rakotondravony, P., Radespiel, U., and Zimmermann, E. (2003). Seasonality in general activity, body mass and reproduction of the golden brown mouse lemur (*Microcebus ravelobensis*) in northwestern Madagascar and the brown mouse lemur (*M. rufus*) in eastern Madagascar. *Primates* 44:321–331.

Rasmussen, M. A. (1999). Ecological influences on activity cycle in two cathemeral primates, the mongoose lemur (*Eulemur mongoz*) and the common brown lemur (*Eulemur fulvus fulvus*) [PhD diss.]. Duke University, Durham, NC.

Rasoazanabary, E. A. (2004). Preliminary study of mouse lemurs in the Bezamahafaly Special Reserve, southwest Madagascar. *Lemur News* 9:4–7.

Rasoloarison, R. M., Goodman, S. M., and Ganzhorn, J. U. (2000). Taxonomic revision of mouse lemurs (*Microcebus*) in the western portions of Madagascar. *Int. J. Primatol.* 21:963–1019.

Rasoloharijaona, S. (1998). Vocal communication in a nocturnal lemur in northwestern Madagascar. XVIIth Congress of the International Primatological Society, Antananarivo, Madagascal. Abstract 170.

Rasoloharijaona, S. B., Rakotosamimanana, B., Randrianambinina, B., and Zimmermann, E. (2003). Pair-specific usage of sleeping sites and their implications for social organization in a nocturnal Malagasy primate, the Milne Edwards sportive lemur (*Lepilemur edwardsi*). *Am. J. Phys. Anthropol.* 122:251–258.

Ratsimbazafy, J. H. (2001). On the brink of extinction and the process of recovery: responses of black-and-white ruffed lemurs (*Varecia variegata variegata*) to disturbance in Manombo Forest, Madagascar [PhD diss.]. State University of New York, Stony Brook.

Ratsimbazafy, J. H. (2002). How do black and white ruffed lemurs still survive in a highly disturbed habitat? *Lemur News* 7:7–10.

Ratsirarson, J., Anderson, J., Warter, S., and Rumpler, Y. (1987). Notes on the distribution of *Lepilemur septentrionalis* and *L. mustelinus* in northern Madgascar. *Primates* 28:119–122.

Ratsirarson, J., and Rumpler, Y. (1988). Contribution à l'étude comparée de l'écoéthologie de deux espèces de lémuriens, *Lepilemur mustelinus* (I. Geoffroy 1850), *Lepilemur septentrionalis* (Rumpler and Albignac 1975). In: Rakotovao, L., Barre, V., and Sayer, J. (eds.), *L'Equilibre des Ecosystèmes forestiers à Madagascar. Actes d'un Séminaire International.* IUCN, Gland, Switzerland. pp. 100–102.

Raxworthy, C. J., and Rakotondraparany, F. (1988). Mammals report. In: Quansah, N. (ed.), *Manongarivo Special Reserve (Madagascar), 1987/88 Expedition Report.* Madagascar Environmental Research Group, London.

Razanahoera-Rakotomalala, M. (1981). Les adaptations alimentaires comparées de deux lémuriens folivores sympatriques: *Avahi Jourdan*, 1834—*Lepilemur* I. Geoffroy 1851 [PhD thesis]. University of Madagascar, Antananarivo.

Reimann, E. W. (2002). Coexistence and feeding ecology in female grey and golden-brown mouse lemurs (*Microcebus murinus* and *M. ravelobensis*) in north-western Madagascar [diss.]. Tierärztliche Hochschule, Hannover.

Richard, A. F. (1973). Social organisation and ecology of *Propithecus verreauxi*, Grandidier 1867 [PhD diss.]. University of London, London.

Richard, A. F. (1974). Intra-specific variation in the social organization and ecology of *Propithecus verreauxi*. *Folia Primatol.* 22:178–207.

Richard, A. F. (1978). *Behavioral Variation: Case Study of a Malagasy Lemur.* Associated University Press, Princeton, NJ.

Richard, A. F. (1984). Lemurs. In: MacDonald, D. (ed.), *The Encyclopedia of Mammals*, vol. 1. George Allen and Unwin, London. pp. 330–331.

Richard, A. F. (1985). *Primates in Nature.* W. H. Freeman, New York. pp. 98–99.

Richard, A. F. (1987). Malagasy prosimians: female dominance. In: Smuts, B. B., Cheney, D. L., Seyfarth, R. M., Wrangham, R. W., and Struhsaker, T. T. (eds.), *Primate Societies.* University of Chicago Press, Chicago. pp. 227–239.

Richard, A. F., and Dewar, R. E. (1991). Lemur ecology. *Annu. Rev. Ecol. Syst.* 22:145–175.

Richard, A. F., Dewar, R. E., Schwartz, M., and Ratsirarson, J. (2002). Life in the slow lane: demography and life histories of male and female sifaka (*Propithecus verreauxi verreauxi*). *J. Zool.* 256:421–436.

Richard, A. F., Rakotomanga, P., and Schwartz, M. (1993). Dispersal by *Propithecus verreauxi* at Beza Mahafaly, Madagascar: 1984–1991. *Am. J. Primatol.* 30:1–20.

Rigamonti, M. M. (1993). Home range and diet in red ruffed lemurs (*Varecia variegata rubra*) on the Masoala Peninsula, Madagascar. In: Kappeler, P. M., and Ganzhorn, J. U. (eds.), *Lemur Social Systems and Their Ecological Basis.* Plenum Press, New York. pp. 25–39.

Russell, R. J. (1977). The behavior, ecology, and environmental physiology of a nocturnal primate, *Lepilemur mustelinus* (Strepsirhini, Lemuriformes, Lepilemuridae) [PhD thesis]. Duke University, Durham, NC.

Russell, R. J. (1980). The environmental physiology and ecology of *Lepilemur ruficaudatus* (*L. leucopus*) in arid southern Madagascar. *Am. J. Phys. Anthropol.* 52:273–274.

Sauther, M. L. (1991). Reproductive behavior of free-ranging *Lemur catta* at Beza Mahafaly Special Reserve, Madagascar. *Am. J. Phys. Anthropol.* 84:463–477.

Sauther, M. L. (1992). The effect of reproductive state, social rank and group size on resource use among free-ranging ringtailed lemurs (*Lemur catta*) of Madagascar [PhD diss.]. Washington University, St. Louis.

Sauther, M. L. (1993). Resource competition in wild populations of ringtailed lemurs (*Lemur catta*): implications for female dominance. In: Kappeler, P. M., and Ganzhorn, J. U. (eds.), *Lemur Social Systems and Their Ecological Basis*. Plenum Press, New York. pp. 135–152.

Sauther, M. L. (1994). Wild plant use by pregnant and lactating ringtailed lemurs; Implications for ringtailed lemur conservation. *Folia Primatol.* 69(suppl. 1):309–320.

Sauther, M. L. (1998). The interplay of phenology and reproduction in ringtailed lemurs: implications for ringtailed lemur conservation. *Folia Primatol.* 69(suppl. 1):309–320.

Sauther, M. L., and Sussman, R. W. (1993). A new interpretation of the social organization and mating system of the ringtailed lemur (*Lemur catta*). In: Kappeler, P. M., and Ganhorn, J. U. (eds.), *Lemur Social Systems and Their Ecological Basis*. Plenum Press, New York. pp. 111–122.

Sauther, M. L., Sussman, R. W., and Gould, L. (1999). The myth of the typical *Lemur catta*: 30 years of research on the ringtailed lemur. *Evol. Anthropol.* 8:120–132.

Schmelting, B., Ehresmann, P., Lutermann, H., Randrianambinina, B., and Zimmermann, E. (2000). Reproduction of two sympatric mouse lemur species (*Microcebus murinus* and *Microcebus ravelobensis*) in north-west Madagascar: first results of a long term study. In: Lourenco, W. R., and Goodmann, S. M. (eds.), *Diversity and Endemism in Madagascar*. Mémoires de la Société de Biogéographie, Paris. pp. 165–175.

Schmid, J. (1998). Tree holes used for resting by gray mouse lemurs (*Microcebus murinus*) in Madagascar: insulation capacities and energetic consequences. *Int. J. Primatol.* 19:797–809.

Schmid, J., and Kappeler, P. M. (1994). Sympatric mouse lemurs (*Microcebus* spp.) in western Madagascar. *Folia Primatol.* 63:162–170.

Schulke, O., and Kappeler, P. M. (2003). So near and yet so far: territorial pairs but low cohesion between pair partners in a nocturnal lemur, *Phaner furcifer*. *Anim. Behav.* 65:331–343.

Schwab, D. (2000). A preliminary study of the social and mating system of pygmy mouse lemurs (*Microcebus myoxinus*). *Am. J. Primatol.* 51:41–60.

Simons, E. L., Burney, D. A., Chatrath, P. S., Godfrey, L. R., Jungers, W. L., and Rakotosamimanana, B. (1995). AMS ^{14}C dates for extinct lemurs from caves in the Ankarana Massif, northern Madagascar. *Quat. Res.* 43:249–254.

Sterling, E. J. (1992). Timing the reproduction of aye-ayes (*Daubentonia madagascariensis*) in Madagascar. *Am. J. Primatol.* 27:59–60.

Sterling, E. J. (1993). Patterns of range use and social organization in aye-ayes (*Daubentonia madagascariensis*) on Nosy Mangabe. In: Kappeler, P. M., and Ganzhorn, J. U. (eds.), *Lemur Social Systems and Their Ecological Basis*. Plenum Press, New York. pp. 1–10.

Sterling, E. J., Dierenfeld, E. S., Ashbourne, C. G., and Feistner, A. T. C. (1994). Dietary intake, food composition and nutrient intake in wild and captive populations of *Daubentonia madagascariensis*. *Folia Primatol.* 62:115–124.

Sterling, E. J., and Richard, A. F. (1995). Social organization in the aye-aye (*Daubentonia madagascariensis*) and the perceived distinctiveness of nocturnal primates. In: Alterman, L., Doyle, G. A., and Izard, M. K. (eds.), *Creatures of the Dark: The Nocturnal Prosimians*. Plenum Press, New York. pp. 439–451.

Sussman, R. W. (1972). An ecological study of two Madagascan primates: *Lemur fulvus rufus* Audebert and *Lemur catta* Linnaeus [PhD diss.]. Duke University, Durham, NC.

Sussman, R. W. (1974). Ecological distinctions between two species of lemur. In: Martin, R. D., Doyle, D. A., and Walker, C. (eds.), *Prosimian Biology*. Duckworth, London. pp. 75–108.

Sussman, R. W. (1977). Socialization, social structure and ecology of two sympatric species of lemur. In: Chevalier-Skolnikoff, S., and Poirier, F. E. (eds.), *Primate Bio-Social Development: Biological, Social, and Ecological Determinants*. Garland, New York. pp. 515–528.

Sussman, R. W. (1991). Demography and social organization of free-ranging *Lemur catta* in the Beza Mahafaly Reserve, Madagascar. *Am. J. Phys. Anthropol.* 84:43–58.

Sussman, R. W. (1992). Male life histories and inter-group mobility among ringtailed lemurs (*Lemur catta*). *Int. J. Primatol.* 13:395–413.

Sussman, R. W. (1999). *Primate Ecology and Social Structure. Lorises. Lemurs and Tarsiers*, vol. 1. Pearson, Needham Heights, MA.

Sussman, R. W., and Richard, A. F. (1986). Lemur conservation in Madagascar: the status of lemurs in the south. *Primate Conserv.* 7:85–92.

Sussman, R. W., and Tattersall, I. (1976). Cycles of activity, group composition and diet of *Lemur mongoz* Linnaeus 1766 in Madagascar. *Folia Primatol.* 26:270–283.

Tan, C. L. (1998). Group composition, home range size, and diet of three sympatric bamboo lemur species (Genus *Hapalemur*) in Ranomafana National Park. *Int. J. Primatol.* 20:547–563.

Tan, C. L. (2001). Behavior and ecology of three sympatric bamboo lemur species (genus *Hapalemur*) in Ranomafana National Park, Madagascar [PhD diss.]. State University of New York, Stony Brook.

Tattersall, I. (1976). Group structure and activity rhythm in *Lemur mongoz* (Primates, Lemuriformes) on Anjouan and Mohéli Islands, Comoro Archipelago. *Anthropol. Papers Am. Mus. Nat. Hist.* 53:257–261.

Tattersall, I. (1977a). Distribution of the Malagasy lemurs, part 1. The lemurs of northern Madagascar. *Ann. N.Y. Acad. Sci.* 293:160–169.

Tattersall, I. (1977b). The lemurs of the Comoro Islands. *Oryx* 13:445–448.

Tattersall, I. (1982). *The Primates of Madagascar*. Columbia University Press, New York.

Tattersall, I. (2004). Comments and conclusions. The historical biogeography of the strepsirhini: understanding the colonization of Madagascar. 2004 IPS Symposium. *Folia Primatol.* 75(suppl. 1):120.

Terranova, C. J., and Coffman, B. S. (1997). Body weights of wild and captive lemurs. *Zoo Biol.* 16:17–30.

Thalmann, U. (2001). Food resource characteristics in two nocturnal lemurs with different social behaviour: *Avahi occidentalis* and *Lepilemur edwardsi. Int. J. Primatol.* 22:287–324.

Thalmann, U., and Geissmann, T. (2000). Distributions and geographic variation in the western woolly lemur (*Avahi occidentalis*) with description of a new species (*A. unicolor*). *Int. J. Primatol.* 21:915–941.

Van Schaik, C. P., and Kappeler, P. M. (1996). The social systems of gregarious lemurs: lack of convergence with anthropoids due to evolutionary disequilibrium? *Ethology* 102:915–941.

Vargas, A., Jimenez I., Palomares, F. and Palacios, M. J. (2002). Distribution, status, and conservation needs of the golden-crowned sifaka (*Propithecus tattersalli*). *Biol. Conserv.* 108:325–334.

Vasey, N. (2000). Niche separation in *Varecia variegata rubra* and *Eulemur fulvus albifrons*: I. Interspecific patterns. *Am. J. Phys. Anthropol.* 112:411–431.

Vasey, N. (2002). Niche separation in *Varecia variegata rubra* and *Eulemur fulvus albifrons*: II. Intraspecific patterns. *Am. J. Phys. Anthropol.* 118:169–183.

Walker A. (1967). Patterns of extinction among the subfossil Madagascan lemuroids. In: Martin, P. S., and Wright, H. E., Jr. (eds.), *Pleistocene Extinctions: The Search for a Cause.* Yale University Press, New Haven, CT. pp. 425–432.

Warren, R. (1994). Lazy leapers: a study of the locomotor ecology of two species of saltatory nocturnal lemur in sympatry at Ampijoroa, Madagascar [PhD thesis]. University of Liverpool, Liverpool.

Weidt, A., Hagenah, N., Randrianambinina, B., Radespiel, U., and Zimmermann, E. (2004). Social organization of the golden brown mouse lemur (*Microcebus ravelobensis*). *Am. J. Phys. Anthropol.* 123:40–51.

White, F. J. (1989). Diet, ranging behavior and social organization of the black and white ruffed lemur, *Varecia variegata variegata*, in southeastern Madagascar [Abstract]. *Am. J. Phys. Anthropol.* 78:323.

Wilson, J. M., Stewart, P. D., Ramangason, G. S., Denning, A. M., and Hutchings, M. S. (1989). Ecology and conservation of the crowned lemur, *Lemur coronatus*, at Ankarana, N. Madagascar. *Folia Primatol.* 53:1–26.

Winn, R. M. (1989). The aye-ayes, *Daubentonia madagascariensis*, at the Paris Zoological Gardens: maintenance and preliminary behavioural observations. *Folia Primatol.* 52:109–123.

Wright, P. C. (1993). The evolution of female dominance and biparental care among non-human primates. In: Miller, B. (ed.), *Sex and Gender Hierarchies.* Cambridge University Press, Cambridge. pp. 127–147.

Wright, P. C. (1995). Demography and life history of free-ranging *Propithecus diadema edwardsi* in Ranomafana National Park, Madagascar. *Int. J. Primatol.* 16:835–854.

Wright, P. C. (1999). Lemur traits and Madagascar ecology: coping with an island environment. *Ybk. Phys. Anthropol.* 42:31–72.

Wright, P. C., Daniels, P. S., Meyers, D. M., Overdorff, D. J., and Rabesoa, J. (1987). A census and study of *Hapalemur* and *Propithecus* in southeastern Madagascar. *Primate Conserv.* 8:84–88.

Wright, P. C., Heckscher, S. K., and Dunham, A. E. (1997). Predation on Milne-Edwards' sifaka (*Propithecus diadema edwardsi*) by the fossa (*Cryptoprocta ferox*) in the rain forest of southeastern Madagascar. *Folia primatol.* 68:34–43.

Wright, P. C., and Martin, L. B. (1995). Predation, pollination and torpor in two nocturnal prosimians: *Cheirogaleus major* and *Microcebus rufus* in the rain forest of Madagascar. In: Alterman, L., Doyle, G. A., and Izard, M. K. (eds.), *Creatures of the Dark: The Nocturnal Prosimians.* Plenum Press, New York. pp. 45–60.

Wroegemann, D., and Zimmermann, E. (2001). Aspects of reproduction in the eastern rufous mouse lemur (*Microcebus rufus*) and their implications for captive management. *Zoo Biol.* 20:157–167.

Yoder, A. D., Burns, M. M., Zehr, S., Delefosse, T. Veron, G., Goodman, S. M., and Flynn, J. J. (2003). Single origin of Malagasy carnivora from an African ancestor. *Nature* 421:734–737.

Yoder, A. D., Rasoloarison, R. M., Goodman, S. M., Irwin, J. A., Atsalis, S., Ravosa, M. J., and Ganzhorn, J. U. (2000). Remarkable species diversity in Malagasy mouse lemurs (Primates, *Microcebus*). *Proc. Natl. Acad. Sci. USA* 97:11325–11330.

Young, A. L., Richard, A. F., and Aiello, L. C. (1990). Female dominance and maternal investment in strepsirhine primates. *Am. Nat.* 135:473–488.

Zimmermann, E. (1998). Nachtgeister im Tropenwald: die nachtaktiven Lemuren Madagaskars. *Biol. Zeit.* 28:294–303.

Zimmerman, E., Cepok, S., Rakotoarison, N., Zietemann, V., and Radespiel, U. (1998). Sympatric mouse lemurs in north-west Madagascar: a new rufous mouse lemur species. *Folia Primatol.* 69:106–114.

5

Tarsiiformes

Sharon Gursky

TAXONOMY

Tarsiers first became known to Western scientists through a description given by the missionary J. G. Camel, based on his observations of an individual from the island of Luzon in the Philippines (Hill 1955). It is important to note that the Philippine tarsiers are restricted to the southern Philippine islands and, consequently, this individual tarsier could not have been from Luzon. In his description, Camel described the tarsier as a small monkey with big, round eyes that rarely closed; skinny, hairless ears; and a tail and hindfeet of similar length to the rest of the body. Petiver published Camel's description in 1705 and named the animal *Cercopithecus luzonis minimus* due to its primate-like features. Based on Petiver's account of Camel's discovery, Linnaeus (1758) coined the name *Saimiri syrichta*. In 1765, Buffon also described a juvenile tarsier from an unknown locale that he gave the name *Didelphis macrotarsus: Didelphis* because he believed that tarsiers were related to opossums (or perhaps jerboas, small rodents from the Dipodidae family) and *macrotarsus* because of the extremely long tarsal bones exhibited by this animal. Interestingly, in a later edition of his *Systema naturae*, Linnaeus recognized Buffon's tarsier as *D. macrotarsus*, also regarding it as an opossum. He did not apparently associate it with his own *S. syrichta*, based on Camel's animal. Erxleben (1777) was the first scientist to link tarsiers with prosimian primates and gave them the name *Lemur tarsier*. Several years later, Storr (1780) recognized the numerous differences between lemurs and tarsiers and argued for their generic separation. He coined the genus name *Tarsius* for these unusual primates that possessed exceptionally long tarsal bones.

Since 1780, when tarsiers were finally recognized as primates (Storr 1780), they have been the focus of numerous controversies. Most notably, there has been considerable debate over the classification of the tarsiers within the order Primates (Pocock 1918, Wolin and Massoupust 1970, Cave 1973, Luckett 1976, Cartmill and Kay 1978, Rosenberger and Szalay 1980, Aiello 1986). There are two major schools of thought. The first is based on the work of Pocock (1918), who believed that tarsiers are more similar to the simian primates than to the lemurs and lorises. To express this relationship, Pocock established two infraorders within the

order Primates: the Haplorhini, which included the suborders Tarsoidea, Ceboidea, Cercopithecoidea, and Hominoidea, and the Strepsirhini, which included the lemurs and lorises. Pocock believed tarsiers were more similar to the anthropoid primates because of numerous shared derived characteristics. Among the characteristics shared by tarsiers and anthropoid primates are hemochorial placentation, dry rhinarium, absence of a tapetum lucidum, presence of a fovea centralis, lack of a toothcomb, reduced olfactory bulbs, a well-developed promontory artery, and lack of an attached upper lip (Pocock 1918, Wolin and Massoupust 1970, Cave 1973, Luckett 1976, Cartmill and Kay 1978, Rosenberger and Szalay 1980, Aiello 1986, MacPhee and Cartmill 1986, Fleagle 1999). The second school of classification is based on the idea that lemurs, lorises, and tarsiers are grouped together in the suborder Prosimii, with the simian primates all grouped together in the Anthropoidea (Simpson 1945, Cartmill and Kay 1978, Rosenberger and Szalay 1980, Fleagle 1999). This grouping pattern is based on numerous ancestral characteristics, such as the tarsier's nocturnal habits, its small body size, and its parenting strategy (infant parking and oral transport). This scheme represents a gradistic classification compared to the cladistic classification proposed by Pocock (1918). Anatomical characteristics that are shared by tarsiers and other prosimian primates include an unfused mandibular symphysis, a grooming claw, multiple pairs of mammae, and a bicornuate uterus (Pocock 1918, Wolin and Massoupust 1970, Cave 1973, Luckett 1976, Cartmill and Kay 1978, Rosenberger and Szalay 1980, Aiello 1986, MacPhee and Cartmill 1986, Fleagle 1999).

Molecular data have not been useful in elucidating the phylogenetic position of tarsiers. Specifically, a study of the apolipoprotein b gene (*APOB*) showed strong support for an association of tarsiers with anthropoids, thus supporting the haplorhine–strepsirhine dichotomy (Amrine-Madsen et al. 2003). On the other hand, the substitution rate of the cytochrome *b* gene has increased approximately twofold along lineages leading to simian primates in comparison to tarsiers, other prosimians, and nonprimate mammalian species, thus supporting the prosimian–anthropoid dichotomy (Andrews et al. 1998). Resulting from the large amounts of conflicting data and interpretations, the question of tarsier phylogeny is tremendously complicated and unlikely to be resolved anytime soon.

There has also been substantial disagreement concerning the number of tarsier species (Hill 1955; Niemitz 1984; Musser and Dagosto 1987; Niemitz et al. 1991; Nietsch and Kopp 1998; Nietsch 1999; Groves 1998, 2001a,b; Shekelle 2003). At present, five species are formally recognized: *Tarsius syrichta* (Linnaeus 1758), the Philippine tarsier; *T. bancanus* (Horsfield 1821), the Bornean tarsier; *T. pumilus* (Miller and Hollister 1921, Musser and Dagosto 1987), the pygmy tarsier from central Sulawesi; *T. spectrum* (Pallas 1778), the spectral tarsier from Sulawesi; and *T. dianae* (Niemitz et al. 1991), another central Sulawesi lowland tarsier species.

According to Hill (1955), Cabrera suggested that Buffon's tarsier was actually an eastern tarsier (*T. spectrum*) and not a Philippine tarsier (*T. syrichta*). Despite the loss of this specimen but based on Buffon's written description, Groves (2001b) and Shekelle (2003) have reaffirmed this opinion. The nomenclatural implication of this error is that the name *L. tarsier* (Erxleben 1777) has seniority over *L. spectrum* (Pallas 1778). Thus, *T. spectrum* would be *T. tarsier*. I continue to use the name *T. spectrum* but recognize that the correct name might be *T. tarsier*.

All tarsier species differ morphologically from one another in a variety of ways. They differ in their body weight (Table 5.1), absolute orbit size, absolute tooth size, proportion of tail that is covered by hair, and limb proportions. In terms of absolute orbit size and absolute tooth size, *T. bancanus* possesses the largest orbits and largest teeth of any of the species, followed by *T. syrichta*, *T. spectrum*, and then *T. pumilus* (Musser and Dagosto 1987).

There is also a cline in the proportion of the tail that is covered by hair for each species. *T. spectrum* and the other Sulawesian tarsier species possess the most tail hair, followed by *T. bancanus* and *T. syrichta*, whose tail is often described as "naked" due to the sparse, fine hairs that cover it. Niemitz (1984) also recorded numerous differences in limb proportions between the three major tarsier species (*T. spectrum*, *T. bancanus*, and *T. syrichta*). *T. bancanus* has the longest hindlimbs and hands, followed by *T. syrichta*, whose hindlimbs and hands are intermediate in length, and then *T. spectrum*, whose hindlimbs and hands are shortest in length. The recent work by Dagosto and colleagues (2001) has also demonstrated a cline in the locomotor patterns of tarsiers. They found that *T. bancanus* is the most specialized, using vertical clinging postures the most frequently, followed by *T. syrichta*, and then the Sulawesian tarsiers *T. spectrum* and then *T. dianae*.

According to Nietsch and Kopp (1998), two of the Sulawesian tarsiers (*T. spectrum* and *T. dianae*) can be distinguished from one another based on their vocalizations. Using playbacks of the vocal duets sung each morning, Nietsch and Kopp found that each species responded only to the vocalizations of their own species. Neither species responded to the vocalizations of the other Sulawesian tarsier species, while avidly responding to the vocalizations of their own species. Nietsch and Kopp also found that the tarsier population on Togian Island, adjacent to Sulawesi, did not respond to the playback vocalizations of *T. spectrum* or *T. dianae* and proposed that the Togian tarsier population may also represent a distinct species, *T. togianensis*. Specifically, Nietsch (1999) found that the degree of separation of the acoustic characteristics of the Togian tarsier

Table 5.1   Body Weights of Four Different Tarsier Species Separated by Sex

SPECIES	BODY WEIGHT (G)	SEX	N	REFERENCES
Tarsius bancanus	128	M	—	Niemitz 1984
	117	F	—	Niemitz 1984
	128	M	5	Crompton and Andau 1987
	118	F	2	Crompton and Andau 1987
T. dianae	110	F	1	Niemitz et al. 1991
	105	F	1	Muskita cited in Sussman 1999
	104	M	1	Muskita cited in Sussman 1999
	100	F	4	Shekelle 2003
	119	M	5	Shekelle 2003
	130	M	1	Gursky 1997
	110	F	1	Gursky 1997
T. spectrum	113	F	1	Niemitz et al. 1991
	110	M	1	Niemitz et al. 1991
	104	F	21	Shekelle 2003
	115	M	11	Shekelle 2003
	107	F	8	Gursky 1999
	126	M	5	Gursky 1999
T. syrichta	136	M	4	Neri Arboleda 2001
	120	F	6	Neri Arboleda 2001
	141	M	4	Dagosto et al. 2001
T. pumilus	58		1	Maryanto and Yani in press

population from the *T. spectrum* population was substantial enough to warrant their designation as a distinct species.

Two additional Sulawesian tarsier species, *T. pelengensis* and *T. sangirensis*, have also been tentatively proposed and are awaiting additional studies utilizing larger sample sizes (Shekelle et al. 1997; Groves 1998, 2001a; Shekelle 2003). Some scientists naturally question whether there really are so many distinct tarsier species in Sulawesi or whether they are in fact subspecies. Based on preliminary genetic data (Shekelle 2003) as well as my own preliminary observations of behavior, vocalizations, and morphology, I believe that these populations, as well as several others in Sulawesi, are distinct and represent new species. It is clear that the continued use of vocalizations is integral for the proper identification of species boundaries when dealing with cryptic primates such as the tarsier (Bearder 1999).

Another taxonomic argument relating to tarsiers concerns whether the Sulawesian tarsiers are distinct enough to warrant their own genus name. Groves (1998) argues in his analyses of the three historically recognized species (*T. bancanus*, *T. syrichta*, and *T. spectrum*) that *T. spectrum* is distinct enough from *T. bancanus* and *T. syrichta* to warrant its own genus name, *Rabienus*. At present, however, even Groves (2001b) continues to use the genus name *Tarsius* for the Sulawesian tarsiers (*T. spectrum*, *T. dianae*, and *T. pumilus*).

Interestingly, although there are only minor differences in morphology, there are substantial chromosomal differences between some tarsier species. *T. bancanus* and *T. syrichta* are each reported to have a karotype of 80 chromosomes (Klinger 1963, Poorman et al. 1985, Dutrillaux and Rumpler 1988), whereas *T. dianae* is reported to have only 46 chromosomes (Niemitz et al. 1991). The karotypes for both *T. spectrum* and *T. pumilus* are still unknown. Dutrillaux and Rumpler (1988) report that the chromosomes of *T. syrichta* and *T. bancanus* are distinct from over 100 other primate species and other mammals. It has been suggested therefore that the tarsiers do not have a primitive karotype but have undergone a large number of chromosomal rearrangements (Dutrillaux and Rumpler 1988). A large number of fissions have cer-

tainly occurred since the diploid chromosome number of 80 is among the highest in mammals.

ECOLOGY

Despite the great interest in the phylogenetic relationships of tarsiers and their taxonomic nomenclature, only recently have primatologists begun to study their behavior, ecology, and conservation status. In part, this is a reflection of the difficulty of observing a small, nocturnal primate. In addition, tarsiers do not possess a *tapetum lucidum* (a reflective layer behind the eye), which helps observers locate most other nocturnal mammals in the dark. Over the last decade, several scientists have come to recognize that the tarsier's behavioral repertoire is relatively unusual and deserves equal attention. Behavioral and ecological studies of wild, free-ranging, and semicaptive tarsiers have concentrated on four species (Table 5.2): *T. bancanus* (Fogden 1974; Niemitz 1972, 1974, 1979, 1984; Crompton and Andau 1986, 1987), *T. spectrum* (MacKinnon and MacKinnon 1980; Gursky 1995, 1997, 1998a,b, 1999, 2000, 2002a,b, 2003), *T. syrichta* (Dagosto and Gebo 1998, Dagosto et al. 2001, Neri-Arboleda 2001), and *T. dianae* (Niemitz et al. 1991, Tremble et al. 1993, Gursky 1998b). These studies represent the primary behavioral and ecological literature for the remainder of this chapter, which will compare the differences between the member species of the genus *Tarsius* with respect to their behavior. No field study has been conducted on *T. pumilus*.

ACTIVITY PATTERNS

Tarsiers are nocturnal and begin their activity around sunset. Although they are active throughout the night, they are known to have peaks of activity in the early evening and again just before sunrise. Niemitz (1984) found that there was a minor peak in activity early and a major peak late in the evening in two *T. bancanus* individuals. Crompton and Andau (1987) observed more travel in the early evening in

Table 5.2 Major Studies of Wild Tarsier Species and the Locations Where They Were Conducted

SPECIES	STUDY LOCATION	YEAR	REFERENCES
Tarsius bancanus	Semongok, Sarawak	1970–1985	Niemitz 1972, 1974, 1979, 1984
T. bancanus	Semongok, Sarawak	1970	Fogden 1974
T. bancanus	Sepilok, Sabah	1985	Crompton and Andau 1987
T. spectrum	Tangkoko, Sulawesi	1978–1979	MacKinnon and MacKinnon 1980
T. spectrum	Tangkoko, Sulawesi	1994–2003	Gursky 1995, 1997, 1998a,b, 1999, 2000, 2002a,b, 2003
T. dianae	Kamarora, Sulawesi	1992	Tremble et al. 1993
T. dianae	Kamarora and Posangke, Sulawesi	1997	Gursky 1997, 1998b
T. dianae	Kamarora, Sulawesi	1990	Niemitz et al. 1991
T. syrichta	Mt. Pangasugan, Leyte	1995–2003	Dagosto and Gebo 1998, Dagosto et al. 2001
T. syrichta	Corella, Bohol	2000	Neri-Arboleda 2001

the same species. In *T. spectrum*, the most energetic periods occurred during the first half-hour of activity and then again during the last half-hour of activity (MacKinnon and MacKinnon 1980, Gursky 1997). In *T. syrichta*, there is a big peak in activity early in the evening and then three minor peaks throughout the night (Neri-Arboleda 2001).

In *T. spectrum*, males spent approximately 54% of their night foraging, 30% traveling, 11% resting, and 5% socializing; females spent approximately 53% of their night foraging, 26% traveling, 15% resting, and 6% socializing (Gursky 1997). *T. dianae* spent 48% of the night foraging, 27% traveling, 21% resting, and 7% socializing (Tremble et al. 1993). *T. bancanus* spent approximately 62% of the night foraging, 27% traveling, and 11% resting (Crompton and Andau 1986). The similarity in time spent foraging and traveling is quite remarkable given that all these species live in very different habitats. The only observable difference in activity patterns across tarsier species is that *T. dianae* spent nearly twice as much time resting in comparison with the other species. No quantifiable data have been obtained for *T. syrichta* or *T. pumilus* on nightly activity patterns.

DIET

Tarsiers are also unusual among nocturnal prosimians in that they consume exclusively animal food (arthropods and vertebrates). At present, there are no observations in the literature of any tarsier species consuming fruit, leaves, or gum (MacKinnon and MacKinnon 1980, Niemitz 1984, Crompton and Andau 1987, Tremble et al. 1993, Gursky, 2000). Niemitz (1984) has occasionally observed *T. bancanus* bite a leaf but never chewing or digesting them. The same pattern has also been observed in *T. spectrum* (S. Gursky personal observation). Niemitz's observations (Niemitz 1979, 1984) on semiwild *T. bancanus* indicated that their diet includes beetles (35%), ants (21%), locusts (16%), cicadas (10%), cockroaches (8%), and other miscellaneous vertebrates (11%: birds, smaller fruit bats, spiders, and poisonous snakes).

My own research on *T. spectrum* has shown that this species also prefers moths (Lepidoptera) as well as crickets and cockroaches (Orthoptera). These two arthropod orders comprised more than 50% of the study population's diet. I noted though that during the dry season, when these types of insect were less abundant, *T. spectrum* substantially increased their consumption of beetles (Coleoptera) and ants (Hymenoptera) (Gursky 2000). The mean size of insects captured by *T. spectrum* was 1.9 cm and ranged 1–6 cm. The majority of the insects consumed either were located on or underneath a leaf (43%) or were flying (35%). A smaller percentage of the insects were located on branches (13%) or on the ground (9%).

Tremble et al.'s (1993) observation on the dietary items of wild *T. dianae* included moths, crickets, and a lizard. At present, nothing is known about the diet of wild *T. syrichta* or *T. pumilus*.

SLEEPING TREES

There is a large amount of variation among tarsier species with regard to sleeping site preferences. This has important implications for determination of social groups and mating systems since tarsier groups are often identified by which animals sleep together. The majority of sleeping trees utilized by *T. spectrum* are *Ficus spp.*, with *F. caulocarpa* being used much more frequently than all other species (Gursky 1997, MacKinnon and MacKinnon 1980). *T. spectrum* primarily utilized one major sleeping site, although most groups had one alternate sleeping site as well. The mean circumference of these strangler fig sleeping trees was 287 cm, ranging 30–700 cm. The mean height of the sleeping trees was 20.17 m, ranging 6–38 m. MacKinnon and MacKinnon (1980) also observed several *T. spectrum* groups utilizing vine tangles and grass platforms when groups were found in secondary forest and grassland areas.

In comparison, data from Crompton and Andau (1986) on *T. bancanus* indicate that the average height of sleeping trees was 4.0–5.5 m but occasionally as low as 2.5 m. Vine tangles and naturally formed platforms of creepers were the most common sleeping sites. Thus, the sleeping sites utilized by *T. bancanus* were much lower in the forest canopy and more open compared to those used by *T. spectrum*.

Niemitz et al. (1991), in their description of *T. dianae*, state that this species does not return to the same nest each night. This description agrees with MacKinnon and MacKinnon's (1980) observation that tarsiers in central Sulawesi do not use the same sleeping site each night. Tremble et al. (1993), however, found that *T. dianae* did return to the same nest/tree each night. At two sites (Kamarora and Posangke), Gursky (1997, 1998b) found that *T. dianae* also returned to the same nest each morning. Nest sites for *T. dianae* include vine tangles, tree cavities, and fallen logs (Tremble et al. 1993). The type of sleeping site used might account for this variation in whether or not groups return to their sleeping site.

T. bancanus sleeps singly unless it is a mother and her infant (Niemitz 1984, Crompton and Andau 1987). Niemitz found that this population of *T. bancanus* sleeps clinging to vertical branches at 2 m or lower. In another population of *T. bancanus*, Crompton and Andau (1987) found that individuals sleep in vine tangles 4–5.5 m above the ground and on 50–90 degree supports. They also noted that the sleeping sites tend to cluster at the edge of a tarsier's home range, in areas of overlap with neighbors of the opposite sex. This contrasts with *T. spectrum*, whose sleeping sites tend to be located near the center of the group's home range.

Dagosto and Gebo (1998) report that *T. syrichta* in Leyte always sleeps singly, never in groups, except for mother–infant pairs; and this observation has been confirmed by Neri-Arboleda (2001) with another population. *T. syrichta* in Leyte utilized three or four different sleeping sites. They were primarily *Arctocarpus*, *Pterocarpus*, and *Ficus*, low to the ground and surrounded by very dense vegetation. *T. syrichta* on the island of Bohol generally utilized one

Table 5.3   Reported Home Range Sizes, Nightly Path Length (NPL), Mean Group Size, and Dispersal Distance in Wild Tarsier Species

SPECIES	HOME RANGE (HA)	NPL (M)	MEAN SOCIAL GROUP SIZE	DISPERSAL DISTANCE (M)
Tarsius bancanus	Males and females = 2.5–3.0 Males and females = 1.0–2.0 Males and females = 4.5–11.25	Males = 2,082 Females = 1,448	1	N/A
T. spectrum	Males and females = 1.6–4.1	Males = 791 Females = 448	3	Males = 641 Females = 255
T. dianae	Males and females = 0.5–0.8	N/A	2	N/A
T. syrichta	Males and females = 0.6–2.0 Males = 6.86 Females = 2.76 Subadult = 13.4	Males = 1,636 Females = 1,118	1	N/A

sleeping site but had several alternate sleeping sites. Males utilized seven or eight sleeping sites, while females tended to limit their sleeping sites to three or four different locations over the entire duration of the study. Five of the 14 tree species used as sleeping sites were *Ficus spp.* The average tree height of sleeping sites was 7.53 m. The sleeping sites of the Bohol *T. syrichta* were highly variable. They ranged from rock crevices near the ground (3%) to vine tangles (10%), dense thickets of *Pandanus* palms (15%), tree trunks (45%), and fork branches of trees (27%).

HOME RANGE

The early work of Fogden (1974) on *T. bancanus* suggested that the Bornean tarsier's home range is 2.5–3.0 ha (Table 5.3). This study was conducted over a 2-year period on 26 individuals. Niemitz (1979, 1984) conducted his own observations on *T. bancanus* under semiwild conditions (within a 90 m² cage enclosure). He utilized urine and epigastric (abdominal) gland marking sites to determine the home range of the animal. His results indicated that the species' home range size varied from 1 to 2 ha and was remarkably stable. Niemitz (1984) found that the ranges were larger for males than females and that they overlapped extensively.

Crompton and Andau (1987) conducted the first study of tarsiers involving radiotelemetry in Sabah, Borneo. They collected 120 hr of observation on four individuals. Males had a home range of 8.75–11.25 ha and females, 4.5–9.5 ha. Individuals traveled a mean distance of 1,800 m per night; males moved significantly farther (2,082 m) than females (1,448 m). Crompton and Andau (1987) report that males utilized 50%–75% of their total home range and females utilized 66%–100% of their range per night.

In a long-term radiotelemetry study of *T. spectrum*, Gursky (1998a) found that the home range size varied 1.6–4.1 ha, with males having larger ranges (mean 3.07 ha) than females (mean 2.32 ha). This is notably larger than the 1 ha range size previously estimated by MacKinnon and MacKinnon (1980) based on occasional sightings. The mean

nightly path length (NPL) for males was 791 m (*n* = 5), while that for females was 448 m (*n* = 8).

Dagosto and Gebo (1998) radiocollared four male *T. syrichta* on the island of Leyte during a 6-week study. Male home range size was 0.6–2.0 ha. An 8-month study of *T. syrichta* was conducted by Neri-Arboleda (2001), who found that the home range was 6.86 ha for males (*n* = 4) and 2.76 ha for females (*n* = 6). Interestingly, Neri-Arboleda (2001) also had one home range of a subadult that averaged 13.4 ha, slightly larger than Crompton and Andau's (1987) 11.25 ha for male *T. bancanus*. Neri-Arboleda (2001) noted that the incremental plot of home range never reached an asymptote (flattening out), suggesting that the home range would be even larger with additional data collection.

The mean NPL for *T. syrichta* females was 1,118 m, ranging 820–1,304 m (Neri-Arboleda 2001). The mean NPL for males was 1,636 m. While these figures are comparable with those of *T. bancanus,* they are substantially greater than those observed for *T. spectrum* (Gursky 1998) or the Leyte population of *T. syrichta*. Dagosto and Gebo's (1998) preliminary study of tarsiers in Leyte suggests that the mean NPL was 301 m per night for two males.

It is apparent that tarsiers, in general, have relatively small home ranges. However, not all home range sizes are readily comparable because they were collected using different techniques (trap–retrap vs. scent-marked trees vs. radiotracking) and on different sample sizes in studies of varying duration.

REPRODUCTIVE PARAMETERS

The gestation period for captive *T. bancanus* is 178 days (Izard et al. 1985), which is comparable to the mean gestation period of 191 days estimated (Gursky 1997) for wild *T. spectrum* (Table 5.4). There are no published estimates of gestation length in either wild or captive *T. syrichta*, *T. dianae*, or *T. pumilus*.

Litter size for tarsiers has been reported to be strictly limited to one offspring; that is, there have been no reports of any intraspecific variation in litter size (Catchpole and

Table 5.4   Life History Variables for Captive and Wild Tarsier Species

SPECIES	MEAN BIRTH WEIGHT (G)	LITTER SIZE	GESTATION LENGTH (DAYS)	METHOD OF INFANT TRANSPORT	BIRTH SEASONALITY	LACTATION LENGTH (DAYS)	INTERBIRTH INTERVAL (MONTHS)
Captive							
Tarsius bancanus	25	1	178	Fur/oral	No	49–82	–
T. spectrum	–	–	–	–	–	–	–
T. dianae	–	–	–	–	–	–	–
T. syrichta	26	1	180	Oral	No	60	–
Wild							
T. bancanus	–	1	–	–	Yes	–	–
T. spectrum	24	1	191	Oral	Yes	78	13
T. dianae	–	1	–	Oral	–	–	–
T. syrichta	37	1	–	–	Yes	78	–

Fulton 1943, Ulmer 1963, Niemitz 1984, Crompton and Andau 1987, Haring and Wright 1989, Roberts and Kohn 1993, Roberts 1994, Gursky 1997). The lack of variation in litter size in *Tarsius spp.* is surprising for two reasons. First, it is surprising because tarsiers are morphologically capable of producing twins; that is, they possess a bicornuate uterus, enabling two offspring to be produced (Luckett 1976, Fleagle 1999). Second, all tarsiers have two or three pairs of mammary glands (Schultz 1948, Niemitz 1984, S. Gursky, personal observation) such that they could nurse more than one infant without any difficulty. There is a well-known relationship between nipple number and litter size, where litter size is usually one-half of nipple number (Gilbert 1986, Lagreca et al. 1992, Sherman et al. 1999). Yet, tarsiers have not been observed producing litters of two or three offspring despite being morphologically capable. Instead, they exhibit a life history strategy that minimizes the number of offspring produced each year, while at the same time producing a single large infant (25% of adult weight at birth) over an exceptionally long gestation period, with rapid postnatal growth (Wright 1990). The presence of the bicornuate uterus and multiple mammary glands represents evidence that historically tarsiers produced multiple offspring and not a single large offspring as they do today.

BIRTH SEASONALITY

Although Neri-Arboleda (2001) did not have data for an entire year on *T. syrichta* (March–October), the data she does have are certainly suggestive of birth seasonality in this species. Of the 10 births that she observed over the 8-month study, all were between April and July: one in April, two in May, four in June, and three in July. Captive *T. syrichta* has not been reported to be a seasonal breeder (Catchpole and Fulton 1943, Haring and Wright 1989). Fogden (1974) reported that wild *T. bancanus* has a sharply defined breeding season, with mating in October through December and births in January through March. However, it was observed to be a nonseasonal breeder in semiwild conditions (Niemitz 1984) and in captivity (Wright et al. 1988, Roberts and Kohn

1993). Wild *T. spectrum* also has been reported to be a seasonal breeder, with two distinct breeding seasons within a year (MacKinnon and MacKinnon 1980, Gursky 1997): April–May and November–December.

Lactation length for wild tarsiers is known only for *T. spectrum* and *T. syrichta*. Gursky (1997) and Neri-Arboleda (2001) report a lactation length of approximately 78 days. For *T. syrichta*, weaning was determined when infants no longer shared a sleeping site with their mother. In captive tarsiers, lactation length is reportedly much more variable, 49–120 days (Niemitz 1984, Haring and Wright 1989, Roberts and Kohn 1993, Roberts 1994).

PARENTAL CARE

One of the most unusual reproductive behaviors of the tarsier concerns its parenting strategy. Tarsier mothers do not continually transport their infants on their bodies (ventrally or dorsally). Historically, tarsiers were thought to be rather nonchalant parents, parking their infants through the entire night and coming back to get them at dawn (Charles-Dominique 1977, MacKinnon and MacKinnon 1980, Niemitz 1984). However, a study of *T. spectrum* infants has demonstrated that *T. spectrum* are not negligent parents but are optimizing their behavior given numerous constraints. Specifically, Gursky (1997) found that although infants are parked throughout the territory, the mean length of each parking bout was a mere 27 min. Also, in a given night, the mean number of locations that the infant was parked was 11 per night and that the average distance between a mother and her parked infant was 4 m (standard deviation 3.9). In fact, a *T. spectrum* mother more often than not foraged in the same tree where her infant was parked. These results together indicate that *T. spectrum* are not inattentive parents. The mother caches the infant in one tree where she forages and then carries the infant (in her mouth) into another tree, parks it, and then begins foraging again within the tree where she parked her infant. Thus, "cache and carrying" better describes the infant caretaking system of *T. spectrum* than does "infant parking."

SOCIAL ORGANIZATION

In *T. spectrum*, group size is highly variable, ranging from two individuals to as many as eight individuals per sleeping site. The mean observed group size is three, but composition varies from two adult individuals of the opposite sex to one adult male and two adult females and their numerous offspring. Approximately 15% of groups contain two adult females. Groups with two adult females and two young infants suggest that there may be several breeding females in a group. Nietsch and Niemitz's (1992) preliminary observations produced similar results but with one group containing three adult females. Whereas MacKinnon and MacKinnon (1980) observed *T. spectrum* groups with multiple adult males, Gursky (1997) reports only groups with one adult male, although some groups also contained a subadult male. The variation in group size and composition observed in *T. spectrum* is inconsistent with the traditional interpretation of a strictly monogamous social structure for this species (MacKinnon and MacKinnon 1980). Instead, "facultative polygyny" more aptly describes the social structure of this nocturnal primate. In a study of *T. dianae* in central Sulawesi, Gursky (1998b) censused nine groups containing a total of 22 individuals. The mean group size was two. Six groups contained two individuals, two groups contained three individuals, and one group contained four individuals.

In contrast to *T. spectrum*, *T. bancanus* and *T. syrichta* do not live in groups but have overlapping home ranges (Fogden 1974, Niemitz 1984, Crompton and Andau 1987). Thus, they are traditionally considered to exhibit a dispersed social organization, where a male's range overlaps the ranges of several females. However, Fogden (1974) reports a male and female *T. bancanus* traveling together on eight occasions. In one situation, a male was seen within a few days with two different females. Niemitz (1984) reports that *T. bancanus* possibly live in pairs in that they live synterritorially for long periods. As the above-mentioned studies were conducted not only at different field sites but also with different groups, it is possible that this tarsier species exhibits both male–female pair bonds (monogamy) as well as a dispersed social system.

According to Neri-Arboleda (2001), the Bohol *T. syrichta* are also solitary. They foraged and slept solitarily except for females carrying their infants. She never observed an individual in close proximity (<10 m) with a conspecific during nocturnal activity. A single observation was made of a male and female having sleeping sites 5 m apart. The Leyte *T. syrichta* are also solitary (Dagosto et al. 2001). All of the tarsiers captured were alone at their sleep sites, and three of the four males were never within close proximity to another tarsier. However, Dagosto et al. (2001) did observe one male sharing a sleep site with another tarsier (sex unknown) on three out of eight nights of observation. At the sleep site, the two tarsiers were never closer than 1 m from each other.

SOCIALITY

Recent studies of *T. spectrum* have demonstrated that this species exhibits substantial amounts of gregariousness, relative to its sister species as well as to other nocturnal prosimian primates (Gursky 2002a). For example, in *T. spectrum* groups, approximately 28% of nightly forays outside the sleeping tree were spent at distances from one another of ≤10 m. Almost 40% of the night was spent at <20 m from one another, including 11% of their time spent in actual physical contact. These observations of encounters with other group members are not the result of chance given the home range size, NPL, and travel speed of this species (Gursky 2002b). Gursky compared the observed encounter frequency with that expected based on Waser's random gas model (Waser 1976, Jolly et al. 1993, Holenweg et al. 1996, Gursky 2002b). Waser's (1976) random gas model is based on the idea that if group members' movements are random and independent, then their territory can be considered a two-dimensional gas of tarsier individuals. Using this model, Gursky (2002b) found that *T. spectrum* individuals encountered one another more frequently than expected by chance alone for distances of <50 m. This is especially true at closer distances. *T. spectrum* also spent more time foraging when close to another group member (≤10 m) compared to when individuals were 50 m or 100 m apart. However, individuals foraging close to another adult group member had lower insect capture rates compared to more isolated foraging individuals. The decrease in foraging efficiency experienced by *T. spectrum* when foraging close to other group members may explain the corresponding increase in time allocated to foraging.

This result naturally raises the following question: If *T. spectrum* individuals experience substantial intragroup competition over food resources when foraging in close proximity to other group members, then why do they continue to forage in gregarious situations? Group living is predicted to occur only when the benefits outweigh the costs incurred from group living (Krebs and Davies 1984, Dunbar 1988, Kappeler and Ganzhorn 1993). If living in a group is costly to *T. spectrum*, then why do they not actively avoid one another while traveling throughout their territory? One possibility is that *T. spectrum* forages in proximity to other adult group members only when the benefits of being near another group member outweigh the costs of intraspecific food competition (lower insect capture rates when foraging in proximity). A few situations under which it might benefit *T. spectrum* to be gregarious are (*1*) when predation pressure is high, (*2*) when prospects of meeting up with potentially infanticidal males are high, and (*3*) when females are sexually receptive. Currently, only data addressing the first potential situation are available.

In an experimental study of the determinants of gregariousness for *T. spectrum*, Gursky (2002a) found evidence to support the importance of predation pressure. In response to artificially high predator pressure, *T. spectrum* individuals

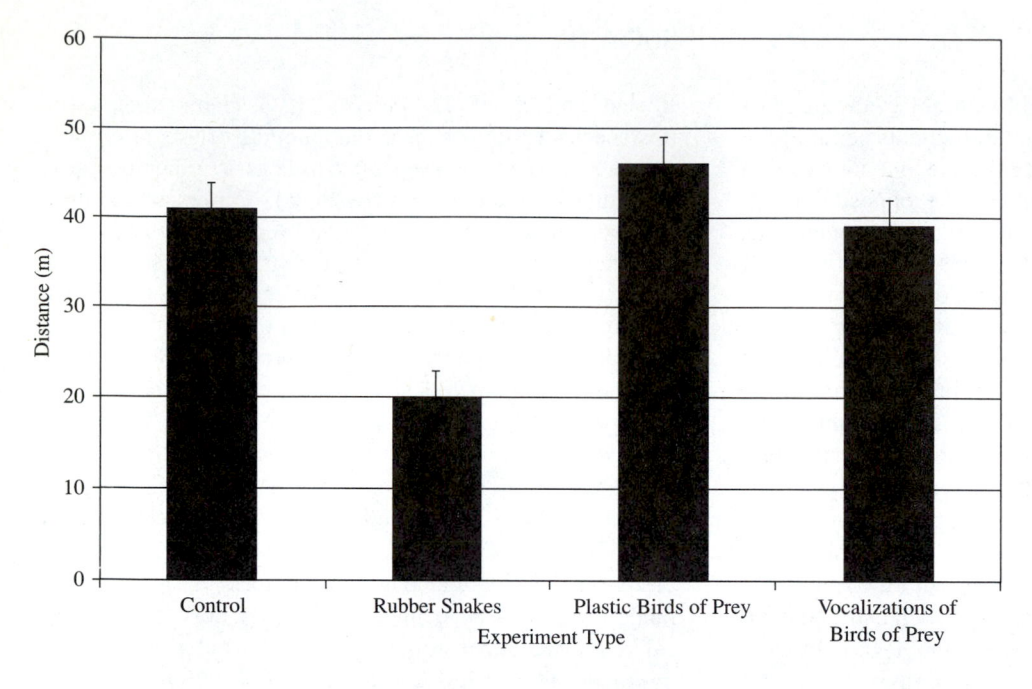

Figure 5.1 Mean distance (m) between adult group members on nights when rubber snakes, plastic birds of prey, or vocalizations of birds of prey were distributed/played throughout the focal group's territory. The mean distance between adult group members when no experiments were conducted is represented by the control.

were observed to spend more time together, often meeting up to try and scare away the predator. Adult group members were significantly closer on nights when predation pressure experiments were conducted (snake models, birds of prey models, or vocalizations of birds of prey distributed or played back) than on nights when no experiments were conducted (Fig. 5.1). In particular, on 60% of the nights that rubber snakes were placed in a group's territory, *T. spectrum* were observed mobbing the rubber snake. Mobbing involved all group members traveling to the area of the individual that initially located the snake. Group members began vocalizing loudly and then lunging and retreating at the snake. On three occasions, an adult male *T. spectrum* was observed literally biting the rubber snake (Gursky 2002a). In each instance, it was the group's adult male. In addition, the mean distance between group members was twice as far on nights when no rubber snakes were present in the territory as on experimental nights with rubber snakes (Gursky 2002a).

An interesting behavioral observation that lends support to the importance of predation pressure for *T. spectrum* sociality is this species' response to bright moonlight (Gursky 1999, 2003). *T. spectrum* were observed to increase the amount of time spent foraging and to decrease the amount of time spent resting during full moons relative to other moon phases (Fig. 5.2a,b). Similarly, they were also observed to increase the number of loud calls emitted, territorial disputes, and intragroup encounters, while they decreased the number of scent marks compared to nights with a new moon (Gursky 2003). These behavioral responses are quite surprising because nearly all other nocturnal mammals appear to avoid bright moonlight. Numerous studies of nocturnal mammals, including many nocturnal primates, have consistently shown that they restrict their foraging activity, restrict their movement, reduce their use of open space, reduce the duration of the activity period, or switch their activity to

darker periods of the night in response to bright moonlight (Morrison 1978, Wolfe and Summerlin 1989). The only consistent exceptions to this pattern have been observed in three primates, the night monkey (*Aotus*), the southern lesser galago (*Galago moholi*), and the Mysore slender loris (*Loris tardigradus lydekkerianus*) (Trent et al. 1977, Erkert and Grober 1986, Nash 1986, Bearder et al. 2002). Instead of reducing their activity during the full moon, all three species are reported to increase their activity levels during periods of moonlight compared to periods without moonlight.

The lunar phobia exhibited by most nocturnal mammals is believed to be a form of predator avoidance (Morrison 1978, Wolfe and Summerlin 1989, Daly et al. 1992, Kotler et al. 1993, Bearder et al. 2002). This implies that *T. spectrum* and a few other nocturnal prosimians increase their exposure to predators when they increase their activity during full moons. This raises two equally intriguing questions. First, are the benefits of foraging during the full moon so high that they outweigh the costs of increased predation pressure? Second, how do *T. spectrum* deal with the increasing predator pressure during full moons? Concerning the first question, a comparison of *foraging efficiency* (defined as the number of insects captured per unit time) during full moons relative to the other moon phases suggests that the benefits of foraging during the full moon are in fact tremendous. Gursky (2003) observed that *T. spectrum* captured three times as many insects during full moons than during moonless nights (new moon). These numbers do not simply reflect a bias in observation conditions because this pattern was also observed in terms of the numbers of insects captured in insect traps during the different moon phases (Gursky 1997, 2000, 2003). Thus, for *T. spectrum*, the benefits of foraging during full moons are substantial.

To deal with the potential increase in predation during a full moon (when diurnal and nocturnal predators can take

Figure 5.2 A: The mean percent time male (black bars) and female (dotted bars) spectral tarsiers allocated to foraging during various phases of the lunar cycle. B: The mean percent time male (black bars) and female (dotted bars) spectral tarsiers allocated to resting during various phases of the lunar cycle.

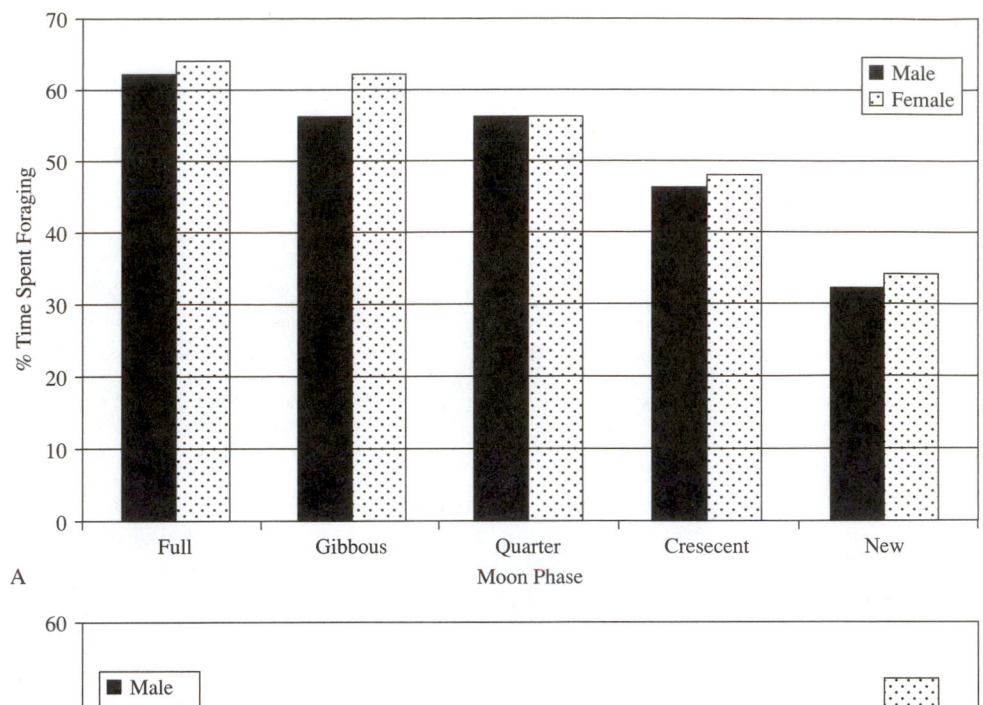

A

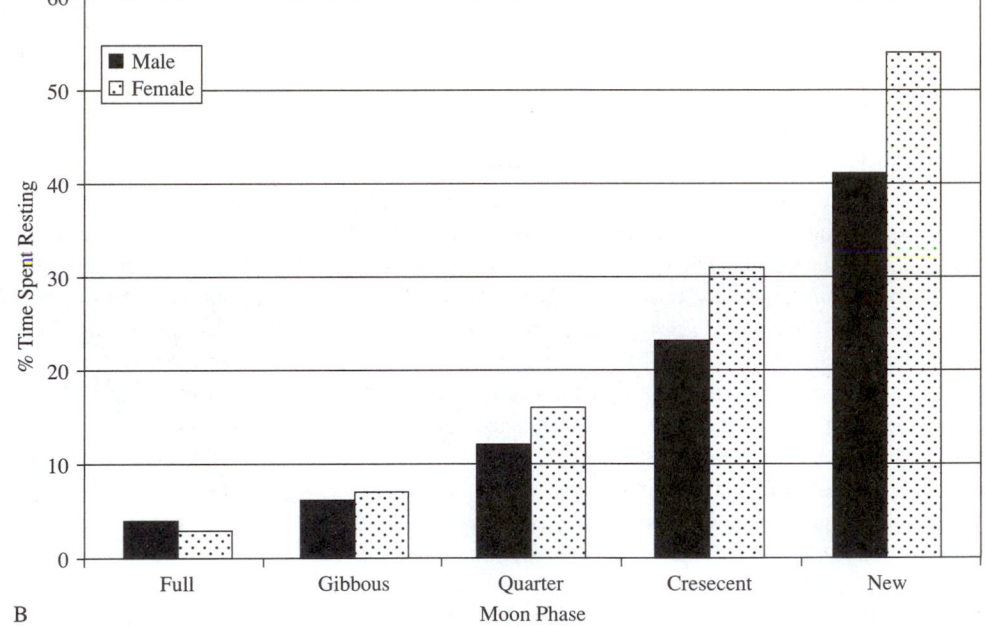

B

advantage of the additional light), *T. spectrum* modified other aspects of its behavioral repertoire. Most interestingly, they increased the frequency that group members traveled together. In particular, the frequency that any two group members were observed together increased substantially during full moons. It is believed that the more individuals there are in the group to scan for predators, the less time each individual will have to spend in vigilance (Krebs and Davies 1984, Dunbar 1988, Kappeler and Ganzhorn 1993). However, living in a group can be costly in terms of intragroup foraging competition (Janson 1988). *T. spectrum* overcomes this costly behavior by consuming insects that are abundant in moonlight.

In summary, *T. spectrum*, like three other nocturnal primates, does not exhibit lunar phobia. Instead, they increase their activity during full moons and decrease their activity during new moons. The increased activity during full moons may increase their exposure to predators. To counter this putative increase risk of predation, *T. spectrum* increase the frequency that they travel in groups. *T. spectrum* also increase consumption of insects that are abundant during full moons to counter the within-group competition for insects that may result from traveling in groups.

CONSERVATION

Gursky's study of the population density of *T. spectrum* indicates that they are relatively abundant (Table 5.5). The population density was calculated as 156 individuals within

Table 5.5   Reported Population Density and Conservation Status of Wild Tarsier Species

SPECIES	IUCN CONSERVATION STATUS[1]	REVISED CONSERVATION STATUS[2]	POPULATION DENSITY (INDIV. KM²)	REFERENCES
Tarsius bancanus	LR lc	LR cd	80, 15–20	Niemitz 1984, Crompton and Andau 1987
T. spectrum	DD	VU	30–100, 156	MacKinnon and MacKinnon 1980, Gursky 1998a
T. dianae	DD	VU	129	Tremble et al. 1993, Gursky 1998b
T. syrichta	LR cd	VU	57	Neri-Arboleda 2001
T. pumilus	DD	DD	—	

[1] World Conservation Union conservation status was taken from Appendix 1 in Cowlishaw and Dunbar (2000). LR 1c, low risk, least concern; LR cd, low risk, conservation-dependent; DD, data deficient.
[2] The revised conservation status is based on a workshop attended by M. Shekelle, A. Nietsch, and S. Gursky.

a 100 ha study area, and the number of groups estimated within the study area was 56 (Gursky 1998b). This estimate is one-half of the lowest estimate calculated by Bearder (1987) (300–1,000 individuals/km² over 100 ha), based on density estimates determined by MacKinnon and MacKinnon (1980) of 3–10 individuals/ha in north Sulawesi.

In comparison, studies of *T. bancanus* have also yielded varying estimates of population density. Crompton and Andau (1987) determined that there are 14–20 individual tarsiers/km². On the other hand, Niemitz (1984) reports an estimate of 80 individuals/km². Given the significantly larger home range of individuals observed by Crompton and Andau, the reduced number of individuals in a given area can be better understood. Neri-Arboleda (2001) estimated the density of the Bohol *T. syrichta* as approximately 57 individuals/km². Gursky (1998b) also estimated the population density of Dian's tarsier at Kamarora as 129 individuals/km². No data on population density have been obtained for the pygmy tarsier.

These values are somewhat deceptive because tarsier population density is known to change substantially with both altitude and habitat type. For example, the density of both *T. spectrum* and *T. dianae* was significantly lower at higher altitudes than lower altitudes (Gursky 1998b). Their population density was also significantly higher in the secondary disturbed forest than in the primary habitat.

The conservation status of the five accepted tarsier species is presently categorized as either low risk (least concern), low risk (conservation-dependent), or data deficient. Following a workshop organized by Myron Shekelle and the Pusat Primata Schmutzer in Jakarta, February 2002, Shekelle, Gursky, and Nietsch compiled and reanalyzed data on each species' distribution, abundance, and conservation status. Using one of the World Conservation Union criteria for calculating conservation status (estimates of the amount of land within each species' distribution), we determined that *T. bancanus* should be listed as low risk, conservation-dependent; *T. syrichta* as vulnerable; *T. spectrum* as vulnerable; *T. dianae* as vulnerable; and *T. pumilus* as data deficient.

Neri-Arboleda (2001) conducted a population viability analysis of *T. syrichta* using a metapopulation model analysis

of the likelihood of extinction. Given the values of the life history and environmental parameters used for the analysis (Neri-Arboleda 2001) and in the absence of environmental catastrophe, the minimum habitat area that could attain a <5% probability of extinction within 100 years is 60 ha. Therefore, 60 ha is considered the minimum viable habitat area for *T. syrichta*. Each tarsier has a home range of approximately 2.5 ha, with little overlap; thus, the minimum viable habitat of 60 ha contains 24 female individuals. Neri-Arboleda (2001) notes that the model's output is very sensitive to changes in adult mortality. That is, slight increases in adult mortality made the population very unstable and caused substantial increases in the probability of extinction. Given numerous observations of predation attempts (Gursky 2003) and the relatively high rate of predation in *T. spectrum*, it is fair to assume that adult female mortality is relatively high. This estimate of 60 ha takes into consideration that the forested area is fully occupied, which in the wild is most often not the case. This estimate is also based on life history parameters derived from one 8-month study on the island of Bohol in the Philippines.

DIRECTIONS FOR FUTURE RESEARCH

Directions for future research on tarsiers are numerous. There is the question of the phylogenetic relationship of tarsiers to other primates, both living and fossil. Numerous primate evolutionary biologists have been working on this problem (Beard 1998, Ross 1994, Beard and Wang 1990, Ginsburg and Mein 1986, Simons and Bown 1985). There is also the question for taxonomists concerning the number of tarsier species. The use of vocalizations for the identification of distinct tarsier species is a method that needs to be systematically applied to all tarsier populations. It has already proved quite successful on galagos in Africa (Bearder 1999) and is beginning to be used in Sulawesi. Its use in Borneo, the Philippines, and southern Sumatra is needed. Using this method, I anticipate that several tarsier species will be identified in Borneo. While Crompton and Andau (1986, 1987) found that *T. bancanus* at Sepilok Forest Reserve regularly have calling concerts, Niemitz (1984) reported that *T. bancanus* at Semongok, Sarawak, rarely vocalize. These vocalization differences may be the result of niche differences, thus representing different dialects. However, according to Bearder (1999), these observations suggest that these two populations are not members of the same species. Studies of the number of tarsier species as well as their distribution are critical if we are to develop any conservation schemes for the tarsiers. Thus, an avenue for future research would be an island-wide census of tarsiers on Borneo, Sumatra, Sulawesi, and several islands within the Philippines while controlling for altitude and forest type.

No studies of the population density of *T. pumilus* have been conducted. Recently, two Indonesian scientists trapping rats in the highlands of central Sulawesi also trapped a pygmy tarsier on the ground (Maryanto and Yani, in press). It is imperative that additional systematic attempts be made to locate this species in the wild, such as a systematic survey using 100 m increments in altitude on the mountains in central Sulawesi, where the museum specimens that were utilized to identify this as a new species (Musser and Dagosto 1987) were obtained. Unfortunately, the region is very hostile due to fighting between Muslim and Christian populations, making research in the area quite dangerous.

Another extremely important avenue for future research involves theoretically based studies. Most research on prosimians, including tarsiers, is generally descriptive and naturalistic. Only within the last 10 years have primatologists begun to ask theoretical questions of the nocturnal primates, trying to understand what factors pattern their behavior in space and time. One theoretical question that stands out concerns the variation in social systems among the different tarsier species. Is it due to ecology or perhaps phylogeny? Why are *T. spectrum* so much more social than the other tarsier species? Is it a reflection of their different habitat? *T. bancanus* and *T. syrichta* live in dipterocarp forests that fruit on 7-year cycles, whereas the Sulawesian tarsiers live in forests with very few dipterocarps and higher proportions of figs that fruit all year round. As the tarsiers consume insects that are attracted to the fruit of these trees, this may account for some of the differences. Note that the Sulawesian tarsiers are east of Wallace's line, while the other tarsier species are west of Wallace's line. Crompton (1989) believes it may have to do with the different types of insect that each species consumes: *T. bancanus* eats large ground-dwelling insects, while *T. spectrum* consumes more flying and colonial insects. Research is needed to determine which, if any, of these explanations can account for the different social systems in these sister species.

REFERENCES

Aiello, L. (1986). The relationship of the Tarsiiformes: a review of the case for the Haplorhini. In: Wood, B. A., Martin, L. B., and Andrews, P. J. (eds.), *Major Topics in Primate and Human Evolution*, Cambridge University Press, Cambridge.

Amrine-Madsen, H., Koepfli, K. P., Wayne, R. K., and Soringer, M. S. (2003). A new phylogenetic marker APOB provides compelling evidence for eutherian relationships. *Mol. Phylogenet. Evol.* 28:225–240.

Andrews, T. D., Jermiin, L. S., and Easteal, S. (1998). Accelerated evolution of cytochrome *b* in Simian primates: adaptive evolution in concert with other mitochondrial proteins? *J. Mol. Evol.* 47:249–257.

Beard, K. (1998). A new genus of Tarsiidae (Mammalia; Primates) from the middle Eocene of Shanxi Province, China with notes on the historical biogeography of tarsiers. *Bull. Carnegie Mus. Nat. Hist.* 34:260–277.

Beard, K., and Wang, B. (1990). Phylogenetic and biogeographic significance of tarsiiform primate *Asiomomys changbaicus* from the Eocene of Jilin Province, People's Republic of China. *Am. J. Phys. Anthropol.* 85:159–166.

Bearder, S. (1999). Physical and social diversity among nocturnal primates: a new view based on long-term research. *Primates* 40:267–282.

Bearder, S., Nekaris, K., and Buzzell, C. (2002). Dangers in the night: are some nocturnal primates afraid of the dark? In: Miller, L. (ed.), *Eat or Be Eaten: Predator Sensitive Foraging*. Cambridge University Press, Cambridge. pp. 21–40.

Bearder, S. K. (1987). Lorises, bushbabies, and tarsiers: diverse societies in solitary foragers. In: Smuts, B., Cheney, D., Seyfarth, R., Wrangham, R., and Struhsaker, T. (eds.), *Primate Societies*, University of Chicago Press, Chicago.

Buffon, G. (1765). *Histoire Naturelle, Generale et Particuliere*, vol. 13. L'Imprimerie du Roi Paris. pp. 87–91.

Cartmill, M., and Kay, R. (1978). Cranio-dental morphology, tarsier affinities, and primate suborders. In: Chivers, D., and Joysey, K. (eds.), *Recent Advances in Primatology. Evolution*, vol. 3. Academic Press, London. pp. 205–214.

Catchpole, H., and Fulton, J. (1943). The oestrus cycle in *Tarsius*. *J. Mammal.* 24:90–93.

Cave, A. J. (1973). The primate nasal fossa. *Biol. J. Linnean Soc.* 5:377–387.

Charles-Dominique, P. (1977). *Ecology and Behavior of Nocturnal Primates*. Columbia University Press, New York.

Cowlishaw, G., and Dunbar R. I. M. (2000). Primate Conservation Biology. Chicago University Press, Chicago, Appendix.

Crompton, R. (1989). Mechanisms for speciation in *Galago* and *Tarsius*. *Hum. Evo.* 4:105–116.

Crompton, R., and Andau, P. (1986). Locomotion and habitat utilization in free ranging *Tarsius bancanus*: a preliminary report. *Primates* 27:337–355.

Crompton, R., and Andau, P. (1987). Ranging, activity rhythms, and sociality in free-ranging *Tarsius bancanus*: preliminary report. *Int. J. Primatol.* 8:43–71.

Dagosto, M., and Gebo, D. (1998). A preliminary study of the Philippine tarsier (*Tarsius syrichta*) in Leyte. *Am. J. Phys. Anthropol.* 26(suppl.):73.

Dagosto, M., Gebo, D., and Dolino, C. (2001). Positional behavior and social organization of the Philippine tarsier (*Tarsius syrichta*). *Primates* 42:233–243.

Daly, M., Behrends, P., Wilson, M., and Jacobs, L. (1992). Behavioral modulation of predation risk: moonlight avoidance and crepuscular compensation in a nocturnal desert rodent, *Dipodomys merriami*. *Anim. Behav.* 44:1–9.

Dunbar, R. (1988). *Primate Social Systems*. Cornell University Press, Ithaca.

Dutrillaux, B., and Rumpler, Y. (1988). Absence of chromosomal similarities between tarsiers (*Tarsius syrichta*) and other primates. *Folia Primatol.* 50:130–133.

Erkert, H., and Grober, J. (1986). Direct modulation of activity and body temperature of owl monkeys (*Aotus lemurinus griseimenbra*) by low light intensities. *Folia Primatol.* 47:171–188.

Erxleben, J. (1777). *Systema Regni Animalis.* Impensis Weygandianis, Liepzig.

Fleagle, J. (1999). *Primate Adaptation and Evolution.* Academic Press, New York.

Fogden, M. L. (1974). A preliminary field study of the western tarsier, *Tarsius bancanus* Horsfield. In: Martin, R. D., Doyle, G. A., and Walker, A. C. (eds.), *Prosimian Biology.* Duckworth Press, London.

Gilbert, A. (1986). Mammary number and litter size in Rodentia: the one-half rule. *Proc. Nat. Acad. Sci. USA* 83:4828–4830.

Ginsburg, L., and Mein, P. (1986). *Tarsius thailandica nov. sp.,* Tarsiidae (Primates, Mammalia) fossile d'Asie. *C. R. Acad. Sci.* 304:1213–1215.

Groves, C. (1998). Systematics of tarsiers and lorises. *Primates* 39:13–27.

Groves, C. (2001a). *Primate Taxonomy.* Smithsonian Institution Press, Washington DC.

Groves, C. (2001b). Getting to know the tarsiers: yesterday, today, and tomorrow. Paper presented at the International Society of Primatologists, 14th congress, Adelaide, Australia, January 7–12, 2001.

Gursky, S. (1995). Group size and composition. *Trop. Biodivers.* 3:57–62.

Gursky, S. (1997). Modeling maternal time budgets: the impact of lactation and gestation on the behavior of the spectral tarsier, *Tarsius spectrum* [PhD diss.]. State University of New York, Stony Brook.

Gursky, S. (1998a). Conservation status of the spectral tarsier *Tarsius spectrum*: population density and home range size. *Folia Primatol.* 69:191–203.

Gursky, S. (1998b). The conservation status of two Sulawesian tarsier species: *Tarsius spectrum* and *Tarsius dianae. Primate Conserv.* 18:88–91.

Gursky, S. (1999). The effect of moonlight on the behavior of spectral tarsiers. *Am. J. Phys. Anthropol.* 18(suppl.):142.

Gursky, S. (2000). The effect of seasonality on the behavior of an insectivorous primate, *Tarsius spectrum. Int. J. Primatol.* 21:477–495.

Gursky, S. (2002a). Determinants of gregariousness in the spectral tarsier (Prosimian: *Tarsius spectrum*). *J. Zool.* 256:1–10.

Gursky, S. (2002b). The behavioral ecology of the spectral tarsier. *Evol. Anthropol.* 11:226–234.

Gursky, S. (2003). The effect of moonlight on a nocturnal prosimian primates. *Int. J. Primatol.* 24:351–367.

Haring, D., and Wright, P. (1989). Hand raising a Philippine tarsier, *Tarsius syrichta. Zoo Biol.* 8:265–274.

Hill, W. C. (1955). *Primates Haplorhini, Tarsoidea.* Interscience, New York.

Holenweg, A., Noe, R., and Schabel, M. (1996). Waser's gas model applied to associations between diana monkeys and colobus monkeys in Tai National Park, Ivory Coast. *Folia Primatol.* 67:125–136.

Horsfield, J. (1821). *Zoological Research in Java and in the Neighboring Islands.* Kingsbury, Parbury and Allen, London.

Izard, M., Wright, P., and Simons, E. (1985). Gestation length in *Tarsius bancanus. Am. J. Primatol.* 9:327–331.

Janson, C. (1988). Intraspecific food competition and primate social structure: a synthesis. *Behaviour* 105:1–17.

Jolly, A., Rasamimanana, H., Kinnaird, M., O'Brien, T., Crowley, H., Harcourt, C., Gardner, S., and Davidson, J. (1993). Territoriality in *Lemur catta* groups during the birth season at Berenty, Madagascar. In: Kappeler, P., and Ganzhorn, J. (eds.), *Lemur Social Systems and Their Ecological Basis.* Plenum Press, New York.

Kappeler, P., and Ganzhorn, J. (1993). *Lemur Social Systems and Their Ecological Basis.* Plenum Press, New York.

Klinger, H. (1963). The somatic chromosomes of some primates: *Tupaia glis, Nycticebus coucang, Tarsius bancanus, Cercocebus aterrimus, Symphalangus syndactylus. Cytogenetics* 2:140–151.

Kotler, B., Brown, J., and Hasson, O. (1993). Factors affecting gerbil foraging, behavior and rates of owl predation. *Ecology* 72:2249–2260.

Krebs, J., and Davies, N. (1984). *Behavioral Ecology.* Blackwell Scientific, Oxford.

Lagreca, L., Marotta, E., and Vales, L. (1992). Relationship between sow's teat number and the litter size at birth and at weaning. *An. Vet. Murica* 1992:863–867.

Linneaus, C. (1758). *Systema Naturae per Regna Tria Naturae, Secundum Classes, Ordines, Genera, Species, cum Characteribus et Differentiis,* 10th ed. L. Salvii, Holmiae.

Luckett, W. (1976). Comparative development and evolution of the placenta in primates. *Contrib. Primatol.* 3:142–234.

MacKinnon, J., and MacKinnon, K. (1980). The behavior of wild spectral tarsiers. *Int. J. Primatol.* 1:361–379.

MacPhee, R., and Cartmill, M. (1986). Basicranial structures and primate systematics. In: Swindler, D., and Erwin, J. (eds.), *Comparative Primate Biology, Systematics, Evolution and Anatomy,* vol. 1. Alan R. Liss, New York. pp. 219–275.

Maryanto, I., and Yani, M. (in press). The third record of pygmy tarsier (*Tarsius pumilus*) from Lore Lindu National Park, central Sulawesi, Indonesia. *Trop. Biodivers.*

Miller, G., and Hollister, F. (1921). A review of *Tarsius spectrum* from Celebes. *Proc. Biol. Soc. Wash.* 34:103–228.

Morrison, D. (1978). Lunar phobia in a neotropical fruit bat, *Artibeus jamaicensis* (Chiroptera, Phyllostomidae). *Anim. Behav.* 26:852–855.

Musser, G., and Dagosto, M. (1987). The identity of *Tarsius pumilus,* a pygmy species endemic to montane mossy forests of central Sulawesi. *Am. Mus. Novit.* 2867:1–53.

Nash, L. (1986). Influence of moonlight level on traveling and calling patterns in two sympatric species of *Galago* in Kenya. In: Taub, D., and King, F. (eds.), *Current Perspectives in Primate Social Dynamics.* Van Nostrand Reinhold, New York. pp. 357–367.

Neri-Arboleda, I. (2001). Ecology and behavior of *Tarsius syrichta* in Bohol, Philippines: implications for conservation [MSc thesis]. University of Adelaide, Adelaide.

Niemitz, C. (1972). Puzzle about *Tarsius. Sarawak Mus. J.* 20:329–337.

Niemitz, C. (1974). A contribution to the postnatal behavioral development of *Tarsius bancanus,* studied in two cases. *Folia Primatol.* 21:250–276.

Niemitz, C. (1979). Results of a field study on the western tarsier in Sarawak. *Sarawak Mus. J.* 27:171–228.

Niemitz, C. (1984). *Biology of Tarsiers*. Gustav Fischer, Stuttgart.

Niemitz, C., Nietsch, A., Warter, S., and Rumpler, Y. (1991). *Tarsius dianae*: a new primate species from central Sulawesi (Indonesia). *Folia Primatol.* 56:105–116.

Nietsch, A. (1999). Duet vocalizations among different populations of Sulawesi tarsiers. *Int. J. Primatol.* 20:567–583.

Nietsch, A., and Kopp, M. (1998). The role of vocalization in species differentiation of Sulawesi tarsiers. *Folia Primatol.* 69:371–378.

Nietsch, A., and Niemitz, C. (1992). Indication for facultative polygamy in free-ranging *Tarsius spectrum*, supported by morphometric data. In: *International Primatological Society Abstracts*. International Primatological Society, Strasbourg. p. 318.

Pallas, P. S. (1778). *Novae Species quad e Glirium Ordinae cum Illustrationibus Variis Complurium ex hoc Ordinae Animalium*, W. Walther, Erlangen.

Petiver, J. (1705). *Gazophylacii Naturae et Artis*, London.

Pocock, R. I. (1918). On the external characteristics of the lemurs and of *Tarsius*. *Proc. Zool. Soc. Lond.* 1918:19–53.

Poorman, P., Cartmill, M., and MacPhee, R. (1985). The G banded karotype of *Tarsius bancanus* and its implications for primate phylogeny. *Am. J. Phys. Anthropol.* 66:215.

Roberts, M. (1994). Growth, development and parental care patterns in western tarsiers, *Tarsius bancanus* in captivity: evidence for a slow life history and non-monogamous mating system. *Int. J. Primatol.* 15:1–28.

Roberts, M., and Kohn, F. (1993). Habitat use, foraging behavior and activity patterns in reproducing western tarsiers, *Tarsius bancanus*, in captivity: a management synthesis. *Zoo Biol.* 12:217–232.

Rosenberger, A., and Szalay, F. (1980). On the tarsiiform origins of the Anthropoidea. In: Ciochon, R., and Chiarelli, A. (eds.), *Evolutionary Biology of the New World Monkeys and Continental Drift*. Plenum Press, New York. pp. 139–157.

Ross, C. (1994). The craniofacial evidence for anthropoid and tarsier relationships. In: Fleagle, J., and Kay, R. (eds.), *Anthropoid Origins*. Plenum Press, New York. pp. 469–547.

Schultz, A. (1948). The number of young at a birth and the number of nipples in primates. *Am. J. Phys. Anthropol.* 6:1–23.

Shekelle, M. (2003). *Taxonomy and biogeography of eastern Tarsiers* [PhD thesis]. Washington University, St. Louis.

Shekelle, M., Leksono, S., Ischwan, L., and Masala, Y. (1997). The natural history of the tarsiers of north and central Sulawesi. *Sulawesi Prim. News.* 4:4–11.

Sherman, P., Braude, S., and Jarvis, J. (1999). Litter sizes and mammary numbers of naked mole-rats: breaking the one-half rule. *J. Mammal.* 80:720–733.

Simons, E., and Bown, T. (1985). *Afrotarsius chatrathi*, the first tarsiiform primate (Tarsiidae) from Africa. *Nature* 313:4750–477.

Simpson, G. G. (1945). The principles of classification and a classification of mammals. *Bull. Am. Mus. Nat. Hist.* 85:1–350.

Storr, G. (1780). *Prodromus methodi mammalium*. Tubingen.

Tremble, M., Muskita, Y., and Supriatna, J. (1993). Field observations of *Tarsius dianae* at Lore Lindu National Park, central Sulawesi, Indonesia. *Trop. Biodivers.* 1:67–76.

Trent, B., Tucker, M., and Lockard, J. (1977). Activity changes with illumination in slow loris *Nycticebus coucang*. *Appl. Anim. Ethnol.* 3:281–286.

Ulmer, F. A. (1963). Observations on the tarsier in captivity. *Zool. Garten* 27:106–121.

Waser, P. (1976). *Cercocebus albigena*: site attachment, avoidance and intergroup spacing. *Am. Nat.* 110:911–917.

Wolfe, J., and Summerlin, C. (1989). The influence of lunar light on nocturnal activity of the old field mouse. *Anim. Behav.* 37:410–414.

Wolin, L., and Massoupust, L. (1970). Morphology of the primate retina. In: Noback, C. R., and Montagna, W. (eds.), *The Primate Brain*. Appleton-Century-Crofts, New York. pp. 1–27.

Wright, P. C. (1990). Patterns of paternal care in primates. *Int. J. Primatol.* 11:89–102.

Wright, P. C., Toyama, L., and Simons, E. (1988). Courtship and copulation in *Tarsius bancanus*. *Folia Primatol.* 46:142–148.

6

Callitrichines
The Role of Competition in Cooperatively Breeding Species
Leslie J. Digby, Stephen F. Ferrari, and Wendy Saltzman

INTRODUCTION

The callitrichines are best known for their suite of reproductive and behavioral characteristics that are unusual or unique among the primates. Social suppression of reproduction, postpartum ovulation, twinning, cooperative care of young, and flexible mating systems make this a useful group of animals for testing hypotheses about the evolution of reproductive strategies and social systems. In addition, these species are characterized by claw-like nails on all digits but the hallux, two molars instead of the three typical of platyrrhines (2.1.3.2 dental formula, except in *Callimico*), small body size (from 120 g in *Cebuella* to 650 g in *Leontopithecus*), and dramatic variation in coloration, ear tufts, and even "mustaches" (detailed physical descriptions in Eisenberg and Redford 1999). When reviewed by Goldizen (1987a), information on these species in the wild was limited to five long-term studies, most of which focused on the genus *Saguinus* (e.g., Dawson 1978, Neyman 1978, Terborgh and Goldizen 1985). The past two decades have witnessed a surge of work on the subfamily Callitrichinae, including field studies of several species for which little or no data were available 15 years ago (e.g., *Callimico goeldii*, Porter et al. 2001, *Callithrix geoffroyi*, Passamani and Rylands 2000, *Leontopithecus* spp., Kleiman and Rylands 2002, *Saguinus tripartitus*, Kostrub 2003). Nevertheless, intensive long-term ecological and behavioral studies are still restricted to just over a dozen species (e.g., Rylands 1993a, Kleiman and Rylands 2002), and the majority of taxa, including the 10 newly described species and subspecies (Rylands et al. 2000), are known primarily from general surveys or initial species descriptions.

Field studies of callitrichines have investigated topics as diverse as cognitive mapping (Bicca-Marques and Garber 2001), the influence of color vision polymorphism on foraging behavior (Caine et al. 2003, Smith et al. 2003), reproductive endocrinology (Savage et al. 1997, French et al. 2003), foraging biomechanics (Hourani et al. 2003, Vinyard et al. 2003), nutritional content of food (Heymann and Smith 1999, Smith 2000), and genetics (Nievergelt et al. 2000, Faulkes et al. 2003). Field data have been complemented by laboratory studies on energetics (Genoud et al. 1997, Power et al. 2003), nutrition and digestion (Power et al. 1997, Ullrey et al. 2000), phylogenetics (Barroso et al. 1997, Seuánez et al. 2002), neuroendocrine and behavioral control of reproduction (Abbott et al. 1998, Saltzman 2003), and ovulation and pregnancy using ultrasonography (Jaquish et al. 1995, Oerke et al. 2002), as well as many biomedically oriented studies of reproduction, immune response, and disease (reviewed in Mansfield 2003).

Our goal here is to summarize the most recent information available on the behavior, ecology, and reproduction of callitrichine primates, with an emphasis on comparisons among genera. We then examine the extent to which competition plays a role in social interactions and reproductive success in these cooperatively breeding primates.

TAXONOMY AND DISTRIBUTION

Taxonomy

The systematics of the Platyrrhini has undergone extensive revision at all levels since the classic review of Hershkovitz (1977), and the callitrichines have been among the most controversial taxa. Because it combines morphological, genetic, and ecological perspectives, the recent revision by

Rylands et al. (2000, see also Groves 2001) is perhaps most representative of the current consensus and is followed here (see Appendix 6.1). According to these authors, the subfamily Callitrichinae is a monophyletic group containing six genera [*Callimico* (Goeldi's monkey), *Callithrix* (Atlantic marmosets), *Cebuella* (pygmy marmoset), *Leontopithecus* (lion tamarins), *Mico* (Amazonian marmosets), and *Saguinus* (tamarins)], with a total of 60 species and subspecies. We will also include the newly proposed genus name *Callibella* for the dwarf marmoset (van Roosmalen and van Roosmalen 2003).

Callimico goeldii has posed the major problem for callitrichine taxonomists. Its small body size and claw-like nails are characteristic of callitrichines, but its third molar and singleton births are typical of the larger-bodied platyrrhines. This led to the monospecific genus being placed at various times within the Callitrichidae (Hill 1957, Napier and Napier 1967), the Cebidae (Cabrera 1958, Simons 1972), or even its own family, the Callimiconidae (Hershkovitz 1977). The current, widely held consensus is that *Callimico* is a true, albeit atypical, callitrichine. This is strongly supported by a number of recent molecular studies (Schneider and Rosenberger 1996, Tagliaro et al. 1997, Canavez et al. 1999, von Dornum and Ruvolo 1999) that clearly place *Callimico* as a sister group of the marmosets (*Callithrix*, *Mico*, *Callibella*, and *Cebuella*).

The marmosets have also been the subject of recent taxonomic revisions. Rosenberger (1981) proposed that *Cebuella* should be included within the genus *Callithrix*. This view has been supported by many genetic studies (e.g., Canavez et al. 1999, Tagliaro et al. 2000) but never widely accepted. Schneider and Rosenberger (1996) and Rylands et al. (2000) not only excluded this proposition but instead reinstated the genus *Mico*, which is equivalent to Hershkovitz's (1977) Amazonian *Callithrix argentata* group. The inclusion of *Mico* avoids paraphyly and is consistent with differences in the dental morphology of *Callithrix* and *Mico* as well as with their allopatric distribution. Recent studies of Amazonian marmosets (Corrêa et al. 2002, Gonçalves et al. 2003) have adopted the new arrangement.

The newest genus to be added to the callitrichine subfamily is *Callibella* (van Roosmalen and van Roosmalen 2003). Although it was originally included in *Mico*, detailed genetic and morphological studies of the recently described black-crowned dwarf marmoset (intermediate in size between *Cebuella* and *Mico*) (van Roosmalen et al. 1998) support the creation of the new genus (Aguilar and Lacher 2003, van Roosmalen and van Roosmalen 2003).

Distribution

Current knowledge of the zoogeography of the callitrichine genera is little changed from that reviewed by Hershkovitz (1977). Perhaps the most significant alteration has been the extension of the southern limit of the range of *Leontopithecus*, following the discovery of *Leontopithecus*

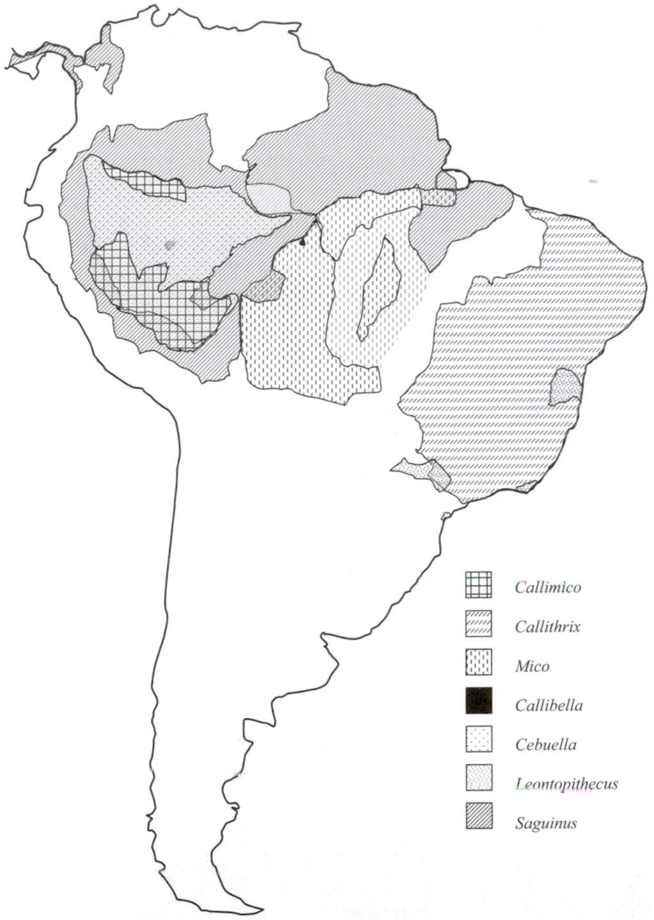

Figure 6.1 Distribution map of the Callitrichinae (based on Rylands et al. 1993, Ferrari 1993, Eisenberg and Redford 1999, van Roosmalen and van Roosmalen 2003, Infonatura 2004). Lighter shades indicate probable but unconfirmed presence in that area.

Legend:
- Callimico
- Callithrix
- Mico
- Callibella
- Cebuella
- Leontopithecus
- Saguinus

caissara in the southern Brazilian state of Paraná (Lorini and Persson 1990). *Leontopithecus* species are now found in four distinct areas, corresponding to the geographic ranges of the four known species (Fig. 6.1).

Two genera, *Callimico* and *Cebuella*, have roughly equivalent geographic ranges in western Amazonia. Recent studies (van Roosmalen and van Roosmalen 1997, Ferrari et al. 1999) have confirmed that both genera range as far east as the left bank of the Madeira, where they are potentially parapatric with *Mico*.

Saguinus is sympatric with *Callimico* and *Cebuella* throughout their distributions but also ranges much farther north and east. There is now confirmation of sympatric zones between *Saguinus* and *Mico* on the upper Madeira River (Schneider et al. 1987) as well as the lower Toncantins–Xingu interfluviam (Ferrari and Lopes Ferrari 1990).

Marmosets of the genus *Mico* are found in the southern Amazon Basin between the Madeira and the Tocantins Rivers and as far south as northeastern Paraguay (Hershkovitz 1977). The newly described *Callibella* also inhabits the area west of the Rio Aripuanã and overlaps with at least one *Mico* species. *Callithrix* ranges farther south and east than *Mico*. This includes sympatry with *Leontopithecus chrysomelas*

and *L. rosalia* in some regions (Rylands 1989), though preferences for high or low elevation may limit range overlap.

ECOLOGY

Habitat

As might be expected from their wide geographic distribution, callitrichines inhabit a wide variety of neotropical habitats, with correspondingly variable patterns of occupation. There are some general ecological differences among the genera, with *Callithrix* and *Mico* tending to form larger groups, to occupy smaller home ranges, and consequently, to have higher population densities than *Saguinus* and *Callimico*. However, home range size and population density vary by up to two orders of magnitude not only within the Callitrichinae but also within some genera (Table 6.1). Consequently, it is difficult to identify genus-specific or even species-specific patterns, especially where data are based on observations of a single group or population.

The key factor determining differences between marmosets and tamarins appears to be the marmosets' morphological specializations for the dietary exploitation of plant exudates (e.g., gums and saps) (Ferrari 1993). The ability to exploit exudates systematically as a substitute for fruit throughout the year allows marmosets to inhabit resource-poor or highly seasonal habitats in which tamarins may be unable to survive. Such habitats include not only the wooded ecosystems of the Cerrado, Caatinga, and Chaco biomes and Amazonian savannas but also forests that have suffered intense anthropogenic disturbance. Distributed throughout tropical Brazil south of the Amazon, marmosets —in particular *Callithrix jacchus* and *C. penicillata*—can thrive in such unlikely habitats as city parks, backyards, and even coconut plantations (Rylands and de Faria 1993, L. J. D. and S. F. F., personal observation). *Callibella* takes this ability to make use of disturbed habitats to an extreme and may be dependent on human occupation of a habitat (e.g., the presence of orchards and gardens) (van Roosmalen and van Roosmalen 2003).

Cebuella, in contrast to *Callithrix* and *Mico*, appears to be a habitat specialist. It may reach extremely high ecological densities in riparian forest but is usually absent from neighboring areas of *terra firma* forest (Soini 1988). This unusual distribution pattern appears to be related to *Cebuella*'s specialization for exudativory (groups may inhabit a single gum tree for long periods) and the avoidance of competition with sympatric callitrichines. In this context, the apparent specialization of *Cebuella* can be interpreted as an accentuated preference for a marginal habitat type, which may be relatively unsuitable for other callitrichines.

Ranging Patterns

With regard to both home range size and population density, the major division between *Callithrix* and *Mico* appears to

Table 6.1   Habitat, Home Range, and Daily Path Length

SPECIES AND LOCATION	HABITAT[1]	HOME RANGE (HA)	DAILY PATH LENGTH (M/DAY)	HOME RANGE OVERLAP (%)	REFERENCES
Callimico goeldii					
Pando, Bolivia	SC	30–150	Approx. 2,000		Pook and Pook 1981, Porter 2001a
Virazon, Bolivia	RV	45–50	-		Christen 1999
Callithrix jacchus					
João Pessoa, Brazil	AF (2)	2–5	1,300		Maier et al. 1982, Alonso and Langguth 1989
Nisia Floresta, Brazil	AF (2)	0.7–5.2	912–1,243	46–86	Digby and Barreto 1996, Castro 2000
Dois Irmãos, Brazil	AF (2)	4.11		23–99	Mendes Pontes and Monteiro da Cruz 1995
C. penicillata					
Brasilia, Brazil	G	3.5			de Faria 1986
C. aurita					
Cunha, Brazil	AF (2)	35.3	958.8	15	Ferrari et al. 1996
Fazenda Lagoa, Brazil	SC	16.5	986		Martins 1998
C. flaviceps					
Caratinga, Brazil	AF (2)	33.86–35.5	883.8–1,222.5	80	Ferrari 1988, Ferrari et al. 1996, Corrêa et al. 2000, Guimarães 1998
C. kuhlii					
Lemos Maia, Brazil	WS	10	830–1,120	50	Rylands 1989
Cebuella pygmaea					
Peru	V	0.1–0.5	280–300	0	Soini 1982, 1988; Heymann and Soini 1999
Ecuador	TF	0.4–1.09			de la Torre et al. 2000
Mico intermedius					
Mato Grosso, Brazil	ED	22.1	772–2,115	22	Rylands 1986a
M. argentatus					
Tapajós, Brazil	WS	4–24			Albernaz and Magnusson 1999
Caxiuanã, Brazil	AM	35			Veracini 2000
Saguinus fuscicollis					
Manu, Peru	AM (L)	30–100[2]	1,220		Terborgh 1983, Goldizen 1987a
Rio Blanco, Peru	AM (L)	40	1,849	23	Garber 1988
Rio Urucu, Brazil	TF	149	1,150–2,700	76	Peres 2000
S. f. weddelli					
Cachoeira Samuel, Brazil	TF	44+	1,312		Lopes and Ferrari 1994
S. imperator					
Manu, Peru	AM (L)	30–100[2]	1,420		Terborgh 1983, Goldizen 1987a
S. mystax					
Rio Blanco, Peru	AM (L)	40[2]	1,946	23	Garber 1988
Quebrada Blanco, Peru	AM	41–45	1,500–1,720		Heymann 2000, 2001
S. niger					
Caxiuanã, Brazil	AM	35			Veracini 2000
S. midas midas					
French Guiana	AM (H)	31.1–42.5			Day and Elwood 1999
S. tripartitus					
Tiputini, Ecuador	AM	16–21	500–2,300		Kostrub 2003
Leontopithecus rosalia					
Poço das Antas, Brazil	AF (2)	21.3–73	955–2,405	61	Dietz et al. 1997, Peres 2000
Fazenda União, Brazil	AF	65–229	1,873–1,745		Kierulff et al. 2002
L. caissara					
Superagüi, Brazil	AF	125.5–300	1,082–3,398		Prado 1999a,b as cited in Kierulff et al. 2002
L. chrysomelas					
Lemos Maia, Brazil	WS	36	1,410–2,175	7	Rylands 1989
Una, Brazil	AF	66–130.4	1,684–2,044	Approx 10–14	Dietz et al. 1994b, Keirulff et al. 2002, Raboy and Dietz 2004
L. chrysopygus					
Morro do Diabo, Brazil	AF (1)	113–199	1,362–2,088		Valledares-Padua 1993, Valledares-Padua and Cullen 1994 as cited in Kierulff et al. 2002
Caetetus, Brazil	SC	276.5–394	1,164–3,103		Passos 1997 as cited in Kierulff 2002, Passos 1998

[1] AF, Atlantic Forest; AF (2), secondary Atlantic Forest; AF (1), interior Atlantic Forest; SC, sandy clay forest; RV, riverine forest; G, gallery/Cerrado forest; TF, terra firma forest; V, Varzea; ED, evergreen dryland forest; AM, Amazonian forest; AM (L), Amazonian lowland forest; AM (H), Amazonian highland forest; WS, white-sand forest.
[2] Home range of mixed groups (two species) of tamarins.

lie not between the two genera but between patterns characteristic of *C. jacchus* and *C. penicillata* and the remaining species. Typically, *C. jacchus* and *C. penicillata* occupy home ranges of less than 10 ha, with correspondingly high population densities. Other marmosets (Table 6.1) tend to occupy much larger home ranges (up to 35 ha) with correspondingly lower population densities. However, the data available for *Mico argentatus* indicate that this species may be more similar to *C. jacchus* and *C. penicillata*, at least under equivalent ecological conditions (i.e., forest patches in savanna ecosystems and anthropogenic fragments) (Albernaz and Magnusson 1999, Corrêa et al. 2002, Gonçalves et al. 2003). With additional data on the ecology of these species, we may find that observed differences are more closely related to habitat characteristics than to taxon-specific variables.

These findings implicate habitat quality as a primary determinant of home range size and, consequently, population density. For marmosets, the key factors are likely to be the availability of exudate sources and arthropod abundance. Because many gum-producing plants (e.g., Leguminosae, Vochysiaceae) are abundant in the habitats favored by marmosets, arthropod abundance is probably the limiting factor. This may account for the relatively large home ranges recorded for *C. aurita* and *C. flaviceps*, which occur in comparatively seasonal ecosystems in which arthropods may be relatively scarce.

Home range sizes of *Saguinus* and *Leontopithecus* are highly variable, with some species and populations showing small home ranges similar to those of the marmosets and others extending to over 100 ha (Table 6.1). The larger home ranges and correspondingly lower population densities are consistent with the tamarins' relatively high degree of frugivory (Tables 6.1 and 6.2). *Callimico* shows a similar pattern of large home range and relatively small group size (Tables 6.1 and 6.3), in spite of its ability to utilize fungus during periods of fruit scarcity (Porter 2001b).

Population densities of both *Callithrix* and *Leontopithecus* decrease farther south, where climate and, presumably, resource abundance are far more seasonal. Once again, differences are disproportionate relative to differences in body size, with home ranges reported for *L. caissara* and *L. chrysopygus* being the largest for any callitrichine (Passos 1998, Prado 1999b as cited in Kierulff et al. 2002). It is interesting to note, however, that *L. rosalia* may be relatively abundant in anthropogenic forest patches, with home range sizes similar to those recorded for *C. aurita* and *C. flaviceps* (Table 6.1).

Territoriality

Callitrichines typically respond to intergroup exchanges with vocalizations, chases, and occasionally physical aggression (Hubrecht 1985; Garber 1988, 1993a; Peres 1989, Lazaro-Perea 2001; but see also van Roosmalen and van Roosmalen 2003). While such intergroup aggression can be interpreted as territoriality (e.g., Peres 1989, Lazaro-Perea 2001), the high degree of overlap between home ranges in some species (e.g., up to 80%–90% in *C. jacchus*; Mendes Pontes and Monteiro da Cruz 1995, Digby and Barreto 1996) (Table 6.1) may indicate that these behaviors are being used primarily for mate defense or defense of a specific set of currently fruiting trees or exudate sources (e.g., Garber 1988, 1993a; Peres 2000). Intergroup interactions may also serve a second function in allowing individuals to monitor the composition and status of neighboring groups and to assess possible mating opportunities (Digby 1992, Ferrari and Diego 1992, Goldizen et al. 1996, Schaffner and French 1997, Lazaro-Perea 2001, Kostrub 2003; see below).

Mixed–Species Troops

The presence of *mixed-species troops* (two or more species associating in a nonrandom fashion, often coordinating activities; e.g., Pook and Pook 1982) plays an important role in the ecology of several callitrichine species. To date, all mixed-species troops include *Saguinus fuscicollis* interacting with either *S. mystax, S. imperator, S. labiatus,* or *Mico emiliae* (reviewed in Heymann and Buchanan-Smith 2000). At some sites, two of the *Saguinus* species also associate with *Callimico* (e.g., Porter 2001b). Though expected to generate costs due to niche overlap, mixed-species troops appear to limit competition via differential use of forest strata and foraging techniques (Terborgh 1983, Heymann and Buchanan-Smith 2000). Potential benefits include increased protection from predators, increased foraging efficiency (including increased insect capture rates), and resource defense (reviewed in Heymann and Buchanan-Smith 2000).

Foraging Behavior

Exudates

In the wild, callitrichines are known to exploit a wide variety of food types, avoiding only non-reproductive plant parts such as leaves and bark (Garber 1993a,b, Rylands and de Faria 1993, Digby and Barreto 1998, Heymann and Buchanan-Smith 2000, Smith 2000, Porter 2001b, Kierulff et al. 2002) (Table 6.2). The principal feature of callitrichine diets is their variety, and the only clear taxon-specific pattern is the marmosets' use of plant exudates (primarily gums, with some sap) as a dietary staple. All callitrichines eat some exudates, but *Callithrix, Mico,* and *Cebuella* are morphologically specialized for the systematic harvesting and digestion of gum and are thus able to sustain high levels of exudativory throughout the year (Ferrari 1993).

The marmosets' specializations for exudativory include elongated, chisel-like lower incisors and a wide jaw gape that permit them to gouge through the bark of gum-producing plants, thus provoking exudate flow (Hershkovitz 1977; Vinyard et al. 2001, 2003). Gums contain complex polysaccharides, a potentially high-energy source but one which cannot be broken down enzymatically by other mammals (reviewed in Power and Oftedal 1996, Heymann and Smith

Table 6.2   Diet

SPECIES AND LOCATION	% TIME FEEDING	% TIME FEEDING ON				
		REPRODUCTIVE PLANT PARTS[1]	EXUDATES	ANIMAL PREY	OTHER	REFERENCES
Callimico goeldii						
Pando, Bolivia	15	29	1	33	29 (fungus), 27 (unknown)	Porter 2001b, 2004
Callithrix jacchus						
Joao Pessoa, Brazil	27	18.1	76.4	5.4	12.9 (fungus)	Alonso and Langguth 1989
Nisia Floresta, Brazil	43[2]	23	68	9		Digby (unpublished data)
C. aurita						
Cunha, Brazil	17	40.5	30	29.5	2 (fungus)	Corrêa et al. 2000
Fazenda Lagoa, Brazil	6	11	50.5	38.5		Martins and Setz 2000
C. flaviceps						
Caratinga, Brazil (1)		14.4	65.7	19.9		Ferrari et al. 1996
Caratinga, Brazil (2)	11.8	2.0	83.2	14.7		Corrêa et al. 2000
C. kuhlii						
Lemos Maia, Brazil	23	58.2	28.3	13.5		Rylands 1989, Corrêa et al. 2000
C. geoffroyii						
Espírito Santo, Brazil	21	15	68.6	15.4		Passamani 1998, Passamani and Rylands 2000
Cebuella pygmaea						
Rio Nanay, Peru	48[2]	Minor	67	33		Ramirez et al. 1977
Mico intermedius						
Mato Grosso, Brazil		74.9	15.5	9.6		Rylands 1982 as cited in Corrêa et al. 2000
M. argentatus						
Caxiuanã, Brazil		36	59	5		Veracini 1997 as cited in Corrêa et al. 2000
Saguinus fuscicollis						
Quebrada Bl., Peru		64.2	30.3	5.8		Knogge and Heymann 2003
Rio Blanco, Peru	12.9	39.2	7.6	53.1		Garber 1988
Pando, Bolivia		63	12	26	6 (unknown)	Porter 2001b
S. f. weddelli						
Cachoeira Samuel, Brazil	9.8	69.6	15.8	7.3		Lopes and Ferrari 1994
S. labiatus						
Pando, Bolivia		73	8	11	8 (unknown)	Porter 2001b
S. mystax						
Quebrada Blanco Peru		77.6	19.8	2.7		Knogge and Heymann 2003
Rio Blanco, Peru	13.1	50.6	1.5	47.8		Garber 1988
S. niger						
Fazenda Vitória, Brazil		87.5	3.1	9.4		Oliveira and Ferrari 2000
Caxiuanã, Brazil	17.6	71.1	23.81	4.5		Veracini 2000
S. tripartitus						
Tiputini, Ecuador		62	12	21	5 (unknown)	Kostrub 2003
Leontopithecus rosalia						
Poço das Antas, Brazil	18.5	83.5	1.4	14.9		Dietz et al. 1997
União, Brazil	10.4	84.6	0	15.4		Kierulff 2000 as cited in Kierulff et al. 2002
L. caissara						
Superagüi, Brazil	29.4[2]	75.5	1.3	10.3	12.9 (fungus)	Prado 1999b as cited in Kierulff et al. 2002
L. chrysomelas						
Lemos Maia, Brazil	27	74–89,[3] 11 flowers	3–11[3]	13–15		Rylands 1989
L. chrysopygus						
Morro do Diabo, Brazil	6–10	78.5	7.8	13.5		Valladares-Padua 1993 as cited in Kierulff et al. 2002
Caetetus, Brazil	29.9[2]	74.7	15.2	10.1		Passos 1999 as cited in Kierulff et al. 2002

[1] Includes fruit, seeds, flowers, and nectar.
[2] Includes foraging.
[3] Percent of plant records only; flowers made up 0%–20% of monthly diet (percent overall diet not reported).

Table 6.3 Group Size, Composition, and Mating Patterns

SPECIES	GROUP SIZE	ADULT MALES	ADULT FEMALES	MATING PATTERNS[1]	REFERENCES
Callimico goeldii	4–12	1–3	1–3	M/PG/PA	Reviewed in Porter 2001a
Callibella humilis	6–8			M?/PG	van Roosmalen and van Roosmalen 2003
Callithrix jacchus	3–16	2–7	2–6	M/PG/PGA	Barreto 1996, Digby and Barreto 1996, Lazaro-Perea et al. 2000, Faulkes et al. 2003
C. penicillata	3–13	3	2		Reviewed in de Faria 1986
C. aurita	4–11	1–2	2–3	M/PG	Ferrari et al. 1996, Martins and Setz 2000
C. flaviceps	5–20	3–5	1–6	M/PG	Ferrari and Diego 1992, Coutinho and Corrêa 1995, Guimarães 1998
C. geoffoyi	2–8	2	1		Passamani 1998, Chiarello and de Melo 2001, Price et al. 2002
C. kuhlii	5	2	1		Rylands 1989
Cebuella pygmaea	2–9	1–2	1–2	M/PG?[2]	Soini 1982, 1988; Heymann and Soini 1999; de la Torre et al. 2000
Mico intermedius	9–15	2–3	2–5	M/PA	Rylands 1986b
M. argentatus	6–10	1–3	1–2		Albernaz and Magnusson 1999, Tavares and Ferrari 2002
Saguinus fuscicollis	2–10	1–4	1–2	M/PA/ PG/PGA	Goldizen 1987a,b 2003; Garber 1988; Porter 2001a; Heymann 2001
S. f. weddelli	4–11	1–2	1–2	M/PG	Garber and Leigh 2001, Buchanan–Smith et al. 2000, Lopes and Ferrari 1994
S. mystax	3–11	1–4	1–2	M/PG[3]	Garber 1988, Heymann 2000, Smith et al. 2001
S. niger	5–7	3–4	1		Oliveira and Ferrari 2000
S. labiatus labiatus	2–13	2	1		Buchanan-Smith et al. 2000, Garber and Leigh 2001
S. oedipus	2–10			M/PG[3]	Savage et al. 1996a
S. tripartitus	2–9	1–4	1–2	PA	Heymann et al. 2002, Kostrub 2003
Leontopithecus rosalia	2–11	0–5	0–4	M/PA/PG	Baker et al. 1993, Dietz and Baker 1993
L. caissara	4–7				Prado 1999a
L. chrysomelas	3–10	1–3	1–3	M/PG	Dietz et al. 1994b, Baker et al. 2002, Raboy and Dietz 2004
L. chrysopygus	2–7	≤2	≤2		Passos 1994, reviewed in Kierulff et al. 2002

[1] Designations are based on reports from the field (direct observation of copulations and/or births unless otherwise noted). M, monogamy; PG, polygyny; PA, polyandry; PGA, polygynandry.
[2] PG based on short interbirth interval.
[3] PG based on presence of pregnant females.

1999). Marmosets are able to digest gums more efficiently than other callitrichines because their intestines have a comparatively enlarged and complex cecum (e.g., Ferrari and Martins 1992, Ferrari et al. 1993), allowing for relatively slow gut passage rates and microbial fermentation (Power and Oftedal 1996). Once digested, plant gums provide not only carbohydrates but also minerals (particularly calcium) and proteins (Garber 1984, Smith 2000).

Saguinus, Leontopithecus, and *Callimico* are opportunistic gummivores. Though exudates may provide as much as half of the diet during some periods (Porter 2001b), this is invariably a temporary phenomenon and exudates are never a dietary staple. *Saguinus* has been observed parasitizing gouge holes made by *Cebuella* (Soini 1988), but most exudate sources are either rare (broken branches) or seasonal (e.g., insect bore holes, *Parkia pendula* fruit pods; Garber 1993a). In addition, the digestive tracts of *Saguinus* and *Leontopithecus* are not specialized for gum consumption, decreasing their digestive efficiency (Ferrari and Martins 1992, Power and Oftedal 1996). One compensatory strategy

may be to concentrate gum feeding in the late afternoon, which allows longer retention time (overnight) and, presumably, better absorption of nutrients (Heymann and Smith 1999).

Fruits, Nectar, and Fungus

Fruit is the primary food item for many callitrichine species, especially among *Saguinus* and *Callimico*. Callitrichines typically exploit relatively small fruit patches, characteristic of the lower forest strata, and prefer disturbed and edge habitats (thus reducing competition with larger-bodied primates). Fruit can be a highly seasonal resource, especially in habitats of low diversity. Because *Saguinus* and *Leontopithecus* are unable to compensate systematically for fruit scarcity by gouging exudate sources, they are especially vulnerable to such seasonality. Small body size limits the potential for the exploitation of nonreproductive plant parts, as indicated by the lack of reports of folivory (Table 6.2). As a last resort under extreme seasonal fruit scarcity, *Saguinus* may turn to resources such as nectar

(Terborgh and Goldizen 1985). Although relatively nutritious, nectar is normally available in quantities too small to be harvested adequately by vertebrates as large as *Saguinus*.

Another potential alternative resource is fungus, which may contribute significantly to the diet of some species (*Callimico*, Porter 2001b; *C. aurita*, Corrêa et al. 2000) during certain parts of the year. It is interesting to note that, while fungi (including types of jelly and bamboo fungus) are a dietary staple of *Callimico* in some months, they are not exploited by sympatric tamarins (*S. fuscicollis* and *S. labiatus*), reinforcing the role of divergent diets in reducing competition in these species (Porter 2001b).

Availability of fruit tends to correlate with arthropod abundance, with both resources relatively scarce during the dry season. While seasonal fluctuations in resource abundance are faced by all primates, their effects may be relatively severe for callitrichines because of their small body size and high metabolic rate. Goldizen et al. (1988), for example, found that resource scarcity in the dry season resulted in potentially deleterious weight loss in *S. fuscicollis*. Thus, specialization for gum feeding among the marmoset genera may be the decisive factor permitting them to reproduce at 5- to 6-month intervals (Ferrari 1993, Ah-King and Tullberg 2000), in contrast with other callitrichines, which normally produce only one litter per year (but see Smith 2000).

Prey

Animal material, predominantly arthropods, is the third major component of callitrichine diets. Typical arthropod prey species are large-bodied and mobile and generally depend on camouflage as a predator-avoidance strategy. The characteristic callitrichine foraging behavior is the stealthy "scan and pounce" technique (Soini 1988, Ferrari 1993, Rylands and de Faria 1993, Porter 2001b). In addition, callitrichines will often pursue disturbed prey that has fallen to the forest floor (Rylands and de Faria 1993).

Extractive foraging is an important strategy for both *S. fuscicollis* and *Leontopithecus* species. In both cases, the animals spend a relatively large proportion of foraging time investigating concealed hiding places manually, rather than visually. For *S. fuscicollis*, the manipulative investigation of substrates, such as bark crevices, on vertical supports in the lower strata of the forest appears to be a key factor in niche separation with sympatric tamarin species (Terborgh 1983, Heymann and Buchanan-Smith 2000). Extractive foraging is also thought to contribute to more complex cognitive abilities (e.g., Gibson 1986, Day et al. 2003).

In addition to arthropods, vertebrates (e.g., nestlings, small lizards, and frogs) are included in callitrichine diets (Ferrari 1988, Digby and Barreto 1998, Smith 2000, Porter 2001b, Kierulff et al. 2002) (Table 6.2). All large prey, but especially vertebrates, are highly valued food items; and their capture almost invariably provokes solicitation from younger group members, resulting in a variety of social interactions ranging from passive food transfer to agonistic behavior (Digby and Barreto 1998).

REPRODUCTION

Reproductive Potential

Callitrichines are characterized by a collection of traits that results in a high reproductive potential for some females while at the same time restricting breeding opportunities for others. All species studied, with the exception of *Callimico goeldii* and possibly *Callibella humilis*, typically ovulate multiple ova and produce litters of two or more infants. In captivity, 50%–80% of litters comprise dizygotic twins, with the remainder comprising singletons, triplets, or, less frequently, quadruplets (Ziegler et al. 1990a, Baker and Woods 1992, Tardif et al. 2003, De Vleeschouwer et al. 2003). Only rarely, however, do more than two infants survive from a single litter. Ovulation number correlates with maternal body mass and thus may vary in response to maternal nutritional status (e.g., *C. jacchus*, Tardif et al. 2003).

Callitrichines typically ovulate and may conceive within 2–4 weeks after giving birth (Heistermann and Hodges 1995, French et al. 2002, Tardif et al. 2003). Unlike other primates, ovulation in these species is not inhibited by lactation, although postpartum ovulation may be delayed slightly in females nursing more than one infant (Ziegler et al. 1990b, Baker and Woods 1992) (Table 6.4). Combined with gestation lengths ranging from 125 days in *Leontopithecus* spp. to 184 days in *Saguinus oedipus* (Ziegler et al. 1987a, French et al. 2002) (Table 6.4), this pattern allows breeding females to produce litters at approximately 5- to 6-month intervals. Most species frequently produce two litters per year in captivity, and most genera (*Callithrix*, *Mico*, *Cebuella*, and *Callimico*) also do so in the wild (Stevenson and Rylands 1988, Soini 1993, Digby and Ferrari 1994, Porter 2001a). *Leontopithecus* and *Saguinus*, in contrast, typically breed only once per year in the wild, with most births clustered during the first half of the rainy season, corresponding with the period of maximal fruit availability (Snowdon and Soini 1988, Ferrari and Lopes Ferrari 1989) (Table 6.4).

The reproductive potential of callitrichines is further enhanced by rapid maturation in both males and females. Females are typically capable of ovulation and conception by 12–17 months of age (Table 6.4), while males produce sperm by 13–18 months and can sire infants by 15–25 months (Abbott and Hearn 1978, Epple and Katz 1980, Ginther et al. 2002). However, the onset of sexual maturity, especially in females, is often obscured by social suppression of reproductive function (see below). Ovarian cycles last from approximately 19 days in *L. rosalia* to 28 days in *C. jacchus* (Harlow et al. 1983, French and Stribley 1985) (Table 6.4). Like other platyrrhines, callitrichines do not menstruate and exhibit no conspicuous external signs of ovulation. Sexual behavior can occur throughout the ovarian cycle and pregnancy but is most common during the periovulatory period (Kendrick and Dixson 1983, Converse et al. 1995, Digby 1999, De Vleeschouwer et al. 2000a).

Table 6.4   Reproductive Parameters (Based on Captive Animals, Unless Otherwise Noted; Average ± Standard Error)

SPECIES	GESTATION LENGTH (DAYS)	INTERBIRTH INTERVAL (DAYS)	AGE AT FEMALE MATURITY (MONTHS)[1]	CYCLE LENGTH (DAYS)	PEAK BIRTH PERIOD	POSTPARTUM OVULATION[2]	REFERENCES
Callimico goeldii	151–152		Approx. 13	23.9 ± 0.4	Sept–Nov (Bolivia)	22–23	Dettling 2002, Dettling and Pryce 1999, Pook and Pook 1981
Callithrix jacchus	143–144	162	Approx. 13	28.6 ±1.0	Weakly bi-modal, most Oct–Feb	10–20	Abbott and Hearn 1978, Digby and Barreto 1993, Harlow et al. 1983, Lazaro-Perea et al. 2000, Tardif et al. 2003, Torii et al. 1987
C. kuhlii	143.1 ± 1.6	156.3 ± 2.9	Approx. 12–15	24.9 ± 0.6		13.6 ± 1.2	French et al. 1996, Smith et al. 1997
Cebuella pygmaea	141.9	212.7 ± 122.3	15–17	28.6 ± 4.1	May–Jun, Oct–Jan (NE Peru)	15.6 ± 4.1	Ziegler et al. 1990a, Carlson et al. 1997, Spurlock 2002
Saguinus fuscicollis	149.8 ± 2.4	185	13	25.7 ± 1.0	Nov–Feb (SE Peru)	17 ± 3.4	Epple and Katz 1980, Tardif et al. 1984, Goldizen et al. 1988, Heistermann and Hodges 1995, Kuederling and Heistermann 1997
S. oedipus	183.7 ± 1.1	240–267[3] 332.9 ± 53.6 (field)	15–17	23.6 ± 1.2	Mar–Jun (Colombia)	16.5 ± 1.6 to 30.8 ± 5.2[3]	French et al. 1983; Ziegler et al. 1987a,b, 1990b; Baker and Woods 1992; Savage et al. 1997
Leontopithecus rosalia	125.0	182–215[3] 311 ± 11.2 (field)	12–17	18.5 ± 0.3	Sept–Nov (SE Brazil)	Non-conceptive	French and Stribley 1985; French et al. 1989, 2002; Baker and Woods 1992; Dietz et al. 1994a; Monfort et al. 1996
L. chrysomelas	125.3 ± 3.0	160.6–257.8[3]	17	21.5 ± 2.5	Oct–Apr (E Brazil)	17.3 ± 3.5, non-conceptive	De Vleeschouwer et al. 2000a,b, 2003; Bach et al. 2001; French et al. 2002
L. chrysopygus		242		23.0 ± 2.0		Non-conceptive	Wormell and Price 2001, French et al. 2002

[1] Age at which females first undergo ovulatory cycles or reproductive hormone elevations, especially when housed in the absence of a dominant female.
[2] Days from parturition.
[3] Depending on lactation, litter size, and/or origin of dam (see text).

Reproduction can continue into old age; however, ovulatory cycles may become irregular or cease, and reproductive output may decline in the oldest females (Tardif and Ziegler 1992, Tardif et al. 2002). Thus, while callitrichines have the highest annual reproductive potential of any anthropoid primate, actual lifetime reproductive output is limited by their relatively short reproductive tenure, relatively high infant mortality rates, and suppression of reproduction in socially subordinate individuals (Tardif et al. 2003).

Mechanisms of Reproductive Suppression

One of the most striking features of callitrichine reproduction is the monopolization of breeding by a single, behaviorally dominant female in most social groups. Although similar breeding patterns are found throughout the callitrichine subfamily (French 1997), the underlying mechanisms, particularly the relative contributions of physiological suppression and behavioral inhibition, appear to differ across genera. In captive *Leontopithecus* and *Callimico*, for example, eldest daughters living with their natal families typically undergo ovulatory cycles indistinguishable from those of breeding females (Dettling and Pryce 1999, French et al.

2002), suggesting that inhibition of sexual behavior is the primary cause of reproductive failure. Among captive *Saguinus*, in contrast, adult daughters routinely fail to ovulate while living with their natal families (Epple and Katz 1984, Ziegler et al. 1987b, Kuederling et al. 1995). Captive *Callithrix* females appear to be intermediate, with up to 50% or more of eldest daughters ovulating while living with their natal families (Saltzman et al. 1997, Smith et al. 1997). Endocrine studies of free-living callitrichines, however, have not consistently supported findings from captivity. For example, data on wild *L. rosalia* indicate that periods of ovarian insufficiency occur in adult daughters living with their natal families (French et al. 2003), whereas both ovulatory cyclicity and pregnancy have been detected in wild *S. oedipus* daughters (Savage et al. 1997).

The physiological, sensory, and behavioral determinants of ovulation suppression have been investigated in several species. Circulating or excreted concentrations of cortisol, a stress-responsive hormone from the adrenal cortex, are similar in dominant and subordinate females or, in some cases, lower in subordinates (Saltzman et al. 1994, 1998; Ziegler et al. 1995; Smith and French 1997), suggesting that ovulation suppression cannot be attributed to generalized stress (Abbott et al. 1997). Instead, anovulation appears to result from a specialized neuroendocrine mechanism activated by specific social cues (Abbott et al. 1997). In *C. jacchus*, anovulation in subordinate females is mediated by suppression of luteinizing hormone (LH) secretion from the anterior pituitary, which is associated with enhanced negative feedback and diminished positive feedback effects of estrogen. Hypothalamic secretion of gonadotropin-releasing hormone (GnRH) does not appear to be altered by social subordination, suggesting that pituitary responsiveness to GnRH may be dampened (Abbott et al. 1997). Olfactory cues from dominant females have been implicated in the initiation and maintenance of ovulation suppression in several species but may play a redundant role with other cues (Epple and Katz 1984; Savage et al. 1988; Barrett et al. 1990; Abbott et al. 1993, 1998).

Reproductive failure in subordinate females can occur in response to either intrasexual (i.e., rank-related suppression) or intersexual (i.e., inbreeding avoidance) influences. The specific roles of these two factors differ among species. Captive *S. oedipus* females require cohabitation with an unrelated male in order to commence ovulatory cyclicity, even after removal from the natal family (Widowski et al. 1990, 1992). Among captive *C. jacchus* females, in contrast, ovulation suppression is determined by intrasexual dominance relationships: daughters living with their families frequently ovulate even in the absence of unrelated males but only if they are not behaviorally subordinate to another female (Saltzman et al. 2004). Nonetheless, *C. jacchus* daughters do not normally engage in sexual behavior unless they have access to an unrelated male (Saltzman 2003, Saltzman et al. 2004).

Male callitrichines, like females, engage in little or no intersexual copulatory behavior while living with their natal families. This appears to reflect inbreeding avoidance rather than intrasexual, rank-related suppression and is not generally associated with suppression of testosterone or LH concentrations (French et al. 1989, Baker et al. 1999, Ginther et al. 2001).

SOCIAL ORGANIZATION

Group Composition

Group composition in callitrichines varies from two to 20 individuals, with *Mico* and *Callithrix* species tending to have larger groups than *Saguinus*, *Callimico*, or *Leontopithecus* (Table 6.3). *Cebuella* tends to have the smallest groups, with most containing a single breeding pair and young (Soini 1982, 1988; de la Torre et al. 2000) (Table 6.3). Solitary animals have been noted in several populations, and male–female pairs appear to be rare and not always successful at raising young (Terborgh and Goldizen 1985, Goldizen 2003; see also Porter 2001a for *Callimico*). Differences in group size across species are

likely tied to increased recruitment rates through biannual births in marmosets and to the diet and habitat preferences of the different genera (Koenig 1995, Heymann 2000, Goldizen 2003; see Ecology above). Intriguing preliminary data on *Callibella* note that, while group size falls within the typical callitrichine range (six to eight individuals), aggregations of up to 30 individuals sometimes form (van Roosmalen and van Roosmalen 2003).

Although some early reports indicated that group membership could be dynamic, with individuals frequently moving in and out of groups (e.g., Dawson 1978, Neyman 1978, Scanlon et al. 1988), more recent studies based on direct follows of animals indicate more stable, extended family groups (Ferrari and Digby 1996, Goldizen et al. 1996, Nievergelt et al. 2000, Baker et al. 2002; but see also Garber et al. 1993). Typical of family groups, both sexes may emigrate as individuals mature and as groups grow in size (e.g., Goldizen et al. 1996, Porter et al. 2001, Baker et al. 2002). In *L. rosalia*, some 60% of individuals born into a group will have dispersed by 3 years of age and 90% by 4 years of age (Baker et al. 2002). Of those individuals that remain with their natal group, most become breeders by the time they are 4 years old (Baker et al. 2002). Similar patterns have been described for *S. fuscicollis* (Goldizen et al. 1996). Although both sexes may disperse, males have a higher probability than females of entering an established group (Ferrari and Diego 1992; Baker et al. 1993, 2002; Goldizen et al. 1996). Females may "float" as solitary individuals waiting for a breeding vacancy in an already established group, or they may form a new group with other recent emigrants (Lazaro-Perea et al. 2000, Baker et al. 2002). Breeding takeovers by individuals within their natal group are more likely to occur when unrelated mates are available (Ferrari and Diego 1992, Goldizen et al. 1996, Saltzman et al. 2004). The general pattern of extended family groups with occasional immigration (typically, but not always, into a breeding position) has been supported by studies of genetic relatedness within *C. jacchus* groups (Nievergelt et al. 2000), but periods of instability can result in groups of mixed parentage (Faulkes et al. 2003).

Social Relationships

Groups are typically cohesive, and individuals often rest in physical contact with one another and allogroom (Ferrari 1988, Alonso and Langguth 1989, Digby 1995b, Heymann 1996). Allogrooming has been described as asymmetrical in at least some wild groups, with females typically receiving more grooming than they perform and breeding individuals of both sexes being favored grooming partners (Goldizen 1989, Digby 1995b, Heymann 1996, Kostrub 2003, Lazaroa-Perea et al. 2004). Breeding females may be performing a "service" that entices nonbreeding females to remain in the group (Lazaro-Perea et al. 2004). There is no evidence that breeding females preferentially groom one potential sexual partner over another (Heymann 1996, Kostrub 2003).

Aggressive behavior in callitrichines is relatively rare (e.g., fewer than 0.1 acts/hr in *S. mystax*; Heymann 1996, Garber 1997) and mild, typically consisting of cuffs, "arch walks," piloerection, chases, avoidance, and submissive vocalizations; it occurs most often in feeding contexts (Goldizen 1989, Baker et al. 1993, Digby 1995b, Garber 1997, Kostrub 2003). Rates of aggression can increase following changes in group composition, for example, the loss of a breeding female (Lazaro-Perea et al. 2000). Experimental exposure to unfamiliar "intruders" or changes in membership of captive groups can also elicit physical attacks (reviewed in Anzenberger 1993, Caine 1993). The pattern of aggression suggests that it is used to limit access to mates and to control group membership (Anzenberger 1993, Lazaro-Perea et al. 2000, Baker et al. 2002). Where intersexual dominance has been described, either there is no clear pattern of male or female dominance (e.g., Digby 1995b, Kostrub 2003) or males are able to displace females at feeding sites (Baker and Dietz 1996). Intrasexual dominance can have profound implications for reproductive success in these species (Baker et al. 1993, Digby 1995a,b; Lazaro-Perea et al. 2000) and will be discussed in detail below.

Mating Systems

Callitrichine mating systems have been described as monogamous, polyandrous, polygynous, and polygynandrous, with variation occurring both within and between groups and populations (reviewed in Garber 1997, Baker et al. 2002, Goldizen 2003, Saltzman 2003) (Table 6.3) (note: we use these terms to describe patterns of copulation only). Such flexible mating strategies are linked to changes in both group composition and social relationships, and specific patterns appear to be more typical of some genera than others.

The presence of more than one reproductively active male (polyandry and polygynandry) has been described for a number of species but appears to be most prevalent in *Saguinus* and *Leontopithecus* (Table 6.3). In *L. rosalia*, "potentially polyandrous" groups (based on the presence of two or more potentially reproductive males) were noted in 46% of monthly censuses, but direct observations of the timing of copulatory behavior together with information on probable fertile periods indicated that most groups were "genetically" monogamous, including six of seven potentially polyandrous groups (Baker et al. 1993). Other species exhibit little or no competition over access to breeding females (e.g., Kostrub 2003, Schaffner and French 2004). Only a handful of polyandrous groups have been noted among *Mico* (*M. humeralifer*, Rylands 1986b) and *Callithrix* (*C. jacchus*, two males copulating with two females in a newly formed group, Lazaro-Perea et al. 2000, see also Schaffner and French 2004 for captive *C. kuhlii*). The limited data available for *Cebuella* (Soini 1988) and *Callimico* suggest that groups typically contain a single breeding male. Possible reasons for differences across genera may involve differential costs of infant care: species using larger home ranges and having smaller overall group sizes may benefit more from infant care shared by two "potential" fathers (Heymann 2000, Goldizen 2003).

Female reproductive strategies also vary across marmoset and tamarin species. Although breeding is typically restricted to a single female, the presence of multiple breeding females in some groups demonstrates that reproductive suppression is not absolute. It is notable that even when suppression is relaxed, breeding appears to be restricted to no more than two females, even in groups containing additional females of breeding age. Polygyny has been documented in multiple wild groups of *Callithrix*, *Callibella*, *Saguinus*, and *Leontopithecus* and inferred (based on a short interval between births) in *Cebuella* (Table 6.3). Additional groups have been documented to contain two pregnant/lactating females, although tenure in the group was unknown (i.e., a female may have immigrated into a group while pregnant) or one of the females failed to raise young (*S. mystax*, Garber et al. 1993, *S. oedipus*, Savage et al. 1996a; *Callimico goeldii*, Porter 2001a; plus additional groups of *S. fuscicollis*, Goldizen et al. 1996, and *L. rosalia*, Baker et al. 2002). Only in *Mico* have multiple breeding females not been reported. The presence of two breeding females and their resulting young would presumably increase average costs of infant care in terms of time spent carrying and food sharing (see below). Some species or populations may be better able to accommodate these costs because of larger group sizes, more steady food supply (e.g., exudates), and smaller home ranges that allow caretakers to reduce travel while maintaining contact with their group (Digby and Barreto 1996, Goldizen 2003).

A further complication in describing the mating system of callitrichines is the occurrence of extragroup copulations in some species. In *C. jacchus*, for example, both reproductive and nonreproductive males and females have been observed participating in extragroup copulations, often during or just after aggressive intergroup interactions (Digby 1999, Lazaro-Perea et al. 2000). Thus, even some groups described as "monogamous" based on within-group copulation patterns may show more complex patterns of potential and actual paternity.

Infant Care

The high cost of infant care (due to twinning, high infant/maternal weight ratios, and frequent overlap of lactation and pregnancy in some species) is thought to be a key determinant of many aspects of callitrichine behavior, most notably the high level of cooperative care of young (reviewed in Tardif et al. 1993, 2002). Captive studies have provided detailed information on the proximate mechanisms and energetics of infant carrying (e.g., Tardif et al. 1993, Tardif 1997, Nievergelt and Martin 1999) and have demonstrated several species differences in patterns of infant care, including latency to onset of alloparental behavior, degree of maternal involvement, and overall time spent carrying (Tardif et al. 1993).

In both *L. rosalia* and *Callimico*, mothers are typically the sole caretakers for up to the first 3 weeks of an infant's life (Schradin and Anzenberger 2001, Tardif et al. 2002, see also de Oliveira et al. 1999 for an exception in *L. chrysomelas*); but *Saguinus*, *Callithrix*, *Mico*, and *Cebuella* adult males and other group members may begin carrying the infant as early as the day of birth (reviewed in Tardif et al. 2002). The mother's social status can also influence the onset of allomaternal care, with subordinate breeding females avoiding potential helpers for the first 10 days postpartum (Digby 1995a). In some tamarin species, males may actually care for infants more than the mother (e.g., *S. oedipus*, Savage et al. 1996b; *S. fuscicollis*, Goldizen 1987a,b), whereas mothers act either as primary caretakers or as equal partners in *Callithrix*, *Mico*, and *Callimico* (*C. jacchus*, Digby 1995a, Yamamoto and Box 1997; *M. humeralifer*, Rylands 1986b; *Callimico*, Schradin and Anzenberger 2001). The extent of maternal care can also be influenced by group size and composition, litter size, and maternal weight (reviewed in Tardif et al. 1993, Bales et al. 2002). Both captive and field data confirm the general pattern that marmosets (e.g., *C. jacchus* and *M. argentata*) carry less frequently and for a shorter period of time than do tamarins (e.g., *S. fuscicollis* and *S. oedipus*), with *L. rosalia* being intermediate (Tardif et al. 1993, 2002; Savage et al. 1996b). Differences in carrying patterns are once again likely to be linked to daily path length, home range size, and ultimately differences in diet and habitat use (Tardif et al. 1993, Goldizen 2003, see above).

Several hypotheses have been proposed to explain why group members other than the parents (alloparents) participate in infant care in the callitrichines: (*1*) enhancing direct fitness by gaining experience in caretaking behaviors, (*2*) increasing inclusive fitness, (*3*) maintaining group membership while waiting for a breeding spot to become vacant, and (*4*) caretaking as a courtship strategy for breeding males (reviewed in Tardif 1997, Bales et al. 2000). While there is some evidence that caretaking experience and overall number of helpers increase infant survival, evidence for a connection between helping behavior and future reproductive opportunities (either inheritance of breeding position or as a courting strategy) is currently lacking (Tardif 1997, Bales et al. 2000).

CONSERVATION STATUS

The Callitrichinae include not only some of the world's most critically endangered species but also species such as *C. jacchus* that thrive under disturbed conditions and are unlikely to be threatened with extinction. A characteristic shared by all endangered callitrichine species is a relatively small geographic range combined with critical levels of anthropogenic habitat alteration. As a group, *Leontopithecus* faces the most serious threat, with two species listed as critically endangered (*L. caissara* and *L. chrysopygus*) and two listed as endangered (*L. rosalia* and *L. chrysomelas*,

Rylands and Chiarello 2003). Only four other species are allocated endangered status by the World Conservation Union (*S. oedipus*, *S. bicolor*, *C. flaviceps*, and *C. aurita*; IUCN Species Survival Commission 2003), but numerous species are classified as data deficient, including several of the newly described species of *Mico*. While the more endangered species of callitrichines receive official protection and are the subjects of conservation-oriented research projects (e.g., the Golden Lion Tamarin Project), their survival still depends on careful metapopulation management (e.g., chapters in Kleiman and Rylands 2002).

COMPETITION IN COOPERATIVELY BREEDING SPECIES

Cooperatively breeding species are characterized, in part, by their unusual propensity to share in the rearing of offspring. In this system, parents, older siblings, members of the extended family, and in some cases unrelated individuals participate in resource and territory defense, infant carrying, food sharing, babysitting, and even allonursing (chapters in Solomon and French 1997). Such behaviors, along with low rates of physical aggression, are typical of the callitrichines (Garber 1997, Schaffner and Caine 2000). The cooperative nature of some aspects of callitrichine behavior, however, belies the fact that typically only two (or sometimes three) individuals receive the majority of the benefits from such a "cooperative" system. Unlike communal breeding (Price and Evans 1991), in which there is shared parentage of the young, most callitrichines invest time and energy into rearing the offspring of other group members. Though helpers may gain indirect benefits in the form of inclusive fitness or caretaking experience (reviewed in Tardif 1997), direct benefits accrue primarily to those individuals that are able to breed. This reproductive skew is expected to give rise to intense, if sometimes subtle, reproductive competition.

Control or manipulation of breeding opportunities can occur prior to conception as well as after birth. Competition can be manifest in acquisition and maintenance of social status, physiological suppression of ovulation, interference in the feeding and care of young, and, in extreme cases, infanticide. Below, we outline some of the means by which callitrichines compete for reproductive opportunities and examine potential causes for variation in these mechanisms across genera.

Social Status and Reproduction

Within groups, social status can play a profound role in determining the reproductive opportunities and reproductive success of both males and females. Breeding individuals (both males and females) are typically those that are socially dominant over all others within the group (Baker et al. 1993, Dietz and Baker 1993, Digby 1995a,b, Goldizen et al. 1996).

The mechanisms by which females' social status results in differential reproductive success include both inhibition of sexual behavior and suppression of ovulation (see Mechanisms of Reproductive Suppression, above). There has been some debate on whether subordinate females' failure to reproduce results from direct manipulation by the dominant female (e.g., dominant control model; Snowdon 1996) or whether it is more appropriately interpreted as a type of self-inhibition on the part of the subordinate female in an attempt to reduce wasted reproductive effort (e.g., conceiving or giving birth to infants that are unlikely to survive—self-restraint model; Snowdon 1996; see also Wasser and Barash 1983). In either scenario, socially dominant females are able to maintain reproductive sovereignty in most cases and thus also gain the benefits of a relatively higher reproductive success. Even in groups where subordinate females are able to conceive, the dominant female will typically produce more infants and have higher infant survival rates (Digby 1995a, Goldizen et al. 1996). In *L. rosalia*, for example, dominant females had twice the reproductive success of subordinate breeding females (Dietz and Baker 1993); and in *C. jacchus*, dominant breeding females gave birth to twice as many infants and had nearly twice the overall survival rate (Digby 1995a).

For males in potentially polyandrous groups, dominance may determine the likelihood of paternity. As noted above, dominance plays little or no role in access to females during the periovulatory period in some species (*S. mystax*, Garber et al. 1993, *S. tripartitus*, Kostrub 2003, *S. fuscicollis*, Goldizen 1989, but note that some consort behavior has been reported). In contrast, Baker et al. (1993) were able to demonstrate that the more dominant of the two sexually active males in *L. rosalia* groups was responsible for 94% of sexual behavior during periods when the female was most likely to conceive. In one of these groups, direct aggression was used by a dominant male to prevent the subordinate male's access to the breeding female. Mate guarding by dominant males has also been observed in *C. jacchus* (Digby 1999) and *Cebuella* (Soini 1987).

In summary, both subordinate males and females may be subject to interference or inhibition of sexual behavior, and subordinate females may also undergo physiological suppression of reproduction. With both polyandrous and polygynous groups occurring in several genera, however, it is clear that these mechanisms are not always successful in maintaining reproductive sovereignty for the dominant male–female pair. When subordinates do breed, postpartum reproductive competition may play a role in determining which infants survive.

Infanticide

Perhaps the most extreme form of reproductive competition in callitrichines is the killing of infants by females other than the mother. Infanticide in primates is typically associated with the killing of infants by unrelated males that have recently joined or taken over a group (e.g., langurs, *Presbytis entellus*, Hrdy 1979; gorillas, *Gorilla gorilla beringei*,

Watts 1989). However, the sexual selection hypothesis (Hrdy 1979) put forth to explain these cases of infanticide is not applicable to most callitrichine incidents because the killing of dependent young will not necessarily bring females back into estrus more quickly (due to the limited influence of lactation on ovulation in these species). Instead, competition for access to alloparents or other resources in polygynous groups could lead to enforced neglect and/or killing of infants (Digby 1995a, 2000). Infanticide by females has now been reported in several free-ranging groups and populations of *C. jacchus* (Digby 1995a, Yamamoto et al. 1996, Roda and Mendes Pontes 1998, Lazaro-Perea et al. 2000), several captive groups of *C. jacchus* (reviewed by Saltzman 2003), and at least one captive group of *L. chrysomelas* (De Vleeschouwer et al. 2001). In five of the six cases where social status was known, the dominant female was observed or strongly suspected of killing the offspring born to the socially subordinate female (in the sixth case, the perpetrator became dominant following the infanticide) (Roda and Mendes Pontes 1998). In at least three cases, the perpetrator gave birth within days or weeks after the infanticide (Digby 1995a, Roda and Mendes Pontes 1998, Lazaro-Perea et al. 2000). In some cases, subordinate females that had lost their infants (either to infanticide or to unknown causes) subsequently carried and occasionally nursed infants born to the dominant female (Digby 1995a, Roda and Mendes Pontes 1998).

The harassment of subordinate females with young, the protective rearing strategies used by subordinate mothers (e.g., the extended period before alloparental care is tolerated), and the increased chance of infant loss when the births of dominant and subordinate females are closely spaced support the hypothesis that infanticide by females is a response to resource competition (Digby 1995a, Saltzman 2003). The intensity of competition between females is likely to vary across callitrichine genera due in part to such factors as typical group size (the smaller groups of *Saguinus* and *Cebuella* would result in less severe reproductive skew within the population), population density (higher densities will likely result in fewer reproductive vacancies within the population), and diet (with less seasonally influenced foods such as gum allowing for smaller home range sizes and, thus, lower infant care costs). Specifically, the marmosets' ability to maintain large groups in high-density areas may allow for a higher proportion of polygynous groups in these species. Once reproductive suppression is relaxed, however, infanticide by females may provide an alternative strategy by which dominant females may maintain their reproductive sovereignty (Digby 2000, Hager and Johnstone 2004).

Infanticide by females other than the mother is unusual among primate species but has been well documented in several other cooperatively breeding species (e.g., wild dogs, *Lycaon pictus*, Frame et al. 1979, black-tailed prairie dogs, *Cynomys ludovicianus*, Hoogland 1985, meerkats, *Suricata suricatta*, Clutton-Brock et al. 1998). In addition to the reproductive state of the perpetrator (either in the late stages of pregnancy or lactating), the other striking pattern among these species is that the perpetrator is likely to be closely related to the victim. Given the potential loss of inclusive fitness in these cases, we can only assume that the benefits to the infanticidal female must be greater than the costs, suggesting particularly intense reproductive competition. This hypothesis remains to be tested.

CONCLUSION

Although social tolerance and cooperation are important aspects of callitrichine social organization, it is important to consider the reproductive skew that often results from this type of social system. Only a portion of the population directly benefits from the behavior of helpers, who may delay their own reproduction for many years or forfeit it altogether (e.g., Goldizen et al. 1996). As a result, intense competition over the limited number of reproductive positions may occur. This competition may be subtle (suppression of ovulation and inhibition of sexual behavior) or overt (infanticide), but it nonetheless has a profound impact on the reproductive success of these animals. Ultimately, our understanding of callitrichine social organization and reproductive strategies will need to balance the cooperative aspects of their social interactions with the inevitable reproductive competition inherent in all social systems.

REFERENCES

Abbott, D. H., Barrett, J., and George, L. M. (1993). Comparative aspects of the social suppression of reproduction in female marmosets and tamarins. In: Rylands, A. B. (ed.), *Marmosets and Tamarins: Systematics, Behaviour, and Ecology*. Oxford University Press, Oxford. pp. 152–163.

Abbott, D. H., and Hearn, J. P. (1978). Physical, hormonal and behavioural aspects of sexual development in the marmoset monkey, *Callithrix jacchus. J. Reprod. Fertil.* 53:155–166.

Abbott, D. H., Saltzman, W., Schultz-Darken, N. J., and Smith, T. E. (1997). Specific neuroendocrine mechanisms not involving generalized stress mediate social regulation of female reproduction in cooperatively breeding marmoset monkeys. *Ann. N.Y. Acad. Sci.* 807:219–238.

Abbott, D. H., Saltzman, W., Schultz-Darken, N. J., and Tannenbaum, P. L. (1998). Adaptations to subordinate status in female marmoset monkeys. *Comp. Biochem. Physiol.* 119:261–274.

Aguilar, J. M., and Lacher, T. E., Jr. (2003). On the morphological distinctiveness of *Callithrix humilis* van Roosmalen et al., 1998. *Neotrop. Primates* 11:11–17.

Ah-King, M., and Tullberg, B. S. (2000). Phylogenetic analysis of twinning in Callitrichinae. *Am. J. Primatol.* 51:135–146.

Albernaz, A. L., and Magnusson, W. E. (1999). Home-range size of the bare-ear marmoset (*Callithrix argentata*) at Alter do Chão, Central Amazonia, Brazil. *Int. J. Primatol.* 20:665–677.

Alonso, C., and Langguth, A. (1989). Ecologia e comportamento de *Callithrix jacchus* (Primates: Callithrichidae) numa ilha de foresta Atlântica. *Rev. Nordestina Biol.* 6:107–137.

Anzenberger, G. (1993). Social conflict in two monogamous New World primates: pairs and rivals. In: Mason, W. A., and Mendoza, S. P. (eds.), *Primate Social Conflict*. State University of New York Press, New York. pp. 291–329.

Bach, A., Raboy, B., and Dietz, J. (2001). Birth seasonality in wild golden-headed lion tamarins (*Leontopithecus chrysomelas*) in Una Reserve, Bahia state, Brazil. *Am. J. Primatol.* 54:69.

Baker, A. J., Bales, K., and Dietz, J. M. (2002). Mating system and group dynamics in lion tamarins. In: Kleiman, D. G., and Rylands, A. B. (eds.), *Lion Tamarins: Biology and Conservation*. Smithsonian Institution Press, Washington DC. pp. 188–212.

Baker, A. J., and Dietz, J. M. (1996). Immigration in wild groups of golden lion tamarins (*Leontopithecus rosalia*). *Am. J. Primatol.* 38:47–56.

Baker, A. J., Dietz, J. M., and Kleiman, D. G. (1993). Behavioural evidence for monopolization of paternity in multi-male groups of golden lion tamarins. *Anim. Behav.* 46:1091–1103.

Baker, A. J., and Woods, F. (1992). Reproduction of the emperor tamarin (*Saguinus imperator*) in captivity, with comparisons to cotton-top and golden lion tamarins. *Am. J. Primatol.* 26:1–10.

Baker, J. V., Abbott, D. H., and Saltzman, W. (1999). Social determinants of reproductive failure in male common marmosets housed with their natal family. *Anim. Behav.* 58:501–513.

Bales, K., French, J. A., and Dietz, J. M. (2002). Explaining variation in maternal care in a cooperatively breeding mammal. *Anim. Behav.* 63:453–461.

Bales, K. L., Dietz, J. M., Baker, A. J., Miller, K., and Tardif, S. D. (2000). Effects of allocare-givers on fitness of infants and parents in callitrichid primates. *Folia Primatol.* 71:27–38.

Barreto, C. E. (1996). Comportamento de fêmeas reproductivas em grupos poligínicos de *Callithrix jacchus* (Primates: Callithrichidae) no ambiente natural: perfil das interaçãos afiliativas, agonísticas e da marcação de cheiro [masters thesis]. Universidade Federal do Rio Grande do Norte.

Barrett, J., Abbott, D. H., and George, L. M. (1990). Extension of reproductive suppression by pheromonal cues in subordinate female marmoset monkeys, *Callithrix jacchus. J. Reprod. Fertil.* 90:411–418.

Barroso, C. M. L., Schneider, H., Schneider, M. P. C., Sampaio, I., Garada, M. L., Czelusniak, J., and Goodman, M. (1997). Update of the phylogenetic systematics of New World monkeys: further DNA evidence for placing the pygmy marmoset (*Cebuella*) within the genus *Callithrix. Int. J. Primatol.* 18:651–674.

Bicca-Marques, J. C., and Garber P. A. (2001). Cognitive ecology and within-patch foraging decisions in tamarins. *Am. J. Primatol.* 32(suppl.):40.

Buchanan-Smith, H. M., Hardie, S. M., Caceres, C., and Prescott, M. J. (2000). Distribution and forest utilization of *Saguinus* and other primates of the Pando department, northern Bolivia. *Int. J. Primatol.* 21:353–379.

Cabrera, A. (1958). Catálogo de los mamíferos de America del Sur. *Revista del Museo Argentino de Ciencias Naturales "Bernardino Rivadavia"* 26:1–307.

Caine, N. G. (1993). Flexibility and co-operation as unifying themes in *Saguinus* social organization and behaviour: the role of predation pressures. In: Rylands, A. B. (ed.), *Marmosets and*

Tamarins: Systematics, Behaviour, and Ecology. Oxford University Press, Oxford. pp. 200–219.

Caine, N. G., Surridge, A. K., and Mundy, N. I. (2003). Dichromatic and trichromatic Callithrix geoffroyi differ in relative foraging ability for red–green color camouflaged and non-camouflaged food. Int. J. Primatol. 24:1163–1174.

Canavez, F. C., Moreira, M. A. M., Simon, F., Parham, P., and Seuánez, H. N. (1999). Phylogenetic relationships of the Callitrichinae (Platyrrhini, Primates) based on beta2-microglobin DNA sequences. Am. J. Primatol. 48:225–236.

Carlson, A. A., Ziegler, T. E., and Snowdon, C. T. (1997). Ovarian function of pygmy marmoset daughters (Cebuella pygmaea) in intact and motherless families. Am. J. Primatol. 43:347–355.

Castro, C. (2000). Ecology and behavior of the common marmoset, Callithrix jacchus. Neotrop. Primates 8:50–51.

Chiarello, A. G., and de Melo, F. R. (2001). Primate population densities and sizes in Atlantic forest remnants of northern Espirito Santo, Brazil. Int. J. Primatol. 22:379–396.

Christen, A. (1999). Survey of Goeldi's monkeys (Callimico goeldii) in northern Bolivia. Folia Primatol. 70:107–111.

Clutton-Brock, T. H., Brotherton, P. N. M., Smith, R., McIlrath, G. M., Kansky, R., Gaynor, D., O'Riain, M. J., and Skinner, J. D. (1998). Infanticide and expulsion of females in a cooperative mammal. Proc. R. Soc. Lond. B Biol. Sci. 265:2291–2295.

Converse, L. J., Carlson, A. A., Ziegler, T. E., and Snowdon, C. T. (1995). Communication of ovulatory state to mates by female pygmy marmosets, Cebuella pygmaea. Anim. Behav. 49:615–621.

Corrêa, H. K. M., Coutinho, P. E. G., and Ferrari, S. F. (2000). Between-year differences in the feeding ecology of highland marmosets (Callithrix aurita and Callithrix flaviceps) in southeastern Brazil. J. Zool. 252:421–427.

Corrêa, H. K. M., Coutinho, P. E. G., and Ferrari, S. F. (2002). Dieta de grupos de Mico argentatus em fragmentos naturais de Alter do Chão, Santarém, Pará. Livro de Resumos do X Congresso Brasileiro de Primatologia, p. 46.

Coutinho, P. E. G., and Corrêa, H. K. M. (1995). Polygyny in a free-raning group of buffy-tufted-ear marmosets, Callithrix aurita. Folia Primatol. 65:25–29.

Dawson, G. A. (1978). Composition and stability of social groups of the tamarin, Saguinus oedipus geoffroyi, in Panama: ecological and behavioral implications. In: Kleiman, D. G. (ed.), The Biology and Conservation of the Callitrichidae. Smithsonian Institution Press, Washington DC. pp. 23–37.

Day, R. L., Coe, R. L., Kendal, J. R., and Laland, K. N. (2003). Neophilia, innovation and social learning: a study of intergeneric differences in callitrichid monkeys. Anim. Behav. 65:559–571.

Day, R. T., and Elwood, R. W. (1999). Sleeping site selection by the golden-handed tamarin Saguinus midas midas: the role of predation risk, proximity to feeding sites, and territorial defense. Ethology 105:1035–1051.

de Faria, D. S. (1986). Tamanho, composição de um grupo social e área de vivência (home-range) do sagüi Callithrix jacchus penicillata na mata ciliar do córrego capetinga, Brasília, DF. In: Thiago de Mello, M. (ed.), A Primotologia no Brasil, vol. 2. pp. 87–105.

de la Torre, S., Snowdon, C. T., and Bejarana, M. (2000). Effects of human activities on wild pygmy marmosets in Ecuadorian Amazonia. Biol. Conserv. 94:153–163.

Dettling, A., and Pryce, C. R. (1999). Hormonal monitoring of age at sexual maturation in female Goeldi's monkeys (Callimico goeldii) in their family groups. Am. J. Primatol. 48:77–83.

Dettling, A. C. (2002). Reproduction and development in Goeldi's monkey (Callimico goeldii). Evol. Anthropol. 1(suppl.):207–210.

de Oliveira, M. S., Lopes, F. A., Alonso, C., and Yamamoto, M. E. (1999). The mother's participation in infant carrying in captive groups of Leontopithecus chrysomelas and Callithrix jacchus. Folia Primatol. 70:146–153.

De Vleeschouwer, K., Heistermann, M., Van Elsacker, L., and Verheyen, R. F. (2000a). Signaling of reproductive status in captive female golden-headed lion tamarins (Leontopithecus chrysomelas). Int. J. Primatol. 21:445–465.

De Vleeschouwer, K., Leus, K. K., and Van Elsacker, L. (2003). Characteristics of reproductive biology and proximate factors regulating seasonal breeding in captive golden-headed lion tamarins (Leontopithecus chrysomelas). Am. J. Primatol. 60:123–137.

De Vleeschouwer, K., Van Elsacker, L., Heistermann, M., and Leus, K. (2000b). An evaluation of the suitability of contraceptive methods in golden-headed lion tamarins (Leontopithecus chrysomelas), with emphasis on melengestrol acetate (MGA) implants: (II) endocrinological and behavioural effects. Anim. Welf. 9:385–401.

De Vleeschouwer, K., Van Elsacker, L., and Leus, K. (2001). Multiple breeding females in captive groups of golden-headed lion tarmarins (Leontopithecus chrysomelas): causes and consequences. Folia Primatol. 72:1–10.

Dietz, J. M., and Baker, A. J. (1993). Polygyny and female reproductive success in golden lion tamarins, Leontopithecus rosalia. Anim. Behav. 46:1067–1078.

Dietz, J. M., Baker, A. J., and Miglioretti, D. (1994a). Seasonal variation in reproduction, juvenile growth, and adult body mass in golden lion tamarins (Leontopithecus rosalia). Am. J. Primatol. 34:115–132.

Dietz, J. M., de Sousa, N. F., and da Silva, J. R. (1994b). Population structure and territory size in golden-headed lion tamarins, Leontopithecus chrysomelas. Neotrop. Primates 2(suppl.):21–23.

Dietz, J. M., Pere, C. A., and Pinder, L. (1997). Foraging ecology and use of space in wild golden lion tamarins (Leontopithecus rosalia). Am. J. Primatol. 41:289–305.

Digby, L. (1992). Intruders in the wild: intergroup encounters in a natural population of Callithrix jacchus. Abstracts of the XIVth Congress of the International Primatological Society Strasbourg, France, p. 293.

Digby, L. J. (1995a). Infant care, infanticide, and female reproductive strategies in polygynous groups of common marmosets (Callithrix jacchus). Behav. Ecol. Sociobiol. 37:51–61.

Digby, L. J. (1995b). Social organization in a wild population of Callithrix jacchus. Part II: Intragroup social behavior. Primates 36:361–375.

Digby, L. J. (1999). Sexual behavior and extra-group copulations in a wild population of common marmosets (Callithrix jacchus). Folia Primatol. 70:136–145.

Digby, L. J. (2000). Infanticide by female mammals: implications for the evolution of social systems. In: van Schaik, C. P., and Janson, C. H. (eds.), Infanticide by Males and Its Implications. Cambridge University Press, Cambridge. pp. 423–446.

Digby, L. J., and Barreto, C. E. (1993). Social organization in a wild population of *Callithrix jacchus*. Part I: Group composition and dynamics. *Folia Primatol.* 61:123–134.

Digby, L. J., and Barreto, C. E. (1996). Activity and ranging patterns in common marmosets (*Callithrix jacchus*): implications for reproductive strategies. In: Norconk, M., Rosenberger, A., and Garber, P. A. (eds.), *Adaptive Radiations of Neotropical Primates*, Plenum Press, New York, pp. 173–185.

Digby, L. J., and Barreto, C. E. (1998). Vertebrate predation in common marmosets. *Neotrop. Primates* 6:124–126.

Digby, L. J., and Ferrari, S. F. (1994). Multiple breeding females in free-ranging groups of *Callithrix jacchus*. *Int. J. Primatol.* 15:389–397.

Eisenberg, J. F., and Redford, K. H. (1999). *Mammals of the Neotropics: The Central Neotropics. Ecuador, Peru, Bolivia, Brazil*, vol. 3. University of Chicago Press, Chicago.

Epple, G., and Katz, Y. (1980). Social influences on first reproductive success and related behaviors in the saddle-back tamarin (*Saguinus fuscicollis*, Callitrichidae). *Int. J. Primatol.* 1:171–183.

Epple, G., and Katz, Y. (1984). Social influences on estrogen excretion and ovarian cyclicity in saddle back tamarins (*Saguinus fuscicollis*). *Am. J. Primatol.* 6:215–227.

Faulkes, C. G., Arruda, M. F., and Monteiro da Da Cruz, A. O. M. (2003). Matrilineal genetic structure within and among populations of the cooperatively breeding common marmoset, *Callithrix jacchus*. *Mol. Ecol.* 12:1101–1108.

Ferrari, S. F. (1988). The behaviour and ecology of the buffy-headed marmoset, *Callithrix flaviceps* (O. Thomas, 1903) [PhD thesis]. University College, London.

Ferrari, S. F. (1993). Ecological differentiation in the Callitrichidae. In: Rylands, A. B. (ed.), *Marmosets and Tamarins: Systematics, Ecology and Behaviour*. Oxford University Press, Oxford. pp. 314–328.

Ferrari, S. F., Corrêa, H. K. M., and Coutinho, P. E. G. (1996). Ecology of the "southern" marmosets (*Callithrix aurita* and *Callithrix flaviceps*). In: Norconk, M. A., Rosenberger, A. L., and Garber, P. A. (eds.), *Adaptive Radiations of Neotropical Primates*. Plenum Press, New York. pp. 157–171.

Ferrari, S. F., and Diego, V. H. (1992). Long-term changes in a wild marmoset group. *Folia Primatol.* 58:215–218.

Ferrari, S. F., and Digby, L. J. (1996). Wild *Callithrix* groups: stable extended families? *Am. J. Primatol.* 38:19–27.

Ferrari, S. F., Iwanaga, S., Ramos, E. M., Messias, M. R., Ramos, P. C. S., and Cruz Neto, E. H. (1999). Expansion of the known distribution of Goeldi's monkey (*Callimico goeldii*) in southwestern Brazilian Amazonia. *Folia Primatol.* 70:112–116.

Ferrari, S. F., Lopes, M. A., and Krause, E. A. K. (1993). Gut morphology of *Callithrix nigriceps* and *Saguinus labiatus* from western Brazilian Amazonia. *Am. J. Phys. Anthropol.* 90:487–493.

Ferrari, S. F., and Lopes Ferrari, M. A. (1989). A re-evaluation of the social organization of the Callitrichidae, with references to the ecological differences between genera. *Folia Primatol.* 52:132–147.

Ferrari, S. F., and Lopes Ferrari, M. A. (1990). A survey of primates in central Pará. *Bol. Mus. Para. Emílio Goeldi Série Zool.* 6:169–179.

Ferrari, S. F., and Martins, E. S. (1992). Gummivory and gut morphology in two sympatric callitrichids (*Callithrix emiliae*

and *Saguinus fuscicollis weddelli*) from western Brazilian Amazonia. *Am. J. Phys. Anthropol.* 88:97–103.

Frame, L. H., Malcolm, J. R., Frame, G. W., and van Lawick, H. (1979). Social organization of African wild dogs (*Lycaon pictus*) on the Serengeti Plains, Tanzania 1967–78. *Z. Tierpsychol.* 50:225–249.

French, J. A. (1997). Proximate regulation of singular breeding in Callitrichid primates. In: Solomon, N. G., and French, J. A. (eds.), *Cooperative Breeding in Mammals*. Cambridge University Press, Cambridge. pp. 34–75.

French, J. A., and Stribley, J. A. (1985). Patterns of urinary oestrogen excretion in female golden lion tamarins (*Leontopithecus rosalia*). *J. Reprod. Fertil.* 75:537–546.

French, J. A., Abbott, D. H., Scheffler, G., Robinson, J. A., and Goy, R. W. (1983). Cyclic excretion of urinary oestrogens in female tamarins (*Saguinus oedipus*). *J. Reprod. Fertil.* 68:177–184.

French, J. A., Bales, K. L., Baker, A. J., and Dietz, J. M. (2003). Endocrine monitoring of wild dominant and subordinate female *Leontopithecus rosalia*. *Int. J. Primatol.* 24:1281–1300.

French, J. A., Brewer, K. J., Schaffner, C. M., Schalley, J., Hightower-Merritt, D. L., Smith, T. E., and Bell, S. M. (1996). Urinary steroid and gonadotropin excretion across the reproductive cycle in female Wied's black tufted-ear marmosets (*Callithrix kuhli*). *Am. J. Primatol.* 40:231–245.

French, J. A., De Vleeschouwer, K., Bales, K., and Heistermann, M. (2002). Lion tamarin reproductive biology. In: Kleiman, D. G., and Rylands, A. B. (eds.), *Lion Tamarins: Biology and Conservation*. Smithsonian Institution Press, Washington DC. pp. 133–156

French, J. A., Inglett, B. J., and Dethlefs, T. M. (1989). The reproductive status of nonbreeding group memebers in captive golden lion tamarin social groups. *Am. J. Primatol.* 18:73–86.

Garber, P. A. (1984). Proposed nutritional importance of plant exudates in the diet of the Panamanian tamarin, *Saguinus Oedipus geoffroyi*. *Int. J. Primatol.* 5:1–15.

Garber, P. A. (1988). Diet, foraging patterns, and resource defense in a mixed species troop of *Saguinus mystax* and *Saguinus fuscicollis* in Amazonian Peru. *Behaviour* 105:18–34.

Garber, P. A. (1993a). Feeding ecology and behaviour of the genus *Saguinus*. In: Rylands, A. B. (ed.), *Marmosets and Tamarins: Systematics, Behaviour, and Ecology*. Oxford University Press, Oxford. pp. 273–295.

Garber, P. A. (1993b). Seasonal patterns of diet and ranging in two species of tamarin monkeys: stability versus variability. *Int. J. Primatol.* 14:145–166.

Garber, P. A. (1997). One for all and breeding for one: cooperation and competition as a tamarin reproductive strategy. *Evol. Anthropol.* 5:187–222.

Garber, P. A., Encarnacion, F., Moya, L., and Pruetz, J. D. (1993). Demographic and reproductive patterns in moustached tamarin monkeys (*Saguinus mystax*): implications for reconstructing platyrrhine mating systems. *Am. J. Primatol.* 29:235–254.

Garber, P. A., and Leigh, S. R. (2001). Patterns of positional behavior in mixed species groups of *Callimico goeldii, Saguinus labiatus,* and *Saguinus fuscicollis* in northwestern Brazil. *Am. J. Primatol.* 55:17–31.

Genoud, M., Martin, R. D., and Glaser, D. (1997). Rate of metabolism in the smallest simian primate, the pygmy marmoset (*Cebuella pygmaea*). *Am. J. Primatol.* 41:229–245.

Gibson, K. R. (1986). Cognition, brain size, and the extraction of embedded food resources. In: Else, J. G., and Lee, P. C. (eds.), *Primate Ontogeny, Cognition, and Social Behavior*. Cambridge University Press, Cambridge. pp. 93–103.

Ginther, A. J., Carlson, A. A., Ziegler, T. E., and Snowdon, C. T. (2002). Neonatal and pubertal development in males of a cooperatively breeding primate, the cotton-top tamarin (*Saguinus oedipus oedipus*). *Biol. Reprod.* 66:282–290.

Ginther, A. J., Ziegler, T. E., and Snowdon, C. T. (2001). Reproductive biology of captive male cottontop tamarin monkeys as a function of social environment. *Anim. Behav.* 61:65–78.

Goldizen, A. W. (1987a). Tamarins and marmosets: communal care of offspring. In: Smuts, B. B., Cheney, D. L., Seyfarth, R. M., Wrangham, R. W., and Struhsaker, T. T. (eds.), *Primate Societies*. University of Chicago Press, Chicago. pp. 34–43.

Goldizen, A. W. (1987b). Facultative polyandry and the role of infant-carrying in wild saddle-back tamarins (*Saguinus fuscicollis*). *Behav. Ecol. Sociobiol.* 20:99–109.

Goldizen, A. W. (1989). Social relationships in a cooperatively polyandrous group of tamarins (*Saguinus fuscicollis*). *Behav. Ecol. Sociobiol.* 24:79–89.

Goldizen, A. W. (2003). Social monogamy and its variations in callitrichids: do these relate to the costs of infant care? In: Reichard, U. H., and Boesch, C. (eds.), *Monogamy: Mating Strategies and Partnerships in Birds, Humans, and Other Mammals*. Cambridge University Press, Cambridge. pp. 232–247.

Goldizen, A. W., Mendelson, J., van Vlaardingen, M., and Terborgh, J. (1996). Saddle-back tamarin (*Saguinus fuscicollis*) reproductive strategies: evidence from a thirteen-year study of a marked population. *Am. J. Primatol.* 38:57–84.

Goldizen, A. W., Terborgh, J., Cornejo, F., Porras, D. T., and Evans, R. (1988). Seasonal food shortage, weight-loss and the timing of births in saddle-back tamarins (*Saguinus fuscicollis*). *J. Anim. Ecol.* 57:893–901.

Gonçalves, E. C., Ferrari, S. F., Silva, A., Coutinho, P. E. G., Menezes, E. V., and Schneider, M. P. C. (2003). Effects of habitat fragmentation on the genetic variability of silvery marmosets, *Mico argentatus*. In: Marsh, L. K. (ed.), *Primates in Fragments*. Kluwer Academic, New York. pp. 17–28.

Groves, C. P. (2001). *Primate Taxonomy*. Smithsonian Institution Press, Washington DC.

Guimarães, A. (1998). Ecology and social behavior of buffy-headed marmosets, *Callithrix flaviceps*. *Neotrop. Primates* 6:51–52.

Hager, R., and Johnstone, R. A. (2004). Infanticide and control of reproduction in cooperative and communal breeders. *Anim. Behav.* 67:941–949.

Harlow, C. R., Gems, S., Hodges, J. K., and Hearn, J. P. (1983). The relationship between plasma progesterone and the timing of ovulation and early embryonic development in the marmoset monkey (Callitrichidae). *J. Zool.* 201:273–282.

Heistermann, M., and Hodges, J. K. (1995). Endocrine monitoring of the ovarian cycle and pregnancy in the saddle-back tamarin (*Saguinus fuscicollis*) by measurement of steroid conjugates in urine. *Am. J. Primatol.* 35:117–127.

Hershkovitz, P. (1977). *Living New World Monkeys (Platyrrhini), with an Introduction to Primates*. vol. 1. Chicago University Press, Chicago.

Heymann, E. W. (1996). Social behavior of wild moustached tamarins, *Saguinus mystax*, at the Estación Biológica Quebrada Blacno, Peruvian Amazonia. *Am. J. Primatol.* 38:101–113.

Heymann, E. W. (2000). The number of adult males in callitrichine groups and its implications for callitrichine social evolution. In: Kappeler, P. M. (ed.), *Primate Males: Causes and Consequences of Variation in Group Composition*. Cambridge University Press, Cambridge. pp. 64–71.

Heymann, E. W. (2001). Interspecific variation of scent-marking behaviour in wild tamarins, *Saguinus mystax* and *Saguinus fuscicollis*. *Folia Primatol.* 72:253–267.

Heymann, E. W., and Buchanan-Smith, H. (2000). The behavioral ecology of mixed-species troops of callitrichine primates. *Biol. Rev.* 75:169–190.

Heymann, E. W., Encarnación, F., and Canaquin, J. E. (2002). Primates of the Río Curaray, northern Peruvian Amazon. *Int. J. Primatol.* 23:191–201.

Heymann, E. W., and Smith, A. C. (1999). When to feed on gums: temporal patterns of gummivory in wild tamarins, *Saguinus mystax* and *Saguinus fuscicollis* (Callitrichinae). *Zoo Biol.* 18:459–471.

Heymann, E. W., and Soini, P. (1999). Offspring number in pygmy marmosets, *Cebuella pygmaea*, in relation to group size and number of adult males. *Behav. Ecol. Sociobiol.* 46:400–404.

Hill, W. C. O. (1957). *Primates, Comparative Anatomy and Taxonomy. Hapalidae*, vol. III. University of Edinburgh Press, Edinburgh.

Hoogland, J. L. (1985). Infanticide in prairie-dogs: lactating females kill offspring of close kin. *Science* 230:1037–1040.

Hourani, P. E., Vinyard, C. J., and Lemelin, P. (2003). Forelimb forces during gouging and other behaviors on vertical substrates in common marmosets. *Am. J. Phys. Anthropol.* 36(suppl.):118.

Hrdy, S. B. (1979). Infanticide among animals—review, classification, and examination of the implications for the reproductive strategies of females. *Ethol. Sociobiol.* 1:13–40.

Hubrecht, R. C. (1985). Home range size and use and territorial behavior in the common marmoset, *Callithrix jacchus jacchus*, at the Tapacura Field Station, Brazil. *Int. J. Primatol.* 6:533–550.

InfoNatura (2004). *Birds, Mammals, and Amphibians of Latin America*, version 3.2. Arlington, Virginia, NatureServe, http://www.natureserve.org/infonatura (accessed July 1, 2004).

IUCN Species Survial Commission (2003). *IUCN Red List*. IUCN, Gland, Switzerland, http://www.redlist.org/.

Jaquish, C. E., Toal, R. L., Tardif, S. D., and Carson, R. L. (1995). Use of ultrasound to monitor prenatal growth and development in the common marmoset (*Callithrix jacchus*). *Am. J. Primatol.* 36:259–275.

Kendrick, K. M., and Dixson, A. F. (1983). The effect of the ovarian cycle on the sexual behaviour of the common marmoset (*Callithrix jacchus*). *Physiol. Behav.* 30:735–742.

Kierulff, M. C. M. (2000). Ecology and behaviour of translocated groups of golden lion tamarins (*Leontopithecus rosalia*) [PhD diss.]. University of Cambridge, Cambridge.

Kierulff, M. C. M., Raboy, B. E., Procópio de Oliveira, P., Miller, K., Passos, F. C., and Prado, F. (2002). Behavioral ecology of lion tamarins. In: Kleiman, D. G., and Rylands, A. B. (eds.), *Lion Tamarins: Biology and Conservation*. Smithsonian Institution Press, Washington DC. pp. 157–187.

Kleiman, D. G., and Rylands, A. B. (2002). *Lion Tamarins: Biology and Conservation*. Smithsonian Institution Press, Washington DC.

Knogge, C., and Heymann, E. W. (2003). Seed dispersal by sympatric tamarins, *Saguinus mystax* and *Saguinus fuscicollis*: diversity and characteristics of plant species. *Folia Primatol.* 74:33–47.

Koenig, A. (1995). Group size, composition, and reproductive success in wild common marmosets (*Callithrix jacchus*). *Am. J. Primatol.* 35:311–317.

Kostrub, C. E. (2003). The social organization and behavior of golden-mantled tamarins, *Saguinus tripartitus*, in eastern Ecuador [PhD diss.]. University of California, Davis.

Kuederling, I., Evans, C. S., Abbott, D. H., Pryce, C. R., and Epple, G. (1995). Differential excretion of urinary oestrogen by breeding females and daughters in the red-bellied tamarin (*Saguinus labiatus*). *Folia Primatol.* 64:140–145.

Kuederling, I., and Heistermann, M. (1997). Ultrasonographic and hormonal monitoring of pregnancy in the saddle back tamarin, *Saguinus fuscicollis*. *J. Med. Primatol.* 26:299–306.

Lazaro-Perea, C. (2001). Intergroup interactions in wild common marmosets, *Callithrix jacchus*: territorial defense and assessment of neighbors. *Anim. Behav.* 62:11–21.

Lazaro-Perea, C., Arruda, M. F., and Snowdon, C. T. (2004). Grooming as a reward? Social function of grooming between females in cooperatively breeding marmosets. *Anim. Behav.* 67:627–636.

Lazaro-Perea, C., Castro, C. S. S., Harrison, R., Araujo, A., Arruda, M. F., and Snowdon, C. T. (2000). Behavioral and demographic changes following the loss of the breeding female in cooperatively breeding marmosets. *Behav. Ecol. Sociobiol.* 48:137–146.

Lopes, M. A., and Ferrari, S. F. (1994). Foraging behaviour of a tamarin group (*Saguinus fuscicollis weddelli*), and interactions with marmosets (*Callithrix emiliae*). *Int. J. Primatol.* 15:373–387.

Lorini, V. G., and Persson, M. L. (1990). Uma nova espécie de *Leontopihecus* Lesson, 1840, do sul do Brasil (Primates, Callitrichidae). *Bol. Mus. Nac. Rio Janeiro Zool.* 338:1–14.

Maier, W., Alonso, C., and Langguth, A. (1982). Field observations on *Callithrix jacchus jacchus*. *Z. Saugetierk.* 47:334–346.

Mansfield, K. (2003). Marmoset models commonly used in biomedical research. *Comp. Med.* 53:383–392.

Martins, M. M. (1998). Feeding ecology of *Callithrix aurita* in a forest fragment of Minas Gerais. *Neotrop. Primates* 6:126–127.

Martins, M. M., and Setz, E. Z. F. (2000). Diet of buffy tufted-eared marmosets (*Callithrix aurita*) in a forest fragment in southeastern Brazil. *Int. J. Primatol.* 21:467–476.

Mendes Pontes, A. R., and Monteiro da Cruz, M. A. O. (1995). Home range, intergroup transfers, and reproductive status of common marmosets, *Callithrix jacchus*, in a forest fragment in north-eastern Brazil. *Primates* 36:335–347.

Monfort, S. L., Bush, M., and Wildt, D. E. (1996). Natural and induced ovarian synchrony in golden lion tamarins (*Leontopithecus rosalia*). *Biol. Reprod.* 55:875–882.

Napier, J. R., and Napier, P. H. (1967). *A Handbook of Living Primates*. Academic Press, New York.

Neyman, P. F. (1978). Aspects of the ecology and social organization of free-ranging cotton-top tamarins (*Saguinus oedipus*) and the conservation status of the species. In: Kleiman, D. G. (ed.), *The Biology and Conservation of the Callitrichidae*. Smithsonian Institution Press, Washington DC. pp. 39–71.

Nievergelt, C. M., Digby, L. J., Ramakrishnan, U., and Woodruff, D. S. (2000). Genetic analysis of group composition and breeding system in a wild common marmoset (*Callithrix jacchus*) population. *Int. J. Primatol.* 21:1–20.

Nievergelt, C. M., and Martin, R. D. (1999). Energy intake during reproduction in captive common marmosets (*Callithrix jacchus*). *Physiol. Behav.* 65:849–854.

Oerke, A. K., Heistermann, M., Kuderling, I., Martin, R. D., and Hodges, J. K. (2002). Monitoring reproduction in Callitrichidae by means of ultrasonography. *Evol. Anthropol.* 11(suppl. 1):183.

Oliveira, A. C. M., and Ferrari, S. F. (2000). Seed dispersal by black-handed tamarins, *Saguinus midas niger* (Callitrichinae, Primates): implications for the regeneration of degraded forest habitats in eastern Amazonia. *J. Trop. Ecol.* 16:709–716.

Passamani, M. (1998). Activity budget of Geoffroy's marmoset (*Callithrix geoffroyi*) in an Atlantic forest in southeastern Brazil. *Am. J. Primatol.* 46:333–340.

Passamani, M., and Rylands A. B. (2000). Feeding behavior of Geoffroy's marmoset (*Callithrix geoffroyi*) in an Atlantic Forest fragment of south-eastern Brazil. *Primates* 41:27–38.

Passos, F. C. (1994). Behavior of the black lion tamarin, *Leontopithecus chrysopygus*, in different forest levels in the Caetetus Ecological Station, São Paulo, Brazil. *Neotrop. Primates* 2(suppl.):40–41.

Passos, F. C. (1997). Padrão de atividades, dieta e uso do espaço em um grupo de mico-leão-preto (*Leontopithecus chrysopygus*) na Estação Ecológica do Caetetus, SP [PhD thesis]. Univerdade Federal de São Carlos, São Carlos.

Passos, F. C. (1998). Ecology of the black lion tamarin. *Neotrop. Primates* 6:128–129.

Passos, F. C. (1999). Dieta de um grupo de mico-leão-preto, *Leontopithecus chrsyopygus* (Mikan) (Mammalia, Callitrichidae), na Estação Ecológica do Caetetus, São Paolo. *Rev. Brasil. Zool.* 16(suppl. 2):269–278.

Peres, C. A. (1989). Costs and benefits of territorial defense in wild golden lion tamarins, *Leontopithecus rosalia*. *Behav. Ecol. Sociobiol.* 25:227–233.

Peres, C. A. (2000). Territorial defense and the ecology of group movements in small-bodied neotropical primates. In: Boinski, S., and Garber, P. A. (eds.), *On the Move: How and Why Animals Travel in Groups*. University of Chicago Press, Chicago. pp. 100–123.

Pook, A. G., and Pook, G. (1981). A field study of the status and socioecology of Goeldi's monkey (*Callimico goeldii*) in northern Bolivia. *Folia Primatol.* 35:288–312.

Pook, A. G., and Pook, G. (1982). Polyspecific association between *Saguinus fuscicollis*, *Saguinus labiatus*, *Callimico goeldii* and other primates in north-western Bolivia. *Folia Primatol.* 38:196–216.

Porter, L. M. (2001a). Social organization, reproduction and rearing strategies of *Callimico goeldii*: new clues from the wild. *Folia Primatol.* 72:69–79.

Porter, L. M. (2001b). Dietary differences among sympatric Callitrichinae in northern Bolivia: *Callimico goeldii*, *Saguinus fuscicollis* and *S. labiatus*. *Int. J. Primatol.* 22:961–992.

Porter, L. M. (2004). Forest use and activity patterns of *Callimico goeldii* in comparison to two sympatric tamarins, *Saguinus fuscicollis* and *Saguinus labiatus*. *Am. J. Phys. Anthropol.* 124:139–153.

Porter, L. M., Hanson, A. M., and Becerra, E. N. (2001). Group demographies and dispersal in a wild group of Goeldi's monkeys (*Callimico goedii*). *Folia Primatol.* 72:108–110.

Power, M. L., and Oftedal, O. T. (1996). Differences among captive callitrichids in the digestive responses to dietary gum. *Am. J. Primatol.* 40:131–144.

Power, M. L., Oftedal, O. T., Savage, A., Blumer, E. S., Soto, L. H., Chen, T. C., and Holick, M. F. (1997). Assessing vitamin D status of callitrichids: baseline data from wild cotton-top tamarins (*Saguinus oedipus*) in Colombia. *Zoo Biol.* 16:39–46.

Power, M. L., Tardif, S. D., Power, R. A., and Layne, D. G. (2003). Resting energy metabolism of Goeldi's monkey (*Callimico goeldi*) is similar to that of other callitrichids. *Am. J. Primatol.* 60:57–67.

Prado, F. (1999a). Ecology and behavior of black-faced lion tamarins. *Neotrop. Primates* 7:137.

Prado, F. (1999b). Ecologia, comportamento e conservação do mico-leão-de-cara-preta (*Leontopithecus caissara*) no Parque Nacional de Superagüi, Guaraqueçaba Paraná [Master's. thesis]. Universidade Estadual Paulista, Botucatu.

Price, E. C., and Evans, S. (1991). Terminology in the study of callitrichid reproductive strategies. *Anim. Behav.* 34:31–41.

Price, E. C., Piedade, H. M., and Wormell, D. (2002). Population densities of primates in a Brazilian Atlantic forest. *Folia Primatol.* 73:54–56.

Raboy, B. E., and Dietz, J. M. (2004). Diet, foraging, and use of space in wild golden-headed lion tamarins. *Am. J. Primatol.* 63:1–15.

Ramirez, M. F., Freese, C. H., and Revilla, J. C. (1977). Feeding ecology of the pygmy marmoset, *Cebuella pygmaea*, in northeastern Peru. In: Kleiman, D. G. (ed.), *The Biology and Conservation of the Callitrichidae*. Smithsonian Institution Press, Washington DC. pp. 91–104.

Roda, S. A., and Mendes Pontes, A. R. (1998). Polygyny and infanticide in common marmosets in a fragment of Atlantic Forest of Brazil. *Folia Primatol.* 69:372–376.

Rosenberger, A. L. (1981). Systematics of the higher taxa. In: Coimbra-Filho, A. F., and Mittermeier, R. A. (eds.), *Ecology and Behavior of Neotropical Primates*, vol. 1. Academia Brasileira de Ciências, Rio de Janeiro. pp. 9–27.

Rylands, A. B. (1982). Behaviour and ecology of three species of marmosets and tamarins (Callitrichidae, Primates) in Brazil [PhD thesis]. Unviersity of Cambridge, Cambridge.

Rylands, A. B. (1986a). Ranging behaviour and habitat preference of a wild marmoset group, *Callithrix humeralifer* (Callitrichidae, Primates). *J. Zool. Lond.* (A) 210:489–514.

Rylands, A. B. (1986b). Infant-carrying in a wild marmoset group, *Callithrix humeralifer*: evidence for a polyandrous mating system. In: Thiago de Mello, M. (ed.), *A Primatologia No Brasil*, vol. 2. Sociedade Brasileira de Primatologia, Brasilia. pp. 131–144.

Rylands, A. B. (1989). Sympatric Brazilian callitrichids—the black-tufted-ear marmoset, *Callithrix kuhli*, and the golden-headed lion tamarin, *Leontopithecus chrysomelas*. *J. Hum. Evol.* 18:679–695.

Rylands, A. B. (1993a). *Marmosets and Tamarins: Systematics, Behaviour, and Ecology*. Oxford University Press, Oxford.

Rylands, A. B. (1993b). The ecology of the lion tamarins, *Leontopithecus*: some intrageneric differences and comparisons with other callitrichids. In: Rylands, A. B. (ed.), *Marmosets and Tamarins: Systematics, Behaviour, and Ecology*. Oxford University Press, Oxford. pp. 296–313.

Rylands, A. B., and Chiarello, A. G. (2003). Official list of Brazilian fauna threatened with extinction—2003. *Neotrop. Primates* 11:43–49, http://www.mma.gov.br/port/sbf/fauna/index.cfm.

Rylands, A. B., and de Faria, D. S. (1993). Habitats, feeding ecology, and home range size in the genus *Callithrix*. In: Rylands, A. B. (ed.), *Marmosets and Tamarins: Systematics, Behaviour, and Ecology*. Oxford University Press, Oxford. pp. 262–272.

Rylands, A. B., Schneider, H., Langguth, A., Mittermeier, R. A., Groves, C. P., and Rodríguez-Luna, E. (2000). An assessmentof the diversity of New World primates. *Neotrop. Primates* 8:61–93.

Saltzman, W. (2003). Reproductive competition among female common marmosets (*Callithrix jacchus*): proximate and ultimate causes. In: Jones, C. (ed.), *Sexual Selection and Reproductive Competition in Primates: New Perspectives and Directions*. American Society of Primatologists, Norman, OK. pp. 197–229.

Saltzman, W., Pick, R. R., Salper, O. J., Liedl, K. J., and Abbott, D. H. (2004). Onset of plural cooperative breeding in common marmoset families following replacement of the breeding male. *Anim. Behav.* 68:59–73.

Saltzman, W., Schultz-Darken, N. J., and Abbott, D. H. (1997). Familial influences on ovulatory function in common marmosets (*Callithrix jacchus*). *Am. J. Primatol.* 41:159–177.

Saltzman, W., Schultz-Darken, N. J., Scheffler, G., Wegner, F. H., and Abbott, D. H. (1994). Social and reproductive influences on plasma cortisol in female marmoset monkeys. *Physiol. Behav.* 56:801–810.

Saltzman, W., Schultz-Darken, N. J., Wegner, F. H., Wittwer, D. J., and Abbott, D. H. (1998). Suppression of cortisol levels in subordinate female marmosets: reproductive and social contributions. *Horm. Behav.* 33:58–74.

Savage, A., Giraldo, L. H., Soto, L. H., and Snowdon, C. T. (1996a). Demography, group composition, and dispersal in wild cotton-top tamarins (*Saguinus oedipus*) groups. *Am. J. Primatol.* 38:85–100.

Savage, A., Shideler, S. E., Soto, L. H., Causado, J., Giraldo, L. H., Lasley, B. L., and Snowdon, C. T. (1997). Reproductive events of wild cotton-top tamarins (*Saguinus oedipus*) in Colombia. *Am. J. Primatol.* 43:329–337.

Savage, A., Snowdon, C. T., Giraldo, L. H., and Soto, L. H. (1996b). Parental care patterns and vigilance in wild cotton-top tamarins (*Saguinus oedipus*). In: Norconk, M. A., Rosenberger, A. L., and Garber, P. A. (eds.), *Adaptive Radiations of Neotropical Primates*. Plenum Press, New York, pp. 187–199.

Savage, A., Ziegler, T. E., and Snowdon, C. T. (1988). Sociosexual development, pair bond formation, and mechanisms of fertility suppression in female cotton-top tamarins (*Saguinus oedipus oedipus*). *Am. J. Primatol.* 14:345–359.

Scanlon, C. E., Chalmers, N. R., and Monteiro da Cruz, M. A. O. (1988). Changes in the size, composition and reproductive condition of wild marmoset groups (*Callithrix jacchus jacchus*) in north east Brazil. *Primates* 29:295–305.

Schaffner, C. M., and Caine, N. G. (2000). The peacefulness of cooperatively breeding primates. In: Aureli, F., and de Waal, F. B. M. (eds.), *Natural Conflict Resolution*. University of California Press, Berkeley. pp.155–169.

Schaffner, C. M., and French, J. A. (1997). Group size and aggression: "recruitment incentives" in a cooperatively breeding primate. *Anim. Behav.* 54:171–180.

Schaffner, C. M., and French, J. A. (2004). Behavioral and endocrine responses in male marmosets to the establishment of multimale breeding groups: evidence for non-monopolizing facultative polyandry. *Int. J. Primatol.* 25:709–732.

Schneider, H., Martins, E. S., and Leão, V. (1987). Syntopy and troops association between *Callithrix* and *Saguinus* from Rondônia. *Int. J. Primatol.* 8:527.

Schneider, H., and Rosenberger, A. L. (1996). Molecules, morphology and platyrrhine systematics. In: Norconk, M. A., Rosenberger, A. L., and Garber, P. A. (eds.), *Adaptive Radiations of Neotropical Primates.* Plenum Press, New York. pp. 3–19.

Schradin, C., and Anzenberger, G. (2001). Infant carrying in family groups of Goeldi's monkeys (*Callimico goeldii*). *Am. J. Primatol.* 53:57–67.

Seuánez, H. N., Di Fiore, A., Moreira, Â. M., da Silva Almeida, C. A., and Canavez, F. C. (2002). Genetics and evolution of lion tamarins. In: Kleiman, D. G., and Rylands, A. B. (eds.), *Lion Tamarins: Biology and Conservation.* Smithsonian Institution Press, Washington DC. pp. 117–132.

Simons, E. L. (1972). *Primate Evolution: An Introduction to Man's Place in Nature*, MacMillan, New York.

Smith, A. C. (2000). Composition and proposed nutritional importance of exudates eaten by saddleback (*Saguinus fuscicollis*) and mustached (*Saguinus mystax*) tamarins. *Int. J. Primatol.* 21:69–83.

Smith, A. C., Buchanan-Smith, H. M., Surridge, A. K., and Mundy, N. I. (2003). Leaders of progressions in wild mixed-species troops of saddleback (*Saguinus fuscicollis*) and mustached tamarins (*S. mystax*), with emphasis on color vision and sex. *Am. J. Primatol.* 61:145–157.

Smith, A. C., Tirado Herrara, E. R., and Buchanan-Smith, H. M. (2001). Multiple breeding females and allo-nursing in a wild group of moustached tamarins (*Saguinus mystax*). *Neotrop. Primates* 9:67–69.

Smith, T. E., and French, J. A. (1997). Social and reproductive conditions modulate urinary cortisol excretion in black tufted-ear marmosets (*Callithrix kuhli*). *Am. J. Primatol.* 42:253–267.

Smith, T. E., Schaffner, C. M., and French, J. A. (1997). Social and developmental influences on reproductive function in female Wied's black tufted-ear marmosets (*Callithrix kuhli*). *Horm. Behav.* 31:159–168.

Snowdon, C. T. (1996). Infant care in cooperatively breeding species. *Adv. Study Behav.* 25:648–689.

Snowdon, C. T., and Soini, P. (1988). The tamarins, genus *Saguinus*. In: Mittermeier, R. A., Rylands, A. B., Coimbra-Filho, A. F., and da Fonseca, G. A. B. (eds.), *Ecology and Behavior of Neotropical Primates*, vol. 2. World Wildlife Fund, Washington DC. pp. 223–298.

Soini, P. (1982). Ecology and population dynamics of the pygmy marmoset, *Cebuella pygmaea. Folia Primatol.* 39:1–21.

Soini, P. (1987). Sociosexual behavior of a free-ranging *Cebuella pygmaea* Callitrichidae, Platyrrhini troop during post-partum estrus of its reproductive female. *Am. J. Primatol.* 13:223–230.

Soini, P. (1988). The pygmy marmoset, genus *Cebuella*. In: Mittermeier, R. A., Rylands, A. B., Coimbra-Filho, A. F., and da Fonseca, G. A. B. (eds.), *Ecology and Behavior of Neotropical Primates*, vol. 2. World Wildlife Fund, Washington DC. pp. 79–129.

Soini, P. (1993). The ecology of the pygmy marmoset, *Cebuella pygmaea*: some comparisons with two sympatric tamarins. In: Rylands, A. B. (ed.), *Marmosets and Tamarins: Systematics, Behaviour, and Ecology.* Oxford University Press, Oxford. pp. 257–261.

Solomon, N. G., and French, J. A. (1997). *Cooperative Breeding in Mammals.* Cambridge University Press, Cambridge.

Spurlock, L. B. (2002). Reproductive suppression in the pygmy marmoset *Cebuella pygmaea. Diss. Abst. Int.* B62:3872.

Stevenson, M. F., and Rylands, A. B. (1988). The marmosets, genus *Callithrix.* In: Mittermeier, R. A., Rylands, A. B., Coimbra-Filho, A. F., and da Fonseca, G. A. B. (eds.), *Ecology and Behavior of Neotropical Primates*, vol. 2. World Wildlife Fund, Washington DC. pp. 131–222.

Tagliaro, C. H., Schneider, M. P. C., Schneider H., Sampaio, I. C., and Stanhope, M. J. (1997). Marmoset phylogenetics, conservation perspectives, and evolution of the mtDNA control region. *Mol. Biol. Evol.* 14:674–684.

Tagliaro, C. H., Schneider, M. P. C., Schneider, H., Sampaio, I., and Stanhope, M. J. (2000). Molecular studies of *Callithrix pygmaea* (Primates, Platyrrhini) based on transferrin intronic and ND1 regions: implications for taxonomy and conservation. *Genet. Mol. Biol.* 23:729–737.

Tardif, S. D. (1997). The bioenergetics of parental behavior and the evolution of alloparental care in marmosets. In: Solomon, N. G., and French, J. A. (eds.), *Cooperative Breeding in Mammals.* Cambridge University Press, Cambridge. pp. 11–33.

Tardif, S. D., Harrison, M. L., and Simek, M. A. (1993). Communal infant care in marmosets and tamarins: relation to energetics, ecology, and social organization. In: Rylands, A. B. (ed.), *Marmosets and Tamarins: Systematics, Behaviour, and Ecology.* Oxford University Press, Oxford. pp. 220–234.

Tardif, S. D., Richter, C. B., and Carson, C. L. (1984). Reproductive performance of three species of Callitrichidae. *Lab. Anim. Sci.* 34:272–275.

Tardif, S. D., Santos, C. V., Baker, A. J., van Elsacker, L., Ruiz-Miranda, C. R., Moura, A. C. A., Passos, F. C., Price, E. C., Rapaport, L. G., and De Vleeschouwer, K. (2002). Infant care in lion tamarins. In: Kleiman, D. G., and Rylands, A. B. (eds.), *Lion Tamarins: Biology and Conservation.* Smithsonian Institution Press, Washington DC. pp. 213–232.

Tardif, S. D., Smucny, D. A., Abbott, D. H., Mansfield, K., Schultz-Darken, N., and Yamamoto, M. E. (2003). Reproduction in captive common marmosets (*Callithrix jacchus*). *Comp. Med.* 53:364–368.

Tardif, S. D., and Ziegler, T. E. (1992). Features of female reproductive senescence in tamarins (*Saguinus* spp.), a New World primate. *J. Reprod. Fertil.* 94:411–421.

Tavares, L. I., and Ferrari, S. F. (2002). Diet of the silvery marmoset (*Callithrix argentata*) at CFPn: seasonal and longitudinal variation. In: Lisboa, P. L. B. (ed.), *Caxiuanã, Populações Tradicionais, Meio Físico e Diversidade Biológica.* Belém. pp. 707–719.

Terborgh, J. (1983). *Five New World Primates: A Study in Comparative Ecology.* Princeton University Press, Princeton, NJ.

Terborgh, J., and Goldizen, A. W. (1985). On the mating system of the cooperatively breeding saddle-backed tamarin (*Saguinus fuscicollis*). *Behav. Ecol. Sociobiol.* 16:293–299.

Torii, R., Koizumi, H., Tanioka, Y., Inaba, T., and Mori, J. (1987). Serum LH, progesterone and estradiol-17β levels throughout the ovarian cycle, during the early stage of pregnancy and after the parturition and the abortion in the common marmoset, *Callithrix jacchus. Primates* 28:229–238.

Ullrey, D. E., Bernard, J. B., Peter, G. K., Lu, Z., Chen, T. C., Sikarski, J. G., and Holick, M. F. (2000). Vitamin D intakes by cotton-top tamarins (*Saguinus oedipus*) and associated serum 25-hydroxyvitamin D concentrations. *Zoo Biol.* 18:473–480.

Valladares-Padua, C. (1993). The ecology, behavior and conservation of the black lion tamarin (*Leontopithecus chrysopygus*, Mikan, 1823) [PhD thesis]. University of Florida, Gainesville.

Valladares-Padua, C., and Cullen, L., Jr. (1994). Distribution, abundance and minimum viable metapopulation of the black lion tamarin (*Leontopithecus chrysopygus*). *Dodo J. Wildl. Preserv. Trusts* 30:80–88.

van Roosmalen, M. G. M., and van Roosmalen, T. (1997). An eastern extension of the geographical range of the pygmy marmoset, *Cebuella pygmaea*. *Neotrop. Primates* 5:3–6.

van Roosmalen, M. G. M., and van Roosmalen, T. (2003). The description of a new marmoset genus, *Callibella* (Callitrichinae, Primates), including its molecular phylogentic status. *Neotrop. Primates* 11:1–10.

van Roosmalen, M. G. M., van Roosmalen, T., Mittermeier, R. A., and Rylands, A. B. (1998). A new and distinctive species of marmoset (Callitrichidae, Primates) from the lower Rio Aripuanã, state of Amazonas, central Brazilian Amazonia. *Goeldiana Zool.* 22:1–27.

Veracini, C. (1997). O comportamento alimentar de *Callithrix argentata* (Linnaeus 1771) (Primata Callitrichinae). In: Lisboa, P. L. B. (ed.) *Caxiuana* MCT/CNPq, Belem. pp. 437–446.

Veracini, C. (2000). Dado preliminaries sobre a ecologia de *Saguinus niger* na Estação Científica Ferreira Penna, Caxiuaná, Brasil. *Neotrop. Primates* 8:108–113.

Vinyard, C. J., Wall, C. E., Williams, S. H., and Hylander, W. L. (2003). Comparative function of skull morphology of tree-gouging primates. *Am. J. Phys. Anthropol.* 120:153–170.

Vinyard, C. J., Wall, C. E., Williams, S. H., Schmitt, D., and Hylander, W. L. (2001). A preliminary report on the jaw mechanics during tree gouging in common marmosets (*Callithrix jacchus*). In: Brook, A. (ed.), *Dental Morphology: 12th International Symposium on Dental Morphology*. Sheffield Academic Press, Sheffield. pp. 283–297.

von Dornum, M., and Ruvolo, M. (1999). Phylogenetic relationships of the New World monkeys (Primates, Platyrrhini) based on nuclear G6PD DNA sequences. *Mol. Phylogenet. Evol.* 11:459–476.

Wasser, S. K., and Barash, D. P. (1983). Reproductive suppression among female mammals: implications for biomedicine and sexual selection theory. *Q. Rev. Biol.* 58:513–538.

Watts, D. P. (1989). Infanticide in mountain gorilla—new cases and a reconsideration of the evidence. *Ethology* 81:1–18.

Widowski, T. M., Porter, T. A., Ziegler, T. E., and Snowdon, C. T. (1992). The stimulatory effect of males on the initiation but not the maintenance of ovarian cycling in cotton-top tamarins (*Saguinus oedipus*). *Am. J. Primatol.* 26:97–108.

Widowski, T. M., Ziegler, T. E., Elowson, A. M., and Snowdon, C. T. (1990). The role of males in the stimulation of reproductive function in female cotton-top tamarins, *Saguinus o. oedipus*. *Anim. Behav.* 40:731–741.

Wormell, D., and Price, E. (2001). Reproduction and management of black lion tamarin, *Leontopithecus chrysopygus* at Jersey Zoo. *Dodo J. Wildl. Preserv. Trusts* 37:34–40.

Yamamoto, M. E., and Box, H. O. (1997). The role of non-reproductive helpers in infant carrying in captive *Callithrix jacchus*. *Ethology* 103:760–771.

Yamamoto, M. E., Box, H. O., Albuquerque, F. S., and Arruda, M. F. (1996). Carrying behavior in captive and wild marmosets (*Callithrix jacchus*): a comparison between two colonies and a field site. *Primates* 37:297–304.

Ziegler, T. E., Bridson, W. E., Snowdon, C. T., and Eman, S. (1987a). Urinary gonadotropin and estrogen excretion during the postpartum estrus, conception, and pregnancy in the cotton-top tamarin (*Saguinus oedipus oedipus*). *Am. J. Primatol.* 12:127–140.

Ziegler, T. E., Savage, A., Scheffler, G., and Snowdon, C. T. (1987b). The endocrinology of puberty and reproductive functioning in female cotton-top tamarins (*Saguinus oedipus*) under varying social conditions. *Biol. Reprod.* 37:618–627.

Ziegler, T. E., Scheffler, G., and Snowdon, C. T. (1995). The relationship of cortisol levels to social environment and reproductive functioning in female cotton-top tamarins, *Saguinus oedipus*. *Horm. Behav.* 29:407–424.

Ziegler, T. E., Snowdon, C. T., and Bridson, W. E. (1990a). Reproductive performance and excretion of urinary estrogens and gonadotropins in the female pygmy marmoset (*Cebuella pygmaea*). *Am. J. Primatol.* 22:191–203.

Ziegler, T. E., Widowski, T. M., Larson, M. L., and Snowdon, C. T. (1990b). Nursing does affect the duration of the post-partum to ovulation interval in cotton-top tamarins (*Saguinus oedipus*). *J. Reprod. Fertil.* 90:563–570.

Appendix 6.1   Species and Subspecies Names for the Callitrichines

SPECIES AND SUBSPECIES	COMMON NAME
Callimico goeldii	Goeldi's monkey
Callithrix aurita	Buffy-tufted-ear marmoset
Callithrix flaviceps	Buffy-headed marmoset
Callithrix geoffroyi	Geoffroy's tufted-ear marmoset
Callithrix jacchus	Common or white-tufted-ear marmoset
Callithrix kuhlii	Wied's black-tufted-ear marmoset
Callithrix penicillata	Black-tufted-ear or black-pencilled marmoset
Cebuella pygmaea pygmaea	Pygmy marmoset
Cebuella pygmaea niveiventris	White-bellied pygmy marmoset
Callibella humilis	Black-crowned dwarf marmoset
Mico acariensis	Rio Acarí marmoset
Mico argentatus	Silvery marmoset
Mico chrysoleucus	Golden-white tassel-ear marmoset
Mico emiliae	Snethlage's marmoset
Mico humeralifer	Black and white tassel-ear marmoset
Mico intermedius	Aripuanã marmoset
Mico leucippe	Golden-white bare-ear marmoset
Mico manicorensis	Manicoré marmoset
Mico marcai	Marca's marmoset
Mico mauesi	Maués marmoset
Mico melanurus	Black-tailed marmoset
Mico nigriceps	Black-headed marmoset
Mico saterei	Sateré marmoset
Saguinus bicolor	Pied bare-face tamarin
Saguinus fuscicollis avilapiresi	Ávila Pires' saddle-back tamarin
Saguinus fuscicollis crandalli	Crandall's saddle-back tamarin
Saguinus fuscicollis cruzlimai	Cruz Lima's saddle-back tamarin
Saguinus fuscicollis fuscicollis	Spix's saddle-back tamarin
Saguinus fuscicollis fuscus	Lesson's saddle-back tamarin
Saguinus fuscicollis illigeri	Illiger's saddle-back tamarin
Saguinus fuscicollis lagonotus	Red-mantle saddle-back tamarin
Saguinus fuscicollis leucogenys	Andean saddle-back tamarin
Saguinus fuscicollis melanoleucus	White saddle-back tamarin
Saguinus fuscicollis nigrifrons	Geoffroy's saddle-back tamarin
Saguinus fuscicollis primitivus	Saddle-back tamarin
Saguinus fuscicollis weddelli	Weddell's saddle-back tamarin
Saguinus geoffroyi	Geoffroy's tamarin
Saguinus graellsi	Graell's black-mantle tamarin
Saguinus imperator imperator	Black-chinned emperor tamarin
Saguinus imperator subgrisescens	Bearded emperor tamarin
Saguinus inustus	Mottled-face tamarin
Saguinus labiatus labiatus	Red-bellied tamarin
Saguinus labiatus rufiventer	Red-bellied tamarin
Saguinus labiatus thomasi	Thomas' mustached tamarin
Saguinus leucopus	Silvery-brown bare-face tamarin
Saguinus martinsi martinsi	Martin's bare-face tamarin
Saguinus martinsi ochraceus	Ochraceous bare-face tamarin
Saguinus midas	Golden-handed or Midas tamarin
Saguinus mystax mystax	Spix's mustached tamarin
Saguinus mystax pileatus	Red-cap mustached tamarin
Saguinus mystax Pluto	White-rump mustached tamarin
Saguinus niger	Black-handed tamarin
Saguinus nigricollis hernandezi	Hernández-Camacho's black-mantle tamarin
Saguinus nigricollis nigricollis	Spix's black-mantle tamarin
Saguinus oedipus	Cotton-top tamarin
Saguinus tripartitus	Golden-mantle saddle-back tamarin
Leontopithecus caissara	Black-faced lion tamarin
Leontopithecus chrysomelas	Golden-headed lion tamarin
Leontopithecus chrysopygus	Black or golden-rumped lion tamarin
Leontopithecus rosalia	Golden lion tamarin

Source: Adapted from Rylands et al. (2000), van Roosmalen and van Roosmalen (2003).

7

The Cebines
Toward an Explanation of Variable Social Structure
Katharine M. Jack

INTRODUCTION

Capuchins (genus *Cebus*) and squirrel monkeys (genus *Saimiri*), which together make up the subfamily Cebinae, are among the most widely recognized species of New World primates. Some may recognize capuchins as the hat-tipping, money-collecting partner of the organ grinder or may have seen them in movies or television ads spreading a deadly strain of the ebola virus or as the "spokesperson" for any number of products, both of which have little to do with being a capuchin or a primate. Although perhaps not quite as recognizable as their capuchin cousins, squirrel monkeys are also well known because of their prominence in the illegal pet trade and their use in biomedical research; they are one of the most commonly used laboratory primates, second only to rhesus macaques (*Macaca mulatta*) (Kinzey 1997). Although these two primate genera have much in common and can often be observed in the same forests in Central and South America, they comprise a very diverse subfamily, particularly in respect to their morphology and behavioral ecology.

TAXONOMY AND DISTRIBUTION

The taxonomic classification of the platyrrhines has seen a long history of debate, and the exact placement of the genera *Cebus* and *Saimiri* within the infraorder has been particularly problematic (see Janson and Boinski 1992 for review). Ford and Davis (1992: 494) sum the problem up succinctly, stating that "*Cebus* and *Saimiri* are in some ways enigmatic" and "[t]heir potential relationships to each other and to other New World monkeys remain unclear" (see also Tyler 1991). They argue that morphological and life history data indicate that *Cebus* may have separated from the other platyrrhines as many as 20 million years ago and, accordingly, they should be placed within their own subfamily. Groves (2001) agrees, placing *Cebus* in the subfamily Cebinae and *Saimiri* in Chrysotrichinae (also Saimirinae; see Rosenberger's interpretation in Schneider and Rosenberger 1996), although he does group these two neotropical primates together in

the family Cebidae, accompanied only by the small squirrel-like marmosets and tamarins (*Callithrix*, *Callimico*, *Leontopithecus*, and *Saguinus*; subfamily Hapalinae). Contrary to these classification schemes that place *Cebus* and *Saimiri* in different subfamilies, other taxonomists argue that current phylogenetic analyses of both morphological and molecular data support the placement of these two genera together within the Cebinae subfamily (Schneider et al. 1993, for recent reviews see Rylands et al. 2000 and Schneider and Rosenberger 1996). The question of whether or not the marmosets and tamarins, and perhaps owl monkeys (*Aotus*), should accompany *Cebus* and *Saimiri* in the Cebinae subfamily is still highly debated; and here I follow Rylands et al. (2000) in considering *Cebus* and *Saimiri* to be the exclusive members of Cebinae and use the term *cebine* to refer solely to these two genera.

Four species of *Cebus* (*C. apella*, *C. albifrons*, *C. capucinus*, and *C. olivaceus*) and more than 30 subspecies are traditionally recognized (Ford and Hobbs 1996). A fifth species, *C. kaapori*, discovered in 1992 (Queiroz 1992; cf. Masterson 1995), is increasingly included in this list. Recently, Groves (2001) and Rylands et al. (2000) have called for the recognition of *C. libidinosus*, *C. xanthosternos*, and *C. nigritus* as distinct species rather than subspecies of *C. apella*, as they have previously been classified (see Fragaszy et al. 2004b for review). The four better-known *Cebus* species are divided into two main groups according to the presence or absence of tufts on the top of the head. The tufted group contains only one species, *C. apella*, while the untufted group contains *C. albifrons*, *C. capucinus*, and *C. olivaceus* (Hershkovitz 1949 based on Elliot 1913).

The number of species included within the genus *Saimiri* has been widely investigated, and molecular studies conducted in the last decade have shed much light on the topic. Historically, the genus was divided into only two species based on geographic distribution: *S. oerstedii* in Central America and *S. sciureus* in South America. In some accounts, *S. oerstedii* is listed as an offshoot of *S. sciureus* that was thought to have been introduced to Central America by humans in pre-Columbian times (Hershkovitz 1969; cf. Cropp and Boinski 2000, Costello et al. 1993; see

Boinski 1999 for review of *Saimiri* taxonomy); however, genetic analyses have shown this not to be the case, and it is now known to be a distinct species (Cropp and Boinski 2000). Taxonomists are now in general agreement on the division of *Saimiri* into five species, with as many as 12 subspecies: *S. oerstedii*, *S. sciureus*, *S. boliviensis*, *S. ustus*, and *S. vanzolinii* (Groves 2001, Rylands et al. 2000). This division is based on a variety of molecular, morphological, and behavioral data (Boinski and Cropp 1999, see also Ayres 1985, Groves 2001). It should be noted that the previous practice of classifying all South American squirrel monkeys together as *S. sciureus* does lead to some confusion when trying to decipher early reports of squirrel monkey taxonomy and behavior. In particular, most of our current information on *S. boliviensis* comes from Mitchell's (e.g., 1990) study at Manu, Peru, where the species was formerly referred to as *S. sciureus* (see Boinski 1999). The species designator *sciureus* is now reserved for squirrel monkey populations in the northeastern Amazon Basin (Colombia, Ecuador, and Venezuela) and the Guyana Shield (Boinski 1999, Boinski and Cropp 1999).

The geographic distribution of the cebines overlaps extensively; indeed, the two genera are often found in sympatry and commonly form mixed-species associations (see Ecology, below). The geographic range of *Cebus* is more extensive than that of *Saimiri* and, among New World monkeys, is second only to that of *Alouatta* (howler monkeys) (Sussman 2000). *Cebus* range throughout much of Central and South America, from Honduras in the north to Argentina in the south (Eisenberg 1989) (see Fig. 7.1 for map of *Cebus* distribution). With the exception of *C. capucinus*, which ranges from Honduras through the northwestern coast of Ecuador (Rowe 1996), all *Cebus* species are indigenous to South America (see Table 7.1 for species

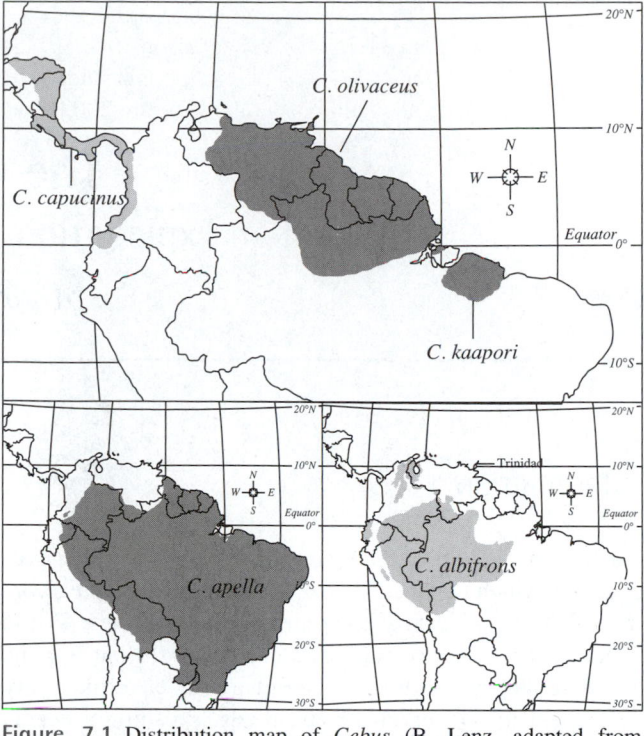

Figure 7.1 Distribution map of *Cebus* (B. Lenz, adapted from Fragaszy et al. 2004a).

distributions and common names). Despite the large geographic distribution of this genus, long-term field studies have been limited, with the majority of data coming from six main field sites: *C. capucinus* in Santa Rosa and Lomas Barbudal, Costa Rica; *C. olivaceus* in Hato Masaguaral, Venezuela; *C. apella* in Iguazu, Argentina; Manu, Peru; and a study begun in 2000 at Raleighvallen, Suriname (see Fragaszy et al. 2004b for a complete overview of field

Table 7.1   Species, Common Names, and Geographic Distribution of the Most Commonly Recognized Cebines

SCIENTIFIC NAME	COMMON NAME	GEOGRAPHIC DISTRIBUTION
Cebus albifrons	White-fronted capuchin	Brazil, Colombia, Ecuador, Peru, Venezuela, northern Bolivia, and Trinidad
Cebus apella[1]	Tufted, black-capped, or brown capuchin	Northern and central South America from Colombia and Ecuador (west of the Andes) through coastal Brazil
Cebus capucinus	White-faced, white-headed, or white-throated capuchin	Honduras through northwestern Ecuador
Cebus olivaceus (also *C. nigrivittatus*)	Weeper or wedge-capped capuchin	Northeastern Brazil, Guyana, French Guiana, Suriname, and Venezuela
Cebus kaapori	Ka'apor capuchin	Brazil
Saimiri oerstedii	Red-backed or Central American squirrel monkey	Central Pacific coast of Costa Rica through the Pacific coast of Panama
Saimiri sciureus	Common or South American squirrel monkey	North–central South America (Brazil, Guyana, French Guiana, Suriname, Venezuela, Ecuador, and Colombia)
Saimiri ustus	Golden-backed or bare-eared squirrel monkey	Northwestern Brazil
Saimiri boliviensis	Bolivian or black-capped squirrel monkey	Brazil, Bolivia, Peru
Saimiri vanzolinii	Black squirrel monkey; many classify with *S. ustus*	Northwestern Brazil

[1] *C. libidinosus*, *C. xanthosternos*, and *C. nigritus* are now considered by many taxonomists to be distinct species, rather than subspecies of *C. apella*. The above list follows more traditional taxonomy.
Sources: *Cebus*, Groves 2001, Rowe 1996, Fragaszy et al. 2004b; *Saimiri*, Groves 2001, Boinski 1999, Rowe 1996.

studies of *Cebus*). To date, only short-term studies of *C. albifrons* have been undertaken in Peru (Janson 1984, 1986) and Colombia (Defler 1979a,b, 1982), although studies of this species in Trinidad (Phillips and Abercrombi 2003) and Ecuador (K. Jack, unpublished data) are currently under way. Despite this apparent paucity of studies focused on this genus in the wild, particularly in comparison to the long-term field studies of many Old World monkey and ape species, *Cebus* is one of the most extensively studied genera of New World primates, with studies of *C. capucinus* in Costa Rica now entering their third decade.

Saimiri are found in two geographically separated pockets. A single species, *S. oerstedii*, is isolated in Central America in a relatively small area of lowland forests that extend from the central Pacific coastline of Costa Rica through the Pacific coast of western Panama. The remaining species occur in the lowland forests of the Amazon Basin of South America from Guyana through Paraguay (Boinski et al. 2002) (Fig. 7.2 and Table 7.1). To date, long-term studies of *Saimiri* have been limited: *S. oerstedii* in Corcovado, Costa Rica (e.g., Boinski 1987c, 1988), *S. boliviensis* in Manu, Peru (Mitchell 1990, 1994; Boinski 1991, 1994), and *S. sciureus* in Raleighvallen, Suriname (Boinski 1999), which is currently the only long-term investigation being undertaken on this genus (see Boinski et al. 2002 for a summary of shorter studies).

ECOLOGY

Morphological Adaptations to Habitat and Diet

Cebus are medium-sized primates, with females weighing 1.4–3.4 kg (mean = 2.3 kg) and males ranging 1.3–4.8 kg (mean = 3.0 kg) depending on the species under consideration (see Table 7.2). *Cebus* display moderate levels of sexual dimorphism, with males weighing 19.5%–27% more than females (mean = 24%, Table 7.2) and possessing canine

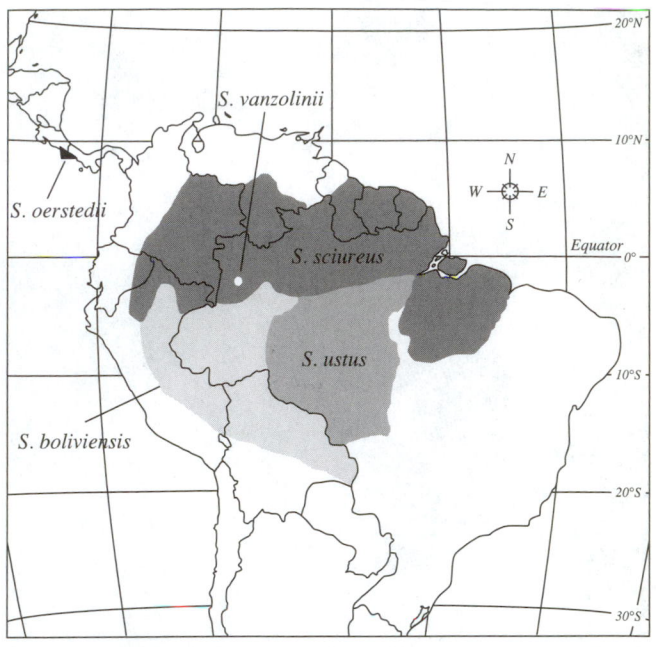

Figure 7.2 Distribution map of *Saimiri* (B. Lenz, adapted from Boinski 1999).

teeth that are 16–22% larger than female canines (Kay et al. 1988). *Saimiri* are considerably smaller than *Cebus*, with male weights ranging 620–1,200 g (mean = 884 g) and female weights ranging 600–880 g (mean = 700 g). However, the degree of sexual dimorphism in body size is comparable across the two genera (male *Saimiri* are on average 19.7% larger than females) (Table 7.2) (see also Janson 1984). Both *Cebus* and *Saimiri* show large brain to body weight ratios, particularly for primates of their size, a trend that may be an adaptation to their relatively large home ranges, variable diets (Stephen et al. 1988; see Terborgh 1983 for home range comparisons), and/or complex foraging patterns (e.g., Fragaszy 1990, Parker and Gibson 1977, Terborgh 1983) (see Chapter 41).

Table 7.2   Adult Weights and Sexual Dimorphism of the Cebines

GENUS	SPECIES	MALE BODY WEIGHT RANGE (MEAN) (G)	FEMALE BODY WEIGHT RANGE (MEAN) (G)	DEGREE OF SEXUAL DIMORPHISM
Cebus	*albifrons*	1,700–3,260 (2,480)	1,400–2,228 (1,814)	27%
	apella	1,350–4,800 (3,050)	1,758–3,400 (2,385)	22%
	capucinus	3,765–3,970 (3,668)	2,610–2,722 (2,666)	27%
	olivaceus	1,447–4,500 (2,974)	1,589–3,200 (2,395)	19.5%
Mean		**3,043**	**2,315**	**24%**
Saimiri	*boliviensis*	(992)	(751)	24.3%
	oerstedii	750–950 (829)	600–800 (695)	16%
	sciureus	(740)	(635)	14%
	ustus	620–1,200 (910)	710–880 (795)	13%
	vanzolinii	(950)	(650)	31.5%
Mean		**884**	**705**	**19.7%**

Sources: Ford and Davis 1992 (data based on wild individuals only). Boinski (1999); with the exception of those weights for female *S. oerstedii*, sexual dimorphism for *S. boliviensis*, male *S. oerstedii*, and *S. sciureus* has been calculated based on weights provided in Boinski (1999) rather than Ford and Davis (1992) because of earlier practices of lumping numerous species under *S. sciureus*.

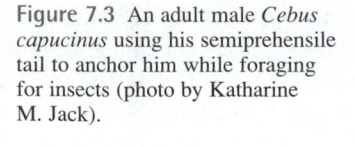
Figure 7.3 An adult male *Cebus capucinus* using his semiprehensile tail to anchor him while foraging for insects (photo by Katharine M. Jack).

In addition to differences in absolute body size, *Cebus* and *Saimiri* display considerable morphological variability. For example, although both genera have skeletons adapted for quadrupedal locomotion, they differ markedly in their relative limb proportions. Unlike other platyrrhines, *Cebus* tend to have forelimbs and hindlimbs of fairly equal proportions, while *Saimiri* have comparatively longer hindlimbs (Fleagle 1999). This is likely an adaptation for the more frequent leaping that characterizes *Saimiri* locomotion, while the equal limb lengths of capuchins may be an adaptation for their more terrestrial locomotor patterns (Janson and Boinski 1992). Overall, *Cebus* postcranial anatomy is more similar to that of terrestrial Old World monkeys than it is to that of other platyrrhines, which is likely because capuchins, particularly the untufted species, spend more time on the ground than other platyrrhines. Capuchins also have a semiprehensile tail that, unlike their ateline cousins (see Chapter 10), is fully covered in fur and unable to sustain the full weight of an adult. Instead, the tail provides support and balance and is commonly used to anchor an individual to trees while foraging (see Fig. 7.3). Squirrel monkeys locomote quadrupedally along the surfaces of thin branches and lianas, often leap between them (Boinski et al. 2002), and rarely descend to the ground to forage. Although they are born with prehensile tails, the grasping ability is lost with age and the rather large tail of adults assists only in balance (Boinski 1989a) (Fig. 7.4).

The dentition of *Cebus* and *Saimiri* is also variable and appears to reflect dietary differences between the two genera. Although cebines are considered the most omnivorous of the platyrrhines (Fleagle 1999) and both *Cebus* and *Saimiri* are traditionally categorized as frugivore–insectivores

(Robinson and Janson 1987), *Saimiri* display a relatively heavier reliance on insects, while *Cebus* rely more on fruit (see Janson and Boinski 1992 for review). The cebines spend upward of 75% of their day engaged in active foraging (Robinson and Janson 1987), and *Saimiri* can spend up to 80% of their time foraging just for insects. The diets of both genera are extremely flexible, and individuals are able to alter their diet to become very specialized in response to changing environmental conditions. For example, during the peak of the dry season, *Saimiri* are able to become complete insectivores for up to a week at a time, while *Cebus*, which can survive in very marginal habitats, can rely almost completely on bromeliads (a plant in the family that includes pineapples) or become seed predators in times of food stress (see Sussman 2000 for review). *Cebus* are very adept at capturing small vertebrate prey, including birds (and eggs), lizards, squirrels (Fig. 7.5), and coatis, although such animal prey form only a small portion of their diet (~0.5%–2.5%) (Sussman 2000). The hunting behavior displayed by *Cebus* is opportunistic rather than cooperative in nature (see Perry and Rose 1994). Interestingly, at least among *C. capucinus*, the individual that captures the prey, regardless of age or sex class, is able to eat it with only minimal harassment from other group members (K. M. Jack personal observation). Although food sharing does occur in this genus, it is indirect in that bits of food are dropped or left behind for other group members to eat (Perry and Rose 1994).

Cebus are often referred to as "habitat generalists" (Chapman et al. 1989) in that they occupy a great diversity of habitats, including primary forests of all types (rain, cloud, dry, deciduous, etc.), highly disturbed and fragmented areas, as well as swamp and seasonally flooded forests (Freese and

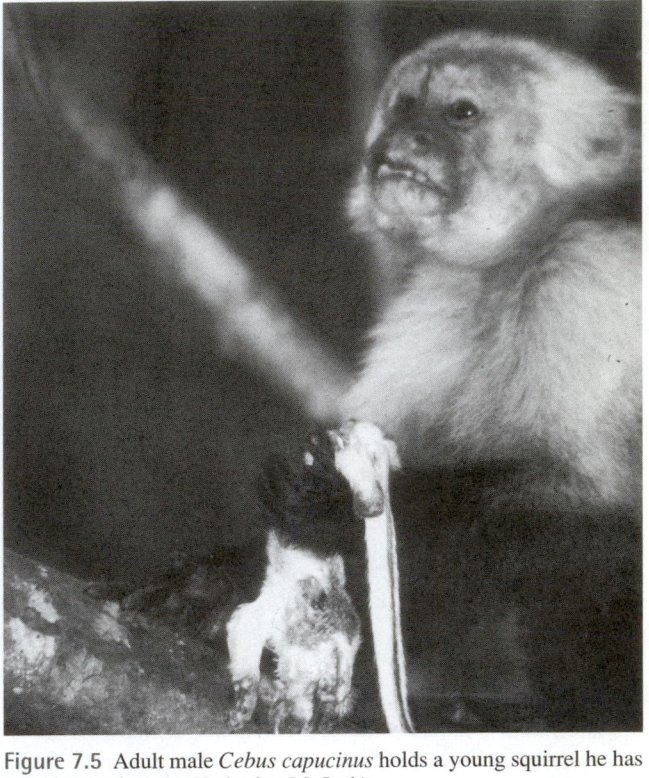

Figure 7.4 An adult female *Saimiri oerstedii* with 6-month-old infant. Note the lack of prehensile tails (photo by Katharine M. Jack).

Figure 7.5 Adult male *Cebus capucinus* holds a young squirrel he has just killed (photo by Katharine M. Jack).

Oppenheimer 1981). Members of the genus *Cebus* are the most dextrous of the platyrrhine primates. They have shortened fingers, "pseudo-opposable thumbs," and the ability to move all digits independently of the others (Janson and Boinski 1992). This dexterity, accompanied by what seems to be an innate curiosity, is readily visible in their foraging habits. Capuchins are extractive foragers (Parker and Gibson 1977) and appear to specialize in consuming foods that "fight back"—foods that have hard protective coverings, toxins, and other predatory defenses or insects and vertebrates that bite and/or sting (Fragaszy et al. 2004b). In order to access these foods, wild capuchins will pound hard husked or shelled objects onto a substrate, and they frequently rid food items of spines or poisonous hairs by rubbing them against branches to rid them of their noxious substances (Panger et al. 2002). In some cases, these food items are wrapped in leaves prior to rubbing, perhaps as a way of directly protecting the hands from the toxins they are attempting to remove (Panger et al. 2002). Fragaszy et al. (2004a) recently described a wild group of *C. apella* in Brazil that are regularly observed to crack open palm nuts using rocks in a hammer and anvil-type fashion, further demonstrating the dexterity and ingenuity of these small New World primates. Capuchins have been observed using a variety of other tools in the wild, including a club to attack a venomous snake, leaves as containers, and probing tools to access imbedded foods; among the nonhuman primates, capuchins and the great apes are considered to be the most adept and varied tool users (see Panger et al. 2002). Another characteristic that *Cebus* has evolved as a means of gaining

access to protected foods (e.g., palm nuts) is their thick tooth enamel; in fact, when tooth size is taken into account, *Cebus* shows the thickest tooth enamel of any nonhuman primate (Kay 1981, see Janson and Boinski 1992 for review).

Despite their large geographic distribution, the habitat preferences of *Saimiri* show extreme ecological consistency (see Janson and Boinski 1992). *Saimiri* show a preference for tropical lowland rain forests (Boinski et al. 2002), with most activities concentrated in secondary forests and relatively little time in primary and late successional forests (Boinski 1987c; see Sussman 2000 for review). Although, in general, the cebines are more manually dextrous than other platyrrhines, *Saimiri* do not possess the manipulative abilities that are so characteristic of capuchins. Squirrel monkeys do not have independently mobile digits, nor do they possess opposable thumbs, a combination that prohibits them from generating a strong grip between the fingers and thumb (Janson and Boinski 1992). Although both *Saimiri* and *Cebus* rarely go after prey that is in motion, *Saimiri*, are more prone to hunt potentially mobile prey that are visible on substrate surfaces (branches, leaves, tree trunks) than are *Cebus*. Squirrel monkeys also specialize in finding insects by unrolling both dead and living leaves to reveal invertebrate prey imbedded within them (Fig. 7.6) (Janson and Boinski 1992).

Predation: Susceptibility and Behavioral Adaptations

Both *Cebus* and *Saimiri* are at risk of being taken by similar types of predator, including raptors, felids, constricting and

Figure 7.6 Adult male *Saimiri boliviensis* unrolling a dead leaf in search of imbedded invertebrate prey (photo by B. Lenz).

venomous snakes, coyotes, and tayras (an arboreal mammal that is part of the weasel family, *Mustelidae*) (Freese 1983, van Schaik and van Hooff 1983, Terborgh 1983, Chapman 1986, Boinski et al. 2002). However, given that an adult capuchin outweighs an adult squirrel monkey by as much as seven times, it is not surprising that *Saimiri* are much more susceptible to predation than are *Cebus* (Fedigan et al. 1996). Raptors are the main predators observed to kill *Saimiri*, and Boinski (1987b) reports that 50% of all infants born to *S. oerstedii* do not survive to 6 months of age due to confirmed or suspected predation by avian predators. Boinski et al. (2002) report similar predation rates on *S. boliviensis*, while *S. sciureus* of Suriname appear to be at lower risk despite an intact predator community. Perhaps as a response to their susceptibility to predation, *Saimiri* form groups that are usually between three and six times the size of *Cebus* troops (see Social Organization, below); and although both genera are known for being highly vocal primates that display a great repertoire of call types, *Saimiri* have more calls that serve as contact calls than do *Cebus*. While these vocalizations do not directly function in predator avoidance or detection, they are believed to function as a type of "security blanket," to make up for the visual separation that occurs while squirrel monkeys forage and enable group members to keep alerted to the presence and location of their group mates (see Fedigan et al. 1996 for review).

The South American populations of *Saimiri* have developed another way of combating their high risk of predation: in all areas where they occur sympatrically, *Saimiri* parasitize on the predator-detecting abilities of *Cebus* by joining them in their daily activities (Waser 1986). These mixed-species associations may last several days, and for the most part the two genera will interact peacefully. However, it is not uncommon for an adult male capuchin to chase away an entire group of squirrel monkeys, although the displaced squirrel monkeys usually quickly return to their capuchin comrades. Data from South American cebine populations, particularly those coming from Manu, Peru (e.g., Terborgh 1983, Terborgh and Janson 1986), have shown that it is *Saimiri* rather than *Cebus* that maintain these mixed-species associations. Terborgh (1983) reports that, in Manu, *Saimiri* actually seek out *Cebus* and that although *Saimiri* often appear to be leading the movements and direction of the mixed-species foraging excursions, they will usually backtrack if the capuchins lag behind or fail to follow them. The maintenance of these associations with capuchins is thought to benefit squirrel monkeys in terms of both increased foraging efficiency and predation avoidance. For example, when they are not in association with *Cebus*, squirrel monkeys travel twice as far in an hour. They also appear to benefit from the more detailed knowledge that *Cebus* have of fruiting tree locations during times of resource scarcity. In addition, *Saimiri* gain access to fruits dropped by *Cebus* that they would otherwise be unable to open (e.g., tough husked palm nuts). Despite the obvious foraging advantages that these associations provide, the real advantage for *Saimiri* seems to lay in the increased predator detection skills of *Cebus* (Terborgh 1983; see also Sussman 2000 for review). *Saimiri* appear to benefit greatly from the ever-vigilant capuchin males and the alarm calls that they emit; in fact, squirrel monkeys respond more readily to *Cebus* alarm calls than they do to the alarm calls of their own group members (Terborgh 1983). *Cebus*, on the other hand, do not seem to benefit much from these associations; they pay little attention to *Saimiri* alarm calls, and they travel up to 40% more when in the company of squirrel monkeys. It is no wonder, then, that capuchins do not wait for sidetracked squirrel monkeys to catch up when they lag behind.

Interestingly, in Central America, sympatrically occurring *Cebus* and *Saimiri* do not form mixed-species associations more frequently than predicated by chance. Given that *S. oerstedii* are under severe predation pressure, particularly from raptors (Boinski 1987b), would they not also benefit from the predator detection skills of *C. capucinus*? Boinski (1989b) reports that in Corcovado National Park, Costa Rica, the two cebine genera spent only approximately 6% of their time in association with one another, which greatly contrasts with the 90% rate reported for *S. boliviensis* with *C. apella* and *C. albifrons* in Peru. Boinski (1989b) suggests that the two Central American species do not form mixed troops simply because the presence of *C. capucinus* would impose very high foraging costs on *S. oerstedii*. Her study showed that *Saimiri* would have little to gain by associating with *Cebus* even in terms of increased predator detection and avoidance. Boinski concluded that, although *Cebus* in Costa Rica are susceptible to predation, their vigilance does not appear to function for detecting predators. Several studies focusing on male vigilance in *C. capucinus* have shown that while they are alert to potential predators, as evidenced by their elaborate alarms calls that differ according to predator types, the majority of their vigilance is directed at detecting extragroup males (Rose and Fedigan 1995) or monitoring the activities of their coresident males (Jack 2001). The foraging costs and limited benefits associated with detecting predators appear to have precluded the formation of mixed-species groups between the Costa Rican cebines, although this topic requires further detailed investigation.

REPRODUCTIVE PARAMETERS

Saimiri are highly seasonal breeders, with all matings being confined to an annual 2-month period and births being even more tightly synchronized (Table 7.3). For example, *S. oerstedii* are reported to have the most restricted birth season of any primate species, with group females giving birth within the same 1–week period (Kinzey 1997, Boinski 1987b). *C. apella* in Argentina is the only capuchin population studied thus far that displays a true birth season, which in turn reflects mating seasonality (Di Bitetti and Janson 2000, 2001). While the other *Cebus* species do show a yearly birth peak during which the majority of infants are born, they are not strict seasonal breeders as births do occur throughout all months of the year (Freese and Oppenheimer 1981, Robinson and Janson 1987, Fedigan and Rose 1995). Like most New World monkeys, cebine birth peaks or seasons generally coincide with seasonal peaks in resource availability (Di Bitetti and Janson 2000). *Cebus* interbirth intervals range 18–26.4 months (mean = 22 months), while *Saimiri* interbirth intervals, with the exception of *S. boliviensis*

where females give birth every other year, tend to be shorter and average 12 months (see Table 7.3).

Compared with other mammals, where life history variables strongly correlate with body size, primates show very long life histories, particularly in their tendency toward extreme delays in reproductive maturity (Promislow and Harvey 1990, Purvis et al. 2003). If we restrict the examination of life history variables and body size to the order Primates, the general pattern holds, with larger primates showing slower life histories than smaller ones (Eisenberg 1979). However, comparing published data available for New World species, *Cebus* do not fit this general pattern. Rather, their maturation rates and overall life history variables are more comparable with members of the subfamily Atelinae (spider monkeys and woolly monkeys), which generally weigh about four times as much as the average capuchin (see data in Robinson and Janson 1987, Rowe 1996, Kinzey 1997). Compare, for example, the largest of the New World monkeys, the muriqui (*Brachyteles spp*), with *Saimiri* and *Cebus*. Female muriquis weigh about 9.5 kg (Rowe 1996) and are 7.5 years of age when they give birth to their first offspring (Strier 1997). *Saimiri* females weigh on average 700 g and give birth to their first infant at 2.5 years, while *Cebus* females weigh 2.3 kg and are 7–8 years of age when they give birth for the first time (Tables 7.2 and 7.3). Even male *Cebus* show slow maturation rates, with most studies estimating sexual maturity to occur after 7 years of age; and in several species, males do not attain full adult body size until after 10 years of age (Jack and Fedigan 2004a,b). Interestingly, the small size and long life history patterns characteristic of *Cebus* continue to stand out even when we extend comparisons to include the Old World monkeys.

These slow maturation rates of *Cebus* may be related to their extremely long life spans, which average about

Table 7.3   Cebine Reproductive Parameters and Life History Variables

GENUS	SPECIES	BIRTH SEASON	INTERBIRTH INTERVAL (MONTHS)	FEMALE AT FIRST BIRTH (YEARS)	MALE ADULT AGE (YEARS, REPRODUCTIVE MATURITY)	LIFE SPAN (CAPTIVE)
Cebus	*albifrons*	Peak[1]	18[3]	N/A	N/A	44[15]
	apella	Peak–Peru Season–Argentina[1]	19.4[4]	7 (wild)[1]	4.5[1], 7[11,12] (captive)	45.1[15]
	capucinus	Peak[1]	26.4[5]	7 (wild)[1]	10 (wild)[13]	54.8[15]
	olivaceus	Peak[1]	26[6]	6 (wild)[1]	15 (wild)[13]	41[15]
Saimiri	*boliviensis*	Yes (2 months)[2]	~24[7]	2.5 (wild)[10]	6 (wild)[7]	N/A
	oerstedii	Yes (1 week)[2]	~12[8]	2.5 (wild)[10]	5 (2.5, wild)[14]	N/A
	sciureus	Yes (1 week)[2]	~12[9]	N/A	N/A	21[16]

Sources:
[1] Fragaszy et al. 2004b.
[2] Boinski 1987b.
[3] Kappeler and Pereira 2003.
[4] Di Bitetti and Janson 2000.
[5] Fedigan and Rose 1995.
[6] Robinson and Janson 1987.
[7] Mitchell 1994.
[8] Boinski 1999.
[9] Rowe 1996.
[10] Baldwin and Baldwin 1981.
[11] Patino et al. 1996.
[12] Nagle and Denari 1982.
[13] Jack and Fedigan 2004b.
[14] Boinski 1994.
[15] Hakeem et al. 1996.
[16] Harvey et al. 1987.

41 years in captivity (the record is nearly 55 years; Hakeem et al. 1996). In stark contrast, the smaller *Saimiri* are reported to live only 21 years and most capuchin-sized (and larger) primates live approximately 30 years (see data presented in Rowe 1996). It is also possible that delayed maturation in *Cebus* is related to their unusually slow postnatal brain growth and motor skill development (see Hartwig 1996). *Cebus* display more postnatal brain growth and development than any of the other platyrrhines, while *Saimiri* neonates are more precocial in terms of both brain growth and motor skill development (Hartwig 1995).

GROUP STRUCTURE

Both *Cebus* and *Saimiri* form mixed-sex groups, but the two genera show much diversity in group size and composition (see Cebine Dispersal Patterns, below). Although the mean size of *Cebus* groups does not vary greatly across the four better-known species (range = 16.4–21, mean = 18.8), there is considerable variation in adult sex ratios within groups. For example, *C. olivaceus* show the most skewed sex ratios, with approximately one adult male for every two adult females in a group, while *C. albifrons* show the most equitable sex ratios, with 1.08 adult males per female (Table 7.4). The

size of *Saimiri* groups is much more variable across species, ranging from 15 to 75 individuals. *S. boliviensis* form the largest groups (mean = 54), while *S. sciureus* display the smallest mean group sizes, at 23 individuals per group. Several researchers have reported *Saimiri* groups sizes of up to 300 individuals; however, in all instances, it appeared that these large groups were actually temporary unions of multiple groups (see Sussman 2000 for review). Sex ratios also vary across species, with *S. oerstedii* showing one male per 1.6 adult females and *S. boliviensis* showing a much more skewed ratio, with one male per 2.5 adult females (data are not available for *S. sciureus*) (see Table 7.4). Interestingly, *S. boliviensis* is also the only species of *Saimiri*, indeed of the platyrrhines, reported to form all-male groups; and even within mixed-sex groups, males of this species are generally peripheral and have little to no interactions with group females outside of the mating season (Boinski 1999).

MATING SYSTEMS

Similar to the other platyrrhines, the cebines do not show external cues that signal ovulation or pregnancy like those characteristic of many of the catarrhine primates, making it extremely difficult to discern reproductive from

Table 7.4   Cebine Social Organization and Social Structure

SPECIES	MEAN GROUP SIZE	MEAN ADULT SEX RATIO (M:F)	DISPERSAL PATTERNS	DOMINANCE
Cebus albifrons[1]	19.8	1.08	Male	Both sexes display linear hierarchies, and males are individually dominant over females; males well-integrated into the group
C. apella[1]	18	0.85	Male	Both sexes display linear hierarchies, and males are individually dominant over females; alpha female may rank below alpha male; subordinate males are often peripheral group members
C. capucinus[1]	16.4	0.71	Male: parallel dispersal is common and lasts through multiple emigration events; female dispersal is rare, but it does occur	Both sexes display linear hierarchies, and males are individually dominant over females; alpha female may rank below alpha male; female coalitions can displace alpha male
C. olivaceus[1]	21	0.53	Male	Both sexes display linear hierarchies, and males are individually dominant over females; alpha female may rank below alpha male
Mean	18.8	0.79		
Saimiri oerstedii	35–65 (41)[2]	0.63 (1:1.6)[3]	Females disperse prior to first mating season/flexible male philopatry or dispersal[5]	Egalitarian; no female dominance hierarchy or coalitions; male hierarchies only evident during mating season[5]
S. boliviensis (Peru)	45–75 (54)[2]	0.40 (1:2.5)[4]	Male:[2] parallel dispersal is common and lasts through multiple emigration events[4]	Females dominant (matrilineal hierarchies and frequent coalitions); both sexes form stable linear dominance hierarchies;[5,6] males are peripheral group members
S. sciureus (Suriname)	15–50 (23)[2]	NA	Both sexes thought to disperse; female dispersal is flexible (may spend first mating season in natal group)[5]	Males dominant; both males and females form stable linear hierarchies[5]

Sources:
[1] All *Cebus* data are from Fragaszy et al. 2004b.
[2] Boinski et al. 2003.
[3] Boinski 1987a.
[4] Mitchell 1994.
[5] Boinski et al. 2002.
[6] Mitchell 1990.

nonreproductive states (Dixson 1983; see also Chapter 25). In *Cebus*, this lack of obvious fertility signaling, accompanied by the fact that members of the genus show limited reproductive seasonality and engage in frequent non-conceptive copulations (complete with copulatory displays, see below), makes it nearly impossible for observers to discern fertile versus nonfertile matings (e.g., see Manson et al. 1997). Given that *Saimiri* are highly seasonal breeders, with all mating activity being confined to a 2-month period each year, and that individual females are sexually active for only about a 2-day period throughout these short mating seasons (Boinski 1992) (Table 7.3), determining fertile vs. nonfertile matings is not quite as challenging.

Mating behavior in *Cebus*, both conceptive and non-conceptive, is preceded, at least in some species, by elaborate courtship rituals. The most detailed data available to date come from studies of captive *C. apella*. In this species, a female begins the courtship ritual by gazing toward the male of her choice (usually the alpha male), often tilting her head from side to side and raising her eyebrows at him in an attempt to catch his gaze (see Fragaszy et al. 2004b). She slowly decreases the distance between herself and the male and, when she is close enough, reaches out, quickly touches him, and then runs away. Throughout this process, the female continuously rubs her chest and emits soft vocalizations. In these initial stages, her attempts to gain the male's attention seem futile; he appears utterly and completely disinterested. Such solicitation by a female can go on for hours, but the male eventually relents and begins to reciprocate her movements and vocalizations, adding a mutual gaze and a grin. At this point, the movements of the pair become mirrored in a coordinated dance display. One member of the pair advances, while the other retreats, all the while maintaining their locked gaze. Pirouettes of 180 degrees are added, with the dancer always coming back to face the partner. At this point in the courtship, the participants appear oblivious to what is going on around them. The pair eventually comes together and mating ensues. During copulation, the male emits a loud vocalization that is unique to the mating context, and following ejaculation, the pair remain together, continuing their dance display as if nothing has happened and beginning the process anew.

C. capucinus also show a similarly elaborat pattern of courtship and mating, although the eyebrow raising and grinning characteristic of *C. apella* are exchanged for a "duck face," which involves a protrusion of the lips in the direction of the partner (Manson et al. 1997). During their dance displays, *C. capucinus* males and females coordinate their pirouettes and often face away from each other, only to look at their partner either over the shoulder or through the legs. A unique vocalization also accompanies mating in this species, and like the patterns described for *C. apella*, individuals involved in the mating and dance display do not make any attempts to hide their activities. Data on mating in *C. olivaceus* are completely absent from the published literature. The mating behavior and rituals of *C. albifrons*, although not well described do, not appear to be as elaborate as those of either *C. apella* or *C. capucinus*, but data for this species are very limited.

Unlike *Cebus*, *Saimiri* do not show any sort of preparatory behaviors that precede copulation; there are no special vocalizations, displays, or behaviors that indicate readiness or willingness to mate (Mendoza and Mason 1994). That said, *Saimiri* do display one of the most unique reproductive systems reported for nonhuman primates. Prior to and during the short annual mating season, male *S. boliviensis* and *S. oerstedii* undergo dramatic physiological and morphological changes; they gain up to 222 g, a shocking 22% of their body weight (Dumond and Hutchinson 1967) (see Table 7.3; there are no data available for other *Saimiri* species). Males begin accumulating "fat" prior to the start of the annual mating season, and this fattened state is maintained throughout the season, with males reaching their largest size during the same months that the majority of conceptions occur (Schiml et al. 1996). This "seasonal enhancement of male body size" (Boinski 1998: 174) results not from the accumulation of fat, as has been reported for rhesus macaques and other mammalian species (see Berkovitch 1992), but from the deposition of water between the skin and muscles, particularly along the shoulders, back, and arms; and it has been likened to the water retention–induced swelling that women often experience before the onset of menstrual cycles (Boinski 1998). This swelling of males does not occur because they are eating more or doing less. On the contrary, given the high level of reproductive synchrony observed among group females, males are extremely active during the mating season. Male "fattening" is also not related to a restricted seasonal ability to produce sperm. Mendoza et al. (1978) have shown that, although males do experience an increase in testes size and sperm production during the brief mating season, they are fertile year-round. Instead, the fattening appears to be in response to changing hormone levels within individual males, and males that become the fattest are those with the highest testosterone levels (Schiml et al. 1996).

During the nonmating season, there are no discernable size differences among coresident male *Saimiri*; however, with the onset of the mating season, some males become significantly larger than others and one male stands out as the largest. It is this male, the fattest male in a group, that is selectively preferred as a mating partner by group females. Indeed, Boinski (1992) reports that the largest male in her study group of *S. oerstedii* participated in over 70% of all observed copulations and the less swollen males were successful only in gaining solicitations from females after they had been rejected by the largest male. For *S. boliviensis*, the dominant male becomes the most fattened and apparently suppresses the reproductive potential of less swollen subordinate males, although it remains to be determined by exactly what means this suppression occurs (DuMond and Hutchison 1967). Outside of the mating season, *S. boliviensis* males usually occupy peripheral positions within the group and rarely, if ever, interact with the females that dominate them. However, during the mating season, all males become more central within the group and interact regularly with females (DuMond and Hutchinson 1967). Within groups of *S. oerstedii*, there are no discernable dominance hierarchies either within or between the sexes, even during the mating season. Therefore, rather than dominance status, male enlargement in this species appears to be linked with the length of time a male has spent in a particular group, with long-term residents (>3 years) being the most enlarged and, therefore, the most popular during the mating season (Boinski 1992). At present, it is unclear whether all populations and species of *Saimiri* undergo these seasonal changes; data from additional field sites are needed to address these issues.

The proximate trigger for these morphological changes is still unknown; however, it is possible that males respond to subtle pheromonal or behavioral cues from group females. Male *S. oerstedii* frequently perform genital inspections of females, and this has been interpreted as a means of assessing female reproductive status (Boinski 1987a). Although these inspections occur throughout the year, during the 2 months prior to the commencement of the annual mating season, which is when males begin to fatten, there is a notable increase in the frequency of male coalitionary mobbing of females in order to perform such inspections. Boinski's (1987a) study showed that, at their peak, these aggressive inspections occur approximately twice per hour and may include up to 16 males. Boinski even describes one female being wounded during such a mobbing (see Boinski et al. 2002). It is possible that these close inspections trigger the onset of seasonal swelling in males, although additional comparative data of intrasexual hormonal states are needed to test this suggestion. However, the fact that males are at their most swollen at the same time that group females achieve peak fertility (see Boinski 1992, Schiml et al. 1996, 1999) is a good indication that pheromones are at work.

Dominance appears to play a very strong role in the reproductive success of male *Cebus*. In all three species for which data on wild groups are available, a significant link between male dominance rank and reproductive success has been found, with alpha males siring the majority of infants born (*C. apella*, Escobar-Páramo 2000; *C. capucinus*, Jack and Fedigan 2003, in press; *C. olivaceus*, Valderrama et al. 2000). Given the rather divergent mating systems seen across the genus, the consistency of the positive correlation between male dominance rank and reproductive success is surprising. Although a distinctive male dominance hierarchy is discernable in *C. capucinus*, male–male intragroup competition for access to mates has not been documented. The alpha male does not monopolize matings, nor does he appear to be the exclusive target of female choice (Fedigan 1993); and all males within the group experience some degree of mating success (e.g., Rose 1998). However, paternity analysis in the Santa Rosa study groups has demonstrated that alpha males sire the majority of infants born into their groups (Jack and Fedigan 2003, in press). Consequently, regardless of the apparently egalitarian mating system of this

species, reproduction is not shared among group males (see Jack and Fedigan in press). A recent study of *C. capucinus* by Carnegie et al. (in press) has shown that matings between females and alpha males occur during fertile periods, while those between females and subordinate males tend to occur during nonfertile times (when females are pregnant, lactating, or in the postovulatory phase). However, it is still unclear which sex is responsible for the timing of fertile matings, that is, whether females select to mate with alpha males during their fertile times or if a more subtle form of mating competition occurs and alpha males somehow exclude subordinates from mating with fertile females.

The two other well-studied species of *Cebus* (*C. apella* and *C. olivaceus*) show tense relationships among groups males, and although both species reside in multimale groups, they are described as being unimale, with the dominant male monopolizing matings and subordinate males being peripheral group members (e.g., *C. apella*, Janson 1985, 1986; *C. olivaceus*, Robinson 1988, O'Brien 1991). Reproductive success within groups of these two species largely reflects the observed social patterns, with alpha males monopolizing reproductive success (*C. apella*, Escobar-Páramo 2000; *C. olivaceus*, Valderrama et al. 2000), although at least for *C. apella*, beta males also sire offspring, particularly with the daughters of alpha males (Escobar-Páramo 2000).

CEBINE DISPERSAL PATTERNS: TOWARD AN EXPLANATION OF VARIABLE SOCIAL STRUCTURE

The dispersal of one or both sexes from the birth group is a pattern common to all social mammals and birds, and many species show a bias toward the dispersal of one sex over the other (Greenwood 1980). The majority of Old World cercopithecines and prosimians are characterized by male-biased dispersal and female *philopatry* (the tendency to remain in the birth group for life), while New World monkeys and apes tend toward bisexual dispersal (e.g., *Alouatta* and *Gorilla*) or female-biased dispersal (e.g., *Ateles* and *Pan*) (see Pusey and Packer 1987 for review, Strier 1999). Dispersal is said to be among "the most important life history traits involved in both species persistence and evolution" (Clobert et al. 2001); however, despite its importance and the fact that it is one of the most studied phenomena, it remains among the most poorly understood issues in ecology and evolutionary biology (Clobert et al. 2001). Without a doubt, dispersal patterns have a profound influence on the type and nature of social relationships within groups. For example, in species characterized by male-biased dispersal, females within groups are related, most often show close bonds, and regularly form matrilineally based dominance hierarchies (e.g., *Macaca mulatta*, *M. fuscata*, and *Papio cynocephalus*). Within female philopatric groups, males are generally considered to be unrelated and their social relationships are usually characterized by very low levels of affiliation and cooperation. In general, the opposite pattern of social relationships is observed when males are philopatric (e.g., chimpanzees and muriquis, although bonobos are an exception to this pattern), while those species characterized by bisexual dispersal most often show much looser bonds among same-sexed individuals and closer bonds between the sexes (e.g., gorillas) (for reviews see Pusey and Packer 1987 and van Hooff 2000).

Unlike most platyrrhine species, *Cebus* are characterized by female philopatry and male dispersal from the birth group; in this respect, they more closely resemble Old World monkeys. In general, *Cebus* are considered female-bonded (Wrangham 1980) or resident nepotistic (Sterck et al. 1997) in that (*1*) they exhibit female-biased philopatry and male dispersal, (*2*) affiliative bonds among females are generally stronger than they are between the sexes or among males, (*3*) females develop dominance hierarchies, and (*4*) females appear to be responsible for the direction of group movement (Fragaszy et al. 2004b; but see Phillips and Newlon 2000 on *C. albifrons trinitatis*). In addition, field studies are shedding light on the importance of kinship in the formation of female–female relationships and, although it may not be as decisive a factor as observed among the Old World cercopithecines, this does appear to play a prominent role in at least some species of *Cebus* (*C. capucinus*, Rose 1998, Perry 1996; *C. olivaceus*, O'Brien and Robinson 1991). In terms of relationships between the sexes, males are generally individually dominant over adult females for all species; however, at least for the three better-studied species (*C. apella*, *C. capucinus*, and *C. olivaceus*), an alpha female will often rank directly below the alpha male and dominate individual subordinate males within the group (Fragaszy et al. 2004b). In addition, females will often form coalitions that enable them to displace even the alpha male from feeding trees, and at least for *C. capucinus*, females tend to direct more threats toward males than they receive from them (Perry 1997, Rose 1998). Overall, this results in somewhat egalitarian relationships between males and females; and across the genus, these relationships are best characterized as being affiliative in nature, with little physical aggression being exchanged between the sexes.

Throughout the genus, social relationships among group members are maintained through the exchange of frequent grooming, the maintenance of proximity, and the frequent formation of coalitions with preferred partners (Fragaszy et al. 2004b). Although social relationships among females and between the sexes across the better-studied *Cebus* species are remarkably consistent, male capuchins display extensive variation in their relationships with one another. Male–male relationships within the genus range from despotic (*C. apella*, Janson 1986; *C. capucinus*, Perry 1998b; *C. olivaceus*, Robinson 1988) to highly affiliative and cooperative (*C. albifrons*, Janson 1986; *C. capucinus*, Jack 2003a), and at least for *C. apella* and *C. capucinus*, which have been studied the most extensively and at multiple sites, this variability in male–male relationships exists both within and among species. For example, the first published accounts of

male relationships in *C. apella* in Peru reported that alpha males directed high rates of aggression toward subordinate males, which remained on the periphery of the group, avoiding interactions with the alpha male and resident females (Janson 1985, 1986). However, additional studies on *C. apella* at multiple sites have yielded a more diverse view of male relationships within this species. In Colombia (Izawa 1980, 1994), Brazil (Lynch et al. 2002), and Suriname (Kauffman et al. 2004), males are reported to display low rates of aggression and frequently interact affiliatively, while in Argentina (Janson 1998 and personal communication), male *C. apella* exhibit greater levels of cooperation in resource defense, although relationships between dominants and subordinates are still described as agonistic.

Similarly, male relationships in *C. capucinus* have been reported as highly variable, and this variability occurs not only across study sites but also among groups within the same study population (e.g., Santa Rosa National Park in Costa Rica, where studies have been ongoing since 1983; see Jack 2003a). In terms of access to mates and resources, relationships among males are fairly egalitarian, with little aggression being exchanged among group males. Although a distinct alpha male is discernable within groups, it is the extent to which he asserts his dominance that appears to be most variable. For example, some alpha males actively, and often aggressively, disrupt affiliative interactions among subordinate coresident males and spend the majority of their time affiliating with groups females (Perry 1998a,b), while other alphas form close affiliative relationships with subordinates and, in some groups, spend more time interacting with them than with group females (Jack 2003a).

This diversity in male relationships across *Cebus* likely reflects the behavioral plasticity characteristic of the genus (see Fragaszy et al. 1990), and within species, male relationships are apt to change in response to ecological and/or social pressures (e.g., Janson 1998, Di Bitetti and Janson 2001). Ecological factors appear to play an important role in determining male–male relationships in *C. apella*, which has now been studied in five countries throughout South America (Argentina, Brazil, Colombia, Peru, and most recently Suriname). Janson (1986) suggests that it is the distribution of food resources and a male's ability to monopolize access to them that most profoundly influences primate mating systems, which in turn dictate the types of social relationship among group males. Janson (1986) argues that if a single male is able to dominate individual access to resources, it is likely that females will choose to mate with him to ensure resource access for themselves and their infants. Under these circumstances, the mating system essentially becomes unimale. Such a skewed mating system results in rather tense male–male relationships, with males tending to be less inclined to cooperate in group and resource defense when there is little reward for their assistance (i.e., little mating activity) (see van Hooff 2000). If, on the other hand, resources are distributed in such a way that they cannot be monopolized by a single individual (e.g., they occur in large patches), females need not choose to mate exclusively with the alpha male and a more egalitarian mating system may result. This mating system may lead to increased male cooperation in resource defense (Janson 1984), although it may also result in increased within-group mating competition and does not necessarily trigger affiliative relationships among group males.

Such differences in ecological factors, namely the defensibility of food resources, do not, however, explain the intergroup variability in relationships among coresident male *C. capucinus* that is reported to occur within the same study populations (see Jack 2003a; S. Perry, personal communication). It appears that within groups of *C. capucinus*, male familiarity, and perhaps kinship, best explains the observed variability (Jack 2003a). Although *C. capucinus*, like all *Cebus* species, is characterized by the emigration of males from their birth group, high rates of parallel dispersal can, in the absence of philopatry, promote the retention of kinship among group males (van Hooff 2000). Parallel dispersal can occur through the coordinated emigration of male siblings or the movement of males toward groups that contain familiar, previously dispersed males. This type of coordinated male dispersal has been reported for several primate species, although in most species it appears to be limited to particular life phases and to be most common among immature males (see Jack 2003b for review). However, studies of *C. capucinus* in Santa Rosa National Park, Costa Rica, found that parallel dispersal occurs at extremely high levels (67%–80% of all emigrations depending on male age class), lasts through multiple migrations, and persists at high rates even among adult males (Jack and Fedigan 2004a,b). In addition, analysis of male–male interactions in this species has shown that familiar males do display more affiliative relationships (Jack 2003a), and Janson (personal communication) suspects that male familiarity and/or relatedness may well account for the more cooperative relationships he has observed among male *C. apella* in Argentina. *C. apella* groups in Argentina contain more males than those Janson first studied in Peru, which may enable parallel dispersal to occur more readily. Relatedness among group males in both *C. apella* in Argentina and *C. capucinus* in Costa Rica, and perhaps in *C. olivaceus* (although very little is known about the dispersal patterns of this species), may explain the high degree of cooperation among group males in the face of the unimale reproductive system each of these species exhibits (see Mating Systems, above). We eagerly await genetic analyses of male kinship within groups so that these predictions can be tested.

Despite the remarkable similarity that *Saimiri* species display in terms of morphology and general habitat preference, the genus shows an even greater diversity in social relationships than we see in *Cebus*. However, across the three species for which sufficient data are available (*S. oerstedii*, *S. boliviensis*, and *S. sciureus*), it is the social relationships among group females and between males and females that are extremely variable, while relationships among group males are surprisingly consistent (Table 7.4). Boinski, who has studied this genus extensively over the past two decades, claims that "squirrel monkeys arguably exhibit the most geographically variable social organization of any set of closely related primate populations" (1998: 179). The variability that we see across the genus appears to directly reflect the extremely divergent dispersal patterns across species, which are likely determined by the nature and distribution of resources each species exploits. Interestingly, the formation and maintenance of social bonds throughout this genus are not based on grooming interactions, as is the case for many nonhuman primates. In fact, *Saimiri* are among the few primate species where social grooming is almost completely absent. Instead, social relationships are based on proximity patterns, tolerance around feeding sources, the frequent exchange of vocalizations, and the formation of alliances in some species (see Sussman 2000 for review; Boinski 1999).

S. oerstedii in Costa Rica are characterized by female emigration from the birth group (Boinski and Mitchell 1994) and a flexible pattern of male philopatry. A natal male will generally remain in his birth group until he reaches sexual maturity (4–5 years), after which he will either take up one of the few reproductive positions within the group or disperse with members of his age cohort and attempt to take over breeding positions in another group (Boinski 1998). No matter which pattern is followed, it is thought that resident males of this species are related as a result of parallel dispersal (see above). This dispersal pattern leads to strong affiliative bonds among group males, while female bonds are described as weak (Boinski 1999, Boinski and Mitchell 1994) and relationships between males and females are described as being extremely egalitarian, with neither males nor females being dominant over the other (Boinski 1987a, 1988). With the exception of the 2-month period prior to the mating season, when males are observed to form coalitions and frequently mob females to perform genital inspections (see Mating Systems, above), aggression within and between the sexes very rarely occurs. Even intergroup interactions are described as neutral, although avoidance is generally practiced (Boinski, 1987a, 1988).

In contrast, *S. boliviensis* in Peru display female philopatry; and accordingly, although perhaps more so than any other species of platyrrhine primate, females form tight-knit matrilineal relationships that cooperate in resource acquisition and defense (Mitchell 1990, 1994; Mitchell et al. 1991). Intragroup resource competition is described as occurring at moderate levels within this species, and it occurs both among and between the sexes (Boinski 1999). Interestingly, despite the fact that this is the most sexually dimorphic of the *Saimiri* species, with males being 24% larger than females (Table 7.2), females form stable linear dominance hierarchies and actually dominate males, aggressively forcing them to occupy peripheral positions within the group (Mitchell 1990, 1994). Within these peripheral subgroups, males also form stable linear dominance hierarchies and, although social aggression is described as commonly occurring (Boinski 1999), affiliative coalitions that are often maintained through multiple emigrations (i.e., parallel dispersal) and may work to ensure kinship among group males (Mitchell 1994). *S. boliviensis* groups have home ranges that overlap extensively, and intergroup interactions are nonaggressive; in areas where fruit patch size allows, it is not uncommon to see groups foraging together (Mitchell 1990).

S. sciureus presents yet another configuration of dispersal patterns, social interactions, and dominance structures for the genus. Both male and female *S. sciureus* appear to disperse from their birth group, although Boinski et al. (2002) describe the dispersal patterns as being flexible in that the timing of female natal emigration may occur before or after the first mating season. *S. sciureus* display the highest level of intragroup aggression reported among the three *Saimiri* species. The majority of this aggression occurs within the context of resource acquisition, and wounding of females and immatures is common, something which is rarely seen in the other two species (Boinski et al. 2002). Overall, *S. sciureus* is best described as being male-dominant; and although both males and females form stable linear dominance hierarchies, males are reported to be more closely bonded in that they form more frequent cooperative coalitions with moderate to close affiliative relationships (Boinski 1999). Unlike the other two species of *Saimiri*, which do not display tense interactions with neighboring groups, *S. sciureus* in Suriname are described as territorial. Their home ranges show minimal overlap with other groups, and intergroup interactions are described as highly agonistic (Boinski 1999).

In an attempt to explain the variable patterns of female relationships seen across this genus, Boinski et al. (2002) provide an extensive examination of the detailed ecological data available for the three species of *Saimiri* discussed above. Following Janson (1986) and his explanation of variable relationships among *Cebus* males (see above), Boinski et al. (2002) suggest that the extreme variability in dispersal patterns and the corresponding social relationships among group members (at least among group females) that characterize *Saimiri* are largely attributed to ecological factors associated with the defensibility of food resources, namely fruit patch size and abundance. As mentioned above, *Saimiri* are classified as insectivore–frugivores, and insects form the bulk of their diets, with all species spending an average of 75%–80% of their time foraging for insects (Terborgh 1983, Boinski 1988). The distribution of arthropods does not enable a single individual, or even groups of individuals, to effectively monopolize the food supply; therefore, direct competition (contest competition) for this resource is absent (Boinski 1988, Mitchell 1990). However, fruit is also a key resource exploited by *Saimiri*; and even though the habitats of the three well-known species are quite uniform in terms of relative abundance and distribution of arthropods, these habitats do display extensive variability in fruit production.

A comparison of fruit patch size and abundance demonstrates that, overall, Peruvian forests have the highest levels of fruit abundance and the lowest fluctuations in seasonal availability of fruit, followed by Costa Rican forests, while Surinamese forests show the lowest abundance and greatest seasonal fluctuations (Boinski et al. 2002). Hence, it is not surprising that in Peruvian forests, where food patches are too large to be effectively monopolized by a single individual, females cooperate with kin to defend access to these resources and exclude males. Cooperation is necessary because males are larger and, therefore, it is beneficial for females to remain philopatric and cooperate with their kin to control access to resources (Boinski et al. 2002; see also Sterck et al. 1997). In Costa Rica, where fruit is distributed in such a way that it cannot be monopolized by either a group or a single individual (it can be eaten on the spot on a first-come, first-served basis), it is not advantageous for females to remain in their birth group, a condition which most often leads to male philopatry. In the Surinamese forests, fruit patches are small and in relatively low abundance, making an individual patch easily monopolized by single individuals and of great benefit for those that can gain access to them, namely dominants. This type of distribution leads to very strong contest competition; therefore, it makes sense for a female to disperse so that she is not directly competing with kin for access to necessary resources. According to Boinski et al. (2002), this type of situation will not promote stable coalitions among individuals because cheating (i.e., reneging on sharing resources with a coalition partner) would provide great benefits to dominant individuals.

Boinski and Cropp (1999) state that, across the three species, relationships among male squirrel monkeys are rather consistent—males show affiliative bonds—while female–female bonds vary from strong to weak and male–female relationships also vary from species to species. They provide a very good example of how even slight differences in ecological factors (e.g., quality, size, and defensibility of food patches) can have profound effects on primate social structure and how these factors influence the variable nature of female–female and female–male relationships across the genus. They do not, however, offer an explanation for the consistency of male–male relationships in the face of the vast ecological variability that can be seen across the genus. Familiarity and/or relatedness among group males likely explains the relative consistency in male–male relationships across the genus *Saimiri*. It seems likely that in all three species males within groups maintain some degree of relatedness to one another. In *S. oerstedii*, males either are philopatric or disperse together in cohorts, which also appears to be the case for male dispersers in *S. sciureus* and *S. boliviensis*. In *S. boliviensis*, where male emigration is the norm, relatedness among group males also appears to be maintained through parallel dispersal. Mitchell (1994) reports that males form migration alliances whose composition often remains constant over multiple emigration events, and this may act to ensure the relatedness of males in the face of dispersal. These data show that, regardless of ecological pressures and female dispersal patterns, males in the genus appear to have evolved a mechanism for retaining residence with their kin, regardless of dispersal patterns.

Additional comparative studies are needed to further our understanding of ecological pressures on social relationships, mating systems, and dispersal patterns. In addition, genetic data on relatedness of group members (both males and females) and more detailed reports on dispersal patterns and the fates of dispersing individuals in additional species are required in order to attain a more complete overview of the dynamics and interactions of these processes on primate behavior.

REFERENCES

Ayres, J. M. (1985). On a new species of squirrel monkey, genus *Saimiri*, from Brazilian Amazonia (Primates, Cebidae). *Pap. Avul. Zool.* 36:147–164.

Baldwin, J. D., and Baldwin, J. L. (1981). The squirrel monkeys, genus *Saimiri*. In: Coimbra-Filho, A. F., and Mittermeir, R. A. (eds.), *Ecology and Behavior of Neotropical Primates*. Academis Brasileira de Ciencias, Rio de Janeiro. pp. 277–330.

Berkovitch, F. R. (1992). Estradiol concentrations, fat deposits, and reproductive strategies in male rhesus macaques. *Horm. Behav.* 26:272–282.

Boinski, S. (1987a). Mating patterns in squirrel monkeys (*Saimiri oerstedii*): implications for seasonal sexual dimorphism. *Behav. Ecol. Sociobiol.* 21:13–21.

Boinski, S. (1987b). Birth synchrony in squirrel monkeys (*Saimiri oerstedii*): a strategy to reduce neonatal predation. *Behav. Ecol. Sociobiol.* 21:393–400.

Boinski, S. (1987c). Habitat use by squirrel monkeys (*Saimiri oerstedii*) in Costa Rica. *Folia Primatol.* 49:151–167.

Boinski, S. (1988). Sex differences in foraging behavior of squirrel monkeys: ecological implications. *Behav. Ecol. Sociobiol.* 21:177–186.

Boinski, S. (1989a). The positional behavior and substrate use of squirrel monkeys: ecological implications. *J. Hum. Evol.* 18:659–677.

Boinski, S. (1989b). Why don't *Saimiri oerstedii* and *Cebus capucinus* for mixed-species groups? *Int. J. Primatol.* 10:103–114.

Boinski, S. (1991). The coordination of spatial position: a field study of the vocal behavior of adult female squirrel monkeys. *Anim. Behav.* 41:89–120.

Boinski, S. (1992). Monkeys with inflated sex appeal. *Nat. Hist.* 101:42–49.

Boinski, S. (1994). Affiliation patterns among male Costa Rican squirrel monkeys. *Behaviour* 130:191–209.

Boinski, S. (1998). Monkeys with inflated sex appeal. In: Ciochon, R. L., and Nisbett, R. A. (eds.), *The Primate Anthology: Essays on Primate Behavior, Ecology, and Conservation from Natural History*. Prentice Hall, Englewood Cliffs, NJ. pp. 174–179.

Boinski, S. (1999). The social organization of squirrel monkeys: implications for ecological models of social evolution. *Evol. Anth.* 8:101–112.

Boinski, S., and Cropp, S. (1999). Disparate data sets resolve squirrel monkey (*Saimiri*) taxonomy: implications for behavioral ecology and biomedical usage. *Int. J. Primatol.* 20:237–256.

Boinski, S., Kauffman, L., Westoll, A., Stickler, C. M., Cropp, S., and Ehmke, E. (2003). Are vigilance, risk from avian predators

and group size consequences of habitat structure? A comparison of three species of squirrel monkey (*Saimiri oerstedii, S. boliviensis*, and *S. sciureus*). *Behaviour* 140:1421–1467.

Boinski, S., and Mitchell, C. L. (1994). Male resident and association patterns in Costa Rican squirrel monkeys (*Saimiri oerstedii*). *Am. J. Primatol.* 34:157–169.

Boinski, S., Sughrue, K., Selvaggi, L., Quatrone, R., Henry, M., and Cropp, S. (2002). An expanded test of the ecological model of primate social evolution: competitive regimes and female bonding in three species of squirrel monkeys (*Saimiri oerstedii, S. boliviensus*, and *S. sciureus*). *Behaviour* 139:227–261.

Carnegie, S. D., Fedigan, L. M., and Ziegler, T. E. (in press). Postconceptive mating in white-faced capuchins, *Cebus capucinus*: hormonal and sociosexual patterns of cycling, non-cycling and pregnant females. In: Estrada, A., Garber, P., Luecke, L., and Pavelka, M. S. M. (eds.), *New Perspectives in the Study of Mesoamerican Primates: Distribution, Ecology, Behavior, and Conservation*. Kluwer Academic/Plenum, New York.

Chapman, C. A. (1986). Boa constrictor predation and group response in white-faced *Cebus* monkeys. *Biotropica* 18:171–172.

Chapman, C. A., Chapman, L. J., and Glander, K. E. (1989). Primate population in northwestern Costa Rica: potential for recovery. *Primates* 10:37–44.

Clobert, J., Danchin, E., Dohndt, A. A., and Nichols, J. D. (2001). *Dispersal*. Oxford University Press, Oxford.

Costello, R. K., Dickinson, C., Rosenberger, A. L., Boinski, S., and Szalay, F. S. (1993). Squirrel monkey (genus *Saimiri*) taxonomy: a multidisciplinary study of the biology of species. In: Kimbell, W. H., and Martin, L. B. (eds.), *Species, Species Concepts, and Primate Evolution*. Plenum Press, New York. pp. 177–210.

Cropp, S., and Boinski, S. (2000). The Central American squirrel monkey (*Saimiri oerstedii*): introduced hybrid or endemic species? *Mol. Phylogenet. Evol.* 16:350–365.

Defler, T. R. (1979a). On the ecology and behavior of *Cebus albifrons* in eastern Colombia: I. Ecology. *Primates* 20:475–490.

Defler, T. R. (1979b). On the ecology and behavior of *Cebus albifrons* in eastern Colombia: II. Behavior. *Primates* 20:491–502.

Defler, T. R. (1982). A comparison of intergroup behavior in *Cebus albifrons* and *C. apella*. *Primates* 23:385–392.

Di Bitetti, M. S., and Janson, C. H. (2000). When will the stork arrive? Patterns of birth seasonality in neotropical primates. *Am. J. Primatol.* 50:109–130.

Di Bitetti, M. S., and Janson, C. H. (2001). Reproductive socioecology of tufted capuchins (*Cebus apella nigritus*) in northeastern Argentina. *Int. J. Primatol.* 22:127–142.

Dixson, A. F. (1983). Observations on the evolution and behavioral significance of "sexual skin" in female primates. *Adv. Study Behav.* 13:63–106.

DuMond, F. V., and Hutchinson, T. C. (1967). Squirrel monkey reproduction: the "fatted" male phenomenon and seasonal spermatogenesis. *Science* 158:1067–1070.

Eisenberg, J. F. (1979). Habitat, economy, and society: some correlations and hypotheses for the neotropical primates. In: Bernstein, I. S., and Smith, E. O. (eds.), *Primate Ecology and Human Origins: Ecological Influences on Social Organization*. Garland STPM Press, New York. pp. 215–262.

Eisenberg, J. F. (1989). *Mammals of the Neotropics. The Northern Neotropics Panamá, Colombia, Venezuela, Guyana, Suriname, French Guiana*, vol. 1. University of Chicago Press, Chicago.

Elliot, D. G. (1913). *A Review of the Primates. Monograph Series*, vol. II. American Museum of Natural History, New York.

Escobar-Páramo, P. (2000). *Inbreeding avoidance and the evolution of male mating strategies* [PhD thesis]. State University of New York, Stony-brook.

Fedigan, L. M. (1990). Vertebrate predation in *Cebus capucinus*: meat eating in a neotropical monkey. *Folia Primatol.* 54:196–205.

Fedigan, L. M. (1993). Sex differences and intersexual relations in adult white-faced capuchins (*Cebus capucinus*). *Int. J. Primatol.* 14:853–877.

Fedigan, L. M., and Rose, L. M. (1995). Interbirth interval variation in three sympatric species of neotropical monkey. *Am. J. Primatol.* 37:9–24.

Fedigan, L. M., Rosenberger, A. L., Boinski, S., Norconk, M. A., and Garber, P. A. (1996). Critical issues in cebine evolution and behavior. In: Norconk, M. A., Rosenberger, A. L., and Garber, P. A. (eds.), *Adaptive Radiations of Neotropical Primates.* Plenum Press, New York. pp. 219–228.

Fleagle, J. G. (1999). *Primate Adaptation and Evolution*, 2nd ed. Academic Press, San Diego.

Ford, S. M., and Davis, L. C. (1992). Systematics and body size: implications for feeding adaptations in New World monkeys. *Am. J. Phys. Anthropol.* 88:415–468.

Ford, S. M., and Hobbs, D. G. (1996). Species definition and differentiation as seen in the postcranial skeleton of *Cebus*. In: Norconk, M. A., Rosenberger, A. L., and Garber, P. A. (eds.), *Adaptive Radiations of Neotropical Primates.* Plenum Press, New York. pp. 229–249, 540–541.

Fragaszy, D. (1990). Sex and age differences in the organization of behavior in wedge-capped capuchins, *Cebus olivaceus. Behav. Ecol.* 1:81–94.

Fragaszy, D., Izar, P., Visalberghi, E., Ootóni, E. B., and de Oliveira, M. G. (2004a). Wild capuchin monkeys use anvils and stone pounding tools. *Am. J. Primatol.* 64:359–366.

Fragaszy, D., Vesalberghi, E., and Fedigan, L. (2004b). *The Complete Capuchin: The Biology of the Genus* Cebus. Cambridge University Press, Cambridge.

Fragaszy, D. M., Visalberghi, E., and Robinson, J. G. (1990). Variability and adaptability in the genus *Cebus. Folia Primatol.* 54:114–118.

Freese, C. H. (1983). *Cebus capucinus*, mono cara blanca, white-faced capuchin. In: Janzen, D. H. (ed.), *Costa Rica Natural History*. University of Chicago Press, Chicago. pp. 458–460.

Freese, C. H., and Oppenheimer, J. R. (1981). The capuchin monkey, genus *Cebus*. In: Coimbra-Filho, A. F., and Mittermeier, R. H. (eds.), *Ecology and Behaviour of Neotropical Primates*, vol. 1. Academia Brasilia, Rio de Janeiro. pp. 331–390.

Greenwood, P. J. (1980). Mating systems, philopatry, and dispersal in birds and mammals. *Anim. Behav.* 28:1140–1162.

Groves, C. P. (2001). *Primate Taxonomy*. Smithsonian Institution Press, Washington DC.

Hakeem, A., Sandoval, R. G., Jones, M., and Allman, J. (1996). Brain and life span in primates. In: Birren, J. E., and Schaie, K. W. (eds.), *Handbook of the Psychology of Aging*, 4th ed. Academic Press, San Diego. pp. 78–104.

Hartwig, W. C. (1995). Effect of life history on the squirrel monkey (Platyrrhini, *Saimiri*) cranium. *Am. J. Phys. Anthropol.* 97:435–449.

Hartwig, W. C. (1996). Perinatal life history traits in New World monkeys. *Am. J. Primatol.* 40:99–130.

Harvey, P. H., Martin, R. D., and Clutton-Brock, T. H. (1987). Life histories in comparative perspective. In: Smuts, B. B., Cheney, D. L., Seyfarth, R. M., Wrangham, R. W., and Struhsaker, T. T. (eds.), *Primate Societies*. University of Chicago Press, Chicago. pp. 181–196.

Hershkovitz, P. (1949). Mammals of northern Colombia. Preliminary report no. 4. Monkeys (Primates) with taxonomic revisions of some forms. *Proc. U.S. Natl. Mus.* 98:323–427.

Hershkovitz, P. (1969). The recent mammals of the neotropical region: a zoogeographic and ecological review. *Q. Rev. Biol.* 44:1–70.

Izawa, K. (1980). Social behavior of the wild black-capped capuchin (*Cebus apella*). *Primates* 20:57–76.

Izawa, K. (1994). Group division of wild black-capped capuchins. *Field Studies of the New World Monkeys. La Macarena, Colombia*, 9:5–14.

Jack, K. (2003a). Affiliative relationships among male white-faced capuchins (*Cebus capucinus*): evidence of male bonding in a female bonded species. *Folia Primatol.* 74:1–16.

Jack, K. (2003b). Males on the move: evolutionary significance of secondary dispersal in male nonhuman primates. *Primate Rep.* 67:61–84.

Jack, K. M. (2001). Effect of male emigration on the vigilance behavior of coresident males in white-faced capuchins (*Cebus capucinus*). *Int. J. Primatol.* 22:715–732.

Jack, K. M., and Fedigan, L. M. (in press). Why be alpha? Dominance and reproductive success in wild white-faced capuchins (*Cebus capucinus*). In: Estrada, A., Garber, P., Luecke, L., and Pavelka, M. S. M. (eds.), *New Perspectives in the Study of Mesoamerican Primates: Distribution, Ecology, Behavior, and Conservation*. Kluwer Academic/Plenum, New York.

Jack, K., and Fedigan, L. M. (2003). Male dominance and reproductive success in white-faced capuchins (*Cebus capucinus*). *Am. J. Phys. Anthropol.* 36(suppl.):121–122.

Jack, K. M., and Fedigan, L. M. (2004a). Male dispersal patterns in white-faced capuchins (*Cebus capucinus*). Part 1: Patterns and causes of natal emigration. *Anim. Behav.* 67:761–769.

Jack, K. M., and Fedigan, L. M. (2004b). Male dispersal patterns in white-faced capuchins (*Cebus capucinus*). Part 2: Patterns and causes of secondary dispersal. *Anim. Behav.* 67(4):771–782.

Janson, C. H. (1984). Female choice and mating system of the brown capuchin monkey (*Cebus apella*) (Primates: Cebidae). *Z. Tierpyschol.* 65:177–200.

Janson, C. H. (1985). Aggressive competition and individual food consumption in wild brown capuchin monkeys (*Cebus apella*). *Behav. Ecol. Sociobiol.* 18:125–138.

Janson, C. H. (1986). The mating system as a determinant of social evolution in capuchin monkeys (*Cebus*). In: Else, J. G., and Lee, P. C. (eds.), *Primate Ecology and Conservation*. Cambridge University Press, Cambridge. pp. 169–179.

Janson, C. H. (1998). Capuchin counterpoint. In: Ciochon, R. L., and Nisbett, R. A. (eds.), *The Primate Anthology. Essays on Primate Behavior, Ecology, and Conservation from Natural History*. Prentice Hall, Englewood Cliffs, NJ. pp. 153–159.

Janson, C. H., and Boinski, S. (1992). Morphological and behavioral adaptations for foraging in generalist primates: the case of the cebines. *Am. J. Phys. Anthropol.* 88:483–498.

Kappeler, P. M., and Pereira, M. E. (2003). *Primate Life Histories and Socioecology*. University of Chicago Press, Chicago.

Kauffman, L., Ehmke, E., and Boinski, S. (2004). Increased male–male cooperation among brown capuchin monkeys (*Cebus apella*) in Suriname. *Am. J. Phys. Anthropol.* 23:123.

Kay, R. F. (1981). The nut-crackers—a new theory of the adaptations of the Ramapithecinae. *Am. J. Phys. Anthropol.* 55:141–151.

Kay, R. F., Plavcan, J. M., Glander, K. E., and Wright, P. C. (1988). Sexual selection and canine dimorphism in New World monkeys. *Am. J. Phys. Anthropol.* 77:385–397.

Kinzey, W. G. (1997). *New World Primates. Ecology, Evolution and Behavior.* Aldine Press, New York.

Lynch, J. W., Ziegler, T. E., and Strier, K. B. (2002). Individual and seasonal variation in fecal testosterone and cortisol levels in wild male tufted capuchin monkeys, *Cebus apella nigritus.* *Horm. Behav.* 41:275–287.

Manson, J. H., Perry, S., and Parish, A. R. (1997). Nonconceptive sexual behavior in bonobos and capuchins. *Int. J. Primatol.* 18:767–786.

Masterson, T. J. (1995). Morphological relationships between the Ka'apor capuchin (*Cebus kaapori* Queiroz 1992) and other male *Cebus* crania: a preliminary report. *Neotrop. Primates* 3:165–169.

Mendoza, S. P., Lowe, E. L., Davidson, J. M., and Levine, S. (1978). Annual cyclicity in the squirrel monkey (*Saimiri sciureus*): the relationship between testosterone, fatting, and sexual behavior. *Horm. Behav.* 11:295–303.

Mendoza, S. P., and Mason, W. A. (1994). Constitution and context: the social modulation of temperament. In: Thierry, B., Anderson, J. R., Roeder, J. J., and Herrenschmidt, N. (eds.), *Current Primatology. Social Development, Learning and Behaviour,* vol II. University Louis Pasteur, Strasbourg, Germany. pp. 251–256.

Mitchell, C. L. (1990). The ecological basis for female social dominance: a behavioral study of the squirrel monkeys (*Saimiri sciureus*) [PhD thesis]. Princeton University, Princeton, NJ.

Mitchell, C. L. (1994). Migration alliances and coalitions among adult male South American squirrel monkeys (*Saimiri sciureus*). *Behavior* 130:169–190.

Mitchell, C. L., Boinski, S., and van Schaik, C. P. (1991). Competitive regimes and female bonding in two species of squirrel monkeys (*Saimiri oerstedii* and *S. sciureus*). *Behav. Ecol. Sociobiol.* 28:55–60.

Nagle, C. A., and Denari, J. H. (1982). The reproductive biology of capuchin monkeys. *Int. Zoo Yrbk.* 22:143–150.

O'Brien, T. C. (1991). Female–male social interactions in wedge-capped capuchin monkeys: benefits and costs of group living. *Anim. Behav.* 41:555–567.

O'Brien, T. C., and Robinson, J. G. (1991). Allomaternal care by female wedge-capped capuchin monkeys: effects of age, rank and relatedness. *Behaviour* 119:30–50.

Panger, M. A., Perry, S., Rose, L., Gros-Louis, J., Vogel, E., Mackinnon, K. C., and Baker, M. (2002). Cross-site differences in foraging behavior of white-faced capuchins. *Am. J. Phys. Anthropol.* 119:52–66.

Parker, S. T., and Gibson, K. R. (1977). Object manipulation, tool use and sensorimotor intelligence as feeding adaptations in *Cebus* monkeys and great apes. *J. Hum. Evol.* 6:623–641.

Patino, E., Borda, J. T., and Ruiz, J. C. (1996). Sexual maturity and seasonal reproduction in captive *Cebus apella.* *Lab. Primate Newsl.* 35:8–10.

Perry, S. (1996). Female–female social relationships in wild white-faced capuchin monkeys, *Cebus capucinus.* *Am. J. Primatol.* 40:167–182.

Perry, S. (1997). Male–female social relationships in wild white-faced capuchins (*Cebus capucinus*). *Behaviour* 134:477–510.

Perry, S. (1998a). A case report of a male rank reversal in a group of wild white-faced capuchins (*Cebus capucinus*). *Primates* 39:51–70.

Perry, S. (1998b). Male–male social relationships in wild white-faced capuchins, *Cebus capucinus.* *Behaviour* 135:139–172.

Perry, S., and Rose, L. M. (1994). Begging and transfer of coati meat by white-faced capuchin monkeys, *Cebus capucinus.* *Primates* 35:409–415.

Phillips, K. A., and Abercrombie, C. L. (2003). Distribution and conservation status of the primates of Trinidad. *Primate Conserv.* 19:19–22.

Phillips, K. A., and Newlon, K. (2000). Female–female social relationships in white-fronted capuchins (*Cebus albifrons*): testing hypotheses about resource size and quality. *Am. J. Primatol.* 51(suppl.):81.

Promislow, D. E. L., and Harvey, P. H. (1990). Living fast and dying young: a comparative analysis of life-history variation among mammals. *J. Zool.* 220:417–437.

Purvis, A., Webster, A. J., Agapow, P. M., Jones, K. E., and Isaac, N. J. B. (2003). Primate life histories and phylogeny. In: Kappeler, P. M., and Pereira, M. E. (eds.), *Primate Life Histories and Socioecology.* University of Chicago Press, Chicago. pp. 24–40.

Pusey, A. E., and Packer, C. (1987). Dispersal and philopatry. In: Smuts, B. B., Cheney, D. L., Seyfarth, R. M., Wrangham, R. W., and Struhsaker, T. T. (eds.), *Primate Societies.* University of Chicago Press, Chicago. pp. 250–266.

Queiroz, H. L. (1992). A new species of capuchin monkey, genus *Cebus* Erxleben 1977 (Cebidae, Primates), from eastern Brazil Amazonia. *Goeldiana Zool.* 15:1–3.

Robinson, J. G. (1988). Demography and group structure in wedge-capped capuchin monkeys, *Cebus olivaceus.* *Behaviour* 104:202–232.

Robinson, J. G., and Janson, C. H. (1987). Capuchins, squirrel monkeys, and atelines: socioecological convergence with Old World primates. In: Smuts, B. B., Cheney, D. L., Seyfarth, R. M., Wrangham, R. W., and Struhsaker, T. T. (eds.), *Primate Societies.* University of Chicago Press, Chicago. pp. 69–82.

Rose, L. M. (1998). Behavioral ecology of white-faced capuchins (*Cebus capucinus*) in Costa Rica [PhD diss.]. Washington University, St. Louis.

Rose, L. M., and Fedigan, L. M. (1995). Vigilance in white-faced capuchins, *Cebus capucinus,* in Costa Rica. *Anim. Behav.* 49:63–70.

Rowe, N. (1996). *The Pictorial Guide to the Living Primates.* Ponganias Press, New York.

Rylands, A. B., Schneider, H., Langguth, A., Mittermeier, R. A., Groves, C. P., and Rodríguez-Luna, E. (2000). An assessment of the diversity of New World primates. *Neotrop. Primates* 8:61–93.

Schiml, P. A., Mendoza, S. P., Saltzman, W., Lyons, D. M., and Madon, W. A. (1996). Seasonality in squirrel monkeys (*Saimiri sciureus*): social facilitation by females. *Physiol. Behav.* 60:1105–1113.

Schiml, P. A., Mendoza, S. P., Saltzman, W., Lyons, D. M., and Madon, W. A. (1999). Annual physiological changes in individually housed squirrel monkeys (*Saimiri sciureus*). *Am. J. Primatol.* 47:93–103.

Schneider, H., Canavez, F. C., Sampaio, I., Moreira, M. A. M., Tagliaro, C. H., and Seuanez, H. N. (2001). Can molecular data place each neotropical monkey in its own branch? *Chromosoma* 109:515–523.

Schneider, H., and Rosenberger, A. L. (1996). Molecules, morphology and platyrrhine systematics. In: Norconk, M. A., Rosenberger, A. L., and Garber, P. A. (eds.), *Adaptive Radiations of Neotropical Primates*. Plenum Press, New York. pp. 3–19.

Schneider, H., Schneider, M. P. C., Sampaio, I., Harada, M. L., Barroso, Stanhopes, M., Czelusniak, J., and Goodman, M. (1993). Molecular phylogeny of the New World monkeys (Platyrrhini, Primates). *Mol. Phylogenet Evol.* 2:225–242.

Stephen, H., Baron, G., and Frahm, H. D. (1988). Comparative size of brains and brain components. In: Steklis, H. D., and Erwin, J. (eds.), *Comparative Primate Biology. Neurosciences*, vol. 4. Wiley-Liss, New York. pp. 1–38.

Sterck, E. H. M., Watts, D. P., and van Schaik, C. P. (1997). The evolution of female social relationships in nonhuman primates. *Behav. Ecol. Sociobiol.* 41:291–309.

Strier, K. B. (1997). Mate preference of wild muriqui monkeys (*Brachyteles arachnoids*): reproductive and social correlates. *Folia Primatol.* 68:120–133.

Strier, K. B. (1999). Why is female kin bonding so rare? Comparative sociality of neotropical primates. In: Lee, P. C. (ed.), *Comparative Primate Socioecology*. Cambridge University Press, Cambridge. pp. 300–319.

Strier, K. B. (2001). Reproductive ecology of New World monkeys. In: Ellison, P. T. (ed.), *Reproductive Ecology and Human Evolution*. Aldine de Gruyter, New York. pp. 351–367.

Sussman, R. W. (2000). *Primate Ecology and Social Structure. New World Monkeys*, vol. 2. Pearson Custom Publishing, Boston.

Terborgh, J. W. (1983). *Five New World Primates: A Study in Comparative Ecology*. Princeton University Press, Princeton, NJ.

Terborgh, J. W., and Janson, C. H. (1986). The socioecology of primate groups. *Annu. Rev. Ecol. System.* 17:111–136.

Tyler, D. E. (1991). The evolutionary relationships of *Aotus. Folia Primatol.* 56:50–52.

Valderrama, X., Robinson, J. G., and Melnick, D. J. (2000). Females control male reproductive success in wedge-capped capuchins, based on genetic and behavioral data. *Am. J. Phys. Anthropol.* Suppl. 30:308.

van Hooff, J. A. R. A. M. (2000). Relationships among non-human primate males: a deductive framework. In: Kappeler, P. M. (ed.), *Primate Males: Causes and Consequences of Variation in Group Composition*. Cambridge University Press, Cambridge. pp. 183–191.

van Schaik, C. P., and van Hooff, J. A. R. A. M. (1983). On the ultimate causes of primate social systems. *Behaviour* 85:91–117.

Waser, P. (1986). Interactions among primate species. In: Smuts, B. B., Cheney, D. L., Seyfarth, R. M., Wrangham, R. W., and Struhsaker, T. T. (eds.), *Primate Societies*. University of Chicago Press, Chicago. pp. 210–226.

Wrangham, R. (1980). An ecological model of female-bonded primate groups. *Behaviour* 75:262–299.

8

Sakis, Uakaris, and Titi Monkeys

Behavioral Diversity in a Radiation of Primate Seed Predators

Marilyn A. Norconk

INTRODUCTION

The pitheciines are a cohesive group of platyrrhines phylogenetically and ecologically but exhibit a range of variation in group size and social dynamics. At one end of the continuum, titi monkeys (*Callicebus* spp.) form small, cohesive, pair-bonded groups that in many ways represent the "classic monogamous" pattern (Fuentes 1999, van Schaik and Kappeler 2003). Pairs are generally territorial, adults are monomorphic in body size and color, adults exhibit social and physiological mechanisms that promote and reinforce attachment between mates, and males are strongly paternalistic (Mason 1968, 1971; Fragaszy et al. 1982; Kinzey 1981;

Menzel 1986; Mendoza and Mason 1986a,b; Mason and Mendoza 1998; Schradin et al. 2003).

Bearded sakis (*Chiropotes* spp.) and uakaris (*Cacajao* spp.) are at the opposite end of the continuum from titi monkeys. They form large, more loosely structured groups that may fission into smaller feeding parties (Ayres 1986; Norconk and Kinzey 1994; Kinzey and Cunningham 1994; Defler 1999, 2003a). Groups travel through large home ranges and day ranges as long as those of any platyrrhine (Ayres 1981, 1986; Norconk and Kinzey 1994; Aquino 1998; Boubli 1999; Defler 1999; Peetz 2001; Barnett et al. 2002). Males do not take an active part in infant care, and there is little sexual dimorphism in body mass (Ford 1994) and minimal difference in pelage color (for uakaris, Hershkovitz 1987a). Uakari males do exhibit sex-specific traits (enlarged frontal and parietal areas, Fontaine 1981), but sex-specific cues in bearded sakis are minimal. Both male and female bearded sakis exhibit well-developed beards upon sexual maturity and colorful, relatively large external genitalia (pink scrotum and enlarged, pink vaginal lips) (van Roosmalen et al. 1981, Peetz 2001).

Members of the genus *Pithecia* are intermediate between titis and bearded sakis/uakaris. Groups are usually reported as being small (Table 8.1), and many, but not all, conform to the "two-adult group" suggested by Fuentes (1999). White-faced sakis (*Pithecia pithecia*) exhibit aggressive intergroup behavior both in the wild and in captivity (Homburg 1997, Savage et al. 1992, Shideler et al. 1994, Norconk et al. 2003). Mothers are primary care-givers (Brush and Norconk 1999), but adult male interest increases as infants mature; and adult males have been observed to play and share food with older infants (Buzzell and Brush 2000). Ryan (1995) reported that white-faced saki males carried infants, but Homburg (1997) did not find that to be the case. *Pithecia* adult males vary from subtle sexual dichromatism in the western (Amazonian) species to striking pelage differences in the eastern (Guiana Shield) *P. pithecia* ssp. (Hershkovitz 1987b, Gerald 2003). Small group size, small body mass, and territorial behavior ally *Pithecia* with *Callicebus*, whereas diet and dental anatomy, relatively low level of male infant care, and bushy tails ally them with *Chiropotes* and *Cacajao*.

SYSTEMATICS AND GEOGRAPHIC DISTRIBUTION OF THE PITHECIINES

Callicebus (Titis)

In the process of preparing the second volume of *Living New World Monkeys (Platyrrhini)*, Hershkovitz wrote six taxonomic reviews of nonprehensile-tailed platyrrhines including the four genera reviewed here (Hershkovitz 1985, 1987a,b, 1990). Of the four genera of the Pitheciinae, *Callicebus* is the largest and the most complex group. Ten taxa distributed in two species were recognized in an earlier review (Hershkovitz 1963), but in 1990, after examining close to 1,200 specimens, he raised the number of taxa in the genus *Callicebus* to 25, which he distributed into 13 species (Hershkovitz 1990). Van Roosmalen et al. (2002) reexamined the genus, named two new species, raised all subspecific taxa to species level, and divided the genus into five groups—for a total of 28 species. Thus, *Callicebus* has become the second most speciose genus of platyrrhines after *Saguinus*.

Titi monkey species are largely distinguished by pelage color, small differences in body size (Table 8.2), and chromosome number (2n = 20 in *Callicebus torquatus*, 46 in *C. cupreus*, 48 in *C. brunneus* and *C. moloch*, and 50 in *C. dubius donacophilus*) (Hershkovitz 1990: Table 12). Molecular studies place *Callicebus* (as represented by *C.*

torquatus and *C. moloch*) as the sister group of the sakis/ uakaris, supporting the view that the four genera represent a single clade (Schneider and Rosenberger 1996; but also see Ford 1986, Kay 1990, Marroig and Cheverud 2001).

Callicebus species groups are distributed throughout the western and southern Amazon Basin (*cupreus, donacophilus, moloch* groups) and southern Orinoco River Basin (*torquatus* group), with a disjunct distribution between the Atlantic coastal forests of southeastern Brazil (*personatus* group) and central Bolivia (Ferrari et al. 2000, van Roosmalen et al. 2002 and distribution maps therein). They are absent from the northern Guiana Shield forests. Amazonian and Orinoco Basin titis are sympatric with one or more species of the larger pitheciines (see below).

Some species of titis, as well as uakaris, have been called "habitat specialists." Kinzey and Gentry (1979) proposed that *C. torquatus* was a white-sand specialist in Peru, but this hypothesis was reviewed and rejected by Defler, working in Colombia. Defler (1994, 2003b) found that two species, *C. cupreus (moloch)* and *C. torquatus*, were broadly sympatric in Colombia. Each exhibited habitat preferences, the former in low forests along streams and the latter in tall, well-stratified forests; but neither was restricted to these habitats. This view conforms with the recent review of *Callicebus* spp. by van Roosmalen et al. (2002). They suggest that *moloch* and *cupreus* groups are tolerant of habitat disturbance, due to both human activity and seasonal flooding. These two groups are comprised of species that are ecologically very similar and allopatric, but both species are broadly sympatric with the *C. torquatus* group in the sense suggested by Defler above (van Roosmalen et al. 2002). *C. donacophilus* may express a preference for grassland habitats in the state of Rondônia, Brazil (Ferrari et al. 2000).

Pithecia (Sakis)

Pithecia spp., like *Callicebus*, have a broad geographic distribution and occupy a range of habitats from tropical wet to tropical dry forests throughout the Amazon Basin and north into the Guianas and eastern Venezuela (approximately to 9E° N to 14E° S latitudes; see Table 8.1). Hershkovitz (1987b) revised the systematics of *Pithecia* and divided them into two groups based on geographic distribution: a Guianan group (two taxa) and an Amazonian group (six taxa). The distinction between these groups is partly based on pelage color. A recent study of cranial morphology advocates rasing all members of the Amazonian (Monacha) group to species status and leaving the Guianan (Pithecia) group as subspecies (Marroig and Cheverud 2004).

Guianan sakis (white-faced and pale-faced sakis) found north of the Amazon River and ranging into the Guianas are strongly sexually dichromatic: male body pelage is entirely black with a white or yellowish face; females are gray–brown with white or off-white facial markings above the brow and stripes on the side of the face. Amazonian males found south of the Amazon River (Brazil) and in the western Amazon

Table 8.1   Group Size and Composition and Use of Space by the Pitheciines

SPECIES[1]	GROUP SIZE (N)	GROUP SIZE, RANGE (N)	ADULT MALES	ADULT FEMALES	JUVENILES + SUBADULTS	INFANTS	HOME RANGE SIZE (HA)	DAY RANGE (KM)	POPULATION DENSITY (INDIVIDUALS/KM²) OR INDIVIDUAL SIGHTING RATE (SR)/10 KM	SOURCE
Callicebus brunneus, Peru (c. 11°S) ST		2–5 (6)	1(–2?)	1	0–3		1.4			Ferrari et al. 2000, Lawrence 2003
C. brunneus, Bolivia (10°35'S to 11°31'S) C	3	1–5 (39)								Buchanan-Smith et al. 2000
C. brunneus, Brazil (10°–12°S) C	2.25 ± 0.97 (118 sightings)	max = 5								Ferrari et al. 2000
C. caligatus, Brazil (c. 10°S) C	2.20 ± 0.84 (5 sightings)	max = 4								Ferrari et al. 2000
C. cinerascens, Brazil (c. 12°S) C	1.67 ± 0.33 (3 sightings)	max = 2								Ferrari et al. 2000
C. cupreus, Peru (5°35'S) C	3.0–3.73	2–7 (18)							14.6	Bennett et al. 2001
C. donacophilus, Brazil (c. 13°S) C	2.0 ± 1.0	max = 3								Ferrari et al. 2000
C. moloch, Brazil (c. 12°S) C	2.47 ± 0.92 (15 sightings)	max = 4								Ferrari et al. 2000
C. moloch, Peru (11°52'S) LT	4.1 (2 sightings)	2–5 (2)	1	1	1–2	0–1	6–8	0.55 ± 0.1 and 0.67 ± 0.2	20–26/km²	Wright 1984, 1985, 1986
C. moloch, Peru (11°52'S) LT	4.0	2–7					11.5 (6–18)			Bossuyt 2002
C. ornatus, Colombia (c. 4°N) ST	3.2 (9)	2–4 (9)	1	1	0–1	0–1	4.4 (3.2–5.1)	0.57 (0.3–0.87)		Mason 1968
C. ornatus, Colombia (5°N) LT							3.5–14			Defler 1994
C. personatus melanochir, Brazil (15°18'S) LT		2–6[2]	1	1	1–4		24	1.0		Müller 1996
C. p. personatus, Brazil (19°S) ST		6 (1)	1	1	3	1	4.7	0.69 ± 0.04 (0.52–0.80)		Kinzey 1981, Kinzey and Becker 1983
C. p. personatus, Brazil (c. 20°S) ST		3–5 (2)	1(–2?)	1(–2?)	1	1	10.7–12.3	1.0 ± 0.2 (0.8–1.3)		Price and Piedade 2001
C. torquatus torquatus, Peru (4°S) LT		3–5[2]	1	1	1–2	1	29 (4–30)	0.8 ± 0.04 (0.5–1.4)	16	Kinzey 1978, 1981
C. t. lugens, Colombia (1°5.55'S) RC	4.8 (10)	3–5 (10)	1	1	0–1	1	14.2 (9–22)		6.08 (8 groups)	Defler 1983, 2003a
Cacajao calvus calvus, Brazil (3°22'N) LT		30–48 (3)	16	16	13	3	500–550	2.5–5.0	7–8	Ayres 1986, 1989
C. c. ucayalii, Peru (4°23'S and 4°30'S) ST	41.9 ± 16.7 (21)[3]	8–70								Aquino 1998, Aquino and Encarnación 1999
C. c. rubicundus, Peru (5°35'S) C	15.8–33.1	2–55 (18)							7.44 and 25.78	Bennett et al. 2001
C. melanocephalus melanocephalus, Brazil (00°24'N) LT		c. 70						4.4		Boubli 1999
C. m. ouakary, Colombia (1°5.55'S) RC	20–30	1– >108							4.15 overall, 12.0 in igapó habitat	Defler 2001
Chiropotes albinasus, Brazil (10°10'S) LT		19–26 (4)	8	9	8		250–350	2.5–3.5	10–11	Ayres 1981, 1989
C. albinasus, Brazil (9–13°S) C	4.2 ± 3.1 (7)[4]								1.6–2.5 (SR)	Ferrari et al. 1999

Table 8.1 (*cont'd*)

SPECIES[1]	GROUP SIZE (N)	GROUP SIZE, RANGE (N)	ADULT MALES	ADULT FEMALES	JUVENILES + SUBADULTS	INFANTS	HOME RANGE SIZE (HA)	DAY RANGE (KM)	POPULATION DENSITY (INDIVIDUALS/KM²) OR INDIVIDUAL SIGHTING RATE (SR)/10 KM	SOURCE
C. satanas chiropotes, Suriname (4°41′N) and Brazil (2°N) LT		8–27+ (4)	8	9	5	2–3	200–250	2.5		van Roosmalen et al. 1981, Ayres 1981
C. s. chiropotes, Suriname (4°41′N) ST	9 and 13 (2)							3.2 ± 1.1		Norconk and Kinzey 1994
C. s. chiropotes, Venezuela (7°21′N) LT		15–22[2] (1)	1–2	8–10	3–5	1–5	180	1.6 (0.5–2.7)		Norconk 1996, Peetz 2001
C. s. chiropotes, Suriname (5°01′N) C	32.7 (3)	22–44 (3)							37.4 (SR)	Norconk et al. 2003
C. satanas, Brazil (multiple sites, 1–5°S) C									1.8–10.08, 0.3 (SR)	Ferrari and Lopes 1996
Pithecia aequatorialis, Peru (2°S) C		1–7 (4)								Heymann et al. 2002
P. albicans, Brazil (4°51′S) RC	4.6 ± 1.5 (5)	3–7					172.4 (147–204)		4.1	Peres 1993
P. hirsuta, Peru (c. 5°S) RC	3.8	2–8	1–3	1–2			24.9 (9.7–42)		12.8	Soini 1986
P. irrorata, Brazil (9°–13°S) C	2.68 ± 1.38								1.3 (SR)	Ferrari et al. 1999
P. irrorata, Bolivia (10°35′S to 11°24′S) C	3.5	2–5 (6)								Buchanan-Smith et al. 2000
P. monachus, Peru (5°35′S) C	3.75	2–5 (16)							9–17.2	Bennett et al. 2001
P. pithecia chrysocephala, Brazil (2°25′S) LT	6 (1)	4–7[2]	1	1–3	0–2	0–2				Setz and Gaspar 1997, Gilbert and Setz 2001
P. p. chrysocephala, Brazil (c. 2°S) ST	2.6 ± 0.5	2–3	0–2	0–2	0–1					Oliveira et al. 1985
P. p. pithecia, French Guiana (4°N) LT	2.8 ± 1.0 (4)	1–4								Kessler 1998
P. pithecia Guyana (various) C	4.8 ± 2.4 (21)	2–12	2.0	1.8	1.0 (m) 1.3 (f)	1.0				Lehman et al. 2001
P. p. pithecia, Guyana (various) C	3.3 ± 1.7 (10)	1–5								Muckenhirn et al. 1975
P. p. pithecia, Suriname (4°41′N) RC	2.7 ± 0.8	9	1–2	1	0–1					Mittermeier 1977
P. p. pithecia, Suriname (5°01′N) C	3.7 (10)	2–6	1–2	1–3	1		10.3		14.1 (SR)	Norconk et al. 2003
P. p. pithecia, French Guiana (5°04′N) ST	2.3 (35)	1–5						1.88	0.64	Vie et al. 2001
P. p. pithecia, Venezuela (7°21′N) LT	9	5–9	3	2	2	2	15[5]			Homburg 1997
P. p. pithecia, Venezuela (7°21′N) LT	6.9 ± 1.4 (1)[2]	5–9	1–4	2–3	0–2 (m) 0–2 (f)	0–2	12.8[5]	1.5		Norconk, in press

[1] In addition to species name, data in column one include country of study, latitude, and study length. These data are not repeated for Tables 8.2 and 8.3. C, census; RC, repeated census; ST, short term (<1 year); LT, long term (≥1 year).
[2] Range = the change in group size of one group.
[3] Lower number of range estimates used to calculate mean ± standard deviation. The count (100–120) of multiple groups was excluded.
[4] An average of mean group sizes is reported (*n* = 7 locations).
[5] Island population.

Table 8.2 Social and Reproductive Characteristics of the Pitheciines

SPECIES	MALE BODY MASS (G)	FEMALE BODY MASS (G)	GROUP SEX RATIO	AGE AT SEXUAL MATURITY (MONTHS)/ INTERBIRTH INTERVAL (MONTHS)	BIRTH PEAK (MONTHS) AND SEASON (WET OR DRY)	PATERNAL CARE	TERRITORIAL INTERACTIONS WITH CONSPECIFIC GROUPS	VOCALIZATIONS	SOURCE
Callicebus brunneus				c. 36/12		Males carry, share food, and play with infants	7 encounters in 15 months of observation (♀♀ participated in 2:7)	90% calls before 0900 (calls on 15 mornings a month)	Wright 1985, 1986
C. ornatus	1,178	1,163	1.0				Calling, chasing, piloerection, tail lashing, chest rubbing (5%–7% HR overlap). Rate = 1.67/day	Regular dawn calls and during ITEs	Mason 1968, Hershkovitz 1990
C. discolor	935	1,075		44.4 ± 15.6 /11.8	Winter: Dec–Mar (Davis, CA)				Hershkovitz 1990, Valeggia et al. 1999
C. moloch	1,016	877							Hershkovitz 1990
C. ornatus	845	850	1.0		Nov–March	Males carry infants			Mason 1968, Hershkovitz 1990
C. personatus	1,270	1,378			Sept–Oct (dry)		Vocal responses on 3:6 occasions (no other groups in forest patch)	6:16 mornings, dawn calls given in sleeping trees	Hershkovitz 1990, Kinzey and Becker 1983
C. p. personatus							Rare: vocalizations only; no chasing, physical contact, or threat displays	97% before 1,000, not daily	Price and Piedade 2001
C. torquatus torquatus	1,110	1,310		c. 36	Nov–March	Males carry infants	Vocalizations and movement away from intruders (playbacks), occasional active encounters (10% overlap of HR)	Solo male and duets, from sleeping trees	Kinzey 1981, Hershkovitz 1990, Kinzey and Robinson 1983, Easley and Kinzey 1986
C. t. lugens					Jan (dry)		Not daily, duetting 10–20 m apart from stable position		Defler 1983
Cacajao calvus	3,450	2,880	1.20		Oct–Nov (dry)				Ayres 1986
Chiropotes albinasus	3,170	2,520	1.26						Ayres 1981, 1989
C. satanas chiropotes	2,880	2,660	1.08						Ayres 1981
C. s. chiropotes					wet season				van Roosmalen et al. 1981
C. s. chiropotes				c. 36/24+	Dec–April (late wet to dry)		Only one group in forest patch		Peetz 2001
Pithecia albicans	3,000								Peres 1993
P. irrorata	2,010 (2,920)	1,875 (1,980–2,160)							Hershkovitz 1987b, Ford 1994
P. hirsuta				? / 24–36	Sept–Dec (late dry to early wet)		Rare		Soini 1986
P. monachus	2,795 (2,500–3,100)	1,900 (1,300–2,000)							Hershkovitz 1987b, Ford 1994
P. pithecia	1,732 (1,380–1,866)	1,515 (1,347–1,875)		Regular cycles (birth)/23.2 (15–34)	Nov–April (dry)	Play starting c. 4 months of age	Calling (roaring), chasing, piloerection, neck rubbing, urine marking. Rate: 9/100 hr	During travel and ITEs	Hershkovitz 1987b, Ford 1994, Norconk in press
P. pithecia						Play and share food with older infants	Aggressive toward same-sex intruders		Shideler et al. 1994, Savage et al. 1992
P. p. chrysocephala							Neck rubbing unrelated to territorial behavior		Setz and Gaspar 1997

HR, home range; ITE, inter-troop encounters.

Basin (Colombia, Ecuador, Peru) exhibit a range of variation in facial color but are more subtly marked than Guianan males. The range of variation may have taxonomic significance (L. Marsh personal communication, A. Rylands personal communication), but there is not yet sufficient evidence to make such an assessment. Females of all *Pithecia* spp. are very similar in appearance (Hershkovitz 1987b).

According to Ford (1994), sexual dimorphism in body mass ranges from 1.1 to 1.4 for three species of *Pithecia* sakis (see Table 8.2). *P. albicans* nearly bridges the body mass gap between *Pithecia* and larger sakis/uakaris: *Chiropotes/Cacajao*. Peres' (1993) research on *P. albicans* in central Amazonia suggests that relatively larger body size compared with other *Pithecia* spp., in addition to a preference for higher canopy travel and larger home ranges, may reflect competitive release of *P. albicans* in the absence of both *Cacajao* and *Chiropotes*.

Chiropotes (Bearded Sakis)

Bearded sakis (*Chiropotes* spp.) are found primarily in upland, nonflooded habitats in eastern Amazonia, both north and south of the Amazon River (Hershkovitz 1985, Walker 1996, Auricchio 1995). In the Guianas, they are absent from the region west of the Essequibo River in Guyana and their presence in Venezuela may have been due to their ability to follow the right bank of the Orinoco River from Brazil into southern Venezuela (state of Amazonas) and then east into the state of Bolívar. Their present eastern boundary in Venezuela appears to be the left bank of the Caroní River (Norconk et al. 1996). In Brazil, they range north and south of the Amazon, east of Rio Madeira and throughout eastern Amazonia (66° to 44° W latitude) (Ferrari and Lopes 1996 and distribution maps therein). The Rio Tocantins provides a boundary between two subspecies of *Chiropotes satanas*, *Ch. s. satanas* and *Ch. s. utahicki*.

It is unclear why bearded sakis are absent from western Guyana and why their distribution is spotty in French Guiana. They are not found at the Nouragues Research Station in French Guiana (Bongers et al. 2001) despite geological similarities to sites in Suriname and Guyana and high species diversity of one of their most important plant food families, the Lecythidaceae (Mori 1989). de Granville's (1982) description of the forests of southern French Guiana as xeric and scrubby may constitute a barrier to the northern migration of bearded sakis from Brazil. If vegetation or riverine barriers do exist for *Chiropotes*, they have not limited *Pithecia* dispersal into either Guyana or French Guiana. Having a wider tolerance for seasonally dry habitats, *P. pithecia* is apparently broadly distributed across the entire region.

Hershkovitz (1985) reviewed the systematics of *Chiropotes*, dividing the genus into two species, *C. satanas* (with three subspecies) and *C. albinasus*, but the group was reevaluated recently by Bonvicino et al. (2003). Based on pelage coloration and karyotypic analysis, they suggested that the most westerly group could be a different species

and recommended raising the present subspecies to species status. If accepted, this would increase the number of *Chiropotes* species to five (Bonvicino et al. 2003) and perhaps more if the Guianan bearded sakis (from Venezuela and the Guianas) are included in future genetic analyses.

Cacajao (Uakaris)

Hershkovitz (1987a) summarized data from collection localities for uakaris in the Orinoco and western Amazon Basins, ranging from southern Venezuela and western Brazil to eastern Colombia and Peru. He designated two species, black-headed uakaris (*C. melanocephalus*) with two subspecies and bare-headed uakaris (*C. calvus*) with three subspecies. *Chiropotes* and *Cacajao* are generally found to be allopatric, with *Cacajao* inhabiting the western Amazon Basin and *Chiropotes* inhabiting the eastern Amazon Basin (Hershkovitz 1985, 1987a; Auricchio 1995 and distribution maps therein). A permeable species boundary apparently exists on the eastern edge of Pico de Neblina National Park, Brazil (c. 65° W latitude), where Boubli (2002) found a few *Chiropotes* individuals in an area also occupied by *Cacajao melanocephalus*.

Uakaris appear to range widely on a daily and seasonal basis but particularly inhabit areas that flood seasonally along white-water rivers (*várzea*) and black-water rivers (*igapó*). These habitats are flooded up to 9 months of the year to a depth of 6–20 m (Ferreira and Prance 1998) and support fewer primate species year-round than terra firma forests occupied by bearded sakis. Ayres et al. (1999) found that both uakaris and squirrel monkeys were endemic to the *igapó* habitats of the Mamirauá Reserve in central Amazonia but only uakaris traveled through the entire extent of the reserve, moving deeply into flooded forest.

Low concentrations of suspended nutrients and a low pH of black-water rivers result in relatively lower plant species diversity in *igapó*, although plant diversity is still much higher in flooded tropical forests than in temperate forests (Junk 1989). Plant strategies related to growth and reproduction differ in *várzea* and *igapó* forests. Parolin (2000, 2001) found that habitats flooded by white-water rivers originating from the Andes (*várzea*) have rich soils due to a high load of suspended sediments. Selection for fast-growing stems allows young plants to reach sufficient height and to survive seasonal floods. Rapid stem growth is replaced by high maternal investment in large seeds in the habitats drained by black-water streams (*igapó*) (Parolin 2000, 2001). It is not yet understood how these differences in plant growth rates and maternal investment might translate into habitat and dietary differences among populations of uakaris.

Várzea and *igapó* habitats may provide reliable resources for primate seed predators with few competitors, but long day ranges, large home ranges, and reports of low population densities of uakaris suggest that something about these resources is limiting. Population density estimates of bearded sakis are also relatively low. Stevenson (2001) found that fruit production (as estimated by fruit traps) did not predict either pitheciine biomass or number of species found at 30 neotropical field sites (he excluded *Callicebus* in this analysis). Indeed, "the most striking result [of the study] was the association between the abundance of pitheciine species and *Eschweilera* trees" (Stevenson 2001:172). This observation accords well with research on the biogeography of the family Lecythidaceae. Mori (1989) suggested that many Lecythidaceae species had their origin in the ancient Guiana Shield forests and have recently migrated into the alluvial habitats of the Amazon Basin. *Eschweilera* and other Lecythidaceae genera are now abundant in *várzea* forests, with *E. turbinata* documented as the most abundant tree species by Ayres (1986). The geographic distribution of *Cacajao* and possibly *Chiropotes* may be more dependent on and limited by seed availability of specific plant families than either *Pithecia* or *Callicebus*.

PHYSICAL CHARACTERISTICS OF THE PITHECIINES

Pitheciines (including *Callicebus*) are small- to medium-sized platyrrhines, ranging in body size from about 850 g to 3,500 g. Titis are the smallest-bodied of the pitheciines, ranging in size from approximately 800 to 1,300 g (see Table 8.2). The sexual dimorphism ratio in body mass ranges 0.85–1.16. Adult females are heavier than males in four of six species for which body mass data exist, although Hershkovitz (1990:37) remarked that there were "no appreciable morphological differences between the sexes at comparable ages." In the other three genera, males are slightly larger than females (1.08–1.26 in *Chiropotes* and *Cacajao*, n = 3 species, and 1.07–1.47 in *Pithecia*, n = 3 species). Sexual dichromatism was discussed above for *Pithecia* spp.; is absent in *Callicebus* spp., *Chiropotes* spp., and *Cacajao calvus*; but is expressed to varying degrees in *Cacajao melanocephalus* (Hershkovitz 1987b, Gerald 2003).

Similarities in dental anatomy unite the sakis/uakaris and separate them from titis. Titis lack the highly derived incisor/canine complex of the larger pitheciines but exhibit (with *Aotus*) tall incisors and enlarged incisor roots that suggest heavy use of anterior dentition during food acquisition or processing (Kinzey 1992).

Dental adaptations in the sakis/uakaris are strongly correlated with a high incidence of seed predation (see Table 8.3). Kinzey and Norconk (1990) described them as "sclerocarpic" foragers, specialists in opening mechanically protected (i.e., hard and/or thick-husked) fruit. The protected characteristics of fruit exocarp contrasts with the relatively soft seeds, particularly if they are young seeds. Mechanically protected fruits are breached using robust, widely flaring canines; but the canines of *P. pithecia* are also used very precisely to open small, multiloculed fruit, like many species of the Euphorbiaceae, to extract tiny seeds (M. A. Norconk, personal observation). Procumbent incisors are used to scrape adherent mesocarp from the inside of fruit husks.

Compared with anterior dentition, saki/uakari molars appear to be unspecialized—low-crowned with thin enamel. However, two specializations in the enamel have been noted. First, Kinzey (1992) described the well-crenulated enamel of molars, particularly in *Chiropotes* and *Cacajao*, and proposed that the uneven surface of the tooth facilitated positioning seeds during mastication. Second, Martin et al. (2003) examined the microstructure of saki/uakari enamel and found it to be infused with Hunter-Schreger bands, which enable the teeth to resist the propagation of cracks. This finding, they believe, fits well with a diet that requires mastication of "tough, pliable, and generally soft seeds, rather than hard food items" (Martin et al. 2003:360–361). The lack of crack-resistant properties in the enamel of *Callicebus* correlates with the lower proportion of seeds in their diet (Martin et al. 2003).

Both bearded sakis and uakaris are above-branch quadrupeds (pronograde clamberers) and leapers, dropping between tree crowns (Walker 1996). White-faced sakis are vertical clingers and leapers and above-branch quadrupeds. Body proportions and use of the tail as a rudder to "direct turning of the body in the mid-air phase" of vertical clinging and leaping distinguish *Pithecia* from bearded sakis and uakaris (Walker 1996:346).

All sakis/uakaris have bushy tails—long and bushy in *Pithecia* and *Chiropotes*, short and bushy in *Cacajao*. Despite the reduced tail length in uakaris, tail wagging occurs in both bearded sakis and uakaris (Fontaine 1981, Fernandes 1993, Defler 2003a). Tail wagging, whether below branch or arched over the head (in bearded sakis) and accompanied by vocalizations, occurs in a variety of contexts from mild to severe agitation (e.g., predator sightings and in response to alarm calls, reunion of group members, and group reorganization after rest or feeding periods) (Fernandes 1993, Walker and Ayres 1996, Peetz 2001, Defler 2003a). Tail wagging is absent in *Pithecia* and titi monkeys, but tail twining among group members is common in titis. Rather than providing a medium-distance visual cue, as it may in bearded sakis and uakaris, titi monkey tail twining provides a tactile cue "which, it seems reasonable to suppose, contribute[s] to the formation and maintenance of the bond between male and female" (Mason 1974:7).

Kinzey (1986:136) reviewed the available literature on scent marking in platyrrhines and observed that scent marking "plays a major role in regulating social behavior in the marmosets and tamarins, and almost all species of platyrrhines utilize specialized skin scent glands and/or urine for chemical communication." To date, however, there is little information on the function of scent marking in pitheciines. White-faced sakis (*P. pithecia*) possess scent glands in the gular, sternal, and anogenital areas and at times combine scent marking with urine washing (Brumloop et al. 1994, Setz and Gaspar 1997, Gleason 1998). Adult males were scored more often than adult females in scent-marking activities by all of these observers. Gleason (1998) found that the frequency of scent marking peaked in an overlap

Table 8.3   Feeding Ecology of the Pitheciines[1]

SPECIES AND COUNTRY OF RESEARCH	FEEDING GROUP SIZE (RANGE)	% FEEDING					TOP PLANT FAMILIES	SOURCE
		SEEDS	FLESHY FRUIT	FLOWERS	LEAVES	INSECTS		
Callicebus brunneus, Peru	Cohesive group				23–66, varies seasonally		Moraceae, Leguminosae, Annonaceae	Wright 1986
C. personatus personatus, Brazil	Entire group	21.9	54.7		17.2		Myrtaceae, Sapotaceae, Moraceae (51.4%)	Müller 1996
C. p. personatus, Brazil			81[3]	1	18		Sapotaceae (41%)	Kinzey and Becker 1983
C. torquatus torquatus, Peru			71[3]		4	20	Moraceae, Guttiferae, Leguminosae, Euphorbiaceae, Convolvulaceae, Palmae (52%)	Kinzey 1978, 1981
C. t. lugens, Colombia							Euphorbiaceae	Defler 1983, 2003b
Cacajao calvus calvus, Brazil	8.95 (1–50), small subgroups	66.9	18.4		3.3	5.2	Lecythidaceae, Moraceae, Hippocrateaceae, Sapotaceae, Annonaceae (62.4%)	Ayres 1986, 1989; Walker and Ayres 1996
C. c. ucayalii, Peru		46.0	50.0				Sapotaceae, Leguminosae, Apocynaceae (38%)	Aquino 1998, Aquino and Encarnación 1999
C. melanocephalus melanocephalus, Brazil	Dispersed	67.0	28.8	5.0	4.0	2.0	Euphorbiaceae (24%), Caesalpinoidea, Fabaceae, Lecythidaceae, Sapotaceae	Boubli 1999
C. m. ouakary, Colombia	Fission–fusion						Lecythidaceae	Defler 1999, 2003a
Chiropotes albinasus, Brazil	22.5 ± 3.5	35.9	53.9	3.0	7.2[4]		Palmae, Sapotaceae, Leguminosae, Caryocaraceae, Moraceae (54%)	Ayres 1981, 1989
C. satanas chiropotes, Brazil	2.88	63.3	9.3	11.4	16.1[4]		Moraceae, Leguminosae, Lecythidaceae, Sapotaceae (62%)	Ayres 1981
C. s. chiropotes, Suriname	Fission–fusion (especially when fruit trees are within 100 m)	66.4	27.6	4.6				van Roosmalen et al. 1981
C. s. chiropotes, Venezuela[2]		50.7 (4–60)	0–52.0	0–8.1	0–1.5	1–21	Sapotaceae, Loranthaceae, Moraceae (76%)	Peetz 2001
C. s. chiropotes, Venezuela[2]	Travel together, fission when in the vicinity of several feeding trees	74.8	21.6	0.4	0.2	0.5	Sapotaceae, Loranthaceae, Moraceae, Meliaceae (74.5%)	Norconk 1996, Kinzey and Norconk 1993
Pithecia albicans, Brazil	Often fragmenting	46.2	28.6	6.5	9.5	0.4	Sapotaceae (21%), Leguminosae (20%)	Peres 1993
P. hirsuta, Peru		38.0	55.0	3.0	4.0		Lecythidaceae, Leguminosae, Annonaceae	Soini 1986
P. pithecia chrysocephala, Brazil		26–31		15.9	4.0–18.4			Setz 1993
P. p. pithecia, Venezuela[2]		53.3	31.0	2.0	10.4	3.7	Connaraceae, Erythroxylaceae, Rubiaceae, Chrysobalanaceae (53.7)	Homburg 1997
P. p. pithecia, Venezuela[2]	Entire group	60.6	27.8	2.2	7.1	2.3	Connaraceae, Lecythidaceae, Loganiaceae, Leguminosae, Erythroxylaceae (57%)	Kinzey and Norconk 1993, Norconk 1996

[1] Percentage feeding on various resources was taken directly from sources; no attempt was made to total the "% feeding" resources to equal 100%.
[2] Studies at the same site, overlapped in time.
[3] "Fruit" did not specify seed eating.
[4] "Other" catagory included leaves, insects, bark.

zone between two groups, and most of his samples occurred while the sakis were traveling. Setz and Gaspar (1997) concluded that scent marking was related to sexual behavior, but there was only one group at their site, a forest fragment in central Amazonia. Neither Gleason (1998) nor Setz and Gaspar (1997) found sakis to scent mark while feeding.

In addition to scent marking branches, Gleason (personal communication) noted that adult males huddled, rubbed their chests against each other, and possibly exchanged scent just prior to some inter-troop encounters. Group huddles were seen only in a year in which the study group had four adult males, an unusual composition for most wild white-faced saki groups.

Bearded sakis do not scent mark branches, but they may exchange scent through body contact. Peetz (2001:146) observed "ritualized behavior patterns of hugging and lining up." The behaviors were not limited to adult males. Of 44 incidences of hugging, half involved four or more individuals. Peetz treated hugging and lining up as separate activities, but both involved body contact during which scent may have been exchanged. Titi monkeys also "line up," but they engage in lengthy periods of body contact during resting periods (see above), unlike bearded sakis whose contact periods are brief and active and often take on characteristics of a "reunion" (M. A. Norconk, personal observation).

FEEDING ECOLOGY AND DIET

Fleshy fruit comprises the largest component of titi monkey diets, particularly fruit of the Moraceae, Leguminosae, and Sapotaceae families. Leaves are the second highest food category, but leaf composition of the diet ranges 4%–66% depending on the season and titi monkey species (Table 8.3). The *moloch* and *cupreus* groups of titis appear to have a higher proportion of leaves in the diet than the *torquatus* and *personatus* groups (Table 8.3). Insect eating represented 20% of the diet of *C. torquatus* in a study by Kinzey (1978, 1981).

Only a few studies have examined *Callicebus* diets in a long-term, comprehensive manner (Wright 1986, Müller 1996); but titis do not appear to ingest a high proportion of seeds (Table 8.3). In contrast, seeds are often found to comprise a third of saki/uakari diets, sometimes more than two-thirds (see Table 8.3). Sakis and uakaris frequently ingest unripe, dull-colored (green or brown) fruit that has a woody or well-protected exocarp (Ayres 1986, 1989; Peres 1993; Norconk et al. 1998; Boubli 1999). Mature seeds and fruit pulp are also eaten, but they usually make up a smaller proportion of the diet.

Leaves, flowers or nectar, insects, bark, pith, termite nests, and wasp nests are secondary resources for sakis and uakaris but may be important seasonally. Ayres (1989), Norconk (1996), and Boubli (1999) found that these secondary resources make up only about 10% of the annual diet; but Peetz (2001) found that insects made up over 20% of the diet of bearded sakis seasonally. B. Urbani (personal communication) suggests that secondary resources provide an intermittent nutritional boost to the white-faced saki diet. *Polistes* spp. wasp nests, taken opportunistically whether or not the wasps had deserted the nest, were higher in crude protein (10.9% dry matter, DM) than most other resources ingested by white-faced sakis. Grasshoppers (*Tropidacris* spp.) had the highest crude protein value (58.4% DM) and were relatively high in lipids (8.0% DM). Nothing is known about the specific mineral requirements of wild sakis, but iron and manganese were found to be in significantly higher concentrations in termite nests than in fruit and leaves ingested by the sakis (B. Urbani, personal communication). Thus, incidental items are likely to provide important nutritional diversity to saki/uakari diets.

Venezuelan white-faced sakis ingest a diet rich in seeds, but they also ingest young leaves daily and insects and fleshy fruit seasonally (Norconk and Conklin-Brittain 2004). This combination of items provided an intake that was calculated, primarily from fruit, to be seasonally rich in lipids (11.4%–27.5% estimated DM basis), high in total dietary fiber (25.4%–40.8% DM), and seasonally low in both free simple sugars (4.0%–21.3% DM) and crude protein (4.0%–12.6% DM) (Norconk and Conklin-Brittain 2004).

Bearded sakis and uakaris may also have a high intake of and preference for lipid-rich seeds. Ayres (1986:191–192) found that lipid value was higher in large seeds than small seeds. Large seeds also had a significantly higher ratio of lipids plus protein content to condensed tannin plus acid detergent fiber content. This finding correlates well with the high proportion of large seeds found in *igapó* habitats, where plant maternal investment is high and compensates for poor nutrient availability in the soils of black-water river basins (Parolin 2001). Seed eating is similar to leaf eating with regard to dietary fiber intake so that (white-faced) sakis and uakaris may be forced to accept a trade-off between high lipids and high fiber as part of their seed-eating strategy (Ayres 1986, Norconk et al. 2002, Norconk and Conklin-Brittain 2004).

The major advantage of a diet rich in seeds may lie in the ability to reduce or shift the impact of seasonal fruit shortages relative to that experienced by other platyrrhine frugivores. For both bearded sakis and white-faced sakis inhabiting the tropical dry forests of Lago Guri, Venezuela, seeds represented more than 60% of the diet in the early and late dry seasons. Both species shifted to fleshy fruits in the late wet season, and their seed intake fell (Norconk 1996). The period of food shortage as measured by low food species diversity seems to occur at the end of the wet season and beginning of the dry season in Lago Guri. At this time, many seeds are small, still early in their maturation, and fleshy fruits have declined in abundance. Ayres (1986:206) provided support for this observation by noting that *Cacajao calvus* "seem to accumulate extensive fat tissues, comparable to that reported in *Saimiri*" at a time when other frugivorous primates have difficulty finding fleshy fruits in the white-water flooded forests of Lake Teiú, Brazil.

A second advantage of seed eating is the duration of availability of at least some seed species. Boubli (1999) and Norconk (1996) noted that uakaris and sakis, respectively, can gain access to seeds in well-protected fruits that have slowly maturing seeds. Third, these pitheciines have the dental and gnathic strength to break open large, woody young pods of the Bignoniaceae, a family of primarily wind-dispersed seeds, fruiting in the dry season. Winged seeds are largely ignored by other primate frugivores and as such may be an important fallback food for sakis (Norconk and Conklin-Brittain 2004).

GROUP SIZE, USE OF SPACE, AND INTERGROUP RELATIONS

Titis and *Pithecia* sakis form small social groups, have relatively small home ranges, and often exhibit behaviors that are associated with defense of territories (Tables 8.1 and 8.2). Group sizes, ranging from two to seven individuals in titis, conform to the expectation of pair-bonded primates (see Table 8.1), but *Pithecia* groups are more variable. A number of censuses have reported *Pithecia* group sizes between two and five individuals, consistent with a pair-bonded primate (e.g., Ferrari et al. 2000, Buchanan-Smith et al. 2000, Bennett et al. 2001). However, Lehman et al.'s (2001) extensive survey in Guyana documented a wide range of variation in group size. They observed 21 groups, of which one had 12 individuals. Only five groups had the expected one adult male:one adult female ratio typical of pair-bonded primates, but slow dispersal of young adults may account for the "extra" adults in the group. Studies exceeding a year in length reported group sizes as high as nine individuals, with stable compositions of multiple males and multiple females (Peres 1993, Setz 1993, Norconk 1996); but the Setz and Norconk studies were in habitat fragments with limited dispersal opportunities. In the absence of genetic data on paternity and long-term studies in intact (nonfragmented) habitats, it is best to interpret *Pithecia* social groups as usually small but flexible and responsive to variables such as population density, food distribution, and perhaps sympatry with the two larger-bodied pitheciines, *Chiropotes* and *Cacajao* (see Ferrari et al. 1999).

Using playbacks, Robinson (1979) and Kinzey (1981, 1997) defined two distinctive patterns of home range use and defense by *Callicebus* spp. *C. (moloch) ornatus* gave early-morning calls regularly followed by intertroop encounters depending on the proximity of callers to their territorial boundary (Robinson 1979, Robinson et al. 1987). In contrast, a *C. torquatus* group moved away from playbacks of both adult males and male–female pairs using, what Kinzey and Robinson (1983) called, "proximity-dependent avoidance." *C. personatus* seems to be closer to the *C. torquatus* pattern of group dispersion (Ferrari et al. 2000), but other behavioral and spatial use differences may exist among these species to suggest that we have not exhausted all of the habitat use strategies of species in this very widespread and diverse genus.

A few studies suggest that white-faced sakis in Venezuela and Suriname are also territorial (Norconk et al. 2003; M. A. Norconk personal observation), and Shideler et al. (1994) have reported aggressive interactions with same-sex intruders among captive sakis. Territorial activities that consist of behavior-specific vocalizations and chasing are strikingly similar to Mason's (1968) description of behaviors associated with territoriality in *C. ornatus* and Fernandez-Duque's (see Chapter 9) description of territoriality in *Aotus*.

Chiropotes and *Cacajao* live in relatively large groups and are as far-ranging as any platyrrhine. Daily travel distance appears to be driven by the density and dispersion of seed crops, and some authors have reported extensive group fissioning and wide variations in feeding group sizes seasonally (Ayres 1986, Defler 1999). Most authors report minimum estimates of home range instead of defining use of space in precise terms. Ayres (1986) estimated white uakari home ranges to be 500–550 ha, with daily path lengths up to 5 km. Path lengths averaged 4.4 km for black-headed uakaris in Brazil (Boubli 1999) and 2.5–4.0 km for bearded sakis in Suriname (Norconk and Kinzey 1994).

The difficulty of tracking fast-moving, widely ranging bearded saki and uakari groups, as well as the paucity of studies on captive groups, has resulted in little information on how *Chiropotes* and *Cacajao* groups are organized socially. Ayres (1981) suggested that the organization of bearded saki groups may be based on an underlying structure of multiple male–female units. This hypothesis might provide information on how groups fission during feeding and travel, but logistical difficulties have thus far precluded the ability to address the question. To my knowledge, there have been no studies of wild bearded sakis or uakaris with individually identified animals.

Defler (2003b) documented seasonal variation in group size for black-headed uakaris in Colombia, but Boubli (1999) found them to be more cohesive in Brazil. Van Roosmalen et al. (1988) and Norconk and Kinzey (1994) found that bearded sakis often fissioned when multiple feeding trees were within about 100 m, but they traveled cohesively between feeding sites. That view may have to be modified as recent work in Suriname suggests that there may be seasonal variation in group patterns for bearded sakis (M. A. Norconk personal observation) similar to what Defler (2003b) noted for uakaris.

To cautiously summarize the socioecology of pitheciines using available data from 9–15 studies (see Tables 8.1–8.3 for raw data), there is a strong positive correlation between group size and day range (Fig. 8.1: *tau* = 0.66, *p* < 0.01 two-tailed, *n* = 14) and between group mass and home range size (Fig. 8.2: *tau* = 0.69, *p* < 0.01, *n* = 18). Diet and home range size were not correlated, either in terms of the percentage of seeds in the diet (Fig. 8.3: *tau* = 0.42, *p* not significant, *n* = 9) or as a percentage of all fruit resources in the diet (Fig. 8.4:

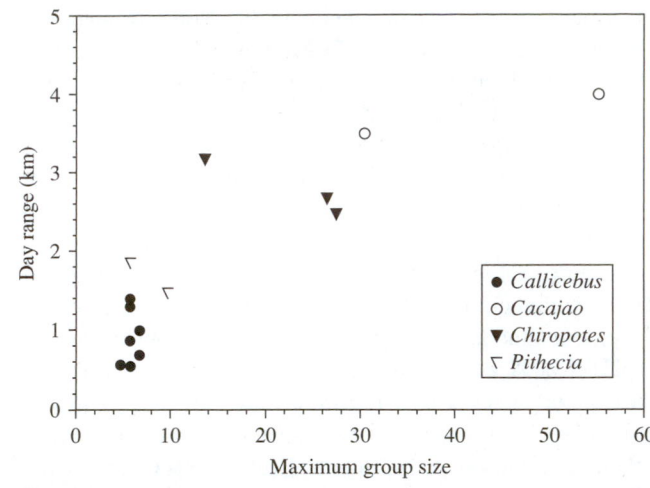

Figure 8.1 Maximum group size is plotted against estimated day range. The maximum group size was used since group sizes are underestimated for many of the pitheciine census samples. Home ranges are also estimates, particularly for the larger-bodied species. *Pithecia* and *Callicebus*, whose group size is smaller than 10 individuals, have smaller home ranges (<40 ha) than *Chiropotes* and *Cacajao* (except *P. albicans*). Data taken from Tables 8.1 and 8.2.

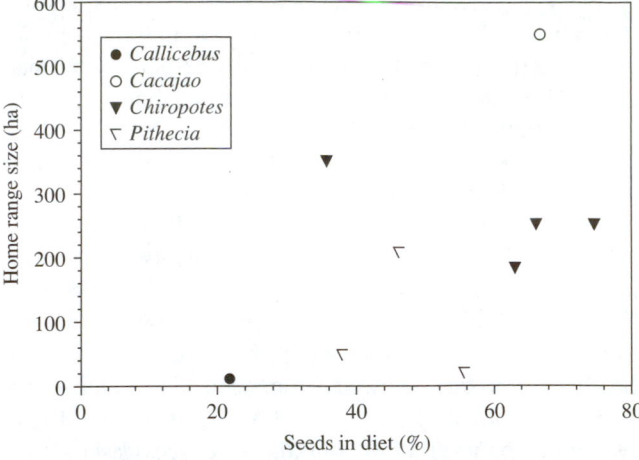

Figure 8.3 Percentage of seeds in diet is plotted against estimated home range size. Data on seed eating are available for only one species of *Callicebus* (*C. personatus*), but seed eating does not increase with larger home ranges for those populations with sufficient data.

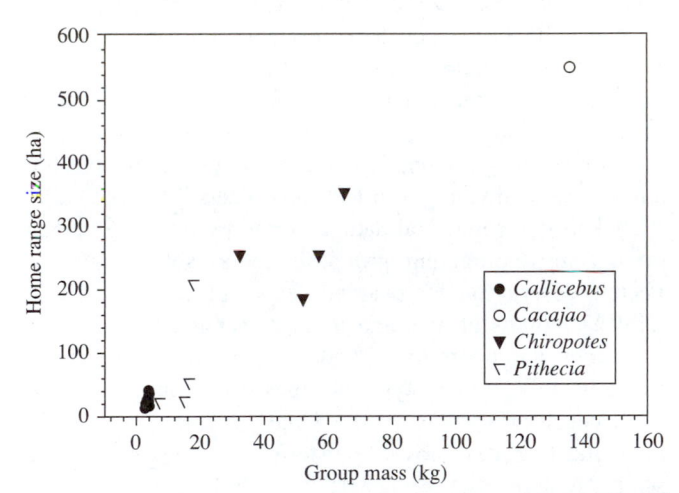

Figure 8.2 Estimated group mass is plotted against estimated home range size. Calculation of group mass: [(male body mass (kg) $*$ n males) + (female mass $*$ n females) + (n immatures $*$ (0.75) female body mass]. *Pithecia* sakis are intermediate between *Callicebus* and the two larger-bodied sakis. Sample sizes are low for all taxa but particularly for *Cacajao*. The outlier for *Pithecia* is *P. albicans* (see Tables 8.1 and 8.2).

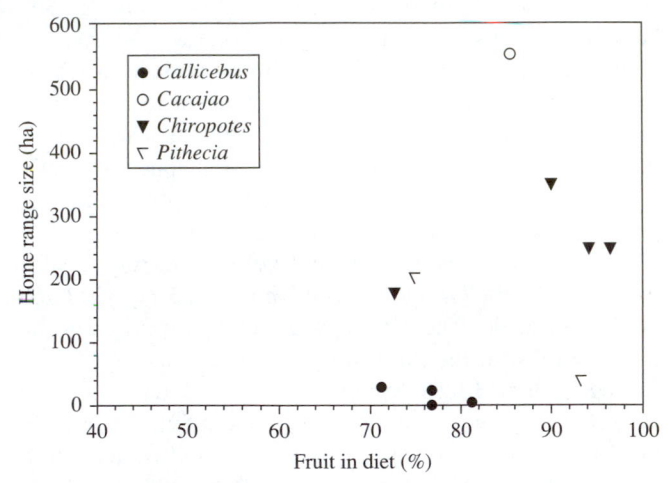

Figure 8.4 Percentage of fruit (including seeds) in diet is plotted against estimated home range size. All pitheciines have a total fruit intake of >70% of their diets.

$tau = 0.28$, p = not significant, $n = 11$). The latter analysis is influenced by the striking reliance on fruit resources (>70%) by pitheciines.

REPRODUCTION AND DEVELOPMENT

Pitheciine female reproductive physiology has been studied for the smaller members of the group: white-faced sakis (*P. pithecia*, Savage et al. 1992, Shideler et al. 1994) and titis (*C. cupreus*, Hoffman et al. 1995, Valeggia et al. 1999). White-faced sakis have a 16- to 17-day ovarian cycle, gestation of 150 days, and approximately 195 days of lacta-

tional amenorrhea. The interbirth interval in captivity was 15.1 months, and sexual maturity occurred at about 30 months. Interbirth intervals are longer in wild sakis (21.5 ± 9 months, $n = 6$), and first birth occurred at the age of 5 years (M. A. Norconk and S. E. Shideler unpublished data; Norconk in press; see Table 8.2).

The length of titi monkey ovarian cycles is very similar to that of white-faced sakis (*C. cupreus*, 17.2 ± 1.5 days), but gestation length and interbirth intervals were shorter (128.6 ± 4.4 days and 11.8 months, respectively) (Valeggia et al. 1999). Titi monkey females housed with an unfamiliar male had their first infants at the age of 3.7 years (range 2–6 years) (Valeggia et al. 1999).

Mason, Mendoza, and colleagues conducted a number of elegant studies on the behavior and physiology of filial attachment in *C. (moloch) cupreus* (e.g., Mason 1971, 1974; Mendoza and Mason, 1986a,b; Mason and Mendoza 1998; Hennessy et al. 1995; Hoffman et al. 1995). They found that titi monkey infants exhibit preferences for fathers (Mason and Mendoza 1998) and that separation from the father, but not the mother, elicited a strong cortisol response (Hoffman 1998). Male caretaking peaked in the second month and continued until the sixth month (Fragaszy et al. 1982).

In a series of experiments, Hoffman (1998) studied physiological and behavioral parameters in the interaction between nearly mature titi monkeys and their parents. His experimental studies on captive *C. cupreus* "dispersal" accord well with data collected by Bossuyt (2002) on wild *C. moloch*. Hoffman's data suggested that sons were more reticent to leave the family group than daughters, but both daughters and sons showed strong behavioral attachment to parents until they were 3.5–4 years of age (Hoffman 1998:92). Bossuyt (2002), working in Cocha Cashu, Peru, found that both daughters and sons dispersed from their natal groups between 3 and 4 years of age, that daughters tended to disperse earlier than sons, and that, in light of high juvenile mortality, parents may benefit from retaining subadults in the group as individuals that could provide "critical aid" to siblings. Existing data on dispersal patterns in titis appear to be similar to preliminary data on white-faced sakis: both sons and daughters appear to leave their natal group but do not do so until they are mature (at least 3 years of age). Like titis, white-faced sakis appear to retain affiliative relations within their natal group prior to dispersal (M. A. Norconk, personal observation).

Homburg (1997) documented developmental patterns for the first 5 months of life for wild white-faced sakis in Lago Guri, Venezuela. The mother was the only carrier for the first 8 weeks. In the third month, two other adult females carried the infant for 15–85 min once or twice a day (Homburg 1997:131). Distance between mother and infant increased in the fourth month, and solitary play was observed in the fifth month. Infants made brief contact with males, but no male infant carrying was observed (but see Ryan 1995). A wild white-faced saki infant was observed to spend about 50% of its time off the mother's back by the fifteenth week of life (Buzzell and Brush 2000). Captive infants showed an accelerated level of independence and were on the mother in only 3% of the samples at the end of the twelfth week of life. Carrying by other group members was seen during week 12 in the wild but was not seen in the captive sample (Buzzell and Brush 2000). Whereas white-faced saki infants become self-locomoting by their fourth month, both bearded saki and uakari infants were carried by the mother for at least 9 months (see below). Larger body mass may account for the slower developmental pattern in bearded sakis and uakaris, but longer daily travel paths for these species may also influence the duration of maternal carrying.

The transition to the white face and black body pelage characteristic of adult male white-faced sakis (*P. pithecia*) appears to follow a prolonged but variable developmental path. Some young males in Lago Guri, Venezuela, exhibited the dark body pelage of adult males within a few months, and others maintained the orange–brown agouti coloration

of females into their third year (M. A. Norconk, personal observation). Homburg (1997) estimated that body pelage changed at about 20 months of age but body hair was not entirely black until the third year. Development of the white facial mask appears to proceed at a variable rate with respect to changes in body pelage in Guianan sakis (M. A. Norconk, personal observation), but nothing is known about the development of pelage differences in Amazonian sakis.

Peetz (2001) provides the only information available on the development of wild bearded sakis, but Hick provided developmental data on captive sakis (cited in van Roosmalen et al. 1981). According to Peetz, bearded saki births peaked in the dry season (December to April); however, van Roosmalen et al. (1981) correlated the birth season with the beginning of the wet season in Suriname. Peetz (2001) estimated a minimum interbirth interval of 2 years, that infants were carried ventrally for the first 2 months of life and dorsally through the fifth month, and that they were still carried in the ninth month whenever mothers lept between widely spaced tree crowns. Older juveniles traveled independently by age 10–13 months (Peetz 2001) but continued to suckle into the second year. Testes descended at about 3 years of age.

Peetz (2001) did not document any courtship behavior pre- or postcopulation. After copulation, both males and females resumed precopulation activities (Peetz 2001). Adult male care of infants is not as extensive in bearded sakis; it was limited to grooming and playful interactions between adult males and young bearded sakis (Peetz 2001).

Uakari developmental data, reported by Fontaine (1981) on a semi-free-ranging group, is remarkably similar to Peetz's description for bearded sakis. Suckling continued until 22 months of age, and mothers persisted in carrying 12-month-old infants across wide gaps between tree crowns (Fontaine 1981). Phenotypic changes from young uakari to adult began at about 2 years of age: "a darkly pigmented glandular field develops in the sternal area, alopecia of the scalp develops, and the general body pelage of the dorsal torso and lateral limb surfaces fills out to form a mantle" (Fontaine 1981:457). Sexually dimorphic traits develop slowly in red uakaris as they do in white-faced sakis. Infant red uakaris are born with only a trace of pink in the face, and color change begins gradually in the juvenile period (3–12 months). Over a period of 2 years, Fontaine (1981) observed young male uakaris to increase in body mass and musculature relative to females and to develop the charateristic paired muscle masses overlying the frontal and parietal areas of the skull. *P. pithecia* saki males complete the developmental process by about 3 years of age, when uakaris are just beginning it. Fontaine (1981) also noted that the developmental process in male red uakaris is remarkably plastic, as it appears to be in male white-faced sakis.

CONCLUSIONS

I opened this chapter with the view that the pitheciines represent a closely related group of species arrayed along a continuum of smaller to larger body mass, with group compositions ranging from pair-bonded and territorial to multimale and relatively nomadic. *Pithecia*, *Chiropotes*, and *Cacajao* are clearly a very cohesive group from the perspective of feeding adaptations. Body size aside, they share the same dental adaptations. *Pithecia*, the smallest of the three, may have a broader diet that includes leaves on a regular basis as well as fewer habitat restrictions. Tropical dry and savanna habitats probably limit the distribution of bearded sakis and uakaris but do not seem to be a barrier to *Pithecia* dispersal, particularly for *P. pithecia*, the white-faced sakis. This view is complicated by some diversity within the small radiation of *Pithecia* species, as suggested by Peres (1993) and Walker (1996). The smaller-bodied, vertical clinging and leaping white-faced sakis of the Guianas are well adapted to moving through low- to middle-canopy levels compared with the Amazonian sakis.

There also appears to be a size-related continuum among titi monkeys, with the smaller ones (*C. moloch, donacophilus, cupreus*) inhabiting small, well-defended territories using voice and movement toward territorial boundaries and *C. torquatus* and *C. personatus* using voice to announce their presence in a larger home range. *Callicebus* contrasts with the other three pitheciines in aspects of feeding ecology, but *Pithecia* is closer to *Callicebus* in aspects of body size, group size, and perhaps home range size.

The larger sakis (bearded sakis and uakaris) are very similar in diet and dentition, ranging patterns, group size, and slow developmental patterns. Their primary difference lies in habitat preferences: bearded sakis are found more often in nonflooded forests and uakaris, while not limited to flooded forests, use those habitats extensively.

One can make the case that, for most primates, more wild studies will improve the resolution with which we interpret their social behavior, but this is particularly true for the pitheciines. Long-term studies are rare even for *Callicebus*, even though wild studies began in the 1960s (Mason 1968). No long-term free-ranging studies have been conducted on *Pithecia*, and our understanding of social behavior in the bearded sakis and uakaris is hindered not only by the difficulty of studying social behavior in these relatively nomadic primates but also by the few opportunities to study them in captivity. We must begin to shift from the more easily gathered data on feeding ecology and develop methods to study behavioral mechanisms that serve to coordinate large groups of bearded sakis and uakaris, to explore ecological and social differences that may underlie the diversity of *Pithecia* populations, and to further investigate the seemingly cohesive pattern of behavior in the 28 species of *Callicebus*.

REFERENCES

Aquino, R. (1998). Some observations on the ecology of *Cacajao calvus ucayalii* in the Peruvian Amazon. *Primate Conserv.* 18:21–24.

Aquino, R., and Encarnación, C. F. (1999). Observaciones preliminares sobre la dieta de *Cacajao calvus ucayalii* en el Nor-Oriente Peruano. *Neotrop. Primates* 7:1–5.

Auricchio, P. (1995). *Primatas do Brasil*. Terra Brasilis Comercio de Material Didático e Editora Letda, São Paulo.

Ayres, J. M. (1981). Observações sobre a ecologia e o comportamento dos cuxiús (*Chiropotes albinasus* e *Chiropotes satanas*, Cebidae: Primates). Fundação Universidade do Amazonas (FUA), Brazil.

Ayres, J. M. (1986). Uakaris and Amazonian flooded forest [PhD diss.]. Cambridge University, Cambridge.

Ayres, J. M. (1989). Comparative feeding ecology of the uakari and bearded saki, *Cacajao* and *Chiropotes. J. Hum. Evol.* 18:697–716.

Ayres, J. M., Alves, A. R., de Queiroz, H. L., Marmontel, M., Moura, E., de Magalhães Lima, D., Azevedo, A., Reis, M., Santos, P., da Silveira, R., and Masterson, D. (1999). Mamirauá: the conservation of biodiversity in an Amazonian flooded forest. In: Padoch, C., Ayres, J. M., Pinedo-Vasquez, M., and Henderson, A. (eds.), *Várzea: Diversity, Development, and Conservation of Amazonia's Whitewater Floodplains*. New York Botanical Garden Press, Bronx. pp. 203–216.

Barnett, A. A., Borges, S. H., de Castilho, C. V., Neri, F. M., and Shapley, R. L. (2002). Primates of the Jaú National Park, Amazonas, Brazil. *Neotrop. Primates* 10:65–70.

Bennett, C. L., Leonard, S., and Carter, S. (2001). Abundance, diversity, and patterns of distribution of primates on the Tapiche River in Amazonian Peru. *Am. J. Primatol.* 54:119–126.

Bongers, F., Charles-Dominique, P., Forget, P.-M., and Théry, M. (2001). *Nouragues: Dynamics and Plant–Animal Interactions in a Neotropical Rainforest*. Kluwer Academic Publishers, Dordrecht.

Bonvicino, C. R., Boubli, J. P., Otazú, I. B., Almeida, F. C., Nascimento, F. F., Coura, J. R., and Seuánez, H. N. (2003). Morphologic, karyotypic, and molecular evidence of a new form of *Chiropotes* (Primates, Pitheciinae). *Am. J. Primatol.* 61:123–133.

Bossuyt, F. (2002). Natal dispersal of titi monkeys (*Callicebus moloch*) at Cocha Cashu, Manu National Park, Peru. *Am. J. Phys. Anthropol.* 34(suppl):47.

Boubli, J. P. (1999). Feeding ecology of black-headed uacaris (*Cacajao melanocephalus melanocephalus*) in Pico de Neblina National Park, Brazil. *Int. J. Primatol.* 20:719–749.

Boubli, J. P. (2002). Western extension of the range of bearded sakis: a possible new taxon of *Chiropotes* sympatric with *Cacajao* in the Pico da Neblina National Park, Brazil. *Neotrop. Primates* 10:1–4.

Brumloop, A., Homburg, I., Peetz, A., and Riehl, R. (1994). Gular scent glands in adult female white-faced saki, *Pithecia pithecia pithecia*, and field observations on scent-marking behaviour. *Folia Primatol.* 63:212–215.

Brush, J. A., and Norconk, M. A. (1999). Early behavioral development in a wild white-faced saki monkey (*Pithecia pithecia*). *Am. J. Phys. Anthropol.* 28(suppl.):99.

Buchanan-Smith, H. M., Hardie, S. M., Caceres, C., and Prescott, M. J. (2000). Distribution and forest utilization of *Saguinus* and other primates of the Pando Department, northern Bolivia. *Int. J. Primatol.* 21:353–373.

Buzzell, C. A., and Brush, J. A. (2000). Ontogeny of independence in wild and captive white-faced saki monkeys (*Pithecia pithecia*). *Am. J. Primatol.* 51(suppl. 1):49–50.

Defler, T. R. (1983). Some population characteristics of *Callicebus torquatus lugens* (Humboldt, 1812) (Primates: Cebidae) in eastern Colombia. *Lozania (Acta Zool. Colomb.)* 38:1–9.

Defler, T. R. (1994). *Callicebus torquatus* is not a white-sand specialist. *Am. J. Primatol.* 33:149–154.

Defler, T. R. (1999). Fission–fusion in the black-headed uacari (*Cacajao melanocephalus*) in eastern Colombia. *Neotrop. Primates* 7:5–8.

Defler, T. R. (2001). *Cacajao melanocephalus ouakary* densities on the lower Apaporis River, Colombian Amazon. *Primate Rep.* 61:31–36.

Defler, T. R. (2003a). *Primates de Colombia*. Conservation International, Bogotá. Defler, T. R. (2003b). Densidad de especies y organización espacial de una comunidad de primates: Estación Biológica Caparú, Deparamento de Vaupés, Colombia. In: Pereira-Bengoa, V., Nassar-Montoya, F., and Savage, A. (eds.), *Primatología del Nuevo Mundo: Biología, Medicina, Manejo y Conservación*. Centro de Primatología Araguatos, Bogotá. pp. 23–39.

de Granville, J.-J. (1982). Rain forest and xeric flora refuges in French Guiana. In: Prance, G. T. (ed.), *Biological Diversification in the Tropics*. Columbia University Press, New York. pp. 159–181.

Easley, S. P., and Kinzey, W. G. (1986). Territorial shift in the yellow-handed titi monkey (*Callicebus torquatus*). *Am. J. Primatol.* 11:307–318.

Fernandes, M. E. B. (1993). Tail-wagging as a tension relief mechanism in pithecines. *Folia Primatol.* 61:52–56.

Ferrari, S. F., Iwanaga, S., Coutinho, P. E. G., Messias, M. R., Cruz Neto, E. H. D., Ramos, E. M., and Ramos, P. C. S. (1999). Zoogeography of *Chiropotes albinasus* (Platyrrhini, Atelidae) in southwestern Amazonia. *Int. J. Primatol.* 20:995–1004.

Ferrari, S. F., Iwanaga, S., Messias, M. R., Ramos, E. M., Ramos, P. C. S., Cruz Neto, E. H. D., and Coutinho, P. E. G. (2000). Titi monkeys (*Callicebus* spp., Atelidae: Platyrrhini) in the Brazilian state of Rondônia. *Primates* 41:229–234.

Ferrari, S. F., and Lopes, M. A. (1996). Primate populations in eastern Amazonia. In: Norconk, M., Rosenberger, A. L., and Garber, P. A. (eds.), *Adaptive Radiations of Neotropical Primates*. Plenum Press, New York. pp. 53–67.

Ferreira, L. V., and Prance, G. T. (1998). Structure and species richness of low-diversity floodplain forest on the Rio Tapajós, eastern Amazonia, Brazil. *Biodivers. Conserv.* 7:585–596.

Fontaine, R. (1981). The uakaris, genus *Cacajao*. In: Coimbra-Filho, A. F., and Mittermeier, R. A. (eds.), *Ecology and Behavior of Neotropical Primates*, vol. 1. Academia Brasileira de Ciências, Rio de Janeiro. pp. 443–493.

Ford, S. M. (1986). Systematics of the New World monkeys. In: Swindler, D. R., and Erwin, J. (eds.), *Comparative Primate Biology. Systematics, Evolution, and Anatomy*, vol. 1. Alan R. Liss, New York. pp. 73–135.

Ford, S. M. (1994). Evolution of sexual dimorphism in body weight in platyrrhines. *Am. J. Primatol.* 34:221–244.

Fragaszy, D. M., Schwarz, S., and Shimosaka, D. (1982). Longitudinal observations of care and development of infant titi monkeys (*Callicebus moloch*). *Am. J. Primatol.* 2:191–200.

Fuentes, A. (1999). Re-evaluating primate monogamy. *Am. Anthropol.* 100:890–907.

Gerald, M. S. (2003). How color may guide the primate world: possible relationships between sexual selection and sexual dichromatism. In: Jones, C. B. (ed.), *Sexual Selection and Reproductive Competition in Primates: New Perspectives and Directions*, vol. 3. American Society of Primatologists, Norman, OK. pp. 141–171.

Gilbert, K. A., and Setz, E. Z. (2001). Primates in a fragmented landscape. Six species in central Amazonia. In: Bierregaard, R. O., Jr., Gascon, C., Lovejoy, T. E., and Mesquita, R. (eds.), *Lessons from Amazonia: The Ecology and Conservation of a Fragmented Forest*. Yale University Press, New Haven. pp. 262–270.

Gleason, T. M. (1998). The ecology of olfactory communication in Venezuelan white-faced sakis. *Am. J. Primatol.* 45:183.

Hennessy, M. D., Mendoza, S. P., Mason, W. A., and Moberg, G. P. (1995). Endocrine sensitivity to novelty in squirrel monkeys and titi monkeys: species differences in characteristic modes of responding to the environment. *Physiol. Behav.* 57:331–338.

Hershkovitz, P. (1963). A systematic and zoogeographic account of the monkeys of the genus *Callicebus* (Cebidae) of the Amazonas and Orinoco River basins. *Mammalia* 27:1–80.

Hershkovitz, P. (1985). A preliminary taxonomic review of the South American bearded saki monkeys genus *Chiropotes* (Cebidae, Platyrrhini), with the description of a new subspecies. *Fieldiana* 27[1363], 1–46. Field Museum of Natural History, Chicago.

Hershkovitz, P. (1987a). Uacaries, New World monkeys of the genus *Cacajao* (Cebidae, Platyrrhini): a preliminary taxonomic review with the description of a new subspecies. *Am. J. Primatol.* 12:1–53.

Hershkovitz, P. (1987b). The taxonomy of South American sakis, genus *Pithecia* (Cebidae, Platyrrhini): a preliminary report and critical review with the description of a new species and a new subspecies. *Am. J. Primatol.* 12:387–468.

Hershkovitz, P. (1990). Titis, New World monkeys of the genus *Callicebus* (Cebidae, Platyrrhini): a preliminary taxonomic review. *Fieldiana* [55], 1–109. Chicago, Field Museum of Natural History. Zoology, New Series.

Heymann, E. W., Encarnación, C. F., and Canaquin, Y. J. E. (2002). Primates of the Río Curaray, northern Peruvian Amazon. *Int. J. Primatol.* 23:191–201.

Hick, U. (1968). Erstmalig gelungene Zucht eines Bartsakis (Vater: Rotrückensaki) *Chiropotes chiropotes* (Humboldt 1811). Mutter: Weissnasensaki, *Chiropotes albinasus* (Geoffroy et Deville 1848) im Kölner Zoo. *Freunde des Kölner Zoo* 11(2):35–41.

Hoffman, K. A. (1998). Transition from juvenile to adult stages of development in titi monkeys (*Callicebus moloch*) [PhD diss.]. University of California, Davis.

Hoffman, K. A., Valeggia, C. R., Mensoza, S. P., and Mason, W. A. (1995). Responses to mature titi monkey daughters following a one month separation from their family groups. *Am. J. Primatol.* 36:128.

Homburg, I. (1997). Ökologie and sozialverhalten einer gruppe von weißgesicht-sakis (*Pithecia pithecia pithecia* Linnaeus 1766) im Estado Bolívar, Venezuela [PhD diss.]. Universität Bielefeld, Bielefeld.

Junk, W. J. (1989). Flood tolerance and tree distribution in central Amazonian floodplains. In: Holm-Nielsen, L. B., Nielsen, I. C., and Balslev, H. (eds.), *Tropical Forests: Botanical Dynamics, Speciation and Diversity*. Acadmic Press, London. pp. 47–73.

Kay, R. F. (1990). The phyletic relationships of extant and fossil Pitheciinae (Platyrrhini, Anthropoidea). *J. Hum. Evol.* 19:175–208.

Kessler, P. (1998). Primate densities in the natural reserve of Nourague, French Guiana. *Neotrop. Primates* 6:45–46.

Kinzey, W. G. (1978). Feeding behaviour and molar features in two species of titi monkey. In: Chivers, D. J., and Herbert, J. (eds.), *Recent Advances in Primatology*, vol. 1. Academic Press, London. pp. 373–385.

Kinzey, W. G. (1981). The titi monkeys, genus *Callicebus*. In: Coimbra-Filho, A. F., and Mittermeier, R. A. (eds.), *Ecology and Behavior of Neotropical Primates*, vol. 1. Academia Brasileira de Ciências, Rio de Janeiro. pp. 241–276.

Kinzey, W. G. (1986). New World primate field studies: what's in it for anthropology? *Annu. Rev. Anthropol.* 15:121–148.

Kinzey, W. G. (1992). Dietary and dental adaptations in the Pitheciinae. *Am. J. Phys. Anthropol.* 88:499–514.

Kinzey, W. G. (1997). *Callicebus*. In: Kinzey, W. G. (ed.), *New World Primates: Ecology, Evolution and Behavior*. Aldine de Gruyter, New York. pp. 213–221.

Kinzey, W. G., and Becker, M. (1983). Activity pattern of the masked titi monkey, *Callicebus personatus*. *Primates* 24:337–343.

Kinzey, W. G., and Cunningham, E. P. (1994). Variability in platyrrhine social organization. *Am. J. Primatol.* 34:185–198.

Kinzey, W. G., and Gentry, A. H. (1979). Habitat utilization in two species of *Callicebus*. In: Sussman, R. W. (ed.), *Primate Ecology: Problem-Oriented Field Studies*. John Wiley & Sons, New York. pp. 89–100.

Kinzey, W. G., and Norconk, M. A. (1990). Hardness as a basis of fruit choice in two sympatric primates. *Am. J. Phys. Anthropol.* 81:5–15.

Kinzey, W. G., and Norconk, M. A. (1993). Physical and chemical properties of fruit and seeds eaten by *Pithecia* and *Chiropotes* in Surinam and Venezuela. *Int. J. Primatol.* 14:207–227.

Kinzey, W. G., and Robinson, J. G. (1983). Intergroup loud calls, range size, and spacing in *Callicebus torquatus*. *Am. J. Phys. Anthropol.* 60:539–544.

Lawrence, J. M. (2003). Preliminary report on the natural history of brown titi monkeys (*Callicebus brunneus*) at the Los Amigos Research Station, Madre de Dios, Peru. *Am. J. Phys. Anthropol.* 36(suppl.):136.

Lehman, S. M., Prince, W., and Mayor, M. (2001). Variations in group size in white-faced sakis (*Pithecia pithecia*): evidence for monogamy or seasonal cogregations? *Neotrop. Primates* 9:96–101.

Marroig, G., and Cheverud, J. M. (2001). A comparison of phenotypic variation and covariation patterns and the role of phylogeny, ecology, and ontogeny during cranial evolution of New World monkeys. *Evolution* 55:2576–2600.

Marroig, G., and Cheverud, J. M. (2004). Cranial evolution in sakis (*Pithecia*, Platyrrhini) I: interspecific differentiation and allometric patterns. *Am. J. Phys. Anthropol.* 125:266–278.

Martin, L. B., Olejniczak, A. J., and Maas, M. C. (2003). Enamel thickness and microstructure in pitheciin primates, with comments on dietary adaptations of the middle Miocene hominoid *Kenyapithecus*. *J. Hum. Evol.* 45:351–367.

Mason, W. A. (1968). Use of space by *Callicebus* groups. In: Jay, P. C. (ed.), *Primates: Studies in Adaptation and Variability*. Holt, Rinehart and Winston, New York. pp. 200–216.

Mason, W. A. (1971). Field and laboratory studies of social organization in *Saimiri* and *Callicebus*. In: Rosenblum, L. A. (ed.), *Primate Behavior*, vol. 2. Academic Press, New York. pp. 107–137.

Mason, W. A. (1974). Comparative studies of social behavior in *Callicebus* and *Saimiri*: behavior of male–female pairs. *Folia Primatol.* 22:1–8.

Mason, W. A., and Mendoza, S. P. (1998). Generic aspects of primate attachments: parents, offspring and mates. *Psychoneuroendocrinology* 23:765–778.

Mendoza, S. P., and Mason, W. A. (1986a). Parental division of labour and differentiation of attachments in a monogamous primate (*Callicebus moloch*). *Anim. Behav.* 34:1336–1347.

Mendoza, S. P., and Mason, W. A. (1986b). Parenting within a monogamous society. In: Else, J. G., and Lee Phyllis, C. (eds.), *Primate Ontogeny, Cognition and Social Behaviour*. Cambridge University Press, Cambridge. pp. 255–266.

Menzel, C. R. (1986). An experimental study of territory maintenance in captive titi monkeys (*Callicebus moloch*). In: Else, J. G., and Lee, P. C. (eds.), *Primate Ecology and Conservation*. Cambridge University Press, Cambridge. pp. 133–143.

Mittermeier, R. A. (1977). Distribution, synecology and conservation of Surinam monkeys [PhD diss.]. Harvard University, Boston.

Mori, S. A. (1989). Diversity of Lecythidaceae in the Guianas. In: Holm-Nielsen, L. B., Nielsen, I. C., and Balslev, H. (eds.), *Tropical Forests: Botanical Dynamics, Speciation and Diversity*. Academic Press, London. pp. 319–322.

Muckenhirn, N. A., Mortensen, B. K., Vessey, S., Fraser, C. E. O., and Singh, B. (1975). *Report on a Primate Survey in Guyana, 1975*. Pan American Health Organization, Washington DC. pp. 1–49.

Müller, K.-H. (1996). Diet and feeding ecology of masked titis (*Callicebus personatus*). In: Norconk, M. A., Rosenberger, A. L., and Garber, P. A. (eds.), *Adaptive Radiations of Neotropical Primates*. Plenum Press, New York. pp. 383–401.

Norconk, M. A. (1996). Seasonal variation in the diets of white-faced and bearded sakis (*Pithecia pithecia* and *Chiropotes satanas*) in Guri Lake, Venezuela. In: Norconk, M. A., Rosenberger, A. L., and Garber, P. A. (eds.), *Adaptive Radiations of Neotropical Primates*. Plenum Press, New York. pp. 403–423.

Norconk, M. A. (in press). A long-term study of venezuelan white-faced sakis (*Pithecia pithecia*): group dynamics and female reproduction. *Int. J. Primatol.*

Norconk, M. A., and Conklin-Brittain, N. L. (2004). Variation on frugivory: the diet of Venezuelan white-faced sakis. *Int. J. Primatol.* 25:1–26.

Norconk, M. A., Grafton, B. W., and Conklin-Brittain, N. L. (1998). Seed dispersal by neotropical seed predators. *Am. J. Primatol.* 45:103–126.

Norconk, M. A., and Kinzey, W. G. (1994). Challenge of neotropical frugivory: travel patterns of spider monkeys and bearded sakis. *Am. J. Primatol.* 34:171–183.

Norconk, M. A., Oftedal, O. T., Power, M. L., Jakubasz, M., and Savage, A. (2002). Digesta passage and fiber digestibility in captive white-faced sakis (*Pithecia pithecia*). *Am. J. Primatol.* 58:23–34.

Norconk, M. A., Raghanti, M. A., Martin, S. K., Grafton, B. W., Gregory, L. T., and DiDijn, B. P. E. (2003). Primates of Brownsberg Natuurpark, Suriname, with particular attention to the pitheciins. *Neotrop. Primates* 11:94–100.

Norconk, M. A., Sussman, R. W., and Phillips-Conroy, J. E. (1996). Primates of Guyana Shield forests. In: Norconk, M., Rosenberger, A. L., and Garber, P. A. (eds.), *Adaptive*

Radiations of Neotropical Primates. Plenum Press, New York. pp. 69–83.

Oliveira, J. M. S., Lima, M. C., Bonvincino, C., Ayres, J. M., and Fleagle, J. G. (1985). Preliminary notes on the ecology and behavior of the Guianan saki (*Pithecia pithecia*, Linnaeus 1766; Cebidae, Primate). *Acta Amazonica* 15:249–263.

Parolin, P. (2000). Seed mass in Amazonian floodplain forests with contrasting nutrient supplies. *J. Trop. Ecol.* 16:417–428.

Parolin, P. (2001). Seed germination and early establishment of 12 tree species from nutrient-rich and nutrient-poor central Amazonian floodplains. *Aqu. Bot.* 70:89–103.

Peetz, A. (2001). Ecology and social organization of the bearded saki *Chiropotes satanas chiropotes* (Primates: Pitheciinae) in Venezuela. *Ecotrop. Monogr.* 1:1–170.

Peres, C. A. (1993). Notes on the ecology of buffy saki monkeys (*Pithecia albicans*, Gray 1860): a canopy seed-predator. *Am. J. Primatol.* 31:129–140.

Price, E. C., and Piedade, H. M. (2001). Ranging behavior and intraspecific relationships of masked titi monkeys (*Callicebus personatus personatus*). *Am. J. Primatol.* 53:87–92.

Robinson, J. G. (1979). Vocal regulation of use of space by groups of titi monkeys, *Callicebus moloch*. *Behav. Ecol. Sociobiol.* 5:1–15.

Robinson, J. G., Wright, P. C., and Kinzey, W. G. (1987). Monogamous cebids and their relatives: intergroup calls and spacing. In: Smuts, B. B., Cheney, D. L., Seyfarth, R. M., Wrangham, R. W., and Struhsaker, T. T. (eds.), *Primate Societies*. University of Chicago Press, Chicago. pp. 44–53.

Ryan, K. M. (1995). Preliminary report on social structure and alloparental care on an island in Guri Reservoir. *Am. J. Phys. Anthropol.* 20(suppl.):187.

Savage, A., Shideler, S. E., Moorman, E. A., Ortuño, A. M., Whittier, C. A., Casey, K. K., and McKinney, J. (1992). The reproductive biology of the white-faced saki (*Pithecia pithecia*) in captivity. Abstracts of the XIVth Congress of the International Primatological Society, Strasbourg, France. pp. 59–60.

Schneider, H., and Rosenberger, A. L. (1996). Molecules, morphology, and platyrrhine systematics. In: Norconk, M. A., Rosenberger, A. L., and Garber, P. A. (eds.), *Adaptive Radiations of Neotropical Primates*. Plenum Press, New York. pp. 3–19.

Schradin, C., Reeder, D. M., Mendoza, S. P., and Anzenberger, G. (2003). Prolactin and paternal care: comparison of three species of monogamous New World monkeys (*Callicebus cupreus, Callithrix jacchus*, and *Callimico goeldii*). *J. Comp. Psychol.* 117:166–175.

Setz, E. Z. F. (1993). Ecologia alimentar de um grupo de parauacus (*Pithecia pithecia chrysocephala*) em um fragmento florestal na Amazônia central. Ciências Biológicas (Ecologia) [PhD diss.]. Instituto Biologica, UNICAMP, Campinas, Brazil.

Setz, E. Z. F., and Gaspar, D. A. (1997). Scent-marking behaviour in free-ranging golden-faced saki monkeys, *Pithecia pithecia chrysocephala*: sex differences and context. *J. Zool. Lond.* 241:603–611.

Shideler, S. E., Savage, A., Ortuño, A. M., Moorman, E. A., and Lasley, B. L. (1994). Monitoring female reproductive function by measurement of fecal estrogen and progesterone metabolites in the white-faced saki (*Pithecia pithecia*). *Am. J. Primatol.* 32:95–108.

Soini, P. (1986). A synecological study of a primate community in Pacaya-Samiria National Reserve, Peru. *Primate Conserv.* 7:63–71.

Stevenson, P. R. (2001). The relationship between fruit production and primate abundance in neotropical communities. *Biol. J. Linn. Soc.* 72:161–178.

Valeggia, C. R., Mendoza, S. P., Fernandez-Duque, E., Mason, W. A., and Lasley, B. (1999). Reproductive biology of female titi monkeys (*Callicebus moloch*) in captivity. *Am. J. Primatol.* 47:183–195.

van Roosmalen, M. G. M., Mittermeier, R. A., and Fleagle, J. G. (1988). Diet of the northern bearded saki (*Chiropotes satanas chiropotes*): a neotropical seed predator. *Am. J. Primatol.* 14:11–36.

van Roosmalen, M. G. M., Mittermeier, R. A., and Milton, K. (1981). The bearded sakis, genus *Chiropotes*. In: Coimbra-Filho, A. F., and Mittermeier, R. A. (eds.), *Ecology and Behavior of Neotropical Primates*, vol. 1. Academia Brasileira de Ciências, Rio de Janeiro. pp. 419–441.

van Roosmalen, M. G. M., van Roosmalen, T., and Mittermeier, R. A. (2002). A taxonomic review of the titi monkeys, genus *Callicebus* Thomas, 1903, with the description of two new species, *Callicebus bernhardi* and *Callicebus stephennashi* from Brazilian Amazonia. *Neotrop. Primates* 10(suppl.):1–52.

van Schaik, C. P., and Kappeler, P. M. (2003). The evolution of social monogamy in primates. In: Reichard, U. H., and Boesch, C. (eds.), *Monogamy: Mating Strategies and Partnerships in Birds, Humans and Other Mammals*. Cambridge University Press, Cambridge. pp. 59–80.

Vie, J.-C., Richard-Hansen, C., and Fournier-Chambrillon, C. (2001). Abundance, use of space, and activity patterns of white-faced sakis (*Pithecia pithecia*) in French Guiana. *Am. J. Primatol.* 55:203–221.

Walker, S. E. (1996). The evolution of positional behavior in the saki-uakaris (*Pithecia, Chiropotes*, and *Cacajao*). In: Norconk, M. A., Rosenberger, A. L., and Garber, P. A. (eds.), *Adaptive Radiations of Neotropical Primates*. Plenum Press, New York. pp. 335–367.

Walker, S. E., and Ayres, J. M. (1996). Positional behavior of the white uakari (*Cacajao calvus calvus*). *Am. J. Phys. Anthropol.* 101:161–172.

Wright, P. C. (1984). Biparental care in *Aotus trivirgatus* and *Callicebus moloch*. In: Small, M. (ed.), *Female Primates: Studies by Women Primatologists*. Alan R. Liss, New York. pp. 59–75.

Wright, P. C. (1985). The costs and benefits of nocturnality for *Aotus trivirgatus* (the night monkey) [PhD diss.]. City University of New York, New York.

Wright, P. C. (1986). Ecological correlates of monogamy in *Aotus* and *Callicebus*. In: Else, J. G., and Lee, P. C. (eds.), *Primate Ecology and Conservation*. Cambridge University Press, Cambridge. pp. 159–167.

9

Aotinae
Social Monogamy in the Only Nocturnal Haplorhines
Eduardo Fernandez-Duque

OVERVIEW

The two most salient features of owl monkeys are their nocturnal habits and their monogamous social organization. Owl monkeys concentrate their activities during the dark portion of the 24 hr cycle, with peaks of activity at dawn and dusk. Interestingly, our understanding of the evolution of nocturnality in the genus is further challenged by the existence of at least one owl monkey species that shows some remarkable temporal plasticity in its activity patterns. Differently from all owl monkey species in the tropics, which are strictly nocturnal, *Aotus azarai azarai* of the South American Gran Chaco is cathemeral, showing activity during the day as well as during the night (Fig. 9.1).

Owl monkeys are also one of the few socially monogamous primates in the world. They live in small groups that include only one pair of reproducing adults, one infant, one or two juveniles, and sometimes a subadult. Males show intense care of the infants. For a long time, the difficulties of studying a small arboreal and nocturnal primate limited our understanding of the evolution and maintenance of monogamy in this taxon. More recently, studies on the cathemeral owl monkeys of the Gran Chaco have begun to offer some insights into their social organization.

Figure 9.1 An owl monkey (*Aotus azarai azarai*) social group sun-basking during a cold morning in the Argentinean Chaco.

DISTRIBUTION AND TAXONOMY

Distribution

Owl monkeys range from Panama to northern Argentina and from the foothills of the Andes to the Atlantic Ocean (Aquino and Encarnación 1988, 1994; Aquino et al. 1992b; Barnett et al. 2002; Bennett et al. 2001; de Sousa e Silva and Nunes 1995; Defler 2003, 2004; Fernandes 1993; Ford 1994; García and Tarifa 1988; Hernández-Camacho and Cooper 1976; Hernández-Camacho and Defler 1985; Hershkovitz 1983; Kinzey 1997a; Peres 1993; Villavicencio Galindo 2003; Wright 1981; Zunino et al. 1985). They inhabit a variety of forests of both primary and secondary growth, sometimes up to 3,200 m above sea level (Defler 2003, Hernández-Camacho and Cooper 1976). At the southern end of their range, owl monkeys are distributed across the South American Gran Chaco of Argentina, Bolivia, and Paraguay, where they can be found in dry forests that receive only 500 mm of annual rainfall (Brooks 1996, Fernandez-Duque et al. 2002, Lowen et al. 1996, Neris et al. 2002, Stallings et al. 1989, Wright 1985) (Fig. 9.2).

Taxonomy

Owl monkeys, also known as night monkeys, douroucoulis or *mirikinás*, belong in the genus *Aotus*. The genus name is derived from combining the latin words *a*, meaning "without," and *otis*, meaning "ear," making reference to the inconspicuous earlobes usually hidden by dense fur. Several taxonomic issues remain largely unsettled, including the classification of the genus at the family and subfamily levels and the number of recognized species and subspecies within it.

Regarding its suprageneric classification, the genus was for many years placed alternatively in the Atelidae or in the Cebidae families. Based on morphological data, owl monkeys were considered to be closely related to the pithecines within the atelids (Rosenberger 1981, Schneider and Rosenberger 1996). A close affinity between owl monkeys and the atelines has also been suggested based on dental morphology (Tejedor 1998, 2001). On the other hand,

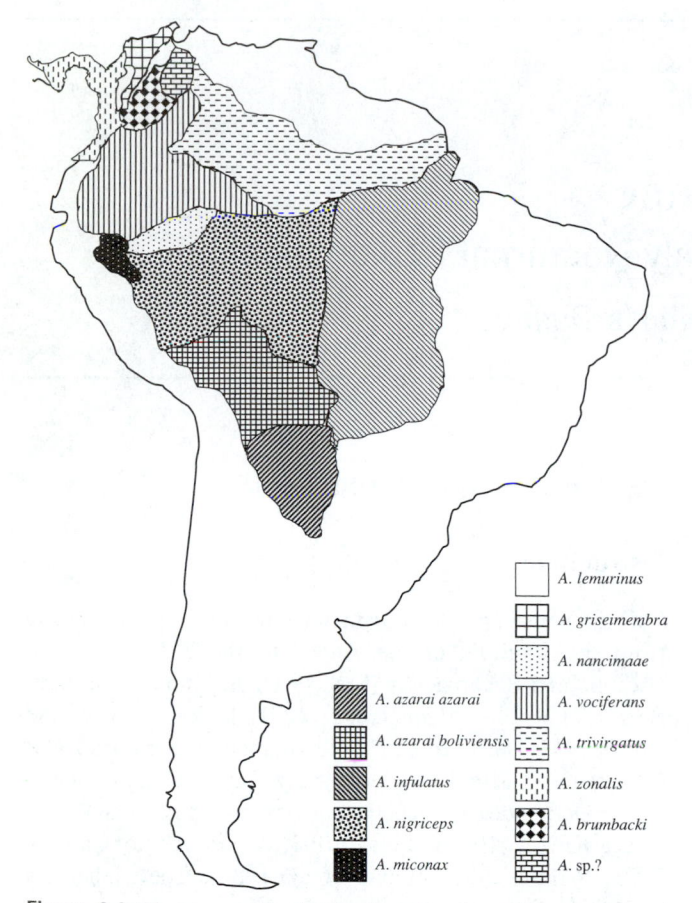

A. lemurinus

A. griseimembra

A. nancimaae

A. azarai azarai

A. azarai boliviensis

A. vociferans

A. infulatus

A. trivirgatus

A. nigriceps

A. zonalis

A. miconax

A. brumbacki

A. sp.?

Figure 9.2 The geographic distribution of owl monkey species and subspecies. The map has been redrawn based on Defler (2003), Erkert (1999), Ford (1994), and Rylands (unpublished data).

analysis of molecular genetic data led researchers to place the genus within the cebids (Porter et al. 1997; Schneider and Rosenberger 1996; Schneider et al. 1993, 1996). More recently, *Aotus* was placed in its own separate family, Aotidae (Defler 2003, Groves 2001, Rylands et al. 2000).

The genus only included the species *A. trivirgatus* when first described. Following the discovery of various karyotypes (Brumback 1973, 1974; Brumback et al. 1971; Ma 1981; Ma et al. 1976a,b, 1977, 1978, 1985), Hershkovitz (1983) divided the genus into nine species organized in two groups based on their karyotypes, coloration of the neck, and susceptibility to *Plasmodium falciparum*, one of the pathogens of human malaria: the gray-necked group occurs to the north and the red-necked group to the south of the Amazon River. Later on, an independent evaluation of craniodental measures and color pelage and patterns led Ford (1994) to accept at least five, and possibly seven, of the nine species identified by Hershkovitz.

However, because some of the owl monkey species in the gray-necked group are part of a sibling species complex, reliance on phenotypic traits for species recognition may be problematic (Defler 2003, 2004; Defler and Bueno 2003; Defler et al. 2001). In fact, extensive and systematic research using both phenotypes and karyotypes (Defler and Bueno

2003, Defler et al. 2001, Giraldo et al. 1986, Torres et al. 1998) led researchers to recognize up to 11 species (Defler 2003, Defler and Bueno 2003, Defler et al. 2001, Galbreath 1983, Rylands et al. 2000). There is enough information to recognize at least the following seven species in the gray-necked group: *A. lemurinus*, *A. zonalis*, *A. brumbacki*, *A. griseimembra*, *A. vociferans*, *A. trivirgatus*, and an undescribed eastern Colombian species. In turn, the red-necked group includes four species: *A. miconax*, *A. nancymaae* (the ending should be with *e*, not *i*), *A. nigriceps*, and *A. azarai* (the ending should be with *i*, not *e*) (Mudry de Pargament et al. 1984, Mudry et al. 1990, Pieczarka et al. 1993) (Table 9.1).

The lumping until so recently of all owl monkey diversity as *A. trivirgatus* (e.g., Robinson et al. 1987) should be carefully considered when evaluating the published literature. Given the existing knowledge on significant differences among owl monkey species in body size and body mass (Smith and Jungers 1997), activity patterns (Fernandez-Duque 2003; Wright 1989, 1994a, 1996), and canine sexual dimorphism (Hershkovitz 1983), published data from *A. trivirgatus* should be checked against the geographic origin of the sample or the locality where the study was conducted before they are assumed to be from any particular species.

ECOLOGY

Population Density and Group Size

Few reliable estimates of population densities exist for most owl monkey species (Table 9.2). The most comprehensive population studies have been done on *A. vociferans* and *A. nancymaae* in Peru, where numerous researchers have evaluated population densities over a long period of time and in a large number of different localities and habitats (Aquino and Encarnación 1986b, 1988, 1989, 1994; Aquino et al. 1990, 1992a,b, 1993; Moya et al. 1990).

Only a handful of studies have specifically evaluated the density of owl monkeys during the time when they are most active. On the other hand, several studies have reported densities estimated from sightings of nocturnal animals during the day while censusing diurnal primates. Unless the time of activity of the specific owl monkey species is considered, the estimates may be of little value (Aquino and Encarnación 1994, Weisenseel et al. 1993). For example, nocturnal censuses must consider moon phase since the probability of detection of owl monkeys is severely affected by available moonlight (Fernandez-Duque 2003, Fernandez-Duque and Erkert, 2006, Wallace et al. 2000). Given these serious limitations in some of the published data on owl monkey densities, there are no adequate data to evaluate the potential influence of habitat quality, predators, or competitors in determining population parameters across species. The value of discussing differences among estimates is highly questionable unless one has access to the raw sighting data obtained by the investigators (Peres 1999). Furthermore, the

Table 9.1   Taxonomy, Body Mass, and Chromosome Numbers of Owl Monkey Species and Subspecies

SPECIES	CAPTIVE (C) OR WILD (W)	ADULT MALE BODY MASS (KG) AVERAGE	RANGE	NUMBER OF INDIVIDUALS	ADULT FEMALE BODY MASS (KG) AVERAGE	RANGE	NUMBER OF INDIVIDUALS	CHROMOSOME 2N[1]	SOURCE
Aotus azarai azarai	W	1,254 ± 118	990–1,580	40	1,246 ± 114	1,010–1,450	39	m 49, f 50	Fernandez-Duque 2004
A. a. boliviensis	W	1,180		4	1,230		8	m 49, f 50	Smith and Jungers 1997
A. a. boliviensis	C	1,091		7	1,141		7		Dixson 1983
A. a. infulatus	W	1,190		1	1,240		1	m 49, f 50	Fernandes 1993
A. brumbacki	W	875 (unsexed)		1	455		1	50	Hernández-Camacho and Defler 1985
A. griseimembra	C	1,009 ± 200		20	923 ± 63		16	52, 53, 54	Dixson 1983
A. griseimembra	C	925 (median)	800–1,080	12					Dixson et al. 1980
A. lemurinus	W	920.7 ± 79.9	608–1,150	7	859 ± 87.6	578–1,050	6	58	Hernández-Camacho and Defler 1985
A. miconax								Unknown	
A. nancymaae	C				930 (median)[2]	1,190–760	16		Málaga et al. 1991
A. nancymaae	W	794		32	780		24	54	Aquino and Encarnación 1986b
A. nancymaae	C	946.5 ± 142.3	750–1,077	4	905.6 ± 123.6	706–1,055	6		S. Evans (unpublished)
A. nigriceps	W	875		1	1,040		2	m 51, f 52	Peres 1993
A. trivirgatus	W	813		20	736		17	Unknown	Smith and Jungers 1997
A. trivirgatus	W	1,200		1	1,000		1		Fernandes 1993
A. vociferans	C	708		20	698		20		Montoya et al. 1995
A. vociferans	W	697.5 ± 24	568–800	4				46, 47, 48	Hernández-Camacho and Defler 1985
A. zonalis	W	889		6	916		11	55, 56	Crile and Quiring 1940

[1] Chromosome numbers from Defler 2003, Erkert 1999, and Santos-Mello 1986.
[2] The median was visually estimated from the original data depicted in Figure 1 of Malaga et al. 1991.

estimates presented in Table 9.2 span 30 years and were collected by a wide range of researchers using different methods. Before differences among densities are attributed to ecological factors, the possible effects of different methodologies and temporal fluctuations in population values must be at least considered.

All reports agree that owl monkey group size ranges between two and six individuals (Aquino and Encarnación 1986b, Brooks 1996, Brown and Zunino 1994, Fernandes 1993, Fernandez-Duque and Bravo 1997, Fernandez-Duque et al. 2001, García and Braza 1989, García and Tarifa 1988, Peres 1993, Rathbun and Gache 1980, Stallings et al. 1989, Wright 1994a). In every locality, owl monkeys were found in small groups generally composed of an adult heterosexual pair, one infant, and one or two individuals of smaller size (Table 9.2). In an *A. a. azarai* population in Formosa, Argentina, many groups also included 3- and 4-year-old subadults (Fernandez-Duque and Huntington 2002).

At least in one population of *A. a. azarai*, a significant number of adults (25%–30%) do not belong to a social group but range solitarily (Fernandez-Duque 2004, Fernandez-Duque and Rotundo 2003). These individuals are either young adults that have recently emigrated from their natal groups or relatively old adults that have been evicted from their groups by incoming adults. A recent survey of *Aotus* spp. in northern Colombia also found a very significant number of solitary individuals (Villavicencio Galindo 2003). It seems likely that the less conspicuous solitary individuals

will be detected in other populations and species as more long-term studies of identified individuals are conducted.

Ranging and Territoriality

Owl monkeys are territorial, each group occupying a range that overlaps only slightly with the area used by neighboring groups (Table 9.3).

Groups regularly encounter other groups at range boundaries (García and Braza 1987; Garcia Yuste 1989; Robinson et al. 1987; Schwindt et al. 2004; Solano 1995; Wright 1978, 1985, 1994a), vocalizing and chasing each other (Wright 1978, 1985). Confrontations last from a few minutes to almost half an hour (Aquino and Encarnación 1986a, Wright 1985). Vocal confrontations include resonant whooping by both groups (Moynihan 1964, Wright 1981). In the strictly nocturnal species, encounters are more likely when the moon is full or directly overhead, whereas the cathemeral *A. a. azarai* has intergroup encounters during the night as well as during daylight hours (E. Fernandez-Duque, personal observation).

The function of territories and the behavioral mechanisms that maintain them remain unknown. It has been suggested that the animals defend feeding sources (Wright 1985), and it is also possible that at least some of the reported encounters are related to social or reproductive interactions with other groups or floaters. For example, in a population of *A. a. azarai*, some of the most aggressive interactions

Table 9.2 Population Density and Group Size of Owl Monkey Species

TAXON	STUDY SITE	GROUP DENSITY (G/KM²)	INDIVIDUAL DENSITY (KM²)	AVERAGE GROUP SIZE	RANGE	NUMBER OF GROUPS SIGHTED	TYPE OF STUDY[1]	SOURCE
Aotus azarai azarai	Guaycolec, Formosa, Argentina	14[2]	32.3	—	—	—	c	Zunino et al. 1985
A. a. azarai	Guaycolec, Formosa, Argentina	8.0	25.0	3.3	—	47	c	Arditi and Placci 1990
A. a. azarai	Guaycolec, Formosa, Argentina	16.0	64.0	4.0	2–6	11	be	Fernandez-Duque et al. 2001
A. a. azarai	Formosa Province, Argentina	5.5	12.8	2.3	—	12	c	Zunino et al. 1985
A. a. azarai	Formosa Province, Argentina		15.0	3.2	—	6	c	Brown and Zunino 1994
A. a. azarai	Formosa Province, Argentina	10.0	29.0	2.9	1–4	25	c	Rathbun and Gache 1980
A. a. azarai	Teniente Enciso, Paraguay	3.3	8.9	2.7	—	6	ce	Stallings et al. 1989
A. a. azarai	Agua Dulce, Paraguay	4.7	14.4	3.1	—	21	ce	Stallings et al. 1989
A. a. azarai	Presidente Hayes, Paraguay	—	—	—	2–4	2	be	Wright 1985
A. a. boliviensis	Departamento Beni, Bolivia	68.9[2]	242.5[2]	3.5	—	21	be	García and Braza 1989
A. a. boliviensis	Departamento Beni, Bolivia	—		3.6	2–5	23	c	García and Tarifa 1988
A. brumbacki	Río Duda, Tinigua NP, Colombia	—	—	3	—	1	be	Solano 1995
A. griseimembra	Northern Colombia	—	150[3]	—	2–4	—	c	Heltne 1977
A. nancymaae	Isla Iquitos, Peru	10.0	—	—	—	—	c	Soini 1976 as cited in Aquino and Encarnación 1989
A. nancymaae	Río Tahuayo, Peru	7.5	29.0	3.4	—	42	c	Aquino and Encarnación 1986b
A. nancymaae	Northeastern Peru	11.3	46.3	4.1	—	75	c	Aquino and Encarnación 1988
A. nancymaae	Northeastern Peru	5.9	24.2	—	—	23	c	Aquino and Encarnación 1988
A. nancymaae	Ríos Marañón-Amazonas, Peru	—	—	4	2–6	142	c	Aquino et al. 1990
A. nigriceps	Cocha Cashu, Manú NP, Peru	10[4]	36–40	4.1	2–5	9	be	Wright 1985
A. trivirgatus[5]	Departmento Bolivar, Colombia	0.5	1.5	2.5	—	8	c	Green 1978
A. trivirgatus[6]	Pto. Bermúdez, Dept. Pasco, Peru	—	—	3.3	2–4	3	be	Wright 1978
A. vociferans	Ríos Nanay and Napo, Peru	—	—	3.3	2–5	82	c	Aquino et al. 1990
A. vociferans	Northeastern Peru	10.0	33.0	3.3	—	22	c	Aquino and Encarnación 1988
A. vociferans	Northeastern Peru	2.4	7.9	—	—	11	c	Heltne 1977

[1] c, census; be, behavior and ecology study; ce, census and ecology study.
[2] Data are based on censusing an island of forest of 0.33 ha. Thus, estimates of density are based on repeated sampling of a very small area corresponding to the territory of one group.
[3] Data were collected in a forest remnant that may have served as a refuge, thus explaining the very unusual high density of 150 individuals/km².
[4] Number of groups was not reported, so it was estimated by dividing the density of individuals by the reported average group size.
[5] Only eight sightings while censusing during daylight. Source refers to *A. trivirgatus*, but it should be *A. griseimembra*.
[6] Source refers to *A. trivirgatus*, but it should be *A. nigriceps*, given its locality.

Table 9.3 Home Range Size and Ranging Patterns

TAXON	STUDY SITE	HOME RANGE SIZE (HA)	NUMBER OF GROUPS	AVERAGE DAY RANGE (M)	AVERAGE NIGHT RANGE (M)	SAMPLING EFFORT	SOURCE
Aotus azarai azarai	Presidente Hayes, Paraguay	5	1	199 (100–400)	420 (120–600)	10 day and 10 night follows during 2 weeks in July–August (n = 1 group)	Wright 1989, 1994a
A. a. azarai	Guaycolec, Formosa, Argentina	4–10	15	—	—	80 day and 40 night follows during 1 year (n = 5 groups)	Schwindt et al. 2004, E. Fernandez-Duque (unpublished)
A. a. azarai	Guaycolec, Formosa, Argentina	12	—	—	—	12 months, 262 hr	Arditi 1992
A. a. boliviensis	Departamento Beni, Bolivia	10.33[1]	1	—	337 (153–440)	10 nights during 5 months, only during full moon nights (n = 1 group)	García and Braza 1987
A. brumbacki	Río Duda, Tinigua NP, Colombia	17.5	1	—	837.3[2]	53 night follows during 5 months (n = 1 group)	Solano 1995
A. nigriceps	Cocha Cashu, Manú NP, Peru	7–14	3	—	708 (340–1,025)	60 night follows during 1 year (n = 1 group)	Wright 1989, 1994a
A. trivirgatus[3]	Pto. Bermúdez, Dept. Pasco, Peru	3.1	1	—	252 (60–450)	34 night follows during 9 weeks (n = 1 group)	Wright 1978

[1] The group occupied an island forest of 0.33 ha.
[2] Source refers to "average daily ranging," it is assumed it referred to night ranging.
[3] Source refers to *A. trivirgatus*, but it should be *A. nigriceps*, given its locality.

Table 9.4 Diet Composition (Percent)

SPECIES AND SUBSPECIES	LOCATION	FRUIT	NECTAR AND FLOWERS	LEAVES	FUNGUS	INSECTS	LENGTH OF STUDY (MONTHS)	SOURCE
Aotus azarai azarai	Presidente Hayes, Paraguay	16	33	40		[1]	2	Wright 1985
A. a. azarai	Guaycolec, Formosa, Argentina	45	14	41		[1]	12	Arditi 1992
A. a. azarai	Guaycolec, Formosa, Argentina	66	1	15	[2]	[1]	2	Giménez and Fernandez-Duque 2003, Giménez 2004
A. brumbacki	Río Duda, Tinigua NP, Colombia	59	13			28	6	Solano 1995
A. nigriceps	Cocha Cashu, Manú NP, Peru	60	[3]	[4]		[1]	15	Wright 1985
A. vociferans	Puerto Huamán y Mishana, Peru	83	17			–	9	Puertas et al. 1992
A. zonalis	Burro Colorado Island, Panama	65		30		5		Hladik et al. 1971[5]

[1] No quantitative estimate of insect consumption.
[2] On several occasions, they have been observed ingesting an unidentified fungus.
[3] Nectar was mainly ingested in July and August; the average of those 2 months was 67%.
[4] No quantitative data available; the author considered it not extensive since there were no leaf veins or leaf refuse in 36 fecal samples examined.
[5] Stomach content of "several" individuals.

involve the resident group and a dispersing animal, possibly in search of reproductive opportunities (Fernandez-Duque 2003, 2004; Fernandez-Duque and Huntington 2002).

Ranging is strongly influenced by available moonlight in all examined species. The distance traveled was significantly longer during full-moon nights than during new-moon nights in *A. a. boliviensis* in Bolivia (García and Braza 1987), *A. a. azarai* in Argentina (Fernandez-Duque 2003) and Paraguay (Wright 1985, 1989), and *A. nigriceps* in Peru (Wright 1978, 1981, 1985). In *A. a. azarai*, ranging distance was also influenced by ambient temperature. When the moon was full, animals traveled more during warm nights than during cold ones, but there were no effects of temperature during the new moon (Fernandez-Duque 2003).

Diet

There are still no satisfactory quantitative estimates of diet composition and foraging for any of the strictly nocturnal species (Table 9.4). Wright's work (1986, 1994a) on *A. nigriceps* in Manú National Park, Peru, is the only thorough attempt at quantifying diet in one of the nocturnal species, but the problems of obtaining quantitative estimates in a nocturnal primate were numerous. Cathemeral owl monkeys (*A. a. azarai*) have provided opportunities to examine the diet in some detail during daylight hours (Arditi 1992, Giménez and Fernandez-Duque 2003, Giménez 2004, Wright 1985), but determining their foraging habits during the night remains a challenge.

Owl monkeys are primarily frugivorous. Fruits are the most consumed item in *A. nigriceps* (Wright 1985, 1986, 1994a), *A. a. azarai* (Arditi 1992, Fernandez-Duque et al.

2002, Giménez and Fernandez-Duque 2003, Wright 1985), and *A. vociferans* (Puertas et al. 1992). *Ficus* spp. fruits are a highly preferred item in all examined species.

Leaf and insect eating is virtually impossible to quantify during the night (Arditi 1992, Wright 1985). The absence of leaf veins or leaf refuse in 36 fecal samples and the fact that she observed *A. nigriceps* eating a vine leaf only occasionally led Wright (1985) to conclude that leaf eating may not be important in *A. nigriceps*. Still, it remains possible that leaf eating in strictly nocturnal owl monkeys occurs but remains undetected.

On the other hand, leaf consumption has been regularly observed in the cathemeral *A. a. azarai*. A study of the *A. a. azarai* diet in Paraguay (Ganzhorn and Wright 1994, Wright 1985) and two studies in Argentina (Arditi 1992, Giménez and Fernandez-Duque 2003) found a significant consumption of leaves (Table 9.4). Arditi and Placci (1990) reported that the diet of diurnal and crepuscular owl monkeys in Argentina was more than 40% leaves. Giménez (2004) found that the consumption of leaves reached 30% for a group inhabiting a patch of thorn forest, whereas it was lower (15%) for a group inhabiting a type of forest where fruits are more abundant.

Although insects are undoubtedly part of the diet, quantifying their representation in the diet has proven so far impossible. Most authors have observed owl monkeys eating insects (Arditi 1992, Giménez and Fernandez-Duque 2003, Moynihan 1964, Puertas et al. 1992, Wright 1985), but none has béen able to obtain quantitative estimates of its prevalence. Wright (1985:60) observed them eating lepidopterans (moths 2–8 cm), coleopterans (beetles, 2–4 cm), and spiders; but she "could not collect quantitative data on insect-foraging." Fernandez-Duque (unpublished data) observed *A. a. azarai* eating cicadas a few times during the daylight hours, and it is frequently reported that pet owl monkeys keep houses clean of spiders.

Flowers may be an important food item for *A. a. azarai* during certain times of the year in the very seasonal Paraguayan and Argentinean Chaco (Ganzhorn and Wright 1994, Giménez and Fernandez-Duque 2003, Wright 1985). In August and September, when fruit availability is lower (Fernandez-Duque et al. 2002), *A. a. azarai* routinely feeds on trumpeter tree flowers (*Tabebuia ipe*).

REPRODUCTIVE PARAMETERS

The reproductive biology of some species has been well documented in captive animals (Table 9.5). In captivity, *A. lemurinus* males enter puberty at a surprisingly early age (Dixson 1994, 1998; Dixson et al. 1980), with testosterone first increasing when individuals were between 7 and 11 months of age ($n = 6$ males). The timing of puberty does not seem to be affected by the social context since there are no differences in levels of testosterone or glandular growth between males caged together with the family or caged alone.

Achievement of full sexual maturation of males in *A. lemurinus* takes place at approximately 2 years of age, as indicated by measures of body mass, growth of the subcaudal scent gland, and circulating reproductive hormones (Dixson 1982, 1983, 1994; Dixson and Gardner 1981; Dixson et al. 1980; Hunter et al. 1979). The timing of first reproduction in captive *A. vociferans* and *A. nancymaae* females occurred between 3 and 4 years of age.

Sexually mature captive males of *A. lemurinus* show partially arrested spermatogenesis with a very low sperm count

and small testis size (Dixson 1986, Dixson and Gardner 1981, Dixson and George 1982, Dixson et al. 1979). The median testis volume was 514 mm³ (range 378–673, $n = 12$ males). The average testis volume of 14 free-ranging adult *A. a. azarai* was 190 mm³ (range 86–308) (Fernandez-Duque, unpublished data). Although estimates in the wild were obtained by applying the same methodology used with the captive animals, it seems reasonable that the differences in volume between species may be research artifacts.

The knowledge of reproductive parameters in wild individuals is much more limited. A physical examination of 115 individuals of *A. a. azarai* showed that they did not reach adult body mass or exhibit a fully developed subcaudal gland until they were approximately 4 years of age (Fernandez-Duque 2004, Juárez et al. 2003). In this population, age at first reproduction was at least 5 years as indicated by age at dispersal, the average delay from pairing to parturition (see below), and the observation of one female of known date of birth who reproduced for the first time when she was 58 months old (Table 9.5).

Mating is very infrequent in owl monkeys, although it is extremely likely that unobserved mating behavior takes place during the night. Over the course of 3 years and more than 2,000 hr of observations, mating behavior was recorded only on eight occasions ($n = 5$ groups) (Fernandez-Duque et al. 2002). Further evidence for the low frequency of mating comes from studies in captivity. Dixson (1994) recorded it on only 19 occasions during 278 hr of observation. Mating has been observed during pregnancy in free-ranging *A. a. azarai* and in captive *A. lemurinus* (Dixson 1994, Hunter et al. 1979).

Owl monkey females' reproductive cycle lasts approximately 16 days (range 13–19 days, *A. lemurinus*) (Dixson et al. 1980), and females produce one infant per year. In captivity, twinning occurred in one of 169 births (Gozalo and

Table 9.5   Life History and Reproductive Parameters

SPECIES	CAPTIVE (C) OR WILD (W)	AVERAGE INTERBIRTH INTERVAL	GESTATION LENGTH (DAYS)	BIRTH SEASONALITY	FEMALE AGE AT FIRST REPRODUCTION (MONTHS)	SOURCE
Aotus griseimembra	C	253 (166–419), 36 IBIs		No		Dixson et al. 1980, Bonney et al. 1979, Dixson 1983
A. griseimembra	C	271 (median = 258), 48 IBIs	133	No		Hunter et al. 1979
A. trivirgatus	C		126, 138, and 148–159			Meritt, pers. comm. as cited in Hunter et al. 1979, Hall and Hodgen 1979, Elliot et al. 1976 as cited in Dixson 1983
A. vociferans	C	12.8 ± 6.5 ($n = 110$ IBIs w), 10.2 ± 3.2 ($n = 19$ IBIs c)	146, 132, and 151 ($n = 3$)	58.8% (87/148) December–March, 75.7% (112/148) December–May	48 ± 12 ($n = 15$)	Montoya et al. 1995
A. nancymaae	C	12.72 ± 5.72 ($n = 75$ IBIs)		52.3% (66/126) October–January	40.56 ± 7.82 ($n = 9$)	Gozalo and Montoya 1990
A. nigriceps	W			100% (9/9) August–February		Wright 1985
A. nancymaae	C		122–141	No		Málaga et al. 1997
A. nancymaae	W			December–March		Aquino et al. 1990
A. vociferans	W			November–January		Aquino et al. 1990
A. azarai azarai	W	Median 370 (345–426), 13 IBIs		88% (24/27) October–November	5 years ($n = 1$)	Fernandez-Duque et al. 2002

Montoya 1990) and in one of 287 births (Málaga et al. 1997) in *A. nancymaae*. In the wild, it has been reported only once in *A. vociferans* (Aquino et al. 1990). Two newborns, weighing 125 and 150 g, were removed from the male's back when the group was captured.

Survival tends to be higher among infants than juveniles, both in captivity and in the wild. Infant survival in captivity was 93.8% during the first week of life (Gozalo and Montoya 1990, Málaga et al. 1997) and 96% (26 of 27 infants) during the first 6 months of life in *A. a. azarai* in the wild (Fernandez-Duque et al. 2002). On the other hand, survival was lower for 1-year-old captive juveniles of *A. vociferans* (76.9%), (Montoya et al. 1995) and *A. nancymaae* (85.8%) (Gozalo and Montoya 1990). Further evidence for lower survival among juveniles comes from a wild population of *A. nancymaae*, where juveniles were the least numerous age category (6%–7%) in 14 captured and 23 partially captured groups (Aquino and Encarnación 1989).

There is a strong environmental influence on the timing of births in free-ranging individuals. All nine births recorded in four groups of *A. nigriceps* in Peru occurred between August and February (Wright 1985). Births were estimated to occur between December and March in *A. nancymaae* based on the presence of dependent and independent offspring in 75 captured groups in northeastern Peru (Aquino et al. 1990). In the Argentinean Chaco, most births (88%, $n = 24$) took place during an 8-week period between late September and late November (Fernandez-Duque et al. 2002).

In captivity, the tropical species *A. lemurinus* and *A. nancymaae* bred throughout the year when photoperiod was kept constant (Dixson 1994, Málaga et al. 1997) but *A. nancymaae* and *A. vociferans* adjusted their reproduction accordingly when housed under conditions of natural photoperiod (Gozalo and Montoya 1990, Montoya et al. 1995). Finally, when owl monkeys that had been living in indoor facilities were housed in an outdoor facility at 25° of latitude in the Northern Hemisphere, they became increasingly more seasonal the longer they lived outdoors (Holbrook et al. 2004). Very interestingly, most births were confined to the March–June period, a 6-month difference from the birth season in the Argentinean Chaco *A. a. azarai* population, which is also located at 25° of latitude but in the Southern Hemisphere.

SOCIAL ORGANIZATION

Serial Monogamy

Owl monkeys are one of the few socially monogamous primates in the world (Fuentes 1999; Kappeler and Van Schaik 2002; Kinzey 1997a,b; Moynihan 1964, 1976; Robinson et al. 1987; Wright 1981). They live in small groups (two to six individuals), which include an adult heterosexual pair, one infant, one or two juveniles, and sometimes a subadult (Table 9.2). There is never more than one reproducing female in the group or more than one adult male caring for the offspring. Still, data from a long-term study of wild *A. a. azarai* suggest that the social system of owl monkeys may be more dynamic than commonly assumed, in good agreement with cumulating evidence from other socially monogamous primates (Brockelman et al. 1998, Palombit 1994, Reichard 1995). In the aforementioned *A. a. azarai* population, turnover of resident adults is very frequent: 14 of 15 pairs had at least one of the mates replaced during a 3-year period. Males and females were similarly likely to be replaced. In almost all cases, mate replacement occurred as a consequence of an intruding adult expelling the same-sex resident individual and not because of desertion of one of the pairmates (Fernandez-Duque 2004).

Biparental Care

The intensive involvement of the male in infant care is one of the most fascinating and unique aspects of the social organization of the genus. It is comparable only to the pattern described in *Callicebus* spp., where there is intensive paternal care but no involvement of the siblings in the care of infants (Hoffman et al. 1995, Mendoza and Mason 1986, Wright 1984). Sibling care has been recorded infrequently in captive groups of *A. lemurinus* and *A. a. boliviensis* (Dixson and Fleming 1981, Jantschke et al. 1996, Wright 1984), and it does not seem to be prevalent in free-ranging *A. a. azarai* (Rotundo et al. 2005).

There have been several detailed studies of parental behavior and infant development in captive groups (Dixson and Fleming 1981, Jantschke et al. 1996, Málaga et al. 1997, Meritt 1980, Robinson et al. 1987, Wright 1984). Although studies of free-ranging populations have been limited by the nocturnal habits of owl monkeys (Wright 1984, 1985), studies of the cathemeral *A. a. azarai* have confirmed the findings in captivity (Juárez et al. 2003; Rotundo et al. 2002, 2005; Schwindt et al. 2004; Wright 1981, 1984, 1986, 1994a).

Newborns are carried in a distinctive ventrolateral position, in the flexure of the mother's thigh during the first 2 postnatal weeks. In captivity, the mother is the main carrier of the single offspring only during the first week in *A. lemurinus* (Dixson and Fleming 1981) and during the first 2 weeks of life in *A. a. boliviensis* (Jantschke et al. 1996); thereafter, the male takes over the role. After the first week of life, free-ranging *A. a. azarai* males carry the infant 84% of the time (Juárez et al. 2003, Rotundo et al. 2002). Rather than providing comfort and extended periods of physical contact with their infants, it appears that female owl monkeys have evolved mechanisms to reject the developing offspring once suckling is complete and to encourage it to transfer to the adult male (Dixson and Fleming 1981, Wright 1984). For example, 40 of 146 suckling bouts were followed by maternal rejection in captive *A. lemurinus* (Dixson 1994).

The male not only carries the infant most of the time but also plays, grooms, and shares food with the infant. Food sharing between weaned infants and other group members occurs only rarely under natural conditions (Rotundo et al. 2005). On the other hand, food sharing was prevalent among *A. nancymaae* and *A. lemurinus* in captivity and more frequently done by the male than the female (Feged et al. 2002, Wright 1984). The closer relationship of the male with the infant continues as the infant approaches maturity (Dixson 1983). Then, juveniles maintain more frequent social proximity with the male than the female (Juárez et al. 2003) and may even disperse from their natal group following eviction of their putative father by an incoming male (E. Fernandez-Duque, personal observation). The strong male–infant attachment is also manifested in the preference of the infant to stay in the group with the male following the eviction of the mother by an incoming female (Fernandez-Duque 2004), instead of dispersing with the mother. These observations suggest that, in owl monkeys, the infant may be primarily attached to the putative father, similar to what has been described in titi monkeys (Hoffman et al. 1995, Mendoza and Mason 1986).

The unquestionable intensive male care notwithstanding, the evolution and maintenance of paternal care in owl monkeys remain largely unexplained. It has been hypothesized that paternal care may be adaptive because it increases offspring survival or offers increased foraging opportunities for the lactating female (Tardif 1994; Wright 1984, 1986). Data from free-ranging *A. a. azarai* do not support the "increased foraging" hypothesis since the adult carrying the infant traveled most frequently in the middle of the group, sometimes first but rarely last as predicted by the hypothesis (Rotundo et al. 2005). At present, there is no evidence suggesting that male care reduces the risk of infanticide in free-ranging *A. a. azarai*. A third possibility is that male care could function as mating effort. This possibility is somewhat supported by observations of three intruding *A. a. azarai* males that, following eviction of the putative father, took care of the infant in a manner that could not be distinguished behaviorally from the care provided by the putative father (E. Fernandez-Duque, unpublished data).

Dispersal

Based on the species-specific social group size, it has usually been assumed that natal dispersal might occur when individuals are 2 or 3 years of age and have reached adult size. Additional inferences on age at dispersal were made from changes in group composition (Fernandez-Duque and Huntington 2002, Huntington and Fernandez-Duque 2001, Wright 1985). More recently, the first data on dispersal from identified individuals of known sex and age have been obtained from free-ranging radiocollared *A. a. azarai*.

Both sexes disperse at various ages in *A. a. azarai* (Fernandez-Duque 2004, Juárez et al. 2003). Animals disperse from their natal groups before reaching full sexual maturation, around the time of sexual maturation, and well after it but never before they are 26 months old. Dispersal of relatively young animals seems to be associated with some major changes in group composition, like eviction of the resident male or female. Although most animals disperse when they are approximately 3 years of age, some stay until they are 4 or even 5 years old (Fernandez-Duque and Huntington 2002, Juárez et al. 2003). Aquino et al. (1990) also found more than two adults in captured groups, suggesting that non-reproductive adult offspring may be tolerated for a while. Thus, the available evidence suggests that age at dispersal may be flexible and dependent on the social context in which the pre-dispersing individual finds itself.

There are no data on free-ranging populations revealing whether the observed peripheralization of subadults that precedes dispersal is triggered by aggression from within the group. In captivity, maturing offspring are tolerated by their parents and no increase in agonistic behavior occurs between pubertal monkeys and their parents (Dixson 1983). Two- or 3-year-old *A. a. azarai* individuals sometimes lag behind the rest of the group or sleep in different trees, suggesting that some peripheralization may start taking place around that time (E. Fernandez-Duque, personal observation). Some animals leave the group for a few days and then return. The lack of unoccupied areas of the forest where dispersing animals could establish their own territory may lead offspring to postpone dispersal in this population of *A. a. azarai*.

Most dispersal events in *A. a. azarai* occurr immediately before (August–September, $n = 3$) or during (October–December, $n = 5$) the birth season. The concentration of these events around the birth season raises the possibility that births within the group may trigger the process of dispersal. On the other hand, it is also possible that dispersal is timed to take place in anticipation of the May–June mating season. Finally, given that dispersed individuals need to range solitarily for a few weeks to many months before moving into a new group, it is possible that dispersal takes place during the spring and summer months when temperatures are not as harsh and food resources are more abundant (Fernandez-Duque et al. 2002).

Mate Choice and Pair Formation

A. a. azarai individuals appear to disperse from their natal groups to find reproductive opportunities. There is no indication that they may find reproductive opportunities within their natal territory, as has been described for gibbons (Brockelman et al. 1998, Palombit 1994). Following dispersal, young adults range alone for various periods of time, in apparent search of a reproductive opportunity within a social group. Both male and female floaters replace resident adults through a process that may take a few days and involves aggressive interactions (Fernandez-Duque 2004). Frequently, mating takes place as soon as the new adult is accepted into the group. The observation of mating following pair formation agrees well with results from pair-testing experiments in captivity. *A. lemurinus* females tested daily with the same previously unfamiliar male copulated more frequently than females paired with their mates at all stages of the cycle, not only during the periovulatory phase (Dixson 1994).

New pairs of free-ranging *A. a. azarai* take at least 1 year until the female produces offspring, even if the pair is formed during the mating season. In captivity, the latency to reproduce could be taken as an indirect indicator of successful pairbonding. Montoya et al. (1995) found that, on average, a captive reproducing *A. vociferans* female took 26 months to reproduce after pairing. Another indirect indication of an improved pairbond with time could be the shortening of interbirth intervals in multiparous pairs. *A. nancymaae* captive females took, on average, 11 months between the first and second births but only 8.8 months following the third one (Málaga et al. 1997).

Intrasexual Competition and Aggression

Social groups sometimes interact aggressively with each other at territory boundaries. Interactions between members of social groups and floaters can also be aggressive. Both sexes participate in these interactions (Fernandez-Duque 2004, Fernandez-Duque and Rotundo 2003). Most resident adults in 15 social groups of *A. a. azarai* had broken or missing canines (78% of males and 56% of females), and approximately one-third had damaged earlobes (28% of males and 32% of females). Given the lack of observed intragroup aggression, the wounds are most likely the consequence of the severe fights that take place during encounters with other groups or with floaters trying to take over (Fernandez-Duque 2004). Fights have also been reported in *A. nigriceps* (Wright 1985) and *A. nancymaae* (Aquino and Encarnación 1989). In the latter case, when the authors captured the observed group, they confirmed that the adult male had wounds in the forelimb and earlobe due to bites.

The prevalence of aggressive interactions is supported by data from experiments in captivity. Most pairs (seven of eight) of confronting *A. lemurinus* males fought vigorously during testing so that four tests had to be terminated to avoid serious injuries to the animals (Hunter 1981, Hunter and Dixson 1983). Similarly, contact aggression occurred in five of eight pairs of confronting females. In 71% of male pairs and 40% of female pairs, the resident monkey became the aggressor.

Communication

Vocal, olfactory, and visual communication are undoubtedly important for owl monkeys (Robinson et al. 1987; Wright 1985, 1989). Their vocal repertoire can be divided into eight different categories (Moynihan 1964). The two most salient calls are resonant whoops and hoots. Resonant whoops are usually produced by both sexes during intergroup encounters and occur together with visual displays like swaying or arching (Moynihan 1964, Wright 1985).

Hoots are low-frequency calls given by one individual in the social group or by a solitary individual that convey information over long distances. Playback of these calls elicits responses from animals in the area, both groups and solitaries (E. Fernandez-Duque personal observation).

Olfactory cues also play an important role. Owl monkeys use both urine and cutaneous secretions in their scent-marking behaviors (Hill et al. 1959). Olfaction plays a prominent role in sexual recognition and aggression (Dixson 1994, Hunter 1981, Hunter and Dixson 1983, Hunter et al. 1984). During the confrontations of *A. lemurinus*, contact aggression between same-sex individuals was always preceded by some form of olfactory communication. Blocking olfactory input led to a reduction in intermale aggression (Hunter and Dixson 1983). In captivity, owl monkeys have been observed self-annointing with plants and millipedes (Zito et al. 2003).

Temporal Plasticity in Activity Patterns

Owl monkeys concentrate most of their activity during the dark portion of the 24 hr cycle, as indicated by observational studies of free-ranging *A. nigriceps* (Wright 1978, 1985, 1989, 1994a,b, 1996), *A. a. boliviensis* (García and Braza 1987, 1993; Garcia Yuste 1989), and *A. a. azarai* (Arditi 1992; Fernandez-Duque 2003; Fernandez-Duque and Erkert 2004; E. Fernandez-Duque and H. G. Erkert, unpublished; Rotundo et al. 2000; Wright 1994a,b, 1996) (Table 9.6). The nocturnal activity of these owl monkey species is strongly influenced by available moonlight. Activity is maximal during full-moon nights and minimal when there is no moonlight. *A. nigriceps* spent approximately half of the night active, but maximal levels of activity were recorded when the full moon was near the meridian (Wright 1985, 1989). *A. nancymaae* and *A. a. boliviensis* also showed intensive nocturnal activity during nights of full moon (Aquino and Encarnación 1986a; García and Braza 1987; Wright 1978, 1989). Captive *A. lemurinus* in Colombia showed marked lunar periodic variations in activity patterns under natural lighting conditions (Erkert 1974, 1976).

The strictly nocturnal owl monkeys of Colombia (*A. griseimembra*) were the focus of a series of laboratory experiments analyzing circadian rhythms of locomotor activity as well as their entrainment and masking by light (Erkert 1976, 1991; Erkert and Grober 1986; Erkert and Thiemann-Jager 1983; Rappold and Erkert 1994; Rauth-Widmann et al. 1991). These studies replicated the patterns observed under natural lighting conditions in the controlled setting of the laboratory (Erkert and Grober 1986, Erkert and Thiemann-Jager 1983). When new-moon conditions are simulated by means of continuous low light intensity, there is suppression of locomotor activity, indicating that low light intensity or brightness changes during the dark phase causes particularly strong masking of the light/dark-entrained circadian activity rhythm. Under nonmasking lighting conditions, *A. lemurinus* females show an increase in locomotor activity every 2

Table 9.6 Activity Patterns of Free-Ranging Owl Monkeys (Percentage of Sampling Points when Group Was Active)

SPECIES NAME	MOON PHASE OR SEASON	TIME OF DAY																									SOURCE
		1 hr	2 hr	3 hr	4 hr	5 hr	6 hr	7 hr	8 hr	9 hr	10 hr	11 hr	12 hr	13 hr	14 hr	15 hr	16 hr	17 hr	18 hr	19 hr	20 hr	21 hr	22 hr	23 hr	24 hr	
Aotus azarai azarai	Full moon	100	100	50	40	50	100	0	0	0	0	0	0	0	100	0	0	0	75	75	100	80	60	90	90	Wright 1985, 1989
A. a. azarai	New moon	0	0	40	40	10	50	40	0	30	40	0	60	0	0	0	0	80	100	90	0	50	0	30	10	Wright 1985, 1989
A. a. azarai	Annual mean	50	50	45	40	30	0	20	0	15	20	0	30	0	0	50	0	78	100	83	50	65	0	45	50	Wright 1985, 1989
A. a. azarai	Summer	–	–	–	–	–	92	65	30	50	20	5	0	5	5	8	0	10	10	55	10	5	–	–	–	Arditi 1992
A. a. azarai	Fall	–	–	–	–	–	90	90	65	40	15	15	5	5	0	3	15	45	92	100	–	–	–	–	–	Arditi 1992
A. a. azarai	Winter	–	–	–	–	–	100	100	88	30	32	28	38	25	18	20	35	70	95	100	–	–	–	–	–	Arditi 1992
A. a. azarai	Spring	–	–	–	–	–	95	55	25	60	50	5	0	0	0	0	0	10	80	95	–	–	–	–	–	Arditi 1992
A. a. azarai	Annual mean	–	–	–	–	–	94	78	52	45	29	13	11	7.5	5.8	7.8	13	34	69	88	10	5	–	–	–	Arditi 1992
A. a. azarai	Annual mean	45	48	37	43	51	70	70	47	29	24	14	12	16	11	12	18	46	73	87	68	45	42	45	34	Fernandez-Duque 2003
A. a. boliviensis	Full moon	60	38	35	25	38	40	0	–	–	–	–	–	–	–	–	–	–	12	85	92	50	50	65	45	Garcia and Braza 1987
A. nigriceps	Full moon	80	100	100	70	70	0	0	0	0	0	0	0	0	0	0	0	0	60	60	90	90	75	0	25	Wright 1985, 1989
A. nigriceps	New moon	90	0	0	75	75	0	0	0	0	0	0	0	0	0	0	0	0	60	100	100	70	80	80	70	Wright 1985, 1989
A. nigriceps	Mean	85	50	50	88	73	0	0	0	0	0	0	0	0	0	0	0	0	60	80	95	80	78	40	48	Wright 1985, 1989

weeks, which corresponds approximately to the ovarian cycle length (Rauth-Widmann et al. 1996).

Despite a preference for being active during the night, at least one owl monkey species is also active during daylight. *A. a. azarai* of the Argentinean and Paraguayan Chaco shows peaks of activity during the day as well as during the night (Arditi 1992, Erkert 2004, Fernandez-Duque 2003, Fernandez-Duque and Bravo 1997, Fernandez-Duque and Erkert 2006, Rotundo et al. 2000, Wright 1985). The results from observational studies were confirmed by fitting owl monkeys with accelerometer collars that allow uninterrupted activity recordings over a span of 6 months (Erkert 2004, Fernandez-Duque and Erkert 2006. Owl monkeys showed a clearly bimodal pattern of motor activity, with two main peaks of activity from 1800 to 2100 hr and from 0500 to 0800 hr (Fig. 9.3, bottom panel). During the night,

significant amounts of activity occurred regularly only during the moonlit parts of the night (Fig. 9.3, top panel).

An understanding of the mechanisms regulating cathemerality in this species is emerging (Erkert 2004, Fernandez-Duque 2003, Fernandez-Duque and Erkert 2006). The monkeys show marked lunar periodic and seasonal modulations of their activity pattern. At full moon, they are active throughout the night and show reduced activity during the day. With a new moon, activity decreases during the dark portion of the night, peaks during dawn and dusk, and extends over the bright morning hours. Waxing and waning moon induces a significant increase in activity during the first and second halves of the night, respectively. During the cold winter months, the monkeys display twice as much activity throughout the warmer bright part of the day than during the other months. These findings suggest that *A. a. azarai* is mainly a dark-active species but still able to shift a considerable portion of activity into the bright part of the day if unfavorable lighting and/or temperature conditions prevail during the night.

On the other hand, understanding the evolution of cathemerality remains a challenge. There have been several hypotheses proposed and evaluated to explain cathemerality in primates (Curtis and Rasmussen 2002; Curtis et al. 1999; Donati et al. 2001; Ganzhorn and Wright 1994; Morland 1993; Overdorff 1996; Overdorff and Rasmussen 1995; Tattersall 1987; van Schaik and Kappeler 1993, 1996; Warren and Crompton 1997; Wright 1989, 1999). One of them suggests that cathemerality may result from unusually harsh climatic conditions, whereas a second one poses that cathemerality may be the consequence of a pronounced seasonality in resource availability (Engqvist and Richard 1991, Ganzhorn and Wright 1994, Overdorff and Rasmussen 1995).

If cathemerality in owl monkeys is, at least partially, a response to food availability and digestive constraints, diurnal activity should increase during months of less fruit and insect availability. Owl monkeys do not show changes in activity patterns that can be interpreted this way. The total amount of diurnal activity in *A. a. azarai* in the Argentinean Chaco (Fernandez-Duque 2003, Fernandez-Duque and Erkert 2004, E. Fernandez-Duque and H. G. Erkert, unpublished) remained fairly constant throughout the year despite seasonal changes in food availability, temperature, and rainfall.

Instead, owl monkeys adjust their periods of activity to changes in ambient temperature, although the effects of temperature are contingent on moon phase. Owl monkeys apparently have a very narrow thermoneutral zone that may range 28°–30°C and a relatively low resting metabolic rate (Le Maho et al. 1981, Morrison and Simoes 1962). In captivity, *A. lemurinus* was most active when ambient temperature was 20°C and least active when it was 30°C (Erkert 1991). Still, the existing characterization of the thermoneutral zone and the relatively low basal metabolic rate should be interpreted with special attention to the geographic origins of the specimens studied in view of the various activity patterns observed in the genus.

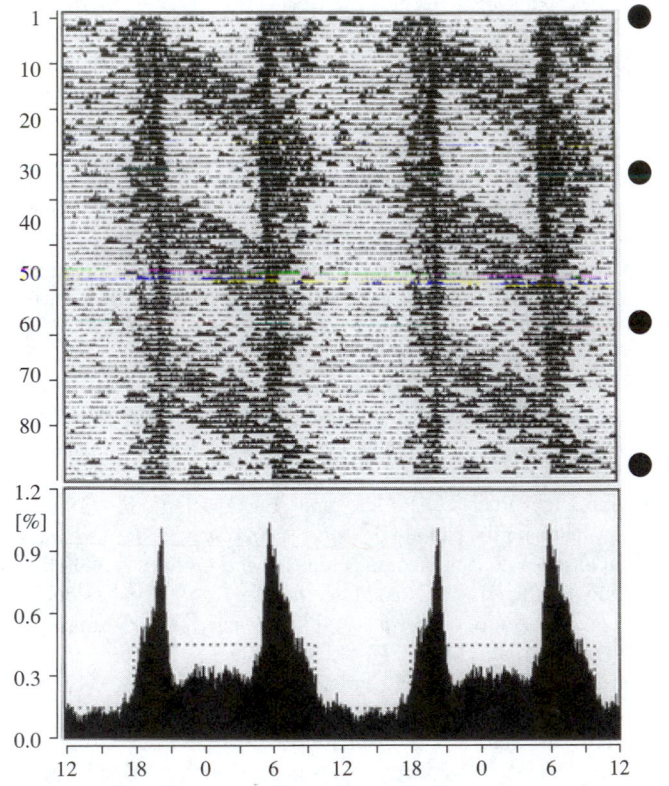

F: 1040903.011 C: 11 D: 23-11-03 12:05 - 21-02-04 12:00 R: 3 - 100 (ACTI)

Figure 9.3 *Top*: Double plot of original activity recordings in a male *Aotus azarai azarai* over a period of 3 months from November 2003 to February 2004 (top to bottom). Line spaces between the histograms correspond to 100 counts per 5 min interval. Please note the pronounced lunar periodic course of nighttime activity and that around the new moon (indicated by black circles on the right margin); only little motor activity is produced throughout the night, while much more activity occurs throughout the morning hours than at the other phases of the moon. *Bottom*: Double plotted bimodal activity pattern of this owl monkey as averaged over the whole recording period shown above. During the nighttime (21:00–06:00), the average level of activity is considerably higher than during the bright hours of the day. *Abscissa*: Argentinean official time, which is about 1 hr advanced in relation to the local time at the study site. *Ordinate*: relative activity, in percent of the mean 24 hr total as averaged over the whole recording period.

It is also reasonable to hypothesize that the risk of predation may influence the activity patterns of owl monkeys.
Wright (1989) suggested that a release from diurnal predation pressure may have been a selective force favoring
cathemerality in *A. a. azarai*. Unfortunately, no adequate
data exist on predation risk, predator activity patterns, or
predation events. There is some anecdotal evidence that runs
counter to the hypothesis that the absence of large diurnal
raptors may have favored cathemerality in owl monkeys.
At El Beni in Bolivia, where harpy eagles are present, owl
monkeys (*A. a. boliviensis*) showed diurnal activity when
the climate was unusually cold (Mann 1956).

CONCLUSIONS

Our understanding of owl monkeys' behavior, ecology,
and evolution remains severely limited. Although a picture
is emerging about the social organization, behavior, and
ecology of the southernmost taxon (*A. a. azarai*), several
intriguing aspects of this subspecies will need to be examined in other species before any broad generalizations can
be made for the genus as a whole.

Although owl monkeys are undoubtedly socially monogamous, the unexpected fast rate of adult replacement in the
A. a. azarai population suggests that serial monogamy may be
the norm. Thus, the long-held assumption of stable, lasting
pair bonds in monogamous primates will need to be once more
revised. As a consequence, it may also become necessary to
reevaluate many of our assumptions about the evolutionary
forces leading to monogamy. For example, if it is confirmed
that putative and nonputative males provide similar care to
infants, the function of the intensive care provided by owl
monkey males will have to be reevaluated. Logically, a
thorough evaluation of the costs and benefits of intensive
male care will not be complete until there is an understanding of the genetic structure of the different social groups.

An ongoing study of genetic relatedness among all individuals of 20 social groups in one *A. a. azarai* population (E. Fernandez-Duque unpublished data, Lau et al.
2004, Lau 2002, Sharma et al. 2003) will hopefully shed
light on the genetic aspects of serial monogamy. Preliminary
analyses have indicated a very low level of genetic diversity
in the population, with a relatively high degree of relatedness
among individuals in different groups. Is it possible that the
potential costs of serial monogamy are attenuated through
kin selection effects resulting from neighboring groups
being formed by closely related individuals.

The function of territoriality in owl monkeys will also
need careful examination. To successfully identify some of
the relevant factors driving and maintaining territoriality, it
will be necessary to develop a semiexperimental approach to
examine some of the unresolved issues. For example, playback experiments to simulate intruders or food-provisioning
experiments to manipulate available food resources will
need to be considered and implemented.

Finally, advancing our understanding of the evolutionary
forces favoring monogamy in owl monkeys will require a
comparative approach that considers some of the other more
tropical, strictly nocturnal owl monkey species as well as
some of the other socially monogamous primates. Recent
efforts to begin a long-term research program on the behavior and ecology of *A. vociferans*, *Pithecia monachus*,
and *Callicebus cupreus* in Yasuní National Park, Ecuador
(Schwindt et al. 2004), are one step toward accomplishing a
more general and broad description of owl monkey ecology
and behavior.

REFERENCES

Aquino, R., and Encarnación, F. (1986a). Characteristics and use
of sleeping sites in *Aotus* (Cebidae: Primates) in the Amazon
lowlands of Perú. *Am. J. Primatol.* 11:319–331.

Aquino, R., and Encarnación F. (1986b). Population structure of
Aotus nancymai (Cebidae: Primates) in Peruvian Amazon
lowland forest. *Am. J. Primatol.* 11:1–7.

Aquino, R., and Encarnación, F. (1988). Population densities and
geographic distribution of night monkeys (*Aotus nancymai* and
Aotus vociferans) (Cebidae: Primates) in northeastern Perú.
Am. J. Primatol. 14:375–381.

Aquino, R., and Encarnación, F. (eds.) (1989). *Aspectos de la
dinámica poblacional de* Aotus nancymai *(Cebidae: Primates)*.
IUCN/CSE-World Wildlife Fund, Washington DC.

Aquino, R., and Encarnación, F. (1994). Owl monkey populations
in Latin America: field work and conservation. In: Baer, J. F.,
Weller, R. E., and Kakoma, I. (eds.), Aotus: *The Owl Monkey*.
Academic Press, San Diego. pp. 59–95.

Aquino, R., Puertas, P. E., and Encarnación, F. (1990). Sup-
plemental notes on population parameters of northeastern
Peruvian night monkeys, genus *Aotus* (Cebidae). *Am. J.
Primatol.* 21:215–221.

Aquino, R., Puertas, P. E., and Encarnación, F. (1992a).
Evaluacion post-captura de *Aotus vociferans* y *Aotus nancymae*
en bosques de la amazonia peruana. *Folia Amazon.* 4:141–151.

Aquino, R., Puertas, P. E., and Encarnación, F. (1992b). Las poblaciones de *Aotus vociferans* y *Aotus nancymae* en la Amazonía
Peruana. *Alma Mater* 3:77–82.

Aquino, R., Puertas, P. E., and Encarnación, F. (1993). Effects
of cropping on the *Aotus nancymae* population in a forest of
Peruvian Amazonia. *Primate Rep.* 37:31–40.

Arditi, S. I. (1992). Variaciones estacionales en la actividad y dieta
de *Aotus azarae* y *Alouatta caraya* en Formosa, Argentina. *Bol.
Primatol. Latinoam.* 3:11–30.

Arditi, S. I., and Placci, G. L. (1990). Hábitat y densidad de *Aotus
azarae* y *Alouatta caraya* en Riacho Pilagá, Formosa. *Bol.
Primatol. Latinoam.* 2:29–47.

Barnett, A. A., Borges, S. H., Castilho, C. V. D., Neri, F. M., and
Shapley, R. L. (2002). Primates of the Jaú National Park,
Amazonas, Brazil. *Neotrop. Primates* 10:65–70.

Bennett, C. L., Leonard, S., and Carter, S. (2001). Abundance,
diversity, and patterns of distribution of primates on the Tapiche
River in the Amazonian Perú. *Am. J. Primatol.* 54:119–126.

Bonney, R. C., Dixson, A. F., and Fleming, D. (1979). Cyclic
changes in the circulating and urinary levels of ovarian steroids
in adult owl monkeys. *J. Reprod. Fertil.* 56:271–280.

Brockelman, W. Y., Reichard, U., Treesucon, U., and Raemaekers, J. J. (1998). Dispersal, pair formation and social structure in gibbons (*Hylobates lar*). *Behav. Ecol. Sociobiol.* 42:329–339.

Brooks, D. M. (1996). Some observations of primates in Paraguay. *Neotrop. Primates* 4:15–19.

Brown, A. D., and Zunino, G. E. (1994). Hábitat, densidad y problemas de conservación de los primates en Argentina. *Vida Silv. Neotrop.* 3:30–40.

Brumback, R. A. (1973). Two distinctive types of owl monkeys (*Aotus*). *J. Med. Primatol.* 2:284–289.

Brumback, R. A. (1974). A third species of the owl monkey (*Aotus*). *J. Hered.* 65:321–323.

Brumback, R. A., Staton, R. D., Benjamin, S.A., and Lang, C. M. (1971). The chromosomes of *Aotus trivirgatus* Humboldt 1812. *Folia Primatol.* 15:264–273.

Crile, G., and Quiring, D. P. (1940). A record of the body weight and certain organ and gland weights of 3690 animals. *Ohio. JSci.* XL:219–241.

Curtis, D. J., and Rasmussen, M. A. (2002). Cathemerality in lemurs. *Evol. Anthropol.* 1(suppl.):83–86.

Curtis, D. J., Zaramody, A., and Martin, R. D. (1999). Cathemerality in the mongoose lemur, *Eulemur mongoz. Am. J. Primatol.* 47:279–298.

Defler, T. R. (2003). *Primates de Colombia*. Conservación Internacional, Bogotá.

Defler, T. R. (2004). *Primates of Colombia*. Conservación Internacional, Bogotá.

Defler, T. R., and Bueno, M. L. (2003). Karyological guidelines for *Aotus* taxonomy. *Am. J. Primatol.* 60:134–135.

Defler, T. R., Bueno, M. L., and Hernández-Camacho, J. I. (2001). Taxonomic status of *Aotus hershkovitzi*: its relationship to *Aotus lemurinus lemurinus*. *Neotrop. Primates* 9:37–52.

de Sousa e Silva, J., Jr., and Nunes, A. (1995). Geographic distribution of night monkeys, *Aotus*, in northern Brazil: new data and a correction. *Neotrop. Primates* 3:72–74.

Dixson, A. F. (1982). Some observations on the reproductive physiology and behaviour of the owl monkey. *Int. Zoo Ybk.* 22:115–119.

Dixson, A. F. (1983). The owl monkey (*Aotus trivirgatus*). In: Hearn, J. P. (ed.), *Reproduction in New World Primates. New Models in Medical Sciences*. International Medical Publishers, Lancaster. pp. 69–113.

Dixson, A. F. (1986). Plasma testosterone concentrations during postnatal development in the male common marmoset. *Folia Primatol.* 47:166–170.

Dixson, A. F. (1994). Reproductive biology of the owl monkey. In: Baer, J. F., Weller, R. E., and Kakoma, I. (eds.), Aotus: *The Owl Monkey*. Academic Press, San Diego. pp. 113–132.

Dixson, A. F. (1998). *Primate Sexuality. Comparative Studies of the Prosimians, Monkeys, Apes, and Human Beings*. Oxford University Press, Oxford.

Dixson, A. F., and Fleming, D. (1981). Parental behaviour and infant development in owl monkeys (*Aotus trivirgatus griseimembra*). *J. Zool. Lond.* 194:25–39.

Dixson, A. F., and Gardner, J. S. (1981). Diurnal variations in plasma testosterone in a male nocturnal primate, the owl monkey (*Aotus trivirgatus*). *J. Reprod. Fertil.* 62:87–92.

Dixson, A. F., Gardner, J. S., and Bonney, R. C. (1980). Puberty in the male owl monkeys (*Aotus trivirgatus griseimembra*): a study of physical and hormonal development. *Int. J. Primatol.* 1:129–139.

Dixson, A. F., and George, L. (1982). Prolactin and parental behaviour in a male New World primate. *Nature* 299:551–553.

Dixson, A. F., Martin, R. D., Bonney, R. C., and Fleming, D. (1979). Reproductive biology of the owl monkey, *Aotus trivirgatus griseimembra*. In: *7th Congress of the International Primatological Society*. Karger, Bangalore. pp. 61–68.

Donati, G., Lunardini, A., Kappeler, P. M., and Borgognini Tarli, S. M. (2001). Nocturnal activity in the cathemeral red-fronted lemur (*Eulemur fulvus rufus*), with observations during a lunar eclipse. *Am. J. Primatol.* 53:69–78.

Engqvist, A., and Richard, A. (1991). Diet as a possible determinant of cathemeral activity patters in primates. *Folia Primatol.* 57:169–172.

Erkert, H. G. (1974). Der einflub des mondlichtes auf die aktivitatsperiodik nachtaktiver saugetiere. *Oecologia* 14:269–287.

Erkert, H. G. (1976). Lunarperiodic variation of the phase-angle difference in nocturnal animals under natural zeitgeber-conditions near the equator. *Int. J. Chronobiol.* 4:125–138.

Erkert, H. G. (1991). Influence of ambient temperature on circadian rhythms in Colombian owl monkeys, *Aotus lemurinus griseimembra*. In: Ehard, A., Kimura, T., Takenaka, O., and Iwamoto, M. (eds.), *Primatology Today*. Elsevier Science, Amsterdam. pp. 435–438.

Erkert, H. G. (1999). Owl monkeys. In: Poole, T. (ed.), *The UFAW Handbook on the Care and Management of Laboratory Animals*. Blackwell Science, London. 574–590.

Erkert, H. G. (2004). Chronobiological background to cathemerality. *Int. J. Primatol.* 75(suppl. 1):65.

Erkert, H. G., and Grober, J. (1986). Direct modulation of activity and body temperature of owl monkeys (*Aotus lemurinus griseimembra*) by low light intensities. *Folia Primatol.* 47:171–188.

Erkert, H. G., and Thiemann-Jager, A. (1983). Dark switch in the entrainment of circadian activity rhythms in night monkeys, *Aotus trivirgatus* Humboldt. *Comp. Biochem. Physiol.* 74A:307–310.

Feged, A., Wolowich, C., and Evans, S. (2002). Food sharing in owl monkeys. *Am. J. Primatol.* 57(suppl. 1):26.

Fernandes, M. E. B. (1993). New field records of night monkeys, genus *Aotus*, in northern Brazil. *Neotrop. Primates* 1:6–7.

Fernandez-Duque, E. (2003). Influences of moonlight, ambient temperature and food availability on the diurnal and nocturnal activity of owl monkeys (*Aotus azarai*). *Behav. Ecol. Sociobiol.* 54:431–440.

Fernandez-Duque, E. (2004). High levels of intrasexual competition in sexually monomorphic owl monkeys. *Folia Primatol.* 75(suppl. 1):260.

Fernandez-Duque, E., and Bravo, S. (1997). Population genetics and conservation of owl monkeys (*Aotus azarai*) in Argentina: a promising field site. *Neotrop. Primates* 5:48–50.

Fernandez-Duque, E., and Erkert, H. G. (2006). Cathemerality and lunarphilia in owl monkeys of the Argentinean Chaco. 76(1–2). *Folia Primatol.*

Fernandez-Duque, E., and Huntington, C. (2002). Disappearances of individuals from social groups have implications for understanding natal dispersal in monogamous owl monkeys (*Aotus azarai*). *Am. J. Primatol.* 57:219–225.

Fernandez-Duque, E., and Rotundo, M. (2003). Are female owl monkeys imposing social monogamy on males? *Rev. Etol.* 5(suppl.):92.

Fernandez-Duque, E., Rotundo, M., and Ramírez-Llorens, P. (2002). Environmental determinants of birth seasonality in owl monkeys (*Aotus azarai*) of the Argentinean Chaco. *Int. J. Primatol.* 23:639–656.

Fernandez-Duque, E., Rotundo, M., and Sloan, C. (2001). Density and population structure of owl monkeys (*Aotus azarai*) in the Argentinean Chaco. *Am. J. Primatol.* 53:99–108.

Ford, S. M. (1994). Taxonomy and distribution of the owl monkey. In: Baer, J. F., Weller, R. E., and Kakoma, I. (eds.), Aotus: *The Owl Monkey*. Academic Press, San Diego. pp. 1–57.

Fuentes, A. (1999). Re-evaluating primate monogamy. *Am. Anthropol.* 100:890–907.

Galbreath, G. J. (1983). Karyotypic evolution in *Aotus*. *Am. J. Primatol.* 4:245–251.

Ganzhorn, J. U., and Wright, P. C. (1994). Temporal patterns in primate leaf eating: the possible role of leaf chemistry. *Folia Primatol.* 63:203–208.

García, J. E., and Braza, F. (1987). Activity rhythms and use of space of a group of *Aotus azarae* in Bolivia during the rainy season. *Primates* 28:337–342.

García, J. E., and Braza, F. (1989). Density comparisons using different analytic-methods in *Aotus azarae*. *Primate Rep.* 25:45–52.

Garcia, J. E., and Braza, F. (1993). Sleeping sites and lodge trees of the night monkey (*Aotus azarae*) in Bolivia. *Int. J. Primatol.* 14:467–477.

García, J. E., and Tarifa, T. (1988). Primate survey of the Estación Biológica Beni, Bolivia. *Primate Conserv.* 9:97–100.

Garcia Yuste, J. E. (1989). Patrones etológicos y ecológicos del mono nocturno, *Aotus azarae boliviensis*. [*Diss. Abstr. Int.*] 50:473–C.

Giménez, M., and Fernandez-Duque, E. (2003). Summer and winter diet of night monkeys in the gallery and thorn forests of the Argentinean Chaco. *Rev. Etol.* 5(suppl.):164.

Giménez, M. C. (2004). Dieta y comportamiento de forrajeo en verano e invierno del mono mirikiná (*Aotus azarai azarai*) en bosques secos y húmedos del Chaco Argentino. Undergraduate thesis, Biology Department, University of Buenos Aires, Buenos Aires.

Giraldo, A., Bueno, M. L., Silva, E., Ramírez, J., Umaña, J., and Espinal, C. (1986). Estudio citogenético de 288 *Aotus* Colombianos. *Biomédica* 6:5–13.

Gozalo, A., and Montoya, E. (1990). Reproduction in the owl monkey (*Aotus nancymai*) (Primates: Cebidae) in captivity. *Am. J. Primatol.* 21:61–68.

Green, K. M. (1978). Primate censusing in northern Colombia: a comparison of two techniques. *Primates* 19:537–550.

Groves, C. (2001). *Primate Taxonomy*. Smithsonian Institution Press, Washington DC.

Hall, R. D., and Hodgen, G. D. (1979). Pregnancy diagnosis in owl monkeys (*Aotus trivirgatus*): Evaluation of the hemaglutination inhibition test for urinary chorionic gonadotropin. Laboratory Animal Science 29:345–348.

Heltne, P. G. (1977). *Census of Aotus in Northern Colombia*. Panamerican Health Organization, Washington DC. pp. 1–11.

Hernández-Camacho, J., and Cooper, R. W. (1976). The non-human primates of Colombia. *Neotropical Primates: Field Studies and Conservation*. In: Thorington, R. W., and Heltne, J. P. G. (eds.). National Academy of Sciences, Washington DC. 35–69.

Hernández-Camacho, J., and Defler, T. R. (1985). Some aspects of the conservation of non-human primates in Colombia. *Primate Conserv.* 6:46–50.

Hershkovitz, P. (1983). Two new species of night monkeys, genus *Aotus* (Cebidae: Platyrrhini): a preliminary report on *Aotus* taxonomy. *Am. J. Primatol.* 4:209–243.

Hill, W. C. O., Appleby, H. M., and Auber, L. (1959). The specialised area of skin glands in *Aotus* Humboldt (Simiae Platyrrhini). *Trans. R. Soc. Edinburgh* 63:535–551.

Hladik, C. M., Hladik, A., Bousset, J., Valdegbouze, P., Viroben, G., and Delort'Laval, J. (1971). Le régime alimentaire del primates de l'oele de Barro Colorado (Panamá): résultats des analyses quantitatives. *Folia Primatol.* 16:85–122.

Hoffman, K. A., Mendoza, S. P., Hennessy, M. B., and Mason, W. A. (1995). Responses of infant titi monkeys, *Callicebus moloch*, to removal of one or both parents: evidence for paternal attachment. *Dev. Psychobiol.* 28:399–407.

Holbrook, G. D., Chambers, C. M., and Evans, S. (2004). Seasonal breeding in captive owl monkeys. *Folia Primatol.* 75:276.

Hunter, A. J. (1981). *Chemical Communication, Aggression and Sexual Behavior in the Owl Monkey* (Aotus trivirgatus griseimembra). [PhD diss.] University of London, London.

Hunter, A. J., and Dixson, A. F. (1983). Anosmia and aggression in male owl monkeys (*Aotus trivirgatus*). *Physiol. Behav.* 30:875–879.

Hunter, A. J., Fleming, D., and Dixson, A. F. (1984). The structure of the vomeronasal organ and nasopalatine ducts in *Aotus trivirgatus* and some other primate species. *J. Anat.* 138:217–225.

Hunter, J., Martin, R. D., Dixson, A. F., and Rudder, B. C. C. (1979). Gestation and interbirth intervals in the owl monkey (*Aotus trivirgatus griseimembra*). *Folia Primatol.* 31:165–175.

Huntington, C., and Fernandez-Duque, E. (2001). Natal dispersal in the monogamous owl monkey (*Aotus azarai*) of Formosa, Argentina. *Am. J. Phys. Anthropol.* 32(suppl.):83–84.

Jantschke, B., Welker, C., and Klaiber-Schuh, A. (1996). Rearing without paternal help in the Bolivian owl monkey *Aotus azarae boliviensis*: a case study. *Folia Primatol.* 69:115–120.

Juárez, C., Rotundo, M., and Fernandez-Duque, E. (2003). Behavioral sex differences in the socially monogamous night monkeys of the Argentinean Chaco. *Rev. Etol.* 5(suppl.):174.

Kappeler, P. M., and Van Schaik, C. P. (2002). Evolution of primate social systems. *Int. J. Primatol.* 23:707–741.

Kinzey, W. G. (1997a). *Aotus*. In: Kinzey, W. G. (ed.), *New World Primates: Ecology, Evolution and Behavior*. Aldine de Gruyter, New York. pp. 186–191.

Kinzey, W. G. (ed.) (1997b). *New World Primates. Ecology, Evolution and Behavior*. Aldine de Gruyter, New York.

Lau, J. (2002). Characterization of microsatellite loci in three owl monkey species (*Aotus azarai, A. lemurinus, A. nancymaae*). [MSc. thesis] University of California, San Diego.

Lau, J., Fernandez-Duque, E., Evans, S., Dixson, A. F., and Ryder, O. A. (2004). Heterologous amplification and diversity of microsatellite loci in three owl monkey species (*Aotus azarai, A. lemurinus, A. nancymaae*). Conservation Genetics. 5(5):727–731.

Le Maho, Y., Goffart, M., Rochas, A., Felbabel, H., and Chatonnet, J. (1981). Thermoregulation in the only nocturnal simian: the night monkey *Aotus trivirgatus*. *Am. J. Physiol.* 240:R156–R165.

Lowen, J. C. L., Bartrina, C. R. P., and Tobias, J. A. (1996). Biological surveys and conservation priorities in eastern Paraguay. CSB Conservation Publications, Cambridge.

Ma, N. S. F. (1981). Chromosome evolution in the owl monkey, *Aotus*. *Am. J. Phys. Anthropol.* 54:293–303.

Ma, N. S. F., Aquino, R., and Collins, W. E. (1985). Two new karyotypes in the Peruvian owl monkey (*Aotus trivirgatus*). *Am. J. Primatol.* 9:333–341.

Ma, N. S. F., Elliot, M. W., Morgan, L. M., Miller, A. C., and Jones, T. C. (1976a). Translocation of Y chromosome to an autosome in the Bolivian owl monkey, *Aotus. Am. J. Phys. Anthropol.* 45:191–202.

Ma, N. S. F., Jones, T. C., Bedard, M. T., Miller, A. C., Morgan, L. M., and Adams, E. A. (1977). The chromosome complement of an *Aotus* hybrid. *J. Hered.* 68:409–412.

Ma, N. S. F., Jones, T. C., Miller, A., Morgan, L., and Adams, E. (1976b). Chromosome polymorphism and banding patterns in the owl monkey (*Aotus*). *Lab. Anim. Sci.* 26:1022–1036.

Ma, N. S. F., Rossan. R. N., Kelley, S. T., Harper, J. S., Bedard, M. T., and Jones, T. C. (1978). Banding patterns of the chromosomes of two new karyotypes of the owl monkey, *Aotus*, captured in Panamá. *J. Med. Primatol.* 7:146–155.

Málaga, C. A., Weller, R. E., and Buschbom, R. L. (1991). Twinning in the karyotype I night monkey (*Aotus nancymai*). *J. Med. Primatol.* 20:370–372.

Málaga, C. A., Weller, R. E., Buschbom, R. L., Baer, J. F., and Kimsey, B. B. (1997). Reproduction of the owl monkey (*Aotus* spp.) in captivity. *J. Med. Primatol.* 26:147–152.

Mann, G. F. (1956). Efecto del frío en mamíferos Amazónicos. *Invest. Zool. Chil.* III:155.

Mendoza, S. P., and Mason, W. A. (1986). Parental division of labour and differentiation of attachments in a monogamous primate (*Callicebus moloch*). *Anim. Behav.* 34:1336–1347.

Meritt, D. A. J. (1980). Captive reproduction and husbandry of the Douroucouli and the titi monkey. *Int. Zoo Ybk.* 20:52–59.

Montoya, E. G., Moro, J., Gozalo, A., and Samame, H. (1995). Reproducción de *Aotus vociferans* (Primates: Cebidae) en cautiverio. Revista de Investigaciones Pecuarias IVITA 7(2): 122–126.

Morland, H. S. (1993). Seasonal behavorial variation and its relationship to thermoregulation in ruffed lemurs. In: Kappeler, P. M., and Ganzhorn, J. U. (eds.), *Lemur Social Systems and Their Ecological Basis.* Plenum Press, New York. pp. 193–203.

Morrison, P., and Simoes, J. (1962). Body temperatures in two Brazilian primate. *Zoologia* 24:167–178.

Moya, L., Encarnación, F., Aquino, R., Tapia, J., Ique, C., and Puertas, P. (1990). El estado de las poblaciones naturales de primates y los beneficios de las cosechas sostenidas. In: Arámbulo III, P. Encarnación, F., Estupinan, J., Samame, H., Watson, C. R., and Weller, R. E. (eds.), *Primates de las Américas.* Battelle Press, Columbus. Ohio pp. 173–192.

Moynihan, M. (1964). Some behavior patterns of playtyrrhine monkeys. I. The night monkey (*Aotus trivirgatus*). *Smith. Misc. Coll.* 146:1–84.

Moynihan, M. (1976). *The New World Primates. Adaptive Radiation and the Evolution of Social Behavior, Languages, and Intelligence.* Princeton University Press, Princeton, NJ.

Mudry, M. D., Slavutsky, I., and Labal de Vinuesa, M. (1990). Chromosome comparison among five species of platyrrhini (*Alouatta caraya, Aotus azarae, Callithrix jacchus, Cebus apella, Saimiri sciureus*). *Primates* 31:415–420.

Mudry de Pargament, M., Colillas, O. J., and de Salum, S. B. (1984). The *Aotus* from northern Argentina. *Primates* 25:530–537.

Neris, N., Colmán, F., Ovelar, E., Sukigara, N., and Ishii, N. (2002). *Guía de Mamíferos Medianos y Grandes del Paraguay. Distribución, Tendencia Poblacional y Utilización.* Secretaría del Ambiente. Asunción, Paraguay.

Overdorff, D. J. (1996). Ecological correlates to activity and habitat use of two prosimian primates: *Eulemur rubriventer* and *Eulemur fulvus rufus* in Madagascar. *Am. J. Primatol.* 40:327–342.

Overdorff, D. J., and Rasmussen, M. A. (1995). Determinants of nightime activity in "diurnal" lemurid primates. In: Alterman, L. (eds.), *Creatures of the Dark: The Nocturnal Prosimians,* Plenum Press, New York. pp. 61–74.

Palombit, R. (1994). Dynamic pair bonds in hylobatids: implications regarding monogamous social systems. *Behavior* 128:65–101.

Peres, C. (1993). Notes on the primates of the Juruá River, western Brazilian Amazonia. *Folia Primatol.* 61:97–103.

Peres, C. A. (1999). Effects of subsistence hunting and forest types on the structure of Amazonian primate communities. In: Fleagle, J. G., Janson, C. H., and Reed, K. E. (eds.), *Primate Communities.* Cambridge University Press, Cambridge. pp. 268–283.

Pieczarka, J. C., de Souza Barros, R. M., de Faria, F. M., and Nagamachi, C. Y. (1993). *Aotus* from the southwestern Amazon region is geographically and chromosomally intermediate between *A. azarae boliviensis* and *A. infulatus. Primates* 34:197–204.

Porter, C. A., Page, S. L., Czelusniak, J., Schneider, H., Schneider, M. P. C., Sampaio, I., and Goodman, M. (1997). Phylogeny and evolution of selected primates as determined by sequences of the e-globin locus and 5′ flanking regions. *Int. J. Primatol.* 18:261–295.

Puertas, P. E., Aquino, R., and Encarnación, F. (1992). Uso de alimentos y competición entre el mono nocturno *Aotus vociferans* y otros mamíferos, Loreto, Perú. *Folia Amazon.* 4:135–144.

Rappold, I., and Erkert, H. G. (1994). Re-entrainment, phase-response and range of entrainment of circadian rhythms in owl monkeys (*Aotus lemurinus*) of different age. *Biol. Rhythm Res.* 25:133–152.

Rathbun, G. B., and Gache, M. (1980). Ecological survey of the night monkey, *Aotus trivirgatus*, in Formosa Province, Argentina. *Primates* 21:211–219.

Rauth-Widmann, B., Fuchs, E., and Erkert, H. G. (1996). Infradian alteration of circadian rhythms in owl monkeys (*Aotus lemurinus griseimembra*): an effect of estrus? *Physiol. Behav.* 59:11–18.

Rauth-Widmann, B., Thiemann-Jager, A., and Erkert, H. G. (1991). Significance of nonparametric light effects in entrainment of circadian rhythms in owl monkeys (*Aotus lemurinus griseimembra*) by light–dark cycles. *Chronobiol. Int.* 8:251–266.

Reichard, U. (1995). Extra-pair copulations in a monogamous gibbon (*Hylobates lar*). *Ethology* 100:99–112.

Robinson, J. G., Wright, P. C., and Kinzey, W. G. (1987). Monogamous cebids and their relatives: intergroup calls and spacing. In: Smuts, B. B., Cheney, D. L., Seyfarth, R. M., Wrangham, R. W., and Struhsaker, T. (eds.), *Primate Societies.* University of Chicago Press, Chicago. pp. 44–53.

Rosenberger, A. L. (1981). Systematics: the higher taxa. In: Coimbra-Filho, A. F., and Mittermeier, R. A. (eds.), *Ecology and Behavior of Neotropical Primates.* Academia Brasileira de Ciencias, Rio de Janeiro. pp. 9–27.

Rotundo, M., Fernandez-Duque, E., and Dixson, A. F. (2005). Infant development and parental care in free-ranging groups of owl monkeys (*Aotus azarai azarai*) in Argentina. *Int. J. Primatol.* 26(6).

Rotundo, M., Fernandez-Duque, E., and Giménez, M. (2002). Cuidado biparental en el mono de noche (*Aotus azarai azarai*) de Formosa, Argentina. *Neotrop. Primates* 10:70–72.

Rotundo, M., Sloan, C., and Fernandez-Duque, E. (2000). Cambios estacionales en el ritmo de actividad del mono mirikiná (*Aotus azarai*) en Formosa Argentina. In: Cabrera, E., Mércolli, C., and Resquin, R. (eds.), *Manejo de Fauna Silvestre en Amazonía y Latinoamérica*. Fundación Moise's Bertoni, Asunción, Paraguay. pp. 413–417.

Rylands, A. B., Schneider, H., Langguth, A., Mittermeier, R. A., Groves, C. P., and Rodriguez-Luna, E. (2000). An assessment of the diversity of New World primates. *Neotrop. Primates* 8:61–93.

Santos-Mello, R., and Thiago de Mello, M. (1986). Cariótipo de *Aotus trivirgatus* (Macaco-da-Noite) das proximidades de Manaus, Amazonas. Nota preliminar. In: Thiago de Mello, M. (ed.), *A Primatologia no Brasil*, vol. 2, p. 388.

Schneider, H., and Rosenberger, A. L. (1996). Molecules, morphology and platyrrhine systematics. In: Norconk, M. A., Rosenberger, A. L., and Garber, P. A. (eds.), *Adaptive Radiations of Neotropical Primates*. Plenum Press, New York. pp. 3–19.

Schneider, H., Sampaio, I., Harada, M. L., Barroso, C. M. L., Schneider, M. P. C., Czelusniak, J., and Goodman, M. (1996). Molecular phylogeny of the New World monkeys (Platyrrhini, Primates) based on two unlinked nuclear genes: IRBP intron 1 and epsilon-globin sequences. *Am. J. Phys. Anthropol.* 100:153–179.

Schneider, H., Schneider, M. P. C., Sampaio, I., Harada, M. L., Stanhope, M., Czelusniak, J., and Goodman, M. (1993). Molecular phylogeny of the New World primates (Platyrrhini, Primates). *Mol. Phylogenet. Evol.* 2:225–242.

Schwindt, D. M., Carrillo, G. A., Bravo, J. J., Di Fiore, A., and Fernandez-Duque, E. (2004). *Comparative Socioecology of Monogamous Primates in the Amazon and Gran Chaco. Int. J. Primatol.* 75(suppl. 1):412.

Sharma, N. E., Fernandez, E., and Lawrance, S. K. (2003). The major histocompatibility complex of the owl monkey, *Aotus azarai*: a model system for the study of mate choice. *Ohio J. Sci.* 1031:A–9.

Smith, R. J., and Jungers, W. L. (1997). Body mass in comparative primatology. *J. Hum. Evol.* 32:523–559.

Solano, P. (1995). Patrón de actividad y área de acción del mico nocturno *Aotus brumbacki hershkovitz*, 1983 (Primates: Cebidae) Parque Nacional Natural Tinigua, Meta, Colombia. [PhD diss.]. Pontificia Universidad Javeriana, Bogotá.

Stallings, J. R., West, L., Hahn, W., and Gamarra, I. (1989). Primates and their relation to habitat in the Paraguayan Chaco. In: Redford, K. H., and Eisenberg, J. F. (eds.), *Advances in Neotropical Mammalogy*. Sandhill Crane Press, Gainesville, FL. pp. 425–442.

Tardif, S. D. (1994). Relative energetic cost of infant care in small-bodied neotropical primates and its relation to infant-care patterns. *Am. J. Primatol.* 34:133–143.

Tattersall, I. (1987). Cathemeral activity in primates: a definition. *Folia Primatol.* 49:200–202.

Tejedor, M. (1998). La posición de *Aotus* y *Callicebus* en la filogenia de los primates platirrinos. *Bol. Primatol. Latinoam.* 7:13–29.

Tejedor, M. (2001). *Aotus* y los atelinae: nuevas evidencias en la sistemática de los primates platirrinos. *Mastozool. Neotrop.* 8:41–57.

Torres, O. M., Enciso, S., Ruiz, F., Silva, E., and Yunis, I. (1998). Chromosome diversity of the genus *Aotus* from Colombia. *Am. J. Primatol.* 44:255–275.

van Schaik, C. P., and Kappeler, P. M. (1993). Life history, activity period and lemur social systems. In: Kappeler, P. M., and Ganzhorn, J. U. (eds.), *Lemur Social Systems and Their Ecological Basis*. Plenum Press, New York. pp. 241–260.

van Schaik, C. P., and Kappeler, P. M. (1996). The social systems of gregarious lemurs: lack of convergence with anthropoids due to evolutionary disequilibrium? *Ethology* 102:915–941.

Villavicencio Galindo, J. M. (2003). Distribución geográfica de los primates del género *Aotus* en el Departamento Norte de Santander, Colombia. In: Pereira-Bengoa, V., Nassar-Montoya, F., and Savage, A. (eds.), *Primatología del Nuevo Mundo*. Centro de Primatología Araguatos, Bogotá.

Wallace, R. B., Painter, L. E., Rumiz, D. I., and Taber, A. B. (2000). Primate diversity, distribution and relative abundance in the Rios Blanco y Negro Wildlife Reserve, Santa Cruz Department, Bolivia. *Neotrop. Primates* 8:24–28.

Warren, R. D., and Crompton, R. H. (1997). A comparative study of the ranging behaviour, activity rhythms and sociality of *Lepilemur edwardsi* (Primates, Lepilemuridae) and *Avahi occidentalis* (Primates, Indriidae) at Ampijoroa, Madagascar. *J. Zool. Lond.* 243:397–415.

Weisenseel, K., Chapman, C. A., and Chapman, L. J. (1993). Nocturnal primates of the Kibale Forest: effects of selective logging on prosimian densities. *Primates* 34(4):445–45.

Wright, P. C. (1978). Home range, activity patterns, and agonistic encounters of a group of night monkeys (*Aotus trivirgatus*) in Perú. *Folia Primatol.* 29:43–55.

Wright, P. C. (1981). The night monkeys, genus *Aotus*. In: Coimbra-Filho, A., and Mittermeier, R. A. (eds.), *Ecology and Behavior of Neotropical Primates*. Academia Brasileira de Ciencias, Rio de Janeiro. pp. 211–240.

Wright, P. C. (1984). Biparental care in *Aotus trivirgatus* and *Callicebus moloch*. In: Small, M. (ed.), *Female Primates: Studies by Women Primatologists*. Alan R. Liss, New York. pp. 59–75.

Wright, P. C. (1985). *The Costs and Benefits of Nocturnality for* Aotus trivirgatus (*the Night Monkey*). [PhD diss.] City University of New York, New York.

Wright, P. C. (1986). Ecological correlates of monogamy in *Aotus* and *Callicebus*. In: Else, J. G., and Lee, P. C. (eds.), *Primate Ecology and Conservation*. Cambridge University Press. New York. pp. 159–167.

Wright, P. C. (1989). The nocturnal primate niche in the New World. *J. Hum. Evol.* 18:635–658.

Wright, P. C. (1994a). The behavior and ecology of the owl monkey. In: Baer, J. F., Weller, R. E., and Kakoma, I. (eds.), *Aotus: The Owl Monkey*. Academic Press, San Diego. pp. 97–112.

Wright, P. C. (1994b). Night watch on the Amazon. *Nat. Hist.* 103:45–51.

Wright, P. C. (1996). The neotropical primate adaptation to nocturnality. In: Norconk, M. A., Rosenberger, A. L., and Garber, P. A. (eds.), *Adaptive Radiations of Neotropical Primates*. Plenum Press, New York. pp. 369–382.

Wright, P. C. (1999). Lemur traits and Madagascar ecology: coping with an island environment. *Ybk. Phys. Anthropol.* 42:31–72.

Zito, M., Evans, S., and Weldon, P. (2003). Owl monkeys (*Aotus* spp.) self-annoint with plants and millipedes. *Folia Primatol.* 74:159–161.

Zunino, G. E., Galliari, C. A., and Colillas, O. J. (1985). Distribución y conservación del mirikiná (*Aotus azarae*), en Argentina: resultados preliminares. A. Primatologia No. Brasil. S. B. d. Primatologia. Campinas 2:305–316.

10

The Atelines
Variation in Ecology, Behavior, and Social Organization
Anthony Di Fiore and Christina J. Campbell

GENERAL DESCRIPTION

Primates of the subfamily Atelinae are the largest monkeys in the New World and include the howler monkeys (genus *Alouatta*), spider monkeys (genus *Ateles*), woolly monkeys (genus *Lagothrix*), and muriquis (genus *Brachyteles*); additionally, according to some authors, the yellow-tailed woolly monkey is sufficiently distinct from other woolly monkeys (and other atelines) to warrant its elevation to a separate genus, *Oreonax* (e.g., Groves 2001). Atelines are a monophyletic group (descended from a common ancestor), constituting one of the three major platyrrhine radiations recognized in modern molecular systematic studies (e.g., Schneider 2000, Schneider et al. 2001). Morphologically, atelines are characterized by having a muscular prehensile tail, which, in all genera, is commonly used to support the full weight of the body during feeding and as an additional support during locomotion (Fig. 10.1).

In two genera (*Ateles*, *Brachyteles*), the use of rapid suspensory, semibrachiating locomotion is particularly common. Both genera show dramatic modifications of the hand (e.g., a reduced thumb and elongation of the remaining digits) and shoulder (e.g., an elongated and dorsally positioned scapula), as well as elongation of the tail and limbs relative to trunk length, as adaptations to this distinctive locomotor pattern (Hill 1962, Erickson 1963). Adult body size in atelines ranges from just over 3 kg to an estimated 15 kg (Table 10.1). In howler monkeys and woolly monkeys, males are considerably larger than females, while in spider monkeys and muriquis there is little sexual dimorphism in body size (Table 10.1).

DISTRIBUTION, TAXONOMY, AND PHYLOGENY

Alouatta is the most widely distributed ateline genus, with a geographic range extending from the dry, deciduous forests of the northern Argentine Chaco to Central America as far north as eastern Mexico. *Ateles*, too, is distributed widely in moist tropical and deciduous forests throughout South and Central America, from northern Bolivia to the coastal regions of southern Mexico and the Yucatan peninsula.

Figure 10.1 An ateline primate (*Ateles belzebuth belzebuth*) hanging from its prehensile tail. (D. Schwindt)

Lagothrix shows a broad distribution in rain forest habitats throughout the western Amazon and upper Orinoco Basins, while *Oreonax* is restricted to only small areas of moist, evergreen forest on the eastern Andean slopes in northern Peru. The geographic range of *Brachyteles* is limited to several small remnants of Brazilian Atlantic Forest (Fig. 10.2).

Table 10.1 Currently Recognized Genera and Species of Ateline Primates Based on Recent Molecular and Morphological Studies[1]

GENUS	SPECIES[2]	COMMON NAME[2]	ADULT MALE BODY WEIGHT (G)[3]			ADULT FEMALE BODY WEIGHT (G)[3]			DIMORPHISM[4]	BODY WEIGHT SOURCE	IUCN RED LIST STATUS 2003[5]
			AVERAGE	RANGE	NUMBER OF INDIVIDUALS	AVERAGE	RANGE	NUMBER OF INDIVIDUALS			
Alouatta	*belzebul*	Red-handed howler monkey	7,270	6,540–8,000	27	5,525	4,850–6,200	26	1.32	Ford and Davis 1992	CR
A.	*caraya*	Brown (black) howler monkey	6,800	4,000–9,600	≥19	4,605	3,800–5,410	13	1.48	Ford and Davis 1992	Not listed
			6,420		58	4,330		117	1.48	Rumiz 1990	
A.	*guariba*[6]	Red and black (brown) howler monkey	6,175	5,200–7,150	4	4,550	4,100–5,000	3	1.36	Ford and Davis 1992	NT, CR
			6,730		4	4,350		5	1.55	Smith and Jungers 1997	
A.	*macconnelli*	Guyanan red howler monkey	7,585	7,170–8,000	3	5,000	5,000	2	1.52	Ford and Davis 1992	Not listed
A.	*nigerrima*	Amazon black howler monkey									Not listed
A.	*palliata*[7]	Mantled howler monkey	7,150	4,500–9,800	≥56	5,350	3,100–7,600	≥67	1.34	Ford and Davis 1992	VU, CR
A.	*pigra*	Guatemalan black howler monkey	11,352	11,113–11,590	2	6,434	6,290–6,577	4	1.76	Ford and Davis 1992	EN
A.	*sara*	Bolivian red howler monkey	7,200	5,400–9,000	61	5,600	4,200–7,000	61	1.29	Ford and Davis 1992	Not listed
			5,617		31	4,034		29	1.39	Braza et al. 1983	VU
A.	*seniculus*	Venezuelan red howler monkey	7,540	5,000–12,500	8	6,298	4,000–10,000	9	1.20	Hernández-Camacho and Defler 1985	VU
Ateles	*belzebuth*[8]	White-fronted (white-bellied) spider monkey	6,306		64	4,670		46	1.35	Rodríguez and Boher 1988	VU, EN
			8,260		10	7,880		16	1.05	Smith and Jungers 1997	
A.	*geoffroyi*[9]	Geoffroy's (black-handed) spider monkey	8,532	7,264–9,800	12	8,112	5,824–10,400	15	1.05	Ford and Davis 1992	VU, EN, CR
			8,160	7,600–8,600	5	8,275	7,800–8,750	2	0.99	Karesh et al. 1998	
			8,210	7,420–9,000	56	7,700	6,000–9,400	≥101	1.07	Ford and Davis 1992	
A.	*hybridus*	Brown or variegated spider monkey	8,250	7,875–8,625	2	9,151	7,500–10,500	7	0.90	Hernández-Camacho and Defler 1985[10]	CR
A.	*paniscus*	Red-faced (red-faced black) spider monkey	9,110		20	8,440		42	1.08	Smith and Jungers 1997	Not listed
Brachyteles	*arachnoides*	Southern muriqui or woolly spider monkey	7,335	5,470–9,200	8	8,750	6,500–11,000	12	0.84	Ford and Davis 1992	EN
			10,200	10,200	1	8,500	8,500	1	1.20	Lemos de Sá and Glander 1993	
B.	*hypoxanthus*	Northern muriqui or woolly spider monkey	9,416	9,250–9,600	3	8,333	6,900–9,300	3	1.13	Lemos de Sá and Glander 1993	CR
Lagothrix	*cana*	Gray (Geoffroy's) woolly monkey	13,800	13,800	1	13,500	~12,000–15,000[11]	1	—	Peres 1994a,b	NT
L.	*lagotricha*	Brown (Humboldt's) woolly monkey	9,493	8,930–10,200	3	7,650	7,650	1	1.24	Lemos de Sá and Glander 1993	NT
			9,000	8,000–10,000	3	5,750	5,000–6,500	6	1.57	Ford and Davis 1992	
L.	*lugens*	Colombian woolly monkey	—			6,000	6,000	1	—	Ford and Davis 1992	VU
L.	*poeppigii*	Silvery (Poeppig's) woolly monkey				4,533		9	1.57	Lu 1999	NT
Oreonax	*flavicauda*	Yellow-tailed or Hendee's woolly monkey	7,100		6	~10,000[12]			—	Leo Luna 1984	CR

1 Taxonomy follows that of Rylands et al (2000) and Groves (2001) with recent modifications based on molecular data for *Alouatta* (Cortés-Ortiz et al. 2003) and *Ateles* (Collins and Dubach 2000b).

2 First name noted is that listed in Groves (2001). Additional common names, listed in parentheses, come from Rylands et al. (2000).

3 Body weight averages and ranges are based only on free-ranging animals or those presumed in the cited source to be free-ranging. Where Ford and Davis's (1992) compilation of platyrrhine body weights is cited as the data source, we recalculated the appropriate range of weights based on the taxonomy adopted here and calculated average weights as the mid-value of the range reported. The mid-value of Ruschi's (1964) estimated size for female northern muriquis is also used as an estimate of average female body weight for that species. For all other sources, average weights represent the arithmetic mean of individual body weights either reported in the cited source or calculated from the original data. Where both Ford and Davis (1992) and Smith and Jungers (1997) have compiled data on a taxon from some of the same original sources, care was taken to include those data only once in the summary presented here.

4 Dimorphism calculated as the ratio of average male to average female body weight.

5 Data compiled from the World Conservation Union (IUCN) online resource at www.redlist.org according to either 2001 or 1994 criteria. Listings apply either to entire species or to one or more recognized subspecies; thus, more than one listing is possible for a taxon. For many species, conservation status has been assessed only for a subset of currently recognized subspecies. CR, critically endangered; EN, endangered; VU, vulnerable; NT, near threatened.

6 *Alouatta guariba* was formerly called *A. fusca* and appears in many earlier publications as such.

7 *A. palliata* includes *A. coibensis* recognized by Rylands et al. (2000), Groves (2001), and other authors.

8 *Ateles belzebuth* includes *A. chamek* and *A. marginatus* recognized by Rylands et al. (2000), Groves (2001), and other authors.

9 *A. geoffroyi* includes *A. fusciceps* recognized by Groves (2001) and other authors.

10 Source refers to taxon as *A. paniscus brunneus*.

11 Estimated range for the species. This estimate and reference are included here only for completeness of the data set.

12 The only body weight available for *O. flavicauda* is an estimate for which the sex is not specified. This estimate and reference are included here only for completeness of the data set.

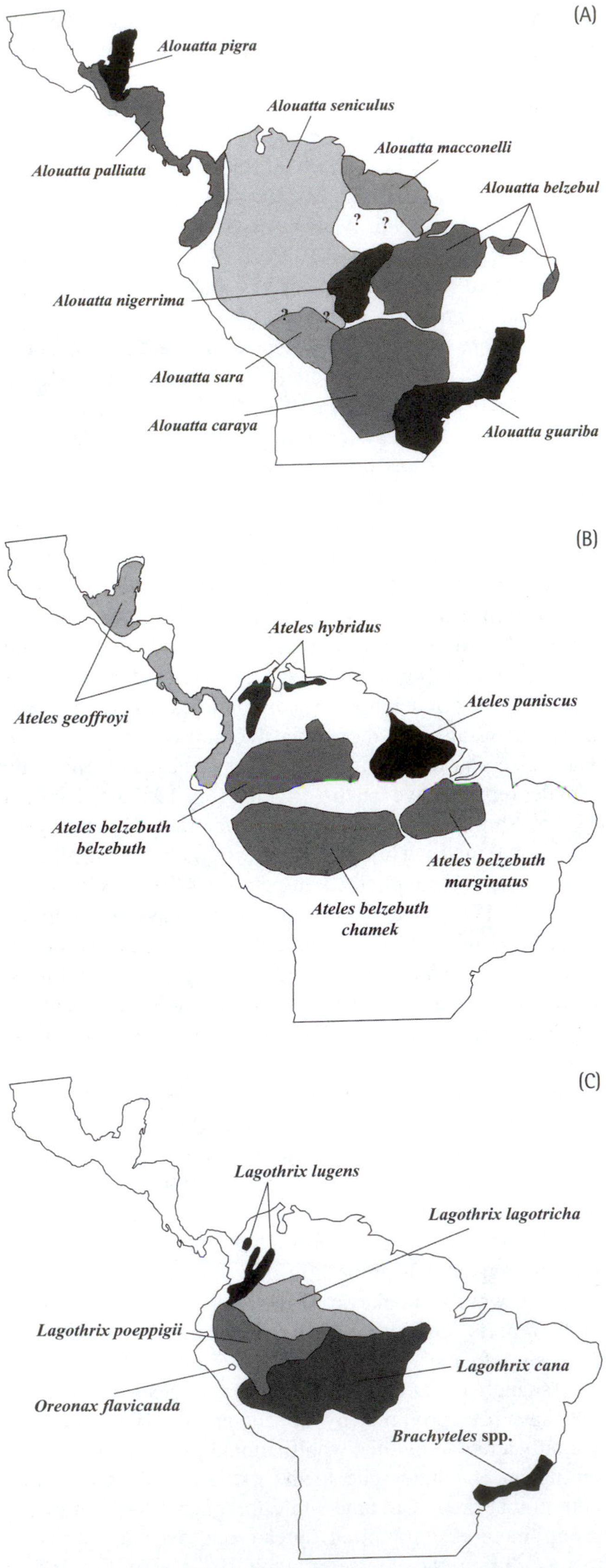

(A)

Alouatta pigra

Alouatta seniculus

Alouatta macconelli

Alouatta palliata

Alouatta belzebul

Alouatta nigerrima

? ?

? ?

Alouatta sara

Alouatta caraya

Alouatta guariba

(B)

Ateles hybridus

Ateles geoffroyi

Ateles paniscus

Ateles belzebuth belzebuth

Ateles belzebuth marginatus

Ateles belzebuth chamek

(C)

Lagothrix lugens

Lagothrix lagotricha

Lagothrix poeppigii

Lagothrix cana

Oreonax flavicauda

Brachyteles spp.

Figure 10.2 The geographic distributions of extant atelines. (A) *Alouatta*. (B) *Ateles*. (C) *Brachyteles*, *Lagothrix*, and *Oreonax*. Maps are redrawn based on Fooden (1963), Defler and Defler (1996), Kinzey (1997), Collins and Dubach (2000a), and Cortés-Ortiz et al. (2003).

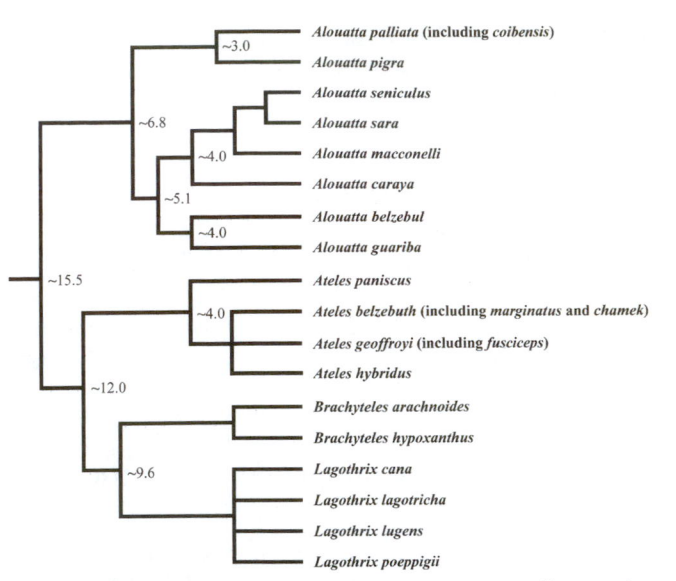

Figure 10.3 Phylogenetic relationships among extant ateline species (excluding *Alouatta nigerrima* and *Oreonax flavicauda*) and selected approximate crown group ages in millions of years. Branching order and dates are derived from Meireles et al. (1999), Collins and Dubach (2000a,b), Collins (2001), and Cortés-Ortiz et al. (2003). Where multiple estimates of the age of a crown group were available from different sources, the midpoint of the range of estimated dates is presented. Note that branch lengths are not drawn to scale.

Most modern taxonomic treatments divide the atelines into two groups, one comprising the various species of *Alouatta* and one containing the remaining genera, which had a common ancestor roughly 15–16 million years ago (Goodman et al. 1998, Schneider 2000) (Fig. 10.3). These groups are typically separated at the level of the tribe (Alouattini and Atelini, Schneider et al. 1993, 1996; Schneider and Rosenberger 1996; Goodman et al. 1998; Rylands et al. 2000), although Groves (2001) assigns the howler monkeys to their own subfamily. Here, we adopt the former and more commonly used tribe-level division, and we thus use the term *atelin* to refer to the non-alouattin genera *Brachyteles*, *Ateles*, *Lagothrix*, and *Oreonax*.

Both morphological and molecular phylogenetic studies consistently place *Alouatta* as the basal genus within the ateline clade, but the relationships among the remaining genera are more controversial. Morphological studies have tended to place *Brachyteles* and *Ateles* or *Brachyteles* and *Lagothrix* as sister taxa (e.g., Ford 1986); nonetheless, Kay (1990) has noted a possible sister relationship between *Brachyteles* and *Alouatta* on the basis of dental characteristics, and cranial features suggest a closer relationship between *Lagothrix* and *Ateles* (Groves 2001). Rosenberger and Strier (1989) argue that a sister grouping of *Brachyteles* and *Ateles* is most likely on the basis of shared morphological and behavioral features associated with their extensive use of suspensory locomotor behavior and feeding postures. In contrast, several recent molecular studies (Schneider et al. 1993, 1996; von Dornum and Ruvolo 1999; Canavez et al. 1999; Meireles et al. 1999) have linked *Brachyteles* and *Lagothrix* as sister taxa, although Collins (2003) argues that the molecular evidence is inconclusive and favors instead an unresolved trichotomy among the atelins. No molecular studies to date have examined *Oreonax*, but on the basis of limited morphological data, Groves (2001) suggests it may be linked to *Ateles* rather than occupying a basal position within the woolly monkey clade. Several of the atelines have been the subject of recent molecular analyses concerned with elucidating species-level phylogenies and the biogeographic histories of the genera (*Alouatta*, Meireles et al. 1999, Cortés-Ortiz et al. 2003; *Ateles*, Collins and Dubach 2000a,b; *Brachyteles*, Lemos de Sá et al. 1990, Pope 1998b) (Fig. 10.3).

GROUP COMPOSITION AND SOCIAL STRUCTURE

One of the most remarkable characteristics of atelines is the diversity of grouping patterns and social systems represented in the clade (Table 10.2). Howler monkeys live in cohesive social groups usually containing one or a small number of adult males and several adult females, although occasionally groups with just one reproductive-age individual of each sex have been reported. In most species of howler monkeys (e.g., *Alouatta seniculus*, *A. caraya*, *A. guariba*, *A. pigra*), group size is limited to fewer than 10–15 animals and there is commonly only one adult male per group and seldom more than three. In *A. palliata*, however, groups are much larger and typically contain three or more adult males and up to nine or more adult females; groups of this species containing over 40 members have also been reported (Fedigan et al. 1985, Chapman 1988; see also review in Neville et al. 1988). In this genus, individuals of both sexes typically disperse from their natal social groups prior to first reproduction, although some males and females may inherit breeding positions in their natal groups (Rudran 1979; Glander 1980, 1992; Crockett 1984; Clarke and Glander 1984; Pope 1992, 2000; Rumiz 1990; Crockett and Pope 1993; Calegaro-Marques and Bicca-Marques 1996; Brockett et al. 2000a). In general, females appear more likely to disperse than males, and they move farther distances than males do upon dispersal (Gaulin and Gaulin 1982; Crockett 1984; Glander 1992; Pope 1989, 1992; Crockett and Pope 1993). In this respect, the dispersal pattern can be characterized as "female-biased." Males that disperse typically try to take over existing social groups by evicting the resident dominant male, sometimes forming coalitions with one another to do so (Rudran 1979, Clarke 1983, Pope 1990, Agoramoorthy and Rudran 1993). Among red howler monkeys (*A. seniculus*), dispersing females have often been ousted from their natal groups by aggression from older females and are seldom able to integrate themselves into established social groups. Instead, they must form new social groups with other dispersers and successfully defend a home range against other groups before they begin breeding (Crockett 1984; Crockett and Pope 1988, 1993; Pope 2000). The situation is somewhat different

in mantled howler monkeys (*A. palliata*), where female dispersers sometimes succeed in joining established groups (Jones 1980b, Glander 1992).

Spider monkeys (*Ateles* spp.) live in *fission–fusion* societies, in which individual animals from a large community associate on a daily basis in small, flexible parties that change size and membership frequently (Klein 1972, Cant 1977, van Roosmalen 1985, McFarland 1986, Ahumada 1989, Symington 1990, Chapman 1990a). Fission–fusion sociality is thought to represent an adaptation to mitigate the costs of direct competition over food items when high-quality resources are scarce and presented in patches too small to support all of the individuals in a feeding party (Klein and Klein 1977; Wrangham 1980, 1987; Symington 1990). Alternatively, such systems may mitigate the increased daily travel costs that large parties would incur under these same conditions of resource abundance and distribution (Chapman et al. 1995). Supporting these hypotheses is the fact that the size of foraging and traveling parties is positively correlated with the habitat-wide abundance of preferred, high-value food items, such as ripe fruits (Symington 1987a, 1988b; Chapman et al. 1995). When intragroup feeding competition is reduced due to an extensive home range and lack of neighboring communities, spider monkeys appear to forage as a more cohesive unit (Campbell 2002). Party size and composition in *Ateles* is also influenced by female reproductive status (Symington 1987a, Chapman 1990a, Shimooka 2003). Dispersal in *Ateles* is largely or solely by females, while males mature and begin breeding in their natal communities (Symington 1987b, 1988a, 1990; Ahumada 1989). The fission–fusion social system of spider monkeys can also be characterized as sex-segregated in that adult females and their dependent offspring often forage and travel independently of the group's adult males (Fedigan and Baxter 1984, Symington 1988a, Ahumada 1989, Chapman 1990a).

Woolly monkey (*Lagothrix* spp.) group size is large (18–45 individuals), and groups contain multiple reproductive-age animals of both sexes (Ramirez 1980, 1988; Nishimura 1990a; Peres 1994a; Stevenson et al. 1994; Defler 1995, 1996; Di Fiore 1997). Little is known of the natural history of yellow-tailed woolly monkeys (*Oreonax*), but published reports suggest that group size may be much smaller than in other woolly monkeys (Leo Luna 1980, 1982). Some early reports suggested that woolly monkeys might live in fission–fusion societies like those of spider monkeys (Kavanaugh and Dresdale 1975). This suggestion, however, may have been prompted by the fact that observers often can see only a few individual woolly monkeys at any given time during an encounter due to the extremely diffuse spatial pattern that groups assume while foraging. Woolly monkey group members are often spread out over hundreds of meters during the course of their daily activities (Peres 1996, Di Fiore 1997 and unpublished data); nonetheless, groups appear to maintain social cohesion through the frequent use of contact vocalizations. Social groups of woolly

Table 10.2 Population Characteristics and Group Composition in Atelines Based on Both Population Censuses and Long-Term Behavioral or Ecological Studies[1]

GENUS AND SPECIES	STUDY SITE	POPULATION DENSITY (INDIVIDUALS/KM²)	AVERAGE GROUP SIZE (RANGE)	GROUP STRUCTURE[2]						NUMBER OF GROUPS	% UNIMALE GROUPS	ADULT SEX RATIO WITHIN GROUPS (FEMALES/MALE)	DATES OF STUDY	SOURCE
				AM	AF	SM	SF	JUV	INF					
Alouatta belzebul	Sapé, Paraíba, Brazil	—	7.4 (5–14)	1.2 (1–2)	—	—	—	—	—	5	80%	—	1985–1986	Bonvicino 1989
A. belzebul	Paranaíta, Mato Grosso, Brazil	—	7	1	2	0–1	1	—	0–3	1	—	1.5	1999–2000	Pinto and Setz 2004
A. caraya	Río Riachuelo, Corrientes, northeastern Argentina	81	6.4 (3–10)	1.4 (1–2)	2.0 (1–3)	0.5 (0–3)	—	1.9 (0–4)	0.6 (0–2)	11	64%	1.4	1982	Rumiz 1990
A. caraya	Río Riachuelo, Corrientes, northeastern Argentina	102	8.4 (5–13)	1.4 (1–3)	2.3 (1–3)	0.9 (0–2)	—	2.4 (0–5)	1.4 (0–3)	11	73%	1.6	1984	Rumiz 1990
A. caraya	Isla Guascára, Corrientes, northeastern Argentina	280	10.2 (5–15)	2.7 (1–4)	3.8 (2–6)	—	2.5 (0–5)		1.2 (0–3)	11	27%	1.4	1981	Rumiz 1990
A. caraya	Río Riachuelo, Corrientes, northeastern Argentina	—	7.1 (2–12)	1.7 (0–4)	2.4 (1–5)	0.3 (0–2)	0.3 (0–2)	1.6 (0–4)	0.8 (0–3)	24	54%	1.4	1997	Agoramoorthy and Lohmann 1999
A. caraya	Ríos Paraná and Paraguay, northeastern Argentina	—	18.5 (16–21)	2 (2–2)	5.5 (5–6)	1 (1–1)	0.5 (0–1)	6 (5–7)	3.5 (3–4)	2	0%	2.8	1998–2000	Bravo and Sallenave 2003
A. guariba	Cantareira Reserve, São Paulo, Brazil	48–113	5.8 (2–11)	1.8 (1–3)	2.4 (1–4)	—	—	1.2 (0–3)	0.4 (0–1)	25	36%	1.3	1979	da Silva 1981
A. guariba	Estação Biológica Caratinga, Minas Gerais, Brazil	117	6.8	1.2 (1–2)	2.3 (1–3)	0.4 (0–1)	—	2.0 (0–4)	0.9 (0–2)	19	84%	1.9	1983–1984	Mendes 1989
A. guariba	Estação Biológica Caratinga, Minas Gerais, Brazil	—	7.0 (4–11)	1.1 (1–2)	1.9 (1–3)	0.5 (0–1)	—	3.4 (2–6)		10	90%	1.7	2000	Strier et al. 2001
A. palliata	Santa Rosa National Park, Guanacaste, Costa Rica	4.9	13.6	22%	44%	—	—	20%	14%	25	—	2.0	1984	Fedigan et al. 1985, Fedigan 1986
A. palliata	Santa Rosa National Park, Guanacaste, Costa Rica[3]	7.9	16.3	24%	41%	—	—	16%	19%	34	—	1.7	1992	Fedigan et al. 1998
A. palliata	Santa Rosa National Park, Guanacaste, Costa Rica	—	12.1	22%	39%	—	—	23%	16%	45	—	1.8	1999	Fedigan and Jack 2001
A. palliata	Cabo Blanco, Costa Rica	—	14.9 (8–19)	2.5 (1–4)	7.8 (4–12)	—	—	2.4 (0–7)	2.3 (1–4)	8	12.5%	3.1	1987–1988	Lippold 1988, 1989
A. palliata	Various locations, Guanacaste, Costa Rica	—	21.8 (13–35)	3.1 (2–6)	10.2 (6–15)	—	—	5.8 (3–12)	2.7 (1–6)	11	0.0%	3.3	not specified	Jones 1996

Table 10.2 (cont'd)

GENUS AND SPECIES	STUDY SITE	POPULATION DENSITY (INDIVIDUALS/KM²)	AVERAGE GROUP SIZE (RANGE)	GROUP STRUCTURE[2]						NUMBER OF GROUPS	% UNIMALE GROUPS	ADULT SEX RATIO WITHIN GROUPS (FEMALES/MALE)	DATES OF STUDY	SOURCE
				AM	AF	SM	SF	JUV	INF					
A. palliata	Hacienda la Pacifica, Guanacaste, Costa Rica[4]	—	15.5 (4-29)	2.6	7.6	—	—	4.3	—	15	—	2.9	1974-1976	Clarke et al. 1986
A. palliata	Hacienda la Pacifica, Guanacaste, Costa Rica	—	15.7 (4-31)	2.1	8.9	—	—	4.0	—	16	6%	4.2	1984	Clarke et al. 1986
A. palliata	Hacienda la Pacifica, Guanacaste, Costa Rica	26	12.6 (2-45)	2.5	6.8	—	—	3.1		27	26%	2.7	1991	Clarke and Zucker 1994[5]
A. palliata	Hacienda la Pacifica, Guanacaste, Costa Rica	30	10.2 (4-28)	1.6	4.8	—	—	1.5	2.6	34	44%	3.0	1998	Clarke et al. 2002
A. palliata	Finca Taboga, Guanacaste, Costa Rica	—	11.5 (2-39)	21%	49%	—	—	21%	10%	7-22	12-50%	2.3	1966-1971	Heltne et al. 1976[6]
A. palliata	La Selva Biological Reserve, Heredia, Costa Rica	7-15	11 (6-15)	3.3 (1-5)	4.0 (3-6)	2.3 (0-4)	—	—	1.3 (0-2)	7	14%	1.2	1990	Stoner 1994
A. palliata	Barro Colorado Island, Panama[7]	—	19.4 ± 6.3 SD	3.1 ± 1.2 SD	8.6 ± 3.1 SD	—	—	2.5 ± 1.5 SD	5.3 ± 2.4	—	—	2.8	1977-1993	Milton 1996[8]
A. palliata	Barro Colorado Island, Panama	—	21.3 ± 4.5 SD (16-28)	3.9 ± 1.1 SD	8.4 ± 2.7 SD	—	—	2.6 ± 1.3 SD	7.0 ± 1.3 SD	13	—	2.2	1980	Milton 1982[8]
A. palliata	Inland lowland forest, Chiriqui, Panama	—	18.9 (7-28)	3.9 (2-6)	8 (3-13)	0	—	3.8 (2-6)	3.3 (0-6)	8	0%	2.1	1970-1971	Baldwin and Baldwin 1976
A. pigra	Los Tuxtlas, Veracruz, Mexico	23.3	9.1 (5-16)	3.0 (1-5)	4.1 (2-6)	—	—	0.9 (0-3)	1.2 (0-3)	17	6%	1.4	1978-1981	Estrada 1982, 1984
A. pigra	Palenque, Chiapas, Mexico	23	7.0 (2-12)	2.0 (1-4)	1.9 (1-4)	—	—	1.0 (0-4)	1.0 (0-3)	20	35%	1.0	2000	Estrada et al. 2002a
A. pigra	Calakmul, Campeche, Mexico	15.2	7.5 (4-9)	2.5 (1-3)	2.2 (1-4)	—	—	1.8 (0-3)	1.0 (0-2)	8	25%	0.9	2001-2002	Estrada et al. 2004
A. pigra	Yaxchilán, Chiapas, Mexico	12.8	6.6 (4-10)	2.8 (1-5)	2.0 (1-3)	—	—	0.9 (0-2)	1.3 (0-2)	8	25%	0.7	2001-2002	Estrada et al. 2002b, 2004
A. pigra	Muchukux Forest, Quintana Roo, Mexico	16.5	3.2 (1-8)	—	—	—	—	—	—	25	—	—	1995	Gonzalez-Kirchner 1998
A. pigra	Tikal National Park, Guatemala	5	6.3	—	—	—	—	—	—	4	—	—	1973	Coelho et al. 1976a,b
A. pigra	Tikal National Park, Guatemala	17.8	8.7 (6-12)	2.2 (1-3)	2.9 (1-4)	—	—	1.8 (0-3)	1.8 (1-3)	10	10%	1.3	2002	Estrada et al. 2004
A. pigra	Bermuda Landing, Gulf Coast, Belize	—	6.8 (4-10)	1.7 (1-3)	2.1 (1-3)	0.3 (0-1)	0.2 (0-1)	1.4 (0-3)	0.4 (0-1)	9	44%	1.3	1981	Horwich and Gebhard 1983
A. pigra	Bermuda Landing, Gulf Coast, Belize	8.1	4.4 (2-7)	1.1 (1-2)	1.2 (1-2)	—	—	1.3 (0-4)	0.7 (0-2)	13	92%	1.1	1978-1979	Bolin 1981

A. pigra	Community Baboon Sanctuary, Gulf Coast, Belize	31.9–178.2	6.0 (5.2–6.9)	1.6 (1.3–1.8)	2.1 (1.8–2.4)	—	—	—	10–36	—	1.3	1985–2001	Horwich et al. 2000[9]
A. pigra	Community Baboon Sanctuary, Gulf Coast, Belize	47–257	5.8 (3–16)	1.5 (1–3)	2.0 (1–4)	—	—	2.2 (0–10)	38	58%	1.3	1994–1997	Ostro et al. 2000[10]
A. pigra	Monkey River, Gulf Coast, Belize	102	6.4 (2–9)	—	—	—	—	—	8	—	—	1999–2001	Pavelka et al. 2003
A. seniculus	Bush Bush Forest, Trinidad	—	8.5 (8, 9)	3.5 (4, 3)	2.5 (2, 3)	—	—	1 (1, 1)	2	0%	0.7	1968	Neville 1972a
A. seniculus	Nourague Station, French Guiana	—	6–8	1	1–2	0–2	—	—	1	—	1.2	1988–1990	Julliot 1996
A. seniculus	Ríos Tuparro and Tomo, Vichada, eastern Colombia	23–27	6.3 (3–9)	1.9 (1–3)	2.4 (1–4)	0.1 (0–1)	—	1.0 (0–2)	10	30%	1.3	1977–1978	Defler 1981[11]
A. seniculus	Río Peneya, Meta, Colombia	—	7.6 (5–11)	1.8 (1–3)	3.0 (2–4)	—	—	1.4 (1–3)	5	40%	1.7	1971–1974	Izawa 1976
A. seniculus	La Macarena National Park, Meta, Colombia	—	9 (9, 9)	2 (2, 2)	2.5 (3, 2)	0 (0, 0)	0 (0, 0)	3 (3, 3)	2	0	1.3	1986	Izawa and Nishimura 1988
A. seniculus	Finca Merenberg, Huila, Colombia	—	7.5 (2–13)	2.5 (1–5)	2.5 (1–5)	0.6 (0–2)	0.6 (0–2)	1.4 (0–4)	8	50%	1.7	1975	Gaulin and Gaulin 1982
A. seniculus	Hato El Frío, Apure State, Venezuela	25	5.5 (3–11)	1.2 (1–3)	1.6 (1–4)	0.1 (0–1)	0.5 (0–2)	1.3 (0–3)	29	87%	1.3	1975–1976	Braza et al. 1981[12]
A. seniculus	Hato Masaguaral, Guárico State, Venezuela	M: 112	10.5 (6–16)	1.6	2.9	—	—	—	27	—	1.8	1978–1981	Crockett 1984[13]
A. seniculus	Hato Masaguaral, Guárico State, Venezuela	G: 36	7.7 (4–12)	1.2	2.5	—	—	—	25	—	2.1		
A. seniculus	Hato Masaguaral, Guárico State, Venezuela	58–223	UM: 7.9 (5.9–9.6) MM: 9.1 (8.5–10.5)	UM: 1 MM: 2.2	UM: 2.5 MM: 2.7	UM: 1.3 MM: 1.2	UM: 1.0 MM: 1.1	UM: 7.1 MM: 7.7	15–36	36–93%	UM: 2.5 MM: 1.2	1976–1999	Rudran and Fernandez-Duque 2003[14]
Ateles belzebuth chamek	Cocha Cashu, Manu National Park, Peru	25–31	38.5 (37, 40)	5 (5, 5)	15.5 (15, 16)	4 (4, 4)	14 (13, 15)	—	2	—	3.1	1983–1986	Symington 1987a, 1988b
A. belzebuth	La Macarena National Park, Meta, Colombia	11.6–15.4	23.5 (20, 27)	4 (3, 5)	11.5 (11, 12)	4 (3, 5)	4 (3, 5)	—	2	—	2.9	1968	Klein 1972, Klein and Klein 1976[15]
A. belzebuth	La Macarena National Park, Meta, Colombia	—	16	4	5	0	1	3	1	—	1.3	Mid 1980s	Izawa and Nishimura 1988, Ahumada 1989
A. belzebuth	Maracá Ecological Station, Roraima, Brazil	—	19–23	6	8	5–9	—	1	1	—	1.3	1987–1989	Nunes 1995
A. belzebuth	Yasuní National Park, Ecuador	11.5	~28	3	10–11	~15	—	—	1	—	3.5	1995–1996	Dew 2001
A. geoffroyi	Santa Rosa National Park, Guanacaste, Costa Rica	—	42	4	17–18	1–2	3–4	5–8	1	—	4.4	1983–1988	Chapman 1990a,b
A. geoffroyi	Barro Colorado Island, Panama	2.1–2.3	21–24	4–5	7–10	0–3	0–2	5	1	—	1.9	1991–1998	Campbell 2000

Table 10.2 *(cont'd)*

GENUS AND SPECIES	STUDY SITE	POPULATION DENSITY (INDIVIDUALS/KM²)	AVERAGE GROUP SIZE (RANGE)	GROUP STRUCTURE[2] AM	AF	SM	SF	JUV	INF	NUMBER OF GROUPS	% UNIMALE GROUPS	ADULT SEX RATIO WITHIN GROUPS (FEMALES/MALE)	DATES OF STUDY	SOURCE
A. geoffroyi	Reserva Punta Laguna, Quintana Roo, Mexico	89.5, 6.3	28.5 (16, 41)	4.5 (3, 6)	10 (5, 15)	—	—	7 (4, 10)	7 (4, 10)	2	—	2.2	1997–1998	Ramos-Fernández and Ayala-Orozco 2003[16]
A. geoffroyi	Calakmul, Campeche, Mexico	17.2	—	22%	45%	—	—	21%	12%	—	—	2.1	2001–2002	Estrada et al. 2004
A. geoffroyi	Yaxchilán, Chiapas, Mexico	17.0	—	35%	29%	—	—	18%	18%	—	—	0.8	2001–2002	Estrada et al. 2004
A. geoffroyi	Nuchukux Forest, Quintana Roo, Mexico	27.1	—	—	—	—	—	—	—	—	—	2.6	1995, 1997	Gonzalez-Kirchner 1999
A. geoffroyi	Najil Tucha Forest, Quintana Roo, Mexico	14.5	—	—	—	—	—	—	—	—	—	—	—	—
A. geoffroyi	Tikal National Park, Guatemala	28	—	18%	32%	—	—	37%	12%	—	—	1.8	1975–1976	Cant 1977
A. geoffroyi	Tikal National Park, Guatemala	45	—	15%	33%	—	—	30%	22%	—	—	2.2	1973	Coelho et al. 1976a,b
A. paniscus	Raleighvallen-Voltzberg Nature Reserve, Surinam	8.2	18	3	8	0	1	4	2	1	—	2.7	1976–1977	van Roosmalen 1985
A. geoffroyi	Tikal National Park, Guatemala	56.4	—	27%	43%	—	—	17%	13%	1	—	1.6	2002	Estrada et al. 2004
Brachyteles arachnoides	Serra de Paranapiacaba, São Paulo, Brazil	2.3	7.0	—	—	—	—	—	—	8	—	—	1998	González-Solís et al. 2001
B. arachnoides	Fazenda Barreiro Rico, São Paulo, Brazil	—	~13	4	3 or 4	—	—	2	3	1	—	0.9	1979	Torres de Assumpção 1983
B. arachnoides	Parque Estadual de Carlos Botelho, São Paulo, Brazil	2.0–3.3	11.3	—	—	—	—	—	—	—	—	—	1985–1986	Paccagnella 1991
B. hypoxanthus	Estação Biológica Caratinga, Minas Gerais, Brazil	—	22–27	6	8	0–2	0–2	2–7	1–6	1	—	1.3	1982–1984	Strier 1991a
B. hypoxanthus	Estação Biológica Caratinga, Minas Gerais, Brazil	—	32–33	7–8	9	0	1–2	9–13	1–6	1	—	1.2	1986–1987	Strier 1991a
B. hypoxanthus	Estação Biológica Caratinga, Minas Gerais, Brazil	—	31–43	6–8	10–12	0–2	3–6	8–15	2–7	1	—	1.6	1988–1990	Strier 1991a
B. hypoxanthus	Estação Biológica Caratinga, Minas Gerais, Brazil	—	52	7	17	—	—	28	—	1	—	2.4	1994	Strier 1996b
B. hypoxanthus	Estação Biológica Caratinga, Minas Gerais, Brazil	—	~64–~73	~13–~20	~19–~28	—	—	~11–~17	~5–~14	2	—	1.4	August 1999	Strier et al. 2002

Species	Site/location	Group size	AM	AF	SM	SF	JUV	INF	Sex ratio	Census period	Source
B. hypoxanthus	Estação Biológica Caratinga, Minas Gerais, Brazil	~70~80	~19~21	~20~24	~14~18	~14	2	—	—	September 2002	Strier et al. 2002[17]
Lagothrix cana	Rio Urucu, Amazonas, Brazil	44–49	7	12–14	2	15–18	5–8	1	1.9	1988–1989	Peres 1996
L. lagotricha	Estación Biológica Caparú, Colombia	24	4	11	0	0	3	6	2.8	1984–1987	Defler 1996
L. lagotricha	Rio Peneya, Meta, Colombia	42–43	11	15	3	2	5–6	1	1.4	1971–1972	Izawa 1976
L. lagotricha	Rio Peneya, Meta, Colombia	29 (13, 45)	5.5 (4, 7)	7.5 (3, 12)	2 (1, 3)	3 (3, 3)	7 (2, 12)	4 (0, 8)	1.4	1975–1976	Nishimura 1990a
L. lugens	Tinigua National Park, Colombia	~21 (~12~33)	5 (4–8)	6.5 (4–10)	1.0 (0–2)	1.3 (0–4)	4.1 (2–6)	2.0 (1–4)	1.3	1988–1991	Stevenson et al. 1994, Izawa 1988[18]
L. poeppigii	Yasuní National Park, Ecuador	31+									
L. poeppigii	Yasuní National Park, Ecuador	23 (22, 24)	3 (4, 2)	9.5 (8, 11)	5 (5, 5)	—	4.5 (4, 5)	1.0 (1, 1)	3.2	1995–1996	Di Fiore 1997[19]
Oreonax flavicauda	East Andean Forests, northern Peru	4–14	1–3	—	—	—	—	—	11	1978–1980	Leo Luna 1980, 1982

1 Crockett and Eisenberg 1987, Neville et al. 1988, and Chapman and Balcomb 1998 provide excellent overviews of population characteristics for *Alouatta* based on data published up to 1985–1987, including longitudinal data for several well-studied populations. Instead of repeating all of that information, we concentrate here on data published since 1985 and those not included in previous reviews and direct the reader to earlier surveys. For less well-studied populations of *Alouatta* we include data from pre-1985 publications that may also have been cited in earlier reviews in order to provide a thorough overview of the variation in *Alouatta* population characteristics. Far fewer population surveys have been published for other atelines; thus, we include all available data on group size and composition, including those gleaned from studies of only a small number of groups.

2 AM, adult male; AF, adult female; SM, subadult male; SF, subadult female; JUV, juvenile; INF, infant; —, data not available in the cited source. We report data as presented in the original sources and make no attempt to standardize the definitions of various age sex classes across studies. Wherever possible, the average and number of individuals per class are given, though some studies report only the percentage of the population belonging to each age sex class. Where two groups from the same population were studied, the average and actual number of individuals per group (in parentheses, separated by commas) are noted, rather than the range. In general, few howler monkey population surveys have explicitly recognized "subadults" as a distinct age class, while many studies of other atelines have. Sex ratio calculation based on average number of adult females and males per group, as given in the table.

3 Data from an earlier population survey at this site are presented in Freese 1976.

4 Data from an earlier population survey at this site are presented in Heltne et al. 1976.

5 Another three groups in this population had no adult males and do not contribute to the summary values included in the table. Results are presented for the more extensive "repeat" rather than "initial" survey at this site.

6 Values noted in table represent the mean, across years, of average group size and proportion of individuals in each age-sex class reported within years. Ranges span the minimum to maximum values recorded across years.

7 Data from earlier population surveys at this site are presented in Carpenter 1934, 1965, Collias and Southwick 1952, Chivers 1969, Mittermeier 1973, and Smith 1977 and are summarized in Milton 1982.

8 Source reports average number of individuals in each age sex class (± standard deviation [SD]) per group during the census period but does not provide the range.

9 Values in table represent averages from across the 14-year census period. Ranges given in the group size and age sex composition columns represent the minimum to maximum average yearly values recorded across 11 census years within the census period. The total range of observed group sizes across the census period was 2–16 individuals.

10 Summary of surveys conducted at five local sites within the same general region.

11 Summary of surveys conducted at two local sites within the same general region.

12 Estimate of population density comes from a larger survey of 159 groups with an average group size of 6.3 individuals and range of sizes from 3 to 13.

13 Data from earlier population surveys at this site are presented in Neville 1972a, 1976 and Rudran 1979. Group size and composition data are separated for two habitat types: open shrub woodland or mata (M) and gallery forest (G).

14 In the original source, group size and composition data are separated for unimale (UM) and multimale (MM) groups and are thus presented separately here. Values in table represent averages from across the ~30 year census period. Ranges given in the group size column represent the minimum to maximum average yearly group size recorded across years of the census. The total range of observed group sizes across the census period was 2–18 individuals.

15 Population density calculation for this population is taken from van Roosmalen 1985.

16 Population densities reported for two types of forest: old and successional.

17 Population in 2002 also included about eight additional individuals not assigned to a particular age sex class.

18 Two of these groups were censused in each of several years. For these groups, averages were taken for each age sex class for the two most complete censuses (Stevenson et al. 1994), and these were then averaged with censuses of other groups to find the mean number of individuals in each age sex class. Ranges reported cover minimum to maximum numbers of individuals recorded in each age sex class across all groups.

19 Counts taken at end study. Some individuals assigned to adult female age sex class perhaps are better classified as subadults.

monkeys occasionally do split into "dispersed subunits" (Defler 1996) or independently traveling subgroups that remain apart for periods of several hours to several days (Ramirez 1980, 1988; Di Fiore 1997 and unpublished data). In addition, individual animals sometimes visit other social groups for variable periods of time (Nishimura 1990a, 2003). In this sense, woolly monkey grouping patterns and social organization can be considered relatively flexible (Di Fiore and Strier 2004). Nonetheless, members of the same group forage and travel as a socially cohesive unit the vast majority of the time; although group subdivision does occasionally occur, it happens with nowhere near the frequency nor the associated small party sizes seen in the fission–fusion system of spider monkeys. Observed cases of transfer among woolly monkey groups suggest that dispersal is predominantly by females (Nishimura 1990a, 2003; Stevenson et al. 1994; Stevenson 2002), and genetic studies confirm that the level of female transfer is substantial (Di Fiore 2002, Di Fiore and Fleischer 2005). Nonetheless, solitary males, including adults, have been seen in at least one species (*Lagothrix poeppigii*, Di Fiore 2002 and unpublished data), suggesting some degree of male transfer as well.

Grouping patterns among muriquis (*Brachyteles* spp.) are quite flexible. Among southern muriquis, at least two populations have been characterized as living in the same kinds of fission–fusion society as spider monkeys (Torres de Assumpção 1983, Milton 1984a, de Moraes et al. 1998, Coles and Talebi 2004). Northern muriquis, by contrast, were initially reported to live in large, cohesive, multimale-multifemale groups (Strier 1986a, 1992c). Later research on the same population, however, showed that these groups began fissioning into smaller foraging units as group size increased (Strier et al. 1993). Nonetheless, these subgroups tend to be much larger, on average, than the foraging parties of spider monkeys. As in spider monkeys, females are the dispersing sex, while males mature and begin breeding in their natal groups (Strier 1987b, 1990, 1991a).

ECOLOGY AND BEHAVIOR

Diet and Ecological Strategies

Broadly speaking, all atelines have diverse and seasonally variable diets, and even the most folivorous atelines (*Alouatta*, *Brachyteles*) eat considerable amounts of fruit and other plant parts (Table 10.3). Immature and, to a lesser extent, mature leaves are the predominant component in the diets of both *Brachyteles* and *Alouatta*, and both genera show morphological adapatations for folivory in the teeth (e.g., high, shearing crests on the molars) (Zingeser 1973, Kay 1990). While howler monkeys lack extreme modifications in gut morphology (Milton 1998), they do show very long retention times for digesta in the alimentary canal (Milton 1984b, Crissey et al. 1990). This allows the gut bacteria to more thoroughly break down the structural carbohydrates common in leaves (Milton 1998, Lambert 1998). *Brachyteles*

gut passage rates are much faster than those of howler monkeys and more in line with those of other atelins (Milton 1984a,b), suggesting an important difference between the folivorous dietary strategies of howler monkeys and muriquis. Whereas howler monkeys rely on slow and efficient digestion to extract energy locked up in leaf structural carbohydrates, muriquis appear to rely on faster intake, processing, and elimination of a greater volume of less thoroughly digested material (Milton 1984a). Not surprisingly, howler monkeys are highly selective feeders and focus their feeding on immature, more easily digested leaves with a high ratio of protein to fiber (Milton 1979, 1980) and on plant species and leaf parts containing lower concentrations of certain plant secondary compounds (Glander 1978, 1982).

The remaining atelines are predominantly frugivorous. Spider monkeys are considered "ripe fruit specialists" (Cant 1977, 1990; Klein and Klein 1977; van Roosmalen and Klein 1988; Kinzey 1997; Dew 2001), and at all sites where they have been studied long-term, more than 70% of the annual diet consists of fruits, predominantly ripe ones (Table 10.3) (Russo et al. 2005). Nonetheless, in some months at some sites (e.g., Chapman 1987, 1988; Chapman et al. 1995), immature leaves and other plant parts comprise the majority of the diet. The diet of *Lagothrix* also consists predominantly of ripe fruit, and throughout their geographic range this genus focuses almost as heavily on ripe fruits as *Ateles* (Soini 1986, 1990; Stevenson 1992; Stevenson et al. 1994; Peres 1994a; Defler and Defler 1996; Di Fiore 1997, 2004; Dew 2001). However, in some populations of *Lagothrix*, foraging for animal prey, particularly insects, is also clearly an important component of the ecological strategy (Stevenson 1992, Stevenson et al. 1994, Di Fiore and Rodman 2001, Di Fiore 2004). Where *Ateles* and *Lagothrix* occur sympatrically, there is considerable overlap in the set of plant species contributing to the fruit diet of the two genera. Partitioning of the ripe fruit frugivore niche appears to take place, at least in part, along a dimension of fruit phytochemical composition, with *Ateles* focusing more on lipid-rich, or "fatty," fruits and *Lagothrix*, on fruits containing high concentrations of easily digestible sugars (van Roosmalen 1985; Castellanos and Chanin 1996; Di Fiore 1997, 2004; Dew 2001).

Strier (1992a) has suggested that the two groups of atelines exemplify contrasting ecological strategies, i.e., that *Alouatta* follows a strategy of energy minimization, moving little and resting for long periods of time each day, thereby allowing them to subsist predominantly on a diet of hard-to-digest leaves, while *Lagothrix*, *Oreonax*, and *Brachyteles* typically adopt a strategy of traveling widely and rapidly between patches of higher-quality, more easily digested resources (e.g., fruits, flowers) to maximize energy intake. Under this scenario, the more folivorous diet of *Brachyteles* relative to the other atelins, its larger body size, and its convergence with *Alouatta* in craniodental features are interpreted as convergent adaptations that permit intensive leaf eating during periods of scarcity of higher-quality foods (Strier 1991b, 1992a).

Table 10.3 Diets of Wild Ateline Primates Based on Studies Lasting 6 Months or More[1]

GENUS AND SPECIES	LOCATION	PERCENT FRUIT[2] MEAN	RANGE	PERCENT LEAVES[3] MEAN	RANGE	PERCENT PREY[4] MEAN	RANGE	PERCENT FLOWERS[5] MEAN	RANGE	PERCENT OTHER[6] MEAN	RANGE	STUDY LENGTH[7]	SOURCE
Alouatta belzebul	Brazil	55.6 (55.0)	25–80	24.8	11–54	–	–	5.7	0–13	14.0	4–26	10	Pinto 2002a,b, Pinto and Setz 2004
A. belzebul	Brazil	59.0	43–92	13.3	8–15	0	0	27.6	0–41	0	0	13	Bonvicino 1989[8]
A. caraya	Brazil	28.5	~6–~63	67.3	~37–~86	0	0	2.7	~0–~13	1.6	~0–~7	12	Bicca-Marques and Calegaro-Marques 1994[9]
A. caraya	Argentina	19 (18)	~0–~60 (~0–~60)	68	~36–~95	0	0	12	0–44	1	~0–~4	17	Bravo and Sallenave 2003[10]
A. caraya	Argentina	24	7–27	76	72–91	0	0	0	0	0	0	6	Zunino and Rumiz 1986[8]
A. guariba	Brazil	15.6	1–30	76.0	64–88	0	0	8.4	6–11	0	0	7	Mendes 1989[8]
A. guariba	Brazil	5.2	~0–~12	73.0	~56–~92	0	0	11.7	~0–~27	10.1	~3–~20	12	Chiarello 1994
A. palliata	Mexico	40.6 (34.8)	3–87	54.4	17–87	0	0	0.7	0–3	4.3	0–11	12	Estrada et al. 1999
A. palliata	Mexico	49.9 (41.4)	~0–~80 (~0–~66)	49.3	~20–~100	0	0	<1	–	0	0	12	Estrada 1984
A. palliata	Panama	42.1	~10–~66	48.2	~26–~84	0	0	9.6	~0–~24	–	–	10	Milton 1980
A. palliata	Costa Rica	12.5	~9–~16	69.3	~64–~72	0	0	18.2	~17–~21	0	0	12	Glander 1978[8]
A. palliata	Costa Rica	23.0	–	68.5	–	0	–	8.5	–	0	–	15	Stoner 1996[11]
A. palliata	Costa Rica	28.5	0–55	49.0	0–95	0	0	22.5	–	0	0	24	Chapman 1987, 1988
A. palliata	Nicaragua	34.8 (26.6)	16–48	55.8	~32–~82	0	0	7.9	0–28	1.6	~0–~8	13	Williams-Guillén 2003
A. pigra	Belize	40.8	11–65	45.1	~32–~66	0	0	10.6	–	3.4	–	14	Silver et al. 1998
A. seniculus	French Guiana	25.5 (21.5)	~0–~74 (0–51)	57.0	~27–~79	–	–	12.6	0–39	~5	~0–~17	19	Julliot and Sabatier 1993
A. seniculus	Colombia	52.3	~17–~76	35.3	~25–~69	0	0	1.1	~0–~11	11.3	~0–~22	10	Palacios and Rodriguez 2001
A. seniculus	Colombia	42.3 (28.4)	–	52.1	–	0	–	5.4	–	<1	–	10	Gaulin and Gaulin 1982
Ateles belzebuth belzebuth	Colombia	83	78–100	7	0–22	0	–	<1	–	10	0–18	10	Klein and Klein 1977
A. belzebuth belzebuth	Colombia	73.0	41–96	13	3–25	1.4	0–9	12	1–36	1.2	0–4	12	A. Link (unpublished data)
A. belzebuth belzebuth	Brazil	91.7 (88.5)	~74–~100	8.3	~0–26	0	0	0	0	0	0	12	Nunes 1998
A. belzebuth belzebuth	Ecuador	87	64–100	9	0–23	<1	–	1	~0–~6	~3	~0–~17	12	Dew 2001
A. belzebuth belzebuth	Ecuador	78.8	52–92	7.7	0–22	0	0	3.2	0–24	10.3	0–24	17	Suarez 2003[12]
A. belzebuth chamek	Peru	74.7	54–99	15.5	<1–38	–	–	4.5	~0–~22	4.5	~0–~21	12	Symington 1987a, 1988b[13]
A. geoffroyi	Guatemala	54.9	31–84	15.1	1–34	1.9	0–17	6.3	0–22	21.8	1–66	9	Cant 1977[8]
A. geoffroyi	Costa Rica	71.4	14–100	12.5	0–86	2.1	0–30	14	–	0	0	38	Chapman et al. 1995[14]
A. geoffroyi	Panama	82.2	69–91	17.2	6–32	0.6	0–2	1.0	0–9	0	0	14	Campbell 2000
A. paniscus	Surinam	79.8	54–92	7.9	1–23	<1	–	6.4	1–28	5.6	1–17	26	van Roosmalen 1985
Brachyteles arachnoides	Brazil	20.8	4–59	67.1	41–93	0	0	12.2	0–34	0	0	10	Milton 1984a[14]
B. hypoxanthus	Brazil	26.8 (14.8)	11–66	50.9	28–78	0	0	10.9	0–33	11.4	2–33	14	Strier 1991b
B. hypoxanthus	Brazil	19	–	51	–	–	–	28	–	–	–	7	Lemos de Sá 1988[14,15]
Lagothrix cana	Brazil	67.4 (66.6)	~6–~96 (2–95)	16.2	~2–~47	<1	–	3.1	0–8	13.4	~2–~59	11	Peres 1994a[16]
L. lagotricha	Colombia	78.9	–	11.4	–	4.9	–	0.1	–	4.7	–	12	Defler and Defler 1996
L. lugens	Colombia	53	–	13	–	25	–	2	–	7	–	36	Stevenson 2002[17]
L. lugens	Colombia	60	~37–~78	~16	~5–~54	23	~4–~36	~1	–	<1	–	13	Stevenson et al. 1994[18,19]
L. lugens	Colombia	78	72–84	9	5–18	12	9–22	–	–	<1	–	6	Stevenson 1992[19,20]
L. poeppigii	Peru	77	69–79	7	4–20	0	–	2	1–5	14	5–24	72	Soini 1990[21]
L. poeppigii	Ecuador	75.5 (71.3)	64–89 (61–86)	9.8	5–17	9.3	4–16	3.5	0–11	1.9	0–12	12	Di Fiore 1997, 2004
L. poeppigii	Ecuador	73	47–88	10	0–20	6	1–10	5	~0–~19	~6	~0–~17	12	Dew 2001

[1] Means and ranges are taken from or calculated based on data in original sources. Approximate means and ranges are based on extrapolation of values from figures presented in original sources when raw data are not given. —, values not available or could not be calculated or estimated from the original source.

[2] Wherever possible, the overall percentage and range of the total diet comprising "ripe" or "mature" fruit is separated out and noted in parentheses.

[3] Includes leaves and other plant vegetative parts (shoots, leaf buds, petioles, meristem of epiphytes).

[4] Includes both invertebrate and vertebrate prey, although the latter is negligible.

[5] Includes nectar where distinguished in original source.

[6] Includes seeds, seed pod exudates, bark, fungi, aquatic plants, moss, termitaria soil, ground soil, and undetermined food items.

[7] The number of months during which systematic dietary data were collected or the overall length of the study in months not including those periods when data collection was suspended, if that information was available from the original source.

[8] The listed range of values for each food category is based on averages for different seasons (e.g., "dry" and "wet") noted in original source. The actual range across the years should be somewhat greater.

[9] In the original source, seeds were included within the category "fruit." Based on reported values for seed consumption averaged across four seasonal samples, we recalculated dietary proportions including seeds in "other."

[10] Original study does not specify whether "buds" consumed were from leaves or flowers. They are included here in "leaves."

[11] Means reported represent the average of means for two study groups.

[12] Based on 10 follows of up to 2 weeks spread among three focal individuals.

[13] Means reported represent the average of means for two study groups, while ranges cover the minimum to maximum percentages noted in both groups. Some animal prey (caterpillers) were consumed by members of one study group in 2 months of the year but are included in "other" as the original source does not allow these to be separated out into "prey."

[14] Some seeds are included in the "fruit" category but cannot be separated out given how the data are presented in the cited source.

[15] Original source not available, but the values included were taken from multiple secondary sources citing this original.

[16] In the original source, the author includes seeds with "fruit." Here, percentages have been recalculated including seeds as "other" for comparability across studies.

[17] Values included are from Stevenson 2002, which incorporates most of the data also presented in Stevenson et al. 1994 but explicitly excludes seeds from the category of "fruit."

[18] Authors of original source included leaves, flowers, and other plant vegetative parts in same dietary category. Here, we have pulled out the approximate percentage of the diet comprising flowers based on information provided in the original source.

[19] Some seeds are included in the "fruit" category but cannot be separated out accurately given how the data are presented in the cited source. For Stevenson et al. 1994, this is presumed to be about 5%.

[20] Author of original source included leaves, flowers, and other plant vegetative parts in the same dietary category. Here, all of these items are assigned to the category "leaves" as leaves constituted the vast majority of the items consumed, and no information was available in the original source that allowed us to estimate the contribution of "flowers" to this total.

[21] Diet presented is based on periodic observations of feeding accumulated over a period of 7 years.

Activity Patterns and Ranging Behavior

Consistent with their ecological strategy, howler monkeys spend a large portion of their day (66%–80%) resting—more, on average, than any of the other atelines (Table 10.4). *Brachyteles* and *Ateles*, too, are typically inactive for more than half of the day, while *Lagothrix* are the most active atelines, spending upward of 60% of their daily activity either moving or engaged in subsistence activity (Stevenson et al. 1994, Defler 1996, Di Fiore and Rodman 2001). The greater activity of *Lagothrix* compared to the other atelines has several likely explanations. First, the large size of their social groups, combined with their high degree of frugivory,

means that, all else being equal, *Lagothrix* groups need to visit more food patches per day to meet the nutritional requirements of group members than do those of other atelines. Moreover, woolly monkeys tend to use slower and more deliberate quadrupedal locomotion and less suspensory locomotion than do spider monkeys (Defler 1999, Cant et al. 2001) and presumably muriquis, and they may take longer, less direct routes to traverse tree crowns (Cant et al. 2003). As a result, they may commonly spend more time and/or energy covering the same distance as these other atelins.

As in other features of their natural history, there is considerable variation within and between genera of atelines in home range size and ranging behavior (Table 10.5), and

Table 10.4 Percent of Daily Time Budget Spent in Various Activities for Wild Ateline Primates Based on Studies Lasting 6 Months or More[1]

GENUS AND SPECIES	ACTIVITY CATEGORY				STUDY LENGTH (MONTHS)[2]	SOURCE
	EAT + FORAGE	MOVE	REST	SOCIAL + OTHER		
Alouatta belzebul	20.0	18.2	58.7	3.1[3]	10	Pinto 2002a,b
A. belzebul	7.5	19.3	55.9	16.9	13	Bonvicino 1989[4]
A. caraya	15.9	17.6	61.6	4.9	12	Bicca-Marques 1993
A. guariba	17.3	11.0	71.8	0	13	Mendes 1989
A. guariba	19.0	18.8	57.6	4.6[3]	12	de Marques 1995, 1996
A. guariba	18.7	13.2	63.7	4.4	12	Chiarello 1993a[5]
A. palliata	16.2	10.2	65.5	8.1[3]	10	Milton 1980
A. palliata	17.3	2.2	79.7	0.9	12	Estrada et al. 1999
A. palliata	18.1	7.8	72.5	1.6[3]	49[6]	Teaford and Glander 1996
A. palliata	~24.5	~14	~56	~5.5	15	Stoner 1996[7]
A. palliata	13.6	27.4	57.0	2.1	13	Williams-Guillén 2003
A. pigra	24.4	9.8	61.9	3.8	14	Silver et al. 1998
A. seniculus	12.7	6.2	78.5	2.5	10	Gaulin and Gaulin 1982
Ateles belzebuth belzebuth	22.2	9.7	61	7.1	10	Klein and Klein 1977
A. b. belzebuth	18	36	45	1[3]	12	Nunes 1995
A. b. belzebuth	16.7	24.9	58.2	0.1	17	Suarez 2003[8]
A. b. chamek	29	26	45	<1	12	Symington 1988a
A. b. chamek	18.9	29.7	45.5	5.9	11	Wallace 2001
A. belzebuth	50.5	18.1	23.7	7.7	~16	Castellanos 1995 cited in Wallace 2001
A. geoffroyi	33.5	32.6	24.1	9.8[3]	36	Chapman et al. 1989a
Brachyteles arachnoides	27.8	9.7	61.0	1.6	10	Milton 1984a
B. hypoxanthus	18.8	29.4	49.3	2.5	14	Strier 1987a
Lagothrix lagotricha	25.8	38.8	29.9	5.5	12	Defler 1995
L. lugens	36	26	35	3	36	Stevenson 2002[9]
L. lugens	36	24	36	4	13	Stevenson et al. 1994
L. poeppigii	36.2	34.5	23.2	6.1	12	Di Fiore and Rodman 2001

[1] Values are taken from or calculated based on data in original sources. Approximate values are based on extrapolation from figures presented in original sources when raw data are not given.
[2] The number of months during which systematic behavioral data were collected, or the overall length of study in months not including those periods when data collection was suspended, if that information was available from the original source.
[3] Value calculated by subtracting the sum of the other category percentages, reported in the original source, from 100%.
[4] Budget calculated by averaging dry and wet season budgets for each of two groups sampled for very different periods of time and then taking the mean of the group budget, weighted by sampling time.
[5] Time budget calculated here as the mean of four seasonal budgets presented in the original source. The total yearly budget presented therein, while similar to that reported here, did not sum to sufficiently close to 100%.
[6] Activity budget is based on four groups sampled periodically over this time period.
[7] Values reported are the average of means for two study groups, extrapolated from a figure presented in the original source.
[8] Based on 10 follows of up to 2 weeks spread among three focal individuals.
[9] Incorporates data from Stevenson et al. 1994.

Table 10.5 Range Size and Range Use Patterns for Wild Ateline Primates Based on Studies Lasting 6 Months or More

TAXON	GROUP SIZE	GROUP OR COMMUNITY RANGE SIZE (HA)	SEX DIFFERENCES IN RANGING[1]	SPACE USED (HA/INDIVIDUAL)[2]	RANGE OVERLAP	AVERAGE DAY RANGE (M)	MAX. DAY RANGE (M)	SOURCE
Alouatta belzebul	7–9	63.2	—	7.9	—	761	—	Pinto 2002a,b
A. belzebul	6–8	9.5	—	1.4	11%	—	—	Bonvicino 1989
A. belzebul	6–8	4.7+	—	0.7	21%	—	—	Bonvicino 1989
A. caraya	15–17	2	—	0.1	—	454	893	Bicca-Marques 1994
A. caraya	16	1.7	—	0.1	—	513	—	Bravo and Sallenave 2003
A. caraya	21	2.2	—	0.1	—	602	1,515	Bravo and Sallenave 2003
A. caraya	5.5	5.5	—	1.0	<20%	345	—	Zunino and Rumiz 1986[3]
A. guariba	6	4.1	—	0.7	33%	467	808	Chiarello 1993b
A. guariba	7	7.9	—	1.1	30%	364	540	Mendes 1989
A. pallitata	40	108	—	2.7	—	—	—	Chapman 1988
A. seniculus	8	25	—	3.1	—	—	—	Izawa and Nishimura 1988
A. seniculus	6–8	45	—	~6.4	—	—	1,620	Julliot 1996
A. seniculus	9	~22	—	~2.4	—	706	—	Gaulin and Gaulin 1982[4]
A. seniculus	—	25	—	—	—	542	655	Crockett and Eisenberg 1987
A. seniculus	9.5	5.1	—	0.5	63%	375	610	Sekulic 1982a[5]
A. seniculus	11.5	7.4	—	0.6	64%	355	790	Sekulic 1982a[5]
A. seniculus	9.5	3.9	—	0.4	56%	340	520	Sekulic 1982a[5]
A. seniculus	10.5	5.8	—	0.6	28%	445	840	Sekulic 1982a[5]
A. seniculus	7–8	182.0	—	24.3	—	1,150	2,200	Palacios and Rodriguez 2001
A. palliata	13	9.9	—	0.8	—	596	1,261	Glander 1978
A. palliata	18	31.7	—	1.8	Extensive	392	792	Milton 1980
A. palliata	18	31.1	—	1.7	100%	488	—	Milton 1980
A. palliata	14	60	—	4.3	—	123	503	Estrada 1984
A. palliata	26	21.9	—	0.8	Minimal	752	—	Williams-Guillén 2003
A. palliata	20	21.3	—	1.1	Minimal	572	—	Williams-Guillén 2003
Average		**28**		**2.7**		**526**	**968**	
Ateles belzebuth belzebuth	12–16	300	—	21.4	—	3,311	6,039	Suarez 2003[6]
A. belzebuth belzebuth	15–16	80–90	—	5.5	—	—	—	Ahumada 1989
A. belzebuth belzebuth	19–23	316	M: 120, F: 88	15.0	—	1,750	~2,600	Nunes 1995
A. belzebuth belzebuth	15–18	259–388	—	19.6	20%–30%	—	4,000+	Klein 1972, Klein and Klein 1976
A. belzebuth chamek	37	231	M "core": 57, F "core": 28	6.2	10%	1,977	4,070	Symington 1988a
A. belzebuth chamek	40	153		3.8	16%			Symington 1988a
A. geoffroyi	20–24	963	—	43.8	—	2,055	4,500	Campbell 2000
A. geoffroyi	42	170	M: 81, F: 55	4.0	—	1,297	—	Chapman et al. 1989a, Chapman 1990a
A. geoffroyi	—	—	M: 93, F: 50	—	—	—	—	Fedigan et al. 1988[7]
A. geoffroyi	16	95	M "core": 15, F "core": 8	5.9	Minimal	2,302	3,872	Ramos-Fernández and Ayala-Orozco 2003
A. geoffroyi	41	166	M "core": 21, F "core": 12	4.0	Minimal			Ramos-Fernández and Ayala-Orozco 2003
A. paniscus	12	—	—	—	Minimal	2,300	—	Norconk and Kinzey 1994
A. paniscus	18	255	M > F	14.2	0%	—	5,000	van Roosmalen 1985
Average		**278**		**13.9**		**2,142**	**4,297**	
Brachyteles arachnoides	7–10	70+	M > F	~8.2+	—	630	~1,350	Milton 1984a[8]
B. hypoxanthus	26	168	—	6.5	46%	1,283	3,403	Strier 1987b
B. hypoxanthus	34	184	—	5.4	—	—	—	Strier 1991a
B. hypoxanthus	62–63	309	—	~4.9	—	1,313	2,835	Dias and Strier 2003
B. hypoxanthus	15–18	40	—	~2.4	—	—	—	Lemos de Sá 1988[9]
Average		**154**		**5.5**		**1,075**	**2,529**	
Lagothrix cana	44–49	1,021+	—	~22.0+	—	—	—	Peres 1996
L. lagotricha	13	350	—	26.9	65%–100%	—	—	Nishimura 1990a
L. lagotricha	45	450	—	10.0	50%–100%	—	—	Nishimura 1990a
L. lagotricha	24	760	—	31.7	90%–100%	2,880	3,582	Defler 1989, 1996[10]
L. lugens	17–19	169	—	~9.4	100%	1,633	1,853	Stevenson et al. 1994[11]
L. lugens	14	—	—	—	—	2,500	3,450	Stevenson and Castellanos 2000[12]
L. lugens	17–33	—	—	—	—	1,750	2,200	Stevenson and Castellanos 2000[12]
L. lugens	17–21	—	—	—	—	1,900	2,250	Stevenson and Castellanos 2000[12]
L. lugens	27–32	—	—	—	—	2,450	3,200	Stevenson and Castellanos 2000[12]
L. poeppigii	24–25	124	—	~5.1	45%	1,792	2,738	Di Fiore 2003b[13]
L. poeppigii	23	108	—	4.7	47%	1,878	2,859	Di Fiore 2003b[13]
L. poeppigii	17–23	350	—	~17.5	—	540	950	Soini 1986[14]
L. poeppigii	14	400+	—	28.6+	—	—	—	Ramirez 1980
L. poeppigii	10	250+	—	25+	—	—	—	Ramirez 1980
Average		**398**		**18.1**		**1,925**	**2,565**	

[1] Differences in the size, in hectares, of the home range used by males (M) versus females (F) are noted for populations in which males and females often range separately. For some populations, data are not available on actual range sizes but the source indicates the relative size of the ranges of males versus females.

[2] Calculated by dividing the group or community range size by group size. If a range of values is given for group size or home range size, the midpoint of the range is used in the calculation.

[3] Based on average values reported for two groups studied for 6 months in a forest fragment.

[4] Day range estimated by dividing the total horizontal distance traversed throughout the study by total observation time, normalized by hour of the day.

[5] Similar home range sizes were reported by Neville (1972a) for a larger number of groups studied for a shorter period of time in the same population. Average day ranges reported in Sekulic (1982a) are medians rather than means.

[6] Based on 10 follows of up to 2 weeks spread among three focal individuals.

[7] Based on radiotelemetry data from two males and five females. Same population as studied by Chapman et al. (1989a) and Chapman (1990a).

[8] Mean day range calculated as the grand mean of the average day ranges recorded during 12 ~1-week sampling periods across 10 months. Maximum day range estimated as the largest 1-week average (814 m) plus two standard deviations for that period.

[9] Original source not available, but the values included were taken from multiple secondary sources citing this original.

[10] Table entries reflect the grand mean and maximum of average monthly day ranges. Without first averaging within months, the overall mean day range for this population was 2,670 m and the maximum distance traveled in a day was 5,130 m.

[11] The maximum day range value reflects the largest of four trimester means.

[12] Mean and maximum day range values estimated from Figure 3 of the original source.

[13] Day ranges calculated for period from 0700 to 1700 only.

[14] Day ranges calculated as the straight line displacement between sleeping sites on successive days.

generalizations about the range use patterns characteristic of any one taxon are thus difficult to make. In howler monkeys, for example, variation in home range size is extensive, and the variation within species is greater than that between species (Crockett and Eisenberg 1987). Similarly, the degree of range overlap between groups of howler monkeys at the same study site varies from slight to complete. A cogent explanation for the variation in range size and range overlap among howler monkeys remains elusive. Although howler monkey groups do not appear to defend exclusive territories (Milton 1980), groups are site-faithful, occupying roughly the same range from year to year (Crockett 1996, Fedigan and Jack 2001, Clarke et al. 2002, Pavelka et al. 2003). Interactions between adjacent groups in the best-studied species (*Alouatta palliata*, *A. seniculus*, and *A. guariba*) are typically agonistic and often involve loud, coordinated vocal displays that may either influence or reflect the eventual outcome of the encounter (Baldwin and Baldwin 1972, Sekulic 1982a,b, Chiarello 1995b). Day ranges in howler monkeys typically average 700 m or less (Crockett and Eisenberg 1987) and are much shorter than those of other atelines (Table 10.5).

Among different species of *Lagothrix*, too, there is considerable variation in range size (Stevenson et al. 1994, Defler 1996, Peres 1996, Di Fiore 2003b) (Table 10.5). This is probably due, in part, to the fact that a larger supplying area is needed to support larger group biomass (Milton and May 1976, Dunbar 1988, Isbell 1991, Janson and Goldsmith 1995). No single species of woolly monkey has yet been studied well enough to address the extent of range size variation within species. Despite variation among the three species of woolly monkeys for which data are available, average day range length is clearly much longer for *Lagothrix* than for *Alouatta* or *Brachyteles* and is comparable to, or greater than, that seen for *Ateles* (Table 10.5). Di Fiore (2003b) has suggested that the day ranges of western Amazonian *L. poeppigii* may be associated with the high level of insect prey foraging noted in this species or with the monitoring of potential feeding trees. As with home range size and day range length, the degree of range overlap between adjacent groups varies within *Lagothrix* but can be extensive. Although groups do not defend exclusive areas from other groups, interactions between groups occasionally occur in areas of range overlap. Sometimes these encounters are peaceful, with groups associating passively with one another for up to several hours (Nishimura 1990a, Di Fiore 1997). At other times, these encounters can be quite hostile, accompanied by chases and loud vocal behavior (Di Fiore 1997). Males seem to be the more active participants, and aggressive intergroup encounters are one of the few instances where adult males from the same social group show either affiliative or cooperative behavior toward one another (Di Fiore personal observation).

Among spider monkeys, females from the same community typically range within separate but overlapping areas that comprise but a fraction of the entire community range (McFarland 1986, Symington 1988a, Fedigan et al. 1988,

Chapman 1990a). Symington (1988a), for example, found the average area used by female spider monkeys in two adjacent communities at Cocha Cashu, Peru to be only 34% and 75% of their respective community home ranges; and the "core" areas used by these females were considerably smaller, averaging just 21% and 33% of their community ranges, respectively. On Barro Colorado Island, Panama, females do not appear to use core areas but, rather, travel in larger, more cohesive groups throughout the community's range; this is probably a result of the fact that they have no conspecific competitors on the island (Campbell 2000, 2002). Male spider monkeys typically spend more time moving, range farther each day, and use a larger overall area than do most females (Fedigan et al. 1988, Symington 1988a, Chapman 1990a, Nunes 1995, Ramos-Fernández and Ayala-Orozco 2003) (Table 10.5). Moreover, dominant individuals tend to have larger individual ranges than subordinate individuals in at least one species of spider monkey, *Ateles geoffroyi* (Chapman 1990a). Although average day range length in spider monkeys is broadly comparable to that of woolly monkeys, which are also highly frugivorous, *Ateles* sometimes range much farther in a given day than the longest day range lengths reported for *Lagothrix*. The overlap in home range between adjacent communities of spider monkeys is far less than that seen in woolly monkeys, and intercommunity interactions in spider monkeys are typically aggressive. Males are the main participants in intergroup encounters, and some of the ranging behavior of males appears to be associated specifically with territorial defense as males band together and patrol over larger areas and spend more time in border areas than do females (Fedigan and Baxter 1984, Chapman 1990a).

For one southern muriqui population studied for 10 months, Milton (1984a) found range use patterns very comparable to those that characterize *Ateles*, with small groups of females occupying predictable core areas and "itinerant" males that ranged over a much larger area, sometimes joining females and other times traveling independently or in parties with other males. In early studies of a population of northern muriquis, however, females were not found to maintain individual core areas (Strier et al. 1993). Moreover, in this species, large social groups with multiple adult males and females remained socially cohesive and ranged as a single unit at lower population densities (Strier 1989). As group size has increased in this population, groups have begun to split into several independent foraging subunits, which are nonetheless considerably larger than those seen in spider monkeys (Dias and Strier 2003). *Brachyteles* day ranges tend to be longer, on average, than those of *Alouatta* but are considerably shorter than those of the more frugivorous *Lagothrix* and *Ateles* (Strier 1987c). For one group of northern muriquis monitored over several years, home range size increased as the community grew in size (Strier 1987c, Strier et al. 1993, Dias and Strier 2003). Interestingly, however, even though recently observed subgroups in this population were, on average, larger than the original community size, average day range length did not change appreciably, suggesting that factors other than intragroup feeding competition may also influence subgroup size (Dias and Strier 2003). Dias and Strier (2003) suggest that the shift to a more fluid subgrouping pattern in this growing community while day range length has remained relatively stable may actually reflect the existence of an upper limit to the number of individuals that can coordinate their movements rather than a response to increased intragroup feeding competition. Intergroup interactions among muriquis are similar to those among spider monkeys and woolly monkeys: males are the principal participants in encounters between groups and cooperate with one another against male intruders into their group's range (Strier 1994a, Strier et al. 2002). Depending on the local population density, home range overlap between adjacent groups can be slight or extensive (de Moraes et al. 1998).

Kin Structure and Social Behavior

All atelines are characterized by a high incidence of female dispersal and, in the atelin genera at least, by a greater degree of male than female philopatry (Strier 1990, 1994a,b). Under conditions of female-biased dispersal, males within social groups and within local populations are expected to be, on average, more closely related to one another than are females (Morin et al. 1994, Di Fiore 2003a). This, theoretically, should have important implications for the patterning of social behavior within ateline groups. Indeed, despite considerable variation among ateline genera in the expression of affiliative and agonistic social relationships, these relationships are, broadly speaking, consistent with expectations based on kin structure patterns. However, the presumption of greater average male–male versus female–female relatedness in atelines has been tested genetically only for *Lagothrix* and shown to be true for several (but not all) social groups (Di Fiore and Fleischer 2005). A comparative overview of some aspects of social relationships in atelines is presented in Table 10.6.

Among howler monkeys, animals of both sexes disperse (Clarke and Glander 1984; Crockett and Pope 1993; Rumiz 1990; Glander 1980, 1992); thus, many individuals are likely to live a good portion of their adult lives in groups without close kin. In most well-studied populations, both males and females are reported to form dominance hierarchies, although in contrast to the pattern commonly seen among many Old World primates, rank is inversely related to age among males in *Alouatta palliata* and among females in both *A. palliata* and *A. seniculus* (Glander 1980; Jones 1980a, 1981; Crockett 1984; Zucker and Clarke 1998). Moreover, for several populations where the relationship has been studied, both mating and reproductive success appear to be related to dominance rank among males (Clarke 1983, Jones 1985, Pope 1990) and among females dominant individuals commonly direct aggression toward subordinates and their offspring (Crockett 1984, Clarke et al. 1998). However, in one large group of *A. palliata*, Wang and Milton (2003)

could not discern dominance relationships among the adult males, apart from the existence of an alpha male.

In general, intrasexual social relationships in howler monkeys are weak and affiliation is rare (Zucker and Clarke 1998, Wang and Milton 2003). Grooming, when it occurs, is frequently between the sexes, most often from females to adult males; and proximity partners among adults tend to be opposite-sex individuals (Neville 1972b, Chiarello 1995a, Zucker and Clarke 1998, Wang and Milton 2003). Kin networks are generally not as developed in howler monkeys as in primate species where dispersal is predominantly by members of one sex. Despite these observations, howler monkeys provide some of the clearest examples yet documented in primates of the actual fitness benefits that may accrue as a result of kin-directed nepotism (Pope 1998a, 2000). For example, in howler monkeys, males often form coalitions to take over groups of females from other males in order to begin breeding (Pope 1990, Horwich et al. 2000). Pope (1990) has demonstrated that in *A. seniculus*, coalitions comprising male kin last, on average, 3.6 times longer than coalitions made up of unrelated males. Similarly, females in new groups are generally unrelated individuals that join together with one or more emigrating males and begin defending a range. As groups become more established, however, founding females try to recruit their own daughters as breeders and to force the emigration of other females' daughters (Crockett 1984, Pope 2000). As a result, the degree of relatedness among females within groups of red howler monkeys tends to increase over time, as does females' net reproductive output, indicating a clear fitness benefit for kin-directed recruitment (Pope 2000).

Among spider monkeys and muriquis, females disperse while males are philopatric. Not surprisingly, in these two genera, social interactions among males are more affiliative and cooperative than those among females or between the sexes. In spider monkeys, males are each other's most common associates and interact affiliatively with each other much more often than do females. For example, Symington (1990) found that "association indices" for male–male dyads were much higher, on average, than indices for either male–female or female–female dyads. Similarly, for grooming bouts not occurring between mothers and offspring, Symington (1987a, 1990) noted that the most common grooming dyads involved pairs of males rather than pairs of females or mixed-sex pairs. Fedigan and Baxter (1984) also found that males engaged in more affiliative behaviors than did females and directed these more toward other males than toward females. For example, males "greet" each other significantly more than females with an embrace in which pectoral scent glands are often inspected (Eisenberg and Kuehn 1966, Fedigan and Baxter 1984). Male *Ateles* also cooperate with one another to patrol and defend a large community range (Fedigan and Baxter 1984; Symington 1987a, 1988a; Chapman 1990a).

Grooming (auto- and allo-) is absent from the *Brachyteles* behavioral repertoire (Strier 1994a); however, the same

Table 10.6   Comparative Social and Reproductive Behavior of Ateline Primates

	ALOUATTA	ATELES	LAGOTHRIX	BRACHYTELES
Female dispersal	Yes	Yes	Yes	Yes
Male dispersal	Yes	No	Possible	No
Intersexual dominance relationships	Males > females	Males > females	Males > females	Codominant
Female competition over group membership				
Female response to immigrants	Intense hostility	Harass immigrants	Indifferent	Harass immigrants
Females force emigration of maturing females	Usually	No?	No?	No
Female dominance relationships within groups	Hierarchical	Hierarchical	Hierarchical?	Egalitarian
Male competition over group membership				
Male response to immigrants	Intense hostility	–	Tolerant to hostile	–
Male reaction to males in other groups	Hostile	Hostile	Tolerant to hostile	Hostile
Male dominance relationships within groups	Hierarchical and competitive	Hierarchical and tolerant?	Hierarchical and tolerant?	Egalitarian and tolerant
Affiliative bonds among adult females	Weak	Weak	Weak	Weak?
Affiliative bonds among adult males	Weak	Strong	Weak	Strong
Intersexual affiliation	Strong	Weak	Strong?	Weak
Strongest grooming and proximity relationships	Male–female	Male–male	Male–female and adult male–subadult male	Male–male
Contexts of male–male cooperation	Related or unrelated males may form coalitions with each other to take over a group and oust resident male	Males of a community may jointly patrol community range boundaries, aggress against intruders, and cooperate in intergroup encounters	Males from the same social group may cooperate during intergroup encounters; at other times, are tolerant of but not affiliative with each other	Males of a community jointly aggress against intruders, cooperate in intergroup encounters, spend considerable time in proximity with one another, and are overtly affiliative with each other
Females solicit copulations from males?	Yes	Yes	Yes	Yes
Females copulate outside of periovulatory period?	Yes	Yes	Yes	Yes
Females mate with multiple males?	Occasionally	Yes	Yes, both within and across receptive periods, sometimes on the same day	Yes, both within and across receptive periods, sometimes on the same day and in quick succession
Social context of copulations	Dominant male may form consortship with receptive female and keep other males from mating	Often take place during consortships or in seclusion, away from other group members	Take place in exposed positions, in view of other group members, males tolerant of mating by other males	Take place in exposed positions, in view of other group members, males tolerant of mating by other males
Differential male mating success?	Yes, related to dominance rank or tenure as alpha male	Unknown	Some, may be related to male age, dominance, or size	Yes, based on female preferences
Importance of female choice in mating system	Weak	Strong	Strong	Strong
Direct female competition over matings?	Yes	No	Yes	No

general pattern seen among male spider monkeys is apparent for other affiliative behaviors, at least among the well-studied northern muriqui. In this species, for instance, males are the most common proximity and embrace partners of other males (Strier 1992b, 1994a, 1997a), and they cooperate to defend group females from males of neighboring groups (Strier 1992c, 1994a). Female social relationships in muriquis and spider monkeys are less well described. In both taxa, females do interact affiliatively with one another, although, at least in *Ateles*, the frequency with which affinitive behavior is seen is far less among females than among males. Northern muriqui females spend more time in proximity to other females than they do to males, and common proximity partners tend to be common embracing partners as well (Strier 1990, 1992b); however, the relative strength of female–female versus male–male bonds has not been addressed. Among southern muriquis, the pattern appears to be somewhat different, with inter-sexual relationships apparently more important than same-sex ones in at least one population (Coles and Talebi 2004). Although low in both taxa, the rate of intragroup aggression is much higher in *Ateles* than *Brachyteles*, in which very little aggression has ever been seen (Fedigan and Baxter 1984, Strier 1992c). Aggression in spider monkeys is typically directed from males toward females (Fedigan and Baxter 1984, Campbell 2003). The context of this aggression is unclear but does not appear to be directly related to sexual encounters. It may, however, allow males to maintain dominance over equally sized females (Strier 1994a, Campbell 2003). Unlike in *Ateles*, the sexes are codominant in *Brachyteles* and the relationships between males and females appear to be more egalitarian (Strier 1994a).

Grooming and other affiliative behaviors are rarely observed in studies of *Lagothrix* social behavior. When grooming does occur, adult males are the most common recipients, being groomed more often by subadult males than by females; adult males rarely, if ever, groom one another, and grooming among adult females is also uncommon (Nishimura 1990a, 1994; Di Fiore 1997; Di Fiore and Fleischer 2005). Male bonds are not nearly so well developed in woolly monkeys as in the other atelins, and adult males from the same social group rarely interact with one another except during hostile intergroup encounters. Nonetheless, while not overtly affiliative, adult male woolly monkeys are extremely tolerant of one another, even in the context of mating (see below). In general, the most developed affiliative social interactions among nonjuvenile woolly monkeys appear to be those between subadult and adult males (but not among adults) and between males and females (Di Fiore and Fleischer 2005); interestingly, relationships among females seem to be even less developed than in other atelins. Nishimura (1990a, 1994) suggests that male woolly monkeys can be ranked into a dominance hierarchy that correlates with age and body size; while likely to be true, interactions among males are sometimes too infrequent to preclude determination of a hierarchy (Di Fiore 1997 and unpublished data).

REPRODUCTIVE STRATEGIES AND LIFE HISTORY

Among the atelines, mating patterns are varied and the two ateline tribes demonstrate behaviorally divergent reproductive strategies. In howler monkeys, reproduction appears to be strongly skewed toward socially dominant males. In the best-studied case of red howler monkeys (*A. seniculus*) living in the llanos region of central Venezuela, the single resident male in one-male groups, and the dominant male in multimale groups, is responsible for most of the observed mating and sires all of the infants born in the group during his tenure (Pope 1990). Additionally, aggression among red howler monkey males for breeding positions is intense, and infanticide has been reported when a new male takes over a group (Crockett and Sekulic 1984, Agoramoorthy and Rudran 1995), with 44% of infant mortality presumably due to infanticide in one well-studied population (Crockett and Rudran 1987b). Infanticide has also been reported in three other howler monkey species living in groups with one or a small number of adult males (e.g., *A. caraya*, Zunino et al. 1986, Rumiz 1990, Calegaro-Marques and Bicca-Marques 1996; *A. guariba*, Galetti et al. 1994; *A. pigra*, Brockett et al. 1999, Knopff et al. 2004). Consistent with the sexual selection hypothesis, which interprets infanticide as an adaptive male reproductive strategy (Hrdy 1977, 1979), red howler monkey females are able to breed more quickly following the loss of an infant, and they mate with infanticidal males following group takeovers (Clarke 1983, Agoramoorthy and Rudran 1995, Crockett and Sekulic 1984, Crockett and Rudran 1987b).

Overt male–male competition is less extensive in mantled howler monkeys (*A. palliata*) where larger, multimale groups are common. In some populations of mantled howlers, males can be ranked into dominance hierarchies, male mating success is associated with rank, and group takeovers and subsequent infanticidal attacks by new males have been observed (Jones 1980a, 1985; Clarke 1983; Clarke and Glander 1984; Glander 1980, 1992; Clarke et al. 1994). However, in one large group of mantled howler monkeys on Barro Colorado Island, Panama, all six adult males, regardless of rank, mated with receptive females with no overt competition and the highest-ranking male did not copulate the most (Wang and Milton 2003). Taken together, these results suggest a general importance of direct reproductive competition among males across howler species, the intensity of which appears to decrease with increasing male group size. Among female howler monkeys, as among males, direct competition over group membership and/or breeding positions is fierce. In *A. seniculus*, maturing females are often forced from their natal groups by resident females and rarely are able to immigrate into existing groups, instead securing breeding opportunities by establishing new groups with other emigrants (Crockett 1984, Crockett and Sekulic 1984, Crockett and Pope 1993, Pope 2000). Female *A. palliata*, by contrast, are able to immigrate successfully into established groups, where they either fight their way to the top of the female dominance hierarchy or, if unsuccessful, disperse again and try to enter a different group (Jones 1980b, Glander 1992).

Within the atelins, observed mating is not nearly so skewed toward specific males as it appears to be in *Alouatta*, nor is direct intrasexual competition over mating opportunities nearly so overt. Nonetheless, mating behavior and the intensity of male–male competition are qualitatively different in the different atelin genera. In all three genera, females have been observed to mate promiscuously with multiple males during a given fertile period, sometimes even on the same day (*Ateles*, Campbell 2000; *Brachyteles*, Aguirre 1971, Milton 1985a,b, Strier 1986b; *Lagothrix*, Nishimura 1988, 1990b, Di Fiore 1997 and unpublished data). Moreover, in *Brachyteles* and *Lagothrix*, males are extremely tolerant of mating by other males (Strier 1993, 1994a; Nishimura 1994; Di Fiore 1997, Di Fiore and Fleischer 2005; Strier et al. 2000); and in muriquis, multiple males have even been observed mating with the same female in quick succession (Aguirre 1971, Milton 1985a,b). Among muriquis, certain males have been recorded mating more often than others, but relative mating success does not appear to be related to male dominance rank in any way; indeed, no dominance hierarchy is discernable among muriqui males, among which very few agonistic interactions have ever been seen (Strier 1990). Rather, this differential mating pattern has been attributed to female preferences for particular males or classes of males (Strier 1994a, 1997b). Although various males of all ages mate with females among woolly monkeys (Nishimura 1990b, Di Fiore 1997), Nishimura (1990b, 1994) has suggested that mating success may be skewed toward older, perhaps more socially dominant, males. Observations in captivity (Williams 1967) also suggest that socially dominant males may monopolize copulations during a female's periovulatory period, implying a somewhat more dominance-based but promiscuous mating system for *Lagothrix* than for *Brachyteles* (see also Strier 1996a). Very little is known of mating patterns in *Ateles* as mating tends to take place in seclusion, away from other group members (van Roosmalen 1985, Campbell 2000, Di Fiore personal observation). However, females have also been reported to mate with multiple males, sometimes on the same day (van Roosmalen 1985, Robinson and Janson 1987), and perhaps to form consortships with specific males (Symington 1987a, Campbell 2000). No molecular studies have yet examined paternity patterns in any of the atelins.

Unlike the situation in most populations of howler monkeys, overt male–male competition appears to be of little importance in atelin societies. In fact, as noted above, among spider monkeys and muriquis, social relationships among males are overtly affiliative and the rate of aggressive interactions among males is extremely low. No cases of aggression have been reported in the context of copulation for either species (Fedigan and Baxter 1984; Strier 1986a, 1990, 1992b; Strier et al. 2000), although for spider monkeys this may be due, in part, to the secluded nature of copulation.

Among woolly monkeys, too, adult males are very tolerant of one another, especially in the context of mating. Males will often mate within sight of other males without receiving overt aggression from them (Di Fiore 1997, Di Fiore and Fleischer 2005), although Nishimura (1990b) has observed copulation by a subadult male being interrupted by an adult male, and Stevenson (1997) has reported that the only case of "strong" aggression observed between adult males of the same group during a year-long study occurred within half an hour of a copulation involving one of the males.

Instead of competing overtly over reproductive opportunities, male atelins (and perhaps male howler monkeys in the largest groups of *A. palliata*) appear to compete more subtly. Atelins in general are characterized by having very long intromission times relative to howler monkeys (Table 10.7), suggesting that mate guarding may be an important component in the male reproductive strategy. Similarly, at least in muriquis, males have large testes relative to their body size, an indication that competition among males over fertilizations may be relegated largely to competition among sperm (Milton 1985b; Strier 1992b,c; Strier et al. 2002; see also Chapter 25). Also, female choice appears to be an extremely important component of the mating systems of all atelin primates (Strier 1990, Di Fiore 1997 and unpublished data). In spider monkeys, woolly monkeys, and muriquis, females commonly (and sometimes insistently) solicit copulation from males and may advertise their receptivity broadly to males through other behavioral signals (e.g., by manually stimulating male genitals, *Ateles geoffroyi*; Campbell 2000) as well as olfactory cues (e.g., Williams 1967, Klein and Klein 1971, Milton 1987, Ramirez 1988, Di Fiore 1997, Campbell 2000). Based on tracking ovarian cycles using fecal steroid assays, Strier and Ziegler (1997) and Campbell (2004) found that female northern muriquis (*Brachyteles hypoxanthus*, previously named *B. arachnoides hypoxanthus*) and black-handed spider monkeys (*A. geoffroyi*), respectively, often mate outside of their periovulatory periods. In captivity, female *Lagothrix* also have been reported to mate outside of their 3- to 4-day periovulatory period (Williams 1967, 1974), but this has not been confirmed in the wild. Interestingly, female woolly monkeys have been observed to compete directly over reproductive opportunities. It is not uncommon for female woolly monkeys to harass copulating pairs, to the point where mating is interrupted prior to ejaculation by the male, suggesting that direct female mate competition, in addition to female choice, may be a significant part of the mating system of *Lagothrix* (Di Fiore 1997, Di Fiore and Fleischer 2005).

Life history parameters are rather poorly studied for ateline primates, even in captivity. Nonetheless, it is clear that spider monkeys, woolly monkeys, and muriquis have "slow" life histories compared to other platyrrhines and other nonhominoid primates generally, with a late age at first reproduction and a long interbirth interval for their body size (Fig. 10.4). Howler monkeys, in contrast, begin reproducing somewhat earlier and have shorter interbirth intervals than

Table 10.7   Reproductive and Life History Parameters for Ateline Primates[1]

			TAXON			
PARAMETER	*ALOUATTA*	SPECIES	*ATELES*	SPECIES	*LAGOTHRIX*	*BRACHYTELES*
Female cycle length (days)	16.3 ± 3.5	*palliata* (La Pacifica)	~25	*paniscus* (Raleighvallen)	~18[2]	21.0 ± 5.4
			22.7	*geoffroyi* (Barro Colorado)		
Copulatory or receptive period (days)	2–4	*palliata* (La Pacifica)	8–10	*paniscus* (Raleighvallen)	3.1 ± 2.4	2.1 ± 1.2
Number of cycles or months to conception	8–15 months	*palliata* (La Pacifica)	At least 3–8 cycles 3–6 cycles	*paniscus* (Raleighvallen) *geoffroyi* (Barro Colorado)	~12 cycles[3]	2–7 cycles
Average intromission time	<50 sec 41 ± 11 sec (max 70 sec)	*palliata* (La Pacifica) *caraya* (Brazil)	10 min 19 min 14.7 min (max 23 min)	*paniscus* (Raleighvallen) *geoffroyi* (Barro Colorado) *belzebuth* (Manu)	6.6 min (max >20 min)	6.5 min (max 18 min)[4] 4.1 min (max 8.3 min)[5]
Gestation length (days)	186 ± 6 184–194 152–195	*palliata* (La Pacifica) *seniculus* (Hato Masaguaral) *caraya* (Brazil)	226–232	*geoffroyi* (captive, Barro Colorado)	~210–225	216.4 ± 1.5
Sex ratio at birth (M:F)	1:0.88 1:1.67 1:0.94	*caraya* (NE Argentina) *palliata* (La Pacifica) *seniculus* (Hato Masaguaral)	1:2.67 1:1.00	*belzebuth* (Manu) *geoffroyi* (Santa Rosa)	1:1.18	1:1.76
Birth seasonality	Dry season No seasonality No seasonality No seasonality No seasonality Dry season Early dry to early wet season Little seasonality but more births in dry season	*caraya* (NE Argentina, mainland gallery forest) *caraya* (NE Argentina, island flooded forest) *guariba* (EBC) *palliata* (La Pacifica) *palliata* (Barro Colorado) *palliata* (Santa Rosa) *pigra* (Belize) *seniculus* (Hato Masaguaral)	Late wet/early dry season Late wet/early dry season Dry season Late dry/early wet season Little seasonality but more births in early wet season	*belzebuth* (La Macarena) *geoffroyi* (Barro Colorado) *geoffroyi* (Santa Rosa) *paniscus* (Raleighvallen) *belzebuth* (Manu)	Late wet/early dry season	Dry season
Age at weaning (months)	12 11–14	*palliata* (La Pacifica) *seniculus* (Hato Masaguaral)	26+ 24+ ~36	*geoffroyi* (Barro Colorado) *geoffroyi* (Santa Rosa) *paniscus* (Raleighvallen)	—	18–24
Sex and age at emigration (years)	M: 1–2, F: 2–3 M: 4–6, F: 2–4	*palliata* (La Pacifica) *seniculus* (Hato Masaguaral)	F: 4–5	*belzebuth* (Manu)	F: 6.0 ± 0.4	F: 6.1 ± 0.6
Age at menarche (years)	~2.8	*palliata* (La Pacifica)	Earliest at ~6.5	*geoffroyi* (Barro Colorado)	—	~6
Female age at first reproduction (years)	~3.6 5.2 (4–7)[6]	*palliata* (La Pacifica) *seniculus* (Hato Masaguaral)	Earliest at ~7.1	*geoffroyi* (Barro Colorado)	9.0 ± 0.7 9.0 (captive, wild-born)	9.0 ± 1.8, earliest at 7.25 for a single philopatric female
Male age at sexual maturity (years)	3.5 7 (6–8)[6]	*palliata* (La Pacifica) *seniculus* (Hato Masaguaral)	Earliest behavioral correlates of sexual maturity at 4 to 5	*geoffroyi* (Barro Colorado)	First copulation without ejaculation at 3.5, first "true" copulation at 5.5 to 6	Earliest observed mating at 5.5
Interval from female immigration to first copulation (months)	3–4	*palliata* (La Pacifica)	—	—	Typically less than 1 week, else female does not remain in the group	11.2 ± 2.2
Interval from immigration to first birth (months)	19.7	*palliata* (Santa Rosa)	—	—	22.8 ± 7.2	33.8 ± 7.3
Interval from parturition to next conception (months)	12.8 13.7	*palliata* (La Pacifica) *palliata* (Santa Rosa)	27.1	*geoffroyi* (Santa Rosa)	23.4	—
Interbirth interval (months)	15.9 21.2 22.5 19.9 17.0	*caraya* (NE Argentina) *guariba* (EBC) *palliata* (La Pacifica) *palliata* (Santa Rosa) *seniculus* (Hato Masaguaral)	34.5 ± 5.8 31.9 ± 3.0 34.7 46–50	*belzebuth* (Manu) *geoffroyi* (Barro Colorado) *geoffroyi* (Santa Rosa) *paniscus* (Raleighvallen)	36.7 ± 4.7	36.4 ± 4.3

Table 10.7   *(cont'd)*

			TAXON			
PARAMETER	*ALOUATTA*	SPECIES	*ATELES*	SPECIES	*LAGOTHRIX*	*BRACHYTELES*
Infant survivorship to 12 months	74%+ M: 53%, F: 100% 70% 80%	*guariba* (EBC) *palliata* (La Pacifica) *palliata* (Santa Rosa) *seniculus* (Hato Masaguaral)	67%	*belzebuth* (Manu)	93.7% (captive)	Close to 100%
Females cycle synchronously?	–	–	Yes	*paniscus* (Raleighvallen)	No	No
% of matings with extragroup males	None reported	None reported			None reported	13%
Average male life span (years)	16.6	*palliata* (Barro Colorado)	–	–	–	–
Average female life span (years)	15.5	*palliata* (Barro Colorado)	–	–	F: 13 (captive)	–
Maximum life span (years)	M: >20.5, F: >19 F: >16 4.2–6.2	*palliata* (Barro Colorado) *palliata* (La Pacifica) *seniculus* (Hato Masaguaral)	M: >24, F: >22	*geoffroyi* (Barro Colorado)	F: >30 (captive)	–
Intrinsic rate of natural increase	0.18 0.18	*palliata* *seniculus*	0.11 0.09	*geoffroyi* *paniscus*	0.16[7]	

[1] Data compiled from the sources listed below. In some cases, we have converted from time in months to time in years or vice versa. —, parameters for which data are not available. Where data on both wild and captive animals are available for a taxon, only those from wild animals are reported. If only captive data are available, then these are reported in lieu of —.

[2] Calculated as the sum of the average copulation period plus the average length of the interval between periods.

[3] Calculated as the number of cycles in 7.2 months, the average number of months from resumption of sexual activity postpartum to next conception.

[4] *Brachyteles hypoxanthus* at the Estação Biológica Caratinga (EBC), Minas Gerais.

[5] *Brachyteles arachnoides* at Fazenda Barreiro Rico, São Paulo.

[6] Average value reported is median rather than mean.

[7] Species of *Lagothrix* for which this value was derived was not specified.

Sources: *Alouatta*, Neville 1972a, Glander 1980, Froelich et al. 1981, Crockett and Sekulic 1982, Milton 1982, Crockett 1984, Clarke and Glander 1984, Jones 1985, Crockett and Rudran 1987a,b, Pope 1990, Rumiz 1990, Ross 1991, Calegaro-Marques and Bicca-Marques 1993, Crockett and Pope 1993, Fedigan and Rose 1995, Zucker et al. 1997, Fedigan et al. 1998, Brockett et al. 2000b, Strier et al. 2001, Kowaleski and Zunino 2004; *Ateles*, Eisenberg 1973; Milton 1981; van Roosmalen 1985; Symington 1987a,b, 1988a; Chapman et al. 1989b; Ross 1991; Campbell 2000; Izawa 2000 cited in Nishimura 2003; *Lagothrix*, Ross 1988; Nishimura 1988, 1990b, 2003; Nishimura et al. 1992; Stevenson 1997; Di Fiore 1997; Mooney and Lee 1999; Di Fiore and Fleischer 2005; *Brachyteles*, Milton 1985a; Strier 1986b, 1991a, 1992b, 1996b, 1997b; Strier and Ziegler 1997, 2000; Strier et al. 2003; Martins and Strier 2004.

Figure 10.4 Ateline life history in a comparative context. Relationship between female age at first reproduction and female body weight (log-transformed) for (A) atelines relative to other platyrrhines and (B) atelines and hominoids relative to other primates. Relationship between interbirth interval and female body weight (log-transformed) for (C) atelines relative to other platyrrhines and (D) atelines and hominoids relative to other primates. In all cases, the atelines (spider monkeys, woolly monkeys, and muriquis) fall substantially above the regression line, indicating they are characterized by having long prenatal and postnatal developmental periods compared to other primates of comparable size. Similar life history features characterize the hominoids. Data on female body weight, age at first birth, and interbirth interval come from Ross and Jones (1999) and Nunn and Barton (2001), except that data from Table 10.1 are used in lieu of the ateline data presented in those sources.

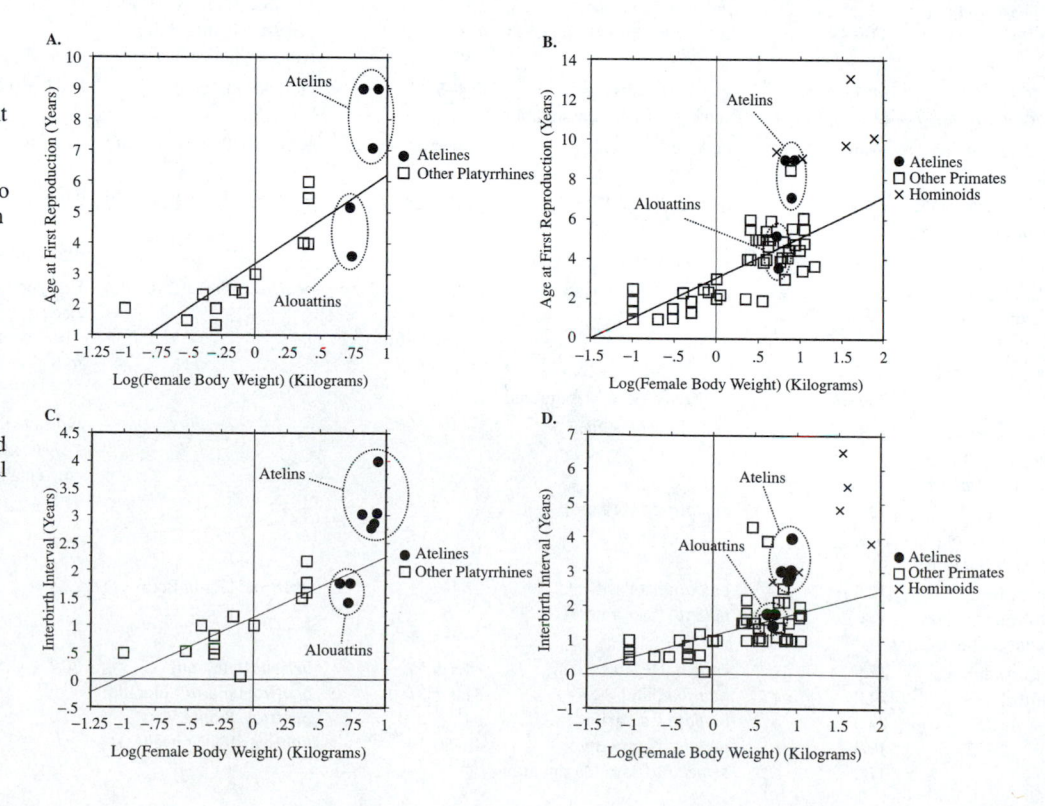

might be predicted on the basis of body size, and these features of life history may help explain their widespread distribution and persistence in regenerating secondary habitats. While female howler monkeys may begin breeding even before they reach 4 years of age and thereafter produce offspring at 1.5- to 2-year intervals, female atelins begin breeding much later and typically have interbirth intervals of 3 years or more; woolly monkey and muriqui females, for example, do not typically begin breeding until about age 9 (Table 10.7). For all of the atelins, too, realized female fertility over the reproductive life span is low.

CONSERVATION

Throughout their geographic ranges, atelines are the New World monkeys most susceptible to the negative impacts of anthropogenic activities. Atelines are the preferred and most common mammals in the diets of many indigenous groups in South America and are among the first primates to go locally extinct in the face of human hunting pressure (Peres 1990, 1991; Mena et al. 2000). In fact, throughout the neotropics, the number of ateline primates killed by subsistence hunting alone is staggering. For example, in lowland Ecuador, Yost and Kelley (1983) recorded 562 woolly monkeys and 146 howler monkeys being killed in a 275-day period by hunters in three villages of indigenous Huaorani. Similarly, Peres (1991) noted that during an 18-month period

in the mid-1980s a single family of rubber tappers in rural Brazil killed about 200 woolly monkeys, 100 spider monkeys, and 80 howlers; and Mena et al. (2000) reported that 395 woolly monkeys, 85 howler monkeys, and 10 spider monkeys were killed during an 11-month period in yet another Huaorani community in Ecuador in the mid-1990s. On a regional scale, Peres (2000) has estimated that 1.1–2.6 million individual ateline primates (corresponding to a biomass of 6,812–16,710 tons) are harvested each year by the low-income rural population of the Brazilian Amazon.

Such a large harvest of ateline primates is clearly not sustainable and can dramatically influence primate community structure in local areas. In a study, in part, of the density of ateline primates at 25 western Amazonian sites subject to varying degrees of hunting pressure, Peres (2000) found that population densities were nearly 10 times lower at sites subject to moderate or heavy hunting pressure compared to sites where little or no hunting occurred. Combined with their susceptibility to hunting, the slow reproductive rates of atelines can make it difficult for populations to recover even after hunting in a local area has been curtailed. Tellingly, two of the world's 25 most endangered primates identified in 2000 by Conservation International, the yellow-tailed woolly monkey, *Oreonax (Lagothrix) flavicauda*, and the northern muriqui, *B. hypoxanthus*, are atelines (Mittermeier et al. 2000); and the northern muriqui was listed again in 2002 (Konstant et al. 2002). Moreover, the majority of the ateline species listed in Table 10.1 for which data are

Table 10.8   Socioecological Convergence of Atelines and African Apes

	ALOUATTA AND *GORILLA GORILLA*	*ATELES* AND *PAN TROGLODYTES*	*BRACHYTELES* AND *PAN PANISCUS*
Ecological strategy	Largely folivorous	Frugivorous, focus on ripe fruits	Both frugivorous and folivorous, but both taxa rely heavily on plant vegetative parts at times
Dispersal pattern	Dispersal by both females and males	Dispersal by females	Dispersal by females
Social organization	Unimale, age-graded male, and multimale social groups	Fission–fusion communities	Both cohesive social groups and fission–fusion communities in muriquis, only the latter reported for bonobos
Male–male relations	Intense male competition over group membership and thus breeding opportunities, age-related male dominance hierarchies	Strong male–male affiliation in both taxa, male dominance hierarchies present among chimpanzees but unknown among spider monkeys	Tolerant to affiliative relations among within-group males, no dominance hierarchies among males
Infanticide	Reported	Reported in chimpanzees, not reported in spider monkeys	Not reported for either taxon
Male intergroup relations	Resident males aggressive toward males in other groups and solitaries	Males cooperate in patrolling and defense of community range against other groups of males	Male muriquis cooperate against extragroup males, unknown for bonobos
Male–female dominance	Males dominant to females	Males dominant to females	Males and females are codominant
Ranging patterns	Groups are cohesive, sexes range together	Average subgroup size is small, individual females utilize different core areas within community range, male ranges larger and up to size of community range, males and females often range separately	Subgroup size is large, individual females do not have own core areas, societies not sex-segregated
Mating patterns	Male mating success is largely dominance-based	Consortships reported in both species, some promiscuous and dominance-based mating in chimpanzees	Promiscuous matings and tolerant male–male relationships

available are considered vulnerable, endangered, or critically endangered by the World Conservation Union (2004), and all are included in either Appendix I or II of the Convention on International Trade in Endangered Species of Wild Flora and Fauna, IUCN 2004.

SOCIOECOLOGICAL CONVERGENCE: ATELINES AND THE AFRICAN GREAT APES

One of the interesting aspects about ateline primates is their convergence with some of the African great apes (subfamily Homininae) in various aspects of their morphology, ecology, social organization, life history, and behavior. For example, as discussed above, a high level of female dispersal and at least some degree of male philopatry characterize all atelines; and these features likewise characterize all hominines (see Chapters 18 and 19). Similarly, both atelines and African great apes show marked morphological adaptations for more upright, suspensory locomotion and feeding postures, including a shortened lumbar portion of the vertebral column, more dorsally placed scapulae, and shoulder joints capable of full rotation. Finally, at least some members of both clades take to extremes the primate tendency of having "slow" life histories relative to other mammals, with long periods of juvenile development, late age at first reproduction, long interbirth intervals, and low realized lifetime fertility.

Several specific examples further illustrate the marked socioecological convergences between atelines and distantly related hominines. The first of these highlights the similarities in foraging ecology, social structure, and mating system between the various howler monkey species (*Alouatta*) and gorillas (*Gorilla gorilla*); the second of these considers the convergence of spider monkeys (*Ateles*) and muriquis (*Brachyteles*) with common chimpanzees (*Pan troglodytes*) and bonobos (*Pan paniscus*), respectively, in light of contemporary socioecological theory.

Howler monkeys (*Alouatta* spp.) and gorillas (*G. gorilla*) are more folivorous than their closest relatives and, depending on the species, subspecies, or population being considered, may live in groups containing one or multiple reproductive-age males. Red, brown, and red-and-black howler monkeys (*A. seniculus*, *A. caraya*, and *A. guariba*), like mountain gorillas (*G.g. beringei*), typically live in groups containing one or a few adult males; and group membership is essential for male reproduction. Competition among males for breeding positions is strong, and infanticide has been reported when a new male enters a group of females (*Alouatta*, Crockett and Sekulic 1984, Zunino et al. 1986, Crockett and Rudran 1987b, Rumiz 1990, Galetti et al. 1994, Agoramoorthy and Rudran 1995, Calegaro-Marques and Bicca-Marques 1996; *Gorilla*, Fossey 1984, Watts 1989). Within groups of mantled howlers (*A. palliata*) and western lowland gorillas (*G.g. gorilla*) (and in some mountain gorilla groups), there may be multiple breeding males and male reproductive suc-

cess is likely to be correlated with male competitive ability. Additionally, in both *Alouatta* and *Gorilla*, intrasexual relationships tend to be weak and adult males appear to be the primary focus of social attention from other group members.

Spider monkeys (*Ateles* spp.) and chimpanzees (*Pan troglodytes*) are ripe fruit specialists that typically live in fission–fusion social systems characterized by the formation of subgroups whose composition is relatively fluid (*Ateles*, Klein 1972, Cant 1977, van Roosmalen 1985, McFarland 1986, Symington 1990, Chapman 1990a; *Pan*, Nishida 1968, Wrangham 1977, Wrangham and Smuts 1980, Goodall 1986). For both taxa, subgroup and feeding party size closely track habitat-wide measures of the availability of ripe fruits (Symington 1987a, 1988c; Chapman et al. 1995), suggesting that fission–fusion sociality may represent an adaptation for reducing the level of intragroup feeding competition experienced during times of scarcity. Moreover, in both taxa, the sexes often range separately (Wrangham and Smuts 1980, Symington 1988a, Fedigan et al. 1988, Chapman 1990a, Nunes 1995); and males in each taxon are affiliative and cooperative with one another (Nishida 1979, Fedigan and Baxter 1984, Goodall 1986, Watts 1998), commonly direct high levels of aggression toward females (Goodall 1986, Campbell 2002), and join with other males to patrol community boundaries and defend these areas against males from other groups (Bygott 1979, Watts and Mitani 2001). Additionally, mating strategies of both chimpanzees and spider monkeys include the formation of consortships (Tutin 1979, van Roosmalen 1985, Symington 1987a, Goodall 1986, Campbell 2000), although the overt promiscuity seen occasionally in chimpanzees is not reported for *Ateles*. Both within and across periovulatory periods, females of both species mate with multiple males; however, copulation usually does not take place in the presence of other adult males in spider monkeys (Campbell 2000). Chimpanzee males may become very possessive of copulatory partners close to the presumed time of ovulation (Tutin 1979).

In contrast, muriquis (*Brachyteles* spp.) and bonobos (*Pan paniscus*) feed more on leafy material found in larger patches. Probably as a result of the more uniform distribution of their food, the social system of these two primate taxa differs subtly from that of *Ateles* and *P. troglodytes*. While both bonobos and muriquis may be found in fission–fusion societies, at least in some populations (bonobos, Kuroda 1979, Kano 1982, Badrian and Badrian 1984, Nishida and Hiraiwa-Hasegawa 1987, White 1986, 1988; muriquis, Milton 1984a, Strier et al. 1993), subgroup size tends to be much larger than that seen in comparably sized communities of spider monkeys or chimpanzees (Dias and Strier 2003). Social life in these two species is likewise dramatically different from that of chimpanzees and spider monkeys. Neither muriqui nor bonobo society is considered sex-segregated to the extent seen in spider monkeys and chimpanzees: females do not typically maintain individual core areas within the community range, and foraging parties are far more likely to contain individuals of both sexes. Additionally, relationships between males and females are more egalitarian, and aggressive behaviors, whether within or between the sexes, are observed only rarely (bonobos, Furuichi 1997; muriquis, Strier 1992c). Finally, sexual behavior in both taxa is often overtly promiscuous and often takes place in the presence of other adult males (Milton 1985a,b, Furuichi 1992).

Convergences such as these are clearly interesting as they imply that similar ecological pressures have resulted in the evolution of similar social systems and similar mating patterns in widely divergent taxa. Nonetheless, it is important to stress that the atelines should not be thought of as simply New World equivalents of the African great apes (nor, for that matter, should the African apes be thought of as merely "Old World atelines"). Rather, the atelines represent a unique radiation of primates that fill important roles in the ecology of contemporary neotropical forests and that have social lives we are only just beginning to understand.

REFERENCES

Agoramoorthy, G., and Lohmann, R. (1999). Population and conservation status of the black-and-gold howler monkeys, *Alouatta caraya*, along the Rio Riachuelo, Argentina. *Neotrop. Primates* 7:43–44.

Agoramoorthy, G., and Rudran, R. (1993). Male dispersal among free-ranging howler monkeys (*Alouatta seniculus*) in Venezuela. *Folia Primatol.* 61:92–96.

Agoramoorthy, G., and Rudran, R. (1995). Infanticide by adult and subadult males in free-ranging red howler monkeys, *Alouatta seniculus*, in Venezuela. *Ethology* 99:75–88.

Aguirre, A. C. (1971). *O Mono* Brachyteles arachnoides. Academia Brasiliera de Ciencias, Rio de Janeiro.

Ahumada, J. A. (1989). Behavior and social structure of free ranging spider monkeys (*Ateles belzebuth*) in La Macarena. *Field Studies New World Monkeys La Macarena Colombia* 2:7–31.

Badrian, A., and Badrian, N. (1984). Social organization of *Pan paniscus* in the Lomako Forest, Zaire. In: Susman, R. L. (ed.), *The Pygmy Chimpanzee: Evolutionary Biology and Ecology.* Plenum Press, New York. pp. 325–346.

Baldwin, J. D., and Baldwin, J. I. (1972). Population density and use of space by howling monkeys (*Alouatta villosa*) in southwestern Panama. *Primates* 13:371–379.

Baldwin, J. D., and Baldwin, J. I. (1976). Primate populations in Chiriqui, Panama. In: Thorington, R. W., Jr., and Heltne, P. G. (eds.), *Neotropical Primates: Field Studies and Conservation.* National Academy of Sciences, Washington DC. pp. 20–31.

Bicca-Marques, J. C. (1993). Padrão de atividades diárias do bugio-preto *Alouatta caraya* (Primates: Cebidae): uma análise temporal e bioenergética. *Primatol. Brasil* 4:35–49.

Bicca-Marques, J. C. (1994). Padrão de utilização de uma ilha de mata por *Alouatta caraya* (Primates: Cebidae). *Rev. Brasil. Biol.* 54:161–171.

Bicca-Marques, J. C., and Calegaro-Marques, C. (1994). Feeding behavior of the black howler monkey (*Alouatta caraya*) in a seminatural forest. *Acta Biol. Leopold.* 16:69–84.

Bolin, I. (1981). Male parental behavior in black howler monkeys (*Alouatta palliata pigra*) in Belize and Guatemala. *Primates* 22:349–360.

Bonvicino, C. R. (1989). Ecologia e comportamento de *Alouatta belzebul* (Primates: Cebidae) na mata Atlântica. *Rev. Nordest. Biol.* 6:149–179.

Bravo, S. P., and Sallenave, A. (2003). Foraging behavior and activity patterns of *Alouatta caraya* in the northeastern Argentinean flooded forest. *Int. J. Primatol.* 24:825–846.

Braza, F., Alvarez, F., and Azcarate, T. (1981). Behaviour of the red howler monkey (*Alouatta seniculus*) in the llanos of Venezuela. *Primates* 22:459–473.

Braza, F., Alvarez, F., and Azcarate, T. (1983). Feeding habits of the red howler monkey (*Alouattta seniculus*) in the llanos of Venezuela. *Mammalia* 42:205–215.

Brockett, R. C., Horwich, R. H., and Jones, C. B. (1999). Disappearance of infants following male takeovers in the Belizean black howler monkey (*Alouatta pigra*). *Neotrop. Primates* 7:86–88.

Brockett, R. C., Horwich, R., and Jones, C. B. (2000a). Female dispersal in the Belizean black howling monkey (*Alouatta pigra*). *Neotrop. Primates* 8:32–34.

Brockett, R. C., Horwich, R. H., and Jones, C. B. (2000b). Reproductive seasonality in the Belizean black howling monkey (*Alouatta pigra*). *Neotrop. Primates* 8:136–138.

Bygott, J. D. (1979). Agonistic behavior, dominance, and social structure in wild chimpanzees of the Gombe National Park. In: Hamburg, D. A., and McCown, E. R. (eds.), *The Great Apes.* Benjamin/Cummings, Menlo Park, CA. pp. 405–427.

Calegaro-Marques, C., and Bicca-Marques, J. C. (1993). Reprodução de *Alouatta caraya* Humboldt, 1812 (Primates, Cebidae). *Primatol. Brasil* 4:51–66.

Calegaro-Marques, C., and Bicca-Marques, J. C. (1996). Emigration in a black howling monkey group. *Int. J. Primatol.* 17:229–237.

Campbell, C. J. (2000). The reproductive biology of black-handed spider monkeys (*Ateles geoffroyi*): integrating behavior and endocrinology [PhD thesis]. University of California, Berkeley.

Campbell, C. J. (2002). The influence of a large home range on the social structure of free ranging spider monkeys (*Ateles geoffroyi*) on Barro Colorado Island, Panama. *Am. J. Phys. Anthropol.* S34:51–52.

Campbell, C. J. (2003). Female-directed aggression in free-ranging *Ateles geoffroyi*. *Int. J. Primatol.* 24:223–237.

Campbell, C. J. (2004). Patterns of behavior across reproductive states of free-ranging female black-handed spider monkeys (*Ateles geoffroyi*). *Am. J. Phys. Anthropol.* 124:166–176.

Canavez, F. C., Moreira, M. A. M., Ladasky, J. J., Pissinatti, A., Parham, P., and Seuánez, H. N. (1999). Molecular phylogeny of New World primates (Platyrrhini) based on β_2–microglobulin DNA sequences. *Mol. Phylogenet. Evol.* 12:74–82.

Cant, J. G. H. (1977). Ecology, locomotion, and social organization of spider monkeys (*Ateles geoffroyi*) [PhD thesis]. University of California, Davis.

Cant, J. G. H. (1990). Feeding ecology of spider monkeys (*Ateles geoffroyi*) at Tikal, Guatemala. *Hum. Evol.* 5:269–281.

Cant, J. G. H., Youlatos, D., and Rose, M. D. (2001). Locomotor behavior of *Lagothrix lagothricha* and *Ateles belzebuth* in Yasuní National Park, Ecuador: general patterns and nonsuspensory modes. *J. Hum. Evol.* 41:141–166.

Cant, J. G. H., Youlatos, D., and Rose, M. D. (2003). Suspensory locomotion of *Lagothrix lagothricha* and *Ateles belzebuth* in Yasuní National Park, Ecuador. *J. Hum. Evol.* 44:685–699.

Carpenter, C. R. (1934). A field study of the behavior and social relations of howling monkeys. *Comp. Psych. Monogl.* 10:1–168.

Carpenter, C. R. (1965). The howlers of Barro Colorado Island. In: DeVore, I. (ed.), *Primate Behavior: Field Studies of Monkeys and Apes.* Holt, Rinehart, and Winston, New York. pp. 250–291.

Castellanos, H. G. (1995). Feeding behaviour of *Ateles belzebuth* E. Geoffroy, 1806 (Cebidac: Atelinae) in Tawadu Forest Southern Venezuela [D. Phil. thesis] University of Exeter, UK.

Castellanos, H. G., and Chanin, P. (1996). Seasonal differences in food choice and patch preference of long-haired spider monkeys (*Ateles belzebuth*). In: Norconk, M. A., Rosenberger, A. L., and Garber, P. A. (eds.), *Adaptive Radiations of Neotropical Primates.* Plenum Press, New York. pp. 451–466.

Chapman, C. (1987). Flexibility in diets of three species of Costa Rican primates. *Folia Primatol.* 49:90–105.

Chapman, C. (1988). Patterns of foraging and range use by three species of neotropical primates. *Primates* 29:177–194.

Chapman, C. A. (1990a). Association patterns of spider monkeys: the influence of ecology and sex on social organization. *Behav. Ecol. Sociobiol.* 26:409–414.

Chapman, C. A. (1990b). Ecological constraints on group size in three species of neotropical primates. *Folia Primatol.* 55:1–9.

Chapman, C. A., and Balcomb, S. R. (1998). Population characteristics of howlers: ecological conditions or group history. *Int. J. Primatol.* 19:385–403.

Chapman, C. A., Chapman, L. J., and McLaughlin, R. L. (1989a). Multiple central place foraging by spider monkeys: travel consequences of using many sleeping sites. *Oecologia* 79:506–511.

Chapman, C. A., Fedigan, L. M., Fedigan, L., and Chapman, L. J. (1989b). Post-weaning resource competition and sex ratios in spider monkeys. *Oikos* 54:315–319.

Chapman, C. A., Wrangham, R. W., and Chapman, L. J. (1995). Ecological constraints on group size: an analysis of spider monkey and chimpanzee subgroups. *Behav. Ecol. Sociobiol.* 36:59–70.

Chiarello, A. G. (1993a). Activity pattern of the brown howler monkey, *Alouatta fusca*, Geoffroy 1812, in a forest fragment of southeastern Brazil. *Primates* 34:289–293.

Chiarello, A. G. (1993b). Home range of the brown howler monkey, *Alouatta fusca*, in a forest fragment of southeastern Brazil. *Folia Primatol.* 60:173–175.

Chiarello, A. G. (1994). Diet of the brown howler monkey *Alouatta fusca* in semi-deciduous forest fragment of southeastern Brazil. *Primates* 35:25–34.

Chiarello, A. G. (1995a). Grooming in brown howler monkeys, *Alouatta fusca. Am. J. Primatol.* 35:73–81.

Chiarello, A. G. (1995b). Role of loud calls in brown howlers, *Alouatta fusca. Am. J. Primatol.* 36:213–222.

Chivers, D. J. (1969). On the daily behavior and spacing of howling monkey groups. *Folia Primatol.* 10:48–102.

Clarke, M. R. (1983). Infant-killing and infant disappearance following male takeovers in a group of free-ranging howling monkeys (*Alouatta palliata*) in Costa Rica. *Am. J. Primatol.* 5:241–247.

Clarke, M. R., Crockett, C. M., Zucker, E. L., and Zaldivar, M. (2002). Mantled howler population of Hacienda la Pacifica, Costa Rica, between 1991 and 1998: effects of deforestation. *Am. J. Primatol.* 56:155–163.

Clarke, M. R., and Glander, K. E. (1984). Female reproductive success in a group of free-ranging howling monkeys (*Alouatta palliata*) in Costa Rica. In: Small, M. F. (ed.), *Female Primates: Studies by Women Primatologists.* Alan R. Liss, New York. pp. 111–126.

Clarke, M. R., Glander, K. E., and Zucker, E. L. (1998). Infant–nonmother interactions of free-ranging mantled howlers (*Alouatta palliata*) in Costa Rica. *Int. J. Primatol.* 19:451–472.

Clarke, M. R., and Zucker, E. L. (1994). Survey of the howling monkey population at La Pacifica: a seven-year follow-up. *Int. J. Primatol.* 15:61–73.

Clarke, M. R., Zucker, E. L., and Glander, K. E. (1994). Group takeover by a natal male howling monkey (*Alouatta palliata*) and associated disappearance and injuries of immatures. *Primates* 35:435–442.

Clarke, M. R., Zucker, E. L., and Scott, N. J., Jr. (1986). Population trends of the mantled howler groups of La Pacifica, Guanacaste, Costa Rica. *Am. J. Primatol.* 11:79–88.

Coehlo, A. M., Jr., Bramblett, C. A., Quick, L. B., and Bramblett, S. S. (1976a). Resource availability and population density in primates: a socio-bioenergetic analysis of energy budgets of Guatemalan howler and spider monkeys. *Primates* 17:63–80.

Coehlo, A. M., Jr., Coehlo, L. S., Bramblett, C. A., Bramblett, S. S., and Quick, L. B. (1976b). Ecology, population characteristics, and sympatric association in primates: a socio-bioenergetic analysis of howler and spider monkeys in Tikal, Guatemala. *Ybk. Phys. Anthropol.* 20:96–135.

Coles, R., and Talebi, M. G. (2004). Spatial relations in wild southern muriquis (*Brachyteles arachnoides*): choices of nearest neighbour and proximity. *Folia Primatol.* 75 (suppl. 1):365–366.

Collias, N., and Southwick, C. (1952). A field study of population density and social organization in howling monkeys. *Proc. Am. Phil. Soc.* 96:143–156.

Collins, A. C. (2001). The importance of sampling for reliable assessment of phylogenetics and conservation among neotropical primates: a case study in spider monkeys (*Ateles*). *Primate Rep.* 61:9–30.

Collins, A. C. (2003). Atelinae phylogenetic relationships: the trichotomy revisited? *Am. J. Phys. Anthropol.* S36:78.

Collins, A. C., and Dubach, J. M. (2000a). Biogeographic and ecological forces responsible for speciation in *Ateles. Int. J. Primatol.* 21:421–444.

Collins, A. C., and Dubach, J. M. (2000b). Phylogenetic relationships of spider monkeys (*Ateles*) based on mitochondrial DNA variation. *Int. J. Primatol.* 21:381–420.

Cortés-Ortiz, L., Bermingham, E., Rico, C., Rodríguez-Luna, E., Sampaio, I., and Ruiz-García, M. (2003). Molecular systematics and biogeography of the neotropical monkey genus, *Alouatta. Mol. Phylogenet. Evol.* 26:64–81.

Crissey, S. D., Oftedal, O. T., Currier, J. A., and Rudran, R. (1990). Gastro-intestinal tract capacity, food passage rates and the possible role of fiber in diets fed to captive red howler monkeys (*Alouatta seniculus*) in Venezuela. *Am. Assoc. Zoo Vet. Annu. Proc.* 1990:81–86.

Crockett, C. M. (1984). Emigration by female red howler monkeys and the case for female competition. In: Small, M. F. (ed.), *Female Primates: Studies by Women Primatologists.* Alan R. Liss, New York. pp. 159–173.

Crockett, C. M. (1996). The relation between red howler monkey (*Alouatta seniculus*) troop size and population growth in two habitats. In: Norconk, M. A., Rosenberger, A. L., and Garber, P. A. (ed.), *Adaptive Radiations of Neotropical Primates*. Plenum, New York. pp. 489–510.

Crockett, C. M., and Eisenberg, J. F. (1987). Howlers: variation in group size and demography. In: Smuts, B. B., Cheney, D. L., Seyfarth, R. M., Wrangham, R. W., and Struhsaker, T. T. (eds.), *Primate Societies*. University of Chicago Press, Chicago. pp. 54–68.

Crockett, C. M., and Pope, T. R. (1988). Inferring patterns of aggression from red howler monkey injuries. *Am. J. Primatol.* 14:1–21.

Crockett, C. M., and Pope, T. R. (1993). Consequences of sex differences in dispersal for juvenile red howler monkeys. In: Pereira, M. E., and Fairbanks, L. A. (eds.), *Juvenile Primates: Life History, Development, and Behavior*. Oxford University Press, New York. pp. 104–118.

Crockett, C. M., and Rudran, R. (1987a). Red howler monkey birth data. I: Seasonal variation. *Am. J. Primatol.* 13:347–368.

Crockett, C. M., and Rudran, R. (1987b). Red howler monkey birth data. II: Interannual, habitat, and sex comparisons. *Am. J. Primatol.* 13:369–384.

Crockett, C. M., and Sekulic, R. (1982). Gestation length in red howler monkeys. *Am. J. Primatol.* 3:291–294.

Crockett, C. M., and Sekulic, R. (1984). Infanticide in red howler monkeys (*Alouatta seniculus*). In: Hausfater, G., and Hrdy, S. B. (eds.), *Infanticide: Comparative and Evolutionary Perspectives*. Aldine, New York. pp. 173–191.

da Silva, E. C., Jr. (1981). A preliminary survey of brown howler monkeys (*Alouatta fusca*) at the Catareira Reserve, São Paulo, Brazil. *Rev. Brasil. Biol.* 41:897–909.

Defler, T. R. (1981). The density of *Alouatta seniculus* in the eastern llanos of Colombia. *Primates* 22:564–569.

Defler, T. R. (1989). Recorrido y uso del espacio en un grupo de *Lagothrix lagotricha* (Primates: Cebidae) mono lanudo churuco en la Amazonia Colombiana. *Trianea* 3:183–205.

Defler, T. R. (1995). The time budget of a group of wild woolly monkeys (*Lagothrix lagotricha*). *Int. J. Primatol.* 16:107–120.

Defler, T. R. (1996). Aspects of the ranging pattern in a group of wild woolly monkeys (*Lagothrix lagotricha*). *Am. J. Primatol.* 38:289–302.

Defler, T. R. (1999). Locomotion and posture in *Lagothrix lagotricha*. *Folia Primatol.* 70:313–327.

Defler, T. R., and Defler, S. B. (1996). Diet of a group of *Lagothrix lagothricha lagothricha* in southeastern Colombia. *Int. J. Primatol.* 17:161–190.

de Marques, A. A. B. (1995). O Bugio-Ruivo *Alouatta fusca clamitans* (Cabrera, 1940) (Primates: Cebidae) na Estação Ecológica de Aracuri, RS; Variações Sazonias de Forregeamento [MSc thesis]. Pontifícia Universidade Católica do Rio Grande do Sul, Porto Alegre.

de Marques, A. A. B. (1996). Ecology and behavior of brown howlers in Araucaria Pine Forest, southern Brazil. *Neotrop. Primates* 4:90–91.

de Moraes, P. L. R., Oswaldo de Carvalho, J., and Strier, K. B. (1998). Population variation in patch and party size in muriquis (*Brachyteles arachnoides*). *Int. J. Primatol.* 19:325–337.

Dew, J. L. (2001). Synecology and seed dispersal in woolly monkeys (*Lagothrix lagotricha poeppigii*) and spider monkeys (*Ateles belzebuth belzebuth*) in Parque Nacional Yasuní, Ecuador [PhD thesis]. University of California, Davis.

Dias, L. G., and Strier, K. B. (2003). Effects of group size on ranging patterns in *Brachyteles arachnoides hypoxanthus*. *Int. J. Primatol.* 24:209–221.

Di Fiore, A. (1997). Ecology and behavior of lowland woolly monkeys (*Lagothrix lagotricha poeppigii*, Atelinae) in eastern Ecuador [PhD thesis]. University of California, Davis.

Di Fiore, A. (2002). Molecular perspectives on dispersal in lowland woolly monkeys (*Lagothrix lagotricha poeppigii*). *Am. J. Phys. Anthropol.* S34:63.

Di Fiore, A. (2003a). Molecular genetic approaches to the study of primate behavior, social organization, and reproduction. *Ybk. Phys. Anthropol.* 46:62–99.

Di Fiore, A. (2003b). Ranging behavior and foraging ecology of lowland woolly monkeys (*Lagothrix lagotricha poeppigii*) in Yasuní National Park, Ecuador. *Am. J. Primatol.* 59:47–66.

Di Fiore, A. (2004). Diet and feeding ecology of woolly monkeys in a western Amazonian rain forest. *Int. J. Primatol.* 24:767–801.

Di Fiore, A., and Fleischer, R. C. (2005). Social behavior, reproductive strategies, and population genetic structure of *Lagothrix lagotricha poeppigii*. *Int. J. Primatol.* 26:1137–1173.

Di Fiore, A., and Rodman, P. S. (2001). Time allocation patterns of lowland woolly monkeys (*Lagothrix lagotricha poeppigii*) in a neotropical terra firma forest. *Int. J. Primatol.* 22:449–480.

Di Fiore, A., and Strier, K. B. (2004). Flexibility in social organisation in ateline primates. *Folia Primatol.* 75 (suppl. 1):140–141.

Dunbar, R. I. M. (1988). *Primate Social Systems*. Cornell University Press, Ithaca, NY.

Eisenberg, J. (1973). Reproduction in two species of spider monkeys, *Ateles fusciceps* and *Ateles geoffroyi*. *J. Mammal.* 54:955–957.

Eisenberg, J. F., and Kuehn, R. E. (1966). The behavior of *Ateles geoffffroyi* and related species. *Smithsonian Misc. Coll.* 151:1–63.

Erickson, G. E. (1963). Brachiation in New World monkeys and in anthropoid apes. *Symp. Zool. Soc. Lond.* 10:135–164.

Estrada, A. (1982). Survey and census of howler monkeys (*Alouatta palliata*) in the rain forest of "Los Tuxtlas," Veracruz, Mexico. *Am. J. Primatol.* 2:363–372.

Estrada, A. (1984). Resource use by howler monkeys (*Alouatta palliata*) in the rain forest of Los Tuxtlas, Veracruz, Mexico. *Int. J. Primatol.* 5:105–131.

Estrada, A., Castellanos, L., Garcia, Y., Franco, B., Muñoz, D., Ibarra, A., Rivera, A., Fuentes, E., and Jimenez, C. (2002a). Survey of the black howler monkey, *Alouatta pigra*, population at the Mayan site of Palenque, Chiapas, Mexico. *Primates* 43:51–58.

Estrada, A., Juan-Solano, S., Ortíz Martínez, T., and Coates-Estrada, R. (1999). Feeding and general activity patterns of a howler monkey (*Alouatta palliata*) troop living in a forest fragment at Los Tuxtlas, Mexico. *Am. J. Primatol.* 48:167–183.

Estrada, A., Luecke, L., van Belle, S., French, K., Muñoz, D., García, Y., Castellanos, L., and Mendoza, A. (2002b). The black howler monkey (*Alouatta pigra*) and spider monkey (*Ateles geoffroyi*) in the Mayan site of Yaxchilán, Chiapas, Mexico: a preliminary survey. *Neotrop. Primates* 10:89–95.

Estrada, A., Luecke, L., van Belle, S., Barrueta, E., and Meda, M. R. (2004). Survey of black howler (*Alouatta pigra*) and spider (*Ateles geoffroyi*) monkeys in the Mayan sites of Calakmul and Yaxchilán, Mexico and Tikal, Guatemala. *Primates* 45:33–39.

Fedigan, L. M. (1986). Demographic trends in the *Alouatta palliata* and *Cebus capucinus* populations of Santa Rosa National Park,

Costa Rica. In: Else, J. G., and Lee, P. C. (eds.), *Primate Ecology and Conservation*. Cambridge University Press, Cambridge. pp. 285–293.

Fedigan, L. M., and Baxter, M. J. (1984). Sex differences and social organization in free-ranging spider monkeys (*Ateles geoffroyi*). *Primates* 25:279–294.

Fedigan, L. M., Fedigan, L., and Chapman, C. (1985). A census of *Alouatta palliata* and *Cebus capucinus* in Santa Rosa National Park, Costa Rica. *Brenesia* 23:309–322.

Fedigan, L. M., Fedigan, L., Chapman, C., and Glander, K. E. (1988). Spider monkey home ranges: a comparison of telemetry and direct observation. *Am. J. Primatol.* 16:19–29.

Fedigan, L. M., and Jack, K. (2001). Neotropical primates in a regenerating Costa Rican dry forest: a comparison of howler and capuchin population patterns. *Int. J. Primatol.* 22:689–713.

Fedigan, L. M., and Rose, L. M. (1995). Interbirth interval variation in three sympatric species of neotropical monkey. *Am. J. Primatol.* 37:9–24.

Fedigan, L. M., Rose, L. M., and Avila, R. M. (1998). Growth of mantled howler groups in a regenerating Costa Rican dry forest. *Int. J. Primatol.* 19:405–432.

Fooden J. (1963). A revision of the woolly monkeys (genus *Lagothrix*). *J. Mammal.* 44:213–247.

Ford, S. M. (1986). Systematics of the New World monkeys. In: Swindler, D. R., and Erwin, J. (eds.), *Systematics, Evolution, and Anatomy*. Alan R. Liss, New York. pp. 73–135.

Ford, S. M., and Davis, L. C. (1992). Systematics and body size: implications for feeding adaptations in New World monkeys. *Am. J. Phys. Anthropol.* 88:415–468.

Fossey, D. (1984). Infanticide in mountain gorillas (*Gorilla gorilla berengei*) with comparative notes on chimpanzees. In: Hausfater, G., and Hrdy, S. B. (eds.), *Infanticide: Comparative and Evolutionary Perspectives*. Aldine, Hawthorne, NY. pp. 217–235.

Freese, C. (1976). Censusing *Alouatta palliata*, *Ateles geoffroyi*, and *Cebus capucinus* in the Costa Rican dry forest. In: Thorington, R. W., Jr., and Heltne, P. G. (eds.), *Neotropical Primates: Field Studies and Conservation*. National Academy of Sciences, Washington DC. pp. 4–9.

Froelich, J. W., Thorington, R. W., and Otis, J. S. (1981). The demography of howler monkeys (*Alouatta palliata*) on Barro Colorado Island, Panamá. *Int. J. Primatol.* 2:207–236.

Furuichi, T. (1992). The prolonged oestrus of females and factors influencing mating in a wild group of bonobos (*Pan paniscus*) in Wamba, Zaire. In: Itoigawa, N., Sugiyama, Y., Sackett, G. P., and Thompson, R. K. R. (eds.), *Topics in Primatology*, vol. 2. University of Tokyo Press, Tokyo. pp. 179–190.

Furuichi, T. (1997). Agonistic interactions and matrifocal dominance rank of wild bonobos. *Int. J. Primatol.* 18:855–875.

Galetti, M., Pedroni, F., and Paschoal, M. (1994). Infanticide in the brown howler monkey, *Alouatta fusca*. *Neotrop. Primates* 2:6–7.

Gaulin, S. J. C., and Gaulin, C. K. (1982). Behavioral ecology of *Alouatta seniculus* in Andean cloud forest. *Int. J. Primatol.* 3:1–32.

Glander, K. E. (1978). Howling monkey feeding behavior and plant secondary compounds: a study of strategies. In: Montgomery, G. G. (ed.), *The Ecology of Arboreal Folivores*. Smithsonian Institution Press, Washington DC. pp. 561–573.

Glander, K. E. (1980). Reproduction and population growth in free-ranging mantled howling monkeys. *Am. J. Phys. Anthropol.* 53:25–36.

Glander, K. E. (1982). The impact of plant secondary compounds on primate feeding behavior. *Ybk. Phys. Anthropol.* 25:1–18.

Glander, K. E. (1992). Dispersal patterns in Costa Rican mantled howling monkeys. *Int. J. Primatol.* 13:415–436.

Gonzalez-Kirchner, J. P. (1998). Group size and population density of the black howler monkey (*Alouatta pigra*) in Muchukux Forest, Quintana Roo, Mexico. *Folia Primatol.* 9:260–265.

González-Solís, J., Guix, J. C., Mateos, E., and Lorrens, L. (2001). Population density of primates in a large fragment of the Brazilian Atlantic rainforest. *Biodiv. Conserv.* 10:1267–1282.

Goodall, J. (1986). *The Chimpanzees of Gombe: Patterns of Behavior*. Belknap Press, Cambridge, MA.

Goodman, M., Porter, C. A., Czelusniak, J., Page, S. L., Schneider, H., Shoshani, J., Gunnell, G., and Groves, C. P. (1998). Toward a phylogenetic classification of primates based on DNA evidence complemented by fossil evidence. *Mol. Phylogenet. Evol.* 9:585–598.

Groves, C. (2001). *Primate Taxonomy*. Smithsonian Institution Press, Washington DC.

Heltne, P. G., Turner, D. C., and Scott, N. J., Jr. (1976). Comparison of census data on *Alouatta palliata* from Costa Rica and Panama. In: Thorington, R. W., Jr., and Heltne, P. G. (eds.), *Neotropical Primates: Field Studies and Conservation*. National Academy of Sciences, Washington DC. pp. 10–19.

Hernandez-Camacho, J., and Defler, T. R. (1985). Some aspects of the conservation of non-human primates in Colombia. *Primate Conserv.* 42–50.

Hill, W. C. O. (1962). *Primates: Comparative Taxonomy and Anatomy V: Cebidae B*. Edinburgh University Publications, Edinburgh.

Horwich, R., and Gebhard, K. (1983). Roaring rhythms in black howler monkeys (*Alouatta pigra*) of Belize. *Primates* 24:290–296.

Horwich, R. H., Brockett, R. C., and Jones, C. B. (2000). Alternative male reproductive behaviors in the Belizean black howler monkey (*Alouatta pigra*). *Neotrop. Primates* 8:95–98.

Hrdy, S. B. (1977). Infanticide as a primate reproductive strategy. *Am. Sci.* 65:40–49.

Hrdy, S. B. (1979). Infanticide among animals: a review, classification and examination of the implications for the reproductive strategies of females. *Ethol. Sociobiol.* 1:1–13.

Isbell, L. A. (1991). Contest and scramble competition: patterns of female aggression and ranging behavior among primates. *Behav. Ecol.* 2:143–155.

IUC 2004. 2004 IUCN Red List of Threatened SpeciesN. www.redlist.org.

Izawa, K. (1976). Group sizes and compositions of monkeys in the upper Amazon Basin. *Primates* 17:367–399.

Izawa, K., and Nishimura, A. (1988). Primate fauna at the study site, La Macarena, Colombia. *Field Studies of New World Monkeys, La Macarena, Colombia* 1:5–11.

Janson, C. H., and Goldsmith, M. L. (1995). Predicting group size in primates: foraging costs and predation risks. *Behav. Ecol.* 6:326–336.

Jones, C. B. (1980a). The functions of status in the mantled howler monkey, *Alouatta palliata* Gray: intraspecific competition for group membership in a folivorous neotropical primate. *Primates* 21:389–405.

Jones, C. B. (1980b). Seasonal parturition, mortality, and dispersal in the mantled howler monkey, *Alouatta palliata* Gray. *Brenesia* 17:1–10.

Jones, C. B. (1981). The evolution and socioecology of dominance in primate groups: a theoretical formulation, classification, and assessment. *Primates* 22:70–83.

Jones, C. B. (1985). Reproductive patterns in mantled howler monkeys: estrus, mate choice and copulation. *Primates* 26:130–142.

Jones, C. B. (1996). Relative reproductive success in the mantled howler monkey: implications for conservation. *Neotrop. Primates* 4:21–23.

Julliot, C. (1996). Fruit choice by red howler monkeys (*Alouatta seniculus*) in a tropical forest. *Am. J. Primatol.* 40:261–282.

Julliot, C., and Sabatier, D. (1993). Diet of the red howler monkeys (*Alouatta seniculus*) in French Guiana. *Int. J. Primatol.* 14:527–550.

Kano, T. (1982). The social group of pygmy chimpanzees. *Primates* 23:171–188.

Karesh, W. B., Wallace, R. B., Painter, R. L. E., Rumiz, D., Braselton, W. E., Dierenfeld, E. S., and Puche, H. (1998). Immobilization and health assessment of free-ranging black spider monkeys (*Ateles paniscus chamek*). *Am. J. Primatol.* 44:107–123.

Kavanaugh, M., and Dresdale, L. (1975). Observations on the woolly monkey (*Lagothrix lagothricha*) in northern Colombia. *Primates* 16:285–294.

Kay, R. F. (1990). The phyletic relationships of extant and fossil Pitheciinae (Platyrrhini, Anthropoidea). *J. Hum. Evol.* 19:175–208.

Kinzey, W. G. (1997). *New World Primates: Ecology, Evolution, and Behavior*. Aldine de Gruyter, New York.

Klein, L. L. (1972). The ecology and social behavior of the spider monkey, *Ateles belzebuth* [PhD thesis]. University of California, Berkeley.

Klein, L., and Klein, D. (1971). Aspects of social behavior in a colony of spider monkeys *Ateles geoffroyi* at the San Francisco Zoo. *Int. Zoo Ybk.* 11:175–181.

Klein, L. L., and Klein, D. J. (1976). Neotropical primates: aspects of habitat usage, population density, and regional distribution in La Macarena, Colombia. In: Thorington, R. W., Jr., and Heltne, P. G. (eds.), *Neotropical Primates: Field Studies and Conservation*. National Academy of Sciences, Washington DC. pp. 70–78.

Klein, L. L., and Klein, D. J. (1977). Feeding behavior of the Colombian spider monkey, *Ateles belzebuth*. In: Clutton-Brock, T. H. (ed.), *Primate Ecology: Studies of Feeding and Ranging Behaviour in Lemurs, Monkeys, and Apes*. Academic Press, London. pp. 153–181.

Knopff, K. H., Knopff, A. R. A., and Pavelka, M. S. M. (2004). Observed case of infanticide committed by a resident male Central American black howler monkey (*Alouatta pigra*). *Am. J. Primatol.* 63:239–244.

Konstant, W. R., Mittermeier, R. A., Rylands, A. B., Butynski, T. M., Eudey, A. A., Ganzhorn, J., and Kormos, R. (2002). The world's 25 most endangered primates–2002. *Neotrop. Primates* 10:128–131.

Kowaleski, M., and Zunino, G. E. (2004). Birth seasonality in *Alouatta caraya* in Northern Argentina. *Int. J. Primatol.* 25:283–400.

Kuroda, S. (1979). Grouping of pygmy chimpanzees. *Primates* 20:161–183.

Lambert, J. E. (1998). Primate digestion: interactions among anatomy, physiology and feeding ecology. *Evol. Anthropol.* 7:8–20.

Lemos de Sá, R. M. (1988). Situação de uma população de Mono Carvoeiro, *Brachyteles arachnoides*, em um fragmento de Mata Atlântica (M.G.) e implicações para sua conservação [MSc thesis]. Universidade de Brasília, Brasília.

Lemos de Sá, R. M., and Glander, K. E. (1993). Capture techniques and morphometrics for the woolly spider monkey, or muriqui (*Brachyteles arachnoides* E. Geoffroy 1806). *Am. J. Primatol.* 29:145–153.

Lemos de Sá, R. M., Pope, T. R., Glander, K. E., Struhsaker, T. T., and da Fonseca, G. A. B. (1990). A pilot study of genetic and morphological variation in the muriqui (*Brachyteles arachnoides*). *Primate Conserv.* 11:26–30.

Leo Luna, M. (1980). First field study of the yellow-tailed woolly monkey. *Oryx* 15:386–389.

Leo Luna, M. (1982). Conservation of the yellow-tailed woolly monkey. *Int. Zoo Ybk.* 22:47–52.

Leo Luna, M. (1984). The effects of hunting, selective logging and clear-cutting on the conservation of the yellow-tailed woolly monkey (*Lagothrix flavicauda*) [MA thesis]. University of Florida, Gainesville.

Lippold, L. K. (1988). A census of primates in Cabo Blanco Absolute Nature Reserve, Costa Rica. *Brenesia* 29:101–105.

Lippold, L. K. (1989). Primates in Cabo Blanco Absolute Nature Reserve, Costa Rica. *Primate Conserv.* 10:23–25.

Lu, F. E. (1999). Changes in subsistence patterns and resource use of the Huaorani Indians in the Ecuadorian Amazon [PhD thesis]. University of North Carolina, Chapel Hill.

Martins, W. P., and Strier, K. B. (2004). Age at first reproduction in philopatric female muriquis (*Brachyteles arachnoides hypoxanthus*). *Primates* 45:63–67.

McFarland, M. J. (1986). Ecological determinants of fission–fusion sociality in *Ateles* and *Pan*. In: Else, J. G., and Lee, P. C. (eds.), *Primate Ecology and Conservation*. Cambridge University Press, Cambridge. pp. 181–190.

Meireles, C. M., Czelusniak, J., Schneider, M. P. C., Muniz, J. A. P. C., Brigido, M. C., Ferreira, H. S., and Goodman, M. (1999). Molecular phylogeny of ateline New World monkeys (Platyrrhini, Atelinae) based on β-globin gene sequences: evidence that *Brachyteles* is the sister group of *Lagothrix*. *Mol. Phylogenet. Evol.* 12:10–30.

Mena, P., Stallings, J. R., Regalado, B. J., and Cueva, L. R. (2000). The sustainability of hunting practices by the Huaorani. In: Robinson, J. G., and Bennett, E. L. (eds.), *Hunting for Sustainability in Tropical Forests*. Columbia University Press, New York. pp. 57–78.

Mendes, S. L. (1989). Estudo ecológico de *Alouatta fusca* (Primates: Cebidae) na Estaçao Biológica de Caratinga, MG. *Rev. Nordest. Biol.* 6:71–104.

Milton, K. (1979). Factors affecting leaf choice by howler monkeys: a test of some hypotheses of food selection by generalist herbivores. *Am. Nat.* 114:362–378.

Milton, K. (1980). *The Foraging Strategy of Howler Monkeys: A Study in Primate Economics*. Columbia University Press, New York.

Milton, K. (1981). Estimates of reproductive parameters for free-ranging *Ateles geoffroyi*. *Primates* 22:574–579.

Milton, K. (1982). Dietary quality and demographic regulation in a howler monkey population. In: Leigh, E. G., Jr., Rand, A. S., and Windsor, D. M. (eds.), *The Ecology of a Tropical Forest: Seasonal Rhythms and Long-Term Changes*. Smithsonian Institution Press, Washington DC. pp. 273–289.

Milton, K. (1984a). Habitat, diet, and activity patterns of free-ranging woolly spider monkeys (*Brachyteles arachnoides* E. Geoffroy 1806). *Int. J. Primatol.* 5:491–514.

Milton, K. (1984b). The role of food processing factors in primate food choice. In: Rodman, P. S., and Cant, J. G. H. (eds.), *Adaptations for Foraging in Nonhuman Primates: Contributions to an Organismal Biology of Prosimians, Monkeys and Apes.* Columbia University Press, New York. pp. 249–279.

Milton, K. (1985a). Mating patterns of woolly spider monkeys, *Brachyteles arachnoides*: implications for female choice. *Behav. Ecol. Sociobiol.* 17:53–59.

Milton, K. (1985b). Multimale mating and absence of canine tooth dimorphism in woolly spider monkeys (*Brachyteles arachnoides*). *Am. J. Phys. Anthropol.* 68:519–523.

Milton, K. (1987). Mating behaviors in woolly spider monkeys (*Brachyteles arachnoides*). *Int. J. Primatol.* 8:460.

Milton, K. (1996). Effects of bot fly (*Alouattamyia baeri*) parasitism on a free-ranging howler monkey (*Alouatta palliata*) population in Panama. *J. Zool. Lond.* 239:39–63.

Milton, K. (1998). Physiological ecology of howlers (*Alouatta*): energetic and digestive considerations and comparison with the Colobinae. *Int. J. Primatol.* 19:513–548.

Milton, K., and May, M. L. (1976). Body weight, diet, and home range area in primates. *Nature* 259:459–462.

Mittermeier, R. A. (1973). Group activity and population dynamics of the howler monkey on Barro Colorado Island. *Primates* 14:1–19.

Mittermeier, R. A., Konstant, W. R., and Rylands, A. B. (2000). The world's top 25 most endangered primates. *Neotrop. Primates* 8:49.

Mooney, J. C., and Lee, P. C. (1999). Reproductive parameters in captive woolly monkeys (*Lagothrix lagotricha*). *Zoo Biol.* 18:421–427.

Morin, P. A., Moore, J. J., Chakraborty, R., Jin, L., Goodall, J., and Woodruff, D. S. (1994). Kin selection, social structure, gene flow, and the evolution of chimpanzees. *Science* 265:1193–1201.

Neville, M. K. (1972a). The population structure of red howler monkeys (*Alouatta seniculus*) in Trinidad and Venezuela. *Folia Primatol.* 17:56–86.

Neville, M. K. (1972b). Social relations within troops of red howler monkeys (*Alouatta seniculus*). *Folia Primatol.* 18:47–77.

Neville, M. (1976). The population and conservation of howler monkeys in Venezuela and Trinidad. In: Thorington, R. W., Jr., and Heltne, P. G. (eds.), *Neotropical Primates: Field Studies and Conservation.* National Academy of Sciences, Washington DC. pp. 101–109.

Neville, M. K., Glander, K. E., Braza, F., and Rylands, A. B. (1988). The howling monkeys, genus *Alouatta.* In: Mittermeier, R. A., Rylands, A. B., Coimbra-Filho, A. F., and da Fonseca, G. A. B. (eds.), *Ecology and Behavior of Neotropical Primates,* vol 2. World Wildlife Fund, Washington DC. pp. 349–453.

Nishida, T. (1968). The social group of wild chimpanzees in the Mahale Mountains. *Primates* 9:167–224.

Nishida, T. (1979). The social structure of chimpanzees of the Mahale Mountains. In: Hamburg, D. A., and McCown, E. R. (eds.), *The Great Apes.* Benjamin/Cummings, Menlo Park, CA. pp. 73–121.

Nishida, T., and Hiraiwa-Hasegawa, M. (1987). Chimpanzees and bonobos: cooperative relationships among males. In: Smuts, B. B., Cheney, D. L., Seyfarth, R. M., Wrangham, R. W., and

Struhsaker, T. T. (eds.), *Primate Societies.* University of Chicago Press, Chicago. pp. 165–178.

Nishimura, A. (1988). Mating behavior of woolly monkeys, *Lagothrix lagotricha*, at La Macarena, Colombia. *Field Studies of New World Monkeys, La Macarena, Colombia* 1:19–27.

Nishimura, A. (1990a). A sociological and behavioral study of woolly monkeys, *Lagothrix lagotricha*, in the Upper Amazon. *Sci. Eng. Rev. Doshisha Univ.* 31:87–121.

Nishimura, A. (1990b). Mating behavior of woolly monkeys (*Lagothrix lagotricha*) at La Macarena, Colombia (II): mating relationships. *Field Studies of New World Monkeys, La Macarena, Colombia* 3:7–12.

Nishimura, A. (1994). Social interaction patterns of woolly monkeys (*Lagothrix lagotricha*): a comparison among the atelines. *Sci. Eng. Rev. Doshisha Univ.* 35:236–254.

Nishimura, A. (2003). Reproductive parameters of wild female *Lagothrix lagotricha*. *Int. J. Primatol.* 24:707–722.

Nishimura, A., Wilches, A. V., and Estrada, C. (1992). Mating behaviors of woolly monkeys, *Lagothrix lagotricha*, at La Macarena, Colombia (III): reproductive parameters viewed from a longterm study. *Field Studies of New World Monkeys, La Macarena, Colombia* 7:1–7.

Norconk, M. A., and Kinzey, W. G. (1994). Challenge of neotropical frugivory: travel patterns of spider monkeys and bearded sakis. *Am. J. Primatol.* 34:171–183.

Nunes, A. (1995). Foraging and ranging patterns in white-bellied spider monkeys. *Folia Primatol.* 65:85–99.

Nunes, A. (1998). Diet and feeding ecology of *Ateles belzebuth belzebuth* at Maracá Ecological Station, Roraima, Brazil. *Folia Primatol.* 69:61–76.

Nunn, C. L., and Barton, R. A. (2001). Comparative methods for studying primate adaptation and allometry. *Evol. Anthropol.* 10:81–98.

Ostro, L. E. T., Silver, S. C., Koontz, F. W., Horwich, R. H., and Brockett, R. (2001). Shifts in social structure of black howler (*Alouatta pigra*) groups associated with natural and experimental variation in population density. *Int. J. Primatol.* 22:733–748.

Paccagnella, S. G. (1991). Censo da população de monos (*Brachyteles arachnoides*) do Parque Estadual Carlos Botelho, Estado de São Paulo. *Primatol. Brasil* 3:225–233.

Pavelka, M. S. M., Brusselers, O. T., Nowak, D., and Behie, A. M. (2003). Population reduction and social disorganization in *Alouatta pigra* following a hurricane. *Int. J. Primatol.* 24:1037–1055.

Peres, C. A. (1990). Effects of hunting on western Amazonian primate communities. *Biol. Conserv.* 54:47–59.

Peres, C. A. (1991). Humboldt's woolly monkeys decimated by hunting in Amazonia. *Oryx* 25:89–95.

Peres, C. A. (1994a). Diet and feeding ecology of gray woolly monkeys (*Lagothrix lagotricha cana*) in central Amazonia: comparisons with other atelines. *Int. J. Primatol.* 15:333–372.

Peres, C. A. (1994b). Which are the largest New World monkeys? *J. Hum. Evol.* 26:245–249.

Peres, C. A. (1996). Use of space, spatial group structure, and foraging group size of gray woolly monkeys (*Lagothrix lagotricha cana*) at Urucu, Brazil. In: Norconk, M. A., Rosenberger, A. L., and Garber, P. A. (eds.), *Adaptive Radiations of Neotropical Primates.* Plenum Press, New York. pp. 467–488.

Peres, C. A. (2000). Effects of subsistence hunting on vertebrate community structure in Amazonian forests. *Conserv. Biol.* 14:240–253.

Pinto, L. P. (2002a). Diet, activity, and home range of *Alouatta belzebul discolor* (Primates: Atelidae) in Paranaíta, northern Mato Grosso. *Neotrop. Primates* 10:98–99.

Pinto, L. P. (2002b). Dieta, padrão de atividades e área de vida de *Alouatta belzebul discolor* (Primates: Atelidae) em Paranaíta, norte do Mato Grosso [MSc thesis]. Universidade Estadual de Campinas (UNICAMP), São Paulo.

Pope, T. R. (1989). The influence of mating systems and dispersal patterns on the genetic structure of red howler monkey populations [PhD thesis]. University of Florida, Gainesville.

Pope, T. R. (1990). The reproductive consequences of male cooperation in the red howler monkey: paternity exclusion in multi-male and single-male troops using genetic markers. *Behav. Ecol. Sociobiol.* 27:439–446.

Pope, T. R. (1992). The influence of dispersal patterns and mating systems on genetic differentiation within and between populations of the red howler monkey (*Alouatta seniculis*). *Evolution* 46:1112–1128.

Pope, T. R. (1998a). Effects of demographic change on group kin structure and gene dynamics of populations of red howling monkeys. *J. Mammal.* 79:692–712.

Pope, T. R. (1998b). Genetic variation in remnant populations of the woolly spider monkey (*Brachyteles arachnoides*). *Int. J. Primatol.* 19:95–109.

Pope, T. R. (2000). Reproductive success increases with degree of kinship in cooperative coalitions of female red howler monkeys (*Alouatta seniculus*). *Behav. Ecol. Sociobiol.* 27:439–446.

Ramirez, M. (1980). Grouping patterns of the woolly monkey, *Lagothrix lagothricha*, at the Manu National Park, Peru. *Am. J. Phys. Anthropol.* 52:269.

Ramirez, M. (1988). The woolly monkeys, genus *Lagothrix*. In: Mittermeier, R. A., Rylands, A. B., Coimbra-Filho, A. F., and da Fonseca, G. A. B. (eds.), *Ecology and Behavior of Neotropical Primates,* vol. 2. World Wildlife Fund, Washington DC. pp. 539–575.

Ramos-Fernández, G., and Ayala-Orozco, B. (2003). Population size and habitat use of spider monkeys at Punta Laguna, Mexico. In: March, L. K. (ed.), *Primates in Fragments: Ecology and Conservation.* Kluwer Academic/Plenum, New York. pp. 191–209.

Robinson, J. G., and Janson, C. H. (1987). Capuchins, squirrel monkeys, and atelines: socioecological convergence with Old World primates. In: Smuts, B. B., Cheney, D. L., Seyfarth, R. M., Wrangham, R. W., and Struhsaker, T. T. (eds.), *Primate Societies.* University of Chicago Press, Chicago. pp. 69–82.

Rodríguez, G. A. C., and Boher, S. (1988). Notes on the biology of *Cebus nigrivittatus* and *Alouatta seniculus* in northern Venezuela. *Primate Conserv.* 9:61–66.

Rosenberger, A. L., and Strier, K. B. (1989). Adaptive radiation of the ateline primates. *J. Hum. Evol.* 18:717–750.

Ross, C. (1988). The intrinsic rate of natural increase and reproductive effort in primates. *J. Zool. Lond.* 214:199–219.

Ross, C. (1991). Life history patterns of New World monkeys. *Int. J. Primatol.* 12:481–502.

Ross, C., and Jones, K. E. (1999). Socioecology and the evolution of primate reproductive rates. In: Lee, P. C. (ed.), *Comparative Primate Socioecology.* Cambridge University Press, Cambridge. pp. 73–110.

Rudran, R. (1979). The demography and social mobility of a red howler (*Alouatta seniculus*) population in Venezuela. In: Eisenberg, J. F. (ed.), *Vertebrate Ecology of the Northern Neotropics.* Smithsonian Institution Press, Washington DC. pp. 107–126.

Rudran, R., and Fernandez-Duque, E. (2003). Demographic changes over thirty years in a red howler population in Venezuela. *Int. J. Primatol.* 24:925–947.

Rumiz, D. I. (1990). *Alouatta caraya*: population density and demography in northern Argentina. *Am. J. Primatol.* 21:279–294.

Ruschi, A. (1964). Macacos do estado do Espírito Santo. *Bol. Mus. Mello-Leitão* 23:1–18.

Russo, S. E., Campbell, C. J., Dew, J. L., Stevenson, P. R., and Suarez, S. (2005). A multi-site comparison of dietary preferences and seed dispersal by spider monkeys (*Ateles* spp.). *Int. J. Primatol.* 26:1017–1037.

Rylands, A. B., Schneider, H., Langguth, A., Mittermeier, R. A., Groves, C. P., and Rodríguez-Luna, E. (2000). An assessment of the diversity of New World primates. *Neotrop. Primates* 8:61–93.

Schneider, H. (2000). The current status of the New World monkey phylogeny. *Ann. Acad. Brasil. Ciênc.* 72:165–172.

Schneider, H., Canavez, F. C., Sampaio, I., Moreira, M. Â. M., Tagliaro, C. H., and Seuánez, H. N. (2001). Can molecular data place each neotropical monkey in its own branch? *Chromosoma* 109:515–523.

Schneider, H., and Rosenberger, A. L. (1996). Molecules, morphology, and platyrrhine systematics. In: Norconk, M. A., Rosenberger, A. L., and Garber, P. A. (eds.), *Adaptive Radiations of Neotropical Primates.* Plenum Press, New York. pp. 3–19.

Schneider, H., Sampaio, I., Harada, M. L., Barroso, C. M. L., Schneider, M. P. C., Czelusniak, J., and Goodman, M. (1996). Molecular phylogeny of the New World monkeys (Platyrrhini, Primates) based on two unlinked nuclear genes: IRBP intron 1 and β-globin sequences. *Am. J. Phys. Anthropol.* 100:153–179.

Schneider, H., Schneider, M. P. C., Sampaio, M. I. C., Harada, M. L., Stanhope, M., Czelusniak, J., and Goodman, M. (1993). Molecular phylogeny of the New World monkeys (Platyrrhini, Primates). *Mol. Phylogenet. Evol.* 2:225–242.

Sekulic, R. (1982a). Daily and seasonal patterns of roaring and spacing in four red howler (*Alouatta seniculus*) troops. *Folia Primatol.* 39:22–48.

Sekulic, R. (1982b). The function of howling in red howler monkeys (*Alouatta seniculus*). *Behaviour* 81:38–54.

Shimooka, Y. (2003). Seasonal variation in association patterns of wild spider monkeys (*Ateles belzebuth belzebuth*) at La Macarena, Colombia. *Primates* 44:83–90.

Silver, S. C., Ostro, L. E. T., Yeager, C. P., and Horwich, R. (1998). Feeding ecology of the black howler monkey (*Alouatta pigra*) in northern Belize. *Am. J. Primatol.* 45:263–279.

Smith, C. C. (1977). Feeding behavior and social organization in howling monkeys. In: Clutton-Brock, T. H. (ed.), *Primate Ecology: Studies of Feeding and Ranging Behaviour in Lemurs, Monkeys, and Apes.* Academic Press, London. pp. 97–126.

Smith, R. J., and Jungers, W. L. (1997). Body mass in comparative primatology. *J. Hum. Evol.* 32:523–559.

Soini, P. (1986). A synecological study of a primate community in the Pacaya-Samiria National Reserve, Peru. *Primate Conserv.* 7:63–71.

Soini, P. (1990). Ecologia y dinámica poblacional del "choro" (*Lagothrix lagotricha*, Primates) en el Rio Pacaya, Peru. In: *La Primatologia en el Perú.* Proyecto Peruano de Primatologia, Lima. pp. 382–396.

Stevenson, P. (1997). Notes on the mating behavior of woolly monkeys (*Lagothrix lagotricha*) at Tinigua National Park, Colombia. *Field Studies of Fauna and Flora at La Macarena, Colombia* 10:13–15.

Stevenson, P. R. (1992). Diet of woolly monkeys (*Lagothrix lagotricha*) at La Macarena, Colombia. *Field Studies of New World Monkeys, La Macarena, Colombia* 6:3–14.

Stevenson, P. R. (2002). Frugivory and seed dispersal by woolly monkeys at Tinigua National Park, Colombia [PhD thesis]. State University of New York, Stony Brook.

Stevenson, P. R., and Castellanos, M. C. (2000). Feeding rates and daily path range of the Colombian woolly monkeys as evidence for between- and within-group competition. *Folia Primatol.* 71:399–408.

Stevenson, P. R., Quiñones, M. J., and Ahumada, J. A. (1994). Ecological strategies of woolly monkeys (*Lagothrix lagotricha*) at Tinigua National Park, Colombia. *Am. J. Primatol.* 32:123–140.

Stoner, K. E. (1994). Population density of the mantled howler monkey (*Alouatta palliata*) at La Selva Biological Reserve, Costa Rica: a new technique to analyze census data. *Biotropica* 26:332–340.

Stoner, K. E. (1996). Habitat selection and seasonal patterns of activity and foraging in mantled howling monkeys (*Alouatta palliata*) in northeastern Costa Rica. *Int. J. Primatol.* 17:1–30.

Strier, K. B. (1986a). The behavior and ecology of the woolly spider monkey, or muriqui (*Brachyteles arachnoides* E Geoffroy 1806) [PhD thesis]. Harvard University, Cambridge, MA.

Strier, K. B. (1986b). Reprodução de *Brachyteles arachnoides* (Primates: Cebidae). *Primatol. Brasil* 2:163–175.

Strier, K. B. (1987a). Activity budgets of woolly spider monkeys, or muriquis (*Brachyteles arachnoides*). *Am. J. Primatol.* 13:385–395.

Strier, K. B. (1987b). Demographic patterns in one group of muriquis. *Primate Conserv.* 8:73–74.

Strier, K. B. (1987c). Ranging behavior of woolly spider monkeys, or muriquis, *Brachyteles arachnoides*. *Int. J. Primatol.* 8:575–591.

Strier, K. B. (1989). Effects of patch size on feeding associations in muriquis (*Brachyteles arachnoides*). *Folia Primatol.* 52:70–77.

Strier, K. B. (1990). New World primates, new frontiers: insights from the woolly spider monkey, or muriqui (*Brachyteles arachnoides*). *Int. J. Primatol.* 11:7–19.

Strier, K. B. (1991a). Demography and conservation of an endangered primate, *Brachyteles arachnoides*. *Conserv. Biol.* 5:214–218.

Strier, K. B. (1991b). Diet in one group of woolly spider monkeys, or muriquis (*Brachyteles arachnoides*). *Am. J. Primatol.* 23:113–126.

Strier, K. B. (1992a). Atelinae adaptations: behavioral strategies and ecological constraints. *Am. J. Phys. Anthropol.* 88:515–524.

Strier, K. B. (1992b). Causes and consequences of nonaggression in the woolly spider monkey, or muriqui (*Brachyteles arachnoides*). In: Silverberg, J., and Gray, J. P. (eds.), *Aggression and Peacefulness in Humans and Other Primates*. Oxford University Press, New York. pp. 100–116.

Strier, K. B. (1992c). *Faces in the Forest: The Endangered Muriqui Monkeys of Brazil*. Oxford University Press, New York.

Strier, K. B. (1993). Growing up in a patrifocal society: sex differences in the spatial relations of immature muriquis. In: Pereira, M. E., and Fairbanks, L. A. (eds.), *Juvenile Primates:*

Life History, Development, and Behavior. Oxford University Press, New York. pp. 138–147.

Strier, K. B. (1994a). Brotherhoods among atelins: kinship, affiliation, and competition. *Behaviour* 130:151–167.

Strier, K. B. (1994b). The myth of the typical primate. *Ybk. Phys. Anthropol.* 37:233–271.

Strier, K. B. (1996a). Male reproductive strategies in New World primates. *Hum. Nat.* 7:105–123.

Strier, K. B. (1996b). Reproductive ecology of female muriquis (*Brachyteles arachnoides*). In: Norconk, M. A., Rosenberger, A. L., and Garber, P. A. (eds.), *Adaptive Radiations of Neotropical Primates*. Plenum Press, New York. pp. 511–352.

Strier, K. B. (1997a). Subtle cues of social relations in male muriqui monkeys (*Brachyteles arachnoides*). In: Kinzey, W. G. (ed.), *New World Primates: Ecology, Evolution, and Behavior*. Aldine de Gruyter, New York. pp. 109–118.

Strier, K. B. (1997b). Mate preferences of wild muriqui monkeys (*Brachyteles arachnoides*): reproductive and social correlates. *Folia Primatol.* 68:120–133.

Strier, K. B., Carvalho, D. S., and Bejar, N. O. (2000). Prescription for peacefulness. In: Aureli, F., and de Waal, F. B. M. (eds.), *Natural Conflict Resolution*. University of California Press, Los Angeles. pp. 315–317.

Strier, K. B., Dib, L. T., and Figueria, J. E. C. (2002). Social dynamics of male muriquis (*Brachyteles arachnoides hypoxanthus*). *Behaviour* 139:315–342.

Strier, K. B., Lynch, J. W., and Ziegler, T. E. (2003). Hormonal changes during the mating and conception seasons of wild northern muriquis (*Brachyteles arachnoides hypoxanthus*). *Am. J. Primatol.* 61:85–99.

Strier, K. B., Mendes, F. D. C., Rimoli, J., and Rimoli, A. O. (1993). Demography and social structure of one group of muriquis (*Brachyteles arachnoides*). *Int. J. Primatol.* 14:513–526.

Strier, K. B., Mendes, S. L., and Santos, R. R. (2001). Timing of births in sympatric brown howler monkeys (*Alouatta fusca clamitans*) and northern muriquis (*Brachyteles arachnoides*). *Am. J. Primatol.* 55:87–100.

Strier, K. B., and Ziegler, T. E. (1997). Behavioral and endocrine characteristics of the reproductive cycle in wild muriqui monkeys, *Brachyteles arachnoides*. *Am. J. Primatol.* 42:299–310.

Strier, K. B., and Ziegler, T. E. (2000). Lack of pubertal influences on female dispersal in muriqui monkeys, *Brachyteles arachnoides*. *Anim. Behav.* 59:849–860.

Suarez, S. A. (2003). Spatio-temporal foraging skills of white-bellied spider monkeys (*Ateles belzebuth belzebuth*) in the Yasuní National Park, Ecuador [PhD thesis]. State University of New York, Stony Brook.

Symington, M. M. (1987a). Ecological and social correlates of party size in the black spider monkey, *Ateles paniscus chamek* [PhD thesis]. Princeton University, Princeton, NJ.

Symington, M. M. (1987b). Sex ratio and maternal rank in wild spider monkeys: when daughters disperse. *Behav. Ecol. Sociobiol.* 20:421–425.

Symington, M. M. (1988a). Demography, ranging patterns, and activity budgets of black spider monkeys (*Ateles paniscus chamek*) in the Manu National Park, Peru. *Am. J. Primatol.* 15:45–67.

Symington, M. M. (1988b). Environmental determinants of population densities in *Ateles*. *Primate Conserv.* 9:74–79.

Symington, M. M. (1988c). Food competition and foraging party size in the black spider monkey (*Ateles paniscus chamek*). *Behaviour* 105:117–132.

Symington, M. M. (1990). Fission–fusion social organization in *Ateles* and *Pan. Int. J. Primatol.* 11:47–61.

Teaford, M. F., and Glander, K. E. (1996). Dental microwear and diet in a wild population of mantled howling monkeys (*Alouatta palliata*). In: Norconk, M. A., Rosenberger, A. L., and Garber, P. A. (eds.), *Adaptive Radiations of Neotropical Primates*. Plenum Press, New York. pp. 433–449.

Torres de Assumpção, C. (1983). Ecological and behavioural information on *Brachyteles arachnoides*. *Primates* 24:584–593.

Tutin, C. E. G. (1979). Mating patterns and reproductive strategies in a community of wild chimpanzees (*Pan troglodytes schweinfurthii*). *Behav. Ecol. Sociobiol.* 6:29–38.

van Roosmalen, M. G. M. (1985). Habitat preferences, diet, feeding strategy and social organization of the black spider monkey (*Ateles paniscus paniscus* Linnaeus 1758) in Surinam. *Acta Amazon.* 15:1–238.

van Roosmalen, M. G. M., and Klein, L. L. (1988). The spider monkeys, genus *Ateles*. In: Mittermeier, R. A., Rylands, A. B., Coimbra-Filho, A. F., and da Fonseca, G. A. B. (eds.), *Ecology and Behavior of Neotropical Primates*. World Wildlife Fund, Washington DC. pp. 455–537.

von Dornum, M., and Ruvolo, M. (1999). Phylogenetic relationships of the New World monkeys (Primates, Platyrrhini) based on nuclear G6PD DNA sequences. *Mol. Phylogenet. Evol.* 11:459–476.

Wallace, R. B. (2001). Diurnal activity budgets of black spider monkeys, *Ateles chamek*, in a southern Amazonian tropical forest. *Neotrop. Primates* 9:101–107.

Wang, E., and Milton, K. (2003). Intragroup social relationships of male *Alouatta palliata* on Barro Colorado Island, Republica of Panama. *Int. J. Primatol.* 24:1227–1243.

Watts, D. (1989). Infanticide in mountain gorillas: new cases and a reconsideration of the evidence. *Ethology* 81:1–18.

Watts, D. (1998). Coalitionary mate-guarding by male chimpanzees at Ngogo, Kibale National Park, Uganda. *Behav. Ecol. Sociobiol.* 44:43–55.

Watts, D., and Mitani, J. (2001). Boundary patrols and intergroup encounters among wild chimpanzees. *Behaviour* 138:299–327.

White, F. J. (1986). Behavioral ecology of the pygmy chimpanzee [PhD thesis]. State University of New York, Stony Brook.

White, F. J. (1988). Party composition and dynamics in *Pan paniscus. Int. J. Primatol.* 9:179–193.

Williams, L. (1967). Breeding Humboldt's woolly monkey *Lagothrix lagotricha* at Murrayton Monkey Sanctuary. *Int. Zoo Ybk.* 7:86–89.

Williams, L. (1974). *Monkeys and the Social Instinct*. Monkey Sanctuary Publications, Looe, England.

Williams-Guillén, K. (2003). The behavioral ecology of mantled howling monkeys (*Alouatta palliata*) living in a Nicaraguan shade coffee plantation [PhD thesis]. New York University, New York.

Wrangham, R. W. (1977). Feeding behaviour of chimpanzees in Gombe National Park, Tanzania. In: Clutton-Brock, T. H. (ed.), *Primate Ecology: Studies of Feeding and Ranging Behaviour in Lemurs, Monkeys, and Apes*. Academic Press, London. pp. 504–538.

Wrangham, R. W. (1980). An ecological model of female-bonded primate groups. *Behaviour* 75:262–300.

Wrangham, R. W. (1987). Evolution of social structure. In: Smuts, B. B., Cheney, D. L., Seyfarth, R. M., Wrangham, R. W., and Struhsaker, T. T. (eds.), *Primate Societies*. University of Chicago Press, Chicago. pp. 282–296.

Wrangham, R. W., and Smuts, B. B. (1980). Sex differences in the behaviour ecology of chimpanzees in the Gombe National Park, Tanzania. *J. Reprod. Fertil.* S28:13–31.

Yost, J. A., and Kelley, P. M. (1983). Shotguns, blowguns, and spears: the analysis of technological efficiency. In: Hames, R. B., and Vickers, W. T. (eds.), *Adaptive Responses of Native Amazonians*. Academic Press, New York. pp. 189–224.

Zingeser, M. R. (1973). Dentition of *Brachyteles arachnoides* with reference to alouattine and ateline affinities. *Folia Primatol.* 20:351–390.

Zucker, E. L., and Clarke, M. R. (1998). Agonistic and affiliative relationships of adult female howlers (*Alouatta palliata*) in Costa Rica over a 4-year period. *Int. J. Primatol.* 19:433–449.

Zucker, E. L., Clarke, M. R., and Glander, K. E. (1997). Latencies to first births by immigrating adult female howling monkeys (*Alouatta palliata*) in Costa Rica. *Am. J. Primatol.* 42:158.

Zunino, G. E., Chalukian, S. C., and Rumiz, D. I. (1986). Infanticide and infant disappearance related to male takeover in groups of *Alouatta caraya*. *Primatol. Brasil* 2:185–190.

Zunino, G. E., and Rumiz, D. I. (1986). Observaciones sobre el comportamiento territorial del mono aullador negro (*Alouatta caraya*). *Bol. Primatol. Arg.* 4:36–52.

11

The Asian Colobines
Diversity Among Leaf-Eating Monkeys
R. Craig Kirkpatrick

INTRODUCTION

Colobines are known as "leaf-eating monkeys." This is because they have physiological adaptations—most notably, a multichambered stomach with fermenting bacteria—that help them digest leaves. Yet, this generalization masks the remarkable diversity within colobines.

Colobines are found throughout Asia, in a broad range of habitat types. These monkeys also have a broad range of diet and social organization. The diets of Asian colobines are often seasonal, with a preference for young leaves supplemented by seeds and fruits. Relative to cercopithecines, groups of Asian colobines are usually small, with just several females, their offspring, and one male. As might be expected, these groups have relatively small home ranges and short daily travel distances, too.

Knowledge about Asian colobines is important to our understanding of primates in general. Although Asian colobines have modal patterns in diet, social organization, and range use, there are variations both within and among species. This makes Asian colobines useful in the study of one of primatology's enduring themes: relationships among ecological variables, feeding strategies, range use, and social organization. Asian colobines play a key role in primatology's quest to better understand food choice, such as trade-offs between food availability and food quality. Asian colobines also are central to investigations of social behavior, such as male–male competition and alloparental care.

TAXONOMY AND DISTRIBUTION

The taxonomy of the Asian colobines has been, and continues to be, an exciting area of debate. All taxonomies are hypotheses, revised as new information becomes available (Groves 2001). An example of this comes from the "odd-nosed colobines." Discussion of whether these species should be lumped in the two genera *Nasalis* (proboscis monkeys) and *Pygathrix* (doucs) or split into four with the addition of *Rhinopithecus* (snub-nosed monkeys) and *Simias* (pig-tailed langurs) took place over 25 years (see Groves 1970,

Jablonski 1998). Another example comes from discussion over the past 15 years of whether species historically grouped under the genus *Presbytis* are better represented as two genera, *Presbytis* (surilis) and *Trachypithecus* (lutongs).

In the absence of a unified framework, researchers have used idiosyncratic and varying models, leading to competing taxonomies and a literature that at times resembles the Tower of Babel. Groves (2001) provides comprehensive structure, guiding a path through recent debates (Table 11.1). Groves's taxonomy is in overall accord with comprehensive solutions proposed by others (e.g., Brandon-Jones et al. 2004) and is used in this chapter.

Asian colobines are divided into seven genera: *Nasalis*, *Simias*, *Pygathrix*, *Rhinopithecus*, *Presbytis*, *Trachypithecus*, and *Semnopithecus*. The genus *Semnopithecus* (gray langurs) has seven species, previously described as one, *Presbytis entellus*. These langurs are distributed widely in South Asia, from the temperate Himalayas to tropical Sri Lanka; growing analysis suggests that different populations of the former *P. entellus* are better represented by separate species. There is also a strong case for separating *Presbytis* and *Trachypithecus* on the basis of both dental and genital morphology. The remaining genera—*Nasalis*, *Simias*, *Pygathrix*, and *Rhinopithecus*—are grouped as the "odd-nosed colobines," although questions remain as to the phyletic relations among this clade.

Semnopithecus is generally found in western Asia, from Pakistan through India into Nepal, Bhutan, and Sri Lanka. *Presbytis* and *Trachypithecus* are broadly sympatric in Indonesia and Malaysia, with *Trachypithecus* extending farther north into Indochina, Burma, and Bangladesh. The genera of the odd-nosed group are rather dispersed in distribution, with *Rhinopithecus* in southern China and northern Vietnam; *Pygathrix* in Vietnam, Laos, and Cambodia; *Nasalis* on the island of Borneo; and *Simias* only on the Mentawai Islands of Indonesia.

Some species of Asian colobines are widely distributed. *Semnopithecus dussumieri* (southern plains gray langur) is found throughout northern India, *Trachypithecus cristatus* (silver leaf monkey) across Indonesia, and *Presbytis rubicunda* (maroon leaf monkey) throughout Borneo. More

Table 11.1   Taxonomy of Asian Colobines

LATIN NAME	COMMON NAME	NOTES
Semnopithecus	**Gray langur or Hanuman langur**	
Semnopithecus entellus	Northern plains gray langur	Formerly *Presbytis entellus*
S. schistaceus	Nepal gray langur	Formerly in *S. entellus*
S. ajax	Kashmir gray langur	Formerly in *S. entellus*
S. hector	Tarai gray langur	Formerly in *S. entellus*
S. hypoleucos	Black-footed gray langur	Formerly in *S. entellus*
S. dussumieri	Southern plains gray langur	Formerly in *S. entellus*
S. priam	Tufted gray langur	Formerly in *S. entellus*
Trachypithecus	**Lutong**	
Vetulus group		
Trachypithecus vetulus	Purple-faced langur	Formerly *Presbytis senex*
T. johnii	Nilgiri langur	Formerly *Presbytis johnii*
Cristatus group		
T. auratus	Javan lutong	Formerly in *T. cristatus*
T. cristatus	Silver leaf monkey	Formerly *Presbytis cristatus*
T. germaini	Indochinese lutong	Formerly in *T. cristatus*
T. barbei	Tenasserium lutong	Formerly in *T. cristatus*
Obscurus group		
T. obscurus	Dusky, or spectacled, leaf monkey	Formerly *Presbytis obscurus*
T. phayrei	Phayre's leaf monkey	Formerly in *T. obscurus*
Pileatus group		
T. pileatus	Capped langur	Formerly *Presbytis pileatus*
T. shortridgei	Shortridge's langur	Formerly in *T. pileatus*
T. geei	Gee's golden langur	Formerly *Presbytis geei*
Francoisi group		
T. francoisi	Francois's langur	Formerly *Presbytis francoisi*
T. hatinhesis	Hatinh langur	Formerly in *T. francoisi*
T. poliocephalus	White-headed langur	Formerly in *T. francoisi*
T. laotum	Laotian langur	Formerly in *T. francoisi*
T. delacouri	Delacour's langur	Formerly in *T. francoisi*
T. ebenus	Indochinese black langur	Formerly in *T. auratus*
Presbytis	**Surili**	
Presbytis melalophos	Sumatran surili	
P. femoralis	Banded surili	Formerly in *P. melalophos*
P. chrysomelas	Sarawak surili	Formerly in *P. femoralis*
P. siamensis	White-thighed surili	Formerly in *P. femoralis*
P. frontata	White-fronted langur	
P. comata	Javan surili	Formerly *P. aygula*
P. thomasi	Thomas's langur	Formerly in *P. aygula*
P. hosei	Hose's langur	Formerly in *P. aygula*
P. rubicunda	Maroon leaf monkey	
P. potenziani	Mentawai langur	
P. natunae	Natuna Islands surili	Formerly in *P. femoralis*
Pygathrix	**Douc**	
Pygathrix nemaeus	Red-shanked douc	
P. nigripes	Black-shanked douc	Formerly in *Py. nemaeus*
P. cinerea	Gray-shanked douc	Formerly in *Py. nemaeus*
Rhinopithecus	**Snub-nosed monkey**	
Rhinopithecus roxellana	Golden snub-nosed monkey	Formerly *Py. roxellanae*
R. bieti	Black snub-nosed monkey	Formerly *Py. bieti*
R. brelichi	Gray snub-nosed monkey	Formerly *Py. brelichi*
R. avunculus	Tonkin snub-nosed langur	Formerly *Py. avunculus*
Nasalis		
Nasalis larvatus	Proboscis monkey	
Simias		
Simias concolor	Pig-tailed langur	Formerly *Nasalis concolor*

Source: Groves (2001).

often, however, species have restricted ranges, such as *Trachypithecus poliocephalus* (white-headed langur) in southern China, *T. barbei* (Tenasserium lutong) in the northern peninsular areas of Burma and Thailand, and *Pygathrix cinerea* (gray-shanked douc) in central Vietnam. Although deforestation in Asia has severely constricted habitat, the limited ranges are not simply the result of human action. The restricted ranges of Asian colobines and the discontinuous distributions of those species with larger ranges may be an artifact of habitat refugia resulting from climate fluctuations during the Pleistocene (Brandon-Jones 1996).

ECOLOGY

Habitats and Range Use

In keeping with their distribution throughout Asia, the colobines of Asia are found in many types of habitat (Table 11.2). Most species live in tropical or subtropical forests, expected habitat for monkeys. Less conventional habitats are home to some species, too. These include the temperate Himalayan forests of Nepal (*Semnopithecus schistaceus* [Nepal gray langur]) and southwest China (*Rhinopithecus bieti* [black snub-nosed monkey]), the peat swamps of Borneo (*Nasalis larvatus* [proboscis monkey]), and the dry scrub deserts of western India (*S. dussumieri*). Perhaps most striking, members of the Franciosi group of *Trachypithecus* live in forests on vertical cliffs of karst limestone in Vietnam and China, regularly using caves as sleeping sites (e.g., *T. poliocephalus*, Huang et al. 2003).

The population densities of Asian colobines vary by an order of magnitude (Table 11.2). The Chinese species of *Rhinopithecus*, for example, have relatively low densities of about 10 individuals/km^2 (Kirkpatrick 1998), perhaps because they live in temperate forests of low productivity. At the other extreme, *Trachypithecus johnii* (Nilgiri langur) reach densities of over 70 individuals/km^2 (Oates et al. 1980) and *Presbytis siamensis* (white-thighed surili) reach densities of over 100 individuals/km^2 (Davies 1994). Population densities vary not only among species but also among different populations of the same species. Densities of *S. dussumieri* span from under 5/km^2 (Orcha, India; Dolhinow 1972) to over 45/km^2 (Kanha, India; Newton 1992); the habitats of Orcha and Kanha are both moist deciduous forest. Similarly, in both *N. larvatus* and *Trachypithecus geei* (Gee's golden langur), densities vary substantially between populations, from about 10 to over 60 individuals/km^2, without notable differences in habitat (see Table 11.2).

The range sizes used by groups of Asian colobines vary widely (Table 11.2). Most Asian colobines have range sizes below 1 km^2. Some species, however, such as *S. schistaceus* in the Himalayas and *R. bieti* in southwest China, have ranges that are substantially larger, over 10 km^2 (Bishop 1979, Kirkpatrick 1998). In general, species with lower average densities have larger average range sizes (Yeager and

Kirkpatrick 1998), suggesting that resource availability may regulate the number of individuals in an area.

Colobines typically have daily travel distances of under 1 km (Table 11.2). Even the Asian colobines with large home ranges do not have a correspondingly large increase in the length of their daily travel distances. Daily travel distances appear to be related to food resources. *P. rubicunda*, *P. siamensis*, and *S. dussumieri* all travel farther in months when diet is based on fruits and flowers rather than leaves, for example (Davies 1984, Bennett 1986, Newton 1992). Similarly, groups of *Trachypithecus pileatus* (capped langur) move their farthest distances when feeding on fruits and their shortest distances when feeding on mature leaves (Stanford 1991a).

Diet

Colobines are widely known for their physiological adaptations to eat leaves. These adaptations include sharp molars to chew leaves, enlarged salivary glands that help degrade them, and a multichambered stomach with symbiotic microbes that break down leaf fibers (Kay and Davies 1994).

Yet, their generalization as "leaf-eaters" does injustice to the diversity of diets found among the Asian colobines. The feeding adaptations that aid the digestion of leaves also aid the digestion of other food types. This capacity for diet variability is reflected in the ability of Asian colobines to live in diverse habitats and to endure seasonality in those habitats (Kirkpatrick 1999).

Certainly, many Asian colobines subsist primarily on leaves (Table 11.2). Most of these leaves are young, easier to digest than mature leaves, and often quite seasonal. Asian colobines also typically get a quarter of their diet from seeds and fruits. This reaches an extreme in *P. rubicunda*, whose diet is over 80% seeds in some months (Davies 1991). The most extreme examples of diet diversity come from *Rhinopithecus roxellana* (golden snub-nosed monkey) and *R. bieti*. *R. roxellana* at Baihe, China, switch from leaves in summer to lichens and the green bark and buds of dicot plants in winter (Kirkpatrick et al. 1999); at Wuyapiya, China, *R. bieti* feeds almost exclusively on lichens (Kirkpatrick 1996). Animal matter is rarely a major part of the diet for Asian colobines, although caterpillars and other insects can provide up to a quarter of the diet of *S. dussumieri* at Kanha, India, during the monsoons, when insects are plentiful (Newton 1992).

The diets of Asian colobines are often marked by seasonality. *T. pileatus* at Madhupur, Bangladesh, eats over 80% mature leaves from November to March but switches to over 50% young leaves or fruit when these foods are available (Stanford 1991b). *N. larvatus* at Kinabatangan, Malaysian Borneo, typically feeds on young leaves, switching to fruits when seasonally available (Boonratana 1994). Switching between mature leaves, young leaves, and seeds or fruits is associated with availability (e.g., *T. johnii*, Oates et al. 1980; *N. larvatus*, Yeager 1989; *T. pileatus*, Stanford 1991b; *S. dussumieri*, Newton 1992) showing preference for these foods.

Table 11.2 Range Use and Diet in Asian Colobines[1]

SPECIES	SITE	HABITAT	RANGE SIZE (HA)	OVERLAP (%)	DAY RANGE (M)	POPULATION DENSITY (INDIVIDUALS/KM²)	L (%)	YL (%)	ML (%)	S/F (%)	FL (%)	B/B (%)	L/F (%)	O/U (%)	REFERENCE
Semnopithecus															
Semnopithecus dussumieri	Orcha	Moist deciduous forest	~375	Extensive		3–6									Dolhinow 1972
	Kaukori	Dry scrub forest	~775			3									Dolhinow 1972
	Kanha	Moist deciduous forest	74			46	39	4	35	24	9	—	—	3	Newton 1992, 1994
	Gir	Dry deciduous forest	20	0		121									Starin 1978
	Dhawar (open)	Tropical broadleaf forest	149		1,083	11									Sugiyama 1964
	Dhawar (closed)	Tropical broadleaf forest	19			76	58	7		29	2	—	—	4	Sugiyama 1964
S. schistaceus	Junbesi	Conifer/broadleaf forest													Curtin 1975
S. ajax	Rannagar	Tropical broadleaf forest	~1,275			~2									Borries et al. 2001
S. priam	Polonnaruwa	Semievergreen forest				26	48	27	21	45	7	—	—	—	Hladik 1977
Trachypithecus															
Vetulus group															
Trachypithecus vetulus	Polonnaruwa	Dry deciduous forest	2	Considerable			60	20	40	28	12	—	—	—	Hladik 1977
T. johnii	Nilgiri district	Moist forest	65–250												Poirier 1970
	Kakachi	Evergreen forest	24	10		71	52	25	27	25	9	—	—	6	Oates et al. 1980
Cristatus group															
T. auratus	Pangandaran	Secondary forest			<200	56	56	46	<10	32	14	—	—	8	Kool 1992, 1993
T. cristatus	Rantau Panjang	Mangrove forest													Furuya 1961
Obscurus group															
T. obscurus	Kuala Lompat	Dipterocarp forest	17–33	3	560	87	58	36	22	35	7	—	—	—	Curtin 1976, 1980
Pileatus group															
T. pileatus	Madhupur	Tropical moist forest	22	84	325	52	53	11	42	34	7	4	—	1	Stanford 1991b,c
	Manas	Tropical moist forest	64	~60		46									Mukherjee 1978
T. geei	Ripu	Tropical moist forest				64									Srivastava et al. 2001
	Chirrang	Tropical moist forest				46									Srivastava et al. 2001
	Manas 01	Tropical moist forest				20									Srivastava et al. 2001
	Manas 02	Tropical moist forest				8									Srivastava et al. 2001

Table 11.2 (cont'd)

SPECIES	SITE	HABITAT	RANGE SIZE (HA)	OVERLAP (%)	DAY RANGE (M)	POPULATION DENSITY (INDIVIDUALS/KM²)	L (%)	YL (%)	ML (%)	S/F (%)	FL (%)	B/B (%)	L/F (%)	O/U (%)	REFERENCE
Presbytis															
Presbytis femoralis	Perawang	Rubber plantation	22	25	935	42	29	26	3	58	<1	8	—	5	Megantara 1989
P. siamensis	Kuala Lompat	Dipterocarp forest	27	~80	750	36	36	28	8	49	12	—	—	3	Davies et al. 1988, Bennett 1983
P. comata	Kuala Lompat	Dipterocarp forest	21	33	750	35	35	24	11	56	6	—	—	2	Curtin 1980
	Kutai	Dipterocarp forest				20	65			27	—	—	—	8	Rodman 1978
P. thomasi	Kamojang	Montane forest	38	10	500	11	65	59	6	14	7	1	4	7	Ruhiyat 1983
	Ketambe	Primary rain forest													Steenbeek et al. 2000
P. hosei	Bohorok	Rubber plantation	14		683	21	36			53	8	—	—	3	Gurmaya 1986
	Lipad	Dipterocarp forest	35		740	~26	78	45	5	19	3	—	—		Mitchell 1994
P. rubicunda	Sepilok	Dipterocarp forest	67	10	890		37	36	1	49	11	—	2	2	Davies 1984, Davies et al. 1988
	Tanjung Puting	Dipterocarp forest and peat swamp	33–99			36	36			52	12	—	—	—	Supriatna et al. 1986
P. potenziani	Betumonga	Dipterocarp forest	34	~40	540	11	55			32	—	—	—	13	Fuentes 1994, Watanabe 1981
	Sarabua	Dipterocarp forest	13			14									
Pygathrix															
Pygathrix nemaeus	Son Tra	Tropical moist forest													Lippold 1977
Rhinopithecus															
Rhinopithecus roxellana	Shennongjia	Conifer/broadleaf forest	2,600	100		8									Ren et al. 1998
R. bieti	Wuyapiya	Conifer forest	2,500	100	1,310	7	>6						86	8	Kirkpatrick et al. 1998, Kirkpatrick 1996
R. brelichi	Fanjingshan	Temperate broadleaf forest	3,500	100	1,290	11	71		<1	15	7	—	0.2	6	Bleisch et al. 1993, Bleisch and Xie 1998
R. avunculus	Ta Ke/Nam Trang	Tropical broadleaf forest	>1,000			<8	38			62					Boonratana and Le 1998
Nasalis															
Nasalis larvatus	Kinabatangan-Sukai	Peat swamp	220	~100	910	34	74	73	<1	8	—	—	—	8	Boonratana 1994
	Tanjung Puting	Peat swamp	130	96		63	52	42	10	40	3	—	—	5	Yeager 1989, 1990
	Samunsam	Mangrove/heath forest	770	87	485	6	41	38	3	58					Bennett and Sebastian 1988
Simias															
Simias concolor	Pagai Islands	Swamp and lowland forest	7–20			21									Tenaza and Fuentes 1995
	Sarabua	Dipterocarp forest	13	~5		8									Watanabe 1981
	Grukna	Dipterocarp forest	3.5			220									Watanabe 1981

[1] L, leaf; YL, young leaf (a subset of leaf); ML, mature leaf (a subset of leaf); S/F, seed/fruit; FL, flower; B/B, buds and bark; L/F, lichens and fungus; O/U, other or unknown.

Although leaves may be ubiquitous, they are often defended by toxins (e.g., phenols) and digestion-inhibiting materials (e.g., fiber). Asian colobines must therefore be selective in the plant parts they choose and in the plant species from which they choose. This can be quantified with "selectivity ratios," comparing the representation of a species in the habitat to the representation of that species in the diet. Such selectivity ratios show that Asian colobines are picky eaters. This is, of course, variable between species and sites. *P. rubicunda* at Sepilok in Malaysian Borneo, for example, derives only 7% of its diet from the five most common tree families (83% of basal area), while *P. siamensis* at Kuala Lompat in Peninsular Malaysia derives 46% of its diet from the five most common tree families (49% of basal area) (Davies et al. 1988).

Sharp molars, enlarged salivary glands, and multi-chambered stomachs are common to all colobines. Species differences in diet may come from physiological special-izations within these general colobine adaptations (Chivers and Hladik 1980, Caton 1999). *Trachypithecus* may be better adapted to leaf eating than is *Presbytis*, for example (Bennett and Davies 1994, Yeager and Kool 2000). The molars of *Trachypithecus* are better at shearing than the molars of *Presbytis*, and the forestomachs of *Trachypithecus* are larger than the forestomachs of *Presbytis* (Chivers and Hladik 1980, Davies 1991, Chivers 1994). Both *T. obscura* (dusky leaf monkey) and *T. pileatus* eat more mature leaves than any known *Presbytis*; *T. obscura* also eats more leaves than *P. siamensis* at Kuala Lompat, where they are sympatric (Curtin 1980). If indeed *Trachypithecus* is better adapted to leaf eating than *Presbytis*, this may support niche separation in sympatric Asian colobines and explain, in part, the broader distribution of *Trachypithecus* relative to *Presbytis*. Although this hypothesis is tantalizing, there is some evidence that contradicts it. *T. auratus* (Javan lutong) does not eat more leaves than the broadly sympatric *Presbytis comata* (Javan surili), for example (Rodman 1978, Ruhiyat 1983, Kool 1993). Kool's study of *T. auratus* covered only 8 months, however, possibly underrepresenting the leaf portion of the annual diet.

Predation

While the feeding behavior of Asian colobines is well documented, information on predation rates remains elusive. Documented cases of predation have been isolated and opportunistically observed. Examples of these rare events include a female *N. larvatus* preyed upon by a crocodilian during a river crossing at Tanjung Puting in Indonesian Borneo (Yeager 1991a), a juvenile *T. pileatus* preyed upon by jackals at Madhupur in Bangladesh (Stanford 1991a), and a juvenile *R. roxellana* preyed upon by a goshawk at Qinling in China (Zhang et al. 1999).

Predation pressure is best measured by the importance of a species to a predator's diet. This combined with the density of the predator population gives an indication of the frequency with which individuals are at risk from predation

events. Using frequency in feces as an estimate of pro-portion in diet, *S. dussumieri* provided 6% and 27% of the diet of leopards and tigers, respectively, at Kanha in India (Dolhinow 1972). *R. roxellana* provided 2% of the diet for leopards at Wolong in China from 1981 to 1983 and 0% from 1984 to 1987 (Johnson et al. 1993).

At least relative to African colobines, predation rates for Asian colobines generally seem low. Predation is a significant risk for Asian colobines, nonetheless, as evidenced by the evolution of an array of antipredator behaviors. Leopards evoke alarm calls from *S. dussumieri* (Starin 1978), and birds of prey evoke alarm calls in *R. bieti* (Cui 2003). *N. larvatus* sit and scan rivers before crossing, possibly to detect predators (Yeager 1991a). *Presbytis thomasi* (Thomas's langur) have more neighbors when feeding on the ground, presumably an antipredator strategy (Sterck 2002).

REPRODUCTIVE PARAMETERS AND SOCIAL BEHAVIOR

Social and Sexual Behavior

As measured by grooming, Asian colobines are an unsocial lot. They typically spend no more than 2% of their time in social grooming (Table 11.3). Some species, such as *T. johnii* and *P. siamensis*, have an almost complete absence of social grooming (Poirier 1970, Bennett 1983). *S. dussumieri* and *R. bieti* are relatively social, however, with grooming occupy-ing about 6% of their time (Newton 1994, Kirkpatrick et al. 1998).

Social grooming in Asian colobines is primarily a female affair. Males are generally peripheral to social interaction, whether measured by lack of involvement in grooming (e.g., *Presbytis femoralis* [banded surili], Megantara 1989) or in greater distances between males and the other members of a group (e.g., *P. rubicunda*, Davies 1984). Since males apparently do not regulate their relationships with females through social grooming, it is perhaps surprising that male *herding* (i.e., aggressively maintaining proximity of females, particularly in the presence of extragroup males) is reported only for *T. pileatus* (Stanford 1991c).

Almost universally, Asian colobine sex is initiated by females. Females use species-typical behavior patterns, such as the "head shake" in *S. dussumieri* and the "crouch" in *R. roxellana*, to signal sexual proceptivity and encourage male mounting. With the head shake solicitation, the female directs her hindquarters toward the male while "frantically shuddering" her head (Hrdy 1977:49). The crouch is similar, although there is no shaking of the head and the female's forearms are bent to lower the upper body (Ren et al. 1995). Copulations are almost always preceded by female solici-tation, although not all solicitations result in copulation (e.g., *Semnopithecus ajax* [Kashmir gray langur], Borries et al. 2001). *Simias concolor* (pig-tailed langur) is the only Asian colobine in which sexual swellings indicate receptivity (Tenaza 1989).

Table 11.3   Social Behavior in Asian Colobines

SPECIES	SITE	BIRTH SEASON	WEANING AGE (MONTHS)	INTERBIRTH INTERVAL (YEARS)	ALLOPARENTAL CARE?	MALE–INFANT RELATIONS	SOCIAL TIME (%)	REFERENCE
Semnopithecus								
Semnopithecus dussumieri	Orcha	Weak	11–15	~2	Yes			Dolhinow 1972
	Kaukori	Apr–May			Yes			Dolhinow 1972
	Kanha						6	Newton 1994
	Jodhpur	Mar (peak)	13	1.4			6.5	Sommer and Rajpurohit 1989, Sommer et al. 1992, Borries et al. 1994
	Gir						<2	Starin 1978
	Dhawar (open)					Violent during male replacements		Sugiyama 1964
S. schistaceus	Melemchi	Feb–Apr						Bishop 1979
S. ajax	Ramnagar	Jan–Jun (47% Mar)	25	2.4	To 5 months	At times protective, at times violent		Borries et al. 1999a, 2001
Trachypithecus								
Vetulus group								
Trachypithecus vetulus	Polonnaruwa	May–Aug (40% Jun)	7–8	2				Rudran 1973
	Horton Plains			1.3				Rudran 1973
T. johnii	Nilgiri district	May–Jun	10–11	~1.75	Mostly at 3–7 weeks	Passively tolerant	0.1	Poirier 1970
Pileatus group								
T. pileatus	Madhupur	Dec–May		2	Yes	Passively tolerant	0.4	Stanford 1991a,c
Presbytis								
Presbytis femoralis	Perawang	No	9–10		To 3 months		1.5	Megantara 1989
P. siamensis	Kuala Lompat				Rare		Rare	Bennett 1983
P. comata	Kamojang		>12					Ruhiyat 1983
P. thomasi	Bohorok	No						Gurmaya 1986
P. rubicunda	Sepilok				Yes		Rare	Davies 1984
P. potenziani	Betumonga				No		<2	Fuentes 1994
Rhinopithecus								
Rhinopithecus bieti	Wuyapiya	Mar/Apr		~3	Yes	Passively tolerant	6.1%	Kirkpatrick et al. 1998
Nasalis								
Nasalis larvatus	Kinabatangan/Sukai	Weak			Yes		2.2	Boonratana 1994
	Tanjung Puting						Rare	Yeager 1990
	Samunsam				Yes		Rare	Rajanathan and Bennett 1990

Gestation in Asian colobines typically lasts 6–7 months (e.g., *Trachypithecus vetulus* [purple-faced langur], Rudran 1973; *R. roxellana*, Qi 1986; *Semnopithecus dussumieri*, Sommer et al. 1992; *S. ajax*, Borries et al. 2001). Births— and, presumably, copulations—are often concentrated in "seasons" lasting 2–6 months (Table 11.3). It is unclear what regulates birth seasonality, although it is possibly based on nutrition (e.g., *S. ajax*, Ziegler et al. 2000). Seasonality in copulations and births is typically retained by captives, how- ever (e.g., *R. roxellana*, Zhang et al. 2000), so regulatory mechanisms are probably not directly tied to nutritional intake. In general, interbirth intervals are around 2 years,

sometimes less (Table 11.3). Interbirth intervals often vary between different populations of the same species. *Trachypi- thecus vetulus* in Sri Lanka, for example, has an interbirth interval of 2 years at Polonarruwa but only 1.3 years at Horton Plains (Rudran 1973).

Infants and Adults

Asian colobines often wean their infants within the first year (e.g., *T. johnii*, Poirier 1970; *T. vetulus*, Rudran 1973; *P. femoralis*, Megantara 1989). Some studies have reported suckling beyond 1 year, however (e.g., *P. comata*, Ruhiyat

1983; *S. dussumieri*, Dolhinow 1972; Sommer et al. 1992). Weaning may be based on nutritional constraints, with monkeys at poorer sites weaning later. The longest reported lactation period is for *S. ajax* at Ramnagar, Nepal, a relatively poor site with long interbirth intervals. Females at Ramnagar typically wean their infants only after again becoming pregnant, an average of 25 months (Borries et al. 2001).

Nonmaternal caretakers, or *alloparents*, are found in virtually all Asian colobines for which we have good records (Table 11.3). The alloparent may hold, groom, or carry the infant. Alloparents are usually adult females, at times juveniles (mainly female, sometimes male). Alloparenting can occur soon after birth and provide a substantial portion of parental care. In the first 4 weeks of life, *P. femoralis* infants are held about a third of the time by individuals other than the mother (Megantara 1989). In *R. bieti*, juveniles provide about a quarter of infant grooming and 15% of infant carriage during long-distance travel (Kirkpatrick et al. 1998).

Adult male Asian colobines are typically "passively tolerant" of infants (Whitten 1987). Males are alloparents only on the rarest occasions. More common are reports of adult males being "protective" of infants, giving alarm calls or standing between the infant and harm. Male protection is context-specific, however, and may depend on relatedness. While certain males (i.e., probable fathers) may protect infants at times, other males (i.e., nonfathers) may be extremely aggressive toward them. At Ramnagar in Nepal, for example, *S. ajax* males in multimale groups are principal defenders of infants against attack but only if they are the genetic father or were resident when the infant was conceived (Borries et al. 1999a).

The question of infanticide in Asian colobines has generated heated debate (e.g., Sussman et al. 1995, Hrdy et al. 1995). The social pathology hypothesis suggests that infanticide results from crowded living conditions. The sexual selection hypothesis ties primates into the larger body of literature on infanticide in mammals, suggesting that males will kill unrelated infants if this increases their reproductive success. An infanticidal male may benefit his reproductive success if (*1*) the female losing the infant becomes sexually receptive sooner than if she had not lost her infant and (*2*) the infanticidal male gains sexual access to her.

As with other types of predation, infanticide is rarely observed. In Asian colobines, it is reported primarily in some populations of *Semnopithecus* (e.g., *S. dussumieri* of Dharwar and of Abu, both in India, Sugiyama 1964, Hrdy 1977, *S. ajax* of Ramnagar in Nepal, Borries et al. 1999a). Infanticide is estimated to cause 20%–30% of infant mortality at Ramnagar (*S. ajax*) and at Kanha and Jodhpur (both sites of *S. dussumieri* in India) (Borries 1997). Data from both *S. ajax* and *S. dussumieri* tend to support the sexual selection hypothesis. In both species, a mother losing an infant to infanticide has a shorter interbirth interval than a mother not losing an infant (Sommer et al. 1992, Borries 1997). Further, genetic analysis of *S. ajax* at Ramnagar clearly shows that a male will not attack his own infant and

that, when a female loses her infant, the male that attacked the infant has a higher chance of siring the female's next offspring (Borries et al. 1999b).

Whatever the pressures leading to infanticide, they are variable among sites; many studies of *Semnopithecus* have not reported infanticide. Also, for all the discussion of infanticide in Asian colobines, it is perhaps surprising that the only other Asian colobines for which there are eyewitness reports in the wild are *T. cristatus* (Wolf and Fleagle 1977) and *P. thomasi* (Sterck 1995). Infanticide has also been suggested for *T. vetulus* at Horton Plains in Sri Lanka, although this was inferred only because births were synchronous within groups, not among groups (Rudran 1973).

SOCIAL ORGANIZATION

The modal pattern of mixed-sex groups in Asian colobines is the one-male unit (OMU). In such groups, one male associates with multiple females and their offspring. In many Asian colobines, over 50% of the males in mixed-sex groups are in OMUs, suggesting that this is the primary strategy (Table 11.4). There is much variation around this central theme, however. In some species, mixed-sex groups are mainly multimale, with two or, rarely, three or more males. In some species, OMUs have overlapping ranges and a second level of social organization, the "band." In such species, OMUs have relatively nonaggressive relationships with a subset of OMUs in the surrounding area, at times traveling together. *T. pileatus* males in OMUs at Madhupur, Bangladesh, for example, respond less aggressively to familiar OMUs than to unfamiliar OMUs (Stanford 1991c). The band form of social organization reaches an extreme in *Rhinopithecus*, with OMUs having complete range overlap and traveling together as a single entity of 200 or more individuals in ranges of over 20 km^2 (Kirkpatrick 1998).

As in most Old World monkeys, male Asian colobines are the "immigrating sex." Male departures typically occur before physical maturity (e.g., *N. larvatus*, Boonratana 1994; *P. thomasi*, Steenbeek et al. 2000). Adult, subadult, and juvenile males that are not part of mixed-sex groups may be solitary or found in all-male units (AMUs). Juvenile females are present in AMUs at times (e.g., *N. larvatus*, *T. auratus*, *T. johnii*, *P. thomasi*, Yeager and Kool 2000, Steenbeek et al. 2000), suggesting that *AMU* is an imprecise term that masks more complex demographic and life history processes. Heterosexual sex has not been reported for AMUs, however.

OMUs vary widely in their size and in the number of females within them (Table 11.4). The OMUs of *T. pileatus* at Madhupur, Bangladesh, are typical of Asian colobines as a whole, with up to 13 individuals and up to five adult females (Stanford 1991a). In contrast, the OMUs of *S. dussumieri* at Kanha, India, are relatively large, with up to 34 individuals and up to 15 adult females (Newton 1987). At the other extreme, *Simias concolor* and *Presbytis potenziani* (Mentawai langur) typically, though not exclusively, have OMUs with

Table 11.4　Social Organization in Asian Colobines

SPECIES	SITE	AVERAGE GROUP SIZE (MSG)	% MALES IN MSG IN OMUS	AM/AF RATIO (MSG)	ADULT/ IMMATURE RATIO (MSG)	I/AF RATIO	INTERGROUP RELATIONS	LENGTH OF MALE TENURE	REFERENCE
Semnopithecus									
Semnopithecus entellus	Orcha	19	0	1:1.6	1:0.9	1:1.4	Avoidance, nonaggressive		Dolhinow 1972
	Kaukori	54	0	1:3.2	1:1.2	1:1.4			Dolhinow 1972
	Kanha (meadow)	22	87	1:7.9	1:1.1	1:1.7		8.7 years	Newton 1994 (data for 1982)
	Gir	28	33	1:6.0	1:0.6	1:2.0			Starin 1978 (data for April)
	Jodhpur	38	100	1:4.9				2.2 years	Sommer and Rajpurohit 1989, Sommer et al. 1992
	Abu							28–31 months	Hrdy 1977
	Dhawar	15	42	1:3.1	1:1.1			27–46 months	Sugiyama 1964, Newton 1994
	Raipur	29	23	1:5.7	1:0.6	1:2.6			Sugiyama 1964
S. schistaceus	Junbesi			1:1.6			Rare (once in 16-month study with 243 days of contact)		Curtin 1975
S. ajax	Ramnagar	18	38	1:2.6					Borries 1997, Borries et al. 2001
Trachypithecus									
Vetulus group									
Trachypithecus vetulus	Horton Plains		100	1:3.4	1:1.0	1:1.4			Rudran 1973
T. johnii	Nilgiri district	9	35	1:1.2	1:1.0	1:1.4	Primarily vocal		Poirier 1970
	Kakachi	17							Oates et al. 1980
Cristatus group									
T. auratus	Pangandaran	14	100	1:6.5	1:0.7	No infants			Kool 1993
T. cristatus	Rantau Panjang	35	0	1:4.6	1:1.4	1:2.6			Furuya 1961
Obscurus group									
T. obscurus	Kuala Lompat	17	0	1:2.4	1:1.0	1:1.7			Curtin 1980
Pileatus group									
T. pileatus	Madhupur	8	100	1:3.6	1:0.8	1:1.4	~1 encounter per day, OMUs with variable aggression based on familiarity, no male loud call	At least to 26 months	Stanford 1991c
T. geei	Manas	10	71	1:3.3	1:0.9	1:1.2			Mukherjee 1978
	Ripu	7		1:2.2	1:0.6	1:2.4			Srivastava et al. 2001
	Chirrang	9		1:1.5	1:0.7	1:2.3			Srivastava et al. 2001
	Manas 01	11		1:2.2	1:0.8	1:1.9			Srivastava et al. 2001
	Manas 02	7		1:3.8	1:0.6	1:4.6			Srivastava et al. 2001
	Jam Duar	10	71	1:3.4	1:1.0	1:1.7	Common, nonaggressive		Mukherjee 1978, Mukherjee and Saha 1974
Presbytis									
Presbytis femoralis	Perawang	11	71	1:4.8	1:0.8		Males aggressive		Megantara 1989
P. siamensis	Kuala Lompat	15	100	1:7.8	1:0.7	1:2.9	~1 encounter per day		Bennett 1983
	Kuala Lompat	16	11	1:2.6	1:1.2	1:3.3	Mostly mutual avoidance, mediated vocally, relations typically aggressive		Curtin 1980
	Kuala Lompat	14	38	1:3.1	1:1.2	1:1.7			Curtin 1976

Table 11.4   (*cont'd*)

SPECIES	SITE	AVERAGE GROUP SIZE (MSG)	% MALES IN MSG IN OMUS	AM/AF RATIO (MSG)	ADULT/ IMMATURE RATIO (MSG)	I/AF RATIO	INTERGROUP RELATIONS	LENGTH OF MALE TENURE	REFERENCE
P. comata	Kamojang	8	100	1:3.0	1:1.0	1:2.4	Males aggressive, ~1 encounter every 30 hr of observation		Ruhiyat 1983
P. thomasi	Ketambe						Both neutral and with male chases	4.9 years (0.3–7.7)	Steenbeek et al. 2000
	Bohorok	8	70	1:3.0	1:0.7	1:2.3	~1 encounter every 130 hr of observation, typically aggressive		Gurmaya 1986
P. rubicunda	Sepilok	6	100	1:2.0	1:1.0	1:2.0	~1 encounter every 20 days		Davies 1984, 1991
	Tanjung Puting		100	1:2.6	1:0.7	1:1.5			Supriatna et al. 1986
P. potenziani	Betumonga		100	1:1.0	1:0.8	1:3.3	Mostly mutual avoidance, mediated vocally, relations typically aggressive		Fuentes 1994
	Sarabua/ Grukna	3	100	1:1.0	1:0.6	1:2.3			Watanabe 1981
Pygathrix									
Pygathrix nemaeus	Son Tra	9	50	1: 2.7	1: 0.9	1:1.6			Lippold 1977
Rhinopithecus									
Rhinopithecus roxellana	Shennongjia	12 (B: 340)	"Most"				OMUs travel together as band		Su et al. 1998
R. bieti	Wuyapiya	<15 (B: 175)	100	1:3.1	1:1.0	1:2.3	OMUs travel together as band		Kirkpatrick et al. 1998
R. brelichi	Fanjingshan	6 (B: 225)	100	1:2.2	1:0.9	1:2.4	OMUs travel together as band, OMUs "rarely aggressive" to one another		Bleisch et al. 1993, Bleisch and Xie 1998
R. avunculus	Ta Ke/ Nam Trang	15 (B: 65)					OMUs occasionally travel together as band		Boonratana and Le 1998
Nasalis									
Nasalis larvatus	Kinabatangan/ Sukai	17	100	1:7.5	1:1.1	1:2.3	Associations among some OMUs		Boonratana 1994
	Tanjung Puting	12	100	1:3.6	1:0.8	1:1.3	Associations among some OMUs, commonly seen at riverbanks, typically neutral encounters with low-intensity displays by males		Yeager 1990
	Samunsam	9	100	1:3.8	1:0.9	1:1.9	Associations among some OMUs, commonly seen at riverbanks, male displays when OMUs approach one another		Bennett and Sebastian 1988
Simias									
Simias concolor	Pagai Islands	4	95	1:1.8	1:0.4	1:3.2	Apparently rare		Tenaza and Fuentes 1995
	Sarabua/ Grukna	5 (2–20)	100	1:1.7	1:0.8	1:3.2	Rare, both neutral and aggressive		Watanabe 1981

MSG, mixed-sex group; OMU, one-male group; B, band (i.e., an association of OMUs); AM, adult male; AF, adult female; I, infant.

only one female, giving the appearance of monogamous pairs (Watanabe 1981, Tenaza and Fuentes 1995).

Multimale groups of Asian colobines generally have neither more nor fewer females than do OMUs (Sterck and van Hooff 2000). This suggests that multimale groups are not a strategy to maintain access to more females. In many cases, multimale groups may be relatively transitory, signaling the end of a resident male's tenure, to be replaced by either a maturing son or, more commonly, a male that has recently entered the OMU. The dynamic, unstable nature of multimale groups in most species of Asian colobines is brought about by the reproductive interests of individual males at different times in their lives and is reviewed more fully below (see Population Consequences of Individual Strategies).

At times, the home ranges of Asian colobines do not overlap much. In these cases, physical encounters between mixed-sex groups are rare. When range overlap is greater, encounters are more frequent. Usually, encounters are mediated by loud calls given by males. In some species, mixed-sex groups are aggressive when they meet one another, with chases by the males of differing groups. In other species, mixed-sexed groups are tolerant of one another, even traveling together (*N. larvatus*, Yeager 1991b; *T. pileatus*, Stanford 1991c; *T. geei*, Mukherjee and Saha 1974). Although OMUs in these species may travel together, they apparently maintain integrity and are still at times aggressive to one another (e.g., *Rhinopithecus brelichi* [gray snub-nosed langur], Bleisch et al. 1993).

This striking combination—spatial overlap of groups with maintenance of OMU boundaries—has led to the hypothesis that intergroup encounters are based on mate defense by males and not, as widely argued, on food defense (van Schaik et al. 1992). Groups do not fight over food; rather, males defend their access to females. Two examples of this are *P. siamensis* and *T. pileatus*. In *P. siamensis*, two OMUs will at times come together in a single feeding tree, apparently not using contest competition for access to food, and intergroup aggression is a male affair (Bennett 1983). In *T. pileatus*, OMUs do not defend food resources and only males, not females, are involved in aggressive intergroup encounters (Stanford 1991a).

POPULATION CONSEQUENCES OF INDIVIDUAL STRATEGIES

In making decisions about whether to join or leave a group—or to contest the entry of or exit by another in one's group—different individuals have different interests. For example, male and female interests may overlap but they are not identical. Further, individual interests will change over time—as a young adult becomes fully mature, for example.

Population structure emerges from the choices made by individuals within that population. This is seen most clearly, perhaps, in changes of group composition over time, including the establishment of new groups. As with other primates,

groups of Asian colobines are dynamic. In addition to births and deaths, group change is driven by reproductive decisions: by male–male competition for access to females, often seen in aggressive contests, and by female choice, with females leaving their current group to join another group or to join a lone male, forming a new group.

Male–Male Competition

If an adult male is to have offspring, he needs residence in an OMU or high rank in a multimale group. For *S. dussumieri* at Jodhpur, India, the 25% of males that are never adult residents of OMUs—and have a longer life span because of this—share only 5% of conceptions (Sommer and Rajpurohit 1989). In multimale groups of *S. ajax* at Ramnagar, Nepal, male dominance rank correlates with number of matings and, presumably, offspring (Borries 1997).

An adult male generally will tolerate other males in his group only if the potential fitness benefits are greater than the costs. Theoretically, benefits can come from improved predator protection, enhanced intergroup competition for food, or support against infanticidal assaults by outside males. There is little reason to believe that an additional male provides more or better predator protection than an additional female. This is particularly true in Asian colobines without sexual dimorphism (e.g., *Presbytis* spp.). Also, Asian colobines do not have aggressive intergroup competition for food. More likely, multimale groups result when more than one male has a fitness interest in the group, either because one male cannot monopolize the females in a group or because a male has reached maturity within his natal group. Although some multimale groups are stable over time, many are not, with the males in these groups engaging in either sudden contests or gradual transitions that result in OMUs.

Stable multimale groups are best known for *Semnopithecus*. Groups of *S. schistaceus* are relatively large, possibly because they live in low-productivity environments (Bishop 1979). A single male cannot monopolize the females in these groups. Multimale groups of *S. ajax* at Ramnagar have up to five adult males, with a hierarchy of sexual access to females (Borries et al. 1999a). Low rank may cause some males to transfer; new resident males in groups at Ramnagar are former residents of adjacent groups about a third of the time (Borries et al. 1999b).

Although some multimale groups are relatively stable, most are in transition. When there are two or more males in a group, their relations can become quite aggressive, to the point of group split or one of the males being expelled. This transition can happen slowly, as with most age-graded multimale groups, or quickly, as when extragroup males aggressively join OMUs.

When the dominant male of an OMU is tolerant of his sons, allowing them to stay after maturation, the OMU becomes an age-graded, multimale group. This occurs about half the time in *P. thomasi* at Ketambe, Indonesia (Steenbeek et al.

2000). These young adult males are aggressive with extragroup males, both from mixed-sex groups and all-male bands, keeping them away from females and infants. Age-grade groups in *P. thomasi* merely delay immigration; young adult males rarely replace their fathers as the sexually active male of the group (Steenbeek et al. 2000, cf. Gurmaya 1986). This is variable among species, however. In *T. johnii*, age-graded groups may split, with both the older male and younger male resident in the resulting OMUs (Hohmann 1989).

The transitory multimale groups of Asian colobines are not always age-graded. In *T. johnii* and *P. femoralis*, extragroup males at times join OMUs with surprisingly little aggression (Poirier 1970, Hohmann 1989, Megantara 1989). Over a period of months, however, these groups split due to gradually increasing aggression between the males. Alternately, the transition from multimale group to OMU or group split can happen quite quickly. Violence may start immediately after an extragroup male joins a group, continue for days or weeks, and lead to replacement of the OMU's resident male. The multimale phase in some populations of *S. dussumieri*, for example, is generally short-lived and seems to occur only in the context of aggressive contests between males (e.g., at Dharwar, Sugiyama 1964; at Abu, Hrdy 1977; at Kanha, Newton 1987—all three sites in India). At Dharwar, male invasions from AMUs will split a group; aggression between the resident male and invading males can result in two OMUs, with females divided between them (Sugiyama 1964). Although age-graded multimale groups are common in *P. thomasi*, violent change in resident males of OMUs occurs as well (Steenbeek et al. 2000).

Female Choice

Females are active participants in decisions about group size and structure. In mammals, variance in reproductive success is generally lower for females than for males, leading to female philopatry and male immigration. In primates, females generally resist movement between groups because foods and predators are unknown in the range of the new group and there is less chance to find related individuals with which to form alliances for contest competition (Wrangham 1980). Yet, colobine females may have lower resistance to changing groups because foods are relatively ubiquitous and, for Asian colobines, predation pressure is low. Scramble competition, rather than contest competition, is more relevant. Particularly in species or populations in which groups have overlapping ranges, female resistance to movement may be negligible. Although males are consistently the immigrating sex in colobines (i.e., the sex that almost invariably changes group residency prior to becoming sexually active), females may also transfer (i.e., change group residency either before or after becoming sexually active).

Female transfer is common in Asian colobines (Yeager and Kool 2000, Sterck and van Hooff 2000), and female mate choice is important in determining group size (e.g., *T. pileatus*, Stanford 1991a). Females interact with lone males following established OMUs, at times leaving their current group to join a lone male or transferring to a small, newly formed group (*T. pileatus*, Stanford 1991c; *P. thomasi*, Steenbeek et al. 2000). In *T. johnii*, *P. siamensis*, and *N. larvatus*, even females with infants will transfer (Poirier 1970, Bennett 1983, Bennett and Sebastian 1988). By inference, this suggests that the risk of infanticide is low.

Female choice is important during division of multimale groups as well. In *T. johnii* and *P. femoralis*, division of multimale groups is presaged by consistent membership in subgroups—seen during feeding, for example (Hohmann 1989, Megantara 1989). Females make choices about which male to join and, therefore, are agents in deciding whether the resident male will be replaced or whether the group will split (e.g., *S. dussumieri*, Sugiyama 1964; *T. johnii*, Hohmann 1989).

What do females need from males? A group's male defends females from the harassment of males from other mixed-sex groups and from all-male units. In some cases, Asian colobine females may also compete for sperm (e.g., *S. dussumieri*, Sommer et al. 1992). In larger groups, one male may not provide enough sperm to impregnate all receptive females. If so, species with tightly constrained breeding seasons should have the smallest groups; females choose smaller groups to ensure access to sperm during the brief window for impregnation.

The male characters subject to female choice remain unclear, as are the mechanisms by which Asian colobine males maintain the allegiance, or forbearance, of females. Male Asian colobines are unsocial, are minor participants in grooming networks, and do not "herd" females. When adult females transfer between existing groups, they may be evaluating female companions as much as male characteristics. Given the general consistency in size of groups of different species of Asian colobines, it seems likely that females balance mate choice with relative group size.

Stability and Change in Social Organization

Population structure results from the interactions of strategies of different individuals. Further, different processes are at work in the same group at different times. Since group changes are based on individual decisions, it is unsurprising that the speed and extent of group change vary over time and across both populations and species. In both *N. larvatus* and *P. thomasi*, for example, some groups remain stable, while others change radically from year to year (Rajanathan and Bennett 1990, Yeager 1990, Sterck 1995). Groups of *T. johnii* also change substantially over varying periods of time, following unpredictable and idiosyncratic patterns (Poirier 1970, Hohmann 1989).

Notwithstanding social change within groups, social organization at the level of the population can be relatively constant. In *N. larvatus*, for example, multitiered social organization—OMUs associating in bands—is consistent among the three sites with long-term studies (Bennett and

Sebastian 1988, Yeager 1991b, Boonratana 1994). In some species, however, there is variation in social organization among sites. *S. dussumieri* is the best-studied Asian colobine and is found only in OMUs at Jodhpur, only in multimale groups at Orcha, and in a variable mix of the two at Kanha, at Gir, and at other sites (Sommer and Rajpurohit 1989, Dolhinow 1972, Newton 1987, Starin 1978; see Newton and Dunbar 1994 for a comprehensive review).

Grooming relationships and the lack of male herding behavior suggest that female decisions maintain internal group structure, including the number of females in a group. Intergroup relations are the province of both males and females, with male–male aggression creating distance between groups and female transfer possibly bringing groups closer together. *Rhinopithecus* provides an instructive example. Given the size of bands of *Rhinopithecus*, in which over 200 individuals travel together, it is perhaps surprising that these are not multimale groups. Males attach to and defend subgroups of females within the band. This is possibly based on phylogenetic constraints whereby colobine males are relatively ineffective at forming the cooperative relationships necessary to jointly exclude other males from a larger group.

As we build a greater understanding of Asian colobines, it is worthy and justifiable to look for attributes at the level of genus and species. We must be wary of an "essentialist" approach that dismisses variation between studies, however. Observations of the most widely studied Asian colobine, *S. dussumieri*, show that the more information we have, the more diversity we see. Similarly, the studies that occur over the longest time spans, such as those for *P. thomasi* at Ketambe, Indonesia, show that individual behavior is flexible and opportunistic, leading to much range around species means. This diversity on common themes is the central lesson of the study of Asian colobines.

REFERENCES

Bennett, E. L. (1983). The banded langur: ecology of a colobine in west Malaysian rain-forest [PhD diss]. Cambridge University, Cambridge.

Bennett, E. L. (1986). Environmental correlates of ranging behaviour in the banded langur, *Presbytis melalophus*. *Folia Primatol*. 47:26–38.

Bennett, E. L., and Davies, A. G. (1994). The ecology of Asian colobines. In: Davies, A. G., and Oates, J. F. (eds.), *Colobine Monkeys: Their Ecology, Behaviour and Evolution*. Cambridge University Press, Cambridge. pp. 129–172.

Bennett, E. L., and Sebastian, A. C. (1988). Social organization and ecology of proboscis monkeys (*Nasalis larvatus*) in mixed coastal forest in Sarawak. *Int. J. Primatol*. 9:233–255.

Bishop, N. H. (1979). Himalayan langurs: temperate colobines. *J. Hum. Evol*. 8:251–281.

Bleisch, W. V., Cheng, A. S., Ren, X. D., and Xie, J. H. (1993). Preliminary results from a field study of wild Guizhou snub-nosed monkeys (*Rhinopithecus brelichi*). *Folia Primatol*. 60:72–92.

Bleisch, W. V., and Xie, J. H. (1998). Ecology and behavior of the Guizhou snub-nosed langur (*Rhinopithecus [Rhinopithecus] brelichi*), with a discussion of socioecology in the genus. In: Jablonski, N. G. (ed.), *The Natural History of the Doucs and Snub-Nosed Monkeys*. World Scientific Press, Singapore. pp. 217–240.

Boonratana, R. (1994). The ecology and behaviour of the proboscis monkey (*Nasalis larvatus*) in the Lower Kinabatangan, Sabah [PhD diss]. Mahidol University, Bangkok.

Boonratana, R., and Le, X. C. (1998). Preliminary observations of the ecology and behavior of Tonkin snub-nosed monkey (*Rhinopithecus [Presbytiscus] avunculus*) in northern Vietnam. In: Jablonski, N. G. (ed.), *The Natural History of the Doucs and Snub-Nosed Monkeys*. World Scientific Press, Singapore. pp. 207–216.

Borries, C. B. (1997). Infanticide in seasonally breeding multimale groups of Hanuman langurs (*Presbytis entellus*) in Ramnagar (south Nepal). *Behav. Ecol. Sociobiol*. 41:139–150.

Borries, C. B., Koenig, A., and Winkler, P. (2001). Variation of life history traits and mating patterns in female langur monkeys (*Semnopithecus entellus*). *Behav. Ecol. Sociobiol*. 50:391–402.

Borries, C. B., Laumhardt, K., Epplen, C., Epplen, J. T., and Winkler, P. (1999a). Males as infant protectors in Hanuman langurs (*Presbytis entellus*) living in multimale groups—defense pattern, paternity and sexual behavior. *Behav. Ecol. Sociobiol*. 46:350–356.

Borries, C. B., Laumhardt, K., Epplen, C., Epplen, J. T., and Winkler, P. (1999b). DNA analyses support the hypothesis that infanticide is adaptive in langur monkeys. *Proc. R. Soc. Lond. B* 266:901–904.

Borries, C. B., Sommer, V., and Srivastava, A. (1994). Weaving a tight social net: allogrooming in free-ranging female langurs (*Presbytis entellus*). *Int. J. Primatol*. 15:421–443.

Brandon-Jones, D. (1996). The Asian colobinae (Mammalia: Cercopithecidae) as indicators of Quaternary climatic change. *Biol. J. Linn. Soc*. 59:327–350.

Brandon-Jones, D., Eudey, A. A., Geissmann, T., Groves, C. P., Melnick, D. J., Morales, J. C., Shekelle, M., and Stewart, C. B. (2004). Asian primate classification. *Int. J. Primatol*. 25:97–164.

Caton, J. M. (1999). Digestive strategy of the Asian colobine genus *Trachypithecus*. *Primates* 40:311–325.

Chivers, D. J. (1994). The gastrointestinal tract. In: Davies, A. G., and Oates, J. F. (eds.), *Colobine Monkeys: Their Ecology, Behaviour and Evolution*. Cambridge University Press, Cambridge. pp. 205–227.

Chivers, D. J., and Hladik, C. M. (1980). Morphology of the gastrointestinal tract in primates: comparisons with other mammals in relation to diet. *J. Morphol*. 166:337–386.

Cui, L. W. (2003). A note on an interaction between *Rhinopithecus bieti* and a buzzard at Baima Snow Mountain. *Folia Primatol*. 74:51–53.

Curtin, R. A. (1975). The socioecology of the common langur, *Presbytis entellus*, in the Nepal Himalayas [PhD diss]. University of California, Berkeley.

Curtin, S. H. (1976). Niche separation in sympatric Malaysian leaf-monkeys (*Presbytis obscura* and *Presbytis melalophos*). *Ybk. Phys. Anthropol*. 20:421–439.

Curtin, S. H. (1980). Dusky and banded leaf monkeys. In: Chivers, D. J. (ed.), *Malayan Forest Primates*. Plenum Press, New York. pp. 107–145.

Davies, A. G. (1984). An ecological study of the red leaf monkey (*Presbytis rubicunda*) in the dipterocarp forest of northern Borneo [PhD diss]. Cambridge University, Cambridge.

Davies, A. G. (1991). Seed-eating by red leaf monkeys (*Presbytis rubicunda*) in dipterocarp forest of northern Borneo. *Int. J. Primatol.* 12:119–144.

Davies, A. G. (1994). Colobine populations. In: Davies, A. G., and Oates, J. F. (eds.), *Colobine Monkeys: Their Ecology, Behaviour and Evolution*. Cambridge University Press, Cambridge. pp. 285–310.

Davies, A. G., Bennett, E. L., and Waterman, P. G. (1988). Food selection by two South-east Asian colobine monkeys (*Presbytis rubicunda* and *Presbytis melalophos*) in relation to plant chemistry. *Biol. J. Linn. Soc.* 34:33–56.

Dolhinow, P. J. (1972). The north Indian langur. In: Dolhinow, P. J. (ed.), *Primate Patterns*. Holt, Rinehart and Winston, New York. pp. 181–238.

Fuentes, A. (1994). The socioecology of the Mentawai Island langur [PhD diss]. University of California, Berkeley.

Furuya, Y. (1961). The social life of silvered leaf monkeys. *Primates* 3:41–60.

Groves, C. P. (1970). The forgotten leaf-eaters, and the phylogeny of the Colobinae. In: Napier, J. R., and Napier, P. H. (eds.), *Old World Monkeys: Evolution, Systematics, and Behavior*. Academic Press, New York. pp. 555–587.

Groves, C. P. (2001). *Primate Taxonomy*. Smithsonian Institution Press, Washington DC.

Gurmaya, K. J. (1986). Ecology and behavior of *Presbytis thomasi* in northern Sumatra. *Primates* 27:151–172.

Hladik, C. M. (1977). A comparative study of the feeding strategies of two sympatric species of leaf monkeys: *Presbytis senex* and *Presbytis entellus*. In: Clutton-Brock, T. H. (ed.), *Primate Ecology: Studies of Feeding and Ranging Behaviour in Lemurs, Monkeys and Apes*. Academic Press, New York. pp. 323–353.

Hohmann, G. (1989). Group fission in Nilgiri langurs (*Presbytis johnii*). *Int. J. Primatol.* 10:441–454.

Hrdy, S. B. (1977). *The Langurs of Abu: Female and Male Strategies of Reproduction*. Harvard University Press, Cambridge, MA.

Hrdy, S. B., Janson, C., and van Schaik, C. (1995). Infanticide: let's not throw out the baby with the bath water. *Evol. Anthropol.* 3:151–154.

Huang, C. M., Wei, F. W., Li, M., Li, Y. B., and Sun, R. Y. (2003). Sleeping cave selection, activity pattern and time budget of white-headed langurs. *Int. J. Primatol.* 24:813–824.

Jablonski, N. G. (1998). The evolution of the doucs and snub-nosed monkeys and the question of the phyletic unity of the odd-nosed colobines. In: Jablonski, N. G. (ed.), *The Natural History of the Doucs and Snub-Nosed Monkeys*. World Scientific Press, Singapore. pp. 13–52.

Johnson, K. G., Wang, W., Reid, D. G., and Hu, J. C. (1993). Food habits of Asiatic leopards (*Panthera pardus fusea*) in Wolong Reserve, Sichuan, China. *J. Mamm.* 74:646–650.

Kay, R. N. G., and Davies, A. G. (1994). Digestive physiology. In: Davies, A. G., and Oates, J. F. (eds.), *Colobine Monkeys: Their Ecology, Behaviour and Evolution*. Cambridge University Press, Cambridge. pp. 229–250.

Kirkpatrick, R. C. (1996). Ecology and behavior of the Yunnan snub-nosed langur (*Rhinopithecus bieti*, Colobinae) [PhD diss]. University of California, Davis.

Kirkpatrick, R. C. (1998). Ecology and behavior of the snub-nosed and douc langurs. In: Jablonski, N. G. (ed.), *The Natural History*

of the Doucs and Snub-Nosed Monkeys. World Scientific Press, Singapore. pp. 155–190.

Kirkpatrick, R. C. (1999). Colobine diet and social organization. In: Dolhinow, P., and Fuentes, A. (eds.), *The Nonhuman Primates*. Mayfield, Palo Alto, CA. pp. 93–105.

Kirkpatrick, R. C., Gu, H. J., and Zhou, X. P. (1999). A preliminary report on the ecology of *Rhinopithecus roxellana* at Baihe Nature Reserve. *Folia Primatol.* 70:117–120.

Kirkpatrick, R. C., Long, Y. C., Zhong, T., and Xiao, L. (1998). Social organization and range use in the Yunnan snub-nosed langur *Rhinopithecus bieti*. *Int. J. Primatol.* 17:13–51.

Kool, K. M. (1992). Food selection by the silver leaf monkey, *Trachypithecus auratus sondaicus*, in relation to plant chemistry. *Oecologia* 90:527–533.

Kool, K. M. (1993). The diet and feeding behavior of the silver leaf monkey (*Trachypithecus auratus sondaicus*) in Indonesia. *Int. J. Primatol.* 14:667–700.

Lippold, L. K. (1977). The douc langur: a time for conservation. In: Rainier, H. S. H. P., and Bourne, G. H. (eds.), *Primate Conservation*. Academic Press, New York. pp. 513–538.

Megantara, E. N. (1989). Ecology, behavior and sociality of *Presbytis femoralis* in east central Sumatra. *Comp. Primatol. Monogr.* 2:171–301.

Mitchell, A. H. (1994). Ecology of Hose's langur, *Presbytis hosei*, in mixed logged and unlogged dipterocarp forest of northeast Borneo [PhD diss]. Yale University, New Haven.

Mukherjee, R. P. (1978). Further observations on the golden langur (*Presbytis geei* Khajuria, 1956), with a note to capped langur (*Presbytis pileatus* Blyth, 1843) of Assam. *Primates* 19:737–747.

Mukherjee, R. P., and Saha, S. S. (1974). The golden langurs (*Presbytis geei* Khajuria, 1956) of Assam. *Primates* 15:327–340.

Newton, P. N. (1987). The social organization of forest hanuman langurs (*Presbytis entellus*). *Int. J. Primatol.* 8:199–232.

Newton, P. N. (1992). Feeding and ranging patterns of forest hanuman langurs (*Presbytis entellus*). *Int. J. Primatol.* 13:245–285.

Newton, P. N. (1994). Social stability and change among forest Hanuman langurs (*Presbytis entellus*). *Primates* 35:489–498.

Newton, P. N., and Dunbar, R. I. M. (1994). Colobine monkey society. In: Davies, A. G., and Oates, J. F. (eds.), *Colobine Monkeys: Their Ecology, Behaviour and Evolution*. Cambridge University Press, Cambridge. pp. 311–346.

Oates, J. F., Waterman, P. G., and Choo, G. M. (1980). Food selection by the south Indian leaf-monkey, *Presbytis johnii*, in relation to leaf chemistry. *Oecologia* 45:45–56.

Poirier, F. E. (1970). The Nilgiri langur (*Presbytis johnii*) of south India. In: Rosenblum, R. A. (ed.), *Primate Behavior: Developments in Field and Laboratory Research*, vol. 1. Academic Press, New York. pp. 251–383.

Qi, J. F. (1986). The Chinese golden monkey: husbandry and reproduction. In: Benirschke, K. (ed.), *Primates: The Road to Self-Sustaining Populations*. Springer-Verlag, New York. pp. 837–843.

Rajanathan, R., and Bennett, E. L. (1990). Notes on the social behavior of wild proboscis monkeys (*Nasalis larvatus*). *Malay. Nat. J.* 44:35–44.

Ren, R. M., Su, Y. J., Yan, K. H., Li, J. J., Yin, Z., Zhu, Z. Q., Hu, Z. L., and Hu, Y. F. (1998). Preliminary survey of the social organization of *Rhinopithecus [Rhinopithecus] roxellana* in Shennongjia National Natural Reserve, Hubei, China. In: Jablonski, N. G. (ed.), *The Natural History of the Doucs* and *Snub-Nosed Monkeys*. World Scientific Press, Singapore. pp. 269–278.

Ren, R. M., Yan, K. H., Su, Y. J., Qi, H. J., Liang, B., Bao, W. Y., and de Waal, F. B. M. (1995). The reproductive behavior of golden monkeys in captivity (*Rhinopithecus roxellana roxellana*). *Primates* 36:135–143.

Rodman, P. S. (1978). Diets, densities, and distributions of Bornean primates. In: Montgomery, G. G. (ed.), *The Ecology of Arboreal Folivores*. Smithsonian Institution Press, Washington DC. pp. 465–478.

Rudran, R. (1973). The reproductive cycles of two subspecies of purple-faced langurs (*Presbytis senex*) with relation to environmental factors. *Folia Primatol*. 19:41–60.

Ruhiyat, Y. (1983). Socio-ecological study of *Presbytis aygula* in west Java. *Primates* 24:344–359.

Sommer, V., and Rajpurohit, L. S. (1989). Male reproductive success in harem troops of Hanuman langurs (*Presbytis entellus*). *Int. J. Primatol*. 10:293–000.

Sommer, V., Srivastava, A., and Borries, C. (1992). Cycles, sexuality, and conception in free-ranging langurs (*Presbytis entellus*). *Am. J. Primatol*. 28:1–27.

Srivastava, A., Biswas, J., Das, J., and Bujarbarua, P. (2001). Status and distribution of golden langurs (*Trachypithecus geei*) in Assam, India. *Am. J. Primatol*. 55:15–23.

Stanford, C. B. (1991a). *The Capped Langur in Bangladesh: Behavioral Ecology and Reproductive Tactics*. Karger, New York.

Stanford, C. B. (1991b). The diet of the capped langur (*Presbytis pileata*) in a moist deciduous forest in Bangladesh. *Int. J. Primatol*. 12:199–216.

Stanford, C. B. (1991c). Social dynamics of intergroup encounters in the capped langur (*Presbytis pileata*). *Am. J. Primatol*. 25:35–47.

Starin, E. D. (1978). A preliminary investigation of home range use in the Gir Forest langur. *Primates* 19:551–568.

Steenbeek, R., Sterck, E. H. M., de Vries, H., and van Hooff, J. A. R. A. M. (2000). Costs and benefits of the one-male, age-graded, and all-male phases in wild Thomas's langur groups. In: Kappeler, P. M. (ed.), *Primate Males: Causes and Consequences of Variation in Group Composition*. Cambridge University Press, Cambridge. pp. 130–145.

Sterck, E. H. M. (1995). Females, foods and fights: a socioecological comparison of the sympatric Thomas langur and long-tailed macaque [PhD diss.]. University of Utrecht, Utrecht.

Sterck, E. H. M. (2002). Predator sensitive foraging in Thomas langurs. In: Miller, L. E. (ed.), *Eat or Be Eaten: Predator Sensitive Foraging Among Primates*. Cambridge University Press, Cambridge. pp. 74–91.

Sterck, E. H. M., and van Hooff, J. A. R. A. M. (2000). The number of males in langur groups: monopolizability of females or demographic processes? In: Kappeler, P. M. (ed.), *Primate Males: Causes and Consequences of Variation in Group Composition*. Cambridge University Press, Cambridge. pp. 120–129.

Su, Y. J., Ren, R. M., Yan, K. H., Li, J. J., Yin, Z., Zhu, Z. Q., Hu, Z. L., and Hu, Y. F. (1998). Preliminary survey of the home range and ranging behavior of golden monkeys (*Rhinopithecus [Rhinopithecus] roxellana*) in Shennongjia National Natural Reserve, Hubei, China. In: Jablonski, N. G. (ed.), *The Natural History of the Doucs and Snub-Nosed Monkeys*. World Scientific Press, Singapore. pp. 255–268.

Sugiyama, Y. (1964). Group composition, population density and some sociological observations of hanuman langurs (*Presbytis entellus*). *Primates* 5:7–37.

Supriatna, J., Manullang, B. O., and Soekara, E. (1986). Group composition, home range, and diet of the maroon leaf monkey (*Presbytis rubicunda*) at Tanjung Puting Reserve, central Kalimantan, Indonesia. *Primates* 27:185–190.

Sussman, R. W., Cheverud, J. M., and Barlett, T. Q. (1995). Infant killing as an evolutionary strategy: reality or myth. *Evol. Anthropol*. 3:149–151.

Tenaza, R. R. (1989). Female sexual swellings in the Asian colobine *Simias concolor*. *Am. J. Primatol*. 17:81–86.

Tenaza, R. R., and Fuentes, A. (1995). Monandrous social organization of pigtailed langurs (*Simias concolor*) in the Pagai Islands, Indonesia. *Int. J. Primatol*. 16:295–310.

van Schaik, C. P., Assink, P. R., and Salafsky, N. (1992). Territorial behavior in Southeast Asian langurs: resource defense or mate defense? *Am. J. Primatol*. 26:233–242.

Watanabe, K. (1981). Variations in group composition and population density of the two sympatric Mentawaian leaf-monkeys. *Primates* 22:145–160.

Whitten, P. L. (1987). Infants and adult males. In: Smuts, B. B., Cheney, D. L., Seyfarth, R. M., Wrangham, R. W., and Struhsaker, T. T. (eds.), *Primate Societies*. Chicago University Press, Chicago. pp. 343–357.

Wolf, K., and Fleagle, J. G. (1977). Adult male replacement in a group of silvered leaf monkeys (*Presbytis cristata*) at Kuala Selangor, Malaysia. *Primates* 18:949–955.

Wrangham, R. W. (1980). An ecological model of female-bonded primate groups. *Behaviour* 75:262–300.

Yeager, C. P. (1989). Feeding ecology of the proboscis monkey (*Nasalis larvatus*). *Int. J. Primatol*. 10:497–530.

Yeager, C. P. (1990). Proboscis monkey (*Nasalis larvatus*) social organization: group structure. *Am. J. Primatol*. 20:95–106.

Yeager, C. P. (1991a). Possible antipredator behavior associated with river crossings by proboscis monkeys (*Nasalis larvatus*). *Am. J. Primatol*. 24:61–66.

Yeager, C. P. (1991b). Proboscis monkey (*Nasalis larvatus*) social organization: intergroup patterns of association. *Am. J. Primatol*. 23:73–86.

Yeager, C. P., and Kirkpatrick, R. C. (1998). Asian colobine social structure: ecological and evolutionary constraints. *Primates* 39:147–155.

Yeager, C. P., and Kool, K. (2000). The behavioral ecology of Asian colobines. In: Whitehead, P. F., and Jolly, C. J. (eds.), *Old World Monkeys*. Cambridge University Press, Cambridge. pp. 496–521.

Zhang, S., Liang, B., and Wang, L. (2000). Seasonality of matings and births in captive Sichuan golden monkeys (*Rhinopithecus roxellana*). *Am. J. Primatol*. 51:265–269.

Zhang, S., Ren, B., and Li, B. (1999). A juvenile Sichuan golden monkey (*Rhinopithecus roxellana*) predated by a goshawk (*Accipiter gentiles*) in the Qinling Mountains. *Folia Primatologica* 70:175–176.

Ziegler, T., Hodges, K., Winkler, P., and Heistermann, M. (2000). Hormonal correlates of reproductive seasonality in wild female Hanuman langurs (*Presbytis entellus*). *Am. J. Primatol*. 51:119–134.

12

African Colobine Monkeys
Patterns of Between-Group Interaction
Peter J. Fashing

INTRODUCTION

The Colobinae are a subfamily of monkeys found in Africa and Asia that includes some of the most striking and attractive taxa in the primate order. Colobines are perhaps best known for their multichambered stomachs that permit them to digest cellulose and detoxify secondary compounds, thereby enabling them to exploit leaves to a greater extent than most other primates (Chivers 1994, Kay and Davies 1994). Though this morphological adaptation offers colobines an ecological advantage over other primates and may help account for the unusually high contribution of colobines to the primate biomass at many locations (Davies 1994), it does not constrain them to a strictly folivorous diet. Many colobines consume large quantities of seeds, whole fruits, or even lichens as well as leaves (Kirkpatrick 1999). In addition to their stomach morphology, colobines vary from other primates in that their thumbs are reduced, and the feet of some species are unusually large (Struhsaker and Leland 1987, Oates and Davies 1994).

The colobines of Africa differ from their Asian counterparts (see Chapter 11) in several ways. First, due to Africa's more contiguous geography and the lower overall ecological diversity of its forests, less taxonomic divergence has taken place among African colobines than among those in Asia (Oates 1994). Second, while all African colobines are arboreal, there is one Asian colobine genus, *Semnopithecus* or the Hanuman langur, that is predominantly terrestrial. Lastly, there are several minor morphological differences between African and Asian colobines, including a more pronounced reduction of the thumb in African taxa (Szalay and Delson 1979, Oates et al. 1994).

TAXONOMY AND DISTRIBUTION

African colobine taxonomy has long been in a state of flux (Lydekker 1905, Schwartz 1929, Rahm 1970, Szalay and Delson 1979, Struhsaker 1981, Oates and Trocco 1983, Groves 2001, Grubb et al. 2003). Conservative taxonomists split African colobines into only two genera, *Colobus* (black and white colobus) and *Procolobus* (red and olive colobus),

with five species of *Colobus* and two of *Procolobus* (Oates et al. 1994). A recent taxonomy and the one adopted for this chapter (Groves 2001), however, splits the African colobines into three genera, *Colobus* (black and white), *Procolobus* (olive), and *Piliocolobus* (red). It argues that there are five species of *Colobus*, one of *Procolobus*, and nine of *Piliocolobus* (Table 12.1). Based on means of adult male and female weights, *Colobus* spp. are the heaviest at an average of 8.9 kg, *Piliocolobus* spp. are intermediate at 7.6 kg, and *Procolobus verus* is the lightest at 4.5 kg (Oates et al. 1994).

Most species can be identified by differences in their pelage. Four of the five species of *Colobus* feature differing black and white pelage patterns, while the fifth is entirely black (*C. satanas*). *Piliocolobus* exhibits greater color variation than the other genera with pelages of the different forms incorporating various combinations of red, orange, brown, black, gray, and white. The single species of *Procolobus*, *P. verus*, has a dull olive-brown pelage (Groves 2001).

The African colobines inhabit many of the forested regions of equatorial and West Africa (Table 12.1). *Colobus* occur from Ethiopia to Guinea and are allopatric, with the exception of one area of sympatry between *C. angolensis* and *C. guereza* in the Ituri Forest, Democratic Republic of the Congo (DRC; Oates 1994, Bocian 1997). *Piliocolobus* are entirely allopatric and occur from Zanzibar to Senegal. *Procolobus* are limited to West Africa, occurring from Nigeria to Sierra Leone (Oates et al. 1994). At some rain forest locations in West Africa, all three African colobine genera are sympatric (Davies et al. 1999, Korstjens 2001).

ECOLOGY

Habitat Requirements

African colobines live in a variety of forested habitats, though some taxa exhibit greater habitat flexibility than others (Table 12.1). Olive colobus are restricted to lowland rain forests and seem to gravitate toward disturbed areas where they can feed and move stealthily through dense vegetation in the lower and middle canopy (Booth 1957; Oates 1988, 1994; Korstjens 2001). Most forms of red colobus also

Table 12.1   The African Colobines and Their Distributions, Habitats, Major Study Sites, and Active Long-Term Researchers

LATIN NAME	COMMON NAME	GENERAL DISTRIBUTION	HABITATS	MAJOR STUDY SITES	ACTIVE LONG-TERM RESEARCHERS
Colobus angolensis	Angolan colobus	East and central Africa	Lowland rain, coastal, gallery, and montane forests	Ituri (DRC), Nyungwe (Rwanda), Salonga (DRC)	None
C. guereza	Guereza	East and central Africa	Lowland rain, gallery, and montane forests	Ituri (DRC), Kakamega (Kenya), Kibale (Uganda)	C. Chapman, P. Fashing, T. Harris
C. polykomos	King colobus	West Africa	Lowland rain and gallery forests	Taï (Ivory Coast), Tiwai (Sierra Leone)	None
C. satanas	Black colobus	Central Africa	Lowland rain and montane forests	Douala-Edea (Cameroon), Fôret des Abeilles (Gabon), Lopé (Gabon)	None
C. vellerosus	Ursine colobus	West Africa	Lowland rain, dry, and gallery forests	Bia (Ghana), Boabeng-Fiema (Ghana)	P. Sicotte
Piliocolobus badius	Western red colobus	West Africa	Lowland rain forests	Abuko (Gambia), Taï (Ivory Coast), Tiwai (Sierra Leone)	None
P. foai	Central African red colobus	Central Africa	Lowland rain forests	None	None
P. gordonorum	Udzungwa red colobus	Tanzania	Lowland rain and montane forests	Udzungwa Mountains (Tanzania)	T. Struhsaker
P. kirkii	Zanzibar red colobus	Zanzibar	Dry coastal forests	Jozani (Zanzibar)	K. Siex
P. pennantii	Pennant's colobus	Central and West Africa	Lowland rain and montane forests	Gbanraun (Nigeria)	None
P. preussi	Preuss's red colobus	Cameroon and Nigeria	Lowland rain forests	None	None
P. tephrosceles	Ugandan red colobus	East Africa	Lowland rain forests	Gombe (Tanzania), Kibale (Uganda)	C. Chapman, S. Teelen
P. rufomitratus	Tana River red colobus	Kenya	Gallery forests	Tana River (Kenya)	D. Mbora
P. tholloni	Thollon's red colobus	DRC	Lowland rain forests	Salonga (DRC)	None
Procolobus verus	Olive colobus	West Africa	Lowland rain forests	Taï (Ivory Coast), Tiwai (Sierra Leone)	None

Taxonomy based on Groves (2001). Abbreviation for all tables: DRC, Democratic Republic of Congo. Distribution given as general region unless the species occupies two countries or fewer, in which case the individual countries are given.

appear to be limited to lowland rain forests (Oates 1994), though some are capable of occupying montane forests (*Piliocolobus pennantii*, Gonzalez-Kirchner 1997), dry coastal forests (*P. kirkii*, Siex 2003), or gallery forests (*P. badius*, Gatinot 1976; *P. rufomitratus*, Mbora and Meikle 2004a). Unlike olive colobus, red colobus tend to spend most of their time in the upper and middle forest canopy (Struhsaker and Oates 1975, Oates 1994), though some species are known to occasionally descend to the ground to cross forest gaps or feed on aquatic plants, soil, or charcoal (Gatinot 1976, Galat-Luong and Galat 1979, Starin 1991, Struhsaker et al. 1997). Red colobus also tend to be more averse to disturbed areas than olive colobus and may require many decades to recover from population declines caused by human disturbance (Chapman et al. 2000; Mbora and Meikle 2004b) (Fig. 12.1).

Black and white colobus generally occupy a wider variety of habitats than the other African colobines, with *C. angolensis* and *C. guereza* exhibiting particular habitat lability (Table 12.1). *C. guereza* is unusual among the African colobines in that it seems to thrive most in fragmented and secondary forests (Oates 1977a, 1994). Guereza densities are typically low in primary rain forest blocks (Bocian 1997, Poulsen et al. 2001), intermediate in disturbed rain forest blocks (Oates 1974, Fashing and Cords 2000), and high in forest fragments (Schenkel and Schenkel-Hulliger 1967, Leskes and Acheson 1971, Dunbar 1987, Krüger et al. 1998). In fact, *C. guereza* is one of the few species of primates known to reach higher densities in logged than unlogged forest (Skorupa 1986, Plumptre and Reynolds 1994, Chapman et al. 2000). Though they spend considerable amounts of time in the canopy, black and white colobus tend to occupy a lower strata than red colobus at sites where they are sympatric (Struhsaker and Oates 1975, Korstjens 2001) and are the African colobines that most often descend to ground level to cross forest gaps or to feed on soil, swamp plants, climbers, or terrestrial herbaceous vegetation (Booth 1956; Dunbar and Dunbar 1974; McKey 1978a,b; Oates 1978; Harrison and Hladik 1986; Fashing 2001b; Fimbel et al. 2001; Fashing et al. in press).

Figure 12.1 *Piliocolobus kirkii* inhabit the dry coastal forests of Zanzibar and, with fewer than 2,000 individuals remaining, are among the world's most endangered primates. (Siex and Struhsaker 1999a, photo by Kirstin Siex)

Predators

The natural predators of African colobines are crowned-hawk eagles (*Stephanoaetus coronatus*), chimpanzees (*Pan troglodytes*), and leopards (*Panthera pardus*) (Oates 1996, Schultz et al. 2004). In the case of *C. guereza*, for example, the primary predator appears to be crowned-hawk eagles (Fashing and Oates in press). Studies in nearby forest compartments at Kanyawara (Kibale Forest, Uganda) found that guerezas accounted for 39% and 10%, respectively, of the prey of crowned-hawk eagles (Skorupa 1987, Struhsaker and

Leakey 1990). In comparison, the rate of chimpanzee predation on *C. guereza* at another Kibale site (Ngogo) proved to be quite low, with guerezas accounting for only 4% of the chimpanzee's mammalian prey items (Watts and Mitani 2002). Leopards also prey on black and white colobus at low rates. *C. angolensis* and *C. guereza* combined to account for 1% of prey items in leopard scats at Ituri Forest, DRC (Hart et al. 1996), and *C. polykomos* remains were found in 2%–8% of leopard scats at Taï Forest, Ivory Coast (Hoppe-Dominik 1984, Zuberbühler and Jenny 2002).

Unlike black and white colobus, red colobus are often preyed upon heavily by chimpanzees (Fig. 12.2). At the four sites where chimpanzee hunting behavior has been studied intensively (Taï, Mahale [Tanzania], Gombe [Tanzania], Kibale), red colobus account for 53%–88% (mean = 76%) of the chimpanzee's mammalian prey items (Boesch and Boesch 1989, Uehara 1997, Stanford 1998, Watts and Mitani 2002). Stanford (1998, 2002) provided compelling evidence that chimpanzee hunting patterns exert a major influence on red colobus population size, demography, and behavior at Gombe. He found that mean annual mortality due to chimpanzee predation was 18% for the Gombe red colobus, with immatures disproportionately represented among the individuals killed. Once chimpanzees were detected by red colobus, the latter often engaged in long bouts of alarm calling and males banded together to counterattack chimpanzee hunters that approached their group (Stanford 1998). The more red colobus males that counterattacked, the less likely a hunt was to be successful (Stanford 1998). At Taï, where the canopy is more closed than at Gombe, male red colobus do not counterattack but rather retreat quietly along with other group members when chimpanzees are detected (Bshary and Nöe 1997a). Another strategy employed by Taï red colobus to reduce risk of predation by chimpanzees is to form polyspecific associations with a guenon species (*Cercopithecus diana*) that is particularly vigilant and utters alarm calls when chimpanzees are detected (Bshary and Nöe 1997a,b).

Though chimpanzees are often their main predator, crowned-hawk eagles also consume red colobus. In the two aforementioned studies of eagles at Kibale, red colobus accounted for 18% and 20% of eagle prey items, respectively (Skorupa 1987, Struhsaker and Leakey 1990). Red colobus are also eaten by leopards, and their remains were found in 4%–11% of leopard scats at Ituri and Taï (Hoppe-Dominik 1984, Hart et al. 1996, Zuberbühler and Jenny 2002). Little is known about the predators of the secretive olive colobus. However, it has been speculated that the small group sizes, tendency to form polyspecific associations with *Cercopithecus diana*, and cryptic habits of this species are responses to the threat of chimpanzee predation (Oates and Whitesides 1990, Korstjens 2001).

Diet

In all African colobine species studied, leaves make up a substantial portion (26%–92%) of the diet (Table 12.2).

Figure 12.2 Western red colobus (*Piliocolobus badius*) are the primary targets of both human and chimpanzee hunters in Taï Forest, Ivory Coast (photo courtesy of www.florianmoellers.com).

Though their diet has been investigated at only two sites, olive colobus may be the most consistently folivorous (74%–85%) of the African colobine genera. At both sites, they consumed far less fruit than sympatric *C. polykomos* and *Piliocolobus badius*. Most forms of red colobus are also primarily folivorous, with an apparent preference for young leaves (e.g., Struhsaker 1975, Marsh 1981a, Werre 2000, Usongo and Amubode 2001). *P. tephrosceles* appears to be especially folivorous and at one site, Kibale, has exhibited an increasing reliance on leaves over the past several decades. *P. badius* populations in West Africa are unusual in that they tend to consume nearly as much fruit and/or seeds as leaves. Many red colobus populations differ from other African colobines in that flowers comprise ≥10% of their diet.

Black and white colobus appear to have the most variable diets of the African colobines (Table 12.2). While most are primarily folivorous, numerous populations, especially those of *C. satanas*, feed heavily on seeds when they are available. This granivory often occurs at central and West African sites where leguminous seed-producing tree species are abundant and is not a response to scarcity of high-quality foliage (Oates et al. 1990). Studies in Nyungwe Forest, Rwanda, show that *C. angolensis* diets vary widely over time even among groups occupying the same general area and, in some instances, include large quantities of lichen, an unusual food for primates. Even *C. guereza*, which early studies suggested was among the most folivorous colobines, appears to be quite flexible dietarily. Guerezas at three different sites across their range include large quantities of whole fruits and/or seeds in their annual diets (Budongo, Uganda, A. J. Plumptre unpublished data; Dia, Cameroon, Poulsen et al. 2002; Kakamega, Kenya, Fashing 2001b), with those in Kakamega Forest, Kenya, clearly selecting

whole fruits over young leaves when both are available (Fashing 2001b).

While young leaves, seeds, or whole fruits generally appear to be their preferred foods, most red as well as black and white colobus are capable of switching their focus to lower-quality mature leaves in times of preferred food scarcity (Struhsaker 1975, Oates 1977a, Marsh 1981b, McKey et al. 1981, Dasilva 1994, Fashing 2004). However, even at times of preferred food scarcity, olive colobus appear limited in their ability to exploit mature leaves (Oates 1988). This limitation may result from the small body size of olive colobus, which limits their ability to efficiently engage in the forestomach fermentation necessary to extract nutrients from mature leaves (Oates 1988, Kay and Davies 1994).

When feeding on leaves, African colobines select those that are high in protein (*C. angolensis*, Maisels et al. 1994; *C. guereza*, Bocian 1997; *Piliocolobus preussi*, Usongo and Amubode 2001; *P. tholloni*, Maisels et al. 1994), low in fiber (*C. angolensis*, Bocian 1997; *Procolobus verus*, Oates 1988), or both (*C. guereza*, Chapman et al. 2004; *C. satanas*, McKey et al. 1981; *Piliocolobus rufomitratus*, Mowry et al. 1996). Some colobines also avoid leaves high in secondary compounds (*C. guereza*, Oates et al. 1977; P. J. Fashing et al. unpublished data; *C. polykomos*, Dasilva 1994; *Procolobus verus*, Oates 1988), whereas others appear undeterred by these compounds (*C. angolensis* and *Piliocolobus tholloni*, Maisels et al. 1994; *P. rufomitratus*, Mowry et al. 1996; *P. tephrosceles*, Chapman and Chapman 2002). The consumption of secondary compounds by colobines is probably facilitated by the fermentative abilities of their forestomachs, which may be capable of neutralizing the effects of some plant toxins before they are absorbed into the stomach (Kay and Davies 1994).

Table 12.2 Diets of African Colobine Monkeys (9-Month Study Minimum)[1]

SPECIES	STUDY SITE	% COMPOSITION OF DIET BY FOOD PART								NUMBER OF SPECIES IN DIET	REFERENCES
		YL	ML	UL	TL	FL	FR	(SD)	OT		
Colobus angolensis	Ituri, D.R. Congo	26	2	22	50	7	28	(22)	15	37	Bocian 1997
C. angolensis	Nyungwe, Rwanda[2] (1987–88)	30	7	1	38	1	23	(20)	37[3]	45+	Vedder and Fashing 2002
C. angolensis	Nyungwe, Rwanda (1993)	25	40	7	72	5	17	—	6	59+	Fimbel et al. 2001
C. guereza	Kibale (Kanyawara), Uganda (1971–72)	65	13	3	81	2	15	—	2	43	Oates 1977a, 1994
C. guereza	Kibale (Kanyawara), Uganda (1998–99)	81	5	1	86	1	5	—	8	—	Wasserman and Chapman 2003
C. guereza	Kibale (fragment), Uganda	65	14	1	80	6	12	—	2	—	Wasserman and Chapman 2003
C. guereza	Kibale (logged), Uganda	78	5	3	87	3	10	—	0	—	Wasserman and Chapman 2003
C. guereza	Ituri, DRC	30	4	24	58	3	25	(22)	15	31	Bocian 1997
C. guereza[4]	Kakamega, Kenya	24	7	23[5]	54	1	39	(1)	8	37+	Fashing 2001b
C. guereza	Budongo (unlogged), Uganda	—	—	63	63	6	29	(12)	2	—	Plumptre unpub. data
C. guereza	Budongo (logged), Uganda	—	—	51	51	7	40	(11)	2	—	Plumptre unpub. data
C. polykomos	Tiwai, Sierra Leone	30	26	2	58	3	35	(32)	3	46+	Dasilva 1989
C. polykomos	Taï, Ivory Coast	28	20	0	49	3	48	—	1	—	Korstjens et al. in press
C. satanas	Douala-Edea, Cameroon	21	18	0	39	3	53	(53[6])	5	84+	McKey et al. 1981
C. satanas	Lopé, Gabon	23	3	0	26	5	64	(60)	4	65	Harrison personal communication cited in Oates 1994
C. satanas	Fôret des Abeilles, Gabon	35	3	0	38	12	50	(41)	0	109	Gautier-Hion et al. 1997, Fleury and Gautier-Hion 1999
Piliocolobus badius	Fathala, Senegal	42	5	0	47	9	36	(19)	9	39	Gatinot 1978, Oates 1994
P. badius	Abuko, Gambia	35	12	0	47	9	42	(3)	3	89+	Starin 1991
P. badius	Taï, Ivory Coast (1992–93)	24	7	0	31	30	37	—	2	153	Wachter et al. 1997
P. badius	Taï, Ivory Coast (1996–98)	46	4	0	50	20	29	—	1	—	Korstjens et al. in press
P. badius	Tiwai, Sierra Leone	32	20	0	52	16	31	(25)	1	51	Davies et al. 1999
P. kirkii[4,7]	Jozani (forest), Zanzibar (1980–81)	50	7	7	63	8	32	—	0	63	Mturi 1993
P. kirkii[8]	Jozani (forest), Zanzibar (1999)	51	9	4	64	5	26	(2)	6	21	Siex 2003
P. kirkii[9]	Jozani (*shamba*), Zanzibar (1999)	55	7	6	67	6	5	(0)	22[10]	27	Siex 2003
P. pennantii	Gbanraun, Nigeria	56	10	0	66	9	16	(12)	9	19+	Werre 2000
P. preussi[11]	Korup, Cameroon	89	0	0	89	10	1	—	0	17	Usongo and Amubode 2001
P. rufomitratus	Tana River (Mchelelo), Kenya (1973–74)	52	11	1	65	6	25	(1)	4	22	Marsh 1981b
P. rufomitratus[12]	Tana River (Mchelelo), Kenya (1986–88)	61	2	0	63	13	22	—	2	28	Decker 1994
P. rufomitratus[13]	Tana River (Baomo S.), Kenya	46	1	0	47	27	26	—	1	26	Decker 1994
P. tephrosceles[14]	Gombe, Tanzania	35	44	0	79	7	11	—	3	58+	Clutton-Brock 1975a
P. tephrosceles	Kibale (Kanyawara), Uganda (1972–75)	42	24	8	73	16	6	(1)	5	57	Struhsaker 1978
P. tephrosceles	Kibale (Kanyawara), Uganda (1998–99)	70	10	7	87	2	8	—	3	—	Wasserman and Chapman 2003
P. tephrosceles	Kibale (fragment), Uganda	60	22	2	84	2	7	—	7	—	Wasserman and Chapman 2003
P. tephrosceles	Kibale (logged), Uganda	79	7	6	92	2	6	—	1	—	Wasserman and Chapman 2003
P. tholloni	Salonga, DRC	54	6	0	61	1	38	(31)	0	84	Maisels et al. 1994
Procolobus verus[15]	Tiwai, Sierra Leone	59	11	4	74	7	19	(14)	0	50+	Oates 1988
P. verus	Taï, Ivory Coast	83	1	1	85	4	8	—	3	—	Korstjens et al. in press

[1] YL, young leaves; ML, mature leaves; UL, unidentified leaves; TL, total leaves; FL, flowers; FR, fruit; SD, seeds; OT, other.
[2] Different group studied several years earlier than the Fimbel et al. (2001) study.
[3] Includes the 32% of the annual diet composed of lichen.
[4] Mean diet for two groups.
[5] Believed to be mostly mature leaves.
[6] Includes primarily seeds but some records of seeds and pericarp.
[7] Totals exceed 100%.
[8] Mean for three groups.
[9] Mean for four groups.
[10] Includes the 15% of the annual diet composed of herbs.
[11] Based on 17-month study.
[12] Based on 24-month study.
[13] Based on 19-month study.
[14] Based on 9 months of data between August 1969 and June 1970.
[15] Based on 11 months of data between June 1983 and October 1985.

Table 12.3   Ranging Behavior of African Colobine Monkeys (6-Month Study Minimum). (Home Range Sizes Based on Number of 50 × 50 m Grid Cells Entered Unless Otherwise Noted)

SPECIES	STUDY SITE	DENSITY (IND/KM²)	HOME RANGE SIZE (HA)	RANGE OVERLAP (%)	DAILY PATH LENGTH (M) MEAN	RANGE (M)	NUMBER OF GROUPS	CORE AREA	REFERENCES
Colobus angolensis	Ituri, DRC	3	371[1]	100	983	312–1,914	1	No	Bocian 1997
C. angolensis	Nyungwe, Rwanda	–	2,440[2]	–	–	–	1	Yes	Fashing et al. in press
C. guereza	Kibale, Uganda	100	28	74	535	288–1,004	1	Yes	Oates 1977a,c
C. guereza	Ituri, DRC	17	100[1]	≥22	609	268–1,112	1	Yes	Bocian 1997
C. guereza	Kakamega, Kenya	150	18[3]	67	588	166–1,360	5	Yes	Fashing 2001a,c
C. guereza	Entebbe, Uganda	63[4]	8	+	386	62–1,036	3	–	Grimes 2000
C. guereza	Budongo, Uganda (1964–65)	–	14	Minimal	–	–	5	–	Marler 1969
C. guereza	Budongo, Uganda (1967–73)	49	14[5]	Minimal	–	–	25	–	Suzuki 1979
C. guereza	Budongo, Uganda (1994–95)	20–50	10–33	–	–	–	6	–	Plumptre unpub. data
C. polykomos	Tiwai, Sierra Leone	67	24	≥20[6]	832	350–1,410	1	Yes	Dasilva 1989, Oates 1994
C. polykomos	Taï, Ivory Coast	47	77[7]	>21[8]	617	241–1,341	2	Yes	Korstjens 2001
C. satanas	Douala-Edea, Cameroon	38	60	+	459	<100->800	1	Yes	McKey 1978a,b; McKey and Waterman 1982
C. satanas	Lopé, Gabon	11[9]	184	–	510	40–1,100	1	–	Harrison 1986, Harrison personal communication cited in Oates 1994, White 1994
C. satanas	Fôret des Abeilles, Gabon	8	573[10]	≥65	852	20–1,983	1	No[11]	Fleury and Gautier-Hion 1999
C. vellerosus	Bia, Ghana	–	48	?	307	75–752	1	–	Olson 1986, Oates 1994
C. vellerosus	Boabeng-Fiema, Ghana	105	9	≥33	300	–	1	–	Sicotte and MacIntosh 2004
Piliocolobus badius	Fathala, Senegal	–	20	–	–	–	1	–	Gatinot 1977
P. badius	Abuko, Gambia	124	34	60	–	–	1	–	Starin 1991
P. badius	Tiwai, Sierra Leone	66	53	10–20	–	–	1	–	Davies et al. 1999
P. badius	Taï, Ivory Coast (1992–93)	–	64[10]	–	967	–	1	–	Holenweg et al. 1996
P. badius	Taï, Ivory Coast (1992–93)	–	65[10]	+	–	–	2	–	Höner et al. 1997
P. badius	Taï, Ivory Coast (1996–98)	158	58[7]	>6[8]	872	300–1,532	2	–	Korstjens 2001
P. kirkii	Jozani (forest), Zanzibar (1980–81)	100	60[10]	70	1,044	–	2	–	Mturi 1991, 1993
P. kirkii	Jozani (forest), Zanzibar (1999)	176	25	31–65	565	50–1,270	3	Yes	Siex 2003
P. kirkii	Jozani (*shamba*), Zanzibar (1999)	784	13	70–95	310	0–690	4	Yes	Siex 2003
P. pennantii	Gbanraun, Nigeria	120	73	>42	1,040	450–1,900	1	–	Werre 2000
P. rufomitratus	Tana River (Mchelelo), Kenya (1973–74)	253	9[12]	36	603	≤200-<1,100	1	Yes	Marsh 1979a,b, 1981c; Marsh 1978b
P. rufomitratus	Tana River (Mchelelo), Kenya (1986–88)	56	12	–[13]	532	210–870	1	–	Decker 1994
P. rufomitratus	Tana River (Baomo S.), Kenya	33	13	–[13]	461	180–780	1	–	Decker 1994
P. tephrosceles	Gombe, Tanzania	–	114[14]	–	–	–	1	–	Clutton-Brock 1975b
P. tephrosceles	Kibale (Kanyawara), Uganda (1970–71)	≥260	65	99	649	223–1,185	1	Yes	Struhsaker 1975
Procolobus verus	Tiwai, Sierra Leone	11	28	–	–	–	1	–	Oates and Whitesides 1990
P. verus	Taï, Ivory Coast	14	56[7]	>14[8]	1,212	482–2,105	2	–	Korstjens 2001

[1] Based on number of 1.56 ha cells entered.
[2] Based on the 95% utilization distribution derived from fixed kernel analysis.
[3] Mean for two groups.
[4] Calculated from number of individuals in the 35 ha study area.
[5] Based on mean home range size of 25 groups, though the number of hours of observation for each group was not reported.
[6] Calculated based on map of intergroup encounter locations from Dasilva (1989:305).
[7] Based on number of 1 ha cells entered per year.
[8] Based on 2 years of data, measures only overlap with one of the multiple groups that overlap study groups' home ranges.
[9] Based on mean for five sites at Lopé.
[10] Based on number of 1 ha cells entered.
[11] Certain areas were intensively used but were not contiguous.
[12] Based on 12 months.
[13] No other red colobus groups were adjacent to Decker's study groups.
[14] Based on number of 100 × 100 yard grid cells.

Ranging

Colobines typically travel shorter distances each day and occupy smaller home ranges than sympatric cercopithecines and apes (Clutton-Brock and Harvey 1977, Chapman and Chapman 2000). It is unusual for an African colobine group to have a home range >100 ha or to travel >1,000 m/day (Table 12.3). Even within their generally small ranges, many colobines spend most of their time in a "core area" that covers only a small portion of their entire range. These limited ranging patterns likely result from the relative abundance and even distribution of colobine food sources as well as the energy conservation strategies of some species (Dasilva 1992, 1993; Kay and Davies 1994).

As the most selective feeders among the African colobines, it is not surprising that olive colobus have the longest daily path lengths (mean = 1,212 m, n = 1). The generally larger group sizes of red colobus probably explain why they travel longer distances per day (mean = 920 m, n = 5) than most black and white colobus (mean = 582 m, n = 12) despite their relatively similar diets. Groups of C. satanas in Fôret des Abeilles, Gabon, and of C. angolensis in Nyungwe Forest, Rwanda, are unusual in that they adopt semi-nomadic lifestyles involving the occupation of very large ranges that continue to expand in size even after years of study. In Fôret des Abeilles, C. satanas' seminomadism may be explained by the irregular fruiting patterns of the Caesalpinaceae that dominate this site and provide an important component of their diet (Fleury and Gautier-Hion 1999). In Nyungwe, however, food is relatively abundant and of very high quality (Fimbel et al. 2001), suggesting that it is the unusually large group sizes (>300 individuals) of C. angolensis at this site that force them to range so widely (Fashing et al. in press).

Several colobine species exhibit a tendency for home range size to become compressed as population density increases (Dunbar 1987, Newton and Dunbar 1994). For example, in a comparison of nine East African C. guereza populations, Fashing (2001a) found a significant negative correlation between population density and mean home range size. Moreover, the mean range size of Piliocolobus kirkii groups in Jozani Forest decreased from 60 ha to 25 ha over a recent 19-year period during which population density increased from 100 individuals/km² to 176 individuals/km² (Mturi 1993, Siex 2003). Consistent with this pattern is the fact that range size averaged only 13 ha in nearby shambas (farms) at Jozani, where P. kirkii density reached an astounding 784 individuals/km² in 1999 (Siex 2003).

Like many other primates, most African colobines exhibit seasonal variation in ranging patterns corresponding with changes in the consumption of major dietary items and/or food availability (Clutton-Brock 1975b, Marsh 1981c, McKey and Waterman 1982, Dasilva 1989, M. Harrison personal communication cited in Oates 1994, Bocian 1997, Fleury and Gautier-Hion 1999). However, for several populations, ranging behavior is not clearly linked to these primary food-related variables (Struhsaker 1974, Oates 1977a, Bocian 1997, Fashing 2001a). In C. guereza, for example, temporal variability in ranging patterns appears to be more related to the distribution of infrequently eaten food items that are spatially rare yet nutritionally important, like swamp plants or eucalyptus bark (Oates 1978, Fashing 2001a and unpublished data). Olive colobus and some red colobus (e.g., Piliocolobus badius at Taï) are also unusual in that their ranging patterns appear to be dictated primarily by the movements of the guenon (Cercopithecus diana) groups with which they form near-permanent polyspecific associations as an apparent antipredation strategy (Oates and Whitesides 1990, Holenweg et al. 1996, Bshary and Nöe 1997a, Korstjens 2001).

Activity Patterns

Colobines are notorious for their lethargic lifestyles. Among the African colobine genera, activity level appears to be inversely related to body size: mean time spent resting is 54% for nine Colobus populations, 47% for eight Piliocolobus populations, and 40% for one Procolobus population (Table 12.4). This pattern may result from the greater intake of mature leaves by Colobus (mean = 14%, n = 13; Table 12.1) than by the other two genera (Piliocolobus, mean = 11%, n = 15; Procolobus, mean = 6%, n = 2) and the concomitant greater reliance by Colobus on foregut fermentation, which requires large amounts of resting time to facilitate digestion. Though some have suggested that colobines have abnormally low metabolic rates (McNab 1978, Müller et al. 1983), Dasilva (1992, 1993) argues that most available evidence contradicts this contention. Instead, she provides evidence that the low activity levels of one species, C. polykomos, are part of a behavioral, rather than a physiological, strategy of conserving energy. C. polykomos' strategy of behavioral thermoregulation involves actions like reducing energy expenditure at times of preferred food scarcity and adopting hunched positions during rainstorms and when the temperature is low. Procolobus do not appear to have this option due to both their small body size, which requires them to maintain a higher-quality diet, and their need to move a great deal to keep up with the guenons with whom they typically travel (Oates 1994).

Competition over Food

Because most food items in their often largely folivorous diets are assumed to be relatively evenly dispersed, African colobines are typically expected to experience scramble, but not contest, competition over food within groups (van Schaik 1989). Intragroup scramble competition is typically inferred from a positive correlation between group size and daily path length within a primate population (Isbell 1991, Wrangham et al. 1993; see Gillespie and Chapman 2001 for a critique of this technique). However, at all three sites where three or more groups of the same African colobine

Table 12.4   Activity Budgets of African Colobines from Studies of ≥4 Months (Based on Scan Sampling of the First Behavior Performed for ≥5 sec Continuously Unless Otherwise Noted)

SPECIES	STUDY SITE	REST	FEED	MOVE	SOCIAL	OTHER	REFERENCES
Colobus angolensis[1]	Ituri, DRC	43	27	24	5	1	Bocian 1997
C. angolensis[2,3]	Nyungwe, Rwanda	32	42	20	5	1	Fashing et al. in press
C. guereza	Kibale, Uganda	57	20	5	11	7	Oates 1977a
C. guereza[1]	Ituri, DRC	52	19	22	5	2	Bocian 1997
C. guereza[1,4]	Kakamega, Kenya	63	26	2	7	2	Fashing 2001a
C. guereza[2,5]	Entebbe, Uganda	58	20	12	10	1	Grimes 2000
C. polykomos[6]	Tiwai, Sierra Leone	61	28	9	1	1	Dasilva 1989, 1992
C. satanas	Douala-Edea, Cameroon	60[7]	23	4	14[8]	0	McKey and Waterman 1982
C. vellerosus[6,9]	Boabeng-Fiema, Ghana	59	24	15	3	0	Teichroeb et al. 2003
Piliocolobus badius	Abuko, Gambia	52	21	13	13	1	Starin 1991
P. badius	Tiwai, Sierra Leone	55	37	5	—	3	A. G. Davies personal communication cited in Oates 1994, Davies et al. 1999
P. kirkii	Jozani (forest), Zanzibar	47	29	12	7	5	Siex 2003
P. kirkii	Jozani (*shamba*), Zanzibar	44	29	6	15	7	Siex 2003
P. pennantii[10]	Gbanraun, Nigeria	33	37	25	6	0	Werre 2000
P. rufomitratus	Tana River (Mchelelo), Kenya (1973–74)	55	30	7	8	0	Marsh 1978a
P. rufomitratus[10]	Tana River (Mchelelo), Kenya (1986–88)	48	29	22	2	0	Decker 1994
P. rufomitratus[10]	Tana River (Baomo S.), Kenya	50	23	24	3	0	Decker 1994
P. tephrosceles[10]	Gombe, Tanzania	54	25	8	9	5	Clutton-Brock 1974
P. tephrosceles	Kibale (Kanyawara), Uganda	38	45	9	8	0	Struhsaker 1975
Procolobus verus	Tiwai, Sierra Leone	40	27	25	—	8	Oates 1994, Davies et al. 1999

[1] Recorded first behavior performed for ≥3 sec continuously.
[2] Conducted instantaneous scan samples.
[3] Based on data collected primarily between 10 A.M. and 3 P.M.
[4] Mean of activity budgets of two groups.
[5] Mean of activity budgets for three groups studied for 6 months each. Values recalculated after discarding 18.6% of 5,790 total scans, which were categorized as "out of sight."
[6] Recorded first behavior performed 5 sec after spotting an individual.
[7] Total of time spent sitting, lying, and clinging.
[8] Authors included time spent self-cleaning and grooming as one category, so time spent social (playing and grooming) is overestimated here.
[9] Mean of activity budgets of three groups studied for 4 months each.
[10] Scanning method not described.

species have been studied, there is no correlation between group size and daily path length (*C. guereza*, Fashing 2001a; *Piliocolobus kirkii*, Siex 2003; *P. tephrosceles*, Struhsaker and Leland 1987). Only in unusually large groups of black and white colobus does scramble competition perhaps begin to appear (*C. guereza*, Fashing 1999, 2001a; *C. vellerosus*, Teichroeb et al. 2003).

Direct studies of contest competition within African colobine groups have seldom been conducted, probably due to the rarity of intragroup aggression in most populations (Struhsaker and Leland 1979, Dunbar 1987, Fashing 2001a). A study of sympatric *C. polykomos* and *P. badius* at Taï provides a notable exception (Korstjens et al. 2002). In this study, *C. polykomos* were more frugivorous and consumed foods that required greater handling time and longer food patch residence than *P. badius*. Rates of intragroup aggression were also over three times higher in *C. polykomos*, suggesting that intragroup contest competition is more intense for this species than for *P. badius* at Taï (Korstjens et al. 2002).

Because the *C. polykomos* at Taï are unusually frugivorous for an African colobine (Table 12.2) and often consume seeds that may incite competition owing to their long handling time, they may be one of the few African colobine populations experiencing such intense contest competition. Competition between groups will be discussed in detail in the final section of this chapter.

Biomass and Its Determinants

The biomass of African colobines varies widely among sites (Fashing and Cords 2000). At Kibale's Kanyawara study site, for example, two colobine species (*Piliocolobus badius* and *C. guereza*) combine to reach a biomass of 2,386 kg/km², or 82% of the total primate biomass (Struhsaker 1975, 1997). In contrast, at Lomako, DRC, the only colobine species (*C. angolensis*) occurs at a biomass of 57 kg/km², or 6% of the total primate biomass (McGraw 1994). While it has been suggested that a variety of biotic and abiotic factors probably contribute to this intersite variation in colobine biomass, the ratio of protein to fiber in the mature foliage at a site is the strongest predictor of colobine biomass (Oates et al. 1990). The predictive powers of this ratio have proven to be remarkably robust and have held not only between forests (Oates et al. 1990, Davies 1994) but also among sites within the same forest (Chapman et al. 2002a, 2004; Wasserman and Chapman 2003). The link between the protein:fiber ratio and colobine biomass is founded on the assumption that since mature leaves are the fallback foods of many colobines, their quality plays a critical role in determining the density of the colobine populations that can be supported through lean periods (McKey 1978a,b; Davies 1994).

Another major influence on colobine biomass is human disturbance through habitat destruction and hunting (Starin 1989; Oates 1977b, 1996, 1999; Struhsaker 1997). Logging has had adverse effects on red colobus populations across Africa (Skorupa 1986, Davies 1994). Furthermore, forest clearance by local people has led to a drastic reduction in density and mean group size of *Piliocolobus rufomitratus* over the past several decades at Tana River, Kenya (Decker 1994, Wieczkowski and Mbora 1999/2000, Mbora and Meikle 2004b). Hunting of colobines is especially prevalent in central and West Africa, where the wild meat trade is rampant. Davies (1994) noted that in Sierra Leone commercial hunting led to major declines in *P. badius* populations in accessible forests. Sadly, one red colobus subspecies, *P. b. waldronae*, formerly of Ghana and Ivory Coast, appears to have been hunted to near extinction during the twentieth century (Oates et al. 2000a, McGraw and Oates 2002, McGraw 2005).

REPRODUCTIVE PARAMETERS

The reproductive parameters of African colobines have not been studied extensively (Table 12.5). Though rare in the Asian colobines, periodic swellings of the sex skin occur in all *Piliocolobus* and *Procolobus* spp. as well as in *C. angolensis* and *C. satanas* (Bocian 1997, Dixson 1998). Among *Piliocolobus badius* in Abuko Forest, The Gambia, males direct most of their sexual interest at inflating and maximally tumescent females (Starin 1991). It is generally assumed that ovulation occurs at some point while a female is swollen, though hormonal studies are still needed to test this idea (Oates 1994).

Little is known about the copulatory behavior of most African colobines. Both sexes solicit copulations in *C. guereza*, *C. vellerosus*, *Piliocolobus badius*, *P. tephrosceles*, and *Procolobus verus*, while only males have been observed soliciting copulations in *C. angolensis* and *C. polykomos* (Table 12.5). All three species for which mounting behavior has been described are multiple mounters. Among African colobine genera, only female red colobus give copulation calls (Struhsaker 1975, Korstjens 2001). These calls are audible up to 400 m away in *P. badius* at Taï and may function to attract other males (Korstjens 2001). Copulatory harrassment by immatures has been described for *C. guereza*, *P. badius*, and *P. tephrosceles* (Oates 1977c, Struhsaker and Leland 1979, Starin 1991, Harris and Monfort 2003). In *P. badius* and *P. tephrosceles*, copulating couples are also harassed by adult males (Struhsaker and Leland 1979, Starin 1991, Firos 2001). Consort pairs, in which a male and a female leave their group for 1 day or longer and mate exclusively with one another, have been reported only for *P. badius* and *P. rufomitratus* (Starin 1991, Mbora and McGrew 2002).

Gestation lengths have been estimated at 5.5–6 months for African colobines (Table 12.5). Olive colobus are the only African colobine known to exhibit strict birth seasonality at multiple sites. In red colobus, the typical pattern is for births to occur throughout the year but to peak during certain periods. *C. polykomos* on Tiwai Island, Sierra Leone, are unusual among black and white colobus populations in that they have a strict birth season, whereas another *C. polykomos* population at Taï and two populations of *C. guereza* feature births throughout the year with no obvious peaks. The possibility of intragroup birth synchrony has rarely been considered for African colobines (Struhsaker and Leland 1987), though it does appear to occur among *C. guereza* at Kakamega and *P. badius* at Abuko. Few African colobine studies have been of sufficient duration to record interbirth intervals for a large number of individually recognized females. An exception is provided by Struhsaker and Leland (1987), who found that based on over 20 births, *P. tephrosceles* interbirth intervals averaged 25.5 ± 5.1 months at Kibale. Two 1-year studies of *P. kirkii* on Zanzibar conducted 7 years apart suggest that interbirth intervals can change dramatically within a population over time (Siex and Struhsaker 1999b, Siex 2003). Estimated interbirth interval in Siex's (2003) *shamba*-dwelling *P. kirkii* population decreased from 55 months in 1992–1993 to 26 months in 1999. Dunbar (1987) inferred that interbirth intervals are significantly longer in multimale groups than in one-male groups of *C. guereza*, perhaps due to female reproductive suppression resulting from the stress created by competition among males in multimale groups.

Infants of all *Colobus* spp. except *C. satanas* are born with entirely white coats which contrast strikingly with those of adults (Oates 1994). Red colobus newborns also differ from adults in coat color, though less strikingly. In olive colobus, the coats of newborns are darker than, though essentially the same color as, those of adults. During the first few months of life, when black and white colobus infants remain flamboyantly colored, they are very attractive to, and are often held by, females other than their mothers (Horwich and Manski 1975, Oates 1977c, Dasilva 1989, Korstjens 2001). Similar "allomothering" behavior is rarely tolerated in red colobus (though *P. kirkii* provides an interesting exception: [Siex 2003]) and has never been observed in olive colobus (Oates 1994). African colobine males typically exhibit little interest in infants, though preliminary observations suggest that *C. angolensis* may be different in this

Table 12.6   (*cont'd*)

| SPECIES | STUDY SITE | HABITAT | GROUP SIZE | | N | GROUP COMPOSITION | | | | | | REFERENCES |
| | | | | | | ADULT MALES | | ADULT FEMALES | | IMMATURES | | |
			MEAN	RANGE		MEAN	RANGE	MEAN	RANGE	MEAN	RANGE	
P. badius	Taï, Ivory Coast	Lowland moist primary forest	52.3	41–64	4	10.5	6–15	18.3	14–22	23.5	20–27	Korstjens 2001
P. foai	Badane, C. A. R.	Seasonally flooded forest	11.7	3–18	3	–	–	–	–	–	–	Galat-Luong and Galat 1979
P. gordonorum	Magombera, Tanzania	Disturbed groundwater forest	24.8	21–33	4	2.8	1–5	9.3	7–11	12.8[8]	–	Struhsaker and Leland 1987, T. T. Struhsaker, personal communication cited in Oates 1994
P. kirkii	Jozani (forest), Zanzibar	Secondary groundwater forest	31.1	23–36	3	3.9	–	12.0	–	13.7	–	Siex 2003
P. kirkii	Jozani (*shamba*), Zanzibar	Cultivated areas and regenerating forest	37.5	20–65	4	4.7	–	14.9	–	17.7	–	Siex 2003
P. pennantii	Gbanraun, Nigeria	Lowland disturbed marsh forest	~46	~15–~80	4[9]	7	7	26	26	27	27	Werre 2000
P. preussi	Korup, Cameroon	Lowland moist primary forest	>47.0	>24–≥80	7	–	–	–	–	–	–	Struhsaker 1975
P. rufomitratus	Tana River, Kenya	Lowland disturbed gallery forest	11.2	4–24	9	1.1	1–2	5.6	2–11	4.5	1–11	Decker 1994
P. tephrosceles	Kibale, Uganda	Moist medium-altitude mixed forest	≥44	9–80	20[10]	–	≥2–≥7	–	≥3–≥16	–	?–≥21	Struhsaker 1975, 2000
P. tephrosceles	Gombe, Tanzania	Disturbed mixed forest	82	–	1	11	–	24	–	47[8]	–	Clutton-Brock 1975a
P. tholloni	Salonga, DRC	Lowland moist primary forest	–	>60	1	–	–	–	–	–	–	Maisels et al. 1994
Procolobus verus	Taï, Ivory Coast	Lowland moist primary forest	7.1	2–12	10[11]	1.5	1–3	3.0	1–6	2.6	1–4	Korstjens 2001, Korstjens and Schippers 2003
P. verus	Tiwai, Sierra Leone	Lowland moist secondary forest	8.5	3–11	1	2.0	1–2	3.5	1–5	3.0	1–4	Oates 1988, Whitesides et al. 1988, Davies et al. 1999

[1] Group composition based on the seven groups for which composition data were complete.
[2] Group composition data based on only two groups since age/sex class of some individuals in the other groups could not be determined.
[3] Group composition based on the four groups for which composition data were complete.
[4] Group composition based on only 26 groups.
[5] Group size and composition data based on values at start of study.
[6] Excludes two groups for which widely variable sizes and no information on group composition were reported.
[7] Authors describe forest as primary despite selective logging there 3 years before their study.
[8] Includes both immatures and individuals of indeterminate age/sex class.
[9] Group composition data based on only one group.
[10] Group composition based on only 14 groups.
[11] Group composition data based on only six groups since age/sex class of some individuals in the other groups could not be determined.

adult females, and their dependent offspring, are typical of some taxa, such as olive colobus and *C. guereza*. Group sizes in several other black and white colobus, including *C. polykomos*, *C. satanas*, and *C. vellerosus*, tend to be larger, averaging ~15 individuals, including multiple males, at most sites. *C. angolensis* has a remarkably labile social organization, with group sizes ranging from fewer than 10 individuals at several sites to more than 300 individuals at Nyungwe. Most red colobus species live in large multimale groups averaging 25–50 individuals.

Habitat type, logging history, predation risk, and feeding competition all appear to influence African colobine group sizes. For example, groups of red colobus in marginal, strongly seasonal habitats, like *Piliocolobus rufomitratus* in the gallery forests at Tana River and *P. foai* in the seasonally flooded forests at Badane, Central African Republic, are often only one-fourth to one-half the size of red colobus groups in rain forest habitats (Table 12.6). Similarly, among *C. guereza* populations across East Africa, mean group size tends to be lower in gallery or scrub forest than in rain forest (Dunbar 1987). Evidence that logging history influences group size comes from Kibale, where *C. guereza* groups in lightly logged forest are 25% smaller than those in unlogged forest (Struhsaker 1997). The influence of predation risk on group size can be inferred by comparing colobine group sizes at locations with and without certain predators.

Struhsaker (2000) has shown that on Bioko, where crowned-hawk eagles are absent, groups of *C. satanas* and *Piliocolobus pennantii* are distinctly smaller than in forests on the African mainland, where crowned-hawk eagles occur. Furthermore, Stanford (1998) has noted that red colobus groups tend to be smaller in forests where chimpanzees are absent. Finally, there is evidence that feeding competition may set the upper limit of group size in some African colobines, especially black and white colobus (see Ecology, above).

The composition of African colobine groups is also influenced by several factors. As in many other primates (Mitani et al. 1996), the number of adult females appears to have a strong influence on the number of adult males within some African colobine populations (e.g., *P. verus*, Korstjens 2001). Furthermore, adult sex ratios in red colobus populations appear to be influenced by the presence or absence of chimpanzees. At sites where red colobus co-occur with chimpanzees, ratios of adult red colobus males to females tend to be much higher than at sites where chimpanzees are absent (Stanford 1998). Crowned-hawk eagles do not, however, appear to have a similar influence on red colobus sex ratios (Struhsaker 2000).

Mating systems have been identified for only a handful of African colobines and are related largely to group composition. One-male groups are typified by polygynous mating, in which the group male monopolizes copulations with females in his group (e.g., some *C. guereza* at Ituri and Kakamega; Bocian 1997, Fashing unpublished data). Multimale groups are characterized by promiscuous mating (Ituri *C. angolensis*, Bocian 1997; Tiwai *C. polykomos*, Dasilva 1989; Abuko *Piliocolobus badius*, Starin 1991; Kibale *P. tephrosceles*, Struhsaker 1975; some Taï *Procolobus verus*, Korstjens and Noë 2004), though often the dominant male does most of the copulating with the females in his group (Struhsaker 1975, Starin 1991, Korstjens and Noë 2004). Occasional extragroup copulations have been reported (Kakamega *C. guereza*, Fashing 2001c; Taï *P. verus*, Korstjens and Noë 2004) but are not known to be common in any species of African colobine.

Olive colobus are unusual in that group membership appears to be remarkably fluid (Oates 1994, Korstjens and Schippers 2003) (Fig. 12.3). Juveniles and adults of both sexes are known to frequently disperse to new groups at Taï, with females being the sex that disperses most often (Korstjens and Schippers 2003). Costs of dispersal appear to be low for olive colobus because even while in-between groups, individuals typically associate with a diana monkey group that provides protection against predation. Adult female olive colobus appear to disperse as a means of reducing feeding competition, whereas adult males seem to disperse as a means of improving their mating opportunities (Korstjens and Schippers 2003). Dispersal is believed to be far more common among females than males in most red colobus taxa (e.g., *Piliocolobus badius*, Starin 1994; *P. rufomitratus*, Marsh 1979a,b; Decker 1994; *P. tephrosceles*,

Figure 12.3 Group membership in olive colobus (*Procolobus verus*) is unusually fluid, with juveniles and adults of both sexes regularly dispersing to new groups (photo by Melanie Krebs).

Struhsaker and Leland 1987). At Kibale, for example, several females immigrated into Struhsaker and Leland's (1987) *P. tephrosceles* study groups each year, whereas male immigration occurred roughly only once every 9 years. Female *P. badius* at Abuko also transfer more often than males, though most juveniles of both sexes typically disperse from their natal groups in this population (Starin 1994). Studies of *P. rufomitratus* at Tana River show that rates of female transfer can vary widely over time; female migrations were 10 times more common in 1973–1975 (11.6/year; Marsh 1979b) than in 1986–1988 (1.1/year; Decker 1994). Patterns of dispersal in the *P. kirkii* population inhabiting *shambas* at Jozani are unusual for red colobus in that males transfer more often than females (Siex 2003). The unusual degree of female philopatry in the *shambas* may have arisen in response to the intense intergroup feeding competition resulting from the extraordinarily high population density at this site (Siex 2003).

Transfer patterns among black and white colobus remain poorly known, with most long-term studies having reported only a few observed or inferred instances of dispersal. Based on the limited data available, it appears that members of both sexes transfer at least occasionally in *C. guereza, C. polykomos, C. satanas,* and *C. vellerosus* (Dasilva 1989, M. Harrison personal communication cited in Oates 1994, Sicotte and MacIntosh 2004, Fashing unpublished data). Considering the number of bachelor males living solitarily or in all-male bands, it is likely that dispersal is biased toward males in *C. guereza* and *C. vellerosus* (Sicotte and MacIntosh 2004, Fashing unpublished data).

SOCIAL BEHAVIOR

African colobines vary widely among species and among populations of the same species in the amount of time they devote to social activities (Table 12.4). Though patterns of social behavior remain largely unknown for most species, below I discuss two of the better-studied topics in African colobine social behavior: grooming and aggression.

Grooming

Grooming is the predominant social behavior in most African colobines (Oates 1977c, Struhsaker and Leland 1979, Dasilva 1989, Starin 1991, Grimes 2000, Werre 2000, Fashing 2001a, Siex 2003). In black and white colobus, adult females are the primary groomers while adult males rarely groom others (*C. guereza,* Leskes and Acheson 1970, Oates 1977c, Fashing unpublished data; *C. polykomos,* Dasilva 1989, Korstjens 2001). The primary recipients of grooming are typically adult females and/or adult males, though the exact pattern varies across black and white colobus study sites (Oates 1977c, Dasilva 1989, Grimes 2000).

Red colobus grooming patterns exhibit considerable interpopulational variability and are often strikingly different from those of black and white colobus. In *Piliocolobus pennantii* at Gbanraun, grooming occurred far more often among adult males than among adult females (Werre 2000). Similarly, at Kibale, adult male *P. tephrosceles* groomed each other more than expected by chance, while adult females primarily directed their grooming at adult males rather than each other (Struhsaker and Leland 1979). Strushsaker and Leland (1979) argue that the male-centered grooming patterns of *P. tephrosceles* are probably an outcome of the fact that males are philopatric and females disperse in this species. Among *P. badius* at Abuko, where both sexes usually disperse, adult males rarely groomed each other and females mostly groomed their offspring and adult males (Starin 1991). Similarly, among *P. badius* at Taï, where dispersal patterns are unknown, adult males were no more likely to groom other males than they were to groom females, though females were still more likely to groom males than other females (Korstjens 2001). At Jozani, where dispersal is male-biased, *P. kirkii* females still exhibit the typical red colobus pattern of directing much of their grooming at adult males, while adult males rarely groom

each other as at Abuko and Taï (Siex 2003). Grooming patterns have been described for only one population of olive colobus. Oates (1994) reported that adult male *Procolobus verus* at Tiwai primarily groomed adult females, while adult females were more equitable in whom they groomed, albeit with some bias toward adult males. Oates (1994) also noted that male–female grooming bouts frequently occurred before or, particularly, after copulation.

Aggression

Aggression typically occurs at relatively low rates within African colobine groups (Marler 1972, Oates 1977c, Dunbar 1987, Dasilva 1989, Werre 2000, Fashing 2001a, Saj and Sicotte 2005). As in most primates, males tend to be the more aggressive sex (Oates 1977c, Dasilva 1989, Werre 2000, Siex 2003). In an Ethiopian population of *C. guereza,* for example, Dunbar and Dunbar (1976) found that rates of aggression were higher in multimale groups than in one-male groups, due primarily to aggression among males. Most aggression in *Piliocolobus tephrosceles* groups at Kibale is directed by adult males at other adult males, often in the form of harassment during copulation (Struhsaker and Leland 1979, 1987). Most aggression by adult male *P. badius* at Abuko and Taï occurs in the context of sex or feeding (Starin 1991, Korstjens 2001). In most African colobine groups where multiple males coexist, one male is clearly dominant to all others (*C. guereza,* Dunbar and Dunbar 1976; Bocian 1997; *P. badius,* Starin 1991), and in some cases, a linear dominance hierarchy is apparent (*C. polykomos,* Dasilva 1989; *P. tephrosceles,* Struhsaker and Leland 1979). *C. angolensis* may be unusual in that two adult males were described as co-dominant, with three other adult males occupying indistinguishable ranks below them in a group at Ituri (Bocian 1997, Oates et al. 2000b).

Patterns of adult female aggression have been described for several populations of African colobines. Among *P. badius* at Abuko, most aggression by females occurs in the contexts of protecting their offspring and weaning (Starin 1991). Struhsaker and Leland (1979) contend that aggression by female *P. tephrosceles* at Kibale may also be related to restricting access to their infants and possibly to the weaning process. Female aggression occurs primarily in the feeding context in both *C. polykomos* and *P. badius* at Taï (Korstjens et al. 2002). Strong linear dominance hierarchies exist among female *C. polykomos* at Taï and *P. badius* at Abuko (Starin 1991, Korstjens et al. 2002). Only a weakly developed dominance hierarchy has been detected for female *P. tephrosceles* at Kibale (Struhsaker 1975), and no dominance hierarchy appears to exist among female *P. badius* at Taï (Korstjens et al. 2002). The benefits of high rank in the populations that feature female hierarchies are as yet unclear. Starin (1991) found no correlation between female rank and age at first parturition, birth rate, or total number of surviving offspring. No information is yet available on aggression and dominance in olive colobus.

INFANTICIDE

Though infanticide was first described among primates in the Asian colobine genus *Semnopithecus* (Hrdy 1974) and the strongest evidence to date for the hypothesis that infanticide by males is an adaptive strategy resulting from sexual selection comes from this genus (Borries et al. 1999), infanticide has rarely been reported for African colobines (Table 12.7). In fact, four of the five confirmed cases of infanticide or attempted infanticide in African colobines occurred in a single forest, Kibale. The incidents at Kibale consisted of two successful infanticide attempts in *C. guereza* (Onderdonk 2000, Harris and Monfort 2003) and one successful and one unsuccessful attempt in *P. tephrosceles* (Struhsaker and Leland 1985). An additional unsuccessful infanticide attempt was recently observed in *C. vellerosus* at Boabeng-Fiema, Ghana (Saj and Sicotte 2005). In all cases in which the identity of the perpetrator could be determined during attacks on infants (*n* = 4), the infanticidal individual was an adult or subadult male (Struhsaker and Leland 1985, Onderdonk 2000, Harris and Monfort 2003, Saj and Sicotte 2005). However, in only one instance (*P. tephrosceles*) was the infanticidal male observed to copulate with the mother whose infant he killed subsequent to the attack, as would be expected by the sexual selection hypothesis that is generally accepted as the most plausible explanation for infanticide by males as an adaptive behavior (Struhsaker and Leland 1985, van Schaik and Janson 2000). Whether the fact that infanticide has primarily been observed at Kibale is due to unusual features of this forest (e.g., extremely high densities of colobines [Fashing and Cords 2000]) or the longer time span that colobines have been monitored at Kibale relative to other forests remains to be determined. At the very least, however, the combined tens of thousands of hours of observation at most other sites without reliable observations of infanticide suggest that the strategy may not be widespread among African colobine males (Table 12.7).

INTERGROUP RELATIONSHIPS

Interactions between individuals in different groups are less common and more difficult to observe than those occurring among members of the same group. As a result, far more is known about relationships within than between primate groups (Cheney 1987). Our limited knowledge of intergroup relationships is unfortunate considering that most models of primate socioecology feature predictions about intergroup interactions that have not been well tested (e.g., Wrangham 1980, van Schaik 1989, Isbell 1991, Sterck et al. 1997). Nevertheless, intergroup relationships have been surprisingly well studied for African colobines relative to other forest primates. The relatively small home range sizes and often predictable ranging patterns, as well as the small group sizes and spatial cohesion, of many African colobines have facilitated this unusually intensive investigation into intergroup

relationships. Patterns of intergroup relations in African colobines have been found to vary at several levels; among different species (Struhsaker and Oates 1975), among different populations of the same species (Dasilva 1989, Korstjens 2001), and even among different subpopulations of the same species in a single forest (Siex 2003). Here, I review the patterns of intergroup relations among the African colobines and evaluate how consistent they are with the predictions of recent models of primate socioecology.

Loud Calls

Though encounters between colobine groups are usually defined as occasions when two or more groups are within 50 m of each other (e.g., Oates 1977c, Stanford 1991), the vocal repertoire of adult male African colobines allows groups to remain in contact over much larger distances. Male black and white colobus utter low-frequency, high-amplitude loud calls, referred to as "roars," which in *C. guereza* have been estimated to be audible up to 1.6 km from the source (Marler 1969, Oates and Trocco 1983, Oates et al. 2000b). Though black and white colobus roars appear to play roles in male–male competition and predator intimidation, there is little doubt they also serve as intergroup spacing mechanisms, especially in *C. guereza* (Marler 1969, 1972; Oates 1994). Lacking a similar low-frequency, high-amplitude loud call, red colobus males utter several other calls (e.g., *bark*, *chist*, *nyow*) that carry over distances of >300 m and facilitate intergroup communication (Struhsaker 1975, Oates 1994). The loudest call uttered by olive colobus males appears to be their "laughing" call (Oates 1994), which carries ~150 m (M. Korstjens unpublished data cited in Wich and Nunn 2002). With their travel patterns largely dictated by the movements of the diana monkey group with which they associate, olive colobus are probably less likely to use their loud calls to mediate intergroup spacing than the other African colobine taxa. In fact, laughing calls have primarily been mentioned as occurring when olive colobus groups are already in proximity (e.g., during intergroup encounters involving neighboring olive colobus–diana monkey associations [Oates and Whitesides 1990]).

Territoriality and Other Spacing Systems

The word *territory* is often used (and misused) when discussing intergroup relationships (Maher and Lott 1995). It has been employed to describe a variety of patterns of range use and defense, the most common of which are *(1)* an area defended by a group, *(2)* an area exclusively used by a group, or *(3)* an area in which a group exhibits site-specific dominance. Since use of a common terminology is critical for making comparisons of spacing systems across studies (Maher and Lott 1995), I conceptually and operationally define as *territorial* any populations of African colobines in which groups consistently defend large parts of their home

Table 12.7 Long-Term (≥6 Months) Studies of Intergroup Encounters (IEs) in African Colobines

SPECIES	STUDY SITE	STUDY LENGTH (MONTHS)	HOME RANGE SIZE (HA)	RANGE OVERLAP (%)	DAILY PATH LENGTH (M) MEAN	NUMBER OF GROUPS	CORE AREA	D INDEX	TERRITORIAL	NUMBER OF IEs	IEs/HR	% IEs AGGRESSIVE[22] MALE(S)	FEMALE(S)	ANYONE	DISPERSING SEX[1]	INFANTICIDE REPORTED	REFERENCES
Colobus angolensis	Ituri, D. R. Congo	11	371[2]	100	983	1	No	0.45	No	≥25	>0.03	+		+	M?,F?	No	Bocian 1997; Oates et al. 2000b
C. guereza	Budongo, Uganda (1964–65)	8	14	minimal		5			Maybe			*		+		No	Marler 1969
C. guereza	Kibale, Uganda	12	28	74	535	1	Yes	0.90	No	65	0.09	+	*	+	M	Yes[3]	Oates 1977a,c
C. guereza	Ituri, D. R. Congo	11	100[2]	≥22	609	1	Yes	0.54	No	2	0.003	+		+	M?,F?	No	Bocian 1997
C. guereza	Kakamega, Kenya	12	18[4]	67	588	5	Yes	1.23	No	136	0.07	62	18	70	M?,F?	No	Fashing 2001a,c
C. guereza	Entebbe, Uganda	6	7.5	+	386	3	+	1.25		12[5]		+	–	+	M	No	Grimes 2000
C. polykomos	Tiwai, Sierra Leone	12	24	≥20[6]	832	1	Yes	1.50	Maybe	9	0.010	67	0	67	M,F	No	Dasilva 1989, Oates 1994
C. polykomos	Taï, Ivory Coast	42	77[7]	>21[8]	617	2	Yes	0.62	No	99	0.012	79	>40[9]	79	M,F	No	Korstjens 2001, pers. comm.
C. satanas	Douala-Edea, Cameroon	11	60	+	459	1	Yes	0.53	Yes	4	0.006	+	–	+		No	McKey 1978a,b; McKey and Waterman 1982
C. satanas	Forêt des Abeilles, Gabon	18	573[10]	≥65	852	1	No[11]	0.32	No	19[12]	0.02	+	–	+		No	Fleury and Gautier-Hion 1999
C. vellerosus	Boabeng-Fiema, Ghana	6	9	≥33	300	1	Yes	2.61		47	0.12	83	9	83	M,F	Yes	Sicotte and MacIntosh 2004, Saj and Sicotte 2005
Piliocolobus badius	Abuko, Gambia	51	34	60		1			No	≥64[13]		+	*	+	F,M	Suspected	Starin 1991, 1994
P. badius	Taï, Ivory Coast (1996–98)	35	58[7]	>6[8]	872	2		1.01	Maybe			+	*	+	F	No	Korstjens 2001, pers. comm.
P. kirkii	Jozani (forest), Zanzibar (1999)	12	25	31–65	565	3	Yes	1.00	No			+[14]		+	M,F	No	Siex 2003
P. kirkii	Jozani (shamba), Zanzibar (1999)	12	13	70–95	310	4	Yes	0.76	No			+	+	+	M,F	No	Siex 2003
P. pennantii	Gbanraun, Nigeria	12	73	>42	1,040	1	No	1.08	No	6	0.008	17[15]	17	17		No	Werre 2000
P. rufomitratus	Tana River (Mchelelo), Kenya (1986–88)	15	9[16]	36	603	1	No	1.78	No	8[17]	rare	+?[18]	50	50	F,M	Suspected	Marsh 1979a
P. tephrosceles	Kibale (Kanyawara), Uganda (1970–71)	12	65	99	649	1	Yes	0.71	No	≥64[19]		≤70	≤2	≤70	F,M	Yes	Struhsaker 1975; Struhsaker and Leland 1985, 1987
Procolobus verus	Tiwai, Sierra Leone	8[20]	28			1			No	10[21]		–	–	–	F,M	No	Oates and Whitesides 1990
P. verus	Taï, Ivory Coast	21	56[7]	>14[8]	1,212	2	Maybe	1.43	Maybe			+	–	+	F,M	No	Korstjens 2001, pers. comm.

1 Bold lettering indicates cases in which one sex clearly disperses more often than the other. 2 Based on number of 1.56 ha cells entered.
3 Infanticide suspected by Oates (1977c), but definitive reports of infanticide for this population come from Onderdonk (2000) and Harris and Monfort (2003). 4 Mean for two groups.
5 Encounters appear to have been defined by visual contact between groups rather than by groups being within 50 m of each other.
6 Calculated based on map of intergroup encounter locations from Dasilva (1989:305).
7 Based on number of 1 ha cells entered per year.
8 Based on 2 years of data, measures only overlap with one of the multiple groups that overlap study groups' home ranges.
9 Calculated from data in Table 2.1 in Korstjens (2001).
10 Based on number of 1 ha cells entered.
11 Certain areas were intensively used but were not contiguous.
12 Encounters defined as when two groups were within 200 m of each other.
13 Starin (1994) gives only the number of aggressive encounters (n = 64).
14 Rate of adult male aggressive acts was >3 times higher than the rate for adult females.
15 In the one of six encounters involving aggression, males were more persistently aggressive than females.
16 Based on 12 months.
17 Does not include 15 encounters with solitaries and small parties since a larger cut-off distance (<300 m) appears to have been used to define these encounters relative to those with bisexual groups (<50 m).
18 Does not say which sex(es) participated in intergroup aggression but mentions male aggression (chasing) during 4 of 15 interactions with solitaries and small parties.
19 Struhsaker (1975) notes that encounters (aggressive or tolerant) occurred on 49 of 83 study days and that on 30 of these days, 45 aggressive encounters occurred.
20 Based on 33 days during 8 months over a 29-month period.
21 Encounters defined by loud ("laughing") call bouts between different groups rather than distances between groups.
22 +, often or usually aggressive; *, rarely to occasionally aggressive or clearly less aggressive than members of the opposite sex; –, never aggressive.

ranges against conspecific intruders and maintain ≥80% of their ranges as areas of exclusive use. This definition seems appropriate since it implies that a group actively defends its range and has considerable success keeping out intruders, two features that are at the heart of the concept of territoriality (Maher and Lott 1995).

Mitani and Rodman (1979) developed a simple yet powerful model that predicts which primate populations are capable of being territorial. This model treats home ranges as circular and assumes that only those groups that travel distances equivalent to at least the length of the diameter of their home range on a typical day can be territorial. The formula for the model is $D = d/\sqrt{(4a/\pi)}$, where D represents defendability, a represents home range area, and d represents mean daily path length. Mitani and Rodman found that only primate groups with $D \geq 1$ maintain a territory, though some groups with $D \geq 1$ are not territorial.

Although according to the defendability index, about half the African colobine populations studied to date are theoretically capable of territoriality, very few, if any, are actually territorial (Table 12.7). Based on the 20 long-term studies summarized in Table 12.7, only four populations, *C. guereza* at Budongo, *C. polykomos* at Tiwai, *Procolobus verus* at Taï, and *Piliocolobus badius* at Taï, qualify as possibly territorial, though there is some uncertainty as to the amount of home range overlap that occurs in these populations. It is possible that the energetic constraints placed on colobines by their specialized digestive systems make it difficult for them to be territorial even when they are theoretically capable of doing so. Because many colobines seem to require long resting periods after meals to allow time for bacterial fermentation, they may simply not be energetically capable of evicting intruders from their home range at all times.

Colobine populations in which groups at least sometimes behave aggressively toward one another but are not territorial can be categorized as exhibiting *(1)* core area defense, *(2)* intergroup dominance, or *(3)* "supertroop" formation.

Core area defense is less intensive than territoriality in that it involves defending only frequently used portions of the home range. It occurs in *C. guereza* populations at Kibale, Kakamega, and Entebbe and is the most common spacing system for this species (Oates 1977a,c; Fashing 1999; Grimes 2000). *C. polykomos* at Taï also appear to feature core area defense (Korstjens 2001).

Intergroup dominance has been described for two red colobus populations, *Piliocolobus tephrosceles* at Kibale and *P. badius* at Abuko (Struhsaker and Leland 1979, Starin 1991), and appears to play a secondary role to core area defense in determining the outcome of encounters among *C. guereza* groups at Kibale (Oates 1977c). Three red colobus groups at Kibale shared completely overlapping home ranges, and the outcomes of encounters between them were based not on the locations of encounters but, rather, on the number of adult males in each group, their fighting abilities, and their inclination to involve themselves in

encounters (Struhsaker and Leland 1979). Among red colobus at Abuko, the number of adult males did not influence intergroup dominance, though changes in the identities and physical condition of adult males did appear to exert an influence on the outcome of encounters (Starin 1991). Female dispersal from one group to another also appeared to influence the balance of power between groups at Abuko (Starin 1991). Among *C. guereza* at Kibale, core area defense usually determined the outcome of encounters, though groups seemed to have differentiated dominance relationships that occasionally influenced the outcome of encounters instead (Oates 1977c).

Supertroops (or "associations") occur when two or more groups rest, feed, or travel while intermingled for periods ranging from >1 hr to >1 day. Among the African colobines, these peculiar associations appear to be unique to *C. angolensis* and have been reported for populations at Diani Beach, Kenya; Sango Bay, Uganda; and Ituri Forest, DRC (Moreno-Black and Bent 1982, Oates 1994, Bocian 1997). Despite the intermingling of group members that occurs when supertroops form, there have been no reports of affiliative behaviors between members of different groups. In contrast, intergroup aggression escalating to the level of grappling and even biting among males has been reported within supertroops at Ituri (Bocian 1997). The largest supertroop yet reported was observed at Sango Bay and included at least 51 individuals (Oates 1994). The fascinating >300-member groups of *C. angolensis* at Nyungwe appear to be permanent associations rather than supertroops (A. Vedder, personal communication), and nothing has been written about intergroup relationships in this population.

Male and Female Strategies During Intergroup Encounters

As noted before, encounters between African colobine groups range in intensity from peaceful comminglings to chases resulting in physical attacks and even infanticide (Bocian 1997, Fashing 2001c, Harris and Monfort 2003). When encounters are aggressive (i.e., when individuals in one group threaten, chase, and/or physically attack members of another group), not all individuals participate equally in the aggression. A comparison across African colobine populations reveals that adult males are nearly always the primary aggressors during intergroup encounters (Table 12.7). Among the 18 populations for which adult participation in intergroup encounters has been reported, males exhibit intergroup aggression in 17 and females in only 11. Furthermore, in the six populations for which data on the percentage of encounters in which males and females behaved aggressively are available, males (64 ± 10%) were aggressive during a far greater percentage of encounters than females (14 ± 6%). In fact, regular female participation in intergroup aggression appears to occur in only three of the populations studied to date (*C. polykomos*, Korstjens 2001; *Piliocolobus badius*, Starin 1991; *shamba*-dwelling

P. kirkii, Siex 2003). More often, females are described as only occasional aggressive participants, while males are regarded as regular aggressive participants, in encounters.

Predictions about the behavior of adults of each sex during intergroup encounters play important roles in models of primate socioecology (Wrangham 1980, van Schaik 1989, Isbell 1991, Sterck et al. 1997). Though the specifics of the various models differ, all agree that if food is limiting and distributed in such a way that it can (or must) be defended against other groups, females will spend their lives in their natal groups and form coalitions with female relatives to defend food sources against other groups. The rationale here is that as the sex that makes the greatest energetic investment in their offspring, females are more dependent on food and high-quality nutrition for enhancing their lifetime reproductive success than males (Trivers 1972). Conversely, as a sex whose reproductive success is limited more by access to mates than by access to food, males are expected to focus primarily on defending access to females during intergroup encounters.

Studies of intergroup encounters in primates have often used Trivers' (1972) differential investment theory as a means of inferring the behavioral strategies adopted by males and females during encounters (Cheney 1987, van Schaik et al. 1992). Female participation in intergroup aggression has typically been assumed to reflect food defense, whereas male participation has generally been assumed to reflect mate defense. Only recently have researchers begun to empirically test these assumptions by examining patterns of male and female participation in intergroup encounters with regard to the ecological and social contexts in which encounters occur. Several such intensive studies have now been conducted on African colobines, and the results offer evidence that the strategies adopted during encounters may

be more complex than predicted by current models of primate socioecology.

Recent studies of *C. guereza* at Kakamega, *C. vellerosus* at Boabeng-Fiema, and *C. polykomos* at Taï have all produced evidence consistent with the surprising conclusion that males, not females, may be the primary defenders of food resources during intergroup encounters in these taxa (Fashing 2001c, Korstjens 2001, Sicotte and MacIntosh 2004) (Fig. 12.4). In all three populations, males are more frequent aggressive participants in encounters than females, with females commonly participating in intergroup aggression only in *C. polykomos* at Taï (Table 12.7). Furthermore, in *C. guereza* at Kakamega, levels of male, but not female, aggression during encounters were positively related to the amount of time group members spent feeding that month in encounter grid cells (Fashing 2001c). Among *C. vellerosus* at Boabeng-Fiema, intergroup encounters were significantly more likely to occur in the feeding context than predicted by chance alone and it was common for males to direct aggression toward females as well as males during encounters (Sicotte and MacIntosh 2004). In *C. polykomos* at Taï, aggressive encounters (in which males always participated) occurred more often than expected by chance in "important" grid cells (i.e., where the study group spent large amounts of time), while peaceful encounters occurred less often than expected by chance in these important cells (Korstjens 2001).

Taken collectively, these results suggest that males tend to play a larger role in food defense than females among black and white colobus monkeys. Fashing (2001c) has suggested that male resource defense in *C. guereza* may be related to an indirect form of mate defense in which males that consistently successfully defend food sources during intergroup encounters are preferred as mates by females.

Figure 12.4 Females, like the ones pictured here, rarely behave aggressively during encounters between guereza (*Colobus guereza*) groups. Instead, males appear to play the primary role in intergroup food defense in this and other black and white colobus species (photo by Peter Fashing).

This hypothesis may be relevant to other taxa in which males defend resources as well, though further studies will be necessary to test it. If indirect mate defense is indeed a regular strategy of black and white colobus males, they appear to adopt it in addition to the strategy of direct mate defense predicted by traditional socioecological models. For each of the studies discussed above, in addition to the evidence suggesting that males play an important role in food defense, there was also (not mutually incompatible) evidence that males were involved in directly defending mates (e.g., chasing away bachelor males) during at least some encounters (Fashing 2001c, Korstjens 2001, Sicotte and MacIntosh 2004).

Still, based on the available evidence, recent primate socioecological models do a poor job of explaining patterns of between-group competition in the several African colobines for which intergroup relationships have been intensively investigated. In particular, the assumption that when food resources must be defended, females will be the sex that takes the lead role comes into question. It appears that, at least among black and white colobus, males are primarily responsible for resource defense during intergroup encounters, perhaps as a form of indirect mate defense. Furthermore, it seems unlikely that the male resource defense strategy is exclusive to colobines (Fashing 2001c). Indeed, there is evidence that males defend food during intergroup encounters in other primate taxa as varied as *Hapalemur griseus*, *Cebus olivaceus*, *Chlorocebus aethiops*, *Macaca radiata*, and *Pan troglodytes* (Harrison 1983, Robinson 1988, Nievergelt et al. 1998, Mutschler et al. 2000, Cooper et al. 2004, Williams et al. 2004). With the growing evidence that males play a prominent role in between-group feeding competition in many primate taxa, it may ultimately prove necessary to reconsider existing socioecological models whose foundations rest on the importance of female bonds and female intergroup food defense.

REFERENCES

Bocian, C. M. (1997). Niche separation of black-and-white colobus monkeys (*Colobus angolensis* and *C. guereza*) in the Ituri Forest [PhD thesis]. City University of New York, New York.

Boesch, C., and Boesch, H. (1989). Hunting behavior of wild chimpanzees in the Taï National Park. *Am. J. Phys. Anthropol.* 78:547–573.

Booth, A. H. (1956). The distribution of primates in the Gold Coast. *J. West Afr. Sci. Assoc.* 2:122–133.

Booth, A. H. (1957). Observations on the natural history of the olive colobus monkey, *Procolobus verus* (van Beneden). *Proc. Zool. Soc. Lond.* 129:421–430.

Borries, C., Launhardt, K., Epplen, C., Epplen, J. T., and Winkler, P. (1999). DNA analyses support the hypothesis that infanticide is adaptive in langur monkeys. *Proc. R. Soc. Lond. B* 266:901–904.

Brugière, D., Gautier, J.-P., Moungazi, A., and Gautier-Hion, A. (2002). Primate diet and biomass in relation to vegetation composition and fruiting phenology in a rain forest in Gabon. *Int. J. Primatol.* 23:999–1024.

Bshary, R., and Noë, R. (1997a). Anti-predation behaviour of red colobus monkeys in the presence of chimpanzees. *Behav. Ecol. Sociobiol.* 41:321–333.

Bshary, R., and Noë, R. (1997b). Red colobus and diana monkeys provide mutual protection against predators. *Anim. Behav.* 54:1461–1474.

Chapman, C. A., Balcomb, S. R., Gillespie, T. R., Skorupa, J. P., and Struhsaker, T. T. (2000). Long-term effects of logging on African primate communities: a 28-year comparison from Kibale National Park, Uganda. *Conserv. Biol.* 14:207–217.

Chapman, C. A., and Chapman, L. J. (2000). Determinants of group size in primates: the importance of travel costs. In: Boinski, S., and Garber, P. A. (eds.), *On the Move: How and Why Animals Travel in Groups*. University of Chicago Press, Chicago. pp. 24–42.

Chapman, C. A., and Chapman, L. J. (2002). Foraging challenges of red colobus monkeys: influence of nutrients and secondary compounds. *Comp. Biochem. Physiol.* 133:861–875.

Chapman, C. A., Chapman, L. J., Bjorndal, K. A., and Onderdonk, D. A. (2002a). Application of protein-to-fiber ratios to predict colobine abundance on different spatial scales. *Int. J. Primatol.* 23:283–310.

Chapman, C. A., Chapman, L. J., and Gillespie, T. R. (2002b). Scale issues in the study of primate foraging: red colobus of Kibale National Park. *Am. J. Phys. Anthropol.* 117:349–363.

Chapman, C. A, Chapman, L. J., Naughton-Treves, L., Lawes, M. J., and McDowell, L. R. (2004). Predicting folivorous primate abundance: validation of a nutritional model. *Am. J. Primatol.* 62:55–69.

Cheney, D. L. (1987). Interactions and relationships between groups. In: Smuts, B. B., Cheney, D. L., Seyfarth, R. M., Wrangham, R. W., and Struhsaker, T. T. (eds.), *Primate Societies*. University of Chicago Press, Chicago. pp. 267–281.

Chivers, D. J. (1994). Functional anatomy of the gastrointestinal tract. In: Davies, A. G., and Oates, J. F. (eds.), *Colobine Monkeys: Their Ecology, Behavior and Evolution*. Cambridge University Press, Cambridge. pp. 205–227.

Clutton-Brock, T. H. (1974). Activity patterns of red colobus (*Colobus badius tephrosceles*). *Folia Primatol.* 21:161–187.

Clutton-Brock, T. H. (1975a). Feeding behaviour of red colobus and black and white colobus in East Africa. *Folia Primatol.* 23:165–207.

Clutton-Brock, T. H. (1975b). Ranging behavior of red colobus (*Colobus badius tephrosceles*) in the Gombe National Park. *Anim. Behav.* 23:706–722.

Clutton-Brock, T. H., and Harvey, P. H. (1977). Species differences in feeding and ranging behaviour in primates. In: Clutton-Brock, T. H. (ed.), *Primate Ecology: Studies of Feeding and Ranging Behavior in Lemurs, Monkeys and Apes*. Academic Press, New York. pp. 557–584.

Cooper, M. A., Aureli, F., and Singh, M. (2004). Between-group encounters among bonnet macaques (*Macaca radiata*). *Behav. Ecol. Sociobiol.* 56:217–227.

Dasilva, G. L. (1989). The ecology of the western black and white colobus (*Colobus polykomos polykomos* Zimmerman 1780) on a riverine island in south-eastern Sierra Leone [PhD thesis]. University of Oxford, Oxford.

Dasilva, G. L. (1992). The western black-and-white colobus as a low-energy strategist: activity budgets, energy expenditure and energy intake. *J. Anim. Ecol.* 61:79–91.

Dasilva, G. L. (1993). Postural changes and behavioural thermoregulation in *Colobus polykomos*: the effect of climate and diet. *Afr. J. Ecol.* 31:226–241.

Dasilva, G. L. (1994). Diet of *Colobus polykomos* on Tiwai Island: selection of food in relation to its seasonal abundance and nutritional quality. *Int. J. Primatol.* 15:655–680.

Davies, A. G. (1994). Colobine populations. In: Davies, A. G., and Oates, J. F. (eds.), *Colobine Monkeys: Their Ecology, Behavior and Evolution*. Cambridge University Press, Cambridge. pp. 285–310.

Davies, A. G., Oates, J. F., and Dasilva, G. L. (1999). Patterns of frugivory in three West African colobine monkeys. *Int. J. Primatol.* 20:327–357.

Decker, B. S. (1994). Effects of habitat disturbance on the behavioral ecology and demographics of the Tana River red colobus (*Colobus badius rufomitratus*). *Int. J. Primatol.* 15:703–737.

Dixson, A. F. (1998). *Primate Sexuality: Comparative Studies for the Prosimians, Monkeys, Apes, and Human Beings*. Oxford University Press, Oxford.

Dunbar, R. I. M. (1987). Habitat quality, population dynamics, and group composition in a colobus monkey (*Colobus guereza*). *Int. J. Primatol.* 8:299–329.

Dunbar, R. I. M., and Dunbar, E. P. (1974). Ecology and population dynamics of *Colobus guereza* in Ethiopia. *Folia Primatol.* 21:188–208.

Dunbar, R. I. M., and Dunbar, E. P. (1976). Contrasts in social structure among black-and-white colobus monkey groups. *Anim. Behav.* 24:84–92.

Fashing, P. J. (1999). The behavioral ecology of an African colobine monkey: diet, range use, and patterns of intergroup aggression in eastern black and white colobus monkeys (*Colobus guereza*) [PhD thesis]. Columbia University, New York.

Fashing, P. J. (2001a). Activity and ranging patterns of guerezas in the Kakamega Forest: intergroup variation and implications for intragroup feeding competition. *Int. J. Primatol.* 22:549–577.

Fashing, P. J. (2001b). Feeding ecology of guerezas in the Kakamega Forest, Kenya: the importance of Moraceae fruit in their diet. *Int. J. Primatol.* 22:579–609.

Fashing, P. J. (2001c). Male and female strategies during intergroup encounters in guerezas (*Colobus guereza*): evidence for resource defense mediated through males and a comparison with other primates. *Behav. Ecol. Sociobiol.* 50:219–230.

Fashing, P. J. (2002). Population status of black and white colobus monkeys (*Colobus guereza*) in the Kakamega Forest, Kenya: are they really on the decline? *Afr. Zool.* 37:119–126.

Fashing, P. J. (2004). Mortality trends in the African cherry (*Prunus africana*) and the implications for colobus monkeys (*Colobus guereza*) in Kakamega Forest, Kenya. *Biol. Conserv.* 120:449–459.

Fashing, P. J., and Cords, M. (2000). Diurnal primate densities and biomass in the Kakamega Forest: an evaluation of census methods and a comparison with other forests. *Am. J. Primatol.* 50:139–152.

Fashing, P. J., Mulindahabi, F., Gakima, J.-B., Masozera, M., Mununura, I., Plumptre, A. J., and Nguyen, N. (in press). Activity and ranging patterns of Angolan black-and-white-colobus (*Colobus angolensis ruwenzorii*) in Nyungwe Forest, Rwanda: possible costs of large group size. *Int. J. Primatol.*

Fashing, P. J., and Oates, J. F. (in press). *Colobus guereza*. In: Kingdon, J., Happold, D., and Butynski, T. (eds.), *The Mammals of Africa*. Academic Press, London.

Fimbel, C., Vedder, A., Dierenfeld, E., and Mulindahabi, F. (2001). An ecological basis for large group size in *Colobus angolensis* in the Nyungwe Forest, Rwanda. *Afr. J. Ecol.* 39:83–92.

Firos, S. (2001). Absence of intragroup coalitions in adult male red colobus (*Colobus badius tephrosceles*) in the Kibale National Park, Uganda. *Folia Primatol.* 72:54–56.

Fleury, M. C., and Gautier-Hion, A. (1999). Seminomadic ranging in a population of black colobus (*Colobus satanas*) in Gabon and its ecological correlates. *Int. J. Primatol.* 20:491–509.

Galat-Luong, A., and Galat, G. (1979). Quelques observations sur l'écologie de *Colobus pennanti oustaleti* en Empire Centrafricain. *Mammalia* 43:309–312.

Gatinot, B. L. (1976). Les milieux fréquentés par le colobe bai d'Afrique de l'ouest (*Colobus badius temmincki*, Kuhn 1820) en Sénégambie. *Mammalia* 40:1–12.

Gatinot, B. L. (1977). Le régime alimentaire du colobe bai au Séngéal. *Mammalia* 41:373–402.

Gatinot, B. L. (1978). Characteristics of the diet of the West African red colobus. In: Chivers, D. J., and Herbert, J. (eds.), *Recent Advances in Primatology*, vol. 1. Academic Press, London. pp. 253–255.

Gautier-Hion, A., Gautier, J.-P., and Moungazi, A. (1997). Do black colobus in mixed-species groups benefit from increased foraging. *C. R. Acad. Sci. Paris Sci. Vie* 320:67–71.

Gillespie, T. R., and Chapman, C. A. (2001). Determinants of group size in the red colobus monkey (*Procolobus badius*): an evaluation of the generality of the ecological-constraints model. *Behav. Ecol. Sociobiol.* 50:329–338.

Gonzalez-Kirchner, J. P. (1997). Behavioural ecology of two sympatric colobines on Bioko Island, Equatorial Guinea. *Folia Zool.* 46:97–104.

Grimes, K. H. (2000). Guereza dietary and behavioural patterns at the Entebbe Botanical Gardens [PhD thesis]. University of Calgary, Calgary.

Groves, C. P. (1973). Notes on the ecology and behaviour of the Angola colobus (*Colobus angolensis* P. L. Sclater 1860) in N. E. Tanzania. *Folia Primatol.* 20:12–26.

Groves, C. P. (2001). *Primate Taxonomy*. Smithsonian Institution Press, Washington DC.

Grubb, P. J., Butynski, T. M., Oates, J. F., Bearder, S. K., Disotell, T. R., Groves, C. P., and Struhsaker, T. T. (2003). Assessment of the diversity of African primates. *Int. J. Primatol.* 24:1301–1357.

Harris, T. R., and Monfort, S. L. (2003). Behavioral and endocrine dynamics associated with infanticide in a black and white colobus monkey (*Colobus guereza*). *Am. J. Primatol.* 61:135–142.

Harrison, M. J. S. (1983). Territorial behavior in the green monkey, *Cercopithecus sabaeus*: seasonal defense of local food supplies. *Behav. Ecol. Sociobiol.* 12:85–94.

Harrison, M. J. S. (1986). Feeding ecology of black colobus, *Colobus satanas*, in central Gabon. In: Else, J. G., and Lee, P. C. (eds.), *Primate Ecology and Conservation*. Cambridge University Press, Cambridge. pp. 31–37.

Harrison, M. J. S., and Hladik, C. M. (1986). Un primate granivore: le colobe noir dans la forêt du Gabon; potentialité d'évolution du comportement alimentaire. *Rev. Ecol. (Terre Vie)* 41:281–298.

Hart, J. A., Katembo, M., and Punga, K. (1996). Diet, prey selection and ecological relations of leopard and golden cat in the Ituri Forest, Zaire. *Afr. J. Ecol.* 34:364–379.

Holenweg, A.-K., Noë, R., and Schabel, M. (1996). Waser's gas model applied to associations between red colobus and diana monkeys in the Taï National Park, Ivory Coast. *Folia Primatol.* 67:125–136.

Höner, O. P., Leumann, L., and Noë, R. (1997). Dyadic associations of red colobus and diana monkey groups in the Taï National Park, Ivory Coast. *Primates* 38:281–291.

Hoppe-Dominik, B. (1984). Étude du spectre des proies de la panthère, *Panthera pardus*, dans le Parc National de Taï en Côte d'Ivoire. *Mammalia* 48:477–487.

Horwich, R. H., and Manski, D. (1975). Maternal care and infant transfer in two species of *Colobus* monkeys. *Primates* 16:49–73.

Hrdy, S. B. (1974). Male–male competition and infanticide among the langurs (*Presbytis entellus*) of Abu, Rajasthan. *Folia Primatol.* 22:19–58.

Isbell, L. A. (1991). Contest and scramble competition: patterns of female aggression and ranging behavior among primates. *Behav. Ecol.* 2:143–155.

Kanga, E. M. (2001). Survey of black and white colobus monkeys (*Colobus angolensis palliatus*) in Shimba Hills National Reserve and Maluganji Sanctuary, Kenya. *Am. Soc. Primatol. Bull.* 25:8–9.

Kanga, E. M., and Heidi, C. M. (1999/2000). Survey of the Angolan black-and-white colobus monkey *Colobus angolensis palliatus* in the Diani forests, Kenya. *Afr. Primates* 4:50–54.

Kay, R. N. B., and Davies, A. G. (1994). Digestive physiology. In: Davies, A. G., and Oates, J. F. (eds.). *Colobine Monkeys: Their Ecology, Behaviour and Evolution.* Cambridge University Press, Cambridge. pp. 229–249.

Kirkpatrick, R. C. (1999). Colobine diet and social organization. In: Dolhinow, P., and Fuentes, A. (eds.), *The Nonhuman Primates.* Mayfield, Mountain View, CA. pp. 93–105.

Korstjens, A. H. (2001). The mob, the secret sorority, and the phantoms: an analysis of the socio-ecological strategies of the three colobines at Taï [PhD thesis]. University of Utrecht, Utrecht.

Korstjens, A. H., and Noë, R. (2004). Mating system of an exceptional primate, the olive colobus (*Procolobus verus*). *Am. J. Primatol.* 62:261–273.

Korstjens, A. H., and Schippers, E. Ph. (2003). Dispersal patterns among olive colobus in Taï National Park. *Int. J. Primatol.* 24:515–539.

Korstjens, A. H., Schippers, E. P., Nijssen, E. C., van Oirschot, B. M. A., Krebs, M., Bergman, K., Deffernez, C., and Paukert, C. (in press). The influence of food on the social organisation of three colobine species. In: Noë, R., McGraw, S., and Zuberbühler, K. (eds.), *Monkeys of the Taï Forest: An African Primate Community.* Cambridge University Press, Cambridge.

Korstjens, A. H., Sterck, E. H. M., and Noë, R. (2002). How adaptive or phylogenetically inert is primate social behavior? A test with two sympatric colobines. *Behaviour* 139:203–225.

Krüger, O., Affeldt, E., Brackmann, M., and Milhahn, K. (1998). Group size and composition of *Colobus guereza* in Kyambura Gorge, southwest Uganda, in relation to chimpanzee activity. *Int. J. Primatol.* 19:287–297.

Leskes, A., and Acheson, N. H. (1971). Social organization of a free-ranging troop of black and white colobus monkeys (*Colobus abyssinicus*). In: Kummer, H. (ed.), *Proceedings of the Third International Congress of Primatology, Zurich 1970. Behavior,* vol. 3. S. Karger, Basel. pp. 22–31.

Lydekker, R. (1905). Colour evolution in guereza monkeys. *Proc. Zool. Soc. Lond.* 2:325–329.

Maher, C. R., and Lott, D. F. (1995). Definitions of territoriality used in the study of variation in vertebrate spacing systems. *Anim. Behav.* 49:1581–1597.

Maisels, F., Gautier-Hion, A., and Gautier, J.-P. (1994). Diets of two sympatric colobines in Zaire: more evidence on seed-eating in forests on poor soils. *Int. J. Primatol.* 15:681–701.

Marler, P. (1969). *Colobus guereza*: territoriality and group composition. *Science* 163:93–95.

Marler, P. (1972). Vocalizations of East African monkeys II: black and white colobus. *Behaviour* 42:175–197.

Marsh, C. (1978a). Comparative activity budgets of red colobus. In: Chivers, D. J., and Herbert, J. (eds.), *Recent Advances in Primatology,* vol. 1. Academic Press, London. pp. 249–251.

Marsh, C. (1978b). Ecology and Social Organization of the Tana River Red Colobus (Colobus badius rufomitratus). [PhD thesis]. University of Bristol, Bristol.

Marsh, C. W. (1979a). Comparative aspects of social organization in the Tana River red colobus, *Colobus badius rufomitratus. Z. Tierpsychol.* 51:337–362.

Marsh, C. W. (1979b). Female transference and mate choice among Tana River red colobus. *Nature* 281:568–569.

Marsh, C. W. (1981a). Diet choice among red colobus (*Colobus badius rufomitratus*) on the Tana River, Kenya. *Folia Primatol.* 35:147–178.

Marsh, C. W. (1981b). Ranging behaviour and its relation to diet selection in Tana River red colobus (*Colobus badius rufomitratus*). *J. Zool. Lond.* 195:473–492.

Marsh, C. W. (1981c). Time budget of Tana River red colobus. *Folia Primatol.* 35:30–50.

Mbora, D. N. M., and McGrew, W. C. (2002). Extra-group sexual consortship in the Tana River red colobus (*Procolobus rufomitratus*)? *Folia Primatol.* 73:210–213.

Mbora, D. N. M., and Meikle, D. B. (2004a). Forest fragmentation and the distribution, abundance and conservation of the Tana River red colobus (*Procolobus rufomitratus*). *Biol. Conserv.* 118:67–77.

Mbora, D. N. M., and Meikle, D. B. (2004b). The value of unprotected habitat in conserving the critical endangered Tana River red colobus (*Procolobus rufomitratus*). *Biol. Conserv.* 120:91–99.

McGraw, S. (1994). Census, habitat preference, and polyspecific associations of six monkeys in the Lomako Forest, Zaire. *Am. J. Primatol.* 34:295–307.

McGraw, W. S. (2005). Update on the search for Miss Waldron's red colobus monkey (*Procolobus badius waldroni*). *Int. J. Primatol.* 26:605–620.

McGraw, W. S., and Oates, J. F. (2002). Evidence for a surviving population of Miss Waldron's red colobus. *Oryx* 36:223.

McKey, D. B. (1978a). Plant chemical defenses and the feeding and ranging behavior of colobus monkeys in African rainforests [PhD thesis]. University of Michigan, Ann Arbor.

McKey, D. B. (1978b). Soils, vegetation, and seed-eating by black colobus monkeys. In: Montgomery, G. G. (ed.), *The Ecology of Arboreal Folivores.* Smithsonian Institution Press, Washington DC. pp. 423–437.

McKey, D. B., Gartlan, J. S., Waterman, P. G., and Choo, G. M. (1981). Food selection by black colobus monkeys (*Colobus satanas*) in relation to food chemistry. *Biol. J. Linn. Soc.* 16:115–146.

McKey, D. B., and Waterman, P. G. (1982). Ranging behavior of a group of black colobus (*Colobus satanas*) in the Douala-Edea Reserve, Cameroon. *Folia Primatol.* 39:264–304.

McNab, B. (1978). The energetics of arboreal folivores: the problem of feeding on a ubiquitous food source. In: Montgomery, G. G. (ed.), *The Ecology of Arboreal Folivores*. Smithsonian Institution Press, Washington DC. pp. 153–162.

Mitani, J. C., Gros-Louis, J., and Manson, J. H. (1996). Number of males in primate groups: Comparative tests of competing hypotheses. *Am. J. Primatol.* 38:315–332.

Mitani, J. C., and Rodman, P. S. (1979). Territoriality: the relation of ranging pattern and home range size to defendability, with an analysis of territoriality among primate species. *Behav. Ecol. Sociobiol.* 5:241–251.

Moreno-Black, G. S., and Bent, E. F. (1982). Secondary compounds in the diet of *Colobus angolensis*. *Afr. J. Ecol.* 20:29–36.

Mowry, C. B., Decker, B. S., and Shure, D. J. (1996). The role of phytochemistry in dietary choices of Tana River red colobus monkeys (*Procolobus badius rufomitratus*). *Int. J. Primatol.* 17:63–84.

Mturi, F. A. (1991). The feeding ecology and behavior of the red colobus monkey (*Colobus badius kirkii*). [PhD thesis]. Dares Salaam, University of Dares Salaam.

Mturi, F. A. (1993). Ecology of the Zanzibar red colobus monkey, *Colobus badius kirkii* (Gray, 1968), in comparison with other red colobines. In: Lovett, J. C., and Wasser, S. K. (eds.), *Biogeography and Ecology of the Rain Forests of Eastern Africa*. Cambridge University Press, Cambridge. pp. 243–266.

Müller, E. F., Kamau, J. M. Z., and Maloiy, G. M. O. (1983). A comparative study of basal metabolism and thermoregulation in a folivorous (*Colobus guereza*) and an omnivorous (*Cercopithecus mitis*) primate species. *Comp. Biochem. Physiol.* 74:319–322.

Mutschler, T., Nievergelt, C. M., and Feistner, A. T. C. (2000). Social organization of the Alaotran gentle lemur (*Hapalemur griseus alaotrensis*). *Am. J. Primatol.* 46:251–258.

Newton, P. N., and Dunbar, R. I. M. (1994). Colobine monkey ciety. In: Davies, A. G., and Oates, J. F. (eds.), *Colobine Monkeys: Their Ecology, Behavior and Evolution*. Cambridge University Press, Cambridge. pp. 311–346.

Nievergelt, C. M., Mutschler, T., and Feistner, A. T. C. (1998). Group encounters and territoriality in wild Alaotran gentle lemurs (*Hapalemur griseus alaotrensis*). *Am. J. Primatol.* 50:9–24.

Oates, J. F. (1974). The ecology and behaviour of the black-and-white colobus monkey (*Colobus guereza* Ruppell) in East Africa [PhD thesis]. University of London, London.

Oates, J. F. (1977a). The guereza and its food. In: Clutton-Brock, T. H. (ed.), *Primate Ecology: Studies of Feeding and Ranging Behavior in Lemurs, Monkeys and Apes*. Academic Press, New York. pp. 275–321.

Oates, J. F. (1977b). The guereza and man. In: Prince Ranier III and Bourne, G. H. (eds.), *Primate Conservation*. Academic Press, London. pp. 419–467.

Oates, J. F. (1977c). The social life of a black and white colobus monkey, *Colobus guereza*. *Z. Tierpsychol.* 45:1–60.

Oates, J. F. (1978). Water-plant and soil consumption by guereza monkeys (*Colobus guereza*): a relationship with minerals and toxins in the diet? *Biotropica* 10:241–253.

Oates, J. F. (1988). The diet of the olive colobus monkey, *Procolobus verus*, in Sierra Leone. *Int. J. Primatol.* 9:457–478.

Oates, J. F. (1994). The natural history of African colobines. In: Davies, A. G., and Oates, J. F. (eds.), *Colobine Monkeys: Their Ecology, Behavior and Evolution*. Cambridge University Press, Cambridge. pp. 75–128.

Oates, J. F. (1996). Habitat alteration, hunting and the conservation of folivorous primates in African forests. *Aust. J. Ecol.* 21:1–9.

Oates, J. F. (1999). *Myth and Reality in the Rain Forest: How Conservation Strategies Are Failing in West Africa*. University of California Press, Berkeley.

Oates, J. F., Abedi-Lartey, M., McGraw, W. S., Struhsaker, T. T., and Whitesides, G. H. (2000a). Extinction of a West African red colobus monkey. *Conserv. Biol.* 14:1526–1532.

Oates, J. F., Bocian, C. M., and Terranova, C. J. (2000b). The loud calls of black-and-white colobus monkeys: new information and a reappraisal of their phylogenetic and functional significance. In: Whitehead, P. F., and Jolly, C. J. (eds.), *Old World Monkeys*. Cambridge University Press, Cambridge. pp. 431–452.

Oates, J. F., and Davies, A. G. (1994). What are the colobines? In: Davies, A. G., and Oates, J. F. (eds.), *Colobine Monkeys: Their Ecology, Behavior and Evolution*. Cambridge University Press, Cambridge. pp. 1–9.

Oates, J. F., Davies, A. G., and Delson, E. (1994). The diversity of living colobines. In: Davies, A. G., and Oates, J. F. (eds.), *Colobine Monkeys: Their Ecology, Behavior and Evolution*. Cambridge University Press, Cambridge. pp. 45–73.

Oates, J. F., Swain, T., and Zantovska, J. (1977). Secondary compounds and food selection by colobus monkeys. *Biochem. Syst. Ecol.* 5:317–321.

Oates, J. F., and Trocco, T. F. (1983). Taxonomy and phylogeny of black-and-white colobus monkeys: inferences from an analysis of loud call variation. *Folia Primatol.* 40:83–113.

Oates, J. F., and Whitesides, G. H. (1990). Association between olive colobus (*Procolobus verus*), diana guenons (*Cercopithecus diana*), and other forest monkeys in Sierra Leone. *Am. J. Primatol.* 21:129–146.

Oates, J. F., Whitesides, G. H., Davies, A. G., Waterman, P. G., Green, S. M., Dasilva, G. L., and Mole, S. (1990). Determinants of variation in tropical forest primate biomass: new evidence from West Africa. *Ecology* 71:328–343.

Olson, D. K. (1980). Male interactions and troop split among black-and-white colobus monkeys (Colobus polykomosvellerosus). Paper presented at the Eighth Congress of the International Primatological Society, Florence, Italy.

Olson, D. K. (1986). Determining range size for arboreal monkeys: methods, assumptions, and accuracy. In: Taub, D. M., and King, F. A. (eds.), *Current Perspectives in Primate Social Dynamics*. Van Nostrand Reinhold, New York. pp. 212–227.

Onderdonk, D. A. (2000). Infanticide of a newborn black-and-white colobus monkey (*Colobus guereza*) in Kibale National Park, Uganda. *Primates* 41:209–212.

Onderdonk, D. A., and Chapman, C. A. (2000). Coping with forest fragmentation: the primates of Kibale National Park, Uganda. *Int. J. Primatol.* 21:587–611.

Plumptre, A. J., and Reynolds, V. (1994). The effect of selective logging on the primate populations in the Budongo Forest Reserve, Uganda. *J. Appl. Ecol.* 31:631–641.

Poulsen, J. R., Clark, C. J., and Smith, T. B. (2001). Seed dispersal by a diurnal primate community in Dja Reserve, Cameroon. *J. Trop. Ecol.* 17:787–808.

Poulsen, J. R., Clark, C. J., Connor, E. F., and Smith, T. B. (2002). Differential resource use by primates and hornbills: implications for seed dispersal, *Ecology* 83:228–240.

Rahm, U. H. (1970). Ecology, zoogeography and systematics of some African forest monkeys. In: Napier, J. R., and Napier, P. M. (eds.), *Old World Monkeys: Evolution, Systematics and Behavior*. Academic Press, New York. pp. 589–626.

Robinson, J. G. (1988). Group size in wedge-capped capuchin monkeys *Cebus olivaceus* and the reproductive success of males and females. *Behav. Ecol. Sociobiol.* 23:187–197.

Rose, M. D. (1978). Feeding and associated positional behavior of black and white colobus monkeys (*Colobus guereza*). In: Montgomery, C. G. (ed.), *The Ecology of Arboreal Folivores*. Smithsonian Institution Press, Washington DC. pp. 253–262.

Rowell, T. E., and Richards, S. M. (1979). Reproductive strategies of some African monkeys. *J. Mammal.* 60:58–69.

Saj, T. L., and Sicotte, P. (2005). Male takeover in *Colobus vellerosus* at Boabeng-Fiema Monkey Sanctuary, central Ghana. *Primates* 46:211–214.

Saj, T. L., Teichroeb, J. A., and Sicotte, P. (2005). The population status and habitat quality of the Geoffroy's pied colobus (*Colobus vellerosus*) at Boabeng-Fiema sacred grove, Ghana. In: Paterson, J. D. (ed.), *Commensalism and Conflict: Human–Primate Interface*. American Society of Primatology, Norman, OK. pp. 264–287.

Schenkel, R., and Schenkel-Hulliger, L. (1967). On the sociology of free-ranging colobus (*Colobus guereza caudatus* Thomas 1885). In: Starck, D., Schneider, R., and Kuhn, H. J. (eds.), *Progress in Primatology*. Karger, Basel. pp. 185–194.

Schultz, S., Noë, R., McGraw, W. S., and Dunbar, R. I. M. (2004). A community-level evaluation of the impact of prey behavioural and ecological characteristics on predator diet composition. *Proc. R. Soc. Lond. B* 271:725–732.

Schwarz, E. (1929). On the local races and distribution of black and white colobus monkeys. *Proc. Zool. Soc. Lond.* 1929:585–598.

Sicotte, P., and MacIntosh, A. J. (2004). Inter-group encounters and male incursions in *Colobus vellerosus* in central Ghana. *Behaviour* 141:533–553.

Siex, K. S. (2003). Effects of population compression on the demography, ecology, and behavior of the Zanzibar red colobus monkey (*Procolobus kirkii*) [PhD thesis]. Duke University, Durham, NC.

Siex, K. S., and Struhsaker, T. T. (1999a). Colobus monkeys and coconuts: a study of perceived human-wildlife conflicts. *J. Appl. Ecol.* 36:1009–1020.

Siex, K. S., and Struhsaker, T. T. (1999). Ecology of the Zanzibar red colobus monkey: demographic variability and habitat stability. *Int. J. Primatol.* 20:163–192.

Skorupa, J. P. (1986). Responses of rainforest primates to selective logging in Kibale Forest, Uganda: a summary report. In: Benirschke, K. (ed.), *Primates: The Road to Self-Sustaining Populations*. Springer-Verlag, New York. pp. 57–70.

Skorupa, J. P. (1987). Crowned eagles *Stephanoaetus coronatus* in rainforest: observations on breeding chronology and diet at a nest in Uganda. *Ibis* 131:294–298.

Stanford, C. B. (1991). The capped langur in Bangladesh: behavioral ecology and reproductive tactics. *Contrib. Primatol.* 26:1–179.

Stanford, C. B. (1998). *Chimpanzee and Red Colobus: The Ecology of Predator and Prey*. Harvard University Press, Cambridge, MA.

Stanford, C. B. (2002). Avoiding predators: expectations and evidence in primate antipredator behavior. *Int. J. Primatol.* 23:741–757.

Starin, E. D. (1988). Gestation and birth-related behaviors in Temminck's red colobus. *Folia Primatol.* 51:161–164.

Starin, E. D. (1989). Threats to the monkeys of the Gambia. *Oryx* 23:208–214.

Starin, E. D. (1991). Socioecology of the red colobus monkey in the Gambia with particular reference to female–male differences and transfer patterns [PhD thesis]. City University of New York, New York.

Starin, E. D. (1994). Philopatry and affiliation among red colobus. *Behaviour* 130:253–270.

Sterck, E. H. M., Watts, D. P., and van Schaik, C. P. (1997). The evolution of female social relationships in nonhuman primates. *Behav. Ecol. Sociobiol.* 41:291–309.

Struhsaker, T. T. (1974). Correlates of ranging behavior in a group of red colobus monkeys (*Colobus badius tephrosceles*). *Am. Zool.* 14:177–184.

Struhsaker, T. T. (1975). *The Red Colobus Monkey*. University of Chicago Press, Chicago.

Struhsaker, T. T. (1978). Food habits of five monkey species in the Kibale Forest, Uganda. In: Chivers, D. J., and Herbert, J. (eds.), *Recent Advances in Primatology*, vol. 1. Academic Press, New York. pp. 225–248.

Struhsaker, T. T. (1981). Vocalizations, phylogeny and palaeogeography of red colobus monkeys (*Colobus badius*). *Afr. J. Ecol.* 19:265–283.

Struhsaker, T. T. (1997). *Ecology of an African Rain Forest: Logging in Kibale and the Conflict Between Conservation and Exploitation*. University Press of Florida, Gainesville.

Struhsaker, T. T. (2000). The effects of predation and habitat quality on the socioecology of African monkeys: lessons from the islands of Bioko and Zanzibar. In: Whitehead, P. F., and Jolly, C. J. (eds.), *Old World Monkeys*. Cambridge University Press, Cambridge. pp. 393–430.

Struhsaker, T. T., Cooney, D. O., and Siex, K. S. (1997). Charcoal consumption by Zanzibar red colobus monkeys: its function and its ecological and demographic consequences. *Int. J. Primatol.* 18:61–72.

Struhsaker, T. T., and Leakey, M. (1990). Prey selectivity by crowned-hawk eagles on monkeys in the Kibale Forest, Uganda. *Behav. Ecol. Sociobiol.* 26:435–443.

Struhsaker, T. T., and Leland, L. (1979). Socioecology of five sympatric monkey species in the Kibale Forest, Uganda. *Adv. Stud. Behav.* 9:159–228.

Struhsaker, T. T., and Leland, L. (1985). Infanticide in a patrilineal society of red colobus monkeys. *Z. Tierpsychol.* 69:89–132.

Struhsaker, T. T., and Leland, L. (1987). Colobines: infanticide by adult males. In: Smuts, B. B., Cheney, D. L., Seyfarth, R. M., Wrangham, R. W., and Struhsaker, T. T. (eds.), *Primate Societies*. University of Chicago Press, Chicago. pp. 83–97.

Struhsaker, T. T., and Oates, J. F. (1975). Comparison of the behavior and ecology of red colobus and black-and-white colobus monkeys in Uganda: a summary. In: Tuttle, R. H. (ed.), *Socioecology and Psychology of Primates*. Mouton, The Hague. pp. 103–123.

Suzuki, A. (1979). The variation and adaptation of social groups of chimpanzees and black and white colobus monkeys. In: Bernstein, I. S., and Smith, E. O. (eds.), *Primate Ecology and Human Origins*. Garland STPM Press, New York. pp. 153–173.

Szalay, F. S., and Delson, E. (1979). *Evolutionary History of the Primates*. Academic Press, New York.

Teichroeb, J. A., Saj, T. L., Paterson, J. D., and Sicotte, P. (2003). Effect of group size on activity budgets of *Colobus vellerosus* in Ghana. *Int. J. Primatol.* 24:743–758.

Trivers, R. L. (1972). Parental investment and sexual selection. In: Campbell, B. (ed.), *Sexual Selection and the Descent of Man.* Aldine, Chicago. pp. 136–179.

Uehara, S. (1997). Predation on mammals by chimpanzees (*Pan troglodytes*). *Primates* 38:193–214.

Usongo, L. I., and Amubode, F. O. (2001). Nutritional ecology of Preuss's red colobus monkey (*Colobus badius preussi* Rahm 1970) in Korup National Park, Cameroon. *Afr. J. Ecol.* 39:121–125.

van Schaik, C. P. (1989). The ecology of social relationships amongst female primates. In: Standen, V., and Foley, R. A. (eds.), *Comparative Socioecology: The Behavioral Ecology of Humans and Other Mammals.* Blackwell, Oxford. pp. 195–218.

van Schaik, C. P., Assink, P. R., and Salafsky, N. (1992). Territorial behavior in Southeast Asian langurs: resource defense or mate defense? *Am. J. Primatol.* 26:233–242.

van Schaik, C. P., and Janson, C. H. (2000). *Infanticide by Males and Its Implications.* Cambridge University Press, Cambridge.

Vedder, A., and Fashing, P. J. (2002). Diet of a 300-member Angolan colobus monkey (*Colobus angolensis*) supergroup in the Nyungwe Forest, Rwanda. *Am. J. Phys. Anthropol.* 117(suppl. 34):159–160.

Wachter, B., Schabel, M., and Noë, R. (1997). Diet overlap and polyspecific associations of red colobus and diana monkeys in the Taï National Park, Ivory Coast. *Ethology* 103:514–526.

Wasserman, M. D., and Chapman, C. A. (2003). Determinants of colobine monkey abundance: the importance of food energy, protein and fibre content. *J. Anim. Ecol.* 72:650–659.

Watts, D. P., and Mitani, J. C. (2002). Hunting behavior of chimpanzees at Ngogo, Kibale National Park, Uganda. *Int. J. Primatol.* 23:1–28.

Werre, J. L. R. (2000). Ecology and behavior of the Niger Delta red colobus (*Procolobus badius epieni*) [PhD thesis]. City University of New York, New York.

White, L. J. T. (1994). Biomass of rain forest mammals in the Lopé Reserve, Gabon. *J. Anim. Ecol.* 63:499–512.

Whitesides, G. H., Oates, J. F., Green, S. M., and Kluberdanz, R. P. (1988). Estimating primate densities from transects in a West African rain forest: a comparison of techniques. *J. Anim. Ecol.* 57:345–367.

Wich, S. A., and Nunn, C. L. (2002). Do male "long-distance calls" function in mate defense? A comparative study of long-distance calls in primates. *Behav. Ecol. Sociobiol.* 52:474–484.

Wieczkowski, J., and Mbora, D. N. M. (1999/2000). Increasing threats to the conservation of endemic endangered primates and forests of the lower Tana River, Kenya. *Afr. Primates* 4:32–40.

Williams, J. M., Oehlert, G. W., Carlis, J. V., and Pusey, A. E. (2004). Why do male chimpanzees defend a group range? *Anim. Behav.* 68:523–532.

Wrangham, R. W. (1980). An ecological model of female-bonded primate groups. *Behaviour* 75:262–300.

Wrangham, R. W., Gittleman, J. L., and Chapman, C. A. (1993). Constraints on group size in primates and carnivores: population density and day range as assays of exploitation competition. *Behav. Ecol. Sociobiol.* 32:199–209.

Zuberbühler, K., and Jenny, D. (2002). Leopard predation and primate evolution. *J. Hum. Evol.* 43:873–886.

13

The Macaques
A Double-Layered Social Organization

Bernard Thierry

INTRODUCTION

The genus *Macaca* is one of the most successful primate radiations. Its representatives are present in both Asia and Africa, and it has the widest geographical range of primates after *Homo* (Fooden 1982) (Fig. 13.1). Whereas many macaques live at the heart of the tropics, others inhabit northern regions where snowfall occurs during winter. In spite of their spread, macaques remain interfertile (Bernstein and Gordon 1980). All possess cheek pouches, allowing them to harvest food and minimize feeding time. Their social organization is characterized by both a profound unity and a great diversity, which may be best described as variations on the same theme (Thierry et al. 2004). On the one hand, macaques share the same basic patterns of organization. They form multimale/multifemale groups. Neighboring groups have overlapping home ranges. Females are philopatric, consequently forming kin-bonded subgroups within their natal group. Most males disperse and periodically transfer from one group to another. On the other hand, macaques widely nepotism govern all macaque societies (Matsumura 2001). Though there were early hints that some macaques might be "nicer" than others (Rosenblum et al. 1964), two decades passed before interspecific contrasts in the conciliatory dispositions of macaques were directly addressed (Thierry 1986, 2000; de Waal and Luttrell 1989). A number of contrasts are now documented in the social relationships of macaques, making the genus an especially interesting taxon to compare homologous social features and investigate their determinism. In what follows, I will successively review the evolution, ecology, life history, reproduction, and social behaviors of macaques, with the aim of unraveling the two layers that structure their social organization: social styles and female defensibility by males.

EVOLUTION AND BIOGEOGRAPHY

Presently, 21 species of macaques are recognized (Table 13.1) (Fooden 1976, Groves 2001, Sinha et al. 2005a),

Table 13.1   Species and Phyletic Lineages of the Genus *Macaca*

SPECIES

***Silenus–sylvanus* lineage**

  M. sylvanus (Barbary macaque)

  M. silenus (lion-tailed macaque)

  M. nigra (crested macaque)

  M. nigrescens (Gorontalo macaque)

  M. hecki (Heck's macaque)

  M. tonkeana (Tonkean macaque)

  M. maurus (moor macaque)

  M. ochreata (booted macaque)

  M. brunnescens (Muna-Butung macaque)

  M. pagensis (Mentawai macaque)[1]

  M. nemestrina (pig-tailed macaque)[2]

***Sinica–arctoides* lineage**

  M. sinica (toque macaque)

  M. radiata (bonnet macaque)

  M. assamensis (Assamese macaque)

  M. munzala (Arunachal macaque)[3]

  M. thibetana (Tibetan macaque)

  M. arctoides (stump-tailed macaque)

***Fascicularis* lineage**

  M. fascicularis (long-tailed macaque)

  M. mulatta (rhesus macaque)

  M. fuscata (Japanese macaque)

  M. cyclopis (Taiwanese macaque)

Source: Fooden (1976, 1982), Delson (1980), Groves (2001).
[1] Mentawai macaques may represent two different species, *M. pagensis* and *M. siberu* (Roos et al. 2003).
[2] Pigtailed macaques may be split into *M. nemestrina* and *M. leonina* (Groves 2001).
[3] The discovery of Arunachal macaques is recent, dating from 2004 (Sinha et al. 2005a).

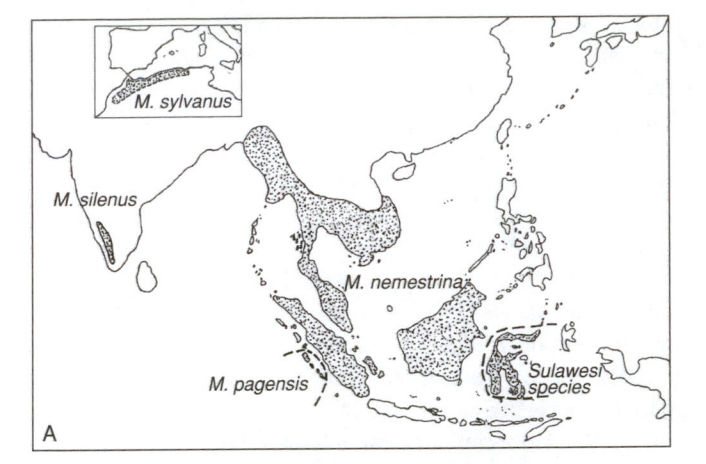

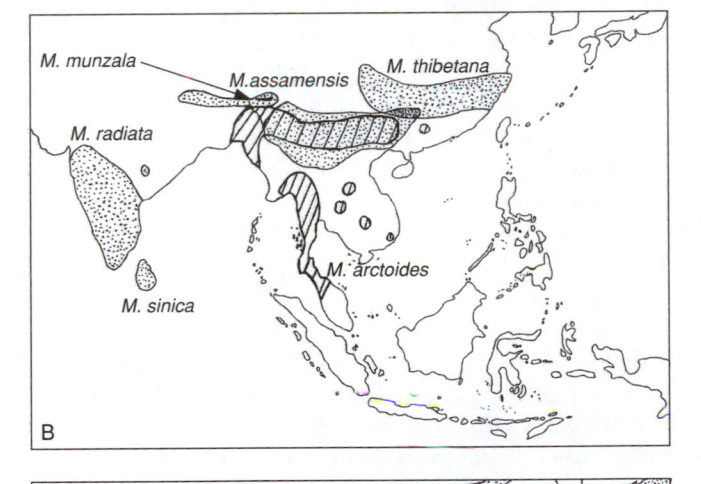

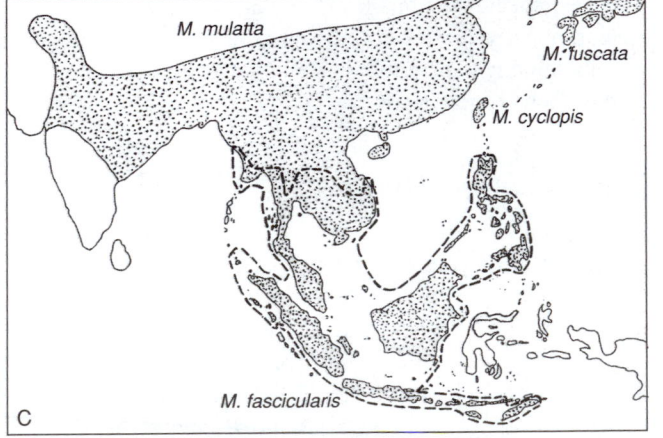

Figure 13.1 Present geographical distribution of macaques (from Fooden 1982). (A) *Silenus–sylvanus* lineage, (B) *sinica–arctoides* lineage, (C) *fascicularis* lineage. The arrow indicates the narrow range of *Macaca munzala* (in black), within the western distribution of Assamese macaques.

differ in their patterns of mating, aggression, conciliation, dominance, and nepotism. Studies of *Macaca fuscata* and *M. mulatta* have shown that their social relationships are characterized by harsh competition and that the dominance status of individuals mainly depends on the support of their kin (Kawai 1958, Kawamura 1958, Missakian 1972, Sade 1972). For a time, this led to the belief that hierarchy and representing a monophyletic group (Delson 1980, Morales and Melnick 1998). The fossil record indicates that they diverged from other papionins about 7 million years ago in northern Africa. Macaques invaded Eurasia about 5.5 million years ago, probably via the Near East. They then branched into several phyletic lineages that have been identified from morphological (Fooden 1976, Delson 1980) and molecular (Hoelzer and Melnick 1996) evidence.

Three main phyletic groups of extant macaques are distinguished, which correspond to three dispersal waves in Asia (Fooden 1976) (Table 13.1). The *silenus* lineage has the most disjunct geographical distribution (Fig. 13.1). Its 10 species are situated far away from each other, and most are represented by relict distributions, indicating an early dispersal. Only *M. nemestrina* has a large range. *M. silenus* is found in the evergreen forests of south India. The other species of the lineage inhabit the evergreen forests of the Sulawesi and Mentawai Islands. The *sinica* lineage has a moderately fragmented distribution in south Asia; it is thought to be the second to have dispersed (Fig. 13.1). *M. radiata*, *M. assamensis*, *M. munzala*, and *M. thibetana* live in tropical and subtropical continental areas; only *M. sinica* is present on an island. The *fascicularis* lineage has the most broadly continuous distribution; it is likely the third to have dispersed (Fig. 13.1). *M. fascicularis* inhabits equatorial and tropical regions. The other species occur in subtropical and temperate Asia; while *M. fuscata* and *M. cyclopis* are confined to islands, *M. mulatta* occurs from Pakistan to China. Species from different lineages may be found sympatrically in many areas: for example, *M. silenus* and *M. radiata* in south India and *M. mulatta*, *M. assamensis*, *M. nemestrina*, and *M. arctoides* in the evergreen forests of the Brahmaputra Valley in Assam.

The taxonomic position of two further species is still debated. *M. sylvanus* lives in the montane forests of North Africa; it is likely the most ancient taxon of the genus and in fact its last African representative (Fig. 13.1). It is alternatively classified as either being the only member of its own species group or belonging to the *silenus–sylvanus* lineage. *M. arctoides* inhabits south Asian broadleaf evergreen forests (Fig. 13.1); it is either ascribed to its own species group or included in the *sinica-arctoides* lineage (see Fooden 1976, Delson 1980, Deinard and Smith 2001).

The replacement of older lineages by younger ones was for some time explained by the occurrence of interspecific competition (Fooden 1976, Delson 1980). The most recent colonizers were supposed to have been better adapted to the new environmental conditions. Only geographical barriers would have prevented a later lineage from invading some refuges where remnant populations of an earlier lineage still survived. It has been pointed out, however, that the climatic changes of the Pleistocene significantly affected the distribution of primates (Eudey 1980, Brandon-Jones 1996). The aridity associated with glaciations eliminated species from their distribution ranges and split up areas where relict populations were able to survive. A reappraisal of the cir-

cumstances of the deployment of macaques in Southeast Asia indicates that the disappearance of the first lineage from large areas was caused by glacially induced deforestation (Abegg and Thierry 2002a). Several populations took refuge in relict moist habitats. When forests recovered during wetter and warmer periods, some species could not disperse due to geographical obstacles and oceanic gaps whereas others were lucky enough to find migration routes and colonize a wealth of habitats devoid of primates. Historical contingencies likely played a major role in the deployment and present distribution of macaque species.

With the lowering of sea levels in recent geological time, certain species have become marginally sympatric and low rates of hybridization may occur at their boundary. A narrow contact zone extends 2,000 km across the Indochinese peninsula between the ranges of *M. mulatta* and *M. fascicularis* (Fooden 1997). Hybrids displaying an intermediate morphology are found on this border, but their number remains limited. Such a sharp transition between species may result from the removal of a biogeographical barrier that was present during a Pleistocene glacial period. Genetic evidence indicates that gene flow goes from *M. mulatta* into *M. fascicularis* (Tosi et al. 2002). Another case of secondary intergradation is provided by the seven parapatric species inhabiting Sulawesi. This island was subdivided into an archipelago during the Pleistocene, inducing allopatric speciation (Fooden 1969). A recent uplifting of the lands has reunited the sister species. As a consequence, contemporary hybridization is observed in all the contact zones between Sulawesi species (Camperio Ciani et al. 1989). Gene flow appears mostly limited to narrow contact zones; however, the morphology of parental species changes abruptly on either side of these zones (Bynum et al. 1997, Evans et al. 2001).

HABITAT AND ECOLOGY

Macaques are semiterrestrial primates that inhabit a wide range of habitats, including evergreen, deciduous, and coniferous forests; grasslands; swamps; semideserts; and areas settled by humans. Ecological preferences do not relate directly to phylogeny (Fooden 1982, Richard et al. 1989). Species like *M. mulatta*, *M. fascicularis*, and *M. nemestrina* have extensive distributions, which include diverse types of habitat. Other species have a more limited geographical distribution, yet they may live in diverse habitats. *M. maurus*, for instance, is found both in dipterocarp and tropical seasonal forests of south Sulawesi from sea level to near 1,000 m elevation (Supriatna et al. 1992). Alterations of environment induced by human activities have further increased the diversity of ecological niches. *M. mulatta*, *M. fascicularis*, *M. radiata*, and *M. sinica* may live commensally with humans and are able to cope with degraded habitats and even urbanized areas (Richard et al. 1989). Other species have a narrower ecological range. Fooden (1976) stresses that some are restricted to primary broadleaf evergreen forests (e.g.,

Table 13.2   Diet of Nonprovisioned Macaques: Percentage of Feeding Time

SPECIES (REFERENCE AND STUDY SITE)	FRUITS	SEEDS	LEAVES AND BUDS	ROOTS, FLOWERS, AND HERBS	INVERTEBRATES	OTHERS
Macaca sylvanus (Ménard and Vallet 1996: Akfadou, Algeria)	0.8	32.2	8.8	28.9	10.5	18.8
M. sylvanus (Ménard and Vallet 1996: Djurdjura, Algeria)	4.3	26.7	13	44.4	5.6	6.0
M. nemestrina (Caldecott 1986b: Pasoh, Malaysia)	74.2	0	11.1	1.1	12.2	1.4
M. fascicularis (Yeager 1996: Tanjung Puting, Kalimantan)	66.7	0	17.2	8.9	4.1	3.1
M. fascicularis (Wheatley 1980: Kutai, Kalimantan)	87.0	0	1.6	5.4	4.0	2.0
M. mulatta (Goldstein and Richard 1989: Murree Hills, Pakistan)	8.5	0	84.4	5.9	0	1.1
M. fuscata (Hill 1997: Yakushima, Japan)	30.2	13.2	35.0	5.5	10.3	5.8
M. fuscata (Agetsuma 1995: Yakushima, Japan)	28.6	28.2	22.0	4.9	8.9	7.4
M. fuscata (Agetsuma and Nakagawa 1998: Kinkazan, Japan)	10	44	14	18	2	12
M. cyclopis (Su and Lee 2001: Jentse, Taiwan)	53.8	0	29.1	7.2	9.8	0

Source: Ménard (2004).

M. silenus, M. nemestrina, M. arctoides). When macaque species are sympatric, they often exploit different ecological niches or habitats (Fooden 1982). In Borneo, Sumatra, and Malaysia, *M. nemestrina* prefers primary montane forests whereas *M. fascicularis* inhabits riverine and secondary forests at lower altitudes. Both species additionally use different forest strata; *M. nemestrina* travel more often on the ground between more dispersed food resources (Rodman 1979, Crockett and Wilson 1980).

Although macaques are usually considered to be mainly frugivorous animals, the actual picture is more complex (Ménard 2004). Most species show a highly flexible diet with marked annual and inter-annual variations. Fruits represent a major proportion of the feeding time in tropical species like *M. nemestrina, M. fascicularis*, and *M. cyclopis*. These appear to be the most frugivorous among known species (Table 13.2). During periods of fruit shortage, however, they switch to a diet that is mainly leaves, and *M. cyclopis* may even become completely folivorous (Su and Lee 2001). *M. silenus, M. nigra, M. radiata*, and *M. assamensis* are also mainly frugivorous, but insects represent a significant part of their diet (O'Brien and Kinnaird 1997, Ménard 2004). In *M. fuscata*, populations from warm-temperate areas like Yakushima are more frugivorous than those inhabiting cool-temperate areas like Kinkazan, which are mainly granivorous (Table 13.2). Food availability is highly variable from month to month and from one season to another; in northern regions, the diet reduces to bark and buds during winter. Fruit production is limited in temperate regions, and the animals must rely on other major food sources. *M. thibetana*

is primarily a leaf eater (Zhao 1996). *M. sylvanus* appears to be either mainly granivorous or folivorous depending on the habitat; the proportion of herbaceous plants in their diet is consistently elevated (Ménard and Vallet 1996) (Table 13.2). Lastly, *M. mulatta* appears to be mainly folivorous in the deciduous forests of the Himalayan foothills (Table 13.2), but it may be largely frugivorous on occasion; in many parts of its range, it relies heavily on agricultural crops and food from people (Richard et al. 1989, Fooden 2000). Like *M. fuscata* and *M. sylvanus, M. mulatta* should be considered a non-specialized feeder (Ménard 2004).

The daily feeding and foraging time usually ranges between 2 and 6 hr (Ménard 2004). Comparisons between seasons and species indicate that feeding and foraging time increases when the proportion of leaves rises and that more frugivorous diets are associated with more time spent moving. Animals travel along greater areas when food is clumped (e.g., Agetsuma 1995). The average daily path length ranges from 100 m to 3 km according to the dispersion of resources. Foraging efficiency is also affected by group size; larger groups travel more than others as a result of feeding competition (van Schaik et al. 1983, O'Brien and Kinnaird 1997). Home range size usually fluctuates between some dozen hectares and some square kilometers, but minima and maxima may be as extreme as a few hectares and 22 km² in a single species like *M. mulatta* (Caldecott 1986a; Fooden 1986, 1995, 2000; Kurup and Kumar 1993; Fa and Lindburg 1996; O'Brien and Kinnaird 1997; Yamagiwa and Hill 1998). The degree of arboreality represents a further source of diversity. Many macaques spend more time on the ground than in the trees during the day. Species like *M. arctoides* and *M. thibetana* are mostly terrestrial, foraging on the forest floor with their heavy bodies and short tails. However, other species are primarily arboreal. *M. fascicularis* and *M. silenus* are gracile animals whose long tails serve as rudders during leaping; they may spend more than 90% of their time in the trees.

The availability of food resources is a main factor determining the size of populations. For example, a population of *M. sinica* declined by 15% during an 11-month drought (Dittus 2004). When animals are artificially fed, groups grow to unusual sizes. Provisioned groups of *M. fuscata* may reach several hundred individuals (Asquith 1989). In undisturbed habitats, groups rarely exceed 100. Wide intraspecific variations are reported for group size. The mean size is quite comparable across macaque species, most groups ranging between 15 and 50 individuals. Maximal group size before fission, however, may differ between species. Social groups usually do not go above 30–40 individuals in *M. radiata*, *M. sinica*, and *M. silenus*, whereas groups of 70–90 are not rare in some habitats for *M. mulatta*, *M. fuscata*, and *M. sylvanus* (Fooden 1986, 1995, 2000; Yamagiwa and Hill 1998; Ménard 2004).

In species like *M. nemestrina* and *M. fascicularis*, groups may temporarily split into smaller foraging units that gather at night (Caldecott 1986a, van Schaik and van Noordwijk 1986). Smaller parties may allow individuals to maximize feeding efficiency and cope with the diverse nutritional requirements of individuals. The facultative partitioning documented in *M. sylvanus* and *M. fuscata* supports this argument. The latter two species are characterized by cohesive groups, but instances of temporary partitioning have been reported in conditions of increased food competition because of either overcrowding (Fukuda 1989) or limited food resources (Ménard and Vallet 1996).

It should be added that knowledge about the impact of disease and predation on the size of populations and groups is limited. No information exists about how acute and chronic diseases affect adults, juveniles, and infants. It is known that moderate intestinal parasitism may not induce a heightened mortality, but there are no quantified data about its toll on food intake (Dittus 2004). Cases of predation by leopards, tigers, eagles, crocodiles, and pythons have been reported (Fooden 1986, 1995, 2000). Groups of *M. fascicularis* appear smaller when predators are absent (van Noordwijk and van Schaik 1985). Nevertheless, it is not possible to provide accurate estimations of predation rates in the wild because a majority of studies have been carried out in habitats where large predators had been eliminated (Dittus 2004). Actually, feral dogs and humans represent the primary threat for most macaque populations.

LIFE HISTORY AND REPRODUCTIVE BIOLOGY

The range of adult body weight is quite large in the genus (Table 13.3). Patterns of growth and reproduction are heavily dependent on food intake. In general, food-enhanced females grow faster, mature earlier, survive longer, and produce more offspring (Lyles and Dobson 1988, Asquith 1989). Reproduction starts prior to full body growth, with females reaching sexual maturity between 2 and 5 years of age depending on species and nutritional condition (Bercovitch and Harvey 2004). The first birth usually occurs between 4 and 6 years (Table 13.3), but it may be delayed beyond 10 years of age when food is scarce. Females nurse infants until 6–12 months of age. Mean interbirth interval varies from 1 to 2 years. Gestation length averages 24 weeks, with 1 further week for heavier species and 1 less week for lighter ones (Table 13.3). Female reproduction ceases around the age of 20–25 years. Though maximum life span ranges 25–35 years, in free-ranging conditions few females live long enough to become post-reproductive (Bercovitch and Harvey 2004; see also Chapter 26).

In some macaques, females display cyclical changes in coloration and swelling of the sexual skin in the anogenital region and the rump (see Chapter 25). The genus is notable for its diversity regarding these changes (Table 13.3). In all species of the *silenus–sylvanus* lineage, adult females display an enlarged anogenital swelling, which develops gradually during the follicular stage of the ovarian cycle. This characteristic has been lost to various degrees in the other groups. In the *fascicularis* lineage, the perineum reddens during estrus but only adolescent females display some swelling of the sexual skin. In most species of the *sinica–arctoides* lineage, no changes of the female appearance are visible during the cycle (Dixson 1998, Bercovitch and Harvey 2004). The occurrence of sexual swellings in a species is generally associated with the formation of multimale groups, its arousing effect on males potentially raising competition among them. The absence of sexual swellings in some lineages remains unexplained, however.

An extension of the follicular period might favor the development of signals advertising the female reproductive state (Aujard et al. 1998). Females show a long-lasting ovarian cycle in species displaying voluminous swellings (Table 13.3). Macaque females may display a second conspicuous sexual signal, the estrus call. This staccato vocalization is triggered by copulation in some macaque species; it additionally occurs outside the mating context in others, and it is absent in a third category of species (Table 13.3). The rise of female sexual displays is tracked to an increase of estrogen levels. Several functions have been proposed for their occurrence: signaling fertility, indicating female quality, promoting intermale competition, facilitating paternity confidence or paternity confusion (Dixson 1998, Soltis 2004). In several species, the climax of sexual displays corresponds with the periovulatory period of adult females, actually providing males with reliable cues about female conception time (Aujard et al. 1998, Bercovitch and Harvey 2004, Engelhardt et al. 2004).

Puberty starts in males at around 3–4 years of age, at the time of testicular enlargement; rates of agonistic interactions

Table 13.3   Life History Traits of Macaques

SPECIES	MALE BODY WEIGHT (KG)[1]	FEMALE BODY WEIGHT (KG)[1]	SEXUAL DIMORPHISM[2]	SOCIONOMIC SEX RATIO (F/M)[3]	FIRST BIRTH (YEARS)[4]	GESTATION LENGTH (DAYS)[4]	OVARIAN CYCLE LENGTH (DAYS)[4]	SEX SKIN SWELLING[4]	ESTRUS CALL[5]	COPULATION PATTERNS[6]	BIRTH SEASONALITY[7]
Macaca sylvanus	16.0	11.0	0.37	1.1	5	165		Large	C, NC	SM, MM[8]	Discrete
M. silenus	8.9	6.1	0.38	9.9	3.9	172	32	Large	C, NC	MM	Year round
M. nigra	9.9	5.5	0.59	3.5		176	36	Large	C	MM	Year round
M. nigrescens								Large	None		
M. hecki								Large			
M. tonkeana	14.9	9.0	0.50		5	176	37.4	Large	C, NC	MM	Year round
M. maurus	9.7	6.1	0.46	4.0	6–7	176	36.2	Large	None	MM	Year round, birth peak
M. ochreata								Large			
M. brunnescens								Large			
M. nemestrina	11.2	6.5	0.54	6.3	3.9	171	38	Large	C, NC	MM	Year round, birth peak
M. pagensis								Large			
M. sinica	5.7	3.2	0.58	2.4	5	168		None	None	SM	Birth peak
M. radiata	6.7	3.9	0.54	1.7	4	168	27.8	None	None	SM	Discrete
M. assamensis	11.3	6.9	0.49	2.4			32.0	Adolescence[9]		SM	Seasonal
M. thibetana	18.3	12.8	0.36	1.9	5.5		26.4	None	None	MM	Discrete
M. arctoides	12.2	8.4	0.37	5.7	4.9[10]	175	29.5	None	None	SM	Year round
M. fascicularis	5.4	3.6	0.41	4.8	5.2	163	33.5	Adolescence[9]	C	SM, MM	Birth peak
M. mulatta	11.0	8.8	0.22	2.9	4	167	28.3	Adolescence[9]	None	MM	Discrete
M. fuscata	11.0	8.0	0.32	1.3	5	173	26.5	Adolescence[9]	NC	MM	Discrete
M. cyclopis	6.0	4.9	0.20	1.7	4–6	163	29.4	Adolescence[9]		MM	Discrete

[1] From Smith and Jungers (1997).
[2] Weight dimorphism is calculated as the natural logarithm of the male body weight minus that of the female body weight.
[3] From Singh and Sinha (2004).
[4] From Bercovitch and Harvey (2004).
[5] From Oi (1996), Aujard et al. (1998), Thierry et al. (2000a). C, copulatory call; NC, noncopulatory call.
[6] From Soltis (2004). SM, single mount; MM, multiple mount.
[7] From Bercovitch and Harvey (2004): birth peak means a period of the year in which a high proportion of births, but not all births, are concentrated, contrary to discrete periods, in which all births are confined.
[8] From Ménard (personal communication).
[9] Swelling occurs in adolescent females only.
[10] From Fooden (1990).

with other males increase, and in the next 2 years adolescents undergo a rise in weight and testosterone levels associated with an elevation in dominance rank (e.g., Nieuwenhuijsen et al. 1988, Bercovitch 1993). Like females, growth and sexual maturation in males are both linked to feeding and social conditions. In *M. mulatta*, a high dominance status provided by maternal protection accelerates maturation (Bercovitch 1993). In *M. fuscata*, the age at testicular descent may be delayed by 1 or 2 years when food supply is limited (Nigi et al. 1989). Age at reproduction is significantly delayed relative to sexual maturation in males. Most have to wait until 7–11 years of age to reach full body size and compete with other adult males for access to reproductive females.

Macaque species differ by their patterns of copulation. In some, males need only one mount to ejaculate, whereas others show multimount ejaculatory patterns (Table 13.3). *M. arctoides* is featured by a further peculiarity: after ejaculation, the male sits and brings the female onto his lap while the intromission continues over the next 1–2 min

(Fooden 1990). The significance of such variation remains ill-understood (see Soltis 2004).

Other differences among macaques relate to the periodicity of reproduction. In species from temperate regions, breeding is annual (Table 13.3). A discrete mating season centered on fall produces a discrete birth season around spring. In *M. mulatta* and *M. fuscata*, testis volume, testosterone levels, and spermatogenesis increase at the onset of the mating season and the face of males and their scrotal area redden (Nigi et al. 1989, Lindburg 1983, Bercovitch and Harvey 2004). Females are seasonally polyestrous, several of them coming simultaneously into estrus. No male can monopolize all of them, and mate guarding is limited. Associations between males and females typically last some hours, but this may reduce to momentary encounters or, on the contrary, extend to several days. Both males and females mate with multiple partners. Dominant males can interfere in the mounts of subordinates and chase them away from females. Nevertheless, males do not fight for access to females. The latter may express mate choice, favoring some males and rejecting others; and sneaky copulations are not rare (Lindburg 1983, Huffman 1991, Manson 1992, Bercovitch 1997, Soltis et al. 2001). A similar diversity in sexual relationships is documented in *M. thibetana* and *M. radiata*, albeit with a somewhat higher degree of tolerance among males (Glick 1980, Zhao 1996). In *M. sylvanus*, the lowest-ranking males have to attract females in remote places to mate, but medium-ranking males may copulate in full view of the highest-ranking males; associations with females are brief, the latter commonly shifting from one partner to another every quarter- or half-hour (Taub 1980). Priority of access and furtive mating are not the only tactics open to males. Forming coalitions to displace higher-ranking rivals is a tactic reported in *M. sylvanus* (Taub 1980). Males also compete by endurance rivalry. High energetic expenditures over extended periods of time lead some males to forego mating opportunities. In *M. mulatta*, males increase their fat reserves before the mating season; they may experience a loss of 10% of their body weight during the mating season (Bercovitch 1997; see also Cooper et al. 2004). Some authors assume that sperm competition also occurs, but no strong evidence has been provided yet (see Soltis 2004).

In species living in the tropics, reproduction occurs year-round (Table 13.3). When there is no more than one sexually receptive female at the time, males and females form long-lasting consortships in which one male typically follows and mates with the female during days or weeks, excluding other males from reproduction. In such a case, the close control exerted by the male leaves little room for female mate choice, as observed in a group of *M. tonkeana* (Aujard et al. 1998, Thierry and Gachot-Neveu 2004). In this species, male dominance rank functions as a queue for mating opportunities; top-ranking males have priority of access to fertile females and exclude rivals from reproduction. When the number of males is especially high, however, some of them form coalitions against top-ranking males and the queuing system breaks down (Thierry and Gachot-Neveu 2004).

The *operational sex ratio*—that is, the ratio of adult males to fertile females—appears to be a determinant factor in the choice of mating tactics. In several species, mounts are more frequent during a particular period of the year, and limited peaks of births are reported (Table 13.3); there can be one or several fertile females depending on time, female defensibility and the extent of consortship exclusivity being consequently variable. In *M. fascicularis* and *M. nemestrina*, the duration of associations between males and females may be brief or long; consorting females may copulate with further partners if the attention of the male is diverted by another female or if he is displaced by a higher-ranking male (van Noordwijk 1985, Caldecott 1986b, Fooden 1995, Oi 1996, Engelhardt et al. 2004). In captive *M. arctoides*, subordinate males can mount fertile females only when out of sight of the higher-ranking male, but neither consortships nor reproductive seasonality has been reported (Nieuwenhuijsen et al. 1986).

DEMOGRAPHIC STRUCTURE AND GENE FLOW

Macaques live in bisexual social groups where the socionomic sex ratio is in favor of females (Table 13.3). The ratio is elevated in *M. nemestrina* and *M. silenus*, for example, whereas it is quite balanced in *M. sylvanus*. Although sexual dimorphism is commonly held to arise from mating competition, there is no direct relation between sex ratio and sexual dimorphism (see Plavcan 2001). The level of competition may, however, differ between species according to the extent of monopoly of males over females (Oi 1996, Thierry et al. 1996, Paul 2004). In general, the number of offspring sired by males weakly correlates with their dominance ranks in seasonally breeding macaques. This should not come as a surprise given the diversity of mating tactics that rivals may use to break down the dominants' priority of access to females. When the number of females cycling simultaneously is limited, alternative mating tactics are less effective than overt competition and higher-ranking males gain more paternity of offspring (Soltis et al. 2001). The variance in the reproductive success of males should be especially strong when mating occurs year-round. In three groups of tropics-living species, a positive correlation was indeed found between dominance and paternity (wild *M. fascicularis*, de Ruiter et al. 1994; captive *M. arctoides*, Bauers and Hearn 1994; captive *M. tonkeana*, Gachot-Neveu et al. 2004). If future studies confirm that priority of access best applies to year-round breeders, the genetic structure of macaque groups would then be contrasted. In year-round breeders, each age cohort would represent a separate paternal sibship, contrary to seasonal breeders, in which offspring may have several fathers at each generation.

Competition and dispersal make the variance in reproductive success greater in males than in females. The latter usually remain and reproduce in their natal group throughout life—notwithstanding some notable exceptions, for example, *M. radiata* (Sinha et al. 2005b). The lifetime reproductive success of females primarily depends on their longevity, and social rank has no significant impact on it (in provisioned groups at least) (Fedigan et al. 1986, Bercovitch and Berard 1993, Paul and Kuester 1996; see also van Noordwijk and van Schaik 1999 for limited effects of dominance in wild *M. fascicularis*). By contrast, most males break their social ties and emigrate. Young males commonly disperse at sexual maturation; they often leave in the company of same-aged peers. In *M. fuscata*, *M. mulatta*, and *M. cyclopis*, a majority of males have left their natal group by 5 years of age; they may join small all-male bands or be solitary for months or even years (Drickamer and Vessey 1973, Sugiyama 1976, Hsu and Lin 2001). In other species, first dispersal occurs at least 1 or 2 years later, all-male bands are rarely reported, and males may directly transfer into another group within 1 day or spend months solitary or semisolitary (*M. fascicularis*, van Noordwijk and van Schaik 1985; *M. radiata*, Simonds 1992; *M. sinica*, Dittus 1975; *M. sylvanus*, Ménard and Vallet 1996; *M. thibetana*,

Zhao 1996; *M. maurus*, Okamoto et al. 2000). Dispersal is not limited to younger individuals; males continue to shift groups during their whole life. This secondary dispersal may occur every 2–5 years. Though a drop in rank or a lack of mating opportunities in the home group sometimes prompts a male to leave, even high-ranking males regularly transfer between groups (Sugiyama 1976, Zhao 1996, Berard 1999).

Males often immigrate into groups adjacent to their original group, and it is not rare that they find ex-groupmates in the new group (Sugiyama 1976, Melnick et al. 1984, van Noordwijk and van Schaik 1985, Mehlman 1986, Zhao 1996). Two tactics are open to newcomers: they may either settle at the bottom of the male hierarchy and increase rank with tenure or challenge the highest-ranking males and directly attain the top of the hierarchy. The first tactic is unobtrusive; it is used by males of any age. The second one is risky, and it is mainly reported among full-grown males at their prime (van Noordwijk and van Schaik 1985, Sprague 1992). Tactical choices are commanded by the strength of individuals but also by the demographic context, a main parameter of which may be the operational sex ratio. Males usually enter as subordinates into groups of *M. mulatta* and *M. fuscata*; this might be typical of seasonal breeders (see Drickamer and Vessey 1973; Sprague 1992, 1998; Berard 1999). By contrast, aggressively outranking the top-ranking males is the only way to reproduce when priority of access to females represents the prevailing mating tactic. Rates of challenge appear higher in smaller groups of *M. fuscata*, in which the dominant males could control females more effectively (see Sprague 1992). The relatively high rate of outranks in *M. fascicularis* may be explained by the fact that the top-ranking males sire the majority of offspring (de Ruiter et al. 1994). Take-overs also seem to be the rule in groups of *M. silenus*, which usually contain no more than one adult male (Kumar et al. 2001). *M. tonkeana* provides one further case. In a captive group of this species, the highest-ranking male most of the time had sexual monopoly over fertile females. To obtain sexual access to females, other males had no choice but to challenge him directly (Thierry and Gachot-Neveu 2004). If the occurrence of aggressive outranking among male rivals depends on the extent of female defensibility, it should be more frequent in nonseasonal breeders than in seasonal ones.

Solitary life and immigration increase the likelihood of succumbing to predation, starvation, and injuries. Such costs should then be offset by reproductive benefits (Dittus 2004, Bercovitch and Harvey 2004). It is likely that dispersal increases mating opportunities. As a general finding, females prefer to mate with novel males, which contributes to maintaining genetic diversity (Huffman 1991, Bercovitch 1997, Sprague 1998, Berard 1999, Jack 2003; but see Manson 1995). Regardless of the dominance rank at which they join a group, newcomers display elevated rates of copulation. Correlatively, the reproductive activity of males decreases with increasing tenure length. This may be an effect of a lower sexual attractiveness of familiar males when female mate choice matters. It may also be the outcome of an age-related decline in fighting ability whenever overt competition is the main determinant of reproductive success.

A significant amount of gene flow may additionally occur through extragroup copulations. In all species documented so far, mating occurs between females and male visitors (e.g., Dittus 1975, Mehlman 1986, Simonds 1992, Sprague 1992, Oi 1996). The visitors may be either bachelors or males that move to other groups for some hours or days, then return to their original group. They are mainly attracted by sexually active females. Among seasonal breeders, the mobility of males peaks at the mating season. In *M. fuscata*, it sometimes happens that nonresident males invade groups during this season and become involved in as much as one-third of the resident females' copulations (Sprague 1992). Nonresident males actually sire a significant number of offspring (Soltis et al. 2001).

Whereas male transfer tends to homogenize the genetic structure of populations, the differential reproduction of males is liable to increase within-group relatedness (de Ruiter and Geffen 1998). Another phenomenon liable to augment intergroup genetic differentiation is group fission. This is a rare but important event, usually preceded by an increase in group size. The division of a large group into smaller groups follows genealogical lines, maternal relatives staying in the same groups. Such nonrandom splits create daughter groups, which are genetically distinct. By contrast, patrilineal relatives are randomly distributed into the new groups (Gachot-Neveu and Ménard 2004). In many cases, females promote fission; lower-ranking kin subgroups join in one group, whereas higher-ranking kin subgroups form another group. In *M. maurus*, however, the dominance status of females does not influence group membership (Okamoto and Matsumura 2001).

The past history of groups and the regular transfer of individuals make their relationships complex. Individuals meeting after a transfer may be strangers or ex-groupmates. It is understandable that encounters between groups have quite variable outcomes among macaques, going from peaceful to neutral or agonistic, as in many other primates. When conflicts occur, subadult and young adult males represent the main category of individuals involved (Cooper et al. 2004). Bigger groups usually supplant smaller ones, which may have significant effects regarding access to the food supply (Dittus 2004). In *M. nigra*, the rate of intergroup conflicts was found to increase when fruit availability decreased (Kinnaird and O'Brien 2000).

SOCIAL BEHAVIORS AND RELATIONSHIPS

Dominance and kinship relations structure macaque societies. Decades of study have documented this point in *M. mulatta* and *M. fuscata* (Kawamura 1958, Kawai 1958, Missakian 1972, Sade 1972, Kurland 1977, Chapais 1988, Datta 1992). They form strong hierarchies in which higher-ranking individuals may be recognized by their steady walk and up-held tail carriage. By contrast, lower-ranking individuals are prompt to flee and avoid confrontations and use a particular facial expression to submit, the silent bared-teeth display; by retracting the lips and exposing the teeth, they formally acknowledge their lower dominance status relative to higher-ranking individuals (de Waal and Luttrell 1985). Dominance asymmetry determines which members may inter-act with which others. It affects how an individual chooses partners for proximity, affiliation, or play and whether the distribution of choices is skewed in favor of higher-ranking individuals. Dominance status partly rests upon individual power, that is, on factors like physical strength, personality, and experience. Differences in dominance status also depend on variations in social power, that is, the strength of group-mates that an individual can summon to support him or her (Kawai 1958, Chapais 1988).

Third-party interventions in conflicts have a decisive role in social relationships. Since individuals are all the more prone to help mates in conflicts as they are nearer relatives, this results in the formation of alliance networks based on kinship. These networks constitute subgroups of kin-related individuals called *matrilines* that descend from one female and represent the core of macaque social groups. In this female-bonded system, mothers, daughters, and sisters main-tain strong ties during their lifetime; they are often found together, frequently groom each other, and offer mutual support in conflicts. Strict rules of inheritance determine the acquisition of dominance rank within matrilines (Kawamura 1958, Datta 1992). By adulthood, females achieve a domi-nance rank just below their mothers and rarely outrank them. Because females choose to support their youngest relatives in contests, dominance ranks are ordered inversely to age within matrilines, and younger daughters dominate their elder sisters. Matrilines are themselves ordered along a linear hierarchy. Chapais (1988) has experimentally shown that the ranks of matrilines depend on the number of kin able to support each other. It should be added that to apply, rank inheritance needs the occurrence of strong matrilines. In small groups and/or when the demographical rates do not allow females to have many adult relatives, rules of rank inheritance become less reliable (Datta 1992, Hill and Okayasu 1995).

The picture drawn from *M. mulatta* and *M. fuscata* does not strictly apply to other macaque species. In several species, social relationships are characterized by higher levels of tol-erance between dominants and subordinates, and the record provides regular exceptions to the previous rules of rank inheritance (Angst 1975, Silk et al. 1981, Nieuwenhuijsen et al. 1988). In some species, the exceptions are so numerous that the rules are no longer valid. In *M. sylvanus*, daughters often outrank older mothers and females are usually sub-ordinate to their older sisters (Paul and Kuester 1987, Prud'homme and Chapais 1993). Whereas the society re-mains based on mutual support between kin-related partners, matrilines are more open and the gradient of dominance is

weaker in species like *M. sylvanus*, *M. arctoides*, *M. radiata*, *M. maurus*, and *M. tonkeana*; this favors social tolerance and the development of alliances even between nonrelatives (e.g., de Waal and Luttrell 1989, Thierry et al. 1994, Chapais 1995). In *M. sylvanus* and *M. silenus*, the silent, bared-teeth display may have a positive meaning or express subordi-nation according to the context (Preuschoft 1995). In Sulawesi macaques (e.g., *M. nigra*, *M. tonkeana*, *M. maurus*), the same display does not have a communicative function about dominance status; it signals the sender's peaceful intentions and is used like a smile to initiate affiliative interactions (Thierry et al. 1989, Petit and Thierry 1992).

The degree of kin bias in social relationships is a main variable of macaque societies. According to Chapais and colleagues (1997), there is a relatedness threshold for kin preference. In *M. fuscata*, maternal relatives are favored until a degree of relatedness equal to 0.25, that is, those cor-responding to half-sibling and grandmother–granddaughter bonds. This number also applies to the avoidance of sexual interactions between relatives. The authors propose that the individuals' ability to recognize maternal kin only would not allow them to go beyond the 0.25 threshold. In *M. sylvanus*, however, mating inhibition occurs in niece–uncle and aunt–nephew dyads (maternal relatedness = 0.125) and even between cousins (maternal relatedness = 0.063) (Kuester et al. 1994). This discrepancy may reflect a different kin bias in the relationships of *M. fuscata* and *M. sylvanus*, the former being more nepotistic than the latter. It is worth adding that the social networks of mothers and offspring show a significant degree of similarity in *M. mulatta* and *M. fuscata* (de Waal 1996, Berman and Kapsalis 1999, Schino et al. 2004). On the one hand, this points to the possible occurrence of an intergenerational transmission of social preferences. On the other hand, relatives spend much time in proximity to each other in these species, and their association may account as well for the similarity of mothers and offspring in partner preferences. In fact, no shared partner preferences were found in less nepotistic species for which grooming associations were studied (Schino et al. 2004).

Adult social relationships influence the socialization of immatures. In macaque species typified by strong kin bias and strict hierarchies, mothers appear quite protective; they frequently retrieve their infants and restrict their interactions mostly to relatives. In species displaying more balanced relationships, the mother is quite permissive, alloparental care is common, and even nonkin females may handle and carry infants from an early age (Thierry 2004). Adult males rarely take care of infants in macaques, but there are some noteworthy exceptions. In *M. sylvanus* and *M. thibetana* in particular, males display a strong commitment toward them, they provide active support and frequently handle infants. They use infants to regulate their relations: typically, an adult male picks and carries an infant so as to approach another male and engage in an affiliative interaction (Deag 1980, Ogawa 1995). No correlation was found between male caretaking and paternity in *M. sylvanus*. The behavior of males cannot be held as paternal investment; neither is it a successful strategy which would increase the chances of mating with the mother (Paul et al. 1996, Ménard et al. 2001). The interest of males for infants may be a side effect of their exploitation in males' interactions (Paul et al. 1996, Thierry 2000).

Affinitive associations between adult males and females may develop outside the mating season. Such "friendships" are characterized by frequent proximity and grooming between partners. It is reported in *M. fuscata* (Takahata 1982), *M. mulatta* (Chapais 1986, Manson 1994), and *M. sylvanus* (Paul et al. 1996, Ménard et al. 2001). There is little correspondence between male–female associations and paternity, however. This finding is consistent with the fact that males do not care for mothers' infants as part of their mating effort and that individuals preferentially choose sex partners among less familiar conspecifics (see above). Support in conflicts may be the main benefit drawn by males and females.

SOCIAL STYLES AND COVARIATION

The comparative study of agonistic interactions in captive groups of macaques has yielded marked contrasts between species (Thierry 2000). In *M. mulatta* and *M. fuscata*, most conflicts are unidirectional, high-intensity aggression is common, and reconciliations are not frequent, the conciliatory tendency (see Chapter 36) among unrelated individuals scoring 4%–12%. These results depart from those found in Sulawesi macaques (*M. nigra*, *M. tonkeana*, and *M. maurus*), in which a majority of conflicts are bidirectional; that is, most aggressive acts induce protests or counterattack. Aggression is generally of low intensity, and measuring rates of reconciliation yields especially high values: conciliatory tendencies score around 50% among unrelated partners. Other macaques are intermediate between the previous species.

M. fascicularis and *M. nemestrina* are more similar to *M. mulatta* and *M. fuscata*, whereas *M. arctoides*, *M. sylvanus*, and *M. silenus* tend toward the Sulawesi macaques. Behavioral constraints may account for the covariation of aggression patterns and reconciliation rates at a functional level. On one side, asymmetric conflicts and increased risk of injury may inhibit the occurrence of affiliative contacts between opponents, while approximate symmetry and uncertainty about outcomes create room for negotiation (Silk 1997). On the other side, conciliatory behaviors may reduce the probability of conflict escalation by facilitating information exchange between adversaries (de Waal 1986, Thierry 1986, Aureli et al. 2002).

Macaque species may be set along a 4-grade scale based on their patterns of aggression and reconciliation (Table 13.4) (Thierry 2000). Species from grades 3 and 4 display high rates of affiliative contacts likes clasps and embraces; they are also characterized by the development of behaviors that reduce social tension, such as the use of infants in *M. sylvanus*, ritualized biting in *M. arctoides*, or peaceful interventions by third parties in conflicts among *M. tonkeana* and *M. nigra*. Such behavior patterns are absent in species from grade 1 (*M. mulatta* and *M. fuscata*) and occur at low rates in species from grade 2 (*M. fascicularis* and *M. nemestrina*).

The coupling between social characters may stem from a link between the various traits of the individual, either at the genetic level (e.g., gene pleiotropy and linkage disequilibrium) or the phenotypic level (e.g., trade-offs and structural constraints). It may also arise through the mutual influence of individuals at the social level. This induces constraints which favor some behaviors while forbidding others, as specified for agonistic interactions. In view of the numerous constraints liable to link the characters of a social organization, the covariation hypothesis states that any significant variation in a single character induces a set of correlated changes (Thierry 2004). The contrasts observed in aggression and conciliation patterns extend to a consistent suite of

Table 13.4 Tentative Scaling of Macaque Social Styles[1]

GRADE 1	GRADE 2	GRADE 3	GRADE 4
Macaca mulatta	M. fascicularis	M. arctoides	M. tonkeana
M. fuscata	M. nemestrina	M. sylvanus	M. maurus
(M. cyclopis)		M. silenus	M. nigra
		M. radiata	M. brunnescens[2]
		(M. sinica)	(M. ochreata)
		(M. thibetana)[3]	(M. hecki)
		(M. assamensis)[3]	(M. nigrescens)
			(M. siberu)[4]

Source: Thierry (2000).
[1] Species are ordered mainly based on conciliatory tendency and social tolerance, which increase from the left (grade 1) to right (grade 4), and based on asymmetry of conflicts, dominance gradient, and kin bias, which decrease from left to right. For the least known species (indicated in parentheses), location on grade is predicted from only a few behavioral characteristics.
[2] From Slater (2002).
[3] Recent studies indicate that *M. assamensis* and *M. thibetana* might be located in grade 2 or even grade 1 (Cooper and Bernstein 2002, Berman et al. 2004).
[4] From Abegg and Thierry (2002b).

interspecific variations. Many of the differences previously reported among macaques at the level of dominance asymmetry, the degree of kin bias, the modes of socialization, and the patterns of male dispersal make sense from such a viewpoint (Table 13.4).

At the level of the individual, interspecific variations in temperament have the potential to explain many differences in social behaviors. Consistent differences have been found between species in response to stress and novelty when measured by arousal, alarm, and exploration behaviors; corticosteroid levels; and heart rates. *M. mulatta* and *M. fascicularis* are less explorative with regard to their physical environment compared with *M. silenus* or *M. tonkeana* (Thierry et al. 1994, Clarke and Boinski 1995). As a general rule, species from grades 3 and 4 are less easily aroused than species from grades 1 and 2 (de Waal 1989, Clarke and Boinski 1995). It is likely that macaque species differ in a number of psychobiological variables. For instance, lower serotoninergic activity is associated with higher aggression intensity in *M. mulatta* compared with *M. nemestrina* (Westergaard et al. 1999).

At the social level, the linkage between the degree of nepotism and the gradient of dominance may be accounted for by the frequent occurrence of coalitions within groups of macaques (Thierry 2004). When most alliances involve relatives, the dominance status of individuals primarily depends on the power of the kin subgroup to which they belong. This increases rank differences between non-relatives and further develops kin alliances, generating group structures based on strong hierarchies. Conversely, when kin bias is less pronounced, coalitions involving nonrelatives are more common, dominance appears to be more a question of individual attributes, and the individual retains some degree of freedom with regard to power networks. Dominance relationships remain balanced among group members, and close ties exist even between nonrelatives. Comparisons between species support the view that dominance and nepotism among females are connected by positive feedback (Aureli et al. 1997, Demaria and Thierry 2001).

A number of characters thus appear interrelated: individual temperament, rate of conciliation behaviors, level of aggression asymmetry, gradient of dominance, degree of nepotism, degree of mother permissiveness, amount of female alloparental care, patterns of female rank inheritance, and male dispersal. Each species is featured by a given set of characters that may label its dominance style (de Waal and Luttrell 1989) or more generally its social style (Thierry 2004). Direct comparisons have been made only for a limited sample of species, groups, and characters; and intraspecific variability calls for caution (Thierry 2000, Ménard 2004). Rating species along a discrete and bipolar scale is an idealization. Although each species is assigned to one grade, a more accurate picture would represent the various study populations of each species using a cluster of points centered on one modal location and would allow for overlaps with other clusters centered on neighboring locations.

PHYLOGENETIC INERTIA AND THE SOCIOECOLOGICAL MODEL

The finding that macaque social styles represent covariant sets of characters indicates that they belong to a single family of forms. By limiting the changes possible to social organizations, interconnections between characters act as constraints that channel evolutionary processes and allow only a subset of organizations to arise. Thus, it should not come as a surprise that the three phyletic lineages of macaques have a different distribution on the 4-grade scale (compare Tables 13.1 and 13.4). Variations in the social styles of macaques significantly correlate with their phylogeny (Matsumura 1999; Thierry et al. 2000b). By tracing each of the characters on the phylogenetic tree of macaques, it is possible to recognize their most ancient states and reconstruct their typical ancestral organization. The resulting set of characters closely matches grade 3 on the scale, which may be tentatively considered the ancestral state (Thierry et al. 2000b). The fact that *M. sylvanus* and *M. silenus* are grade 3 reinforces the previous finding since both species come closer to the root of the phylogenetic trees established from morphological and molecular data. The other species of the first lineage have diverged either on grade 4 (Sulawesi macaques) or on grade 2 (*M. nemestrina*). Members of the second lineage have remained on grade 3 (with the possible exceptions of *M. assamensis* and *M. thibetana*, which could have diverged on grade 2). The third lineage evolved toward grades 1 and 2. The location of every lineage is mostly restricted to two grades in the scale. The good match between phylogeny and the 4-grade scale of macaques points to the occurrence of phylogenetic inertia in their social patterns, indicating that the core of the species-specific systems of interconnections underwent limited changes during several hundred thousands or even millions of years in several taxa.

The powerful action of internal attractors pointed out by phylogenetic inertia does not mean that external factors play no role in the determinism of social relationships. Attempts have been made to correlate the social styles of macaques with the main ecological features of their habitats, predation risk and food distribution (de Waal and Luttrell 1989, van Schaik 1989, Sterck et al. 1997; see also Caldecott 1986b). According to a basic ecological argument, animals live in groups to protect themselves from predators, whereas group living in turn induces competition between individuals and groups (see Chapters 28 and 34). In mainly frugivorous species like macaques, females should constitute kin-bonded coalitions to face overt competition that arises for resources within and between groups. If predation risks are high, the costs of leaving the group are elevated for subordinates. As a consequence, dominants take the lion's share and the relationships between unrelated group members are despotic. On the contrary, when predation risks are low, subordinates are not forced to remain in the group. Dominants in this case benefit from subordinates' cooperation against external threats and so must accept a relatively equal exploitation of resources. This condition would produce rather tolerant relationships even among nonkin. Available data on the ecology of macaques do not fit the expectations of the model, however. No correspondence has been found between social styles and between-group competition, food patchiness, or the diet of macaques (Matsumura 1999, Ménard 2004) (see Table 13.2). The ecological tolerance of *M. mulatta* would single-handedly challenge any ecological theory of macaque social organization.

Several reasons may be proposed to explain the failure of the socioecological model. Some relevant ecological factors may still go unidentified. It could also be that social styles have been selected under past ecological conditions and that phylogenetic inertia has maintained them in spite of subsequent ecological changes (Sterck et al. 1997, Ménard 2004). This argument of anachronism is not satisfactory, however. By forbidding any possibility of falsification, it makes the hypothesis of ecological determinism a dogma. Socioecological models commonly envision social organizations as sets of coadapted characters that result from trade-offs between the selective advantages of each of them. Future models should integrate the structural coupling of characters, liable to produce phylogenetic inertia. Evolutionary changes depend on a balance between the action of selective processes and the degree of entrenchment of characters. The important role of historical contingencies in the deployment of macaque species (Abegg and Thierry 2002a) may help explain the lack of correlation between ecology and social style.

CONCLUSION

The study of macaque societies reveals a double-layered organization. On one side, individual behaviors are linked through multiple influences that produce sets of covarying characters. On the other side, patterns of competition and reproduction heavily depend on the more or less seasonal nature of the climate. By influencing the extent of female defensibility by males, the latter is the ecological variable that most directly affects macaque social organization. Both layers are disconnected. A first consequence is that knowing the typical social relationships of a species is not sufficient to predict the outcome of intermale competition. The dominance status has more influence on the reproductive success of males in nonseasonal breeders with limited dominance asymmetry than in species in which asymmetry is marked but where females' fertility is synchronous. A second consequence is that layers may vary at different speeds. Macaque populations have experienced more than once a change between warm and temperate climate during recent geological time (Eudey 1980, Richard et al. 1989). Switching between seasonal and aseasonal reproduction may represent an easy modular change regarding physiological mechanisms. On the contrary, modifying the social style—and by the way the whole system of social relationships and the individual characters on which they rest—may be more difficult to achieve at the evolutionary level.

REFERENCES

Abegg, C., and Thierry, B. (2002a). Macaque evolution and dispersal in insular Southeast Asia. *Biol. J. Linn. Soc.* 75:555–576.

Abegg, C., and Thierry, B. (2002b). The phylogenetic status of Siberut macaques: hints from the bared-teeth display. *Primate Rep.* 63:63–72.

Agetsuma, N. (1995). Foraging synchrony in a group of Yakushima macaques (*Macaca fuscata yakui*). *Folia Primatol.* 64:167–179.

Agetsuma, N., and Nakagawa, N. (1998). Effects of habitat differences on feeding behaviors of Japanese monkeys: comparison between Yakushima and Kinkazan. *Primates* 39:275–289.

Angst, W. (1975). Basic data and concepts on the social organization of *Macaca fascicularis*. In: Rosenblum, L. A. (ed.), *Primate Behavior: Developments in Field and Laboratory Research*, vol. 4. Academic Press, New York. pp. 325–388.

Asquith, P. J. (1989). Provisioning and the study of free-ranging primates: history, effects, and prospects. *Ybk. Phys. Anthropol.* 32:129–158.

Aujard, F., Heistermann, M., Thierry, B., and Hodges, J. K. (1998). Functional significance of behavioral, morphological, and endocrine correlates across the ovarian cycle in semifree ranging female Tonkean macaques. *Am. J. Primatol.* 46:285–309.

Aureli, F., Cords, M., and van Schaik, C. P. (2002). Conflict resolution following aggression in gregarious animals: a predictive framework. *Anim. Behav.* 64:325–343.

Aureli, F., Das, M., and Veenema, H. C. (1997). Differential kinship effect on reconciliation in three species of macaques (*Macaca fascicularis*, *M. fuscata*, and *M. sylvanus*). *J. Comp. Psychol.* 111:91–99.

Bauers, K. A., and Hearn, J. P. (1994). Patterns of paternity in relation to male social rank in the stumptailed macaque, *Macaca arctoides*. *Behaviour* 129:149–176.

Berard, J. D. (1999). A four-year study of the association between male dominance rank, residency status, and reproductive activity in rhesus macaques (*Macaca mulatta*). *Primates* 40:159–175.

Bercovitch, F. B. (1993). Dominance rank and reproductive maturation in male rhesus macaques (*Macaca mulatta*). *J. Reprod. Fertil.* 99:113–120.

Bercovitch, F. B. (1997). Reproductive strategies of rhesus macaques. *Primates* 38:247–263.

Bercovitch, F. B., and Berard, J. D. (1993). Life history costs and consequences of rapid reproductive maturation in female rhesus macaques. *Behav. Ecol. Sociobiol.* 32:103–109.

Bercovitch, F. B., and Harvey, N. (2004). Reproductive life history. In: Thierry, B., Singh, M., and Kaumanns, W. (eds.), *Macaque Societies: A Model for the Study of Social Organization*. Cambridge University Press, Cambridge. pp. 61–80.

Berman, C. M., Ionica, C. S., and Li, L. (2004). Dominance style among *Macaca thibetana* on Mt. Huangshan, China. *Int. J. Primatol.* 25:1283–1312.

Berman, C. M., and Kapsalis, E. (1999). Development of kin bias among rhesus monkeys: maternal transmission or individual learning? *Anim. Behav.* 58:883–894.

Bernstein, I. S., and Gordon, T. P. (1980). Mixed taxa introductions, hybrids and macaque systematics. In: Lindburg, D. G. (ed.),

The Macaques: Studies in Ecology, Behavior and Evolution. van Nostrand Reinhold, New York. pp. 125–147.

Brandon-Jones, D. (1996). The Asian Colobinae (Mammalia: Cercopithecidae) as indicators of Quaternary climatic changes. *Biol. J. Linn. Soc.* 59:327–350.

Bynum, E. L., Bynum, D. Z., and Supriatna, J. (1997). Confirmation and location of the hybrid zone between wild populations of *Macaca tonkeana* and *Macaca hecki* in central Sulawesi, Indonesia. *Am. J. Primatol.* 43:181–209.

Caldecott, J. O. (1986a). *An Ecological and Behavioral Study of the Pig-Tailed Macaque.* Karger, Basel.

Caldecott, J. O. (1986b). Mating patterns, societies and the ecogeography of macaques. *Anim. Behav.* 34:208–220.

Camperio Ciani, A., Stanyon, R., Scheffrahn, W., and Sampurno, B. (1989). Evidence of gene flow between Sulawesi macaques. *Am. J. Primatol.* 17:257–270.

Chapais, B. (1986). Why do adult male and female rhesus monkeys affiliate during the birth season? In: Rawlins, R. G., and Kessler, M. J. (eds.), *The Cayo Santiago Macaques: History, Behavior and Biology.* State University of New York Press, Albany. pp. 173–200.

Chapais, B. (1988). Rank maintenance in female Japanese macaques: experimental evidence for social dependency. *Behaviour* 104:41–59.

Chapais, B. (1995). Alliances as a means of competition in primates: evolutionary, developmental and cognitive aspects. *Ybk. Phys. Anthropol.* 38:115–136.

Chapais, B., Gauthier, C., Prud'homme, J., and Vasey, P. (1997). Relatedness threshold for nepotism in Japanese macaques. *Anim. Behav.* 53:1089–1101.

Clarke, A. S., and Boinski, S. (1995). Temperament in nonhuman primates. *Am. J. Primatol.* 37:103–125.

Cooper, M. A., and Bernstein, I. S. (2002). Counter aggression and reconciliation in Assamese macaques (*Macaca assamensis*). *Am. J. Primatol.* 56:215–230.

Cooper, M. A., Chaitra, M. S., and Singh, M. (2004). Effect of dominance, reproductive state, and group size on body mass in *Macaca radiata. Int. J. Primatol.* 25:165–178.

Crockett, C. M., and Wilson, W. L. (1980). The ecological separation of *Macaca nemestrina* and *M. fascicularis* in Sumatra. In: Lindburg, D. G. (ed.), *The Macaques: Studies in Ecology, Behavior and Evolution.* Van Nostrand Reinhold, New York. pp. 148–181.

Datta, S. B. (1992). Effects of availability of allies on female dominance structure. In: Harcourt, A. H., and de Waal, F. B. M. (eds.), *Coalitions and Alliances in Humans and Other Animals.* Oxford University Press, Oxford. pp. 61–82.

Deag, J. M. (1980). Interactions between males and unweaned Barbary macaques: testing the agonistic buffering hypothesis. *Behaviour* 75:54–81.

Deinard, A., and Smith, D. G. (2001). Phylogenetic relationships among the macaques: evidence from the nuclear locus *NRAMP1. J. Hum. Evol.* 41:45–59.

Delson, E. (1980). Fossil macaques, phyletic relationships and a scenario of deployment. In: Lindburg, D. G. (ed.), *The Macaques: Studies in Ecology, Behavior and Evolution.* Van Nostrand Reinhold, New York. pp. 10–30.

Demaria, C., and Thierry, B. (2001). A comparative study of reconciliation in rhesus and Tonkean macaques. *Behaviour* 138:397–410.

de Ruiter, J. R., and Geffen, E. (1998). Relatedness of matrilines, dispersing males and social groups in long-tailed macaques (*Macaca fascicularis*). *Proc. R. Soc. Lond. B* 265:79–87.

de Ruiter, J., van Hooff, J. A. R. A. M., and Scheffrahn, W. (1994). Social and genetic aspects of paternity in wild long-tailed macaques (*Macaca fascicularis*). *Behaviour* 129:203–224.

de Waal, F. B. M. (1986). The integration of dominance and social bonding in primates. *Q. Rev. Biol.* 61:459–479.

de Waal, F. B. M. (1989). *Peacemaking among Primates.* Harvard University Press, Cambridge, MA.

de Waal, F. B. M. (1996). Macaque social culture: development and perpetuation of affiliative networks. *J. Comp. Psychol.* 110:147–154.

de Waal, F. B. M., and Luttrell, L. M. (1985). The formal hierarchy of rhesus monkeys: an investigation of the bared-teeth display. *Am. J. Primatol.* 9:73–85.

de Waal, F. B. M., and Luttrell, L. M. (1989). Toward a comparative socioecology of the genus *Macaca*: different dominance styles in rhesus and stumptail macaques. *Am. J. Primatol.* 19:83–109.

Dittus, W. P. J. (1975). Population dynamics of the toque monkey, *Macaca sinica*. In: Tuttle, R. H. (ed.), *Socioecology and Psychology of Primates.* Mouton, The Hague. pp. 125–151.

Dittus, W. P. J. (2004). Demography: a window to social evolution. In: Thierry, B., Singh, M., and Kaumanns, W. (eds.), *Macaque Societies: A Model for the Study of Social Organization.* Cambridge University Press, Cambridge. pp. 87–112

Dixson, A. F. (1998). *Primate Sexuality: Comparative Studies of the Prosimians, Monkeys, Apes, and Human Beings.* Oxford University Press, Oxford.

Drickamer, L. C., and Vessey, S. H. (1973). Group changing in free ranging male rhesus monkeys. *Primates* 14:359–368.

Engelhardt, A., Pfeifer, J. B., Heistermann, M., Niemitz, C., van Hooff, J. A. R. A. M., and Hodges, J. K. (2004). Assessment of female reproductive status by male long-tailed macaques, *Macaca fascicularis*, under natural conditions. *Anim. Behav.* 67:915–924.

Eudey, A. A. (1980). Pleistocene glacial phenomena and the evolution of Asian macaques. In: Lindburg, D. G. (ed.), *The Macaques: Studies in Ecology, Behavior and Evolution.* Van Nostrand Reinhold, New York. pp. 52–83.

Evans, B. J., Supriatna, J., and Melnick, D. J. (2001). Hybridization and population genetics of two macaque species in Sulawesi, Indonesia. *Evolution* 55:1986–1702.

Fa, J. E., and Lindburg, D. G. (eds.) (1996). *Evolution and Ecology of Macaque Societies.* Cambridge University Press, Cambridge.

Fedigan, L. M., Fedigan, L., Gouzoules, S., Gouzoules, H., and Koyama, N. (1986). Lifetime reproductive success in female Japanese macaques. *Folia Primatol.* 47:143–157.

Fooden, J. (1969). *Taxonomy and Evolution of the Monkeys of Celebes.* Karger, Basel.

Fooden, J. (1976). Provisional classification and and key to the living species of macaques (Primates: *Macaca*). *Folia Primatol.* 25:225–236.

Fooden, J. (1982). Ecogeographic segregation of macaque species. *Primates* 23:574–579.

Fooden, J. (1986). Taxonomy and evolution of the sinica group of macaques: 5. Overview of natural history. *Fieldiana Zool.* 29:1–22.

Fooden, J. (1990). The bear macaque, *Macaca arctoides*: a systematic review. *J. Hum. Evol.* 19:607–686.

Fooden, J. (1995). Systematic review of Southeast Asian longtailed macaques, *Macaca fascicularis* (Raffles, [1821]). *Fieldiana Zool.* 81:1–206.

Fooden, J. (1997). Tail length variation in *Macaca fascicularis* and *M. mulatta. Primates* 38:221–231.

Fooden, J. (2000). Systematic review of the rhesus macaque *Macaca mulatta* (Zimmermann, 1780). *Fieldiana Zool.* 96:1–179.

Fukuda, F. (1989). Habitual fission–fusion and social organization of the Hakone troop T of Japanese macaques in Kanagawa Prefecture, Japan. *Int. J. Primatol.* 10:419–439.

Gachot-Neveu, H., Fazio, G., and Thierry, B. (2004). Evolution of the genetic structure in a group of Tonkean macaques: consequences of the mating system [abstract]. *Folia Primatol.* 75:170.

Gachot-Neveu, H., and Ménard, N. (2004). Gene flow, dispersal patterns and social organization. In: Thierry, B., Singh, M., and Kaumanns, W. (eds.), *Macaque Societies: A Model for the Study of Social Organization.* Cambridge University Press, Cambridge. pp. 117–131.

Glick, B. B. (1980). Ontogenetic and psychobiological aspects of the mating activities of male *Macaca radiata.* In: Lindburg, D. G. (ed.), *The Macaques: Studies in Ecology, Behavior and Evolution.* Van Nostrand Reinhold, New York. pp. 345–369.

Groves, C. (2001). *Primate Taxonomy.* Smithsonian Institution Press, Washington DC.

Hill, D. A., and Okayasu, N. (1995). Absence of "youngest ascendancy" in the dominance relations of sisters in wild Japanese macaques (*Macaca fuscata yakui*). *Behaviour* 132:367–379.

Hoelzer, G. A., and Melnick, D. J. (1996). Evolutionary relationships of the macaques. In: Fa, J. E., and Lindburg, D. G. (eds.), *Evolution and Ecology of Macaque Societies.* Cambridge University Press, Cambridge. pp. 3–19.

Hsu, M. J., and Lin, J. F. (2001). Troop size and structure in free-ranging Formosan macaques (*Macaca cyclopis*) at Mt Longevity, Taiwan. *Zool. Stud.* 40:49–60.

Huffman, M. A. (1991). Mate selection and partner preferences in female Japanese macaques. In: Fedigan, L. M., and Asquith, P. J. (eds.), *The Monkeys of Arashiyama: Thirty-Five Years of Research in Japan and the West.* State University of New York Press, Albany. pp. 101–122.

Jack, K. (2003). Males on the move: evolutionary explanations of secondary dispersal by male primates. *Primate Rep.* 67:61–83.

Kawai, M. (1958). On the system of social ranks in a natural troop of Japanese monkeys: I, II [in Japanese]. *Primates* 1/2:111–148. English translation in Imanishi, K., and Altmann, S. A. (eds.) (1965). *Japanese Monkeys: A Collection of Translations.* Emory University, Atlanta. pp. 66–104.

Kawamura, S. (1958). Matriarchal social ranks in the Minoo-B troop: a study of the rank system of Japanese macaques [in Japanese]. *Primates* 1:149–156. English translation in Imanishi, K., and Altmann, S. A. (eds.) (1965). *Japanese Monkeys: A Collection of Translations.* Emory University, Atlanta. pp. 105–112.

Kinnaird, M. F., and O'Brien, T. G. (2000). Comparative movement patterns of two semiterrestrial cercopithecine primates: the Tana River crested mangabey and the Sulawesi crested black macaque. In: Boinski, S., and Garber, P. A. (eds.), *On the Move: How and Why Animals Travel in Groups.* University of Chicago Press, Chicago. pp. 327–350.

Kuester, J., Paul, A., and Arnemann, J. (1994). Kinship, familiarity and mating avoidance in Barbary macaques, *Macaca sylvanus. Anim. Behav.* 48:1183–1194.

Kumar, A., Singh, M., Kumara, H. N., Sharma, A. K., and Bertsch, C. (2001). Male migration in lion-tailed macaques. *Primate Rep.* 59:5–17.

Kurland, J. A. (1977). *Kin Selection in the Japanese Monkey.* Karger, Basel.

Kurup, G. U., and Kumar, A. (1993). Time budget and activity patterns of the lion-tailed macaque (*Macaca silenus*). *Int. J. Primatol.* 14:27–39.

Lindburg, D. G. (1983). Mating behavior and estrus in the Indian rhesus monkey. In: Seth, P. K. (ed.), *Perspectives in Primate Biology.* Today & Tomorrow, New Delhi. pp. 45–61.

Lyles, A. M., and Dobson, A. P. (1988). Dynamics of provisioned and unprovisioned primate populations. In: Fa, J. E., and Southwick, C. H. (eds.), *Ecology and Behavior of Food-Enhanced Primate Groups.* Alan R. Liss, New York. pp. 167–198.

Manson, J. H. (1992). Measuring female mate choice in Cayo Santiago rhesus macaques. *Anim. Behav.* 44:405–416.

Manson, J. H. (1994). Mating patterns, mate choice, and birth season heterosexual relationships in free-ranging rhesus macaques. *Primates* 35:417–433.

Manson, J. H. (1995). Do female rhesus monkeys choose novel males? *Am. J. Primatol.* 37:285–296.

Matsumura, S. (1999). The evolution of "egalitarian" and "despotic" social systems among macaques. *Primates* 40:23–31.

Matsumura, S. (2001). The myth of despotism and nepotism: dominance and kinship in matrilineal societies of macaques. In: Matsuzawa, T. (ed.), *Primate Origins of Human Cognition and Behavior.* Springer, Tokyo. pp. 441–462.

Mehlman, P. (1986). Male intergroup mobility in a wild population of the Barbary macaque (*Macaca sylvanus*), Ghomaran Rif Mountains, Morocco. *Am. J. Primatol.* 10:67–81.

Melnick, D. J., Pearl, M. C., and Richard, A. F. (1984). Male migration and inbreeding avoidance in wild rhesus monkeys. *Am. J. Primatol.* 7:229–243.

Ménard, N. (2004). Do ecological factors explain variation in social organization? In: Thierry, B., Singh, M., and Kaumanns, W. (eds.), *Macaque Societies: A Model for the Study of Social Organization.* Cambridge University Press, Cambridge. pp. 237–262.

Ménard, N., and Vallet, D. (1996). Demography and ecology of Barbary macaques (*Macaca sylvanus*) in two different habitats. In: Fa, J. E., and Lindburg, D. G. (eds.), *Evolution and Ecology of Macaque Societies.* Cambridge University Press, Cambridge. pp. 106–131.

Ménard, N., von Segesser, F., Scheffrahn, W., Pastorini, J., Vallet, D., Gaci, B., Martin, R. D., and Gautier-Hion, A. (2001). Is male–infant caretaking related to paternity and/or mating activities in wild Barbary macaques (*Macaca sylvanus*)? *C. R. Acad. Sci. Paris Life Sci.* 324:601–610.

Missakian, E. A. (1972). Genealogical and cross-genealogical dominance relations in a group of free-ranging rhesus monkeys (*Macaca mulatta*) on Cayo Santiago. *Primates* 13:169–180.

Morales, J. C., and Melnick, D. J. (1998). Phylogenetic relationships of the macaques (Cercopithecidae: *Macaca*), as revealed by high resolution restriction site mapping of mitochondrial ribosomal genes. *J. Hum. Evol.* 34:1–23.

Nieuwenhuijsen, K., Bonke-Jansen, M., Broekhuijzen, E., de Neef, K. J., van Hooff, J. A. R. A. M., van der Werff ten Bosch, J. J., and Slob, A. K. (1988). Behavioral aspects of puberty in group-living stumptail monkeys (*Macaca arctoides*). *Physiol. Behav.* 42:255–264.

Nieuwenhuijsen, K., de Neef, K. J., and Slob, A. K. (1986). Sexual behaviour during ovarian cycles, pregnancy and lactation in group-living stumptail macaques (*Macaca arctoides*). *Hum. Reprod.* 1:159–169.

Nigi, H., Hayama, S. I., and Torii, R. (1989). Rise in age of sexual maturation in male Japanese monkeys in Takasakiyama in relation to nutritional conditions. *Primates* 30:571–575.

O'Brien, T. G., and Kinnaird, M. F. (1997). Behavior, diet, and movements of the Sulawesi crested black macaque (*Macaca nigra*). *Int. J. Primatol.* 18:321–351.

Ogawa, H. (1995). Bridging behavior and other affiliative interactions among male Tibetan macaques (*Macaca thibetana*). *Int. J. Primatol.* 16:727.

Oi, T. (1996). Sexual behaviour and mating system of the wild pig-tailed macaque in West Sumatra. In: Fa, J. E., and Lindburg, D. G. (eds.), *Evolution and Ecology of Macaque Societies.* Cambridge University Press, Cambridge. pp. 342–368.

Okamoto, K., and Matsumura, S. (2001). Group fission in moor macaques (*Macaca maurus*). *Int. J. Primatol.* 22:481–493.

Okamoto, K., Matsumura, S., and Watanabe, K. (2000). Life history and demography of wild moor macaques (*Macaca maurus*): summary of ten years of observations. *Am. J. Primatol.* 52:1–11.

Paul, A. (2004). Dominance and paternity. In: Thierry, B., Singh, M., and Kaumanns, W. (eds.), *Macaque Societies: A Model for the Study of Social Organization.* Cambridge University Press, Cambridge. pp. 131–134.

Paul, A., and Kuester, J. (1987). Dominance, kinship and reproductive value in female Barbary macaques (*Macaca sylvanus*) at Affenberg Salem (FRG). *Behav. Ecol. Sociobiol.* 21:323–331.

Paul, A., and Kuester, J. (1996). Differential reproduction in male and female Barbary macaques. In: Fa, J. E., and Lindburg, D. G. (eds.), *Evolution and Ecology of Macaque Societies.* Cambridge University Press, Cambridge. pp. 293–317.

Paul, A., Kuester, J., and Arnemann, J. (1996). The sociobiology of male–infant interactions in Barbary macaques, *Macaca sylvanus. Anim. Behav.* 51:155–170.

Petit, O., and Thierry, B. (1992). Affiliative function of the silent bared-teeth display in moor macaques (*Macaca maurus*): further evidence for the particular status of Sulawesi macaques. *Int. J. Primatol.* 13:97–105.

Plavcan, J. (2001). Sexual dimorphism in primate evolution. *Ybk Phys. Anthropol.* 44:25–53.

Preuschoft, S. (1995). "Laughter" and "smiling" in macaques: an evolutionary perspective [PhD thesis]. University of Utrecht, Utrecht.

Prud'homme, J., and Chapais, B. (1993). Rank relations among sisters in semi-free ranging Barbary macaques (*Macaca sylvanus*). *Int. J. Primatol.* 14:405–420.

Richard, A. F., Goldstein, S. J., and Dewar, R. E. (1989). Weed macaques: the evolutionary implications of macaques feeding ecology. *Int. J. Primatol.* 10:569–594.

Rodman, P. S. (1979). Skeletal differentiation of *Macaca fascicularis* and *Macaca nemestrina* in relation to arboreal and terrestrial quadrupedalism. *Am. J. Phys. Anthropol.* 51:51–62.

Roos, C., Ziegler, T., Hodges, J., Zischler, H., and Abegg, C. (2003). Molecular phylogeny of Mentawai macaques: taxo-nomic and biogeographic implications. *Mol. Phylogenet. Evol.* 29:139–150.

Rosenblum, L. A., Kaufman, I. C., and Stynes, A. J. (1964). Individual distance in two species of macaque. *Anim. Behav.* 12:338–342.

Sade, D. S. (1972). Sociometrics of *Macaca mulatta*. I. Linkages and cliques in grooming matrices. *Folia Primatol.* 18:196–223.

Schino, G., Aureli, F., Ventura, R., and Troisi, A. (2004). A test of the cross-generational transmission of grooming preferences in macaques. *Ethology* 110:137–146.

Silk, J. B. (1997). The function of peaceful post-conflict contacts among primates. *Primates* 38:265–279.

Silk, J. B., Samuels, A., and Rodman, P. (1981). Hierarchical organization of female *Macaca radiata* in captivity. *Primates* 22:84–95.

Simonds, P. E. (1992). Primate behavioral dynamics. In: Seth, P. K., and Seth, S. (eds.), *Perspectives in Primate Biology.* Today & Tomorrow, New Dehli. pp. 95–109.

Singh, M., and Sinha, A. (2004). Life-history traits: ecological adaptations or phylogenetic relics? In: Thierry, B., Singh, M., and Kaumanns, W. (eds.), *Macaque Societies: A Model for the Study of Social Organization.* Cambridge University Press, Cambridge. pp. 80–83.

Sinha, A., Datta, A., Madhusudan, M. D. & Mishra, C. (2005a). The Arunachal macaque *Macaca munzala*: a new species from western Arunachal Pradesh, northeastern India. *Int. J. Primatol.* 26:977–989.

Sinha, A., Mukhopadhyay, K., Datta-Roy, A., and Ram, S. (2005b). Ecology proposes, behaviour disposes: ecological variability in social organisation and male behavioural strategies among wild bonnet macaques. *Curr. Sci.* 89:1166–1179.

Slater, K. (2002). Dominance style, grooming and biological markets in the Buton macaque (*Macaca ochreata brunnescens*) [MSc thesis]. University of Liverpool, Liverpool.

Smith, R. C., and Jungers, W. L. (1997). Body mass in comparative primatology. *J. Hum. Evol.* 32:523–559.

Soltis, J. (2004). Mating systems. In: Thierry, B., Singh, M., and Kaumanns, W. (eds.), *Macaque Societies: A Model for the Study of Social Organization.* Cambridge University Press, Cambridge. pp. 135–151.

Soltis, J., Thomsen, R., and Takenaka, O. (2001). The interaction of male and female reproductive strategies and paternity in wild Japanese macaques, *Macaca fuscata. Anim. Behav.* 62:485–494.

Sprague, D. S. (1992). Life history and male intertroop mobility among Japanese macaques (*Macaca fuscata*). *Int. J. Primatol.* 13:437–454.

Sprague, D. S. (1998). Age, dominance rank, natal status, and tenure among male macaques. *Am. J. Phys. Anthropol.* 105:511–521.

Sterck, E. H. M., Watts, D. P., and van Schaik, C. P. (1997). The evolution of female social relationships in nonhuman primates. *Behav. Ecol. Sociobiol.* 41:291–309.

Su, H. H., and Lee, L. L. (2001). Food habits of Formosan rock macaques (*Macaca cyclopis*) in Jentse, northeastern Taiwan, assessed by fecal analysis and behavioral observation. *Int. J. Primatol.* 22:359–377.

Sugiyama, Y. (1976). Life history of male Japanese macaques. *Adv. Stud. Behav.* 7:255–284.

Supriatna, J., Froehlich, J. W., Erwin, J. M., and Southwick, C. H. (1992). Population, habitat and conservation status of *Macaca maurus*, *Macaca tonkeana* and their putative hybrids. *Trop. Biodiv.* 1:31–48.

Takahata, Y. (1982). Social relations between adult males and females of Japanese monkeys in the Arashiyama B troop. *Primates* 23:1–23.

Taub, D. M. (1980). Female choice and mating strategies among wild Barbary macaques (*Macaca sylvanus* L.). In: Lindburg, D. G. (ed.), *The Macaques: Studies in Ecology, Behavior and Evolution.* Van Nostrand Reinhold, New York. pp. 287–344.

Thierry, B. (1986). A comparative study of aggression and response to aggression in three species of macaque. In: Else, J. G., and Lee, P. C. (eds.), *Primate Ontogeny, Cognition, and Social Behaviour.* Cambridge University Press, Cambridge. pp. 307–313.

Thierry, B. (2000). Covariation of conflict management patterns across macaque species. In: Aureli, F., and de Waal, F. B. M. (eds.), *Natural Conflict Resolution.* University of California Press, Berkeley. pp. 106–128.

Thierry, B. (2004). Social epigenesis. In: Thierry, B., Singh, M., and Kaumanns, W. (eds.), *Macaque Societies: A Model for the Study of Social Organization.* Cambridge University Press, Cambridge. pp. 267–290.

Thierry, B., Anderson, J. R., Demaria, C., Desportes, C., and Petit, O. (1994). Tonkean macaque behaviour from the perspective of the evolution of Sulawesi macaques. In: Roeder, J. J., Thierry, B., Anderson, J. R., and Herrenschmidt, N. (eds.), *Current Primatology,* vol. 2. Université Louis Pasteur, Strasbourg. pp. 103–117.

Thierry, B., Bynum, E. L., Baker, S., Kinnaird, M. F., Matsumura, S., Muroyama, Y., O'Brien, T. G., Petit, O., and Watanabe, K. (2000a). The social repertoire of Sulawesi macaques. *Primate Res.* 16:203–226.

Thierry, B., Demaria, C., Preuschoft, S., and Desportes, C. (1989). Structural convergence between silent bared-teeth display and relaxed open-mouth display in the Tonkean macaque (*Macaca tonkeana*). *Folia Primatol.* 52:178–184.

Thierry, B., and Gachot-Neveu, H. (2004). The balance of power between males influences mating tactics in Tonkean macaques [abstract]. *Folia Primatol.* 75 (suppl. 1):345.

Thierry, B., Heistermann, M., Aujard, R., and Hodges, J. K. (1996). Long-term data on basic reproductive parameters and evaluation of endocrine, morphological, and behavioral measures for monitoring reproductive status in a group of semifree-ranging Tonkean macaques (*Macaca tonkeana*). *Am. J. Primatol.* 39:47–62.

Thierry, B., Iwaniuk, A. N., and Pellis, S. M. (2000b). The influence of phylogeny on the social behaviour of macaques (Primates: Cercopithecidae, genus *Macaca*). *Ethology* 106:713–728.

Thierry, B., Singh, M., and Kaumanns, W. (eds.) (2004). *Macaque Societies: A Model for the Study of Social Organization.* Cambridge University Press, Cambridge.

Tosi, A. J., Morales, J. M., and Melnick, D. J. (2002). Y-chromosome and mitochondrial markers in *Macaca fascicularis* indicate introgression with Indochinese *M. mulatta* and a biogeographic barrier in the Isthmus of Kra. *Int. J. Primatol.* 23:161–178.

van Noordwijk, M. A. (1985). Sexual behaviour of Sumatran long-tailed macaques (*Macaca fascicularis*). *Z. Tierpsychol.* 70:277–296.

van Noordwijk, M., and van Schaik, C. P. (1999). The effects of dominance rank and group size on female lifetime reproductive success in wild long-tailed macaques, *Macaca fascicularis.* *Primates* 40:105–130.

van Noordwijk, M. A., and van Schaik, C. P. (1985). Male migration and rank acquisition in wild long-tailed macaques (*Macaca fascicularis*). *Anim. Behav.* 33:849–861.

van Schaik, C. P. (1989). The ecology of social relationships amongst female primates. In: Standen, V., and Foley, R. A. (eds.), *Comparative Socio-ecology: The Behavioral Ecology of Humans and Other Animals.* Blackwell, Oxford. pp. 195–218.

van Schaik, C. P., and van Noordwijk, M. A. (1986). The hidden costs of sociality: intra-group variation in feeding strategies in Sumatran long-tailed macaques (*Macaca fascicularis*). *Behaviour* 99:296–315.

van Schaik, C. P., van Noordwijk, M. A., de Boer, R. J., and den Tonkelaar, I. (1983). The effect of group size on time budgets and social behaviour in wild long-tailed macaques (*Macaca fascicularis*). *Behav. Ecol. Sociobiol.* 13:173–181.

Westergaard, G. C., Suomi, S. J., Higley, J. D., and Mehlman, P. T. (1999). CSF 5-HIAA and aggression in female macaque monkeys: species and interindividual differences. *Psychopharmacology* 146:440–446.

Yamagiwa, J., and Hill, D. A. (1998). Intraspecific variation in the social organization of Japanese macaques: past and present scope of field studies in natural habitats. *Primates* 39:257–273.

Zhao, Q. K. (1996). Etho-ecology of Tibetan macaques at Mount Emei, China. In: Fa, J. E., and Lindburg, D. G. (eds.), *Evolution and Ecology of Macaque Societies.* Cambridge University Press, Cambridge. pp. 263–289.

14

Baboons, Mandrills, and Mangabeys
Afro-Papionin Socioecology in a Phylogenetic Perspective
Clifford J. Jolly

The baboons, mandrills, and mangabeys of Africa (subtribe Papionina; the informal term *Afro-papionins* is used here since there is no vernacular contraction specific to subtribes —*papionin* implies membership of the tribe Papionini, which also includes Macacina) share many features in ecology, behavior, anatomy, and physiology but are also diverse enough to justify a comparative study. Most, but not all, forage mostly on the ground. Some inhabit regions with sparse, seasonal rainfall and little or no natural tree cover; others live in moist tropical rain forest, and some are able to exploit both these environments and all the woodlands and savannas between them. Their behavioral diversity makes them an excellent group in which to develop and test theories about the connections between ecology and society in long-lived, highly social animals. Afro-papionins fall into five genera: *Papio* ("common" baboons), *Theropithecus* (gelada), *Lophocebus* ("baboon-mangabeys") (Kingdon 1996), *Mandrillus* (drills and mandrills), and *Cercocebus* ("drill-mangabeys"). Genetic evidence (Barnicot and Hewett-Emmett 1972, Disotell 2000) strongly supports two subclades: *Theropithecus–Lophocebus–Papio* and *Mandrillus–Cercocebus*.

A recent taxonomic review (Grubb et al. 2003) is generally followed here, except that the *Cercocebus* mangabeys are split at the species, rather than the subspecies, level (Kingdon 1996, Rowe 1996). *Theropithecus* includes only the gelada of highland Ethiopia. *Lophocebus* includes the gray-cheeked mangabey (*L. albigena*), with a range extending from Nigeria to Uganda; the black mangabey (*L. aterrimus*), found south and west of the Congo and Lualaba Rivers; and the highland mangabey (*L. kipunji*) (Jones et al. 2005), apparently confined to a few montane forest fragments in Tanzania.

There is less agreement about the species of *Papio*. The many geographical variants have ranges that never overlap but often adjoin, and interbreeding occurs where they do. They can be called subspecies of a single species (*P. hamadryas*) or recognized as full species. This account follows Grubb et al. (2003), recognizing five species: chacma, anubis or olive, hamadryas, Guinea or red, and yellow baboons (*P. ursinus*, *P. anubis*, *P. hamadryas*, *P. papio*, and *P. cynocephalus*, respectively). Between them, they occupy most of sub-Saharan Africa and southwest Arabia (Fig. 14.1).

The *Mandrillus* subclade includes two living genera. *Cercocebus* divides into two species groups (formerly species). The *galeritus* group includes the Tana River (*C. galeritus*), agile (*C. agilis*), golden-bellied (*C. chrysogaster*), and Sanje (*C. sanjei*) mangabeys. They are distributed in forest blocks and fragments from Gabon to the Kenyan coast. The *torquatus* group includes the sooty (*C. atys*), capped (*C. torquatus*), and white-collared (*C. lunulatus*) mangabeys. It is entirely West African, its species extending along the Atlantic coast from Upper Guinea to Angola. The subclade's second genus, *Mandrillus*, includes the mandrill (*M. sphinx*) and the drill (*M. leucophaeus*), which have adjoining ranges within the equatorial rain forest of the coast of west–central Africa.

Excellent illustrations of each species and summary accounts of their general biology are to be found in Kingdon (1996) and Rowe (1996). Following are some topics relevant to understanding the socioecology of species in an evolutionary context.

SEXUAL DIMORPHISM, MATURATION, AND REPRODUCTION

Table 14.1 includes median body mass estimates for some Afro-papionin samples (Delson et al. 2000, Olupot 2000, J. E. Phillips-Conroy and C. Jolly unpublished data). Even if measured in the wild, mass is affected by environmental quality (e.g., Olupot 2000), pregnancy, and even the state of the individual's digestive tract, so these data must be treated as approximations. Even so, they show two broad size categories. *Lophocebus* and *Cercocebus* are medium-sized, comparable to macaques, the subtribe's sister taxon. *Papio*, *Theropithecus gelada*, and *Mandrillus* are larger, the smallest females barely overlapping the largest male mangabeys. There is considerable variation in body mass within *Papio*, even within most species (*P. anubis*, *P. cynocephalus*, *P. ursinus*, and probably *P. hamadryas*). The heaviest living papionins are probably male mandrills, although some chacma (*P. ursinus ursinus*) and anubis baboons come close.

Figure 14.1 Distribution of the major geographical forms of *Papio* baboons. *1*, Guinea baboon, *P. papio*; *2*, Anubis baboon, *P. anubis*; *3*, Hamadryas baboon, *P. hamadryas*; *4*, "large" yellow baboons, *P. cynocephalus cynocephalus* and *P. cynocephalus ibeanus*; *5*, "small" yellow baboon, *P. cynocephalus kindae*; *6*, cape chacma baboon, *P. ursinus ursinus*; *7*, gray-footed chacma baboon, *P. ursinus griseipes*; *8*, desert chacma baboon, *P. ursinus ruacana*.

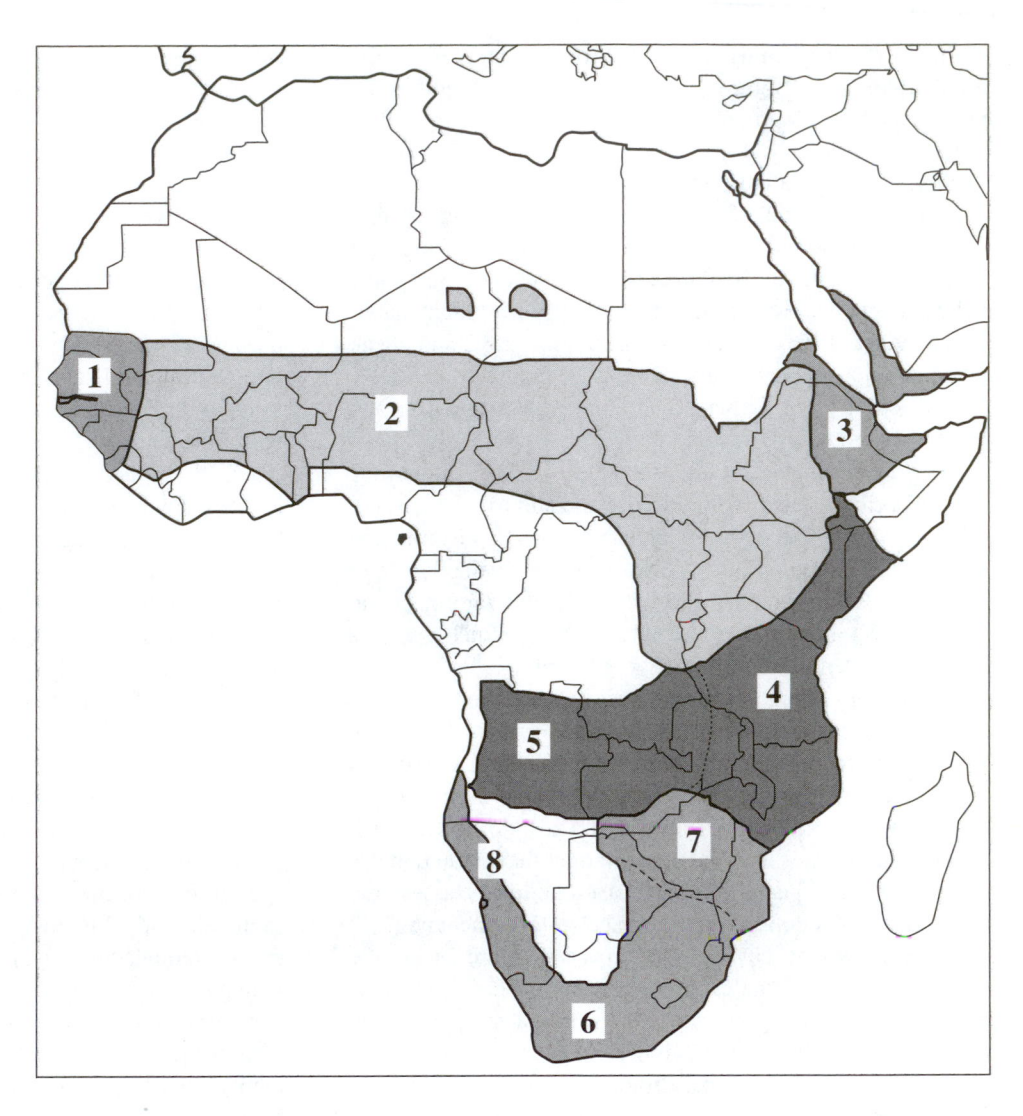

Table 14.1   Body Mass and Sexual Dimorphism in Some Papionin Samples

	FEM. MASS	MALE MASS	MID MASS	F/M	M/F	RES–MID
Lophocebus	5.7	8.7	7.2	65.5	152.6	0.3
Theropitheecus	11	18.5	14.8	59.5	168.2	−10.7
Papio (yellow)	12.3	22.6	17.5	54.4	183.7	−4.6
Papio (Kinda)	10	16	13.0	62.5	160.0	−12.7
P. anubis (small)	12.5	22.8	17.7	54.8	182.4	−6.7
P. anubis (large)	16	31	23.5	51.6	193.8	−16.0
P. hamadryas	12	20.8	16.4	57.7	173.3	−11.4
Papio (chacma)	16	28	22.0	57.1	175.0	−29.4
Cercocebus	5.8	10.3	8.1	56.3	177.6	22.3
Mandrillus sphinx	12	33	22.5	36.4	275.0	68.8

Medians for each sex and sexes combined derived from Delson et al. (2000), Phillips-Conroy and Jolly unpublished data. Mass, body mass (kg); Fem., female; F/M, female mass as percentage of male mass; M/F, male mass as percentage of female mass; Res-Mid, unstandardized residual from linear regression of M/F on mid mass.

All Afro-papionins are sexually dimorphic in body size, but the ratio of male to female body mass is much greater in *Mandrillus* than any of the other taxa (Setchell and Dixson 2002). Absolute size is quite strongly and positively correlated with size dimorphism. Within the subclades, *Mandrillus* is much more dimorphic than *Cercocebus* and *Papio*, more than *Lophocebus*. A grade shift apparently separates the subclades: both *Cercocebus* and *Mandrillus* have much higher, positive residuals from the regression than the members of the baboon subclade.

In all Afro-papionins, the sexes differ in external appearance. Males usually have more luxuriant coats, with more pronounced development of crests, manes, and capes, than females and young. At maturity, male geladas, Guinea baboons, and olive baboons develop a much longer, more conspicuous version of the shoulder cape of waved hair that is seen in females and young. Similarly, the fringes of long straight hair on the flanks, seen in chacmas and some yellow baboons, become more pronounced in adult males. In hamadryas baboons, the male's shoulder cape and cheek whiskers not only become longer at maturity but also change color from brown to light gray and the face and buttocks become bright red.

In *Mandrillus*, the multicolored skin of the hindquarters and the striking facial coloration of the adult males (especially *M. sphinx*) are much brighter versions of juvenile and female coloration. Males also develop distinctive hair patterns at sexual maturity (Setchell and Dixon 2002) and an odorous, glandular area on the chest, which secretes copiously when the male is actively breeding.

Though developmental trajectories in craniodental features vary, all papionins, especially the larger forms, exhibit strong, positive allometry in facial length (Collard and O'Higgins 2001). Variation in absolute body size is therefore accompanied by marked differences in facial profile. In each species, adult males have relatively longer faces than females and young. Any papionin can thus immediately assess the age and sex of a conspecific in a face-to-face encounter.

Male papionins have higher-crowned, more blade-like canine teeth than females, especially in the maxilla. These are effective as weapons and as signals during "yawning" displays. Mandrills have the highest male canine crowns and the most pronounced sexual dimorphism in canine size. In large juvenile and younger subadult male papionins, the encrypted maxillary canines form distinct bulges along either side of the muzzle. In most species, the bulges recede as the canines erupt, but in *Mandrillus*, they remain distinct and (in the mandrill) become bosses covered with specialized, brightly colored skin.

Considering their diversity in ecology, size, and social structure, Afro-papionins appear remarkably uniform in reproductive parameters and the timing of maturation (though some subtle but meaningful differences might well emerge with larger samples and data sets) (Jolly and Phillips-Conroy 2003). The ovulation–menstruation cycle of 4–5 weeks includes visible menses. All except *T. gelada* have prominent, highly visible perineal swellings that occur regularly around ovulation. In geladas, which sit most of the time while foraging, ovulation is signaled by blister-like vesicles bordering the hairless chest patch seen in both sexes (Matthews 1956).

Lophocebus, *Papio*, and *Theropithecus* are year-round breeders. Although weak birth peaks occur in some populations, there is no defined mating period when social behavior of the whole group changes. Mandrills (Setchell and Wickings 2004, Abernethy et al. 2002) and at least some populations of *Cercocebus* (Benneton and Noe 2004) are more obviously seasonal breeders, with changes in social behavior in the breeding season.

Gestation in all Afro-papionins is about 6 months (170–180 days), with no convincing evidence for intertaxon variation. Remarkably, larger species do not have longer gestation than smaller ones. A convenient reproductive landmark is the first periovulatory swelling of a mother of a surviving infant. This seems generally to occur about the middle of the second year after the last infant's birth, although intervals as short as 9 months have been recorded in the wild (Swedell in press). Combined with the 6-month gestation, a 15- to 16-month postpartum amenorrhea means that each mother could produce one surviving infant every 2 years. Female mandrills appear to be exceptional, regularly becoming pregnant with a surviving offspring only 7 months old, and therefore producing one offspring per year (Wickings and Dixon 1992). This rate was, however, observed in a provisioned, though free-ranging, colony; and it is not known whether it is matched in the wild.

In most Afro-papionins, periovulatory swellings first occur during a female's fourth or fifth year, and her first infant is usually born during her sixth year (e.g., Altmann et al. 1981, Dunbar 1984). Two species may follow an accelerated program of female maturation. Female mandrills produce their first infant at as young as 2.7 years, and mean age at first birth is 4.5 years in a provisioned colony (Setchell and Wickings 2004). The other case is more circumstantial. Swellings are seen in female hamadryas at younger ages than any female recorded among nearby anubis baboons (J. E. Phillips-Conroy and C. Jolly unpublished data). This presumably indicates that hamadryas experience earlier menarche, though not necessarily younger first parturition.

The male developmental trajectory broadly parallels the female's until subadulthood, when a female's growth slows as her physiology reorients toward reproduction but males continue to add body mass. A growth spurt in testicular size starts about the time of the eruption of the second molars, at 4–5 years of age. Males are probably capable of fertile mating from then on but reach social maturity and full body size much later, at 8–10 years, when canine teeth are fully erupted and secondary sexual characters are fully developed. In male mandrills, full development of skin coloration and testicular size occurs only with high social rank and may be delayed in the presence of a dominant male (Setchell and Dixon 2001a,b). Like the chest patch of the male gelada, the bright red coloration of the male mandrill's facial skin (which is similarly attributed to subcutaneous vascularization; Hill 1969) fades rapidly after loss of breeding status.

ECOLOGY

Each Afro-papionin species has a characteristic range of habitats. Both genera of mangabeys and *Mandrillus* species are confined to mainly closed, moist, evergreen, or semideciduous forest, where most primary productivity occurs above ground level. *Lophocebus* mangabeys are highly arboreal, rarely descending to the ground. Typically inhabitants of continuous evergreen forest (Olupot et al. 1994), they are also able to survive in habitats in which only patches of forest remain (Chalmers 1968). In much of their range, *Cercocebus* mangabeys are associated with riverine and swamp forest, a habitat that might represent a refuge from competition with the more recently evolved and ecologically aggressive forest guenons (Kingdon 1996). Geladas are also habitat specialists, confined to high montane grasslands in which there are few trees and where most productivity occurs in the herb layer.

Papio baboons have the widest geographical and habitat range of the Afro-papionins. They can survive in all tropical, subtropical, and temperate-zone vegetation types within sub-Saharan Africa, from equatorial evergreen forest to semideserts, so long as they find adequate food sources in trees or on the ground, surface water, and night refuges in the form of rock faces or trees. The range of each of the recognized species tends to be centered on a different set of vegetational/climatic zones. Most hamadryas habitats are semi arid; much of the range of yellow baboons consists of miombo woodlands, and most chacma baboon populations occupy southern subtropical or temperate habitats. There are many exceptions to these associations, however. Anubis baboons, for instance, are found in the Ituri Forest, rain forest, the semidesert Tibesti massif in the Sahara, and all habitats between these extremes. Even the Guinea baboon, which has a comparatively restricted range, occupies habitats ranging from dry sahel to moist coastal mangrove (Galat-Luong et al. 2005). In some areas, boundaries between species correspond to ecotones. In Kenya, for example, anubis baboons are "highland" and yellow baboons, "lowland" (Kingdon 1971); but in wider perspective, this contrast breaks down. Habitat diversity among populations within species seems at least as great as interspecific differences, so it seems most likely that local associations between ecological zones and species (as in Kenya) result from historical accident rather than intrinsic, species-specific habitat limitations. Local populations have, however, adapted physically to local conditions. Hamadryas baboons of Ethiopia, for example, have on average shorter and stouter digits than neighboring populations of anubis baboons (Jolly 2001). Their ancestors presumably acquired shorter toes when confined to a habitat that provided cliff faces but few tall trees for sleeping.

In feeding habits, all Afro-papionins are opportunistic omnivores, lacking major specializations of the digestive system and capable of living on whatever high-quality vegetable or animal foods are available. What each population actually consumes, however, is constrained by the quality and distribution of resources in its habitat and by potential competitors, especially other monkeys. All species have specialties in their diet and foraging tactics, which are reflected in adaptations of the masticatory and digestive systems.

The facial shape common to both mangabey genera perhaps provides a mechanically advantageous configuration of masticatory muscles, permitting them to exploit hard-shelled fruits that cannot be opened by other, smaller forest monkeys. In other respects, however, the ecology of the two mangabey genera is rather different. *Lophocebus* feeds mainly in the forest canopy. Where available, figs and other fruits comprise a major part of the diet (Olupot et al. 1997, Freeland 1979), but hard-shelled fruits and seeds are its fall-back resource (Lambert et al. 2004, Poulsen et al. 2001).

Cercocebus mangabeys forage mostly on the ground and have been described as "forest-floor gleaners" (Fleagle and McGraw 1999, 2002), specializing in fallen fruit, fungi, as well as invertebrates.

Mandrills (and probably drills) are also specialized forest-floor gleaners, feeding mostly on unfallen and fallen fruit, especially tough-shelled fruits and seeds; fungi; and small vertebrates and invertebrates. Leaves, stems, and pith make up a smaller percentage of the diet (Hoshino 1985, Lahm 1986). The annual range of mandrill foraging groups is unknown but certainly very large (Abernethy et al. 2002). The forest-floor gleaning adaptation, involving strenuous activities such as log rolling and bark ripping, as well as cracking hard nuts, has been associated with postcranial and dental features common to *Mandrillus* and *Cercocebus* (Fleagle and McGraw 1999, 2002).

The open, high plateau habitat of *T. gelada* provides few tree-derived resources, and those that are present are rarely exploited by the geladas (Crook and Aldrich-Blake 1968), which derive most of their nutrition from low-growing herbaceous plants, such as grasses and clovers, gathered by hand from a sitting position. During the dry season, roots, bulbs, and rhizomes are intensively eaten (Iwamoto 1993). Gelada herds commonly forage on grassy flats close to the lips of the deep gorges that dissect their high-plateau habitat and afford them night refuges. The species exhibits a complex of dental and postcranial adaptations to seated hand grazing (Jolly 1970).

The diet of *Papio* baboons is so eclectic that few generalizations are possible. Baboons frequently feed high in trees, as well as on the ground. In more wooded habitats, fruits such as figs (*Ficus* spp.) are an important dietary component; while in savannas and seasonal woodlands, the flowers, seeds, fruits, and pods of trees and shrubs, such as *Acacia* and *Grewia* spp., are exploited. The leaves of grasses are eaten when young and palatable, and at other times grass and sedge rhizomes are consumed in quantity. *Papio* baboons eat invertebrates and have been reported to capture and consume small to medium-sized vertebrates such as hares, infant ruminants, vervet monkeys, ostrich chicks, and guinea fowl. Besides food and water, the home range of a *Papio* group includes one or more regularly used sleeping places, either a grove of tall trees, often on a watercourse, or a cliff face on which there are narrow ledges. The area needed to provide these necessities varies widely by habitat; it is largest in semidesert and smallest in productive woodland or forest.

AFRO–PAPIONIN SOCIETIES AND THEIR EVOLUTION

In the following account, *Lophocebus*, *Cercocebus*, and *Mandrillus* are each represented principally by their best known species: *L. albigena*, *C. atys*, and *M. sphinx*, respectively. *Papio* species, which are generally better documented, are considered individually.

Some Common Features

There are clear differences in social behavior among the Afro-papionins, but all are variants on a theme. One common element is a group of animals that coordinate their foraging activities and generally sleep at a single site. This basic, permanent, two-sex, multigenerational group is a troop. (In accounts of hamadryas [Kummer 1968] and gelada [Dunbar 1984] society, the equivalent unit has been called a "band"; for consistency, the term *troop* is used here for all species.) Troops acquire new members by birth and immigration. In most Afro-papionin species, almost all females remain in the troop of their birth. Males generally emigrate to another troop before full sexual maturity, and many migrate again during their reproductive career. The major known exception is *P. hamadryas*, in which most males, as well as females, apparently breed in their natal troop.

In some species, troops commonly join into larger, temporary associations, at a favored sleeping place or foraging area. These aggregations in hamadryas and gelada baboons have been called "troops" (Kummer 1968). In this account, they are called "multitroop aggregations" (MTAs).

Within the Afro-papionin troop, patterns of association and interaction among individuals comprise a network of relationships, some transient, others longer-lasting. Where mothers and (female) offspring continue to interact frequently into adulthood, the result is a *matriline*, a cluster of related females. Because most males disperse from the troop of their birth (e.g., Alberts and Altmann 1995), kin-based associations among adult males are rarely seen. The one known exception again is hamadryas baboon society, where observers have distinguished clans, interacting clusters of males, within troops that appear (but have not yet been proved) to be genetically related.

A feature of papionin societies is social dominance —asymmetrical, power-based relationships maintained by force or the threat of it. Being dominant enables an animal to displace others from a resource that both desire— a mating opportunity, a favored food, or sleeping place. Awareness of dominance-defined relationships is also expressed in social vigilance or "attention structure" (Chance 1967). Low-ranking animals spend time and energy monitoring the whereabouts and behavior of dominants, who themselves appear much more oblivious. Where matrilines are spatially and socially coherent, they are usually dominance-ranked so that a maturing animal's rank in the troop is determined by that of its mother and is retained for life. Within matrilines, newly matured females (who have the highest reproductive value to the matriline) outrank their elder siblings. Generally, adults are dominant over juveniles, and subadult and adult males are dominant over females. The dominance hierarchy among same-sex adults is usually more or less linear, but in some species a group of two or more individuals (typically, low-ranking adult males seeking a mating opportunity) may form a coalition that unites to displace an animal that could dominate each of them individually.

Infanticide is the systematic hunting down and killing of a suckling infant, almost always by an adult male. Its advantage to the male is believed to be that the infant's mother returns to breeding condition sooner and can then be mated by the killer. While infanticide is probably in the repertoire of all male Afro-papionins, its actual incidence varies widely from one study to another, even within the same species. Most infants are killed in circumstances where the killer can be reasonably sure that he did not father the victim and where he runs little risk of a damaging confrontation with its defenders. Sometimes regarded as a by-product of social upheaval (Sussman et al. 1994), the threat of infanticide has recently been implicated as a major determinant of social evolution in papionins (Henzi and Barrett 2003, 2005).

Relationships between males and females of reproductive age fall into two distinct categories. *Consortship* is a temporary association between a sexually active female and an adult male. The context of mating and the tense interactions involved in it distinguish consortship from *friendship* (Smuts 1985), in which an adult male enjoys a close, long-term, and typically relaxed and hedonic relationship with one or more females and their immature offspring. In some Afro-papionin societies, consortships and friendships are coincident, the same partners regularly participating in both kinds of relationship. In this case, the resulting long-term clusters are usually designated as "harems" or one-male units (OMUs). Other clusters include *bachelor bands*, coherent associations of unmated adult males. Where birth cohorts are large enough, playgroups of juveniles, typically with their mothers in attendance, may be observable.

In the following hypothetical reconstruction of Afro-papionin behavioral evolution, conventional socioecological theory (e.g., Dunbar 1988, Kappeler and van Schaik 2002) provides most of the components of the explanatory scenarios, but the whole story is hung onto a phylogenetic roadmap derived from molecular data. The society at the phylogenetic root can be reconstructed by subtracting the most obvious derived elements of present-day Afro-papionin societies. The result is not unlike societies of most macaques (see Chapter 13). Females would be philopatric, males would disperse into neighboring groups; it would have no closed OMUs, no obvious MTAs, and probably no bachelor bands. Troop size would vary with habitat, the number of adult males per troop would be proportional to the number of females, and breeding, in a tropical habitat, would probably not be seasonal.

Socioecological Evolution in the Mandrill Subclade

Cercocebus, the less derived genus of this subclade, is quite diverse in the size of troops and foraging ranges. Except where constrained by habitat fragmentation, however, its home range and troop sizes seem to be consistently larger than those of *Lophocebus* (Waser 1984, Homewood 1978, Homewood and Rodgers 1985, Karere et al. 2004, Wieczkowski 2004, Mitani 1989). The largest troops and most complex social behavior so far reported are in *C. atys* living in the Taï Forest, Cote D'Ivoire (Range and Noe 2002, Benneton and Noe 2004). At Taï, at least, breeding is seasonal. Some males are resident within the troop outside the breeding season, while others apparently belong to all-male groups that attempt to enter the troop when females are in breeding condition.

Cercocebus socioecology becomes clearer in the light of recent work on its more derived relative, *M. sphinx*. Studies of both *Mandrillus* species in the wild have been severely handicapped by their dense forest habitat, the nervousness of the animals, and the large distances over which they travel. Observations of adult males associating closely with groups of females coupled with occasional sightings of very large groups led to the view that mandrills (and drills) live in a multilevel society in which permanent, closed OMUs move independently but sometimes congregate in larger troops (e.g., Stammbach 1987). Some observers, however, have postulated more conventional, multimale troops (Kudo 1987).

The behavior of drills in the wild is still largely unknown, but recent work in the Lope National Park, Gabon (Abernethy et al. 2002), has greatly clarified the social behavior of mandrills in the wild and has provided a context for observations on captive groups. The mandrill foraging group includes only females and young, 600–800 in number. It is highly cohesive, although rarely visible as a whole. This group, within which no substructure has yet been identified, has been named a "horde," for its large size; but it seems homologous with the troops of other species. Adult males are apparently solitary except in the breeding season (of about 3 months), when they join the horde and compete intensely for mates. Male–male competition is highly energy-expensive, with non-stop activity, fights and threats, marathon grunting, and scent marking of trees with the sternal gland. Successful competitors have brighter skin colors, known to signal higher testosterone levels (Setchell and Dixon 2001a,b).

Assuming that the horde structure seen at Lope is characteristic of mandrills (and perhaps drills), a hypothetical scenario can be constructed deriving the mandrill pattern from that seen in *Cercocebus* and linking both to forest-floor gleaning. In many respects (troop size, the presence of extra-troop males, sexual dimorphism in body mass), mandrill society shows the derived features of *Cercocebus* carried to a greater extreme. Both *Cercocebus* and mandrills exploit fruit when it is available and will climb trees to find it. Their specialty, however, is resistant seeds that have accumulated on the forest floor and small but nutritious items derived especially from rotting logs and tree trunks—arthropods, fungi, and so on (Fleagle and McGraw 2002, Range and Noe 2002, Lahm 1986). Unlike canopy foods such as leaves and fruit, these are resources that accumulate over time and, once harvested, are slow to regenerate. The large home range of *Cercocebus* troops, the even larger range of mandrill foraging groups, and the fact that females and juveniles travel the circuit as a single, large party enable each area's resources to recover to a harvestable level between visits.

In both, and unusually for Afro-papionins, breeding is seasonal. In mandrills, concentrating breeding in the dry season implies a birth season just before and during the period of maximal rainfall; but it is not clear which, if either, of these facts is ecologically significant. The advantage may lie not in the season so much as synchronicity of births per se. If females whose infants are at the same stage of development tend to determine the pace at which the whole troop moves during long daily foraging marches, those who are encumbered by a larger infant than other females or who are giving birth are likely to be left behind and risk predation.

Whatever its initial advantage, breeding synchronicity allows males to spend most of the year away from the troop. In *C. atys* at Taï, some, but not all, males adopt this strategy and the most dominant, and presumably reproductively successful, male is resident. Among subordinate males, residents and incomers form opposing coalitions (Benneton and Noe 2004). The high reproductive stakes of such contests could explain the somewhat increased sexual size dimorphism in *Cercocebus*.

Mandrills (at least at Lope) appear to have evolved further down the same path, relinquishing the resident male option entirely. Since males develop few, if any, long-term bonds with individual females, their breeding success is determined entirely by hectic, intrasexual competition during the breeding season (Abernethy et al. 2002). Contending for access to the breeding group is an all-or-nothing effort, costly in injury risk as well as energy. It has favored the evolution of extreme body mass and canine size in the male and specialized deposits of body fat to carry him through a period when he has little time to forage. Also important are unambiguous signals of colored skin, which inform a male's potential rivals whether or not he is a serious contender that season. The decision to become a contender seems to be directly influenced by the presence of other males in breeding condition and, in a confined colony at least, may be delayed for years after sexual maturity (Setchell and Dixon 2001a). The analogy with other mammals in which males compete intensely for a female herd suggests that in the wild the breeding career of each male may be limited to a few seasons.

The behavior of male mandrills outside the breeding season is virtually unknown, but no all-male groups have been reported. Large body mass would make travel energetically costly. This suggests that solitary adult males might advantageously follow a shorter daily path and perhaps occupy a smaller home range than the hordes. In fact, once males no longer had to keep pace with females and young, the energetic constraint on male body mass was removed, permitting evolution of the mandrill's uniquely high level of size dimorphism.

Socioecological Evolution in the Baboon Subclade

In the baboon subclade, the macaque-like ancestral pattern has been retained in the highly arboreal *Lophocebus*. Troops consist of up to 30 individuals, including one or more adult males, living in a home range of 50 ha or less (Waser 1977). Males migrate singly between troops, after a period of "visiting" in which they do not mate (Olupot and Waser 2001) but presumably assess breeding prospects.

T. gelada has a more derived multilevel society, in which a troop ("band") of 30–300 animals includes one or more all-male groups and many breeding clusters of close matrilineal kin each with one attached, reproductively active adult male (OMUs) (Dunbar 1984, Kawai et al. 1983). The breeding male remains resident for as long as he can maintain his position against attacks of bachelors. Once deposed, he may stay as a peripheral, nonbreeding member of the OMU, which presumably enables him to protect his young offspring against infanticidal attack. Though infanticidal motivations are not absent in the species (Mori et al. 1997), the act itself is evidently rare in the wild. Female geladas are socially adept; perhaps, acting together, they are able to protect infants by establishing relations with the incoming male and even manipulate the take-over to a time when the fewest vulnerable infants are present. Though a recognizable social unit, the gelada troop is relatively incoherent; OMUs can move from one troop to another to forage, and several troops may gather together at a feeding site, resulting in MTAs of several hundred individuals. Gelada OMUs survive male turnovers, but become vulnerable to take-over or dispersal when they exceed a critical size, which varies with habitat but rarely exceeds 12 individuals (Dunbar 1984).

Like its physical adaptations, the society of *T. gelada*—matriline-based OMUs in a loose troop structure, with large feeding aggregations—seems to be intimately tied to its ecology as a specialized grazer on open, short-grass pastures. The short distance between a group's sleeping site and feeding grounds means that travel time and effort are hardly significant. Grass sward is continuously and evenly distributed in large patches. Its value is realized only with considerable effort spent picking out edible grass blades and rhizomes, so displacing a neighboring troop from its feeding area would produce little, if any, net gain. Similarly, predation risk, even before humans removed most large predators from gelada habitat, would have provided little incentive for troop cohesion. The large feeding aggregations provide maximal vigilance, and the cliff faces below the plateau are close enough to provide refuge. The main factors that could favor an investment in troop cohesion—intertroop conflict, predator detection, and the value of collective memory when traveling between feeding areas—are minimal.

Troop members therefore would lose little by following their own individual path across the feeding grounds, paying little attention to other troop members. Lone mother–infant dyads, however, would be vulnerable to harassment and infanticidal attacks from males. It is therefore advantageous for matriline members, which share inclusive fitness interest in all their infants, to remain spatially cohesive and collectively attempt to manage males' access. In a population that is more or less demographically stable, clusters of closely related females will rarely be large enough to manage multiple males. With only one resident male, which putatively fathers all the infants, unambiguous paternity means that other males will be more inclined to infanticide. Anti-infanticide vigilance, on the part of both the females and the resident male, is strongly favored. The adult male's very small testes (Jolly et al. 1997) suggest that the strategy of excluding other males is so effective that sperm competition is very rare. Since some male turnover is inevitable, the females' need to assess their male's prospects of survival would lead to the evolution of honest signals of his condition. A copious, well-groomed mane signals condition and social standing in the medium term, while the color of the breast patch can convey change in status almost instantaneously (Dunbar 1984).

Any discussion of socioecological variation in *Papio* baboons must start by eliminating a major source of confusion—the fallacy of the "savanna–desert dichotomy." The hamadryas (the desert baboon) is often put into one species (*P. hamadryas*), and all the others (savanna baboons) in another (*P. cynocephalus*). Though widely used (e.g., Smuts et al. 1987), this scheme is unsatisfactory in several ways. For one, the ecologically based labels are misleading. Some hamadryas baboons do not live in deserts (Zinner et al. 2001). Many savanna baboons do not inhabit savannas, and some actually live in deserts. Recognizing just two species with ecological labels promotes unwarranted assumptions about the homogeneity of "savanna baboons" (Williams-Blangero et al. 1990, Henzi and Barrett 2003, 2005). For systematists, however, its fatal flaw is that it misrepresents phylogeny. There is no evidence (*contra*, e.g., Purvis 1995) that the hamadryas is the sister taxon to all other extant baboons. Its morphology and behavior ally it most closely with the Guinea baboon (*P. papio*) of West Africa (Jolly 1993, Jolly and Phillips-Conroy 2003). Rooting the hamadryas phylogenetically among, rather than beside, the savanna baboons makes the inclusive savanna baboon species non-holophyletic and thus unacceptable to most taxonomists. So the two-species desert vs. savanna dichotomy is taxonomically invalid as well as biologically inaccurate and analytically misleading.

Papio baboon societies vary in features such as troop size, the presence or absence of OMUs or corporate matrilines within the troop, and whether MTAs are formed. At the relationship level, they vary in features such as the incidence of infanticide, the duration and exclusivity of consortships, and the formation of coalitions among adult males. Some of this variation clearly stems from environmental factors working directly on behavioral plasticity—the size of the foraging group in all baboons is related to the distribution and size of food patches, for example. Other variation is taxon-, rather than habitat-, specific and can be explained only if evolutionary history is considered along with socio-ecological theory (Henzi and Barrett 2003, 2005). The scenario constructed here follows Henzi and Barrett's (2003, 2005) synthesis, with a few modifications.

Since chacma baboons are closest both geographically (McKee 1993, Broadfield et al. 1994) and phylogenetically (Newman et al. 2004, Wildman et al. 2004) to the original stem of all living *Papio*, species, it is reasonable to suppose that their society is also ancestral. Certainly, it lacks derived features seen in other baboons, notably same-sex coalitions and permanent, male-centered OMUs. Chacma habitats range from near desert (Cowlishaw 1999) to tropical woodlands and savannas (e.g., Cheney et al. 2004), winter-rainfall regimes (Hall 1963), and seasonally frigid montane grasslands (Henzi et al. 1997a,b). In less productive habitats, most troops are too small to attract more than one resident male, and infanticide generally follows when an incoming male takes over. In more productive areas, troops are larger and multimale but infanticide is still frequent because paternity is highly correlated with male rank (Palombit et al. 2000). To guard against infanticide, a female with a dependent infant will form a bond of "friendship" with a male that mated with her at the time of her infant's conception. Alternatively, a cluster of females may evade a potentially infanticidal newcomer by emigrating along with their infants' most probable father (Weingrill et al. 2003).

Chacma society is an appropriate base for the evolution of the somewhat more complex societies of yellow and anubis baboons, known from study sites in the East African savanna–woodland corridor, such as Amboseli (e.g., Altmann et al. 1985), Mikumi (e.g., Rhine et al. 2000, Norton et al. 1987), Gilgil/Isiolo (e.g., Strum 2001), and Gombe Stream (e.g., Ransom 1981). Here, more densely distributed and larger resource patches favor larger troops, which include cohesive matrilines and attract multiple immigrant adult males. The probability of paternity is spread among them by the female strategy of mating serially with most of the high-ranking males and by the males' ability to form coalitions that displace a higher-ranking male from a consortship. Consortships last hours, rather than days, as in chacmas (Henzi and Barrett 2003), so a female is typcally polyandrous in each conception cycle. Though infanticide remains on the males' agenda, it occurs much more rarely because immigrants are outnumbered in large troops by resident males and potential fathers. A male can also benefit from the protection that paternity confusion affords his offspring—provided, of course, that he is indeed their father. In a troop with many potential rivals, a male might gain fitness by investing energy and resources in sperm competition, rather than defending his consortship once he has mated. This might explain why anubis and yellow baboons combine very short consortships and very large testes (Jolly and

Phillips-Conroy 2003). It could also explain another odd behavior. A male anubis will sometimes take an infant from its protesting mother and handle it so roughly that it screams. This behavior is quite distinct from using an infant as a shield or protecting the male's own putative offspring in a dangerous situation. Perhaps it is a way for a male considering infanticide to size up the likely opposition. Meanwhile, with less risk of infanticide, "friendship" seems to have become a way that low ranking males can "engage" a mate for the future while also protecting their existing offspring (Henzi and Barrett 2003).

Hamadryas society has a number of derived features not seen in chacma, yellow, or olive baboons (Kummer 1968, Sigg et al. 1982, Abegglen 1984, Swedell in press). Troops ("bands") are often very large, in spite of the arid habitat. Two or more troops may gather at a single sleeping place (forming a MTA) but remain separate and mutually wary. Large troops can exist because, when necessary, they can split up to forage as OMUs. Hamadryas troops are not just aggregations of OMUs, however. Their membership includes many bachelor males with no OMU affiliation. Moreover, as in most other Afro-papionins, troop members sleep and file out of the sleeping site together. They forage together when conditions permit and collectively threaten other troops and potential predators. As among other Afro-papionins, too, most females stay in their natal troop to breed but some are taken by males from neighboring troops within the MTA (Swedell 2005).

The hamadryas OMU (unlike the matrilineal OMU of geladas) is a cluster of 1–10 females (with their offspring) that the male has personally accumulated. He jealously guards access to them, insisting that they follow his movements closely. If they or their attention wander, they may be rounded up, held down, and bitten. There is little sign of a dominance hierarchy among OMU leaders, which interact in a peculiar ritual called "notifying." This avoids fights that might cause injury or loss of females to a bachelor. An OMU often includes a peripheral subadult or young adult male that is pursuing a "follow-and-wait" reproductive strategy. Although (mostly) excluded from the group's females by the leader's vigilance, he may eventually inherit some of them. More likely, he will adopt one of its prepubertal females as the nucleus of his own OMU (Kummer 1968).

The OMU system explains some physical peculiarities of the hamadryas. The adult male's red bottom is used in "notifying," and his showy mane (like the gelada's) is a signal to his females of his condition and durability. Small adult testes are correlated with low sperm competition, but, perhaps because as followers they can sometimes sneak copulations, juveniles have testes that are somewhat large for their age (Jolly and Phillips-Conroy 2003). Meanwhile, protected from female–female competition (Wasser and Starling 1988) by the male that has adopted her, a young female hamadryas can begin her reproductive career earlier than other baboons.

Unique as it is, the hamadryas OMU clearly has roots in the friendship groups of other baboons and fulfills the same functions, albeit in a modified and exaggerated form. Exclusive mating in the OMU makes for very high paternity certainty and thus infanticide risk (Swedell 2000, Swedell and Tesfaye 2003), so in keeping his OMU close to him, the leader is also protecting his existing offspring. The "follow-and-wait" strategy is a way of establishing ties with future mates, just as in a friendship bond among yellow or anubis baboons (Henzi and Barrett 2003). The difference is that it is initiated earlier in the male's life cycle and has to be abandoned when he becomes an OMU leader himself.

The reason that a young male can initiate his "follow-and-wait" strategy long before he is mature enough to be an OMU leader is that he usually stays to form his OMU in the troop where he was born. This is the truly unusual feature of hamadryas troops—not the minimal outflow of females but the fact that most natal males do not disperse (Abegglen 1984, Sigg et al. 1982, Swedell 2005). Philopatry has some advantages. It enables a young male hamadryas to follow and wait among older male relatives, which have little to gain by opposing him. It also allows OMUs whose leaders are related to form cooperating groups.

Clearly, the hamadryas system, with OMUs that can forage independently, works well in the semideserts of the Horn of Africa, but did it evolve as an adaptation to that environment? Most interpretations (grounded in the erroneous desert–savanna dichotomy) have taken this for granted (Kummer 1968, 1995; Jolly 1963; Henzi and Barrett 2005), but an element of doubt enters when we look at other baboon species. Some chacmas live in deserts at least as forbidding as those inhabited by hamadryas (Cowlishaw 1999, Brain 1990), and as predicted, they have small foraging groups. However, they have not become male-philopatric, nor have they evolved a multilevel, hamadryas-style society. Conversely, the Guinea baboon, *P. papio*, of far western Africa, which lives in many savanna, woodland, and forest habitats (Galat-Luong et al. 2005), apparently does show some very similar behavioral traits.

Guinea baboons share a number of anatomical details with hamadryas, which suggest that the two are probably sister taxa (Jolly 1993). Additionally, the male Guinea baboon has a prominent shoulder mane, though this is not distinctively colored as in hamadryas. As recently summarized (Galat-Luong et al. 2005), Guinea baboon society, too, appears quite hamadryas-like in some ways (Boese 1975, Galat-Luong et al. 2005, Dunbar and Nathan 1972, Sharman 1981). Guinea baboons live in large troops (100–250 strong). Several troops may cluster at night at a favored sleeping place, typically in a grove of large, riverside trees; however, they remain distinct, and each moves out as a separate column in the morning. Within troops are obvious OMUs, consisting typically of a leader, one to three females with their young, and often a subadult male follower. Troops often split while foraging into second-level subgroups of about 30 animals that also nap together in the afternoon heat. Adult males within these groups are very affiliative and relaxed

with each other. Second-level subgroups (Galat-Luong et al. 2005) seem much like hamadryas clans, in both size and the relaxed relations among their adult males (whether these males are actually close kin, as has been speculated, is still unproven). Like hamadryas, also, Guinea baboon OMU leaders have no obvious dominance hierarchies but exhibit stylized greeting behavior (Boese 1975). We do not know whether Guinea OMUs are exclusive mating units in the wild, but it seems likely as adult testes are small (Jolly and Phillips-Conroy 2003), suggesting monandry and lack of sperm competition. Tolerance among males, the presence of subadult males attached to OMUs, and the clans all hint at male philopatry; but much more field work is needed to test this hypothesis. Another unknown is whether the females of an OMU are a matriline or are recruited individually by the male. On the other hand, there are some striking hamadryas features that Guinea baboons do not have. Male Guinea baboons have prominent manes, but no striking change in color signals their transition from subadult to adult; also, they lack the prominent red buttocks the hamadryas uses in male–male greeting. OMUs exist, but as independent foraging units they seem less important than the clan-like second-level subgroups. Though male Guinea baboons, like hamadryas, "herd" females, they have not been observed using the same rough tactics to keep the OMU together. Altogether, Guinea baboons seem behaviorally intermediate between hamadryas and other baboons. Boese (1975) pointed this out long ago, but his insight went unremarked at the time, when Guineas were "just another savanna baboon."

The phylogenetic relationship between Guinea and hamadryas baboons suggests that the derived features their societies share were inherited from their common ancestor. Where, when, and why the ancestral stock aquired the novel pattern are unknown; but given that Guinea baboons today live in habitats ranging from dry steppe to moist evergreen forest, it is hard to find any plausible explanation in ecology alone. (A speculative scenario, beyond the scope of this chapter, ties large troops, OMUs, and male philopatry to demographic structure peculiar to the initial geographical expansion of *Papio* baboons.)

The final transition to the yet more derived hamadryas system, however, can indeed be related to the move from savanna–woodland to semidesert. In the sparser environment, OMUs often foraged independently, rather than in second-level subgroups. Without this support, it became even more urgent for a male to keep his group physically together, to pro-tect his offspring against predation and infanticide; and this favored more insistent herding behavior, including the neck bite. More of the OMU leader's interactions with other males involved comparative strangers in the troop, rather than the familiar members of his clan. This put a higher value on avoiding ambiguity in male–male signaling, favoring both greater stereotypy of signals such as notifying and the evolution of structures specialized to convey them. On the female side, the cost of misjudging a male's viability as a protector was now higher, leading, via sexual selection, to the evolution of even more visible, hormone-sensitive shoulder manes in males.

CONCLUSION

This chapter has examined the evolution of the Afro-papionins from a somewhat novel perspective, more explicitly using phylogeny and population history along with socioecological theory to reconstruct the story. Stretching the fabric of data over the phylogenetic framework, however, has revealed its many holes. In spite of the enormous, painstaking effort that has been expended in studying Afro-papionins, in both the lab and the field, synthesis is only just becoming possible. In particular, for most species, we still need to know how much variation is due to population-specific, inherited factors and how much to simple plasticity. Whole important taxa have yet to be subjected to the kind of sustained, long-term field work needed to tackle interesting evolutionary questions. Behavioral diversity in *Cercocebus* has hardly been sampled, *Mandrillus* is still a challenge, and the highly distinctive Kinda baboon has yet to be watched in the field at all. This chapter is necessarily full of speculations, untested hypotheses, and unproven scenarios. Hopefully, it will provide some stimulus for filling the informational gaps before the opportunity, and the monkeys, disappear forever.

REFERENCES

Abegglen, J. J. (1984). *On Socialization in Hamadryas Baboons: A Field Study*. Lewisburg, PA: Bucknell University Press.

Abernethy, K. A., White, L. J. T., and Wickings, E. J. (2002). Hordes of mandrills (*Mandrillus sphinx*): extreme group size and seasonal male presence. *J. Zool.* 258:131–137.

Alberts, S. C., and Altmann, J. (1995). Balancing costs and opportunities: dispersal in male baboons. *Am. Nat.* 145:279–306.

Altmann, J., and Alberts, S. C. (2005). Growth rates in a wild primate population: ecological influences and maternal effects. *Behav. Ecol. Sociobiol.* 57:490–501.

Altmann, J., Altmann, S., and Hausfater, G. (1981). Physical maturation and age estimates of yellow baboons, *Papio cynocephalus*, in Amboseli National Park, Kenya. *Am. J. Primatol.* 1:389–399.

Altmann, J., Hausfater, G., and Altmann, S. A. (1985). Demography of Amboseli baboons, 1963–1983. *Am. J. Primatol.* 8:113–125.

Barnicot, N. A., and Hewett-Emmett, D. (1972). Red cell and serum proteins of *Cercocebus, Presbytis, Colobus* and certain other species. *Folia Primatol.* 17:442–457.

Benneton, C., and Noe, R. (2004). Reproductive tactics of adult male sooty mangabeys in Taï National Park, Ivory Coast [abstract]. *Folia Primatol.* 75:169.

Boese, G. K. (1975). Social behavior and ecological considerations of West African baboons (*Papio papio*). In: Tuttle, R. H. (ed.), *Socioecology and Psychology of Primates*. Mouton, The Hague, pp. 205–230.

Brain, C. (1990). Spatial usage of a desert environment by baboons (*Papio ursinus*). *J. Arid Environ.* 18:67–73.

Broadfield, D. C., Delson, E., and Atsalis, S. (1994). Cercopithecid fossils from the later Pleistocene of Taung, South Africa [abstract]. *Am. J. Phys. Anthropol.* (Suppl. 18):59–60.

Chalmers, N. R. (1968). Group composition, ecology and daily activities of free living mangabeys in Uganda. *Folia Primatol.* 8:247–262.

Chance, M. R. A. (1967). Attention structure as the basis of primate rank orders. *Man* 2:503–518.

Cheney, D. L., Seyfarth, R. M., Fischer, J., Beehner, J., Bergman, T., Johnson, S. E., Kitchen, D. M., Palombit, R. A., Rendall, D., and Silk, J. B. (2004). Factors affecting reproduction and mortality among baboons in the Okavango Delta, Botswana. *Int. J. Primatol.* 25:401–428.

Collard, M., and O'Higgins, P. (2001). Ontogeny and homoplasy in the papionin monkey face. *Evol. Dev.* 3:322–331.

Cowlishaw, G. (1999). Ecological and social determinants of spacing behaviour in desert baboon groups. *Behav. Ecol. Sociobiol.* 45:67–77.

Crook, J. H., and Aldrich-Blake, P. (1968). Ecological and behavioural contrasts between sympatric ground dwelling primates in Ethiopia. *Folia Primatol.* 8:192–227.

Delson, E., Terranova, C. J., Jungers, W. L., Sargis, E. J., Jablonski, N. G., and Dechow, P. C. (2000). Body mass in Cercopithecidae (Primates, Mammalia): estimation and scaling in extinct and extant taxa. *Anthropol. Pap. Am. Mus. Nat. Hist.* 83:1–159.

Disotell, T. R. (2000). Molecular systematics of the Cercopithecidae. In: Whitehead, P. F., and Jolly, C. J., eds., *Old World Monkeys*. Cambridge University Press, Cambridge. pp. 29–56.

Dunbar, R. I. M. (1984). *Reproductive Decisions: An Economic Analysis of Gelada Baboon Social Strategies*. Princeton University Press, Princeton, NJ.

Dunbar, R. I. M. (1988). *Primate Social Systems*. Cornell University Press, Ithaca.

Dunbar, R. I. M. (1992). Time: a hidden constraint on the behavioural ecology of baboons. *Behav. Ecol. Sociobiol.* 33:35–49.

Dunbar, R. I. M., and Nathan, M. F. (1972). Social organization of the Guinea baboon, *Papio papio. Folia Primatol.* 17:321–334.

Fleagle, J. G., and McGraw, W. S. (1999). Skeletal and dental morphology supports diphyletic origin of baboons and mandrills. *Proc. Natl. Acad. Sci. USA* 96:1157–1161.

Fleagle, J. G., and McGraw, W. S. (2002). Skeletal and dental morphology of African papionins: unmasking a cryptic clade. *J. Hum. Evol.* 42:267–292.

Freeland, W. J. (1979). Mangabey (*Cercocebus albigena*) social organization and population density in relation to food use and availability. *Folia Primatol.* 32:108–124.

Galat-Luong, A., Galat, G., and Hagell, S. (2005). The social and ecological flexibility of Guinea baboons: implications for Guinea baboon social organization and male strategies. In: Swedell, L. and Leigh, S. (eds), Reproduction and Fitness in Baboons: Behavioral, Ecological, and Life History Perspectives. Kluwer, New York.

Grubb, P., Butynski, T. M., Oates, J. F., Bearder, S. K., Disotell, T. R., Groves, C. P., and Struhsaker, T. T. (2003). Assessment of the diversity of African primates. *Int. J. Primatol.* 24:1301–1357.

Hall, K. R. L. (1963). Variations in the ecology of the chacma baboon, *Papio ursinus. Symp. Zool. Soc. Lond.* 10:1–28.

Henzi, S. P., and Barrett, L. (2003). Evolutionary ecology, sexual conflict, and behavioral differentiation among baboon populations. *Evol. Anthropol.* 12:217–230.

Henzi, S. P., and Barrett, L. (2005). The historical socioecology of savanna baboons (*Papio hamadryas*). *J. Zool. Lond.* 265:215–226.

Henzi, S. P., Lycett, J. E., and Weingrill, T. (1997a). Cohort size and the allocation of social effort by female mountain baboons. *Anim. Behav.* 54:1235–1243.

Henzi, S. P., Lycett, J. E., Weingrill, T., Byrne, R., and Whiten, A. (1997b). The effect of troop size on travel and foraging in mountain baboons. *S. Afr. J. Sci.* 93:333–335.

Hill, W. C. O. (1969). The vascular supply of the face in long-snouted primates (*Papio* and *Mandrillus*). Preliminary notes. In: Hofer, H. O. (ed.), *Proceedings of the Second International Congress of Primatology. Recent Advances in Primatology*, vol. 2. S. Karger, Basel. pp. 155–159.

Homewood, K. M. (1978). Feeding strategy of the Tana mangabey (*Cercocebus galeritus galeritus*) (Mammalia: Primates). *J. Zool. Lond.* 186:375–391.

Homewood, K. M., and Rodgers, A. (1985). Tanzania's newest primate. *Anim. King.* 88:14–17.

Horn, A. D. (1987). The socioecology of the black mangabey (*Cercocebus aterrimus*) near Lake Tumba, Zaire. *Am. J. Primatol.* 12:165–180.

Hoshino, J. (1985). Feeding ecology of mandrills (*Mandrillus sphinx*) in Campo Animal Reserve, Cameroon. *Primates* 26:248–273.

Iwamoto, T. (1993). The ecology of *Theropithecus gelada*. In: Jablonski, N. G. (ed.), *Theropithecus: The Rise and Fall of a Primate Genus*. Cambridge University Press, Cambridge. pp. 441–452.

Jolly, C. J. (1963). A suggested case of evolution by sexual selection in primates. *Man* 63:177–178.

Jolly, C. J. (1970). The large African monkeys as an adaptive array. In: Napier, J. R., and Napier, P. H. (eds.), *Old World Monkeys*. Academic Press, New York. pp. 139–174.

Jolly, C. J. (1993). Species, subspecies, and baboon systematics. In: Kimbel, W. H., and Martin, L. B. (eds.), *Species, Species Concepts, and Primate Evolution*. Plenum Press, New York. pp. 67–107.

Jolly, C. J., and Phillips-Conroy, J. E. (2003). Testicular size, mating system, and maturation schedules in wild anubis and hamadryas baboons. *Int. J. Primatol.* 24:125–142.

Jolly, C. J., Woolley-Barker, T., Beyene, S., Disotell, T. R., and Phillips-Conroy, J. E. (1997). Intergeneric hybrid baboons. *Int. J. Primatol.* 18:597–627.

Jones, T., Ehardt, C. L., Butynski, T. M., Davenport, T. R. B., Mpunga, N. E., Machaga, S. J., and De Luca, D. W. (2005). The highland mangabey *Lophocebus kipunji*: a new species of African monkey. *Science* 308:1161–1164.

Kappeler, P. M., and van Schaik, C. P. (2002). Evolution of primate social systems. *Int. J. Primatol.* 23:707–740.

Karere, G. M., Oguge, N. O., Kirathe, J., Muoria, P. K., Moinde, N. N., and Suleman, M. A. (2004). Population sizes and distribution of primates in the lower Tana River forests, Kenya. *Int. J. Primatol.* 25:351–365.

Kawai, M., Dunbar, R., Ohsawa, H., and Mori, U. (1983). Social organization of gelada baboons: social units and definitions. *Primates* 24:13–24.

Kingdon, J. S. (1971). East African Mammals: an atlas of evolution in Africa. Vol. 1. Academic Press, London.

Kingdon, J. (1996). *The Kingdon Field Guide To African Mammals*. Academic Press, San Diego.

Kudo, H. (1987). The study of vocal communication of wild mandrills in Cameroon in relation to their social structure. *Primates* 28:289–308.

Kummer, H. (1968). Social organization of hamadryas baboons. A field study. *Bibl. Primatol.* 6:1–189.

Kummer, H. (1995). *In Quest of the Sacred Baboon: A Scientist's Journey*. Princeton University Press, Princeton, NJ.

Lahm, S. A. (1986). Diet and habitat preference of *Mandrillus sphinx* in Gabon: implications of foraging strategy. *Am. J. Primatol.* 11:9–26.

Lambert, J. E., Chapman, C. A., Wrangham, R. W., and Conklin-Brittain, N. L. (2004). Hardness of cercopithecine foods: implications for the critical function of enamel thickness in exploiting fallback foods. *Am. J. Phys. Anthropol.* 125:363–368.

Matthews, L. H. (1956). The sexual skin of the gelada baboon (*Theropithecus gelada*). *Trans. Zool. Soc. Lond.* 28:543–552.

McKee, J. K. (1993). Taxonomic and evolutionary affinities of *Papio izodi* fossils from Taung and Sterkfontein. *Palaeontol. Afr.* 30:43–49.

Mitani, M. (1989). *Cercocebus torquatus*: adaptive feeding and ranging behaviors related to seasonal fluctuations of food resources in the tropical rain forest of south-western Cameroon. *Primates* 30:307–323.

Mori, A., Iwamoto, T., and Bekele, A. (1997). A case of infanticide in a recently found gelada population in Arsi, Ethiopia. *Primates* 38:79–88.

Newman, T. K., Jolly, C. J., and Rogers, J. (2004). Mitochondrial phylogeny and systematics of baboons (*Papio*). *Am. J. Phys. Anthropol.* 124:17–27.

Norton, G. W., Rhine, R. J., Wynn, G. W., and Wynn, R. D. (1987). Baboon diet: a five-year study of stability and variability in the plant feeding and habitat of the yellow baboons (*Papio cynocephalus*) of Mikumi National Park, Tanzania. *Folia Primatol.* 48:78–120.

Olupot, W. (2000). Mass differences among male mangabey monkeys inhabiting logged and unlogged forest compartments. *Conserv. Biol.* 14:833–843.

Olupot, W., Chapman, C. A., Brown, C. H., and Waser, P. M. (1994). Mangabey (*Cercocebus albigena*) population density, group size, and ranging: a twenty-year comparison. *Am. J. Primatol.* 32:197–205.

Olupot, W., Chapman, C. A., Waser, P. M., and Isabirye-Basuta, G. (1997). Mangabey (*Cercocebus albigena*) ranging patterns in relation to fruit availability and the risk of parasite infection in Kibale National Park, Uganda. *Am. J. Primatol.* 43:65–78.

Olupot, W., and Waser, P. M. (2001). Correlates of intergroup transfer in male grey-cheeked mangabeys. *Int. J. Primatol.* 22:169–187.

Palombit, R. A., Cheney, D. L., Fischer, J., Johnson, S., Rendall, D., Seyfarth, R. M., and Silk, J. B. (2000). Male infanticide and defense of infants in chacma baboons. In: van Schaik, C. P., and Janson, C. H. (eds.). *Infanticide by Males and Its Implications*. Cambridge University Press, Cambridge. pp. 123–152.

Poulsen, J. R., Clark, C. J., and Smith, T. B. (2001). Seasonal variation in the feeding ecology of the grey-cheeked mangabey (*Lophocebus albigena*) in Cameroon. *Am. J. Primatol.* 54:91–105.

Purvis, A. (1995). A composite estimate of primate phylogeny. *Phil. Trans. R. Soc. Lond.* B348:405–421.

Range, F., and Noe, R. (2002). Familiarity and dominance relations among female sooty mangabeys in the Taï National Park. *Am. J. Primatol.* 56:137–153.

Ransom, T. W. (1981). *Beach Troop of the Gombe*. Bucknell University Presses, Lewisburg, PA.

Rhine, R. J., Norton, G. W., and Wasser, S. K. (2000). Lifetime reproductive success, longevity, and reproductive life history of female yellow baboons (*Papio cynocephalus*) of Mikumi National Park, Tanzania. *Am. J. Primatol.* 51:229–241.

Rowe, N. (1996). *The Pictorial Guide to the Living Primates*. Pogonias Press, East Hampton, NY.

Setchell, J. M., and Dixson, A. F. (2001a). Arrested development of secondary sexual adornments in subordinate adult male mandrills (*Mandrillus sphinx*). *Am. J. Phys. Anthropol.* 115:245–252.

Setchell, J. M., and Dixson, A. F. (2001b). Changes in the secondary sexual adornments of male mandrills (*Mandrillus sphinx*) are associated with gain and loss of alpha status. *Horm. Behav.* 39:177–184.

Setchell, J. M., and Dixson, A. F. (2002). Developmental variables and dominance rank in adolescent male mandrills (*Mandrillus sphinx*). *Am. J. Primatol.* 56:9–25.

Setchell, J. M., Lee, P. C., Wickings, E. J., and Dixson, A. F. (2001). Growth and ontogeny of sexual size dimorphism in the mandrill (*Mandrillus sphinx*). *Am. J. Phys. Anthropol.* 115:349–360.

Setchell, J. M., and Wickings, E. J. (2004). Social and seasonal influences on the reproductive cycle in female mandrills (*Mandrillus sphinx*). *Am. J. Phys. Anthropol.* 125:73–84.

Setchell, J. M., and Wickings, E. J. (2005). Dominance, status signals and coloration in male mandrills (*Mandrillus sphinx*). *Ethology* 111:25–50.

Sharman, M. J. (1981). Feeding, ranging, and social behavior of the Guinea baboon [PhD diss.]. University of St. Andrews, fife.

Sigg, H., and Stolba, A. (1981). Home range and daily march in a hamadryas baboon troop. *Folia Primatol.* 36:40.

Sigg, H., Stolba, A., Abegglen, J. J., and Dasser, V. (1982). Life history of hamadryas baboons: physical development, infant mortality, reproductive parameters and family relationships. *Primates* 23:473–487.

Smuts, B. B. (1985). *Sex and Friendship in Baboons*. Aldine, New York.

Smuts, B. B., Cheney, D. L., Seyfarth, R. M., Wrangham, R. W., and Struhsaker, T. T. (eds.) (1987). *Primate Societies*. University of Chicago Press, Chicago.

Stammbach, E. (1987). Desert, forest and montane baboons: multilevel societies. In: Smuts, B. B., Cheney, D. L., Seyfarth, R. M., Wrangham, R. W., and Struhsaker, T. T. (eds.), *Primate Societies*. University of Chicago Press, Chicago. pp. 112–120.

Strum, S. C. (2001). *Almost Human: A Journey into the World of Baboons*. University of Chicago Press, Chicago.

Sussman, R. W., Cheverud, J. M., and Bartlett, T. Q. (1994). Infant killing as an evolutionary strategy: reality or myth? *Evol. Anthropol.* 3:149–151.

Swedell, L. (2000). Two takeovers in wild hamadryas baboons. *Folia Primatol.* 71:169–172.

Swedell, L. (2005). Dispersal by force: residence patterns of wild female hamadryas baboons [abstract]. *Am. J. Phys. Anthropol.* (Suppl. 40):202.

Swedell, L. (in press). *Strategies of Sex and Survival in Hamadryas Baboons: Through a Female Lens*. Pearson Prentice Hall, Upper Saddle River, NJ.

Swedell, L., and Tesfaye, T. (2003). Infant mortality after takeovers in wild Ethiopian hamadryas baboons. *Am. J. Primatol.* 60:113–118.

Waser, P. (1977). Feeding, ranging and group size in the mangabey *C. albigena*. In: Clutton-Brock, T. H. (ed.), *Primate Ecology: Studies of Feeding and Ranging Behavior in Lemurs, Monkeys and Apes*. Academic Press, New York. pp. 183–222.

Waser, P. (1984). Ecological differences and behavioral contrasts between two mangabey species. In: Rodman, P. S., and Cant, J. G. H. (eds.), *Adaptations for Foraging in Non-human Primates*. Columbia University Press, New York. pp. 195–216.

Wasser, S. K., and Starling, A. K. (1988). Proximate and ultimate causes of reproductive suppression among female yellow baboons at Mikumi National Park, Tanzania. *Am. J. Primatol.* 16:97–121.

Weingrill, T., Lycett, J. E., Barrett, L., Hill, R. A., and Henzi, S. P. (2003). Male consortship behaviour in chacma baboons: the role of demographic factors and female conceptive probabilities. *Behaviour* 140:405–427.

Wickings, E. J., and Dixson, A. F. (1992). Development from birth to sexual maturity in a semi-free-ranging colony of mandrills (*Mandrillus sphinx*) in Gabon. *J. Reprod. Fertil.* 95:129–138.

Wieczkowski, J. (2004). Ecological correlates of abundance in the Tana mangabey (*Cercocebus galeritus*). *Am. J. Primatol.* 63:125–138.

Wildman, D. E., Bergman, T. J., al-Aghbari, A., Sterner, K. N., Newman, T. K., Phillips-Conroy, J. E., Jolly, C. J., and Disotell, T. R. (2004). Mitochondrial evidence for the origin of hamadryas baboons. *Mol. Phylogenet. Evol.* 32:287–296.

Williams-Blangero, S., Vandeberg, J. L., Blangero, J., Konigsberg, L., and Dyke, B. (1990). Genetic differentiation between baboon subspecies: relevance for biomedical research. *Am. J. Primatol.* 20:67–81.

Zinner, D., Pelaez, F., and Torkler, F. (2001). Distribution and habitat associations of baboons (*Papio hamadryas*) in central Eritrea. *Int. J. Primatol.* 22:397–413.

15

The Guenons (Genus *Cercopithecus*) and Their Allies

Behavioral Ecology of Polyspecific Associations

Karin L. Enstam and Lynne A. Isbell

INTRODUCTION

This chapter deals with the taxonomy, distribution, ecology, reproduction, and social organization of the guenons and their allies. Strictly speaking, the term *guenon* includes catarrhine monkeys of the genus *Cercopithecus* (Butynski 2002); but in this chapter, we will expand our discussion to include three other closely related genera of the tribe Cercopithecini— *Allenopithecus*, *Erythrocebus*, and *Miopithecus* (Grubb et al. 2003) (Table 15.1)—and refer to all members of the tribe Cercopithecini as "guenons" (see also Butynski 2002). Specific genera and species will be discussed as warranted.

Guenons are small to medium-sized monkeys (average 3.6 kg for females, 5.9 kg for males; Haltenorth and Diller 1988) distributed throughout sub-Saharan Africa. The members of the genus *Cercopithecus* are the most colorful of all primates (Rowell 1988), particularly in the face; and the variation in coloration may be used in intraspecific communication (see Vine 1970, Kingdon 1987) for recognition of individuals, species, and appropriate mates (e.g., sexual selection, see Ryan and Rand 1993). *Allenopithecus* and *Miopithecus* are drab, with coat colors ranging from greenish gray in *Allenopithecus* to varying shades of olive–yellow in *Miopithecus* (Haltenorth and Diller 1988). *Erythrocebus* is unique among guenons with its reddish brown to orange coat color (Haltenorth and Diller 1988).

TAXONOMY AND DISTRIBUTION

Considerable debate still exists regarding the taxonomy of the tribe Cercopithecini (see Butynski 2002), but one thing is clear: the guenons are the largest group of African primates. Recent taxonomic work lists as few as 23 (Grubb et al. 2003) and as many as 36 (Groves 2001) species of guenons (Table 15.1). Within the guenons, the genus *Cercopithecus* is divided into several superspecies groups. According to Groves (2001), there are eight superspecies groups: *C. cephus*, *C. diana*, *C. dryas*, *C. hamlyni*, *C. lhoesti*,

C. mitis, *C. mona*, and *C. neglectus*. Grubb et al. (2003) give a similar, but slightly different, organization of the superspecies groups, dividing the genus into *C. aethiops*, *C. cephus*, *C. diana*, *C. dryas*, *C. hamlyni*, *C. preussi* (*C. lhoesti* group in Groves [2001] taxonomy), *C. nictitans* (*C. mitis* group in Groves [2001] taxonomy), *C. mona*, and *C. neglectus* (see Table 15.1). One should note that there are nine superspecies groups in Grubb et al.'s (2003) taxonomy of *Cercopithecus* because they retain *Cercopithecus* as the genus name for vervet/grivet/green monkeys (*C. aethiops*). Groves (2001) elevates *C. aethiops* to its own genus, *Chlorocebus*, due to certain morphological differences (e.g., shape of the auditory tube and eye orbits, shape of the second upper incisor, etc.) between this species group and the genus *Cercopithecus* (see Groves 1989).

Likewise, there is dispute between these two studies regarding the division within the vervet/grivet/green monkey group. Grubb et al. (2003), following Napier (1981), consider the *C. aethiops* group to be one highly polytypic species partly because of uncertainty about the boundaries between the species and partly because Struhsaker (1970) was unable to distinguish between populations of *C. aethiops* and *C. tantalus* based on their vocalizations. According to Grubb et al. (2003), this group consists of six subspecies of *C. aethiops*, all of which Groves (2001) has elevated to a species within the genus *Chlorocebus* (Table 15.1). Other divisions of the vervet/grivet/green monkey group also exist. Kingdon et al. (in press) retain the *Cercopithecus* genus name for the vervet/grivet/green monkey group but divide the group into five species rather than subspecies.

There is less dispute over the taxonomy of *Allenopithecus*, *Erythrocebus*, and *Miopithecus*. Both *Allenopithecus* and *Erythrocebus* are monotypic, but *Erythrocebus* may contain from zero (Groves 2001, Grubb et al. 2003) to four (Kingdon 1997) subspecies. Isbell (in press) describes three subspecies of *Erythrocebus* (*E. patas patas* in West Africa, *E. p. baumstarki* in Tanzania, and *E. p. pyrrhonotus* in the rest of East Africa) based on differences in facial coloration. *Miopithecus*, once believed to be monotypic, is now believed to contain

Table 15.1 Comparison of Two Most Recent Guenon Taxonomies

SCIENTIFIC NAME	COMMON NAME	GROVES (2001)[1]	GRUBB ET AL. (2003)[1]	DISCREPANCIES[2]
Allenopithecus nigroviridis	Allen's swamp monkey	x	x	
Miopithecus talapoin	Angolan talapoin	x	x	
Miopithecus ogouensis	Gabon talapoin	x	x	
Erythrocebus patas	Patas monkey	x	x	
Cercopithecus aethiops superspecies			X	Groves (2001): *Chlorocebus* (six species)
C. aethiops	Vervet/grivet/green monkey	x	x	*Ch. aethiops* (Grivet)
		x		*Ch. cynosuros* (Malbrouck)
		x		*Ch. djamdjamensis* (Bale Mt. vervet)
		x		*Ch. pygerythrus* (Vervet)
		x		*Ch. sabaeus* (Green monkey)
		x		*Ch. tantalus* (Tantalus monkey)
C. cephus superspecies		X	X	
C. ascanius	Red-tailed monkey	x	x	
C. cephus	Mustached guenon	x	x	
C. erythrogaster	White-throated guenon	x	x	
C. erythrotis	Red-eared guenon	x	x	
C. petuarista	Lesser spot-nosed guenon	x	x	
C. sclateri	Sclater's guenon	x	x	
C. signatus		x		Grubb et al. (2003): not recognized as valid species
C. diana superspecies		X	X	
C. diana	Diana monkey	x	x	
C. roloway	Roloway monkey	x	x	
C. dryas superspecies		X	X	
C. dryas	Dryas guenon	x	x	
C. hamlyni superspecies		X	X	
C. hamlyni-	Owl-faced guenon	x	x	
C. l'hoesti superspecies		X	X	Grubb et al. (2003): *C. pruessi* superspecies
C. lhoesti	L'Hoest's monkey	x	x	
C. preussi	Preuss's monkey	x	x	
C. solatus	Sun-tailed guenon	x	x	
C. mitis superspecies		X	X	Grubb et al. (2003): *C. nictitans* superspecies
C. albogularis	Sykes monkey	x		Grubb et al. (2003): *C. mitis* subspecies
C. doggetti	Silver monkey	x		Grubb et al. (2003): *C. mitis* subspecies
C. kandti	Golden monkey	x		Grubb et al. (2003): *C. mitis* subspecies
C. mitis	Blue monkey	x	x	
C. nictitans	Putty-nosed/greater spot-nosed guenon	x	x	
C. mona superspecies		X	X	
C. campbelli	Campbell's guenon	x	x	
C. denti	Dent's guenon	x		Grubb et al. (2003): *C. pogonias* subspecies
C. lowei	Lowe's guenon	x		Grubb et al. (2003): *C. campbelli* subspecies
C. mona	Mona monkey	x	x	
C. pogonias	Crowned guenon	x	x	
C. wolfi	Wolf's guenon	x		Grubb et al. (2003): *C. pogonias* subspecies
C. neglectus superspecies		X	X	
C. neglectus	De Brazza's monkey	x	x	
Total *Cercopithecus* species		25	20	
Total guenon species		36	24	

[1] An *x* indicates that taxonomy includes the species (or subspecies) listed. An *X* indicates the taxonomy includes the superspecies group listed.
[2] Indicates the differences between the two taxonomies and where the discrepancy lies. For example, Grubb et al. (2003) lists one species (*C. aethiops*) in the *C. aethiops* superspecies group, with six subspecies. Groves (2001) lists six species of *Chlorocebus*.

Table 15.2   Guenon Distributions

SPECIES	DISTRIBUTION
Cercopithecus aethiops	Widespread through sub-Saharan Africa, except Namib Desert in South Africa and dense rain forests of West Africa
C. ascanius	Central Africa: from Angola in the southwest to Central African Republic in the north to Uganda and Zambia in the east
C. campbelli	Coastal northwest Africa from Gambia to Ghana
C. cephus	West Africa: Cameroon in the north to Angola in the south
C. diana	Coastal northwest Africa: Sierra Leone to Ghana
C. dryas	Central Africa: northern Democratic Republic of Congo
C. erythrogaster	West Africa: Benin and Nigeria
C. erythrotis	West Africa: Nigeria, Cameroon, Equatorial Guinea
C. hamlyni	Central Africa: Democratic Republic of Congo and Rwanda
C. lhoesti	Central/East Africa: Democratic Republic of Congo, Rwanda, Uganda, Burundi
C. mitis	Widespread through central and Eastern Africa, Ethiopia in the north, northeast coast of South Africa in the south and Angola in the west
C. mona	Coastal West Africa: Ghana to Cameroon
C. neglectus	Central Africa: Cameroon in the northwest to Uganda and Kenya in the northeast to Angola in the south; small population in Sudan
C. nictitans	West Africa: Libera to Ivory Coast and Nigeria to Angola
C. petaurista	Coastal West Africa: Gambia to Benin
C. pogonias	West Africa: Nigeria to Gabon to Central African Republic
C. preussi	West Africa: Cameroon and Equatorial Guinea
C. sclateri	West Africa: southern Nigeria
C. solatus	West Africa: Gabon
C. wolfi	Central Africa: Democratic Republic of Congo to Uganda
Allenopithecus nigroviridis	Central Africa: Democratic Republic of Congo to Angola
Miopithecus talapoin	West Africa: Angola and Democratic Republic of Congo in south and Cameroon and Gabon in north
Erythrocebus patas	Band just below Sahara: Senegal and Guinea in west to Ethiopia, Kenya, and Tanzania in east

Source: Wolfheim (1983).

two species, *M. talapoin* in the northern part of its range and *M. ogouensis* in the south (Kingdon 1997, Groves 2001, Grubb et al. 2003). In this chapter, we follow the most recent taxonomy of Grubb et al. (2003) (Table 15.1).

All guenons are endemic to sub-Saharan Africa (Table 15.2). Within the genus *Cercopithecus*, the *C. aethiops* superspecies group is by far the most widespread, being found throughout sub-Saharan Africa, except in the Namib Desert in southern Africa and the dense rain forests of the West African nations of Nigeria, Cameroon, Gabon, Congo, and the Democratic Republic of Congo. The rest of the *Cercopithecus* species are found primarily in central and western Africa. The distributions of a few species, including *C. lhoesti*, *C. mitis*, and *C. neglectus*, extend as far east as Rwanda, Uganda, and Kenya. *Allenopithecus nigroviridis* is found in central Africa, from Angola in the southwest to the Democratic Republic of Congo to the northeast of its distribution. *E. patas* is found in a relatively narrow band just south of the Sahara Desert, from Senegal and Guinea in West Africa to Ethiopia, Kenya, and Tanzania in East Africa. *Miopithecus* is found in West Africa, with *M. talapoin* distributed in Cameroon and Gabon in the north and *M. ogouensis* in Angola and the Democratic Republic of Congo in the south.

ECOLOGY

Habitat

All members of the genus *Cercopithecus*, except the vervet/grivet group, occupy varying forest habitats, from primary, secondary, and gallery rain forest (e.g., *C. ascanius* and *C. pogonias*) to bamboo forest (e.g., *C. hamlyni* and *C. neglectus*) to flooded and swamp forest (e.g., *C. ascanius*, *C. neglectus*, and *C. wolfi*) (Table 15.3). Members of the *C. aethiops* superspecies group prefer closed savanna woodland habitat to primary and secondary rain forest, in contrast to other members of their genus, and sleep in trees along rivers or other water courses but also forage away from rivers in closed woodland during the day. Like most members of the *Cercopithecus* genus, *A. nigroviridis* and *M. talapoin* occupy forest habitats. *Miopithecus* is found in a variety of forest types, from primary and secondary forest to swamp forest. *A. nigroviridis* in particular prefer primary lowland swamp forest, which is a likely contributor to the relative lack of information about this species in the wild. *E. patas* differ from the forest guenons, *A. nigroviridis* and *Miopithecus*, but are similar to the *C. aethiops* superspecies group in their preference for more open habitats. *E. patas*

Table 15.3   Guenon Habitats

SPECIES	HABITAT	REFERENCE
Cercopithecus aethiops	Closed riparian savanna woodland	Struhsaker 1967a, Nakagawa 1999, Enstam and Isbell 2002
C. ascanius	Moist deciduous forest	Cords 1986
	Swamp forest	Zeeve 1991
C. campbelli	Primary, secondary swamp forest	Gautier-Hion et al. 1981, Zuberbühler 2001
C. cephus	Primary tropical rain forest	Fleury and Gautier-Hion 1997
C. diana	Tropical, moist forest	Wolters and Zuberbühler 2003
	Secondary forest	Whitesides 1989, Oates and Whitesides 1990
C. dryas	Tropical rain forest	Colyn et al. 1991
C. erythrogaster	Lowland rain forest	Oates 1985
C. erythrotis	Lowland, tropical forest	Struhsaker 2000
C. hamlyni	Bamboo forest, lowland forest	Wolfheim 1983
C. lhoesti	Tropical montane forest, mature lowland rain forest	Struhsaker 1969, Kaplin 2001, Kaplin and Moermond 2000
C. mitis	Moist, semideciduous forest, evergreen rain forest	Cords 1986, Butynski 1990
	Tropical montane forest	Kaplin 2001
C. mona	Mangrove swamp forest	Struhsaker 1969
C. neglectus	Swampy, flooded areas of riverine forest	Wahome et al. 1993
	Montane forest	Zeeve 1991
C. nictitans	Primary tropical rain forest	Fleury and Gautier-Hion 1997
C. petaurista	Primary rain forest	Zuberbühler et al. 1999
C. pogonias	Primary tropical rain forest	Fleury and Gautier-Hion 1997
C. preussi	Primary and secondary forest	Fleury and Gautier-Hion 1997
C. sclateri	Moist, tropical, swamp, riverine forest	Oates et al. 1990
C. solatus	Primary tropical rain forest	Fleury and Gautier-Hion 1997
C. wolfi	Primary, secondary lowland rain forest, swamp forest	Zeeve 1991
Allenopithecus nigroviridis	Primary lowland riverine swamp forest	Gautier 1985, Zeeve 1991
Miopithecus talapoin	Inundated primary riverine forest	Gautier-Hion 1971, 1973
Erythrocebus patas	Open woodland, wooded savanna	Hall 1965, Nakagawa 1999, Chism and Rowell 1988, Enstam and Isbell 2002

occupy non-riverine grassland and open woodland in both western (Cameroon, Nakagawa 1999) and eastern (Kenya, Chism and Rowell 1988, Enstam and Isbell 2002) Africa. *E. patas* and *C. aethiops* avoid dense rainforest habitat (although *C. aethiops* in South Africa can be found in forest habitats [e.g., van der Zee and Skinner 1977]), unlike all other members of the Cercopithecini tribe.

Diet

The guenons are members of the subfamily Cercopithecinae (the cheek-pouch monkeys), which also includes the tribe Papionini (*Papio* [baboons], *Cercocebus* [mangabeys], *Lophocebus* [mangabeys], and *Macaca* [macaques]) (Grubb et al. 2003). Like the other members of the Cercopithecinae, the guenons can be distinguished from the Colobinae (the leaf-eating monkeys) by morphological adaptations to a frugivorous diet (e.g., low and rounded molar cusps, simple stomachs, pouches in their cheeks for storing food) (Fleagle 1999). For the forest guenons (including *A. nigroviridis* and *Miopithecus*), the majority of the diet is composed of fruit

(range 24.5%–91%); but the guenons do display diverse dietary preferences, consuming leaves, flowers, vertebrates, and invertebrates to varying degrees (Table 15.4), even within species. Dietary data for *C. mitis*, for example, are given for seven different sites; and the amount of fruit in the blue monkeys' diet at these sites is highly variable, accounting for the entire range of fruit composition in forest guenon diets stated above (24.5%–91%). Similarly, the composition of fruit in the diets of *C. ascanius* and *C. cephus* across several sites ranges 29.5%–57% and 49%–78%, respectively.

Despite being highly frugivorous, forest guenons also feed on leaves; and in some populations of some species, leaves can make up as much as a third of the diet (e.g, *C. preussi* 41%, Beeson et al. 1996; *C. lhoesti* 35%, Kaplin and Moermond 2000). In only one species, however, do leaves compose over 50% of the diet (e.g., *C. ascanius* in Budongo, up to 74%, Sheppard 2000). Mature leaves are more difficult to digest than immature leaves because of their higher levels of cellulose and potentially higher levels of toxic secondary compounds (Richard 1985, Strier 2003), and since guenons do not possess adaptations for digesting leaves, when leaves

Table 15.4 Guenon Diets

SPECIES	STUDY SITE	LEAVES	SEEDS	FRUIT	FLOWERS	INSECTS	GUM	OTHER	REFERENCE
Cercopithecus aethiops	Amboseli National Park, Kenya (n = 3)[1]	26.60%	2.60%	11.10%	14.30%	7.70%	30.00%	0.2% bark	Wrangham and Waterman 1981
	Saumburu-Isiolo Reserve, Kenya (n = 1)	0.00%	19.60%	5.80%	44.70%	NA	NA	8.3% grass, 1.3% swollen thorns	Whitten 1983
	Segera Ranch, Kenya (n = 1)	2.00%	8.00%	10.00%	8.00%	23.00%	37.00%	2% swollen thorns 8% grass	Isbell et al. 1998
C. ascanius	Kakamega, Kenya (n = 1)	7.20%	0.40%	61.30%	2.00%	25.10%	2.80%	0.7% stems, shoots 0.5% unidentified	Cords 1986
	Kibale, Uganda (n = 5)	21.20% (range 13%–35%)	0.20% (range 0%–1%)	48.20% (range 36%–60%)	8.40% (range 3%–15%)	21.40% (range 15%–31%)	NA	NA	Struhsaker 1978, Chapman et al. 2002
	Budongo, Uganda (n = 2)	58.00% (range 42%–74%)	0.00%	29.50% (range 13%–46%)	7.00% (range 5%–9%)	NA	NA	NA	Sheppard 2000
C. campbelli	Bia National Park, Ghana (n = 1)	9.6%[3]	12.2%[3]	22.6%[3]	7.5%[3]	NA	NA	NA	Lambert 2002
	Lomako, Zaire (n = NA)	32%	5.70%	39.60%	0.00%	22.60%	0.00%	NA	Zeeve 1991
	Not specified (n = NA)	1.00%	Combined with Fruit category	78.00%	0.00%	15.00%	NA	NA	Gautier-Hion 1988a
C. cephus	Makokou, Gabon (n = 1)	6.00%	0.00%	78.00%	1.00%	13.00%	NA	NA	Gautier-Hion 1980
	Lope, Gabon (n = 1)	11.00%	7.00%	67.00%	6.00%	9.00%	NA	0.6% other	Tutin et al. 1997
C. diana	Bia National Park, Ghana (n = 1)	8.00%	24.30%	36.10%	5.90%	25.20%	NA	0.6% other	Curtin 2002
C. dryas	No data[2]								
C. erythrogaster	No data								
C. erythrotis	No data								
C. hamlyni	No data								
C. lhoesti	Nyungwe Forest, Rwanda (n = 1)	35.20%	17.70%	24.50%	4.00%	8.80%	NA	9.8% unspecified	Kaplin and Moermond 2000, Kaplin 2001
C. mitis	Kakamega, Kenya (n = 1)	18.90%	2.50%	54.60%	3.70%	16.80%	NA	1.1% stems, shoots 0.5% unidentified	Cords 1986
	Nyungwe Forest, Rwanda (n = 1)	6.20%	9.30%	47.40%	6.20%	24.90%	NA	6.2% unspecified	Kaplin and Moermond 2000, Kaplin 2001
	Kibale, Uganda (n = 2)	0.00%	20.5%[3]	43.5%[3]	NA	NA	1.9% stems, shoots	0.5% unidentified	Lambert 2002
	Cape Vidal, South Africa (n = 1)	26.00%	0.00%	57.00%	13.00%	6.00%	NA	NA	Lawes 1991
	Zomba Plateau (n = 1)	32.60%	53.50%	10.20%	1.00%	1.00%	NA	2.9% other plant material	Beeson et al. 1996
	Ngoye Forest, South Africa (n = 1)	3.00%	0.00%	91.00%	2.00%	0.00%	NA	NA	Lawes et al. 1990
	Kanwayara, Kibale (high biomass) (n = 4)	33.00% (range 22.4%–35.4%)	NA	27.70% (range 15%–28.9%)	6.90% (range 2.9%–7.8%)	37.70% (range 35.1%–45.4%)	NA	0.6% unspecified (range 0.4%–1.7%)	Butynski 1990
	Ngogo, Kibale (low biomass) (n = 1)	22.80%	NA	30.10%	9.80%	35.90%	NA	1.3% unspecified	Butynski 1990

Species	Study site[1]								Reference
C. mona	No data								
C. neglectus	Mpassa, Gabon (n = 1)	9.10%	Combined with Fruit category	74.40%	3.00%	NA	NA	3.7% soil	Gautier-Hion and Gautier 1978
C. nictitans	Lomako, Zaire (n = NA)	0.00%	0.00%	33.30%	0.00%	0.00%	66.70%		Zeeve 1991
	Makokou, Gabon (n = 1)	17.00%	0.00%	71.00%	1.00%	9.00%	NA		Gautier-Hion 1980
	Lope (continuous), Gabon (n = 1)	16.00%	11.00%	59.00%	3.00%	10.00%	NA		Tutin et al. 1997
	Lope (fragment), Gabon (n = 1)	17.00%	4.00%	44.00%	9.00%	24.00%	NA		Tutin et al. 1997
	Makande, Gabon (n = 1)	10.30%	50.20%	35.50%	4.10%	NA	NA		Brugiere et al. 2002
C. petaurista	Taï, Ivory Coast (n = NA)	6.00%	Combined with Fruit category	77.00%	NA	7.00%	NA		Gautier-Hion 1988a
C. pogonias	Makokou, Gabon (n = 1)	1.00%	3.00%	80.00%	1.00%	16.00%	NA		Gautier-Hion 1980
	Lope Reserve, Gabon (n = 1)	7.00%	9.00%	69.00%	9.00%	7.00%	NA		Tutin et al. 1997
	Makande, Gabon (n = 1)	12.60%	49.80%	26.90%	4.70%	NA	NA		Brugiere et al. 2002
C. preussi	Kilum (n = 1)	41.40%	Combined with Fruit category	51.90%	1.20%	1.00%	NA	4.6% other plant material	Beeson et al. 1996
C. sclateri	No data								
C. solatus	No data								
C. wolfi	Salonga (n = 1)	30.00%	27.00%	32.00%	11.00%	NA	NA		Gautier-Hion et al. 1983
	Lomako, Zaire (n = NA)	26.70%	0.00%	46.70%	6.70%	20.00%	0.00%		Zeeve 1991
Allenopithecus nigroviridis	Zaire (n = NA)	<5%	Combined with Fruit category	81.00%	NA	18.00%	NA		Gautier-Hion 1988a
	Lomako, Zaire (n = NA)	0.00%	0.00%	40.00%	20.00%	0.00%	6.70%	13.3% other plant material	Zeeve 1991
Miopithecus talapoin	Not specified (n = NA)	<5%	Combined with Fruit category	43.00%	35.00%	35.00%	NA		Gautier-Hion 1988a
Erythrocebus patas	Segera Ranch, Kenya (n = 1)	2.00%	Included in Seeds category	10.00%	7.00%	35.00%	37.00%	1% vertebrates	Isbell 1998
	Kala Maloue, Cameroon (n = 1)	6.00%	0.00%	0.00%	65.00%	12.00%	7.00%		Nakagawa 1989

Data for each food type are provided as percent of the total diet. Note that different authors may have used different methods, so the data may not be comparable. NA, authors did not provide data for that food type.

[1] "n" refers to the number of study groups at the given study site for which data were provided. If data were provided for multiple groups, the percentages indicate the average percent of the diet across all study groups, and the range is given. If the author(s) did not indicate the number of study groups on which the data are based, "n = NA" follows the study site.

[2] "No data" indicates that dietary data are not available for that species.

[3] For C. mitis and C. ascanius at Kibale, data by Lambert (2002) are provided only as percent of plant diet.

are eaten, young, immature leaves are preferred (see also Butynski 1988). For example, in a study of *C. mitis* and *C. ascanius* in Kakamega Forest, Kenya, Cords (1987a) found that leaves made up almost 23% and 10% of the plant diet, respectively. However, the majority of the feeding scores for both species were on young leaves (*C. mitis* 20% young, 2.7% mature; *C. ascanius* 9.1% young, 0.5% mature). Similar results have also been reported for two guenon species in Gabon (*C. nictitans* 10.2% young, 0.1% mature; *C. pogonias* 12.5% young, 0.1% mature, Brugiere et al. 2002). Unusually for guenons, seeds comprise the majority of the diet of *C. nictitans* (50.2%) and *C. pogonias* (49.8%) in Makande, Gabon (Brugiere et al. 2002). Although seeds can be difficult for guenons to digest because they can contain toxic compounds (Janson and Chapman 1999), irregular and recurrent food shortages due to a lack of fruit-bearing trees at Makande means *C. nictitans* and *C. pogonias* must sometimes eat less desirable foods (i.e., seeds) (Brugiere et al. 2002).

More than half of the forest guenons (11 of 21, including *A. nigroviridis* and *Miopithecus*) have been observed feeding on arthropods or other invertebrates (Table 15.4; see also Butynski 1982). In some cases, insects make up a large proportion of the diet (e.g., *C. mitis* at Kibale 35.1%–45.4%, *C. cephus* at Lope 35%). The diet of *Miopithecus* consists primarily of fruit and seeds (43%), but a third comes from arthropods (Gautier-Hion 1988a). The high proportion of arthropods in the *Miopithecus* diet is not surprising considering their small body size.

Few studies provide detailed information on the arthropod foraging and capture methods of guenons. However, Cords (1986) found that *C. mitis* and *C. ascanius* at Kakamega Forest, Kenya, capture the majority of their arthropod prey from the surface of mature leaves (61.9% of captures for *C. mitis*, 60.1% of captures for *C. ascanius*). The next most common substrate for insect capture in both species was tree trunks or branches (10.1% of captures for *C. mitis*, 13.2% of captures for *C. ascanius*). Cords (1986) documented 11 capture techniques, all of which were used by both species, to varying degrees. These techniques vary in *(1)* the speed of capture, *(2)* whether the prey item is ingested with or without handling, and *(3)* whether the monkey holds or braces the substrate containing the prey item. Both species used an intermediate speed to capture prey, although blue monkeys preferred to hold the substrate containing the prey item and ingest the insect directly from the substrate without handling it (42.4% of captures). *C. ascanius* also used this method of capture (29.1% of captures) but used their hands to "pick" insect prey from the substrate almost as frequently (28.1% of captures).

Unlike *C. mitis* and *C. ascanius*, which actively hunt for their arthropod prey, the primary arthropod prey of *E. patas* in Kenya is colonially living ants of the genus *Crematogaster*, which *E. patas* obtain by biting open swollen thorns of *Acacia drepanolobium* trees (Isbell 1998). Holding on to the swollen thorn or branch with their hands, *E. patas* extract adult and larval ants with their tongues (K. L. Enstam, personal observation). Since the ants defend their nests and the tree by biting (Young et al. 1997), *E. patas* usually open only one or two swollen thorns per tree (97% of feeding observations) before moving on to the next one (Isbell 1998). *E. patas* will occasionally eat other invertebrates, especially grasshoppers, which they usually capture in the grass, sometimes after chasing them (K. L. Enstam, personal observation).

Gum of *A. drepanolobium* also composes a high proportion (37%) of the *E. patas* diet in Kenya (Isbell 1998). For both *E. patas* and *C. aethiops*, fruit makes up a very small percentage of the diet (10% or less in both species: patas, Nakagawa 1989, Isbell 1998; vervets, Whitten 1983, Isbell et al. 1998), in stark contrast to most of the forest guenons (see Table 15.4). *Acacia* trees are also important food sources for *C. aethiops*. Leaves, seeds, fruits, flowers, gum, and bark of *A. xanthophloea* and *A. tortilis* account for as much as 75% of the diet of *C. aethiops* in Amboseli National Park, Kenya; and *Acacia* gum alone accounts for 30% (Wrangham and Waterman 1981). Swollen thorns, seeds, flowers, leaves, and gum from *A. xanthophloea* and *A. drepanolobium* trees account for 57.2% of the diet of *C. aethiops* at Segera Ranch, Kenya (Pruetz and Isbell 2000). *C. aethiops* at the Samburu-Isiolo Reserve in Kenya also rely primarily on *Acacia* trees, but the majority of their diet (44%) comes from *Acacia* flowers (Whitten 1983).

Ranging

Previous research has shown a link between diet and ranging behavior, particularly home range size and daily travel distance. Frugivores tend to have larger home ranges and longer day ranges than folivores with the same group biomass (Clutton-Brock and Harvey 1977), perhaps because their preferred foods are more patchily distributed (Richard 1985). However, Kaplin (2001) found the opposite to be true of *C. lhoesti* and *C. mitis*. In her study, *C. lhoesti* ate greater amounts of leaves than *C. mitis*, which incorporated more fruit and insects into their diet. However, the more folivorous *C. lhoesti* had larger home ranges and longer day ranges than *C. mitis*. Kaplin (2001) suggests that *C. lhoesti* travel farther because the terrestrial herbs on which they feed are not as densely distributed as other types of foliage. This suggests that resource abundance and distribution may be more central than food type in determining home range size and day range length (see also Isbell 1991).

Ranging behavior varies among the guenons. Home range size varies from as little as 4 ha in *C. neglectus* (Wahome et al. 1993) to as much as 4,000 ha in *E. patas* (Chism and Rowell 1988, Isbell unpublished data), which have the largest home ranges for their group biomass of any primate (Clutton-Brock and Harvey 1977, Isbell et al. 1998) (Table 15.5). Daily travel distance follows a similar pattern, being smallest in *C. neglectus* (530 m; Gautier-Hion and Gautier 1978) and largest in *E. patas* (2,250–4,025 m; Chism and Rowell 1988) (Table 15.5).

Table 15.5 Guenon Home Range Sizes and Day Range Lengths

SPECIES	STUDY SITE	AVERAGE GROUP SIZE	HOME RANGE SIZE (HA)	DAY RANGE LENGTH (M)	RANGE DEFENSE?	INTERGROUP AGGRESSION?	SOURCE
Cercopithecus aethiops	Segera Ranch, Kenya	8	10	1,025[1]	Yes	Yes	Isbell unpublished, Isbell et al. 1999
	Segera Ranch, Kenya	27	40	1,632[1]	Yes	Yes	Isbell unpublished, Isbell et al. 1999
	Amboseli, Kenya	7.3	34	NA	Yes	Yes	Struhsaker 1967b
	Amboseli, Kenya	16.4	96	NA	Yes	Yes	Struhsaker 1967b
	Amboseli, Kenya	17.2	19	NA	Yes	Yes	Struhsaker 1967b
	Amboseli, Kenya	49.9	34	NA	Yes	Yes	Struhsaker 1967b
C. ascanius	Kibale, Uganda	15	28	1,595	Yes	Yes	Struhsaker and Leland 1988
	Kibale, Uganda	45–50	68	1,198	Yes	Yes	Struhsaker and Leland 1988
	Kakamega, Kenya	25	60	NA	Yes	Yes	Cords 1987a
C. campbelli	No data						
C. cephus	Makokou, Gabon	4	18	800	NA	NA	Gautier-Hion and Gautier 1974
	Makokou, Gabon	8	45	NA	NA	NA	Gautier-Hion and Gautier 1974
C. diana	Sierra Leone	20	41	1,019	NA	NA	Whitesides 1989
	Sierra Leone	27	29	1,513	NA	NA	Whitesides 1989
C. dryas	No data						
C. erythrogaster	No data						
C. erythrotis	No data						
C. hamlyni	No data						
C. lhoesti	Nyungwe Forest, Rwanda	29	50[2]	2,092	NA	NA	Kaplin 2001
C. mitis	Ngogo (Kibale) Uganda	18	335	1,406	No	No	Butynski 1990
	Kanwayara (Kibale) Uganda	17	44	1,300	Yes	Yes	Butynski 1990
	Kakamega, Kenya	45	38	NA	Yes	Yes	Cords 1987a
	Nyungwe Forest, Rwanda	27	26[2]	1,307	NE[3]	NE	Kaplin 2001
C. mona	Cameroon, West Africa	8.8	NA	NA	NA	NA	Struhsaker 1969
	Cameroon, West Africa	10.8	NA	NA	NA	NA	Struhsaker 1969
C. neglectus	Kisere Forest, Kenya	11	≥5.0	NA	No	NA	Wahome et al. 1993
	Kisere Forest, Kenya	13	4	NA	No	NA	Wahome et al. 1993
	Makokou, Gabon	3–4	4–10	530	No	No	Gautier-Hion and Gautier 1978
C. nictitans	Makokou, Gabon	38[4]	148	1,825	NA	NA	Gautier-Hion et al. 1983
	Makokou, Gabon	53[5]	119	1,980	NA	NA	Gautier-Hion et al. 1983
C. petaurista	No data						
C. pogonias	Makokou, Gabon	38[4]	148	1,825	NA	NA	Gautier-Hion et al. 1983
	Makokou, Gabon	53[5]	119	1,980	NA	NA	Gautier-Hion et al. 1983
C. sclateri	No data						
C. solatus	No data						
C. wolfi	No data						
Allenopithecus nigroviridis	No data						
Miopithecus talapoin	Makakou, Gabon	64	140	2,323	NA	NA	Gautier-Hion 1971
Erythrocebus patas	Mutara, Kenya	16	2,340	3,830	No	Yes	Chism and Rowell 1988
	Mutara, Kenya	47	3,200	4,220	No	Yes	Chism and Rowell 1988
	Segera, Kenya	45	4,000	3,188[1]	No	Yes	Isbell unpublished, Isbell et al. 1999
	Murchison Falls, Uganda	31	5,200	2,250	No	Yes	Hall 1965

[1] Day range length data based on Isbell et al. (1999) were calculated by multiplying "straight-line distance of groups" (per 30 min) for each group by 22 to obtain an estimate of day range length for an 11 h day.
[2] Home range size is an average of two methods, the minimum convex polygon method and grid-cell analysis.
[3] NE, no encounters, indicating that the researcher did not observer encounters between conspecific groups.
[4] Data on group size, home range size, and day range length for *C. nictitans* and *C. pogonias* are given for a mixed-species group containing 20 *C. nictitans* individuals and 18 *C. pogonias* individuals. No data on *C. nictitans* or *C. pogonias* in a monospecific group were provided.
[5] Data on group size, home range size, and day range length for *C. nictitans* and *C. pogonias* were given for a mixed-species group containing 20 *C. nictitans* individuals, 18 *C. pogonias* individuals, and 15 *C. cephus* individuals. No data on *C. nictitans* or *C. pogonias* in a monospecific group were provided.

Table 15.6   Guenon Predators

SPECIES	BODY WT. (KG)	SOURCE	CONFIRMED[1]	SOURCE	POTENTIAL[2]	SOURCE
Cercopithecus aethiops	m: 5.6, f: 2.79	Haltenorth and Diller 1988, Gevaerts 1992	Leopard, martial eagle, python, yellow baboon, black eagles	Seyfarth et al. 1980, Hausftater 1975, Boshoff et al. 1991	Lion, spotted hyena, African wild cat, serval, black-backed jackal, cheetah, caracal	Isbell and Enstam 2002
C. ascanius	m: 3.7, f: 2.8	Gevaerts 1992, Colyn 1994	Crowned eagle, chimpanzee	Cords 1987a,b, 1990; Struhsaker and Leakey 1990; Skorupa 1989; Sanders et al. 2003; Wrangham and Riss 1990; Mitani and Watts 1999; Mitani et al. 2001		
C. campbelli	m: 4.5, f: 2.7	Oates et al. 1990	Leopard, crowned eagle, chimpanzee	Hoppe-Dominik 1984, Shultz et al. 2004, Shultz 2001, Zuberbühler and Jenny 2002, Zuberbühler 2001		
C. cephus	m: 4.0, f: 2.9	Gautier-Hion and Gautier 1976	Crowned eagle, human	Gautier-Hion et al. 1983, Gautier-Hion and Tutin 1988	Python, golden cat, leopard	Gautier-Hion et al. 1983
C. diana	m: 5.2, f: 3.9	Oates et al. 1990	Chimpanzee, human, leopard, crowned eagle	Zuberbühler et al. 1999, Zuberbühler and Jenny 2002, Shultz 2001, Shultz et al. 2004, Hoppe-Dominik 1984, Zuberbühler et al. 1997		
C. lhoesti	m: 6, f: 3.5	Gevaerts 1992, Colyn 1994	Crowned eagle	Struhsaker and Leakey 1990, Sanders et al. 2003, Skorupa 1989, Mitani et al. 2001	Leopard, golden cat, python	Haltenorth and Diller 1988
C. mitis	m: 5.9–8.9, f: 3.83	Napier 1981, Gevaerts 1992	Crowned eagle, chimpanzee, human	Cords 1987a,b, 1990, 2002a; Struhsaker and Leakey 1990; Skorupa 1989; Wrangham and Riss 1990; Mitani and Watts 1999; Struhsaker 2000		
C. neglectus	m: 7.0, f: 3.5	Gautier-Hion and Gautier 1976, Gevaerts 1992, Colyn 1994	Crowned eagle	Wahome et al. 1993	Leopard, golden cat, python	Haltenorth and Diller 1988, Wahome et al. 1993
C. nictitans	m: 6.6, f: 3.65–4.2, f: 3.65–4.2	Gautier-Hion and Gautier 1976, Gevaerts 1992	Crowned eagle, human	Gautier-Hion et al. 1983, Zuberbühler and Jenny 2002, Gautier-Hion and Tutin 1988	Python, golden cat, leopard	Gautier-Hion et al. 1983
C. petaurista	m: 3.8, f: 2.76–2.9	Oates et al. 1990	Leopard, crowned eagle, chimpanzee	Hoppe-Dominik 1984, Zuberbühler and Jenny 2002, Shultz et al. 2004		
C. pogonias	m: 4.5, f: 3.0	Gautier-Hion and Gautier 1976	Crowned eagle, human	Gautier-Hion et al. 1983, Gautier-Hion and Tutin 1988	Python, golden cat, leopard	Gautier-Hion et al. 1983
C. wolfi	m: 3.8, f: 2.76–2.9	Gevaerts 1992, Colyn 1994	Crowned eagle	Zeeve 1991	Leopard	Zeeve 1991
Erythrocebus patas	m: 10.25, f: 5.75	Haltenorth and Diller 1988	Black-backed jackal, domestic dog	Chism and Rowell 1988, Isbell, unpub. data	Leopard, serval, caracal, African wild cat, lion, spotted hyena, martial eagle, chimpanzee (in West Africa), wild dog, baboon	Chism and Rowell 1988, Isbell and Enstam 2002, Isbell unpub., Chism et al. 1983
Miopithecus talapoin	m: 1.4, f: 1.1	Gautier-Hion and Gautier 1976 Gautier-Hion 1971			Leopard, golden cat, genet, nile monitor, crowned eagle	Haltenorth and Diller 1988

[1] "Confirmed" predators include species that have been observed preying on a specific guenon species, whether the attack was successful or not. Confirmed predators also include species that have left remains of monkeys in their nests or dung.
[2] "Potential" predators are species that researchers have listed as possible predators but have not been observed attacking or attempting to attack the guenon species in question. In general, potential predators are those that co-occur with guenons and are known to take prey of equal or greater size than a guenon, even if they have not been observed preying on guenons.

Data on daily ranging behavior other than daily travel distance are sparse. Cords (1987a) found that *C. mitis* and *C. ascanius* tend not to cross over their path during their daily travels. *C. mitis* tend to move in rather circular courses, covering about one-third of their home range per day. *C. ascanius* also traveled in a circular motion but in a more linear fashion (Cords 1987a). *C. ascanius* covered less of their home range (about one-fourth) than *C. mitis* per day.

E. patas have a more meandering, circuitous ranging pattern, and often recross portions of their path during a single day of ranging (Isbell et al. 1999). *E. patas* ranging behavior also appears to be unique among guenons in that they seem to have a number of "sub-home ranges" in which they forage for various lengths of time (K. L. Enstam, unpublished data). Compared to *E. patas*, *C. aethiops* tend to follow a straighter daily path (Isbell et al. 1999).

Predators

Guenons are susceptible to predation from a large number of mammalian, reptilian, and aerial predators. Confirmed and potential predators of various guenon species are listed in Table 15.6. Although data on predators of guenons are sparse, crowned eagles (*Stephanoaetus coronatus*) appear to be the most pervasive predator species of forest guenons (see Struhsaker and Leakey 1990). Crowned eagles have been documented as predators of eight species (Table 15.6). Among mammals, confirmed predators include chimpanzees (*Pan troglodytes*), leopards (*Panthera pardus*), and humans (*Homo sapiens*). Potential predators of forest guenons include African rock pythons (*Python sebae*) and golden cats (*Profelis aurata*). Given their small size, *Miopithecus* are presumably susceptible to a fairly large number of predators, and potential predators include *S. coronatus*, *P. pardus*, *Profelis aurata*, domestic dogs (*Canis familiaris*), *Python sebae* (Wahome et al. 1993), and Nile monitors (*Varanus niloticus*) (Gautier-Hion et al. 1983).

Guenons living in more open habitats tend to be preyed upon by a different set of predators. Confirmed predators of *E. patas* include black-backed jackals (*Canis mesomelas*) and *C. familiaris* (Chism and Rowell 1988, Enstam and Isbell 2002). *P. pardus* have not been observed to attack *E. patas* but are strongly suspected in overnight disappearances (Chism et al. 1983, L. A. Isbell unpublished data). Lions (*Panthera leo*), spotted hyenas (*Crocuta crocuta*), wild dogs (*Lycaon pictus*), martial eagles (*Polemaetus bellicosus*), olive baboons (*Papio anubis*), and in West Africa *Pan troglodytes* are all potential predators (Enstam and Isbell 2002). Known predators of *C. aethiops* include *P. pardus*, *S. coronatus*, *Polemaetus bellicosus*, *Python sebae*, and yellow baboons (*Papio cynocephalus*) (Struhsaker 1967a, Altmann and Altmann 1970, Hausfater 1975, Seyfarth et al. 1980, Balldelou and Henzi 1992), whereas *P. leo*, *Cr. crocuta*, *Canis mesomelas*, and cheetahs (*Acinonyx jubatus*) are potential predators (Struhsaker 1967a, Seyfarth et al. 1980).

REPRODUCTIVE PARAMETERS

Age at Sexual Maturity

Within the genus *Cercopithecus*, age at sexual maturity averages 64.4 months for males (range 48 months in *C. campbelli* [Hunkeler et al. 1972] to 72 months in *C. ascanius* [M. Cords unpublished data in Cords 1987b]) and 47.6 months for females (range 42 months in *C. neglectus* [Gautier-Hion and Gautier 1976] to 54 months in *C. ascanius* [T. T.

Struhsaker, unpublished data in Cords 1987b]) (Table 15.7). *E. patas* males and females mature faster than the forest guenons, with males maturing at 48 months and females at 30 months (Hall 1965, Chism et al. 1984). Female *M. talapoin* reach maturity at 48 months, which is comparable to females in most *Cercopithecus* species. Data on age at sexual maturity are unavailable for *A. nigroviridis*.

Age at First Birth

Age at first birth for female *Cercopithecus* species averages 51.4 months (range 36 months in *C. campbelli* [Bourlière et al. 1970] and *E. patas* [Chism et al. 1984] to 60.8 months in *C. diana* [Cheney et al. 1988]). Habitat quality is known to affect age at first birth in *C. aethiops*. At Amboseli National Park in Kenya, Cheney *et al.* (1988) found that females in a group without access to permanent water sources and associated flora gave birth for the first time later than females in two groups with access to permanent water and associated flora. Similarly, after the heavy rains of an El Niño weather event, a female in one study group gave birth for the first time at 2.5 years of age (L. A. Isbell and K. L. Enstam unpublished data), apparently because of plentiful resources. This is in contrast to the late age (6 years) at first birth of the two females that matured after the El Niño rains. *Miopithecus* females fall within the range of ages for *Cercopithecus* females, with an average age at first birth of 53.5 months (see Table 15.7). Data on age at first birth are not available for *A. nigrovirdis*.

Gestation Length

There are little data on gestation length for the guenons, but for the six species for which these data are available, average gestation length is 163.2 days (range 140 days in *C. mitis* [Rowell 1970] to 180 days in *C. campbelli* [Bourlière et al. 1970]). Gestation lengths for *E. patas* and *Miopithecus* (167 days [Sly et al. 1983] and 165 days [Gautier-Hion and Gautier 1976], respectively) are well within the range of guenons (Table 15.7). Data on gestation length are not available for *A. nigroviridis*.

Interbirth Interval

The average interbirth interval for *Cercopithecus* species is 25 months (range 16 months in *C. aethiops* to 54 months in *C. mitis*) (Table 15.7). Interbirth intervals in vervet monkeys can also be affected by availability of water and food. In Amboseli, Cheney et al. (1988) found that a *C. aethiops* group that did not have access to permanent water and associated flora had an interbirth interval of almost 2 years, significantly longer than that of the females in two groups with access to water year-round (average 16.3 and 13.8 months for the two groups). *E. patas* (Chism et al. 1984, L. A. Isbell unpublished data) have considerably shorter interbirth intervals of around 12 months. There are no data on interbirth intervals for *A. nigroviridis*.

Table 15.7 Guenon Reproductive Parameters

SPECIES	SITE	SEXUAL MATURITY (MONTHS)	SOURCE	FIRST BIRTH (MONTHS)	SOURCE	GESTATION LENGTH (DAYS)	SOURCE	INTERBIRTH INTERVAL (MONTHS)	SOURCE	MATING/BIRTH SEASON	SOURCE
Cercopithecus aethiops	Amboseli, Kenya	m: 60, f: 48	Cheney et al. 1988	60.8	Cheney et al. 1988	NA[1]		17.1	Cheney et al. 1988	MS: Apr–Oct; BS: Oct–Dec	Andelman 1987; Cheney et al. 1988
	Sumburu Isiologame reserve, Kenya	NA		NA		NA		NA		MS: Mar–Jun; BS: Oct–Jan	Whitten 1984; Whitten 1984
	Austin, Texas	NA		NA		163[2]	Bramblett et al. 1975	NA		NA	
C. ascanius	Kakamega, Kenya	m: 72	Cords unpub. data in Cords 1987b	NA		NA		>24	Cords unpub. data in Cords 1987b	Birth peak: Dec–Mar; MS: Jun–Oct	Cords and Rowell 1987; Cords 2002b
	Zaire	NA		NA		NA		52	Cords and Rowell 1987	Birth peak: Apr–Nov	Gevaerts 1992
	Kibale, Uganda	f: 48–60	Struhsaker unpub. data in Cords 1987b	NA		NA		17.8 (range 11.6–24)	Struhsaker unpub. data in Cords 1987b	Birth peak: Nov–Feb	Struhsaker unpub. data in Cords 1987b
C. campbelli	Ivory Coast	m: 48, f: 36	Hunkeler et al. 1972	36–48	Bourlière et al. 1970	180	Bourlière et al. 1970	12	Hunkeler et al. 1972	MS: Jun–Sept, BS: Nov–Jan	Bourlière et al. 1970
C. cephus	Paimpont, France	NA		60[2]	Cords 1988	NA		27.4[1]	Cords 1988	NA	
C. diana	Bia National Park, Ghana	NA		NA		NA		NA		MS: Feb–Jul, Mating peak: Jul; BS: Nov–Dec, Mar	S. Curtin pers. comm.; S. Curtin pers. comm.
C. dryas	No data[3]										
C. erythrogaster	No data										
C. erythrotis	No data										
C. hamlyni	Zaire	NA		NA		NA		NA		BS: Jun–Nov	Gevaerts 1992
C. lhoesti	Zaire	NA		NA		NA		NA		BS: Apr–Dec	Gevaerts 1992
C. mitis	Kakamega, Kenya	m: 72, f: 48–60	Cords unpub. data in Cords 1987b	NA		140[2]	Rowell 1970	47	Cords and Rowell 1987	Birth peak: Dec–Mar	Cords and Rowell 1987
C. mitis	Makere University, Uganda	NA		NA		NA		NA		NA	
C. mona	Makokou, Gabon	m: 60–72, f: 48	Gautier-Hion and Gautier 1976, 1978	54[5]		182	Haltenorth and Diller 1988	NA		NA	
C. neglectus	Kisere Reserve, Kenya	NA		NA		NA		>24	Wahome et al. 1993	Birth peak, Dec–Mar	Wahome et al. 1993
	Nairobi, Kenya	NA		NA		170[2,4]	Rowell and Richards 1979	NA		NA	

Species	Location	Age at sexual maturity	Ref.	Age at first birth	Gestation	Interbirth interval	Ref.	Birth season	Ref.
C. nictitans	Makokou, Gabon	m: 60–72, f: 48	Gautier-Hion and Gautier 1976	NA	NA	24	Butynski 1988	Birth peak: Dec–May	Butynski 1988
C. petaurista	Paimpont, France	NA		60[2]	NA	NA	Cords 1988	NA	NA
	No data								
C. pogonias	Makokou, Gabon	m: 60–72, f: 48	Gautier-Hion and Gautier 1976	NA	NA	NA		BS: Dec–Apr	Napier 1981, Cords 1988
C. pogonias	Paimpont, France	NA		60[2]	NA	NA	Cords 1988	NA	NA
C. preussi	No data								
C. sclateri	No data								
C. wolfi	Zaire	NA		NA	NA	NA		BS: Jun–Dec	Gevaerts 1992
C. solatus	No data								
Allenopithecus nigroviridis	No data								
Miopithecus talapoin	Makokou, Gabon	f: 48	Gautier-Hion and Gautier 1976	53.5[5]	167	NA	Gautier-Hion and Gautier 1976	BS: Nov–Apr	Gautier-Hion 1973
Miopithecus talapoin	Mbalmayo Reserve, Cameroon	m: 48, f: 48	Rowell and Dixson 1975	NA	NA	NA		MS: Jan–Mar; BS: Jun–Aug	Rowell and Dixson 1975
Erythrocebus patas	Mutara Ranch, Kenya	f: 30	Chism et al. 1984	36	NA	11.8	Chism et al. 1984	MS: Jun–Aug, BS: Dec–Feb	Chism et al. 1983, 1984; Chism and Rowell 1986
	Meloy Laboratories, Maryland	NA		NA	167[2]	NA	Sly et al. 1983	MS: Jul–Sep, BS: Jan–Mar	Enstam et al. 2002, Isbell unpub. data
	Murchison Falls, Uganda	m: 36[6]	Hall 1965	NA	NA	NA		BS: Dec–Feb	Hall 1965
	Waza Reserve, Cameroon	NA		NA	NA	NA		BS: Nov–Jan	Struhsaker and Gartlan 1970
	Kala Maloue, Cameroon	NA		36.5	NA	14.4	Nakagawa et al. 2003	BS: Dec–Feb	Nakagawa et al. 2003

[1] NA, data for that particular reproductive parameter are not available for that species.

[2] Data come from captive animals.

[3] "No data" indicates that there are no data on any reproductive parameters for that species.

[4] Estimated by the authors of the study.

[5] Age at first birth for de Brazza's and talapoin monkeys was estimated by the authors by combining age at sexual maturity and gestation length data.

[6] For patas males, age at sexual dispersal is used as an estimate of age at sexual maturity.

Mating/Birth Seasonality

The vast majority of guenons have distinct mating and birth seasons (Table 15.7). *C. ascanius* is the only guenon species known to give birth year-round (Butynski 1988). Even so, only 3% of births occur outside of the birth peak from April to November. Variation in birth seasons (and therefore mating seasons) tends to coincide with variation in rainfall and resource availability (Butynski 1988). That rainfall and subsequently resource availability are important correlates of guenon reproduction is highlighted by the response of vervet females after a particularly heavy rainy season, during an El Niño weather event. Normally, the birth season of *C. aethiops* on Segera Ranch, Kenya, falls between January and March (L. A. Isbell unpublished data), but two female vervets in one study group unusually gave birth in July, during the normal mating season (L. A. Isbell and K. L. Enstam unpublished data).

SOCIAL ORGANIZATION

Social and Mating Systems

Most guenons live in single-male/multifemale groups (Rowell 1988, Cords 2002b). The exceptions are *C. aethiops* (Struhsaker 1967b), *M. talapoin* (Gautier-Hion 1971, 1973; Rowell 1971; Rowell and Dixson 1975), and *A. nigroviridis* (Gautier 1985; see also Zeeve 1991), which live in multimale/multifemale groups. The existence of apparently single-male/single-female groups of *C. neglectus* (Gautier-Hion and Gautier 1978) is odd given their extreme sexual dimorphism. It is possible that single-male/single-female groups of *C. neglectus* largely occur in declining populations, and other researchers have reported *C. neglectus* living in single-male/multifemale groups (Wahome et al. 1993). In all species for which there are data, males disperse both socially and locationally (see below) at sexual maturity, whereas females for the most part are socially philopatric, living their entire lives in their natal groups. Females also rarely leave their natal home range or group. When they do disperse, it is usually under unusual conditions. For example, female vervet monkeys in Amboseli National Park, Kenya, transferred with juveniles into neighboring groups when the population declined substantially (Hauser et al. 1986, Isbell et al. 1991).

The single-male/multifemale social system can lead to extreme competition among males for access to females. In some guenons, including *C. ascanius*, *C. mitis*, and *E. patas*, heterosexual groups at times experience influxes of multiple extragroup males during the breeding season (Chism and Rowell 1986; Harding and Olson 1986; Cords 1988, 2002b; Ohsawa et al. 1993; Kaplin et al. 1998; Carlson and Isbell 2001; Enstam et al. 2002). In some cases, an influx of multiple males during the breeding season results in the resident male being replaced (e.g., Enstam et al. 2002), while in other cases, the resident male may not be expelled

but may remain in the group after the influx (e.g., Tsingalia and Rowell 1984). Since multimale influxes occur only during *some* breeding seasons, Cords (2002b) was able to divide her study of Kakamega *C. mitis* into "influx" and "noninflux" periods. Multiple males were present in 50%–94% of days during the six influx breeding seasons but in only 4%–28% of days during noninflux breeding seasons (Cords 2002b). Breeding seasons among *E. patas* have also been divided into influx and noninflux periods, based on the number of males observed with the group during the breeding season (Harding and Olson 1986, Carlson and Isbell 2001, Enstam et al. 2002).

Among *C. mitis*, the most likely explanation for multimale influxes appears to be a single male's inability to defend multiple cycling females (Cords 2002b). The presence of two or more females cycling at the same time is highly correlated with the presence of more than one male in a group, apparently because it is too difficult for a single male to maintain exclusive access to multiple receptive females (Cords 2002b). This explanation is supported by data from other species that sometimes experience multimale influxes during the breeding season. Studies of *C. mitis labiatus* (Henzi and Lawes 1987, 1988), *C. ascanius* (Struhsaker 1988), and *E. patas* (Chism and Rowell 1988, Harding and Olson 1986) have shown a positive correlation between the number of males in a heterosexual group during the breeding season and the number of receptive females. The ability of a male to defend reproductively cycling females from extragroup males may be made even more difficult if the ratio of extragroup males to groups of females is high (Cords 2000a). Surveys of the home ranges of *C. mitis* groups routinely turned up more extragroup males during influx breeding seasons than noninflux seasons (Cords 2002b). Similarly, during a study of *E. patas* that spanned three breeding seasons, Carlson and Isbell (2001) found that the one breeding season during which multimale influxes were observed also had the highest number of sightings of extragroup males per month.

Although the number of receptive females or the density of extragroup males appears to influence multimale influxes in the species noted above, it does not explain the year-round existence of multiple males in *C. aethiops* (Struhsaker 1967b, Isbell et al. 2002), *Miopithecus* (Gautier-Hion 1971, 1973; Rowell 1971; Rowell and Dixson 1975), and *A. nigroviridis* (Gautier 1985; see also Zeeve 1991) groups. A recently proposed explanation for the formation of multimale groups in *C. aethiops* is the limited dispersal hypothesis, which states that *C. aethiops* males have limited dispersal abilities because of the configuration of the habitat and the costs of dispersal (Isbell et al. 2002). Unlike forest guenons, where groups can live adjacent to multiple other groups, vervets' restricted use of habitats along water courses in otherwise unsuitable savanna habitats means that *C. aethiops* groups often line up, one after another, along rivers, rather than being scattered throughout the savanna habitat (a notable exception is Amboseli, where vervet groups occupied more swampy habitat; Cheney and Seyfarth 1983). The result is that *C. aethiops* males have limited options when they disperse from their natal groups (i.e., only two directions) and usually disperse into adjacent groups in which other males (including male relatives) already reside (Isbell et al. 2002). Further research into the dispersal patterns of *Miopithecus* and *A. nigroviridis* needs to be conducted to determine whether similar limitations are also responsible for multiple males in these species.

Miopithecus are even more unusual guenons than their multimale/multifemale social system suggests, however. Male and female *M. talapoin* interact only during the breeding season. Outside of the breeding season, *M. talapoin* society is highly segregated, with males interacting and traveling with other males and females only with females (Rowell 1971, Rowell and Dixson 1975). Given the high level of interaction between males, it would be interesting to examine the behavioral genetics of *M. talapoin* to determine the degree of relatedness among males in the same group. Few field studies of *A. nigroviridis* have been conducted, but those that have been indicate that this species lives in multimale/multifemale groups (e.g., Gautier 1985).

Intragroup and Intergroup Relations

Given the fact that most guenon species live in single-male/ multifemale social groups (see above), it is not surprising that the majority of adult social interactions occur between females (Rowell 1988) and between females and the resident male primarily during the mating season (e.g., Chism et al. 1984). In such species, the resident male is socially peripheral to the rest of the group (e.g., Hall 1965, Struhsaker and Leland 1979). Intragroup interactions between females are generally amicable, as expected of female-philopatric species, consisting primarily of grooming (Chism and Rogers 2002) and, in some species, allomaternal care of infants (Chism 1978, Struhsaker and Leland 1979, Zucker and Kaplan 1981). Most guenon species differ from the closely related *Papio* and *Macaca* in lacking strong, linear dominance hierarchies (*C. ascanius*, Cords 1987a; *C. campbelli*, Cords 1987a; *C. mitis*, Cords 2000b; *E. patas*, Isbell and Pruetz 1998). *C. aethiops* appear to be the exception among guenons as they exhibit linear dominance hierarchies for both males and females (Struhsaker 1967b, Cheney et al. 1988).

Conspecific groups of guenons are generally intolerant of one another, with both males and females behaving aggressively in intergroup encounters (see Cheney 1987 for a review). Several species of guenons are known to engage in territorial range defense, including *C. aethiops* (Struhsaker 1967b; Cheney 1981, 1992), *C. diana* (Hill 1994), *C. mitis* (Struhsaker and Leland 1979, Cords 1987a, Butynski 1990), and *C. ascanius* (Struhsaker and Leland 1979, Cords 1987a). In some guenons, intergroup encounters are marked primarily by territorial calling. Among *C. diana*, for example, range boundaries are advertised and defended primarily through intergroup calling and secondarily through aggres-

sive interactions (Hill 1994). In this species, females play an important role in range defense; intergroup calls by females apparently stimulate the group's male to give his intergroup vocalization or loud call (Hill 1994). In other species, such as *C. aethiops*, range defense includes both intergroup vocalizations and active defense at territorial boundaries, in which both males and females participate (Cheney 1981). Territorial encounters are similar in *C. ascanius*. After a large *C. ascanius* group (50 individuals) fissioned into two smaller groups of similar size, intergroup territorial encounters increased significantly as the two daughter groups struggled to establish a new territorial border between them (Windfelder and Lwanga 2002). Adult females were the primary participants during the encounters, which included vocalizations, chases, and physical aggression (Windfelder and Lwanga 2002).

Although other guenons, including *E. patas* and *C. neglectus*, do not defend territorial boundaries (Rowell 1988, Chism 1999), they may still engage in aggressive intergroup encounters. Among *E. patas*, whose home ranges are apparently too large to actively defend, adult females and juveniles chase the members of the other group over long distances (Chism 1999). Although resident male *E. patas* are invariably aggressive toward extragroup males, they are seldom involved in intergroup encounters, except during the breeding season (Struhsaker and Gartlan 1970, Chism 1999).

Although female guenons are philopatric and generally intolerant of conspecifics in other groups (see above), occasionally females do transfer between groups. Sometimes such transfers result in apparently peaceful coexistence between established and new residents of the groups (Hauser et al. 1986, Isbell et al. 1991). However, such transfers by females are not always without incident. In the Cape Vidal Nature Reserve in South Africa, a female *C. mitis*, apparently attempting to transfer into another group, was attacked by the females of the new group and a few hours later died of her injuries (Payne et al. 2003).

BEHAVIORAL ECOLOGY OF POLYSPECIFIC ASSOCIATIONS

Polyspecific associations, or mixed-species groups, are associations between two or more different species that involve behavioral changes by at least one of the participating species (Strier 2003). For example, associations between brown capuchins (*Cebus apella*) and squirrel monkeys (*Saimiri sciureus*) are maintained because *S. sciureus* change their ranging behavior and actively follow the *C. apella* group (Terborgh 1983). Such associations are well known in both New World and Old World monkeys. In some cases, associations between different species are not true ones but occur by chance as a result of high primate biomass (e.g., *Lophocebus albigena* with *C. mitis*, *Procolobus badius*, and *Colobus guereza*, Waser 1982; *Procolobus verus* with *Cercocebus atys*, western black and white colobus [*Colobus* polykomos], and *Procolobus badius*, Oates et al. 1990). In other cases, they are true associations, occurring at frequencies greater than expected (e.g., Goeldi's monkeys [*Callimico goeldii*] with saddleback tamarins [*Saguinus fuscicollis*], Porter 2001; *L. albigena* with *C. ascanius*, Waser 1982; *C. ascanius* with *C. mitis*, Cords 1990; *Procolobus verus* with *C. campbelli*, *C. diana*, and *C. petaurista*, White-sides 1989; Oates et al. 1990), with one or more species gaining real benefits. In some cases, these associations may be nearly permanent (e.g., mustached tamarins [*Saguinus mystax*] with *fuscicollis*, Peres 1992; emperor tamarins [*imperator*] with *Saguinus fuscicollis*, Terborgh 1983).

Within the Cercopithecini, the tendency to form polyspecific associations is more common in the forest guenons, though still variable. *E. patas*, *C. aethiops* (K. L. Enstam personal observation), and *C. neglectus* (Rowell 1988, Mugambi et al. 1997, Zeeve 1991) do not form associations with other species, although sometimes individual vervets or patas join groups of different species (Isbell personal observation) or vice versa (Starin 1993). *C. diana* and *C. lhoesti* spend little time in association with other species (Gautier-Hion 1988b, Whitesides 1989). In contrast, *C. pogonias* and *C. nictitans* spend virtually all of their time in association with each other or other guenon species (Gautier-Hion and Gautier 1974, Gautier-Hion et al. 1983).

Guenons also vary in the species with which they choose to associate (Table 15.8). *C. pogonias* (Gautier-Hion 1988b) and *A. nigroviridis* (Zeeve 1991) have been observed associating only with other guenons. *C. ascanius* (Struhsaker 1981, Cords 1990, Zeeve 1991), *C. mitis* (Struhsaker 1981, Cords 1990), and *C. wolfi* (Zeeve 1991) form associations with both other guenons and nonguenon species. *C. erythrotis* are reported to associate only with a nonguenon species, *L. albigena* (Waser 1980) despite the fact that it is sympatric with *C. preussi* and *C. pogonias* in parts of its range (Oates 1988). Finally, although rare, some guenon species have been observed associating with nonprimate species (e.g., *C. ascanius* with red forest duikers [*Cephalophus natalensis*] in Kibale Forest, Uganda). In this case, the association appears to be maintained by *Ceph. natalensis*, which forage on food dropped by *C. ascanius* (Struhsaker 1981).

In some cases, association between two species, or simply sympatry, can lead to interspecific matings and even fertile hybrids (*C. mitis* and *C. ascanius*, Aldrich-Blake 1968, Struhsaker et al. 1988). However, in the vast majority of cases, polyspecific associations do not result in viable offspring or even interspecific matings, so a primary question is what (nonreproductive) benefits do individuals of different species acquire by forming long-term, stable polyspecific associations?

Explanations for such associations center on benefits individuals acquire from increased group size without suffering the costs associated with traveling in larger groups. The benefits of polyspecific associations can be divided into two broad categories which mirror primary benefits that have

been forwarded to explain the evolution of group living in primates: increased access to foods and enhanced predator detection and avoidance (Table 15.9). The specific benefits acquired will likely differ for each polyspecific association, depending on the species involved, the habitat type, and potential predators, and may involve any combination of benefits listed. Associations between guenon species and between guenons and other catarrhines can involve both categories of benefits.

Increased competition for resources is thought by most to be a primary cost of increased group size (Alexander 1974, van Schaik 1983, Terborgh and Janson 1986, Isbell 1991; but see Isbell 2004). Such a cost may be alleviated to some extent if group size is increased by associating with allospecifics that eat different foods (Oates and Whitesides 1990, Wolters and Zuberbühler 2003). Even if species have similar diets, however, they may still obtain foraging advantages by associating. For example, in a community of *Cercopithecus* species in Makokou, Gabon, polyspecific associations occur between three species: *C. cephus*, *C. nictitans*, and *C. pogonias* (Gautier-Hion et al. 1983). Despite having a high degree of dietary overlap (97%), monkeys in polyspecific associations increased foraging efficiency by reducing their chances of visiting a food source that had already been depleted (Gautier-Hion et al. 1983; see also Isbell 2004 for a similar argument for group living in general).

Studies of associations between *C. mitis* and *C. ascanius* suggest also that one species may use another to locate food sources. In Kakamega Forest, Kenya, *C. ascanius*, which travel farther per day than *C. mitis*, apparently use "local" *C. mitis* groups as guides to food sources and by doing so may avoid feeding at sites that have been recently visited by *C. mitis* (Cords 1987a). A study of polyspecific associations in *C. mitis* and *C. ascanius* in Kibale Forest, Uganda, on the other hand, suggests that *C. mitis* use *C. ascanius* as guides to food sources. In Kibale, unlike Kakamega, *C. ascanius* have smaller home ranges than *C. mitis*, and may have better knowledge of food source locations (Struhsaker 1981; see also Cords 1990).

The other category of benefits obtained from polyspecific associations is increased predator avoidance. Like foraging benefits, anti-predator-related benefits of polyspecific associations come in many forms. Studies of birds have shown that increased group size reduces the amount of time any individual needs to spend scanning because there are more eyes with the potential to detect a predator (e.g., Powell 1974, Kenward 1978). Guenons may gain a similar benefit. In their study of associations between *C. campbelli* and *C. diana*, Wolters and Zuberbühler (2003) found that both species spent less time being vigilant when in association. The freed-up time was apparently used (at least by *C. diana*) to search for food as their time spent foraging increased when in association. Wolters and Zuberbühler (2003) interpret this as a benefit to *C. diana* of associating, but the possibility exists that increased time spent foraging could actually represent a cost. Given that *C. campbelli* are known to be

Table 15.8 Guenon Polyspecific Associations

SPECIES	ASSOCIATES WITH	SOURCES
Cercopithecus aethiops	Does not associate[1]	Enstam pers. obs.
C. ascanius	Red colobus (*Procolobus badius*)	Struhsaker 1981
	Gray-cheeked mangabeys (*Lophocebus albigena*)	Struhsaker 1981
	Blue monkeys	Struhsaker 1981 (Uganda), Cords 1990 (Kenya)
	Black and white colobus (*Colobus guereza*)	Cords 1990 (Kenya)
	Allen's swamp monkeys	Zeeve 1991
	Wolf's guenons	McGraw 1994
	Angolan colobus (*C. angolensis*)	McGraw 1994
C. campbelli	Lesser spot-nosed guenons	Gautier-Hion 1988b
	Olive colobus (*Procolobus verus*)	Oates and Whitesides 1990
	Diana monkeys	Gautier-Hion 1988b
C. cephus	Putty-nosed guenons	Gautier-Hion 1988b
	Crowed guenons	Gautier-Hion 1988b
	White-collared mangabeys (*Cercocebus torquatus*)	Mitani 1991
	Gray-cheeked mangabeys	Mitani 1991
C. diana	Campbell's guenons	Gautier-Hion 1988b, Wolters and Züberhbuhler 2003
	Lesser spot-nosed guenons	Gautier-Hion 1988b
	Olive colobus	Oates and Whitesides 1990
	Red colobus	Oates and Whitesides 1990, Bshary and Noë 1997, Noë and Bshary 1997
C. dryas	No data[2]	Oates 1985
C. erythrogaster	Mona monkeys	Oates 1985
	White-collared mangabeys	Waser 1980
C. erythrotis	Gray-cheeked mangabeys	
C. hamlyni	No data	Gautier-Hion 1988b
C. lhoesti	Rarely associates with other guenons	Struhsaker 1981 (Uganda), Cords 1990 (Kenya)
C. mitis	Red-tailed guenons	Struhsaker 1981
	Red colobus	Cords 1990
	Black and white colobus	Struhsaker 1981
	Gray-cheeked mangabeys	Zeeve 1991
	Angolan colobus	Oates 1985
C. mona	White-throated guenons	Harding 1984
	White-collared mangabeys	Rowell 1988, Zeeve 1991, Wahome et al. 1993
C. neglectus	Does not associate with other primates	Gautier-Hion et al. 1983
C. nictitans	Crowned guenons	Mitani 1991
	Gray-cheeked mangabeys	Gautier-Hion 1988b
	Mustached guenons	Gautier-Hion 1988b
C. petaurista	Campbell's guenon	Gautier-Hion 1988b
	Diana monkeys	Oates and Whitesides 1990
	Olive colobus	Gautier-Hion 1988b
C. pogonias	Mustached guenons	Gautier-Hion 1988b
	Putty-nosed guenons	
C. preussi	No data	
C. sclateri	No data	
C. solatus	No data	Gevaerts 1992, Zeeve 1991
C. wolfi	Black mangabeys (*Lophocebus aterrimus*)	McGraw 1994, Zeeve 1991
	Red-tailed guenons	McGraw 1994, Zeeve 1991
	Angolan colobus	Zeeve 1991
	Allen's swamp monkeys	Gautier 1985, Zeeve 1991
Allenopithecus nigroviridis	Red-tailed guenons	Zeeve 1991
	Wolf's guenons	
Miopithecus talapoin	No data	Enstam pers. obs.
Erythrocebus patas	Does not associate	

[1] "Does not associate" indicates that groups of a specific species have not been observed forming polyspecific associations with other groups. Individuals of these species may associate with allospecifics (see text).
[2] "No data" indicates data on polyspecific association formation are not available for that species.

Table 15.9   Proposed Benefits of Polyspecific Associations

FORAGING EFFICIENCY ADVANTAGES	ANTIPREDATOR ADVANTAGES
Reduced scramble competition	Increased predator detection
Increased detection of resources	Increased predator confusion
Access to otherwise unavailable resources	Decreased chance of "being the victim"
Increased prey capture rate	Increased ability to defend against predators
Increased competitive ability	

largely responsible for the formation and maintenance of associations with *C. diana* (Bshary and Noë 1997), *C. diana*'s increased foraging may be a means of off-setting the cost of increased competition for resources they incur by associating with *C. campbelli*.

The increased chance of detecting a predator when in association also seems to lead to a change in *C. campbelli*'s behavior. When in association with *C. diana*, *C. campbelli* apparently experience a reduced risk of predation and *(1)* spend more time in exposed (i.e., more dangerous) areas of the habitat, *(2)* vocalize more often, and *(3)* display greater inter-individual distances, apparently indicating reduced perceived risk of predation when in association. While all three of these results are expected if perceived risk of predation is reduced, the last result, at least, may also be due to *C. diana* being interspersed among *C. campbelli*, forcing the *C. campbelli* individuals to spread out. In addition, both species apparently benefit from associating by using a broader ecological niche (i.e., both species increase their use of intermediate strata; Wolters and Zuberbühler 2003). Increased use of intermediate strata may be an antipredator benefit of poly-specific associations because of the seeming protection from both terrestrial and arboreal predators.

As with the study of *C. diana–C. campbelli* associations, other researchers have also noted increased used of vertical space as an antipredator benefit because, when in association, individuals are able to use strata that are otherwise unavailable to them. For example, polyspecific associations between *C. cephus*, *C. nictitans*, and *C. pogonias* in Gabon indicate that in association with the other two species, *C. cephus* expand their vertical use of space and are less dependent on low, dense vegetation, perhaps because they are less susceptible to predation from aerial predators (Gautier-Hion et al. 1983). Similarly, when in association, both *Procolobus badius* and *C. diana* changed their habitat use, with *C. diana* spending more time in higher strata and *Procolobus badius*, in lower strata (Bshary and Noë 1997). According to the authors, this change in strata use indicates that *C. diana* experience a reduced risk of predation by eagles when in association (Bshary and Noë 1997). Since *Procolobus badius* and *C. diana* share responsibility or the formation and maintenance of these associations (Bshary and Noë 1997), it is

likely that both species gain some advantage from associating. It is difficult to determine, however, whether individuals in polyspecific associations are forced to change their stratum use because of the association or are freed up to do so.

Increased use of otherwise inaccessible strata is apparently a common result of polyspecific associations, but these examples highlight the difficulty of teasing apart foraging benefits from antipredator benefits. The use of an expanded ecological niche during polyspecific associations may be the result of antipredator benefits, but such habitat use clearly has foraging benefits as species would likely encounter additional foods when they forage in strata they do not usually occupy.

The difficulty of separating predation from feeding as the ultimate cause of polyspecific associations or behaviors while in association can be seen with the association between *C. diana* and *C. campbelli*. In addition to other behavioral changes, Wolters and Zuberbühler (2003) note that both species increased their day range lengths when in association and cite this as an antipredator benefit because moving is a dangerous, attention-grabbing behavior and primates should display adaptations to reducing predation risk while traveling (see also Boinski et al. 1999). On the other hand, increased daily travel distance could be a true cost of polyspecific associations, just as it appears to be for many single-species groups (Isbell 1991, Janson and Goldsmith 1995). Dietary overlap is high in these two species (see Gautier-Hion 1988b), so a larger mixed-species group of *C. diana* and *C. campbelli* may need to travel farther to obtain enough food for all group members than smaller, single-species groups. If this is the case, increased day range lengths and increased (scramble) competition for resources are apparently costs that both species are willing to incur given the other benefits they obtain from associating.

Perhaps the most obvious evidence that polyspecific associations provide antipredator benefits is supplied by alarm-calling behavior. In the associations of *C. cephus*, *C. nictitans*, and *C. pogonias* in Makokou, Gabon, the species had different predator detection roles, depending on the strata they typically occupied (Gautier-Hion et al. 1983). When the potential predator was terrestrial, *C. cephus*, which tend to be closer to the ground, gave the first alarm call 75% of the time. In contrast, in the case of aerial predators, *C. cephus*, which were higher in the canopy, were the first to alarm call 88% of the time and the first caller was invariably the male. In a study of associations between single male *C. pogonias* and black colobus (*Colobus satanas*) groups, *C. pogonias* were better at detecting aerial predators and *Colobus santanas* were better at detecting *Pan troglodytes*, despite the fact that both were found at the same height in the canopy (Fleury and Gautier-Hion 1997).

Similarly, in *Procolobus badius–C. diana* associations, both species benefit from the other's alarm calls at crowned eagles by being alerted to predator presence, and *C. diana* may further reduce their susceptibility to terrestrial predators simply by associating with *Procolobus badius* (Bshary and Noë 1997), even though this seems counterintuitive (i.e., *C. diana* would seem to be in harm's way more often by associating with a preferred prey species). Although *Pan troglodytes* eat *C. diana*, they prefer *Procolobus badius* as prey (Bshary and Noë 1997, Stanford 1998). Given this inclination, *C. diana* may decrease their chances of being preyed upon by associating with *Procolobus badius* because, given the choice, *Pan troglodytes* prefer to hunt and eat *Procolobus badius*, at least in part, it seems, because they are easier to catch (Boesch 1994).

SUMMARY

The guenons are a diverse group of African monkeys in a number of ways, including pelage coloration (ranging from the drab and cryptic *A. nigroviridis* to the brightly colored and conspicuous *C. diana*), diet (ranging from the highly frugivorous *C. cephus* to the highly gummivorous and insectivorous *E. patas*), habitat preferences (dense rain forests of the forest guenons to open woodland occupied by *C. aethiops* and *E. patas*), and social structure (single-male/multifemale in most forest guenons to multimale/multifemale in *C. aethiops* and *Miopithecus* to possible single-male/single-female in some *C. neglectus* populations). In addition, guenons display diversity in their willingness to form associations with other guenon (and nonguenon) species. Polyspecific associations are relatively widespread among the guenons, and benefits resulting from such associations fall into two broad categories: resource acquisition and predator avoidance.

In this chapter, we have provided an overview of the taxonomy, ecology, and behavior of the guenons. This overview has shed light on two areas in particular that require further attention. First, while many guenon species are relatively well known, the behavior and ecology of a number of others largely remain a mystery. We hope this review encourages further research on all species of guenons, but we especially hope some intrepid explorers will take the lead in studying *A. nigroviridis*, *C. erythrogaster*, *C. erythrotis*, *C. sclateri*, *C. solatus*, *C. dryas*, and *C. hamlyni*, for which little data currently exist.

Second, although much work has been conducted on polyspecific associations in guenons, teasing apart the costs and benefits of such associations continues to be a difficult task and will require further attention. The confounding nature of many of the costs and benefits listed in Table 15.9 means that researchers must be careful not to assume that changes in behavior (e.g., feeding behavior, habitat use) are benefits. In order to determine whether such behavioral changes are costs or benefits, it is important to take a number of issues into account, including which species is responsible for forming and maintaining the association. For example, increased time spent feeding in association may indeed be a benefit if the species that increases feeding is responsible for the association. However, if the species that increases feeding in association is not responsible for the

association, this same change in behavior more likely represents a cost in terms of resource competition. In light of this, we encourage future research into the costs and benefits of polyspecific associations, with specific attention paid to the formation and maintenance of these associations.

REFERENCES

Aldrich-Blake, F. P. G. (1968). A fertile hybrid between two *Cercopithecus* spp. in the Budongo Forest, Uganda. *Folia Primatol.* 9:15–21.

Alexander, R. D. (1974). The evolution of social behavior. *Annu. Rev. Ecol. System.* 5:325–383.

Andelman, S. J. (1987). Evolution of concealed ovulation in vervet monkeys (*Cercopithecus aethiops*). *Am. Nat.* 129:785–799.

Altmann, S. A., and Altmann, J. (1970). *Baboon Ecology*. University of Chicago Press, Chicago.

Balldelou, M., and Henzi, S. P. (1992). Vigilance, predator detection and the presence of supernumerary males in vervet monkey troops. *Anim. Behav.* 117:220–241.

Beeson, M., Tame, S., Keeming, E., and Lea, S. E. G. (1996). Food habits of guenons (*Cercopithecus* spp.) in Afro-montane forest. *Afr. J. Ecol.* 34:202–210.

Boesch, C. (1994). Chimpanzee–red colobus monkeys: a predator–prey system. *Behaviour* 47:1135–1148.

Boinski, S., Treves, A., and Chapman, C. A. (1999). A critical evaluation of the influence of predators on primates: effects on group travel. In: Boinski, S., and Garber, P. A. (eds.), *On the Move: How and Why Animals Travel in Groups*. University of Chicago Press, Chicago. pp. 43–72.

Boshoff, A. F., Palmer, N. G., Avery, G., Davies, R. A. G., and Jarvis, M. J. F. (1991). Biogeograpical and topographical variation in the prey of the black eagle in the Cape Province, South Africa. *Ostrich.* 62:59–72

Bourlière, F., Hunkeler, C., and Bertrand, M. (1970). Ecology and behaviour of Lowe's guenon (*Cercopithecus cambelli lowei*) in the Ivory Coast. In: Napier, J. R., and Napier, P. H. (eds.), *Old World Monkeys: Evolution, Systematics, and Behaviour*. Academic Press, New York. pp. 397–350.

Bramblett, C. A., Pejaver, L. D., and Drickman, D. J. (1975). Reproduction in captive vervet and Sykes' monkeys. *J. Mammal.* 56:940–946.

Brugiere, D., Gautier, J. P., Moungazi, A., and Gautier-Hion, A. (2002). Primate diet and biomass in relation to vegetation composition and fruiting phenology in a rain forest in Gabon. *Int. J. Primatol.* 23:999–1024.

Bshary, R., and Noë, R. (1997). Red colobus and diana monkeys provide mutual protection against predators. *Anim. Behav.* 54:1461–1574.

Butynski, T. M. (1982). Vertebrate predation by primates: a review of hunting patterns and prey. *J. Hum. Evol.* 11:421–430.

Butynski, T. M. (1988). Guenon birth seasons and correlates with rainfall and food. In: Gautier-Hion, A., Bourlière, F., Gautier, J. P., and Kingdon, J. (eds.), *A Primate Radiation: Evolutionary Biology of the African Guenons*. Cambridge University Press, New York. pp. 284–322.

Butynski, T. M. (1990). Comparative ecology of blue monkeys (*Cercopithecus mitis*) in high- and low-density subpopulations. *Ecol. Monogr.* 60:1–26.

Butynski, T. M. (2002). The guenons: an overview of diversity and taxonomy. In: Glenn, M. E., and Cords, M. (eds.), *The Guenons: Diversity and Adaptation in African Monkeys*. Kluwer Academic/Plenum Publishers, New York. pp. 3–13.

Carlson, A. A., and Isbell, L. A. (2001). Causes and consequences of single-male and multi-male mating in free-ranging patas monkeys, *Erythrocebus patas*. *Anim. Behav.* 62:1047–1058.

Chapman, C. A., Chapman, L. J. Cords, M., Gathua, J. M., Gautier-Hion, A., Lambert, J. E., Rode, K., Tutin, C. E. G., and White, L. J. T. (2002). Variation in the diets of cercopithecus species: Differences within forests, among forests and across species. In: *The Guenons: Diversity and adaptation in African Monkeys*. Glenn, M. E., and Cords, M. (eds.), Kluwer Academic/Plenum publishers, New York. pp. 325–350.

Cheney, D. L. (1981). Intergroup encounters among free-ranging vervet monkeys. *Folia Primatol.* 35:124–146.

Cheney, D. L. (1987). Interactions and relationships between groups. In: Smuts, B. B., Cheney, D. L., Seyfarth, R. M., Wrangham, R. W., and Struhsaker, T. T. (eds.), *Primate Societies*. University of Chicago Press, Chicago. pp. 267–281.

Cheney, D. L. (1992). Within-group cohesion and inter-group hostility: the relation between grooming distributions and inter-group competition among female primates. *Behav. Ecol.* 3:334–345.

Cheney, D. L., and Seyfarth, R. M. (1983). Non-random dispersal in free-ranging vervet monkeys: social and genetic consequences. *Am. Nat.* 122:392–412.

Cheney, D. L., Seyfarth, R. M., Andelman, S. J., and Lee, P. C. (1988). Reproductive success in vervet monekys. In: Clutton-Brock, T. H. (ed.), *Reproductive Success: Studies of Individual Variation in Contrasting Breeding Systems*. University of Chicago Press, Chicago. pp. 384–402.

Chism, J. (1978). Behavior of group members other than the mother toward captive patas infants. In: Chivers, D., and Herbert, J. (eds.), *Recent Advances in Primatology*, vol. 1. Academic Press, New York. pp. 173–176.

Chism, J. (1999). Intergroup encounters in wild patas monkeys (*Erythrocebus patas*) in Kenya. *Am. J. Primatol.* 49:43.

Chism, J., Olson, D. K., and Rowell, T. E. (1983). Diurnal births and perinatal behavior among wild patas monkeys: evidence of an adaptive pattern. *Int. J. Primatol.* 4:167–184.

Chism, J., and Rogers, W. (2002). Grooming and social cohesion in patas monkeys and other guenons. In: Glenn, M. E., and Cords, M. (eds.), *The Guenons: Diversity and Adaptation in African Monkeys*. Kluwer Academic/Plenum Publishers, New York. pp. 233–244.

Chism, J., and Rowell, T. E. (1986). Mating and residence patterns of male patas monkeys. *Ethology* 72:31–39.

Chism, J., and Rowell, T. E. (1988). The natural history of patas monkeys. In: Gautier-Hion, A., Bourlière, F., Gautier, J. P., and Kingdon, J. (eds.), *Primate Radiation: Evolutionary Biology of the African Guenons*. Cambridge University Press, Cambridge. pp. 412–438.

Chism, J., Rowell, T., and Olson, D. (1984). Life history patterns of female patas monkeys. In: Small, M. F. (ed.), *Female Primates: Studies by Women Primatologists*. Alan R. Liss, New York. pp. 175–190.

Clutton-Brock, T. H., and Harvey, P. H. (1977). Primate ecology and social organization. *J. Zool. Lond.* 183:1–39.

Colyn, M. (1994). Données pondérales sur les primates Cercopithecidae d-Afrique Central (Bassin du Zaire/Congo). *Mammalia* 58:483–487.

Colyn, M., Gautier-Hion, A., and Thys van den Audenaerde, D. (1991). *Cercopithecus dryas* Schwarz 1932 and *C. solango* Thys van den Audenaerde 1977 are the same species with an age-related coat pattern. *Folia Primatol.* 56:167–170.

Cords, M. (1986). Interspecific and intraspecific variation in diet of two forest guenons, *Cercopithecus ascanius* and *C. mitis*. *J. Anim. Ecol.* 55:811–827.

Cords, M. (1987a). *Mixed-Species Association of* Cercopithecus *Monkeys in the Kakamega Forest, Kenya. University of California Publications in Zoology*, vol. 17. University of California Press, Berkeley.

Cords, M. (1987b). Forest guenons and patas monkeys: male–male competition in one-male groups. In: Smuts, B. B., Cheney, D. L., Seyfarth, R. M., Wrangham, R. W., and Struhsaker, T. T. (eds.), *Primate Societies*. University of Chicago Press, Chicago. pp. 98–111.

Cords, M. (1988). Mating systems of forest guenons: a preliminary review. In: Gautier-Hion, A., Bourlière, F., Gautier, J. P., and Kingdon, J. (eds.), *A Primate Radiation: Evolutionary Biology of the African Guenons*. Cambridge University Press, New York. pp. 323–339.

Cords, M. (1990). Mixed-species association of East African guenons: general patterns or specific examples? *Am. J. Primatol.* 21:101–114.

Cords, M. (2000a). The number of males in guenon groups. In: Kappeler, P. M. (ed.) *Primate Males*. Cambridge University Press, Cambridge. pp. 84–96.

Cords, M. (2000b). The agonistic and affiliative relationships of adult females in a blue monkey group. In: Whitehead, P., and Jolly, C. (eds.), *Old World Monkeys*. Cambridge University Press, Cambridge. pp. 453–479.

Cords, M. (2002a). Foraging and safety in adult female blue monkeys in the Kakamega Forest, Kenya. In: Miller, L. E. (ed.), *Eat or be Eaten: Predator Sensitive Foraging among Primates*. Cambridge University Press, Cambridge. pp. 205–221.

Cords, M. (2002b). When are there influxes in blue monkey groups? In: Glenn, M. E., and Cords, M. (eds.), *The Guenons: Diversity and Adaptation in African Monkeys*. Kluwer Academic/Plenum Publishers, New York. pp. 189–201.

Cords, M., and Rowell, T. E. (1987). Birth intervals of *Cercopithecus* monkeys of the Kakamega Forest, Kenya. *Primates* 28:277–281.

Curtin, S. H. (2002). Diet of the Roloway monkey, *Cercopithecus diana roloway* in Bia National Park, Ghana. In: Glenn, M. E., and Cords, M. (eds.), *The Guenons: Diversity and Adaptation in African Monkeys*. Kluwer Academic/Plenum Publishers, New York. pp. 351–371.

Enstam, K. L., and Isbell, L. A. (2002). Comparison of responses to alarm calls by patas (*Erythrocebus patas*) and vervet (*Cercopithecus aethiops*) monkeys in relation to habitat structure. *Am. J. Phys. Anthropol.* 119:3–14.

Enstam, K. L., Isbell, L. A., and de Maar, T. W. (2002). Male demography, female mating behavior, and infanticide in wild patas monkeys (*Erythrocebus patas*). *Int. J. Primatol.* 23:85–104.

Fleagle, J. G. (1999). *Primate Adaptation and Evolution*, 2nd ed. Academic Press, San Diego.

Fleury, M. C., and Gautier-Hion, A. (1997). Better to live with allogenerics than to live alone? The case of single male *Cercopithecus pogonias* in troops of *Colobus santanas*. *Int. J. Primatol.* 18:967–974.

Gautier, J.-P. (1985). Quelques caracteristique ecologiques du singe des marais: *Allenopithecus nigroviridis* Lang 1923. *Terre Vie* 40:331–342.

Gautier-Hion, A. (1971). L'ecologie du talapoin du Gabon. *Terre Vie* 25:427–490.

Gautier-Hion, A. (1973). Social and ecological features of talapoin monkeys—comparisons with sympatric cercopithecines. In: Michael, R. P., and Crook, J. H. (eds.), *Comparative Ecology and Behaviour of Primates*. Academic Press, London, pp. 147–170.

Gautier-Hion, A. (1980). Seasonal variation of diet related to species and sex in a community of *Cercopithecus* monkeys. *J. Anim. Ecol.* 49:237–269.

Gautier-Hion, A. (1988a). Diet and dietary habits of forest guenons. In: Gautier-Hion, A., Bourlière, F., Gautier, J. P., and Kingdon, J. (eds.), *A Primate Radiation: Evolutionary Biology of the African Guenons*. Cambridge University Press, New York. pp. 257–283.

Gautier-Hion, A. (1988b). Polyspecific associations among forest guenons: ecological, behavioural and evolutionary aspects. In: Gautier-Hion, A., Bourlière, F., Gautier, J. P., and Kingdon, J. (eds.), *A Primate Radiation: Evolutionary Biology of the African Guenons*. Cambridge University Press, New York. pp. 452–576.

Gautier-Hion, A., and Gautier, J. P. (1974). Les associations polyspécifiques du plateau de M'passa, Gabon. *Folia Primatol.* 22:134–177.

Gautier-Hion, A., and Gautier, J. P. (1976). Croissance, maturité sociale et sexuelle, reproduction chez les cercopithecinés forestiers arboricoles. *Folia Primatol.* 4:103–118.

Gautier-Hion, A., and Gautier, J. P. (1978). Le singe de Brazza: une stratégie originale. *Z. Tierpsychol.* 46:84–104.

Gautier-Hion, A., and Tutin, C. E. G. (1988). Simultaneous attack by adult males of a polyspecific troop of monkeys against a crowned hawk eagle. *Folia Primatol.* 51:149–151.

Gautier-Hion, A., Gautier, J. P., and Quris, R. (1981). Forest structure and fruit availability as complementary factors influencing habitat use by a troop of monkeys (*Cercopithecus cephus*). *Rev. Ecol.* 35:511–536.

Gautier-Hion, A., Quris, R., and Gautier, J. P. (1983). Monospecific vs. polyspecific life: a comparative study of foraging and anti-predatory tactics in a community of *Cercopithecus* monkeys. *Behav. Ecol. Sociobiol.* 12:325–335.

Gevaerts, H. (1992). Birth seasons of *Cercopithecus, Cercocebus* and *Colobus* in Zaire. *Folia Primatol.* 59:105–113.

Groves, C. P. (1989). *A Theory of Human and Primate Evolution*. Clarendon Press, Oxford.

Groves, C. P. (2001). *Primate Taxonomy*. Smithsonian Institution Press, Washington DC.

Grubb, P., Butynski, T. M., Oates, J. F., Bearder, S. K., Disotell, T. R., Groves, C. P., and Struhsaker, T. T. (2003). Assessment of the diversity of African primates. *Int. J. Primatol.* 24:1301–1357.

Hall, K. R. L. (1965). Behaviour and ecology of the wild patas monkey, *Erythrocebus patas*, in Uganda. *J. Zool.* 148:15–87.

Haltenorth, T., and Diller, H. (1988). *The Collins Field Guide to the Mammals of Africa, Including Madagascar*. Stephen Greene Press, Lexington, MA.

Harding, R. S. O. (1984). Primates of the Killini area, northwest Sierra Leone. *Folia Primatol.* 42:96–114.

Harding, R. S. O., and Olson, D. K. (1986). Patterns of mating among male patas monkeys (*Eyrthrocebus patas*) in Kenya. *Am. J. Primatol.* 11:343–358.

Hauser, M. D., Cheney, D. L., and Seyfarth, R. M. (1986). Group extinction and fusion in free-ranging vervet monkeys. *Am. J. Primatol.* 11:63–77.

Hausfater, G. (1975). Predatory behavior of yellow baboons. *Behaviour* 61:44–68.

Henzi, S. P., and Lawes, M. J. (1987). Breeding influxes and the behaviour of adult male samango monkeys (*Cercopithecus mitis albogularis*). *Folia Primatol.* 48:125–136.

Henzi, S. P., and Lawes, M. J. (1988). Strategic responses of male samango monkeys (*Cercopithecus mitis*) to a decline in the number of receptive females. *Int. J. Primatol.* 9:479–495.

Hill, C. M. (1994). The role of female diana monkeys, *Cercopithecus diana*, in territorial defence. *Anim. Behav.* 47:425–431.

Hoppe-Dominik, B. (1984). Etude du spectre des proies de la panthèra, *Panthera pardus*, dans le Parc National de Taï en Côte d'Ivoire. *Mammalia* 48:477–487.

Hunkeler, C., Bourlière, F., and Bertrand, M. (1972). Le comportement de la Mone de Lowe (*Cercopithecus campbelli lowei*). *Folia Primatol.* 17:218–236.

Isbell, L. A. (1991). Contest and scramble competition: patterns of female aggression and ranging behavior among primates. *Behav. Ecol.* 2:143–155.

Isbell, L. A. (1998). Diet for a small primate: insectivory and gummivory in the (large) patas monkey (*Erythrocebus patas pyrrhonotus*). *Am. J. Primatol.* 45:381–389.

Isbell, L. A. (2004). Is there no place like home? Ecological bases of female dispersal and philopatry and their consequences for the formation of kin groups. In: Chapais, B., and Berman, C. (eds.), *Kinship and Behavior in Primates*. Oxford University Press, New York. pp. 71–108.

Isbell, L. A. (in press). Patas monkeys (*Erythrocebus patas*). In: Kingdon, J., Happold, D., and Butynski, T. (eds.), *The Mammals of Africa*. Academic Press, London.

Isbell, L. A., Cheney, D. L., and Seyfarth, R. M. (1991). Group fusions and minimum group sizes in vervet monkeys (*Cercopithecus aethiops*). *Am. J. Primatol.* 25:57–65.

Isbell, L. A., Cheney, D. L., and Seyfarth, R. M. (2002). Why vervet monkeys (*Cercopithecus aethiops*) live in multimale groups. In: Glenn, M. E., and Cords, M. (eds.), *The Guenons: Diversity and Adaptation in African Monkeys*. Kluwer Academic/Plenum Publishers, New York. pp. 173–187.

Isbell, L. A., and Enstam, K. L. (2002). Predator (in)sensitive foraging in sympatric female vervets (*Cercopithecus aethiops*) and patas monkeys (*Erythrocebus patas*): a test of ecological models of group dispersion. In: Miller, L. E. (ed.), *Eat or Be Eaten: Predator Sensitive Foraging Among Primates*. Cambridge University Press, Cambridge. pp. 154–168.

Isbell, L. A., and Pruetz, J. D. (1998). Differences between vervets (*Cercopithecus aethiops*) and patas monkeys (*Erythrocebus patas*) in agonistic interactions between adult females. *Int. J. Primatol.* 19:837–855.

Isbell, L. A., Pruetz, J. D., Nzuma, B. M., and Young, T. P. (1999). Comparing measures of travel distances in primates: methodological considerations and socioecological implications. *Am. J. Primatol.* 48:87–98.

Isbell, L. A., Pruetz, J. D., and Young, T. P. (1998). Movements of vervets (*Cercopithecus aethiops*) and patas monkeys

(*Eyrthrocebus patas*) as estimators of food resource size, density, and distribution. *Behav. Ecol. Sociobiol.* 42:123–133.

Janson, C. H., and Chapman, C. (1999). Resources and primate community structure. In: Fleagle, J. G., Janson, C., and Read, C. K. (eds.), *Primate Communities*. Cambridge University Press, Cambridge. pp. 237–267.

Janson, C. H., and Goldsmith, M. L. (1995). Predicting group size in primates: foraging costs and predation risks. *Behav. Ecol.* 6:326–336.

Kaplin, B. A. (2001). Ranging behavior of two species of guenons (*Cercopithecus lhoesti* and *C. mitis doggetti*) in the Nyungwe Forest Reserve, Rwanda. *Int. J. Primatol.* 22:521–548.

Kaplin, B. A., and Moermond, T. C. (2000). Foraging ecology of the mountain monkey (*Cercopithecus lhoesti*): implications for its evolutionary history and use of disturbed forest. *Am. J. Primatol.* 50:227–246.

Kaplin, B. A., Munyaligoga, V., and Moermond, T. C. (1998). The influence of temporal changes in fruit availability on diet composition and seed handling in blue monkeys (*Cercopithecus mitis doggetti*). *Biotropica* 30:56–71.

Kenward, R. E. (1978). Hawks and doves: factors affecting success and selection in goshawk attacks on wild pigeons. *J. Anim. Ecol.* 47:449–460.

Kingdon, J. (1987). What are face patterns and do they contribute to reproductive isolation in guenons? In: Gautier-Hion, A., Bourlière, F., Gautier, J. P., and Kingdon, J. (eds.), *A Primate Radiation: Evolutionary Biology of the African Guenons*. Cambridge University Press, New York. pp. 227–245.

Kingdon, J. (1997). *Kingdon Field Guide to the Mammals of Africa*. Princeton University Press, Princeton, NJ.

Kingdon, J., Happold, D., and Butynski, T. (in press). *The Mammals of Africa*. Academic Press, London.

Lambert, J. E. (2002). Resource switching and species co-existence in guenons: a community analysis of dietary flexibility. In: The Guenons: Diversity and Adaptation in African Monkeys. Glenn, M. E., and Cords, M. (eds.), Kluwer Academic/Plenum Press, New York. pp. 309–323.

Lawes, M. J. (1991). Diet of samango monkeys (*Cercopithecus mitis erthrarchus*) in the Cape Vidal dune forest, South Africa. *J. Zool. Lond.* 224:149–173.

Lawes, M. J., Henzi, S. P., and Perrin, M. R. (1990). Diet and feeding behaviour of samango monkeys (*Cercopithecus mitis labiatus*) in Ngoye Forest, South Africa. *Folia Primatol.* 54:57–69.

McGraw, S. (1994). Census, habitat preference and polyspecific association of six monkeys in the Lomako Forest, Zaire. *Am. J. Primatol.* 34:295–307.

Mitani, J. C. (1991). Niche overlap and polyspecific association among sympatric *Cercopithecus* in the Campo Animal Reserve, southwest Cameroon. *Primates* 32:137–151.

Mitani, J. C. and Watts, D. P. (1999). Demographic influences on the hunting behavior of chimpanzees. *Am. J. Phys. Anthropol.* 109:439–545.

Mitani, J. C., Sanders, W. J., Lwanga, J. S., and Windfelder, T. L. (2001). Predatory behavior of crowned hawk-eagles (*Stephanoaetus coronatus*) in Kibale National Park, Uganda. *Behav. Ecol. Sociobiol.* 49:187–195.

Mugambi, K. G., Butynski, T. M., Suleman, M. A., and Ottichilo, W. (1997). The vanishing De Brazza's monkey (*Cercopithecus neglectus*) in Kenya. *Int. J. Primatol.* 18:995–1004.

Nakagawa, N. (1989). Activity budget and diet of patas monkeys in Kala Maloue National Park, Cameroon: a preliminary report. *Primates* 30:27–34.

Nakagawa, N. (1999). Differential habitat utilization by patas monkeys (*Erythrocebus patas*) and tantalus monkeys (*Cercopithecus aethiops tantalus*) living sympatrically in northern Cameroon. *Am. J. Primatol.* 49:243–264.

Nakagawa, N., Ohsawa, H., and Muroyama, Y. (2003). Life-history parameters of a wild group of West African patas monkeys (*Erythrocebus patas patas*). *Primates* 44:281–290.

Napier, P. H. (1981). *Catalogue of Primates in the British Museum (Natural History). Part 2: Family Cercopithecidae, subfamily Cercopithecinae*. British Museum (Natural History), London.

Noë, R., and Bshary, R. (1997). The formation of red colobus–diana monkey associations under predation pressure from chimpanzees. *Proc. R. Soc. Lond.* 264:253–259.

Oates, J. F. (1985). The Nigerian guenon *Cercopithecus erythrogaster*: ecological, behavioural, systematic, and historical observations. *Folia Primatol.* 45:25–43.

Oates, J. F. (1988). The distribution of *Cercopithecus* monkeys in West African forests. In: Gautier-Hion, A., Bourlière, F., Gautier, J. P., and Kingdon, J., (eds.), A *Primate Radiation: Evolutionary Biology of the African Guenons*. Cambridge University Press, Cambridge. pp. 79–103.

Oates, J. F., and Whitesides, G. H. (1990). Association between olive colobus (*Procolobus verus*), diana guenons (*Cercopithecus diana*), and other forest monkeys in Sierra Leone. *Am. J. Primatol.* 21:129–146.

Oates, J. F., Whitesides, G. H., Davies, A. G., Waterman, P. G., Green, S. M., Desilva, G., and Mole, S. (1990). Determinants of variation in tropical forest primate biomass: new evidence from West Africa. *Ecology* 71:328–343.

Ohsawa, H., Inoue, M., and Takenaka, O. (1993). Mating strategy and reproductive success of male patas monkeys (*Erythrocebus patas*). *Primates* 34:533–544.

Payne, H. F. P., Lawes, M. J., and Henzi, S. P. (2003). Fatal attack on an adult female *Cercopithecus mitis erythrarchus*: implications for female dispersal in female-bonded societies. *Int. J. Primatol.* 24:1245–1250.

Peres, C. A. (1992). Prey-capture benefits in a mixed-species group of Amazonian tamarins, *Saguinus fuscicollis* and *S. mystax*. *Behav. Ecol. Sociobiol.* 31:339–347.

Porter, L. M. (2001). Benefits of polyspecific associations for the Goeldi's monkey (*Callimico goeldii*). *Am. J. Primatol.* 54:143–158.

Powell, G. V. N. (1974). Experimental analysis of the social value of flocking by starlings (*Sturnus vulgaris*) in relation to predation and foraging. *Anim. Behav.* 22:501–505.

Pruetz, J. D., and Isbell, L. A. (2000). Correlations of food distribution and patch size with agonistic interactions in female vervets (*Chlorocebus aethiops*) and patas monkeys (*Erythrocebus patas*) living in simple habitats. *Behav. Ecol. Sociobiol.* 49:38–47.

Richard, A. F. (1985). *Primates in Nature*. W. H. Freeman and Co., New York.

Rowell, T. E. (1970). Reproductive cycles of two *Cercopithecus* monkeys. *J. Reprod. Fertil.* 22:321–328.

Rowell, T. E. (1971). Social organization of wild talapoin monkeys. *Am. J. Phys. Anthropol.* 38:593–598.

Rowell, T. E. (1988). The social system of guenons compared with baboons, macaques, and mangabeys. In: Gautier-Hion, A., Bourlière, F., Gautier, J. P., and Kingdon, J. (eds.), *A Primate*

Radiation: Evolutionary Biology of the African Guenons. Cambridge University Press, Cambridge. pp. 439–451.

Rowell, T. E., and Dixson, A. F. (1975). Changes in the social organization during the breeding seasons of wild talapoin monkeys. *J. Reprod. Fertil.* 43:419–434.

Rowell, T. E., and Richards, S. M. (1979). Reproductive strategies of some African monkeys. *J. Mammal.* 60:58–69.

Ryan, M. J., and Rand, A. S. (1993). Species recognition and sexual selection as a unitary problem in animal communication. *Evolution* 47:647–657.

Sanders, W. J., Trapani, J., and Mitanti, J. C. (2003). Taphonomic aspects of crowned hawk-eagle predation on monkeys. *J. Hum. Evol.* 44:87–105.

Shultz, S. (2001). Notes on interactions between monkeys and African crowned eagles in Taï National Park, Ivory Coast. *Folia Primatol.* 72:248–250.

Shultz, S., Noë, R., McGraw, W. S., and Dunbar, R. I. M. (2004). A community-level evaluation of the impact of prey behavioural and ecological characteristics on predator diet composition. *Proceedings of the Royal Society of London*, B. 271:725–732.

Seyfarth, R. M, Cheney, D. L., and Marler, P. (1980). Vervet monkey alarm calls: semantic communication in a free-ranging primate. *Anim. Behav.* 28:1070–1094.

Sheppard, D. J. (2000). Ecology of the Budongo Forest redtail: patterns of habitat use and population density in primary and regenerating forest sites [MSC thesis]. University of Calgary, Alberta.

Skorupa, J. P. (1989). Crowned eagles *Strephanoaetus coronatus* in rainforest: Observations on breeding chronology and diet at a nest in Uganda. *Ibis.* 131:294–298.

Sly, D. L., Harbaugh, S. W., London, W. T., and Rice, J. M. (1983). Reproductive performance of a laboratory breeding colony of patas monkeys (*Erythrocebus patas*). *Am. J. Primatol.* 4:23–32.

Stanford, C. B. (1998). *Chimpanzee and Red Colobus: The Ecology of Predator and Prey.* Harvard University Press, Cambridge.

Starin, E. D. (1993). The kindness of strangers. *Nat. Hist.* 102:44–50.

Strier, K. B. (2003). *Primate Behavioral Ecology*, 2nd ed. Allyn and Bacon, San Francsico.

Struhsaker, T. T. (1967a). Ecology of vervet monkeys (*Cercopithecus aethiops*) in the Masai-Amboseli Game Reserve, Kenya. *Ecology* 48:891–904.

Struhsaker, T. T. (1967b). Social structure among vervet monkeys (*Cercopithecus aethiops*). *Behaviour* 29:83–121.

Struhsaker, T. T. (1969). Correlates of ecology and social organization among African cercopithecines. *Folia Primatol.* 11:80–118.

Struhsaker, T. T. (1970). Phylogenetic implications of some vocalizations of *Cercopithecus* monkeys. In: Napier, J. R., and Napier, P. H. (eds.), *Old World Monkeys: Evolution, Systematics, and Behaviour.* Academic Press, New York. pp. 367–344.

Struhsaker, T. T. (1978). Food habits of five monkey species in the Kibale Forest, Uganda. In: Chivers, D. J., and Herbert, J. (eds.), *Recent Advances in Primatology*, vol. 1, Behaviour. Academic Press, London. pp. 225–248.

Struhsaker, T. T. (1981). Polyspecific associations among tropical rain forest primates. *Z. Tierpsychol.* 57:268–304.

Struhsaker, T. T. (1988). Male tenure, multi-male influxes, and reproductive success in redtail monkeys (*Cercopithecus ascanius*). In: Gautier-Hion, A., Bourlière, F., Gautier, J. P., and Kingdon, J., (eds.), *A Primate Radiation: Evolutionary Biology of the African Guenons.* Cambridge University Press, Cambridge. pp. 340–363.

Struhsaker, T. T. (2000). The effects of predation and habitat quality on the socioecology of African monkeys: lessons from the islands of Bioko and Zanzibar. In: Whitehead, P. F., and Jolly, C. J. (eds.), *Old World Monkeys*, Cambridge University Press, Cambridge, pp. 393–430.

Struhsaker, T. T., Butynski, T. M., and Lwanga, J. S. (1988). Hybridization between redtail (*Cercopithecus ascanius schmidti*) and blue (*C. mitis stuhlmanni*) monkeys in the Kibale Forest, Uganda. In: Gautier-Hion, A., Bourlière, F., Gautier, J. P., and Kingdon, J. (eds.), *A Primate Radiation: Evolutionary Biology of the African Guenons.* Cambridge University Press, Cambridge. pp. 477–497.

Struhsaker, T. T., and Gartlan, J. S. (1970). Observations on the behaviour and ecology of the patas monkey (*Erythrocebus patas*) in the Waza Reserve, Cameroon. *J. Zool. Lond.* 161:49–63.

Struhsaker, T. T., and Leakey, M. (1990). Prey selectivity by crowned hawk-eagles on monkeys in the Kibale Forest, Uganda. *Behav. Ecol. Sociobiol.* 26:435–443.

Struhsaker, T. T., and Leland, L. (1979). Socioecology of five sympatric monkey species in the Kibale Forest, Uganda. In: Rosenblatt, J. S., Hinde, R. A., Beer, C., and Busnel, M. C. (eds.), *Advances in the Study of Behavior.* Academic Press, New York. pp. 159–228.

Strushaker, T. T., and Leland, L. (1988). Group fission in redtail monkeys (*Cercopithecus ascanius*) in the Kibale Forest, Uganda. In: Gautier-Hion, A., Bourlière, F., Gautier, J. P., and Kingdon, J. (eds.), *A Primate Radiation: Evolutionary Biology of the African Guenons.* Cambridge University Press, Cambridge. pp. 364–388.

Terborgh, J. (1983). *Five New World Primates: A Study in Comparative Ecology.* Princeton University Press, Princeton.

Terborgh, J., and Janson, C. H. (1986). The socioecology of primate groups. *Annu. Rev. Ecol. System.* 17:111–135.

Tsingalia, H. M., and Rowell, T. E. (1984). The behavior of adult male blue monkeys. *Z. Tierpsychol.* 64:253–268.

Tutin, C. E. G., Ham, R. M., White, L. J. T., and Harrison, M. J. S. (1997). The primate community of the Lopé Reserve, Gabon: diets, responses to fruit scarcity, and effects on biomass. *Am. J. Primatol.* 42:1–24.

van der Zee, D., and Skinner, J. D. (1977). Preliminary observations on samango and vervet monkeys near Lake Sibayi. *S. Afr. J. Sci.* 73:381–382.

van Schaik, C. P. (1983). Why are diurnal primates living in groups? *Behaviour* 87:120–144.

Vine, I. (1970). Communication by facial visual signals. In: Crook, J. H. (ed.), *Social Behaviour in Birds and Mammals.* Academic Press, New York. pp. 279–354.

Wahome, J. M., Rowell, T. E., and Tsingalia, H. M. (1993). The natural history of de Brazza's monkey in Kenya. *Int. J. Primatol.* 14:445–466.

Waser, P. M. (1980). Polyspecific associations of *Cercocebus albigena*: geographic variation and ecological correlates. *Folia Primatol.* 33:57–76.

Waser, P. M. (1982). Primate polyspecific associations: do they occur by chance? *Anim. Behav.* 30:1–8.

Whitesides, G. H. (1989). Interspecific associations of diana monkeys, *Cercopithecus diana*, in Sierra Leone, West Africa: biological significance or chance? *Anim. Behav.* 37:760–776.

Whitten, P. L. (1983). Diet and dominance among female vervet monkeys (*Cercopithecus aethiops*). *Am. J. Primatol.* 5:139–159.

Whitten, P. L. (1984). Competition among female vervet monkeys. In: Small, M. F. (ed.), *Female Primates: Studies by Women Primatologists.* Alan R. Liss, New York. pp. 127–140.

Windfelder, T. L., and Lwanga, J. S. (2002). Group fission in red-tail monkeys (*Cercopithecus ascanius*) in Kibale National Park, Uganda. In: Glenn, M. E., and Cords, M. (eds.), *The Guenons: Diversity and Adaptation in African Monkeys.* Kluwer Academic/Plenum Publishers, New York. pp. 147–160.

Wolfheim, J. H. (1983). *Primates of the World: Distribution, Abundance, and Conservation.* University of Washington Press, Seattle.

Wolters, S., and Zuberbühler, K. (2003). Mixed-species associations of diana and Campbell's monkeys: the costs and benefits of a forest phenomenon. *Behaviour* 140:371–385.

Wrangham, R. W., and Riss, E. (1990). Rates of predation on mammals by Gombe chimpanzees, 1972–1975. *Primates* 31:157–170.

Wrangham, R. W., and Waterman, P. G. (1981.) Feeding behaviour of vervet monkeys on *Acacia tortillas* and *Acacia xanthophloea*: With special reference to reproductive strategies and tannin production. *J. Anim. Ecol.* 50:715–731.

Young, T. P., Stubblefield, C. H., and Isbell, L. A. (1997). Ants on swollen-thorn acacias: species coexistence in a simple system. *Oecologia* 109:98–107.

Zeeve, S. R. (1991). Behavior and ecology of primates in the Lomako Forest, Zaire [PhD thesis]. State University of New York, Stony Brook.

Zuberbühler, K. (2001). Predator-specific alarm calls in Campbell's monkeys, *Cercopithecus campbelli. Behav. Ecol. Sociobiol.* 50:414–422.

Zuberbühler, K. and Jenny, D. (2002). Leopard predation and primate evolution. *J. Hum. Evol.* 43:873–886.

Zuberbühler, K., Jenny, D., and Bshary, R. (1999). The predator deterrence function of primate alarm calls. *Ethology* 105:477–490.

Zuberbühler, K., Noë, R., and Seyfarth, R. M. (1997). Diana monkey long-distance calls: messages for conspecifics and predators. *Anim. Behav.* 53:589–604.

Zucker, E. L., and Kaplan, J. R. (1981). Allomaternal behavior in a group of free-ranging patas monkeys. *Am. J. Primatol.* 1:57–64.

16

The Hylobatidae
Small Apes of Asia
Thad Q. Bartlett

INTRODUCTION

Members of the family Hylobatidae, known commonly as gibbons, include 12 species of small arboreal ape that inhabit the rain forests of eastern and southeastern Asia. Owing to shared locomotor morphology, gibbons, along with the great apes and humans, make up the superfamily Hominoidea (Simpson 1945, Groves 1972). Yet, gibbons are far smaller than other hominoids, a trait that has given rise to the label *lesser apes* (Preuschoft et al. 1984). Depending on the species, gibbons weigh anywhere from 5 to 15 kg. A related characteristic is that gibbons are almost exclusively arboreal. They move quickly through the trees via a mixture of brachiation, bipedal running, and leaping (Fleagle 1976). On the rare occasions that gibbons do come to the forest floor, they walk bipedally, which contrasts with most great apes that routinely come to the ground to travel and, when they do, walk quadrupedally on either fists (*Pongo* spp.) or knuckles (*Gorilla* spp., *Pan* spp.).

While hylobatids are easily distinguished from other hominoids by their small body size, the label *lesser apes* is perhaps just as much an indication of their perceived relevance to understanding human evolution (Yerkes and Yerkes 1929). For example, gibbons do not display the cognitive abilities typically exhibited by other ape species (Povinelli and Cant 1995). However, while the relevance of gibbons as a model for human evolution may be debated, their unique suite of morphological and behavioral characters makes them of particular interest in the study of primate behavioral ecology.

In addition to their small size, gibbons can be distinguished morphologically from the great apes based on three sets of traits (Groves 1972, 1984). First, gibbons possess a number of ancestral characteristics shared with cercopithecoid monkeys but absent in most great apes. One such trait are *ischial tuberosities*, bony structures on the lower portion of the pelvis that serve as areas of attachment for callus-like sitting pads. This trait is probably linked to the fact that gibbons, like many monkeys, sleep while sitting on bare branches, rather than in nests like other apes. Second, gibbons are distinguished by unique derived traits related to brachiation. While all apes share an *orthograde* (or upright) posture, the pendulum-like swing of gibbon locomotion is unique among the hominoids. Related morphological traits include extremely long forearms, a highly mobile shoulder joint, and long hook-like hands with a deep cleft between the first and second digits. Finally, gibbons are also unique among apes in the morphological similarity of males and females. Gibbon males and females are the same size, and both sexes have long saber-like canines with sectorial premolars to hone the upper canines (Groves 1972, 1984).

Behaviorally, gibbons can be distinguished from other primates based on a complex of characters that include frugivory, territoriality, stable pairbonds, and regular vocal displays or songs. Much of the scientific literature on gibbons has focused on the selective pressures that gave rise to this unique suite of traits (Preuschoft et al. 1984, Leighton 1987), and this subject remains an area of active inquiry (Fuentes 2000, Reichard 2003a, Brockelman 2005).

TAXONOMY

Relative to most other primate families, the Hylobatidae are fairly homogenous as a group in respect to both anatomy and behavior (Chivers 1984). As Brockelman (2005) has observed, "the study of gibbons is in many respects a study in adaptive constraints rather than of adaptive radiation." Yet, in terms of gibbon systematics, four different taxa are readily recognized based on chromosome number (Prouty et al. 1983a,b). Given the overall ecological similarity among gibbons, these four groups have typically been distinguished at the level of the subgenus (Marshall and Sugardjito 1986, Geissmann 1995, Chivers 2001, Groves 2001). However, recent molecular data suggest that the temporal split between gibbon subgenera is as deep as that between chimpanzees and humans (Hayashi et al. 1995, Hall et al. 1998, Roos and

Geissmann 2001). As a result, there is an emerging consensus that the recognized subgenera should be promoted to the genus level (Roos and Geissmann 2001, Geissmann 2002b, Groves 2001, Brandon-Jones et al. 2004, Mootnick and Groves 2005). Nevertheless, investigators continue to differ on how many gibbon species should be recognized. Marshall and Sugardjito (1986) identify nine, Groves (2001) 14, and Geissmann (1995, 2002b) 12. Further complicating matters is the observation by Groves (2001, see also Brandon-Jones et al. 2004) that the type specimen for *Bunopithecus sericus* is not congeneric with other members of the genus. Consequently, the scientific name for the hoolock gibbon has recently been changed to *Hoolock hoolock*. At the species level, the greatest area of disagreement over the taxonomy of the living gibbons concerns the crested group (*Nomascus* spp.). This genus has been studied extensively by Geissmann (1995, 2002b), and his taxonomy is followed here (Table 16.1).

Most gibbon species are very similar in skeletal anatomy and more readily differentiated based on pelage and characteristics of their vocal repertoire (Fooden 1971, Chivers 2001). Hoolock, pileated (*Hylobates pileatus*), and crested gibbons are all sexually dichromatic. The predominant color of males is black, while females are mostly buff-colored. In these species, sexes also differ in the markings of the crown, face, and chest. Sexually dichromatic species are the most northerly distributed, while monochromatic species are restricted to more southerly regions (Fooden 1971). Chivers (2001) speculates that increased visibility in more seasonal forests may increase selection for visible indicators of sex. Lar gibbons (*Hylobates lar*), which have the broadest north–south distribution of all gibbon species, may represent an evolutionarily intermediate condition. This species is asexually dichromatic; that is, both sexes can be either black or buff. Extensive demographic data on lar gibbons indicate that coat color is a simple Mendelian trait, with black dominant to buff (Fooden 1971, Brockelman 2004). Coat color changes more rapidly over the course of evolution than

Table 16.1   Species of the Family Hylobatidae[1]

GENUS	SPECIES	COMMON NAME(S)	WEIGHT[2] (KG)	COAT COLORATION	DUET
Hoolock (diploid number 38)	*H. hoolock*	Hoolock or white-browed gibbon	6.1–6.9	Males black, females buff	Yes
Hylobates (diploid number 44)	*H. agilis*	Dark-handed or agile gibbon	5.5–6.4	Varies, males have white cheeks	Yes
	H. klossii	Kloss's gibbon or biloh	5.8	Both sexes black	No
	H. lar	White-handed or lar gibbon	4.4–7.6	Both sexes black or buff	Yes
	H. moloch	Silvery or Javan gibbon	5.7	Both sexes silver–gray	No
	H. muelleri	Mueller's or Bornean gibbon	5.0–6.4	Both sexes brown to gray	Yes
	H. pileatus	Pileated or capped gibbon	6.3–10.4	Males black, females buff	Yes
Nomascus (diploid number 52)	*N. concolor*	Western black-crested gibbon	4.5–9.0	Males black, females buff	Yes
	N. cf. nasutus	Eastern black-crested gibbon	–	Males black, females buff	Yes
	N. gabriellae	Yellow-cheeked crested gibbon	5.8	Males black, females buff	Yes
	N. leucogenys	White-cheeked crested gibbon	5.6–5.8	Males black, females buff	Yes
Symphalangus (diploid number 50)	*S. syndactylus*	Siamang	10.0–14.7	Both sexes black	Yes

[1] Taxonomy disputed. See text.
[2] See references in Rowe (1996).

other traits and therefore taken alone may be less useful than vocal or genetic traits in reconstructing the phylogenetic relationships between gibbon species (Geissmann 2002b).

Songs

Gibbons exhibit a broad variety of loud calls, or songs, that are structurally complex and can be heard over long distances. Current speculation is that gibbon calls serve both to cement the bond between pairmates and to advertise the location of the gibbon territory (Wickler 1980, Raemaekers et al. 1984). Studies of hybrid gibbons demonstrate that there is a significant genetic component to gibbon songs (Brockelman and Schilling 1984; Geissmann 1984, 1993; Tenaza 1985), and phylogenetic analyses of song characteristics have delineated species by genus (Geissmann 2002b). The kinds of song produced by gibbons can be divided minimally into solos and duets. Among *Hylobates* spp., mated males sing solos frequently, though not exclusively, at or before dawn (Geissmann 2002a). In the remaining genera, solos are absent among mated males. Mated females are reported to sing solos in two species, Kloss (*H. klossii*) and silvery gibbons (*H. moloch*). Notably, these two species lack the highly coordinated duets seen in other taxa (Table 16.1).

Duets differ broadly between species but share many elements of their structure, which typically includes an introductory sequence, a female great call, and the male's reply or coda (Marshall and Marshall 1976, Raemaekers et al. 1984). Male and female contributions are distinctive but often blend together or overlap so that different parts of the song are not apparent to the novice listener. Duets are sung mostly in the morning and may last up to 30 min. Immatures do not customarily sing during duets but do occasionally contribute during the vocal interlude to produce a group chorus. It is not uncommon to hear maturing females, and less often males, sing great calls with their mothers. Groups also sing when disturbed by a predator or the alarm calls of another species. As Raemaekers et al. (1984) observe, calls with apparently different functions blend into one another such that it is difficult to mark objectively the boundary between different classes of vocalization. Calling rates vary seasonally, with the highest rates during periods of resource abundance (Chivers 1974, Bartlett 1999a).

Biogeography

Gibbons are confined to areas of eastern and southeastern Asia (Fig. 16.1). Most gibbon species are distributed allopatrically, with the main exception being siamangs (*Symphalangus syndactylus*), which are sympatric with two smaller species, lar gibbons in northern Sumatra and on the Malay Peninsula and agile gibbons (*H. agilis*) south of Lake Toba on Sumatra (Gittins and Raemaekers 1980). Brockelman and Gittins (1984) describe the known contact zones between species of the genus *Hylobates* at three separate sites, including a small area of sympatry between lar and pileated gibbons in Khao Yai National Park,

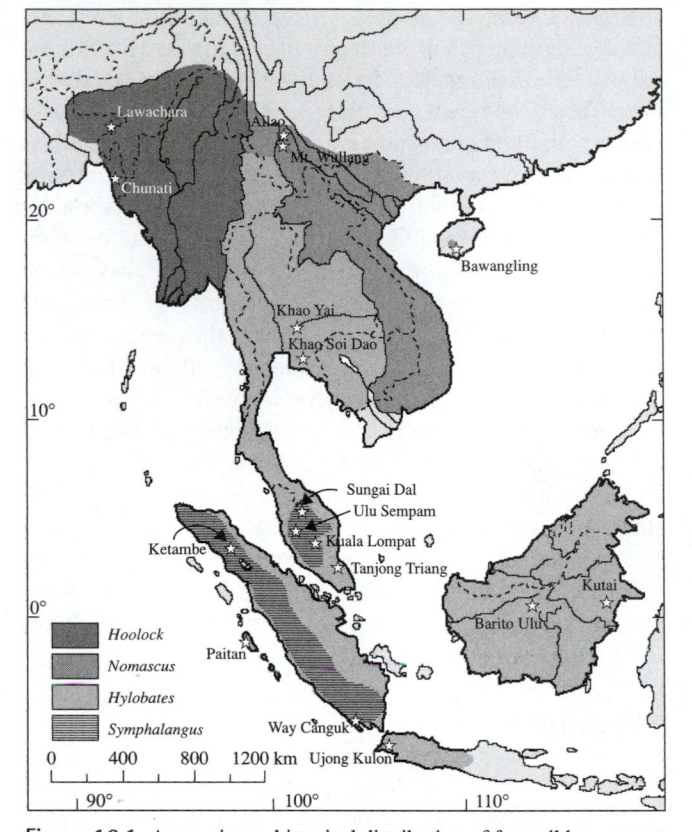

Figure 16.1 Approximate historical distribution of four gibbon genera including major study sites. See text for details (distribution follows Chivers 1977, 1984; Geissmann 2004).

Thailand, and a second between lar and agile gibbons along the Thai–Malay border (Gittins 1978). A third contact area occurs between Mueller's (*H. muelleri*) and agile gibbons in the Barito watershed, central Kalimantan, Borneo, Indonesia. This population, first documented by Marshall and Sugardjito (1986), has been studied intensively by Chivers and colleagues (Mather 1992, Bricknell 1999, McConkey 2000, Chivers 2001). An extensive survey by Mather (1992) showed that the hybrid zone extends over at least 3,500 km². Because no pure-species individuals were documented in the Barito area, it has been argued that the species rank of these two taxa should be reconsidered (see Chivers 2001).

Field Studies

With the exception of some members of the crested group, most gibbon species have now been the subject of studies of at least several months. Studies amounting to several years have been completed on lar gibbons and siamangs at Kuala Lompat, Malaysia (Chivers 1980); agile gibbons in the Gunung Palung Nature Reserve, west Kalimantan, Borneo, Indonesia (Mitani 1990); lar gibbons and siamangs at the Ketambe Research Station, Sumatra, Indonesia (West 1981, Palombit 1992); and hybrid gibbons (*H. muelleri × agilis*) at the Barito Ulu Research Area, central Kalimantan, Borneo, Indonesia (Mather 1992; Bricknell 1999; McConkey 2000; Chivers 2001; McConkey et al. 2002, 2003). Several other long-term projects are on-going, including studies on lar gibbons in Khao Yai National Park, Thailand (Reichard and Sommer 1997, Brockelman et al. 1998, Bartlett 1999a); black gibbons (*Nomascus concolor*) at Mt. Wuliang, Yunnan Province, China (Lan 1993, Lan and Sheeran 1995); and siamangs at the Way Canguk Research Station, Sumatra, Indonesia (O'Brien et al. 2003, 2004). Much of this research has been reviewed previously (Chivers 1977, 2001; Leighton 1987; Bartlett 1999b; Geissmann 2004). The present chapter will emphasize developments in our understanding of the behavioral ecology of gibbons during the last two decades.

ECOLOGY

Because gibbons live in small social groups, are mostly active high in the canopy, and have the ability to move very quickly through the forest, detailed studies of their social behavior have been slow to materialize. Most studies have focused on aspects of ecology, and comparative data are now available for all species except for some members of the crested group (Table 16.2). As a rule, gibbons occupy small, well-defended territories that they travel via regular pathways in search of sources of ripe fruit (Chivers 1984, 2001; Leighton 1987). The limited data available on the ecology of crested gibbons suggest that these species may diverge from this pattern; consequently, crested gibbons are afforded separate attention below.

The Gibbon Feeding Niche

All long-term studies of gibbon feeding behavior have found that they rely heavily on ripe fruit pulp (Chivers 1984, 2001). On average, fruit comprises 58% of the annual diet (Table 16.3). Most comparative studies have revealed that gibbons feed more heavily on fruit than other frugivorous primates, including macaques (*Macaca* spp., Feeroz et al.

Table 16.2 A Summary of Gibbon Feeding and Ranging Behavior

SPECIES	STUDY SITE	DIET (%)				RANGE USE			SOURCE
		FRUIT (FIG[1])	LEAVES	FLOWERS	INSECTS	DAY RANGE (KM)	HOME RANGE (HA)	TERRITORY (HA, %[2])	
Hoolock	Lawachara, Bangladesh	89 (38)	6	5	0	1.2	35	32 (91)	Islam and Feeroz 1992
		77 (44)	12	4	5	1.7	63	51 (81)	Ahsan 2001, Chivers 2001
	Chunati, India	71 (30)	13	9	1	0.9	26	—	Ahsan 2001, Chivers 2001
Hylobates agilis	Sungai Dal, W. Malaysia	58 (17)	39	3	1	1.3	29	22 (76)	Gittins 1982
H. klossii	Paitan, Siberut, Indonesia	72 (26)	2	0	26	1.5	32	21 (66)	Whitten 1984
H. lar	Tanjon Triang, W. Malaysia	67 (—)	33	0	0	—	—	—	Ellefson 1974
	Kuala Lompat, W. Malaysia	50 (22)	29	7	13	1.4	57	—	Raemaekers 1979
		64 (27)	31	1	5	1.8	53	28 (53)	MacKinnon and MacKinnon 1980
	Ketambe, Sumatra, Indonesia	71 (45)	4	1	24	—	—	—	Palombit 1997
	Mo Singto, Khao Yai, Thailand	66 (19)	24	1	9	1.2	23	19 (83)	Bartlett 1999a
	Klong Sai, Khao Yai, Thailand	66 (20)	27	2	7	1.3	43	40 (93)	Suwanvecho 2003
H. moloch	Ujong Kulan, Java, Indonesia	61 (—)	38	1	0	1.4	17	16 (94)	Kappeler 1984
H. muelleri	Kutai, Kalimantan, Indonesia	62 (24)	32	4	2	0.9	44	39 (89)	Leighton 1987
H. muelleri × *agilis*	Barito Ulu, Kalimantan, Indonesia	62 (17)	24	13	1	1.5	53	42 (79)	Bricknell 1999, McConkey et al. 2002
H. pileatus	Khao Soi Dao, Thailand	71 (26)	13	0	15	—	—	—	Srikosamatara 1984
	Klong Sai, Khao Yai, Thailand	64 (21)	27	2	7	1.3	38	37 (96)	Suwanvecho 2003
Nomascus concolor	Mt. Wuliang, Yunnan, China	21 (7)	72	7	0	—	—	—	Lan 1993
	Ailao, Yunnan, China	24 (0)	54	6	14	1.3	87	—	Chen 1995
N. leucogenys	Meng La, Yunnan, China	39 (—)	53	5	4	1.3	—	—	Hu et al. 1989, 1990
Symphalangus syndactylus	Ulu Sempan, W. Malaysia	47 (41)	50	2	1	0.8	15	13 (87)	Chivers 1974
	Kuala Lompat, W. Malaysia	32 (24)	58	9	2	0.9	35	26 (76)	Chivers 1974
		36 (22)	43	6	15	0.7	47	—	Raemaekers 1979
		45 (31)	44	4	8	0.6	28	18 (65)	MacKinnon and MacKinnon 1980
	Ketambe, Sumatra, Indonesia	61 (43)	17	1	21	—	—	—	Palombit 1997

[1] Percent of total diet.
[2] Percent of home range.

Table 16.3   Average Diet and Ranging Parameters in Gibbons

| SPECIES | DIET (%) | | | | RANGE USE | | |
	FRUIT (FIGS[1])	LEAVES	FLOWERS	INSECTS	DAY RANGE (KM)	HOME RANGE (HA)	TERRITORY (HA, %[2])
Hoolock	77 (36)	11	7	2	1.2	38	42 (86)
Hylobates agilis	58 (17)	39	3	1	1.3	29	22 (76)
H. klossii	72 (23)	2	0	25	1.5	32	21 (66)
H. lar	66 (27)	24	1	9	1.4	40	29 (76)
H. moloch	61 (−)	38	1	0	1.4	17	16 (94)
H. muelleri	62 (24)	32	4	2	0.9	44	39 (89)
H. muelleri × agilis	62 (17)	24	13	1	1.5	53	42 (79)
H. pileatus	68 (24)	20	1	11	1.3	38	37 (96)
Nomascus concolor	23 (4)	63	7	7	1.3	87	−
N. leucogenys	39 (−)	53	5	4	−	−	−
Symphalangus syndactylus	49 (37)	38	3	10	0.8	26	18 (79)
Average	58 (23)	31	4	7	1.3	40	29 (82)

[1] Percent of total diet.
[2] Percent of home range.

1994, Ungar 1995) and orangutans (*Pongo pygmaeus*, Ungar 1995). However, quantitative differences are only part of the story and, in fact, do not always apply. For example, MacKinnon and MacKinnon (1980) found that long-tail macaques included as much fruit in the diet as lar gibbons and were much more frugivorous than sympatric siamangs. Where gibbons and sympatric species differ most is in the supplementary foods: macaques have a more diverse diet overall, and orangutans typically make greater use of hard or unripe fruit and seeds (MacKinnon 1977, Feeroz et al. 1994, Ungar 1995). Gibbons may also avoid competition with sympatric primates by feeding higher in the canopy (MacKinnon and MacKinnon 1980, Ungar 1996) and by making greater use of the small terminal branches of the crown margin (Cant 1992). The efficiency with which gibbons exploit the terminal branch niche has long been recognized as an essential component of their adaptive complex (Ellefson 1974).

The remainder of the gibbon diet comes mostly from young leaves, flowers, and insects. The contribution of these additional foods varies considerably between species and study sites. Rates of insectivory may account for up to 26% of feeding time but more typically less than 10%. To date, there has been no attempt to compare insect feeding to insect abundance, but at Khao Yai, Thailand, insect feeding varied seasonally (Bartlett 1999a). It is likely that, as with other foods, the amount of insects in the diet varies in relation to availability. The significance of insectivory to gibbon socioecology remains an important area for future research (Di Fiore and Rodman 2001).

Leaves are the second most common food item for most species, accounting for 31% of the diet on average. Siamangs tend to feed more heavily on leaves than sympatric lar gibbons (MacKinnon and MacKinnon 1980, Palombit 1997),

but very high rates of folivory have been recorded for *Hylobates* spp. where they occur in the absence of larger siamangs (*H. agilis*, 39%; *H. moloch*, 38%). In general, there is a negative correlation between the contribution of leaves and insects in the diet, suggesting that these food items are exploited as alternative sources of protein. Flowers generally contribute least to the gibbon diet; however, this is likely due to the short period of availability. Selection of flowers during periods of abundance is generally high (Bartlett 1999a, Bricknell 1999, McConkey et al. 2003).

Attempts to further define the gibbon feeding niche have focused principally on three areas: fruit size, patch size, and the contribution of figs (*Ficus* spp.) to the diet. Ungar (1995) studied fruit choice by four sympatric primates at Ketambe, Sumatra. He argues that gibbons demonstrate a preference for small fruit (<7.92 cm³). Though gibbons fed equally on fruits of all sizes, they selected small fruit more often than all other sympatric primate species except smaller-bodied macaques (*Macaca fascicularis*). Ungar (1995) speculates that the gibbon preference for small fruit may be related to suspensory foraging and "a need to limit hand use during ingestion." However, research on fruit selection by hybrid gibbons (*H. muelleri × H. agilis*) at Barito Ulu, Borneo, failed to confirm Ungar's findings. According to McConkey et al. (2002), gibbons at Barito Ulu selected medium-sized fruit overall and favored large fruit during periods when fruit abundance was low. Methodological differences between the two studies make it difficult to draw firm conclusions; McConkey et al.'s estimate of fruit size was based on weight rather than volume. In addition, the two study sites differ in terms of the presence of primate competitors. At Ketambe, sympatric primates (e.g., *P. pygmaeus*, *M. fascicularis*) shared many of the same food sources, while at Barito Ulu, primate competitors were largely absent from the study area.

Elsewhere it has been argued that the key aspect of the gibbon niche is a preference for food distributed in small patches. According to Gittins and Raemaekers (1980:98), "the gibbons' feeding niche appears to be the succulent fruit occurring in smaller scarcer sources;" yet, despite the persistence of this view (e.g., Leighton 1987, Chivers 2001), the size and dispersion of gibbon feeding sources are rarely quantified. In fact, available data do not support the characterization of gibbons as small-patch specialists. At Khao Yai, Thailand, for example, lar gibbons routinely fed on the largest tree species in the forest (Bartlett 1999a). Furthermore, the characterization of gibbons as small-patch specialists stands in stark contrast to the fact that gibbons feed heavily on large fig trees throughout most of their range (Gittins and Raemaekers 1980).

Fig Exploiters?

On average, fig fruit makes up a quarter of the gibbon diet, and gibbons feed more heavily on figs than co-occurring macaques or langurs (Kuala Lompat, MacKinnon and MacKinnon 1980; Ketambe, Ungar 1995; Lawachara, Feeroz et al. 1994). Accordingly, MacKinnon and MacKinnon (1980) conclude that gibbons are best described as "fig exploiters" (see also Chivers and Raemaekers 1986, Palombit 1997). It is not clear, however, if fig consumption alone distinguishes gibbons and larger orangutans, which also rely heavily on figs (MacKinnon 1977, Leighton 1993). Both gibbons (*H. meulleri*, Leighton and Leighton 1983; *H. lar*, Bartlett 1999a) and orangutans (*P. pygmaeus*, Leighton and Leighton 1983, Leighton 1993) turn to figs most when preferred foods are unavailable. On the other hand, seasonal fruit shortages may impact gibbons more than orangutans, which are able to exploit foods such as wood pith and bark, for which gibbons lack the necessary anatomical traits to exploit. In any case, high rates of fig feeding are not universal for gibbons. At many sites, rates of fig feeding are less than half that reported for Ketambe (Table 16.2). It is unlikely, therefore, that fig feeding per se defines the gibbon feeding niche.

Range Use and Territoriality

The average home range size for all gibbon species is 40 ha, but estimates differ considerably both within and between species (Table 16.2). In lar gibbons, for example, average home range size at Kuala Lompat, Malaysia (55 ha), is over twice that reported for Mo Singto in Khao Yai, Thailand (23 ha). This range is comparable to that exhibited by the Hylobatidae as a whole. In general, it appears that ranges are larger at sites where at least two hominoid species are present (e.g., *H. lar* and *S. syndactylus* in Kuala Lompat; *H. muelleri* and *P. pygmaeus* in Kutai). However, ranges for hybrid gibbons at Barito Ulu, Borneo, are relatively large despite the fact that orangutans are only rarely sighted. According to McConkey (2000), the forest at Barito Ulu is less productive than other sites. Mather (1992) demonstrated that gibbon biomass is related to the local abundance of gibbon food trees, especially fig trees (see Chivers 2001). Figs contribute relatively little to the gibbon diet at Barito Ulu (17%), which may account for the large home range size at that site.

Territory can be distinguished from home range as the portion of the range actively defended against incursion by conspecifics (Bourliere 1964). On average, gibbon territory size is 82% of home range size (range 53%–96%). Day range is typically long relative to home range size, with the result that gibbons routinely enter all parts of their range. This is likely a prerequisite for territoriality in primates (Mitani and Rodman 1979).

Patterns of range use by gibbons appear to be influenced most by the location of food sources (Chivers 1977, Whitten 1982a), but factors such as patrolling of territorial borders (MacKinnon 1977) or the distribution of sleeping trees (Tenaza 1975) may also be important determinants of gibbon travel routes. Patterns of range use do not differ considerably by season (Whitten 1982b), but several investigators have shown that distance traveled per day does change over the course of the year. At Kuala Lompat, Malaysia, both lar gibbons and siamangs reduced travel during months when the abundance of preferred food was low (Chivers 1974, Raemaekers 1980). Similarly, at Khao Yai, Thailand, lar gibbons reduced travel during the cool dry season when fruit abundance was lowest; but day length was also shortest during the cool dry season and may have influenced daily path length (Bartlett 1999a). Gittins (1982) concludes that day range length for agile gibbons was shortest during the wet season and correlated with activity period.

During directed travel, gibbons move in single file along shared pathways. Studies of siamang (Chivers 1974), Kloss (Tenaza 1975), lar (Reichard and Sommer 1997), and hoolock (Ahsan 2001) gibbons determined that females lead the majority of group travel bouts. Given the greater nutritional stress on females due to pregnancy, lactation, and infant carrying, female primates tend to devote more time to feeding behavior relative to males (Lee 1996, Strier 2003), which may account for female group leadership in gibbons. Alternatively, Tenaza (1975) has suggested that Kloss gibbon males travel behind the group as a form of predator defense. It is possible that both factors influence group leadership in gibbons; however, female group leadership is not universal. Gittins (1979) determined that the male leads the majority of travel bouts in the group of agile gibbons he studied. This raises the possibility that other factors, such as detailed knowledge of resource location, may influence group leadership in gibbons (Brockelman 2005).

Activity Pattern

Unlike most primates, gibbons lack a prolonged mid-day rest period and retire 2–3 hr before sunset (Gittins and Raemaekers 1980). Optimal foraging theory predicts that feeding and travel time should be balanced with other behaviors as a means to optimize energy intake (MacArthur and Pianka 1966, Schoener 1971, Pyke et al. 1977). Most often, this approach views animals as energy maximizers. That is, they use their time in a way that insures maximal energy intake in the available time, where "available time" is assumed to be dawn to dusk. Gibbons, on the other hand, retire early even when resources are abundant. In seasonal environments, activity is further reduced during periods of food scarcity (Chivers 1974, Bartlett 1999a). Thus, gibbons may best be characterized as time minimizers (see Raemaekers 1979, 1980). That is, they are active only for as long as it takes to fulfill their minimum caloric needs. If this were strickly true, however, the overall activity period should be most limited during periods of fruit abundance. This is not the case, but gibbons do have more time for nonsubsistence activities such as grooming and playing during periods when ripe fruit is plentiful (Bartlett 1999a). It is not known if gibbons store calories in the form of fat during periods of fruit abundance as, for example, orangutans do (Knott 1998); but this might account, in part, for longer activity periods when abundance is high. Alternatively, a short activity period relative to sympatric primates may be necessitated by territoriality. In the small territories occupied by gibbons, increased foraging time would translate into much more intensive use of the available resources. Gibbons appear to ensure continued access to highly favored resources, namely ripe fruit, by exploiting their ranges less intensively. Access is nevertheless ensured as a result of territorial defense. Accordingly, comparisons across species suggest that range defense is less intense when high-quality resources are less abundant. At Kuala Lompat, for example, higher rates of folivory by siamangs relative to lar gibbons have been correlated with reduced rates of travel, longer feeding bouts, smaller home range size, and fewer territorial encounters (Chivers 1974, MacKinnon 1977, Raemaekers 1979, Gittins and Raemaekers 1980).

Gibbons and the Forest

The role of gibbons as seed dispersers (see Chapter 31) has been studied only at a few sites, but the evidence indicates that hylobatids are important dispersal agents in Asian forests (Whitington and Treesucon 1991, Corlett 1998). McConkey's (2000) extensive study of the seed shadow generated by hybrid gibbons at Barito Ulu, Borneo, determined that gibbons dispersed up to 81% of the species they consumed, while the seeds of 12% of species were destroyed. Furthermore, germination experiments have shown that gibbons improve the chance of germination of at least some species in their diet (Whitington and Treesucon 1991, McConkey 2000, Ahsan 2001). According to McConkey (2000), gibbons may be important as fig dispersers as many fig seeds will not germinate without prior passage through the gut. Given the loss of primates in many Asian forests, the role of gibbons in maintaining biodiversity is an area of pressing interest.

Predator Pressure

As with all primates, predator pressure on gibbons is difficult to assess. It has been argued that small, two-adult groups are more likely to occur when predator pressure is relaxed (van Schaik and van Hooff 1983). As a result, the vulnerability of gibbons to predators is a topic of considerable interest. According to van Schaik and Dunbar (1990, pp. 40–41) the fact that they are extremely fast and occupy the upper canopy renders gibbons "virtually immune from cursorial and arboreal predators" (see also van Schaik et al. 1983). Yet, it is unlikely that any arboreal animal's key defensive strategy would be to outrun its pursuers. Snakes, to which gibbons respond with vigorous displays, are stealth hunters; and raptors, too, rely on surprise. It is true that gibbons spend considerable time high in the canopy, but they regularly come close enough to the ground to be at risk (Uhde and Sommer 2002). Rabinowitz (1989) found gibbon hair in leopard feces, and Uhde and Sommer (2002) report a case where a group of lar gibbons mobbed a tiger. The reticence of gibbons to come to the ground, even to recover a fallen infant (T. Q. Bartlett personal observation), further indicates the threat posed by terrestrial predators. It is also true that gibbons are relatively large compared to other arboreal primates and may experience reduced predator pressure as a result. However, a small gibbon could certainly be taken by a large snake, and even adults are vulnerable. For example, Schneider (1906) recovered a full-grown siamang from the belly of a python (see Uhde and Sommer 2002).

One possible behavioral response to predator pressure is increased vigilance by adult males. In fact, male gibbons do feed less than females (Bartlett 1999a), which may mean they are better able to detect predators (see Dunbar and Dunbar 1980). However, Uhde and Sommer (2002) were unable to connect reduced activity by male gibbons to increased vigilance (as defined by visual scanning of the environment). In contrast to Tenaza (1975), they concluded that the fact that male gibbons most often travel behind females argues against the view that male gibbons play a major role in predation detection and defense because the first animal in progression is the most likely to encounter a predator.

Ecology and Behavior of Crested Gibbons

To date, comparatively little is known about the feeding and ranging behavior of crested gibbons. These species are critically endangered, and long-term studies of habituated animals have yet to be completed. Nevertheless, available data hint at significant differences between crested gibbons and other gibbon species (Sheeran et al. 1998). Preliminary studies at three sites in Yunnan, China, have revealed comparably very low rates of frugivory (Meng La, 39%; Mt. Wuliang, 21%; Ailao, 24%). Furthermore, Chen (1995) noted the complete absence of figs in the home range of the group he studied, and fruit of any kind was available only during 4 months of the year; leaves accounted for 54% of the annual diet.

Differences in diet appear to correspond to differences in range use. Home range size is larger for crested gibbons than any other species (Bleish and Chen 1991, Chen 1995, Sheeran et al. 1998). In addition, Sheeran et al. (1998) recorded increased travel during the cold, dry season by black crested gibbons at Mt. Wuliang. They reasoned that gibbons increase day range to find more fruit. This contrasts with the apparent energy-minimizing strategy of other gibbon species. In the context of marked drops in resource abundance during which minimum caloric needs are otherwise unmet, increased search time may be the only viable foraging strategy.

REPRODUCTIVE PARAMETERS

Gibbons live in pairbonded social groups that typically comprise two adults and one to three offspring. Mean group size is approximately four individuals and rarely exceeds six animals (MacKinnon 1977, Gittins 1980, Tilson 1981, Mitani 1990, Fuentes 2000, Reichard 2003b). It does not appear that different species consistently differ in terms of group size. Haimoff et al. (1987) estimate that the average group size for crested gibbons in Yunan, China, is seven or eight individuals, and one group included 10 animals. However, subsequent observations of crested gibbons in China have failed to corroborate these findings (Lan 1989, Bleish and Chen 1991, Sheeran et al. 1998).

Typically, five age grades are recognized (Brockelman et al. 1998): infant (0–2 years), juvenile (2–5 years), adolescent (5–8 years), subadult (8+ years), adult (mated with territory). Females give birth to a single young after a gestation length of approximately 7 months (Napier and Napier 1985). Infants are carried ventrally by females until they are weaned sometime during their second year. In siamangs, adult males as well as older juveniles are known to carry infants (Chivers 1974, Dielentheis et al. 1991). Data from both captive (Alberts 1987) and wild (S. Lappan personal communication) populations suggest that infant siamangs seek out male contact beginning in their second year of life. Males in all species may contribute parental care in the form of play, grooming, protection, and territorial defense (Rutberg 1983, Brockelman 2005).

In the wild, gibbon females first give birth at 6–9 years and have an interbirth interval (IBI) of 2–4 years (Leighton 1987, Geissmann 1991). Based on data for agile gibbons at Gunung Palung, Indonesia, Mitani (1990) reports that the average IBI was 3.2 years. In a cross-sectional study, O'Brien et al. (2003) monitored a population of 24–37 siamang groups at Way Canguk, Indonesia, over 4 years and found a mean IBI of 2.6–2.8 years. Comparable data for other species are lacking, but most estimates fall between 3 and 4 years, though there is considerable variation between individuals (Raemaekers and Chivers 1980, Ahsan 1994, Palombit 1995, Brockelman et al. 1998).

Dispersal and Pair Formation

Data from captivity indicate that gibbons reach sexual maturity at 5 years of age or earlier (Geissmann 1991). However, studies in the wild suggest that dispersal from the natal group is often greatly delayed. Brockelman et al. (1998) observed six dispersals by lar subadults over a period of 18 years. They estimate that individuals reached mature body size at about 8 years of age. The five males that emigrated during the observation period did so at 10 years on average, and a single female left her group at approximately 8.6 years. Contrary to previous studies (e.g., Tilson 1981), Brockelman et al. (1998) did not find evidence that subadults are forced out via intrasexual aggression upon reaching sexual maturity. They speculate that extra subadults provide a beneficial service to their parents by acting as social partners for younger infants and contributing to territorial defense. Even when they do disperse, subadults may remain close by. Demographic data for Kloss (Tilson 1981) and lar (Brockelman et al. 1998) gibbons suggest that most subadults transfer into neighboring territories, and preliminary genetic data collected from siamangs at Way Canguk, Indonesia, are consistent with short dispersal distances (one or two territories) for emigrating males (S. Lappan personal communication). At Khao Yai, three lar males transferred into a neighboring group together. Subsequently, the two younger animals transferred again (Brockelman et al. 1998). Codispersal has also been observed in lar gibbons at Ketambe, Indonesia (Palombit 1994a). To date, the majority of data on dispersal reflect emigration by subadult males. The sex difference in dispersal distance remains an important area for future research.

Early considerations of pair formation in gibbons focused on new pairs formed via coincident emigration by maturing subadults (Raemaekers and Chivers 1980, Tilson 1981, Palombit 1994a, Brockelman et al. 1998, Ahsan 2001), but subsequent work has revealed that gibbon pairbonds are less stable than originally thought and that new bonds may form by a number of alternative routes, including replacement of one pairmate by an outside usurper (Brockelman et al. 1998, Ahsan 2001), filling of a vacancy left open by a dead or departed pairmate (Palombit 1994a), or replacement of one pairmate by a maturing offspring (Raemaekers and Chivers 1980, Tilson 1981, Raemaekers and Raemaekers 1984, Palombit 1994a, Ahsan 2001). In fact, it now appears that bonds formed by two neophytes are relatively rare. Long-term research on the population dynamics of a community of lar gibbons at Khao Yai, Thailand, determined that just one of seven documented pair formations was the result of two dispersing subadults jointly entering an unoccupied territory (Brockelman et al. 1998). At Ketambe, Indonesia, just one of six new pairbonds was established in such a manner (Palombit 1994a). At Khao Yai, bonds were formed most often via takeovers during which an established mate was usurped by a younger individual (take-overs were clearly observed in two cases and strongly suspected in three others). While at Ketambe, new bonds resulted most often when an adult or subadult filled a vacancy after the previous pairbond was terminated due to death, disappearance, or desertion.

Though apparently not uncommon, the extent to which mate desertion is a successful reproductive strategy will require further long-term data. In two of the cases reported by Palombit (1994a), the deserting adult returned to its mate; and in two others, the males that left their territories failed to acquire a new mate during the period of the study. On the whole, existing data suggest that abandoning a mate entails considerable risk: lone males may experience considerable delay in attracting a suitable mate (Aldrich-Blake and Chivers 1973, Tilson 1981), lone females may reject the approaches of lone males (Tilson 1981), and even successful pairings may experience considerable delay before first birth (Raemaekers and Chivers 1980).

Pairbonds and Sexual Monogamy

As illustrated above, one means by which a paired gibbon may increase its reproductive access is to abandon its current mate for real or potential opportunities to mate with another adult. However, while the potential for gibbons to practice serial monogamy has long been recognized (Kloss 1908), recent observations of lar gibbons and siamangs demonstrate that an alternative sexual strategy employed by gibbons is to mate opportunistically with already paired neighbors during intergroup encounters. These findings have helped to erode the presumed connection between pairbonds and sexual fidelity in gibbons.

Over a period of 2.5 years, Palombit (1994b) documented five instances of extrapair copulation (EPC) involving a single pairbonded siamang female and three males of a neighboring group. Similarly, Reichard (1995, 2003b) determined that EPC accounted for 9% of the copulations observed in a population of lar gibbons over an 11-year period at Khao Yai, Thailand. As of yet, the genetic consequences of EPCs have not been reported. Palombit (1994b) notes that EPC observed at Ketambe did not result in the birth of an infant. However, in the case of lar gibbons, Reichard (1995) concludes that some of the EPCs took place over the period during which the female is believed to have conceived, though she also copulated with her own mate during this period.

EPC can also take place within the context of the social group itself. Fuentes (2000) observes that at least 10% of gibbon groups across all species contain multiple adults. While extra adults are often thought to be temporary or nonreproductive residents, in some cases polygamous matings have been observed (*H. lar*, T. Q. Bartlett personal observation, W. Y. Brockelman personal communication; *S. syndactylus*, S. Lappan personal communication; *H. agilis* × *H. muelleri*, Bricknell 1999); and in a few cases, observers have reported groups containing multiple females with infants (Haimoff et al. 1987, Brockelman and Srikosamatara

1984, Sommer and Reichard 2000). While demographic data indicate that two-adult groups are the most common pattern in gibbons, frequent exceptions undermine the description of gibbons as sexually monogamous and have stimulated extensive discussion about gibbon sociality and social organization (Palombit 1994a, Brockelman et al. 1998, Fuentes 2000, Sommer and Reichard 2000).

SOCIALITY AND SOCIAL ORGANIZATION

Intragroup Social Behavior

In gibbons, heterosexual pairbonds are signaled by a host of affiliative interactions, including joint defense of territory, duetting, social proximity, and affiliative social interactions (Palombit 1994a, Geissmann and Orgeldinger 2000; but see Reichard and Sommer 1997). In general, data on affiliative social behavior (e.g., grooming and play) have been slow to materialize. According to Leighton (1987), the paucity of data on gibbon sociality is largely due to a lack of social contact in gibbons. Gittins and Raemaekers (1980), for example, concluded, based on a 2-year study, that social behavior was almost wholly lacking in agile gibbons. Yet, low rates of social activity are not universal. At Khao Yai, for example, rates of affiliative social behavior have been found to be comparable to levels reported for primates generally (see Chapter 39). Bartlett (2003) found that social activity accounted for 11% of the activity budget in lar gibbons and that, during periods of fruit abundance, play and grooming accounted for as much as 20% of the activity time. Bouts of mutual grooming between adults tend to parallel juvenile play bouts, though adults, especially adult males, do engage in various forms of play with juveniles (Ellefson 1974, Bartlett 2003).

Available data suggest that the effort contributed to grooming relationships differs between males and females. In lar and hoolock gibbons, males groom females more often than the reverse (Palombit 1996, Ahsan 2001). Among siamangs, however, the time devoted to the grooming relationship is more equal (Gittins and Raemaekers 1980, Palombit 1996). Palombit (1996) argues that differences in grooming effort indicate that lar gibbon males contribute more to the maintenance of the pairbond than females do, whereas in siamangs the effort is more reciprocal. He suggests that higher rates of intragroup feeding competition may indicate a higher cost for bonding in smaller-bodied lar gibbons. Palombit allows that there is a certain amount of intraspecific variability (see Chivers 1974, Fisher and Geissmann 1990); nevertheless, he concludes that past research has overestimated the similarity in the social system of different gibbon species.

Intergroup Encounters

Gibbons actively defend a fixed range or territory via a host of behaviors that include duets and male and female solos, in addition to protracted chases between adults, especially males (Ellefson 1974, Gittins 1984). Disputes are initiated when two groups come into visual contact or when the male of one group approaches the calls of rival gibbons near the area of the territorial boundary. Adult males take the lead role in virtually all disputes; however, subadult males also participate in chases and calls. Though females are not regular participants in disputes, it is not uncommon for neighbor males to approach females over the course of an encounter (Reichard and Sommer 1997, Bartlett 2003). Such incursions are generally repelled by the resident male or by the female herself, though observations of EPC in lar gibbons and siamangs demonstrate that some incursions are tolerated, if not welcomed, by females. During much of the dispute, rival males sit or hang within view of one another. Such standoffs, which may or may not be accompanied by conflict vocalizations, are punctuated by rapid chases back and forth between the two territories. Encounters tend to end rather suddenly, with the males quickly moving to rejoin their respective groups. In rare cases, encounters may be lethal to one of the participants (Palombit 1993, Brockelman et al. 1998, Reichard 2003b).

Comparing data from eight species, Gittins (1984) concludes that gibbon groups sing on average once a day and engage in territorial disputes on average once every 5 days. However, rates of intergroup encounters are subject to numerous complicating factors. For example, the main study group observed by Chivers (1974) had no immediate neighbors and therefore engaged in no territorial disputes. All disputes recorded by Kappeler (1984) were terminated once his presence was revealed due to the fearfulness of the unhabituated neighboring groups. Thus, while territorial activity is one of the most easily recognized gibbon traits, it remains poorly understood (Gittins 1984). Studies have rarely involved simultaneous observations of multiple groups, and as a result, it is often difficult to evaluate the cause of a given encounter or its outcome.

Playback Experiments

One methodology that has been employed to better elucidate the function of territorial encounters is the use of playback experiments. Mitani (1984, 1985) used playbacks of male and female calls to determine the response of mated Mueller's gibbons to simulated intrusions from lone gibbons and neighbors. Responses differed depending on the location of the playback and the sex of the singer. Female solos or duets played at least 100 m inside the group's territory elicited approach, typically led by females, and countersinging from the resident group. Songs played from the territorial boundary elicited duets but were less likely to elicit approaches. In contrast, residents did not sing in response to male solos played inside the group's territory. Instead, males led silent approaches to the position of the playback or approached alone. These findings corroborate early suggestions that territoriality among gibbons is maintained by intrasexual intolerance (Ellefson 1974).

Additional playback experiments conducted on lar (Raemaekers and Raemaekers 1985) and agile (Mitani 1987) gibbons confirm the location-dependent response of mated gibbons to neighbor songs. Gibbons respond most quickly to territorial intruders. In contrast to Mueller's gibbons, however, both lar and agile gibbon males lead the majority of approaches, including those elicited by female solos or duets. Mitani (1987) notes that in several cases in which male Mueller's gibbons led approaches, the female in question carried a young infant. It is likely that the role of female gibbons in territorial defense is mediated in part by their reproductive condition. Females without small infants are more likely to engage in territorial defense (Mitani 1987, Bartlett 1999a).

Agonistic encounters are by their nature conspicuous, but extensive data on lar gibbons at Khao Yai indicate that not all group encounters are agonistic (Reichard and Sommer 1997). According to Bartlett (2003), 17% of encounters occurring over a 12-month period involved affiliative activity between neighboring groups. The majority of affiliative activity included chase play between juveniles, but during two encounters a neighbor male groomed the adolescent and juvenile of the neighboring group. Affiliative encounters between neighboring adults were not observed; nevertheless, neighboring adults do exhibit unappreciated levels of tolerance, which is likely a product of the genetic relatedness of the animals in adjacent groups (Brockelman et al. 1998).

Social Monogamy

While it has long been customary to refer to gibbons as a monogamous species, authors have not always taken care to explain what precisely a monogamous social system entails. Often masked by an ostensibly simple label are hidden assumptions about sexual and genetic relationships of group members. While warnings about the imprecision of the term *monogamy* are not new (Wickler and Seibt 1983), data accumulated over the last two decades have forced a wholesale reevaluation of hylobatid social organization. The most important developments can be summarized as follows. First, observations of EPC reveal that pairbonding neither implies nor depends upon sexual fidelity. Second, long-term studies show that adult membership changes regularly as a result of death, mate desertion, and displacement. Third, when adults or subadults enter new groups, they are occasionally accompanied by one or more immatures, including siblings. Fourth, short dispersal distances mean that the members of neighboring groups may be close relatives. Emerging from these findings is the recognition that during the course of their lives most gibbons will inhabit a nonnuclear family group at some point. In at least one case, this has been confirmed using deoxyribonucleic acid (DNA) analysis (Okra and Takenaka 2001). In addition, it is likely that neighboring family groups frequently consist, in part, of unrelated individuals that have lived in the same social group during a portion of their lives (Reichard and Sommer 1997, Brockelman et al. 1998). Apart from genetic relationships, an individual's knowledge of the other animals within a community is likely to extend beyond those individuals it has seen. Group choruses, duets, and solo vocalizations provide opportunities for gibbons to learn the composition of nonadjacent groups. While it remains to be demonstrated, it is likely that gibbons know individuals they have never seen based on their vocal signature alone (Dallmann and Geissmann 2001).

These findings have stimulated many authors to reject the label *monogamy* in favor of specific models more reflective of the complex realities of gibbon social systems (Fuentes 1998, 2000; Sommer and Reichard 2000; Reichard, 2003b). One of the earliest attempts to accommodate groups with complex origins was that of Quiatt (1985, 1987) and Quiatt and Kelso (1987), who proposed the term *household*, which they defined as a group that travels and exploits territorial resources as a single unit. As described by Quiatt and Kelso (1987:429), the advantage of the term *household* is that membership is less likely to imply either a mating relationship or genes in common with some other members. This analogy can be extended to the level of the neighborhood to describe a nested hierarchy in which several households occupy a shared social and physical landscape (Bartlett 2003). Conceptually, the household model is similar to population-level approaches advocated by Fuentes (2000) and Reichard (2003b). Despite different emphases, most investigators converge on two main points. First, gibbon groups should no longer be portrayed as socioreproductive isolates. Neighboring groups frequently consist of social and sexual partners as well as kin. Second, future discussions of gibbon social organization must separate mating systems from social systems because the conditions that favor one do not necessarily favor the other.

Frugivory, Territoriality, and the Evolution of Gibbon Pairbonds

However labeled, the social organization of gibbons is rare, if not unique, among primates; and it has been the subject of numerous attempts to account for its evolution (Table 16.4). The most widely cited explanation for pairbonding maintains that females represent an "overdispersed resource" (van Schaik and Dunbar 1990) and that males are forced to accept a single mate due to mutual female intolerance (see Emlen and Oring 1977, Wrangham 1979). As detailed by Leighton (1987:144), female antagonism is the product of resource competition due to a feeding preference for ripe fruits that occur in small patches:

> The critical feature of these "small" patches is that they are large enough for one female and her offspring to use regularly, but small enough that if two females and their offspring visited the same patches, the dominant could not be assured a sufficient meal or high feeding rates.

As a result, females distribute themselves in exclusive ranges. Males accept a single mate only because they are unable to successfully defend the ranges occupied by two or more females. While playback experiments offer some support for this model, other investigators have questioned the assumption that gibbon males are unable to simultaneously defend multiple female ranges (van Schaik and Dunbar 1990, Reichard 2003b). van Schaik and Dunbar (1990) argue, for example, that gibbons travel far enough each day to defend the home ranges of as many as seven females, on average. They conclude, therefore, that some other factor must compel male/female spatial cohesion. Based on a comparison of four alternative hypotheses, van Schaik and Dunbar assert that the most likely explanation for pairbonding in primates, including gibbons, is the near universal threat of infanticide posed by extragroup males. While a specific test of the infanticide-protection hypothesis is confounded by the almost complete lack of evidence for infanticide in gibbons, van Schaik and Dunbar argue that the wariness of females during intergroup encounters is clear evidence that infants are at risk.

The case for primate pairbonding as a form of infanticide defense has been repeated many times (e.g., Sommer and Reichard 2000, van Schaik and Kappeler 2003), but as a general explanation for pairbonding in gibbons, infanticide

Table 16.4   Support for Alternative Hypotheses for Pair Bonding in Gibbons

HYPOTHESIS	SELECTED REFERENCE(S)	SUPPORT FOR HYPOTHESIS			
		WITTENBERGER AND TILSON 1980	VAN SCHAIK AND DUNBAR 1990	FUENTES 2000	REICHARD 2003A,B
Male paternal care	Kleiman 1977	−	−	+/−	n.c.
Females as overdispersed resource	Emlen and Oring 1977, Rutberg 1983	+	−	−	−
Mate guarding	Wittenberger and Tilson 1980, Palombit 1999	−	n.c.	+	−
Predation, protection	Tenaza 1975, Dunbar and Dunbar 1980	−	+/−	−	+/−
Exclusive resource defense	Rutberg 1983, Brockelman 2004	n.c.	−	+/−	−
Infanticide defense	Wrangham 1979, van Schaik and Dunbar 1990	n.c.	+	−	+/−

+, hypothesis was rejected; +/−, evaluation of the hypothesis was equivocal; −, hypothesis was rejected; n.c., hypothesis was not considered.

defense faces a number of significant hurdles (Palombit 1999, Brotherton and Komers 2003, Brockelman 2005). First, adult gibbons are monomorphic, and females may therefore be capable of defending infants themselves or making the cost of infanticidal attacks prohibitive. Second, unmated males, those who are expected to pose the greatest risk of infanticide, are rare in most gibbon populations. Third, males and females maintain spatial proximity even when no infants are present. Fourth, infanticide protection does not necessitate that gibbons live in two adult groups. Furthermore, pair living in gibbons is uniformly associated with territoriality and, as Brockelman (2005) has observed, territorial defense and infanticide protection have conflicting aims—territorial defense requires males to leave females and their infants during intergroup encounters, precisely when they are most vulnerable to attack by neighbor males.

As an alternative to the infanticide-protection hypothesis, Palombit (1996, 1999) argues that pairbonding in gibbons is a form of male mate guarding. Given the potential for EPC, it is argued that gibbon males may improve their fitness by limiting access to their female partner. The benefits of mate guarding are realized in two ways. In the short term, mate guarding allows a male to increase paternal certainty. In the long term, it limits a female's ability to evaluate potential sexual partners, thus ensuring the resident male future breeding opportunities. According to Palombit, this may account for the persistence of male–female bonds outside of estrus. For their part, females accept male companionship, not because of any tangible benefits but because the cost of rejecting it is too high (Gowaty 1996). However, the mate-defense hypothesis is vulnerable to some of the same criticisms detailed above. In particular, field observations indicate that EPCs typically occur during territorial encounters (Palombit 1994b, Reichard 1995, Brockelman 2005); thus, males invite EPC whenever they lead group travel toward territorial boundaries.

Pairbonds and Pair Living

The failure of either infanticide defense or mate defense to adequately account for gibbon behavior during intergroup encounters suggests that multiple selection pressures may act together to shape gibbon social organization (Fuentes 2000, 2002; Brockelman 2005). Almost forgotten in the historical shift toward models derived from sexual selection theory (Janson 2000, Paul 2002) is the importance of ecological pressures in limiting group size in primates (Chapman and Chapman 2000). Only recently have investigators refocused attention on the relationship between foraging efficiency and pair living in primates (Kappeler and van Schaik 2002, Isbell 2004, Brockelman 2005). Irrespective of mating strategies, it can be argued that the combination of small foraging parties (i.e., two-adult groups) and territoriality entails benefits to both males and females that would be lost if adults foraged alone. For example, gibbons may increase foraging efficiency by sharing

knowledge about resources (Brockelman 2005) or by ensuring that resources are not unknowingly depleted by adults foraging alone (Terborgh 1983). In fact, limited support for models based on foraging efficiency may be found in the observation that gibbons alter foraging in response to resource availability. Three observations are relevant. First, comparative data demonstrate that home range size is larger in forests where there are two or more hominoid competitors (Chivers 2001) or where the relative abundance of ripe fruit is limited (Mather 1992, McConkey 1999, Chen 1995); second, seasonal comparisons indicate that gibbons reduce levels of activity during periods of fruit scarcity (Chivers 1974, Raemaekers 1980, Bartlett 1999a); and third, longitudinal data from Khao Yai, Thailand, demonstrate that range size in lar gibbons fluctuates in response to group size (Brockelman 2005, T. Q. Bartlett unpublished data). If food is limiting to gibbons, strategies for increasing foraging efficiency should have demonstrable benefits in terms of reproductive rates. In fact, Ellefson (1974) hypothesized about such a relationship over 30 years ago; but despite circumstantial evidence (e.g., O'Brien et al. 2003), essential data on the relationship between resource access and reproductive success are almost completely absent (van Schaik and Dunbar 1990). It is ironic that while the last two decades of field research on wild gibbons have yielded unanticipated insights into the complex social lives of this important primate family, long-standing fundamental questions about the relationship between resource abundance and reproductive success remain unanswered. Fortunately, the current emphasis on population-level studies (e.g., Lappan and Whittaker 2004) promises to finally address these questions.

REFERENCES

Ahsan, M. F. (1994). Feeding ecology of the primates of Bangladesh. In: Thierry, B., Anderson, J. R., Roeder, J. J., and Herrenschmidt, N. (eds.), *Current Primatology. Ecology and Evolution*, vol. I. Université Louis Pasteur, Strasbourg. pp. 79–86.

Ahsan, M. F. (2001). Socio-ecology of the hoolock gibbon (*Hylobates hoolock*) in two forests of Bangladesh. In: *The Apes: Challenges for the 21st Century. Conference Proceedings*. Chicago Zoological Society, Chicago. pp. 286–299.

Alberts, S. (1987). Parental care in captive siamangs (*Hylobates syndactylus*). *Zoo Biol.* 6:401–406.

Aldrich-Blake, F. P. G., and Chivers, D. J. (1973). On the genesis of a group of siamang. *Am. J. Phys. Anthropol.* 38:631–636.

Bartlett, T. Q. (1999a). Feeding and ranging behavior of the white-handed gibbon (*Hylobates lar*) in Khao Yai National Park, Thailand [PhD thesis]. Washington University, St. Louis.

Bartlett, T. Q. (1999b). The gibbons. In: Dolhinow, P., and Fuentes, A. (eds.), *The Nonhuman Primates*. Mayfield, Mountain View, CA. pp. 44–49.

Bartlett, T. Q. (2003). Intragroup and intergroup social interactions in white-handed gibbons. *Int. J. Primatol.* 24:239–259.

Bleisch, W. V., and Chen, N. (1991). Ecology and behavior of wild black-crested gibbons (*Hylobates concolor*) in China with a reconsideration of evidence for polygyny. *Primates* 32:539–548.

Bourliere, F. (1964). *The Natural History of Mammals,* 3rd ed. Knopf, New York.

Brandon-Jones, D., Eudey, A. A., Geissmann, T., Groves, C. P., Melnick, D. J., Morales, J. C., Shekelle, M., and Stewart, C.-B. (2004). Asian primate classification. *Int. J. Primatol.* 25:97–164.

Bricknell, S. J. (1999). Hybridisation and behavioral variation: a socio-ecological study of hybrid gibbons (*Hylobates agilis albibarbis* × *H. muelleri*) in central Kalimantan, Indonesia [PhD thesis]. Australian National University, Canberra.

Brockelman, W. Y. (2004). Inheritance and selective effects of color phase in white-handed gibbons (*Hylobates lar*) in central Thailand. *Mamm. Biol.* 69:73–80.

Brockelman, W. Y. (2005). Ecology and the social system of gibbons. In: Galdikas, B. M. F., Briggs, N., Sheeran, L. K., Shapiro, G. L., and Goodall, J. (eds.), *All Apes Great and Small,* vol. II. Kluwer Academic/Plenum Publishers.

Brockelman, W. Y., and Gittins, S. P. (1984). Natural hybridization in the *Hylobates lar* species group: implications for speciation in gibbons. In: Preuschoft, H., Chivers, D. J., Brockelman, W. Y., and Creel, N. (eds.), *The Lesser Apes: Evolutionary and Behavioural Biology.* Edinburgh University Press, Edinburgh. pp. 498–532.

Brockelman, W. Y., Reichard, U., Treesucon, U., and Raemaekers, J. J. (1998). Dispersal, pair formation and social structure in gibbons (*Hylobates lar*). *Behav. Ecol. Sociobiol.* 42:329–339.

Brockelman, W. Y., and Schilling, D. (1984). Inheritance of stereotyped gibbon calls. *Nature* 312:634–636.

Brockelman, W. Y., and Srikosamatara, S. (1984). Maintenance and evolution of social structure in gibbons. In: Preuschoft, H., Chivers, D. J., Brockelman, W. Y., and Creel, N. (eds.), *The Lesser Apes: Evolutionary and Behavioural Biology.* Edinburgh University Press, Edinburgh. pp. 298–323.

Brotherton, P. N. M., and Komers, P. E. (2003). Mate guarding and the evolution of social monogamy in mammals. In: Reichard, U. H., and Boesch, C. (eds.), *Monogamy: Mating Strategies and Partnerships in Birds, Humans, and Other Mammals.* Cambridge University Press, Cambridge. pp. 42–58.

Cant, J. G. H. (1992). Positional behavior and body size of arboreal primates: a theoretical framework for field studies and an illustration of its application. *Am. J. Phys. Anthropol.* 88:273–284.

Chapman, C. A., and Chapman, L. J. (2000). Determinants of group size in primates: the importance of travel costs. In: Boinski, S., and Garber, P. A. (eds.), *On the Move: How and Why Animals Travel in Groups.* University of Chicago Press, Chicago. pp. 24–42.

Chen, N. (1995). Ecology of the black-crested gibbon (*Hylobates concolor*) in the Ailao Mt. Reserve, Yunnan, China [MA thesis]. Mahidol University, Bangkok, Thailand.

Chivers, D. J. (1974). *The Siamang in Malaya: A Field Study of a Primate in Tropical Rain Forest. Contributions to Primatology.* S. Karger, Basel.

Chivers, D. J. (1977). The lesser apes. In: Prince Rainier of Monaco and Bourne, G. H. (eds.), *Primate Conservation.* Academic Press, New York. pp. 539–598.

Chivers, D. J. (1980). Introduction. In: Chivers, D. J. (ed.), *Malayan Forest Primates. Ten Years' Study in Tropical Rain Forest.* Plenum Press, New York. pp. 1–27.

Chivers, D. J. (1984). Feeding and ranging in gibbons: a summary. In: Preuschoft, H., Chivers, D. J., Brockelman, W. Y., and Creel, N. (eds.), *The Lesser Apes: Evolutionary and Behavioural Biology.* Edinburgh University Press, Edinburgh. pp. 267–281.

Chivers, D. J. (2001). The swinging singing apes: fighting for food and family in the far-east forests. In: *The Apes: Challenges for the 21st Century. Conference Proceedings.* Chicago Zoological Society, Chicago. pp. 1–28.

Chivers, D. J., and Raemaekers, J. J. (1986). Natural and synthetic diets of Malayan gibbons. In: Else, J. G., and Lee, P. C. (eds.), *Primate Ecology and Conservation.* Cambridge University Press, New York. pp. 39–56.

Corlett, R. T. (1998). Frugivory and seed dispersal by vertebrates in the Oriental (Indomalayan) region. *Biol. Rev.* 73:413–448.

Dallmann, R., and Geissmann, T. (2001). Different levels of variability in the female song of wild silvery gibbons (*Hylobates moloch*). *Behaviour* 138:629–648.

Dielentheis, T. F., Zaiss, E., and Geissmann, T. (1991). Infant care in a family of siamangs (*Hylobates syndactylus*) with twin offspring at Berlin Zoo. *Zoo Biol.* 10:309–317.

Di Fiore, A., and Rodman, P. S. (2001). Time allocation patterns of lowland woolly monkeys (*Lagothrix lagotricha poeppigii*) in a neotropical terra firma forest. *Int. J. Primatol.* 22:449–480.

Dunbar, R. I. M., and Dunbar, E. P. (1980). The pairbond in klipspringer. *Anim. Behav.* 28:219–229.

Ellefson, J. (1974). A natural history of white-handed gibbons in the Malayan Peninsula. In: Rumbaugh, D. M. (ed.), *Gibbon and Siamang.* Karger, Basel. 3:1–136.

Emlen, S. T., and Oring, L. W. (1977). Ecology, sexual selection and the evolution of mating systems. *Science* 197:215–223.

Feeroz, M. M., Islam, M. A., and Kabir, M. M. (1994). Food and feeding behaviour of hoolock gibbon (*Hylobates hoolock*) capped langur (*Presbytis pileata*) and pigtailed macaque (*Macaca nemestrina*) of Lawachara. *Bangladesh J. Zool.* 22:123–132.

Fischer, J. O., and Geissmann, T. (1990). Group harmony in gibbons: comparison between white-handed gibbon (*Hylobates lar*) and siamang (*H. syndactylus*). *Primates* 31:481–494.

Fleagle, J. G. (1976). Locomotion and posture of the Malayan siamang and implications for hominoid evolution. *Folia Primatol.* 26:245–269.

Fooden, J. (1971). Color and sex in gibbons. *Bull. Field Mus. Nat. Hist.* 42:2–7.

Fuentes, A. (1998). Re-evaluating primate monogamy. *Am. Anthropol.* 100:890–907.

Fuentes, A. (2000). Hylobatid communities: changing views on pair bonding and social organization in hominoids. *Ybk. Phys. Anthropol.* 43:33–60.

Fuentes, A. (2002). Patterns and trends in primate pair bonds. *Int. J. Primatol* 23:953–978.

Geissmann, T. (1984). Inheritance of song parameters in the gibbon song, analysed in 2 hybrid gibbons (*Hylobates pileatus* × *H. lar*). *Folia Primatol.* 42:216–235.

Geissmann, T. (1991). Reassessment of age of sexual maturity in gibbons. *Am. J. Primatol.* 23:11–22.

Geissmann, T. (1993). Evolution of communication in gibbons (Hylobatidae) [PhD diss.]. Universitaet Zuerich, Zuerich.

Geissmann, T. (1995). Gibbon systematics and species identification. *Int. Zoo News* 42:467–501.

Geissmann, T. (2002a). Duet-splitting and the evolution of gibbon songs. *Biol. Rev.* 77:57–76.

Geissmann, T. (2002b). Taxonomy and evolution of gibbons. *Evol. Anthropol. Suppl.* 1:28–31.

Geissmann, T. (2004). "The Gibbons (Hylobatidae): An Introduction," http://www.tiho-hannover.de/gibbons/main/index.html (accessed September 14, 2004).

Geissmann, T., and Orgeldinger, M. (2000). The relationship between duet songs and pair bonds in siamangs, *Hylobates syndactylus. Anim. Behav.* 60:805–809.

Gittins, S. P. (1978). The species range of the gibbon *Hylobates agilis*. In: Chivers, D. J., and Joysey, K. A. (eds.), *Recent Advances in Primatology. Evolution*, vol. 3. Academic Press, New York. pp. 319–321.

Gittins, S. P. (1979). The behaviour and ecology of the agile gibbon (*Hylobates agilis*) [PhD diss.]. Cambridge University, Cambridge.

Gittins, S. P. (1980). Territorial behavior in the agile gibbon. *Int. J. Primatol.* 1:381–399.

Gittins, S. P. (1982). Feeding and ranging in the agile gibbon. *Folia Primatol.* 38:39–71.

Gittins, S. P. (1984). Territorial advertisement and defense in gibbons. In: Preuschoft, H., Chivers, D. J., Brockelman, W. Y., and Creel, N. (eds.), *The Lesser Apes: Evolutionary and Behavioural Biology*. Edinburgh University Press, Edinburgh. pp. 420–424.

Gittins, S. P., and Raemaekers, J. J. (1980). Siamang, lar and agile gibbons. In: Chivers, D. J. (ed.), *Malayan Forest Primates: Ten Years' Study in Tropical Rain Forest*. Plenum Press, New York. pp. 63–105.

Gowaty, P. A. (1996). Battles of the sexes and origins of monogamy. In: Black, J. M. (ed.), *Partnerships in Birds: The Study of Monogamy*. Oxford University Press, Oxford. pp. 21–52.

Groves, C. P. (1972). Systematics and phylogeny of gibbons. *Gibbon Siamang* 1:1–89.

Groves, C. P. (1984). A new look at the taxonomy and phylogeny of the gibbons. In: Preuschoft, H., Chivers, D. J., Brockelman, W. Y., and Creel, N. (eds.), *The Lesser Apes: Evolutionary and Behavioural Biology*. Edinburgh University Press, Edinburgh. pp. 542–561.

Groves, C. P. (2001). *Primate Taxonomy*. Smithsonian Institution Press, Washington DC.

Haimoff, E. H., Yang, X.-J., He, S.-J., and Chen, N. (1987). Preliminary observations of wild black-crested gibbons (*Hylobates concolor concolor*) in Yunnan Province, People's Republic of China. *Primates* 28:319–335.

Hall, L. M., Jones, D. S., and Wood, B. A. (1998). Evolution of gibbon subgenera inferred from cytochrome *b* DNA sequence data. *Mol. Phylogenet. Evol.* 10:281–286.

Hayashi, S., Hayasaka, K., Takenaka, O., and Horai, S. (1995). Molecular phylogeny of gibbons inferred from mitochondrial DNA sequences: preliminary report. *Mol. Evol.* 41:359–365.

Hu, Y., Xu, H., and Yang, D. (1989). The studies on ecology of *Hylobates leucogenys* [in Chinese]. *Zool. Res.* 10(suppl.):61–66.

Hu, Y., Xu, H., and Yang, D. (1990). Feeding ecology of the white-cheek gibbon (*Hylobates concolor leucogenys*) [in Chinese]. *Acta Ecol. Sin.* 10:155–159.

Isbell, L. A. (2004). Is there no place like home? Ecological bases of female dispersal and philopatry and their consequences for the formation of kin groups. In: Chapais, C., and Berman, C. (eds.), *Kinship and Behavior in Primates*. Oxford University Press, New York. pp. 71–108.

Islam, M. A., and Feeroz, M. M. (1992). Ecology of hoolock gibbon of Bangladesh. *Primates* 33:451–464.

Janson, C. H. (2000). Primate socio-ecology: the end of a golden age. *Evol. Anthropol.* 9:73–86.

Kappeler, M. (1984). Diet and feeding behavior of the moloch gibbon. In: Preuschoft, H., Chivers, D. J., Brockelman, W. Y., and Creel, N. (eds.), *The Lesser Apes: Evolutionary and Behavioural Biology*. Edinburgh University Press, Edinburgh. pp. 228–241.

Kappeler, M., and van Schaik, C. P. (2002). Evolution of primate social systems. *Int. J. Primatol.* 23:707–740.

Kleiman, D. (1977). Monogamy in mammals. *Q. Rev. Biol.* 52:39–69.

Kloss, C. B. (1908). The white-handed gibbon. *Journal of the Straits Branch of the Royal Asiatic Society* 50:79–80.

Knott, C. D. (1998). Changes in orangutan caloric intake, energy balance, and ketones in response to fluctuating fruit availability. *Int. J. Primatol.* 19:1061–1077.

Lan, D. Y. (1989). Preliminary study on the group composition, behavior and ecology of the black gibbons (*Hylobates concolor*) in southwest Yunnan. *Zool. Res.* 10(suppl.):119–126.

Lan, D. Y. (1993). Feeding and vocal behaviours of black gibbons (*Hylobates concolor*) in Yunnan: a preliminary study. *Folia Primatol.* 60:94–105.

Lan, D. Y., and Sheeran, L. K. (1995). The status of black gibbons (*Hylobates concolor jingdongensis*) at Xiaobahe Wuliang Mountains, Yunnan Province, China. *Asian Primates* 5:2–4.

Lappan, S., and Whittaker, D. J. (2004). Trends in the study of gibbon populations, 1940–2004 [abstract]. Presented at the International Primatological Society Congress, Torino, Italy, August 24, 2004.

Lee, P. C. (1996). The meanings of weaning: growth, lactation, and life history. *Evol. Anthropol.* 5:87–96.

Leighton, M. (1987). Gibbons: territoriality and monogamy. In: Smuts, B. B., Cheney, D. L., Seyfarth, R. M., Wrangham, R. W., and Struhsaker, T. T. (eds.), *Primate Societies*. University of Chicago Press, Chicago. pp. 135–145.

Leighton, M. (1993). Modeling dietary selectivity by Bornean orangutans: evidence for integration of multiple criteria in fruit selection. *Int. J. Primatol.* 14:257–313.

Leighton, M., and Leighton, D. R. (1983). Vertebrate responses to fruiting seasonality within a Bornean rain forest. In: Sutton, S. L., Whitmore, T. C., and Chadwick, A. C. (eds.), *Tropical Rain Forests: Ecology and Management*. Blackwell Scientific Publications, Oxford. pp. 181–196.

MacArthur, R. H., and Pianka, E. R. (1966). An optimal use of a patchy environment. *Am. Nat.* 100:603–609.

MacKinnon, J. R. (1977). A comparative ecology of Asian apes. *Primates* 18:747–772.

MacKinnon, J. R., and MacKinnon, K. S. (1980). Niche differentiation in a primate community. In: Chivers, D. J. (ed.), *Malayan Forest Primates: Ten Years Study in Tropical Rain Forest*. Plenum Press, New York. pp. 167–191.

Marshall, J. T., and Marshall, E. R. (1976). Gibbons and their territorial songs. *Science* 193:235–238.

Marshall, J., and Sugardjito, J. (1986). Gibbon systematics. In: Swindler, D. R., and Erwin, J. (eds.), *Comparative Primate Biology. Systematics, Evolution, and Anatomy*, vol. 1. Alan R. Liss, New York. pp. 137–185.

Mather, R. J. (1992). A field study of hybrid gibbons in central Kalimantan, Indonesia [PhD thesis]. University of Cambridge, Cambridge.

McConkey, K. R. (1999). Gibbons as seed dispersers in the rain forests of central Borneo [PhD thesis]. Cambridge University, Cambridge.

McConkey, K. R. (2000). Primary seed shadow generated by gibbons in the rain forests of Barito Ulu, central Borneo. *Am. J. Primatol.* 52:13–29.

McConkey, K. R., Aldy, F., Ario, A., and Chivers, D. J. (2002). Selection of fruit by gibbons (*Hylobates muelleri* × *agilis*) in the rain forests of central Borneo. *Int. J. Primatol.* 23:123–145.

McConkey, K. R., Ario, A., Aldy, F., and Chivers, D. J. (2003). Influence of forest seasonality on gibbon food choice in the rain forests of Barito Ulu, central Kalimantan. *Int. J. Primatol.* 24:19–32.

Mitani, J. C. (1984). The behavioral regulation of monogamy in gibbons (*Hylobates muelleri*). *Behav. Ecol. Sociobiol.* 15:225–229.

Mitani, J. C. (1985). Location-specific responses of gibbons (*Hylobates muelleri*) to male songs. *Z. Tierpsychol.* 70:219–224.

Mitani, J. C. (1987). Species discrimination of male song in gibbons. *Am. J. Primatol.* 13:413–423.

Mitani, J. C. (1990). Demography of agile gibbons (*Hylobates agilis*). *Int. J. Primatol.* 11:411–424.

Mitani, J. C., and Rodman, P. S. (1979). Territoriality: the relation of ranging pattern and home range size to defendability, with an analysis of territoriality among primate species. *Behav. Ecol. Sociobiol.* 5:241–251.

Mootnick, A., and Groves, C. (2005). A new generic name for the hoolock gibbon (Hylobatidae). *Int. J. Primatol.* 26:971–976.

Napier, J. R., and Napier, P. H. (1985). *The Natural History of the Primates.* Cambridge, MA: MIT Press.

O'Brien, T. G., Kinnaird, M. F., Nurcahyo, A., Iqbal, M., and Rusmanto, M. (2004). Abundance and distribution of sympatric gibbons in a threatened Sumatran rain forest. *Int. J. Primatol.* 25:267–284.

O'Brien, T. G., Kinnaird, M. F., Nurcahyo, A., Prasetyaningrum, M., and Iqbal, M. (2003). Fire, demography and the persistence of siamang (*Symphalangus syndactylus*: Hylobatidae) in a Sumatran rainforest. *Anim. Conserv.* 6:115–121.

Okra, T., and Takenaka, O. (2001). Wild gibbons' parentage tested by non-invasive DNA sampling and PCR-amplified polymorphic microsatellites. *Primates* 42:67–73.

Palombit, R. A. (1992). Pair bonds and monogamy in wild siamang (*Hylobates syndactylus*) and white-handed gibbon (*Hylobates lar*) in northern Sumatra [PhD thesis]. University of California, Davis.

Palombit, R. A. (1993). Lethal territorial aggression in a white-handed gibbon. *Am. J. Primatol.* 31:311–318.

Palombit, R. A. (1994a). Dynamic pair bonds in hylobatids: implications regarding monogamous social systems. *Behaviour* 128:65–101.

Palombit, R. A. (1994b). Extra-pair copulations in a monogamous ape. *Anim. Behav.* 47:721–723.

Palombit, R. A. (1995). Longitudinal patterns of reproduction in wild female siamang (*Hylobates syndactylus*) and white-handed gibbons (*Hylobates lar*). *Int. J. Primatol.* 16:739–760.

Palombit, R. A. (1996). Pair bonds in monogamous apes: a comparison of the siamang *Hylobates syndactylus* and the white-handed gibbon *Hylobates lar*. *Behaviour* 133:321–356.

Palombit, R. A. (1997). Inter- and intraspecific variation in the diets of sympatric siamang (*Hylobates syndactylus*) and lar gibbons (*Hylobates lar*). *Folia Primatol.* 68:321–337.

Palombit, R. A. (1999). Infanticide and the evolution of pair bonds in nonhuman primates. *Evol. Anthropol.* 7:117–129.

Paul, A. (2002). Sexual selection and mate choice. *Int. J. Primatol.* 23:877–904.

Povinelli, D. J., and Cant, J. G. H. (1995). Arboreal clambering and the evolution of self-conception. *Q. Rev. Biol.* 70:393–421.

Preuschoft, H., Chivers, D. J., Brockelman, W. Y., and Creel, N. (eds.) (1984). *The Lesser Apes: Evolutionary and Behavioural Biology.* Edinburgh University Press, Edinburgh.

Prouty, L. A., Buchanan, P. D., Pollitzer, W. S., and Mootnick, A. R. (1983a). A presumptive new hylobatid subgenus with 38 chromosomes. *Cytogenet. Cell Genet.* 35:141–142.

Prouty, L. A., Buchanan, P. D., Pollitzer, W. S., and Mootnick, A. R. (1983b). *Bunopithecus*: a genus-level taxon for the hoolock gibbon (*Hylobates hoolock*). *Am. J. Primatol.* 5:83–87.

Pyke, G. H., Pulliam, H. R., and Charnov, E. L. (1977). Optimal foraging: a selective review of theory and tests. *Q. Rev. Biol.* 52:137–154.

Quiatt, D. (1985). The "household" in non-human primate evolution: a basic linking concept. *Anthropol. Contemp.* 3:187–193.

Quiatt, D. (1987). *A-group: one example of gibbon social organization.* Videotape available from the Primate Center Library, Wisconsin Regional Primate Research Center, Madison.

Quiatt, D., and Kelso, J. (1987). The concept of the household: linking behavior and genetic analyses. *Hum. Evol.* 2:429–435.

Rabinowitz, A. (1989). The density and behavior of large cats in a dry tropical forest mosaic in Huai Kha Khaeng Wildlife Sanctuary, Thailand. *Nat. Hist. Bull. Siam Soc.* 37:235–251.

Raemaekers, J. J. (1979). Ecology of sympatric gibbons. *Folia Primatol.* 31:227–245.

Raemaekers, J. J. (1980). Causes of variation between months in the distance traveled daily by gibbons. *Folia Primatol.* 34:46–60.

Raemaekers, J. J., and Chivers, D. J. (1980). Socio-ecology of Malayan forest primates. In: Chivers, D. J. (ed.), *Malayan Forest Primates: Ten Years' Study in Tropical Rain Forest.* Plenum Press, New York. pp. 279–316.

Raemaekers, J. J., and Raemaekers, P. M. (1984). Vocal interactions between two male gibbons, *Hylobates lar*. *Nat. Hist. Bull. Siam Soc.* 32:95–106.

Raemaekers, J. J., and Raemaekers, P. M. (1985). Field playback of loud calls to gibbons (*Hylobates lar*): territorial, sex-specific, and species-specific responses. *Anim. Behav.* 33:481–493.

Raemaekers, J. J., Raemaekers, P. M., and Haimoff, E. H. (1984). Loud calls of the gibbon (*Hylobates lar*): repertoire, organisation and context. *Behaviour* 91:147–189.

Reichard, U. (1995). Extra-pair copulation in a monogamous gibbon (*Hylobates lar*). *Ethology* 100:99–112.

Reichard, U. (2003a). Monogamy: past and present. In: Reichard, U. H., and Boesch, C. (eds.), *Monogamy: Mating Strategies and Partnerships in Birds, Humans, and Other Mammals.* Cambridge University Press, Cambridge. pp. 3–25.

Reichard, U. (2003b). Social monogamy in gibbons: the male perspective. In: Reichard, U. H., and Boesch, C. (eds.), *Monogamy: Mating Strategies and Partnerships in Birds, Humans, and Other Mammals.* Cambridge University Press, Cambridge. pp. 190–213.

Reichard, U., and Sommer, V. (1997). Group encounters in wild gibbons (*Hylobates lar*): agonism, affiliation, and the concept of infanticide. *Behaviour* 134:1135–1174.

Roos, C., and Geissmann, T. (2001). Molecular phylogeny of the major hylobatid divisions. *Mol. Phylogenet. Evol.* 19:486–494.

Rowe, N. (1996). *The Pictorial Guide to the Living Primates.* Pogonias, East Hampton, NY.

Rutberg, A. T. (1983). The evolution of monogamy in primates. *J. Theor. Biol.* 104:93–112.

Schneider, G. (1906). Ergebnisse zoologischer Forschungsreisen in Sumatra. *Zool. Jahr. Abt. Syst. Geogr. Biol. Tiere* 23:1–172.

Sheeran, L. K., Yongzu, Z., Poirier, F. E., and Dehua, Y. (1998). Preliminary report on the behavior of the Jingdong black gibbon (*Hylobates concolor jingdongensis*). *Trop. Biodivers.* 5:113–125.

Shoener T. W. (1971). Theory of feeding strategies. *Annu. Rev. Ecol. Syst.* 2:369–404.

Simpson, G. G. (1945). The Principles of Classification and a Classification of Mammals. Bulletin of the American Museum of Natural History 85:1–350.

Sommer, V., and Reichard, U. (2000). Rethinking monogamy: the gibbon case. In: Kappeler, P. M. (ed.), *Primate Males: Causes and Consequences of Variation in Group Composition.* Cambridge University Press, Cambridge. pp. 159–168.

Srikosamatara, S. (1984). Ecology of pileated gibbons in southeast Thailand. In: Preuschoft, H., Chivers, D. J., Brockelman, W. Y., and Creel, N. (eds.), *The Lesser Apes: Evolutionary and Behavioural Biology.* Edinburgh University Press, Edinburgh. pp. 242–257.

Strier, K. B. (2003). *Primate Behavioral Ecology.* Allyn and Bacon, New York.

Suwanvecho, U. (2003). Ecology and interspecific relations of two sympatric *Hylobates* species (*H. lar* and *H. pileatus*) in Khao Yai National Park, Thailand [PhD thesis]. Mahidol University, Bangkok, Thailand.

Tenaza, R. R. (1975). Territory and monogamy among Kloss' gibbons (*Hylobates klossii*) in Siberut Island, Indonesia. *Folia Primatol.* 24:60–80.

Tenaza, R. R. (1985). Songs of hybrid gibbons (*Hylobates lar* × *H. muelleri*). *Am. J. Primatol.* 8:249–253.

Terborgh, J. (1983). *Five New World Primates.* Princeton University Press, Princeton, NJ.

Tilson, R. L. (1981). Family formation strategies of Kloss's gibbons. *Folia Primatol.* 35:259–287.

Uhde, N. L., and Sommer, V. (2002). Antipredatory behavior in gibbons (*Hylobates lar*, Khao Yai/Thailand). In: Miller, L. E. (ed.), *Eat or Be Eaten: Predator Sensitive Foraging Among Primates.* Cambridge University Press, New York. pp. 268–291.

Ungar, P. S. (1995). Fruit preferences of four sympatric primate species at Ketambe, northern Sumatra, Indonesia. *Int. J. Primatol.* 16:221–245.

Ungar, P. S. (1996). Feeding height and niche separation in sympatric Sumatran monkeys and apes. *Folia Primatol.* 67:163–168.

van Schaik, C. P., and Dunbar, R. I. M. (1990). The evolution of monogamy in large primates: a new hypothesis and some crucial tests. *Behaviour* 115:30–62.

van Schaik, C. P., and Kappeler, P. M. (2003). The evolution of social monogamy in primates. In: Reichard, U. H., and Boesch, C. (eds.), *Monogamy: Mating Strategies and Partnerships in Birds, Humans, and Other Mammals.* Cambridge University Press, Cambridge. pp. 59–80.

van Schaik, C. P., and van Hooff, J. A. R. A. M. (1983). On the ultimate causes of primate social systems. *Behaviour* 85:91–117.

van Schaik, C. P., van Noordwijk, M. A., Warsono, B., and Sutriono, E. (1983). Party size and early detection of predators in Sumatran forest primates. *Primates* 24:211–221.

West, K. (1981). The behavior and ecology of the siamang in Sumatra [MA thesis]. University of California, Davis.

Whitington, C., and Treesucon, U. (1991). Selection and treatment of food plants by white-handed gibbons (*Hylobates lar*) in Khao Yai National Park, Thailand. *Nat. Hist. Bull. Siam Soc.* 39:111–122.

Whitten, A. J. (1982a). Diet and feeding behaviour of Kloss gibbons on Siberut Island, Indonesia. *Folia Primatol.* 37:177–208.

Whitten, A. J. (1982b). Home range use by Kloss gibbons (*Hylobates klossii*) on Siberut Island, Indonesia. *Anim. Behav.* 30:182–198.

Whitten, A. J. (1984). Ecological comparisons between Kloss gibbons and other small gibbons. In: Preuschoft, H., Chivers, D. J., Brockelman, W. Y., and Creel, N. (eds.), *The Lesser Apes: Evolutionary and Behavioural Biology.* Edinburgh University Press, Edinburgh. pp. 219–227.

Wickler, W. (1980). Vocal duetting and the pair bond. *Z. Tierpsychol.* 52:201–209.

Wickler, W., and Seibt, U. (1983). Monogamy: an ambiguous concept. In: Bateson, P. (ed.), *Mate Choice.* Cambridge University Press, Cambridge.

Wittenberger, J. F., and Tilson, R. L. (1980). The evolution of monogamy: hypothesis and evidence. *Annu. Rev. Ecol. Syst.* 11:197–232.

Wrangham, R. W. (1979). On the evolution of ape social systems. *Soc. Sci. Inf.* 18:335–368.

Yerkes, R. M., and Yerkes, A. (1929). *The Great Apes: A Study in Anthropoid Life.* Yale University Press, New Haven.

17

Orangutans in Perspective
Forced Copulations and Female Mating Resistance
Cheryl D. Knott and Sonya M. Kahlenberg

INTRODUCTION

Orangutans (genus *Pongo*) represent the extreme of many biological parameters. Among mammals they have the longest interbirth interval (up to 9 years) and are the largest of those that are primarily arboreal. They are the most solitary of the diurnal primates and the most sexually dimorphic of the great apes. They range over large areas and do not have obviously distinct communities. Additionally, adult male orangutans come in two morphologically distinct types, an unusual phenomenon known as *bimaturism*. Some males also pursue a mating strategy that is rare among mammals: they obtain a large proportion of copulations by force.

Although intriguing, this suite of features has proven difficult for researchers to fully understand. The semi-solitary lifestyle of orangutans combined with their slow life histories and large ranges means that data accumulate slowly. Also, rapid habitat destruction and political instability in Southeast Asia have left only a handful of field sites currently in operation (Fig. 17.1). Despite these difficulties, long-term behavioral studies and recent genetic and hormonal data enable us to begin answering some of the most theoretically interesting questions about this endangered great ape. In this chapter, we summarize what is currently known about the taxonomy and distribution, ecology, social organization, reproductive parameters, and conservation of wild orangutans. We also discuss recent advances in our

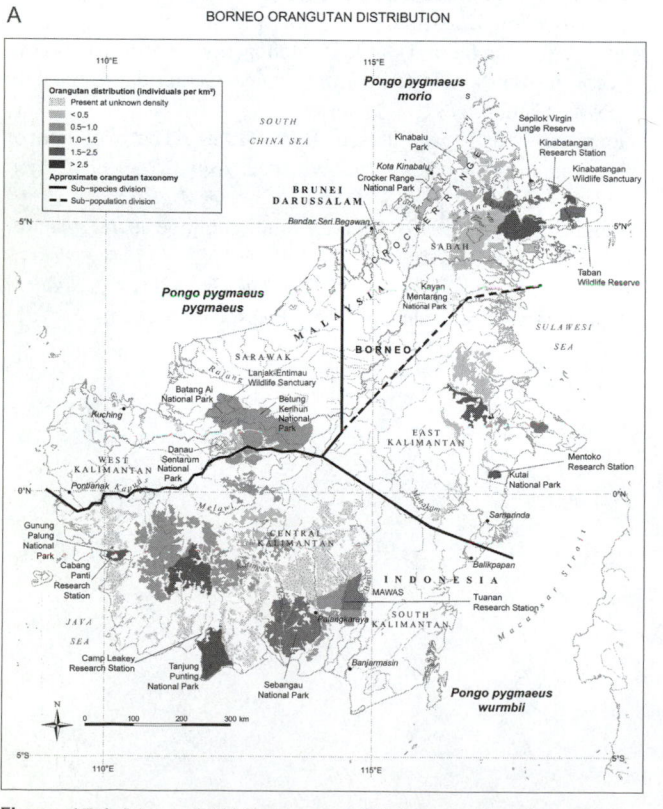

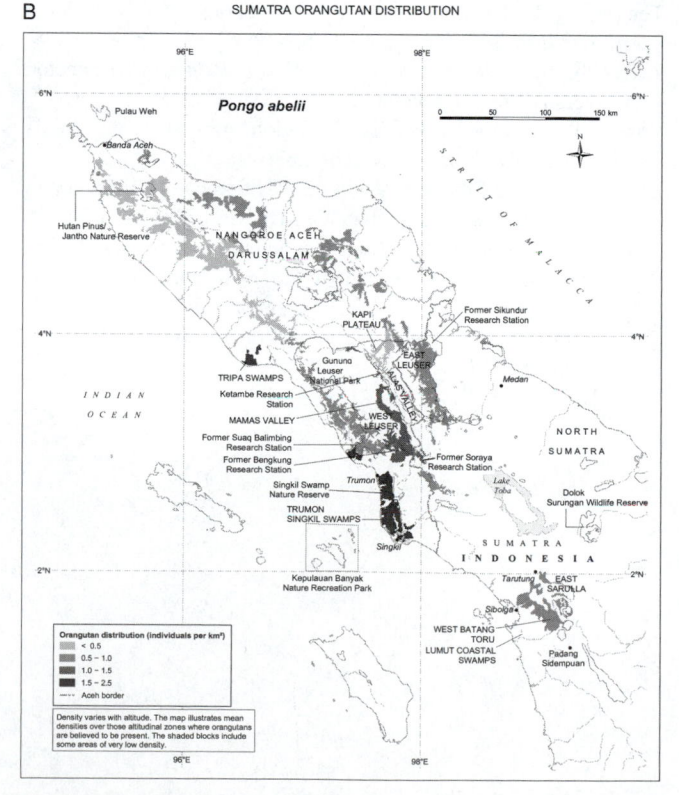

Figure 17.1 Maps of (A) Borneo and (B) Sumatra showing species and sub-species distinctions, study sites, densities and distributions [modified from Caldecott and Miles (2005) and Singleton et al. (2004) by Kee, L. and Hendrickson, K.].

understanding of one distinctive orangutan behavior—forced copulations.

TAXONOMY AND DISTRIBUTION

Evolutionary History and Past and Present Distribution

Orangutans evolved in Asia during the Miocene epoch. *Sivapithecus*, a Miocene ape, was hypothesized to be their putative ancestor, based on similarities in the face and palate (Pilbeam 1982); but later details of its lower jaw (Ward 1997) and postcrania (Pilbeam et al. 1990) indicate it is instead a sister taxon. The teeth of *Lufengpithecus* show similarities to those of orangutans (Kelley 2002), but differences in its face and periorbital region also exclude it from *Pongo* ancestry (Chaimanee et al. 2004). Recently, Chaimanee and colleagues (2004) found a new hominoid, *Khoratpithecus piriyai*, in Thailand from the early Late Miocene (7–9 million years ago), which shares unique, derived traits with orangutans (such as the absence of anterior digastric muscles) and is therefore a likely ancestral candidate.

At the height of their distribution, orangutans ranged throughout the tropical and subtropical regions of eastern and southeastern Asia. Orangutan fossils and subfossils primarily consist of teeth and jaw fragments found in China, Thailand, Vietnam, Sumatra, Java, and possibly Burma and northeastern India (Bacon and Long 2001; Gu et al. 1987; Hooijer 1948, 1960; Kahlke 1972; Pei 1935; Schwartz et al. 1994, 1995). These extinct populations had significantly larger teeth than do modern orangutans (Bacon and Long 2001, Hooijer 1948, Schwartz et al. 1995). The first complete fossil orangutan skeleton was recently described (Bacon and Long 2001). The specimen has a small body and proportionally longer arms, particularly forearms, than legs (Bacon and Long 2001), consistent with arboreality.

Orangutans are now restricted to the islands of Borneo and Sumatra. Both ecological and anthropogenic factors seem to be responsible for their range collapse (Delgado and van Schaik 2000). Tropical and subtropical zones shifted during the Pleistocene (Jablonski 1998, Jablonski et al. 2000), and mean annual temperature rose (Harrison 1999, Medway 1977), leading to habitat constriction. Among other evidence of human influence (Delgado and van Schaik 2000), orangutan remains found in caves in northern Borneo show that they were preyed upon by humans (Hooijer 1960, Medway 1977).

Orangutans live in true wet rain forests with reported average annual rainfall between 2,000 (Galdikas 1988) and 4,500 (Lawrence and Leighton 1996) mm. The primary habitat types they occupy are peat and freshwater swamps and lowland forests. Orangutans have been found living up to an altitude of 1,200 m; however, their distribution declines with increasing elevation (Djojosudharmo and van Schaik 1992).

Taxonomic Status

Formerly, orangutans were classified into two subspecies (Courtenay et al. 1988, Jones 1969). However, orangutan taxonomy has been under recent study and revision, with the majority of molecular and morphological analyses finding the level of differentiation between Sumatran and Bornean populations to be equal to or greater than that between the two species of chimpanzee (Bruce and Ayala 1979, Ferris et al. 1981, Groves 1986, Janczewski et al. 1990, Ruvolo et al. 1994, Ryder and Chemnick 1993, Uchida 1998, Warren et al. 2001, Xu and Arnason 1996, Zhi et al. 1996). Thus, recent taxonomies now elevate the subspecies distinction between Sumatran and Bornean orangutans to the species level (Groves 2001), distinguishing them as *Pongo pygmaeus* in Borneo and *P. abelii* in Sumatra. Warren et al. (2001) estimate a divergence time of approximately 1.1 million years ago. However, dissenting opinions exist (Muir et al. 1995, 1998), and Bornean and Sumatran orangutans interbreed in captivity and produce fertile offspring. Although morphological and behavioral differences are emerging between the two orangutan species (Delgado and van Schaik 2000), the level of observed variation does not approach that witnessed within other hominoid genera, such as *Pan*.

The Bornean species shows further genetic (Warren et al. 2000, 2001; Zhi et al. 1996) and morphological (Groves 1986, Groves et al. 1992, Uchida 1998) variation, with most studies pointing to geographically distinct populations, separated in large part by major rivers. Groves (2001) thus separates the Bornean orangutan into three subspecies: *P. pygmaeus pygmaeus* in Sarawak and northwestern Kalimantan, *P. p. wurmbii* in southwestern and central Kalimantan, and *P. p. morio* in Sarawak and east Kalimantan (Fig. 17.1). Furthermore, analysis of the mitochondrial DNA control region reveals four subpopulations on Borneo that diverged 860,000 years ago (Warren et al. 2001). These populations correspond to the *pygmaeus* and *wurmbii* subspecies, with the *morio* subspecies being further subdivided into Sabah and east Kalimantan populations (Warren et al. 2001).

Morphology

Often referred to as the "red ape," orangutans are covered with thick reddish orange hair. Despite this coloration, orangutans are often hard to see as they spend over 95% of their time high up in the canopy (Knott 2004). Orangutans negotiate the canopy through *quadrumanual clambering*, grasping supports with both their hands and feet (Knott 2004). Morphological adaptations for arboreality include arms, hands, and feet that are longer than those of humans and the other great apes (Fleagle 1999) as well as a shallow hip joint that allows them to hang from any hand–foot combination because legs can be extended over 90 degrees (MacLatchy 1996).

Orangutans are unusual because males appear to have indeterminate growth, meaning that they put on weight

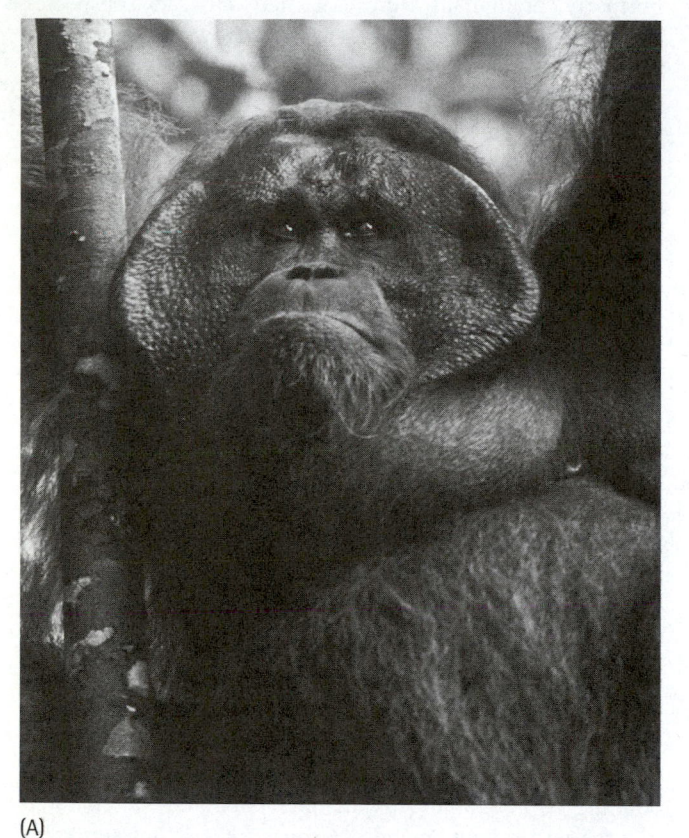

(A)

(B)

Figure 17.2 Photos of (A) flanged and (B) unflanged males from Gunung Palung National Park (photos by Tim Laman).

throughout their lives (Leigh and Shea 1995). This growth pattern has been described only for a handful of other mammals and is thought to be associated with intense inter-male competition. Orangutans are also exceptional for having two morphologically distinct types of adult male (Galdikas 1985a,b; Graham and Nadler 1990; Kingsley 1982, 1988; MacKinnon 1979; Maggioncalda et al. 1999, 2000, 2002; Mitani 1985a,b; Schürmann and van Hooff 1986; te Boekhorst et al. 1990). One type is large, weighing over 80 kg in the wild (Markham and Groves 1990), and possesses secondary sexual characteristics, including projecting cheek pads or "flanges," a pendulous throat pouch, a coat of long hair, and a musky odor (reviewed in Crofoot and Knott in press) (Fig. 17.2A). Several times daily, these so-called flanged males produce loud vocalizations known as long calls, which are audible to humans up to 800 m away (Mitani 1985b). Long calls seem to mediate spacing between flanged males (Mitani 1985b), coordinate community movements (Delgado and van Schaik 2000), and perhaps attract females (Delgado in press, Fox 2002, Galdikas 1983, Utami and Mitra Setia 1995, Utami Atmoko 2000). The second type of male is up to half the size of flanged males and does not have the secondary sexual characteristics described above or produce long calls (Fig. 17.2B). These unflanged males, were originally referred to as *subadults* (e.g., Galdikas 1985a,b), but this

term is misleading since these males can remain in this stage for over 20 years (Utami Atmoko and van Hooff 2004) and are fully capable of siring offspring (Kingsley 1982, 1988; Utami et al. 2002). Endocrinological data from captive orangutans show that the decoupling of fertility and secondary sexual trait development in unflanged males results from gonadotropin and testicular steroid levels being sufficient for spermatogenesis but inadequate for triggering secondary sexual maturation (Kingsley 1982, Maggioncalda et al. 1999).

Wild orangutan females weigh on average 39 kg, which is only 45% of flanged male body size (Markham and Groves 1990). This makes orangutans one of the most sexually dimorphic species on record. The evolution of this extreme dimorphism has been attributed to male–male competition (Rodman and Mitani 1987), female choice (Fox 1998, Utami et al. 2002), and sexual coercion (Smuts and Smuts 1993).

Phenotypic differences between the two orangutan species seem to exist, but no systematic study comparing soft tissue and appearance has yet been completed. Skeletal differences, though, are comparatively well studied. Cranial morphology and dimensions are highly variable in modern *Pongo*; however, there is perhaps more variation within the Bornean species than between *P. pygmaeus* and *P. abelii* (Courtenay et al. 1988). Among these differences are the shape and profile of the face, degree of prognathism, shape of the brain case, and several features of the teeth (Groves 1986, Jacobshagen 1979, Rorhrer-Ertl 1988, van Bemmel 1968).

ECOLOGY

Food Availability

The Southeast Asian rain forest is characterized by dramatic fluctuations in the availability of the orangutan's preferred food—fruit. These fluctuations are particularly pronounced in the lowland forests that were once the predominant habitat type. Bornean and Sumatran forests are dominated by trees in the dipterocarp family that periodically experience a mast fruiting phenomenon in which up to 88% of the trees of this and other plant families flower and fruit in synchrony (Appanah 1985, Ashton 1988, Medway 1972, van Schaik 1986). Mast fruiting occurs every 2–10 years (Ashton 1988) and appears to be driven by climatic events associated with the El Niño weather pattern (Curran et al. 1999), although the El Niño effect may wane farther west (Wich and van Schaik 2000). By fruiting in synchrony, on an unpredictable cue, these plants are able to swamp out seed predators (primarily insects but also vertebrates such as forest pigs and primates), thus ensuring a higher likelihood of seedling survival than if they reproduced at nonsynchronized regular intervals (Curran et al. 1999, Curran and Leighton 2000). Mast fruitings are often followed by periods of extremely low fruit availability (Knott 1998b). Other plant species reproduce at more regular intervals, and smaller fruiting peaks are also observed. Overall, compared to African and South American rain forests, these forests experience much greater inter- and intra-annual variability in fruit production (Fleming et al. 1987). The high species diversity of these forests also leads to more uneven food distribution. Thus, the rain forest habitat for orangutans provides resources that are more patchily distributed in space and time and an overall lower productivity than is found in rain forests of other tropical regions (Knott 2005b). Certain orangutan habitat types, such as peat swamps, are less subject to mast fruiting because of lower dipterocarp density, but fruit availability is still highly variable (Galdikas 1988). It also appears that mast fruiting and fluctuations in fruit availability may be more pronounced on Borneo than on Sumatra (Delgado and van Schaik 2000, Rijksen and Meijaard 1999).

Activity Patterns

Orangutans cope with this overall lower food availability by being primarily solitary foragers. Averaging across several studies, orangutans divide their time among feeding (50.0%), resting (34.5%), traveling (12.9%), and nest building (1.3%) activities, with the remaining time (1.3%) spent socializing, mating, grooming, etc. (Fox et al. 2004, Galdikas 1988, Knott 1999b, Mitani 1989, Rijksen 1978, Rodman 1979). However, these percentages may vary significantly from day to day, and there may be individual variation, even within a site, in

how orangutans react to the same ecological conditions (Knott 1999b, 2005b). Orangutans modify time spent feeding (Galdikas 1988, MacKinnon 1974, Mitani 1989, Rodman 1977), traveling (Galdikas 1988, Knott 1999b, Mitani 1989), and resting (Mitani 1989) during periods of increased sociality. Pronounced changes in activity profiles also occur with fluctuations in fruit availability (Knott 2005a). Data from Gunung Palung in Borneo show that during periods of high fruit availability orangutans spend more time awake and more time feeding, traveling, socializing, and mating (Knott 1999b). When fruit is low, orangutans at Gunung Palung conserve energetic resources by spending less time awake and traveling shorter distances per day (Knott 1998c, 1999b).

Diet and Nutrient Consumption

Orangutans may spend up to 100% of their foraging time eating fruit when it is available (Knott 1998b). Fruits produced during masts are not only more abundant but also higher in calories, carbohydrates, and lipids (Knott 1999b). Orangutans cope with periods of low fruit availability by feeding on leaves and bark as fallback food resources. They also incorporate other foods into their diet, such as pithy vegetation and insects, primarily termites. Their diet during these low-fruit periods is much lower in quality, with a large portion being derived from fiber. Vertebrate animal matter is rarely eaten, but orangutans in Sumatra have been seen opportunistically hunting slow lorises on seven occasions (Utami 1997) and eating a gibbon (Sugardjito and Nurhada 1981). Only one meat-eating observation, of a tree rat, has been reported from Borneo (Knott 1998a). Daily caloric consumption has been computed so far for one population, Gunung Palung in Borneo. These records show dramatic fluctuations associated with fruit availability, with orangutans consuming two to five times more kilocalories during fruit-rich periods compared to fruit-poor periods (Knott 1998b).

Physiologically, orangutans are particularly adapted to fat storage. During periods of high fruit availability, they put on fat stores, as measured by their intake of significantly more calories than they expend (Knott 1998b). This adaptation appears to have evolved as a way to sustain them through extended periods of low fruit availability, when they have been shown, through the production of ketones in urine (Knott 1998b), to metabolize fat reserves. The orangutan propensity for fat storage has long been known in zoos, where they become obese more readily than other captive great apes and show higher rates of diseases associated with obesity, such as diabetes (Gresl et al. 2000).

Interisland Differences in Ecology

Because orangutan dietary composition changes with food availability (Fox et al. 2004, Knott 1998b), it is difficult to compare typical diets between sites as the lengths of studies and food availability during studies are not held constant. Some reviews suggest that Sumatran orangutans are subjected to fewer environmental fluctuations and thus eat more fruit and insects, whereas Bornean orangutans eat more bark (Delgado and van Schaik 2000). However, other reviews do not find a clear interisland difference (Fox et al. 2004). It has been proposed that the Sumatran rain forest has a richer volcanic soil than Bornean forests and thus may be more productive (Wich et al. in press). Site-specific idiosyncrasies exist as well. For example, the majority of research in Sumatra has been done at Ketambe, which is known for its unusually high density of very large crowned fig trees that may provide a buffer during periods of low fruit availability. Comparative research is needed to adequately address whether there are overall interisland differences in ecology and dietary composition. The most substantiated difference is that orangutans at Suaq Balimbing in Sumatra eat twice as many insects as orangutans at other sites, and insect consumption is not dependent on food availability (Fox et al. 2004). Suaq is a rich peat swamp where orangutan group size is larger than at other sites, and individuals often use tools to eat insects. Fox et al. (2004) found that insect abundance was almost twice as high at Suaq compared to nearby Ketambe and suggest that this may contribute to the invention and transmission of tool use there.

SOCIAL ORGANIZATION

Ranging and Philopatry

Orangutan ranging behavior is still poorly understood because the areas that they occupy exceed the boundaries of study sites. Where orangutans have access to multiple habitats, such as at Gunung Palung, they change habitats depending on food availability (Leighton and Leighton 1983). At Suaq Balimbing, Singleton and van Schaik (2001) found that minimum female range size was 850 ha, and flanged and unflanged males used at least 2,500 ha. Orangutans are not territorial in the classic sense of defending a home range. Indeed, using Mitani and Rodman's (1979) defensibility index, their ranges are much larger than can be actively defended (van Schaik and van Hooff 1996).

In contrast to the other great apes, orangutans appear to be female-philopatric (Galdikas 1988, Rijksen 1978, Rodman 1973, Singleton and van Schaik 2002, van Schaik and van Hooff 1996). Adult female ranges show considerable overlap (Galdikas 1988; Knott 1998c; Singleton and van Schaik 2001, 2002; van Schaik and van Hooff 1996), and adolescent females occupy ranges that overlap the ranges of their mothers (Singleton and van Schaik 2002). Although it was initially thought that males defended exclusive ranges (Rodman 1973), it is now known that ranges of flanged males overlap considerably (Knott 1998c, Singleton and van Schaik 2001). Some flanged males remain in a fixed area ("residents"), while others range over greater regions ("nonresidents"). The highest-ranking flanged male appears to occupy a range that overlaps with the ranges of numerous females (Galdikas 1979, 1985a; Mitani 1985a; Utami and Mitra Setia 1995). Patterns of residency are not permanent, however, as resident males are known to leave their areas voluntarily (Galdikas 1979) or to be forced out by nonresident or resident challengers (Utami and Mitra Setia 1995).

Grouping Patterns

Although orangutans are predominantly solitary, three types of grouping can be distinguished (Utami et al. 1997): travel bands, wherein individuals co-feed and travel together during periods of high food availability; temporary aggregations, in which individuals feed together during times of food scarcity but travel independently; and consortships, in which a receptive female travels in a coordinated fashion with a flanged or unflanged male for a period of several hours, days, or weeks. Additionally, a mother and her dependent offspring may travel with an older daughter and her offspring. At some sites, such as Suaq Balimbing in Sumatra, orangutans have been observed to be much more gregarious (van Schaik 1999).

Feeding competition (Wrangham 1980) seems to be important in the low levels of gregariousness seen in orangutans and commonly takes the form of scramble competition, where orangutans compete for patchily distributed food sources. Competition may be more intense for orangutans compared to other apes. Orangutans feed in trees that are significantly smaller than those utilized by chimpanzees and bonobos (Knott 1999a,b), and they experience periods of much sparser food availability (reviewed above). Contest competition also occurs in large fig trees (Sugardjito et al. 1987, Utami et al. 1997). Predation probably plays an insignificant role in regulating grouping patterns since it has been reported only for formerly captive juveniles traveling without a mother's protection (Rijksen 1978).

Social Relationships

Orangutan social interactions are more limited than in many other primates, with some individuals spending days or weeks without contact with conspecifics. Orangutans rarely groom themselves or others, which may reflect less of a need for the social functions of grooming or, alternatively, a lower ecoparasite load because of their more solitary lifestyle.

Despite their semisolitary nature, behavioral and experimental evidence suggests that individualized relationships exist (Delgado and van Schaik 2000, van Schaik and van Hooff 1996). Male long calls appear to act as an important mechanism by which relationships are communicated within the dispersed society (Delgado and van Schaik 2000). One flanged male can typically be recognized as dominant over all other males within the immediate vicinity and gives long calls at the highest rate (Delgado in press, Galdikas 1983, Mitani 1985b, Utami and Mitra Setia 1995). By monitoring the dominant male's long calls, subordinate flanged males avoid altercations with this male (Mitani 1985b), which is totally intolerant of his flanged peers. Sexually receptive females prefer to mate with the dominant male and may also use long calls to find this male during peak fecundity (Delgado in press, Galdikas 1983). Flanged males are invariably dominant to unflanged males, as evidenced by observations of unflanged males fleeing from flanged males (Galdikas 1979, 1985a,b; Mitani 1985a; Utami Atmoko 2000) and being supplanted in fruit trees (Utami et al. 1997). However, flanged males are markedly more tolerant of un-flanged males and often allow them to remain in proximity undisturbed (Galdikas 1985a). This tolerance is not due to males being closely related (Utami et al. 2002), and it wanes when a potentially reproductive female is present (Galdikas 1985a,b; Utami Atmoko 2000); however, aggression between flanged and unflanged males rarely involves physical contact. Unflanged males seldom act aggressively toward one another, even in the presence of females (Galdikas 1985b, Mitani 1985a). Dominance relationships among unflanged males have not been fully described, but observations of approach–avoidance interactions, aggression, and mating interruptions involving these males suggest that dominance relationships exist (Galdikas 1979, 1985b; Utami Atmoko 2000). Little is known about dominance relationships among adult female orangutans. At Ketambe, Utami and colleagues (1997) described a nearly unidirectional pattern of female–female displacements in fig trees, suggesting that a female hierarchy is present.

Tool Use and Culture

Tool use is relatively rare in wild orangutans. This contrasts with reports of captive orangutans habitually making and using tools (Lethmate 1982). However, orangutans at all sites make "leaf umbrellas" to hold over their heads during rainstorms. Leaves are also used in agonistic displays (Galdikas 1982, MacKinnon 1974, Rijksen 1978), for self-cleaning (MacKinnon 1974, Rijksen 1978), as protection in food acquisition (Rijksen 1978), and as drinking vessels (Knott 1999a). Sticks have been used for scratching (Galdikas 1982) and dead wood for opening up durian fruits (Rijksen 1978). In a recent comparative study of orangutan sites, van Schaik and colleagues (2003) determined that many of these behaviors constitute cultural differences between populations that are not explained by ecological circumstances. Other non-tool-use local traditions include using leaves or the hands to produce "kiss–squeak" vocalizations, riding tree snags, exaggerated symmetrical scratching, building bunk nests, constructing nest covers, and peculiar vocalizations emitted routinely during nest building (van Schaik et al. 2003).

The site with the most habitual tool use is Suaq Balimbing in Sumatra. Orangutans there strip sticks and use them to harvest insects and *Neesia* seeds (Fox et al. 1999, van Schaik and Knott 2001, van Schaik et al. 1996). Use of sticks to extract *Neesia* seems to be a cultural difference as orangutans at Gunung Palung in Borneo regularly eat this fruit but do not use tools to extract the seeds (van Schaik and Knott 2001). Orangutans may invent tool use at a fairly constant rate, but the high density of individuals at Suaq may permit more regular transmission and diffusion throughout the population (Fox et al. 2004, van Schaik 2002).

REPRODUCTIVE PARAMETERS

Male Development

For males, the onset of sexual maturity occurs around 14 years of age in the wild (Wich et al. 2004b). The timing of full development of secondary sexual characteristics, however, is highly variable. Some males develop into flanged males immediately following adolescence, while others experience maturational delay that can last for a few or many years. At the extreme, adult males have been known to remain unflanged for more than 20 years in the wild (Utami Atmoko and van Hooff 2004). The variable timing of male development contrasts sharply with the more predictable male development of the other great apes. Maturational delay may be a temporary state, but the possibility that some males never become flanged cannot be ruled out (Crofoot and Knott in press).

Unflanged males can develop secondary sexual characteristics quite rapidly (in less than a year) under both captive and wild conditions (Kingsley 1988, Utami and Mitra Setia 1995). This development has been linked physiologically to a surge in testicular steroids and growth hormones (Maggioncalda et al. 1999, 2000). However, the underlying mechanism involved in triggering or, alternatively, suppressing development of unflanged males remains elusive. It may be that the presence of flanged males inhibits unflanged male development (Graham and Nadler 1990; Kingsley 1982, 1988; Maggioncalda et al. 1999). This hypothesis stems from anecdotal zoo reports that describe unflanged males developing secondary sexual features soon after being separated from their flanged cagemate (e.g., Maple 1980). Field observations from Ketambe also support this view since two unflanged males at this site developed secondary sexual characteristics shortly after the dominant male was overthrown (Utami Atmoko 2000). Against this hypothesis, unflanged males in captivity have fully matured while a flanged male was present (Kingsley 1982, Maggioncalda et al. 1999). In their captive study, Maggioncalda and colleagues (2002) rejected the possibility that "arrested development" in unflanged males is due to chronic stress imposed upon them by flanged males since cortisol, the hormone that when elevated is associated with psychosocial stress, is lower in unflanged than flanged males. They suggest that selection has maintained arrested development as an alternative male reproductive strategy (Maggioncalda et al. 1999). Under this model, males who "play it safe" by postponing the development of secondary sexual features at puberty avoid the high costs associated with being a flanged male (e.g., male–male aggression, immunosuppression caused by high androgen levels) but are less reproductively successful due to female mate choice for flanged males and reduced competitive ability (Maggioncalda et al. 1999: 27). However, males that experience developmental arrest are thought to achieve approximately the same lifetime fitness as males that "take a chance" (i.e., immediately undergo secondary maturation at puberty) because they end up living longer (Maggioncalda et al. 1999: 27). The influence of energetic status, genetic variation, or a combination of these factors has not been considered so far but may also play an important role in male maturational delay (Crofoot and Knott in press, Knott 1999a).

If flanged males do suppress the development of unflanged males, how might this occur? Because wild males are rarely in visual or olfactory contact, some have proposed that long calls given by flanged males may suppress unflanged male development via direct connections between the auditory area of the brain and the hypothalamus (Maggioncalda et al. 1999). It is thought that full maturation is initiated only once the density of calling males is sufficiently low. However, the putative auditory–endocrine mechanism is problematic, given that not all flanged males regularly produce long calls (Utami and Mitra Setia 1995), making long calls an unreliable indicator of flanged male density. Additionally, no auditory–endocrine mechanism has yet been described in primates. Further investigation is necessary, although controlled experiments are not possible in the wild and would be difficult in captivity.

Female Development and Ovarian Function

In the wild, orangutan females reach sexual maturity at approximately 10–11 years (Knott 2001). Like most other primates, they experience a period of adolescent subfecundity (a lower probability of conception in each cycle and/or irregular occurrence of ovulation within cycles), which lasts 1–5 years (Knott 2001). Age at first birth is around 15 years (Knott 2001, Wich et al. 2004b). As in humans, the menstrual cycle averages 28 days (Knott 2005a, Markham 1990, Nadler 1988). Gestation in orangutans is approximately 8 months (Markham 1990). Pregnancy is recognizable to human observers by a slight but conspicuous swelling and lightening in color of the perineal tissues and enlargement of the nipples and eventually the abdominal region (Fox 1998, Galdikas 1981, Schürmann 1981).

Unlike the other great apes (Graham 1981), orangutans exhibit no sexual swellings or other outward indicators of ovulation (Graham-Jones and Hill 1962, Schultz 1938). In the wild, copulations are concentrated around mid-cycle, and ovulation appears to be effectively concealed from males since males rely on female proceptive behavior for information about cycle timing (Fox 1998). Nadler's (1982, 1988) experimental work with captive orangutans also supports this conclusion. In these experiments, when male and female orangutans were freely accessible to each other in a single cage, males forcibly copulated with females on most days of the cycle. However, when females controlled male access, copulations were restricted to mid-cycle and were cooperative, with females exhibiting proceptive behavior.

Orangutans are striking for giving birth only once every 8–9 years, on average (Galdikas and Wood 1990, Wich et al. 2004b). These long interbirth intervals seem to be due to suppressed ovarian function during much of this time (Knott 2001). Knott (1999b) showed that during periods of low fruit availability orangutans have lower levels of estrone conjugates (E_1C) compared to periods of high fruit availability. Levels of E_1C have been shown in captivity to be positively correlated with conception rates (Masters and Markham 1991).

Why should orangutans give birth so rarely? Due to the low fruit availability in the Southeast Asian rain forests, orangutans may spend significant periods in negative energy balance when their ovarian function is suppressed (Knott 1999b). This in combination with postpartum amenorrhea due to the energetics of lactation may shorten the period when orangutans can conceive (Knott 1999a,b). In captivity, where orangutan energy status is greatly improved, females reach sexual maturity faster (Masters and Markham 1991), conceive faster after each birth (Lasley et al. 1980), and have dramatically shorter interbirth intervals (Knott 2001, Markham 1990). Additionally, in captivity, low body weight has been shown to result in amenorrhea, and increased ovarian hormones are linked to weight gain (Masters and Markham 1991). Low food availability may also lead orangutan mothers to postpone weaning until their juveniles can forage on their own (van Noordwijk and van Schaik 2005).

Mating Behavior

Copulation typically occurs face-to-face (Galdikas 1981) and averages 10 min in length (Galdikas 1981, MacKinnon 1979, Mitani 1985a). In the wild, mating strategies differ between unflanged and flanged males. Flanged males predominately use a "consort/combat" tactic whereby they advertise their presence with long calls and wait for females to approach them to establish a consortship (Galdikas 1985a: 20). Mating usually occurs multiple times during a consortship and involves the cooperation of the female (Fig. 17.3). Due to the dispersed nature of orangutan society, dominant males cannot entirely prevent subordinates from mating. Females express strong mating preferences for the locally dominant male and seek out this male around the time of presumed ovulation (Fox 1998, Utami Atmoko 2000). In addition to initiating consortships, females play an active role during mating with the dominant male by aiding with intromission, manually or orally stimulating male genitalia, and performing pelvic thrusts (Fox 1998; Galdikas 1979, 1981; Schürmann 1981, 1982; Utami Atmoko 2000). Females are less proceptive toward subordinate flanged males. These males initiate most of their consortships (Galdikas 1981, Utami Atmoko 2000) and frequently force females to copulate. For example, the unusually high proportion of forced copulations by flanged males at Kutai (Fig. 17.3) is attributable to low-ranking flanged males (Mitani 1985a).

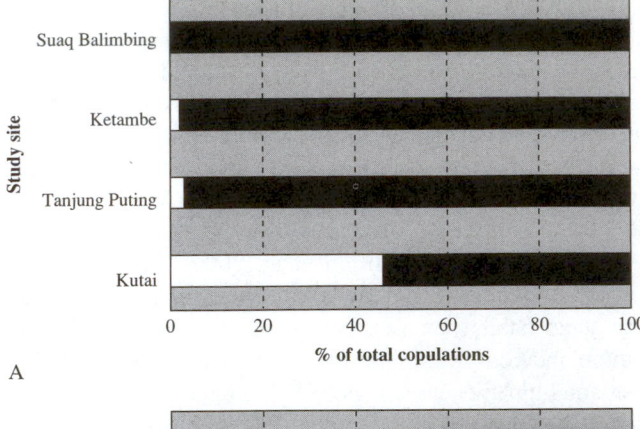

A

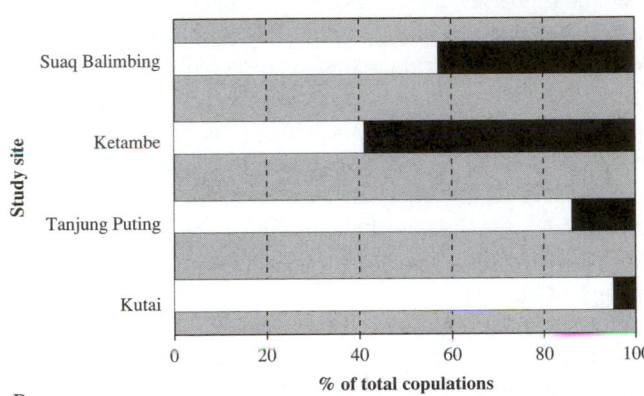

B

Figure 17.3 Proportion of forced and cooperative matings at major orangutan study sites for (A) flanged males and (B) unflanged males. White bars indicate percentage of matings that were forced (i.e. females resisted) and black bars indicate percentage of matings that were co-operative. Data sources include studies with > 20 observed copulations per category. Copulation sample sizes are as follows (unflanged males/flanged males): Ketambe (N = 38/50, 94/70; averaged between two studies): Schürmann & van Hooff 1986, Utami Atmoko, 2000; Kutai (N = 151/28): Mitani 1985a; Tanjung Puting (N = 22/30): Galdikas 1985a,b; Suaq Balimbing (N = 90 with known outcomes/66): Fox, 2002.

While on consort, flanged males closely guard females and react aggressively to the calls or presence of other males, including unflanged males that often follow consorting pairs (Galdikas 1981). If two flanged males meet in the presence of a receptive female, escalated contests often ensue and the dominant male will take over a subordinate male's consortship (Galdikas 1985a). Wounds suffered during male–male fights can be serious (Galdikas 1985a) and even fatal (Knott 1998a). Participating in consortships is costly for flanged males because they must adjust their normal activity budgets to accommodate the longer day ranges of females (Mitani 1989, Utami Atmoko and van Hooff 2004). Flanged males are therefore somewhat selective about the identity of their consorts: they sometimes reject or react passively to nulliparous females but readily accept parous females as consort partners (Galdikas 1981, 1985a, 1995; Schürmann 1981, 1982). Such selectivity is likely adaptive since young females have a high probability of being subfecund.

Unflanged males usually adopt a lower-profile mating strategy in which they do not emit long calls or form consortships but instead actively search for mating opportunities (Galdikas 1985a,b). Females are generally not proceptive toward these males, but unflanged males try to copulate despite female resistance (Table 17.1). The degree of female resistance varies from mild to violent (Fox 1998, Galdikas 1981), but females never or very rarely receive wounds during forced copulations (Galdikas 1981). Female resistance occasionally alerts a nearby flanged male, which will chase away the unflanged male (e.g., Galdikas 1981). However, this scenario is rare as most forced copulations occur in the absence of others.

Based on mating records and conception estimates, field researchers have long assumed that dominant flanged males father the majority of offspring (Fox 1998; Galdikas

Table 17.1 Predictions of and Problems with Hypotheses for the Function of Female Mate Choice in Orangutans (Points of Difference Are Italicized)

INFANTICIDE AVOIDANCE	GOOD GENES
Predictions	**Predictions**
• Females preferably mate with flanged males, especially the dominant male for *protection* from infanticide.	• Females preferably mate with flanged males, especially the dominant male for "*good genes.*"
• During stable rank periods, females resist mating with unflanged males and nondominant flanged males to *concentrate paternity into the dominant male.*	• During stable rank periods, females resist mating with unflanged males and nondominant flanged males *when chance of insemination is high to avoid conception.*
• During unstable periods, both *reproductive and nonreproductive females* mate polyandrously to *confuse paternity.* If females can recognize which male will overthrow the dominant, this male will be targeted.	• During unstable periods, *potentially reproductive* females should mate with new dominant male following a take-over and with up-and-coming males, if they can be recognized, in order to *obtain these males' superior genes.*
Problems	**Problems**
• During stable periods, how does dominant male know if females mated with others in his absence?	• Ketambe paternity results.
• No infanticides have been observed or inferred in orangutans, despite theoretical expectations.	• Why do nonreproductive females mate polyandrously during unstable periods?

1979, 1985a,b; Rijksen 1978; Rodman and Mitani 1987; Schürmann 1982; Schürmann and van Hooff 1986). Relatively lower reproductive success of unflanged males is consistent not only with female preference for flanged males but also the relaxed attitude flanged males display toward unflanged males. In addition, theory predicts that flanged males should outreproduce unflanged males; otherwise, the development of metabolically expensive secondary sexual features would have been selected against (Maggioncalda et al. 1999). However, the first paternity study of wild orangutans found that unflanged males fathered six of the ten offspring genotyped (Utami et al. 2002). Whether or not conceptions occurred during forced copulations is unknown, but the authors suggest that the unexpected success of unflanged males is largely due to changing female strategies when the locally dominant male's position is challenged. During periods of male rank instability (i.e., a single dominant male cannot be recognized), females at Ketambe copulated with unflanged males, particularly residents, more often and with little female resistance (Utami Atmoko 2000). It was during these periods that two-thirds of the offspring sired by unflanged males were conceived (Utami et al. 2002). Whether the Ketambe paternity results will hold up in other orangutan populations awaits further study. Given the current evidence, however, we can rule out the possibility that flanged males enjoy a reproductive monopoly and must take seriously the suggestion that equally successful alternative male reproductive strategies exist (Crofoot and Knott in press, Maggioncalda et al. 1999, Utami et al. 2002, Utami and van Hooff 2004).

CONSERVATION

Orangutans are listed as endangered according to the World Conservation Union red book listing. The most recent orangutan population and habitat viability analysis estimates that only 7,500 orangutans remain in Sumatra (Singleton et al. 2004). Bornean populations are estimated to include 13,600 in Sabah, 4,335 in east Kalimantan, 32,000 in central Kalimantan, and 7,500 in west Kalimantan and Sarawak (Singleton et al. 2004). However, the relevant figure for species viability is not the overall population size but the size of each isolated unit. This presents an even bleaker picture. Only populations of 500 or more individuals are considered to be demographically and genetically stable (Singleton et al. 2004). On Sumatra, there are only four remaining populations of over 500 individuals. Habitat is being lost at an annual rate of 10%–15%; thus, orangutans are projected to be extinct from Sumatra in 50–100 years, except for one population (Singleton et al. 2004). Even within their stronghold in central Kalimantan, orangutan numbers are rapidly decreasing. The largest population in Kalimantan, the Sebangau ecosystem, suffered a 50% decline from 12,000 to 6,000 individuals between 1995 and 2004 (Singleton et al. 2004).

The major cause of orangutan decline is habitat destruction. Despite current conservation efforts, illegal logging has reached epidemic levels, even within national parks (Felton et al. 2003, Jepson et al. 2001, van Schaik et al. 2001). In 1993, it was estimated that over 80% of the orangutan's rain forest habitat had been cut down in the previous 20 years (Tilson et al. 1993). Between 1985 and 2001, over 56% of Kalimantan's lowland rain forests within protected areas were destroyed (Curran et al. 2004). In addition to logging, both legal and illegal, other primary threats to orangutan habitat are fire, conversion of rain forest to oil palm plantations, mining, fragmentation, peat swamp drainage, and human encroachment (Rijksen and Meijaard 1999, Robertson and van Schaik 2001, Singleton et al. 2004). Additionally, the illegal pet trade is responsible for the death of thousands of orangutans each year as mothers are killed in order to capture their infants, which often subsequently die in captivity (Rijksen and Meijaard 1999). Orangutans are also lost through poaching for meat in some areas. The slow growth rate of orangutans indicates that even if habitat is not lost, populations cannot survive with an annual removal rate of 1% (Singleton et al. 2004). This rate is probably exceeded in many populations.

Orangutan conservation relies on accurate assessments of population size, and it is usually not feasible to count actual orangutans due to their low density and solitary roaming. Thus, estimating density from calculations of orangutan nests has become the standard methodology (van Schaik et al. 1995). Refinements to this methodology include double counts of each transect (Johnson et al. 2005), use of soil pH (Buij et al. 2003, Wich et al. 2004a) and nest tree species (Ancrenaz et al. 2004a) to estimate nest decay rate, use of population-specific values for nests built per day (Johnson et al. 2005, Ancrenaz et al. 2004a), and aerial nest count surveys by helicopter (Ancrenaz et al. 2004b). These studies suggest that orangutan densities are related to the proportion of soft pulp fruit in a given area (Djojosudharmo and van Schaik 1992, van Schaik et al. 1995, Buij et al. 2003), the density of large strangler figs (Wich et al. 2004a), and, at least in Sumatra, soil pH (a correlate of strangler fig density) (Buij et al. 2003, Wich et al. 2004a; but see Johnson et al. 2005 for a negative finding from Borneo). Thus, habitat protection is clearly key to orangutan survival.

DISCUSSION: FORCED COPULATIONS IN ORANGUTANS

In this chapter, we highlighted several aspects of orangutan biology that are exceptional among primates. One of the most striking of these features is the regular occurrence of forced copulations. The incidence of forced copulations varies among populations and between Borneo and Sumatra (Fig. 17.3), possibly due to differing costs of grouping for flanged males (Delgado and van Schaik 2000). However, at all sites, a substantial proportion of matings—mostly those involving unflanged males (Fig. 17.3)—are forcibly obtained. Aggression in the mating context is not unique to orangutans as it occurs in many Old World primates (van Schaik et al. 2004). However, forced copulation, the most extreme form of mating aggression, is remarkably rare. Forced copulations can occur only when females resist male mating attempts; thus, understanding why females resist is critical for understanding the evolution of this unusual mating pattern (van Schaik 2004). Here, we review current explanations for female mating resistance in orangutans.

Forced Copulation: Sexual Coercion or Mate Assessment?

It is curious that female orangutans resist male mating attempts since females incur costs by doing so, such as reduced feeding efficiency (Fox 1998). Resistance is also rarely effective. For example, at Kutai in Borneo, Mitani (1985a) observed females struggling free in fewer than 10% of forced mating attempts. It is therefore expected that females must gain some benefit by trying to prevent intromission. What might this benefit be?

Females express clear mating preferences for the dominant flanged male. Therefore, one possibility is that females resist the solicitations of unflanged and subordinate flanged males simply because these males are not preferred. If this is the case, forced copulations can be considered a form of sexual coercion because female mate choice is negated (Smuts and Smuts 1993). Because of their semi-solitary nature, orangutan females lack allies and are therefore particularly susceptible to sexual coercion (Smuts and Smuts 1993). Female resistance does not necessarily indicate that a female is unwilling to mate, however. For instance, in some species, females resist mating in order to gain information about the health and vigor of prospective mates (Smuts and Smuts 1993). This possibility may be especially applicable to orangutans because the sexes have little contact with each other outside of mating. If mate assessment is why orangutan females resist, then stronger, more aggressive males are expected to achieve the highest reproductive success and to be chosen by females for their good genes.

Data on female proceptive behavior suggest that the mate assessment scenario is unlikely. Female orangutans prefer dominant flanged males, and aggression is typically absent when females mate with these males. However, it could be that, when unflanged, these males were the most aggressive in their cohorts and female preference for these males took place during the unflanged stage (Fox 1998). Recent work at Suaq by Fox (1998) provides the only test to distinguish between the sexual coercion and mate assessment hypotheses in orangutans. Fox predicted that if female resistance was due to negative mate choice (i.e., females being unwilling to mate with a male), females would respond in a consistent way to solicitations by particular males. On the other hand, if resistance has to do with mate assessment, females would decrease their resistance to males that had mated aggressively

with them in the past. Using 135 observed copulations, she found that females at this site showed constancy in resistance or variation in resistance that did not support the mate assessment hypothesis (Fox 1998). In sum, given the available data, female orangutans appear to resist mating with unflanged males because these males are not preferred. Forced copulations can therefore be considered a form of male sexual coercion.

The Function of Female Mate Choice

Why do orangutan females exercise such strong negative and positive mate choice? Two main hypotheses, which are not necessarily mutually exclusive, have been proposed: infanticide avoidance and good genes (Fox 1998, 2002; Delgado and van Schaik 2000; van Schaik 2004). Predictions of these hypotheses are outlined in Table 17.1.

Infanticide Avoidance

It is becoming increasingly evident that infanticide is a potential reproductive strategy for primate males (see reviews by Hiraiwa-Hasegawa 1988, Hrdy and Hausfater 1984, van Schaik 2000; but see Bartlett et al. 1993, Sussman et al. 1995). By killing infants they have not sired, males can create a new mating opportunity by interrupting a mother's lactational amenorrhea, causing her to return to a fecund state sooner than would occur otherwise (Hrdy 1979). Given their slow life histories, female orangutans, along with the other great apes, should be among the primates most vulnerable to infanticide. Infanticide is especially common (85% of observed cases in wild primates; van Schaik 2000) when a new male assumes a position of dominance either by successfully challenging the former dominant or through other means. Thus, in orangutans, stranger flanged males that challenge the dominant male or resident unflanged males that quickly rise in rank ("up-and-coming" males) should pose the most serious infanticidal threat (van Schaik 2004).

One way primate females can protect themselves against infanticide is by associating with a protector male (van Schaik and Kappeler 1997). To secure a male's protective services, a female should ensure that he is reasonably certain that he is the father of her offspring. The infanticide-avoidance hypothesis thus sees female orangutan preference for dominant flanged males and against unflanged and subordinate flanged males as a way to concentrate paternity into the male that can offer protection. Regular association between the sexes appears impossible in orangutans, but recent research suggests that males and females maintain loose ties that may offer females protection. At Suaq, Fox (2002) showed that females which associated with a flanged male received less harassment from unflanged males seeking copulation. Females seem to use long calls to locate protector males when a threat is imminent. For example, females were observed traveling rapidly toward long-calling males when pursued by unflanged males (Fox 2002).

However, females do not always resist copulations with nondominant males. If the infanticide-avoidance hypothesis is correct, how can these cooperative matings be explained? Hrdy (1979) originally proposed that in species vulnerable to infanticide, females may use sexual behavior to decrease the risk of attack. Because males do not recognize their offspring, females are expected to mate with potentially infanticidal males whenever possible, including when conception is unlikely, to increase these males' assessments of likely paternity. If a newly dominant male is reasonably uncertain about whether he sired an infant, it pays him, in terms of fitness, to refrain from killing the infant (van Schaik et al. 2004). Thus, if female orangutans can recognize that a dominant male may be overthrown before the birth of their next infant, the infanticide-avoidance hypothesis predicts females should willingly copulate with up-and-coming unflanged males or flanged males that are likely to be successful in challenging the dominant male. Intriguing new observations of orangutan mating behavior at Ketambe in Sumatra are consistent with these predictions. When a single dominant male could not be recognized, females copulated with unflanged males more frequently and with little resistance (Utami Atmoko 2000). Lactating females, which typically do not behave proceptively, also readily engaged in consortships with flanged males (Utami Atmoko 2000). After a new dominant male assumed power, Ketambe females began selectively approaching this male, even when the former dominant was still around (Utami and Mitra Setia 1995). Paternity data also indicate that female orangutans may be able to recognize males rising in rank because one unflanged male at Ketambe which later became dominant fathered 50% of all offspring sired by unflanged males (Utami et al. 2002). These data, although limited, appear to support the predictions of the infanticide-avoidance hypothesis.

There are also important problems with the infanticide-avoidance hypothesis (Table 17.1). First, it is unclear why, if female resistance to male mating attempts during stable periods (i.e., when a single dominant male can be recognized) functions to concentrate paternity into the dominant male, females should resist copulations when the dominant male is absent. How would the dominant male know if females mated with others? Second, and much more damaging to this hypothesis, is that despite nearly four decades of study, successful or attempted infanticides have never been reported in wild orangutans (van Schaik and van Hooff 1996, van Schaik 2000). The absence of infanticide is surprising since numerous infanticides have been recorded in other great apes (chimpanzees, see Chapter 19; gorillas, Watts 1989, see Chapter 18), with the exception of bonobos. Orangutan females are often not seen by researchers for months at a time; however, it is unlikely that infanticides have been missed because infant survivability is remarkably high (Wich et al. 2004b). van Schaik (2004) attributes the discrepancy between theoretical expectations and the lack of infanticides to the successfulness of female counterstrategies. However, so far, no explanation for why orangutan counterstrategies would

be more effective than those of gorilla and chimpanzee females has been offered. Thus, the infanticide-avoidance hypothesis currently remains highly speculative.

Good Genes

Female orangutan mating preferences may also reflect a female's effort to obtain "good genes" (Galdikas 1981, Mitani 1985b). Because female orangutans give birth to only a few offspring, they should, whenever possible, select males with the best genetic makeup to father their young. Thus, since it has been assumed that dominant males sire the majority of offspring, good genes has been a plausible explanation for why females resist nondominant males and prefer to mate with the dominant male. Females do not appear to assess male genetic quality through mating resistance, but they may use male body size, secondary sexual characteristics, or dominance status as indirect indicators of genetic quality (Fox 2002). New data have introduced major challenges to the good-genes hypothesis (Table 17.1). First, the paternity data introduced earlier show that, at least in one population, the dominant male was not the most reproductively successful (Utami et al. 2002). This suggests that female preference for dominant males and against others cannot be due to good genes alone. However, regardless of the paternity data, other problems with the good-genes explanation remain. For example, if female orangutans are strictly interested in receiving good genes, how can we explain the fact that pregnant and lactating females occasionally resist copulations (Fox 1998, Utami Atmoko 2000)? It could be that females never want to risk conception by nondominant males, so they always resist to "play it safe." Alternatively, they may be attempting to avoid other mating costs such as sexually transmitted diseases. No study has evaluated this possibility. Another challenge to the good-genes explanation for female mating preferences comes from observations of nonreproductive females willingly mating during unstable rank periods (Utami Atmoko 2000).

In sum, there is currently no consensus about the functional significance of female mating preferences in orangutans. Neither the infanticide-avoidance nor the good-genes explanation appears to be adequate, given the available data. Future paternity analyses from other populations will help clarify whether the Ketambe results are generalizable to all sites. It is possible that females pursue a mixed strategy: they mate for good genes as well as protection against infanticide. In the meantime, theoretical work should focus on what other factors might explain the available data.

CONCLUSION

After more than 30 years of orangutan research in the wild, extraordinary yet puzzling aspects of orangutan biology are finally being addressed. However, much work remains. Important goals for future research include documenting ecological, morphological, and behavioral differences between and within the two orangutan species, developing tests to elucidate the mechanism underlying arrested male development, and collecting paternity data at multiple sites to understand male development and male and female mating strategies. However, the issue that must move to the forefront of all agendas is conservation—habitat protection, in particular. For without increased efforts to save orangutans, we will lose the opportunity to fully understand and appreciate this unique species.

REFERENCES

Ancrenaz, M., Calaque, R., and Lackman-Ancrenaz, I. (2004a). Orangutan nesting behavior in disturbed forest of Sabah, Malaysia: implications for nest census. *Int. J. Primatol.* 25:983–1000.

Ancrenaz, M., Goossens, B., Gimenez, O., Sawang, A., and Lackman-Ancrenaz, I. (2004b). Determination of ape distribution and population size using ground and aerial surveys: a case study with orang-utans in lower Kinabatangan, Sabah, Malaysia. *Anim. Conserv.* 7:375–385.

Appanah, S. (1985). General flowering in the climax rain forests of South-east Asia. *J. Trop. Ecol.* 1:225–250.

Ashton, P. S. (1988). Dipterocarp biology as a window to the understanding of tropical forest structure. *Annu. Rev. Ecol. Syst.* 19:347–370.

Bacon, A.-M., and Long, V. T. (2001). The first discovery of a complete skeleton of a fossil orang-utan in a cave of the Hoa Binh Province, Vietnam. *J. Hum. Evol.* 41:227–241.

Bartlett, T. Q., Sussman, R. W., and Cheverud, J. M. (1993). Infant killing in primates: a review of observed cases with specific reference to the sexual selection hypothesis. *Am. Anthropol.* 95:958–990.

Bruce, E., and Ayala, F. (1979). Phylogenetic relationships between man and the apes: electrophoretic evidence. *Evol. Hum. Behav.* 33:1040–1056.

Buij, R., Singleton, I., Krakauer, E., and van Schaik, C. P. (2003). Rapid assessment of orangutan density. *Biol. Conserv.* 114:103–113.

Chaimanee, Y., Suteethorn, V., Jintasakul, P., Vidthayanon, C., Marandat, B., and Jaeger, J. J. (2004). A new orang-utan relative from the Late Miocene of Thailand. *Nature* 427:439–441.

Courtenay, J., Groves, C., and Andrews, P. (1988). Inter- or intra-island variation? An assessment of the differences between Bornean and Sumatran orang-utans. In: Schwartz, J. H. (ed.), *Orang-utan Biology*. Oxford University Press, Oxford. pp. 19–29.

Crofoot, M. C., and Knott, C. D. (in press). What we do and don't know about orangutan male dimorphism. In: Galdikas, B. M. F., Briggs, N., Sheeran, L. K., and Shapiro, G. L. (eds.), *Great and Small Apes of the World*. Orangutans and Gibbons, vol. II.

Curran, L. M., Caniago, I., Paoli, G. D., Astiani, D., Kusneti, M., Leighton, M., Nirarita, C. E., and Haeruman, H. (1999). Impact of El Nino and logging on canopy tree recruitment in Borneo. *Science* 286:2184–2188.

Curran, L. M., and Leighton, M. (2000). Vertebrate responses to spatio-temporal variation in seed production by mast-fruiting Bornean Dipterocarpaceae. *Ecol. Monogr.* 70:101–128.

Curran, L. M., Trigg, S. N., McDonald, A. K., Astiani, D., Hardiono, Y. M., Siregar, P., Caniago, I., and Kasischke, E. (2004). Lowland forest loss in protected areas of Indonesian Borneo. *Science* 303:1000–1003.

Delgado, R. A. (in press). Sexual selection in the loud calls of male primates: signal content and function. *Int. J. Primatol.*

Delgado, R. A., and van Schaik, C. P. (2000). The behavioral ecology and conservation of the orangutan (*Pongo pygmaeus*): a tale of two islands. *Evol. Anthropol.* 9:201–218.

Djojosudharmo, S., and van Schaik, C. P. (1992). Why are orangutans so rare in the highlands?: altitudinal changes in a Sumatran forest. *Trop. Biodivers.* 1:11–22.

Felton, A. M., Engstrom, L. M., Felton, A., and Knott, C. D. (2003). Orangutan population density, forest structure and fruit availability in hand-logged and unlogged peat swamp forests in West Kalimantan, Indonesia. *Biol. Conserv.* 114:91–101.

Ferris, S. D., Brown, W. M., Davidson, W. S., and Wilson, A. C. (1981). Extensive polymorphism in the mitochondrial DNA of apes. *Proc. Natl. Acad. Sci. USA* 78:6319–6323.

Fleagle, J. G. (1999). *Primate Adaptation and Evolution*. Academic Press, San Diego.

Fleming, T. H., Breitwisch, R., and Whitesides, G. H. (1987). Patterns of tropical vertebrate frugivore diversity. *Annu. Rev. Ecol. Syst.* 18:91–109.

Fox, E. A. (1998). The function of female mate choice in the Sumatran orangutan (*Pongo pygmaeus abelii*) [PhD thesis]. Duke University, Durham, NC.

Fox, E. A. (2002). Female tactics to reduce sexual harassment in the Sumatran orangutan (*Pongo pygmaeus abelii*). *Behav. Ecol. Sociobiol.* 52:93–101.

Fox, E. A., Sitompul, A. F., and van Schaik, C. P. (1999). Intelligent tool use in wild Sumatran orangutans. In: Parker, S. T., Mitchell, R. W., and Miles, H. L. (eds.), *The Mentalities of Gorillas and Orangutans*. Cambridge University Press, Cambridge. pp. 99–116.

Fox, E. A., van Schaik, C. P., Sitompul, A. F., and Doniele, N. W. (2004). Intra- and interpopulational differences in orangutan (*Pongo pygmaeus*) activity and diet: implications for the invention of tool use. *Am. J. Phys. Anthropol.* 125:162–174.

Galdikas, B. M. F. (1979). Orangutan adaptation at Tanjung Puting Reserve: mating and ecology. In: Hamburg, D. A., and McCown, E. R. (eds.), *The Great Apes*. Benjamin/Cummings, Menlo Park, CA. pp. 194–233.

Galdikas, B. M. F. (1981). Orangutan reproduction in the wild. In: Graham, C. E. (ed.), *Reproductive Biology of the Great Apes: Comparative and Biomedical Perspectives*. Academic Press, New York. pp. 281–300.

Galdikas, B. M. F. (1982). An unusual instance of tool-use among wild orang-utans in Tanjung Puting Reserve, Indonesian Borneo. *Primates* 23:138–139.

Galdikas, B. M. F. (1983). The orangutan long call and snag crashing at Tanjung Puting Reserve. *Primates* 24:371–384.

Galdikas, B. M. F. (1985a). Adult male sociality and reproductive tactics among orangutans at Tanjung Puting. *Folia Primatol.* 45:9–24.

Galdikas, B. M. F. (1985b). Subadult male orangutan sociality and reproductive behavior at Tanjung Puting. *Am. J. Primatol.* 8:87–99.

Galdikas, B. M. F. (1988). Orangutan diet, range, and activity at Tanjung Puting, central Borneo. *Int. J. Primatol.* 9:1–35.

Galdikas, B. M. F. (1995). Social and reproductive behavior of wild adolescent female orangutans. In: Nadler, R. D., Galdikas,

B. M. F., Sheeran, L. K., and Rosen, N. (eds.), *The Neglected Ape*. Plenum Press, New York. pp. 163–182.

Galdikas, B. M. F., and Wood, J. W. (1990). Birth spacing patterns in humans and apes. *Am. J. Phys. Anthropol.* 83:185–191.

Graham, C. E. (1981). Menstrual cycle of the great apes. In: Graham, C. E. (ed.), *Reproductive Biology of the Great Apes: Comparative and Biomedical Perspectives*. Academic Press, New York. pp. 1–43.

Graham, C. E., and Nadler, R. D. (1990). Socioendocrine interactions in great ape reproduction. In: Ziegler, T. E., and Bercovitch, F. B. (eds.), *Socioendocrinology of Primate Reproduction*. Wiley-Liss, New York. pp. 33–58.

Graham-Jones, O., and Hill, W. C. O. (1962). Pregnancy and parturition in a Bornean orangutan. *Proc. Zool. Soc. Lond.* 139:503–510.

Gresl, T. A., Baum, S. T., and Kemnitz, J. W. (2000). Glucose regulation in captive *Pongo pygmaeus abelii, P. p. pygmaeus*, and *P. p. abelii × P. p. pygmaeus* orangutans. 19:193–208.

Groves, C. P. (1986). Systematics of the great apes. In: Swindler, D. R. (ed.), *Comparative Primate Biology*. Alan Liss, New York. pp. 187–217.

Groves, C. P. (2001). *Primate Taxonomy*. Smithsonian Institution Press, Washington DC.

Groves, C. P., Westwood, C., and Shea, B. T. (1992). Unfinished business: mahalanois and a clockwork orang. *J. Hum. Evol.* 22:327–340.

Gu, Y. M., Huang, W. P., Song, F. Y., Guo, X. F., and Chien, D. Y. (1987). The study of some fossil orang-utan teeth from Guangdong and Guangxi. *Acta Anthropol. Sin.* 6:272–283.

Harrison, T. (1999). The palaeoecological context at Niah Cave, Sarawak: evidence from the primate fauna. *Bull. Indo-Pac. Prehist. Assoc.* 1964:521–530.

Hiraiwa-Hasegawa, M. (1988). Adaptive significance of infanticide in primates. *Trends. Ecol.* 3:102–105.

Hooijer, D. A. (1948). Prehistoric teeth of man and of the orang-utan from central Sumatra with notes on the fossil orang-utan from Java and southern China. *Zool. Med.* 29:175–301.

Hooijer, D. A. (1960). The orang-utan in Niah cave pre-history. *Sarawak Mus. J.* 9:408–421.

Hrdy, S. B. (1979). Infanticide among animals: a review, classification, and examination of the implications for the reproductive strategies of females. *Ethol. Sociobiol.* 1:13–40.

Hrdy, S. B., and Hausfater, G. (1984). Comparative and evolutionary perspectives on infanticide: introduction and overview. In: Hausfater, G., and Hrdy, S. B. (eds.), *Infanticide: Comparative and Evolutionary Perspectives*. Aldine de Gruyter, New York. pp. xiii–xxxv.

Jablonski, N., Whitfort, M., Roberts-Smith, J. N., and Qinqi, X. (2000). The influence of life history and diet on the distribution of catarrhine primates during the Pleistocene in eastern Asia. *J. Hum. Evol.* 39:131–157.

Jablonski, N. G. (1998). The response of catarrhine primates to Pleistocene environmental fluctuations in east Asia. *Primates* 39:29–37.

Jacobshagen, B. (1979). Morphometrics studies in the taxonomy of the orang-utan (*Pongo pygmaeus*, L. 1760). *Folia Primatol.* 32:29–34.

Janczewski, D. N., Goldman, D., and O'Brien, S. J. (1990). Molecular genetic divergence of orangutan (*Pongo pygmaeus*) subspecies based on isozyme and two-dimensional gel electrophoresis. *J. Hered.* 81:375–387.

Jepson, P., Jarvie, J. K., MacKinnon, K., and Monk, K. A. (2001). The end for Indonesia's lowland forests? *Science* 292:859–861.

Johnson, A., Knott, C., Pamungkas, B., Pasaribu, M., and Marshall, A. (2005). A survey of the orangutan *Pongo pygmaeus pygmaeus* population in and around Gunung Palung National Park, West Kalimantan, Indonesia based on nest counts. *Biol. Conserv.* 121:495–507.

Jones, M. L. (1969). The geographical races of orangutan. *Proc. 2nd Int. Congress Primatol.* 2:217–223.

Kahlke, H. D. (1972). A review of the Pleistocene history of the orang-utan. *Asian Persp.* 15:5–13.

Kelley, J. (2002). The hominoid radiation in Asia. In: Hartwig, W. C. (ed.), *The Primate Fossil Record*. Cambridge University Press, Cambridge. pp. 369–384.

Kingsley, S. R. (1982). Causes of non-breeding and the development of secondary sexual characteristics in the male orangutan: a hormonal study. In: de Boer, L. E. M. (ed.), *The Orangutan: Its Biology and Conservation*. Dr. W. Junk Publishers, The Hague. pp. 215–229.

Kingsley, S. R. (1988). Physiological development of male orang-utans and gorillas. In: Schwartz, J. H. (ed.), *Orang-utan Biology*. Oxford University Press, New York. pp. 123–131.

Knott, C. D. (1998a). Orangutans in the wild. *Natl. Geogr.* 194:30–57.

Knott, C. D. (1998b). Changes in orangutan diet, caloric intake and ketones in response to fluctuating fruit availability. *Int. J. Primatol.* 19:1061–1079.

Knott, C. D. (1998c). Social system dynamics, ranging patterns and male and female strategies in wild Bornean orangutans (*Pongo pgymaeus*). *Am. J. Phys. Anthropol.* Suppl. 26:140.

Knott, C. D. (1999a). Orangutan behavior and ecology. In: Dolhinow, P., and Fuentes, A. (eds.), *The Nonhuman Primates*. Mayfield Press. Mountain View, CA. pp. 50–57.

Knott, C. D. (1999b). Reproductive, physiological and behavioral responses of orangutans in Borneo to fluctuations in food availability [PhD thesis]. Harvard University, Cambridge, MA.

Knott, C. D. (2001). Female reproductive ecology of the apes: implications for human evolution. In: Ellison, P. (ed.), *Reproductive Ecology and Human Evolution*. Aldine de Gruyter, New York. pp. 429–463.

Knott, C. (2004). Orangutans: the largest canopy dwellers. In: Lowman, M. D., and Rinker, H. B. (eds.), *Forest Canopies*. Academic Press, San Diego. pp. 322–324.

Knott, C. D. (2005a). Radioimmunoassay of estrone conjugates from urine dried on filter paper. *Am. J. Primatol.*

Knott, C. D. (2005b). Energetic responses to food availability in the great apes: implications for hominin evolution. In: Brockman, D. K., and van Schaik, C. P. (eds.), *Primate Seasonality: Implications for Human Evolution*. Cambridge University Press, Cambridge.

Lasley, B., Presley, S., and Czekala, N. (1980). Monitoring ovulation in great apes; urinary immunoreactive estrogen and bioactive lutenizing hormone. *Annu. Proc. Am. Soc. Zoo Vet.* 37–40.

Lawrence, D. C., and Leighton, M. (1996). Ecological determinants of feeding bout length in long-tailed macaques (*Macaca fascicularis*). *Trop. Biodivers.* 3:227–242.

Leigh, S. R., and Shea, B. T. (1995). Ontogeny and the evolution of adult body size dimorphism in apes. *Am. J. Primatol.* 36:37–60.

Leighton, M., and Leighton, D. (1983). Vertebrate responses to fruiting seasonality within a Bornean rain forest. In: Sutton,

S. L., Whitmore, T. C., and Chadwick, A. C. (eds.), *Tropical Rain Forest: Ecology and Management*. Blackwell Scientific Publications, Boston. pp. 181–196.

Lethmate, J. (1982). Tool-using skills of orangutans. *J. Hum. Evol.* 11:49–64.

MacKinnon, J. (1979). Reproductive behavior in wild orangutan populations. In: Hamburg, D. A., and McCown, E. R. (eds.), *The Great Apes*. Benjamin/Cummings, Menlo Park, CA. pp. 257–273.

MacKinnon, J. R. (1974). The behaviour and ecology of wild orang-utans (*Pongo pygmaeus*). *Anim. Behav.* 22:3–74.

MacLatchy, L. M. (1996). Another look at the australopithecine hip. *J. Hum. Evol.* 31:455–476.

Maggioncalda, A. N., Czekala, N. M., and Sapolsky, R. M. (2000). Growth hormone and thyroid stimulating hormone concentrations in captive male orangutans: implications for understanding developmental arrest. *Am. J. Primatol.* 50:67–76.

Maggioncalda, A. N., Sapolsky, R. M., and Czekala, N. M. (1999). Reproductive hormone profiles in captive male orangutans: implications for understanding developmental arrest. *Am. J. Phys. Anthropol.* 109:19–32.

Maggioncalda, A. N., Sapolsky, R. M., and Czekala, N. M. (2002). Male orangutan subadulthood: a new twist on the relationship between chronic stress and developmental arrest. *Am. J. Phys. Anthropol.* 118:25–32.

Maple, T. (1980). *Orangutan Behavior*. van Nostrand Reinhold, New York.

Markham, R., and Groves, C. P. (1990). Weights of wild orangutans. *Am. J. Phys. Anthropol.* 81:1–3.

Markham, R. J. (1990). Breeding orangutans at Perth zoo: twenty years of appropriate husbandry. *Zoo Biol.* 9:171–182.

Masters, A., and Markham, R. J. (1991). Assessing reproductive status in orangutans by using urinary estrone. *Zoo Biol.* 10:197–208.

Medway, L. (1972). Phenology of a tropical rain forest in Malaya. *Biol. J. Linn. Soc.* 4:117–146.

Medway, L. (1977). The Niah excavations and an assessment of the impact of early man on mammals in Borneo. *Asian Persp.* 20:51–69.

Mitani, J. C. (1985a). Mating behaviour of male orangutans in the Kutai Game Reserve, Indonesia. *Anim. Behav.* 33:392–402.

Mitani, J. C. (1985b). Sexual selection and adult male orangutan long calls. *Anim. Behav.* 33:272–283.

Mitani, J. C. (1989). Orangutan activity budgets: monthly variation and the effects of body size, parturition, and sociality. *Am. J. Primatol.* 18:87–100.

Mitani, J. C., and Rodman, P. S. (1979). Territoriality: the relation of ranging pattern and home range size to defendability, with an analysis of territoriality among primate species. *Behav. Ecol. Sociobiol.* 5:241–251.

Muir, C., Galdikas, B. M. F., and Beckenbach, A. T. (1995). Genetic variability in orangutans. In: Nadler, R. D., Galdikas, B. M. F., Sheeran, L. K., and Rosen, N. (eds.), *The Neglected Ape*. Plenum Press, New York. pp. 267–272.

Muir, C., Galdikas, B. M. F., and Beckenbach, A. T. (1998). Is there sufficient evidence to elevate the orangutan of Borneo and Sumatra to separate species? *J. Mol. Evol.* 46:378–379.

Nadler, R. D. (1982). Reproductive behavior and endocrinology of orangutans. In: de Boer, L. E. M. (ed.), *The Orangutan: Its Biology and Conservation*. Dr. W. Junk Publishers, The Hague. pp. 231–248.

Nadler, R. D. (1988). Sexual and reproductive behavior. In: Schwartz, J. H. (ed.), *Orang-utan Biology*. Oxford University Press, New York. pp. 105–116.

Pei, W. C. (1935). Fossil mammals from the Kwangsi caves. *Bull. Geol. Soc. China* 14:413–425.

Pilbeam, D. (1982). New hominoid skull material from the Miocene of Pakistan. *Nature* 295:232–234.

Pilbeam, D. R., Rose, M. D., Barry, J. C., and Shah, S. M. I. (1990). New *Sivapithecus* humeri from Pakistan and the relationship of *Sivapithecus* and *Pongo*. *Nature* 348:237–239.

Rijksen, H. D. (1978). A field study on Sumatran orang-utans (*Pongo pygmaeus abelii*, Lesson 1827): ecology, behavior, and conservation. Veenman, H., and Zonen, B. V. (eds.), Wageningen, The Netherlands. pp. 157–323.

Rijksen, H. D., and Meijaard, E. (1999). *Our Vanishing Relative: The Status of Wild Orang-utans at the Close of the Twentieth Century*. Kluwer Academic Publishers, Boston.

Robertson Yarrow, J. M., and van Schaik, C. P. (2001). Causal factors underlying the dramatic decline of the Sumatran orang-utan. *Oryx* 35:26–38.

Rodman, P. S. (1973). Population composition and adaptive organisation among orangutans of the Kutai Reserve. In: Clutton-Brock, T. H. (ed.), *Primate Ecology: Studies of Feeding and Ranging Behaviour in Lemur, Monkeys and Apes*. Academic Press, London. pp. 171–209.

Rodman, P. S. (1977). Feeding behavior of orangutans in the Kutai Reserve, East Kalimantan. In: Clutton-Brock, T. H. (ed.), *Primate Ecology*. Academic Press: London. pp. 383–413.

Rodman, P. S. (1979). Individual activity profiles and the solitary nature of orangutans. In: Hamburg, D. A., and McCown, E. R. (eds.), *The Great Apes*. Benjamin/Cummings, Menlo Park, CA. pp. 234–255.

Rodman, P. S., and Mitani, J. C. (1987). Orangutans: sexual dimorphism in a solitary species. In: Smuts, B. B., Cheney, D. L., Seyfarth, R. M., Wrangham, R. W., and Struhsaker, T. T. (eds.), *Primate Societies*. University of Chicago Press, Chicago. pp. 146–152.

Rorhrer-Ertl, O. (1988). Research history, nomenclature, and taxonomy of the orang-utan. In: Schwartz, J. H. (ed.), *Orang-utan Biology*. Oxford University Press, New York. pp. 7–18.

Ruvolo, M., Pan, D., Zehr, S., Goldberg, T., and Disotell, T. R. (1994). Gene trees and hominoid phylogeny. *Proc. Natl. Acad. Sci. USA* 91:8900–8904.

Ryder, O. A., and Chemnick, L. G. (1993). Chromosomal and mitochondrial DNA variation in orangutan. *J. Hered.* 84:405–409.

Schultz, A. H. (1938). Genital swelling in the female orang-utan. *J. Mammal.* 19:363–366.

Schürmann, C. L. (1981). Courtship and mating behavior of wild orangutans in Sumatra. In: Chiarelli, A. B., and Corruccini, R. S. (eds.), *Primate Behavior and Sociobiology*. Springer-Verlag, Berlin. pp. 130–135.

Schürmann, C. (1982). Mating behaviour of wild orangutans. In: de Boer, L. E. M. (ed.), *The Orangutan: Its Biology and Conservation*. Dr. W. Junk Publishers, The Hague. pp. 269–284.

Schürmann, C. L., and van Hooff, J. A. R. A. M. (1986). Reproductive strategies of the orang-utan: new data and a reconsideration of existing sociosexual models. *Int. J. Primatol.* 7:265–287.

Schwartz, J. H., Long, V. T., Cuong, N. L., Kha, L. T., and Tattersall, I. (1994). A diverse hominoid fauna from the late Middle Pleistocene brecia cave of Tham Khuyen, Socialist

Republic of Vietnam. *Anthropol. Pap. Am. Mus. Nat. Hist.* 74:1–11.

Schwartz, J. H., Long, V. T., Cuong, N. L., Kha, L. T., and Tattersall, I. (1995). A review of the Pleistocene hominoid fauna of the socialist republic of Vietnam (excluding Hylobatidae). *Anthropol. Pap. Am. Mus. Nat. Hist.* 76:1–24.

Singleton, I., Wich, S. A., Husson, S., Stephens, S., Utami Atmoko, S. S., Leighton, M., Rosen, N., Traylor-Holzer, K., Lacy, R., and Byers, O. (2004). *Orangutan Population and Habitat Viability Assessment.* IUCN/SSC Conservation Breeding Specialist Group, Apple Valley, MN.

Singleton, I. S., and van Schaik, C. P. (2001). Orangutan home range size and its determinants in a Sumatran swamp forest. *Int. J. Primatol.* 22:877–911.

Singleton, I. S., and van Schaik, C. P. (2002). The social organisation of a population of Sumatran orang-utans. *Folia Primatol.* 73:1–20.

Smuts, B. B., and Smuts, R. W. (1993). Male aggression and sexual coercion of females in nonhuman primates and other mammals: evidence and theoretical implications. *Adv. Study Behav.* 22:1–63.

Sugardjito, J., and Nurhada. (1981). Meat-eating behavior in wild orangutan *Pongo pygmaeus. Primates* 22:414–416.

Sugardjito, J., te Boekhorst, I. J. A., and van Hooff, J. A. R. A. M. (1987). Ecological constraints on the grouping of wild orangutans (*Pongo pygmaeus*) in the Gunung Leuser National Park, Sumatra, Indonesia. *Int. J. Primatol.* 8:17–41.

Sussman, R. W., Cheverud, J. M., and Bartlett, T. Q. (1995). Infant killing as an evolutionary strategy: reality or myth? *Evol. Anthropol.* 3:149–151.

te Boekhorst, I. J. A., Schürmann, C. L., and Sugardjito, J. (1990). Residential status and seasonal movements of wild orang-utans in the Gunung Leuser Reserve (Sumatra, Indonesia). *Anim. Behav.* 39:1098–1109.

Tilson, R., Seal, U. S., Soemarna, K., Ramono, W., Sumardja, E., Poniran, S., van Schaik, C., Leighton, M., Rijksen, H., and Eudey, A. (1993). Orangutan population and habitat viability analysis report. Presented at Orangutan Population and Habitat Viability Analysis Workshop, Medan, North Sumatra, Indonesia, October 20, 1993.

Uchida, A. (1998). Variation in tooth morphology of *Pongo pygmaeus. J. Hum. Evol.* 34:71–79.

Utami, S. (1997). Meat-eating behavior of adult female orangutans (*Pongo pygmaeus abelii*). *Am. J. Primatol.* 43:159–165.

Utami, S., and Mitra Setia, T. (1995). Behavioral changes in wild male and female orangutans (*Pongo pygmaeus abelii*) during and following a resident male take-over. In: Nadler, R. D., Galdikas, B. M. F., Sheeran, L. K., and Rosen, N. (eds.), *The Neglected Ape.* Plenum Press, New York. pp. 183–190.

Utami, S. S., Goossens, B., Bruford, M. W., de Ruiter, J. R., and van Hooff, J. A. R. A. M. (2002). Male bimaturism and reproductive success in Sumatran orangutans. *Behav. Ecol.* 13:643–652.

Utami, S. S., Wich, S. A., Sterck, E. H. M., and van Hooff, J. A. R. A. M. (1997). Food competition between wild orangutans in large fig trees. *Int. J. Primatol.* 18:909–927.

Utami Atmoko, S. S. (2000). Bimaturism in orang-utan males: reproductive and ecological strategies [PhD thesis]. University of Utrecht, Utrecht.

Utami Atmoko, S. S., and van Hooff, J. A. R. A. M. (2004). Alternative male reproductive strategies: male bimaturism in orangutans. In: Kappeler, P. M., and van Schaik, C. P. (eds.),

Sexual Selection in Primates: New and Comparative Perspectives. Cambridge University Press, Cambridge. pp. 196–207.

van Bemmel, A. C. V. (1968). Contributions to the knowledge of the geographical races of *Pongo pygmaeus* (Hoppius). *Bijdr. Dierk.* 38:13–15.

van Noordwijk, M. A., and van Schaik, C. P. (2005). Development of ecological competence in Sumatran orangutans. *Am. J. Phys. Anthropol.* 127:79–94.

van Schaik, C. P. (1986). Phenological changes in a Sumatran rainforest. *J. Trop. Ecol.* 2:327–347.

van Schaik, C. P. (1999). The socioecology of fission–fusion sociality in orangutans. *Primates* 40:73–90.

van Schaik, C. P. (2000). Infanticide by males: the sexual selection hypothesis revisited. In: van Schaik, C. P., and Janson, C. H. (eds.), *Infanticide by Males and Its Implications.* Cambridge University Press, Cambridge. pp. 27–60.

van Schaik, C. P. (2002). Fragility of traditions: the disturbance hypothesis for the loss of local traditions in orangutans. *Int. J. Primatol.* 23:527–538.

van Schaik, C. P. (2004). *Among Orangutans: Red Apes and the Rise of Human Culture.* Belknap Press of Harvard University, Cambridge, MA.

van Schaik, C. P., Ancrenaz, M., Borgen, G., Galdikas, B., Knott, C., Singleton, I., Suzuki, A., Utami, S., and Merrill, M. (2003). Orangutan cultures and the evolution of material culture. *Science* 299:102–105.

van Schaik, C. P., Fox, E. A., and Sitompul, A. F. (1996). Manufacture and use of tools in wild Sumatran orangutans. *Naturwissenschaften* 83:186–188.

van Schaik, C. P., and Kappeler, P. M. (1997). Infanticide risk and the evolution of male–female association in primates. *Proc. R. Soc. Lond. B* 264:1687–1694.

van Schaik, C., and Knott, C. D. (2001). Geographic variation in tool use on *Neesia* fruits in orangutans. *Am. J. Phys. Anthropol.* 114:331–342.

van Schaik, C. P., Monk, K., and Yarrow Robertson, J. M. (2001). Dramatic decline in orang-utan numbers in the Leuser Ecosystem, northern Sumatra. *Oryx* 35:14–25.

van Schaik, C. P., Pradhan, G. R., and van Noordwijk, M. A. (2004). Mating conflict in primates: infanticide, sexual harassment, and female sexuality. In: Kappeler, P. M., and van Schaik, C. P. (eds.), *Sexual Selection in Primates: New and Comparative Perspectives.* Cambridge University Press, Cambridge. pp. 131–150.

van Schaik, C. P., Priatna, A., and Priatna, D. (1995). Population estimates and habitat preferences of orangutans based on line transects of nests. In: Nadler, R. D., Galdikas, B. M. F., Sheeran, L. K., and Rosen, N. (eds.), *The Neglected Ape.* Plenum Press, New York. pp. 129–147.

van Schaik, C. P., and van Hooff, J. A. R. A. M. (1996). Towards an understanding of the orangutan's social system. In: McGrew, W. C., Marchant, L. F., and Nishida, T. (eds.), *Great Ape Societies.* Cambridge University Press, Cambridge. pp. 3–15.

Ward, S. (1997). The taxonomy and phylogenetic relationships of *Sivapithecus* revisited. In: Begun, D. R., Ward, C. V., and Rose, M. D. (eds.), *Function, Phylogeny, and Fossils: Miocene Hominoid Evolution and Adaptations.* Plenum Press, New York. pp. 269–290.

Warren, K. S., Nijman, I. J., Lenstra, J. A., Swan, R. A., Heriyanto, and Den Boer, M. H. (2000). Microsatellite DNA variation in Bornean orang-utans, *Pongo pygmaeus. J. Med. Primatol.* 29:57–62.

Warren, K. S., Nijman, I. J., Lenstra, J. A., Swan, R. A., Heriyanto, and Den Boer, M. H. (2001). Speciation and intrasubspecific variation of Bornean orangutans, *Pongo pygmaeus pygmaeus*. *Mol. Biol. Evol*. 18:472–480.

Watts, D. P. (1989). Infanticide in mountain gorillas: new cases and a review of evidence. *Ethology* 81:1–18.

Wich, S., Buij, R., and van Schaik, C. P. (2004a). Determinants of orangutan density in the dryland forests of the Leuser Ecosystem. *Primates* 45:177–182.

Wich, S. A., Geurts, M. L., Mitra Setia, T., and Utami-Atmoko, S. S. (in press). Sumatran orangutan sociality, reproduction and fruit availability. In: Hohmann, G., Robbins, M., and Boesch, C. (eds.), *Feeding Ecology in Apes and Other Primates*. Cambridge University Press, Cambridge.

Wich, S. A., Utami-Atmoko, S. S., Setia, T. M., Rijksen, H. R., Schurmann, C., van Hooff, J. A. R. A. M., and van Schaik, C. P. (2004b). Life history of wild Sumatran orangutans (*Pongo abelii*). *J. Hum. Evol*. 47:385–398.

Wich, S. A., and van Schaik, C. P. (2000). The impact of El Niño on mast fruiting in Sumatra and elsewhere in Malesia. *J. Trop. Ecol*. 16:563–577.

Wrangham, R. W. (1980). An ecological model of female-bonded primate groups. *Behaviour* 75:262–300.

Xu, X., and Arnason, U. (1996). The mitochondrial DNA molecule of Sumatran orangutan and a molecular proposal for two (Bornean and Sumatran) species of orangutan. *J. Mol. Evol*. 43:431–437.

Zhi, L., Karesh, W. B., Janczewski, D. N., Frazier-Taylor, H., Sajuthi, D., Gombek, F., Andau, M., Martenson, S., and O'Brien, S. J. (1996). Genomic differentiation among natural populations of orangutan (*Pongo pygmaeus*). *Curr. Biol*. 6:326–1336.

18

Gorillas
Diversity in Ecology and Behavior
Martha M. Robbins

INTRODUCTION

Gorillas are found in 10 central African countries, in a broad diversity of habitats ranging from coastal lowland forests to high-altitude, Afromontane rain forests (Fig. 18.1). The distribution of gorillas makes them a particularly appealing species to study, both in terms of understanding how they have adapted to such a variety of habitats and by providing an excellent opportunity to test many hypotheses of primate behavioral ecology that assume variation in ecology will lead to variation in behavior and demography (e.g., Sterck et al. 1997; *Papio*, Barton et al. 1996; *Presbytis entellus*, Koenig et al. 1998; *Saimiri*, Boinski et al. 2002). However, perhaps one of the biggest ironies of primatology is that the majority of information on gorillas has come from a very small population studied for over 35 years living at a unique ecological extreme (high altitude; Karisoke Research Center, Rwanda). Fortunately, in recent years, several studies of other gorilla populations have begun to provide us with information that is more representative of the genus as a whole, especially on feeding ecology; but because of the difficulty of habituating lowland gorillas, only a limited number of studies involving direct observations have been conducted and much remains unknown. One obvious conclusion, though, is that due to the large differences in ecology between western and eastern gorillas, we can no longer assume that all knowledge of the behavior and demography of mountain gorillas applies to all gorilla populations.

TAXONOMY AND DISTRIBUTION

The taxonomic classification of gorillas has changed many times over the past century (Groves 2003). For much of the past few decades, gorillas were considered only one species with three subspecies (western lowland gorillas, eastern lowland gorillas, and mountain gorillas). A recent taxonomic reclassification now groups gorillas as two species and four subspecies (Grooves 2001) (Fig. 18.1, Table 18.1). Western gorillas include *Gorilla gorilla gorilla* (western lowland gorilla), found in Equatorial Guinea, Gabon, Angola, Cameroon, Central African Republic, and Republic of Congo, and *G. g. diehli* (Cross River gorilla), found in a handful of small populations in Nigeria and Cameroon. Eastern gorillas

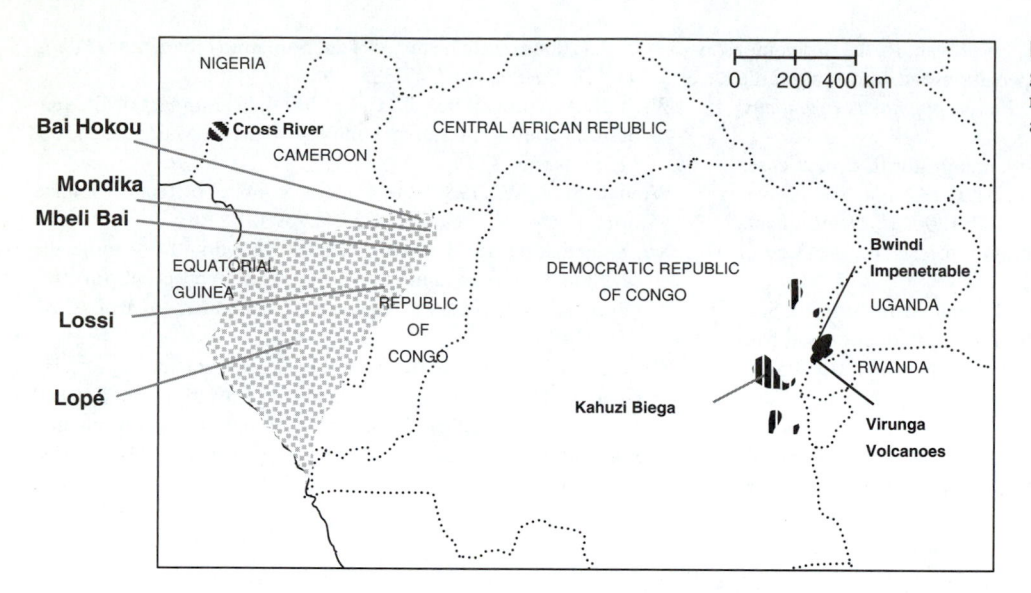

Figure 18.1 Map of distribution of gorillas in Africa. The locations of major field sites discussed in the text are indicated.

Western lowland gorilla (*Gorilla gorilla gorilla*)

Cross River gorilla (*G. g. diehli*)

Grauer's gorilla (*G. beringei graueri*)

Mountain gorilla (*G. beringei beringei*)

Table 18.1   Subspecies of Gorillas, Countries Which They Inhabit, Estimated Area of Habitat, and Estimated Population Size

SPECIES/SUBSPECIES	COUNTRIES FOUND	AREA OF HABITAT (KM²)	ESTIMATED POPULATION SIZE
Eastern gorilla (*Gorilla beringei*)			
Mountain gorilla[a] (*G. b. beringei*)	Rwanda, Uganda, Democratic Republic of Congo	700	700
Grauer's gorilla[b] (*G. b. graueri*)	Democratic Republic of Congo	15,000	5,000–15,000
Western gorilla (*Gorilla gorilla*)			
Cross River gorilla[c] (*G. g. diehli*)	Nigeria, Cameroon	140–200	200–250
Western lowland gorilla[d] (*G. g. gorilla*)	Gabon, Cameroon, Equatorial Guinea, Central African Republic, Angola, Democratic Republic of Congo	445,000	50,000–110,000

[a] McNeilage et al. (2001), Kalpers et al. (2003).
[b] Hall et al. (1998), Plumptre et al. (2003).
[c] Oates et al. (2003).
[d] Butynski (2001), Plumptre et al. (2003), Walsh et al. (2003).

include *G. beringei graueri* (eastern lowland or Grauer's gorilla), found in the east of Democratic Republic of Congo, and *G. b. beringei* (mountain gorilla), found in two small populations at the Virunga volcanoes of Rwanda, Uganda, and Democratic Republic of Congo and the Bwindi Impenetrable National Park of Uganda (Figs. 18.2 and 18.3).

FEEDING ECOLOGY

Gorillas are largely vegetarian, with the only non-vegetative foods in their diet being ants and termites (Williamson et al. 1990; Deblauwe et al. 2003; Doran et al. 2002; Watts 1984,

1989a; Yamagiwa et al. 1991; Ganas and Robbins 2004; Ganas et al. 2004; Rogers et al. 2004). Their large body size, enlarged hindgut, and long hindgut fermentation time enable them to meet their nutritional requirements with a diet high in bulky foods and structural carbohydrates (Watts 1996). Gorillas spend approximately 50% of their daylight hours feeding (Watts 1988). Gorillas eat nonreproductive plant parts (leaves, stems, pith, and bark) as well as fruit. A comparison of the diet of gorillas across Africa is best done by considering an elevation continuum from the lowland forests of central Africa to the Afromontane forests of the Virunga volcanoes (Table 18.2). In general, the degree of frugivory decreases as altitude increases because of reduced

(A)

(B)

(C)

(D)

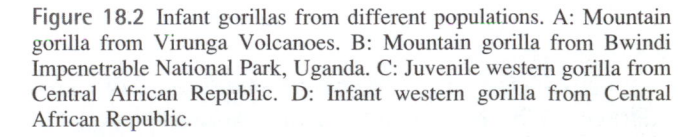

Figure 18.2 Infant gorillas from different populations. A: Mountain gorilla from Virunga Volcanoes. B: Mountain gorilla from Bwindi Impenetrable National Park, Uganda. C: Juvenile western gorilla from Central African Republic. D: Infant western gorilla from Central African Republic.

fruit availability (Doran and McNeilage 2001, Goldsmith 2003, Ganas et al. 2004). Detailed studies on mountain gorillas at Karisoke Research Center initially labeled them as folivore–herbivores that included only a negligible amount of fruit in their diet (Fossey and Harcourt 1977; Watts 1984; Vedder 1984; McNeilage 1995, 2001). However, these studies were conducted in the only gorilla habitat that contains almost no fruit (>2,500 m in elevation). Moving only 30 km away, the mountain gorillas of Bwindi Impenetrable National Park, whose range is between 1,400 and 2,500 m in altitude, eat fruit on approximately 60%–80% of days (Robbins and McNeilage 2003, Ganas et al. 2004). Several

studies of both Grauer's gorillas and western gorillas in lowland habitats reveal an almost daily consumption of fruit (Nishihara 1995; Remis 1997; Doran et al. 2002; Rogers et al. 2004; Yamagiwa et al. 1994, 1996, 2003a). Cross River gorillas consume fruit slightly less frequently than other lowland gorillas (approximately 90% of days), probably because they are faced with the ecological constraint of a very long dry season due to the northern location (5 months/year with less than 100 mm of rainfall compared to 3 months at other sites) (Oates et al. 2003). It is now clear that all gorilla subspecies should be considered folivore–frugivores. However, because nearly all studies have measured fruit consumption

Figure 18.3 Silverback mountain gorilla from Bwindi Impenetrable National Park, Uganda, descending a tree after foraging for fruit.

by analyzing feces for the presence of fruit remains (seeds) and signs of feeding remains along trails made by the gorillas and not with direct observations, the proportion and quantity of the gorilla diet that is fruit versus fibrous vegetation is still unknown.

While much emphasis has been placed on the degree of frugivory by gorillas, all populations do incorporate a significant amount of fibrous food (nonfruit foods including herbaceous vegetation) in their diet (Watts 1984, Ganas et al. 2004, Doran et al. 2002, Oates et al. 2003, Yamagiwa et al. 2003a, Rogers et al. 2004). In particular, during times of fruit scarcity, gorillas rely on fibrous foods as fallback foods. Cross River gorillas likely experience a more drastic shift in diet from fruit to fibrous foods because of the prolonged dry season compared with other lowland gorillas (Oates et al. 2003). There is almost no temporal variability in food availability for Virunga mountain gorillas, which almost entirely lack fruit in their diet (Watts 1984, 1991a, 1998a; Vedder 1984; McNeilage 1995, 2001). The only seasonal food for Virunga mountain gorillas is bamboo shoots (Vedder 1984). While bamboo is considered to be an

important component of mountain gorilla diet because it is high in protein, not all mountain gorillas consume it due to lack of availability (Watts 1984; McNeilage 1995, 2001; Ganas et al. 2004). The number of herb species consumed by gorillas also decreases with increasing altitude, probably as a result of decreased plant diversity as altitude increases.

Gorillas are selective in their feeding behavior (Watts 1984, 1991a; Vedder 1984; Yamagiwa et al. 1996, 2003a; Ganas et al. 2004; Rogers et al. 2004). Their diet is primarily influenced by availability of food resources, but it is clear that gorillas seek out foods with particular nutritional compositions and often forage on rare species. Gorillas select leaves and herbaceous material that are high in protein and fruits that are high in soluble sugars (Waterman et al. 1983, Calvert 1985, Rogers et al. 1990, Watts 1991a, Remis et al. 2001, Remis 2003). Tests with captive gorillas have shown them to be tolerant of relatively high levels of tannins if accompanied by high levels of sugar (Remis and Kerr 2002). Certain foods that are found only in swamps offer high levels of particular minerals (e.g., sodium and potassium), which may influence the use of these areas by gorillas and their ranging patterns (Magliocca and Gautier-Hion 2002, Doran-Sheehy et al. 2004).

A species' ranging patterns are strongly influenced by its diet and the availability of food resources. Karisoke mountain gorillas forage efficiently by preferentially feeding in areas containing a high biomass of food, especially those foods that are high in protein (Vedder 1984, Watts 1991a). Additionally, it is hypothesized that a frugivorous diet should affect travel distances and home range size of primates because fruit resources are typically more dispersed in the environment than herbaceous vegetation (Clutton-Brock and Harvey 1977, Janson and Goldsmith 1995). Because gorillas exhibit such variability in frugivory, they are an interesting species in which to examine the relationship between diet and ranging patterns. Gorillas are not territorial and have overlapping home ranges (Watts 1998b, Tutin 1996, Doran-Sheehy et al. 2004; Ganas and Robbins 2005). Current evidence suggests that gorillas are probably not opportunistic in their frugivory patterns but, rather, fruit "pursuers" that incur costs to forage on more dispersed resources (Rogers et al. 2004; Doran-Sheehy et al. 2004; Ganas and Robbins 2005). Daily travel distance is positively correlated to fruit consumption at several sites (Goldsmith 1999, Yamagiwa et al. 2003a, Doran-Sheehy et al. 2004; J. Ganas and M. M. Robbins 2005). How this variability in diet and travel distance influences net energy gain depending on fruit availability has not yet been examined, but gorillas are predicted to forage efficiently so as to maximize energy gain for distance traveled. Across study sites there is a positive relationship between the degree of frugivory (and decreased density of herbaceous vegetation) and daily travel distance, with Karisoke mountain gorillas traveling only approximately 0.5 km/day, Grauer's and western gorillas traveling 1–2 km/day, and Bwindi mountain gorillas being intermediate (Table 18.2).

Table 18.2   Dietary and Ranging Patterns of Gorillas

SUBSPECIES/STUDY LOCATION	ALTITUDE (M)	NUMBER OF FOOD SPECIES EATEN	NUMBER OF FRUIT SPECIES EATEN	DEGREE OF FRUGIVORY[a]	DAY JOURNEY LENGTH (M)	HOME RANGE SIZE (KM²)
Mountain gorilla (*Gorilla beringei beringei*)						
Virungas–Karisoke[b,c]	2,680–3,710	36	1	<1%*	570	3–15
Virungas–tourist group[c]	2,500–2,800	42	2	<1%**	756	5
Bwindi–Ruhija[d]	2,100–2,500	112	30	16–66%	1,034	21–40
Bwindi–Buhoma[e]	1,450–1,800	140	36	68–89%	547–978	16–22
Grauer's gorilla (*G. b. graueri*)						
Kahuzi-Biega–Kahuzi[f]	1,800–2,600	79	24	96%	850	23–31
Kahuzi-Biega–Itebero[f]	600–1,300	142	67	89%	–	23–31
Cross River gorilla (*G. gorilla diehli*)						
Cross River–Afi[g]	400–1,300	166	83	90%	–	~32
Western lowland gorilla (*G. g. gorilla*)						
Lopé, Gabon[h]	100–700	134	95	98%	1,105	22
Mondika, Central African Republic[i]	<400	100	70	99%	2,014	16
Bai Hokou, Central African Republic[j]	460	138	77	99%	1,580	11

[a] Degree of frugivory, percent of fecal samples that contained seeds as indication of daily fruit consumption, except for Virungas, where it was based on percent time of observation (*) or of feeding remains on trails that contained fruit (**).
[b] Vedder (1984), Watts (1984, 1991a, 1998b).
[c] McNeilage (1995, 2001).
[d] Robbins and McNeilage (2003), Ganas et al. (2004), Nkurunungi (2004), Ganas and Robbins (2005).
[e] Ganas et al. (2004), Ganas and Robbins (2005).
[f] Yamagiwa et al. (1992, 1994, 1996, 2003a).
[g] Oates et al. (2003).
[h] Rogers et al. (1990), Williamson et al. (1990), Tutin (1996).
[i] Doran et al. (2002, 2004).
[j] Remis (1997), Cipolletta (2003, 2004).

The relationship between frugivory and home range size is less clear (Table 18.2). Karisoke mountain gorillas include only a negligible amount of fruit in their diet, their habitat contains a very high herb density, and they have the smallest home ranges of any population studied. Otherwise, comparable values have been observed among the size of home ranges of all other populations examined.

According to the ecological constraints model, one cost of group living is that as group size increases, the amount of food needed collectively by the group also increases and, therefore, the day journey length and home range size should increase accordingly (Clutton-Brock and Harvey 1977, Chapman and Chapman 2000). It has been possible to test this hypothesis with gorillas only at the two mountain gorilla sites because of the lack of more than one group of habituated gorillas at other sites. At both Karisoke and Bwindi, a positive relationship was found between group size, daily travel distance, and home range size (Fossey and Harcourt 1977; McNeilage 1995; Watts 1991a, 1998b; Ganas and Robbins 2005). However, solitary males travel farther per day and have larger home range sizes than necessary to sustain a single individual because their movement patterns are also influenced by mate acquisition strategies (Watts 1994a).

The population density of gorillas is also expected to be correlated with food availability. Studies of western gorillas suggest that their density is highest where particular herbaceous, staple foods are found in high abundance (Rogers et al. 2004, Bermejo 1999). The relationship between the density of seasonal and fallback fruits and gorilla density is not yet known. Additionally, the density of gorillas is likely to be affected by interspecific competition for food resources, particularly with chimpanzees, depending on the degree of niche separation for fruit and herb species (Tutin et al. 1991, Yamagiwa et al. 1996, Stanford and Nkurunungi 2003).

REPRODUCTIVE, DEMOGRAPHIC, AND LIFE HISTORY PARAMETERS

As expected, given their large brain and body size, gorillas have a long maturation time and are long-lived, have long interbirth intervals, and reproduce relatively few times in their lives; but they are faster or shorter in most life history parameters than other ape species (Bentley 1999). Due to the lack of long-term studies of habituated western gorillas, nearly all the information available on reproductive and life history parameters comes from mountain gorillas (but see also Yamagiwa and Kahekwa 2001 and Yamagiwa et al. 2003b for Grauer's gorillas). Age/sex classifications typically used for gorillas include infants (0–3.5 years), juveniles (3.5–6 years), subadults (6–8 years), adult females (>8 years), blackback males (8–12 years), and silverback males (>12 years).

Females become sexually active at approximately age 6 and go through a period of adolescent sterility for usually at least 2 years (Watts 1990a, 1991b). Menstrual cycles are approximately 28 days long, and the length of the proceptive period is typically only 1–2 days (Watts 1990a). Based on hormonal analysis, mating behavior predominantly correlates to the time of ovulation (Czekala and Sicotte 2000). Nulliparous females may have small sexual swellings, but parous females show no swellings or other external signs of ovulation (Czekala and Sicotte 2000). Parous females typically conceive after four to six menstrual cycles (Watts 1990a, 1991b). Females show no obvious signs of being pregnant and usually mate while pregnant (Watts 1990a, 1991b). Gestation length is 8.5 months. The average age at first parturition is approximately 10 years (range 8–13). Typically, only one infant is born at a time, although twinning has been observed rarely (Watts 1988). Females go through lactational amenorrhea for approximately 3 years so that the typical interbirth interval is 4 years (Stewart 1988, Gerald 1995, Watts 1991b). If an infant dies before weaning, the female typically produces another offspring 1 year after the infant death (Watts 1991b, Gerald 1995). Female mountain gorillas that survive to adulthood (60%) have an average reproductive life span of 14 years and produce an average of 4.6 offspring that survive past infancy (Gerald 1995).

The reproductive parameters of wild mountain gorillas are remarkably similar to those observed in captive western gorillas (Czekala and Robbins 2001). However, the question remains whether the differences in ecological conditions of wild Grauer's and western gorillas from mountain gorillas lead to differences in reproductive parameters. In particular, if lowland gorillas face greater ecological constraints, later age at first parturition and longer interbirth intervals are expected (Doran and McNeilage 2001). No such differences have yet been observed for these variables between western, Grauer's, and mountain gorillas, nor are birth rates of western gorillas and mountain gorillas significantly different (range 0.18–0.226 births/adult female/year) (Robbins et al. 2004).

Males experience a longer period until maturity than females and do not go through such well-defined physiological events to mark certain points of maturation. Therefore, it is difficult to determine when males reach sexual maturity. Male gorillas begin the development of secondary sexual characteristics at approximately age 10 years, including a large increase in body size, the silvering of hair on the back and legs, and the formation of a peaked sagittal crest. They are typically considered to be silverbacks by age 12–13 years, but they may not obtain full adult size until 15 years of age (Watts 1991b, Watts and Pusey 1993).

Whereas female reproductive success is constrained by the costs of gestation and lactation, male reproductive success is limited by access to females. Male gorillas exhibit sociosexual behavior as juveniles and have been observed to start copulating at the age of 8–9 years (Watts 1990a; M. M. Robbins personal observation), long before attaining

final body size and secondary sexual characteristics. Captive western gorillas are known to sire offspring by 7 years of age (Beck 1982, Kingsley 1988). In one-male groups, the silverback is assumed to sire all offspring (western gorillas, Bradley et al. 2004) because extragroup copulations have been observed only very rarely (Sicotte 2001). In multimale mountain gorilla groups, dominant males mate more than subordinate silverbacks but subordinates do mate, including at the likely time of conception (Robbins 1999, Watts 1990a). Paternity determination studies at both Karisoke and Bwindi show that subordinate males sire offspring, even while still blackbacks (Bradley et al. 2005, Nsubuga et al. 2001, Nsubuga 2005). Variance in male reproductive success is high because some adult males never reproduce and others may be dominant males of large groups for many years (Robbins 1995; Robbins and Robbins 2005). Male lifetime reproductive success is correlated with the dominance tenure of silverbacks and the number of adult females in the group (Robbins 1995, Watts 2000; Robbins and Robbins 2005). Paternity determination studies will assist in determining the upper limits of male reproductive success. The degree of reproductive skew among males is likely to vary between western gorillas and mountain gorillas because of the variation in the occurrence of multimale groups and is likely to influence the genetic structuring within and between social groups in the two species (Bradley 2003, Bradley et al. 2001, Nsubuga et al. 2001).

Understanding patterns of mortality is one of the greatest challenges for researchers of long-lived species. As for most large-bodied mammals, mortality rates of gorillas are highest for infants and older adults (Gerald 1995). Infant mortality (to age 3 years) of mountain gorillas is 34% (Watts 1991b, Gerald 1995), and preliminary data on Grauer's and western gorillas do not appear to differ significantly (Robbins et al. 2004). Approximately 60% of mountain gorillas survive to age 8 years (Gerald 1995). Mortality rates are very low through adulthood. Despite the expectation that adult mortality would be higher for males than females because of the high level of male–male competition, no sex differences in mortality have yet been observed; but this may be because the majority of available data are for gorillas under the age of 12 years when competition is low (Gerald 1995).

The impact of predation and disease on mortality rates may also vary among gorilla populations. For example, respiratory diseases that are often fatal for mountain gorillas (Watts 1998a, Mudakikwa et al. 2001) have not been reported for western gorillas. Ebola has had devastating effects on certain western gorilla populations (Walsh et al. 2003, Huijbergts et al. 2003, Leroy et al. 2004). Attacks by leopards on mountain gorillas were recorded in the 1950s and 1960s, but none has been noted in recent decades (Schaller 1963, Sholley 1991). Predation by leopards currently is likely to have a greater impact on western gorillas (Fay et al. 1995, Robbins et al. 2004) than on mountain gorillas, which probably have no predators other than humans.

Table 18.3 Reproductive Parameters and Summary of Dispersal Patterns and Transitions of Group Structure Observed in Western Gorillas, Grauer's Gorillas, and Mountain Gorillas

	WESTERN GORILLA	GRAUER'S GORILLA[a]	MOUNTAIN GORILLA
Birth rate (births/adult female/year)	0.19 (Lossi) 0.198 (Mbeli)	Not calculated	0.226[b]
Interbirth interval (surviving births only)	4–6 years (n = 6)	4.6 years (n = 9, range 3.4–6.6)	3.9 years (n = 62)[b]
Infant mortality (to age 3 years)	22.0% (Lossi) 65% (Mbeli)	26.1%	34%[b]
Infanticide	Suspected cases	Observed (n = 3)[i]	Observed (n = 6)[g]
Adult male emigration	Common	Common	<50%[d,e]
Female natal/primary transfer	Common	Common	Common[c]
Female secondary transfer	Common	Common	Common[c]
Multimale groups	Rare	Rare	Common[d,f]
Nonreproductive groups	Observed[j]	Not observed	Observed[d,h]
Group formation by solitary male	Observed (n = 2)	Observed (n = 6)	Observed (n = 2)[d]
Group formation by group fissioning	Not observed	Observed (n = 2)	Observed (n = 5)[f]
Group disintegration	Observed (n = 6)	Observed (n = 3, due to poaching)	Observed (n = 5)[f]
Group takeover by outsider male	Not observed	Not observed	Not observed

Source: Modified from Robbins et al. (2004).
[a] Yamagiwa and Kahekwa (2001).
[b] Gerald (1995).
[c] Sicotte (2001).
[d] Robbins (1995, 2001).
[e] Watts (2000).
[f] Kalpers et al. (2003).
[g] Watts (1989b).
[h] Yamagiwa (1987a,b).
[i] Yamagiwa (personal communication).
[j] Gatti et al. (2004).

GROUP STRUCTURE AND DISPERSAL PATTERNS

Gorillas live primarily in stable, cohesive social units, although adult males may be solitary. Groups typically contain several adult females, their immature offspring, and always at least one silverback. Groups may be one-male, multimale, or all-male (nonreproductive, containing no adult females) (Stewart and Harcourt 1987; Robbins 1995, 2001; Robbins et al. 2004; Yamagiwa 1987a; Yamagiwa et al. 2003b; Gatti et al. 2004).

The variability in the social system of mountain gorillas is due to the following transitions (Robbins 1995, 2001; Yamagiwa 1987a; Yamagiwa and Kahekwa 2001) (Table 18.3). New social groups form when females transfer to lone silverbacks. Such groups remain one-male until male offspring mature into silverbacks, and the group is then multimale. Multimale groups return to a one-male structure if males emigrate, the original adult male dies, or the group fissions. When the silverback of a one-male group dies, the group disintegrates. If a breeding group loses all of its adult females, it becomes an all-male, or nonreproductive, group. All-male groups may also form through a merger of immature males (evicted from a heterosexual group taken over by a new silverback following the death of the previous silverback) and a solitary silverback (Robbins 1995; Yamagiwa 1987a; Gatti et al. 2003, 2004). All-male groups can become heterosexual if females transfer into them. If a dominant male loses all of his group members, he becomes a lone silverback. Group fissions have been reported only with eastern gorillas (Robbins 2001, Kalpers et al. 2003, Yamagiwa

and Kahekwa 2001). Take-overs by outsider males of established groups containing a mature silverback have not been reported for any gorilla population.

The average group size of both western and eastern gorillas is 8–10 individuals (range 2–40+) (Tutin 1996, Parnell 2002a, Kalpers et al. 2003, Yamagiwa et al. 2003b, Gatti et al. 2004). However, groups with more than 20 individuals have been found in several eastern gorilla populations but at only one location of western gorillas, Odzala National Park in the Republic of Congo (Bermejo 1999, Magliocca et al. 1999, Yamagiwa and Kahekwa 2001, Kalpers et al. 2003). Ecological constraints on western gorillas may prevent groups from surpassing a particular size. The percentage of groups in the Virunga population containing over 20 gorillas has been increasing in the past three decades (3.5% in 1978–1979, 9% in 1989, and 18% in 2000), which may be a reflection of increased protection and reduced poaching (Weber and Vedder 1983, Sholley 1991, Kalpers et al. 2003).

It is hypothesized that females form permanent associations with males as a means to avoid infanticide by extragroup males (Watts 1989b, 2000; van Schaik and Kappeler 1997; van Schaik 2000) and to benefit from protection against predators. Males benefit from the long-term associations with females. Males can use the risk of infanticide as an effective strategy to retain mates because of high male–male competition for mates and the long interbirth interval. Most infanticides in mountain gorillas have occurred when the only silverback of a group died and the new male (from outside the group) that took over killed the unweaned offspring (Watts 1989b). In Grauer's gorillas, a silverback was observed to kill three young infants of recent immigrant females, but females with unweaned offspring were also observed without a silverback for extended periods of time and did not suffer from infanticide (Yamagiwa and Kahekwa 2001; J. Yamagiwa personal communication). Infanticide has been inferred in western gorillas twice when the infants disappeared following their mothers' transfer after the death of a silverback; however, infants survived in two other cases in similar circumstances (Stokes et al. 2003). Furthermore, Harcourt and Greenberg (2001) used a mathematical model to show that females traveling unaccompanied by a silverback would be at three times greater risk of infanticide than those in social groups. Infanticide does not need to occur in all possible cases in order for it to be an evolutionarily stable strategy (van Schaik 2000).

Gorillas are one of only a few primate species in which both males and females may be philopatric or disperse (Harcourt et al. 1976, Tutin 1996). Females transfer directly to a solitary male or to another group during interunit encounters (Harcourt et al. 1976, Harcourt 1978). Female primary (natal) and secondary dispersals have been observed in mountain, Grauer's, and western lowland gorillas (Harcourt et al. 1976, Sicotte 2001, Yamagiwa and Kahekwa 2001, Stokes et al. 2003). Female transfer is considered to be an important method of female mate choice (Sicotte 2001). Female primary transfer may occur as early as age 5.5 (Sicotte 2001, Watts 1991b). Females may transfer several times in their lives, but because of the risk of infanticide by extragroup males, the window of opportunity for females to disperse is quite narrow, limited to the 3 to 6-month period per 4-year interbirth interval when the female is neither pregnant nor lactating (Watts 1989b, Sicotte 2001; but see Sicotte 2000 and Yamagiwa and Kahekwa 2001 for cases of transfer with unweaned infants). Transfer possibilities are further dependent on the occurrence of intergroup encounters. The average rate of intergroup encounter per group is approximately one per month at Karisoke, but in some cases many months may pass without any intergroup encounters, which would further constrain the possibility for female transfer (Sicotte 2001).

Currently, more is known about what group conditions influence female transfer decisions than which male traits females prefer (e.g., size, sexually selected characteristics). Female western gorillas appear to prefer smaller groups, presumably because of reduced feeding competition (Tutin 1996, Stokes et al. 2003). In contrast, female mountain gorillas, which face low feeding competition, show no group size preference, yet they do prefer multimale groups over one-male groups and solitary males (Watts 2000). This is likely due to females in multimale groups deriving greater protection against infanticide than those in one-male groups because in the event of one silverback death in a multimale group, the other silverback can assume leadership and prevent group disintegration and take-over by an external male (Robbins 1995, Watts 2000). Females may also transfer to avoid inbreeding or to increase the genetic variability among offspring (Sicotte 2001).

Subordinate silverbacks emigrate to become solitary males at approximately 12–18 years of age (Robbins 1995, Stokes et al. 2003, Yamagiwa and Kahekwa 2001). It is unclear whether males voluntarily emigrate or if they are forcibly evicted, but dispersal decisions are presumably influenced by male–male competition, likelihood of mate acquisition within the natal group versus in neighboring areas, and inbreeding avoidance. Interestingly, in western gorillas, there are reports of immature males voluntarily immigrating into heterosexual groups (Gatti et al. 2004, Robbins et al. 2004). The proportion of males that emigrate is higher in western gorillas than in mountain gorillas, and correspondingly, multimale groups are rare among western gorillas (Tutin 1996, Magliocca et al. 1999, Parnell 2002a, Gatti et al. 2004, Robbins et al. 2004) (Table 18.3) and Grauer's gorillas (Yamagiwa et al. 1993, Yamagiwa and Kahekwa 2001) but common in mountain gorillas (8%–50% of groups; McNeilage et al. 2001, Kalpers et al. 2003). These comparisons suggest that male western gorillas may typically attain higher reproductive success by dispersing and trying to form groups than by queuing for dominant positions within multimale groups, as is the case in mountain gorillas (Robbins 1995, Watts 2000; Robbins and Robbins 2005). This is likely to be influenced by how females distribute themselves in groups. Both male eastern and western gorillas have low success rates in acquiring females after emigration; males have been observed as solitary for 5 years or more (Robbins 1995, Watts 2000, Robbins et al. 2004). The rate of group disintegration appears to be higher in western gorillas than mountain gorillas, and group fissions have not been observed in western gorillas; these findings are to be expected given the rare occurrence of multimale groups in western gorillas (Robbins et al. 2004) (Table 18.3).

INTERGROUP INTERACTIONS

In most cases, intergroup interactions do not appear to be related to defending a group's range or to food resources (but see Bermejo 2004). They are an important time for female choice and male–male competition because these interactions are the only time that female transfers occur (Watts 1989b, Sicotte 2001, Robbins 2003). However, female choice may be limited by males that may herd females away from a neighboring group to prevent transfers (Sicotte 1993). Solitary males have been observed to pursue groups for several days at a time in both mountain and western gorillas, presumably in an attempt to acquire mates (Watts 1994a, Tutin 1996, Bermejo 2004). The frequency of intergroup encounters is likely to vary depending on the density of groups and of solitary males within a particular area.

During intergroup interactions, silverbacks typically perform chest-beating displays at each other, which sometimes elevates to contact aggression. However, interactions may also be peaceful, with group members intermingling. Sicotte

(1993) found that the intensity of intergroup encounters was positively related to the number of potential migrant females. Interestingly, studies of western gorillas have reported that intergroup encounters are often peaceful and involve aggression less frequently than has been observed in mountain gorillas (Parnell 2002b, Doran-Sheehy et al. 2004, Bermejo 2004). Bradley et al. (2004) hypothesized that the lower frequency of aggression during encounters observed in western gorillas may be due to males interacting differently with relatives and nonrelatives occupying neighboring home ranges. However, intergroup male–male competition occurs in western gorillas, based on observations of wounds on silverbacks that were likely to have been caused by other silverbacks (lacerations and bite marks on the head, neck, and back) (Parnell 2002b). Silverbacks have died as a result of serious injuries incurred during encounters in both western and mountain gorillas (Tutin 1996, Robbins 2003). Additional studies at all locations are needed to understand how the level of aggressiveness during intergroup encounters is related to the following variables: relatedness between males, number of silverbacks and potential migrant females in each group, familiarity between groups (e.g., frequency of encounters), competition for food resources, and presence of researchers around unhabituated gorillas, which may inhibit normal behavior.

SOCIAL RELATIONSHIPS

Many current socioecological models predict that the types of social relationship exhibited by primates are most influenced by three variables: distribution of food resources, predation risk, and risk of infanticide (van Schaik 1989, Sterck et al. 1997, Koenig 2002, Isbell and Young 2002). The extensive research on the mountain gorillas of Karisoke Research Center, Rwanda, has contributed significantly to the development of these models (Watts 2003). Given the known variation in ecological conditions and dietary patterns of gorillas, the opportunity exists to use them as a test of these models. Unfortunately, data on social relationships in other populations of gorillas are minimal at this time; therefore, the focus here is on what is known from Karisoke, and comparisons with other populations are made when possible.

Female–Female Social Relationships

Overall, at the Virunga volcanoes, female mountain gorillas have very weak social relationships with one another (Harcourt 1979a; Stewart and Harcourt 1987; Watts 1991c, 1994a,b, 2001, 2003) that are markedly different from those observed among many other primate species. They are classified as being "non-female-bonded" (Wrangham 1979, van Schaik 1989) and "dispersal egalitarian" (Sterck et al. 1997). Food resources are abundant and evenly distributed, which leads to very low levels of contest competition (Watts 1984, 1985, 1994b). Therefore, there are few benefits of

associating with other female kin, and the social and ecological costs of transferring to new habitats are low (Watts 1990b). This allows for female transfer (see above). As a result, females are typically in social groups containing unrelated individuals, although mother–daughter and sister pairs are not uncommon because females do not always transfer and related females may transfer into the same group (Watts 1996, 2001). Levels of aggressive and affiliative interactions among female–female dyads are highly variable and may change over time (Stewart and Harcourt 1987, Watts 2001).

Despite low levels of feeding competition, aggression among females most commonly occurs while feeding, but it also occurs during resting bouts and traveling, when females may be intolerant of close proximity to one another (Watts 1994b). The majority of aggressive interactions among females do not have a clear winner; they are undecided because the recipient ignores the action or retaliates. Submissive behavior (cowering, backing away, or an appeasement vocalization, "grumbling") is rarely exhibited among females (Watts 1994b). Dominance relationships are determined according to patterns of displacement and avoidance (approach–retreat interactions) and are not based on any formal submissive behavior, as is used in many other primate species. While females can be placed in linear dominance hierarchies in some cases, in many groups such dominance relationships are weak or nonexistent (Harcourt 1979a; Harcourt and Stewart 1987, 1989; Watts 1985, 1994b). However, many dominance relationships among females are stable over time (Robbins et al. 2005). Additionally, aggression within individual dyads is frequently bidirectional (both females of a dyad initiate aggression against each other) so that females cannot be placed into linear hierarchies using all agonistic behavior (Watts 1994b). These relationships, described as being egalitarian (Sterck et al. 1997; Watts 2001, 2003), may be explained by two factors which show that there is little benefit of targeting aggression toward particular individuals (Watts 2001). First, there are low gains from agonistic behavior in an environment with highly abundant food resources (Watts 1985, 1994b). Second, males frequently intervene in aggressive contests between females and do not clearly support either female (control intervention) (Harcourt and Stewart 1987, 1989; Watts 1994b, 1997). The fact that relationships among females are of little value is further emphasized by the very low rates of reconciliation following conflicts (Watts 1995).

Even in an environment of high food availability, scramble competition is expected to increase as group size increases. As the number of females per group increased, the rates of aggression among females increased but not significantly (Watts 1985, 2001). While females that immigrate into average-sized groups do not receive high levels of aggression from resident females, five females that immigrated into a large group within a few months of each other received high levels of harassment from the resident females (Watts 1991c, 1994c). However, this may have occurred in this case because several related resident females often formed coalitions against the immigrants or because the dominant male may not have been able to effectively intervene in aggression among so many females. Aggression among females in contexts other than feeding, for example, during rest sessions and traveling, indicates that they may be competing for male services, in particular protection against predators and extragroup males (Watts 1994b, 2001). A male will be limited in his ability to maintain close proximity to all females as group size exceeds a certain number.

Affiliative interactions, particularly grooming and resting in close proximity, occur at higher rates among maternally related individuals than unrelated individuals, with intermediate values for putative paternally related females (Watts 1994c). Grooming is typically reciprocal among dyads, but this is due to the higher occurrence of grooming among relatives. In general, female relatives (mother–daughter and sister dyads) most commonly have "good" relationships with each other (higher than median values of affiliative interactions and lower than median values for aggressive interactions for both partners) and unrelated females typically have "bad" relationships with each other (lower than median values of affiliative interactions and higher than median values for aggressive interactions), though sometimes they do have good relationships (Watts 2001).

Because fruit is usually distributed in a clumped fashion in time and space, the increase in frugivory experienced by all other gorilla populations besides those at the Virunga volcanoes is predicted to lead to higher levels of within-group contest competition and differentiated female–female social relationships that include linear agonistic dominance hierarchies (Doran and McNeilage 1998, 2001). The lack of habituated gorillas at most study sites currently limits tests of this hypothesis to two locations. Female mountain gorillas in Bwindi Impenetrable National Park, Uganda, exhibited higher rates of aggression when foraging on fruit versus herbaceous vegetation; but overall, they exhibited weak dominance relationships, a high degree of bidirectionality in aggression, and low rates of affiliative behavior, as has been observed in Karisoke mountain gorillas (M. M. Robbins unpublished data). Bwindi also has abundant quantities of evenly distributed herbaceous vegetation (Ganas et al. 2004, Nkurunungi 2004), and the gorillas spend only approximately 10% of their foraging time eating fruit, which may limit the effect that frugivory has on their social relationships (M. M. Robbins, unpublished data). Observations of western gorillas at Mbeli Bai, Republic of Congo, also failed to show a noticeable difference in social relationships among females from those observed at Karisoke (Stokes 2004). However, this study provides only a limited view of their social behavior because the gorillas were observed only in a *bai* (swampy clearing) feeding on abundant, evenly distributed aquatic vegetation and the observations do not include any frugivorous behavior and are not representative of the majority of western gorilla daily activity (a group typically spends only 1% of its time in a bai) (Magliocca et al. 1999, Parnell 2002b). Further detailed studies of

habituated western gorillas are necessary to understand the influence that frugivory has on their social relationships.

Male–Female Social Relationships

Social relationships among males and females are the foundation of gorilla groups (Harcourt 1979b; Fossey 1983; Stewart and Harcourt 1987; Watts 1992, 2003). These relationships reflect the costs and benefits of the reproductive strategies of both males and females. Females benefit from the protection against predators and infanticide by extra-group males, and males benefit by permanent association with females as a long-term mating strategy. Male–female social relationships vary depending on the reproductive status of the females, the residency status of females, the number of males in the group, and kinship (Harcourt 1979a; Watts 1992, 2003). Particular male–female dyads may coinhabit a social group for 10 or more years. Male–female relationships may influence female transfer patterns (Watts 1992, 1996, 2003) and, if a multimale group fissions, with which male each female remains (Robbins 2001, Watts 2003).

The high level of affiliation between males and females is emphasized by proximity patterns. Females may spend up to 20% of their feeding time and over 50% of resting time within 5 m of the dominant silverback (Watts 1992). Females increase the amount of time they spend in close proximity to the silverback upon the birth of a new infant (Harcourt 1979b). Females are more responsible than males for maintaining proximity to the silverback in one-male groups (Harcourt 1979b), but in multimale groups males are more responsible for maintaining proximity to females when they are proceptive (Sicotte 1994). Grooming among adult gorillas is most common between males and females. However, there is high variability among groups in whether the silverback does more grooming of adult females or visa versa (Harcourt 1979b, Watts 1992).

Silverbacks are clearly dominant over adult females based on approach–retreat interactions, directionality of aggressive behavior, and submissive behavior (Harcourt 1979b; Watts 1992, 1994b). The majority of aggression by males toward females consists of mild vocalizations and displays (e.g., chest-beating displays) and only rarely results in physical injuries (Watts 1992). Females frequently "grumble" after receiving aggression from males, especially following displays; and this is the only formal submissive behavior observed in gorillas (Watts 1994b). Females frequently reconcile with males following aggressive encounters (Watts 1995).

Given that males are clearly larger in size and dominant over females, why do they need to frequently exhibit aggression toward females? This aggression is not predominantly due to feeding competition, but it likely serves as a means of reasserting their protective abilities as well as courtship aggression or sexual coercion. In fact, only a minority of agonistic behavior that silverbacks direct toward females occurs in the context of feeding, and more than 60%

may be classified as courtship aggression (displays and interventions in sexual interactions) (Watts 1992). Additionally, the levels of aggression directed toward individual females vary depending on their reproductive status, further emphasizing the role of courtship aggression. Recent immigrant females receive higher levels of aggression from males than do resident females in both one-male and multimale groups (Harcourt 1979b, Sicotte 2000, Watts 1992). While in one-male groups, the silverback does not increase the rate of aggression toward females when they are sexually active (Harcourt 1979b); but in multimale groups, both dominant and subordinate males do (Robbins 2003).

Females in multimale groups will commonly mate with more than one male, which may be a result of sexual coercion by the males and/or female choice and may serve to confuse paternity (Robbins 1999, 2003; Sicotte 2001). Subordinate males may use affiliative behavior such as grooming to influence female choice (Watts 1992). Dominant males are more likely than subordinate males to mate with cycling and pregnant females, whereas subordinate males are more likely to mate with subadult females (Robbins 1999). Mate guarding may occur in multimale groups but not in one-male groups (Harcourt et al. 1980, Sicotte 1994, Watts 1996).

Observations of male–female relationships in a multimale group in Bwindi Impenetrable National Park largely corroborate those at Karisoke (M. M. Robbins personal observation), which is unsurprising given that reproductive strategies, not ecological conditions, are likely to have a larger influence on these social relationships. At Mbeli Bai, males exhibited higher rates of aggression toward females than was observed among females, but most of it consisted of mild aggression and could be considered courtship aggression (Stokes 2004). Again, until more detailed studies of western gorillas are conducted, it is not possible to draw conclusions about the differences in male–female social relationships between the species.

Male–Male Social Relationships

In many primate species, male–male social relationships are very weak and characterized by high levels of competition for mates (van Hooff and van Schaik 1994, Kappeler 2000). Gorillas are not an exception to this pattern. In multimale groups, males may be related as father–son or half-brothers (maternal or paternal), but in some cases they may be unrelated (Robbins 1995, Bradley et al. 2005). The coexistence of unrelated males may be caused by immigration of immature or blackback males into social groups (rarely) or lower levels of reproductive skew in multimale groups that lead to cohorts of young being sired by more than one male.

In the majority of social groups that are one-male, the only adult male present has no other males to form social relationships with except for maturing males. Silverback–blackback relationships are weak, based on low rates of close proximity and affiliation and unidirectional aggression

from the silverback to blackback. Silverbacks are always dominant over blackbacks, which spend considerable amounts of time on the periphery of the group, perhaps to avoid conflict with the silverback or adult females. More affiliative relationships may exist between males that are related or more familiar with each other, which may influence dispersal decisions of young adult males (Harcourt 1979c, Harcourt and Stewart 1981). Unfortunately, male behavior shortly before emigration has not been systematically studied to determine if dispersal is related to increased rates of aggression and intolerance by the dominant male. If dominant males want to evict subordinate individuals, it would seem logical that they do it before the younger males become formidable competitors.

Heterosexual social groups that contain at least two fully mature silverbacks may exist for 10 or more years in mountain gorillas (Robbins 1995, 2001). Among the males, dominance hierarchies are obvious, but rank does not always positively correlate with age (Harcourt 1979c, Sicotte 1994, Watts 1992, Robbins 1996). Silverbacks spend little time in close proximity to each other (<5 m), and they rarely affiliate with each other (Sicotte 1994, Robbins 1996). Rates of aggressive behavior between males depend on a variety of factors, including the number and reproductive status of females, the age of males, and the stability of their relationships. Aggression between males may increase when females are sexually active but not always (Sicotte 1994, Robbins 2003). In some cases, dominant males increase aggression toward subordinates; in other cases, subordinate males direct more aggression toward dominant males, or there may be an increase in either direction. Both dominant and subordinate males may harass each other during mating, but dominant males are more successful in stopping copulations than subordinate males (Robbins 1999). This variability may be due to the stability of the dominance relationship among the males and/or the reproductive status of the female (subadult females, cycling adult females, pregnant females). Reconciliation between adult males has not been observed (Watts 1995). In general, when males have stable dominance relationships, they appear to coexist predominantly through avoidance or tolerance rather than frequent, high levels of aggression or by forming affiliative relationships. One benefit of a multimale structure is that males will cooperate during intergroup encounters against extragroup males (Sicotte 1993; Robbins 2001; Watts 1989b, 2000).

Silverbacks may be dominant for 10 or more years. Usurpation of the alpha male by younger silverbacks in multimale groups has been observed three times in three decades at Karisoke (Watts 1990a, 1992; Sicotte 1993; Robbins 1996); however, the transitional process was not well documented, so the strategies that younger males use to attain alpha status or how dominant males resist are not yet understood. Observations in Bwindi suggest that usurpation may involve a lengthy process (several years) of agonistic encounters, rather than a quick turnover (M. M. Robbins, personal observation). Deposed males are usually not evicted and typically retain positive relationships with adult females. If deposed males sire offspring and reduce the direct fitness of the dominant male, the question remains of why these males are tolerated in the group. Eviction may be costly to the dominant male, and/or the benefits of retaining a multi-male group status (e.g., more attractive to females, reduced risk of infanticide, and additional defense during intergroup encounters) may outweigh the costs.

Relationships among males in all-male groups are markedly different from those in heterosexual groups, presumably because the absence of females reduces competitive interactions (Yamagiwa 1987b, Robbins 1996). Dominance relationships are still apparent among the males, but they may be less clear within age classes, especially with blackbacks. While rates of aggression may be higher than those observed in heterosexual groups, most of the aggression consists of vocalizations and displays, and lower incidences of wounding caused by serious fights are observed (Robbins 1996). Males exhibit higher rates of grooming and playing and spend more time in close proximity than in heterosexual groups. Sociosexual behavior occurs but at low rates. Despite the lack of reproductive opportunities, all-male groups may be a better alternative for males than being solitary because they provide a social setting in which males can gain experience in aggressive and affiliative interactions (Robbins 1996, Gatti et al. 2004).

Development and Social Relationships of Immature Gorillas

Gorillas have an extended period of immaturity, during which time they develop their social skills. Infants are completely dependent on their mothers at birth and are unlikely to survive the death of the mother until they reach the age of 2–3 years because of the prolonged lactation period as well as their dependence for locomotion. The first 6 months of an infant's life are spent in constant contact with the mother, and then slowly the infant begins to spend increasing amounts of time farther away so that by the age of 30 months it spends only approximately 50% of its time within 5 m of the mother (Fletcher 2001). Mothers may continue to serve as an important social partner for offspring through immaturity, and relationships with other adult females are usually weak because female dispersal limits the number of female kin within a social group (Fossey 1979, Fletcher 1994).

Play likely serves as one of the most important behaviors for the development of social skills in gorillas. Play begins typically when infants are about 9 months of age, peaks in frequency during juvenility, and then decreases during adolescence (Fletcher 1994). Sex differences in play exist, with males playing more than females, especially during adolescence (Watts and Pusey 1993, Fletcher 1994).

While silverbacks' primary contribution to immature gorillas is to provide protection against infanticide from outsider males (Watts 1989b), there are other positive benefits to silverback–immature associations. As the time spent near

the mother decreases in late infancy, the time spent in close proximity to the silverback increases. Immature gorillas are attracted to the silverback as the focal point of the group and spend considerable time in close proximity to him even when their mothers are not nearby (Watts and Pusey 1993, Fletcher 1994, Stewart 2001). Juveniles groom silverbacks more than vice versa in most cases (Stewart 2001). Silverbacks frequently intervene during aggressive conflicts involving immatures and usually support the younger opponent, which serves to protect immatures from high levels of aggression from older individuals (Harcourt and Stewart 1987, Watts 1997, Stewart 2001). Silverbacks may increase affiliative behavior toward young that have lost their mothers through death or dispersal, including grooming and co-nesting (Watts and Pusey 1993, Stewart 2001).

CONSERVATION

All subspecies of gorillas are considered to be endangered (IUCN 2000). However, the population size and area inhabited vary between subspecies, country, and protected area (Table 18.1). The most critically endangered subspecies are the Cross River gorilla, with only approximately 200–250 individuals remaining in nine small, isolated pockets of forest (Oates et al. 2003), and mountain gorillas, with approximately 700 remaining in two isolated populations (Virunga volcanoes and Bwindi Impenetrable) (McNeilage et al. 2001, Kalpers et al. 2003). While both Grauer's gorillas and western gorillas have larger population sizes, both have declined dramatically in recent years (Hall et al. 1998, Plumptre et al. 2003, Walsh et al. 2003), and their current population estimates should be viewed with caution because the survey methods used have many biases and lack precision, which result in huge estimation errors that render the final numbers not particularly reliable (Walsh et al. 2003 and unpublished data). Accurate population estimates are necessary in order to monitor the changes in populations and assess the effectiveness of conservation strategies.

The major threats to gorillas are habitat destruction, poaching, disease, and political instability; but there is variation in the degree to which each of these threats impacts each gorilla population (Plumptre and Williamson 2001, Plumptre et al. 2003, Walsh et al. 2003). What can be done to ensure that gorillas do not go extinct? Currently, there are many conservation activities that focus on wild gorillas and their habitat, but clearly efforts must be expanded given the evidence of declining populations. Effective conservation of gorillas ultimately depends on the will and efforts of both local and the international communities.

REFERENCES

Barton, R. A., Byrne, R. W., and Whiten, A. (1996). Ecology, feeding competition and social structure in baboons. *Behav. Ecol. Sociobiol.* 38:321–329.

Beck, B. B. (1982). Fertility in North American lowland gorillas. *Am. J. Primatol.* 1(suppl.):7–11.

Bentley, G. R. (1999). Aping our ancestors: comparative aspects of reproductive ecology. *Evol. Anthropol.* 7:175–185.

Bermejo, M. (1999). Status and conservation of primates in Odzala National Park, Republic of the Congo. *Oryx* 33:323–331.

Bermejo, M. (2004). Home range use and inter-group encounters in western gorillas (*Gorilla g. gorilla*) at Lossi Forest, north Congo. *Am. J. Primatol.* 64:223–232.

Boinski, S., Sughure, K., Selvaggi, L., Quatrone, R., Henry, M., and Cropp, S. (2002). An expanded test of the ecological model of primate social evolution: competitive regimes and female bonding in three species of squirrel monkeys (*Saimiri oerstedii*, *S. boliviensis* and *S. sciureus*). *Behaviour* 139(2–3):227–261.

Bradley, B. J. (2003). Molecular ecology of wild gorillas [PhD diss.]. State University of New York, Stony Brook.

Bradley, B. J., Doran-Sheehy, D. M., Lukas, D., Boesch, C., and Vigilant, L. (2004). Dispersed male networks in western gorillas. *Curr. Biol.* 14:510–514.

Bradley, B. J., Robbins, M. M., Williamson, E. A., Boesch., C., and Vigilant, L. (2001). Assessing male reproductive success in wild mountain gorillas using DNA analysis. *Primate Rep.* 60:16–17.

Bradley, B. J., Robbins, M. M., Williamson, E. A., Steklis, H. D., Gerald-Steklis, N., Eckhardt, N., Boesch, C., and Vigilant, L. (2005). Mountain gorilla tug-of-war: silverbacks have limited control over reproduction in multi-male groups. *Proc. Nati. Acade. Science, USA*, 102:9418–9423.

Butynski, T. M. (2001). Africa's great apes. In: Beck, B. B., Stoinski, T. S., Hutchins, M., Maple, T. L., Norton, B., Rowan, A., Stevesn, E. F., and Arluke, A. (eds.), *Great Apes & Humans: The Ethics of Coexistence*. Smithsonian Institute Press, Washington DC. pp. 3–56.

Butynski, T. M., and Kalina, J. (1998). Gorilla tourism: a critical look. In: Milner-Gulland, E. J., and Mace, R. (eds.), *Conservation of Biological Resources*. Blackwell Science, Oxford. pp. 280–300.

Calvert, J. J. (1985). Food selection by western gorillas (*G. g. gorilla*) in relation to food chemistry. *Oecologia* 65:236–246.

Chapman, C. A., and Chapman, L. J. (2000). Determinants of group size in primates: the importance of travel costs. In: Boinski, S., and Garber, P. A. (eds.), *On the Move: How and Why Animals Travel in Groups*. University of Chicago Press, Chicago. pp. 24–42.

Cipolletta, C. (2003). Ranging patterns of a western gorilla group during habituation to humans in the Dzanga-Ndoki National Park, Central African Republic. *Int. J. Primatol.* 24:1207–1226.

Cipolletta, C. (2004). Effects of group dynamics and diet on the ranging patterns of a western gorilla group (*Gorilla gorilla gorilla*) at Bai Hokou, Central African Republic. *Am. J. Primatol.* 64:193–205.

Clutton-Brock, T. H., and Harvey, P. H. (1977). Species differences in feeding and ranging behavior in primates. In: Clutton-Brock, T. H. (ed.), *Primate Ecology: Studies of Feeding and Ranging Behaviour in Lemurs, Monkeys, and Apes*. Academic Press, New York. pp. 557–584.

Czekala, N. M., and Robbins, M. M. (2001). Assessment of reproduction and stress through hormone analysis in gorillas. In: Robbins, M. M., Sicotte, P., and Stewart, K. J. (eds.), *Mountain Gorillas: Three Decades of Research at Karisoke*. Cambridge University Press, Cambridge. pp. 317–339.

Czekala, N. M., and Sicotte, P. (2000). Reproductive monitoring of free-ranging female mountain gorillas by urinary hormone analysis. *Am. J. Primatol.* 51:209–215.

Deblauwe, I., Dupain, J., Nguenang, G. M., Werdenich, D., and Van Elsacker, L. (2003). Insectivory by *Gorilla gorilla gorilla* in southeast Cameroon. *Int. J. Primatol.* 24:493–501.

Doran, D. M., and McNeilage, A. (1998). Gorilla ecology and behavior. *Evol. Anthropol.* 6:120–131.

Doran, D. M., and McNeilage, A. (2001). Subspecific variation in gorilla behavior: the influence of ecological and social factors. In: Robbins, M. M., Sicotte, P., and Stewart, K. J. (eds.), *Mountain Gorillas: Three Decades of Research at Karisoke*. Cambridge University Press, Cambridge. pp. 123–149.

Doran, D. M., McNeilage, A., Greer, D., Bocian, C., Mehlman, P. T., and Shah, N. (2002). Western lowland gorilla diet and resource availability: new evidence, cross-site comparisons, and reflections on indirect sampling methods. *Am. J. Primatol.* 58:91–116.

Doran-Sheehy, D. M., Greer, D., Mongo, P., and Schwindt, D. (2004). Impact of ecological and social factors on ranging in western gorillas. *Am. J. Primatol.* 64:207–222.

Fay, M., Carroll, R. W., Kerbis Peterhans, J. C., and Harris, D. (1995). Leopard attack on and consumption of gorillas in the Central African Republic. *J. Hum. Evol.* 29:93–99.

Fletcher, A. W. (1994). The social development of immature mountain gorillas (*Gorilla gorilla beringei*) [PhD diss.]. Bristol University, Bristol.

Fletcher, A. W. (2001). Development of infant independence from the mother in wild mountain gorillas. In: Robbins, M. M., Sicotte, P., and Stewart, K. J. (eds.), *Mountain Gorillas: Three Decades of Research at Karisoke*. Cambridge University Press, Cambridge. pp. 153–182.

Fossey, D. (1979). Development of the mountain gorilla (*Gorilla gorilla beringei*): the first 36 months. *Perspect. Hum. Evol.* 5:139–184.

Fossey, D. (1983). Gorillas in the Mist. Houghton-Mifflin, Boston.

Fossey, D., and Harcourt, A. H. (1977). Feeding ecology of free ranging mountain gorillas (*Gorilla gorilla beringei*). In: Clutton-Brock, T. H. (ed.), *Primate Ecology: Studies of Feeding and Ranging Behavior in Lemurs, Monkeys and Apes*. Academic Press, London. pp. 539–556.

Ganas, J., and Robbins, M. M. (2004). Intrapopulation differences in ant eating in the mountain gorillas of Bwindi Impenetrable National Park, Uganda. *Primates* 45:275–279.

Ganas, J., and Robbins, M. M. (2005). Ranging behavior of the mountain gorillas (*Gorilla beringei beringei*) in Bwindi Impenetrable National Park, Uganda: a test of the ecological constraints model. *Behav. Ecol. Sociobiol.* 58:277–288.

Ganas, J., Robbins, M. M., Nkurunungi, J. B., Kaplin, B. A., and McNeilage, A. (2004). Dietary variability of mountain gorillas in Bwindi Impenetrable National Park, Uganda. *Int J. Primatol.* 25:1043–1072.

Gatti, S., Levréro, F., Ménard, N., and Gautier-Hion, A. (2004). Population and group structure of western lowland gorillas (*Gorilla gorilla gorilla*) at Lokoué, Republic of Congo. *Am. J. Primatol.* 63:111–123.

Gatti, S., Levréro, F., Ménard, N., Petit, E., and Gautier-Hion, A. (2003). Bachelor groups of western lowland gorillas (*Gorilla gorilla gorilla*) at Lokoué Clearing, Odzala National Park, Republic of Congo. *Folia Primatol.* 74:195–196.

Gerald, C. N. (1995). Demography of the Virunga mountain gorilla (*Gorilla gorilla beringei*) [MSc thesis]. Princeton University, Brinceton, NJ.

Goldsmith, M. L. (1999). Ecological constraints on the foraging effort of western gorillas (*Gorilla gorilla gorilla*) at Bai Hokou, Central African Republic. *Int. J. Primatol.* 20:1–23.

Goldsmith, M. L. (2003). Comparative behavioral ecology of a lowland and highland gorilla population: where do Bwindi gorillas fit? In: Taylor, A. B., and Goldsmith, M. L. (eds.), *Gorilla Biology: A Multidisciplinary Perspective*. Cambridge University Press, Cambridge. pp. 358–384.

Groves, C. (2001). *Primate Taxonomy*. Smithsonian Institute Press, Washington DC.

Groves, C. (2003). A history of gorilla taxonomy. In: Taylor, A. B., and Goldsmith, M. L. (eds.), *Gorilla Biology: A Multidisciplinary Perspective*. Cambridge University Press, Cambridge. pp. 15–34.

Hall, J. S., White, L. J. T., Inogwabini, B. I., Omari, I., Morland, H. S., Williamson, E. A., Saltonstall, K., Walsh, P., Sikubwabo, C., Bonny, D., Kiswele, K. P., Vedder, A., and Freeman, K. (1998). Survey of Grauer's gorillas (*Gorilla gorilla graueri*) and eastern chimpanzees (*Pan troglodytes schweinfurthii*) in the Kahuzi-Biega National Park lowland sector and adjacent forest in eastern Democratic Republic of Congo. *Int. J. Primatol.* 19:207–235.

Harcourt, A. H. (1978). Strategies of emigration and transfer by female primates, with special reference to mountain gorillas. *Z. Tierpsychol.* 48:401–420.

Harcourt, A. H. (1979a). Social relationships among adult female mountain gorillas. *Anim. Behav.* 27:251–264.

Harcourt, A. H. (1979b). Social relationships between adult male and female mountain gorillas. *Anim. Behav.* 27:325–342.

Harcourt, A. H. (1979c). Contrasts between male relationships in wild gorilla groups. *Behav. Ecol. Sociobiol.* 5:39–49.

Harcourt, A. H., Fossey, D., Stewart, K. J., and Watts, D. P. (1980). Reproduction in wild gorillas and some comparisons with chimpanzees. *J. Reprod. Fertil.* 28(suppl.):59–70.

Harcourt, A. H., and Greenberg, J. (2001). Do gorilla females join males to avoid infanticide? A quantitative model. *Anim. Behav.* 62:905–915.

Harcourt, A. H., and Stewart, K. J. (1981). Gorilla male relationships: can differences during immaturity lead to contrasting reproductive tactics in adulthood? *Anim. Behav.* 29:206–210.

Harcourt, A. H., and Stewart, K. J. (1987). The influence of help in contests on dominance rank in primates: hints from gorillas. *Anim. Behav.* 35:182–190.

Harcourt, A. H., and Stewart, K. J. (1989). Functions of alliances in contests within wild gorilla groups. *Behaviour* 109:176–190.

Harcourt, A. H., Stewart, K. J., and Fossey, D. (1976). Male emigration and female transfer in wild mountain gorilla. *Nature* 263:226–227.

Huijbergts, B., De Wachter, P., Ndong Obiang, L. S., and Ella Akou, M. (2003). Ebola and the decline of gorilla (*Gorilla gorilla*) and chimpanzee (*Pan troglodytes*) populations in Minkebe Forest, north-eastern Gabon. *Oryx* 37:437–443.

Isbell, L. A., and Young, T. P. (2002). Ecological models of female social relationships in primates: similarities, disparities, and some directions for future clarity. *Behaviour*, 139:177–202.

IUCN (2000). *Red List of Threatened Species*. Species Survival Commission, IUCN–World Conservation Union, Gland, Switzerland.

Janson, C. H., and Goldsmith, M. L. (1995). Predicting group size in primates: foraging costs and predation risks. *Behav. Ecol.* 6:326–336.

Kalpers, J., Williamson, E. A., Robbins, M. M., McNeilage, A., Nzamurambaho, A., Ndakasi, L., and Mugiri, G. (2003). Gorillas in the crossfire: assessment of population dynamics of the Virunga mountain gorillas over the past three decades. *Oryx* 37:326–337.

Kappeler, P. M. (2000). *Primate Males*. Cambridge University Press, Cambridge.

Kingsley, S. R. (1988). Physiological development of male orang-utans and gorillas. In: Schwartz, J. H. (ed.), *Orang-utan Biology*. Oxford University Press, New York. pp. 123–131.

Koenig, A. (2002). Competition for resources and its behavioral consequences among female primates. *Int. J. Primatol.* 23:759–783.

Koenig, A., Beise, J., Chalise, M. K., and Ganzhorn, J. U. (1998). When females should contest for food-testing hypotheses abut resource density, distribution, size, and quality with hanuman langurs (*Presbytis entellus*). *Behav. Ecol. Sociobiol.* 42:225–237.

Leroy, E. M., Rouquet, P., Formenty, P., Souquière, S., Kilbourne, A., Froment, J. M., Bermejo, M., Smit, S., Karesh, W., Swanepoel, R., Zaki, S. R., and Rollin, P. E. (2004). Multiple ebola virus transmission events and rapid decline of central African wildlife. *Science* 303:387–390.

Magliocca, F., and Gautier-Hion, A. (2002). Mineral content as a basis for food selection by western lowland gorillas in a forest clearing. *Am. J. Primatol.* 57:67–77.

Magliocca, F., Querouil, S., and Gautier-Hion, A. (1999). Population structure and group composition of western lowland gorillas in north-western Republic of Congo. *Am. J. Primatol.* 48:1–14.

McNeilage, A. (1995). Mountain gorillas in the Virunga Volcanoes: ecology and carrying capacity [PhD thesis]. University of Bristol, Bristol.

McNeilage, A. (2001). Diet and habitat use of two mountain gorilla groups in contrasting habitats in the Virungas. In: Robbins, M. M., Sicotte, P., and Stewart, K. J. (eds.), *Mountain Gorillas: Three Decades of Research at Karisoke*. Cambridge University Press, Cambridge. pp. 265–292.

McNeilage, A., Plumptre, A. J., Brock-Doyle, A., and Vedder, A. (2001). Bwindi Impenetrable National Park, Uganda: gorilla census 1997. *Oryx* 35:39–47.

Mudakikwa, A. B., Cranfield, M. R., Sleeman, J. M., and Eilenberger, U. (2001). Clinical medicine, preventive health care and research on mountain gorillas in the Virunga Volcanoes region. In: Robbins, M. M., Sicotte, P., and Stewart, K. J. (eds.), *Mountain Gorillas: Three Decades of Research at Karisoke*. Cambridge University Press, Cambridge. pp. 341–360.

Nishihara, T. (1995). Feeding ecology of western lowland gorillas in the Nouable-Ndoki National Park, Congo. *Primates* 36:151–168.

Nkurunungi, J. B. (2004). The availability and distribution of fruit and non-fruit resources in Bwindi: their influence on gorilla habitat and food choice [PhD thesis]. Makerere University, Makerere, Uganda.

Nsubuga, A. M. (2005). Genetic analysis of the social structure in wild mountain gorillas (*Gorilla beringei beringei*) of Bwindi Impenetrable National Park, Uganda [PhD diss.]. University of Leipzig, Leipzig.

Nsubuga, A. M., Robbins, M. M., Vigilant, L., and Boesch, C. (2001). A non-invasive genetic assessment of the social structure and male reproductive success of mountain gorillas in Bwindi Impenetrable National Park, Uganda. *Primate Eye* 75:21–22.

Oates, J. F., McFarland, K. L., Groves, J. L., Bergl, R. A., Linder, J. M., and Disotell, T. R. (2003). The Cross River gorilla: natural history and status of a neglected and critically endangered subspecies. In: Taylor, A. B., and Goldsmith, M. L. (eds.), *Gorilla Biology: A Multidisciplinary Perspective*. Cambridge University Press, Cambridge. pp. 472–497.

Parnell, R. J. (2002a). Group size and structure in western lowland Gorillas (*Gorilla gorilla gorilla*) at Mbeli Bai, Republic of Congo. *Am. J. Primatol.* 56:193–206.

Parnell, R. J. (2002b). The social structure and behaviour of western lowland gorillas (*Gorilla gorilla gorilla*) at Mbeli Bai, Republic of Congo [PhD thesis]. University of Stirling, Stirling.

Plumptre, A. J., McNeilage, A., Hall, J. S., and Williamson, E. A. (2003). The current status of gorillas and threats to their existence at the beginning of a new millennium. In: Taylor, A. B., and Goldsmith, M. L. (eds.). *Gorilla Biology: A Multidisciplinary Perspective*. Cambridge University Press, Cambridge. pp. 414–431.

Plumptre, A. J., and Williamson, E. A. (2001). Conservation-oriented research in the Virunga region. In: Robbins, M. M., Sicotte, P., and Stewart, K. J. (eds.), *Mountain Gorilla: Three Decades of Research at Karisoke*. Cambridge University Press, Cambridge. pp. 361–389.

Remis, M. J. (1997). Western lowland gorillas (*Gorilla gorilla gorilla*) as seasonal frugivores: use of variable resources. *Am. J. Primatol.* 43:87–109.

Remis, M. J. (2003). Are gorillas vacuum cleaners of the forest floor? The roles of body size, habitat, and food preferences on dietary flexibility and nutrition. In: Taylor, A. B., and Goldsmith, M. L. (eds.), *Gorilla Biology: A Multidisciplinary Perspective*. Cambridge University Press, Cambridge. pp. 385–404.

Remis, M. J., Dierenfeld, E. S., Mowry, C. B., and Carroll, R. W. (2001). Nutritional aspects of western lowland gorilla (*Gorilla gorilla gorilla*) diet during seasons of fruit scarcity at Bai Hokou, Central African Republic. *Int. J. Primatol.* 22:807–836.

Remis, M. J., and Kerr, M. E. (2002). Taste response to fructose and tannic acid among gorillas (*Gorilla gorilla gorilla*). *Int. J. Primatol.* 23:251–261.

Robbins, A. M., and Robbins, M. M. (2005). Fitness consequences of dispersal decisions for male mountain gorillas (*Gorilla beingei beringei*). *Behav. Ecol. Sociobiol.* 58:295–309.

Robbins, M. M. (1995). A demographic analysis of male life history and social structure of mountain gorillas. *Behaviour* 132:21–47.

Robbins, M. M. (1996). Male–male interactions in heterosexual and all-male wild mountain gorilla groups. *Ethology* 102:942–965.

Robbins, M. M. (1999). Male mating patterns in wild multimale mountain gorilla groups. *Anim. Behav.* 57:1013–1020.

Robbins, M. M. (2001). Variation in the social system of mountain gorillas: the male perspective. In: Robbins, M. M., Sicotte, P., and Stewart, K. J. (eds.), *Mountain Gorillas: Three Decades of Research at Karisoke*. Cambridge University Press, Cambridge. pp. 29–58.

Robbins, M. M. (2003). Behavioral aspects of sexual selection in mountain gorillas. In: Jones, C. L. (ed.), *Sexual Selection & Reproductive Competition in Primates: New Perspectives and Directions*. American Society of Primatologists. pp. 477–501.

Robbins, M. M., Bermejo, M., Cipolletta, C., Magliocca, F., Parnell, R. J., and Stokes, E. (2004). Social structure and life history patterns in western gorillas (*Gorilla gorilla gorilla*). *Am. J. Primatol.* 64:145–159.

Robbins, M. M., and McNeilage, A. (2003). Home range and frugivory patterns of mountain gorillas in Bwindi Impenetrable National Park, Uganda. *Int. J. Primatol.* 24:467–491.

Robbins, M. M., Robbins, A. M., Gerald-Steklis, N., and Steklis, H. D. (2005). Long-term dominance relationships in female mountain gorillas: strength, stability and determinants of rank. *Behaviour*. 142:779–809.

Rogers, M. E., Abernethy, K., Bermejo, M., Cipolletta, C., Doran, D., McFarland, K., Nishihara, T., Remis, M., and Tutin, C. E. G. (2004). Western gorilla diet: a synthesis from six sites. *Am. J. Primatol.* 64:173–192.

Rogers, M. E., Maisels, F., Williamson, E. A., Tutin, C. E. G., and Fernandez, M. (1990). Gorilla diet in the Lope Reserve, Gabon: a nutritional analysis. *Oecologia* 84:326–339.

Schaller, G. (1963). *The Mountain Gorilla: Ecology and Behavior*. University of Chicago Press, Chicago.

Sholley, C. (1991). *Conserving gorillas in the Midst of Guerrillas*. American Association of Zoological Parks and Aquariums Annual Conference Proceedings, pp. 30–37.

Sicotte, P. (1993). Inter-group encounters and female transfer in mountain gorillas: influence of group composition on male behavior. *Am. J. Primatol.* 30:21–36.

Sicotte, P. (1994). Effects of male competition on male–female relationships in bi-male groups of mountain gorillas. *Ethology* 97:47–64.

Sicotte, P. (2000). A case of mother–son transfer in mountain gorillas. *Primates* 41:95–103.

Sicotte, P. (2001). Female mate choice in mountain gorillas. In: Robbins, M. M., Sicotte, P., and Stewart, K. J. (eds.). *Mountain Gorillas: Three Decades of Research at Karisoke*. Cambridge University Press, Cambridge. pp. 59–87.

Stanford, C. B., and Nkurunungi, J. B. (2003). Behavioral ecology of sympatric chimpanzees and gorillas in Bwindi Impenetrable National Park, Uganda: diet. *Int. J. Primatol.* 24:901–918.

Sterck, E. H. M., Watts, D. P., and van Schaik, C. P. (1997). The evolution of female social relationships in nonhuman primates. *Behav. Ecol. Sociobiol.* 41:291–309.

Stewart, K. J. (1988). Suckling and lactational anoestrus in wild gorillas (*Gorilla gorilla*). *J. Reprod. Fertil.* 83:627–634.

Stewart, K. J. (2001). Social relationships of immature gorillas and silverbacks. In: Robbins, M. M., Sicotte, P., and Stewart, K. J. (eds.), *Mountain Gorillas: Three Decades of Research at Karisoke*. Cambridge University Press, Cambridge. pp. 183–213.

Stewart, K. J., and Harcourt, A. H. (1987). Gorillas: variation in female relationships. In: Smuts, B. B., Cheney, D. L., Seyfarth, R. M., Wrangham, R. W., and Struhsaker, T. T. (eds.), *Primate Societies*. University of Chicago Press, Chicago. pp. 155–164.

Stokes, E. J. (2004). Within-group social relationships amongst females and adult males in wild western lowland gorillas (*Gorilla gorilla gorilla*). *Am. J. Primatol.* 64:233–246.

Stokes, E. J., Parnell, R. J., and Olejniczak, C. (2003). Female dispersal and reproductive success in wild western lowland gorillas (*Gorilla gorilla gorilla*). *Behav. Ecol. Sociobiol.* 54:329–339.

Tutin, C. E. G. (1996). Ranging and social structure of lowland gorillas in the Lopé Reserve, Gabon. In: McGrew, W. C., Marchant, L. F., and Nishida, T. (eds.), *Great Ape Societies*. Cambridge University Press, Cambridge. pp. 58–70.

Tutin, C. E. G., Fernandez, M., Rogers, M. E., Williamson, E. A., and McGrew, W. C. (1991). Foraging profiles of sympatric lowland gorillas and chimpanzees in the Lope Reserve, Gabon. *Phil. Trans. R. Soc. Lond. B* 334:179–186.

van Hooff, J. A. R. A. M., and van Schaik, C. P. (1994). Male bonds: affiliative relationships among nonhuman primate males. *Behaviour* 130:309–337.

van Schaik, C. P. (1989). The ecology of social relationships amongst female primates. In: Standon, V., and Foley, R. A. (eds.), *Comparative Socioecology: The Behavioural Ecology of Humans and Other Mammals*. Blackwell Scientific Publications, Oxford. pp. 195–218.

van Schaik, C. P. (2000). Vulnerability to infanticide by males: patterns among mammals. In: van Schaik, C., and Janson, C. (eds.), *Infanticide by Males & Its Implications*. Cambridge University Press, Cambridge. pp. 61–71.

van Schaik, C. P., and Kappeler, P. M. (1997). Infanticide risk and the evolution of male–female associations in primates. *Proc. R. Soc. Lond. B* 264:1687–1694.

Vedder, A. L. (1984). Movement patterns of a free-ranging group of mountain gorillas (*Gorilla gorilla beringei*) and their relation to food availability. *Am. J. Primatol.* 7:73–88.

Walsh, P. D., Abernethy, K. A., Bermejo, M., Beyers, R., De Wachter, P., Akou, M. E., Huijbregts, B., Mambounga, D. I., Toham, A. K., Kilbourn, A. M., Lahm, S. A., Latour, S., Maisels, F., Mbina, C., Mihindou, Y., Obiang, S. N., Effa, E. N., Starkey, M. P., Telfer, P., Thibault, M., Tutin, C. E. G., White, L. J. T., and Wilkie, D. S. (2003). Catastrophic ape decline in western equatorial Africa. *Nature* 422:611–614.

Waterman, P. G., Choo, G. M., Vedder, A. L., and Watts, D. (1983). Digestibility, digestion-inhibitors and nutrients of herbaceous foliage and green stems from an African montane flora and comparison with other tropical flora. *Oecologia* 60:244–249.

Watts, D. P. (1984). Composition and variability of mountain gorilla diets in the central Virungas. *Am. J. Primatol.* 7:323–356.

Watts, D. P. (1985). Relations between group size and composition and feeding competition in mountain gorilla groups. *Anim. Behav.* 33:72–85.

Watts, D. P. (1988). Environmental influences on mountain gorilla time budgets. *Am. J. Primatol.* 15:195–211.

Watts, D. P. (1989a). Ant eating behavior of mountain gorillas. *Primates* 30:121–125.

Watts, D. P. (1989b). Infanticide in mountain gorillas: new cases and a reconsideration of the evidence. *Ethology* 81:1–18.

Watts, D. P. (1990a). Mountain gorilla life histories, reproductive competition, and sociosexual behavior and some implications for captive husbandry. *Zoo Biol.* 9:185–200.

Watts, D. P. (1990b). Ecology of gorillas and its relation to female transfer in mountain gorillas. *Int. J. Primatol.* 11:21–44.

Watts, D. P. (1991a). Strategies of habitat use by mountain gorillas. *Folia Primatol.* 56:1–16.

Watts, D. P. (1991b). Mountain gorilla reproduction and sexual behavior. *Am. J. Primatol.* 24:211–225.

Watts, D. P. (1991c). Harassment of immigrant female mountain gorillas by resident females. *Ethology* 89:135–153.

Watts, D. P. (1992). Social relationships of immigrant and resident female mountain gorillas, I. Male–female relationships. *Am. J. Primatol.* 28:159–181.

Watts, D. P. (1994a). The influence of male mating tactics on habitat use in mountain gorillas (*Gorilla gorilla beringei*). *Primates* 35:35–47.

Watts, D. P. (1994b). Agonistic relationships of female mountain gorillas. *Behav. Ecol. Sociobiol.* 34:347–358.

Watts, D. P. (1994c). Social relationships of immigrant and resident female mountain gorillas, II. Relatedness, residence, and relationships between females. *Am. J. Primatol.* 32:13–30.

Watts, D. P. (1995). Post-conflict social events in wild mountain gorillas, 1. Social interactions between opponents. *Ethology* 100:158–174.

Watts, D. P. (1996). Comparative socioecology of gorillas. In: McGrew, W. C., Marchant, L. F., and Nishida, T. (eds.), *Great Ape Societies*. Cambridge University Press, Cambridge. pp. 16–28.

Watts, D. P. (1997). Agonistic interventions in wild mountain gorilla groups. *Behaviour* 134:23–57.

Watts, D. P. (1998a). Seasonality in the ecology and life histories of mountain gorillas (*Gorilla gorilla beringei*). *Int. J. Primatol.* 19:929–948.

Watts, D. P. (1998b). Long term habitat use by mountain gorillas (*Gorilla gorilla beringei*) 1. Consistency, variation, and home range size and stability. *Int. J. Primatol.* 19:651–680.

Watts, D. P. (2000). Causes and consequences of variation in male mountain gorilla life histories and group membership. In: Kappeler, P. M. (ed.), *Primate Males*. Cambridge University Press, Cambridge. pp. 169–179.

Watts, D. P. (2001). Social relationships of female mountain gorillas. In: Robbins, M. M., Sicotte, P., and Stewart, K. J. (eds.), *Mountain Gorillas: Three Decades of Research at Karisoke*. Cambridge University Press, Cambridge. pp. 215–240.

Watts, D. P. (2003). Gorilla social relationships: a comparative overview. In: Taylor, A. B., and Goldsmith, M. L. (eds.), *Gorilla Biology: A Multidisciplinary Perspective*. Cambridge University Press, Cambridge. pp. 302–327.

Watts, D. P., and Pusey, A. E. (1993). Behavior of juvenile and adolescent great apes. In: Pereira, M. E., and Fairbanks, L. A. (eds.), *Juvenile Primates*. Oxford University Press, New York. pp. 148–167.

Weber, A. W., and Vedder, A. (1983). Population dynamics of the Virunga gorillas: 1959–1978. *Biol. Conserv.* 26:341–366.

Williamson, E. A., Tutin, C. E. G., Rogers, M. E., and Fernandez, M. (1990). Composition of the diet of lowland gorillas at Lope in Gabon. *Am. J. Primatol.* 21:265–277.

Wrangham, R. W. (1979). On the evolution of ape social systems. *Soc. Sci. Inform.* 18:335–368.

Yamagiwa, J. (1987a). Male life history and the social structure of wild mountain gorillas (*Gorilla gorilla beringei*). In: Kawano, S., Connell, J. H., and Hidaka, T. (eds.), *Evolution and Coadaptation in Biotic Communities*. University of Tokyo Press, Tokyo. pp. 31–51.

Yamagiwa, J. (1987b). Intra- and inter-group interactions of an all-male group of Virunga mountain gorillas (*Gorilla gorilla berengei*). *Primates* 28:1–30.

Yamagiwa, J., Basabose, K., Kaleme, K., and Yumoto, T. (2003a). Within-group feeding competition and socioecological factors influencing social organization of gorillas in the Kahuzi-Biega National Park, Democratic Republic of Congo. In: Taylor, A. B., and Goldsmith, M. L. (eds.), *Gorilla Biology: A Multidisciplinary Perspective*. Cambridge University Press, Cambridge. pp. 328–357.

Yamagiwa, J., and Kahekwa, J. (2001). Dispersal patterns, group structure, and reproductive parameters of eastern lowland gorillas at Kahuzi in the absence of infanticide. In: Robbins, M. M., Sicotte, P., and Stewart, K. J. (eds.), *Mountain Gorillas: Three Decades of Research at Karisoke*. Cambridge University Press, Cambridge. pp. 89–122.

Yamagiwa, J., Kahekwa, J., and Basabose, A. K. (2003b). Intraspecific variation in social organization of gorillas: implications for their social evolution. *Primates* 44:359–369.

Yamagiwa, J., Maruhashi, T., Yumoto, T., and Mwanza, N. (1996). Dietary and ranging overlap in sympatric gorillas and chimpanzees in Kahuzi-Biega National Park, Zaire. In: McGrew, W. C., Marchant, L. F., and Nishida, T. (eds.), *Great Ape Societies*. Cambridge University Press, Cambridge. pp. 82–98.

Yamagiwa, J., Mwanza, N., Spangenberg, A., Maruhashi, T., Yumoto, T., Fischer, A., and Steinhauer-Bukart, B. (1993). A census of the eastern lowland gorilla (*Gorilla gorilla graueri*) in the Kahuzi-Biegu National Park with reference to mountain gorillas (*G. g. beringei*) in the Virunga region, Zaire. *Biol. Conserv.* 64:83–89.

Yamagiwa, J., Mwanza, N., Yumoto, T., and Maruhashi, T. (1991). Ant eating by eastern lowland gorillas. *Primates* 32:247–253.

Yamagiwa, J., Mwanza, N., Yumoto, T., and Maruhashi, T. (1992). Travel distances and food habits of eastern lowland gorillas: a comparative analysis. In: Itoigawa, N., Sugiyama, Y., and Sackett, G. P. (eds.), *Topics in Primatology. Behavior, Ecology and Conservation*, vol. 2. University of Tokyo Press, Tokyo. pp. 267–281.

Yamagiwa, J., Mwanza, N., Yumoto, T., and Maruhashi, T. (1994). Seasonal change in the composition of the diet of eastern lowland gorillas. *Primates* 35:1–14.

19

Chimpanzees and Bonobos
Diversity Within and Between Species
Rebecca Stumpf

INTRODUCTION

Field research on wild chimpanzees (*Pan troglodytes*) and bonobos (*Pan paniscus*) has been conducted since the 1960s and 1970s. As recently as 1980, virtually everything that was known about wild chimpanzee natural history and behavior came from two long-term field sites in East Africa. Far less was known about wild bonobos. As more and more chimpanzee and, to a lesser extent, bonobo communities were studied in captivity and across equatorial Africa (Table 19.1, Fig. 19.1), a far greater understanding has emerged of the substantial diversity both within and between the two species that comprise the genus *Pan*. Such variation within *Pan* has led to an ongoing debate surrounding the distinctiveness of chimpanzees and bonobos, as well as the remarkable behavioral variation among chimpanzee subspecies. Researchers are continuing their efforts to determine which combination of ecological, cultural, genetic, and/or demographic factors are the primary bases for these differences.

The principal goal of this chapter is to convey what is presently known of the taxonomy, morphology, ecology, and social and sexual behaviors of wild chimpanzees and bonobos and to examine the substantial range of variation both within and between the two species. This chapter is primarily concerned with studies of wild chimpanzees and bonobos. Since tool use and cognition, communication, and culture are reviewed extensively elsewhere in this volume (see Chapters 38, 40, and 41), these areas are not emphasized here.

CHIMPANZEE AND BONOBO RESEARCH: A HISTORICAL PERSPECTIVE

Research began on wild chimpanzees in the 1960s when Jane Goodall initiated a field study at Gombe Stream Reserve, Tanzania. During the same year, a Japanese team led by Toshisada Nishida began field research on chimpanzees in the Mahale Mountains National Park, at a site 200 km from Gombe (see Fig. 19.1), both near the far eastern range of chimpanzee geographic distribution. This initial glimpse into chimpanzee daily life, diet, and intra- and intergroup social interactions laid the foundation for many subsequent studies of wild chimpanzee behavior.

Table 19.1 Chimpanzee and Bonobo Field Projects

SPECIES	FIELD SITE	COUNTRY	PRINCIPAL RESEARCHER(S)	PROJECT DATES	RESEARCH FOCUS	REFERENCES
Chimpanzee (*Pan troglodytes*)						
P. t. schweinfurthii	Gombe National Park[1]	Tanzania	J. Goodall, A. Pusey	1960–present	Social behavior, tool use, reproductive parameters	Goodall (1986), Pusey et al. (1997), Constable et al. (2001)
P. t. schweinfurthii	Mahale Mountains National Park[1]	Tanzania	T. Nishida	1965–present	Social behavior, feeding ecology, zoopharmacognosy, male–male relationships	Nishida (1990)
P. t. schweinfurthii	Ugalla	Tanzania	J. Moore	1985+, intermittent	Feeding ecology, nesting	Moore (1996)
P. t. schweinfurthii	Kanyawara, Kibale National Park[1]	Uganda	R. Wrangham	1987–present	Behavior, social relationships, ecology, physiology, and cognition	Wrangham et al. (1996), Chapman et al. (1994), Goldberg and Wrangham (1997)
P. t. schweinfurthii	Ngogo, Kibale National Park[1]	Uganda	J. Mitani, D. Watts	1994–present	Behavioral ecology, male relationships	Mitani et al. (2002a–c), Muller and Wrangham (2004a,b)
P. t. schweinfurthii	Budongo Forest Reserve[1]	Uganda	V. Reynolds	1990–present	Conservation, feeding ecology, reproductive behavior, male strategies	Reynolds (1992), Newton-Fisher (2002a,b)
P. t. schweinfurthii	Semliki-Toro Wildlife Reserve	Uganda	K. Hunt	1996–present	Ecology, feeding behavior, locomotion, positional behavior	Hunt and McGrew (2002), Hunt et al. (1999)
P. t. schweinfurthii	Kalinzu Forest	Uganda	C. Hashimoto, T. Furuichi, Y. Tashiro	1995–present	Feeding ecology	Furuichi et al. (2001)
P. t. schweinfurthii	Bwindi Impenetrable Forest	Uganda	C. Stanford	1996–present	Sympatric chimpanzee-gorilla behavioral ecology	Stanford and Nkurunungi (2003)
P. t. schweinfurthii	Ishasha River	Uganda, Democratic Republic of Congo	J. Sept	1989+, intermittent	Nesting, feeding ecology	Sept (1998)
P. t. schweinfurthii	Kahuzi-Biega National Park	Democratic Republic of Congo	J. Yamagiwa, A. Basabose	1991–present	Nesting behavior, sympatric chimpanzee-gorilla behavior	Basabose (2002)
P. t. troglodytes	Nouabalé-Ndoki National Park	Congo (Brazzaville)	S. Koruda	1989–present, intermittent	Tool use, feeding ecology	Kuroda (1992), Kuroda et al. (1996)
P. t. troglodytes	Lopé Reserve	Gabon	C. Tutin	1984–present	Sympatric chimpanzee-gorilla feeding ecology, tool use	Tutin and Fernandez (1993)
P. t. troglodytes	Goualougo Triangle	Democratic Republic of Congo	C. Sanz, D. Morgan	1999–present	Behavior, tool use	Morgan and Sanz (2003)
P. t. troglodytes	Ngotto Forest	Central African Republic	R. and D. Fouts	2000–present	Tool use, gestural communication	Hicks et al. (2005)
P. t. vellerosus	Gashaka-Gumti National Park	Nigeria	V. Sommer	2002–present	Behavior, habituation	Sommer et al. (2004)
P. t. verus	Taï National Park[1]	Ivory Coast	C. Boesch	1979–present	Social behavior, reproductive strategies, feeding ecology, cognition, and culture	Boesch and Boesch-Achermann (2000), Boesch (1994), Boesch et al. (1994)
P. t. verus	Bossou National Park, Mt. Nimba[1]	Guinea, Ivory Coast	Y. Sugiyama, T. Matsuzawa	1976–present	Ecology, behavior, cognition, conservation	Sugiyama (1989, 2004)
P. t. verus	Mt. Assirik, Fongoli	Senegal	W. McGrew, J. Pruetz	1976–1979, 2001–present	Tool use, feeding ecology, ape–human habitat use	McGrew et al. (1988), Pruetz et al. (2002)
Bonobo (*Pan paniscus*)						
P. paniscus	Wamba Forest[1]	Democratic Republic of Congo	T. Kano	1973–present, intermittent	Sexual behavior, intergroup interactions	Kano (1992)
P. paniscus	Lomako Forest[1]	Democratic Republic of Congo	A. and N. Badrian, R. Susman, G. Hohmann, F. White	1974–present, intermittent	Socioecology, food sharing, feeding ecology	Susman (1984), Doran (1993), Malenky and Stiles (1991), Thompson-Handler et al. (1984), Fruth and Hohmann (1993), Hohmann and Fruth (1993, 2000, 2002), White (1988)
P. paniscus	Lukuru Forest	Democratic Republic of Congo	J. Thompson	1992–present	Conservation, intersite comparisons	Thompson (2002)
P. paniscus	Salonga National Park	Democratic Republic of Congo	G. Hohmann, B. Fruth, J. Eriksson	2000–present	Socioecology, phylogeography	Hohmann and Fruth (2002)

[1] Denotes a long-term field site (more than 10 years of continuous, habituated chimpanzee or bonobo research).

Figure 19.1   Map of equatorial Africa showing distribution of chimpanzee subspecies and bonobos as well as the major study sites. Predominant study sites follow taxonomic names. Habitats are highly fragmented in many of the chimpanzee ranges, particularly for the Nigerian-Cameroon and western chimpanzees. It is unclear whether chimpanzees west of the Niger River belong to *Pan troglodyte vellerosus* or *P. t. verus*. Figure modified after de Waal and Lanting (1997).

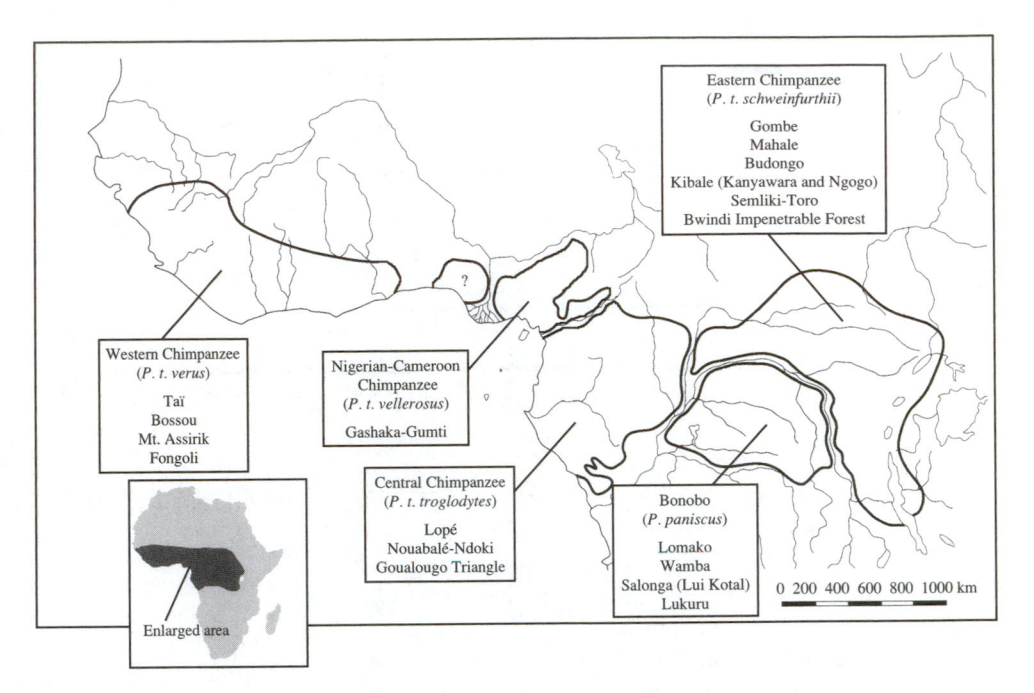

In the late 1970s, Christophe Boesch (in the Taï National Park, Ivory Coast) and Yukimaru Sugiyama (in Bossou, Guinea) began field research on chimpanzees in West Africa. Although many aspects of behavior appeared to be relatively similar across all chimpanzee sites (see below), substantial differences in chimpanzee behavior emerged between the western and eastern populations (Sugiyama and Koman 1979, Boesch and Boesch 1990). During the last two decades, studies in East Africa, such as those in Budongo and Kibale (containing the Kanyawara and Ngogo sites) and West Africa (e.g., Mt. Assirik and Fongoli in Senegal) have added further insight into the complexity and diversity of chimpanzee grouping patterns, behavior, and feeding ecology, even within the same subspecies. At present, little is known about the central (*P. t. troglodytes*) and Nigerian-Cameroon (*P. t. vellerosus*) subspecies, but future data on these two subspecies may provide even more evidence of behavioral variation within chimpanzees.

Bonobos have been studied to a far lesser extent than chimpanzees. This is largely due to their limited distribution and remote location, along with a civil war in Congo which prohibited continuous research throughout most of the 1990s. Current knowledge of bonobo behavior and ecology comes largely from observations in Lomako and Wamba, Congo (see Fig. 19.1), as well as from captive populations. Initial data from both wild and captive bonobos suggested that their behavior differed substantially from that of chimpanzees, particularly in the patterns of dominance, aggression, sexual behavior, and intercommunity interactions (see below; Kuroda 1979, 1980; de Waal 1989, 1998; Kano 1992; Thompson-Handler et al. 1984; White 1996a,b; Wrangham and Peterson 1996; Nishida 1997).

In the last decade, the behavioral disparities between chimpanzees and bonobos have been challenged (Boesch 1996, Fruth 1998, McGrew 1998, Stanford 1998). Both species show considerable behavioral overlap (Stanford 1998, Hohmann and Fruth 2002). Some behaviors observed in chimpanzees that have never been seen in bonobos may be because bonobos have been less studied than chimpanzees (Fruth 1998, McGrew 1998). Some chimpanzee behaviors, like lethal aggression, were seen only after many years of close observation (Goodall 1986). Moreover, behavioral variation across chimpanzee field sites is far greater than previously considered (Wrangham et al. 1994, Boesch 2002). Differences in bonobo behavior between research sites also have been reported (White 1992, 1996b; Hohmann and Fruth 2003a), and more differences are likely to be observed as bonobo research progresses in more diverse habitats. Bonobo studies have begun in more open and drier habitats (Thompson 2002), and such ecological variation may substantially influence bonobo behavioral variation. Thus, the extent of behavioral diversity within chimpanzees or within bonobos may rival the behavioral diversity between the two species. What is presently known of the nature of the similarities and differences within and between these sister taxa will be discussed in more detail below.

CHANGING TAXONOMIC PERSPECTIVES

Debate has long surrounded the taxonomic and evolutionary relationships among all the great apes. Historically, chimpanzees and bonobos were placed with gorillas and orangutans in the family Pongidae, distinct from the family Hominidae, which included only humans and our fossil ancestors. However, recent genetic analyses indicate that chimpanzees and bonobos are more closely related to humans than either are to gorillas or orangutans (Sibley and Alquist 1984, Sarich et al. 1989, Caccone and Powell 1989, Horai et al. 1995, Takahata 1995, Ruvolo 1997, Chen and Li

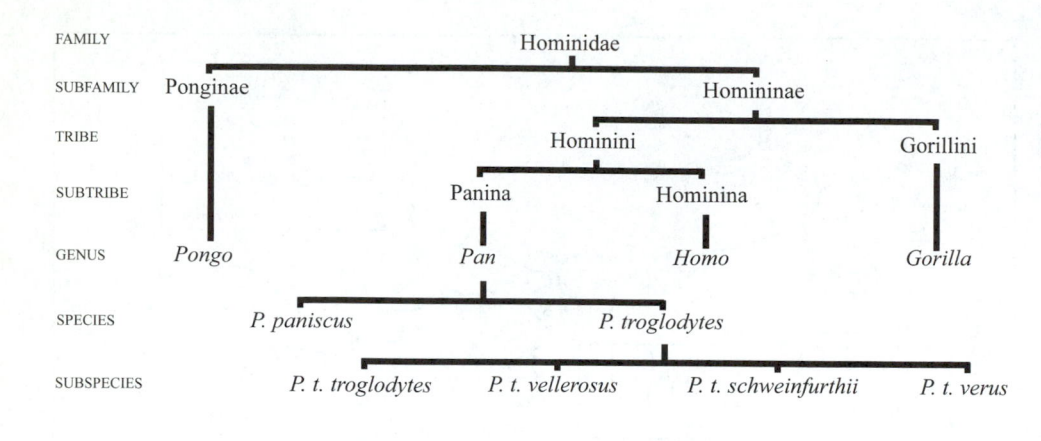

FAMILY — Hominidae
SUBFAMILY — Ponginae / Homininae
TRIBE — Hominini / Gorillini
SUBTRIBE — Panina / Hominina
GENUS — *Pongo* / *Pan* / *Homo* / *Gorilla*
SPECIES — *P. paniscus* / *P. troglodytes*
SUBSPECIES — *P. t. troglodytes* / *P. t. vellerosus* / *P. t. schweinfurthii* / *P. t. verus*

Figure 19.2   Recent taxonomy subsumes humans, the great apes, and their ancestors all within the family Hominidae. The subfamily Ponginae contains only the orangutans and their ancestors. African apes, humans, and our fossil ancestors are grouped within the subfamily Homininae. Chimpanzees, bonobos and humans are united within the tribe Hominini. Chimpanzees and bonobos should be further united within the subtribe Panina (after Groves 2001).

2001). The recent taxonomy proposed by Groves (2001) (Fig. 19.2) subsumes humans, the great apes, and their ancestors all within the family Hominidae. The subfamily Ponginae contains only the orangutans and their ancestors. The African apes along with humans and our fossil ancestors are grouped within the subfamily Homininae. While Groves (2001) does not address subfamilial distinctions within the Homininae, molecular analyses suggest that chimpanzees, bonobos, and humans are united within the tribe Hominini. Chimpanzees and bonobos should be further united within the subtribe Panina.

Bonobos were first formally recognized as different from chimpanzees in 1929, based on skull morphology. Initially considered a chimpanzee subspecies (Schwartz 1929), bonobos were later classified as a separate species (Coolidge 1933) (Fig. 19.2). This classification is supported by additional morphological, histological, and genetic data (Zihlman and Cramer 1978; Zihlman 1984; Susman 1984; Moor-Jankowski et al. 1972, 1975; Ruvolo et al. 1994; Burrows and Ryder 1997; Jenson-Seaman and Li 2003).

Chimpanzees are further divided into four subspecies. These include the East African (*P. t. schweinfurthii*), central African (*P. t. troglodytes*), West African (*P. t. verus*), and the recently recognized fourth subspecies from Nigeria and Cameroon (*P. t. vellerosus*), based on molecular data (Gonder et al. 1997).

Several molecular studies have estimated that chimpanzees and bonobos diverged between 1.7 million years ago (mya) and 2.7 mya (mitochondrial deoxyribonucleic acid [DNA], Morin 1994, Gagneux et al. 1999; X chromosome, Kaessmann et al. 1999; Y chromosome, Stone et al. 2002; random nuclear genome, Yu et al. 2003). The split between western (*P. t. verus* and *P. t. vellerosus*) and central/eastern (*P. t. troglodytes* and *P. t. schweinfurthii*) chimpanzees is estimated at 1.6 mya (Morin 1994, Gagneux et al. 1999). Mitochondrial differentiation of *P. t. verus* compared to the other subspecies led Morin (1994) to recommend species-level distinction, but this remains controversial. The genetic variation within chimpanzee subspecies is greatest within *P. t. troglodytes* (Kaessmann et al. 1999, Gonder 2000). Bonobos have substantially less genetic variation than chimpanzees (Reinartz et al. 2000).

MORPHOLOGICAL VARIATION

Although males of both *Pan* species are larger and have longer canines than females, bonobos show lower levels of craniodental dimorphism than chimpanzees (Table 19.2) (White 1996b). Yet body mass dimorphism can be greater or equal to that of chimpanzees with male bonobos weighing about 15% more than females while male chimpanzees weigh between 5% and 14% more than females (Table 19.2). Of all of the chimpanzee subspecies, *P. t. troglodytes* is the largest and most sexually dimorphic in body mass, while *P. t. schweinfurthii* is the smallest.

Head plus body length ranges 635–940 mm for chimpanzees and 700–830 mm for bonobos. Although bonobos have a more slender build than chimpanzees, body mass of the two species are comparable (Table 19.2), which may be explained by the bonobos' longer thigh bones, heavier thigh muscles (Zihlman 1984, 1996), and longer feet (Fleagle 1999). Due to their relatively longer legs, the bonobo intermembral index—[(humerus + radius)/(femur + tibia)]* 100—of 102.2 is lower than that of chimpanzees at 108.7 (Coolidge 1933, Zihlman and Cramer 1978). The crania and mandible of bonobos are clearly distinct from those of chimpanzees, even when data are adjusted for differences in size (Cramer 1977, Shea 1984, Shea and Coolidge 1988, Shea et al. 1993, Taylor and Groves 2003). Bonobos have a more anteriorly placed vulva (de Waal and Lanting 1997). Bonobo vocalizations are also notably higher in pitch than those of chimpanzees (de Waal 1988).

Table 19.2   Average Adult Body Mass (kg) for Bonobos and Chimpanzee Subspecies

	MALE	FEMALE
Pan paniscus	45	33.2
P. troglodytes schweinfurthii	42.7 (39.0–42.0*)	33.7 (31.3–35.2*)
P. t. troglodytes	59.7	45.8
P. t. verus	46.3	41.6
P. t. vellerosus	?	?

Source: Smith and Jungers (1997) (*wild data from Uehara and Nishida 1987; Pusey et al., 2005).

Facial patterns vary across the chimpanzee subspecies. Most infants of all chimpanzee subspecies are born with light pink faces. *P. t. verus* generally has a darker mask around the eyes and lighter skin tone in the rest of the face, and this pattern often remains into adulthood. The face of *P. t. troglodytes* generally darkens completely to deep black when adult. The pink face of *P. t. schweinfurthii* is often freckled and darkens with age. In contrast to most chimpanzees, bonobo faces are black from birth. The lips of bonobos are pink, and the hair on the head is parted in the middle. While both chimpanzees and bonobos are born with a white tail tuft, only bonobos maintain it into adulthood.

ECOLOGY

Habitat and Distribution

Chimpanzees occupy very diverse habitats across equatorial Africa, including dense lowland tropical rain forest as well as savanna woodland, and are found at altitudes of up to 3,000 m (Suzuki 1969, Kano 1971, McGrew et al. 1979) (Table 19.3, Fig. 19.2). They are distributed as far northwest as Senegal, south to the Congo River, across northern Congo, and east of Lake Tanganyika, Tanzania (see Fig. 19.1). *P. t. schweinfurthii* is thought to have moved into its seasonally dry habitat relatively recently (Goldberg and Ruvolo 1997).

Bonobos live in lowland rain forest, swamp forest, dry forest, and grassland habitats. Bonobo distribution is more limited than that of chimpanzees, being bordered to the north and west by the Congo River, to the east by the Lomami River, and to the south by the Kasai and Sankuru Rivers, completely within the Democratic Republic of the Congo (Groves 2001). Chimpanzee and bonobo ranges are not known to overlap. The formation of the Congo River separating the two species is estimated geologically at 1.5 mya (Beadle 1981), which is also the time of large climatic, vegetation, and habitat changes that may have initiated speciation in other hominins (Vrba 1995).

Diet

Chimpanzees and bonobos are highly frugivorous (Wrangham et al. 1998, Conklin-Brittain et al. 1998) (Table 19.4). Both species obtain substantial amounts of protein from terrestrial herbaceous vegetation (THV), although eastern chimpanzees appear to rely on it more as a fallback food in times of food scarcity (Wrangham et al. 1994, Newton-Fisher 1999a). The range of plant species eaten by chimpanzees varies widely across sites and depends on both local and seasonal availability (Table 19.4). During times of preferred-food scarcity, chimpanzees increase the proportion of leaves and other lower-quality foods in their diet (Gombe, Wrangham 1977, Taï, Doran 1997a) and/or eat a wider range of foods (Gombe, Wallis 1995, Taï, Doran 1997a). At very arid sites, such as Semliki, chimpanzees regularly consume a limited

range of foods (about 45 items), while at Mahale, more than 300 food items are eaten (Hunt and McGrew 2002). Animal protein (e.g., vertebrates, insects, eggs) constitutes approximately 8%–10% of the chimpanzee diet (Goodall 1986), with considerable intersite variation. Approximately 1%–3% of the chimpanzee diet consists of mammalian flesh (McGrew 1992). Where chimpanzees and gorillas overlap, their diets are seasonally similar but diverge during food scarcity (Tutin and Fernandez 1993, Basabose 2002, Stanford and Nkurunungi 2003).

Bonobos also eat a wide variety of foods (over 110 species) (Badrian and Malenky 1984, Kano 1992), including large ripe *Anonidium* (10 kg) and *Treculia* (30 kg) fruits. Bonobos rely heavily on THV and seem to prefer plant foods high in carbohydrates and low in fiber (Conklin-Brittain et al. 2000). Less than 1% of the bonobo diet is derived from animal protein (Kano 1992).

Hunting has been documented at all long-term chimpanzee research sites (Goodall 1986, Boesch and Boesch 1989, Nishida et al. 1979), although there are considerable intersite differences in prey type, habitat characteristics, and hunting methods (Boesch et al. 2002). At Mahale, chimpanzees hunt at least 17 vertebrate species, whereas only about five species are hunted at Taï (Boesch and Boesch 1989). Red colobus monkeys are the favored prey at all study sites where they are present (Nishida *and* Uehara 1983, Boesch 1994, Stanford et al. 1994a), though Gombe chimpanzees prefer immature monkeys while Taï chimpanzees prefer adults (Stanford et al. 1994a, Boesch and Boesch-Achermann 2000).

Over 50% of all hunts are successful across all chimpanzee study sites. However, hunting success appears to depend on the level of forest cover. Hunting success is lower where the canopy is more continuous than in more interrupted forest cover. Presumably, this is because the continuous canopy allows more escape routes for arboreal prey species (Watts and Mitani 2002a). In the east, where forest canopies are more interrupted, individuals hunt more opportunistically, whereas in the taller, more continuous forest canopy of Taï, Ivory Coast, more cooperative hunting may be necessary to successfully trap the prey (Boesch and Boesch 1989, Boesch 1994, Boesch and Boesch-Achermann 2000).

Chimpanzee males hunt more often than females (Stanford et al. 1994a). The frequency of chimpanzee hunting varies throughout the year and depends upon food availability, group composition, and the presence of estrous females. The relative importance of each of these factors differs across sites. For example, at Mahale, hunting is positively associated with the presence of larger groups, estrous females, and higher fruit availability (Mitani and Watts 2001, Watts and Mitani 2002a). However, at Ngogo, estrous female presence does not influence the likelihood of male hunting, nor do males that hunt have higher mating success (Mitani and Watts 2001). Rather, Mitani and Watts (2001) suggest that hunting coincides with the presence of larger parties containing more males, which facilitates hunting

Table 19.3 *Pan Study Sites*

	PAN TROGLODYTES VERUS			*P. T. SCHWEINFURTHII*						*P. T. VELLEROSUS*		*P. PANISCUS*		
	TAÏ[1]	BOSSOU[2]	MT. ASSIRIK[3]	GOMBE[4]	MAHALE[5]	KANYAWARA[6]	NGOGO[7]	SEMLIKI[8]	BUDONGO[9]	KAHUZI-BIEGA[10]	GASHAKA-GUMTI[11]	WAMBA[12]	LOMAKO[13]	LUKURU[14]
Study site location	Ivory Coast	Guinea	Senegal	Tanzania	Tanzania	Uganda	Uganda	Uganda	Uganda	D. R. Congo	Nigeria	D. R. Congo	D. R. Congo	D. R. Congo
Environment	Tropical rain forest	Tropical forest	Savanna woodland	Riverine forest and woodland	Riverine forest and woodland	Moist evergreen forest	Moist evergreen forest	Savanna with riverine forest	Semideciduous moist tropical rain forest	Montane and bamboo forest	Woodland, lowland, gallery forest	Dry, swamp, and secondary forest, cultivated fields	Primary evergreen forest, swamp forest and dry grassland	Evergreen forest and dry grassland
Annual mean rainfall (mm)	1,829	2,230	954	1,775	1,836	1,671	1,671	1,452	1,842	1,619	1,826	2,000	1,844	1,775
Elevation (m)	202	550	100–311	1,137	1,040	1,500	1,500	1,200	1,100	2,200	300–1,000	400	390	564
Dry season months (rainfall <60 mm)	3	3	7	4	4	4	4	4	3	3	4–5	0	0.5	?
Provisioned?	No	Oil palm	No	Until 2000	Until 1987	No	No	No	No	No	No	Yes	No	No
Community size	29–82	16–22	>15	38–60	45–101	44	>140	>29	32–56	22	>35	28	33	?
Average party size	8.3	4	4	4.5	6.1	5.1	10.3	4.8	5.66	4.4	3.7	15	7.1	6.4
Home range size (km²)	13–26	15–20	50	4–24	7–14	16	35	38.3	7	10	26	22	14.7	?
Average day range (m)	2,400	1,000	—	3,450	4,825			?		400	?	2,400	3,035	?
Number of plant species consumed		>200	43	103	198	102		45	83	114	?	>110	>110	?

1 Boesch (1997), Boesch and Boesch-Achermann (2000), Herbinger et al. (2001), Lehmann and Boesch (2003, 2004).
2 Sugiyama (1989, 1994, 2004), Sugiyama and Koman (1992).
3 McGrew et al. (1996, 2004), Tutin et al. (1983).
4 Wrangham (1979a,b), Goodall (1986), Wallis (1997).
5 Nishida et al. (2003), Hasegawa and Hiraiwa-Hasegawa (1983).
6 Chapman and Wrangham (1993), Wrangham et al. (1996).
7 Watts (1998), Watts and Mitani (2001), Mitani et al. (2002a), Mitani (unpublished data cited in Nishida et al. 2003).
8 Hunt and McGrew (2002).
9 Newton-Fisher (1999a, 2003).
10 Basabose (2002), Basabose and Yamagiwa (2002).
11 Sommer et al. (2004).
12 Kano (1992), Badrian and Malenky (1984).
13 Thompson-Handler (1990).
14 Myers Thompson (2002).

Table 19.4   Chimpanzee and Bonobo Dietary Composition

	CHIMPANZEES[1]		BONOBOS[2]	
	(%) OF DIET	RANGE (%)	(%) OF DIET	RANGE (%)
Fruit	64	19–99	55	0–100
Leaves	16	0–56	14	0–28
THV[3]	7	0–27	25	0–100
Bark and Misc.	4	0–41	2	0–11
Prey	4	0–28	2	0–3
Flowers	2	0–14	2	0–7
Seeds	3	0–30	–	–

All dietary data compiled by Conklin-Brittain et al. (in press).
[1] Chimpanzee data from Conklin-Brittain et al. (1998), Galdikas and Teleki (1981), Ghiglieri (1984), van Lawick-Goodall (1968), Hladik (1973, 1977), Newton-Fisher (1999a), Isabirye-Basuta (1989), Kuroda (1992), Kuroda et al. (1996), Matsumoto-Oda and Hayashi (1999), Mc-Grew et al. (1981), Peters and O'Brien (1981), Rodman (1984), Sabater-Pi (1979), Sugiyama and Koman (1987), Tutin and Fernandez (1993), Tutin et al. (1984, 1991, 1997), Wrangham (1977, 1996), and Yamagiwa et al. (1992).
[2] Bonobo data from Badrian and Malenky (1984), Badrian et al. (1981), Hashimoto et al. (1998), Kano (1992), Kano and Mulavwa (1984), and Uehara (1990).
[3] THV, terrestrial herbaceous vegetation.

success. That large parties are able to form during these times suggests that feeding competition does not substantially limit grouping and, thus, that hunting is not purely a subsistence activity.

Compared to chimpanzees, bonobos appear to hunt rarely (Kano 1992, Ingmanson 1996). Duikers (small ungulates) are the more common prey (Fruth and Hohmann 2002), even though colobus monkeys are present in their habitat (Ihobe 1990). Though low, the hunting frequency of bonobos falls within the range of chimpanzees (Fruth and Hohmann 2002) since chimpanzees at some sites hunt only rarely (Mt. Assirik, Budongo, McGrew et al. 1988, Newton-Fisher 1999a). Ecological differences do not adequately account for this variation in hunting propensity among chimpanzees and bonobos: bonobos hunt infrequently despite the apparent similarity of their moist tropical forest habitats to those of chimpanzees with high hunting propensities.

SOCIAL ORGANIZATION AND GROUPING PATTERNS

Primate social organization and grouping patterns are strongly influenced by ecological factors. Chimpanzees in most of the eastern sites experience greater seasonality of rainfall and a greater number of dry season months compared to the west, whereas bonobos have few, if any, dry months per year (Doran et al. 2002) (Table 19.3). This seasonal variation results in substantial spatial and temporal variation in food availability, as well as a greater period of food scarcity (Goodall 1986, Nishida 1990, Wrangham et al. 1996, Doran et al. 2002, Anderson et al. 2002). This section

will discuss how these habitat differences and seasonal variation in food availability in turn appear to affect female grouping patterns and social behaviors across *Pan*.

Community Structure

Chimpanzees and bonobos live in multimale/multi-female communities, which vary greatly in size (van Lawick Goodall 1968) (Table 19.3). Chimpanzee communities range from a handful of individuals to as many as 150 (Watts 2002). Bonobo communities show similar variation in size, ranging from 25 to 75 or perhaps 120 individuals (Furuichi 1989). In both species, males are the philopatric sex, remaining permanently in their natal group (but see Sugiyama 1999). Females generally transfer into a new community after reaching sexual maturity, presumably to avoid inbreeding (Nishida and Kawanaka 1972, Pusey 1979). Upon transfer to a new group, immigrant female chimpanzees establish initial relationships with males. Resident chimpanzee females are at first largely intolerant of immigrant females and sometimes attack them (Pusey 1980, Goodall 1986, Nishida 1989). In contrast, young immigrant female bonobos primarily establish strong bonds with older resident females in order to integrate themselves into the community (Furuichi 1989, Idani 1991).

Although females of both species generally disperse during adolescence, the likelihood of female transfer varies between sites (Table 19.5). At most chimpanzee sites (e.g., Mahale, Ngogo, and Taï), the majority of females do transfer (Nishida et al. 2003, Boesch and Boesch-Achermann 2000). However, only about half of all Gombe females transfer (Pusey et al. 1997), while none is confirmed to have transferred at Bossou (Sugiyama 1999). Intersite differences may be influenced by the extent of community isolation as well as by the potential benefits of remaining in the natal group, such as inheriting the mother's core area and avoiding having to settle in the periphery of a new territory, as most Gombe immigrants do (Williams et al. 2002b). Male transfer has been observed twice at Lomako and potentially in Bossou, Guinea (Sugiyama 1999), suggesting that patterns of male philopatry may be more flexible for bonobos (Hohmann and Fruth 2002) and some chimpanzee populations.

Ranging Patterns, Day Ranges, and Nesting

Chimpanzee home ranges are relatively large in comparison to those of other apes, but there is considerable variation across sites (Table 19.3). Home ranges vary from the smallest (7–10 km[2]) in Gombe and Budongo (Goodall 1986, Newton-Fisher 2003) to 16–30 km[2] in primary forested habitat of Kanyawara and Taï (Herbinger et al. 2001, Wilson 2001) to the largest estimates of over 50 km[2] in more arid habitats such as Mt. Assirik and Fongoli (both in Senegal) as well as Semliki, Uganda (Baldwin et al. 1982, Hunt and McGrew 2002, Pruetz 2004, Hunt 2004). Chimpanzee home range sizes also vary considerably within sites (e.g., Herbinger et al.

Table 19.5   Grouping Patterns of Chimpanzees and Bonobos

	PAN TROGLODYTES VERUS		P. T. SCHWEINFURTHII		P. PANISCUS	
	TAÏ[1]	BOSSOU[2]	GOMBE[3]	MAHALE[4]	WAMBA[5]	LOMAKO[6]
Adult sex ratio (M/F)	1:4	1:2.3	1:1.35	1:3.6	1:3	1:1.2
Mixed parties[7] (%)	52	42	32	52	74	75.6
All-male parties[7] (%)	20	0	10	11	2	4.1
All-female parties[7] (%)	18	49	24	13	5	20.2
Solitary[7] (%)	18	3.5	18	21	6	4
Time spent alone by females[7] (%)	4	10	65	31.8	2	3

[1] Boesch (1997), Boesch and Boesch-Achermann (2000), Herbinger et al. (2001), Lehmann and Boesch (2003, 2004).
[2] Sugiyama (1994, 2004).
[3] Wrangham (1979a,b), Goodall (1986), Wallis (1997).
[4] Nishida (1990), Nishida et al. (2003), Takahata et al. (1996).
[5] Furuichi et al. (1998).
[6] Hohmann and Fruth (2002).
[7] Data compiled by Doran et al. (2002).

2001). This variation may be influenced by a combination of factors such as chimpanzee density and the number of males in the community (Lehmann and Boesch 2003; but see Williams 1999).

Male and female chimpanzees differ substantially in their ranging patterns and day range lengths (Wrangham and Smuts 1980). Male chimpanzees have longer day ranges than anestrous females, whereas estrous females often travel in larger groups with males, so their day ranges are comparable (Goodall 1986). These differences also vary across sites (Table 19.3). West African chimpanzee females associate frequently with both males and females and utilize the entirety of their community's range (Sugiyama 1988, Boesch 1996, Boesch and Boesch-Achermann 2000, Lehmann and Boesch 2004). In contrast, eastern female chimpanzees (Mahale, Gombe, Kibale) are less social than males, spending the majority of their time when not in estrus either alone with dependent offspring or with other mothers and young (Goodall 1986, Pusey 1980, Wrangham and Smuts 1980, Nishida et al. 1985, Pepper et al. 1999). Eastern (but not western) chimpanzee females occupy small, overlapping core areas within the home range and, unlike the males, do not utilize the entire community's territory (Wrangham 1979a; Chapman and Wrangham 1993; Wilson 2001; Williams et al. 2002a; Lehmann and Boesch 2004). Consequently, eastern female chimpanzee association rates are lower than at other chimpanzee sites (Williams et al. 2002a). Each night, both chimpanzees and bonobos sleep in fresh, individual nests interwoven from tree branches. Where trees are scarce, nests are sometimes reused (Fruth and Hohmann 1996).

Grouping Patterns and Variation in Party Size

Ecological factors, such as food abundance and distribution, the presence of estrous females, and predation pressure, are thought to influence primate grouping patterns (Wrangham 1980, van Schaik 1983, Symington 1988, Strier 1989, Chapman et al. 1995, Newton-Fisher et al. 2000). Heavy de-

pendence on fruit presents a challenge for both chimpanzees and bonobos. Both species exhibit a flexible, fission–fusion social system whereby individuals form smaller subgroups (parties) that change in size and composition throughout the day (Nishida 1968, Goodall 1986). This fission–fusion strategy may limit intragroup feeding competition and improve foraging efficiency. Smaller parties form during times of food scarcity and increase in times of food abundance (White 1996a,b; Doran 1997a; Wrangham 2002; Mitani et al. 2002c; Chapman et al. 1995; Anderson et al. 2002).

Party size and composition are also influenced by the number of estrous females. Chimpanzee party sizes increase and numbers of both males and anestrous females increase with the presence of estrous females (Goodall 1986; Wrangham et al. 1992; Matsumoto-Oda et al. 1998; Wrangham 2000; Anderson et al. 2002; Mitani et al. 2002c; Newton-Fisher 2002a,b). Bonobo party sizes seem independent of the number of estrous females, although the proportion of males in the party increases (Hohmann and Fruth 2002).

Predation risk is also thought to affect primate group size, with larger groups providing increased vigilance and decreased individual risk of predation (van Schaik 1983). The relatively large body size of chimpanzees and bonobos may reduce predation risk, permitting the formation of smaller parties during diurnal foraging (Terborgh and Janson 1986). At night, however, both chimpanzee and bonobo party sizes increase (Fruth and Hohmann 1993, Anderson et al. 2002), perhaps as a response to the real threat of nocturnal predation (e.g., leopards; Boesch 1991).

Bonobo parties are generally larger and less variable than those of chimpanzees (Furuichi 1989, Hohmann and Fruth 2002) (Table 19.3), and unlike most chimpanzees, bonobos are very rarely alone (Hohmann and Fruth 2002, Goodall 1986; but see Lehmann and Boesch 2004) (Table 19.5). However, food availability is also an important factor influencing the size of bonobo parties. Though less well understood, their less seasonal habitat, more abundant and larger fruiting trees, and more even distribution of THV may result in lower within-group feeding competition among bonobos and permit larger and more cohesive groups than among most chimpanzees (Chapman et al. 1994, Malenky and Wrangham 1994).

Generally, chimpanzees in West Africa form larger parties than East African chimpanzees (Boesch and Boesch-Achermann 2000, Wrangham 2000) and, like bonobos, are rarely alone. This variation in party size may be affected by ecological differences across sites. Large party sizes may be permitted where food availability and diet diversity are higher, reliance on THV is greater, predation risk is higher, or within-group competition is lower (Boesch 1991, Wrangham 2000, Williams et al. 2002a, Mitani et al. 2002c). Some intersite differences in bonobo grouping patterns have also been reported (White 1992, 1996b). Fissioning is less common in Wamba compared to Lomako (Kano 1992), and party sizes are generally smaller at Lomako (White 1992, 1996b; Hohmann and Fruth 2002).

Party composition also varies considerably between chimpanzees and bonobos, as well as between sites (Table 19.5). Bonobo parties most frequently comprise both sexes, usually containing more females than males (Kano 1982, White 1988, Fruth 1995). All-female parties are more common among bonobos than chimpanzees (Hohmann and Fruth 2002). When food is scarce, male bonobos are more likely to disperse, whereas females remain together (White 1988). Eastern chimpanzee parties contain mostly males, whereas western chimpanzee parties more closely resemble bonobo parties in their more mixed composition (Sugiyama and Koman 1979; Boesch 1991, 1996). Very little is known about grouping patterns of the central and Nigerian-Cameroon chimpanzee subspecies. Intersite differences in bonobo party composition are also apparent: at Wamba, parties contain equal proportions of males and females (Kano 1992), whereas the proportion of females in a party is higher for Lomako bonobos (Fruth 1995, Hohmann et al. 1999).

Intragroup Social Relationships

Male dominance and strong male–male bonds characterize chimpanzees across all sites (Goodall 1986, Wrangham 1986, Nishida and Hiraiwa-Hasegawa 1987, Nishida and Hosaka 1996, Boesch and Boesch-Achermann 2000, Watts 2000b, Arnold and Whiten 2003). In most chimpanzee communities, males associate more with each other than in mixed parties. Male chimpanzees also groom one another more often than they groom females and than females groom other females (Takahata 1990, Watts 2000a). Males groom those with whom they associate (Mitani et al. 2000), reciprocally (Watts 2000a, 2002; Arnold and Whiten 2003), and up the hierarchy, with higher-ranking males being groomed more than lower-ranking males (Newton-Fischer 2002a; but see Takahata 1990). Male chimpanzees join in coalitionary mate guarding (Watts 1998), hunting (Boesch and Boesch-Achermann 2000), and intergroup aggression (Wrangham 1999, Muller 2002). Moreover, prey capture and possession is male-

biased, and males share food predominantly with other males (and some high-ranked females) as opposed to other group members (Boesch 1994).

Since males are philopatric, kin selection was proposed to explain the close relationships among male chimpanzees. However, chimpanzee male–male alliances can change easily, which does not fit well with the predictions of kin selection (Nishida 1983). Furthermore, genetic analyses reveal that chimpanzee male affiliation and bonding are not significantly correlated with maternal genetic relatedness (Goldberg and Wrangham 1997, Mitani et al. 2000), and the levels of relatedness among chimpanzee males are unexpectedly low and no higher than those among females (Vigilant et al. 2001; but see Morin 1994). Since chimpanzee interbirth intervals are long and preadult mortality is high (Goodall 1986, Wallis 1997, Hill et al. 2001), it is unlikely that brothers will reach adulthood together. Thus, male chimpanzees rely on close age and rank cohorts (rather than kin) when forming alliances (Goldberg and Wrangham 1997, Watts 2000b, Mitani et al. 2002). Comparable correlations between male affiliations and paternal relatedness have not yet been performed, so kin selection cannot be excluded.

Male chimpanzee hierarchies are linear, and dominance is very important for males since dominant males generally sire more offspring (Constable et al. 2001, Vigilant et al. 2001). Male coalitions in chimpanzees are very common and important in rank ascension and maintenance (de Waal 1982, Nishida 1983). Male chimpanzees are aggressive more frequently than females (Muller 2002). Though females are often victims of male aggression (Muller 2002, Stumpf 2005; see below), most chimpanzee aggression is between males and is related to male–male competition and rank (Goodall 1986, Nishida and Hosaka 1996, Muller 2002, Wittig and Boesch 2003). The fission–fusion social system and the fluctuating nature of male coalitions may impede a male's ability to monitor other males' social relationships, and frequent aggression may be necessary to maintain or reestablish rank (Muller and Wrangham 2001, 2004a). Lethal within-group aggression has been reported at Budongo (Fawcett and Muhumuza 2000), Mahale (Kitopeni et al. 1995), and Ngogo (Watts 2004) and may function to reduce mating competition (Watts 2004).

In contrast to males, female chimpanzee relationships are traditionally classified as egalitarian and weak (Wrangham 1980, Sterck et al. 1997). This is usually attributed to female emigration from their natal groups, decreased female sociality and female–female affiliative behaviors at most sites, and less research focused on female relationships. However, this view of female–female relationships has recently come into question. Relationships between female primates may be influenced by the level and type of feeding competition (see Chapter 28; Wrangham 1979b, van Schaik 1989). For example, the strength of female bonds and hierarchies is predicted to vary directly with the level of contest competition. When females are able to group more, as in the case of western chimpanzees and bonobos, the increased likelihood of contest competition leads to alliances among unrelated females (Williams et al. 2002a,b, Wittig and Boesch 2003). Western chimpanzee females, in particular, do have higher than expected grooming frequencies and do form alliances and strong bonds (Sugiyama 1988, Boesch 1996, Boesch and Boesch-Achermann 2000). Interestingly, it has recently been suggested that Taï females have a linear dominance hierarchy, implying that while food abundance permits larger subgrouping in western chimpanzees, intragroup contest competition may be considerable (Wittig and Boesch 2003). Recent data from Gombe indicate that hierarchies are important for females since their reproductive success, longevity, infant survivorship, body weight, and interbirth intervals are strongly affected by female rank (Pusey et al. 1997, 2005; Williams et al. 2002). Dominant females travel with other high-ranking females, support each other in food competition, and occupy prime feeding sites, while low-ranked females spend little time with higher-ranked females (Goodall 1986, Boesch and Boesch-Achermann 2000, Williams et al. 2002a,b). When infanticide by females occurs, lower-ranking females are the targets (Goodall 1977, Pusey et al. 1997). Most female–female aggression among chimpanzees occurs over food and defense of offspring (Williams et al. 2002a,b, Wittig and Boesch 2003). Presumably, the positive correlation between female reproductive success and rank is due to the increased access to better feeding sites (Pusey et al. 1997, Williams 1999, Williams et al. 2002). These findings suggest that feeding and, in particular, contest competition are not inconsequential and that female relationships among chimpanzees are not as egalitarian or as weak as originally considered (Williams et al. 2002a,b, Wittig and Boesch 2003).

Relationships within bonobo communities are markedly different from those of chimpanzees. Whereas male–male relations and male hierarchy are most evident among chimpanzees, female bonobos are characterized as dominant (Parish 1996, Kano 1998), codominant (Furuichi 1989, 1997), or dominant at feeding sites (White 1996b) to males. Despite male philopatry, female–female relations are paramount among bonobos (Kano 1992, Furuichi 1992, Hohmann and Fruth 1996, Parish 1996). The lower seasonality, presumed greater food availability, and fewer costs of grouping appear to increase the potential for female bonding. Female bonobos forage together without much competition (White and Wrangham 1988). Females can easily take food from males, more often possess shareable foods (Kuroda 1980, Fruth and Hohmann 2002), and regularly band together to chase away harassing males (Idani 1991, Kano 1992, Furuichi 1992, Hohmann and Fruth 1996). In contrast to chimpanzees, male bonobos are least likely to share food with other males (Kuroda 1980).

Bonobo grooming bouts are most common between the sexes, followed by female–female grooming, then lastly by grooming among males (Badrian and Badrian 1984, Thompson-Handler 1990, Kano 1992). The duration of these bouts suggests a different pattern of affiliation, with male–male grooming bouts lasting longer than either mixed-sex

or female–female bouts (Wamba, Furuichi and Ihobe 1994; Lomako, Hohmann et al. 1999, Kano 1992). Although chimpanzee and bonobo grooming patterns appear to differ, grooming rates may be influenced by the bonobos' female-biased sex ratio, as well as larger and mixed composition parties. Thus, grooming patterns for bonobos show some overlap with those for chimpanzees (Hohmann and Fruth 2002).

Although alpha-male bonobos have higher reproductive success than subordinates (Gerloff et al. 1999), male hierarchies appear to be less important for bonobos than for chimpanzees. Male bonobos do not assist one another during conflicts, and apart from the alpha position, male bonobo ranks are not clearly defined (de Waal and Lanting 1997).

Wild and captive bonobos are characterized by less aggression and higher conciliation rates than are chimpanzees (de Waal 1998, Hohmann and Fruth 2002). When aggression occurs, most fights are either between individual males or directed by females toward males (Parish 1994, de Waal 1998), while males rarely attack females (Wrangham 1993; Parish 1994, 1996; Vervaecke et al. 1999). Male–male aggression increases on mating days (Hohmann and Fruth 2003b). Very few fights are female to female (3.4%, Kano 1992) except in the context of mating (Hohmann and Fruth 2003b, Vervaecke et al. 2003).

In both chimpanzees and bonobos, enduring bonds exist between mothers and their offspring, particularly their sons (Goodall 1986, Boesch 1997). However, the period of mother–son dependence is longer for bonobos than for chimpanzees. Both bonobo and chimpanzee mothers appear to play a role in influencing a son's dominance rank (Furuichi 1989, Goodall 1986, Boesch and Boesh-Achermann 2000; but see Williams et al. 2002a,b).

Due to the variation in chimpanzee and bonobo grouping patterns across sites, three social system models have been applied to *Pan*. Because of male philopatry and the patterns of association, grooming, meat sharing, and boundary patrols at some sites, as well as the relatively weak egalitarian relationships among females (Nishida 1968, Wrangham and Smuts 1980, Boesch 1996, Nishida and Hosaka 1996, Watts 2000a, Boesch and Boesch Achermann 2000, Watts and Mitani 2001), chimpanzee social organization historically has been considered male-bonded (Nishida 1968, Wrangham 1979b, Wrangham and Smuts 1980, Kawanaka 1984). However, because Taï chimpanzee females and males are equally social and maintain high range overlap between the sexes, Boesch (1996) labeled the social system of *P. t. verus* as "bisexually bonded." Bonobos are considered female-bonded because of the female-focused patterns of association and social interactions (Parish 1998).

Intercommunity Interactions

Chimpanzees are very territorial, and intercommunity interactions are usually hostile and sometimes deadly (Goodall 1986, Nishida 1979, Wrangham 1999, Watts and Mitani 2000, Muller 2002, Wilson and Wrangham 2003, Herbinger

2004). Chimpanzees patrol their boundaries about once a month (Herbinger et al. 2001, Anderson et al. 2002) and encounter other communities in 26%–40% of patrols (Boesch and Boesch-Achermann 2000, Wilson and Wrangham 2003). At other times, chimpanzees generally avoid the periphery of their territory (Herbinger et al. 2001). Such patrolling and avoidance of the periphery may reflect the risk of aggression from other communities (Nishida et al. 1985, Watts and Mitani 2000). The prevalence of intergroup encounters varies across chimpanzee sites: 9.5% of follows at Gombe (Goodall 1986), 11.8% at Taï (Boesch and Boesch-Achermann 2000, Herbinger 2004), and far less frequently at Kanyawara (2.8% cited in Wilson and Wrangham 2003). Composition of patrols also varies across sites. All-male patrols are the norm for eastern chimpanzees (Goodall 1986, Watts and Mitani 2001), while mixed-sex patrols are more common among Taï chimpanzees, though females rarely participate in physical attacks (Boesch and Boesch-Achermann 2000). In addition, intercommunity rates of aggression vary substantially (Wilson and Wrangham 2003). Lethal aggression against adults, while observed multiple times in the east, is rare in western chimpanzees. Territoriality and lethal aggression in chimpanzees are linked to improved access to food and females and to improved male and female reproductive success (Nishida et al. 1985, Manson and Wrangham 1991, Williams 1999, Goodall 1986, Wrangham 1999, Boesch and Boesch-Achermann 2000, Pusey 2001, Williams et al. 2004).

Chimpanzee strategies during intercommunity encounters depend upon the perceived balance of power between themselves and their opponents (Manson and Wrangham 1991, Wilson and Wrangham 2003). During intercommunity encounters, males will approach, attack, or retreat depending on the size of their patrol and the size and composition of the foreign group (Wilson et al. 2001). Chimpanzees will usually approach neighboring groups even when they are outnumbered, implying that the benefits of resource defense are substantial (Herbinger 2004). In the case of an extreme imbalance of power, such as finding a strange male traveling alone, lethal aggression may occur (Manson and Wrangham 1991, Wrangham 1999). Females avoid peripheral areas more than do males (Wilson et al. 2001). When nongroup females are encountered, males may tolerate, groom, and mate with them or, depending on the female's reproductive status, may attack her or her infant (Goodall 1986).

Bonobos, like chimpanzees, defend their territories (Kano 1992, Hohmann et al. 1999). However, compared to chimpanzees, bonobos are less territorial, intergroup interactions are less aggressive (Kano 1992, Kano and Mulavwa 1984, Ingmanson 1998), and lethal aggression has never been observed (Stanford 1998, Wrangham 1999). Bonobo intergroup interactions can last for several hours and include varied behaviors such as grooming, genitogenital (g–g) rubbing, mating, as well as some aggression (Idani 1990, Kano 1992, White 1996b). Bonobos have greater home range overlap than do chimpanzees, and male bonobos have not

been observed to patrol borders (Hohmann and Fruth 2002). In addition, while most chimpanzee interactions occur on the periphery, encounters with other bonobo groups can occur in the center of the territory (Hohmann 2001). Differences between chimpanzees and bonobos in territoriality and intergroup aggression may result from lower feeding and mating competition among bonobos compared to chimpanzees.

Infanticide

Infanticide and attacks on juveniles have been observed in the majority of long-term chimpanzee sites, both within (Goodall 1977, 1986; Arcadi and Wrangham 1999; Takahata 1985; Hiraiwa-Hasegawa 1987; Hamai et al. 1992; Kawanaka 1981) and between (Goodall 1986; Watts and Mitani 2000, 2001; Nishida et al. 1979; Newton-Fisher 1999b; Kutsukake and Matsusaka 2002; Wilson et al. 2004) communities. These attacks can have a substantial effect on infant mortality. For example, at Mahale, conspecifics were responsible for 16% of community deaths, and the majority of these were infants killed by mature males (Hamai et al. 1992, Nishida et al. 2003).

Male between-group infanticide may function to eliminate future competitors and drive neighboring females from territorial borders (Muller and Mitani, in press). Infanticide may also be a strategy to obtain higher reproductive success over same-sex conspecifics (Hrdy 1979; but see Sussman et al. 1995). For males, killing dependent infants sired by other males results in a shortening of the female interbirth interval. Females resume cycling sooner and may permit the infanticidal male a possibility to sire the female's next infant.

Although males are responsible for the majority of infanticides observed, two female chimpanzees killed many infants at Gombe (Goodall 1986). Infanticide by females may be more common at Gombe than other chimpanzee sites (Pusey et al. 1997, Arcadi and Wrangham 1999) and may function to diminish within-group competition. Difference between sites may be explained by lower intergroup competition and, for western chimpanzees, increased female–male sociality, which is thought to function as an anti-infanticide strategy (van Schaik 1996). Infanticide has never been observed among bonobos. This lack of evidence may be due to less observation time for bonobos or may be due to the less aggressive nature and more effective female counterstrategies in bonobos.

REPRODUCTIVE PARAMETERS

Differences in Reproductive Parameters Within *Pan*

On average, chimpanzee females give birth only once every 5–6 years (Tutin and McGinnis 1981, Nishida et al. 1990, Boesch and Boesch-Achermann 2000). Interbirth intervals are generally shortest among Bossou chimpanzees, followed by Taï, Mahale, and then Kanyawara (Wrangham et al. 1996)

(Table 19.6). Bonobo interbirth intervals are, on average, slightly shorter than chimpanzees' (Kano 1992), but they vary substantially and do not differ significantly from those of Bossou chimpanzees (Fruth and Hohmann 2002, Hohmann and Fruth 2002). Resumption of cycling after parturition also varies considerably within *Pan*. Bonobos and female Taï chimpanzees often resume sexual swelling 1 year after parturition (Kano 1992, Fruth and Hohmann 2002, Boesch and Boesch-Achermann 2000). However, resumption of swelling among eastern chimpanzees is considerably later (see Table 19.6). Although bonobos have longer swelling phases than chimpanzees, there is also wide variation within chimpanzees, with longer swellings observed among western than eastern chimpanzee females (Wrangham 2002). Both chimpanzees and bonobos continue to have swellings and sexually active phases after conception. However, the period of postconception (pregnancy) swellings is longer for bonobos than for chimpanzees (Kano 1992, Furuichi and Hashimoto 2002).

Ecological constraints influence female fecundity and reproductive parameters. The onset of postpartum cycling is seasonal in chimpanzees at Gombe, Mahale, and Budongo (Wallis 1992, 1995; Nishida 1990; Wallis and Reynolds 1999), coinciding most frequently with the end of the dry season. There is also a strong seasonal influence on chimpanzee sexual activity and conception/birth cycles (Wallis 1995, 2002; Wallis and Reynolds 1999; Boesch and Boesch-Achermann 2000; Anderson et al. 2002). Since party sizes in many chimpanzee communities increase with the presence of estrous females (Anderson et al. 2002, Mitani and Watts

Table 19.6   Reproductive Parameters

	PAN TROGLODYTES VERUS		P. T. SCHWEINFURTHII		P. PANISCUS
	TAÏ[1]	BOSSOU[2]	GOMBE[3]	MAHALE[4]	WAMBA[5]
Onset of menses[6] (age in years)	10	8.5	10.8	10.7	~10
Adolescent sterility (between first max. swelling and birth, years)	2.6	2.4	2.4	2.5	~4
Age at emigration/immigration (years)	11	>8?[a,b]	9–13	11 (median), 11.27 (mean)	8–13
Emigration rate (%)	~90	0–70[b]	50	90	100
Time between immigration and first birth (years)	2.7	n/a		2.7 (median), 3.3 (mean)	
Age at first birth (years)	13.7	13 (median), 10.9 (mean)[c]	13.3	13 (median), 15.6 (mean)	13–15
Estrous cycle	35		34–36	31.5	41.5 (range 33–49)
Days of maximal swelling	12	12	13	12.5	12.9–14.6
Maximal swelling % of whole cycle	34.3		36.1	39.7	40.7 (range 35–48)
Maximal swelling % of interbirth interval	4.9		4.2	6.4	27
Estrous sex ratio (M/max. tumescent F)	1.6		12.3	4.2	2.8
Female copulation rate per hour (within the cycle)	0.43		0.52	0.79	0.11–0.37
Number of pregnancy cycles	1.5		2	2.6	6.5 (months)
Gestation length (months)	8		7.6	7.6	7.6
Infant survival M (%)	60	80	66	36	81.8[7]
Infant survival F (%)	60	80	76	49	81.8[7]
Resumption of swelling after parturition (months)	24.5	36	46.8	55	12
Cycle resumption to conception (months)	5.1		3.6–4.75	8.9	38
Interbirth interval (months)	69.1	61.2	61.8	69.1	57.6
Annual birth rate per adult F	0.2	0.194		0.2	0.18
Age at last birth (years, max. observed)	~43	~41	37	39	~39
Lifetime reproductive success[8]	1.46	4.25	3.32	2.2	

[1] Boesch (1997), Boesch and Boesch-Achermann (2000), Deschner et al. (2003), R. Stumpf (2005), R. Stumpf, unpublished data.
[2] Sugiyama (1994, 1999, 2004). [a]no females are known to have immigrated into Bossou, [b]70% of females disappeared by adolescence but emigration not confirmed, [c]includes females that remained in Bossou.
[3] Tutin (1979), Wrangham (1979a,b), Tutin and McGinnis (1981), Goodall (1983, 1986), Pusey (1990), Pusey et al. (1997), Wallis (1997).
[4] Nishida et al. (1990, 2003), Hasegawa and Hiraiwa-Hasegawa (1983), Hiraiwa-Hasegawa et al. (1984), Furuichi and Hashimoto (2002), Takahata et al. (1996).
[5] Kano (1989, 1992, 1996), Takahata et al. (1996), Furuichi (1987, 1989, 1992), Furuichi et al. (1998), Furuichi and Hashimoto (2002), Kuroda (1989).
[6] First full-sized swelling.
[7] Data for both sexes combined.
[8] Lifetime reproductive success was calculated as [(40 − (age of primiparity/interbirth interval)] + 1 × survival to primiparity (see Sugiyama 2004).

2001), these females would be subjected to higher feeding competition. Females appear to avoid this by cycling in times of food abundance (Anderson et al. 2002). For chimpanzees, increased feeding competition and decreased food availability may constrain the optimal period for resuming cycling and producing new offspring more so than for bonobos (Furuichi and Hashimoto 2002, Fruth and Hohmann 2002). Thus, the shorter interbirth interval and period of lactational amenorrhea among bonobos and some chimpanzees (Taï and Bossou) may be related to the lower costs of grouping and loosened constraints for females in denser forest habitats with more reliable food availability (Wrangham 2002).

Swelling and Ovulation

Female chimpanzees have, on average, a 35-day menstrual cycle. Their sexually active phase (estrus) lasts about 10–15 days and is marked by a large, pink perineal swelling. Ovulation occurs near the end of maximal swelling (Graham 1981, Tutin and McGinnis 1981, Deschner et al. 2003). Female chimpanzees are most sexually receptive during the period of maximal tumescence (Goodall 1986, Wallis 1997, Stumpf 2005), yet much of chimpanzee (and bonobo) sexual behavior is nonreproductive since sperm longevity is short and ovulation and conception occur only within a limited time window (Gomendio and Roldan 1993). In chimpanzees (and possibly bonobos), nonconceptive, multimale mating appears to be a female strategy to avoid infanticide (Stumpf 2005) and strengthen social bonds.

The sexual cycle of female bonobos (averaging 40 days) is somewhat longer than that of chimpanzees, and sexual swelling occurs through most of the cycle (Kano 1996; but see Stanford 1998, Hohmann and Fruth 2000). Because party sizes generally increase with increasing numbers of estrous females, the more lengthy maximal swelling phases of female bonobos may occur because of the more relaxed constraints of feeding competition and lower costs of grouping compared to chimpanzees (e.g., Stanford 1998, see also Wrangham 2002). Swellings in chimpanzees reliably advertise the probability of ovulation (Emery and Whitten 2003, Deschner et al. 2003). Female bonobo ovulation is less reliably tied to the swelling (thus more concealed) and often occurs after detumescence (Reichert et al. 2002).

After menarche and before emigration, adolescent females in both *Pan* species go through a period of subfecundity where conception is delayed. Although they have frequent swellings, they are generally not yet fertile. During this intermediate period, when females are still in their natal group, adolescent subfecundity may protect these females from interbreeding with their male relatives (Pusey 1980).

Sexual Behavior and Copulation Rates

Some aspects of sexual behavior among chimpanzees and bonobos differ substantially. For example, among bonobos, almost all mating attempts are initiated by males, whereas females initiate one-quarter to one-third of all mating attempts among chimpanzees (Takahata et al. 1996, Nishida 1997, Stumpf 2004). Dorso-ventral copulation is the most common pattern for both species. However, ventroventral mating is also relatively common in bonobos (Kano 1980, Thompson-Handler et al. 1984). In bonobos, as in chimpanzees, sex is often nonreproductive but, in contrast to chimpanzees, functions frequently to appease and reduce aggression and tension (de Waal 1987).

Another aspect of sexuality among bonobos that is quite distinct from chimpanzees is the preponderance of same-sex, particularly female–female, sexual behavior. This behavior, g–g rubbing, occurs when females rapidly rub their clitorises together in a ventroventral position. G–g rubbing occurs regularly (Kano 1980, Thompson-Handler et al. 1984, de Waal 1987, Parish 1994, 1996) but is very rare among both wild and captive chimpanzees (R. Stumpf unpublished data; but see Anestis 2004). G–g rubbing may function to regulate tension (White 1988, Hohmann and Fruth 2000), as well to enhance and reconcile female–female relationships (Hohmann and Fruth 2000) and access to preferred food patches (Parish 1994). Rates of g–g rubbing are particularly high when access to food is limited and contested (Hohmann and Fruth 2000). However, White and Lanjouw (1992) found that g–g rubbing occurred more when food was abundant and decreased when food was rare. This suggests that g–g rubbing may depend on the quality of food, type of feeding competition, food distribution and accessibility, and extent of monopolization of the resource (Hohmann and Fruth 2000).

Bonobos are often considered to be more sexually active than chimpanzees. This determination depends upon the measures being considered. The majority of copulations for both species occur during the period of maximal tumescence, though this varies between them (96% for chimpanzees, Goodall 1986, 67%–82% for bonobos, Furuichi 1987, Takahata et al. 1996). Female chimpanzees have higher copulation rates during maximal tumescence than bonobos (Takahata et al. 1996) (Table 19.6). In contrast, female copulation rates between successive births (across the interbirth interval) are higher for bonobos than for chimpanzees (Wamba, 0.14/hr [Kano 1992] Mahale, 0.04/hr [Takahata et al. 1996]). This latter difference can be explained by the fact that bonobos have longer tumescent periods (Kano 1996) and, on average, resume cycling earlier than chimpanzees (Takahata et al. 1996, Furuichi and Hashimoto 2002). Male copulation rates do not differ between chimpanzees and bonobos and range between 0.1 and 0.2/hr for both taxa (Furuichi and Hashimoto 2002).

Male and Female Sexual Strategies

There are at least four types of mating described for chimpanzees (Tutin 1979). Opportunistic mating occurs when females mate with many males in succession, with little outright aggression among the males. Consortship occurs when a male and female leave the group and travel together for a period of days or weeks. Possessive mating occurs when a male guards a female and prevents mating by other males. A type of possessive mating, coalitionary mate guarding, has been described by Watts (1998) and occurs when two or three males unite to guard a female from other males. Finally, extragroup mating (with individuals from outside the community) occurs rarely and results in a small proportion (<10%) of all paternities (Constable et al. 2001, Vigilant et al. 2001). Between sites, the propensity for consortships differs subtantially. Bonobo mating patterns are similar to chimpanzees', although possessive mating and consortships are rare (Kano 1992, Takahata et al. 1996). Among bonobos, extragroup mating is both common and tolerated.

Initial data suggested that most chimpanzee offspring were conceived during consortships (Tutin 1979), and this strategy does appear to be successful for low-ranking males at Gombe (Constable et al. 2001). However, additional data from Taï, Mahale, and Gombe indicate that opportunistic matings are responsible for the majority of conceptions (Hasegawa and Hiraiwa-Hasegawa 1983, 1990; Wallis 1997, Boesch and Boesch-Achermann 2000).

At times, no relationship has been found between male dominance rank and mating success (Tutin 1979, Hasegawa and Hiraiwa-Hasegawa 1990, Nishida 1997, Boesch and Boesch-Achermann 2000, Newton-Fisher 2004). However, mating success for high-ranking males is improved by a stable hierarchy, fewer males, and the likelihood of conception (Goodall 1986, Nishida 1997, Kano 1996, Gerloff et al. 1999, Mitani et al. 2002c). Male interest intensifies around ovulation (Wallis 1992), and primarily high-ranked or coalitionary males respond by increasing mate guarding (Tutin and McGinnis 1981, Watts 1998). High-ranking males have greater mating success during the females' most conceptive phase (Tutin and McGinnis 1981, Nishida 1997, Muller and Wrangham 2001, Stumpf 2004, Deschner et al. 2004) and higher paternity success (Constable et al. 2001, Vigilant et al. 2001), which is similar for bonobos (Gerloff et al. 1999). Although dominant males benefit from increased reproductive success, dominance comes at a cost as high rank in chimpanzees is positively correlated with high cortisol and testosterone levels (Muller and Wrangham 2004b).

Chimpanzees are highly promiscuous in their mating behavior, but in spite of this promiscuity and male dominance, females exhibit preferences and can alter the frequency with which they mate with males (Stumpf 2005). Female chimpanzees may influence paternity by manipulating males during both the nonconceptive and conceptive phases of tumescence. Taï females follow a mixed mating strategy of being selective when conception is likely and more promiscuous when conception is unlikely (Stumpf 2005, see also Nunn 1999).

Female preferences during periovulatory phases correlate positively with male mating success in chimpanzees at Taï (Stumpf 2004). This appears to result in part from females'

successful resistance to solicitations of nonpreferred males, thereby averting copulation. Female choice may be less effective in eastern chimpanzees due to the increased proportion of males, higher male–male competition, female attractivity (Wrangham 2002), more solitary nature of eastern females, and more intense sexual coercion (Wrangham and Smuts 1980, Goodall 1986, Matsumoto-Oda and Oda 1998). For bonobos, increased female cohesiveness and female dominance may afford females more opportunity to implement their mating preferences (Hohmann and Fruth 2002).

CONSERVATION

Both chimpanzees and bonobos are characterized by long interbirth intervals, slow life histories, substantial maternal care, and relatively high infant mortality (Harvey et al. 1987) (Table 19.7). They are slow to reproduce and, therefore, especially vulnerable. Bonobos and chimpanzees are presently classified as endangered by the World Conservation Union and threatened with extinction by the Convention on International Trade in Endangered Species. The total wild chimpanzee population was recently estimated to be 173,000–300,000 (Butynski 2003), and populations have declined by as much as 66% in the last three decades due to increasing human populations, habitat destruction, and infectious diseases such as ebola (Walsh et al. 2003, Butynski 2003). Chimpanzees are also hunted for meat and medicinal and animistic religious purposes (Nishida and van Hooff 2002). In West Africa, chimpanzee populations have declined by 50% in the last two decades (Walsh et al. 2003) and are now extinct in Benin and Togo. Only 38,000 West African chimpanzees are estimated to remain (Butynski 2003). Eastern and central chimpanzees are estimated at

Table 19.7   Life History Parameters (*Pan*)

Female growth complete (stature, age in years)[1]	11–12
Male growth complete (stature, age in years)[2]	13–17
Growth cessation (adult mass reached, age in years)[3]	21 F, 24 M
Age at locomotor independence (years)[4]	>5
Lifetime offspring produced[5]	3.9
Lifetime offspring survive weaning[6]	1.4–2
Longevity (years)[7]	40–55 estimate
Age at weaning[8] (years)	5
Birth weight (g)[9]	1,800,* 1,381**

[1,2] Leigh and Shea (1995), Pusey (1990 and 2004), Pusey et al. (2005), Goodall (1986).
[3] Pusey et al. (2005), Kuroda (1989).
[4] Doran (1997b).
[5] Hill et al. (2001), Nishida et al. (2003).
[6] Nishida et al. (2003).
[7] Hill et al. (2001, cited in de Waal and Lanting 1997).
[8] Pusey (1983), Hiraiwa-Hasegawa (1989), Kuroda (1980).
[9] Graham (1970) (*captive *P. troglodytes*), Thompson-Handler (1990) (**wild *P. paniscus*).

98,000, and 93,000, respectively (Butynski 2003). The estimated number of *P. t. vellerosus* remaining is about 6,000 (Butynski 2003).

In 1998, the bonobo population was estimated at 10,000–20,000 (Bowen-Jones 1998). Habitat destruction and the bushmeat trade are the biggest threats to bonobos (Walsh et al. 2003). Although many local taboos prohibit eating bonobos, hunting bonobos increased markedly after the Congolese civil war. Based on the recent rate of African ape decline, Walsh et al. (2003) recommend changing the status of chimpanzees and bonobos from endangered to critically endangered. Salonga National Park, established in 1970, is presently the only national forest reserve in the Democratic Republic of Congo to effectively conserve both *Pan* species. More information is needed on their distribution, abundance, ecological requirements, threats to survival, and areas of high potential for conservation.

CONCLUSION

Presently, there are over 40 chimpanzee and bonobo research sites, and more are being established. Moreover, within the last 15 years, the inclusion of hormonal and genetic methodologies together with more extensive behavioral studies has further expanded our understanding of relationships and behavioral strategies within *Pan*.

Until recently, chimpanzees and bonobos were thought to be quite behaviorally distinct from one another. However, it is becoming increasingly evident that there is substantial behavioral overlap between the two species, and that the interspecies differences are not as marked as were originally thought (Stanford 1998, Boesch 2002, Hohmann and Fruth 2002, Matsumoto-Oda 2002, Williams et al. 2002a) (Table 19.8). Many behavioral patterns once thought to distinguish chimpanzees and bonobos, such as their propensity for hunt-

Table 19.8 Behavioral Similarities and Distinctions Between Chimpanzees and Bonobos: Whether All of These Interspecies Distinctions are Supported Over Time as More Information Is Gained Across All *Pan* Study Sites Remains to Be Seen

	CHIMPANZEES	BONOBOS
Activity pattern	Diurnal	Diurnal
Social organization	Fission–fusion	Fission–fusion
Community structure	Multimale/multifemale	Multimale/multifemale
Dispersing sex	Adolescent females	Adolescent females
Locomotion	Knuckle walking	Knuckle walking
Substrate use	Arboreal/semiterrestrial	Arboreal/semiterrestrial
Sleep site	Arboreal nests	Arboreal nests
Diet	Primarily frugivorous	Primarily frugivorous
Sympatric gorillas?	Variable, only in some areas	No
Dominant sex	Male	Female or codominant
Intercommunity interactions	Usually antagonistic	Usually tolerant
Male–male association	Very strong	Moderate
Male–female association	Weak in eastern, strong in western	Strong
Female–female association	Very weak in eastern, strong in western	Strong
Resolve tension with	Aggression	Sex
Male–female aggression	Common	Rare
Female–male aggression	Rare	Common
Pant–grunt submissive vocalization	Common	Never observed
Grooming	Male–male most, female–female least	Male–female most, male–male least
Infanticide	Rare but present at all sites	Never observed
Tool use	Habitual	Infrequent
Consorts	Some but not all sites	Rare
Boundary patrolling	Habitual	Never observed
Hunting	Occurs at most sites	Rarely observed
Food sharing	Yes (mainly meat)	Yes (mainly fruit)
Vocalizations	Lower pitch	Higher pitch
Sexual behavior during nonswelling	Very rare	Common (33%)
Copulation posture	Dorsoventral	Dorsoventral (70%–89%), ventroventral (11%–30%)
Female–female sexual behavior	Very rare	Genitogenital rubbing common

Figure 19.3 Young male bonobo (*Pan paniscus*) (photo by Henrike Wolf).

ing, consorts, and tool use as well as party sizes and composition vary substantially across *Pan* sites and thus may not support the distinction between these two taxa. These considerable behavioral variations observed within *Pan* have been attributed to ecological, cultural, genetic and demographic factors. Because of their close phylogenetic relationship, both species are likely to have similar behavioral responses to similar environmental pressures (Boesch 2002). Additionally, the perceived behavioral distinctions may be the product of differing levels of research efforts (Stanford 1998). Perhaps with equivalent observation and more comparative data, other aspects of chimpanzee and bonobo behavior may prove to be more similar than is currently thought.

In spite of the substantial behavioral overlap, it is apparent that many key differences do exist between these two species (de Waal 1998, Kano 1998, Parish 1998, Doran et al. 2002) (Table 19.8). These nontrivial differences include dominance relations between the sexes, levels and patterns of aggression, the nature of intercommunity interactions, and female sociosexual behaviors. These differences may be influenced by variation in ecological factors such as rainfall, food availability and seasonality (Wrangham 1986, Wrangham et al. 1996, White and Wrangham 1988, Malenky 1990, Malenky and Wrangham 1994, Chapman et al. 1994, Doran et al. 2002). Boinski's (1999) work on *Saimiri* has demonstrated that even subtle ecological differences can have substantial effects on social organization. In addition, the divergent evolutionary pathways of the two species may be a factor in their differing responses to comparable ecological pressures. Nevertheless, understanding the specific effects of ecological differences on chimpanzee and bonobo behavior requires more detailed research.

Further attempts to quantify food availability, defensibility, quality, and size of food patches using comparative methods across sites are in progress; and a results may lead to a clearer explication of interspecific differences. With time, we hope to better understand whether the perceived behavioral distinctions between chimpanzees and bonobos will hold or whether the behavioral diversity within *Pan* is simply a continuum.

REFERENCES

Anderson, D., Nordheim, E. V., Boesch, C., and Moermond, T. C. (2002). Factors influencing fission–fusion grouping in chimpanzees in the Taï National Park, Cote d'Ivoire. In: Boesch, C., Hohmann, G., and Marchant, L. F. (eds.), *Behavioral Diversity in Chimpanzees and Bonobos*. Cambridge University Press, Cambridge. pp. 90–101.

Anestis, S. F. (2004). Female genito-genital rubbing in a group of captive chimpanzees. *Int. J. Primatol.* 25:477–488.

Arcadi, A. C., and Wrangham, R. W. (1999). Infanticide in chimpanzees: review of cases and a new within-group observation from the Kanyawara study group in Kibale National Park. *Primates* 40:337–351.

Arnold, K., and Whiten, A. (2003). Grooming interactions among the chimpanzees of the Budongo Forest, Uganda: tests of five explanatory models. *Behaviour* 140:519–552.

Badrian, A., and Badrian, N. R. (1984). Social organization of *Pan paniscus* in the Lomako Forest, Zaire. In: Susman, R. L. (ed.), *The Pygmy Chimpanzee*. Plenum, New York. pp. 325–346.

Badrian, N., Badrian, A., and Susman, R. (1981). Preliminary observations of the feeding behavior of *Pan paniscus* in the Lomako Forest of Central Zaire. *Primates* 22:173–181.

Badrian, N., and Malenky, R. K. (1984). Feeding ecology of *Pan paniscus* in the Lomako Forest, Zaire. In: Susman, R. L. (ed.), *The Pygmy Chimpanzee*. Plenum, New York. pp. 275–299.

Baldwin, P. J., McGrew, W. C., and Tutin, C. E. G. (1982). Wide-ranging chimpanzees at Mt. Assirik, Senegal. *Int. J. Primatol.* 3:367–385.

Basabose, A. K. (2002). Diet composition of chimpanzees inhabiting the montane forest of Kahuzi, Democratic Republic of Congo. *Am. J. Primatol.* 58:1–21.

Basabose, A. K., and Yamagiwa, J. (2002). Factors affecting nesting site choice in chimpanzees at Tshibati, Kahuzi-Biega National Park: influence of sympatric gorillas. *Int. J. Primatol.* 23:263–282.

Beadle, L. C. (1981). *The Inland Waters of Tropical Africa*. Longman, London.

Boesch, C. (1991). The effects of leopard predation on grouping patterns in forest chimpanzees. *Behaviour* 117:220–242.

Boesch, C. (1994). Cooperative hunting in wild chimpanzees. *Anim. Behav.* 48:653–667.

Boesch, C. (1996). Social grouping in Taï chimpanzees. In: McGrew, W., Marchant, L., and Nishida, T. (eds.), *Great Ape Societies*. Cambridge University Press, Cambridge. pp. 101–113.

Boesch, C. (1997). Evidence for dominant wild female chimpanzees investing more in sons. *Anim. Behav.* 54:811–815.

Boesch, C. (2002). Behavioural diversity in *Pan*. In: Boesch, C., Hohmann, G., and Marchant, L. F. (eds.), *Behavioural Diversity in Chimpanzees and Bonobos*. Cambridge University Press, Cambridge. pp. 1–8.

Boesch, C., and Boesch, H. (1989). Hunting behavior of wild chimpanzees in the Taï National Park. *Am. J. Phys. Anthropol.* 78:547–573.

Boesch, C., and Boesch, H. (1990). Tool use and tool making in wild chimpanzees. *Folia Primatol.* 54:86–99.

Boesch, C., and Boesch-Achermann, H. (2000). *The Chimpanzees of the Taï Forest: Behavioral Ecology and Evolution.* Oxford University Press, Oxford.

Boesch, C., Marchesi, P., Marchesi, N., Fruth, B., and Joulian, F. (1994). Is nut cracking in wild chimpanzees a cultural behaviour? *J. Hum. Evol.* 26:325–338.

Boesch C., Uehara, S., and Ihobe, H. (2002). Variations in chimpanzee–red colobus interactions. In: Boesch, C., Hohmann, G., and Marchant, L. (eds.), *Behavioural Diversity in Chimpanzees and Bonobos.* Cambridge University Press, Cambridge. pp. 221–230.

Boinski, S. (1999). The social organization of squirrel monkeys: implications for ecological models of social evolution. *Evol. Anthropol.* 8:101–112.

Bowen-Jones, E. (1998). A review of the commercial bushmeat trade with emphasis on central/West Africa and the great apes. In: *Report for the Ape Alliance.* Fauna & Flora International, Cambridge.

Burrows, W., and Ryder, O. A. (1997). Y-chromosome variation in great apes. *Nature* 385:125–126.

Butynski, T. (2003). The robust chimpanzee, *Pan troglodytes*: taxonomy, distribution, abundance and conservation status. In: Kormos, R., Boesch, C., Bakarr, M. I., and Butynski, T. (eds.), *West African Chimpanzees. Status Survey and Conservation Action Plan.* IUCN/SSC Primate Specialist Group, Gland, Switzerland. pp. 31–39.

Caccone, A., and Powell, J. R. (1989). DNA divergence among hominoids. *Evolution* 43:925–942.

Chapman, C. A., White, F. J., and Wrangham, R. W. (1994). Party size in chimpanzees and bonobos. In: Wrangham, R. W., McGrew, W. C., de Waal, F. B. M., Heltne, P. G., and Marquardt, L. A. (eds.), *Chimpanzee Cultures.* Harvard University Press, Cambridge, MA. pp. 41–57.

Chapman, C., and Wrangham, R. W. (1993). Range use of the forest chimpanzees of Kibale: implications for the understanding of chimpanzee social organization. *Am. J. Primatol.* 31:263–273.

Chapman, C., Wrangham, R. W., and Chapman, L. J. (1995). Ecological constraints on group size: an analysis of spider monkey and chimpanzee subgroups. *Behav. Ecol. Sociobiol.* 36:59–70.

Chen, F. C., and Li, W. H. (2001). Genomic divergences between humans and other hominoids and the effective population size of the common ancestor of humans and chimpanzees. *Am. J. Hum. Genet.* 68:444–456.

Conklin-Brittain, N., Knott, C. D., and Wrangham, R. W. (2001). The feeding ecology of apes. In: *Apes: Challenges for the 21st Century.* Chicago Zoological Society, Brookfield, IL. pp. 167–174.

Conklin-Brittain, N. L., Knott, C. D., and Wrangham, R. W. (in press). Energy intake by wild chimpanzees and orangutans: methodological considerations and a preliminary comparison. In: Hohmann, G., Robbins, M., and Boesch, C. (eds.), Feeding Ecology of Primates. Cambridge University Press, Cambridge.

Conklin-Brittain, N. L., Wrangham, R. W., and Hunt, K. D. (1998). Dietary response of chimpanzees and cercopithecines to seasonal variation in fruit abundance: II. Nutrients. *Int. J. Primatol.* 19:971–987.

Constable, J. L., Ashley, M. V., Goodall, J., and Pusey, A. E. (2001). Noninvasive paternity assignment in Gombe chimpanzees. *Mol. Ecol.* 10:1279–1300.

Coolidge, H. J. (1933). *Pan paniscus*, pygmy chimpanzee from south of the Congo River. *Am. J. Phys. Anthropol.* 18:1–59.

Cramer, D. L. (1977). Craniofacial morphology of *Pan paniscus*: a morphological and evolutionary appraisal. *Contrib. Primatol.* 10:1–64.

Deschner, T., Heistermann, M., Hodges, K., and Boesch, C. (2003). Timing and probability of ovulation in relation to sex skin swelling in wild West African chimpanzees, *Pan troglodytes verus. Anim. Behav.* 66:551–560.

de Waal, F. B. M. (1982). *Chimpanzee Politics: Power and Sex among Apes.* Allen and Unwin, London.

de Waal, F. B. M. (1987). Tension regulation and non-reproductive functions of sex among captive bonobos (*Pan paniscus*). *Natl. Geogr. Res.* 3:318–335.

de Waal, F. B. M. (1988). The communicative repertoire of captive bonobos (*Pan paniscus*), compared to that of chimpanzees. *Behaviour* 106:183–251.

de Waal, F. B. M. (1989). Behavioral contrasts between bonobos and chimpanzees. In: Helte, P. G., and Marquardt, L. A. (eds.), *Understanding Chimpanzees.* Harvard University Press, Cambridge, MA. pp. 154–175.

de Waal, F. B. M. (1998). Reply to Stanford (1998) The social behavior of chimpanzees and bonobos—empirical evidence and shifting assumptions. *Curr. Anthropol.* 39:407–408.

de Waal, F. B. M., and Lanting, F. (1997). *Bonobos: The Forgotten Ape.* University of California Press, Berkeley.

Doran, D. M. (1993). Sex differences in adult chimpanzee positional behavior: the influence of morphology on locomotion. *Am. J. Phys. Anthropol.* 91:99–115.

Doran, D. (1997a). Influence of seasonality on activity patterns, feeding behavior, ranging and grouping patterns in Taï chimpanzees. *Int. J. Primatol.* 18:183–206.

Doran, D. (1997b). Ontogeny of locomotion in mountain gorilla and chimpanzees. *J. Hum. Evol.* 32:323–344.

Doran, D., Jungers, W. L., Sugiyama, Y., Fleagle, J. G., and Heesy, C. P. (2002). Multivariate and phylogenetic approaches to understanding chimpanzee and bonobo behavioral diversity. In: Boesch, C., Hohmann, G., and Marchant, L. (eds.), *Behavioural Diversity in Chimpanzees and Bonobos.* Cambridge University Press, Cambridge. pp. 14–34.

Emery, M. A., and Whitten, P. L. (2003). Size of sexual swellings reflects ovarian function in wild chimpanzees (*Pan troglodytes*). *Behav. Ecol. Sociobiol.* 54:340–351.

Fawcett, K., and Muhumuza, G. (2000). Death of a wild chimpanzee community member: possible outcome if intense sexual competition. *Am. J. Primatol.* 51:243–247.

Fleagle, J. G. (1999). *Primate Adaptation and Evolution,* 2nd ed. Academic Press, New York.

Fruth, B. (1995). *Nests and Nest Groups in Wild Bonobos* (Pan paniscus): *Ecological and Behavioural Correlates.* Ludwig-Maximilians-Universität, München.

Fruth, B. (1998). Reply to Stanford (1998) The social behavior of chimpanzees and bonobos—empirical evidence and shifting assumptions. *Curr. Anthropol.* 39:408–409.

Fruth, B., and Hohmann, G. (1993). Ecological and behavioral aspects of nest-building in wild bonobos (*Pan paniscus*). *Ethology* 94:113–126.

Fruth, B., and Hohmann, G. (1996). Nest building behavior in the great apes: the great leap forward? In: McGrew, W. C., Marchant, L. F., and Nishida, T. (eds.), *Great Ape Societies.* Cambridge University Press, Cambridge. pp. 225–240.

Fruth, B., and Hohmann, G. (2002). How bonobos handle hunts and harvests: why share food? In: Boesch, C., Hohmann, G., and Marchant, L. (eds.), *Behavioural Diversity in Chimpanzees and Bonobos*. Cambridge University Press, Cambridge. pp. 231–243.

Furuichi, T. (1987). Sexual swelling, receptivity, and grouping of wild pygmy chimpanzee females at Wamba Zaire. *Primates* 28:309–318.

Furuichi, T. (1989). Social interactions and the life history of female *Pan paniscus* in Wamba, Zaire. *Int. J. Primatol.* 10:173–197.

Furuichi, T. (1992). The prolonged estrus of females and factors influencing mating in a wild group of bonobos (*Pan paniscus*) in Wamba, Zaire. *Topics Primatol.* 179–190.

Furuichi, T. (1997). Agonistic interactions and matrifocal dominance rank of wild bonobos (*Pan paniscus*) at Wamba. *Int. J. Primatol.* 18:855–875.

Furuichi, T., and Hashimoto, C. (2002). Social relationships between cycling females and adult males in Mahale chimpanzees. In: Boesch, C., Hohmann, G., and Marchant, L. (eds.), *Behavioural Diversity in Chimpanzees and Bonobos*. Cambridge University Press, Cambridge. pp. 156–167.

Furuichi, T., Hashimoto, C., and Tashiro, Y. (2001). Fruit availability and habitat use by chimpanzees in the Kalinzu Forest, Uganda: examination of fallback foods. *Int. J. Primatol.* 22:929–945.

Furuichi, T., Idani, G., Ihobe, H., Kuroda, S., Kitamura, K., Mori, A., Enomoto, T., Okayasu, N., Hashimoto, C., and Kano, T. (1998). Population dynamics of wild bonobos (*Pan paniscus*) at Wamba. *Int. J. Primatol.* 19:1029–1043.

Furuichi, T., and Ihobe, H. (1994). Variation in male relationships in bonobos and chimpanzees. *Behaviour* 130:211–228.

Gagneux, P., Wills, C., Gerloff, U., Tautz, D., Morin, P. A., Boesch, C., Fruth, B., Hohmann, G., Ryder, O. A., and Woodruff, D. S. (1999). Mitochondrial sequences show diverse evolutionary histories of African hominoids. *Proc. Natl. Acad. Sci. U.S.A.* 96:5077–5082.

Gerloff, U., Hartung, B., Fruth, B., Hohmann, G., and Tautz, D. (1999). Intra-community relationships, dispersal patterns and paternity success in a wild living community of bonobos (*Pan paniscus*) determined from DNA analysis of faecal samples. *Proc. R. Soc. Lond.* 266:1189–1195.

Goldberg, T. L., and Ruvolo, M. (1997). The geographic apportionment of mitochondrial genetic diversity in East African chimpanzees, *Pan troglodytes schweinfurthii*. *Mol. Biol. Evol.* 14:976–984.

Goldberg, T. L., and Wrangham, R. W. (1997). Genetic correlates of social behaviour in wild chimpanzees: evidence from mitochondrial DNA. *Anim. Behav.* 54:559–570.

Gomendio, M., and Roldan, E. R. S. (1993). Mechanisms of sperm competition: linking physiology and behavioral ecology. *Trends Ecol. Evol.* 8:95–100.

Gonder, K. M. (2000). Evolutionary genetics of chimpanzees (*Pan troglodytes*) in Nigeria and Cameroon [PhD thesis]. City University of New York, New York.

Gonder, K. M., Oates, J. F., Disotell, T. R., Forstner, M. R., Morales, J. C., and Melnick, D. J. (1997). A new West African chimpanzee subspecies? *Nature* 388:337.

Goodall, J. (1977). Infant killing and cannibalism in free-living chimpanzees. *Folia Primatol.* 28:259–282.

Goodall, J. (1986). *The Chimpanzees of Gombe: Patterns of Behavior*. Harvard University Press, Cambridge, MA.

Graham, C. E. (1970). Reproductive physiology of the chimpanzee. In: Bourne, G. H. (ed.), *The chimpanzee*, vol. 3. Karger, Basel. pp. 183–220.

Graham, C. E. (1981). Menstrual cycle of the great apes. In: Graham, C. E. (ed.), *Reproductive Biology of the Great Apes. Comparative and Biomedical Perspectives*. Academic Press, New York. pp. 1–41.

Groves, C. P. (2001). *Primate Taxonomy*. Smithsonian Institution Press, Washington, DC.

Hamai, M., Nishida, T., Takasaki, H., and Turner, L. (1992). New records of within-group infanticide and cannibalism in wild chimpanzees. *Primates* 33:151–162.

Harvey, P. H., Martin, R. D., and Clutton-Brock, T. H. (1987). Life histories in the comparative perspective. In: Smuts, B. B., Cheney, D. L., Seyfarth, R. M., Wrangham, R. W., and Struhsaker, T. T. (eds.), *Primate Societies*. University of Chicago Press, Chicago. pp. 181–196.

Hasegawa, T., and Hiraiwa-Hasegawa, M. (1983). Opportunistic and restrictive matings among wild chimpanzees in the Mahale Mountains, Tanzania. *J. Ethol.* 1:75–85.

Hasegawa, T., and Hiraiwa-Hasegawa, M. (1990). Sperm competition and mating behavior. In: Nishida, T. (ed.), *The Chimpanzees of the Mahale Mountains: Sexual and Life History Strategies*. University of Tokyo Press, Tokyo. pp. 115–132.

Hashimoto, C., Tashiro, Y., Kimura, D., Enomoto, T., Ingmanson, E. J., Idani, G., and Furuichi, T. (1998). Habitat use and ranging of wild bonobos (*Pan paniscus*) at Wamba. *Int. J. Primatol.* 19:1045–1060.

Herbinger, I. (2004). Inter-group interactions among three communities of wild West African chimpanzees (*Pan troglodytes verus*) [PhD thesis]. University of Leipzig, Leipzig.

Herbinger, I., Boesch, C., and Rothe, H. (2001). Territory characteristics among three neighboring chimpanzee communities. *Int. J. Primatol.* 22:143–167.

Hicks, T. C., Fouts, R. S., and Fouts, D. H. (2005). Chimpanzee (*Pan troglodytes troglodytes*) tool use in the Ngotto Forest, Central African Republic. *Am. J. Primatol.* 65:221–237.

Hill, K., Boesch, C., Goodall, J., Pusey, A., Williams, J., and Wrangham, R. (2001). Mortality rates among wild chimpanzees. *J. Hum. Evol.* 40:437–450.

Hiraiwa-Hasegawa, M. (1987). Infanticide in primates and a possible case of male-biased infanticide in chimpanzees. In: Ito, Y., Brown, J. L., and Kikkawa, J. (eds.), *Animal Societies: Theories and Facts*. Japan Scientific Societies Press, Tokyo. pp. 125–139.

Hiraiwa-Hasegawa, M. (1989). Sex differences in the behavioral development of chimpanzees at Mahale. In: Heltne, P. G., and Marquardt, L. A. (eds.), *Understanding Chimpanzees*. Harvard University Press, Cambridge, MA. pp. 104–115.

Hiraiwa-Hasegawa, M., Hasegawa, T., and Nishida, T. (1984). Demographic study of a large-sized unit-group of chimpanzees in the Mahale Mountains, Tanzania: a preliminary report. *Primates* 25:401–413.

Hladik, C. (1973). Alimentation et activite d'un groupe de chimpanzes reintroduits en foret Gabonaise. *La Terre et la Vie* 27:343–413.

Hladik, C. (1977). Chimpanzees of Gabon and chimpanzees of Gombe: some comparative data on the diet. In: Clutton-Brock, T. (ed.), *Primate Ecology: Studies of Feeding and Ranging Behaviour in Lemurs, Monkeys and Apes*. Academic Press, London. pp. 481–501.

Hohmann, G. (2001). Association and social interactions between strangers and residents in bonobos (*Pan paniscus*). *Primates* 42:91–99.

Hohmann, G., and Fruth, B. (1993). Field observations on meat sharing among bonobos (*Pan paniscus*). *Folia Primatol.* 60:225–229.

Hohmann, G., and Fruth, B. (1996). Food sharing and status in unprovisioned bonobos. In: Wiessner, P., and Schiefenhoevel, W. (eds.), *Food and the Status Quest*. Berghahn, Oxford. pp. 47–67.

Hohmann, G., and Fruth, B. (2000). Use and function of genital contacts among female bonobos. *Anim. Behav.* 60:107–120.

Hohmann, G., and Fruth, B. (2002). Dynamics in social organization of bonobos (*Pan paniscus*). In: Boesch, C., Hohmann, G., and Marchant, L. (eds.), *Behavioural Diversity in Chimpanzees and Bonobos*. Cambridge University Press, Cambridge. pp. 138–155.

Hohmann, G., and Fruth, B. (2003a). Culture in bonobos? Between-species and within-species variation in behavior. *Curr. Anthropol.* 44:563–571.

Hohmann, G., and Fruth, B. (2003b). Intra- and inter-sexual aggression by bonobos in the context of mating *Behaviour* 140:1389–1413.

Hohmann, G., Gerloff, U., Tautz, D., and Fruth, B. (1999). Social bonds and genetic ties: kinship association and affiliation in a community of bonobos (*Pan paniscus*). *Behaviour* 136:1219–1235.

Horai, S., Hayasaka, K., Kondo, R., Tsugane, K., and Takahata, N. (1995). Recent African origin of modern humans revealed by complete sequences of hominoid mitochondrial DNAs. *Proc. Natl. Acad. Sci. U.S.A.* 92:532–536.

Hrdy, S. B. (1979). Infanticide among animals: a review, classification, and examination of the implications for the reproductive strategies of females. *Ethol. Sociobiol.* 1:13–40.

Hunt, K. D. (2004). Chimpanzee Socioecology in a Hot, Dry Habitat: Implications for Early Hominin Evolution. Conference on African Great Apes: Evolution, Diversity, and Conservation. Kyoto University, Kyoto, March 4, 2004.

Hunt, K. D., Cleminson, A. J. M., Latham, J., Weiss, R. I., and Grimmond, S. (1999). A partly habituated community of dry-habitat chimpanzees in the Semliki Valley Wildlife Reserve, Uganda. *Am. J. Phys. Anthropol. Suppl.* 28:157.

Hunt, K. D., and McGrew, W. (2002). Chimpanzees in the dry habitats of Assirik, Senegal and Semliki Wildlife Reserve, Uganda. In: Boesch, C., Hohmann, G., and Marchant, L. (eds.), *Behavioural Diversity in Chimpanzees and Bonobos*. Cambridge University Press, Cambridge. pp. 35–51.

Idani, G. (1990). Relations between unit-groups of bonobos at Wamba, Zaire: encounters and temporary fusions. *Afr. Study Monogr.* 11:153–186.

Idani, G. (1991). Social relationships between immigrant and resident bonobo (*Pan paniscus*) females at Wamba. *Folia Primatol.* 57:83–95.

Ihobe, H. (1990). Interspecific interactions between wild pygmy chimpanzees (*Pan paniscus*) and red colobus (*Colobus badius*). *Primates* 31:109–112.

Ingmanson, E. J. (1996). Tool-using behavior in wild *Pan paniscus*: social and ecological considerations. In: Russon, A. E., Bard, K. A., and Parker, S. T. (eds.), *Reaching into Thought*. Cambridge University Press, Cambridge. pp. 190–210.

Ingmanson, E. J. (1998). Reply to Stanford (1998) The social behavior of chimpanzees and bonobos—empirical evidence and shifting assumptions. *Curr. Anthropol.* 39:409–410.

Isabirye-Basuta, G. (1989). Feeding ecology of chimpanzees in the Kibale Forest, Uganda. In: Heltne, P., and Marquardt, L. (eds.), *Understanding Chimpanzees*. Harvard University Press, Cambridge. pp. 116–127.

Jensen-Seaman, M. I., and Li, W. H. (2003). Evolution of the hominoid semenogelin genes, the major proteins of ejaculated semen. *J. Mol. Evol.* 57:261–270.

Jungers, W. L., and Susman, R. L. (1984). Body size and skeletal allometry in African apes. In: Susman, R. L. (ed.), *The Pygmy Chimpanzee: Evolutionary Biology and Behavior*. Plenum, New York. pp. 131–177.

Kaessmann, H., Wiebe, V., and Paabo, S. (1999). Extensive nuclear DNA sequence diversity among chimpanzees. *Science* 286:1159–1162.

Kano, T. (1971). The chimpanzees of Filabanga, western Tanzania. *Primates* 12:229–246.

Kano, T. (1980). Social behavior of wild pygmy chimpanzees *Pan paniscus* of Wamba: a preliminary report. *J. Hum. Evol.* 9:243–260.

Kano, T. (1982). The social group of pygmy chimpanzees (*Pan paniscus*) of Wamba. *Primates* 23:171–188.

Kano, T. (1989). The sexual behavior of pygmy chimpanzees. In: Heltne, P. G., and Marquardt, L. (eds.), *Understanding Chimpanzees*. Harvard University Press, Cambridge, MA. pp. 176–183.

Kano, T. (1992). *The Last Ape: Pygmy Chimpanzee Behavior and Ecology*. Stanford University Press, Stanford, CA.

Kano, T. (1996). Male rank order and copulation rate in a unit-group of bonobos at Wamba, Zaire. In: McGrew, W. C., Marchant, L. F., and Nishida, L. (eds.), *Great Ape Societies*. Cambridge University Press, Cambridge. pp. 135–145.

Kano, T. (1998). Reply to Stanford (1998) The social behavior of chimpanzees and bonobos—empirical evidence and shifting assumptions. *Curr. Anthropol.* 39:410–411.

Kano, T., and Mulavwa, M. (1984). Feeding ecology of *Pan paniscus* at Wamba. In: Susman, R. L. (ed.), *The Pygmy Chimpanzee: Evolutionary Biology and Behavior*. Plenum, New York. pp. 233–274.

Kawanaka, K. (1981). Infanticide and cannibalism in chimpanzees, with special reference to the newly observed case in the Mahale Mountains. *Afr. Study Monogr.* 1:69–99.

Kawanaka, K. (1984). Association, ranging, and the social unit in chimpanzees of the Mahale Mountains, Tanzania. *Int. J. Primatol.* 5:411–434.

Kitopeni, R., Kasagula, M., and Turner, L. A. (1995). Ntologi falls? *Pan Afr. News* 2:9–11.

Kuroda, S. (1979). Grouping of the pygmy chimpanzees. *Primates* 20:161–183.

Kuroda, S. (1980). Social behavior of the pygmy chimpanzees. *Primates* 21:181–197.

Kuroda, S. (1992). Ecological interspecies relationships between gorillas and chimpanzees in the Ndoki-Nouabale Reserve, northern Congo. In: Itoigawa, N., Sugiyama, Y., Sackett, G. P., and Thompson, R. K. R. (eds.), *Topics in Primatology. Behavior, Ecology, and Conservation*, vol. 2. Tokyo University Press, Tokyo. pp. 385–394.

Kuroda, S., Nishihara, T., Suzuki, S., and Oko, R. A. (1996). Sympatric chimpanzees and gorillas in the Ndoki Forest, Congo. In: McGrew, M. C., Marchant, L. F., and Nishida, T. (eds.), *Great Ape Societies*. Cambridge University Press, Cambridge. pp. 71–81.

Kutsukake, N., and Matsusaka, T. (2002). Incident of intense aggression by chimpanzees against an infant from another group in Mahale Mountains National Park, Tanzania. *Am. J. Primatol.* 58:175–180.

Lehmann, J., and Boesch, C. (2003). Social influences on ranging patterns among chimpanzees (*Pan troglodytes verus*) in the Taï National Park, Cote d'Ivoire. *Behav. Ecol.* 14:642–649.

Lehmann, J., and Boesch, C. (2004). To fission or to fusion: effects of community size on wild chimpanzee (*Pan troglodytes verus*) social organization. *Behav. Ecol. Sociobiol.* 56:207–216.

Leigh, S. R., and Shea, B. T. (1995). Ontogeny and the evolution of adult body size dimorphism in apes. *Am. J. Primatol.* 36:37–60.

Malenky, R. (1990). Ecological factors affecting food choice and social organization in *Pan paniscus* [PhD diss.]. State University of New York, Stony Brook.

Malenky, R. K., and Stiles, E. W. (1991). Distribution of terrestrial herbaceous vegetation and its consumption by *Pan paniscus* in the Lomako Forest, Zaire. *Am. J. Primatol.* 23:153–169.

Malenky, R. K., and Wrangham, R. W. (1994). A quantitative comparison of terrestrial herbaceous food-consumption by *Pan paniscus* in the Lomako Forest, Zaire, and *Pan troglodytes* in the Kibale Forest, Uganda. *Am. J. Primatol.* 32:1–12.

Manson, J. H., and Wrangham, R. W. (1991). Intergroup aggression in chimpanzees and humans. *Curr. Anthropol.* 32:369–390.

Matsumoto-Oda, A. (2002). Social relationships between cycling females and adult males in Mahale chimpanzees. In: Boesch, C., Hohmann, G., and Marchant, L. (eds.), *Behavioural Diversity in Chimpanzees and Bonobos*. Cambridge University Press, Cambridge. pp. 168–180.

Matsumoto-Oda, A., and Hayashi, Y. (1999). Nutritional aspects of fruit choice by chimpanzees. *Folia Primatol.* 70:154–162.

Matsumoto-Oda, A., Hosaka, K., Huffman, M. A., and Kawanaka, K. (1998). Factors affecting party size in chimpanzees of the Mahale Mountains. *Int. J. Primatol.* 19:999–1011.

Matsumoto-Oda, A., and Oda, R. (1998). Changes in the activity budget of cycling female chimpanzees. *Am. J. Primatol.* 46:157–166.

McGrew, W. (1992). *Chimpanzee Material Culture: Implications for Human Evolution*. Cambridge University Press, Cambridge.

McGrew, W. (1998). Reply to Stanford (1998) The social behavior of chimpanzees and bonobos—empirical evidence and shifting assumptions. *Curr. Anthropol.* 39:411.

McGrew, W., Baldwin, P. J., and Tutin, C. E. (1979). Chimpanzees, tools and termites: cross-cultural comparisons of Senegal, Tanzania and Rio Muni. *Man* 14:185–214.

McGrew, W., Baldwin, P., and Tutin, C. E. G. (1981). Chimpanzees in a hot, dry and open habitat, Mt. Assirik, Senegal, West Africa. *J. Hum. Evol.* 10:227–244.

McGrew, W., Baldwin, P. J., and Tutin, C. E. (1988). Diet of wild chimpanzees (*Pan troglodytes verus*) at Mt. Assirik, Senegal: I. Composition. *Am. J. Primatol.* 16:213–226.

McGrew, W. C., Ensminger, A. L., Marchant, L., Pruetz, J., and Vigilant, L. (2004). Genotyping aids field study of unhabituated wild chimpanzees. *Am. J. Primatol.* 63:87–93.

McGrew, W. C., Marchant, L., and Nishida, T. (1996). *Great Ape Societies*. Cambridge University Press, Cambridge.

Mitani, J., and Watts, D. (2001). Why do chimpanzees hunt and share meat? *Am. J. Phys. Anthropol.* 109.

Mitani, J., Watts, D., and Lwanga, J. S. (2002a). Ecological and social correlates of chimpanzee party size and composition. In: Boesch, C., Hohmann, G., and Marchant, L. F. (eds.), *Behavioural Diversity in Chimpanzees and Bonobos*. Cambridge University Press, Cambridge. pp. 102–111.

Mitani, J. C., Merriwether, D. A., and Zhang, C. (2000). Male affiliation, cooperation and kinship in wild chimpanzees. *Anim. Behav.* 59:885–893.

Mitani, J. C., Watts, D. P., and Muller, M. N. (2002b). Recent developments in the study of wild chimpanzee behavior. *Evol. Anthropol.* 11:9–25.

Mitani, J. C., Watts, D. P., Pepper, J. W., and Merriwether, D. A. (2002c). Demographic and social constraints on male chimpanzee behaviour. *Anim. Behav.* 64:727–737.

Moor-Jankowski, J., Wiener, A. S., Socha, W. W., Gordon, E. B., and Mortelmans, J. (1972). Blood groups of the dwarf chimpanzee (*Pan paniscus*). *J. Med. Primatol.* 1:90–101.

Moor-Jankowski, J., Wiener, A. S., Socha, W. W., Gordon, E. B., Mortelmans, J., and Sedgwick, C. J. (1975). Blood groups of pygmy chimpanzees (*Pan paniscus*): human-type and simian-type. *J. Med. Primatol.* 4:262–267.

Moore, J. (1996). Savanna chimpanzees, referential models and the last common ancestor. In: McGrew, W. C., Marchant, L., and Nishida, T. (eds.), *Great Ape Societies*. Cambridge University Press, Cambridge. pp. 275–292.

Morgan, D., and Sanz, C. (2003). Naive encounters with chimpanzees of the Gouaougo Triangle, Republic of Congo. *Int. J. Primatol.* 24:369–381.

Morin, P. A. (1994). Decoding chimp genes and lives. *Science* 265:1172–1173.

Muller, M. N. (2002). Agonistic relations among Kanyawara chimpanzees. In: Boesch, C., Hohmann, G., and Marchant, L. (eds.), *Behavioural Diversity in Chimpanzees and Bonobos*. Cambridge University Press, Cambridge. pp. 112–123.

Muller, M., and Mitani, J. C. (in press). Conflict and cooperation in wild chimpanzees. In: Slater, P. J. B., Rosenblatt, J., Snowdon, C., Roper, T., and Naguib, M. (eds.), *Advances in the Study of Behavior*. Academic Press, New York.

Muller, M. N., and Wrangham, R. (2001). The reproductive ecology of male hominoids. In: Ellison, P. (ed.), *Reproductive Ecology and Human Evolution*. Aldine de Gruyter, New York. pp. 397–427.

Muller, M. N., and Wrangham, R. W. (2004a). Dominance, aggression and testosterone in wild chimpanzees: a test of the "challenge hypothesis." *Anim. Behav.* 67:113–123.

Muller, M. N., and Wrangham, R. W. (2004b). Dominance, cortisol and stress in wild chimpanzees (*Pan troglodytes schweinfurthii*). *Behav. Ecol. Sociobiol.* 55:332–340.

Myers Thompson, J. (2002). Bonobos of the Lukuru Wildlife Research Project. In: Boesch, C., Hohmann, G., and Marchant, L. (eds.), *Behavioural Diversity in Chimpanzees and Bonobos*. Cambridge University Press, Cambridge. pp. 61–70.

Newton-Fisher, N. E. (1999a). The diet of chimpanzees in the Budongo Forest Reserve, Uganda. *Afr. J. Ecol.* 37:344–354.

Newton-Fisher, N. E. (1999b). Infant killers of Budongo. *Folia Primatol.* 70:167–169.

Newton-Fisher, N. E. (2002a). Relationships of male chimpanzees in the Budongo Forest, Uganda. In: Boesch, C., Hohmann, G., and Marchant, L. F. (eds.), *Behavioural Diversity in Chimpanzees and Bonobos*. Cambridge University Press, Cambridge. pp. 125–137.

Newton-Fisher, N. E. (2002b). Ranging patterns of male chimpanzees in the Budongo Forest, Uganda: range structure and individual differences. In: Harcourt, C. S., and Sherwood, B.

(eds.), *New Perspectives in Primate Evolution and Behaviour.* Westbury Academic & Scientific Publishing, London. pp. 287–308.

Newton-Fisher, N. E. (2003). The home range of the Sonso community of chimpanzees from the Budongo Forest, Uganda. *Afr. J. Ecol.* 41:150–156.

Newton-Fisher, N. E., Reynolds, V., and Plumptre, A. (2000). Food supply and chimpanzee (*Pan troglodytes schweinfurthii*) in the Budongo Forest Reserve, Uganda. *Int. J. Primatol.* 21:615–628.

Nishida, T. (1968). The social group of wild chimpanzees in the Mahale Mountains. *Primates* 9:167–224.

Nishida, T. (1979). The social structure of chimpanzees of the Mahale Mountains. In: Hamburg, D. A., and McCown, E. R. (eds.), *The Great Apes.* Benjamin/Cummings, Menlo Park, CA. pp. 72–121.

Nishida, T. (1983). Alpha status and agonistic alliance in wild chimpanzees (*Pan troglogytes schweinfurthii*). *Primates* 24:318–336.

Nishida, T. (1989). Social interactions between resident and immigrant female chimpanzees. In: Heltne, P. G., and Marquardt, L. A. (eds.), *Understanding Chimpanzees.* Harvard University Press, Cambridge, MA. pp. 68–89.

Nishida, T. (1990). A quarter century of research in the Mahale Mountains: an overview. In: Nishida, T. (ed.), *The Chimpanzees of the Mahale Mountains: Sexual and Life History Strategies.* University of Tokyo Press, Tokyo. pp. 3–36.

Nishida, T. (1997). Sexual behavior of adult male chimpanzees of the Mahale Mountains National Park, Tanzania. *Primates* 38:379–398.

Nishida, T., Corp, N., Hamai, M., Hasegawa, T., Hiraiwa-Hasegawa, M., Hosaka, K., Hunt, K. D., Itoh, N., Kawanaka, K., Matsumoto-Oda, A., Mitani, J. C., Nakamura, M., Norikoshi, K., Sakamaki, T., Turner, L., Uehara, S., and Zamma, K. (2003). Demography, female life history, and reproductive profiles among the chimpanzees of Mahale. *Am. J. Primatol.* 59:99–121.

Nishida, T., and Hiraiwa-Hasegawa, M. (1987). Chimpanzees and bonobos: cooperative relationships among males. In: Smuts, B. B., Cheney, D. L., Seyfarth, R. M., Wrangham, R. W., and Struhsacker, T. T. (eds.), *Primate Societies.* University of Chicago Press, Chicago.

Nishida, T., Hiraiwa-Hasegawa, M., Hasegawa, T., and Takahata, Y. (1985). Group extinction and female transfer in wild chimpanzees in the Mahale Mountains. *Z. Tierpsychol.* 67:284–301.

Nishida, T., and Hosaka, K. (1996). Coalition strategies among adult male chimpanzees of the Mahale Mountains, Tanzania. In: McGrew, W. C., Marchant, L. F., and Nishida, T. (eds.), *Great Ape Societies.* Cambridge University Press, Cambridge. pp. 114–134.

Nishida, T., and Kawanaka, K. (1972). Inter-unit-group relationships among wild chimpanzees of the Mahale Mountains. *Kyoto Univ. Afr. Stud.* 7:131–169.

Nishida, T., Takasaki, H., and Takahata, Y. (1990). Demography and reproductive profiles. In: Nishida, T. (ed.), *The Chimpanzees of the Mahale Mountains: Sexual and Life History Strategies.* University of Tokyo Press, Tokyo. pp. 63–97.

Nishida, T., and Uehara, S. (1983). Natural diet of chimpanzees (*Pan troglodytes schweinfurthii*): long-term record from the Mahale Mountains, Tanzania. *Afr. Study Monogr.* 3:109–130.

Nishida, T., Uehara, S., and Nyundo, R. (1979). Predatory behavior among wild chimpanzees of the Mahale Mountains. *Primates* 20:1–20.

Nishida, T., and Van Hooff, J. (2002). UNESCO World Heritage status for the great apes. Caring for Primates. Abstracts of the XIXth Congress, International Primatological Society, Beijing, Mammalogical Society of China, August, 2002. p. 194.

Nunn, C. L. (1999). The evolution of exaggerated sexual swellings in primates and the graded-signal hypothesis. *Anim. Behav.* 58:229–246.

Parish, A. R. (1994). Sex and food control in the uncommon chimpanzee: how bonobo females overcome a phylogenetic legacy of male-dominance. *Ethol. Sociobiol.* 15:157–179.

Parish, A. R. (1996). Female relationships in bonobos (*Pan paniscus*)—evidence for bonding, cooperation, and female dominance in a male-philopatric species. *Hum. Nat.* 7:61–96.

Parish, A. R. (1998). Reply to Stanford (1998) The social behavior of chimpanzees and bonobos—empirical evidence and shifting assumptions. *Curr. Anthropol.* 39:413–414.

Pepper, J., Mitani, J., and Watts, D. (1999). General gregariousness and specific social preferences among wild chimpanzees. *Int. J. Primatol.* 20:613–632.

Peters, C., and O'Brien, E. (1981). The early hominid plant-food niche: insights from an analysis of plant exploitation by Homo, Pan, and Papio in eastern and southern Africa. *Curr. Anthropol.* 22:127–140.

Pruetz, J. D. (2004). Feeding and Ranging Ecology of Chimpanzees in an Open Environment at Fongoli, Senegal. Feeding Ecology Conference, Max Planck Institute for Evolutionary Anthropology, Leipzig, Germany, Oct. 2004.

Pruetz, J. D., Marchant, L. F., Arno, J., and McGrew, W. C. (2002). Survey of savanna chimpanzees (*Pan troglodytes verus*) in southeastern Senegal. *Am. J. Primatol.* 58:35–43.

Pusey, A. E. (1979). Inter-community transfer of chimpanzees in Gombe National Park. In: Hamburg, D. A., and McCown, E. R. (eds.), *The Great Apes.* Benjamin/Cummings, Menlo Park, CA. pp. 465–479.

Pusey, A. E. (1980). Inbreeding avoidance in chimpanzees. *Anim. Behav.* 28:543–582.

Pusey, A. E. (1990). Behavioural changes at adolescence in chimpanzees. *Behaviour* 115:203–246.

Pusey, A. E. (2001). Of genes and apes: chimpanzee social organization and reproduction. In: de Waal, F. (ed.), *The Tree of Origin.* Harvard University Press, Cambridge, MA. pp. 9–37.

Pusey, A. E., Oehlert, G., Williams, J., and Goodall, J. (2005). Influence of ecological and social factors on body mass of wild chimpanzees. *Int. J. Primatol.* 26:3–31.

Pusey, A. E., Williams, J., and Goodall, J. (1997). The influence of dominance rank on the reproductive success of female chimpanzees. *Science* 277:828–831.

Reichert, K. E., Heistermann, M., Hodges, J. K., Boesch, C., and Hohmann, G. (2002). What females tell males about their reproductive status: are morphological and behavioural cues reliable signals of ovulation in bonobos (*Pan paniscus*)? *Ethology* 108:583–600.

Reinartz, G. E., Karron, J. D., Phillips, R. B., and Weber, J. L. (2000). Patterns of microsatellite polymorphism in the range-restricted bonobo (*Pan paniscus*): considerations for interspecific comparison with chimpanzees (*P. troglodytes*). *Mol. Ecol.* 9:315–328.

Reynolds, V. (1992). Chimpanzees in the Budongo Forest, 1962–1992. *J. Zool.* 228:695–699.

Rodman, P. (1984). Foraging and social systems of orangutans and chimpanzees. In: Rodman, P., and Cant, J. (eds.), *Adaptations for Foraging in Nonhuman Primates.* Columbia University Press, New York. pp. 135–156.

Ruvolo, M. (1997). Molecular phylogeny of the hominoids: inferences from multiple independent DNA sequence data sets. *Mol. Biol. Evol.* 14:248–265.

Ruvolo, M., Pan, D., Zehr, S., Goldberg, T., Disotell, T. R., and von Dornum, M. (1994). Gene trees and hominoid phylogeny. *Proc. Natl. Acad. Sci. U.S.A.* 91:8900–8904.

Sabater-Pi, J. (1979). Feeding behaviour and diet of chimpanzees (*Pan troglodytes troglodytes*) in the Okorobik Mountains of Rio Muni (West Africa). *Z. Tierpsychol.* 50:265–281.

Sarich, V. M., Schmid, C. W., and Marks, J. (1989). DNA hybridization as a guide to phylogenies: a critical analysis. *Cladistics* 5:3–32.

Schwartz, E. (1929). Das vorkommen des Schimpansen auf den linken Kongo-Ufer. *Rev. Zool. Bot. Afr.* 16:425–433.

Sept, J. (1998). Shadows on a changing landscape: comparing nesting patterns of hominids and chimpanzees since their last common ancestor. *Am. J. Primatol.* 46:85–101.

Shea, B. T. (1984). An allometric perspective on the morphological and evolutionary relationships between pygmy (*Pan paniscus*) and common (*Pan troglodytes*) chimpanzees. In: Susman, R. L. (ed.), *The Pygmy Chimpanzee. Evolutionary Biology and Behavior.* Plenum, New York. pp. 89–130.

Shea, B. T., and Coolidge, H. J. (1988). Craniometric differentiation and systematics in the genus *Pan. J. Hum. Evol.* 17:671–685.

Shea, B. T., Leigh, S. R., and Groves, C. P. (1993). Multivariate craniometric variation in chimpanzees: implications for species identification in paleoanthropology. In: Kimbel, W. H., and Martin, L. B. (eds.), *Species, Species Concepts, and Primate Evolution.* Plenum, New York. pp. 265–296.

Sibley, C. G., and Alquist, J. E. (1984). The phylogeny of the hominid primates as indicated by DNA–DNA hybridization. *J. Mol. Evol.* 20:2–15.

Smith, R. J., and Jungers, W. L. (1997). Body mass in comparative primatology. *J. Hum. Evol.* 32:523–559.

Sommer, V., Adanu, J., Faucher, I., and Fowlera, A. (2004). Nigerian chimpanzees (*Pan troglodytes vellerosus*) at Gashaka: two years of habituation efforts. *Folia Primatol.* 75:295–316.

Stanford, C. B. (1998). The social behavior of chimpanzees and bonobos—empirical evidence and shifting assumptions. *Curr. Anthropol.* 39:399–420.

Stanford, C. B., Wallis, J., Matama, H., and Goodall, J. (1994a). Patterns of predation by chimpanzees on red colobus monkeys in Gombe-National-Park, 1982–1991. *Am. J. Phys. Anthropol.* 94:213–228.

Stanford, C. B., and Nkurunungi, J. B. (2003). Behavioral ecology of sympatric chimpanzees and gorillas in Bwindi Impenetrable National Park, Uganda: diet. *Int. J. Primatol.* 24:901–918.

Sterck, E. H. M., Watts, D. P., and van Schaik, C. P. (1997). The evolution of female social relationships in non-human primates. *Behav. Ecol. Sociobiol.* 41:291–309.

Stone, A. C., Griffiths, R. C., Zegura, S. L., and Hammer, M. F. (2002). High levels of Y-chromosome nucleotide diversity in the genus *Pan. Proc. Natl. Acad. Sci. U.S.A.* 99:43–48.

Strier, K. B. (1989). Effects of patch size on feeding associations in muriquis (*Brachyteles arachnoides*). *Folia Primatol.* 52:70–77.

Stumpf, R. M. (2004). Female reproductive strategies of chimpanzees of the Taï National Park, Cote d'Ivoire [PhD thesis]. State University of New York, Stony Brook.

Stumpf, R. M. (2005). Does promiscuity preclude choice? Female sexual strategies and mate preferences in chimpanzees of the Taï National Park, Cote d'Ivoire. *Behav. Ecol. Sociobiol.* 57:511–524.

Sugiyama, Y. (1988). Grooming interactions among adult chimpanzees at Bossou, Guinea, with special reference to social structure. *Int. J. Primatol.* 9:393–407.

Sugiyama, Y. (1989). Population dynamics of chimpanzees at Bossou, Guinea. In: Heltne, P. G., and Marquardt, L. G. (eds.), *Understanding Chimpanzees.* Harvard University Press, Cambridge, MA. pp. 134–145.

Sugiyama, Y. (1994). Age-specific birth rate and life time reproductive success of Chimpanzees at Bossou, Guinea. *Am. J. Primatol.* 32:311–318.

Sugiyama, Y. (1999). Socio-ecological factors of male chimpanzee migration at Bossou, Guinea. *Primates* 40:61–68.

Sugiyama, Y. (2004). Demographic parameters and life history of chimpanzees at Bossou, Guinea. *Am. J. Phys. Anthropol.* 124:154–165.

Sugiyama, Y., and Koman, J. (1979). Social structure and dynamics of wild chimpanzees at Bossou, Guinea. *Primates* 20:323–339.

Sugiyama, Y., and Koman, J. (1992). The flora of Bossou: its utilization by chimpanzees and humans. *Afr. Study Monogr.* 13:127–169.

Susman, R. L. (ed.) (1984). *The Pygmy Chimpanzee: Evolutionary Biology and Behavior.* Plenum, New York.

Sussman, R. W., Cheverud, J. R., and Bartlett, T. Q. (1995). Infant killing as an evolutionary strategy: reality or myth? *Evol. Anthropol.* 3:149–151.

Suzuki, A. (1969). An ecological study of chimpanzees in a savanna woodland. *Primates* 10:103–148.

Symington, M. M. (1988). Demography ranging patterns and activity budgets of black spider monkeys (*Ateles paniscus chamek*) in the Manu National Park Peru. *Am. J. Primatol.* 15:45–67.

Takahata, Y. (1985). Adult male chimpanzees kill and eat a male newborn infant: newly observed intragroup infanticide and cannibalism in Mahale National Park Tanzania. *Folia Primatol.* 44:161–170.

Takahata, Y. (1990). Social relationships among adult males. In: Nishida, T. (ed.), *The Chimpanzees of the Mahale Mountains: Sexual and Life History Strategies.* University of Tokyo Press, Tokyo. pp. 149–170.

Takahata, Y. (1995). A genetic perspective on the origin and history of humans. *Annu. Rev. Ecol. Syst.* 26:343–372.

Takahata, Y., Ihobe, H., and Idani, G. (1996). Comparing copulations of chimpanzees and bonobos: do females exhibit proceptivity or receptivity? In: McGrew, W. C., Marchant, L. F., and Nishida, T. (eds.), *Great Ape Societies.* Cambridge University Press, Cambridge. pp. 146–155.

Taylor, A. B., and Groves, C. P. (2003). Patterns of mandibular variation in *Pan* and *Gorilla* and implications for African ape taxonomy. *J. Hum. Evol.* 44:529–561.

Terborgh, J. W., and Janson, C. H. (1986). The socioecology of primate groups. *Annu. Rev. Ecol. Syst.* 17:111–135.

Thompson, J. (2002). Bonobos of the Lukuru Wildlife Research Project. In: Boesch, C., Hohmann, G., and Marchant, L. (eds.), *Behavioural Diversity in Chimpanzees and Bonobos.* Cambridge University Press, Cambridge. pp. 61–70.

Thompson-Handler, N., Malenky, R. K., and Badrian, N. (1984). Sexual behavior of *Pan paniscus* under natural conditions in the Lomako Forest, Zaire. In: Susman, R. L. (ed.), *The Pygmy Chimpanzee: Evolutionary Biology and Behavior*. Plenum, New York. pp. 347–368.

Thompson-Handler, N. E. (1990). The pygmy chimpanzee: sociosexual behaviour, reproductive biology and life history [PhD diss.]. Yale University, New Haven.

Tutin, C. E. G. (1979). Mating patterns and reproductive strategies in a community of wild chimpanzees (*Pan troglodytes schweinfurthii*). *Behav. Ecol. Sociobiol.* 6:29–38.

Tutin, C. E. G., and Fernandez, M. (1993). Composition of the diet of chimpanzees and comparisons with that of sympatric lowland gorillas in the Lope Reserve, Gabon. *Am. J. Primatol.* 30:195–211.

Tutin, C. E. G., and McGinnis, P. R. (1981). Chimpanzee reproduction in the wild. In: Graham, C. E. (ed.), *Reproductive Biology of the Great Apes*. Academic Press, New York. pp. 239–264.

Tutin, C. E. G., McGrew, W. C., and Baldwin, P. J. (1983). Social organization of savanna-dwelling chimpanzees, *Pan troglodytes verus* at Mt. Assirik, Senegal. *Primates* 24:154–173.

Tutin, C. E., Fernandes, M., Pierce, A. H., and Williamson, E. A. (1984). Foods consumed by sympatric populations of *Gorilla g. gorilla* and *Pan t. troglodytes* in Gabon. *Int. J. Primatol.* 6:27–43.

Tutin, C., Fernandez, M., Rogers, M., Williamson, E., and McGrew, W. (1991). Foraging profiles of sympatric lowland gorillas and chimpanzees in the Lope Reserve, Gabon. *Phil. Trans. R. Soc. Lond.* B334:179–186.

Tutin, C., Ham, R., White, L., and Harrison, M. (1997). The primate community of the Lope Reserve, Gabon: diets, responses to fruit scarcity, and effects on biomass. *Am. J. Primatol.* 42:1–24.

Uehara, S. (1990). Utilization patterns of a marsh grassland within the tropical rain-forest by the bonobos (*Pan-paniscus*) of Yalosidi, Republic of Zaire. *Primates* 31:311–322.

Uehara, S., and Nishida, T. (1987). Body weights of wild chimpanzees in the Mahale Mountains National Park, Tanzania. *Am. J. Phys. Anthropol.* 72:315–321.

van Lawick-Goodall, J. (1968). The behaviour of free-living chimpanzees in the Gombe Stream Reserve. *Anim. Behav. Monogr.* 1:165–311.

van Schaik, C. P. (1983). Why are diurnal primates living in groups? *Behaviour* 87:120–144.

van Schaik, C. P. (1989). The ecology of social relationships amongst female primates. In: Standen, V., and Foley, R. A. (eds.), *Comparative Socioecology*. Blackwell, Oxford. pp. 195–218.

van Schaik, C. P. (1996). Social evolution in primates: the role of ecological factors on male behavior. In: Runciman, W. G., Smith, J. M., and Dunbar, R. I. M. (eds.), *Evolution of Social Behaviour Patterns in Primates and Man*. Oxford University Press, Oxford. pp. 9–31.

Vervaecke, H., De Vries, H., and Van Elsacker, L. (1999). An experimental evaluation of the consistency of competitive ability and agonistic dominance in different social contexts in captive bonobos. *Behaviour* 136:423–442.

Vervaecke, H., Stevens, J., and Van Elsacker, L. (2003). Interfering with others: female–female reproductive competition in *Pan paniscus*. In: Jones, C. P. (ed.), *Sexual Selection and Reproductive Competition in Primates: New Perspectives and Directions*. American Society of Primatologists, Norman, OK. pp. 231–253.

Vigilant, L., Hofreiter, M., Siedel, H., and Boesch, C. (2001). Paternity and relatedness in wild chimpanzee communities. *Proc. Natl. Acad. Sci. U.S.A.* 98:12890–12895.

Vrba, E. (1995). The fossil record of African antelopes (Mammalia, Bovidae) in relation to human evolution and paleoclimate. In: Vrba, E., Denton, G., Burckle, L., and Partridge, T. (eds.), *Paleoclimate and Evolution with Emphasis on Human Origins*. Yale University Press, New Haven. pp. 385–424.

Wallis, J. (1992). Socioenvironmental effects on timing of first postpartum cycles in chimpanzees. In: Nishida, T., McGrew, W. C., Marler, P., Pickford, M., and de Waal, F. M. (eds.), *Topics in Primatology*. University of Tokyo Press, Tokyo. pp. 119–130.

Wallis, J. (1995). Seasonal influence on reproduction in chimpanzees of Gombe National Park. *Int. J. Primatol.* 16:435–451.

Wallis, J. (1997). A survey of reproductive parameters in the free-ranging chimpanzees of Gombe National Park. *J. Reprod. Fertil.* 109:297–307.

Wallis, J. (2002). Seasonal aspects of reproduction and sexual behavior in two chimpanzee populations: a comparison of Gombe (Tanzania) and Budongo (Uganda). In: Boesch, C., Hohmann, G., and Marchant, L. (eds.), *Behavioural Diversity in Chimpanzees and Bonobos*. Cambridge University Press, Cambridge. pp. 181–191.

Walsh, P. D., Abernethy, K. A., Bermejo, M., Beyersk, R., De Wachter, P., Akou, M. E., Huljbregis, B., Mambounga, D. I., Toham, A. K., Kilbourn, A. M., Lahm, S. A., Latour, S., Maisels, F., Mbina, C., Mihindou, Y., Obiang, S. N., Effa, E. N., Starkey, M. P., Telfer, P., Thibault, M., Tutin, C. E. G., White, L. J. T., and Wilkie, D. S. (2003). Catastrophic ape decline in western Equatorial Africa. *Nature* 422:611–614.

Watts, D. P. (1998). Coalitionary mate guarding by male chimpanzees at Ngogo, Kibale national Park, Uganda. *Behav. Ecol. Sociobiol.* 44:43–55.

Watts, D. P. (2000a). Grooming between male chimpanzees at Ngogo, Kibale National Park. I. Partner number and diversity and grooming reciprocity. *Int. J. Primatol.* 21:189–210.

Watts, D. P. (2000b). Grooming between male chimpanzees at Ngogo, Kibale National Park. II. Influence of male rank and possible competition for partners. *Int. J. Primatol.* 21:211–238.

Watts, D. P. (2002). Interchange of grooming and coalitionary support by wild male chimpanzees. *Am. J. Phys. Anthropol. Suppl.* 34:162.

Watts, D. P. (2004). Intracommunity coalitionary killing of an adult male chimpanzee at Ngogo, Kibale National Park, Uganda. *Int. J. Primatol.* 25:507–521.

Watts, D. P., and Mitani, J. C. (2000). Infanticide and cannibalism by male chimpanzees at Ngogo, Kibale National Park, Uganda. *Primates* 41:357–365.

Watts, D. P., and Mitani, J. C. (2001). Boundary patrols and intergroup encounters in wild chimpanzees. *Behaviour* 138:299–327.

Watts, D. P., and Mitani, J. C. (2002a). Hunting behavior of chimpanzees at Ngogo, Kibale National Park, Uganda. *Int. J. Primatol.* 23:1–28.

White, F. J. (1988). Party composition and dynamics in *Pan paniscus*. *Int. J. Primatol.* 9:179–193.

White, F. J. (1992). Pygmy chimpanzee social organization: variation with party size and between study sites. *Am. J. Primatol.* 26:203–214.

White, F. J. (1996a). Comparative socio-ecology of *Pan paniscus*. In: McGrew, W. C., Marchant, L. F., and Nishida, T. (eds.), *Great Ape Societies*. Cambridge University Press, Cambridge. pp. 29–41.

White, F. J. (1996b). *Pan paniscus* 1973 to 1996: twenty-three years of field research. *Evol. Anthropol.* 5:11–17.

White, F. J., and Lanjouw, A. (1992). Feeding competition in Lomako bonobos: variation in social cohesion. *Topics Primatol.* 1:67–79.

White, F. J., and Wrangham, R. W. (1988). Feeding competition and patch size in the chimpanzee species, *Pan paniscus* and *Pan troglodytes*. *Behaviour* 105:148–163.

Williams, J. (1999). Female strategies and the reasons for territoriality: lessons from three decades of research at Gombe [PhD diss.]. University of Minnesota, Minneapolis.

Williams, J., Liu, H.-Y., and Pusey, A. (2002a). Costs and benefits of grouping for female chimpanzees at Gombe. In: Boesch, C., Hohmann, G., and Marchant, L. (eds.), *Behavioural Diversity in Chimpanzees and Bonobos*. Cambridge University Press, Cambridge. pp. 192–203.

Williams, J., Oehlert, G., Carlis, J. V., and Pusey, A. (2004). Why do male chimpanzees defend a group range? *Anim. Behav.* 68:523–532.

Williams, J. M., Pusey, A. E., Carlis, J. V., Farms, B. P., and Goodall, J. (2002b). Female competition and male territorial behaviour influence female chimpanzees' ranging patterns. *Anim. Behav.* 63:347–360.

Wilson, M. L. (2001). Imbalances of power: how chimpanzees respond to the threat of intergroup aggression (*Pan troglodytes*) [PhD diss.]. Harvard University, Cambridge, MA.

Wilson, M. L., Hauser, M. D., and Wrangham, R. W. (2001). Does participation in inter-group conflict depend on numerical assessment, range location, or rank for wild chimpanzees? *Anim. Behav.* 61:1203–1216.

Wilson, M. L., Wallauer, W. R., and Pusey, A. E. (2004). New cases of inter-group violence among chimpanzees in Gombe National Park, Tanzania. *Int. J. Primatol.* 25:523–549.

Wilson, M. L., and Wrangham, R. W. (2003). Intergroup relations in chimpanzees. *Annu. Rev. Anthropol.* 32:363–392.

Wittig, R. M., and Boesch, C. (2003). Food competition and linear dominance hierarchy among female chimpanzees of the Taï National Park. *Int. J. Primatol.* 24:847–867.

Wrangham, R. (2000). Why are male chimpanzees more gregarious than mothers? A scramble competition hypothesis. In: Kappeler, P. M. (ed.), *Primate Males*. Cambridge University Press, Cambridge. pp. 248–258.

Wrangham, R. (2002). The cost of sexual attraction: is there a trade-off in female *Pan* between sex appeal and received coercion? In: Boesch, C., Hohmann, G., and Marchant, L. (eds.), *Behavioural Diversity in Chimpanzees and Bonobos*. Cambridge University Press, Cambridge. pp. 204–215.

Wrangham, R. W. (1977). Feeding behavior of chimpanzees in Gombe National Park, Tanzania. In: Clutton-Brock, T. H. (ed.), *Primate Ecology*. London: Academic Press. pp. 504–538.

Wrangham, R. W. (1979a). Sex differences in chimpanzee dispersion. In: Hamburg, D. A., and McCown, E. R. (eds.), *The Great Apes*. Benjamin/Cummings, Menlo Park, CA. pp. 481–489.

Wrangham, R. W. (1979b). On the evolution of ape social systems. *Soc. Sci. Inf.* 18:335–368.

Wrangham, R. W. (1980). An ecological model of female-bonded primate groups. *Behaviour* 75:262–300.

Wrangham, R. W. (1986). Ecology and social relationships in two species of chimpanzee. In: Rubenstein, D. I., and Wrangham, R. W. (eds.), *Ecological Aspects of Social Evolution: Birds and Mammals*. Princeton University Press, Princeton, NJ. pp. 352–378.

Wrangham, R. W. (1993). The evolution of sexuality in chimpanzees and bonobos. *Hum. Nat.* 4:47–79.

Wrangham, R. W. (1999). Evolution of coalitionary killing. *Ybk. Phys. Anthropol.* 42:1–30.

Wrangham, R. W., Chapman, C. A., Clark-Arcadi, A. P., and Isabirye-Basuta, G. (1996). Social ecology of Kanyawara chimpanzees: implications for understanding the costs of great ape groups. In: McGrew, W. C., Marchant, L., and Nishida, T. (eds.), *Great Ape Societies*. Cambridge University Press, Cambridge. pp. 45–57.

Wrangham, R. W., Clark, A. P., and Isabirye-Basuta, G. (1992). Female social relationships and social organization of Kibale Forest chimpanzees. In: Nishida, T., McGrew, W. C., Marler, P., Pickford, M., and de Waal, F. B. M. (eds.), *Topics in Primatology, Human Origins*. vol. I. University of Tokyo Press, Tokyo. pp. 81–98.

Wrangham, R. W., Conklin-Brittain, N. L., and Hunt, K. D. (1998). Dietary response of chimpanzees and cercopithecines to seasonal variation in fruit abundance: I. Antifeedants. *Int. J. Primatol.* 19:949–970.

Wrangham, R. W., McGrew, W., de Waal, F., and Heltne, P. G. (1994). *Chimpanzee Cultures*. Harvard University Press, Cambridge, MA.

Wrangham, R. W., and Peterson, D. (1996). *Demonic Males*. New York: Houghton Mifflin.

Wrangham, R. W., and Smuts, B. B. (1980). Sex difference in the behavioral ecology of chimpanzees in the Gombe National Park, Tanzania. *J. Reprod. Fertil. Suppl.* 28:13–31.

Yamigawa, J. (1999). Socioecologic factors influence population structure of gorillas and chimpanzees. *Primates* 40:87–101.

Yu, N., Jensen-Seaman, M. I., Chemnick, L., Kidd, J. R., Deinard, A. S., Ryder, O., Kidd, K. K., and Li, W. H. (2003). Low nucleotide diversity in chimpanzees and bonobos. *Genetics* 164:1511–1518.

Zihlman, A. (1984). Body build and tissue composition in *Pan paniscus* and *Pan troglodytes* with comparison to other hominoids. In: Susman, R. (ed.), *The Pygmy Chimpanzee*. Plenum, New York. pp. 179–200.

Zihlman, A. (1996). Reconstructions reconsidered: chimpanzee models and human evolution. In: McGrew, W. C., Marchant, L. F., and Nishida, T. (eds.), *Great Ape Societies*. Cambridge University Press, Cambridge. pp. 293–304.

Zihlman, A. L., and Cramer, D. L. (1978). A skeletal comparison between pygmy (*Pan paniscus*) and common chimpanzees (*Pan troglodytes*). *Folia Primatol.* 29:86–94.

PART THREE

Methods

20

Research Questions

Elsworth Ray

INTRODUCTION

In the beginning, there is the question. It all begins with the question.

Before research can be conducted, a question must be asked. Sometimes the research question is very basic or general in nature, particularly when little is known about a species. In this situation, primatologists seek basic information: which species are present in a region, what are the environmental conditions, what plants or animals are used as food resources, how many individuals live in the social groups, what are the sex ratios and age ranges? At other times, when basic ecological and social information about a species is available, primatologists ask more specific questions: how do primates respond to the stresses and seasonal changes in their environments, such as food shortages or temperature, or how do individuals integrate themselves into a new social group when they disperse from their natal group at maturity? Some research questions might require the collection of multiple types of data to gain an answer: what is the association between circulating cortisol levels and patterns of aggression, or what are the relationships between proceptive behaviors, estrogen, and food resources? Whatever the focus is, researchers need to create a well-thought-out question, or series of questions, as the research question establishes nearly everything about the research to be conducted.

The questions researchers pose are usually derived from prior research and theory. Previous research, having provided insights into the lives of nonhuman primates, establishes a foundation for new questions to gain a deeper understanding of nonhuman primate life. Theories of animal behavior, and the models derived from those theories, guide researchers as they investigate the processes which create and maintain the behavior of nonhuman primates. A thorough familiarization with the pertinent scientific literature is the starting point for all research. Chapais (1988), for example, investigated four differing hypotheses, obtained from previously published research, to determine which of these explanations best described the processes involved in the maintenance of intergenerational rank stability among macaque matrilines. It is important to keep in mind that theories, and the models derived from theoretical perspectives, are conceptualizations created by us, human observers. Although these conceptualizations may be logical and are very seldom capricious, they may not represent the actual processes functioning in specific social or ecological conditions. We need to continually ask new questions.

Each species is unique. Every species has evolved its own specific set of features and adaptations, enabling individuals to be successful in their environments, to grow, to maintain themselves, and to reproduce successfully. Each species has its own suite of sensory and cognitive mechanisms, ecological capabilities, and constraints. Although the cognitive system and sensory mechanisms of humans are very similar to those of other primate species, due to our shared evolutionary heritage, they are not identical. Many species of primates rely far more heavily on olfaction to gain information on their environments than we humans do. Our cognitive abilities are far beyond those of other primates. Our technological and symbolic world is far more complicated than are their worlds.

Our shared histories raise an important issue requiring attention when developing a research program and when analyzing the information gathered from research. There is a serious danger of *anthropomorphism*, the attribution of human characteristics to nonhuman entities. The features we share with other primate species, due to our common evolutionary heritage, allow us insight into the lives and actions of the other members of our mammalian order. However, they are others, alien beings, and we need to pay attention to the differences as well as the similarities. Despite the differences between ourselves and other primate species, it is possible to investigate other primates and to gain perspective into how they operate within their worlds. All that is needed is a good question requiring an answer and a carefully considered project designed to provide that answer. By questioning and testing new and existing notions, we advance our understanding of the world of which they, and we, are parts.

STUDY DESIGN

Once a research question is formulated, researchers design their studies considering all phases of a project from beginning to end. There are a series of steps in any research project: collecting data, establishing a formal hypothesis, testing the data, analyzing the outcome of the tests, and finally publishing the results of the study. Researchers design studies based upon the procedure(s) needed to gain a viable result, an answer to their question(s). In modern primatology, answers are determined by the rejection or acceptance of a null hypothesis. A *null hypothesis* is a statement refuting the formal hypothesis of the study, the explanation for the phenomenon under investigation. As research questions are, or become, formalized statistical hypotheses, testing and analysis cannot be ignored until some late stage in the project. During the testing phase of the research project, both the hypothesis and the null hypothesis will be transformed into mathematical formulae for testing. Consequently, they need to be very specific statements. It is important for researchers to consider testing procedures when determining the data collection techniques they will use. Otherwise, the data gathered may be inappropriate or inadequate for obtaining a valid result to the research question. Although each phase in a research project is distinct, they are pieces of an integrated whole.

The testing procedure establishes the standard for the amount of data required (sample size). The testing procedure also determines the characteristics and size of the sample population needed to obtain accurate results. A *sample population* is the assembly of individuals from which the data to be analyzed are drawn. The sample population needs to represent all constituents of the group under investigation, to avoid sampling bias. Biased samples invalidate the results of statistical tests as they do not represent the real world. Once researchers determine the testing procedure, they determine which method of data collection best provides the information needed.

Considerations of data collection include very basic issues, such as from which species the information should be gathered, which data collection technique should be used, or the research conditions that will best provide the data needed. Research conditions can affect the behavior and activities of the subject animals. Researchers need to consider whether the information is best gathered in laboratories, under captive conditions, from free-ranging populations, or in the field from naturally occurring primate groups. The primary factor in deciding between different data collection techniques is the type(s) of data required for testing the hypothesis once the data are collected. The data must be accurate and without bias. To compile accurate records, researchers must become proficient at the technique(s) they will use in collecting data and be able to habituate the subjects to the presence of the observer. Pilot studies can be very useful in determining the viability of different data collection techniques.

In the process of research design, researchers are usually aware of the conditions under which data will be gathered, the type of information to be collected, and the method(s) required for data analysis. For example, if the research question is "What is the correlation between dominance and reproductive success for males in a given species?" the data collected would be different from the data collected if the research question were "How do females establish and/or maintain matrilineal societies?" or "What are the responses of the animals to different predation threats, and how successful are these different responses?" Each question requires different kinds of information to obtain an answer.

The first question requires construction of dominance rankings by the researcher (in and of itself a thorny issue, see Bernstein 1981) and intergenerational data on paternity to show a correspondence between the constructed rankings and reproductive success. Statistical analysis of the data would require regression testing. The second question would require information on the interaction patterns of matriline members with each other and their interactions with individuals not of their matriline. To answer this question, researchers may well need to examine data sequentially, determine the responses to particular actions, and establish how time-dependent the responses are or need to be. The last question requires the gathering of information on predators themselves, their hunting techniques, and success rates. Alternatively, the answer may require analyzing the predator's scat as well as collecting information on the alarm calls given in response to the presence of predators, the flight patterns of the primate group, or the effectiveness of mobbing predators.

There are times when research is not driven by hypothesis testing. Sometimes, researchers gather basic information about a particular species and report their findings descriptively. These studies, general in their design, provide starting points for future research. The purpose of generalized research projects is to provide basic ecological and social information on species that have not been well studied. General studies continue to be important in primatology. We have very little information on many primate species. Often, our knowledge is limited to a single study group living in one locality. We have little understanding of how representative the information for that species is across its geographic distribution or the extent of variation displayed by the species in differing circumstances. Although these studies do not provide in-depth insights on the ecological or social dynamics of a primate species, they are essential for providing basic information regarding the ecological niche of a species. General studies are also useful in establishing conservation programs. Without basic information on the requirements of different primate species, efforts to establish and maintain conservation projects can be futile. Pressures on nonhuman primate habitats continue to mount in the face of the increasing demands of our growing human population. Demands for logging, agricultural fields, pasturage for grazing livestock, and cooking fuel are impacting and/or destroying the habitats of many primates around the globe.

As conducting research can impact the lives of the study subjects, another fundamental consideration required during research design is the well-being of the investigative subjects. The well-being of the research subjects not only requires deliberation in the design of any research project but also needs to be paramount throughout the research project. The research must, and is required by law in most, if not all, countries, incorporate consideration of the short-term and long-term effects upon the animals. This is true for field studies and for laboratory studies. As Williamson and Feistner (2003) have noted, simply habituating primates to the presence of observers can have serious consequences for the very individuals we wish to study. Many of the diseases experienced by people can be contracted by other species of primates due to our shared evolutionary histories. Our presence can transmit the diseases to the study groups with devastating effects. The trails we cut allow access into the forests, enabling others to come into contact with the study groups. The animals having become accustomed to the presence of humans in their environments can become easy prey for hunters, pet collectors, and the bushmeat trade. The Animal Behavior Society has periodically published guidelines (Dawkins and Gosling 1991). Become familiar with these and all regulations which have been established. Incorporate them into the research project. We primatologists gather a great deal of knowledge from our subjects, and we owe them the common decency of ethical consideration and treatment.

DATA COLLECTION

Two major and deeply integrated spheres of primatological research are primate ecology and primate sociality. Primates live in many types of environment: savannas, rain forests, deciduous forests, mountains, cities, etc. Primates also live in many types of social group with multiple factors contributing to the structure and maintenance of these groups. To understand the ecological requirements of primates and how primates use their immediate habitats to obtain their needs, researchers systematically study the environments of the animals. Ecological factors can influence the composition of primate social groups and the interaction patterns of the individual primates within groups as each individual acts to obtain its needs. The social dynamics of a primate group are not determined solely by ecological factors however, so researchers also examine the behavior of the individuals to understand the many variables affecting primate social life.

Estimating Population Densities

Information on population densities, habitat structure and composition, resource seasonality, and diet provides insight into the lives of our closest relatives. It is the most basic information primatologists must gather in order to gain such insight. This knowledge enables researchers to determine which resources are available in a habitat and the demands being placed on the resources available within a habitat. A common method of establishing reliable estimates of population density is the transect line survey (Whitesides et al. 1988). In this sampling system, researchers travel along mapped routes, counting the number of animals encountered.

Researchers also record the location, distance, and direction of the animals from the transect line. Other information collected during line surveys can include, if possible, the size, sexes, or ages of the individuals encountered. Initially, the ages of the animals are often recorded according to basic life history stages: infants, juveniles, and adults. The responses of the animals to the presence of the researchers are also recorded. Do they give alarm calls or flee from researchers, and if so, in what direction do they flee? Do they engage in cryptic behaviors, cautiously observing the researchers, or simply ignore them? Response information, as when the animals flee from the presence of the observers, can be important in determining the population density. Researchers can record individuals only once or the results of the census will be inaccurate. Vocalizations can be used also to determine the presence of some species in an area (Brockelman and Srikosamatara 1993).

Several formulae have been developed to estimate population density (Brower et al. 1990). Essentially, the data obtained during the census are converted to a ratio using the census numbers, the length of the transect route, and the distances of the animals from the transect line. This conversion provides an estimation of the population density in a given area. The reported standard of measurement is the number of individuals per square hectare. Transect sampling needs to be replicated several times and in different seasons to ensure accurate results and to find out what changes in population density may occur as animals adjust to the constantly changing availability of resources in their habitats.

To determine which of the several techniques that have been proposed for determining population densities from transect surveys provides the most accurate estimation of primate densities, Fashing and Cords (2000) compared data on four known primate populations and examined the alternative methods which have been proposed to determine population density. Their results indicate that the system developed by Whitesides et al. (1988), wherein the ranges of primate groups are treated as if shaped in circles, provided the most accurate estimation of the techniques tested.

Other techniques for obtaining population density estimates exist. Using a point survey system is a common alternative to conducting a transect survey. In this system, researchers go to specific predetermined locations within the research area. These sight locations, or points, are determined from a grid overlaid on a map of the research area to create adequate coverage of a region. At the site locations, observers spend a preestablished period of time (e.g., 30 min or 1 hr) counting all animals observed during the observation period and recording the distance(s) and direction(s) from the observer. Again, the census data are converted to a ratio, number of animals per unit of space, to estimate the population density. Other techniques used to census the population of animals in a region can include road counts, wherein researchers travel preestablished routes such as roads or fire trails and record the animals they observe. Trapping animals may be appropriate for some studies (Brower et al.

1990, Krebs 2000). Counting scat and/or resting nests to determine gorilla populations has a long history (Schaller 1963, Tutin et al. 1995) and may be a useful method for counting nonhabituated populations.

Habitat and Resources

As with determining population density, transect and plot surveys are basic systems for the study of primate habitat—its structure, composition, and available food resources. Researchers overlay a grid on an area map and randomly designate plots and or transect lines to census the plant life in a region. All plants within a plot are identified and located on a map (using the Global Positioning System may be very useful, see below). Within each plot, researchers label with an identification tag all plants whose trunks are greater than 10 cm in diameter at breast height. Researchers identify each plant species and collect samples from the plants to identify unknown species. Samples should, if possible, include flowers, fruit, or nuts as well as leaves to ensure proper identification of the plant species. All ecological samples should be preserved for identification, by experts, if necessary. Techniques for the preservation of plant, animal, or fecal samples can be found in many texts on ecological sampling.

As each species of plant possesses its own phenological cycle, it is important to regularly census sampling plots to record when the different species put out new leaves, flowers, and fruits. Researchers also record basic information on variables such as rainfall and temperature, which may affect the phenological cycles. This information can be important as the young leaves, flowers, fruits, and nuts of plants are primary food sources for most primate species. Insects and small vertebrates are also components of many species' diets. Throughout the study, researchers should collect samples of the food items in primate diets (Ganzhorn 2003). Whenever possible, the *refuse* (dropped or discarded portions of food items) should be gathered and the portions actually used by the primates should be recorded. Often, primates use only a portion or a specific part of a particular plant in their diet. This can be important information in understanding the ecological niche developed by each species.

Global Positioning System

Two technologies with growing importance to primatological research are the Global Positioning System (GPS) and Geographic Information Systems (GIS, discussed under Data Analysis, below), which are often used in conjunction with one another. GPS provides accurate location data. These data can be used to demarcate important locations in research studies. It is possible to precisely record the boundaries of home ranges, or core areas, allowing greater accuracy to studies interested in the ecological parameters of primates, their ranging patterns, and the dynamics of primate energy use. GPS data are of particular importance to effective conservation efforts. By compiling location data, researchers can demonstrate the areas most important to the animals and local peoples, accurately establishing those locations where protection is most vital.

GPS was developed, and is operated, by the United States Department of Defense. GPS uses signals sent between Earth-based GPS devices and satellites in low Earth orbit to triangulate location in space: longitude, latitude, and altitude. One issue of importance affecting the use of GPS is the degradation or blocking of the signal between transmitters and receivers (Phillips et al. 1998). Degraded signals can and do produce inaccurate data. For primatological research, one factor affecting signal transmission, and therefore affecting the applicability of GPS technology, is the need for a clear line of sight between the ground-based receiver and the satellites for the system to function properly. Local conditions require consideration. Research in forested regions or in sites with deep valleys can be difficult. The forest canopy scatters, or blocks, the transmission signal from being relayed to and from the satellites. Steep valley walls in mountainous terrain can block access to the satellites.

These problems are being dealt with as the technology develops. GPS devices using multiple channels are available. Some possible solutions, such as using collapsible or specialized antennas or using a base receiver as a geographical reference point coordinated to roving handheld devices, might be practical in some field conditions. Some suggestions are given in Johnston (1998) and Hughes (2003). Despite issues with obtaining accurate data under some conditions, GPS is an increasingly useful technology, adding to the tools of primatological research. One important feature of GPS is the ability to download the location data directly into GIS programs, which in turn are used for mapping and spatial analyses.

BEHAVIOR SAMPLING

Before it is possible to record the activities of their subject animals, researchers need to develop a standardized, well-defined repertoire of behaviors. This considered list of behavioral definitions is called an *ethogram*. Animals do many things, and to accurately record these actions, researchers must construct a well-defined inventory of their behavior before a study is conducted. This is a vital step in the process of research design, whether the research focus is ecological or social. The behavioral definitions should be as noninterpretive and non-idiosyncratic as possible.

A *behavior* can be defined as any activity an animal is engaged in doing. The activity can be solitary, such as simply sitting; interactive, as when one individual approaches another; or participatory, as when two or more individuals engage in a behavior together. Definitions should be descriptive and exclude, to the best of our abilities, inferences as to the motivations of the animals engaging in any particular action. To repeat, as this is very important for later analysis and interpretation, definitions of the behaviors should be neutral. They should be simple and precise statements of motor patterns (e.g., Bertrand 1969, Dolhinow 1978). By following these steps in defining the specific actions of the animals, researchers can most accurately record the activities in the animals' lives. The process of establishing well-defined behaviors results in data allowing researchers to make accurate comparisons between the activities from one site to another or between species. Ethograms allow researchers to establish a common "language" with which they can explicitly communicate with one another, avoiding confusion.

Since primates cannot communicate through the use of language, they use their bodies to communicate with one another. They also use vocalizations and other sensorial mechanisms, such as olfaction, in communication with one another. Paterson (2001) states that 90% of all behaviors can be identified within 100 hr of observation. This may be true. I would estimate the most commonly recorded behaviors fall into just a few basic categories: sit, rest, look at, locomote, and feed. However, it is often the case that less frequently occurring behaviors, such as reproductive or agonistic interactions, are extremely important to understanding the overall dynamics and patterns of primate life. While many behaviors may be self-evident, many others are subtle, simple variations of body posture. Because they are subtle, these behaviors can be difficult to distinguish from one another until a researcher gains experience observing a particular species or social group. For example, one common behavior many species of primates perform is called a *present*. A present occurs when one individual stands, directing its hindquarters toward another individual. Presents can occur in many social contexts. In many species, females use a slight variation of the basic present posture when inviting a male to mate, the hindlimbs might be held more rigidly, slightly elevating the female's hindquarters. By orientating the body differently, another variation of the present posture is an invitation to groom the presenting individual.

An important aspect of behavioral observation requiring consideration in the process of data recording and later, when compiling those data for an analysis of the types of behavior observed, is the context within which a behavior occurs. An individual may approach another individual for many reasons. It might seek to groom the individual it is approaching or to sit near that individual. It might also approach another with the intent of committing an agonistic act, such as displacing the other from a specific location in order to occupy that space itself. It could also approach another without any intent to interact and merely pass by as it moves to another locality. Researchers should not assume intent but simply note an approach and then record the interaction, if any, between the approaching individual and the receiver of the approach.

A part of the process of constructing an ethogram, or repertoire, for a study is the realization that not all research questions require the use of the full repertoire of a species' behaviors. If researchers are investigating a specific subset of behavioral patterns, they may establish, for the purpose of a specific study, a delimited or specialized ethogram. For example, mountain gorillas routinely feed on herbaceous plants which have evolved protective mechanical defenses such as thorns, hooks, casings, etc. To examine the techniques used by the gorillas to process these food items, Byrne and Byrne (1993) established six different categories of "skill elements." Within each skill category, between three and six distinct actions were identified. Specialized sets of behaviors like this allow researchers to examine in finer detail the variation and expression of a particular class or type of behavior.

Habituation and Recognition

Primates, by and large, are social animals living in groups of different sizes and composed of smaller social units such as matrilineal networks, coalitions, cohorts, and preferential relationships. These smaller social units reflect the social dynamics, kinship relationships, and particular histories of a group and its members. To be able to gather accurate data on social processes, researchers need to be able to distinguish the individuals of the social group from one another. This process, of accurately identifying the individuals of a group and subsequently recording their behaviors, takes time and effort on the part of researchers.

Individual recognition requires the habituation of the animals to the continued presence of researchers and the researchers' actions. The term *habituation* refers to the animals' acceptance of the researchers at a distance that allows the researchers to observe the primates' actions. The amount of time required to habituate primate groups is variable, depending upon a number of factors including the species being studied, the presence or absence of hunting or logging pressures on the animals, and the actions of the observers as they study the animals (Williamson and Feistner 2003). During the habituation process, researchers can and do gather important information on the animals. Often, this is the period during which initial ecological surveys on population densities and habitat composition are conducted and the identification of individuals occurs.

Until a researcher is able to identify the individuals of a group, it is not possible to accurately collect specific data on their personal interactions. It is also impossible to collect data in a systematic method. As a result, the data that are accrued will be biased. Some group members will be sampled frequently and others infrequently or not at all. Consequently, the information collected will be a distorted representation of the animals' activity patterns. The results from any analysis based upon biased data sets will be meaningless as the data used in the analysis are not representative of a normal population.

Observational Sampling Techniques

As the fields of animal behavior studies in general and primatology more specifically have developed, different techniques for collecting data on the behavior of animals have been devised. Researchers use different techniques to gather the information which allows them to examine primate social and ecological dynamics. In 1974, Jeanne Altmann published a landmark study addressing the advantages and problems of many sampling techniques. This paper remains not only influential but vital for students and researchers of animal behavior. It is not possible to reiterate her analysis in its entirety. I will, however, describe a few of the more common data collection techniques used to gather information on primate behavior. For more detailed descriptions of these and other techniques, the Altmann paper and various workbooks, such as those by Paterson (2001) and Martin and Bateson (1986), provide in-depth discussion.

No technique of data collection is perfect. Each method of data collection has advantages and drawbacks. It is important to remember this when designing a research project. Researchers choose the method most appropriate for their project. It should also be mentioned that there are software programs available for collecting data. These programs do not alter the fundamentals of data collection but can greatly ease the transformation of observational records into databases and spreadsheets for later analysis. I describe the basics of two common and important techniques herein, but be advised: these are not the only methods developed for the recording of behaviors. For novices, it is probably best to master these data collection techniques by hand, using paper and pencil, even if planning to use machines to gather data during the research project. I suggest this out of practical concerns. Machines and software malfunction and batteries go dead, often at the most inopportune moments. Proficiency in manual collection techniques allows researchers to circumvent what could be a disaster for a research project.

Instantaneous Scan Sampling

One of the most widely used, and simplest, methods for gathering information on primates is instantaneous scan sampling. In this system, the observer notes at preestablished time intervals the actions of the individuals of a primate group and/or the spatial relationships of the individuals to one another. This can be done either with a checklist or by entering the identities of individuals into the cells of a matrix. Scan sampling provides a coarse approximation of the activities or the spatial relationships of the research subjects.

Imagine the interval established for conducting each scan sample is 10 min. Then, every 10 min, the researcher enters information as to what activities the individuals of the group are engaged in at that moment: who is grooming whom, who is sleeping, or who is feeding. If spatial relationships are the focus of the study, the researcher would enter at each 10 min interval the distance between each individual from the others of the group, (e.g., 1, 5, or 20 m). Every hour, six sample sets consisting of one sample on every individual in the study group would be acquired. In a short period of time, a great deal of information would be collected. Scan samples provide a general estimation of the actions of primates. Researchers can establish basic patterns of daily life with these data. Is grooming a common activity? Do males and females socialize affiliatively within the group, or do they ignore one another? Do the juveniles play with all group members or interact only with the associates of their mothers?

It is important to note, however, that while this method provides a simple and direct system of data collection, it has a serious drawback. Scan sampling cannot provide information on the duration of behaviors. All behavioral actions must be treated as events (i.e., instantaneous occurrences) when the data are analyzed. Scan samples show only whether or not specific behaviors occur. Scan sample data cannot be transformed into rates of occurrence or establish the amount of time a given behavior takes, its *duration*. As data are collected at specific intervals, there is no information as to when behaviors begin or end. Scan samples consequently are inappropriate for use in answering questions requiring temporal analysis. Another drawback of scan sampling is that these data give limited, or inaccurate, information on behaviors that are rare or instantaneous. It takes but an instant for one primate to give a threat display to another. The odds of such a display occurring at the exact moment designated for recording a sample are low. Other limitations of this method include an inability to use the data for establishing the sequence of events or the responses of one individual to another's actions. For these kinds of data, different techniques are required.

Focal Animal Sampling

Focal animal sampling focuses upon the activities of a single individual at a time, giving the method its name. Over a predetermined period of time, the researcher documents all of the activities in which the focal animal is engaged. The researcher records what the focal animal does, as an actor or participant, and what happens to the focal animal. In addition, the researcher notes the activity of the focal animal when each minute changes. By noting these occurrences, a complete record of a single, "focal" individual's behavior during the sampling time interval is obtained. After each sample is collected, the researcher begins a new sample on a different individual, rotating sample collection through all members of the group. By following this procedure, a sample set is acquired consisting of one focal animal sample on each group member.

Focal animal sampling is the only technique that creates a record allowing analysis of both behavioral events and states of behavior (Altmann 1974). A *behavioral event* is simply any occurrence of a behavior, while a *state* refers to the persistence of a behavior for some duration. One major advantage of focal animal sampling is that it provides researchers with data to establish rates of the occurrence of behaviors and to analyze how much time is devoted to different behaviors.

The technique does have the disadvantage of forcing the observer to ignore activities occurring between other group members while the focal sample is being recorded. As samples are recorded for set intervals of time, it is not uncommon for behavioral episodes to continue after a sample is completed. One problem with this method arises when visibility is difficult. It is not unusual for the focal animal to disappear altogether, or for some time, from the observer's view. As researchers are not allowed to imagine the activities of the focal, the disappearance invalidates the sample. That sample must be rejected and a new one begun. Another drawback of focal animal sampling is that it is labor-intensive. Not only does the technique require considerable time to master, but it is also easy for an observer to make simple recording errors and invalidate a sample.

Supplemental Data: Field Notes and Anecdotes

As noted previously, systematic data collection requires observers to know and to be able to identify all group members in order to obtain accurate records of behavioral transactions. The process of learning the group requires time, and research time is valuable. As researchers obtain the ability to distinguish group members from one another, they are not inactive. They record a great deal of information on habitat and on the daily events affecting the study group, such as contact with other groups or predators, in their general field notes. Data gathered at will or outside of a preestablished methodology are referred to as *ad libitum notes*.

Important social or ecological events can and do occur outside of the established sampling intervals or are performed by animals which are not the focus of a sample. Animals are unconcerned with research projects. They conduct their lives regardless of sampling protocols. Consequently, researchers record important and/or rare interactions and events as they occur regardless of the sampling system. These records of events are reported and incorporated, as appropriate, within the researchers' larger analysis of the dynamics of the group. For example, a female may give birth during a study. Another may be injured in a fight or by a fall from a tree. These would be important events affecting the dynamics of the group and require inclusion in the larger analysis of the animals' lives.

Ad libitum notes are often valuable or insightful for the researchers. However, these data cannot be included in statistical evaluations as they are not systematically gathered. Drickamer (1974) showed how biased ad libitum data can be in his study of macaque reproduction. Lower-ranked individuals and their actions were overlooked in favor of higher-ranked and more observable individuals in study groups. This does not mean researchers cannot use ad libitum notes. Researchers can and do. They simply inform other researchers of the data's origin so that other researchers are aware of the nature and possible limitations of this information.

Variables: Dependent and Independent

For some research questions, the data collection techniques available for field research are not sufficient or appropriate. When research is directed toward understanding sensory responses, neurological mechanisms, endocrinological states, physiological or anatomical variation, embryonic development, disease, genetic conditions, cognitive processes, etc., researchers may have to conduct their investigations in laboratories or with captive animals. These areas of research often require controlled manipulations of the circumstances in which the animals are embedded and technology that cannot be, or is not yet, available for data collection in the field or that may entail procedures which are invasive to the subjects. Although researchers seek to understand the natural world, it is often the case that in order to understand the actual processes involved in the manifestation of specific behaviors or complex situations, researchers must isolate the variables operating to create these behavioral complexes or the situation's dynamics. These types of study require the ability to impose controls on the animals or on the variables of interest to the researcher.

Researchers establish control by manipulating the conditions of the animals and recording the exhibited responses. The manipulated factor is termed the *independent variable*. By changing the independent variable, researchers can determine whether there is or is not a response to that variable. They can also establish the extent of the response to the altered conditions. The response under investigation is itself known as the *dependent variable* because the extent of observed change relies, or depends, upon manipulation of the independent variable by the researchers.

Control of research variables is not the only reason researchers might conduct studies within tightly controlled circumstances. It is often necessary to work out and establish data collection techniques under laboratory conditions before they are implemented in field studies to demonstrate the viability of the study's methodology or to establish baseline or comparative data. Captive populations also provide a means to observe animal behavior in detail not possible under some field conditions as the habitat of the animals may be too dense to allow good observation or the forest canopy creates conditions where continuous observation is nearly impossible.

DATA ANALYSIS

Statistics are the common language of current scientific investigation. Without a solid background in statistical procedures, it is impossible not only to conduct serious behavioral research but also to seriously evaluate the research of others. Without an understanding of statistics, one cannot properly judge whether the results of a study should be taken seriously, accepted with caution under very restricted circumstances, or considered as preliminary results requiring further research and verification. This statement should not be inferred as suggesting that all researchers need to obtain a degree in statistics or applied mathematics, but students of animal behavior do need to develop the fundamental ability to carry out statistical procedures and to understand the assumptions and limits of statistical testing.

Descriptive Statistics

The most basic type of statistics is descriptive statistics. *Descriptive statistics* depict the patterns of behavior actually observed by researchers. A descriptive statistic can be a simple statement of the total number of behavioral events (e.g., there were 33 grooming episodes observed or the group ate figs on 68 different occasions). These statements may be accurate but do not provide much information that allows researchers to compare and evaluate behaviors to patterns observed in other groups. The statements lack context. In order for researchers to make useful comparisons, raw data need to be transformed into standardized measurements of how often the behavior occurs or into a breakdown of the percentages of behavioral events. Do 33 bouts of grooming show the behavior is common or rare? If the 33 bouts are the total number of grooming events seen in a group of 20 individuals over the course of a 1-year study, then it is reasonable to conclude grooming is a fairly rare pattern of interaction. One needs to incorporate group size and the dimension of time, however, to reach that conclusion. Consequently, descriptive statistics are most commonly reported as *frequencies* (the number of bouts per interval of time) or as proportions.

Frequencies are ratios, giving standardized measures of how often a given behavior occurs. The rates are standardized by a specific interval of "sample time." *Sample time* refers only to those periods of time during which data were being systematically collected. The interval of sample time provides the standard for comparison and might be given in hours, days, or whatever measure the researcher deems appropriate for a behavior. Again, only those behaviors recorded during the sampling process can be transformed into frequencies. Behaviors used for statistical analysis cannot include episodes recorded and observed outside of the sampling process as there is no temporal control for these events. By standardizing how often events occur, it is possible to compare the behavioral patterns of different individuals, different social groups, or even species.

Proportions are given as percentages and are based upon the number of times the study animals engage in a behavior. Researchers divide the number of episodes performed by a specific individual or subgroup of individuals, such as all adult females, by the total sum of the measure for a specific behavior or behavior class recorded for the study group during sampling. This procedure provides a means to identify which individuals or subgroups perform which activities or to represent the importance of a given activity, such as traveling between feeding sights, to the animals' daily or annual routine.

Researchers are often interested in providing an account of their study group's general patterns of behavior over the course of the entire study, to provide a broad picture of how the animals live. One way to do this in a standardized system is to establish a time budget. A *time budget* is an overview of the major behavioral activities of the study animals. Large categories of state behaviors, such as rest, locomotion, feeding, etc., are presented. Researchers sum the total number of minutes for each behavior for all animals in the group and divide these sums by the total number of sample minutes. Following this procedure provides a standardized and useful representation of the group's normal activity pattern.

Inferential Statistics

As many, perhaps most, questions of interest to primatologists cannot be answered through the use of descriptive statistics, researchers need to use a different kind of statistical analysis, *inferential statistics*, to obtain answers. Inferential statistics are based upon probabilities. Conducting an inferential statistical test begins by converting an explanation for the patterns of behavior observed and under investigation into a specific mathematical hypothesis or series of alternative hypotheses. Once an explicit mathematical hypothesis is created, researchers construct a null hypothesis for the purpose of conducting a statistical test. Most often, a null hypothesis is based upon the concept of the phenomenon under investigation being without function, spread equally among all members of a population, or a random event.

All statistical tests rely upon a number of assumptions or requirements in order to provide a reliable result. Failure to obtain data of the kind required for a test or in inadequate amounts will invalidate the results of a test. Among the most basic assumptions concerns the population from which the data are gathered. The source population from which the data are gathered must be representative of normal populations, conditions, or situations. Another requirement is that all data used in an analysis must be collected under the same conditions using the same methodology. This is the reason that researchers cannot incorporate ad libitum notes when conducting hypothesis tests. A third factor concerns sample size, size counts. Statistical tests are sensitive to the amounts of data being analyzed. Generally speaking, the likelihood of a statistical test returning an accurate, verifiable result increases with the use of more data. Small sample size is an issue of serious concern in many primatological studies.

Inferential statistical analyses return results based upon probability values, or *p* values. The *p* value represents the probability, or likelihood, that the result of a hypothesis test is accurate and not the erroneous acceptance or rejection of the null hypothesis. The general standard used in most research for the *p* value is 0.05. This number represents a measure of reliability; that is, the results of the test can be accepted as valid with a 95% degree of certainty. There is only a 5% chance that the results and conclusions drawn from the results are due to error. More specifically, the *p* value refers to the probability of a Type I, or alpha, error being committed in the refutation of the null hypothesis. A Type I error occurs when the null hypothesis is rejected on the basis of the result of a statistical test when in reality the null hypothesis should have been accepted. When the result of a statistical test is reported as significant, or nonsignificant, the researchers are simply stating whether or not the predetermined standard of the *p* value has been met.

There have been great advances in the field of statistics throughout the last century. The field of statistics is dynamic, with new statistical tests and techniques being developed. There is a growing body of dynamic modeling and complexity analyses. Data mining techniques are growing in importance, as are *meta-analyses*, techniques which allow the combining of the results from many studies. Another important development has been in the analysis of *sequential data*, the examination of the sequence of behavioral responses among multiple individuals.

Even with continuing advances, there are some standard statistical tests deserving brief mention as their use is very common. Chi-squared tests can determine whether or not an observed pattern of behavior is in fact distinctive and not simply a random set of events or evenly distributed among all individuals. Is the pattern of grooming within a group distinctive, with specific individuals grooming each other more often than they groom with other individuals? The analysis of variation (ANOVA) and the multiple analysis of variation (MANOVA) allow researchers to determine the source(s) of variation. Is the distinctive grooming pattern observed a result of females grooming each other, grooming males, or grooming primarily their young? Regression analyses allow researchers to examine to what extent variables change in response to specific factors affecting the subjects' lives or to detect the association or interaction between differing factors of primate life. Each statistical test is a distinct tool. Each has its own set of assumptions and requirements. The decision as to which statistical test should be used is determined by the research question being asked.

Geographic Information Systems

Use of GIS analyses is increasingly common in environmental studies and developing into an important technique for data analysis in many academic fields from geology to demography. GIS are computer database programs used for spatial analyses. Consequently, GIS programs can be used to create topographical maps or depictions of ranging patterns of one or multiple groups in a region.

As with all database programs, the key to using a GIS is in the data itself. The data must be collected or tabulated in a manner compatible for use by the software programs. GIS analyses are created from three basic types of data: spatial data, imaged data, and tabular data. *Spatial data* consist of information referencing specific locations, regions, or segments within an area. GPS devices providing longitude and latitude of a site's location are a common source of spatial data. Common sources for *imaged data* are maps or aerial and satellite photographs. Images of geographical space need to be digitally transformed for computation. *Tabular data* generally refer to some aspect or feature of a particular site. The type of food resource available at a specific location and the number of primate groups and species living in a region are examples of tabular data.

The data compiled in GIS databases are analyzed using two different types of modeling program: vector models and raster models. Vector modeling combines discrete locations or events to create maps of a region. These maps can include tabular data to show site usage by the study group, creating profiles of the ranging behavior of the primates. Raster modeling uses nondiscrete or continuous data. Raster modeling converts data to produce three-dimensional representations of an area. It can be useful for topographical mapping or depicting the strata of forest structure.

Once the data are entered into a database, however, GIS representations are not limited to showing the extant conditions of an environment. They can be used to model or predict changes, such as the responses of primates to alterations in their habitat. Although GIS software has been in use for over 20 years, researchers in primatology are just beginning to recognize its utility and incorporate it in their analyses. The potential of GIS programs is enormous and not limited to range use studies. M'Kirera and Unger (2003) used GIS software to show distinctive molar wear patterns due to the differences in the diets of gorillas and chimpanzees. As technological and theoretical advances continue, the possibilities inherent in spatial analyses may establish GIS as a common analytical tool for many aspects of primatological research.

CLOSING REMARKS

Unfortunately, it is not possible in a single chapter to cover all of the research protocols and techniques used in, or available to, primatological research. The field of primatology encompasses too many areas of focus to make that possible. Technology advances far too quickly for the printed page to keep up with changes that might be exceptionally valuable for researchers. Consequently, I have focused on a few fundamental aspects of research here. Those interested in more detailed explications of the various aspects of conducting primatological research can find many useful texts and guidelines to assist them. Search out reviews of techniques and the descriptions of new data collection methods in the pertinent journals. Martin and Bateson's text (1986) provides a useful summary of behavioral research, as does Bakeman and Gottman's (1986) text on the sequential analyses of behavior. Currently, those individuals seeking greater detail on ecological research methods can find them in Krebs (2000) and Henderson (2003). Many useful statistical procedures are outlined by Sokal and Rohlf (1994) and Krebs (2000). The edited volume of Setchell and Curtis (2003) offers specific essays on many of the techniques used by primatologists, as well as useful references on each topic.

Whatever the specific focus of the research might be, the process of conducting research relies on a few basic procedures. Establish a well-thought-out project, considering as many details as possible, including the well-being of the subjects to be studied. Determine the techniques to be used. Become proficient with these techniques. Test and analyze the data collected. Be cautious with the conclusions drawn from the research, but be creative in asking questions and in designing the study. Advances in understanding come from questioning established ideas and explanations, not through the simple acceptance of prevailing conceptualizations. In the end, there is always more to know, new questions to ask.

REFERENCES

Altmann, J. (1974). Observational study of behaviour: sampling methods. *Behaviour* 49:227–265.

Bakeman, R., and Gottman, J. (1986). *Observing Interaction: An Introduction to Sequential Analysis*. Cambridge University Press, Cambridge.

Bernstein, I. S. (1981). Dominance: the baby and the bathwater. *Behav. Brain Sci.* 4:419–457.

Bertrand, M. (1969). *The Behavioral Repertoire of the Stumptail Macaque. Bibliotheca Primatologica*, 11. S. Karger, Basel.

Brockelman, W., and Srikosamatara, S. (1993). Estimation of density of gibbon groups by use of loud songs. *Am. J. Primatol.* 29:93–108.

Brower, J., Zar, J., and von Ende, C. (1990). *Field and Laboratory Methods for General Ecology*. Wm. C. Brown Publishers, Dubuque.

Byrne, R., and Byrne, J. (1993). Complex leaf-gathering skills of mountain gorillas (*Gorilla g. beringei*): variability and standardization. *Am. J. Primatol.* 31:241–261.

Chapais, B. (1988). Experimental matrilineal inheritance of rank in female Japanese macaques. *Anim. Behav.* 36:1025–1037.

Dawkins, M., and Gosling, M. (1991). *Ethics in Research on Animal Behaviour*. Academic Press, London.

Dolhinow, P. (1978). A behavior repertoire for the Indian langur monkey (*Presbytis entellus*). *Primates* 19:449–472.

Drickamer, L. C. (1974). Social rank, observability and sexual behavior of rhesus monkeys (*Macaca mulatta*). *J. Reprod. Fertil.* 37:117–120.

Fashing, P., and Cords, M. (2000). Diurnal primate densities and biomass in the Kakamega Forest: an evaluation of census methods and a comparison with other forests. *Am. J. Primatol.* 50:139–152.

Ganzhorn, J. (2003). Habitat description and phenology. In: Curtis, D., and Setchell, J. (eds.), *Field and Laboratory Methods in Primatology*. Cambridge University Press, Cambridge. pp. 40–56.

Henderson, P. (2003). *Practical Methods in Ecology*. Blackwell Publishing, Malden.

Hughes, K. (2003). The global positioning system, geographical information systems and remote sensing. In: Curtis, D., and Setchell, J. (eds.), *Field and Laboratory Methods in Primatology*. Cambridge University Press, Cambridge. pp. 57–73.

Johnston, C. (1998). *Geographic Information Systems in Ecology*. Blackwell Science, Malden.

Krebs, C. (2000). *Ecology: The Experimental Analysis of Distribution and Abundance*, 5th ed. Pearson Addison Wesley, Boston.

Martin, P., and Bateson, P. (1986). *Measuring Behaviour: An Introductory Guide*. Cambridge University Press, Cambridge.

M' Kirera, F., and Unger, P. (2003). Occlusal relief changes with molar wear in *Pan t. troglodytes* and *Gorilla g. gorilla*. *Am. J. Primatol.* 60:31–41.

Paterson, J. (2001). *Primate Behavior: An Exercise Workbook*, 2nd ed. Waveland Press, Prospect Heights, IL.

Phillips, K., Elvey, R., and Abercrombie, C. (1998). Applying GPS to the study of primate ecology: a useful tool? *Am. J. Primatol.* 46:167–172.

Schaller, G. (1963). *The Mountain Gorilla: Ecology and Behavior*. University of Chicago Press, Chicago.

Setchell, D., and Curtis, J. (eds.) (2003). *Field and Laboratory Methods in Primatology*. Cambridge University Press, Cambridge.

Sokal, R., and Rohlf, F. J. (1994). *Biometry: The Principles and Practice of Statistics in Biological Research*, 5th ed. W. H. Freeman, New York.

Tutin, C., Parnell, R., White, L., and Fernandez, M. (1995). Nest building by lowland gorillas in the Lope Reserve, Gabon: environmental influences and implications for censusing. *Int. J. Primatol.* 16:53–76.

Whitesides, G., Oates, J., Green, S., and Kluberdanz, R. (1988). Estimating primate densities from transects in a West African rain forest: a comparison of techniques. *J. Anim. Ecol.* 57:345–367.

Williamson, E., and Feistner, A. (2003). Habituating primates: processes, techniques, variables and ethics. In: Curtis, D., and Setchell, J. (eds.), *Field and Laboratory Methods in Primatology*. Cambridge University Press, Cambridge. pp. 25–39.

21

Advances in the Understanding of Primate Reproductive Endocrinology

Bill L. Lasley and Anne Savage

INTRODUCTION

The field of primate reproductive endocrinology has grown exponentially from initial studies that focused on collecting samples using chemical immobilization and/or physical restraint to study reproductive functioning to a new era where non-invasive means are used to collect samples. Today, technology plays an integral role in how we manage animals in captivity and allows us to study various aspects of reproduction in free-ranging animals. This technology has provided us with information that is essential for (*1*) understanding the basic biology of reproduction in captive and wild primates, (*2*) developing husbandry strategies that maximize successful production of offspring, (*3*) diagnosing departures from normal reproduction, and (*4*) developing assisted reproductive technologies when necessary. The development of reliable methods for monitoring reproductive events such as ovulation and pregnancy has allowed for increased reproductive performance but also has provided the basis for studies designed to accelerate reproduction using assisted reproductive technologies. This chapter will review the history of measuring hormones in primates and discuss the current technology available to assist in monitoring reproductive function in captive and wild primates.

HORMONE MEASUREMENT

Two assumptions of comparative reproductive physiology have been used to develop general strategies that allow us to monitor hormone levels in all mammals (Lasley and Kirkpatrick 1991, Lasley and Shideler 1993). The first assumption is that gonadal steroid hormones have the same molecular structure throughout the animal kingdom. While secretion and metabolism of steroids may vary among species, it is universally accepted that testosterone, estradiol, and progesterone are the primary bioactive sex steroids and that they reflect gonadal function in all mammals. The second assumption is that all steroid hormones are excreted rapidly through the kidneys or gut and then cleared from the body after undergoing only minor physical change. The resulting metabolites are concentrated in the urine, saliva, or feces and are relatively stable for a prolonged period of time (Lasley and Kirkpatrick 1991, Lasley and Shideler 1993). Although there are commonalities in the reproductive endocrinology of mammals, there are significant differences in both the pattern and levels of hormones produced, making extrapolation of findings from one species to another challenging and potentially misleading. Hormones are present and can be measured in a variety of biological fluids; however, the choice of which to use depends on various factors, such as the type of information desired, the assay techniques to be used, species differences in steroid metabolism and route of excretion, and the ease of sample collection (Hodges 1996).

ENDOCRINE ASSAYS

Endocrine assays are now applied in a wide range of disciplines to study the broadest possible range of species and in every conceivable physical setting (Lasley 1985, Lasley and Kirkpatrick 1991, Lasley and Shideler 1993). Blood, urine, saliva, and feces represent the range of possible biological samples that are collected in laboratories, zoos, forests, and savannas. Some of the earliest assay formats remain in use today, and new methods are constantly being developed. There are few known endocrine analytes that have defied assay development, and assay methods are available to virtually anyone who requires their use.

IMMUNOASSAYS

While a few reports predated the advent of immunoassays, most of the endocrine data that are still in use today were reported after 1965, when the first truly practical steroid hormone assays were developed. The first of these were the protein binding assays. These assays were developed to measure only a few steroid hormones. The methods associated with this particular technology generally used corticobinding protein from the dog to measure either circulating cortisol or progesterone. Tissue extracts were also used for other analytes, such as estradiol. While the method was completely replaced by radioimmunoassay within

5 years, the basic concept remains valid today. Prior to this time, physical–chemical methods, such as thin-layer or paper chromatography followed by fluorescence or gas–liquid chromatography with flame or ion detection, were required. The advantages of protein binding technology were increased sensitivity and reduced sample size preparation. While this technology was limited to progesterone and estradiol as the only reproductive hormones measured, it was sufficient to make it an important contribution since little beyond ovulation and the luteal phase of ovarian cycles was well defined at this time.

PROTEIN HORMONE ASSAYS

The principle of protein binding assays was simple: add insufficient binding protein to permit all of the analyte to bind and add a similar amount of radiolabeled analyte to compete with the endogenous analyte. The resulting competition would permit only a fraction of the label to be bound, and once this bound portion was separated from the original mixture, the amount of label bound would permit calculation of the amount of endogenous analyte (e.g., luteinizing hormone [LH], follicle-stimulating hormone [FSH]) that had competed. With two small exceptions, this basic principle is still the prevailing principle in small analyte detection today. The weaknesses of protein binding assays were (1) the limited number of types of specific protein binding globulin that exist, (2) the difficulty of removing all endogenous analyte in order to obtain protein binding sites for the assays, and (3) separating the "bound" from the "free" components without changing the dynamic equilibrium that had been established. This method was made obsolete, when Bersen et al. (1956) revealed that specific antibodies could be produced to replace the binding globulins that had more specificity and increased binding avidity. Over the next 10 years, immunoassays were developed to measure every known human circulating reproductive hormone. With immunoassays, peptide and protein hormone assays could be developed as easily as steroid hormone assays, once a pure form of the targeted analyte was made available for antibody production. Beginning in the late 1960s, radioimmunoassays were being applied with varied success to a wide range of nonhuman primate species.

STEROID HORMONE ASSAYS

Assays for the steroid hormones have always been highly adaptable because steroid hormones (estrogen, progesterone, testosterone, etc.) are physically identical in all mammalian species. There are species-specific differences, however, in the concentrations and potential interfering analytes and matrices; but in general, steroid hormone immunoassays have been used to measure circulating levels of sex steroids in all well-studied primate species. With the advent of radioimmunoassay, both competitive radiolabels and specific antisera became available for all clinically relevant steroid hormones. Unfortunately, this was not true for the peptide, protein, and glycoprotein hormones. Because peptides and proteins are derived from the genome and because alterations of the genome are the basis of speciation, the physical structure of larger peptides and all protein hormones varies between species. While the physical difference in molecular structure between the same hormone in different species can be modest, if that difference in structure between species changes the immunological properties of the analogous hormone, then assay reagents will not be adaptable between those species. This has been, and remains, the major problem in the application of existing hormone assays to a wide range of nonhuman primate species. This problem led to the creation of bioassays that permit the measurement of peptide/protein hormones in a wide range of species.

BIOASSAYS

The principle underlying bioassays is that while the overall physical structure of the hormone molecule changes, or "evolves," with speciation, the portion of that molecule that actually participates in sending the biological signal remains conserved. In other words, the portion of the molecule that endows it with biological activity is highly conserved and will be capable of exerting that biological activity in target cells of many species. Making use of this principle, target cells from laboratory animals were used to respond to the hormones of other species. The initial assays were in vivo assays and were used for both protein and steroid hormones. This use of live animals for assays was labor-intensive and short-lived. In a short time, steroid hormone assays were replaced by immunoassays and the protein hormones were directed to in vitro bioassays in which the target tissues were incubated with the sample to obtain results. The best example of such an assay is the use of minced mouse testes to respond to the serum of a wide range of species to produce testosterone as a function of the amount of LH present in the serum. The receptors that peptide and protein hormones bind to and transduce their signals have been cloned and transfected into otherwise inert cells. When co-transfected with reporter genes that produce light or other signals, these transfected immortalized cell lines can be used to obtain the same result that was obtained using whole animals or fresh tissues from animals. While this technology is still growing in popularity, it is likely to be replaced by immunoassays as a result of new molecular technologies.

THE NEED TO MEASURE HORMONES

The occurrence and timing of reproductive events such as ovulation, conception, early pregnancy, and losses, which are largely concealed, can be detected and documented only through hormone analysis. Early studies of female reproductive function required repeated blood draws from restrained animals to obtain sufficient data (Eaton and Resko 1974) to answer many of these questions. By their nature and by endocrine definition, the results of assays that measure circulating hormones from blood provide the "gold standard" for all other hormone assay results. While a single blood sample provided useful information, only serial blood samples provided information on hormonal profiles. Repeated capture and restraint to obtain blood samples was often stressful and led to potential behavioral abnormalities that were thought to influence reproductive events (Smith and French 1997a,b). Moreover, when studying small primates, repeated blood sampling can lead to significantly lower hematocrit values (French et al. 1992). Thus, techniques were developed to noninvasively monitor reproductive function in captive and wild primates (Czekala et al. 1983; Hodges et al. 1986; Mitchell et al. 1982; Savage et al. 1997, 1995; Shideler et al. 1994, 1983a,b; Wasser et al. 1988; Ziegler et al. 1997a, 1987a,b), which have revolutionized our ability to understand the complex reproductive biology of primates. Using urine and fecal samples to measure hormones is now commonplace, and the changes in the concentration of specific analytes excreted in urine and feces are thought to accurately reflect the physiological processes and time course of events associated with those analytes.

URINARY ASSAYS

During the early 1970s, the success of immunoassays for measuring circulating hormone levels made it clear that daily measurement of reproductive hormones was required if the reproductive biology of any species was to be completely understood. By 1975, zoological parks and gardens aggressively took steps to study and understand the reproductive biology of endangered species and developed first-rate captive breeding programs (Hodges et al. 1979, Lasley 1985, Lasley and Kirkpatrick 1991, Lasley and Shideler 1993, Lasley et al. 1981a). The zoo community sponsored the effort to reevaluate their current management strategies and asked whether the new technologies developed for investigating reproductive physiology could also improve the ability to obtain useful information from urine-based assays. This new perspective in the field of reproductive biology soon became a major factor in the development of urinary hormone assays.

The majority of steroids in urine are present in the conjugated form. Initial analyses of steroids in urine involved a very labor-intensive process of hydrolysis and solvent extraction prior to assay. With the advent of nonextraction assays, it became possible to directly measure steroid conjugates (see Hodges 1996 for a complete review). Direct assays for steroid conjugates have the distinct advantage of producing a more informative hormonal profile than is possible with extraction methods (Hodges and Eastman 1984, Lasley 1981b, Shideler et al. 1983b). Less complicated assays based on the use of enzymatic procedures are available as an alternative to assays using radioisotopes and are commonly used in most endocrine service laboratories. The advantages of these enzyme immunoassays (EIAs) include low cost, ease of performance, and no use of radioactive material. These assay systems require minimal instrumentation and are more likely to be used in the field. The application of EIAs to reproductive monitoring in primates has been carried out by many researchers (e.g., Czekala et al. 1988, Heistermann et al. 1995b, Hodges 1985, Hodges and Green 1989, Lasley 1985, Lasley and Kirkpatrick 1991, Lasley et al. 1980, Shideler et al. 1990).

Gonadotropic hormones, such as pituitary LH and FSH, are also excreted in urine. However, they have proven to be more challenging to measure by immunoassay methods since the species specificity in the immunological activities of these hormones makes multispecies application of assays difficult (Hodges 1996, Matteri et al. 1987). In contrast, biological activity is less species-specific, and in vitro bioassays have been established for urinary LH (Harlow et al. 1984, Hodges et al. 1979, Ziegler et al. 1993a,b) and FSH (Dahl et al. 1987, for a complete review, see Hodges 1996).

ADVANTAGES OF URINE SAMPLES

Urine has favorable qualities that make it ideal as a sample for monitoring reproduction. It is constantly produced by the accumulation of metabolic products. This insures that the time course of metabolism is preserved in the components of urine, and it is a concentrate of metabolic by-products. In general, urine can be collected without capture or restraint and in relatively large quantities over indefinite periods of time. Because urine is largely water with concentrated analytes in solution, assay formats can use urine directly or, following a simple dilution, in assay buffer. Because the volume of urine is determined by water intake and need, changes in intake will be reflected in urine volume. The potential problem presented by variation in urine volume is usually addressed by indexing the concentration of the targeted analyte to some measure of the urine volume. The specific gravity of urine or urinary creatinine concentration is used as a divisor and produces a quotient that is a normalized concentration of the analyte (Taussky 1954). When daily changes in the concentration of the circulating parent analyte are compared to daily changes in the indexed urinary analyte, the patterns are parallel, thus verifying the validity of this method. Despite certain limitations in using creatinine measurements, good correlations between hormone/creatinine index have been demonstrated for several species (Hodges and Eastman 1984, Shideler and Lasley 1982).

COLLECTING AND STORING URINE SAMPLES

The species and individual subjects usually determine the timing and method of collecting urine samples (Hodges 1985, Lasley 1985). For most studies conducted in captivity, it is relatively easy to condition most primates to urine collection. Primates can be trained to enter individual cages that allow urine to be collected (Anzenberger and Gossweiler 1993, Pryce et al. 1995) and to accept a caretaker collecting a urine sample upon waking (Pryce et al. 1995, Ziegler et al. 1993a), or urine samples may be collected from a clean floor of a cage when the animal has been moved (Bahr et al. 2001, Pryce et al. 1995).

While collection of urine in the field is not commonplace with most small-bodied primates, it has been used successfully with great apes. Urine is collected either upon waking or ad libitum by placing a collecting tube in the stream of urine upon urination (mountain gorillas, Robbins and Czekala 1997) or by pipetting urine found on nearby vegetation or soil (orangutans and chimpanzees, Knott 1997, bonobos, Marshall and Hohmann 2005).

Most species have predictable voiding habits, and these can be used to insure the collection of clean samples. Frequently, portions of early-morning voids are collected because they represent an integrated accumulation of analytes over several hours of minimal physical activity (Czekala et al. 1983; Savage et al. 1995; Shideler et al. 1983a,b; Ziegler et al. 1987a,b). This collection schedule obviates concerns relating to wide changes in fluid intake and minimizes the need for adjustments for differences in analyte concentration due to changes in urine volume production. If intact LH or FSH is to be measured, then preservative (glycerol) needs to be added immediately to prevent dissociation of the subunits. Usually, the entire amount of urine collected is not needed and an aliquot is centrifuged to remove cell debris and other particulate matter prior to storage at a temperature that will insure the sample is frozen solid.

Urine samples can be stored indefinitely if kept frozen and rethawed multiple times without loss of the ability to accurately measure steroid conjugates. Multiple thaws are not recommended if the intact gonadotropins are to be measured. However, recent advances have also shown that small amounts of urine (200 μl) can be saved on absorbent filter paper and stored in airtight plastic containers with silica gel as a viable method to store samples in the field (Marshall and Hohmann 2005, Shideler et al. 1995). The resulting profiles generated by urinary hormone analysis generally parallel those of circulating hormones.

URINARY STEROIDS

The majority of steroids in urine are present in the conjugated form. Early analyses of steroids in urine involved the laborious process of hydrolysis and solvent extraction prior to assay; however, the introduction of nonextraction assays allowing direct measurement of steroid conjugates has now greatly simplified procedures. Direct assays for steroid conjugates have the additional advantage of often generating a more informative hormone profile than previously possible with extraction methods (Hodges and Eastman 1984, Lasley

1985, Shideler et al. 1983b). Even simpler assays based on the use of enzymatic procedures are now available as an alternative to conventional assays using radioisotopes. The advantages of these EIAs, including low cost, ease of performance, convenience, as well as avoidance of the problems associated with the use and disposal of radioactive materials, make them particularly suitable for institutions with limited laboratory facilities. Furthermore, the end point of these tests is a color change that is simple to quantify. These tests require minimal instrumentation and are therefore amenable to use under field conditions. EIAs have been applied to reproductive monitoring in primates by several researchers (e.g., Czekala et al. 1986, 1988; Hodges 1985; Hodges and Green 1989; Shideler et al. 1993a,b).

FECAL ASSAYS

In many captive situations, urine samples can be collected regularly for most primate species; however, in the field, it may be more practical to collect fecal samples (de Sousa et al. 2002, Fujita et al. 2001, Savage et al. 1997, Strier and Ziegler 1997, Wasser et al. 1991, Ziegler et al. 1997a). In addition, there are some species (*Nycticebus coucang*) that have an insignificant about of hormones excreted in the urine, with primary hormonal excretion found in the feces (Perez et al. 1988). In such cases, fecal samples can be collected and analyzed in much the same fashion as urine. The primary differences in fecal analysis are the need to remove solid materials, the limitation of measurements to the metabolites of steroid hormones, and usually the need to measure unconjugated rather than conjugated forms of the steroid metabolites. As with urinary measurements, there is a need to index or adjust the amount of analyte to compensate for the total volume of feces produced. In most cases, this is accomplished by simply adjusting the mass of analyte to the mass of the fecal sample analyzed. While this is clearly less accurate than the adjustment of urine by creatinine, it is currently the only practical method available and has been successfully applied to several primate species.

COLLECTION AND STORAGE OF FECAL SAMPLES

Fecal samples can be collected systematically or randomly, stored in a secure container, and frozen solid until processed (Savage et al. 1997). In the field or in situations where access to refrigeration is limited, fecal samples have been stored in 95% ethanol and 0.2% sodium azide (Wasser 1996, Wasser et al. 1988), dried in gas ovens (Stavisky et al. 1995), or collected and stored on filter paper (Shideler et al. 1995).

SALIVA SAMPLES

Collecting saliva for hormone analysis has been successful for studies investigating stress and cortisol levels

(Tiefenbacher et al. 2003) and has also been used to assess reproductive function in selected species (DiGiano et al. 1992). Saliva is typically collected by enticing animals to bite on a dental rope. The rope is then centrifuged and the saliva removed and stored for analysis (Tiefenbacher et al. 2003). Others have trained animals, such as gorillas, to chew on dental rope and "trade" the dental rope for a food reward (Bettinger et al. 1998).

STUDYING PRIMATE REPRODUCTION IN CAPTIVITY AND THE FIELD

With more than 300 primate species and more than 600 primate taxa, it is surprising how little we know about the basic reproductive biology of primates (see Table 21.1). Less than 20% of all the primate species have had their basic reproductive biology characterized, either wild or in captivity, and even fewer have studies that compare the reproductive biology of wild and captive animals. Ziegler and colleagues have spent many years examining the reproductive biology and behavior of cotton-top tamarins (*Saguinus oedipus*) in captivity (Ziegler et al. 2000b,c; 1997b; 1990c; 1987a,b) and have provided the most complete information on any callitrichid studied in captivity. Savage et al. (1997) compared data from wild caught females to the data published for captive cotton-top tamarins and found some interesting differences. Wild female cotton-top tamarins are seasonal breeders, giving birth every 48 weeks, compared to captive females, which give birth every 28 weeks. Wild cotton-top tamarins have a postpartum suppression of fertility that averages approximately 144 days, while captive females typically conceive within 1 month of parturition. The postpartum period of ovarian inactivity in wild cotton-top tamarin females is influenced by environmental conditions (e.g., drought), and similar to captive females, it is possible to experience a 19-day postpartum ovulation under specific environmental conditions. Interestingly, these studies allow us greater insight into the factors that influence reproduction both in captivity and in the wild and demonstrate how both social and environmental factors may cause females to regulate their own fertility. Wild female tamarins are observed to terminate their pregnancies when the environmental conditions (e.g., drought) are not conducive to a successful outcome. This type of comparative information is critically important when one is trying to develop conservation plans for a species. Understanding the reproductive capacity of a species in addition to a variety of other important factors provides the information necessary when developing models for long-term population viability.

HUSBANDRY

In many captive facilities in the United States and abroad, the hope is to develop self-sustaining captive breeding colonies that are genetically, demographically, and behav-

iorally viable. Attempts to breed animals without a clear understanding of the reproductive biology is often challenging since without hormonal data it is often difficult to document puberty, ovarian cycles, etc., in species that do not exhibit clear physical or behavioral signs of estrus or pregnancy. Early studies in white-faced sakis (*Pithecia pithecia*) estimated that sexual maturation, based on first conception, occurred within 26–32 months (Shoemaker 1982) after pairing with a male. However, it was not until a comprehensive study was conducted on the reproductive biology of the species (Savage et al. 1995, Shideler et al. 1994) that it was demonstrated that puberty occurred at approximately 32 months of age and that it did not appear that puberty could be accelerated in this species by pairing a prepubescent female with an adult male.

The ability to diagnose pregnancy in both captive and field conditions is essential. In captive colonies, managers are tasked with diagnosing pregnancy so that they can make the appropriate husbandry decisions from an animal welfare perspective. While pregnancy confirmation through a collection of serial samples is commonplace for many species, the ease and rapidity with which one can diagnose pregnancy was suggested as a high priority for technology development. Shimizu et al. (2001) developed a noninstrumented enzyme-linked immunosorbant assay (NELISA) for the measurement of urinary monkey chorionic gonadotropin (mCG) for the detection of early pregnancy in macaque monkeys (*Macaca mulatta* and *M. fascicularis*) for use in both the laboratory and the field. The advantage of this technique is that the color change can be detected without the use of a spectrophotometer. The mCG in pregnant monkey samples was indicated by a dark green color change. In contrast, nonpregnant monkey urine samples exhibited no color change. These findings suggest that the simple, economical, and reliable urinary mCG NELISA may be useful for diagnosing early pregnancy in these and related species.

ASSISTED REPRODUCTIVE TECHNOLOGIES

While assisted reproductive technology (ART) has revolutionized the breeding of domestic animals and the treatment of infertile human couples (Trounson 1993), the use of ART to assist with the propagation of threatened and endangered nonhuman primate species has been limited (Bavister and Boatman 1993, Reddy et al. 1992).

While infertility is commonplace with humans, documented infertility in captive and wild primates is rare. In general, there has not been the need to develop ART for most primate species and, as a consequence, that field has progressed slowly. Documented individual infertility in free-ranging primates is either rare or seldom a breeding problem as the relatively complex interrelated social and sexual behaviors seem to override physiological deficiencies. Nonetheless, there is a significant amount of information on

Table 21.1   Overview of Primate Reproductive Profiles

PRIMATES	AUTHORS	MEDIA/HORMONE	PURPOSE
Prosiminians			
Propithecus verrauxi	Brockman et al. (1998)	Fecal testosterone	Hormones and behavior in the field
	Brockman et al. (1996)	Fecal estradiol and progesterone	Hormones and behavior in the field
Microcebus murinus	Perret (1986)	Plasma progesterone	Hormones and behavior in captivity
Eulemur macaco	Lasley et al. (1978)	Serum progesterone, estrogens	Endocrine knowledge in captivity
Lemur catta	Cavigelli and Pereira (2000)	Fecal testosterone	Hormones and behavior in semi-free-ranging animals
	Norman et al. (1978)	Serum luteinizing hormone	Endocrine knowledge in captivity
	Lasley et al. (1978)	Serum progesterone, estrogens	Endocrine knowledge in captivity
Varecia variegatus	Lasley et al. (1978)	Serum progesterone, estrogens	Endocrine knowledge in captivity
	Shideler et al. (1983a,b)	Urinary estrogens	Endocrine knowledge in captivity
Nycticebus pygmaeus	Fitch-Snyder and Jurke (2003)	Fecal testosterone	Hormones and behavior in captivity
Galago crassicaudatus	Van Horn and Eaton (1979)	Plasma progesterone and estradiol	Hormones and behavior in captivity
G. moholi	Lipschitz (1996)	Serum luteinizing hormone, estrogens, progesterone, testosterone	Endocrine knowledge in captivity
New World primates			
Callimico goeldii	Jurzke et al. (1994)	Urinary pregnanediol-3-α-glucuronide	Endocrine knowledge in captivity
	Pryce et al. (1993, 1994)	Fecal estrogen, urinary and plasma progesterone, estrogen	Endocrine knowledge in captivity
	Ziegler et al. (1990b)	Urinary luteinizing hormone, estrone conjugates	Endocrine knowledge in captivity
Callithrix jacchus	Möhle et al. (2002)	Urinary and fecal testosterone	Endocrine knowledge in captivity
	Harding et al. (1982)	Serum progesterone	Endocrine knowledge in captivity
	Abbott et al. (1988)	Serum luteinizing hormone	Hormones and behavior in captivity
	Heistermann et al. (1993)	Fecal estradiol, progesterone, pregnanediol, plasma progesterone	Endocrine knowledge in captivity
	Hodges and Eastman (1984)	Urinary estrogen	Endocrine knowledge in captivity
	Ziegler et al. (1996)	Serum and fecal progesterone, pregnanediol, estradiol, estrone, and serum luteinizing hormone	Endocrine knowledge in captivity
	Boes et al. (1993)	Urinary estradiol	Endocrine knowledge in captivity
C. kuhlii	Nunes et al. (2002)	Urinary testosterone and estradiol	Endocrine knowledge in captivity
C. pygmaea	Ziegler et al. (1990a)	Urinary estrone, luteinizing hormone	Endocrine knowledge in captivity
Leontopithecus chrysomelas	Chaoui and Hasler-Gallusser (1997)	Fecal progesterone	Hormones and behavior in captivity
L. rosalia	French et al. (2003)	Fecal estrone conjugates and pregnanediol-3-glucuronide	Endocrine knowledge in the field
	French et al. (1992)	Urinary and plasma luteinizing hormone, estrogens	Endocrine knowledge in captivity
Saguinus bicolor	Heistermann et al. (1987)	Urinary estrogen and progesterone	Endocrine knowledge in captivity
S. fuscicollis	Heistermann et al. (1995b)	Urinary estrone conjugates and pregnanediol glucuronide, fecal estradiol, progesterone, pregnanediol, plasma progesterone	Endocrine knowledge in captivity
	Epple and Katz (1983)	Urinary progesterone, testosterone, estradiol	Endocrine knowledge in captivity
	Lerchl et al. (1991)	Urinary testosterone	Endocrine knowledge in captivity
S. labiatus	Kuederling et al. (1995)	Urinary estrogens	Endocrine knowledge in captivity
S. oedipus	Savage et al. (1997)	Fecal estrone conjugates and pregnanediol-3-glucuronide	Endocrine knowledge in the field
	Ziegler et al. (1987a,b; 1990c; 1993a,b; 1996; 1997b; 2000c)	Fecal and serum progesterone, estrone, estradiol, luteinizing hormone and urinary estrone conjugates, luteinizing hormone, estrone, estradiol	Endocrine knowledge, hormones and behavior in captivity
Alouatta seniculus	Herrick et al. (2000)	Urinary progesterone	Endocrine knowledge in the field

Table 21.1   *(cont'd)*

PRIMATES	AUTHORS	MEDIA/HORMONE	PURPOSE
Aotus trivirgatus	Dixson (1994)	Urinary and serum testosterone, estrogens, progesterone	Endocrine knowledge in captivity
	McConnell et al. (1990)	Serum progesterone	Endocrine knowledge in captivity
	Setchell and Bonney (1981)	Urinary estrogens, progesterone, cortisol	Endocrine knowledge in captivity
	Bonney et al. (1980)	Plasma estradiol-17β, estrone, progesterone, testosterone	Endocrine knowledge in captivity
	Dixson and Gardner (1981)	Plasma testosterone	Endocrine knowledge in captivity
Ateles geoffroyi	Campbell (2003, 2004)	Fecal estrone conjugates and pregnanediol-3-glucuronide	Hormones and behavior in the field
	Campbell et al. (2001)	Urinary and fecal estrone conjugates and pregnanediol-3-glucuronide	Endocrine knowledge in captivity
	Hernandez-Lopez et al. (1998)	Serum estrogens and progesterone	Endocrine knowledge in captivity
Brachyteles arachnoids	Strier and Ziegler (1997)	Fecal estradiol, progesterone	Hormones and behavior in the field
	Ziegler et al. (1997a)	Fecal estrogens, progesterone	Endocrine knowledge in the field and in captivity
Callicebus moloch	Valeggia et al. (1999)	Urinary estrone conjugates and pregnandiol-3-glucuronide	Endocrine knowledge in captivity
Cebus apella	Lynch et al. (2002)	Fecal testosterone and cortisol	Endocrine knowledge in the field
	DiGiano et al. (1992)	Salivary and plasma progesterone	Endocrine knowledge in captivity
	Nagle et al. (1989)	Serum progesterone	Endocrine knowledge in captivity
Saimiri boliviensis	Yeoman et al. (2000)	Serum follicle-stimulating and luteinizing hormones, estradiol, progesterone	Endocrine knowledge in captivity
Sai. sciureus	Moorman et al. (2002)	Urinary and fecal estradiol and progesterone metabolites	Endocrine knowledge in captivity
	Diamond et al. (1987)	Serum estradiol, progesterone, chorionic gonadotropin, prolactin	Endocrine knowledge in captivity
	Chen et al. (1981)	Plasma testosterone	Endocrine knowledge in captivity
Pithecia pithecia	Savage et al. (1995)	Urinary and fecal estrone conjugates and pregnanediol-3-glucuronide	Endocrine knowledge in captivity
	Shideler et al. (1994)	Fecal estrone conjugates and pregnanediol-3-glucuronide	Endocrine knowledge in captivity
Old World primates			
Cercocebus atys	Whitten and Russell (1996)	Fecal estrogen and progesterone	Endocrine knowledge in captivity
	Mann et al. (1982)	Serum luteinizing hormone, follicle-stimulating hormone	Endocrine knowledge in captivity
	Aidara et al. (1981)	Serum estrogen, progesterone, prolactin, luteinizing hormone, follicle-stimulating hormone	Endocrine knowledge in captivity
	Stabenfeldt and Hendrickx (1973)	Serum progesterone	Endocrine knowledge in captivity
Chlorocebus aethiops	Eley et al. (1989)	Serum estradiol, progesterone, and luteinizing hormone	Endocrine knowledge in captivity
	Andelman et al. (1985)	Urinary pregnanediol-3-α-glucuronide	Hormones and behavior in the field
Erythrocebus patas	Winterer et al. (1985)	Serum estrogen and progesterone	Endocrine knowledge in captivity
Macaca arctoides	Wilks (1977)	Serum estrogen, progesterone, luteinizing hormone, follicle-stimulating hormone	Endocrine knowledge in captivity
M. assamensis	Wehrenberg et al. (1980)	Serum estrogen, progesterone, luteinizing hormone, follicle-stimulating hormone	Endocrine knowledge in captivity
M. fascicularis	Möhle et al. (2002)	Urinary and fecal testosterone	Endocrine knowledge in captivity
	Shideler et al. (1993a,b)	Fecal estrone conjugates and pregnanediol-3-glucuronide; urinary progesterone, estradiol, estrone conjugates and pregnanediol-3-glucuronide	Endocrine knowledge in captivity
	Michael et al. (1987)	Serum testosterone	Endocrine knowledge in captivity
M. fuscata	Fujita et al. (2001)	Urinary and fecal estrone conjugates, pregnanediol-3-glucuronide, and estrone conjugates; plasma estradiol-17β, progesterone, and luteinizing hormone	Endocrine knowledge in captivity and in the field
	Matsubayashi and Enomoto (1983)	Plasma testosterone	Endocrine knowledge in captivity
	Nigi and Torii (1983)	Plasma estradiol and progesterone	Endocrine knowledge in captivity

Table 21.1 (*cont'd*)

PRIMATES	AUTHORS	MEDIA/HORMONE	PURPOSE
M. mulatta	Shideler et al. (1990)	Serum estradiol and progesterone, urinary estrone conjugates, and pregnanediol-3-glucuronide	Endocrine knowledge in captivity
	Monfort et al. (1987)	Serum estradiol, luteinizing hormone, progesterone; urinary estrone conjugates	Endocrine knowledge in captivity
	Plant (1981)	Serum prolactin, testosterone, cortisol	Endocrine knowledge in captivity
M. nemestrina	Risler et al. (1987)	Serum estrogens and progesterone	Endocrine knowledge in captivity
	Eaton and Resko (1974)	Serum estrogen and progesterone	Endocrine knowledge in captivity
M. nigra	Thomas et al. (1992)	Serum estradiol, progesterone, follicle-stimulating hormone, and luteinizing hormone	Endocrine knowledge in captivity
M. radiata	Jagannadha Rao et al. (1984)	Serum chorionic gonadotropin, estradiol-17β, and progesterone	Endocrine knowledge in captivity
	Kholkute et al. (1981)	Serum testosterone	Endocrine knowledge in captivity
M. silenus	Heistermann et al. (2001)	Fecal estradiol, 5α-pregnane-3α-ol-20-one	Endocrine knowledge in captivity
	Shideler et al. (1983a, 1985)	Urinary estrone conjugates and pregnanediol-3-glucuronide	Endocrine knowledge in captivity
M. tonkeana	Theirry et al. (1996)	Urinary estrone conjugates and pregnanediol glucuronide	Endocrine knowledge in captivity
Miopithecus talapoin	Dixson and Herbert (1977)	Serum estrogen, testosterone	Endocrine knowledge in captivity
Papio hamadryas	Khan et al. (1985)	Serum luteinizing hormone	Endocrine knowledge in captivity
	Cekan et al. (1977)	Serum estrogen, progesterone, testosterone	Endocrine knowledge in captivity
P. h. anubis	Fortman et al. (1993)	Serum estradiol, progesterone, chorionic gonadotropin	Endocrine knowledge in captivity
	Hodges et al. (1986)	Urinary estrone conjugates and pregnanediol-3α-glucuronide	Endocrine knowledge in captivity
P. h. cynocephalus	Wasser (1996)	Fecal estrogens and progestins	Endocrine knowledge in the field
	Stavisky et al. (1995)	Fecal estradiol and progesterone	Endocrine knowledge in the field
Theropithecus gelada	Klecha and McCann (2001)	Urinary progesterone	Endocrine knowledge in captivity
Pygathrix bieti	He et al. (2001)	Urinary estrogen and progesterone	Endocrine knowledge in captivity
Semnopithecus entellus	Ziegler et al. (2000a)	Fecal pregnanediol glucuronide	Endocrine knowledge in the field
	Heistermann et al. (1995a)	Urinary and fecal estrone conjugates, pregnanediol glucuronide, estradiol, progesterone, pregnanediol, 20α-hydroxprogesterone	Endocrine knowledge in captivity
	Lohiya, et al. (1988, 1998)	Serum estradiol and progesterone, serum testosterone and cortisol	Endocrine knowledge in captivity
Great apes			
Gorilla gorilla beringei	Robbins and Czekala (1997)	Urinary testosterone and cortisol	Endocrine knowledge in the field
G. g. gorilla	Bahr et al. (2001)	Urinary estrone conjugates and pregnanediol-3α-glucuronide	Endocrine knowledge in captivity
	Miyamoto et al. (2001)	Fecal estrogens and progesterone	Endocrine knowledge in captivity
	Bellem et al. (1995)	Urinary estrogens and progesterone	Endocrine knowledge in captivity
	Czekala et al. (1983, 1987)	Urinary estrone conjugates, estradiol	Endocrine knowledge in captivity
Pan paniscus	Jurke et al. (2000, 2001)	Urinary and fecal estrogens, progesterone, cortisol	Hormones and behavior in captivity
	Heistermann et al. (1996)	Urinary estrone conjugates, pregnanediol glucuronide and estriol, fecal progesterone	Endocrine knowledge in captivity
P. troglodytes	Möhle et al. (2002)	Urinary and fecal testosterone	Endocrine knowledge in captivity
	Nadler et al. (1985)	Serum estrogens, progesterone, luteinizing hormone, follicle-stimulating hormone, testosterone	Endocrine knowledge in captivity
Pongo pygmaeus	Czekala et al. (1983)	Urinary estrogens	Endocrine knowledge in captivity
	Bonney and Kingsley (1982)	Urinary estrogens and progesterone	Endocrine knowledge in captivity

the production of embryos/infants with the great apes (Gould and Dahl 1993, Joslin et al. 1995, Pope et al. 1997), Old World primates (Cranfield et al. 1993, Curnow et al. 2002, Fazleabas et al. 1993, Ogonuki et al. 2003, Torii et al. 2000), and New World primates (Dukelow 1993, Marshall et al. 1999, Morrell et al. 1998).

SUMMARY AND CONCLUSIONS

Assays for monitoring reproductive events in primates continue to evolve as new technological developments make it less expensive, less labor-intensive, and more efficient to provide results. New sophisticated automated platforms will revolutionize the standard assay lab into one that will have the ability to process hundreds of samples in each run with national and international standards of quality control. In the future, participating scientists will be able to compare results without concern for assay differences, and those results will be obtained within weeks, if not days.

With the advent of new and less expensive technology, it is our hope that more scientists will develop projects to assess reproductive functioning in wild and captive primates that have not yet been evaluated. Many of these species are highly endangered and in areas that require protection or assistance with developing self-sustaining captive breeding programs. As part of any long-term conservation program, it is critical that we have an understanding of the reproductive biology of the species to insure its protection for the future.

REFERENCES

Abbott, D. H., Hodges, J. K., and George, L. M. (1988). Social status controls LH secretion and ovulation in female marmoset monkeys (*Callithrix jacchus*). *J. Endocrinol.* 117:329–339.

Aidara, D., Badawi, M., Tahiri-Zagret, C., and Robyn, C. (1981). Changes in concentrations of serum prolactin, FSH, oestradiol and progesterone and of the sex skin during the menstrual cycle in the mangabey monkey (*Cercocebus atys lunulatus*). *J. Reprod. Fertil.* 62:475–481.

Andelman, S. J., Else, J. G., Hearn, J. P., and Hodges, J. K. (1985). The non-invasive monitoring of reproductive events in wild vervet monkeys (*Cercopithecus aethiops*) using urinary pregnanediol-3-alpha-glucuronide and its correlation with behavioural observations. *J. Zool.* A205:467–477.

Anzenberger, G., and Gossweiler, H. (1993). How to obtain individual urine samples from undisturbed marmoset families. *Am. J. Primatol.* 31:223–230.

Bahr, N. I., Martin, R. D., and Pryce, C. R. (2001). Peripartum sex steroid profiles and endocrine correlates of postpartum maternal behavior in captive gorillas (*Gorilla gorilla gorilla*). *Horm. Behav.* 40:533–541.

Bavister, B. D., and Boatman, D. E. (1993). IVF in nonhuman primates: current status and future directions. In: Wolf, D. P., Stouffer, R. L., and Brenner, R. M. (eds.), *In Vitro Fertilization and Embryo Transfer in Primates.* Spring-Verlag, New York. pp. 30–45.

Bellem, A. C., Monfort, S. L., and Goodrowe, K. L. (1995). Monitoring reproductive development, menstrual cyclicity, and pregnancy in the lowland gorilla (*Gorilla gorilla*) by enzyme immunoassay. *J. Zoo Wildlife Med.* 26:24–31.

Berson, S. A., Yalow, R. S., Bauman, A., Rothchild, M. A., and Newerly, K. (1956). Insulin-I[131] metabolism in human subjects: demonstration of insulin binding globulin in the circulation on insulin-treated subjects. *J. Clin. Invest.* 35:170–190.

Bettinger, T., Kuhar, C., Sironen, A., and Laudenslager, M. (1998). Behavior and salivary cortisol in gorillas housed in an all male group. In: *American Zoological Association Annual Conference Proceedings.* pp. 22–246.

Boes, A., Probst, B., and Erkert, H. G. (1993). Urinary estradiol-17-beta excretion in common marmosets, *Callithrix jacchus*: diurnal pattern and relationship between creatinine-related values and excreted amount. *Comp. Biochem. Physiol.* A105:287–292.

Bonney, R. C., Dixson, A. F., and Fleming, D. (1980). Plasma concentrations of oestradiol-17-beta, estrone, progesterone and testosterone during the ovarian cycle of the owl monkey (*Aotus trivirgatus*). *J. Reprod. Fertil.* 60:101–107.

Bonney, R. C., and Kingsley, S. (1982). Endocrinology of pregnancy in the orang-utan (*Pongo pygmaeus*) and its evolutionary significance. *Int. J. Primatol.* 34:431–444.

Brockman, D. K., and Whitten, P. L. (1996). Reproduction in free-ranging *Propithecus verreauxi*: estrus and the relationship between multiple partner matings and fertilization. *Am. J. Phys. Anthropol.* 100:57–69.

Brockman, D. K., Whitten, P. L., Richard, A. F., and Schneider, A. (1998). Reproduction in free-ranging male *Propithecus verreauxi*: the hormonal correlates of mating and aggression. *Am. J. Phys. Anthropol.* 105:137–151.

Campbell, C. J. (2003). Female-directed aggression in free-ranging *Ateles geoffroyi*. *Int. J. Primatol.* 24:223–237.

Campbell, C. J. (2004). Pattterns of behavior across reproductive states of free-ranging black-handed spider monkeys (*Ateles geoffroyi*). *Am. J. Phys. Anthropol.* 124:166–176.

Campbell, C. J., Shideler, S. E., Todd, H. E., and Lasley, B. L. (2001). Fecal analysis of ovarian cycles in female black-handed spider monkeys (*Ateles geoffroyi*). *Am. J. Primatol.* 54:79–89.

Cavigelli, S. A., and Pereira, M. E. (2000). Mating season aggression and fecal testosterone levels in male ring-tailed lemurs (*Lemur catta*). *Horm. Behav.* 37:246–255.

Cekan, S. Z., Goncharov, N., Aso, T., and Diczfalusy, E. (1977). Circulating steroid levels in male and female baboons (*Papio hamadryas*). In: Prasad, M. R. N., and Anand Kumar, T. C. (eds.), *Use of Non-Human Primates in Biomedical Research.* Indian National Science Academy, New Delhi. pp. 234–241.

Chaoui, N. J., and Hasler-Gallusser, S. (1997). Incomplete sexual suppression in *Leontopithecus chrysomelas*: a behavioural and hormonal study in a semi-natural environment. *Folia Primatol.* 70:47–54.

Chen, J. J., Smith, E. R., Gray, G. D., and Davidson, J. M. (1981). Seasonal changes in plasma testosterone and ejaculatory capacity in squirrel monkeys (*Saimiri sciureus*). *Primates* 22:253–260.

Cranfield, M. R., Bavister, B. D., Boatman, D. E., Berger, N. G., Schaffer, N., Kempski, S. E., Ialeggio, D. M., and Smart, J. (1993). Assisted reproduction in the propagation management of the endangered lion-tailed macaque (*Macaca silenus*). In:

Wolf, D. P., Stouffer, R. L., and Brenner, R. M. (eds.), *In Vitro Fertilization and Embryo Transfer in Primates*. Spring-Verlag, New York. pp. 331–348.

Curnow, E. C., Kuleshova, L., Shaw, J. M., and Hayes, E. S. (2002). Comparison of slow- and rapid-cooling protocols for early-cleavage-stage *Macaca fascicularis* embryos. *Am. J. Primatol.* 58:169–174.

Czekala, N. M., Benirschke, K., McClure, H., and Lasley, B. L. (1983). Urinary estrogen excretion during the pregnancy in the gorilla (*Gorilla gorilla*), orangutan (*Pongo pygmaeus*) and the human (*Homo sapiens*). *Biol. Reprod.* 28:289–294.

Czekala, N. M., Gallusser, S., Meier, J. E., and Lasley, B. L. (1986). The development and application of an enzyme immunoassay for urinary estrone conjugates. *Zoo Biol.* 5:1–6.

Czekala, N. M., Mitchell, W. R., and Lasley, B. L. (1987). Direct measurements of urinary estrone conjugates during the normal menstrual cycle of the gorilla (*Gorilla gorilla*). *Am. J. Primatol.* 12:223–229.

Czekala, N. M., Roser, J. F., Mortensen, R. B., Reichard, T., and Lasley, B. L. (1988). Urinary hormone analysis as a diagnostic tool to evaluate the ovarian function of female gorillas (*Gorilla gorilla gorilla*). *J. Reprod. Fertil.* 82:255–261.

Dahl, K. D., and Hseuh, A. J. (1987). Use of the granulosa cell aromatase bioassay for measurement of bioactive follicle-stimulating hormone in urine and serum samples of diverse species. *Steroids.* 50:375–392.

de Sousa, M. B. C., Naciemento Lopes, M. C., and Albuquerque, A. C. S. R. (2002). Fecal collection in free-ranging common marmosets, *Callithrix jacchus. Neotrop. Primates* 10:21–24.

Diamond, E. J., Aksel, S., Hazelton, J. M., Jennings, R. A., and Abee, C. R. (1987). Serum oestradiol, progesterone, chorionic gonadotropin and prolactin concentrations during pregnancy in the Bolivian squirrel monkey (*Saimiri sciureus*). *J. Reprod. Fertil.* 80:373–381.

DiGiano, L., Nagle, C. A., Quiroga, S., Paul, N., Farinati, Z., Torres, M., and Mendizabal, A. F. (1992). Salivary progesterone for the assessment of the ovarian function in the capuchin monkey (*Cebus apella*). *Int. J. Primatol.* 13:113–123.

Dixson, A. F. (1994). Reproductive biology of the owl monkey. In: Baer, J. F., Weller, R. E., and Kakoma, I. (eds.), Aotus: *The Owl Monkey*. Academic Press, San Diego. pp. 113–132.

Dixson, A. F., and Gardner, J. S. (1981). Diurnal variations in plasma testosterone in a male nocturnal primate, the owl monkey (*Aotus trivirgatus*). *J. Reprod. Fertil.* 62:83–86.

Dixson, A. F., and Herbert, J. (1977). Gonadal hormones and sexual behavior in groups of adult talapoin monkeys (*Miopithecus talapoin*). *Horm. Behav.* 8:141–154.

Dukelow, W. R. (1993). Assisted reproduction in New World primates. In: Wolf, D. P., Stouffer, R. L., and Brenner, R. M. (eds.), *In Vitro Fertilization and Embryo Transfer in Primates*. Spring-Verlag, New York. pp. 73–84.

Eaton, G. G., and Resko, J. A. (1974). Ovarian hormones and sexual behavior in *Macaca nemestrina*. *J. Comp. Physiol. Psychol.* 86:919–925.

Eley, R. M., Tarara, R. P., Worthman, C. M., and Else, J. G. (1989). Reproduction in the vervet monkey (*Cercopithecus aethiops*): III. The menstrual cycle. *Am. J. Primatol.* 17:1–10.

Epple, G., and Katz, Y. (1983). The saddle back tamarin and other tamarins. In: Hearn, J. (ed.), *Reproduction in New World Primates: New World Models in Medical Science*. MTP Press, Lancaster. pp. 115–148.

Fazleabas, A. T., Hild-Petito, S., Donnelly, K. M., Mavrogianis, P., and Verhage, H. G. (1993). Interactions between the embryo and uterine endometrium during implantation and early pregnancy in the baboon (*Papio anubis*). In: Wolf, D. P., Stouffer, R. L., and Brenner, R. M. (eds.), *In Vitro Fertilization and Embryo Transfer in Primates*. Spring-Verlag, New York. pp. 169–181.

Fitch-Synder, H., and Jurke, M. (2003). Reproductive patterns in pygmy lorises (*Nycticebus pygmaeus*): behavioral and physiological correlates of gonadal activity. *Zoo Biol.* 22:15–32.

Fortman, J. D., Herring, J. M., Miller, J. B., Hess, D. L., Verhage, H. G., and Fazleabas, A. T. (1993). Chorionic gonadotropin, estradiol, and progesterone levels in baboons (*Papio anubis*) during early pregnancy and spontaneous abortion. *Biol. Reprod.* 49:737–742.

French, J. A., Bales, K. L., Baker, A. J., and Dietz, J. M. (2003). Endocrine monitoring of wild dominant and subordinate female *Leontopithecus rosalia*. *Int. J. Primatol.* 24:1281–1300.

French, J. A., deGraw, W. A., Hendricks, S. E., Wegner, F., and Bridson, W. E. (1992). Urinary and plasma gonadotropin concentrations in golden lion tamarins (*Leontopithecus r. rosalia*). *Am. J. Primatol.* 26:53–59.

Fujita, S., Mitsunaga, F., Sugiura, H., and Shimuzu, K. (2001). Measurement of urinary and fecal steroid metabolites during the ovarian cycle in captive and wild Japanese macaques, *Macaca fuscata. Am. J. Primatol.* 53:167–176.

Gould, K. G., and Dahl, J. F. (1993). Assisted reproduction in the great apes. In: Wolf, D. P., Stouffer, R. L., and Brenner, R. M. (eds.), *In Vitro Fertilization and Embryo Transfer in Primates*. Spring-Verlag, New York. pp. 46–72.

Harding, R. D., Hulme, M. J., Lunn, S. F., Henderson, C., and Aitken, R. J. (1982). Plasma progesterone levels throughout the ovarian cycle of the common marmoset (*Callithrix jacchus*). *J. Med. Primatol.* 11:43–51.

Harlow, C. R., Hearn, J. P., and Hodges, J. K. (1984). Ovulation in the marmoset monkeys: endocrinology, prediction, and detection. *J. Endocrinol.* 103:17–24.

He, Y. M., Pei, Y. J., Zou, R. J., and Ji, W. Z. (2001). Changes of urinary steroid conjugates and gonadotropin excretion in the menstrual cycle and pregnancy in the Yunnan snub-nosed monkey (*Rhinopithecus bieti*). *Am. J. Primatol.* 55:223–232.

Heistermann, M., Finke, M., and Hodges, J. K. (1995a). Assessment of female reproductive status in captive-housed hanuman langurs (*Presbytis entellus*) by measurement of urinary and fecal steroid excretion patterns. *Am. J. Primatol.* 37:275–284.

Heistermann, M., Moestl, E., and Hodges, J. K. (1995b). Non-invasive endocrine monitoring of female reproductive status: methods and applications to captive breeding and conservation of exotic species. In: Ganslosser, U., and Hodges, J. K. (eds.), *Research and Captive Propagation*. Filander Verlag, Fuerth. pp. 36–48.

Heistermann, M., Möhle, U., Vervaecke, H., van Elsacker, L., and Hodges, J. K. (1996). Application of urinary and fecal steroid measurements for monitoring ovarian function and pregnancy in the bonobo (*Pan paniscus*) and evaluation of perineal swelling patterns in relation to endocrine events. *Biol. Reprod.* 55:844–853.

Heistermann, M., Proeve, E., Wolter, H. J., and Mika, G. (1987). Urinary oestrogen and progesterone excretion before and during pregnancy in a pied bare-faced tamarin (*Saguinus bicolor bicolor*). *J. Reprod. Fertil.* 80:635–640.

Heistermann, M., Tari, S., and Hodges, J. K. (1993). Measurement of faecal steroids for monitoring ovarian function in New World primates, Callitrichidae. *J. Reprod. Fertil.* 99:243–251.

Heistermann, M., Uhrigshardt, J., Husung, A., Kaumanns, W., and Hodges, J. K. (2001). Measurement of faecal steroid metabolites in the lion-tailed macaque (*Macaca silenus*): a non-invasive tool for assessing female ovarian function. *Primate Rep.* 59:27–42.

Hernandez-Lopez, L., Mayagoitia, L., Esquivel-Lacroix, C., Rojas-Mayas, S., and Mondragon-Ceballos, R. (1998). The menstrual cycle of the spider monkey (*Ateles geoffroyi*). *Am. J. Primatol.* 44:183–195.

Herrick, J. R., Agoramoorthy, G., Rudran, R., and Harder, J. D. (2000). Urinary progesterone in free-ranging red howler monkeys (*Alouatta seniculus*): preliminary observations of the estrous cycle and gestation. *Am. J. Primatol.* 51:257–263.

Hodges, J. K. (1985). The endocrine control of reproduction. *Sym. Zool. Soc. Lond.* 54:149–168.

Hodges, J. K. (1996). Determining and manipulating female reproductive parameters. In: Kleiman, D. G., Allen, M. E., Thompson, K. V., and Lumpkin, S. (eds.), *Wild Mammals in Captivity.* University of Chicago Press, Chicago. pp. 418–428.

Hodges, J. K., Czekala, N. M., and Lasley, B. L. (1979). Estrogen and luteinizing hormone secretion in diverse primate species from simplified urinary analysis. *J. Med. Primatol.* 3:349–364.

Hodges, J. K., and Eastman, S. A. K. (1984). Monitoring ovarian function in marmosets and tamarins by the measurement of urinary oestrogen metabolites. *Am. J. Primatol.* 6:187–197.

Hodges, J. K., and Green, D. G. (1989). A simplified enzyme immunoassay for urinary pregnanediol-3-glucuronide: applications to reproductive assessment of exotic species. *J. Zool.* 219:89–99.

Hodges, J. K., Tarara, R., Hearn, J. P., and Else, J. G. (1986). The detection of ovulation and early pregnancy in the baboon by direct measurement of conjugated steroids in urine. *Am. J. Primatol.* 10:329–338.

Jagannadha Rao, A., Kotagi, S. G., and Moudgal, N. R. (1984). Serum concentrations of chorionic gonadotropin, oestradiol-17beta and progesterone during early pregnancy in the south Indian bonnet monkey (*Macaca radiata*). *J. Reprod. Fertil.* 70:449–455.

Joslin, J. O., Weissman, W. D., Johnson, K., Forster, M., Wasser, S., and Collins, D. (1995). In vitro fertilization of Bornean orangutan (*Pongo pygmaeus pygmaeus*) gamete followed by embryo transfer into a surrogate hybrid orangutan (*Pongo pygmaeus*). *J. Zoo Wildlife Med.* 26:32–42.

Jurke, M. H., Hagey, L. R., Czekala, N. M., and Harvey, N. C. (2001). Metabolites of ovarian hormones and behavioral correlates in captive female bonobos (*Pan paniscus*). In: Galdikas, B. M. G., Briggs, N. E., Sheeran, L. K., Shapiro, F. L., and Goodall, J. (eds.), *All Apes Great and Small.* African Apes, vol. 1. Kluwer Academic/Plenum, New York. pp. 217–229.

Jurke, M. H., Hagey, L. R., Jurke, S., and Czekala, N. M. (2000). Monitoring hormones in urine and feces of captive bonobos (*Pan paniscus*). *Primates* 41:311–319.

Jurke, M. H., Pryce, C. R., Dobeli, M., and Martin, R. D. (1994). Non-invasive detection and monitoring of pregnancy and the postpartum period in Goeldi's monkey (*Callimico goeldii*) using urinary pregnanediol-3-alpha-glucuronide. *Am. J. Primatol.* 34:319–331.

Khan, S. A., Katzija, G., Lindberg, M., and Dicsfalusy, E. (1985). Radioimmunoassay of luteinizing hormone in the baboon (*Papio hamadryas*). *J. Med. Primatol.* 14:143–157.

Kholkute, S. D., Joseph, R., Joshi, U. M., and Munshi, S. R. (1981). Diurnal variations of serum testosterone levels in the male bonnet monkey (*Macaca radiata*). *Primates* 22:427–430.

Klecha, F., and McCann, C. (2001). A profile of progesterone excretion during pregnancy in gelada baboons (*Theropithecus gelada*). In: Chan, S. D. (ed.), *American Association of Zoo Keepers 27th National Conference.* Columbus Zoo and Aquarium, Powell. pp. 62–71.

Knott, C. D. (1997). Field collection and preservation of urine in orangutans and chimpanzees. *Trop. Biodivers.* 4:95–102.

Kuederling, I., Evans, C. S., Abbott, D. H., Pryce, C. R., and Epple, G. (1995). Differential excretion of urinary oestrogen by breeding females and daughters in the red-bellied tamarin (*Saguinus labiatus*). *Folia Primatol.* 64:140–145.

Lasley, B. L. (1985). Methods for evaluating reproductive function in exotic species. *Adv. Vet. Sci. Comp. Med.* 30:209–228.

Lasley, B. L., Bogart, M. H., and Shideler, S. E. (1978). A comparison of lemur ovarian cycles. In: Alexander, N. J. (ed.), *Animal Models for Research on Contraception and Fertility.* Harper and Row, Hagerstown. pp. 417–424.

Lasley, B. L., Hodges, J. K., and Czekala, N. M. (1980). Monitoring the female reproductive cycle of great apes and other species by estrogen and bioactive luteinizing hormone determinations in small volumes of urine. *J. Reprod. Fertil.* 28:121–129.

Lasley, B. L., and Kirkpatrick, J. F. (1991). Monitoring ovarian function in captive and free-ranging wildlife by means of urinary and fecal steroids. *J. Zoo Wildlife Med.* 22:23–31.

Lasley, B. L., Lindburg, D. G., Robinson, P. T., and Benirschke, K. (1981a). Captive breeding of exotic species. *J. Zoo Anim. Med.* 12:67–73.

Lasley, B. L., Monfort, S. L., Hodges, J. K., and Czekala, N. M. (1981b). Comparison of urinary estrogens during pregnancy in diverse species. In: Novy, M. J., and Resko, J. A. (eds.), *Fetal Endocrinology.* Academic Press, New York. pp. 111–126.

Lasley, B. L., and Shideler, S. E. (1993). Methods for assessing reproduction in nondomestic species. In: Fowler, M. E. (ed.), *Zoo and Wild Animal Medicine: Current Therapy*, 3rd ed., W. B. Saunders, Philadelphia. pp. 79–86.

Lerchl, A., Kuederling, I., and Epple, G. (1991). Diurnal variations of urinary testosterone excretion in the male saddle-back tamarin (*Saguinus fuscicollis*). *Primate Rep.* 30:17–23.

Lipschitz, D. L. (1996). A preliminary investigation of the relationship between ovarian steroids, LH, reproductive behaviour and vaginal changes in lesser bushbabies (*Galago moholi*). *J. Reprod. Fertil.* 107:164–174.

Lohiya, N. K., Sharma, R. S., Manivannan, B., and Anand Kumar, T. C. (1998). Reproductive exocrine and endocrine profiles and their seasonality in male langur monkeys (*Presbytis entellus entellus*). *J. Med. Primatol.* 27:15–20.

Lohiya, N. K., Sharma, R. S., Puri, C. P., David, G. F. X., and Anand Kumar, T. C. (1988). Reproductive exocrine and endocrine profile of female langur monkeys, *Presbytis entellus.* *J. Reprod. Fertil.* 82:485–492.

Lynch, J. W., Zielger, T. E., and Strier, K. B. (2002). Individual and seasonal variation in fecal testosterone and cortisol levels of wild male tufted capuchin monkeys, *Cebus apella nigritus.* *Horm. Behav.* 41:275–287.

Mann, D. R., Blank, M. S., Gould, K. G., and Collins, D. C. (1982). Validation of radioimmunoassays for LH and FSH in the sooty mangabey (*Cercocebus atys*): characterization of LH and the response to GnRH. *Am. J. Primatol.* 2:275–283.

Marshall, A. J., and Hohmann, G. (2005). Urinary testosterone levels of wild male bonobos (*Pan paniscus*) in the Lomako Forest, Democratic Republic of Congo. *Am. J. Primatol.* 65:87–92.

Marshall, V. S., Tannenbaum, P., Browne, M. A., Knowles, L., Kalishman, J. K., and Thomson, J. A. (1999). Assisted reproductive technologies in a New World primate, the common marmoset monkey (*Callithrix jacchus*). *American Association of Zoo Veterinarians Annual Conference Proceedings*. pp. 93–94.

Matsubayashi, K., and Enomoto, T. (1983). Longitudinal studies on annual changes in plasma testosterone, body weight and spermatogenesis in adult Japanese monkeys (*Macaca fuscata fuscata*) under laboratory conditions. *Primates* 24:521–529.

Matteri, R. L., Roser, J. F., Baldwin, D. M., Lipovetsky, V., and Papkoff, H. (1987). Characterization of a monoclonal antibody which detects luteinizing hormone from diverse mammalian species. *Dom. Anim. Endocrinol.* 4:157–165.

McConnell, I. W., Atwell, R., Renquist, D., and Gunnels, R. (1990). Serum concentrations of progesterone and estrogens during the reproductive cycle of owl monkeys (*Aotus trivirgatus*). *Georgia J. Sci.* 48:96–101.

Michael, R. P., Bonsall, R. W., and Zumpe, D. (1987). Testosterone and its metabolites in male cynomolgus monkeys (*Macaca fascicularis*): behavior and biochemistry. *Physiol. Behav.* 40:527–537.

Mitchell, W. R., Presely, S., Czekala, N. M., and Lasley, B. L. (1982). Urinary immunoreactive estrogen and pregnanediol-3-glucuronide during the normal menstrual cycle of the female lowland gorilla (*Gorilla gorilla*). *Am. J. Primatol.* 2:167–175.

Miyamoto, S., Chen, Y., Kurotori, H., Sankai, T., Yoshida, T., and Machida, T. (2001). Monitoring the reproductive status of female gorillas (*Gorilla gorilla gorilla*) by measuring the steroid hormones in fecal samples. *Primates* 42:291–299.

Möhle, U., Heistermann, M., Palme, R., and Hodges, J. K. (2002). Characterization or urinary and fecal metabolites of testosterone and their measurement for assessing gonadal endocrine function in male non-human primates. *Gen. Comp. Endocrinol.* 129:135–145.

Monfort, S. L., Hess, D. L., Shideler, E. S., Samuels, S. J., Hendrickx, A. G., and Lasley, B. L. (1987). Comparison of serum estradiol to urinary estrone conjugates in rhesus macaque (*Macaca mulatta*). *Biol. Reprod.* 37:832–837.

Moorman, E. A., Mendoza, S. P., Shideler, S. E., and Lasley, B. L. (2002). Excretion and measurement of estradiol and progesterone metabolites in the feces and urine of female squirrel monkeys (*Saimiri sciureus*). *Am. J. Primatol.* 57:79–90.

Morrell, J. M., Nubbemeyer, R., Heistermann, M., Rosenbusch, J., Kuederling, I., Holt, W., and Hodges, J. K. (1998). Artificial insemination in *Callithrix jacchus* using fresh or cryopreserved sperm. *Anim. Reprod. Sci.* 52:165–174.

Nadler, R. D., Graham, C. E., Gosselin, R. E., and Collins, D. C. (1985). Serum levels of gonadotropin and gonadal steroids, including testosterone, during the menstrual cycle of the chimpanzee (*Pan troglodytes*). *Am. J. Primatol.* 9:273–284.

Nagle, C. A., Paul, N., Mazzoni, I., Quiroga, S., Torres, M., Mendizabal, A. F., and Farinati, Z. (1989). Interovarian relationship in the secretion of progesterone during the luteal phase of the capuchin monkey (*Cebus apella*). *J. Reprod. Fertil.* 85:389–396.

Nigi, H., and Torii, R. (1983). Periovulatory time courses of plasma estradiol and progesterone in the Japanese monkey (*Macaca fuscata*). *Primates* 24:410–418.

Norman, R. L., Brandt, H., and Van Horn, R. N. (1978). Radioimmunoassay for luteinizing hormone (LH) in the ring-tailed lemur (*Lemur catta*) with antiovine LH and ovine ^{125}I-LH[1]. *Biol. Reprod.* 19:1119–1124.

Nunes, S., Brown, C., and French, J. A. (2002). Variation in circulating and excreted estradiol associated with testicular activity in male marmosets. *Am. J. Primatol.* 546:27–42.

Ogonuki, N., Tsuchiya, H., Hirose, Y., Okada, H., Ogura, A., and Sankai, T. (2003). Pregnancy by the tubal transfer of embryos developed after injection of round spermatids into oocyte cytoplasm of the cynomolgus monkey (*Macaca fascicularis*). *Hum. Reprod.* 18:1273–1280.

Perez, L., Czekala, N. M., Weisenseel, A., and Lasley, B. L. (1988). Excretion of radiolabeled estradiol metabolites in the slow loris (*Nycticebus coucang*). *Am. J. Primatol.* 16:321–330.

Perret, M. (1986). Social influences on oestrous cycle length and plasma progesterone concentrations in the female lesser mouse lemur (*Microcebus murinus*). *J. Reprod. Fertil.* 77:303–311.

Plant, T. M. (1981). Time courses of concentrations of circulating gonadotropin, prolactin, testosterone, and cortisol in adult male rhesus monkeys (*Macaca mulatta*) throughout the 24h light–dark cycle. *Biol. Reprod.* 25:244–252.

Pope, C. E., Dresser, B. L., Chin, N. W., Liu, J. H., Loskutoff, N. M., Behnke, E. J., Brown, C., McRae, M. M., Sinovway, C. E., Campbell, M. K., Cameron, K. N., Owens, O. M., Johnson, C. A., Evans, R. R., and Cedars, M. I. (1997). Birth of a western lowland gorilla (*Gorilla gorilla gorilla*) following in vitro fertilization and embryo transfer. *Am. J. Primatol.* 41:247–260.

Pryce, C. R., Jurke, M., Shaw, H. J., Sandmeier, I. G., and Doebeli, M. (1993). Determination of ovarian cycle in Goeldi's monkey (*Callimico goeldii*) via the measurement of steroids and peptides in plasma and urine. *J. Reprod. Fertil.* 99:427–435.

Pryce, C. R., Schwarzenberger, F., and Doebeli, M. (1994). Monitoring fecal samples for estrogen excretion across the ovarian cycle in Goeldi's monkey (*Callimico goeldii*). *Zoo Biol.* 13:219–230.

Pryce, C. R., Schwarzenberger, F., Doebeli, M., and Etter, K. (1995). Comparative study of oestrogen excretion in female New World monkeys: an overview of non-invasive ovarian monitoring and a new application in evolutionary biology. *Folia Primatol.* 64:107–123.

Reddy, S. M., Ray, A. K., and Gupta, P. (1992). The assisted reproductive technologies (ART) in conserving endangered wildlife: a review. *Proc. Zool. Soc. Calcutta* 45(suppl. A):149–159.

Risler, L., Wasser, S. K., and Sackett, G. P. (1987). Measurement of excreted steroids in *Macaca nemestrina*. *Am. J. Primatol.* 12:91–100.

Robbins, M. M., and Czekala, N. M. (1997). A preliminary investigation of urinary testosterone and cortisol levels in wild male mountain gorillas. *Am. J. Primatol.* 43:51–64.

Savage, A., Lasley, B. L., Vecchio, A. J., Miller, A. E., and Shideler, S. E. (1995). Selected aspects of female white-faced saki (*Pithecia pithecia*) reproductive biology in captivity. *Zoo Biol.* 14:441–452.

Savage, A., Shideler, S. E., Soto, L. H., Causado, J., Lasley, B. L., and Snowdon, C. T. (1997). Reproductive events of wild cotton-top tamarin (*Saguinus oedipus*) in Colombia. *Am. J. Primatol.* 43:329–337.

Setchell, K. D. R., and Bonney, R. C. (1981). The excretion of urinary steroids by the owl monkey (*Aotus trivirgatus*) studied using open tubular capillary column gas chromatography and mass spectrometry. *J. Steroid Biochem.* 14:37–43.

Shideler, S. E., Czekala, N. M., Kasman, L. H., Lindburg, D. G., and Lasley, B. L. (1983a). Monitoring ovulation and implantation in the lion-tailed macaque (*Macaca silenus*) through urinary estrone conjugate evaluations. *Biol. Reprod.* 29:905–911.

Shideler, S. E., and Lasley, B. L. (1982). A comparison of primate ovarian cycles. *Am. J. Primatol.* 1(suppl.):171–180.

Shideler, S. E., Lindburg, D. G., and Lasley, B. L. (1983b). Estrogen–behavior correlates in the reproductive physiology and behavior of the ruffed lemur (*Lemur variegatus*). *Horm. Behav.* 17:249–263.

Shideler, S. E., Mitchell, W. R., Lindburg, D. G., and Lasley, B. L. (1985). Monitoring luteal function in the lion-tailed macaque (*Macaca silenus*) through urinary progesterone metabolite measurements. *Zoo Biol.* 4:65–73.

Shideler, S. E., Munro, C. J., Johl, H. K., Taylor, H. W., and Lasley, B. L. (1995). Urine and fecal sample collection on filter paper for ovarian hormone evaluations. *Am. J. Primatol.* 37:305–315.

Shideler, S. E., Munro, C. J., Tell, L., Owiti, G., Laughlin, L., Chatterton, R., Jr., and Lasley, B. L. (1990). The relationship of serum estradiol and progesterone concentrations to the enzyme immunoassay measurements of urinary estrone conjugates and immunoreactive pregnanediol-3-glucuronide in *Macaca mulatta*. *Am. J. Primatol.* 22:113–122.

Shideler, S. E., Ortuno, A. M., Moran, F. M., Moorman, E. A., and Lasley, B. L. (1993a). Simple extraction and enzyme immunoassays for estrogen and progesterone metabolites in the feces of *Macaca fascicularis* during non-conceptive and conceptive ovarian cycles. *Biol. Reprod.* 48:1290–1298.

Shideler, S. E., Savage, A., Ortuno, A. M., Moorman, E. A., and Lasley, B. L. (1994). Monitoring female reproductive function by measurement of fecal estrogen and progesterone metabolites in the white-faced saki (*Pithecia pithecia*). *Am. J. Primatol.* 32:95–108.

Shideler, S. E., Shackleton, C. H. L., Moran, F. M., Stauffer, P., Lohstroh, P. N., and Lasley, B. L. (1993b). Enzyme immunoassays for ovarian steroid metabolites in the urine of *Macaca fascicularis*. *J. Med. Primatol.* 22:301–312.

Shimizu, K., Lohstroh, P. N., Laughlin, L. S., Gee, N. A., Todd, H., Shideler, S. E., and Lasley, B. L. (2001). Noninstrumented enzyme-linked immunosorbant assay for detection of early pregnancy in macaques. *Am. J. Primatol.* 54:57–62.

Shoemaker, A. H. (1982). Notes on the reproductive biology of the white-faced saki in captivity. *Int. Zoo Ybk.* 22:124–127.

Smith, T. E., and French, J. A. (1997a). Social and reproductive conditions modulate urinary cortisol excretion in black tufted-ear marmosets (*Callithrix kuhli*). *Am. J. Primatol.* 42:253–267.

Smith, T. E., and French, J. A. (1997b). Psychosocial stress and urinary cortisol excretion in marmoset monkeys (*Callithrix kuhli*). *Physiol. Behav.* 62:225–232.

Stabenfeldt, G. H., and Hendrickx, A. G. (1973). Progesterone levels in the sooty mangabey (*Cercocebus atys*) during the menstrual cycle, pregnancy, and parturition. *J. Med. Primatol.* 2:1–10.

Stavisky, R., Russell, E., Stallings, J., Smith, E. O., Worthman, C., and Whitten, P. L. (1995). Fecal steroid analysis of ovarian cycles in free-ranging baboons. *Am. J. Primatol.* 36:285–297.

Strier, K. B., and Ziegler, T. E. (1997). Behavioral and endocrine characteristics of the reproductive cycle in wild muriqui monkeys, *Brachyteles arachnoids*. *Am. J. Primatol.* 42:299–310.

Taussky, H. H. (1954). A microcolorimetric determination of creatinine in urine by the Jaffe reaction. *J. Biol. Chem.* 208:853–861.

Thierry, B., Heistermann, M., Aujard, F., and Hodges, J. K. (1996). Long-term data on basic reproductive parameters and evaluation of endocrine, morphological, and behavioral measures for monitoring reproductive status in a group of semifree-ranging Tonkean macaques (*Macaca tonkeana*). *Am. J. Primatol.* 39:47–62.

Thomas, J. A., Hess, D. L., Dahl, K. D., Iliff-Sizemore, S. A., Stouffer, R. L., and Wolf, D. P. (1992). The Sulawesi crested black macaque (*Macaca nigra*) menstrual cycle: changes in perineal tumescence and serum estradiol, progesterone, follicle-stimulating hormone, and luteinizing hormone levels. *Biol. Reprod.* 46:879–884.

Tiefenbacher, S., Lee, B., Meyer, J. S., and Spealman, R. D. (2003). Noninvasive technique for the repeated sampling of salivary free cortisol in awake, unrestrained squirrel monkeys. *Am. J. Primatol.* 60:69–75.

Torii, R., Hosoi, Y., Masuda, Y., Iritani, A., and Nigi, H. (2000). Birth of the Japanese monkey (*Macaca fuscata*) infant following in-vitro fertilization and embryo transfer. *Primates* 41:39–47.

Trouson, A. (1993). State of the art and future directions in human IVF. In: Wolf, D. P., Stouffer, R. L., and Brenner, R. M. (eds.), *In Vitro Fertilization and Embryo Transfer in Primates*. Spring-Verlag, New York. pp. 3–29.

Valeggia, C. R., Mendoza, S. P., Fernandez-Duque, E., and Mason, W. A. (1999). Reproductive biology of female titi monkeys (*Callicebus moloch*) in captivity. *Am. J. Primatol.* 47:183–195.

Van Horn, R. N., and Eaton, G. G. (1979). Reproductive physiology and behavior in prosimians. In: Doyle, G. A., and Martin, R. D. (eds.), *The Study of Prosimian Behavior*. Academic Press. New York. pp. 79–122.

Wasser, S. K. (1996). Reproductive control in wild baboons measured by fecal steroids. *Biol. Reprod.* 55:393–399.

Wasser, S. K., Monfort, S. L., and Wildt, D. E. (1991). Rapid extraction of faecal steroids for measuring reproductive cyclicity and early pregnancy in free-ranging yellow baboons (*Papio cynocephalus cynocephalus*). *J. Reprod. Fertil.* 92:415–423.

Wasser, S. K., Risler, L., and Steiner, R. A. (1988). Excreted steroids in primate feces over the menstrual cycle and pregnancy. *Biol. Reprod.* 39:862–872.

Wehrenberg, W. B., Dyrenfurth, I., and Ferin, M. (1980). Endocrine characteristics of the menstrual cycle in the Assamese monkey (*Macaca assamensis*). *Biol. of Reprod.* 23:522–525.

Whitten, P. L., and Russell, E. (1996). Information context of sexual swelling and fecal steroids in sooty mangabeys (*Cercocebus torquatus atys*). *Am. J. Primatol.* 40:67–82.

Wilks, J. W. (1977). Endocrine characterization of the menstrual cycle of the stumptailed macaque (*Macaca arctoides*). *Biol. Reprod.* 16:474–478.

Winterer, J., Palmer, A. E., Cicmanec, J., Davis, E., Harbaugh, S., and Loriaux, D. L. (1985). Endocrine profile of pregnancy in the patas monkey (*Erythrocebus patas*). *Endocrinology* 116:1090–1093.

Yeoman, R. R., Wegner, F. H., Gibson, S. V., Williams L. E., Abbot, D. H., and Abee, C. R. (2000). Midcycle and luteal elevations of follicle stimulating hormone in squirrel monkeys (*Saimiri boliviensis*) during the estrous cycle. *Am. J. Primatol.* 52:207–211.

Ziegler, T., Hodges, K., Winker, P., and Heistermann, M. (2000a). Hormonal correlates of reproductive seasonality in wild female hanuman langurs (*Presbytis entellus*). *Am. J. Primatol.* 51:119–134.

Ziegler, T. E., Bridson, W. E., Snowdon, C. T., and Eman, S. (1987a). Urinary gonadotropin and estrogen excretion during the postpartum estrus, conception, and pregnancy in the cotton-top tamarin (*Saguinus oedipus oedipus*). *Am. J. Primatol.* 12:127–140.

Ziegler, T. E., Carlson, A. A., Ginther, A. J., and Snowdon, C. T. (2000b). Gonadal source of testosterone metabolites in urine of male cotton-top tamarin monkeys (*Saguinus oedipus*). *Gen. Comp. Endocrinol.* 118:332–343.

Ziegler, T. E., Matteri, R. L., and Wegner, F. H. (1993a). Detection of urinary gonadotropins in callitrichid monkeys with a sensitive immunoassay based upon a unique monoclonal antibody. *Am. J. Primatol.* 31:181–188.

Ziegler, T. E., Santos, C. V., Pissinatti, A., and Strier, K. B. (1997a). Steroid excretion during the ovarian cycle in captive and wild muriquis, *Brachyteles arachnoids*. *Am. J. Primatol.* 42:311–321.

Ziegler, T. E., Savage, A., Scheffler, G., and Snowdon, C. T. (1987b). The endocrinology of puberty of reproductive functioning in female cotton-top tamarins (*Saguinus oedipus*) under varying social conditions. *Biol. Reprod.* 37:618–627.

Ziegler, T. E., Scheffler, G., Wittwer, E. J., Schultz-Darken, N., Snowdon, C. T., and Abbott, D. H. (1996). Metabolism of reproductive steroids during the ovarian cycle in two species of callitrichids, *Saguinus oedipus* and *Callithrix jacchus*, and estimate of the ovulatory period from fecal steroids. *Biol. Reprod.* 54:91–99.

Ziegler, T. E., and Snowdon, C. T. (1997b). Role of prolactin in paternal care in a monogamous New World primate, *Saguinus oedipus*. *Ann. N. Y. Acad. Sci.* 807:599–601.

Ziegler, T. E., Snowdon, C. T., and Bridson, W. E. (1990a). Reproductive performance and excretion of urinary estrogens and gonadotropins in the female pygmy marmoset (*Cebuella pygmaea*). *Am. J. Primatol.* 22:191–203.

Ziegler, T. E., Snowdon, C. T., Warneke, M., and Bridson, W. E. (1990b). Urinary excretion of oestrone conjugates and gonadotrophins during pregnancy in the Goeldi's monkey (*Callimico goeldii*). *J. Reprod. Fertil.* 89:163–168.

Ziegler, T. E., Wegner, F. H., Carlson, A. A., Lazaro-Perea, C., and Snowdon, C. T. (2000c). Prolactin levels during the periparturitional period in the biparental cotton-top tamarin (*Saguinus oedipus*): interactions with gender, androgen levels, and parenting. *Horm. Behav.* 38:111–122.

Ziegler, T. E., Widowski, T. M., Larson, M. L., and Snowdon, C. T. (1990c). Nursing does affect the duration of the post-partum to ovulation interval in cotton-top tamarins (*Saguinus oedipus*). *J. Reprod. Fertil.* 90:563–570.

Ziegler, T. E., Wittwer, D. J., and Snowdon, C. T. (1993b). Circulating and excreted hormones during the ovarian cycle in the cotton-top tamarin, *Saguinus oedipus*. *Am. J. Primatol.* 31:55–65.

22

Molecular Primatology

Anthony Di Fiore and Pascal Gagneux

THE UTILITY AND PROMISE OF MOLECULAR PRIMATOLOGY

The fundamental aim of molecular primatology is to describe and interpret the patterns of molecular variation found within and between primate taxa. There are a number of areas in contemporary primatology where molecular studies are particularly useful.

First, molecular studies provide a means for quickly and efficiently generating data that can be used for inferring phylogenetic relationships among groups of organisms. One advantage of molecular data is that the alternative states of a molecular character are typically far less ambiguous than for many morphological traits (e.g., a nucleotide base that is "adenine" versus a molar crest that is "high"). Another advantage of molecular over morphological data is that, at least for some loci, reasonably good estimates of the rate of change (mutation rates) for deoxyribonucleic acid (DNA) and amino acid sequences are available. If these rates of change are reasonably consistent over evolutionary time (i.e., approximate a so-called molecular clock), researchers can infer the timing of lineage splitting events within a phylogeny. Over the last decade, molecular phylogenies for extant primates have largely replaced morphology-based ones.

Second, molecular studies provide a direct way to investigate numerous features of primate social systems that are not easily observed in the field. For example, using various kinds of molecular markers, researchers can conduct parentage assessments, document kin relationships among individuals within or between social groups, examine population substructuring, assess levels of inbreeding within and gene flow between populations, and evaluate sex biases in dispersal behavior, allowing inferences to be made about mating systems, social organization, and dispersal patterns that are typically impossible to make based on observational studies alone (Di Fiore 2003).

Third, molecular techniques can contribute significantly to applied primate conservation efforts both in the wild and in captivity. For example, researchers might use genotyping of noninvasively collected samples (e.g., feces, hair from nests) to perform molecular mark–recapture studies for determining population density and sex composition or, if species-specific markers are available, to assay the presence or absence of a particular species in an area. Similarly, with genotype data, captive colony managers and zoo personnel can make more informed decisions about which animals to breed in order to minimize inbreeding and preserve the genetic variation present in captive populations. Routine health monitoring and veterinary care for captive primates benefit from molecular studies as well. A variety of polymerase chain reaction (PCR)–based tests are already available that allow animals to be quickly screened for various hereditary disorders as well as for infection by numerous pathogens.

Fourth, molecular studies of primates are increasingly relevant to questions of human health and disease as well as to the tracking and management of disease outbreaks in wild animal populations. Primates are natural reservoirs and zoonotic sources for a variety of current and emerging human diseases (e.g., ebola, Marburg, simian or human immunodeficiency virus, herpes simiae virus) (Daszak et al. 2000, Hahn et al. 2000). Molecular studies of both pathogen and host populations can lead to better understanding of the source and mode of transmission of these diseases, as well as the factors responsible for differential susceptibility to various pathogens.

Finally, the current era of *genomics* (the study of the complete DNA sequence of the genome of an organism) holds promise for evolutionary biologists to be able to undertake far more complete and sophisticated comparisons of the organization, function, and evolution of genomes across taxa. Nearly complete genome sequences are now known for a variety of organisms, ranging from microbes to invertebrates to mammals, including mice (Mouse Genome Sequencing Consortium 2002), modern humans (International Human Genome Sequencing Consortium 2001, 2004), and our close relative the chimpanzee (Chimpanzee Sequencing

and Analysis Consortium 2005), with the completed cow, dog, rat, and rhesus macaque genomes expected soon. Comparative genomic studies are providing insights into the history of genes associated with important diseases in humans and are shedding light on a variety of candidate genes that have been implicated in the development of many aspects of biology and behavior that appear to distinguish humans from the great apes. For example, despite being extremely closely related genetically and sharing close to 99% of their genomes in coding regions, humans and chimpanzees differ notably in many aspects of life history, physiology, behavior, and, fundamentally, their capacity for symbolic and acoustic communication. Molecular studies are now beginning to reveal some of the important molecular differences between modern humans and chimpanzees (Gagneux and Varki 2001).

BREADTH OF THE FIELD

Current use of the term *molecular primatology* typically equates molecular studies with studies of DNA, the fundamental molecule of genetic heredity. Indeed, studies of primate DNA have burgeoned in the last decade as researchers have increasingly employed genetic techniques (particularly microsatellite genotyping and direct DNA sequencing) to investigate mating systems and social structure (Altmann et al. 1996; Keane et al. 1997; Gagneux et al. 1997b, 1999; Gerloff et al. 1999; Nievergelt et al. 2000; Laundhart et al. 2001; Constable et al. 2001; Vigilant et al. 2001; Utami et al. 2002; Radespiel et al. 2002; Kappeler et al. 2002; Wimmer et al. 2002), dispersal patterns (Morin et al. 1994, Melnick and Hoelzer 1996, Di Fiore 2002, Di Fiore and Fleischer 2005), and within-group social behavior, kin recognition, and relatedness (Goldberg and Wrangham 1997, Alberts 1999, Hohmann et al. 1999, Mitani et al. 2000, Vigilant et al. 2001, Widdig et al. 2001, Di Fiore and Fleischer 2005). Genetic data are also contributing increasingly to attempts to infer phylogenetic relationships among all major radiations of primates (strepsirhines, Yoder et al. 1996, Yoder 1997, Yoder and Yang 2004; New World monkeys, von Dornum and Ruvolo 1999, Schneider et al. 2001; Old World monkeys, Stewart and Disotell 1998, Disotell 2000, hominoids, Arnason et al. 1996, Ruvolo 1997, Satta et al. 2000), as well as to evaluate the demographic (Gagneux et al. 1999; Storz et al. 2002a,b) and biogeographic (Rosenblum et al. 1997a,b; Goldberg and Ruvolo 1997; Evans et al. 1999; Collins and Dubach 2000a,b; Hapke et al. 2001; Gagneux et al. 2001; Tosi et al. 2000, 2003; Cortés-Ortiz et al. 2003; Wildman et al. 2004) histories of various primate groups.

However, the true scope of molecular primatology is much broader than simply the study of DNA sequence-level genetic variation, and the discipline has deep historical roots (Table 22.1). First, there are many other types of biomolecules besides DNA that investigators might use to explore patterns of natural variation within and between primate species,

Table 22.1   Brief Overview of Significant Developments in Molecular Primatology

DEVELOPMENT	REFERENCE(S)
Discovery of human blood groups	Landsteiner (1900)
Early immunological experiments on blood serum cross-reactivity	Nuttal (1904)
First correct assessment of the human chromosome number ($2n = 46$)	Tjio and Levan (1956)
Immunological cross-reactivity experiments to elucidate the phylogeny of hominoid primates: results suggested a much more recent common ancestry for humans and great apes than previously thought	Sarich and Wilson (1967)
Dramatic improvements in cytogenetic techniques for staining metaphase chromosomes	Caspersson et al. (1968)
Phylogenetic studies of many of the major primate radiations using immunological methods	Goodman and Moore (1971), Goodman et al. (1974), Dene et al. (1976a,b, 1980), Baba et al. (1979), Sarich and Cronin (1976)
Early applications of amino acid sequence data to primate phylogenetic studies	Goodman et al. (1975), Goodman (1976), Hewett-Emmett et al. (1976), Romero-Herrera et al. (1976), Tashian et al. (1976)
Allozyme methods applied to studying the population structure, dispersal pattern, and mating system of a variety of wild primates (e.g., baboons, lion tamarins, howler monkeys, rhesus macaques, long-tailed macaques)	Jolly and Brett (1973), Brett et al. (1976), Melnick et al. (1984a,b), Forman et al. (1987), Pope (1990, 1992), de Ruiter et al. (1992, 1994)
Cytogenetic studies of hominoid primates suggest a sister-taxon relationship for humans and chimpanzees to the exclusion of gorillas	Yunis and Prakash (1982)
DNA–DNA hybridization studies of the phylogenetic relationships among African great apes provide additional support for a sister-taxon relationship between humans and chimpanzees	Sibley and Ahlquist (1984, 1987)
Development of PCR, which allows in vitro cloning of specific fragments of the genome from a complex DNA sample	Saiki et al. (1985, 1988), Mullis and Faloona (1987)
Evolutionary relationships among hominoids investigated using 2D allozyme electrophoresis: results provide additional support for a chimpanzee–human clade within the hominoids	Goldman et al. (1987)
RFLP studies of mitochondrial DNA variation suggest a recent African origin for all modern humans	Cann et al. (1987), Vigilant et al. (1991)
Demonstration of the utility of noninvasive sampling (e.g., hair, buccal cells, feces) as a source of DNA for PCR-based genotyping and sequencing studies of primates	Higuchi et al. (1988), Takasaki and Takenaka (1991), Morin and Woodruff (1992), Sugiyama et al. (1993), Constable et al. (1995)
Minisatellite DNA "fingerprinting" applied to paternity testing in captive and semi-free-ranging primates (e.g., ring-tailed lemurs, mandrills, Barbary macaques, rhesus macaques, stump-tailed macaques)	Periera and Weiss (1991), Dixson et al. (1993), Paul et al. (1993), Berard et al. (1993, 1994), Bauers and Hearn (1994)
Widespread use of PCR-based genotyping and sequencing in molecular ecological and phylogenetic studies of primates	1995 to present

PCR, polymerase chain reaction; RFLP, restriction fragment length polymorphism; 2D, two-dimensional.

from various types of ribonucleic acid (RNA) to regulatory enzymes to cell membrane proteins to "classic" serum protein markers. Moreover, there are numerous ways in which many of these molecules can be modified following transcription and translation through posttranslational modifications (e.g., addition of sugars, phosphates, sulfates, fatty acids, etc.). The study of variation among and within taxa in all of these molecules also falls under the rubric of "molecular primatology" and can contribute to our understanding of the evolutionary relationships among primate taxa, the selective pressures which have acted on populations within taxa, and the differences in physiology and behavior observed between taxa. Additionally, advances in the study of chromosomes (e.g., cytogenetic techniques such as chromosome painting through fluorescence in situ hybridization, or FISH) and new computational methods for mining and comparing large

sets of sequence data are providing greater insights into variation among primate taxa in genome structure. Among such findings are the discovery of differences among taxa in the type, frequency, and distribution within the genome of various kinds of repetitive DNA, transposons (i.e., mobile pieces of DNA, or "jumping genes"), and endogenous retroviruses (Gagneux and Varki 2001, Trask 2002). Finally, prompted by insights from comparative genomic and gene expression studies, we have come to realize the significance of variation in both coding and noncoding (e.g., promoter and enhancer binding sites, pre-messenger RNA [mRNA] splicing signals present in introns and at intron/exon boundaries) regions of the genome for producing phenotypic variation, either through their regulation of transcription or through their effects on pre-mRNA splicing (Hastings and Krainer 2001, Modrek and Lee 2002, Cartegni et al. 2002).

As these loci play a central role in the link between genotype and phenotype and likely explain much of the evolutionarily significant phenotypic variation seen among closely related primate taxa, they should be important targets for future research.

In the remainder of this chapter, our aim is to provide a brief but broad summary of modern molecular primatology. We begin by summarizing some of the major laboratory methods that are currently used to analyze molecular variation, dividing our discussion into four sections: variation at the level of the genome, variation at the level of the gene or DNA sequence, variation in gene structure and expression, and variation in other types of non-DNA molecules. We then briefly detail the kinds of samples primatologists should collect in the field in order to be able to carry out these types of molecular studies and provide some methodological suggestions for sample collection and storage. Finally, we point to a number of resources where students and researchers new to this field can find more information on laboratory and analytical methods.

MODERN TECHNIQUES FOR ASSAYING MOLECULAR VARIATION

Depending on the question being addressed, researchers will be interested in different levels of molecular variation. For example, comparative karyotypic or cytogenetic data are useful for broad phylogenetic studies as well as for comparing genome organization across taxa. Data on variation between taxa in a variety of non-DNA molecules (e.g., cell surface proteins) and comparative data on gene expression can provide insights into the fundamental biological differences among nonhuman primate taxa. DNA sequence data, like cytogenetic data, are useful for evaluating the phylogenetic relationships among a group of taxa and can also be used to examine intraspecific variation (e.g., for phylogeographic studies) if the loci under investigation evolve rapidly enough. Finally, for detailed analyses of identity, parentage, and relatedness within populations of a particular species, researchers would be most interested in assaying variation at very rapidly evolving and hypervariable loci such as microsatellites or minisatellites.

Variation at the Level of the Genome

Studying the organization and comparative composition of whole genomes is the purview of the fields of cytogenetics and comparative genomics. Classical cytogenetics uses precipitated metaphase chromosomes, which are harvested from actively dividing cells and then dropped on a glass slide, causing the chromosomes to form a spread or splatter. The chromosomes can then be treated with enzymes (e.g., trypsin) to partially digest the histone proteins around which DNA is coiled and then stained with various dyes to create banding patterns that reflect the different base pair com-

position of particular areas (e.g., GC-rich bands, which contain the majority of active genes). The species-specific number and appearance (morphology) of such chromosome spreads is called a *karyotype*.

Studies of the primate genome classically done by cytogenetic analysis are rapidly evolving to merge with molecular genetic techniques such as fluorescent nucleotide probe hybridization on whole chromosomes (FISH) (Ried et al. 1993; Wienberg and Stanyon 1995, 1997, 1998). FISH studies use PCR to produce chromosome-specific probes from flow cytometry–sorted chromosomes of a particular species. Fluorescent dyes incorporated in the amplified fragments can then be used to "paint" the target chromosomes of other species, where the fluorescent probes will hybridize to complementary sequences found on any of the chromosomes of the species under investigation. This technique allows the detection of major rearrangements in chromosomes between taxa by highlighting where genetic material with similar sequence has been maintained through evolution (Fig. 22.1). The use of reciprocal chromosome painting via FISH has produced insights into chromosome evolution within a number of major primate groups, including strepsirhines (Müller et al. 1997, Stanyon et al. 2002), platyrrhines (Stanyon et al. 2000, 2001; Neusser et al. 2001), colobines (Bigoni et al. 2003, 2004), and hylobatids (Arnold et al. 1996; Koehler et al. 1995a,b; Müller et al. 2003).

At a more detailed level, comparative genomic studies have revealed a surprising degree of variation among taxa in genome organization due to dynamic processes of inter- and intrachromosomal segmental duplications. These duplications, which range in size from a few thousand to a few million base pairs, are widespread within the genome; and their prevalence suggests that the genome is much more plastic than previously thought (Samonte and Eichler 2002, van Geel et al. 2002). While segmental duplications are more common in gene-poor regions of the genome, some duplications also involve regions containing functional genes, and some of these appear to have surprisingly high rates of evolutionary change (e.g., the *morpheus* gene family, Johnson et al. 2001). Recent work is focusing on the relationship between chromosomal rearrangements and the quick pace of sequence evolution in genes located near rearrangement sites (Navarro and Barton 2003, Marquès-Bonet et al. 2004). The jury is still out, however, on the importance of such effects for gene expression evolution and its potential role in the speciation process (Zhang et al. 2004).

Variation at the Level of the Gene or DNA Sequence

Modern techniques for studying DNA sequence-level variation take advantage of a relatively small set of fundamental concepts of molecular biology: the cutting of DNA with restriction enzymes, the joining of DNA fragments with DNA ligase, the separation (denaturing) and joining (annealing) of complementary strands of DNA by heating and cooling, the replication of DNA using DNA polymerases,

Figure 22.1 Fluorescence in situ hybridization (FISH). This technique allows the localization of DNA sequences on chromosomes with the help of fluorescent dyes attached to short DNA probes. It can be used to study chromsome rearrangements (as in reciprocal chromosome painting) or to detect the precise chromosomal location of a known stretch of DNA, such as a fragment of genomic DNA containing a gene of interest.

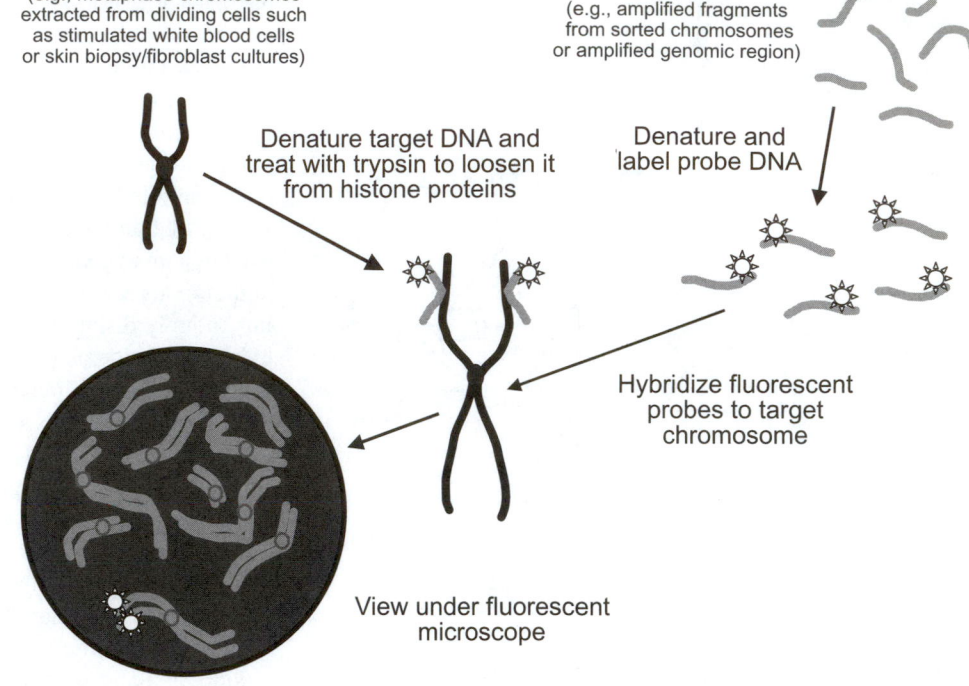

Target DNA
(e.g., metaphase chromosomes extracted from dividing cells such as stimulated white blood cells or skin biopsy/fibroblast cultures)

Probe DNA
(e.g., amplified fragments from sorted chromosomes or amplified genomic region)

Denature target DNA and treat with trypsin to loosen it from histone proteins

Denature and label probe DNA

Hybridize fluorescent probes to target chromosome

View under fluorescent microscope

and the separation of fragments of DNA by size through gel electrophoresis. Since there are many excellent reviews of the basic molecular biological concepts applied in modern genetic analysis, we will not undertake such a task and instead direct the reader to those reviews (Hoelzel 1998, Hillis et al. 1996, Ferraris and Palumbi 1996, Baker 2000). We focus here on only a small set of techniques that are currently used to generate marker data (e.g., genotypes) and DNA sequences.

The workhorse tool of virtually all modern genetic studies is PCR (Saiki et al. 1985, 1988; Mullis and Faloona 1987). PCR is essentially a tool for in vitro cloning, allowing the many-million-fold amplification of a specific target fragment of the genome from the midst of a complex DNA sample. PCR takes advantage of the discovery of various thermostable polymerases whose activity is not degraded by the high temperatures required to denature double-stranded DNA prior to replication. In PCR, template DNA and DNA polymerase are mixed with free nucleotides (deoxynucleotide triphosphates, dNTPs) and *primers*, short oligonucleotides of DNA that are complementary to the sequence flanking the region to be amplified, thus providing the double-stranded template necessary for the polymerase to function properly (Fig. 22.2).

Microsatellite Genotyping

Microsatellites, also known as simple sequence repeats or short tandem repeats, are one class of molecular markers known more generally as variable number of tandem repeat (VNTR) loci. These loci consist of a short 1–6 bp motif repeated in series a variable number of times (Fig. 22.3).

Microsatellite regions tend to evolve far more rapidly than other portions of the genome, presumably due to slippage of

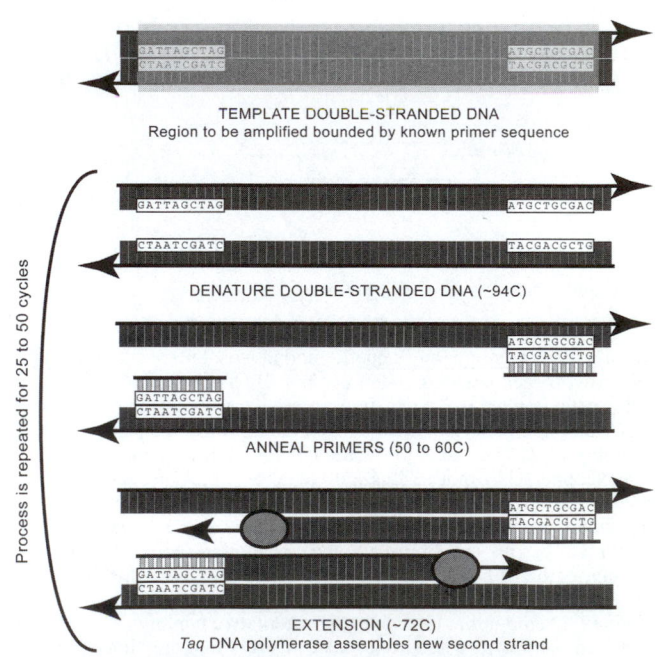

Process is repeated for 25 to 50 cycles

TEMPLATE DOUBLE-STRANDED DNA
Region to be amplified bounded by known primer sequence

DENATURE DOUBLE-STRANDED DNA (~94C)

ANNEAL PRIMERS (50 to 60C)

EXTENSION (~72C)
Taq DNA polymerase assembles new second strand

Figure 22.2 The polymerase chain reaction (PCR). The most common element in almost all modern genetic studies of primates, PCR takes advantage of a heat-stable DNA-replicating enzyme isolated from a hot-spring bacterium (*Thermophilus aquaticus*) from Yellowstone National Park to generate large numbers of identical copies of a given target DNA sequence.

DNA polymerase as DNA is replicated during the process of gametogenesis (Levinson and Gutman 1987, Kruglyak et al. 1998, Toth et al. 2000). The various alleles present in a population at a given microsatellite locus can be assayed through PCR amplification of the locus followed by electrophoretic

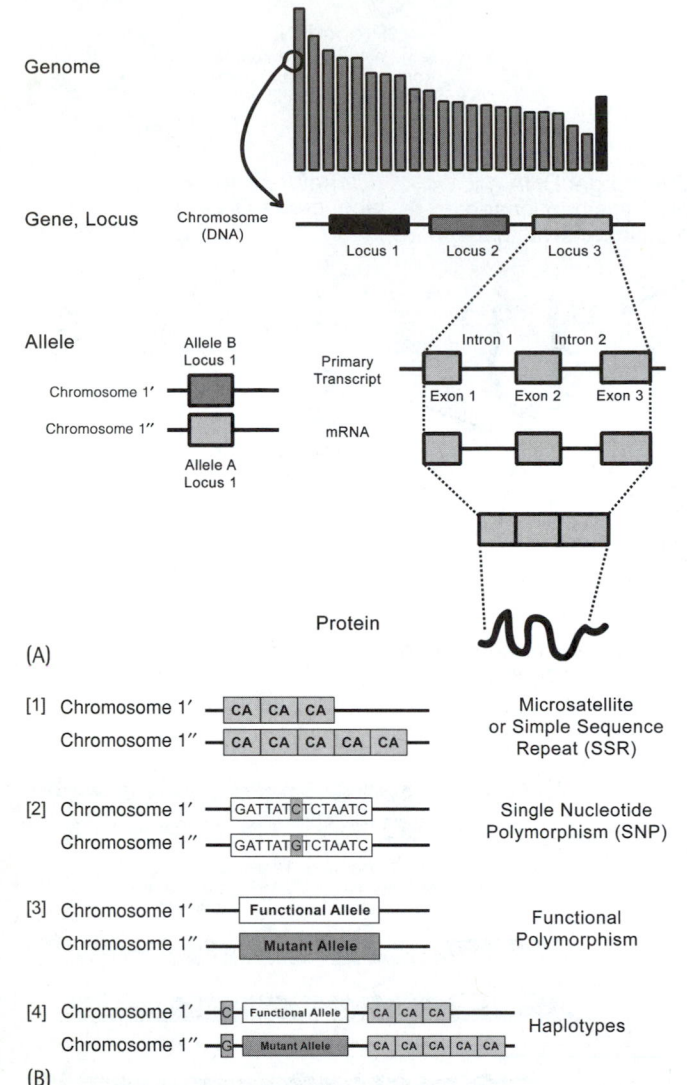

Genome

Gene, Locus

Allele

Protein

(A)

[1] Chromosome 1' |CA|CA|CA| Microsatellite
    Chromosome 1'' |CA|CA|CA|CA|CA| or Simple Sequence
                                             Repeat (SSR)

[2] Chromosome 1' |GATTAT**C**TCTAATC| Single Nucleotide
    Chromosome 1'' |GATTAT**G**TCTAATC| Polymorphism (SNP)

[3] Chromosome 1' | **Functional Allele** | Functional
    Chromosome 1'' | **Mutant Allele** | Polymorphism

[4] Chromosome 1' —C—| **Functional Allele** |—|CA|CA|CA|—
    Chromosome 1'' —G—| **Mutant Allele** |—|CA|CA|CA|CA|CA| Haplotypes

(B)

Figure 22.3 Genome and genetic marker vocabulary. A: Schematic overview of genome and gene structure. B: Various classes of molecular genetic markers. Each line represents a region of homologous DNA on one member of each autosomal chromosome pair. From top to bottom: *(1)* Microsatellite marker locus involving the repeat of a dinucleotide (CA) motif. This individual is a $(CA)_3/(CA)_5$ heterozygote. *(2)* Biallelic single-nucleotide polymorphism, or SNP, site. This individual is an A/G heterozygote. *(3)* Functional polymorphism. This individual carries one "normal" and one mutant copy of a protein-coding region of DNA. An example might be an individual heterozygous for the normal (Hb^A) and sickle (Hb^S) alleles coding for the β chain of a hemoglobin molecule. *(4)* Alternative haplotypes at a locus. The term *haplotype* refers to a set of markers that are "linked" or inherited together most of the time because they are located close to one another on the same chromosome and therefore less likely to undergo recombination during the process of gamete formation.

separation of the *amplicons* (the millions of copies of the same DNA fragment produced by PCR). Because alleles will differ in length due to the presence of different numbers of repeats of the microsatellite motif, these fragments will migrate to different positions in a gel during electrophoresis and individual genotypes can be scored directly, without the need for sequencing (Fig. 22.4A).

In the last decade, PCR-based microsatellite genotyping has been increasingly used to provide insight into various aspects of the behavioral biology of nonhuman primates, including mating systems, dispersal patterns, and population genetic structure (Di Fiore 2003, de Ruiter 2004). One of the limitations of PCR-based microsatellite genotyping, however, is that primers need to be available for amplifying variable microsatellite loci. Many such primer pairs are known for a variety of hominoid and cercopithecoid taxa since primers used in human genotyping and linkage studies often amplify homologous regions in other catarrhines. Researchers working on more distantly related primates, however, often need to invest considerable time and money isolating novel microsatellites either from taxon-specific *genome libraries* (sets of short fragments of DNA covering the entire genome that are inserted into biological vectors and cloned; e.g., Strassman et al. 1996, Hamilton et al. 1999, Paetkau 1999) or from subsets of genome fragments generated through a variety of PCR procedures (e.g., Fisher et al. 1996, Lunt et al. 1999, Zane et al. 2002). An up-to-date database of microsatellite loci used in primate studies has been compiled by Di Fiore (2003, www.nyu.edu/projects/difiore/yearbook/appendix.html), and a similar database is currently under development by INPRIMAT (www.inprimat.org), an international consortium to promote primate molecular biology research.

Single Nucleotide Polymorphism Genotyping

Even faster methods of detecting genetic variation have been developed that do not require separation of PCR products via electrophoresis. Much emphasis has been placed recently on the search for single nucleotide polymorphisms (SNPs, or "snips") in the human genome for use as molecular markers in large-scale linkage studies. SNPs, as their name implies, are single nucleotide positions in the genome that differ between different chromosomes in a population (Fig. 22.3). Most SNPs are biallelic, existing in only two alternative forms. In humans, SNPs occur approximately once every 1,000–2,000 bp in the genome, and over 1.4 million SNPs have already been putatively identified in the roughly 2.85 billion bp comprising the completed sequence of the euchromatic portion of the human haploid genome (International Human Genome Sequencing Consortium 2001, 2004; International SNP Map Working Group 2001).

SNPs have been used extensively in human studies to examine population genetic structure and migration patterns. For example, a number of studies have used Y chromosome SNP data to trace the spread of modern humans around the globe (Underhill et al. 2000, 2001; Hammer and Zegura 2002). As yet, however, SNPs have not been used widely in studies on nonhuman primates, primarily due to the lack of known markers. In the coming years, this will likely change as improved protocols for isolating SNPs are developed (e.g., Aitken et al. 2004).

Because of their utility in linkage studies in humans, a number of high-throughput, automated methods have been derived to directly assay variation at multiple SNP loci

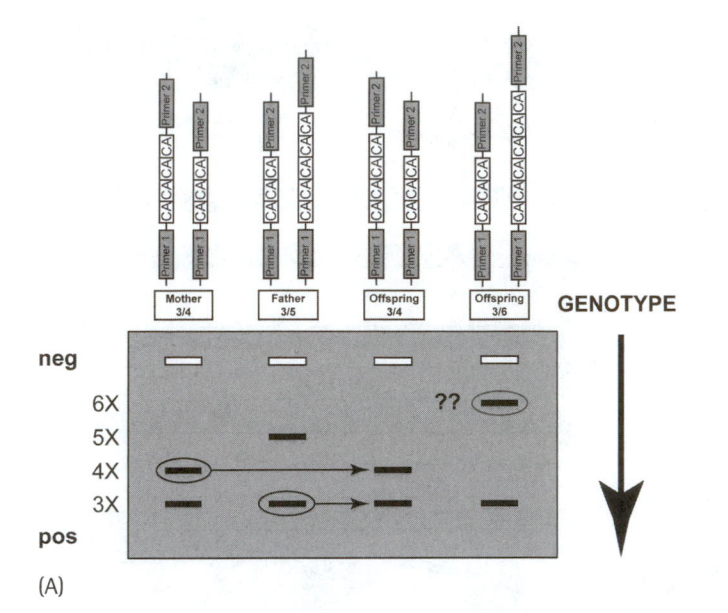

(A)

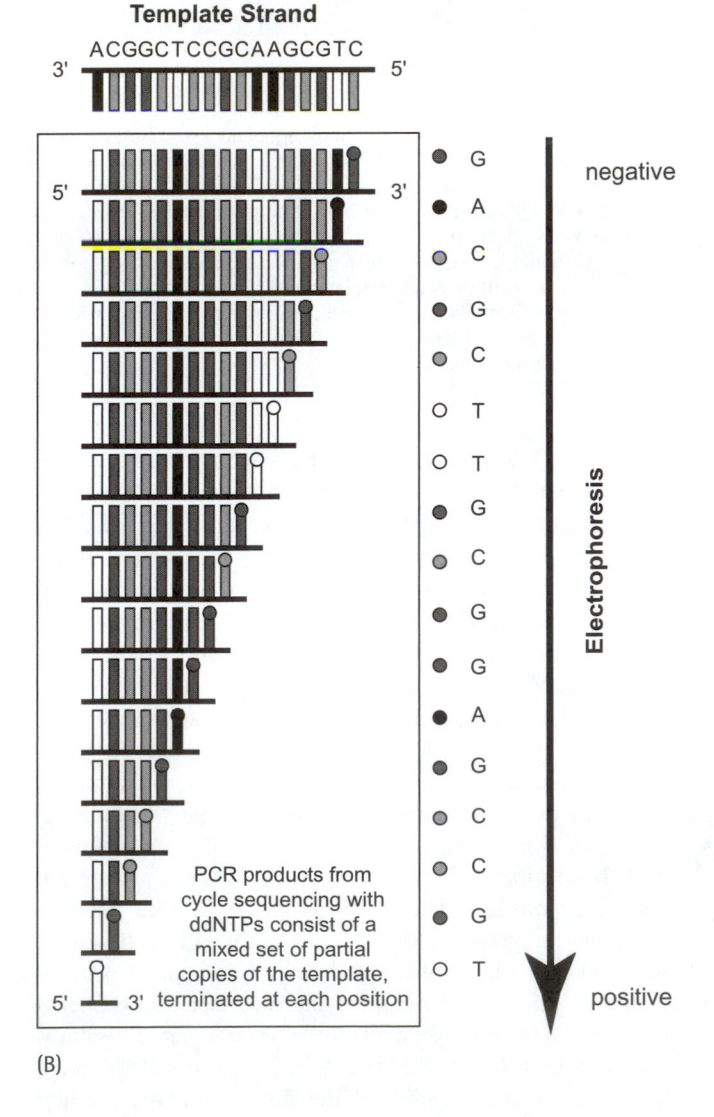

Template Strand

ACGGCTCCGCAAGCGTC

PCR products from cycle sequencing with ddNTPs consist of a mixed set of partial copies of the template, terminated at each position

(B)

simultaneously, including the use of allele-specific extension on DNA microarrays (Pastinen et al. 2000), hybridization of different alleles to complementary oligonucleotides bound on microarrays (Hacia et al. 1999), and the detection of alternative alleles created via PCR using mass spectrometry (Haff and Smirnov 1997, Bray et al. 2001). Reviews of these and other automated SNP typing methods are provided by Gut (2001) and Syvänen (2001).

Direct DNA Sequencing

The most detailed form of genetic variation that molecular primatologists can assay is variation in an actual DNA sequence between individuals or among taxa. Modern phylogenetic studies commonly use direct DNA sequencing to generate DNA sequence data for a particular site (locus) of interest and then compare the sequence composition, the distribution of variable sites, and the relative genetic distance between individuals or groups of sequences. Modern DNA sequencing relies on the controlled replication of a DNA template of interest using PCR. Template DNA is mixed with polymerase and other reaction components, along with free nucleotides (A, T, G, and C) in the form of dNTPs. Some small fraction of these free nucleotides are di-deoxynucleotides (ddNTPs), which, when randomly incorporated into a growing DNA strand, prevent further extension as they lack the 3′-hydroxyl group needed to form a phosphodiester bond with a subsequent nucleotide (Sanger et al. 1977). The result of the sequencing reaction is thus a mixed population of copied DNA fragments differing in length by a single base pair, from 1 base to the full complementary length of the template. These fragments are then separated by electrophoresis, allowing the sequence of DNA bases to be determined based on which ddNTP is incorporated at the end of each fragment (Fig. 22.4b). Studies of sequence variation are useful at multiple levels (e.g., for inferring phylogenetic relationships among higher-order taxa, such as species or genera, as well as for inferring intraspecific phylogenies that can be used in phylogeographic studies).

Figure 22.4 (*left*) Molecular genetics techniques routinely used in primatological studies. A: Microsatellite genotyping by electrophoretic separation of amplied DNA fragments. Fragments are separated by size in an electrophoretic medium such as a polyacrylamide gel. DNA is negatively charged and will migrate toward the positive pole in a gel. Microsatellite alleles are codominant, allowing genotypes to be scored directly. Offspring must inherit one allele from each parent, thus *Offspring 2* in this example appears to have been fathered by an unsampled individual. B: Direct DNA sequencing by the dideoxy method, which causes DNA replication of a template to be terminated when a dideoxynucleotide base analog is incorporated into a growing DNA strand (Sanger et al. 1977). Electrophoresis separates a mixture of partially copied DNA sequence fragments differing in size by one base pair in a medium sensitive enough to resolve these differences. In modern automated sequencing, the four possible dideoxynucleotide triphosphates (ddNTPs; A, T, G, and C) are each tagged with a different fluorescent molecule, allowing the type of nucleotide at the terminal position of a DNA fragment to be resolved by laser detection. From shortest to longest fragments, the order of the passing fragments gives the complementary sequence of the template DNA.

Alternative Methods for Detecting Sequence Variation

Several alternative methods have been developed as shortcuts for rapidly evaluating whether sequence variation exists within a population at a given locus without having to directly sequence every individual sample. These include a variety of techniques—single-strand confirmation polymorphism analysis (SSCP), denaturing gel gradient electrophoresis analysis (DGGE), and conformation-sensitive gel electrophoresis analysis (CSGE)—that take advantage of the fact that fragments of DNA differing by as little as a single base pair will migrate differently through a variety of electrophoretic media. Amplified DNA fragments from a sample can also be screened for variation using the more recently developed technique of denaturing high-performance liquid chromatography (DHPLC) (Xiao and Oefner 2001). All of these methods can thus be used to rapidly screen a population of samples for variation and to identify a set of representative variants for subsequent direct sequencing. They can also be used as less costly alternatives to microarrays and mass spectrometry for rapidly genotyping individuals at well-characterized variable loci, such as SNP sites.

Alternative methods for fast typing of well-characterized sequence variants, such as SNPs, include "real-time" or "quantitative" PCR and flow cytometry. Real-time PCR is a modified form of PCR that, in addition to the two oligonucleotide primers bracketing the target sequence, incorporates a third oligonucleotide that is complementary to a short segment of the target. This third oligonucleotide, or "probe," carries both a fluorescent dye molecule and a "quenching" molecule, which prevents the dye from emitting light as long as the probe is intact. During PCR amplification, the DNA polymerase, which also has exonuclease activity, digests the central probe as it works its way down the template synthesizing a new complementary strand (Fig. 22.5). Digestion of the probe separates the fluorescent dye from the quencher, causing it to emit light when stimulated by a laser. The intensity of the fluorescent emission is proportional to the amount of template DNA present at each cycle; thus, the initial quantity of template in an experimental sample can be evaluated by comparing its amplification profile to that of a standard template of known DNA quantity. A simpler and cheaper form of real-time PCR uses the nucleic acid dye Cyber green and simply measures the change in the total amount of DNA (PCR product) over time, without the need for a highly specific probe. The coupling of allele-specific amplification (e.g., for an SNP locus) with real-time PCR detection allows for fast and efficient screening of a population for sequence variation.

Flow cytometry is a general procedure that can be used for the analysis of a variety of particulate samples, from fragments of DNA to organelles to bacteria to whole eukaryotic cells (Fig. 22.6). The technique relies on samples being kept as a suspension in a "sheath" fluid, the flow of which is controlled such that the particles to be studied flow through a chamber one-by-one at a rate of several hundred

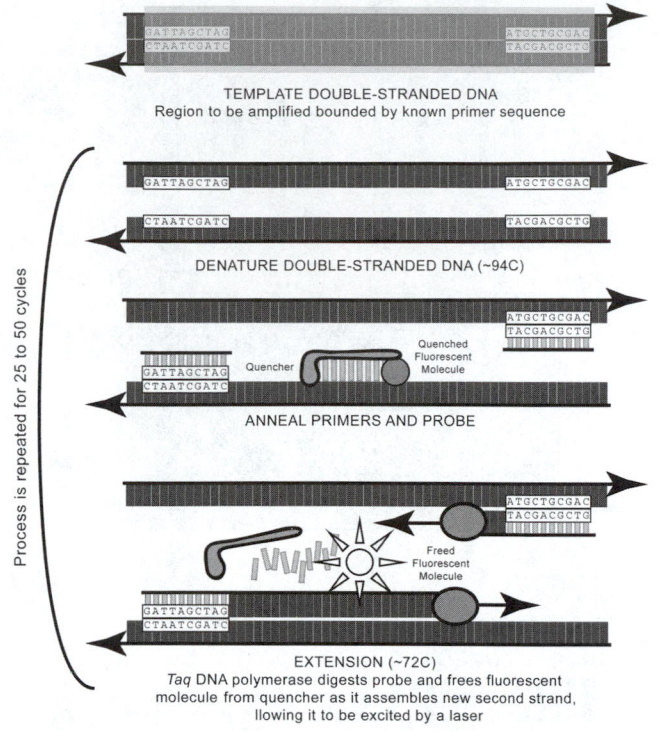

Figure 22.5 Real-time or quantitative polymerase chain reaction (Q-PCR). In addition to flanking PCR primers, a third oligonucleotide is used that is complementary to a portion in the middle of the target region to be amplified and contains both a fluorescent and a "quencher" molecule. As replication proceeds, the DNA polymerase digests this third primer and releases the quencher molecule, allowing the fluorescent molecule to emit detectable light when stimulated by a laser. The intensity of this light can be measured to provide an accurate estimate of the number of amplified copies of a DNA sequence present after each cycle of replication.

per second. Laser light illuminates each passing particle, and a collection of light detectors (photomultipliers) measures the emission on the other side of the chamber. This procedure allows the rapid determination of particle size and granularity and, based on the emission profile, can be used to analyze samples by detecting molecules present on the particles and using fluorescence intensity to quantify their abundance. Methods for SNP genotyping by flow cytometry make use of microspheres that are tagged with a distinct fluorescent color and that have short, allele-specific oligonucleotide capture probes attached to them. Populations of microspheres bearing probes for the alternative alleles at an SNP site are then hybridized to sample DNA subject to allele-specific PCR, which also incorporates a different fluorescent marker. The flow cytometer is then used to determine the presence of each alternative allele, based on the combined microsphere and allelic fluorescent signals recorded. Taking advantage of the fact that microspheres with 100 distinct fluorescent tags are commercially available (Luminex Corp., Austin, TX), Ye et al. (2001) have demonstrated the potential of the flow-cytometric method for multiplex genotyping of DNA samples at up to 15 SNP loci simultaneously.

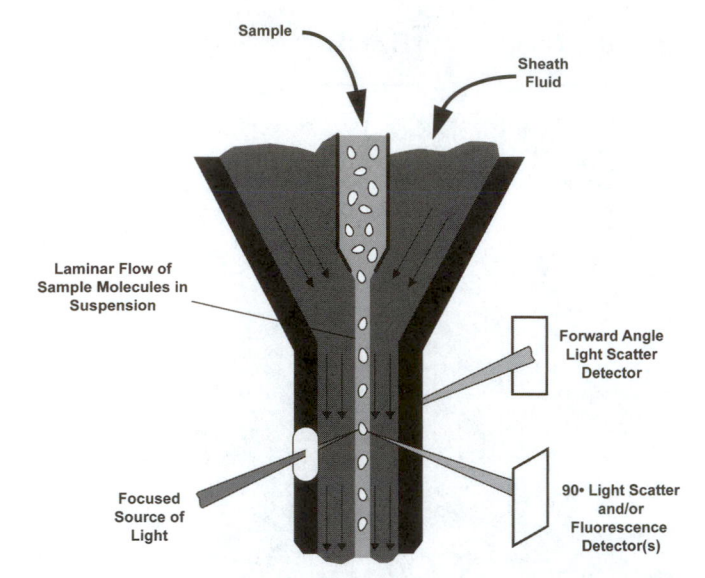

Figure 22.6 Flow cytometry (literally "cell measuring") is a technique used to measure various aspects of individual particles or cells at a rate of several hundred per second. Sample particles in suspension are injected into a sheath fluid (usually an isotonic buffer or medium). By adjusting the pressure on the sheath and sample channels, the sample fluid can be directed to flow through the center of the sheath fluid column without mixing and contain a focused stream of separated molecules. These molecules flow one-by-one past a source of illumination (e.g., a laser). The amount of light scattered in the forward and side directions can be detected and used to characterize the size and granularity of the sample particles. Additionally, particles of interest can be assayed and quantified by measuring the color and amount of light emanating from laser-stimulated fluorescent probes used to stain those particles. Flow cytometry can be applied to many different tasks, from identifying and sorting different types of cell (on the basis of their forward and side scatter illumination profiles) to evaluating what stage of the cell cycle particular cells are in (based on fluorescent quantification of the amount of DNA present in the cell) to sorting chromosomes (based on fluorescent quantification of two different dyes, one that binds preferentially to GC-rich areas of chromosomes and one that binds to AT-rich areas) to multiplex genotyping of a mixture of polymerase chain reaction–amplified DNA fragments (based on the incorporation of allele-specific fluorescent molecules and their binding to similarly tagged microspheres).

Variation in Gene Structure and Expression

Studying the structure and expression of genes requires the copying of large stretches of chromosomal DNA (up to several tens of thousands of base pairs) that contain the full coding sequence as well as the regulatory regions and noncoding exons and introns of a gene. This is because regulation of gene expression is governed by the binding of various transcription factors (mostly proteins) to well-defined binding regions upstream from the gene itself. Variants of basic PCR procedures, such as "long-distance" or "long-range" PCR, make such amplification possible. Amplified stretches of that genomic DNA can then be probed for the presence of known sequence elements (e.g., regulatory sequences). This is done by separating the fragments on agarose gels, transferring the fragments to a membrane by Southern blotting, and screening the membrane with labeled probes for the DNA regions of interest. Such information is crucial for understanding how genes are structured.

Furthermore, many genes can be "read" in more than one way by the process of *alternative splicing* (the joining together of different exonic portions of a primary mRNA transcript in different ways). The DNA sequence at splice sites can provide information on the total number of potential transcripts (mRNAs) generated by alternative ways of transcribing the same gene. Different patterns of mutation in the exons and introns, for example, can be used to infer selection affecting a particular gene or even a particular functional region of the protein (Sonnenburg et al. 2004).

In order to more directly study the nature of expressed genes and the process of alternative splicing of genetic messages to generate more than one product from the same gene, mRNA can be extracted from a tissue of interest, separated by electrophoresis on agarose gels, and then transferred to a membrane. There, it can be probed with single-stranded DNA oligonucleotides labeled with either radioactive or fluorescent biomolecules to determine what genes (and combinations of genes) and in what quantity contribute to the mRNA transcripts expressed in particular tissues. Single-stranded mRNA, which is far less stable than DNA, can also be copied back into DNA via reverse-transcription PCR (RT-PCR), which allows it to be studied more easily. Real time RT-PCR can also be used to quantify the genetic message by providing an estimate of the amount of starting mRNA template isolated from a particular tissue (Cáceres et al. 2003). The term *transcriptomics* is used to describe such studies of the complete collection of genetic messages, or "transcripts," present in a given tissue at a given time.

Increasingly, transcriptomic studies are being carried out using microarray technology to measure the presence and relative quantity of different mRNAs that are expressed in specific tissues or cell types. This involves the hybridization of reverse-transcribed mRNA to arrays of unique short oligonucleotide or complementary DNA (cDNA) probes specific to genes of interest. These arrays may be commercially available "DNA chips" containing thousands of probes or homemade dot plots of a much smaller number of probes bound to a membrane or slide (Fig. 22.7). Karaman et al. (2003) used this procedure to evaluate the relative expression of a large number of genes in gorillas, bonobos, and humans by interrogating Affymetrix (Santa Clara, CA) U95Av2 DNA chips (which contain short probes for over 10,000 human gene transcripts) with mRNA isolated from multiple fibroblast cell lines (cultures of skin cells initiated from biopsy samples) of each of these three taxa. Their analysis identified a number of candidate genes (many involved in metabolic pathways, neural signal transduction, and cell and tissue growth) that were differentially expressed among these species, providing some directions for future research on the underlying causes of phenotypic differentiation within hominoids.

In independent studies using similar DNA chips, Enard et al. (2002) and Cáceres et al. (2003) compared the expression of the same set of human genes in the brains of humans, chimpanzees, orangutans, and macaques and compared

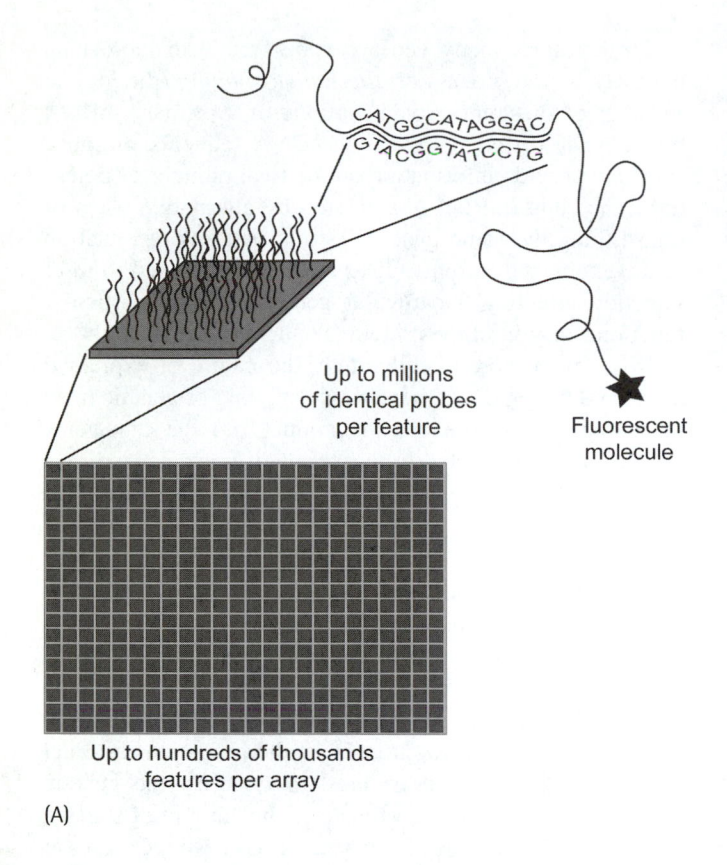

(A)

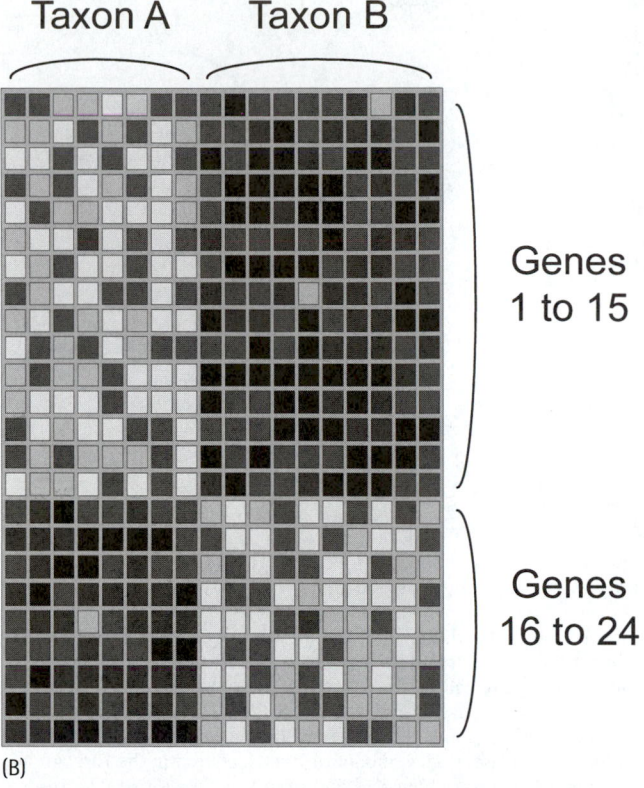

(B)

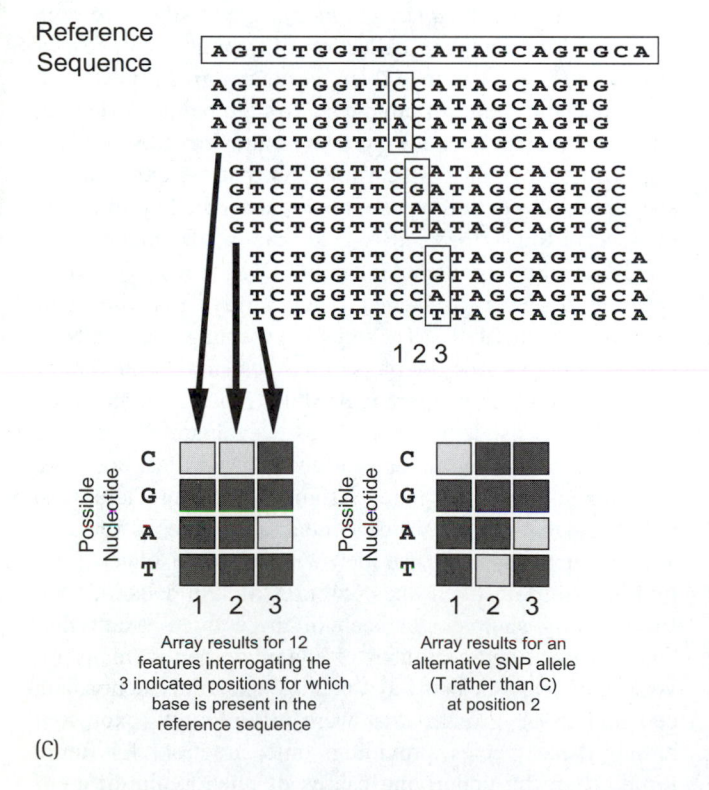

(C)

Figure 22.7 DNA microarrays and gene chips. Microarrays are contraptions allowing miniature hybridization experiments of sample DNA against an array of up to thousands of known DNA probes fixed in a grid. A: Single-stranded DNA oligonucleotides are either blotted onto a small, flat substrate (e.g., a glass slide) in well-defined positions (a "DNA microarray") or synthesized directly on a silica-based chip using photolithography (a "gene chip"). Multiple copies of the same oligonucleotide probe bound to the same region of the microarray or chip constitute a "feature." A single photolithographically produced chip may contain up to several hundred thousand features. The microarray or chip is washed with a denatured nucleic acid sample of interest that has been tagged with a radioactive or fluorescent label. Sample DNA or RNA will hybridize only where a complementary sequence is found and can be detected based on the corresponding luminescent signal. B: As discussed in the text, microarrays and chips can be used in comparative studies of gene expression. As a hypothetical example, a cDNA-based microarray containing probes for 24 genes (*rows*) is used to interrogate fluorescently labeled mRNA extracted from tissue samples of 18 individuals from two different taxa (columns 1–8, taxon A; columns 9–18, taxon B). Lighter coloration for a feature indicates that more of the labeled mRNA sample bound to the probe. In this case, genes 1–15 show greater expression in taxon A, while genes 16–24 show greater expression in taxon B. C: Microarrays and chips can also be used for single nucleotide polymorphism (SNP) genotyping of polymerase chain reaction–amplified samples. In this case, features consist of staggered sets of oligonucleotides that span a known reference sequence containing the SNP site. For each of these sets, there are four features on the array, one corresponding to each potential nucleotide present at a "query" position located in the middle of the oligonucleotide. Amplified samples will hybridize only to the feature containing the appropriate complementary base at each query position, which allows for rapid SNP genotyping. A similar procedure can be used for "resequencing" an entire DNA amplicon.

these results to those for other body tissues. In both studies, nearly 170 genes were expressed differentially in the brains of chimpanzees and humans, with humans showing higher expression levels (upregulation) for the vast majority of these genes, many of which appear to be related to cell growth and maintenance and to neural function. These differences in expression levels, however, did not extend to heart or liver tissue, where only just over half of the genes with different expression levels were upregulated in humans (Cáceres et al. 2003). If real, these differences in expression levels suggest a possible neurological basis for some of the behavioral and cognitive adaptations thought to be associated with the divergence of humans from other African apes. Interestingly, the locations of these upregulated genes appear to be associated with areas of chromosomal rearrangement and segmental duplication between humans and chimpanzees (Khaitovich et al. 2004, Marquès-Bonet et al. 2004). A similar study by another group demonstrated that genes expressed in cortical tissue are upregulated in all of the African great apes relative to macaques, though in humans to a greater extent than in either chimpanzees or gorillas (Uddin et al. 2004). Preuss et al. (2004) provide a comprehensive review of these and other microarray studies of gene expression differences among primates.

One of the important limitations of the microarray approach is that, to date, it is still impossible to detect subtle changes (e.g., less than twofold differences) in gene expression levels (Forster et al. 2003). A further limitation is that these studies are heavily biased toward "prominent" messages (i.e., those genes that are expressed at very high levels, producing high copy numbers of the same mRNA transcript). Other genes with relatively low and much more narrowly localized expression may nonetheless have important phenotypic consequences (e.g., glycosyl transferases and other protein-modifying enzymes).

Assaying Variation in non–DNA Molecules

In addition to nucleic acids (DNA and RNA), the cells of living organisms contain several other classes of important biomolecules, including proteins, lipids, and sugars, which exist in various combinations (Fig. 22.8). Techniques from other fields of molecular biology used to study these other kinds of molecules are starting to find wider application in primatology, even if to a much lesser degree than genetic methods. For example, protein comparisons have become much more sophisticated since the days of early allozyme studies, contributing to the rapidly expanding field of *proteomics* (the study of the complete set of proteins found in a particular type of cell or tissue). Soluble and extracted proteins can be separated in two dimensions via isoelectric focusing and sodium dodecyl sulfate polyacrylamide gel electrophoresis (SDS-PAGE) and then transferred to a membrane and probed with various labeled antibodies to assay the presence of particular proteins as well as their electro-phoretic signatures. An example of this approach is the comparison of soluble proteins in the blood plasma of great apes and humans (Gagneux et al. 2001). Sugar molecules on glycoproteins can be assayed similarly, by probing with carbohydrate-specific lectins and certain antibodies that bind to these molecules (Boenisch 2001). A more recent approach, multidimensional protein identification technology (MudPIT), is based on the partial digestion of proteins followed by mass spectrometry, which allows the identification of membrane proteins and their post-translational modification (Wu et al. 2003).

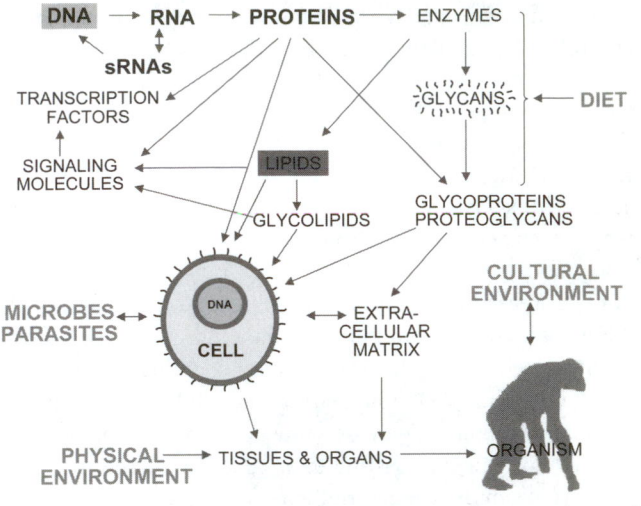

Figure 22.8 Molecules in context. Living organisms are made up of four classes of molecules—nucleic acids (DNA and RNA), proteins, lipids, and carbohydrates/sugars—which exist in various combinations. The assembly, distribution, and abundance of the final molecules are affected by many factors, ranging from genetic control to cellular dialog within the organism to interactions with the biotic and abiotic environment faced during development. DNA (*light gray shading*) is found in the nucleus (and mitchondria, not shown), lipids (*dark gray shading*) are prevalent in the nuclear and cellular membranes, and carbohydrate-containing glycans (*black threads*) are attached to proteins expressed on the surfaces of cells. (Figure kindly provided by A. Varki.)

Variation in soluble or extracted proteins can also be screened immunologically using the enzyme-linked immunosorbent assay (ELISA). In this technique, the presence of a protein or protein variant in a sample is detected via its reactivity with antibodies that are labeled with a *chromogenic* (color-producing) enzyme. In a variant of ELISA, the line immunoassay (LIA), probes consisting of known proteins are blotted on a membrane, over which a sample is then washed. This technique has been used to test for the presence of antibodies against simian immunodeficiency virus in a large number of wild primates in Africa (Peeters et al. 2002). Animals that tested positive for antiviral antibodies in their serum or urine were then further investigated with PCR or RT-PCR to confirm the presence of the virus and to characterize its nucleotide sequence (Santiago et al. 2003).

Flow cytometry is also increasingly being used for studying the comparative cell biology of primates. The process can be characterized as a form of "liquid ELISA" as it allows the study of molecules on and inside living or fixed cells by taking advantage of fluorescently labeled probes that can be bound to the molecule of interest. Flow cytometry has recently been used to document differences among several primate species in the distribution and abundance of immune molecules present on white blood cells and in the ability of these cells to bind certain well-defined ligands (Brinkman-van der Linden et al. 2000). Analyses of primate sperm are also now conducted with the help of flow cytometry as this powerful method allows for the quantitative analysis of large numbers of cells in a very short time (~10,000 per minute) (Shankar et al. 2004).

The detection and visualization of specific molecules in and on cells can also be accomplished using immunohisto-chemical methods. These methods rely on cells left intact in the architecture of a particular tissue or isolated from such tissue by enzymatic digestion, and they are sensitive enough to detect extremely low amounts of target molecules. This sensitivity is critical because certain components of a tissue may be present at very low concentration (and thus relatively underrepresented in a tissue extract) and yet be of the utmost biological importance (e.g., receptors found only on the apical side of a single-cell epithelial cell layer such as hormone receptors, olfactory receptors, endogenous lectins, or glycoconjugates) (Gagneux et al. 2003). Additionally, proteins with known genetic sequence can be cloned in vitro (i.e., by copying and pasting their genes in mRNA or cDNA to form plasmids, which are then introduced into a bacterial vector and replicated) and then expressed in an expression system (e.g., Chinese hamster ovary cells) for comparative functional and/or structural studies.

Fibroblast lines established from skin biopsy samples form precious repositories of live cells, which can be stored indefinitely and regrown as needed as a source of DNA for genetic studies or for use in comparative cytogenetic, transcriptomic, and proteomic studies as well as studies of gene expression in various tissues. With support from the National Science Foundation, numerous institutions, such as the Frozen Zoo of the Zoological Society of San Diego, the Coriell Institute, and Princeton University, are currently collaborating to establish the Integrated Primate Biomaterials and Information Resource (www.ipbir.org) containing a large collection of such cell lines from animals with known provenience and, often, known life histories. Additionally, for hominoid primates, immortal cell lines created from white blood cells form a crucial resource for comparative studies of gene function, particularly for studies of the many important genes of the immune system, such as *MHC* and *KIR* (Adams and Parham 2001, Rajalingam et al. 2001). These lines are created from immune system B cells via transformation using Epstein-Barr virus (EBV). Unfortunately, however, there are no methods of immortalizing B cells from the majority of nonhuman primate species, as EBV has a fairly restricted host range, and other appropriate transforming viruses remain to be identified.

Finally, although little explored to date, interest is growing in the comparative study of the *microbiome* (the complete set of microbial symbionts and parasites living in and on an animal) of humans and other primates. For example, molecular genetic techniques, in combination with classical methods from microbiology such as serology, can be used to identify the types and strains of bacteria inhabiting the bodies of healthy and sick primates, as well as potential adaptations of the host primate genome to their presence (e.g., Stewart et al. 1987, Zhang et al. 2002). This field promises to uncover a vast and brand new realm of primate biodiversity, one with clear significance for understanding aspects of primate evolution relevant to health and diseases.

FIELD METHODS: SAMPLE TYPES AND TECHNIQUES

In order to take advantage of any of these molecular techniques, it is crucial to collect appropriate samples. The collection, exportation, and importation of nonhuman primate samples are regulated by a variety of national and international laws and conventions. For both practical and ethical reasons, it is imperative that researchers comply fully with these regulations. For the United States, permission to import various nonhuman primate samples needs to be obtained from one or more of the following agencies: the Centers for Disease Control and Prevention of the U.S. Public Health Service, the U.S. Fish and Wildlife Service, and the U.S. Department of Agriculture. Also, as many nonhuman primates appear in Appendices I and II of the Convention on International Trade in Endangered Species of Wild Flora and Fauna (CITES), the U.S. Fish and Wildlife Service requires that CITES export permits be obtained from the country of origin for most primate samples. Furthermore, nonhuman primates, by virtue of their close taxonomic proximity to humans, can carry a large number of pathogens of great concern (e.g., herpes simiae and polio viruses in Old World monkeys; simian immuno-deficiency virus, ebola, and anthrax in great apes). Before embarking on a molecular study utilizing primate samples, researchers must be trained in safe laboratory practices and follow special primate protocols that are now required at many research institutions, and they must be vaccinated against a number of infectious agents likely to be encountered in primate tissues or secretions.

Sampling for Genetic Analyses

For genetic analyses of wild primates, a variety of sample types have been used as sources for DNA, including blood, tissue, hair, feces, urine, semen, buccal swabs, and masticated food wadges. There are, of course, advantages and disadvantages associated with the use of different sample types. Blood and tissue are the best sources for large quantities of high-quality genomic DNA, but they are the most difficult samples to acquire. Blood collection requires that animals be immobilized, often under anesthesia, which may be difficult to do in field conditions. Additionally, collectors must have received professional veterinary training in how to draw blood without injuring the animal. Tissue sampling (e.g., collection of ear punches) can also be done on anesthetized animals and requires less training and practice to perform safely. Nonetheless, any such invasive procedures carry risks to the health of both animal and researcher. Such procedures will also be subject to Institutional Animal Care and Use Committee approval, which may require the presence of professional animal-care technicians during sampling. The logistical complexities associated with anesthetizing and collecting blood or tissue from wild animals may preclude the use of these kinds of sample if large numbers of samples are required.

Biopsy darts propelled from a compressed air- or CO_2-powered pistol or rifle are a minimally invasive alternative for retrieving small tissue samples that are suitable sources of large amounts of high-quality genomic DNA. This technique has been successfully used in studies of several primate species, including leaf monkeys (Rosenblum et al. 1997a), woolly monkeys (Di Fiore 2002, Di Fiore and Fleischer 2005), and spider monkeys (Di Fiore unpublished data). Darts can be made by hand from standard 3 cc syringes and 16-gauge needles (e.g., Henry Schein) or purchased from commercial suppliers (e.g., Pneudart, Williamsport, PA).

Some of the earliest field studies of primates to employ molecular techniques used noninvasively collected hair samples as a source of DNA (e.g., Morin et al. 1994; Gagneux et al. 1997b, 1999; Goldberg and Wrangham 1997; Mitani et al. 2000). Unfortunately, it is now clear that shed hairs are generally a very poor source of nuclear DNA, though follicle-containing plucked hairs are more suitable. A variety of creative techniques have been employed to collect fresh hair samples from wild nonhuman primates, including shooting animals with flat-tipped darts that have sticky tape mounted on the front, presenting animals with food bait wrapped in tape, and coating the edges of the entrance to a food corral with tape to capture an animal's hair as it enters (Valderrama et al. 1999).

Fecal samples have also been used as a source of DNA for numerous genetic studies of wild primates. Feces typically contains small numbers of cells sloughed from the epithelial lining of an animal's lower digestive tract as a bolus of fecal material passes through. An extraction of DNA from feces will thus contain a complex mix of DNA from a variety of sources, including the animal itself, as well as gut microbes and plants or insects that the animal consumed. PCR allows the preferential amplification of a specific target sequence present at low copy number (i.e., a piece of the genome of the animal) from amidst this complex sample. The low template copy number in hair and fecal samples poses a potential problem for determining reliable genotypes on the basis of fecal DNA due to allelic dropout (Taberlet et al. 1996, 1999; Gagneux et al. 1997a). However, real-time PCR can be used to prescreen fecal DNA extracts, as well as extracts from any other noninvasive samples, to identify those containing sufficient template for subsequent analysis (Morin et al. 2001). Over the last several years, improved methods for storing (Roeder et al. 2004, Nsubaga et al. 2004) and extracting DNA from fecal samples (e.g., commercially available DNA extraction kits, such as the QIAgen™ DNA Stool Mini Kit; Qiagen, Chatsworth, CA) have been developed. As a result, fecal samples are now routinely used as a source of DNA for molecular analysis in a number of field studies (e.g., Vigilant et al. 2001, Bradley et al. 2004). Coupled with pre-amplification methods (Piggott et al. 2004) and new methods for whole-genome amplification from very limited copy number templates (Dean et al. 2002), fecal samples will probably become even more widely used in future studies.

Although blood, tissue, hair, and feces are the most commonly used samples as sources of DNA for genetic studies, several additional sample types might also be used. For example, in a study of captive Japanese macaques, Hayakawa and Takenaka (1999) demonstrated that urine could be used as a source of DNA for microsatellite genotyping and that it yielded concordant genotypes to fecal samples from the same individuals (see also Valière and Taberlet 2000). More recently, Domingo-Roura et al. (2004) demonstrated the efficacy of amplifying mitochondrial DNA and microsatellite loci from DNA extracted from the semen of wild Japanese macaques. DNA extracted from buccal cells collected either via cheek swabs or from chewed food wadges (Takasaki and Takenaka 1991, Sugiyama et al. 1993, Vigilant et al. 2001) has also been used on occasion for microsatellite genotype and mitochondrial DNA sequencing. Woodruff (2004) and Goossens et al. (2003) provide excellent overviews of the use of non-invasive techniques for sampling primates for genetic studies.

Sampling for Other Kinds of Molecular Study

Depending on the questions being asked, primatologists might be interested in collecting a variety of other samples from trapped or darted animals for studies of other biomolecules. These include saliva, scent gland secretions, milk, nose swabs, vaginal swabs, copulatory plugs, semen, skin biopsies, and microbial samples from the gastrointestinal tract.

Samples for immunohistochemical studies are most valuable when they are rapidly frozen in an appropriate medium (e.g., embedded in optimal cutting temperature [OCT] compound and flash-frozen in a dry ice and 2-methylbutane/isopentane slush). Such treatment will maintain the cellular architecture of the tissue, and subsequent frozen sectioning on a cryostat allows for the detection of proteins and nucleic acids in situ. Immediate fixation in buffered formalin will result in the loss of a large number of protein epitopes as these are denatured by the preservative but will maintain much of the cellular architecture, which can be studied by embedding the fixed samples in paraffin for sectioning at a later point in time. Samples taken for other kinds of biochemical analysis (e.g., proteomics, lipid composition) should ideally be kept and transported frozen. Samples of glandular secretions for chromatographic analysis can be collected with a sterile dry swab and then stored in an airtight glass vial for long periods of time in the field.

Many kinds of sample typically collected for nucleic acid extraction also provide abundant opportunities for the study of other molecules. For example, because certain hormones (e.g., prolactin, testosterone, cortisol, progesterone, estradiol) and metabolic by-products (e.g., ketones from metabolism of fat) are present in feces and urine, these samples can be used for endocrinological studies of female reproductive cycles (reviewed in Hodges and Heistermann 2003), for examining the hormonal correlates of social status (van Schaik et al. 1991, Strier et al. 1999), or for investigating seasonal changes in energy budgets (Knott 1997, 1998). It is important to keep aliquots of the various fractions created during DNA extraction as a potential source for some of these other biomolecules. Unlike larger and more fragile molecules, mono- and oligosaccharides are generally robust and subject to slower decay, unless there is bacterial or fungal contamination of the samples. As such contamination almost always renders the samples useless, an aseptic technique should be followed in the field for preserving these molecules. Skin biopsies, if taken with proper caution (aseptic techniques) and kept in cell culture medium, can survive at room temperature for a week to 10 days and still yield viable fibroblast cultures. Alternatively, if liquid nitrogen is available, skin biopsy samples can be minced and frozen in dimethyl sulfoxide (DMSO) in the field to prevent crystal formation and resulting cell death and then shipped in dry ice (Houck et al. 1995).

Sample Storage

Storing samples while in the field can present challenges as electricity for running refrigerators or freezers for preserving samples is typically not available. However, many sample types can be kept for long periods of time at room temperature without substantial degradation of DNA and other molecules of interest if stored under the proper conditions. For example, all samples potentially containing DNA and RNA can be kept at room temperature for several weeks if

stored in a sufficient volume of RNA*later*® (Ambion, Austin, TX), which is a nonflammable, nontoxic buffer that will maintain RNA and DNA for later isolation and amplification. Table 22.2 summarizes a set of appropriate field storage conditions (most at room temperature) for a range of sample types and includes recipes for a number of several useful field buffers. Whenever possible, these samples should be transferred to appropriate long-term storage temperatures (e.g., 4°C–20°C) as soon as possible.

A BRIEF NOTE ON LABORATORY AND ANALYTICAL METHODS

Because a number of excellent publications present detailed protocols for many of the laboratory methods discussed above—and because very few of these protocols are specific to primates—we saw little need to repeat that information here. Instead, we have compiled a short list of useful references where readers can go for more detail on how to implement some of the general methods discussed above (Table 22.3). In addition to these, Humana Press' *Methods in Molecular Biology* series and a variety of titles from Cold Spring Harbor Laboratory Press provide even more detail on using particular methods (e.g., RT-PCR) for specific applications. Table 22.3 also highlights a set of excellent sources providing overviews of important methods for analyzing molecular data generated using the techniques described above.

Table 22.2   Suggested Field Storage Conditions for a Variety of Sample Types

SAMPLE TYPE	USED FOR	STORAGE MEDIUM	RECIPE	HOW TO USE AND REFERENCE
Blood	DNA	Blood field buffer	0.2M NaCl, 0.1M EDTA, 2% SDS, store at RT	Mix 1 volume whole blood with 5 volumes of buffer and store at RT
	DNA	STE buffer	0.1M NaCl, 10 mM Tris, 1 mM EDTA, adjust pH to 8.0 with HCl and store at RT	Mix whole blood with equal volume of buffer and store at RT (Goossens et al. 2003)
	DNA	Citrate blood buffer	0.48 g citric acid, 1.32 g sodium citrate, 1.47 g glucose, add ddH$_2$O to 100 ml and store at RT	Mix 1 volume whole blood with 6 volumes of buffer and store at RT
	DNA	Filter paper cards (e.g., Whatman FTA® cards)		Place drops of sample on card, allow to air dry, and store in airtight container at RT with desiccant; FTA® cards are impregnated with chemicals that lyse cells, denature proteins, and protect DNA from damage; suitable for long-term storage
	DNA	Queen's lysis buffer	10 mM NaCl, 10 mM Tris, 10 mM EDTA, 1% *n*-lauroylacrosine, adjust pH to 8.0 with HCl and store at RT	Mix whole blood with 20 volumes of buffer (e.g., 50 µl in 1 ml) (Seutin et al. 1991)
	White blood cells, DNA	ACK lysis buffer, followed by storage in field buffer or PBS	150 mM NH$_4$Cl, 1 mM KHCO$_3$, 0.1 mM EDTA, adjust pH to 7.2–7.4 with HCl, sterilize with 0.22 mm filter, and store at RT	Mix 1 volume of pellet with 5 volumes of ACK lysis buffer, vortex, spin, and remove supernatant; repeat until no more red blood cells are visible in the remaining pellet of concentrated white blood cells; resuspend the pellet in 5 volumes of field buffer at RT or frozen in PBS
Blood serum	Proteins (albumin, globulins, hormones)			Collect whole blood into special serum separator or blood collection tubes; allow blood to coagulate at RT for 2–4 hr, then carefully pipet off liquid phase (serum) and store in liquid nitrogen or at −80°C

Table 22.2 (*cont'd*)

SAMPLE TYPE	USED FOR	STORAGE MEDIUM	RECIPE	HOW TO USE AND REFERENCE
Tissue	DNA	NaCl-saturated DMSO	20% (wt/vol) DMSO, 0.25M EDTA, NaCl to saturation, adjust pH to 7.5 with HCl and store at RT	Score tissue with razor to expose more of surface to solution, immerse sample and store at RT (Seutin et al. 1991)
	DNA	90%–100% ethanol		Score tissue with razor to expose more of surface to solution; immerse sample and store at RT (Note: DNA yield will be lower and DNA more degraded [fewer large fragments] using this storage medium)
	DNA	Queen's lysis buffer	See above	Store small tissue biopsies in 1 ml of buffer (Bruford et al. 1998)
	DNA, RNA, proteins	RNA*later*®		Follow manufacturer's protocols for sample preparation and storage
	Histology	OCT (optimal cutting temperature) compound	Available from several commercial suppliers (e.g., Pelco)	Embed sample in OCT compound, flash-freeze in dry ice and 2-methylbutane slush by floating it in a small plastic sample holder/boat, and store in liquid nitrogen or at −80°C
Hair	DNA	Paper or glassine envelopes		Collect sample using flame-sterilized forceps, being careful to include root; store envelopes in airtight container desiccated with silica gel beads
Feces	DNA	RNA*later*® buffer	Available from several commercial suppliers (e.g., Ambion)	Add 1 ml of sample to 5 ml of buffer and shake vigorously to homogenize; buffer should cover sample
	DNA	70%–100% ethanol		Add 1 ml of sample to 5 ml of buffer and shake vigorously to homogenize; buffer should cover sample
	DNA	Desiccating silica gel beads	Available from several commercial suppliers (e.g., Sigma)	Place sample in airtight plastic bag or collection tube and add sufficient silica gel beads to thoroughly desiccate sample (6–8× weight of sample)
	DNA	90% ethanol and desiccating silica gel beads		Place 200–250 mg of feces in 4 ml ethanol for 1 day, then decant ethanol and place sample on filter paper in a collection tube containing 25–30 ml of silica gel beads and store at RT (Roeder et al. 2004)
	DNA	DETs solution	20% (wt/vol) DMSO, 0.25M EDTA, 100 mM Tris, NaCl to saturation, adjust pH to 7.5 with HCl and store at RT	Add ~2 g wet weight sample to 1 ml of DETs solution (Frantzen et al. 1998)
Buccal swabs	DNA	Plastic sealable bag containing desiccating silica beads		Using a sterile foam-tipped swab or nylon cytology brush, rub up and down along the inner cheek of the subject for ~15 sec; allow to air dry for 10–15 min, then remove tip and store in airtight environment with desiccating silica beads
		Filter paper cards (e.g., Whatman FTA® cards)		Collect sample as above and then express directly onto card; allow card to air dry and store in airtight container at RT with desiccant
		Cell lysis buffer (or Queen's lysis buffer or STE buffer)	0.1M NaCl, 0.1M Tris, 10 mM EDTA, 1% SDS, store at RT	Collect sample as above, then swish tip around in buffer thoroughly, express excess fluid against side of collection tube, and store at RT
Skin biopsy	Fibroblast cell culture	Sterile growth medium	Alpha MEM (GIBCO BRL) supplemented with 10% fetal calf serum, 1% penicillin/streptomycin, and 200 mM L-glutamine	Aseptic technique for skin biopsy collection from a properly anesthesized animal: (*1*) Clean area on the skin with alcohol swabs (*2*) Shave area clear of any hair and repeat cleaning with second sterile alcohol pad (*3*) Using sterile forceps, pinch and lift the skin (*4*) Using a new scalpel blade, cut a 5 × 5 mm piece of skin (*5*) Place the cut sample directly into a flask containing sterile medium (*6*) Disinfect cut and suture if necessary
		DMSO for freezing on site of minced tissue	Alpha MEM (GIBCO BRL) supplemented with 10% fetal calf serum, 1% penicillin/streptomycin, 200 mM L-glutamine 10% (wt/vol) DMSO	

RT, room temperature; EDTA, ethylene diamide tetra acetate acid (a chelating agent); SDS, sodium dodecyl sulfate (a detergent for lysing cells); PBS, phosphate-buffered saline; DMSO, dimethyl sulfoxide; STE,—Sodium-Tris-EDTA solution; ACK,—ACK buffer; DETs,—DMSO-EDTA-Tris-Sodium solution; ddH$_2$O, double-distilled water; MEM, minimal essential medium.

Table 22.3   Texts Providing Laboratory Protocols and Overviews of Theoretical and Analytical Methods Useful to Molecular Primatologists

TITLE	REFERENCE	DESCRIPTION
DNA Fingerprinting	Burke et al. (1991)	Provides a thorough introduction to minisatellite DNA fingerprinting and an early overview of DNA typing in general
Genetic Data Analysis II	Weir (1996)	Offers a comprehensive overview of the statistical analysis of population genetic data (evaluating linkage between loci, selection, inbreeding, population structure, etc), also provides an overview of methods used in phylogeny reconstruction
Fundamentals of Molecular Evolution	Graur and Li (2000)	Provides a clearly written overview of many aspects of molecular evolution, including the structure of genes and genomes, genome evolution, population genetics, and methods of phylogeny reconstruction
Inferring Phylogenies	Felsenstein (2004b)	Provides a comprehensive, readable overview of phylogenetic methods
Molecular Cloning	Sambrook and Russell (2001)	Serves as a central source of useful, basic protocols for manipulating and studying DNA in vitro, including DNA extraction, polymerase chain reaction, electrophoresis, visualization, and vector cloning; also a comprehensive source of basic buffer and reagent "recipes"
Molecular Genetic Analysis of Populations, 2nd ed.	Hoelzel (1998)	Provides a short but comprehensive set of laboratory protocols for finding and assaying variation in allozymes and DNA
Molecular Markers, Natural History, and Evolution, 2nd ed.	Avise (2004)	Offers a comprehensive theoretical and practical overview of the various kinds of molecular markers evolutionary biologists might be interested in and of the utility of these markers for investigating many aspects of a taxon's biology and history
Molecular Methods in Ecology	Baker (2000)	Provides a general overview of both basic and newly developed molecular biology techniques that can be useful in behavioral and ecological research that is accessible to nonspecialists
Molecular Systematics	Hillis et al. (1996)	Provides a large set of protocols useful for assaying variation in a variety of molecule types, including allozymes, entire chromosomes, and DNA fragments; also includes thorough overviews of many of the analytical techniques used for inferring phylogenies using molecular data
Molecular Zoology	Ferraris and Palumbi (1996)	Overviews a number of zoological research programs that use molecular techniques as a central research tool and provides a compendium of laboratory protocols relevant to these techniques

One set of laboratory techniques that is more specific to primates involves those used to genetically determine the sex of a DNA sample. A number of methods for sex typing primate DNA have been developed. Most commonly, these involve the coamplification of homologous fragments of the nonrecombining region of the sex chromosomes that show fixed sequence length differences between the X and the Y. For several hominoids, an assay developed by Sullivan et al. (1993), which relies on the presence of a 6 bp difference between the X- and Y-borne copies of the gene for the tooth enamel protein amelogenin, has proved effective for molecular sex typing (Bradley et al. 2001). However, this assay does not work effectively outside of the Hominoidea (Ensminger and Hoffman 2002), nor is it effective for the genus *Pongo* (Steiper and Ruvolo 2003). Wilson and Erlandsson (1998) developed a PCR-based assay effective for a broad range of anthropoid primates; however, it relies on the co-amplification of two fragments of ~700 and ~1,100 bp in length from the X- and Y-borne copies of the zinc finger protein gene, and these are too large to amplify reliably from many types of noninvasively collected sample. Fredsted and Villesen (2004) have designed a similar assay based on the amelogenin locus and involving amplicons of >1,300 bp that appears to be effective in a number of strepsirhines as well as humans. Finally, Di Fiore (in press) has recently developed a novel sex-typing assay that is effective in a broad range of anthropoid primates. This assay uses multiplex PCR to coamplify fragments of the amelogenin gene from the X chromosome and the sex-determining region of the Y gene (*SRY*) from the Y chromosome. The advantage of this assay is that the target fragments are short enough (<200 bp) to reliably amplify from degraded DNA samples.

Finally, there now exists a wealth of software packages useful for analyzing various kinds of molecular data, many of which are publicly available for downloading from the internet. These include packages for multiple sequence alignment, phylogeny inference (using a variety of algorithms, including distance, parsimony, and likelihood-based methods), basic population genetic analyses, and paternity assignment (Table 22.4). Additionally, some journals (e.g., *Molecular Ecology Notes*, *Bioinformatics*) routinely publish notes on new software suitable for particular types of analysis.

CONCLUSIONS

Molecular primatology has already contributed much to our understanding of primate evolution. For the future, the

Table 22.4   Software Packages Useful for the Analysis of Molecular Data

PROGRAM	DESCRIPTION	REFERENCE	PLATFORM AND SOURCE
Sequence alignment			
CLUSTAL	Program for multiple alignment of nucleotide and protein sequence data using a progressive sequence alignment algorithm, can also be used for inferring the phylogenetic relationships among sequences using the neighbor-joining method	Higgins et al. (1996)	Web: www.ebi.ac.uk/clustalw/ Windows, Mac, Linux (via X Windows): ftp-igbmc.u-strasbg.fr/pub/ClustalX/
MALIGN	Program for multiple alignment of nucleotide sequence data that builds phylogenetic trees as it constructs an alignment and uses parsimony criteria to assess the best alignment	Wheeler and Gladstein (1994)	Windows, Linux, Sun: research.amnh.org/scicomp/projects /malign.php
TreeAlign	Program for multiple alignment of nucleotide and protein sequence data that builds phylogenetic trees as it constructs an alignment and uses approximate parsimony criteria to assess the best alignment	Hein (1994)	Web: bioweb.pasteur.fr/seqanal/ interfaces/treealign-simple.html Unix: ftp.ebi.ac.uk/pub/software/unix/
DNASIS	Commercially available program for multiple alignment of nucleotide and protein sequence data using a progressive sequence alignment algorithm, also used for archiving of sequence data and for various kinds of sequence analysis (e.g., protein structure)		Windows, Mac: www.oligo.net/dnasis.htm
Sequencher	Commercially available program for multiple sequence alignment and editing and for archiving of sequence data		Windows, Mac: www.genecodes.com/
Phylogenetic inference			
PAUP*	Very widely used and general-purpose program for phylogenetic analysis; calculates numerous measures of genetic distance between taxa according to a user-specified model of evolution; performs phylogeny estimation using distance, parsimony, and likelihood methods	Swofford (2002)	Windows, Mac, Unix, DOS: paup.csit.fsu.edu/
MEGA	General-purpose program for DNA and protein sequence analysis, calculates various measures of genetic distance between sequences from molecular data according to a user-specified model of evolution, performs phylogeny estimation using distance and parsimony methods	Kumar et al. (2001)	Windows: www.megasoftware.net/
PHYLIP	Large set of programs for phylogeny estimation that can use sequence data, allele frequency data, or character data; performs phylogeny estimation using distance, parsimony, and likelihood methods	Felsenstein (2004a)	Windows, Mac, Linux: evolution.genetics.washington.edu /phylip.html
MrBayes	Program to estimate phylogenetic trees based on Bayesian inference	Ronquist and Huelsenbeck (2003)	Windows, Mac, Unix: morphbank.ebc.uu.se/mrbayes/info.php
PAML	Set of programs for phylogeny estimation using maximum likelihood methods and implementing many different models of sequence evolution	Yang (1997)	Windows, Mac, Linux: abacus.gene.ucl.ac.uk/software/paml.html
TREE-PUZZLE	Program to estimate phylogenetic trees from DNA and protein sequence data using a maximum likelihood quartet-puzzling method	Schmidt et al. (2002)	Windows, Mac, Linux: www.tree-puzzle.de/
Population genetics and analysis of population structure			
GENEPOP	General-purpose software package for population genetic analysis; calculates allele frequencies as well as observed and expected genotype frequencies and evaluates whether genotype data deviate significantly from Hardy-Weinberg equilibrium expectation; also calculates various estimators of population differentiation and gene flow among populations and can be used to test for linkage among loci; web version includes useful routines for converting among file formats used by different population genetics programs	Raymond and Roussett (1995)	Web, DOS: wbiomed.curtin.edu.au/genepop/
FSTAT	Versatile general-purpose program for population genetic analysis; calculates allele frequencies as well as observed and expected genotype frequencies and goodness of fit to Hardy-Weinberg equilibrium expectations; also calculates various estimators of population subdivision and inbreeding as well as estimators of genetic distance and relatedness between populations using allelic data; can also be used to test for linkage among loci and to evaluate potential sex biases in dispersal	Goudet (1995, 2001)	Windows: www.unil.ch/izea/softwares/fstat.html
RSTCALC	Program for analysis of population structure and gene flow specifically designed for use with microsatellite marker data	Goodman (1997)	DOS: helios.bto.ed.ac.uk/evolgen/rst/rst.html

Table 22.4   (*cont'd*)

PROGRAM	DESCRIPTION	REFERENCE	PLATFORM AND SOURCE
ARLEQUIN	Versatile multiplatform program for population genetic analysis; calculates allele frequencies as well as observed and expected genotype frequencies, various estimators of population subdivision and gene flow, and various estimators of genetic distance between populations using allelic, haplotype, or sequence data; conducts tests for goodness of fit to Hardy-Weinberg expectations and can also be used to perform assignment tests and to test for isolation by distance	Schneider et al. (2000)	Windows, Mac, Linux (via Java): lgb.unige.ch/arlequin/
Structure	Program for analysis of population structure using multilocus genotype data from various kinds of genetic markers, can evaluate whether a population is subdivided and can be used to assign individuals to subpopulations or to investigate hybridization and admixture	Pritchard et al. (2000)	Windows: pritch.bsd.uchicago.edu/
GeneClass	Program to detect migrants into a population and to assign individuals to a population based on multilocus genotype data, also useful for converting among several file formats used by different population genetics programs	Piry et al. (2004)	Windows, Linux: www.montpellier.inra.fr/URLB/index.html
Identity and parentage analysis			
RELATEDNESS	Estimates pairwise relatedness between individuals or average pairwise relatedness between groups of individuals using a regression-based estimator of relatedness	Queller and Goodnight (1989)	Mac: gsoft.smu.edu/GSoft.html
KINSHIP	Evaluates user-specified hypotheses about pedigree relationships using likelihood methods and can be used for parentage assignment	Goodnight and Queller (1999)	Mac: gsoft.smu.edu/GSoft.html
CERVUS	Conducts parentage assignments using likelihood-based methods that allow for incomplete sampling of potential parents and for some error in genotyping, also can be used to test for the existence of null alleles at a locus and for evaluating whether genotype data deviate significantly from Hardy-Weinberg expectations	Marshall et al. (1998)	Windows:helios.bto.ed.ac.uk/evolgen /cervus/cervus.htm

promise of combining molecular genetic data with information gleaned from other biomolecules, such as proteins, lipids, and carbohydrates and their various combinations, is likely to reveal additional important insights into primate adaptations. The acute conservation crisis faced by most nonhuman primate species makes it imperative that we use all the tools at our disposal to further the understanding of primate biology and support conservation management decisions with key scientific knowledge. Both molecular and whole-organism primatologists have much to gain from the many possibilities for fertile collaboration.

We conclude this chapter by making one caveat. While the various techniques described here clearly constitute an impressive set of tools for examining and interpreting the molecular variation seen within and among different primate taxa, implementation of any of these techniques often requires substantial investment in time, energy, and resources to become proficient with both the theoretical and practical aspects of a particular kind of analysis. Thus, we caution field workers to beware of succumbing to the idea that it is easy to simply "do genetics on the side." To be sure, molecular data can provide important insights into many aspects of the biology of wild primates that may be difficult to realize through field studies alone, but the rationale for adding a molecular component to any field study must be clearly scientifically motivated and worth the investment. This is especially true when the collection of samples for molecular analyses requires anything but completely noninvasive procedures. Thus, as a point of caution, we urge students starting out in the field to not underestimate the workload involved in the molecular part of field projects, and we suggest that students undertake their laboratory training before taking off to the field to start a project. Likewise, we caution molecular biologists to not underestimate the crucial importance of reliable behavioral and ecological data from the field. Some of the most provocative molecular findings may be meaningless without their context in natural history.

REFERENCES

Adams, E. J., and Parham, P. (2001). Species-specific evolution of MHC class I genes in the higher primates. *Immunol. Rev.* 183:41–64.

Aitken, N., Smith, S., Schwarz, C., and Morin, P. A. (2004). Single nucleotide polymorphism (SNP) discovery and genotyping in mammals: a targeted-gene approach. *Mol. Ecol.* 13:1423–1431.

Alberts, S. C. (1999). Paternal kin discrimination in wild baboons. *Proc. R. Soc. Lond. B* 266:1501–1506.

Altmann, J., Alberts, S. C., Haines, S. A., Bubach, J., Muruthi, P., Coote, T., Geffen, E., Cheesman, D. J., Mututa, R. S., Saiyalel, S. N., Wayne, R. K., Lacy, R. C., and Bruford, M. W. (1996).

Behavior predicts genetic structure in a wild primate group. *Proc. Nat. Acad. Sci. USA* 93:5797–5801.

Arnason, U., Gullberg, A., Janke, A., and Xu, X. (1996). Pattern and timing of evolutionary divergences among hominoids based on analyses of complete mtDNAs. *J. Mol. Evol.* 43:650–661.

Arnold, N., Stanyon, R., Jauch, A., O'Brien, P., and Wienberg, J. (1996). Identification of complex chromosome rearrangements in the gibbon by fluorescent in situ hybridization (FISH) of a human chromosome 2q specific microlibrary, yeast artificial chromosomes, and reciprocal chromosome painting. *Cytogenet. Cell Genet.* 74:80–85.

Avise, J. C. (2004). *Molecular Markers, Natural History, and Evolution*, 2nd ed. Sinauer Associates, Sunderland, MA.

Baba, M. L., Darga, L. L., and Goodman, M. (1979). Immuno-diffusion systematics of the primates. Part V. The Platyrrhini. *Folia Primatol.* 32:207–238.

Baker, A. J. (ed.) (2000). *Molecular Methods in Ecology*. Blackwell Scientific, Oxford.

Bauers, K. A., and Hearn, J. P. (1994). Patterns of paternity in relation to male social rank in the stumptailed macaque, *Macaca arctoides*. *Behaviour* 129:149–176.

Berard, J. D., Nürnberg, P., Epplen, J. T., and Schmidtke, J. (1993). Male rank, reproductive behavior, and reproductive success in free-ranging rhesus macaques. *Primates* 34:481–489.

Berard, J. D., Nürnberg, P., Epplen, J. T., and Schmidtke, J. (1994). Alternative reproductive tactics and reproductive success in male rhesus macaques. *Behaviour* 129:177–201.

Bigoni, F., Houck, M. L., Ryder, O. A., Wienberg, J., and Stanyon, R. (2004). Chromosome painting shows that *Pygathrix nemaeus* has the most basal karyotype among Asian Colobinae. *Int. J. Primatol.* 4:679–688.

Bigoni, F., Stanyon, R., Wimmer, R., and Schempp, W. (2003). Chromosome painting shows that the proboscis monkey (*Nasalis larvatus*) has a derived karyotype and is phylogenetically nested within Asian colobines. *Am. J. Primatol.* 60:85–93.

Boenisch, T. (ed.) (2001). *Immunochemical Staining Methods*, 3rd ed. Dako, Carpinteria, CA.

Bradley, B. J., Chambers, K. E., and Vigilant, L. (2001). Accurate DNA-based sex identification of apes using non-invasive samples. *Conserv. Genet.* 2:179–181.

Bradley, B. J., Doran-Sheehy, D. M., Lukas, D., Boesch, C., and Vigilant, L. (2004). Dispersed male networks in western gorillas. *Curr. Biol.* 14:510–513.

Bray, M. S., Boerwinkle, E., and Doris, P. A. (2001). High-throughput multiplex SNP genotyping with MALDI-TOF mass spectrometry: practice, problems and promise. *Hum. Mutat.* 17:296–304.

Brett, F. L., Jolly, C. J., Socha, W., and Wiener, A. S. (1976). Human-like ABO blood groups in wild Ethiopian baboons. *Ybk. Phys. Anthropol.* 20:276–289.

Brinkman-van der Linden, E. C., Sjoberg, E. R., Juneja, L. R., Crocker, P. R., Varki, N., and Varki, A. (2000). Loss of *N*-glycolylneuraminic acid in human evolution. Implications for sialic acid recognition by siglecs. *J. Biol. Chem.* 275:8633–8640.

Bruford, M. W., Hanotte, O., and Burke, T. (1998). Multi- and single-locus DNA fingerprinting. In: Hoelzel, A. R. (ed.), *Molecular Genetic Analysis of Populations—A Practical Approach*, 2nd ed. ILR Press, Oxford. pp. 225–269.

Burke, T., Dolf, G., Jeffreys, A. J., and Wolff, R. (eds.) (1991). *DNA Fingerprinting: Approaches and Applications*. Birkhäuser Verlag, Basel.

Cáceres, M., Lachuer, J., Zapala, M. A., Redmond, J. C., Kudo, L., Geschwind, D. H., Lockhart, D. J., Preuss, T. M., and Barlow, C. (2003). Elevated gene expression levels distinguish human from non-human primate brains. *Proc. Natl. Acad. Sci. USA* 100:13030–13035.

Cann, R. L., Stoneking, M., and Wilson, A. C. (1987). Mitochondrial DNA and human evolution. *Nature* 325:31–36.

Cartegni, L., Chew, S. L., and Krainer, A. R. (2002). Listening to silence and understanding nonsense: exonic mutations that affect splicing. *Nat. Rev. Genet.* 3:285–298.

Caspersson, T., Farber, S., Foley, G. E., Kudynowski, J., Modest, E. J., Simonsson, E., Wagh, U., and Zeck, L. (1968). Chemical differentiation along metaphase chromosomes. *Exp. Cell Res.* 49: 219–222.

Chimpanzee Sequencing and Analysis Consortium (2005). Initial sequence of the chimpanzee genome and comparison with the human genome. *Nature* 437:69–87.

Collins, A. C., and Dubach, J. M. (2000a). Biogeographic and ecological forces responsible for speciation in *Ateles*. *Int. J. Primatol.* 21:421–444.

Collins, A. C., and Dubach, J. M. (2000b). Phylogenetic relationships of spider monkeys (*Ateles*) based on mitochondrial DNA variation. *Int. J. Primatol.* 21:381–420.

Constable, J. L., Packer, C., Collins, D. A., and Pusey, A. E. (1995). Nuclear DNA from primate dung. *Nature* 373:393.

Constable, J. L., Ashley, M. V., Goodall, J., and Pusey, A. E. (2001). Noninvasive paternity assignment in Gombe chimpanzees. *Mol. Ecol.* 10:1279–1300.

Cortés-Ortiz, L., Bermingham, E., Rico, C., Rodríguez-Luna, E., Sampaio, I., and Ruiz-García, M. (2003). Molecular systematics and biogeography of the neotropical monkey genus, *Alouatta*. *Mol. Phylogenet. Evol.* 26:64–81.

Daszak, P., Cunningham, A. A., and Hyatt, A. D. (2000). Emerging infectious diseases of wildlife—threats to biodiversity and human health. *Science* 287:443–449.

Dean, F. B., Hosono, S., Fang, L., Wu, X., Faruqi, A. F., Bray-Ward, P., Sun, Z., Zong, Q., Du, Y., Du, J., Driscoll, M., Song, W., Kingsmore, S. F., Egholm, M., and Lasken, R. S. (2002). Comprehensive human genome amplification using multiple displacement amplification. *Proc. Natl. Acad. Sci. USA* 99:5261–5266.

Dene, H., Goodman, M., and Prychodko, W. (1980). Immuno-diffusion systematics of the primates. IV. Lemuriformes. *Mammalia* 44:211–223.

Dene, H., Goodman, M., Prychodko, W., and Moore, G. W. (1976a). Immunodiffusion systematics of the primates. III. The Strepsirhini. *Folia Primatol.* 25:35–61.

Dene, H. T., Goodman, M., and Prychodko, W. (1976b). Immuno-diffusion evidence on the phylogeny of the primates. In: Goodman, M., Tashian, R. E., and Tashian, J. H. (eds.), *Molecular Anthropology: Genes and Proteins in the Evolutionary Ascent of the Primates*. Plenum Press, New York. pp. 171–195.

de Ruiter, J. R. (2004). Genetic markers in primate studies: elucidating behavior and its function. *Int. J. Primatol.* 25:1173–1189.

de Ruiter, J. R., Scheffrhan, W., Trommelen, G. J. J. M., Uitterlinden, A. G., and Martin, R. D. (1992). Male social rank and reproductive success in wild long-tailed macaques. In: Martin, R. D., Dixson, A. F., and Wickings, E. J. (eds.), *Paternity in Primates: Genetic Tests and Theories*. Karger, Basel. pp. 175–191.

de Ruiter, J. R., van Hooff, J. A. R. A. M., and Scheffrhan, W. (1994). Social and genetic aspects of paternity in wild long-tailed macaques (*Macaca fascicularis*). *Behaviour* 129:203–223.

Di Fiore, A. (2002). Molecular perspectives on dispersal in lowland woolly monkeys (*Lagothrix lagotricha poeppigii*). *Am. J. Phys. Anthropol.* S34:63.

Di Fiore, A. (2003). Molecular genetic approaches to the study of primate behavior, social organization, and reproduction. *Ybk. Phys. Anthropol.* 46:62–99.

Di Fiore, A. (in press). A rapid genetic method for sex assignment in nonhuman primates. *Conserv. Genet.*

Di Fiore, A., and Fleischer, R. C. (2005). Social behavior, reproductive strategies, and population genetic structure of *Lagothrix lagotricha poeppigii. Int. J. Primatol.* 26:1137–1173.

Disotell, T. R. (2000). Molecular systematics of the Cercopithecidae. In: Whitehead, P. F., and Jolly, C. J. (eds.), *Old World Monkeys.* Cambridge University Press, Cambridge. pp. 29–56.

Dixson, A. F., Bossi, T., and Wickings, E. J. (1993). Male dominance and genetically determined reproductive success in the mandrill (*Mandrillus sphinx*). *Primates* 34:525–532.

Domingo-Roura, X., Marmi, J., Andrés, O., Yamagiwa, J., and Terradas, J. (2004). Genotyping from semen of wild japanese macaques (*Macaca fuscata*). *Am. J. Primatol.* 62:31–42.

Enard, W., Khaitovich, P., Klose, J., Zöllner, S., Heissig, F., Giavalisco, P., Nieselt-Struwe, K., Muchmore, E., Varki, A., Ravid, R., Doxiadis, G. M., Bontrop, R. E., and Pääbo, S. (2002). Intra- and interspecific variation in primate gene expression patterns. *Science* 296:340–343.

Ensminger, A. L., and Hoffman, S. M. G. (2002). Sex identification assay useful in great apes is not diagnostic in a range of other primate species. *Am. J. Primatol.* 56:129–134.

Evans, B. J., Morales, J. C., Supriatna, J., and Melnick, D. J. (1999). Origin of the Sulawesi macaques (Cercopithecidae: *Macaca*) as suggested by mitochondrial DNA phylogeny. *Biol. J. Linn. Soc.* 66:539–560.

Felsenstein, J. (2004a). *PHYLIP (Phylogeny Inference Package)*, version 3.6. Department of Genome Sciences, University of Washington, Seattle (distributed by the author).

Felsenstein, J. (2004b). *Inferring Phylogenies.* Sinauer Associates, Sunderland, MA.

Ferraris, J. D., and Palumbi, S. R. (eds.) (1996). *Molecular Zoology: Advances, Strategies, and Protocols.* Wiley-Liss, New York.

Fisher, P. J., Gardner, R. C., and Richardson, T. E. (1996). Single locus microsatellites isolated using 5′-anchored PCR. *Nucleic Acids Res.* 24:4369–4371.

Forman, L., Kleiman, D. G., Bush, R. M., Dietz, J. M., Ballou, J. D., Phillips, L. G., Coimbra-Filho, A. F., and O'Brien, S. J. (1987). Genetic variation within and among lion tamarins. *Am. J. Phys. Anthropol.* 71:1–11.

Forster, T., Roy, D., and Ghazal, P. (2003). Experiments using microarray technology: limitations and standard operating procedures. *J. Endocrinol.* 178:195–204.

Frantzen, M. A. J., Silk, J. B., Ferguson, J. W. H., Wayne, R. K., and Kohn, M. H. (1998). Empirical evaluation of preservation methods for faecal DNA. *Mol. Ecol.* 7:1423–1428.

Fredsted, T., and Villesen, P. (2004). Fast and reliable sexing of prosimian and human DNA. *Am. J. Primatol.* 64:345–350.

Gagneux, P., Amess, B., Diaz, S., Moore, S., Patel, T., Dillmann, W., Parekh, R., and Varki, A. (2001). Proteomic comparison of human and great ape blood plasma reveals conserved glycosylation and differences in thyroid hormone metabolism. *Am. J. Phys. Anthropol.* 115:99–109.

Gagneux, P., Boesch, C., and Woodruff, D. S. (1997a). Microsatellite scoring errors associated with noninvasive genotyping based on nuclear DNA amplified from shed hair. *Mol. Ecol.* 6:861–868.

Gagneux, P., Boesch, C., and Woodruff, D. S. (1999). Female reproductive strategies, paternity and community structure in wild West African chimpanzees. *Anim. Behav.* 57:19–32.

Gagneux, P., Cheriyan, M., Hurtado-Ziola, N., van der Linden, E. C., Anderson, D., McClure, H., Varki, A., and Varki, N. M. (2003). Human-specific regulation of alpha 2-6-linked sialic acids. *J. Biol. Chem.* 278:48245–48250.

Gagneux, P., Gonder, M. K., Goldberg, T. L., and Morin, P. A. (2001). Gene flow in wild chimpanzee populations: what genetic data tell us about chimpanzee movement over space and time. *Phil. Trans. R. Soc. Lond. B* 356:889–897.

Gagneux, P., and Varki, A. (2001). Genetic differences between humans and great apes. *Mol. Phylogenet. Evol.* 18:2–13.

Gagneux, P., Woodruff, D. S., and Boesch, C. (1997b). Furtive mating by female chimpanzees. *Nature* 387:327–328.

Gerloff, U., Hartung, B., Fruth, B., Hohmann, G., and Tautz, D. (1999). Intracommunity relationships, dispersal pattern, and paternity success in a wild living community of bonobos (*Pan paniscus*) determined from DNA analysis of faecal samples. *Proc. R. Soc. Lond. B* 266:1189–1195.

Goldberg, T. L., and Ruvolo, M. (1997). Molecular phylogenetics and historical biogeography of East African chimpanzees. *Biol. J. Linn. Soc.* 61:301–324.

Goldberg, T. L., and Wrangham, R. W. (1997). Genetic correlates of social behaviour in wild chimpanzees: evidence from mitochondrial DNA. *Anim. Behav.* 54:559–570.

Goldman, D., Giri, P. R., and O'Brien, S. J. (1987). A molecular phylogeny of the hominoid primates as indicated by two-dimensional protein electrophoresis. *Proc. Natl. Acad. Sci. USA* 84:3307–3311.

Goodman, M. (1976). Toward a geneological description of the primates. In: Goodman, M., Tashian, R. E., and Tashian, J. H. (eds.), *Molecular Anthropology: Genes and Proteins in the Evolutionary Ascent of the Primates.* Plenum Press, New York. pp. 321–353.

Goodman, M., Farris, W., Jr., Moore, W., Prychodko, W., Poulik, E., and Sorenson, M. (1974). Immunodiffusion systematics of the primates: II. Findings on *Tarsius*, Lorisidae and Tupaiidae. In: Martin, R. D., Doyle, G. A., and Walker, A. C. (eds.), *Prosimian Biology.* Duckworth, London. pp. 881–890.

Goodman, M., and Moore, G. W. (1971). Immunodiffusion systematics of the primates. I. The catarrhini. *Syst. Zool.* 20:19–62.

Goodman, S. J. (1997). RST CALC: a collection of computer programs for calculating unbiased estimates of genetic differentiation and gene flow from microsatellite data and determining their significance. *Mol. Ecol.* 6:881–886.

Goodnight, K. F., and Queller, D. C. (1999). Computer software for performing likelihood tests of pedigree relationship using genetic markers. *Mol. Ecol.* 8:1231–1234.

Goossens, B., Anthony, N., Jeffrey, K., Johnson-Bawe, M., and Bruford, M. W. (2003). Collection, storage, and analysis of non-invasive genetic material in primate biology. In: Setchell, J. M., and Curtis, D. M. (eds.), *Field and Laboratory Methods in Primatology: A Practical Guide.* Cambridge University Press, Cambridge. pp. 295–308.

Goudet, J. (1995). FSTAT (version 1.2): a computer program to calculate F-statistics. *J. Hered.* 86:485–486.

Goudet, J. (2001). *FSTAT*, a program to estimate gene diversity and fixation indices (version 2.9.3). Institute of Ecology, Laboratory for Zoology, University of Lausanne, Lausanne.

Graur, D., and Li, W.-H. (2000). *Fundamentals of Molecular Evolution*, 2nd ed. Sinauer Associates, Sunderland, MA.

Gut, I. G. (2001). Automation in genotyping of single nucleotide polymorphisms. *Hum. Mutat.* 17:475–492.

Hacia, J. G., Fan, J.-B., Ryder, O., Jin, L., Edgemon, K., Ghandour, G., Mayer, R. A., Sun, B., Hsie, L., Robbins, C. M., Brody, L. C., Wang, D., Lander, E. S., Lipshutz, R., Fodor, S. P. A., and Collins, F. S. (1999). Determination of ancestral alleles for human single nucleotide polymorphisms using high-density oligonucleotide arrays. *Nat. Genet.* 22:164–167.

Haff, L., and Smirnov, I. (1997). Single nucleotide polymorphism identification assays using a thermostable DNA polymerase and MALDI-TOF MS. *Genome Res.* 7:378–388.

Hahn, B. H., Shaw, G. M., De Cock, K. M., and Sharp, P. M. (2000). AIDS as a zoonosis: scientific and public health implications. *Science* 287:607–614.

Hamilton, M. B., Pincus, E. L., Di Fiore, A., and Fleischer, R. C. (1999). Universal linker and ligation procedures for construction of genomic DNA libraries enriched for microsatellites. *Biotechniques* 27:500–507.

Hammer, M. F., and Zegura, S. L. (2002). The human Y chromosome haplogroup tree: nomenclature and phylogeography of its major divisions. *Annu. Rev. Anthropol.* 31:303–321.

Hapke, A., Zinner, D., and Zischler, H. (2001). Mitochondrial DNA variation in Eritrean hamadryas baboons (*Papio hamadryas hamadryas*): life history influences population genetic structure. *Behav. Ecol. Sociobiol.* 50:483–492.

Hastings, M. L., and Krainer, A. R. (2001). Pre-mRNA splicing in the new millennium. *Curr. Opin. Cell Biol.* 13:302–309.

Hayakawa, S., and Takenaka, O. (1999). Urine as another potential source for template DNA in polymerase chain reaction (PCR). *Am. J. Primatol.* 48:299–304.

Hein, J. (1994). TreeAlign. In: Griffin, A. M., and Griffin, H. G. (eds.), *Methods in Molecular Biology. Computer Analysis of Sequence Data, Part II*, vol. 25. Humana Press, Totowa, NJ. pp. 349–364.

Hewett-Emmett, D., Cook, C. N., and Barnicot, N. A. (1976). Old World monkey hemoglobins: deciphering phylogeny from complex patterns of molecular evolution. In: Goodman, M., Tashian, R. E., and Tashian, J. H. (eds.), *Molecular Anthropology: Genes and Proteins in the Evolutionary Ascent of the Primates.* Plenum Press, New York. pp. 257–275.

Higgins, D. G., Thompson, J. D., and Gibson, T. J. (1996). Using CLUSTAL for multiple sequence alignments. *Methods Enzymol.* 266:383–402.

Higuchi, R., von Berholdingen, C. H., Sensabaugh, G. F., and Erlich, H. A. (1988). DNA typing from single hairs. *Nature* 332:543–546.

Hillis, D. M., Moritz, C., and Mable, B. K. (eds.) (1996). *Molecular Systematics*, 2nd ed. Sinauer Associates, Sunderland, MA.

Hodges, J. K., and Heistermann, M. (2003). Field endocrinology: monitoring hormonal changes in free-ranging primates. In: Setchell, J. M., and Curtis, D. M. (eds.), *Field and Laboratory*

Methods in Primatology: A Practical Guide. Cambridge University Press, Cambridge. pp. 282–294.

Hoelzel, A. R. (ed.) (1998). *Molecular Genetic Analysis of Populations—A Practical Approach*, 2nd ed. ILR Press, Oxford.

Hohmann, G., Gerloff, U., Tautz, D., and Fruth, B. (1999). Social bonds and genetic ties: kinship, association, and affiliation in a community of bonobos (*Pan paniscus*). *Behaviour* 136:1219–1235.

Houck, M. L., Ryder, O. A., Kumamoto, A. T., and Benirschke, K. (1995). Cytogenetics of the Rhinocerotidae. *Verhandlungsbericht des 37. Internationalen Symposiums über die Erkrankungen der Zootiere.* Dresden. pp. 25–32.

International Human Genome Sequencing Consortium (2001). Initial sequencing and analysis of the human genome. *Nature* 409:860–921.

International Human Genome Sequencing Consortium (2004). Finishing the euchromatic sequence of the human genome. *Nature* 431:931–945.

International SNP Map Working Group (2001). A map of human genome sequence variation containing 1.42 million single nucleotide polymorphisms. *Nature* 409:928–933.

Johnson, M. E., Viggiano, L., Bailey, J. A., Abdul-Rauf, M., Goodwin, G., Rocchi, M., and Eichler, E. E. (2001). Positive selection of a gene family during the emergence of humans and African apes. *Nature* 413:514–519.

Jolly, C. J., and Brett, F. L. (1973). Genetic markers and baboon biology. *J. Med. Primatol.* 2:85–99.

Kappeler, P. M., Wimmer, B., Zinner, D., and Tautz, D. (2002). The hidden matrilineal structure of a solitary lemur: implications for primate social evolution. *Proc. R. Soc. Lond. B* 269:1755–1763.

Karaman, M. W., Houck, M. L., Chemnick, L. G., Nagpal, S., Chawannakul, D., Sudano, D., Pike, B. L., Ho, V. V., Ryder, O. A., and Hacia, J. G. (2003). Comparative analysis of gene-expression patterns in human and African great ape cultured fibroblasts. *Genome Res.* 13:1619–1630.

Keane, B., Dittus, W. P. J., and Melnick, D. J. (1997). Paternity assessment in wild groups of toque macaques *Macaca sinica* at Polonnaruwa, Sri Lanka using molecular markers. *Mol. Ecol.* 6:267–282.

Khaitovich, P., Muetzel, B., She, X., Lachmann, M., Hellmann, I., Dietzsch, J., Steigele, S., Do, H.-H., Weiss, G., Enard, W., Heissig, F., Arendt, T., Nieselt-Struwe, K., Eichler, E. E., and Pääbo, S. (2004). Regional patterns of gene expression in human and chimpanzee brains. *Genome Res.* 14:1462–1473.

Knott, C. D. (1997). Field collection and preservation of urine in orangutans and chimpanzees. *Trop. Biodivers.* 4:95–102.

Knott, C. D. (1998). Changes in orangutan caloric intake, energy balance and ketones in response to fluctuating fruit availability. *Int. J. Primatol.* 19:1061–1071.

Koehler, U., Arnold, N., Wienberg, J., Tofanelli, S., and Stanyon, R. (1995a). Genomic reorganization and disrupted chromosomal synteny in the siamang (*Hylobates syndactylus*) revealed by fluorescence in situ hybridization. *Am. J. Phys. Anthropol.* 97:37–47.

Koehler, U., Bigoni, F., Wienberg, J., and Stanyon, R. (1995b). Genomic reorganization in the concolor gibbon (*Hylobates concolor*) revealed by chromosome painting. *Genomics* 30:287–292.

Kruglyak, S., Durrett, R. T., Schug, M. D., and Aquadro, C. F. (1998). Equilibrium distribution of microsatellite repeat length resulting from a balance between slippage events and point mutations. *Proc. Natl. Acad. Sci. USA* 95:10774–10778.

Kumar, S., Tamura, K., Jakobsen, I. B., and Nei, M. (2001). MEGA2: Molecular Evolutionary Genetics Analysis software. *Bioinformatics* 17:1244–1245.

Landsteiner, K. (1900). Zur Kenntnis der antifermentativen, lytischen und agglutinierenden Wirkungen des Blutserums und der Lymphe. *Zentralbl. Bakteriol. Parasit. Infekt.* 27:357–362.

Launhardt, K., Borries, C., Hardt, C., Epplen, J. T., and Winkler, P. (2001). Paternity analysis of alternative male reproductive routes among the langurs (*Semnopithecus entellus*) of Ramnagar. *Anim. Behav.* 61:53–64.

Levinson, G., and Gutman, G. A. (1987). Slipped-strand mispairing: a major mechanism for DNA sequence evolution. *Mol. Biol. Evol.* 4:203–221.

Lunt, D. H., Hutchinson, W. F., and Carvalho, G. R. (1999). An efficient method for PCR-based identification of microsatellite arrays (PIMA). *Mol. Ecol.* 8:893–894.

Marquès-Bonet, T., Cáceres, M., Bertranpetit, J., Preuss, T. M., Thomas, J. W., and Navarro, A. (2004). Chromosomal rearrangements and the genomic distribution of gene-expression divergence in humans and chimpanzees. *Trends Genet.* 20:524–529.

Marshall, T. C., Slate, J., Kruuk, L. E. B., and Pemberton, J. M. (1998). Statistical confidence for likelihood-based paternity inference in natural populations. *Mol. Ecol.* 7:639–655.

Melnick, D. J., and Hoelzer, G. A. (1996). The population genetic consequences of macaque social organization and behaviour. In: Fa, J. E., and Lindburg, D. G. (eds.), *Evolution and Ecology of Macaque Societies*. Cambridge University Press, Cambridge. pp. 413–443.

Melnick, D. J., Jolly, C. J., and Kidd, K. K. (1984a). The genetics of a wild population of rhesus monkeys (*Macaca mulatta*): I. Genetic variability within and between social groups. *Am. J. Phys. Anthropol.* 63:341–360.

Melnick, D. J., Pearl, M. C., and Richard, A. F. (1984b). Male migration and inbreeding avoidance in wild rhesus monkeys. *Am. J. Primatol.* 7:229–243.

Mitani, J. C., Merriwether, A., and Zhang, C. (2000). Male affiliation, cooperation and kinship in wild chimpanzees. *Anim. Behav.* 59:885–893.

Modrek, B., and Lee, C. (2002). A genomic view of alternative splicing. *Nat. Genet.* 30:13–19.

Morin, P. A., Chambers, K. E., Boesch, C., and Vigilant, L. (2001). Quantitative polymerase chain reaction analysis of DNA from noninvasive samples for accurate microsatellite genotyping of wild chimpanzees (*Pan troglodytes*). *Mol. Ecol.* 10:1835–1844.

Morin, P. A., Moore, J. J., Chakraborty, R., Jin, L., Goodall, J., and Woodruff, D. S. (1994). Kin selection, social structure, gene flow, and the evolution of chimpanzees. *Science* 265:1193–1201.

Morin, P. A., and Woodruff, D. S. (1992). Paternity exclusion using multiple hypervariable microsatellite loci amplified from nuclear DNA of hair cells. In: Martin, R. D., Dixson, A. F., and Wickings, E. J. (eds.), *Paternity in Primates: Genetic Tests and Theories*. Karger, Basel. pp. 63–81.

Mouse Genome Sequencing Consortium (2002). Initial sequencing and comparative analysis of the mouse genome. *Nature* 420:520–561.

Müller, S., Hollatz, M., and Wienberg, J. (2003). Chromosomal phylogeny and evolution of gibbons (Hylobatidae). *Hum. Genet.* 113:493–501.

Müller, S., O'Brien, P. C. M., Ferguson-Smith, M. A., and Wienberg, J. (1997). Reciprocal chromosome painting between human and prosimians (*Eulemur macaco macaco* and *E. fulvus mayottensis*). *Cytogenet. Cell Genet.* 78:260–271.

Mullis, K., and Faloona, F. (1987). Specific synthesis of DNA in vitro via a polymerase catalyzed chain reaction. *Methods Enzymol.* 155:335–350.

Navarro, A., and Barton, N. H. (2003). Chromosomal speciation and molecular divergence—accelerated evolution in rearranged chromosomes. *Science* 300:321–324.

Neusser, M., Stanyon, R., Bigoni, F., Wienberg, J., and Müller, S. (2001). Molecular cytotaxonomy of New World monkeys (Platyrrhini)—comparative analysis of five species by multicolor chromosome painting gives evidence for a classification of *Callimico goeldii* within the family of Callitrichidae. *Cytogenet. Cell Genet.* 94:206–215.

Nievergelt, C. M., Digby, L. J., Ramakrishnan, U., and Woodruff, D. S. (2000). Genetic analysis of group composition and breeding system in a wild common marmoset (*Callithrix jacchus*) population. *Int. J. Primatol.* 21:1–20.

Nsubuga, A. M., Robbins, M. M., Roeder, A. D., Morin, P. A., Boesch, C., and Vigilant, L. (2004). Factors affecting the amount of genomic DNA extracted from ape faeces and the identification of an improved sample storage method. *Mol. Ecol.* 13:2089–2094.

Nuttal, G. H. F. (1904). *Blood Immunity and Blood Relationship*. Cambridge University Press, Cambridge.

Paetkau, D. (1999). Microsatellites obtained using strand extension: an enrichment protocol. *Biotechniques* 26:690–697.

Pastinen, T., Raitio, M., Lindroos, K., Tainola, P., Peltonen, L., and Syvänen, A.-C. (2000). A system for specific, high-throughput genotyping by allele-specific primer extension on microarrays. *Genome Res.* 10:1031–1042.

Paul, A., Kuester, J., Timme, A., and Arnemann, J. (1993). The association between rank, mating effort, and reproductive success in male Barbary macaques (*Macaca sylvanus*). *Primates* 34:491–502.

Peeters, M., Courgnaud, V., Abela, B., Auzel, P., Pourrut, X., Bibollet-Ruche, F., Loul, S., Liegeois, F., Butel, C., Koulagna, D., Mpoudi-Ngole, E., Shaw, G. M., Hahn, B. H., and Delaporte, E. (2002). Risk to human health from a plethora of simian immunodeficiency viruses in primate bushmeat. *Emerg. Infect. Dis.* 8:451–457.

Pereira, M. E., and Weiss, M. L. (1991). Female mate choice, male migration, and the threat of infanticide in ringtailed lemurs. *Behav. Ecol. Sociobiol.* 28:141–152.

Piggott, M. P., Bellemain, E., Taberlet, P., and Taylor, A. C. (2004). A multiplex pre-amplification method that significantly improves microsatellite amplification and error rates for faecal DNA in limiting conditions. *Conserv. Genet.* 5:417–420.

Piry, S., Alapetite, A., Cornuet, J.-M., Paetkau, D., Baudouin, L., and Estoup, A. (2004). GeneClass2: a software for genetic assignment and first-generation migrant detection. *J. Hered.* 95:536–539.

Pope, T. R. (1990). The reproductive consequences of male cooperation in the red howler monkey: paternity exclusion in multi-male and single-male troops using genetic markers. *Behav. Ecol. Sociobiol.* 27:439–446.

Pope, T. R. (1992). The influence of dispersal patterns and mating systems on genetic differentiation within and between populations of the red howler monkey (*Alouatta seniculis*). *Evolution* 46:1112–1128.

Preuss, T. M., Cáceres, M., Oldham, M. C., and Geschwind, D. H. (2004). Human brain evolution: insights from microarrays. *Nat. Rev. Genet.* 5:850–860.

Pritchard, J. K., Stephens, M., and Donnelly, P. (2000). Inference of population structure using multilocus genotype data. *Genetics* 155:945–959.

Queller, D. C., and Goodnight, K. F. (1989). Estimating relatedness using genetic markers. *Evolution* 43:258–275.

Radespiel, U., dal Secco, V., Drögemüller, C., Braune, P., Labes, E., and Zimmermann, E. (2002). Sexual selection, multiple mating and paternity in grey mouse lemurs, *Microcebus murinus*. *Anim. Behav.* 63:259–268.

Rajalingam, R., Hong, M., Adams, E. J., Shum, B. P., Guethlein, L. A., and Parham, P. (2001). Short *KIR* haplotypes in pygmy chimpanzee (bonobo) resemble the conserved framework of diverse human *KIR* haplotypes. *J. Exp. Med.* 193:135–146.

Raymond, M., and Roussett, M. (1995). GENEPOP (version 1.2), a population genetics software for exact tests and ecumenicism. *J. Hered.* 86:248–249.

Ried, T., Arnold, N., Ward, D. C., and Wienberg, J. (1993). Comparative high-resolution mapping of human and primate chromosomes by fluorescence in situ hybridization. *Genomics* 18:381–386.

Roeder, A. D., Archer, F. I., Poinar, H. N., and Morin, P. A. (2004). A novel method for collection and preservation of faeces for genetic studies. *Mol. Ecol. Notes* 4:761–764.

Romero-Herrera, A. E., Lehmann, H., Joysey, K. A., and Friday, A. E. (1976). Evolution of myoglobin amino acid sequences in primates and other vertebrates. In: Goodman, M., Tashian, R. E., and Tashian, J. H. (eds.), *Molecular Anthropology: Genes and Proteins in the Evolutionary Ascent of the Primates*. Plenum Press, New York. pp. 289–300.

Ronquist, F., and Huelsenbeck, J. P. (2003). MrBayes 3: Bayesian phylogenetic inference under mixed models. *Bioinformatics* 19:1572–1574.

Rosenblum, L. L., Supriatna, J., Hasan, M. N., and Melnick, D. J. (1997a). High mitochondrial DNA diversity with little structure within and among leaf monkey populations (*Trachypithecus cristatus* and *Trachypithecus auratus*). *Int. J. Primatol.* 18:1005–1028.

Rosenblum, L. L., Supriatna, J., and Melnick, D. J. (1997b). Phylogeographic analysis of pigtail macaque populations (*Macaca nemestrina*) inferred from mitochondrial DNA. *Am. J. Phys. Anthropol.* 104:35–45.

Ruvolo, M. (1997). Genetic diversity in hominoid primates. *Annu. Rev. Anthropol.* 26:515–540.

Saiki, R. K., Gelfand, D. H., Stoffel, S., Scharf, S. J., Higuchi, R., Horn, G. T., Mullis, K., and Erlich, H. A. (1988). Primer-directed enzymatic amplification of DNA with a thermostable DNA polymerase. *Science* 239:487–491.

Saiki, R. K., Scharf, S. J., Faloona, F., Mullis, K., Horn, G. T., Erlich, H. A., and Arnheim, N. (1985). Enzymatic amplification of β-globin genomic sequences and restriction site analysis for diagnosis of sickle-cell anemia. *Science* 230:1350–1354.

Sambrook, J., and Russell, D. (2001). *Molecular Cloning: A Laboratory Manual*, 3rd ed. Cold Spring Harbor Laboratory Press, Cold Spring Harbor, NY.

Samonte, R. V., and Eichler, E. E. (2002). Segmental duplications and the evolution of the primate genome. *Nat. Rev. Genet.* 3:65–72.

Sanger, F., Nicklen, S., and Coulson, A. R. (1977). DNA sequencing with chain-terminating inhibitors. *Proc. Natl. Acad. Sci. USA* 74:5463–5467.

Santiago, M. L., Lukasik, M., Kamenya, S., Li, Y., Bibollet-Ruche, F., Bailes, E., Muller, M. N., Emery, M., Goldenberg, D. A., Lwanga, J. S., Ayouba, A., Nerrienet, E., McClure, H. M., Heeney, J. L., Watts, D. P., Pusey, A. E., Collins, D. A., Wrangham, R. W., Goodall, J., Brookfield, J. F., Sharp, P. M., Shaw, G. M., and Hahn, B. H. (2003). Foci of endemic simian immunodeficiency virus infection in wild-living eastern chimpanzees (*Pan troglodytes schweinfurthii*). *J. Virol.* 77:7545–7562.

Sarich, V. M., and Cronin, J. E. (1976). Molecular Systematics of the Primates. In: Goodman, M., Tashian, R. E., and Tashian, J. H. (eds.), *Molecular Anthropology: Genes and Proteins in the Evolutionary Ascent of Primates*. Plenum Press, New York. pp. 141–170.

Sarich, V. M., and Wilson, A. C. (1967). Immunological time scale for hominid evolution. *Science* 179:1144–1147.

Satta, Y., Klein, J., and Takahata, N. (2000). DNA archives and our nearest relative: the trichotomy problem revisited. *Mol. Phylogenet. Evol.* 14:259–275.

Schmidt, H. A., Strimmer, K., Vingron, M., and von Haeseler, A. (2002). TREE-PUZZLE: maximum likelihood phylogenetic analysis using quartets and parallel computing. *Bioinformatics* 18:502–504.

Schneider, H., Canavez, F. C., Sampaio, I., Moreira, M. Â. M., Tagliaro, C. H., and Seuánez, H. N. (2001). Can molecular data place each neotropical monkey in its own branch? *Chromosoma* 109:515–523.

Schneider, S., Roessli, D., and Excoffier, L. (2000). *Arlequin*: a software for population genetics data analysis (version 2.000). Genetics and Biometry Lab, Department of Anthropology, University of Geneva, Geneva.

Seutin, G., White, B. N., and Boag, P. T. (1991). Preservation of avian blood and tissue samples for DNA analysis. *Can. J. Zool.* 69:82–90.

Shankar, S., Mohapatra, B., Verma, S., Selvi, R., Jagadish, N., and Suri, A. (2004). Isolation and characterization of a haploid germ cell specific sperm associated antigen 9 (SPAG9) from the baboon. *Mol. Reprod. Dev.* 69:186–193.

Sibley, C. G., and Ahlquist, J. E. (1984). The phylogeny of the hominoid primates as indicated by DNA–DNA hybridization. *J. Mol. Evol.* 20:2–15.

Sibley, C. G., and Ahlquist, J. E. (1987). DNA hybridization evidence of hominoid phylogeny: results from an expanded data set. *J. Mol. Evol.* 26:99–121.

Sonnenburg, J. L., Altheide, T. K., and Varki, A. (2004). A uniquely human consequence of domain-specific functional adaptation in a sialic acid–binding receptor. *Glycobiology* 14:339–346.

Stanyon, R., Consigliere, S., Bigoni, F., Ferguson-Smith, M., O'Brien, P. C. M., and Wienberg, J. (2001). Reciprocal chromosome painting between a New World primate, the woolly monkey, and humans. *Chromosome Res.* 9:97–106.

Stanyon, R., Consigiliere, S., Müller, S., Moreschalchi, A., Neusser, M., and Wienberg, J. (2000). Fluorescence in situ hybridization (FISH) maps chromosomal homologies between the dusky titi and squirrel monkey. *Am. J. Primatol.* 50:95–107.

Stanyon, R., Koehler, U., and Consigliere, S. (2002). Chromosome painting reveals that galagos have highly derived karyotypes. *Am. J. Phys. Anthropol.* 117:319–326.

Steiper, M. E., and Ruvolo, M. (2003). Genetic sex identification of orangutans. *Anthropol. Anz.* 61:1–5.

Stewart, C.-B., and Disotell, T. R. (1998). Primate evolution—in and out of Africa. *Curr. Biol.* 8:R582–R588.

Stewart, C.-B., Schilling, J. W., and Wilson, A. C. (1987). Adaptive evolution in the stomach lysozymes of foregut fermenters. *Nature* 330:401–404.

Storz, J. F., Beaumont, M. A., and Alberts, S. C. (2002a). Genetic evidence for long-term population decline in a savannah-dwelling primate: inferences from a hierarchical Bayesian model. *Mol. Biol. Evol.* 19:1981–1990.

Storz, J. F., Ramakrishnan, U., and Alberts, S. C. (2002b). Genetic effective size of a wild primate population: influences of current and historical demography. *Evolution* 56:817–829.

Strassmann, J. E., Solís, C. R., Peters, J. M., and Queller, D. C. (1996). Strategies for finding and using highly polymorphic DNA microsatellite loci for studies of genetic relatedness and pedigrees. In: Ferraris, J. D., and Palumbi, S. R. (eds.), *Molecular Zoology: Advances, Strategies, and Protocols*. Wiley-Liss, New York. pp. 163–180.

Strier, K. B., Ziegler, T. E., and Wittwer, D. J. (1999). Seasonal and social correlates of fecal testosterone and cortisol levels in wild male muriquis (*Brachyteles arachnoides*). *Horm. Behav.* 35:125–134.

Sugiyama, Y., Kawamoto, S., Takenaka, O., Kumazaki, K., and Miwa, N. (1993). Paternity discrimination and inter-group relationships of chimpanzees at Bossou. *Primates* 34:545–552.

Sullivan, K., Walton, A., Kimpton, C., Tully, G., and Gill, P. (1993). A rapid and quantitative DNA sex test: fluorescence-based PCR analysis of X–Y homologous gene amelogenin. *Biotechniques* 15:637–641.

Swofford, D. L. (2002). *PAUP*: Phylogenetic Analysis Using Parsimony (and Other Methods)*, 4.0 Beta. Sinauer Associates, Sunderland, MA.

Syvänen, A.-C. (2001). Accessing genetic variation: genotyping single nucleotide polymorphisms. *Nat. Rev. Genet.* 2:930–942.

Taberlet, P., Griffin, S., Goossens, B., Questiau, S., Manceau, V., Escaravage, N., Waits, L., and Bouvet, J. (1996). Reliable genotype of samples with very low DNA quantities using PCR. *Nucleic Acids Res.* 24:3189–3194.

Taberlet, P., Walts, L. P., and Luikart, G. (1999). Noninvasive genetic sampling: look before you leap. *Trends Ecol. Evol.* 14:321–325.

Takasaki, H., and Takenaka, O. (1991). Paternity testing in chimpanzees with DNA amplification from hairs and buccal cells in wadges: a preliminary note. In: Ehara, A., Kimura, T., Takanaka, O., and Iwamoto, M. (eds.), *Primatology Today*. Elsevier Scientific Publishers, Amsterdam. pp. 613–616.

Tashian, R. E., Goodman, M., Ferrell, R. E., and Tanis, R. J. (1976). Evolution of carbonic anhydrase in primates and other mammals. In: Goodman, M., Tashian, R. E., and Tashian, J. H. (eds.), *Molecular Anthropology: Genes and Proteins in the Evolutionary Ascent of the Primates*. Plenum Press, New York. pp. 301–319.

Tjio, H. J., and Levan, A. (1956). The chromosome numbers of man. *Hereditas* 42:1–6.

Tosi, A. J., Morales, J. C., and Melnick, D. J. (2000). Comparison of Y chromosome and mtDNA phylogenies leads to unique inferences of macaque evolutionary history. *Mol. Phylogenet. Evol.* 17:133–144.

Tosi, A. J., Morales, J. C., and Melnick, D. J. (2003). Paternal, maternal, and biparental molecular markers provide unique

windows onto the evolutionary history of macaque monkeys. *Evolution* 57:1419–1435.

Toth, G., Gaspari, Z., and Jurka, J. (2000). Microsatellites in different eukaryotic genomes: survey and analysis. *Genome Res.* 10:967–981.

Trask, B. J. (2002). Human cytogenetics: 46 chromosomes, 46 years and counting. *Nat. Rev. Genet.* 3:769–778.

Uddin, M., Wildman, D. E., Liu, G., Xu, W., Johnson, R. M., Hof, P. R., Kapatos, G., Grossman, L. I., and Goodman, M. (2004). Sister grouping of chimpanzees and humans as revealed by genome-wide phylogenetic analysis of brain gene expression profiles. *Proc. Natl. Acad. Sci. USA* 101:2957–2962.

Underhill, P. A., Passarino, G., Lin, A. A., Shen, P., Lahr, M. M., Foley, R. A., Oefner, P. J., and Cavalli-Sforza, L. L. (2001). The phylogeography of Y chromosome binary haplotypes and the origins of modern human populations. *Ann. Hum. Genet.* 65:43–62.

Underhill, P. A., Shen, P., Lin, A. A., Jin, L., Passarino, G., Yang, W. H., Kauffman, E., Bonné-Tamir, B., Bertranpetit, J., Francalacci, P., Ibrahim, M., Jenkins, T., Kidd, J. R., Mehdi, S. Q., Seielstad, M. T., Wells, R. S., Piazza, A., Davis, R. W., Feldman, M. W., Cavalli-Sforza, L. L., and Oefner, P. J. (2000). Y chromosome sequence variation and the history of human populations. *Nat. Genet.* 26:358–361.

Utami, S. S., Goossens, B., Bruford, M. W., Ruiter, J. R. D., and Hooff, J. A. R. A. M. V. (2002). Male bimaturism and reproductive success in Sumatran orang-utans. *Behav. Ecol.* 13:643–652.

Valderrama, X., Karesh, W. B., Wildman, D. E., and Melnick, D. J. (1999). Noninvasive methods for collecting fresh hair tissue. *Mol. Ecol.* 8:1749–1752.

Valière, N., and Taberlet, P. (2000). Urine collected in the field as a source of DNA for species and individual identification. *Mol. Ecol.* 9:2150–2152.

van Geel, M., Eichler, E. E., Beck, A. F., Shan, Z. H., Haaf, T., van der Maarel, S. M., Frants, R. R., and de Jong, P. J. (2002). A cascade of complex subtelomeric duplications during the evolution of the hominoid and Old World monkey genomes. *Am. J. Hum. Genet.* 70:269–278.

van Schaik, C. P., van Noordwijk, M. A., and van Bragt, T. (1991). A pilot study of the social correlates of levels of urinary cortisol, prolactin, and testosterone in wild long-tailed macaques (*Macaca fascicularis*). *Primates* 32:345–356.

Vigilant, L., Hofreiter, M., Siedel, H., and Boesch, C. (2001). Paternity and relatedness in wild chimpanzee communities. *Proc. Natl. Acad. Sci. USA* 98:12890–12895.

Vigilant, L., Stoneking, M., Harpending, H., Hawkes, K., and Wilson, A. C. (1991). African populations and the evolution of human mitochondrial DNA. *Science* 253:1503–1507.

von Dornum, M., and Ruvolo, M. (1999). Phylogenetic relationships of the New World monkeys (Primates, Platyrrhini) based on nuclear G6PD DNA sequences. *Mol. Phylogenet. Evol.* 11:459–476.

Weir, B. S. (1996). *Genetic Data Analysis II*. Sinauer Associates, Sunderland, MA.

Wheeler, W. C., and Gladstein, D. G. (1994). MALIGN: a multiple sequence alignment program. *J. Hered.* 85:417–418.

Widdig, A., Nürnberg, P., Krawczak, M., Streich, W. J., and Bercovitch, F. B. (2001). Paternal relatedness and age proximity regulate social relationships among adult female rhesus macaques. *Proc. Natl. Acad. Sci. USA* 98:13769–13773.

Wienberg, J., and Stanyon, R. (1995). Chromosome painting in mammals as an approach to comparative genomics. *Curr. Opin. Genet. Dev.* 5:792–797.

Wienberg, J., and Stanyon, R. (1997). Comparative painting of mammalian chromosomes. *Curr. Opin. Genet. Dev.* 7:784–791.

Wienberg, J., and Stanyon, R. (1998). Comparative chromosome painting of primate genomes. *ILAR J.* 39:77–91.

Wildman, D. E., Bergman, T. J., al-Aghbari, A., Sterner, K. N., Newman, T. K., Phillips-Conroy, J. E., Jolly, C. J., and Disotell, T. R. (2004). Mitochondrial evidence for the origin of hamadryas baboons. *Mol. Phylogenet. Evol.* 32:287–296.

Wilson, J. F., and Erlandsson, R. (1998). Sexing of human and other primate DNA. *Biol. Chem.* 379:1287–1288.

Wimmer, B., Tautz, D., and Kappeler, P. M. (2002). The genetic population structure of the gray mouse lemur (*Microcebus murinus*), a basal primate from Madagascar. *Behav. Ecol. Sociobiol.* 52:166–175.

Woodruff, D. S. (2004). Noninvasive genotyping and field studies of free-ranging nonhuman primates. In: Chapais, B., and Berman, C. M. (eds.), *Kinship and Behavior in Primates*. Oxford University Press, Oxford. pp. 46–68.

Wu, C. C., MacCoss, M. J., Howell, K. E., and Yates, J. R., 3rd (2003). A method for the comprehensive proteomic analysis of membrane proteins. *Nat. Biotechnol.* 21:532–538.

Xiao, W., and Oefner, P. J. (2001). Denaturing high-performance liquid chromatography: a review. *Hum. Mutat.* 17:439–474.

Yang, Z. (1997). PAML: a program package for phylogenetic analysis by maximum likelihood. *CABIOS* 13:555–556.

Ye, F., Li, M.-S., Taylor, J. D., Nguyen, Q., Colton, H. M., Casey, W. M., Wagner, M., Weiner, M. P., and Chen, J. (2001). Fluorescent microsphere-based readout technology for multiplexed human single nucleotide polymorphism analysis and bacterial identification. *Hum. Mutat.* 17:305–316.

Yoder, A. D. (1997). Back to the future: a synthesis of strepsirhine systematics. *Evol. Anthropol.* 6:11–22.

Yoder, A. D., Cartmill, M., Ruvolo, M., Smith, K., and Vilgalys, R. (1996). Ancient single origin of Malagasy primates. *Proc. Natl. Acad. Sci. USA* 93:5122–5126.

Yoder, A. D., and Yang, Z. (2004). Divergence dates for Malagasy lemurs estimated from multiple gene loci: geological and evolutionary context. *Mol. Ecol.* 13:757–773.

Yunis, J. J., and Prakash, O. (1982). The origin of man: a chromosomal pictorial legacy. *Science* 215:1525–1530.

Zane, L., Bargelloni, L., and Patarnello, T. (2002). Strategies for microsatellite isolation: a review. *Mol. Ecol.* 11:1–16.

Zhang, J., Wang, X., and Podlaha, O. (2004). Testing the chromosomal speciation hypothesis for humans and chimpanzees. *Genome Res.* 14:845–851.

Zhang, J., Zhang, Y. P., and Rosenberg, H. F. (2002). Adaptive evolution of a duplicated pancreatic ribonuclease gene in a leaf-eating monkey. *Nat. Genet.* 30:411–415.

PART FOUR

Reproduction

23

Life History

Steven R. Leigh and Gregory E. Blomquist

INTRODUCTION

> And then increasingly, it was screaming at me, "These are the most interesting individuals; this has the greatest evolutionary impact; this is where the ecological pressures are". (Jeanne Altmann, interview with D. Haraway on baboon mothers and infants [Haraway 1989:312])

The events, transitions, and phases that occur during the life course are fundamental to the diversity of life (Raff 1996). Primates are especially interesting in this regard, presenting a fascinating array of variation in terms of changes that occur during life. A special class of theory, termed *life history theory*, explains the evolution of changes during the life course by analyzing demography, genetics, behavior, and morphology in a developmental and, typically, quantitative context (Lande 1982; Roff 1992, 2002; Stearns 1992; Charnov 1993). Most broadly, life history theory includes "not only the age-specific fecundity and mortality rates, but the entire sequence of changes through which an organism passes in its development from conception to death" (Lande 1982:608).

THEORETICAL CONTEXT OF PRIMATE LIFE HISTORY

Life history studies have a long tradition in both primatology and biological anthropology, manifested by several distinct research "lineages." The first theoretical focus owes much to Jeanne Altmann's interest in baboon infants and their mothers, with projects centering on long-term field studies such as the Amboseli Baboon Project (J. Altmann 1980, S. Altmann 1998, Altmann and Alberts 2002, Silk et al. 2003), the Cayo Santiago rhesus macaque colony (Sade et al. 1985), and Jane Goodall's (1986) chimpanzee research at Gombe. These studies couple information on vital population parameters (e.g., birth and death rates) with fine-grained behavioral observations on the day-to-day lives of study subjects. A second research area explores human life histories, concentrating on how an apparently distinctive set of human demographic and ontogenetic characteristics might have evolved (Bogin 1999, Hawkes et al. 2002, Hill and Hurtado 1996, Kaplan et al. 2000, Leigh 2001). Studies of human life histories also utilize both demographic and behavioral data, but many studies also consider patterns of somatic growth in relation to demography, behavior, and culture. In this chapter,

we emphasize a third research tradition that highlights interspecific comparisons of ontogenetic and *allometric* (size-related) variation in life history traits. Studies here focus on key events in life histories, such as birth, age at maturation, birth rates, and longevity, to understand how evolutionary forces shape the life courses of primates (Harvey et al. 1987; see also Godfrey et al. 2002, 2004; Harvey and Clutton-Brock 1985; Martin 1983, 1996; Martin and MacLarnon 1990; Ross 1988, 1991, 1992, 2002, 2004; Ross and Jones 1999; Shea 1987, 1990). This perspective departs somewhat from theoretical priorities on genetic and demographic data in traditional life history theory, which emphasize measures of heritability and rates of gene substitution in populations (Stearns 1992). However, these approaches are compatible, and their interrelations have received some theoretical treatment (e.g., Charnov 1993). We stress theoretical developments and current ideas that emerge from interspecific analyses but briefly incorporate advances from other research areas and explore ways in which genetic and demographic insights articulate with interspecific analyses.

HISTORICAL CONSIDERATIONS

Research conducted in the late 1970s through the 1980s marked a transition in studies of primate life histories (Martin 1983, Harvey and Clutton-Brock 1985, Harvey et al. 1987). Prior to this period, such studies focused most on comparative analyses of human growth and development (Gavan and Swindler 1966, Schultz 1969, Watts and Gavan 1982). Adolph Schultz made major contributions to this tradition, encapsulated by his often-repeated line drawing that shows "progressively" extended life stages in primates (1969:Fig. 57). The figure shows unidimensional, evenly proportioned, and progressive increases in each life stage during primate evolution, conveying the erroneous impression that primate life histories follow an orderly, orthogenetic, and unilinear advance. Schultz's view was consistent with bioanthropology's focus on evolutionary trends prior to the 1960s. However, cladistic thinking showed flaws with this and other "trends" by revealing that several "trends" were, in fact, independently evolved similarities (*homoplasies*) (see Leigh 2001).

Unfortunately, advances in biogeography, particularly the concepts of r and K selection, seemed to reinforce Schultz's schematic (MacArthur and Wilson 1967, Pianka 1970). These ideas found rapid deployment in bioanthropology, particularly in studies of human evolution. For example, the late Stephen Jay Gould (1977) promoted the idea that K selection produced a slowing of developmental change during the course of human evolution. Similarly, Owen Lovejoy's (1981) influential model linked early hominin monogamy with life history, explicitly invoking r–K selection theory under Schultz's orthogenetic scheme. Regrettably, these theoretical turns came at the expense of more powerful genetic and demographic concepts, advanced mainly by George C. Williams, that emphasized the unequal force of selection throughout the life span (1957, 1966a; see also Lack 1954; Medawar 1946, 1952). Differences in the effects of selection could explain diversity in the attributes of life stages, especially how genes deleterious late in life, including those producing senescence, could become fixed in populations. Senescence and diminishing reproductive output might occur because genes with tiny advantages during phases of relatively high reproduction can be selectively favored during these phases and, thus, increase in frequency. However, such genes may have detrimental effects later in life, with little consequence from selection because of low reproductive output. This basic insight imbued life history theory with ideas of genetic trade-offs (antagonistic pleiotropy), juvenile phases, and risks of juvenility. The notion of trade-offs is fundamental to life history theory, particularly when formalized by the concepts of reproductive value (Fisher 1930) and residual reproductive value (Williams 1966b). A *trade-off* means that energy invested in one area impacts, and usually limits, expenses in other areas. For example, an investment in a current offspring, like delaying weaning, may limit investment in future offspring.

The ideas of Williams, Fisher, and Medawar seem to have gone unrecognized in early primate life history studies, probably owing both to priorities established by Schultz and to the general appeal of r and K selection. Primates are long-lived relative to other mammals (Austad and Fischer 1992, Flower 1931, Prothero and Jürgens 1987), a finding superficially consistent with r–K ideas. In addition, the suggestion that the brain served as a pace-setter of life histories was consistent with r and K selection. In this model, larger brains were associated with greater longevity by increased "precision of physiological regulation" (Sacher 1959:129, Sacher and Staffeldt 1974; see also Allman and Hasenstaub 1999, Deaner et al. 2002). This suggestion had obvious implications for an order classically defined in part by a trend toward large relative brain size (Le Gros Clark 1959, Martin 1983, Shea 1987), but it relied upon an ageing paradigm uninformed by Williams' ideas.

Influential analyses in the 1980s marked a change in primate life history perspectives, although these analyses worked within a general framework that derived mainly from the traditions of Schultz and biogeography (Harvey et al. 1987). These studies concentrated on questions relating primate life history to brain and body size, but they also presented data enabling researchers to test novel hypotheses and raised awareness of statistical complications posed by phylogeny. Harvey et al. assembled large databases representing estimates of "life history variables" for many primate species. Variables included markers of time points or intervals (gestation length, weaning age, age at maturation, age at first breeding, interbirth interval, and maximum recorded life span) and morphological variables (adult brain size, neonatal brain size, adult body size, and neonatal body size). They interpreted significant statistical variation among species as an outcome of evolutionary changes in body size, given high correlations between mass and other variables. Analyses of these data suggested considerable complexity in the relations of brain size to life history. Specifically, brain size and maturation age seemed to be linked, but further analyses indicated that this correlation reflected prenatal factors to a greater degree than postnatal factors. Essentially, later-maturing species have larger-brained infants, but the majority of brain growth seem to occur prenatally in larger-brained species. The mechanistic bases and implications of these correlations for life history have been difficult to comprehend fully and unambiguously (see below).

POPULATION DYNAMICS AND PRIMATE LIFE HISTORIES

Despite Harvey et al.'s (1987) major advances, several areas of theoretical concern went unaddressed. Perhaps most importantly, while the article drew inspiration from population biology, it neglected questions regarding demography, quantitative genetics, and developmental biology. Fortunately, an influential study, published by Caroline Ross (1988), made a major step in this direction by merging comparative approaches with ideas about population dynamics codified by Cole (1954). Specifically, Ross (following Hennemann 1983, 1984) solved for r, the intrinsic rate of natural population increase (Table 23.1), by measuring key variables (age at maturation, age at death, and birth rate) from published studies of many primate species. She suggested that values for each species provided a measure of maximum possible reproductive output (r_{max}). Ross discovered that species with low body mass tended to have high r_{max} values, while large-bodied species had low values (Pearson product-moment correlation $r = -0.869$). The details of this correlation implied to her that the general idea behind r and K selection may hold for primates. Specifically, given the pattern of high r_{max} residuals from species designated as occupying "unpredictable" habitats, she suggested that such species tended to be r-selected. Covariation between r_{max} values and environments thus seemed to fit basic predictions of r and K selection.

Despite these important findings, some results were tenuous. Our objective in pointing this out is not to diminish Ross' important research. Rather, we wish to recognize limitations of r and K selection for primates and follow Ross in

Table 23.1   Important Life History Equations

Equation 1: Cole's equation, used by Ross (1988) for primates:

$$1 = e^{-r} + be^{-r\alpha} + be^{-r(n+\alpha)} \qquad (1)$$

In this equation, e is the base of the natural logarithm, α is age at first reproduction, b equals the birth rate of female offspring, n is estimated by calculating $\omega - (\alpha + 1)$ (where ω = age at last reproduction), and the value of r is estimated through iteration (Cole 1954:eq. 21).

Equation 2: Euler-Lotka or characteristic equation:

$$\int_0^\infty l(x)m(x)e^{-rx}dx = 1 \qquad (2)$$

This equation summarizes population rates of age-specific survivorship, $l(x)$ and production of female offspring, $m(x)$, for age x and measures the intrinsic rate of population increase (r).

Equation 3: Algebraic expression of the Euler-Lotka equation:

$$\sum_{x=0}^{\omega} l(x)m(x)e^{-rx} = 1 \qquad (3)$$

this regard (Ross and Jones 1999; see also Stearns 1992). First, the measure of habitat predictability employed was imprecise, being derived from a tertiary literature source (see Ross 1992 for improved habitat estimates). Second, the association between higher relative r_{max} values and habitat predictability was modest at best, being represented by positive residual values that overlapped entirely with data points from species found in "predictable" habitats. In fact, a species occupying a predictable habitat seemed to present the highest positive residual r_{max} value. Third, as with most studies at this time, phylogenetic adjustment methods could only roughly counter effects of phylogeny. More recent analyses that include phylogenetic adjustments show "no significant links" between habitat and measures of reproductive rate (Ross and Jones 1999:94).

Ross' original analysis, despite these difficulties, was extremely important because it incorporated Cole's research (1954) and that of demographically oriented theoreticians (Blomquist in press) into considerations of primate life histories. Unfortunately, it seems simultaneously to have reified r and K selection among primatologists. The role of r–K selection is sometimes abstracted by the catchy phrase that primates experience "slow" life histories, with some occupying slower "lanes" than others, with variation among species described along a "fast–slow continuum" (Ross 1992:383, Kelley 2004, Promislow and Harvey 1990). In any case, the inclusion of ideas about population dynamics and life histories significantly advanced the field, sustaining many further theoretical developments and forcing a reconsideration of the brain's role in primate life history.

EMERGENCE OF THEORETICAL DIVERSITY

Ross' studies were complemented by two influential life history theory texts published in 1992 (Roff 1992, Stearns

1992). Charnov's theoretical monograph (1993) and two major edited volumes devoted to primate life histories and juvenility (DeRosseau 1990, Pereira and Fairbanks 1993) supplemented these works, as did quantitative analyses at the population level (Sade 1990, Stucki et al. 1991). Moreover, phylogenetic adjustment techniques advanced tremendously (Felsenstein 1985, Garland and Adolph 1994).

Volumes by Roff (1992, 2002) and Stearns (1992) ground contemporary life history theory firmly in the ideas and methods of demography and quantitative genetics. This foundation derives directly from R. A. Fisher's (1930) interest in the demographic dimensions of the maintenance of genetic variation in populations. As Stearns notes, the main idea is that demographic methods facilitate a concern with "marginal effects of gene substitutions, not with numbers of organisms in a population" (1992:21). Stearns' six-point "rejection of r and K selection" is an exceptionally valuable contribution (1992:206–207). In showing the limitations of this idea beyond biogeographic studies, Stearns advocates a life history paradigm shift from the idea (latent in r–K selection theory) of direct habitat effects on life history (habitat → life history) to that of habitat → mortality regime → life history (1992:208). In a related advance, DeRousseau argued that a life history perspective differs considerably from traditional approaches to evolution, relaxing many of the simplifying assumptions of the evolutionary synthesis by introducing sex and, in particular, age structure to populations (1990; see also Shea 1990).

These advances set the stage for an especially active period of theory construction in life history studies that improve our understanding of primate life history variation. We review key theories and briefly consider the problem of trade-offs and the calibration of trade-offs in primate life histories. While not a theory per se, the problem of trade-offs requires more extensive theoretical and empirical investigation. Trade-offs may be especially important for long-lived organisms such as primates and have considerable relevance for understanding the evolution of human life histories (Hawkes et al. 2002). We use this review to point to new ways of exploring primate life histories.

Charnov and Life History Invariants

Models formulated in the early 1990s are still debated, as are long-standing ideas about learning and the brain as a direct pace-setter of life history (Allman and Hasenstaub 1999, Leigh 2004, Ross and Jones 1999, Sacher 1959, Sacher and Staffeldt 1974). These models have generally undergone initial rounds of empirical testing with primate species (Ross and Jones 1999), but critical evaluation of alternative life history models shows that much research remains to be done. A major theoretical development was the integration of optimality theory and comparative data in Charnov's life history invariants model (Charnov 1991, 1993; Charnov and Berrigan 1993; Berrigan et al. 1993). Charnov argued that primate life histories are impacted by a single trade-off between delaying reproduction to increase body size (an investment offset by higher fecundity) and the risk of mortality during the waiting period prior to reproduction. Primates could grow for a long time, to enjoy larger size and higher fecundity; but the risk of dying limits the degree to which they can delay reproduction. Mathematically, the model proposes that $dM/dt = AM^{0.75}$, where M is the mass of either a growing (female) individual or an adult, dM/dt is the growth rate or annual litter mass, and A describes the height of the production function. The *production function* describes the fraction of total energy put into either growth or reproduction. Growth, by definition, ceases at maturity, diverting energy from adding body mass (growth) into producing offspring. Externally imposed adult mortality rates set the optimal age of maturation, circumscribing an optimal body size. Optimal maturation age is also conditioned by the reproductive benefits of larger size, which translates to higher levels of energy available for production. Finally, juvenile mortality is assumed to be density-dependent and keeps populations stable ($R_o = 1$) (see Mylius and Diekmann 1995 for problems of this assumption).

Empirical analyses of Charnov's model indicate exceptionally low values of A for primates ($A = 0.42$), with $A = 1$ for other mammalian taxa (Charnov 1993). So, for primates, the amount of energy allocated to either growth or reproduction is small. This forces primate females to spend much time growing until they reach sizes large enough to produce offspring efficiently. Consequently, they grow for longer time frames and produce fewer offspring than other mammals. Comparatively low mortality should mean that primates have the opportunity to live longer, achieving sufficient fitness to compensate for lost reproduction during long preadult periods.

A powerful element of Charnov's (1993) model lies in its potential to fuse growth or ontogenetic phenomena to population dynamics. Charnov demonstrates the compatibility between these fields by deriving the value of A through several different means. For example, he shows that the value of A can be calculated directly from data that describe ontogeny, including size at weaning (δ), body mass at maturation, and age at maturation (Charnov 1993:eq. 5.3). However, A also emerges from population dynamics, such as his value of A_3, obtained by estimating the height of a regression line of Ross' (1992) r_{max} against body size (Charnov 1993:Fig. 6.4, Charnov and Berrigan 1993). Articulations between ontogenetic and population dynamic perspectives are significant because they tie ontogenetic patterns, particularly aspects of growth and developmental energetics, directly to population parameters. Consequently, the impact of life history variation on individual fitness can be assessed with respect to population dynamics. Finally, Charnov and Berrigan bluntly question life history models that involve the brain (1993; cf. Deaner et al. 2002, Ross 2002), contrasting with older models (Sacher 1959, Sacher and Staffeldt 1974; see also Allman and Hasenstaub 1999, Kelley 2004).

Empirical tests of Charnov's model among mammals (Purvis and Harvey 1995) and within primates (Ross and Jones 1999) support some of its predictions but reveal a number of deficiencies, at least given available comparative data sets. One particular prediction of Charnov's model lacking empirical support is the invariance of mass at independence (weaning, δ) with adult mass (but see Hawkes et al. 2002). In addition, many of the simplifying assumptions of the original Charnov model have been criticized and reevaluated. The production constant A and constant exponent of about 0.75 permit evolutionary change in body size only through differences in duration of growth and do not allow for change in growth rates. However, primate growth rates vary substantially: some species reach different adult sizes in the same amount of time (Fig. 23.1A,B), but others reach the same size in different amounts of time (Fig. 23.1C) (Pereira and Leigh 2002). Charnov (2001) noted how changes in the exponent affect the scaling of body size and annual fecundity. Jones and MacLarnon (2001), applying Charnov's logic to analyses of bat life histories, found a higher exponent (~1) which still seemed to follow a power function relation with fecundity. Finally, a loose interpretation of Charnov's (1993) model could be construed as a restatement of r–K selection, possibly promoted by the subtitle of Charnov and Berrigan's article "Life in the Slow Lane."

Juvenile Risks

The juvenile risk aversion model, proposed by Janson and van Schaik (1993), answers some questions unaddressed by

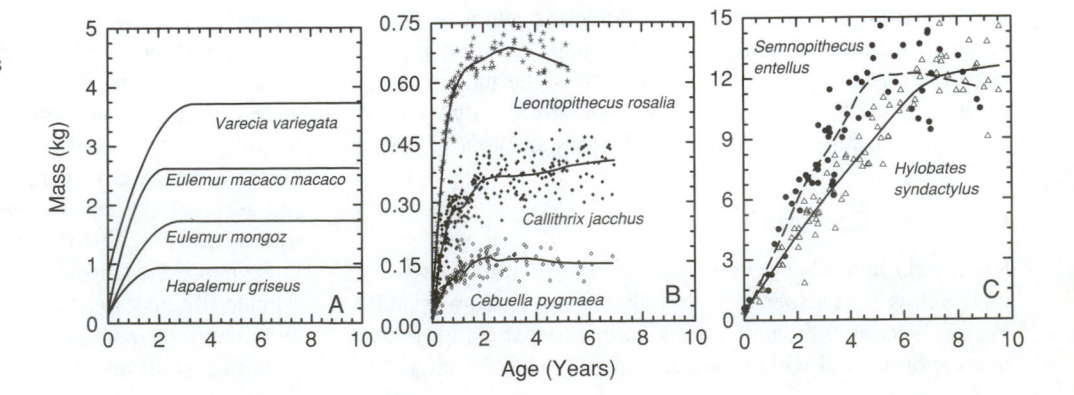

Figure 23.1 Comparative body mass growth curves for strepsirhines (A), platyrrhines (B), and catarrhines (C). Primates can reach the same size through differences in rates of growth, either with no difference in total growth time (A, B) or with a difference in growth duration (C). Data on females are shown for *Semnopithecus entellus*, but sexes are combined in these plots, given the absence of dimorphism in these taxa. In A, lines represent piecewise regressions (Leigh and Terranova 1998), but other lines represent lowest regressions (Leigh 1992).

Charnov's model, particularly growth rate variation among species. Their model also accommodates Williams' (1966a) notions about life stage duration and risk. Janson and van Schaik attribute the long period of primate juvenility to a low growth rate that serves as metabolic risk adaptation during ontogeny. Juvenile primates face a simultaneous trade-off between predation risks and the metabolic costs of feeding competition from conspecifics: they can forage at the center of groups, encountering competition from adult group members, or they can forage at the periphery, increasing their susceptibility to predators. Predators (and infanticide) should select against a low growth rate, favoring juveniles that grow out of harm's way (a constant growth rate is assumed for each species). On the other hand, growing rapidly raises metabolic risks because growth is metabolically costly as a result of protein turnover (Tanner 1978). Thus, feeding competition should favor a slow growth rate, especially if the metabolic expenses of growth mean that juveniles face disproportionately high costs of intra-group feeding competition. Juveniles face extra burdens if they forage less efficiently than adults. In this model, trade-offs involving predation pressure and possibly infanticide, contrasted with metabolic risks of feeding competition, select on growth rate. For primates, low growth rates seem to predominate, resulting in deferred maturation by forcing extended juvenile phases.

This model provides a compelling way to account for growth rate variation among primate species and follows from Williams' recognition that mortality risk conditions the pace of development, affecting either rates of morphological change, duration of high-risk developmental phases, or both. However, the model does not explain growth spurts or growth rate variability within a species (Leigh 1996). Analyses of restricted data sets show mixed support for the model, and field-based evaluations have been slow in coming. For example, growth patterns in leaf-eating anthropoids are consistent with the model (Leigh 1994), and variation among ape species aligns with its predictions (Leigh and Shea 1996). In contrast, neither small-bodied New World monkey growth (Garber and Leigh 1997) nor Malagasy lemur growth (Godfrey et al. 2004) seems to fit the model. A recent field study of *Saimiri sciureus* (squirrel monkeys) shows mixed support for components of the model. Juvenile squirrel monkeys experience negligible intraspecific feeding competition, and few differences in foraging proficiency occur across age classes. In squirrel monkeys, predation pressure seems to influence group dynamics, so elements of the model seem to fit (Stone 2004). Interspecific comparative analyses also show modest support for this model (Ross and Jones 1999).

Body Mass Optimization

Kozłowski and Wiener (1997) modified Charnov's model, interpreting it as a special case of their own. Kozłowski and Weiner's computer simulations indicate that adult body mass is optimized within a species in response to ecological parameters that affect the growth function. These ecological parameters can generate interspecific correlations between life history and morphological and physiological variables of body mass. These interspecific allometries are, however, epiphenomena of selection that act only within species to optimize body size. The model also predicts some recognized empirical patterns of growth rate scaling and species body mass distributions (Kozłowski and Galwelczyk 2002, Kindleman et al. 1999). Most importantly, their research generally supports the idea, following Charnov, that age and size at maturity are strongly influenced by optimal resource allocations to either growth or reproduction. We expect difficulties in formulating critical tests between the Kozłowski-Wiener and Charnov models and note that applications to primates have yet to be undertaken.

Charnov and Sigmoidal Growth

Charnov's (2001) recent model of mammalian life history evolution incorporates a sigmoidal growth law developed by West et al. (2001; but see Ricklefs 2003). In sigmoidal growth, rates of change follow a curvilinear pattern, much like an italicized letter *S*. Along with this somewhat more realistic assumption about growth, the model does not reflect reproductive output as merely diverted self-growth but as some proportion of it. Charnov proposes several new invariant relationships based on adult size and mortality, cellular maintenance costs, offspring production, and growth rate and age at adulthood. This model has not been subjected to empirical tests, and many of its predictions may be untestable until both accurate life tables and growth data are available on a large number of species.

Life History Trade-Offs

The idea of a trade-off is fundamental to life history thinking. This concept requires treatment in the context of the present theoretical review because trade-offs are evident in these models and complicate attempts to test them. The basic problem is that fitness components (variables directly related to fitness like maturation age, interbirth interval, and litter size) cannot be individually maximized, which may lead to negative correlations among them. In other words, organisms face "decisions" of how, when, and in what proportions energy should be allocated to various fitness components and the physiological processes that impact these components. This produces classic patterns like the trade-off of current versus future reproduction or current reproduction versus survival (Fisher 1930, Williams 1966a). Furthermore, trade-offs may not act directly on reproduction and survival but can be mediated through other variables, such as body size (Roff 2002).

Perhaps the most significant problem in explaining primate life histories, especially human life histories, is that researchers have had difficulty in actually finding predicted trade-offs (Hill and Hurtado 1996). For example, humans, and possibly other primates, do not show expected trade-offs between current and future reproduction, so it appears that the cost of current reproduction does not seem to reduce resources for future reproduction. New models are beginning to provide an answer to this paradox, by suggesting that variation in resource allocation and acquisition must be analyzed. Specifically, two conditions may obscure trade-offs. First, resource acquisition may be high enough to obviate trade-offs. Second, and more subtly, trade-offs may not be evident when variation in resource acquisition exceeds variation in allocation (van Noordwijk and de Jong 1986). *Variation in allocation* refers to differences in how organisms devote energy to either survival (e.g., body maintenance, size, condition) or reproduction. If there are huge differences in populations in terms of energy acquired but few differences in how it is allocated, then trade-offs may not be apparent at the population level.

Quantitative genetic models help resolve this, showing that trade-offs among fitness components may be obscured under two conditions. First, more genes may contribute to variation in acquisition than to resource allocation. Second, there could be greater variation at higher levels in an allocation hierarchy (Houle 1991, de Jong and van Noordwijk 1992, de Jong 1993, de Laguerie et al. 1993, Worley et al. 2003). For example, an organism may allocate a certain fraction of resources to reproduction with the remainder going to somatic maintenance, but the fraction allotted to reproduction may then be subdivided among traits that are traded off against one another, such as offspring size and number. Establishing whether or not primates face such trade-offs is a crucial area of future research, requiring ecological, demographic, and quantitative genetic information.

PATTERNS OF LIFE HISTORY VARIATION

This brief theoretical review establishes a context in which we can present summary analyses of life history correlations. Our objective is not to test various models. Instead, these correlations offer empirical experience with the relevant data and show limitations of data used to analyze life histories. These deficiencies reveal a need for new kinds of data for testing and extending life history theories. They also raise questions about precisely how life history analyses should be conducted and interpreted.

Space limitations preclude detailed descriptions of data and methods. However, we use the most up-to-date source for interspecific data known to us (Kappeler and Pereira 2002:Appendix). This compendium includes carefully vetted sources for both adult and neonatal body mass data (Smith and Jungers 1997, Smith and Leigh 1998, respectively). We conduct analyses of both raw and phylogenetically adjusted estimates, with adjustments undertaken with independent contrasts computed through the PDAP module of Mesquite phylogenetic software (Maddison and Maddison 2004). Phylogenies utilized are modified from Smith and Cheverud's (2002) rendition of primate phylogeny. Statistically, species values are not independent data points, but standard statistical techniques can be used if we can measure of the degree to which data points are independent. In essence, phylogenetic adjustments account for the fact that we expect closely related species to share greater similarity than distantly related species. Controlling for phylogenies is essential in order to account for differences in the degree to which species are related. Throughout, we use Pearson product-moment correlations to measure associations on both kinds of data.

Correlation estimates are consistent with numerous previous studies of primate life history variation (Table 23.2). However, phylogenetically adjusted correlations document surprisingly weak associations among certain variables traditionally regarded as fundamentally important (Table 23.2). For example, the correlation between age at reproductive maturation and body size, measured in Harvey et al.'s (1987) study at $r = 0.92$, diminishes to $r = 0.059$ given our phylogenetic adjustment. Ross and Jones (1999) report a higher adjusted value ($r = 0.42$), estimated after a special statistical procedure (trimming contrasts among terminal nodes, which may be problematic).

Table 23.2 A: Correlations Estimated by the Current Study with Values Published by Kappeler and Pereira (2002:Appendix)

	ADULT BODY MASS (G)	ADULT BRAIN MASS (G)	AGE AT FIRST REPRODUCTION (YEARS)	GESTATION LENGTH (DAYS)	INTERBIRTH INTERVAL (MONTHS)	LITTER SIZE	NEONATAL MASS (G)	WEANING AGE (DAYS)	NEONATAL BRAIN MASS (G)
Adult body mass (g)	—	0.961	0.881	0.683	0.821	−0.579	0.960	0.860	0.973
Adult brain mass (g)	0.846	—	0.898	0.707	0.745	−0.465	0.968	0.865	0.989
Age at first reproduction (years)	0.059	0.202	—	0.727	0.829	−0.571	0.914	0.861	0.948
Gestation length (days)	0.484	0.429	0.074	—	0.604	−0.537	0.735	0.733	0.760
Interbirth interval (months)	0.389	0.342	0.345	0.394	—	−0.400	0.779	0.813	0.837
Litter size	−0.149	−0.191	−0.023	−0.198	0.040	—	−0.647	−0.560	−0.633
Neonatal mass (g)	0.798	0.867	0.255	0.386	0.277	−0.280	—	0.879	0.987
Weaning age (days)	0.366	0.207	0.055	0.328	0.436	−0.118	0.404	—	0.872
Neonatal brain mass (g)	0.891	0.972	0.714	0.454	0.402	−0.488	0.853	0.574	—

Estimates above the diagonal represent unadjusted or "species" values, and estimates below the diagonal are adjusted using independent contrasts; raw data concern values for females unless otherwise noted.

Table 23.2   B: Correlations Published by Harvey et al. (1987) (Based on Subfamily Data)

	MALE ADULT BODY MASS (G)	GESTATION LENGTH (DAYS)	INDIVIDUAL NEONATAL MASS (G)	NUMBER OF OFFSPRING PER LITTER	WEANING AGE (DAYS)	AGE AT FIRST BREEDING (YEARS)	AGE AT SEXUAL MATURITY (YEARS)	MAXIMUM RECORDED LIFE SPAN (YEARS)	INTERBIRTH INTERVAL (MONTHS)	AGE AT SEXUAL MATURITY (MALE, YEARS)	NEONATAL BRAIN WEIGHT (G)	ADULT BRAIN WEIGHT (G)
Adult body mass (g)	0.996	0.74	0.97	−0.52	0.91	0.92	0.89	0.78	0.86	0.89	0.95	0.96
Male adult body mass (g)		0.73	0.97	−0.51	0.92	0.92	0.89	0.78	0.85	0.91	0.95	0.96
Gestation length (days)			0.82	−0.61	0.84	0.81	0.81	0.62	0.63	0.84	0.84	0.8
Individual neonatal mass (g)				−0.56	0.94	0.95	0.94	0.8	0.87	0.95	0.99	0.98
Number of offspring per litter					−0.56	−0.49	−0.47	−0.3	−0.41	−0.44	−0.51	−0.5
Weaning age (days)						0.9	0.92	0.7	0.89	0.93	0.89	0.91
Age at first breeding (years)							0.97	0.87	0.88	0.94	0.93	0.96
Age at sexual maturity (years)								0.83	0.85	0.96	0.95	0.94
Maximum recorded life span (years)									0.72	0.78	0.82	0.85
Interbirth interval (months)										0.83	0.86	0.86
Age at sexual maturity (male, years)											0.97	0.96
Neonatal brain weight (g)												0.99

Table 23.2   C: Correlations Published by Ross and Jones (1999) Comparing Body Mass to Other Variables

	Body Mass	
	SPECIES VALUES	INDEPENDENT CONTRASTS
Length of Juvenile Period (years)	0.71	0.60
Birth rate	0.81	0.60
R_{max}	0.89	0.65
Age at first reproduction (years)	0.87	0.42
Average instantaneous adult mortality rate	0.22	0.32
Prereproductive mortality rate	0.41	0.20
Average infant mortality rate	0.37	0.46
Survival to reproductive age	0.16	0.35

RETHINKING PRIMATE LIFE HISTORIES

Our review points to problems with theories, models, and empirical analyses that can be solved by new approaches to life history problems. Both the ontogenetic and population dynamic sides of the field require critical theoretical examination, but we prioritize ontogenetic studies. Recent theoretical advances in developmental biology (Raff 1996) have yet to be considered by life historians but provide important insights into life history problems. More specifically, developmental biology sees animal development as modular, with dissociation among developing parts playing a major role (Raff 1996, Raff and Raff 2000). *Modularity* means that morphological structures or organ systems may vary in the degree to which they are interrelated during development. Therefore, different levels of correlation, integration, and interaction among tissues should be expected during ontogeny. This means that different organs or organ systems may grow over separate age spans and at very different rates (i.e., they may be dissociated during ontogeny).

Dissociation of morphological structures causes the emergence of particular patterns or modes of ontogeny during development. These modes are mediated by adjustments in the growth of metabolically expensive tissues (e.g., the brain) so that there may be various ways of being a juvenile. The "fast versus slow" continuum that has dominated discussion of primate life histories does not account for these different modes and, thus, appears to be inaccurate. The emerging view of primate life history is much more complicated than is revealed by the "fast versus slow" perspective so that understanding life histories requires high-quality developmental data.

In this view, dissociation of developing structures is especially important for primates because of their extended life spans. Primate longevity may be a benefit conferred by life in the trees (or at least, typically, not on the ground [Austad and Fischer 1992]). Bats and birds also have relatively long life spans (Williams 1957), but developing anatomy for flight may actually favor reduced juvenile periods. In terrestrial taxa, selection probably favors a short developmental period coupled with a very rapid shift of energetic resources from growth to reproduction, especially if juvenile mortality is high (Williams 1966a,b). We further expect few opportunities for dissociation and the emergence of life history modes in these taxa, so these species should meet the assumptions of Charnov's model well. In primates, life history modes minimize risks from energetic competition between still-growing structures and offspring production, offering flexibility as to when to grow different structures and when to complete the juvenile period. Reduced primate mortality provides opportunities for the evolution of diverse patterns of development, depending on energetic risks and mortality profiles at different stages. Therefore, instead of a "fast–slow continuum," "modes" and phases characterize primate life histories (see Pereira and Leigh 2002, Leigh and Bernstein in press). We define a *life history mode* as a distinctive pattern or arrangement of ontogeny with respect to the rate and scheduling of growth for various organs, organ systems, or modules.

Unfortunately, the theoretical infrastructure devoted to modularity in life history remains underdeveloped, as lamented by Stearns over a decade ago (1992, see also Watkinson and White 1985). Despite these difficulties, it is important to note that previous primate life history theory tends to assume, either implicitly or explicitly, that primate life histories are tightly integrated, with low dissociation and minimal modularity (e.g., Kelley 2004). High interspecific correlations reported by previous studies but shown to be problematic by the current study are at the heart of this interpretation. In addition, correlation studies virtually always lack true ontogenetic dimensions, blinding them to modularity. Correlation studies assume that all structures cease growth synchronously at the age of reproductive maturation. In contrast, ontogenetic studies accommodate variation in pathways through the juvenile period that arise from modularity. Moreover, an ontogenetic view incorporates ideas about metabolic risks and extrinsic mortality (Janson and van Schaik 1993, Williams 1966a).

We expect that alternative life history modes among primates have evolved in response to combinations of metabolic risks and differences in mortality at various phases of life histories. This idea relies on classic trade-off theory, anticipating that organisms face "decisions" as to how and when to allocate effort to growth and reproduction. We suggest that the situation for primates is more complicated than this, involving hierarchies of trade-offs. Specifically, a relatively lengthy period of primate ontogeny means that primates (including humans) have options (unavailable to other species) regarding energy allocation among modules. Three factors in primate energy allocation may be important. First, energy allocation may follow temporal patterns, involving a simple energetic shift from growth to reproduction, as predicted by classic theory. Second, dissociation may enable trade-offs among modules, with expensive tissues growing during ages of relatively low risk. Third, variation in total body mass growth rates may mediate energy allocation.

Obviously, different combinations of these factors may play roles, but the important point is that the idea of a life history mode captures these different ways of arranging ontogeny and, thus, of "assembling" adults.

THE BRAIN AND MODULAR LIFE HISTORY

A cursory empirical example focusing on the brain illustrates advantages of the life history mode concept (see Leigh 2004, Leigh and Bernstein in press, Pereira and Leigh 2002 for additional examples). Full documentation of modular ontogenies mandates analyses of many species and several organ systems, or structures (modules). However, consideration of body and brain size growth can provide preliminary insights into this idea. First, body mass growth rates vary substantially among species to produce comparably sized organisms (Fig. 23.1). Growth rates also fluctuate (Leigh 1996) so that, when measured by the progression of size, life histories can consist of both "fast" and "slow" phases. Different combinations of these phases imply alternative pathways through the juvenile period. Second, turning to the brain (and mindful of editorial space limitations), we assume the brain is an energetically costly module that is responsive to selection partly on metabolic performance (Aiello and Wheeler 1995).

Brain modularity can be illustrated by summarizing analyses of brain growth in relation to key life history variables. Our example tests the hypothesis that the age at brain growth cessation, adult brain size, and age at reproductive maturation are tightly intercorrelated. Rejection of this hypothesis implies that modularity plays a role in primate life history evolution, while failure to reject it supports models that see the brain as a pace-setter of life histories (see also studies by Deaner et al. 2002; Ross 2002, 2004; Ross and Jones 1999; Sacher 1959; Sacher and Staffeldt 1974). Moreover, this test has significant implications for refining ideas proposed by Martin (1983, 1996; Harvey et al. 1987) that reveal ties between gestation, maternal reproductive or energetic effort, brain size, and life history.

Ontogenetic data for brain size can be gathered for only a handful of species. We follow procedures outlined previously (Leigh 2004), which involve estimating age at brain growth cessation, then compare these estimates against adult brain size and literature-reported estimates of age at reproductive maturation. We present both raw and phylogenetically adjusted correlations. The latter correlations take into account differences in degrees of relatedness among species.

Evaluations of brain growth curves reveal variation independent of reproductive maturation age. An especially obvious case concerns comparisons between squirrel monkeys (*S. sciureus*) and tamarins (*Saguinus fuscicollis*). Brain growth curves show that the former grow brains at higher rates than the latter but over a much shorter age interval (Fig. 23.2). Moreover, squirrel monkey brains and bodies are larger as adults than those of tamarins, with squirrel monkeys reaching reproductive maturation later than tamarins (Garber

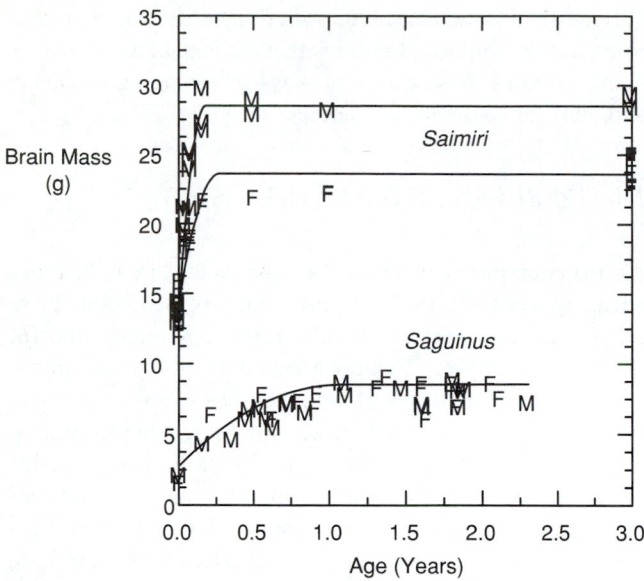

Figure 23.2 New World monkey brain growth trajectories. Squirrel monkeys (*Saimiri sciureus*) cease brain growth earlier than saddle-back tamarins (*Saguinus fuscicollis*). Lines represent best-fit piecewise regressions. M, males; F, females. Sexes are combined in the tamarin sample. See Leigh (2004) for details on data utilized.

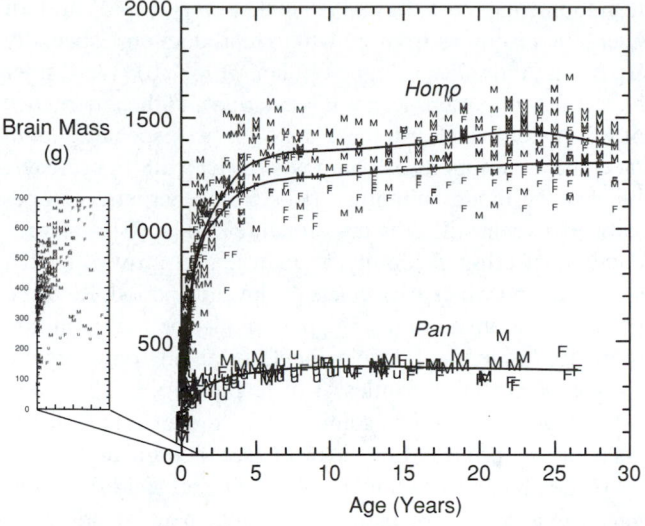

Figure 23.3 Brain mass growth data for humans (*Homo sapiens*) and chimpanzees (*Pan troglodytes*). Lines represent best-fit loess regressions through data. Symbols represent females (F), males (M), or unrecorded (u). The inset shows brain mass growth for each species during the first postnatal year. See Leigh (2004) for details on data utilized.

and Leigh 1997). Chimpanzees and humans provide another clear example, showing major differences in brain size produced by different growth rates (Fig. 23.3). Major distinctions in total time of brain growth are difficult to discern between these species, despite differences in brain size, age at reproductive maturation, age at body mass growth cessation, and adult body size. Both comparisons indicate some level of independence of brain and somatic growth.

Statistical analyses for a larger sample show that adult brain size and age at reproductive maturation are strongly correlated (Fig. 23.4A, $r = 0.93$) and that age at brain growth cessation and age at reproductive maturation are correlated (Fig. 23.4B, $r = 0.64$). Phylogenetic adjustment does not greatly alter the first correlation (brain size and reproductive maturation age, $r_{adj} = 0.70$), but the latter correlation evaporates ($r_{adj} = 0.12$).

Our small sample size presents obvious limitations. Nevertheless, these results refute models that designate the

brain as a direct pace-setter of life histories. The time it takes to grow a brain and the length of the juvenile period are unrelated. However, these results point toward models that see roles for energetics and risk aversion in driving life history variation. Specifically, adult brain size and reproductive maturation age are correlated, no matter how we measure the association. This suggests indirect effects of brain ontogeny on both body size and reproductive maturation age. So, larger, faster-growing brains seem to require larger, later-maturing mothers (Leigh 2004). Species in which mothers invest relatively little in prenatal brain growth seem to produce offspring that grow brains slowly but through much of the postnatal period. In these cases, costs of brain growth are borne by the offspring itself; and in the specific case of tamarins, males and other group members subsidize offspring (Garber and Leigh 1997).

These results suggest a significant role for the brain in primate life history. Species that produce large-brained offspring during the prenatal period invest heavily and

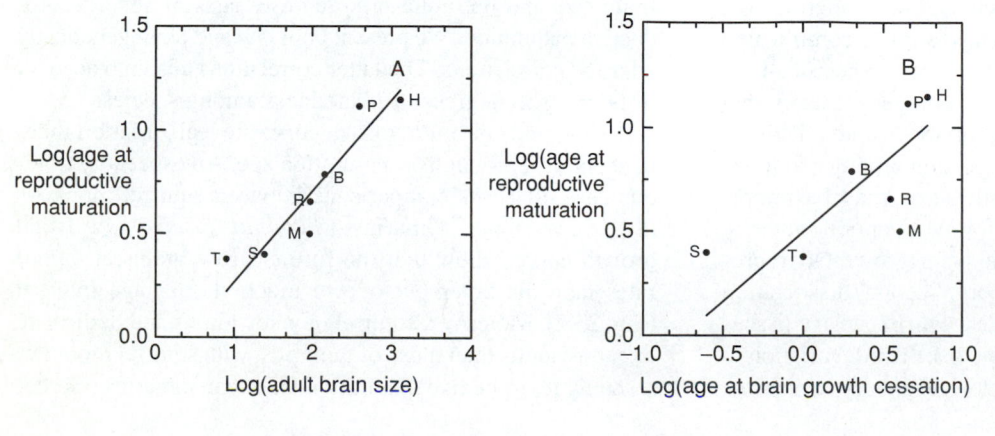

Figure 23.4 Regression analyses of brain attributes and age at reproductive maturation. A: Age at reproduction is highly correlated with adult brain size. B: Age at reproduction is also correlated with age at brain growth cessation, but the relationship is not as strong. Plot abbreviations are as follows: T, *Saguinus fuscicollis*; S, *Saimiri sciureus*; R, *Macaca mulatta*; M, *Lophocebus/Cercocebus* (mangabeys); B, *Papio*; P, *Pan troglodytes*; H, *Homo sapiens*. All values are measured on females.

"single-handedly" in brain size, a situation ameliorated by late maturation and large maternal size (see also Leigh 2004, Leigh and Bernstein in press), while other species mature early in part because of limited investment in prenatal brain growth. These inferences rely on Martin's (1983, 1996) ideas tying brain size to life histories through maternal energetics (see Leigh 2004). Perhaps most importantly, these results suggest the potential for relating ontogeny to demographic views of life histories.

PROSPECTUS

This chapter illustrates several key issues for the future of life history studies. First, we advocate a shift in life history studies toward ontogenetic data. Such data are absolutely essential to furthering our knowledge. We recommend a shift from the kinds of life history data analyzed by traditional studies to data that speak directly to ontogeny and development (see Shea 1990). A longitudinal perspective is likely to be especially valuable in this context (DeRousseau 1990) because it provides an opportunity to measure fitness consequences of events occurring prior to adulthood (see S. Altmann 1998). This perspective readily accommodates ideas about genetic trade-offs. Second, traditional interspecific comparative studies are not likely to add to what we currently know about life history evolution in primates. However, it is clear that such analyses serve as a valuable foundation for large-scale life history theories (e.g., Charnov 1993). These kinds of analysis also provide insights into historical processes (Shea 1987). Third, theoreticians should follow Charnov's lead by investigating ontogenetic dimensions of maternal energetics, metabolic risk aversion, and differences in extrinsic mortality among phases of life history (including infanticide and predation) and the relation of these to population dynamics. This will require approaches that rely on quantitative genetics and demography. Fourth, we are sorely lacking in ontogenetic data from noncaptive populations. Recent analyses are beginning to solve this problem (S. Altmann 1998, Johnson 2003, Stone 2004). Future researchers must be well versed in both ontogeny and population dynamics to productively extend these theories. We can note that ontogenetic studies of captive samples are also necessary, particularly given new imaging technologies that will revolutionize studies of ontogeny. Taken together, these advances present significant opportunities to develop life history theories that accommodate both ontogenetic and population phenomena.

CONCLUSIONS

Applications of life history theory to primates have generated significant insights at many levels. These theories prioritize understanding of how energetic resources are allocated either to growth or to reproduction and seek a general understanding of the mechanisms responsible for patterns of allocation and resulting life histories. Theories reviewed here, including Charnov's ideas, Janson and van Schaik's risk aversion hypothesis, and Kozłowski and Weiner's optimization model, all contribute to our understanding of primate life history evolution. However, this theoretical plurality implies possibilities for a more robust life history theory.

Improvements in life history theory rely on a better understanding of trade-offs and development of models that establish stronger links between ontogenetic processes, quantitative genetics, and demography. Moreover, explicit recognition of the paradigm established by Williams can aid in advancing contemporary primate life history theory. Specifically, greater attention to selection during all phases of life history (e.g., S. Altmann 1998, Johnson 2003) will improve our understanding of how changes during life evolve. Reconceptualizing primate life histories in terms of modes and phases offers new insight into patterns of variation across the primate order by fitting this priority more adequately than traditional approaches. Theories that address modularity of development may be especially valuable in promoting theoretical advances. Of course, the goal of better theory also relies upon generation and analysis of high-quality data from primate species and populations.

REFERENCES

Aiello, L. C., and Wheeler, P. (1995). The expensive tissue hypothesis. *Curr. Anthropol.* 36:199–222.

Allman, J., and Hasenstaub, A. (1999). Brain maturation times, and parenting. *Neurobiol. Aging* 20:447–454.

Altmann, J. (1980). *Baboon Mothers and Infants*. Harvard University Press, Cambridge, MA.

Altmann, J., and Alberts, S. C. (2003). Variability in reproductive success viewed from a life-history perspective in baboons. *Hum. Biol.* 15:401–409.

Altmann, S. A. (1998). *Foraging for Survival*. University of Chicago Press, Chicago.

Austad, S. N., and Fischer, K. E. (1992). Primate longevity: its place in the mammalian scheme. *Am. J. Primatol.* 28:251–262.

Berrigan, D., Charnov, E. L., Purvis, A., and Harvey, P. H. (1993). Phylogenetic contrasts and the evolution of mammalian life histories. *Evol. Ecol.* 7:270–278.

Blomquist, G. E. (in press). Population regulation and the life history studies of LaMont Cole. *J. Hist. Biol.*

Bogin, B. B. (1999). *Patterns of Human Growth*. Cambridge University Press, Cambridge.

Charnov, E. L. (1991). Evolution of life history variation among female mammals. *Proc. Natl. Acad. Sci. USA* 88:1134–1137.

Charnov, E. L. (1993). *Life History Invariants*. Oxford University Press, Oxford.

Charnov, E. L. (2001). Evolution of mammalian life histories. *Evol. Ecol. Res.* 3:521–535.

Charnov, E. L., and Berrigan, D. (1993). Why do female primates have such long lifespans and so few babies? or Life in the slow lane. *Evol. Anthropol.* 1:191–194.

Cole, L. C. (1954). The population consequences of life history phenomena. *Q. Rev. Biol.* 29:103–137.

Deaner, R. O., Barton, R. A., and van Schaik, C. P. (2002). Primate brains and life histories: renewing the connection. In: Kappeler, P. M., and Pereira, M. E. (eds.), *Primate Life Histories and Socioecology*. University of Chicago Press, Chicago. pp. 233–265.

de Jong, G. (1993). Covariances between traits deriving from successive allocations of a resource. *Funct. Ecol.* 7:75–83.

de Jong, G., and van Noordwijk, A. J. (1992). Acquisition and allocation of resources: genetic (co)variances, selection, and life histories. *Am. Nat.* 139:749–770.

de Laguerie, P., Olivieri, I., Atlan, A., and Gouyon, (1993). Analytic and simulation models predicting positive genetic correlations between traits linked by tradeoffs. *Evol. Ecol.* 5:361–369.

DeRosseau, C. J. (1990). Life-history thinking in perspective. In: DeRosseau, C. J. (ed.), *Primate Life History and Evolution*. Wiley-Liss, New York. pp. 1–13.

Felsenstein, J. (1985). Phylogenies and the comparative method. *Am. Nat.* 125:1–15.

Fisher, R. A. (1930). *The Genetical Theory of Natural Selection*. Clarendon Press, Oxford.

Flower, S. S. (1931). Contributions to our knowledge of the duration of life in vertebrate animals. *Proc. Zool. Soc. Lond.* 1931:145–233.

Garber, P. A., and Leigh, S. R. (1997). Ontogenetic variation in small-bodied New World primates: implications for patterns of reproduction and infant care. *Folia Primatol.* 68:1–22.

Garland, T., and Adolph, S. C. (1994). Why not to do two-species comparative studies: limitations on inferring adaptation. *Physiol. Zool.* 67:797–828.

Gavan, J. A., and Swindler, D. R. (1966). Growth rates and phylogeny in primates. *Am. J. Phys. Anthropol.* 24:181–190.

Godfrey, L. R., Samonds, K. E., and Jungers, W. L. (2002). Dental development and primate life histories. In: Kappeler, P. M., and Pereira, M. E. (eds.), *Primate Life Histories and Socioecology*, University of Chicago Press, Chicago. pp. 177–203.

Godfrey, L. R., Samonds, K. E., Jungers, W. L., Sutherland, M. R., and Irwin, M. T. (2004). Ontogenetic correlates of diet in Malagasy lemurs. *Am. J. Phys. Anthropol.* 123:250–276.

Goodall, J. (1986). *Chimpanzees of Gombe*. Harvard University Press, Cambridge, MA.

Gould, S. J. (1977). *Ontogeny and Phylogeny*. Belknap Press, Cambridge.

Haraway, D. J. (1989). *Primate Visions*. Routledge, Chapman, and Hall, New York.

Harvey, P. H., and Clutton-Brock, T. H. (1985). Life history variation in primates. *Evolution* 39:559–581.

Harvey, P. H., Martin, R. D., and Clutton-Brock, T. H. (1987). Life histories in comparative perspective. In: Smuts, B. B., Cheney, D. L., Seyfarth, R. M., Wrangham, R. W., and Struhsaker, T. T. (eds.), *Primate Societies*. University of Chicago Press, Chicago. pp. 181–196.

Hawkes, K., O'Connell, J. F., and Blurton-Jones, N. G. (2002). Human life histories: primate trade-offs, grandmothering socioecology, and the fossil record. In: Kappeler, P. M., and Pereira, M. E. (eds.), *Primate Life Histories and Socioecology*. University of Chicago Press, Chicago. pp. 204–232.

Hennemann, W. W. (1983). Relationship among body mass, metabolic rate, and the intrinsic rate of natural increase in mammals. *Oecologia* 56:419–424.

Hennemann, W. W. (1984). Basal metabolic rate and the intrinsic rate of natural increase: an empirical and theoretical reexamination. *Oecologia* 64:421–423.

Hill, K., and Hurtado, A. M. (1996). *Ache Life History*. Aldine de Gruyter, New York.

Houle, D. (1991). Genetic covariance of fitness correlates: what genetic correlations are made of and why it matters. *Evolution* 45:630–645.

Janson, C. H., and van Schaik, C. P. (1993). Ecological risk aversion in juvenile primates: slow and steady wins the race. In: Pereira, M. E., and Fairbanks, L. A. (eds.), *Juvenile Primates*. Oxford University Press, New York. pp. 57–74.

Johnson, S. E. (2003). Life history and the competitive environment: trajectories of growth, maturation, and reproductive output among chacma baboons. *Am. J. Phys. Anthropol.* 120:83–98.

Jones, K. E., and MacLarnon, A. (2001). Bat life histories: testing models of mammalian life-history evolution. *Evol. Ecol. Res.* 3:465–476.

Kaplan, H., Hill, K., Lancaster, J., and Hurtado, A. M. (2000). A theory of human life history evolution: diet, intelligence, and longevity. *Evol. Anthropol.* 9:156–183.

Kappeler, P. M., and Pereira, M. E. (eds.) (2002). *Primate Life Histories and Socioecology*. University of Chicago Press, Chicago.

Kelley, J. (2004). Life history and cognitive evolution in the apes. In: Russon, A. E., and Begun, D. (eds.), *The Evolution of Thought*. Cambridge University Press, New York. pp. 280–297.

Kindleman, P., Dixon, A. F. G., and Dostalkova, I. (1999). Does body size optimization result in skewed body size distribution on a logarithmic scale? *Am. Nat.* 153:445–447.

Kozłowski, J., Gawelczyk, A. T. (2002). Why are species' body size distributions usually skewed to the right? *Funct. Ecol.* 16:419–432.

Kozłowski, J., and Wiener, J. (1997). Interspecific allometries are by-products of body size optimization. *Am. Nat.* 149:352–380.

Lack, D. (1954). *The Natural Regulation of Animal Numbers*. Oxford University Press, Oxford.

Lande, R. (1982). A quantitative genetic theory of life history evolution. *Ecology* 63:607–615.

Le Gros Clark, W. E. (1959). *The Antecedents of Man*. Harper and Row, New York.

Leigh, S. R. (1994). Ontogenetic correlates of diet in anthropoid primates. *Am. J. Phys. Anthropol.* 94:499–522.

Leigh, S. R. (1996). Evolution of human growth spurts. *Am. J. Phys. Anthropol.* 101:455–474.

Leigh, S. R. (2001). Evolution of human growth. *Evol. Anthropol.* 10:223–236.

Leigh, S. R. (2004). Brain growth, life history, and cognition in primate and human evolution. *Am. J. Primatol.* 62:139–164.

Leigh, S. R., and Bernstein, R. M. (in press). Ontogeny, life history, and maternal reproductive strategies in baboons. In: Swedell, L., and Leigh, S.R. (eds.), *Life History, Reproductive Strategies, and Fitness in Baboons*. Kluwer/Academic Press, New York.

Leigh, S. R., and Shea, B. T. (1996). The ontogeny of body size variation in African apes. *Am. J. Phys. Anthropol.* 99:43–65.

Lovejoy, C. O. (1981). The origin of man. *Science* 211:341–350.

MacArthur, R. H., and Wilson, E. O. (1967). *The Theory of Island Biogeography*. Princeton University Press, Princeton, NJ.

Maddison, W. P., and Maddison, D. R. (2004). *Mesquite*: a modular system for evolutionary analysis, version 1.01, http://mesquiteproject.org.

Martin, R. D. (1983). *Human Brain Evolution in an Ecological Context. 52nd James Arthur Lecture on the Evolution of the Human Brain*. American Museum of Natural History, New York.

Martin, R. D. (1996). Scaling of the mammalian brain: the maternal energy hypothesis. *News Physiol. Sci.* 11:149–156.

Martin, R. D., and MacLarnon A. M. (1990). Reproductive patterns in primates and other mammals: the dichotomy between altricial and precocial offspring. In: DeRousseau, C. J. (ed.), *Primate Life History and Evolution*. Wiley-Liss, New York. pp. 47–80.

Medawar, P. B. (1946). Old age and natural death. *Med Q.* 1:30–56.

Medawar, P. B. (1952). *An Unsolved Problem in Biology*. H. K. Lewis, London.

Mylius, S. D., and Diekmann, O. (1995). On evolutionary stable life histories, optimisation, and the need to be specific about density dependence. *Oikos* 74:218–224.

Pereira, M. E., and Fairbanks, L. A. (eds.) (1993). *Juvenile Primates*. Oxford University Press, New York.

Pereira, M. E., and Leigh, S. R. (2002). Modes of primate development. In: Kappeler, P. M., and Pereira, M. E. (eds.), *Primate Life Histories and Socioecology*. University of Chicago Press, Chicago. pp. 149–176.

Pianka, E. R. (1970). On "r" and "K" selection. *Am. Nat.* 102:592–597.

Promislow, D. E. L., and Harvey, P. H. (1990). Living fast and dying young: a comparative analysis of life-history variation among mammals. *J. Zool.* 220:417–437.

Prothero, J., and Jürgens, K. D. (1987). Scaling of maximal lifespan in mammals: a review. In: Woodhead, A. D., and Thompson, K. H. (eds.), *Evolution of Longevity in Animals*. Plenum Press, New York. pp. 49–74.

Purvis, A., and Harvey, P. H. (1995). Mammal life-history evolution: a comparative test of Charnov's model. *J. Zool.* 237:259–283.

Raff, R. A. (1996). *The Shape of Life*. University of Chicago Press, Chicago.

Raff, E. C., and Raff, R. A. (2000). Dissociability, modularity, evolvability. *Evol. Dev.* 2:235–237.

Ricklefs, R. E. (2003). Is rate of ontogenetic growth constrained by resource supply or tissue growth potential? A comment on West et al.'s model. *Funct. Ecol.* 17:384–393.

Roff, D. A. (1992). *The Evolution of Life Histories*. Chapman and Hall, New York.

Roff, D. A. (2002). *Life History Evolution*, 2nd ed. Sinauer Associates, Sunderland, MA.

Ross, C. (1988). The intrinsic rate of natural increase and reproductive effort in primates. *J. Zool. Lond.* 214:199–219.

Ross, C. (1991). Life history patterns in New World monkeys. *Int. J. Primatol.* 12:481–502.

Ross, C. (1992). Environmental correlates of the intrinsic rate of natural increase in primates. *Oecologia* 90:383–390.

Ross, C. (2002). Life history, infant care strategies, and brain size in primates. In: Kappeler, P. M., and Pereira, M. E. (eds.), *Primate Life Histories and Socioecology*. University of Chicago Press, Chicago. pp. 266–284.

Ross, C. (2004). Life history and the evolution of large brain size in great apes. In: Russon, A. E., and Begun, D. (eds.), *The Evolution of Thought*. Cambridge University Press, New York. pp. 122–139.

Ross, C., and Jones, K. E. (1999). Socioecology and the evolution of primate reproductive rates. In: Lee, P. C. (ed.), *Comparative Primate Socioecology*. Cambridge University Press, Cambridge. pp. 73–110.

Sacher, G. A. (1959). The relation of life span to brain weight and body weight in mammals. In: Wolstenholme, G. E. W., and O'Connor, M. (eds.), *CIBA Foundation Colloquia on Aging. The Lifespan of Animals*, vol. 5. Churchill, London. pp. 115–133.

Sacher, G. A., and Staffeldt, E .F. (1974). Relation of gestation time to brain weight for placental mammals. *Am. Nat.* 108:593–615.

Sade, D. S. (1990). Intrapopulation variation in life history parameters. In: DeRousseau, C. J. (ed.), *Primate Life History and Evolution*. Wiley-Liss, New York. pp. 181–194.

Sade, D. S., Chepko-Sade, B. D., Schneider, J. M., Roberts, S. S., and Richtsmeier, J. T. (1985). *Basic Demographic Observations on Free-Ranging Rhesus Monkeys*. Human Relations Area Files, New Haven.

Schultz, A. H. (1969). *The Life of Primates*. Universe Press, New York.

Shea, B. T. (1987). Reproductive strategies, body size, and encephalization in primate evolution. *Int. J. Primatol.* 8:139–156.

Shea, B. T. (1990). Dynamic morphology: growth, life history, and ecology in primate evolution. In: DeRousseau, C. J. (ed.), *Primate Life History and Evolution*. Wiley-Liss, New York. pp. 325–352.

Silk, J. B., Alberts, S. C., and Altman, J. (2003). Social bonds of female baboons enhance infant survival. *Science* 302:1231–1234.

Smith, R. J., and Cheverud, J. M. (2002). Scaling of sexual dimorphism in body mass: a phylogenetic analysis of Rensch's rule in primates. *Int. J. Primatol.* 23:1095–1135.

Smith, R. J., and Jungers, W. L. (1997). Body mass in comparative primatology. *J. Hum. Evol.* 32:523–559.

Smith, R. J., and Leigh, S. R. (1998). Sexual dimorphism in primate neonatal body mass. *J. Hum. Evol.* 34:173–201.

Stearns, S. C. (1992). *Life History Evolution*. Yale University Press, New Haven.

Stone, A. (2004). Juvenile feeding ecology and life history in a neotropical primate, the squirrel monkey (*Samiri sciureus*). [PhD diss.]. University of Illinois, Urbana.

Stucki, B. R., Dow, M. M., and Sade, D. S. (1991). Variance in intrinsic rates of growth among free-ranging rhesus monkey groups. *Am. J. Phys. Anthropol.* 84:181–192.

Tanner, J. M. (1978). *Foetus into Man*. Harvard University Press, Cambridge, MA.

van Noordwijk, A. J., and de Jong, G. (1986). Acquisition and allocation of resources: their influence on variation in life history tactics. *Am. Nat.* 128:137–142.

Watkinson, A. R., and White, J. (1985). Some life-history consequences of modular construction in plants. *Phil. Trans. R. Soc. Lond. B* 313:31–51.

Watts, E. S., and Gavan, J. A. (1982). Postnatal growth of nonhuman primates: the problem of the adolescent spurt. *Hum. Biol.* 54:53–70.

West, G. B., Brown, J. H., and Enquist, B. J. (2001). A general model for ontogenetic growth. *Nature* 413:628–631.

Williams, G. C. (1957). Pleiotropy, natural selection, and the evolution of senescence. *Evolution* 11:398–411.

Williams, G. C. (1966a). *Adaptation and Natural Selection: A Critique of Some Current Evolutionary Thought*. Princeton University Press, Princeton, NJ.

Williams, G. C. (1966b). Natural selection, the costs of reproduction and a refinement of Lack's principle. *Am. Nat.* 100:687–690.

Worley, A. C., Houle, D., and Barrett, S. C. H. (2003). Consequences of hierarchical allocation for the evolution of life-history traits. *Am. Nat.* 161:153–167.

24

Primate Growth and Development
A Functional and Evolutionary Approach
Debra Bolter and Adrienne Zihlman

INTRODUCTION

Growth and development are of concern for evolutionary studies in two major ways. From a phylogenetic perspective, different growth patterns produce adults that diverge morphologically, even when the early stages show considerable similarity. From a life history perspective, natural selection operates on immature individuals: before any adult primate can reproduce, it must survive long infant and juvenile stages when mortality is high. If a young animal does not reach maturity, reproductive success, longevity, and speciation are irrelevant.

We begin with a brief historical background of a few influential scientists of growth research, the questions they asked, and the methods they used. Early data collection on extant primates generally described one or two body systems neglecting both functional and evolutionary frameworks. More recent studies model growth of different body parts, for example, the face, neurocranium, or long bones. These studies speak of ontogenetic scaling, allometry, heterochrony, and neoteny; but meanings and usages vary among the researchers (cf. Shea 2000). Such studies do not seem relevant to field primatologists and behavioral ecologists: conclusions do not consider function or adaptation, and they are cloaked in terminology for specialists only.

Our goal is to present a holistic view of individuals. We incorporate information from multiple systems in order to link the developing body to behaviors observed in the wild: the young primate that gains locomotor independence, the shift from mother's milk to adult diet, social integration, and the approach to reproductive maturity. We hope to convey the sense of flow between the immature and the adult and the ways that growth affects species adaptation. We draw on available data from several somatic systems that mature at different rates—for example, dental, neurological, skeletal, and muscular—and from a range of primate species, including lemurs, tamarins, vervet monkeys, baboons, macaques, gorillas, and chimpanzees.

HISTORICAL PERSPECTIVE: NINETEENTH AND EARLY TWENTIETH CENTURIES

An evolutionary framework is fundamental to the interpretation of immaturity as a developmental stage, with the adult as "end product." The individual—the whole individual—survives and reproduces, and the challenge is to integrate functionally the parts and the whole.

Charles Darwin's Contributions

In formulating the mechanism of natural selection, Charles Darwin discussed in *On the Origin of Species* (Chapter 13) the physical traits and adaptations of immature individuals. Darwin considered that his thoughts on embryology and development in *Origin* were among his most original (Darwin 1876:125). Very early stages of life were not under heavy selection pressure and, therefore, embryos of related groups of animals looked similar (Darwin 1859:449). Darwin argued that selection acted on an immature individual and subsequent features were modulated only when the young animal must provide for its own survival (Darwin 1859, 1860, 1876:125).

Adolph Schultz's Contributions

Adolph Schultz contributed some of the earliest systematic and comparative primate studies well before fieldwork on socioecology had begun (e.g., 1924, 1927, 1930). He gathered information on fetuses, neonates, and other immature individuals. Schultz, who was trained in human biology, drew on museum collections of wild-shot primates and on captive animals from zoos and breeding colonies.

Because dental eruption can be evaluated in both living primates and their skeletal remains, Schultz used dentition as the life stage marker: he wanted a dental "comparable age,"a baseline for growth and development. He determined an order of eruption of the deciduous (milk, temporary or "baby") teeth and the timing of eruption of the first permanent molar (M1) from zoo records (e.g. for chimpanzees, *Pan troglodytes*,

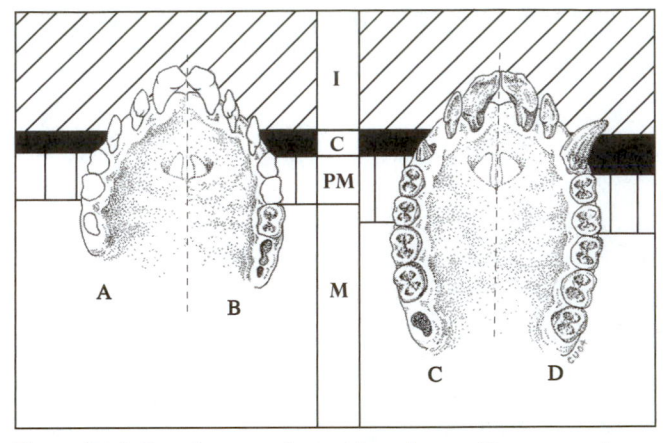

Figure 24.1 Dental stages of catarrhine primates. Drawn to scale. A: Infant, deciduous dentition. B: Juvenile 1, deciduous and permanent dentition. C: Juvenile 2, permanent teeth, not all erupted. D: Adult, all permanent teeth erupted. Abbreviations: I, incisors; C, canines; PM, premolars, M, molars. (Drawing by Carol Underwood.)

1940; for orangutans, *Pongo pygmaeus*, 1941). Based on dental eruption, Schultz defined four life stages that are still used: infant, juvenile 1, juvenile 2, adult (e.g., 1940; cf. Krogman 1969). Infants have a partial or full set of deciduous teeth; juvenile 1 has a combination of deciduous and permanent teeth; juvenile 2 has only permanent teeth but not all of them erupted; and adults have all the permanent teeth erupted (Fig. 24.1). Schultz's dental definition of the adult life stage with all permanent dentition erupted had no reference to skeletal growth or joint fusion. Based on these comparative data, he illustrated primate life stages, which remains one of the most widely recognized of his self-penned diagrams (e.g., Bogin 1999) (see Fig. 24.2).

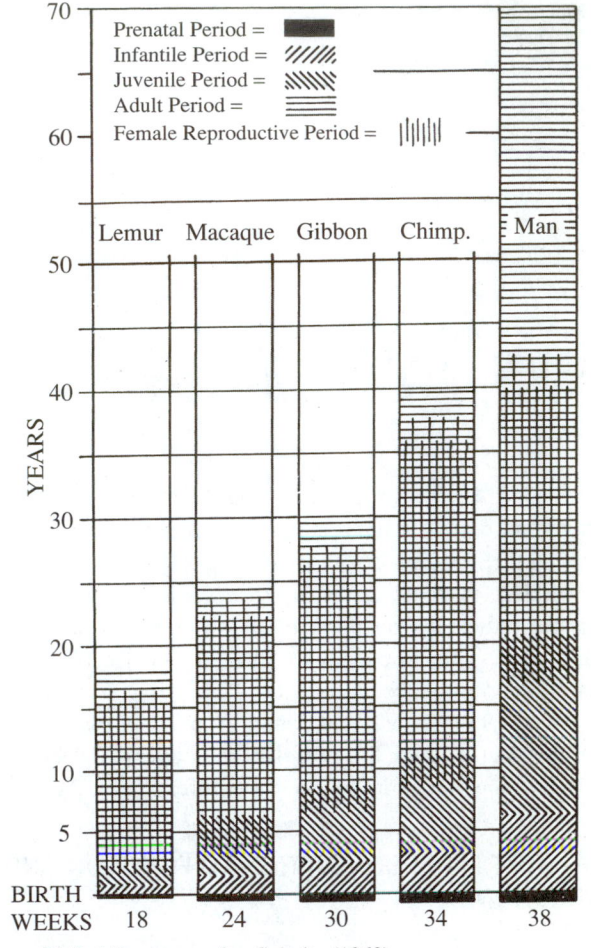

Figure 24.2 Life stages, after Schultz (1963).

MID-TWENTIETH CENTURY: THE MODERN SYNTHESIS AND NEW APPROACHES TO GROWTH

New possibilities for studies of growth and development came with the "rediscovery" of Darwin's ideas about the individual. This came from biologists, the bridge builders, who integrated information from several fields: genetics and the importance of mutation as a source of variation, the concept of surviving and reproducing individuals as the focus for natural selection, and the population—its size, gene pool composition, and migration patterns. All these processes contribute to speciation and therefore to evolutionary change (see reviews in Mayr 1980, 1997, 2004).

Connecting the study of growth to the modern synthesis, G. G. Simpson, a mammalian paleontologist and one of the bridge builders, emphasized the individual as four-dimensional. Although usually conceptualized as a three-dimensional adult, Simpson maintained that an individual is not a "combination of tissues arrested at a moment in time, but a whole sequence of states and forms through which one animal passes from fertilized ovum to death" (1941:9). Thus, growth is the dynamic element which gives the individual its fourth dimension (see Fig. 24.3).

In order to introduce evolutionary ideas into the study of human evolution, Sherwood Washburn and Theodosius Dobzhansky organized the "Origin of Man" conference at Cold Spring Harbor, New York, in 1950. Washburn (1951a) urged that anthropologists move away from mere description of bones, teeth, and traits and include information from living and fossil primates, in the same way that he emphasized the integration of bone, muscle, and joints into functional regions in his research rather than describing isolated bones or taking nonbiological measurements (1946a, 1946b, 1947a, 1947b, 1951a; Detwiler and Washburn 1943). Washburn's importation of the modern synthesis into physical anthropology in the mid-twentieth century energized the entire field (Zihlman 2000).

Washburn recognized that growth had a role to play in an evolutionary approach. In his article "The New Physical Anthropology," Washburn (1951b) specifically noted growth and development as a main subarea for future research. This evolutionary framework has been slow in acceptance, in part because of the Darwinian ideas about selection pressures on the individual (Bowler 1988, Mayr 2004) and because developmental processes were not fully appreciated by those who supported the modern synthesis.

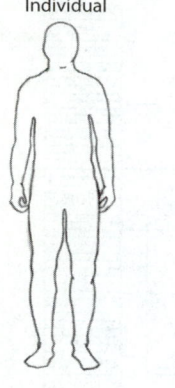

3 Dimension
Individual

4 Dimension Individual

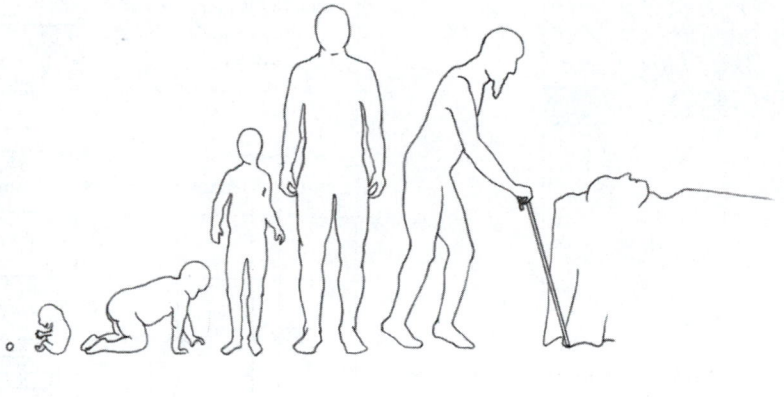

Figure 24.3 Third and fourth dimensions of the individual, after Simpson (1944).

Kenneth Brizzee and William Dunlap (1986) of Tulane University reviewed the literature on primate growth studies through the mid-1980s. This paper highlights the absence of the modern synthesis and illustrates some of the pitfalls of growth studies: narrow description of one anatomical region and neglect of its functional assessment and effect in another anatomical region (see also Watts 1986). For example, they reviewed facial growth and dental eruption in two different sections but missed the correlation between the erupting teeth and the elongation of the face (refer to Fig. 24.1).

THE SHIFT FROM DESCRIPTION TO FUNCTION AND EVOLUTION

Primate Immaturity in the Field

Primate field research began in the 1940s (e.g., Carpenter 1940). Longer-term primatology studies in the late 1950s and early 1960s depended on the identification of individuals and brought new possibilities for studying individuals in Simpson's "fourth dimension" (through their lifetimes). As field studies extended beyond a few months, researchers observed variation between individuals and varied patterns of social organization (Washburn et al. 1965, Crook and Gartlan 1966). In the early studies, the high-profile males received the most attention; they could be readily identified by their size; infants and juveniles were less likely to be named and observed in much detail (Fedigan and Fedigan 1989, Fedigan 1992).

Two early projects focused on immature primates. A study on Indian langur monkeys (*Presbytis entellus*) described infant behavior and mother–infant interactions (Jay 1962, Dolhinow 1972), which later led to behavioral development in captive groups (e.g. Dolhinow and Murphy 1982). Early fieldwork coincided with emerging laboratory observations on infant monkey social development by psychologist Harry Harlow and his colleagues in the United States that revealed the importance of the mother–infant bond (Blum 2002).

In her studies on chimpanzees (*P. troglodytes*) at Gombe National Park beginning in 1960, Jane Goodall named individuals and followed them through their lives. Early on,

she charted the development of newborn infants (Goodall 1971, 1986, 1990). Focus on Gombe infant, juvenile, and adolescent chimpanzees continued in the research of Anne Pusey, Frans Plooij, and Hedwig van de Rijt-Plooij, which described physical and behavioral markers in these early immature stages (Pusey 1978, 1983, 1990; Plooij 1984; van de Rijt-Plooij and Plooij 1987). Other long-term studies have provided parallel life history data on wild chimpanzee populations of the Mahale Mountains and at Bossou (e.g., Nishida et al. 1990, 2003; Sugiyama 1994).

Researchers of baboons, macaques, and langurs also catalogued age classes of wild populations—e.g., infant, juvenile, subadult, adult—based on physical appearance and relative size (*Papio hamadryas*, Kummer 1968, Sigg et al. 1982; *Papio cynocephalus*, J. Altmann et al. 1977; *Macaca fuscata*, Mori 1979; *Pr. entellus*, Dolhinow 1972). This fieldwork on Old World monkeys showed the significance of life stages for understanding complex social behaviors in different age and sex classes. Jeanne Altmann (1980) linked infant development and survival to the mother's energetics in her research on the intense infant–mother relationship in yellow baboons (*Papio cynocephalus*). On this same population, Stuart Altmann (1991, 1998) summarized decades of research on wild infant baboon social and ecological strategies. Phyllis Lee (1984, 1987) investigated how differential maternal investment affected infant vervet social development (*Cercopithecus aethiops*). As research accumulated, a more detailed picture of immature primates began to emerge.

Infancy

Long-term research demonstrates that infancy is quite risky. Provisioned primates may experience only 10% infant mortality, whereas among wild primates, one-half to two-thirds of infants die (*Lemur catta*, Gould et al. 2003; *Papio cynocephalus*, J. Altmann et al. 1977; *M. fuscata*, Mori 1979, Itoigawa et al. 1992; *P. troglodytes*, Boesch and Boesch-Achermann 2000). Infants depend on the mother for food, protection, and companionship and are affected by maternal health and nutrition (Mori 1979). While still dependent, infants must learn foraging techniques, food choices, and social skills that will help them survive as adults (e.g.,

Mason 1979, Walters 1987, S. Altmann 1998). For example, in captive tarsiers (*Tarsius bancanus*), the infancy phase is extremely short compared to other haplorhines. During this accelerated infancy period, the babies perfect their hunting skills; and by 80 days, they are weaned and independent juveniles (Roberts 1994).

In field studies, infants may be easily distinguished from adults in hair and facial skin color. These physical markers signify behavioral transition to weaning, end of dependence, and differential social behaviors from mother and other group members. For example, infant hamadryas baboons (*Papio hamadryas*) differ from other group members by a black neonatal coat; as the infant matures, the color turns into the tan coat of a juvenile, which elicits different kinds of interaction from other group members. The tan coats "get away with less" and become potential food competitors with other juveniles and adults (Kummer 1968, 1995). Crested langur infants (*Trachypithecus cristatus*) are born with neon orange hair but, at about 5 months of age, acquire the silvery coats of the adults (Wolf 1984). Proboscis monkey infants (*Nasalis larvatus*) have blackish coats and blue faces at birth; during the eighth month, faces become the fleshy color of adults and coat color transitions to silvery red (Pournelle 1967, Rüedi 1981).

Juvenility

The juvenile stage is one of marked physical and social change. In recognition of this, Michael Pereira and Lynn Fairbanks (1993a) collected a number of studies on juveniles. Juvenile primates travel and forage independently, while remaining socially connected to their mother and siblings. They are not adult in size; by definition, they are reproductively immature. They are vulnerable because they must compete with other group members for resources. In times of scarcity, only about half of the juveniles may survive (Pereira and Altmann 1985, Watanabe et al. 1992, Pereira and Fairbanks 1993b, Gould et al. 2003).

The risks of increasing the duration of immaturity are balanced by the skills and abilities that they acquire for later benefit. Through play and group activities, immature primates learn complex behaviors and social roles and develop flexibility in coping with stressful situations. They acquire knowledge of food location and territorial boundaries and gain experience in mothering and in motor coordination during play while their bodies gradually transform into the adult form (Mason 1979; Lancaster 1971; Pereira and Altmann 1985; Watts 1985; Walters 1987; Fairbanks 1993, 2000; Fagan 1993; S. Altmann 1998).

Sexual Maturity

As immature individuals shift to sexual maturity, the duration of reproduction varies across species. Some primates have a distinct adolescent stage: marked physical and social behavioral divergence of females and males. This stage has not been well studied except for chimpanzees (*P. troglodytes*). Pusey (1978, 1983, 1990) describes early and late adolescent stages and notes distinct physical and behavioral changes in individuals, as well as in the behavior directed to them by other group members. For example, in early adolescence, a female's first estrus is defined by the response of the adult males, though only in late adolescence does a female conceive (Pusey 1978, 1990). This phase of life, as in the other stages, is marked by changes in physiology, anatomy, physical appearance, and behavior among group members; no one component is independent of the others.

LEVELS OF LIFE HISTORY: WHOLE BODIES, MULTIPLE SYSTEMS

Whether one studies the maturing individual in captivity or in the wild or studies behavior, physiology, or anatomy, the process is time-dependent: the individual—a member of a social group, population, and species—passes time as an infant, a juvenile, and in some species, an adolescent.

Temporal shifts over the course of life are called "life history" and can be studied on three "macro" levels (Fedigan 1997). First, life history may represent a personalized biography of an individual in terms of social rank, relations, and life experiences. The second level of analysis focuses on variation in timing of growth events of individuals within a social group or local population, such as the range of ages at which infants are weaned or when females and males diverge in their development. Third, at the species level, the "average" age for major life events establishes a species pattern, such as age at weaning, locomotor independence, dispersal, sexual maturity, and first reproduction, as well as birth intervals and life span. These variables are among the life history features laid out by Harvey et al. (1987) and expanded in volumes edited by deRousseau (1990) and Morbeck et al. (1997). The personalized biographies, population variation in maturation, and species patterns represent alternate survival and reproductive strategies employed across primate species. These approaches illustrate the different levels of this topic, with growth and development at the center.

At the "micro" level, body systems transform and interact to promote survival at the infant and juvenile stages. The individual must be able to function during day-to-day activities while its body adjusts, changes, and matures in its transformation to adulthood. Anatomical systems underpin behavior and provide physical markers for behavioral expression. Both anatomy and behavior shape survival. Therefore, a functional approach to growth and development integrates multiple lines of information, for example, timing of brain maturity and dental emergence, rate of skeletal growth and fusion, degree of change in body proportions and muscularity, and attainment of adult body mass.

The Part and the Whole

In courses and textbooks on human anatomy, the body is necessarily divided into systems—musculo-skeletal, neural, reproductive, hormonal. These courses often begin with a discussion of cells, move into tissues, then rise up a level to individual systems. Dividing the body into finite pieces provides a way to study a complex machine. Growth and development studies have also followed this path, isolating the body into a series of systems, like dentition, brain size, regions (e.g., pelvis, limbs, facial structure) as well as body mass, body proportions, and composition, and comparing the sexes in these areas. However, the body is integrated and these separations, although an aid in sorting through this dynamic process, oversimplify it.

Different systems in the body mature at different rates, a point illustrated in Scammon's (1930) classic graph of human growth. Physical anthropologists have demonstrated that the timing of physical growth events can also fingerprint non-human primate groups (e.g., Schultz 1927, 1937; Washburn 1943; Tappen and Severson 1971; Grand 1977b, 1992; Watts 1985, 1990; Leigh 1994; Bolter and Zihlman 2002, 2003; Bolter 2003, 2004; King 2004). We elaborate on this theme and offer ideas on relating physical changes to locomotion and other behaviors for future application to laboratory and field research. We begin with the dentition because hard tissues are consistent markers of immature individuals. We move on to the skeleton, body mass and sex differences, body proportions, and body composition. Although preliminary, our findings illustrate functional ways to look at growth.

Dentition

Life stage classifications of individuals, particularly skeletal specimens, are most often assessed based on the timing and order of tooth emergence. In the 1980s, Holly Smith highlighted the importance of dental eruptions in primate life histories, correlating the emergence of the first permanent molar (M1) with 90%–95% brain growth in primates (Smith 1989, 1991, 1994); thus, she connected two systems, dental and neural. Other studies of tooth enamel and the counting of incremental markings showed that crown formation of M1 begins about 2 weeks before birth. These findings make it possible to interpret growth and development in extinct genera by estimations of chronological age at death for immature specimens, extrapolation to the tempo of brain growth in extinct species (especially hominids), and evolution of a childhood stage in the life history pattern of *Homo sapiens* (e.g., Laird 1967, Bromage and Dean 1985, Dean et al. 1986, Bogin and Smith 1996, Leigh and Park 1998, Bogin 1999, Zollikofer et al. 1998, Dean and Reid 2001, Dean et al. 2001, Ponce de Leon and Zollikofer 2001, Kelley and Smith 2003, Ramirez Rozzi and De Castro 2004, Zihlman et al. 2004b, Zihlman and Bolter 2004).

An individual is labeled "infant" when no permanent teeth have yet erupted but there is a set of deciduous (milk) teeth: four incisors, two canines, and four (pre)molars in the lower and upper jaw. Typically, the incisors come in before the canines and premolars (Smith et al. 1994). Lemurs (*L. catta*) have erupted deciduous teeth at birth. Infants do not rely on their milk teeth as they depend on nursing for the bulk of their diet.

The transition from infant to juvenile comes with the eruption of the first permanent (M1) tooth. The M1 is a "new" tooth which does not replace any of the deciduous teeth. Permanent incisors replace deciduous incisors and permanent canines, deciduous canines. In the catarrhines, the permanent premolars replace the deciduous molars (Fig. 24.1). The full eruption of M1 has been interpreted as signifying the timing of weaning and therefore the end of infancy. Weaning may be abrupt due to marked seasonality in reproduction in lemurs (Wright 1999) and in some platyrrhines where twinning is the norm (Garber and Leigh 1997). In catarrhines, weaning is gradual when young animals shift from maternal calories to independently acquired foods (e.g., S. Altmann 1998).

Juveniles have a mixture of deciduous and permanent teeth. In catarrhines, the second molar (M2) emerges while the trunk and long bones continue to grow, and body mass and musculature increase. During this stage, females and males diverge and juvenile males in some species leave their natal group.

The eruption and full occlusion of the last molar (M3s or M2s in callitrichids) serves as only one biological marker of adulthood. Skeleton and body mass are two examples of systems still undergoing considerable growth after M2/M3 occlusion (e.g., platyrrhines, Tappen and Severson 1971; colobines, Schultz 1942, Dirks 2003; cercopithecines, Turner et al. 1997, Bolter and Zihlman 2002; hominoids, Randall 1943).

Data on dental eruption are available for a few species (Smith et al. 1994) (Table 24.1). The species in Table 24.1 represent the major phylogenetic groups and demonstrate that a pattern of increased periods of development does exist, just as Schultz recognized. However, although timing in dental eruption is delayed from prosimians to anthropoids, from platyrrhines to catarrhines, from cercopithecoids to hominoids, body mass of the female does not follow this same trajectory. Timing of dental eruption and weight of adult females are disassociated. The table demonstrates the limited data across species on other aspects of growth. Dental information may be available for an individual, but facts about skeletal fusion stages, body mass, or behavioral markers like weaning age may not be known for the same individual (Watts 1986, Smith et al. 1994, Godfrey et al. 2001).

The Skeleton

The skeleton is the frame that provides a stable structure for body mass. It anchors the muscles and facilitates movement around the joints. Bones add length during development in the *growth plate*, or the area of cartilage that sits between the shaft (diaphysis) and the tip (epiphysis). The new bone pushes the epiphysis away and increases the length of body segments such as the arm, forearm, thigh, and leg. Fusion of the epiphysis to the bone shaft terminates growth and additional length (see Fig. 24.4).

Washburn noted a consistent order of skeletal fusion. For example, in Old World monkeys—crested langurs (*T. cristatus*) and crab-eating macaques (*Macaca fascicularis*)

Table 24.1   Dental Emergence Time (Years) and Approximate Body Mass (kg) in Female Primates (Wild Weights Used from Postcranially Mature Individuals, Where Available)

SPECIES	M1 GINGIVAL EMERGENCE (YEARS)	M3 GINGIVAL EMERGENCE (YEARS)	FEMALE ADULT BODY MASS (KG)
Prosimian			
Lemur catta (ring-tailed lemur)	0.3	1.3	2.5
Anthropoid			
Platyrrhine			
Saguinus nigricollis	0.4	Absent in genera	0.5
Siamiri sciureus	0.4	1.7	0.6
Catarrhine			
Cercopithecoid			
Trachypithecus cristata	1	3.5	6.2
Semnopithecus entellus	1.4[1]	4.7[2]	9.9
Cercopithecus aethiops	1.2	4.4	3.5
Macaca mulatta	1.3[1]	5.8[1]	7.1
Papio hamadryas	1.7	7.7	9.9
Hominoid			
Pan troglodytes	4	13	35.2
Homo sapiens	6	18.5	60

Sources: Long and Cooper (1968), Chase and Cooper (1969), Wolf (1984), Iwamoto et al. (1987), Uehara and Nishida (1987), Harvey et al. (1987), Phillips-Conroy and Jolly (1988), Smith et al. (1994), Zihlman (1997), Vancata et al. (2000), Bolter and Zihlman (2003), Liversidge (2003), Dirks (2003), Zihlman et al. (2004b).
[1] Emergence through the jawbone.

—the skeletal elements comprising the elbow, hip, and knee joints fuse first, then the ankle, wrist, and finally shoulder joints. He emphasized regional union of joints—for example, the elbow joint with several epiphyses in the distal humerus, proximal radius, and proximal ulna—as opposed to a single, isolated bone. Washburn (1942, 1943) implied that variation in fusion patterns between species was due to locomotor specialization.

The timing of skeletal fusion compared to the eruption of teeth can vary within sexually dimorphic species. In rhesus macaques (*Macaca mulatta*) and Japanese macaques (*M. fuscata*), females complete skeletal fusion of the long bones prior to male conspecifics of comparable dental development (van Wagenen and Asling 1958, Cheverud 1981, Kimura and Hamada 1990). In colobine monkeys—crested langurs (*T. cristatus*), maroon leaf monkeys (*Presbytis rubicunda*), and proboscis monkeys (*N. larvatus*)—females tend to fuse pelvic and long bones earlier than males of the same dental age (Bolter 2004).

Body Systems as a Developmental Mosaic in Two Monkey Genera

The following case studies from the callitrichine New World monkeys (platyrrhines) and cercopithecine African monkeys (catarrhines) illustrate that primates do not synchronize maturity in the skeleton, dentition, reproductive anatomy, body mass, and behavioral events (cf. Watts 1990). Furthermore, sexual maturity and reaching adult stage involve a whole pattern, one that differs for females and males (*bimaturation*). This combination of information integrates

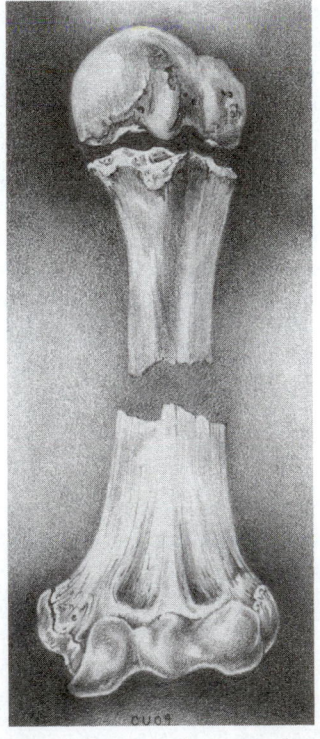

Figure 24.4 Immature chimpanzee humerus, showing proximal and distal ends; proximal end fusion indicates adult stage reached. (Drawing by Carol Underwood.)

and highlights physical and behavioral elements of maturation and reminds fieldworkers of the dynamic anatomy that underlies the behavior of immature primates.

Tamarins. The saddle-back tamarins of South America (*Saguinus fuscicollis*) are among the smallest of anthropoid primates, with minimal sex difference in adult body mass and fast maturation. Tamarins are born with 12 deciduous teeth (eight incisors and four canines); at 3 months, they have all the deciduous molars and are weaned (Glassman 1983, Garber and Leigh 1997). At 5 months, permanent teeth appear; at 12 months, all of them have erupted (12 premolars and eight molars).

Skeletal fusion begins around the third month, at the time of weaning. Elbow and pelvic elements are the first bones to complete growth. Most elements of the hip, knee, and wrist joints (bones that contribute to the joints) fuse by 15 months, after all of the permanent teeth are erupted. The ankle and shoulder joints mature much later, at about 24 months, around the time of social adulthood (Glassman 1983) (Table 24.2).

Sexual maturity of captive saddle-back tamarin females is reached as early as 16 months of age. Pregnancy may occur before the skeleton has finished growing but after the permanent teeth have erupted. The average age at first birth is 24 months, which follows a 5-month pregnancy, and they most commonly give birth to twins (Glassman 1983).

The closely related black-mantled tamarins (*S. nigricollis*) similarly have all of their deciduous teeth at 2 months and wean at age 3 months (Chase and Cooper 1969). In body mass, these young weanlings are only 40% of adult body mass. By 12 months, they have reached 85% of their adult weights and all of their adult linear proportions are achieved. Adult body mass is not reached until 16 months. Females, like their saddle-back relatives, have their first offspring at 24 months (Chase and Cooper 1969).

A "fast track" timing of dental–skeletal maturation compared to other anthropoids may reflect callitrichine-unique reproductive and social strategies: twinning as the norm, with two smaller infants rather than one, heavier infant, as typical in most anthropoids; only one reproductively active female per social group; siblings and adult males that help carry the infants. These species features may influence their accelerated weight gains compared to other anthropoids (Garber and Leigh 1997) and quicker dental–skeletal maturation.

Vervet Monkeys (C. aethiops). These African monkeys exhibit moderate sex difference in adult body mass and illustrate two aspects of growth that characterize many catarrhine primates: sex differences in pattern of development (skeletal fusion, body mass) and the order in which different body systems mature.

In contrast to tamarins, tooth eruption is later. Incisor, canine, and (pre)molar deciduous teeth erupt between 6 and 115 days, first permanent molars between 12 and 14 months, with all permanent teeth erupted by 67 months (Fairbanks *personal communication*, Bolter and Zihlman 2003). Females and males at the same dental age depart in timing of skeletal fusion; for example, female vervet monkeys fuse their pelvic bones (at the ischiopubic ramus and acetabulum) before males (Bolter and Zihlman 2003). Female vervets with erupted canines have completed pelvic growth, whereas males with erupted canines have not.

With all teeth erupted, the limb and trunk lengths and body mass of wild vervet females are within the adult female average for that population, although all elements of the postcrania (shoulder joint) have not fused. Males of comparable dental age have not completed their limb length and trunk growth and have achieved only 80% of the adult body mass (Bolter and Zihlman 2003).

A comparison of immature and adult vervet monkeys from the same population illustrates the mosaic order in which body systems mature (Bolter and Zihlman 2003). This sample with data available on body mass and soft tissue (testicle weights and specific muscles) expands information normally available on growth patterns. As other studies have shown, brain volume matures early; linear growth of trunk and long bones follow, then dental eruption, skeletal fusion, and body mass; muscularity as measured by masticatory muscles is late (see Fig. 24.5).

Body Mass

Body mass, commonly reported in kilograms, is seemingly a straightforward measure of growth. However, it can be unreliable as an assessment of stages of growth for many reasons, some of which are outlined here: (*1*) mass maturity is not synchronized with maturity in particular body systems, (*2*) increase in body mass during growth has a different pattern in males and females, (*3*) body mass fluctuates over a lifetime, seasonally and reproductively and with illness, (*4*) average adult mass varies within a species by population, and (*5*) few weights are available for infants and juvenile individuals for populations where adult weights are available. Each point is considered in brief below.

Nonsynchronicity. As illustrated with the black-mantled tamarins (*S. nigricollis*) and the vervet monkeys (*C. aethiops*), growth in body mass does not predict the level of anatomical

---

Table 24.2   Developmental Events of *Saguinus*[1]

AGE (MONTHS)	EVENT
At birth	12 deciduous teeth
2–3	All deciduous teeth
3	Skeletal fusion—elbow, pelvic elements—weaning complete
4–5	First permanent teeth, body mass 40% of adult
10–12	All permanent teeth (12 premolars/8 molars)
12	Body mass 85%, linear growth complete
15	Skeletal fusion—hip, knee, wrist joints
16	Sexual maturity of female, adult body mass
24	First birth, skeletal fusion—ankle, shoulder—social adults

Sources: Glassman (1983), Chase and Cooper (1969), Garber and Leigh (1997).

[1] Combines information from *Saguinus fuscicollis* and *S. nigricollis*.

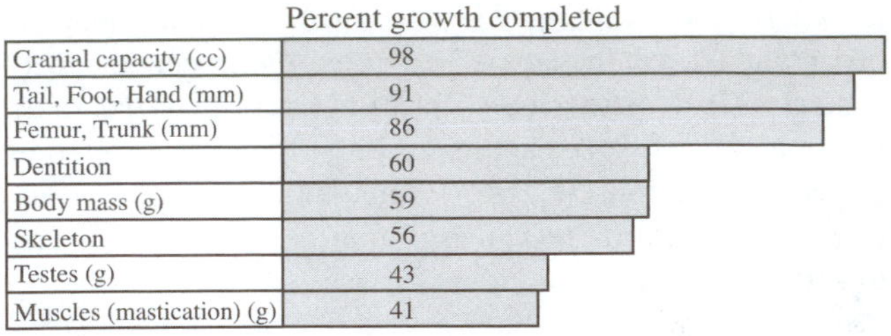

Percent growth completed		
Cranial capacity (cc)	98	
Tail, Foot, Hand (mm)	91	
Femur, Trunk (mm)	86	
Dentition	60	
Body mass (g)	59	
Skeleton	56	
Testes (g)	43	
Muscles (mastication) (g)	41	

Figure 24.5 Percentage of growth completed between immature and mature vervet monkeys (*Cercopithecus aethiops*), after Bolter and Zihlman (2003).

maturity of dental eruption, linear proportions, and reproductive systems. Furthermore, behavioral studies may rely on more easily observable external attributes like nipple size, mantle length, or coat coloration for classification of the individual into an age class. Few studies consider multiple physical and behavioral features when assessing survival strategies of immaturity, and body mass does not translate into the development of particular systems.

Bimaturity. The timing of maturation in body mass in sexually dimorphic catarrhines relates to the second problem. Male adult body mass is consistently one of the last physical markers of adult growth. Female primates may stop growing earlier and shift energetics into nutritional transfer (lactation), infant transport, pregnancy, and heat transport (J. Altmann 1980, 1986; Lee 1987). Larger-bodied males of the species continue growth in the skeleton, canine teeth, and body mass. For example, female olive baboons (*Papio anubis*) and gray-cheeked mangabeys (*Lophocebus albigena*) exhibit a period of declining body mass growth around menarche, while males spurt in growth between 4 and 6 years (Coelho 1985, Deputte 1992). In wild toque macaques (*Macaca sinica*), females finish muscle mass and weight growth at about 8 years of age and males, at age 12 years (Cheverud et al. 1992). In extremely dimorphic species, such as hamadryas baboons (*Papio hamadryas*), wild females cease body mass growth around 5.5 years while males add body weight for another 5 years, by which time mass has doubled (Sigg et al. 1982).

Species Variability. There is a high degree of variability within species, a point that is often neglected (Smith and Jungers 1997; cf. Vancata et al. 2000). Variation in body mass has also been documented within a subspecies and across local populations. For example, *P. troglodytes schweinfurtheii* at Gombe Reserve average less than the Mahale Mountain populations—e.g., adult females at Gombe average 31.4 vs. Mahale 35.2 kg (Pusey et al. 2005, Uehara and Nishida 1987, Smith and Jungers 1997). Such differences may reflect ecological quality or a broader genetic variation.

Seasonal and Lifetime Variability. Information available from field studies demonstrates not only that body mass changes during growth but also that accurate measurements are further complicated because the individual's mass may fluctuate with reproductive status, social status, seasonality,

and time of day (Bercovitch 1987, J. Altmann et al. 1993, Pusey et al. 2005). For example, a full stomach in a colobine leaf-eating monkey can weigh up to 1.4 kg (Washburn unpublished field notes 1937) or between 7% and 18% of the total body weight (Kay and Davies 1994).

Weights from the Wild. Body mass data collected in the wild are a more accurate reflection of age and sex and, therefore, are preferred (and specified in publications) over captive population weights. In most instances, these field weights are recorded during expeditions and often associated with individual specimens in museums (e.g., Allen and Coolidge 1940). These specimens are not always complete and may be represented only by cranial remains. As a result of historical practices in categorizing massey in specimens according to dental eruption status, wild body masses for adults have been underestimated (e.g., Napier and Napier 1967, Harvey and Clutton-Brock 1985, Jungers 1985, Brizzee and Dunlap 1986, Harvey et al. 1987, Smith and Jungers 1997).

For example, when calculating body mass estimates of adult *N. larvatus* from the Asiatic Primate Expedition collection of the Comparative Museum of Zoology at Harvard, Bolter (2004) noted that females who are dental adults but skeletally immature average only 9.1 kg, while dental plus skeletal adults (shoulder joint fused) average 10.4 kg; for males the difference is more extreme, 16.7 vs. 21.0 kg. Reliance on wild-collected, museum body weights brings into doubt studies that depend on these "adult" masses as the size baseline for scaling of other somatic and life history variables through growth (limb stoutness, gestation length, etc) (e.g., Jungers 1985).

A few field researchers have obtained accurate weights, although adults and not infants and juveniles were often the targets of the data collection. Animals may be captured and sedated to obtain weights and other anthropometrics (e.g., Turner et al. 1997), or feral individuals may be enticed to climb onto a scale to reach food or sit for a long enough period to weigh (e.g., J. Altmann et al. 1993, Johnson 2003, Pusey et al. 2005). One notable success is the collection of body weights on Gombe chimpanzees (*P. troglodytes*), measurements that span more than 30 years (Pusey et al. in press). Another notable study on wild toque macaques (*M. sinica*) has tracked body masses and other demographic and genetic variables over several generations of immatures and adults (Dittus 2004). Most often, field researchers do not have weights on their study populations and rarely on immature animals; captive animals provide the majority of known body mass data on immature animals (Brizzee and Dunlap 1986, Leigh 1994).

BEYOND BODY MASS: BODY COMPOSITION OF SOFT TISSUE

Increase in body mass during growth reflects an overall change but masks changes in the components, for example, muscle, bones, skin, brain, and fat. When the measure is only "body mass," the relative proportion of the components is smoothed into one all-encompassing number (cf., Grand 1977a, 1990). This number obscures the functional relationships between the systems at each stage. Differences in the relative proportions of these elements in age and sex classes reflect the functional aspects of survival: motor skills and locomotion, reproduction, perception and cognition.

Brain and Nervous System

An infant macaque (*M. mulatta*) has a relatively heavy head, 20% of body mass compared to the adult's 6% (Grand 1977b). Head mass consists primarily of the brain, which is 10 times heavier (proportional to body mass) than the adult's. A large brain at birth correlates with a young primate's early social interaction, though in physical skills it lags behind other mammals (Grand 1992). Well-developed brains during infancy function early as adaptive centers to process information from the visual, olfactory, auditory, and tactile senses (J. Altmann 1980, Dominy and Lucas 2001). The expanded forebrain, especially the somatosensory, motor, and visual cortices, controls the functions of locomotion, manipulation, visual processing, memory, communication, social interaction, and emotional responses—critical to infant survival into the next life stage. Infant primates divert extensive somatic energy and resources into brain growth, which reaches adult mass around the time they transition to adult foods, as measured by the eruption of their permanent teeth as juveniles (Smith 1989).

Muscular System

Muscularity and proportions of muscle groups shift as newborn primates acquire motor skills. At birth, an infant macaque (*M. mulatta*) consists of 25% muscle, compared to 43% in the adult (Grand 1977b). The hands and feet function for survival in the baby's ability to flex its digits (Hines 1942). This use is reflected in the infant's hand and foot proportions: infant hands are 2.6% of body mass compared to the adult, 1.2%; feet are 3.8% versus the adult's 2.4% (Grand 1977b). An infant is therefore well equipped to grip its mother's hair tightly as she carries, feeds, and protects it.

As the infant achieves locomotor independence, its thigh increases in muscularity and it decreases emphasis on the grasping function of the foot; the thigh provides much of the propulsion in adult quadrupedal locomotion (Grand 1983, Bolter and Zihlman 2002). Thigh mass, particularly muscularity, is the best single indicator of locomotor independence in primates (Grand 1990).

Linear dimensions also reflect proportional changes in the forearm, leg, hands, and feet—the ability of the dependent infant to cling. Schultz (1937) noted that during fetal development the relative lengths of the hands and feet of near-term rhesus macaques (*M. mulatta*) comprise a very large overall proportion of the distal limb segments (90% and 130%, respectively), but in postnatal life both indexes drop

markedly. Two important points are made with Schultz's observations: (*1*) linear skeletal proportions and the percentage of soft tissue mass per body mass give the same information, but (*2*) linear dimensions without consideration of surrounding muscle and locomotor function give an incomplete picture of how the proportions equip the infant to survive.

Body Fat

Body fat, its metabolism and storage, is a mammalian attribute that plays a critical role in reproduction (Pond 1977). In nonhuman primate females, levels of body fat affect ovulation, successful pregnancy, and ability to lactate (McFarland 1997). Body fat differences in two groups of a wild population of yellow baboons (*Papio cynocephalus*) were dramatic: 2% in wild-feeding females compared to 23% in garbage-feeding females, which traveled much less each day (J. Altmann et al. 1993). The greater body fat stores in the garbage eaters correlated with a first conception 1 year earlier and accelerated weaning of infants than leaner females. Little is known about when during development body fat is acquired, though it probably parallels the timing in human females, who add body fat at puberty under hormonal influence (Short 1976, Frisch 2002).

Reproductive Organs

Rate and timing of maturation of reproductive anatomy diverge significantly in females and males of some species. For example, ovaries in yearling vervet monkey females (*C. aethiops*) are mature in structure but are not functional. In contrast, testicular development in vervet monkeys is relatively complete in yearlings and relatively rapid during juvenility in baboons (*Papio anubis* and *hamadryas*), chimpanzees (*P. troglodytes*), and humans (J. Altmann et al. 1981, Bolter and Zihlman 2003, Pusey 1990, Short 1976). The baboon testicle growth spurt during juvenility appears to vary with different forms of these species' social organizations (Jolly and Phillips-Conroy 2003). In colobine leaf monkeys (*T. cristatus*, *P. rubicunda*, and *N. larvatus*), testicles descend into the scrotum before dental, skeletal, and body mass maturity (Bolter 2004). Male cercopithecine, chimpanzee, and human testicles are sexually mature (descend and produce sperm) before reaching adult body size and before males develop the social skills and strategies to mate successfully (Short 1976, Matsubayashi and Mochizuki 1982, Pusey 1990, Dixson 1998).

Case Studies: Body Composition in Tamarins, Gorillas, and Chimpanzees

Golden Lion Tamarins

Golden lion tamarins (*Leontopithecus rosalia*) of South America are small-bodied New World monkeys that produce twins as the norm. Gestation takes 140 days, and newborns weigh about 55 g, or 10% of adult body mass (Grand 1992). Brain mass at birth is about 50% of the adult size, but by 2 months of age the brain is fully grown and indicates the mature level of development of the sensory system. In contrast, muscle development is about half of the adult value at birth and continues to increase for 12 months to reach maturity. The mass and growth pattern indicates relatively delayed physical strength compared to their social responsiveness, which translates to more socially but less physically interactive juveniles (Grand 1992).

Distribution of musculature also differs between infant and adult tamarins. Newborns have heavier forelimb muscles and heavier flexors in the fingers and toes but less back extensor muscle mass—all related to an infant's primary anatomical function: clinging to its mother for transport. As adults, hindlimb and back extensors disproportionately increase, reflecting the shift to quadrupedalism (Grand 1992).

Gorillas and Chimpanzees

Gorilla (*Gorilla gorilla*) and chimpanzee (*P. troglodytes*) infants are born at similar weights, but by 3 years of age captive gorillas are over three times the body mass of chimpanzees (Bolter and Zihlman 2002). Dental development in the gorilla is also on an accelerated schedule compared to chimpanzees (Dean and Reid 2001). In a study of the skin, muscle, bone, and fat among infant, juvenile, adult chimpanzees and gorillas, limb masses increased disproportionately in gorillas compared to chimpanzees. The infant gorilla, infant chimpanzee, and juvenile chimpanzee are about 30% muscle to total body mass. The juvenile gorilla muscle mass is more like an adult chimpanzee ratio at 38% of total body mass. In particular, the juvenile gorilla thigh musculature reaches adult proportions earlier than that of the chimpanzee.

Wild gorillas achieve locomotor independence between 2.5 and 3 years. In wild chimpanzees, independence is achieved much later, around 6 years (Doran 1997). The juvenile gorilla's accelerated maturation of body mass and "adult" composition in thigh muscularity provide a possible functional explanation for the gorilla's accelerated acquisition of quadrupedal knuckle-walking compared to chimpanzee's.

Ecological–Social Aspects: Growth Rates

Captive Versus Wild

Captive primates grow faster than wild primates. Dental development accelerates in captive, food-provisioned *Papio cynocephalus* baboons compared to wild baboon populations (J. Altmann et al. 1981, Phillips-Conroy and Jolly 1988). Vervet females (*C. aethiops*) reach sexual maturation consistently earlier than their free-ranging counterparts (Bramblett 1980, Cheney et al. 1988). Macaques (*M. sinica*) in Sri Lanka prolong growth by several years compared to laboratory populations (Cheverud et al. 1992).

This trend for developmental acceleration in captive chimpanzees has been recently documented (Zihlman et al.

2004a,b). Dental maturity is accelerated up to 3+ years earlier compared to wild populations, which parallels the fact that captive females give birth up to 3 years earlier than their wild relatives (Boesch and Achermann-Boesch 2000). Skeletal elements fuse up to 2 years earlier in captive compared to wild individuals (Watts 1993, Zihlman et al. 2004a). These captive accelerations are attributed to better nutrition, lower levels of predation, daily activity and travel, as well as less social uncertainty and stress; but the actual regulating mechanisms remain unknown. This phenomenon provides another caution about modeling phylogeny on the growth patterns of provisioned animals.

Seasonal Variation

Ring-tailed lemurs (*L. catta*), diurnal prosimians, reach behavioral adulthood at age 3, a younger age than most monkeys and apes (Gould et al. 2003). Three-year-old females begin reproducing seasonally. The schedules for conception, birth, and lactation correspond with optimal food resources (Sauther et al. 1999). The 18-week gestation occurs during the dry season, when fruits (their main staple food) are scarce and, consequently, females have little opportunity to accumulate fat (Sauther et al. 1999). As both mothers and infants rely directly on the resources in the environment and not on stored fat reserves, infant mortality is tied to the seasonal ecology (Sauther et al. 1999, Gould et al. 2003). An infant nurses during the wet season, when increased rainfall promotes an abundant food supply for the mother. Mortality of infants can rise 50%–80% when food is scarce, for example, following a drought year; this is higher than the mortality encountered by anthropoids (Wright 1999, Gould et al. 2003).

Infant ring-tailed lemurs grow quickly and are weaned at 4 months during a second peak of food productivity, when fruit and young leaves are available to the weanling. This behavioral change coincides with the eruption of the first permanent molars (Smith et al. 1994). The permanent teeth complete eruption at about 15 months, when they are behavioral juveniles and their mothers are giving birth to the next offspring (Gould et al. 2003). The "subadult" phase is between 2 and 3 years, when reproductive systems mature and individuals learn adult social skills. Short life stages make the ring-tailed lemurs more vulnerable to rainfall and temperature, yet allow them to begin reproductive life earlier than most of their anthropoid cousins.

Same Ecology, Variation in Strategy

The length of the juvenile stage in primates dramatically affects adult life and survival. Comparison of vervet monkeys (*C. aethiops*) and yellow baboons (*Papio cynocephalus*) illustrates the importance of this stage in learning and practicing skills. These closely related monkeys inhabit the same woodland areas of eastern Africa; they are also similar in dental and digestive anatomy; they differ in body size and in length of the juvenile stage. Vervet monkeys mature about 2 years earlier and, thus, reach adulthood before the larger baboons.

During the juvenile stage, both species learn to feed independently and to choose their foods. Stuart Altmann (1998) documents the adeptness of young baboons at finding and choosing quality foods; they exploit more than 250 species over their lifetimes, far more than vervets. Local vervet monkey populations have become extinct during periods of food scarcity, whereas baboons have survived (Lee and Hauser 1988). Narrow food choices and reliance on seasonally abundant foods limit vervet foraging, especially in times of reduced availability. Baboons, in contrast, are able to exploit less lush and less available plant foods. With a shorter juvenile stage, vervets essentially "give up" time during immaturity for learning, finish growth earlier, and begin reproducing. Attenuated immature stages with more opportunities for acquiring skills and life strategies make young animals more resistant to temperature, drought, and predation, therefore leading to higher survival rates.

Juvenile primates not only acquire ecological knowledge but also learn social skills that contribute to survival and, later in life, to reproductive outcome and form the basis for species-typical social organization. Although the acquisition of social skills may be more difficult to measure than the numbers of utilized plants, the time to learn and practice social behavior and communication is as critical to survival as mastering a complex ecology (Rowell 1988a,b).

SUMMARY AND FUTURE STUDIES

Growth and development began as a descriptive area within physical anthropology. Its emphasis on bones and teeth traces back to Schultz and to his training as a human osteologist. Although Washburn's bone growth and laboratory work demonstrated functional and dynamic aspects of immature animals, research on immature primates and the dynamics of growth processes has tended to remain atheoretical, nonfunctional, and divorced from field research.

New data, approaches, and syntheses offer possibilities for bridging informational and conceptual gaps based on laboratory, museum, captive, and field studies: dental and brain development (e.g., Smith 1989), dental development through incremental lines (e.g., Schwartz et al. 2006), field research tracking individuals through immaturity (Pusey 1978, Pusey et al. 2005), laboratory work on soft tissue (e.g., Grand 1977a, 1990; Bolter and Zihlman 2002, 2003), and long-term demographic field studies (e.g., Cheverud et al. 1992, Boesch and Boesch-Achermann 2000, Nishida 2003, Dittus 2004). Individual case studies that integrate multiple dimensions begin to answer some of our questions about the timing of somatic growth events relative to the social development of a species or genus (e.g., Zihlman et al. 1990, Fragaszy et al. 2004) or about the life stages in fossil species (e.g., Kelley and Smith 2003, Dean and Leakey 2004). These represent only the beginning of a more holistic analysis of primate growth and social as well as physical development. We are entering a new era in which Simpson's vision of primates, both extant and extinct, in the four dimensions of time and space is closing in on reality.

REFERENCES

Allen, G. M., and Coolidge, H. J. (1940). Mammal and bird collections of the Asiatic primate expedition: mammals. *Bull. Mus. Comp. Zool. Harv. Univ.* 87:131–166.

Altmann, J. (1980). *Baboon Mothers and Infants*. Harvard University Press, Cambridge, MA.

Altmann, J. (1986). Adolescent pregnancies in non-human primates: an ecological and developmental perspective. In: Lancaster, J., and Hamburg, B. (eds.), *School-Age Pregnancy and Parenthood: Biosocial Dimensions*. Aldine Press, New York. pp. 247–262.

Altmann, J., Altmann, S., and Hausfater, G. (1981). Physical maturation and age estimates of yellow baboons, *Papio cynocephalus*, in Amboseli National Park, Kenya. *Am. J. Primatol.* 1:389–399.

Altmann, J., Altmann, S., Hausfater, G., and McCuskey, S. (1977). Life history of yellow baboons: physical development, reproductive parameters, and infant mortality. *Primates* 18:315–330.

Altmann, J., Schoeller, D., Altmann, S., Muruthi, P., and Sapolsky, R. (1993). Body size and fatness of free-living baboons reflect food availability and activity levels. *Am. J. Primatol.* 30:149–161.

Altmann, S. (1998). *Foraging for Survival: Yearling Baboons in Africa*. University of Chicago Press, Chicago.

Altmann, S. A. (1991). Diets of yearling female primates (*Papio cynocephalus*) predict lifetime fitness. *Proc. Natl. Acad. Sci. USA* 88:420–423.

Bercovitch, F. B. (1987). Female weight and reproductive condition in a population of olive baboons (*Papio anubis*). *Am. J. Primatol.* 12:189–195.

Blum, D. (2002). *Love at Goon Park: Harry Harlow and the Science of Affection*. Perseus, Cambridge.

Boesch, C., and Boesch-Achermann, H. (2000). *The Chimpanzees of the Taï Forest: Behavioural Ecology and Evolution*. Oxford University Press, Oxford.

Bogin, B. (1999). *Patterns of Human Growth*. Cambridge University Press, Cambridge.

Bogin, B., and Smith, B. H. (1996). Evolution of the human life cycle. *Am. J. Phys. Anthropol.* 8:703–716.

Bolter, D. R. (2003). Anatomical growth and development in hamadryas baboons (*Papio hamadryas*) compared with the closely related vervet monkeys (*Cercopithecus aethiops*). *Am. J. Phys. Anthropol.* 36 (suppl.):70.

Bolter, D. R. (2004). Anatomical growth patterns in colobine monkeys and implications for primate evolution [PhD diss.]. University of California, Santa Cruz.

Bolter, D. R., and Zihlman, A. L. (2002). Growth and development in body tissues and proportions in African apes (*Gorilla gorilla* and *Pan troglodytes*): a preliminary report. *Am. J. Phys. Anthropol.* 34 (suppl.):46.

Bolter, D. R., and Zihlman, A. L. (2003). Morphometric analysis of growth and development in wild-collected vervet monkeys (*Cercopithecus aethiops*) with implications for growth patterns in Old World monkeys, apes and humans. *J. Zool. Lond.* 260:99–110.

Bowler, P. (1988). *The Non-Darwinian Revolution: Reinterpreting a Historical Myth*. Johns Hopkins University Press, Baltimore.

Bramblett, C. A. (1980). Model for development of social behavior in vervet monkeys. *Dev. Psychol*. 13:205–223.

Brizzee, K., and Dunlap, W. (1986). Growth. In: Dukelow, W., and Erwin, J. (eds.), *Comparative Primate Biology: Reproduction and Development*. Alan R Liss, New York. pp. 363–413.

Bromage, T., and Dean, M. C. (1985). Re-evaluation of the age at death of immature fossil hominids. *Nature* 317:525–527.

Carpenter, C. R. (1940). A field study in Siam of the behavior and social relations of the gibbon (*Hylobates lar*). *Comp. Psychol. Monogr*. 16:1–212.

Chase, J. E., and Cooper, R. W. (1969). *Saguinus nigricollis*—physical growth and dental eruption in a small population of captive-born individuals. *Am. J. Phys. Anthropol*. 30:111–116.

Cheney, D., Seyfarth, R., Andelman, S., and Lee, P. (1988). Reproductive success in vervet monkeys. In: Clutton-Brock, T. (ed.), *Reproductive Success: Studies of Individual Variation in Contrasting Breeding Systems*. Chicago University Press, Chicago. pp. 384–402.

Cheverud, J. M. (1981). Epiphyseal union and dental eruption in *Macaca mulatta*. *Am. J. Phys. Anthropol*. 56:157–167.

Cheverud, J. M., Wilson, P., and Dittus, W. (1992). Primate population studies at Polonnaruwa. III. Somatometric growth in a natural population of toque macaques (*Macaca sinica*). *J. Hum. Evol*. 23:51–77.

Coelho, A. M. (1985). Baboon dimorphism: growth in weight, length and adiposity from birth to 8 years of age. In: Watts, E. S. (ed.), *Nonhuman Primate Models for Human Growth and Development*. Alan R. Liss, New York. pp. 125–159.

Crook, J. H., and Gartlan, J. S. (1966). Evolution of primate societies. *Nature* 210:1200–1203.

Darwin, C. (1859). *On the Origin of Species by Means of Natural Selection*. John Murray, London. Reprinted 1964, Harvard University Press, Cambridge, MA.

Darwin, C. (1860). Letter written to Asa Gray from Down on September 10.

Darwin, C. (1876). *The Autobiography of Charles Darwin 1809–1882*. Reprinted 1969, W. W. Norton and Company, New York.

Dean, C., Stringer, C., and Bromage, T. (1986). Age at death of the Neanderthal child from Devil's Tower, Gibraltar and the implications for studies of general growth and development in Neanderthals. *Am. J. Phys. Anthropol*. 70:301–309.

Dean, M. C., and Leakey, M. G. (2004). Enamel and dentine development and the life history profile of *Victoriapithecus macinnesim* from Maboko Island, Kenya. *Ann. Anat*. 186:405–412.

Dean, M. C., Leakey, M. G., Reid, D., Schrenk, F., Schwartz, G. T., Stringer, C., and Walker, A. (2001). Growth processes in teeth distinguish modern humans from *Homo erectus* and earlier hominins. *Nature* 414:627–631.

Dean, M. C., and Reid, D. J. (2001). Perikymata spacing and distribution on hominid anterior teeth. *Am. J. Phys. Anthropol*. 116:209–215.

Deputte, B. L. (1992). Life history of captive gray-cheeked mangabeys: physical and sexual development. *Int. J. Primatol*. 13:509–531.

deRousseau, C. J. (ed.) (1990). *Primate Life History and Evolution*. Wiley-Liss, New York.

Detwiler, S. R., and Washburn, S. L. (1943). An experiment bearing on the problems of physical anthropology. *Am. J. Phys. Anthropol*. 1:171–190.

Dirks, W. A. (2003). Effect of diet on dental development in four species of catarrhine primates. *Am. J. Primatol*. 61:29–40.

Dittus, W. (2004). Demography: a window to social evolution. In: Thierry, B., Singh, M., and Kaumanns, W. (eds.), *Macaque Societies: A Model for the Study of Social Organization*. Cambridge University Press, New York. pp. 87–112.

Dixson, A. (1998). *Primate Sexuality*. Oxford University Press, Oxford.

Dolhinow, P. (1972). *Primate Patterns*. Holt, Rinehart and Winston, New York.

Dolhinow, P., and Murphy, G. (1982). Langur monkeys (*Presbytis entellus*) development: the first 3 months of life. *Folia Primatol*. 39:305–331.

Dominy, N., and Lucas, P. W. (2001). Ecological importance of trichromatic vision to primates. *Nature* 410:363–366.

Doran, D. (1997). Ontogeny of locomotion in mountain gorillas and chimpanzees. *J. Hum. Evol*. 32:323–344.

Fagan, R. (1993). Primate juveniles and primate play. In: Pereira, M., and Fairbanks, L. (eds.), *Juvenile Primates*. Oxford University Press, New York. pp. 182–196.

Fairbanks, L. (1993). Juvenile vervet monkeys: establishing relationships and practicing skills for the future. In: Pereira, M., and Fairbanks, L., (eds.), *Juvenile Primates*. Oxford University Press, Oxford. pp. 211–227.

Fairbanks, L. (2000). The developmental timing of primate play: a neural selection model. In: Parker, S. T., Langer, J., and McKinney, M. L. (eds.), *Biology, Brains and Behavior: The Evolution of Human Development*. SAR Press, Santa Fe. pp. 131–158.

Fedigan, L. (1992). *Primate Paradigms: Sex Roles and Social Bonds*. Chicago University Press, Chicago.

Fedigan, L. (1997). Changing views of female life histories. In: Morbeck, M. E., Galloway, A., and Zihlman, A. L. (eds.), *The Evolving Female: A Life History Perspective*. Princeton University Press, Princeton, NJ. pp. 15–26.

Fedigan, L. M., and Fedigan, L. (1989). Gender and the study of primates. In: Morgan, S. (ed.), *Gender and Anthropology. Critical Reviews for Teaching and Research*. American Anthropological Association, Washington DC. pp. 41–64.

Fragaszy, D. M., Visalberghi, E., and Fedigan, L. M. (2004). *The Complete Capuchin. The Biology of the Genus* Cebus. Cambridge University Press, Cambridge.

Frisch, R. E. (2002). *Female Fertility and the Body Fat Connection*. Chicago University Press, Chicago.

Garber, P., and Leigh, S. R. (1997). Ontogenetic variation in small-bodied New World primates: implications for patterns of reproduction and infant care. *Folia Primatol*. 68:1–22.

Glassman, D. M. (1983). Growth and development in the saddleback tamarin: the sequence and timing of dental eruption and epiphyseal union. *Am. J. Phys. Anthropol*. 5:51–60.

Godfrey, L. R., Samonds, K. E., Junder, W. L., and Sutherland, M. R. (2001). Teeth, brains, and primate life histories. *Am. J. Phys. Anthropol*. 114:192–214.

Goodall, J. (1971). *In the Shadow of Man*. Delta Books, New York.

Goodall, J. (1986). *The Chimpanzees of Gombe: Patterns of Behavior*. Harvard University Press, Cambridge, MA.

Goodall, J. (1990). *Through a Window: My Thirty Years with the Chimpanzees of Gombe.* Houghton Mifflin, Boston.

Gould, L., Sussman, R. W., and Sauther, M. (2003). Demographic and life-history patterns in a population of ring-tailed lemurs (*Lemur catta*) at Beza Mahafaly reserve, Madagascar: a 15-year perspective. *Am. J. Phys. Anthropol.* 120:182–194.

Grand, T. (1977a). Body weight: its relation to tissue composition, segment distribution, and motor function. I. Interspecific comparisons. *Am. J. Phys. Anthropol.* 47:211–239.

Grand, T. I. (1977b). Body weight: its relation to tissue composition, segment distribution, and motor function II. Development of *Macaca mulatta. Am. J. Phys. Anthropol.* 47:241–247.

Grand, T. I. (1983). Body weight: its relationship to tissue composition, segmental distribution of mass, and motor function III. The Didelphidae of French Guiana. *Aust. J. Zool.* 31:299–312.

Grand, T. I. (1990). The functional anatomy of body mass. In: Damuth, J., and MacFadden, B. J. (eds.), *Body Size in Mammalian Paleobiology: Estimation and Biological Implications.* Cambridge University Press, Cambridge. pp. 39–47.

Grand, T. I. (1992). Altricial and precocial mammals: a model of neural and muscular development. *Zoo Biol.* 11:3–15.

Harvey, P., and Clutton-Brock, T. (1985). Life history variation in primates. *Evolution* 39:559–581.

Harvey, P., Martin, R., and Clutton-Brock, T. (1987). Life histories in comparative perspective. In: Smuts, B., Cheney, D., Seyfarth, R., Wrangham, R., and Struhsaker, T. T. (eds.), *Primate Societies.* Chicago University Press, Chicago. pp. 181–196.

Hines, M. (1942). The development and regression of reflexes, postures and progression in the young macaque. *Contrib. Embryol.* 30:153–209.

Itoigawa, N., Tanaka, T., Ukai, N., Fujii, H., Kurokawa, T., Koyama, T., Ando, A., Watanabe, Y., and Imakawa, S. (1992). Demography and reproductive parameters of a free-ranging group of Japanese macaques (*Macaca fuscata*) at Katsuyama. *Primates* 33:49–68.

Iwamoto, M., Watanabe, T., and Hamada, Y. (1987). Eruption of permanent teeth in Japanese monkeys (*Macaca fuscata*). *Primate Res.* 3:18–28.

Jay, P. C. (1962). Aspects of maternal behavior among langurs. *Ann. N. Y. Acad. Sci.* 102:468–476.

Jay, P. C. (1968). Primate field studies and human evolution. In: Jay, P. C. (ed.), *Primates: Studies in Adaptation and Variability.* Holt, Rinehart and Winston, New York. pp. 487–519.

Johnson, S. (2003). Life history and the competitive environment: trajectories of growth, maturation, and reproductive output among chacma baboons. *Am. J. Phys. Anthropol.* 120:83–98.

Jolly, C. J., and Phillips-Conroy, J. (2003). Testicular size, mating system, and maturation schedules in wild anubis and hamadryas baboons. *Int. J. Primatol.* 24:125–142.

Jungers, W. L. (1985). Body size and scaling in limb proportions in primates. In: Jungers, W. L. (ed.), *Size and Scaling in Primate Biology.* Plenum Press, New York. pp. 345–381.

Kay, R. N. B., and Davies, A. G. (1994). Digestive physiology. In: Davies, A. G. and Oates, J. F. (eds.), *Colobine monkeys.* Cambridge University Press, Cambridge. pp. 229–250.

Kelley, J., and Smith, T. (2003). Age at first molar emergence in early Miocene *Afropithecus turkanensis* and life-history evolution in the Hominoidea. *J. Hum. Evol.* 44:307–329.

Kimura, T., and Hamada, Y. (1990). Development of epiphyseal union in Japanese macaques of known chronological age. *Primates* 31:79–93.

King, S. J. (2004). Relative timing of ontogenetic events in primates. *J. Zool. Lond.* 264:267–280.

Krogman, W. M. (1969). Growth changes in skull, face, jaws, and teeth of the chimpanzee. In: Bourne, G. H. (ed.), *The chimpanzee.* University Park Press, Baltimore. pp. 104–164.

Kummer, H. (1968). *Social Organization of Hamadryas Baboons.* University of Chicago Press, Chicago.

Kummer, H. (1995). *In Quest of the Sacred Baboon.* Princeton University Press, Princeton, NJ.

Laird, A. K. (1967). Evolution of the human growth curve. *Growth* 31:345–355.

Lancaster, J. (1971). Play-mothering: the relations between juvenile females and young infants among free-ranging vervet monkeys (*Cercopithecus aethiops*). *Folia Primatol.* 15:161–182.

Lee, P. (1984). Early infant development and maternal care in free-ranging vervet monkeys. *Primates* 25:36–47.

Lee, P. C. (1987). Nutrition, fertility and maternal investment in primates. *J. Zool. Lond.* 213:409–422.

Lee, P. C., and Hauser, M. D. (1988). Long-term consequences of changes in territory quality on feeding and reproductive strategies of vervet monkeys. *J. Anim. Ecol.* 67:347–358.

Lee, P. C., Majluf, P., and Gordon, I. (1991). Growth, weaning and maternal investment from a comparative perspective. *J. Zool. Lond.* 225:99–114.

Leigh, S. R. (1994). Ontogenetic correlates of diet in anthropoid primates. *Am. J. Phys. Anthropol.* 94:499–522.

Leigh, S. R., and Park, P. B. (1998). Evolution of human growth prolongation. *Am. J. Phys. Anthropol.* 107:331–350.

Liversidge, H. (2003). Variation in modern human dental development. In: Thompson, J. L., Krovitz, G. E., and Nelson, A. J. (eds.), *Patterns of Growth and Development in the Genus Homo.* Cambridge University Press, Cambridge. pp. 73–113.

Long, J. O., and Cooper, R. W. (1968). Physical growth and dental eruption in captive-bred squirrel monkeys, *Saimiri sciureus* (letica, colombia). In: Rosenblum, L. A., and Cooper, R. W. (eds.), *The Squirrel Monkey.* Academic Press, New York. pp. 193–205.

Mason, W. (1979). Ontogeny of social behavior. In: Marler, P., and Vandenbergh, J. G. (eds.), *Handbook of Behavioral Neurobiology.* Plemium Press, New York. pp. 1–28.

Matsubayashi, K., and Mochizuki, K. (1982). Growth of male reproductive organs with observation of their seasonal morphological changes in Japanese monkeys (*Macaca fuscata*). *Jpn. J. Vet.* 44:891–902.

Mayr, E. (1980). Prologue: some thoughts on the history of the evolutionary synthesis. In: Mayr, E., and Provine, W. B. (eds.), *The Evolutionary Synthesis. Perspectives on the Unification of Biology.* Harvard University Press, Cambridge, MA. pp. 1–48.

Mayr, E. (1997). *This Is Biology: The Science of the Living World.* Harvard University Press, Cambridge, MA.

Mayr, E. (2004). *What Makes Biology Unique? Considerations on the Autonomy of a Scientific Discipline.* Cambridge University Press, Cambridge.

McFarland, R. (1997). Primate females: fit or fat. In: Morbeck, M. E., Galloway, A., and Zihlman, A. L. (eds.), *The Evolving Female: A Life-History Perspective.* Princeton University Press, Princeton, NJ. pp. 163–175.

Morbeck, M. E., Galloway, A., and Zihlman, A. L. (eds.) (1997). *The Evolving Female: A Life-History Perspective.* Princeton University Press, Princeton, NJ.

Mori, A. (1979). Analysis of population changes by measurement of body weight in the Koshima troop of Japanese monkeys. *Primates* 20:371–397.

Napier, J. R., and Napier, P. H. (1967). *A Handbook of Living Primates*. Academic Press, London.

Nishida, T. (1990). *The Chimpanzees of the Mahale Mountains: Sexual and Life History Strategies*. University of Tokyo Press, Tokyo.

Nishida, T., Corp, N., Hamai, M., Hasegawa, T., Hiraiwa-Hasegawa, M., Hosaka, K., Hunt, K., Itoh, N., Kawanaka, K., Matsumoto-Oda, A., Mitani, J., Nakamura, M., Norikoshi, K., Sakamaki, T., Turner, L., Uehara, S., and Zamma, K. (2003). Demography, female life history, and reproductive profiles among the chimpanzees of Mahale. *Am. J. Primatol.* 59:99–121.

Nishida, T., Takasaki, H., and Takahata, Y. (1990). Demography and reproductive profiles. In: Nishida, T. (ed.), *The Chimpanzees of the Mahale Mountains*. Tokyo University Press, Tokyo, pp. 63–97.

Pereira, M., and Fairbanks, L. (1993a). *Juvenile Primates*. Oxford University Press, Oxford.

Pereira, M., and Fairbanks, L. (1993b). What are juvenile primates all about? In: Pereira, M., and Fairbanks, L. (eds.), *Juvenile Primates*. Oxford University Press, New York. pp. 3–12.

Pereira, M. E., and Altmann, J. (1985). Development of social behavior of free-living nonhuman primates. In: Watts, E. S. (ed.), *Nonhuman Primate Models for Human Growth and Development*. Alan R Liss, New York. pp. 217–309.

Phillips-Conroy, J., and Jolly, C. J. (1988). Dental eruption schedules of wild and captive baboons. *Am. J. Primatol.* 15:17–29.

Plooij, F. (1984). *The Behavioral Development of Free-Living Chimpanzee Babies and Infants*. Ablex Publishing, Norwood.

Ponce de Leon, M. S., and Zollikofer, C. P. E. (2001). Neanderthal cranial ontogeny and its implications for late hominid diversity. *Nature* 412:534–538.

Pond, C. M. (1977). The significance of lactation in the evolution of mammals. *Evolution* 31:177–199.

Pournelle, G. H. (1967). Observations on reproductive behaviour and early postnatal development of the proboscis monkey (*Nasalis larvatus orientalis*). *Int. Zoo Ybk.* 7:90–92.

Pusey, A. (1978). The physical and social development of wild adolescent chimpanzees (*Pan trolgodytes schweinfurthii*) [PhD thesis]. Stanford University, Palo Alto, CA.

Pusey, A. (1983). Mother–offspring relationships in chimpanzees after weaning. *Anim. Behav.* 31:363–377.

Pusey, A. (1990). Behavioural changes at adolescence in chimpanzees. *Behaviour* 115:203–246.

Pusey, A., Oehlert, G. W., Williams, J. M., and Goodall, J. (2005). Influence of ecological and social factors on body mass of wild chimpanzees. *Int. J. Primatol.* 26:3–31.

Ramirez Rozzi, F. V., and De Castro, J. M. B. (2004). Surprisingly rapid growth in Neanderthals. *Nature* 428:936–939.

Randall, F. E. (1943). The skeletal and dental development and variability of the gorilla. *Hum. Biol.* 15:236–254.

Roberts, M. (1994). Growth, development and parental care in the western tarsier (*Tarsius bancanus*) in captivity: evidence for a "slow" life history and nonmonogamous mating system. *Int. J. Primatol.* 15:1–28.

Rowell, T. (1988a). The social system of guenons, compared with baboons, macaques and mangabeys. In: Gautier-Hion, A., Bourliere, F., and Gautier, J. (eds.), *A Primate Radiation: Evolutionary Biology of the African Guenons*. Cambridge University Press, Cambridge. pp. 439–451.

Rowell, T. (1988b). Beyond the one-male group. *Behaviour* 104:189–201.

Rüedi, D. (1981). Hand-rearing and reintegration of a caesarian-born proboscis monkey *Nasalis larvatus* at Basle Zoo. *Int. Zoo Ybk.* 21:225–229.

Sauther, M., Sussman, R. W., and Gould, L. (1999). The socio-ecology of the ringtailed lemur: thirty-five years of research. *Evol. Anthropol.* 8:120–132.

Scammon, R. E. (1930). The measurement of the body in childhood. In: Harris, J. A., Jackson, C. M., Paterson, D. G., and Scammon, R. E. (eds.), *The Measurement of Man*. University of Minnesota Press, Minneapolis. pp. 174–215.

Schultz, A. H. (1924). Observations on colobus fetuses. *Bull. Am. Mus. Nat. Hist.* XLIX:443–457.

Schultz, A. H. (1927). Studies on the growth of gorilla and of other higher primates with special reference to a fetus of gorilla, preserved in the Carnegie Museum. *Mem. Carnegie Mus.* 11:1–87.

Schultz, A. H. (1930). The skeleton of the trunk and limbs of higher primates. *Hum. Biol.* II:303–438.

Schultz, A. H. (1937). Growth and development. In: Hartman, C. G., and Straus, W. L. (eds.), *The Anatomy of the Rhesus Macaque*. Hafner Publishing, New York. pp. 10–27.

Schultz, A. H. (1940). Growth and development of the chimpanzee. *Contrib. Embryol.* 170:3–63.

Schultz, A. H. (1941). Growth and development of the orang-utan. *Contrib. Embryol.* 182:57–110.

Schultz, A. H. (1942). Growth and development of the proboscis monkey. *Bull. Mus. Comp. Zool.* 89:279–323.

Schultz, A. H. (1963). Age changes, sex differences, and variability as factors in the classification of primates. *Viking Fund publ. Anthropol.* 37:85–115.

Schwartz, G. T., Reid, D. J., Dean, M. C., and Zihlman, A. L. (2006). A faithful record of stress life events in the dental developmental record of a juvenile gorilla. *Int. J. Primatol.* 27(3).

Shea, B. T. (2000). Current issues in the investigation of evolution by heterochrony, with emphasis on the debate over human neoteny. In: Taylor Parker, S., Langer, J., and McKinney, M. L., (eds.), *Biology, Brains and Behavior: The Evolution of Human Development*. School of American Research Press, Santa Fe. pp. 181–213.

Short, R. V. (1976). The evolution of human reproduction. *Proc. R. Soc. Lond.* 195:3–24.

Sigg, H., Stolba, A., Abegglen, J.-J., and Dasser, V. (1982). Life history of hamadryas baboons: physical development, infant mortality, reproductive parameters and family relationships. *Primates* 23:473–487.

Simpson, G. G. (1944). Paleontology. The role of the individual in evolution. *J. Wash. Acad. Sci.* 31:1–20.

Smith, B. H. (1989). Dental development as a measure of life history in primates. *Evolution* 43:683–688.

Smith, B. H. (1991). Dental development and the evolution of life history in Hominidae. *Am. J. Phys. Anthropol.* 86:157–174.

Smith, B. H. (1994). Patterns of dental development in *Homo*, *Australopithecus*, *Pan* and *Gorilla*. *Am. J. Phys. Anthropol.* 94:307–325.

Smith, B. H., Crummett, T. L., and Brandt, K. L. (1994). Age of eruption of primate teeth: a compendium for aging individuals and comparing life histories. *Ybk. Phys. Anthropol.* 37:177–231.

Smith, R. J., and Jungers, W. L. (1997). Body mass in comparative primatology. *J. Hum. Evol.* 32:523–559.

Sugiyama, Y. (1994). Age-specific birth rate and lifetime reproductive success of chimpanzees at Bossou, Guinea. *Am. J. Primatol.* 32:311–318.

Tappen, N. C., and Severson, A. (1971). Sequence of eruption of permanent teeth and epiphyseal union in New World monkeys. *Folia Primatol.* 15:293–312.

Turner, T. R., Anapol, F., and Jolly, C. J. (1997). Growth, development, and sexual dimorphism in vervet monkeys (*Cercopithecus aethiops*) at four sites in Kenya. *Am. J. Phys. Anthropol.* 103:19–35.

Uehara, S., and Nishida, T. (1987). Body weights of wild chimpanzees (*Pan troglodytes schweinfurthii*) of the Mahale Mountains National Park, Tanzania. *Am. J. Phys. Anthropol.* 72:315–321.

Vancata, V., Vancatova, M., Chalyan, V., and Meishvili, N. (2000). Longitudinal study of growth and body mass changes in ontogeny in captive rhesus macaques (*Macaca mulatta*) from the Institute of Medical Primatology. *Sochi. Var. Evol.* 8:51–81.

van de Rijt-Plooij, H. H. C., and Plooij, F. X. (1987). Growing independence, conflict and learning in mother–infant relations in free-ranging chimpanzees. *Behaviour* 101:1–86.

van Wagenen, G., and Asling, C. W. (1958). Roentgenographic estimation of bone age in the rhesus monkey (*Macaca mulatta*). *Am. J. Anat.* 103:163–185.

van Wagenen, G., and Catchpole, H. R. (1956). Physical growth of the rhesus monkey (*Macaca mulatta*). *Am. J. Phys. Anthropol.* 14:245–273.

Walters, J. R. (1987). Transition to adulthood. In: Smuts, B., Cheney, D., Seyfarth, R., Wrangham, R., and Struhsaker, T. T. (eds.), *Primate Societies*. University of Chicago Press, Chicago. pp. 358–369.

Washburn, S. L. (1942). Skeletal proportions of adult langurs and macaques [abridged version of doctoral diss., Harvard University]. *Hum. Biol.* 14:444–472.

Washburn, S. L. (1943). The sequence of epiphyseal union in Old World monkeys. *Am. J. Anat.* 72:339–360.

Washburn, S. L. (1946a). The effect of facial paralysis on the growth of the skull of rat and rabbit. *Anat. Rec.* 94:163–168.

Washburn, S. L. (1946b). The effect of removal of the zygomatic arch in the rat. *J. Mammal.* 27:121–124.

Washburn, S. L. (1947a). The relation of the temporal muscle to the form of the skull. *Anat. Rec.* 99:239–248.

Washburn, S. L. (1947b). The biological basis of measurement. *Am. J. Phys. Anthropol.* n.s. 5:237.

Washburn, S. L. (1951a). The analysis of primate evolution with particular reference to the origin of man. In: Washburn, S. L., and Dobshansky, T. (eds.), *Cold Spring Harbor Symposia on Quantitative Biology, Origin and Evolution of Man*, vol. XV.

Cold Spring Harbor Laboratory Press, Cold Spring Harbor, NY. pp. 67–78.

Washburn, S. L. (1951b). The new physical anthropology. *Trans. N. Y. Acad. Sci.* II:298–304.

Washburn, S. L., Jay, P. C., and Lancaster, J. (1965). Field studies of old world monkeys and apes. *Science* 150:1541–1547.

Watanabe, K., Mori, A., and Kawai, M. (1992). Characteristic features of the reproduction of Koshima monkeys, *Macaca fuscata fuscata*: a summary of thirty-four years of observation. *Primates* 33:1–32.

Watts, E. S. (1985). Adolescent growth and development of monkeys, apes and humans. In: Watts, E. S. (ed.), *Nonhuman Primate Models for Human Growth and Development*. Alan R. Liss, New York. pp. 41–65.

Watts, E. S. (1986). Skeletal development. In: Dukelow, W., and Erwin, J. (eds.), *Comparative Primate Biology: Reproduction and Development*. Alan R. Liss, New York. pp. 415–439.

Watts, E. (1990). Evolutionary trends in primate growth and development. In: DeRousseau, C. J. (ed.), *Primate Life History and Evolution*. Wiley-Liss, New York. pp. 89–104.

Watts, E. S. (1993). Epiphyseal union in captive chimpanzees. *Am. J. Phys. Anthropol.* 16 (suppl.):206.

Wolf, K. E. (1984). *Reproductive Competition among Co-resident Male Silvered Leaf Monkeys* (Presbytis cristata). Yale University Press, New Haven.

Wright, P. (1999). Lemur traits and Madagascar ecology: coping with an environment. *Ybk. Phys. Anthropol.* 42:31–72.

Zihlman, A. L. (1997). Natural history of apes: life history features in females and males. In: Morbeck, M. E., Galloway, A., and Zihlman, A. L. (eds.), *The Evolving Female*. Princeton University Press, Princeton, NJ. pp. 86–104.

Zihlman, A. L. (2000). A skeletal survey of physical anthropology in the U.S. *Riv. Anthropol.* (Roma) 78:57–66.

Zihlman, A. L., and Bolter, D. R. (2004). Mammalian and primate roots of human sociality. In: Sussman, R. W., and Chapman, A. R. (eds.), *The Origins and Nature of Sociality*. Aldine de Gruyter, New York. pp. 23–52.

Zihlman, A. L., Bolter, D. R., and Boesch, C. (2004a). Skeletal and dental development in wild chimpanzees from the Taï National Forest, Ivory Coast, and Gombe Reserve, Tanzania. *Am. J. Phys. Anthropol.* 38(suppl.):215.

Zihlman, A. L., Bolter, D. R., and Boesch, C. (2004b). Wild chimpanzee dentition and its implications for assessing life history in immature hominin fossils. *Proc. Natl. Acad. Sci. USA* 101:10541–10543.

Zihlman, A. L., Morbeck, M. E., and Goodall, J. (1990). Skeletal biology and individual life history of Gombe chimpanzees. *J. Zool. Lond.* 221:37–61.

Zollikofer, C. P. E., Ponce de Leon, M. S., and Martin, R. (1998). Computer-assisted paleoanthropology. *Evol. Anthropol.* 6:41–54.

25

Primate Sexuality and Reproduction

Christina J. Campbell

INTRODUCTION

The title of this chapter suggests that the terms *sexuality* (sexual behavior) and *reproduction* (the production of offspring) refer to different avenues of primatological inquiry. Indeed, for many animal species, the terms refer to synonymous events; that is, sexual activity takes place only in the context of reproduction (although this by no means implies that every sexual encounter results in conception). Within the primate order, strepsirhine primates (lemurs and lorises) resemble nonprimate mammals in that sexual activity is largely restricted to times when conception is most likely (Shideler et al. 1983, Dixson 1998, Koyama et al. 2001, although see Brockman and Whitten 1996 for an exception). The sexual behavior of tarsiers remains largely unstudied, but from the limited information available, they also appear to restrict sexual behavior to brief periods of time associated with midcycle ovulation (Wright et al. 1986). Many species of monkey and ape (anthropoid primates), however, deviate from this pattern and engage in sexual activity at times when it is unlikely or even impossible for them to conceive (Loy 1987, Dixson 1998). For these primates, sexuality therefore is not necessarily synonymous with reproduction and may have been co-opted for functions other than the production of offspring. Mountings and indeed copulations have multiple functions, including dominant–subordinate interactions, appeasement, stress reduction, and the confirmation of social relationships. In reality, for many primate species, it may be extremely difficult for researchers to determine whether the function of a sexual encounter is reproductive or social in nature. Much of the primatological literature, however, treats sexual behavior as a reproductive event, often without strong evidence of it being so. My aim in this chapter is to discuss patterns and trends in primate sexuality. Separating out sociosexual and reproductive sexual behaviors in many cases, however, is beyond the scope of this chapter.

Biases in Research

In addition to the difficulty in consistently correlating sexual behavior with reproduction in many primate species, much of our knowledge of the sexual and reproductive biology of nonhuman primates comes from research on captive animals.

Maintaining captive primates for medical and biological research requires extensive knowledge of their sexual and reproductive biology for successful breeding. Such knowledge is increasingly important as more and more species become endangered in the wild (see Chapter 30), forcing captive programs to breed all their animals "in house." In addition, captive situations lend themselves to studying many facets of sexual and reproductive biology far more than field settings (see Chapter 20). Studies investigating the interaction between female ovarian cycles and sexual behavior provide a good example. In the wild, such studies are extremely difficult due to the fact that females rarely cycle (i.e., they are usually acyclic as a result of pregnancy or lactation). In contrast, captive conditions allow researchers to prevent pregnancy, a condition that can result in prolonged periods of ovarian cyclicity. Detailed information on the sexual behavior and reproduction of species in more natural settings however, is a fundamental requirement to our full understanding of evolution and behavioral ecology of primate species. Given the precarious natural existence of many species, understanding their reproductive and sexual biology in the wild is also becoming more vital, to ensure the success of conservation endeavors.

Additional biases of research into primate sexuality and reproduction stem from the fact that the focus of early primatological inquiry was largely restricted to terrestrial species of Old World monkeys, for example, various species of macaques (*Macaca* spp.) and baboons (*Papio* spp.). These early studies provided extensive and highly useful information regarding the sexuality of these species (e.g., Carpenter 1942, Conaway and Koford 1965, Kauffman 1965). Recently, it has become clear, however, that the data from these early studies and the theories regarding primate sexual behavior garnered from them were largely taxon-specific and did not reflect the great diversity in the reproductive and sexual morphology, physiology, and behavior of the primate order. For example, the taxonomic ubiquity of certain morphological features, such as sexual skins (see Sexual Skins, and Female Genitalia, below), was overemphasized in the past due to a dearth of studies on New World monkeys and prosimians. In addition, differences in the sexual behavior of human and nonhuman primate females tended to, and often continue to, be overemphasized because of the continued use of incorrect terminology in the description of monkey and ape sexuality (see The Question of Estrus, below).

The taxonomic imbalance in data concerning sexuality and reproduction stemmed from disparities in general research on certain primate groups but also from the fact that the ease of studying these topics specifically varies widely from species to species. For example, rhesus monkeys (*Macaca mulatta*) are largely terrestrial (making them easier to follow than an arboreal species), females have sexual skin coloration changes (allowing for visual determination of ovulation) (Zuckerman et al. 1938), and individuals copulate frequently during a defined breeding season and, in fact, within a given day (Manson 1992). As a result of these characteristics, researchers of rhesus monkey sexual behavior and reproduction can record large numbers of sexual encounters and visually estimate ovulation from sexual skin changes. Such forms of data allow investigations into a wide variety of topics, such as the relationship of sexual behavior to physiology (e.g., Herbert 1970, Michael and Zumpe 1970, Dierschke et al. 1974, Chambers and Phoenix 1987), mate choice (Manson 1992, see Chapter 27), and reproductive success (Bercovitch and Nürnberg 1996). In contrast, the primates that are the main focus of my research, spider monkeys (genus *Ateles*), spend nearly all of their time high in the forest canopy (Youlatus 2002) and often copulate in seclusion from other group members (Campbell in press), making it highly unlikely that researchers will actually witness a copulatory event. In addition, female spider monkeys have no visual cue to ovulation, so this can only be determined from hormonal data (Hernández-López et al. 1998, Hodges et al. 1981, Campbell et al. 2001) and vaginal swab histology (McDaniel et al. 1993, Goodman and Wislocki 1935). Recent advances in noninvasive forms of collecting hormonal information (see Chapter 21) are slowly changing this disparity in research on certain taxonomic groups. However, the sexual and reproductive biology of many species remains largely unstudied in either captive or natural settings.

FEMALE MORPHOLOGY AND PHYSIOLOGY

Ovarian and Menstrual Cycles

After puberty, primate females experience cyclical changes in the excretion of certain hormones (e.g., estrogen and progesterone) as a result of changing reproductive states. Alternating periods of ovarian cycling (including ovulation), pregnancy, and lactation (in this order) each bring a unique hormonal and, in some cases, behavioral profile. When a female is lactating, most species do not experience ovarian cycles and, consequently, do not ovulate, usually preventing females from conceiving when the nutritional requirements of their current offspring are still great (Valeggia and Ellison 2001). An exception to this general pattern is found among female callitrichids and some lorisiforms (e.g., *Arctocebus calabarensis*), where ovulation and conception may occur within a few days of giving birth, that is, when they are lactating (Dixson 1992, Bearder 1987). In all primates, ovarian cycling involves the cyclical build-up and regression of the uterine lining of blood capable of supporting a fertilized egg upon conception. In many primate species, uterine blood is reabsorbed during regression, and thus, no external bleeding is observed (Strassman 1996). Some New World monkey species (e.g., *Ateles geoffroyi*, Goodman and Wislocki 1935, Campbell et al. 2001) experience some level of external blood loss, but this is highly variable both within cycles and between females. In many species of Old World anthropoids, including *Homo sapiens*, copious externally visible bleeding occurs; and in such cases, these cycles are referred to as menstrual cycles.

The evolution of overt menstrual bleeding has received much attention in the primatological literature. Profet (1993) proposed that menstruation is a means of removing pathogens transported to the uterus via sperm. This hypothesis carries with it three predictions. First, pathogens should be more prevalent before, rather than after, menstrual bleeding. Second, menstruation should occur during life stages when pathogen loads are likely to be highest. Third, in species where promiscuous breeding is common, and hence the likelihood of exposure to multiple pathogen sources is increased, menstruation should be more prevalent. Strassman (1996) tested each of these predictions with data from the primatological literature and found that there is little to no evidence to support the pathogen-removal hypothesis. Alternatively, the regression and rebuilding of the lining is proposed to be energetically less expensive than maintaining it constantly (Strassman 1996). Strassman also hypothesizes that overt cyclical bleeding, as opposed to absorption of blood, may be a by-product of the different endometrial microvasculature in Old World anthropoids; that is, levels of endometrial blood in these primates may be too great to be reabsorbed effectively and, thus, are partially shed.

Sexual Skins and Female Genitalia

Female reproductive morphology is highly variable in the primate order. As already noted, early studies of many Old World monkeys emphasized sexual skins (swelling of the skin surrounding the anogenital region) in the signaling of ovulation to adult males. Where sex skins were absent, ovulation was considered "concealed," and various hypotheses were put forward to explain this concealment in both nonhuman primate (Burley 1979, Andelman 1987, Datta 1987) and human (Benshoff and Thornhill 1979, Lovejoy 1981, Strassman 1981, Daniels 1983, Turke 1984) females. Although a highly conspicuous event (Fig. 25.1), the importance of these sexual skins in the reproductive and sexual biology of the primate order as a whole has been overemphasized as the trait is by no means widely distributed. Today, it is recognized that this characteristic has evolved at least three times, twice in the Cercopithecoidea and once in the Hominoidea. Only 25 species of Old World monkey and two species of ape (*Pan troglodytes* and *P. paniscus*)

Figure 25.1 Chimpanzee female (*Pan troglodytes*) with a sexual swelling (photo by Ilka Herbinger).

show exaggerated cyclical swelling or coloration changes of the perineal skin (Dixson 1983). Given the limited distribution of this trait within the primate order, focus should, and in some cases has, changed to asking the question of why such elaborate signals are present (e.g., Burt 1992, Nunn 1999), rather than why they are not, as a lack of strong visual signaling of female reproductive state appears to be the "norm" in the primate order.

Sexual skin swelling (*tumescence*) results from cyclical water retention in the skin surrounding the anogenital region in association with fluctuating hormone levels, particularly estrogen (Bullock et al. 1972, Wildt et al. 1977, Dixson 1983a, Thomson et al. 1992). Swellings can be so exaggerated that they may result in significant changes in female body weight; for example, in some subspecies of red colobus monkeys (*Piliocolobus badius*), female body weight may increase by 25% when fully swollen (Struhsaker 1975). In addition to swelling, the skin often changes color, becoming bright pink or red as a result of increases in blood supply to the region (Collings 1926). In most species, maximal tumescence is reached close to or on the day of ovulation after being stimulated by increasing estrogen levels during the follicular phase of the cycle (the stage associated with development of the ovum prior to ovulation). This is by no means a perfect association, however, and females may actually ovulate within a number of days surrounding maximal tumescence (e.g., *Macaca tonkeana*, Thierry et al. 1996; *Cercocebus torquatus atys*, Whitten and Russell 1996). After ovulation, the area detumesces, the color intensity reduces, and the cycle begins again (Wildt et al. 1977, Thomson et al. 1992).

Researchers have long suggested that sex skins act as strong visual stimulants to males, which find fully swollen and/or colored females highly attractive, and experimental work appears to lend strong support to this hypothesis (Bielert and Girolami 1986, Girolami and Bielert 1987). In a series of experiments involving captive chacma baboons (*Papio ursinus*), males showed a 20-fold increase in seminal ejaculations when shown an ovariectomized female with a false sexual swelling attached to her rump. A lack of ovarian function and invitational behavior in the experimental females strongly suggests that changes in male behavior resulted solely from the visual cue of the fake swelling (Girolami and Bielert 1987). Further experiments showed that the red coloration of a sex skin is also important in stimulating male sexual arousal (Bielert et al. 1989).

Several hypotheses have been proposed to explain the adaptive function of sex skins. Clutton-Brock and Harvey (1976) suggest that the visual stimulus of sex skins prompts adult males to compete for females, thus allowing females to choose "the best male" from among the competitors. Hrdy (1981) proposes that, rather than instigating male–male competition, sex skins attract "multiple males," all of which may be potential mating partners (see also Hrdy and Whitten 1987, Wallis 1992). Hamilton (1984) hypothesizes that sex skins are true indicators of the time of ovulation ("obvious ovulation"), thus allowing males to pinpoint the best time to copulate with a female. Van Noordwijk (1995) elaborates on the previous hypothesis by suggesting that the "obvious ovulation" indicator allows females to manipulate the behavior of males in a way that provides them protection from predators and conspecifics during consortships. Such consortships are more likely to occur around the time of maximal swelling. Alternatively, Pagel (1994) suggests that sex skins provide information to males regarding the reproductive "quality" of the female, allowing males to predictably choose females that are likely to be successful mothers. Nunn (1999) hypothesizes that sex skins act as a "graded signal" of the probability of ovulation. This hypothesis proposes that the imperfect association of maximal swelling with ovulation was selected for as it allows females to limit paternity to the more dominant males (which are more likely to mate-guard at maximum swelling) while at the same time attracting multiple males outside of peak swelling in order to confuse paternity (Nunn 1999). Zinner and Deschner (2000) suggest that sexual swellings occurring in female hamadryas baboons (*Papio hamadryas*) after a male take-over may act as a deceptive signal in order to prevent infanticide by the new breeding male. All of these hypotheses are at least partially supported by data concerning sex skins and behavior; however, there is little consensus among authors as to whether any one of them truly explains the evolution of this trait. It also appears unlikely, given the discontinuous taxonomic distribution of the trait, that one hypothesis will sufficiently explain its function across all species.

There is general lack of exaggerated physical change in the appearance of external female genital regions in the prosimians and New World monkeys. That said, females of many species without exaggerated sex skins show some swelling and color changes in their genitalia during the periovulatory phase of the cycle. Female ring-tailed lemurs (*Lemur catta*) show moderate swelling and reddening of the external genitalia during estrus (Jolly 1966), and spider monkey (*Ateles* spp.) females may show a small amount of swelling of the clitoris; however, this is highly variable within cycles and between individuals (Goodman and Wislocki 1935, C. J. Campbell personal observation). Strier (1996)

believes that in an arboreal habitat (as inhabited by all New World monkeys) visual cues of reproductive state would be of little use as visibility is often obscured. Thus, chemical or olfactory cues may be more important signals of female reproductive status for New World monkeys and prosimians than for the Old World anthropoids (Epple 1976).

While there is little evidence of exaggerated external changes in the genital morphology of female strepsirhines, some species show extreme internal changes in association with changing hormonal levels and ovulation. Vaginal closure by means of a "vaginal closure membrane" is associated with seasonal shortages in food (Dixson 1998) and has been documented in various species of galagos and lemurs (e.g., *Galago senegalensis*, Manley 1966, Butler 1974; *Microcebus murinus*, Peter-Rousseaux 1964; *Cheirogaleus major*, Peter-Rousseaux 1964; *Varecia variegata*, Shideler et al. 1983). Such a feature means that copulation is physically impossible in these species except for around the time of ovulation when the vaginal closure membrane regresses.

The clitoris shows great variation in shape and size within the primate order. A hypertrophied clitoris is present in many species of New World monkey, and in some cases (e.g., members of the Atelini tribe) it is pendulous (Fig. 25.2) and larger than the flaccid penis of the male (Wislocki 1936). The clitoris is pendulous from birth and can thus be employed to successfully determine the sex of newborn monkeys. It has been proposed that such a clitoris may aid in communication of the sexual state via chemical cues in the urine (Klein 1971). However, olfactory cues are successfully employed in many other primate species without such an extreme morphological feature. Alternatively, I have observed that the dorsal surface of the clitoris in *A. geoffroyi* rubs along the inferior surface of the erect penis during

Figure 25.2 Hypertrophied clitoris of a juvenile female spider monkey, *Ateles geoffroyi* (photo by author).

intromission attempts and copulation and may act as a "guide" to direct the large penis of this species into the vaginal canal (Campbell 2000). The clitoris of capuchin monkeys (*Cebus* spp.) and some strepsirhine species actually contains a small bone known as the *os clitoris* (Hill 1972, Carosi et al. 2001); however, the function of this bone, if any, remains unknown. The clitoris of some anthropoid species is capable of a certain level of erection (e.g., *Cercopithecus* spp., Hill 1966; *Macaca fuscata*, Wolfe 1984). However, these species are not reported to have an *os clitoris*, so it is unlikely that this bone functions in a similar way to the male baculum (see below). Unlike in studies of the male penis (see below), there has been little to no attempt to explain the variety in primate clitoral morphology in terms of ecological or social environments.

In many species of New World monkey, the labia are pseudoscrotal. This is extreme in some species of howler monkey (*Alouatta* spp.), where they are of similar color and location as the scrota of males. However, *Alouatta caraya* females do not show any genital mimicry or exaggeration (Thorington et al. 1984). Thus, even within a genus there can be considerable variation in reproductive morphology.

Female Orgasm

Long thought to be a unique feature of human female sexuality, there is now strong evidence for the presence of orgasms (as evidenced by the presence of uterine muscle contractions, increases in heartbeat rate, and climax faces) in female nonhuman primates (Goldfoot et al. 1980, Phoebus 1982, Slob et al. 1986, Triosi and Carosi 1998). The function of the female orgasm remains unclear however. Early work stressed its function in maintaining a pairbond in humans by making sex with one partner enjoyable for the human female (Alexander 1979). The presence of an orgasm in nonhuman primates, especially in species with multimale/multifemale mating systems, appears to negate this possibility. It has been postulated that the female orgasm functions in one of two ways to ensure conception. Firstly, muscle contractions may enhance the movement of the seminal fluids and hence sperm up the vaginal canal (Morris 1967). Studies imaging the movement of seminal fluid injected with radioactive opaque materials, however, suggest that this is not the case (Masters and Johnson 1966). Secondly, female orgasm may aid in sperm retention (Morris 1967, Baker and Bellis 1993), and there is some evidence in humans that this may be the case (Baker and Bellis 1993). Female orgasm is not necessary for spermatic transport, however, and thus is unlikely to have evolved as an adaptation solely to ensure conception (Smith 1984, Baker and Bellis 1993). Work with Japanese macaques (*M. fuscata*) suggests that social stimuli may be important in the promotion of female orgasm in this species; that is, low-ranking females were more likely to experience orgasm than their high-ranking counterparts. In addition, females of all rank were more likely to orgasm when copulating with a high-ranking male (Triosi and Carosi 1998).

The authors of this study suggest their data provide evidence that the female orgasm may have evolved as a mate choice adaptation. Alternatively, the function of the female orgasm may be far more basic in that it may simply provide enjoyment for the female in order to ensure her participation in sexual encounters. Testing this possibility will prove difficult, however, as measuring "enjoyment" is beyond our current capabilities.

MALE MORPHOLOGY AND PHYSIOLOGY

Penile Morphology

Like the female clitoris, penis length and complexity are highly variable. Species that typically exhibit pairbonded or polygynous breeding systems tend to have relatively shorter penises with relatively simple distal ends (Dixson 1987a). In contrast, where females typically mate with multiple males, penis length is on average longer and the distal end of the penis tends to be more complex in terms of general shape and in the presence of specialized features such as penile spines (Dixson 1987a). Some descriptions of penile morphology in the literature may be misleading, however, as most come from inspections of the flaccid penis. For example, all references to the penis of male spider monkeys (*Ateles* spp.) come from one citation (Hill 1962) that describes the penis as being elongated with a relatively simple or undifferentiated distal end. However from my observations of spider monkeys, it is clear that the distal end of the erect penis is actually mushroom-shaped (Fig. 25.3), as was described by Hill (1962) for the closely related muriquis (*Brachyteles* spp.) and woolly monkeys (*Lagothrix* spp.).

The penis in primates typically contains a bone (*os penis*, *os priapi*, or baculum), as it does in several other orders of mammals. This bone is completely absent, however, from tarsiers, the atelini primates, and humans (Hill 1972, Dixson

Figure 25.3 Erect penis of an adult male spider monkey (*Ateles geoffroyi*) soon after copulation (redrawn by Jody Biggers from photo by author). Note the distinctive shape of the distal end of the penis.

1987b). Bacula size, relative to body size, varies significantly (see Dixson 1987b, 1998 for a full list of bacula lengths), and this variation in length can be largely explained by taxonomy and copulatory pattern. In general, for their body size, there is great variation in baculum length in the strepsirhines, being greatest in the lorises. New World monkeys in general have smaller bacula than their Old World counterparts and, indeed, as mentioned above, in some species of New World monkey this tendency toward size reduction is so extreme that the baculum is now absent. Among the Old World monkeys, the colobines have shorter bacula lengths in relation to body size than the cercopithecines. Hominoids have smaller relative bacula lengths than most of the Old World monkeys (Dixson 1987b).

Once differences in phylogeny have been controlled for, species that show prolonged intromission times tend to have longer relative bacula length (Dixson 1987a,b). It has been suggested that the bacula may assist in copulation when initial penetration is difficult either as a result of the male being much larger than the female or because of a complex penile morphology (Long and Frank 1968). Alternatively, Romer (1962) proposed that the role of the baculum during copulation might be as a supporting rod for the penis, allowing for the maintenance of erection during a prolonged intromission period. However, the relationship between bacula length and copulatory pattern is by no means concrete as the atelini monkeys show extended copulation lengths but lack a baculum entirely.

The "lock-and-key" hypothesis has been proposed to explain the evolution of genital morphology in the primate order. This hypothesis suggests that penile morphology has evolved to complement female genitalia and eventually forms a species-isolating mechanism. In general, however, there is little support for this hypothesis. Penis morphology can often vary significantly within a genus, and the classic example of such intrageneric variation is found within the macaques, *Macaca* spp. (Fooden 1976). The lock-and-key hypothesis predicts that different species within this genus would not be able to hybridize as the vaginal morphology of one species would not match the penile morphology of another. In fact, this is not the case, and many species have been demonstrated to hybridize in captivity (Bernstein and Gordon 1980). One species of macaque, *M. arctoides*, does appear to show complementary evolution of male and female genitalia (Fooden 1967), suggesting that, in isolated cases, the lock-and-key hypothesis may have some validity. The increased complexity of penile morphology (length, spines, shape, and baculum length) in species where females typically mate with multiple males supports Eberhard's (1985) hypothesis of sexual selection by female choice.

Testes Size

Schultz (1938) first documented the wide variation of testes size relative to body size in the primate order. Much later,

Parker (1970) hypothesized that male insects might compete by means of the quality or quantity of their sperm in species where females mate with multiple males. Primatological data lend support to this "sperm competition" hypothesis in that relative testes size is generally larger in males of species where the females mate with multiple males and smallest in those species that typically exhibit pairbonded or polygynous mating systems (Short 1979, Harcourt et al. 1981a, Dixson 1987a, Jolly and Phillips-Conroy 2003). Larger testes produce greater quantities of sperm (Møller 1988), a greater percentage of motile sperm (Møller 1988), and longer sperm (Dixson 1993) that tend to have higher motility rates (Gomendio and Roldan 1991). Males with larger testes are proposed to experience an increased likelihood of siring the offspring of females with which they copulate. Investigations into the potential role of sperm competition and the evolution of testes morphology are typically comparative in nature, that is, comparing testes size and mating strategies over different species. Concrete evidence supporting the hypothesis, however, can come only from investigations within a species, that is, evidence showing that males with larger testes produce more sperm than males of the same species with smaller testicles. Empirical evidence of this kind is, however, entirely lacking. In addition, it must be shown that males with larger testes have higher reproductive success than those males of the same species with smaller testes. Some evidence to support this comes from Pullen et al. (2000), who have shown that male lesser galagos (*Galago moholi*) with the largest testes during the mating season have the highest mating success, but no data are available to determine whether this translates to greater reproductive success.

Copulatory Plugs

The semen of male primates coagulates after ejaculation to varying degrees. In the most extreme cases, the ejaculate forms a solid plug-like structure in the female vaginal canal. The adaptive function of copulatory plugs has traditionally been explained by one of several hypotheses, the two most common of which involve the prevention of sperm leakage out of the vaginal canal (Michener 1984) and a form of male–male competition by creating a physical barrier to subsequent matings (Parker 1970, Schwab 2000). Behavioral observations from various primate species, however, show that intromission by subsequent males may easily remove a copulatory plug from the female's vaginal canal (e.g., *Macaca nemestrina*, Busse and Estep 1984; *Brachyteles arachnoides*, Milton 1985; *L. catta*, Parga 2003; *A. geoffroyi*, Campbell in press). Such observations suggest that copulatory plugs may be relatively ineffective at preventing subsequent matings. In ring-tailed lemurs (*L. catta*), removal of a copulatory plug appears to be a viable mating strategy for males that are not the first to mate with a female during her receptive period. Males that first mate with females during this time are more likely to practice postejaculatory mate guarding (Sauther 1991, Parga 2003), thus reducing the possibility of subsequent males from mating with the female. Once mate guarding by the first mating male is terminated, the distinct penile morphology of this species (enlarged distal end with penile spines, Dixson 1987a) is thought to help non-first-mating males to remove copulatory plugs from a female's vaginal canal (Sauther 1991, Parga 2003). Evidence showing that copulatory plugs function to prevent leakage of semen (and consequently sperm) from the vaginal canal in primates is lacking.

SEXUAL BEHAVIOR

Patterns of Sexual Behavior Across the Female Ovarian Cycle

The Question of Estrus

In both seasonal and aseasonal breeders (see below), the sexual behavior of female mammals is often limited to a short period of time coinciding with ovulation, making conception more likely to occur (Heape 1900). This period of female sexual activity is known widely as "heat" and/or estrus and was first defined by Heape (1900:6) as "the special period of sexual desire of the female; it is during oestrus and only at that time, the female is willing to receive the male and fruitful coition rendered possible in most, if not in all, mammals." Today, the term *estrus* is widely used by scientists and laypeople alike and is applicable to most mammal species. Like many nonprimate mammals, sexual behavior in many lemur and loris species is strictly limited by hormonally controlled mechanisms, such as vaginal closure and the presence of lordotic postures (positional changes necessary for copulation). As such, sexual behavior in these species is seen only during limited periods of time in association with ovulation, for example, *L. catta* (Evans and Goy 1968) and *V. variegata* (Shideler et al. 1983). In some lemur species (e.g., *Propithecus verreauxi*), the pattern is not so clear-cut and sexual behavior may take place under certain social conditions regardless of female hormonal status (Brockman and Whitten 1996). Indeed, Whitten and Brockman (2001) suggest that social factors may impact lemur and loris sexuality more than previously thought.

It is unclear whether or not the concept of estrus, given Heape's original definition, is applicable at all to the anthropoid primates. Female anthropoids of many species copulate at times when conception is not likely or possible, that is, throughout the ovarian cycle, when lactating, or when pregnant (e.g., *A. geoffroyi*, Campbell 2004; *B. arachnoides*, Strier and Ziegler 1997; *Cercocebus mitis*, *C. aethiops*, *C. neglectus*, Booth 1962; *Miopithecus talapoin*, Rowell and Dixson 1975; *M. mulatta*, Michael and Zumpe 1970; *Papio ursinus*, Saayam 1970; *P. troglodytes*, Tutin 1980; *P. paniscus*, Thompson-Handler et al. 1984). Such a major deviation from Heape's (1900) definition has resulted in a plethora of

publications attempting to redefine the concept of "estrus" to fit the sexual behavior of the anthropoid primates. Heape (1900) himself noted that sexual behavior does occur during pregnancy in some species and termed such behavior "abnormal estrus". Other authors have coined terms such as *postconception estrus* (Conaway and Koford 1965), *perimenstrual estrus* (Loy 1970, Wolfe 1979), *pseudoestrus* (Hrdy 1977), and *periabortive estrus* (Wolfe 1979) to describe the sexual behavior of female primates during times other than the time close to ovulation, or the periovulatory period.

In addition, various authors differ in the factors they include in a definition of estrus. Loy (1970) states that the term *ovulation* must be removed from the definition. He coined the term *perimenstrual estrus* to describe increases in copulation frequency in rhesus macaques (*M. mulatta*) near menstruation. He noted that hormonally this sexual activity is similar to estrus in that it is associated with heightened levels of estrogen. Thus, he stresses the importance of endocrinology over histology. Other authors have suggested that researchers may be stressing the histological and endocrinological components of estrus at the expense of the behavioral component (e.g., Carpenter 1942). Rowell (1972) shows that there is often little correlation between sexual behavior and stage of the menstrual cycle of Old World monkeys. She goes on to suggest that there are no histological components to estrus. Likewise, Hrdy and Whitten (1987) remove any relationship with ovulation from the concept of estrus.

Resulting from this confusion over how to define estrus in anthropoid primates, other authors have suggested that the term should not be used when discussing the sexual behavior of female monkeys and apes (Loy 1987, Dixson 1998, Campbell 2004). Removal of the term allows for a more detailed understanding of female sexuality. A clear example of how this term has clouded our vision of female sexuality relates to its ubiquitous use in reports of female sexual behavior of monkeys and apes but not of humans. By realizing that, in fact, monkey and ape females do not exhibit true estrous cycles, we can begin to understand that human female sexuality and its seeming lack of association with hormonal fluctuations may not be as different from that of other anthropoids as has long been postulated in the literature.

Alternatives to the Estrus Concept

Alternatives to the concept of estrus in female primates are available. Beach (1976) proposed a model that separates female sexual behaviors into three characteristics: proceptivity, receptivity, and attractivity. In the same year, Keverne (1976) proposed that female behaviors should be investigated in terms of hormonal influences on attractivity and receptivity. Behaviors and physiological conditions that are necessary and sufficient for copulation to take place indicate *receptivity*. For example, female black and white ruffed lemurs (*V. variegata*) are not receptive outside of the periovulatory period as their vaginal canal is closed, lordosis is not inducible, and therefore copulation is physically impossible (Shideler et al. 1983). *Proceptive* behaviors are those that a female engages in to indicate her motivation to mate, such as presentation of the anogenital region, facial displays, vocalizations, etc. *Attractivity* refers to the stimulus effects of the female on the male, that is, how attractive she is to an interested male. Any changes in the attractiveness of a female primate can be observed by changes in the behavior of adult males toward the female in question, often in relation to olfactory (urinary scents) and visual (sexual swellings) cues. Changes in the frequency of sexual behaviors in anthropoids are usually attributed to changes in attractivity and proceptivity, rather than receptivity (Beach 1976, Keverne 1976); that is, monkey and ape females have no physical barrier to copulation at any time. These two models deal solely with the behavioral component of female sexuality, but they do provide a framework to investigate the association of these behavioral changes with underlying histological and endocrinological changes. While these terms are becoming more widely used in the primatological literature, they are not without their own problems. The most persistent of these is that it can often be difficult to determine whether certain behaviors are related to attractivity, proceptivity, or receptivity. For example, it is likely that a proceptive female is more attractive to a male because of the mere fact that she is, by definition, motivated to copulate. Thus, these behavioral categories are not necessarily independent.

Copulatory Patterns

Specific copulatory sequences and behaviors within the primate order vary widely and are beyond the scope of this chapter. All primates, however, can be classified as having one of three copulatory patterns (Dewsbury and Pierce 1989, Dixson 1998). These patterns are typically distinguished by whether each copulatory bout involves single or multiple *intromissions* (insertion of the penis into the vagina). In the case of multiple intromission patterns, such as that found in the rhesus monkey (*M. mulatta*), ejaculation occurs only after a series of brief intromissions, each merely a few seconds long (Carpenter 1942). Single intromissions are classified as brief (e.g., mantled howler monkeys, *Alouatta palliata*, approximately 30 sec; C. J. Campbell personal observation) or prolonged (e.g., black-handed spider monkeys, *A. geoffroyi*, 12–24 min; Campbell 2000, in press). Additional features, such as the presence or absence of a copulatory lock, are often used to distinguish among mammals in general; but it is unclear as to whether or not this phenomenon actually exists within the primate order (Dewsbury and Pierce 1989, Dixson 1998). Many species appear to maintain intromission past male ejaculation. This can be anywhere from a few minutes or less in species such as the black-handed spider monkey (*A. geoffroyi*, Campbell in press) to up to several hours in thick-tailed bushbabies (*Galago crassicaudatus*) (Eaton et al. 1973). Whether the male and female are actually locked together and unable to

Figure 25.4 Dorsoventral copulatory posture with male foot clasp in the Japanese macaque, *Macaca fuscata* (photo by Linda Wolfe).

physically separate, as is the case in dogs, is not known however (Dewsbury and Pierce 1989, Dixson 1998).

Copulatory Postures

Primate copulatory postures are somewhat stereotyped, with a dorsoventral position being the "norm"; that is, the dorsal surface of the female meets with the ventral surface of the male (Fig. 25.4). Variations on this dorsoventral posture are related to different ways in which the animals are oriented in their environment and ways in which males and females hold on to each other. Members of the Lorisinae are usually suspended underneath a branch while copulating in an otherwise standard dorsoventral posture (Schulze and Meier 1995). In many species of Old World monkey (e.g., baboons and macaques) and in some species of New World monkey (e.g., capuchins and squirrel monkeys) and strepsirhines (e.g., galagos), the dorsoventral position is modified slightly by the male "climbing" on the female and grasping her lower hindlegs with his feet, also referred to as a "double foot clasp" (Dixson 1998) (Fig. 25.4). Members of the Atelini tribe (see Chapter 10) vary from the standard posture in that the male does not mount the female but, rather, sits upright on a branch and the female then backs onto him (Fig. 25.5). Probably to stabilize the pair during extended copulation times, the male wraps his legs over those of the female in what has become known as a "leg-lock" (Klein 1971, Milton 1985, Nishimura 1988). Copulatory postures in the apes appear to be more varied, especially in the bonobo, where ventroventral copulations are often observed (26% of copulations in one study, Thompson-Handler et al. 1984). In the other apes, face-to-face copulations are observed less frequently and the more standard dorsoventral posture

Figure 25.5 Dorsoventral copulatory posture of spider monkeys, *Ateles geoffroyi*. Note the "leg lock" behavior of the male seated behind the female (photo by author).

predominates (Harcourt et al. 1981b, Goodall 1986), although the arboreal orangutan is often seen copulating in a variety of suspensory positions (Nadler 1977).

Sociosexual Behavior

Homosexuality

While homosexual relationships are present in the human species, there has been little to no evidence to date of a consistent choice for same-sex sexual partners among any nonhuman primate species (Dixson 1998, Kirkpatrick 2000). Indeed, there is some disagreement among authors as to how to define homosexuality, and this probably adds to the confusion over whether homosexuality is a phenomenon specific to humans or one that applies to primates in general. Some define any behavioral patterns performed by members of the same sex that are common to heterosexual copulations as evidence for homosexuality, e.g., mountings, presentations, and genital manipulation (Nadler 1990, Wolfe 1991, Kano 1992, Vasey 1995, Fox 2001). Such behaviors are common throughout the primate order and often function in the formation and/or maintenance of social relationships, alliances, etc. Perhaps the most widely publicized example of such behavior among the nonhuman primates is the so-called genitogenital (g–g) rubbing of bonobos, where females rub their genitals together (Kano 1992). This behavior appears to act both as a tension reducer and as a social bonding mechanism among adult females (Kano 1992). In contrast, Wallen and Parsons (1997) and Dixson (1998) stipulate that homosexuality in humans is not defined by mere behavioral patterns but by the presence of a "preference" for members of the same sex as sexual partners. Thus, they define homosexuality as a preference for sexual encounters with members of the same sex. Using this definition, evidence for homosexuality in the nonhuman primates is generally lacking.

Masturbation

Among the Old World anthropoids, self-genital stimulation by males, which may or may not result in ejaculation, appears to be relatively common: e.g., *Piliocolobus badius* (Struhsaker 1975), *M. mulatta* (Carpenter 1942, Lindburg 1971), *Papio anubis* (Ransom 1981), *P. troglodytes* (Goodall 1986). Reports of female masturbation are less common—e.g., *Papio anubis* (Ransom 1981) and *P. troglodytes* (Goodall 1986)—and this may be partially due to the fact that male masturbation is likely to be more visually obvious to human observers. Among the New World monkeys, reports of masturbation are less common. Female spider monkeys (*A. geoffroyi*) are known to rub or hold their own clitoris, and both of these behaviors appear to occur more frequently in females that are reproductive cycling (Campbell 2004). There does not, however, appear to be a relationship of this behavior with any particular stage of the ovarian cycle (Campbell 2004). In addition, it is difficult to determine whether or not females are sexually stimulating themselves. A possible alternative explanation is that they are reacting to the presence of irritating insects such as mosquitoes that are likely to be attracted to the large hairless clitoris. Masturbation in prosimian primates has not been reported. Much remains unknown about nonhuman primate masturbation. For example, why the prosimians and New World monkeys differ from the Old World anthropoids in the occurrence of this behavior is unclear. In addition, whether or not masturbation has any adaptive function is unknown, and it is more likely that such behavior is a side effect of the increased tactile sensitivity of genitalia in response to their copulatory function.

REPRODUCTION

Conception

Young female primates typically go through a period of adolescent sterility or subfecundity (Knott 2001) where they do not ovulate during ovarian cycling or where seemingly normal hormonal cycling coincides with a complete lack of sexual behavior, a lack of interest by adult males, and a lack of conception: e.g., *Papio cynocephalus* (Altmann et al. 1977), *Gorilla gorilla* (Harcourt et al. 1980, Watts 1991). Although adult female primates in natural conditions rarely cycle (i.e., they are usually pregnant or lactating), data suggest that multiparous females (those who have previously reproduced) usually cycle for a few months before conception occurs (e.g., *A. geoffroyi*, three to six cycles, Campbell et al. 2001; *B. arachnoides*, three to six cycles, Strier and Ziegler 1997; *Papio cynocephalus*, 5 months, Altmann et al. 1977; *Papio hamadryas*, one to four cycles, Zinner and Deschner 2000). Females are sexually active during these non-conceptive cycles, and to date, there are no data available to explain why sexual encounters during these cycles do not result in conception. Given the huge number of variables related to a successful conception (effective sperm motility, ovulation, fertilization, implantation, etc.), it is not surprising that the majority of copulatory events are unsuccessful in this regard.

Paternity

When indeed a female does conceive, it is often of interest to primatologists to know which male sired the offspring. Historically, paternity in primates has been inferred by behavioral data; that is, males that were seen to copulate with a female around the presumed time of conception were considered to be likely fathers. With the increased availability of deoxyribonucleic acid (DNA) technology (see Chapter 22), reliance on this highly imperfect source of paternal information is no longer acceptable. Indeed, many early concepts in primatology, such as the exclusive mating rights of dominant males, have since been challenged by genetic information showing that subordinate males indeed sire offspring: e.g., *P. troglodytes* (Vigilant et al. 2001), *Papio cynocephalus* (Altmann et al. 1996). However, male reproductive success does usually correlate with dominance rank, and even though dominant males do not appear to have exclusive mating rights, they often do sire the majority of offspring (Pope 1990, Altmann et al. 1996, Gachot-Neveu et al. 1999, Vigilant et al. 2001). The level of exclusive mating access experienced by top-ranking males is likely related to the number of additional males present in the group and the level of synchronicity in female fertile periods. Higher male-to-female ratios and synchronized female fertility potentially make it more difficult for dominant males to monopolize fertile females. In such situations, subordinate males are hypothesized to be more likely to mate with females and even to sire offspring.

Seasonality

Not only are many species of nonhuman primate seasonal breeders but sexual behavior is also usually limited to this season. Mating seasons can be a matter of days (e.g., ring-tailed lemurs, *L. catta*; Jolly 1966) to a few months (e.g., *M. fuscata*; Kawai et al. 1967). In such species, births are also strictly limited in their occurrence throughout the year. Some species may even exhibit multiple distinct birthing periods within a year, such as lesser galagos, *G. moholi* (Pullen et al. 2000). Alternatively, in other species, there is no strict reproductive seasonality; thus, sexual behavior and conception can take place during most months of the year, although "birth peaks" may be recorded (e.g., *Ateles* spp, Milton 1981, Symington 1987, Chapman and Chapman 1990, Campbell 2000; *Alouatta fusca*, Strier et al. 2001).

Most researchers agree that seasonal reproduction is a response to seasonal changes in food supply (triggered by changes in rainfall patterns and day length); however, species do differ in how they react to these changes. In gen-

eral, it is thought that reproduction is timed so that the most energetically expensive stage of reproduction for females occurs when there is the most food available (Knott 2001). Sichuan snub-nosed monkeys (*Rhinopithecus roxellana*) time their births so that females begin to lactate when food becomes more abundant (Zhang et al. 2000, Bao-Ping et al. 2003). Squirrel monkeys (*Saimiri* spp.) are the most seasonal of the neotropical primates (Di Bitetti and Janson 2000), and their short birthing season appears to serve two functions. First, females give birth during the period of maximal food availability; second, the tightly synchronized birthing season may act to swamp the predator community (Boinski 1987). For species with longer interbirth intervals, it is difficult to time any one reproductive stage to food availability (Bronson and Heideman 1988, Knott 2001). In such cases, seasonal variation in births is most likely related to female conditional changes in response to fluctuating food availability. A lack of conception during certain months is almost certainly due to poor female condition as a result of low food availability (Knott 1997, Ziegler et al. 2000, Strier et al. 2001). During such months, a reduction in food intake can negatively impact hormonal levels, leading to a cessation of ovulation and an inability to conceive (Bercovitch 1987, Ziegler et al. 2000). More folivorous primates, such as howler monkeys (*Alouatta* spp.) and mountain gorillas (*Gorilla gorilla beringei*), may be less likely to show seasonal peaks in births or conceptions as their ability to digest mature leaves prevents major seasonal changes in food availability (Watts 1998, Di Bitetti and Janson 2000, Knott 2001, Strier et al. 2001). Thus, in species with shorter reproductive parameters (e.g., gestation, lactation), breeding can be timed to coincide with a certain level of food production and seasonal breeding is usually present. In species with longer reproductive parameters (usually larger species), any seasonal pattern is more likely to be an immediate interaction between food availability, available energy, and female reproductive condition ("ecological energetics hypothesis," Knott 2001).

Males of some primate species undergo spermatogenesis during the breeding season only (Gordon et al. 1976, Wiebe et al. 1984, Rostal et al. 1986). For example, in the squirrel monkey (*Saimiri* spp.), a seasonal cycle of spermatogenesis is correlated with a seasonal fatted appearance of the male (Kinzey 1971; Baldwin and Baldwin 1981; Boinski 1987; see also Chapter 7). During this time, the male is heavy with subcutaneous fat in the upper torso, shoulders, and arms. This appearance is gradually lost at the end of the breeding season and is absent for much of the year (Kinzey 1971). Male mouse lemurs (*Microcebus murinus*) also gain weight prior to the mating season, probably in response to testicular tissue development and steroid action (Schmid and Kappeler 1998). In other species, spermatogenesis may continue throughout the year but sperm quality may differ. In male spider monkeys (*A. geoffroyi*), sperm quality appears (in terms of motility and percent of live sperm) to be better during months when females are more likely to conceive

(Hernández-López et al. 2002), suggesting that declining female body condition may not be the only factor leading to reduced conception rates.

CONCLUSIONS

The information presented in this chapter briefly reviews the sexual and reproductive biology of the nonhuman primates. It should be obvious to the reader that the topics presented here are diverse and that the primate order also shows a great deal of diversity in many of the areas covered. Indeed, entire books have been written on the topics presented here, and I would refer anyone interested in further information to sources such as Dixson (1998) for a more extensive review. Future studies of primate sexuality and reproduction will benefit greatly from advancements in endocrinological and genetic techniques such as those discussed in Chapters 21 and 22 of this volume. In addition, an important avenue of future research is the increasing tendency within the primatological community to study species about which we currently know little. Data across all taxonomic categories will allow us to perform more complete comparative studies and help to broaden our knowledge in this fascinating field of study.

REFERENCES

Alexander, R. D. (1979). Sexuality and sociality in humans and other primates. In: Katchadourian, A. (ed.), *Human Sexuality: A Comparative and Developmental Perspective*. University of Californian Press, Berkeley. pp. 81–97.

Altmann, J., Alberts, S. C., Haines, S. A., Dubach, J., Muruthi, P., Coote, T., Geffen, E., Cheeseman, D. J., Mututua, R. S., Saiyalele, S. N., Wayne, R. K., Lacy, R. C., and Bruford, M. W. (1996). Behavior predicts genetic structure in a wild primate group. *Proc. Natl. Acad. Sci. USA* 93:5797–5801.

Altmann, J., Altmann, S. A., Hausfater, G., and McCuskey, S. A. (1977). Life history of yellow baboons: physical development, reproductive parameters, and infant mortality. *Primates* 18:315–330.

Andelman, S. J. (1987). Evolution of concealed ovulation in vervet monkeys (*Cercopithecus aethiops*). *Am. Nat.* 129:785–789.

Baker, R. R., and Bellis, M. A. (1993). Human sperm competition: ejaculate manipulation by females and a function for the female orgasm. *Anim. Behav.* 46:887–909.

Baldwin, J. D., and Baldwin, J. I. (1981). The squirrel monkeys, genus *Saimiri*. In: Coimbra-Filho, A. F., and Mittermeier, R. A. (eds.), *Ecology and Behavior of Neotropical Primates*, vol. 1. Academia Brasileira de Ciências, Rio de Janeiro. pp. 277–330.

Bao-Ping, R., Shu-Yi, Z., Shu-Zhong, X., Qing-Fen, L., Bing, L., and Min-Qiang, L. (2003). Annual reproductive behavior of *Rhinopithecus roxellana*. *Int. J. Primatol.* 24:575–589.

Beach, F. A. (1976). Sexual attractivity, proceptivity and receptivity in female mammals. *Horm. Behav.* 7:105–138.

Bearder, S. K. (1987). Lorises, bushbabies, and tarsiers: diverse societies in solitary foragers. In: Smuts, B. B., Cheney, D. L., Seyfarth, R. M., Wrangham, R. W., and Strushsaker, T. T. (eds.), *Primate Societies*. University of Chicago Press, Chicago. pp. 11–24.

Benshoff, L., and Thornhill, R. (1979). The evolution of monogamy and concealed ovulation in humans. *J. Soc. Biol.* 2:95–106.

Bercovitch, F. B. (1987). Female weight and reproductive condition in a population of olive baboons (*Papio cynocephalus anubis*). *Am. J. Primatol.* 12:189–195.

Bercovitch, F. B., and Nürnberg, P. (1996). Socioendocrine and morphological correlates of paternity in rhesus macaques. *J. Reprod. Fertil.* 107:59–68.

Bernstein, I. S., and Gordon, T. P. (1980). Mixed taxa introduction, hybrids and macaques systematics. In: Lindburg, D. G. (ed.), *The Macaques: Studies in Ecology, Behavior and Evolution*. Van Nostrand Reinhold, New York. pp. 125–147.

Bielert, C., and Girolami, C. (1986). Experimental assessments of behavioral and anatomical components of female chacma beboon (*Papio ursinus*) sexual attractiveness. *Psychoneuroendocrinology* 11:75–90.

Bielert, C., Girolami, L., and Jowell, S. (1989). An experimental examination of the colour component in visually mediated sexual arousal of the male chacma baboon (*Papio ursinus*). *J. Zool.* 219:569–579.

Boinski, S. (1987). Birth synchrony in squirrel monkeys (*Saimiri oerstedi*): a strategy to reduce neonatal predation. *Behav. Ecol. Sociobiol.* 21:393–400.

Booth, C. (1962). Some observations on behaviour of *Cercopithecus* monkeys. *Ann. N. Y. Acad. Sci.* 102:477–487.

Brockman, D. K., and Whitten, P. L. (1996). Reproduction in free-ranging *Propithecus verrauxi*: estrus and the relationship between multiple partner matings and fertilization. *Am. J. Phys. Anthropol.* 100:57–69.

Bronson, F. H., and Heideman, P. D. (1988). Seasonal regulation of reproduction in mammals. In: Knobil, E., and Neill, J. D. (eds.), *The Physiology of Reproduction*. Raven Press, New York. pp. 541–582.

Bullock, D. W., Paris, C. A., and Goy, R. W. (1972). Sexual behavior, swelling of the sex skin and plasma progesterone in the pigtail macaque. *J. Reprod. Fertil.* 31:225–236.

Burley, N. (1979). The evolution of concealed ovulation. *Am. Nat.* 114:835–858.

Burt, A. (1992). "Concealed ovulation" and sexual signals in primates. *Folia Primatol.* 58:1–6.

Busse, C. D., and Estep, D. Q. (1984). Sexual arousal in mail pigtailed monkeys (*Macaca nemestrina*); effects of serial matings by two males. *J. Comp. Psychol.* 98:227–231.

Butler, H. (1974). Evolutionary trends in primate sex cycles. In: *Contributions to Primatology*, vol. 3. Karger, Basel. pp. 2–35.

Campbell, C. J. (2000). The reproductive biology of the black-handed spider monkey: integrating behavior and endocrinology [PhD thesis]. University of California, Berkeley.

Campbell, C. J. (2004). Patterns of behavior across reproductive states of free-ranging female black-handed spider monkeys (*Ateles geoffroyi*). *Am. J. Phys. Anthropol.* 124(2):166–176.

Campbell, C. J. (in press). Copulatory behavior of free ranging black-handed spider monkeys (*Ateles geoffroyi*). *Am. J. Primatol.*

Campbell, C. J., Shideler, S. E., Todd, H. E., and Lasley, B. L. (2001). Fecal analysis of ovarian cycles in female black-handed spider monkeys (*Ateles geoffroyi*). *Am. J. Primatol.* 54:79–89.

Carosi, M., Ulland, A. E., Gerald, M. S., and Suomi, S. J. (2001). Male-like external genitalia in female tufted capuchins (*Cebus apella*) and the presence of a clitoral bone (baubellum): a cross sectional study. *Folia Primatol.* 72:149.

Carpenter, C. R. (1942). Sexual behavior of free ranging rhesus monkeys (*Macaca mulatta*). I. Specimens, procedures and behavioral characteristics of estrus. *J. Comp. Psychol.* 33:118–142.

Chambers, K. C., and Phoenix, C. H. (1987). Differences among ovariectomized female rhesus macaques in the display of sexual behavior without and with estradiol treatment. *Behav. Neurosci.* 101:303–308.

Chapman, C. A., and Chapman, L. J. (1990). Reproductive biology of captive and free-ranging spider monkeys. *Zoo Biol.* 9:1–9.

Clutton-Brock, T. H., and Harvey, P. H. (1976). Evolutionary rules and primate societies. In: Bateson, P. P. G., and Hinde, R. A. (eds.), *Growing Points in Ethology*. Cambridge University Press, Cambridge. pp. 195–237.

Collings, M. R. (1926). A study of the cutaneous reddening and swelling about the genitalia of the monkey *Macacus rhesus*. *Anat. Rec.* 33:271–287.

Conaway, C. H., and Koford, C. B. (1965). Estrous cycles and mating behavior in a free-ranging band of rhesus monkeys. *J. Mammal.* 45:577–588.

Daniels, D. (1983). The evolution of concealed ovulation and self-deception. *Ethol. Sociobiol.* 4:69–87.

Datta, S. (1987). The evolution of concealed ovulation in vervet monkeys. *TREE* 2:323–324.

Dewsbury, D. A., and Pierce, J. D. (1989). Copulatory patterns of primates as viewed in a broad mammalian perspective. *Am. J. Primatol.* 15:51–72.

Di Bittetti, M. S., and Janson, C. H. (2000). When will the stork arrive? Patterns of birth seasonality in neotropical primates. *Am. J. Primatol.* 50:109–130.

Dierschke, D. S., Weiss, G., and Knobil, E. (1974). Sexual maturation in the female rhesus monkeys and the development of estrogen-induced gonadotropic hormone release. *Endocrinology* 94:198–206.

Dixson, A. F. (1983). Observations on the evolution and behavioral significance of "sexual" skin in female primates. *Adv. Stud. Behav.* 13:63–106.

Dixson, A. F. (1987a). Observations on the evolution of the genitalia and copulatory behaviour in male primates. *J. Zool. Lond.* 213:423–443.

Dixson, A. F. (1987b). Baculum length and copulatory behavior in primates. *Am. J. Primatol.* 13:51–60.

Dixson, A. F. (1992). Observations on postpartum changes in hormones and sexual behavior in callitrichid primates: do females exhibit post partum "estrus"? In: Itoigawa, N., Sugiyama, Y., Sackett, G. P., and Thompson, R. K. R. (eds.), *Topics in Primatology*, vol. 2. University of Tokyo Press, Tokyo. pp. 141–149.

Dixson, A. F. (1993). Sexual selection, sperm competition and the evolution of sperm length. *Folia Primatol.* 61:221–227.

Dixson, A. F. (1998). *Primate Sexuality: Comparative Studies of the Prosimians, Monkeys, Apes, and Human Beings*. Oxford University Press, Oxford.

Eaton, G. G., Slob, A., and Resko, J. A. (1973). Cycles of mating behaviour, oestrogen and progesterone in the thick-tailed bushbaby (*Galago crassicaudatus*) under laboratory conditions. *Anim. Behav.* 21:309–315.

Eberhard, W. G. (1985). *Sexual Selection and Animal Genitalia*. Harvard University Press, Cambridge, MA.

Epple, G. (1976). Chemical communication and reproductive processes in nonhuman primates. In: Doty, R. L. (ed.), *Mammalian Olfaction, Reproductive Processes, and Behavior*. Academic Press, New York. pp. 257–282.

Evans, C. S., and Goy, R. W. (1968). Social behaviour and reproductive cycles in captive ring-tailed lemurs (*Lemur catta*). *J. Zool. Lond.* 156:181–197.

Fooden, J. (1967). Complementary specialization of male and females reproductive structures in the bear macaque, *Macaca arctoides*. *Nature* 214:939–941.

Fooden, J. (1976). Provisional classification and key to living species of macaques (primates: *Macaca*). *Folia Primatol.* 25:225–236.

Fox, E. A. (2001). Homosexual behavior in wild Sumatran orangutans (*Pongo pygmaeus abelii*). *Am. J. Primatol.* 55:177–181.

Gachot-Neveu, H., Petit, M., and Roeder, J. J. (1999). Paternity determination in two groups of *Eulemur fulvus mayottensis*: implications for understanding mating strategies. *Int. J. Primatol.* 20:107–119.

Girolami, C., and Bielert, C. (1987). Female perineal swelling and its effects on male sexual arousal: on apparent sexual releaser in the chacma baboon (*Papio ursinus*). *Int. J. Primatol.* 8:651–661.

Goldfoot, D. A., Westerborg-Van loon, H., Groeneveld, W. and Slob, A. K. (1980). Behavioral and physiological evidence of sexual climax in the female stump-tailed macaque (*Macaca arctoides*). *Science* 208:1477–1479.

Gomendio, M., and Roldan, E. R. S. (1991). Sperm competition influences sperm size in mammals. *Proc. R. Soc. Lond. B* 243:181–185.

Goodall, J. (1986). *The Chimpanzees of Gombe: Patterns of Behavior*. Belknap Press, Harvard.

Goodman, L. M., and Wislocki, G. B. (1935). Cyclical uterine bleeding in a New World monkey (*Ateles geoffroyi*). *Anat. Rec.* 61:379–387.

Gordon, T. P., Rose, R., and Bernstein, I. S. (1976). Seasonal rhythm in plasma testosterone levels in the rhesus monkey (*Macaca mulatta*): a three-year study. *Horm. Behav.* 7:229–243.

Hamilton, W. J. III (1984). Significance of paternal investment by primates to the evolution of adult male-female associations. In: Taub, D. M. (ed.), *Primate Paternalism*. Van Nostrand Reinhold, New York. pp. 309–335.

Harcourt, A. H., Fossey, D., Stewart, K. J., and Watts, D. P. (1980). Reproduction in wild gorillas and some comparisons with chimpanzees. *J. Reprod. Fertil.* 28(suppl.):59–70.

Harcourt, A. H., Harvey, P. H., Larson, S. G., and Short, R. V. (1981a). Testis weight, body weight and breeding system in primates. *Nature* 293:55–57.

Harcourt, A. H., Stewart, K. J., and Fossey, D. (1981b). Gorilla reproduction in the wild. In: Graham, C. E. (ed.), *Reproductive Biology of the Great Apes: Comparative and Biomedical Perspectives*. Academic Press, New York. pp. 265–318.

Heape, W. (1900). The "sexual season" of mammals and the relation of the "pro-oestrum" to menstruation. *Q. J. Microsc. Sci.* 44:1–70.

Herbert, J. (1970). Hormones and reproductive behavior in rhesus and talapoin monkeys. *J. Reprod. Fertil.* 11:119–140.

Hernández-López, L., Mayagoita, L., Esquivel-Lacroix, C., Rojas-Maya, S., and Mondragon-Ceballos, R. (1998). The menstrual cycle of the spider monkey (*Ateles geoffroyi*). *Am. J. Primatol.* 44:183–195.

Hernández-López, L., Parra, G. C., Cerda-Molina, A. L., Pérez-Bolaños, S. C., Sánchez, V. D., and Mondragón-Ceballos, R.

(2002). Sperm quality differences in captive black-handed spider monkeys (*Ateles geoffroyi*). *Am. J. Primatol.* 57:35–41.

Hill, W. C. O. (1962). *Primates, Comparative Anatomy and Taxonomy. Cebidae*, part B, vol. 4. Edinburgh University Press, Edinburgh.

Hill, W. C. O. (1966). *Primates, Comparative Anatomy and Taxonomy. Catarrhini, Cercopithecoidea, Cercopithecus*, vol. 6. Edinburgh University Press, Edinburgh.

Hill, W. C. O. (1972). *Evolutionary Biology of the Primates.* Academic Press, London.

Hodges, J. K., Gulick, B. A., Czekala, N. M., and Lasley, B. L. (1981). Comparison of urinary oestrogen excretion in South American primates. *J. Reprod. Fertil.* 61:83–90.

Hrdy, S. B. (1977). *The Langurs of Abu: Female and Male Strategies of Reproduction.* Harvard University Press, Cambridge, MA.

Hrdy, S. B. (1981). *The Women that Never Evolved.* Harvard University Press, Cambridge, MA.

Hrdy, S. B., and Whitten, P. L. (1987). Patterning of sexual activity. In: Smuts, B. B., Cheney, D. L., Seyfarth, R. M., Wrangham, R. W., and Struhsaker, T. T. (eds.), *Primate Societies.* University of Chicago Press, Chicago. pp. 370–384.

Jolly, A. (1966). *Lemur Behavior.* University of Chicago Press, Chicago.

Jolly, C. J., and Phillips-Conroy, J. E. (2003). Testicular size, mating system, and maturation schedules in wild anubis and hamadryas baboons. *Int. J. Primatol.* 24:125–142.

Kano, T. (1992). *The Last Ape: Pygmy Chimpanzee Behavior and Ecology.* Stanford University Press, Stanford, CA.

Kauffman, J. H. (1965). A three-year study of mating behavior in a freeranging band of rhesus monkeys. *Ecology* 46:500–512.

Kawai, M., Azuma, S., and Yoshiba, K. (1967). Ecological studies of reproduction in Japanese monkeys (*Macaca fuscata*). 1. Problems of the birth season. *Primates* 8:35–74.

Keverne, E. B. (1976). Sexual receptivity and attractiveness in the female rhesus monkey. *Adv. Study Behav.* 7:155–196.

Kinzey, W. G. (1971). Male reproductive system and spermatogenesis. In: Hafez, E. S. E. (ed.), *Comparative Reproduction of Nonhuman Primates.* Charles C. Thomas, Springfield, IL. pp. 85–114.

Kirkpatrick, R. C. (2000). The evolution of human homosexual behavior. *Curr. Anthropol.* 41:385–413.

Klein, L. L. (1971). Observations on copulation and seasonal reproduction of two species of spider monkey, *Ateles belzebuth* and *A. geoffroyi. Folia Primatol.* 15:233–248.

Knott, C. D. (1997). Interactions between energy balance, hormonal patterns, and mating behavior in wild Borean orangutans (*Pongo pygmaeus*). *Am. J. Primatol.* 42:124.

Knott, C. (2001). Female reproductive ecology of the apes. In: Ellison, P. T. (ed.), *Reproductive Ecology and Human Evolution.* Aldine de Gruyter, New York. pp. 429–464.

Koyama, N., Nakamichi, M., Oda, R., Miyamoto, N., Ichino, S., and Takahata, Y. (2001). A ten year summary of reproductive parameters for ring-tailed lemurs at Berenty, Madagascar. *Primates* 42:1–14.

Lindburg, D. G. (1971). The rhesus monkey in north India: an ecological and behavioral study. In: Rosenblum, L. A. (ed.), *Primate Behavior*, vol. 2. Academic Press, New York. pp. 1–106.

Long, C. A., and Frank, T. (1968). Morphometric function and variation in the baculum with comments on the correlation of parts. *J. Mammal.* 49:32–43.

Lovejoy, O. (1981). The origins of man. *Science* 211:241–250.

Loy, J. (1970). Peri-menstrual sexual behavior among rhesus monkeys. *Folia Primatol.* 13:286–297.

Loy, J. (1987). The sexual behavior of African monkeys and the question of estrus. In: Zucker, E. L. (ed.), *Comparative Behavior of African Monkeys.* Alan R. Liss, New York. pp. 175–195.

Manley, G. H. (1966). Reproduction in lorisoid primates. In: Rowlands, I. W. (ed.), *Comparative Biology of Reproduction in Mammals.* Academic Press, New York. pp. 493–509.

Manson, J. H. (1992). Measuring female mate choice in Cayo Santiago rhesus macaques. *Anim. Behav.* 44:405–416.

Masters, W. H., and Johnson, V. A. (1966). *Human Sexual Response.* J. and A. Churchill, London.

McDaniel, P. S., Janzow, F. T., Porton, I., and Asa, C. S. (1993). The reproductive and social dynamics of captive *Ateles geoffroyi* (black-handed spider monkey). *Am. Zool.* 33:173–179.

Michael, R. P., and Zumpe, D. (1970). Rhythmic changes in the copulatory frequencies of rhesus monkeys (*Macaca mulatta*) in relation to the menstrual cycle, and a comparison with the human cycle. *J. Reprod. Fertil.* 21:199–201.

Michener, G. R. (1984). Copulatory plugs in Richardsons' ground squirrels. *Can. J. Zool.* 62:267–270.

Milton, K. (1981). Estimates of reproductive parameters for free-ranging *Ateles geoffroyi. Primates* 22:574–579.

Milton, K. (1985). Mating patterns of woolly spider monkeys, *Brachyteles arachnoides*: implications for female choice. *Behav. Ecol. Sociobiol.* 17:53–59.

Møller, A. P. (1988). Ejaculate quality, testes size and sperm competition in primates. *J. Hum. Evol.* 17:479–488.

Morris, D. (1967). *The Naked Ape: A Zoologist's Study of the Human Animal.* Johnathan Cape, London.

Nadler, R. D. (1977). Sexual behavior of captive orang-utans. *Arch. Sex. Behav.* 6:457–475.

Nadler, R. D. (1990). Homosexual behavior in nonhuman primates. In: McWhirter, D., Sanders, P., and Reinisch, J. (eds.), *Homosexuality/Heterosexuality: Concepts of Sexual Orientation.* Oxford University Press, New York. pp. 138–170.

Nishimura, A. (1988). Mating behavior of woolly monkeys *Lagothrix lagotricha*, at La Macarena, Colombia. *Field Stud. New World Monkeys* 1:19–27.

Nunn, C. L. (1999). The evolution of exaggerated sexual swellings in primates and the graded-signal hypothesis. *Anim. Behav.* 58:229–246.

Pagel, M. (1994). The evolution of conspicuous prestrous advertisement in Old World monkeys. *Anim. Behav.* 27:1–36.

Parga, J. A. (2003). Copulatory plug displacement evidences sperm competition in *Lemur catta. Int. J. Primatol.* 24:889–899.

Parker, G. A. (1970). Sperm competition and its evolutionary consequences in the insects. *Biol. Rev.* 45:525–567.

Peter-Rousseaux, A. (1964). Reproductive physiology and behaviour of the Lemuroidea. In: Buettner-Janusch, J. (ed.), *Evolutionary and Genetic Biology of Primates.* Academic Press, New York. pp. 92–132.

Phoebus, E. C. (1982). Primate female orgasm. *Am. J. Primatol.* 2:223–224.

Pope, T. R. (1990). The reproductive consequences of male cooperation in the red howler monkey: paternity exclusion using genetic markers. *Behav. Ecol. Sociobiol.* 27:439–446.

Profet, M. (1993). Menstruation as a defense against pathogens transported by sperm. *Q. Rev. Biol.* 68:335–386.

Pullen, S. L., Bearder, S. K., and Dixson, A. F. (2000). Preliminary observations on sexual behavior and the mating system in free-ranging lesser galagos (*Galago moholi*). *Am. J. Primatol.* 51:79–88.

Ransom, T. W. (1981). *Beach Troop of the Gombe*. Bucknell University Press, Lewisburg.

Rostal, D. C., Glick, B. B., Eaton, G. G., and Resko, J. A. (1986). Seasonality of adult male Japanese macaques (*Macaca fuscata*): androgens and behavior in a confined troop. *Horm. Behav.* 20:452–462.

Rowell, T. E. (1972). Female reproduction cycles and social behavior in primates. In: Lehrmen, D. S., Hinde, R. A., and Shaw, E. (eds.), *Advances in the Study of Behavior*, vol. 4. Academic Press, New York. pp. 69–105.

Rowell, T. E., and Dixson, A. F. (1975). Changes in the social organization during the breeding season of wild talapoin monkeys. *J. Reprod. Fertil.* 43:419–434.

Saayman, G. S. (1970). The menstrual cycle and sexual behaviour in a troop of free-ranging chacma baboons (*Papio ursinus*). *Folia Primatol.* 12:81–110.

Sauther, M. L. (1991). Reproductive behavior of free-ranging *Lemur catta* at Beza Mahafaly Special Reserve, Madagascar. *Am. J. Phys. Anthropol.* 84:463–477.

Schmid, J., and Kappeler, P. M. (1998). Fluctuating sexual dimorphism and differential hibernation by sex in a primate, the grey mouse lemur (*Microcebus murinus*). *Behav. Ecol. Sociobiol.* 43:125–132.

Schultz, A. H. (1938). The relative weights of the testes in primates. *Anat. Rec.* 72:387–394.

Schulze, H. G., and Meier, B. (1995). Behavior of captive *Loris tardigradus nordicus*: a qualitative description, including some information about morphological bases of behavior. In: Doyle, G. A., and Izard, M. K. (eds.), *Creatures of the Dark, the Nocturnal Prosimians*. Plenum Press, New York. pp. 221–249.

Schwab, D. (2000). A preliminary study of spatial distribution and mating system of pygmy mouse lemurs (*Microcebus myoxinus*). *Am. J. Primatol.* 51:41–60.

Shideler, S. E., Lindburg, D. G., and Lasley, B. L. (1983). Estrogen–behavior correlates in the reproductive physiology of the ruffed lemur (*Lemur variegatus*). *Horm. Behav.* 17:249–263.

Short, R. V. (1979). Sexual election and its component parts, somatic and genital selection, as illustrated by man and the great apes. *Adv. Study Behav.* 9:131–158.

Smith, R. L. (1984). Human sperm competition. In: Smith, R. L. (ed.), *Sperm Competition and the Evolution of Animal Mating Systems*. Wildwood House, London. pp. 601–660.

Strassmann, B. T. (1981). Sexual selection, paternal care, and concealed ovulation in humans. *Ethol. Sociobiol.* 2:31–40.

Strassmann, B. T. (1996). The evolution of endometrial cycles and menstruation. *Q. Rev. Biol.* 71:181–220.

Strier, K. B. (1996). Reproductive ecology of female muriquis. In: Norconk, M. A., Rosenberger, A. L., and Garber, P. E. (eds.), *Adaptive Radiations of Neotropical Primates*. Plenum Press, New York. pp. 511–532.

Strier, K. B., Mendes, S. L., and Santos, R. R. (2001). Timing of births in sympatric brown howler monkeys (*Alouatta fusca clamitans*) and northern muriquis (*Brachyteles arachnoides hypoxanthus*). *Am. J. Primatol.* 55:87–100.

Strier, K. B., and Ziegler, T. E. (1997). Behavioral and endocrine characteristics of the reproductive cycle in wild muriqui monkeys, *Brachyteles arachnoides*. *Am. J. Primatol.* 42:299–310.

Struhsaker, T. T. (1975). *The Red Colobus Monkey*. University of Chicago Press, Chicago.

Symington (1987). Ecological and social correlates of party size in the black spider monkey, *Ateles paniscus chamek* [PhD thesis]. Princeton University, Princeton, NJ.

Thierry, B., Heistermann, M., Aujard, F., et al. (1996). Long-term data on basic reproductive parameters and evaluation of endocrine, morphological, and behavioral measures for monitoring reproductive status in a group of semifree-ranging Tonkean macaques (Macaca tonkeana). *Am. J. Primatol.* 39:47–62.

Thompson-Handler, N., Malenky, R. K., and Badrian, N. (1984). Sexual behavior of *Pan paniscus* under natural conditions in the Lomako Forest, Equateur, Zaire. In: Susman, R. L. (ed.), *The Pygmy Chimpanzee: Evolution, Biology, and Behavior*. Plenum Press, New York. pp. 347–368.

Thomson, J. A., Hess, D. L., Dahl, K. D., Illiff-Sizemore, S. A., Stouffer, R. L., and Wolfe, D. P. (1992). The Sulawesi crested black macaque menstrual cycle; changes in perineal tumescence and serum estradiol, progesterone, follicle-stimulating hormone, and lutenizing hormone levels. *Biol. Reprod.* 46:879–884.

Thorington, R. W., Jr., Ruiz, J. C., and Eisenberg, J. F. (1984). A study of a black howling monkey (*Alouatta caraya*) population in Northern Argentina. *Am. J. Primatol.* 6:357–366.

Triosi, A., and Carosi, M. (1998). Female orgasm rate increases with male dominance in Japanese macaques. *Anim. Behav.* 56:1261–1266.

Turke, P. W. (1984). Effects of ovulatory concealment and synchrony on protohominid mating systems and parental roles. *Ethol. Sociobiol.* 5:33–44.

Tutin, C. E. G. (1980). Reproductive behaviour of wild chimpanzees in the Gombe National Park, Tanzania. *J. Reprod. Fertil.* 28(suppl.):43–57.

Valeggia, C. R., and Ellison, P. T. (2001). Lactation, energetics, and postpartum fecundity. In: Ellison, P. T. (ed.), *Reproductive Ecology and Human Evolution*. Aldine de Gruyter, New York. pp. 85–106.

Van Noordwijk, M. A. (1985). Sexual behavior of Sumatran long-tailed macaques (*Macaca fascicularis*). *Z. Tierpsychol.* 70:277–296.

Vasey, P. L. (1995). Homosexual behavior in primates: a review of evidence and theory. *Int. J. Primatol.* 61:173–204.

Vigilant, L., Hofreiter, M., Seidel, H., and Boesch, C. (2001). Paternity and relatedness in wild chimpanzee communities. *Proc. Natl. Acad. Sci. USA* 98:12890–12895.

Wallen, K., and Parsons, W. A. (1997). Sexual behavior in same-sexed nonhuman primates: is it relevant to understanding human homosexuality. *Ann. Revs. Sex. Res.* 8:195–224.

Wallis, J. (1992). Chimpanzee genital swelling and its role in the pattern of sociosexual behavior. *Am. J. Primatol.* 28:101–113.

Watts, D. P. (1991). Mountain gorilla reproduction and sexual behavior. *Am. J. Primatol.* 24:211–225.

Watts, D. P. (1998). Seasonality in the ecology and life histories of mountain gorillas (*Gorilla gorilla berengei*). *Int. J. Primatol.* 19:929–948.

Whitten, P. L., and Brockman, D. K. (2001). Strepsirhine reproductive ecology. In: Ellison, P. T. (ed.), *Reproductive Ecology and Human Evolution*. Aldine de Gruyter, New York. pp. 321–350.

Whitten, P. L., and Russell, E. (1996). Information content of sexual swellings and fecal steroids in sooty mangabeys (*Cercocebus torquatus atys*). *Am. J. Primatol.* 40:67–82.

Wiebe, R. H., Diamond, E., Askel, S., Liu, M. D., and Abee, C. R. (1984). Diurnal variation in androgens in sexually mature Bolivian squirrel monkeys (*Saimiri sciureus*) studied longitudinally. *Acta Endocrinol.* 87:424–433.

Wildt, D. E., Doyle, L. L., Stone, S. C., and Harrison, R. M. (1977). Correlation of perineal swelling with serum ovarian hormone levels, vaginal cytology, and ovarian follicular development during the baboon reproductive cycle. *Primates* 18:261–270.

Wislocki, G. B. (1936). The external genitalia of the simian primates. *Hum. Biol.* 8:309–347.

Wolfe, L. (1979). Behavioral patterns of estrous females of the Arashiyama West troop of Japanese macaques (*Macaca fuscata*). *Primates* 20:525–534.

Wolfe, L. D. (1984). Japanese macaque female sexual behavior: a comparison of Arashiyama East and West. In: Small, M. F. (ed.), *Female Primates: Studies by Women Primatologists*. Alan Liss, New York. pp. 141–157.

Wolfe, L. D. (1991). Human evolution and the sexual behavior of female primates. In: Loy, J., and Peters, C. B. (eds.), *Understanding Behavior: What Primate Studies Tell Us About Human Behavior*. Oxford University Press, New York. pp. 121–151.

Wright, P. C., Toyama, T. M., and Simons, E. I. (1986). Courtship and copulation in *Tarsius bancanus*. *Folia Primatol.* 46:142–148.

Youlatus, D. (2002). Positional behavior of black spider monkeys (*Ateles paniscus*) in French Guiana. *Int. J. Primatol.* 23:1071–1094.

Zhang, S., Liang, B., and Wang, L. (2000). Seasonality of matings and births in captive Sichuan golden monkeys (*Rhinopithecus roxellana*). *Am. J. Primatol.* 51:265–269.

Ziegler, T., Hodges, K., Winkler, P., and Heistermann, M. (2000). Hormonal correlates of reproductive seasonality in wild female hanuman langurs (*Presbytis entellus*). *Am. J. Primatol.* 51:119–134.

Zinner, D., Deschner, T. (2000). Sexual swellings in female hamadryas baboons after male take-overs: "Deceptive" swellings as a possible female counter-strategy against infanticide. *Am. J. Primatol.* 52:157–168.

Zuckerman, S., Van Wagenen, G., and Gardner, R. H. (1938). The sexual skin of the rhesus monkey. *Proc. Zool. Soc. Lond.* 108:385–401.

26

Reproductive Cessation in Female Primates
Comparisons of Japanese Macaques and Humans

Linda Marie Fedigan and Mary S. M. Pavelka

THE EVOLUTION OF MENOPAUSE

In the past 25 years or so, menopause has emerged from the private domain of aging women and into public view. As the postwar bulge in the populations of developed nations (the "baby boomers") reached middle age, science and the media alike began to take an increasing interest in understanding why human females universally cease to experience menstrual cycles and fertility around the age of 50. A recent check of a popular search engine with the key word *menopause* garnered close to 9 million results on the worldwide web, and databases for journal articles indicate that more than 5,000 papers have been published on this subject in the biological literature alone. Clearly, this is no longer a topic that is discussed only rarely and guardedly among older women.

The vast majority of the scientific and popular literature on menopause concerns the treatment of its side effects. However, evolutionary biologists and physical anthropologists have also begun to address the issues of why menopause originated, what maintains it, and whether it is unique to humans or common in mammalian species. It was to be expected that the search for an evolutionary context and an animal model of menopause would turn to the accumulating data on life histories and reproductive senescence in the nonhuman primates, our closest relatives.

In this chapter, we first describe the competing explanations for the origin and maintenance (selection) of menopause in humans and then turn to the search for a larger understanding of reproductive senescence in female primates. As a life history characteristic of human females, menopause is universal, it happens halfway through the maximum life span of our species, and it consistently occurs at the average age of 50–52 years in different populations around the globe and throughout history. There is some debate about whether menopause is a discrete event or merely the end point of a long process of follicular depletion (Wood et al. 2001), whether the term might also apply to reproductive decline in men (Peccei 2001), whether individual variation in age at menopause is affected by genetic and/or environmental factors (Brambilla and McKinlay 1989), and whether increasing life expectancies in developed countries indicate that menopause is a recent artifact of modernization (Sherman 1998, Leidy 1999). We have described/addressed these debates in prior publications (e.g., Pavelka and Fedigan 1991). In this chapter, we focus on competing explanations for the distinctive pattern called menopause in the human female, that is, the pattern of permanent and universal cessation of menstrual cycles in females who live to the age of 60, an aging effect that occurs in middle age, long before the senescence of other somatic systems.

There are three primary evolutionary explanations for menopause, which, as noted by Peccei (2001), are not mutually exclusive. These can be referred to as the adaptation explanation (the grandmother or maternal investment hypothesis) and the two epiphenomenon explanations, antagonistic pleiotropy (the trade-off hypothesis) and the physiological constraint model (the prolonged life hypothesis).

Taking these in reverse order, the *physiological constraint model* argues that during the course of human evolution there was strong selection for a longer life span but that the reproductive system of the human female could not keep pace (Weiss 1981). Judging by brain-to-body ratios in fossil hominids, the maximum human life span increased from about 50 years in early hominids to approximately 120 years in *Homo sapiens* (Bogin and Smith 1996, Hammer and Foley 1996). However, the age at which human females cease to experience ovarian cycles appears to have remained stationary at approximately 50 years, which is not much different from the age at which great ape females cease to cycle (Graham 1979, Nishida et al. 2003). Why would this be the case? One answer is that human females are born with all the oocytes and primordial follicles they will ever possess (approximately 1 million), a mammalian pattern that is referred to as *semelgametogenesis*. It is widely held that oocytes and follicles are depleted throughout a woman's life until they reach a minimum threshold below which hormonal signals fail, at which point cycling ceases (Armstrong 2001). However, there is some recent evidence from mice that it is actually the germline stem cells that deplete and that female mice may be able to produce new oocytes during their lifetimes (Johnson et al. 2004). Nonetheless, this explanation holds that, due to the physiological constraints of what we might refer to as the "shelf life" of eggs, it may simply not be possible to select for longer and longer reproductive life spans in mammalian females. Thus, menopause is viewed as an epiphenomenon or side effect of selection for a prolonged life span.

However, the epiphenomenon explanation that is more commonly endorsed by evolutionary biologists is known as *antagonistic pleiotropy*, or the *trade-off hypothesis* (e.g., Williams 1957, Rose 1991, Gosden and Faddy 1998). According to this model, patterns that have high adaptive value early in the life course (intense reproductive output) will be selected for even if they result in reduced fitness (follicular depletion) later in the life course. Certain genes, such as those governing semelgametogenesis, may be selected if they have pleiotropic fitness effects, that is, beneficial effects at early ages and deleterious effects at later ages. In the wider research on somatic aging, it is commonly accepted that antagonistic pleiotropy is a good explanation for the evolution of senescence (Wood et al. 2001).

In a sense, pleiotropy is an adaptive explanation for menopause since it argues that there is an adaptive benefit to follicular depletion early in life that outweighs its later deleterious effects. However, the *grandmother hypothesis*, developed by several theorists (e.g., Lancaster and King 1985; Hawkes et al. 1989, 1997; Alvarez 2000; Hawkes 2003) is the only one of the three explanations that argues for a direct benefit of midlife cessation of fertility in the human female and is widely regarded as the *adaptive explanation*. This hypothesis proposes that selection has favored cessation of reproduction midway through the life span because as human females age there are greater benefits from investing care in existing children and grandchildren than from continuing to produce additional offspring. In other words, older women will be more reproductively successful if they put their energies toward ensuring the survival of their offspring than they would be by producing yet more children. Within this directly adaptive school of thought, there are several hypotheses related through the underlying argument that a postreproductive life exists in order to facilitate enhanced parental investment. Peccei (2001) distinguished between a "mother hypothesis," focused on improving survivorship and fertility of the first generation of offspring, and a "grandmother hypothesis," directed toward increased survival rates in grandchildren. However, in terms of inclusive fitness, the benefits to females and their children and grandchildren are clearly interrelated. If a female protects or provisions her grandchildren, she may enhance their survival as well as the reproductive success of their parents (her children) and, thus, her own inclusive fitness. Hence, the *grandmother hypothesis* is often used as the general and more memorable term for postreproductive investment in any existing progeny, even though *enhanced maternal investment hypothesis* might be a more global and accurate name.

There is a growing body of literature that addresses the grandmother hypothesis in humans. A few studies have examined what are considered to be costs of old-age reproduction in human females, such as age-related risks of dying in childbirth, risks to fetuses and neonates resulting from old-age pregnancies, risks to siblings of maternal death during childbirth, and fertility costs to offspring from sibling competition. According to Peccei (2001), the only clearly established cost of old-age reproduction is increased risk of fetal loss, stillbirths, and birth defects. However, Mace (2000) modeled sibling competition and concluded that competition between siblings for food, status, territory, breeding opportunities, or kin support can greatly increase the cost of maintaining a mother's fertility beyond the age of 50 years.

Other studies have looked for fitness benefits of ceasing to reproduce, with mixed results. For example, Hill and Hurtado (1996) addressed the question of whether the presence of living, helpful grandmothers among the Ache of South America significantly enhanced the fertility of their children and the survival of grandchildren. These authors concluded that the presence of postreproductive mothers and grandmothers did not enable their children to raise more offspring. Rogers (1993) developed a theoretical model that came to the same conclusion. However, Hawkes and collaborators (1989, 1997, 2003) provided evidence that postmenopausal Hadza grandmothers supply sufficient surplus calories and babysitting services to allow their daughters to successfully raise more offspring. Blurton-Jones et al. (1999) concluded that grandmothering facilitated the evolution of earlier weaning in hominids, and Mace (2000) showed that babies in The Gambia are more likely to survive if their maternal grandmother is alive. Most recently, Lahdenpera et al. (2004) documented that premodern Finnish and Franco-Canadian women with a prolonged postreproductive life span have more grandchildren. There is scattered evidence from a variety of societies that older women are a substantial help to their progeny, but whether or not this has a significant effect on inclusive fitness is still a matter of debate (e.g., compare Kaplan et al. 2000 to Hawkes 2003, 2004).

There is further divergence of opinion between those who see the grandmother hypothesis as applicable only to the case of human females (Lancaster and King 1985, Hawkes et al. 1989) and those who would extend this adaptive explanation to other species of primates (Hrdy 1981, Sommer et al. 1992, Paul et al. 1993). Although universal midlife termination of female reproduction is not known to occur in any primate species other than humans, there are several reports of individual female monkeys and apes living part of their lives in a postreproductive state. This has generated interest in the possibility that some form of early reproductive termination may have been selected for in nonhuman primate females.

PRIMATE COMPARISONS: THE SEARCH FOR PARALLELS IN OLD-AGE REPRODUCTION

Endocrine Studies of Reproductive Senescence in Captive Female Primates

If menopause is not unique to humans but rather shared with the nonhuman primates as part of our phylogenetic heritage, then we do not need a special explanation for its existence in *Homo sapiens*. Furthermore, if we could find a nonhuman primate model suitable for laboratory research on aging and for testing various clinical aspects of ovarian decline, then we could learn more about this phenomenon in our own species. Thus, it is not surprising that biomedical primatologists spent a good part of the 1970s and 1980s examining cases in which captive female primates lost ovarian function.

Macaques (genus *Macaca*) are the most common type of primate kept in biomedical labs, and several examinations of

colony records showed that particular aged female macaques cease to experience ovarian cycles around the age of 22 years, if they live to be that old (e.g., Hodgen et al. 1977, Graham et al. 1979, Graham 1986). Maximum life span in captive macaques is about 30–35 years. Chimpanzees (genus *Pan*) are the nonhuman primates most closely related to humans, and examination of reproductive decline in captive chimpanzees also turned up a few cases of very old female chimpanzees that ceased to cycle close to their deaths at 48–50 years of age, although the majority died while still cycling (Graham 1979, Gould et al. 1981).

The few studies available on the hormonal profiles of aging female monkeys indicate that the decline of ovarian function in nonhuman primates parallels the hormonal events associated with menopause in women—prolonged follicular phases, failure to ovulate, breakthrough bleeding, high plasma luteinizing hormone concentrations, low estrogen levels and/or lack of patterned estradiol/pregnonediol-3-glucuronide dynamics, etc. (e.g., Tardif and Ziegler 1992, Nozaki et al. 1995, Gilardi et al. 1997, Bellino and Wise 2003). A study of aged tamarin females indicated there may be some differences in ovarian ageing between neotropical and Old World primates (Tardif and Ziegler 1992), and we clearly have much more to learn on this topic; however, the general pattern of ovarian hormonal senescence (progression to cycle termination) in human and nonhuman female primates appears similar.

Thus, from an endocrinological perspective, reproductive decline may well follow a similar pattern in all primates, and we could use cases of individual postreproductive monkeys and apes as clinical models of the physiological basis for menopause in women. However, from an evolutionary perspective, these studies fail to demonstrate similarity between reproductive senescence in nonhuman primates and menopause in the human female. Instead, they highlight the critical differences: female macaques and chimpanzees that cease to cycle are very close to age at death, whereas human females cease to cycle in middle age; female macaques and chimpanzees cease to cycle on an idiosyncratic basis, whereas human females universally cease to cycle at the average age of approximately 50 years.

Although a recent book on aging in primates (Erwin and Hof 2002) argues that there is a pressing need for endocrine monitoring of the oldest female primates in captivity to document patterns of reproductive senescence, there has been surprisingly little published on this subject in the past decade (for exceptions, see Tardif and Ziegler 1992, Nozaki et al. 1995, Walker 1995, Gilardi et al. 1997, Bellino and Wise 2003, Coleman and Kemnitz 1998).

Demographic and Behavioral Studies of Reproductive Cessation in Socially Living Primates

In Table 26.1, we summarize the available data on reproductive cessation in primates. There are three problems with this data set. The first is the dearth of studies; especially problematic is the relative lack of information from

Table 26.1 Demographic and Behavioral Estimates of Life History Variables Related to Reproductive Cessation in Primates

SPECIES	FREE-RANGING/ CAPTIVE[1]	ESTIMATED LIFE SPAN (YEARS)	OLDEST AGE AT LAST BIRTH (YEARS)	% LIFE SPAN COMPLETED AT TIME OF LAST BIRTH	SAMPLE SIZE	% SAMPLE REACHING REPRODUCTIVE CESSATION	LENGTH OF PRLS (YEARS)	METHOD FOR CALCULATING PRLS[2]
Pan troglodytes[3]	FP	50.0	40	80.0	39	41.0	4.75	2
Papio anubis[4]	FP	27.0	25	92.6	392	?	?	3
Semnopithecus entellus[5]	FP	34.0	32	94.1	19	21.0	5.1	1
Macaca fuscata[6]	FP	32.7	25	67.3	32	25.0	4.5	4
M. mulatta[7]	SF	34.0	25	73.5	285	10	>2.0	5
M. sylvanus[8]	SF	30.0	28	93.3	12	58.0	5.7	1
Pan troglodytes[9]	C	60.0	36	60.0	15	60.0	9.25	6
Pongo pygmaeus[10]	C	58.7	40	68.1	53	31.9	7.08	6
Gorilla gorilla[11]	C	54.0	28	51.9	12	40.0	4.54	6
Papio anubis/cynocephalus hybrids[12]	C	33.4	17	50.9	13	10.0	3.45	6
M. mulatta[13]	C	30.0	20	66.7	38	13.2	2.58	6
M. nemestrina[14]	C	28.9	20	69.2	209	25.6	4.02	6
M. radiata[15]	C	28.9	19	65.7	13	3.8	6.70	6
Chlorocebus aethiops[16]	C	25.4	17	66.9	12	0	N/A	6
Saimiri sciureus[17]	C	21.0	19	90.5	28	32.1	3.28	6
Saguinus fuscicollis[18]	C	15.8	13	82.3	6	20.0	4.02	6
Leontopithecus rosalia[19]	C	24.7	12	48.6	21	47.4	3.87	6
Callithrix jacchus[20]	C	16.2	10	61.7	14	36.4	2.11	6
Lemur spp.[21]	C	28.7	22	76.7	30	44.8	3.56	6
Homo sapiens[22]	N/A	100–120	50	41.7–50.0	106	99.1	29.26	6

Sources: (source for life span value listed first, source for other life history values in this species listed second):

[1] FP, free-ranging but provisioned; SF, semi-free-ranging and provisioned; C, breeding colonies (usually caged).
[2] PRLS, postreproductive life span, calculated as follows: 1, age at death − age at last birth; 2, D − LP − 5 years; 3, cycling (or cessation thereof) inferred from perineal swellings; 4, D − LP − 1.5 years; 5, >25 years of age + no births for 2 years; 6, LP − D/(xIBI + 2SD). D, age at death; LP, age at last parturition; IBI, interbirth interval; SD, standard deviation.
[3] Nishida et al. (2003), Nishida et al. (2003).
[4] Packer et al. (1998), Packer et al. (1998).
[5] Sommer et al. (1992), Sommer et al. (1992).
[6] Takahata et al. (1995), Takahata et al. (1995).
[7] Walker (1995), Johnson and Kapsalis (1998).
[8] Paul et al. (1993), Paul et al. (1993).
[9] Judge and Carey (2000), Caro et al. (1995).
[10] Judge and Carey (2000), Caro et al. (1995).
[11] Kaplan et al. (2000), Caro et al. (1995).
[12] Judge and Carey (2000), Caro et al. (1995).
[13] Walker (1995), Caro et al. (1995).
[14] Judge and Carey (2000), Caro et al. (1995).
[15] Judge and Carey (2000), Caro et al. (1995).
[16] Judge and Carey (2000), Caro et al. (1995).
[17] Judge and Carey (2000), Caro et al. (1995).
[18] Judge and Carey (2000), Caro et al. (1995).
[19] Judge and Carey (2000), Caro et al. (1995).
[20] Judge and Carey (2000), Caro et al. (1995).
[21] Judge and Carey (2000), Caro et al. (1995).
[22] Kaplan et al. (2000), Caro et al. (1995).

free-ranging and semi-free-ranging primates, from which it is very difficult to obtain data on old age and reproduction. Apart from an anecdotal report of one postreproductive *Cercocebus albigena* female (Waser 1978), there are no studies at all of reproductive senescence in nonprovisioned, free-ranging primates. Second, sample sizes are not large, even from the few studies we have. Again, the reason is that it is difficult to obtain data on the reproductive status and patterns of very old primates of known age. The third problem is that there is very little consistency in how reproductive cessation was determined and how postreproductive life span (PRLS) was calculated in these studies. Some researchers (e.g., Sommer et al. 1992, Paul et al. 1993) treat the time lag between last parturition and maternal death as the postreproductive period, when in fact these mothers may simply have died before they could have another infant. Others (e.g., Takahata et al. 1995, Nishida et al. 2003) subtract a value equal to the weaning age of an infant as part of the calculation so that age at death minus age at last parturition minus weaning age is considered the postreproductive period. Altogether, six different ways of calculating PRLS are presented in this table. In our view, the only method that adequately takes account of the mother's own reproductive history is the formula presented by Caro et al. (1995):

$$LP - D/(\bar{x}IBI + 2\,SD)$$

where *LP* is age at last parturition; *D* is age at death; $\bar{x}IBI$ is mean length of a female's interbirth intervals across her lifetime, and SD is standard deviation.

In order for a female to be classified as postreproductive, this method requires that a mother live (without producing an infant) significantly longer than her own average inter-birth interval. Caro's formula reduces the likelihood of categorizing females as reproductively terminated simply because they died before having another baby, and it allows researchers to explore questions about variation within the population at the age at which reproductive termination can occur.

With these shortcomings of the data set in mind, we briefly review here what is presently known about reproductive cessation from demographic and behavioral studies of female primates. Estimated life span in the six free-ranging and semi-free-ranging populations is reported to be 27 years (*Papio anubis*) to 50 years (*Pan troglodytes*). For the 13 examples of captive primates, the estimated life spans range from 16 years (*Saguinus fuscicollis*) to 60 years (*P. troglodytes*). Oldest age at birth in the free-ranging and semi-free-ranging primates is from 25 years (*Macaca fuscata*) to 40 years (*P. troglodytes*), and in the captive populations, oldest age at birth ranges from 10 years (*Callithrix jacchus*) to 40 years (*Pongo pygmaeus*). The length of the postreproductive period ranges 2–9 years, and anywhere from 0% (*Clorocebus aethiops*) to 60% (captive *P. troglodytes*) of the adult females sampled are reported to experience reproductive cessation ($\bar{x} = 28.9\%$, SD = 17.7).

Most nonhuman primate females experience reproductive termination when a large proportion of their life span is already completed (49%–94% of life span completed at time of final birth; $\bar{x} = 72\%$, SD = 14.3). In contrast, human females typically experience reproductive cessation when only 42% of their life span is completed (Table 26.1). Caro et al. (1995) report the example of a human population (eighteenth-century Germans) in which 99.1% of the adult females sampled reached reproductive termination, and they lived an average of more than 29 years in a postreproductive state. As we have previously described, it is universal in human populations for all women who reach the age of 60 to have experienced complete reproductive termination and at a point that is only halfway through the life span. Although life expectancy values are always lower than life spans and are highly variable across time and space, it is nonetheless typical for women to live a substantial proportion of their lives in a postreproductive state.

Our conclusion is that reproductive cessation in human females (menopause) is quite distinct from the patterns of reproductive senescence and termination found in the nonhuman primates. It was the apparent difference between menopause and other forms of primate reproductive termination and the lack of any substantial data on a large sample of nonhuman primate females that led us to begin a series of analyses on old-age reproduction in the Arashiyama West population of Japanese macaque females. In the rest of this chapter, we will outline our ongoing examination of reproductive cessation in these monkeys and how it compares to the human case.

CASE STUDY: REPRODUCTIVE TERMINATION IN JAPANESE MONKEYS

To address some of the problems of earlier cross-species comparisons, we used a large sample ($n = 95$) of completed lives, for which exact ages and reproductive histories are known, to explore a series of questions regarding reproductive termination and postreproductive life in female Japanese macaques. Data for this study were collected on the Arashiyama West population of Japanese monkeys. These monkeys were studied in Japan from 1954 to 1972, then transplanted as an entire group to a large ranch in south Texas, where they were studied in semi-free-ranging conditions until 1996. (For more information on group history, management, demography, and environment, see Fedigan and Asquith 1991, Pavelka 1993). Genealogical records were maintained between 1954 and 1996 on each individual born into the group, which includes date of birth, reproductive history, and date of death or disappearance. Females first mate around 4.5 years of age and can produce their first infant at 5 years of age. Japanese macaques are seasonal breeders, with a fall mating season and a spring birth season (Fedigan and Griffin 1996). The youngest female in our sample died at the age of 5 years, and the oldest lived to 32.6 years.

Reproductively terminated females were identified by Caro et al.'s (1995) interbirth interval criterion, when $LP - D/(\bar{x}IBI + 2 \text{ SD}) > 1.0$; that is, when the time lag between last parturition and the death of the mother ($LP - D$) exceeds two standard deviations of the female's own mean lifetime interbirth interval. This criterion for identifying reproductive termination does not provide direct evidence for the cessation of reproductive capabilities because it relies on externally observable events to infer internal states. However, it is a good estimate of reproductive termination for populations in which hormonal profiles, ovarian histology, and direct measures of menstrual activity are not available. Out of the sample of 95 females, 70 individuals had given birth to at least three infants (the minimum required to calculate a mean interbirth interval and standard deviation).

Reproductive Termination and Age

From the sample of 70 females that had given birth to at least three infants, 20 (28.5%) were identified as having terminated reproduction using the $LP - D/(\bar{x}IBI + 2 \text{ SD}) > 1.0$ criterion. These 20 females ranged from 14.5 to 32.7 years of age at death, with a mean of 24.6 years. Females that continued to reproduce ranged in age at death from 8.8 to 25.7 years, with a mean of 17.3 years. The difference in the mean age at death of the reproductive and reproductively terminated females was statistically significant (Pavelka and Fedigan 1999). No females under the age of 10 were identified as reproductively terminated. While possible, it is unlikely that a female under the age of 10 would meet the criterion for reproductive termination since she would need to produce three infants, then live significantly longer than her own interbirth interval, and then die all within 5 years. Reproductive termination was found in approximately 12% of the females between the ages of 10 and 20. Of particular interest are the females aged 20–25 years. Twenty years of age is uniformly regarded as aged for Japanese macaques (e.g., Pavelka 1991, Takahata et al. 1995) since it represents the beginning of the third trimester of the life span for this species. Yet, 81% of these females were still reproductive, with only 19% showing termination. Thus, old females between the ages of 20 and 25 that become reproductively terminated are not representative of their age class. Continued reproduction in these old females is the norm.

After age 25, however, reproductive termination appears to become universal, with 86% of subjects over the age of 25 (12 out of 14 females) becoming reproductively terminated. Neither of the two subjects over age 25 that were classified as reproductive gave birth after age 25. In fact, they are false-negatives in that they gave birth at 23 and 25 years of age and then quickly thereafter died within the normal range of their interbirth interval. Given that there are no records of any female in this (or any Japanese macaque) population ever giving birth after the age of 25, we appear to be dealing with a biologically meaningful cut-off point in the reproductive lives of female Japanese monkeys. This parallels the findings of Walker (1995) and Johnson and Kapsalis (1998), who report reproductive termination after age 25 for the closely related rhesus macaque.

How Old Is a 25-Year-Old Japanese Monkey in Human Years?

Our finding that reproductive termination occurs at a low frequency in female Japanese monkeys between the ages of 10 and 25, becoming a certainty after age 25, suggests some similarities between Japanese macaque and human females. Women too may stop producing infants and even occasionally experience full-blown menopause in the decades before age 50, with termination becoming universal and certain for women who live into their fifties. However, before we can conclude that Japanese monkeys at age 25 experience something equivalent to the human female menopause, other factors must be considered. Menopause, the complete cessation of ovulation, menstruation, and reproductive capabilities at 50 years of age is universal—not idiosyncratic—among women. Further, it occurs only half way through the species life span of 100 years. This termination of reproductive capabilities in women does not occur in association with extreme old age or with advanced deterioration of the organism as it approaches the maximum life span of the species. Thus, we are led to the following question: How old is a 25-year-old Japanese monkey in human years, and how does she compare to a 50-year-old woman?

One way to draw such a comparison would be to use survivorship values to compare our monkey data with a human population whose survivorship values are unlikely to have been affected by the forces of modernization. For example,

Howell (1979) reports a 2% survivorship to age 85 for the Dobe !Kung, a value which is comparable to the 2.3% survivorship to age 26 for Japanese macaque females (Pavelka and Fedigan 1999). Based on this comparison, one might argue that in Japanese monkeys reproductive termination is unlikely to occur before the equivalent human age of 85 years. However, the use of survivorship values to make comparisons between species is problematic, due to the fact that survivorship values are environmentally dependent and vary widely among populations of the same species.

The question of how old a 25-year-old monkey is compared to a 50-year-old woman might be better approached using the following equation:

$$\frac{\text{Species A Age at Reproductive Termination}}{\text{Species A Maximum Life Span}} = \frac{\text{Species B Age at Reproductive Termination}}{\text{Species B Maximum Life Span}}$$

There are no reports of Japanese monkeys living longer than the oldest female in our own sample; thus, age 32 is considered the maximum life span of Japanese monkeys. The oldest woman of documented age at death lived to be 122 years (*Detroit News*, 4 August 1997). Using these maximum life span values in the above equation, we could argue that a 25-year-old monkey is the equivalent human age of 95 years. Undoubtedly, the 122-year value represents an exceptionally rare outlier for human maximum life span. Nonetheless, this is the value for the oldest known individual, and there is no other agreed-upon value for maximum life span of humans. Estimates range from 90 (Weiss 1981) to 122 years. Using more modest values of 100 years for the maximum human life span and 30 years for the maximum Japanese macaque life span, we would argue that a 25-year-old monkey is the equivalent human age of 83 years.

Although it is difficult to obtain exact figures, it is clear that population-wide reproductive termination in Japanese macaques occurs very late in the life course. Reproductive termination at age 25 is much later in the life course of Japanese monkeys than is menopause in women at age 50. Japanese monkeys are widely regarded as old when they reach age 20 (only 7.9% of our population lived to this age), yet those aged 20–25 years are unlikely to experience reproductive termination: 81% of this age group is still reproductive.

Is Reproductive Termination Adaptive in Japanese Macaques?

Female macaques living in matrilineal societies are good candidates for an investigation of the grandmother hypothesis since they frequently engage in kin-directed affiliative behaviors. Pavelka (1991) found that old females in the Arashiyama West population continue to have active social lives and to interact affiliatively with their offspring, relatives, and friends. Grandmother macaques that have ceased to produce infants of their own would be in a position to offer a variety of supportive behaviors to their children and grandchildren, including carrying, retrieving when lost, predator protection, interventions in agonistic interactions, and alloparental supervision when the mother is absent. They may also lactate longer than usual for their final infant. In theory, these caregiving behaviors could be sufficient to advantageously influence survivorship of descendents. We investigated the possible adaptive value of reproductive termination in Japanese macaques by comparing the survivorship of the descendents of our postreproductive females to those that continued reproducing until death. We compared the reproductive ($n = 50$) and postreproductive ($n = 20$) females for three measures of offspring survivorship: (1) mean survival of all offspring, (2) survival of final offspring, and (3) survival of daughters' offspring (matrilineal grandchildren). Survival analyses did not reveal any significant differences between these two groups (Fedigan and Pavelka 2001). Infant survival rates to age 1 were remarkably similar for those offspring born to postreproductive and reproductive females (85% versus 83%) and were not significantly different for survival to age 5 (71% versus 79%). Survivorship of the final infant was greater for those born to postreproductive females than to reproductive females (85% versus 72% survival to age 1 and 80% versus 67% to age 5); however, this difference was not statistically significant. Finally, we found that the survival rates of daughters' offspring did not differ between postreproductive and reproductive grandmothers (86% versus 83% survival of matrilineal grandoffspring to age 1 and 80% versus 77% to age 5). In this sample of Japanese macaque females, the cessation of reproduction before death does not result in greater survivorship of immediate offspring or of daughters' offspring, and these tests failed to support the grandmother hypothesis. In spite of the fact that female Japanese macaques do direct differential caregiving behaviors to their descendents and other matrilineal relatives, these behaviors do not confer fitness benefits in the form of greater survivorship of descendents.

Other Differences Between Reproductive and Postreproductive Females

Next, we attempted to determine if there are other traits with possible adaptive value that distinguish females that cease to reproduce before death from those that do not and found that reproductive and postreproductive females were not significantly different in dominance, matriline affiliation, body weight, infant sex ratios, age at first birth, or lifetime reproductive success (number of infants surviving to reproductive age). They were, however, significantly different in age at death, the length of time between last parturition and death, cause of death, fecundity (number of infants produced) and reproductive life span (Fedigan and Pavelka 2001). Those 20 females classified as postreproductive lived on average 7 years longer (24.6 compared to 17.4) and specifically lived five times longer after last parturition (60.0 versus 12.1 months). These older postreproductive females were

more likely to disappear, whereas reproductive (younger) females were more likely to succumb to infectious diseases, in line with a previous study with a much larger sample of mortality causes which found that death from infectious disease is more common in younger Japanese macaques (Fedigan and Zohar 1997).

Postreproductive females also had more infants (9.7 compared to 7.7) and had a longer reproductive life span (age at final birth – age at first birth). Postreproductive females had a reproductive span of 13.8 years compared to 10.1 years for those that died while still reproducing. Thus, postreproductive females experienced a significantly greater number of years during which they produced infants than did reproductive females. This means that postreproductive females not only were longer-lived in total life spans but also had longer reproductive life spans than did those who died while still reproducing. Since age at first birth varied little, postreproductive females must have acquired these extra reproductive years at the end of the reproductive phase of their lives. A linear regression of fecundity against age at death showed that age at death is highly predictive of fecundity. We concluded that postreproductive females produced more offspring because, in spite of experiencing some postreproductive years at the end of their lives, they still lived through more years in which to give birth than did reproductive females, which died younger.

Why did postreproductive females have higher fecundity but not significantly greater lifetime reproductive success than reproductive females? Lifetime reproductive success was calculated as the number of offspring to reach breeding age at 5 years. We suggest that because living long enough to become postreproductive was not associated with greater survivorship of one's offspring, the greater fecundity of postreproductive females did not translate into significantly higher lifetime reproductive success.

Based on these analyses, we concluded that reproductive cessation in Japanese monkeys is adaptive only in its indication of a set of females that have reached very old age and have lived long enough to produce many infants. However, the grandmother hypothesis predicts that postreproductive females are able to direct additional care to their grandchildren; and in the above analysis, we investigated all postreproductive females, without specifying which were grandmothers. Thus, we undertook an investigation of the availability and adaptive value of reproductive and postreproductive mothers and grandmothers (Pavelka et al. 2002). This allowed us to target not just postreproductive females in general (many of which may not have living adult daughters or grandchildren) but also the theoretically important subcategories of postreproductive mothers and grandmothers in particular. Eight of the 70 females that could be categorized as reproductive or postreproductive did not produce any daughters (and hence grandchildren that were known to us), so this analysis is based on 62 grandmothers and their 175 daughters and 905 grandchildren. For the analyses in which we needed the daughters' death dates, we were able to use only 88 of the 175 daughters, and only 74 of these produced infants. Of the 905 grandchildren born, we have complete information on 886 individuals that were included in the survival analyses. Of these, 504 survived to age 5. The probability of survival to age 5 is calculated based on the survivorship of all of the 886 individuals in the sample.

Availability of Reproductive and Postreproductive Mothers and Grandmothers

Of the 175 Japanese macaque daughters in our sample, 64 (36.6%) had a deceased mother, 106 (60.5%) had a living reproductive mother, and only 5 (2.8%) had a living postreproductive mother at the time they first gave birth. Therefore, nearly two-thirds of daughters had their mother still alive at the time of their first birth—a mother that could potentially contribute to her daughter's production of infants or to the survivorship of those infants—but the vast majority of these grandmothers were themselves still producing infants. Very few daughters had a postreproductive mother available to them when they began to reproduce.

The mean length of reproductive life for the sample of 74 daughters was 9.71 years. Over half of the reproductive years of the daughters in our sample (5.26 years, 54.2%) were lived without a mother present at all. During the other half (4.45 years or 46% of the daughter's reproductive life), the grandmother was available, but the vast majority of these years were spent with a grandmother that was herself still reproducing. Only 4.2% of the reproductive life span of the daughter was spent with a postreproductive grandmother available to help the daughter. This represents less than 5 months for the average female in the population.

This pattern is further reflected in the next generation in the time available for a postreproductive grandmother to have an impact on her grandchildren's survival—the essence of the grandmother hypothesis. Most of the grandchildren's first 5 years of life was spent without a living grandmother (3.6 years, or 72%). Grandchildren that survived to age 5 had a living grandmother available to them for only 1.4 years, or 28% of that time; and during most of that time, the grandmother had an infant of her own. Only 0.2 years, or 4%, of the first 5 years of life of the grandchildren in our sample—that is, 2 months on average—was spent with a postreproductive grandmother available to offer the extra caregiving behavior predicted by the grandmother hypothesis.

Thus, grandmother Japanese macaques proved to be less common than we had expected. This study showed that while nearly two-thirds of newly adult females have a living mother when they first give birth, the mother will likely remain alive for less than half of her daughter's reproductive life. Furthermore, postreproductive grandmothers are exceedingly rare. Only 5 out of 175 daughters had a postreproductive mother available to them when they reached reproductive age, and the average adult daughter had a postreproductive mother available to her for less than 6 months of her reproductive years. The average grandchild had a postreproductive

grandmother available to it for less than 2 months. This surprisingly restricted availability of postreproductive grandmothers, even in a long-lived, multigenerational, provisioned population such as the Arashiyama West Japanese macaques, must limit opportunity for the classic maternal investment models to operate.

Adaptive Value of Reproductive and Postreproductive Mothers and Grandmothers

In the Arashiyama West population, most Japanese macaque females do not have their mothers around and free to help them with their infants because the living grandmothers are still reproducing themselves. Likewise, the vast majority of grandchildren do not have a postreproductive grandmother around to help take care of them. Still, it is interesting to explore the possible effects that a female (grandmother) can have on the reproduction of her daughter. Does having a living mother improve any aspect of a daughter's reproduction? For example, do females whose mother is alive start to give birth earlier, or do they have shorter interbirth intervals? Do the grandchildren survive better if the grandmother is alive, especially in the rare case of a grandmother that is alive and postreproductive? In these Japanese macaque data, having a mother alive did appear to be associated with improvements in some of the daughter's reproductive parameters. Females whose mother was alive were more likely to give birth at age 5 rather than age 6, although it did not matter if that mother was reproductive or postreproductive. The presence of a mother also appears to be of some benefit to females in terms of shortening their interbirth intervals from 19.2 months for females whose mother was dead to 18.1 months for females whose mother was alive. Moreover, if the mother was alive and postreproductive, the interbirth intervals of the daughters dropped to 16 months, although this trend is not quite statistically significant (Pavelka et al. 2002).

Thus, despite being less common than expected, we found evidence that the presence of a living mother—most of whom were still reproductive—is advantageous to the reproduction of their daughters. Females with a living mother were more likely to begin reproducing at age 5 rather than age 6. This may be because the presence of the mother improves a young female's chance of having a "successful" first fertile cycle (one that results in a conception) at age 4.5. The first proceptive period for female Japanese monkeys requires that these young females venture out of the tight female kinship units for the first time to establish contact with unrelated adult males—animals with which they would have had little need or opportunity to interact previously. The inexperienced behavior of these young Japanese monkey females increases the likelihood that adult males will target them for aggression (McDonald 1985). In vervet monkeys, adult daughters whose mothers were still living in the group received less aggression and were defended more often than were young adult females whose mothers had died (Fairbanks 1988). Thus, the mother of the pubescent female, through the agonistic support she provides to her daughter, may help to increase confidence on the part of the daughter and/or reduce the frequency and intensity of serious aggression from adult males, thus increasing the likelihood of the daughter forming a successful consortship.

The essence of the grandmother hypothesis is that grandmothers, specifically postreproductive grandmothers, are able to enhance the survivorship of their grandchildren. Do the grandchildren in this Japanese macaque population survive better if the grandmother is alive and especially if she is alive and no longer producing infants of her own? Interestingly, given the initial survival analysis above, we found that they do: there is a significant difference in survivorship to age 1 depending on the status of the grandmother. Specifically, grandchildren with a living postreproductive grandmother were significantly more likely to live to age 1 than were grandchildren with either a dead grandmother or a live one that was herself still reproducing. There are no differences in the survivorship of grandchildren to age 5 based on whether the grandmother was dead, alive and reproductive, or alive and postreproductive.

Our earlier test of all 20 reproductively terminated females compared with all 50 reproductive females (irrespective of whether these females were mothers or grandmothers) found no difference in the survivorship of first- or second-generation descendants. Reproductive termination was characteristic of only a small cohort of very old females, and postreproductive females were significantly older at time of death than were reproductive ones, supporting the conclusion that reproductive termination in this population is a by-product of selection favoring longevity. Yet, in the present study, in which we were able to target the theoretically important postreproductive mothers and grandmothers, we are seeing some tantalizing evidence for the adaptive value of postreproduction. Those few grandchildren that had a postreproductive grandmother present during their first year of life were significantly more likely to survive (95%) than were those who had a dead grandmother (85%) or a living and reproducing grandmother (89%). The first year of life is critical as most mortality of immatures occurs in this period (Fedigan and Zohar 1997) and most infants are weaned at 1 year. However, the presence of a postreproductive grandmother did not enhance the survival of grandchildren to age 5, so the effects appear to be restricted to the infant or preweaning stage of life. Recall, however, that by this stage almost 75% of the grandchildren no longer have a grandmother alive. Grandmothers have a much better opportunity to affect survival to age 1 than to age 5 because they are more available during the earlier years of the grandchildren's lives.

SUMMARY

Human females are unique among the primates in experiencing menopause—the universal permanent cessation of menstrual cycles and reproductive capability at only halfway through the maximum life span for the species. Biological anthropologists and evolutionary biologists interested in the evolution of this peculiar trait (generally speaking, natural selection should favor reproducing phenotypes) have developed two categories of explanation. Either menopause is a by-product of increased life span or of traits with high selective value early in life or menopause is directly adaptive because non-reproducing middle-aged women can better maximize their inclusive fitness by aiding their daughters and grandchildren in reproduction and survival than by continuing to reproduce themselves. A small number of individual monkeys and apes that have ceased to reproduce and that show endocrine changes similar to menopausal women have been identified in captivity; however, in addition to being rare individuals, they are very old and close to the known maximum life span for their species. Reproductive cessation has also been reported for a few free-ranging nonhuman primates, although the problem of distinguishing reproductive termination from death in an interbirth interval plagues these reports.

We investigated reproductive termination in a large sample of free-ranging Japanese macaques by using a female's own reproductive history to determine when she had lived significantly longer than her own average lifetime interbirth interval without reproducing. We found that reproductive termination was uncommon in individuals under 25 years of age but universal after that. A 25-year-old Japanese monkey, however, is a rare creature as only just over 2% of the population will live to this age. A 25-year-old Japanese monkey is estimated to be equivalent to an 85-year-old woman. There was no difference in the survivorship of the grandchildren of these females that lived long enough to become postreproductive; however, this group did produce significantly more infants than those that died while reproducing because fecundity and longevity are positively correlated. For a Japanese monkey female in our sample, the key to having many infants was living a long life, even though this meant living a few years beyond reproductive ability before death in very old age. In the very rare cases when a reproductively terminated female also had a live grandchild to which she could direct caregiving behaviors, those grandchildren did have an improved chance of surviving to age 1. We conclude that the reproductive termination that we have documented in female Japanese macaques of the Arashiyama West population probably occurs too late in life, with too few females reaching and remaining in this stage for any substantial proportion of their descendants' lives, to have sufficient inclusive fitness effects to compensate for the loss of the grandmother's direct reproductive output, as theorized by the grandmother hypothesis. Our findings here suggest that increased survivorship of grandchildren during their first year of life and more rapid production of young by their daughters may be a secondary benefit enjoyed by those few postreproductive females that find themselves without an unweaned infant of their own but with a grandchild available to benefit from their free time and energy.

Thus far, our investigation of Japanese monkeys has focused on the benefits that might be associated with reproductive termination. However, the grandmother hypothesis is based on the assertion that the benefits of ceasing to reproduce will outweigh the costs of continuing to reproduce. Thus, we are continuing this investigation by turning to the other side of the equation and considering the possible costs for older females of continuing to reproduce.

REFERENCES

Alvarez, H. P. (2000). Grandmother hypothesis and primate life histories. *Am. J. Phys. Anthropol.* 113:435–450.

Armstrong, D. T. (2001). Effects of maternal age on oocyte developmental competence. *Theriogeneology* 55:1303–1322.

Bellino, F. L., and Wise, P. M. (2003). Nonhuman primate models of menopause workshop. *Biol. Reprod.* 68:10–18.

Blurton-Jones, N., Hawkes, K., and O'Connell, J. F. (1999). Some current ideas about evolution of the human life history. In: Lee, P. C. (ed.), *Comparative Primate Socioecology.* Cambridge University Press, Cambridge. pp. 140–166.

Bogin, B., and Smith, B. H. (1996). Evolution of the human life cycle. *Am. J. Hum. Biol.* 8:703–716.

Brambilla, D. J., and McKinlay, S. M. (1989). A prospective study of factors affecting age at menopause. *J. Clin. Epidemiol.* 42:1031–1039.

Caro, T. M., Sellen, D. W., Parish, A., Frank, R., Brown, D. M., Voland, E., and Borgerhoff Mulder, M. (1995). Termination of reproduction in nonhuman and human female primates. *Int. J. Primatol.* 16:205–220.

Coleman, R. J., and Kemnitz, J. W. (1998). Aging experiments using nonhuman primates. In: Yu, B. P. (ed.), *Methods in Aging Research.* CRC Press, Boca Raton. pp. 249–267.

Erwin, J. M., and Hof, P. R. (eds.) (2002). *Aging in Nonhuman Primates. Interdisciplinary Topics in Gerontology*, vol. 31. S. Karger, Basel.

Fairbanks, L. A. (1988). Vervet monkey grandmothers: effects on mother–infant relationships. *Behaviour* 104:176–188.

Fedigan, L. M., and Asquith, P. J. (eds.) (1991). *The Monkeys of Arashiyama: Thirty-Five Years of Research in Japan and the West.* State University of New York, Albany. pp. 54–73.

Fedigan, L. M., and Griffin, L. (1996). Determinants of reproductive seasonality in the Arashiyama West Japanese macaques. In: Fa, J. E., and Lindburg, D. G. (eds.), *Evolution and Ecology in Macaque Societies.* Cambridge University Press, Cambridge. pp. 369–388.

Fedigan, L. M., and Pavelka, M. S. M. (2001). Is there adaptive value to reproductive termination in Japanese macaques? A test of maternal investment hypotheses. *Int. J. Primatol.* 22:109–125.

Fedigan, L. M., and Zohar, S. (1997). Sex differences in mortality of Japanese macaques: twenty-one years of data from the Arashiyama West population. *Am. J. Phys. Anthropol.* 102:161–175.

Gilardi, K. G., Shideler, S. E., Valverde, C., Roberts, J. A., and Lasley, B. L. (1997). Characterization of the onset of menopause in the rhesus macaque. *Biol. Reprod.* 57:335–340.

Gosden, R. G., and Faddy, M. J. (1998). Biological bases of premature ovarian failure. *Reprod. Fertil. Dev.* 10:644–661.

Gould, K. G., Flint, M., and Graham, C. E. (1981). Chimpanzee reproductive senescence: a possible model for evolution of the menopause. *Maturitas* 3:157–166.

Graham, C. E. (1979). Reproduction function in aged female chimpanzees. *Am. J. Phys. Anthropol.* 50:291–300.

Graham, C. E. (1986). Endocrinology of reproductive senescence. In: Dukelow, W. R., and Erwin, J. (eds.), *Comparative Primate Biology. Reproduction and Development*, vol. 3. Alan R. Liss, New York. pp. 93–99.

Graham, C. E., Kling, O. R., and Steiner, R. A. (1979). Reproductive senescence in female nonhuman primates. In: Bowden, D. J. (ed.), *Aging in Nonhuman Primates.* Van Nostrand Reinhold, New York. pp. 183–209.

Hammer, M. L. A., and Foley, R. A. (1996). Longevity and life history in hominid evolution. *Hum. Evol.* 11:61–66.

Hawkes, K. (2003). Grandmothers and the evolution of human longevity. *Am. J. Hum. Biol.* 15:380–400.

Hawkes, K. (2004). The grandmother effect. *Nature* 428:128–129.

Hawkes, K., O'Connell, J. F., and Blurton Jones, N. G. (1989). Hardworking Hadza grandmothers. In: Standon, V., and Foley, R. (eds.), *Comparative Socioecology of Mammals and Man.* Blackwell, London. pp. 341–366.

Hawkes, K., O'Connell, J. F., and Blurton-Jones, N. G. (1997). Hadza women's time allocation, offspring provisioning, and the evolution of long postmenopausal life Spans. *Curr. Anthropol.* 18:551–577.

Hill, K., and Hurtado, A. M. (1996). *Ache Life History. The Ecology and Demography of a Foraging People.* Aldine de Gruyter, New York.

Hodgen, G. D., Goodman, A. L., O'Conner, A., and Johnson, D. K. (1977). Menopause in rhesus monkeys: model for study of disorders in the human climacteric. *Am. J. Obstet. Gynecol.* 127:581–584.

Howell, N. (1979). *The Demography of the Dobe !Kung.* New York, Academic Press.

Hrdy, S. B. (1981). Nepotists and altruists: the behavior of old females among macaques and langur monkeys. In: Amoss, P. T., and Harrell, S. (eds.), *Other Ways of Growing Old: Anthropological Perspectives.* Stanford University Press, Stanford, CA. pp. 59–96.

Johnson, J., Canning, J., Kaneko, T., Pru, J. K., and Tilly, J. L. (2004). Germline stem cells and follicular renewal in the postnatal mammalian ovary. *Nature* 428:145–150.

Johnson, R. L., and Kapsalis, E. (1998). Menopause in free-ranging rhesus macaques: estimated incidence, relation to body condition, and adaptive significance. *Int. J. Primatol.* 19:751–765.

Judge, D. S., and Carey, J. R. (2000). Postreproductive life predicted by primate patterns. *J. Gerontol.* 55A:B201–B209.

Kaplan, H., Hill, K., Lancaster, J., and Hurtado, A. M. (2000). A theory of human life history evolution: diet, intelligence and longevity. *Evol. Anthropol.* 9:156–185.

Lahdenpera, M., Lummaa, V., Helle, S., Tremblay, M., and Russell, A. F. (2004). Fitness benefits of prolonged post-reproductive lifespan in women. *Nature* 428:178–181.

Lancaster, J. B., and King, B. J. (1985). An evolutionary perspective on menopause. In: Borwn, J. K., and Kerns, V. (eds.), *In Her Prime. A New View of Middle-Aged Women.* Bergin and Garvey Press, S. Hadley, MA. pp. 13–20.

Leidy, L. E. (1999). Menopause in evolutionary perspective. In: Trevathan, W. R., Smith, E. O., and McKenna, J. T. (eds.), *Evolutionary Medicine.* Oxford University Press, New York. pp. 407–427.

Mace, R. (2000). Evolutionary ecology of human life history. *Anim. Behav.* 59:1–10.

McDonald, M. (1985). Courtship behavior of female Japanese monkeys. *Can. Rev. Phys. Anthropol.* 4:67–75.

Nishida, T., Corp, N., Hamai, M., Hasegawa, T., Hiraiwa-Hasegawa, M., Hosaka, K., Hunt, K. D., Itoh, N., Kawanaka, K., Matsumoto-Oda, A., Mitani, J. C., Nakamura, M., Norikoshi, K., Sakamaki, T., Turner, L., Uehara, S., and Zamma, K. (2003). Demography, female life history, and reproductive profiles among the chimpanzees of Mahale. *Am. J. Primatol.* 59:99–121.

Nozaki, M., Mitsunaga, F., and Shimizu, K. (1995). Reproductive senescence in female Japanese monkeys (*Macaca fuscata*): age and season-related changes in hypothalamic–pituitary–ovarian functions and fecundity rates. *Biol. Reprod.* 52:1250–1257.

Packer, C., Tatar, M., and Collins, A. (1998). Reproductive cessation in female mammals. *Nature* 392:807–811.

Paul, A., Kuester, J., and Podzuweit, D. (1993). Reproductive senescence and terminal investment in female Barbary macaques (*Macaca sylvanus*) at Salem. *Int. J. Primatol.* 14:105–124.

Pavelka, M. S. M. (1991). Sociability in old female Japanese monkeys, human versus nonhuman primate aging. *Am. Anthropol.* 93:588–598.

Pavelka, M. S. M. (1993). *Monkeys of the Mesquite: The Social Life of the South Texas Snow Monkey.* Kendall/Hunt Publishing, Dubuque.

Pavelka, M. S. M., and Fedigan, L. M. (1991). Menopause: a comparative life history perspective. *Ybk. Phys. Anthropol.* 34:13–38.

Pavelka, M. S. M., and Fedigan, L. M. (1999). Reproductive termination in female Japanese monkeys: a comparative life history perspective. *Am. J. Phys. Anthropol.* 109:455–464.

Pavelka, M. S. M., Fedigan, L. M., and Zohar, S. (2002). Availability and adaptive value of reproductive and postreproductive Japanese macaque mothers and grandmothers. *Anim. Behav.* 64:407–414.

Peccei, J. S. (2001). Menopause: adaptation or epiphenomenon? *Evol. Anthropol.* 10:43–57.

Rogers, A. R. (1993). Why menopause? *Evol. Ecol.* 7:406–420.

Rose, M. R. (1991). *Evolutionary Biology of Aging.* Oxford University Press, Oxford.

Sherman, P. (1998). The evolution of menopause. *Nature* 392:759–761.

Sommer, V., Srivastava, A., and Borries, C. (1992). Cycles, sexuality and conception in free-ranging langurs (*Presbytis entellus*). *Am. J. Primatol.* 28:1–27.

Takahata, Y., Koyama, N., and Suzuki, S. (1995). Do the old aged females experience a long post-reproductive life span? The cases of Japanese macaques and chimpanzees. *Primates* 36:169–180.

Tardif, S. D., and Ziegler, T. E. (1992). Features of female reproductive senescence in tamarins (*Saguinus* spp), a New World primate. *J. Reprod. Fertil.* 94:411–421.

Walker, M. L. (1995). Menopause in female rhesus monkeys. *Am. J. Primatol.* 35:59–71.

Waser, P. M. (1978). Postreproductive survival and behavior in a free-ranging female mangabey. *Folia Primatol.* 29:142–160.

Weiss, K. M. (1981). Evolutionary perspectives on human aging. In: Amoss, P. T., and Harrell, S. (eds.), *Other Ways of Growing Old: Anthropological Perspectives.* Stanford University Press, Stanford, CA. pp. 25–58.

Williams, G. C. (1957). Pleiotropy, natural selection, and the evolution of senescence. *Evolution* 11:398–411.

Wood, J. W., O'Conner, K. A., Holman, D. J., Brindle, E., Barson, S. H., and Grimes, M. A. (2001). The evolution of menopause by antagonistic pleiotropy. Working paper 01-04. Center for Studies in Demography and Ecology, University of Washington, Seattle http://csde.washington.edu/pubs/wps/01-04.pdf.

27

Mate Choice

Joseph H. Manson

HISTORY AND THEORY

Background and Definitions

During the first few decades of behavioral primatology, mate choice among nonhuman primates was given almost no formal attention. As in biology generally, female mate choice was dismissed as unlikely or unimportant despite Darwin's (1871) and Fisher's (1930) theoretical arguments for its likely role in producing exaggerated male ornaments (Cronin 1991). Male mate choice was regarded as even less plausible than female mate choice. The rise of evolutionary social theory (Hamilton 1964; Williams 1966; Trivers 1971, 1972; Maynard Smith 1974; Wilson 1975) and the growing prominence of female primatologists (Hrdy 1986, Fedigan 1982) put female primate mate choice on the map as a research topic by the 1980s. The past 10 years have witnessed the steady production of systematic studies of primate female mate choice, some of them accompanied by genetic paternity data that have greatly increased researchers' ability to test hypotheses. Few systematic studies of male mate choice have been carried out, but some inferences are possible based on available data. In this chapter, I will review relevant theoretical concepts and use these as a framework within which to examine findings about primate mate choice. I will focus on studies conducted since 1990 because (*1*) few rigorous studies focusing on mate choice were conducted before then and (*2*) reviews by Fedigan (1982: 282–285), Smuts (1987), Small (1989), and Keddy-Hector (1992) summarize the largely anecdotal information available at that time.

Mate choice may be defined as "any pattern of behavior, shown by members of one sex, that leads to their being more likely to mate with certain members of the opposite sex than others" (Halliday 1983: 4). "Mating" here refers to the actual production of offspring, although many studies have used the occurrence or frequency of copulation as a proxy for reproduction. The strength of Halliday's definition is that it focuses attention on the evolutionary causes and consequences of repeated behavior patterns (i.e., their effects on the fitness of both the choosing and the chosen sex), rather than on the internal states (preferences, desires) of putative choosers, which are difficult to infer and which in any case have evolutionary consequences only insofar as they are manifest in overt behavior. In contrast to mate *choice*, mate *preference* "denotes a disposition or propensity that an individual possesses whether or not it is exercised" (Heisler et al. 1987: 99).

Sexual Selection Theory and Mate Choice

Intrasexual and Intersexual Selection

Darwin (1871) developed the concept of sexual selection to explain certain traits that apparently decrease their bearers' probability of survival, the prototypical example being the peacock's tail, which presumably increases vulnerability to predation. He proposed that such traits confer an advantage in acquiring mates that compensates for their negative effects on survival. Sexually selected traits may be weapons (e.g., antlers, large body size) used in contests over mates (*intrasexual selection*) or ornaments that increase their bearers' attractiveness to the opposite sex (*intersexual selection*). Darwin (1871) noted that, where the sexes differed in their endowments of weapons and/or ornaments (*sexual dimorphism*), it was usually the males that had more elaborate versions of sexually selected traits. Although intrasexual selection was accepted by biologists, female mate choice for male ornaments was, for the most part, decisively rejected (e.g., Huxley 1938), even after Fisher (1930) argued that an initially modest male ornament and a slight female preference for it could lead to the "runaway" evolution of a highly exaggerated ornament.

Parental Investment and Sexual Selection

Trivers (1972) played a major role in rehabilitating Darwin's theory of intersexual selection by proposing, as the source of the usual sex difference in weapons and ornaments, sex differences in parental investment. He defined *parental investment* as any investment in time, energy, or risk by a parent in an offspring that enhances the offspring's reproductive success (RS) and that detracts from the parent's ability to invest in other offspring. In this model, access to members of the more heavily investing sex (usually females, particularly in mammals) becomes the resource limiting the RS of the less heavily investing sex. In species with the typical sex difference in parental investment, males are then forced to compete against each other for access to females, either directly or indirectly through contests for dominance, producing selection for weaponry (in excess of female weaponry) and possibly for ornaments. Variance in male RS is hypothesized to exceed variance in female RS, and males try to acquire large numbers of mates, whereas females (whose RS cannot be increased by mating with additional males) focus on choosing mates that offer either genes that enhance offspring survivorship (possibly signaled by elaborate ornaments, see below) or parental investment to supplement the mother's contribution (see Chapter 37 for additional discussion).

Although this model remains a cornerstone of sexual selection research, recent theoretical and empirical work has eroded the sharp distinction between "choosy" females and "promiscuous" males. Females may, in fact, increase their RS by mating with multiple males for a number of reasons: some males may be infertile, or a single male may not be able to produce enough sperm to impregnate all his mates (Small 1988, Smith 1984); a more genetically diverse set of offspring may experience higher survivorship on average than the offspring of a single male (Parker 1970, Keller 1994, Olsson et al. 1994, Williams 1975); and multiple males may be induced to invest in an offspring, or at least to refrain from harming it, by providing them with some probability of paternity (Hrdy 1981; but see Soltis and McElreath 2001). On the other hand, males do not always benefit by indiscriminate mating, even in species where they provide little parental investment. Ejaculates may be costly to produce (Dewsbury 1982), male–male competition may be potentially damaging enough that it pays males to choose their contests carefully (Pagel 1994), and females may vary in their reproductive and parenting capacities and, hence, in the likely reproductive payoff from mating with them (Altmann 1997).

Mate Choice for Genes

Theorists generally agree that when males provide females with resources, parental care, or other phenotypic benefits, females should choose males on the basis of their ability and willingness to provide these resources (Andersson 1994). However, controversy continues regarding the adaptive value of choosing mates on the basis of their ornamental features such as bright plumage or elaborate displays. In Fisher's (1930, Lande 1981) formulation, sexually selected ornaments are arbitrary and the only benefit to females of preferring ornamented males is that their sons will inherit the ornament and therefore be attractive to females. Zahavi

(1975) argued, in contrast, that by decreasing their bearers' survival prospects, inherited ornaments send the unfakeable message that their bearers' *other* genes increase viability: they tell prospective mates, in effect, "Having survived to breeding age with this ornament, I must be of superior genetic quality." This hypothesis (the "handicap principle") was overwhelmingly rejected by behavioral ecologists until Grafen (1990) produced a formal model that illuminated two aspects of the idea that Zahavi's verbal arguments had left unclear. First, an individual's level of ornamentation must be adjusted facultatively. Second, low-quality individuals must suffer larger viability costs than high-quality individuals for producing the same level of ornamentation. Under these conditions, an evolutionarily stable state is possible in which males produce costly ornaments proportionately to their underlying quality and females use these ornaments as reliable indicators of male genetic quality. A large number of empirical studies in various animal taxa have supported predictions from the handicap principle (e.g., Petrie 1994, Kose et al. 1999), though support for Fisher's "runaway" process has also been found (e.g., Brooks 2000).

"Good genes" models, such as Zahavi's (1975), also faced the problem of explaining how heritable variation in fitness could be maintained over evolutionary time. Would selection not quickly eliminate this variation and thereby eliminate the basis of mate choice? Hamilton (1980, Hamilton and Zuk 1982) proposed that parasites, being selected to circumvent hosts' immune systems and having much shorter generation times than their hosts, create cyclically varying selection pressures, such that the most fit host genotypes change fairly rapidly. A large number of studies have found that sexually selected ornaments accurately display parasite resistance (e.g., Duffy and Ball 2003, Doucet and Montgomerie 2003). A related line of research has discovered evidence that heterozygotes, particularly at the loci of the major histocompatibility complex (MHC), which affects immune function, are better able to resist pathogens than homozygotes (Potts and Wakeland 1993, Brown 1997). Choice for genetically dissimilar or otherwise genetically compatible mates could potentially explain (*1*) apparently idiosyncratic mates choices as well as (*2*) systematic inbreeding avoidance, which has long been a hypothesized mate choice criterion because of the greatly increased probability that offspring of closely related mates will be homozygous for rare deleterious recessive alleles (e.g., Ralls and Ballou 1982). Because of females' greater investment in each offspring, they are predicted to be more averse to inbreeding than males (Clutton-Brock and Harvey 1976). Some evidence suggests that overly outbred offspring also suffer depressed fitness and that there is an optimal level of outbreeding (Bateson 1983).

Conflict Between the Sexes

Behavioral ecologists have recognized for some time that conflicts between the sexes may arise about mating decisions (Parker 1979). Smuts and Smuts (1993: 2–3) proposed, as a third form of sexual selection, *sexual coercion*, which they defined as "the use by a male of force, or the threat of force, that functions to increase the chances that a female will mate with him at the time when she is likely to be fertile, and to decrease the chances that she will mate with other males, at some cost to the female." Sexual coercion includes sexually selected infanticide (Hrdy 1974, 1977), in which males are hypotheszed to kill unweaned infants that were sired by other males to hasten the victims' mothers' return to ovulatory cycling and thereby gain quicker sexual access to them. Although debate continues about the validity of this hypothesis (Sussman et al. 1995, van Schaik 2000), many researchers use the concept of sexually selected infanticide as a platform from which to investigate various aspects of social and sexual behavior (van Schaik and Janson 2000). Most relevant to the focus of this chapter, Hrdy (1981) proposed that females may seek out multiple mates, particularly newly arrived males, to confuse paternity and thereby reduce the likelihood of infanticide.

Sexual dialectics theory (Gowaty 1997) holds that conflict between the sexes is ubiquitous and that its resolution has profound effects on a species' evolution. Mating partners' genetic compatibility could affect the number and/or quality of the offspring they produce, and individuals may be able to assess prospective mates' genetic compatibility with themselves. If so, then social constraints imposed by males on free mate choice will decrease female fitness even as they benefit the competitively successful males. Some of a successful male's mates would have higher RS if they were allowed to mate with a preferred but competitively less successful male. In fruit flies (*Drosophila pseudoobscura*) (Gowaty et al. 2002) and house mice (*Mus domesticus*) (Drickamer et al. 2000, 2003), both quality and quantity of offspring are enhanced when individuals are mated with preferred partners.

Postcopulatory Sexual Selection

Neither male–male competition nor female mate choice necessarily ends with the initiation of copulation. Parker (1970) first drew attention to the possibility of *sperm competition*, in which sperm from different males compete to fertilize a female's egg(s). *Cryptic female choice*, or *postcopulatory choice* (Thornhill and Alcock 1983, Eberhard 1996), is the process whereby a female's anatomy, physiology, or behavior selectively alters the probability of paternity among her copulatory partners. No studies have sought direct evidence for cryptic female choice in primates, but there are numerous theoretical and empirical reasons for suspecting that it occurs (Reeder 2003).

SOURCES OF EVIDENCE ABOUT MATE CHOICE

Observational Studies

The simplest but least precise kind of information about mate choice is a statistical description for a particular study group or population of who has copulated with whom or of which male–female pairs have associated closely while the female was sexually receptive ("consortships," see Manson 1997 for a critique of this concept). Because mate choice is defined as behavior that affects the relative probabilities of mating among different pairs, data describing these probabilities tell us something about mate choice, particularly when infants' paternities are determined using genetic methods (see Chapter 22). However, observed patterns of mating and parentage reflect the operation of several processes (see also Small 1989): female mate choice, male mate choice, male–male competition, female–female competition, and possibly male sexual coercion. Ascribing any observed pattern in the distribution of copulations or parentage among dyads (e.g., high-ranking females conceive more than expected by chance with high-ranking males) to just one or two of these processes is problematic. Furthermore, neither mating nor paternity data alone reveal anything about the mechanisms of mate choice—i.e., the behaviors that alter dyads' likelihood of reproduction.

Some behaviors have strong *face validity*, or *apparent validity*, as forms of mate choice. If a sexually attractive/proceptive (see Chapter 25 for definitions) female persistently approaches one adult male or a small subset of her group's adult males, we may infer that she is positively affecting her likelihood of mating with him or them. If she presents sexually to a male or engages in other species-typical sexual solicitation behaviors, we have even stronger grounds for regarding her behavior as a form of mate choice. Chimpanzee (*Pan troglodytes*) consortships, in which a cycling female and a male range together apart from the rest of their community for up to several weeks, are regarded as dependent on female mate choice because a female can vocalize to summon other males to break up an incipient consortship if she does not desire it (Tutin 1979). In female-transfer species (e.g., mountain gorillas, Sicotte 2001), females' transfer decisions are almost certainly a form of mate choice. However, many putative mate choice behaviors are subject to alternative interpretations. If a fertile female persistently avoids a particular male, is she engaging in negative mate choice or is she testing his ability to continue following her? Strong evidence for mate choice is produced when a putative mate choice behavior is linked statistically with an altered likelihood of parentage or, at least, an altered likelihood of copulation during the female's fertile period. On the other hand, the same behavior pattern may affect dyads' mating probabilities in some social or demographic contexts but not in others within the same species (see below).

Experimental Studies

As in any area of behavioral research, experimental studies of mate choice permit stronger inferences about causality at some cost in confidence about ecological validity, particularly when experimental subjects are captive animals (see Chapter 20). Experimental subjects (usually fertile females) are typically given a choice between two or more prospective mating partners that differ along a theoretically relevant dimension. The differences themselves may be natural (e.g., Schneider 1998) or manipulated (e.g., Waitt et al. 2003). Experiments involving pairs or trios of animals have particular value in the study of female mate choice because of the possibility that female behavior in normal group settings is constrained by male sexual coercion (Buck 1998, Schneider 1998).

FEMALE MATE CHOICE

Table 27.1 summarizes studies of primates that have produced data about female mate choice.

Female Choice Can Undermine the Effects of Male–Male Competition

Studies using genetic paternity assignment technologies have revealed that in multimale group-living primates, dominant males usually produce more offspring than subordinates and almost never produce fewer offspring than subordinates (e.g., Pope 1990, de Ruiter 1992, Altmann et al. 1996, Bercovitch and Nurnberg 1997, Gerloff et al. 1999, Constable et al. 2001). However, dominant males often fail to monopolize reproduction to the extent predicted by Altmann's (1962) "priority-of-access" model (e.g., Alberts et al. 2003). Traditionally, it was assumed that females would choose the winners of male–male competition (see Fedigan 1982: 273–277 for a critique). However, evidence from several primate taxa indicate that female choice for multiple mates or for male qualities uncorrelated with dominance rank may be partly responsible for observed deviations from the priority-of-access model.

Cercopithecines

Female aversion to mating with high-ranking males is particularly well documented in Japanese and rhesus macaques (*Macaca fuscata* and *M. mulatta*). Males frequently disrupt the consortships of lower-ranking males, usually by chasing or approaching the female of the pair; but females usually return to their previous consort partner within a short time and only rarely join the more dominant male (Huffman 1987, Manson 1992, Perloe 1992, Soltis et al. 1997b). Studies that quantified spatial proximity maintenance have found that greater female responsibility for proximity maintenance is sometimes associated with higher copulation rates during the female's fertile period (Manson 1992) or a greater likelihood of producing offspring (Soltis et al. 1997b). Sexually attractive/proceptive females have been observed to maintain proximity preferentially to lower-ranking males (Takahata 1982b, Manson 1992, Perloe 1992, Soltis et al. 2000), though this result was not always statistically significant. As a result of these female choice behaviors, alpha and other high-ranking males are sometimes virtually excluded from

Table 27.1 Studies of Female Mate Choice in Primates

SPECIES	TYPE OF EVIDENCE[1]	EFFICACY OF FEMALE CHOICE RELATIVE TO EFFECTS OF MALE–MALE COMPETITION	CHOICE CRITERIA	SOURCES
Microcebus murinus (gray mouse lemur)	Premating behavior *Dist. copulations; Paternity	No reproductive advantage to dominant males	Inbreeding avoidance	Radespiel et al. (2002), Radespiel and Zimmermann (2003)
Propithecus verreauxi (sifaka)	Premating behavior; Dist. copulations	Subordinates sometimes chosen	Inbreeding avoidance	Richard (1992), Brockman (1999)
Eulemur fulvus rufus (red-fronted lemur)	Premating behavior	No male dominance relationships	Prior affiliative relationship (?)	Pereira and McGlynn (1997)
Lemur catta (ring-tailed lemur)	Premating behavior *Dist. copulations *Paternity	Subordinates chosen if unfamiliar	Inbreeding avoidance	Pereira and Weiss (1991)
	Premating behavior; Dist. copulations	Males of all ranks chosen	Inbreeding avoidance	Sauther (1991)
Cebus apella (brown capuchin)	Premating behavior *Dist. copulations	Alpha male chosen, especially if food is monopolizable	Alpha male chosen, especially if food is monopolizable; inbreeding avoidance	Janson (1984, 1994), Di Bitetti and Janson (2001)
Saimiri oerstedii (Costa Rican squirrel monkey)	Premating behavior; Dist. copulations	No male dominance relationships	Largest (seasonally enlarged) male chosen	Boinski (1987)
Brachyteles arachnoides (muriqui)	Dist. copulations	No male dominance relationships	Inbreeding avoidance; sorting of mates by age	Strier (1997)
Cercocebus torquatus (sooty mangabey)	Premating behavior *Dist. copulations	Female choice affects male copulatory success independent of male rank	?	Buck (1998)
Erythrocebus patas (patas monkey)	Premating behavior †Dist. copulations	Neither female solicitations nor male fighting success affects male copulatory success	None found	Chism and Rogers (1997)
Cercopithecus aethiops (vervet)	Experimental	Alpha male chosen	Alpha male chosen	Keddy (1986)
	Premating behavior; Dist. copulations	Dominants have no mating advantage with fertile females	Older males chosen	Andelman (1987)
Macaca cyclopis (Formosan macaque)	Premating behavior †Dist. copulations	Dominance rank, but not female proximity maintenance, enhances male mating success	Females maintain proximity preferentially to lower-ranking and extra-group males	Birky (2002)
Macaca sylvanus (Barbary macaque)	Premating behavior †Dist. copulations	Females incited male–male contests; males most successful at coalition formation had highest mating success	Inbreeding avoidance by both sexes; No other criteria	Paul and Kuester (1985), Kuester and Paul (1992), Kuester et al. (1994)
Macaca fuscata (Japanese macaque)	Premating behavior *Dist. copulations	No reproductive advantage to dominant males	Particular middle-ranking young adult males preferred	Huffman (1987, 1992)
	Premating behavior; Dist. copulations	Mating advantage to dominant males, though females maintained proximity most to low-rankers	Possible preference for males likely to retain or attain high rank in the future	Perloe (1992)
Captive group	Premating behavior *Dist. copulations and Premating behavior *Paternity	Female choice affects male copulatory success independent of male rank	Inbreeding avoidance; frequency of displays; idiosyncratic choices	Soltis et al. (1997a,b, 1999)
Wild group	Premating behavior †Dist. copulations and Premating behavior †Paternity	Reproductive advantage to dominant males, though female proximity maintenance was unrelated to male rank	Females seeking multiple mates	Soltis et al. (2001)
Macaca nigra (Sulawesi crested black macaques)	Premating behavior; Dist. copulations	Dominant males preferred	Dominant males preferred	Reed et al. (1997)
Macaca mulatta (rhesus macaques)	Premating behavior *Dist. copulations	Female choice affects male copulatory success independent of male rank	Inbreeding avoidance; idiosyncratic choices	Manson (1992, 1994a,b, 1995, 1996), Manson and Perry (1993)
	Dist. copulations	?	Choice for males that give copulation calls	Hauser (1993)
	Experimental	N.A.	Females preferred to look at red male faces	Waitt (2003)

Table 27.1 *(cont'd)*

SPECIES	TYPE OF EVIDENCE[1]	EFFICACY OF FEMALE CHOICE RELATIVE TO EFFECTS OF MALE–MALE COMPETITION	CHOICE CRITERIA	SOURCES
Papio hamadryas anubis (anubis or olive baboon)	Premating behavior *Dist. consortships	Dominant male preferred	Dominant male preferred	Seyfarth (1978)
	Dist. consortships	No mating advantage to dominant males	Males with established affiliative relationships (Friends) preferred	Smuts (1985)
	Premating behavior *Dist. consortships	Dominant males preferred	Young, high-ranking recent immigrants preferred; multiple mates sought	Bercovitch (1991, 1995)
	Experimental	Mixed evidence for relative roles of males and females in consortship formation	Preference for frequent grooming partner	Schneider (1998)
Papio hamadryas hamdryas (hamadryas baboons)	Experimental	Males inhibited from "taking" females from preferred subordinate males	?	Bachmann and Kummer (1980)
Papio hamdryas hamadryas × *P. h. anubis* (hamadryas /anubis baboon hybrid)	Premating behavior *Dist. copulations	?	Preference for males of own phenotype	Bergman and Beehner (2003)
Pongo pygmaeus (orangutan)	Premating behavior; Dist. fertile copulations	Locally dominant male preferred	Protection from harassment by subordinate males	Galdikas (1981); Schürmann and van Hooff (1986); Fox (1998, 2002)
Gorilla beringei beringei (mountain gorillas)	Premating behavior; Dist. copulations[2]	Dominant not always preferred	Inbreeding avoidance	Sicotte (2001)
Pan troglodytes (chimpanzees) Gombe	Dist. of consortships	Consortships not used by dominant males	Inbreeding avoidance; choice for males that groom and share meat with the choosing female	Tutin (1979); Pusey (1980)
Mahale	Premating behavior *Dist. copulations	Dominant males preferred (?)	Dominant males preferred (?)	Matsumoto-Oda (1999)
Tai	Premating behavior *Dist. fertile copulations	Alpha and omega males preferred in one group; alpha disfavored in a another group	Inbreeding avoidance; age sorting; idiosyncratic preferences	Stumpf and Boesch (2005)

[1] Premating behavior, female choice inferred from premating behaviors (e.g., sexual solicitations, rejections of mount attempts) that have face validity as forms of mate choice; Dist. copulations, distribution of copulations among dyads; Dist. consortships, distribution of consortships among dyads; Paternity: paternity determined using genetic technology; * indicates a statistical association between the occurrence of a female behavior pattern and a measure of mating outcome (either distribution of copulations or genetically assigned paternity), † indicates that statistical tests failed to find an association between the occurrence of a female behavior pattern and a measure of mating outcome (When "Premating behavior" is separated from "Dist. copulations" and/or "Paternity" by a semicolon, these sources of evidence were used but no attempt was made to find a statistical association between female behavior and mating outcomes); Experimental, female given a choice of males in an experimental setting.

[2] In multimale groups.

mating (Huffman 1992, Perloe 1992, Manson 1992). Statistical comparisons of the effects of male dominance and attractiveness to females on male RS have sometimes shown that the two variables have independent significant effects (Manson 1992, Soltis et al. 1997a). Although these macaque results must be qualified by the consideration that they were obtained in provisioned or captive groups (in which, e.g., males may remain longer than they would under natural conditions), they at least show the potential for female choice to undermine the effects of male–male competition. Among captive sooty mangabeys (*Cercocebus torquatus*), female sexual solicitations and acceptance of male courtship positively affected male mating success independently of the effects of male dominance rank (Buck 1998).

Female choice also affects putative male RS, independently of the effects of male strategies, in a hybrid anubis/

hamadryas baboon (*Papio anubis* × *P. hamadryas*) population studied by Bergman and Beehner (2003). Males varied in their employment of a hamadryas-like social strategy (i.e., spending time in close proximity to nonfecundable females; see Chapter 14). Hamadryas-like social behavior was positively correlated with mating success. However, after statistically controlling for male social strategy, males that received more grooming and hindquarter presents from females experienced greater mating success. In a classic experimental study of hamadryas baboons, Bachmann and Kummer (1980) showed that males are inhibited from challenging each other for the "possession" of a female if they have seen that the female prefers her current partner, even if he is subordinate to the would-be challenger.

Some studies have quantified putative female mate choice behaviors in cercopithecines and found that they had little effect on male RS. Soltis et al. (2001), using the same methods that revealed the effectiveness of female mate choice in a captive group of Japanese macaques, found that females of a wild group maintained proximity preferentially to males of various dominance ranks, yet high-ranking males sired most of the offspring that were sired by troop males. Proceptive female Formosan macaques (*M. cyclopis*) maintained proximity preferentially to lower-ranking and extragroup males, yet male copulatory success was positively correlated with dominance rank (Birky 2002). Among wild anubis baboons studied by Bercovitch (1991, 1995), rates of grooming and hindquarter presentations by sexually swollen females affected whether dyads formed consortships but these and other female behaviors had little effect on male sexual access to a female after they had formed a consortship. Persistent female refusal to allow a male to copulate did not decrease consortship duration, nor was mount rejection rate correlated with male ejaculatory success. Among wild patas monkeys (*Erythrocebus patas*), male copulatory success was correlated neither with rate of female sexual solicitations received nor with success in male–male fights (Chism and Rogers 1997). Female barbary macaques (*M. sylvanus*) reject almost no copulation attempts (Small 1990), although they may incite male–male competition by leaving a mating partner and approaching another male (Kuester and Paul 1992). Because mating success was concentrated among those males that successfully passed these female-initiated tests, female behavior in this species reinforces the effects of male–male competition.

Lemuroids

Perhaps because female lemuroids are usually dominant to males (see Chapter 4), this taxon provides several examples of female choice partially or completely negating the predicted effects of male dominance on RS. In one study, female ring-tailed lemurs (*Lemur catta*) preferentially solicited newly immigrated males that were lower-ranking than most long-term residents, and the newcomers sired a disproportionately large share of the offspring (Pereira and Weiss 1991). Among wild Verreaux's sifakas (*Propithecus verreauxi*), females generally prefer resident and older dominant males but sometimes vigorously resist dominants' mating attempts and successfully conceive with subordinates (Brockman 1999) or even incite male–male fights and then copulate with a third, uninvolved male (Richard 1992). In a captive population of gray mouse lemurs (*Microcebus murinus*), Radespiel and colleagues (Radespiel and Zimmermann 2003, Radespiel et al. 2002) found that (*1*) females terminated 40% of all copulations, always by biting, slapping, or fighting the male; (*2*) females sometimes failed to conceive at all during a mating season by refusing to mate; and (*3*) dominant males sired no more offspring than would be expected by chance.

Chimpanzees

In a study of proceptivity and sexual resistance by female wild chimpanzees in the Taï Forest, the females of one group showed the most resistance and the least proceptivity toward the highest ranking among three males, whereas in the other group they preferred both the highest and the lowest ranking of four males (R. M. Stumpf and C. Boesch 2005). Among male–female dyads, female proceptivity rate was positively correlated with copulation frequency throughout female ovarian cycles, whereas female sexual rejection rate was negatively correlated with copulation frequency only during periovulatory periods.

Female Choice May Contribute to Discrepancies Between Grouping Pattern and Mating System

Early accounts of primate social organization (e.g., Kummer 1971: 31) equated species-typical grouping pattern with species-typical mating system. Females' potential mates were assumed to be limited to the male members of her social group. Long-term observation and paternity assignment studies have led primatologists to reject this assumption for many species, just as ornithologists earlier discovered that extrapair copulations (EPCs) are common in many socially monogamous avian species (e.g., Westneat 1987). EPCs and pair dissolutions are common in socially monogamous siamangs (*Hylobates syndactylus*) (Palombit 1994a,b), white-handed gibbons (*H. lar*) (Reichard 1995, Sommer and Reichard 2000), fat-tailed dwarf lemurs (*Cheirogaleus medius*) (Fietz et al. 2000), and possibly titi monkeys (*Callicebus molloch*) (Mason 1966). Seasonal influxes of males apparently prevent resident males from monopolizing reproduction in guenon (*Cercopithecus*) groups (Cords 1987). In multimale groups, females sometimes mate with extragroup males (Agoramoorthy and Hsu 2000), and such males may sire offspring (e.g., Berard et al. 1993, Gerloff et al. 1999), sometimes via "sneaked" matings unseen by observers (Soltis et al. 2001). Although there are few systematic data quantifying the importance of female choice in producing these discrepancies between social system and mating system, some evidence suggests that female behavior plays a major role in these phenomena. Proceptive females of many species have been observed soliciting matings from extragroup males, sometimes moving considerable distances from their own group to do so (Hrdy 1977, Henzi and Lucas 1980, Brereton 1981, Burton and Fukuda 1981, Richard 1985, Wolfe 1986, Sprague 1991, Manson 1995, Strier 1997, Brockman 1999).

Mate Choice Can Be Costly to Females

If females incur costs by acting in ways that alter the probability of mating with a particular male, it is reasonable to infer that mate choice confers compensating benefits on females—if it did not, then selection would favor female indifference to a mate's identity. Furthermore, costs of female mate choice play a major role in models that attempt to determine the conditions under which Fisherian "runaway," as opposed to Zahavian "good genes," selection processes will prevail (e.g., Pomiankowski et al. 1991).

By seeking out matings with low-ranking, peripheral, and/or extragroup males, primate females may incur costs in the form of increased aggression from dominant males of their own group (Smuts and Smuts 1993). I found (Manson 1994a) that among free-ranging rhesus macaques, sexually attractive/proceptive females experienced a higher rate of male escalated aggression (chases and bites) while in proximity to low-ranking males than while in proximity to high-ranking males. Aggressors in most cases were males dominant to the male that was accompanying the female. The potential costs entailed by such aggression are illustrated dramatically by Lindburg's (1971) account of a male fatally wounding a proceptive female that was accompanying a lower-ranking male in a wild rhesus macaque group.

Influences on the Efficacy of Female Mate Choice

Although the number of systematic studies of primate female mate choice is still too small to permit quantitative comparisons among species or populations, it is possible that female behavior has more effect (relative to male–male competition) on male reproductive success when more females are sexually receptive simultaneously (Soltis et al. 2001). This hypothesis is supported by comparisons between the two Japanese macaque populations studied by Soltis and colleagues (1997a, 2001) and by Paul's (1997) finding that breeding seasonality is associated, in cross-species comparisons, with a weakened association between male dominance and RS. Variation among species in degree of sexual dimorphism may also affect females' ability to influence the distribution of copulations among males (Barenstain and Wade 1983, Smuts and Smuts 1993, Kuester and Paul 1992; but see Bercovitch 1995).

Female Mate Choice Criteria

Choice for Dominant Males

Female choice for subordinate males, as discussed above, is conspicuous to observers because of (*1*) the resultant clash with "conventional wisdom" and (*2*) the high drama created by events such as prolonged refusals to mate with alpha males and attacks by dominant males on courting pairs including subordinate males. However, several studies have clearly documented female choice for high-ranking males (Hausfater 1975, Seyfarth 1978, Janson 1984, Bercovitch 1991, Fox 2002). Sexually swollen nonconsort female anubis baboons directed more grooming toward higher-ranking males, which also tended to be young and to be new immigrants (Bercovitch 1991). Interestingly, these males did not experience an overall mating advantage (Bercovitch 1986). In an experimental setting, female vervets (*Cercopithecus aethiops*) consistently chose to mate with their group's alpha male (Keddy 1986). Female Sulawesi crested black macaques (*Macaca nigra*) directed more sexual solicitations toward dominant than toward subordinate males (Reed et al. 1997). Among wild chimpanzees at Mahale, copulation rate

was correlated with male dominance rank during females' periovulatory periods and female grooming rate was positively correlated with male mating rate in comparisons among dyads; however, it is unclear whether female grooming rate was correlated with male dominance rank (Matsumoto-Oda 1999).

Access to Resources

A possible benefit to females of choosing dominant males is to provide females and/or their offspring with access to reproductively valuable resources. Few empirical examples of this process have been discovered in primates, presumably because individuals rarely have exclusive control over food. In the clearest case, Janson (1984) found that female brown capuchins at Manu National Park in Peru showed more proceptive behavior toward the alpha male than other males, that females were largely responsible for maintaining proximity to the alpha male, and that the alpha male monopolized copulations during females' likely fertile periods. At this site, preferred food sources were easily monopolizable, so the alpha male could exclude juveniles that he was unlikely to have sired from access to valuable feeding sites. At another brown capuchin site, in contrast, preferred food was less monopolizable and females showed a weaker mating preference for the alpha male (Janson 1994). In one study of wild chimpanzees, the most preferred consort partners were males that shared meat most readily with females (Tutin 1979).

Inbreeding Avoidance

The most widely documented primate female mate choice criterion is avoidance of genetic relatives. Sex-biased dispersal usually separates adult opposite-sexed close kin into different social groups (Pusey and Packer 1987); however, when relatives are potential mates, they copulate much less than expected by chance, and females are usually primarily responsible for this pattern. Female chimpanzees resist their sons' and brothers' courtship attempts (Tutin 1979, Pusey 1980, Goodall 1986) and sometimes avoid high-ranking male relatives while sexually swollen by engaging in consortships with unrelated males. These tactics usually prevent incestuous conceptions (Constable et al. 2001). Among free-ranging rhesus macaques, males breeding in their natal group experienced copulatory success equal to that of immigrant males, but they received greater proximity maintenance and fewer sexual refusals from unrelated attractive/proceptive females than from related attractive/proceptive females (Manson and Perry 1993). These behavioral patterns are probably responsible for the observed low frequency of incestuous conceptions in captive rhesus macaques (Smith 1995). Soltis et al.'s (1999) study of Japanese macaques also found that attractive/proceptive females avoided proximity to their male matrilineal kin and that related dyads copulated less than expected by chance and never during females' periovulatory periods. Female ring-tailed lemurs showed proceptivity toward unrelated males and aggression toward close relatives that courted them and avoided being impregnated by kin (Pereira and Weiss 1991, Taylor and Sussman 1985, Sauther 1991). Similar results have been obtained from other lemuroids (Richard 1992, Brockman 1999, Radespiel et al. 2002, Radespiel and Zimmermann 2003). A sex difference in aversion to inbreeding is suggested by Chapais and Mignault's (1991) analysis of heterosexual and homosexual interactions in Japanese macaques. Female–female consortships were frequent, yet never involved close kin, even though (1) female kin preferred each other for nonsexual interactions and (2) occasional incestuous heterosexual consortships were observed. In barbary macaques, in contrast, inbreeding avoidance appears to show no obvious sex difference in intensity: matings and conceptions occurred at much lower frequencies than expected in maternally related dyads, but females did not resist the few copulations between kin that were observed (Paul and Kuester 1985, Kuester et al. 1994). Matings and conceptions between paternally related kin did not occur at lower than expected frequencies. However, mating was inhibited in male–female dyads in which the male had had a close ("caretaking") relationship with the female when she was an infant. Taken together, these results support the hypothesis that individuals avoid inbreeding by using early social familiarity (which occurs among maternal, but not necessarily paternal, kin in this species) as a cue to genetic relatedness (Paul and Kuester 2004), a hypothesis originally proposed to account for human inbreeding avoidance (Westermarck 1891). However, yellow baboon (*Papio cynocephalus*) paternal half-siblings engage in less affiliative and sexual behavior while in consort than do nonsiblings (Alberts 1999), and dyads of similar age (which are statistically likely to be paternal half-sibs) engage in consortships at lower frequency than dyads of dissimilar age. These results suggest that in species in which dominant males temporarily monopolize paternity, selection can favor natal individuals' use of age proximity (in addition to early social familiarity) as a cue of relatedness.

Unfamiliarity

Females are predicted to choose unfamiliar males for five reasons. First, familiarity may a reliable cue of relatedness, as discussed above. Second, altered sociodemographic conditions (e.g., provisioning leading to very large group size and extended male tenures) may cause unrelated or distantly related dyads to experience enough familiarity that they behave, maladaptively, as if they were kin (Takahata 1982a). Third, choice for unfamiliar males presumably leads to the production of a genetically diverse set of offspring, and this in turn may increase females' long-term RS in temporally or spatially variable environments (Parker 1970, Williams 1975). Fourth, choice for unfamiliar males increases the number of potential sires of the choosing female's subsequent offspring and, thus, reduces the number of potentially infanticidal males (Hrdy 1981), assuming that males refrain from killing infants that they might have sired, as documented in a recent review (van Schaik 2000). Fifth, choice for unfamiliar

males could increase the number of potentially investing males (Hrdy 1981; but see Soltis and McElreath 2001).

Female primates frequently approach and solicit matings from extragroup, peripheral, and newly immigrated males (Cords et al. 1986, Strier 1997, Hrdy 1977, Brockman 1999, Henzi and Lucas 1980, Richard 1985, Brereton 1981, Burton and Fukuda 1981, Wolfe 1986, Manson 1995, Sprague 1991, Packer 1979). Bercovitch (1991) found that newly immigrated male anubis baboons (that were also usually high-ranking, thus confounding the variables of rank and novelty) received more grooming from sexually swollen nonconsort females compared to long-term resident males. These behaviors almost certainly increase these males' chances of siring offspring (e.g., Berard et al. 1993, Gerloff et al. 1999, Soltis et al. 2001), and, thus, qualify as mate choice for novelty. However, interpreting these observations as evidence for female preference for novel males (Small 1993) is less straightforward (Manson 1995). Consider a case in which females prefer males independently of their novelty (i.e., using a criterion that is uncorrelated with novelty). Because achieving copulations with an extragroup or peripheral male probably requires more effort than mating with a central male of the female's own group, females will need to direct more choice behavior toward preferred novel males than toward preferred familiar males.

Protection of Self and/or Infants

Females might choose mates that provide protection from male aggression against themselves or their infants, either in courtship contexts only or at other times as well. This choice criterion is thought to be central to the phenomenon of male–female friendships in anubis baboons (Ransom and Ransom 1971, Strum 1983, Smuts 1985). These relationships consist of preferential proximity and grooming while the female is not sexually receptive, agonistic support of the female by the male, and preferential mating while the female has a sexual swelling. Smuts (1985) showed that some friend dyads form their relationships before mating and argued that females prefer to mate with males with which they already have an affiliative relationship. Other studies (e.g., Palombit et al. 1997, Bercovitch 1991, Busse and Hamilton 1981, Packer 1980, Noe and Sluijter 1990, Weingrill 2000) indicate that baboon friendships form only after consortships during females' fertile periods and function to enable the male to protect his likely offspring from the danger of infanticide. Bercovitch (1991) argues that females choose young recent immigrants as mates because they are more likely than established residents to remain in the troop and provide protection to infants. However, most baboon researchers have not examined, as Smuts (1985) did, whether males experience greater consorting success with friends than would be expected given their overall level of consorting activity. Macaque male–female friendships do not include preferential mating access (Fedigan and Gouzoules 1978, Enomoto 1978, Hill 1990, Manson 1994b). Among redfronted lemurs (*Eulemur fulvus rufus*), one of the few non-female-dominant lemuroids, male–female friendships involve spatial association and mutual agonistic support (Pereira and McGlynn 1997). Although completed copulations were rarely observed in this study, all four female subjects accepted copulation attempts from male friends on likely days of ovulation, and two subjects aggressively rejected mating attempts by nonfriends on those same days.

Orangutans (*Pongo pygmaeus*) are unique among nonhuman primates in that forced copulations are frequent (MacKinnon 1974, Galdikas 1981). Subadult males and males using the arrested development strategy (see Chapter 17) are more likely than full-sized (flanged) males to encounter female sexual resistance and to force copulations on females (Galdikas 1985, Mitani 1985a). Fox (1998, 2002) found that (*1*) female resistance to male mating attempts did not consistently decrease over time, suggesting that resistance is "honest" rather than a test of male health and vigor, and (*2*) females that maintained proximity to a full-sized male gained some protection from harassment by unflanged males in that a smaller proportion of unflanged males' copulation attempts were successful, although the overall rate of unflanged male copulations, including forced copulations, was not affected by the presence of a full-sized male.

A hypothesized function of female dispersal, in the species in which it occurs, is to enable females to transfer to a male that is better able to protect infants from infanticide. Transfer by female mountain gorillas has been linked to infanticide (Watts 1989), and Smuts (1995) has hypothesized that males kill infants during intergroup encounters to induce the victims' mothers to transfer to them. The logic of this argument is that a male that can successfully kill an infant is a better protector than a male that was unable to prevent the killing of his infant. However, in most of the cases for which complete information is available, either the female's transfer preceded the infanticide or she did not transfer into the killer's group (Sicotte 2001).

Traits that May Signal "Good Genes"

In contrast to researchers of other taxa (e.g., insects, Moore 1994), primatologists have gathered only a small number of data sets suggesting female choice for genes, have not conclusively demonstrated choice for genes in any study population, and have made almost no progress in determining whether any particular cases conform better to Fisherian or Zahavian models of female choice. Four studies have addressed whether male macaque copulation calls (loud vocalizations given exclusively while copulating) are costly, reliable (i.e., Zahavian) indicators of male quality. Specifically, they have tested the hypothesis that calling males incite aggression against themselves and thereby demonstrate their ability to tolerate harassment and wounds. Hauser (1993) reported that, among male rhesus macaques, (*1*) the occurrence and frequency of calling were positively associated with the frequency of male aggression suffered and (*2*) calling males experienced higher mating success than noncallers. However, studying the same population, I

found that (*1*) males were almost never attacked during or immediately after a copulation and their likelihood of being attacked was not affected by whether they called and (*2*) there was no evidence from analysis of female choice behavior (proximity maintenance, cooperation with mount attempts) to suggest that calling males were more attractive than silent males (Manson 1996). In one study of Formosan macaques, the occurrence of calling was variable among males yet males were never harassed or attacked during or shortly following copulations (Hsu et al. 2002). Another study of the same species found no evidence that calling males were more attractive to females then silent males (Birky 2002). Thus, the adaptive significance of macaque male copulation calls remains unclear.

In an experimental setting, estrous female gray mouse lemurs (*Microcebus murinus*) showed a strong trend toward preferring males that produced trill calls at higher rates (Craul et al. 2004). Because trill call rate is correlated with dominance rank (Zimmermann and Lerch 1993) and males provide no paternal care, this result suggests female choice for genes.

Male macaques engage in courtship displays consisting of shaking branches, leaping, and other conspicuous behaviors. Modahl and Eaton (1977) found that male Japanese macaque display rates did not correlate with dominance rank but did correlate with rate of completed copulations, suggesting that females choose males that display more frequently. Consistent with this result, Soltis et al. (1999) report that periovulatory female Japanese macaques maintained proximity most actively to males that had displayed most frequently before the female's first periovulatory period of the mating season. Orangutan male long calls are clearly products of sexual selection, but field experiments reveal that they function to repel rival males rather than to attract females (Mitani 1985b).

Several dozen primate species are sexually dichromatic, raising the possibility that females could use male color as a signal of genetic quality (Gerald 2003). Very little research has addressed this possibility, although some evidence indicates that male color variation plays a role in male–male competition (Wickings and Dixson 1992, Gerald 2001, Setchell and Dixson 2001). Using digitized images in an experimental setting, Waitt et al. (2003) found that female rhesus macaques spent more time looking at images of reddened male faces than at otherwise identical pale faces. Among mandrills (*Mandrillus sphinx*), the most strikingly sexually dichromatic primate, semi-free-ranging periovulatory females directed more proceptive and receptive behavior toward more brightly colored males, independent of the effects of male dominance rank (Setchell 2005). In the wild, male mandrills may lack the long-term social bonds of other male cercopithecines and may be forced to rely on displays of color to establish dominance over rival males and to induce females to choose them (Abernethy et al. 2002).

Boinski's (1987) classic study of mate choice in wild Costa Rican squirrel monkeys (*Saimiri oerstedii*) showed that receptive females preferentially follow and solicit copulations from males that have attained the largest body size via a seasonal, steroid-mediated "fattening" process. Boinski suggests that this unusual dimorphic trait could have evolved by Fisherian ("runaway") sexual selection, but available data cannot distinguish between this and a Zahavian ("good genes") hypothesis.

"Choice for genes" includes choice for mates whose genotypes are compatible with (i.e., optimally dissimilar to) that of the chooser, in the sense that such matings would produce offspring with higher fitness, most likely by increasing heterozygosity (Brown 1997). No study has sought evidence that female primates choose mates on the basis of genetic compatibility, although this phenomenon has been abundantly documented for rodents (e.g., Potts et al. 1991). However, two studies of macaques have found that MHC genotypic dissimilarity increases offspring fitness. Among captive pig-tail macaques (*Macaca nemestrina*), parental sharing of certain MHC antigens significantly predicted pregnancy wastage (Knapp et al. 1996); and among free-ranging rhesus macaques, males that were heterozygous at an MHC locus sired significantly more offspring than homozygotes, even though they did not live significantly longer (Sauermann et al. 2001). The authors of the latter study claim that, because mother–sire couples were no more likely than expected by chance to share zero or one allele at this locus, their data falsify the hypothesis that females use this locus as a basis of mate choice. However, patterns of parentage reflect other processes besides female choice (i.e., male–male competition and male sexual coercion) (Gowaty 1997), and behavioral evidence from this population indicates that attractive/proceptive females do not always copulate with the males to which they maintain proximity preferentially (Manson 1992). Thus, it remains an open question whether female macaques choose mates on the basis of MHC genotypes. The finding that female rhesus macaques maintained proximity preferentially to the same males in consecutive years, even when the two animals were not observed to interact at all during the intervening birth season (Manson 1992), suggests idiosyncratic choice possibly based on genetic characteristics.

"Choice" for Multiple Mates

Hrdy (1986) and Small (1989) have cautioned against confusing female mate choice with generalized female sexual assertiveness, which could be an adaptation to prevent infanticide (Hrdy 1981) or a non-adaptive by-product of sexual motivation (Small 1993). Several studies of multimale group-living primates show that as the duration of attractive/proceptive periods, or focal observation time on a female, increases, the number of her copulatory partners increases also (Small 1990, Manson 1992, Soltis et al. 2001), indicating that females seek out multiple mates. However, with rare exceptions, such as barbary macaques (Small 1990, Kuester and Paul 1992), females of multimale primate groups reject the sexual advances of some males, as reviewed above. Even

in muriquis, which do not form dominance hierarchies and almost never engage in aggression, copulations are not distributed evenly among males (Strier 1997). Furthermore, some studies (Soltis et al. 1999, Stumpf and Boesch 2005) show that females become more selective when they are most likely to be ovulating. These results suggest that females of most group-living primate species pursue a mixed mating strategy, incorporating elements of both selectivity and promiscuity. A similar view is supported by comparative analyses of sexual swellings (Nunn 1999).

MALE MATE CHOICE

Conditions likely to favor male discrimination among potential mates include (*1*) high costs of male–male competition (Pagel 1994, Domb and Pagel 2001), (*2*) high variation in female fertility and/or offspring survivorship (Altmann 1997), and (*3*) synchronized breeding that prevents even the most dominant or attractive males from mating with multiple females (Kuester and Paul 1996). Hypothesized male mate choice criteria include female dominance rank, age, and heritable quality.

Choice for Female Rank

Dominant females in multifemale primate groups often experience reproductive advantages such as shorter interbirth intervals and higher rates of infant survival (Silk 1993, Packer et al. 1995). Thus, males should be expected to choose high-ranking mates. Evidence that they do so (reviewed by Robinson 1982, Berenstain and Wade 1983, Silk and Boyd 1983) consists largely of patterns of assortative mating in which high-ranking females tend to mate with high-ranking or otherwise competitively successful males, leading to the inference that males compete most strongly for high-ranking females. For example, old male barbary macaques are more successful than younger males at forming coalitions and thereby experience increased RS (Kuester and Paul 1996). Old males mated significantly more frequently with high-ranking than with low-ranking females, whereas this correlation was not found for young males. Because sons of high-ranking females outreproduced sons of low-ranking females, a preference for high-ranking females could increase a male's production of grandoffspring.

Choice for Female Age

Adolescent female primates are typically (*1*) not as likely to conceive as adults and (*2*) less likely than adults to successfully raise an offspring to weaning age (see Chapter 25). Presumably for this reason, male primates find sexually receptive adolescents less attractive than adult females in a wide range of cercopithecines, colobines, and apes (Anderson 1986, Anderson and Bielert 1994, Perry and Manson 1995), in both natural and experimental settings (Bielert et al. 1986). Even in species in which adolescents display exaggerated versions of sexual signals (e.g., sexual swellings), males disfavor them (Anderson and Bielert 1994). Male choice for older females, like male choice for high-ranking females, is often inferred on the basis of mating patterns. For example, Robbins (1999) found that adult female mountain gorillas were more likely to mate with the dominant male than with subordinate males, whereas subadult females were equally likely to mate with dominant and subordinate males.

Sexual Swellings: A Signal of Heritable Quality?

The function(s) of primate female sexual swellings remains a puzzle (see Chapter 25). Most hypotheses proposed to account for them (Clutton-Brock and Harvey 1976, Hrdy 1981, Hamilton 1984, Nunn 1999) assume that sexual swellings signal some probability that the female is ovulating. These hypotheses differ in the proposed benefit to females, but all agree that sexual swellings enable choosing males to make within-female, across-time comparisons only. Pagel (1994), in contrast, has hypothesized that female sexual swellings are Zahavian signals of quality and that males use them to assess interfemale differences in fertility. Among anubis baboons, (*1*) females with larger swellings reached sexual maturity sooner and produced more surviving offspring per year than females with smaller swellings and (*2*) males consorting with females with larger swellings received more aggression than males consorting with females with smaller swellings, suggesting that males are more attracted to the former (Domb and Pagel 2001). However, swelling size was not corrected for body size in this study (Zinner et al. 2002). Furthermore, a comparative study by Nunn et al. (2001) found no evidence that exaggerated sexual swellings are associated with measures of intensity of female–female mate competition such as the adult sex ratio and estimated female mating synchrony.

CONCLUSIONS

Over 20 years ago, Fedigan (1982: 275–276) asked whether primatologists, like behavioral biologists generally, had inadvertently reduced female mate choice to Hobson's choice, that is, not a real choice but mere validation of the results of male–male competition. Theory and data generated since then have partially corrected the biases identified by Fedigan. Female choice sometimes undermines the effects of male–male competition, though we do not yet have the comparative data to determine the conditions under which this occurs. Inbreeding avoidance is a strong and widespread primate mate choice criterion, as in almost all mammals. Unsurprisingly, given the great diversity of primate ecological niches, social organizations, and cognitive abilities, other female choice criteria vary throughout the order. Smuts' (1985) proposal that nonsexual "friendships" affect primate mate choice has enjoyed only limited empirical validation, although it has not been tested properly as frequently

as it deserves. Potential effects of genetic quality, including genetic compatibility, on male attractiveness have only recently begun to attract attention from primatologists. We may anticipate that in the coming years primatological data will begin to catch up with theory and with data from other taxa regarding these issues.

REFERENCES

Abernethy, K. A., White, L. J. T., and Wickings, E. J. (2002). Hordes of mandrills (*Mandrillus sphinx*): extreme group size and seasonal male presence. *J. Zool.* 258:131–137.

Agoramoorthy, G., and Hsu, M. J. (2000). Extragroup copulation among wild red howler monkeys in Venezuela. *Folia Primatol.* 71:147–151.

Alberts, S., Watts, H. E., and Altmann, J. (2003). Queuing and queue-jumping: long-term patterns of reproductive skew in male savannah baboons, *Papio cynocephalus. Anim. Behav.* 65:821–840.

Alberts, S. C. (1999). Paternal kin discrimination in wild baboons. *Proc. R. Soc. Lond. B* 266:1501–1506.

Altmann, J. (1997). Mate choice and intrasexual reproductive competition: contributions to reproduction that go beyond acquiring more mates. In: Gowaty, P. A. (ed.), *Feminism and Evolutionary Biology: Boundaries, Intersections, and Frontiers*. Chapman and Hall, New York. pp. 320–333.

Altmann, J., Alberts, S. C., Haines, S. A., Dubach, J., Muruthi, P., Coote, T., Geffen, E., Cheesman, D. J., Mututua, R. S., Saiyalel, S. N., Wayne, R. K., Lacy, R. C., and Bruford, M. W. (1996). Behavior predicts genetic structure in a wild primate group. *Proc. Natl. Acad. Sci. USA* 93:5797–5801.

Altmann, S. A. (1962). A field study of the sociology of rhesus monkeys, *Macaca Mulatta. Ann. N. Y. Acad. Sci.* 102:338–435.

Andelman, S. J. (1987). Evolution of concealed ovulation in vervet monkeys (*Cercopithecus aethions*). *Am. Nat.* 129:785–799.

Anderson, C. M. (1986). Female age–male-preference and reproductive success in primates. *Int. J. Primatol.* 7:305–326.

Anderson, C. M., and Bielert, C. F. (1994). Adolescent exaggeration in female catarrhine primates. *Primates* 35:283–300.

Andersson, M. B. (1994). *Sexual Selection*. Princeton University Press, Princeton, NJ.

Bachmann, C., and Kummer, H. (1980). Male assessment of female choice in hamadryas baboons. *Behav. Ecol. Sociobiol.* 6:315–321.

Barenstain, L., and Wade, T. D. (1983). Intrasexual selection and male mating strategies in baboons and macaques. *Int. J. Primatol.* 4:201–235.

Bateson, P. (1983). Optimal outbreeding. In: Bateson, P. (ed.), *Mate Choice*. Cambridge University Press, Cambridge. pp. 257–278.

Berard, J. D., Nurnberg, P., Epplen, J. T., and Schmidtke, J. (1993). Male rank, reproductive behavior, and reproductive success in free-ranging rhesus macaques. *Primates* 34:481–489.

Bercovitch, F. B. (1986). Male rank and reproductive activity in savanna baboons. *Int. J. Primatol.* 7:533–550.

Bercovitch, F. B. (1991). Mate selection, consortship formation, and reproductive tactics in adult female savanna baboons. *Primates* 32:437–452.

Bercovitch, F. B. (1995). Female cooperation, consortship maintenance, and male mating success in savanna baboons. *Anim. Behav.* 50:137–149.

Bercovitch, F. B., and Nurnberg, P. (1997). Genetic determination of paternity and variation in male reproductive success in two populations of rhesus macaques. *Electrophoresis* 18:1701–1705.

Berenstain, L., and Wade, T. D. (1983). Intrasexual selection and male mating strategies in baboons and macaques. *Int. J. Primatol.* 4:201–235.

Bergman, T. J., and Beehner, J. C. (2003). Hybrid zones and sexual selection: insights from the Awash baboon hybrid zone (*Papio hamdryas anubis* × *P. h. hamadryas*). In: Jones, C. B. (ed.), *Sexual Selection and Reproductive Competition in Primates: New Perspectives and Directions.* American Society of Primatologists, Norman, OK. pp. 503–537.

Bielert, C., Girolami, L., and Anderson, C. (1986). Male chacma baboon (*Papio ursinus*) sexual arousal—studies with adolescent and adult females as visual stimuli. *Dev. Psychobiol.* 19:368–383.

Birky, W. A. (2002). Mating patterns and social structure in a wild group of Formosan macaques [PhD thesis]. Rutgers University, Piscataway, NJ.

Boinski, S. (1987). Mating patterns in squirrel monkeys (*Saimiri oerstedi*): implications for seasonal sexual dimorphism. *Behav. Ecol. Sociobiol.* 21:13–21.

Brereton, A. (1981). Intertroop consorting by a free-ranging female rhesus monkey. *Primates* 22:417–423.

Brockman, D. K. (1999). Reproductive behavior of female *Propithecus verreauxi* at Beza Mahafaly, Madagascar. *Int. J. Primatol.* 20:375–398.

Brooks, R. (2000). Negative genetic correlation between male sexual attractiveness and survival. *Nature* 406:67–70.

Brown, J. L. (1997). A theory of mate choice based on heterozygosity. *Behav. Ecol.* 8:60–65.

Buck, M. R. (1998). Female mate choice in sooty mangabeys: social constraints on mating behavior [PhD thesis]. Emory University, Atlanta, GA.

Burton, F. D., and Fukuda, F. (1981). On female mobility: the case of the Yugawara-T group of *Macaca fuscata*. *J. Hum. Evol.* 10:381–386.

Busse, C., and Hamilton, W. J. (1981). Infant carrying by male chacma baboons. *Science* 212:1281–1283.

Chapais, B., and Mignault, C. (1991). Homosexual incest avoidance among females in captive Japanese macaques. *Am. J. Primatol.* 23:171–183.

Chism, J., and Rogers, W. (1997). Male competition, mating success and female choice in a seasonally breeding primate (*Erythrocebus patas*). *Ethology* 103:109–126.

Clutton-Brock, T. R., and Harvey, P. H. (1976). Evolutionary rules and primate societies. In: Bateson, P. P. G., and Hinde, R. A. (eds.), *Growing Points in Ethology.* Cambridge University Press, Cambridge. pp. 195–237.

Constable, J. L., Ashley, M. V., Goodall, J., and Pusey, A. E. (2001). Noninvasive paternity assignment in Gombe chimpanzees. *Mol. Ecol.* 10:1279–1300.

Cords, M. (1987). Forest guenons and patas monkeys: male–male competition in one-male groups. In: Smuts, B. B., Cheney, D. L., Seyfarth, R. M., Wrangham, R., and Struhsaker, T. T. (eds.), *Primate Societies.* University of Chicago Press, Chicago. pp. 98–111.

Cords, M., Mitchell, B. J., Tsingalia, H. M., and Rowell, T. E. (1986). Promiscuous mating among blue monkeys in the Kakamega Forest, Kenya. *Ethology* 72:214–226.

Craul, M., Zimmermann, E., and Radespiel, J. (2004). First experimental evidence for female mate choice in a nocturnal primate. *Primates* 45:271–274.

Cronin, H. (1991). *The Ant and the Peacock: Altruism and Sexual Selection from Darwin to Today.* Cambridge University Press, Cambridge.

Darwin, C. (1871). *The Descent of Man and Selection in Relation to Sex.* John Murray, London.

de Ruiter, J. R. (1992). Male social rank and reproductive success in wild long-tailed macaques: paternity exclusions by blood protein analysis and DNA-fingerprinting. In: Martin, R. D., Dixson, A. F., and Wickings, E. J. (eds.), *Paternity in Primates: Genetic Tests and Theories.* Karger, Basel. pp. 175–191.

Dewsbury, D. A. (1982). Ejaculate cost and male choice. *Am. Nat.* 119:601–610.

DiBitetti, M. S., and Janson, C. H. (2001). Reproductive socioecology of tufted capuchins (*Cebus apella*) in northeastern Argentina. *Int. J. Primatol.* 22:127–142.

Domb, L. G., and Pagel, M. (2001). Sexual swellings advertise female quality in wild baboons. *Nature* 410:204–206.

Doucet, S. M., and Montgomerie, R. (2003). Multiple sexual ornaments in satin bowerbirds: ultraviolet plumage and bowers signal different aspects of male quality. *Behav. Ecol.* 14:503–509.

Drickamer, L. C., Gowaty, P. A., and Holmes, C. M. (2000). Free female mate choice in house mice affects reproductive success and offspring viability and performance. *Anim. Behav.* 59:371–378.

Drickamer, L. C., Gowaty, P. A., and Wagner, D. M. (2003). Free mutual mate preferences in house mice affect reproductive success and offspring performance. *Anim. Behav.* 65:105–114.

Duffy, D. L., and Ball, G. F. (2003). Song predicts immunocompetence in male European starlings (*Sturnus vulgaris*). *Proc. R. Soc. Biol. Sci. B* 269:847–852.

Eberhard, W. G. (1996). *Female Control: Sexual Selection by Cryptic Female Choice.* Princeton University Press, Princeton, NJ.

Enomoto, T. (1978). On social preference in sexual behavior of Japanese monkeys (*Macaca fuscata*). *J. Hum. Evol.* 7:283–293.

Fedigan, L. M. (1982). *Primate Paradigms: Sex Roles and Social Bonds.* Eden Press, Montreal.

Fedigan, L. M., and Gouzoules, H. (1978). The consort relationship in a troop of Japanese monkeys. In: Chivers, D. J., and Herbert, J. (eds.), *Recent Advances in Primatology.* Academic Press, New York. pp. 493–495.

Fietz, J., Zischler, H., Schwiegk, C., Tomiuk, J., Dausmann, K. H., and Ganzhorn, J. U. (2000). High rates of extra-pair young in the pair-living fat-tailed dwarf lemur, *Cheirogaleus medius*. *Behav. Ecol. Sociobiol.* 49:8–17.

Fisher, R. A. (1930). *The Genetical Theory of Natural Selection.* Clarendon Press, Oxford.

Fox, E. A. (1998). The function of female mate choice in the Sumatran orangutan (*Pongo pygmaeus abelii*) [PhD thesis]. Duke University, Durham, NC.

Fox, E. A. (2002). Female tactics to reduce sexual harassment in the Sumatran orangutan (*Pongo pygmaeus abelii*). *Behav. Ecol. Sociobiol.* 52:93–101.

Galdikas, B. M. F. (1981). Orangutan reproduction in the wild. In: Graham, D. E. (ed.), *Reproductive Biology of the Great Apes.* Academic Press, New York. pp. 281–300.

Galdikas, B. M. F. (1985). Adult male sociality and reproductive tactics among orangutans at Tanjung-Puting. *Folia Primatol.* 45:9–24.

Gerald, M. S. (2001). Primate colour predicts social status and aggressive outcome. *Anim. Behav.* 61:559–566.

Gerald, M. S. (2003). How color may guide the primate world: possible relationships between sexual selection and sexual dichromatism. In: Jones, C. B. (ed.), *Sexual Selection and Reproductive Competition in Primates: New Perspectives and Directions.* American Society of Primatologists, Norman, OK. pp. 141–171.

Gerloff, U., Hartung, B., Fruth, B., Hohmann, G., and Tautz, D. (1999). Intracommunity relationships, dispersal pattern and paternity success in a wild living community of bonobos (*Pan paniscus*) determined from DNA analysis of faecal samples. *Proc. R. Soc. Lond. B* 266:1189–1195.

Goodall, J. (1986). *The Chimpanzees of Gombe: Patterns of Behavior.* Harvard University Press, Cambridge, MA.

Gowaty, P. A. (1997). Sexual dialectics, sexual selection, and variation in reproductive behavior. In: Gowaty, P. A. (ed.), *Feminism and Evolutionary Biology: Boundaries, Intersections, and Frontiers.* Chapman and Hall, New York. pp. 351–384.

Gowaty, P. A., Steinichen, R., and Anderson, W. W. (2002). Mutual interest between the sexes and reproductive success in *Drosophila pseudoobscura. Evolution* 56:2537–2540.

Grafen, A. (1990). Sexual selection unhandicapped by the Fisher process. *J. Theor. Biol.* 144:473–516.

Halliday, T. R. (1983). The study of mate choice. In: Bateson, P. (ed.), *Mate Choice.* Cambridge University Press, Cambridge. pp. 3–22.

Hamilton, W. D. (1964). The genetical evolution of social behavior. *J. Theor. Biol.* 7:1–51.

Hamilton, W. D. (1980). Sex versus non-sex versus parasite. *Oikos* 35:282–290.

Hamilton, W. D., and Zuk, M. (1982). Heritable true fitness and bright birds: a role for parasites? *Science* 218:384–387.

Hamilton, W. J. (1984). Significance of paternal investment by primates to the evolution of male–female associations. In: Taub, D. M. (ed.), *Primate Paternalism.* Van Nostrand Reinhold, New York. pp. 309–355.

Hauser, M. D. (1993). Rhesus monkey copulation calls: honest signals for female choce? *Proc. R. Soc. Lond. B* 254:93–96.

Hausfater, G. (1975). *Dominance and Reproduction in Baboons* (Papio cynocaphalus). Karger, Basel.

Heisler, L., Andersson, M. B., Arnold, S. J., Boake, C. R., Borgia, G., Hausfater, G., Kirkpatrick, M., Lande, R., Maynard Smith, J., O'Donald, P., Thornhill, R., and Weissing, F. J. (1987). The evolution of mating preferences and sexually selected traits. In: Bradbury, J. W., and Andersson, M. B. (eds.), *Sexual Selection: Testing the Alternatives.* John Wiley and Sons, Chichester. pp. 97–118.

Henzi, S. P., and Lucas, J. W. (1980). Observations on the intertroop movement of adult vervet monkeys (*Cercopithecus aethiops*). *Folia Primatol.* 33:220–235.

Hill, D. A. (1990). Social relationships between adult male and female rhesus macaques: II. Non-sexual affiliative behavior. *Primates* 31:33–50.

Hrdy, S. B. (1974). Male–male competition and infanticide among the langurs (*Presbytis entellus*) of Abu, Rajasthan. *Folia Primatol.* 22:19–58.

Hrdy, S. B. (1977). *The Langurs of Abu: Female and Male Strategies of Reproduction.* Harvard University Press, Cambridge, MA.

Hrdy, S. B. (1981). *The Woman that Never Evolved.* Harvard University Press, Cambridge, MA.

Hrdy, S. B. (1986). Empathy, polyandry, and the myth of the coy female. In: Bleier, R. (ed.), *Feminist Approaches to Science.* Pergamon Press, New York. pp. 119–146.

Hsu, M. J., Lin, J. F., Chen, L. M., and Agoramoorthy, G. (2002). Copulation calls in male Formosan macaques: honest signals of male quality? *Folia Primatol.* 73:220–223.

Huffman, M. A. (1987). Consort intrusion and female mate choice in Japanese macaques (*Macaca fuscata*). *Ethology* 75:221–234.

Huffman, M. A. (1992). Influences of female partner preference on potential reproductive outcome in Japanese macaques. *Folia Primatol.* 59:77–88.

Huxley, J. (1938). Darwin's theory of sexual selection and the data subsumed by it, in light of recent research. *Am. Nat.* 72:416–433.

Janson, C. H. (1984). Female choice and mating system of the brown capuchin monkey *Cebus apella* (Primates: Cebidae). *Z. Tierpsychol.* 65:177–200.

Janson, C. H. (1994). Comparison of mating system across two populations of brown capuchin monkeys. *Am. J. Primatol.* 33:217.

Keddy, A. C. (1986). Female mate choice in vervet monkeys (*Cercopithecus aethiops sabaeus*). *Am. J. Primatol.* 10:125–134.

Keddy-Hector, A. C. (1992). Mate choice in non-human primates. *Am. Zool.* 32:62–70.

Keller, L. (1994). Rewards of promiscuity. *Nature* 372:229–230.

Knapp, L. A., Ha, J. C., and Sackett, G. P. (1996). Parental MHC antigen sharing and pregnancy wastage in captive pigtailed macaques. *J. Reprod. Immunol.* 32:73–88.

Kose, M., Maend, R., and Møller, A. P. (1999). Sexual selection for white tail spots in the barn swallow in relation to habitat choice by feather lice. *Anim. Behav.* 58:1201–1205.

Kuester, J., and Paul, A. (1992). Influence of male competition and female mate choice on male mating success in Barbary macaques (*Macaca sylvanus*). *Behaviour* 120:192–217.

Kuester, J., and Paul, A. (1996). Female–female competition and male mate choice in Barbary macaques (*Macaca sylvanus*). *Behaviour* 133:763–790.

Kuester, J., Paul, A., and Arnemann, J. (1994). Kinship, familiarity and mating avoidance in Barbary macaques, *Macaca sylvanus. Anim. Behav.* 48:1183–1194.

Kummer, H. (1971). *Primate Societies: Group Techniques of Ecological Adaptation.* Harlan Davidson, Arlington Heights, IL.

Lande, R. (1981). Models of speciation by sexual selection on polygenic traits. *Proc. Natl. Acad. Sci. USA* 78:3721–3725.

Lindburg, D. G. (1971). *The Rhesus Monkey in Northern India: An Ecological and Behavioral Study.* Academic Press, New York.

MacKinnon, J. (1974). The behaviour and ecology of wild orangutans (*Pongo pygmaeus*). *Anim. Behav.* 22:3–74.

Manson, J. H. (1992). Measuring female mate choice in Cayo Santiago rhesus macaques. *Anim. Behav.* 44:405–416.

Manson, J. H. (1994a). Male aggression: a cost of female mate choice in Cayo Santiago rhesus macaques. *Anim. Behav.* 48:473–475.

Manson, J. H. (1994b). Mating patterns, mate choice, and birth season heterosexual relationships in free-ranging rhesus macaques. *Primates* 35:417–433.

Manson, J. H. (1995). Do female rhesus macaques choose novel males? *Am. J. Primatol.* 37:285–296.

Manson, J. H. (1996). Rhesus macaque copulation calls: re-evaluating the "honest signal" hypothesis. *Primates* 37:145–154.

Manson, J. H. (1997). Primate consortships: a critical review. *Curr. Anthropol.* 38:353–374.

Manson, J. H., and Perry, S. E. (1993). Inbreeding avoidance in rhesus macaques: whose choice? *Am. J. Phys. Anthropol.* 90:335–344.

Mason, W. A. (1966). Social organization of the South American monkey, *Callicebus moloch*: a preliminary report. *Tulane Stud. Zool.* 13:23–28.

Matsumoto-Oda, A. (1999). Female choice in the opportunistic mating of wild chimpanzees (*Pan troglodytes schweinfurthii*) at Mahale. *Behav. Ecol. Sociobiol.* 46:258–266.

Maynard Smith, J. (1974). The theory of games and the evolution of animal conflicts. *J. Theor. Biol.* 47:209–221.

Mitani, J. C. (1985a). Mating behaviour of male orangutans in the Kutai Game Reserve, Indonesia. *Anim. Behav.* 33:392–402.

Mitani, J. C. (1985b). Sexual selection and adult male orangutan long calls. *Anim. Behav.* 33:272–283.

Modahl, K. B., and Eaton, G. G. (1977). Display behavior in a confined troop of Japanese macaques. *Anim. Behav.* 25:525–535.

Moore, A. J. (1994). Genetic evidence for the "good genes" process of sexual selection. *Behav. Ecol. Sociobiol.* 35:235–241.

Noe, R., and Sluijter, A. A. (1990). Reproductive tactics of male savanna baboons. *Behaviour* 113:117–170.

Nunn, C. L. (1999). The evolution of exaggerated sexual swellings in primates and the graded-signal hypothesis. *Anim. Behav.* 58:229–246.

Nunn, C. L., van Schaik, C. P., and Zinner, D. (2001). Do exaggerated sexual swellings function in female mating competition in primates? A comparative test of the reliable indicator hypothesis. *Behav. Ecol.* 12:646–654.

Olsson, M., Madsen, T., Shine, R., Gullberg, A., and Tegelström, H. (1994). Can female adders multiply? *Nature* 369:528.

Packer, C. (1979). Inter-troop transfer and inbreeding avoidance in *Papio anubis*. *Anim. Behav.* 27:1–36.

Packer, C. (1980). Male care and exploitation of infants in *Papio anubis*. *Anim. Behav.* 28:512–520.

Packer, C., Collins, D. A., Sindimwo, A., and Goodall, J. (1995). Reproductive constraints on aggressive competition in female baboons. *Nature* 373:60–63.

Pagel, M. (1994). The evolution of conspicuous oestrous advertisement in Old World monkeys. *Anim. Behav.* 47:1333–1341.

Palombit, R. (1994a). Dynamic pair bonds in hylobatids: implications regarding monogamous social systems. *Behaviour* 128:65–101.

Palombit, R. A. (1994b). Extra-pair copulations in a monogamous ape. *Anim. Behav.* 47:721–723.

Palombit, R. A., Seyfarth, R. M., and Cheney, D. L. (1997). The adaptive value of "friendships" to female baboons: experimental and observational evidence. *Anim. Behav.* 54:599–614.

Parker, G. A. (1970). Sperm competition and its evolutionary consequences in insects. *Biol. Rev. Cambr. Philos. Soc.* 45:525–567.

Parker, G. A. (1979). Sexual selection and sexual conflict. In: Blum, M. S., and Blum, N. A. (eds.), *Sexual Selection and Reproductive Competition in Insects*. Academic Press, New York. pp. 123–166.

Paul, A. (1997). Breeding seasonality affects the association between dominance and reproductive success in non-human male primates. *Folia Primatol.* 68:344–349.

Paul, A., and Kuester, J. (1985). Intergroup transfer and incest avoidance in semifree-ranging barbary macaques (*Macaca sylvanus*) at Salem (FRG). *Am. J. Primatol.* 8:317–322.

Paul, A., and Kuester, J. (2004). The impact of kinship on mating and reproduction. In: Chapais, B., and Berman, C. M. (eds.), *Kinship and Behavior in Primates*. Oxford University Press, Oxford. pp. 271–291.

Pereira, M. E., and McGlynn, C. A. (1997). Special relationships instead of female dominance for redfronted lemurs, *Eulemur fulvus rufus*. *Am. J. Primatol.* 43:239–258.

Pereira, M. E., and Weiss, M. L. (1991). Female mate choice, male migration, and the threat of infanticide in ringtailed lemurs. *Behav. Ecol. Sociobiol.* 28:141–152.

Perloe, S. I. (1992). Male mating competition, female choice and dominance in a free ranging group of Japanese macaques. *Primates* 33:289–304.

Perry, S., and Manson, J. H. (1995). A comparison of the mating behavior of adolescent and adult female rhesus macaques (*Macaca mulatta*). *Primates* 36:27–39.

Petrie, M. (1994). Improved growth and survival of offspring of peacocks with more elaborate trains. *Nature* 371:598–599.

Pomiankowski, A., Iwasa, Y., and Nee, S. (1991). The evolution of costly mate preferences: 1. Fisher and biased mutation. *Evolution* 45:1422–1430.

Pope, T. R. (1990). The reproductive consequences of male cooperation in the red howler monkey: paternity exclusion in multi-male and single-male troops using genetic markers. *Behav. Ecol. Sociobiol.* 27:439–446.

Potts, W. K., Manning, C. J., and Wakeland, E. K. (1991). Mating patterns in seminatural populations of mice influenced by MHC genotype. *Nature* 352: 619–621.

Potts, W. K., and Wakeland, E. K. (1993). Evolution of MHC genetic diversity: a tale of incest, pestilence and sexual preference. *Trends Genet.* 9:408–412.

Pusey, A., and Packer, C. (1987). Dispersal and philopatry. In: Smuts, B. B., Cheney, D. L., Seyfarth, R. M., Wrangham, R., and Struhsaker, T. T. (eds.), *Primate Societies*. University of Chicago Press, Chicago. pp. 250–266.

Pusey, A. E. (1980). Inbreeding avoidance in chimpanzees. *Anim. Behav.* 28:543–552.

Radespiel, U., Dal Secco, V., Drögemüller, C., Braune, P., Labes, E., and Zimmermann, E. (2002). Sexual selection, multiple mating and paternity in grey mouse lemurs, *Microcebus murinus*. *Anim. Behav.* 63:259–268.

Radespiel, U., and Zimmermann, E. (2003). The influence of familiarity, age, experience, and female mate choice on pregnancies in captive grey mouse lemurs. *Behaviour* 140:301–318.

Ralls, K., and Ballou, J. (1982). Effects of inbreeding on infant mortality in captive primates. *Int. J. Primatol.* 3:491–505.

Ransom, T. W., and Ransom, B. S. (1971). Adult male–infant relations among baboons (*Papio anubis*). *Folia Primatol.* 16:179–195.

Reed, C., O'Brien, T. G., and Kinnaird, M. F. (1997). Male social behavior and dominance hierarchy in the Sulawesi crested black macaque (*Macaca nigra*). *Int. J. Primatol.* 18:247–260.

Reeder, D. M. (2003). The potential for cryptic female choice in primates: behavioral, anatomical, and physiological considerations. In Jones, C. B. (ed.), *Sexual Selection and Reproductive Competition in Primates: New Perspectives and Directions*. American Society of Primatologists, Norman, OK. pp. 255–303.

Reichard, U. (1995). Extra-pair copulations in a monogamous gibbon (*Hylobates lar*). *Ethology* 100:99–112.

Richard, A. F. (1985). Social boundaries in a Malagasy prosimian, the sifaka (*Propithecus verreauxi*). *Int. J. Primatol.* 6:553–568.

Richard, A. F. (1992). Aggressive competition between males, female-controlled polygyny and sexual monomorphism in a Malagasy primate, *Propithecus verreauxi*. *J. Hum. Evol.* 22:395–406.

Robbins, M. M. (1999). Male mating patterns in wild multimale mountain gorilla groups. *Anim. Behav.* 57:1013–1020.

Robinson, J. G. (1982). Intrasexual competition and mate choice in primates. *Am. J. Primatol.* 131–144.

Sauermann, U., Nurnberg, P., Bercovitch, F. B., Berard, J. D., Trefilov, A., Widdig, A., Kessler, M., Schmidtke, J., and Krawczak, M. (2001). Increased reproductive success of MHC class II heterozygous males among free-ranging rhesus macaques. *Hum. Genet.* 108:249–254.

Sauther, M. L. (1991). Reproductive behavior of free-ranging *Lemur catta* at Beza Mahafaly Special Reserve, Madagascar. *Am. J. Phys. Anthropol.* 84:463–477.

Schneider, R. D. (1998). Female social preferences and mating behavior in captive group-living baboons (*Papio cynocephalus anubis*): an experimental study [PhD thesis]. University of Michigan, Ann Arbor, MI.

Schurmann, C. L., and van Hooff, J. (1986). Reproductive strategies of the orang-utan: new data and a reconsideration of existing sociosexual models. *Int. J. Primatol.* 7:265–287.

Setchell, J. M. (2005). Do female mandrills (*Mandrillus sphinx*) prefer brightly colored males? *Int. J. Primatol.* 26:715–735.

Setchell, J. M., and Dixson, A. F. (2001). Changes in the secondary sexual adornments of male mandrills (*Mandrillus sphinx*) are associated with gain and loss of alpha status. *Horm. Behav.* 39:177–184.

Seyfarth, R. M. (1978). Social relationships among adult male and female baboons. I. Behaviour during sexual consortship. *Behaviour* 64:204–226.

Sicotte, P. (2001). Female mate choice in mountain gorillas. In: Robbins, M. M., Sicotte, P., and Stewart, K. J. (eds.), *Mountain Gorillas: Three Decades of Research at Karisoke*. Cambridge University Press, Cambridge. pp. 59–88.

Silk, J. B. (1993). The evolution of social conflict among female primates. In: William, A., and Mason, S. P. M. (eds.), *Primate Social Conflict*. State University of New York Press, Albany. pp. 49–83.

Silk, J. B., and Boyd, R. (1983). Cooperation, competition, and mate choice in matrilineal macaque groups. In: Wasser, S. K. (ed.), *Social Behavior of Female Vertebrates*. Academic Press, New York. pp. 315–347.

Small, M. F. (1988). Female primate sexual behavior and conception: are there really sperm to spare? *Curr. Anthropol.* 29:81–100.

Small, M. F. (1989). Female choice in nonhuman primates. *Ybk. Phys. Anthropol.* 32:103–127.

Small, M. F. (1990). Promiscuity in Barbary macaques (*Macaca sylvanus*). *Am. J. Primatol.* 20:267–282.

Small, M. F. (1993). *Female Choices: Sexual Behavior of Female Primates*. Cornell University Press, Ithaca.

Smith, D. G. (1995). Avoidance of close consanguineous inbreeding in captive groups of rhesus macaques. *Am. J. Primatol.* 35:31–40.

Smith, R. (1984). Human sperm competition. In: Smith, R. (ed.), *Sperm Competition and the Evolution of Animal Mating Systems*. Academic Press, New York. pp. 601–659.

Smuts, B., and Smuts, R. W. (1993). Male aggression and sexual coercion of females in nonhuman primates and other mammals: evidence and theoretical implications. *Adv. Study Behav.* 22: 1–63.

Smuts, B. B. (1985). *Sex and Friendship in Baboons*. Aldine, New York.

Smuts, B. B. (1987). Sexual competition and mate choice. In: Smuts, B. B., Cheney, D. L., Seyfarth, R. M., Wrangham, R. W., and Struhsaker, T. T. (eds.), *Primate Societies*. University of Chicago Press, Chicago. pp. 385–399.

Smuts, B. B. (1995). The evolutionary origins of patriarchy. *Hum. Nat.* 6:1–32.

Soltis, J., and McElreath, R. (2001). Can females gain extra paternal investment by mating with multiple males? A game theoretic approach. *Am. Nat.* 158:519–529.

Soltis, J., Mitsunaga, F., Shimizu, K., Nozaki, M., Yanagihara, Y., DomingoRoura, X., and Takenaka, O. (1997a). Sexual selection in Japanese macaques: 2. Female mate choice and male–male competition. *Anim. Behav.* 54:737–746.

Soltis, J., Mitsunaga, F., Shimizu, K., Yanagihara, Y., and Nozaki, M. (1997b). Sexual selection in Japanese macaques: 1. Female mate choice or male sexual coercion? *Anim. Behav.* 54:725–736.

Soltis, J., Mitsunaga, F., Shimizu, K., Yanagihara, Y., and Nozaki, M. (1999). Female mating strategy in an enclosed group of Japanese macaques. *Am. J. Primatol.* 47:263–278.

Soltis, J., Thomsen, R., and Takenaka, O. (2000). Reproductive strategies and paternity in wild Japanese macaques on Yakushima Island, Japan. *Am. J. Phys. Anthropol.* (suppl. 30): 287.

Soltis, J., Thomsen, R., and Takenaka, O. (2001). The interaction of male and female reproductive strategies and paternity in wild Japanese macaques, *Macaca fuscata. Anim. Behav.* 62:485–494.

Sommer, V., and Reichard, U. (2000). Rethinking monogamy: the gibbon case. In: Kappeler, P. (ed.), *Primate Males: Causes and Consequences of Variation in Group Composition*. Cambridge University Press, Cambridge. pp. 159–168.

Sprague, D. S. (1991). Mating by nontroop males among the Japanese macaques of Yakushima Island. *Folia Primatol.* 57:156–158.

Strier, K. B. (1997). Mate preferences of wild muriqui monkeys (*Brachyteles arachnoides*): reproductive and social correlates. *Folia Primatol.* 68:120–133.

Strum, S. C. (1983). Use of females by male olive baboons (*Papio anubis*). *Am. J. Primatol.* 5:93–109.

Stumpf, R. M., and Boesch, C. (2005). Does promiscuous mating preclude female choice? Female sexual strategies and mate preferences in chimpanzees of the Taï Forest, Côte d'Ivoire. *Behav. Ecol. Sociobiol.* 57:511–524.

Sussman, R. W., Cheverud, J. M., and Bartlett, T. Q. (1995). Infant killing as an evolutionary strategy: reality or myth? *Evol. Anthropol.* 3:149–151.

Takahata, Y. (1982a). Social relations between adult males and females of Japanese monkeys in the Arashiyama B troop. *Primates* 23:1–23.

Takahata, Y. (1982b). The socio-sexual behavior of Japanese monkeys. *Z. Tierpsychol.* 59:89–108.

Taylor, L., and Sussman, R. W. (1985). A preliminary study of kinship and social organization in a semi-free-ranging group of *Lemur catta. Int. J. Primatol.* 6:601–614.

Thornhill, R., and Alcock, J. (1983). *The Evolution of Insect Mating Systems*. Harvard University Press, Cambridge, MA.

Trivers, R. L. (1971). The evolution of reciprocal altruism. *Q. Rev. Biol.* 46:35–57.

Trivers, R. L. (1972). Parental investment and sexual selection. In: Campbell, B. (ed.), *Sexual Selection and the Descent of Man.* Aldine, Chicago. pp. 136–179.

Tutin, C. E. G. (1979). Mating patterns and reproductive strategies in a community of wild chimpanzees. *Behav. Ecol. Sociobiol.* 6:29–38.

van Schaik, C. P. (2000). Infanticide by male primates: the sexual selection hypothesis revisited. In: van Schaik, C. P., and Janson, C. H. (eds.), *Infanticide by Males and Its Implications.* Cambridge University Press, Cambridge. pp. 27–60.

van Schaik, C. P., and Janson, C. H. (2000). *Infanticide by Males and Its Implications.* Cambridge University Press, Cambridge.

Waitt, C., Little, A. C., Wolfensohn, S., Honess, P., Brown, A. P., Buchanan-Smith, H. M., and Perrett, D. I. (2003). Evidence from rhesus macaques suggests that male coloration plays a role in female primate mate choice. *Proc. R. Soc. Biol. Sci. B* 270 (suppl.): 5144–5146.

Watts, D. P. (1989). Infanticide in mountain gorillas: new cases and a reconsideration of the evidence. *Ethology* 81:1–18.

Weingrill, T. (2000). Infanticide and the value of male–female relationships in mountain chacma baboons. *Behaviour* 138:337–359.

Westermarck, E. (1891). *The History of Human Marriage.* Macmillan, London.

Westneat, D. F. (1987). Extra-pair copulations in a predominantly monogamous bird: observations of behaviour. *Anim. Behav.* 35:865–876.

Wickings, E. J., and Dixson, A. F. (1992). Testicular function, secondary sexual development and social status in male mandrills (*Mandrillus sphinx*). *Physiol. Behav.* 52:909–916.

Williams, G. C. (1966). *Adaptation and Natural Selection: A Critique of Some Current Evolutionary Thought.* Princeton University Press, Princeton, NJ.

Williams, G. C. (1975). *Sex and Evolution.* Princeton University Press, Princeton, NJ.

Wilson, E. O. (1975). *Sociobiology: The New Synthesis.* Harvard University Press, Cambridge, MA.

Wolfe, L. D. (1986). Sexual strategies of female Japanese macaques (*Macaca fuscata*). *Hum. Evol.* 1:267–275.

Zahavi, A. (1975). Mate selection: a selection for a handicap. *J. Theor. Biol.* 205:14.

Zimmermann, E., and Lerch, C. (1993). The complex acoustic design of an advertisement call in male mouse lemurs (*Microcebus murinus*, Prosimii, Primates) and sources of its variation. *Ethology* 93:211–224.

Zinner, D., Alberts, S., Nunn, C. L., and Altmann, J. (2002). Significance of primate sexual swellings. *Nature* 420:142–143.

PART FIVE

Ecology

28

The New Era of Primate Socioecology
Ecology and Intersexual Conflict
Deborah Overdorff and Joyce Parga

Primatology is not something over there. It is not a logically conceived body of knowledge and techniques. Primatology is us, and we are deeply immersed in a particular European history. That history determined our education and set our social rewards; the greatest contribution to the study of the nonhuman primates might be to free us from some of the traditional limitations and points of view. (Washburn 1973:180)

INTRODUCTION

Historically, primate socioecology has been the study of how environmental variables (usually food, its seasonality, distribution, and abundance) influence primate group size, composition, and social dynamics. Our understanding of primate socioecology largely comes from long-term work at well-established research sites on such anthropoid genera as *Pan, Macaca, Papio,* and other related catarrhines. As more studies on these species and others are conducted in a variety of habitats, researchers have recognized a wide range of behavioral and dietary variation, even within closely related taxa such as *Semnopithecus* spp. (Koenig and Borries 2001), *Papio* spp. (Barton et al. 1996), *Colobus* spp. (Chapman et al. 2002), *Pan troglodytes* and *P. paniscus* (Boesch 2002, Mitani et al. 2002, Stanford 1998), *Pongo* spp. (Delgado and van Schaik 2000), and *Gorilla* spp. (Doran and McNeilage 1998). Work on less well-studied groups, most notably platyrrhines and strepsirhines, has further underscored the potential for inter- and intraspecific social diversity (i.e., *Saimiri* spp., Boinski et al. 2002; *Cebus,* Panger et al. 2002; *Callimico* and *Saguinus,* Porter 2004; Porter and Garber 2004; *Pithecia,* Norconk and Conklin-Brittain 2004; *Hapalemur,* Grassi 2002; *Eulemur,* Overdorff and Johnson 2003; *Lemur* and *Propithecus,* Yamashita 2002).

Researchers now realize that not all primates live in large groups, maintain rigid dominance hierarchies, show male dominance over females, or exhibit female philopatry (Strier 1994). Strepsirhines, for example (Fig. 28.1), are highly nonconvergent with haplorhines based on the prevalence of *(1)* small group sizes which converge on those documented for platyrrhines, *(2)* female dominance, and *(3)* nocturnal and cathemeral activity cycles (Curtis 2004, Curtis and Zaramody

(A)

(B)

Figure 28.1 (Upper) Ring-tailed lemurs (*Lemur catta*) feeding on terrestrial foliage (photo by Joyce Parga) and (lower) a red-fronted lemur female (*Eulemur fulvus rufus*) feeding on *Ficus* sp. fruits.

1999, Overdorff and Erhart 2001, Pereira et al. 1990, Rasmussen 1999, Tattersall 1987, Richard 1987). Consequently, identifying a common set of selection pressures that can explain such diversity has become more challenging.

Although primatology has come a long way theoretically since its inception in the 1940s and 1950s (Jolly 2000, Strum and Fedigan 2000), there are still several obstacles that primatologists must overcome. One inherent tendency for primatologists is to assume that nonhuman primates are "special"—set apart from other social mammals and birds due to their complex sociality, exceptional cognitive abilities, and close taxonomic affinity to humans. While no one would question that these are indeed uniquely primate traits, it has led to a general ignorance in our field of the value of a broader comparative perspective (see Harcourt 1998, Rowell 2000). Because primates are long-lived, reproduce slowly, and demonstrate extraordinary behavioral flexibility compared to similar-sized mammals, researchers must be extraordinarily patient, collect a much larger volume of data, and/or commit to data collection for a longer period of time when looking for emergent patterns. Primatologists also tend to place great value on studying their subjects under natural conditions; this devotion to naturalism poses its own set of problems experimentally. Despite years of study, small sample sizes for certain behaviors or age groups are common. This is exacerbated by the fact that wild subjects will die of unknown causes—or, worse, will mysteriously disappear—leaving the explanation for such events up to the researcher's discretion. Finally, the degree of historical or ongoing anthropogenic interference may also influence behavior in unpredicted ways, leading researchers to wonder if current conditions actually represent the original evolutionary context of the observed behavior (Janson 2000a).

As a result, primate socioecological studies have passed through various stages of increasing sophistication since the 1950s (see reviews by Dunbar 1988, Janson 2000a, Kappeler and van Schaik 2002, Strum and Fedigan 1999, Terborgh and Janson 1986). One constant throughout these stages has been the acceptance that the main selection pressures that influence primate behavior are predation and feeding ecology. Although these two variables go hand in hand, this chapter will focus on the explanations offered to interpret primate behavior based on feeding ecology. We will describe briefly each of the historical stages, discuss the pros and cons of existing theoretical approaches, describe where primatology is heading, and discuss the newest approaches which, ironically, consider selection pressures other than feeding ecology (specifically, intersexual conflict).

IN THE BEGINNING

Primatological studies in their infancy (Carpenter 1942, 1954; Zuckerman 1932) were largely descriptive and not based on models or hypothesis testing. Subsequent studies attempted to synthesize broad categories of primate behavior and searched for correlations between primate sociality and elements of their ecology. Two such studies (Crook and Gartlan 1966, Eisenberg et al. 1972) provided the first major contributions that helped move primatology away from being a solely descriptive discipline by attempting to find common ground between different primate species.

Crook and Gartlan's (1966) approach was to develop specific dietary and ecological categories and then place primates into these categories based on activity, type of habitat used, and group size. Using this technique, they found broad consistencies within groups, such as the fact that solitary primates tended to be nocturnal and savanna groups tended to live in large social units. Eisenberg et al. (1972) expanded their analysis by including additional categories such as diet and reproductive units (mating system). Despite these additions, it was still an exercise in assigning primates to categories with little adaptive rationale. Nevertheless, each of these studies laid the groundwork for the idea that there existed an evolutionary grade among social systems as primates evolved from a more ancestral (nocturnal and solitary) social pattern to a more derived, socially complex pattern (diurnal and group living). This assumption still forms the basis for more recent socioecological models (van Schaik and Kappeler 1996).

Though ambitious, these first theoretical papers reflected work with only a small number of well-studied species and did not take into account phylogeny, which can explain the presence of many similar traits found in closely related taxa (Rendall and Di Fiore 1994, but see Korstjens et al. 2002). They also did not account for the possible causal mechanisms that might shape behavior, although Crook's later work (1972, Crook et al. 1976) made a more significant attempt to place the behaviors in an adaptive and comparative (nonprimate) framework.

At approximately the same time, a more inductive and comparative approach documented more specific correlations between ecological and behavioral variables within primates (Clutton-Brock 1974, 1977, 1989; Clutton-Brock and Harvey 1980; Milton and May 1976). The implementation of more standardized behavioral data collection methods (Altmann 1974) and the increase in available data on a greater number of primate species also made such an approach more productive. The relationship between activity, diet, and ranging patterns was the main focus in these papers; and researchers relied on a more quantitative, statistical approach to their analyses. Other previously disregarded variables were included, such as daily ranging patterns, body size, and sex ratio. Some trends emerged during this time that are now taken for granted: nocturnal primates are smaller than their diurnal counterparts, frugivores range farther on a daily and long-term basis compared to folivores, smaller groups are more cryptic than larger groups, and sexual dimorphism increases with increasing group size where females outnumber males. Ongoing studies of the behavioral ecology of sympatric, congeneric species further confirmed that these have been consistent trends among primates (Crockett and

Wilson 1980; Gautier-Hion 1978; Hladik 1977; Kinzey and Gentry 1979; Milton 1981; Mittermeier and Roosmalen 1981; Struhsaker and Oates 1975; Sussman 1972, 1974). One of the main assumptions in the study of sympatric congeners was that habitat and evolutionary forces remain broadly constant so that finer details of social behavior can emerge. However, as with other approaches, there were still drawbacks: *(1)* confounding variables were problematic (e.g., frugivory and monogamy often coexist), *(2)* which independent variable should be considered the most important (group size, resource size, social interactions, activity, etc.) was not well defined or agreed upon, and *(3)* the causal mechanisms behind the observed patterns were vaguely articulated.

An additional weakness during this comparative/correlation phase was that the potential costs and benefits of behaviors were not really considered. There were also no generalized models that could produce testable hypotheses to guide future study. Conversely, this was not the situation for other taxonomic groups being studied concurrently. Ornithologists and mammalogists were developing internally consistent, evolutionarily based models that were used to explain sociality in nonprimate mammals and birds, which were firmly grounded in a sociobiological approach (Alexander 1974, Orians 1969, Wittenberger 1980, Wittenberger and Tilson 1980). Although ecological variables were considered important (i.e., the degree of territoriality and habitat quality), there was more focus on the role that sexual selection might play (especially male intrasexual competition for mates). More consideration was also given to kin selection and the role that inclusive fitness plays in determining social and mating systems. These papers ultimately suggested that grouping patterns among animals are a result of *(1)* specific selection pressures (ecological conditions, predation levels, etc.) and *(2)* the interplay between the costs and the benefits (in the form of reproductive success) of certain behaviors to males and females.

Primatologists were not completely ignorant of such considerations. In a paper not often cited, Denham (1971) detailed a model that broadly linked three variables into a predictive framework: energy acquisition, predation pressure

(what he called "defensive response"), and reproductive strategies. He laid out several key ecological variables, such as food predictability, food density, and antipredator strategies—all are concepts that would play a central role in later, more high-profile models. Shortly thereafter, Altmann (1974) published a more specific test of the energetics of baboon movements in which he considered these same variables.

Wrangham's (1979, 1980, 1982, 1987) work comprises one of the most pivotal and often-cited set of papers and demonstrates a significant turning point in primatological theory. In these papers, he developed the first ecological model for primates that provided for testable predictions and a clear direction for future research. Intergroup competition in this model is considered the most important factor in the evolution of primate grouping patterns. The roles of kin selection and inclusive fitness (which were essential elements in nonprimate models at the time) were also considered to be integral.

Generally, Wrangham proposed a strong link between the defendability of food resources, the degree of female bonding, and female dispersal patterns (Fig. 28.2). Food was thought to be the primary limiting factor for female reproductive success, and the number of mates was considered to be the primary limiting factor for male reproductive success. It is worthwhile noting that the foundation for this premise was laid 30 years earlier in experimental work on sexual selection using *Drosophila* as a model (Bateman 1948, Tang-Martinez 2000). Wrangham's own work (1980), concurrent platyrrhine studies (Robinson 1988, Terborgh 1983, Terborgh and Janson 1986), and more recent work back up the general premise of his model, although the picture has become more complex as more independent variables are considered (Dunbar 2000, Eberle and Kappeler 2002).

COMPETING COMPETITION MODELS

A good model should stimulate others to challenge and improve on previous work, and Wrangham's model did just that. The two major points of dispute that guided the next

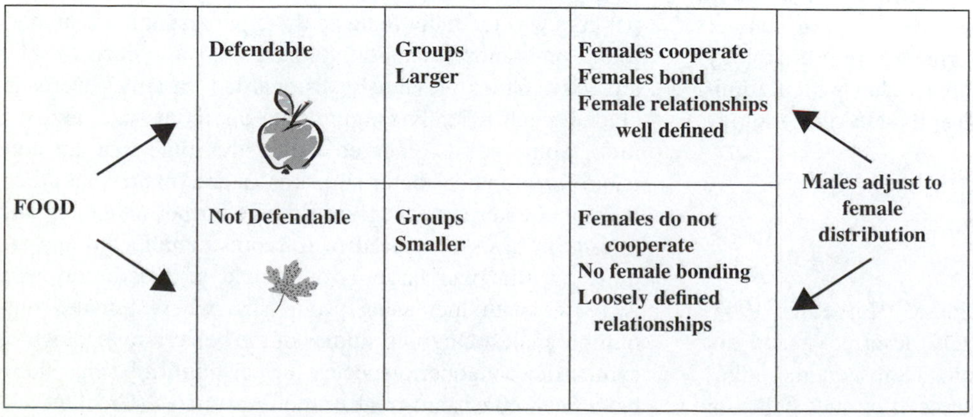

	Defendable	**Groups Larger**	**Females cooperate** **Females bond** **Female relationships well defined**
FOOD			**Males adjust to female distribution**
	Not Defendable	**Groups Smaller**	**Females do not cooperate** **No female bonding** **Loosely defined relationships**

Figure 28.2 A summary of Wrangham's initial model used to explain the relationship between types of food, food defendability, and female and male grouping patterns.

generation of modeling focused on the role predation avoidance should play (in conjunction with feeding ecology) and whether group size and social organization (how individuals interact within groups) function as independent variables. One such example is van Schaik and van Hooff's model (1983), which included predation pressure along with other ecological factors (e.g., food distribution and density). The combined influences of net food intake and predation risk will ultimately produce the optimal group size and social organization for a particular primate species. In a more rigorous test of this premise, van Schaik (1983) examined birth rates (which he used as a proxy for foraging efficiency) as a function of group size. By comparing the observed slopes to the expected slopes for two conditions (predation with intragroup feeding competition and intragroup feeding competition alone), he concluded that the lower limit to group size is set by predation pressure, while the upper limit is influenced by the degree of feeding competition. This paper demonstrated the need for direct, measurable variables to better test hypotheses.

Subsequent papers addressed this issue by including an explicit list of measurable variables (mostly related to feeding) that could be used to test predictions. Like their predecessors, Terborgh and Janson (1986) introduced a model for optimal group size that considered both predation and foraging efficiency but also introduced the idea that food resource size (or patch size) was an integral piece of the puzzle. Janson (1988a) subsequently produced a more complete list of variables related to net food intake that should be considered (e.g., encounter rate, amount of food consumed during each feeding bout, nutritional content of food, and time spent foraging).

Another pivotal moment occurred when primatologists finally integrated two concepts that had already been incorporated into group size models in other taxa: *(1)* the two main forms of competition and *(2)* that competition levels can differ within and between groups. Individuals can aggressively compete over food which is monopolizable (known as *contest* or *interference competition*), or they can indirectly compete by exploiting nonmonopolizable food on a "first come–first served" basis (known as *scramble competition*). Either of these forms of competition can occur within or between groups. Of note, avian models at this time emphasized that both the abundance and the divisibility of food resources were important variables (Elgar 1986, Mangel 1990), an idea featured prominently in the next round of seminal primate papers.

In 1988, a special issue of the journal *Behaviour* was dedicated to the question of which selection pressures most profoundly affect primate foraging and competitive regimes. These papers aptly demonstrated that it was possible to document differing levels of competition in a range of primate species (Chapman 1988, Garber 1988, Isabirye-Basuta 1988, Janson 1988a,b, Rowell 1988, Symington 1988, van Schaik and van Noordwijk 1988, White and Wrangham 1988, Whitten 1988). They also underscored the fact that researchers lacked a uniform methodological approach with which to investigate feeding competition from an evolutionary perspective. Accordingly, Janson and van Schaik (1988) emphasized the need for a more standardized methodology to measure the degree of competition for food within and between primate groups and provided a conceptual model that could more readily predict when different types of competition might occur. Using the *Behaviour* collection of articles as their starting point, they synthesized a comprehensive list of data that should be collected in future studies. This list included many previously published variables as well as a more sophisticated estimate for energy or gain rates while feeding. They also suggested that researchers should pay more attention to collecting data on food handling time, energy ingestion rate, and the amount of energy expended in different activity states. Although their intention was to challenge primatologists to develop and use more precise and quantitative methods in future competition studies, few studies (if any) have since incorporated *all* of the variables suggested. Of these variables, patch size continues to be commonly used, yet there remains little uniformity across studies as to how it is measured (Chapman 1990, Chapman et al. 1995, Overdorff 1996, Phillips 1995, Rodrigues de Moraes et al. 1998, Strier 1989, Wich et al. 2002, White 1989).

Subsequent revisions of the competition model refined predictions regarding intergroup dynamics, most often relying on previously published data. van Schaik (1989) revisited the concepts of between- and within-group competition and developed a more complex list of competitive regimes that included the possible strength of scramble and contest competition. Using these possible combinations, he made specific predictions about female intrasexual relationships based on food density, population density, and levels of between-group contest competition.

Isbell's (1991) variation on this approach also considered intra- and intergroup contest and scramble competition, but her model made more specific predictions about which ecological factors would influence competition. To test her model, she used data on day range and home range size for 20 primate species as proxies for food distribution and food abundance. After considering the variation in group size and ranging patterns, she concluded that food abundance shapes between-group relationships, while food distribution (clumped versus dispersed) shapes within-group relationships. One significant factor that sets her model apart from others is that outcomes are not dependent on levels of predation pressure. It is interesting to note that predation pressure becomes more or less a silent partner in many of the subsequent models: either it is dismissed as an unimportant influence or it is recognized as important but not considered any further in deference to behavior (competition) and diet.

Sterck et al. (1997) offered one more refinement to the competition model by considering how competition levels vary along three discrete dimensions: *(1)* egalitarian versus despotic (see Vehrencamp 1983), *(2)* individualistic versus nepotistic (kin), and *(3)* the degree of social tolerance (as measured by the severity of aggression and amount of reconciliation observed). By including female dispersal patterns, they then created four categories in which primates can be grouped based on the degree of within- and between-group competition, which could be strong (high) or low (weak). These categories were dispersal-egalitarian, resident-egalitarian, resident-nepotistic (but also despotic), and resident-nepotistic-tolerant. Using an updated comparative mammalian database, the authors demonstrated that the predicted categories were supported by empirical data (e.g., Barton et al. 1996, Boinski et al. 2002).

There were, of course, some exceptions to predicted patterns, and two alternative explanations were provided to guide future work. One was that habitat fragmentation or saturation may lead to nonadaptive behaviors within populations (*behavioral disequilibrium*). This has subsequently been proposed as a possible explanation for the unique grouping patterns of Malagasy lemurs (van Schaik and Kappeler 1996, but see Kay and Kirk 2000). Another consideration is the role that sexual coercion (Smuts and Smuts 1993) or, more explicitly, infanticide plays in shaping female relationships. Sterck et al. (1997) hypothesized that infanticide avoidance may act as a strong selective pressure on female social behavior and may promote varying counter-strategies depending on the degree of perceived threat (see data from *Gorilla*, Watts 1989, 1990; *Presbytis*, Sterck 1999, Sterck and Steenbeek 1997).

FINE TUNING

Although the "competition modeling" phase has provided primatologists with a more unified perspective on primate social behavior, there remain several major drawbacks (Isbell and Young 2002, Koenig 2002). One problem is the inconsistency in the primatological lexicon regarding social behavior categories. Despite early attempts at defining social systems, different researchers mean different things when they refer to "social structure," "social organization," and "mating system" (see Chapter 37). A second problem is that there is still a lack of consensus among primatologists with respect to which ecological variables should be considered, how they are measured, and which function as independent variables. Finally, there is still considerable focus on the proximate causes of behavior and not enough consideration of the ultimate, or evolutionary, causes.

The Importance of Words

Clear, operational definitions are a necessity when using the scientific method (Lehner 1996, Martin and Bateson 1993), yet primatologists over the years have been inconsistent in their use of definitions for varying aspects of social behavior and the ecological variables used within models. In earlier models, there was a tendency to portray social systems as a set of Russian dolls; social organization (group size)

determined social structure (social dynamics within groups), which in turn influenced mating system (van Schaik and van Hooff 1983). However, long-term work on populations with known individuals coupled with the recent ability to obtain paternity information through deoxyribonucleic acid (DNA) testing demonstrates that this strictly linear approach is no longer appropriate. In many species, group size and composition vary seasonally; therefore, the subsequent mating system may not always be predicted based on group composition. Both monogamous and polygynous mating systems, for example, have been observed in "solitary" strepsirhines (Bearder 1999, Fietz 1999, Mueller 1999). In other species, group composition and mating patterns can vary seasonally (Cords 2002, Cords et al. 1986) or considerably between groups within populations, such as in *Propithecus* spp., where pairbonded, unimale, and multimale groups are commonly observed (Pochron and Wright 2003, D. J. Overdorff unpublished data).

Independent Variables

Authors also have used more subjective terms or categories when classifying social and feeding patterns. Primatologists seem to have a penchant for forcing variability into dichotomous categories. For example, food sources can be clumped or dispersed, social dominance can be strong or weak, and food patches can be small or large. Lack of consistent use of these terms between researchers and the less than quantitative way that these terms are used severely limit their application (Isbell and Young 2002). Additionally, there is a growing consensus that primatologists have relied on variables that are easy to measure (e.g., patch size) to the exclusion of others which may be harder to measure but are more important to our understanding of the ultimate mechanisms at work (Kappeler and van Schaik 2002, Koenig 2002). Energy exchange rates (as first proposed by Janson and van Schaik 1988) and actual reproductive success are key variables that primatologists need to measure more precisely. The "patchiness" of food not only is hard to determine but may not universally influence primate feeding patterns. Other feeding variables that could be more rigorously quantified, such as food depletion time or even the size of individual food items, may be more appropriate (Isbell and Young 2002, Overdorff 1996, Pruetz and Isbell 2000, Vogel and Janson 2003). Primatologists also should consider duplicating some of the more sophisticated methods that have been used to examine energy output in captive primates during reproduction as reflected by milk output and quality (Edwards 1990, Tilden 1993, Tilden and Oftedal 1995).

Another general consensus is that the standard set of available methods and models that have been used by primatologists should be used more effectively (Dunbar 2002, Isbell and Young 2002, Koenig 2002), and entirely new directions need to be considered (Janson 2000a). The benefits of a more rigorous approach are that researchers are forced to be more explicit with their assumptions and to make clear distinctions between which variables are dependent and independent. This process may highlight variables that have not previously been considered important and has the potential of teasing out the evolutionary consequences of a behavior (as opposed to the evolutionary cause).

Innovative Use of Modeling

Some of the more recent and novel approaches to understanding the complexity of primate behavior include the application of economic principles to behavior and virtual modeling. For example, a better understanding of the complexities of individual interactions within groups might be achieved by applying collective action models (Nunn 2000, Nunn and Lewis 2001). While Nunn (2000) does not stray far from Wrangham's initial model, he adds another dimension for consideration: that of social rules and interactions, effects which can be illustrated by identifying collective action problems within groups. Nunn (2000) showed, using loud calls in macaques, that at some point males choose to cooperate to exclude additional males from the group, to reduce the incidence of "free-riders" that take advantage of services provided by other group members without contributing themselves. Such decisions have an obvious impact on optimal group size and may contribute to the degree of variation observed in social systems.

Other concepts commonly used by economists and sociologists, such as power and leverage (Fedigan 1983, 1992; Hand 1986), can be used to further refine the role dominance plays in dyadic and group interactions and can take primatologists beyond the more vague categories of "strong" versus "weak" dominance hierarchies. Lewis (2002) suggests that the concept of power is one way to address the current deficiencies of the dominance concept. In her model, power is considered the outcome of dominance relations and an individual's potential for leverage over another. *Dominance* retains its traditional definition: what an individual can gain based on force or the threat of force (i.e., one's ability to coerce, fight, and/or compete) and leverage is defined as "power based on inalienable resources," such as eggs or services. This approach is promising in terms of making progress toward a less subjective and more uniform technique to examine dyadic relationships and their impact on competitive abilities within groups.

Hemelrijk (2002a,b) has also explored the variation in dominance hierarchies from a unique perspective using virtual models. In a simplified primate world dubbed "DomWorld," Hemelrijk recreates interactions among primate "agents" that have only two tendencies: to group and to perform dominance interactions. By manipulating single variables, such as food, she demonstrates how despotic societies can arise from egalitarian ones as the result of individual selection, self-organizing principles, and group selection. Admittedly, there are drawbacks to such a model: it overly simplifies real-world dynamics, does not really consider what causes social grouping to begin with, and contains some controversial perspectives (i.e., group selection) which historically have had little support from primatologists (Bradley 1999). Nevertheless, such approaches may jumpstart primatologists into using such models to generate testable hypotheses in wild populations (e.g., the evolution of female dominance in pygmy chimpanzees, Hemelrijk 2002b).

ECOLOGY AND BEYOND

Janson (2000a) suggested that the business of primate socioecology as it has been conducted needs to be reassessed. By considering new paradigms and methodologies to think "outside of the box," some of the inconsistencies in primate social patterns might be better elucidated. To this end, the most recent wave of inquiry focuses on what have been traditionally considered rare occurrences or weak selection pressures, such as dispersal, predation, parasites, disease, and infanticide.

Dispersal

One of the basic assumptions behind the earlier competition models is that there is a strong association between female philopatry and female intergroup aggression. However, recent work with some species, especially colobines (Korstjens et al. 2002), indicates otherwise. Cooperation can occur in the absence of kinship. Also, female dispersal is more common in some species (especially platyrrhines) than previously thought. Consequently, Isbell (2004) proposed the "dispersal/foraging model" to explain variation seen in dispersal patterns not previously accommodated in other models. Instead of focusing on the inclusive benefits accrued by kin defending food sources, she explores how the locational (travel to unfamiliar home range) and social (harassment, see Pope 1992, Watts 1991) costs of dispersal influence formation of kin groups. Foraging efficiency remains the main selection pressure, but success in intergroup competition is no longer the main benefit of living with kin. How well this new model holds up as a replacement for its predecessors depends on more ad hoc tests, which must include data on travel patterns (wandering versus goal-oriented), travel rates (slow versus fast), mother–infant tolerance, and dispersal patterns.

Predation

Although the topic of predation and its influence on primate social behavior is reviewed in more detail in this volume (see Chapter 32) and others (Miller 2002), we will add that some of the more recent attempts to examine the influence of predation on primate social behavior are much more sophisticated in addressing previous gaps (Boinski and Chapman 1995, Isbell 1994). Improvements in understanding the dynamic between predators and their primate prey have included experimental studies of prey response to predator presence (i.e., playback calls, Gursky 2003, Karpanty and Grella 2001) and studies that focus on predators. In Madagascar, it was believed that current predator pressure was minimal given the low diversity and small body size of potential extant predators. It had even been suggested that some of the unique behaviors exhibited by lemuriform primates, such as increasingly large group size in diurnal species and cathemeral activity, are the result of the recent extinction of a giant eagle (*Aquila*, Csermely 1996, Goodman 1994). However, recent predator studies coupled with long-term studies on marked primate populations indicate that predation pressure is still very much an issue for lemuriforms (Britt et al. 2001, Brockman 2003, Dollar 1999, Goodman et al. 1993, Karpanty and Goodman 1999, Wright 1998). The real future of understanding the interaction between predators and prey lies in studying both populations simultaneously.

Parasites

In the late 1970s, Freeland (1976) published a non-food-driven model to explain primate sociality that has been relatively overlooked until recently. His hypothesis was that primate sociality can be explained as an adaptation that minimizes the risk of acquiring new pathogens and limits the pathogenicity of diseases already harbored, thereby maximizing individuals' reproductive fitness. In Freeland's view, group size is limited by food availability and by infant mortality rates, which can increase as a result of the higher disease density brought about by the presence of too many group individuals. Group stability and fidelity to a home range are influenced by perceived disease risk. More stable groups with little turnover in membership are less likely to have new pathogens introduced. Site fidelity to a home range is also less likely to introduce members to new diseases. When primates do disperse to avoid inbreeding (and/or to counterbalance increased infant mortality due to higher disease density), the rate of dispersal and the acceptance of immigrants are influenced by perceived disease risk. Freeland also suggests that female mate choice and, more notably, promiscuity may play a strong role in regulating the numbers of males and their success upon entering a group. For example, males that survive the stress experienced during intense mating competition are likely to be "superior" (i.e., rapid, aggressive take-overs of unimale groups). In other cases, individuals attempting to enter a new group more subtly may be observed by resident females across a long transitional period, during which time their overall health can be assessed.

One of the drawbacks to this hypothesis is in devising ways to separate out the influence of food versus perceived pathogen risk; any one of the behavioral patterns described above could be parsimoniously explained by patterns of food abundance, density, and/or seasonality. Steps toward testing hypotheses regarding pathogen influences must include understanding the evolution of the primate immune system, identifying proxies that can be used to assess pathogen risk, and actual collection of parasites from feces, blood, and or corpses (Stuart and Strier 1995). Nunn and colleagues have attempted to tackle the first two issues (Nunn 2002, 2003a,b; Nunn et al. 2000). Working with the assumption that group size, population density, and number of mating partners are likely to increase disease transmission, Nunn et al. (2000) examined white blood cell counts in 41 primate species. While they found no correlations with group size, population density, terrestriality, or body mass, they did find a positive correlation with the number of males with which a female mated. Females that mate with higher numbers of males should be more at risk of contracting sexually transmitted diseases, and this risk may be a significant driving force behind variation in the primate immune system.

So far, there has been little success in identifying proxies that can be used to assess disease risk in nonhuman primates. Spleen size, which has been a useful predictor of disease risk in birds and other nonprimate mammals, does not appear to correlate with the presumed disease risks to primates such as sociality (larger groups with higher population densities are likely to have higher disease density), terrestriality (exposure to soil and feces located on the ground poses a higher degree of pathogen exposure than arboreal habitats), or lifehistory (longer life spans allow for more exposure to disease vectors) (Nunn 2002). Many of the presumed behavioral defenses against disease exposure, such as genital inspection and postcopulatory urination, do not correlate well with the degree of promiscuity observed in primate populations (Nunn 2003a). Given the positive correlations observed in other animal taxa (birds and carnivores), this line of investigation may not be entirely closed for primates. Current research is hampered by poor statistical power due to small sample sizes (a common problem with primate studies) and the fact that there may be other as yet unidentified anatomical and physiological defenses employed by primates to counter disease risk (Nunn 2000). A better understanding of the potential coevolutionary or symbiotic relationship between primates and parasites will be achieved once a more comparative field database is available (Stuart and Strier 1995).

Infanticide and Intersexual Conflict

With an increasing number of long-term studies and with a wider range of species being studied, the number of observed infanticide cases by adult males and females has increased. Male infanticide has been observed or inferred in a number of species in all primate radiations (van Schaik and Janson 2000). Consequently, many primatologists have taken the stance that infanticide (or sexual coercion) exerts a strong selection pressure on behavior. This perspective has its supporters and its detractors and has become a divisive issue (Sommer 2000). Much of the controversy centers on whether or not one considers infanticide to be a product of selection.

Skeptics have taken aim at the infanticide database, which, in their view, is still considered weak evidence for sexual selection (Bartlett et al. 1993, Sussman et al. 1995). One of the major deficiencies is that there remains a small number of directly observed infanticides that fit the "sexual selection profile." Additionally, infanticide is often inferred when infants disappear, and inference should not be substituted for direct observation. The prevalence of observations in some species (such as *Presbytis*) and the rarity of instances in others (such as *Eulemur*) are also cited as evidence against considering behavioral responses to infanticide as a widespread primate adaptation. Also cited as problematic is the lack of substantial data documenting infanticidal counterstrategies. One final point considers the issue of how such behaviors are passed from one generation to the next (Sussman et al. 1995, Dolhinow 1999). For a behavior to be "selected for" and maintained in a population, it must be heritable and inherited by progeny. If infanticidal behavior were favored, such a scenario would lead to a form of runaway selection with increasingly aggressive males, which would eventually be a maladaptive strategy within a society. Instead, they suggest that infanticide is a "by-product" of generalized male aggression.

Responses to these criticisms emphasize that the database is, in fact, more robust than suggested (Borries and Koenig 2000, Crockett and Janson 2000, Palombit et al. 2000, Steenbeek 2000). Sommer (2000) points to a large number of incidences in *Presbytis*, *Pan*, and *Gorilla* that he considers support infanticide as an adaptive strategy (while other researchers do not), leading him to conclude that "What counts as unequivocal evidence is obviously largely in the eye of the beholder" (Sommer 2000:23). Therefore, it is up to the beholder to be explicit in his or her description of what happened and to which individual it happened and to be clear about why he or she interprets the behavior as infanticidal (or not). This is one of the benefits of the dialogue between those in favor of infanticide hypotheses or those opposed: it forces both sides to be more specific about actual observations and predictions.

Such dialogue also challenges primatologists to test predictions regarding infanticide in creative ways, such as identifying proxies for a relatively rare behavior. One example uses the ratio of lactation length to gestation length to estimate the vulnerability of infants to infanticide: when lactation is longer than gestation, infanticide should be more likely to occur (van Schaik 2000). Using a broader mammalian database, van Schaik concluded that infanticide does appear to occur more often in species whose lifehistory trajectory is such that a nonreproductive male would benefit by jump-starting a female's reproductive cycle via infant killing. Finally, actual documentation of changes in infant mortality and birth rate within populations as the result of infant killing as well as an individual infanticidal male's lifetime reproductive success compared to that of noninfanticidal males must be achieved to fully test the infanticide/sexual coercion hypothesis (Fedigan 2003). It is equally important to note that negative results that provide evidence against the presence of sexual coercion must also be equally weighted (e.g., *Ateles*, Campbell 2003).

Evidence for possible infanticidal counterstrategies is also mounting. Paternity confusion by females mating with multiple males (Hrdy 1979) has long been thought to be a viable counterstrategy to infanticide (see recent work by Borries et al. 1999, Enstam et al. 2002, Heistermann et al. 2001, van Schaik et al. 2000, Soltis 2002, Soltis et al. 2000). Additionally, Treves (2000) speculates that there may be some self-protection strategies implemented by infants themselves, such as age deception, concealment, or simply stranger avoidance. Strong affiliative bonds between males and females in some species may be the result of protection from infanticide (Palombit 1996, 1999, 2000). Female dispersal (Sterck and Korstjens 2000, Isbell 2004), female alliances (Treves and Chapman 1996), and male–infant relationships (Paul et al. 2000) also may be shaped by infanticide risk.

One of the biggest drawbacks to the infanticide perspective is the risk of it becoming a "megaparadigm" (see Sommer 2000). Infanticide prevention currently seems to be the stock explanation for behavioral patterns when all others fail. Pairbonding in gibbons (van Schaik and Dunbar 1990) and the presence of male–female dyads within several other primate taxa, such as *Lemur* and *Eulemur* spp., are such examples (Pereira and McGlynn 1997, van Schaik and Kappeler 1996).

Clearly, more data are needed to adequately test the infanticide protection hypothesis to address the following: *(1)* What are the actual fitness costs and benefits to males performing infanticide across their lifetime? *(2)* How does male presence in pairbonds actually influence infanticide risk? *(3)* How effective are counterstrategies to infanticide, and can these in fact be most parsimoniously explained by infanticide/sexual coercion hypotheses?

REMODELING

One of the continued obstacles primatologists face is determining ways in which to tease out the complicated effects of *(1)* easily documented (but not rigorously or systematically quantified) behaviors such as feeding, *(2)* rare and difficult to observe behaviors (due to observer affect) such as infanticide and predation, and *(3)* aspects of social living such as disease risk, power relationships, and degree of relatedness that are difficult to quantify. In any case, the challenge to the current and next generations of primatologists is to devise creative and technologically savvy methods to tackle these issues. These methods should *(1)* incorporate experimentation under field conditions; *(2)* integrate experimental work from captive populations more effectively into the theory developed from fieldwork; *(3)* bring technology to the field; *(4)* continue to think "outside the box" and incorporate new perspectives; *(5)* implement studies which truly attempt to tease out the effects of differing selection pressures such as foraging, predation, and intersexual conflict; and *(6)* acknowledge the potential for political or gender bias.

Experimental Fieldwork

Many workers have begun to develop more controlled, experimental studies under field conditions to better understand the social and economic rules that primates follow when making foraging decisions. Such studies have involved designing feeding platforms and then altering their arrangement and/or the type of food offered to better understand how primates form cognitive maps, what cues are implemented during foraging, the costs and benefits to specific kinds of foraging choice, and the pay-offs under specific ecological conditions that certain individuals might exact based on their dominance status or sex (Bicca-Marques and Garber 2003; Boinski and Garber 2000; Di Bitetti and Janson 2001; Dominy et al. 2003; Janson 2000b, 2001; Janson and Di Bitetti 1997). Although field experimentation does not occur under the pristinely controlled conditions that can be achieved in captivity, such work allows for insight into how social relationships shape foraging choices and can inform on cognitive abilities as well.

Captive Contributions

Many fieldworkers have a bias against captive work as the resultant behavior "isn't natural" or does not occur under adaptive conditions. However, many unsolved problems and limitations faced by field studies could be overcome in captive, semi-free-ranging, or free-ranging populations in novel environments where animals may be more accessible visually and physically, genealogies are usually well documented, and conditions are carefully controlled. Kinship, dominance interactions, mate choice, and other aspects of reproduction are areas that can in some cases be documented more carefully in captivity (Carosi and Visalberghi 2002; Chapais 1988a,b; Chapais and Lecomte 1995; Chapais et al. 1995; Craul et al. 2003; Fisher et al. 2003; Gerald 2001; Keddy 1986; Parga 2003; Pereira and Weiss 1991; Soltis et al. 1997; Taylor 1986; Radespiel and Zimmermann 2001; Vervaecke et al. 2000; de Waal 1982, 1986; Zinner et al. 1994). Work done in such novel environments can add to the existing knowledge base gathered from studies in the wild and may reveal entirely new information that would be difficult or impossible to document under natural field conditions. Additionally, captive studies can elucidate the behavioral plasticity inherent in the primate repertoire that might otherwise go undetected, especially for species having much condition-dependent behavioral flexibility (i.e., Sauther et al. 1999).

Field Technology

One of the long-standing criticisms of evolutionary paradigms is the fact that hypotheses regarding social strategies hinge on how a strategy ultimately influences male and female reproductive success (Fedigan 1983, 1992; Richard and Schulman 1982). While it is usually easy to make assumptions about mother–offspring relatedness based on association, it is often virtually impossible to know which male fathered offspring, even within monogamous primate groups (Palombit 1994a,b). The advent of molecular methods that can be implemented in the field makes the possibility of establishing infant paternity a very real goal (Andres et al. 2003, Cowlishaw and O'Connell 1996, Di Fiore 2003, Henzi 1996, Pazol 2003, Pullen et al. 2000, Weiss 2000, Yamane et al. 2003). Ultimately, these data can inform more fully on the impacts that intersexual conflict, infanticide, dominance, and within-group competition have on individual reproductive success. Long-term study of populations of known individuals over their lifetime to establish patterns of lifelong reproductive success and/or age-specific fecundity patterns is another important element that is being achieved across the order *Gorilla* (Doran and McNeilage 1998, Watts 1990), *Pongo* (Utami et al. 2002), *Pan* (Mitani et al. 2002, Nishida et al. 2003, Pusey et al. 1997), *Papio* (Altmann et al. 1996), *Macaca* (Koyama et al. 1992, Sugiyama and Ohsawa 1982), *Semnopithecus* (Koenig and Borries 2001), *Brachyteles* (Strier 1992), *Alouatta* (Glander 1992), and a variety of strepsirhines (Gould et al. 2003, Jolly et al. 2002, Koyama et al. 2001, Overdorff et al. 1999, Richard et al. 2002, Wright 1995).

Another area in which technology has greatly enhanced the type, amount, and quality of data gathered in the field is in the use of radiotelemetry for locating primate troops and tracking the movement of individuals (Campbell and Sussman 1994, Fedigan et al. 1988); satellite tracking is yet another research tool now available (Honess and MacDonald 2003). Global information system analysis of habitat (Dominy and Duncan 2001) is also being used more often, which greatly augments the explanatory power of studies beyond the traditional usage of global positioning system coordinates to estimate home range size (Sprague et al. 2004).

Out of the Box

As mentioned above, some of the most recent paradigm breakthroughs are the result of theorists daring to go "outside the box." The wealth of new, recent research directions (intersexual conflict, disease risk, dispersal, role of behaviors not necessarily related to aggression or competition) have evolved as primatologists challenge the basic assumptions made by earlier models regarding the role that food or competition plays in shaping primate dynamics, the nature of female relationships (Barrett and Henzi 2002), and the acceptance that there is not a typical or universal primate pattern (Strier 1994). Sussman et al. (2005) recently questioned the assumption that competition alone drives primate sociality. They proposed that cooperative and affiliative behaviors are equally likely to influence sociality given the higher proportion of time primates devote to these behaviors. Although some of these research avenues have generated heated debate (infanticide being one of the best examples), this is still the scientific process at its very best; paradigm shifts will not occur unless established paradigms are challenged forcefully (Kuhn 1962).

Confounding Variables

In the same vein, primatologists also must develop new ways to meaningfully tease apart the influence of differing selection pressures. Cowlishaw (1997) elegantly examined the potential trade-offs for wild mountain baboons between foraging and predation risk and found that his study animals made choices to minimize predation risk to the detriment of feeding. In a comparative study, Nunn and van Schaik (2000) tackled the broader question of whether ecological pressure or intersexual conflict drives variability in primate social behavior. Rather than focus on observed rates of behavior, which can vary simply due to observation conditions or number of contact hours, they used estimates of the intrinsic risk for each of their categories (predation, female intrasexual competition, infanticide, harassment). In the end, their results upheld the traditional assumption that ecological forces in the form of predation pressure and competition within groups can explain more of the variation in social behavior observed across species. While this comparative perspective is one of the innovative ways by which primatologists can tackle this question, Wolfe (1995) argues that a truly comparative approach cannot be realized until primatologists can have access to all available data. She advocates the formation of a database similar in function to that of the Human Resource Area Files, which would allow primatologists to access, retrieve, and organize information more effectively.

Bias

As scientists, we would like to believe that we are objective and unbiased in our approach to research questions. However, paradigms can be swayed by cultural influences and political agendas: this may be exacerbated by the fact that primatology has an anthropological ancestry rather than strictly evolutionary biological roots. Primatology is often considered a "feminist" science (Bleier 1986, Haraway 1986, Sperling 1991) given the high proportion of female researchers relative to other related fields and the emphasis on gender roles (Fedigan 2000). Primatology is also steeped in a decidedly Western perspective (Strum and Fedigan 2000), often ignoring other non-Western approaches (Asquith 2000). The marginalization of the Japanese perspective on primate sociality is a case in point. The Kyoto school's easy acceptance of anthropomorphism and dismissal of sociobiological interpretations sets it apart from the historical perspective outlined in this chapter (Takasaki 2000). This does not mean that primate studies are any less scientific (Strum 2000) or that our science is "bad." Rather, by recognizing our potential biases and acknowledging new ways to interpret behavior (outside of the cultural and social paradigms to which we are accustomed), we may in fact generate new questions and new ways of exploring those questions (see also Nakagawa and Okamoto 2003).

Primatology may be at the end of its "golden age" (Janson 2000a), which is often considered the first and best age of the world, represented by a time of ideal happiness, prosperity, and innocence. While it is true that the past decades have been prosperous in terms of the numbers and quality of studies conducted and the models proposed, the best age of primatology is yet to come. In keeping with the analogy, primatological theory in some senses has already embraced aspects of the second, "silver," age, characterized by opulence and loss of devoutness. Opulence in this case is reflected in the increasing number of comparative studies on the same species and the addition of new ones to the database. The third age of primatology, the "bronze" age, which is characterized by warfare and violence, is being realized as new researchers begin to aggressively challenge the devout following of particular paradigms. Exploring alternative hypotheses beyond those traditionally considered to be important provides opportunities for new perspectives to emerge, although it may result in creation of tremendous controversy (e.g., the infanticide debate). Thus, the study of primate socioecology is certainly not dead but in an exciting state of flux and entering a new age of its own definition.

REFERENCES

Alexander, R. D. (1974). The evolution of social behavior. *Annu. Rev. Ecol. Syst.* 5:325–383.

Altmann, J. (1974). Observational study of behavior: sampling methods. *Behaviour* 49:227–265.

Altmann, J., Alberts, S. C., Haines, S. A., Dubach, J., Muruthi, P., Coote, T., Geffen, E., Cheesman, D. J., Mututua, R. S., Saiyalel, S. N., Wayne, R. K., Lacey, R. C., and Bruford, M. W. (1996). Behavior predicts genetic structure in wild primate group. *Proc. Natl. Acad. Sci. USA* 93:5797–5801.

Altmann, S. A. (1974). Baboons, space, time, and energy. *Am. Zool.* 14:221–248.

Andres, M., Solignac, M., and Perret, M. (2003). Mating system in mouse lemurs: theories and facts, using analysis of paternity. *Folia Primatol.* 74:355–366.

Asquith, P. (2000). Negotiating science: internationalization and Japanese primatology. In: Strum, S. C., and Fedigan, L. M. (eds.), *Primate Encounters: Models of Science, Gender, and Society.* University of Chicago Press, Chicago. pp. 296–309.

Barrett, L., and Henzi, P. (2002). Constraints on relationship formation among female primates. *Behaviour* 139:263–289.

Bartlett, T. Q., Sussman, R. W., and Cheverud, J. M. (1993). Infant killing in primates: a review of observed cases with specific reference to the sexual selection hypothesis. *Am. Anthropol.* 95:958–990.

Barton, R. A., Byrne, R. W., and Whiten, A. (1996). Ecology, feeding competition and social structure in baboons. *Behav. Ecol. Sociobiol.* 38:321–329.

Bateman, A. J. (1948). Intra-sexual selection in *Drosophila. Heredity* 2:349–368.

Bearder, S. K. (1999). Physical and social diversity among nocturnal primates: a new view based on long term research. *Primates* 40:267–282.

Bicca-Marques, J. C., and Garber, P. A. (2003). Experimental field study of the relative costs and benefits to wild tamarins (*Saguinus imperator* and *Saguinus fuscicollis*) of exploiting contestable food patches as single- and mixed-species troops. *Am. J. Primatol.* 60:139–153.

Bleier, R. (1986). *Feminist Approaches to Science*. Pergamon Press, New York.

Boesch, C. (2002). Behavioural diversity in *Pan*. In: Boesch, C., Hohmann, G., and Marchant, L. F. (eds.), *Behavioural Diversity in Chimpanzees and Bonobos*. Cambridge University Press, New York. pp. 1–8.

Boinski, S., and Chapman, C. A. (1995). Predation on primates: where are we and what's next? *Evol. Anthropol.* 4:1–3.

Boinski, S., and Garber, P. A. (2000). *On the Move: How and Why Animals Travel in Groups*. University of Chicago Press, Chicago.

Boinski, S., Sughrue, K., Selvaggi, L., Quatrone, R., Henry, M., and Cropp, S. (2002). An expanded test of the ecological model of primate social evolution: competitive regimes and female bonding in three species of squirrel monkeys (*Saimiri oerstedii, S. boliviensis* and *S. sciureus*). *Behaviour* 139:227–261.

Borries, C., and Koenig, A. (2000). Infanticide in hanuman langurs: social organization, male migration, and weaning age. In: van Schaik, C. P., and Janson, C. H. (eds.), *Infanticide by Males and Its Implications*. Cambridge University Press, Cambridge. pp. 99–122.

Borries, C., Launhardt, K., Epplen, C., Epplen, J. T., and Winkler, P. (1999). DNA analyses support the hypothesis that infanticide is adaptive in langur monkeys. *Proc. R. Soc. Lond. B* 266:901–904.

Bradley, B. J. (1999). Levels of selection, altruism, and primate behavior. *Q. Rev. Biol.* 74:171–194.

Britt, A., Welch, C., and Katz, A. (2001). The impact of *Cryptoprocta ferox* on the *Varecia variegata* reinforcement project at Betampona. *Lemur News* 6:35–37.

Brockman, D. K. (2003). *Polybroides radiatus* predation attempts on *Propithecus verreauxi*. *Folia Primatol.* 74:71–74.

Campbell, A. F., and Sussman, R. W. (1994). The value of radio tracking in the study of neotropical rain forest monkeys. *Am. J. Primatol.* 32:291–301.

Campbell, C. J. (2003). Female-directed aggression in free-ranging *Ateles geoffroyi*. *Int. J. Primatol.* 24:223–237.

Carosi, M., and Visalberghi, E. (2002). Analysis of tufted capuchin (*Cebus apella*) courtship and sexual behavior repertoire: changes throughout the female cycle and female interindividual differences. *Am. J. Phys. Anthropol.* 118:11–24.

Carpenter, C. R. (1942). Characteristics of social behavior in non-human primates. *Trans. N. Y. Acad. Sci.* 2:248–258.

Carpenter, C. R. (1954). Tentative generalizations on the grouping behavior of non-human primates. *Hum. Biol.* 26:269–276.

Chapais, B. (1988a). Experimental matrilineal inheritance of rank in female Japanese macaques. *Anim. Behav.* 36:1025–1037.

Chapais, B. (1988b). Rank maintenance in female Japanese macaques: experimental evidence for social dependency. *Behaviour* 104:41–59.

Chapais, B., Gauthier, C., and Prud'homme, J. (1995). Dominance competition through affiliation and support in Japanese macaques: an experimental study. *Int. J. Primatol.* 16:521–536.

Chapais, B., and Lecomte, M. (1995). Induction of matrilineal rank instability by the alpha male in a group of Japanese macaques. *Am. J. Primatol.* 36:299–312.

Chapman, C. (1988). Patch use and patch depletion by the spider and howling monkeys of Santa Rosa National Park, Costa Rica. *Behaviour* 105:99–116.

Chapman, C. A. (1990). Ecological constraints on group size in three species of neotropical primates. *Folia Primatol.* 55:1–9.

Chapman, C. A., Chapman, L. J., and Gillespie, T. R. (2002). Scale issues in the study of primate foraging: red colobus of Kibale National Park. *Am. J. Phys. Anthropol.* 117:349–363.

Chapman, C. A., Wrangham, R. W., and Chapman, L. J. (1995). Ecological constraints on group size: an analysis of spider monkey and chimpanzee subgroups. *Behav. Ecol. Sociobiol.* 36:59–70.

Clutton-Brock, T. H. (1974). Primate social organisation and ecology. *Nature* 250:539–542.

Clutton-Brock, T. H. (1977). Some aspects of intraspecific variation in feeding and ranging behaviour in primates. In: Clutton-Brock, T. H. (ed.), *Primate Ecology*. Academic Press, London. pp. 539–556.

Clutton-Brock, T. H. (1989). Mammalian mating systems. *Proc. R. Soc. Lond. B* 236:339–372.

Clutton-Brock, T. H., and Harvey, P. H. (1980). Primates, brains and ecology. *J. Zool. Lond.* 190:309–323.

Cords, M. (2002). When are there influxes in blue monkey groups? In: Glenn, M. E., and Cords, M. (eds.), *The Guenons: Diversity and Adaptation in African Monkeys*. Kluwer Academic/Plenum Press, New York. pp. 189–201.

Cords, M., Mitchell, B. J., Tsingalia, H. M., and Rowell, T. E. (1986). Promiscuous mating among blue monkeys in the Kakamega Forest, Kenya. *Ethology* 72:214–226.

Cowlishaw, G. (1997). Trade-offs between foraging and predation risk determine habitat use in a desert baboon population. *Anim. Behav.* 53:667–686.

Cowlishaw, G. O., and O'Connell, S. M. (1996). Male–male competition, paternity certainty and copulation calls in female baboons. *Anim. Behav.* 51:235–238.

Craul, M., Zimmermann, E., and Radespiel, U. (2003). Experimental evidence for female mate choice in the grey mouse lemur (*Microcebus murinus*). *Folia Primatol.* 74:187.

Crockett, C. M., and Janson, C. H. (2000). Infanticide in red howlers: female group size, male membership and a possible link to folivory. In: van Schaik, C. P., and Janson, C. H. (eds.), *Infanticide by Males and Its Implications*. Cambridge University Press, Cambridge. pp. 75–98.

Crockett, C. M., and Wilson, W. L. (1980). The ecological separation of *Macaca nemestrina* and *M. fasicularis* in Sumatra. In: Lindburg, D. (ed.), *The Macaques. Studies in Ecology, Behavior, and Evolution*. van Nostrand Rheinhold, New York. pp. 148–181.

Crook, J. H. (1972). Sexual selection, dimorphism, and social organization in the primates. In: Campbell, B. (ed.), *Sexual Selection and the Descent of Man*. Aldine, Chicago. pp. 231–281.

Crook, J. H., Ellis, J. E., and Goss-Custard, J. D. (1976). Mammalian social systems. *Anim. Behav.* 24:261–274.

Crook, J. H., and Gartlan, J. S. (1966). Evolution of primate societies. *Nature* 210:1200–1203.

Csermely, D. (1996). Antipredator behavior in lemurs: evidence of an extinct eagle on Madagascar or something else? *Int. J. Primatol.* 17:349–354.

Curtis, D. J. (2004). Diet and nutrition in wild mongoose lemurs (*Eulemur mongoz*) and their implications for the evolution of female dominance and small group size in lemurs. *Am. J. Phys. Anthropol.* 124:234–247.

Curtis, D. J., and Zaramody, A. (1999). Social structure and seasonal variation in the behaviour of *Eulemur mongoz*. *Folia Primatol.* 70:79–96.

Delgado, R. A., and van Schaik, C. P. (2000). The behavioral ecology and conservation of the orangutan (*Pongo pygmaeus*): a tale of two islands. *Evol. Anthropol.* 9:201–218.

Denham, W. W. (1971). Energy relations and some basic properties of primate social organization. *Am. Anthropol.* 73:77–95.

de Waal, F. (1982). *Chimpanzee Politics: Power and Sex Among Apes.* Johns Hopkins University Press, Baltimore.

de Waal, F. B. M. (1986). The integration of dominance and social bonding in primates. *Q. Rev. Biol.* 61:459–479.

Di Bitetti, M. S., and Janson, C. H. (2001). Social foraging and the finder's share in capuchin monkeys, *Cebus apella*. *Anim. Behav.* 62:47–56.

Di Fiore, A. D. (2003). Molecular genetic approaches to the study of primate behavior, social organization, and reproduction. *Ybk. Phys. Anthropol.* 46:62–99.

Dolhinow, P. (1999). A mystery: explaining behavior. In: Strum, S. C., Lindburg, D. G., and Hamburg, D. (eds.), *The New Physical. Anthropology: Science, Humanism, and Critical Reflection.* Prentice Hall, Upper Saddle River, NJ. pp. 119–131.

Dollar, L. (1999). Preliminary report on the status, activity cycle, and ranging of *Cryptoprocta ferox* in the Malagasy rain forest, with implications for conservation. *Sm. Carniv. Conserv.* 20:7–10.

Dominy, N., Garber, P. A., Bicca-Marques, J. C., and Azevedo-Lopes, M. A. O. (2003). Do female tamarins use visual cues to detect fruit rewards more successfully than do males? *Anim. Behav.* 66:829–837.

Dominy, N. J., and Duncan, B. (2001). GPS and GIS methods in an African rain forest: applications to tropical ecology and conservation. *Conserv. Ecol.* 5:6, http://www.consecol.org/vol5/iss2/art6.

Doran, D. M., and McNeilage, A. (1998). Gorilla ecology and behavior. *Evol. Anthropol.* 6:120–131.

Dunbar, R. I. M. (1988). *Primate Social Systems.* Cornell University Press, Ithaca, NY.

Dunbar, R. I. M. (2000). Male mating strategies: a modeling approach. In: Kappeler, P. M. (ed.), *Primate Males.* Cambridge University Press, Cambridge. pp. 259–268.

Dunbar, R. I. M. (2002). Modeling primate behavioral ecology. *Int. J. Primatol.* 23:785–819.

Eberle, M., and Kappeler, P. M. (2002). Mouse lemurs in space and time: a test of the socioecological model. *Behav. Ecol. Sociobiol.* 51:131–139.

Edwards, J. E. (1990). Validity of the doubly labeled water method of measuring energy expenditure with rest and exercise in nonhuman primates (*Erythrocebus patas*). [Ph.D diss.]. Indiana University, Bloomington.

Eisenberg, J. F., Muckenhirn, N. A., and Rudran, R. (1972). The relation between ecology and social structure in primates. *Science* 176:863–874.

Elgar, M. A. (1986). House sparrows establish foraging flocks by giving chirrup calls if the resources are divisible. *Anim. Behav.* 34:169–174.

Enstam, K. L., Isbell, L. A., and De Maar, T. W. (2002). Male demography, female mating behavior, and infanticide in wild patas monkeys (*Erythrocebus patas*). *Int. J. Primatol.* 23:85–104.

Fedigan, L. M. (1983). Dominance and reproductive success in primates. *Ybk. Phys. Anthropol.* 26:91–129.

Fedigan, L. M. (1992). *Primate Paradigms: Sex Roles and Social Bonds.* Eden Press, Montreal.

Fedigan, L. M. (2000). Gender encounters. In: Strum, S. C., and Fedigan, L. M. (eds.), *Primate Encounters: Models of Science, Gender and Society.* University of Chicago Press, Chicago. pp. 498–520.

Fedigan, L. M. (2003). Impact of male takeovers on infant deaths, births, and conceptions in *Cebus capucinus* at Santa Rosa, Costa Rica. *Int. J. Primatol.* 24:723–741.

Fedigan, L. M., Fedigan, L., Chapman, C., and Glander, K. E. (1988). Spider monkey home ranges: a comparison of radio telemetry and direct observation. *Am. J. Primatol.* 16:19–29.

Fietz, J. (1999). Monogamy as a rule rather than exception in nocturnal lemurs: the case of the fat-tailed dwarf lemur, *Cheirogaleus medius. Ethology* 105:259–272.

Fisher, H. S., Swaisgood, R. R., and Fitch-Snyder, H. (2003). Countermarking by male pygmy lorises (*Nycticebus pygmaus*): do females use odor cues to select mates with competitive ability? *Behav. Ecol. Sociobiol.* 53:123–130.

Freeland, W. J. (1976). Pathogens and the evolution of primate sociality. *Biotropica* 8:12–24.

Garber, P. A. (1988). Diet, foraging patterns, and resource defense in a mixed species troop of *Saguinus mystax* and *Saguinus fuscicollis* in Amazonian Peru. *Behaviour* 105:18–34.

Gautier-Hion, A. (1978). Food niches and coexistence in sympatric primates in Gabon. In: Chivers, D. J., and Herbert, J. (eds.), *Recent Advances in Primatology.* Academic Press, London. pp. 269–286.

Gerald, M. S. (2001). Primate colour predicts social status and aggressive outcome. *Anim. Behav.* 61:559–566.

Glander, K. E. (1992). Dispersal patterns in Costa Rican mantled howling monkeys. *Int. J. Primatol.* 13:415–436.

Goodman, S. M. (1994). The enigma of antipredator behavior in lemurs: evidence of a large extinct eagle on Madagascar. *Int. J. Primatol.* 15:129–134.

Goodman, S. M., O'Connor, S., and Langrand, O. (1993). A review of predation on lemurs: implications for the evolution of social behavior in small, nocturnal primates. In: Kappeler, P. M., and Ganzhorn, J. U. (eds.), *Lemur Social Systems and Their Ecological Basis.* Plenum Press, New York. pp. 51–66.

Gould, L., Sussman, R. W., and Sauther, M. L. (2003). Demographic and life-history patterns in a population of ring-tailed lemurs (*Lemur catta*) at Beza Mahafaly Reserve, Madagascar: a 15-year perspective. *Am. J. Phys. Anthropol.* 120:182–194.

Grassi, C. (2002). The behavioral ecology of *Hapalemur griseus griseus*: the influences of microhabitat and population density on this small-bodied folivore (Madagascar) [diss.]. University of Texas, Austin.

Gursky, S. (2003). Predation experiments on infant spectral tarsiers (*Tarsius spectrum*). *Folia Primatol.* 74:272–284.

Hand, J. L. (1986). Resolution of social conflicts: dominance, egalitarianism, spheres of dominance and game theory. *Q. Rev. Biol.* 61:201–220.

Haraway, D. (1986). Primatology is politics by other means. In: Bleier, R. (ed.), *Feminist Approaches to Science.* Pergamon Press, New York. pp. 77–117.

Harcourt, A. H. (1998). Does primate socioecology need non-primate socioecology? *Evol. Anthropol.* 7:3–7.

Heistermann, M., Ziegler, T., van Schaik, C. P., Launhardt, K., Winkler, P., and Hodges, J. K. (2001). Loss of oestrus, concealed ovulation and paternity confusion in free-ranging hanuman langurs. *Proc. R. Soc. Lond. B* 268:2445–2451.

Hemelrijk, C. K. (2002a). Self-organization and natural selection in the evolution of complex despotic societies. *Biol. Bull.* 202:283–288.

Hemelrijk, C. K. (2002b). Self-organizing properties of primate social behavior: a hypothesis for intersexual rank overlap in chimpanzees and bonobos. *Evol. Anthropol.* 11:91–94.

Henzi, S. P. (1996). Copulation calls and paternity in chacma baboons. *Anim. Behav.* 51:233–234.

Hladik, C. M. (1977). A comparative study of the feeding strategies of two species of leaf monkeys: *Presbytis entellus* and *Presbytis senex*. In: Clutton-Brock, T. H. (ed.), *Primate Ecology: Studies of Feeding and Ranging Behaviour in Lemurs, Monkeys, and Apes*. Academic Press, London. pp. 323–353.

Honess, P. E., and MacDonald, D. W. (2003). Marking and radio-tracking primates. In: Setchell, J. M., and Curtis, D. J. (eds.), *Field and Laboratory Methods in Primatology: A Practical Guide*. Cambridge University Press, Cambridge. pp. 158–173.

Hrdy, S. B. (1979). Infanticide among mammals: a review, classification, and examination of the implications for the reproductive strategies of females. *Ethol. Sociobiol.* 1:13–40.

Isabirye-Basuta, G. (1988). Food competition among individuals in a free-ranging chimpanzee community in Kibale Forest, Uganda. *Behaviour* 105:135–147.

Isbell, L. A. (1991). Contest and scramble competition: patterns of female aggression and ranging behavior among primates. *Behav. Ecol.* 2:143–155.

Isbell, L. A. (1994). Predation on primates: ecological patterns and evolutionary consequences. *Evol. Anthropol.* 3:61–71.

Isbell, L. A. (2004). Is there no place like home? Ecological bases of female dispersal and philopatry and their consequences for the formation of kin groups. In: Chapais, C., and Berman, C. (eds.), *Kinship and Behavior in Primates*. Oxford University Press, New York. pp. 71–108.

Isbell, L. A., and Young, T. P. (2002). Ecological models of female social relationships in primates: similarities, disparities, and some directions for future clarity. *Behaviour* 139:177–202.

Janson, C. H. (1988a). Intra-specific food competition and primate social structure: a synthesis. *Behaviour* 105:1–17.

Janson, C. H. (1988b). Food competition in brown capuchin monkeys (*Cebus apella*): quantitative effects of group size and tree productivity. *Behaviour* 105:53–76.

Janson, C. H. (2000a). Primate socio-ecology: the end of a golden age. *Evol. Anthropol.* 9:73–86.

Janson, C. H. (2000b). Spatial movement strategies: theory, evidence and challenges. In: Boinski, S., and Garber, P. A. (eds.), *On The Move: How and Why Animals Travel in Groups*. Chicago University Press, Chicago. pp. 165–203.

Janson, C. H. (2001). Field experiments in primate ecology: the monkeys are always right. *Am. J. Primatol.* 54:107.

Janson, C. H., and Di Bitetti, M. S. (1997). Experimental analysis of food detection in capuchin monkeys: effects of distance, travel speed, and resource size. *Behav. Ecol. Sociobiol.* 41:17–24.

Janson, C. H., and van Schaik, C. P. (1988). Recognizing the many faces of primate food competition: methods. *Behaviour* 105:165–186.

Jolly, A. (2000). The bad old days of primatology? In: Strum, S. C., and Fedigan, L. M. (eds.), *Primate Encounters: Models of Science, Gender, and Society*. University of Chicago Press, Chicago. pp. 71–84.

Jolly, A., Dobson, A., Rasaminmanana, H., Walker, J., O'Connor, S., Solberg, M., and Perel, V. (2002). Demography of *Lemur catta* at Berenty Reserve, Madagascar: effects of troop size, habitat and rainfall. *Int. J. Primatol.* 23:327–353.

Kappeler, P. M., and van Schaik, C. P. (2002). Evolution of primate social systems. *Int. J. Primatol.* 23:707–740.

Karpanty, S. M., and Goodman, S. M. (1999). Diet of the Madagascar harrier-hawk *Polyboroides radiatus*, in southeastern Madagascar. *J. Rapt. Res.* 33:313–316.

Karpanty, S. M., and Grella, R. (2001). Lemur responses to diurnal raptor calls in Ranomafana National Park, Madagascar. *Folia Primatol.* 72:100–103.

Kay, R. F., and Kirk, E. C. (2000). Osteological evidence for the evolution of activity pattern and visual acuity in primates. *Am. J. Phys. Anthropol.* 113:235–262.

Keddy, A. C. (1986). Female mate choice in vervet monkeys (*Cercopithecus aethiops sabaeus*). *Am. J. Primatol.* 10:125–134.

Kinzey, W. G., and Gentry, A. H. (1979). Habitat utilization in two species of *Callicebus*. In: Sussman, R. W. (ed.), *Primate Ecology: Problem Oriented Field Studies*. John Wiley and Sons, New York. pp. 89–100.

Koenig, A. (2002). Competition for resources and its behavioral consequences among female primates. *Int. J. Primatol.* 23:759–783.

Koenig, A., and Borries, C. (2001). Socioecology of hanuman langurs: the story of their success. *Evol. Anthropol.* 10:122–137.

Korstjens, A. H., Sterck, E. H. M., and Noe, R. (2002). How adaptive or phylogenetically inert is primate social behaviour? A test with two sympatric colobines. *Behaviour* 139:203–225.

Koyama, N., Nakamichi, M., Oda, R., Miyamoto, N., Ichino, S., and Takahata, Y. (2001). A ten-year summary of reproductive parameters for ring-tailed lemurs at Berenty, Madagascar. *Primates* 42:1–14.

Koyama, N., Takahata, Y., Huffman, M. A., Norikoshi, K., and Suzuki, H. (1992). Reproductive parameters of female Japanese macaques: thirty years data from the Arashiyama troops, Japan. *Primates* 33:33–47.

Kuhn, T. S. (1962). *The Structure of Scientific Revolutions*. University of Chicago Press, Chicago.

Lehner, P. N. (1996). *Handbook of Ethological Methods*. Cambridge University Press, Cambridge.

Lewis, R. J. (2002). Beyond dominance: the importance of leverage. *Q. Rev. Biol.* 77:149–164.

Mangel, M. (1990). Resource divisibility, predation and group formation. *Anim. Behav.* 39:1163–1172.

Martin, P., and Bateson, P. (1993). *Measuring Behavior*. Cambridge University Press, Cambridge.

Miller, L. E. (2002). *Eat or Be Eaten: Predator Sensitive Foraging Among Primates*. Cambridge University Press, New York.

Milton, K. (1981). Food choice and digestive strategies of two sympatric primate species. *Am. Nat.* 117:496–505.

Milton, K., and May, M. L. (1976). Body weight, diet and home range area in primates. *Nature* 259:459–462.

Mitani, J., Watts, D. P., and Muller, M. N. (2002). Recent developments in the study of wild chimpanzee behavior. *Evol. Anthropol.* 11:9–25.

Mittermeier, R. A., and Roosmalen, M. G. M. (1981). Preliminary observations on habitat utilization and diet in eight Surinam monkeys. *Folia Primatol.* 36:1–39.

Mueller, A. E. (1999). Social organization of the fat-tailed dwarf lemur (*Cheirogaleus medius*) in northwestern Madagascar. In: Rakotosamimanana, B., Rasaminmanana, H., and Ganzhorn, J. U. (eds.), *New Directions in Lemur Studies*. Plenum Press, New York. pp. 139–157.

Nakagawa, N., and Okamoto, K. (2003). Van Schaik's socioecological model: developments and problems. *Reichorui Kenkyu/Primate Res.* 19:243–264.

Nishida, T., Corp, N., Hamai, M., Hasegawa, T., Hiraiwa-Hasegawa, M., Hosaka, K., Hunt, K. D., Itoh, N., Kawanaka, K., Matsumoto-Oda, A., Mitani, J. C., Nakamura, M., Norikoshi, K., Sakamaki, T., Turner, L., Uehara, S., and Zamma, K. (2003). Demography, female life history and reproductive profiles among the chimpanzees of Mahale. *Am. J. Primatol.* 59:99–121.

Norconk, M. A., and Conklin-Brittain, N. L. (2004). Variation on frugivory: the diet of Venezuelan white-faced sakis. *Int. J. Primatol.* 25:1–26.

Nunn, C. L. (2000). Collective benefits, free-riders and male extra-group conflict. In: Kappeler, P. M. (ed.), *Primate Males: Causes and Consequences of Variation in Group Composition*. Cambridge University Press, Cambridge. pp. 192–204.

Nunn, C. L. (2002). Spleen size, disease risk, and sexual selection: a comparative study in primates. *Evol. Ecol. Res.* 4:91–107.

Nunn, C. L. (2003a). Behavioural defenses against sexually transmitted diseases in primates. *Anim. Behav.* 66:37–48.

Nunn, C. L. (2003b). Sociality and disease risk: a comparative study of leukocyte counts in primates. In: de Waal, F. B. M., and Tyack, P. L. (eds.), *Animal Social Complexity: Intelligence Culture, and Individualized Societies*. Harvard University Press, Cambridge, MA. pp. 26–31.

Nunn, C. L., Gittleman, J. L., and Antonovics, J. (2000). Promiscuity and the primate immune system. *Science* 290:1168–1170.

Nunn, C. L., and Lewis, R. J. (2001). Cooperation and collective action in animal behaviour. In: Noe, R., van Hooff, J. A. R. A. M., and Hammerstein, P. (eds.), *Economics in Nature: Social Dilemmas, Mate Choice, and Biological Markers*. Cambridge University Press, Cambridge. pp. 42–66.

Nunn, C. L., and van Schaik, C. P. (2000). Social evolution in primates: the relative roles of ecology and intersexual conflict. In: van Schaik, C. P., and Janson, C. H. (eds.), *Infanticide by Males and Its Implications*. Cambridge University Press, Cambridge. pp. 388–419.

Orians, G. H. (1969). On the evolution of mating systems in birds and mammals. *Am. Nat.* 103:589–603.

Overdorff, D. J. (1996). Ecological correlates to social structure in two lemur species in Madagascar. *Am. J. Phys. Anthropol.* 100:487–506.

Overdorff, D. J., and Erhart, E. M. (2001). Social and ecological influences on female dominance in day-active prosimian primates. *Am. J. Phys. Anthropol.* 32(suppl.):116.

Overdorff, D. J., and Johnson, S. E. (2003). *Eulemur*, true lemurs. In: Goodman, S. M., and Benstead, J. P. (eds.), *The Natural History of Madagascar*. University of Chicago Press, Chicago. pp. 1320–1324.

Overdorff, D. J., Merelender, A. M., Talata, P., Telo, A., and Forward, Z. (1999). Life history of *Eulemur fulvus rufus* from 1988–1998 in southeastern Madagascar. *Am. J. Phys. Anthropol.* 108:295–310.

Palombit, R. A. (1994a). Dynamic pair bonds in hylobatids: implications regarding monogamous social systems. *Behaviour* 128:65–101.

Palombit, R. A. (1994b). Extra-pair copulations in a monogamous ape. *Anim. Behav.* 47:721–723.

Palombit, R. A. (1996). Pair bonds in monogamous apes: a comparison of the siamang *Hylobates syndactylus* and the white-handed gibbon *Hylobates lar*. *Behaviour* 133:321–356.

Palombit, R. A. (1999). Infanticide and the evolution of pair bonds in nonhuman primates. *Evol. Anthropol.* 7:117–129.

Palombit, R. A. (2000). Infanticide and the evolution of male–female bonds in animals. In: van Schaik, C. P., and Janson, C. H. (eds.), *Infanticide by Males and Its Implications*. Cambridge University Press, Cambridge. pp. 239–268.

Palombit, R. A., Cheney, D. L., Fischer, J., Johnson, S., Rendall, D., Seyfarth, R. W., and Silk, J. B. (2000). Male infanticide and defense of infants in chacma baboons. In: van Schaik, C. P., and Janson, C. H. (eds.), *Infanticide by Males and Its Implications*, Cambridge University Press, Cambridge. pp. 123–152.

Panger, M. A., Perry, S., Rose, L., Gros-Louis, J., Vogel, E., Mackinnon, K. C., and Baker, M. (2002). Cross-site differences in foraging behavior of white-faced capuchins (*Cebus capuchinus*). *Am. J. Phys. Anthropol.* 119:52–66.

Parga, J. A. (2003). Copulatory plug displacement evidences sperm competition in *Lemur catta*. *Int. J. Primatol.* 24:889–899.

Paul, A., Preuschoft, S., and van Schaik, C. P. (2000). The other side of the coin: infanticide and the evolution of affiliative male–infant interactions in Old World primates. In: van Schaik, C. P., and Janson, C. H. (eds.), *Infanticide by Males and Its Implications*. Cambridge University Press, Cambridge. pp. 269–292.

Pazol, K. (2003). Mating in the Kakamega Forest blue monkeys (*Cercopithecus mitis*): does female sexual behavior function to manipulate paternity assessment? *Behaviour* 140:473–499.

Pereira, M. E., and McGlynn, C. A. (1997). Special relationships instead of female dominance for redfronted lemurs, *Eulemur fulvus rufus*. *Am. J. Primatol.* 43:239–258.

Pereira, M. E., Kaufman, R., Kappeler, P. M., and Overdorff, D. J. (1990). Female dominance does not characterize all of the Lemuridae. *Folia Primatol.* 55:96–103.

Pereira, M. E., and Weiss, M. L. (1991). Female mate choice, male migration, and the threat of infanticide in ringtailed lemurs. *Behav. Ecol. Sociobiol.* 28:141–152.

Phillips, K. A. (1995). Resource patch size and flexible foraging in white-faced capuchins (*Cebus capucinus*). *Int. J. Primatol.* 16:509–519.

Pochron, S. T., and Wright, P. C. (2003). Variability in adult group compositions of a prosimian primate. *Behav. Ecol. Sociobiol.* 54:285–293.

Pope, T. R. (1992). The influence of dispersal patterns and mating system on genetic differentiation within and between populations of the red howler monkey (*Alouatta seniculus*). *Evolution* 46:1112–1128.

Porter, L. M. (2004). Forest use and activity patterns of *Callimico goeldii* in comparison to two sympatric tamarins *Saguinus fuscicollis* and *Saguinus labiatus*. *Am. J. Phys. Anthropol.* 124:139–153.

Porter, L. M., and Garber, P. A. (2004). Goeldi's monkey: a primate paradox. *Evol Anthropol.* 13:104–115.

Pruetz, J. D., and Isbell, L. A. (2000). Correlations of food distribution and patch size with agonistic interactions in female vervets (*Chlorocebus aethiops*) and patas monkeys (*Erythrocebus patas*) living in simple habitats. *Behav. Ecol. Sociobiol.* 49:38–47.

Pullen, S. L., Bearder, S. K., and Dixson, A. F. (2000). Preliminary observations on sexual behavior and the mating systems in free-ranging lesser galagos (*Galago moholi*). *Am. J. Primatol.* 51:79–88.

Pusey, A., Williams, J., and Goodall, J. (1997). The influence of dominance rank on reproductive success in female chimpanzees. *Science* 277:828–831.

Radespiel, U., and Zimmermann, E. (2001). The influence of familiarity, age, experience and female mate choice on pregnancies in captive grey mouse lemurs. *Behaviour* 140:301–318.

Rasmussen, M. A. (1999). Ecological influences on activity cycle in two cathemeral primates, the mongoose lemur (*Eulemur mongoz*) and the common brown lemur (*Eulemur fulvus rufus*) [diss.]. Duke University, Durham, NC.

Rendall, D., and Di Fiore, A. D. (1994). Evolution of social organization: a reappraisal for primates by using phylogenetic methods. *Proc. Natl. Acad. Sci. USA* 91:9941–9945.

Richard, A., Dewar, R. E., Schwartz, M., and Ratsirarson, J. (2002). Life in the slow lane? Demography and life histories of male and female sifaka (*Propithecus verreauxi verreauxi*). *J. Zool. Lond.* 256:421–436.

Richard, A., and Schulman, S. R. (1982). Sociobiology: primate field studies. *Annu. Rev. Anthropol.* 11:231–255.

Richard, A. F. (1987). Malagasy prosimians: female dominance. In: Smuts, B. B., Cheney, D. L., Seyfarth, R. M., Wrangham, R. W., and Struhsaker, T. T. (eds.), *Primate Societies*. University of Chicago Press, Chicago. pp. 25–33.

Robinson, J. G. (1988). Group size in wedge-capped capuchin monkeys *Cebus olivaceus* and the reproductive success of males and females. *Behav. Ecol. Sociobiol.* 23:187–197.

Rodrigues de Moraes, P. L., de Carvalho, O., and Strier, K. B. (1998). Population variation in patch and party size in muriquis (*Brachyteles arachnoides*). *Int. J. Primatol.* 19:325–337.

Rowell, T. E. (1988). Beyond the one-male group. *Behaviour* 105:189–201.

Rowell, T. E. (2000). A few peculiar primates. In: Strum, S. C., and Fedigan, L. M. (eds.), *Primate Encounters: Models of Science, Gender, and Society*. University of Chicago Press, Chicago. pp. 57–70.

Sauther, M. L., Sussman, R. W., and Gould, L. (1999). The socioecology of the ringtailed lemur: thirty-five years of research. *Evol. Anthropol.* 8:120–132.

Smuts, B. B., and Smuts, R. W. (1993). Male aggression and sexual coercion of females in nonhuman primates and other mammals: evidence and theoretical implications. *Adv. Stud. Behav.* 22:1–63.

Soltis, J. (2002). Do female primates gain nonprocreative benefits by mating with multiple males? Theoretical and empirical considerations. *Evol. Anthropol.* 11:187–197.

Soltis, J., Mitsunaga, F., Shimizu, K., Yanagihara, Y., and Nozaki, M. (1997). Sexual selection in Japanese macaques. I: Female mate choice or male sexual coercion? *Anim. Behav.* 54:725–736.

Soltis, J., Thomsen, R., Matsubayashi, K., and Takenaka, O. (2000). Infanticide by resident males and female counter-strategies in wild Japanese macaques (*Macaca fuscata*). *Behav. Ecol. Sociobiol.* 48:195–202.

Sommer, V. (2000). The holy wars about infanticide. Which side are you on? And why? In: van Schaik, C. P., and Janson, C. H. (eds.), *Infanticide by Males and Its Implications*. Cambridge University Press, Cambridge. pp. 9–26.

Sperling, S. (1991). Baboons with briefcases vs. langurs in lipstick: feminism and functionalism in primate studies. In: Leonardo, M. (ed.), *Gender at the Crossroads of Knowledge: Feminist Anthropology in the Postmodern Era*. University of California Press, Berkeley. pp. 204–234.

Sprague, D. S., Kabaya, H., and Hagihara, K. (2004). Field testing a global positioning system (GPS) collar on a Japanese monkey: reliability of automatic GPS positioning in a Japanese forest. *Primates* 45:151–154.

Stanford, C. B. (1998). The social behavior of chimpanzees and bonobos: empirical evidence and shifting assumptions. *Curr. Anthropol.* 39:399–420.

Steenbeek, R. (2000). Infanticide by males and female choice in wild Thomas's langurs. In: van Schaik, C. P., and Janson, C. H. (eds.), *Infanticide by Males and Its Implications*. Cambridge University Press, Cambridge. pp. 153–177.

Sterck, E. H. M. (1999). Variation in langur social organization in relation to the socioecological model, human habitation alteration, and phylogenetic constraints. *Primates* 40:199–213.

Sterck, E. H. M., and Korstjens, A. H. (2000). Female dispersal and infanticide avoidance in primates. In: van Schaik, C. P., and Janson, C. H. (eds.), *Infanticide by Males and Its Implications*. Cambridge University Press, Cambridge. pp. 293–321.

Sterck, E. H. M., and Steenbeek, R. (1997). Female dominance relationships and food competition in the sympatric Thomas langur and long-tailed macaque. *Behaviour* 134:749–774.

Sterck, E. H. M., Watts, D. P., and van Schaik, C. P. (1997). The evolution of female social relationships in nonhuman primates. *Behav. Ecol. Sociobiol.* 41:291–309.

Strier, K. B. (1989). Effects of patch size on feeding associations in muriquis (*Brachyteles arachnoides*). *Folia Primatol.* 52:70–77.

Strier, K. B. (1992). *Faces in the Forest: The Endangered Muriqui Monkeys of Brazil*. Oxford University Press, New York.

Strier, K. B. (1994). Myth of the typical primate. *Ybk. Phys. Anthropol.* 37:233–271.

Struhsaker, T. T., and Oates, J. F. (1975). Comparison of the behaviour and ecology of red colobus and black-and-white colobus monkeys in Uganda: a summary. In: Tuttle, R. H. (ed.), *Socioecology and Psychology of Primates*. Mouton, The Hague. pp. 103–123.

Strum, S. C. (2000). Science encounters. In: Strum, S. C., and Fedigan, L. M. (eds.), *Primate Encounters: Models of Science, Gender, and Society*. University of Chicago Press, Chicago. pp. 475–497.

Strum, S. C., and Fedigan, L. M. (1999). Theory, method, gender, and culture: what changes our views of primate sociality? In: Strum, S. C., Lindburg, D. G., and Hamburg, D. (eds.), *The New Physical Anthropology: Science, Humanism, and Critical Reflection*. Prentice Hall, Upper Saddle River, NJ. pp. 67–105.

Strum, S. C., and Fedigan, L. M. (2000). Changing views of primate society: a situated North American view. In: Strum, S. C., and Fedigan, L. M. (eds.), *Primate Encounters: Models of Science, Gender, and Society*. University of Chicago Press, Chicago. pp. 1–49.

Stuart, M. D., and Strier, K. B. (1995). Primates and parasites: a case for a multidisciplinary approach. *Int. J. Primatol.* 16:577–593.

Sugiyama, Y., and Ohsawa, Y. (1982). Population dynamics of Japanese monkeys with special reference to the effect of artificial feeding. *Folia Primatol.* 39:238–263.

Sussman, R. W. (1972). An ecological study of two Madagascan primates: *Lemur fulvus rufus* Audebert and *Lemur catta* Linnaeus [diss.]. Duke University, Durham, NC.

Sussman, R. W. (1974). Ecological distinctions in sympatric species of *Lemur*. In: Martin, R. D., Doyle, G. A., and Walker, A. C. (eds.), *Prosimian Biology*. Duckworth, London. pp. 75–108.

Sussman, R. W., Cheverud, J. M., and Bartlett, T. Q. (1995). Infant killing as an evolutionary strategy: reality or myth? *Evol. Anthropol.* 3:149–151.

Sussman, R. W., Garber, P. A., and Cheverud, J. M. (2005). Importance of cooperation and affiliation in the evolution of primate sociality. *Am. J. Phys. Anthropol.* 1281:84–97.

Symington, M. M. (1988). Food competition and foraging party size in the black spider monkey (*Ateles paniscus chamek*). *Behaviour* 105:117–134.

Takasaki, H. (2000). Traditions of the Kyoto school of field primatology in Japan. In: Strum, S. C., and Fedigan, L. M. (eds.), *Primate Encounters: Models of Science, Gender, and Society*. University of Chicago Press, Chicago. pp. 151–164.

Tang-Martinez, Z. (2000). Paradigms and primates: Bateman's principle, passive females, and perspectives from other taxa. In: Strum, S. C., and Fedigan, L. M. (eds.), *Primate Encounters: Models of Science, Gender, and Society*. University of Chicago Press, Chicago. pp. 261–274.

Tattersall, I. (1987). Cathemeral activity in primates: a definition. *Folia Primatol.* 49:200–202.

Taylor, L. (1986). Kinship, dominance, and social organization in a semi-free ranging group of ringtailed lemurs (*Lemur catta*) [diss.]. Washington University, St. Louis.

Terborgh, J. (1983). *Five New World Primates: A Study in Comparative Ecology*. Princeton University Press, Princeton, NJ.

Terborgh, J., and Janson, C. H. (1986). Socioecology of primate groups. *Annu. Rev. Ecol. Syst.* 17:111–135.

Tilden, C. (1993). Reproductive energetics of prosimian primates [diss.]. Duke University, Durham, NC.

Tilden, C. D., and Oftedal, O. T. (1995). The bioenergetics of reproduction in prosimian primates: is it related to female dominance? In: Izard, M. K., Alterman, L., and Doyle, G. A. (eds.), *Creatures of the Dark: The Nocturnal Prosimians*. Plenum Press, New York. pp. 119–131.

Treves, A. (2000). Prevention of infanticide: the perspective of infant primates. In: van Schaik, C. P., and Janson, C. H. (eds.), *Infanticide by Males and Its Implications*. Cambridge University Press, Cambridge. pp. 223–238.

Treves, A., and Chapman, C. A. (1996). Conspecific threat, predation avoidance, and resource defense: implications for grouping in langurs. *Behav. Ecol. Sociobiol.* 39:43–53.

Utami, S. S., Goossens, B., Bruford, M. W., de Ruiter, J. R., and van Hooff, J. A. R. A. M. (2002). Male bimaturism and reproductive success in orang-utans. *Behav. Ecol.* 13:643–652.

van Schaik, C. P. (1983). Why are diurnal primates living in groups? *Behaviour* 87:120–143.

van Schaik, C. P. (1989). The ecology of social relationships amongst female primates. In: Standen, V., and Foley, R. A. (eds.), *Comparative Socioecology: The Behavioral Ecology of Humans and Other Mammals*. Blackwell, Oxford. pp. 195–218.

van Schaik, C. P. (2000). Vulnerability to infanticide by males: patterns among mammals. In: van Schaik, C. P., and Janson, C. H. (eds.), *Infanticide by Males and Its Implications*. Cambridge University Press, Cambridge. pp. 61–71.

van Schaik, C. P., and Dunbar, R. I. M. (1990). The evolution of monogamy in large primates: a new hypothesis and some crucial tests. *Behaviour* 115:30–62.

van Schaik, C. P., Hodges, J. K., and Nunn, C. L. (2000). Paternity confusion and the ovarian cycles of female primates. In: van Schaik, C. P., and Janson, C. H. (eds.), *Infanticide by Males and Its Implications*. Cambridge University Press, Cambridge. pp. 361–387.

van Schaik, C. P., and Janson, C. H. (2000). *Infanticide by Males and Its Implications*. Cambridge University Press, Cambridge.

van Schaik, C. P., and Kappeler, P. M. (1996). The social systems of gregarious lemurs: lack of convergence with anthropoids due to evolutionary disequilibrium? *Ethology* 102:915–941.

van Schaik, C. P., and van Hooff, J. A. R. A. M. (1983). On the ultimate causes of primate social systems. *Behaviour* 85:91–117.

van Schaik, C. P., and van Noordwijk, M. A. (1988). Scramble and contest in feeding competition among female long-tailed macaques (*Macaca fascicularis*). *Behaviour* 105:77–98.

Vehrencamp, S. L. (1983). A model for the evolution of despotic versus egalitarian societies. *Anim. Behav.* 31:667–682.

Vervaecke, H., de Vries, H., and van Elsacker, L. (2000). Dominance and its behavioral measures in a captive group of bonobos (*Pan paniscus*). *Int. J. Primatol.* 21:47–68.

Vogel, E. R., and Janson, C. H. (2003). The role of food patches in primate socioecology: a monkey's eye view. *Am. J. Phys. Anthropol.* 60 (suppl. 1):43–44.

Washburn, S. L. (1973). The promise of primatology. *Am. J. Phys. Anthropol.* 38:177–182.

Watts, D. P. (1989). Infanticide in mountain gorillas: new cases and a reconsideration of the evidence. *Ethology* 81:1–18.

Watts, D. P. (1990). Mountain gorilla life histories, reproductive competition, and sociosexual behavior and some implications for captive husbandry. *Zoo Biol.* 9:185–200.

Watts, D. P. (1991). Harassment of immigrant female mountain gorillas by resident females. *Ethology* 89:135–153.

Weiss, M. K. (2000). A view on the science: physical anthropology at the millennium. *Am. J. Phys. Anthropol.* 111:295–299.

White, F. J. (1989). Ecological correlates of pygmy chimpanzee social structure. In: Standen, V., and Foley, R. A. (eds.), *Comparative Socioecology: The Behavioral Ecology of Humans and Other Mammals*. Blackwell, Oxford. pp. 151–164.

White, F. J., and Wrangham, R. W. (1988). Feeding competition and patch size in the chimpanzee species *Pan paniscus* and *Pan troglodytes*. *Behaviour* 105:148–164.

Whitten, P. L. (1988). Effects of patch quality and feeding subgroup size on feeding success in vervet monkeys (*Cercopithecus aethiops*). *Behaviour* 105:5–52.

Wich, S. A., Fredriksson, G., and Sterck, E. H. M. (2002). Measuring fruit patch size for three sympatric Indonesian primate species. *Primates* 43:19–27.

Wittenberger, J. F. (1980). Group size and polygamy in social mammals. *Am. Nat.* 115:197–222.

Wittenberger, J. F., and Tilson, R. L. (1980). The evolution of monogamy: hypotheses and evidence. *Annu. Rev. Ecol. Syst.* 11:197–232.

Wolfe, L. D. (1995). Current research in field primatology. In: Boaz, N. T., and Wolfe, L. D. (eds.), *Biological Anthropology: The State of the Science*. International Institute for Human Evolutionary Research, Bend, OR. pp. 149–157.

Wrangham, R. W. (1979). On the evolution of ape social systems. *Soc. Sci. Info.* 18:335–368.

Wrangham, R. W. (1980). An ecological model of female-bonded primate groups. *Behaviour* 75:262–300.

Wrangham, R. W. (1982). Mutualism, kinship, and social evolution. In: *Current Problems in Sociobiology.* Cambridge University Press, Cambridge. pp. 270–289.

Wrangham, R. W. (1987). Evolution of social structure. In: Smuts, B. B., Cheney, D. L., Seyfarth, R., Wrangham, R. W., and Struhsaker, T. T. (eds.), *Primate Societies.* Chicago University Press, Chicago. pp. 282–297.

Wright, P. C. (1995). Demography and life history of free-ranging *Propithecus diadema edwardsi* in Ranomafana National Park, Madagascar. *Int. J. Primatol.* 16:835–854.

Wright, P. C. (1998). Impact of predation risk on the behaviour of *Propithecus diadema edwardsi* in the rain forest of Madagascar. *Behaviour* 135:483–512.

Yamane, A., Shotake, T., Mori, A., Boug, A. I., and Iwamoto, T. (2003). Extra-unit paternity of hamadryas baboons (*Papio hamadryas*) in Saudi Arabia. *Ethol. Ecol. Evol.* 15:379–387.

Yamashita, N. (2002). Diets of two lemur species in different microhabitats in Beza Mahafaly Special Reserve, Madagascar. *Int. J. Primatol.* 23:1025–1051.

Zinner, D., Schwibbe, M. H., and Kaumanns, W. (1994). Cycle synchrony and probability of conception in female hamadryas baboons *Papio hamadryas. Behav. Ecol. Sociobiol.* 35:175–183.

Zuckerman, S. (1932). *The Social Life of Monkeys and Apes.* Routledge, London.

29

Primate Nutritional Ecology
Feeding Biology and Diet at Ecological and Evolutionary Scales
Joanna E. Lambert

The whole of nature is a conjugation of the verb to eat, in the active and the passive. (Inge 1927, p. 4)

INTRODUCTION

Given its critical role in all aspects of an animal's biology, it is not surprising that research on feeding has occupied a central position in the history of primate studies. In the landmark volume *Primate Societies*, Oates (1987) provided an excellent overview of the ecological aspects of primate feeding and paid particular attention to the patterns in which primate foraging is influenced by the spatial and temporal distribution of food resources. I offer a follow-up in this chapter with the benefit of an almost two-decade period in which we have seen remarkable progress in our understanding of diet and feeding in wild primates and its implications for animal physiology (Ross 1992, Messier and Stewart 1997, Knott 1998), morphology (e.g., McGraw 1998, Vinyard et al. 2003), ontogeny, growth and development (Leigh 1994, Dirks 2003), and ecology (Oates et al. 1990, Milton 1996, Shultz and Noe 2002) at individual (Altmann 1998, Curtis 2004, Garber 2004), population (e.g., Isbell 1991, Chapman

1990, White 1998), and community (Fleagle et al. 1999, Brugiere et al. 2002) scales.

An understanding of the nutritional content of different food types is necessary for gaining insight into an animal's energetic requirements which can, in itself, provide a direct link to the complexities of feeding adaptations in primates. To wit: food in its various forms is the only source of energy for animals, and we can use an understanding of energy to define the oft-used term *dietary quality*. Moreover, as I discuss below, energetic requirements scale with body size; since body size correlates with a number of other ecological and behavioral variables, we can integrate our understanding of nutrients and diet to explain other components of an animal's behavior and biology.

However, nutrition and dietary quality do not exist in a void. Indeed, primates are confronted with an array of feeding challenges that influence the quality of food. These challenges can be intrinsic—representing some inherent chemical, nutritional, or structural feature of the food—or extrinsic—a function of availability of that food and the costs (e.g., competition) associated with feeding on it as a consequence of that availability. Like all biological phenomena, questions regarding feeding challenges may be evaluated at two intersecting scales: proximate (ecological) and ultimate (evolutionary). *Proximate* mechanisms are short-term responses within the lifetime of an individual to immediate circumstances. For example, food availability influences which foods an individual eats, which can then, as discussed below, influence such responses as increases in day range length or production of endogenous enzymes and gastrointestinal nutrient transporters depending on what food an animal finds and consumes. If an animal's response both has a heritable component and is under enough selective pressure (which presumably many components of diet and feeding would be, given its centrality in an animal's biology), then we might think of the answers at an *ultimate* scale and evaluate these responses in terms of evolved strategies and species' features. Obviously, these scales are synergistic and intrinsically related, and neither is the more correct scale at which to discuss and evaluate behavior and biology. They are, instead, complementary and facilitate answering both "how" questions regarding an animal's response to immediate conditions and "why" questions concerning the reasons that animals have evolved mechanisms to carry out these responses (Alcock 1989). The interplay between the proximate and ultimate is extraordinarily complex, as are the nexi of responses resulting in the diversity of feeding adaptations observed in primates. It is the job of primate nutritional ecologists to determine the nature of feeding challenges confronted by primates in acquiring their requisite nutrients and energy and to unravel the solutions exhibited by primate species.

I cannot give the spectrum of foraging, feeding, diet, and nutrition full justice in this chapter. Instead, I focus on those aspects not largely discussed by Oates (1987), namely nutritional biology. I start by evaluating feeding categories, primate nutritional requirements, and what wild primate diets offer with respect to micro- and macronutrients. After discussing metabolism, energy, and dietary quality, I outline the remainder of the discussion in terms of feeding challenges to primates and provide several examples of solutions to these challenges. I conclude by arguing that questions regarding diet and nutrition are not just academic concerns but in fact have important implications for the conservation of wild species.

PRIMATE DIET: FEEDING CATEGORIES AND NUTRITIONAL REQUIREMENTS

Feeding Categories

The nutrients required by vertebrates are divided into six broad classes: carbohydrates, protein, lipids, vitamins, minerals, and water. The first three (carbohydrates, protein, lipids) are commonly referred to as *macronutrients* and are required in large quantities for the energy (typically measured in calories) for growth and maintenance, while minerals and vitamins, or *micronutrients*, are not used for energy per se but are instead vitally important for innumerable physiological pro-

cesses (Leonard 2000). The Council on Food and Nutrition has defined nutrition as "the science of food, the nutrients and substances therein, their action, interaction, and balance in relation to health and disease, and the process by which the organism ingests, absorbs, transports, utilizes, and excretes food substances" (cited in Wardlaw et al. 2004, p. 3). Nutritional ecology, then, can be defined as the broader field of investigation into the means by which animals procure these nutrients from their habitat.

Evaluating how primate nutrition per se relates to a species' nutritional ecology can be vexing, largely because of the disjuncture between what we know from studies of primates in captivity and from those of primates in the wild. On the one hand, we have a decent understanding of micro- and macronutritional requirements in a few commonly captive species (e.g., *Macaca* spp.) as a consequence of work by the National Research Council (NRC). The nutritional requirements of nonhuman primates were first considered and outlined at the behest of the NRC's Committee on Animal Nutrition in 1972; this report was later expanded (1978). The current *Nutrient Requirements of Nonhuman Primates* represents a second revised edition (National Research Council 2003). The current NRC report is distinct from other animal nutrition reports in that it is not based on dietary requirements of a single domesticated species but, rather, on an entire order of wild species, exhibiting an extraordinary array of diet-related adaptations. Much of the information in the NRC report hence focuses on a few model species. Regardless of these caveats, we now have updated nutritional recommendations for primates for energy, fiber, protein, lipids, minerals, and vitamins (National Research Council 2003).

On the other hand, primate biologists today have both intensive knowledge on several groups of a few well-known wild species in well-studied communities and extensive but more superficial knowledge on most, if not all, wild species. For example, we have specific information regarding the diets of several groups of *Lophocebus albigena* and *Cercopithecus ascanius* living in the Kanyawara study area of Kibale National Park, Uganda, and can calculate dietary particulars for these groups for an almost 30-year period (Waser 1977, Struhsaker 1978, Olupot et al. 1998, Lambert 2002a). Moreover, in a few cases of particularly well-studied primate populations, data are available for the nutritional components of these foods as well. For example, Conklin-Brittain et al. (1998) have described Kanyawara *C. ascanius* and *L. albigena* foods with regard to their macronutritional content, finding few differences with regard to levels of crude lipid, crude protein, soluble carbohydrates, and fiber in the diets of these two species. Such long-term comparisons of particular primate communities were not possible until fairly recently with continuing research in given populations.

The challenge to primate biologists is to link data on feeding and foraging in wild populations with knowledge regarding nutrition in captive primates. For example, despite our understanding of what a particular group (or even several groups) of wild primate species is eating at a given time, it is difficult to know whether these foods are or are not meeting the basic vitamin, mineral, and macronutrient needs of that wild population. This is in part a consequence of the difficulties of measuring the wild diet in the first place (Oftedal and Allen 1996). Primate diet is typically expressed as time spent foraging and feeding, although the quantity of food (e.g., g/min) actually ingested can vary greatly depending on food type, age/sex, food availability, etc. (Hladik 1977, Milton 1984, Oftedal 1991, Oftedal and Allen 1996). Moreover, details of ingestion can be extremely difficult to observe in arboreal primates living in tall, closed-canopy forest. Integrating these bodies of information (i.e., data on natural diet and nutritional requirements) is where fieldworkers and nutritional biologists working together have the most to offer.

Adding to the complexity of evaluating primate nutrition is the plethora of terms associated with feeding. Many of these terms are used in slightly different ways by different researchers. Animals that consume from only one trophic level are labeled either *herbivores* (plants only) or *carnivores* (animals only). Organisms that consume foods from more than one trophic level (e.g., both plants and animals) are technically *omnivores*, although in some cases authors use the term *omnivory* to suggest flexibility in some sense or another. With very few exceptions (e.g., *Tarsius* spp., *Loris* spp.; see Chapters 3 and 5), primates derive a majority of their energetic and nutritional requirements from plants; indeed, there are reports of virtually all components of plant anatomy consumed by different primate species around the world, including fruit (both ripe and unripe), seeds (immature and mature), leaves (all developmental phases from buds to mature leaves), petioles, corms (base of stem that gives rise to vegetative plant growth), rhizome (underground stems), bark, flowers (buds and petals), nectar, sap, and gum. Some primates are technically best defined as herbivores (e.g., colobines and indriids), although most primates consume both plants and animals. Harding (1981), for example, undertook a review of the diet of 131 species of wild primates and found that 90% of these species ate fruit, 79% ate soft foliar parts (i.e., buds, shoots), 69% consumed mature leaves, 41% consumed seeds, 65% ate invertebrates (primarily insects), and 37% ate other vertebrate prey (small vertebrates, eggs).

Omnivory is often taken to be synonymous with *generalist* or *flexible*. This is not necessarily always the case, and in fact, this term masks the extreme diversity of feeding strategies in primates; indeed, within this large category, a species may focus on some particular category of food, hence terms such as *frugivory* (fruit), *folivory* (leaves), *gummivory* (exudates), *gramnivory* (seeds), *insectivory* (insects), and *faunivory* (animals), etc. However, these labels have not been rigorously defined and can mask variability. For example, it is common to encounter the idea that "frugivores" should behave in a particular way and "folivores" in another. Such categorization can obfuscate the complexity of feeding, and many assumptions we make regarding what a frugivore or folivore should do are not so much a result of what these animals actually

feed on but more a result of the labels that we apply. Related to these complexities are issues such as the point at which an animal, population, or species should actually be considered, say, frugivorous and how this is determined (actual food intake, time spent feeding, anatomical adaptations of the teeth and/or gut, etc.). Different populations of the same species can differ markedly in their diet, as can different members of the same group and even the same individual at different times, depending on myriad factors such as microhabitat, seasonal shifts in availability, community structure, age/rank, etc. Indeed, it is rarely, if ever, possible to define what is "normal" for a species with regard to the foods they consume or the nutrients they ingest (Oftedal and Allen 1996), suggesting that, at the very least, we must get a handle on intraspecific variation and how it relates to interspecific patterns (Chapman et al. 2003, Strier 2003).

Anyone perusing primary literature on primate feeding biology will inevitably encounter the food category of "other," which may include such foods as bark, fungus, sap, flowers, and petioles, among others. While this food category tends to receive the least attention by researchers, there is the potential that, of all food categories, it may be the most important. For example, comments regarding minerals and primate population density are largely anecdotal despite the fact that in other literature phosphorus has been demonstrated to limit populations of large herbivores (e.g., *Equus asinus* in Australia; Freeland and Choquenot 1990). Janson and Chapman (1999) have thus suggested that this trace and little-studied primate mineral requirement may in fact limit folivorous primate populations. This example highlights the fact that, with regard to nutritional factors that may influence population density and evolution of feeding traits, it is not so much what is required most of the time but what food resources are required during critical periods. For example, myself and colleagues (Lambert et al. 2004) have recently argued that it is the difference in the mechanical properties of fallback foods during critical periods that may have served as the selective pressure for thick molar enamel in *Lophocebus albigena* relative to thinner-enameled cercopithecines (e.g., *Cercopithecus* spp.). The fact that for the most part *L. albigena* consumes similar food to that fed on by sympatric cercopithecines (e.g., soft, fleshy fruit) but also has a diet comprising a small (but probably very important) percentage of mechanically tough foods, such as bark and seeds, suggests that the thick enamel of *L. albigena* serves a critical function (Rosenberger 1992, Kinzey 1978). That is, thick dental enamel was selected for because of its benefits during periods when preferred—and softer—foods were unavailable.

Nutritional Requirements

Regardless of the feeding and food categories we employ to describe primate diet, primates require the full suite of nutrients required by most mammals in general (45–47 in total of amino acids, fatty acids, vitamins, and minerals;

Oftedal and Allen 1996) from the principal classes of carbohydrates, protein, lipids, vitamins, and minerals. Here, I describe features of these basic nutritional classes and evaluate the major dietary sources of these nutrients for wild primate species.

Carbohydrates

Carbohydrates take the form of monosaccharide sugars, disaccharides, and polysaccharides. Monosaccharides (simple sugars) such as glucose or fructose are readily absorbed by the body and utilized directly along ordinary metabolic pathways. Disaccharides consist of two monosaccharide units bonded together; sucrose, for example, is a common fruit sugar comprising the monosaccharides glucose and fructose. Along with monosaccharides, the disaccharides are often referred to as the "soluble sugars." Disaccharides must be hydrolyzed into simple sugars (in the small intestine) before they can be absorbed and utilized (Schmidt-Nielsen 1997). The most complex sugars are the polysaccharides, polymers of monosaccharides, which can be divided into the starch or starch-like polysaccharides (storage polysaccharides) and the nonstarch polysaccharides. The starch and starch-like polysaccharides are essentially plant energy reserves and can be broken down by the consumer. The enzyme amylase, for example, is secreted in the pancreas and saliva of humans and other primates, and its function is to reduce polysaccharide starch to disaccharide sugars that can then be hydrolyzed in the small intestine; this process is initiated by amylase directly in the mouth.

The nonstarch polysaccharides are the fiber components and can be further divided into the soluble nonstarch carbohydrates (soluble fiber) and the insoluble nonstarch polysaccharides (insoluble fiber). The insoluble, nonstarch polysaccharides comprise the structural components of the plant cell walls (hence, "structural polysaccharides") and include hemicellulose, cellulose, and lignin. Cellulose is the most abundantly distributed carbohydrate in the world (Sharon 1980) and represents a large proportion of the available energy content of plant foods (Blaxter 1962, Alexander 1993), although no vertebrate has the cellulose-digesting enzyme (cellulase) for breaking down this carbohydrate, an interesting evolutionary question given that primates have relied on plants heavily throughout their evolutionary history (Milton 1987, 1993). The structural carbohydrates must instead be broken down with the assistance of protozoans or, more commonly, bacterial symbionts (Lambert 1998). Like cellulose, hemicellulose cannot be digested enzymatically but, instead, requires fermentation. There is however, some evidence that hemicellulose can be partially hydrolyzed in a low pH stomach; *Pan troglodytes* (and *Homo sapiens*), for example, ferment hemicellulose more completely than cellulose (Milton and Demment 1987, National Research Council 2003). The nonstructural, non-starch polysaccharides include some components that, like starch, represent sources and storage of energy for plant metabolism. Although they are not as digestible as starch, they are completely fermentable and include substances in plants known as fructans. Pectin (a component of plant cell walls) is not a plant energy resource per se but is associated with the plant cell wall; it is closely associated with hemicellulose and is completely fermentable.

In the diets of wild primates, fruit pulp is an excellent source of carbohydrate energy, providing about 50–100 kcal/g (Leonard 2000). In comparison to cultivated fruit, the sugar of wild fruits is hexose-dominated (i.e., some fructose and glucose) (Milton 1999). Fruits can also have more pectin than leaves, although gastrointestinal anaerobic bacteria very effectively ferment such pectin (National Research Council 2003). The volatile fatty acids thereby produced can provide considerable energetic benefits. A number of vegetative plant parts can also be fermented for their energy, including leaves, petioles, and bark. Exudates, too, are exploited by primates, although there are chemical and nutritional differences among them (Bearder and Martin 1980, Nash 1986). For example, gum is water-soluble and high in complex carbohydrates composed of nonstarch, multibranched polysaccharides (Power 1996, Caton et al. 1996). Saps, on the other hand, are typically high in relatively easy-to-digest, simple, and water-soluble carbohydrates (Nash 1986).

Protein

Protein provides energy and is critical for growth and replacement of tissue in the body (Leonard 2000). Proteins are composed of amino acids; primates require 20 amino acids for body maintenance, nine of which are considered essential because they are not synthesized endogenously—i.e., they must be derived from dietary sources (Leonard 2000). Protein requirements are greatest during growth and reproduction and can increase by as much as 30% during such critical periods (Oftedal 1991). Oftedal (1991) has argued that protein deficiency is probably not common as protein requirements in the order generally are low, likely because of the slow life history that characterizes primates, particularly when compared to carnivorous mammals (Oftedal 1991, Oftedal and Allen 1996).

Approximately 75% of protein synthesis comes from protein breakdown in the body; the other 25% comes from dietary input. As indicated in isotope labeling techniques, there is a very high turnover rate of proteins in the body, and virtually all "marked" proteins are eventually incorporated throughout the body (Schmidt-Nielsen 1997). The major sources of dietary protein in primate diets are leaves, insects, and other animal matter. With exceptions, immature leaves offer more protein than mature leaves since as they age, structural components increase and protein is allocated to other parts of the plant (storage organs, seeds). Leaves in general offer approximately 10–20 kcal/g (Wu Lueng 1968, Eaton and Konner 1985, Leonard 2000). Overall, fruit pulp is not high in protein, and few mammals and birds exist on an exclusively frugivorous diet even when this resource is abundant (Whiten et al. 1991, Leighton 1993). Indeed, it has been argued that fruit availability generally limits fitness and primate population density (Whiten et al. 1991, Plumptre and Reynolds 1994, Gupta and Chivers 1999), although others have suggested that there are behavioral mechanisms for the selection of rare but protein- and mineral-rich resources during critical periods of growth and reproduction (Cords 1986, Sourd and Gautier-Hion 1986, Butynski 1988). For example, young primates or pregnant/lactating females often increase their intake of leaves and/or insects, and Cords (1986) has found that the size of juvenile *Cercopithecus ascanius* is proportional to the percentage of leaves and insects in the diet.

Lipids (or Fats)

Lipids (or fats) are the body's most highly concentrated source of energy, providing more than two times the energy per unit weight as either carbohydrates or proteins. Fats are divided into three main classes: simple, compound, and derived. The simple fats comprise triglycerides that are made up of two molecules: glycerol and fatty acids. Fatty acids are divided into either saturated fatty acids or unsaturated fatty acids, based on differences in the bonding between carbon atoms (Leonard 2000). The major sources of fat for primates are insects and other animal matter, seeds, and the arils of some fruit species (e.g., *Virola* spp.). Insects, for example, provide upward of 100–200 kcal/g. We know very little about the consumption of lipids by primates, although wild plant foods tend to show a fairly equal balance between saturated and unsaturated fats (Chamberlain 1993). Milton (1999) has analyzed the fatty acid composition of wild plant foods consumed by *Alouatta palliata* and found that the most common fatty acids are palmitic (30%), linoleic (23%), α-linolenic (16%), and oleic (15%). Conklin-Brittain et al. (1998) report that lipid intake by anthropoids in Kibale National Park, Uganda, is marginal. As lipids influence neurotransmitter levels that regulate reproductive hormones, insufficient lipid intake can result in developmental and reproductive problems in mammals (Robbins 1993).

Minerals

Minerals perform functions essential to life and are generally divided into the macrominerals (calcium, phosphorus, magnesium, potassium, sodium, chlorine, sulfur; typically expressed as a percentage) and trace elements (iron, copper, manganese, zinc, iodine, selenium, chromium, cobalt; typically expressed as parts per million) (National Research Council 2003). Minerals are inorganic elements that serve as structural components in many biological molecules (e.g., iron in hemoglobin), as activators in hormonal and enzymatic processes, as regulators of cell activity, and as components of body fluids (National Research Council 2003). Collectively, minerals make up approximately 4% of body weight and are critical for animal health and for maintaining physiological function (Leonard 2000, Robbins 1993). Calcium and sodium, for example, are very important for lactating females as insufficient intake can result in slower growth and higher

infant mortality (Buss and Cooper 1970, Power et al. 1999). We have few data on the distribution of important dietary minerals in natural habitats, although a number of studies have indicated that primates select specific plant foods for their mineral content and that the highest levels of minerals are in foliar material, followed by bark and fruit (Hladik 1978, Oates 1978, Nagy and Milton 1979, Yeager et al. 1997, Waterman 1984, Milton 1999). Overall, there is high variability in the distribution of these minerals (Milton 1999). Rode et al. (2003) found that ripe and unripe fruits in Kibale National Park, Uganda, are lower in minerals than other primate plant foods and that rarely consumed foods, such as bark, petioles, and caterpillars, had the highest mineral levels. *Geophagy* (i.e., soil consumption) is quite common in primates, and the role of soil as a mineral supplement has been suggested by a number of authors (see review by Krishnamani and Mahaney 2000). Primates are likely to encounter the same mineral-related problems as other, better-studied primary consumers (Rode et al. 2003). Oates (1978), for example, has suggested that the presence of swampy areas containing sodium-rich plants partly explains the relatively high density of *Colobus guereza* in the Kanyawara area of Kibale National Park, Uganda. Rode et al. (2003) have found that sodium is low in primate foods generally in Kibale and that the iron and copper content of primate foods is marginal. These are very similar findings to those of Nagy and Milton (1979), who found that *Alouatta palliata* in Panama had a natural diet low in sodium and copper.

Vitamins

Vitamins, like minerals, facilitate the use of energy rather than providing energy per se (Leonard 2000). Vitamins are divided into those that are water-soluble (B, C) and those that are fat-soluble (A, D, E, K). In contrast to fat-soluble vitamins, primates should generally consume water-soluble vitamins daily because they cannot be stored in the body; any excess is excreted in urine. Very little is known about the vitamin requirements of nonhuman primates, and even less is known about the distribution of vitamins in primate diets, although it is thought that virtually all vitamins except for B_{12} can be met by a plant diet. B_{12} is provided by the activity of microbes in the gut as well as animal resources, including insect larvae found in fruit pulp (Milton 1999). E, K, and B-complex (niacin, thiamine, riboflavin, pyrodoxine) are widespread and found in seeds, nuts, and leaves; A is found mostly in leaves and C in fruit flesh, leaves, and seeds (Waterman 1984). Vitamin C (ascorbic acid) is of particular interest because, like other highly herbivorous mammals (e.g., some rodents, lagomorphs, and bats), primates cannot synthesize this important vitamin themselves (Oftedal and Allen, 1996, Milton 1999). While there can be significant intra- and interspecific variation, the pulp of fleshy tropical fruit can be extremely high in vitamins. For example, Milton and Jenness (1987) have found that on average diets of *Alouatta palliata* and *Ateles geoffryi* contain upwards of ten times the daily recommendation for *Homo sapiens*. Given the important health consequences of vitamins and minerals, we clearly have much more work to undertake on the variability in mineral and vitamin content of plants (Milton 1999, 2000).

METABOLISM AND ENERGY, OR WHAT IS A "QUALITY" DIET?

With the exception of minerals and vitamins (i.e., the micronutrients), all nutrients (macronutrients) are involved in the balance of energy in the body. Determinants of any primate's energy requirements are dietary *thermogenesis* (i.e., the energy required to process, absorb, and store nutrients), growth, and reproduction, as well as basal metabolism, or basic life processes such as respiration and regulation of body temperature (Krebs 1972, Begon et al. 1990, Schmidt-Nielsen 1997, Leonard 2000). Energy is required for all other components of a primate's behavioral repertoire as well, such as traveling, foraging, social interactions in all forms, and predator avoidance. The energetic costs associated with food handling (i.e., the thermic effect of food) make up a relatively small proportion of daily energy expenditure and are influenced by amount consumed and the composition of the diet (e.g., high-protein meals elevate dietary thermogenesis) (Leonard 2000).

To link nutrition with an animal's feeding adaptations, we must define dietary quality; to do so, we need an understanding of metabolism. *Metabolism* refers to all the chemical reactions of the body and represents a balancing act between two processes: anabolic, or synthetic, processes and catabolic, or degradative, processes (Schmidt-Nielsen 1997, Tortora and Anagnostakos 1987). *Anabolic* processes are chemical reactions that combine simple substances into more complex molecules, for example, building proteins out of amino acids. Such synthesis requires energy. *Catabolic* processes are those chemical reactions in the body that break down complex organic compounds into simple ones. Catabolic reactions release energy by breaking bonds between atoms of food molecules, for example, breaking down disaccharides to monosaccharides. These processes provide energy for the body. Thus, dietary quality is best defined in terms of its capacity to yield energy.

Among mammals, there are important and reasonably predictable relationships among metabolism (energetic needs), body mass, and dietary quality. The relationship between body mass and basal metabolism is a negatively allometric one. Smaller mammals thus require relatively more energy to maintain endothermic homeothermy and larger mammals, relatively less (Kleiber 1961; Bell 1971; Jarman 1974; Parra 1978; Gaulin 1979; Schmidt-Nielsen 1984, 1997; Harvey et al. 1987; Martin 1990). This negative relationship can have important implications for the quality of foods (as defined above by a food's nutrient density, or capacity to yield

energy) primates consume (Bell 1971, Jarman 1974, Gaulin 1979). Most importantly, the relatively greater metabolic requirements of smaller mammals must be met by a diet with a capacity to yield energy readily, either as a function of how easily it is digested or by its energetic density or both. However, since they do not have large absolute requirements for food, smaller primates can afford to consume foods that are spatially and temporally rare (Kleiber 1961, Parra 1978, Schmidt-Nielsen 1984, Van Soest 1982, Gaulin 1979, Kay 1985 Harvey et al. 1987, Martin 1990). This is in contrast to larger-bodied mammals that have lower energy requirements per unit body weight, and although they require absolutely more food and thus cannot rely on rare resources, they can afford to consume either more food with lower nutrient density or foods that require lengthy processing (via bacterial fermentation) (Gaulin 1979, Van Soest 1982). According to these generalizations, larger mammals theoretically have more options than smaller primates in that they experience fewer metabolic constraints; that is, they can be large and still maintain a high-quality diet (and just eat more of it) or rely on foods that require lengthy processing times (Milton 1984). According to these predictions, the latter option is not open to smaller-bodied primates, which are more metabolically constrained and must instead rely on foods that yield more readily accessible energy (e.g., insects, nectar, ripe fruit).

HOW TO EVALUATE DIETARY QUALITY: FEEDING CHALLENGES AND WAYS AROUND THEM

Ecological Challenges

The initial challenge to all animal consumers is to find food in the first place, a not insignificant task given the extreme spatial and temporal variability in food availability (Oates 1987, White 1998). There are differences among broad food types however (Isbell 1991). Certainly, "a leaf is not a leaf is not a leaf" (Janson and Chapman 1999: 246); however, on average within a given tropical habitat, leaves do tend to be a more abundant and predictably available resource relative to ripe fruit, exudates, and insects. Milton (1980) for example quantified resource availability on Barro Colorado Island, Panama, and found that plant vegetative parts (e.g., leaves, petioles) are more evenly distributed than plant reproductive parts (e.g., fruit). In contrast, fruit as a primate food resource can be highly unpredictable. This is a function of the fact that most tropical forests are characterized by tree species with differences in seasonal and annual fruit production, and by trees that produce fruit either in small quantities or in abundance over widely scattered individuals (Janzen 1967, Frankie et al. 1974, Whitmore 1990, van Schaik et al. 1993).

Primates exhibit a diversity of responses to the challenges of locating food. At an ecological or more proximate scale, one means by which to deal with the vagaries of food availability is to shift those features that are more labile than others, for example, the number of individuals in a foraging group. The relationship between foraging group size and food availability, particularly fruit, has been a focus of recent research, particularly as it may shed light on patterns of intraspecific feeding competition (Gillespie and Chapman 2001). While there is disagreement over the degree to which group living and group size can be explained by avoidance of predators, conspecific threat, or foraging advantages, it is generally held that increased feeding competition with more group members can be a cost and that the size of a primate social group mitigates the costs of predation at the expense of increasing feeding competition (Struhsaker 1981, Terborgh and Janson 1986, Dunbar 1988, Garber 1988, Janson and van Schaik 1988, van Schaik 1989, Isbell 1991, Janson 1992, van Schaik and Kappeler 1993, Chapman et al. 1994, Janson and Goldsmith 1995, Treves and Chapman 1996, Sterck et al. 1997, Boinski et al. 2000, Koenig 2002). The "ecological constraints model" posits that as group size increases, the food resources required by that group will also increase, a primary consequence being that day range length, and hence home range length, will also increase (Chapman 1990, Gillespie and Chapman 2001, Chapman and Chapman 2002). Different primate species, and different populations for that matter (because of differences in habitat), will experience these patterns to varying degrees. For example, it has been suggested that primates that rely heavily on fruit will exhibit a stronger relationship between group size and day range because of the temporally and spatially patchy nature of fruit distribution in forests. Also, primates that rely on fruit patches are seen to incur greater costs of contest competition because fruit is more monopolizable than other resources (e.g., leaves). Thus, fruit-eating primates are seen to have greater limitations on group size than primates that are not so reliant on fruit. Indeed, although there are exceptions it has been demonstrated that more frugivorous primates in larger groups travel farther than those in smaller groups (van Schaik et al. 1983, Chapman 1990, Wrangham et al. 1993, Chapman et al. 1995, O'Brien and Kinnaird 1997).

Less labile traits, such as morphological adaptations that have evolved over time (evolutionary, ultimate scale), may also represent adaptations to the variations in food availability. For example, cheek pouches are found ubiquitously within the subfamily Cercopithecinae (Hill 1966, Murray 1973, Fleagle 1999). Most recently, I (Lambert 2005) have investigated the function of cheek pouches in closed-canopy cercopithecines and found that cheek pouches were used more commonly by *Cercopithecus ascanius* and *Lophocebus albigena* when feeding on fruit and when the number of conspecifics increased in a feeding patch. For both species, cheek pouch use increased in larger food patches, a result that corroborates those of Pruetz and Isbell (1999), who found that the size of the food patch has a direct effect on the potential for competition. In their work on *Erythrocebus patas* and *Cercopithecus aethiops* in Kenya, these authors found that small food sources may be used up too quickly to result in feeding competition since the finder of the food

essentially uses up that food before other foragers begin exploiting the patch. Using cheek pouches may be more important in those settings where the patch size is large enough to engender competition but not large enough to accommodate all group members.

Digestive strategies may also mitigate the costs of interspecific feeding competition over foods that vary in availability. Cercopithecines in particular are noted for their diverse and eclectic diet (Rudran 1978, Struhsaker 1978, Gautier-Hion 1988, Beeson 1989, Richard et al. 1989, Maisels 1993, Chapman et al. 2002). *Papio cynocephalus* in Kenya, for example, ingest over 200 food types by the age of 1 year (Altmann 1998). Such dietary flexibility is almost certainly related to their digestive strategy: very long mean retention times relative to body size in conjunction with a simple, unspecialized stomach (Lambert 1998, 2002a,b). If a flexible digestive and dietary strategy means that more items can be used for food, then resource switching may provide a means to minimize intra- and interspecific contest, with switchers able to access a diversity of foods over a wider area (Lambert 2002a). An ability to consume many food types can facilitate larger feeding spheres, which may allow a species to avoid pushing forward (van Schaik et al. 1983, Chapman and Chapman 2000). That is, for species with digestive flexibility, the likelihood of encroaching on another individual's feeding space is decreased because a given area essentially holds greater food richness. Such a mechanism (digestive and dietary flexibility) may help to explain the coexistence of closely related cercopithecines in similar trophic space (Lambert 2002a).

Chemical Challenges

Many plants, as well as some insects, defend themselves from animal feeding by arming themselves with chemical compounds. Because they are the result of secondary processes involved in defense, rather than the primary metabolism important for basic plant processes such as reproduction and growth, these compounds are collectively known as "secondary metabolites" (Harborne 1988). Primates are thus confronted with the challenge of selecting a nutritionally balanced diet while minimizing their intake of plant secondary compounds which are toxic or inhibit digestion (chemical challenges). To date, approximately 12,000 secondary metabolites have been identified (Levin 1971, 1976; Freeland and Janzen 1974; Rosenthal and Janzen 1979; Harborne 1988; McKey 1974; McKey et al. 1978; Gartlan et al. 1980), falling roughly into two broad categories: *(1)* digestion inhibitors, which interfere with the efficiency with which nutrients are obtained by the animal, and *(2)* toxins, which interfere with normal physiology (Feeny 1976, Waterman and Kool 1994).

However, as with ecological challenges, primates exhibit various ways around chemical defense. For example, one of the best-studied digestion inhibitors is the group of polyphenolic compounds known as tannins (both condensed and hydrolyzable forms). These compounds can interfere with protein uptake by the animals because they have the capacity to bind with plant proteins, forming insoluble complexes. Several studies have demonstrated that the presence of condensed tannins in plant material is negatively related to primate feeding behavior (e.g., Oates et al. 1977, Glander 1981, McKey et al. 1981, Wrangham and Waterman 1981). However, hydrolyzable tannins tend to be more rare than condensed tannins, and much less is known of their effect (Marks et al. 1988, *Macaca mulatta*). Yet, the degree to which the presence of tannins in plant food is problematic for animal feeding remains uncertain. *Surfactant* substances (i.e., agents that lower surface tension) found in the gastrointestinal tract of some insects and some mammals can interfere with the formation of tannin–protein complexes, and in some cases tannins have the potential to enhance the rate at which protein is hydrolyzed via the regeneration of denatured protein(s) (Mole and Waterman 1985, Waterman and Kool 1994). In addition, some primates, including humans, have "proline-rich" proteins in their saliva known as tannin-binding salivary proteins (TBSPs). These TBSPs have a higher than average affinity for binding with tannins, allowing for uptake of plant proteins in the presence of tannins (Foley and McArthur 1994, Waterman and Mole 1994).

Toxic compounds such as alkaloids and cyanogenic glycosides are typically absorbed through the gastrointestinal tract and have a specific toxic effect on the consumer (Foley and McArthur 1994). As first clarified by Freeland and Janzen (1974), there are two major ways in which mammals deal with toxic secondary metabolites: microbial activity and microsomal enzymes. In herbivorous mammals with specialized stomachs, such as the Colobinae, some toxic secondary metabolites may be broken down via a diverse bacterial and protozoan microflora supported in the anaerobic, alkaline stomach environment. Essentially, secondary compounds are degraded during fermentation before they are absorbed by the animal (Freeland and Janzen 1974, Kay and Davies 1994). For primates without such stomach specializations, it is probable that a complex detoxification system is employed that relies heavily on the microsomal enzymatic activity that takes place after digesta has left the stomach. Microsomal enzymes are produced in mammalian liver cells, and their major function is the degradation of foreign chemicals, with most of their activity concentrated in the liver and kidneys (Freeland and Janzen 1974). While there is very little known about microsomal detoxification systems in primates, information stemming from research on domestic mammal species suggests that the intestinal mucosa may be an important site for detoxification of plant allelochemicals. For example, in sheep and cattle, the activity of one important allelochemical enzyme (uridine diphosphate glucuronosyltransferase) was found to be three times as great in the intestinal wall as it was in the liver (Foley and McArthur 1994). As suggested by Foley and McArthur (1994), this finding may be significant, given that the gastrointestinal tract is the first organ exposed to plant secondary compounds.

In addition, rates of enzymatic activity scale negatively with mammal body size (Walker 1978, Freeland 1991). Freeland (1991) thus suggests that smaller mammals are at an advantage for detoxifying plant secondary metabolites and that the larger the mammal, the greater the preference for foods with low amounts of toxic plant metabolites. These assertions rest largely on results by Walker (1978), who has expressed rates of enzyme activity in rats as a function of liver mass relative to body mass. Thus, small body size in some primates may facilitate the consumption of plant foods high in secondary metabolites (Lambert 2002b), which may explain why cercopithecines are able to consume plant foods with greater secondary metabolite loads than plants consumed by sympatric chimpanzees (Waser 1977, Wrangham et al. 1998, Lambert 2000). Indeed, Wrangham et al. (1998) reported that *Cercopithecus ascanius*, *C. mitis*, and *Lophocebus albigena* have absolutely more condensed tannins, monoterpenoids, and triterpenoids than chimpanzees in their annual diet.

In addition to toxins and digestion inhibitors, primates must deal with other nonnutritive components of plants (e.g., fiber, especially lignin) in order to maximize nutritional intake. For example, the ratio of protein and fiber in leaves is known to be an important selection criterion for primates and may serve as an excellent, quantifiable index for dietary quality (McKey 1978, Milton 1979, Chapman and Chapman 2002). Several studies have demonstrated a positive correlation between folivorous primate biomass and protein:fiber ratios, including colobines in Africa and Asia (Waterman et al. 1988, Oates et al. 1990, Davies 1994, Chapman and Chapman 2002) and lemurs in Madagascar (Ganzhorn 1992). Chapman and Chapman (2002) found that *Procolobus badius* (a.k.a. Piliocolobus badius) in Kibale National Park, Uganda, selected young leaves over mature leaves and that young leaves had more protein, were more digestible, and had higher protein:fiber ratios than mature leaves. Selection of leaves was not related to levels of secondary metabolites as these authors found no difference in secondary metabolite load in young versus mature leaves. Indeed, one of the most preferred trees (*Prunus africana*) had the highest levels of toxins (cyanogenic glycosides and saponins) and yet was among the most commonly consumed plant species with high protein content in young leaves.

Mechanical Challenges

Feeding primates are presented with an array of mechanical challenges as well. In many cases, although not all, these are plant mechanisms for protecting against predation such as thick exo- or pericarps for protecting seeds (see Chapter 41 for a discussion of tooluse to access foods). Primate anatomical adaptations for getting around such plant mechanisms tend to be less labile than responses to shifts in food availability (group size shift) or dietary selectivity to avoid secondary metabolites or fiber. For example, Kinzey and Norconk (1990, 1993) evaluated the responses of *Ateles paniscus* and *Chiropotes satanus* in Suriname to both chemical and mechanical challenges and found that these two species, while exhibiting a high degree of overlap in their fruit diet, had very different means to deal with these challenges. They found that the outer coverings of the fruit (pericarp) that *C. satanus* consume are significantly harder than those consumed by *A. paniscus*, while the seed covering (exocarp) consumed by *A. paniscus* was significantly harder. *C. satanus* consume the young seeds of hard fruits and, by doing so, avoid the secondary metabolites of mature seeds but are confronted with the mechanical challenge of difficult-to-penetrate hard pericarps. Hence, the evolution of small, procumbant lower incisors and larger, laterally splayed canines, which work together to crack open and gouge hard foods. *A. paniscus*, on the other hand, have no such dental adaptations. Also, while the pericarps of the fruits they consume are significantly softer, the seeds of consumed fruit are significantly harder. *A. paniscus* swallow fruits whole to remove adhesive pulp and defecate the nonmasticated seeds, thereby avoiding this mechanical challenge.

In another example of ways around mechanical challenges, Lambert (1999) studied the variable methods that *Pan troglodytes* and *Cercopithecus ascanius* employ to deal with the mechanical challenges of what to do with seeds when pulp flesh is the food and not seeds. The seeds of tropical fruits range in size from less than 1 mm in length to large palm nuts that can be greater than 30 cm in width and can account for more than half of the weight of fruits (van Roosmalen 1984, Garber 1986, Howe and Westley 1988, Whitmore 1990). Swallowed seeds can thus represent a significant cost to a fruit pulp consumer in that they not only increase an animal's body mass but may also displace more readily processed, nutritious digesta from the gut (Snow and Snow 1988, Fleming 1988, Corlett and Lucas 1990, Levey and Grajal 1991, Leighton 1993, Levey and Karasov 1994). The seeds of fruits are literally an unwanted mass, and the adhesive pulp must be removed and the seeds discarded in some way. Lambert (1999) found that *P. troglodytes* swallow most seeds while *C. ascanius* spit most seeds and argues that these seed handling differences can be attributed to differences in digestive retention times, oral anatomy, and alternative mechanisms by which to avoid the mechanical cost of seed ballast. Cercopithecine monkeys have among the longest (both absolutely and relatively) digestive retention times across the Primate order (Clemens and Phillips 1980; Clemens and Maloiy 1981; Maisels 1993; Lambert 1997, 1998, 2002b; see Warner 1981 for definitions of digestive passage times). Seed spitting is important for cercopithecines because if large-seeded fruit were swallowed whole to remove pulp, then the energetic cost of indigestible seed ballast would be incurred for an absolutely and relatively long time. Moreover, it could severely limit space for incoming food. On the other hand, *P. troglodytes* lack the oral adaptations of the Cercopithecinae (e.g., large postcanine teeth, high-crested bilophodont molars, and cheek pouches). In addition, they are large primates that exhibit the more typical emphasis of relatively rapid digestive processing common to frugivores

(Milton 1984; Milton and Demment 1987; Levey and Grajal 1991; Lambert 1998, 2002b). Because chimpanzees move seeds through the gastrointestinal tract relatively faster than cercopithecines, they are also less likely to incur such a cost in terms of seed ballast and displacement of nutritious digesta. Moreover, being a large animal, swallowed seeds will represent a smaller mass relative to chimpanzee body size. A potential result of this is that natural selection has been less intense in hominoids than in cercopithecines for evolving behavioral mechanisms to avoid the cost of seeds in the gut.

CONCLUSIONS AND CONSERVATION IMPLICATIONS

As I have endeavored to illustrate in this chapter, diet and nutrition are tied to virtually every aspect of an animal's biology, and understanding the means by which primates solve feeding challenges can lend invaluable insight into the ecology and evolution of species. However, such understanding is more than academic. Insight into nutrition and its myriad consequences for animal biology is essential in at least three ways to an increasingly critical goal: conservation of wild primate populations. The sobering reality is that 96 of the world's extant 250 primate species are presently in the World Conservation Union's critically endangered or endangered category and could continue to decline to extinction within the next century (Struhsaker 1999, Wright and Jernvall 1999; see Chapter 30). Population size directly influences the intrinsic components of population decline and is hence critically related to species vulnerability (Cowlishaw and Dunbar 2000). Since food is commonly argued to be among the most important biotic variables limiting population size, an understanding of the direct and causal relationships among food resource availability, diet, nutrition, and population size should be part of any sound in situ conservation plan (Struhsaker 1976, Milton 1982, Gautier-Hion 1988, Cowlishaw and Dunbar 2000, Chapman et al. 2003). Yet, few studies have directly linked specific nutrients to changes in population size, despite the fact that such information is key for the implementation of conservation tactics (Chapman et al. 2003). For example, as suggested by Chapman et al. (2003), if it could be demonstrated that a wild primate population is limited by x nutrient, tree species rich in this nutrient could be left standing in selective logging operations. With careful planning, post-felling population decline could be lower and/or recovery may be faster.

Information on diet and nutrition is critical for the successful ex situ management of endangered species in captivity as well. One response on the part of conservation managers to the decline of wild populations is to maintain endangered taxa in captive settings. *Procolobus badius waldronii* (a.k.a. *Piliocolobus badius waldronii*) was declared extinct in 2000 (Oates et al. 2000); yet, while some colobines (e.g., *Colobus guereza*, *Presbytis* spp.) do reasonably well in captivity, maintaining *P. badius* in captivity has proven extremely difficult. Why this is the case is not clear; however, *P. badius* has a quadripartite stomach and complex digestive ecology, and the biology of maintaining a healthy community of anaerobic gastric microbes in a delicately balanced stomach pH, along with unknown mineral/vitamin requirements, may be implicated.

Finally, nutritional information can also lend insight into dietary flexibility, which too can inform our decisions as conservation managers. For example, dietary flexibility and a capacity to fall back on low-quality foods may facilitate animal persistence in degraded landscapes and forest fragments. Are these then the same species whose populations are less impacted overall by anthropogenic change? Should they in turn receive greater conservation priority because they are likely to tolerate future human-induced change or less attention because they are presently tolerating anthropogenic stress? Such questions are central to conservation tactics, and we are yet to get a handle on what flexibility actually means in terms of nutritional requirements. These examples clearly highlight the fact that questions regarding diet and nutrition represent more than academic concerns.

REFERENCES

Alcock, J. (1989). *Animal Behavior: An Evolutionary Approach*. Sinauer Associates, Sunderland, MA.

Alexander, R. M. (1993). The relative merits of foregut and hindgut fermentation. *J. Zool. Lond.* 231:391–401.

Altmann, S. A. (1998). *Foraging for Survival: Yearling Baboons in Africa*. University of Chicago Press, Chicago.

Bearder, S. K., and Martin, R. D. (1980). Acacia gum and its use by bushbabies, *Galago senegalensis* (Primates: Lorisidae). *Int. J. Primatol.* 1:103–128.

Beeson, M. (1989). Seasonal dietary stress in a forest monkey (*Cercopithecus mitis*). *Oecologia* 78:565–570.

Begon M., Harper, J. L., and Townsend, C. R. (1990). *Ecology: Individuals, Populations, and Communities*, 2nd ed. Blackwell Scientific, New York.

Bell, R. H. V. (1971). A grazing ecosystem in the Serengeti. *Sci. Am.* 225:86–93.

Blaxter, K. L. (1962). *The Energy Metabolism of Ruminants*. Hutchison, London.

Boinski, S., Treves, A., and Chapman, C. A. (2000). A critical evaluation of the influence of predators on primates: effects on group travel. In: Boinski, S., and Garber, P. A. (eds.), *On the Move: How and Why Animals Travel in Groups*. University of Chicago Press, Chicago. pp. 491–518.

Brugiere, D., Gautier, J. P., Moungazi, A., and Gautier-Hion, A. (2002). Primate diet and biomass in relation to vegetation composition and fruiting phenology in a rain forest Gabon. *Int. J. Primatol.* 23:999–1024.

Buss, D. H., and Cooper, R. W. (1970). Composition of milk from talapoin monkeys. *Folia Primatol.* 13:196–206.

Butynski, T. M. (1988). Guenon birth seasons and correlates with rainfall and food. In: Gautier-Hion, A., Bourliere, F., and Gautier, J.-P. (eds.), *A Primate Radiation: Evolutionary Biology of the African Guenons*. Cambridge University Press, Cambridge. pp. 284–322.

Caton, J. M., Hill, D. M., Hume, I. D., and Crook, G. A. (1996). The digestive strategy of the common marmoset, *Callithrix jacchus. Comp. Biochem. Physiol.* A114:1–8.

Chapman, C. A. (1990). Ecological constraints on group size in three species of neotropical primates. *Folia Primatol.* 55:1–9.

Chapman, C. A., and Chapman, L. J. (2000). Constraints on group size in red colobus and red-tailed guenons: examining the generality of the ecological constraints model. *Int. J. Primatol.* 21:565–585.

Chapman, C. A., and Chapman, L. J. (2002). Foraging challenges of red colobus monkeys: influence of nutrients and secondary compounds. *Comp. Biochem. Physiol.* A133:861–875.

Chapman, C. A., Chapman, L. J., Gautier-Hion, A., Lambert, J. E., Rode, K., Tutin, C. E. G., and White, L. J. T. (2002). Variation in the diet of *Cercopithecus* monkeys: differences within forests, among forests, and across species. In: Glenn, M., and Cords, M. (eds.), *The Guenons: Diversity and Adaptation in African Monkeys*. Kluwer Academic, New York. pp. 319–344.

Chapman, C. A., Chapman, L. J., Rode, K. D., Hauck, E. M., and McDowell, L. R. (2003). Variation in the nutritional value of primate foods: among trees, time periods, and areas. *Int. J. Primatol.* 24:317–333.

Chapman, C. A., White, F. J., and Wrangham, R. W. (1994). Party size in chimpanzees and bonobos: a reevaluation of theory based on two similarly forested sites. In: Wrangham, R. W., McGrew, W. C., deWaal, F. B., and Heltne, P. G. (eds.), *Chimpanzee Cultures*. Harvard University Press, Cambridge, MA. pp. 41–58.

Chapman, C. A., Wrangham, R. W., and Chapman, L. J. (1995). Ecological constraints on group sizes: an analysis of spider monkey and chimpanzee subgroups. *Behav. Ecol. Sociobiol.* 36:59–70.

Clemens, E. T., and Moloiy, G. M. O. (1981). Organic acid concentrations and digesta movement in the gastrointestinal tract of the bushbaby (*Galago crassicaudatus*) and vervet monkey (*Cercopithecus aethiops*). *J. Zool. Lond.* 193:487–497.

Clemens, E. T., and Phillips, B. (1980). Organic acid production and digesta movement in the gastrointestinal tract of the baboon and sykes monkey. *Comp. Biochem. Physiol.* 66:529–532.

Conklin-Brittain, N. L., Wrangham, R. W., and Hunt, R. D. (1998). Dietary response of chimpanzees and cercopithecines to seasonal variation in fruit abundance. II. Macronutrients. *Int. J. Primatol.* 1998:971–998.

Cords, M. (1986). Interspecific and intraspecific variation in diet of two forest guenons, *Cercopithecus ascanius* and *C. mitis. J. Anim. Ecol.* 55:811–827.

Corlett, R. T., and Lucas, P. W. (1990). Alternative seed-handling strategies in primates: seed-spitting by long-tailed macaques (*Macaca fascicularis*). *Oecologia* 82:166–171.

Cowlishaw, G., and Dunbar, R. (2000). *Primate Conservation Biology*. University of Chicago Press, Chicago.

Curtis, D. J. (2004). Diet and nutrition in wild mongoose lemurs (*Eulemur mongoz*) and their implications for the evolution of female dominance and small group size in lemurs. *Am. J. Phys. Anthropol* 124:234–247.

Davies, A. G. (1994). Colobine populations. In: Davies, A. G., and Oates, J. F. (eds.), *Colobine Monkeys: Their Ecology, Behaviour and Evolution*. Cambridge University Press, Cambridge. pp. 11–43.

Dirks, W. (2003). Effect of diet on dental development in four species of catarrhine primates. *Am. J. Primatol.* 61:29–40.

Dunbar, R. I. M. (1988). *Primate Social Systems*. Cornell University Press, Ithaca.

Eaton, S. B., and Konner, M. (1985). Paleolithic nutrition: a consideration of its nature and current implications. *N. Engl. J. Med.* 312:283–289.

Feeny, P. (1976). Plant apparency and chemical defense. *Recent Adv. Phytochem.* 10:1–40.

Fleagle, J. F. (1999). *Primate Adaptation and Evolution*. Academic Press, New York.

Fleagle, J. F., Janson, C. H., and Reed, K. E. (1999). *Primate Communities*. Cambridge University Press, Cambridge.

Fleming, T. H. (1988). *The Short-Tailed Fruit Bat: A Study in Plant–Animal Interactions*. University of Chicago Press, Chicago.

Foley, W. J., and McArthur, C. (1994). The effects and costs of allelochemicals for mammalian herbivores: an ecological perspective. In: Chivers, D. J., and Langer, P. (eds.), *The Digestive System of Mammals*. Cambridge University Press, Cambridge. pp. 370–391.

Frankie, G. W., Baker, H. G., and Opler, P. A. (1974). Comparative phenological studies of trees in tropical wet and dry forests in the lowlands of Costa Rica. *Int. J. Ecol.* 62:881–919.

Freeland, W. J. (1991). Plant secondary metabolites. Biochemical coevolution with herbivores. In: Palo, R. T., and Robbins, C. T. (eds.), *Plant Defenses Against Mammalian Herbivory*. CRC Press, Boca Raton. pp. 61–82.

Freeland, W. J., and Choquenot, D. (1990). Determinants of herbivore carrying capacity: plants, nutrients and *Equus asinus* in northern Australia. *Ecology* 701:589–597.

Freeland, W. J., and Janzen, D. H. (1974). Strategies in herbivory by mammals: the role of plant secondary compounds. *Am. Nat.* 108:269–289.

Ganzhorn, J. U. (1992). Leaf chemistry and the biomass of folivorous primates in tropical forests: tests of a hypothesis. *Oecologia* 91:540–547.

Garber, P. A. (1986). The ecology of seed dispersal in two species of callitrichid primates (*Saguinus mystax* and *Saguinus fuscicollis*). *Am. J. Primatol.* 10:155–170.

Garber, P. A. (1988). Diet, foraging patterns, and resources defense in a mixed species troop of *Saguinus mystax* and *Saguinus fuscicollis* in Amazonian Peru. *Behaviour* 105:18–34.

Garber, P. A. (2004). New perspectives in primate cognitive ecology. *Am. J. Primatol.* 62:133–137.

Gartlan, J. S., McKey, D. B., Waterman, P. G., Mbi, C. N., and Struhsaker, T. T. (1980). A comparative study of the chemistry of two African rain forests. *Biochem. Syst. Ecol.* 8:401–422.

Gaulin, S. J. C. (1979). A Jarman/Bell model of primate feeding niches. *Hum. Ecol.* 7:1–20.

Gautier-Hion, A. (1988). The diet and dietary habits of forest guenons. In: Gautier-Hion, A., Bourliere, F., and Gautier, J. P. (eds.), *A Primate Radiation: Evolutionary Biology of the African Guenons*. Cambridge University Press, Cambridge. pp. 257–283.

Gillespie, T. R., and Chapman, C. A. (2001). Determinants of group size in the red colobus monkey (*Procolobus badius*): an evaluation of the generality of the ecological-constraints model. *Behav. Ecol. Sociobiol.* 50:329–338.

Glander, K. E. (1981). Feeding patterns in mantled howling monkeys. In: Kamil, A. C., and Sargent, T. D. (eds.), *Foraging Behavior: Ecological, Ethological, and Psychological Approaches*. Garland STPM Press, New York. pp. 231–257.

Gupta, A. K., and Chivers, C. J. (1999). Biomass and use of resources in south and southeast Asian primate communities.

In: Fleagle, J. G., Janson, C., and Reed, K. E. (eds), *Primate Communties*. Cambridge University Press, Cambridge. pp. 38–54.

Harborne, J. B. (1988). *Introduction to Ecological Biochemistry*. Academic Press, New York.

Harding, R. S. O. (1981). An order of omnivores: nonhuman primate diets in the wild. In: Harding, R. S. O., and Teleki, G. (eds.), *Omnivorous Primates: Gathering and Hunting in Human Evolution*. Columbia University Press, New York. pp. 191–214.

Harvey, P. H., Martin, R. D., and Clutton-Brock, T. H. (1987). Life histories in comparative perspective. In: Smuts, B. B., Cheyney, D. L., Seyfarth, R. M., Wrangham, R. W., and Struhsaker, T. T. (eds.), *Primate Societies*. University of Chicago Press, Chicago. pp. 181–196.

Hill, W. C. O. (1966). *Primates: Comparative Anatomy and Taxonomy. Catarrhini, Cercopithecoidea, Cercopithecinae*, vol. VI. University Press, Edinburgh.

Hladik, C. M. (1977). Chimpanzees of Gabon and chimpanzees of Gombe: some comparative data on the diet. In: Clutton-Brock, T. H. (ed.), *Primate Ecology: Studies of Feeding and Ranging Behavior in Lemurs, Monkeys and Apes*. Academic Press London. pp. 481–503.

Hladik, C. N. (1978). Adaptive strategies of primates in relation to leaf-eating. In: Montgomery, G. G. (ed.), *The Ecology of Arboreal Folivores*. Smithsonian Institution Press, Washington DC. pp. 373–395.

Howe, H. F., and Westley, L. C. (1988). *Ecological Relationships of Plants and Animals*. Oxford University Press, New York.

Inge, W. R. (1927). *Confessio fidei: Outspoken Essays*, 2nd ser. Longmans, Green, New York.

Isbell, L. A. (1991). Contest and scramble competition: patterns of female aggression and ranging behavior among primates. *Behav. Ecol.* 2:143–155.

Janson, C. H. (1992). Evolutionary ecology of primate social structure. In: Smith, E. A., and Winterhalder, B. (eds.), *Evolutionary Ecology and Human Behavior*. Aldine de Gruyter, New York. pp. 95–130.

Janson, C. H., and Chapman, C. A. (1999). Resources and primate community structure. In: Fleagle, J. G., Janson, C., and Reed, K. E., (eds.), *Primate Communities*. Cambridge University Press, Cambridge. pp. 237–260.

Janson, C. H., and Goldsmith, M. L. (1995). Predicting group size in primates: foraging costs and predation risks. *Behav. Ecol.* 6:326–336.

Janson, C. H., and van Schaik, C. P. (1988). Recognizing the many faces of primate food competition: methods. *Behaviour* 105:165–186.

Janzen, D. H. (1967). Synchronization of sexual reproduction of trees within the dry season in Central America. *Evolution* 21:620–637.

Jarman, P. J. (1974). The social organization of antelope in relation to their ecology. *Behavior* 58:215–267.

Kay, R. N. B. (1985). Comparative studies of food propulsion in ruminants. In: Ooms, L. A. A., Degryse, A. D., and van Miert, A. S. J. A. M. (eds.), *Physiological and Pharmacological Aspects of the Reticulo-Rumen*. Martinus Nijhoff, Dordrecht. pp. 155–170.

Kay, R. N. B., and Davies, A. G. (1994). Digestive physiology. In: Davies, A. G., and Oates, J. F. (eds.), *Colobine Monkeys: Their Ecology, Behavior and Evolution*, Cambridge University Press, Cambridge. pp. 229–259.

Kinzey, W. G. (1978). Feeding behaviour and molar features in two species of titi monkey. In: Chivers, D. J., and Herbert, J.

(eds), *Recent Advances in Primatology*, vol. 1. Academic Press, New York. pp. 372–375.

Kinzey, W. G., and Norconk, M. A. (1990). Hardness as a basis of fruit choice in two sympatric primates. *Am. J. Phys. Anthropol.* 81:5–15.

Kinzey, W. G., and Norconk, M. A. (1993). Physical and chemical properties of fruit and seeds eaten by *Pithecia* and *Chiropotes* in Suriname and Venezuela. *Int. J. Primatol.* 14:207–228.

Kleiber, M. (1961). *The Fire of Life: An Introduction to Animal Energetics*. John Wiley & Sons, New York.

Knott, C. D. (1998). Changes in orangutan caloric intake, energy balance, and ketones in response to fluctuating fruit availability. *Int. J. Primatol.* 19:1061–1079.

Koenig, A. (2002). Competition for resources and its behavioral consequences among female primates. *Int. J. Primatol.* 23:759–783.

Krebs, C. J. (1972). *Ecology: the Experimental Analysis of Distribution and Abundance*. Harper and Row, New York.

Krishnamani, R., and Mahaney, W. C. (2000). Geophagy among primates: adaptive significance and ecological consequences. *Anim. Behav.* 59:899–815.

Lambert, J. E. (1997). Digestive strategies, fruit processing, and seed dispersal in the chimpanzees (*Pan troglodytes*) and redtail monkeys (*Cercopithecus ascanius*) of Kibale National Park, Uganda [PhD diss.]. University of Illinois, Champaign-Urbana.

Lambert, J. E. (1998). Primate digestion: interactions among anatomy, physiology, and feeding ecology. *Evol. Anthropol.* 7:8–20.

Lambert, J. E. (1999). Seed handling in chimpanzees (*Pan troglodytes*) and redtail monkeys (*Cercopithecus ascanius*): implications for understanding hominoid and cercopithecine fruit processing strategies and seed dispersal. *Am. J. Phys. Anthropol.* 109:365–386.

Lambert, J. E. (2000). Urine drinking in wild Cercopithecus ascanius, evidence of nitrogen balancing? *Afr. J. Ecol.* 389:360–363.

Lambert, J. E. (2002a). Resource switching in guenons: a community analysis of dietary flexibility. In: Glenn, M., and Cords, M. (eds.), *The Guenons: Diversity and Adaptation in African Monkeys*. Kluwer Academic, New York. pp. 303–317.

Lambert, J. E. (2002b). Digestive retention times in forest guenons with reference to chimpanzees. *Int. J. Primatol.* 26:1169–1185.

Lambert, J. E. (2005). Competition, predation and the evolution of the cercopithecine cheek pouch: the case of *Cercopithecus* and *Lophocebus*. *Am. J. Phys. Anthropol.* 126:183–192.

Lambert, J. E., Chapman, C. A., Wrangham, R. W., and Conklin-Brittain, N. L. (2004). Hardness of mangabey and guenon foods: implications for the critical function of enamel thickness in exploiting fallback foods. *Am. J. Phys. Anthropol.* 125:363–368.

Leigh, S. R. (1994). Ontogenetic correlates of diet in anthropoid primates. *Am. J. Primatol.* 94:499–522.

Leighton, M. (1993). Modeling dietary selectivity by Bornean orangutans: evidence for integration of multiple criteria in fruit selection. *Int. J. Primatol.* 14:257–314.

Leonard, W. R. (2000). Human nutritional evolution. In: Stinson, S., Bogin, B., Huss-Ashmore, R., and O'Rourke, D. (eds.), *Human Biology: An Evolutionary and Biocultural Approach*. Wiley-Liss, New York. pp. 295–343.

Levey, D. J., and Grajal, A. (1991). Evolutionary implications of fruit-processing limitations in cedar waxwings. *Am. Nat.* 138:171–189.

Levey, D. J., and Karasov, W. H. (1994). Gut passage of insects by European starlings and comparison with other species. *Auk* 111:478–481.

Levin, D. A. (1971). Plant phenolics: an ecological perspective. *Am. Nat.* 105:157–181.

Levin, D. A. (1976). Alkaloid-bearing plants: an ecogeographic perspective. *Am. Nat.* 110:261–284.

Maisels, F. (1993). Gut passage rate in guenons and mangabeys: another indicator of a flexible dietary niche? *Folia Primatol.* 61:35–37.

Marks, D. L., Swain, T., Goldstein, S., Richard, A., and Leighton, M. (1988). Chemical correlates of rhesus monkey food choice: the influence of hydrolyzable tannins. *J. Chem. Ecol.* 14:213–235.

Martin, R. D. (1990). *Primate Origins and Evolution: A Phylogenetic Reconstruction*. Princeton University Press, Princeton, NJ.

McGraw, W. S. (1998). Comparative locomotion and habitat use of six monkeys in the Taï Forest, Ivory Coast. *Am. J. Phys. Anthropol.* 105:493–510.

McKey, D. B. (1974). Adaptive patterns in alkaloid physiology. *Am. Nat.* 108:305–320.

McKey, D. B. (1978). Soils, vegetation, and seed-eating by black colobus monkeys. In: Montgomery, G. G. (ed.), *The Ecology of Arboreal Folivores*. Smithsonian Institution Press, Washington DC. pp. 423–437.

McKey, D. B., Gartlan, J. S., Waterman, P. G., and Choo, G. M. (1981). Food selection by black colobus monkeys (*Colobus satanus*) in relation plant chemistry. *Biol. J. Linn. Soc.* 16:115–146.

McKey, D. B., Waterman, P. G., Gartlan, J. S., and Struhsaker, T. T. (1978). Phenolic content of vegetation in two African rain forests: ecological implications. *Science* 202:61–64.

Messier, W., and Stewart, C. B. (1997). Episodic adaptive evolution of primate lysozymes. *Nature* 385:151–154.

Milton, K. (1979). Factors influencing leaf choice by howler monkeys: a test of some hypotheses of food selection by generalist herbivores. *Am. Nat.* 114:362–378.

Milton, K. (1980). *The Foraging Strategy of Howler Monkeys: A Study in Primate Economics*. Columbia University Press, New York.

Milton, K. (1982). Dietary quality and demographic regulation in a howler monkey population. In: Leigh, E. G., Rand, A. S., and Windsor, D. M. (eds.), *The Ecology of a Tropical Forest: Seasonal Rhythms and Long-Term Changes*. Smithsonian Institution Press, Washington DC. pp. 273–289.

Milton, K. (1984). The role of food processing factors in primate food choice. In: Rodman, P. S., and Cant, J. G. H. (eds.), *Adaptations for Foraging in Nonhuman Primates: Contributions to an Organismal Biology of Prosimians, Monkeys and Apes*. Columbia University Press, New York. pp. 249–279.

Milton, K. (1987). Primate diets and gut morphology: implications for hominid evolution. In: Harris, M., and Ross, E. B. (eds.), *Food and Evolution: Toward a Theory of Human Food Habits*. Temple University Press, Philadelphia. pp. 93–115.

Milton, K. (1993). Diet and primate evolution. *Sci. Am.* 269:86–93.

Milton, K. (1996). Effects of bot fly (*Alouattamyia baeri*) parasitism on a free-ranging howler monkey (*Alouatta palliata*) population in Panama. *J. Zool.* 239:39–63.

Milton, K. (1999). Nutritional characteristics of wild primate foods: do the diets of our closest living relatives have lessons for us? *Nutrition* 15:488–498.

Milton, K. (2000). Back to basics: why foods of wild primates have relevance for modern human health. *Nutrition* 16:480–483.

Milton, K., and Demment, M. W. (1987). Digestion and passage kinetics of chimpanzees fed high and low fiber diets and comparison with human data. *J. Nutr.* 118:1082–1088.

Milton, K., and Jenness, R. (1987). Ascorbic acid content of neotropical plant parts available to wild monkeys and bats. *Experientia* 43:339–342.

Mole, S., and Waterman, P. G. (1985). Stimulatory effects of tannins and cholic acid on tryptic hydrolysis of proteins: ecological implications. *J. Chem. Ecol.* 11:1323–1332.

Murray, P. (1973). The anatomy and adaptive significance of cheek pouches (bursae buccales) in the Cercopithecinae, Cercopithecoidea [PhD thesis]. University of Chicago, Chicago.

Nagy, K. A., and Milton, K. (1979). Aspects of dietary quality, nutrient assimilation and water balance in wild howler monkeys (*Alouatta palliata*). *Ecology* 60:475–480.

Nash, L. T. (1986). Dietary, behavioral, and morphological aspects of gummivory in primates. *Ybk. Phys. Anthropol.* 29:113–137.

National Research Council (2003). *Nutrient Requirements of Nonhuman Primates.* National Academies Press, Washington DC.

Oates, J. F. (1978). Water-plant and soil consumption by guereza monkeys (*Colobus guereza*): a relationship with minerals and toxins in the diet? *Biotropica* 10:241–253.

Oates, J. F. (1987). Food distribution and foraging behavior. In: Smuts, B. B., Cheyney, D. L., Seyfarth, R. M., Wrangham, R. W., and Struhsaker, T. T. (eds.), *Primate Societies.* University of Chicago Press, Chicago. pp. 197–209.

Oates, J. F., Abedi-Lartey, M., McGraw, W. S., Struhsaker, T. T., and Whitesides, G. H. (2000). Extinction of a West African red colobus monkey. *Conserv. Biol.* 14:1526–1532.

Oates, J. F., Swain, T., and Zantovska, J. (1977). Secondary compounds and food selection by colobus monkeys. *Biochem. Syst. Ecol.* 5:317–321.

Oates, J. F., Whitesides, G. H., Davies, A. G., Waterman, P. G., Green, S. M., Dasilva, G. L., and Mole, S. (1990). Determinants of variation in tropical forest primate biomass: new evidence from West Africa. *Ecology* 71:328–343.

O'Brien, T. G., and Kinnaird, M. F. (1997). Behavior, diet and movements of the Sulawesi crested black macaque (*Macaca nigra*). *Int. J. Primatol.* 18:321–351.

Oftedal, O. T. (1991). The nutritional consequences of foraging in primates: the relationship of nutrient intakes to nutrient requirements. *Philos. Trans. R. Soc. Lond. B* 334:161–170.

Oftedal, O. T., and Allen, A. E. (1996). The feeding and nutrition of omnivores with emphasis on primates. In: Kleiman, D. G., Allen, M. E., Thompson, K. V., Lumpkin, S., and Harris, H. (eds.), *Wild Mammals in Captivity: Principles and Techniques.* University of Chicago Press, Chicago. pp. 148–157.

Olupot, W., Waser, P. M., and Chapman, C. A. (1998). Fruit finding by mangabeys (*Lophocebus albigena*): are monitoring of fig trees and use of sympatric frugivore calls possible strategies? *Int. J. Primatol.* 19:339–353.

Parra, R. (1978). Comparison of foregut and hindgut fermentation in herbivores. In: Montgomery, G. G. (ed.), *The Ecology of Arboreal Folivores.* Smithsonian Institution Press, Washington DC. pp. 205–229.

Plumptre, A. and Reynolds, V. (1994). The effect of selective logging on the primate populations in the Budongo Forest Reserve, Uganda. *J. Appl. Ecol.* 31:631–641.

Power, M. L. (1996). The other side of callitrichine gummivory: digestibility and nutritional value. In: Norconk, M. A., Rosenberger, A. L., and Garber, P. A. (eds.), *Adaptive Radiations of Neotropical Primates.* Plenum Press, New York. pp. 97–110.

Power, M. L., Tardif, S. D., Layne, D. G., and Schulkin, J. (1999). Ingestion of calcium solutions by common marmosets (*Callithrix jacchus*). *Am. J. Primatol.* 47:255–261.

Pruetz, J. D., and Isbell, L. A. (1999). What makes a food contestable? Food properties and contest competition in vervets and patas monkeys in Laikipia, Kenya. *Am. J. Phys. Anthropol.* 28(suppl.):225–226.

Richard, A. F., Goldstein, S. J., and Dewar, R. E. (1989). Weed macaques: the evolutionary implications of macaque feeding ecology. *Int. J. Primatol.* 10:569–594.

Robbins, C. T. (1993). *Wildlife Feeding and Nutrition.* Academic Press, New York.

Rode, K. D., Chapman, C. A., Chapman, L. J., and McDowell. L. R. (2003). Mineral resource availability and consumption by colobus in Kibale National Park, Uganda. *Int. J. Primatol.* 24:541–573.

Rosenberger, A. L. (1992). Evolution of feeding niches in New World monkeys. *Am. J. Phys. Anthropol.* 88:525–562.

Rosenthal, G. A., and Janzen, D. H. (1979). *Herbivores: Their Interaction with Secondary Plant Metabolites.* Academic Press, New York.

Ross, C. (1992). Basal metabolic rate, body weight and diet in primates: an evaluation of the evidence. *Folia Primatol.* 58:7–23.

Rudran, R. (1978). *Socioecology of the Blue Monkeys of the Kibale Forest, Uganda. Smithsonian Contributions to Zoology, 249.* Smithsonian Institution Press, Washington DC.

Schmidt-Nielsen, K. (1984). *Scaling: Why Is Animal Size so Important?* Cambridge University Press, Cambridge.

Schmidt-Nielsen, K. (1997). *Animal Physiology: Adaptation and Environment.* Cambridge University Press, Cambridge.

Sharon, M. (1980). Carbohydrates. *Sci. Am.* 243:90–116.

Shultz, S., and Noe, R. (2002). The consequences of crowned eagle central-place foraging on predation risk in monkeys. *Proc. R. Soc. Lond. B* 269:1797–1802.

Snow, B. K., and Snow, D. W. (1988). *Birds and Berries.* T&AD Poyser Publishers, Calton, UK.

Sourd, C., and Gautier-Hion, A. (1986). Fruit selection by a forest guenon. *J. Anim. Ecol.* 55:235–244.

Sterck, E. H. M., Watts, D. P., and van Schaik, C. P. (1997). The evolution of female social relationships in nonhuman primates. *Behav. Ecol. Sociobiol.* 41:291–309.

Strier, K. B. (2003). Primatology comes of age; 2002 AAPA luncheon address. *Ybk Phys. Anthropol.* 46:2–13.

Struhsaker, T. T. (1976). A further decline in numbers of Amboseli vervet monkeys. *Biotropica* 8:211–214.

Struhsaker, T. T. (1978). Food habits of five monkey species in the Kibale Forest, Uganda. In: Chivers, D. J., and Herbert, J. (eds.), *Recent Advances in Primatology. Conservation,* vol. 2. Academic Press, London. pp. 87–94.

Struhsaker, T. T. (1981). Polyspecific associations among tropical rain-forest primates. *Z. Tierpsychol.* 57:268–304.

Struhsaker, T. T. (1999). Primate communities in Africa: the consequence of long-term evolution or the artifact of recent hunting? In: Fleagle, J. G., Janson, C., and Reed, K. E. (eds.), *Primate Communities.* Cambridge University Press, Cambridge. pp. 289–294.

Terborgh, J., and Janson, C. H. (1986). The socioecology of primate groups. *Annu. Rev. Ecol. Syst.* 17:111–136.

Tortora, G. J., and Anagnostakos, N. P. (1987). *Principles of Physiology*. Harper and Row, Cambridge.

Treves, A., and Chapman, C. A. (1996). Conspecific threat, predation avoidance, and resource defense: implications for grouping in langurs. *Behav. Ecol. Sociobiol.* 39:43–53.

van Roosmalen, M. G. M. (1984). Subcategorizing foods in primates. In: Chivers, D. J., Wood, B. A., and Bilsborough, A. (eds.), *Food Acquisition and Processing in Primates*. Plenum Press, New York. pp. 167–176.

van Schaik, C. P. (1989). The ecology of social relationships amongst female primates. In: Standen, V., and Foley, R. A. (eds.), *Comparative Socioecology: The Behavioral Ecology of Humans and Other Mammals*. Blackwell, Oxford. pp. 195–218.

van Schaik, C. P., and Kappeler, P. M. (1993). Life history, activity period and lemur social systems In: Kappeler, P. M., and Ganzhorn, J. U. (eds.), *Lemur Social Systems and Their Ecological Basis*. Plenum, New York. pp. 241–260.

van Schaik, C. P., Terborgh, J. W., and Wright, S. J. (1993). The phenology of tropical forests: adaptive significance and consequences for primary consumers. *Annu. Rev. Ecol. Syst.* 24:353–377.

van Schaik, C. P., van Noordwijk, M. A., de Boer R. J., and den Tonkelaar, I. (1983). The effect of group size on time budgets and social behaviour in wild long-tailed macaques (*Macaca fascicularis*). *Behav. Ecol. Sociobiol.* 13:173–181.

Van Soest, P. J. (1982). *Nutritional Ecology of the Ruminant*. O & B Books, Corvallis, OR.

Vinyard, C. J., Wall, C. E., Williams, S. H., and Hylander, W. L. (2003). Comparative functional analysis of skull morphology of tree-gouging primates. *Am. J. Phys. Anthropol.* 120:153–170.

Walker, C. H. (1978). Species differences in microsomal monooxygenase activity and their relationship to biological half-lives. *Drug Metab. Rev.* 7:295–310.

Wardlaw, G. M., Hampl, J. S., and DiSilvestro, R. A. (2004). *Perspectives in Nutrition*. McGraw-Hill, Boston.

Warner, A. C. I. (1981). Rate of passage of digesta through the gut of mammals and birds. *Nutr. Abstr. Rev. B* 51:789–820.

Waser, P. (1977). Feeding, ranging and group size in the mangabey *Cercocebus albigena*. In: Cluttton-Brock, T. H. (ed.), *Primate Ecology: Studies of Feeding and Ranging Behavior in Lemurs, Monkeys and Apes*. Academic Press, New York. pp. 183–222.

Waterman, P. G. (1984). Food acquisition and processing as a function of plant chemistry. In: Chivers, D. J., Wood, B. A., and Bilsborough, A. (eds.), *Food Acquisition and Processing in Primates*. Plenum, New York. pp. 177–212.

Waterman, P. G., and Kool, K. (1994). Colobine food selection and plant chemistry. In: Davies, A. G., and Oates, J. K. (eds.), *Colobine Monkeys: Their Ecology, Behavior and Evolution*. Cambridge University Press, Cambridge. pp. 251–284.

Waterman, P. G., and Mole, K. M. (1994). Colobine food selection and plant chemistry. In: Davies, G., and Oates, J. F. (eds.), *Colobine Monkeys: Their Ecology, Behavior and Evolution*. Cambridge University Press, Cambridge. pp. 251–284.

White, F. J. (1998). Seasonality and socioecology: the importance of variation in fruit abundance to bonobo sociality. *Int. J. Primatol.* 19:1013–1027.

Whiten, A., Byrne, R. W., Barton, R. A., Waterman, P. G., and Henzi, S. P. (1991). Dietary and foraging strategies of baboons. *Philos. Trans. R. Soc. Lond. B* 344:187–197.

Waterman, P. G., Ross, J. A. M., Bennett, E. L., and Davies, A. G. (1988). A comparison of the floristics and leaf chemistry of the tree flora in two Malaysian rain forests and the influence of leaf chemistry on populations of colobine monkeys in the Old World. *Biol. J. Linn. Soc.* 34:1–32.

Whitmore, T. C. (1990). *An Introduction to Tropical Rain Forests*. Clarendon Press, Oxford.

Wrangham, R. W., Conklin, N. L., Etot, G., Obua, J., Hunt, K. D., Hauser, M. D., and Clark, A. P. (1993). The value of figs to chimpanzees. *Int. J. Primatol.* 14:243–256.

Wrangham, R. W., Conklin-Brittain, N. L., and Hunt, K. D. (1998). Dietary response of chimpanzees and cercopithecines to seasonal variation in fruit abundance. I. Antifeedants. *Int. J. Primatol.* 19:949–970.

Wrangham, R. W., and Waterman, P. G. (1981). Feeding behavior of vervet monkeys on *Acacia tortilis* and *Acacia xanthphloea*: with special reference to reproductive strategies and tannin productions. *J. Anim. Ecol.* 50:715–731.

Wright, P. C., and Jernvall, J. (1999). The future of primate communities: a reflection of the present? In: Fleagle, J. G., Janson, C., and Reed, K. E. (eds.), *Primate Communities*. Cambridge University Press, Cambridge. pp. 38–54.

Yeager, C. P., Silver, S. C., and Dierenfeld, E. S. (1997). Mineral and phytochemical influences on foliage selection by the proboscis monkey (*Nasalis larvatus*). *Am. J. Primatol.* 41:117–128.

30

Conservation

Karen B. Strier

INTRODUCTION

Of more than 600 species and subspecies of primates recognized today, nearly one-third are in grave danger of going extinct (Mittermeier et al. 2000, Konstant et al. 2002). This is an alarming increase over estimates from less than a decade ago, and although some of the increase can be attributed to the rarity of taxa that have recently been discovered or re-classified (Groves 2001, Mittermeier et al. 2004), it also exemplifies the rapidity with which extinction risks can rise (Konstant et al. 2002). Some of the primates that are threatened with extinction can still be saved if efforts to protect them and their habitats succeed. Others, however, may already be reduced to such small populations and confined to such limited or altered habitats that the chances for their long-term survival are slim (Cowlishaw 1999). One West African primate, known as Miss Waldron's red colobus monkey (*Pilocolobus badius waldronae*), since its scientific discovery in 1933, may already be extinct (Oates et al. 2000, but see McGraw and Oates 2002, McGraw 2005). Primate conservation is about preserving the diversity of species and subspecies whose futures are endangered and preventing those taxa that are still relatively secure from becoming endangered themselves.

This chapter is structured around a review of the steps that are involved in identifying conservation priorities and developing and implementing informed conservation programs. Informed conservation requires an understanding of the major threats to primates and how the dynamics of small populations affect different taxa in different ways depending on their degree of ecological and behavioral specialization and intrinsic reproductive rates. All conservation efforts call for establishing and enforcing prohibitions on unsustainable hunting, protecting remaining habitats, and targeted research and education; but the relative importance of each of these tactics and the feasibility of implementing them depend on the taxa and the local conditions that affect them. In the final section, I consider some of the many challenging choices that need to be made about the role of zoos and other captive breeding facilities in conserving primates and whether wild populations and disturbed habitats can, or should, be actively managed and restored. These choices raise important questions about the trade-offs involved in meeting conservation priorities and will ultimately determine the long-term prognosis for many of the world's most endangered primates.

IDENTIFYING PRIORITIES

The first step in conserving primates is to identify priorities, which requires an assessment of each taxon's current status in the wild. The Primate Specialist Group of the World Conservation Union (IUCN) regularly updates these assessments using the Red List criteria, which are based on quantitative estimates of the total number of individuals, the size and distribution of their populations, the rate at which their occurrence has declined, and the causes of these declines (Hilton-Taylor 2002). These kinds of data generally come from surveys and censuses conducted by experienced fieldworkers, but the quality of the data can still vary greatly depending on the accessibility of the sites, the sampling methods employed, the history of human contact, and the habits of the primates themselves. Nonetheless, they represent the best, and in many cases the only, available sources of information about the abundance and distribution of the world's extant primates; and there are still a considerable number of taxa for which data on status in the wild are deficient (Hilton-Taylor and Rylands 2002).

Recent evaluations of primate extinction risks have contributed to revisions in primate taxonomy, with the result that many previously recognized species or subspecies have been split into two or more distinct taxa. These revisions include at least four newly recognized genera and various newly recognized species and subspecies (Mittermeier et al. 2000, Groves 2001, Brandon-Jones et al. 2004, Grubb et al. 2005). Although some of these revisions are still subject to debate, there are merits to recognizing primate diversity at the taxonomic level for the purpose of conservation because calling attention to variant forms insures that none is overlooked in assessments of conservation priorities (Rylands et al. 1995, Purvis and Hector 2000, Groves 2001, Agapow et al. 2004, Mace 2004).

The *2002 IUCN Red List of Threatened Species* (Hilton-Taylor 2002) includes 195 primate taxa that have suffered especially severe population and habitat losses in recent years. The degree and rate of these losses determine whether a taxon is classified as critically endangered, endangered, or vulnerable (Table 30.1). All primates that meet the criteria

Table 30.1   World Conservation Union (IUCN) Criteria for Threatened Taxa

Critically Endangered (CR): Facing an extremely high risk of extinction in the wild due to any of the following:

A. Reduction in population size of 80%–90% over the last 10 years or three generations depending on the causes and reversibility of the reductions

B. Extent of occurrence <100 km² or area of occupancy <10 km²

C. Population size estimated to number fewer than 250 mature individuals and to be declining or unevenly distributed

D. Population size estimated to number fewer than 50 mature individuals

E. Probability of extinction within 10 years or three generations at least 50%

Endangered (EN): Facing a very high risk of extinction in the wild due to any of the following:

A. Reduction in population size of 50%–70% over the last 10 years or three generations depending on the causes and reversibility of the reductions

B. Extent of occurrence <5,000 km² or area of occupancy <500 km²

C. Population size estimated to number fewer than 2,500 mature individuals and to be declining or unevenly distributed

D. Population size estimated to number fewer than 250 mature individuals

E. Probability of extinction at least 20% within 20 years or three generations

Vulnerable (VU): Facing a high risk of extinction in the wild due to any of the following:

A. Reduction in population size of 30%–50% over the last 10 years or three generations depending on the causes and reversibility of the reductions

B. Extent of occurrence <20,000 km² or area of occupancy <2,000 km²

C. Population size estimated to number fewer than 1,000 mature individuals and to be declining or unevenly distributed

D. Population size estimated to number fewer than 1,000 mature individuals or restricted area of occupancy or number of locations

E. Probability of extinction at least 10% within 100 years or three generations

Source: Simplified and condensed from http://www.redlist.org/info/categories_criteria2001.html.

for any of these three categories are considered threatened, but the 195 endangered and critically endangered taxa face the highest extinction risks of all. Asian primates now account for nearly 45% of endangered and critically endangered taxa, followed by African (22%), neotropical (17%), and Malagasy (16%) taxa (Konstant et al. 2002). More than 86% of the world's endangered and critically endangered primates occur in just 10 of the 49 countries that support them (Fig. 30.1). More than 70% of these primates occur in just six of the world's biodiversity "hotspots," which have been identified on the basis of their high levels of plant endemism and the severity of habitat losses and degradation (Myers et al. 2000, Konstant et al. 2002).

Taxa with the highest extinction risks are justifiably top priorities for conservation efforts at local, national, and regional scales. Nonetheless, assessments of these extinction risks apply the same population and habitat criteria equally to all taxa regardless of their phylogenetic distinctiveness or evolutionary uniqueness. Indices of evolutionary uniqueness take the number of closely related species or subspecies and their phylogenetic distance from one another into account to determine the amount of evolutionary history that would be lost if the taxon were to go extinct (Nee and May 1997, Hacker et al. 1998, Owens and Bennett 2000, Purvis and Hector 2000).

Assuming similar rates of evolution (Cowlishaw and Dunbar 2000), the extinction of a species belonging to a monotypic genus (or higher taxonomic unit) would result in a greater loss of evolutionary history than the extinction of a species belonging to a polytypic genus (Purvis et al. 2000). Consider, for example, the aye-aye (*Daubentonia madagascariensis*), the sole surviving species and genus of the family Daubentoniidae, which is classified in its own infraorder, Chiromyiformes (Groves 2001). Extinction of the aye-aye would represent the loss of the distinctive evolutionary history of an entire infraorder. Although the extinction of a taxon such as Miss Waldron's red colobus monkey would

Figure 30.1 Distribution of endangered and critically endangered taxa by countries. Adapted from data in Mittermeier et al. (1999).

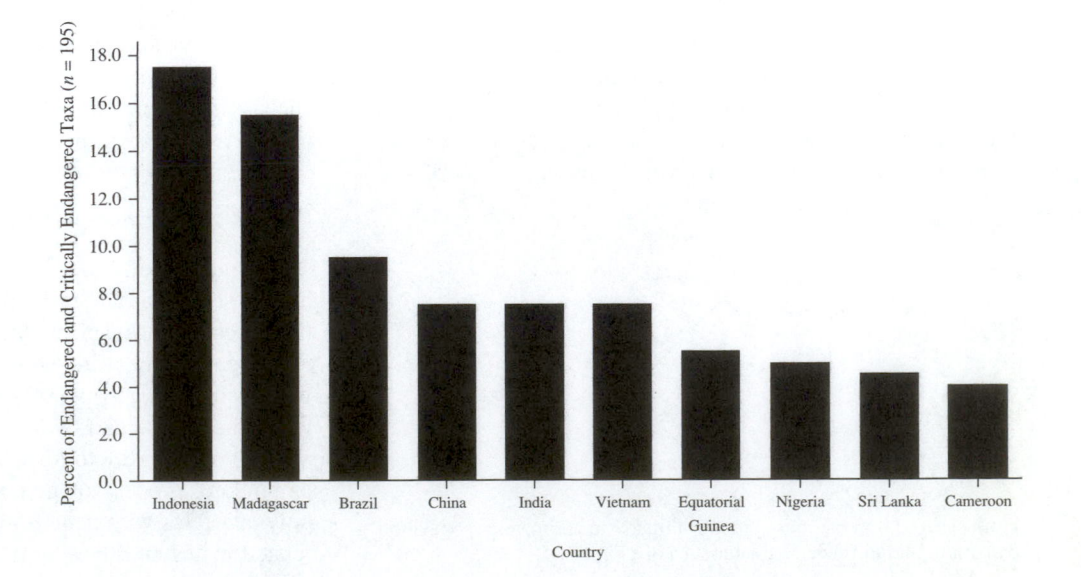

obviously result in the loss of biodiversity, it is also the case that less evolutionary history would be lost because the phylogenetic distance between this taxon and other species and subspecies of red colobus is shorter than that between the aye-aye and its closest living relatives in other infraorders.

Primate evolutionary history is also more heavily concentrated in biodiversity hotspots than would be expected from the numbers of endemic primate taxa, or the *taxon richness*, that these areas support (Sechrest et al. 2002). Prioritizing these hotspots in conservation efforts may therefore be justified based on both the number of endemic taxa and the amount of evolutionary history they support.

MAJOR THREATS TO PRIMATES

Independent of whether conservation priorities emphasize preserving the greatest number of taxa or the most evolutionary history, developing informed conservation plans also requires an understanding of why primates become endangered in the first place and why these threats impact different taxa in different ways. The major threats to primates come from human activities that have reduced the size and viability of local populations. Hunting and other forms of harvesting are serious problems for many primates, but it is the destruction, degradation, and fragmentation of their habitats that pose the most ubiquitous threats of all. Overhunting and harvesting directly deplete local prey populations, while the destruction and deterioration of habitats affect the distribution of primates and further exacerbate local population declines. The impact of these human activities varies widely in different regions of the world, and primates vary in their susceptibility to different kinds of threat. It is the interactions among the major threats, dynamics of small populations, and intrinsic characteristics of primates that affect their extinction risks (Fig. 30.2).

Population and Habitat Dynamics

The IUCN's criteria for assessing extinction risks emphasize the status of populations and their habitats because extinction risks for small or rapidly declining populations often reflect declines in the total area and quality of habitat. Expanding human populations have increased the scope and

intensity of human activities that negatively impact primates, particularly in the tropics where most of the world's endangered primates are found.

Hunting represents one of the most serious threats to wild primate populations in many regions of the world, but other forms of overharvesting due to poaching or live captures can also severely deplete local populations. For example, rhesus macaques (*Macaca mulatta*) were nearly exterminated in some parts of India as a result of overharvesting to meet international demands for experimental subjects in biomedical research. It was not until the 1970s, when India introduced strict prohibitions on these live exports, that the depleted populations began to show signs of recovering (Southwick and Siddiqi 1988). Live captures for zoos, circuses, and other public exhibitions were also routine until strong international restrictions on wildlife trade were established. The Convention on International Trade in Endangered Species of Wild Fauna and Flora, widely known as CITES, went into effect in 1975; and there are now some 160 participating countries. Yet, even with strict laws and enforcement efforts, mortality rates during the illegal capture and transport of live primates continue to be unconscionably high (Cowlishaw and Dunbar 2000). By one estimate from Sierra Leone, five chimpanzees (*Pan troglodytes*) may be killed for every infant captured, and only one of five captives is likely to survive to reach its destination alive (Teleki 1989).

Habitat loss restricts the total geographic distribution of primates and often leads to the fragmentation of whatever habitat remains. Fragmentation is an insidious problem because it isolates populations from one another and thereby disrupts natural dispersal processes and the corresponding gene flow that contributes to the maintenance of genetic variation within populations. Small, isolated populations may not be able to recover without human intervention, in the form of reintroductions from captive stock or translocations from other wild populations. Even then, habitat fragments may be too small to sustain viable populations of primates for very long. Forest fragments also suffer from additional ecological damage due to wind-induced tree falls and colonization from invasive species around their edges and from the elimination of other important members of their ecosystems, such as large predators and pollinators or seed dispersers (Marsh 2003, Ferraz et al. 2003, Strier et al. in press).

Causes and Consequences of Population Declines

The dynamics of small, isolated populations fuel what has been described as an "extinction vortex" because of the negative feedback mechanisms that inevitably ensue (Lacy 1993). Small populations are equivalent to founder populations, which are likely to possess only a fraction of the genetic variation represented by an entire taxon. Small populations are also more likely to lose variation through the stochastic, or random, process of genetic drift. Limited dispersal not only interferes with gene flow but also reduces access to unrelated mates and thereby increases the risks of deleterious

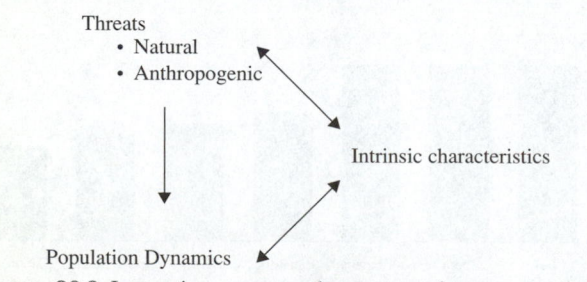

Figure 30.2 Interactions among threats to primates, population dynamics, and intrinsic characteristics of primates.

close inbreeding and compounds the loss of genetic variation.

There are many examples of dramatic population decline due to natural phenomena, such as hurricanes, droughts, and epidemics. Recently, for example, researchers studying black howler monkeys (*Alouatta pigra*) along the coast of Belize have documented the decimation of the population following Hurricane Iris (Pavelka et al. 2003). Before the hurricane struck the site in October 2001, the 53 black howler monkeys in this study area were distributed among eight stable social groups. After the hurricane, which completely defoliated the canopy and uprooted many trees, the howler monkey population was reduced by 49% and the established social groups had disintegrated.

Other striking population declines due to prolonged droughts, storms, or epidemics have been documented in a wide diversity of primates, ranging from mantled howler monkeys (*Alouatta palliata*) on Barro Colorado Island, Panama (Collias and Southwick 1952, Milton 1996), to red howler monkeys (*A. seniculus*) in Venezuela (Rudran and Fernandez-Duque 2003) to ring-tailed lemurs (*Lemur catta*) in Madagascar (Gould et al. 1999) and brown lemurs (*Eulemur fulvus fulvus*) on the Indian Ocean island of Mayotte (Tarnaud and Simmen 2002) to eastern chimpanzees at Gombe National Park, Tanzania (Goodall 1986), to both western and central chimpanzees as well as lowland gorillas (*Gorilla gorilla*) (Leroy et al. 2004). Although the ecological and demographic damage caused by natural catastrophes can be severe, it is also acute and tends to be geographically localized. Human impact, by contrast, tends to be much greater in both its temporal persistence and spatial scale.

The impact of hunting and harvesting on primate populations also differs from that of natural catastrophes in at least three important respects. First, in contrast to droughts and disease, which generally take their greatest tolls on the youngest and weakest members of a population, human hunters either fail to discriminate among their targets or preferentially aim for the largest, most robust adults. They also often target adult females carrying infants, which are then kept or sold as pets if they survive the mother's death (Mittermeier and Cheney 1987). Entire mountain gorilla (*Gorilla beringei*) families have been killed for the illegal capture of a single infant (Fossey 1983), and the persistent availability of infant orangutans (*Pongo pygmaeus*) and other primates through illegal channels is an inevitable sign that their mothers were killed to procure them (Galdikas 1995). The number of reproductive females in a population directly determines the potential and rate of population growth in a way that the number of adult males does not. Consequently, the loss of adult females from hunting and poaching reduces the potential for population recovery. In one computer simulation for northern muriquis (*Brachyteles hypoxanthus*), the harvesting of a single adult female per year was shown to significantly reduce the population's growth rate and, therefore, its ability to recover to a viable size (Rylands et al. 1998). A second factor that makes hunting such a serious threat is that hunters typically live on or near their hunting grounds and therefore exert chronic pressures on local populations of their prey. By one calculation, a single family of hunters in the Brazilian Amazon was estimated to have killed up to 500 large primates, including woolly monkeys (*Lagothrix*), spider monkeys (*Ateles*), and howler monkeys (*Alouatta*), within a 2-year period (Peres 1990). In the Peruvian Amazon, local populations of large primates such as these are especially vulnerable when hunters use shotguns instead of bows (Alvard et al. 1997). The effects of subsistence hunting are even more severe in small, isolated forest fragments than in larger, more contiguous Amazonian forests because there are no opportunities for immigrants to "rescue" an isolated population that has been depleted by local hunters (Peres 2001). A third distinction that makes hunting such a serious threat is that it often extends beyond a local family or community's subsistence needs to satisfy commercial markets (Naughton-Treves et al. 2003). Demands for bushmeat are increasing wherever local human populations are expanding, especially where international logging and development operations have opened new roads into remote areas that were previously inaccessible to hunters (Rose 2002). Bushmeat hunting may be responsible for the recent extinction or near extinction of Miss Waldron's red colobus monkey (Oates et al. 2000), and unsustainable levels of bushmeat hunting of other taxa have been documented in many parts of western and central Africa (Fa et al. 1995), Milner-Gulland and Akçakaya 2001, Rose 2002).

Although population size is a critical predictor of long-term viability, in most taxa only a fraction of individuals in a population reproduce. The number of reproductive members of a population, or its effective population size (N_e), is a more accurate predictor of the population's viability than its total size because N_e describes the number of individuals contributing genes to the next generation (Franklin 1980). Estimates of minimum viable population sizes have ranged from 2,000 to 5,500 individuals necessary to achieve a minimum viable N_e of 200–500 breeding individuals, and more recent analyses suggest that as many as 7,000 breeding-age adults may be necessary for a 99% probability of persistence for 40 generations (Reed et al. 2003). Critically endangered taxa have much smaller total numbers, and even smaller populations, than this. For example, fewer than 500 northern muriquis are estimated to survive in a few dozen forest fragments, and the single largest population includes only about 220 individuals (Strier et al. in press, updated from Strier et al. 2002). Similarly, the largest population of mountain gorillas includes some 350 individuals in all (Weber and Vedder 2001). Although such small total populations and correspondingly lower N_e sizes do not inevitably lead to extinction, the probability of their persistence declines steeply over time.

Both northern muriquis and mountain gorillas rank among the world's top 25 most endangered primate taxa (Table 30.2) because neither their total numbers nor their largest populations meet the IUCN's criteria for long-term viability (Konstant et al. 2002). The largest populations of

Table 30.2 Top 25 Most Endangered Primates, 2004–2006

COMMON NAME	LATIN NAME	COUNTRY
Greater bamboo lemur	*Prolemur simus*	Madagascar
White-collared lemur	*Eulemur albocollaris*	Madagascar
Silky sifaka	*Propithecus candidus*	Madagascar
Perriers sifaka	*Propithecus perrieri*	Madagascar
Mt. Rungwe galago	Undescribed	Southwestern Tanzania
Bioko red colobus	*Procolobus pennantii pennantii*	Equatorial Guinea
White-napped mangabey	*Cercocebus atys lunulatus*	Côte d'Ivoire, Ghana
Tana River red colobus	*Procolobus rufomitratus*	Kenya
Sanje mangebey	*Cercocebus sanjei*	Tanzania
Eastern gorillas (Bwindi, mountain, and Grauer's)	*Gorilla beringei*	Rwanda, Uganda, Democratic Republic of Congo
Cross River gorilla	*Gorilla gorilla diehli*	Cameroon, Nigeria
Horton Plains slender loris	*Loris tardigradus nycticeboides*	Sri Lanka
Pig-tailed langur	*Simias concolor*	Indonesia
Miller's grizzled surili	*Presbytis hosei canicrus*	Indonesia
Delacour's langur	*Trachypithecus delacouri*	Vietnam
Golden-headed langur	*Trachypithecus poliocephalus poliocephalus*	Vietnam
Western purple-faced langur	*Semnopithecus vetulus nestor*	Sri Lanka
Gray-shanked douc	*Pygathrix nemaeus cinerea*	Vietnam
Tonkin snub-nosed monkey	*Rhinopithecus avunculus*	Vietnam
Hainan black-crested gibbon	*Nomascus* sp., cf. *Nasutus hainanus*	China
Sumatran orangutan	*Pongo abelli*	Indonesia
Black-faced lion tamarin	*Leontopithecus caissara*	Brazil
Buff-headed tufted capuchin	*Cebus xanthosternos*	Brazil
Brown spider monkey	*Ateles hybridus brunneus*	Colombia
Northern muriqui	*Brachyteles hypoxanthus*	Brazil

Source: Mittermeier et al. (2004).

each of these taxa have been the subjects of intensive long-term field studies. They have also been the beneficiaries of intensive conservation efforts, which have succeeded in reducing hunting pressures and protecting their remaining habitats. Increases in their populations have been documented during the decades they have been monitored, but it is still not clear whether these protected populations are sufficiently large to rebound from population declines caused by stochastic demographic processes or unpredictable catastrophic environmental events.

Habitat loss and fragmentation compound the problems of small populations because the areas that remain may not be large enough to support viable populations. In the Atlantic Forest of Brazil, for example, the area required to sustain viable populations of endangered species, such as northern muriquis, has been estimated to be at least 20,000 ha in size (Chiarello 2003). This is more than twenty times the size of the forest at the Estação Biológica de Caratinga in Minas Gerais, where the largest known population of northern muriquis resides. There are other tracts of Atlantic Forest within the northern muriquis' geographic distribution that do meet these minimum size estimates, but illegal hunting is difficult to eliminate in these federally or state-protected parks and the population densities of northern muriquis and

other sympatric primates found in them are correspondingly low (Pinto et al. 1993, Strier and Fonseca 1996/1997).

Causes and Consequences of Habitat Loss and Fragmentation

Human demands for timber and land for agricultural and development purposes are the major causes of habitat loss and fragmentation. By 1985, only about a third of eastern Madagascar's rain forests remained, and recent assessments of southern and southwestern forests there show similar, if not faster, rates of loss today than in the recent past (Green and Sussman 1990, Sussman et al. 2003). During the twentieth century alone, more than 90% of the Brazilian Atlantic Forest was destroyed (Câmara 2003). Although the Amazon Basin still has over 50% of its rain forest, it suffers from the highest absolute rate of deforestation in the world (Laurence et al. 2002). The vast Indonesian forests of Borneo and Sumatra, like those in western and central Africa, are now seriously threatened by logging concessions. Even protected areas, such as Borneo's vast Gunung Palung National Park, are suffering irreversible losses and fragmentation due to logging for timber and clearing for oil palm plantations, which devastated the 10 km buffer zone surrounding the park and then began encroaching on the so-called protected forest within the park itself. Since 1999, nearly 10% of the forest within the park's boundaries has been destroyed each year (Curran et al. 2004).

Selective logging involves the harvesting of valuable species of trees instead of clear-cutting entire forest tracts. While the disruptive effects of selective logging may not be as immediately apparent as clear-cutting, they nonetheless cause long-term ecological damage. Selectively logged forests in Indonesia, for example, are more susceptible to damage from fires caused by droughts associated with El Niño Southern Oscillation events (Siegert et al. 2001). There and elsewhere many of the trees being logged are often the same mature, fruit-producing species that serve as important food sources for frugivorous primates, including orangutans. Because many primates, such as chimpanzees, are also responsible for dispersing a considerable number of seeds from the species of fruits they ingest, reductions in the densities of frugivorous primates can have deleterious effects on forest recovery and regeneration (Chapman and Chapman 1996, Chapman and Onderdonk 1998; see Chapter 31).

More than 20 years after selective logging ceased in Kibale National Park, Uganda, the densities of frugivorous primates are still significantly lower in the previously logged areas than in areas that had not been disturbed (Chapman et al. 2000). Male gray-cheeked mangabeys (*Lophocebus albigena*) were also found to be significantly heavier, and presumably healthier, in unlogged forests than in logged areas (Olupot 2000). At the same time, however, the densities of folivorous taxa, such as black and white colobus monkeys (*Colobus guereza*), in these selectively logged forests do not appear to have declined and may even be elevated relative to undisturbed areas. The apparent ability of folivorous primates to thrive in disturbed forests here and elsewhere may be attributed to the higher protein-to-fiber ratios of foliage found in the vegetation that colonizes open gaps within the forests and forest edges (Ganzhorn 1995, Ganzhorn et al. 1999, Oates 1996).

The noise and disturbances caused by logging operations can also cause primates to shift their established ranges into neighboring areas, where encounters with hostile conspecifics can have lethal consequences. The decline of one chimpanzee population in Gabon has been directly attributed to the lethal territorial clashes that occurred when chimpanzees fleeing from logging disturbances in their own ranges were attacked by the residents of the surrounding communities (White and Tutin 2001). By the time the loggers had left the area, the density of chimpanzees, as estimated from nest counts, had declined.

Like the folivorous taxa in the logged forests in Kibale National Park, local populations of western lowland gorillas did not fare as badly as those of sympatric chimpanzees during the logging operations in Gabon. One reason for this difference may be that gorillas are not territorial and are therefore more tolerant toward conspecific refugees that move into their ranges. The ability of gorillas to rely heavily on foliage may also contribute to their tolerance of conspecifics at higher densities (Yamagiwa 1999).

Like human harvesting of primates, anthropogenic impacts on primate habitats tend to have longer-term and farther-reaching consequences than natural ecological processes, such as the annual or super-annual fluctuations in rainfall that affect the availability of food and water. Nonetheless, the short-term impact of natural disasters that affect primate habitats cannot be ignored. For example, extended droughts can reduce standing sources of water, forcing primates, such as vervet monkeys or grivets (*Chlorocebus aethiops*) at Ambosli National Park, Kenya, during the 1970s, to make risky incursions into one another's ranges and exacting high mortality on low-ranking and less competitive individuals (Wrangham 1981). Droughts also reduce food availability, which negatively impacts survival and fertility. One population of ring-tailed lemurs that suffered high mortality due to a 2-year drought did not show signs of recovering until 4 years after the drought had ended (Gould et al. 1999), and smaller, more isolated populations may be unable to recover so quickly from these natural ecological fluctuations (Gould et al. 2003).

Many tropical plant species reproduce on super-annual cycles, or masting events, such as the 4-year fruiting cycles of the Dipterocarpaceae trees in Indonesia, which have, at least until recently, coincided with El Niño Southern Oscillation events (Curran et al. 2004). Orangutan energy balance and fertility are strongly influenced by the timing of these mast fruiting events, with both energy intake and fertility declining in the intervening years when the lipid-rich fruits that many dipterocarps produce are scarce (Knott 1998).

Historical cycles of mast fruiting events or of the unpredictable, harsh climatic conditions on islands such as Madagascar are ones to which primates inhabiting these areas have had generations to adapt (Wright 1999). Human impacts on primate habitats, however, represent new selection pressures; and the complete destruction of habitats is the most extreme and irreversible type. Forests can never be restored to their original conditions after they have been repeatedly cleared for planting crops or grazing livestock, flooded during the construction of hydroelectric dams, or asphalted for urban areas. The viability of primates in forest fragments is directly related to the size of the fragments and their distance from contiguous habitat, and small fragments can lose significant components of their biodiversity within a few decades of being isolated (Marsh 2003). Acute disturbances, such as those caused by selective logging or other forest product extractions, can also have long-term deleterious impacts on habitat quality because these kinds of activity are so targeted in their scopes.

Variation in Primate Vulnerabilities to Different Threats

Pressures from human hunting and habitat destruction are not evenly distributed throughout the world, nor do they impact all primates to similar degrees. Chimpanzees, for example, are protected in some areas by local taboos against hunting animals so similar to humans, whereas sympatric gorillas and other primates are considered edible meat (Sabater-Pi and Groves 1972).

Habitat disturbances also affect different primates in different ways, depending on their degree of ecological specialization. Habitat alterations almost always have negative ecological impacts on specialized niches, with devastating consequences on the primates that occupy them.

Densities and Ecology

Sympatric taxa live at very different population densities, even in relatively undisturbed habitats. These differences reflect a combination of extrinsic factors, such as levels of mortality from predators and infectious disease and competition for food and other essential resources from other species, including nonprimates, and intrinsic factors, such as their respective energetic and nutritional requirements. Food and other critical resources are thought to determine the carrying capacities of different habitats for different taxa, but we still know very little about the carrying capacities at which most primates can live (Cowlishaw and Dunbar 2000).

Many primates have historically restricted geographic distributions and have evolved specializations to survive in narrow, highly specialized ecological niches. Others occur much more widely and can survive in a diversity of habitat types by adjusting their diets, ranging, and grouping patterns in response to the distribution and availability of their food resources. Geographic barriers that isolate populations are responsible for historical speciation events, such as the divergence between bonobos (*Pan paniscus*) and chimpanzess that occurred approximately 2.2 million years ago, when the Congo River became a barrier defining the boundaries of the bonobo's geographic distribution. Chimpanzees occupy a much greater diversity of habitat types throughout sub-Saharan Africa, from rain forest to semideciduous forest to woodland to arid savanna, than either gorillas, which occur sympatrically with chimpanzees (but not bonobos) in some parts of their ranges, or Indonesia's orangutans, which are restricted to the rain forests of Borneo and Sumatra.

Baboons (*Papio* spp.) and macaques are sympatric with these great apes in many African and Asian forests, respectively. However, unlike these apes, baboons and macaques have much more generalized adaptations, and the wider diversity of habitats they occupy is consistent with their more flexible ecological niches. Forest-dwelling baboons in Africa and macaques in Asia are much more frugivorous than conspecific or congeneric populations found in more open or seasonal habitats (Rowell 1966, Richard et al. 1989), just as central lowland gorillas are more arboreal and frugivorous than mountain gorillas, whose present high-altitude habitats necessitate the inclusion of more foliage in their diets (Remis 1997, Doran and McNeilage 1998).

Primates that can subsist on regenerating vegetation, which tends to occur in disturbed forests and along forest edges, may also be more adept at raiding human crops. Rhesus macaques have been described as just such a "weed species" because they appear to thrive in the altered habitats associated with human agricultural expansion (Richard et al. 1989). Brown capuchin monkeys (*Cebus apella*) may be the New World ecological equivalent of a weed species because of their extractive foraging techniques and skills at crop raiding (Fragaszy et al. 2004).

The greater sensitivity of ecological specialists, such as frugivorous spider monkeys or frugivorous–insectivorous blue monkeys (*Cercopithecus mitis*) and redtail monkeys (*C. ascanius*), to habitat disturbances cannot be decoupled from the roles these primates typically play within their ecological communities. Because specialists tend to be so sensitive to disturbances, they are more likely to suffer local extinctions when their populations are isolated in disturbed fragments of habitat (Brashares 2003). The loss of such primates can also negatively impact the plant species they eat, as appears to be the case with spider monkeys and the mimosoids (*Inga ingoides*) that rely on these monkeys for dispersing their seeds (Pacheco and Simonetti 2000).

At a more global scale, recent analyses have indicated that the diverse Malagasy lemurs play a much larger and more unique set of roles in their respective ecosystems than primates in South America or continental Africa, where other groups of animals such as birds, bats, and small or medium-sized mammals contribute to their ecological communities. The extinction of endangered lemurs from Madagascar would leave their ecosystems more severely compromised than the extinction of endangered primates elsewhere (Jernvall and Wright 1998).

BEHAVIOR AND THE COMPOSITION OF GROUPS AND POPULATIONS

The age–sex structure and the degree of relatedness among members of social groups and populations also significantly affect the viability of populations and the ability of small populations to recover to viable sizes. Most primates spend most, if not all, of their lives as members of social groups. The composition of these groups is affected by sex-biased or bisexual dispersal patterns, which also affect the demographic composition of populations. Although models in primate behavioral ecology have tended to focus on the size and composition of social groups, genetic and conservation priorities require focusing on the size and composition of populations as well (Strier 1997, 2003a,b).

Populations with unbalanced age structures grow faster or slower than those with stable age distributions depending on the relative proportion of adults and immatures. Similarly, those that contain more males than females will have slower rates of growth than those with more females. The near quadrupling of the northern muriqui population at the Estação Biológica de Caratinga has been attributed to the 2- to 1-female bias in infant births that occurred in one large group over more than a decade (Strier 1993/1994, 2000a). An increase in the number of male births and survivorship relative to females in recent years has led to projections for slower, if not negative, population expansion in the future (Strier et al. in press).

The composition of large, relatively stable populations can fluctuate due to stochastic demographic processes, such as sex-biased infant sex ratios or mortality rates within groups. The impacts of these fluctuations on population growth rates, genetic structure, and even behavior may not become evident until members of one or more birth cohorts reach reproductive age. Slow maturation rates can result in lags of more than a decade before the demographic consequences of these events are observed. Dispersal can buffer the effects of fluctuations on group demography by the redistribution of individuals across populations, which is another reason that small, isolated populations with limited or no dispersal opportunities are so vulnerable to stochastic demographic events.

Sex-biased or bisexual dispersal patterns also directly affect the degrees of relatedness among and between groups, with potentially important consequences on behavior. For example, the unusually high levels of male–male competition and mate guarding observed in the Ngogo chimpanzees at Kibale National Park, Uganda, have been associated with the disproportionately high number of males in this community (Watts 1998, Mitani and Watts 1999). No one knows why this chimpanzee community is so large or has so many males, but the behavior patterns of its members cannot be compared to those of other chimpanzees without controlling for the effects of its unusual demography, which must be at least partially due to historically male-biased sex ratios or survivorship in their male-philopatric society (Strier 2000b).

Levels of male–male competition vary across different taxa as well as among populations of the same taxa, and in both cases they affect the degree to which fertilizations can be monopolized by individual males or are more equitably distributed among different males (Strier 2003b). While dispersal permits males of many species, such as ring-tailed lemurs (Sussman 1992) and vervet monkeys (Cheney and Seyfarth 1983), to adjust their reproductive opportunities by transferring between groups, male philopatry reduces the ability of males to escape from unfavorable demographic conditions in their natal groups (Strier 2000b). Although the transfers of two adult male bonobos from their natal community into an adjacent community with fewer males may be an indication that even philopatric males respond to population-wide demographic conditions when opportunities to disperse exist (Hohmann 2001), it is also the case that philopatric tendencies affect the genetic composition of local groups.

Reproductive Rates

Compared to most other mammals, primates have slow reproductive rates relative to body size. Taxa with rapid reproductive rates can sustain greater harvesting pressures than those with slower reproductive rates, which also tend to be larger in body size within each of the major clades and therefore preferentially hunted for their meat (Oates 1996). Sustainability implies that population losses due to hunting or harvesting can be offset by replacements from births or immigrations, but when net losses exceed gains, populations decline and extinction risks rise. Comparing the prevalence of different taxa in bushmeat markets provides a way of estimating their minimum harvest rates, which can then be weighed against their respective reproductive rates to evaluate whether hunting pressures are occurring at sustainable or unsustainable rates (Fa et al. 1995). This method works well only for taxa that are still relatively plentiful, however, because hunting success also declines as populations are depleted (Cowlishaw and Dunbar 2000).

Reproductive rates determine the potential for small populations to recover once the pressures that contributed to their declines are relaxed. Consider the difference in the reproductive rates of most marmosets (*Callithrix* spp.), which begin reproducing at about 2 years of age and give birth to twins at roughly 6-month intervals, versus orangutans, which begin reproducing at 8–10 years of age and give birth to single infants at 9-year intervals. The two decades it takes a female orangutan to produce two surviving offspring is longer than the life span of a female marmoset, which can produce two infants in a single pregnancy.

Comparing marmoset and orangutan reproductive rates is an extreme and distorted contrast because it ignores the strong relationship between reproductive rates, on the one hand, and body size and phylogeny, and to a lesser degree ecology, on the other hand (Ross and Jones 1999). Among closely related taxa, intrinsic reproductive rates tend to decline and generation lengths increase as body size increases, and relative to body size, reproductive rates also tend to decline with increasing brain size. Forest-dwelling primates also tend to reproduce more slowly than closely related taxa that occur in more seasonal habitats or are subject to less predictable ecological conditions (Ross 1998, Ross and Jones 1999, Wright 1999).

Slow reproductive rates and long generations represent life history trade-offs for increased infant survivorship (Charnov and Berrigan 1993). The relatively fast maturation rates and 1- to 2-year birth intervals of most lemurs, for example, have been attributed to selection pressures favoring fast life histories under the unpredictable and harsh seasonal environmental conditions that characterize much of Madagascar (Wright 1999, Gould et al. 2003). Taxa that have evolved slower life histories under more stable and predictable environmental conditions in their evolutionary past may be initially more resistant to environmental perturbations because they are more likely to forfeit reproduction than to die but ultimately more vulnerable to being drawn into the extinction vortex if their populations decline below minimal sizes because of their slower intrinsic reproductive rates.

APPROACHES TO CONSERVATION

Conservationists have long realized that their efforts must proceed at multiple levels depending on whether the major threats to primates occur at local, regional, national, or international scales. There is also increasing sensitivity to the necessity of working within the political, economic, and social systems in place at each level. Solutions vary depending on whether the primates are at risk from hunters, habitat destruction, or combinations of the two; but even then, protective policies are almost always much easier to establish than they are to enforce.

Primate conservation is a global concern, and most conservation efforts involve international collaborations between nongovernmental organizations (NGOs), policy makers, law enforcers, researchers, and local people whose lives intersect with those of primates. Most of the conservation NGOs in the United States have international offices or cooperative arrangements with NGOs based in the countries that primates inhabit. The various NGOs that are involved in primate conservation efforts also fill different niches based on their respective commitments to lobbying governmental agencies in support of conservation legislation, securing protected lands, or supporting research and educational activities. However, whatever their particular emphases and target groups, all conservation tactics for primates include protecting remaining populations and their habitats and, when appropriate, active management efforts on their behalf.

The hunting and poaching of primates represent quite different kinds of problem because the markets for meat versus live primates or primate products are so varied. International conservation initiatives, such as those regulated by CITES, and local enforcement and education can help to curtail poaching and illegal trade; but prohibitions on the hunting of primates for their meat require that alternative sources of protein are available to the people who need them (Rose 2002).

Even with controls on hunting, primates generally suffer when humans live at high densities nearby. The rapid emergence and spread of infectious diseases, such as ebola and human immunodeficiency virus disease, have been attributed to the encroachment of humans into primate habitats (Daszak et al. 2000, Walsh et al. 2003). As of 1995, nearly 20% of all humans lived within the world's biodiversity hotspots, and human population growth rates in these areas are higher than elsewhere (Cincotta et al. 2000).

The setting aside of protected areas as sanctuaries for remaining populations of endangered primates has been a major achievement in many parts of the world, and there is evidence that these protected areas can fulfill their functions of protecting primates and their habitats when there are sufficient financial resources for local enforcement and economic alternatives to hunting and habitat destruction are available for the local people (Bruner et al. 2001). Yet, there are still many officially "protected" areas throughout the tropics that are little more than "parks on paper." Like the lowland forests of Indonesia, which may be gone within decades despite their protected status (Jepson et al. 2001), human encroachment on national parks represents a serious and urgent problem in many other parts of the world. For example, the 1,000 km² buffer zone surrounding a 2,300 km² national park in Madagascar has already been penetrated by subsistence rice farmers, who are projected to reach the park's boundaries within a few years if their rate of expansion persists (Kremen et al. 2000).

Other protected areas in Madagascar, such as the Ranomafama National Park, have more promising futures. When Ranomafama National Park was inaugurated in 1991, it was only the fourth national park in the country (Wright 1992). International funds and cooperative programs for sustaining ecotourism and research have contributed to the success of Ranomafama National Park, which is now a World Heritage Site recognized by the United Nations.

Ecotourism can provide economic incentives to increase the participation of local people in conservation. The long-term economic value of ecotourism to visit the mountain gorillas in Rwanda is believed to have contributed to the gorillas' survival during a recent civil war that forced foreign researchers to evacuate the area (Weber and Vedder 2001). Profit sharing with local communities has also been critical to the success of ecotourism in Nepal (Bookbinder et al. 1998). Similarly, increasing the local market value of forest products would provide incentives for local communities in Honduras to conserve their surrounding forests because of the higher returns they would gain by conserving instead of exploiting these products (Godoy et al. 2000).

Research can also play a key role in conservation efforts, both through the critical information that researchers obtain about the primates and ecological communities they study and through the local and international attention that their scientific discoveries attract (Strier in press). Long-term studies provide invaluable insights into ecological, demographic, and reproductive processes; and it is often these data, derived from the small number of taxa that have been studied over multiple decades, that provide the parameters for modeling extinction risks for other, less well-studied taxa. Computer programs "known as population viability analyses", or PVAs, are now widely used to simulate extinction risks based on known or estimated demographic and reproductive variables that affect population dynamics (Brook et al. 2000). However, surveys and more rapid assessments are still essential for evaluating the status of the majority of wild populations that have not been the subjects of long-term studies and about which little else is known.

CHALLENGES AND CHOICES

In addition to protecting primate populations and their habitats from irreversible losses, there are a number of more active, hands-on initiatives that conservationists can adopt. Examples include captive breeding programs and various ways of managing wild populations to increase and maintain their viability and of restoring altered habitats through the planting or maintenance of corridors. Debates about how much time and how many resources should be dedicated to these kinds of activity raise as many questions about the assumptions underlying conservation priorities as they do about which conservation plans should be implemented and how well they are likely to fare.

Captive Breeding and Reintroduction Programs

The stated priority of the Primate Specialist Group is the in situ preservation of species and communities (Cowlishaw and Dunbar 2000). In situ preservation requires protected habitats and sufficient population sizes, which may already be threatened or no longer exist. In such cases, the maintenance of viable captive breeding populations to preserve the gene pools of endangered taxa can serve as insurance against their extinction in the wild. Nonetheless, genetic and phenotypic changes occur in the necessarily small populations that can be maintained in captivity, and the most successful captive breeding programs are those that can reinforce in situ conservation efforts to maintain and restore wild populations (Snyder et al. 1996, Earnhardt et al. 2001).

The most successful example of a captive breeding program for an endangered primate involves the golden lion tamarin (*Leontopithecus rosalia*), many of which have been reintroduced into protected areas of their native Brazilian Atlantic Forest. By 1975, fewer than 200 golden lion tamarins were estimated to survive in just a few small patches of Atlantic Forest in the state of Rio de Janeiro; but by 2000, their numbers in the wild were estimated at about 1,000, with another 500 or so housed in zoos (Kleiman and Rylands 2002). Approximately 40% of all golden lion tamarins living in the wild today either were born in captivity and then reintroduced or are the descendents of captive-born tamarins that subsequently reproduced after being released in the wild. The first captive-born tamarins were reintroduced to the Poço das Antas Biological Reserve, a 5,200 ha forest established to protect the largest remaining wild population of this species. Reintroductions began there in 1984, and management efforts since 1994 have expanded to include the translocation of some 120 individuals from 16 groups from small, isolated forests to larger, protected areas (Kierulff et al. 2002).

The translocated individuals had higher survival rates and adapted more rapidly than the reintroduced monkeys, although the success rates for reintroduced tamarins increased as prerelease training and reintroduction procedures were refined (Kleiman 1996). Translocations were also much more economical, costing only about half as much as the reintroductions from the time of release (Kierulff et al. 2002).

Space limitations and high housing and maintenance costs impose limits on the number of animals that captive facilities can maintain (Snyder et al. 1996). Small captive populations also require careful management because they are susceptible to the same genetic risks that threaten small populations in the wild. The Species Survival Commission (SSC), Captive Breeding Specialist Group (CBSG), and societies such as the American Association of Zoological Parks and Aquariums (AAZPA) maintain records of all registered individuals for management purposes. Transferring individuals from one facility to another can reduce the genetic risks of inbreeding in the captive "stock," and often separate housing facilities or birth-control measures are necessary to keep the captive population at a manageable size.

The success of the Golden Lion Tamarin Project can be attributed to a number of factors, including the self-sustaining population that could be maintained in captivity due to the small size and high reproductive rates of these monkeys and the natural tendencies of both sexes to disperse from their natal groups. Translocating entire family groups together and pairing captive-born individuals with wild-born mates proved to be highly effective procedures. There were also still suitable available habitats into which golden lion tamarins could be released. Protection of these habitats and accompanying research and education efforts have been essential to the ongoing conservation efforts on behalf of this species.

Managing Wild Populations and Habitats

Although less expensive than reintroductions, translocations of primates from unprotected to protected habitats, or among small and isolated populations to increase gene flow, are not entirely risk-free. As with reintroductions, care must be taken with translocations to prevent the spread of infectious diseases and pathogens and to minimize social and ecological disruptions in the communities into which the individuals or groups are being moved (Yeager and Silver 1999).

An increasingly attractive alternative to translocating individuals from isolated populations is to establish habitat corridors between fragments so that the animals can move safely through them on their own. The creation of corridors between isolated populations assumes that natural dispersal processes will increase gene flow, thereby decreasing the deleterious effects of small population size (Lens et al. 2002). Genetic studies indicate that one small population of Scandinavian gray wolves (*Canis lupus*) was "rescued" by the immigration of a single, reproductively successful individual (Vila et al. 2002), suggesting that it may not require many such dispersal events for small populations to recover.

Corridors that connect isolated populations also pose potential risks, such as the spread of infectious diseases (Stoner 1996) or the synchronization of population dynamics that can lead to simultaneous extinctions (Earn et al. 2000). Nonetheless, establishing corridors to increase connectivity among isolated populations can at least partially offset the effects of habitat loss. Many African countries are predicted to lose up to 30% of their primate fauna due to the "extinction debts" caused by the long time lags between habitat losses and population extinctions, and restoring connectivity among fragments habitats is recommended before the populations show signs of collapsing (Cowlishaw 1999).

Prioritizing Charismatic Taxa or Habitats

Most conservation education programs focus on highly charismatic taxa, which then serve as flagships for attracting attention and resources to conservation efforts that benefit other fauna and flora in the process. Primates, along with other mega-fauna, such as the World Wildlife Fund's famous panda emblem, hold a special appeal. Among primates, the great apes have the unique distinction of being most closely related to ourselves, and this has contributed to the establishment of special initiatives, such as the Great Ape Conservation Act of 2000.

Nonetheless, the challenges of establishing conservation priorities extend beyond particular charismatic taxa or their endangered status and evolutionary uniqueness because the largest populations of the most endangered primates do not always persist in the most pristine habitats. For example, some of the highest densities of endangered primates endemic to Brazil's Atlantic Forest are found in small, degraded forest fragments, instead of in the few tracts of large protected forests that remain. Should the protection of these large populations in disturbed habitats take priority over that of smaller populations in undisturbed habitats? Alternatively, should the most pristine ecosystems be given priority, even if they support smaller populations of endangered taxa living at lower densities?

Comparative analyses of the biogeography of endangered species suggest that although populations are often most fragmented at the extremes of their geographic ranges, these may also be the populations with the best prospects for survival because they are more removed from the disturbances responsible for population losses in the first place (Channell and Lomolino 2000). Higher rates of evolution may also occur at the edges of a species' distribution because of the more extreme ecological conditions to which they are subjected (Thomas et al. 2001), and these changes are expected to become more pronounced from the alterations of habitats due to global warming (Pounds and Puschendorf 2004, Thomas et al. 2004). Indeed, some conservationists have already posed the uncomfortable question of whether the parks and reserves established to protect endangered primates today will continue to provide suitable habitat as global climatic conditions change (Cowlishaw and Dunbar 2000).

THE FUTURE OF PRIMATE BIODIVERSITY

Between the inevitable effects of global warming on the world's endangered ecosystems and the ongoing expansion of human populations in and around the world's biodiversity hotspots, it is difficult to foresee how the future of primates —and other animals—that are threatened with extinction can be protected. Global losses of biodiversity and ecosystem changes are predicted to occur by 2050, and major primate extinctions may occur even sooner than this because rates of deforestation in countries such as Indonesia and Madagascar are so high (Jenkins 2003).

There is no question that human pressures are accelerating the extinction risks for many primate taxa. Whether through direct actions, such as unsustainable hunting and habitat destruction, or indirect activities, such as the far-reaching effects of atmospheric pollution on global climate, the impact of humans on other primates today is much greater than it has been in the past (Fuentes and Wolfe 2002). Yet, despite the depressing forecast for primates, increased awareness about the status of the world's endangered primates has fueled intensified international conservation efforts. It is too soon to tell whether these efforts will ultimately succeed in securing the futures of all endangered taxa, but there is no doubt that they are helping gain essential time in what for many primates is now an urgent race against extinction.

REFERENCES

Agapow, P. M., Bininda-Emonds, O. R. P., Crandall, K. A., Gittleman, J. L., Mace, G. M., Marshall, J. C., and Purvis, A. (2004). The impact of species concept on biodiversity studies. *Q. Rev. Biol.* 79:161–179.

Alvard, M. S., Robinson, J. G., Redford, K. H., and Kaplan, H. (1997). The sustainability of subsistence hunting in the neotropics. *Conserv. Biol.* 11:977–982.

Bookbinder, M. P., Dinerstein, E., Rijal, A., Cauley, H., and Rajouria, A. (1998). Ecotourism's support of biodiversity conservation. *Conserv. Biol.* 12:1399–1404.

Brandon-Jones, D., Eudey, A. A., Geismann, T., Groves, C. P., Melnick, D. J., Morules, J. C., Shekelle, M., and Steward, C. B. (2004). Asian primate classification. *Int. J. Primatol.* 25:97–164.

Brashares, J. S. (2003). Ecological, behavioral, and life-history correlates of mammalian extinctinos in West Africa. *Conserv. Biol.* 17:733–743.

Brook, B. W., O'Grady, J. J., Chapman, A. P., Burgman, M. A., Akçakayas, H. R., and Frankham, R. (2000). Predictive accuracy of population viability analysis in conservation biology. *Nature* 404:385–387.

Bruner, A. G., Gullison, R. E., Rice, R. E., and Fonseca, G. A. B. (2001). Effectiveness of parks in protecting tropical biodiversity. *Science* 291:125–128.

Câmara, I. G. (2003). Brief history of conservation in the Atlantic Forest. In: Galindo-Leal, C., and Câmara, I. G. (eds.), *The Atlantic Forest of South America.* Island Press, Washington DC. pp. 31–42.

Channell, R., and Lomolino, M. V. (2000). Dynamic biogeography and conservation of endangered species. *Nature* 403:84–86.

Chapman, C. A., Balcomb, S. R., Gillespie, T. R., Skorupa, J. P., and Struhsaker, T. T. (2000). Long-term effects of logging on African primate communities: a 28-year comparison from Kibale National Park, Uganda. *Conserv. Biol.* 14:207–217.

Chapman, C. A., and Chapman, L. J. (1996). Frugivory and the fate of dispersed and non-dispersed seeds of six African tree species. *J. Trop. Ecol.* 12:491–504.

Chapman, C. A., and Onderdonk, D. A. (1998). Forests without primates: primate/plant codependency. *Am. J. Primatol.* 45:127–141.

Charnov, E. L., and Berrigan, D. (1993). Why do female primates have such long lifespans and so few babies? Or life in the slow lane. *Evol. Anthropol.* 1:191–194.

Cheney, D. L., and Seyfarth, R. M. (1983). Non-random dispersal in free-ranging vervet monkeys: social and genetic consequences. *Am. Nat.* 122:392–412.

Chiarello, A. G. (2003). Primates of the Brazilian Atlantic Forest: the influence of forest fragmentation on survival. In: Marsh, L. K. (ed.), *Primates in Fragments*. Kluwer Academic/Plenum, New York. pp. 99–121.

Cincotta, R. P., Wisnewski, J., and Engelman, R. (2000). Human population in the biodiversity hotspots. *Nature* 404:990–992.

Collias, N., and Southwick, C. (1952). A field study of population density and social organization in howling monkeys. *Proc. Am. Philos. Soc.* 96:143–156.

Cowlishaw, G. (1999). Predicting the decline of African primate diversity. *Conserv. Biol.* 13:1183–1193.

Cowlishaw, G., and Dunbar, R. I. M. (2000). *Primate Conservation Biology*. University of Chicago Press, Chicago.

Curran, L. M., Trig, S. N., McDonald, A. K., Astiani, D., Hardiono, Y. M., Siregar, P., Caniago, I., and Kasischke, E. (2004). Lowland forest loss in protected areas of Indonesian Borneo. *Science* 303:1000–1003.

Daszak, P., Cunningham, A. A., and Hyatt, A. D. (2000). Emerging infectious diseases of wildlife—threats to biodiversity and human health. *Science* 287:443–449.

Doran, D. M., and McNeilage, A. (1998). Gorilla ecology and behavior. *Evol. Anthropol.* 6:120–131.

Earn, D. J. D., Levin, S. A., and Rohani, P. (2000). Coherence and conservation. *Science* 290:1360–1364.

Earnhardt, J. M., Thomson, S. D., and Marhevsky, E. A. (2001). Interactions of target population size, population parameters, and program management on viability of captive populations. *Zoo Biol.* 20:169–183.

Fa, J. E., Juste, J., Perez del Val, J., and Castroviejo, J. (1995). Impact of market hunting on mammal species in Equatorial Guinea. *Conserv. Biol.* 9:1107–1115.

Ferraz, G., Russell, G. J., Stouffer, P. C., Bierregaard, R. O. J., Pimm, S. L., and Lovejoy, T. E. (2003). Rates of species loss from Amazonian forest fragments. *Proc. Natl. Acad. Sci. USA* 100:14069–14073.

Fossey, D. (1983). *Gorillas in the Mist*. Houghton Mifflin, Boston.

Fragaszy, D. M., Visalberghi, E., and Fedigan, L. M. (2004). *The Complete Capuchin Monkey*. Cambridge University Press, Cambridge.

Franklin, I. R. (1980). Evolutionary change in small populations. In: Soulé, M. E., and Wilcox, B. A. (eds.), *Conservation Biology: An Evolutionary–Ecological Perspective*. Sinauer Associates, Sunderland, MA. pp. 135–150.

Fuentes, A., and Wolfe, L. D. (eds.) (2002). *Primates Face to Face*. Cambridge University Press, Cambridge.

Galdikas, B. M. F. (1995). *Reflections of Eden: My Year with the Orangutans of Borneo*. Little, Brown, Boston.

Ganzhorn, J. U. (1995). Low-level forest disturbance effects on primary production, leaf chemistry, and lemur populations. *Ecology* 76:2084–2096.

Ganzhorn, J. U., Wright, P. C., and Ratsimbazafy, J. (1999). Primate communities: Madagascar. In: Fleagle, J. G., Janson, C. H., and Reed, K. E. (eds.), *Primate Communities*. Cambridge University Press, Cambridge. pp. 75–89.

Godoy, R., Wildke, D., Overman, H., Cubas, A., Cubas, G., Demmer, J., McSweeney, K., and Brokaw, N. (2000). Valuation of consumption and sale of forest goods from a Central American rain forest. *Nature* 406:62–63.

Goodall, J. (1986). *The Chimpanzees of Gombe: Patterns of Behavior*. Harvard University Press, Cambridge, MA.

Gould, L., Sussman, R. W., and Sauther, M. L. (1999). Natural disasters and primate population: the effects of a 2-year drought on a naturally occurring population of ring-tailed lemurs (*Lemur catta*) in southwestern Madagascar. *Int. J. Primatol.* 20:69–84.

Gould, L., Sussman, R. W., and Sauther, M. L. (2003). Demographic and life-history patterns in a population of ring-tailed lemurs (*Lemur catta*) at Beza Mahafaly Reserve Madagascar: a 15-year perspective. *Am. J. Phys. Anthropol.* 120:182–194.

Green, G., and Sussman, R. W. (1990). Deforestation history of the eastern rain forests of Madagascar from satellite images. *Science* 248:212–215.

Groves, C. P. (2001). *Primate Taxonomy*. Smithsonian Institution Press, Washington DC.

Grubb, P., Butynski, T. M., Oates, J. F., Bearder, S. K., Disotell, T. R., Groves, C. P., and Struhsaker, T. T. (2003). Assessment of the diversity of African Primates. *Int. J. Primatol.* 24:1301–1357.

Hacker, J. E., Cowlishaw, G., and Williams, P. H. (1998). Patterns of African primate diversity and their evaluation for the selection of conservation areas. *Biol. Conserv.* 84:251–262.

Hilton-Taylor, C. (2002). *2002 IUCN Red List of Threatened Species*. World Conservation Union, Species Survival Commission, Gland, Switzerland.

Hilton-Taylor, C., and Rylands, A. B. (2002). The 2002 IUCN Red List of threatened species. *Neotrop. Primates* 10:149–153.

Hohmann, G. (2001). Association and social interactions between strangers and residents in bonobos (*Pan paniscus*). *Primates* 42:91–99.

Jenkins, M. (2003). Prospects for biodiversity. *Science* 302:1175–1188.

Jepson, P., Jarvie, J. K., MacKinnon, K., and Monk, K. A. (2001). The end for Indonesia's lowland forest? *Science* 292:859–861.

Jernvall, J., and Wright, P. C. (1998). Diversity components of impending primate extinctions. *Proc. Natl. Acad. Sci. USA* 95:11279–11283.

Kierulff, M. C. M., Oliveira, P. P., Beck, B. B., and Martins, A. (2002). Reintroduction and translocation as conservation tools for golden lion tamarins. In: Kleiman, D. G., and Rylands, A. B. (eds.), *Lion Tamarins*. Smithsonian Institution Press, Washington DC. pp. 271–282.

Kleiman, D. G. (1996). Reintroduction programs. In: Kleiman, D. G., Allen, M. E., Thompson, K. V., and Lumpkin, S. (eds.), *Wild Mammals in Captivity*. University of Chicago Press, Chicago. pp. 297–305.

Kleiman, D. G., and Rylands, A. B. (2002). *Lion Tamarins*. Smithsonian Institution Press, Washington DC.

Knott, C. D. (1998). Changes in orangutan caloric intake, energy balance, and ketones in response to fluctuating fruit availability. *Int. J. Primatol.* 19:1061–1079.

Konstant, W. R., Mittermeier, R. A., Rylands, A. B., Butynski, T. M., Eudey, A. A., Ganzhorn, J., and Kormos, R. (2002). The world's top 25 most endangered primates–2002. *Neotrop. Primates* 10:128–131.

Kremen, C., Niles, J. O., Dalton, M. G., Daily, G. C., Ehrlich, P. R., Fay, J. P., Grewal, D., and Guillery, R. P. (2000). Economic incentives for rain forest conservation across scales. *Science* 288:1828–1832.

Lacy, R. C. (1993). VORTEX: a computer simulation model for population viability analysis. *Wildl. Res.* 20:45–65.

Laurance, W. F., Lovejoy, T. E., Vanconcelos, H. L., Bruna, E. M., Didham, R. K., Souffer, P. C., Gascon, C., Bierregaard, R. O., Laurance, S. G., and Sampaio, E. (2002). Ecosystem decay of Amazonian forest fragments: a 22-year investigation. *Conserv. Biol.* 16:605–618.

Lens, L., Van Dongen, S., Norris, K., Githiru, M., and Matthysen, E. (2002). Avian persistence in fragmented rainforest. *Science* 298:1236–1238.

Leroy, E. M., Rouquet, P., Formenty, P., Souquiere, S., Kilbourne, A., Fromment, J.-M., Bermejo, M., Smit, S., Karesh, W., Swanepoel, R., Zaki, S. F., and Rollin, P. E. (2004). Multiple ebola virus transmission events and rapid decline of central African wildlife. *Science* 303:387–390.

Mace, G. M. (2004). The role of taxonomy in species conservation. *Philos. Trans. R. Soc. Lond. B* 359:711–719.

Marsh, L. K. (2003). *Primates in Fragments*. Kluwer Academic/Plenum, New York.

McGraw, W. S. (2005). Update on the search for Miss Waldron's red colobus monkey. *Int. J. Primatol.* 26:605–619.

McGraw, W. S., and Oates, J. F. (2002). Evidence for a surviving population of Miss Waldron's red colobus. *Oryx* 36:223.

Milner-Gulland, E. J., and Akçakaya, H. R. (2001). Sustainability indices for exploited populations. *Trends Ecol. Evol.* 16:686–692.

Milton, K. (1996). Effects of bot fly (*Alouattamyia baeri*) parasitism on a free-ranging howler monkey (*Alouatta palliata*) population in Panama. *J. Zool. Lond.* 239:39–63.

Mitani, J. C., and Watts, D. P. (1999). Demographic influences on the hunting behavior of chimpanzees. *Am. J. Primatol.* 109:439–454.

Mittermeier, R. A., and Cheney, D. L. (1987). Conservation of primates and their habitats. In: Smuts, B. B., Cheney, D. L., Seyfarth, R. M., Wrangham, R. W., and Struhsaker, T. T. (eds.), *Primate Societies*. University of Chicago Press, Chicago. pp. 477–490.

Mittermeier, R. A., Rylands, A. B., and Konstant, W. R. (1999). Introduction. In: Nowak, R. M. (ed.), *Walker's Primates of the World*. Johns Hopkins University Press, Baltimore. pp. 1–52.

Mittermeier, R. A., Rylands, A. B., Konstant, W. R., Eudey, A., Butynski, T., Ganzhorn, J. U., and Rodíguez-Luna, E. (2000). Primate specialist group. *Species* 34:82–88.

Mittermeier, R. A., Rylands, A. B., Konstant, W. R., Kormos, R., Eudey, A. A., Walker, S., and Aguiar, J. M. (2004). *IUCN/SSC Primate Specialist Group Report 2001–2004*. Conservation International, Washington DC.

Myers, N., Mittermeier, R. A., Mittermeier, C. G., Fonseca, G. A. B., and Kent, J. (2000). Biodiversity hotspots for conservation priorities. *Nature* 403:853–858.

Naughton-Treves, L. M., Men, J. L., Treves, A., Alvarez, N., and Radeloff, V. C. (2003). Wildlife survival beyond park boundaries: the impact of slash-and-burn agriculture and hunting on mammals in Tambopata, Peru. *Conserv. Biol.* 17:1106–1117.

Nee, S., and May, R. M. (1997). Extinction and the loss of evolutionary history. *Science* 278:692–694.

Oates, J. F. (1996). Habitat alteration, hunting and the conservation of folivorous primates in African forests. *Aust. J. Ecol.* 21:1–9.

Oates, J. F., Abedi-Lartey, M., McGraw, W. S., Struhsaker, T. T., and Whitesides, G. H. (2000). Extinction of a West African red colobus monkey. *Conserv. Biol.* 14:1526–1532.

Olupot, W. (2000). Mass differences among male mangabey monkeys inhabiting logged and unlogged forest compartments. *Conserv. Biol.* 14:833–843.

Owens, I. P. F., and Bennett, P. M. (2000). Quantifying biodiversity: a phenotypic perspective. *Conserv. Biol.* 14:1014–1022.

Pacheco, L., and Simonetti, J. (2000). Genetic structure of a mimosoid tree deprived of its seed disperse, the spider monkey. *Conserv. Biol.* 14:1766–1775.

Pavelka, M. S. M., Brusselers, O. T., Nowak, D., and Behie, A. M. (2003). Population reduction and social disorganization in *Alouatta pigra* following a hurricane. *Int. J. Primatol.* 24:1037–1055.

Peres, C. A. (1990). Effects of hunting on western Amazonian primate communities. *Biol. Conserv.* 54:47–59.

Peres, C. A. (2001). Synergistic effects of subsistence hunting and habitat fragmentation on Amazonian forest vertebrates. *Conserv. Biol.* 15:1490–1505.

Pinto, L. P. S., Costa, C. M. R., Strier, K. B., and da Fonseca, G. A. B. (1993). Habitats, density, and group size of primates in the Reserva Biologica Augusto Ruschi (Nova Lombardia), Santa Teresa, Brazil. *Folia Primatol.* 61:135–143.

Pounds, J. A., and Puschendorf, R. (2004). Clouded futures. *Nature* 427:107–108.

Purvis, A., Agapow, P., Gittleman, J. L., and Mace, G. M. (2000). Nonrandom extinction and the loss of evolutionary history. *Science* 288:328–330.

Purvis, A., and Hector, A. (2000). Getting the measure of biodiversity. *Nature* 405:212–219.

Reed, D. H., O'Grady, J. J., Brook, B. W., Ballou, J. D., and Frankham, R. (2003). Estimates of minimum viable population sizes for vertebrates and factors influencing those estimates. *Biol. Conserv.* 113:23–34.

Remis, M. J. (1997). Western lowland gorillas (*Gorilla gorilla gorilla*) as seasonal frugivores: use of variable resources. *Am. J. Primatol.* 43:87–109.

Richard, A. F., Goldstein, S. J., and Dewar, R. E. (1989). Weed macaques: the evolutionary implications of macaque feeding ecology. *Int. J. Primatol.* 10:569–594.

Rose, A. L. (2002). Conservation must pursue human–nature biosynergy in the era of social chaos and bushmeat commerce. In: Fuentes, A., and Wolfe, L. D. (eds.), *Primates Face to Face*. Cambridge University Press, Cambridge. pp. 208–239.

Ross, C. (1998). Primate life histories. *Evol. Anthropol.* 6:54–63.

Ross, C., and Jones, K. E. (1999). Socioecology and the evolution of primate reproductive rates. In: Lee, P. C. (ed.), *Comparative Primate Socioecology*. Cambridge University Press, Cambridge. pp. 73–110.

Rowell, T. E. (1966). Forest living baboons in Uganda. *J. Zool. Soc. Lond.* 147:344–364.

Rudran, R., and Fernandez-Duque, E. (2003). Demographic changes over thirty years in a red howler population in Venezuela. *Int. J. Primatol.* 24:925–947.

Rylands, A., Strier, K., Mittermier, R., Borovansky, J., and Seal, U. S. (1998). *Population and Habitat Viability Assessment Workshop for the Muriqui* (Brachyteles arachnoides). Conservation Breeding Specialist Group, Apple Valley, MN.

Rylands, A. B., Mittermeier, R. A., and Luna, E. R. (1995). A species list for the New World primates (Platyrrhini): distribution by country, endemism, and conservation status according to the Mace-Land system. *Neotrop. Primates* 3(suppl.):113–160.

Sabater-Pi, J., and Groves, C. (1972). The importance of higher primates in the diet of the Fang of Rio Muni. *Man* 7:239–243.

Sechrest, W., Brooks, T. M., Fonseca, G. A. B., Konstant, W. R., Mittermeier, R. A., Purvis, A., Rylands, A. B., and Gittleman, J. L. (2002). Hotspots and the conservation of evolutionary history. *Proc. Natl. Acad. Sci. USA* 99:2067–2071.

Siegert, F., Ruecker, G., Hinrichs, A., and Hoffmann, A. (2001). Increased damage from fires in logged forests during droughts caused by El Niño. *Nature* 414:437–440.

Snyder, N. F. R., Derrickson, S. F., Beissinger, S. R., Wiley, J. W., Smith, T. B., Toone, W. D., and Miller, B. (1996). Limitation of captive breeding in endangered species recovery. *Conserv. Biol.* 10:338–348.

Southwick, C. H., and Siddiqi, M. F. (1988). Partial recovery and a new population estimate of rhesus monkey populations in India. *Am. J. Primatol.* 16:187–197.

Stoner, K. E. (1996). Prevalence and intensity of intestinal parasites in mantled howling monkeys (*Alouatta palliata*) in northeastern Costa Rica: implications for conservation biology. *Conserv. Biol.* 10:539–546.

Strier, K. B. (1993/1994). Viability analyses of an isolated population of muriqui monkeys (*Brachyteles arachnoides*): implications for primate conservation and demography. *Primate Conserv.* 14–15(1993–1994):43–52.

Strier, K. B. (1997). Behavioral ecology and conservation biology of primates and other animals. *Adv. Study Behav.* 26:101–158.

Strier, K. B. (2000a). Population viabilities and conservation implications for muriquis (*Brachyteles arachnoides*) in Brazil's Atlantic Forest. *Biotropica* 32:903–913.

Strier, K. B. (2000b). From binding brotherhoods to short-term sovereignty: the dilemma of male Cebidae. In: Kappeler, P. M. (ed.), *Primate Males: Causes and Consequences of Variation in Group Composition*. Cambridge University Press, Cambridge. pp. 72–83.

Strier, K. B. (2003a). Primate behavioral ecology: from ethnography to ethology and back. *Am. Anthropol.* 105:16–27.

Strier, K. B. (2003b). Demography and the temporal scale of sexual selection. In: Jones, C. B. (ed.), *Sexual Selection and Reproductive Competition in Primates*. American Society of Primatologists, Norman, OK. pp. 45–63.

Strier, K. B. (in press). *Primate Behavioral Ecology*, 3rd ed. Allyn & Bacon, Boston.

Strier, K. B., Boubli, J. P., Guimarães, V. O., and Mendes, S. L. (2002). The muriqui population at the Estação Biológica de Caratinga, Minas Gerais, Brasil: updates. *Neotrop. Primates* 10:115–119.

Strier, K. B., Boubli, J. P., Possami, C. B., and Mendes, S. L. (in press). Population demography of northern muriquis (Brachyteles hypoxanthus) at the Estação Biológica de Caratinga/Reserva Patrimônio Particular Natural-Feliciano Miguel Abdula, Miras Gerais, Brazil. *Am. J. Phys. Anthropol.*

Strier, K. B., and Fonseca, G. A. B. (1996/1997). The endangered muriquis in Brazil's Atlantic Forest. *Primate Conserv.* 17:131–137.

Sussman, R. W. (1992). Male life history and intergroup mobility among ringtailed lemurs (*Lemur catta*). *Int. J. Primatol.* 13:395–413.

Sussman, R. W., Green, G. M., Porton, I., Andrianasolondraibe, O. L., and Ratsirarson, J. (2003). A survey of the habitat of *Lemur catta* in southwestern and southern Madagascar. *Primate Conserv.* 19:32–57.

Tarnaud, L., and Simmen, B. (2002). A major increase in the population of brown lemurs on Mayotte since the decline reported in 1987. *Oryx* 36:297–300.

Teleki, G. (1989). Population status of wild chimpanzees (*Pan troglodytes*) and threats to survival. In: Heltne, P. G., and Marquardt, L. A. (eds.), *Understanding Chimpanzees*. Harvard University Press, Cambridge, MA. pp. 312–353.

Thomas, C. D., Bodsworth, E. J., Wilson, J. R., Simmons, A. D., Davies, Z. G., Musche, M., and L., Conradt (2001). Ecological and evolutionary processes at expanding range margins. *Nature* 411:577–581.

Thomas, C. D., Cameron, A., Green, R. E., Bakkenes, M., Beaumont, L. J., Collingham, Y. C., Erasmus, B. F. N., Ferreira de Siqueira, M., Grainger, A., Hannah, L., Hughes, L., Huntley, B., van Jaarsveld, A. S., Midgley, G. F., Miles, L., Ortega-Huerta, M. A., Townsend Peterson, A., Phillips, O. L., and Williams, S. E. (2004). Extinction risk from climate change. *Nature* 427:145–148.

Vila, C., Sundqvist, A., Flagstad, O., Sedden, J., Bjornerfeldt, S., Kojola, I., Casulli, A., Sand, H., Wabakken, P., and Ellegren, H. (2002). Rescue of a severely bottlenecked wolf (*Canis lupus*) population by a single immigrant. *Proc. R. Soc. Lond. B* 270:91–97.

Walsh, P. D., Abernethy, K. A., Bermejo, M., Beyers, R., Wachter, P. D., Akou, M. E., Huijbregts, B., Mambounga, D. I., Toham, A. K., Kilbourn, A. M., Lahm, S. A., Latour, S., Maisels, F., Mbina, C., Mihindou, Y., Oblang, S. N., Effa, E. N., Starkey, M. P., Telfer, P., Thibault, M., Tutin, C. E. G., White, L. J. T., and Wildie, D. S. (2003). Catastrophic ape decline in western equatorial Africa. *Nature* 422:611–614.

Watts, D. P. (1998). Coalitionary mate guarding by male chimpanzees at Ngogo, Kibale National Park, Uganda. *Behav. Ecol. Sociobiol.* 44:43–55.

Weber, B., and Vedder, A. (2001). Afterword. Mountain gorillas at the turn of the century. In: Robbins, M. M., Sicotte, P., and Stewart, K. J. (eds.), *Mountain Gorillas*. Cambridge University Press, Cambridge. pp. 413–423.

White, L. J. T., and Tutin, C. E. G. (2001). Why chimpanzees and gorillas respond differently to logging: a cautionary tale from Gabon. In: Weber, B., White, L. J. T., Vedder, A., and Simons-Morland, H. (eds.), *African Rain Forest Ecology and Conservation*. Yale University Press, New Haven. pp. 449–462.

Wrangham, R. W. (1981). Drinking competition in vervet monkeys. *Anim. Behav.* 29:904–910.

Wright, P. C. (1992). Primate ecology, rainforest conservation, and economic development: building a national park in Madagascar. *Evol. Anthropol.* 1:25–33.

Wright, P. C. (1999). Lemur traits and Madagascar ecology: coping with an island environment. *Ybk. Phys. Anthropol.* 42:31–72.

Yamagiwa, J. (1999). Socioecological factors influencing population structure of gorillas and chimpanzees. *Primates* 40:87–104.

Yeager, C. P., and Silver, S. C. (1999). Translocation and rehabilitation as primate conservation tools: are they worth the cost? In: Dolhinow, P., and Fuentes, A. (eds.), *The Nonhuman Primates*. Mayfield, Mountain View, CA. pp. 164–169.

31

Primate Seed Dispersal

Linking Behavioral Ecology with Forest Community Structure

Colin A. Chapman and Sabrina E. Russo

INTRODUCTION

The role of animals in seed dispersal is well recognized. As many as 75% of tropical tree species produce fruits presumably adapted for animal dispersal (Frankie et al. 1974, Howe and Smallwood 1982), and animals are estimated to move more than 95% of tropical seeds (Terborgh et al. 2002). Some vertebrate groups may be particularly important seed dispersers. Primates, for example, comprise between 25% and 40% of the frugivore biomass in tropical forests (Chapman 1995), eat large quantities of fruit, and defecate or spit large numbers of viable seeds (Lambert 1999). Primate frugivory and seed dispersal have been quantified by studies in South America (Garber 1986, Julliot 1996, Stevenson 2000, Dew 2001, Vulinec 2002), Central America (Estrada and Coates-Estrada 1984, 1986; Chapman 1989a), Africa (Gautier-Hion 1984; Gautier-Hion et al. 1985; Wrangham et al. 1994; Chapman and Chapman 1996; Kaplin and Moermond 1998; Lambert 1999; Voysey et al. 1999a,b), and Asia (Corlett and Lucas 1990, Davies 1991, Leighton 1993, Lucas and Corlett 1998, McConkey 2000). This research has illustrated that primates disperse significant numbers of seeds. For example, on Borneo, a single gibbon group (*Hylobates mulleri × agilis*) dispersed a minimum of 16,400 seeds · km^{-2} · year^{-1} of 160 species; and since the survival rate of seeds to 1 year was 8%, a group of gibbons effectively dispersed 13 seedlings · ha^{-1} · year^{-1} (McConkey 2000).

Although it is clear that primates play an important role in dispersing many seeds throughout tropical forests, the ecological and evolutionary significance of these activities is not well understood. The objective of this chapter is to review the role that primate seed dispersal plays in shaping the ecology of tropical forests and to shed light on inconsistencies in the literature in order to point to new directions for research. We first evaluate how variation among primate species in traits such as digestive anatomy, body size, social structure, movement patterns, and diet influences the spatial distribution of dispersed seeds (the *seed shadow*, App. 31.1) and thereby produces a diversity of seed shadows. We then consider how traits of fruiting species, including patch characteristics and fruit and seed traits, influence the seed shadow that different primates generate. Few studies have

quantified primate-generated seed shadows. However, to understand their effect on plants, it is important to consider primate seed shadows in light of plant population and community ecological theory. To this end, we next review current theories and concepts and evaluate the role that primate seed dispersal plays in plant ecological processes. This analysis of variation in seed shadows points to two critical areas in which our understanding of primates' roles is currently lacking: the significance of primates dispersing seeds in clumped versus scattered patterns and the consequences of primate seed dispersal to different distances away from the parent tree. We conclude by discussing potential conservation implications related to changes in the nature of primate–plant interactions.

VARIATION IN SEED SHADOWS GENERATED BY PRIMATES

Dispersal of seeds by primates results in a seed shadow (App. 31.1). The shape of the seed shadow is defined by the dispersal processes that produce it, namely, what proportion of seeds are dispersed from parent tree crowns, which depends on the visitation rate and the number of seeds dispersed per visit of a dispersal agent, how far those seeds are moved from the parent (*seed dispersal distance*, App. 31.1), and the locations and densities at which seeds are deposited (*dispersal sites* and *seed density*, App. 31.1). The interaction between traits of primates and of the fruit-producing plant species that they feed upon influences these dispersal processes, thereby shaping the seed shadow. The shape of the seed shadow is important because it influences the survival of seeds to the seedling stage (Janzen 1970, McCanny 1985), which is a critical demographic process in the life cycle of a plant, and may ultimately influence the spatial distribution of plants in later life stages (Nathan and Muller-Landau 2000, Wang and Smith 2002).

Traits of Primates

As seed dispersal agents, frugivorous primates have high functional diversity in traits influencing dispersal processes, such as digestive anatomy, body size, social structure, movement patterns, and diet, all of which generate heterogeneity in seed shadows. Few studies have quantified how variation in these factors affects seed deposition patterns. However, data are becoming available that identify trends in these relationships, which should be further evaluated in future studies.

Primates largely disperse seeds by *endozoochory* (App. 31.1), but once an animal has located and acquired fruit, there remains the challenge of what to do with the seeds. Protective seed coats are typically difficult to digest, and seeds themselves can also account for more than half of the weight of fruits consumed by primates (van Roosmalen 1984, Waterman and Kool 1994). Swallowed seeds can thus represent a significant cost in that they not only increase an animal's body mass but may also displace more readily processed, nutritious digesta from the gut. Given these constraints, it is somewhat surprising that seed swallowing is by far the most common means of primate seed dispersal in the Neotropics (Estrada and Coates-Estrada 1984, Chapman 1989a, Andresen 1999). In the Paleotropics, many primates also swallow seeds (Lieberman et al. 1979, Corlett and Lucas 1990, Wrangham et al. 1994, Lambert 1999, Kaplin and Lambert 2002); however, seed spitting is common in African and Asian cheek-pouched monkeys (Cercopithecinae, Gautier-Hion 1980; Rowell and Mitchell 1991; Kaplin and Moermond 1998; Lambert 1999, 2000). Cheek pouches of cercopithecines have nearly the same capacity as their stomachs (Fleagle 1999) and allow these monkeys to store multiple fruits and extract the pulp without incurring the costs of ingested seeds (Lambert 1999). They may process the fruits and spit out seeds at or away from parent trees.

Other primates consistently prey upon seeds (App. 31.1) as ingested seeds do not appear in feces or are fragmented. These include black and white colobus (*Colobus guereza*, Poulsen et al. 2001), saki monkeys (*Pithecia pithecia* and *Chiropotes satanas*, Kinzey and Norconk 1993), and red leaf monkeys (*Presbytis rubicunda*, Davies 1991). The extent to which primates act as seed predators is likely underreported since some authors lump seed predation with the ingestion of fruit. Primates' seed-handling strategies depend on the interactions between their digestive anatomy and the traits of the fruiting species (Norconk et al. 1998).

Variation in seed-handling strategies, social structure, and movement patterns shapes the seed shadows primates generate (Fig. 31.1). Some of these characters are predictably related to body size. Primates range in body mass from the 30 g pygmy mouse lemur (*Microcebus myoxinus*) to the 200 kg gorilla (*Gorilla gorilla*, Fleagle 1999). Larger primates typically have larger gut capacities, longer digesta passage times, and longer day ranges than smaller primates, although substantial variation in these relationships exists (Kay and Davies 1994, Lambert 1998, Nunn and Barton 2000). In particular, similarly sized species may have different ranging patterns due to different social systems. These characters, in turn, influence the quantity of seeds dispersed as well as their sizes, passage times, dispersal distances, and

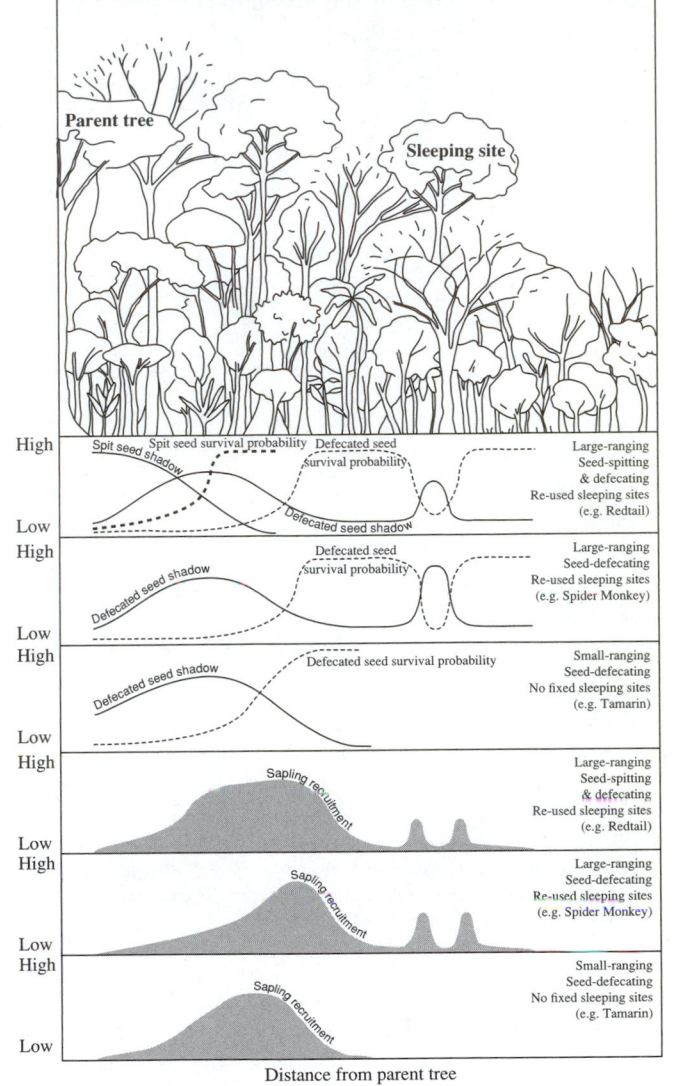

Figure 31.1 The predicted patterns of seed deposition, seed survival, and sapling recruitment for a hypothetical tree species dispersed by primates with different seed-handling strategies and ranging patterns. For many forest tree species, including the hypothetical one depicted here, some habitats may be more suitable for recruitment than others. Hence, the shape of the survival curve will depend on the plant species, but this source of variation is not depicted in the figure. Actual recruitment probabilities are entirely dependent on the actual seed densities relative to realized per-capita-survival probabilities, but in this schematic we show possible qualitative relationships. The relative difference in the height of the seed shadow curve and the survival curve will determine whether clumped seed deposition patterns remain clumped as individuals age. This will partly be determined by the strength of distance- and density-dependent mortality at the seed and seedling stages, as well as other sources of mortality that are nonrandom in space or time, relative to the spatial scale and strength of clumping of seed deposition.

spatial distribution (Table 31.1, App. 31.1). However, based on the limited available data (Table 31.1), we found no statistically significant relationship between either the mean or the maximum seed dispersal distance of primate species and body size (Spearman rank correlation $r = 0.009$, $p = 0.979$, $n = 11$, mean distance; $r = 0.108$, $p = 0.818$, $n = 7$, maximum distance).

Table 31.1 Estimates of the Straight-Line Distance (m) that Primate Species Disperse Seeds Away From the Parent Plant

SPECIES	WEIGHT[1] (KG)	SOCIAL STRUCTURE[2]	PASSAGE TIME[3] (HR)	DISTANCE (SD) (M)	MODE (M)	RANGE (M)	REFERENCE
Seeds swallowed							
Saguinus spp.	0.5	1	3.3	–	100–200	34–513	Garber (1986)
Cebus apella	2.5	3	3.5	390 (215)	–	–	Zhang and Wang (1995)
Cebus capucinus[4]	3.3	3	3.5	–	–	200–1,000	Rowell and Mitchell (1991)
Cebus capucinus[4]	3.3	3	3.5	208 (113)	–	20–844	Wehncke et al. (2003)
Alouatta palliata[4]	6.2	1	20.4	–	–	10–811	Estrada and Coates-Estrada (1984)
Alouatta seniculus	7.3	1	35.0	225 (129)	–	0–550	Julliot (1996)
Ateles paniscus	6.2	2	5.3	151 (241)	–	11–1,119	Russo (2003a)
Ateles paniscus	6.2	2	5.3	254 (145)	–	–	Zhang and Wang (1995)
Ateles belzebuth	6.0	2	–	245	–	50–500	Dew (2001)
Lagothrix lagotricha	6.3	2	6.8	300	–	0–390	Stevenson (2000)
Lagothrix lagotricha	6.3	2	6.8	354 (213)	300–400	0–989	Stevenson (2000)
Lagothrix lagotricha	6.3	2	6.8	245	–	50–500	Dew (2001)
Cercopithecus ascanius	3.6	3	20.2	1,178			Lambert (1997)
Hylobates mulleri × agilis	5.9	1	–	220 (215)		0–1,250	McConkey (2000)
Pan troglodytes	36.4	2	23.2	3,000			Lambert (1997)
Seeds spit out							
Cercopithecus ascanius	3.6	3	20.2	2	10	0–100	Lambert (1999)
Cercopithecus mitis	6.0	3	21.3	–	–	30–50	Rowell and Mitchell (1991)
Pan troglodytes	36.4	2	23.2	4	1	0–20	Lambert (1997, 1999)

[1] Average of male and female weight (from Harvey et al. 1987).
[2] Social structures are 1, cohesive; 2, fission–fusion; 3, dispersed foraging groups.
[3] Average passage time through the digestive tract (from Lambert 1998).
[4] Estimates from passage time and path length.

Social structure can also affect primate seed shadows. Primates with more fluid social structures, such as the fission–fusion societies of spider monkeys (*Ateles* spp.) and chimpanzees (*Pan troglodytes*) and the multi-species foraging groups of capuchins (*Cebus* spp.), cover large areas in their daily ranges; and individuals in groups are often widely dispersed (Terborgh 1983, McFarland 1986). Thus, one might expect primates with fluid social structures to have a more scattered pattern of seed dispersal than species with more cohesive ones. For example, Wehncke et al. (2003) demonstrated that white-faced capuchins (*Cebus capucinus*) on Barro Colorado Island, Panama, spend only 10 min feeding in individual trees, travel between 1 and 3 km a day, and defecate seeds in small clumps throughout the day. Repeated use of the same sleeping site may alter this relationship (Fig. 31.1). Seeds defecated diurnally by spider monkeys (*Ateles paniscus*) in Peru were more widely scattered and had longer dispersal distances than did those defecated at sleeping sites, which were closer to parent trees and more clumped (Russo and Augspurger 2004). No data are available to compare the extent of clumping in seed shadows generated by primates with fluid versus cohesive social structure. However, based on the data in Table 31.1, there is no difference in the mean dispersal distances of species with fluid (e.g., spider monkeys) or dispersed (e.g., redtail monkeys) groups in

comparison to primates whose social structures are more cohesive (e.g., howler monkeys; $t = 1.024$, $p = 0.333$, $n = 12$). Furthermore, in species that do not have cohesive group structures, sex and age may influence seed shadows. In spider monkeys and woolly monkeys (*Lagothrix lagotricha*), males have greater day ranges and tend to disperse seeds longer distances than do females (Symington 1987b, Stevenson 2000, Dew 2001).

The interaction between diurnal variation in movement patterns and passage time affects both dispersal distance and the spatial distribution of seeds. For example, seeds ingested by spider monkeys (*Ateles* spp.) and woolly monkeys (*Lagothrix lagotricha*) in the morning or early afternoon tend to be dispersed the same day. In contrast, seeds ingested later in the afternoon tend to be dispersed at sleeping sites at night or the next morning (Stevenson 2000, Dew 2001, Russo 2003a). In addition, foraging movements in the morning and early afternoon tend to be faster than those in the late afternoon (Chapman and Chapman 1991). As a result, seeds ingested early in the day are likely dispersed longer distances than those ingested later (Stevenson 2000, Dew 2001, Russo 2003a).

Variation in diet can also influence seed shadows, not only by determining which fruiting species are dispersed but also because the quantity of nonfruit foods in the diet typically increases seed passage time (Milton 1981, Kay and Davies 1994, Lambert 1998). Primates that consume substantial quantities of leaves, such as howler monkeys (*Alouatta* spp.), have greater colon volume than do similarly sized close relatives (e.g., spider monkeys [*Ateles* spp.]; Milton 1981, 1986). As a result, howler monkeys have longer passage times than do spider monkeys (Milton 1981, 1986). However, this difference does not necessarily translate into longer seed dispersal distances by howler monkeys as they tend to move slower and have smaller day ranges than do spider monkeys (Milton 1981).

Primate Responses to Traits of Fruiting Species

Primate species respond differently to traits of fruiting species, and this response generates variation in seed shadows. These traits range from those at the community level, such as the abundance and distribution of fruit resources, to those at the individual plant or patch level, such as fruit production, to those at the level of a single fruit or seed, such as seed size. Primates' responses to fruiting plant traits are mediated by many factors, including their foraging decisions, digestive anatomy, seed-handling strategies, social structure, and ranging behavior. Taken together, these plant–primate interactions determine the variation in the seed shadows that primates generate, which in turn influences plant demography and spatial distribution. Here, we consider how variations in *(1)* the distribution and abundance of fruit resources, *(2)* the characteristics of fruit patches, and *(3)* fruit and seed traits influence primates' seed shadows.

Distribution and Abundance of Fruit Resources

Primates' foraging decisions, and hence visitation and seed dispersal of fruiting species, are sensitive to the community-wide abundance and distribution of fruit resources (Agetsuma and Noma 1995, Garber and Paciulli 1997, Janson 1998). Fruit is a spatially and temporally patchy resource (Frankie et al. 1974, van Schaik et al. 1993). As such, any particular fruiting species may not be reliable, especially given the rarity of tropical plant species in general and their tendency to have supra-annual fruiting schedules (Chapman et al. 1999, in press). Such variability in fruit resources likely contributes to the often observed pattern of dietary plasticity and helps explain the great diversity of plant species that primates disperse (Chapman 1987, 1995; Chapman et al. 2002; Russo et al. in press). A comparative study of diets in spider monkeys (*Ateles* spp.) indicated that, despite their catholic diets, spider monkeys appear to prefer relatively abundant fruiting species that consistently produce annual crops (Russo et al. 2005).

The community-wide abundance of fruit may also affect dispersal distances and the frequency of seed predation. During periods of community-wide fruit scarcity, daily path lengths of spider and woolly monkeys decreased and the proportion of time spent resting increased relative to periods of fruit abundance (Symington 1987b, Stevenson 2000). As a result, dispersal distances may decrease during periods of fruit scarcity. In addition, the consumption of unripe fruit by some primates, which likely represents seed predation, often increases during periods of fruit scarcity (Kaplin et al. 1998, Stevenson 2000). Cercopithecine monkeys alternate between acting predominately as seed spitters, seed predators, and seed defecators depending on fruit resource availability (Kaplin et al. 1998).

Since the vast majority of primate field studies are relatively short-term (i.e., 1 year or less), it is difficult to assess the importance of periods of fruit scarcity. Yet, these times may be important periods of selection for both the primate dispersers and the trees. Fruit availability has been quantified in Kibale National Park, Uganda, for over 12 years (Chapman et al. in press). Over this period, temporal variation in fruit availability was high; the proportion of trees per month with ripe fruit varied from 0.1% to 15.9%. In addition, there was dramatic interannual variation in fruit availability: in 1990, on average only 1.1% of trees bore ripe fruit each month, while in 1999 an average of 6.7% of trees bore fruit each month (Fig. 31.2). If a month of fruit scarcity is defined as one with <1% of monitored trees bearing ripe fruit, there is considerable interannual variation in how often frugivores experienced food shortages (Fig. 31.2). For example, 9 of the 12 months in 1990 had <1% of the trees with fruit, while in 2000 no month had <1% of trees fruiting. This level of variability means that for a tree that fruits during a period of general community-wide fruit scarcity, the probability of its seeds being dispersed is likely very different from that in a subsequent year when a greater number of species are attempting to attract the services of dispersers.

Characteristics of Fruit Patches

Characteristics of fruit resource patches, such as size and density of fruits, affect visitation by primates (Leighton 1993). Greater numbers of ripe fruits in the nutmeg tree (*Virola calophylla*, Myristicaceae) increased both visitation and the number of seeds ingested by spider monkeys (*Ateles paniscus*,

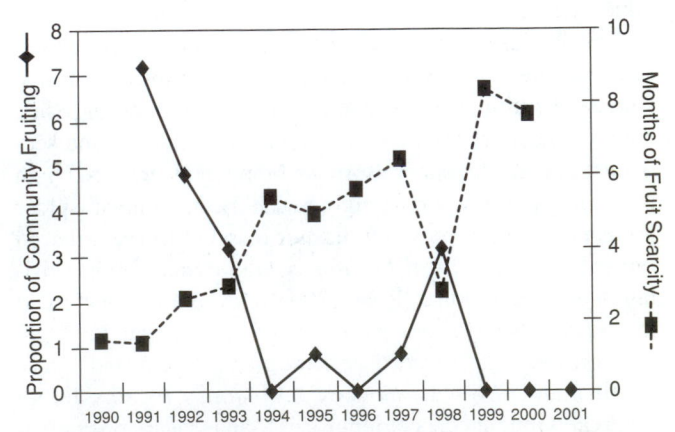

Figure 31.2 The average monthly proportion of the tree community that fruited each year in Kibale National Park, Uganda, and the number of months of fruit scarcity, defined as months with <1% of the entire population bearing fruit.

Russo 2003b). Similarly, orangutans (*Pongo pygmaeus*) preferred to forage in trees with large crop size (Leighton 1993). Furthermore, as the size of primate foraging parties increases, they tend to deplete resource patches completely, thereby increasing travel between patches (Symington 1987a, Chapman and Chapman 2000, Gillespie and Chapman 2001). Seed dispersal distances may therefore increase, all else being equal, as foraging party size increases.

Fruit and Seed Traits

Despite dietary plasticity, primates forage selectively based on traits of fruits and seeds, and these traits have been the basis of describing a primate seed dispersal "syndrome" (van der Pijl 1982, Gautier-Hion et al. 1985). Suites of traits (e.g., color, size, protection) have been interpreted as coadapted features of plants that govern the choice of fruits by dispersers. While these syndromes may be broadly meaningful, responses of different primate species to these traits are nonetheless highly variable, likely due in part to variability in the floristic composition of their habitats and interspecific interactions (Gautier-Hion et al. 1993, Chapman and Chapman 2002, Russo et al. 2005). Fruits considered, based on their traits, to be primarily bird-dispersed can be extensively dispersed by primates (Julliot 1996, Russo 2003b). In fact, it may be more difficult for a plant to exclude primates as dispersers, for example, compared to birds since primates are better able to manipulate even fruits protected by hard outer coverings. Gautier-Hion et al. (1985) went so far as to say that these suites of traits exist despite an outstanding lack of specificity between fruit and consumer species. This lack of specificity has led some to question the generality of such syndromes, and they have pointed out large differences in how syndromes have been defined (Fischer and Chapman 1993). For example, based on observations in Peru, Janson (1983) described a primate syndrome as fruits that are large, yellow, orange, brown, or green with a woody outer covering, whereas based on observations in Gabon, Gautier-Hion et al. (1985) described the primate syndrome as fruits intermediate in size, bright red, orange, or yellow and either dehiscent with arillate seeds or succulent and fleshy.

Despite this controversy over the generality of syndromes, primates are obviously selective in their choice of fruits. Choice of fruiting species can be based on morphology (i.e., color, protection, size) as well as nutrient, caloric, and secondary chemical content. Bornean orangutans preferred high pulp mass per diaspore (Leighton 1993). In contrast, spider monkeys (*Ateles paniscus*) did not respond to the aril:seed ratio of the fruits of *Virola calophylla* in terms of either visitation or seed removal (Russo 2003b). Similarly, long-tailed macaques (*Macaca fascicularis*) did not appear to select fruit on the basis of seed or fruit size (Corlett and Lucas 1990), although howler monkeys (*Alouatta seniculus*, Julliot 1996) and four species of lemurs in Madagascar appeared to select fruits in part based on size (Dew and Wright 1998).

The hardness of the fruit's outer covering and its seed plays an important role in the foraging decisions of primates,

thereby governing which primate species disperse or prey upon seeds of a particular species. When sympatric primates compete for fruit resources, these axes of trait variation may provide means of niche separation. In Suriname and Venezuela, the bearded saki monkey (*Chiropotes satanas*) consumes significantly harder fruits than does the spider monkey (*Ateles belzebuth*, Kinzey and Norconk 1990). Furthermore, seeds masticated by saki monkeys are softer than those dispersed by spider monkeys (Kinzey and Norconk 1990). The fruit consumed by redtail monkeys (*Cercopithecus ascanius*) and mangabeys (*Lophocebus albigena*) do not differ in dietary hardness during periods of fruit abundance, but mangabeys exploited a diet more resistant to puncture than redtail monkeys during periods of fruit scarcity (Lambert et al. 2004). This suggests that it is the differences in the mechanical properties of fallback foods during periods of resource scarcity that may facilitate niche separation.

Many primates pass ingested seeds, resulting in few, if any, negative impacts on germination (Estrada and Coates-Estrada 1984, Dew and Wright 1998, McConkey 2000, Lambert 2001, Poulsen et al. 2001, Stevenson et al. 2002). However, even among primate species that rarely prey upon seeds, seed predation may nonetheless occur, albeit at low frequency, and properties of fruits and seeds influence the likelihood of seed predation. Seeds having relatively soft seed coats or that provide no pulp or aril appear more likely to be preyed upon (Corlett and Lucas 1990, Russo et al. 2005). The shape and size of the seed may also influence seed handling (Chapman 1995). Primates tend to ingest and disperse more smaller than larger seeds (Corlet and Lucas 1990, Lucas and Corlet 1998, Lambert 1999, Oliveira and Ferrari 2000) and more ovoid than round seeds (Garber 1986, McConkey 2000). When pulp or aril is easily removed from the seed, gibbons (*Hylobates mulleri × agilis*) and long-tailed macaques (*Macaca fascicularis*) tend to spit seeds under the parent tree (McConkey 2000), although some observations indicate this effect may be overridden by the effect of seed size (Corlett and Lucas 1990).

The effect of fruit and seed traits on the probability of being ingested has consequences for seed shadows because seeds that are ingested by primates usually are dispersed longer distances than seeds that are spit-dispersed (Lambert 1999, 2001) (Table 31.1). Furthermore, spit or dropped seeds are likely to be deposited on the forest floor singly as fruits are processed one by one (Lambert 1999). However, swallowed seeds can be deposited either in a high-density clump or as just a few seeds, depending on animal size and position in the canopy, defecation size, and the intensity of the feeding bout during which the seeds were swallowed (Wrangham et al. 1994, Kaplin and Moermond 1998, Andresen 1999, Lambert 1999). In Cameroon, for example, Poulsen et al. (2001) found that gorillas and chimpanzees averaged 18 and 41 large (>2 cm) seeds per fecal sample, respectively, while average numbers of large seeds per fecal sample for four frugivorous monkeys were less, ranging 1.0–2.1 seeds. Some primates that largely ingest seeds, however, do not always disperse them in clumps (e.g., *Cebus capucinus*; Wehncke et al. 2003). In addition, seeds ingested by spider monkeys (*Ateles paniscus*) that are defecated diurnally tend to be dispersed at lower densities than those that are defecated at sleeping sites (Russo and Augspurger 2004). A similar pattern has been observed for woolly monkeys (Stevenson 2000) and gorillas (Rogers et al. 1998).

Although seed size can influence the passage time of ingested seeds, this effect may not necessarily translate into dispersal distances, which also vary with seed size (Garber 1986, Stevenson 2000; but see McConkey 2000). In fact, in some primates, passage time correlates poorly with dispersal distance, primarily because movements are rarely unidirectional and because seeds dispersed at sleeping sites are retained in the gut overnight (Stevenson 2000), although in primates the relationship is positive (Garber 1986).

Because fruit pulp attracts animals that may disperse a plant's seed, it generally presents less of a challenge to ingest and digest than other primate foods (Janson and Chapman 1999). In general, most fruits are high in sugars but low in fats and protein. As a result, protein content seems to have little to do with fruit choice (Leighton 1993) since the animals often obtain protein from other sources. Immature fruits are often chemically defended against insect or mammalian herbivory, and some species continue to be defended even when the pulp is ripe (Fenner 1992, Cipollini and Levey 1997). Some fruits have evolved chemical traits that appear to restrict dispersal of seeds to a fraction of the frugivore community. For example, the chemicals that make red peppers spicy to humans and other mammals are not perceived by birds (Janson and Chapman 1999). An interesting avenue for future research would involve investigations of whether some of these apparently defensive chemicals can alter gut passage times in primates and thus dispersal distances, as in birds (Murray et al. 1994).

SEED DISPERSAL MODELS

We have a relatively underdeveloped understanding of both seed-deposition patterns created by primates and the relationship between these patterns and the primate movements and behaviors that produce them. This lack of understanding results partly from the difficulty of quantifying complex movement patterns that often occur over large spatial scales. It does, however, hamper our quantification of the effect of primate dispersal agents on plant population and community dynamics. Modeling approaches have proven useful in describing seed-dispersal curves (App. 31.1), as well as the seed shadows generated by primates.

Two modeling approaches, inverse and mechanistic models, have made fundamental contributions to understanding seed dispersal (Nathan and Muller-Landau 2000). The inverse-modeling approach estimates parameter values for dispersal functions that result in the best fit to seed-dispersal data from seed traps, given a model of dispersal, or *dispersal kernel* (App. 31.1), describing the seed shadow of individual, mapped parent trees (Ribbens et al. 1994, Clark et al. 1999). One limitation of inverse modeling is that it does not directly incorporate information about the behavior of dispersal agents and therefore does not provide an explicit link between seed-deposition patterns and disperser behavior. Furthermore, seed dispersal by more terrestrial primates cannot be assessed using seed traps. In addition, there are logistic difficulties in applying this approach to primates that range over large spatial scales (e.g., chimpanzees). Thus, inferences based on inverse modeling of the role of primate dispersers in plant population and community dynamics are limited.

Mechanistic models predict seed-deposition patterns directly from the traits of plants and their dispersal agents (Nathan and Muller-Landau 2000). Basic mechanistic models predict seed dispersal distances based on the passage time and movement rates of primates (Wehncke et al. 2003). However, such models do not incorporate ecological data on the directionality of movements and behaviors of primates that often cause them to disperse seeds repeatedly in the same locations, such as sleeping sites. Hence, models based on seed passage or handling times likely underestimate clumping of primate-dispersed seeds. To understand the link between primate activity and seed-deposition patterns, more complex mechanistic models need to be developed. These models should be structured based on spatially explicit field observations of the behavior and movement patterns of primates combined with data on the fecundity of mapped parent trees. One such model has been developed for spider monkeys (*Ateles paniscus*) dispersing *Virola calophylla* in Peru (Russo 2003a). It demonstrated that the seed deposition pattern of *V. calophylla* displayed considerable spatial heterogeneity as a direct result of spider monkeys' behavior of dispersing some seeds in a scattered pattern and others in a highly clumped pattern. This heterogeneity was reflected in the multimodality of the seed-dispersal curve, which was long-tailed and did not display a monotonic decrease with distance from the source tree. Thus, the shape of the seed-dispersal curve was a direct consequence of primate dispersal processes, highlighting the need to incorporate primate behavior directly in seed-dispersal models.

The inter- and intraspecific variation in primate traits combined with their interactions with traits of fruiting species suggests that primates produce a diversity of seed shadows. Few studies have quantified primate-generated seed shadows in relation to these characters, but it is this linkage that will illuminate the effect of primate seed dispersal on plant populations and communities.

SEED DISPERSAL, PLANT POPULATION, AND COMMUNITY ECOLOGY: A REVIEW OF THEORY AND CONCEPTS

Seed dispersal influences plant ecological processes in many ways, primarily because it results in colonization of potential recruitment sites and establishes the initial template of off-spring spatial distribution. First, the spatial extent of seed dispersal restricts the suite of potential sites for recruitment (Howe and Smallwood 1982). Second, recruitment of seedlings and saplings often depends on the initial spatial pattern of seed dispersal, such as when survival depends on the distance from a conspecific adult or density of seeds, seedlings, and/or saplings (Janzen 1970, Connell 1971). Such non-random survival produces spatial patterns of juvenile and adult plants that differ from what would result simply from random or uniform thinning of the initial template of dispersed seed (Augspurger 1983). As a result, the spatial pattern of dispersed seeds and the consequences of that pattern for seed and seedling survival affect the density and spatial distribution of recruits (Hubbell 1980, McCanny 1985). Furthermore, theory suggests that, when survival is negatively density-dependent, spatial variation in density among subpopulations can modulate population fluctuations at larger scales (Chesson 1996, 1998). In the case of seed dispersal, as seeds become more clumped, the number of seeds surviving to the seedling stage decreases, with negative consequences for seedling population growth rate (P. Chesson and S. E. Russo unpublished data). Large-scale spatial variation in seed densities resulting from seed dispersal by primates can thereby influence the demographic stability of plant populations. Thus, studies of primate seed dispersal should be designed to estimate variance in seed density as well as the dispersal processes that produce it. Third, the availability and spatial distribution of sites suitable for establishment combined with the interaction between the environment and a plant's *regeneration niche* (Box 31.1) (Grubb 1977) can determine the spatial pattern of seedling and sapling recruitment. This is likely particularly critical for small-seeded and light-demanding species that recruit best in gaps (Dalling and Hubbell 2002, Dalling et al. 2002) but also for larger-seeded, shade-tolerant species that have, for example, elevation- or soil-related habitat requirements (Webb and Peart 2000).

At the community level, the spatial distribution of individual plants mediates intra- and interspecific interactions, such as competition and density-dependent mortality; and the balance of these interactions affects species coexistence (Pacala and Tilman 1994, Chesson 2000). Limitations on seed dispersal, for example, may promote species coexistence if the inability of a better competitor to disperse to a particular site permits poorer competitors to establish there instead (Levin 1974, Tilman 1994, Hurtt and Pacala 1995). If such dispersal limitation occurs on a community-wide basis, then competitive exclusion and loss of species from a community may be slowed, thereby contributing to the maintenance of plant species diversity (Hurtt and Pacala 1995). In addition, theory has suggested and simulations have demonstrated that the spatial scale of seed dispersal affects properties of plant communities, in particular species richness and the shapes of species accumulation curves and rank abundance distributions at community and regional scales (Hubbell 2001, Chave et al. 2002). Thus, understanding the role of seed dispersal in the development of spatial pattern in plant populations is critical to explanations of the structure of plant communities (Levin 1974, Tilman 1994, Hurtt and Pacala 1995, Chesson 2000).

At landscape and larger scales, long-distance seed dispersal mediates the colonization of new habitats (Howe and Smallwood 1982; Cain et al. 2000, 2003). Although long-distance dispersal is notoriously difficult to document (Nathan et al. 2003), it nonetheless has important implications for migration of plant species under global climate change (Cain et al. 2000, Clark et al. 2003), rates of population spread of invasive species (Clark et al. 2003), and recolonization of fragmented or degraded habitats (Duncan and Chapman 1999, Kaplin and Lambert 2002). Long-distance seed dispersal by primates could occur with wide-ranging species like chimpanzees (*Pan troglodytes*) and mandrills (*Mandrillus sphinx*), when young animals disperse to new groups and carry seeds in their guts from their natal home range, and during seasonal movements.

Seed dispersal at all scales has consequences for population genetics of plants. Ecological mechanisms affecting gene movement play an important role in the development of genetic structure; and in seed plants, gene movement is accomplished by pollen and seed dispersal (Loveless and Hamrick 1984, Sork et al. 1999). Although gene flow from pollen movement can be extensive (Loveless and Hamrick 1984), genetic structure nonetheless may develop within populations as a result of nonrandom patterns of seed dispersal (Epperson and Alvarez-Buylla 1997, Schnabel et al. 1998, Hu and Ennos 1999). Genetic structure has important evolutionary consequences for plants, including effects on viability selection through herbivore- or pathogen-mediated effects on survival (Augspurger and Kelly 1984, Schmitt and Antonovics 1986, Sork et al. 1993), levels of inbreeding depression (Ellstrand and Elam 1993), and effective population size, which influences the dynamics of genetic change (Hedrick 2000).

Population genetic theory predicts that the more spatially restricted seed dispersal is, the more likely plant populations are to develop local genetic differentiation (Wright 1943, Loveless and Hamrick 1984). Many primates disperse seeds long distances, and those that are large-bodied can move large quantities of seeds (McConkey 2000, Stevenson 2000, Dew 2001, Poulsen et al. 2001, Russo 2003a). Such gene flow would homogenize plant genetic variation at large scales and reduce the probability of local differentiation of plant populations, thus decreasing the probability of the formation of tight coevolutionary relationships (Herrera 2002).

As detailed above, primates have high functional diversity in terms of the seed-dispersal services they provide. The diversity and heterogeneity of seed shadows they generate means that primate–plant interactions have the potential to influence nearly every aspect of plant population and community ecology.

CONSEQUENCES OF VARIATION IN SEED-DEPOSITION PATTERNS

As demonstrated thus far, many factors influence the initial seed shadows that primates generate. The spatial distribution of dispersed seeds has important consequences for plants in terms of demography and the spatial pattern of recruitment. Two aspects of seed dispersal that warrant special consideration are the scatter versus clump dispersal dichotomy (Howe 1989) and the distances that seeds are dispersed from parent trees.

Scatter Versus Clump Dispersal

Because some primates spit or defecate seeds in low-density seed piles and other primate species defecate seeds into high-density seed clumps, primates lend themselves to an evaluation of the scatter versus clump dispersal hypothesis suggested by Howe (1989) for all fruit-eating animals. Howe (1989) proposed that many tree species are scatter-dispersed by small frugivores that regurgitate, spit, or defecate seeds singly. These offspring recruit as isolated individuals and are unlikely to experience selection for resistance to herbivores, pathogens, or other sources of density-dependent seed or seedling mortality. In contrast, other species are dispersed by large frugivores that deposit large numbers of seeds in a single location. Howe (1989) proposed that these clump-dispersed seeds germinate in close proximity to one another and thus evolve chemical or morphological defenses against seedling predators, pathogens, and herbivores that act in a density-dependent fashion. These processes should therefore be reflected in the spatial distribution of adults, with scatter-dispersed species being random or widely dispersed and clump-dispersed species being highly aggregated.

Howe's (1989) hypothesis rests on one critical assumption: that the initial seed-deposition pattern persists after the seed stage so that density-dependent factors can play a role in determining the spatial pattern of recruitment. Given that processes acting after dispersal may alter patterns generated by the dispersers (Herrera et al. 1994, Jordano and Herrera 1995, Schupp and Fuentes 1995, Kollman et al. 1998, Rey and Alcantara 2000, Balcomb and Chapman 2003), this assumption should be evaluated.

Postdispersal seed predation and secondary dispersal (App. 31.1) can dramatically affect seed survival and, ultimately, seedling recruitment and the dynamics of plant demography (Feer and Forget 2002). In the Paleotropics, Lambert (1997, 2001, 2002) experimentally determined that most seeds do not remain at the site of deposition and that postdispersal processes obscure the initial seed-deposition pattern. Numerous studies cite the importance of rodent seed predation on the fate of dispersed seeds and tree regeneration. For example, DeSteven and Putz (1984) documented the influence of mammalian seed predation on the recruitment of a tropical canopy tree (*Dipteryx panamensis*, Leguminosae) on Barro Colorado Island, Panama. They found that predation of unprotected *D. panamensis* seeds and seedlings exceeded 90% and suggested that predation by seed-eating mammals can be so extensive that even dispersed seeds have little chance of escape (DeSteven and Putz 1984). At Santa Rosa National Park, Costa Rica, 98% of the seeds placed at experimental stations were removed or killed within 70 days (Chapman 1989a). In Peru, 99% of seeds of *Virola calophylla* that had been naturally dispersed by spider monkeys (*Ateles paniscus*) or had fallen below the parent tree were preyed upon within 15 months (Russo and Augspurger 2004).

Initial seed deposition patterns by primates can also be altered by secondary seed dispersal by dung beetles and rodents that cache seeds. Andresen (1999) investigated the fate of monkey-dispersed seeds in Peru and found that 27 species of dung beetle visited the dung of spider monkeys (*Ateles paniscus*) and howler monkeys (*Alouatta seniculus*) and buried 41% of the seeds they encountered. In Uganda, Shepherd and Chapman (1998) documented that dung beetles buried 31% of the seeds placed in experimental primate dung. Clumps of seeds found in chimpanzee defecations rarely remain at the site of deposition after a single day (C. A. Chapman personal observation).

Secondary seed dispersal by rodents, which cache seeds for future consumption, may also alter the initial seed-deposition pattern. However, an important consideration is that the relatively short distances that seeds are secondarily dispersed may not result in substantial modification of the primary seed shadow (Wenny 1999, Brewer and Webb 2001, Russo 2005). Larger rodents, such as agouchies (*Myoprocta agouchy*), may secondarily disperse seeds up to 124 m away from their original location, although most distances were 20 m or less (Jansen et al. 2002). The dominant effect of secondary dispersal by dung beetles and rodents may be mediated by the burial that can increase seed survival (Andresen 2001, Brewer 2001), rather than the actual movement of the seeds. Andresen and Levey (2004) documented that burial of seeds by dung beetles decreased the probability of seed predation by rodents by threefold and increased the probability of establishment by two-fold. Further quantification of how secondary dispersal modifies primate seed shadows in terms of seed distribution and survival are clearly needed; particularly, more emphasis needs to be placed on distinguishing rodent seed predation from secondary dispersal.

Some experimental studies suggest that whether seeds are scatter- or clump-dispersed may not always predictably influence seed fate. Forget and Milleron (1991) investigated the fate of experimentally dispersed *Virola surinamensis* seeds on Barro Colorado Island, Panama. Using thread-marking methods, these authors observed that agoutis (*Dasyprocta punctata*, a large rodent) scatter-hoarded *V. surinamensis* seeds that they found both singly and in clumps. Seed-removal and seed-burial rates were strongly affected by features of forest habitats, such as *V. surinamensis* tree abundance and/ or forest age, but not by seed-dispersal treatment (scattered versus clumped). Predation of unburied seeds by weevils was independent of habitat and dispersal treatment.

Although some experimental studies suggest that clump-dispersed seeds may not remain in a clumped distribution, a different picture emerges when the fate of seeds that were naturally dispersed by primates is linked to the spatial distributions of plants in later life stages. In a study of *Virola calophylla* in Peru, a clumped pattern of seed deposition was generated by the key dispersal agent, the spider monkey (*Ateles paniscus*) (Russo and Augspurger 2004), which dispersed 92% of all dispersed seeds of this tree species (Russo 2003b). The initially clumped seed-deposition pattern was largely maintained through recruitment to the sapling stage. The density of seedlings and saplings was highest where seed fall was greatest, that is, underneath *V. calophylla* parents and at the sleeping sites of spider monkeys. This clumped pattern was maintained even though per capita seed survival to 15 months at these two dispersal site types was the lowest, relative to seeds that spider monkeys dispersed diurnally at low density. In addition, per capita seed survival was negatively dependent on the density of *V. calophylla* seeds and positively dependent on their distance from the nearest female *V. calophylla* tree. Therefore, the clumped recruitment pattern of seedlings and saplings arose despite considerable distance- and density-dependent seed mortality. Thus, spatially aggregated seed dispersal by spider monkeys appeared to play a dominant role in the spatial structuring of this tropical tree population. The interesting challenge that remains is to determine under what conditions the initial seed-deposition pattern generated by primates will persist and what conditions will cause its disruption. The balance likely lies in the strength of distance- and density-dependent mortality at the seed and seedling stages relative to the spatial scale and strength of seed clumping and in the propensity of primates to disperse seeds to habitats consistent with a plant species' regeneration niche, in addition to interactions with a plant species' competitors and mutualists.

Given that many primates use one or a small number of sleeping sites repeatedly over a number of years (Anderson 1984, Chapman 1989b, Chapman et al. 1989, Julliot 1997, Russo and Augspurger 2004), these sites are clearly areas of high seed deposition. A similar pattern of repeated defecation may also exist for feeding trees or any other frequently used location (McConkey 2000, Schupp et al. 2002). Such patterns have led some researchers to speculate that defecation at such sites may alter the distribution of adult trees, depending on how many other dispersers contribute to the seed shadow of a tree species. For example, Milton (1980) suggested that the clumped pattern of seeds dispersed by howler monkeys (*Alouatta palliata*) could account for the patchy distribution of food trees in the animals' home ranges. Lieberman et al. (1979) suggested that by defecating near sleeping sites, baboons (*Papio anubis*) increase the concentration of food plants within their core area. However, these studies have not examined the long-term fate of seeds dispersed to sleeping sites, particularly after the seedling and sapling stages. Thus,

speculating on the impact of these activities with respect to adult distributions is premature. A study done near hornbill nests, large fruit-eating African birds, that examined the long-term fate of seeds deposited near nests found that directed seed dispersal to nests over a decade did not substantially alter sapling community composition at these sites (Paul 2001).

Seed-Dispersal Distance

Janzen (1970) and Connell (1971) suggested that dispersal away from the parent plant enhances survival by removing offspring from mortality factors (e.g., competition, predation, pathogens) acting in a distance- or density-dependent fashion. In one community, negative density-dependent effects on seedling recruitment have been found for every one of the 53 species examined (Harms et al. 2000), but evidence supporting distance- and density-dependent effects is mixed (Clark and Clark 1984, Condit et al. 1992, Hille Ris Lambers et al. 2002, Hyatt et al. 2003). These ideas have led to investigations examining the consequences of primates dispersing seed different distances away from the parent plant (Stevenson 2000, Dew 2001).

Primates disperse seeds to a range of distances, depending in part on seed handling. Seeds that are dispersed through spitting are on average moved only a few meters from the parent tree, whereas those defecated are often moved hundreds of meters from the parent (Fig. 31.1, Table 31.1). However, the effect of seed-dispersal distance on the probability of a seed germinating and surviving will likely be variable and species-specific. For example, several species-specific studies examining seedling survival under parent trees have found little or no recruitment under parent trees (Augspurger 1983, 1984), whereas other studies reveal relatively small differences in the probability of survival between seeds under parent trees and those dispersed farther away (Condit et al. 1992; DeSteven and Putz 1984; Chapman and Chapman 1995, 1996). Howe et al. (1985) found nearly complete mortality (99.96%) within only 12 weeks for *Virola surinamensis* fruit that dropped under the parent. Similarly, Schupp (1988) documented only 7% survival of *Faramea occidentalis* seeds under the crown over 30 weeks in comparison to 24% survival 5 m away from the parent tree. In contrast, Chapman and Chapman (1996) investigated primate-dispersed trees in Kibale National Park, Uganda, and found that *Uvariopsis congensis* experienced 56% more seed predation when dispersed away from parent trees versus directly under a parent tree.

Hence, dispersal by primates away from parent trees may not always guarantee a higher probability of seed survival. Per capita survival of *Virola calophylla* seeds at the sleeping sites where spider monkeys disperse large numbers of seeds is just as low as it is underneath parents (Russo and Augspurger 2004). Although sleeping sites were distant from the nearest female *V. calophylla* tree, the high seed densities that occur there appear to attract seed predators, indicating that distance- and density-dependent mortality can have independent effects on seed survival. In fact, the effect of seed density on seed survival was stronger than that of distance. Thus, for seeds being dispersed to sleeping sites, the value of being dispersed from a parent is reduced in this system.

Understanding whether the strength of distance- or density-dependent survival varies with scale is critical for evaluating the significance of variation among primate species in seed-dispersal distance, particularly that associated with different seed-handling strategies. At least one study has found that the strength of distance- and density-dependent survival drops off dramatically once seeds are beyond the parent crown (Schupp 1988). Furthermore, seed predators, the agents of distance- and density-dependent survival, are likely to respond to different scales of variation of their food resources, depending on their own natural history (Schupp 1992).

Evidence points to substantial interspecific variation in the degree to which distance and density dependence influence seed fate, the scale at which they do so, and the degree to which these factors act independently. Studies should not only determine whether density and distance have separate, significant effects but also quantify and compare the strength of each. Possibly a more significant implication of variation in dispersal distances is a seed's increasing ability to colonize new habitats with increasing dispersal distance (Howe and Smallwood 1982). Newly colonized habitats provide either favorable or unfavorable conditions for establishment, so the underlying spatial arrangement of habitats in a landscape and the degree to which primates cross ecotone boundaries may influence whether longer dispersal distances result in increased seed survival and seedling establishment.

CONSERVATION IMPLICATIONS

Cumulatively, countries with primate populations are losing approximately 125,000 km^2 of forest annually (Chapman and Peres 2001). Other populations are affected by forest degradation (logging and fire), habitat fragmentation, and hunting (Peres 1990; Chapman et al. 2003a,b). Significant reductions in population densities of primates are likely to have far-reaching consequences for plant populations and communities (Redford 1992, Chapman 1995, Chapman and Chapman 1995). Although there have been few empirical documentations of seed dispersal–mediated effects of hunting and fragmentation on plant recruitment, existing evidence points to altered patterns of seed dispersal in hunted compared to nonhunted populations (Pacheco and Simonetti 2000, Wright et al. 2000) and to reduced seed dispersal following habitat fragmentation (Chapman et al. 2003a,b; Cordeiro and Howe 2003). Persistence of primates even in fragmented habitats can, however, facilitate regeneration, particularly when primate dispersers cross habitat boundaries and disperse seeds to a habitat different from that in which they were ingested, as has been documented in black-

handed tamarins (*Saguinus midas niger*) (Oliveira and Ferrari 2000). Chapman et al. (2003a) found that there was movement of seeds among forest fragments in western Uganda by large-bodied dispersers, particularly chimpanzees and hornbills (*Ceratogymna subcylindricus*).

Available evidence also suggests that, although there can be overlap in the assemblages of fruiting species that different agents disperse, this overlap may not be enough to promote redundancy when a group of dispersers, such as primates, is lost as a result of hunting or other disturbance (Poulsen et al. 2002). Furthermore, density compensation by smaller primates when population densities of larger primates fall, if it occurs (Peres and Dolman 2000), is unlikely to make up for the decline in functional diversity of dispersal services. Thus, with the loss of primate dispersal agents, rates of seed dispersal may inevitably decrease and seed shadows may become more stereotyped, to the extent that decreases in species richness correlate with decreases in the functional diversity of disperser services. Whether this results in changes in plant populations and communities depends on the dynamics of postdispersal survival underneath parent trees versus at sites where primates would have dispersed seeds. For example, in Uganda and Bolivia, reduced numbers of large-bodied primates were correlated with lower seedling densities of large-seeded forest trees species (Chapman and Onderdonk 1998, Pacheco and Simonetti 2000) and greater seedling aggregation around parent trees (Pacheco and Simonetti 2000). In contrast, in Mexico and Panama, seedling densities were higher in areas with depleted mammalian communities (Dirzo and Miranda 1991, Wright et al. 2000).

Several authors have argued that the role of primates as seed dispersers is probably particularly important for large-seeded or hard-husked fruit species, which may be inaccessible to smaller, arboreal taxa (Andresen 2000, Kaplin and Lambert 2002). The conservation of primates is therefore argued to be critical to maintaining effective seed dispersal of such species (Andresen 2000). However, in most situations, variation in postdispersal seed fate makes it very difficult to predict accurately how a particular tree species will respond to the removal of its primate dispersers. Thus, if models are developed to predict changes in plant populations and communities with the loss of primate dispersal services, they must account not only for changes in the seed shadow but also for the resulting alterations in the transition probabilities between seed, seedling, sapling, and adult stages; the spatial component of these transitions; and the consequences for interspecific interactions among plants.

REFERENCES

Agetsuma, N., and Noma, N. (1995). Rapid shifting of foraging pattern by Yakushima macaques (*Macaca fuscata yakui*) in response to heavy fruiting of *Myrica rubra*. *Int. J. Primatol.* 16:247–260.

Anderson, J. R. (1984). Ethology and ecology of sleep in monkeys and apes. *Adv. Study Behav.* 14:166–229.

Andresen, E. (1999). Seed dispersal by monkeys and the fate of dispersed seeds in a Peruvian rain forest. *Biotropica* 31:145–158.

Andresen, E. (2000). Ecological roles of mammals: the case of seed dispersal. In: Entwistle, A., and Dunstone, N. (eds.), *Priorities for the Conservation of Mammalian Diversity.* Cambridge University Press, Cambridge. pp. 2–26.

Andresen, E. (2001). Effects of dung presence, dung amount and secondary dispersal by dung beetles on the fate of *Micropholis guyanensis* (Sapotaceae) seeds in central Amazonia. *J. Trop. Ecol.* 17:61–78.

Andresen, E., and Levey, D. J. (2004). Effects of dung and seed size on secondary dispersal, seed predation, and seedling establishment of rain forest trees. *Oecologia* 139:45–54.

Augspurger, C. K. (1983). Offspring recruitment around tropical trees: changes in cohort distance with time. *Oikos* 40:189–196.

Augspurger, C. K. (1984). Seedling survival of tropical tree species: interactions of dispersal distance, light-gaps, and pathogens. *Ecology* 65:1705–1712.

Augspurger, C. K., and Kelly, C. K. (1984). Pathogen mortality of tropical tree seedlings: experimental studies of the effects of dispersal distance, seedling density, and light conditions. *Oecologia* 61:211–217.

Balcomb, S. R., and Chapman, C. A. (2003). Bridging the seed dispersal gap: consequences of seed deposition for seedling recruitment in primate-tree interactions. *Ecol. Monogr.* 73:625–642.

Brewer, S. W. (2001). Predation and dispersal of large and small seeds of a tropical palm. *Oikos* 92:245–255.

Brewer, S. W., and Webb, M. A. H. (2001). Ignorant seed predators and factors affecting the seed survival of a tropical palm. *Oikos* 93:32–41.

Cain, M. L., Milligan, B. G., and Strand, A. E. (2000). Long-distance seed dispersal in plant populations. *Am. J. Bot.* 87:1217–1227.

Cain, M. L., Nathan, R., and Levin, S. A. (2003). Long-distance dispersal. *Ecology* 84:1943–1944.

Chapman, C. A. (1987). Flexibility in diets of three species of Costa Rican primates. *Folia Primatol.* 49:90–105.

Chapman, C. A. (1989a). Primate seed dispersal: the fate of dispersed seeds. *Biotropica* 21:148–154.

Chapman, C. A. (1989b). Spider monkey sleeping sites: use and availability. *Am. J. Primatol.* 18:53–60.

Chapman, C. A. (1995). Primate seed dispersal: coevolution and conservation implications. *Evol. Anthropol.* 4:74–82.

Chapman, C. A., Balcomb, S. R., Gillespie, T. R., Skorupa, J. P., and Struhsaker, T. T. (1999). Long-term effects of logging on African primate communities: a 28-year comparison from Kibale National Park, Uganda. *Conserv. Biol.* 14:207–217.

Chapman, C. A., and Chapman, L. J. (1991). The foraging itinerary of spider monkeys: when to eat leaves? *Folia Primatol.* 56:162–166.

Chapman, C. A., and Chapman, L. J. (1995). Survival without dispersers: seedling recruitment under parents. *Conserv. Biol.* 9:675–678.

Chapman, C. A., and Chapman, L. J. (1996). Frugivory and the fate of dispersed and non-dispersed seeds in six African tree species. *J. Trop. Ecol.* 12:491–504.

Chapman, C. A., and Chapman, L. J. (2000). Determinants of group size in primates: the importance of travel costs. In: Boinski, S., and Garber, P. A. (eds.), *On the Move: How and Why Animals Travel in Groups.* University of Chicago Press, Chicago. pp. 24–42.

Chapman, C. A., and Chapman, L. J. (2002). Plant–animal coevolution: is it thwarted by spatial and temporal variation in animal foraging? In: Levey, D. J., Silva, W. R., and Galetti, M. (eds.), *Seed Dispersal and Frugivory: Ecology, Evolution, and Conservation.* CABI Publishing, New York. pp. 275–290.

Chapman, C. A., Chapman, L. J., and Gillespie, T. R. (2002). Scale issues in the study of primate foraging: red colobus of Kibale National Park. *Am. J. Phys. Anthropol.* 117:349–363.

Chapman, C. A., Chapman, L. J., and McLaughlin, R. L. (1989). Multiple central place foraging by spider monkeys: travel consequences of using many sleeping sties. *Oecologia* 79:506–511.

Chapman, C. A., Chapman, L. J., Struhsaker, T. T., Zanne, A. E., Clark, C. J., and Poulsen, J. R. (in press). A thirty-year evaluation of fruit phenology: importance of climate change. In: Boubli, J., Dew, J. L., and Milton, K. (eds.), *Floristics, Phenology and Frugivore Communities: A Pan Tropical Comparison.* Kluwer Press, New York.

Chapman, C. A., Chapman, L. J., Vulinec, K., Zanne, A. E., and Lawes, M. J. (2003a). Fragmentation and alteration to seed dispersal processes: an initial evaluation of dung beetles, seed fate, and seedling diversity. *Biotropica* 35:382–393.

Chapman, C. A., Lawes, M. J., Naughton-Treves, L., and Gillespie, T. R. (2003b). Primate survival in community-owned forest fragments: are metapopulation models useful amidst intensive use? In: Marsh, L. K. (ed.), *Primates in Fragments: Ecology and Conservation.* Kluwer Academic/Plenum, New York. pp. 63–78.

Chapman, C. A., and Onderdonk, D. A. (1998). Forests without primates: primate/plant codependency. *Am. J. Primatol.* 45:127–141.

Chapman, C. A., and Peres, C. A. (2001). Primate conservation in the new millennium: the role of scientists. *Evol. Anthropol.* 10:16–33.

Chapman, C. A., Wrangham, R. W., Chapman, L. J., Kennard, D. K., and Zanne, A. E. (1999). Fruit and flower phenology at two sites in Kibale National Park, Uganda. *J. Trop. Ecol.* 15:189–211.

Chave, J., Muller-Landau, H. C., and Levin, S. A. (2002). Comparing classical community models: theoretical consequences for patterns of diversity. *Am. Nat.* 159:1–23.

Chesson, P. (1996). Matters of scale in the dynamics of populations and communities. In: Floyd, R. B., Sheppard, A. W., and De Barro, P. J. (eds.), *Frontiers in Population Ecology.* CSIRO Publishing, Melbourne. pp. 353–368.

Chesson, P. (1998). Spatial scales in the study of reef fishes: a theoretical perspective. *Aust. J. Ecol.* 23:209–215.

Chesson, P. (2000). General theory of competitive coexistence in spatially-varying environments. *Theor. Popul. Biol.* 58:211–237.

Cipollini, M. L., and Levey, D. J. (1997). Secondary metabolites of fleshy vertebrate-dispersed fruits: adaptive hypotheses and implications for seed dispersal. *Am. Nat.* 150:346–372.

Clark, D. A., and Clark, D. B. (1984). Spacing dynamics of a tropical rain forest tree: evaluation of the Janzen-Connell model. *Am. Nat.* 124:769–788.

Clark, J. S., Lewis, M., McLachlan, J. S., and HilleRisLambers, J. (2003). Estimating population spread: what can we forecast and how well? *Ecology* 84:1979–1988.

Clark, J. S., Silman, M., Kern, R., Macklin, E., and HilleRisLambers, J. (1999). Seed dispersal near and far: patterns across temperate and tropical forests. *Ecology* 80:1475–1494.

Condit, R., Hubbell, S. P., and Foster, R. B. (1992). Recruitment near conspecific adults and the maintenance of tree and shrub diversity in a neotropical forest. *Am. Nat.* 140:261–286.

Connell, J. H. (1971). One the role of natural enemies in preventing competitive exclusion in some marine animals and rain forest trees. In: dem Boer, P. J., and Gradwell, G. R. (eds.), *Dynamics of Populations*. Centre for Agricultural Publishing and Documentation, Wageningen. pp. 298–312.

Cordeiro, N. J., and Howe, H. F. (2003). Low recruitment of trees dispersed by animals in African forest fragments. *Conserv. Biol.* 15:1733–1741.

Corlett, R. T., and Lucas, P. W. (1990). Alternative seed-handling strategies in primates: seed-spitting by long-tailed macaques (*Macaca fascicularis*). *Oecologia* 82:166–171.

Dalling, J. W., and Hubbell, S. P. (2002). Seed size, growth rate and gap microsite conditions as determinants of recruitment success for pioneer species. *J. Ecol.* 90:557–568.

Dalling, J. W., Muller-Landau, H. C., Wright, S. J., and Hubbell, S. P. (2002). Role of dispersal in the recruitment limitation of neotropical pioneer species. *J. Ecol.* 90:714–727.

Davies, G. (1991). Seed-eating by red leaf monkeys (*Presbytis rubicunda*) in dipterocarp forest of northern Borneo. *Int. J. Primatol.* 12:119–144.

DeSteven, D., and Putz, F. E. (1984). Impact of mammals on early recruitment of a tropical canopy tree, *Dipteryx panamensis*, in Panama. *Oikos* 43:207–216.

Dew, J. L. (2001). Synecology and seed dispersal in woolly monkeys (*Lagothrix lagotricha poeppigii*) and spider monkeys (*Ateles belzebuth belzebuth*) in Parque Nacional Uasuni, Ecuador [PhD diss.]. University of California, Davis.

Dew, J. L., and Wright, P. (1998). Frugivory and seed dispersal by four species of primates in Madagascar's eastern rain forest. *Biotropica* 30:425–437.

Dirzo, R., and Miranda, A. (1991). Altered patterns of herbivory and diversity in the forest understory: a case study of the possible consequences of contemporary defaunation. In: Price, P., Lewinsohn, T. M., Fernandes, G. W., and Benson, W. W. (eds.), *Plant–Animal Interactions: Evolutionary Ecology in Tropical and Temperate Regions*. John Wiley and Sons, New York. pp. 273–287.

Duncan, R. S., and Chapman, C. A. (1999). Seed dispersal and potential forest succession in abandoned agriculture in tropical Africa. *Ecol. App.* 9:998–1008.

Ellstrand, N. C., and Elam, D. R. (1993). Population genetics of small population size: implications for plant conservation. *Annu. Rev. Ecol. Syst.* 24:217–242.

Epperson, B. K., and Alvarez-Buylla, E. R. (1997). Limited seed dispersal and genetic structure in life stages of *Cecropia obtusifolia*. *Evolution* 51:275–282.

Estrada, A., and Coates-Estrada, R. (1984). Fruit eating and seed dispersal by howling monkeys (*Alouatta palliata*) in the tropical rain forest of Los Tuxtlas, Mexico. *Am. J. Primatol.* 6:77–91.

Estrada, A., and Coates-Estrada, R. (1986). Frugivory in howling monkeys (*Alouatta palliata*) at Los Tuxtlas, Mexico: dispersal and fate of seeds. In: Estrada, A., and Fleming, T. H. (eds.), *Frugivores and Seed Dispersers*. Dr. W. Junk Publishers, Dordrecht. pp. 94–104.

Feer, F., and Forget, P. M. (2002). Spatio-temporal variation in post-dispersal seed fate. *Biotropica* 34:555–566.

Fenner, M. (1992). *Seeds: The Ecology of Regeneration in Plant Communities*. CAB International, Wallingford, UK.

Fischer, K., and Chapman, C. A. (1993). Frugivores and fruit syndromes: differences in patterns at the genus and species levels. *Oikos* 66:472–482.

Fleagle, J. G. (1999). *Primate Adaptation and Evolution*. Academic Press, New York.

Forget, P. M., and Milleron, T. (1991). Evidence for secondary seed dispersal by rodents in Panama. *Oecologia* 87:596–599.

Frankie, G. W., Baker, H. G., and Opler, P. A. (1974). Comparative phenological studies of trees in tropical wet and dry forests in the lowlands of Costa Rica. *J. Ecol.* 62:881–919.

Garber, P. A. (1986). The ecology of seed dispersal in two species of callitrichid primates (*Saguinus mystax* and *Saguinus fuscicollis*). *Am. J. Primatol.* 10:155–170.

Garber, P. A., and Paciulli, L. M. (1997). Experimental field study of spatial memory and learning in wild capuchin monkeys (*Cebus capucinus*). *Folia Primatol.* 68:236–253.

Gautier-Hion, A. (1980). Seasonal variation of diet related to species and sex in a community of *Cercopithecus* monkeys. *J. Anim. Ecol.* 49:237–269.

Gautier-Hion, A. (1984). La dissemination des graines par les cercopithecides forestiers Africains. *Terre Vie* 39:159–165.

Gautier-Hion, A., Duplantier, J. M., Quris, R., Feer, F., Sourd, C., Decous, J. P., Doubost, G., Emmons, L., Erard, C., Hecketsweiler, P., Moungazi, A., Roussilhon, C., and Thiollay, J. M. (1985). Fruit characters as a basis of fruit choice and seed dispersal in a tropical forest vertebrate community. *Oecologia* 65:324–337.

Gautier-Hion, A., Gautier, J.-P., and Maisels, F. (1993). Seed dispersal versus seed predation: an intersite comparison of two related African monkeys. *Vegetation* 107/108:237–244.

Gillespie, T. R., and Chapman, C. A. (2001). Determinants of group size in the red colobus monkey (*Procolobus badius*): an evaluation of the generality of the ecological-constraints model. *Behav. Ecol. Sociobiol.* 50:329–338.

Grubb, P. J. (1977). The maintenance of species-richness in plant communities: the importance of the regeneration niche. *Biol. Rev.* 52:107–145.

Harms, K. E., Wright, S. J., Calderon, O., Hernandez, A., and Herre, E. A. (2000). Pervasive density-dependent recruitment enhances seedling diversity in a tropical forest. *Nature* 404:493–495.

Harvey, P., Martin, R. D., and Clutton-Brock, T. H. (1987). Life histories in comparative perspective. In: Smuts, B. B., Cheney, D. L., Seyfarth, R. M., Wrangham, R. W., and Struhsaker, T. T. (eds.), *Primate Societies*. University of Chicago Press, Chicago. pp. 181–196.

Hedrick, P. W. (2000). *Genetics of Populations*. Jones and Bartlett, Sudbury.

Herrera, C. M. (2002). Seed dispersal by vertebrates. In: Herrera, C. M., and Pellmyr, O. (eds.), *Plant–Animal Interactions: An Evolutionary Approach*. Blackwell Science, Oxford. pp. 185–210.

Herrera, C. M., Jordano, P., Lopez-Soria, L., and Amat, J. A. (1994). Recruitment of a mast-fruiting, bird-dispersed tree: bridging frugivore activity and seedling establishment. *Ecol. Monogr.* 64:315–344.

Hille Ris Lambers, J., Clark, J. S., and Beckage, B. (2002). Density-dependent mortality and the latitudinal gradient in species diversity. *Nature* 417:732–735.

Howe, H. F. (1989). Scatter- and clump-dispersal and seedling demography: hypothesis and implications. *Oecologia* 79:417–426.

Howe, H. F., Schupp, E. W., and Westley, L. C. (1985). Early consequences of seed dispersal for a neotropical tree (*Virola surinamensis*). *Ecology* 66:781–791.

Howe, H. F., and Smallwood, J. (1982). Ecology of seed dispersal. *Annu. Rev. Ecol. Syst.* 13:201–228.

Hu, X. S., and Ennos, R. A. (1999). Impacts of seed and pollen flow on population genetic structure for plant genomes with three contrasting modes of inheritance. *Genetics* 152:441–450.

Hubbell, S. P. (1980). Seed predation and the coexistence of tree species in tropical forests. *Oikos* 35:214–229.

Hubbell, S. P. (2001). *The Unified Neutral Theory of Biodiversity and Biogeography*. Princeton University Press, Princeton, NJ.

Hurtt, G. C., and Pacala, S. W. (1995). The consequences of recruitment limitation: reconciling chance, history, and competitive differences between plants. *J. Theor. Biol.* 176:1–12.

Hyatt, L. A., Rosenberg, M. S., Howard, T. G., Bole, G., Fang, W., Anastasia, J., Brown, K., Grella, R., Hunman, K., Kurdziel, J. P., and Gurevitch, J. (2003). The distance dependence prediction of the Janzen-Connell hypothesis: a meta-analysis. *Oikos* 103:590–602.

Jansen, P. A., Bartholomeus, M., Bongers, F., Elzinga, J. A., den Ouden, J., and Van Wieren, S. E. (2002). The role of seed size in dispersal by a scatter-hoarding rodent. In: Levey, D. J., Silva, W. R., and Galetti, M. (eds.), *Seed Dispersal and Frugivory: Ecology, Evolution, and Conservation*. CABI Publishing, New York. pp. 209–225.

Janson, C. H. (1983). Adaptation of fruit morphology to dispersal agents in a neotropical forest. *Science* 219:187–189.

Janson, C. H. (1998). Experimental evidence for spatial memory in foraging wild capuchin monkeys, *Cebus apella*. *Anim. Behav.* 55:1229–1243.

Janson, C. H., and Chapman, C. A. (1999). Resources and the determination of primate community structure. In: Fleagle, J. G., Janson, C. H., and Reed, K. (eds.), *Primate Communities*. Cambridge University Press, Cambridge. pp. 237–267.

Janzen, D. H. (1970). Herbivores and the number of tree species in tropical forests. *Am. Nat.* 104:501–528.

Jordano, P., and Herrera, C. M. (1995). Shuffling the offspring: uncoupling and spatial discordance of multiple stages in vertebrate seed dispersal. *Ecoscience* 2:230–237.

Julliot, C. (1996). Seed dispersal by red howling monkeys (*Alouatta seniculus*) in the tropical rain forest of French Guiana. *Int. J. Primatol.* 17:239–258.

Julliot, C. (1997). Impact of seed dispersal by red howler monkeys *Alouatta seniculus* on the seedling population in the understory of tropical rain forest. *J. Ecol.* 85:431–440.

Kaplin, B. A., and Lambert, J. E. (2002). Effectiveness of seed dispersal by *Cercopithecus* monkeys: implications for seed input into degraded areas. In: Levey, D. J., Silva, W. R., and Galetti, M. (eds.), *Seed Dispersal and Frugivory: Ecology, Evolution, and Conservation*. CABI Publishing, New York. pp. 351–364.

Kaplin, B. A., and Moermond, T. C. (1998). Variation in seed handling by two species of forest monkeys in Rwanda. *Am. J. Primatol.* 45:83–101.

Kaplin, B. A., Munyaligoga, V., and Moermond, T. C. (1998). The influence of temporal changes in fruit availability on diet composition and seed handling in blue monkeys (*Cercopithecus mitis doggetti*). *Biotropica* 30: 56–71.

Kay, R. N., and Davies, A. G. (1994). Digestive physiology. In: Davies, A. G., and Oates, J. F. (eds.), *Colobine Monkeys: Their Ecology, Behaviour and Evolution*. Cambridge University Press, Cambridge. pp. 229–249.

Kinzey, W. G., and Norconk, M. A. (1990). Hardness as a basis of fruit choice in two sympatric primates. *Am. J. Phys. Anthropol.* 81:5–15.

Kinzey, W. G., and Norconk, M. A. (1993). Physical and chemical properties of fruit and seeds eaten by *Pithecia* and *Chiropotes* in Suriname and Venezuela. *Int. J. Primatol.* 14:207–227.

Kollmann, J., Coomes, D. A., and White, S. M. (1998). Consistencies in post-dispersal seed predation of temperate fleshy-fruited species among seasons, years, and sites. *Funct. Ecol.* 12:683–690.

Lambert, J. E. (1997). Digestive strategies, fruit processing, and seed dispersal in the chimpanzees (*Pan troglodytes*) and redtail monkeys (*Cercopithecus ascanius*) of Kibale National Park, Uganda [PhD diss.]. University of Illinois, Urbana-Champaign.

Lambert, J. E. (1998). Primate digestion: interactions among anatomy, physiology, and feeding ecology. *Evol. Anthropol.* 7:8–20.

Lambert, J. E. (1999). Seed handling in chimpanzees (*Pan troglodytes*) and redtail monkeys (*Cercopithecus ascanius*): implications for understanding hominoid and cercopithecine fruit-processing strategies and seed dispersal. *Am. J. Phys. Anthropol.* 109:365–386.

Lambert, J. E. (2000). The fate of seeds dispersed by African apes and cercopithecines. *Am. J. Phys. Anthropol.* (suppl.) 30:204.

Lambert, J. E. (2001). Red-tailed guenons (*Cercopithecus ascanius*) and *Strychnos mitis*: evidence for plant benefits beyond seed dispersal. *Int. J. Primatol.* 22:189–201.

Lambert, J. E. (2002). Exploring the link between animal frugivory and plant strategies: the case of primate fruit-processing and post-dispersal seed fate. In: Levey, D. J., Silva, W. R., and Galetti, M. (eds.), *Seed Dispersal and Frugivory: Ecology, Evolution and Conservation*. CABI Publishing, New York. pp. 365–379.

Lambert, J. E., Chapman, C. A., Wrangham, R. W., and Conklin-Brittain, N. L. (2004). The hardness of cercopithecine foods: implications for the critical function of enamel thickness in exploiting fallback foods. *Am. J. Phys. Anthropol.* 125:363–368.

Leighton, M. (1993). Modeling dietary selectivity by Bornean orangutans: evidence for integration of multiple criteria in fruit selection. *Int. J. Primatol.* 14:257–313.

Levin, S. A. (1974). Dispersion and population interactions. *Am. Nat.* 108:207–228.

Lieberman, D., Hall, J. B., Swaine, M. D., and Lieberman, M. (1979). Seed dispersal by baboons in the Shai Hills, Ghana. *Ecology* 60:65–75.

Loveless, M. D., and Hamrick, J. L. (1984). Ecological determinants of genetic structure in plant populations. *Annu. Rev. Ecol. Syst.* 15:65–95.

Lucas, P. W., and Corlett, R. T. (1998). Seed dispersal by long-tailed macaques. *Am. J. Primatol.* 45:29–44.

McCanny, S. J. (1985). Alternatives in parent–offspring relationships in plants. *Oikos* 45:148–149.

McConkey, K. R. (2000). Primary seed shadow generated by gibbons in the rain forests of Barito Ulu, central Borneo. *Am. J. Primatol.* 52:13–29.

McFarland, M. J. (1986). Ecological determinants of fission–fusion sociality in *Ateles* and *Pan*. In: Else, J. G., and Lee, P. C.

(eds.), *Primate Ecology and Conservation*. Cambridge University Press, Cambridge. pp. 181–190.

Milton, K. (1980). *The Foraging Strategy of Howler Monkeys: A Study in Primate Economics*. Columbia University Press, New York.

Milton, K. (1981). Food choice and digestive strategies by two sympatric primate species. *Am. Nat.* 117:496–505.

Milton, K. (1986). Digestive physiology in primates. *News Phys. Sci.* 1:76–79.

Murray, K. G., Russell, S., Picone, C. M., Winnett Murray, K., Sherwood, W., and Kuhlmann, M. L. (1994). Fruit laxatives and seed passage rates in frugivores: consequences for plant reproductive success. *Ecology* 75:989–994.

Nathan, R., and Muller-Landau, H. C. (2000). Spatial patterns of seed dispersal, their determinants and consequences for recruitment. *TREE* 15:278–285.

Nathan, R., Perry, G., Cronin, J. T., Strand, A. E., and Cain, M. L. (2003). Methods for estimating long-distance dispersal. *Oikos* 103:261–273.

Norconk, M. A., Grafton, B. W., and Conklin-Brittain, N. L. (1998). Seed dispersal by neotropical seed predators. *Am. J. Primatol.* 45:103–126.

Nunn, C. L., and Barton, R. A. (2000). Allometric slopes and independent contrasts: a comparative test of Kleiber's law in primate ranging patterns. *Am. Nat.* 156:519–533.

Oliveira, A. C. M., and Ferrari, S. F. (2000). Seed dispersal by black-handed tamarins, *Saguinus midas niger* (Callitrichinae, Primates): implications for the regeneration of degraded forest habitats in eastern Amazonia. *J. Trop. Ecol.* 16:709–716.

Pacala, S. W., and Tilman, D. (1994). Limiting similarity in mechanistic and spatial models of plant competition in heterogeneous environments. *Am. Nat.* 143:222–257.

Pacheco, L. F., and Simonetti, J. A. (2000). Genetic structure of a mimosoid tree deprived of its seed disperser, the spider monkey. *Conserv. Biol.* 14:1766–1775.

Paul, J. R. (2001). Patterns of seed dispersal by animals: influence on sapling composition in a tropical forest [PhD thesis]. University of Florida, Gainesville.

Peres, C. A. (1990). Effects of hunting on western Amazonian primate communities. *Biol. Conserv.* 54:47–59.

Peres, C. A., and Dolman, P. M. (2000). Density compensation in neotropical primate communities: evidence from 56 hunted and nonhunted Amazonian forests of varying productivity. *Oecologia* 122:175–189.

Poulsen, J. R., Clark, C. J., Connor, E. F., and Smith, T. B. (2002). Differential resource use by primates and hornbills: implications for seed dispersal. *Ecology* 83:228–240.

Poulsen, J. R., Clark, C. J., and Smith, T. B. (2001). Seed dispersal by a diurnal primate community in the Dja Reserve, Cameroon. *J. Trop. Ecol.* 17:787–808.

Redford, K. H. (1992). The empty forest. *Bioscience* 42:412–422.

Rey, P. J., and Alcantara, J. M. (2000). Recruitment dynamics of a fleshy-fruited plant (*Olea europaea*): connecting patterns of seed dispersal to seedling establishment. *J. Ecol.* 88:622–633.

Ribbens, E., Silander, J. A., and Pacala, S. W. (1994). Seedling recruitment in forests: calibrating models to predict patterns of tree seedling dispersion. *Ecology* 75:1794–1806.

Rogers, M. E., Voysey, B. C., McDonald, K. E., Parnell, R. J., and Tutin, C. E. G. (1998). Lowland gorillas and seed dispersal: the importance of nest sites. *Am. J. Primatol.* 45:45–68.

Rowell, T. E., and Mitchell, B. J. (1991). Comparison of seed dispersal of guenons in Kenya and capuchins in Panama. *J. Trop. Ecol.* 7:269–274.

Russo, S. E. (2003a). Linking spatial patterns of seed dispersal and plant recruitment in a neotropical tree, *Virola calophylla* (Myristicaceae) [PhD thesis]. University of Illinois, Urbana.

Russo, S. E. (2003b). Responses of dispersal agents to tree and fruit traits in *Virola calophylla* (Myristicaceae): implications for selection. *Oecologia* 136:80–87.

Russo, S. E. (2005). Linking seed fate to natural dispersal patterns: factors affecting predation and scatter-hoarding of *Birola calophylla* seeds in Peru. *J. Trop. Ecol.* 21:243–253.

Russo, S. E., and Augspurger, C. K. (2004). Aggregated seed dispersal by spider monkeys limits recruitment to clumped patterns in *Virola calophylla*. *Ecol. Lett.* 7:1058–1067.

Russo, S. E., Campbell, C. J., Dew, J. L., Stevenson, P. R., and Suarez, S. (2005). A multi-site comparison of dietary preferences and seed dispersal by spider monkeys (*Ateles* spp.). *Int. J. Primatol.* 26:1017–1037.

Schmitt, J., and Antonovics, J. (1986). Experimental studies of the evolutionary significance of sexual reproduction: III. Maternal and paternal effects during seedling establishment. *Evolution* 40:817–829.

Schnabel, A., Nason, J. D., and Hamrick, J. L. (1998). Understanding the population genetic structure of *Gleditsia triacanthos* L.: seed dispersal and variation in female reproductive success. *Mol. Ecol.* 7:819–832.

Schupp, E. W. (1988). Seed and early seedling predation in the forest understory and in treefall gaps. *Oikos* 51:71–78.

Schupp, E. W. (1992). The Janzen-Connell model for tropical tree diversity: population implications and the importance of spatial scale. *Am. Nat.* 140:526–530.

Schupp, E. W., and Fuentes, M. (1995). Spatial patterns of seed dispersal and the unification of plant-population ecology. *Ecoscience* 2:267–275.

Schupp, E. W., Milleron, T., and Russo, S. E. (2002). Dissemination limitation and the origin and maintenance of species-rich tropical forests. In: Levey, D. J., Silva, W. R., and Galetti, M. (eds.), *Seed Dispersal and Frugivory: Ecology, Evolution, and Conservation*. CABI Publishing, New York. pp. 19–34.

Shepherd, V. E., and Chapman, C. A. (1998). Dung beetles as secondary seed dispersers: impact on seed predation and germination. *J. Trop. Ecol.* 14:199–215.

Sork, V. L., Nason, J., Campbell, D. R., and Fernandez, J. F. (1999). Landscape approaches to historical and contemporary gene flow in plants. *TREE* 14:219–224.

Sork, V. L., Stowe, K. A., and Hochwender, C. (1993). Evidence for local adaptation in closely adjacent subpopulations of northern red oak (*Quercus rubra* L.) expressed as resistance to leaf herbivores. *Am. Nat.* 142:928–936.

Stevenson, P. R. (2000). Seed dispersal by woolly monkeys (*Lagothrix lagothricha*) at Tinigua National Park, Colombia: dispersal distance, germination rates, and dispersal quantity. *Am. J. Primatol.* 50:275–289.

Stevenson, P. R., Castellanos, M. C., Pizarro, J. C., and Garavito, M. (2002). Effects of seed dispersal by three ateline monkey species on seed germination at Tinigua National Park, Colombia. *Int. J. Primatol.* 23:1187–1204.

Symington, M. M. (1987a). Demography, ranging patterns, and activity budgets of black spider monkeys (*Ateles paniscus chamek*) in the Manú National Park, Perú. *Am. J. Primatol.* 15:45–67.

Symington, M. M. (1987b). Ecological and social correlates of party size in the black spider monkey, *Ateles paniscus chamek* [PhD thesis]. Princeton University, Princeton, NJ.

Terborgh, J. (1983). *Five New World Primates*. Princeton University Press, Princeton, NJ.

Terborgh, J., Pitman, N., Silman, M., Schichter, H., and Nunez, P. V. (2002). Maintenance of tree diversity in tropical forests. In: Levey, D. J., Silva, W. R., and Galetti, M. (eds.), *Seed Dispersal and Frugivory: Ecology, Evolution and Conservation*. CABI Publishing, New York. pp. 351–364.

Tilman, D. (1994). Competition and biodiversity in spatially structured habitats. *Ecology* 75:2–16.

van der Pijl, L. (1982). *Principles of Dispersal in Higher Plants*. Springer-Verlag, Berlin.

van Roosmalen, M. G. M. (1984). Subcategorizing foods in primates. In: Chivers, D. J., Wood, B. A., and Bilborough A. (eds.), *Food Acquisition and Processing in Primates*. Plenum Press, New York. pp. 167–176.

van Schaik, C. P., Terborgh, J. W., and Wright, S. J. (1993). The phenology of tropical forests: adaptive significance and consequences for primary consumers. *Annu. Rev. Ecol. Syst.* 24:353–377.

Voysey, B. C., McDonald, K. E., Rogers, M. E., Tutin, C. E. G., and Parnell, R. J. (1999a). Gorillas and seed dispersal in the Lope Reserve, Gabon. I: Gorilla acquisition by trees. *J. Trop. Ecol.* 15:23–38.

Voysey, B. C., McDonald, K. E., Rogers, M. E., Tutin, C. E. G., and Parnell, R. J. (1999b). Gorillas and seed dispersal in the Lope Reserve, Gabon. II: Survival and growth of seedlings. *J. Trop. Ecol.* 15:39–60.

Vulinec, K. (2002). Dung beetle communities and seed dispersal in primary forest and disturbed land in Amazonia. *Biotropica* 34:297–309.

Wang, B. C., and Smith, T. B. (2002). Closing the seed dispersal loop. *TREE* 17:379–385.

Waterman, P. G., and Kool, K. (1994). Colobine food selection and plant chemistry. In: Davies, G., and Oates, J. F. (eds.), *Colobine Monkeys: Their Ecology, Behaviour and Evolution*. Cambridge University Press, Cambridge. pp. 251–284.

Webb, C. O., and Peart, D. R. (2000). Habitat associations of trees and seedlings in a Bornean rain forest. *J. Ecol.* 88:464–478.

Wehncke, E. V., Hubbell, S. P., Foster, R. B., and Dalling, J. W. (2003). Seed dispersal patterns produced by white-faced monkeys: implications for the dispersal limitation of neotropical tree species. *J. Ecol.* 91:677–685.

Wenny, D. G. (1999). Two-stage dispersal of *Guarea glabra* and *G. kunthiana* (Meliaceae) in Monteverde, Costa Rica. *J. Trop. Ecol.* 15:481–496.

Wrangham, R. W., Chapman, C. A., and Chapman, L. J. (1994). Seed dispersal by forest chimpanzees in Uganda. *J. Trop. Ecol.* 10:355–368.

Wright, S. (1943). Isolation by distance. *Genetics* 28:114–138.

Wright, S. J., Zeballos, H., Dominguez, I., Gallardo, M. M., Moreno, M. C., and Ibanez, R. (2000). Poachers alter mammal abundance, seed dispersal, and seed predation in a neotropical forest. *Conserv. Biol.* 14:227–239.

Zhang, S. Y., and Wang L. X. (1995). Fruit consumption and seed dispersal of *Ziziphus cinamomum* (Rhamnaceae) by two species of primates (*Cebus apella* and *Ateles paniscus*) in French Guiana. *Biotropica* 27:397–401.

APPENDIX 31.1. GLOSSARY

Dispersal kernel: A probability density function describing the probability of a seed being deposited per unit area at locations measured with respect to the parent plant. See also Nathan and Muller-Landau (2000).

Endozoochory: Seed dispersal in which seeds are transported inside an animal. For primates, this can occur when seeds are ingested into the gut and passed (*defecation dispersal*) or when seeds are processed in cheek pouches and later spat (*spit dispersal*).

Regeneration niche: A term coined by Grubb (1977) referring to the biotic and abiotic requirements that ensure a high probability of success in the replacement of one mature individual of a plant species by another mature individual of that species. When seeds of a species are dispersed to a habitat outside of its regeneration niche, these seeds have a low probability of germinating, establishing seedlings, or reaching adulthood.

Seed density: The number of seeds in a unit area. Seed density is often used as an indicator of the extent to which dispersed seeds are clumped (higher seed density in a small area) or scattered (lower seed density in that area). See Howe (1989).

Seed dispersal: The movement of seeds away from a parent plant. *Primary seed dispersal* for animal-dispersed seeds refers to the movement of the seed away from the parent by the first dispersal agent, aside from gravity. *Secondary dispersal* refers to subsequent movement of seeds after they have been dispersed by primary dispersal. Secondary dispersal is often by dung beetles (Scarabeidae). When primary dispersal agents defecate seeds, dung beetles accidentally roll seeds up in the dung balls in which they lay their eggs, moving the seeds short distances and burying them in the process. Secondary dispersal also occurs by rodents that cache seeds under the leaf litter or soil (scatter-hoarding and larder-hoarding) for later consumption (see *seed predation*). However, some seeds cached by rodents escape predation because they are not relocated and are, thus, dispersed.

Seed-dispersal curve: The frequency distribution of distances that seeds are moved from the parent plant. Seed-dispersal curves may also be expressed in terms of the density of seeds in annuli (rings) of a particular width surrounding a parent plant, which accounts for the area over which seeds are deposited at each distance from a parent.

Seed-dispersal distance: The straight-line distance a seed travels away from its parent.

Seed passage time: The amount of time that a seed takes to pass through the gut of its animal dispersal agent.

Seed predation: The consumption of seeds, in particular the endosperm, cotyledons, or embryo, rendering the seed, by definition, nonviable for germination. Seed predation is often by either arboreal or terrestrial birds and mammals (particularly rodents) but also by insects that lay eggs inside

seeds, where the larvae develop and consume the seed, or by pathogens, such as fungi.

Seed shadow: The spatial distribution of seeds dispersed from a single plant. Seed shadows from multiple individuals may be summed over an area to represent the spatial distribution of seeds for a population of plants

32

Predation on Primates

Past Studies, Current Challenges, and Directions for the Future

Lynne E. Miller and Adrian Treves

INTRODUCTION

Predation has long been thought to be a major selective pressure shaping the morphology and behavior of primates (Alexander 1974, Cheney and Wrangham 1987, Clutton-Brock and Harvey 1977, Crook and Gartlan 1966, Eisenberg et al. 1972, Goss-Custard et al. 1972, Rodman 1973, van Schaik 1983, Struhsaker 1969, Terborgh and Janson 1986). However, the many challenges of studying primate predator–prey interactions have limited our ability to make rigorous tests of hypothetical relationships. Several reviews have noted the paucity of reliable data on patterns of predation (Boinski and Chapman 1995; Boinski et al. 2000; Isbell 1994; Janson 1998; Treves 1999, 2000). Certainly, more data become available every year, but they still often represent anecdotal accounts rather than systematic studies of interacting variables; therefore, this chapter will again be a call for further attention to this important area of research. It is important because our efforts to understand current primate behavior and past primate evolution will benefit from understanding the subtle nuances of how and when predators exert directional selection and when predation is minor relative to other selective pressures.

This chapter begins with a brief review of the reports of predation on primates, along with some caveats about using such data for quantitative analyses. We then move on to methodological issues, including the challenges of studying predator–primate interactions, and experimental protocols that might allow researchers to tease apart the impact of multiple

selective forces. From there, we will discuss the growing body of data regarding the behaviors associated with predation risk, including the formation of social groups; manipulation of foraging, sleeping, and ranging patterns; and vigilance and alarm calling. This discussion will include reference to the variables that may mediate an individual's vulnerability to predation, which may in turn influence its willingness to take risks. By this approach, we hope to encourage multifactorial analyses of predator–primate interactions and, thus, begin to develop a more realistic model of this complex relationship. The chapter will close with a brief discussion of the role of humans as predators of nonhuman primates, addressing the extent to which humans represent a special category of predators. Although hard data on predator–primate interactions are few, there is an enormous body of literature on the myriad factors related to predation. In a chapter of this length, it is impossible to review all of these studies; thus, our aim is to provide an overview of current issues, along with a review of classic and recent publications in each area.

ACCOUNTS OF PREDATOR–PRIMATE INTERACTIONS

There is a substantial body of reports of predatory attacks on primates. Recent publications have summarized these accounts (Boinski et al. 2000, Treves 1999). Table 32.1 presents an updated list of prey and predator taxa. We have included accounts of successful and unsuccessful attacks, along with a few data based upon scat and nest debris (though we recognize that these remains may represent scavenging behavior rather

Table 32.1   Reports of Primates as Prey

PREY GENUS	PREDATOR SPECIES	REFERENCES
Lemur	Red-fronted lemur (*Eulemur fulvus rufus*)	Jolly et al. (2000)
	Fossa (*Cryptoprocta ferox*)	Goodman et al. (1993)
	Civet (*Civettictis civetta*)	Goodman et al. (1993)
	Feral cat (unknown)	Goodman et al. (1993)
Eulemur	Henst's goshawk (*Accipiter henstii*)	Goodman et al. (1998), Karpanty (2003), Overdorff and Strait (1995), Schwab (1999)
	Fossa (*Cryptoprocta ferox*)	Wright et al. (1998)
Hapalemur	Madagascar harrier-hawk (*Polyboroides radiatus*)	Karpanty (2003)
	Henst's goshawk (*Accipiter henstii*)	Karpanty (2003)
	Boa constrictor (*Boa manditra*)	Rakotondravony et al. (1998)
	Fossa (*Cryptoprocta ferox*)	Wright et al. (1998)
Avahi	Henst's goshawk (*Accipiter henstii*)	Goodman et al. (1998), Karpanty (2003)
	Madagascar buzzard (*Buteo* sp.)	Wright et al. (1998)
Propithecus	Madagascar harrier-hawk (*Polyboroides radiatus*)	Brockman (2003)
	Boa constrictor (*Acrantophis madagascariensis*)	Burney (2002)
	Fossa (*Cryptoprocta ferox*)	Wright et al. (1997), Wright et al. (1998)
Lepilemur	Madagascar harrier-hawk (*Polyboroides radiatus*)	Schulke and Ostner (2001)
Cheirogaleus	Madagascar harrier-hawk (*Polyboroides radiatus*)	Gilbert and Tingay (2001)
	Henst's goshawk (*Accipiter henstii*)	Karpanty (2003)
	Madagascar buzzard (*Buteo* sp.)	Wright et al. (1998)
	Madagascar boa (*Sanzinia madagascariensis*)	Wright et al. (1998)
	Ring-tailed mongoose (*Galidia elegans*)	Wright et al. (1998)
Microcebus	Madagascar harrier-hawk (*Polyboroides radiatus*)	Karpanty (2003), Wright et al. (1998)
	Henst's goshawk (*Accipiter henstii*)	Karpanty (2003)
	Ring-tailed mongoose (*Galidia elegans*)	Wright et al. (1998)
Galago	Genet (*Genetta* sp.)	Andrews and Evans (1979)
	Chimpanzee (*Pan troglodytes*)	Gaspersic and Pruetz (2004)
Nycticebus	Reticulated python (*Python reticulatus*)	Wiens and Zitzmann (1999)
	Orangutan (*Pongo pygmaeus*)	Utami and van Hooff (1997)
Tarsius	Reticulated python (*Python reticulatus*)	Gursky (2002)
Callithrix	Pit viper (*Bothrops jararaca*)	Correa and Coutinho (1997)
Saguinus	Ornate hawk eagle (*Spizaeus ornatus*)	Terborgh (1983)
	Bicolored hawk (*Accipiter bicolor*)	Terborgh (1983)
	Crested eagle (*Morphnus guianensis*)	Oversluijs Vasques and Heymann (2001)
	Anaconda (*Eunectes* sp.)	Heymann (1987)
	Boa constrictor (*Boa constrictor*)	Tello et al. (2002)
	Ocelot (*Felis pardalis*)	Heymann (1990)
	Tayra (*Eira barbara*)	Buchanan-Smith (1990), Galef et al. (1976)
Saimiri	Harpy eagle (*Harpia harpyja*)	Terborgh (1983)
	Guianan crested eagle (*Morphnus guianensis*)	Terborgh (1983)
	Ornate hawk eagle (*Spizaeus ornatus*)	Terborgh (1983)
	Black and white hawk eagle (*Spizastur melanoleucos*)	Terborgh (1983)
	Slate-colored hawk (*Leucopternis shistacea*)	Terborgh (1983)
	Bicolored hawk (*Accipiter bicolor*)	Terborgh (1983)
	Tayra (*Eira barbara*)	Galef et al. (1976)
Callicebus	Guianan crested eagle (*Morphnus guianensis*)	Terborgh (1983)
	Ornate hawk eagle (*Spizaeus ornatus*)	Terborgh (1983)
	Tufted capuchin (*Cebus apella*)	Sampaio and Ferrari (2005)

Table 32.1 (*cont'd*)

PREY GENUS	PREDATOR SPECIES	REFERENCES
Cebus	Harpy eagle (*Harpia harpyja*)	Fowler and Cope (1964), Izor (1985), Terborgh (1983)
	Guianan crested eagle (*Morphnus guianensis*)	Terborgh (1983)
	Boa constrictor (*Boa constrictor*)	Chapman (1986)
	Tayra (*Eira barbara*)	Defler (1980), Phillips (1985)
Pithecia	Harpy eagle (*Harpia harpyja*)	Rettig (1978)
	Crested eagle (*Morphnus guianensis*)	Gilbert (2000)
Ateles	Crested eagle (*Morphnus guianensis*)	Julliot (1994)
	Jaguar (*Panthera onca*)	Emmons (1987)
	Puma (*Felis concolor*)	Chinchilla (1997)
Brachyteles	Jaguar (*Panthera onca*)	Olmos (1994)
Alouatta	Harpy eagle (*Harpia harpyja*)	Eason (1989), Fowler and Cope (1964), Izor (1985), Peres (1990a), Rettig (1978), Sherman (1991), Terborgh (1983)
	Jaguar (*Panthera onca*)	Chinchilla (1997, cited in Di Fiore 2002), Cuaron (1997), Peetz et al. (1992)
	Puma (*Felis concolor*)	Chinchilla (1997, cited in Di Fiore 2002)
	Tayra (*Eira barbara*)	Phillips (1985)
Cercopithecus (*Chlorocebus*)	African crowned eagle (*Stephanoaetus coronatus*)	Brown (1962, 1971), Maisels et al. (1993), Sanders and Mitani (2000), Sanders et al. (2003), Schultz et al. (2004), Struhsaker and Leakey (1990)
	Leopard (*Panthera pardus*)	Cheney et al. (1981), Isbell (1990), Zuberbuhler and Jenny (2002)
	Chimpanzee (*Pan troglodytes*)	Gaspersic and Pruetz (2004), Goodall (1986), Mitani and Watts (1999), Newton-Fisher et al. (2002), Nishida et al. (1992)
Cercocebus	African crowned eagle (*Stephanoaetus coronatus*)	Sanders and Mitani (2000), Schultz et al. (2004), Struhsaker and Leakey (1990)
	Yellow baboon (*Papio cynocephalus*)	Hausfater (1976)
	Leopard (*Panthera pardus*)	Zuberbuhler and Jenny (2002)
Papio	Black eagle (*Ictinaetus malayensis*)	Boshoff et al. (1991)
	Spotted hyena (*Crocuta crocuta*)	Stelzner and Strier (1981)
	Leopard (*Panthera pardus*)	Busse (1980)
	Lion (*Panthera leo*)	Busse (1980), Condit and Smith (1994)
Macaca	Crocodile (*Crocodylus porosus*)	Galdikas and Yeager (1984)
	Leopard (*Panthera pardus*)	Ramakrishnan et al. (1999), Seidensticker (1983)
Colobus	Crowned eagle (*Stephanoaetus coronatus*)	Sanders and Mitani (2000), Struhsaker and Leakey (1990)
	Leopard (*Panthera pardus*)	Zuberbuhler and Jenny (2002)
	Chimpanzees (*Pan troglodytes*)	Mitani and Watts (1999) Newton-Fisher et al. (2002)
Procolobus	Crowned eagle (*Stephanoaetus coronatus*)	Sanders and Mitani (2000)
	Leopard (*Panthera pardus*)	Zuberbuhler and Jenny (2002)
	Chimpanzee (*Pan troglodytes*)	Boesch and Boesch (1989), Boesch (1994), Goodall (1986), Mitani and Watts (1999), Nishida et al. (1992), Stanford et al. (1994), Wrangham and van Zinnicq Bergmann Riss (1990)
Presbytis (*Semnopithecus*) (*Trachypithecus*)	Python (*Python reticulatus*)	Shine et al. (1998)
	Jackal (*Canis aureus*)	Stanford (1989)
	Leopard (*Panthera pardus*)	Ramakrishnan et al. (1999), Seidensticker (1983)
Nasalis	Clouded leopard (*Neofelis nebulosa*)	Jackson and Nowell (1996, cited in Uhde and Sommer 2002)
Hylobates	Leopard (*Panthera pardus*)	Rabinowitz (1989, cited in Uhde and Sommer 2002)
Symphalangus	Python (*Python sp.*)	Schneider (1906, cited in Uhde and Sommer 2002)
Pan	Leopard (*Panthera pardus*)	Boesch (1991), Furuichi (2000), Zuberbuhler and Jenny (2002)
	Lion (*Panthera leo*)	Tsukahara (1993)
Gorilla	Leopard (*Panthera pardus*)	Fay et al. (1995)

than predation). We have not included accounts of attacks among conspecifics (e.g., accounts of intraspecific infanticide) for these fall outside of our proposed discussion; however, we have included those cases where primates have preyed upon members of other primate species, for example, chimpanzees (*Pan troglodytes*) preying upon red colobus monkeys (*Procolobus badius, a.k.a. Piliocolobus badius*). We have chosen not to include human predators in this summary, postponing this issue for the end of the chapter. Besides the accounts of attacks summarized here, many additional publications have described other types of interaction between primates and potential predators, interactions that might include flight, mobbing, alarm calling, etc. Those are not included in the table but are discussed below.

We hope that this summary table is useful to our readers; it certainly demonstrates that diverse primate taxa (from *Microcebus* to *Gorilla*) are attacked by a variety of predators (especially large raptors, felids, and reptiles). However, while such a summary may serve to generate hypotheses, there are major problems in using it to test such hypotheses. First, the data set is surely incomplete because we must have missed some relevant publications. Second, while it summarizes the genera of primates and their predators, it does not indicate the number of predatory incidents for any given genus because such data are often lacking. Third, it only weakly represents the actual distribution of predators across primate taxa because of the rarity of observation of predation (particularly for rare, arboreal, or nocturnal primates and those facing predators wary of humans) and the fact that more predation events are likely to have been observed for those primate taxa that have been the subject of long-term observational research. For example, a glance at the table would suggest that *Alouatta* is much more vulnerable than *Callithrix* or *Tarsius*, as indicated by the relative numbers of predation accounts; however, this disparity more probably represents a difference in observer hours in the field. Thus, we must be very cautious about drawing conclusions from anecdotal data.

A fourth major difficulty with analyses based upon anecdotal accounts is the lack of contextual data. Such accounts rarely include details about the predator's attack strategy or relevant information about the age, sex, social status, microhabitat, or behavior of the victims and/or the survivors; therefore, these accounts shed little light on the relative vulnerability of the victims or survivors. Furthermore, even a perfectly observed and fully documented event might reveal nothing about the selective pressure imposed by predation because the event might be random with regard to the genotype or phenotype of the prey. Only with large samples of predator encounters can we hope to understand how predation imposes directional selection on a population and therefore might shape the evolution of behavior. Even studies that have generated adequate sample sizes (such as Stanford's work with red colobus predation by chimpanzees: Stanford 1995, 1998, 2002) often provide no more than a glimpse of whether a particular behavior predisposed certain individuals to predation or loss of offspring, prerequisites for selection to be acting. Overall, collections of anecdotal data are unlikely to contribute significantly to our understanding of the selective role of predation.

Despite these problems, several authors have used published data on rates of predation and/or predator encounters in selected populations to produce cross-taxonomic analyses, particularly to isolate and elucidate the role of variables like body size and group size (Anderson 1986, Hill and Dunbar 1998, Hill and Lee 1998, Isbell 1994, Lycett et al. 1998; see also Boinski and Chapman 1995 for a critique of these approaches). These attempts have produced some conflicting results. One factor that may explain conflicting results is the relative selective impact of predation *rate* versus predation risk (Cowlishaw 1997b, Dunbar 1988, Hill and Dunbar 1998, Janson 1998, Young 1994). Predation *rate* is the frequency of successful predatory attacks for a given group, population, species, etc. Predation *risk* represents the rate of encounters or unsuccessful attacks. Predation rates are hugely variable, across taxa and over time. Cheney and Wrangham (1987) estimated that predation rates vary from 0% to 15%, with an average of 3%, though Isbell (1990) witnessed a period in which 45% of vervets were preyed upon by leopards and Stanford (1995) documented 25% mortality caused by predation in a population of red colobus. However, even where rates of successful predation are low, frequent encounters will increase an individual's perception of risk, and this alone might stimulate significant behavioral response. In addition to predation rate and perceived risk, individual vulnerability must also be considered. *Vulnerability* reflects the individual's experience, wariness, and a variety of biological and social variables not necessarily shared by associates (Miller 2002a). There is an element of chance in predator encounters such that any individual can potentially fall prey but vulnerable ones are more likely to be taken. It is likely that the significant impact predation can have on the inclusive fitness of slowly reproducing animals serves to make primates highly responsive to predators (Abrams 1994, Altmann 1974, Hill and Dunbar 1998, Janson 1992, Lima 1998, Lima and Dill 1990).

THE STUDY OF PREDATOR–PRIMATE INTERACTIONS

Predation is notoriously difficult to observe. Events happen quickly, and many predators are active at night when observation conditions are poor. In addition, human observers may have a significant influence on predator behavior, deterring approach and attack (Isbell and Young 1993a). Another challenge to studying predation and its impact on primate evolution is the modern, altered environments in which we are making our observations; these often differ significantly from the environments in which primates evolved. Human transformation of primate habitats and active persecution of many predators (Berger et al. 2001, McDougal 1987, Treves and Naughton-Treves 1999) have surely altered patterns of predation. Human modification of habitats or depletion of predator populations can raise, lower, or leave unchanged the rate and risk of predation. For example, in South Asia leopards preyed less on primates when the dominant tiger was eliminated by human activities (Seidensticker 1983). Humans sometimes remove virtually all the predators of primates, leading to widespread reduction in risk (Southwick et al. 1965), while at other times, humans replace wild predators and impose unprecedented rates of predation on primate populations (Naughton-Treves et al. 2003). Given such variation, uniform generalizations about all primate habitats will not do. We must seek a more thorough and nuanced understanding of predator–prey ecology if we wish to understand the evolutionary role of predation. At the same time, we cannot assume that morphological or behavioral characteristics that exist today, even those that clearly serve antipredator functions, were originally selected to serve this purpose (Byers 1997, Rodman 1988). Our understanding of predation as a selective pressure in the evolution of primate behavior and morphology continues to be limited by our data and our understanding of ecosystem complexity and integrity.

Accurate accounts of predator–primate encounters, even those that result in no mortality, are very useful. In order to augment our understanding of predator–prey dynamics, primatologists should be encouraged to make detailed observations of predator attack patterns; microhabitats in which they occur; the age, sex, condition, rank, social context, and spatial location of as many victims and survivors as possible; long-lasting changes in the behavior of the survivors, such as avoidance of that portion of the range and/or increased group cohesion, vigilance, or alarm calling; and how such postattack behaviors diminish over time. Several projects have amassed important data on predation of certain primate taxa (Cheney et al. 1988 on vervets, Wright et al. 1998 on prosimians, Newton-Fisher et al. 2002 and Stanford et al. 1994 on predation by chimpanzees on other primate species). Such data can help to fill the gaps in our understanding of the behavioral avoidance of predation.

The insights gained by careful observation of primates as prey can be enhanced enormously by study of the predators that threaten them; thus, we encourage field researchers to follow and observe the predators themselves, rather than limiting study to the prey. These data can be augmented by studies of raptor nest remains and carnivore scat (though such remains could represent scavenging rather than predation). Recent efforts include those by Karpanty (2003), Karpanty and Goodman (1999), Mitani et al. (2001), Sanders et al. (2003), Shultz and Noe (2002), and Shultz et al. (2004). Such studies have been able to estimate the percentage of a predator's diet represented by primates (Rettig 1978, Skorupa 1989, Struhsaker and Leakey 1990), the proportion of a primate population annually consumed by predators (Goodman et al. 1993, 1998; Struhsaker and Leakey 1990), and even the correlation between predator rates and prey morphology and behavior (Shultz et al. 2004). These quantitative approaches are an excellent starting point for population-specific tests of theory and the formulation of new hypotheses; they have the potential to revolutionize our understanding of predation on primates.

Quasi-experimental studies are those in which the research fortuitously coincides with predator introduction or removal. Many such cases have been studied to understand prey responses among nonprimate taxa (Berger et al. 2001, Hunter and Skinner 1998, Terborgh et al. 2002). In some areas, this is made possible by recovering carnivore populations (Anderson 1989, 1992) or by transplanted primate groups (Gouzoules et al. 1975). However, care must be taken to assure that the prey are responding to the return of predators and not some other change in the environment, social or physical, and that predators actually kill the primates where they did not before. Thus, extensive baseline data are essential to understand the impacts of predators in such quasi-experimental situations.

True experimental manipulation of predation risk offers a unique opportunity to isolate the role of predators in prey behavior. Simply put, the observer controls the timing and perhaps the location of simulated encounters. These studies need controls that present nonpredatory stimuli in a similar modality or they may fail to discriminate antipredatory behavior from xenophobic or neophobic behavior. There has been ample research on this topic, including the landmark work of Cheney and Seyfarth (1990). More recent studies include those by Bshary and Noe (1997a,b), Chapman and Chapman (1996), Gebo et al. (1994), Gil-da-Costa et al. (2003), Treves (1999), Zuberbuhler (2000), and Zuberbuhler et al. (1997). Predator simulations can be time-consuming and expensive, and they must be used cautiously to avoid the confounding effects of social learning by the subjects; however, simulations have the greatest potential for explicit discrimination of hypotheses about current selective pressures or the adaptive function of behavior.

BEHAVIORS RELATED TO PREDATOR PRESSURE

Predator–prey interaction may involve a series of moves and countermoves as the predator seeks to capture and consume prey and the prey seeks to evade consumption (Bednekoff and Lima 1998, Kerfoot and Sih 1987, Sih 1987). However, there is no reason to presuppose that a given predator will employ the same attack strategy for all prey species in its range (Lima 2002). Consequently, it is unlikely that two prey species will respond identically to the same predator. Effective responses will depend on a suite of biological characteristics (such as body size or visual acuity) and ecological factors (such as distribution of refuges), which will vary significantly across species, populations, groups, and even individuals. By the same token, there is no reason to predict that one primate will respond similarly to two different predators (Cheney and Seyfarth 1990, Goodman et al. 1993, Gursky 2003, Macedonia and Evans 1993, van Schaik and Kappeler 1996, Treves 1999). Effective prey responses will depend in part on characteristics of the predator's attack pattern; clearly, an animal will behave differently to avoid an aerial raptor versus a large, terrestrial felid. Prey species with many potential predators may have developed an extensive repertoire of antipredator behaviors or a few generalized responses that are optimally effective. Some of these will be more labile than others (e.g., alarm calling can be used when effective and stifled when ineffective, but a rapid change of group size is generally less easily accomplished). Thus, we encourage primatologists to consider the multiple variables that are relevant to predator–prey interactions and to continue to conduct multifactorial analyses that are appropriate to this complex puzzle.

Predation and Social Structure

Predation has long been thought to play a major role in the formation of social groups (Anderson 1986, Busse 1977, Jarman 1974, Pulliam and Caraco 1984, van Schaik 1983, van Schaik and van Hooff 1983). Individuals in larger social groups are expected to be safer than those in smaller groups because of dilution effects (Hamilton 1971); the added vigilance of many eyes, leading to increased predator detection (Elgar 1989, Pulliam 1973, Rodman 1973, Triesman 1975); or the improved success of mobbing (Altmann 1956, Curio 1978). While many have asserted a strong relationship, others have questioned the role of predation in the formation of primate groups (van Schaik and Kappeler 1996, Smuts 1985, Treves 2000, Treves and Chapman 1996, Wrangham 1979, Wright 1998).

Scientists working with nonprimate species have produced elegant tests of the hypothesis that predation results in aggregation (Caraco et al. 1980, Cresswell 1994, Lindstrom 1989, Wrona and Dixon 1991). On the other hand, some studies, including a few on primates, have reported disaggregation following an increase in predation risk (Anderson 1992, Fitzgibbon 1990, Lima 1993, Treves 1999, Wolf 1985). In any case, the data for primates are largely circumstantial; that is, many studies have explored the extent to which group size is related to antipredator behaviors, such as vigilance, but it has been difficult to isolate the relationship between group size and predation itself. Classic studies include those comparing primate populations in areas of high versus low predator density (Marsh 1979, van Schaik and Horstermann 1994, van Schaik and van Noordwijk 1985, Struhsaker 2000) or the response to inferred changes in predator pressure (Isbell 1990, Isbell and Young 1993b). Further attempts to demonstrate a link between predation and group size include cross-taxonomic analyses of published data, some of which have found the predicted positive correlation (Dunbar 1988, Hill and Lee 1998, Janson and Goldsmith 1995, van Schaik 1983) while others have not (Hill and Dunbar 1998, Isbell 1994). For all such meta-analyses, confounding variables (such as the specific array of predators or the distribution of food resources) often cloud

these patterns. The differences often depend on methods of assessing predation rate, predation risk, and the types of taxa included in the analysis. In certain socioecological settings, larger groups may actually impose greater predation risk and rate (Wright 1998). In one of the most clearly documented investigations of predator–prey interaction, Stanford (1998, 2002) has shown that larger groups of red colobus monkeys are attacked more frequently by chimpanzees and that high rates of successful predation upon immatures actually depress group size for this prey species (Stanford 1995, Kruger et al. 1998). On the other hand, Shultz et al. (2004) demonstrated that, for one African primate community, the relationship between group size and predation rate depended heavily on the type of predator and the specific prey habitat. Thus, we must avoid overgeneralizations. It is inappropriate to conclude that predator pressure is always positively correlated with group size or that large groups are solely attributable to predator pressure. We must develop a more sophisticated model of predator–prey relationships as they pertain to the adaptive value of social aggregations.

A clearer (though not undisputed) relationship exists between predation risk and the formation of mixed-species groups. Mixed-species groups have been observed among various primate taxa, most notably the callitrichids and the cercopithecids. This phenomenon has commonly been attributed to foraging efficiency and/or to predation risk (Caine 1993; Chapman and Chapman 1996; Cords 1990a,b; Garber 1988; Gautier-Hion et al. 1983; Hardie 1998; Hardie and Buchanan-Smith 1997; Heymann 1990; Heymann and Buchanan-Smith 2000; Norconk 1990; Peres 1993; Struhsaker 1981, 2000; Terborgh 1990; Waser 1987). Garber and Bicca-Marques (2002) have argued that the literature on tamarins reveals no consistent differences in predation risk for members of single- vs. mixed-species troops. However, Bshary and Noe (1997b) have demonstrated experimentally that red colobus monkeys are more likely to initiate and maintain associations with diana monkeys (*Cercopithecus diana*) when under conditions of heightened predation risk (see also Noe and Bshary 1997, McGraw and Bshary 2002). In addition, Sauther (2002) found that smaller groups of ring-tailed lemurs (*Lemur catta*) were more likely to join groups of Verreaux's sifaka (*Propithecus Verreauxi*), particularly when vulnerable infants were present. Thus, some primates may achieve safety in numbers by joining with members of other species. However, the variability described above may reflect variation in the predators' methods of prey selection, the vulnerability of prey in different habitats, and the behavioral tactics used for attack and defense (Fitzgibbon 1990).

Similarly, early works proposed that predation pressure played a role in the formation of multimale groups (Anderson 1986, Crook and Gartlan 1966, Devore and Washburn 1963, Hamilton et al. 1978). Males may serve an important role in vigilance and defense (Anderson 1986, Baldellou and Henzi 1992, Chapman and Chapman 1996, Gautier-Hion and Tutin 1988, Gursky 2005, Rose 1994, Rose and Fedigan 1995, van Schaik and van Noodwijk 1989, Stanford 1995). However, there is still debate about the relationship between predation risk and the proportion of males in a group. Cheney and Wrangham (1987) found no evidence that predation rate influenced the number of males in groups of vervets (*Cercopithecus aethiops*) or red colobus; Struhsaker (2000) showed the same for certain populations of African forest monkeys. van Schaik and Horstermann (1994) attributed multimale groups of colobus and howlers to eagle predation and the unimale groups of langurs to relaxed predation; however, many other ecological variables could provide alternative explanations for this disparity. Hill and Lee (1998) did a multivariate analysis of 121 populations of 39 primate species and found that, for multimale groups, the proportion of males was positively correlated with predation risk. Stanford (1998) found a similar pattern in a handful of red colobus populations. Mitani et al. (1996) concluded, from a cross-species analysis, that the number of males in a group was most closely related to the number of females but that this, in turn, may be influenced by multiple ecological factors including, but not limited to, predator risk. Group composition is influenced by many ecological and social variables (Andelman 1986, Dunbar 1988, Hill and Lee 1998, Mitani et al. 1996, Treves and Chapman 1996). Isolating and evaluating the role of just one of these variables is unlikely without experimental controls.

The view that predation has influenced primate social evolution is persistent despite vexing difficulties in producing convincing scientific tests. We share the view that primates associate for safety, but we encourage appropriate tests of several basic assumptions. First, do larger primate groups detect predators earlier? This is by no means clear for nocturnal primates or for the large groups of arboreal primates found across the world, where vigilant animals are continually distracted by associates and where visual obstructions from the trees often occlude clear lines of sight (Treves 2002). An effective test of this basic relationship will require inventive field experiments because we cannot equate detection with flight (Ydenberg and Dill 1986). Other measures of threat detection, such as individual heart rate or pupil dilation, will be more convincing. Second, are alarm calls produced more often or more quickly by larger groups, and do they improve the safety of listeners? The first proposition should be easy to test, but the trick will always be to distinguish antipredator alarms from anticonspecific alarms or false alarms. Furthermore, determining if alarm signals improve the rate of escape will require sophisticated field experiments, comparing situations where an alarm is given with situations where danger is equal but no alarm is given, perhaps requiring the videotaped responses of many individuals. Third, do larger groups actually attract predators? Because many primatologists can find large primate groups more easily than small ones from their noise or other signs, it has long been assumed that predators do the same. Collaboration with predator researchers will be needed to assess this relationship for any given population. Fourth, does per capita predation rate decline with group size, and does it do so for all age–sex classes similarly (Treves 2000)? This question may only be testable in a population suffering high rates of predation until we devise methods of experimentally simulating the predatory targeting of wild primates.

Risk, Vulnerability, and Predator-Sensitive Behaviors

Predator-sensitive behaviors are, in effect, antipredator behaviors in the absence of a predator. Individuals may alter their patterns of foraging, sleeping, and ranging to reduce the probability of encountering a predator (reducing risk). The extent to which they do so depends on multiple biological and social variables that relate to individual vulnerability. The interplay between risk and vulnerability produces a complex set of adaptive behaviors.

Certain environmental variables are clearly linked to risk, including abundance of refuges, height and density of grass, the presence of protective versus obstructive cover, and the intensity of moonlight (Cheney and Wrangham 1987; Cowlishaw 1997a; Dunbar 1988, 1996; Gursky 2003; Isbell 1994; Treves 2002). As individuals move through the microhabitats of their ranges, their risk may change throughout the day.

Many variables have been hypothetically linked to vulnerability, but few have been unequivocally linked to predation rate. For example, it has long been suggested that smaller-bodied species are taken by predators more often than larger-bodied species; however, recent collaborative work on a primate–predator system by Zuberbuhler and Jenny (2002) presents data to the contrary. Similarly, immatures are expected be more vulnerable than adults; this pattern has been supported by data on red colobus monkeys (Stanford 2002, Struhsaker and Leakey 1990), but studies show no such bias for other primate species (Cheney et al. 1981, Mitani et al. 2001, Struhsaker and Leakey 1990). Females may be more vulnerable than males, in part because of smaller body size; Cowlishaw (1997a) showed greater risk to female baboons (genus *Papio*) but Mitani et al. (2001) showed that males more than females were represented in crowned eagle nest remains (see also Olupot and Waser 2001). Females may also be vulnerable because of their proximity to and encumbrance by immatures (Busse 1984, Collins 1984), but so far only one study supports this hypothesis (Boinski 1987, 1992). Social variables such as rank may influence vulnerability. Cheney et al. (1988) showed that high-ranking male vervets were more likely to be eaten at waterholes than were their lower-ranking conspecifics. Group size and composition may also affect vulnerability, but convincing quantitative data are lacking (see above). An individual's vulnerability may change over the course of its lifetime, and predator-sensitive decisions are expected to shift accordingly.

Foraging is one behavior that is predicted to be predator-sensitive; that is, animals may sacrifice foraging success in order to minimize predation risk, and more vulnerable animals might show greater predator sensitivity (Abrams 1994; Altmann 1974; Lima 1987; Lima and Dill 1990; Schoener

1971; Sih 1987, 1992). Many studies of nonprimate taxa have shown clear correlations between foraging patterns and specific variables of vulnerability (see Miller 2002a for a review). The data for primates are less conclusive but still compelling. Cowlishaw (1997a) demonstrated that baboons left refuges principally when traveling and feeding; additional work (Cowlishaw 1997b) showed that baboons opted for low-risk but low-quality feeding sites, rather than exploiting high-quality but high-risk patches. In contrast, Prescott and Buchanan-Smith (2002) showed that red-bellied tamarins (*Saguinus labiatus*) in a captive setting foraged closer to the ground (which was assumed to represent higher risk) for a greater food reward. Cheney et al. (1988) showed that waterholes were visited out of necessity despite the high risk of predation. Such trade-offs are the central issue in studies of predator-sensitive foraging. Numerous anecdotal reports support the existence of such a trade-off. For example, Uhde and Sommer (2002) reported that, in the wake of a tiger encounter, gibbons (*Hylobates lar*) clustered together and delayed foraging for several hours. While many studies have inferred predator impact upon foraging behavior and morphology (Cords 2002, Hill and Cowlishaw 2002, Lambert 2005, Miller 2002b, Overdorff et al. 2002, Sauther 2002), few have examined the trade-off experimentally to control the multiple potentially confounding factors that affect daily behavior. As primatologists increase their attention to these complex patterns, more fine-tuned testing of the specific hypotheses may be achieved.

Primates may also seek to reduce encounters with predators by altering their sleeping habits. Predator pressure has been said to affect sleeping positions (Anderson 1984, 1998); the timing of sleep (Caine 1987, Dawson 1979, Gibbons and Menzel 1980, Hankerson and Caine 2004); the amount of vigilance and noise before retirement (Caine 1987, Heymann 1995, Peres 1993, Terborgh 1983); the choice of sleeping site, with regard to height, cover, and/or access (Caine et al. 1992; Cowlishaw 1994; Day and Elwood 1999; Fruth and Hohmann 1993; Ramakrishnan and Coss 2001a,b; van Schaik et al. 1996; Setz 1999; Tenaza and Tilson 1985; Wright 1998); the choice of path to the sleeping site (Ferrari and Lopes-Ferrari 1990); and the use of sentinels before sleeping (Zullo and Caine 1988). Observations of predatory attacks during primate sleep are rare (Busse 1980, Wright 1998, Wright and Martin 1995). Experimental design, as seen in several of the studies cited here, helps to elucidate the nuances of predator-sensitive sleeping.

Predation is also expected to influence primate ranging (reviewed in Boinski et al. 2000). Surely, primates will avoid moving through areas known to be dangerous, unless there is considerable incentive to do otherwise. This topic has been reviewed by Treves (2002) for arboreal primates and by Cowlishaw (1997a,b) for terrestrial baboons. On a smaller scale, primates are expected to modify their group cohesion in response to perceived risk (Altmann and Altmann 1970, Hamilton 1971, Vine 1971); cohesion might, in turn, impact other behaviors, such as vigilance and foraging (Cords 2002,

Cowlishaw 1998, Robinson 1981, Treves 1998). Predator-related cohesion has been shown in various primate species (Boinski et al. 2000; Byrne et al. 1989; Gursky 2003; Sigg 1980; Stanford 1995, 2002; Tutin et al. 1983; Uhde and Sommer 2002); however, this is not a universal response (Isbell 1990, Treves 1999), nor can cohesion be related solely to predation pressure (Cowlishaw 1999, Isbell and Enstam 2002). There is no reason to assume a priori that clumping is the best strategy for avoiding a predator. In some cases, it might be better to split up and thereby confuse the predator or hide oneself at the expense of associates. Finally, individual primates may change their location within the group, which might affect not only risk but also foraging success. Hypothetically, the periphery of a group presents greater predation risk than do central locations (Devore 1965, Hamilton 1971, Vine 1971). The command of "safer" locations may depend on sex and/or size (Wright 1998), reproductive state (Busse 1984, Collins 1984 on lactating females), and/or social variables such as rank (Gould et al. 1997; Hall and Fedigan 1997; Janson 1990a,b; Robinson 1981; Ron et al. 1996). Individuals, especially those that perceive themselves to be more vulnerable, may opt for safer locations despite a probable drop in foraging success (Chapman and Chapman 2000, Janson 1990a). Again, these trade-offs are the central focus of predator-sensitive behaviors.

In sum, primates have at their disposal a wide variety of options when it comes to predator-sensitive behaviors, and univariate models do little to convey the complexity of these patterns. Moreover, without direct observations of predator encounters to confirm elevated risk, simple observations of behavior consistent with predator avoidance generate equivocal conclusions about antipredator behavior. For example, individuals must respond not only to the threat of predation but also to the distribution of mates, food resources, and shelter from climatic extremes. The relative power of these sometimes competing selective pressures is hard to assess by simple observation of the prey species (Boinski et al. 2000, Chapman and Chapman 1999, Treves et al. 2003, Ydenberg and Dill 1986). Experimental or quasi-experimental manipulation of predation risk is required. We need more observations of verified encounters and experimental simulations.

Antipredator Behaviors

The most fundamental antipredator behavior is to avoid the initial encounter. Predator avoidance may be improved by cryptic habits, including nocturnality, forming smaller and quieter groups, and foraging alone; but few studies have examined these as antipredator strategies because controls are so difficult. Such strategies may be easier for individuals with smaller bodies, populations that can hide effectively by day, and those that have traditions or evolved patterns of cooperation in small group settings (Horrocks and Hunte 1986, Pages-Feuillade 1988, Savage et al. 1996, Wahome et al. 1993, Watanabe 1981, Wright 1998, Wright and Martin

1995). Large-bodied, diurnal species may also behave cryptically when risk is perceived to be high (Fay et. al. 1995 on gorillas, *Gorilla gorilla*; Slatkin and Hausfater 1976 on an adult male baboon).

Predator avoidance is useful, but it does not eliminate chance encounters with predators or prevent ambush at sites used repeatedly. From the perspective of prey, there is a great deal of uncertainty about the location, number, ability, and hunger of predators. This uncertainty can be resolved in part by gathering information using all the senses. Virtually nothing is known about olfactory or auditory vigilance in primates (but see Jaenicke and Ehrlich 1982). However, we know a great deal about visual information gathering or vigilance (for review, see Treves 2000). First, long-term observational studies indicate that vigilance is intermittent and frequent throughout the foraging period of primates (Cowlishaw et al. 2004, Jack 2001, Treves et al. 2001). Studies of sleeping birds indicate that vigilance may also recur during sleep (Lendrem 1983, 1984). This is reasonable as information gathering must be repeated to compensate for unpredictable movements of predators. Correspondingly, one expects that prey vigilance will increase when entering unfamiliar areas. This prediction has never been tested explicitly with time-sensitive and spatially explicit information on the movements of primates into areas that vary in familiarity. We encourage a test of this prediction with real or simulated predator encounters. The key variable will be the length of time elapsing between revisits to the same site as longer intervals imply greater uncertainty and, hence, greater risk of predator encounter.

Measures of vigilance without evidence for the target or the severity of risk are of little use for inferring antipredator function and utility of the vigilance. Indeed, vigilance serves a very important social purpose, which can conflict with antipredator vigilance (Biben et al. 1989; Brick 1998, Treves 1999, 2000; Treves and Pizzagalli 2002). Our own human experience reveals the multifunctional and flexible nature of vigilance such that it need not be intended to detect a predator to nevertheless detect one (Treves 2000), Treves and Pizzagalli 2002). With this in mind, we have found a few vigilance studies that experimentally simulated and varied the risk of predation or related vigilance to real predator encounters (reviewed in Bshary and Noe 1997a, Cheney and Seyfarth 1990, Treves 1999). The studies are concordant: vigilance rises when predation risk is elevated, presumably as primates attempt to gather visual information that will reduce individual vulnerability.

Primates and many other animals use the early detection of predators to escape or signal this detection. Several brilliant studies have documented the use of signals in the predator context. The best-documented research concerns vocal alarms (Cheney and Seyfarth 1990, Macedonia and Evans 1993, Struhsaker 1967, Zuberbuhler et al. 1997, Zuberbuhler 2000), but visual signals have also been noted (Horrocks and Hunte 1986, Savage et al. 1996). To summarize these findings, some primate populations show elaborate signaling following the detection of predators, and associates often use the semantic information encoded in the signals to respond appropriately. In addition, some vocal signals warn predators that they have been detected, thereby potentially deterring ambush predators from further efforts (Zuberbuhler et al. 1997, Zuberbuhler 2000).

Antipredator signals could make groups advantageous—beyond the benefit of dilution—via collective detection (Bednekoff and Lima 1998) if signals improve the efficiency of escape by non-detectors. On the other hand, if signals such as vocal alarms predominantly function as deterrents—warnings to predators that they have been detected and attack will fail (Zuberbuhler et al. 1997, Zuberbuhler 2000)—then the role of the group and the benefit of alarm signals to associates is less clear. For example, Chapman (1986), Ross (1993), Srivastava (1991), and many other authors have reported protracted vocalizations following encounters or even following successful attacks in which the predator no longer poses a danger. Persistent vocalizations seem ill-suited to collective detection—that is, improving associates' chances of escape—because they could attract additional predators or permit the initial attacker to monitor the prey at a distance. Persistent calls are better suited to warning the predator or distant animals, neither of which can be easily tied to the evolution of group living. We advocate that primatologists devote attention to whether and when alarm signals improve escape, to distinguish those alarm signaling systems whose primary role and evolutionary history reflect predator warnings from those signaling systems whose function and evolution reflect collective detection for gregarious prey. Primatologists wishing to investigate these questions might learn a great deal from studies done on birds (Lima 1994, 1995a,b; Lima and Zollner 1996).

Regarding the use of alarms to help associates escape more quickly, Bednekoff and Lima (1998) emphasized that inefficient transmission and reception of predator-related information can nullify the benefits of collective detection for group-living animals. An alarm signal produced too late or responded to slowly offers little benefit to associates, but the efficiency of escape following alarm signals has never been measured in primates to our knowledge. This may be because primatologists have often assumed that production of an alarm signal is instantaneously followed by appropriate responses. This assumption may be false. Collective detection also lessens in value when predators are highly selective of prey (Bednekoff and Lima 1998). In effect, selective predators reduce the number of effective alarm signalers to that subset that are vulnerable themselves or are sensitive to the vulnerability of others. The latter condition may hold in kin-based primate groups and those involving high levels of reciprocity, but this conjecture remains to be tested.

Vigilance and collective detection serve to uncover and share information, respectively, about predators which themselves often strive to remain undetected. However, not all predators rely on surprise. Coursing predators, such as wild dogs, hyenas, and chimpanzees, may pursue despite detection;

and the predators themselves may gain valuable information by observing the flight of their quarry (Fanshawe and Fitzgibbon 1993, Fitzgibbon and Lazarus 1995, Holekamp et al. 1997, Treves 2000). Some raptors may behave in this way as well, but the data are not as clear (Lima 1993). When confronted by coursing predators, prey require rapid escape or refuges that are inaccessible to their pursuers; vigilance and alarm calling are of limited use against such predators. Thus, the use of vocal signals in the predator context is immensely complex and requires further study before we can conclude that the evolution of antipredator signals is tied to the evolution of group living. For individuals, alarm calls usually assist in escape and avoidance of predators.

To escape from predators, all primates seek refuge (with the possible exception of resident adult male patas monkeys; Chism and Rowell 1988, Hall 1965). Refuges for terrestrial taxa include trees and cliffs, while arboreal forms rapidly change level (reviewed in Treves 2002). Prey species might also perform evasive behaviors such as coordinated attempts to confuse the predator; this has been observed in some fish, birds, and open-country ungulates (Edmunds 1974, Lima 1993) but not in primates.

More rarely, primates stand their ground to counterattack or mob predators. Of the two forms, mobbing appears to be the less dangerous for the prey and is more common among smaller primates. It involves two or more prey animals making repeated advances on a predator, usually while vocalizing and displaying in a conspicuous fashion. The predator is often distracted or repelled by persistent approaches (Baeninger et al. 1977, Eason 1989). Adult males, acting alone or in small parties, are more likely to attack predators than are other classes of individuals (Chapman and Chapman 1996, Cowlishaw 1994). The likelihood of counterattack by primates appears to depend on the size difference between predator and prey (Bshary and Noe 1997a, Stanford 1995).

If all the antipredator strategies fail (avoidance, vigilance-assisted early detection and collective detection, alarm signals, escape, counter-attack), the prey is usually doomed. However, close spatial cohesion of prey may allow one individual to interpose an associate between itself and a predator at the last moment, as discussed in the section on grouping. Another last-ditch struggle is the distress call of a captured prey animal. Distress calls in birds are proposed as a tactic to startle the predator into dropping the captured prey, rather than alert associates (Conover 1994). Similar startle responses may exist in lorisids, along with toxic secretions that may deter the final attack (see Chapter 3). Primates have evidently evolved a wide repertoire of morphologies and behaviors to avoid, escape, and otherwise elude their predators.

HUMANS AS PREDATORS, PAST AND PRESENT

Paleontological evidence suggests that humans have been predators of nonhuman primates for hundreds of thousands of years. For example, fossil accumulations from Europe include cercopithecoid remains that may be attributed to hominid activity (Freeman 1994, Klein 1987). Human hunting has long been implicated in early Holocene extinction of North and South American megafauna (Kurten and Anderson 1980, Martin 1984), along with similar patterns in Africa (Klein 1984) and Eurasia (Vereshchagin and Baryshnikov 1984; see also Janzen 1983). Perhaps the most extreme example of human impact on primate populations comes from Madagascar. Fossil data indicate that, approximately 2,000 years ago humans arrived on this island and rapidly drove some 15 primate species into extinction (Richard and Dewar 1991). Perihistoric hunting may also account for much of today's distribution of primates (Peres 1990b, Struhsaker 1999). In addition, ethnographic research has shown that primates represent a small but significant part of the diet for human cultures around the world (Alvard 1993, Bailey and Peacock 1988, de Garine 1993, Fitzgibbon et al. 1995, Hart 1978, Hill and Padwe 2000). From this point of view, primates have long had to adapt to humans as predators.

On the other hand, it could be argued that humans represent a fundamentally unique type of predator. Humans armed with projectile weapons differ from other predators because they can kill at a distance, approximating or even exceeding the detection distance of their prey. Thus, a human armed with a projectile is like an ambush predator with an enormous striking range, especially if the human uses bait or traps. However, unlike other ambush predators, humans are also like coursing predators in that they are rarely deterred by alarm calls and are hard to avoid (see Treves and Palmqvist in press for a review of different predator tactics). Like coursing predators, humans with projectiles target not only the infirm and the unwary but also prime adults (Brain 1981, Faraizl and Stiver 1996, Hill and Hurtado 1995). Hence, humans with projectiles share characteristics of both ambush and coursing predators, making them especially deadly. Moreover, human intelligence and social transmission of cultural knowledge may give these predators greater capacity to study the habits of their prey over time and circumvent prey defenses. In combination, these predatory traits are evolutionary novelties, imposing unique selective pressures on prey species. It is possible that primates have developed specialized morphologies and behaviors to avoid human predators, but we know of no studies exploring this phenomenon; it is probable that no such specialized adaptations exist. While humans have long represented a part of primate ecological communities, in more recent millennia they may have become a selective pressure superseding the prey species' capacity to adapt; thus, today's primate morphology and behavior are likely only weakly related to humans as predators.

Clearly, this has become the case in recent decades. The impact of human hunting on primate populations has reached record proportions in recent decades, assisted by logging roads and four-wheel-drive vehicles, which give access to previously unexploited tracts of forest, and high-powered rifles, which can easily kill even the largest of primates. The bushmeat trade has been noted for many years (Anadu et al. 1988,

Chardonnet et al. 1995, Martin 1983, Njiforti 1996); however, it is a growing problem, reaching unsustainable levels in many parts of the globe, particularly in major regions of Africa (Walsh et al. 2003, Wilkie and Carpenter 1999). Primates are not the only taxa in peril, but they are especially vulnerable due to their large size (relative to many other forest species) and their tendency to form conspicuous social groups (Bodmer et al. 1988, Struhsaker 1999). Although tribal custom prevents some human cultures from hunting primates, the economics of the commercial bushmeat trade is causing rapid change in long-standing traditions (Struhsaker 1999). There is a great deal of recent debate about the possibility and even value of sustainable hunting (Bodmer et al. 1988, 1994; Freese 1997; Hart 2000; Hill and Padwe 2000; Hill et al. 2003; Robinson 2000; Robinson and Bennett 2000; Robinson and Redford 1994; Slade et al. 1998), but few can argue with the fact that current levels of bushmeat hunting are leading many primate species toward extinction (Naughton-Treves et al. 2003, Walsh et al. 2003, Wilkie and Carpenter 1999).

CONCLUSION

In closing, primates show remarkable diversity in their morphological and behavioral adaptations, and surely some of these relate to predator avoidance. Even where successful predation is rare, a single predatory event may limit an individual's lifetime reproductive success in a profound way; thus, there is likely to be considerable selective pressure to avoid becoming prey. Although we recognize the compelling logic behind predation as an influence on primate evolution, we also see the challenge of demonstrating this impact unequivocally and exploring alternatives. The next generation of primatologists can learn from the previous one by conducting rigorous tests of mutually exclusive hypotheses and by wariness toward hasty conclusions based on a few variables. This will demand primatologists who can study both primates and predators with the sophistication of specialists so that we may understand predator behavior in relation to prey vulnerability. It will also take primatologists with facility in experimental design, who can simulate predator attacks in innovative ways and measure antipredator behavior unambiguously. These are not easy tasks, but we are encouraged by the energy, determination, and creativity of the new generation of primatologists. Finally, our chapter calls attention to the plight of primates subject to human predation. In some areas, the first battle must be conservation lest we lose populations or entire species; basic research, though fascinating, is a luxury we must earn.

REFERENCES

Abrams, P. A. (1994). Should prey overestimate the risk of predation? Am. Nat. 144:317–328.

Alexander, R. D. (1974). The evolution of social behavior. Annu. Rev. Ecol. Syst. 5:324–383.

Altmann, S. A. (1956). Avian mobbing behavior and predator recognition. Condor 58:241–253.

Altmann, S. A. (1974). Baboons: space, time, and energy. Am. Zool. 14:221–248.

Altmann, S. A., and Altmann, J. (1970). Baboon Ecology. University of Chicago Press, Chicago.

Alvard, M. S. (1993). Testing the "ecologically noble savage" hypothesis: interspecific prey choice by Piro hunters of Amazonian Peru. Hum. Ecol. 21:355–387.

Anadu, P. A., Elamah, P. O., and Oates, J. F. (1988). The bushmeat trade in southwestern Nigeria: a case study. Hum. Ecol. 16:199–208.

Andelman, S. J. (1986). Ecological and social determinants of cercopithecine mating patterns. In: Rubenstin, D. I., and Wrangham, R. W. (eds.), Ecological Aspects of Social Evolution: Birds and Mammals. Princeton University Press, Princeton, NJ. pp. 201–216.

Anderson, C. M. (1986). Predation and primate evolution. Primates 27:15–39.

Anderson, C. M. (1989). The spread of exclusive mating in a chacma baboon population. Am. J. Phys. Anthropol. 78:355–360.

Anderson, C. M. (1992). Male investment under changing conditions among chacma baboons at Suikerbosrand. Am. J. Phys. Anthropol. 87:479–495.

Anderson, J. R. (1984). Ethology and ecology of sleeping in monkeys and apes. Adv. Stud. Behav. 14:165–228.

Anderson, J. R. (1998). Sleep, sleeping sites, and sleep-related activities: awakening to their significance. Am. J. Primatol. 46:63–75.

Andrews, P., and Evans, E. M. (1979). Small mammal bone accumulations produced by mammalian carnivores. Palaeobiology 9:289–307.

Baeninger, R., Estes, R., and Baldwin, S. (1977). Anti-predator behaviour of baboons and impalas toward a cheetah. East Afr. Wildl. J. 15:327–329.

Bailey, R. C., and Peacock, N. R. (1988). Efe pygmies of northeast Zaire: subsistence strategies in the Ituri Forest. In: de Garine, I., and Harrison, G. A. (eds.), Uncertainty in the Food Supply. Cambridge University Press, Cambridge. pp. 88–117.

Baldellou, M., and Henzi, P. (1992). Vigilance, predator detection and the presence of supernumerary males in vervet monkey troops. Anim. Behav. 43:451–461.

Bednekoff, P. A., and Lima, S. L. (1998). Re-examining safety in numbers: interactions between risk dilution and collective detection depend upon predator targeting behaviour. Proc. R. Soc. Lond. B 265:2021–2026.

Berger, J., Stacey-Peter, B., Bellis, L., and Johnson, M. P. (2001). A mammalian predator–prey imbalance: grizzly bear and wolf extinction affect avian neotropical migrants. Ecol. Appl. 11:947–960.

Biben, M., Symmes, D., and Bernhards, D. (1989). Vigilance during play in squirrel monkeys. Am. J. Primatol. 17:41–49.

Bodmer, R. E., Fang, T. G., and Moya Ibanez, L. (1988). A comparison of susceptibility to hunting. Primate Conserv. 9:79–83.

Bodmer, R. E., Fang, T. G., Moya Ibanez, L., and Gill, R. (1994). Managing wildlife to conserve Amazonian forests: population biology and economic considerations of game hunting. Biol. Conserv. 67:29–35.

Boesch, C. (1991). The effects of leopard predation on grouping patterns in forest chimpanzees. Behaviour 117:220–242.

Boesch, C. (1994). Chimpanzees–red colobus monkeys: a predator–prey system. *Anim. Behav.* 47:1135–1148.

Boesch, C., and Boesch, H. (1989). Hunting behavior of wild chimpanzees in the Taï National Park. *Am. J. Phys. Anthropol.* 78:547–573.

Boinski, S. (1987). Birth synchrony in squirrel monkeys (*Saimiri oerstedi*) a strategy to reduce neonatal predation. *Behav. Ecol. Sociobiol.* 21:393–400.

Boinski, S. (1992). Monkeys with inflated sex appeal. *Nat. Hist.* 7:42–51.

Boinski, S., and Chapman, C. A. (1995). Predation on primates: where are we and what's next? *Evol. Anthropol.* 4:1–3.

Boinski, S., Treves, A., and Chapman, C. A. (2000). A critical evaluation of the influence of predators on primates: effects on group travel. In: Boinski, S., and Garber, P. A. (eds.), *On the Move: How and Why Animals Travel in Groups*. University of Chicago Press, Chicago. pp. 43–72.

Boshoff, A. F., Palmer, N. G., Avery, G., Davies, A. G., and Jarvis, M. J. (1991). Biogeographical and topographical variation in the prey of the black eagle in the Cape Province, South Africa. *Ostrich* 62:59–72.

Brain, C. K. (1981). *The Hunters or the Hunted? An Introduction to African Cave Taphonomy*. University of Chicago Press, Chicago.

Brick, O. (1998). Fighting behaviour, vigilance and predation risk in the cichlid fish, *Namacara anomala*. *Anim. Behav.* 56:309–317.

Brockman, D. K. (2003). *Polyboroides radiatus* predation attempts on *Propithecus verreauxi*. *Folia Primatol.* 74:71–74.

Brown, L. H. (1962). The prey of the crowned hawk eagle *Stephanoaetus coronatus* in central Kenya. *Scopus* 6:91–94.

Brown, L. H. (1971). The relations of the crowned eagle, *Stephanoaetus coronatus* and some of its prey animals. *Ibis* 113:240–243.

Bshary, R., and Noe, R. (1997a). Anti-predation behaviour of red colobus monkeys in the presence of chimpanzees. *Behav. Ecol. Sociobiol.* 41:321–333.

Bshary, R., and Noe, R. (1997b). Red colobus and diana monkeys provide mutual protection against predators. *Anim. Behav.* 54:1461–1474.

Buchanan-Smith, H. M. (1990). Polyspecific association of two tamarin species, *Saguinus labiatus* and *Saguinus fuscicollis*, in Bolivia. *Am. J. Primatol.* 22:205–214.

Burney, D. A. (2002). Sifaka predation by a large boa. *Folia Primatol.* 73:144–145.

Busse, C. D. (1977). Chimpanzee predation as a possible factor in the evolution of red colobus monkey social organization. *Evolution* 31:907–911.

Busse, C. D. (1980). Leopard and lion predation upon chacma baboons living in the Moremi Wildlife Reserve. *Botswana Notes Rec.* 12:15–20.

Busse, C. D. (1984). Spatial sructure of chacma baboon groups. *Int. J. Primatol.* 5:247–261.

Byers, J. A. (1997). *American Pronghorn: Social Adaptations and the Ghosts of Predators Past*. University of Chicago Press, Chicago.

Byrne, R. W., White, A., and Henzi, S. P. (1989). Social relationships of mountain baboons: leadership and affiliation in a non-female-bonded monkey. *Am. J. Primatol.* 18:191–207.

Caine, N. G. (1987). Vigilance, vocalizations, and cryptic behavior at retirement in captive groups of red-bellied tamarins (*Saguinus labiatus*). *Am. J. Primatol.* 12:241–250.

Caine, N. G. (1993). Flexibility and cooperation as unifying themes in *Saguinus* social organization and behaviour: the role of predation pressures. In: Rylands, A. B. (ed.), *Marmosets and Tamarins: Systematics, Behavior, and Ecology*. Oxford University Press, Oxford. pp. 200–219.

Caine, N. G., Potter, M. P., and Mayer, K. E. (1992). Sleeping site selection by captive tamarins (*Saguinus labiatus*). *Ethology* 90:63–71.

Caraco, T., Martindale, S., and Pulliam, H. R. (1980). Avian flocking in the presence of a predator. *Nature* 285:400–401.

Chapman, C. A. (1986). *Boa constrictor* predation and group response in white faced *Cebus* monkeys. *Biotropica* 18:171–172.

Chapman, C. A., and Chapman, L. J. (1996). Mixed-species primate groups in the Kibale Forest: ecological constraints on association. *Int. J. Primatol.* 17:31–50.

Chapman, C. A., and Chapman, L. J. (1999). Implications of small scale variation in ecological conditions for the diet and density of red colobus monkeys. *Primates* 40:215–231.

Chapman, C. A., and Chapman, L. J. (2000). Determinants of group size in primates: the importance of travel costs. In: Boinski, S., and Garber, P. A. (eds.), *On the Move: How and Why Animals Travel in Groups*. University of Chicago Press, Chicago. pp. 24–42.

Chardonnet, P., Fritz, H., Zorzi, N., and Feron, E. (1995). Current importance of traditional hunting and major contrasts in wild meat consumption in sub-Saharan Africa. In: Bissonette, J. A., and Krausman, P. R. (eds.), *Integrating People and Wildlife for a Sustainable Future*. Wildlife Society, Bethesda. pp. 304–307.

Cheney, D. L., Lee, P. C., and Seyfarth, R. M. (1981). Behavioral correlates of non-random mortality among free-ranging female vervet monkeys. *Behav. Ecol. Sociobiol.* 9:153–161.

Cheney, D. L., and Seyfarth, R. M. (1990). *How Monkeys See the World: Inside the Mind of Another Species*. University of Chicago Press, Chicago.

Cheney, D. L., Seyfarth, R. M., Andelman, S. J., and Lee, P. C. (1988). Reproductive success in vervet monkeys. In: Clutton-Brock, T. H. (ed.), *Reproductive Success: Studies of Individual Variation in Contrasting Breeding Systems*. University of Chicago Press, Chicago. pp. 384–402.

Cheney, D. L., and Wrangham, R. W. (1987). Predation. In: Smuts, B. B., Cheney, D. L., Seyfarth, R. M., Wrangham, R. W., and Struhsaker, T. T. (eds.), *Primate Societies*. University of Chicago Press, Chicago. pp. 227–239.

Chinchilla, F. A. (1997). La dieta del Jaguar (*Panthera onca*), el Puma (*Felis concolor*) yel manigordo (*Felis pardalis*) (*Carnivoria: Felidae*) en el Parque Nacional Corcovado, Costa Rica. *Rev. Biol. Trop.* 45:1223–1229.

Chism, J., and Rowell, T. E. (1988). The natural history of patas monkeys. In: Gautier-Hion, F. B. A., Gautier, J.-P., and Kingdon, J. (eds.), *A Primate Radiation: Evolutionary Biology of the African Guenons*. Cambridge University Press, Cambridge. pp. 412–438.

Clutton-Brock, T. H., and Harvey, P. H. (1977). Primate ecology and social organization. *J. Zool.* 183:1–39.

Collins, D. A. (1984). Spatial pattern in a troop of yellow baboons (*Papio cynocephalus*) in Tanzania. *Anim. Behav.* 32:536–553.

Condit, V. K., and Smith, E. O. (1994). Predation on a yellow baboon (*Papio cynocephalus cynocephalus*) by a lioness in the Tana River National Primate Reserve, Kenya. *Am. J. Primatol.* 33:57–64.

Conover, M. R. (1994). Stimuli eliciting distress calls in adult passerines and response of predators and birds to their broadcast. *Behaviour* 131:19–37.

Cords, M. (1990a). Mixed-species association of East African guenons: general patterns or specific examples? *Am. J. Primatol.* 21:101–114.

Cords, M. (1990b). Vigilance and mixed-species associations of some East African forest monkeys. *Behav. Ecol. Sociobiol.* 26:297–300.

Cords, M. (2002). Foraging and safety in adult female blue monkeys in the Kakamega Forest, Kenya. In: Miller, L. E. (ed.), *Eat or Be Eaten: Predator Sensitive Foraging Among Primates.* Cambridge University Press, Cambridge. pp. 205–221.

Correa, H. K. M., and Coutinho, P. E. G. (1997). Fatal attack of a pit viper, *Bothrops jararaca*, on an infant buffy-tufted ear marmoset (*Callithrix aurita*). *Primates* 38:215–217.

Cowlishaw, G. (1994). Vulnerability to predation in baboon populations. *Behaviour* 131:293–304.

Cowlishaw, G. (1997a). Refuge use and predation risk in a desert baboon population. *Anim. Behav.* 54:241–253.

Cowlishaw, G. (1997b). Trade-offs between foraging and predation risk determine habitat use in a desert baboon population. *Anim. Behav.* 53:667–86.

Cowlishaw, G. (1998). The role of vigilance in the survival and reproductive strategies of desert baboons. *Behaviour* 135:431–452.

Cowlishaw, G. (1999). Ecological and social determinants of spacing behaviour in desert baboon groups. *Behav. Ecol. Sociobiol.* 45:67–77.

Cowlishaw, G., Lawes, M. J., Lightbody, M., Martin, A., Pettifor, R., and Rowcliffe, J. M. (2004). A simple rule for the costs of vigilance: empirical evidence from a social forager. *Proc. R. Soc. Lond. B* 271:27–33.

Cresswell, W. (1994). Flocking is an effective anti-predation strategy in redshanks, *Tringa totanus. Anim. Behav.* 47:433–442.

Crook, J. H., and Gartlan, J. S. (1966). Evolution of primate societies. *Nature* 210:1200–1203.

Cuaron, A. D. (1997). Conspecific aggression and predation: costs for a solitary mantled howler monkey. *Folia Primatol.* 68:100–105.

Curio, E. (1978). The adaptive significance of avian mobbing. *Z. Tierpsychol.* 48:175–183.

Dawson, G. A. (1979). The use of time and space by the Panamanian tamarin, *Saguinus oedipus. Folia Primatol.* 31:253–284.

Day, R. T., and Elwood, R. W. (1999). Sleeping site selection by the golden-handed tamarin *Saguinus midas midas*: the role of predation risk, proximity to feeding sites, and territorial defense. *Ethology* 105:1035–1051.

Defler, T. R. (1980). Notes on interactions between the tayra (*Eira barbara*) and the white fronted capuchin (*Cebus albifrons*). *J. Mammal.* 61:156.

de Garine, I. (1993). Food resources and preferences in the Cameroonian forest. In: Hladik, C. M., Hladik, A., Linares, O. F., Pagezy, H., Semple, A., and Hadley, M. (eds.), *Tropical Forests, People and Food: Biocultural Interactions and Applications to Development.* UNESCO, Paris. pp. 561–574.

DeVore, I. (1965). Male dominance and mating behavior in baboons. In: Beach, F. A. (ed.), *Sex and Behavior.* John Wiley and Sons, New York. pp. 266–289.

DeVore, I., and Washburn, S. L. (1963). Baboon ecology and human evolution. In: Howell, F. C., and Bourliere, F. (eds.), *African Ecology and Human Evolution.* Aldine, Chicago. pp. 335–367.

Dunbar, R. I. M. (1988). *Primate Social Systems.* Cornell University Press, Ithaca.

Dunbar, R. I. M. (1996). Determinants of group size in primates: a general model. *Proc. Br. Acad.* 88:33–57.

Eason, P. (1989). Harpy eagle attempts predation on adult howler monkey. *Condor* 91:469–470.

Edmunds, M. (1974). *Defence in Animals: A Survey of Anti-Predator Defences.* Longman, London.

Eisenberg, J. F., Muckenhirn, N., and Rudran, R. (1972). The relationship between ecology and social structure in primates. *Science* 176:863–874.

Elgar, M. (1989). Predator vigilance and group size in mammals and birds: a critical review of the empirical evidence. *Biol. Rev.* 64:13–33.

Emmons, L. H. (1987). Comaparative feeding ecology of felids in a neotropical rainforest. *Behav. Ecol. Sociobiol.* 20:271–283.

Fanshawe, J. H., and Fitzgibbon, C. D. (1993). Factors influencing the hunting success of an African wild dog pack. *Anim. Behav.* 45:479–490.

Faraizl, S. D., and Stiver, S. J. (1996). A profile of depredating mountain lions. *Proc. Vert. Pest Conf.* 17:88–90.

Fay, J. M., Carroll, R., Kerbis Peterhans, J. C., and Harris, D. (1995). Leopard attack and consumption of gorillas in the Central African Republic. *J. Hum. Evol.* 29:93–99.

Ferrari, S. F., and Lopes-Ferrari, M. A. (1990). Predator avoidance behaviour in the buffy-headed marmoset, *Callithrix flaviceps. Primates* 31:323–338.

Fitzgibbon, C. D. (1990). Why do hunting cheetahs prefer male gazelles? *Anim. Behav.* 40:837–845.

Fitzgibbon, C. D., and Lazarus, J. (1995). Antipredator behavior of Serengeti ungulates: individual differences and population consequences. In: Sinclair, A. R. E. (ed.), *Serengeti II: Dynamics, Management and Conservation of an Ecosystem.* University of Chicago Press, Chicago. pp. 274–296.

Fitzgibbon, C. D., Mogaka, H., and Fanshawe, J. H. (1995). Subsistence hunting in Arabuko-Sokoke Forest, Kenya and its effects on mammal populations. *Conserv. Biol.* 9:1116–1126.

Fowler, J. M., and Cope, J. B. (1964). Notes on the harpy eagle in British Guiana. *Auk* 81:257–273.

Freeman, L. G. (1994). Torralba and Ambrona: a review of discoveries. In: Corruccini, R. S., and Chiochon, R. L. (eds.), *Integrative Paths to the Past: Paleoanthropological Advances in Honor of F. Clark Howell.* Prentice Hall, Englewood Cliffs, NJ. pp. 597–637.

Freese, C. H. (1997). *Harvesting Wild Species: Implications for Biodiversity Conservation.* Johns Hopkins University Press, Baltimore.

Fruth, B., and Hohmann, G. (1993). Ecological and behavioral aspects of nest building in wild bonobos (*Pan paniscus*). *Ethology* 94:113–126.

Furuichi, T. (2000). Possible case of predation on a chimpanzee by a leopard in the Petit Loango Reserve, Gabon. *PanAfr. News* 7:21–23.

Galdikas, B. M. F., and Yeager, C. P. (1984). Crocodile predation on a crab-eating macaque in Borneo. *Am. J. Primatol.* 6:49–51.

Galef, B. G., Mittermeir, J. R. A., and Bailey, R. C. (1976). Predation by the tayra (*Eira barbara*). *J. Mammal.* 57:760–761.

Garber, P. A. (1988). Diet, foraging patterns, and resource defense in a mixed species troop of *Saguinus mystax* and *Saguinus fuscicollis* in Amazonian Peru. *Behaviour* 105:18–34.

Garber, P. A., and Bicca-Marques, J. C. (2002). Evidence of predator sensitive foraging in traveling in single- and mixed-species tamarin troops. In: Miller, L. E. (ed.), *Eat or Be Eaten: Predator Sensitive Foraging Among Primates.* Cambridge University Press, Cambridge. pp. 138–153.

Gaspersic, M., and Pruetz, J. D. (2004). Predation on a monkey by savanna chimpanzees at Fongoli, Senegal. *PanAfr. News* 11:8–10.

Gautier-Hion, A., Quris, R., and Gautier, J. P. (1983). Monospecific vs. polyspecific life: a comparative study of foraging and antipredatory tactics in a community of *Cercopithecus* monkeys. *Behav. Ecol. Sociobiol.* 12:325–335.

Gautier-Hion, A., and Tutin, C. E. G. (1988). Simultaneous attack by adult males of a polyspecific troop of monkeys against a crowned hawk eagle. *Folia Primatol.* 51:149–151.

Gebo, D. L., Chapman, C. A., Chapman, L. J., and Lambert, J. (1994). Locomotoer reesponse to predator threat in red colobus monkeys. *Primates* 35:219–223.

Gibbons, E. F., Jr., and Menzel, E. W. (1980). Rank orders or arising, eating, and retiring in a family group of tamarins (*Saguinus fuscicollis*). *Primates* 21:44–52.

Gilbert, K. A. (2000). Attempted predation on a white-faced saki in the central Amazon. *Neotrop. Primates* 8:103–104.

Gilbert, M., and Tingay, R. E. (2001). Predation of a fat-tailed dwarf lemur (*Cheirogaleus medius*) by a Madagascar Harrier-hawk (*Polyboroides radiatus*): an incidental observation. *Lemur News* 6:6.

Gil-da-Costa, R., Palleroni, A., Hauser, M. D., Touchton, J., and Kelley, J. P. (2003). Rapid acquisition of an alarm response by a neotropical primate to a newly introduced avian predator. *Proc. R. Soc. Lond. B* 270:605–610.

Goodall, J. (1986). *The Chimpanzees of Gombe: Patterns of Behaviour.* Belknap Press, Cambridge.

Goodman, S. M., O'Connor, S., and Langrand, O. (1993). A review of predation on lemurs: implications for the evolution of social behavior in small, nocturnal primates. In: Kappeler, P. M., and Ganzhorn, J. U. (eds.), *Lemur Social Systems and Their Ecological Basis.* Plenum Press, New York. pp. 51–66.

Goodman, S. M., Rene de Roland, L. A., and Thorstrom, R. (1998). Predation on the eastern woolly lemur (*Avahi laniger*) and other vertebrates by Henst's goshawk (*Accipiter henstii*). *Lemur News* 3:14–15.

Goss-Custard, J. D., Dunbar, R. I. M., and Aldrich-Blake, F. P. G. (1972). Survival, mating and rearing strategies and the evolution of primate social structure. *Folia Primatol.* 17:1–19.

Gould, L., Fedigan, L. M., and Rose, L. M. (1997). Why be vigilant? The case of the alpha animal. *Int. J. Primatol.* 18:401–414.

Gouzoules, H., Fedigan, L. M., and Fedigan, L. (1975). Responses of a transplanted troop of Japanese macaques (*Macaca fuscata*) to bobcat (*Lynx rufus*) predation. *Primates* 16:335–349.

Gursky, S. (2002). Predation on a wild spectral tarsier (*Tarsius spectrum*) by a snake. *Folia Primatol.* 73:60–62.

Gursky, S. (2003). Lunar philia in a nocturnal primate. *Int. J. Primatol.* 24:351–367.

Gursky, S. (2005). Predator mobbing in *Tarsius spectrum*. *Int. J. Primatol.* 26:207–221.

Hall, C. L., and Fedigan, L. M. (1997). Spatial benefits afforded by high rank in white-faced capuchins. *Anim. Behav.* 53:1069–1082.

Hall, K. R. L. (1965). Behavior and ecology of the wild patas monkey (*Erythrocebus patas*), in Uganda. *J. Zool.* 148:15–87.

Hamilton, W. D. (1971). Geometry for the selfish herd. *J. Theor. Biol.* 31:295–311.

Hamilton, W. J., III, Buskirk, R. E., and Buskirk, W. H. (1978). Omnivory and utilization of food resources by chacma baboons, *Papio ursinus*. *Am. Nat.* 112:911–924.

Hankerson, S. E. J., and Caine, N. G. (2004). Pre-retirement predator encounters alter the morning behavior of captive marmosets, *Callithrix geoffroyi*. *Am. J. Primatol.* 63:75–85.

Hardie, S. M. (1998). Mixed-species tamarin groups (*Saguinus fuscicollis* and *Saguinus labiatus*) in northern Bolivia. *Primate Rep.* 50:39–62.

Hardie, S. M., and Buchanan-Smith, H. M. (1997). Vigilance in single- and mixed-species groups of tamarins (*Saguinus labiatus* and *Saguinus fuscicollis*). *Int. J. Primatol.* 18:217–234.

Hart, J. (2000). Impact and sustainability of indigenous hunting in the Ituri Forest, Congo-Zaire: a comparison of unhunted and hunted duiker populations. In: Robinson, J., and Bennett, E. (eds.), *Hunting for Sustainability in Tropical Forests.* Columbia University Press, New York. pp. 106–153.

Hart, J. A. (1978). From subsistence to market: a case study of the Mbuti net hunters. *Hum. Ecol.* 6:32–53.

Hausfater, G. (1976). Predatory behavior of yellow baboons. *Behaviour* 56:44–68.

Heymann, E. W. (1987). Field observation of predation on a moustached tamarin (*Saguinus mystax*) by an anaconda. *Int. J. Primatol.* 8:193–195.

Heymann, E. W. (1990). Reactions of wild tamarins *Saguinus mystax* and *Saguinus fuscicollis* to avian predators. *Int. J. Primatol.* 11:327–337.

Heymann, E. W. (1995). Sleeping habits of tamarins, *Saguinus mystax* and *Saguinus fuscicollis* (Mammalia; Primates; Callitrichidae), in north-eastern Peru. *J. Zool.* 237:211–226.

Heymann, E. W., and Buchanan-Smith, H. M. (2000). The behavioural ecology of mixed-species troops of callitrichine primates. *Biol. Rev.* 75:169–190.

Hill, K., and Padwe, J. (2000). Sustainability of Ache hunting in the Mbaracayu Reserve, Paraguay. In: Robinson, J. G., and Bennett, E. L. (eds.), *Hunting for Sustainability in Tropical Forests.* Columbia University Press, New York. pp. 79–105.

Hill, K., McMillan, G., and Farina, R. (2003). Hunting-related changes in game encounter rates from 1994 to 2001 in the Mbaracayu Reserve, Paraguay. *Conserv. Biol.* 17:1312–1323.

Hill, K. R., and Hurtado, A. M. (1995). *Ache Life History: the Ecology and Demography of a Foraging People.* Aldine de Gruyter, New York.

Hill, R. A., and Cowlishaw, G. (2002). Foraging female baboons exhibit similar patterns of antipredator vigilance across two populations. In: Miller, L. E. (ed.), *Eat or Be Eaten: Predator Sensitive Foraging Among Primates.* Cambridge University Press, Cambridge. pp. 187–204.

Hill, R. A., and Dunbar, R. I. M. (1998). An evaluation of the roles of predation rate and predation risk as selective pressures on primate grouping behaviour. *Behaviour* 135:411–430.

Hill, R. A., and Lee, P. C. (1998). Predation risk as an influence on group size in cercopithecoid primates: implicatoins for social structure. *J. Zool. Lond.* 245:447–456.

Holekamp, K. E., Smale, L., Berg, R., and Cooper, S. M. (1997). Hunting rates and hunting success in the spotted hyena (*Crocuta crocuta*). *J. Zool. Lond.* 242:1–15.

Horrocks, J. A., and Hunte, W. (1986). Sentinel behaviour in vervet monkeys: who sees whom first? *Anim Behav* 34:1566–1567.

Hunter, T. B., and Skinner, J. D. (1998). Vigilance behaviour in African ungulates: the role of predation pressure. *Behaviour* 135:195–211.

Isbell, L. A. (1990). Sudden short-term increase in mortality of vervet monkeys (*Cercopithecus aethiops*) due to leopard predation in Amboseli National Park, Kenya. *Am. J. Primatol.* 21:41–52.

Isbell, L. A. (1994). Predation on primates: ecological patterns and evolutionary consequences. *Evol. Anthropol.* 3:61–71.

Isbell, L. A., and Enstam, K. L. (2002). Predator (in)sensitive foraging in sympatric female vervets (*Cercopithecus aethiops*) and patas monkeys (*Erythrocebus patas*): a test of ecological models of group dispersion. In: Miller, L. E. (ed.), *Eat or Be Eaten: Predator Sensitive Foraging Among Primates*. Cambridge University Press, Cambridge. pp. 154–168.

Isbell, L. A., and Young, T. P. (1993a). Human presence reduces predation in a free-ranging vervet monkey population in Kenya. *Anim. Behav.* 45:1233–1235.

Isbell, L. A., and Young, T. P. (1993b). Social and ecological influences on activity budgets of vervet monkeys and their implications for group living. *Behav. Ecol. Sociobiol.* 32:377–385.

Izor, R. J. (1985). Sloths and other mammalian prey of the harpy eagle. In: Montgomery, G. G. (ed.), *The Evolution and Ecology of Armadillos, Sloths, and Vermilinguas*. Smithsonian Institution Press, Washington DC. pp. 343–346.

Jack, K. M. (2001). Effects of male emigration on the vigilance behavior of coresident males in white-faced capuchins (*Cebus capucinus*). *Int. J. Primatol.* 22:715–732.

Jackson, P., and Nowell, K. (1996). Wild Cats: Status Survey and Conservation Action Plan. Gland: IUCN.

Jaenicke, C., and Ehrlich, A. (1982). Effects of animate vs. inanimate stimuli on curiosity behavior in greater galago and slow loris. *Primates* 23:95–104.

Janson, C. H. (1990a). Ecological consequences of individual spatial choice in foraging groups of brown capuchin monkeys, *Cebus apella*. *Anim. Behav.* 40:922–934.

Janson, C. H. (1990b). Social correlates of individual spatial choice in foraging groups of brown capuchin monkeys, *Cebus apella*. *Anim. Behav.* 40:910–921.

Janson, C. H. (1992). Evolutionary ecology of primate social structure. In: Smith, E. A., and Winterhalder, B. (eds.), *Evolutionary Ecology and Human Behavior*. Aldine de Gruyter, New York. pp. 95–130.

Janson, C. H. (1998). Testing the predation hypothesis for vertebrate sociality: prospects and pitfalls. *Behaviour* 135:389–410.

Janson, C. H., and Goldsmith, M. L. (1995). Predicting group size in primates: foraging costs and predation risks. *Behav. Ecol.* 6:326–336.

Janzen, D. H. (1983). The Pleistocene hunters had help. *Am. Nat.* 121:598–599.

Jarman, P. J. (1974). The social organisation of antelope in relation to behaviour. *Behaviour* 48:215–267.

Jolly, A., Caless, S., Cavigelli, S., Gould, L., Pereira, M. E., Pitts, A., Pride, R. E., Rabenandrasana, H. D., Walker, J. D., and Zafison, T. (2000). Infant killing, wounding, and predation in *Eulemur* and *Lemur*. *Int. J. Primatol.* 21:21–40.

Julliot, C. (1994). Predation of a young spider monkey (*Ateles paniscus*) by a crested eagle (*Morphnus guianensis*). *Folia Primatol.* 63:75–77.

Karpanty, S. M. (2003). Rates of predation by diurnal raptors on the lemur community of Ranomafana National Park, Madagascar. *Am. J. Phys. Anthropol.* 36(suppl.):126–127.

Karpanty, S. M., and Goodman, S. M. (1999). Diet of the Mdagascar Harrier-hawk, *Polyboroides radiatus*, in southeastern Madagascar. *J. Raptor Res.* 33:313–316.

Kerfoot, W. C., and Sih, A. (1987). Introduction. In: Kerfoot, W. C., and Sih, A. (eds.), *Predation*. University Press of New England, Hanover. pp. vii–viii.

Klein, R. G. (1984). Mammalian extinctions and Stone Age people in Africa. In: Martin, P. S., and Klein, R. G. (eds.), *Quaternary Extinctions: A Prehistoric Revolution*. University of Arizona Press, Tucson. pp. 553–573.

Klein, R. G. (1987). Problems and prospects in understanding how early people exploited animals. In: Nitecki, M. H., and Nitecki, D. V. (eds.), *The Evolution of Human Hunting*. Plenum Press, New York. pp. 11–45.

Kruger, O., Affeldt, E., Brackmann, M., and Milhahn, K. (1998). Group size and composition of *Colobus guereza* in Kyambura Gorge, southwest Uganda, in relation to chimpanzee activity. *Int. J. Primatol.* 19:287–297.

Kurten, B., and Anderson, E. (1980). *Pleistocene Mammals of North America*. Columbia University Press, New York.

Lambert, J. E. (2005). Competition, predation, and the evolutionary significance of the cercopithecine cheek pouch: the case of *Cercopithecus* and *Lophocebus*. *Am. J. Phys. Anthropol.* 126:183–192.

Lendrem, D. W. (1983). Sleeping and vigilance in birds I. Field observations of the mallard (*Anas platyrhynchos*). *Anim. Behav.* 31:532–538.

Lendrem, D. W. (1984). Sleeping and vigilance in birds II. An experimental study of the barbary dove (*Streptopelia risoria*). *Anim. Behav.* 32:243–248.

Lima, S. L. (1987). Vigilance while feeding and its relation to the risk of predation. *J. Theor. Biol.* 125:303–316.

Lima, S. L. (1993). Ecological and evolutionary perspectives on escape from predatory attack: a survey of North American birds. *Wilson Bull.* 105:1–47.

Lima, S. L. (1994). On the personal benefits of anti-predatory vigilance. *Anim. Behav.* 48:734–736.

Lima, S. L. (1995a). Back to the basics of anti-predatory vigilance: the group-size effect. *Anim. Behav.* 49:11–20.

Lima, S. L. (1995b). Collective detection of predatory attack by social foragers: fraught with ambiguity? *Anim. Behav.* 50:1097–1108.

Lima, S. L. (1998). Stress and decision making under the risk of predation: recent developments from behavioral, reproductive, and ecological perspectives. *Adv. Study Behav.* 27:215–290.

Lima, S. L. (2002). Putting predators back into behavioral predator–prey interactions. *Trends Ecol. Evol.* 17:70–75.

Lima, S. L., and Dill, L. M. (1990). Behavioral decisions made under the risk of predation: a review and prospectus. *Can. J. Zool.* 68:619–640.

Lima, S. L., and Zollner, P. A. (1996). Anti-predatory vigilance and the limits to collective detection: visual and spatial separation between foragers. *Behav. Ecol. Sociobiol.* 38:355–363.

Lindstrom, A. (1989). Finch flock size and risk of hawk predation at a migratory stopover site. *Auk* 106:225–232.

Lycett, J. E., Henzi, S. P., and Barrett, L. (1998). Maternal investment in mountain baboons and the hypothesis of reduced care. *Behav. Ecol. Sociobiol.* 42:49–56.

Macedonia, J. M., and Evans, C. S. (1993). Variation among mammalian alarm call systems and the problem of meaning in animal signals. *Ethology* 93:177–197.

Maisels, F. G., Gautier, J. P., Cruickshank, A., and Bosefe, J. P. (1993). Attacks by crowned hawk eagles (*Stephanoaetus coronatus*) on monkeys in Zaire. *Folia Primatol.* 61:157–159.

Marsh, C. W. (1979). Comparative aspects of social organization in the Tana River red colobus, *Colobus badius rufomitratus*. *Z. Tierpsychol.* 51:337–362.

Martin, G. H. G. (1983). Bushmeat in Nigeria as a natural resource with environmental implications. *Environ. Conserv.* 10:125–134.

Martin, P. S. (1984). Prehistory overkill: the global model. In: Martin, P. S., and Klein, R. G. (eds.), *Quaternary Extinctions: A Prehistoric Revolution*. University of Arizona Press, Tucson. pp. 354–403.

McDougal, C. (1987). The man-eating tiger in geographical and historical perspective. In: Tilson, A. U. S. S. R. L. (ed.), *Tigers of the World*. Park City, Noyes. pp. 435–448.

McGraw, W. S., and Bshary, R. (2002). Association of terrestrial mangabeys (*Cercocebus atys*) with arboreal monkeys: experimental evidence for the effects of reduced ground predator pressure on habitat use. *Int. J. Primatol.* 23:311–325.

Miller, L. E. (2002a). An introduction to predator sensitive foraging. In: Miller, L. E. (ed.), *Eat or Be Eaten: Predator Sensitive Foraging Among Primates*. Cambridge University Press, Cambridge. pp. 1–17.

Miller, L. E. (2002b). The role of group size in predator sensitive foraging decisions for wedge-capped capuchin monkeys (*Cebus olivaceus*). In: Miller, L. E. (ed.), *Eat or Be Eaten: Predator Sensitive Foraging Among Primates*. Cambridge University Press, Cambridge. pp. 95–106.

Mitani, J. C., Gros-Louis, J., and Manson, J. H. (1996). Number of males in primate groups: comparative tests of competing hypotheses. *Am. J. Primatol.* 38:315–332.

Mitani, J. C., Sanders, W. J., Lwanga, J. S., and Windfelder, T. L. (2001). Predatory behavior of craned hawk-eagles (*Stephanoaetus coronatus*) in Kibale National Park, Uganda. *Behav. Ecol. Sociobiol.* 49:187–195.

Mitani, J. C., and Watts, D. P. (1999). Demographic influences on the hunting behavior of chimpanzees. *Am. J. Phys. Anthropol.* 109:439–454.

Naughton-Treves, L., Mena, J. L., Treves, A., Alvarez, N., and Radeloff, V. C. (2003). Wildlife survival beyond park boundaries: the impact of swidden agriculture and hunting on mammals in Tambopata, Peru. *Conserv. Biol.* 17:1106–1117.

Newton-Fisher, N. E., Notman, H., and Reynolds, V. (2002). Hunting of mammalian prey by Budongo Forest chimpanzees. *Folia Primatol.* 73:281–283.

Nishida, T., Hasegawa, T., Hayaki, H., Takahata, Y., and Uehara, S. (1992). Meat-sharing as a coalition strategy by an alpha male chimpanzee? In: Nishida, T., McGrew, W., Marler, P., Pickgord, M., and deWall, F. (eds.), *Topics in Primatology. Human Origins*, vol. 1. Tokyo University Press, Tokyo. pp. 159–174.

Njiforti, H. L. (1996). Preference and present demand for bushmeat in northern Cameroon: some implications for wildlife conservation. *Environ. Conserv.* 23:149–155.

Noe, R., and Bshary, R. (1997). The formation of red colobus–diana monkey associations under predation pressure from chimpanzees. *Proc. R. Soc. Lond. B* 264:253–259.

Norconk, M. A. (1990). Introductory remarks: ecological and behavioural correlates of polyspecific primate troops. *Am. J. Primatol.* 21:81–85.

Olmos, F. 1994. Jaguar preation on muriqui *Brachyteles arachnoides*. *Neotrop. Primates* 2:16.

Olupot, W., and Waser, P. M. (2001). Activity patterns, habitat use and mortality risks of mangabey males living outside social groups. *Anim. Behav.* 61:1227–1235.

Overdorff, D. J., and Strait, S. C. (1995). Life history and predation in *Eulemur rubriventer* in Madagascar. *Am. J. Phys. Anthropol.* 100:487–506.

Overdorff, D. J., Strait, S. G., and Seltzer, R. G. (2002). Species differences in feeding in Milne Edwards sifakas (*Propithecus diadema edwardsi*), rufus lemurs (*Eulemur fulvus rufus*), and red-bellied lemurs (*Eulemur rubriventer*) in southern Madagascar: implications for predator avoidance. In: Miller, L. E. (ed.), *Eat or Be Eaten: Predator Sensitive Foraging Among Primates*. Cambridge University Press, Cambridge. pp. 126–137.

Oversluijs Vasques, M. R., and Heymann, E. W. (2001). Crested eagle (*Morphnus guianensis*) predation on infant tamarins (*Saguinus mystax* and *Saguinus fuscicollis*, Callitrichinae). *Folia Primatol.* 72:301–303.

Pages-Feuillade, E. (1988). Spatial distribution and inter-individual relationships in a nocturnal Malagasy lemur, *Microcebus murinus*. *Folia Primatol.* 50:204–220.

Peetz, A., Norconk, M. A., and Kinzey, W. G. (1992). Predation by jaguar on howler monkyes (*Alouatta seniculus*) in Venezuela. *Am. J. Primatol.* 28:223–228.

Peres, C. A. (1990a). A harpy eagle successfully captures an adult male red howler monkey. *Wilson Bull.* 102:560–561.

Peres, C. A. (1990b). Effects of hunting on western Amazonian primate communities. *Biol. Conserv.* 54:47–59.

Peres, C. A. (1993). Anti-predation benefits in a mixed-species group of Amazonian tamarins. *Folia Primatol.* 61:61–76.

Phillips, K. (1985). Differing responses to a predator (*Eira barbara*) by *Alouatta* and *Cebus*. *Neotrop. Primates* 3:45–46.

Prescott, M. J., and Buchanan-Smith, H. M. (2002). Predation sensitive foraging in captive tamarins. In: Miller, L. E. (ed.), *Eat or Be Eaten: Predator Sensitive Foraging Among Primates*. Cambridge University Press, Cambridge. pp. 44–57.

Pulliam, H. R. (1973). On the advantages of flocking. *J. Theor. Biol.* 38:419–422.

Pulliam, H. R., and Caraco, T. (1984). Living in groups: is there an optimal group size? In: Krebs, K. R., and Davies, N. (eds.), *Behavioural Ecology*. Blackwell Scientific, Oxford. pp. 122–147.

Rakotondravony, D., Goodman, S. M., and Soarimalala, V. (1998). Predation on *Hapalemur griseus griseus* by *Boa manditra* (Boidae) in the littoral forest of eastern Madagascar. *Folia Primatol.* 69:405–408.

Ramakrishnan, U., and Coss, R. G. (2001a). A comparison of the sleeping behavior of three sympatric primates. *Folia Primatol.* 72:51–53.

Ramakrishnan, U., and Coss, R. G. (2001b). Strategies used by bonnet macaques (*Macaca radiata*) to reduce predation risk while sleeping. *Primates* 42:193–206.

Ramakrishnan, U., Coss, R. G., and Pelkey, N. W. (1999). Tiger decline caused by the reintroduction of large ungulate prey: evidence from a study of leopard diets in southern India. *Biol. Conserv.* 89:113–120.

Rettig, N. L. (1978). Breeding behavior of the harpy eagle (*Harpia harpyja*). *Auk* 95:629–643.

Richard, A. F., and Dewar, R. E. (1991). Lemur ecology. *Annu. Rev. Ecol. Syst.* 22:145–175.

Robinson, J. G. (1981). Spatial structure in foraging groups of wedge-capped capuchin monkeys, *Cebus nigrivittatus. Anim. Behav.* 29:1036–1056.

Robinson, J. G. (2000). Calculating maximum sustainable harvests and percentage offtakes. In: Robinson, J. G., and Bennett, E. L. (eds.), *Hunting for Sustainability in Tropical Forests.* Columbia University Press, New York. pp. 521–524.

Robinson, J. G., and Bennett, E. L. (eds.) (2000). *Hunting for Sustainability in Tropical Forests.* Columbia University Press, New York.

Robinson, J. G., and Redford, K. H. (1994). Community-based approaches to wildlife conservation in neotropical forests. In: Western, D., Wright, R. M., and Strum, S. C. (eds.), *Natural Connections: Perspectives in Community-Based Conservation.* Island Press, Washington DC. pp. 300–319.

Rodman, P. S. (1973). Population composition and adaptive organization among orang-utans of the Kutai Reserve. In: Michael, R. P., and Crook, J. H. (eds.), *Comparative Ecology and Behaviour of Primates.* Academic Press, London. pp. 171–209.

Rodman, P. S. (1988). The ecology of social behavior. In: Slobodchikoff, C. N. (ed.), *Resources and Group Sizes of Primates.* Academic Press, San Diego. pp. 83–108.

Ron, T., Henzi, S. P., and Motro, U. (1996). Do female chacma baboons compete for a safe spatial position in a southern woodland habitat? *Behaviour* 133:475–490.

Rose, L. M. (1994). Benefits and costs of resident males to females in white-faced capuchins, *Cebus capucinus. Am. J. Primatol.* 32:235–248.

Rose, L. M., and Fedigan, L. M. (1995). Vigilance in white-faced capuchins (*Cebus capucinus*) in Costa Rica. *Anim. Behav.* 49:63–70.

Ross, C. (1993). Predator mobbing by an all-male band of hanuman langurs (*Presbytis entellus*). *Primates* 34:105–107.

Sampaio, D. T., and Ferrari, S. F. (2005). Predation of an infant titi monkey (*Callicebus moloch*) by a tufted capuchin (*Cebus apella*). *Folia Primatol.* 76:113–115.

Sanders, W. J., and Mitani, J. C. (2000). Taphonomic aspects of ealge predation of primates from Kibale Forest, Uganda. *Am. J. Phys. Anthropol.* 30(suppl.):267–268.

Sanders, W. J., Trapani, J., and Mitani, J. C. (2003). Taphonomic aspects of crowned hawk-eagle predation on monkeys. *J. Hum. Evol.* 44:87–105.

Sauther, M. L. (2002). Group size effects on predation sensitive foraging in wild ring-tailed lemurs (*Lemur catta*). In: Miller, L. E. (ed.), *Eat or Be Eaten: Predator Sensitive Foraging Among Primates.* Cambridge University Press, Cambridge. pp. 107–125.

Savage, A., Snowdon, C. T., Giraldo, L. H., and Soto, L. H. (1996). Parental care patterns and vigilance in wild cotton-top tamarins (*Saguinus oedipus*). In: Norconk, A. R. M., and Garber, P. (eds.), *Adaptive Radiation of Neotropical Primates.* Plenum Press, New York. pp. 187–199.

Schoener, T. W. (1971). Theory of feeding strategies. *Annu. Rev. Ecol. Syst.* 2:369–404.

Schulke, O., and Ostner, J. (2001). Predation on *Lepilemur* by a Harrier hawk and implications for sleeping site quality. *Lemur News* 6:5.

Schwab, D. (1999). Predation on *Eulemur fulvus* by *Accipiter henstii* (Henst's goshawk). *Lemur News* 4:34.

Seidensticker, J. (1983). Predation by *Panthera* cats and measures of human influence in habitats of south Asian monkeys. *Int. J. Primatol.* 4:323–326.

Setz, E. Z. F. (1999). Sleeping habits of the golden-faced sakis in a forest fragment in the central Amazon. *Am. J. Primatol.* 49(suppl.):100.

Sherman, P. T. (1991). Harpy eagle predation on a red howler monkey. *Folia Primatol.* 56:53–56.

Shine, R., Harlow, P. S., Keogh, J. S., and Boeadi (1998). The influence of sex and body size on food habits of a giant tropical snake, *Python reticulatus. Funct. Ecol.* 12:248–258.

Shultz, S., and Noe, R. (2002). The consequences of crowned eagle central-place foraging on predation risk in monkeys. *Proc. R. Soc. Lond. B* 269:1797–1802.

Shultz, S., Noe, R., McGraw, W. S., and Dunbar, R. I. M. (2004). A community-level evaluation of the impact of prey behavioural and ecological characteristics on preator diet composition. *Proc R. Soc. Lond. B* 271:725–732.

Sigg, H. (1980). Differentiation of female positions in hamadryas one-male-units. *Z. Tierpsychol.* 53:265–302.

Sih, A. (1987). Predators and prey lifestyles: an evolutionary and ecological overview. In: Kerfoot, W. C., and Sih, A. (eds.), *Predation.* University Press of New England, Hanover. pp. 203–224.

Sih, A. (1992). Prey uncertainty and the balancing of antipredator and feeding needs. *Am. Nat.* 139:1052–1069.

Skorupa, J. P. (1989). Crowned eagles (*Stephanoaetus coronatus*) in rainforest: observations on breeding chronology and diet at a nest in Uganda. *Ibis* 131:294–298.

Slade, N. A., Gomulkiewicz, R., and Alexander, H. M. (1998). Alternatives to Robinson and Redford's method for assessing overharvest from incomplete demographic data. *Conserv. Biol.* 12:148–155.

Slatkin, M., and Hausfater, G. (1976). A note on the activities of a solitary male baboon. *Primates* 17:311–322.

Smuts, B. B. (1985). *Sex and Friendship in Baboons.* Aldine, Hawthorne.

Southwick, C. H., Beg, M. A., and Siddiqi, M. R. (1965). Rhesus monkeys in north India. In: DeVore, I. (ed.), *Primate Behavior: Field Studies of Monkeys and Apes.* Holt, Rinehart and Winston, New York. pp. 111–162.

Srivastava, A. (1991). Cultural transmission of snake-mobbing in free-ranging hanuman langurs. *Folia Primatol.* 56:117–120.

Stanford, C. B. (1989). Predation on capped langurs (*Presbytis pileata*) by cooperatively hunting jackals (*Canis aureus*). *Am. J. Primatol.* 19:53–56.

Stanford, C. B. (1995). The influence of chimpanzee predation on group size and antipredator behavior in red colobus monkeys. *Anim. Behav.* 49:577–587.

Stanford, C. B. (1998). Predation and male bonds in primate societies. *Behaviour* 135:513–533.

Stanford, C. B. (2002). Avoiding predators: expectations and evidence in primate antipredator behavior. *Int. J. Primatol.* 23:741–757.

Stanford, C. B., Wallis, J., Matama, H., and Goodall, J. (1994). Patterns of predation by chimpanzees on red colobus monkeys in Gombe National Park, Tanzania, 1982–1991. *Am. J. Phys. Anthropol.* 94:213–228.

Stelzner, J., and Strier, K. (1981). Hyena predation on an adult male baboon. *Mammalia* 45:269–260.

Struhsaker, T. T. (1967). Auditory communication among vervet monkeys (*Cercopithecus aethiops*). In: Altmann, S. A. (ed.),

Social Communication Among Primates. University of Chicago Press, Chicago. pp. 238–324.

Struhsaker, T. T. (1969). Correlates of ecology and social organization among African cercopithecines. *Folia Primatol.* 11:80–118.

Struhsaker, T. T. (1981). Polyspecific associations among tropical rain-forest primates. *Z. Tierpsychol.* 57:268–304.

Struhsaker, T. T. (1999). Primate communities in Africa: the consequence of long-term evolution or the artifact of recent hunting? In: Fleagle, J. G., Janson, C. H., and Reed, K. E. (eds.), *Primate Communities.* Cambridge University Press, Cambridge. pp. 289–294.

Struhsaker, T. T. (2000). The effects of predation and habitat quality on the socioecology of African monkeys: lessons from the islands of Bioko and Zanzibar. In: Whitehead, P. F., and Jolly, C. J. (eds.), *Old World Monkeys.* Cambridge University Press, Cambridge. pp. 393–430.

Struhsaker, T. T., and Leakey, M. (1990). Prey selectivity by crowned hawk-eagles on monkeys in Kibale Forest, Uganda. *Behav. Ecol. Sociobiol.* 36:435–443.

Tello, N. S., Huck, M., and Heymann, E. W. (2002). Boa constrictor attack and successful group defense in moustached tamarins, *Saguinus mystax. Folia Primatol.* 73:146–148.

Tenaza, R., and Tilson, R. L. (1985). Human predation and Kloss's gibbon (*Hylobates klossii*) sleeping trees in Siberut Island, Indonesia. *Am. J. Primatol.* 8:299–308.

Terborgh, J. (1983). *Five New World Primates: A Study of Comparative Ecology.* Princeton University Press, Princeton, NJ.

Terborgh, J. (1990). Mixed flocks and polyspecific associations: costs and benefits of mixed groups to birds and monkeys. *Am. J. Primatol.* 21:87–100.

Terborgh, J., and Janson, C. H. (1986). The socioecology of primate groups. *Annu. Rev. Ecol. Syst.* 17:111–135.

Terborgh, J., Lopez, L., Nu-ez, P., Rao, M., Shahabudin, G., Orihuela, G., Riveros, M., Ascanio, R., Adler, G. H., Lambert, T. D., and Balbas, L. (2002). Ecological meltdown in predator-free forest fragments. *Science* 294:1923.

Treves, A. (1998). The influence of group size and near neighbors on vigilance in two species of arboreal primates. *Behaviour* 135:453–482.

Treves, A. (1999). Has predation shaped the social systems of arboreal primates? *Int. J. Primatol.* 20:35–67.

Treves, A. (2000). Theory and method in studies of vigilance and aggregation. *Anim. Behav.* 60:711–722.

Treves, A. (2002). Predicting predation risk for foraging, arboreal monkeys. In: Miller, L. E. (ed.), *Eat or Be Eaten: Predator Sensitive Foraging Among Primates.* Cambridge University Press, Cambridge. pp. 222–241.

Treves, A., and Chapman, C. A. (1996). Conspecific threat, predation avoidance and resource defense: implications for grouping in langurs. *Behav. Ecol. Sociobiol.* 39:43–53.

Treves, A., Drescher, A., and Ingrisano, N. (2001). Vigilance and aggregation in black howler monkeys (*Alouatta pigra*). *Behav. Ecol. Sociobiol.* 50:90–95.

Treves, A., Drescher, A., and Snowdon, C. T. (2003). Maternal watchfulness in black howler monkeys (*Alouatta pigra*). *Ethology.*

Treves, A., and Naughton-Treves, L. (1999). Risk and opportunity for humans coexisting with large carnivores. *J. Hum. Evol.* 36:275–282.

Treves, A., and Palmqvist, P. (in press). Reconstructing the paleo-ecology of hominids and their mammalian predators (6.0–1.8 Ma). In: Nekaris, K. A. I., and Gursky, S. L. (eds.), *Primates and Their Predators.* Springer, New York.

Treves, A., and Pizzagalli, D. (2002). Vigilance and perception of social stimuli: views from ethology and social neuroscience. In: Bekoff, C. A. M., and Burghardt, G. (eds.), *The Cognitive Animal: Empirical and Theoretical Perspectives on Animal Cognition.* MIT Press, Cambridge, MA. pp. 463–469.

Triesman, M. (1975). Predation and the evolution of gregariousness. I. Models for concealment and evasion. *Anim. Behav.* 23:799–900.

Tsukahara, T. (1993). Lions eat chimpanzees: the first evidence of predation by lions on wild chimpanzees. *Am. J. Primatol.* 29:1–11.

Tutin, C. E. G., McGrew, W. C., and Baldwin, P. J. (1983). Social organization of savanna-dwelling chimpanzees, *Pan troglodytes verus,* at Mt. Assirik, Senegal. *Primates* 24:154–173.

Uhde, N. L., and Sommer, V. (2002). Antipredatory behavior in gibbons (*Hylobates lar,* Khao Yai/Thailand). In: Miller, L. E. (ed.), *Eat or Be Eaten: Predator Sensitive Foraging Among Primates.* Cambridge University Press, Cambridge. pp. 268–291.

Utami, S. S., and van Hooff, J. A. R. A. M. (1997). Mean-eating by adult female Sumatran orangutans (*Pongo pygmaeus abelii*). *Am. J. Primatol.* 43:159–165.

van Schaik, C. P. (1983). Why are diurnal primates living in groups? *Behaviour* 87:120–144.

van Schaik, C. P., and Horstermann, M. (1994). Predation risk and the number of adult males in a primate group: a comparative test. *Behav. Ecol. Sociobiol.* 35:261–272.

van Schaik, C. P., and Kappeler, P. M. (1996). The social systems of gregarious lemurs: lack of convergence with anthropoids due to evolutionary disequilibrium? *Ethology* 102:915–941.

van Schaik, C. P., van Amerongen, A., and van Noordwijk, M. A. (1996). Riverine refuging by wild Sumatran long-tailed macaques (*Macaca fascicularis*). In: Fa, J. E., and Lindburg, D. G. (eds.), *Evolution and Ecology of Macaque Societies.* Cambridge University Press, Cambridge. pp. 160–181.

van Schaik, C. P., and van Hooff, J. A. R. A. M. (1983). On the ultimate causes of primate social systems. *Behaviour* 85:91–117.

van Schaik, C. P., and van Noordwijk, M. A. (1985). Evolutionary effect of the absence of felids on the social organization of the macaques on the island of Simeulue (*Macaca fascicularis,* Miller 1903). *Folia Primatol.* 44:138–147.

van Schaik, C. P., and van Noordwijk, M. A. (1989). The special role of male *Cebus* monkeys in predation avoidance and its effects on group composition. *Behav. Ecol. Sociobiol.* 24:265–272.

Vereshchagin, N. K., and Baryshnikov, G. G. (1984). Quaternary mammalian extinctions in northern Eurasia. In: Martin, P. S., and Klein, R. G. (eds.), *Quaternary Extinctions: A Prehistoric Revolution.* University of Arizona Press, Tucson. pp. 483–516.

Vine, I. (1971). Risk of visual detection and pursuit by a predator and the selective advantage of flocking behaviour. *J. Theor. Biol.* 30:405–422.

Wahome, J. M., Rowell, T. E., and Tsingalia, H. M. (1993). The natural history of de Brazza's monkeys in Kenya. *Int. J. Primatol.* 14:445–466.

Walsh, P. D., Abernethy, K. A., Bermejos, M., Beyers, R., de Wachter, P., Akou, M. E., Huijbregts, B., Mambounga, D. I., Toham, A. K., Kilbourn, A. M., Lahm, S. A., Latour, S., Maisels, F., Mbina, C., Mihindou, Y., Obiang, S. N., Effa, E.

N., Starkey, M. P., Telfer, P., Thibaul, M., Tutin, C. E. D., White, L. J. T., and Wilkie, D. S. (2003). Catastrophic ape decline in western equatorial Africa. *Nature* 422:611–614.

Waser, P. M. (1987). Interactions among primate species. In: Smuts, B. B., Cheney, D. L., Seyfarth, R. M., Wrangham, R. W., and Struhsaker, T. T. (eds.), *Primate Societies*. University of Chicago Press, Chicago. pp. 210–226.

Watanabe, K. (1981). Variations in group composition and population density of two sympatric Mentawaian leaf-monkeys. *Primates* 22:145–160.

Wiens, F., and Zitzmann, A. (1999). Predation on a wild slow loris (*Nycticebus coucang*) by a reticulated python (*Python reticulatus*). *Folia Primatol.* 70:362–364.

Wilkie, D. S., and Carpenter, J. F. (1999). Bushmeat hunting in the Congo Basin: an assessment of impacts and options for mitigation. *Biodiv. Conserv.* 8:927–955.

Wolf, N. G. (1985). Odd fish abandon mixed-species groups when threatened. *Behav. Ecol. Sociobiol.* 17:47–52.

Wrangham, R. W. (1979). On the evolution of ape social systems. *Soc. Sci. Inf.* 18:334–368.

Wrangham, R. W., and van Zinnicq Bergmann Riss, E. (1990). Rates of predation on mammals by Gombe chimpanzees, 1972–1975. *Primates* 31:157–170.

Wright, P., Heckscher, S., and Dunham, A. (1997). Predation on a Milne Edwards sifaka (*Propithecus diadema edwardsi*) by the fossa (*Cryptoprocta ferox*) in the rain forest of southeastern Madagascar. *Folia Primatol.* 68:34–43.

Wright, P. C. (1998). Impact of predation risk on the behaviour of *Propithecus diadema edwardsi* in the rain forest of Madagascar. *Behaviour* 135:483–512.

Wright, P. C., Heckscher, K., and Dunham, A. (1998). Predation on rain forest prosimians in Ranomafana National Park, Madagascar. *Folia Primatol.* 69(suppl. 1):401.

Wright, P. C., and Martin, L. B. (1995). Predation, pollination and torpor in two nocturnal prosimians: *Cheirogaleus major* and *Microcebus rufus* in the rainforest of Madagascar. In: Alterman, G. A. D. L., and Izard, M. K. (eds.), *Creatures of the Dark: The Nocturnal Prosimians*. Plenum Press, New York. pp. 45–60.

Wrona, F. J., and Dixon, R. W. J. (1991). Group size and predation risk: a field analysis of encounter and dilution effects. *Am. Nat.* 137:186–201.

Ydenberg, R. C., and Dill, L. M. (1986). The economics of fleeing from predators. *Adv. Study Behav.* 16:229–251.

Young, T. P. (1994). Predation risk, predation rate, and the effectiveness of anti-predator traits. *Evol. Anthropol.* 3:67.

Zuberbuhler, K. (2000). Interspecies semantic communication in two forest primates. *Proc. R. Soc. Lond. B* 267:713–718.

Zuberbuhler, K., and Jenny, D. (2002). Leopard predation and primate evolution. *J. Hum. Evol.* 43:873–886.

Zuberbuhler, K., Noe, R., and Seyfarth, R. M. (1997). Diana monkey long-distance calls: messages for conspecifics and predators. *Anim. Behav.* 53:589–604.

Zullo, J., and Caine, N. (1988). The use of sentinels by captive red-bellied tamarins. *Am. J. Primatol.* 14:455.

33

Primate Locomotor Behavior and Ecology

Paul A. Garber

INTRODUCTION

No group of vertebrates shows a wider range of locomotor adaptations to arboreality than do living primates. (Fleagle 1978:243)

Primates represent a successful radiation of primarily arboreal, tropical forest mammals that vary considerably in body mass, limb proportions, postcranial anatomy, and positional adaptations. These differences reflect important evolutionary distinctions in the manner in which species exploit their environment, obtain access to food resources, and avoid predators as well as in the energetic costs of travel. The study of primate locomotion is central to an understanding of primate adaptive diversity because major changes in the ability of primate lineages to exploit their environment are associated with evolutionary changes in positional behavior and positional morphology. These include the evolution of claw-like nails and vertical clinging in tamarins and marmosets (subfamily Callitrichinae); elongated forelimbs and suspensory behaviors in subfossil sloth-lemurs (Palaeopropithecidae), lesser apes (*Hylobatidae*), and great apes; prehensile tails in atelines and capuchins; and hindlimb elongation and vertical clinging and leaping in some prosimians (Niemitz 1984; Cant 1986, 1990; Oxnard et al. 1990, Hunt 1991; Garber 1992; Anemone 1993; Crompton et al. 1993; Davis 1996; Terranova 1996; Larson 1998a; Godfrey and Jungars 2003). In the case of human evolution, for example, adaptations associated with bipedal locomotion have been associated with fundamental changes in hominid social organization, food sharing, tool use, hunting, and patterns of habitat utilization (Lovejoy 1981, Ruff 1987, Foley and Elton 1998, Hunt 1998).

Early in their evolution, the ancestors of modern primates (euprimates) evolved a unique suite of locomotor traits associated with exploiting an arboreal environment. These include grasping feet and hands, an opposable big toe (hallux) bearing a flattened nail (and eventually nails replacing claws on most or all digits), and changes in the visual system (orbital convergence, color vision, visual images from the retinae crossing at the optic chiasm, enhanced depth perception; see Martin 1990). Several theories have been proposed to account for the evolution of these traits, including the possibility that the ancestors of modern primates were visual predators of insects in the low shrub layer of the forest canopy (Cartmill 1972, 1974, 1992), exploited resources found on terminal branches in the periphery of the tree crown (Charles-Dominique and Martin 1970, Rasmussen 1990), or that early primates played an important role in the evolution of angiosperms and tropical forests by serving as agents of pollination and seed dispersal (Sussman and Raven 1977, Sussman 1991). Each of these theories highlights the fact that the evolutionary history of modern primates is closely linked to their ability to travel on, grasp, and exploit small, flexible arboreal supports. In fact, recent fossil evidence (*Carpolestes simpsoni*) dated at 56 million years ago (Sargis 2002) indicates that changes in the grasping capabilities of the feet and hands of primate ancestors (nail replacing a claw on the big toe) preceded changes in visual acuity. These initial primate positional adaptations provide a framework for examining subsequent changes in primate locomotor anatomy and behavior (Schmitt and Lemelin 2002). Derived trait complexes such as prehensile tails, divergent and opposable thumbs, mobile shoulder girdle, elongated forelimbs, the evolution of an elongated tarsus and tibia–fibula fusion, and forms of hindlimb-dominated locomotion (i.e., leaping, hindlimb grasping, bipedalism) appear to represent functional and adaptive solutions enabling modern primates to obtain resources and exploit environments in new and novel ways. As expressed by Jolly and Plog (1979:114), "almost every major adaptive radiation within the primate order has involved changes in locomotor patterns that opened up a new array of econiches."

In the remainder of this chapter, I examine diversity in patterns of posture and locomotion among living primates and the challenges that primates of different body mass, limb proportions, and diet face in exploiting resources in arboreal and terrestrial environments. Specifically, I focus on distinctions between posture and locomotion, definitions of positional categories, the role of the primate tail in balance and weight support, the effects of body mass on patterns of substrate preference and positional behavior, and evidence for seasonal, site-specific, and sex-based differences in locomotion and posture. In general, within a given species, patterns of positional behavior during travel are more conservative than during feeding. This may reflect seasonal changes in diet and foraging strategies that characterize most primate taxa. Although smaller- and larger-bodied primates face different challenges in maintaining balance and weight support and in the cost of travel, body mass is not a strong predictor of positional behavior. This may reflect the fact that both both smaller- and larger-bodied primates have evolved particular morphological adaptations that enable them to exploit similar sets of resources in different ways.

POSTURE AND LOCOMOTION

The study of primate locomotion is more accurately described as the study of primate positional behavior (Prost 1965). Positional behavior consists of two components, posture and locomotion. Posture is defined as a set of behaviors in which an individual's center of mass remains relatively stable and the limb segments proper, rather than the entire skeleton, are being manipulated for some end (Prost 1965). Postural activity tends to dominate the positional repertoire during feeding, resting, and various socially directed behaviors (Rose 1974, Garber 1984, Hunt 1992, Walker 1996). Sitting, hanging, and standing represent common postural behaviors. In most studies of primates in the wild, postural behavior accounts for a greater proportion of the positional repertoire than does locomotor behavior (Rose 1974, McGraw 1998a).

Locomotor behaviors differ from postural behaviors in that "body mass (as opposed to limb mass) is grossly displaced relative to its physical surroundings" (Prost 1965:1200). Locomotor behaviors are associated with such common activities as traveling, foraging, and predator avoidance and represent a primary component of a species' adaptive pattern. Locomotion is a means to an end—for example, traveling to a food patch or resting site, fleeing a predator, or stalking insect prey—and, as such, feeds back onto other social/ maintenance activities (Ripley 1967). Running, climbing, and leaping are common locomotor behaviors.

In mechanical terms, postural and locomotor modes may represent integrated points on a behavioral continuum, for example, when quadrupedal standing transitions into quadrupedal walking or quadrupedal running (Prost 1965, Kinzey 1976). During postural behavior, many of the forces acting directly on the musculoskeletal system may be analogous to those involved in locomotion. Conversely, the skeleton may be exposed to forces that are infrequent but extreme during locomotion and not commonly encountered during postural behavior. In the case of many vertical clinging and leaping prosimians, the high take-off and landing forces generated when leaping to and from vertical trunks (Peters and Preuschoft 1984; Preuschoft 1985; Crompton et al. 1993; Demes et al. 1995; Warren and Crompton 1998a,b) have no biomechanical counterpart during postural behavior. Thus, distinctions between posture and locomotion are important because each of these sets of positional behaviors may select for alternative forms of musculoskeletal design.

Positional behavior is activity- and context-specific. That is, certain positional behaviors occur more frequently during particular activities (feeding vs. traveling vs. resting) and on supports of certain diameter, orientation, and weight-bearing capacity (Ripley 1967; Kinzey 1967; Garber 1980, 1998; Bergeson 1998). The selection of particular substrates may be a response to biomechanical factors (i.e., due to forces of torque and gravity, it is more stable to walk on top of a relatively large horizontal branch than it is to walk on top of a relatively small horizontal branch; Grand 1972, Cartmill and Milton 1977), as well as the physical structure and geometry of the forest canopy (Dunbar and Badam 2000). Many fruit, flower, leaf, and insect resources exploited by primates are located on thin, flexible branches in the periphery of the tree crown (Sussman 1991). Access to these resources commonly is associated with prehensile, grasping, and below-branch suspensory postures (Grand 1972, Meldrum et al. 1997). For example, gibbons (*Hylobates lar*) take advantage of the downward displacement of a branch during forelimb suspensory behavior to expand their feeding sphere and gain access to below-branch resources (Grand 1972). Other foods exploited by primates may be associated with a large branch niche. It has been argued that the evolution of claw-like nails in trunk-foraging primates in Africa (*Euoticus*, needle-clawed galago), Madagascar (*Phaner*, forked-tooth lemur) and Central and South America (*Cebuella, Callithrix, Mico, Saguinus, Leontopithecus, Callimico*; tamarins, marmosets, and Goeldi's monkey) represents parallel postural adaptations, enabling these small-bodied taxa to obtain resources such as exudates, insects, and fungi by clinging to large vertical supports that would otherwise be too large to be spanned by their diminutive hands and feet (Charles-Dominique 1977; Garber 1980, 1992). If one assumes that individuals preferentially adopt particular positional modes and select particular substrates that offer them increased efficiency and stability in an arboreal environment, then factors such as body mass and ontogenetic changes in motor control and limb and body proportions should play an important role in a species' positional repertoire (Cartmill and Milton 1977, Fleagle 1978, Fleagle and Mittermeier 1980). A major goal of the study of positional behavior is to understand the relationship between form and function in a species' natural environment (Bock and von Wahlert 1965).

CHALLENGES IMPOSED BY THE ARBOREAL AND TERRESTRIAL ENVIRONMENTS

Primates exploiting an arboreal environment face different challenges from primates exploiting a terrestrial environment. These include exposure to different types of predator, different strategies for escape, alternative conditions of temperature and sunlight, and access to different types and distributions of food resource (McGraw and Bshary 2002). In terms of positional behavior, terrestrial environments present foragers with a more uniform and stable set of substrates, as well as greater opportunity for continuous pathways of travel than do arboreal environments. Several primate lineages (African apes, savanna baboons, mandrills, geladas, patas monkeys, macaques, Asian colobines, ring-tailed lemurs, humans) commonly travel and feed on the ground (there is no evidence of terrestrial adaptations in any extant New World monkey). Patterns of terrestrial travel in Asian great apes (orangutan, fist-walking), African great apes (knuckle-walking in gorillas, chimpanzees, and bonobos), and humans (bipedalism) are quite distinctive from those of Old World monkeys and represent alternative solutions to problems of balance and weight bearing. For example, African apes exhibit several derived features of the distal radius that extend and stabilize the wrist during knuckle-walking as well as permit climbing and suspensory behavior in the trees (Richmond and Strait 2000). In the case of human ancestors, evidence of bipedal adaptations appears in fossil hominids by at least 4 million years ago (Leakey et al. 1995). However, despite an expanding fossil record, their remains no consensus on whether human ancestors shared a knuckle-walking ancestry with African apes or whether human bipedalism evolved from an arboreal ancestor that emphasized climbing and other suspensory positional behaviors (Richmond and Strait 2000).

In contrast to terrestrial environments, the arboreal canopy is characterized by substrates that are discontinuous, mobile, contoured in a cylindrical shape, orientated at various angles to the ground, and variable in size and weight-bearing capacity (Cartmill 1974, 1985; Grand 1984; Cant 1992). Moreover, tree branches become smaller and more unstable as one approaches the outer or food-bearing zones of the canopy; therefore, animals traveling in an arboreal environment face significant challenges of maintaining balance and support on branches that bend, sway, and possibly break under their weight (Cartmill and Milton 1977, Cant 1992). Arboreal primates attempt to solve these problems by distributing their body mass over a range of different supports, lowering their center of gravity nearer to the support, adopting flexed and abducted limb postures, and maintaining stability by engaging in suspensory or below-branch postures (Fig. 33.1). Given that primates from the size of orangutans to mouse lemurs travel and feed principally in the trees, species differences in body mass are expected to be an important factor in primate positional behavior and musculoskeletal design (Fleagle 1976, Cartmill and Milton 1977, Fleagle and Mittermeier 1980, Jungars 1985, Fleagle and Meldrum 1988, Demes and Gunther 1989, Jungers and Burr 1994, Preuschoft et al. 1996, Polk, 2002).

PRIMATE QUADRUPEDALISM

Virtually all primates walk quadrupedally during slow horizontal progression on the ground or in the trees, and for many species quadrupedal walking is the most common mode of travel (Tables 33.1–33.6). Kay et al. (1997) argue that the earliest anthropoids emphasized quadrupedal walking over

Figure 33.1 Patterns of locomotion and posture among a community of primates in Suriname. From Fleagle (1988).

leaping or other forms of progression. Primates are distinctive among mammals in their use of a *diagonal sequence–diagonal couplet gait* (i.e., a gait sequence in which movement of the left hindfoot is followed by movement of the right forelimb, followed by right hindfoot and left forelimb) when walking quadrupedally (Hildebrand 1967, Larson 1998b). In addition, during quadrupedal walking, the primate forelimb adopts a more protracted orientation at touchdown (primate forelimb is angled at >90° relative to the axis of the body), and peak vertical substrate reaction forces acting on the forelimb are reduced relative to those acting on the hindlimb (Demes et al. 1994, Schmitt and Lemelin 2002). According to Larson (1998b:166), "the reduction in the use of forelimb propulsive muscles is a component of the basic adaptation of reaching out with the clawless grasping hand in an arboreal setting that arose early in the evolution of

primates. . . ." Thus, there appears to have been a major shift in primates from a primarily weight-bearing role of the forelimb to a greater role in grasping, exploration, and object manipulation. Greater forelimb mobility and enhanced cortical control of forelimb movement in primates are consistent with the mechanical demands of exploiting resources in a fine-branch setting (Larson 1998b; Schmitt 1998, 2003; Schmitt and Lemelin 2002).

Relative to most mammals, primates are characterized by elongated forelimbs. During quadrupedal walking, arboreal primates (Fig. 33.2, upper) adopt a crouching or flexed-elbow gait that places their center of mass nearer to the substrate. This serves to improve balance and stability, especially on small or flexible branches (Schmitt 2003). A crouched position also may facilitate rapid changes in travel direction and substrate use and facilitate opportunistic leaping (Schmitt 1998). In contrast, primates that walk quadrupedally on the ground (Fig. 33.2, lower) tend to have their limbs more adducted (nearer to the body) and elbows extended and to experience vertical reaction forces more equally distributed between forelimbs and hindlimbs (Larson 1998b, Polk 2002, Schmitt and Lemelin 2002). According to Fleagle (1988:247), compared to their more arboreal relatives, terrestrial Old World monkeys are characterized by "a narrow, deep trunk and relatively long limbs, designed for long strides and speed, and their tails are often short or absent." These changes in anatomical design facilitate rapid travel in a terrestrial environment (Fleagle 1988).

DEFINITION OF POSITIONAL CATEGORIES

One of the major challenges facing the study of primate positional behavior is constructing precise and functionally meaningful categories of posture and locomotion. There has been a tendency for researchers to group what may be a set of biomechanically diverse behaviors into a small number of discrete categories (Walker 1998). For example, many studies (Tables 33.2 and 33.3) collapse a species' entire locomotor repertoire into four or five activities, such as climb, leap, quadrupedalism, tail–arm suspend, or some other form of bimanual or suspensory locomotor behavior. Clearly, a species' positional repertoire is considerably more diverse, and the inclusion of behaviors that are biomechanically

Table 33.1   Intraspecific Variation in Positional Behavior[1] in Common Chimpanzees, *Pan troglodytes schweinfurthi* and *Pan troglodytes verus*

FIELD SITE	QUAD	CLIMB/SCRAMBLE	SUSPENSORY	BIPEDAL	LEAP	SIT/LIE	STAND	ARM-HANG	N
Gombe	16.9	0.9	0	0.4	0	73.2	3.6	4.8	2,700
Mahale	16.1	0.9	0.3	0.3	0	76.8	1.4	4.4	11,393
Taï Forest	12.1	1.5	0.2	0.2	0.1	79.5	5.0	1.3	10,079

Sources: Adapted from Doran (1993), Hunt (1992).
[1] Data represent percentage of time engaged in each positional behavior. Hunt (1992) collected data using a 2 min instantaneous point sampling technique. Doran (1993) collected data using a 1 min instantaneous point sampling technique.

Table 33.2 Positional Behavior in Seven Sympatric Primate Species in Suriname[1]

SPECIES	BODY MASS[2]	CLIMB	LEAP	QUAD	TAIL–ARM SUSPEND	N
Travel						
Ateles paniscus	7.7	31.1	4.2	25.4	38.6	743
Alouatta seniculus	6.0	16.0	4.0	80.0	0	818
Cebus apella	3.0	5.0	10.0	84.0	0	363
Chiropotes satanas	3.0	Rare	18.0	80.0	0	452
Pithecia pithecia	1.8	Unknown	75.0	Unknown	0	40
Saimiri sciureus	0.63	4.0	42.0	55.0	0	224
Saguinus midas	0.45	0	24.0	76.0	0	625
Feed/forage						
Ateles paniscus	7.7	44.9	2.6	24.8	27.7	274
Alouatta seniculus	6.0	41.0	0	59.0	0	194
Cebus apella	3.0	8.0	4.0	88.0	0	438
Chiropotes satanas	3.0	2.0	9.0	88.0	0	282
Pithecia pithecia	1.8	Unknown	Unknown	Unknown	0	0
Saimiri sciureus	0.63					
Feed		2.0	11.0	87.0	0	158
Forage		3.0	22.0	75.0	0	463
Saguinus midas	0.45	2.0	12.0	87.0	0	128

Sources: From Fleagle and Mittermeier (1980).
[1] Data represent percentage of time engaged in each positional behavior.
[2] Adult female body mass in kilograms.

Table 33.3 Patterns of Positional Behavior and Substrate Use[1] in Taï Forest Monkeys During all Activities

		LOCOMOTOR BEHAVIOR					
SPECIES	BODY MASS (KG)	ARM SWING	CLIMB	LEAP	WALK	RUN	N
Colobus polykomos	8.3	0	14.3	14.5	41.8	29.4	918
Colobus badius	8.2	3.9	17.0	17.8	53.1	8.2	1,466
Cercocebus atys	6.2	0	12.5	1.0	80.7	5.7	466
Colobus verus	4.2	0	12.0	20.4	45.2	22.4	508
Cercopithecus diana	3.9	<1	19.4	10.4	59.3	10.8	1,553
Cercopithecus petaurista	2.9	0	18.8	10.1	61.3	9.8	512
Cercopithecus campbelli	2.7	0	14.5	5.2	72.6	7.7	596

		POSTURAL BEHAVIOR					
SPECIES	BODY MASS (KG)	SIT	QUAD STAND	SPRAWL	LIE	STAND/FORAGE	N
Colobus polykomos	8.3	89.5	<1	2.9	6.3	<1	2,495
Colobus badius	8.2	87.0	1.4	4.2	6.1	<1	2,690
Cercocebus atys	6.2	80.9	15.7	0	0	1.9	854
Colobus verus	4.2	90.7	1.3	3.8	3.7	<1	1,081
Cercopithecus diana	3.9	59.6	22.1	<1	<1	15.2	1,908
Cercopithecus petaurista	2.9	81.9	12.9	0	<1	3.1	1,530
Cercopithecus campbelli	2.7	70.9	21.9	<1	<1	4.8	838

Sources: McGraw (1998a,b, 2000).
[1] Data represent percentage of time engaged in each positional behavior.

Table 33.4   Positional Behavior[1] in Five Species of Primates in Kibale Forest, Uganda

SPECIES	BODY MASS (KG)	CLIMB	LEAP	QUAD	QUAD STAND	BIMANUAL	VERTICAL BOUND	N
Travel								
Colobus badius	8.2	29	30	34	<1	3	3	3,204
Colobus guereza	8.0	11	44	39	<1	<1	5	2,056
Cercocebus albigena	6.4	31	21	46	0	1	<1	2,328
Cercopithecus mitis	3.5	29	18	51	<1	<1	1	1,367
Cercopithecus ascanius	3.0	27	25	41	0	<1	7	1,124
Feed								
Colobus badius	8.2	37	16	4	<1	2	1	1,745
Colobus guereza	8.0	18	33	43	0	1	4	2,087
Cercocebus albigena	6.4	40	11	48	<1	1	<1	1,583
Cercopithecus mitis	3.5	38	6	55	0	<1	<1	2,046
Cercopithecus ascanius	3.0	50	11	38	0	<1	71	2,529

Sources: Gebo and Chapman (1995a).
[1] Data represent percentage of time engaged in each positional behavior.

Table 33.5   Positional Behavior[1] in Malagasy Prosimians in Ranomafana National Park

SPECIES	BODY MASS (KG)	QUAD	LEAP	CLIMB	OTHER	N	SIT	STAND	SUSPEND	VERTICAL CLING	LIE
Propithecus diadema (site Talatakely)	6.5										
Dry season		0.8	88.6	8.9	1.7	1,795	72.3	0.5	4.1	17.8	0.7
Wet season		0.5	89.1	8.6	1.7	1,220	76.4	0.4	1.8	21.0	0.4
Varecia variegata (site Vatoharanana)	3.8										
Dry season		29.0	55.5	12.5	0.0	1,001	83.5	6.6	4.9	0.2	4.8
Wet season		44.0	48.0	6.5	0.0	666	80.1	8.6	2.7	0.7	8.0
Eulemur fulvus (site Talatakely)	2.5										
Dry season		18	73.3	7.1	0.7	597	74.5	3.3	0.1	1.4	20.6
Wet season		25.8	62.7	8.5	1.9	978	80.4	10.5	0.2	1.6	7.3
Eulemur rubriventer (site Talatakely)	2.3										
Dry season		18.0	69.3	12.1	0.7	583	91.9	4.2	0.1	3.8	0.2
Wet season		27.3	59.7	11.3	2.0	1,505	79.6	15.1	0.1	2.6	1.7
Eulemur rubriventer (site Vatoharanana)	2.3										
Dry season		27.5	53.5	15.5	4.0	583	62.0	0.3	0.3	1.1	36.4
Wet season		20.0	64.0	15.5	1.0	1,505	78.9	9.4	0.1	7.5	4.2

Sources: Adapted from Dagosto (1995).
[1] Data represent percentage of time engaged in each positional behavior.

Table 33.6   Positional Behavior[1] (Travel, Feed, and Forage) in Nine Species of Primates in Ecuador

SPECIES	BODY MASS (KG)	QWALK	QBOUND	CLIMB/CLAMBER	BRIDGE	SUSPEND	LEAP	SCANSORIAL	N
Lagothrix lagotricha[2]	5.7–8.7		28.7	44.5	10.2	11.4[3]	3.9	0.0	3,926
Ateles belzebuth[2]	7.7		20.8	40.7	10.7	23.1[3]	2.3	0.0	3,760
Ateles paniscus	7.7		20.1	28.1	8.8	34.2[3]	2.6	0.0	1,089
Cebus albifrons	2.8	41.5	7.0	17.7	5.7	0.8	27.3	0.0	243
Pithecia monachus	2.4	38.4	8.4	22.9	1.5	0.4	28.4	0.0	449
Callicebus moloch	1.0	45.2	8.9	26.7	1.2	0.3	17.7	0.0	325
Saimiri sciureus	0.7	37.9	7.2	25.2	4.4	0.9	25.4	0.0	569
Saguinus tripartitus	0.4	26.1	14.2	9.7	0.3	0.4	33.7	15.6	626
Cebuella pygmaea	0.1	14.8	12.3	5.2	0.3	0.3	24.4	42.7	709

Sources: Youlatos (1999, 2002), and Cant et al. (2001).
QWalk, quadrupedal walk; QBound, quadrupedal bound; scansorial, vertical quadrupedal locomotion on tree trunks in which claw-like nails are embedded in the bark.
[1] Data represent percentage of time engaged in each positional behavior.
[2] Data from Cant et al. (2001).
[3] Primarily arm–tail suspension.

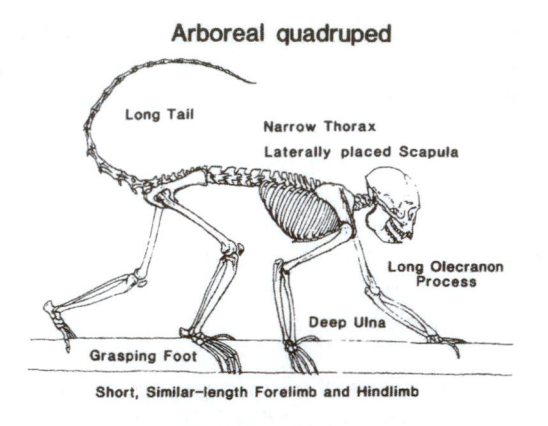

Arboreal quadruped

Long Tail

Narrow Thorax
Laterally placed Scapula

Long Olecranon
Process

Deep Ulna

Grasping Foot

Short, Similar-length Forelimb and Hindlimb

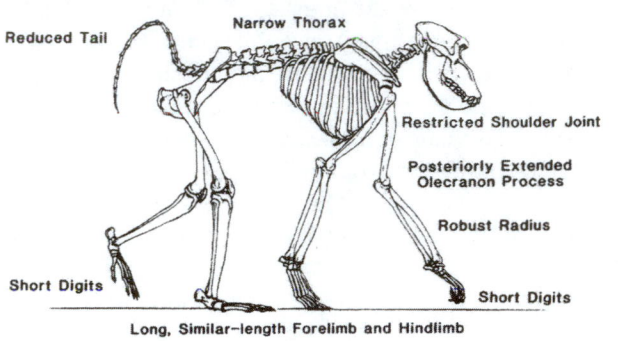

Terrestrial quadruped

Reduced Tail

Narrow Thorax

Restricted Shoulder Joint

Posteriorly Extended
Olecranon Process

Robust Radius

Short Digits

Short Digits

Long, Similar-length Forelimb and Hindlimb

Figure 33.2 Positional behavior and anatomical features in arboreal (upper) and terrestrial (lower) primates (see text for a discussion of these features). From Fleagle (1988).

distinct and exert different forces on the skeleton into a single category is problematic (Walker 1998). For example, the category quadrupedalism may include three distinct behaviors: quadrupedal walking (which in most primates is associated with a slow diagonal sequence–diagonal couplet gait with footfall placements at regular intervals), quadrupedal running (a more rapid form of pronograde travel which is mechanically similar to quadrupedal walking), and bounding (a hindlimb-dominated, asymmetrical form of progression in which the hindlimbs provide powerful and rapid extension and the forelimbs may act as struts, providing stability and orientation in redirecting the line of progression). Bounding also is distinct from running in that it includes a brief in-air phase of stride. Lumping these modes into a single category obscures potentially important species, sex, ontogenetic, seasonal, and biomechanical differences in positional behavior. Similarly, the category of leaping is often inadequately defined. In general, long hindlimbs are advantageous to leapers because they extend the duration of time over which the body can be accelerated during take-off (Niemitz 1984, Emerson 1985, Anemone 1993, Crompton et al. 1993). However, the same primate species may leap to and from large noncompliant (rigid) vertical trunks in the undercanopy, which requires powerful hindlimb extension and mechanisms for absorbing significant force on impact (vertical cling and leap); leap from compliant supports located in the periphery of the canopy onto a mass of terminal branches; leap by

dropping down into the lower levels of the canopy, which requires no hindlimb propulsion; or leap as a transition from high-speed quadrupedal progression requiring powerful and rapid extension of the hindlimbs (Garber 1991, Garber and Pruetz 1995, Terranova 1996). Leaping should be divided into a range of distinct positional categories based on support compliancy (e.g., when taking off from compliant supports, a leaper may be able to take advantage of elastic rebound to generate additional power; Demes et al. 1995), branch orientation and size, distance traversed, forces acting on the forelimbs and hindlimbs during take-off and landing, and adjustments of the body, tail, and center of mass (Dunbar 1988, Demes et al. 1995, Garber and Leigh 2001, Garber et al. 2004). Finally, the locomotor category climbing often has been used as a "catch-all" for a range of prehensile locomotor behaviors involving various combinations of limbs and tail (McGraw 2000). For example, Fleagle and Mittermeier (1980:302) define climbing as "progression along continuous supports using various combinations of three or more limbs, frequently in irregular, opportunistic sequences." The description of a lateral-sequence walking gait (Dunbar and Badam 2000) used by juvenile bonnet macaques when feeding and foraging on small-diameter branches also may fit this definition. Others have restricted climbing to various modes of vertical progression only (Hunt et al. 1996) or use the term *clamber* to describe cautious movements in which "the body is orthograde with the head superior, and various combinations of all the four appendages attach to substrates in different ways, including suspension by the forelimbs from above" (Cant 1992:279).

Hunt and colleagues (1996) called for a standardized method for describing primate positional behavior. These authors highlighted the fact that we have been limited in our ability to identify relationships between positional behavior and positional anatomy because "important behavioral information is lost in a categorical system wherein a critical positional mode may be conflated with one set of behaviors in one study and a different set in another study" (Hunt et al. 1996:365). Clearly, it is difficult to develop a standard set of positional behaviors that is consistent with the behavior of primate species that vary greatly in body mass (from 35 g in the mouse lemur to 180,000 g in an adult male gorilla); are characterized by significant variation in limb and body proportions; have prehensile tails, unspecialized grasping tails, or no tails; and have an opposable thumb, a nonopposable thumb, or a vestigial thumb. Moreover, two primate species of equal body size, one whose lineage has undergone an evolutionary increase in size and one whose lineage has undergone an evolutionary decrease in size, may be characterized by very different allometric growth trajectories, different body proportions, and different ways to locomote (Godfrey et al. 1998). Nevertheless, Hunt et al. (1996) have defined 126 positional behaviors (52 postural and 74 locomotor) that characterize living primates. Although this inventory will need to be amended as new patterns of positional behavior are defined, the goal of standardization is valid and represents an important step forward in comparative studies linking positional behavior with positional anatomy.

THE PRIMATE TAIL

In several species of nonhuman primate, the tail has come to serve an important role in balance, weight support, and arboreal locomotion (Meldrum 1998). Prehensile tails have evolved independently in at least 15 families of mammals, including once in the common ancestor of spider monkeys (*Ateles*), howlers (*Alouatta*), woolly spider monkeys (*Brachyteles*), and woolly monkeys (*Lagothrix*) (subfamily Atelinae) and a second time in capuchins (*Cebus*, subfamily Cebinae, Emmons and Gentry 1983, Rosenberger and Strier 1989, Lemlin 1995, Garber and Rehg 1999). Given significant differences in diet, body mass, and ranging patterns that characterize prehensile-tailed platyrrhines, the prehensile tail likely solves a range of locomotor and postural problems encountered in an arboreal setting (Fig. 33.1).

The ateline prehensile tail is distinct from the capuchin prehensile tail in several ways (see Garber and Rehg 1999: Table 1). Data from Grand (1977) indicate that in *Alouatta* and *Ateles*, the tail accounts for 5.5%–7.8% of total body mass. In capuchins, the prehensile tail accounts for 5.4% of body mass. In addition, the ateline tail is elongated, possesses a hairless sensitive pad on the ventrodistal surface containing specialized tactile receptors (Meissner's corpuscles), and has an expanded caudal sensorimotor area compared to capuchins (Falk 1980, Niemitz 1990, Bergeson 1996, Garber and Rehg 1999). Despite differences in tail musculoskeletal anatomy, both atelines and capuchins employ their prehensile tails in a variety of postural and locomotor behaviors and can support their full body weight hanging by the tail alone (albeit briefly in capuchins). In *Ateles*, tail–arm suspensory locomotion (tail-assisted brachiation) represents a form of high-speed locomotor behavior enabling an individual to rapidly travel long distances following relatively straight-line pathways across gaps in the forest canopy (Fig. 33.1). Youlatos (2002) reports that black spider monkeys (*Ateles paniscus*) use their prehensile tails during 61.5% of travel bouts and 54.2% of locomotor feeding bouts.

In contrast, tail–arm suspensory locomotion is absent from the *Alouatta* positional repertoire (Fleagle and Mittermeier 1980, Gebo 1992, Bergeson 1996). Despite having an ateline prehensile tail, howler monkeys locomote principally using slow, cautious, above-branch quadrupedal walking augmented by climbing and bridging (Fleagle and Mittermeier 1980, Gebo 1992, Bergeson 1996; Fig. 33.1 and Table 33.2). During quadrupedal walking, the howler prehensile tail serves an important role in balance and in providing support by grasping onto branches located above the plane of travel.

Cebus is distinct from the atelines in that it engages in more rapid forms of quadrupedal travel, leaping, and climbing. Garber and Rehg (1999) report that during travel leaping accounted for 24.7% and climbing 9.4% of the white-faced capuchin (*Cebus capucinus*) positional repertoire. During travel, the capuchin prehensile tail served a weight-bearing role 13.3% of the time. However, during feeding and foraging, tail-assisted suspensory postures accounted for 40.6% of capuchin behavior (Garber and Rehg 1999). White-faced capuchins also have been observed to adopt a crouched bipedal posture with the tail anchored around a moderate-sized support when exploiting palm fruits and during destructive insect foraging. Garber and Rehg (1999:335) state "the prehensile tail in *C. capucinus* serves a broad adaptive role in both above and below-branch foraging providing access to resources located on small and medium sized supports in the perimeter and central regions of the tree crown."

In a discussion of the primate tail, Meldrum (1998:147) highlights the fact that several species of Old World monkeys, New World monkeys, and prosimians have been observed to use their tails for "seizing and grasping" arboreal supports. In the case of the bearded saki monkey (*Chiropotes satanas*) and the red-ruffed lemur (*Varecia variegata*), Meldrum (1998) argues that the tail serves an important function during tail-assisted hindlimb suspensory behavior. In this context, their relatively long and fluffy tail is draped over a branch and acts as a brace while the forager hangs vertically downward. Meldrum (1998:154) hypothesizes that "tail bracing and twining, in conjunction with hind limb suspension selected for greater caudal robusticity and prehensile capabilities in some primate lineages, especially ancestral atelines and *Cebus*." However, many primate taxa engage in hindlimb suspension without tail assistance (Meldrum et al. 1997). Moreover, given major differences in *Cebus* and ateline behavior and ecology, it is likely that evolution of a prehensile tail in these lineages is the result of alternative selective pressures.

Finally, many primates use their tail as a counterbalance during quadrupedal walking and leaping. For example, Dunbar and Badam (2000) report that in juvenile bonnet macaques (*Macaca radiata*) the tail may be hooked around a branch to control descent when traveling on small-diameter branches. Moreover, in several species of prosimians, trunk-to-trunk leaping requires mechanical adjustments of limbs, tail, upper torso, and center of gravity during the take-off, in-air, and landing phases of travel. In particular, movement of the tail plays an important role, along with aerial rotation, enabling the hindlimbs to be brought forward and to strike the landing platform well in advance of the forelimbs (Dunbar 1988). Why hominoids lost an external tail, why some primate lineages have a reduced tail, or why no lineage of Old World primate (prosimian or higher primate) evolved a prehensile tail remains unclear. Abitbol (1987) has identified differences between hominoids and Old World monkeys in the development and weight-bearing role of the sacrum that may provide some insights into why ape ancestors lost their tails. He argues that "the tendency of the hominoid sacrum to incorporate the last lumbar vertebrae and to widen markedly . . . offers a firm base of support for the trunk

... and stabilization of the body during bipedal posture and locomotion" (Abitbol 1987:65). Moreover, Pilbeam (2004:254) notes that whereas cercopithecoids are relatively conservative in the number of vertebral elements and appear to be subject to strong stabilizing selection to maintain a fixed pattern, hominoids are characterized by considerably greater axial skeleton variability and "tolerate a broader range of phenotypes." The habitual use of vertically oriented (orthograde) postures associated with increased weight bearing via forelimb suspension and/or hindlimb compression may have resulted in a reorientation and revised function of lumbar, sacral, and coccygeal vertebrae in early hominoids. This argument, however, remains speculative.

EFFECTS OF BODY MASS ON PRIMATE POSITIONAL BEHAVIOR AND HABITAT UTILIZATION

Based on biomechanical principles, larger- and smaller-bodied primates are expected to encounter different problems of weight support and progression in the forest canopy (Cartmill and Milton 1977, Fleagle and Mittermeier 1980, Cant 1992). Body mass and limb and body proportions are critical aspects of a species' adaptive pattern, affect among other things basal metabolic rate and locomotor costs, and have significant biomechanical implications for terrestrial and arboreal progression (Larson 1998b, Schmitt 1998, Polk 2002). Firstly, in terms of the arboreal environment, it has been hypothesized that smaller-bodied animals encounter more gaps or discontinuities in the forest "that could only be crossed by leaping," whereas "a larger animal, by virtue of its longer dimensions, would see a relatively greater number that could be crossed by bridging or climbing" (Fleagle and Mittermeier 1980:309). Secondly, it is expected that "a larger animal must find it easier to hang below a relatively small branch than balance atop it" (Cartmill and Milton 1977:269). Thirdly, larger animals are expected to be more restricted in their use of smaller supports than are smaller-bodied animals and, in order to maintain balance, distribute their weight over a greater number of supports (Fleagle and Mittermeier 1980). Finally, given their relatively shorter

forelimbs (Jungars 1985), smaller-bodied primates are expected to face greater difficulty than larger-bodied primates when grasping and climbing on large vertical or sharply inclined supports (Cant 1992).

In this regard, Fleagle and Mittermeier (1980) conducted a study of positional behavior in seven species of sympatric primates in Suriname (Table 33.2). These primates varied in adult body mass from 500 to 8,000 g. Studying a set of closely related taxa in the same forest represents an important methodological tool in controlling for phylogeny and ecology in order to isolate the potential effects that species differences in body size have on positional behavior. Although these authors concluded that the empirical evidence on positional behavior in neotropical primates was consistent with theoretical predictions, there were many exceptions. For example, *Pithecia pithecia* (white-faced saki) leaped more than expected for its body mass, *Saguinus midas* (golden-handed tamarin) leaped less than expected for its body mass, and *Chiropotes satanas* (bearded saki) climbed less than expected for its body mass (Table 33.2). With respect to support size, *S. midas* was found to use smaller supports less frequently than expected and *Ateles paniscus* (black spider monkey) was found to use smaller supports more frequently than predicted for its body mass. In black spider monkeys, elongated forelimbs and a prehensile tail facilitate arm–tail suspensory behavior on relatively small supports. In the golden-handed tamarin, the evolution of claw-like nails enables these primates to exploit larger vertical supports than expected based on body size alone. Thus, although body mass is an important factor in studies of positional behavior and substrate preference, many lineages have evolved behavioral and anatomical adaptations that enable them to override the predicted effects of body size in exploiting food resources and locomotor pathways in an arboreal environment.

In Tables 33.3–33.7, additional data are presented on body mass and patterns of positional behavior in several primate communities. McGraw (1998a,b, 2000) conducted extensive research on positional behavior in seven species of Old World monkeys inhabiting the Taï Forest, Ivory Coast. As indicated in Table 33.3, larger-bodied species did

Table 33.7 Substrate Utilization[1] (Travel, Feed, and Forage) in Seven Species of Sympatric Primates in Ecuador[2]

SPECIES	BODY MASS (KG)	SMALL	MEDIUM	LARGE	VERY LARGE	N
Ateles belzebuth	7.7	41.4	40.6	16.0	1.8	1,355
Cebus albifrons	2.8	45.7	44.9	2.9	1.6	243
Pithecia monachus	2.4	24.4	54.7	16.3	2.6	449
Callicebus moloch	1.0	32.0	63.7	2.7	1.2	325
Saimiri sciureus	0.7	52.2	27.9	17.2	0.7	569
Saguinus tripartitus	0.4	31.1	46.6	10.5	11.2	626
Cebuella pygmaea	0.1	14.1	54.5	17.2	13.1	709

Sources: Youlatos (1999, 2002).
[1] Substrate size in diameter: small, ≤2 cm; medium, >2 but ≤10 cm; large, >10 but ≤20 cm; very large, >20 cm.
[2] Data represent percentage of time supports of particular size were utilized.

not consistently leap less, climb more, or exploit larger arboreal supports than did smaller-bodied forms. Similar results have been reported in studies of positional behavior in primate communities in Uganda (Gebo and Chapman 1995a, Table 33.4), Madagascar (Dagosto 1994, Table 33.5), and South America (Youlatos 1999, 2002; Cant et al. 2001; Tables 33.6 and 33.7). Overall, patterns of positional behavior and substrate utilization in prosimians, monkeys, and apes appear to be more closely related to species-specific differences in diet (and associations between the physical structure of the canopy and access to particular food resources), foraging strategies, and morphological adaptations than solely to body mass. (Ripley 1967, Fleagle 1978, Rose 1978, Garber 1984, Meldrum et al. 1997, Dunbar and Badam 2000, Schmitt 2003).

HOW VARIABLE IS POSITIONAL BEHAVIOR?

Studies of the relationship between positional behavior and postcranial anatomy in living primates provide important models for identifying evolutionary and ecological relationships in extinct lineages. If a goal of primatology is to reconstruct the lifestyle of fossil primates, then major challenges to the study of locomotion and posture include establishing functional relationships between behavior and anatomy, identifying elements of the positional repertoire that are stable from those that are more variable, and determining more precisely how positional behavior is influenced by habitat, physiology, and the mechanical costs of movement (Garber 1998). It is generally assumed that locomotor patterns are strongly constrained by anatomy. If, however, patterns of positional behavior are found to be highly variable within a species, then functional anatomists will need to incorporate concepts of behavioral plasticity into models of adaptation and anatomy (Dagosto 1995, Gebo and Chapman 1995a, Garber 1998). Figure 33.3 outlines a set of physiological, nutritional, ecological, behavioral, and social factors that are likely to influence the variability in positional behavior, habitat utilization, and substrate preference within a species.

EVIDENCE OF SEX-BASED DIFFERENCES IN POSITIONAL BEHAVIOR

Given male–female differences in body mass, the nutritional costs of reproduction, social dominance, access to resources, and the role of adult males and females in infant care and protection, significant sex-based differences in positional behavior are expected. However, among many primate taxa, male–female differences in positional behavior are extremely subtle. For example, in the case of common chimpanzees (*Pan troglodytes*, males weigh approximately 30% more than females), Doran (1993:102) found "no significant sex differences in the frequency of locomotor activities performed during either feeding or travel" (Table 33.8). This occurred in part because the dominant locomotor activity of both

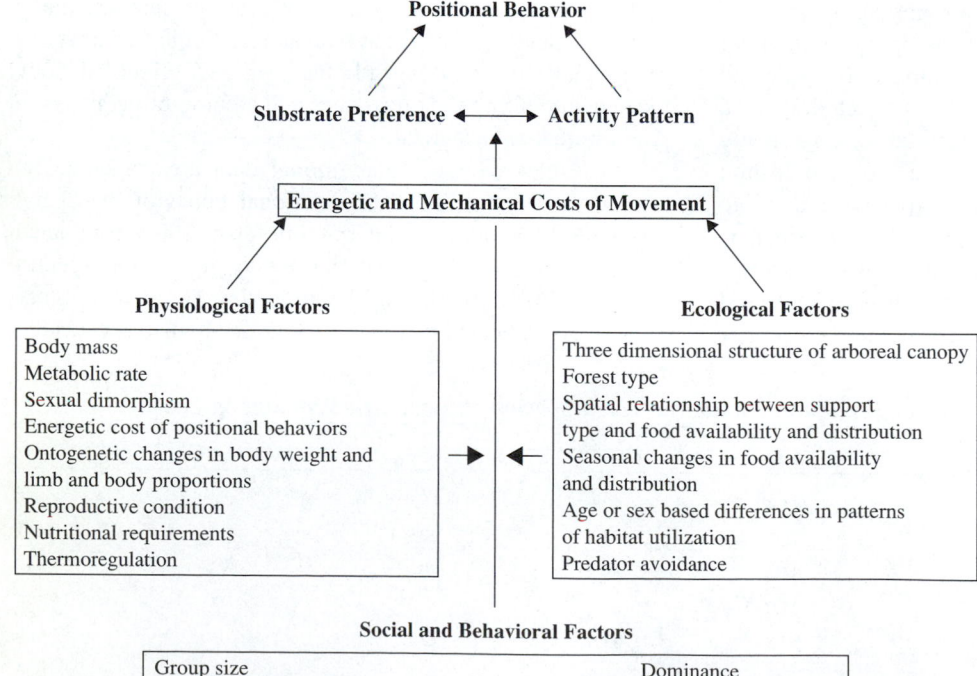

Figure 33.3 Factors influencing patterns of within-species variability in substrate preference and positional behavior.

male and female chimpanzees was quadrupedal travel (knuckle-walking) on the ground. Doran (1993) does report, however, that males and females differ in arboreal feeding locomotion, with males characterized by an increase in climbing and females characterized by an increase in quadrupedal locomotion. However, arboreal feeding locomotion accounted for only 2.2% of the chimpanzee positional repertoire; thus, in general, sex-based differences in chimpanzee positional behavior were minimal.

In Table 33.9, data are presented on arboreal locomotion in western lowland gorillas (*Gorilla gorilla gorilla*). Despite the fact that adult males weigh over twice as much as adult females and females spend several years nursing and caring for young, patterns of positional behavior were virtually identical (Remis 1998). Climbing was the most common locomotor behavior for both males and females, followed by scramble/clamber, quadrupedal travel, and suspension. Male and female lowland gorillas also had a similar positional

Table 33.8 Evidence of Sex-Based Differences in Positional Behavior in Taï Forest Chimpanzees[1] (*Pan troglodytes verus*)

	BODY MASS (KG)	All Locomotion					
		QUAD	CLIMB/SCRAMBLE	SUSPENSORY	BIPEDAL	LEAP	N
Males	46	86.6	11.1	1.1	1.2	0	732
Females	34	85.6	10.9	1.4	1.2	0.6	685

	Arboreal Locomotion					
	QUAD	CLIMB/SCRAMBLE	SUSPENSORY	BIPEDAL	LEAP	N
Males	11.7	76.7	5.8	5.8	0.0	103
Females	30.3	59.8	7.4	0.8	1.6	122

	All Postures				
	SIT	STAND	LIE	ARM-HANG	N
Males	70.8	6.1	22.1	1.0	4,293
Females	80.6	5.6	11.6	2.1	4,367

	Arboreal Postures				
	SIT	STAND	LIE	ARM-HANG	N
Males	82.0	1.6	13.5	2.8	1,524
Females	83.3	1.0	12.4	3.6	2,383

Source: Adapted from Doran (1993).
[1] Data represent percentage of time engaged in each positional behavior.

Table 33.9 Evidence of Sex-Based Differences in Patterns of Arboreal Positional Behavior in Western Lowland[1] Gorillas (*Gorilla gorilla*)[2]

SEX	BODY MASS (KG)	Arboreal Locomotion						
		QUAD	CLIMB	SCRAMBLE/CLAMBER	SUSPEND	BIPED	OTHER	N
Females	71	23.2	40.0	24.0	4.2	4.2	4.2	95
Males	169	21.5	42.1	23.7	7.9	2.6	7.9	38

SEX	BODY MASS (KG)	Arboreal Feeding Postures							
		SIT	SQUAT	SUSPEND	BIPED	TRIPED	QUAD STAND	LIE	N
Females	71	62.0	27.0	3.0	3.4	1.5	3.8	0.4	269
Males	169	54.6	36.1	1.3	0.8	1.1	1.9	0.3	366

Sources: Adapted from Remis (1998).
[1] Lowland gorillas are from Bai Hokou, Central African Republic (Remis 1998).
[2] Data represent percentage of time engaged in each positional behavior.

profile during arboreal feeding postures. The primary difference was that, although sitting was the most common feeding posture for both sexes (Table 33.9), males were found to squat more frequently than females. Overall, females did exploit smaller supports and spend more time in the periphery of the tree crown than did adult males. However, this difference appeared to relate more to social interactions and social role than to mechanical problems associated with body mass and weight support (Remis 1998).

Orangutans (*Pongo pygmaeus*) are the largest habitually arboreal mammal, with males being approximately 2.2 times heavier than females (84 vs. 38 kg). Cant (1987b) conducted a study of positional behavior in male and female Sumatran orangutans exploiting two large fig trees over the course of 11 days. He found that males tended to exploit larger supports than females but that, contrary to theoretical expectations, smaller-bodied females engaged in suspensory postures more frequently than did larger-bodied males (35% vs. 23%). Suspension most commonly occurred on small supports, but even when controlling for support size, females suspended more than males. Males were found to sit and stand more during feeding.

Tables 33.10–33.12 provide a comparison of male and female patterns of positional behavior in five species of Old World monkeys (Gebo and Chapman 1995b) and two species of New World monkeys (Gebo 1992). In the case of arboreal cercopithecids, Gebo and Chapman (1995b) found virtually no differences in postural behavior, locomotor behavior, or substrate utilization (size and orientation) between males and females, even in highly dimorphic species

(Table 33.10). These authors conclude that, either within or between species, "changes in body size do not correspond closely to changes in locomotor frequency" (Gebo and Chapman 1995a:71). A generally similar pattern was found in New World primates (Gebo 1992). Taken together, data on apes, Old World monkeys, and New World monkeys provide only limited evidence for significant sex-based differences in locomotion, posture, and substrate utilization. This supports the contention that for many primate species patterns of positional behavior are highly conservative, at least among adult group members.

EVIDENCE OF SEASONAL AND SITE-SPECIFIC DIFFERENCES IN POSITIONAL BEHAVIOR

Gebo and Chapman (1995a) examined intraspecific variability in positional behavior of red colobus monkeys (*Piliocolobus badius* a.k.a. *Procolobus badius*) in three forests in Uganda. These authors found that the red colobus positional repertoire was consistent from site to site and that common postural and locomotor behaviors at one site also were common at other sites. They note, however, that the frequency of positional behaviors was more variable during feeding than during traveling. This also has been reported in several other primate species (Crompton 1984, Dagosto 1995, Garber 1998, McGraw 1998c). Marked consistency in patterns of positional behavior during travel is likely to reflect the fact that when moving between feeding sites individuals may reuse particular arboreal pathways and/or

Table 33.10 Evidence of Sex-Based Differences in Positional Behavior[1] in Five Species of Primates in Kibale Forest, Uganda

				Travel				
SPECIES	BODY MASS (KG)	CLIMB	LEAP	QUAD	QUAD STAND	BIMANUAL	VERTICAL BOUND	*N*
Colobus guereza								
Males	10.1	11	44	39	<1	<1	6	751
Females	8.0	10	45	39	0	<1	5	1,252
Cercocebus albigena								
Males	8.9	33	21	46	0	<1	<1	821
Females	6.4	31	22	46	0	<1	<1	1,491
Colobus badius								
Males	8.2	29	31	35	0	3	2	1,194
Females	8.2	29	30	33	<1	3	4	1,995
Cercopithecus mitis								
Males	6.0	29	18	50	<1	1	2	576
Females	3.5	29	18	51	0	<1	1	789
Cercopithecus ascanius								
Male	4.1	27	26	42	0	<1	5	313
Females	3.0	27	24	41	0	<1	7	808

Source: Gebo and Chapman (1995b).
[1] Data represent percentage of time engaged in each positional behavior.

Table 33.11   Evidence of Sex-Based Differences in Positional Behavior[1] in *Alouatta palliata* in Costa Rica

	BODY MASS (KG)	Locomotor Behavior						
		QUAD	CLIMB	LEAP	QUAD STAND	BRIDGE	BIMANUALISM	N
Males	6–7	48	35	5	2	9	1	2,400
Females	4–5	45	38	4	2	10	1	2,577

	BODY MASS (KG)	Postural Behavior						
		SIT	RECLINE	QUAD SUSPEND	HINDLIMB SUSPEND+TAIL	QUAD STAND	BIPED STAND	N
Males	6–7	53	12	2	9	21	1	1,338
Females	4–5	52	12	3	11	19	2	1,274

Source: Adapted from Gebo (1992).
[1] Data represent percentage of time engaged in each positional behavior.

Table 33.12   Evidence of Sex-Based Differences in Positional Behavior[1] in *Cebus capucinus* in Costa Rica

	BODY MASS (KG)	Locomotor Behavior						
		QUAD	CLIMB	LEAP	QUAD STAND	BRIDGE	BIMANUALISM	N
Males	3.7	58	22	17	<1	2	<1	454
Females	2.5	49	29	15	<1	4	2	546

	BODY MASS (KG)	Postural Behavior						
		SIT	RECLINE	QUAD SUSPEND	HINDLIMB SUSPEND+TAIL	QUAD STAND	BIPED STAND	N
Males	3.7	38	19	1	2	37	4	189
Females	2.5	50	8	4	5	26	6	219

Source: Gebo (1992).
[1] Data represent percentage of time engaged in each positional behavior.

select substrates that offer them safety and efficiency. A higher degree of variability in the positional repertoire during feeding may be associated with seasonal changes in diet, foraging strategies, and the availability and distribution of feeding sites (Grand 1972; Charles-Dominique 1977; Garber 1980, 1984; Crompton 1984; Cant 1987a; Boinski 1989; McGraw 1998a).

Table 33.13 presents data on site-specific differences in positional behavior in two species of spider monkey (Youlatos 2002). *Ateles paniscus* was studied by Youlatos (2002) in French Guiana and by Mittermeier (1978) in Suriname. Despite differences in methods of data collection (instantaneous vs. bout sampling) and definitions of behavioral categories, there was remarkable consistency between studies. At both sites, tail–arm suspension was the most common positional behavior. At both sites, leaping and bridging were the least common positional behaviors. The greatest difference found between sites was in the frequency of clamber/climbing behavior (28.1% in French Guiana and 17.0 in Suriname).

In the case of *Ateles geoffroyi* (black-handed spider monkey), however, site-specific differences in positional behavior were more marked. In four studies, the frequency

of quadrupedal progression ranged from 22.0% to 52.0% and climbing varied from 6% to 24%. Although some of this variation is likely related to differences in methodology, a study by Bergeson (1996) on *A. geoffroyi* using the same methodology at two different study sites (a dry tropical forest and a rain forest) found considerable within-species variability during gap crossings. In the wetter and taller forest of La Selva, Costa Rica, *A. geoffroyi* was found to use bridging significantly more and tail–arm suspension significantly less than in the drier forest of Santa Rosa, Costa Rica. All other forms of positional behavior were used at similar frequencies. Although the orientations and sizes of substrates were generally similar between sites, at La Selva spider monkeys descended oblique supports at greater frequency, ascended oblique supports at lower frequency, and tended to use smaller supports more frequently than at Santa Rosa (Bergeson 1996).

Finally, Dagosto (1995) provides data on seasonal and site-specific patterns of positional behavior in Malagasy prosimians (Table 33.5). Based on her study of *Eulemur fulvus* (brown lemur), *Eulemur rubriventer* (red-bellied lemur), *Propithecus diadema* (diademed sifaka), and *Varecia variagata* (ruffed lemur), Dagosto (1995:811) concluded that

Table 33.13   Evidence of Within- and Between-Species Variation in Positional Behavior[1] During Travel in Spider Monkeys

SPECIES	Locomotor Behavior					Substrate Size			
	QUAD	CLAMBER/CLIMB	BRIDGE	TAIL–ARM SUSPEND	LEAP	SMALL	MEDIUM	LARGE	N
Ateles paniscus (Youlatos 2002)	20.1	28.1	8.8	34.2	2.6	41.9	40.3	17.8	1,089
Ateles paniscus (Mittermeier 1978)	25.4	17.0	8.6	38.6	4.2	38.2	49.5	12.3	743
Ateles geoffroyi (Mittermeier 1978)	22.0	24.2	4.6	25.7	10.9	28.0	51.6	20.4	1,000
Ateles geoffroyi (Fontaine 1990)	51.0	6.0	1.6	21.9	10.9	16.5	60.7	22.8	1,962
Ateles geoffroyi (Cant 1986)	52.0	19.0	4.0	25.0	1.0	50.0	42.0	8.0	1,685
Ateles geoffroyi (Bergeson 1996)	36.1	18.2	3.0	36.3	2.5	42.2	35.7	22.1	647

Source: Adapted from Youlatos (2002).

[1] Data represent percentage of time engaged in each positional behavior and exploitation of supports of particular size.

"locomotion during travel appears to be fairly conservative while differences during feeding contribute substantially to the overall differences observed."

CONCLUSION

> Throughout its development the study of primate locomotion has been characterized by its interdisciplinary focus combining ecology and behavioral observations with physiological and morphometric studies to understand both how and why primates use different patterns of locomotion and posture. (Fleagle 1998:109)

Evidence presented in this chapter on primate positional behavior points to several important conclusions. First, closely related primates living in the same ecological community are often characterized by significant interspecific differences in patterns of positional behavior, body mass, and positional anatomy. Second, although smaller- and larger-bodied arboreal primates face different challenges in maintaining balance and weight support, body mass appears not to be as strong a predictor of positional behavior as is often assumed. Individual species of different body mass and body size are characterized by particular behavioral and morphological adaptations that enable them to exploit similar sets of resources in different ways. Third, patterns of positional behavior within a species are generally conservative, especially during travel, and vary little among adult males and adult females. In contrast, within-species patterns of positional behavior during feeding are more variable and likely to reflect seasonal changes in diet, day range, foraging strategies, activity pattern, and social interactions. In this regard, Dagosto (1995:832) cautions that "differences of as much as 15–20% between species in any particular positional behavioral category may not be meaningful for interpreting morphology since differences of such magnitude may be observed within a single species and a result of temporal or geographic variation."

A major challenge for the study of primate positional behavior is to continue to develop new methodologies in both field and laboratory settings for identifying functional relationships between patterns of movement, habitat utilization, and anatomical structure. The expanded use of digital video and other imaging technologies, standardized and biomechanically meaningful definitions of positional categories, and more detailed information on the ecological and social referents of positional behavior are critical to this effort. Morphological design sets limits or boundaries on the range and mechanical efficiency of particular locomotor and postural patterns. For a particular species, these limits may be broad or narrow depending on phylogenetic, ontogenetic, ecological, and behavioral factors. We know very little about the ontogeny of positional behavior in primates and the degree to which age-related changes in motor skills, limb and body proportions, and body mass affect an individual's ability to exploit its environment (Doran 1997, Bezanson 2003). In addition, there is evidence that in some primate species neuromuscular development associated with locomotor skills may have become dissociated from neuromuscular development required for fine motor control and object manipulation. For example, many primate species attain locomotor independence within a few months after birth but still lack the fine motor skills required for extractive foraging and prey manipulation. In the case of tamarin monkeys, individuals reach locomotor independence by 3 months of age but are still provisioned with insects by adult caretakers at 9 months of age (Garber and Leigh 1997). Similarly, chimpanzees attain an adult-like locomotor profile at approximately 5 years of age (Doran 1997). However, according to Boesch and Boesch-Achermann (2000), it was not until age 7 that juveniles reached a level of efficiency in cracking *Panda* nuts such that energy intake for cracking exceeded energy intake by sharing nuts opened by their mothers. Moreover, adult levels of *Panda* nut cracking were not reached until age 11. In other primate species, locomotor skills and fine manipulative skills appear to develop at approximately the same time (i.e., *Saimiri oerstedii*; Boinski and Fragaszy 1989). In this regard, studies of primate positional behavior should be placed within the context of primate life history strategies and evaluated in terms of patterns of somatic and neural growth and development, age-related survivorship, and the requirements of efficiently exploiting particular resources and forest habitats.

Finally, greater attention must be paid to identifying the energetic costs of locomotion and posture. Many models of primate socioecology and behavioral ecology assume high energetic costs of increased travel (Janson and Goldsmith 1995, Chapman and Chapman 2000). However, recent physiological studies suggest that for both arboreal and terrestrial primates, the distance traveled each day represents only 3.5%–10% of daily energy expenditure (Steudel 2000).

REFERENCES

Abitbol, M. M. (1987). Evolution of the sacrum in hominoids. *Am. J. Phys. Anthropol.* 74:65–82.

Anemone, R. L. (1993). The functional anatomy of the hip and thigh in primates. In: Gebo, D. L. (ed.), *Postcranial Adaptation in Nonhuman Primates.* Northern Illinois University Press, Dekalb. pp. 150–174.

Bergeson, D. J. (1996). The positional behavior and prehensile tail use of *Alouatta palliata, Ateles geoffroyi,* and *Cebus capucinus* [PhD thesis]. Washington University, St. Louis.

Bergeson, D. J. (1998). Patterns of suspensory feeding in *Alouatta palliata, Ateles geoffroyi,* and *Cebus capucinus.* In: Strasser, E., Fleagle, J. G., Rosenberger, A., and McHenry, H. (eds.), *Primate Locomotion. Recent Advances.* Plenum Press, New York. pp. 45–60.

Bezanson, M. F. (2003). Patterns of positional behavior in juvenile and adult white-faced capuchins (*Cebus capucinus*). *Am. J. Phys. Anthropol.* 120:66.

Bock, W., and von Wahlert, G. (1965). Adaptation and the form–function complex. *Evolution* 19:269–299.

Boesch, C., and Boesch-Achermann, H. (2000). *The Chimpanzees of the Taï Forest: Behavioural Ecology and Evolution.* Oxford University Press, New York.

Boinski, S. (1989). The positional behavior and substrate use of squirrel monkeys: ecological implications. *J. Hum. Evol.* 18:659–677.

Boinski, S., and Fragaszy, D. M. (1989). The ontogeny of foraging in squirrel monkeys, *Saimiri oerstedii. Anim. Behav.* 37:415–428.

Cant, J. G. H. (1986). Locomotion and feeding postures of spider and howling monkeys: field study and evolutionary interpretation. *Folia Primatol.* 46:1–14.

Cant, J. G. H. (1987a). Positional behavior of female Bornean orangutans. *Am. J. Primatol.* 12:71–90.

Cant, J. G. H. (1987b). Effects of sexual dimorphism in body size on feeding postural behavior of Sumatran orangutans (*Pongo pygmaeus*). *Am. J. Phys. Anthropol.* 74:143–148.

Cant, J. G. H. (1990). Feeding ecology of spider monkeys (*Ateles geoffroyi*) at Tikal, Guatemala. *Hum. Evol.* 5:269–281.

Cant, J. G. H. (1992). Positional behavior and body size of arboreal primates: a theoretical framework for field studies and an illustration of its application. *Am. J. Phys. Anthropol.* 88:273–283.

Cant, J. G. H., Youlatos, D., and Rose, M. D. (2001). Locomotor behavior of *Lagothrix lagotricha* and *Ateles belzebuth* in Yasuní National Park, Ecuador: general patterns and suspensory modes. *J. Hum. Evol.* 41:141–166.

Cartmill, M. (1972). Arboreal adaptations and the origin of the order Primates. In: Tuttle, R. (ed.), *The Functional and Evolutionary Biology of Primates.* Aldine, Chicago. pp. 97–122.

Cartmill, M. (1974). Rethinking primate origins. *Science* 184:436–443.

Cartmill, M. (1985). Climbing. In: Hildebrand, M., Bramble, D. M., Liem, K. F., and Wake, D. B. (eds.), *Functional Vertebrate Morphology.* Belknap Press, Cambridge, MA. pp. 73–88.

Cartmill, M. (1992). New views on primate origins. *Evol. Anthropol.* 2:105–111.

Cartmill, M., and Milton, K. (1977). The lorisiform wrist joint and the evolution of "brachiating" adaptations in the Hominoidea. *Am. J. Phys. Anthropol.* 47:249–272.

Chapman, C. A., and Chapman, L. J. (2000). Determinants of group size in primates: the importance of travel costs. In: Boinski, S., and Garber, P. A., (eds.), *On the Move: How and Why Animals Travel in Groups.* University of Chicago Press, Chicago. pp. 24–42.

Charles-Dominique, P. (1977). *Ecology and Behavior of Nocturnal Primates.* Columbia University Press, New York.

Charles-Dominique, P., and Martin, R. D. (1970). Evolution of lorises and lemurs. *Nature* 227:257–260.

Crompton, R. H. (1984). Foraging, habitat structure, and locomotion in two species of *Galago.* In: Rodman, P. S., and Cant, J. G. H. (eds.), *Adaptations for Foraging in Nonhuman Primates.* Columbia University Press, New York. pp. 73–111.

Crompton, R. H., Sellers, W. I., and Gunther, M. M. (1993). Energetic efficiency and ecology as selective factors in saltatory adaptation of prosimian primates. *Proc. R. Soc. Lond. B* 254:41–45.

Dagosto, M. (1994). Testing positional behavior of Malagasy lemurs: a randomization approach. *Am. J. Phys. Anthropol.* 94:189–202.

Dagosto, M. (1995). Seasonal variation in positional behavior of Malagasy lemurs. *Int. J. Primatol.* 16:807–834.

Davis, L. C. (1996). Functional and phylogenetic implications of ankle morphology in Goeldi's monkey (*Callimico goeldii*). In: Norconk, M., Rosenberger, A. L., and Garber, P. A. (eds.), *Adaptive Radiations of Neotropical Primates.* Plenum Press, New York. pp. 133–156.

Demes, B., and Gunther, M. M. (1989). Biomechanics and allometric scaling in primate locomotion and morphology. *Folia Primatol.* 53:125–141.

Demes, B., Jungers, W. L., Gross, T. S., and Fleagle, J. G. (1995). Kinetics of leaping primates: influence of substrate orientation and compliance. *Am. J. Phys. Anthropol.* 96:419–430.

Demes, B., Larson, S. G., Stern, J. T., Jr., Jungers, W. L., Biknevicius, A. R., and Schmitt, D. (1994). The kinetics of "hind limb drive" reconsidered. J. Hum. Evol. 26:353–374.

Doran, D. (1993). Sex differences in adult chimpanzee positional behavior: the influence of body size on locomotion and posture. *Am. J. Phys. Anthropol.* 91:99–116.

Doran, D. (1997). Ontogeny of locomotion in mountain gorillas and chimpanzees. *J. Hum. Evol.* 32:323–344.

Dunbar, D. C. (1988). Aerial maneuvers of leaping lemurs: the physics of whole body rotations while airborne. *Am. J. Primatol.* 16:291–303.

Dunbar, D. C., and Badam, G. L. (2000). Locomotion and posture during terminal branch feeding. *Int. J. Primatol.* 21:649–669.

Emerson, S. B. (1985). Jumping and leaping. In: Hildebrand, M. H., Bramble, D. M., Liem, K. R., and Wake, D. B. (eds.),

Functional Vertebrate Morphology. Belknap Press of Harvard University, Cambridge, MA. pp. 59–72.

Emmons, L. H., and Gentry, A. (1983). Tropical forest structure and the distribution of gliding and prehensile-tailed vertebrates. *Am. Nat.* 12:513–524.

Falk, D. (1980). Comparative study of the endocranial casts of New and Old World monkeys. In: Ciochon, R., and Chiarelli, A. (eds.), *Evolutionary Biology of the New World Monkeys and Continental Drift*. Plenum Press, New York. pp. 275–292.

Fleagle, J. G. (1976). Locomotor behavior and skeletal anatomy of sympatric Malaysian leaf-monkeys (*Presbytis obscura* and *Presbytis melalophos*). *Ybk. Phys. Anthropol.* 20:440–453.

Fleagle, J. G. (1978). Locomotion, posture, and habitat utilization in two sympatric, Malaysian leaf-monkeys (*Presbytis obscura* and *Presbytis melalophos*). In: Montgomery, G. G. (ed.), *The Ecology of Arboreal Folivores*. Smithsonian Institution Press, Washington DC. pp. 243–252.

Fleagle, J. G. (1988). *Primate Adaptations and Evolution*. Academic Press, New York.

Fleagle, J. G. (1998). Introduction to part II. In: Strasser, E., Fleagle, J. G., Rosenberger, A., and McHenry, H. (eds.), *Primate Locomotion: Recent Advances*. Plenum Press, New York. pp. 109–110.

Fleagle, J. G., and Meldrum, D. J. (1988). Locomotor behavior and skeletal morphology of two sympatric pitheciine monkeys, *Pithecia pithecia* and *Chiropotes satanas*. *Am. J. Primatol.* 16:227–249.

Fleagle, J. G., and Mittermeier, R. A. (1980). Locomotor behavior, body size, and comparative ecology of seven Surinam monkeys. *Am. J. Phys. Anthropol.* 52:301–314.

Foley, R. A., and Elton, S. (1998). Time and energy: the ecological context for the evolution of bipedalism. In: Strasser, E., Fleagle, J. G., Rosenberger, A., and McHenry, H. (eds.), *Primate Locomotion: Recent Advances*. Plenum Press, New York. pp. 419–433.

Garber, P. A. (1980). Locomotor behavior and feeding ecology of the Panamanian tamarin (*Saguinus oedipus geoffroyi*, Callitrichidae, Primates). *Int. J. Primatol.* 1:185–201.

Garber, P. A. (1984). Use of habitat and positional behavior in a neotropical primate, *Saguinus oedipus*. In: Rodman, P. S., and Cant, J. G. H. (eds.), *Adaptations for Foraging in Nonhuman Primates: Contributions to an Organismal Biology of Prosimians, Monkeys, and Apes*. Columbia University Press, New York. pp. 112–133.

Garber, P. A. (1991). A comparative study of positional behavior in three species of tamarin monkeys. *Primates* 32:219–230.

Garber, P. A. (1992). Vertical clinging, small body size, and the evolution of feeding adaptations in the Callitrichinae. *Am. J. Phys. Anthropol.* 88:469–482.

Garber, P. A. (1998). Within- and between-site variability in moustached tamarin (*Saguinus mystax*) positional behavior during food procurement. In: Strasser, E., Fleagle, J. G., Rosenberger, A., and McHenry, H. (eds.), *Primate Locomotion: Recent Advances*. Plenum Press, New York. pp. 61–78.

Garber, P. A., Blomquist, G. E., and Anzenberger, G. (2004). Kinematic analysis of trunk-to-trunk leaping in Goeldi's monkey (*Callimico goeldii*). *Int. J. Primatol.* 26:217–234.

Garber, P. A., and Leigh, S. R. (1997). Ontogenetic variation in small-bodied New World primates: implications for patterns of reproduction and infant care. *Folia Primatol.* 68:1–22.

Garber, P. A., and Leigh, S. R. (2001). Patterns of positional behavior in mixed-species troops of *Callimico goeldii*, *Saguinus*

labiatus, and *Saguinus fuscicollis* in northwestern Brazil. *Am. J. Primatol.* 54:17–31.

Garber, P. A., and Pruetz, J. D. (1995). Positional behavior in moustached tamarin monkeys: effects of habitat on locomotor variability and locomotor stability. *J. Hum. Evol.* 28:411–426.

Garber, P. A., and Rehg, J. A. (1999). The ecological role of the prehensile tail in white-faced capuchins (*Cebus capucinus*). *Am. J. Phys. Anthropol.* 110:325–339.

Gebo, D. (1992). Locomotor and postural behavior in *Alouatta palliata* and *Cebus capucinus*. *Am. J. Primatol.* 26:277–290.

Gebo, D. L., and Chapman, C. A. (1995a). Positional behavior in five sympatric Old World monkeys. *Am. J. Phys. Anthropol.* 97:49–76.

Gebo, D. L., and Chapman, C. A. (1995b). Habitat, annual and seasonal effects on positional behavior in red colobus monkeys. *Am. J. Phys. Anthropol.* 96:73–82.

Godfrey, L. R., and Jungars, W. L. (2003). The extinct sloth lemurs of Madagascar. *Evol. Anthropol.* 12:252–263.

Godfrey, L. R., King, S. J., and Sutherland, M. R. (1998). Heterochronic approaches to the study of locomotion. In: Strasser, E., Fleagle, J. G., Rosenberger, A., and McHenry, H. (eds.), *Primate Locomotion: Recent Advances*. Plenum Press, New York. pp. 277–307.

Grand, T. I. (1972). A mechanical interpretation of terminal feeding. *J. Mammal.* 53:198–201.

Grand, T. I. (1977). Body weight: its relationships to tissue composition, segment distribution, and motor function. I. Interspecific comparisons. *Am. J. Phys. Anthropol.* 47:211–240.

Grand, T. I. (1984). Motion economy within the canopy: four strategies for mobility. In: Rodman, P., and Cant, J. G. H. (eds.), *Adaptations for Foraging in Nonhuman Primates*. Columbia University Press, New York. pp. 54–72.

Hildebrand, M. (1967). Symmetyrical gaits in primates. *Am. J. Phys. Anthropol.* 26:119–130.

Hunt, K. D. (1991). Positional behavior in the Hominoidea. *Int. J. Primatol.* 12L:95–118.

Hunt, K. D. (1992). Positional behavior of *Pan troglodytes* in the Mahale Mountains and Gombe Stream National Parks, Tanzania. *Am. J. Phys. Anthropol.* 87:83–105.

Hunt, K. D. (1998). Ecological morphology of *Australopithecus afarensis*; traveling terrestrially, eating arboreally. In: Strasser, E., Fleagle, J. G., Rosenberger, A., and McHenry, H. (eds.), *Primate Locomotion: Recent Advances*. Plenum Press, New York. pp. 387–418.

Hunt, K. D., Cant, J. G. H., Gebo, D. L., Rose, M. D., Walker, S. E., and Youlatos, D. (1996). Standardized description of primate locomotor and postural modes. *Primates* 37:363–387.

Janson, C. H., and Goldsmith, M. (1995). Predicting group size in primates: foraging costs and predation risks. *Behav. Ecol.* 6:326–336.

Jolly, C. J., and Plog, F. (1979). *Physical Anthropology and Archaeology*, 2nd ed. Alfred A. Knopf, New York.

Jungars, W. L. (1985). Body size and scaling of limb proportions in primates. In: Jungars, W. L. (ed.), *Size and Scaling in Primate Biology*. Plenum Press, New York. pp. 345–381.

Jungars, W. L., and Burr, D. B. (1994). Body size, long bone geometry, and locomotion in quadrupedal monkeys. *Z. Morphol. Anthropol.* 80:89–97.

Kay, R. F., Ross, C., and Williams, B. A. (1997). Anthropoid origins. *Science* 275:797–804.

Kinzey, W. G. (1967). Preface. *Am. J. Phys. Anthropol.* 26:115–118.

Kinzey, W. G. (1976). Positional behavior and ecology in *Callicebus torquatus*. *Ybk. Phys. Anthropol.* 20:468–480.

Larson, S. G. (1998a). Parallel evolution in the hominoid trunk and forelimb. *Evol. Anthropol.* 6:87–99.

Larson, S. G. (1998b). Unique aspects of quadrupedal locomotion in nonhuman primates. In: Strasser, E., Fleagle, J. G., Rosenberger, A., and McHenry, H. (eds.), *Primate Locomotion: Recent Advances*. Plenum Press, New York. pp. 157–173.

Leakey, M. C., Fiebel, C. S., McDougall, I., and Walker, A. (1995). New four million year old hominid species from Kanapoi and Allia Bay, Kenya. *Nature* 376:565–571.

Lemlin, P. (1995). Comparative and functional myology of the prehensile tail in New World monkeys. *J. Morphol.* 224:1–18.

Lovejoy, C. O. (1981). The origin of man. *Science* 211:341–350.

Martin, R. D. (1990). *Primate Origins and Evolution: A Phylogenetic Reconstruction*. Princeton University Press, Princeton, NJ.

McGraw, W. S. (1998a). Posture and support use of Old World monkeys (Cercopithecideae): the influence of foraging strategies, activity patterns, and the spatial distribution of preferred food items. *Am. J. Primatol.* 46:229–250.

McGraw, W. S. (1998b). Comparative locomotion and habitat use of six monkeys in the Taï Forest, Ivory Coast. *Am J. Phys. Anthropol.* 105:493–510.

McGraw, W. S. (1998c). Locomotion, support use, maintenance activities, and habitat structure: the case of the Taï Forest cercopithecids. In: Strasser, E., Fleagle, J. G., Rosenberger, A., and McHenry, H. (eds.), *Primate Locomotion: Recent Advances*. Plenum Press, New York. pp. 79–94.

McGraw, W. S. (2000). Positional behavior of *Cercopithecus petaurista*. *Int. J. Primatol.* 21:157–182.

McGraw, W. S., and Bshary, R. (2002). Association of terrestrial mangabeys (*Cercocebus atys*) with arboreal monkeys: experimental evidence for the effects of reduced ground predator pressure on habitat use. *Int. J. Primatol.* 23:311–326.

Meldrum, D. J. (1998). Tail-assisted hind limb suspension as a transitional behavior in the evolution of the platyrrhine prehensile tail. In: Strasser, E., Fleagle, J. G., Rosenberger, A., and McHenry, H. (eds.), *Primate Locomotion: Recent Advances*. Plenum Press, New York. pp. 145–156.

Meldrum, D. J., Dagosto, M., and White, J. (1997). Hindlimb suspension and hind foot reversal in *Varecia variegata* and other arboreal mammals. *Am J. Phys. Anthropol.* 103:85–102.

Mittermeier, R. A. (1978). Locomotion and posture in *Ateles geoffroyi* and *Ateles paniscus*. *Folia Primatol.* 30:161–193.

Niemitz, C. (1984). Locomotion and posture of *Tarsius bancanus*. In: Niemitz, C. (ed.), *Biology of Tarsiers*. Gustav Fischer Verlag, New York. pp. 191–226.

Niemitz, C. (1990). The evolution of primate skin structures in relation to gravity and locomotor patterns. In: Jouffroy, F., Stack, M., and Niemitz, C. (eds.), *Gravity, Posture, and Locomotion in Primates*. Editrice Il Sedicesimo, Florence. pp. 129–156.

Oxnard, C. E., Crompton, R. H., and Liebermann, S. S. (1990). *Animal Lifestyles and Anatomies: The Case of the Prosimian Primates*. Washington University Press, Seattle.

Peters, A., and Preuschoft, H. (1984). External biomechanics of leaping in *Tarsius* and its morphological and kinematic consequences. In: Niemitz, C. (ed.), *Biology of Tarsiers*. Gustav Fischer Verlag, New York. pp. 227–255.

Pilbeam, D. (2004). The anthropoid postcranial axial skeleton: comments on development, variation, and evolution. *J. Exp. Zool.* 302B:241–267.

Polk, J. D. (2002). Adaptive and phylogenetic influences on musculoskeletal design in cercopithecine primates. *J. Exp. Biol.* 205:3399–3412.

Preuschoft, H. (1985). On the quality and magnitude of mechanical stresses in the locomotor system during rapid movement. *Z. Morphol. Anthropol.* 75:245–262.

Preuschoft, H., Witte, H., Christian, A., and Fischer, M. (1996). Size influences primate locomotion and body shape, with special emphasis on the locomotion of "small mammals." *Folia Primatol.* 66:93–112.

Prost, J. H. (1965). A definitional system for the classification of primate locomotion. *Am. Anthropol.* 67:1198–1214.

Rasmussen, D. T. (1990). Primate origins: lessons from a neotropical marsupial. *Am. J. Primatol.* 22:263–277.

Remis, M. (1998). The gorilla paradox: the effects of body size and habitat on the positional behavior of lowland and mountain gorillas. In: Strasser, E., Fleagle, J. G., Rosenberger, A., and McHenry, H. (eds.), *Primate Locomotion: Recent Advances*. Plenum Press, New York. pp. 95–106.

Richmond, B. G., and Strait, D. S. (2000). Evidence that humans evolved from a knuckle-walking ancestor. *Nature* 404:382–385.

Ripley, S. (1967). The leaping of langurs: a problem in the study of locomotor adaptation. *Am. J. Phys. Anthropol.* 26:149–170.

Rose, M. D. (1974). Postural adaptations in New and Old World monkeys. In: Jenkins, F. A., Jr. (ed.), *Primate Locomotion*. Academic Press, New York. pp. 201–222.

Rose, M. D. (1978). Feeding and associated positional behavior of black and white colobus monkeys (*Colobus guereza*). In: Montgomery, G. G. (ed.), *The Ecology of Arboreal Folivores*. Smithsonian Institution Press, Washington DC. pp. 253–262.

Rosenberger, A. L., and Strier, K. B. (1989). Adaptive radiation of the ateline primates. *J. Hum. Evol.* 18:717–750.

Ruff, C. B. (1987). Sexual dimorphism in human lower limb bone structure: relationship to subsistence strategy and sexual division of labor. *J. Hum. Evol.* 16:391–416.

Sargis, E. J. (2002). Primate origins nailed. *Science* 298:1564–1565.

Schmitt, D. (1998). Forelimb mechanics during arboreal and terrestrial quadrupedalism in Old World monkeys. In: Strasser, E., Rosenberger, A., McHenry, H., and Fleagle, J. G. (eds.). *Primate Locomotion: Recent Advances*. Plenum, New York. pp. 175–200.

Schmitt, D. (2003). Substrate size and primate forelimb mechanics: implications for understanding the evolution of primate locomotion. *Int. J. Primatol.* 24:1023–1036.

Schmitt, D., and Lemelin, P. (2002). The origins of primate locomotion: gait mechanics of the woolly opossum. *Am. J. Phys. Anthropol.* 118:231–238.

Steudel, K. (2000). The physiology and energetics of movement: effects on individuals and groups. In: Boinski, S., and Garber, P. A. (eds.), *On the Move: How and Why Animals Travel in Groups*. University of Chicago Press, Chicago. pp. 9–23.

Sussman, R. W. (1991). Primate origins and the evolution of angiosperms. *Am. J. Primatol.* 23:209–223.

Sussman, R. W., and Raven, P. H. (1977). Pollination by lemurs and marsupials: an archaic coevolutionary system. *Science* 200:731–736.

Terranova, C. J. (1996). Variation in the leaping of lemurs. *Am. J. Primatol.* 40:145–165.

Walker, S. E. (1996). Evolution of positional behavior in the saki/uakaris (*Pithecia, Chiropotes, Cacajao*). In: Norconk, M. A., Rosenberger, A. L., and Garber, P. A. (eds.), *Adaptive Radiations of Neotropical Primates*. Plenum Press, New York. pp. 335–367.

Walker, S. E. (1998). Fine-grained differences within positional categories: a case study of *Pithecia* and *Chiropotes*. In: Strasser, E., Fleagle, J. G., Rosenberger, A., and McHenry, H. (eds.), *Primate Locomotion: Recent Advances*. Plenum Press, New York. pp. 31–44.

Warren, R. D., and Crompton, R. H. (1998a). Diet, body size, and energy costs of locomotion in saltatory primates. *Folia Primatol*. 69:86–100.

Warren, R. D., and Crompton, R. H. (1998b). Locomotor ecology of *Lepilemur edwardsi* and *Avahi occidentalis*. *Am. J. Phys. Anthropol*. 104:471–486.

Youlatos, D. (1999). Comparative locomotion of six sympatric primates in Ecuador. *Ann. Sci. Nat*. 20:161–168.

Youlatos, D. (2002). Positional behavior of black spider monkeys (*Ateles paniscus*) in French Guiana. *Int. J. Primatol*. 23:1071–1094.

PART SIX

Social Behavior and Intelligence

34

Social Mechanisms in the Control
of Primate Aggression

Irwin S. Bernstein

INTRODUCTION

Adolph Schultz (1969) said that the special adaptation of the primates was to be understood by recognizing three attributes, none of which was exclusive to the order. First, primates are "smart"; that is, they have the ability to modify their behavior based on past experience. Second, they are social. Third, they have extremely long periods of biological dependence for their body size. (The period of biological dependence is considered to be the time that an individual cannot survive on its own and depends on others for necessities such as nourishment, thermoregulation, etc. Although the end of this period is not clearly marked, infants and young juveniles are thought to be biologically dependent on others for survival.)

This combination means that *(1)* primates will learn as a function of experience, *(2)* much of this learning will be social learning, and *(3)* due to their long periods of biological dependence, young primates will have the opportunity to learn from parents and caregivers, thus setting up the possibility of nongenetic means of transmission of information from generation to generation. Information need not be discovered anew each generation and can accumulate over time. This permits the development of traditions and even culture (see Chapter 40).

Sociality in primates means that they not only establish and maintain social bonds but use social techniques in response to ecological problems, that is, joint action on the environment (both physical and social). This is what Kummer (1971) considered to be their only special ecological adaptation. By "joint action on the environment," he meant that, rather than respond to an environmental challenge individually, primates usually enlist or join others to face a challenge. If a predator threatens, they vocalize, look to see what others are doing, and then may jointly face the predator. If the challenge is temperature, they will huddle with others. If it requires searching for resources, they will join others rather than forage individually. If they are threatened by a conspecific (the social environment), they enlist others to aid them to meet the threat.

Not all primates show the same degree of sociality, but as mammals, all show at least the obligate level of sociality

required for mating and care of the young. Long developmental periods mean that even the least sociable primates typically spend many years in association with at least one other, be it parent, siblings, or offspring. The more sociable primates spend all of their lives as members of a social group, although members of one or both sexes may spend brief periods in transit between groups. Even when moving between two heterosexual groups, many will form unisexual bands or even travel in company with another species (Bernstein 1967).

DEFINING THE SOCIAL UNIT

What does it mean to be a member of a social unit as opposed to simply being in the same place as another, as in an aggregation or some other such grouping? Although different investigators may emphasize different points, I believe that most at least discuss the following five criteria:

1. *Spatial.* If animals were never together, we would have little reason to ask if they were members of a social unit. What constitutes "proximity," however, is a matter of judgment, and some may require physical contact or distances within 1 m or 5 or 10. Others stress being within arm's reach or one body length or one pace (biological rather than physical units).
2. *Temporal.* If proximity is fleeting and not regular, we might think that the individuals were like "ships that pass in the night," with no special attachment or attraction. Again, however, there is no consensus as to how much time they must spend together, nor how often. Some are together nearly constantly, whereas others may part and reunite regularly but within a specified community of others.
3. *Recognition of Group Members.* If animals treat all conspecifics in the same way, it would perhaps suggest general sociality but not specific social units. Social units require boundaries such that group members are treated differently from non-group members. In many cases, behavior directed toward non-group members is described as "xenophobic"

(Scott 1955). This label was suggested when it was noted that the difference between the way group members and nonmembers are treated is largely a matter of the frequency of expressions of agonistic behavior in encounters. Agonistic responses are much more likely in encounters with non-group members.

4. *Communication and Coordination.* Social units will require coordination of activities such that individuals cannot do whatever they want whenever they want. To have each animal travel, eat, drink, rest, etc. without regard to others would quickly result in the spatial dispersal of the group. Coordination requires communication, including social facilitation where the activity of one stimulates the same activity in others. (It is the similarity of the behavior of initiator and others that marks "social facilitation" compared to other interactions in which an observer may respond in a variety of ways to the behavior of another.) Active communication, where animals exchange information and each modifies the behavior of the other, is also required. Indeed, Altmann (1965) suggested that an operational definition of social units could be based on the measurement of frequencies of communication: the area of high density of communication is the social unit and is bounded by areas of much lower density of communication.

5. *Division of Labor.* Not every activity in the group will be displayed by every individual. Of course, we might expect individual, age, and sex differences; but when individuals of the same class show consistent differences in behavioral patterns in specified situations, we suspect a division of labor (or social roles). Sarbin (1954) defined *society* as a network of interrelated social roles. Wilson (1975) suggested that social complexity might be measured by the principle of minimum specification—that is, the number of individuals that have to be studied to see all of the major patterns of behavior in a social group. The term *division of labor* may be construed to be broader than social roles in that social roles usually imply a direct set of social interactions and division of labor simply means that some actions seen in a group are seen only by some individuals (although it is implied that such actions serve beneficial functions (see Chapter 37 for a further discussion of what it means to be social).

SELECTIVE PRESSURES THAT ACT ON SOCIAL LIVING

Why are animals social? Although most primates are social, many other animals live more solitary lives. We suspect that the variation in degrees of sociality reflects adaptations to specific ecological conditions. Clumped and locally abundant food might favor social living since there is enough in one place to feed multiple individuals and the group can jointly fend off other conspecifics trying to displace them at the food source. Moreover, if food clumps are widely scattered, a social unit has many individuals to search for the next food clump. Sharing information upon finding clumped food would not cost an individual if the clump were large enough to feed multiple group members; all would benefit. On the other hand, if food were sparse or the clumps small, then sociality would not be favored because of intragroup competition. If food were evenly scattered (or locally superabundant) but slowly renewing, social feeding might be favored in that groups would not return to a feeding area too soon after exploiting it, whereas individuals feeding alone would likely crisscross each other's paths, wasting time in already visited patches. This coordinated foraging might promote large herds rather than social groups because, whereas there is a benefit in joint feeding, there is no benefit to be obtained by excluding others from coordinating their feeding with you.

Social living will permit larger groups to displace smaller groups at resources, but larger groups will experience more intragroup competition, theoretically requiring greater travel distances to obtain sufficient resources for all. Attempts to empirically demonstrate what seems a logical and obvious conclusion have met with only mixed success (Chapman and Chapman 2000), but it may well be that group size influences travel distances in a more complex fashion. For example, one might assume that baboons (*Papio cynocephalus*) might travel farther to find food when food is scarce, but Altmann and Altmann (1970) found that they used more of their home range in the wet season, when food was more abundant, than in the dry season. Travel in search of food in the dry season was constrained by the necessity to stay within the area of the home range that contained permanent water sources.

Social living provides many sensory systems to detect predators and the possibility of joint defense against predators (mobbing), but it also makes it difficult to be cryptic in that all are betrayed by the most noticeable individual. Predators may also find it profitable to continually track a large group so that, even if an individual benefits by being unlikely to be the one attacked, predator attacks may become regular and frequent.

Finding a potential mate is facilitated by living in heterosexual social groups. In contrast, solitaries must search and advertise for mates, thereby exposing themselves to predators as well as potential mates. The same activities that attract mates can also attract predators, or predators can use mate-advertisement signals to attract prey. On the other hand, social living means that rivals as well as potential mates may already be present.

Thermoregulation may be possible in social groups huddling together, but close social proximity can also favor the transmission of disease and parasites. Sociality will permit ready transfer of information such that the location of resources known to one individual is available to all. In this way, long-term storage of information is based on the memories of the oldest individuals in the group. The sharing of such knowledge insures both the simultaneous use of resources and the presence of potential rivals for those resources. On the other hand, group living means sacrificing personal agendas concerning when to go where to exploit what. Individuals will have to delay satisfaction of some needs or travel to and consume resources at times other than they would have based on personal desires in order to maintain continuous contact with the group. Efforts must be made to coordinate the satisfaction of one's own needs with the satisfaction of those same needs in others. Some social units do temporarily divide in the pursuit of resources, but social living requires the development of means to reassemble or continuously monitor the location of others not in immediate proximity. The information provided on the location of others could also be used by predators searching for prey.

Social living may foster alloparental care and facilitate adoption of orphaned young, but it may also facilitate infanticidal breeding strategies. Social living may facilitate the synchrony of breeding using social communication when there is an advantage to coordinated production of young (as in "swamping" predators by producing all of the infants at one time so that a predator will not have a year-round supply of vulnerable infants to exploit). Breeding synchrony, however, may mean that a female cannot wait until she is in optimal condition to breed. Communication requirements may be increased to facilitate interactions in large groups, and within-group competition may result in agonistic contests that would disrupt the group if not countered by other mechanisms.

Although not exhaustive, the above list indicates that social living will be influenced by multiple selective pressures and that specific ecological conditions will not only select for or against social living but also influence the specific organization of social units when sociality is favored. Various investigators have focused on the different selective pressures that operate on social units and the problems generated by social living, but acknowledge that sociality is influenced by all. (see Chapter 28 for a historical analysis of theories of socioecology).

CONFLICT, COMPETITION, AND AGGRESSION

My research has focused on the expression and control of agonistic behavior (defined as aggression and the usual responses to aggression; Scott 1955) in social groups. Although the expression or control of aggression is important in maintaining social groups, it need not be the central principle behind the organization of social units. Sociality cannot be understood based on the study of any one factor or selective pressure (see Altmann 1974 for a discussion of the multiple pressures that shape baboon activity profiles and energy budgets). Social living can indeed promote conflicts, competition, and aggression; and these can be understood as three separate problems (Mason 1993). Conflicts (two mutually incompatible goals or demands within the same individual or between two or more individuals) are inevitable when multiple individuals want to do different things at the same time and yet all want to remain together. A conflict exists even in situations where one partner wants to groom and the other wants to play, and yet we do not expect such conflicts to be resolved by aggression. Competition (defined to exist whenever a necessary resource is available in insufficient supply for all who require it) is inevitable when a necessary resource is limited and an individual is in proximity to others with the same requirement. Whereas animals may fight over a limited resource, there is little point in fighting over access to each kernel of food when it is widely scattered in small packets, like grains of rice on the ground, and scramble competition is likely to prevail. Although neither conflicts nor competitions need to be resolved using aggressive contests, some conflicts will end in agonistic encounters and some competitions will be resolved by aggressive contests. Aggression can also be generated in social groups, despite the absence of conflict or competition, when individuals under attack by one redirect to another or come to the defense of others by attacking an animal with which they neither had an immediate conflict nor were in competition at the moment. Close proximity can also generate the appearance of conflict or competition when there is none. Hall (1964) stated that the most common cause of aggression in primate groups was a perceived violation of the social code. For example, if an individual does not tolerate others in immediate proximity while feeding and another walks too close in passing, it might be perceived as approaching the food of the feeding individual. This perceived violation of the social code (the expectation that others would not come into proximity during feeding) could result in aggression to repel the violator and bring it into conformity with expectations.

HOW AGGRESSION IS EXPRESSED AND CONTROLLED IN SOCIAL GROUPS

Theoreticians have argued at great length concerning the causes of aggression and even as to how to define aggression (Bernstein 1991a). Pain- or fear-induced or irritable aggression is clearly due to a different cause than instrumental aggression used to achieve a particular goal. Aggression is also often identified by function, such as predatory and antipredatory, defensive, sexual, territorial, and weaning aggression. Parental punishment and even moralistic aggression have been listed, and some believe that aggression may be motivated by striving for dominance (but see Mason 1993). Although aggression has many causes, merely living in a social unit will generate conflicts and competitions, some of which will be resolved by aggressive contests. In these cases, aggression is being used as an instrumental act to achieve a goal.

Since aggression is costly (not only in terms of energy expenditure and social relationships but also in terms of the injuries that can be sustained by resistance or retaliation), natural selection should favor aggression only in those cases where the value of the incentive outweighs the costs.

Moreover, since there can be, at most, one victor in an aggressive contest (either one individual in a dyadic contest or one side when multiple individuals fight with multiple others), individuals will be selected for who can accurately assess not only the relative value of the incentive to the cost but also the likelihood of victory in a particular agonistic encounter. Since agonistic encounters impose costs on both the aggressor and victim, natural selection will favor both victims and aggressors that minimize the frequency and severity of agonistic contests but simultaneously favor individuals that "correctly" use aggressive contests to gain access to resources or to resolve conflicts in their favor.

The decision to initiate aggression then will involve assessing the value of the incentive to be obtained, the costs of the aggression used to obtain the incentive, and the likelihood that the incentive will be obtained. The cost of aggression is not an absolute value but dependent on the relative abilities of the opponent (as is the likelihood of successfully obtaining the incentive); agonistic encounters should, therefore, include a period for assessment of the opponent. It is in the interest of both parties to provide the other maximum information about their aggressive abilities and their willingness to use those abilities in the contest situation. Natural selection will favor individuals that do not respond to false signals and respond appropriately to honest signals. Similarly, individuals that can deter resistance by advertising their abilities and their determination to use those abilities will be favored over individuals that attack immediately and always incur the costs of combat. Each will, therefore, benefit if it can repel a rival without resorting to combat or yield to a rival that it cannot defeat in combat. Each will be subject to selection that favors appropriate responding to honest signals. Some signals, like piloerection to increase apparent body size, may be basically deceptive but do, in fact, convey accurate information about body size. If all individuals use the same device, relative sizes are the same as they would be if no individual did so. Piloerection will be favored only because of the miscommunication of size that would result in animals that failed to do so. A miscommunication may mean that individuals that would have retreated from an individual of a particular size will not do so, requiring escalation to more costly forms of contest. Even if one wins such a contest after escalation, there has been an additional cost, and the loser has also needlessly paid the costs of combat in a losing fight.

The onset of an aggressive encounter will, therefore, include a period of signaling and assessment. If initial information on size, strength, and determination (as well as number of allies and their abilities and willingness to participate) is insufficient to convince either that they are likely to lose, then additional information will be required, which can be obtained by further escalation of signals. Each escalation will be more energy-intensive and closer to actual physical combat. Even the onset of physical combat is an opportunity to gain further information on the aggressive powers of the opponent, and at any time that a contestant is convinced that the cost of further combat will exceed the value of the incentive or is unlikely to result in obtaining the incentive, it is in the interest of both parties to terminate the contest. The loser does not gain the incentive but incurs no additional costs, and the winner does obtain the incentive but would incur additional costs with no further benefit if the fight were prolonged beyond the point at which the rival yielded and indicated that it would no longer contest the outcome. Continuing such a fight due to "anger" or "rage" or a failure to respond to a signal indicating the intention of the rival to break off the contest and retreat may prolong a fight with additional costs and no gain to either contestant. Clear signals of such a decision and appropriate cessation of further hostilities will be favored for both winners and losers, and we may expect that vocalizations, facial expressions, bodily postures, and as many other devices as possible will be used to communicate submission in addition to the simple act of physical withdrawal. Individuals capable of detecting early signs that they cannot win a contest will suffer fewer injuries and defeats, and those willing to allow opponents to withdraw will suffer fewer injuries inflicted by opponents defending themselves against attack. Aggression is thus likely to be ritualized in that agonistic responses will be favored that convey maximum information with minimum costs (including injuries). We thus see the marvelously ritualized combats between animals in the mating season, where few injuries are actually incurred despite the lethal potential of the weaponry used. Primates do not have antlers, horns, and other specializations of many animals but, males especially, may have canine teeth that rival those of the largest carnivores. Nonhuman primate combat, however, rarely results in deaths, and the most vulnerable anatomical sites (throat and ventrum) are injured far less frequently than would be expected based on their relative sizes (Bernstein and Gordon 1974, Owens 1975, Ruehlmann et al. 1987).

DOMINANCE

Inasmuch as primates are smart, repeated fights between the same opponents will be influenced by learning and memories of previous fight outcomes. Recognition of individuals allows primates to learn which individuals have defeated them in the past and which individuals they have defeated. Knowledge of previous fight outcomes can be used to predict the outcome of the next fight, but appropriate action will still depend on evaluation of the incentive and assessment of the determination of the opponent to contest that incentive using aggressive means. Only the fighting ability of the opponent can be stored in memory; the value of an incentive and the determination of the rival must be assessed each time. I use the word *dominance* to describe relationships between individuals where decisions to fight or retreat are largely based on having learned about the outcomes of previous encounters with the same opponent and where the subordinate immediately displays those submissive signals that it displayed at the end of the last agonistic encounter (Bernstein 1981). Note that the loser of a previous fight does not automatically yield every incentive to the past winner but may still decide that the incentive is so valuable that the costs of fighting are overcome (such as a mother defending her baby against an opponent that has defeated her in past fights) or that the opponent is not signaling a high level of motivation to contest the incentive (as when the incentive is food and the rival is already sated) and that it will not be met by aggressive challenge if it takes the incentive. In such cases, a subordinate may successfully obtain the incentive despite the presence of another that has been victorious in the past. If a dominant individual gives no sign of interest in an incentive or objection to an activity, then a subordinate can obtain the incentive or engage in the activity. Priority of access to incentives thus correlates with dominance but is not synonymous with the learned relationship that predicts which of two equally motivated opponents will pursue aggressive contests and which will submit.

If at the very start of an agonistic encounter one individual withdraws, we can invoke the concept of dominance to explain why agonistic signals were unidirectional instead of symmetrical. Such an explanation infers the existence of learning based on a past history of encounters and an assessment of the current situation that the subordinate uses to predict the likelihood of an agonistic encounter and its outcome. At times, however, it appears as if individuals that have never previously met already have a dominance relationship in that one submits to the other virtually immediately. The assessment that results in submission in such a case does not involve knowledge of a past history of agonistic encounters with that opponent, but the rapidity of the decision to submit must mean that sufficient information is available at the start concerning the agonistic potential of the opponent. This need not involve a self concept, such that an individual compares its own size and strength to an opponent, but may be based on generalization. Whereas each animal may be recognized as a distinct individual, some classification and generalizations are to be expected. If a new opponent seems to share those characteristics that are common to the individuals that defeated one in the past, then one can generalize and assume that this individual will also be capable of winning. Socially sophisticated individuals with sufficient experience with others may develop concepts regarding the characteristics of opponents that they can and cannot defeat. Such generalizations may make it appear as if dominance was an attribute of an individual that others could recognize immediately or that individuals had self concepts that they could use to compare against any new opponent, but both explanations seem unlikely. Since all, except the most dominant and most subordinate, are dominant to some and subordinate to others, being dominant (or subordinate) cannot be an absolute attribute of an individual. Moreover, since it is possible for A to dominate B and B to dominate C and still have C dominant to A (as in the child's game of "paper–scissors–rock"), dominance is not even a

property that can always be plotted on a linear scale. The possibility of comparing an opponent to oneself seems unlikely when we realize that monkeys do not seem to have well-developed self concepts in that they fail to recognize that their mirror images are a representation of self (Gallup 1982, 1987).

It has been suggested that primates may infer their dominance relationship to another without direct experience by observing the outcomes of fights between other individuals and then reasoning that if one defeats another that has defeated them, then the first can also defeat them (Cheney and Seyfarth 1990). Surely, such a comparison would yield valuable data and would reduce the number of direct contests required, but evidence for such abilities is lacking. Chapais (1992) has demonstrated that when a younger member of a high-ranking matriline is isolated with all of the members of a lower-ranking matriline, the individual will lose to the intact family and assume subordinate status. The introduction of an adult member of the isolated individual's matriline does not immediately restore dominance, suggesting that merely witnessing submission of the resident matriline to one's relative is not enough. Status is restored to the subject only when its relative comes to its aid against the residents and it personally experiences successful aggression against the other matriline. This suggests that direct experience, rather than passive observation, is required to establish new learned relationships in this domain.

Primate sociality often results in complex agonistic interactions with multiple parties. Complex interactions, however, can be seen as resulting from individual interactions. They add and multiply, but the same principles apply. Fights may be context-specific, and winning and losing may be as well. Dominance, however, is an intervening variable and should not be context-specific. Winning one year is not a permanent condition, but the past influences the future when animals can learn and remember. Even dominance can change as learning is ongoing and new information can change a relationship. If it changes so readily that it allows no prediction, then the intervening variable does not allow prediction and it is pointless to invoke it.

FIGHT INTERFERENCE

Agonistic contests in nonhuman primates often involve more than two individuals. Animals with strong social bonds will use social techniques in facing any challenge, be it from the physical or social environment. At the onset of an agonistic encounter, we should expect animals to seek help from allies and to assess the likelihood that others will aid their opponent. Alliances and coalitions may be in the mutual interest of allies that repel the same opponent (as when groups repel other groups at resources), or individuals with strong social bonds may habitually aid one another regardless of whether they have a conflict or are in competition with the opponent (as when mothers aid their offspring). Such joint action may be balanced or not (a mother may aid her infant more often than her offspring aids her, Kaplan 1978) but it is the social bond that dictates aid rather than any exact reciprocation of services. Whereas reciprocal altruism can be invoked to explain some instances of one animal aiding another, a market analogy (see Noe 2001) may be misleading (e.g., if one thinks that mothers are aiding their young in expectation of payback). Although fight interference on behalf of kin can be explained in terms of classical and inclusive fitness benefits, closely bonded individuals need not be kin. Unrelated animals may aid, groom, and socially interact based on social bonding rather than on elaborate mental bookkeeping of the value of services and favors and ideas of equality (Matheson and Bernstein 2000). Inasmuch as there is very little evidence that primates have a distinct concept of kinship (e.g., Alberts 1999, Silk 2002) but rather end up preferentially associating with kin based on mechanisms such as familiarity that correlate with kinship (Bernstein 1988, 1991b, 1999; Bernstein et al. 1993; Fredrickson and Sackett 1984; Sackett and Fredrickson 1987), it might be expected that behavior toward kin that could be explained based on inclusive fitness benefits might also be directed toward unrelated individuals that are socially bonded for the same reasons (e.g., long-term association in the same social unit). Martin (1997) has demonstrated that the degree of association among members of a captive pig-tailed monkey (*Macaca nemestrina*) group was correlated nearly perfectly with years of association and the intensity of that association. Even though all sisters were genetically equally related to one another on average, those that were of more similar age spent more years dependent on the same mother and were more closely bonded. Sisters with a living mother were also more closely bonded than orphaned sisters, and old matriarchs spent as much time with each other, although not genetically related, as they spent with younger relatives that had been in the group for only a few years.

CONTROL ROLE BEHAVIOR

Aggression has the potential to inflict injuries on group members and thereby degrade their abilities to contribute to group defense against external threats such as predators and rival groups. It is therefore advantageous to every animal in the group to limit the damage done in intragroup fighting. Whereas ritualization of agonistic encounters will reduce actual fighting, there will be times when equally matched individuals must escalate encounters to the most intense levels to decide which will be the winner or loser. (The "fairest" fights are often the most dangerous.) Fights may also occur due to miscommunications and due to redirection and displacement. Regardless of the cause, fights between group members represent a threat to all members of the group. Fight interference may involve family alliances (subgroups within the group), but such interference has as much potential to escalate as to terminate the encounter. Fight interference by a clearly superior group member will likely terminate the encounter as all opponents yield to the interfering individual. I have (Bernstein 1964a, 1966a) previously referred to this as "control role behavior" and indicated that the interfering individual will usually be dominant to the contestants but need not be the most dominant animal in the group and that several individuals may show the same role behavior. Moreover, the control animal may attack either the victim or the aggressor with equal effect in that its aggression will disrupt the initial fight and both combatants will leave the scene on the approach of an aggressor dominant to them both. In fact, the control animal can attack an uninvolved third party in the immediate vicinity of the combatants with the same result; the two combatants (as well as the innocent third party) will flee the vicinity of the aggressive dominant individual. Attacks on uninvolved third parties may be so common and directed to specific classes of individuals that they may selectively socialize individuals. Bernstein and Ehardt (1986a) have suggested that adult male rhesus monkeys (*Macaca mulatta*) interfere in fights between adult females or juveniles by attacking the nearest subordinate male or adolescent male. Such consistent behavior results in subordinate males avoiding the vicinity of noisy agonistic signals produced by females and immature animals and thus curbs their own aggression toward females and immature animals that respond to aggression with noisy agonistic signals. Fight interference, however, is not only seen by animals dominant to the combatants; subordinates that interfere can also be effective at limiting aggression between group members (Boehm 1994).

Whereas the control role is a social technique to control aggression in the group, control role behavior is also seen when individuals protect group members against external sources of disturbance (Bernstein 1966b). Although the same individual may do both, these may be two independent roles (Wilson 1975). Other mechanisms also exist to control the social damage done by intragroup agonistic encounters, and a rich literature on "peacemaking" (de Waal 1989) has developed (see Chapter 36 for a review of reconciliation).

Contests with other social units may be expected whenever social units come into conflict or are in direct competition with one another. Whereas a control animal may take a position between an external threat and potential sources of disturbance, conflicts and competitions are resolved by joint action of as many group members as possible. Again, whereas size and strength may be important in deciding such contests, the sheer number of individuals on each side may be decisive in that few primates can successfully fight multiple opponents simultaneously. In addition, the coordination of individual support and the determination of group members may influence the outcome of an intergroup agonistic encounter. Each individual in a group will then have to assess the opposing group, the relative strength and determination of its own group, and the value of the incentive. This is a more demanding task than assessment of individuals, but primates do learn to recognize other groups and to contest other groups, or not, based on a history of past encounters as well as current relative determination and incentive values. Recognition is unlikely to include recognition of every member of a rival group but may be based on recognition of a few significant group members. Assessment of relative numbers and determination may be based on the behavior of those individuals in one's immediate vicinity and, therefore, somewhat imperfect; but the nature of intergroup fights suggests that these assessments are continual during fights and that the changing course of combat reflects new information provided by which individuals join or flee the fight and the successes and failures of key encounters. Once again, it is in the interest of both groups to engage in fighting only when the incentive is worth the cost and when the incentive is likely to be obtained by contest. Nevertheless, it is still in the interest of both groups to minimize the costs of combat. As long as the opponent still has the potential to inflict damage in self-defense, it is best to let an opponent yield an incentive without further combat when a clear signal is provided.

ANIMAL DENSITIES APPROXIMATE RESOURCE DENSITIES

In this section, I consider intergroup encounters that seem to disperse groups in space roughly in proportion to the density of resources. These encounters usually represent the results of competitions for resources, but several different patterns may be identified, although all serve to prevent all groups from congregating at the richest resources. The four patterns that I have identified are territoriality, intergroup aggression, intergroup dominance, and intergroup avoidance.

Territoriality

Territoriality is defined as the defense of a specific geographic area against a specified class of others (modified from Burt 1943 to permit joint defense of a territory by mixed-species groups as described by Garber 1988). This may result in the defense of resources (but note that males defending an arena or a territory in a lek are not defending any identifiable "resource") but will not result in exclusive use of the territory inasmuch as without invasion there would be no evidence of defense. Territorial defense requires that intruders be detected before they can be repelled, so some intruders may succeed in temporary incursions if undetected. The defense of a specific geographic area by a group (or subclass of members in the group) will usually mean that the class of others being defended against will not have regular access to the resources in the territory. Sequential use of the same resources is thus precluded in animals that maintain territories.

Territorial defense is feasible only when the area being defended is small enough to make it possible to detect intruders and when the number of intruders is not so great that the time and energy required in repelling them exceed the value of the territory to the defenders. If territorial behavior evolves in a species, then there will be an advantage to all that hold and defend territories and individuals or groups not occupying a territory will try to establish one. They will likely go to the richest and most suitable areas first, but if such areas are already occupied, they will be repelled by the occupants. They will then proceed to the next most suitable areas until they find an attractive location without the negative of an opposing occupying group. When all suitable territories are claimed, there will be no recourse but to try to invade an already occupied territory. The choice of area to attempt invasion will be based on the value of the area and the ability (strength and determination) of the occupants to resist. If there is relatively little difference in defensive abilities among groups, but large differences in the quality of areas, then we would expect the richest areas to be the most attractive. Given that there are no alternative areas available, the invading group can lose and possibly die without a suitable territory; win and displace the residents, which will then have to become invaders elsewhere; or partition the territory with the defenders. Partitioning will be possible only if the resources contained are at least sufficient to support both groups. Thus, only the richest territories can be partitioned as partitioning poor territories, barely able to support one group, will result in two nonviable groups. With partitioning, large and rich territories become smaller and more easily defended but contain lesser absolute quantities of resources. The end result of successive invasions of nonestablished groups should be a series of territories each of the size necessary to contain sufficient total resources to sustain the residents. This will mean that areas with the densest resources will have the smallest territories, and the density of animals will be roughly proportional to the density of resources. A "rich" territory will be rich not because of the total quantity of resources available but because the necessary resources are found in a small area. Richer or poorer areas all end up with animals having access to quantities of resources that closely match their needs. Territories will be maintained by active defense, but once groups have learned the boundaries that other groups defend, little fighting need actually occur. Intruders that know the boundaries may still intrude but will withdraw beyond the boundaries whenever challenged by the territory holders. This differs from dominance in that, although it is a learned relationship that influences the directionality of agonistic encounters, it is specific to a location. Neighboring groups can, in fact, take turns chasing each other in a zigzag pattern at a territorial boundary (Bernstein 1968).

Aggressive Contest

Not every habitat is suited to the establishment of a system of territories. If resources must be obtained from scattered areas such that a group cannot defend all of the sectors required to sustain it, then the group will not be able to prevent others from using resources in areas that they do not currently occupy; that is, they will not be able to prevent others from using resources in a sector that is far away and where they cannot detect intrusion. Sequential use of resources will now be possible, but simultaneous use of a resource may still be precluded. Groups may fight one another when they meet at a resource so that the loser must leave; dominant groups may repel subordinates that have learned, based on past fights, that resistance is futile against this particular rival; or groups may avoid the immediate proximity of any group, thereby avoiding the costs of fighting to obtain resources.

When groups fight over resources, one would expect that each group would first be attracted to the most desirable area. If such an area is already occupied, then the two groups will contest the area and the loser will be forced to withdraw and seek an alternative area. It will be expected to go to the next most desirable area and to either use the resource there, if the area is unoccupied, or contest the resource if the area if occupied. After each defeat, the losing group will move to the next most desirable resource until it either finds an available resource or there are no alternatives left. If defeated at all available sites, a group has no choice but to return to a site where it had been defeated and wait for the occupying group to move. They should thus choose a site that is most attractive, with regard to resources, and least aversive due to the group already there. Since any group that defeats another is aversive, the decision will be largely based on the richness of the resource. If there are many more groups than resource sites, then more than one group will be in waiting. The decision as to where to wait will be based on the value of the resource and the number of groups in the area that successfully defend it. Groups will therefore congregate at the richest resources with the fewest groups present that can defeat them. This will result in the distribution of groups in space roughly in proportion to the density of resources; that is, there will be more groups waiting at the areas of richest resources and fewer willing to wait at less attractive areas occupied by groups feeding or waiting that can defeat them. Such a system works only when groups must abandon resources in one area to obtain different resources in another area (e.g., leave a feeding area to travel to a source of water some distance away).

Intergroup Dominance

If groups are capable of recognizing other groups and remembering the outcomes of past agonistic encounters, then rather than fight a group at a resource to learn if it can displace that group or not, a group can decide to contest or yield based on its established relative dominance relationships (i.e., the members will not contest groups that they can identify and remember having lost to in previous contests). Groups will still be attracted to the most attractive resources first and repelled by groups present that are dominant. When all sites are occupied, groups will be most attracted to the richest sites with the fewest groups feeding or waiting to which they are subordinate. Once again, the density of animals

in the vicinity of a resource will be roughly proportional to the density of the resource (i.e., rich areas will have more groups waiting). Sequential use is possible as long as the users cannot permanently occupy the site. If all resources are concentrated in a small area, then animals can remain at the same site permanently and territoriality may result. When the various needs of a group must be met by traveling distances that preclude the defense of all sites that the group utilizes, then sequential use of resources is possible, as is contest at those sites for use of the resources. Since animals have multiple needs, a group cannot permanently remain at a food site when it needs water or shelter or protection from the elements more than it needs food. If it must travel elsewhere to satisfy these other needs, then groups whose need for food is greater than their needs for the other resources will have an opportunity to feed when the most dominant group departs. The order of feeding will be based primarily on relative dominance, and there will be less fighting than if the order was determined by a fight on each encounter. Dominance relationships, however, require repeated contacts with the same groups and sufficient learning ability to identify groups and store memories of past encounters with those groups.

Avoidance

Finally, animals can entirely escape the costs of aggressive competition if they always avoid other groups. In this case, the group travels to the richest resource first but withdraws if the site is occupied and moves to the next richest resource. It will keep moving until it finds an unoccupied site or all sites are found to be occupied. It will now be equally repelled by a group at every site but most attracted to the richest site and may be expected to return to that location and wait for the occupying group to vacate the area. If other groups are already waiting, then their presence is also aversive; thus, a group finding the richest site already occupied with one or more other groups also waiting in the area may move to a less attractive site but one that has fewer waiting groups. There will be a balance between the attraction of the site and the aversion to the number of waiting groups such that animals will once again be distributed in space roughly in proportion to the density of resources. Note that if groups can detect the location of other groups without actually traveling to the site, then considerable time and energy may be saved. Loud vocalizations or prominent visual displays that advertise location may thus benefit all groups as they will not congregate at the same locations and can move directly to sites that no group currently occupies.

All four systems will result in animals being distributed in space roughly in proportion to the density of resources, but only in the case of territoriality is sequential use of resources prohibited. Obviously, there will be ecological conditions that will favor one over another. I have already commented on the conditions that might favor territoriality. Aggressive contests may occur if groups do not regularly meet the same opponents and thus cannot establish dominance relationships. Although aggressive encounters each time are more costly, dominance relationships require an opportunity to learn or make relative assessments without resorting to combat. This will require learning ability, the ability to identify other groups, memory of past encounters, and the ability to assess groups in each situation. Avoiding all others is the least costly in terms of aggressive consequences but is likely to evolve only when resources are so widely distributed that a group can always move to an alternative site and when the quality of each site is roughly similar to all other sites such that no location is sufficiently superior so that the extra cost of aggressive competition is compensated for by the increase in quality. This results in scramble, rather than contest, competition. It is my impression (Bernstein 1964b) that the howlers (*Alouatta palliata*) of Barro Colorado live in such an ecological situation and demonstrate scramble competition based on avoidance, but others have suggested that their spacing is based on territoriality or aggressive competitions and even dominance. It should be possible to empirically test these alternatives.

CONCLUDING COMMENTS

In this chapter, I have focused on the expression and control of aggression that might be expected to be provoked by social living. Such aggression may be triggered between individuals in a social unit or between social units, but the means of expression and control are similar in both cases. Whereas aggression can function to maintain the integrity of groups (defense against external and internal sources of disturbance) and integrate individuals into a group (as in socialization processes where punishment of unacceptable behavior results in individuals learning group standards), uncontrolled aggression has the potential to destroy individuals and groups (Bernstein and Gordon 1974). Parental punishment, defense, and the socialization of newcomers may be essential to the maintenance of social units. Nonhuman primates, lacking the language skills to explain what behavior (or absence of what behavior) resulted in a reward, may rely heavily on punishment of unacceptable behavior in socialization. The behavior of infant rhesus monkeys (*Macaca mulatta*) is modified by frequent punishment of unacceptable behavior on the part of mothers and other kin and close associates (Bernstein and Ehardt 1986b). The behavior of new group members is also modified by selective aggressive responses to their behavior. Aggression may be essential to social life for primates, but it must also be controlled. I have taken the position that natural selection will favor the control of aggressive contests in socially living primates and that such control is essential to sociality but does not define social systems. Dominance need not be a universal attribute of social units nor does the expression of dominance necessarily characterize every aspect of the social unit. On the other hand, every social unit may be expected to experience

aggression and to have evolved mechanisms to control aggression and its consequences.

REFERENCES

Alberts, S. (1999). Paternal kin discrimination in wild baboons. *Proc. R. Soc. Lond. B* 266:1501–1506.

Altmann, S. (1965). Sociobiology of rhesus monkeys. II. Stochastics of social communication. *J. Theor. Biol.* 8:490–522.

Altmann, S. (1974). Baboons, space, time and energy. *Am. Zool.* 14:221–248.

Altmann, S., and Altmann, J. (1970). Baboon ecology. *Bibl. Primatol.* 12:1–220.

Bernstein, I. (1964a). Role of the dominant male rhesus in response to external challenges to the group. *J. Comp. Physiol. Psychol.* 57:404–406.

Bernstein, I. (1964b). A field study of the activities of howler monkeys. *Anim. Behav.* 12:84–91.

Bernstein, I. (1966a). Analysis of a key role in a capuchin (*Cebus albifrons*) group. *Tulane Stud. Zool.* 13:49–54.

Bernstein, I. (1966b). An investigation of the organization of pigtail monkey groups through the use of challenges. *Primates* 7:471–480.

Bernstein, I. (1967). Intertaxa interactions in a primate community. *Folia Primatol.* 7:198–207.

Bernstein, I. (1968). The lutong of Kuala Selangor. *Behaviour* 32:1–16.

Bernstein, I. (1981). Dominance: the baby and the bathwater. *Behav. Brain Sci.* 4:419–457.

Bernstein, I. (1988). Kinship and behavior in nonhuman primates. *Behav. Genet.* 18:511–524.

Bernstein, I. (1991a). Aggression. In: Dubecco, R. (ed.), *Encyclopedia of Human Biology*, Academic Press, San Diego. pp. 113–118.

Bernstein, I. (1991b). The correlation between kinship and behavior in non-human primates. In: Hepper, P. (ed.), *Kin Recognition*. Cambridge University Press, Cambridge. pp. 6–29.

Bernstein, I. (1999). Kinship and the behavior of nonhuman primates. In: Dolhinow, P. (ed.), *The Nonhuman Primates*. Manfield, Mountain View, CA. pp. 202–205.

Bernstein, I., and Ehardt, C. (1986a). Selective interference in rhesus monkey (*Macaca mulatta*) intragroup agonistic episodes by age–sex class. *J. Comp. Psychol.* 100:380–384.

Bernstein, I., and Ehardt, C. (1986b). The influence of kinship and socialization on aggressive behaviour in rhesus monkeys (*Macca mulatta*). *Anim. Behav.* 34:739–747.

Bernstein, I., and Gordon, T. (1974). The function of aggression in primate societies. *Am. Sci.* 62:304–311.

Bernstein, I., Judge, P., and Ruehlmann, T. (1993). Kinship, association, and social relationships in rhesus monkeys (*Macaca mulatta*). *Am. J. Primatol.* 31:41–45.

Boehm, C. (1994). Pacifying interventions at Arnhem Zoo and Gombe. In: Wrangham, R., McGrew, W., and de Waal, F. (ed.), *Chimpanzee Cultures*. Harvard University Press, Cambridge, MA. pp. 211–226.

Burt, W. (1943). Territoriality and home range concepts as applied to mammals. *J. Mammal.* 24:346–352.

Chapais, B. (1992). The role of alliances in social inheritance of rank among female primates. In: de Waal, F., and Harcourt, A. (ed.), *Coalitions and Alliances in Humans and Other Animals*. Oxford University Press, Oxford. pp. 29–59.

Chapman, C., and Chapman L. (2000). Constraints on group size in red colobus and red tailed guenons: examining the generality of the ecological constraints model. *Int. J. Primatol.* 21:565–585.

Cheney, D., and Seyfarth, R. (1990). *How Monkeys See the World. Inside the Mind of Another Species*. University of Chicago Press, Chicago.

de Waal, F. (1989). *Peacemaking Among Primates*. Harvard University Press, Cambridge, MA.

Fredrickson, W., and Sackett, G. (1984). Kin preferences in primates (*Macaca nemestrina*): relatedness or familiarity? *J. Comp. Psychol.* 98:29–34.

Gallup, G. (1982). Self-awareness and the emergence of mind in primates. *Am. J. Primatol.* 2:237–248.

Gallup, G. (1987). Self-awareness. In: Mitchell, G., and Erwin, J. (eds.), *Comparative Primate Biology. Behavior, Cognition, and Motivation*, vol. 2, part B. Alan R. Liss, New York. pp. 3–16.

Garber, P. (1988). Diet, foraging patterns, and resources defense in a mixed species troop of *Saguinus mystax* and *Saguinus fuscicollis* in Amazonian Peru. *Behaviour* 105:18–34.

Hall, K. (1964). Aggression in monkey and ape societies. In: Carthy, J., and Ebling, F. (eds.), *The Natural History of Aggression*. Academic Press, London. pp. 51–64.

Kaplan, J. (1978). Fight interference and altruism in rhesus monkeys. *Am. J. Phys. Anthropol.* 49:241–249.

Kummer, H. (1971). *Primate Societies. Group Techniques of Ecological Adaptation*. Aldine-Atherton, Chicago. pp. 160.

Martin, D. (1997). *Kinship bias: a function of familiarity in pigtailed macaques (Macaca nemestrina)* [diss.]. University of Georgia, Athens.

Mason, W. (1993). The nature of social conflict: a psychoethological perspective. In: Mason, W., and Mendoza, S. (ed.),

Primate Social Conflict. State University of New York Press, Albany. pp. 13–48.

Matheson, M., and Bernstein, I. (2000). Grooming, social bonding, and agonistic aiding in rhesus monkeys. *Am. J. Primatol.* 51:177–186.

Noe, R. (2001). Biological markets: partner choice as the driving force behind the evolution of mutualisms. In: Noe, R., van Hooff, J., and Hammerstein, P. (eds.), *Economics in Nature: Social Dilemmas, Mate Choice and Biological Markets*. Cambridge University Press, Cambridge. pp. 93–118.

Owens, N. W. (1975). Comparison of aggressive play and aggression in free living baboons (*Papio anubis*). *Anim. Behav.* 23:757–765.

Ruehlmann, T., Bernstein, I., Gordon, T., and Balcaen, P. (1987). Wounding patterns in three species of captive macaques. *Am. J. Primatol.* 14:125–134.

Sackett, G., and Fredrickson, W. (1987). Social preferences by pigtailed macaques: familiarity versus degree and type of kinship. *Anim. Behav.* 35:603–606.

Sarbin, T. (1954). Role theory. In: Gardner, L. (ed.), *Handbook of Social Psychology*, vol. I. Addison-Wesley, Cambridge, MA.

Schultz, A. (1969). *The Life of Primates*. The Weidenfeld & Nicolson Natural History. Weidenfeld & Nicolson, London. pp. 281.

Scott, J. (1955). *Animal Behavior*. University of Chicago Press, Chicago.

Silk, J. (2002). Kin selection in primate groups. *Int. J. Primatol.* 23:849–875.

Wilson, E. (1975). *Sociobiology: The New Synthesis*. Belknap Press of Harvard University Press, Cambridge, MA.

35

Social Beginnings
The Tapestry of Infant and Adult Interactions

Katherine C. MacKinnon

INTRODUCTION

Among mammals, primates are characterized by having extended life history stages, particularly infancy. While there is much variation within the Primate order, from an infancy period of 2–3 months in some prosimians to 4+ years in some great apes, all primates are born altricial and require an extended period of care. The relationships that form during

this time are critical for a young primate's survival. Immature primates need not only help with transportation, feeding, and protection but also social stimulation and physical contact (explicitly demonstrated by Harlow's cloth-mother experiments in the 1950s and 1960s; Harlow 1959, Harlow and Harlow 1962).

In most primate taxa, animals live in social groups that are stable through time, individuals know one another well, and this familiarity is a major factor in facilitating and maintaining strong social bonds. Immature animals take an active role in learning about their physical and social environments and often are the determining force behind not only the nature of the social interactions themselves but also how others in the group perceive them (Dolhinow 1991).

For at least the first few months of life, the infant primate is the object of great interest and attention from others in the group. The earliest interactions, apart from those with its mother, involve those individuals with which the mother associates. In many species, all group members at some point approach, stare at, sniff, touch, and inspect the infant; but certain individuals are allowed longer and closer contact, depending on their relationship with the infant's mother and the species in question. Generally, young infants tend to reflect the social preferences of their mothers. Thus, in those species where females show marked preferences for matrilineal kin (e.g., Japanese macaques [*Macaca fuscata*], Kurland 1977), infants show such preferences as well (Berman 1982b, Thierry 1985; and see Berman 2004 for review as well as Chapter 13). This also holds true in species where females do not have strong leanings toward matrilineal kin (e.g., sooty mangabeys [*Cercocebus* spp.], Ehardt 1988, Gust and Gordon 1994).

Once an infant is off of its mother, its social world changes. It can decide whom to approach and when to leave, in addition to being the recipient of invitations for play and other affiliative contacts (Dolhinow and Bishop 1970). Toward the end of infancy, individual foraging increases, play is common, strong bonds among peers develop, and preferential relationships with certain older animals are formed. Crucial to the way infants maneuvre through this change from dependence to independence are their relationships with others in the group (their mothers, adult males, other females, and immatures). Older infants are able to seek out and form relationships with others that do not necessarily accord with their mothers' affiliative preferences. Infants are able to modify their relationships once they are more independent, and thus, complex variables are at work in determining the bonds younger animals may form with older members of the social group.

In this chapter, I will examine infant–adult female and infant–adult male interactions and highlight the extreme variability that exists within the Primate order. Theoretical explanations are presented to better illuminate why primate infants have such a long period of dependence on their mothers and caregivers and why immatures and nonmother adults in the group interact socially. The majority of examples and discussion will come from infant and adult interactions across primate taxa in natural settings.

HISTORY OF INFANT STUDIES

No other dynamic interaction demonstrates the importance of social attachments to primate survival more than the complex bonds infants form with others in the group. Most notably, the mother–infant bond has long been studied in primates, in both captivity and field settings. Through early studies, such as Harlow's work on deprivation (e.g., Harlow 1959, Harlow and Harlow 1962; and see Blum 2002), we see that having the basics of food, water, and protection is simply not enough to ensure an infant's survival. Strong social bonds must be established and maintained in order for an infant to develop well physically, cognitively, and socially. Because of close attention, our knowledge of behavior in infant primates has grown considerably in the past 50 years.

From the time of Carpenter's descriptions of howler monkey infancy in 1934 and Ladygina-Kohts' descriptions of an infant chimpanzee first published in 1935 (reissued in 2002), the development of the infant primate has been scrupulously detailed for decades. While captive studies were the first sources of information, field studies on infants in the wild increased dramatically during the 1960s, 1970s, and 1980s. Table 35.1 includes a partial list of some of the notable scientists that have shed light on how young primates mature and interact with others in natural settings. Due to the legacy of such fieldworkers (and many more not mentioned here), we have a richer understanding of infant sociality. The collective conclusion from these primatologists working on diverse taxa is that primate infant behavior and development is highly variable within and across species.

PRIMATE INFANT SOCIAL DEVELOPMENT

Primates are mammals, and they possess many traits associated with this taxonomic class. Among these is an enlarged brain given body size, especially in the cerebral cortex (Harvey and Clutton-Brock 1985). Another is having offspring born in a dependent state. Unlike the offspring of reptiles, amphibians, fish, and some birds, the young of virtually all mammals are born needing an extended period of constant care, sometimes lasting months or even years. This latter trait is especially striking in primates (see Chapter 24).

Long before an infant primate is independent of its mother, its social development has begun. Particularly, individuals the mother feels comfortable around are allowed to approach, sniff, and touch the young infant. The processes of infant socialization have been well studied in a wide variety of primates (e.g., Hall 1968, Hinde 1974, Hinde et al. 1964, Poirier 1972, Ransom and Rowell 1972, Fragaszy and Mitchell 1974, Chevalier-Skolnikoff 1977, Sussman 1977, McKenna 1979b, Mason 1986, Clarke 1990, Suomi 2002, Parker 2004). An obvious component of socialization is play, and an infant's first play partners (after its mother) are usually other immatures in the group. Such early exposure to group members facilitates social learning. Dolhinow and Murphy (1982) describe the 3-month-old infant langur (*Semnopithecus* [formerly *Presbytis*] *entellus*) as an "active strategist" as it typically has numerous choices for social partners and behavioral actions and ventures away from its mother to explore its social world whenever possible. Many aspects of behavior that are crucial for survival are also

Table 35.1   Some Earlier (Pre–1990) Field Studies that Include Infant Social Behavior and/or Development in Wild (Unless Otherwise Noted) Primate Populations

	SCIENTIFIC NAME	COMMON NAME	LOCATION OF STUDY	SOURCES
Apes	*Gorilla gorilla beringei*	Mountain gorilla	Virunga Volcanoes: Uganda, Rwanda, Democratic Republic of Congo	Schaller (1963), Fossey (1979)
	Pan troglodytes schweinfurthii	Common chimpanzee	Gombe Stream Nature Reserve, Tanzania; Mahale, Tanzania	van Lawick-Goodall (1967), Goodall (1986), Nishida (1968, 1988)
	Pongo pygmaeus	Orangutan	Tanjung Putang, Indonesia	Horr (1977), Galdikas (1981a,b)
Old World monkeys	*Semnopithecus (formerly Presbytis) entellus*	Indian langur monkey	Several sites across north and central India	Jay (1963, 1965)
	Cercopithecus aethiops	Vervet monkey	Zambia	Lancaster (1971)
	Erythrocebus patas	Patas monkey	Uganda	Hall (1965)
	Papio cynocephalus	Yellow baboon	Amboseli Nature Reserve, Kenya	Altmann, 1980
	Papio anubis	Olive baboon	Uganda	Ransom and Rowell, 1972; Nash, 1978
	Macaca mulatta	Rhesus macaque	Cayo Santiago, Puerto Rico (provisioned free-ranging groups)	Sade, 1965, 1980; Berman, (1982a,b, 1984)
	Macaca fuscata	Japanese macaque	Japan and United States (wild and provisioned free-ranging groups)	Itani, 1959; Fedigan, 1976; Hiraiwa, 1981; Tanaka, 1989
New World monkeys	*Alouatta palliata*	Mantled howler monkey	Barro Colorado Island, Panama; Chiriqui, Panama; La Pacifica, Costa Rica	Carpenter, 1934; Baldwin and Baldwin, 1973; Clarke, 1990
	Alouatta seniculus	Red howler monkey	Trinidad and Venezuela	Neville, 1972
	Brachyteles arachnoides	Woolly spider monkey	Fazenda Montes Claros, Brazil	Strier, 1986
	Ateles geoffroyi	Spider monkey	Barro Colorado Island, Panama	Carpenter, 1935
	Callicebus moloch	Titi monkey	Manu National Park, Peru	Wright, 1984
	Aotus trivirgatus	Owl monkey	Barro Colorado Island, Panama; Manu National Park, Peru	Moynihan, 1964; Wright, 1984
	Saimiri sciureus	Squirrel monkey	Panama	Baldwin, 1969; Baldwin and Baldwin, 1972
Prosimians	*Lemur catta*	Ring-tailed lemur	Berenty Reserve, Madagascar; Beza-Mahafaly Reserve, Madagascar	Jolly, 1966; Sussman, 1977; Gould, 1990
	Eulemur fulvus	Brown lemur	Beza-Mahafaly Reserve, Madagascar	Sussman, 1977
	Varecia variegata	Black and white ruffed lemur	Nosy Mangabe, Madagascar	Morland, 1990

learned in a social context: what to eat and where to find it, the group's home range, and what is safe and dangerous in the environment.

Primates display protracted life history stages relative to size compared to other mammals (see Chapter 23). These include extended periods of infant dependence with increased parental care, a delayed maturation to sexual maturity, longer gestation lengths, and longer life spans. Motor and cognitive abilities are not refined until long after birth, and social development continues into adolescence in many species. Therefore, the call for long-term studies in the wild still needs to be sounded. While we have excellent data on several species (e.g., *Pan troglodytes*, *Macaca* spp., *Papio* spp., *Lemur catta*), we still sorely lack longitudinal data on the social development parameters for most primates in natural settings.

INTERACTIONS BETWEEN INFANTS AND MOTHERS

Perhaps no other area of primate behavior has been as closely examined as the interactions between infants and their mothers. With such a long period of dependence and development, the relationship between infants and mothers was an obvious topic for early fieldwork, especially projects that concentrated on easily seen, well-known individuals living in large groups (e.g., yellow baboons [*Papio cynocephalus*], hanuman langurs [*Semnopithecus entellus*], and common chimpanzees [*Pan troglodytes*]). The mother–infant pair is a highly attractive duo in the troop (e.g., Richard 1976, Nakamichi and Koyama 2000) and one that is readily identifiable by group members (and researchers). For example, we see a typical scenario in wild moor macaques (*Macaca maurus*) and white-faced capuchin monkeys (*Cebus capucinus*): females with infants are approached more often than those without infants, and mothers receive more grooming (and do less grooming themselves) when they are holding their infants (Matsumura 1997, Manson et al. 1999).

As Dolhinow (1984:67) notes, "The mother–infant relationship is an elegant and complex system of biologically motivated as well as socially learned interactions." A deep interconnectedness lasts throughout the period of infancy and well into the juvenile and subadult stages in some species.

Figure 35.1 Infant white-faced capuchins (*Cebus capucinus*), like all young primates, first learn about the environment in a social context with their mothers (Photo by Katherine C. MacKinnon).

Indeed, in some primates such as common chimpanzees (*Pan troglodytes*), bonobos (*P. paniscus*), and macaques (e.g., *Macaca fuscata*), the mother–offspring connection is so strong that it is customary for adult individuals and their mothers to have dynamic, complex, and tightly bonded relationships across their entire lives. In bonobos, for example, mothers and sons support each other in conflicts throughout their lives and are intensely close (de Waal 1997).

Mother primates are able to provide situations in which socially mediated learning can occur (see Fig. 35.1). For example, young chimpanzees first learn how to use sticks for anting, wands for termite fishing, and appropriate stones for nutcracking from their mothers (Goodall 1986, Biro et al. 2003, Hirata and Celli 2003). Thus, one of the advantages of group living is that an individual animal, particularly an immature, does not have to "rediscover" certain behaviors for itself: it can learn from others in the group. Social learning, as Box (1984, 1994) and many others have noted, also allows for innovation by young and mature animals alike. That is, newly developed patterns of behavior might, over time, become established parts of a group's repertoire (see Perry et al. 2003 for cultural traditions in wild capuchins [*Cebus capucinus*] and Whiten et al. 2001 for cultural traditions in wild chimpanzees [*Pan troglodytes*]).

Great ape immatures show the longest period of attachment: infants are dependent on their mothers for nourishment, travel, and support in social interactions for the first several years of life (see Fig. 35.2). Old World monkey infants are usually dependent for the first year of life, with a few exceptions (e.g., the fast-maturing patas monkeys [*Erythrocebus patas*]). In many New World monkey species, infants rely on their mothers for locomotion, food, warmth, and social

Figure 35.2 Among the apes, orangutans are more solitary as adults; but for at least the first 5 years of life, infants are in the constant presence of their mothers, siblings, and occasionally other females with offspring. Here, an orangutan (*Pongo pygmaeus pygmaeus*) female relaxes with her dependent infant and older juvenile in Gunung Palung National Park, Indonesia (photo by Tim Laman).

Figure 35.3 Although prosimians generally have shorter life history stages than anthropoid primates, infants are still dependent on their mothers for the first few months of life. At 2 months of age, this infant ring-tailed lemur (*Lemur catta*) relies on its mother for nourishment, transportation, social interaction, and learning which foods to eat and which to avoid (photo by Renee Bauer).

contact for generally the first 6 months and up to 1 year (e.g., capuchins [*Cebus* spp.], bearded sakis [*Chiropotes satanas*]); however, the small-bodied callitrichines mature faster, while some of the larger atelines mature more slowly. Prosimians have the shortest period of infant dependence (generally 1–4 months), but this is still an extended life history stage when compared to similar-sized nonprimate mammals (see Fig. 35.3).

Mothering "Styles" and Infant "Temperaments"

Mothering styles are critical not only to whether or not an infant survives but also to the quality of its early life (Dolhinow 1991). As Dolhinow and DeMay (1982:392) note, "For the infant the mother is the filter to the world surrounding both of them, and her aptitudes or ineptitudes set a measure for the likelihood of infant success and even survival." Adult females can be permissive, restrictive, attentive, and detached in their mothering styles, and these need not be consistent modes throughout an individual female's life (Dolhinow and Krusko 1984). In langurs (*Semnopithecus*, *Presbytis*) and baboons (*Papio*), younger and more inexperienced mothers may be more restrictive, while older females may be more relaxed (Altmann 1980, Dolhinow and Krusko 1984, Dolhinow 1991).

Thus, whether or not an infant is a female's first offspring can have effects on the nature of the interactions (see Table 35.2). In blue monkeys (*Cercopithecus mitis*), infants spend more time in contact with primiparous versus multiparous mothers, and multiparous females tend to reject infants at a later age than first-time mothers (Förster and Cords 2002) In observations on the Mysore slender loris (*Loris tardigradus*), primiparous females "parked" their infants 2 weeks earlier than multiparous mothers (at 2 versus 4 weeks of

age). The difference suggests lack of experience for first-time mothers (Kappeler 1998) as 2 weeks is quite young compared to the majority of observations. Maestripieri (1998) found that previous maternal experience has an important influence on mothering styles in pig-tail macaques (*Macaca nemestrina*). He also investigated behavioral similarities in cross-fostered rhesus macaque females and their biological mothers and found similarities in rates of social contact and aggression between fostered infants and their biological mothers but no clear similarities between infants and their foster mothers. He suggests that there is heritable variation in female social behaviors that may be maintained by natural selection. This variability then leads to adaptation to different sociological niches within macaque populations (Maestripieri 2003).

It is important to remember that the infants themselves are not passive "blank slates" but are individuals that take an active role in their own social development. For example, in western lowland gorillas (*Gorilla gorilla gorilla*), infants encourage their mothers to share food, play, or follow them, not vice versa. Other than encouraging motor skills like walking (see Maestripieri 1995 and 1996 for discussion of this in captive rhesus and pig-tail macaques), gorilla mothers do not show strong encouragement of their infants in other contexts. This suggests that gorilla infants are more active than their mothers in generating situations that increase knowledge acquisition (Maestripieri et al. 2002). Similarly, infant chimpanzees have demonstrated that they are often more active than their mothers in initiating food-sharing bouts, especially those involving higher-quality foods (Ueno and Matsuzawa 2004).

Through recent hormonal studies in captivity, we are now gaining a clearer understanding of the role certain physiological stressors play in infant–mother interactions and temperament formation in young primates. A study of

Table 35.2   Four Variables and Their Possible Effects on Infant–Mother Relationships

VARIABLE	SCIENTIFIC NAME	COMMON NAME	COMMENTS	SOURCES
Primiparity	*Macaca mulatta*	Rhesus macaque	Primiparous mothers less restrictive than multiparous mothers	Berman (1984)
	Papio anubis	Olive baboon	Primiparous mothers rejected infants earlier than multiparous mothers	Ransom and Rowell (1972)
Multiparity	*Gorilla gorilla beringei*	Mountain gorilla	Multiparous mothers less restrictive than primiparous mothers	Fletcher (2001)
	Cercopithecus mitis	Blue monkey	Multiparous females tend to reject infants at a later age than first-time mothers	Förster and Cords (2002)
Dominance rank	*Papio cynocephalus*	Yellow baboon	Low-ranking females are much more restrictive of young infants than high-ranking females	Altmann (1980)
Sex of infant	*Gorilla gorilla beringei*	Mountain gorilla	Females invest more in their male infants	Fletcher (2001)

pig-tail macaques (*Macaca nemestrina*) demonstrates that maternal responsiveness develops during pregnancy and is probably hormonally regulated, instead of being solely determined by social and experiential factors (Maestripieri and Wallen 1995). In captive rhesus macaques (*Macaca mulatta*), Cleveland et al. (2003) found that low serotonin activity and high stress (measured by elevated plasma cortisol) are correlated with patterns of high maternal restrictedness in young adult females. Blood cortisol levels are a reliable indicator of psychosocial stress in primates, and infants do show a cortisol increase after being separated from their mothers (Nakamichi et al. 1994). However, the *social* context into which an infant is placed after maternal separation has an effect on how well they deal with that stress (Nakamichi et al. 1994, Bardi et al. 2003).

Detailed explorations of controlled variables are practically impossible in wild populations of primates; captive work on physiological processes is thus crucial in furthering our understanding of the bidirectional pathways between hormones and behavior (e.g., see Abbott et al. 2003; Sapolsky 1993, 1999, 2000). However, we must be conscious of discerning the effects of captivity and aware that these results may not always be reliable indicators of similar physiological processes in wild individuals.

There is much variability in care patterns among female primates and their offspring, and we know that general flexible strategies of maternal care are one of the hallmarks of primate evolution. Not only do mothering styles vary among individual females in a group, but each female may show different mothering techniques over the course of her life span. Thus, females have the ability to change their mothering styles throughout their lives based on temperament, individual experiences, and the dynamics of the social group in which they live (Dolhinow and Krusko 1984, Berman et al. 1997).

Infant Abuse and Infanticide by Mothers

The warm and tender image of a primate mother cuddling her vulnerable infant is shattered when we examine the maternal tendencies of some primate females. While relatively uncommon, some mothers ignore or abuse their offspring for reasons that are often unclear. The role of a hormonal influence has been examined by Maestripieri and Megna (2000a,b), who found the estradiol and progesterone profiles of abusive and nonabusive rhesus macaque (*Macaca mulatta*) mothers to be similar, with minor differences in other behavioral patterns between the two groups. Maestripieri (1999) also found that while abusive macaque mothers are highly attracted to infants in general, abuse is a phenomenon specific to their own offspring. He concluded that infant abuse is not an accidental by-product of infant handling or the result of maternal inexperience. Rather, he posits it is likely related to reproductive competition among lactating females (Maestripieri 1999). That is, reproductively experienced females may harass or kidnap other females' infants as a form of competition (Silk 1980; Maestripieri 1994, 1999). However, primate females also abuse their own infants due to inexperience or ecological factors or perhaps because something is wrong with the infant.

Infanticide by females toward their own offspring is uncommon in primates, but it occasionally occurs. The phenomenon is better known in some rodents, lagomorphs, mongoose, and lions (e.g., mongoose, Clutton-Brock et al. 1998, 1999; rodents, Blumstein 2000); but when this behavior occurs in any taxa, it is usually in response to an environmental stressor such as drought, shortage of food, or disruption of the habitat (e.g., Andrews 1998 for *Eulemur macaco*) or in response to offspring being directly disturbed. For example, if captive loris (*Loris tardigradus nordicus*) mothers are bothered during the first few days of infants being born, they will sometimes kill their offspring (Schulze and Meier 1995, Digby 2000). The majority of cases of female infanticide are by nonmother females and often for indeterminate reasons (see next section).

INTERACTIONS BETWEEN INFANTS AND NONMOTHER FEMALES

As most primates grow up in a dynamic social setting with multiple members, infants have the opportunity to interact with a variety of others from a young age. Young adult

Table 35.3   Characteristics of Allomaternal Care by Females in Several Primate Species

VARIABLE	SPECIES	SOURCES
Kinship an important factor	*Macaca fuscata*	Schino et al. (2003), Hiraiwa (1981)
	Alouatta seniculus	Crockett and Pope (1993)
	Papio cynocephalus	Cheney (1978)
	Cercopithecus aethiops	Lee (1983)
Kinship not an important factor	*Semnopithecus entellus*	Hrdy (1977), Dolhinow and Murphy (1982)
Young infants preferred	*Alouatta palliata*	Clarke (1990)
	Cercopithecus aethiops	Lee (1983)
Infants of high-ranking mothers receive more allocare (abuse rare)	*Cebus capucinus*	Manson, 1999
	Papio cynocephalus	Cheney, 1978
	Cercopithecus aethiops	Lee, 1983
Infants of low-ranking mothers receive more allocare (often with abuse)	*Macaca radiata*	Silk, 1980
Dominance rank is not an important factor in terms of a female's success with infant handling	*Papio cynocephalus*	Bentley-Condit *et al.*, 2001
	Cebus capucinus	Manson, 1999
Adult (multiparous) females do most allocare	*Lemur catta*	Jolly, 1966; Gould, 1992
Subadult or juvenile (nulliparous) females do most allocare	*Tarsius spectrum*	Gursky, 2000
	Macaca fuscata	Schino et al., 2003; Hiraiwa, 1981
	Cercopithecus aethiops	Lancaster, 1971
No difference in allocare rates of nulliparous vs. multiparous females	*Cebus capucinus*	Manson, 1999
	Erythrocebus patas	Nakagawa, 1995

females in particular are usually intrigued by infants and often sit close to, stare at, touch, and attempt to hold (with varied results) the latest additions to their social worlds.

Allomothering

Allomothering, or taking care of an infant that is not one's own, is common in several primate species; but the phenomenon varies widely (McKenna 1979a, 1987; see Table 35.3). Across taxa we find that nulliparous females are generally more attracted to infants and do more of the allomothering than multiparous females (Nicholson 1987; but see Manson 1999). Given that many species exhibit female philopatry, where females remain in their natal group for many years and often life, allomothering can function to provide younger females with experience taking care of a dependent infant. In species that are not female-bonded (e.g., spider monkeys [genus *Ateles*] and common chimpanzees [*Pan troglodytes*]), younger females can also benefit from allomothering before or after they transfer groups. In either scenario, allomothering provides the dependent infant with socialization opportunities other than those with its mother (see Fig. 35.4).

In some colobine monkeys, such as langurs (e.g., *Semnopithecus entellus*), infants are passed around from the first day of life, with no objections from the mother (Dolhinow and Murphy 1982, Dolhinow and Krusko 1984). Females groom, inspect genitals, nuzzle, touch, and huddle with the infants (Jay 1965, Dolhinow and Murphy 1982). Across

cercopithecine species, we generally see two patterns: patas (*Erythrocebus patas*) and guenon (*Cercopithecus* spp.) mothers permit extensive contact with young infants, while most baboon (*Papio* spp., *Theropithecus* spp.) and some macaque (*Macaca* spp.) mothers restrict early contact (Chism 2000). The genus *Macaca* in particular offers a striking view of allomothering diversity: Japanese (*M. fuscata*), long-tailed (*M. fascicularis*), rhesus (*M. mulatta*), and pig-tailed (*M. nemestrina*) macaque mothers restrict the amount of allocare provided by other females, while Barbary (*M. sylvanus*), lion-tailed (*M. silenus*), Tonkean (*M. tonkeana*), bonnet (*M. radiata*), crested (*M. nigra*), and moor (*M. maura*) macaque mothers are permissive and allocare is common, with many females handling and carrying infants from a young age (Thierry 2000, 2004; see also Chapter 13). Generally, across cercopithecines, early allomothering is most widespread in species with distinct seasonal breeding and relaxed social relationships among adult females (Chism 2000).

In New World primates, allomothering is taken to the extreme in the callitrichines (e.g., *Saguinus* spp., *Callithrix* spp.), titi monkeys (*Callicebus* spp.), and owl monkeys (*Aotus* spp.), with nonbreeding females spending up to 60% of their time caring for other females' offspring (e.g., Chapters 6 and 9). By comparison, we see lower rates of allomothering by female howler monkeys (*Alouatta* spp.) and squirrel monkeys (*Saimiri* spp.), yet in capuchins (e.g., *Cebus capucinus*), infants are often carried and sometimes nursed by nonmother females in the group (Manson 1999, Perry 1996, MacKinnon 2002).

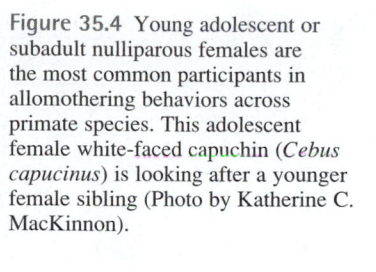

Figure 35.4 Young adolescent or subadult nulliparous females are the most common participants in allomothering behaviors across primate species. This adolescent female white-faced capuchin (*Cebus capucinus*) is looking after a younger female sibling (Photo by Katherine C. MacKinnon).

In prosimians, allomothering occurs in the diurnal (and thus better-studied) species (e.g., ring-tailed lemurs [*Lemur catta*], Jolly 1966, Gould 1992, Nakamichi and Koyama 2000; sifakas [*Propithecus verreauxi*], Richard 1976). In ring-tailed lemurs, other mothers alloparent each other's infants more than do subadult females (Gould 1992 and personal communication), which is contrary to the typical scenario found in most primate species. Allomothering is also an aspect of nocturnal prosimian behavioral patterns. In slender loris (*Loris tardigradus* and *L. lydekkerianus*), parked infants (see Fig. 35.5) are approached and played with by others in the group (Nekaris 2003), while in the spectral tarsier (*Tarsius spectrum*), female subadults share food, groom, play, and carry infants more than other age/sex classes (Gursky 2000; see also Chapter 5).

As with mothering styles, the variability we see in individual females that allomother can reflect experience, temperament, dominance status, the relationship with the infant's mother, and age (McKenna 1987; see Table 35.3). The proclivity for some females in a group to allomother, while others are uninterested, is sometimes difficult to explain. Relatedness may play a role in some species. For example, an older daughter may carry her mother's infant more than other females (e.g., red howlers [*Alouatta seniculus*], Crockett and Pope 1993), while a mother's female siblings may show more allomothering behavior than unrelated individuals (e.g., Japanese macaques [*Macaca fuscata*], Hiraiwa 1981; yellow baboons [*Papio cynocephalus*], Cheney 1978; vervets [*Cercopithecus aethiops*], Lee 1983). In other cases, unrelated females are given access to a female's infant with the same frequency as related females (e.g., hanuman langurs [*Semnopithecus entellus*], Hrdy 1977, Dolhinow and Krusko 1984).

Figure 35.5 In nocturnal prosimians, infants are often "parked" in trees while adults forage. This young slender loris (*Loris lydekkerianus nordicus*) is still nursing, does not yet move efficiently on its own, and will be parked for hours until its mother returns (photo by K. A. I. Nekaris).

Infants of dominant females sometimes receive more allomothering attention than infants of lower-ranking females (Cheney 1978, Lee 1983), while the opposite is occasionally true (Silk 1980). In yellow baboons (*Papio cynocephalus*), Bentley-Condit et al. (2001) found that the dominance rank of females did not seem to offer any privileges in terms of access to other infants; however, Silk et al. (2003) found that while female baboons appeared to be equally attracted to all infants, they had greater access to offspring of their relatives and subordinate females. In langur monkeys (*Semnopithecus entellus*), allomothering females show a wide variety of behavioral differences in their interactions with infants. Some prefer older or younger infants, and these preferences can change throughout an individual female's life (Dolhinow and Krusko 1984). Infants also have their preferences for certain females as caregivers, and degree of relatedness is unreliable when trying to predict the level of interest in infants by females and vice versa (Dolhinow and Krusko 1984). In Japanese macaques (*Macaca fuscata*), due to a social structure with distinct matrilines, infants are handled mainly by related individuals, usually juveniles and nulliparous subadult females (Schino et al. 2003). Maternal responses when infants are being handled are typically quite similar toward kin and nonkin alike but show a great deal of interindividual variability based on age, rank, experience, and relations with others in the group. In white-faced capuchins (*Cebus capucinus*), Manson (1999) suggests infant handling might test bonds between females. He found that females tend to handle the infants of females with which they groom and form coalitions more frequently. He also found little support for other hypotheses (e.g., "learning to mother").

If a dependent infant is rejected early on by its mother or loses her to predation, illness, or other factors, it may be adopted by another individual in the social group. In many species, adoptions are initiated and maintained by the infants themselves, but which female they choose as their new caregiver is not easily predicted or understood. For example, Dolhinow and DeMay (1982) found no preference for related females by infant langur monkeys, and neither was past social interaction an indicator of which female an infant would actively seek out if it lost its mother. In that study, infants chose adult females that were the least approached beforehand; a similar scenario is reported by Clarke et al. (1998) for mantled howler monkeys (*Alouatta palliata*).

What can account for the amazing variability we see in allocare patterns? One of the defining primate characteristics that has been selected for in evolution is a flexibility in behavioral actions. Such plasticity allows variable responses to a wide range of potential outcomes given unpredictable variation in group structure and composition, ecological factors, development patterns, and experiences in individual lives (McKenna 1979a,b, 1987). Yet, the following question remains: Why invest time and energy in someone else's offspring (Trivers 1972)? Several reasons have been proposed to explain why allomothering behavior evolved: (*1*) benefits to the mother, in terms of reproductive competition among females; (*2*) benefits to the infant, in terms of increased survival rates; and (*3*) benefits to the allomother, in terms of enhancing acquisition of maternal skills (Lancaster 1971, Quiatt 1979, Nicholson 1987, Mitani and Watts 1997). For example, Ross and MacLarnon (2000) found that among anthropoid primates, species with high frequencies of allocare grow rapidly and wean their infants at a younger age than species of the same body size with lower allocare levels. Thus, mothers can increase their reproductive output if they allow allocare of their infants (Mitani and Watts 1997).

However, there is still no consensus about the evolutionary significance of allomothering in primates. Silk (1999) reviewed the function of infant handling in bonnet macaques (*Macaca radiata*) and suggested that attraction to dependent infants may have evolved in response to the time period during which maternal care is most critical for infant survival. Females that are highly responsive to infants (usually) make good mothers, so "strong attraction" may be favored by natural selection. Her data on this species of macaque did not support the "learning to mother" hypothesis (i.e., age was not a determining factor of female interest), nor did they support the "reproductive competition" hypothesis (i.e., infant handling did not correspond to patterns of harassment). Rather, her data fit the hypothesis that allomothering is a by-product of selection for proper maternal care (Silk 1999).

In an examination of infant handling by non-mothers in three semi-free-ranging groups of Barbary macaques (*Macaca sylvanus*), Paul and Kuester (1996) also found poor evidence for a functional basis for the behavior. First, experience with infant handling by nulliparous females did not affect the survival rate of their own firstborn offspring. Second, infant handling frequency did not improve infant survival or female reproductive success. Third, while "aunting to death" (Hrdy 1977; and see following section) occurred, the hypothesis that infant handling serves to reduce the fitness of competitors was not supported (Paul and Kuester 1996). They instead found more support for the hypothesis that infant handling evolved as a nonadaptive by-product of selection for mother–offspring bonding. They came to this conclusion because rates of infant handling were highest among females that experienced early infant loss; females that had their own infants handled other infants significantly less often than females without dependent offspring, and both "aunting to death" and a successful adoption occurred regardless of kinship (Paul and Kuester 1996).

In white-faced capuchins (*Cebus capucinus*), Manson (1999) found that both nulliparous and parous females had similar rates of infant handling, which were unrelated to the dominance ranks of the mother and the handler. Rather, females tended to handle the infants of females with which they groomed and formed aggressive coalitions more often; and he suggests that infant handling might serve to test bonds between such females. He found no support for the "learning to mother" hypothesis. Nonetheless, there is evidence that young, nulliparous females do gain experience through allomothering (e.g., Lancaster 1971, Clarke 1978, Caine and Mitchell 1980, Hiraiwa 1981, Nicholson 1987, Paul and Kuester 1996). We also know from captive and wild populations that young females that do not have adequate contact with infants often display poor mothering abilities toward their own offspring (van Lawick-Goodall 1971, Altmann 1980, Suomi 1982). The fact that allomothering occurs far more commonly in nulliparous females suggests this behavior has some sort of developmental role (Lancaster 1971, Nicholson 1987).

A Note on Neonatal Coat Color

In most primates, infants are born with a coat color that is slightly different from the adult shade. Typically, infants are darker or more monochrome in pelage; but in the Old World subfamily Colobinae, we see some striking contrasts (i.e., "conspicuous natal coats," or CNCs). For example, in black and white colobus monkeys (*Colobus guereza*), infants are solid white, while in Francois' leaf monkeys (*Trachypithecus francoisi*), they are a bright pumpkin orange. Coat color typically changes to the adult pattern after a few months, but why is there such a conspicuous distinction at birth? Hrdy (1976) suggests that CNCs evolved to facilitate handling of infants by attracting other females. Similarly, Gouzoules and Gouzoules (2002) suggest that CNCs might be adaptive in situations where predation pressure is just high enough to warrant increased protection by mothers and allomothers. Treves (1997) proposes a different scenario: paternity concealment, such that promiscuous mating systems with multiple unrelated males in a group should favor selection of CNCs. Bright coloring could thereby act as a "paternity cloak" and serve to warn nongroup males from approaching (Treves 1997). However, he notes that "we do not yet have the necessary direct evidence to test whether infants are obscuring paternity or deterring the approach of infanticidal males" (Treves 2000:232–233). A lack of agreement on an adaptive reason for sharply distinct neonatal coat color remains (see also Ross and Regan 2000, Gouzoules and Gouzoules 2002.

Infant Abuse and Infanticide by Nonmother Females

Although abuse of infants by adult females is rare across primate taxa, it occurs more frequently than abuse by adult males (Digby 2000). This phenomenon takes a variety of forms: it can range from mild aggression to "aunting to death" to directed infanticide. Additionally, the behavioral hallmark of infants—insatiable curiosity—can get them into trouble. Even before infants are independent and off their mothers for extended periods of time, they can, and do, get underfoot of bigger, older animals in the group. Attention-grabbing activities such as juveniles playing, adults mating, and agonistic encounters are all worthy of investigation by infants. In many studies, infants have been observed to put themselves in harm's way and, in doing so, are at increased risk of being stepped on, swatted at, or bitten. Occasionally, they are the unfortunate recipients of misplaced or misdirected aggression when two or more individuals are fighting (Dolhinow 1999).

In some species, allomothering can take a turn for the worse, when dominant females take the infants of subordinates and keep them away from their lactating mothers for so long that they die from dehydration or starvation (Quiatt 1979, Kohda 1985, Thierry and Anderson 1986, Maestripieri 1994, Paul and Kuester 1996, Digby 2000). In rhesus macaques (*Macaca mulatta*), females are very attracted to young infants and often approach, hold, and groom immatures that are not theirs. In these scenarios, unskilled handling, abusive behavior, or long periods away from the mother that keep dependent youngsters from nursing are potentially significant risks for infants (Collins et al. 1984; Maestripieri and Carroll 1998, 2000; Digby 2000; Jovanovic and Gouzoules 2001). In 1 year of an intensive long-term field study of patas monkeys (*Erythrocebus patas*), three infants (of eight born) died from inept handling by females that "kidnapped" them from their mothers (Enstam et al. 2002).

Infanticide by nonmother female mammals is likely a more widespread phenomenon than male infanticide and has been documented in over 50 species, in five mammalian orders, with over 300 recorded cases (Digby 2000). In primates, one of the most bizarre cases of female infanticide and cannibalism was documented by Jane Goodall. A mother chimpanzee and her daughter killed and ate at least three (and up to seven inferred) infants of other females in the same community over a 2-year period (Goodall 1977, 1986). That it never happened again after the mother died and that something of this odd nature has never been witnessed at any of the other long-term chimpanzee study sites suggest a truly aberrant case. A different scenario occurs in the Callitrichinae, where female infanticide has also been reported (e.g., *Callithrix jacchus*, Digby 1995). In these cases, it appears that females that are pregnant or lactating perform most of the infanticides and are typically closely related to the targets (see Chapter 6). The (as yet untested) hypothesis of intense reproductive competition is offered as a possible explanation.

INTERACTIONS BETWEEN INFANTS AND ADULT MALES

Across primate species, infants and adult males interact socially, and the majority of interactions are affiliative in nature. Most primate males are not involved in direct caretaking or protective behavior toward infants, but they are usually quite tolerant of the immature members of their social groups. In many species, certain adult males allow certain infants greater access and participation in social interactions. There are likely proximate mechanisms such as temperament factors and relationship histories that determine the attractiveness of a certain adult male to a younger animal and vice versa (Clarke and Boinski 1995; Byrne and Suomi 1995, 1998). Additional theoretical explanations for such interactions will be discussed below.

The Range of Male Involvement

While primate infants are in continual contact with their mothers from the moment they are born, their relations with adult males are, with a few exceptions, much less intensive and consistent. However, infants and adult males do interact, and if we look across the ~250 species in the Primate order (Groves 2001), we see a great range in social patterns (see

Whitten 1987). On the neutral and affiliative side of the spectrum, we see ignorance, tolerance, occasional interest, active interest, and direct caretaking of infants. On the agonistic side, we see mild and harsh aggression, abuse, and infanticide. By far the most common patterns are for adult males to tolerate or take occasional to active interest in young individuals within their social groups (see Table 35.4). Unfortunately, we are still lacking such behavioral data for many wild species.

Among the apes, mountain gorilla males are extremely tolerant and patient with infants in their groups (Fossey 1983), and silverbacks will even nest with youngsters that have lost their mothers (Stewart 2001). Chimpanzee males will occasionally play with immatures and will also adopt orphaned infants (Goodall 1977, 1983). Gibbon males (*Hylobates*) do not perform the intense caretaking duties that characterize some primate males living in pairbonded associations, but they are quite tolerant of infants, including those potentially sired via extrapair copulations or in polyandrous social structures (e.g., white-handed gibbons [*Hylobates lar*], Brockelman et al. 1998, Reichard 2003, and personal communication). Reichard (personal communication) notes that he has never observed an adult male display agonistic behavior toward an immature less than 5 years of age. Siamang males will occasionally carry infants, while males of the other gibbon species generally do not (Chivers 1974). However, all species display male parental care by way of playing, grooming, and protection (Brockelman 2004).

Old World monkey (e.g., *Papio*, *Macaca*) males show varying levels of interest in infants and often play with them and provide transportation. Barbary macaque (*M. sylvanus*) males in particular show strong interest in infants (kin and nonkin alike), and males are tolerated well by adult females in such contexts. Depending on the relationship with the infant's mother, a subadult or adult male may be allowed to take and hold or carry an infant for extended periods of time (Taub 1984; see Fig. 35.6). Male care of offspring in these macaques is not strongly tied to paternity, and the most active male caregivers tend to be subadults that mate with adult females less often (Taub 1984) (see below, *Agonistic Buffering and Triadic Interactions*, for continued discussion of Old World monkeys in this context).

In New World monkeys, brown capuchin (*Cebus apella*) infants older than 2 months are often left with the dominant male, which is tolerant as they play or huddle next to him (Valenzuela 1994). White-faced capuchin (*C. capucinus*) males are also highly tolerant of infants (see Fig. 35.7) and occasionally carry, play with, and sit in contact with immatures (MacKinnon 2002). Interactions between infant and adult male howler monkeys have been documented (e.g., mantled howlers [*Alouatta palliata*], Clarke et al. 1998), and generally, howler males show a passing interest in infants and sometimes "kidnap" them for short periods of time until the infant cries for its mother and is retrieved. In the earliest report of wild howler monkey behavior, Carpenter (1934:128) described infants watching, following, crawling on, and

playing with adult males and noted "In general males behave indifferently to young animals, but they may assist in retrieving fallen infants, protect them from predatory animals, and in unusual situations care for an infant in somewhat the same manner as a mother." Squirrel monkey (*Saimiri* spp.) infants rarely, if ever, interact socially with adult males in the group (Fragaszy et al. 1991). Infants travel independently early on and might not have as much exposure to adult males via their mothers. Squirrel monkeys also spend more time foraging alone than closely related species (e.g., *Cebus* spp.), and social interaction rates overall are quite low.

In sharp contrast, owl monkey (*Aotus*) and titi monkey (*Callicebus*) males are actively involved in parental care (Fragaszy et al. 1982; Mendoza and Mason 1986a,b; Dixson 1994; Wright 1984; see Chapter 9). They also play with, groom, and share food with infants (e.g., Rotundo et al. 2000). In owl monkeys, independent infants return to the adult male if frightened or disturbed (Dixson 1994), and infants are much more attached to their putative fathers than to their mothers (Fernandez-Duque 2004; see also Chapter 9). Similarly, in titi monkeys (*Callicebus*), infants seem to be much more "tuned into" their male caregivers and show increased cortisol levels (indicating stress) upon separation from fathers but not mothers (Mason and Mendoza 1998, Hoffman 1998). Why has such intense male care evolved in these primates? The question remains largely unanswered, although Fernandez-Duque (Chapter 9) suggests that male care in these species may function as a mating effort, especially as other hypotheses (e.g., creating increased foraging opportunities for females) are largely unsupported. Concomitantly, while the exact mechanisms are unknown, paternal care might have influenced the evolution of social monogamy at different stages and to varying degrees (Wittenberger and Tilson 1980, Fuentes 2000, Reichard 2003).

Finally, it bears mention that in some nocturnal prosimians, long believed (incorrectly) to be completely solitary in the wild, males also interact socially with immatures. For example, Nekaris (2003) documented males approaching, grooming, and playing with parked infants throughout the night in three subspecies of slender loris (*Loris lydekkerianus lydekkerianus*, *L. l. nordicus*, and *L. tardigradus tardigradus*) in Sri Lanka and India. It is unknown whether or not these males were related to the infants. The results from this study are an excellent example of how additional fieldwork on the lesser-known members of the Primate order will help to expand our understanding of the social complexities in all gregarious species.

Callitrichinae and Cooperative Infant Care

The small-bodied marmosets and tamarins in the subfamily Callitrichinae (see Chapter 6) are perhaps best known for the extensive care males give to infants. About 80%–90% of Callitrichinae births are twin offspring, and young may be born twice a year (there is no lactational amenorrhea in most species). Infants are relatively large, with both twins collectively weighing 14%–25% of the mother's weight at birth

Table 35.4   Range of Infant and Adult Male Interactions in Several Primate Species (Data Primarily from Wild Studies)

	SCIENTIFIC NAME	COMMON NAME	TYPE[1]	COMMENTS	SOURCES
Apes	*Gorilla gorilla beringei*	Mountain gorilla	MA,T	Infants gain closer proximity to silverback as weaning occurs; silverback increases affiliative behavior toward orphaned young	Stewart (2001)
	Pan troglodytes schweinfurthii	Common chimpanzee	MA,T	Males are generally tolerant of young	Goodall (1977, 1986)
	Pongo pygmaeus	Orangutan	A/T	Males rarely interact with infants	Galdikas (1981b)
	Hylobates lar	White-handed gibbon	MA,T	Males generally interact very little with young infants but are tolerant of them; males will play with older infants but do not carry them	Brockelman *et al.* (1998), Reichard, (2003), Bartlett (2003)
	Hylobates syndactylus	Siamany	MA	Males will sometimes carry young	Chivers (1974), Dielentheis *et al.* (1991)
Old World monkeys	*Semnopithecus (formerly Presbytis) entellus*	Indian langur monkey	MA,T,U/A	Wide range of behavioral interactions reported across several sites (one-male vs. multimale groups)	Boggess (1982), Jay (1965), Hrdy (1977), Borries *et al.* (1999)
	Presbytis potenziani	Mentawai leaf monkey	T,A/I	Males generally do not interact socially with infants	Fuentes, 1995
	Erythrocebus patas	Patas monkey	MA,T	No protective behavior of infants but some play	Hall, 1965; Enstam *et al.*, 2002
	Papio cynocephalus	Yellow baboon	SA,MA,T, A/I,UA	Much variation occurs across groups and sites, with males showing broad range of affiliative behaviors with infants; infants also used by males for agonistic buffering or in more mutualistic interactions	Altmann, 1980; Stein, 1984; Strum, 1984; Stein and Stacey, 1981
	Papio anubis	Olive baboon	SA,MA,T, A/I,UA	Variation occurs between one-male and multimale groups; males often show broad range of affiliative behaviors with infants; infants also used by males for agonistic buffering or in more mutualistic interactions	Strum, 1984; Smuts, 1985
	Theropithecus gelada	Gelada baboon	MA,U/A	Infants used by males for agonistic buffering or in more mutualistic interactions	Dunbar, 1984
	Macaca fuscata	Japanese macaque	SA,MA,T	Males may form special protective relations with particular infants	Itani, 1959
	Macaca sylvanus	Barbary macaque	SA,U/A	Males often groom, handle, and carry infants; infants also used by males in mutualistic male-male interactions (typically without "abuse")	Taub (1980, 1984)
New World monkeys	*Alouatta palliata*	Mantled howler monkey	T,A/I	Some males show a passing interest in infants but are generally indifferent	Clarke *et al.*, 1998
	Cebus apella	Brown capuchin	T	Males are tolerant of young infants left near them	Valenzuela, 1994
	Cebus capucinus	White-faced capuchin	MA,T	Males occasionally carry and play with infants	MacKinnon, 2002
	Callicebus moloch	Titi monkey	IC	Strong attachment between infants and male in group (putative father)	Wright, 1984; Mendoza and Mason, 1986a
	Aotus trivirgatus	Owl monkey	IC	Male carries infant most of the time; much play, grooming, and food sharing	Wright, 1984, 1994; Fernandez-Duque, (2004), (Chapter 9, this volume)
	Saimiri sciureus	Squirrel monkey	A/I	Infants and adult males rarely interact	Fragaszy *et al.*, 1991; Boinski and Cropp, 1999
	Saguinus fuscicollis	Saddle-back tamarin	IC	Males provide extensive allocare	Goldizen, 1987
	Saguinus oedipus	Cotton-top tamarin	IC	Males provide extensive allocare	Savage *et al.*, 1996
	Pithecia pithecia	White-faced saki	MA	Interest in infants by adult males increases as the infant matures; males play and share food with 4-month-old infants	(Chapter 8, this volume)
Prosimians	*Lemur catta*	Ring-tailed lemur	SA,MA	Most males show interest in infants and occasionally participate in allocare; dominance rank or ageclass of male does not affect rates of affiliative behavior with immatures	Gould, 1997
	Eulemur mongoz	Mongoose lemur	IC	Adult males frequently carry the infant after it is 2 weeks old	Curtis and Zaramody, 1997
	Eulemur rubriventer	Red-bellied lemur	IC	Males have been seen carrying infants up to 100 days	Overdorff, 1993; Mittermeier et al., 1994
	Loris lydekkerianus	Slender loris	MA	Males occasionally approach, groom, and play with parked infants	Nekaris, 2003
	Tarsius spectrum	Spectral tarsier	MA	Adult and subadult males occasionally groom and play with infants	Gursky, 2000

[1] IC, intense caretaking: adult males are involved in the majority of caretaking behaviors (excluding nursing), maintain close proximity to infants, and spend much time with them; SA, strong affiliation: males are involved with some caretaking behaviors and often spend time with infants and maintain proximity to them; MA, mild affiliation: males are involved with very few or no direct caretaking behaviors, yet on occasion socially interact with infants; T, tolerance: males do not actively engage infants in social interactions but do not avoid them if approached, and agonistic behaviors occur; A/I, avoidance/indifference: males do not respond to infants, do not engage them in social interactions, and might actively avoid them—no agonistic behaviors occur; U/A, use/abuse: males engage infants in mutualistic exchanges with other males, use them for agonistic buffering, or harm them (excluding infanticide.)

Figure 35.6 In many species, male primates play an active role in the development of infants. This subadult male Barbary macaque (*Macaca sylvanus*) carries a young infant in a species-typical display of male allocare (photo by Katherine C. MacKinnon).

Figure 35.7 Even in species where adult males are not actively involved in direct caregiving, they still play a role in the social and physical development of immatures. Here, an adult male white-faced capuchin (*Cebus capucinus*) carries an unrelated dependent infant. (photo by Katherine C. MacKinnon).

and up to 50% by the time they are weaned (Goldizen 1987). Only one adult female in a group breeds, and there are varying suppression mechanisms at work on the reproductive endocrinology of the other females (Abbott and Hearn 1978, Abbott et al. 1993; see also Chapter 6). Generally speaking, the mother nurses the infants but then passes them on to other members of the group. Thus, multiple individuals take care of infants, and adult males play an active role in the physical and social development of young via carrying, playing, and sharing food (Goldizen and Terborgh 1986, Goldizen 1987; see also Chapter 6). This caregiving system also helps with infant survival (see Bales et al. 2000 for review). For example, in the moustached tamarin monkey, *Saguinus mystax*, Garber et al. (1984) found a statistically significant relationship between the number of adult males in a group and surviving offspring.

Interestingly, there are high levels of the hormone prolactin in "helping" males in many bird and rodent species as well as in several primates (e.g., *Callithrix jacchus* and *Saguinus oedipus*) (Ziegler 2000). Prolactin, a protein hormone secreted from the anterior pituitary gland, is usually associated with stimulating breast development, milk production, and maternal behavior in females but is also produced in males. In a study of paternal care in three species of New World monkeys, the titi monkey (*Callicebus cupreus*), common marmoset (*Callithrix jacchus*), and Goeldi's monkey (*Callimico goeldii*), species differences in the patterns of prolactin levels were apparent and reflected varying paternal roles (Schradin et al. 2003). Mota and Sousa (2000) note that extra prolactin is produced in response to physical contact, may be associated with carrying behavior, and may be related to learning parental skills in some New World monkeys.

When examining the evolution of paternal care in primates, the reproductive effort of both males and females has to be taken into account. Key and Aiello (2000) note that when male reproductive costs are less than female reproductive costs, males cooperate with females even when females do not reciprocate. They suggest that such nonreciprocal altruism is akin to male investment in a female and her offspring.

Agonistic Buffering and Triadic Interactions

Baboons and some macaques are well known for their "use" of infants in a wide variety of contexts, which can involve a tense situation between a male and an infant (e.g., gelada baboons [*Theropithecus gelada*], Dunbar 1984), agonistic situations between two or more adult males (e.g., yellow baboons [*Papio cynocephalus*], Strum 1984), or a more mutualistic interaction that strengthens bonds between males and/or between a male and an infant (e.g., Barbary macaques [*Macaca sylvanus*], Taub 1980; yellow baboons, Strum 1984). In the literature, this sort of interaction is labeled variously as "infant use" (Strum 1984), "triadic male–infant interactions" (Taub 1980), and "agonistic buffering" (Deag and Crook 1971). If a high-ranking baboon male threatens a lower-ranking male, that male might approach a female with an infant, pick up the infant, carry it toward the threatening male, and sit close to him while still holding the infant (Strum 1984). The threatening male then typically leaves; the infant is usually compliant during this exchange. In Barbary macaques, two males will often face each other with an infant in between them, clutch at each other while teeth-chattering, and then chatter to the infant. They will also nuzzle, groom, inspect, and touch the infant, which usually shows no distress at being intensely examined (Taub 1980, 1984). In both baboons and macaques, if infants do show signs of stress, mothers (and others) will attack and chase off the male (Stein 1984, Strum 1984).

Male choice of an infant is specific and not random and does not reflect paternity. In fact, several studies have documented no male preference for infants that are likely their offspring (e.g., olive baboons [*Papio anubis*], Strum 1984; yellow baboons, Stein 1984). These males do not actively avoid their own offspring, nor do they actively seek them. In Barbary macaques, male handling of an infant does not result in increased mating opportunities with the infant's mother (Taub 1980); the benefits males gain seem to be in the form of male–male negotiations, interactions, and alliances (Taub 1980, 1984; Riechelmann et al. 1994). Across species there is a strong affiliation between the male and the infant before, during, and after such interactions (e.g., Dunbar 1984); males that routinely care for infants increase the chance that those infants will cooperate in future triadic interactions (Strum 1984, Stein 1984).

In some species, males that interact with infants gain access to females for future consortships and mating opportunities or merely strengthen already existing relationships with certain females (e.g., olive baboons [*Papio anubis*], Smuts 1985). In ring-tailed lemurs (*Lemur catta*), males might gain benefits from interacting with immatures in the form of achieving more spatial centrality in the social group, which can then lead to increased chances to develop affiliative relationships with females (Gould 1997). Young lemurs can also benefit from these interactions in the form of

enhanced predator detection, allocare, and increased opportunities to develop social skills (Gould 1997).

Adult males sometimes play a role in protecting infants; and in one group of hanuman langurs (*Semnopithecus entellus*), Borries et al. (1999) found that only genetic fathers, or males that were in the group when an infant was conceived, were seen to protect infants. However, in many species, there are reports of males adopting and/or caring for infants that are not their offspring (e.g., chimpanzees [*Pan troglodytes*], Goodall 1977, 1986; Japanese macaques [*Macaca fuscata*], Wolfe 1981; owl monkeys [*Aotus*], see Chapter 9).

Infant interactions with adult males occur across a wide range of contexts. Possible reasons for this type of social interaction are plural and not mutually exclusive. Variables such as adult male dominance rank and the infant's familiarity with certain males seem to be important components in the formation of such relationships. Intense bonds develop between specific infants and specific males in many species, and kinship is not always a consistent predictor of the degree of affiliative interactions. The male's relationship with the infant's mother and his previous history with the infant are more reliable indicators of the degree of contact and interaction the two will share.

Infant Abuse and Infanticide by Males

Overall, the rates of abuse toward infants by adult males are extremely low. As mentioned, agonistic buffering may occur in some species in order to diffuse a tense situation or so that an individual can gain favor with higher-ranking individuals in the group. Such use of infants usually does not result in death, although infants may be indirectly injured. Generally, when fatal aggression by males toward infants is inferred, it is within the context of infanticide.

The majority of reported cases of male infanticide are concentrated in three mammalian orders: carnivores, rodents, and primates (see van Schaik 2000a for review). In many primate species, infanticide has been reported and described as an evolved reproductive strategy for adult males (e.g., hanuman langurs [*Semnopithecus entellus*], Hrdy 1974, 1977, 1979; Hrdy and Hausfater 1984; Sommer 1987, 1994; Borries et al. 1999; Borries and Koenig 2000; Koenig and Borries 2001; chacma baboons [*Papio hamadryas ursinus*], Palombit et al. 2000; howler monkeys [genus *Alouatta*], Crockett 2003; white-faced capuchins [*Cebus capucinus*], Fedigan 2003; and for general review, van Schaik and Janson 2000). In certain cercopithecoid primates (e.g., *Semnopithecus entellus*), infanticide is assumed to be so common (or was common in the recent past) that other aspects of sociality, including the social system itself, have evolved in response to the potential of infanticide occurring (van Schaik and Kappeler 1997, Kappeler and van Schaik 2002).

The sexual selection hypothesis has become the dominant theoretical framework for interpreting infant deaths caused by males in primate groups. This hypothesis is based on the assumption that males practice infanticidal strategies to enhance their reproductive fitness. Other adaptive reasons for infanticide have been proposed (e.g., resource competition, exploitation, parental manipulation; Hrdy 1979). Nonadaptive hypotheses (e.g., social pathology, generalized aggression) have also been put forth to explain infant deaths resulting from male behavior. These alternatives view the majority of infanticides as epiphenomena of heightened male aggression toward other group members (e.g., Curtin and Dolhinow 1978; and see Bartlett et al. 1993 for review). The importance of considering the context of such an event when inferring adaptive reasons (e.g., sexual selection hypothesis) for infanticide by adult males cannot be overstated as the number of infanticides actually witnessed (over several decades and hundreds of thousands of hours of field observations) is still quite low (~70). In some directly observed cases, it is clear that the aberrant behavior of the infant's low-ranking mother placed it unequivocally in harm's way (e.g., MacKinnon 2003). In other cases, where cannibalism has followed infanticide, Melo et al. (2003) hypothesize that this could be a result of competition for scarce resources and the need for animal protein, brought on by forest degradation.

Infanticide by adult males remains a subject of debate in primatology (Bartlett et al. 1993, Sussman et al. 1995, Fuentes 1999, Dolhinow 1999, Sommer 2000, van Schaik 2000b). The main crux of the disagreement is not whether males occasionally kill infants—they do—but whether this behavior is an evolved reproductive strategy: a behavior (or set of behaviors) that has afforded certain males higher reproductive success over the course of their lifetimes and that has been selected for via the mechanism of natural selection (see Cheverud 2004 for a discussion of the natural selection of heritable variation and its application to social behaviors).

SUMMARY

From birth, a young primate's world is complex and constantly changing. Across the Primate order we see a remarkable variability in the characteristics of behavioral interactions between infants and the adult members of their social environments. We also see a degree of plasticity in development not found in any other group of mammals. A prolonged period of physical and emotional dependence by infants on others and a multifaceted social landscape result in an amazing array of infant–adult primate behaviors. With the ongoing development of molecular techniques more applicable in natural settings, future genetic work will elucidate some of the underlying kinship factors that may or may not be at play in the formation and maintenance of immature–adult relationships. We desperately need long-term fieldwork on young primates in their natural environments. Such data coupled with results from captive work, where variables can be more easily manipulated, will give us a comprehensive understanding of the complexities inherent in young primate lives.

REFERENCES

Abbott, D. H., Barrett, J., and George, L. M. (1993). Comparative aspects of the social suppression of reproduction in female marmosets and tamarins. In: Rylands, A. B. (ed.), *Marmosets and Tamarins: Systematics, Behaviour, and Ecology*. Oxford University Press, Oxford. pp. 152–163.

Abbott, D. H., and Hearn, J. P. (1978). Physical, hormonal and behavioural aspects of sexual development in the marmoset monkey (*Callithrix jacchus*). *J. Reprod. Fertil.* 53:155–166.

Abbott, D. H., Keverne, E. B., Bercovitch, F. B., Shively, C. A., Mendoza, S. P., Saltzman, W., Snowdon, C. T., Ziegler, T. E., Banjevic, M., Garland, T., Jr., and Sapolsky, R. M. (2003). Are subordinates always stressed? A comparative analysis of rank differences in cortisol levels among primates. *Horm. Behav.* 43: 67–82.

Altmann, J. (1980). *Baboon Mothers and Infants*. Harvard University Press, Cambridge, MA.

Andrews, J. (1998). Infanticide by a female black lemur, *Eulemur macaco*, in disturbed habitat on Nosy Be, north-western Madagascar. *Folia Primatol.* 69(suppl. 1):14–17.

Baldwin, J. D. (1969). The ontogeny of social behavior of squirrel monkeys (*Saimiri sciureus*) in a seminatural environment. *Folia Primatol.* 11:35–79.

Baldwin, J. D., and Baldwin, J. (1972). The ecology and behavior of squirrel monkeys (*Saimiri oerstedi*) in a natural forest in western Panama. *Folia Primatol.* 18:161–184.

Baldwin, J. D., and Baldwin, J. (1973). Interactions between adult female and infant howling monkeys (*Alouatta palliata*). *Folia Primatol.* 20:27–71.

Bales, K., Dietz, J., Baker, A., Miller, K., and Tardiff, S. D. (2000). Effects of allocare-givers on fitness of infants and parents in callitrichid primates. *Folia Primatol.* 71:27–38.

Bardi, M., Shimizu, K., Barrett, G. M., Borgognini-Tarli, S. M., and Huffman, M. A. (2003). Peripartum cortisol levels and mother–infant interactions in Japanese macaques. *Am. J. Phys. Anthropol.* 120:298–304.

Bartlett, T. Q. (2003). Intragroup and intergroup social interactions in white-handed gibbons. *Int. J. Primatol.* 24:239–259.

Bartlett, T. Q., Sussman, R. W., and Cheverud, J. M. (1993). Infant killing in primates: a review of observed cases with specific reference to the sexual selection hypothesis. *Am. Anthropol.* 95:958–990.

Bentley-Condit, V. K., Moore, T., and Smith, E. O. (2001). Analysis of infant handling and the effects of female rank among Tana River adult female yellow baboons (*Papio cynocephalus cynocephalus*) using permutation/randomization tests. *Am. J. Primatol.* 55:117–130.

Berman, C. M. (1982a). The social development of an orphaned rhesus infant on Cayo Santiago: male care, foster mother–orphan interaction and peer interaction. *Am. J. Primatol.* 3:131–141.

Berman, C. M. (1982b). The ontogeny of social relationships with group companions among free-ranging infant rhesus monkeys. I. Social networks and differentiation. *Anim. Behav.* 30:149–162.

Berman, C. M. (1984). Variation in mother–infant relationships: traditional and nontraditional factors. In: Small, M. F. (ed.), *Female Primates: Studies by Women Primatologists*. Alan R. Liss, New York. pp. 17–36.

Berman, C. M. (2004). Developmental aspects of kin bias in behavior. In: Chapais, B., and Berman, C. M. (eds.), *Kinship and Behavior in Primates*. Oxford University Press, New York. pp. 317–346.

Berman, C. M., Rasmussen, K. L. R., and Suomi, S. J. (1997). Size, infant development and social networks in free-ranging rhesus monkeys. *Anim. Behav.* 53:405–421.

Biro, D., Inoue-Nakamura, N., Tonooka, R., Yamakoshi, G., Sousa, C., and Matsuzawa, T. (2003). Cultural innovation and transmission of tool use in wild chimpanzees: evidence from field experiments. *Anim. Cogn.* 6:213–223.

Blum, D. (2002). *Love at Good Park: Harry Harlow and the Science of Affection*. Perseus Publishing, Cambridge.

Blumstein, D. T. (2000). The evolution of infanticide in rodents: a comparative analysis. In: van Schaik, C. P., and Janson, C. H. (eds.), *Infanticide by Males and Its Implications*. Cambridge University Press, Cambridge. pp.

Boggess, J. (1982). Immature male and adult male interactions in bisexual langur (*Presbytis entellus*) troops. *Folia Primatol.* 38:19–38.

Boinski, S., and Cropp, S. J. (1999). Disparate data sets resolve squirrel monkey (*Saimiri*) taxonomy: implications for behavioural ecology and biomedical usage. *Int. J. Primatol.* 20:237–256.

Borries, C., and Koenig, A. (2000). Infanticide in hanuman langurs: social organization, male migration, and weaning age. In: van Schaik, C. P., and Janson, C. H. (eds.), *Infanticide by Males and Its Implications*. Cambridge University Press, Cambridge. pp. 99–122.

Borries, C., Launhardt, K., Epplen, C., Epplen, J. T., and Winkler, P. (1999). Males as infant protectors in hanuman langurs (*Presbytis entellus*) living in multimale groups—defense pattern, paternity and sexual behaviour. *Behav. Ecol. Sociobiol.* 46:350–356.

Box, H. O. (1984). *Primate Behaviour and Social Ecology*. Chapman and Hall, London.

Box, H. O. (1994). Comparative perspectives in primate social learning: new lessons for old traditions. In: Roeder, J. J., Thierry, B., Anderson, J. B., and Herrenschmidt, N. (eds.), *Current Primatology. Social Development, Learning and Behaviour*, vol. II. Université Louis Pasteur, Strasbourg. pp. 321–327.

Brockelman, W. Y. (2004). Ecology and the social system of gibbons. In: Galdikas, B. M. F., Briggs, N., Sheeran, L. K., Shapiro, G. L., and Goodall, J. (eds.), *All Apes Great and Small*, vol. II. Kluwer Academic/Plenum Publishers, New York.

Brockelman, W. Y., Reichard, U., Treesucon, U., and Raemaekers, J. J. (1998). Dispersal, pair formation and social structure in gibbons (*Hylobates lar*). *Behav. Ecol. Sociobiol.* 42:329–339.

Byrne, G., and Suomi, S. J. (1995). Development of activity patterns, social interactions, and exploratory behavior in infant tufted capuchins (*Cebus apella*). *Am. J. Primatol.* 35:255–270.

Byrne, G., and Suomi, S. J. (1998). Relationship of early infant state measures to behavior over the first year of life in the tufted capuchin monkey (*Cebus apella*). *Am. J. Primatol.* 44:43–56.

Caine, N. G., and Mitchell, G. (1980). Species differences in the interest shown in infants by juvenile female macaques (*Macaca radiata* and *M. mulatta*). *Int. J. Primatol.* 1:323–332.

Carpenter, C. R. (1934). A field study of the behavior and social relations of howling monkeys. *Comp. Psychol. Monogr.* 10:1–168.

Carpenter, C. R. (1935). Behavior of red spider monkeys in Panama. *J. Mammal.* 16:171–180.

Cheney, D. L. (1978). Interactions of immature male and female baboons with adult females. *Anim. Behav.* 26:389–408.

Chevalier-Skolnikoff, S. (1977). A piagetian model for describing and comparing socialization in monkey, ape and human infants. In: Chevalier-Skolnikoff, S., and Poirier, F. E. (eds.), *Primate Bio-Social Development: Biological, Social, and Ecological Determinants*. Garland Publishing, New York. pp. 159–187.

Cheverud, J. M. (2004). Darwinian evolution by the natural selection of heritable variation: definition of parameters and application to social behaviors. In: Sussman, R. W., and Chapman, A. R. (eds.), *Origins and Nature of Sociality*. Aldine de Gruyter, Hawthorne. pp. 140–157.

Chism, J. (2000). Allocare patterns among cercopithecines. *Folia Primatol.* 71:55–66.

Chivers, D. J. (1974). The siamang in Malaya: a field study of a primate in tropical rain forest. In: Contributions to Primatology, Vol. 4. Karger, Basel.

Clarke, A. S., and Boinski, S. (1995). Temperament in nonhuman primates. *Am. J. Primatol.* 37:103–125.

Clarke, M. R. (1978). Social interactions of juvenile female bonnet monkeys, *Macaca radiata*. *Primates* 19:517–524.

Clarke, M. R. (1990). Behavioral development and socialization of infants in a free-ranging group of howling monkeys (*Alouatta palliata*). *Folia Primatol.* 54:1–15.

Clarke, M. R., Glander, K. E., and Zucker, E. L. (1998). Infant–nonmother interactions of free-ranging mantled howlers (*Alouatta palliata*) in Costa Rica. *Int. J. Primatol.* 19:451–472.

Cleveland, A., Westergaard, G. C., Trenkle, M. K., and Higley, J. D. (2003). Physiological predictors of reproductive outcome and mother–infant behaviors in captive rhesus macaque females (*Macaca mulatta*). *Neuropsychopharmacology* 29:891–910.

Clutton-Brock, T. H., Brotherton, P. N. M., Smith, R., McIlrath, G. M., Kansky, R., Gaynor, D., O'Riain, M. J., and Skinner, J. D. (1998). Infanticide and expulsion of females in a cooperative mammal. *Proc. R. Soc. Lond. B* 265:2291.

Clutton-Brock, T. H., Gaynor, D., McIlrath, G. M., Maccoll, A. D. C., Kansky, R., Chadwick, P., Manser, M., Skinner, J. D., and Brotherton, P. N. M. (1999). Predation, group size and mortality in a cooperative mongoose, *Suricata suricatta*. *J. Anim. Ecol.* 68:672–683.

Collins, D. A., Busse, C. D., and Goodall, J. (1984). Infanticide in two populations of savanna baboons. In: Hrdy, S. B., and Hausfater, G. (eds.), *Infanticide: Comparative and Evolutionary Perpectives*. Aldine de Gruyter, New York. pp. 193–215.

Crockett, C. M. (2003). Re-evaluating the sexual selection hypothesis for infanticide by *Alouatta* males. In: Jones, C. B. (ed.), *Selection and Reproductive Competition in Primates: New Perspectives and Directions*. American Society of Primatologists, Norman, OK. pp. 327–365.

Crockett, C. M., and Pope, T. R. (1993). Consequences of sex differences in dispersal for juvenile red howler monkeys. In: Periera, M. E., and Fairbanks, L. A. (eds.), *Juvenile Primates: Life History, Development, and Behavior*. Oxford University Press, New York. pp. 104–118.

Curtin, R., and Dolhinow, P. (1978). Primate social behavior in a changing world. *Am. Sci.* 66:468–475.

Curtis, D. J., and Zaramody, A. (1997). Monogamy and mate monopolization by females in *Eulemur mongoz* [abstract]. *Primate Rep.* 48:16–17.

Deag, J. M., and Crook, J. H. (1971). Social behaviour and "agonistic buffering" in the wild Barbary macaque *Macaca sylvana* L. *Folia Primatol.* 15:183–200.

de Waal, F. B. M. (1997). *Bonobo: The Forgotten Ape*. University of California Press, Berkeley.

Dielentheis, T. F., Zaiss, E., and Geissmann, T. (1991). Infant care in a family of siamangs (*Hylobates syndactylus*) with twin offspring at Berlin Zoo. *Zoo Biol.* 10:309–317.

Digby, L. J. (1995). Infant care, infanticide, and female reproductive strategies in polygynous groups of common marmosets (*Callithrix jacchus*). *Behav. Ecol. Sociobiol.* 37:51–61.

Digby, L. J. (2000). Infanticide by female mammals: implications for the evolution of social systems. In: van Schaik, C. P., and Janson, C. H. (eds.), *Infanticide by Males and Its Implications*. Cambridge University Press, Cambridge. pp. 423–446.

Dixson, A. F. (1994). Reproductive biology of the owl monkey. In: Baer, J. F., Weller, R. E., and Kakoma, I. (eds.), Aotus: *The Owl Monkey*. Academic Press, San Diego. pp. 113–132.

Dolhinow, P. (1984). The primates: age, behavior, and evolution. In: Kertzer, D. I., and Keith, J. (eds.), *Age and Anthropological Theory*. Cornell University Press, Ithaca. pp. 45–81.

Dolhinow, P. (1991). Tactics of primate immaturity. In: Robinson, M. H., and Tiger, L. (eds.), *Man and Beast Revisited*. Smithsonian Institution Press, Washington DC. pp. 139–157.

Dolhinow, P. (1999). A mystery: explaining behavior. In: Strum, S. C., and Lindburg, D. G. (eds.), *The New Physical Anthropology: Science, Humanism, and Critical Reflection*. Prentice Hall, Englewood Cliffs, NJ. pp. 119–132.

Dolhinow, P., and DeMay, M. G. (1982). Adoption: the importance of infant choice. *J. Hum. Evol.* 11:391–420.

Dolhinow, P., and Krusko, N. (1984). Langur monkey females and infants: the female's point of view. In: *Female Primates: Studies by Women Primatologists*. Alan R. Liss, New York. pp. 37–57.

Dolhinow, P., and Murphy, G. (1982). Langur monkey (*Presbytis entellus*) development. The first 3 months of life. *Folia Primatol.* 39:305–331.

Dolhinow, P. J., and Bishop, N. (1970). The development of motor skills and social relationships among primates through play. *Minn. Symp. Child Psychol.* 4:141–198.

Dunbar, R. I. M. (1984). Infant-use by male gelada in agonistic contexts: agonistic buffering, progeny protection, or soliciting support? *Primates* 25:28–35.

Ehardt, C. L. (1988). Absence of strongly kin-preferential behavior by adult female sooty mangabeys (*Cercocebus atys*). *Am. J. Phys. Anthropol.* 76:233–243.

Enstam, K. L., Isbell, L. A., and de Maar, T. W. (2002). Male demography, female mating behavior, and infanticide in wild patas monkeys (*Erythrocebus patas*). *Int. J. Primatol.* 23:85–104.

Fedigan, L. M. (1976). A study of roles in the Arashiyama West troop of Japanese monkeys (*Macaca fuscata*). *Contrib. Primatol.* 9:1–95.

Fedigan, L. M. (2003). Impact of male takeovers on infant deaths, births and conceptions in *Cebus capucinus* at Santa Rosa, Costa Rica. *Int. J. Primatol.* 24:723–741.

Fernandez-Duque E. (2004). High levels of intrasexual competition in sexually monomorphic owl monkeys. *Int. J. Primatol.* 75(Suppl. 1):260.

Fletcher, A. (2001). Development of infant independence from the mother in wild mountain gorillas. In: Robbins, M. M., Sicotte, P., and Stewart, K. J. (eds.), *Mountain Gorillas: Three Decades of Research at Karisoke*. Cambridge University Press, Cambridge. pp. 153–182.

Förster, S., and Cords, M. (2002). Development of mother–infant relationships and infant behavior in wild blue monkeys (*Cercopithecus mitis stuhlmanni*). In: Glenn, M., and Cords, M. (eds.), *The Guenons: Diversity and Adaptation in African Monkeys*. Kluwer Academic/Plenum Publishers, New York. pp. 245–272.

Fossey, D. (1979). Development of the mountain gorilla (*Gorilla gorilla beringei*): the first thirty-six months. In: Hamburg, D. A., and McCown, E. R. (eds.), *Perspectives on Human Evolution. The Great Apes*, vol. 5. Benjamin/Cummings, Menlo Park, CA. pp. 139–184.

Fossey, D. (1983). *Gorillas in the Mist*. Houghton Mifflin, Boston.

Fragaszy, D. M., Baer, J., and Adams-Curtis, L. (1991). Behavioral development and maternal care in tufted capuchins (*Cebus capucinus*) and squirrel monkeys (*Saimiri sciureus*) from birth through seven months. *Dev. Psychobiol.* 24:375–393.

Fragaszy, D. M., and Mitchell, G. (1974). Infant socialization in primates. *J. Hum. Evol.* 3:563–574.

Fragaszy, D. M., Schwarz, S., and Shimosaka, D. (1982). Longitudinal observations of care and development of infant titi monkeys (*Callicebus moloch*). *Am. J. Primatol.* 2:191–200.

Fuentes, A. (1995). The socioecology of the Mentawai Island langur (*Presbytis potenziani*) [PhD diss.]. University of California, Berkeley.

Fuentes, A. (1999). Variable social organization: what can looking at primate groups tell us about the evolution of plasticity in primate societies? In: Dolhinow, P., and Fuentes, A. (eds.), *The Nonhuman Primates*. Mayfield Press, Mountain View, CA. pp. 183–188.

Fuentes, A. (2000). Hylobatid communities: changing views on pair bonding and social organization in hominoids. *Ybk. Phys. Anthropol.* 43:33–60.

Galdikas, B. M. F. (1981a). Wild orangutan studies at Tanjung Puting Reserve, central Indonesian Borneo, 1971–1977. *Nat. Geogr. Soc. Res. Rep.* 13:1–10.

Galdikas, B. M. F. (1981b). Orangutan reproduction in the wild. In: Graham, C. E. (ed.), *Reproductive Biology of the Great Apes: Comparative and Biomedical Perspectives*. Academic Press, New York. pp. 281–300.

Garber, P. A., Moya, L., and Malaga, C. (1984). A preliminary field study of the moustached tamarin monkey (*Saguinus mystax*) in northeastern Peru: questions concerned with the evolution of a communal breeding system. *Folia Primatol.* 42:17–32.

Goldizen, A. W. (1987). Tamarins and marmosets: communal care of offspring. In: Smuts, B. B., Cheney, D. L., Seyfarth, R. M., Wrangham, R. W., and Struhsaker, T. T. (eds.), *Primate Societies*. University of Chicago Press, Chicago. pp. 34–43.

Goldizen, A. W., and Terborgh, J. (1986). Cooperative polyandry and helping behavior in saddle-backed tamarins (*Saguinus fuscicollis*). In: Else, J. G., and Lee, P. C. (eds.), *Primate Ecology and Conservation*. Cambridge University Press, New York. pp. 191–198.

Goodall, J. (1977). Infant killing and cannibalism in free-living chimpanzees. *Folia Primatol.* 28:259–282.

Goodall, J. (1983). Population dynamics during a 15 year period in one community of free-living chimpanzees in the Gombe National Park, Tanzania. *Z. Tierpsychol.* 61:1–60.

Goodall, J. (1986). *The Chimpanzees of Gombe: Patterns of Behavior*. Harvard University Press, Cambridge, MA.

Gould, L. (1990). The social development of free-ranging infant *Lemur catta* at Berenty Reserve, Madagascar. *Int. J. Primatol.* 11:297–318.

Gould, L. (1992). Alloparental care in free-ranging *Lemur catta* at Berenty Reserve, Madagascar. *Folia Primatol.* 58:72–83.

Gould, L. (1997). Affiliative relationships between adult males and immature group members in naturally occurring ringtailed lemurs (*Lemur catta*). *Am. J. Phys. Anthropol.* 103:163–171.

Gouzoules, H., and Gouzoules, S. (2002). Primate communication: by nature honest, or by experience wise? *Int. J. Primatol.* 23:821–848.

Groves, C. P. (2001). *Primate Taxonomy.* Smithsonian Institution Press, Washington DC.

Gursky, S. (2000). Allocare in a nocturnal primate: data on the spectral tarsier, *Tarsius spectrum. Folia Primatol.* 71:39–54.

Gust, D. A., and Gordon, T. P. (1994). The absence of a matrilineally based dominance system in sooty mangabeys (*Cercocebus torquatus atys*). *Anim. Behav.* 47:589–594.

Hall, K. R. L. (1965). Behaviour and ecology of the wild patas monkey, *Erythrocebus patas*, in Uganda. *J. Zool.* 148:15–87.

Hall, K. R. L. (1968). Social learning in monkeys. In: Jay, P. C. (ed.), *Primates. Studies in Adaptation and Variability.* Holt, Rinehart and Winston, New York. pp. 383–397.

Harlow, H. F. (1959). Love in infant monkeys. *Sci. Am.* 200:68–74.

Harlow, H. F., and Harlow, M. K. (1962). Social deprivation in monkeys. *Sci. Am.* 207:136–146.

Harvey, P., and Clutton-Brock, T. (1985). Life history variation in primates. *Evolution* 39:559–581.

Hinde, R. A. (1974). Mother/infant relations in rhesus monkeys. In: White, N. F. (ed.), *Ethology and Psychiatry.* University of Toronto Press, Toronto. pp. 29–46.

Hinde, R. A., Rowell, T. E., and Spencer-Booth, Y. (1964). Behaviour of socially living rhesus monkeys in their first six months. *Proc. Zool. Soc. Lond.* 143:609–649.

Hiraiwa, M. (1981). Maternal and alloparental care in a troop of free-ranging Japanese monkeys. *Primates* 22:309–329.

Hirata, S., and Celli, M. L. (2003). Role of mothers in the acquisition of tool-use behaviours by captive infant chimpanzees. *Anim. Cogn.* 6:235–244.

Hoffman, K. A. (1998). Transition from juvenile to adult stages of development in titi monkeys (*Callicebus moloch*) [PhD diss.]. University of California, Davis.

Horr, D. A. (1977). Orang-utan maturation: growing up in a female world. In: Chevalier-Skolnikoff, S., and Poirier, F. E. (eds.), *Primate Bio-Social Development: Biological, Social, and Ecological Determinants.* Garland Publishing, New York. pp. 289–321.

Hrdy, S. B. (1974). Male–male competition and infanticide among the langurs (*Presbytis entellus*) of Abu, Rajasthan. *Folia Primatol.* 22:19–58.

Hrdy, S. B. (1976). The care and exploitation of non-human primate infants by conspecifics other than the mother. In: Rosenblatt, J. S., Hinde, R. A., Shaw, E., and Beer, C. (eds.), *Advances in the Study of Behaviour*, vol. 6. Academic Press, New York. pp. 101–158.

Hrdy, S. B. (1977). *The Langurs of Abu: Female and Male Strategies of Reproduction.* Harvard University Press, Cambridge, MA.

Hrdy, S. B. (1979). Infanticide among animals: a review, classification, and examination of the implications for the reproductive strategies of females. *Ethol. Sociobiol.* 1:13–40.

Hrdy, S. B., and Hausfater, G. (eds.) (1984). *Infanticide: Comparative and Evolutionary Perpectives.* Aldine de Gruyter, New York.

Itani, J. (1959). Paternal care in the wild Japanese monkey, *Macaca fuscata fuscata. Primates* 2:61–93.

Jay, P. (1963). Mother–infant relations in langurs. In: Rheingold, H. R. (ed.), *Maternal Behavior in Mammals.* John Wiley & Sons, New York. pp. 282–304.

Jay, P. (1965). The common langur of north India. In: DeVore, I. (ed.), *Primate Behavior: Field Studies of Monkeys and Apes.* Holt, Reinhart and Winston, New York. pp. 197–249.

Jolly, A. (1966). *Lemur Behavior: A Madagascar Field Study.* University of Chicago Press, Chicago.

Jovanovic, T., and Gouzoules, H. (2001). Effects of nonmaternal restraint on the vocalizations of infant rhesus monkeys (*Macaca mulatta*). *Am. J. Primatol.* 53:33–45.

Kappeler, P. M. (1998). Nests, tree holes, and the evolution of primate life histories. *Am. J. Primatol.* 46:7–33.

Kappeler, P. M., and van Schaik, C. P. (2002). Evolution of primate social systems. *Int. J. Primatol.* 23:707–740.

Key, C., and Aiello, L. C. (2000). A prisoner's dilemma model of the evolution of paternal care. *Folia Primatol.* 71:77–92.

Koenig, A., and Borries, C. (2001). Socioecology of hanuman langurs: the story of their success. *Evol. Anthropol.* 10:122–137.

Kohda, M. (1985). Allomothering behaviour of New and Old World monkeys. *Primates* 26:28–44.

Kurland, J. A. (1977). *Kin Selection in the Japanese Monkey.* Karger, Basel.

Ladygina-Kohts, N. N. (2002). *Infant Chimpanzee and Human Child: A Classic 1935 Comparative Study of Ape Emotions and Intelligence.* Oxford University Press, New York.

Lancaster, J. (1971). Play-mothering: the relations between juvenile females and young infants among free-ranging vervet monkeys (*Cercopithecus aethiops*). *Folia Primatol.* 15:161–182.

Lee, P. C. (1983). Caretaking of infants and mother–infant relationships. In: Hinde, R. A. (ed.), *Primate Social Relationships: An Integrated Approach.* Sinauer Associates, Sunderland, MA. pp. 145–151.

MacKinnon, K. C. (2002). Social development of wild white-faced capuchin monkeys (*Cebus capucinus*) in Costa Rica: an examination of social interactions between immatures and adult males [PhD diss.]. University of California, Berkeley.

MacKinnon, K. C. (2003). The context of an observed infant-killing event in *Cebus capucinus* at Santa Rosa National Park, Costa Rica. *Am. J. Primatol.* 60(suppl. 1):60–61.

Maestripieri, D. (1994). Social structure, infant handling, and mothering styles in group-living Old World monkeys. *Int. J. Primatol.* 15:531–553.

Maestripieri, D. (1995). First steps in the macaque world: do rhesus mothers encourage their infants' independent locomotion? *Anim. Behav.* 49:1541–1549.

Maestripieri, D. (1996). Maternal encouragement of infant locomotion in pigtail macaques, *Macaca nemestrina. Anim. Behav.* 51:603–610.

Maestripieri, D. (1998). Social and demographic influences on mothering style in pigtail macaques. *Ethology* 104:379–385.

Maestripieri, D. (1999). Fatal attraction: interest in infants and infant abuse in rhesus macaques. *Am. J. Phys. Anthropol.* 110:17–25.

Maestripieri, D. (2003). Similarities in affiliation and aggression between cross-fostered rhesus macaque females and their biological mothers. *Dev. Psychobiol.* 43:321–327.

Maestripieri, D., and Carroll, K. A. (1998). Risk factors for infant abuse and neglect in group-living rhesus monkeys. *Psychol. Sci.* 9:143–145.

Maestripieri, D., and Carroll, K. A. (2000). Causes and consequences of infant abuse and neglect in monkeys. *Aggr. Viol. Behav.* 5:245–254.

Maestripieri, D., and Megna, N. L. (2000a). Hormones and behavior in rhesus macaque abusive and nonabusive mothers. 1. Social interactions during late pregnancy and early lactation. *Physiol. Behav.* 71:35–42.

Maestripieri, D., and Megna, N. L. (2000b). Hormones and behavior in rhesus macaque abusive and nonabusive mothers. 2. Mother–infant interactions. *Physiol. Behav.* 71:43–49.

Maestripieri, D., Ross, S. K., and Megna, N. L. (2002). Mother–infant interactions in western lowland gorillas (*Gorilla gorilla gorilla*): spatial relationships, communication, and opportunities for social learning. *J. Comp. Psychol.* 116:219–227.

Maestripieri, D., and Wallen, K. (1995). Interest in infants varies with reproductive condition in group-living female pigtail macaques (*Macaca nemestrina*). *Physiol. Behav.* 57:353–358.

Manson, J. (1999). Infant handling in wild *Cebus capucinus*: testing bonds between females? *Anim. Behav.* 57:911–921.

Manson, J. H., Rose, L. M., Perry, S., and Gros-Louis, J. (1999). Dynamics of female–female relationships in wild *Cebus capucinus*: data from two Costa Rican sites. *Int. J. Primatol.* 20:679–706.

Mason, W. A. (1986). Early socialization. In: Benirschke, K. (ed.), *Primates: The Road to Self-Sustaining Populations*. Springer-Verlag, New York. pp. 321–329.

Mason, W. A., and Mendoza, S. P. (1998). Generic aspects of primate attachments: parents, offspring and mates. *Psychoneuroendocrinology* 23:765–778.

Matsumura, S. (1997). Mothers in a wild group of moor macaques (*Macaca maurus*) are more attractive to other group members when holding their infants. *Folia Primatol.* 68:77–85.

McKenna, J. J. (1979a). The evolution of allomothering behavior among colobine monkeys: function and opportunism in evolution. *Am. Anthropol.* 81:818–840.

McKenna, J. J. (1979b). Aspects of infant socialization, attachment, and maternal caregiving patterns among primates: a cross-disciplinary review. *Ybk. Phys. Anthropol.* 22:250–286.

McKenna, J. J. (1987). Parental supplements and surrogates among primates: cross-species and cross-cultural comparisons. In: Lancaster, J., Altmann, J., and Rossi, A. (eds.), *Parenting Across the Life Span: Biosocial Dimensions*. Aldine de Gruyter, New York. pp. 143–184.

Melo, L., Mendes Pontes, A. R., and Monteiro da Cruz, M. A. (2003). Infanticide and cannibalism in wild common marmosets. *Folia Primatol.* 74:48–50.

Mendoza, S. P., and Mason, W. A. (1986a). Parental division of labour and differentiation of attachments in a monogamous primate (*Callicebus moloch*). *Anim. Behav.* 34:1336–1347.

Mendoza, S. P., and Mason, W. A. (1986b). Parenting within a monogamous society. In: Else, J. G. and Lee, P. C. (eds.), *Primate Ontogeny, Cognition and Social Behaviour*. Cambridge University Press, Cambridge. pp. 255–266.

Mitani, J. C., and Watts, D. (1997). The evolution of non-maternal caretaking among anthropoid primates: do helpers help? *Behav. Ecol. Sociobiol.* 40:213–220.

Mittermeier, R. A., Tattersall, I., Konstant, W. R., Meyers, D. M., and Mast, R. B. (1994). *Lemurs of Madagascar. Conservation International Tropical Field Guide Series*. Conservation International, Washington DC.

Morland, H. S. (1990). Parental behavior and infant development in ruffed lemurs (*Varecia variegata*) in a northeast Madagascar rain forest. *Am. J. Primatol.* 20:253–265.

Mota, M. T., and Sousa, M. B. C. (2000). Prolactin levels of fathers and helpers related to alloparental care in common marmosets, *Callithrix jacchus*. *Folia Primatol.* 71:22–26.

Moynihan, M. (1964). Some behavior patterns of platyrrhine monkeys. I. The night monkey (*Aotus trivirgatus*). *Smiths. Misc. Coll.* 146:1–84.

Nakagawa, N. (1995). A case of infant kidnapping and allomothering by members of a neighbouring group in patas monkeys. *Folia Primatol.* 64:62–8.

Nakamichi, M., Kanazawa, T., and Terao, K. (1994). Behavior and serum cortisol in infant Japanese monkeys following maternal separation and grouping of infants. In: Roeder, J. J., Thierry, B., Anderson, J. B., and Herrenschmidt, N. (eds.), *Current Primatology. Social Development, Learning and Behaviour*, vol. II. Université Louis Pasteur, Strasbourg. pp. 257–263.

Nakamichi, M., and Koyama, N. (2000). Intra-troop affiliative relationships of females with newborn infants in wild ring-tailed lemurs (*Lemur catta*). *Am. J. Primatol.* 50:187–203.

Nash, L. T. (1978). The development of the mother–infant relationship in wild baboons (*Papio anubis*). *Anim. Behav.* 26:746–759.

Nekaris, K. A. I. (2003). Observations of mating, birthing and parental behaviour in three subspecies of slender loris (*Loris tardigradus* and *Loris lydekkerianus*) in India and Sri Lanka. *Folia Primatol.* 74:312–336.

Neville, M. K. (1972). Social relations within troops of red howler monkeys (*Alouatta seniculus*). *Folia Primatol.* 18:47–77.

Nicholson, N. A. (1987). Infants, mothers and other females. In: Smuts, B. B., Cheney, D. L., Seyfarth, R. M., Wrangham, R. W., and Struhsaker, T. T. (eds.), *Primate Societies*. University of Chicago Press, Chicago. pp. 330–342.

Nishida, T. (1968). The social group of wild chimpanzees in the Mahali Mountains. *Primates* 9:167–224.

Nishida, T. (1988). Development of social grooming between mother and offspring in wild chimpanzees. *Folia Primatol.* 50:109–123.

Overdorff, D. J. (1993). Ecological and reproductive correlates to range use in red-bellied lemurs (*Eulemur rubriventer*) and rufous lemurs (*Eulemur fulvus rufus*). In: Kappeler, P. M., and Ganzhorn, J. U. (eds.), *Lemur Social Systems and Their Ecological Basis*. Plenum Press, New York. pp. 167–178.

Palombit, R. A., Cheney, D. L., Fischer, J., Johnson, S., Rendall, D., Seyfarth, R. M., and Silk, J. B. (2000). Male infanticide and the defense of infants in chacma baboons. In: van Schaik, C. P., and Janson, C. H. (eds.), *Infanticide by Males and Its Implications*. Cambridge University Press, Cambridge. pp. 123–151.

Parker, S. T. (2004). The cognitive complexity of social organization and socialization in wild baboons and chimpanzees: guided participation, socializing interactions, and event representation. In: Russon, A. E., and Begun, D. R. (eds.), *The Evolution of Thought: Evolutionary Origins of Great Ape Intelligence*. Cambridge University Press, Cambridge. pp. 45–60.

Paul, A., and Kuester, J. (1996). Infant handling by female Barbary macaques (*Macaca sylvanus*) at Affenberg Salem: testing functional and evolutionary hypotheses. *Behav. Ecol. Sociobiol.* 39:133–145.

Perry, S. (1996). Female–female social relationships in wild white-faced capuchin monkeys, *Cebus capucinus*. *Am. J. Primatol.* 40:167–182.

Perry, S., Baker, M., Fedigan, L. M., Gros-Louis, J., Jack, K., MacKinnon, K. C., Manson, J. H., Panger, M., Pyle, K., and Rose, L. M. (2003). Social conventions in wild white-faced capuchins: evidence for traditions in a neotropical primate. *Curr. Anthropol.* 44:241–268.

Poirier, F. E. (ed.) (1972). *Primate Socialization.* Random House, New York.

Quiatt, D. (1979). Aunts and mothers: adaptive implications of allomaternal behaviour of non-human primates. *Am. Anthropol.* 81:311–319.

Ransom, T. W., and Rowell, T. E. (1972). Early social development of feral baboons. In: Poirier, F. (ed.), *Primate Socialization.* Random House, New York. pp. 105–144.

Reichard, U. H. (2003). Social monogamy in gibbons: the male perspective. In: Reichard, U. H., and Boesch, C. (eds.), *Monogamy: Mating Strategies and Partnerships in Birds, Humans and Other Mammals.* Cambridge University Press, Cambridge. pp. 190–213.

Richard, A. F. (1976). Preliminary observations on the birth and development of *Propithecus verreauxi* to the age of six months. *Primates* 17:357–366.

Riechelmann, C., Hultsch, H., and Todt, D. (1994). Early development of social relationships in Barbary macaques (*Macaca sylvanus*): trajectories of alloparental behaviour during an infant's first three months of life. In: Roeder, J. J., Thierry, B., Anderson, J. B., and Herrenschmidt, N. (eds.), *Current Primatology. Social Development, Learning and Behaviour,* vol. II. Université Louis Pasteur, Strasbourg. pp. 279–286.

Ross, C., and MacLarnon, A. (2000). The evolution of non-maternal care in anthropoid primates: a test of the hypotheses. *Folia Primatol.* 71:93–113.

Ross, C., and Regan, G. (2000). Allocare, predation risk, social structure and natal coat colour in anthropoid primates. *Folia Primatol.* 71:67–76.

Rotundo, M., Sloan, C., and Fernandez-Duque, E. (2000). Cambios estacionales en el ritmo de actividad del mono mirikiná (*Aotus azarai*) en Formosa Argentina. In: Cabrera, E., Mércolli, C. and Resquin, R. (eds.), *Manejo de Fauna Silvestre en Amazonía y Latinoamérica.* Asunción, Paraguay, pp. 413–417.

Sade, D. S. (1965). Some aspects of parent–offspring and sibling relations in a group of rhesus monkeys, with a discussion of grooming. *Am. J. Phys. Anthropol.* 23:1–17.

Sade, D. S. (1980). Population biology of free-ranging rhesus monkeys on Cayo Santiago, Puerto Rico. In: Cohen, M. N., Malpass, R. S., and Klein, H. G. (eds.), *Mechanisms of Population Regulation.* Yale University Press, New Haven. pp. 171–187.

Sapolsky, R. M. (1993). The physiology of dominance in stable versus unstable social hierarchies. In: Mason, W. A., and Mendoza, S. P. (eds.), *Primate Social Conflict.* State University of New York Press, Albany. pp. 171–204.

Sapolsky, R. M. (1999). Hormonal correlates of personality and social contexts: from non-human to human primates. In: Panter-Brick, C., and Worthman, C. M. (eds.), *Hormones, Health, and Behavior: A Socio-Ecological and Lifespan Perspective.* Cambridge University Press, New York. pp. 18–46.

Sapolsky, R. M. (2000). Physiological correlates of individual dominance style. In: Aureli, F., and de Waal, F. B. M. (eds.), *Natural Conflict Resolution.* University of California Press, Berkeley. pp. 114–116.

Savage, A., Snowdon, C. T., Giraldo, L. H., and Soto, L. H. (1996). Parental care patterns and vigilance in wild cotton-top tamarins (*Saguinus oedipus*). In: Norconk, M. A., Rosenberger, A. L., and Garber, P. A. (eds.), *Adaptive Radiations of Neotropical Primates.* Plenum Press, New York. pp. 187–199.

Schaller, G. B. (1963). *The Mountain Gorilla: Ecology and Behavior.* University of Chicago Press, Chicago.

Schino, G., Speranza, L., Ventura, R., and Troisi, A. (2003). Infant handling and maternal response in Japanese macaques. *Int. J. Primatol.* 24:627–639.

Schradin, C., Reeder, D. M., Mendoza, S. P., and Anzenberger, G. (2003). Prolactin and paternal care: comparison of three species of monogamous new world monkeys (*Callicebus cupreus, Callithrix jacchus,* and *Callimico goeldii*). *J. Comp. Psychol.* 117:166–175.

Schulze, H., and Meier, B. (1995). Behaviour of captive *Loris tardigradus nordicus*: a qualitative description, including some information about morphological bases of behaviour. In: Alterman, L., Doyle, G. A., and Izard, M. K. (eds.), *Creatures of the Dark: The Nocturnal Prosimians.* Plenum Press, New York. pp. 221–250.

Silk, J. B. (1980). Kidnapping and female competition among captive bonnet macaques. *Primates* 21:100–110.

Silk, J. B. (1999). Why are infants so attractive to others? The form and function of infant handling in bonnet macaques. *Anim. Behav.* 57:1021–1032.

Silk, J. B., Rendall, D., Cheney, D. L., and Seyfarth, R. M. (2003). Natal attraction in adult female baboons (*Papio cynocephalus ursinus*) in the Moremi Reserve, Botswana. *Ethology* 109:627–644.

Smuts, B. B. (1985). *Sex and Friendship in Baboons.* Aldine, New York.

Sommer, V. (1987). Infanticide among free-ranging langurs (*Presbytis entellus*) at Jodhpur (Rajasthan/India): recent observations and a reconsideration of hypotheses. *Primates* 28:163–197.

Sommer, V. (1994). Infanticide among langurs of Jodhpur: testing the sexual selection hypothesis with a long-term record. In: Parmigiani, S., and vom Saal, F. S. (eds.), *Infanticide and Parental Care.* Harwood, New York. pp. 155–198.

Sommer, V. (2000). The holy wars about infanticide. Which side are you on? And why? In: van Schaik, C. P., and Janson, C. H. (eds.), *Infanticide by Males and Its Implications.* Cambridge University Press, Cambridge. pp. 9–26.

Stein, D. M. (1984). *The Sociobiology of Infant and Adult Male Baboons.* Albex, Norwood.

Stein, D. M., and Stacey, P. B. (1981). A comparison of infant–adult male relations in a one-male group with those in a multi-male group for yellow baboons (*Papio cynocephalus*). *Folia Primatol.* 36:264–276.

Stewart, K. J. (2001). Social relationships of immature gorillas and silverbacks. In: Robbins, M. M., Sicotte, P., and Stewart, K. J. (eds.), *Mountain Gorillas: Three Decades of Research at Karisoke.* Cambridge University Press, New York. pp. 183–213.

Strier, K. B. (1986). The behavior and ecology of the woolly spider monkey, or muriqui (*Brachyteles arachnoides*, E. Geoffroy 1806) [PhD diss.]. Harvard University, Cambridge, MA.

Strum, S. C. (1984). Why males use infants. In: Taub, D. M. (ed.), *Primate Paternalism.* van Nostrand Reinhold, New York. pp. 146–185.

Suomi, S. J. (1982). The development of social competence by rhesus monkeys. *Ann. Ist. Sup. Sanita* 18:193–202.

Suomi, S. J. (2002). Parents, peers, and the process of socialization in primates. In: Borkowski, J. G., Ramey, S. L., and Bristol-Power, M. (eds.), *Parenting and the Child's World: Influences on Academic, Intellectual, and Social-Emotional Development.* Erlbaum, Mahwah, NJ. pp. 265–279.

Sussman, R. W. (1977). Socialization, social structure, and ecology of two sympatric species of lemur. In: Chevalier-Skolnikoff, S., and Poirier, F. E. (eds.), *Primate Bio-Social Development: Biological, Social, and Ecological Determinants.* Garland Publishing, New York. pp. 515–528.

Sussman, R. W., Cheverud, J. M., and Bartlett, T. Q. (1995). Infant killing as an evolutionary strategy: reality or myth? *Evol. Anthropol.* 3:149–151.

Tanaka, I. (1989). Variability in the development of mother–infant relationships among free-ranging Japanese macaques. *Primates* 30:477–491.

Taub, D. M. (1980). Testing the "agonistic buffering" hypothesis. I. The dynamics of participation in the triadic interaction. *Behav. Ecol. Sociobiol.* 6:187–197.

Taub, D. M. (1984). Male caretaking behaviour among wild Barbary macaques (*Macaca sylvanus*). In: Taub, D. M. (ed.), *Primate Paternalism.* Van Nostrand Reinhold, New York. pp. 20–55.

Thierry, B. (1985). Social development in three species of macaque (*Macaca mulatta, M. fascicularis, M. tonkeana*): a preliminary report on the first ten weeks of life. *Behav. Proc.* 11:89–95.

Thierry, B. (2000). Covariation of conflict management patterns across macaque species. In: Aureli, F., and de Waal, F. B. M. (eds.), *Natural Conflict Resolution.* University of California Press, Berkeley. pp. 106–128.

Thierry, B. (2004). Social epigenesis. In: Thierry, B., Singh, M., and Kaumanns, W. (eds.), *Macaque Societies: A Model for the Study of Social Organization.* Cambridge University Press, New York. pp. 267–290.

Thierry, B., and Anderson, J. R. (1986). Adoption in anthropoid primates. *Int. J. Primatol.* 7:191–216.

Treves, A. (1997). Primate natal coats: a preliminary analysis of distribution and function. *Am. J. Phys. Anthropol.* 104:47–70.

Treves, A. (2000). Protection of infanticide: the perspective of infant primates. In: van Schaik, C. P., and Janson, C. H. (eds.), *Infanticide by Males and Its Implications.* Cambridge University Press, Cambridge. pp. 223–238.

Trivers, R. L. (1972). Parental investment and sexual selection. In: Campbell, B. (ed.), *Sexual Selection and the Descent of Man, 1871–1971.* Aldine, Chicago. pp. 136–179.

Ueno, A., and Matsuzawa, T. (2004). Food transfer between chimpanzee mothers and their infants. *Primates* 45:231–239.

Valenzuela, N. (1994). Early behavioral development in three wild infant *Cebus apella* in Colombia. In: Roeder, J. J., Thierry, B., Anderson, J. B., and Herrenschmidt, N. (eds.), *Current Primatology. Social Development, Learning and Behaviour,* vol. II. Université Louis Pasteur, Strasbourg. pp. 297–302.

van Lawick-Goodall, J. (1967). Mother–offspring relationships in free-ranging chimpanzees. In: Morris, D. (ed.), *Primate Ethology.* Weidenfeld and Nicolson, London. pp. 287–346.

van Lawick-Goodall, J. (1971). Some aspects of mother–infant relationships in a group of wild chimpanzees. In: Schaffer, H. R. (ed.), *The Origins of Human Social Relations.* Academic Press, London. pp. 115–128.

van Schaik, C. P. (2000a). Vulnerability to infanticide by males: patterns among mammals. In: van Schaik, C. P., and Janson, C. H. (eds.), *Infanticide by Males and Its Implications.* Cambridge University Press, Cambridge. pp. 61–71.

van Schaik, C. P. (2000b). Infanticide by male primates: the sexual selection hypothesis revisited. In: van Schaik, C. P., and Janson, C. H. (eds.), *Infanticide by Males and Its Implications.* Cambridge University Press, Cambridge. pp. 27–60.

van Schaik, C. P., and Janson, C. H. (2000). *Infanticide by Males and Its Implications.* Cambridge University Press, Cambridge.

van Schaik, C. P., and Kappeler, P. M. (1997). Infanticide risk and the evolution of male–female association in primates. *Proc. R. Soc. Lond. B* 264:1687–1694.

Whiten, A., Goodall, J., McGrew, W. C., Nishida, T., Reynolds, V., Sugiyama, Y., Tutin, C. E. G., Wrangham, R. W., and Boesch, C. (2001). Charting cultural variation in chimpanzees. *Behaviour* 138:1481–1516.

Whitten, P. L. (1987). Infant and adult males. In: Smuts, B. B., Cheney, D. L., Seyfarth, R. M., Wrangham, R. W., and Struhsaker, T. T. (eds.), *Primate Societies.* University of Chicago Press, Chicago. pp. 343–359.

Wittenberger, J. F., and Tilson, R. L. (1980). The evolution of monogamy: hypotheses and evidence. *Annu. Rev. Ecol. Syst.* 11:197–232.

Wolfe, L. D. (1981). A case of male adoption in a troop of Japanese monkeys (*Macaca fuscata fuscata*). In: Chiarelli, A. B., and Corruccini, R. S. (eds.), *Primate Behavior and Sociobiology.* Springer-Verlag, New York. pp. 156–160.

Wright, P. C. (1984). Biparental care in *Aotus trivirgatus* and *Callicebus moloch.* In: Small, M. (ed.), *Female Primates: Studies by Women Primatologists.* Alan R. Liss, New York. pp. 59–75.

Wright, P. C. (1994). The behavior and ecology of the owl monkey. In: Baer, J. F., Weller, R. E., and Kakoma, I. (eds.), *Aotus: The Owl Monkey.* Academic Press, San Diego. pp. 97–112.

Ziegler, T. (2000). Hormones associated with non-maternal infant care: a review of mammalian and avian studies. *Folia Primatol.* 71:6–21.

36

Postconflict Reconciliation

Kate Arnold and Filippo Aureli

INTRODUCTION

Conflicts of interest between group members occur in many contexts. The most common contexts are competition for access to limited resources, such as mates or food (Darwin 1871, Trivers 1972, Clutton-Brock 1989, van Schaik 1989), and disagreement about decisions, such as the allocation of time to different activities or the direction of travel (van Schaik and van Noordwijk 1986, Menzel 1993, Boinski 2000). Conflicts of interest may escalate into aggression and may compromise the benefits of group living if they induce the loser to leave the group (Janson 1992). Even without such a consequence, aggression may jeopardize future cooperation between the former opponents (de Waal 1986). Thus, behavioral mechanisms that mitigate conflict, prevent aggressive escalation, and resolve disputes should have been strongly selected in group-living animals (Aureli and de Waal 2000).

Although behavioral mechanisms for conflict management before aggressive escalation are of paramount importance, little theoretical and empirical research on nonhuman primates has focused on this aspect (Aureli and Smucny 1998, Cords and Killen 1998, Aureli and de Waal 2000). Most systematic research has dealt with the mechanisms of conflict resolution that occur after aggressive escalation, especially friendly postconflict reunions between former opponents (Kappeler and van Schaik 1992, Silk 1997, de Waal 2000a, Aureli et al. 2002, Judge 2003). Descriptive accounts of such reunions have been reported for various primate species (see de Waal 2000b for a review), but their systematic study started when de Waal and van Roosmalen (1979) showed that, contrary to the common view, chimpanzee (*Pan troglodytes*) opponents actually spent more time in close proximity and exchanging friendly behavior in the aftermath of an aggressive conflict than beforehand. They called these reunions "reconciliations", a heuristic term that clearly implies a specific function (see below). Their pioneering work was instrumental in the shift from simple descriptions of friendly post-conflict reunions to the systematic study of conflict resolution in primates.

In this chapter, we review the knowledge accumulated on postconflict friendly reunions in primates in the last 25 years. Most of the studies used the term *reconciliation*, so we have also adopted it and discuss its functional implications. We start with reviewing methodological issues regarding collection and analysis of postconflict data. We then review the evidence for reconciliation across the Primate order, the benefits of reconciliation in reducing the costs of aggressive conflict, the variation and ontogeny of reconciliation patterns, and the various hypotheses put forward to explain its function and distribution across individuals, groups, and species.

MEASURING RECONCILIATION

Following the first study of reconciliation in chimpanzees, which was based only on postconflict observations (de Waal and van Roosmalen 1979), a second study introduced control observations allowing a direct comparison of postconflict behavior with behavior exhibited under "baseline" conditions (i.e., not affected by a recent conflict) (de Waal and Yoshihara 1983). This procedure, known as the *PC–MC method* (see below), has become the standard methodology and has been used in the majority of postconflict studies (for reviews see Kappeler and van Schaik 1992, Cords 1993, Veenema et al. 1994, Veenema 2000).

Post-conflict observations (PCs) begin immediately after the end of the conflict and usually consist of 10 min continuous focal observations on either the aggressor or victim, during which particular attention is paid to the nature and timing of social interactions. For each PC, a corresponding matched-control observation (MC) of the behavior of the same individual is recorded. This is usually carried out on the next possible day and at the same time as the PC in order to control for seasonal and diurnal activity patterns and to reflect the current social situation, which may change over time. To ensure that MCs are not influenced by a recent conflict, they are postponed until the next day if aggression involving the subject of the MC closely precedes the planned start time. This procedure results in an equal number of PCs and MCs, which are then compared with respect to the timing of the first friendly contact between former opponents.

Further controls have been used in some studies to ensure that PCs and MCs are comparable in other respects. The distance between former opponents at the beginning of corresponding observations has been taken into account in a number of captive studies (de Waal and Ren 1988, York and Rowell 1988, Kappeler 1993, Swedell 1997, Call 1999, Rolland and Roeder 2000). Some studies of wild populations have employed additional controls, selecting MCs which are comparable to PCs in terms of former opponent distance, group activity, and subgroup composition (Aureli 1992, Watts 1995, Matsumura 1996, Arnold and Whiten 2001, Cooper and Bernstein 2002).

If former opponents make friendly contact earlier in the PC than in the corresponding MC, they are considered to be attracted to one another as a result of the conflict. In the reverse situation, where former opponents contact each other earlier (or only) in the MC, the conflict is considered to have resulted in the dispersal of the former opponents. Thus, PC–MC pairs are labeled "attracted" or "dispersed." PC–MC pairs in which no contact between former opponents occurs during either observation or contact occurs at the same time are labeled "neutral" (de Waal and Ren 1988). If post-conflict contacts between two individuals are randomly distributed throughout PCs (as would be expected in MCs), then the ratio of attracted to dispersed pairs should be 1:1 (de Waal and Yoshihara 1983). If this ratio is significantly higher in favor of attracted pairs, then it is concluded that a high proportion of affiliative contacts between former opponents occurred sooner after a conflict than at other times and that reconciliation is demonstrated.

Variations on this method have been used in some studies, where MCs were taken from the period immediately preceding the conflict (Cheney and Seyfarth 1989) or latency to first affiliative contact was calculated as the mean interval between consecutive affiliative interactions from a number of baseline focal observations (Wittig and Boesch 2003). Other studies have compared the rate, instead of the first occurrence, of affiliative behaviors between former opponents during PCs with that during MCs (the *rate method*) (Judge 1991), with MC rates calculated using standard next-day observations (Judge 1991), baseline data collected throughout the study period (de Waal 1987), or the period immediately preceding the conflict (Silk et al. 1996).

A second and complementary approach to demonstrating the occurrence of reconciliation, known as the *time rule*, was developed by Aureli et al. (1989). This method compares the cumulative distribution of first affiliative contacts during all PCs and MCs over time (Fig. 36.1) and identifies the time period during which the frequencies of contact are significantly elevated during PCs compared to MCs. Within this time window, affiliative contacts can be operationally defined as reconciliation and used to study functional aspects by comparing PCs with and those without reconciliation. The relative advantages and disadvantages of the PC–MC method, the rate method, and the time rule are discussed in detail elsewhere (Veenema et al. 1994, Veenema 2000).

de Waal and Yoshihara (1983) also developed a measure of conciliatory tendency (CT) for use in comparative analyses (comparisons between species, groups, or subsets

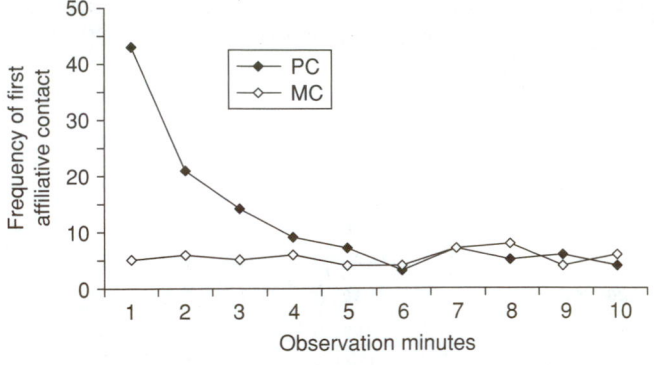

Figure 36.1 The frequency of the first affiliative contact between former opponents in the 10 min of postconflict (PC) and matched control (MC) observations in long-tailed macaques. Modified from Aureli et al. (1989).

of individuals; e.g., kin vs. non-kin). They defined the CT as the proportion of the total number of PC–MC pairs that were attracted. This measure was later refined to fully correct for baseline levels of affiliation by removing a proportion of attracted pairs, which might have occurred by chance (Veenema et al. 1994). The revised measure, known as the corrected conciliatory tendency (CCT), is defined as (number of attracted pairs − number of dispersed pairs)/total number of PC–MC pairs.

EVIDENCE FOR RECONCILIATION AS A WIDESPREAD PHENOMENON IN PRIMATES

Almost all studies of primates have demonstrated that friendly interactions between former opponents occur sooner after aggression than during control periods (Table 36.1). Exceptions are red-bellied tamarins (*Saguinus labiatus*), which are highly cooperative and have very secure social bonds (Schaffner and Caine 2000), and ring-tailed lemurs (*Lemur catta*) and black lemurs (*Eulemur macaco*), whose relationships are either very hostile or very friendly to the point that they are unlikely to be disrupted by conflict (Kappeler 1993, Roeder et al. 2002).

Many studies have also shown that former opponents are selectively attracted to one another as opposed to simply seeking contact in an indiscriminate fashion, thus strengthening the case for such contacts performing a specific function that is directly related to the recent conflict (see Table 36.1).

The vast majority of studies (77%, Table 36.1) have been carried out in captivity, where space is often limited. It could be argued that reconciliation might be an artifact of an artificially intense social environment with no possibility of escape. This has not proven to be the case since a number of studies have demonstrated reconciliation in large enclosures where prolonged avoidance was possible (de Waal and van Roosmalen 1979, Kappeler 1993, Aureli et al. 1994) and in the wild where the option to leave the group, at least temporarily, was available (Cheney and Seyfarth 1989, Aureli 1992, Watts 1995, Matsumura 1996, Silk et al. 1996, Castles

Table 36.1 Studies in which the Occurrence of Reconciliation was Investigated in Primates and the Associated Corrected Conciliatory Tendencies (CCTs) Were Given in the Study or Data Were Available to Calculate Them

SPECIES	LOCATION	RECONCILIATION	SELECTIVE ATTRACTION[1]	CCT/*MEAN CCT* %[2]	SOURCE
Prosimians					
Lemur catta	Captive	No	✓	—	Kappeler (1993)
	Captive	Yes[3]		—	Rolland and Roeder (2000)
Eulemur fulvus rufus	Captive	Yes		6.4	Kappeler (1993)
	Captive	Yes		26.5	Roeder et al. (2002)
Eulemur macaco	Captive	No		—	Roeder et al. (2002)
New World monkeys					
Callithrix jaccus	Captive	Yes	✓	—	Westlund et al. (2000)
Sanguinus labiatus	Captive	No		—	Schaffner and Caine (2000)
Cebus apella	Captive	Yes/No	✓	14.3 (non-food-related conflicts only)	Verbeek and de Waal (1997)
	Captive	Yes		25.0–50.0 (immatures–unrelated adults)	Weaver and de Waal (2003)
Cebus capucinus	Captive	Yes	✓	18.5	Leca et al. (2002)
Saimiri sciureus	Captive	Yes		77.0	Pereira et al. (2000)
Old World monkeys					
Cercopithecus aethiops	Wild	Yes		—	Cheney and Seyfarth (1989)
Erythrocebus patas	Captive	Yes[3]	✓	—	York and Rowell (1988)
Cercocebus torquatus atys	Captive	Yes		—	Gust and Gordon (1993)
Macaca assamensis	Free-ranging[4]	Yes	✓	*11.2*	Cooper and Bernstein (2002)
Macaca arctoides	Captive	Yes	✓	—	de Waal and Ren (1988)
	Captive	Yes		—	Perez-Ruiz and Mondragon-Ceballos (1994)
	Captive	Yes		34.9	Call et al. (1999)
Macaca fascicularis	Captive	Yes		—	Cords (1988)
	Captive	Yes	✓	13.9	Aureli et al. (1989)
	Wild	Yes	✓	—	Aureli (1992)
Macaca fuscata	Captive	Yes	✓	—	Aureli et al. (1993)
	Captive	Yes[3]		22.4	Chaffin et al. (1995)
	Captive	Yes		*7.9–30.3*	Petit et al. (1997)
	Captive	Yes		8.4	Schino et al. (1998)
	Free ranging[4]	Yes		*26.8*	Koyama (2001)
	Free ranging[4]	Yes	✓	*14.0*	Kutsukake and Castles (2001)
	Captive	Yes		—	Abegg et al. (2003)
Macaca maurus	Wild	Yes		40.0	Matsumura (1996)
Macaca mulatta	Captive	Yes	✓	—	de Waal and Yoshihara (1983)
	Captive	Yes		8.1 (low density), *7.1* (high)	Call et al. (1996)
	Captive	Yes		*7.1*	Call (1999)
	Captive	Yes		—	Demaria and Thierry (2001)
Macaca nemestrina	Captive	Yes		—	Judge (1991)
	Captive	Yes		*41.9, 20.4* (two groups)	Castles et al. (1996)
Macaca nigra	Captive	Yes	✓	40.4	Petit and Thierry (1994)
Macaca silensus	Captive	Yes	✓	—	Abegg et al. (1996)
Macaca sylvanus	Captive	Yes	✓	—	Aureli et al. (1994)
Macaca tonkeana	Captive	Yes		—	Demaria and Thierry (2001)
Theropithecus gelada	Captive	Yes		29.8	Swedell (1997)
Papio anubis	Wild	Yes	✓	15.6	Castles and Whiten (1998)
Papio papio	Captive	Yes		27.0	Petit and Thierry (1994)

Table 36.1 *(cont'd)*

SPECIES	LOCATION	RECONCILIATION	SELECTIVE ATTRACTION[1]	CCT/*MEAN CCT* %[2]	SOURCE
Papio ursinus	Wild	Yes		–	Silk et al. (1996)
Rhinopithecus roxellanae	Captive	Yes		43.4	Ren et al. (1991)
Trachypithecus obscura	Captive	Yes	✓	*51.3, 41.3* (two groups)	Arnold and Barton (2001)
Semnopithecus entellus	Wild	Yes/No[5]		–	Sommer et al. (2002)
Colobus guereza	Captive	Yes	✓	45.1	Bjornsdotter et al. (2000)
Great apes					
Gorilla gorilla beringei	Wild	Yes/No[6]	✓	–	Watts (1995)
Pan paniscus	Captive	Yes		–	de Waal (1987)
	Wild	Yes		–	Hohmann and Fruth (2000)
	Captive	Yes		35.6	Palagi et al. (2004)
Pan troglodytes	Captive	Yes		34.7	de Waal and van Roosmalen (1979)
	Wild	Yes	✓	12.3	Arnold and Whiten (2001)
	Captive	Yes	✓	41.2	Preuschoft et al. (2002)
	Captive	Yes		17.3	Fuentes et al. (2002)
	Wild	Yes		–	Wittig and Boesch (2003)

[1] ✓, selective attraction was tested. Selective attraction was demonstrated in every case.
[2] Numbers represent the revised measure of conciliatory tendency at the group level (Veneema et al. 1994). Numbers in italics are the mean CCTs across individuals in a group. Other measures of the tendency to reconcile are not included in order to avoid confusion in making comparisons.
[3] Proximity alone is included as conciliatory behavior.
[4] Free-ranging is used for wild populations that are food provisioned.
[5] Only female–female conflicts were reconciled, although statistical demonstration was not carried out.
[6] Only male–female conflicts were reconciled.

and Whiten 1998a, Hohmann and Fruth 2000, Arnold and Whiten 2001, Koyama 2001, Kutsukake and Castles 2001, Cooper and Bernstein 2002, Wittig and Boesch 2003). Aureli's (1992) study of wild long-tailed macaques (*Macaca fascicularis*) reported very similar results to those of his captive study (Aureli et al. 1989), suggesting that the availability of space may not affect reconciliatory behavior. Studies of free-ranging, food-provisioned Japanese macaques (*Macaca fuscata*) support this view (Koyama 2001, Kutsukake and Castles 2001), although in a study of wild chimpanzees the conciliatory tendency was much lower than reported in most captive studies (Arnold and Whiten 2001, but see Fuentes et al. 2002), possibly due to their fission–fusion social organization.

In some species, particularly those with high conciliatory tendencies, reconciliation can involve conspicuous behavior patterns (Fig. 36.2) that are rarely used in other contexts (i.e., *explicit reconciliation*), while others employ affiliative behaviors that are regularly observed in other contexts, such as grooming or making body contact (i.e., *implicit reconciliation*) (de Waal 1993) (Table 36.2). However, there is some variation in this respect where species reported as having explicit behaviors in their reconciliatory repertoire have not been shown to do so in all groups (pig-tail macaques [*Macaca nemestrina*], Castles et al. 1996; chimpanzees, Arnold and Whiten 2001, Fuentes et al. 2002). Noncontact behaviors might also be sufficient to restore relationships in species that do not have an extensive repertoire of friendly

behavior. Sooty mangabeys (*Cercocebus torquatus atys*) reconcile by merely presenting their hindquarters (Gust and Gordon 1993), and York and Rowell (1988) found reconciliation in patas monkeys (*Erythrocebus patas*) only when they included close proximity as a friendly postconflict interaction between former opponents. Proximity was also found to have a conciliatory function in experiments with pairs of long-tailed macaques, a species that usually uses affiliative behaviors involving body contact (Cords 1993; also see Call 2000). Vocal signals also appear to facilitate friendly interactions following conflicts in chacma baboons (*Papio ursinus*, Cheney et al. 1995, Silk et al. 1996).

Comparisons of species characteristics such as the intensity of aggression, the degree of symmetry in contests, affiliation rates, and conciliatory tendencies have allowed classification of species according to their "dominance styles" (de Waal 1989a, Thierry 2000). For example, long-tailed and rhesus macaques (*Macaca mulatta*), which are relatively aggressive, have steep power gradients and reconcile few conflicts, leading to their classification as "despotic." By contrast "tolerant" Tonkean (*Macaca tonkeana*) and stump-tailed (*Macaca arctoides*) macaques exhibit low-intensity aggression with frequent bidirectional conflicts and high levels of affiliation and reconciliation (de Waal and Luttrell 1989, Thierry 1990, Petit et al. 1997). Only tolerant species have been observed using explicit gestures during reconciliation, whereas despotic species do not. Covariation between reconciliation frequencies and type of gestures occurs even

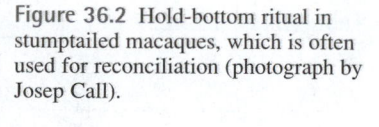
Figure 36.2 Hold-bottom ritual in stumptailed macaques, which is often used for reconciliation (photograph by Josep Call).

Table 36.2   Species that Have Been Shown to Reconcile Conflicts "Explicitly" or "Implicitly" and the Behaviors Observed in This Context

SPECIES	BEHAVIORS USED	SOURCE
Explicit		
Cebus capucinus	Hold-bottom	Leca et al. (2002)
Macaca arctoides	Hold-bottom	de Waal and Ren (1988)
Macaca nemestrina	Standing grasp	Castles et al. (1996)
Macaca tonkeana	Clasp, mount	Demaria and Thierry (2001)
Rhinopithecus roxellanae	Hold lumbar	Ren et al. (1991)
Trachypithecus obscura	Ventro-ventro embrace	Arnold and Barton (2001a)
Colobus guereza	Low-intensity mouth opening	Björnsdotter et al. (2000)
Gorilla gorilla beringei, male–female dyads	Embrace, grumble, reach toward	Watts (1995)
Pan paniscus	Genital massage	de Waal (1987)
Pan troglodytes	Kiss	de Waal and van Roosmalen (1979)
Implicit	e.g., grooming, gentle touch, body contact	
Macaca fascicularis		Cords (1988), Aureli et al. (1989)
Macaca fuscata		Aureli et al. (1993)
Macaca maurus		Matsumura (1996)
Macaca mulatta		de Waal and Yoshihara (1983)
Macaca nigra		Petit and Thierry (1994a)
Theropithecus gelada		Swedell (1997)
Papio anubis		Castles and Whiten (1998a)
Papio papio		Petit and Thierry (1994b)

between groups of the same species: studies of pig-tailed macaques (Castles et al. 1996) and chimpanzees (de Waal and van Roosmalen 1979, Arnold and Whiten 2001, Fuentes et al. 2002) reported that, while one group reconciled often and with explicit gestures, another reconciled less frequently and without the use of specific gestures.

THE COSTS OF AGGRESSIVE CONFLICT AND THE BENEFITS OF RECONCILIATION

Although conflicts of interest may be managed without escalation (Aureli and de Waal 2000), they often result in aggression and the loss of access to a contested resource by one of the participants. Similarly, aggression may serve to establish or reinforce dyadic dominance relationships between individuals. In both cases, the cost to the loser of the conflict is evident. In addition, conflicts themselves are energetically costly and can result in injury. Once aggression is instigated, one or more of these costs are almost impossible to avoid. There are also less obvious costs that can be incurred in the aftermath of conflict. Recipients of aggression are more likely to be the subject of renewed attacks either by the original aggressor (York and Rowell 1988, Aureli and van Schaik 1991, Aureli 1992, Cords 1992, Watts 1995, Silk et al. 1996, Castles and Whiten 1998b, Das et al. 1998, Schino et al. 1998, Kutsukake and Castles 2001) or by third parties (Aureli et al. 1989, Aureli and van Schaik 1991, Kutsukake and Castles 2001). In wild long-tailed macaques, recipients of aggression have also been shown to forage less in postconflict situations than during control periods, perhaps because they then spend more time monitoring the social environment in anticipation of further aggression (Aureli 1992).

Both recipients of aggression and, perhaps more surprisingly, aggressors appear to experience increased levels of anxiety following conflicts. A growing body of pharmacological evidence shows that elevated levels of anxiety result in increased rates of self-directed behaviors (e.g., scratching, self-grooming) (Ninan et al. 1982; Schino et al. 1991, 1996; Barros et al. 2000). This together with circumstantial evidence suggests that self-directed behavior can be used as a noninvasive tool for quantifying levels of anxiety in primates (Maestripieri et al. 1992). A number of studies have demonstrated increased rates of self-directed behaviors following conflicts in one or both opponents (Aureli et al. 1989, 1992; Aureli and van Schaik 1991; Aureli 1997; Castles and Whiten 1998b; Das et al. 1998; Schino et al. 1998; Kutsukake and Castles 2001). Although a prolonged stress response in recipients of aggression is likely to be adaptive in that they are primed for action in the event of further attacks or more predisposed to behave submissively (Leshner 1983), it can have negative physiological and developmental consequences over the long term (Henry 1982, Von Holst 1985, Kaplan 1986). However, anxiety following aggression appears to affect both former opponents, and it has been proposed that this is due to uncertainty about relationships that may have been damaged by a conflict (Aureli and van Schaik 1991, see below).

For both parties, a reduction in the potential for an ongoing, mutually beneficial relationship would be a highly negative mid- and long-term consequence of aggressive conflict, so the ability to approach or to be approached by a former opponent for future affiliation is crucial. In a study of free-ranging Japanese macaques, in which former opponents were monitored for 10 days after conflicts, grooming, time spent in proximity, and approaches by the victim were significantly reduced relative to baseline levels after nonreconciled conflicts (Koyama 2001).

Reconciliation serves to reduce the costs of conflict, both directly and indirectly. For the victim of aggression, the risk of further attacks by the former aggressor or other group members is significantly decreased after reconciliation (Aureli and van Schaik 1991, Cords 1992, de Waal 1993, Watts 1995, Silk et al. 1996, Castles and Whiten 1998b, Koyama 2001, Kutsukake and Castles 2001). In Koyama's (2001) study of the long-term effects of reconciliation, the rate of aggression between former opponents was lower during the 10 days following reconciled conflicts than during baseline or control observations. The anxiety experienced by former opponents following conflicts, expressed by increased rates of self-directed behaviors, returns to baseline levels once reconciliation has taken place (Aureli and van Schaik 1991, Castles and Whiten 1998b, Das et al. 1998, Kutsukake and Castles 2001). Heart rates of former opponents, which increase after a conflict, are also reduced following reconciliation (Smucny et al. 1997, Aureli and Smucny 2000). Interestingly, reconciliation is more effective than affiliative contact with other group members at reducing self-directed behaviors and heart rate (Smucny et al. 1997, Das et al. 1998, Aureli and Smucny 2000).

Cords (1992) provided strong evidence that the function of reconciliation is relationship repair. In a series of experiments, she showed that after conflicts subordinate long-tailed macaques were deterred from approaching dominants in order to drink simultaneously from closely adjacent containers and received more aggression from them than in control periods. Thus, conflicts appeared to have had negative consequences for their relationships. However, when the pair made postconflict affiliative contact before being given access to the containers, the effect of the conflict was nullified in that the pair then accessed the containers with no further avoidance or aggressive behaviors. The results of this study demonstrated that tolerance was restored by reconciliation and implied that the former opponents were no longer uncertain about their relationships. Further support that reconciliation reduces uncertainty between former opponents is provided by playback experiments involving friendly grunts in wild baboons (Cheney and Seyfarth 1997). After playbacks of grunts from the former opponent, recipients of aggression approached former aggressors and tolerated their approaches more often than during postconflict periods without such playbacks.

THE VARIATION IN PATTERNS OF RECONCILIATION

Although reconciliation is a widespread phenomenon in primates, only a proportion of all conflicts are reconciled (Aureli et al. 2002). A number of factors have been considered that might affect the probability that reconciliation will occur, and most studies have investigated one or more of these (see Table 36.3).

Conflict Characteristics

Conflict Intensity

Conflicts vary in intensity and can range from threats to full contact aggression, sometimes resulting in wounding. It seems

Table 36.3   Factors Influencing the Probability of Reconciliation

SPECIES	KINSHIP	FRIENDLINESS	SEX	AGE	RANK	SUPPORT	INTENSITY	COUNTER AGGRESSION	INITIATION	
			Relationship Characteristics				Conflict Characteristics			
Cebus capucinus	Kin						nd	nd	agg/d	Leca et al. (2002)
Saimiri sciureus	Kin	✓	f–f most				High most			Pereira et al. (2000)
Cercopithecus aethiops	Kin									Cheney and Seyfarth (1989)
Erythrocebus patas	nd								nd	York and Rowell (1988)
Cercocebus torquatus atys									vict	Gust and Gordon (1993)
Macaca assamensis			m–m most				nd		agg/s	Cooper and Bernstein (2002)
Macaca arctoides	Kin								agg/s	de Waal and Ren (1988)
	Kin			Complex	Complex					Perez-Ruiz and Mondragon-Ceballos (1994)
	Kin	✓		nd	nd		nd	nd	nd	Call et al. (1999)
Macaca fascicularis	Nonkin								agg/s	Cords (1988)
	Kin	✓						nd	vict	Aureli et al. (1989)
									vict	Aureli (1992)
									nd	Cords (1993)
	nd	✓	jf–af > jm–af + jm–jm > jf–jf	nd		nd	nd	nd	vict/s	Cords and Aureli (1993)
	Kin									Aureli et al. (1997)
Macaca fuscata	Kin								nd	Aureli et al. (1993)
	Kin									Aureli et al. (1997)
			m–m most							Petit et al. (1997)
	Kin	✓	m–f least	j most			Low most			Schino et al. (1998)
	Kin									Kutsukake and Castles (2002)
						✓				Koyama (in prep)
Macaca maurus			m > f						vict	Matsumura (1996)
Macaca mulatta	Kin									de Waal and Yoshihara (1983)
	Kin									Call et al. (1996)
	Kin								agg	Demaria and Thierry (2001)
Macaca nemestrina	Kin				✓				vict	Judge (1991)
	Kin	✓					nd	nd	nd, agg/s	Castles et al. (1996)
Macaca nigra							nd	✓		Petit and Thierry (1994)
			nd							Petit et al. (1997)
Macaca sylvanus	nd									Aureli et al. (1997)
Macaca tonkeana	nd							nd	nd	Demaria and Thierry (2001)
Theropithecus gelada									nd	Swedell (1997)
Papio anubis							nd	nd	nd	Castles and Whiten (1998a)
Papio papio							nd	nd		Petit and Thierry (1994)
			m–f most							Petit et al. (1997)
Papio ursinus	Kin				Closely ranked + high vict rank				Agg	Silk et al. (1996)
Trachypithecus obscura	nd	✓		a–j least				nd	vict/s, agg/s	Arnold and Barton (2001a)
Colobus guereza									vict	Bjornsdottir et al. (2000)
Gorilla gorilla berengei	Kin	✓	m–f only							Watts (1995a)
Pan paniscus							High most		agg	de Waal (1997)
		✓	f–f most						nd	Palagi et al. (2004)
Pan troglodytes			m–m > m–f > f–f	nd					nd	de Waal and van Roosmalen (1979)
		✓	m–m, m–f > f–f	a–j less	nd		Low most (trend)	nd	f/d + vict/d	Arnold and Whiten (2001)
			f–f most, m–m least	nd						Fuentes et al. (2002)
		✓				nd				Preuschoft et al. (2002)
			m–f most, f–f least				nd	Low most	Social, nd Food, vict	Wittig and Boesch (2004)

✓, significantly increases the probability of reconciliation; f, female; m, male; a, adult; j, juvenile; nd, has no effect on the probability of reconciliation; agg, aggressor; vict, victim; s/d, same/opposite pattern observed in nonconflict contexts; a comma separates findings concerning different groups of the same species reported in the same study.

reasonable to assume that the more intense the conflict, the more disturbing to the relationship it will be. If so, conflicts of high intensity should be reconciled more often than those of low intensity. However, this appears not to be the case. Of the 13 studies in which the effect of conflict intensity has been investigated, only bonobos and squirrel monkeys have a higher conciliatory tendency following high-intensity conflicts. The remainder reconciled low-intensity conflicts more often or exhibited no difference in their tendency to reconcile conflicts of different intensities (Table 36.3). These results are difficult to explain, although it is likely that a number of factors are involved. The effect of conflict intensity on resultant inter-opponent distances was investigated in wild chimpanzees; as might be expected, low-intensity conflicts had a less dispersive effect on opponents than high-intensity conflicts, and these conflicts were also reconciled more often (Arnold and Whiten 2001). Similarly, conflicts among rhesus macaques often involve chases over 2 m and are rarely reconciled, whereas stump-tailed macaques tend to remain stationary while slapping or threatening their opponents and have a high conciliatory tendency (Call 1999, Call et al. 1999); however, the cause–effect relation between postconflict distance and likelihood of reconciliation has not been directly examined. Victims of more severe conflicts might be deterred from contacting former opponents (Petit et al. 1997), especially if the likelihood of renewed attacks after a high-intensity conflict is greater, although this has not yet been established. However, if their relationship is valuable to one or both former opponents, then they should still be motivated to communicate their interest in maintaining the relationship by attempting to reconcile the conflict despite the potential risks (Aureli et al. 1989). Based on this view and on the contradictory findings, conflict intensity may not be a major factor affecting the likelihood of reconciliation.

Counteraggression and Conflict Outcome

In general, the proportion of conflicts that involve direct retaliation (or counteraggression) is higher in more tolerant species. Weak power asymmetries allow a wide distribution of affiliative behaviors across group members (Thierry et al. 1990, 1994; Petit et al. 1992). Familiarity with the former opponent and the ability to approach dominants without negative consequences in other contexts (de Waal and Luttrell 1989) combined with the low cost of conflict is likely to increase the probability of reconciliation, especially if good relationships within the group are relatively common. Counteraggression either has not been shown to have any effect on the probability of reconciliation (stump-tailed macaques, Call et al. 1999; Tonkean macaques, Demaria and Thierry 2001; spectacled leaf monkeys, *Trachypithecus obscura*, Arnold and Whiten 2001; white-faced capuchins, *Cebus capucinus* Leca et al. 2002) or has resulted in higher conciliatory tendencies (black macaques [*Macaca nigra*], Petit and Thierry 1994a) in more tolerant species.

The outcome of a conflict can also have an effect on the likelihood of following reconciliation. Aureli et al. (1989)

found that undecided conflicts (i.e., those involving counteraggression and no clear signs of submission) in long-tailed macaques were reconciled more often than decided conflicts and that the initiative to reconcile was taken equally often by the aggressor and victim. They suggested that the uncertainty associated with undecided conflicts resulted in a greater motivation to reconcile in both opponents, whereas decided conflicts rarely resulted in aggressor-initiated reconciliation. As expected, counteraggression was rare in this despotic species but accounted for 25% of conflicts in a study of wild olive baboons (*Papio anubis*, Castles and Whiten 1998a), which have a similar dominance style. In this case, unidirectional conflicts and those involving counteraggression (also called *bidirectional conflicts*) were reconciled equally often. A very similar pattern was reported for Guinea baboons (*Papio papio*, Petit and Thierry 1994b). Castles and Whiten (1998a) suggested that bidirectional conflicts might not result in greater uncertainty if counteraggression is common. By contrast, Wittig and Boesch (2003) found that wild chimpanzees rarely reconciled undecided conflicts and tended to renew aggression instead despite the fact that undecided conflicts were uncommon.

Relationship Characteristics

Kinship

The importance of kinship in determining patterns of social interactions has been intensively studied and shown to be central in many primate species (Gouzoules and Gouzoules 1987). Accordingly, the influence of kinship on patterns of reconciliation has been widely investigated. Species in which cooperative and affiliative relationships occur most often within matrilines reconcile more conflicts with their kin than with unrelated opponents (e.g., *Macaca* spp., see Table 36.3). Kinship was not found to influence reconciliation in species with more open relationship networks (patas monkeys, York and Rowell 1988; spectacled leaf monkeys, Arnold and Barton 2001a), in immature long-tailed macaques (Cords 1988, Cords and Aureli 1993), and in macaque species whose social systems are less kin-biased (Tonkean macaques, Demaria and Thierry 2001; Barbary macaques [*Macaca sylvanus*], Aureli et al. 1997).

Friendliness

Dyads that engage in high levels of affiliation are said to be "friendly" (Cords 1997). Studies that have compared conciliatory tendencies among dyads which are more friendly and less friendly than average have found, without exception, that the most friendly dyads are more conciliatory (see Table 36.3). Primates living in kin-biased societies tend to direct most of their affiliation toward close relatives. In such cases, the influence of friendliness can be confounded by kinship, and a number of studies have overcome this problem by analyzing data on kin and nonkin dyads separately (de Waal and Yoshihara 1983, Aureli et al. 1989) or by using multivariate analyses, which identify variables that independently contribute to differences in reconciliation in dyads of differing types (Schino et al. 1998, Call et al. 1999).

Sex

Sex differences in the occurrence of reconciliation should depend on the nature of the society of the species under study and the role of the sexes within that society. For example, chimpanzees live in *male-philopatric* societies, in which males remain in their natal community and females emigrate (Goodall 1986). Consequently, most males are related to some degree and have lifelong social relationships with other males in the community. Males also cooperate with one another in community defense and in coalition formation. Females have few kin relations within the community and are relatively solitary. These broad sex differences in bondedness were reflected in postconflict patterns of a large captive group, with reconciliation being most common between males and least common between females (de Waal 1986); but this was not the case in a small group in which the only male–male dyad had the lowest conciliatory tendency (Fuentes et al. 2002). In a wild study, male–male conflicts were reconciled most frequently, although not significantly more than male–female conflicts, whereas female–female conflicts were never reconciled (Arnold and Whiten 2001). Reconciliation between females was observed in another study of wild chimpanzees, although rarely, and male–female conflicts were reconciled far more than any other sex combination in a community containing very few males (Wittig and Boesch 2003). In mountain gorillas (*Gorilla gorilla beringei*), bonds between females and between males are weak and reconciliation within these classes of dyads could not be demonstrated (Watts 1995). However, female mountain gorillas rely on males for protection and agonistic support (Watts, 1991, 1994). As a consequence, reconciliation between males and females was common. Among female-bonded species, conflicts between females were reconciled more frequently than male–female conflicts in some cases (rhesus macaques, Thierry 1986; Japanese macaques, Schino et al. 1998; squirrel monkeys [*Saimiri sciureus*], Pereira et al. 2000; Assamese macaques [*Macaca assamensis*], Cooper and Bernstein 2002), but in other cases male–male conflicts were reconciled at least as often as female–female conflicts (Tonkean macaques, Thierry 1986; juvenile long-tailed macaques, Cords and Aureli 1993; Japanese macaques, Petit et al. 1997, Schino et al. 1998; Assamese macaques, Cooper and Bernstein 2002).

Age

The effect of age on the likelihood of reconciliation has been investigated in some studies. Conflicts between adults and juveniles appear to be reconciled less often than conflicts within age classes (Schino et al. 1998, Arnold and Barton 2001a, Arnold and Whiten 2001). In their study of juvenile long-tailed macaques, Cords and Aureli (1993) found that juvenile females reconciled more conflicts with

unrelated adult females than juvenile males did and attributed this difference to the life histories of females, which remain in their natal troops in this species and have long-term relationships with other females. Males, on the other hand, emigrate to other groups and should have less interest in maintaining relationships with unrelated adult females. Perez-Ruiz and Mondragon-Ceballos (1994) found that in stump-tail macaques juvenile females that attacked infants often initiated reconciliation with them. They suggested that this might serve to appease adult females and reduce the probability of being attacked themselves, particularly by mothers.

Dominance Rank

The effect of dominance rank on reconciliation has received little attention to date. When the roles of aggressor and victim were not distinguished, the influence of rank and rank distance on conciliatory tendencies has not been consistent. In chacma baboons, individuals that were close in rank reconciled conflicts more often than individuals that were more distantly ranked (Silk et al. 1996). Close kin often occupy adjacent ranks in this species, and when the confounding effect of kinship was removed, this effect disappeared. However, the effect did remain in pig-tailed macaques when kin were removed from the analysis (Judge 1991). Similar analyses did not reproduce these findings in other species (Call et al. 1999, Arnold and Barton 2001a, Arnold and Whiten 2001). In a study in which the roles of aggressor and victim were distinguished, reconciliation was more frequent in cases where lower-ranking pig-tailed macaques attacked higher-ranking individuals than the reverse (Judge 1991). This finding supports Aureli et al.'s (1989) suggestion that greater uncertainty associated with the outcome of conflict promotes reconciliation between former opponents.

Support

Intuitively, one of the most valuable relationships a socially living primate can have is a coalition partnership (Harcourt and de Waal 1992). A reliable supporter would be a great asset in competitions for any resource, so individuals should be highly motivated to restore such relationships after potentially damaging conflicts. Remarkably, the early studies that investigated this aspect found that individuals did not reconcile more often with individuals that supported them (captive chimpanzees, Preuschoft et al. 2002; juvenile long-tailed macaques, Cords and Aureli 1993). Cords and Aureli (1993) pointed out that supporters of juveniles may not necessarily have good relationships with them or benefit from the relationship in the short term. Their primary motivation might, in some cases, be to direct aggression at the opponent rather than to aid the victim. They suggested that reconciliation should be most common where both former opponents perceive the relationship to be beneficial and less likely where asymmetries exist. Preuschoft et al. (2002) also suggested that their negative result might be due to the fact that there was little overlap between affiliation and support networks and that support relationships were likely to be highly asymmetrical. Wittig and Boesch (2003) found that dyads that supported one another and shared food reconciled more often than other types of dyad, although analyses of the effect of either of these factors alone did not yield significant results. In a recent attempt, Cooper et al. (2005) did find higher rates of reconciliation among regular supporters than among dyads that never supported one another in wild female Assamese macaques (*Macaca assamensis*). A similar result was found in Japanese macaques even when the effect of kinship was controlled (Koyama unpublished data). Contrary to previous studies, there was a significant overlap between dyadic relationships that were highly affiliative and those in which agonistic support was given.

Other Factors

Subgroup Association

In fission–fusion societies, the formation of temporary subgroups means that group members are rarely all together (Goodall 1986, Nishida and Hiraiwa-Hasegawa 1987, Chapman et al. 1993). Thus, subgroup association preferences could be reflected in conciliatory tendencies. If chimpanzees choose to associate with certain individuals over others, it might be expected that reconciliation following conflicts with these individuals should be relatively common. The current evidence is mixed. Whereas Wittig and Boesch (2003) found that chimpanzees that often associate with one another are more likely to reconcile than those that associate rarely, Arnold and Whiten (2001) did not find such a pattern.

Mating Seasonality

Certain species have a defined mating season (Dixson 1998). Aureli and colleagues (1993, 1994) confirmed the occurrence of reconciliation in both mating and non-mating seasons in Japanese and Barbary macaques. The mating season is often a time of increased levels of aggression (Schino et al. 1998, Abegg et al. 2003). Schino et al. (1998) found that Japanese macaques reconciled fewer conflicts during this period and suggested that this was due to a general deterioration of social relationships as a result of increased competition and tension. Other studies have failed to find seasonal differences in conciliatory tendencies (Kutsukake and Castles 2001, Abegg et al. 2003). Arnold and Whiten (2001) also detected no difference in conciliatory tendencies after male–female conflicts involving females that were sexually receptive as opposed to conflicts involving females that were not.

Foraging

The majority of studies that have compared the occurrence of reconciliation in feeding and non-feeding contexts have found that monkeys do not reconcile conflicts over food as often as conflicts in other contexts (de Waal 1984, Aureli 1992, Matsumura 1996, Verbeek and de Waal 1997, Castles and Whiten 1998a, Westlund et al. 2000; but see Cooper and Bernstein 2002). It is likely that many non-food-related conflicts have a more directly social dimension, and it has been suggested that social relationships might not be disturbed by food contest competition (Aureli 1992). Verbeek and de Waal (1997) predicted that chimpanzees might not show the same pattern of reduced reconciliation in feeding contexts as they employ a wide range of reassurance and appeasement behaviors at clumped food sources (Goodall 1986, de Waal 1989a), which suggests that social relationships are at stake. Data on wild populations of chimpanzees support this prediction as reconciliation was at least as common during foraging as it was in other contexts (Arnold and Whiten 2001, Wittig and Boesch 2003).

INITIATIVE TO RECONCILE

Whether it is the aggressor or recipient of aggression that is responsible for the initiation of reconciliation should depend on a number of factors. Most relationships are asymmetrical to some extent, and one party will have more to gain from restoring the relationship than the other (Cords and Aureli 2000). Therefore, the individual that has most to gain should be the one that initiates conciliatory contact. However, approaching a former opponent can be risky, especially for the victim. In despotic species, approaching a dominant individual can have negative consequences, even outside of agonistic contexts, and is relatively rare. For this reason, victims of aggression may be deterred from attempting reconciliation with a higher-ranking former opponent even when it is in their interest to do so. Aggressors may have less to gain from reconciliation than the victim, but they must still maintain their social networks and so should also be motivated to resolve at least some of their conflicts. Both parties should also benefit from the stress-reducing effect of reconciliation and the elimination of uncertainty regarding their future relationship, but again the relative benefit is likely to be different for each opponent.

In addition to methodological difficulties in detecting genuine attempts to reconcile (Aureli and Smucny 2000, Cords and Aureli 2000), the combination of variation in the motivation to reconcile and the risks involved in approaching a former opponent is reflected in inconclusive findings regarding the initiation of reconciliation (Table 36.3). For example, aggressors were primarily responsible for reconciliation in one group of rhesus macaques (Demaria and Thierry 2001), a species in which it might be expected that victims would be reluctant to approach former opponents, and in more tolerant species such as stump-tailed macaques (de Waal and Ren 1988) and bonobos (*Pan paniscus*, de Waal 1987). Variations in patterns of initiation have also been reported for different groups of the same species (e.g., long-tailed macaques, Aureli et al. 1989, Aureli 1992, Cords 1993). A number of other studies have reported no difference in patterns of initiation where biases in favor of one type of opponent might be expected according to their dominance styles (Table 36.3).

The majority of studies that have considered the identities of individuals that take the initiative to reconcile within any particular dyad have found that the individuals that initiate reconciliation are also primarily responsible for friendly contacts in general (Cords 1988, de Waal and Ren 1988, Aureli et al. 1993, Cords and Aureli 1993, Castles et al. 1996, Arnold and Barton 2001a, Cooper and Bernstein 2002). In a few cases, the postconflict pattern was different. In white-faced capuchins, aggressors initiated reconciliation more than affiliation in MCs (Leca et al. 2002), whereas in long-tailed macaques and a newly established group of pig-tailed macaques, victims initiated reconciliation more often than affiliation in MCs (Aureli et al. 1989, Castles et al. 1996). Wild chimpanzee victims were shown to initiate a little more than half of their affiliative contacts with former opponents in normal social contexts, but this figure was significantly elevated in postconflict conditions (Arnold and Whiten 2001). Such changes in patterns of behavior in postconflict situations can be viewed as an indication of the relative importance of re-establishing relationships and the differential motivation to do so between opponents.

Almost all studies have defined the initiator of reconciliation as the individual that approaches the former opponent. Silk and colleagues have usefully pointed out that in some cases chacma baboons grunt before approaching former opponents and suggest that this might serve as reassurance that subsequent interactions will not be antagonistic (Silk 1996, 1997; Silk et al. 1996). A number of species produce vocalizations during reconciliation (stump-tailed macaques, de Waal and Ren 1988, mountain gorillas, Watts 1995, brown capuchins [*Cebus apella*], Verbeek and de Waal 1997, spectacled leaf monkeys, Arnold and Barton 2001a, white-faced capuchins, Leca et al. 2002), although details of whether or not these vocalizations begin prior to the act of reconciliation and which individuals produce such signals have not been provided except in the case of chacma baboons. Other forms of more subtle signaling, such as changes in body posture or facial expressions, may also have been overlooked by researchers (Cords and Aureli 2000). It is possible that the responsibility for initiating reconciliation could have been ascribed wrongly in some cases if such signals are involved in soliciting contact or communicating nonaggressive intentions and are therefore instrumental in bringing about reconciliation (Arnold and Barton 2001b).

THE ONTOGENY OF RECONCILIATION

The interspecific and interindividual variation in conciliatory tendencies and conciliatory styles can be quite marked, as illustrated above. While some of this variation is likely to result from differences in genetically based dispositions, there is good evidence that the social environment has an important role in influencing reconciliation behavior. The absence of social experience can result in an inability to reconcile at all. Rhesus macaques which had been removed

from their mothers before weaning, hand-reared, and housed singly for extended periods did not show any form of reconciliation, in contrast to individuals that after separation from the mother were housed in a peer group at an early age (Ljungberg and Westlund 2000). That early socialization seems sufficient to develop conciliatory behavior is supported by a study on chimpanzees (Spijkerman et al. 1997) where adolescents reared in peer groups did not differ in reconciliation rates from those reared with mothers in a zoo group.

Variation in reconciliation was found between two groups of long-tailed macaques, which were formed by dividing a group according to the female dominance hierarchy: one group contained the higher-ranking females and the other, the lower-ranking females (Butovskaya et al. 1996). Reconciliation was more common in the low-ranking group, suggesting that the difference could have been due to the exposure to a more tolerant social environment during development or to genetic predisposition given the relatedness of the females. Similarly, when juvenile rhesus macaques were cohoused with juvenile stump-tailed macaques over a period of 6 weeks, the conciliatory tendency of the rhesus macaques rose gradually until it matched the much higher conciliatory tendency typical of the more tolerant stump-tailed macaques (de Waal and Johanowicz 1993). Interestingly, the rhesus macaques did not extend their social learning to the specific patterns of behavior, such as hold-bottom, typically used by stump-tailed macaques during reconciliation (Table 36.2, Fig. 36.2), which suggests that some aspects of reconciliation might not be modifiable.

A second line of research has explored the influence of the mother–offspring relationship in the development of reconciliatory behavior. Infant, weanling, and juvenile brown capuchin monkeys were classified as having either secure or insecure relationships with their mothers depending on the relative rate of affiliation and agonism observed between them (Weaver and de Waal 2000, 2003). Reconciliation began to emerge in infancy irrespective of the security of the mother–offspring relationship. The quality of the mother–offspring relationship did, however, affect the way in which reconciliation was accomplished. Secure offspring were not intimidated by their mother's aggression and were similarly relaxed after conflicts with unrelated adults. They performed calm appeasement behaviors toward unrelated adult aggressors and were often approached for reconciliation by former opponents. Insecure offspring, by contrast, were often noisy and agitated after conflicts. They had a higher conciliatory tendency with unrelated adults than secure offspring, but this was a result of a high proportion of offspring-initiated reconciliation. Insecure offspring were more inclined to seek reconciliation that involved directly comforting behaviors involving body contact and rarely had contact with their mothers in post-conflict periods. It appears that the style of individual reconciliation behavior develops from the mother–infant attachment behavior system in this species (Weaver and de Waal 2000, 2003).

HYPOTHESES CONCERNING THE FUNCTION AND DISTRIBUTION OF RECONCILIATION

Since 1979, a large amount of comparative data have been accumulated which has enabled the generation of a number of hypotheses to explain why reconciliation is such a widespread feature of primate social life and to account for the form and distribution of reconciliation within and between species. Early hypotheses (reviewed by Kappeler and van Schaik 1992) include the *minimum cognitive capacity hypothesis* (Gallup 1982), which states that the ability to reconcile requires self-awareness and the capacity to empathize and attribute mental states to others. In fact, there is no reason to assume that such high-order cognitive capacities are necessary since individual recognition, memory, and an affiliative disposition in postconflict situations are sufficient (de Waal and Yoshihara 1983, Castles 2000). Thus, reconciliation is expected to be a widespread phenomenon among social animals. Operational reconciliation has been recently demonstrated in captive bottlenose dolphins (*Tursiops truncatus*, Weaver 2003), domestic goats (*Capra hircus*, Schino 1998), spotted hyenas (*Crocuta crocuta*, Hofer and East 2000, Wahaj et al. 2001), and even between cleaner fish and their clients (*Labroides dimidiatus*, Bshary and Würth 2001).

The *social evolution hypothesis* (de Waal 1989b) states that reconciliation is necessary to maintain group cohesion in the face of the dispersive effects of aggression and should be found in all species which live in social groups. Kappeler's (1993) early failure to demonstrate reconciliation in ring-tailed lemurs appeared not to support this hypothesis, but a later study suggested that reconciliation may occur in this species (Rolland and Roeder 2000). Since then, reconciliation has been reported to be absent in black lemurs and red-bellied tamarins (Roeder et al. 2002, Schaffner and Caine 2000), but explanations have been suggested which do not undermine this hypothesis (Aureli et al. 2002).

de Waal (1986) proposed the *reconciled hierarchy hypothesis*, which is based on the notion that dominance relationships promote peaceful coexistence and that reconciliation is dependent on the subordinate signaling its status to the satisfaction of the dominant following a conflict. Kappeler and van Schaik (1992) interpreted this hypothesis as predicting that reconciliation should be more easily achieved in species with formalized dominance hierarchies where ritualized submission on the part of one opponent can terminate conflict and promote friendly interactions. This has proven not to be the case as such species have among the lowest reported conciliatory tendencies (Kappeler and van Schaik 1992). Furthermore, Arnold and Whiten (2001) found that only about 50% of conciliatory contacts were preceded by submissive "pant-grunts" in wild chimpanzees, which suggests that status signaling is not necessary in more tolerant species at least.

One of the most robust hypotheses proposed to date is the *valuable relationships hypothesis* (de Waal and Aureli 1997). This has been developed from earlier hypotheses which stressed the importance of relationship quality in determining which kinds of dyad are most likely to reconcile (de Waal and Yoshihara 1983, Aureli et al. 1989, Kappeler and van Schaik 1992, Cords and Aureli 1993). This hypothesis assumes the relationship repair function of reconciliation and predicts that conflicts within dyads that have a relationship which has important positive fitness benefits for both parties are more likely to be reconciled than conflicts within dyads that have less valuable relationships (see Cords and Aureli 2000 for a review). Support for this hypothesis is abundant and can be found in the sections above.

Cords and Aureli (1993, 2000) emphasized three dimensions of relationship quality that are likely to influence the probability that reconciliation will occur. The first is *value*, i.e., what an individual can potentially offer, how disposed the individual is to offer it, and how accessible that individual is as a social partner. Value is not an easily defined or measurable quality. It is made easier if it is based on behavioral exchanges that result in clear fitness advantages. A coalition partner fits this description, and although direct evidence that reconciliation is more frequent among coalition partners is currently mixed (Cords and Aureli 1993, Preuschoft et al. 2002, Wittig and Boesch 2003, Cooper et al. 2005, Koyama unpublished data), indirect evidence comes from studies that show high rates of reconciliation within dyads that are known to be the most likely to offer coalitionary support. For example, in species in which matrilineal kin are most often supporters, reconciliation is kin-biased. In chimpanzees, males typically form coalitions with one another for within- and between-group conflicts and male–male conflicts are reconciled more often than female–female conflicts (de Waal 1986, Goodall 1986, Arnold and Whiten 2001, Wittig and Boesch 2003). Direct evidence that value influences reconciliation comes from an experimental study in which pairs of long-tailed macaques were trained to cooperate with one another to gain access to food (Cords and Thurnheer 1993). The conciliatory tendencies of these dyads rose after their value was increased during the training phase.

The second dimension of relationship quality is *compatibility*, defined as the general tenor of social interactions between individuals which derives from both the temperaments of the partners and their shared history of social interactions (Cords and Aureli 1993, 2000). This is straightforward to measure as the relative frequencies of friendly interactions such as grooming, contact sitting, or approaches leading to affiliative behavior compared to aggressive interactions. Reconciliation has been always found to be more common within dyads that have friendly relationships than within those that do not, even after controlling for their higher base-line levels of affiliation (Table 36.3). Additionally, groups with a wide distribution of compatible or friendly partners have higher overall conciliatory tendencies than groups which are more cliquish and restrict affiliation to a small subset of group members (Call et al. 1996, Castles et al. 1996). Similarly, species that show high rates of baseline affiliation are particularly conciliatory (de Waal and Luttrell 1989, Thierry 2000).

The third dimension of relationship quality that should influence patterns of reconciliation is *security*, defined as the perceived probability that the relationship will not change for the worse and measured by the consistency of the partner's behavioral response (Cords and Aureli 1993, 2000). The more secure a relationship is, the less likely it is to need restoration via reconciliation. The influence of relationship security has not yet been directly investigated, but indirect evidence highlights its potential importance. Peer groups of juvenile long-tailed macaques reconciled with nonkin more often than with kin, and this difference was interpreted as reflecting greater uncertainty about these relationships in a situation where both kin and nonkin were similar in their value as social partners (Cords and Aureli 1993). Immature brown capuchin monkeys that generalized their insecure relationships with their mothers to other unrelated adults were more conciliatory than their more secure counterparts (Weaver and de Waal 2000, 2003).

Silk (1996) offers an alternative view of the function of reconciliation. In chacma baboons, dominant former aggressors often approach their victims and grunt to them (Silk et al. 1996). Grunting facilitates friendly contact and the ability of former aggressors to handle the infants of their victims in particular. Silk (1996) views grunts as signals of "benign intent" that communicate a friendly attitude in the former aggressor and serve to reduce anxiety in the victim. She views reconciliation not as a mechanism for relationship repair but as facilitation of friendly interactions in the short term, leading to the fulfillment of immediate goals (Silk 1996, 1997). This distinction is a subtle one but rides on the lack of evidence that relationships are improved following reconciliation compared to preconflict periods, which Silk (1996) believes is the claim of the relationship repair hypothesis. However, as Cords and Aureli (1996) point out, the relationship repair hypothesis predicts a change in the relationship but only in so far as it restores the relationship to its preconflict state, which will have consequences for the maintenance of the relationship into the future but might not necessarily improve it. There is clearly evidence for such restoration (see above for the benefits of reconciliation).

A second line of research has focused on the short-term benefits of reconciliation in reducing stress levels in former opponents. As mentioned previously, the rate of self-directed behaviors increases in both aggressors and victims after conflicts and is reduced to baseline levels when reconciliation is achieved. Following the first studies of this phenomenon, Aureli and collaborators (Aureli et al. 1989, Aureli and van Schaik 1991) proposed the *uncertainty-reduction hypothesis*, which predicts that self-directed behaviors should decrease after reconciliation and interprets this reduction as a decrease in uncertainty in the former opponents about the status of the relationship. This hypothesis has been supported by all subsequent studies of cercopithecines that have investigated the effects of reconciliation on postconflict anxiety (Aureli et al. 1989, Aureli and van

Schaik 1991, Castles and Whiten 1998b, Das et al. 1998, Kutsukake and Castles 2001). In studies that examined the behavior of aggressors and the nature of the conflicts, the effectiveness of reconciliation in reducing anxiety has not been consistent. In one study of long-tailed macaques, the rates of self-directed behaviors of aggressors were restored to baseline levels following reconciliation (Das et al. 1998). However, Castles and Whiten (1998b) found that although reconciliation following bidirectional conflicts reduced rates of self-directed behaviors in aggressors, reconciliation did not have that effect when conflicts were unidirectional. However, partitioning of opponent and conflict type resulted in small sample sizes, so these findings should be treated as preliminary.

The most recent development integrates the valuable relationships hypothesis and the uncertainty-reduction hypothesis in order to take into account the variation and flexibility of reconciliation. The *integrated hypothesis* (Aureli 1997) suggests that postconflict anxiety is due to uncertainty about the future of the relationship between the former opponents and predicts that levels of postconflict anxiety mediate the occurrence of reconciliation. Higher levels of postconflict anxiety are expected after conflicts between valuable partners because of the potential loss of benefits. Thus, conflicts that disturb valuable relationships are expected to result in higher rates of self-directed behavior and then reconciliation. In support of this hypothesis, Aureli (1997) found that long-tailed macaque victims scratched more after conflicts with individuals with which they shared strong affiliative bonds than with other group members; they also reconciled more with such individuals (Aureli et al. 1989). In a recent test of this hypothesis, Kutsukake and Castles (2001) found that conflicts among kin resulted in higher rates of self-directed behavior and reconciliation than conflicts among nonkin. In both studies, the risk of renewed attacks was not associated with such relationship-dependent variation; thus, the higher anxiety levels were attributed to the disturbance of more valuable relationships. Such emotional mediation could serve as a proximate mechanism at the basis of the variation and flexibility of reconciliation depending on the quality of the relationships between former opponents (Aureli and Smucny 2000, Aureli et al. 2002).

CONCLUSIONS

To date, over 50 empirical studies have investigated post-conflict reconciliation in nonhuman primates (Table 36.1). After demonstrating the phenomenon, many of them reported on various factors affecting reconciliation patterns (Table 36.3). Given the overall similarity in the methodology across studies, it has been possible to make meaningful inferences from comparisons of the available evidence (de Waal 1989b, 2000a; Thierry 1990, 2000; Kappeler and van Schaik 1992; Cords and Killen 1998; Silk 1997; Aureli and de Waal 2000; Aureli et al. 2002; Judge 2003). In contrast, only a minority of studies have examined the function of reconciliation. This is certainly a shortcoming as reconciliation may function differently in different species and under different conditions. The next generation of studies on reconciliation should focus on functional aspects, checking first whether the aggressive conflict actually disturbed the social relationship between former opponents rather than simply assuming the need for reconciliation under any circumstance (Aureli et al. 2002).

Another aspect that should receive more attention is the patterning of behavioral exchanges that lead to reconciliation. It is well established that certain species, but not others, use explicit forms of reconciliation characterized by conspicuous affiliative patterns rarely displayed in other contexts. However, variation in the use of these patterns within the same species has been reported (Castles at al. 1996, Arnold and Whiten 2001), and new studies need to focus on the reasons behind the variation. Furthermore, the role each opponent plays in achieving reconciliation is still poorly understood. Empirical data are needed to test the various predictions based on asymmetries between opponents in partner value and in postconflict negative consequences. Better empirical evidence for such testing requires more attention to subtle signals and vocalizations (Silk et al. 1996), but measuring initiative in reconciliation will be challenging because of the difficulty in recognizing subtle cues (Cords and Aureli 2000). Further research may use indicators of emotional responses, such as self-directed behavior, or attention to former opponents to examine how the opponents themselves perceive the postconflict situation and to infer the opponents' relative interest, motivation, and expected initiative in reconciliation (Aureli and Smucny 2000).

A critical aspect of the study of postconflict behavioral exchanges between former opponents is to establish whether reconciliation attempts are successful or not. Reconciliation is a dyadic phenomenon in which both partners need to cooperate. Only attempts that are accepted by the receiver are expected to achieve the beneficial effects implied in the term *reconciliation*. No study on nonhuman primates has explicitly addressed this issue, probably due to the difficulty in determining the actual acceptance of a friendly overture (but see Ljungberg et al. 1999 for a child study). More effort should be put in examining the responses of the receiver of the conciliatory attempt. For example, attempts that are reciprocated by the partner (e.g., mutual grooming) or that require direct participation from the partner (e.g., hold-bottom in stump-tailed macaques, Fig. 36.2) might be more effective in terms of restoring the relationship than attempts that are characterized only by unidirectional friendly contacts.

Our review of the various hypotheses put forward to explain the function and distribution of reconciliation shows that there is a general consensus in favor of the valuable relationships hypothesis. The uncertainty-reduction hypothesis has received even more support, but more research on the emotional responses of former aggressors is needed. These data would help in producing decisive tests for the integrated hypothesis, which combines the valuable relationship hypothesis and the uncertainty-reduction hypothesis to explain the variation and flexibility of reconciliation.

Although more studies are needed to gather the detailed data to fully understand the phenomenon of reconciliation in nonhuman primates, much has been accomplished since the pioneering work of de Waal and van Roosmalen (1979). The wealth of knowledge accumulated so far can be used to predict patterns of postconflict behavior in other primate and nonprimate species, to facilitate more systematic investigation, and to develop a framework for the evolution of conflict resolution in animal societies (Aureli et al. 2002).

REFERENCES

Abegg, C., Petit, O., and Thierry, B. (2003). Variability in behavior frequencies and consistency in transactions across seasons in captive Japanese macaques (*Macaca fuscata*). *Aggress. Behav.* 29:81–93.

Arnold, K., and Barton, R. A. (2001a). Postconflict behavior of spectacled leaf monkeys (*Trachypithecus obscurus*). I. Reconciliation. *Int. J. Primatol.* 22:243–266.

Arnold, K., and Barton, R. A. (2001b). Postconflict behavior of spectacled leaf monkeys (*Trachypithecus obscurus*). II. Contact with third parties. *Int. J. Primatol.* 22:267–286.

Arnold, K., and Whiten, A. (2001). Post-conflict behaviour of wild chimpanzees (*Pan troglodytes schweinfurthii*) in the Budongo Forest, Uganda, *Behaviour* 138:649–690.

Aureli, F. (1992). Post-conflict behaviour among wild long-tailed macaques (*Macaca fascicularis*). *Behav. Ecol. Sociobiol.* 31:329–337.

Aureli, F. (1997). Post-conflict anxiety in nonhuman primates: the mediating role of emotion in conflict resolution. *Aggress. Behav.* 23:315–328.

Aureli, F., Cords, M., and van Schaik, C. P. (2002). Conflict resolution following aggression in gregarious animals: a predictive framework. *Anim. Behav.* 64:325–343.

Aureli, F., Cozzolino, R., Cordischi, C., and Scucchi, S. (1992). Kin oriented redirection among Japanese macaques: an expression of a revenge system. *Anim. Behav.* 44:283–291.

Aureli, F., Das, M., and Veenema, H. C. (1997). Differential kinship effect on reconciliation in three species of macaques (*Macaca fascicularis, M. fuscata, M. sylvanus*). *J. Comp. Psychol.* 111:91–99.

Aureli, F., Das, M., Verleur, D., and van Hooff, J. A. R. A. M. (1994). Post-conflict social interactions among Barbary macaques (*Macaca sylvanus*). *Int. J. Primatol.* 15:471–485.

Aureli, F., and de Waal, F. B. M. (eds) (2000). *Natural Conflict Resolution*. California University Press, Berkeley and Los Angeles.

Aureli, F., and Smucny, D. A. (1998). New directions in conflict resolution research. *Evol. Anthropol.* 6:115–119.

Aureli, F., and Smucny, D. A. (2000). The role of emotion in conflict and conflict resolution. In: Aureli, F., and de Waal, F. B. M. (eds.), *Natural Conflict Resolution*. University of California Press, Berkeley. pp. 199–224.

Aureli, F., and van Schaik, C. P. (1991). Post-conflict behaviour in long-tailed macaques (*Macaca fascicularis*): I. The social events. *Ethology* 89:89–100.

Aureli, F., van Schaik, C. P., and van Hooff, J. A. R. A. M. (1989). Functional aspects of reconciliation among captive long-tailed macaques (*Macaca fascicularis*). *Am. J. Primatol.* 19:39–51.

Aureli, F., Veenema, H. C., van Panthaleon van Eck, C. J., and van Hooff, J. A. R. A. M. (1993). Reconciliation, consolation, and redirection in Japanese macaques (*Macaca fuscata*). *Behaviour* 124:1–21.

Barros, M., Boere, V., Huston, J. P., and Toaz, C. (2000). Measuring fear and anxiety in the marmoset (*Callithrix penicillata*) with a novel predator confrontation model: effects of diazepam. *Behav. Brain Res.* 108:205–211.

Bjornsdotter, M., Larsson, L., and Ljungberg, T. (2000). Postconflict affiliation in two captive groups of black-and-white guereza (*Colobus guereza*). *Ethology* 106:289–300.

Boinski, S. (2000). Social manipulation within and between troops mediates primate group movement. In: Boinski, S., and Garber, P. A. (eds.), *On the Move—How and Why Animals Travel in Groups*. University of Chicago Press, Chicago. pp. 421–446.

Bshary, R., and Würth, M. (2001). Cleaner fish (*Labroides dimidiatus*) manipulate client reef fish providing tactile stimulation. *Proc. R. Soc. Lond. B* 286:1495–1501.

Butovskaya, M., Kozintsev, A., and Welker, C. (1996). Conflict and reconciliation in two groups of crab-eating monkeys differing in social status by birth. *Primates* 37:261–270.

Call, J. (1999). The effect of inter-opponent distance on the occurrence of reconciliation in stumptail (*Macaca arctoides*) and rhesus macaques (*Macaca mulatta*). *Primates*. 40:515–523.

Call, J. (2000). Distance regulation in macaques: a form of implicit reconciliation? In: Aureli, F., and Waal, F. B. M. (eds.), *Natural Conflict Resolution*. University of California Press, Berkeley and Los Angeles. pp. 191–193.

Call, J., Aureli, F., and de Waal, F. B. M. (1999). Reconciliation patterns among stumptail macaques: a multivariate approach. *Anim. Behav.* 58:165–172.

Call, J., Judge, P. G., and de Waal, F. B. M. (1996). Influence of kinship and spatial density on reconciliation and grooming in rhesus monkeys. *Am. J. Primatol.* 39:35–45.

Castles, D. L. (2000). Triadic versus dyadic resolutions. In: Aureli, F., and Waal, F. B. M. (eds.), *Natural Conflict Resolution*. University of California Press, Berkeley and Los Angeles. pp. 289–291.

Castles, D. L., Aureli, F., and de Waal, F. B. M. (1996). Variation in conciliatory tendency and relationship quality across groups of pigtail macaques. *Anim. Behav.* 52:389–403.

Castles, D. L., and Whiten, A. (1998a). Post-conflict behaviour of wild olive baboons. I. Reconciliation, redirection and consolation. *Ethology* 104:126–147.

Castles, D. L., and Whiten, A. (1998b). Post-conflict behaviour of wild olive baboons. II. Stress and self-directed behaviour. *Ethology* 104:148–160.

Chapman, C. A., White, F. J., and Wrangham, R. W. (1993). Defining subgroup size in fission–fusion societies. *Folia Primatol.* 61:31–34.

Cheney, D. L., and Seyfarth, R. M. (1989). Redirected aggression and reconciliation among vervet monkeys, *Cercopithecus aethiops*. *Behaviour* 110:258–275.

Cheney, D. L., and Seyfarth, R. M. (1997). Reconciliatory grunts by dominant female baboons influence victim's behaviour. *Anim. Behav.* 54:409–418.

Cheney, D. L., Seyfarth, R. M., and Silk, J. B. (1995). The role of grunts in reconciling opponents and facilitating interactions among adult female baboons. *Anim. Behav.* 50:249–257.

Clutton-Brock, T. H. (1989). Mammalian mating systems. *Proc. R. Soc. Lond. B* 236:339–372.

Cooper, M. A., and Bernstein, I. S. (2002). Counter aggression and reconciliation in Assamese macaques (*Macaca assamensis*). *Am. J. Primatol.* 56:215–230.

Cooper, M. A., Bernstein, I. S., and Hemelrijk, C. K. (2005). Reconciliation and relationship quality in Assamese macaques (*Macaca assamensis*). *Am. J. Primatol.* 65:269–282.

Cords, M. (1988). Resolution of aggressive conflicts by immature long-tailed macaques, *Macaca fascicularis*. *Anim. Behav.* 36:1124–1135.

Cords, M. (1992). Postconflict reunions and reconciliation in long-tailed macaques. *Anim. Behav.* 44:57–61.

Cords, M. (1993). On operationally defining reconciliation. *Am. J. Primatol.* 29:255–267.

Cords, M. (1997). Friendships, alliances, reciprocity and repair. In: Whiten, A., and Byrne, R. W. (eds.), *Machiavellian Intelligence II: Evaluations and Extensions*. Cambridge University Press, Cambridge. pp. 24–49.

Cords, M., and Aureli, F. (1993). Patterns of reconciliation among juvenile long-tailed macaques. In: Pereira, M. E., and Fairbanks, L. A. (eds.), *Juvenile Primates: Life History, Development and Behavior*. Oxford University Press, New York. pp. 271–284.

Cords, M., and Aureli, F. (1996). Reasons for reconciling. *Evol. Anthropol.* 5:42–45.

Cords, M., and Aureli, F. (2000). Reconciliation and relationship qualities. In: Aureli, F., and de Waal, F. B. M. (eds.), *Natural Conflict Resolution*. University of California Press, Berkeley. pp. 177–198.

Cords, M., and Killen, M. (1998). Conflict resolution in human and nonhuman primates. In: Langer, J., and Killen, M. (eds.), *Piaget, Evolution, and Development*. Lawrence Erlbaum, Mahwah, NJ. pp. 193–219.

Cords, M., and Thurnheer, S. (1993). Reconciling with valuable partners by long-tailed macaques. *Ethology* 93:315–325.

Darwin, C. (1871). *The Descent of Man, and Selection in Relation to Sex*. John Murray, London.

Das, M., Penke, Z., and van Hooff, J. A. R. A. M. (1998). Postconflict affiliation and stress-related behavior of long-tailed macaque aggressors. *Int. J. Primatol.* 19:53–71.

Demaria, C., and Thierry, B. (2001). A comparative study of reconciliation in rhesus and Tonkean macaques. *Behaviour* 138:397–410.

de Waal, F. B. M. (1984). Coping with social tension: sex differences in the effect of food provision to small rhesus monkey groups. *Anim. Behav.* 32:765–773.

de Waal, F. B. M. (1986). The integration of dominance and social bonding in primates. *Q. Rev. Biol.* 61:459–479.

de Waal, F. B. M. (1987). Tension regulation and nonreproductive functions of sex in captive bonobos (*Pan paniscus*). *Natl. Geogr. Res.* 3:318–335.

de Waal, F. B. M. (1989a). Food sharing and reciprocal obligations among chimpanzees. *J. Hum. Evol.* 18:433–459.

de Waal, F. B. M. (1989b). *Peacemaking Among Primates*. Harvard University Press, Cambridge, MA.

de Waal, F. B. M. (1993). Reconciliation among primates: a review of empirical evidence and unresolved issues. In: Mason, W. A., and Mendoza, S. P. (eds.), *Primate Social Conflict*. State University of New York Press, Albany. pp. 111–144.

de Waal F. B. M. (2000a). Primates—a natural heritage of conflict resolution. *Science*. 289:586–590.

de Waal, F. B. M. (2000b). The first kiss: foundations of conflict resolution research in animals. In: Aureli, F., and de Waal, F. B. M. (eds.), *Natural Conflict Resolution*. University of California Press, Berkeley. pp. 15–33.

de Waal, F. B. M., and Aureli, F. (1997). Conflict resolution and distress alleviation in monkeys and apes. In: Carter, C. S.,

Kirkpatrick, B., and Lenderhendler, I. (eds.), *The Integrative Neurobiology of Affiliation*. Annals of the New York Academy of Sciences, New York. pp. 317–328.

de Waal, F. B. M., and Johanowicz, D. L. (1993). Modification of reconciliation behavior through social experience—an experiment with 2 macaque species. *Child Dev.* 64:897–908.

de Waal, F. B. M., and Luttrell, L. M. (1989). Toward a comparative socioecology of the genus *Macaca*: different dominance styles in rhesus and stumptail monkeys. *Am. J. Primatol.* 19:83–109.

de Waal, F. B. M., and Ren, R. M. (1988). Comparison of the reconciliation behavior of stumptail and rhesus macaques. *Ethology* 78:129–142.

de Waal, F. B. M., and van Roosmalen, A. (1979). Reconciliation and consolation among chimpanzees. *Behav. Ecol. Sociobiol.* 5:55–66.

de Waal, F. B. M., and Yoshihara, D. (1983). Reconciliation and redirected affection in rhesus monkeys. *Behaviour* 85:224–241.

Dixon, A. F. (1998). *Primate Sexuality: Comparative Studies of the Prosimians, Monkeys, Apes, and Humans*. Oxford University Press, Oxford.

Fuentes, A., Malone, N., Sanz, C., Matheson, M., and Vaughan, L. (2002). Conflict and post-conflict behavior in a small group of chimpanzees. *Primates* 43:223–235.

Gallup, G. G. J. (1982). Self-awareness and the emergence of mind in primates. *Int. J. Primatol.* 2:237–248.

Goodall, J. (1986). *The Chimpanzees of Gombe: Patterns of Behavior*. Harvard University Press, Cambridge, MA.

Gouzoules, S., and Gouzoules, H. (1987). Kinship. In: Smuts, B. B., Cheney, D. L., Seyfarth, R. M., Wrangham, R. W., and Struhsaker, T. T. (eds.), *Primate Societies*. University of Chicago Press, Chicago. pp. 299–305.

Gust, D. A., and Gordon, T. P. (1993). Conflict resolution in sooty mangabeys. *Anim. Behav.* 46:685–694.

Harcourt, A. H., and de Waal, F. B. M. (1992). *Coalitions and Alliances in Humans and Other Animals*. Oxford University Press, Oxford.

Henry, J. P. (1982). The relation of social to biological processes in disease. *Soc. Sci. Med.* 16:369–380.

Hofer, H., and East, M. L. (2000). Conflict management in female-dominated spotted hyenas. In: Aureli, F., and de Waal, F. B. M. (eds.), *Natural Conflict Resolution*. University of California Press, Berkeley. pp. 232–234.

Hohmann, G., and Fruth, B. (2000). Use and function of genital contacts among female bonobos. *Anim. Behav.* 60:107–120.

Janson, C. H. (1992). Evolutionary ecology of primate social structure. In: Smith, E. A., and Winterhalader, B. (eds.), *Evolutionary Ecology and Human Behavior*. Aldine de Gruyter, Hawthorne, NY. pp. 95–130.

Judge, P. G. (1991). Dyadic and triadic reconciliation in pigtail macaques (*Macaca nemestrina*). *Am. J. Primatol.* 23:225–237.

Judge, P. G. (2003). Conflict resolution. In: Maestripieri, D. (ed.), *Primate Psychology*. Harvard University Press, Cambridge, MA. pp. 41–68.

Kaplan, J. R. (1986). Psychological stress and behavior in nonhuman primates. In: Mitchell, G., and Erwin, J. (eds.), *Comparative Primate Biology*. Alan R. Liss, New York. pp. 455–492.

Kappeler, P. M. (1993). Reconciliation and post-conflict behaviour in ringtailed lemurs, *Lemur catta*, and redfronted lemurs, *Eulemur fulvus rufus*. *Anim. Behav.* 45:901–915.

Kappeler, P. M., and van Schaik, C. P. (1992). Methodological and evolutionary aspects of reconciliation among primates. *Ethology* 92:51–69.

Koyama, N. F. (2001). The long-term effects of reconciliation in Japanese macaques (*Macaca fuscata*). *Ethology* 107:975–987.

Kutsukake, N., and Castles, D. L. (2001). Reconciliation and variation in post-conflict stress in Japanese macaques (*Macaca fuscata fuscata*): testing the integrated hypothesis. *Anim. Cogn.* 4:259–268.

Leca, J. B., Fornasieri, I., and Petit, O. (2002). Aggression and reconciliation in *Cebus capucinus*. *Int. J. Primatol.* 23:979–998.

Leshner, A. I. (1983). Pituitary–adrenocortical effects on intermale agonistic behavior. In: Svare, S. B. (ed.), *Hormones and Aggressive Behavior*. Plenum Press, New York. pp. 27–38.

Ljungberg, T., and Westlund, K. (2000). Impaired reconciliation in rhesus macaques with a history of early weaning and disturbed socialization. *Primates* 41:79–88.

Ljungberg, T., Westlund, K., and Lindqvist Forsberg, A. J. (1999). Conflict resolution in 5-year-old boys: does postconflict affiliative behaviour have a reconciliatory role? *Anim. Behav.* 58:1007–1016.

Maestripieri, D., Schino, G., Aureli, F., and Troisi, A. (1992). A modest proposal: displacement activities as an indicator of emotions in primates, *Anim. Behav.* 44:967–979.

Matsumura, S. (1996). Postconflict affiliative contacts between former opponents among wild moor macaques (*Macaca maurus*). *Am. J. Primatol.* 38:211–219.

Menzel, C. R. (1993). Coordination and conflict in *Callicebus* social groups. In: Mason, W. A., and Mendoza, S. P. (eds.), *Primate Social Conflict.* State University of New York Press, New York. pp. 253–290.

Nishida, T., and Hiraiwa-Hasegawa, M. (1987). Chimpanzees and bonobos: cooperative relationships among males. In: Smuts, B. B., Cheney, D. L., Seyfarth, R. M., Wrangham, R. W., and Struhsaker, T. T. (eds.), *Primate Societies*. University of Chicago Press, Chicago. pp. 165–177.

Ninan, P. T., Insel, T. M., Cohen, R. M., Cook, J. M., Skolnick, P., and Paul, S. M. (1982). Benzodiazepine receptor–mediated experimental "anxiety" in primates. *Science* 218:1332–1334.

Palagi, E., Tommaso, P., and Borgognini Tarli, S. (2004). Reconciliation and consolation in captive bonobos (*Pan paniscus*). *Am. J. Primatol.* 62:15–30.

Pereira, M. E., Schill, J. L., and Charles, E. P. (2000). Reconciliation in captive Guyanese squirrel monkeys (*Saimiri sciureus*). *Am. J. Primatol.* 50:159–167.

Perez-Ruiz, A. L., and Mondragon-Ceballos, R. (1994). Rates of reconciliatory behaviors in stumptail macaques: effects of age, sex, rank and kinship. In: Roeder, J. J., Thierry, B., Anderson, J. R., and Herrenschmidt, N. (eds.), *Current Primatology. Social Development, Learning and Behavior*, vol. II. Presses de l'Université Louis Pasteur, Strasbourg. pp. 147–155.

Petit, O., Abegg, C., and Thierry, B. (1997). A comparative study of aggression and conciliation in three cercopithecine monkeys (*Macaca fuscata, Macaca nigra, Papio papio*). *Behaviour* 134:415–432.

Petit, O., Desportes, C., and Thierry, B. (1992). Differential probability of "coproduction" in two species of macaque (*Macaca tonkeana, M. mulatta*). *Ethology* 90:107–120.

Petit, O., and Thierry, B. (1994a). Reconciliation in a group of black macaques. *Dodo J. Wildl. Preserv. Trusts* 30:89–95.

Petit, O., and Thierry, B. (1994b). Reconciliation in a group of Guinea baboons. In: Roeder, J. J., Thierry, B., Anderson, J. R., and Herrenschmidt, N. (eds.), *Current Primatology. Social Development, Learning and Behavior*, vol. II. Presses de l'Université Louis Pasteur, Strasbourg. pp. 137–145.

Preuschoft, S., Wang, X., Aureli, F., and de Waal, F. B. M. (2002). Reconciliation in captive chimpanzees: a reevaluation with controlled methods. *Int. J. Primatol.* 23:29–50.

Ren, R. M., Yan, K. H., Su, Y. J., Qi, H. J., Liang, B., Bao, W. Y., and de Waal, F. B. M. (1991). The reconciliation behavior of golden monkeys (*Rhinopithecus roxellanae roxellanae*) in small breeding groups. *Primates* 32:321–327.

Roeder, J. J., Fornasieri, I., and Gosset, D. (2002). Conflict and postconflict behaviour in two lemur species with different social organizations (*Eulemur fulvus* and *Eulemur macaco*): a study on captive groups. *Aggress. Behav.* 28:62–74.

Rolland, N., and Roeder, J. J. (2000). Do ringtailed lemurs (*Lemur catta*) reconcile in the hour post-conflict? A pilot study. *Primates* 41:223–227.

Schaffner, C. M., and Caine, N. G. (2000). The peacefulness of co-operatively breeding primates. In: Aureli, F., and de Waal, F. B. M. (eds.), *Natural Conflict Resolution*. University of California Press, Berkeley. pp. 155–169.

Schino, G. (1998). Reconciliation in domestic goats. *Behaviour* 135:343–356.

Schino, G., Perretta, G., Taglioni, A. M., Monaco, V., and Troisi, A. (1996). Primate displacement activities as an ethopharmacological model of anxiety. *Anxiety* 22:186–191.

Schino, G., Rosati, L., and Aureli, F. (1998). Intragroup variation in conciliatory tendencies in captive Japanese macaques. *Behaviour* 135:897–912.

Schino, G., Troisi, A., Perretta, G., and Monaco, V. (1991). Measuring anxiety in nonhuman primates: effects of lorazepam on macaque scratching. *Pharmacol. Biochem. Behav.* 38:889–891.

Silk, J. B. (1996). Why do primates reconcile? *Evol. Anthropol.* 5:39–42.

Silk, J. B. (1997). The function of peaceful post-conflict contacts among primates. *Primates* 38:265–279.

Silk, J. B., Cheney, D. L., and Seyfarth, R. M. (1996). The form and function of post-conflict interactions between female baboons. *Anim. Behav.* 52:259–268.

Smucny, D. A., Price, C. S., and Byrne, E. A. (1997). Post-conflict affiliation and stress reduction in captive rhesus macaques. *Adv. Ethol.* 32:157.

Spijkerman, R. P., van Hooff, J. A. R. A. M., Dienske, H., and Jens, W. (1997). Differences in subadult behaviors of chimpanzees living in peer groups and in a family group. *Int. J. Primatol.* 18:439–454.

Swedell, L. (1997). Patterns of reconciliation among captive gelada baboons (*Theropithecus gelada*): a brief report. *Primates* 38:325–330.

Thierry, B. (1986). A comparative study of aggression and response to aggression in three species of macaque. In: Else, J. G., and Lee, P. C. (eds.), Primate *Ontogony, Cognition and Social Behaviour*. Cambridge University Press, Cambridge. pp. 307–313.

Thierry, B. (1990). Feedback loop between kinship and dominance: the macaque model. *J. Theor. Biol.* 145:511–521.

Thierry, B. (2000). Covariation of conflict management patterns across macaque species. In: Aureli, F., and de Waal, F. B. M. (eds.), *Natural Conflict Resolution*. University of California Press, Berkeley. pp. 106–128.

Thierry, B., Anderson, J. R., Demaria, C., Desportes, C., and Petit, O. (1994). Tonkean macaque behaviour from the perspective of the evolution of Sulawesi macaques. In: Roeder, J. J., Thierry, B., Anderson, J. R., and Herrenschmidt, N. (eds.), *Current Primatology. Social Development, Learning and Behavior*, vol. II. Presses de l'Université Louis Pasteur, Strasbourg.

Thierry, B., Gauthier, C., and Peignot, P. (1990). Social grooming in Tonkean macaques (*Macaca tonkeana*). *Int. J. Primatol.* 11:357–375.

Trivers, R. L. (1972). Parental investment and sexual selection. In: Campbell, B. (ed.), *Sexual Selection and the Descendent of Man*. Aldine, Chicago. pp. 136–179.

van Schaik, C. P. (1989). The ecology of social relationships amongst female primates. In: Standen, V., and Foley, R. (eds.), *Comparative Socioecology: The Behavioural Ecology of Humans and Other Mammals*. Blackwell, Boston. pp. 195–218.

van Schaik, C. P., and van Noordwijk, M. A. (1986). The hidden costs of sociality: intra-group variation in feeding strategies in Sumatran long-tailed macaques (*Macaca fascicularis*). *Behaviour.* 99:296–315.

Veenema, H. C. (2000). Methodological progress in post-conflict research. In: Aureli, F., and de Waal, F. B. M. (eds.), *Natural Conflict Resolution*. University of California Press, Berkeley. pp. 21–23.

Veenema, H. C., Das, M., and Aureli, F. (1994). Methodological improvements for the study of reconciliation. *Behav. Proc.* 31:29–37.

Verbeek, P., and de Waal, F. B. M. (1997). Post-conflict behavior of captive brown capuchins in the presence and absence of attractive food. *Int. J. Primatol.* 18:703–725.

Von Holst, D. (1985). Coping behavior and stress physiology in male tree shrews (*Tupaia belangeri*). *Fortschr. Zool.* 31:461–470.

Wahaj, S. A., Guse, K. R., and Holekamp, K. E. (2001). Reconciliation in the spotted hyena (*Crocuta crocuta*). *Ethology* 107:1057–1074.

Watts, D. P. (1991). Harrassment of immigrant female mountain gorillas by resident females. *Ethology* 89:135–153.

Watts, D. P. (1994). Agonistic relationships among female mountain gorillas. *Behav. Ecol. Sociobiol.* 34:347–358.

Watts, D. P. (1995). Post-conflict social events in wild mountain gorillas. II. Redirection, side direction, and consolation. *Ethology* 100:158–174.

Weaver, A. (2003). Conflict and reconciliation in captive bottlenose dolphins, *Tursiops truncatus*. *Mar. Mammal Sci.* 19:836–846.

Weaver, A., and de Waal, F. B. M. (2003). The mother–offspring relationship as a template in social development: reconciliation in captive brown capuchins (*Cebus apella*). *J. Comp. Psychol.* 117:101–110.

Weaver, A. C., and de Waal, F. B. M. (2000). The development of reconciliation in brown capuchins. In: Aureli, F., and de Waal, F. B. M. (eds.), *Natural Conflict Resolution*. University of California Press, Berkeley. pp. 216–218.

Westlund, K., Ljungberg, T., Borefelt, U., and Abrahamsson, C. (2000). Post-conflict affiliation in common marmosets (*Callithrix jacchus jacchus*). *Am. J. Primatol.* 52:31–46.

Wittig, R. A., and Boesch, C. (2003). The choice of post-conflict interactions in wild chimpanzees (*Pan troglodytes*). *Behaviour* 140:1527–1559.

York, A. D., and Rowell, T. E. (1988). Reconciliation following aggression in patas monkeys, *Erythrocebus patas*. *Anim. Behav.* 36:502–509.

37

Social Organization
Social Systems and the Complexities in Understanding the Evolution of Primate Behavior
Agustín Fuentes

INTRODUCTION

Over 30 years ago, Phyllis Dolhinow proposed that we "focus on those aspects of life common to most primates and in so doing, see to what extent primate behavior is variable" (Dolhinow 1972:352). Today, as primatology begins its second century, the result of Dolhinow's call is a growing awareness of the complexity and variation in intra- and interspecific behavior across the primate order. This complexity and heightened awareness of variation presents itself as a major challenge to our abilities to model and envision primate patterns (Kappeler and van Schaik 2002; Strier 1994, 2003). In their influential review of the evolution of social systems, Kappeler and van Schaik (2002:708) state that "We therefore need a theoretical framework that relates fitness-relevant behavior of individuals, such as foraging, predator avoidance, mating and parental care, to the defining characters of a social system." In this review, Kappeler and van Schaik use the socioecological model (Crook 1970, Emlen and Oring 1977, Terborgh and Janson 1986) as the basal framework in their functional quest to link the individual (as unit of selection) to the higher-level facets of societies such as groups and populations. However, they also recognize that the "social organization and demographic conditions *created* by individual behaviors also impose constraints on the behavioral options of these same individuals leading to complex feedback loops" (Kappeler and van Schaik 2002:708, italics mine). The evident complexity in primate social systems does call for the intensive focus and assessment that Kappeler and van Schaik propose; however, their focus exclusively on functional aspects of behavior, while powerful, may overlook some salient factors in our ongoing quest for effective assessments of primate social evolution (e.g., Isbell 2004). In this chapter, I hope to complement that seminal work by Kappeler and van Schaik, modify some of their definitions, underscore and emphasize their call for more evolutionary theory in primatology, and attempt to contextualize our investigations of social systems into a broader network of selection and other modes of change.

The purpose of this chapter is threefold: (*1*) to refine our concept of primates' social systems or, as I will term it, social organization; (*2*) to contextualize both dramatic variation in individual primates' behavior and observable, characteristic patterns of social behavior in primate groups and species; and (*3*) to expand our horizon regarding the processes of evolution and their impact on variation and patterns in primates as individuals, as members of groups and populations, and at the level of their social organization.

SOCIAL ORGANIZATION AND SOCIAL SYSTEMS

Social organization and its sometime synonyms *society, social system*, and *social structure* are the focus of numerous books and major articles in primatology (Clutton-Brock 1974, Crook and Gartlan 1966, DiFiore and Rendall 1994, Dunbar 1988, Hall 1965, Kummer 1971, Rowell 1993, Dolhinow 1993, Kappeler and van Schaik 2002, van Schaik and van Hoof 1983, Wrangham 1980, 1987, Zuckerman 1932). Various definitions of *social organization* have emerged over time (Table 37.1), and it remains a concept with variable uses in primatology (Dolhinow 1993, Rowell 1993, Dunbar 1988, Fuentes 1999a) and the broader arena of animal behavior (Tinbergen 1953, Wilson 1975, McFarland 1987).

In their overview, Kappeler and van Schaik (2002) use the term *social system*, to reflect the conflux of factors that we see as the collective facet of primate societies. For the purpose of this chapter, I will replace their *social system* with the term *social organization* and, following Hinde and Dolhinow, propose that we envision social organization as an *emergent property* with patterned flexibility in primates (Hinde 1976, Dolhinow 1993). Rowell (1993) cautioned against reifying emergent formations, such as social organization, arguing that we are in danger of creating a monolithic, static representation of dynamic sets of interindividual relationships that hamper our ability to accurately study primate behavior. Dolhinow (1993), however, responded that social systems (*social organization* here) are valuable for describing the "complex and variable behavior patterns we see in primate social

Table 37.1 Examples of Previous Definitions for *Social Organization, Social System,* and *Society*

Social organization

"The totality of all social relationships among members of a group" (McFarland 1987)

"Multi-layered sets of coalition based on relationships that differ in intensity, character or function" (Dunbar 1988)

"A set of 34 traits encompassing several major dimensions of primate social life" (DiFiore and Rendall 1994)

"The structural elements of social groups" (Dolhinow and Fuentes 1999)

Social system

"Society (= social unit; social system) is the set of conspecific animals that interact regularly more so with each other than with members of other such societies" (Kappeler and van Schaik 2002, derived from Struhsaker 1969)

Society

"A group of individuals belonging to the same species and organized in a cooperative manner. The diagnostic criterion is reciprocal communication of a cooperative nature, extending beyond mere sexual activity" (Wilson 1975)

Comparison between Kappeler and van Schaik's (2002) terminology and those presented here

KAPPELER AND VAN SCHAIK	THIS CHAPTER
Mating system	*Mating and rearing patterns*
Social organization	*Demography/grouping patterns*
Social structure	*Intra- and intergroup behavior*
Social system	*Social organization*

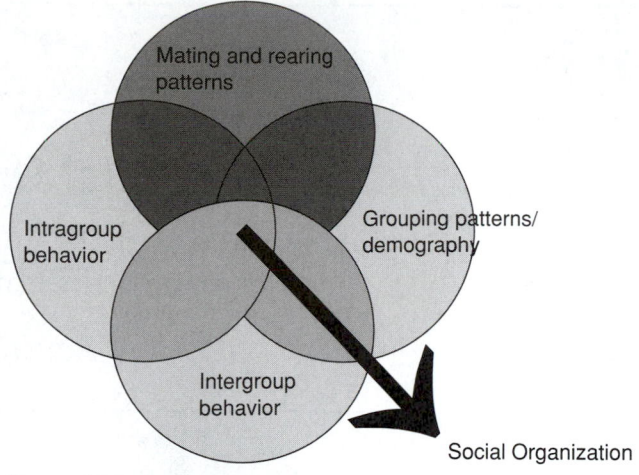

Figure 37.1 Social organization emerges from the conflux of society's constituent components.

behavior" and argued effectively for the retention of approaches that maintain a perspective of primate behavior from the individual, through the group, and population levels. Here, I will use the term *social organization* to mean the network of behaviors and patterns that emerges from the conflux of mating and rearing patterns (Kappeler and van Schaik's *mating system*), grouping patterns/demography (Kappeler and van Schaik's *social organization*), and intragroup and intergroup behavior (combined as Kappeler and van Schaik's *social structure*) (Fig. 37.1). By combining these facets into *social organization*, I intend to emphasize the social interactivity that predominates in primate lives. The term *organization* refers to elements arranged together to provide some form of overview, whereas *system* can imply a set of laws or rules that result in a specific set or pattern. Rather than being a particular system that has specific characteristics unto itself (Rowell 1993), social organization is merely the organization, at a given time, that emerges from the conflux of the structural elements noted above (Dolhinow 1993). I suggest that by using the term *social organization* and by envisioning it as the emergent result from the conflux of observable and measurable features of the social lives of primates, we can avoid deterministic approaches to higher-level patterns and at the same time place social organization in the context of measurable variables. In doing this, we can compare broad-scale patterns across and within taxa in terms of both constituent elements and overarching trends and patterns.

What is the benefit of employing a metaconcept such as *social organization*? Why not simply discuss and measure its constituent factors? Keeping in mind this emergent property of primates' social existence helps us remember that we are in a dual undertaking: examining the behavior and ecology of individuals during their lifetimes (or some segment thereof) but hypothesizing as to the impacts of natural selection and other forces of evolution across time on groups and populations. That is, we examine individual behavior and ecologies but attempt to construct explanations for the observed behavioral patterns by invoking and/or constructing evolutionary histories (past action of selection, adaptations; Williams 1966, Kappeler and van Schaik 2002). The concept of *social organization* allows us to maintain our focus on the individuals while at the same time emphasizing that they do not exist outside of the broader group and population. Behavioral patterns and interactions of individuals create the social environment in which primates exist, and social organization emerges from these interactions. Thus, the behavior exhibited by an individual primate reflects adaptive strategies, stochastic occurrences, and specific responses to the social context in which the individual finds itself. Social organization, then, changes over time as its constituent factors change across generational time in response to selection, genetic drift, gene flow, chance, and historical events. Individual primates do not exist outside of their social context, and understanding that social context and the patterns of social organization in a primate group, population, or species allows us to better attempt broad-scale explanations for the occurrence of observed behavior.

THE CONSTITUENT FACTORS OF SOCIAL ORGANIZATION

In order to effectively assess individual primates' contexts for exhibiting behavior, we require a clear understanding of the constituent factors that make up the social organization in which a primate exists (Fig. 37.1). Each of these factors is composed of a set of variable elements, and understanding the patterns of variation in these elements will facilitate a better envisioning of the social parameters under which primates operate and thus enhance our ability to describe and model the range and expression of primate behaviors.

Extant Primate Grouping Patterns

A primate group can be defined as that collection of individuals that engage in regular, patterned social contact over extended periods of time and use the same general spatial areas (Kummer 1971; see Chapter 34). Rather than seeing a group as a static entity, long-term research on a diverse array of species suggests that groups change over time along demographic lines as individuals die, are born, and migrate and as groups fission or merge. However, despite this variation in individual composition, most species do exhibit one or more basic general demographic patterns in their social groups (Crook 1970, Dolhinow 1972). Table 37.2 describes the basic demographic compositions for primate groups. Some form of multiple adults of both sexes is the most common grouping pattern. There is substantial variation within this category, with differential representation of the variants

Table 37.2   Grouping Types (Demographic Labels) of Extant Primates

Multifemale/multimale

The group consists of more than one adult female and more than one adult male and immatures

Variants: *cohesive group*, the majority of individuals within the group occupy the same general areas at the same time, generally within vocal and/or visual range for the majority of active time

Fission–fussion, the overall group breaks into smaller subgroups of variable composition but uses similar range; subgroups can fuse into larger groups, or the entire group can occasionally occupy the same area at the same time; fissioning or fusion may be seasonally and/or behaviorally driven

Community, similar to fission–fusion; however, subgroups are more dispersed and less likely to coalesce into full group

Unimale/multifemale

A cohesive group consisting of one adult male and more than one adult female and young

Unifemale/multimale

A cohesive group consisting of one adult female and more than one adult male and young

Two-adult group

A cohesive group consisting of one adult female, one adult male, and immatures

Dispersed sociality

Individual primates with individual ranges that overlap extensively; interactions between individuals within the overlapping ranges can be frequent or infrequent, vocal and/or visual and/or tactile; occasionally multi-individual subgroups may form

across the families and subfamilies of Primates. One-adult male and two-adult groups are less common but present in both anthropoids and prosimians. One-adult female groups are rather uncommon and found primarily among the callitrichid primates. There remains some debate as to whether dispersed sociality is truly a group-living situation; however, most, or all, individuals in dispersed social systems, be they apes (*Pongo pygmaeus*) or prosimians, interact with one another through visual, olfactory, vocal, or other contexts (see Chapters 3 and 17). That is, they engage in patterned social interactions in the same general spatial areas, even though they spend less time in spatial proximity. Therefore, we can safely say that the majority of primates live in social groups and that the composition of the social group (of various kinds) is the demography of social organization.

Given that there are general demographic patterns (age/sex compositions) that characterize primate groups but that these patterns vary, it is relatively important to recognize the patterns of intraspecific variation in grouping. Table 37.3 provides examples of intraspecific variation in grouping patterns. At the most basic level, this variation in grouping patterns is a result of ecology and history. Different environments are going to have basic impacts on primate groups (see Chapters 28 and 34). The size and dispersal of food sources, the presence or absence of predators, and the impact of climatological factors will provide differential pressures impacting overall group size and issues of seasonality in reproduction, which in turn impact demographic factors (Kummer 1971, Dolhinow 1972, Dunbar 1988, Sussman 1999, Isbell 2004). Variation in group size can also be due to stochastic occurrences. Chance events can impact the sex ratio, age structure, and ranging patterns of groups (Kappeler and van Schaik 2002). If such events occurred prior to the initiation of observations on a group, then the observers will not have access to the information and may have difficulty explaining current demography (and related issues such as ranging and other behaviors). In these cases, chance events and other historical factors (e.g., a predation event in a particular location leading to members of the group avoiding that location) can have social impacts. That is to say, current social behavior can be impacted by the demography of a group and that demography can be impacted by both ecological and historical factors, not all of which are discernible to the researchers examining the group (Sapolsky and Share 2004, Wasser et al. 2004). For example, historical changes in the group sex ratio that increase or decrease the number of adult females will impact intrasexual competition, mating opportunities, size of peer groups and so on (see Isbell 2004). In other words, social strategies displayed by individual primates can vary depending on the demographic patterning in which they find themselves.

Extant Primate Mating and Rearing Patterns

Patterns of mating and rearing behavior are obviously core to primate (and all mammal) social lives. In primates, there is

Table 37.3   Examples of Intraspecific Variation in Grouping Patterns

SPECIES	GROUP COMPOSITION	STUDY
Propithecus diadema edwardsi	Unimale/multifemale = 9 groups	Pochron and Wright (2003)
	Unifemale/multimale = 11 groups	
	Two-adult group = 12 groups	
Semnopithecus entellus	Unimale/multifemale = 27 groups	Treves and Chapman (1996)
	Multifemale/multimale = 229 groups	
	All male = 4 groups	
Hylobates lar	Two-adult group = 3 groups	Brockelman et al. (1998)
	Unifemale/multimale = 1 group	
Gorilla gorilla beringei	7 groups observed:	Robbins (1995)
	Unimale/multifemale = 41.1% of observation time	
	Multimale/multifemale = 36.8% of observation time	
	All male = 22.1% of observation time	
Saguinus mystax	Two adult group = 27 groups	Sussman and Garber (1987)
	Unimale/multifemale = 30 groups	
	Unifemale/multimale = 52 groups	
	Multifemale/multimale = 163 groups	

a wide array of variability surrounding individual behavior patterns related to mating, with more than one mating pattern frequently being found in a given species, population, or even group (Knott 1999, Alberts et al. 2003, Pochron and Wright 2003). These patterns, taken together, can be seen as a mating system for a species or population and, thus, the reproductive facet of social organization. There is abundant evidence that demographic factors are not the sole determinants, or necessarily good indicators, of the mating patterns of individual primates (Palombit 1995, Digby 1999, Borries 2000, Cords 2000, Sommer and Reichard 2000, Pochron and Wright 2003). Grouping pattern does not equal mating pattern for most primates. Table 37.4 identifies and defines the general mating patterns found in primates. One or more of these patterns can be found in primate species, with polygamy (polygynandry) being the most common. True polyandry is relatively rare and found predominantly in species where an adult female can cause reproductive suppression in other females in the group or where there is only one adult female in the group (as in some callitrichids; Rylands 1993, Garber 1997). Absolute monogamy is also quite rare, but monogamy as the primary mating pattern

Table 37.4   Mating Patterns

Monogamy

One female and one male mate exclusively during mating period

Polygamy (Polygynandry)

Females and males mate with multiple partners during mating periods

Polyandry

One female mates with multiple males during mating period

Polygyny

One male mates with multiple females during mating period

is frequently found in primates that exist mainly in two-adult groups (Fuentes 1999b, 2002). Absolute polygyny is fairly rare as well, except in some highly successful mate-guarding species or species where there is a preponderance of one-male groups. The most important distinction between these mating patterns is their relative potentials for variance in reproductive success. In polygynous and polygamous/polygynandrous patterns where male–male competition for access to females is high, males have a much higher variance due to this competition and the subsequent unequal distribution of fertilizations. Polyandrous mating patterns also tend to result in high variance for both males and females as mating is usually limited by the alpha female and involves only a few males at most (Garber 1997). In monogamous patterns, variance will be relatively equal for both sexes depending on the amount of polygyny or polyandry that is also present. The relative distribution of the different mating patterns within a population can then lead to differential variances in reproductive success, thus impacting individuals' behavior patterns. Here, it is also evident that demographic variables can directly impact the mating patterns via availability of potential mates.

Demographic patterns, phylogenetic influences, and mating patterns also impact the manner in which primates rear their young. Relationships between males and females and between related individuals within a group can impact the manner in which young are treated. In most primates, adult females carry and nurse their offspring; however, some also "park" infants for substantial portions of their active periods. Many species engage in variable levels of *alloparenting*, where infants/immatures are carried and protected for long periods by females (or other age/sex classes) that are not necessarily their mothers (or related). There is also substantial variation in how much assistance in infant carrying, guarding, grooming, and other contact behaviors mothers receive from group mates.

As mating and rearing systems are a prime factor in social organization, it behooves us to briefly discuss current theory and ongoing debates as to why these patterns exist and what the variance in reproductive access might mean (see Chapters 27 and 28). Bateman (1948) and subsequently Trivers (1972, 1974) and others have identified *anisogamy* (differential investment in gametes by males and females) combined with the greater investment by females in rearing offspring as the main driving force behind differences in male and female mating strategies. Recent theoretical discourse (Kokko and Jennions 2003, Borgerhoff-Mulder 2004) recognizes the important, and often overlooked, concept that patterns and contexts of sexual selection drive sexual investment patterns, not merely baseline anisogamy (Tang-Martinez 2000). Bateman (1948) established a basic premise for evolutionary models: females with their large, costly eggs have greater cost per offspring and should therefore invest more in each offspring, whereas males, with their (hypothetically) inexpensive sperm, should invest minimally in individual offspring and maximize their mating opportunities. Trivers (1972, 1974) and others translated this into models of parental investment and created behavioral predictions for many organisms including primates. Kokko and Jennions (2003) suggest that the initial assumptions propagated by Trivers and many others overlook the fact that all offspring have a mother and a father, so if all males and females are equally successfully mating, then their reproductive variance is equal and males should invest in the rearing of offspring. This suggests that anisogamy, in and of itself, does not create inherent differences in reproductive success. However, if there is variance in reproductive potential introduced into the system, by female choice, male–male competition, or demographic factors, then varying investment patterns would be favored. In a sense, then, one can envision sexual selection, not inherent anisogamy, as driving potential selective pressures for male and female investment differences. Behavioral patterns and demographic contexts resulting in sexual selection are malleable and vary across time; therefore, the relationship between the sexes is not necessarily one of strict competition based on differential potential variance and investment but, rather, can be envisioned as one of negotiated interactions (and potentially differential investment patterns) resulting in the production and rearing of offspring.

Given these factors influencing mating and rearing patterns, the impact of mating and rearing systems; the specifics of where, when, and how primate females and males achieve reproduction within a given demographic context; and the manner in which infants are reared are going to have a direct and pronounced effect on the emergent social organization. Given the wide array of ecological and historical conditions that impact primate groups, it is not surprising that we would expect to see a variety of mixed mating and rearing strategies across different local conditions, especially in long-term studies.

Intragroup Behavior Is the Social Structure of Social Organization

The majority of the primatological literature focuses on dyadic interactions to resolve issues of dominance, conflict negotiation, affiliative patterns, etc. in offering descriptive and quantitative analyses of primate social lives (Hinde 1976). The within-group patterns of interaction form the daily nexus for the activity of any given primate. As individuals move across age and sex classes, they encounter different facets of their group's demographic and mating structures and are forced to learn a series of patterned but malleable relationship characteristics that construct their day-to-day dyadic interactions (Morbeck 1997, Bernstein 1999). In the context of assessing social organization, these day-to-day patterns of behavior form the social structure, the interindividual framework, in which an individual behaves. Therefore, variations at the individual level impact the social structure of the group, which can be reflected at the level of social organization. That is, different groups are going to be composed of distinct individuals that will vary in aspects of their behavior. Thus, within the same demographic, ecological, and mating parameters, two groups may appear to exhibit different mean, or emergent, patterns of behavior due to the individual variation in their members (Dolhinow 1972, Fuentes 1999a, Sapolsky and Share 2004). These differences in individual patterns can be attributable to different behavioral strategies (expressed adaptations), stochastic events, and/or historical facets. Obviously, the issue of *personality*, idiosyncratic behavior patterns that are relatively consistently expressed by individuals, may be relevant here. However, measuring personality per se is exceedingly difficult, and there is substantial debate as to its practicality as a concept in the study of nonhuman organisms; thus, for purposes of achieving measurable, quantitative assessments, one must refer to individual variation in measurable behavior rather than to some aspect of primate psychology. By examining the behavior exhibited in a group both as an emergent entity (the mean patterns) and as a collection of individuals (a conglomerate of slightly variant patterns), we are provided with two distinct and complementary ways to envision and analyze the social structure of primate groups.

Intergroup Behavior is the Population Structure of Social Organization

Group behavioral profiling is a strong tool in primatology. To achieve a temporally robust picture of a group's habitat use and relationships with other groups, one can characterize a given group using mean ranging, feeding, activity, etc. patterns. In this sense, for the purposes of comparison across groups within a population, the individual is subsumed into the social structure in which it exists (e.g., Treves and Chapman 1996). These intergroup interactions are the population structure from which aspects of social organization emerge. Demography, mating and rearing, and individual variation impact population-level patterns. However, at this level of analysis, ignoring microvariations in (such as individuals) previous levels provides us with the power of comparison across groups, which is a less refined but robust temporal and spatial tool.

SOCIAL ORGANIZATION AND EVOLUTIONARY PROCESSES: THE CONTEXTS OF CHANGE

The impact of variation in the constituent factors of demography, mating systems, and intra- and intergroup behavior is reflected at the emergent level, the broad patterns we refer to here as *social organization*. By comparing social organizations of different species or populations and the patterns of their constituent factors, we can potentially identify the patterns of action for particular evolutionary processes across primate populations. All four of the evolutionary processes—natural selection, gene flow, genetic drift, and mutation—impact social organization by affecting changes in the patterns of the constituent factors. However, none of these factors is truly independent, thus making social organization an important tool for viewing broad evolved patterns and changes in those patterns.

A substantial component of our attempts to understand primates' lives lies in our ability to model the patterns and processes of evolutionary histories and ongoing evolutionary forces on primate populations. Taking the approach outlined here, examining the individual primate in the context of its demographic characteristics, mating and rearing patterns, inter and intragroup behavior, and membership in a broader social organization facilitates a diverse array of perspectives wherein we can attempt to elucidate evolutionary and other patterns of change.

Natural Selection: Adaptation Shapes Social Organization

> Functional labels should not substitute for a full analysis of the causes of behaviour. (Krebs and Davies 1997:11)

The current primatological approach is couched in the context of behavioral ecology and continues to derive lines of inquiry from Tinbergen's four "why" questions (Table 37.5)

Table 37.5   Four "Why" Questions Regarding a Behavior: A Behavior Has the Following Facets and Each Can Be Approached in Order to Elucidate the Particular Aspect of the Behavior in Question

A. Evolutionary cause underlying a behavior
B. Developmental or ontogenetic cause underlying the behavior
C. Proximate stimulus eliciting the behavior
D. Functional, or fitness-related, impact of the behavior

(Tinbergen 1963, Krebs and Davies 1997, Bernstein 1999). However, despite the recognized and affirmed importance of addressing all of the "why" elements of a behavior, all too frequently our focus in primatology is on the role of natural selection and our quest is solely for functional explanations. Many of our arguments for the evolution of specific traits (be they behavioral or morphological) rely on demonstrating the potential strength of selection pressures on the trait in question. Underlying most estimates of the power of natural selection on a trait is the assumption that there is a causal link between fitness (or lifetime reproductive success) and the trait in question. Given the variable factors that influence any given primate's life (above), one must tread with caution when invoking a functional explanation for an observed behavior (see Chapter 39). It is at the level of the population and via examining changes in the constituent facets of social organization that we can most effectively assess actual evolutionary patterns. The focus on trait commonalities among individuals can provide us with insight, but the inherent variation at the levels of demographics, mating and rearing patterns, and interindividual interactions makes it difficult to always effectively measure the actual impact on the reproductive success of any given heritable behavior or behavioral pattern. This problem of assessing fitness is compounded as it has become apparent that, in at least some cases, observed fitness variation is associated with the environmental component of a trait, not the trait itself. Kruuk et al. (2003). present an example wherein individuals (in a species of bird) that happen to be very healthy can breed early; early breeders can also produce large, healthy broods. This differential birth outcome could result in an erroneous assumption of a positive fitness relationship between the behavior of early breeding and the resulting brood size, leading some to argue that there has been selection of early breeding as a higher fitness behavior. However, there is an important intervening variable in this process: resource availability, which affects bird size and condition (health). In better years, more birds will be healthier and breed earlier than in resource-stressed years. We need not assume that early breeding is an adaptation, the result of a history of selective pressures. Thus, our assumptions about fitness-enhancing values and the "adaptiveness" of certain primate behaviors may overlook the influence of environmental variation and possible stochastic confounding factors. This can mislead us into assuming that a behavior is an adaptation. This may be a reason that natural selection rarely strictly follows the modeled trajectories (as in optimality models; see Dunbar 2002 for an overview of modeling in primatology). These trajectories are often constructed from the readily controlled and/or modeled responses to artificial selection or a proposed ideal and relatively static environmental system. As Kinji Imanishi pointed out in 1941, this may be an important caution when making simple Darwinian evolutionary predictions: artificial selection (or controlled optimal models) may present an overly powerful ideal that is seldom realized in natural selection.

However, despite the importance of caution against assumptions of the universality of natural selection as an explanation for traits, there is a broad and robust body of theory and multiple hypotheses that clearly outline how selection can impact the social organization of primates. Here, we can summarize ideas of how selection impacts individuals, groups, and populations, all of which can be visualized as affecting the emergent social organization.

Impact of Selection on Individuals

While we cannot directly measure selection on individuals, we can model how particular behaviors may contribute to or detract from an individual's reproductive success and thus relative fitness. Selection does not truly impact the individual, but factors that affect the individual's ability to reproduce affect the representation of that individual's unique genetic component in subsequent generations and are of interest in models of evolutionary patterns. Of course, this focus on the individual is relevant to selection models only if there is some heritable facet to the behavior. Alternatively, focus on the individual can also be very useful if the behaviors in question are learned/socially transmitted with no direct biologically heritable facet (see The Social Evolution of Social Organization, below).

Most researchers agree on a basic set of environmental and hypothesized social pressures that can result in differential reproductive success for individuals (Wilson 1975; Rubenstein and Wrangham 1988; Kappeler and van Schaik 2002; see Chapter 28). Additionally, in their overview of the evolution of social systems, Kappeler and van Schaik (2002) follow the dominant socioecological paradigm and treat males and females differently in their explanations of selective pressures on individuals. They do this because under the general model male and female fitness (or relative potential reproductive success) is limited by different factors. That is, "the model assumes that the distribution of females is primarily determined by the distribution of risks and resources in the environment, whereas males distribute themselves primarily in response to the temporal and spatial distribution of receptive females" (Kappeler and van Schaik 2002:708). This orientation follows directly from the anisogamy argument of Bateman (1948) and its subsequent expansion and elaboration into models of parental and reproductive investment (Trivers 1972, 1974; Clutton-Brock 1991). However, for this discussion, I will present the general patterns of pressures on individuals rather than detailed distinctions regarding possible differential pressures on males and females (see Chapters 27 and 28).

Traditionally, biologists and ecologists present the following suite of pressures that face any organism: avoiding predation, finding and ingesting sufficient food and water, finding mates, and achieving effective thermoregulation. Obviously, each of these pressures will be different for organisms in distinct environments. For example, large mammals such as primates may, in some environments, suffer a lower risk of predation than other smaller mammals, or the same species of primate living in two different environments may suffer very different predation risks (see Chapter 32). For primates, living in a group adds to these pressures via the potential for competition within the group and its effects on changing the costs of foraging (generally scramble competition). At the same time, group living may also act to ameliorate the pressures of predation and intergroup foraging competition (generally contest competition) and provide ready access to mates (van Schaik 1989, Sterck et al. 1997, Strier 2000; see Chapter 28 for a review).

Included in Kappeler and van Schaik's (2002) model is the hypothesis that sexually selected infanticide, the killing of infants by unrelated males as an adaptive strategy, is a significant pressure on individual primates (especially females). While being relatively popular among primatologists, this hypothesis remains contested in the scope of its impact and the factors involved in its recording and assessment (see Chapter 28 for a review of the perspectives on sexually selected male infanticide, see also Chapter 6 for female infanticide). If assertions of the prevalence of this type of infanticide are correct (as argued in Janson and van Schaik 2002), then the behavioral pattern of infanticide will be expressed differentially under differing demographic and mating conditions. That is, if this behavioral pattern is common due to past increased reproductive success by those males that carried heritable biological variants that facilitate the appearance of infanticide, then its expression will be dependent on certain demographic, mating, and rearing variables that make it a beneficial endeavor for current males with those variants (van Schaik 2002). In this case, the frequency and effectiveness of infanticidal events will impact on individual primates; the overall pattern should then be apparent at the level of the population, and anti-infanticide behaviors should also become evident. However, the assertion that infanticide is truly widespread across primate taxa and represents an important pressure on primates remains contentious. The majority of infanticidal events are reported in a few species and only at some of the study sites for those species (van Schaik 2002). Therefore, the actual impact of infant killing by males as a potential element of selection on the majority of individual primates in most populations and species remains unknown.

Hamilton (1964) proposed kin selection and Wilson (1975) further invoked the notion of reciprocal alliances between related individuals as core facets of group-living organisms. In most primate species, one sex is philopatric and, thus, has access to kin throughout their lives. In the context of individual primates then, being the philopatric sex alters the potential impact of reproductive success relative to the dispersing sex. Individuals that are around kin are able to enhance their inclusive fitness in a manner that dispersing individuals are not (see Isbell 2004). This implies that the relative impact of reduced individual reproductive success can be greater in the dispersing sex than in the philopatric sex, which may be able to offset individual reductions via assistance provided to relatives (if it results in enhancing successful reproduction).

While Darwin (1871) proposed sexual selection and its result, differential mating success in individuals, subsequent research has clarified aspects of individual patterns and possible reproductive effects (e.g., the handicap principle; Zahavi 1975, 1977). In species or populations where there is strong choice by, and/or intrasexual competition in, one sex, the potential resultant differential reproductive success rates will impact on individuals' behavior patterns. Variable demographic facets will impact the patterns of sexual choices, and variants in individuals' behavior and morphology will result in differential mating behaviors. This is relevant in terms of examining the mating behavior of males and females as it is the pattern of sexual selection that in turn affects investment behavior in offspring and the mating effort of individuals (Kokko and Jennions 2003).

Impact of Selection on Groups

Williams' (1966) criticism and rejection of group selection coupled with Hamilton's (1964) arguments for kin selection as an explanation of apparently altruistic behavior formed the basis for Trivers' (1972) and Wilson's (1975) emphasis on examining the individual as the focal point for understanding selection and subsequent adaptations. However, recent work by Wilson and Sober (1994) argues for envisioning multiple levels in the action of selection (see also Wilson 2004). Groups themselves cannot be units of selection; however, as primates are primarily found in social groups for all, or a majority, of their lives, impacts on groups can have impacts on individuals. Interactions between groups, whether they be direct contests for resources or active partitioning of the environment, can affect the reproductive success of individuals within the groups. Therefore, interactions between groups can have positive or negative fitness results for the individuals in those groups. For example, behavior by a few individuals in a group may result in that group's increased access to contested resources, which in turn may enhance the reproductive opportunities or efficiency of all members in the group. This is not to say that the individuals acting did so "for" the group but, rather, that intergroup interactions can have costs and benefits even for those group members that do not engage in the contests. Obviously, the demographics of a population will have great impact on the ways in which groups interact, and the individual behavioral patterns of group members will also influence these relations. Living in a group also increases the potential for disease transmission. Here, the current demographic composition of a group may dramatically alter the impact that a given pathogen may have on the individuals within the group. In fact, disease itself can produce dramatic demographic shifts which alter the ways the remaining individuals can interact with their environment, other groups, and themselves. For example, Sapolsky and Share (2004) review a case where a tuberculosis outbreak killed the majority of

males in a baboon group. The males that died happened to be highly aggressive, and prior to the disease event, all group males displayed a negative correlation between rank and physiological indicators of stress. The males that survived were atypically non-aggressive, and the overall pattern of behavior in the group was significantly different after the die-off. Over a decade later, the patterns of behavior in this group remain relatively nonaggressive, and there is no correlation between rank and physiological indicators of stress. A disease event changed the demography and the behavior and physiology of this group. Disease may be a major, if somewhat understudied, factor of selection on primates (Kappeler and van Schaik 2002).

Impact of Selection on Populations

Changes in ecological/climatological patterns can cause changes in pressures on individuals and their groups such that changes are readily seen at the population level (Dunbar 1988). Individuals live, die, and reproduce; however, changes at the individual level can alter broader demographics (groups and populations), which can in turn feed back to and alter individual patterns and contexts. In this sense, selection (resulting in differential reproductive success) affects the population and the social organization exhibited in that population.

Examining the impact of selection at the individual, group, and population levels provides a broad, complex picture of the facets of social organization. The three levels of the individual, the group, and the population are interconnected but also can be examined, in a quantitative format, independently to assess distinct patterns and trends (Fig. 37.2).

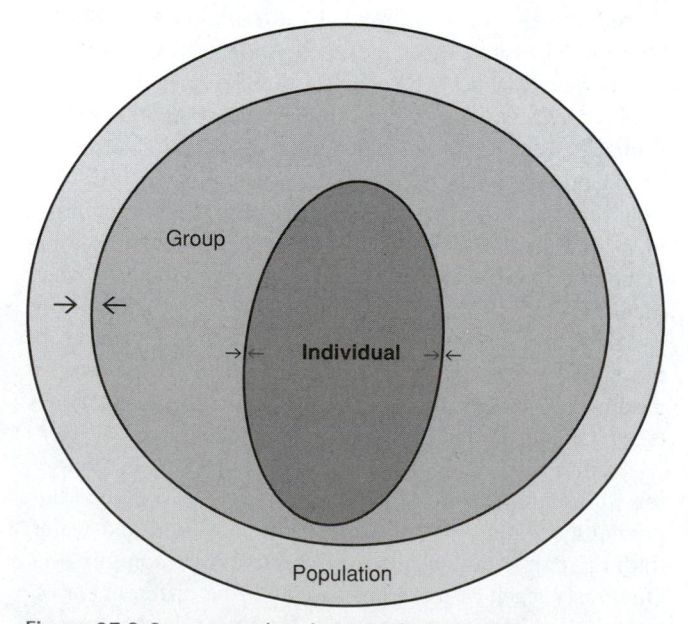

Figure 37.2 Interconnections between the individual, the group, and the population.

Historical and Developmental Facets of Social Organization

Many of the behaviors and patterns we observe today are related not only to the results of selection but also to a wide array of historical and developmental factors that may or may not be reflected as adaptations. *Phylogenetic inertia*, or the presence of ancestral traits that no longer may be directly adaptive or responsive to current selection pressures, is a potentially common confounding factor in understanding extant behavior (DiFiore and Rendall 1994). This suggests that some of what we see at the individual, group, or population level may reflect past adaptations or past events that no longer act as pressures on the extant forms. For example, Dunbar (1992) proposed that cortical/cerebral development may limit group size related to the ability of primates to effectively interact in complex social ways with a limited number of individuals. The ecological structure of the habitat and the climatological patterns in which a primate lives can also have an impact on locomotory patterns, ranging patterns, and the type and frequency of particular behaviors (Dunbar 1988). Finally, gene flow and lack thereof (isolation) can affect the behavior of an individual primate via either the introduction of particular biologically based variants of traits or simply exposure (or lack of exposure) to other members of their species. Gene flow also has implications for the probabilities of pathogen pressure and for the patterns of demographic change or stability in a population.

The Social Evolution of Social Organization?

Numerous primatologists have shown great interest in the possibilities of nonhuman primate "culture" (McGrew 1998, Whiten et al. 1999, Perry and Manson 2003, van Schaik et al. 2003; see Chapter 40). The existence of behaviors, functional and nonfunctional, that are acquired via social learning and transmitted across generations within some social groups and not others is becoming an increasingly common aspect of primatological natural histories. These behaviors vary from patterns of grooming to tool use to greeting behaviors and are reported for a number of primate species (McGrew 1998, Perry and Manson 2003). The debate surrounding whether or not nonhuman primates have the same sort of culture as humans is well covered in other venues (McGrew 1998, see Chapter 40), so I will not review it here. However, the relevance to this discussion lies in the fact that there is little contention that many species of primates do show variation in behavioral patterns, frequently apparently nonfunctional patterns, across sites, groups, and populations. This behavioral variation, or *social tradition*, can be seen as an indication that primates are fairly flexible, at the level of the individual, in their behavioral potential and that living in social groups can facilitate the appearance and spread of social patterns that may have no direct impact on fitness. In other words, primate behavior can change over time because of social innovation and learning without a

direct relationship to the impact on reproductive success of individuals (e.g., Sapolsky and Share 2004). Selection and other evolutionary factors mold the genetic variation in primate populations; however, day-to-day behavioral interactions among primates also actively shape the social environment. When asking questions about a primate's behavior and how that behavior impacts the various aspects of its social organization (at the level of the individual, group, or population), we must consider aspects of behavioral flexibility and plasticity and their potential for leading to social innovation and/or change.

The chapters in this volume effectively demonstrate that a major component of the environment for the majority of primates is the social environment. There is an emerging recognition that *phenotypic plasticity*, continuous and reversible transformations in behavior, physiology, and morphology in response to rapid environmental fluctuations, can be seen as an adaptation. Piersma and Drent (2003:28) suggest that "when environments change over shorter timescales than a lifetime, individuals that can show continuous, but reversible, transformations in behavior, physiology or morphology, might incur a selective advantage." It is also the case that organisms are not only impacted by their immediate environments but also, in part, shape those environments and thus the selection pressures that they face. In a sense, we can see organisms engaged in a certain amount of niche construction (Laland et al. 2001, Stamps 2003). The *niche* is best characterized as a "multidimensional hypervolume" (Hutchinson 1957) and the way that organisms make a living. If niche construction is ongoing, then we can envision the social network (group and population) as a major component of the primate niche. This is a flexible and malleable environment, and thus, we might expect that individual primates are able to exploit a variety of behavioral patterns and strategies in order to effectively negotiate it (i.e., successfully interact on a day-to-day basis with their social partners). Obviously, some behavioral patterns are clearly the result of direct selection resulting in successful variants (adaptations). However, it is also possible that the myriad variable elements involved in the social environment of primates have favored a level of flexibility in the parameters of primate social evolution. That is, many of the behaviors exhibited by primates may reflect an adaptation, but that adaptation may be the broad neural and physiological patterns that allow for the expression of a diverse array of social behaviors. In other words, social complexity in primates may itself be an adaptation, the result of existing in a variable social environment. If this is the case, then we must include aspects of behavioral flexibility, including potentially nonfunctional behavioral patterns, into our assessments of primate social organization. It remains possible that local social variants can change a group's behavior in ways that have an effect on their demographics, mating, and patterns of interaction with one another or with other groups. In short, the major factor in the variation in primate social organization may indeed be the variation in behavior found in individual primates. Documenting social and behavioral variation across the life span is as important as measuring other life history variables (Kappeler and Pereira 2003). Primate social organization can change over time in response to both functional and nonfunctional shifts in primate behavior. Nonfunctional variants can also change the social environment such that they may become functional or may alter aspects of interactions such that there are fitness impacts for group members.

MODERN EVOLUTIONARY BIOLOGY AND PRIMATE SOCIAL ORGANIZATION: THE IMPORTANCE FOR ONGOING ENGAGEMENT WITH EVOLUTIONARY THEORY

Kappeler and van Schaik (2002) express a concern that primatology and evolutionary biology are not as engaged with one another as they should be. They suggest that this may be because of differences in primate mating systems relative to many other animals, the inability to engage in much controlled experimental study with primates (for ethical and practical reasons), and the fact that primate life histories are different from those of many other animals and thus potentially not pertinent to some major questions in evolutionary biology (Kappeler and van Schaik 2002). These points emphasize the differences between primates and many of the avian or insect models that make up the bulk of the behavioral ecological work that is mainstream in evolutionary biology. However, I suggest that these differences are a strong reason to continue to expand and integrate ongoing issues in evolutionary theory into the study of primates, especially at the level of social organization. Primates represent a mammalian taxon that is characterized by high levels of social and behavioral flexibility, with long periods of physical and social development. These patterns do not always easily fit into current socioecological models that have been developed with other organisms. In fact, it is the complexity inherent in primate social organization that requires a diffuse and multi-faceted approach in our attempts to understand its evolution.

Emergent evolutionary biology theory and models allow for a broadening of our inquiry into evolutionary patterns and processes affecting primates. In addition to our basal behavioral ecological model that allows us to assess and model the impacts of natural selection and ecological parameters on primates, we can integrate other prospects, such as phenotypic and developmental plasticity, niche construction, and social traditions, into our attempts to understand and model the complexity and variation in intra- and interspecific behavior across the primate order. These additional evolutionary perspectives can be encompassed under the rubric of developmental systems theory (DST) (Oyama et al. 2001). DST posits *joint determination by multiple causes*, suggesting that every trait is produced by the interaction of many developmental resources and the gene/environment

dichotomy is only one of the many ways to divide up these interactants. Therefore, functional explanations, while forceful and important, are not the only venues for inquiry into primate behavior. *Context sensitivity and contingency* also plays a role, where the significance of any one cause of change is contingent on the state of the rest of the system. This aspect of DST emphasizes the integrated nature of primate social lives. For primates, the DST concept of *extended inheritance*, wherein an organism inherits a wide range of resources that interact to construct that organism's life cycle, is especially relevant. These resources include the memory and experience of other group members, the range in which the group lives, and the patterns of interaction extant in that population. Social traditions can also be a component of the resources inherited by a primate. The assertion of *development as construction*, with neither traits nor representations of traits being transmitted to offspring but, rather, traits being made (reconstructed) in development, is also relevant. Primate life histories are extended relative to those of many animals, and the social environment in which they exist requires much learning and is behaviorally negotiated. Finally, the main DST thrust of viewing *evolution as construction* (i.e., that evolution is not a matter of organisms or populations being molded by their environments but of organism–environment systems changing over time) provides a strong perspective in which to contextualize variability in primate social organization. The emergent property that is referred to as "social organization" in this chapter is constantly being constructed via its constituent elements of demography, mating and rearing systems, and intra- and intergroup behavior patterns. This constant construction, due to the variation in its constituent elements, is a major reason for the variation we see in primate societies at any given time. Integrating our current reliance on socioecological models with the emerging complexities in behavioral ecology and the proposals of DST may facilitate a greater and increasingly successful engagement between primatology and evolutionary biology.

COMPLEXITY DOES NOT MEAN IMPOSSIBILITY: THE CONTINUED IMPORTANCE OF NATURAL HISTORY DATA AND HYPOTHESIS TESTING IN INVESTIGATING SOCIAL ORGANIZATION

The perspective presented here is meant as a complement, not an alternative, to the eloquent and effective paradigms in behavioral ecology currently driving most primatological investigations (Krebs and Davies 1997, Kappeler and van Schaik 2002; see Chapter 28). In combining the different paradigms of approach in primatology, we have a richer context from which to ask questions about the intra- and interspecific patterns of behavior of our subjects. Kappeler and van Schaik (2002) divide primatology into three phases, the early "descriptive inventory" phase, the second "correlational approach" phase, and the current, third phase dominated by "hypothetical–deductive" approaches. They counter Rodman's (1993) and Sussman's (1999) call for caution in hypothesis formulation in the absence of natural history data and suggest that in the absence of relevant data hypotheses are erected to provide focal points for field researchers to collect those data. The evident variability in primates' individual behavior, and thus emergent social organizations, provides an arena for both cautious and aggressive approaches. Extant information does facilitate the deductive creation of hypotheses that can result in specific a priori research projects. However, the variation, especially intraspecific variation, evident from our studies also emphasizes the urgent need for continued and expanded natural history and baseline data collection and publication as many of our hypotheses are currently derived from, or supported/refuted by, data sets limited to a few social groups within a species and rarely from different populations within that species. While Kappeler and van Schaik see the collection of natural history data as a beneficial by-product of hypothesis testing research, I would argue that in and of themselves descriptive research projects and protocols are required for a basal understanding of the patterns and processes in primate evolution. The complexity evident in primate social organization at the level of the individual, the group, and the population in mating, ranging, and foraging patterns is best tackled by a multifaceted program interweaving descriptive, correlational, and hypothetical–deductive approaches, all conducted with an ongoing integration into a dynamic evolutionary biology theory. Rather than being three distinct phases in primatology, I prefer to think that we learn form our past, look toward the future, and retain those features of our endeavors which act to benefit our "fitness" as a discipline.

Finally, some may argue that the focus on variability and emergent properties presented here negates the ability to make specific predictions about patterns and processes. I do not think this is the case. Acknowledging complexity in social patterns does not deny the potential for specific adaptive underlying structures that facilitate the appearance of said variability. Nor does it mean that all facets of primates' lives are best seen as hypervariable. It is in fact the comparison between aspects of behavior that are most variable and those that are less so that maximizes our ability to focus adaptive hypotheses and model the evolution of behavioral strategies, thus connecting the individual and its behavior to the patterns and trends seen at the level of the population. The data to perform these broad-scale assessments are discovered both through comparative natural history data sets and through the specific testing of hypotheses on groups and populations.

REFERENCES

Alberts, S. C., Watts, H. E., and Altmann, J. (2003). Queuing and queue-jumping: long-term patterns of reproductive skew in male savannah baboons, *Papio cynocephalus*. *Anim. Behav.* 65:821–824.

Bateman, A. J. (1948). Intra-sexual selection in *Drosophila*. *Heredity* 2:349–368.

Bernstein, I. (1999). The study of behavior. In: Dolhinow, P., and Fuentes, A. (eds.), *The Nonhuman Primates*. Mayfield Publishing Company, Mountain View, CA. pp. 176–180.

Borgerhoff-Mulder, M. (2004). Are men and women really so different? *Trends Ecol. Evol.* 19:3–6.

Borries, C. (2000). Male dispersal and mating influxes in hanuman langurs living in multimale groups. In: Kappeler, P. M. (ed.), *Primate Males*. Cambridge University Press, Cambridge. pp. 146–158.

Brockelman, W. Y., Reichard, U., Treesucon, U., and Raemaekers, J. J. (1998). Dispersal, pair formation and social structure in gibbons (*Hylobates lar*). *Behav. Ecol. Sociobiol.* 42:329–339.

Clutton-Brock, T. H. (1974). Primate social organization and ecology. *Nature* 250:539–542.

Clutton-Brock, T. H. (1991). *The Evolution of Parental Care*. Princeton University Press, Princeton, NJ.

Cords, M. (2000). The number of males in guenon groups. In: Kappeler, P. M. (ed.), *Primate Males*. Cambridge University Press, Cambridge. pp. 84–96.

Crook, H. H. (1970). The socio-ecology of primates. In: Crook, J. H. (ed.), *Social Behaviour in Birds and Mammals*. Academic Press, Cambridge. pp. 84–96.

Crook, J. H., and Gartlan, J. C. (1966). Evolution of primate societies. *Nature* 210:1200–1203.

Darwin, C. (1871). *The Descent of Man and Selection in Relation to Sex*. Random House, New York.

DiFiore, A., and Rendall, D. (1994). Evolution of social organization: a reappraisal for primates by using phylogenetic methods. *Proc. Natl. Acad. Sci. USA* 91:9941–9945.

Digby, L. (1999). Sexual behavior and extragroup copulations in a wild population of common marmosets (*Callithrix jacchus*). *Folia Primatol.* 70:136–145.

Dolhinow, P. (1972). Primate patterns. In: Dolhinow, P. (ed.), *Primate Patterns*. Holt, Rinehart, and Winston, New York. pp. 352–392.

Dolhinow, P. (1993). Social systems and the individual. *Evol. Anthropol.* 3:73–74.

Dolhinow, P. D., and Fuentes, A. (1999). *The Nonhuman Primates*. Mayfield-McGraw-Hill, New York.

Dunbar, R. I. M. (1988). *Primate Social Systems*. Cornell University Press, Ithaca.

Dunbar, R. I. M. (1992). Neocortex size as a constraint on group size in primates. *J. Hum. Evol.* 28:287–296.

Dunbar, R. I. M. (2002). Modelling primate behavioral ecology. *Int. J. Primatol.* 23:785–819.

Emlen, S. T., and Oring, L. W. (1977). Ecology, sexual selection, and the evolution of mating systems. *Science* 197:215–223.

Fuentes, A. (1999a). Variable social organization in primates: what can looking at primate groups tell us about the evolution of plasticity in primate societies? In: Dolhinow, P., and Fuentes, A. (eds.), *The Nonhuman Primates*. Mayfield Publishing Company, Mountain View, CA. pp. 183–189.

Fuentes, A. (1999b). Re-evaluating primate monogamy. *Am. Anthropol.* 100:890–907.

Fuentes, A. (2002). Patterns and trends in primate pair bonds *Int. J. Primatol.* 23:953–978.

Garber, P. A. (1997). One for all and breeding for one: cooperation and competition as a tamarin reproductive strategy. *Evol. Anthropol.* 5:187–199.

Hall, K. R. L. (1965). Social organization of the Old World monkeys and apes. *Symp. Zool. Soc. Lond.* 14:265–289.

Hamilton, W. D. (1964). The genetical evolution of social behaviour. I, II. *J. Theor. Biol.* 7:1–52.

Hinde, R. A. (1976). Interactions, relationships and social structure. *Man* 11:1–17.

Hutchinson, G. E. (1957). Concluding remarks. *Cold Spring Harb. Symp. Quant. Biol.* 22:415–427.

Imanishi, K. (1941). *A Japanese View of Nature: The World of Living Things*, Asquith, P., Kawakatsu, H., Yagi, S., and Takasaki, H. (trans.). Reprinted 2002, Routledge Curzon, New York.

Isbell, L. A. (2004). Is there no place like home? Ecological bases of female dispersal and philopatry and their consequences for the formation of kin groups. In: Chapias, C., and Berman, C. (eds.), *Kinship and Behavior in Primates*. Oxford University Press, New York. pp. 71–108.

Jansen, C. H., and van Schaik, C. P. (2002). The behavioral ecology of infanticide by males. In: van Schaik, C. P., and Jansen, C. H. (eds.), *Infanticide by Males and Its Implications*. Cambridge University Press, Cambridge. pp. 469–494.

Kappeler, P. M., and Pereira, M. E. (eds.) (2003). *Primate Socioecology and Life History*. University of Chicago Press, Chicago.

Kappeler, P. M., and van Schaik, C. P. (2002). Evolution of primate social systems. *Int. J. Primatol.* 23:707–740.

Knott, C. (1999). Orangutan behavior and ecology. In: Dolhinow, P., and Fuentes, A. (eds.), *The Nonhuman Primates*. Mayfield-McGraw-Hill, New York. pp. 50–57.

Kokko, H., and Jennions, M. (2003). It takes two to tango. *Trends Ecol. Evol.* 18:103–104.

Krebs, J. R., and Davies, N. B. (eds.) (1997). *Behavioural Ecology: An Evolutionary Approach*. Blackwell Science, Oxford.

Kruuk, L. E. B., Merila, J., and Sheldon, B. C. (2003). When environmental variation short-circuits natural selection. *Trends Ecol. Evol.* 18:207–209.

Kummer, H. (1971). *Primate Societies: Group Techniques of Ecological Adaptation*. Aldine, Chicago.

Laland, K. N., Odling-Smee, F. J., and Feldman, M. W. (2001). Niche construction, ecological inheritance, and cycles of contingency in evolution. In: Oyama, S., Griffiths, P. E., and Gray, R. D. (eds.), *Cycles of Contingency: Developmental Systems and Evolution*, MIT Press, Cambridge, MA. pp. 117–126.

McFarland, D. (ed.) (1987). *The Oxford Companion to Animal Behavior*. Oxford University Press, Oxford.

McGrew, W. C. (1998). Culture in nonhuman primates? *Annu. Rev. Anthropol.* 27:301–328.

Morbeck, M. E. (1997). Life history, the individual and evolution. In: Morbeck, M. E., Galloway, A., and Zihlman, A. L. (eds.), *The Evolving Female*. Princeton University Press, Princeton, NJ. pp. 3–14.

Oyama, S., Griffiths, P. E., and Gray, R. D. (2001). Introduction. What is developmental systems theory? In: Oyama, S., Griffiths, P. E., and Gray, R. D. (eds.), *Cycles of Contingency: Developmental Systems and Evolution*. MIT Press, Cambridge, MA. pp. 1–12.

Palombit, R. A. (1995). Longitudinal patterns of reproduction in wild female siamangs (*Symphalangus syndactylus*) and white-handed gibbons (*Hylobates lar*). *Int. J. Primatol.* 16:739–760.

Perry, S., and Manson, J. (2003). Traditions in monkeys. *Evol. Anthropol.* 12:71–81.

Piersma, T., and Drent, J. (2003). Phenotypic flexibility and the evolution of organismal design. *Trends Ecol. Evol.* 18:228–233.

Pochron, S. T., and Wright, P. C. (2003). Variability in adult group compositions of a prosimian primate. *Behav. Ecol. Sociobiol.* 54:285–293.

Robbins, M. M. (1995). A demographic analysis of male life history and social structure of mountain gorillas. *Behaviour* 132:21–47.

Rodman, P. (1993). The human origins program and evolutionary ecology in anthropology today. *Evol. Anthropol.* 2:215–224.

Rowell, T. E. (1993). Reification of social systems. *Evol. Anthropol.* 2:215–224.

Rubenstein, D. I., and Wrangham, R. W. (1988). *Ecological Aspects of Social Evolution: Birds and Mammals*. Princeton University Press, Princeton, NJ.

Rylands, A. B. (1993). *Marmosets and Tamarins: Systematics, Behaviour, and Ecology*. Oxford University Press, Oxford.

Sapolsky, R. M., and Share, L. J. (2004). A pacific culture among wild baboons: its emergence and transmission. *PLoS Biol.* 2:534–541.

Sommer, V., and Reichard, U. (2000). Rethinking monogamy: the gibbon case. In: Kappeler, P. M. (ed.), *Primate Males*. Cambrdige University Press, Cambridge. pp. 159–168.

Stamps, J. (2003). Behavioural processes affecting development: Tinbergen's fourth question comes of age. *Anim. Behav.* 66:1–13.

Sterck, E. H. M., Watts, D. P., and van Schaik, C. P. (1997). The evolution of female social relationships in nonhuman primates. *Behav. Ecol. Sociobiol.* 41:291–310.

Strier, K. B. (1994). The myth of the typical primate. *Ybk. Phys. Anthropol.* 37:233–271.

Strier, K. B. (2000). *Primate Behavioral Ecology*. Allyn and Bacon, Boston.

Strier, K. B. (2003). Primatology comes of age: 2002 AAPA luncheon address. *Ybk. Phys. Anthropol.* 46:2–13.

Struhsaker, T. T. (1969). Correlates of ecology and social organization among African cercopithecines. *Folia Primatol.* 11:80–118.

Sussman, R. W. (1999). *Primate Ecology and Social Structure. Lorises, Lemurs and Tarsiers*, Vol. 1. Pearson Custom Publishing, Needham Heights, MA.

Sussman, R. W., and Garber, P. (1987). A new interpretation of the social organization and mating system of the Callitrichidae. *Int. J. Primatol.* 8:73–92.

Tang-Martinez, Z. (2000). Paradigms and primates: Bateman's principle, passive females, and perspectives from other taxa. In: Strum, S. C., and Fedigan, L. M. (eds.), *Primate Encounters: Models of Science, Gender, and Society*. University of Chicago, Press, Chicago. pp. 261–274.

Terborgh, J., and Janson, C. H. (1986). The socioecology of primate groups. *Annu. Rev. Ecol. Syst.* 17:111–135.

Tinbergen, N. (1953). *Social Behaviour in Animals*. Chapman and Hall, New York.

Tinbergen, N. (1963). On aims and methods of ethology. *Z. Tierpsychol.* 20:410–433.

Treves, A., and Chapman, C. A. (1996). Conspecific threat, predation avoidance, and resource defense: implications for grouping in langurs. *Behav. Ecol. Sociobiol.* 39:43–53.

Trivers, R. L. (1972). Parental investment and sexual selection. In: Campbell, B. (ed.), *Sexual Selection and the Descent of Man 1871–1971*. Aldine, Chicago. pp. 136–179.

Trivers, R. L. (1974). Parent–offspring conflict. *Am. Zool.* 14:249–264.

van Schaik, C. P. (1989). The ecology of social relationships amongst female primates. In: Standen, V., and Foley, R. A. (eds.), *Comparative Socioecology.* Blackwell, Oxford. pp. 195–218.

van Schaik, C. P. (2002). Infanticide by primate males: the sexual selection hypothesis revisted. In: van Schaik, C. P., and Jansen, C. H. (eds.), Infanticide by Males and Its Implications. Cambridge University Press, Cambridge. pp. 27–60.

van Schaik, C. P., Ancrenaz, M., Borgen, G., Galdikas, B., Knott, C. D., Singleton, I., Suzuki, A., Utami, S. S., and Merril, M. (2003). Orangutan cultures and evolution of material culture. *Science* 299:102–105.

van Schaik, C. P., and van Hoof, J. A. R. A. M. (1983). On the ultimate causes of primate social systems. *Behaviour* 85:91–117.

Wasser, S. K., Norton, G. W., Kleindorfer, S., and Rhine, R. J. (2004). Population trend alters the effects of maternal dominance rank of lifetime reproductive success in yellow baboons (*Papio cynocephalus*). *Behav. Ecol. Sociobiol.* 56:338–345.

Whiten, A., Goodall, J., McGrew, W. C., Nishida, T., Reynolds, V., Sugiyama, Y., Tutin, C. E. G., Wrangham, R. W., and Boesch, C. (1999). Cultures in chimpanzees. *Nature* 399:682–685.

Williams, G. C. (1966). *Adaptation and Natural Selection.* Princeton University Press, Princeton, NJ.

Wilson, D. S. (2004). What is wrong with individual fitness? *TREE* 19:245–248.

Wilson, D. S., and Sober, E. (1994). Reintroducing group selection to the human behavioral sciences. *Behav. Brain Sci.* 17:585–608.

Wilson, E. O. (1975). *Sociobiology.* Belknap Press, Cambridge, MA.

Wrangham, R. W. (1980). An ecological model of female-bonded primate groups. *Behaviour* 75:262–300.

Wrangham, R. W. (1987). Evolution of social structure. In: Smuts, B. B., Cheney, D. L., Seyfarth, R. M., Wrangham, R. W., and Struhsaker, T. T. (eds.), *Primate Societies.* University of Chicago Press, Chicago. pp. 159–168.

Zahavi, A. (1975). Mate selection—a selection for a handicap. *J. Theor. Biol.* 53:205–214.

Zahavi, A. (1977). The cost of honesty (further remarks on the handicap principle). *J. Theor. Biol.* 67:603–605.

Zuckerman, S. (1932). *The Social Life of Monkeys and Apes.* Harcourt, Brace, New York.

38

The Conundrum of Communication

Harold Gouzoules and Sarah Gouzoules

INTRODUCTION

Communication underpins social behavior: without it, only the most basic forms of sociality are possible. Species that show complex social interactions are likely to have relatively rich communication repertoires (reviewed in Hauser 1996:644, Blumstein and Armitage 1997), and many primate taxa appear to follow this pattern. Important as it is, however, communication is not easily defined. Bradbury and Vehrencamp (1998:5), in an authoritative 882-page overview, dauntingly construe the book's entirety as a thorough definition of animal communication. Minimally, communication has taken place when a sender emits a signal that is detected by a receiver whose behavior might consequently change. Signals, which can take many and varied forms, have two sources: they can evolve over time (i.e., be the direct target of natural selection) or they can be acquired through learning and convention. Most researchers hold that

signals convey information to receivers (Bradbury and Vehrencamp 1998:2), although definitions of what constitutes information vary (Dusenbery 1992). An ongoing issue to which we will return is the question of what goes on following signal detection by receivers: do they decode or cognitively process information, or is this level of analysis anthropomorphic and unnecessary for nonhuman primate communication?

Primates rely on four major modalities (visual, auditory, olfactory, and tactile) used by most mammals for communication. Species vary in terms of relative emphasis of each form (Hauser 1996). To date, vocal communication has received the most attention from researchers; nonetheless, it is still the case that not even a single species' vocal repertoire has been adequately and fully documented, so much basic research still needs to be done. Early studies of nonhuman primate communication focused on issues such as cataloging the repertoires of various species (Altmann 1962,

Chevalier-Skolnikoff 1974, Marler 1965, Rowell and Hinde 1962, Struhsaker 1967), establishing the neurological bases of communication (Jürgens 1979, 1988; Müller-Preuss 1988; Ploog 1967), comparisons with human language using perceptual and psycholinguistic approaches (Snowdon 1982), the socioecological functions of communication systems (Gautier and Gautier 1977, Green 1975, Marler 1968, Struhsaker 1967), the impact of habitat on signal structure (Waser and Waser 1977), and the capacity of apes to acquire artificial language (reviewed in Wallman 1992). While many of the studies were guided by evolutionary theory in a broad comparative sense (i.e., the view that nonhuman primates were especially appropriate animals to provide insights into human behavior and evolution), very few examined communication systems with respect to narrower, more specific evolutionary hypotheses. Exceptions include studies that examined the ecological and social influences on signal structure (Waser 1982, Waser and Waser 1977, Waser and Brown 1984) and a few that searched for homologies among species (e.g., van Hooff 1967).

Space constraints prevent a comprehensive review of the large literature on primate communication published over the last 40 years. Such an inventory-taking summary is perhaps not the most instructive approach to take in order to assess the progress of and future directions for research. Instead, we examine the literature selectively, with an emphasis on material relevant to current evolutionary issues, and evaluate the available data as they relate to these ideas. We emphasize vocal communication in discussing empirical research, but in many cases, the evolutionary theories to which these studies relate have relevance to other modalities as well. Certain areas that have been recently reviewed and thoroughly debated, for example, laboratory studies of the linguistic abilities of great apes (Wallman 1992, Savage-Rumbaugh et al. 1998, Savage-Rumbaugh and Fields 2000), are not considered here. Other areas neglected here, such as the ontogeny of primate communication (e.g., Seyfarth and Cheney 1997, Snowdon and Hausberger 1997, Snowdon 1999) and facial and gestural communication (e.g., Preuschoft and van Hooff 1995, 1997; van Hooff and Preuschoft 2003; Maestripieri 1999), are also treated elsewhere.

THE ADAPTIVE NATURE OF SIGNAL STRUCTURE

Signal Structure and the Ecological Environment

Identification of the factors involved in the evolution of signal design and form is a critical area in animal communication research, including that on primates. Ecological and biological conditions that impact the physical nature of signals include the distance over which communication must be effective, the energetic costs of communicating, body size and structure, physical properties of the habitat, the timing of communication, the number and phylogenetic relationships of sympatric species, the presence and behavior of both conspecific competitors and those of other species,

and costs imposed by predators (reviewed in Gerhardt 1983, Wiley 1983, Brown and Waser 1988, Brown et al. 1995, Hauser 1996, Bradbury and Vehrencamp 1998).

For some conditions, the impact on design structure has been more significant for signals used over relatively large distances. The diverse habitats of primates both shape and constrain the design of communication systems (Waser and Brown 1984, 1986; Brown and Gomez 1992). Habitats may vary in the level of background noise and in how signals produced in them are modified or distorted.

Sound transmission, for example, is frequency-dependent, and the nature of the habitat will affect the distance over which calls will be effective (Marten and Marler 1977; Marten et al. 1977; Piercy et al. 1977; Waser and Waser 1977; Michelsen 1978; Wiley and Richards 1978; Waser and Brown 1984, 1986). Different habitats can also distort important components of vocal signals, including the frequency, time, and amplitude domains (Brown and Waser 1988, Brown and Gomez 1992). Brown et al. (1995) noted that behavioral motor patterns that govern signal features such as duration, rise and fall time, bandwidth, rate and slope of frequency modulation, amplitude variation, or pulse repetition rate can be adjusted to minimize distortion associated with specific habitat types. They conducted field experiments to test Gish and Morton's (1981) "local adaptation hypothesis," which predicts that the characteristics of the environment in which a signal is normally broadcast have imposed selection pressures on signals resulting in designs that reduce distortion. The hypothesis was tested for calls broadcast in "appropriate" and "inappropriate" habitats of blue monkeys (*Cercopithecus mitis*) and gray-cheeked mangabeys (*Cercocebus albigena*), both of which are rain forest species, and two savanna species, vervet monkeys (*Cercopithecus aethiops*) and yellow baboons (*Papio cynocephalus*). Exemplars of common call categories in each species' repertoire were broadcast and re-recorded at both short and long distances for comparisons between point of origin and transmitted signals that involved acoustical analyses of digitized calls. Overall, the results supported the idea that the general form of the entire repertoire of the vocalizations of forest monkeys had been selected to reduce the significance of distortion. In contrast, the intensity of selection for resistance to distortion posed by habitat acoustics appears to have been relaxed for savanna monkey calls. In the savanna, signalers and receivers can communicate visually, and it is possible that the accessibility of an unobstructed visual channel has resulted in fewer constraints being imposed by habitat acoustics.

Brown and colleagues (1995:958) point out that it is not known whether or not the physical distortion of calls that is measured in their analyses maps faithfully to perceived distortion by the monkeys. As is the case for many studies of primate vocal communication that rely on acoustical analyses, whether or not statistically significant acoustical differences (or the lack of differences) are biologically relevant to the animals themselves is often not known.

Signal Structure and "Motivational–Structural Rules"

In theory, signal structure for shorter-distance communication will be less influenced by selection imposed by the problems of production, effective transmission, and detection in natural habitats. Nonetheless, shorter-range signals are also thought to take forms that may be understood from an evolutionary perspective; such thinking dates back to Darwin (1872). Following Darwin's ideas, Morton (1977, 1982) outlined a set of "motivational–structural rules" for vocalizations after observing that harsh (broadband), lower-frequency vocalizations are widely used in hostile and aggressive contexts, while tonal, higher-frequency calls tend to be employed in appeasing or friendly situations. Morton (1977) noted that the ability to produce harsh sounds of lower frequency is linked to body size and that, in many species, size tends to determine the outcome of aggressive encounters. Owings and Morton (1998:111) have further suggested that more complicated use of vocalizations evolved when the direct and largely fixed sound symbolism of amphibian vocalizations was modified by motivational factors in warm-blooded vertebrates. Vocalizations came to reflect differences in motivation with harsh, low-frequency sounds linked to aggressive states. Appeasing, submissive, or friendly vocalizations, either through selection based on the principle of antithesis or, perhaps, because of an association between the production of higher-pitched sounds by infants, tend to be tonal and of higher frequency. Primate studies have provided general support for Morton's rules (Hauser 1993, 1996). Hauser (1993) looked at several hundred vocalizations from 43 primate species and found patterns consistent with Morton's predictions. However, exceptions to the predictions of motivational–structural rules have been noted (Gouzoules and Gouzoules 2000). That some primate vocalizations in certain contexts deviate from Morton's motivational–structural rules is noteworthy but perhaps not unexpected if the rules are viewed as evolutionary "starting points" from which species might depart and deviate depending on new selective pressures. For example, species in which alliances and coalitions have a greater influence on dominance status than does body size (see Harcourt and de Waal 1992) might rely less on "expressive sound symbolism."

Signal Structure and "Honest Communication"

Zahavi (1975, 1977, 1980, 1987, 1993) and Zahavi and Zahavi (1997) have presented arguments for yet a different perspective on signal evolution that has received much attention. They contend that there should be selection for receivers to ignore signals that are not "honest" (i.e., accurate and reliable); receivers are therefore the "ultimate" determiners of how evolution shapes communication. They reasoned that honest signals could evolve if they took forms that required considerable cost to produce, a condition that would result in ineffective communication if the sender could not bear that cost. They referred to these costly signals as "handicaps." The more extreme the expression of these expensive signals is, the more costly they are; and this expenditure is relatively greater for less fit or poorer-condition individuals. Costliness, according to Zahavi's theory, is essential to the evolution of honesty.

Gouzoules and Gouzoules (2002) reviewed the nonhuman primate literature with respect to Zahavi's ideas on costly signaling. Female sexual swellings, copulation calls, rhesus monkey grunts and girneys, and a number of other possible examples of costly signaling were considered. Game theory models and the empirical studies reviewed suggest that costly signals are not often likely to be the basis for honest communication in nonhuman primates. Costly signals are not evolutionarily obligatory in many contexts where communication occurs among related animals (Bergstrom and Lachmann 1997, 1998; Lachmann and Bergstrom 1998), a prominent feature of many nonhuman primate species. Another condition where inexpensive signals are possible is when sender and receiver both benefit from coordinated interactions (Farrell and Rabin 1996). Also, if individuals interact repeatedly and can use past interactions to assess the honesty of signals and modify future response to signals, then low-cost signals can evolve (Silk et al. 2000). Clearly, much of the communication that nonhuman primates perform occurs under these circumstances.

There are contexts in which costly signaling might evolve in primates, however, and these would include overtly competitive interactions where unrelated individuals have limited access to information about one another: such situations would center around animals with little or no history of interaction and would involve a restricted set of signals. An example might be the long vocal exchanges that male howler monkeys (*Alouatta* spp.) perform when they encounter other troops and solitary males. Sekulic (1982) and Sekulic and Chivers (1986) suggested that the primary function of male howler roaring is to deter outside males from attempting to gain entrance into their groups. They proposed that a high roaring rate might be costly and indicate fighting ability to rivals. While some of the evidence on male howler roaring is consistent with this hypothesis, there has been no direct test examining whether male qualities affect roaring and whether the calls are effective in contests with challengers.

Another possible case of costly signaling is the "red-white-and-blue" display of male vervet monkeys. During dominance interactions, adult males perform a genital display that involves approaching a subordinate with tail erect, then exposing the red perianal region, bright blue scrotum, and the white hair between the scrotum and perineum. Gerald (2001) conducted an experiment on the relationship between scrotal color and social status in adult males. She analyzed the social interactions of paired unfamiliar males, matched for size but differing in scrotal color. Difference in scrotal color predicted eventual dominance status: males with darker scrota came to dominate those with paler ones. Experimental modification of scrotal color revealed that this signal alone did not signal status; males with scrota of similar color were more antagonistic than were those with dissimilar color. Scrotal color pales under stressful conditions (Gartlan and Brain 1968, Isbell 1995) and, thus, may reflect male condition. Furthermore, the display, which involves exposure of the genitals to a potential aggressor, might qualify as risky and, thus, costly. Males transferring into new groups might rely more heavily on scrotal color as a way to evaluate rivals.

In another possible example, coloration might represent a costly signal in the striking natal coats that infants of some nonhuman primate species bear during the first months of life. Ross and Regan (2000) investigated possible relationships between conspicuous natal coats and factors that have been thought to be linked to its evolution—namely, allocare, predation risk, and type of mating system. They point out that a conspicuous natal coat (CNC) very likely enhances visibility to predators and may also advertise an infant's vulnerability, age, and lack of experience, which would also be detrimental where infanticide occurs. Five hypotheses were tested with data from 82 species (28 of which have prominent natal coats). CNC was examined with respect to (1) the degree of allocare, (2) the level of competition among infants for allocare, (3) the risk of predation on infants, (4) terrestrial habitats, and (5) the type of social structure as measured by relative size of the testes (which is presumed to correlate with the level of competition among males for females). Ross and Regan (2000) concluded none of the hypotheses clearly accounted for the distribution of CNCs in anthropoid primates, so the evolution of this feature remains to be explained fully. While acknowledging the likely costs associated with CNCs, Ross and Regan did not consider the implications of costly signal theory in their hypotheses. Gouzoules and Gouzoules (2002) suggested that CNCs may be adaptive for the infants that bear them under conditions where the threat of predation is significant enough to require increased protection and attention from mothers and other group members but not so great as to represent too extravagant a cost. Zahavi and Zahavi (1997:121) consider distinctive and prominent colors or markings of the infants of various nonprimate species as costly handicaps that can be interpreted as a form of "extortion" of parents by infants: in effect, an infant's gambit in a specific example of Trivers' (1974) theory of parent–offspring conflict. Thinking about primate CNCs in this light might result in several of the hypotheses examined by Ross and Regan becoming complementary rather than competing.

Primate olfactory communication might also be an area where costly honest signals are used (Kappeler 1998). Gosling and Roberts (2001) note that while there has been little formal investigation of the costs associated with any olfactory signal, some forms, such as scent marking, are likely to be energetically expensive and, in some nonprimate mammals, have been shown to attract predators and parasites. Scent-marking behavior is very common in some primate species, such as *Lemur catta*, where its use varies by age, sex, and social status (Kappeler 1990). It is used at the boundaries of group home ranges and among group members. Costly signaling theory would predict that in order for scents deposited at the boundaries of territories to effectively deter intruders, they would have to be reliable indictors of a signaler's quality, a likely relationship due to the fact that such signals are linked to physiological conditions (Kappeler 1998).

HOW PRIMATES DEAL WITH DECEPTION

Several major overviews of the empirical data on primate deception are available, as are various theoretical interpretations of deceptive behavior (de Waal 1986; Byrne and Whiten 1988; Whiten and Byrne 1988; Cheney and Seyfarth 1990a,b, 1991; Hauser 1996, 1997; Tomasello and Call 1997). Our concern here is with deception as a problem for honest communication: if nonhuman primates only rarely rely on costly signaling to ensure honesty, what evidence is there that they instead use past interactions to assess the honesty of signals and modify future responses to signals, as predicted by the models discussed above (e.g., Silk et al. 2000)? We briefly summarize the kinds of deception with which nonhuman primates are confronted and then discuss two possible mechanisms that have been suggested as solutions to the problem of dishonesty: punishment (Clutton-Brock and Parker 1995) and skeptical responding (Moynihan 1982, Smith 1986).

Whiten and Byrne (1988) proposed the term *tactical deception* for short-term changes between honest and deceptive behaviors exhibited by an animal in the context of its social group. Key features of tactical deception are that the behavior itself must be part of the normal "honest" repertoire and the deceptive behavior must be rare, subtle, and used in different contexts from those in which it would ordinarily appear. Accounts of functionally deceptive behavior range from the withholding of information (alarm calls, Cheney and Seyfarth 1990a,b; food calls, Hauser 1992, 1997; excitement calls, Goodall 1986, Boesch and Boesch 1989; body postures and facial signals, de Waal 1986) to the use of signals in apparently inappropriate circumstances (alarm calls Byrne and Whiten 1991, Cheney and Seyfarth 1990a,b; screams, Byrne and Whiten 1988).

Evidence from nonhuman primates for punishment in response to deception is scarce, despite theoretical arguments concerning its importance in animal societies. Experiments conducted on Cayo Santiago rhesus monkeys suggested the occurrence of punishment when information was withheld (Hauser 1992; Hauser and Marler 1993a,b). Individuals discovering food (coconut or monkey chow) provided by the researchers gave "food calls" in fewer than half of the trials, and the calling behavior of discoverers was the most significant predictor of aggression they received. Those that produced at least one food call received a lower total of aggressive acts and a shorter duration of aggression. Hauser

(1996:582) has argued that this evidence suggests that rhesus monkeys that withhold information about food are punished. While these data constitute the most formal test of punishment as a response to deception to date, stronger support would be results showing that individuals that call are more likely than silent monkeys not to receive any aggression at all. If callers receive any aggression, as they apparently do, other factors must also be involved. Additional evidence that punishment for withholding information was occurring would be data that revealed the consequences of receiving aggression: are animals that failed to call and were targets of aggression more likely to call subsequently when they encounter food?

One context in which punishment in nonhuman primates has been described is parent–offspring conflict, especially during the weaning period (Altmann 1980). Begging signals that young of various species use to elicit feeding from parents have received considerable attention within the framework of honest communication (Hauser 1986, Kilner and Johnstone 1997). This is a particularly interesting context in which to examine nonhuman primate communication: the individuals involved are related and kinship, as previously noted, is one factor that may promote honest signaling (Bergstrom and Lachmann 1997, 1998; Lachmann and Bergstrom 1998). However, a potential conflict of interest between the infant and parent exists as well. Trivers (1974) suggested that weaning behavior in mammals resulted from parent–offspring conflict over the amount of investment (usually milk) that the offspring should receive. In nonhuman primates, however, it appears that conflicts are related to the timing of investment in the offspring. Conflicts occur as a female restricts her infant's suckling to those times that do not interfere with her ability to engage in other activities, including mating (e.g., baboons, Altmann 1980; Japanese macaques, Worlein et al. 1988; rhesus macaques, Gomendio 1991; gelada baboons, Barrett et al. 1995). At these times, negative behavior directed toward the infants tends to increase as do maternal rejections and punishments. Concurrently, the infants tend to display regressive behavior and signs of distress, including vigorous tantrums. No studies to date have examined the exact coordination of these interactions with respect to the signals used by infants.

Skeptical responding is another mechanism for dealing with deception. Any communication system open to deception will be inherently unstable due to selection pressure on both the deceptive signaler and the receiver to become more proficient (Smith 1986, 1991). In the case of deceptive use of some signals, for example, alarm calls by competent adults, the false alarm is likely to be identical, or nearly so, in terms of its acoustic features to honest alarms. Under these conditions, selection on receivers might promote enhanced capacity to assess the reliability of the signaler, rather than the signal itself. The solution, Moynihan (1982:9) suggested, is that animals "must assess the information proffered by companions and rivals of their own species with some considerable degree of skepticism," an ability labeled "skeptical responding" by Smith (1986). Signalers that misinform sufficiently often may become devalued as sources of information.

Moynihan (1982) noted that skepticism, and any comparisons involved in assessing reliability, cost the receiver in terms of delays and energy expenditures. However, skepticism may be less costly if animals are not equally skeptical of all conspecific signals. An ability to recognize individuals (and their signals) and recall past encounters with them allows skepticism to be restricted to those animals most likely to be deceptive. Playback experiments with vocalizations have been used to assess how monkeys respond to inaccurate or false information (Cheney and Seyfarth 1990a,b; Gouzoules et al. 1996). Not surprisingly, monkeys habituate to these calls: they become less responsive over time. Interestingly, with respect to the skeptical responding hypothesis, their decreased response is specific to particular contexts and individuals. For example, a study of vervet monkeys (Cheney and Seyfarth 1988) used playback of false calls (intergroup *wrrs* and *chutters*, eagle and leopard alarm calls) in habituation experiments (calls repeated eight times at 20 min intervals). Receivers treated calls from a particular individual as unreliable in that referential context but not in others. In other words, skepticism regarding the caller was restricted to the context relevant for the call used in the experiment.

In another study of skeptical responding, Gouzoules et al. (1996) played tape-recorded alarm calls of high- and low-ranking female rhesus monkeys to their groups in a feeding context. False alarm calls from low-ranking monkeys had been previously observed: in each instance, monkeys ran and climbed into trees, except for the caller, which continued to eat. Initial response to alarm playback was strong for the calls of both high- and low-ranking animals. By the eighth day, response had diminished greatly; however, habituation was faster and more extensive to the calls of low-ranking monkeys, those presumably with the most to gain through deception in this context (Gouzoules et al. 1996).

PRIMATE COMMUNICATION, LINGUISTIC MODELS, AND THE EVOLUTION OF LANGUAGE

Exactly what studies of natural nonhuman primate communication and the ape language projects reveal about the evolution of human language has been debated extensively (e.g., Cheney and Seyfarth 1990a, Bickerton 1990, Wallman 1992, Hauser 1996, Deacon 1997, King 1999, Snowdon 2001) and will certainly not be resolved by us here. The continuing debates center on what qualifies as language (in ape linguistic capacity studies, e.g., Savage-Rumbaugh and Fields 2000) and what should be considered as evolutionary precursors to language (in the case of natural communication). There is no doubt, however, that over the last two decades many studies have pointed to the vital role vocal communication plays in primates' lives (reviewed in Cheney and Seyfarth 1990a, Hauser 1996, Seyfarth and Cheney 2003).

Recently, however, some researchers have expressed concerns that some of these results have been over-interpreted, in part due to the use of what they consider to be anthropomorphic and anthropocentric linguistic models that have come to dominate animal communication research, in particular that on nonhuman primates (Owren and Rendall 1997, 2001). They have questioned the view of communication as a process of conveying information to other individuals and have suggested alternative models and concepts; in effect, they have advocated a major paradigm shift in the study of animal communication (Owren and Rendall 1997, 2001; Owings and Morton 1997, 1998; Thompson 1997). A rejection of linguistic models in nonhuman primate communication research effectively eliminates the search for precursors to language. In this section, we assess concerns about the "informational perspective" in animal communication (Owren and Rendall 1997, 2001; Owings and Morton 1998) and some proposed alternatives as they apply to examples of referential vocalizations in primate vocal communication. We begin by reviewing some of the conclusions and remaining uncertainties concerning the best-known example of calls with external referents, the alarm calls of vervet monkeys. We then briefly review another example of putative external reference, macaque agonistic screams. Screams are among the vocalizations that opponents of the informational perspective have targeted for alternative interpretations. Finally, we will come back to the issue of precursors to language in the communication systems of nonhuman primates and argue that recent pessimism about evidence is unwarranted.

Vervet Alarm Calls and External Reference: Evaluating the Evidence

What data have been used to document external reference in primate vocalizations and how do they meet the criteria proposed as necessary to demonstrate this property in animal vocal communication? Only after Seyfarth et al. (1980) focused attention on the alarm calls of vervet monkeys did studies of additional call classes in this and other primate species begin to explore the possibility of more widespread occurrence of externally referential signals. Early on, researchers overemphasized differences between the majority of calls, thought to reflect an individual's internal state or emotions, and a smaller subset of calls that were considered to have referents external to the signaler (e.g., Marler 1984; Gouzoules et al. 1985). This emphasis was not meant to suggest that internal and external referents were in any sense antithetical but, instead, to promote awareness and thinking about the existence of the latter. Recently, more care is taken to be clear that animal vocalizations are more likely to fall along a continuum, with signals at the extremes mostly reflecting the internal state of the sender at one end and referential signals, with relatively little emotional content, at the other (e.g., Marler et al. 1992). Over a similar period, the question of what evidence is sufficient for calls to be considered referential has also been approached in

more complex and sophisticated ways (Marler et al. 1992, Macedonia and Evans 1993). Criteria center on both the calling behavior of the signaler and the effects of the vocalizations on conspecifics.

Evans (1997), for example, contends that, with respect to evidence on calling behavior, referential signals should both be structurally discrete (i.e., belong to an acoustically distinct class) and have a degree of stimulus specificity, meaning that eliciting stimuli should be members of a distinct category. He reasons that one would not expect the same class of referential signal to be generated in response to stimuli that are members of qualitatively different categories. Importantly, it is from the animal's perspective, and not the investigator's, that these criteria should hold: what constitutes a structurally discrete class of calls for a researcher who has examined selected acoustical variables and determined uniformity on the basis of statistical patterns might not correspond to the animal's perception of the same vocalizations; relevant data for the animal's perspective are rare (e.g., Le Prell et al. 2002). The prerequisite that eliciting stimuli should be members of a distinct category requires investigators to have essentially complete knowledge about how the animals categorize their world; here, too, the data are unfortunately lacking, something to which we return below for further comment. The next step in demonstrating external reference for a vocalization has been to establish that calls are capable of eliciting an appropriate response in a receiver deprived of any additional real-time information about the putative referent that might be available from the external environment (Evans et al. 1993, Macedonia and Evans 1993, Evans 1997). (This leaves open the likely possibility that past experience with contextually specific information is available through memory and can influence response.) Evidence of this kind is most convincing for the alarm calls of vervet monkeys (summarized in Cheney and Seyfarth 1990a). Through field experiments in which tape recordings of the three types of alarm call were broadcast from hidden speakers, Seyfarth et al. (1980) determined that the vocalizations were sufficient to elicit adaptive, predator-specific escape responses. Evidence of this sort demonstrates what Marler et al. (1992) and Macedonia and Evans (1993) have labeled "functional reference." Nothing is revealed about the nature of the mental representation that the vervets are using. It is possible that upon hearing a leopard alarm a vervet conjures up a mental image of a large spotted cat armed with sharp teeth and claws and runs for the nearest tree. On the other hand, these playback data alone would also be consistent with the idea that the representation of the alarm call in an animal's auditory system simply might directly trigger motor neurons that control escape behavior (Lloyd 1989, Bickerton 1990). However, the playback data, combined with observations that the monkeys vary their responses to a specific alarm call depending on the setting in which it is heard, suggest that the simpler explanation is unlikely.

Typically, researchers decide what constitutes an appropriate response to an experimentally broadcast vocalization on the basis of contextual analyses or interpretations of the situations as they occur in nature. The alarm calls of vervet monkeys are excellent examples of calls that have demonstrably adaptive and functional responses associated with them. The nature of any stored representation linked to such vocalizations is not easily established through such an approach, however. The exact knowledge or information that either the signaler possesses or the receiver acquires by hearing the call is difficult to assess because this, too, requires the investigator to have a thorough understanding of how the species classifies its world (Cheney and Seyfarth 1990a). Seyfarth et al. (1980) were able to establish, through alarm call playback experiments, that vervet monkeys react differently to calls given in response to leopards and those in response to martial eagles, but such data reveal nothing about what features are critical for the monkeys to distinguish a leopard from an eagle or a leopard from a lion; in fact, such information would be next to impossible to obtain from animals under natural conditions. Thus, Cheney and Seyfarth (1990a:171) noted that, from such evidence, "the external referents of primate vocalizations are best defined in relation to the external referents of other calls within the animals' repertoire."

In additional experiments designed to test whether vervets are able to assess calls according to the things they denote, Cheney and Seyfarth (1988) obtained results supporting the existence of semanticity in what they argue is a stronger sense. Two sets of acoustically different vocalizations were used in a habituation/dishabituation procedure. One pair of calls included the leopard and eagle alarms, while the other pair was composed of two calls the monkeys use during encounters with neighboring groups of vervets. *Wrrs* and *chutters* are thought to be acoustically different calls. (Note that Evans' [1997:120] concern about whether the monkeys "partition the acoustic space" identically to the investigators is an important issue regarding the interpretation of the results of these experiments.) *Wrrs* are typically given when another group has been spotted in the distance, and *chutters* are used during aggressive contact with a different group. Baseline responses of 10 subjects to recordings of both the *chutters* and eagle alarms of one of the group's females were measured. The subjects were then presented with two separate series of eight repetitions of other recorded calls of that same female. In one series, the female's *wrrs* were played and in the other, her leopard alarms. Following the eighth presentation, the other member of the pair of calls was tested: the repeated *wrrs* were followed by a *chutter*, while after the leopard sequence, an eagle alarm was broadcast. Subjects rapidly habituated to both of the repeated calls. Habituation was transferred in the case of the subsequently played *chutter* but not to the eagle alarm that followed the leopard alarms. Cheney and Seyfarth concluded that these results were consistent with the view that the monkeys were assessing calls on the basis of their referents rather than their acoustical structure. They (Cheney and Seyfarth 1990a:154–155) suggest that transfer of habituation occurred in the case of

the *wrr–chutter* pair because these calls had "roughly the same referent," while eagle and leopard alarms denoted significantly different things. Interestingly, when the *chutters* of one animal were tested in a similar fashion with a sequence of *wrrs* from a different monkey, habituation was not transferred to the *chutter*. Thus, call meaning and the identity of the signaler were important determinants of the subjects' responses (see also Owren 1990).

Wrrs and *chutters* seem to be members of a superordinate categorical class that links events concerning other vervet groups, though the precise nature of the similarities is not clear. That habituation does not transfer from an individual monkey's leopard alarm to its eagle alarm suggests that these two basic categories are not members of a superordinate class, even though leopard alarms are occasionally given to stooping eagles about to grasp a victim in its talons (Cheney and Seyfarth 1990a:107). Leopard alarms are sometimes also given to attacking snakes (Macedonia and Evans 1993). Macedonia and Evans (1993:183) have interpreted the relative lack of production specificity for leopard alarms as an indication that the calls are "less specific and hence might be placed further from the referential pole of a 'motivational-to-referential continuum' (Marler et al. 1992) than eagle and snake alarms." Another possible explanation for the more varied use of leopard calls might be that the call's referent is, in fact, primarily external but what is denoted is something shared by, but not restricted to, the spotted cat that is the most frequent elicitor of the vocalization.

In another series of experiments, Seyfarth and Cheney (1990) conducted habituation/dishabituation tests using vervet eagle and leopard alarms paired with the terrestrial predator and raptor alarms of the sympatric superb starling (*Spero superbus*) to test whether the monkeys assessed starling alarm calls on the basis of their meaning. If this were the case, vervet habituation to the raptor (or terrestrial predator) alarm call of one species should transfer to the corresponding call of the other. The starling's alarms, especially its terrestrial predator call, are given to a much wider range of species than are vervet alarm calls, though among the eliciting species are those of concern to the vervets. Thus, Seyfarth and Cheney suspected that transfer of habituation might occur more readily with the raptor/eagle pairing.

Results of the starling/vervet alarm call playback experiments suggested that the monkeys respond to starling raptor calls as relatively precise signals that are similar in meaning to their own eagle alarms. Subjects transferred habituation from one species' alarm call to the other's when the calls broadcast had the same referent. Transfer of habituation also occurred for tests pairing starling terrestrial predator calls and vervet leopard alarms, though the response disparity between control and test calls was not as great as it was for the avian predator pairing. Seyfarth and Cheney (1990) interpret this difference as an indication that, while vervets treat starling terrestrial predator calls and their own leopard alarms as similar, the calls are judged to be less synonymous than starling raptor calls and their own eagle alarms.

Finally, one other interesting outcome of these experiments deserves attention because it reveals something about how the information from calls may vary in different situations. Recall that for the experiment involving habituation to *wrrs* transfer of habituation extended to *chutters* when the identity of the caller was the same for both vocalizations but this did not occur when one monkey's *wrr* was tested with another individual's *chutter*. Cheney and Seyfarth (1990a:201) note that subjects "did not transfer their skepticism to *different* individual's calls, suggesting that doubts about one individual's reliability did not generalize to include other group members." (Here Cheney and Seyfarth use the habituation phase as a "cry wolf" experiment, where a monkey was made to appear unreliable to group members by repeating, through playbacks, unfounded warnings about danger.) In the case of the starling/vervet alarm call tests, however, habituation was transferred, even though members of two different species were involved. The monkeys appear not to attend to information about the identity of the caller when generalizing across species. Why the monkeys should show skepticism to the alarm call of a group member that has not shown itself to be unreliable following exposure to false starling alarms is not clear.

Zuberbühler (2003) employed similar playback techniques to extend the general conclusions of the vervet monkey research to other species in the genus *Cercopithecus*. Nonetheless, the exact nature of the external referents for vervet alarm calls and, certainly, other primate calls remains an elusive but important issue to which we shall return after illustrating this point further with another set of primate vocalizations.

External Reference and Macaque Agonistic Screams: Navigating a Tangled Web

In many species of primates, screams are a prominent feature of aggressive interactions. Studies of agonistic aiding in free-ranging rhesus (Kaplan 1977, 1978) and numerous other species (reviewed in Walters and Seyfarth 1987, Cheney and Seyfarth 1990a:25–44) revealed a complex behavioral system in which alliance formation and recruitment is critical to both the development and maintenance of dominance status in social groups. Vocalizations were frequently noted to play a significant role in eliciting support from allies within the group. Successful recruitment might occur because of vocal recognition of the caller and information provided by the scream about the caller's emotional state or arousal level. Another hypothesis is that screams have external referents and make available more specific information about the agonistic event, information that might be of use to spatially distant allies in making decisions about intervention tactics. While these two hypotheses are not mutually exclusive, our research has focused primarily on this latter possibility.

Gouzoules et al. (1984) found that in rhesus monkeys on Cayo Santiago, victims of attack usually made use of one of five acoustically distinct screams. The major scream types were associated with a particular class of opponent, defined in terms of relative dominance rank and matrilineal relatedness to the signaler, as well as the severity of the aggression, differentiated as to whether or not physical contact had occurred. Noteworthy here are "noisy" and "arched" screams. Acoustically atonal noisy screams were most often given when the opponent was a higher-ranking group member and the incident involved contact aggression such as biting. Arched screams were typically given to lower-ranking opponents during non-contact aggression. They were highly unlikely to occur during serious aggression with higher-ranking opponents or during aggression from matrilineal kin. The different types of scream did not provide a good prediction of the caller's future behavior. Screams can be heard over considerable distances, and this led us to reason that selection had promoted the evolution of call features that enable messages to be conveyed to remote group members; given the energy costs and potential danger in attracting predators, such features would not be predicted for these calls if the primary receivers were opponents at close range (but see Alternative Perspectives, below).

Playback experiments (Gouzoules et al. 1984) revealed that mothers responded most strongly to noisy screams, which are associated with contact aggression from higher-ranking opponents. Responses were measured in terms of latency to look to, and the duration of looking in the direction of, a concealed speaker. Such measures reflect most directly the ability of the calls to attract the attention of the subject. Naturally occurring screams frequently elicit approach and intervention by an ally, and for playback experiments, this might lead to habituation. Thus, short exemplars of calls were used to minimize the likelihood that a subject would approach the concealed speaker. Competing hypotheses based on the probable information made available by the vocalizations were used to predict subjects' interest in the broadcast calls.

Arched screams, associated with noncontact aggression from lower-ranking opponents, elicited the next strongest response from mothers. This finding was pivotal because serious aggression with a risk of injury (i.e., fights involving contact with the opponent) was extremely rare when these screams were produced, and it did not seem likely that caller fear levels should be high. Thus, the physical consequences for a victim of attack from a lower-ranking opponent were minimal, and the high level of response elicited from callers' mothers could not be explained by concern over protecting their offspring from injury. Instead, preservation of the matrilineal dominance rank seems a more probable basis for the allies' reactions to the arched screams.

Additional playback experiments were used to examine whether rhesus monkeys responded differently to calls of dependent offspring as opposed to those from comparably aged collateral kin (e.g., sisters, nieces, and cousins) (Gouzoules et al. 1986). The results supported our earlier conclusions that, for allies, the key information provided by the screams and influencing the likelihood of intervention concerns the nature of the aggression and opponent. While subjects responded more strongly to the noisy screams of their offspring compared to those of more distantly related juveniles, no such difference emerged from a comparison of playbacks involving arched screams from the same callers. When responses to only collateral kin were examined, the same patterns were found: relatedness influenced response to noisy screams but not to arched screams. It may be that the intervention on behalf of a more distant relative whose opponent is higher-ranking (noisy screams) is more costly. We also found that the sex of the caller influenced the intensity of response during playback experiments with the noisy and arched screams of collaterally related juveniles. While there were no differences in response strength to noisy screams of male and female kin, arched screams from young female kin elicited significantly stronger responses than did those from male relatives. This sex difference may be related to male and female life history patterns: pubescent males generally emigrate from the natal group, whereas females remain for life and are dependent on members of their matriline for rank preservation.

Interestingly, rhesus macaque screams would fail a strict application of the production criterion for external reference suggested by Macedonia and Evans (1993) and discussed earlier. Screams are often used in agonistic contexts other than those with which they are primarily associated. However, there are several sources of variation in the use of calls that can account for an apparent lack of specificity and yet allow the possibility of external reference. For example, our analyses of pig-tailed macaque screams (Gouzoules and Gouzoules 1989) suggested that juvenile monkeys under 3 years of age are far more likely than older individuals to make mistakes in the contextual use of calls, but improvement comes with experience and age (Gouzoules and Gouzoules 1995). Production specificity, then, is significantly greater for adult pig-tailed macaques.

Another factor, mentioned above, that might also account for an apparent lack of production specificity is a failure by researchers to accurately describe or recognize the objects or events that are indexed by calls (Gouzoules and Gouzoules 1995, Gouzoules et al. 1998). What might initially appear to be different eliciting stimuli could well have common and relevant features to the monkeys. A further complicating issue is that many studies have approached the problem as though the animals classify items with a logic system having only two truth values, 1 or 0 (true and false). Investigators typically have followed the "essential first step" (Marler et al. 1992) for determining the meaning of a signal: assess correspondence between vocalizations and putative referents. As we have seen, a subsequent stage in demonstrating external reference has been to show that a call is capable of eliciting, in a receiver denied additional information about the referent, a response appropriate to the referent. Both of these steps require a decision by the researcher about the nature of the category to which a

particular referent might belong; often, as a starting point, these presumptive categories have been very simple ones. For example, in our studies of macaque agonistic screams (e.g., Gouzoules and Gouzoules 1989), we classified aggressors in agonistic encounters as either of higher rank or of lower rank than the victim and the intensity of the conflict as involving contact aggression or noncontact aggression. Similarly, Cheney and Seyfarth (1982) differentiated vervet grunts as to whether the recipient was dominant or subordinate to the caller. Studies of primate food calls, which have also been proposed as referential, have categorized food as high-quality or low-quality, divisible or not, preferred or not (e.g., Dittus 1984; Benz 1993; Hauser and Marler 1993a,b; Hauser et al. 1993). Although many of these studies have established associations between different vocalizations and referents employing this dichotomous approach, the mapping between call and referent has been rudimentary. This bisecting of the animal's world was done for logistical and analytical reasons, of course, and we are not suggesting that any investigator has viewed it as reflecting reality from the animal's perspective. Such bivalent logic is probably not the actual basis for the categorization of many things relevant to, and communicated about, by monkeys. For some aspects of the nonhuman primate world, a multivalued or fuzzy logic system, with more than two truth values, might provide a more appropriate and realistic way to conceptualize this problem (Gouzoules et al. 1995, Gouzoules et al. 1998). A fuzzy set approach to categorizing the referents of primate calls—the dominance of opponents, the severity of aggression, aspects of discovered food, or even a predator—would hold that members of the set of objects or events are members to some degree, in contrast to classical set theory where membership is all or none. Thus, for example, an attack from the alpha male or alpha female of the group for a female rhesus monkey usually would have a truth value of close to 1 for the primate proposition "attack from a higher-ranking opponent." The truth value for such a proposition might be considerably less for an incident involving a different monkey that ranks closer to the victim on the ordinal form of dominance hierarchy typically constructed by observers to assess dominance relationships.

Predictions can be made as to how vocal behavior might be reflected by this more complex classification. If, for example, particular calls have as referents the relative dominance status of another group member, as has been argued for vervet monkey grunts (Cheney and Seyfarth 1982) and macaque screams (Gouzoules et al. 1984; Gouzoules and Gouzoules 1989, 1995), those calls should be more reliably used during interactions with individuals that are good members of that set, a prediction that stems from the human psychological literature on fuzzy categories (Mervis and Rosch 1981, Lakoff 1987) and has been supported empirically (Gouzoules et al. 1998). Thus, a strict application of the production criterion for external reference suggested by Macedonia and Evans (1993) would perhaps not be appropriate for call systems with fuzzy categories.

Alternative Perspectives: Criticisms of the Informational Perspective and a Reexamination of Macaque Screams

For some researchers, the issue of reference itself has become problematic (Owren 2000). Owren and Rendall (1997, 2001) were the first to rebel against the view that animal calls provide information of any sort and have proposed an affect-conditioning (or "affect-induction") model of nonhuman primate vocal communication in which they argue that a framework involving roles for both Pavlovian and operant conditioning could account for important aspects of how calls function to influence the behavior of conspecific receivers. They, with Owings and Morton (1998), have moved away from the informational perspective common in communication studies, that signals make available information to which a receiver may respond. The arguments marshaled against the idea that animal displays and signals provide information to receivers take a number of forms. At a general evolutionary level, linguistically based notions about information are discounted as anthropocentric and anthropomorphic; they are viewed as being at odds with, or at least ignoring, two "conceptual pillars of causality in evolutionary analysis," form and function (Owren and Rendall 2001:60). An example of a "form and function" account of animal communication would be Morton's motivational–structural rules discussed above. The stance that animals provide information through communication is also suggested to be group-selectionist and not consistent with modern "selfish gene" (Darwinian) perspectives.

These arguments against reference and information, purportedly based in modern evolutionary theory, are not entirely sound, however. Evolutionary theory readily accommodates cooperation, reciprocity, altruism, and game theoretic approaches, all of which can help explain conditions under which selection would favor providing others with information (Bradbury and Vehrencamp 1998). The anti-information position also claims that the concept of reference implies that senders produce signals specifically to provide information to receivers. This depends on psychological processing in receivers that is conceptual in nature, paralleling mental activity that occurs with meaning extraction in humans. Again, the critics believe this to be anthropomorphic and unnecessary for adequate accounts of animal communication. Instead, the affect-conditioning model holds that individuals vocalize to influence the behavior of others (not to inform them); conditioning determines how sound might have an impact. It should be noted, however, that, with respect to these concerns about the cognitive requirements of referential communication, it remains unclear exactly what level of cognition is necessary for reference to be possible.

Owren and Rendall's model attempts to account for the basic design and function of primate vocalizations by viewing the sounds as stimuli that senders use to elicit simple affective responses in receivers. They envisage roles for both unconditioned and conditioned responses in the manner by which behaviors are elicited by vocalizations, but with respect to monkey screams which are singled out for attention, the former class of learning is suggested to be strongest. They reason that, due to the general properties of the mammalian auditory system, a call can elicit negative or positive reactions; negative reactions are thought to be associated with high amplitude and overall noisiness in calls. They suggest primate screams as an example of calls that produce unconditioned, negative responses in receivers, something they argue would discourage impending or ongoing aggression from a higher-ranking attacker. A subordinate monkey's most effective vocal solution to the problem of attack from a higher-ranking opponent would therefore be to "use sheer magnitude and raw features of acoustic signals for inducing aversive unconditioned responses in opponents" (Owren and Rendall 1997:330).

While a comprehensive evaluation of Owren and Rendall's model for understanding animal vocalizations is not possible within the constraints of this chapter, a few pertinent observations can be made (and others have been offered by Seyfarth and Cheney 2003, Zuberbühler 2003, Gouzoules 2005). One issue relevant to evaluating Owren and Rendall's hypothesis with respect to macaque screams is that of general acoustic structure of the calls, as is clear from the last paragraph. Although not a direct test of their ideas, some of our comparative research on macaque screams has yielded data that seem to be inconsistent with their predictions. For example, Gouzoules and Gouzoules (2000) compared screams of four species of macaques (rhesus monkey, *Macaca mulatta*; pig-tailed monkey, *M. nemestrina*; Sulawesi crested black macaque, *M. nigra*; stump-tailed macaque, *M. arctoides*) by looking at calls produced by victims of attack that involved contact aggression (pulling, pushing, slapping, grappling, and biting) from a higher-ranking opponent. For each macaque species, we digitized 100 screams from adult females and measured acoustic features for each call. Discriminant function analysis was used to determine whether the 400 vocalizations could be assigned to the correct caller species on the basis of their acoustic structure. Calls were assigned to the correct species at a significantly higher rate (93.5%) than expected by chance. These analyses revealed that each of the four macaque species used acoustically distinct screams in a shared context, receiving contact aggression from a higher-ranking opponent (Fig. 38.1). Additional analyses as well as previous work on rhesus and pig-tailed macaques (Gouzoules et al. 1984, 1998; Gouzoules and Gouzoules 1989) suggested that each of the four species is physically capable of producing screams that are similar to those found in the repertoires of the other species, which is not surprising given the similarity among members of the genus in the mechanisms underlying phonation (Hauser 1996:479).

That the four macaque species we examined use distinctly different screams is a finding at odds with Owings and Morton's (1998:115) observation that animal screams often lack species-typical attributes. Their argument was based on the fact that screams in many species are produced

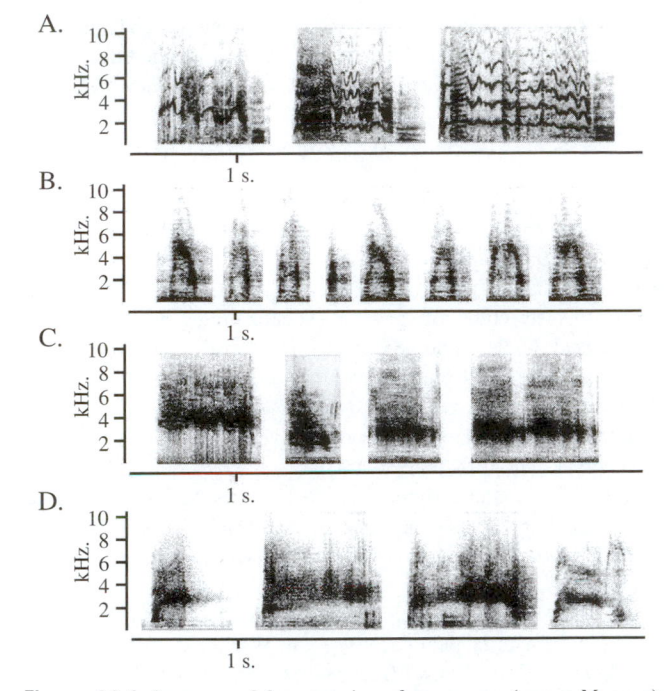

Figure 38.1 Screams of four species of macaques (genus *Macaca*). A. pigtailed macaque, B. black crested macaque, C. rhesus macaque, D. stumptailed macaque

in response to attacks from predators that are usually of a different species. It might be adaptive for such calls to be of a similar acoustic structure either because they serve to startle the predator, attract others (conspecifics or not) to mob the predator, or lure an even larger predator that might chase the first one away.

A significant difference, then, for macaque screams is that they are most prominent and common during intra-specific aggression, and suggestions for understanding their acoustic structure must relate to the nature of this largely within-group conflict. Although fear and pain are likely consequences for the victims of severe aggression in each macaque species, there appears to be no simple correspond-ence among the four species in the acoustic structure of their screams. While all four macaque species' screams are loud, the variation in call structure among species does not appear to be explained by Owren and Rendall's affect-conditioning model. Based on the starting assumption of their model, that general properties of the mammalian auditory system pre-dispose the monkeys to react negatively to certain call pro-perties, it is not clear why the four closely related macaque species we studied should have evolved such acoustically distinct screams if they serve as unconditioned stimuli that are naturally repugnant to aggressors.

Concluding Remarks on Language Precursors

In a recent paper in *Science* that will likely have a substan-tial impact on the study of primate communication, Hauser et al. (2002:1576) reviewed the topic of language evolution and concluded the following:

Without pursuing the matter here, it appears that many of the elementary properties of words, including those that enter into referentiality, have only weak analogs or homologs in natural animal communication systems, with only slightly better evidence from the training studies with apes and dolphins. Future research must therefore provide either stronger support for the precursor position, or abandon this hypothesis.

These conclusions are very reminiscent of those expressed in the 1960s (e.g., Lancaster 1968). Yet, in their review, Hauser and colleagues acknowledge that the empirical re-search in this area over the past 35 years, some of which we have discussed in the present chapter, supports the following conclusions:

1. Monkeys produce certain acoustically distinctive calls in response to functionally significant contexts.
2. The acoustic morphology of the signal, which can be arbitrary in terms of its association with a particular context, is sufficient to enable listeners to respond appropriately without requiring any other contextual information.
3. The number of such signals in the repertoire is small; referents are restricted to objects and events experienced in the present, with no evidence of creative production of new sounds for new situations.
4. The acoustic morphology of the calls is more or less fixed, appearing early in development, with experience playing a larger role in refining the range of objects or events that elicit the calls.
5. While calling is under voluntary control, there is no evidence that it is intentional in the sense of taking into account what other individuals believe, want, or need to know.

In contrast to the conclusions Hauser and colleagues draw, these findings seem to be strong evidence in support of limited forms of referentialty in primate vocal communi-cation that might indeed represent broad precursors to human symbolic capacities. The symbolic capacities of our own species are sometimes hard to define, recognize, and demonstrate, especially in young children (Tomasello 1999, 2003; Namy 2005), something that those working with non-human primates perhaps should keep in mind.

GENERAL CONCLUSIONS

Knowledge concerning nonhuman primate communication has increased dramatically since 1980. Repertoires used in communication are clearly much larger than previously imagined, and studies have revealed a radically different view of the role communication plays in nonhuman primates' lives. In the 1960s, Jane Lancaster (1968:442) summarized the prevalent view for vocal communication as follows:

field and laboratory workers have emphasized that vocalizations do not carry the major burden of meaning in most social interactions, but function instead either to call visual attention to the signaler or to emphasize or enhance the effect of visual or tactile signals. In other words, a blind monkey would be greatly handicapped in his social interactions whereas a deaf one would probably be able to function almost normally.

Our suspicion is that the complexity and richness of non-human primate communication is still underestimated. The degree to which multimodal communication regulates and modifies social interactions is, without doubt, an area that needs more exploration (Partan and Marler 1999, Partan 2002, Parr 2004). Newer technologies, including brain imaging, when combined with socially and evolutionarily relevant hypotheses are likely to be useful in addressing the remaining controversies such as those that surround the issue of language precursors (e.g., Poremba et al. 2004). While linguistic models are useful and should not be rejected out-right as anthropomorphic, their use should not dominate to the point of excluding other evolutionary approaches and hypotheses, especially those that, for nonhuman primates, have received relatively little attention, for example, signal evolution through receiver sensory bias or exploitation (Burley and Symanski 1998). Consistent with Bradbury and Vehrencamp's (1998:5) extended definition of animal communication, continued progress in our understanding of nonhuman primate communication will require a very broad range of empirical approaches and theoretical perspectives.

REFERENCES

Altmann, J. (1980). *Baboon Mothers and Infants*. Harvard University Press, Cambridge, MA.

Altmann, S. A. (1962). A field study on the sociobiology of rhesus monkeys (*Macaca mulatta*). *Ann. N. Y. Acad. Sci.* 102:338–435.

Barrett, L., Dunbar, R. I. M., and Dunbar, P. (1995). Mother–infant contact as contingent behaviour in gelada baboons. *Anim. Behav.* 49:805–810.

Benz, J. J. (1993). Food-elicited vocalizations in golden lion tamarins: design features for representational communication. *Anim. Behav.* 45:443–455.

Bergstrom, C. T., and Lachmann, M. (1997). Signaling among relatives. I. When is signaling too costly? *Philos. Trans. R. Soc. Lond. B* 352:609–617.

Bergstrom, C. T., and Lachmann, M. (1998). Signaling among relatives. III. Talk is cheap. *Proc. Natl. Acad. Sci. USA* 95:5100–5105.

Bickerton, D. (1990). *Language and Species*. University of Chicago Press, Chicago.

Blumstein, D. T., and Armitage, K. B. (1997). Does sociality drive the evolution of communicative complexity? A comparative test with ground-dwelling sciurid alarm calls. *Am. Nat.* 150:179–200.

Boesch, C., and Boesch, H. (1989). Hunting behavior of wild chimpanzees in the Taï National Park. *Am. J. Phys. Anthrop.* 78:547–573.

Bradbury, J. W., and Vehrencamp, S. L. (1998). *Principles of Animal Communication*. Sinauer Associates, Sunderland, MA.

Brown, C. H., and Gomez, R. (1992). Functional design features in primate vocal signals: the acoustic habitat and sound distortion. In: Nishida, T., McGrew, W. C., Marler, P., Pickford, M., and de Waal, F. B. M. (eds.), *Topics in Primatology. Human Origins*, vol. 1. University of Tokyo Press, Tokyo. pp. 177–198.

Brown, C. H., Gomez, R., and Waser, P. M. (1995). Old World monkey vocalizations: adaptation to the local habitat? *Anim. Behav.* 50:945–961.

Brown, C. H., and Waser, P. M. (1988). Environmental influences on the structure of primate vocalizations. In: Todt, D., Goedeking, P., and Symmes, D. (eds.), *Primate Vocal Communication*. Springer-Verlag, Berlin. pp. 51–68.

Burley, N. T., and Symanski, R. (1998). "A taste for the beautiful": latent aesthetic mate preferences for white crests in two species of Australian grassfinches. *Am. Nat.* 152:792–802.

Byrne, R. W., and Whiten, A. (1988). Tactical deception of familiar individuals in baboons. In: Bryne, R. W., and Whiten, A. (eds.), *Machiavellian Intelligence*. Clarendon Press, Oxford. pp. 201–210.

Byrne, R. W., and Whiten, A. (1991). Computation and mindreading in primate tactical deception. In: Whiten, A. (ed.), *Natural Theories of Mind*. Basil Blackwell, Oxford. pp. 127–142.

Cheney, D. L., and Seyfarth, R. M. (1982). How vervet monkeys perceive their grunts: field playback experiments. *Anim. Behav.* 30:739–751.

Cheney, D. L., and Seyfarth, R. M. (1988). Assessment of meaning and the detection of unreliable signals by vervet monkeys. *Anim. Behav.* 36:477–486.

Cheney, D. L., and Seyfarth, R. M. (1990a). *How Monkeys See the World*. University of Chicago Press, Chicago.

Cheney, D. L., and Seyfarth, R. M. (1990b). The representation of social relations by monkeys. *Cognition* 37:167–196.

Cheney, D. L., and Seyfarth, R. M. (1991). Reading minds or reading behaviour? Tests for a theory of mind in monkeys. In: Whiten, A. (ed.), *Natural Theories of Mind*. Basil Blackwell, Oxford. pp. 175–194.

Cheney, D. L., and Seyfarth, R. M. (1992). Meaning, reference, and intentionality in the natural vocalizations of monkeys. In: Nishida, T., McGrew, W. C., Marler, P., Pickford, M., and de Waal, F. B. M. (eds.), *Topics in Primatology. Human Origins*, vol. 1. University of Tokyo Press, Tokyo.

Chevalier-Skolnikoff, S. (1974). The ontogeny of communication in the stumptail macaque (*Macaca arctoides*). In: *Contributions to Primatology*. Karger Press, Basel.

Clutton-Brock, T. H., and Parker, G. A. (1995). Punishment in animal societies. *Nature* 373:209–216.

Darwin, C. (1872). *The Expression of the Emotions in Man and Animals*. J. Murray, London. (reprinted: University of Chicago Press, Chicago, 1965)

Deacon, T. W. (1997). *The Symbolic Species: The Co-evolution of Language and the Brain*. W. W. Norton, New York.

de Waal, F. B. M. (1986). Deception in the natural communication of chimpanzees. In: Mitchell, R. W., and Thompson, N. S. (eds.), *Deception: Perspectives on Human and Nonhuman Deceit*. State University of New York Press, Albany. pp. 221–244.

Dittus, W. P. J. (1984). Toque macaque food calls: semantic communication concerning food distribution in the environment. *Anim. Behav.* 32:470–477.

Dusenbery, D. B. (1992). *Sensory Ecology*. Freeman and Company, New York.

Evans, C. S. (1997). Referential signals. In: Owings, D. H., Beecher, M., and Thompson, N. (eds.), *Perspectives in Ethology. Communication*, vol. 12. Plenum Press, New York. pp. 99–143.

Evans, C. S., Macedonia, J. M., and Marler, P. (1993). Effects of apparent size and speed on the response of chickens, *Gallus gallus*, to computer-generated simulations of aerial predators. *Anim. Behav.* 46:1–11.

Farrell, J., and Rabin, M. (1996). Cheap talk. *J. Econ. Perspect.* 10:110–118.

Gartlan, J. S., and Brain, C. K. (1968). Ecology and social variability in *Cercopithecus aethiops* and *Cercopithecus mitis*. In: Jay, P. C. (ed.), *Primates, Studies in Adaptation and Variability*. University of California Press, Berkeley. pp. 253–292.

Gautier, J.-P., and Gautier, A. (1977). Communication in Old World monkeys. In: Sebeok, T. E. (ed.), *How Animals Communicate*. Indiana University Press, Bloomington. pp. 890–964.

Gerald, M. S. (2001). Primate color predicts status and aggressive outcome. *Anim. Behav.* 61:559–566.

Gerhardt, H. C. (1983). Communication and environment. In: Halliday, T. R., and Slater, P. J. B. (eds.), *Animal Behaviour: Communication*. W. H. Freeman, New York. pp. 82–113.

Gish, S. L., and Morton, E. S. (1981). Structural adaptations to local habitat acoustics in Carolina wren songs. *Z. Tierpsychol.* 56:74–84.

Gomendio, M. (1991). Parent–offspring conflict and maternal investment in rhesus macaques. *Anim. Behav.* 42:993–1005.

Goodall, J. (1986). *The Chimpanzees of Gombe*. Harvard University Press, Cambridge, MA.

Gouzoules, H. (2005). Monkeying around with symbolism: are vocalizations simple symbols . . . or more like cymbals? In: Namy, L. (ed.) *Symbolic Use* and *Symbolic Representation*. Erlbaum, Mahwah, NJ. pp. 245–263.

Gouzoules, H., and Gouzoules, S. (1989). Design features and developmental modification of pigtail macaque, *Macaca nemestrina*, agonistic screams. *Anim. Behav.* 37:383–401.

Gouzoules, H., and Gouzoules, S. (1995). Recruitment screams of pigtail monkeys (*Macaca nemestrina*): ontogenetic perspectives. *Behaviour* 132:431–450.

Gouzoules, H., and Gouzoules, S. (2000). Agonistic screams differ among four species of macaques: the significance of motivation–structural rules. *Anim. Behav.* 59:501–512.

Gouzoules, H., and Gouzoules, S. (2002). Primate communication: by nature honest, or by experience wise? *Int. J. Primatol.* 23:821–848.

Gouzoules, H., Gouzoules, S., and Marler, P. (1985). External reference and affective signaling in mammalian vocal communication. In: Zivin, G. (ed.), *The Development of Expressive Behavior (Biology–Environment Interactions)*. Academic Press, New York. pp. 77–101.

Gouzoules, H., Gouzoules, S., and Miller, K. (1996). Skeptical responding in rhesus monkeys (*Macaca mulatta*). *Int. J. Primatol.* 17:549–568.

Gouzoules, H., Gouzoules, S., and Tomaszycki, M. (1998). Agonistic screams and the classification of dominance relationships: are monkeys fuzzy logicians? *Anim. Behav.* 55:51–60.

Gouzoules, S., Gouzoules, H., and Marler, P. (1984). Rhesus monkey (*Macaca mulatta*) screams: representational signalling in the recruitment of agonistic aid. *Anim. Behav.* 32:182–193.

Gouzoules, S., Gouzoules, H., and Marler, P. (1986). Vocal communication: a vehicle for the study of social relationships.

In: Rawlins, R. G., and Kessler, M. J. (eds.), *The Cayo Santiago Macaques*. State University of New York Press, Albany. pp. 111–129.

Green, S. (1975). Communication by a graded vocal system in Japanese monkeys. In: Rosenblum, L. A. (ed.), *Primate Behavior*, vol. 4. Academic Press, New York. pp. 1–102.

Harcourt, A. H., and de Waal, F. B. M. (eds.) (1992). *Coalitions and Alliances in Humans and Other Animals*. Oxford University Press, New York.

Hauser, M. D. (1986). Parent–offspring conflict: care elicitation behaviour and the "cry-wolf" syndrome. In: Else, J. G., and Lee, P. C. (eds.), *Primate Ontogeny, Cognition and Social Behaviour*. Cambridge University Press, Cambridge. pp. 193–203.

Hauser, M. D. (1992). Costs of deception: cheaters are punished in rhesus monkeys (*Macaca mulatta*). *Proc. Natl. Acad. Sci. USA* 89:12137–12139.

Hauser, M. D. (1993). The evolution of non-human primate vocalizations: effects of phylogeny, body-weight, and social context. *Am. Nat.* 142:528–542.

Hauser, M. D. (1996). *The Evolution of Communication*. MIT Press, Cambridge, MA.

Hauser, M. D. (1997). Minding the behaviour of deception. In: Whiten, A., and Byrne, R. W. (eds.), *Machiavellian Intelligence II*. Cambridge University Press, Cambridge. pp. 112–143.

Hauser, M. D., Chomsky, N., and Tecumseh Fitch, W. T. (2002). The faculty of language: what is it, who has it, and how did it evolve? *Science* 298:1569–1579.

Hauser, M. D., and Marler, P. (1993a). Food-associated calls in rhesus macaques (*Macaca mulatta*): I. Socioecological factors. *Behav. Ecol.* 4:194–205.

Hauser, M. D., and Marler, P. (1993b). Food-associated calls in rhesus macaques (*Macaca mulatta*): II. Costs and benefits of call production and suppression. *Behav. Ecol.* 4:206–212.

Hauser, M. D., Teixidor, P., Field, L., and Flaherty, R. (1993). Food-elicited calls in chimpanzees: effects of food quantity and divisibility. *Anim. Behav.* 45:817–819.

Isbell, L. A. (1995). Seasonal and social correlates of changes in hair, skin, and scrotal condition in vervet monkeys (*Cercopithecus aethiops*) of Amboseli National Park, Kenya. *Am. J. Primatol.* 36:61–70.

Jürgens, U. (1979). Neural control of vocalization in non-human primates. In: Steklis, H. D., and Raleigh, M. J. (eds.), *Neurobiology of Social Communication in Primates*. Academic Press, New York. pp. 11–44.

Jürgens, U. (1988). Central control of monkey calls. In: Todt, D., Goedeking, P., and Symmes, D. (eds.), *Primate Vocal Communication*. Springer-Verlag, Berlin. pp. 162–167.

Kaplan, J. (1977). Patterns of fight interference in free-ranging rhesus monkeys. *Am. J. Phys. Anthropol.* 47:279–288.

Kaplan, J. R. (1978). Fight interference and altruism in rhesus monkeys. *Am. J. Phys. Anthropol.* 49:241–250.

Kappeler, P. M. (1990). Social status and scent-marking behaviour in *Lemur catta*. *Anim. Behav.* 40:774–776.

Kappeler, P. M. (1998). To whom it may concern: the transmission and function of chemical signals in Lemur catta. *Behav. Ecol. Sociobiol.* 42:411–421.

Kilner, R., and Johnstone, R. A. (1997). Begging the question: are offspring solicitation behaviours signals of need? *Trends Ecol. Evol.* 12:11–15.

King, B. J. (ed.) (1999). *The Origins of Language*. School of American Research Press, Santa Fe, NM.

Lachmann, M., and Bergstrom, C. T. (1998). Signaling among relatives. II. Beyond the Tower of Babel. *Theor. Popul. Biol.* 54:146–160.

Lakoff, G. (1987). *Women, Fire, and Dangerous Things: What Categories Reveal about the Mind.* University of Chicago Press, Chicago.

Lancaster, J. (1968). Primate communications and the emergence of human language. In: Jay, P. C. (ed.), *Primates, Studies in Adaptation and Variability.* Holt, Rinehart and Winston, New York. pp. 439–457.

Le Prell, C. G., Hauser, M. D., and Moody, D. B. (2002). Discrete or graded variation within rhesus monkey screams? Psychophysical experiments on classification. *Anim. Behav.* 63:47–62.

Lloyd, L. B. (1989). *Simple Minds.* MIT Press, Cambridge, MA.

Macedonia, J. M., and Evans, C. S. (1993). Variation among mammalian alarm call systems and the problem of meaning in animal signals. *Ethology* 93:177–197.

Maestripieri, D. (1999). Primate social organization, gestural repertoire size, and communication dynamics: a comparative study of macaques. In: King, B. J. (ed.), *The Origins of Language: What Nonhuman Primates Can Tell Us.* School of American Research Press. pp. 55–78.

Marler, P. (1965). Communication in monkeys and apes. In: Devore, I. (ed.), *Primate Behavior.* Holt, Rinehart, and Winston, New York. pp. 544–584.

Marler, P. (1968). Aggregation and dispersal: two functions in primate communication. In: Jay, P.C. (ed.), *Primates, Studies in Adaptation and Variability.* University of California Press, Berkeley. pp. 420–438.

Marler, P. (1984). Animal communication: affect or cognition? In: Scherer, K. R., and Ekman, P. (eds.), *Approaches to Emotion.* Erlbaum, Hillsdale, NJ. pp. 345–365.

Marler, P., Evans, C. S., and Hauser, M. D. (1992). Animal signals: motivational, referential, or both? In: Papousek, H., and Jürgens, U. (eds.), *Nonverbal Communication: Comparative and Developmental Approaches.* Cambridge University Press, Cambridge.

Marten, K., and Marler, P. (1977). Sound transmission and its significance for animal vocalization. I. Temperate habitats. *Behav. Ecol. Sociobiol.* 2:271–290.

Marten, K., Quine, D., and Marler, P. (1977). Sound transmission and its significance for animal vocalization. II. Tropical forest habitats. *Behav. Ecol. Sociobiol.* 2:291–302.

Mervis, C., and Rosch, E. (1981). Categorization of natural objects. *Annu. Rev. Psychol.* 32:89–115.

Michelsen, A. (1978). Sound reception in different environments. In: Ali, B. A. (ed.), *Perspectives in Sensory Ecology.* Plenum Press, New York. pp. 345–373.

Morton, E. S. (1977). On the occurrence and significance of motivation–structural rules in some bird and mammal sounds. *Am. Nat.* 111:855–869.

Morton, E. S. (1982). Grading, discreteness, redundancy, and motivation–structural rules. In: Kroodsma, D. E., and Miller, E. H. (eds.), *Acoustic Communication in Birds*, vol. 1. Academic Press, New York. pp. 183–212.

Moynihan, M. (1982). Why is lying about intentions rare during some kinds of contests? *J. Theor. Biol.* 97:7–12.

Müller-Preuss, P. (1988). Neural basis of signal detection. In: Todt, D., Goedeking, P., and Symmes, D. (eds.), *Primate Vocal Communication.* Springer-Verlag, Berlin. pp. 154–161.

Namy, L. (ed.) (2005). *Symbolic Use and Symbolic Representation*, Erlbaum, Mahwah, NJ.

Owings, D. H., and Morton, E. S. (1997). The role of information in communication: an assessment/management approach. In: Owings, D. H., Beecher, M., and Thompson, N. (eds.), *Perspectives in Ethology. Communication*, vol. 12. Plenum Press, New York.

Owings, D. H., and Morton, E. S. (1998). *Animal Vocal Communication: A New Approach.* Cambridge University Press, Cambridge.

Owren, M. J. (1990). Acoustic classification of alarm calls by vervet monkeys (*Cercopithecus aethiops*) and humans (*Homo sapiens*): I. Natural calls. *J. Comp. Psychol.* 104:20–28.

Owren, M. J. (2000). Standing evolution on its head: the uneasy role of evolutionary theory in comparative cognition and communication. *Rev. Anthropol.* 29:55–69.

Owren, M. J., and Rendall, D. (1997). An affect-conditioning model of nonhuman primate vocal signaling. In: Owings, D. H., Beecher, M., and Thompson, N. (eds.), *Perspectives in Ethology. Communication*, vol. 12. Plenum Press, New York. pp. 299–346.

Owren, M. J., and Rendall, D. (2001). Sound on the rebound: bringing form and function back to the forefront in understanding nonhuman primate vocal signaling. *Evol. Anthropol.* 10:58–71.

Parr, L. A. (2004). Perceptual biases for multimodal cues in chimpanzee (*Pan troglodytes*) affect recognition. *Anim. Cogn.* 7:171–178.

Partan, S., and Marler, P. (1999). Communication goes multimodal. *Science* 283:1272–1273.

Partan, S. R. (2002). Single and multichannel signal composition: facial expressions and vocalizations of rhesus macques (*Macaca mulatta*). *Behaviour* 139:993–1027.

Piercy, J. E., Embleton, T. F. W., and Sutherland, L. C. (1977). Review of noise propagation in the atmosphere. *J. Acoust. Soc. Am.* 61:143–148.

Ploog, D. (1967). The behavior of squirrel monkeys (*Saimiri sciureus*) as revealed by sociometry, bioacoustics, and brain stimulation. In: Altmann, S. A. (ed.), *Social Communication Among Primates.* University of Chicago Press, Chicago. pp. 149–184.

Poremba, A., Malloy, M., Saunders, R. C., Carson, R. E., Herscovitch, P., and Mishkin, M. (2004). Species-specific calls evoke asymmetric activity in the monkey's temporal poles. *Nature* 427:448–451.

Preuschoft, S., and van Hooff, J. A. R. A. M. (1995). Homologizing primate facial displays: a critical review of methods. *Folia Primatol.* 65:121–137.

Preuschoft, S., and van Hooff, J. A. R. A. M. (1997). The social function of "smile" and "laughter": variations across primate species and societies. In: Segerstrale, U., and Molnar, P. (eds.), *Nonverbal Communication: Where Nature Meets Culture.* Erlbaum, Hove, UK. pp. 171–190.

Ross, C., and Regan, G. (2000). Allocare, predation risk, social structure and natal coat color in anthropoid primates. *Folia Primatol.* 71:67–76.

Rowell, T. E., and Hinde, R. A. (1962). Vocal communication by the rhesus monkey (*Macaca mulatta*). *Proc. Zool. Soc. Lond.* 138:279–294.

Savage-Rumbaugh, E. S., and Fields, W. M. (2000). Linguistic, cultural and cognitive capacities of bonobos (*Pan paniscus*). *Cult. Psychol.* 6:131–153.

Savage-Rumbaugh, S., Shanker, S. G., and Taylor, T. J. (1998). *Apes, Language, and the Human Mind.* Oxford University Press, New York.

Sekulic, R. (1982). The function of howling in red howler monkeys (*Alouatta seniculus*). *Behaviour* 81:38–54.

Sekulic, R., and Chivers, D. J. (1986). The significance of call duration in howler monkeys. *Int. J. Primatol.* 7:183–190.

Seyfarth, R. M., and Cheney, D. L. (1990). The assessment by vervet monkeys of their own and another species' alarm calls. *Anim. Behav.* 40:754–764.

Seyfarth, R. M., and Cheney, D. L. (2003). Signalers and receivers in animal communication. *Annu. Rev. Psychol.* 54:145–173.

Seyfarth, R. M., Cheney, D. L., and Marler, P. (1980). Vervet monkey alarm calls: semantic communication in a free-ranging primate. *Anim. Behav.* 28:1070–1094.

Silk, J. B., Kaldor, E., and Boyd, R. (2000). Cheap talk when interests conflict. *Anim. Behav.* 59:423–432.

Smith, W. J. (1986). An "informational" perspective on manipulation. In: Mitchell, R. W., and Thompson, N. S. (eds.), *Deception: Perspectives on Human and Nonhuman Deceit.* State University of New York Press, Albany. pp. 71–86.

Smith, W. J. (1991). Animal communication and the study of cognition. In: Ristau, C. A. (ed.), *Cognitive Ethology: The Minds of Other Animals (Essays in Honor of Donald R. Griffin).* Erlbaum, Hillsdale, NJ. pp. 209–230.

Snowdon, C. T. (1982). Linguistic and psycholinguistic approaches to primate communication. In: Snowdon, C. T., Brown, C. H., and Petersen, M. R. (eds.), *Primate Communication.* Cambridge University Press, Cambridge. pp. 212–238.

Snowdon, C. T. (1999). An empiricist view of language evolution and development. In: King, B. J. (ed.), *The Origins of Language: What Nonhuman Primates Can Tell Us.* School of American Research Press, Santa Fe, NM. pp. 79–114.

Snowdon, C. T. (2001). From primate communication to human language. In: de Waal, F. (ed.), *Tree of Origin: What Primate Behavior Can Tell Us About Human Social Evolution.* Harvard University Press, Cambridge, MA. pp. 193–227.

Struhsaker, T. T. (1967). Auditory communication among vervet monkeys (*Cercopithecus aethiops*). In: Altmann, S. A. (ed.), *Social Communication Among Primates.* University of Chicago Press, Chicago. pp. 281–324.

Thompson, N. S. (1997). Communication and natural design. In: Owings, D. H., Beecher, M., and Thompson, N. (eds.), *Perspectives in Ethology,* vol. 12. Plenum Press, New York. pp. 391–415.

Tomasello, M. (1999). *The Cultural Origins of Human Cognition.* Harvard University Press, Cambridge, MA.

Tomasello, M. (2003). *Constructing a Language: A Usage-Based Theory of Language Acquisition.* Harvard University Press, Cambridge, MA.

Tomasello, M., and Call, J. (1997). *Primate Cognition.* Oxford University Press, New York.

Trivers, R. L. (1974). Parent–offspring conflict. *Am. Zool.* 14:249–264.

van Hoof, J. A. R. A. M. (1967). The facial displays of the catarrhine monkeys and apes. In: Morris, D. (ed.), *Primate Ethology.* Anchor Books, New York. pp. 9–88.

van Hooff, J. A. R. A. M., and Preuschoft, S. (2003). Laughter and smiling: the intertwining of nature and culture. In: de Waal, F., and Tyack, P. (eds.), *Animal Social Complexity: Intelligence, Culture, and Individualized Societies.* Harvard University Press, Cambridge, MA. pp. 260–287.

Wallman, J. (1992). *Aping Language.* Cambridge University Press, New York.

Walters, J. R., and Seyfarth, R. M. (1987). Conflict and cooperation. In: Smuts, B. B., Cheney, D. L., Seyfarth, R. M., Wrangham, R. W., and Struhsaker, T. T. (eds.), *Primate Societies.* University of Chicago Press, Chicago. pp. 306–317.

Waser, P. M. (1982). The evolution of male loud calls among mangabeys and baboons. In: Snowdon, C. T., Brown, C. H., and Petersen, M. R. (eds.), *Primate Communication.* Cambridge University Press, Cambridge. pp. 117–143.

Waser, P. M., and Brown, C. H. (1984). Is there a sound window for primate communication? *Behav. Ecol. Sociobiol.* 15:73–76.

Waser, P. M., and Brown, C. H. (1986). Habitat acoustics and primate communication. *Am. J. Primatol.* 10:135–156.

Waser, P. M., and Waser, M. S. (1977). Experimental studies of primate vocalization: specializations for long-distance propagation. *Z. Tierpsychol.* 43:239–263.

Whiten, A., and Byrne, R. W. (1988). Tactical deception in primates. *Behav. Brain Sci.* 11:233–273.

Wiley, R. H. (1983). The evolution of communication: information and manipulation. In: Halliday, T. R., and Slater, P. J. B. (eds.), *Animal Behaviour: Communication.* W. H. Freeman, New York. pp. 156–189.

Wiley, R. H., and Richards, D. G. (1978). Physical constraints on acoustical communication in the atmosphere: implications for the evolution of animal vocalizations. *Behav. Ecol. Sociobiol.* 3:69–94.

Worlein, J. M., Eaton, G. G., Johnson, D. F., and Glick, B. B. (1988). Mating season effects on mother–infant conflict in Japanese macaques, *Macaca fuscata. Anim. Behav.* 36:1472–1481.

Zahavi, A. (1975). Mate selection: a selection for a handicap. *J. Theor. Biol.* 53:205–214.

Zahavi, A. (1977). The cost of honesty (further remarks on the handicap principle). *J. Theor. Biol.* 67:603–605.

Zahavi, A. (1980). Ritualization and the evolution of movement signals. *Behaviour* 72:77–81.

Zahavi, A. (1987). The theory of signal selection and some of its implications. In: Delfino, V. P. (ed.), *Proceedings of the International Symposium on Biological Evolution, Bari, Italy.* Adriatica Editrica, Rome. pp. 305–327.

Zahavi, A. (1993). The fallacy of conventional signaling. *Philos. Trans. R. Soc. Lond. B* 338:227–230.

Zahavi, A., and Zahavi, A. (1997). *The Handicap Principle.* Oxford University Press, New York.

Zuberbühler, K. (2003). Referential signaling in non-human primates: cognitive precursors and limitations for the evolution of language. *Adv. Study Behav.* 33:265–307.

39

Cooperation and Competition in Primate Social Interactions

Robert W. Sussman and Paul A. Garber

INTRODUCTION

Virtually all species of diurnal primates live in bisexual social groups composed of individuals of all age classes. Although changes in group size and composition are affected by a variety of events, including immigration, emigration, births, deaths, and fissioning, it is common for the same set of related and unrelated individuals to remain in a group for periods of months, years, and in some cases decades. Group living requires that individuals form predictable social relationships and develop affiliative social bonds. As an order, primates are characterized by an extremely broad range of social grouping patterns, including nuclear families, extended matrilines, extended patrilines, multifemale/multimale units, single male-multifemale groups, and fission–fusion groups. In many primate species, group size, composition, and cohesiveness are highly flexible and vary in response to local demographic, ecological, and historical factors.

WHY BE SOCIAL?

Several theories exist concerning the costs and benefits to individuals of interacting in cooperative ways: traveling together, exploiting a common set of resources, defending a common range, and forming alliances. These include sociobiological explanations of kinship and inclusive fitness as well as models of individual selection, mutualism, and multilevel selection (van Schaik and van Hooff 1983, Dugatkin 1997, Wilson 2000, Clutton-Brock 2002, Cheverud 2004; see Chapter 34). In general, it has been argued that the primary benefits of group living include enhanced predator detection, increased access to sexual partners, opportunities for information sharing, avoidance of male infanticide, enhanced food sharing, and cooperative behavior (Terborgh and Janson 1986, van Schaik and Horstermann 1994, Chapman and Chapman 2000, Wilson 2000). The primary costs of sociality are assumed to be increased levels of feeding competition and aggressive interactions, especially with increasing group size (van Schaik 1983, 1989; Janson and Goldsmith 1995; Sterck et al. 1997; Chapman and Chapman

2000; Janson 2000; Wittig and Boesch 2003; see Chapter 34). In this regard, the idea that competition and aggression over access to food and sexual partners are central to an understanding of the origins of group living and sociality in human and nonhuman primates remains a dominant theory in primatology and behavioral ecology. In part, this may relate to a commonly held idea that most social interactions represent a zero-sum game in which all benefits accrue to a single winner. Using this paradigm, competitive and aggressive interactions are expected to be widespread among members of neighboring social groups as well as among members of the same group (Wrangham 1980, van Schaik 1989, Sterck et al. 1997, Wittig and Boesch 2003).

In most primate species, however, cooperative, coordinated, and affiliative interactions are considerably more common than agonistic interactions and may play a more formative role in structuring behavioral interactions. For example, after 15 years of studying baboons (*Papio cynocephalus*), Strum (2001:158) observed that "aggression was not as pervasive or important an influence in evolution as had been thought, and that social strategies and social reciprocity were extremely important." Similarly, Sussman et al. (2003) reports that adult female ring-tailed lemurs (*Lemur catta*) spent on average less than 1 min per day in agonistic interactions. This pattern of extremely low frequencies of aggressive behaviors and high frequencies of affiliative behaviors characterizes individuals in most primate species (Sussman and Garber 2004). Furthermore, there can be considerable advantages to both kin and nonkin group members in developing dyadic, polyadic, and group-level affiliative and cooperative behaviors in which several participants receive collective or positive benefits (Dugatkin 1997, Clutton-Brock 2002, Johnson et al. 2002, Korstjens et al. 2002, Stephens et al. 2002, Silk et al. 2003, Cheverud 2004, Snowdon 2004). Thus, one goal of this chapter is to more clearly examine the relative roles and social functions of agonistic behavior and affiliative behavior in primate sociality. Agonistic, competitive, and affiliative behaviors are factors of social life. That they exist requires no particular explanation. How these behaviors function in promoting individual fitness, group cohesion, and social strategies does require explanation. We argue that primate sociality and affiliative, cooperative, and agonistic behaviors are best understood in terms of the individual and collective advantages of living in a functional social unit.

QUESTIONS CONCERNING THE CONTEXT OF PRIMATE SOCIAL INTERACTIONS

Data on the contexts, functions, and effectiveness of affiliative and agonistic behaviors in wild primates are needed to test theories of primate sociality. For example, how much time do different primate species spend in overt social interactions? How much of this behavior is friendly, and how much is agonistic? How variable are rates of affiliation and agonism among different groups of the same species? How

do group size and composition affect rates of social interactions? Are highly sexually dimorphic species or species that live in larger social groups more agonistic or do they experience higher rates of feeding competition than other species? What are the differences in the frequency and quality of social interactions among kin, friends, and nonkin? Although at the present time it is not possible to answer these questions fully, we begin by addressing a more basic question. How much time do diurnal, social-living prosimians, New World monkeys, Old World monkeys, and apes spend in social behavior; how much of this time is affiliative; and how much of this time is agonistic?

We present data from field studies on the frequency, rate, and nature of social behavior in almost 60 species of primates (Table 39.1). We acknowledge several limitations of the data. For example, researchers have used a variety of alternative

Table 39.1 Activity Budget and Rates of Agonism in Diurnal Primates

SPECIES	% TIME SOCIAL	% AFFILIATIVE	AGONISTIC	REFERENCE
Diurnal prosimians				
Varecia variegata	<1.0	Rare	Rare	Vasey (1997)
Eulemur fulvus	<1.0	Rare	Rare	Vasey (1997)
Eulemur fulvus	1.0	1.0	Rare	Overdorff (1991)
Eulemur rubriventer	2.0	2.0	Rare	Overdorff (1991)
Lemur catta	2.6	2.1	0.5%	Gould (1994)
Lemur catta	2.6	1.9	0.7%	Sussman (unpublished)
Eulemur fulvus	2.8	2.5	0.3%	Sussman (unpublished)
Varecia variegata	3.1	3.1	0.02/hr male	Morland (1991)
			0.17/hr female	
Eulemur fulvus	3.5	3.5	Rare	Tattersall (1977)
Propithecus verreauxi	3.8	n.d.	0.35/hr	Richard (1978)
Eulemur mongoz	4.0	4.0	ND	Curtis (1997)
Eulemur coronatus	4.3	4.0	0.3%	Freed (1996)
Propithecus diadema	5.0	4.5	0.12/hr	Hemingway (1995)
Eulemur fulvus	5.8	5.4	0.4%	Freed (1996)
Eulemur fulvus	8.5	7.7	0.8%	Sussman (unpublished)
Lemur catta	8.6	7.1	1.4%	Sussman (unpublished)
Overall mean	3.7 (± 2.3, mean weighted by species = 3.68)			
New World monkeys				
Alouatta palliata	0.8			Estrada et al. (1999)
Alouatta palliata	0.9			Smith (1977)
Brachyteles arachnoides	0.9			Milton (1984)
Callicebus torquatus	0.9			Kinzey (1981)
Ateles paniscus	0.9			Symington (1988a)
Alouatta palliata	1.0			Stoner (1996)
Cebus olivaceus	1.3			Miller (1992, 1996)
Alouatta seniculus	1.7			Gaulin and Gaulin (1982)
Saguinus mystax	1.8	1.4	0.41%	Castro (1991)
Alouatta palliata	1.9		0.003/(ind · hr)	Larose (1996)
Callicebus torquatus	1.9		0.0006/(ind · hr)	Easley (1982)
Alouatta palliata	2.0			Stoner (1996)
Cebus olivaceus	2.1			Miller (1992, 1996)
Saguinus fuscicollis	2.2	1.9	0.35%	Castro (1991)

Table 39.1 (*cont'd*)

SPECIES	% TIME SOCIAL	% AFFILIATIVE	AGONISTIC	REFERENCE
Brachyteles arachnoides	2.7		0.0006/(ind · hr)	Strier (1986, 1987)
Saguinus fuscicollis	2.8			Peres (1991)
Callithrix geoffroyi	2.8			Passamani (1998)
Saimiri sciureus	2.9		0.0047/(ind · hr)	Mitchell (1990)
Leontopithecus rosalia	3.1			Dietz et al. (1997)
Saguinus mystax	3.1			Peres (1991)
Saguinus mystax	3.5		0.20%	Garber (1997)
			0.0066/(ind · hr)	
Alouatta pigra	3.9		0.10%	Silver et al. (1998)
Lagothrix lagotricha	4.0	3.1	0.60%	Stevenson (1998)
Lagothrix lagotricha	4.8			Defler (1995)
Alouatta caraya	4.9		0.38%	Bicca-Marques (1993)
			0.019/(ind · hr)	
Callithrix humeralifer	5.0	4.2		Rylands (1982)
Leontopithecus rosalia	6.1			Peres (1986)
Leontopithecus rosalia	7.0			Dietz et al. (1997)
Leontopithecus chrysomelas	9.1	8.5		Rylands (1982)
Cebus capucinus	9.9		0.016/(ind · hr)	Mitchell (1989)
Cebus apella	9.9			Zhang (1995)
Cebus capucinus	12.5	7.7	0.034/(ind · hr)	Fedigan (1993)
Saimiri oerstedii	13.9		0.000009/(ind · hr)	Boinski (1986)
Callithrix jacchus	14.0	11.1	0.051/(ind · hr)	Digby (1994)
Ateles geoffroyi	17.0		0.18%	Fedigan and Baxter (1984)
Ateles geoffroyi	22.0		0.92%	Fedigan and Baxter (1984)
Alouatta palliata			0.007/(ind · hr)	Jones (1980)
Ateles geoffroyi			0.0043/(ind · hr)	Klein and Klein (1977)
Cebus apella			0.042/(ind · hr)	Janson (1984, 1988)
Cebus apella			0.007/(ind · hr)	Janson (1984, 1988)
Leontopithecus rosalia			0.0012/(ind · hr)	Baker (1991)
Saguinus fuscicollis			0.01/(ind · hr)	Goldizen et al. (1996)
Saguinus nigricollis			0.20%	de la Torre et al. (1995)
Overall mean	5.1% (± 5.1, mean weighted by species = 5.76)			
Old World monkeys				
Cercopithecus diana	1.2			McGraw (1998)
Colobus badius	1.9	1.8		Decker (1994)
Presbytis pontenziani	1.9			Fuentes (1996)
Colobus polykomos	2.0			Teichroeb et al. (2003)
Colobus vellerosus	2.0			Teichroeb et al. (2003)
Colobus vellerosus	2.0			Teichroeb et al. (2003)
Macaca silenus	2.4			Kurup and Kumar (1993)
Colobus badius	2.6	2.5		Decker (1994)
Cercopithecus campbelli	2.8			McGraw (1998)
Macaca silenus	3.4			Menon and Poirer (1996)
Colobus vellerosus	4.0			Teichroeb et al. (2003)
Cercopithecus petaurista	4.3			McGraw (2000)
Colobus vellerosus	5.0			Teichroeb et al. (2003)
Colobus polykomos	5.3			McGraw (1998)
Colobus guereza	5.9	5.7	0 in 7,793 scans	Fashing (2001)
Colobus badius	6.3			McGraw (1998)
Colobus verus	6.7			McGraw (1998)
Presbytis entellus	7.4	6.7		Fashing (2001)

Table 39.1   (*cont'd*)

SPECIES	% TIME SOCIAL	% AFFILIATIVE	AGONISTIC	REFERENCE
Cercocebus atys	7.9			McGraw (1998)
Colobus badius	8.0	7.6		Decker (1994)
Colobus badius	8.2	7.9		Decker (1994)
Colobus guereza	8.3	8.3	0 in 8,917 scans	Newton (1992)
Macaca silenus	8.4			Menon and Poirer (1996)
Colobus badius	8.5	8.5		Decker (1994)
Macaca sylvanus	10.0			Menard and Vallet (1997)
Papio anubis	10.4	5.2		Harding (1980)
Cercopithecus mitis	10.4			Kaplin and Moermond (2000)
Cercopithecus l'hoesti	11.4			Kaplin and Moermond (2000)
Macaca sylvanus	11.5			Menard and Vallet (1997)
Colobus satanas	13.0			Teichroeb et al. (2003)
Rhinopithecus bieti	13.1	9.8		Kirkpatrick et al. (1998)
Macaca nigra	18.7			O'Brien and Kinnaird (1997)
Macaca fuscata	21.7	18.9		Agetsuma (1995)
Macaca nigra	23.1			O'Brien and Kinnaird (1997)
Macaca nigra	23.5			O'Brien and Kinnaird (1997)
Presbytis francoisi	27.9	27.5		Burton et al. (1995)
Presbytis entellus			0.084/(ind · hr) females	Borries et al. (1991)
Presbytis entellus			0.01/(ind · hr) males	Laws and Laws (1984)
Papio cynocephalus			0.14/(ind · hr) males	Strum (1982)
Papio cynocephalus			0.11/(ind · hr) males	Noe and Sluijter (1995)
Papio cynocephalus			0.079/(ind · hr) males	Noe and Sluijter (1995)
Papio cynocephalus			0.14/(ind · hr) males	Noe and Sluijter (1995)
Cercopithecus aethiops			0.0007/(ind · hr)	Pruetz (1999)
Erythrocebus patas			0.0007/(ind · hr)	Pruetz (1999)
Papio anubis			0.084/(ind · hr) males	Harding (1980)
Papio cynocephalus			0.037/(ind · hr) males	Smuts (1985)
Overall mean	8.6 % (± 6.8, mean weighted by species = 9.38)			
Apes				
Pongo pygmaeus	1.6[1]			Rodman (1973)
Hylobates lar	3.0			Gittens and Raemaekers (1980)
Gorilla gorilla	3.6			Watts (1988)
Hylobates muelleri	4.0			Leighton (1987)
Pongo pygmaeus	5.4[1]			Galdikas (1985)
Gorilla gorilla	7.0	6.7	0.3%	Olejniczak (personal communication)
Hylobates lar	11.0		0.009/hr	Bartlett (1999)
Hylobates syndactylus	15.0		0.15/hr	Chivers (1974)
Pan troglodytes	22.0	9.0% (groom) (includes resting)	0.067/(ind · hr) males	Boesch and Boesch-Achermann (2000)
Pan troglodytes	24.9	16.8 (groom)		Teleki (1981)
Gorilla gorilla		0.39/hr	0.20/hr	Schaller (1963)
Pan troglodytes			0.03/hr	Bygott (1974)
Pan troglodytes			0.05/hr	Ghigliari (1984)
Pan troglodytes			0.016/hr males	Goodall (1986)
			0.009/hr females	Goodall (1986)
Overall mean apes	9.7% (± 8.2, mean weighted by species = 9.7)			

%, percent of total activity budget; ND, no data are provided.
[1] Orangutan data include any proximity between adults (76) or all individuals (77), not all of which would include active social interaction.

definitions for similar behavioral categories and have collected data using different sampling procedures (i.e., focal animal sampling, scan sampling, ad libitum or all-occurrence data; see Chapter 20). Moreover, different species and different individuals of the same species are likely to vary considerably in the expression and conspicuousness of social interactions. Thus, conspicuous behaviors (loud, visual) may be overrepresented in the data and more subtle forms of social interactions may be underrepresented (displacement or avoidance). Similarly, studies may focus on only certain age or sex classes and, therefore, can bias the results (i.e., in a species in which females groom more than males, a concentration on female behavior may overrepresent the overall frequency and nature of social interactions). Finally, most individual primates spend the vast majority of their day in peaceful and close proximity to conspecifics. Although in some species, spatial proximity may represent the strongest measure of social affiliation (Crockett and Eisenberg 1987), it is often not included in studies of primate sociality.

With this in mind, we view published percentages and rates of social interaction as general values that are likely to have considerable variance. However, we have confidence that the data represent reasonable approximations of the true activity patterns for these primates. Furthermore, since these data were collected using widely different methods and trait definitions, we do not attempt to apply complex meta-analysis statistics to them but, rather, use them simply for comparative and illustrative purposes. Using these data, our main purpose is to show that group-living primates generally spend 5%–10% of their day in active social interactions and that extremely few of these social interactions involve agonistic behavior. We hope this chapter will stimulate further research on these questions and the development of more comparable methodologies. However, given the patterns we observe in our large data set, we predict that our general conclusions will stand.

TIME SPENT IN AGONISTIC BEHAVIOR

Overall, we found that diurnal prosimians (3.7% ± 2.3%), New World monkeys (5.1% ± 5.1%), Old World monkeys (8.6% ± 6.8%), and apes (9.7% ± 8.2%) devoted less than 10% of their daily activity budget to direct social interactions.

The overwhelming majority of these interactions were affiliative and cooperative behaviors, such as grooming, food sharing, huddling, and alliance formation (Tables 39.1 and 39.2). In contrast, agonism, which included mild spats, displacements, threats, stares, and fighting, was rare and episodic, typically accounting for less than 1% of all social interactions. For example, even among sexually dimorphic species of Old World monkeys, rates of agonism and aggression were extremely low. In 10 studies of social behavior in hanuman langurs (*Semnopithecus entellus*, formerly *Presbytis entellus*), baboons (*Papio* spp.), vervet monkeys (*Cercopithecus aethiops*), and patas monkeys (*Erythrocebus patas*), hourly rates of agonism varied from 0.14 episodes per individual to 0.0007 per individual (Table 39.1). Although Old World monkeys spent more time than prosimians and New World monkeys engaged in social interactions, this was fully accounted for by an increased level of affiliative behaviors such as grooming.

As in prosimians and monkeys, data on ape social interactions are extremely variable. This is likely to reflect significant differences in social structure among all species of living apes. Gibbons and siamangs (lesser apes, *Hylobates* spp.) principally live in small pairbonded groups. In contrast, mountain gorillas (*Gorilla gorilla beringei*) live in cohesive multifemale/one male and multifemale/multimale social groups, common chimpanzees (*Pan troglodytes*) and bonobos (*Pan paniscus*) live in fission–fusion communities, and orangutans (*Pongo pygmaeus*) tend to be more solitary (except for mother–offspring pairs) and only rarely meet in the wild. Time spent in social interactions for apes ranged from 3.6% of the activity budget of the mountain gorilla to 25% in some studies of chimpanzees. Similar to prosimians and monkeys, rates of agonism in apes are low, ranging from 0.009–0.016 events per hour in chimpanzees to 0.20 per hour in gorillas (Table 39.1).

Based on the expectation that within-group feeding and reproductive competition increases with increasing group size, where possible, we corrected the rate of agonism by the number of potential interactions in the group. Although it is important to acknowledge that not all group members are equally involved in aggressive interactions, there is reason to assume that the number of aggressive events occurring per hour of observation is influenced by the number of potential interactants. One can use the analogy of billiard balls on a

Table 39.2 Summary of Agonistic and Affiliative Social Interactions Across All Activities in a Sample of 41 Species of Diurnal Primates

TAXA	% MEAN TIME SOCIAL	% MEAN TIME AFFILIATIVE	AGONISM (EVENTS/HR)	AGONISM (INDIVIDUAL EVENTS/WEEK)
Diurnal prosimians (10 species)	3.7 (± 2.3)	91.7	0.16	0.002
New World monkeys (21 species)	5.1 (± 5.1)	80.0	0.60	3.6
Old World monkeys (14 species)	10.7 (± 7.1)	89.9	0.58	6.3
Lesser and great apes (6 species)	9.7 (± 8.2)	81.6	0.09	0.0001

Source: Adapted from Sussman and Garber (2004).

table. If there are 15 billiard balls on a table and they are set in random motion, the likelihood of some number of them coming into contact or occupying the same space is much greater than if only two balls were set in motion on the table. We feel this analogy also is informative because it highlights the fact that some agonistic interactions that occur among group members are likely to be random—that is, not the result of a particular social strategy but simply of the fact that two individuals cannot jointly occupy the same space at the same time.

Our review of the literature indicates that overall rates of agonism per adult group member averaged less than one per day in Old World monkeys to approximately one every 2 days in New World monkeys, and such events were extremely rare in apes and prosimians (on average less than one per month; Table 39.2). The highest frequency of agonism (mild spats, displacements, and threats, as well as more severe forms of violent physical aggression) per individual group member per week was 10–11 times in adult male *Papio cynocephalus*.

Quantitative data on the costs and benefits of social cooperation, mutualism, and affiliative behaviors other than grooming, playing, or huddling are not commonly reported in the literature, although qualitative accounts of these behaviors are common. In part, this may reflect difficulties in quantifying both the short-term and long-term benefits and costs of cooperative behaviors. It also may reflect the fact that researchers have tended to underestimate the importance of various forms of mutual and cooperative behavior in primate sociality. Notable examples of affiliative and cooperative behaviors in primates include communal infant care (Garber 1997), food sharing (Boesch and Boesch-Achermann 2000), shared vigilance and other behaviors associated with predator protection and defense (Treves and Chapman 1996, Treves 1998, Miller 2002), alliance and friendship formation (Smuts 1985, de Waal 2000, Silk et al. 2003), coordinated hunting (Boesch 1994, Boesch and Boesch-Achermann 2000, Mitani and Watts 2001), and group-level range and resource defense (Wrangham 1980, Peres 2000).

We found that in prosimians affiliative interactions were eight times more common than agonistic interactions. In New World monkeys and Old World monkeys, affiliative behaviors commonly accounted for over 80%–90% of direct social interactions. In a study of lowland gorillas (*Gorilla gorilla gorilla*), C. Olejniczak (personal communication) reports that over 95% of all social interactions were affiliative. Clearly, affiliative interactions represent the overwhelming majority of primate social interactions and form the basis of individual social bonds. For example, data presented by Smuts (1985) on male and female anubis baboon social interactions and spatial proximity strongly support the idea that affiliative social interactions by newly resident males play a critical role in whether they are able to integrate themselves into a troop or must try their luck elsewhere.

Even in species in which social interactions typically account for only 2%–4% of the activity budget and adult group members are not related (i.e., mantled howler monkeys, *Alouatta palliata*), individuals are found to exhibit consistent partner preferences from year to year (Bezanson et al. 2002). These preferences are based on patterns of spatial proximity and affiliation, enabling individuals to feed together in the same food patch and to develop social and mating bonds.

NEW PERSPECTIVES ON COSTS AND BENEFITS OF GROUP LIVING

Traditional models of primate sociality have focused on issues of feeding competition and the costs to individuals of living in social groups (Wrangham 1980; van Schaik 1983, 1989). In general, it has been suggested that individuals should live in groups that are small enough to avoid the costs of aggression at feeding sites and large enough to benefit from co-operative predator detection and avoidance or infanticide (van Schaik 1983, 1989; Janson 1988; Isbell 1991; Crockett and Janson 2000). These models assume that groups in the population maintain maximal or optimal size and, therefore, the addition of one or a small number of new group members results in significant costs in feeding competition, travel, increased group spread, and aggression (van Schaik 1983, Janson and Goldsmith 1995, Chapman and Chapman 2000). Fewer studies have been designed to identify non-predation benefits of group living, such as allomaternal care, cooperative range and resource defense, social learning, friendships, locating feeding sites, and other forms of cooperative behavior.

SOCIOECOLOGICAL MODEL: FEEDING COMPETITION

Several researchers have argued that feeding competition is a pervasive cost of social group living and has played a determining role in structuring the size, composition, dominance hierarchy, and social system of group-living primates (van Schaik 1989, Janson and Goldsmith 1995, Sterk et al. 1997). Under conditions in which food is limited, the frequency of agonistic interactions, especially among females, is expected to be high, with low-ranking females experiencing costs in reduced fitness. This may result from a decrease in nutrition, fertility, or infant survivorship; an increase in energy costs associated with traveling to more distant feeding sites; or increased vulnerability to predators by foraging apart from other group members (Boinski et al. 2000). This perspective is referred to as the *socioecological model*.

Two forms of feeding competition have been identified, contest competition and scramble competition. *Contest competition* is a direct form of social interaction in which a single or small number of individuals displace or aggressively exclude other individuals from a feeding site. This may take the form of high-cost behaviors such as attacking, chasing, and intimidating displays or more subtle behaviors such as staring, displacement, and avoidance. Contest competition is expected to occur at high-quality, small, or defensible food patches, especially among females, and to result in significant fitness gains for the winner and significant fitness costs for the loser. Given the highly visual and vocal nature of agonistic interactions, they are unlikely to go unnoticed in primate field studies. *Scramble competition* is a more subtle form of social interaction in which an individual that locates a feeding site first is likely to gain a finder's advantage and, in this way, outcompete individuals that might otherwise exclude the finder from a food patch. The finder's advantage is expected to be high when resource patches are depletable, high-quality, ephemeral, and scattered (Ranta et al. 1996). The extent of the finder's advantage is expected to be low when resources are found in intermediate or large clumps or are of low quality (Ranta et al. 1996). However, scramble competition is extremely difficult to document (Chapman and Chapman 2000).

> there is no empirical basis for predicting if a later arriving animal would have obtained full, partial, or no access to the resource, or in fact whether it would have visited the feeding site at all. Given that, in most cases, scramble competition can only be inferred, its role in the evolution of individual foraging and social strategies remains unclear. (Sussman and Garber 2004:162–163)

Finally, Crockett and Janson (2000:76) acknowledge that

> some primate populations actually experiencing little food competition may be vulnerable to increased mortality of infants in larger groups due to factors unrelated to food, a pattern erroneously suggesting relations between group size and female fecundity.

They argue that in the case of red howlers (*Alouatta seniculus*) infanticide and male mating competition are likely to be greater in groups with four adult females than in groups with fewer than four adult females.

It has been hypothesized that due to their increased reproductive requirements, adult females engage in food-related contests more than adult males (van Schaik 1989, Sterck et al. 1997). In particular, van Schaik (1989) has suggested that in species with linear, despotic female dominance hierarchies, aggression and displacements at feeding sites are frequent. However, even in species exhibiting a linear dominance hierarchy, evidence of resource-based aggression among females is extremely low (Table 39.3). For example, in a group of 53–75 Peruvian squirrel monkeys (*Saimiri sciureus*), Mitchell (1990:116) reported that only 9% of "dyadic agonistic interactions involved food resources." Despite the presence of 23 adult female group members in the troop, over the course of 12 months only 28 cases of female–female food-related agonistic interactions were observed. Similarly, in a study of feeding competition among a community of 14 female chimpanzees (*Pan troglodytes verus*), Wittig and Boesch (2003) reported only 103 food conflicts over 1,028 hr of observation. Although these authors argue that "Taï female chimpanzees used aggression to keep resources or to gain them from other females" (Wittig and Boesch 2003:869) and were characterized by despotic

Table 39.3   Rate of Female–Female Food Conflicts per Hour of Feeding in Five Species of Diurnal Primates

SPECIES	CONFLICT RATE NUMBER FEMALES IN GROUP NUMBER CONTESTS DOMINANCE
Cercocebus atys (Range and Noe 2002)	0.058/female 24 N/R Linear
Erythrocebus patas (Pruetz and Isbell 2000)	0.004/female dyad 13–14 75 Egalitarian
Cercopithecus aethiops (Pruetz and Isbell 2000)	0.007/female dyad 8–9 55 Linear
Saimiri sciureus (Mitchell 1990)	0.013/female 23 28 Linear
Pan troglodytes (Wittig and Boesch 2003)	0.007/female 14 103 Linear

dominance relationships and a linear hierarchy, the individual rate of food-related agonism was 0.007/hr. Assuming a 12 hr pattern of daily activity and an activity budget in which 25% of the day is devoted to feeding and foraging, this translates into one food-related agonistic interaction per female every 2–3 months. Moreover, in 35% of these conflicts, the dominant individual did not obtain access to the food reward.

In Table 39.3, we present published data on food conflicts in five primate species. The data represent food conflicts per female per hour of feeding (not over all activities) and indicate extremely low rates of contests even among species forming linear dominance hierarchies. Taken together these data indicate that the frequency of food-related agonism among females occurred at a rate that ranged on average from once per 17 hr of feeding (once every 3–5 days) in mangabeys to once per 143 hr of feeding (once every 1–2 months) in chimpanzees and vervet monkeys. Thus, food-related agonism among females may not be as important a factor in shaping daily social interactions and individual relationships as commonly assumed; therefore, the socioecological model of primate sociality needs to be reevaluated.

ECOLOGICAL CONSTRAINTS MODEL: TRAVEL COSTS

Chapman et al. (1995) and Chapman and Chapman (2000) have proposed an ecological constraints model to evaluate the costs of increased group size on within-group feeding competition in primates. This model builds on an analysis by Janson and Goldsmith (1995) which suggested that a decrease in resource availability or quality or an increase in group size will result in greater travel distance in order to maintain a constant rate of food intake. Individuals in larger groups, therefore, are expected to experience increased travel costs in searching over greater distances to locate a sufficient number of feeding sites to satisfy their energetic needs. Thus, even in the absence of contests at feeding sites, individuals may suffer high costs of scramble competition by increased travel. In this regard, Janson and Goldsmith (1995) reported a positive relationship between group size and daily path length in 16 of 21 primate species studied. The effect was greatest in frugivorous primate species and least or absent in more folivorous species. However, data provided by Chapman and Chapman (2000) and Dias and Strier (2003) indicate that, across a range of different primate species, individuals in larger groups did not travel greater distances than individuals in smaller groups and that even in those cases when travel distance did increase, the additional distance traveled was generally less than 200 m. These data are highlighted in Table 39.4. Although factors other than group size (i.e., seasonal or yearly changes in food availability, range defense) can affect day range, the data indicate that in 50% (8/16) of cases individuals in larger groups of a given species had a day range that was less than that of individuals of the same species living in smaller social groups. In only 18% (3/16) of cases did individuals in larger groups have a mean day range that exceeded by 200 m or more that of individuals of the same species living in a smaller group. In one of these cases, a group of chacma baboons (*Papio hamadryas ursinus*) consisting of 45 individuals traveled an average distance of 4,670 m/day, whereas a group of 47 chacma baboons traveled a distance of 10,460 m/day. However, it is unlikely that a doubling of day range in this species was the direct consequence of adding two additional group members (Table 39.4).

Costs of Additional Travel

Travel costs can be in the form of increased time and energy moving, reduced time feeding and resting, increased conspicuousness to predators, or injury. If these costs are high and individuals in larger groups are required to travel greater distances to obtain resources, then we may expect a decrease in individual fitness. Recent physiological studies of the cost of locomotion in primates indicate that the cost of additional travel represents only a small fraction of an individual's daily energy budget (Warren and Crompton 1998, Steudel 2000). For example, Steudel (2000) calculated that in traveling 6.1 km/day, a female yellow baboon expends 6% of her daily energy budget. Similarly, hamadryas baboons travel approximately 9.5 km/day, and this accounts for 9.1%–9.5% of their daily energy budget. Thus, increases in day range of 100 m, 200 m, or more add relatively little to the energetic costs of travel in these primates. Similarly, Warren and

Table 39.4 Relationship Between Group Size and Distance Traveled per Day Within a Primate Species

SPECIES	GROUP SIZE	DISTANCE TRAVELED (M)	DIFFERENCE (M)
Saguinus fuscicollis	4.7	1,370	
Saguinus fuscicollis	6.5	1,220	−150
Callicebus moloch	3.2	570	
Callicebus moloch	4.2	670	+100
Alouatta palliata	9.1	120	
Alouatta palliata	12.2	600	+480
Alouatta palliata	15.5	440	−160
Alouatta seniculus	7.1	540	
Alouatta seniculus	9.0	710	+170
Alouatta seniculus	9.5	390	−320
Brachyteles arachnoides	24.5	1,280	
Brachyteles arachnoides (Dias and Strier 2003)	57–63	1,313	+33
Procolobus badius	20	600	
Procolobus badius	34	560	−40
Presbytis melalophos	9.3	1,150	
Presbytis melalophos	14	610	−540
Cercopithecus ascanius	26.2	1,540	
Cercopithecus ascanius	32.5	1,450	−90
Cercopithecus mitis	18.7	1,300	
Cercopithecus mitis	32.6	1,140	−160
Erythrocebus patas	20.6	2,250	
Erythrocebus patas	35.5	4,330	+2,080
Papio hamadryas ursinus	45	4,670	
Papio hamadryas ursinus	47.2	10,460	+5,790
Hylobates syndactylus	3.0	740	
Hylobates syndactylus	3.8	930	+190
Hylobates syndactylus	4.0	790	−140
Hylobates syndactylus	5.0	970	+160

Source: Adapted from Chapman and Chapman (2000).

Crompton (1998) calculated the relative cost of locomotion in *Lepilemur edwardsi* (sportive lemur), *Avahi occidentalis* (woolly lemur), *Tarsius bancanus* (Bornean tarsier), *Galago moholi* (South African lesser bushbaby), and *Otolemur crassicaudatus* (large-eared greater bushbaby) based on an estimate of field metabolic rate over a 24 hr period. These prosimians expended 0.9%–3% of their daily metabolic energy budget in locomotion (based on night travel distances, locomotor frequencies, and travel height changes in different locomotor modes). Similarly, in mantled howler monkeys, Nagy and Milton (1979) calculated that the relative cost of locomotion was 2% of daily energy, and Steudel (2000) reported that adult Costa Rican squirrel monkeys (*Saimiri oerstedii*) traveling 3,500 m/day expend only 4.4% of their total daily energy on locomotion. Based on these data, added travel distance is likely to exert only a minor constraint on the energetic costs of primate foraging.

In summary, the empirical data on the frequency and nature of primate social interactions; relationships between group size, day range, and the energetic costs of travel; and female feeding competition in species with linear dominance hierarchies are best interpreted as follows:

1. Agonism occurs at very low frequencies among diurnal primates living in the same social group. This does not fit the predictions of the socioecological model.
2. Agonism at feeding sites does not occur more frequently among female primates that form linear dominance hierarchies than among females that form egalitarian social relationships. This does not fit the feeding competition model.
3. Individuals in larger groups of primates do not necessarily travel greater distances than individuals of the same species living in smaller groups. This does not fit the ecological constraints model.
4. In those cases when individuals in larger groups experience increased travel, the energetic cost of travel for arboreal or terrestrial primates is expected to be relatively low. This does not fit the ecological constraints model.

ALTERNATIVE MODEL

We present an alternative model of primate sociality and group size which we feel is better supported by the available data (Fig. 39.1). This model is based on three principles.

First, resources in tropical forests are found in dispersed, heterogeneous patches. This is supported by recent information on the distribution of plant species in Amazonian (Tuomisto et al. 2003) and Central American (Wehncke et al. 2003) forests, as well as theoretical models of variance in spatiotemporal changes of food availability and resource distributions (Kotliar and Wiens 1990, Johnson et al. 2002). According to Johnson et al. (2002:564), the existence of both nearby and distant food patches of various size and quality scattered across the landscape limits opportunity for feeding competition and increases the probability that several individuals have access to feeding sites, which "shifts the cost–benefit balance towards group living."

Second, compared to many groups of animals, primates are characterized by tremendous dietary breadth in the types of food exploited (insects, fruits, flowers, leaves, seeds, small vertebrates) and the number of plant, insect, and vertebrate species consumed (Garber 1987). It is not uncommon for primates to exploit resources from 50 to over 100 plant species over the course of a single year. Finally, in many primate taxa, grouping patterns are highly flexible and individuals may form subgroups as a facultative response to local ecological and social conditions (Symington 1988a,b, 1990; Chapman 1990a,b; Kinzey and Cunningham 1994; Chapman et al. 1995; Jamieson 1998; Garber et al. 1999; Dias and Strier 2003). The ability to temporarily subgroup in response to changes in the availability, density, and distribution of food resources minimizes opportunity for contest competition at feeding sites. We feel this type of social flexibility is common among a range of primate species and not simply restricted to those taxa exhibiting a fission–fusion system.

Predation is a critical component of all models of sociality. Predation risk in our model is depicted as a U-shaped curve (Fig. 39.1, bold). Predation risk is assumed to be high at low group size, low and relatively constant at intermediate group sizes, and high at larger group size. Although data on the effects of predation on primate sociality are improving (Miller 2002, see Chapter 32), given the limited information on predation rates and predation success (Hill and Dunbar 1998, Zuberbuhler 2004; but see Hart 2000, Hart and Sussman 2005), there is little empirical basis at present to critically evaluate this assumption. However, once groups attain large size, it is likely that (1) a large group may attract a greater number of predators and (2) if large groups are less cohesive due to increased levels of feeding competition and social stress, then individuals that forage apart from the main body of a large group may become more vulnerable to predator attack (Janson 1988). Our model predicts that, across a relatively wide range of group sizes, feeding competition (Fig. 39.1, dotted line) and the associated aggression among conspecifics at feeding sites are expected to be low and to have minimal fitness consequences and levels of predator detection are expected to be high (Fig. 39.1, crosshatched area). This occurs in response to the heterogeneous distribution of food patches in tropical forests, the behavioral flexibility primates show in short-term changes in diet and subgrouping patterns, and the relatively low energetic cost of increased travel. We expect that within-group feeding competition principally becomes a significant problem to individual fitness only at larger group size or among group members living in marginal or degraded habitats.

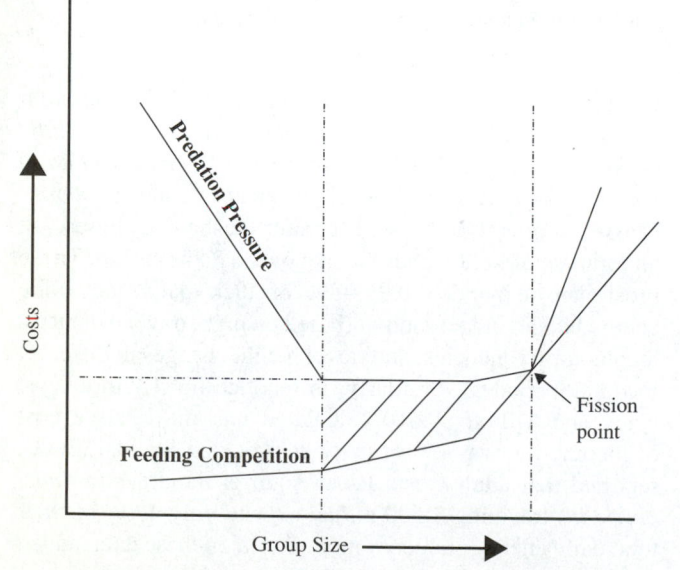

Figure 39.1 Predation pressure, feeding competition, and group size in primates.

The third part of the model involves the question of group demography and the process by which groups increase and decrease in size. A basic assumption of the competition models is that most groups in a population are at or approaching their maximal size. Therefore, within-group feeding competition is expected to be relatively high, and the inclusion of additional individuals results in significant costs in feeding competition, aggression, and additional travel costs. However, there are reasons to suggest that groups in a population are commonly below their maximal size and, therefore, that the effects of feeding competition are low across a range of group sizes. If new groups typically form by a process of fissioning, in which a set of individuals bud off from an established group, then we expect relatively few large groups (they are unstable and fission over a relatively limited period of time), many small groups (newly fissioned elements), and a majority of intermediate-sized groups (those in which a small set of animals have recently left). There is evidence of group fissioning among a wide range of primate taxa, including diurnal lemurs (Sussman 1992, Gould et al. 2003, *Lemur catta*), New World monkeys (Garber 1988, *Saguinus mystax*; Fedigan et al. 1998, *Alouatta palliata*), Old World monkeys (Altmann 1974, *Papio anubis*; Hrdy 1977, *Presbytis entellus*; Duggleby 1977, Cheverud et al. 1978, *Macaca mulatta*; Cords and Rowell 1986, *Cercopithecus mitis*; Dunbar 1988, *Theropithecus gelada*; Bennett and Davies 1994, *Trachypithecus vetulus*; Oates 1994, *Colobus vellerosus*; Windfelder and Lwanga 2002, *Cercopithecus ascanius*), and apes (Watts 1998, *Gorilla gorilla*). For example, Fedigan et al. (1998) report that over the course of a 6-year period, four of nine new mantled howler (*A. palliata*) groups formed by fissioning. Simiarly, Sussman (1992:408–409) argues that in ring-tailed lemurs (*L. catta*) "group fission may be more important than male transfer in adjusting population size over larger areas." It is likely that the dynamics of social living, patterns of group fission, opportunities for subgrouping, and flexibility in the size, composition, and structure of many primate social groups limit the set of conditions under which feeding competition results in significant fitness costs to individuals residing in the same social group. Moreover, individuals that reside together in the same social groups may receive critical fitness benefits by interacting in ways that enhance social cohesion, predator protection, social learning and information exchange, and access to and defense of resources desired by individuals in other social groups.

It is important to point out that we are not suggesting that group size is unrelated to the quality of resources within a home range or that feeding competition is not a factor in primate behavioral ecology and individual fitness. For example, Chapman et al. (2004) found evidence of a significant relationship between the protein-to-fiber ratio of mature leaves and the biomass of African colobines. What we are suggesting is that, in general, rain forest habitats are characterized by food patches that vary considerably in size, richness, and spatial–temporal distribution (Bacon et al. 1991). As species exploit more broad-based diets (increase in both the number of plant species and food types exploited), the potential number of food patches available to group members increases dramatically and is sufficient to support individuals living in groups that vary across a range of sizes (Johnson et al. 2002). Thus, at small and intermediate group sizes, feeding competition is expected to play a minimal role in primate social interactions, whereas in large groups it is likely to result in an increase in short-term rates of agonism and eventual fissioning (group size categories being different for each species and habitat).

BENEFITS OF GROUP LIVING

If the costs of feeding competition and group living are less than previously thought, then researchers need to focus on the benefits of affiliation, cooperation, and mutualism in understanding the evolution of primate sociality. Several recent studies of primate socioecology have highlighted the role of cooperation and affiliation in determining the benefits to individuals of forming groups or subgroups of a particular size and composition (Clutton-Brock 2002, Silk 2002a, Silk et al. 2003, Sussman and Garber 2004).

Cooperation and affiliation represent behavioral tactics that can be used by individual group members to obtain resources, maintain or enhance social position, or increase reproductive opportunities (Garber 1997). Many affiliative or cooperative behaviors can be explained by individual actions that also may benefit several group members. In acts of cooperation, both participants may receive immediate benefits. Coordinated behaviors such as joint resource defense, range defense, cooperative hunting, alliance formation, cooperative food harvesting, and predator vigilance can be explained in terms of immediate benefits to participating individuals. Acts that appear to benefit recipients may also benefit actors. These benefits need not be equal for each individual. If the cost to the actors of affiliative behavior is low, even if the rewards are low and/or variable, we should expect affiliation and cooperation to be common. Thus, many types of social interaction may be best understood in terms of a non-zero-sum game with multiple winners. According to Watts (2002:366), "Formation of low-risk coalitions in which all participants stand to make immediate net gains is widespread in primates and may even incorporate much presumed 'altruism' among kin." Taken together, these data may help to explain observations that nonhuman primates live in relatively stable, cohesive social groups and solve the problems of everyday life in a generally cooperative fashion.

NEUROENDOCRINE MECHANISMS FOR COOPERATION

Animals of many kinds are social. . . . It has often been assumed that animals were in the first place rendered social, and that they feel as a consequence uncomfortable when separated from each other, and comfortable whilst together, but it is a more probable view that these sensations were first developed, in order that those animals which would profit by living in society, should be induced to live together, in the same manner as the sense of hunger and the pleasure of eating were, no doubt, first acquired in order to induce animals to eat. The feeling of pleasure from society is probably an extension of the parental or filial affections, since the social instinct seems to be developed by the young remaining for a long time with their parents; and this extension may be attributed in part to habit, but chiefly to natural selection. (Darwin 1874:97, 102)

As we have stressed, most nonhuman primates are highly social and are members of social groups throughout all or nearly all of their long lives. Investigations into the evolution of primate sociality have tended to focus on aggression and processes by which primates reconcile their conflicts with one another to retain their allies' support, emphasizing competition instead of cooperation. However, many results from behavioral, hormonal, and brain imaging studies offer a new perspective about primates and their proclivities for cooperation, sociality, and peace.

For example, in a recent paper based on 16 years of data on the behavior and ecology of wild savanna baboons, Silk et al. (2003:1231) conclude as follows:

> Our data indicate that social integration has positive effects on the reproductive performance of female baboons. Females who had more social contact with other adult group members and were more fully socially integrated into their groups were more likely than other females to rear infants successfully. These effects were independent of female dominance rank and variation in ecological conditions. . . . These data provide direct empirical evidence that sociality enhances the fitness of nonhuman primate females.

Recently, researchers have identified a set of mechanisms that might lead to cooperative behavior among related and nonrelated individuals that do not necessitate selfish genes, complex calculations of kin recognition or relationships, or accurate cost–benefit predictions of future reciprocity. In experiments using magnetic resonance imaging (MRI, a technique similar to X-ray imaging that uses radio waves, rather than radiation, to elucidate deep structure), mutual cooperation has been associated with consistent activation in two areas of the brain that have been linked with reward processing (the anteroventral striatum and the orbitofrontal cortex). It has been proposed that activation of this neural network positively reinforces cooperative social interactions (Rilling et al. 2002). Even more compelling, the strength of the neural response increases with the persistence of mutual cooperation over successive trials. This activation of the brain's reward center may account for why we tend to feel good when we cooperate and provides a neural mechanism for understanding the positive reinforcement that sustains cooperative social relationships.

Both the anteroventral striatum and the orbitofrontal cortex are rich in neurons that respond to dopamine, the neurotransmitter known for its role in addictive behaviors.

The dopamine system evaluates rewards—both those that flow from the environment and those conjured up in the brain. When the stimulus is positive, dopamine is released. "The dopamine system works unconsciously and globally, providing guidance for making decisions when there is not time to think things through" (Blakesee 1999:347). In experiments with rats, for example, in which electrodes are placed in the striatum, the animals continue to press a bar to stimulate the electrodes, apparently receiving such pleasurable feedback that they will starve to death rather than stop pressing the bar. With these systems, investigators believe that they have identified a pattern of neural activation "that may be involved in sustaining cooperative social relationships, perhaps by labeling cooperative social interactions as rewarding" (Rilling et al. 2002:403).

Another physiological mechanism related to affiliation and nurturing is the neuroendocrine circuitry associated with maternal responses in mammals. Orchestrating the broad suite of these biobehavioral responses is the hormone oxytocin. Oxytocin has been related to every type of animal bonding: parental, fraternal, sexual, and even the capacity to soothe one's self (Angier 1999, Carter 1999, Taylor et al. 2000, Carter and Cushing 2004). Research conducted primarily on monogamous mammals, capable of pairbond formation and biparental care, has revealed that, along with other chemicals, oxytocin influences parental behavior, general sociality, and the capacity to form social bonds. These neurochemicals and their receptors are regulated in early life by epigenetic factors, such as early social experiences, producing lifelong behavioral responses, which may help to account for individual and species-typical variations in social interactions (Bales et al. 2004). *Epigenesis* is the study of heritable changes in gene function that occur without changes in the sequence of nuclear deoxyribonucleic acid (DNA), including how environmental factors affecting a parent can result in changes in gene expression in the offspring (Waterland and Jirtle 2003). In the case of oxytocin, it has been suggested that although its primary role may have been in forging the mother–infant bond, its ability to influence brain circuitry may have been coopted to serve other affiliative purposes that allowed the formation of alliances and partnerships, thus facilitating the evolution of cooperative behaviors (Carter 1999, Taylor et al. 2000).

Results from both behavioral endocrinology and functional MRI studies on cotton-top tamarin monkeys (*Saguinus oedipus*) reveal other hormonal mechanisms critical to cooperation and affiliative behavior (Snowdon 2004). In these primates, males and other nonmaternal helpers provide essential infant care (see Chapter 6). Elevated levels of the hormone prolactin, usually associated with lactation, have been implicated in predisposing individuals to exhibiting nonmaternal infant care. New findings also indicate that prolactin can be considered a "relation hormone" because its levels fluctuate in response to the quality of a male's relationship with his mate. Furthermore, the same regions of the brain associated with positive emotional states (such as mutual cooperation) in humans are also activated in male monkeys when exposed to cues from familiar mates and unfamiliar but ovulating females (Snowdon 2004).

If cooperation and spatial proximity among group-living animals are rewarding in a variety of environmental and social circumstances and if physiological and neurological feedback systems reinforce social tolerance and cooperative behavior, then social group living can persist in the absence of any conscious recognition that material gains might also flow from mutual cooperation. Social animals appear to be wired to cooperate and to reduce stress by seeking each others' company (Carter 1999, Taylor et al. 2000, Rilling et al. 2002, Carter and Cushing 2004, Snowdon 2004). Social affiliation and co-operative behaviors provide psychological, physiological, and ecological benefits to social primates; are reinforced by hormonal and neurological systems; and provide internal positive reinforcement to the participants.

Sociality has evolved independently in many diverse groups of animals. Among primates, sociality may have its origin in the general benefits of mutual cooperation, strong mother–infant bonds, and the evolution of an extended juvenile period in which developing individuals are dependent on other group members. Specifically, neurological, hormonal (oxytocin), and endogenous opioid mechanisms may be at the core of innate cooperative social responses (Carter 1999, Taylor et al. 2000). This could explain the evolution not only of cooperation among non-relatives but also of "nonselfish" altruistic behavior. Again, we acknowledge the important role of aggression and competition in understanding primate social interactions. Our perspective, however, is that affiliation, cooperation, and social tolerance associated with the long-term benefits of mutualism form the core of social group living. In most instances, aggression and competition are better understood as social tactics and individual adjustments to the immediate and ephemeral conditions of particular social situations.

Finally, we highlight the importance of collecting systematic data on the frequency and context of social behavior to better understand the mechanisms that govern everyday interactions within social groups. We must better understand who does what to whom, how often, and when. As stressed by Silk (2002b:440), "we need to pay more attention to methodological details, such as how we should interpret information about the content, frequency, quality and patterns of social interactions." Furthermore, since active social behavior generally takes up less than 10% of an individual's time, social interactions must be understood within a wider context, such as general activity pattern, life history of the individual, group and population demography, and recent perturbations to the ecosystem that may affect the individual, group, or population. Until we have a better understanding of these mechanisms, hypotheses concerning evolutionary explanations of cooperation, agonism, and sociality may be misleading. In any case, we agree with Clutton-Brock (2002:72):

if mutualism proves to be important in maintaining cooperative animal societies, the benefits of cooperation in animals may be more similar to those of cooperation in humans than has been previously supposed. In humans, unrelated individuals commonly assist each other . . . [and] generalized reciprocity appears to be important in maintaining many social institutions . . . [these] trends appear to have close parallels in other cooperative animals.

REFERENCES

Agetsuma, N. (1995). Foraging strategies of Yakushima macaques (*Macaca fuscata yakui*). *Int. J. Primatol.* 16:595–610.

Altmann, S. A. (1974). Baboons, space, time, and energy. *Am. Zool.* 14:221–248.

Angier, N. (1999). Illuminating how bodies are built for sociality. In: Sussman, R. W. (ed.), *The Biological Basis of Human Behavior: A Critical Review*. Prentice Hall, Upper Saddle River, NJ. pp. 350–352.

Bacon, P. J., Ball, F., and Blackwell, P. (1991). A model for territory and group formation in a heterogeneous habitat. *J. Theor. Biol.* 148:445–468.

Baker, A. J. (1991). Evolution of the social system of the golden lion tamarin (*Leontopithecus rosalia*): mating system, group dynamics and cooperative breeding [PhD thesis]. University of Maryland, College Park.

Bales, K. L., Pfeifer, L. A., and Carter, S. C. (2004). Sex differences and developmental effects of manipulations of oxytocin on alloparenting and anxiety in prairie voles. *Dev. Psychobiol.* 44:123–131.

Bartlett, T. Q. (1999). Feeding and ranging behavior of the white-handed gibbon (*Hylobates lar*) in Khao Yai National Park, Thailand [PhD thesis]. Washington University, St. Louis.

Bennett, E. L., and Davies, A. G. (1994). The ecology of Asian colobines. In: Davies, A. G., and Oates, J. F. (eds.), *Colobine Monkeys: Their Ecology, Behaviour and Evolution*. Cambridge University Press, Cambridge. pp. 129–171.

Bezanson, M., Garber, P. A., Rutherford, J. R., and Cleveland, A. (2002). Patterns of subgrouping, social affiliation and social networks in Nicaragua mantled howler monkeys (*Alouatta palliata*). *Am. J. Phys. Anthropol.* 32(suppl.):44.

Bicca-Marques, J. C. (1993). Padrao de atividades diarias do bugio-preto *Alouatta caraya* (Primates, Cebidae): uma analise temporal e bioenergetica. In: Yamamoto, M. E., and Sousa, M. B. C. (eds.), *A Primatologia no Brasil*, vol. 4. Editora Universitaria-UFRN, Natal. pp. 35–49.

Blakeslee, S. (1999). How brain uses a simple dopamine system. In: Sussman, R. W. (ed.), *The Biological Basis of Human Behavior: A Critical Review*. Prentice Hall, Upper Saddle River, NJ. pp. 347–349.

Boesch, C. (1994). Hunting strategies of Gombe and Taï chimpanzees. In: Wrangham, R. W., McGrew, W. C., de Waal, F. B. M., and Heltne. P. G. (eds.), *Chimpanzee Cultures*. Harvard University Press, Cambridge, MA. pp. 77–91.

Boesch, C., and Boesch-Achermann, H. (2000). *The Chimpanzees of the Taï Forest: Behavioural Ecology and Evolution*. Oxford University Press, New York.

Boinski, S., Treves, A., and Chapman, C. A. (2000). A critical evaluation of the influence of predators on primates: effects on group travel. In: Boinski, S., and Garber, P. A. (eds.), *On the*

Move: How and Why Animals Travel in Groups. University of Chicago Press, Chicago. pp. 43–72.

Borries, C., Sommer, V., and Srivastave, A. (1991). Dominance, age, and reproductive success in free-ranging female hanuman langurs (*Presbytis entellus*). *Int. J. Primatol.* 12:231–258.

Burton, F. D., Snarr, K. A., and Harrison, S. E. (1995). Preliminary report on *Presbytis francoisi leucocephalus. Int. J. Primatol.* 16:311–327.

Bygott, J. D. (1974). Agonistic behaviour and dominance in wild chimpanzees [PhD thesis]. Cambridge University, Cambridge.

Carter, S. (ed.) (1999). *Hormones, Brain and Behavior: Integrative Neuroendocrinology of Affiliation.* MIT Press, Boston.

Carter, S., and Cushing, B. S. (2004). Proximate mechanisms regulating sociality and social monogamy, in the context of evolution. In: Sussman, R. W., and Chapman, A. R. (eds.), *Origins and Nature of Sociality.* Aldine de Gruyter, New York. pp. 99–121.

Castro, N. R. (1991). Behavioral ecology of two coexistent tamarin species (*Saguinus fuscicollis nigrifrons* and *Saguinus mystax mystax*, Callitrichidae, Primates) in Amazonian Peru [PhD thesis]. Washington University, St. Louis.

Chapman, C. A. (1990a). Ecological constraints on group size in three species of neotropical primates. *Folia Primatol.* 55:1–9.

Chapman, C. A. (1990b). Association patterns of spider monkeys: the influence of ecology and sex on social organization. *Behav. Ecol. Sociobiol.* 26:409–414.

Chapman, C. A., and Chapman, L. J. (2000). Determinants of group size in primates: the importance of travel costs. In: Boinski, S., and Garber, P.A. (eds.), *On the Move: How and Why Animals Travel in Groups.* University of Chicago Press, Chicago. pp. 24–42.

Chapman, C. A., Chapman, L. J., Naugthon-Treves, L., Lawes, M. J., and McDowell, L. R. (2004). Predicting folivorous primate abundance: validation of a nutritional model. *Am. J. Primatol.* 62:55–69.

Chapman, C. A., Wrangham, R. W., and Chapman, L. J. (1995). Ecological constraints on group size: an analysis of spider monkey and chimpanzee subgroups. *Behav. Ecol. Sociobiol.* 36:59–70.

Cheverud, J. M. (2004). Darwinian evolution by natural selection of heretible variation: definitions of parameters and application to social behavior. In: Sussman, R. W., and Chapman, A. R. (eds.), *Origins and Nature of Sociality.* Aldine de Gruyter, New York. pp. 140–157.

Cheverud, J. M., Buettner-Janusch, J., and Sade, D. (1978). Social group fission and the origin of intergroup genetic differentiation among the rhesus monkeys of Cayo Santiago. *Am. J. Phys. Anthropol.* 49:449–456.

Chivers, D. J. (1974). The siamang in Malaya: a field study of a primate in tropical rain forest. *Contrib. Primatol.* 4:1–335.

Clutton-Brock, T. (2002). Breeding together: kin selection and mutualism in cooperative vertebrates. *Science* 296:69–72.

Cords, M., and Rowell, T. E. (1986). Group fission in blue monkeys of the Kakamega Forest, Kenya. *Folia Primatol.* 46:70–82.

Crockett, C. M., and Eisenberg, J. F. (1987). Howlers: variations in group size and demography. In: Smuts, B. B., Cheney, D. L., Seyfarth, R. M., Wrangham, R. W., and Struhsaker, T. T. (eds.), *Primate Societies.* University of Chicago Press, Chicago. pp. 54–68.

Crockett, C. M., and Janson, C. H. (2000). Infanticide in red howlers: female group size, male membership, and a possible link to folivory. In: van Schaik, C. P., and Janson, C. H. (eds.), *Infanticide by Males and Its Implications.* Cambridge University Press, Cambridge. pp. 75–98.

Curtis, D. J. (1997). The mongoose lemur (*Eulemur mongoz*): a study in behavior and ecology [PhD thesis]. University of Zurich, Zurich.

Darwin, C. (1874). *The Descent of Man and Selection in Relation to Sex*, 2nd ed. Henneberry Company, Chicago.

Decker, B. S. (1994). Effects of habitat disturbance on the behavioral ecology and demographics of the Tana River red colobus (*Colobus badius rufomitratus*). *Int. J. Primatol.* 15:703–737.

Defler, T. R. (1995). The time budget of a group of wild woolly monkeys (*Lagothrix lagotricha*). *Int. J. Primatol.* 16:107–120.

de la Torre, S., Campos, F., and de Vries, T. (1995). Home range and birth seasonality of *Saguinus nigricollis* in Ecuadorian Amazonia. *Am. J. Primatol.* 37:39–56.

de Waal, F. B. M. (2000). The first kiss: foundations of conflict resolution research in animals. In: Aureli, F., and de Waal, F. B. M. (eds.), *Natural Conflict Resolution.* University of California, Berkeley. pp. 15–33.

Dias, L. G., and Strier, K. B. (2003). Effects of group size on ranging patterns in *Brachyteles arachnoides hypoxanthus. Int. J. Primatol.* 24:209–221.

Dietz, J. M., Peres, C. A., and Pinder, L. (1997). Foraging ecology and use of space in wild golden lion tamarins (*Leontopithecus rosalia*). *Am. J. Primatol.* 41:289–305.

Dugatkin, L. A. (1997). *Cooperation Among Animals: An Evolutionary Perspective.* Oxford University Press, New York.

Duggleby, C. R. (1977). Blood group antigens and the population genetics of *Macaca mulatta* on Cayo Santiago. II Effects of social group division. *Ybk Phys. Anthropol.* 20:263–271.

Dunbar, R. I. M. (1988). *Primate Social Systems.* Cornell University Press, Ithaca.

Easley, S. P. (1982). Ecology and behavior of *Callicebus torquatus*: Cebidae, Primates [PhD thesis]. Washington University, St. Louis.

Estrada, A., Juan-Solano, S., Martinz, T. O., and Coates-Estrada, R. (1999). Feeding and general activity patterns of a howler monkey (*Alouatta palliata*) troop living in a forest fragment at Los Tuxtlas, Mexico. *Am. J. Primatol.* 48:167–183.

Fashing, P. J. (2001). Activity and ranging patterns of guerezas in the Kakamega Forest: intergroup variation and implications for intragroup feeding competition. *Int. J. Primatol.* 22:549–578.

Fedigan, L. M., and Baxter, M. J. (1984). Sex differences and social organization in free-ranging spider monkeys (*Ateles geoffroyi*). *Primates* 25:279–294.

Fedigan, L. M., Rose, L. M., and Avila, R. M. (1998). Growth of mantled howler groups in a regenerating Costa Rican dry forest. *Int. J. Primatol.* 19:405–432.

Freed, B. Z. (1996). Co-occurrence among crowned lemurs (*Lemur coronatus*) and Sanford's lemurs (*Lemur fulvus sanfordi*) of Madagascar [PhD thesis]. Washington University, St. Louis.

Fuentes, A. (1996). Feeding and ranging in the Mentawai Island langur (*Presbytis potenziani*). *Int. J. Primatol.* 17:525–548.

Galdikas, B. M. F. (1985). Orangutan sociality in Tanjung Putting. *Am. J. Primatol.* 9:101–119.

Garber, P. A. (1987). Foraging strategies among living primates. *Annu. Rev. Anthropol.* 16:339–364.

Garber, P. A. (1988). Diet, foraging patterns, and resource defense in a mixed species troop of *Saguinus mystax* and *Saguinus fuscicollis* in Amazonian Peru. *Behaviour* 105:18–33.

Garber, P. A. (1997). One for all and breeding for one: cooperation and competition as a tamarin reproductive strategy. *Evol. Anthropol.* 5:187–199.

Garber, P. A., Pruetz, J. D., Lavallee, A. C., and Lavallee, S. G. (1999). A preliminary study of mantled howling monkey (*Alouatta palliata*) ecology and conservation on Isla de Ometepe, Nicaragua. *Neotrop. Primates* 7:113–117.

Gaulin, S. J. C., and Gaulin, C. K. (1982). Behavioral ecology of *Alouatta seniculus* in Andean Cloud Forest. *Int. J. Primatol.* 3:1–32.

Ghigliari, M. P. (1984). *The Chimpanzees of Kibale Forest*. Columbia University Press, New York.

Gittens, G. P., and Raemaekers, J. J. (1980). Siamang, lar and agile gibbons. In: Chivers, D. J. (ed.), *Malayan Forest Primates*. Plenum Press, New York. pp. 63–105.

Goldizen, A. W., Mendelson, J., van Vlaardingen, M., and Terborgh, J. (1996). Saddle-back tamarin (*Saguinus fuscicollis*) reproductive strategies: evidence from a thirteen-year study of a marked population. *Am. J. Primatol.* 38:57–83.

Goodall, J. (1986). *Chimpanzees of Gombe*. Harvard University Press, Cambridge, MA.

Gould, L. (1994). Patterns of affiliative behavior in adult male ringtailed lemurs (*Lemur catta*) at the Beza-Mahafaly Reserve, Madagascar [PhD thesis]. Washington University, St. Louis.

Gould, L., Sussman, R. W., and Sauther, M. L. (2003). Demographic and life-history patterns in a population of ring-tailed lemurs (*Lemur catta*) at Beza Mahafaly Reserve, Madagascar: a 15-year perspective. *Am. J. Phys. Anthropol.* 120:182–194.

Harding, R. S. O. (1980). Agonism, ranking, and the social behavior of adult male baboons. *Am. J. Phys. Anthropol.* 53:203–216.

Hart, D. L. (2000). Primates as prey: ecological, morphological and behavioral relationships between primate species and their predators [PhD thesis]. Washington University, St. Louis.

Hart, D. L., and Sussman, R. W. (2005). *Man the Hunted: Primates, Predators, and Human Evolution*. Westview Press, Boulder.

Hemingway, C. A. (1995). Feeding and reproductive strategies of the Milne-Edwards sifaka, *Propithecus diadema edwardsi* [PhD thesis]. Duke University, Durham, NC.

Hill, R. A., and Dunbar, R. I. M. (1998). An evaluation of the roles of predation rate and predation risk as selective pressures on primate grouping behaviour. *Behaviour* 135:411–430.

Hrdy, S. B. (1977). *The Langurs of Abu: Female and Male Strategies of Reproduction*. Harvard University Press, Cambridge, MA.

Isbell, L. A. (1991). Contest and scramble competition: patterns of female aggression and ranging behaviour among primates. *Behav. Ecol.* 2:143–155.

Jamieson, R. W. (1998). The effects of seasonal variation in fruit availability on social and foraging behavior in *Macaca fascicularis* in Mauritius [PhD thesis]. Washington University, St. Louis.

Janson, C. H. (1984). Female choice and mating system of the brown capuchin monkey, *Cebus apella* (Primates, Cebidea). *Z. Tierpsychol.* 65:177–200.

Janson, C. H. (1988). Food competition in brown capuchin monkeys (*Cebus apella*): quantitative effects of group size and tree productivity. *Behaviour* 105:53–76.

Janson, C. H. (2000). Primate socio-ecology: the end of a golden age. *Evol. Anthropol.* 9:73–86.

Janson, C. H., and Goldsmith, M. (1995). Predicting group size in primates: foraging costs and predation risks. *Behav. Ecol.* 6:326–336.

Johnson, D. P., Kays, R., Blackwell, P. G., and Macdonald, D.W. (2002). Does the resource dispersion hypothesis explain group living?. *Trends Ecol. Evol.* 17:563–570.

Jones, C. B. (1980). The functions of status in the mantled howler monkey, *Alouatta palliata* Gray: intraspecific competition for group membership in a folivorous neotropical primate. *Primates* 21:389–405.

Kaplin, B. A., and Moermond, T. C. (2000). Foraging ecology of the mountain monkey (*Cercopithecus l'hoesti*): implications for its evolutionary history and use of disturbed forest. *Am. J. Primatol.* 50:227–246.

Kinzey, W. G. (1981). The titi monkeys, genus *Callicebus*. In: Coimbra-Filho, A. F., and Mittermeier, R. A. (eds.), *Ecology and Behavior of Neotropical Primates*, vol. 1. Academia Brasileira de Ciencias, Rio de Janeiro. pp. 241–276.

Kinzey, W. G., and Cunningham, E. P. (1994). Variability in platyrrhine social organization. *Am. J. Primatol.* 34:185–198.

Kirkpatrick, R. C., Long, Y. C., Zhong, T., and Xiao, L. (1998). Social organization and range use in the Yunnan snub-nosed. *Int. J. Primatol.* 19:13–52.

Klein, L. L., and Klein, D. J. (1977). Feeding behaviour of the Colombian spider monkey *Ateles belzebuth*. In: Clutton-Brock, T. H. (ed.), *Primate Ecology*. Academic Press, London. pp. 153–181.

Korstjens, A. H., Sterck, E. H. M., and Noe, R. (2002). How adaptive or phylogenetically inert is primate social behaviour? A test with two sympatric colobines. *Behaviour* 139:203–225.

Kotliar, N. B., and Wiens, J. A. (1990). Multiple scales of patchiness and patch structure: a hierarchical framework for the study of heterogeneity. *Oikos* 59:253–260.

Kurup, G. U., and Kumar, A. (1993). Time budget and activity patterns of lion-tailed macaque (*Macaca silenus*). *Int. J. Primatol.* 14:27–39.

Larose, F. (1996). Foraging strategies, group size, and food competition in the mantled howler monkey, *Alouatta palliata* [PhD thesis]. University of Alberta, Edmonton.

Laws, J. W., and Laws, J. V. H. (1984). Social interactions among adult male langurs (*Presbytis entellus*) at Rajaji Wildlife Sanctuary. *Int. J. Primatol.* 5:31–50.

Leighton, D. R. (1987). Gibbons: territoriality and monogamy. In: Smuts, B., Cheney, D., Seyfarth, R., Wrangham, R., and Struhsaker, T. (eds.), *Primate Societies*. University of Chicago Press, Chicago. pp. 135–145.

McGraw, W. S. (1998). Posture and support use of Old World monkeys (Cercopithecidae): the influence of foraging strategies, activity patterns, and the spatial distribution of preferred food items. *Am. J. Primatol.* 46:229–250.

McGraw, W. S. (2000). Positional behavior of *Cercopithecus petaurista*. *Int. J. Primatol.* 21:151–182.

Menard, N., and Vallet, D. (1997). Behavioral responses of Barbary macaques (*Macaca sylvanus*) to variations in environmental conditions in Algeria. *Am. J. Primatol.* 43:285–304.

Menon, S., and Poirer, F. E. (1996). Lion-tailed macaques (*Macaca silenus*) in a disturbed forest fragment: activity patterns and time budget. *Int. J. Primatol.* 17:969–985.

Miller, L. E. (1992). Socioecology of the wedge-capped capuchin monkey (*Cebus olivaceus*) [PhD thesis]. University of California, Davis.

Miller, L. E. (1996). The behavioral ecology of wedge-capped capuchin monkeys (*Cebus olivaceus*). In: Norconk, M. A., Rosenberger, A. L., and Garber, P. A. (eds.), *Adaptative Radiations of Neotropical Primates*. Plenum Press, New York. pp. 271–288.

Miller, L. E. (2002). An introduction to predator sensitive foraging. In: Miller, L. E. (ed.), *Eat or Be Eaten: Predator Sensitive Foraging Among Primates*. Cambridge University Press, Cambridge. pp. 1–20.

Milton, K. (1984). Habitat, diet, and activity patterns of free-ranging woolly spider monkeys (*Brachyteles arachnoides*, E. Geoffroy 1806). *Int. J. Primatol.* 5:491–514.

Mitani, J., and Watts, D. (2001). Why do chimpanzees hunt and share meat. *Anim. Behav.* 61:915–924.

Mitchell, B. J. (1989). Resources, group behavior, and infant development in white-faced capuchin monkeys, *Cebus capucinus* [PhD thesis]. University of California, Berkeley.

Mitchell, C. L. (1990). The ecological basis for female social dominance: a behavioral study of the squirrel monkey (*Saimiri sciureus*) in the wild [PhD thesis]. Princeton University, Princeton, NJ.

Morland, H. S. (1991). Preliminary report on the social organization of ruffed lemurs (*Varecia variegata variegata*) in a northeast Madagascar rainforest. *Folia Primatol.* 56:157–161.

Nagy, K. A., and Milton, K. (1979). Energy metabolism and food consumption by wild howler monkeys (*Alouatta palliata*). *Ecology* 60:475–480.

Newton, P. (1992). Feeding and ranging patterns of forest hanuman langurs (*Presbytis entellus*). *Int. J. Primatol.* 13:245–286.

Noe, R., and Sluijter, A. A. (1995). Which adult male savanna baboons form coalitions. *Int. J. Primatol.* 16:77–106.

Oates, J. F. (1994). The natural history of African colobines. In: Davies, A. G., and Oates, J. F. (eds.), *Colobine Monkeys: Their Ecology, Behaviour and Evolution*. Cambridge University Press, Cambridge. pp. 75–128.

O'Brien, T. G., and Kinnaird, M. G. (1997). Behavior, diet, and movements of the Sulawesi crested black macaque (*Macaca nigra*). *Int. J. Primatol.* 18:321–351.

Overdorff, D. J. (1991). Ecological correlates to social structure in two prosimian primates: *Eulemur fulvus rufus* and *Eulemur rubriventer* in Madagascar [PhD thesis]. Duke University, Durham, NC.

Passamani, M. (1998). Activity budget of the Geoffroy's marmoset (*Callithrix geoffroyi*) in an Atlantic Forest in southeastern Brazil. *Am. J. Primatol.* 46:333–340.

Peres, C. A. (1986). Costs and benefits of territorial defense in golden lion tamairns (*Leontopithecus rosalia*) [MA thesis]. University of Florida, Gainesville.

Peres, C. A. (1991). Ecology of mixed-species groups of tamarins in Amazonian terra firme forests [PhD thesis]. Cambridge University, Cambridge.

Peres, C. A. (2000). Territorial defense and the ecology of group movements in small-bodied neotropical primates. In: Boinski, S., and Garber, P. A. (eds.), *On the Move: How and Why Animals Travel in Groups*. University of Chicago Press, Chicago. pp. 100–123.

Pruetz, J. D. (1999). Socioecology of adult female vervet (*Cercopithecus aethiops*), and patas monkeys (*Erythrocebus patas*) in Kenya: food availability, feeding competition, and dominance relationships [PhD thesis]. University of Illinois, Urbana.

Pruetz, J. D., and Isbell, L. A. (2000). Correlations of food distribution and patch size with agonistic interactions in female vervets (*Chlorocebus aethiops*) and patas monkeys (*Erythrocebus patas*) living in simple habitats. *Behav. Ecol. Sociobiol.* 49:38–47.

Range, F., and Noe, R. (2002). Familiarity and dominance relations amoung female sooty mangabeys in the Taï National Park. *Am. J. Primatol.* 56:137–153.

Ranta, E., Peuhkuri, N., Laurila, A., Rita, H., and Metcalfe, N. B. (1996). Producers, scroungers and foraging group structure. *Anim. Behav.* 51:171–175.

Richard, A. F. (1978). *Behavioral Variation: Case Study of a Malagasy Lemur*. Bucknell University Press, Lewisburg.

Rilling, J. K., Gutman, D. A., Zeh, T. R., Pagnoni, G., Berns, G. S., and Kilts, D. (2002). A neural basis for social cooperation. *Neuron* 35:395–405.

Rodman, P. S. (1973). Population composition and adaptative organization among orang-utans of the Kutai Reserve. In: Michael, R. P., and Crook, J. H. (eds.), *Comparative Ecology and Behavior of Primates*. Academic Press, New York. pp. 171–209.

Rylands, A. B. (1982). The behaviour and ecology of three species of marmosets and tamarins (Callitrichidae, Primates) in Brazil [PhD diss.]. University of Cambridge, Cambridge.

Schaller, G. B. (1963). *The Mountain Gorilla*. University of Chicago Press, Chicago.

Silk, J. (2002a). Introduction. What are friends for? The adaptive value of social bonds in primate groups. *Behaviour* 139:173–176.

Silk, J. (2002b). Using the "F"-word in primatology. *Behaviour* 139:421–446.

Silk, J. B., Alberts, S. C., and Altmann, J. (2003). Social bonds of female baboons enhance infant survival. *Science* 302:1231–1234.

Silver, S. C., Ostro, L. E. T., Yeager, C. P., and Horwich, R. (1998). Feeding ecology of the black howler monkey (*Alouatta pigra*) in northern Belize. *Am. J. Primatol.* 45:263–279.

Smith, C. C. (1977). Feeding behavior and social organization in howling monkeys. In: Clutton-Brock, T. H. (ed.), *Primate Ecology*. Academic Press, San Francisco. pp. 97–129.

Smuts, B. B. (1985). *Sex and Friendship in Baboons*. Aldine de Gruyter, New York.

Snowdon, C. (2004). Affiliative processes and male primate social behavior. Paper presented at the Annual Meeting of the American Association for the Advancement of Science, Seattle, WA, February 2004.

Stephens, D. W., McLinn, C. M., and Stevens, J. R. (2002). Discounting and reciprocity in an iterated prisoner's dilemma. *Science* 298:2216–2218.

Sterck, E. A., Watts, D. P., and van Schaik, C. P. (1997). The evolution of female social relationships in primates. *Behav. Ecol. Sociobiol.* 41:291–310.

Steudel, K. (2000). The physiology and energetics of movement: effects on individuals and groups. In: Boinski, S., and Garber, P. A. (eds.), *On the Move: How and Why Animals Travel in Groups*. University of Chicago Press, Chicago. pp. 9–23.

Stevenson, P. R. (1998). Proximal spacing between individuals in a group of woolly monkeys (*Lagothrix lagotricha*) in Tinigua National Park, Colombia. *Int. J. Primatol.* 19:299–312.

Stoner, K. E. (1996). Habitat selection and seasonal patterns of activity and foraging of mantled howling monkeys (*Alouatta palliata*) in northeastern Costa Rica. *Int. J. Primatol.* 17:1–30.

Strier, K. B. (1986). The behavior and ecology of the woolly spider monkey, or muriqui (*Brachyteles arachnoides* E. Goeffroy 1806) [PhD thesis]. Harvard University, Cambridge, MA.

Strier, K. B. (1987). Ranging behavior of woolly spider monkeys. *Int. J. Primatol.* 8:575–591.

Strum, S. C. (1982). Agonistic dominance in male baboons: an alternative view. *Int. J. Primatol.* 3:175–202.

Strum, S. C. (2001). *Almost Human: A Journey into the World of Baboons.* University of Chicago Press, Chicago.

Sussman, R. W. (1992). Male life history and intergroup mobility among ringtailed lemurs (*Lemur catta*). *Int. J. Primatol.* 13:395–413.

Sussman, R. W., Andrianasolondraibe, O., Soma, T., and Ichino, S. (2003). Social behavior and aggression among ringtailed lemurs. *Folia Primatol.* 74:168–172.

Sussman, R. W., and Garber, P. A. (2004). Rethinking sociality: cooperation and aggression among primates. In: Sussman, R. W., and Chapman, A. (eds.), *The Origin and Nature of Sociality.* Aldine de Gruyter, New York. pp. 161–190.

Symington, M. M. (1988a). Demography, ranging patterns and activity budgets of black spider monkeys (*Ateles paniscus chamek*) in the Manu National Park, Peru. *Am. J. Primatol.* 15:45–67.

Symington, M. M. (1988b). Food competition and foraging party size in the black spider monkey (*Ateles paniscus chamek*). *Behaviour* 105:117–134.

Symington, M. M. (1990). Fission–fusion social organization in *Ateles* and *Pan. Int. J. Primatol.* 11:47–61.

Tattersall, I. (1977). Ecology and behavior of *Lemur fulvus mayottensis* (Primates, Lemuriformes). *Anthropol. Pap. Am. Mus. Nat. Hist.* 52:195–216.

Taylor, S. E., Cousino, L., Klein, B., Gruenewals, T. L., Gurung, R. A. R., and Updegraff, J. A. (2000). Biobehavioral responses to stress in females: tend-and-befriend, nor fight-or-flight. *Psychol. Rev.* 107:411–429.

Teichroeb, J. A., Saj, T. L., Paterson, J. D., and Sicotte, P. (2003). Effect of group size on activity budgets of *Colobus vellerosus* in Ghana. *Int. J. Primatol.* 24:743–758.

Teleki, G. (1981). The omnivorous diet and eclectic feeding habits of chimpanzees in Gombe National Park, Tanzania. In: Harding, R. S. O., and Teleki, G. (eds.), *Omnivorous Primates: Gathering and Hunting in Human Evolution.* Columbia University Press, New York. pp. 303–343.

Terborgh, J., and Janson, C. H. (1986). The socioecology of primate groups. *Annu. Rev. Ecol. Syst.* 17:111–135.

Treves, A. (1998). The influence of group size and near neighbor on vigilance in two species of arboreal primates. *Behaviour* 135:453–482.

Treves, A., and Chapman, C. A. (1996). Conspecific threat, predation avoidance, and resource defense: implications for grouping in langurs. *Behav. Ecol. Sociobiol.* 39:43–53.

Tuomisto, H., Ruokolainen, K., and Yli-Halla, M. (2003). Dispersal, environment, and floristic variation of western Amazonian forests. *Science* 299:241–244.

van Schaik, C. P. (1983). Why are diurnal primates living in groups? *Behaviour* 87:120–122.

van Schaik, C. P. (1989). The ecology of social relationships amongst female primates. In: Standon, V., and Foley, R. A. (eds.), *Comparative Socioecology: The Behavioural Ecology of Humans and Other Animals.* Blackwell Publishing, Oxford. pp. 195–218.

van Schaik, C. P., and Horstermann, M. (1994). Predation risk and the number of adult males in a primate group: a comparative test. *Behav. Ecol. Sociobiol.* 35:261–272.

van Schaik, C. P., and van Hooff, J. A. R. A. M. (1983). On the ultimate causes of primate social systems. *Behaviour* 85:91–117.

Vasey, N. (1997). Community ecology and behavior of *Varecia variegata* and *Lemur fulvus albifrons* on the Masoala Peninsula, Madagascar [PhD thesis]. Washington University, St. Louis.

Warren, R. D., and Crompton, R. H. (1998). Diet, body size and the energy costs of locomotion in saltatory primates. *Folia Primatol.* 69(suppl. 1):86–100.

Waterland, R. A., and Jirtle, R. L. (2003). Transposable elements: targets for early nutritional effects on epigenetic gene regulation. *Mol. Cell. Biol.* 23:5293–5300.

Watts, D. (1988). Environmental influences on mountain gorilla time budgets. *Am. J. Primatol.* 15:195–211.

Watts, D. P. (1998). Chimpanzee male aggression and sexual coercion at Ngogo, Kibale National Park, Uganda. *Am. J. Phys. Anthropol.* 26(suppl.):227.

Watts, D. P. (2002). Reciprocity and interchange in the social relationships of wild male chimpanzees. *Behaviour* 139:343–370.

Wehncke, E. V., Hubbell, S. P., Forester, R. B., and Dalling, J. W. (2003). Seed dispersal patterns produced by white-faced monkeys: implications for the dispersal limitation of neotropical tree species. *J. Ecol.* 91:677–685.

Wilson, D. S. (2000). Animal movement as a group-level adaptation. In: Boinski, S., and Garber, P.A. (eds.), *On the Move: How and Why Animals Travel in Groups.* University of Chicago Press, Chicago. pp. 238–258.

Windfelder, T. L., and Lwanga, J. S. (2002). Group fission in red-tailed monkeys (*Cercopithecus ascanius*) in Kibale National Park, Uganda. In: Glenn, M. E., and Cords, M. (eds.), *The Guenons: Diversity and Adaptation in African Monkeys.* Plenum Press, New York. pp. 147–159.

Wittig, R. M., and Boesch, C. (2003). Food competition and linear dominance hierarchy among female chimps of the Taï National Park. *Int. J. Primatol.* 24:847–868.

Wrangham, R. W. (1980). An ecological model of female-bonded primate groups. *Behaviour* 75:262–300.

Zhang, S. Y. (1995). Activity and ranging patterns in relation to fruit utilization by brown capuchins (*Cebus apella*) in French Guiana. *Int. J. Primatol.* 16:489–508.

Zuberbuhler, K. (2004). In search of the evolutionary impact on predation in primate behavior. *Am. J. Primatol.* 63:37–38.

40

Social Learning in Monkeys and Apes
Cultural Animals?
Christine A. Caldwell and Andrew Whiten

INTRODUCTION

Claims of cultural behaviors in nonhuman primates are becoming ever more common, based on both observational field data and experimental approaches. However, skepticism is still expressed about whether the behavioral variation observed in wild primates is maintained through social learning (Galef 1992; Heyes 1993; Laland and Hoppitt 2003; Tomasello 1990, 1994). Why is such skepticism still widespread in the face of an ever-increasing number of empirical studies on the topic? We address this fundamental question of whether or not there is evidence for what we might label "culture" in apes and monkeys. To do this, we shall review and evaluate several of the more prevalent approaches which have been applied to the investigation of culture in nonhuman primates.

Prior to examining the evidence, however, it is worth emphasizing the significance of the question of whether there is evidence of culture in nonhuman species. To begin with, culture is one of the key traits (alongside others such as language and tool use) that have been proposed, at one time or another, to be unique to humans. To describe an animal as possessing culture challenges traditional preconceptions about humankind's position in relation to the rest of the animal kingdom.

Culture has also fascinated scientists in its role as a nongenetic evolutionary process (Blackmore 1999, Bonner 1980, Mesoudi et al. 2004). Cultural transmission offers another, faster-operating method by which traits can spread, which has to date been much less intensely studied than genetic transmission.

These considerations make the study of culture in any animal important, but there are additional reasons that primates have been the focus of an unusually high proportion of such research. One is simply that some of the earliest, and certainly the most widely cited, examples of apparently cultural behaviors in nonhumans came from primates. Sweet-potato washing by Japanese macaques, *Macaca fuscata* (Itani and Nishimura 1973, Kawai 1965, Kawamura 1959), is the classic example. The macaques on Japan's Koshima Islet were regularly provisioned with sweet potatoes, and one particular individual, Imo, developed a tendency to wash her potatoes prior to consumption. Over several years, this behavior spread

until most of the rest of the group were also regularly washing their potatoes. Shortly afterward, Jane Goodall (1973) suggested that many of the behaviors exhibited by chimpanzees, *Pan troglodytes*, at Gombe in Tanzania were cultural variants, clearly distinct from behaviors exhibited at other field sites. Further, and more subtle, variants have been documented as observations have continued to the present time. Whiten et al. (1999) noted that some of the differences between communities represented alternative versions of otherwise similar patterns. Chimpanzees from the Taï Forest in the Ivory Coast use a short stick to fish ants from their nests, whereas those from Gombe use a much longer stick and a more efficient bimanual technique to achieve the same goal. Through such observations, primatologists have understandably developed a long-standing tradition of their own: a fascination with social learning.

Primates may also be particularly likely, in comparison with other animals, to exhibit culture. We humans are undeniably cultural; and it seems plausible that our closest evolutionary relatives would be the most likely to show some semblance of culture—the question being, to what extent? This close relationship to humans also makes primates particularly good research subjects in terms of providing insight into the evolutionary history of human culture. Monkeys and apes also share with humans several traits which may be associated with a propensity for cultural transmission. A long life span and lengthy period of infancy with high levels of parental investment mean that an ability to learn socially could be particularly valuable in these animals (Roper 1986). Indeed, these features are sometimes proposed to have evolved alongside cultural tendencies because, once a capacity for social learning exists, the fitness of offspring can be increased by maximizing the opportunities for learning from elder relatives (Barrett-Lennard et al. 2001).

Understanding how wild primates develop their full behavioral repertoire may also prove crucial to their protection in an increasingly hostile world. Around half of our 250 primate species are considered to be of conservation concern by the World Conservation Union (IUCN), and recent assessments suggest that over 30% of all primates should be considered seriously threatened with extinction. Although a large number of successful captive breeding programs exist for many endangered species, reintroductions are notoriously unsuccessful (Beck 1994). Captive-bred individuals appear to lack the survival skills possessed by their wild-born counterparts (Beck et al. 2002, Stoinski et al. 2003). If, under natural conditions, a large proportion of learning is dependent on social interaction with skilled conspecifics, then this needs to be taken into account when planning reintroduction programs (Beck 1995, Box 1991, Custance et al. 2002).

DEFINING CULTURE

Culture has been defined in many different ways, ranging from the very broad and inclusive to the extremely narrow and restrictive (Kroeber and Kluckhorn 1963, McGrew 1998, Rendell and Whitehead 2001, Whiten et al. 1999). Analyzing the possible points of reference for culture is beyond the scope of the present chapter (see Whiten et al. 2003 for a fuller discussion), but those interested in culture from a nonhuman perspective typically agree on two central criteria (Boesch and Tomasello 1998, Fragaszy and Perry 2003, Laland and Hoppitt 2003, McGrew 1998, Russell and Russell 1990, Whiten et al. 1999): cultural behaviors are those which are *(1)* specific to members of a group (*group* is used intentionally loosely here as cultural variation could occur on any social

scale) and *(2)* transmitted via some form of social learning. We have adopted this as our definition for this chapter.

The first of the above criteria, group-specific behavior, can be relatively easy to identify. The examples mentioned above of the potato washing of the Koshima macaques and of the chimpanzee tool use behaviors at Taï and Gombe illustrate admirably that primate behavior can show both intriguing uniformity within groups as well as striking variability between them. However, the second criterion is more elusive: how do we know for sure that behaviors have been socially learned? Galef (1990) has pointed out that geographic variation could arise from a number of possible sources, these being: *(1)* heritable differences that influence behavior, *(2)* environmental differences between the habitats of the populations which result in systematic differences in the reinforcement individuals receive for engaging in various behaviors, and *(3)* differences in the behavior of population members that influence the behavioral development of new recruits (i.e., some form of social transmission). Teasing these alternative explanations apart can prove quite problematic.

We shall describe and evaluate three principal approaches that have been taken to the question of whether apparent local traditions really represent evidence for culture in monkeys and apes. These are depicted in Figure 40.1. As is shown in

	Direct manipulation of social learning conditions	Direct study of natural behaviors
Systematic surveys of multiple field sites	No	Yes
Analysis of diffusion data	No	Yes
Experiments	Yes	No (except cross-fostering/ translocation)

	Instrumental behaviors, e.g. foraging techniques, tool use	Non-functional behaviors, e.g. social and play behaviors
Geographically separated groups	Possible genetic confound	Possible genetic confound
	Possible confound from environmental shaping	Confound from environmental conditions unlikely
Sympatric groups	Genetic confound unlikely	Genetic confound unlikely
	No confound with environmental conditions	No confound with environmental conditions

Systematic Field Site Surveys: Behavior patterns are compared between groups either separated geographically or by association patterns. Confounds listed are those possible given basic comparisons of static patterns. Particular examples of these types of research (e.g. Whiten et al., 1999) usually include further information on genetics and environmental influences to help exclude these factors.

	Power to distinguish between alternative mechanisms of social learning	Power to determine capacity to support inter-population differences in behavior
Cross-fostering/ translocation	No	Yes – involves direct study of natural behaviors
Single generation simulation	Yes	No
Multi-generation simulation	Yes	Probably

Social Learning Experiments: Social conditions are directly manipulated either by moving animals to new environments and/or social groups, or by simulating exposure to a novel trait

Figure 40.1 Where is the evidence for culture in primates? A diagrammatic summary of the different approaches to assessing culture. It is difficult to simultaneously study primate social learning and natural behaviors in a direct fashion, as the types of experiments that permit these sorts of manipulation are difficult to carry out in these species. Evidence for culture in primates therefore remains, up to a point, circumstantial, based on the existence of interpopulation differences in behavior that cannot be readily attributable to environmental or genetic factors, and on the results of experiments that show primates can learn from others.

this illustration, the first two approaches, systematic field site surveys and analysis of diffusion patterns, look directly at the naturally expressed behaviors that have been proposed to be socially learned in the wild. The final approach, experimental research, involves direct manipulation of an individual's experiences. Such research allows for definite conclusions to be made about whether or not social learning is involved. However, this type of research may not necessarily reveal much about natural behaviors. Experimental research can focus on natural behaviors, involving transfer of individuals between social groups or alternative environmental conditions; but these manipulations are extremely difficult. Experiments with primates tend only to *simulate* the type of behaviors that might be socially learned in the wild.

SYSTEMATIC SURVEYS OF MULTIPLE FIELD SITES

Ape Field Site Surveys

One of the most influential approaches to date for addressing the issue of primate culture is the systematic field site survey. Such approaches, which compile data from multiple social groups of the same species, have existed in limited form for many years. For example, Goodall's (1973) classic chapter on chimpanzee cultural behavior includes tables detailing the presence or absence of certain behaviors identified at Gombe and proposed by Goodall to be socially learned at several other locations in Africa. McGrew (1992) and Boesch (1996) include ever-larger tables, comparing behaviors observed at different chimpanzee field sites. This method of looking for behavioral differences between groups has been referred to in the literature by a variety of different labels, including *group contrast* (Fragaszy and Perry 2003), *regional contrast* (Dewar 2003), and the *method of elimination* (van Schaik 2003). Such between-group variation provides perhaps the most intuitively compelling evidence of culture in any animal, although as mentioned above, variation could arise from factors other than social learning. The validity of this method therefore relies on the accumulation of data in order to, as far as possible, assess the likelihood of either genetic or environmental influence.

Capitalizing on the potential power of this method, therefore, more recent attempts have taken a more systematic approach. Whiten and collaborators (1999, 2001) pooled data from the seven longest-running chimpanzee field sites. The first stage of this process was to collate a list of proposed cultural variants, that is, behaviors the researchers had observed at their field site that they suspected might be transmitted socially. The leaders of each of the field sites involved were then asked to summarize their knowledge about their chimpanzees' performance of each of these behaviors. For each of the behaviors, there were six possible options: *customary* (observed in all or most able-bodied members of at least one agesex class); *habitual* (not customary but observed repeatedly in several individuals, consistent with some degree of social transmission); *present* (neither customary

nor habitual but clearly identified); *absent* (never recorded); *ecological explanation* (never recorded but absence was explicable because of a local ecological feature); and *unknown* (never recorded but possibly due to an inadequacy of relevant observational opportunities).

Of the 65 original candidate behaviors, only those that were found to be either customary or habitual in at least one site and absent from at least one other (without an apparent ecological explanation) joined the list of behaviors proposed to be cultural variants. A total of 39 different behaviors remained following this process (see Fig. 40.2 for a graphical summary of behaviors found at the different field sites). These spanned a range of contexts, some social, such as grooming and courtship behaviors, and some instrumental, such as foraging behaviors and food-processing skills. It

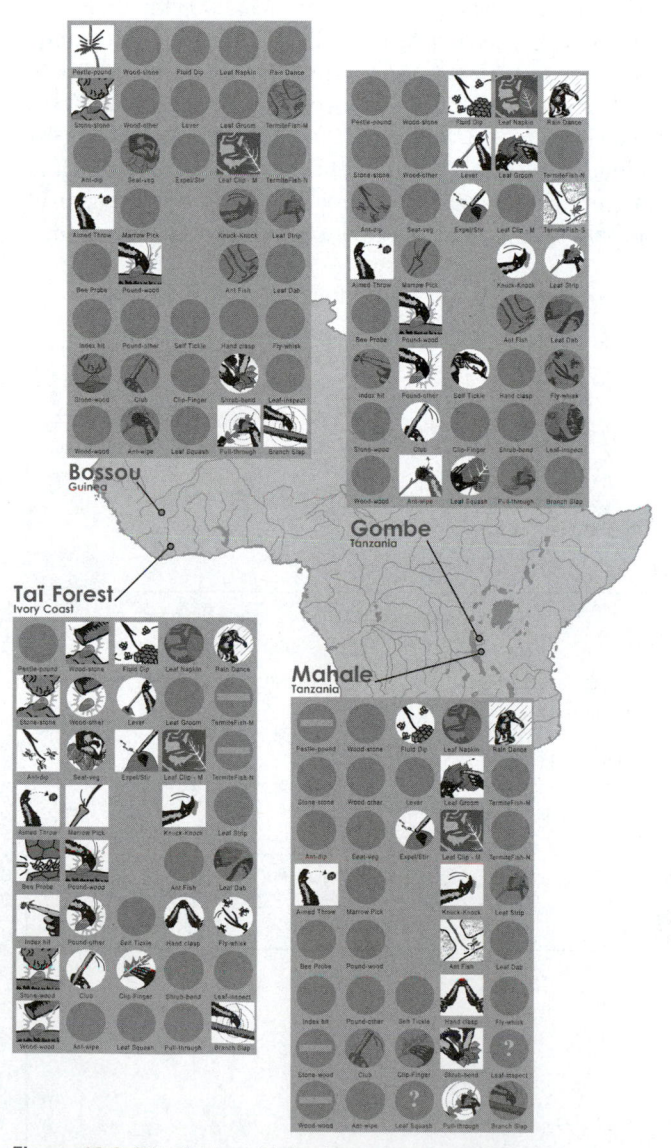

Figure 40.2 Distribution of behavior patterns across four African study sites. Square icons, customary; circular icons on white background, habitual; square and circular icons on dark backgrounds, present; no icon or blank circular icon, absent; horizontal bar, absent with ecological explanation; question mark, answer uncertain (see Whiten et al. 1999).

remains possible that in these geographically separated populations, inconspicuous (possibly extremely subtle) ecological differences could be responsible for the observed variation in some of the food-related behaviors. However, for social behaviors, like handclasp grooming, this is much less of a concern. Furthermore, Whiten et al. (1999) have gone some way toward ruling out genetic influence for some of the behaviors. Nut-cracking behavior, for example, is observed only in the West African field sites, the Sassandra-N'Zo River forming a boundary between those groups that exhibit the behavior and those that do not (Boesch et al. 1994). Given that the very same subspecies, *Pan troglodytes verus*, is found on either side of the river, the abruptness with which the behavior terminates is unlikely to be attributable to divergent genetic makeup.

This same survey methodology was later employed by researchers at six different orangutan (*Pongo pygmaeus*) field sites in Sumatra and Borneo (van Schaik et al. 2003), identifying a total of 24 putative cultural variants. Again, the behaviors were wide-ranging, including social communicatory signals, foraging techniques, and tool-use behaviors. van Schaik et al. (2003) also investigated the relationship between behavioral similarity and geographical proximity. Sites with similar patterns of cultural variation, they hypothesized, would also be ones that were geographically close, supporting a model of innovation followed by social diffusion of the behavior patterns. The cultural difference between any pair of sites was calculated as a percentage of all customary and habitual local variants, and this was found to be significantly correlated with the distance between the sites, suggesting that there was diffusion between localities. This particular finding does not contribute a great deal toward eliminating the possible alternatives of environmental or genetic influence as we would expect ecological and genetic divergence also to correlate with distance. However, the researchers also found that the variety of cultural behaviors was related to the opportunities available for social learning. Repertoire size was measured as the number of customary and habitual variants expressed at each site. This was found to be significantly related to the percentage of time that nondependent animals spent in association. This result seems much less easily dismissed as a potential confound of either environment or genetics. The correlation with association data allowed the researchers to look more directly at social influence on behavior, rather than merely indirectly, through having ruled out the other plausible alternatives.

Monkey Field Site Surveys

Similar methods have now also been applied to the behavior of monkeys as well as apes. Panger et al. (2002) pooled data from three long-term white-faced capuchin (*Cebus capucinus*) field sites in Costa Rica. Their approach (based on the logic that the majority of chimpanzee and orangutan variants involved foraging techniques) was to compile a list of the foods from the diet of this species and compare how the same food was processed by the capuchins from the different sites. Out of the 61 foods that formed part of the diet at more than one site, 20 were found to be processed in a different manner between sites. *Pithecellobium saman* fruits, for example, a species that is broken open by the monkeys for the beetle larvae within, is processed by tapping (rhythmic tapping with the fingertips) at Lomas Barbudal, whereas the capuchins at Palo Verde use a *fulcrum* behavior (the application of force on an object working against a substrate). Individuals from the Santa Rosa site use a variety of techniques with this particular food species, including rubbing and pounding, as well as the fulcrum behavior observed at Palo Verde.

Even more persuasively, data from the Palo Verde field site allowed for an analysis of the relative levels of association of those dyads that shared particular processing techniques. Pairs of animals that had behaviors in common had higher proximity scores than nonmatched pairs. As illustrated in Fig. 40.1, when the groups being compared are sympatric, rather than geographically separated, it makes it considerably less likely that the contrast is produced by environmental shaping alone. Given that the foods being processed are liable to be precisely the same species (possibly even fruits or leaves picked from the very same plant), differences in, for example, hardness or texture are much less likely to be responsible for the differences in behavior within sites than they would be across sites.

It is possible, however, that those individuals that associate closely are typically more closely related compared with pairs that rarely associate. These association measures could therefore be confounded by genetic factors. When dealing with sympatric groups, it should nonetheless be possible, in principle, to assess genetic influence in relation to social factors. If relatedness is known (or can be determined through, e.g., deoxyribonucleic acid [DNA] testing), this can be directly compared with the association data to determine which is more closely correlated with measures of behavioral similarity. If the association data explain a greater amount of the variation in behavior compared with the relatedness data, then a social learning interpretation is supported over a genetic one.

As well as foraging behaviors, a separate analysis (Perry et al. 2003) of possible social traditions in capuchin monkeys was carried out using data from the same three field sites plus one additional site, Curú. As with the previous surveys of this kind, the researchers were able to identify behaviors that were present in certain social groups and absent in others. As discussed in relation to the chimpanzee field site data, social behaviors are less susceptible to the criticism that they could arise as an artifact of environmental conditions. There is nonetheless a slim possibility that, when comparing groups at different field sites, local conditions in some way affect these behaviors (e.g., display behaviors might need to be louder and more conspicuous at more densely forested sites).

However, in addition to having this information about static group differences, the researchers were able to document the patterns of emergence, expansion, and decline of these behaviors within the groups. The three social behaviors under investigation were handsniffing (bouts of sniffing of another individual's hand), body part sucking (sucking of the fingers, toes, ears or tail of another individual), and "games" (which involved the retrieval of an object from another individual's mouth). The dyadic nature of these activities meant that researchers could also chart the probable transfer of the behavior between individuals in the group. When individuals were first observed playing one of the "games," for example, it was assumed that they had acquired the behavior from the particular monkey with which they were interacting. In this way, it was possible to document multiple links in the transmission chain, where former partners of the original innovator went on to engage in game playing with other, previously naive individuals. It is difficult to see how one could explain such diffusion patterns as anything other than some kind of social transmission.

ANALYSIS OF DIFFUSION DATA

When comparing static patterns of group behavior, we are restricted to making post-hoc inferences about the learning processes that must have produced them. However, in some cases, it is possible to record the diffusion process as it is taking place. Although it is generally not possible to determine from exactly which individual an animal has acquired a trait (c.f. Perry et al. 2003 described above), some researchers have used data on the rate of spread of a novel behavior through a population in an attempt to determine whether or not social transmission was likely involved.

The analysis of diffusion data is an approach that has been used by a number of researchers to assess social learning in animals (Galef 1990; Lefebvre, 1995a,b; Reader and Laland 2000; Rendell and Whitehead 2001). The typical assumption is that, if a behavior is socially learned, the number of individuals performing the behavior will increase at an accelerating rate. This is based on the logic that the more potential demonstrators (skilled individuals) there are in a group, the greater the chance that a naive individual will learn the new behavior.

Galef (1990) initially used this reasoning to argue that the potato-washing behavior of the Koshima macaques was unlikely to have been socially learned. He plotted the number of individuals reported to wash potatoes at a given time against the year of data collection. The resulting graph showed a relatively steady increase in the number of potato washers over time (see Fig. 40.3). This led Galef (1990) to suggest that the potato-washing behavior was in fact being individually learned by each monkey through trial and error.

Since Galef's (1990) proposal, more comprehensive mathematical approaches have been used to support similar arguments. Lefebvre (1995a) analyzed the diffusion patterns of a number of behavioral innovations that had been documented in primates. A number of theoretical models of trait diffusion, under conditions of social transmission, exist (e.g., Boyd and Richerson 1985, Cavalli-Sforza and Feldman 1981). Lefebvre (1995a) compared such models against the observed cumulative distributions of the reported behaviors. A total of five theoretical models were utilized, three of which (logistic, hyperbolic sine, and positive exponential equations) featured the accelerating functions assumed to be consistent with social transmission. The other two functions were linear (representing a constant rate of increase of skilled individuals) and logarithmic (featuring a decelerating rate of learning).

Lefebvre (1995a) analyzed data on Japanese macaque innovations. These included the potato-washing behavior as well as seven others, such as the adoption of novel foods into the diet. For four out of these eight behaviors, the best-fit model was one of the accelerating functions. The remaining four were better accounted for by the linear and decelerating equations. Surprisingly, potato washing, the very behavior that had been identified by Galef (1990) as having a distribution inconsistent with social learning, was found by Lefebvre (1995a) to correspond most closely to the (accelerating) logistic function. Lefebvre (1995a) attributes this primarily to his inclusion of a key extra data point, omitted by Galef (1990). The contrasting data sets can be seen in Fig. 40.3.

Although the investigation of cumulative distributions has become popular in the literature, such data should be treated with extreme caution. Quite apart from the fact that

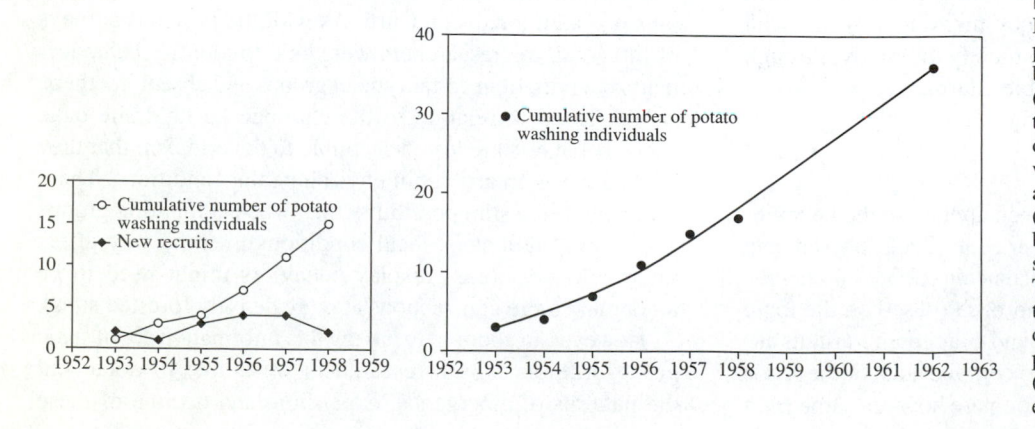

Figure 40.3 Data on the number of Koshima macaques engaging in sweet potato-washing behavior, as plotted by Galef (1990, *left*) and Lefebvre (1995a, *right*). Galef (1990) suggests that the relatively consistent number of new recruits per year to the potato-washing behavior is indicative of asocial learning processes. Lefebvre (1995a) finds an accelerating function best fits the cumulative data, suggesting social learning processes. The contrasting interpretations appear to be primarily attributable to minor differences in the methods used to plot the data reported by Kawai (1965), including the inclusion of an extra data point by Lefebvre (1995a).

analyses may be extremely fragile (dependent even on the inclusion or exclusion of a single data point, as illustrated by the contrasting interpretations of Galef [1990] and Lefebvre [1995a] noted above), there are good reasons why counter intuitive diffusion patterns might arise. Social learning may not necessarily result in accelerating learning curves and, equally, asocial learning may under some circumstances produce accelerating cumulative distributions.

As noted by Lefebvre (1995a) (and more recently by Day et al. 2001 and Laland and Kendal 2003), an S-shaped curve similar to those produced by some acceleratory functions is exactly what would be expected under circumstances of individual learning, given a normal distribution of learning latencies within the population. An accelerating cumulative distribution of learners is therefore not sufficient to conclude that social learning is involved in transmission.

Likewise, the absence of an accelerating curve may not be indicative of an absence of social transmission processes. Laland et al. (1996) have modeled a number of social learning scenarios and note that, under certain circumstances, apparently linear diffusion curves may result. The higher the proportion of skilled individuals in the group, the greater are the opportunities for scrounging (exploiting of food resources found by another individual). Hence, naive individuals may adopt a strategy of scrounging rather than performing the behavior themselves. Opportunities for scrounging behavior would therefore be expected to have a dampening effect on transmission curves. Such effects are certainly liable to operate within primate groups. Fragaszy and Visalberghi (1990) reviewed several of their studies of the learning of novel extractive foraging behaviors within groups of capuchin monkeys (*Cebus apella*) and noted that individuals that scrounged from others typically did not go on to perform the skill themselves.

Furthermore, there may be cases in which a behavior must be learned from a particular demonstrator or demonstrators; otherwise, it will not be learned at all. Identity-dependent social learning such as this would violate the basic assumption (that more demonstrators equal greater opportunities for learning) that underlies the accelerating learning rate models. Some of our own research suggests that such effects may occur in primates. We found that learning of an artificial extractive foraging task in marmosets, *Callithrix jacchus*, was dependent on close interaction with a skilled conspecific demonstrator (Caldwell and Whiten 2003). Subjects that simply observed the demonstration from even a very short distance away were little more successful than those with no pretest experience, as were those that were exposed to the apparatus in a preopened state (equivalent to coming across the products of another individual's efforts). Only those that could interact with the demonstrator had high rates of success in terms of learning the behavior.

This suggests that a tolerant relationship between skilled and naive individuals may be a crucial factor in determining whether or not a particular behavior is transmitted, a notion that has been proposed in some of the previous literature on primate social learning (Coussi-Korbel and Fragaszy 1995, Russon and Galdikas 1995). Such effects also seem consistent with the available field data. Panger et al.'s (2002) research shows that individuals do indeed appear to pick up the particular behavioral variants displayed by their close associates, and the sweet potato-washing behavior is well known to have spread first to Imo's playmates and mother and then to her playmates' mothers (Tomasello and Call 1997).

Currently, therefore, there is little that can be directly determined from cumulative distribution curves as accelerating rates of transmission are not necessarily indicative of social learning processes and, likewise, nonaccelerating diffusion curves may be no indication of asocial learning. Research using this approach to assessing culture will progress only if, at least initially, diffusion data are tied in with other methods of assessing learning processes so that there is some external validation of the meaning of particular cumulative distributions. One suggestion, made by Laland and Kendal (2003), is that techniques using association data (e.g., see Monkey Field Site Surveys, above) could provide a particularly useful metric to verify the presence of social transmission. Such approaches would certainly go a long way toward determining how much cumulative distributions can reveal about learning processes.

EXPERIMENTAL STUDIES OF LEARNING MECHANISM

Experimental Studies of Natural Behaviors

For some group-specific behaviors, it is possible to carry out experiments which can demonstrate conclusively the role of social interaction in acquisition. For example, in Terkel's classic research on the pinecone-stripping behavior of black rats (Aisner and Terkel 1992; Terkel 1995, 1996), it was possible to demonstrate that this skill was maintained in the population through social interaction, by carrying out a cross-fostering experiment.

However, as pointed out by Laland and Hoppitt (2003), under circumstances in which behavioral variation has been found between two geographically separated populations, even cross-fostering experiments tell only half the story. Genetic factors can be conclusively excluded, but the role of environmental shaping is still uncertain. Translocation experiments are required: given two behaviorally distinct populations, each group in its entirety is transferred from its own environment to the former habitat of the other. Truly cultural behaviors would persist within the populations, whereas those that were artifacts of environmental shaping would be expected to die out or, indeed, mutate into the form previously observed in the former residents.

On this basis, Laland and Hoppitt (2003) suggest that the best evidence of culture to date, in any animal, comes from fish as these are the only species with which such experiments have been conducted. Helfman and Schultz (1984) found that French grunts, *Haemulon flavolineatum*, transferred into an existing population adopted the same schooling sites and migration routes as the residents, whereas a control group, introduced after the resident population was removed, did not show these same preferences.

These kinds of experimental design, although extremely informative, are unfortunately of restricted application within the field of primate behavior. Although such studies could certainly be illuminating on a long-lived, large-bodied, and in many cases endangered animal, cross-fostering and translocation not only would be costly and time-consuming but also would involve taking serious risks with an already threatened population. To date, therefore, experimental research on social learning in primates has involved simulations, rather than direct study, of natural behaviors. These experiments allow researchers to establish that social learning mechanisms exist in these animals, and from this we can infer that some of the group-specific behaviors observed in the wild may be supported by such processes.

The Significance of Imitation

Experimental research on social learning in monkeys and apes dates back to the beginning of the century (e.g., Haggerty 1909; and see reviews in Hall 1963, Tomasello and Call 1997, Whiten and Ham 1992), and many primate species have shown clear evidence of social learning effects under experimental conditions. Studies such as those carried out by Tomasello et al. (1987) and Warden and Jackson (1935) with chimpanzees and rhesus macaques (*Macaca mulatta*), respectively, clearly established that subjects that observed a skilled demonstrator performing a simple action (e.g., a lever press for food or raking in of a food reward) acquired the behavior faster than those that had seen no demonstration.

However, it has been proposed by some researchers that not all social learning mechanisms have the capacity to support stable behavioral traditions such as those proposed to exist in the wild. Whiten and Ham (1992) pointed out that many of the early experiments with primates did not have the power to distinguish the mechanism of imitation from other forms of social learning. In particular, results could readily be attributed to stimulus enhancement (Galef 1988, Spence 1937, Whiten and Ham 1992). Stimulus enhancement is said to have occurred when the presence of an individual merely draws an observer's attention to a particular object or part of an object, thus enhancing the observer's opportunity to learn about the object. The behavior itself is therefore not being learned through observation, although similar behavior patterns may often be the result.

Imitation, by contrast, has been defined as "learning to do an act by seeing it done" (Thorndike 1898, Whiten and Ham 1992). Observation of the behavior is therefore both necessary and sufficient for a naive individual to reproduce it through imitation. For this reason, imitation has been singled out by a number of researchers as being particularly suited, among mechanisms of social learning, to support culture (Boyd and Richerson 1985, Dawkins 1976, Galef 1992, Whiten and Ham 1992). As other social learning processes require some

interaction between the naive individual and the environment for learning to take place, it is assumed that these processes would be much more susceptible to interference from environmental shaping. Hence, it is proposed that these mechanisms could not support stable, socially learned, behavioral variation between groups as learners would typically settle on techniques no different from those they would have adopted under conditions of asocial learning (albeit their arrival at these solutions may be made more likely or happen more rapidly).

The evidence from nonprimate species suggest that this assumption is false however. There is no reason to believe that imitative learning was involved in the examples given above of fish migration-route learning and pinecone opening in rats. It seems therefore that stable behavioral traditions can exist without relying on imitation as a transmission mechanism. However, there may be other reasons that imitation is particularly significant to the topic of culture. A number of authors have stressed the importance of cumulative cultural evolution. If by "culture" we refer to a nongenetic evolutionary process, it is crucial that, as well as involving social transmission, the process includes variation and selection, leading to increased complexity. Tomasello (1990, 2000), Tomasello et al. (1993), and Boyd and Richerson (1996) have stressed the significance of cumulative cultural evolution (or "ratcheting," Tomasello 1990) as the resulting behaviors, accumulated over generations of social learning, could not be invented by any single individual. Human societies provide the clearest examples of cumulative cultural evolution: each generation builds on the knowledge, inventions, and achievements of the previous one (Mesoudi et al. 2004). Our present-day technologies exist only thanks to our ability to understand and make use of the imparted knowledge and artifacts of others.

Whether cumulative cultural evolution can exist in the absence of an ability to imitate is presently a moot point. Heyes (1993) sees no reason why imitation should be particularly crucial to the generation of cumulative behavioral change. However, Boyd and Richerson (1996) make a compelling argument to the contrary: since the essence of cumulative cultural evolution is that it produces behaviors more complex than those that an individual could acquire through asocial learning processes, social learning mechanisms other than imitation probably would not be capable of supporting it. As mentioned above, nonimitative social learning, relying as it does on interaction with the environment, is unlikely to produce behaviors that would not ordinarily arise anyway through individual learning alone.

A ratcheting mechanism is not an aspect of culture which features in every definition (indeed, this would considerably narrow the definition given at the beginning of this chapter). However, Galef (1992) argued for defining culture in such a way that it is homologous, rather than simply analogous, to human culture. This entails both a reliance on imitation as a learning process and the capacity for cumulative cultural evolution. Whether or not the behavioral traditions observed in animals can be described as homologous to human culture may be of little importance to some researchers, but to those interested in the question of the evolution of human culture (and primatologists may form a particularly high proportion of this group), this is clearly a significant issue.

A number of reasons have therefore been proposed regarding the significance of imitation to cultural transmission. Fortunately, the kinds of experiment which researchers have been able to carry out with primates are ideally suited to determining social learning mechanisms (a goal that is extremely difficult to achieve in other types of research, which focus on natural behaviors).

Imitation and the Two-Action Method

There have been a large number of studies of imitation in primates, so we are unable to provide an exhaustive review here (see Whiten et al. 2004 and Visalberghi and Fragaszy 2002 for more thorough reviews of the last decade's literature on imitation in apes and monkeys, respectively). A variety of methodologies have been employed, the most successful being "do-as-I-do" ("Simon says") tests and what have become known as "two-action" experiments. Here, we review the two-action tests as these are most relevant to the question of whether the group-specific behaviors observed in wild primates are socially learned. While the do-as-I-do experiments involve training a subject to respond to a cue to imitate (therefore posing the question of whether they can imitate), two-action experiments rely on spontaneous behaviors (thus revealing whether they actually do imitate; Horowitz 2003, Whiten and Custance 1996). A further reason for choosing to focus on two-action studies is that, within imitation research, the two-action design has become something of a methodological gold standard. The majority of scholars in the field, while often disagreeing on other aspects of research design and interpretation, generally agree on this as a preferred methodology (see review in Caldwell and Whiten 2002; but see Byrne 2002 for an alternative view).

In two-action experiments, subjects each observe one of two possible actions on a particular object. If they subsequently produce behaviors more like the version that they saw demonstrated, it is concluded that imitation must be implicated. Two-action designs can control for other social learning effects, such as stimulus enhancement, as these would not be expected to generate differences between the groups in terms of the particular actions displayed.

Experimental Studies of Imitation in Apes

Whiten et al. (1996) used the two-action design to test for imitative effects in chimpanzees. They designed an "artificial fruit," a box containing a food reward, protected by several defenses (see Fig. 40.4). For each of the defensive locks, there were two alternative methods of release. A pair of bolts could be either twisted or poked out, and a handle could be either lifted out or turned around. Chimpanzee subjects were each

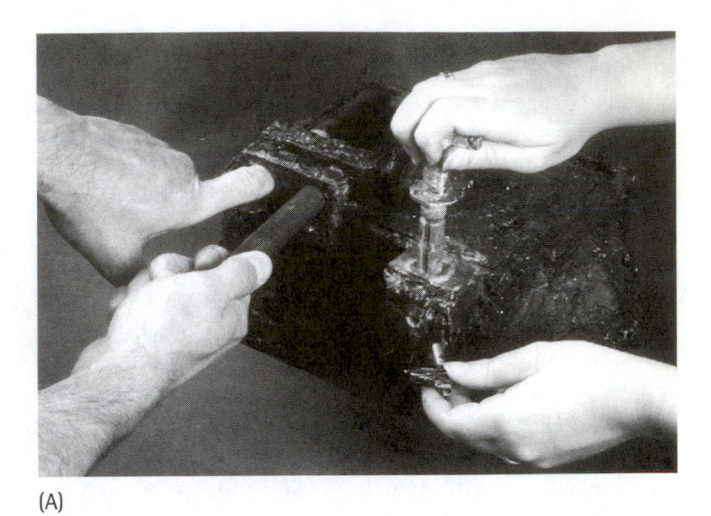

(A)

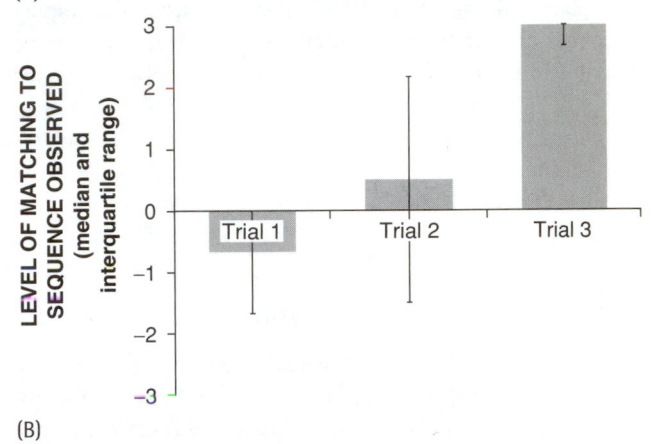

(B)

Figure 40.4 An "artificial fruit" (*above*) with bolts, handle, and pin defenses, which was used to test for imitative tendencies in chimpanzees. The graph shows data from independent coders' ratings (on a 7-point scale) of the closeness of match between the subject's actions and the particular behavior they had seen demonstrated (+3 representing the highest confidence of a sequence that matched what was witnessed and −3 representing the highest confidence of a sequence that was the opposite of what the subject witnessed). Over trials, the chimpanzee subjects showed an increasing tendency to match the behaviors they saw a human demonstrator perform (see Whiten 1998).

shown one of the methods of opening by a human demonstrator. Following the demonstrations, the chimpanzees used actions similar to the versions that they had been shown. This effect showed clear evidence that the chimpanzees were reproducing the details of behaviors, rather than just showing a more generalized stimulus enhancement effect.

The artificial fruit methodology has also been employed to test for imitation in other ape species. Stoinski et al. (2001) found that, as with the chimpanzees, gorilla subjects (*Gorilla gorilla gorilla*) showed a significant tendency to copy the technique they had observed from the human demonstrator. However, in orangutans, results were less clear. Both Custance et al. (2001) and Stoinski and Whiten (2003) used the artificial fruit methodology with orangutan subjects, finding no significant tendency to match the observed behaviors. However, these results contrast strongly with those described by Russon and Galdikas (1993, 1995), who documented many cases of spontaneous imitation in a rehabilitant population of orangutans. Although less systematically collected than is possible in two-action experiments, these observations suggest that the full imitative ability of orangutans may have been underestimated by the current experimental literature.

Research on imitation in apes has also addressed the question of whether the sequence of demonstrated actions is preserved as well as the behaviors themselves. Whiten (1998) was able to test this in chimpanzees using a variation on the artificial fruit methodology. Subjects were again shown one of two different methods of accessing the food reward, although in this version the component elements were ordered differently in each. The chimpanzees showed a significant tendency to unfasten the locks in the order they had seen demonstrated (Fig. 40.4). The imitation of complex sequences of actions may be particularly relevant to the issue of culture in primates as many of the behaviors proposed to be socially learned in the wild are relatively complex, multistage procedures (Byrne and Russon 1998). Experiments that incorporate multiple sequential elements may therefore be extremely revealing in terms of understanding how the natural behaviors are learned.

Experimental Studies of Imitation in Monkeys

Although earlier reviews suggested the possibility of a monkey–ape divide in terms of imitative ability (Visalberghi and Fragaszy 1990, Whiten and Ham 1992), evidence for imitation in monkeys is mounting. Human demonstrators have typically been used in much of the ape imitation research, but this methodology is less suitable for experiments with monkeys, due to the greater discrepancies in body size and morphology. However, recent research has addressed this problem by using trained conspecific demonstrators.

Common marmoset subjects have been tested using the two-action method (Voelkl and Huber 2000). This particular task involved opening photographic film canisters, and two trained marmoset demonstrators were used, one of which removed the lid with its hand while the other used its mouth. Voelkl and Huber (2000) found a significant difference between the two observer groups in the method used to open the canisters: all subjects that saw the hand method used their hands every time, and of the subjects that saw the mouth method, most used their mouth for at least some of the test trials. Similarly, in our research using the artificial fruit methodology (Caldwell and Whiten 2004), marmoset subjects matched some aspects of the actions they saw demonstrated by conspecifics.

The artificial fruit methodology has also been used with capuchin monkeys, *Cebus apella* (Custance et al. 1999). In this experiment, human demonstrations were used as the capuchins were "enculturated" and thus liable to respond well to a human demonstrator. The capuchins produced actions similar to those they saw demonstrated, albeit not perfectly matched. In another two-action study, Fragaszy, Landau,

and Leightly (unpublished, cited in Fragaszy and Visalberghi 2004), also tested for imitation in capuchin monkeys. A juice dispenser which could be operated in two alternative ways (by either turning a wheel or pressing a lever) was introduced to two capuchin groups. Adult animals were provided with the apparatus constrained in such a way that only one method was possible. However, the juveniles, which could enter an adjoining "crèche" area inaccessible to the adults, could access their own version of the apparatus, which could be operated in either fashion. The majority of youngsters in the "wheel" group made use of the wheel and most from the "lever" group used the lever.

CULTURE, CONVERGENCE, AND CONFORMITY

Social Learning and Corruption from Individual Learning

It is clear then, that if it is imitation we are particularly interested in, there is no shortage of evidence from primates. Additionally, the field data clearly show divergent patterns of behavior between different localities and social groups. However, skeptics remain unconvinced that there is evidence for culture in primates (Galef 2003, Laland and Hoppitt 2003). The problem is, essentially, that it is very difficult to show that it is these same mechanisms, which have been identified in experimental research, that support the behavioral variation observed in the field. Field data typically identify social learning only indirectly, through eliminating alternatives, and experimental data (while directly manipulating learning conditions) allow us to conclude only that certain social learning mechanisms are available to a given species. Whether or not these are exploited in the development of natural behaviors is a different matter.

To put forward such an argument might appear to some to be nitpicking to an excessive degree. In the face of such a huge amount of circumstantial evidence, why should one be inclined to draw any other conclusion? In fact, a number of authors have offered persuasive arguments for why interpopulation differences in behavior, such as those observed in chimpanzees, would be unlikely to be the result of the sorts of social learning mechanism identified in experiments.

Galef (1995) makes the point that socially learned behaviors will be either maintained within or extinguished from the repertoire of an individual as a consequence of the reinforcement contingencies that the actor experiences. In other words, asocial learning will interact with the socially learned information, and behaviors will persist only if reinforced. On this basis, therefore, Galef (1995) argues that those behaviors that are maintained in populations (such as those commonly labeled as culture in primates) must be locally adaptive, in the sense that they must result in greater reinforcement relative to alternatives. Such an argument obviously poses a fairly major problem for the interpretation of the contrasting behaviors at different primate field sites. If it is the individually learned reinforcement patterns that determine whether a behavior is maintained in an individual's repertoire (irrespective of whether any form of social learning was involved in the initial acquisition), then the divergent patterns of behavior observed at different field sites almost certainly have their origins in environmental shaping.

Heyes (1993) puts forward a similar argument, proposing that the particular transmission mechanism involved is in fact irrelevant to the question of whether or not a behavior will extinguish. Behaviors learned via imitation, like any other, will be subject to modification from trial-and-error learning. Heyes (1993) cites Galef et al.'s (1986) data from a two-action study on budgerigars (*Melopsittacus undulatus*): subjects that observed conspecifics accessing hidden food using either the beak or the foot tended to match the demonstrated technique for only the first two trials postdemon-stration. In later trials, the difference between the groups disappeared.

What is more pertinent to the issue of culture, proposes Heyes (1993), is a separate insulating mechanism which protects the socially learned information against the influence of trial-and-error learning. Heyes (1993) takes birdsong as an example and notes that socially learned, interpopulation differences in song patterns exist not because they are learned via imitation, but because other mechanisms exist which either prevent exposure to or resist the influence of other song types.

Unfortunately, the question of the longevity of socially learned behaviors in primates has been relatively neglected as a focus of study. However, some illuminating details can be gleaned from some of the experimental literature on transmission mechanisms.

In Bugnyar and Huber's (1997) study of imitation in common marmosets, subjects were provided with a pendulum door that could be either pushed or pulled to gain access to a food reward. These researchers allowed observer monkeys to watch a trained conspecific demonstrator pulling open the door. Some observer monkeys showed a stronger tendency to use the demonstrated opening technique than control (nonobserver) monkeys, whose natural response was to push the door. However, Bugnyar and Huber (1997) report that the initial preference of the observers disappeared over the course of five test trials, and all subjects converged on the simpler alternative preferred by the nonobservers. This effect is consistent with the proposals put forward by Galef (1995) and Heyes (1993).

However, data from Whiten's (1998) study of imitation of a sequence of actions seems to tell a different story. In this experiment, chimpanzee subjects were provided with three opportunities to open the artificial fruit themselves, and only on the third trial was the tendency to match the observed sequence significant. As can be seen from Fig. 40.4, the chimpanzees appeared to show a gradual convergence toward the demonstrated action. This is despite the fact that the chimpanzees had successfully gained the food reward inside (and therefore had received reinforcement) when using alternative sequences.

There are methodological differences between the two studies mentioned above which may be significant. In Bugnyar and Huber's (1997) study, marmosets received three demonstrations on consecutive days, and these were then followed by the five trials, also on consecutive days. In Whiten's (1998) study, however, the subjects also saw repeated demonstrations in between trials. After the first three demonstrations, subjects were presented with the artificial fruit for the first trial. This was followed by a further demonstration and a second trial. Following this, there was a fifth demonstration and then the third and final trial. In fact, this methodology almost certainly corresponds more closely to a naturalistic learning experience, in which the learner's own attempts will often be interspersed with further observations of others' performances.

A mechanism for conformity, whereby individuals typically use the behavioral variants they most regularly see others using (and continue to be influenced by this after the initial learning process has taken place), could easily result in stable interpopulation differences in behavior (Boyd and Richerson 1985, Richerson and Boyd 1998). The type of design used by Whiten (1998) can go some way toward investigating whether primates possess such a mechanism.

Obviously, there is currently so little known regarding the longevity of experimentally induced socially learned behaviors in primates that it is extremely hard to draw any firm conclusions, but the results of this study afford some tantalizing possibilities for future research. It is too early to say whether primates, in addition to being imitative, possess mechanisms for the insulation of socially learned information; but future experiments will no doubt throw a great deal of light onto this issue.

Multigeneration Simulation Experiments

In order to support culture, it is not crucial that a particular behavioral variant remains in an individual's repertoire indefinitely; the real issue is whether or not it persists long enough to be passed on to another individual or individuals. Experiments which simulate social transmission over multiple learner generations are therefore likely to prove particularly illuminating with regard to understanding whether socially learned behaviors can endure within a population (see Fig. 40.1).

Such experiments have so far been carried out with rats (Galef and Allen 1995, Laland and Plotkin 1990), blackbirds (Curio et al. 1978), and guppies (Laland and Williams 1997, 1998), as well as humans (Mesoudi and Whiten 2004). Galef and Allen (1995), for example, induced two different arbitrary food preferences into "founder" rat colonies, with four subjects per colony. These founder individuals were replaced one by one every day by naive rats. After the final founder was removed on day 4, the preferences remained and persisted to the end of testing, which went on for 14 days (three complete learner generations). The social learning of taste preference therefore appears to be sufficiently stable to produce interpopulation differences in diet choice.

Transmission chain experiments such as these are extremely relevant to the question of whether a species possesses the capacity for culture as the goal is essentially to artificially induce cultural variation, that is, interpopulation differences in behavior supported by social learning. Similar methodologics are only just beginning to be exploited within primatology (e.g., Whiten, Horner, and de Waal 2005), but we believe that this will prove to be a particularly fruitful approach for future research on social learning with monkeys and apes.

CONCLUSIONS

We find the case for culture in primates to be a convincing one. Although some skeptics will maintain that the evidence from primate research is less rigorous than that from other species, the literature is full of evidence which suggests that certain nonhuman primates may be more "cultured" than any animal other than ourselves.

From the field research data, the sheer number of apparent cultural variants identified in primates is noteworthy. Although group-specific behaviors have been documented in many different species, from birds to rodents (Lefebvre and Palameta 1988), the variety of such behaviors identified in chimpanzees, orangutans, and capuchin monkeys is striking. Isolated cases may be dependent on relatively inflexible, hard-wired mechanisms; but the multiple cultural variants identified in chimpanzees, and now other primates too, may point to a more general underlying social learning ability or tendency toward adopting the behavior of conspecifics.

All the same, it must be noted that comparable diversity may yet be identified in animals other than monkeys and apes. Whiten et al.'s (1999) survey of chimpanzee culture (once apparently unique in the diversity it revealed) is now echoed to varying extents in Panger et al.'s (2002), Perry et al.'s (2003), and van Schaik et al.'s (2003) studies of capuchin monkeys and orangutans, as well as Rendell and Whitehead's (2001) review of the literature from cetacean field observations and laboratory experiments. Given appropriate research effort, it may be only a matter of time before other phyletic groups are shown to have similarly rich patterns of behavioral variation.

There are still plenty of opportunities for research on this topic however. Approaches such as Panger et al.'s (2002) and van Schaik et al.'s (2003), which make use of association data, will help to rule out the possibilities of environmental or genetic influence. Such data can, in turn, be fed into models of diffusion dynamics in order that more accurate predictions can be generated about expected transmission patterns. There are also scores of experiments waiting to be carried out regarding the longevity of socially learned information, both within the learner's own lifetime and across multiple learner generations.

The case for culture in primates is a strong one. However, there are clearly profitable directions for future research that will enable conclusions about the role of social learning in supporting behavioral variation to be made much more firmly.

REFERENCES

Aisner, R., and Terkel, J. (1992). Ontogeny of pine cone opening behaviour in the black rat, *Rattus rattus*. *Anim. Behav.* 44:327–336.

Barrett-Lennard, L. G., Deecke, V. B., Yurk, H., and Ford, J. K. B. (2001). A sound approach to the study of culture. *Behav. Brain Sci.* 24:325–326.

Beck, B. B. (1994). Reintroduction of captive-born animals. In: Olney, P. J. S., Mace, G. M., and Feistner, A. T. C. (eds.), *Creative Conservation*. Chapman & Hall, London. pp. 265–286.

Beck, B. B. (1995). Reintroduction, zoos, conservation and animal welfare. In: Norton, B. G., M. Hutchins, Stevens, E. F., and Maple, T. L. (eds.), *Ethics on the Ark: Zoos, Animal Welfare, and Wildlife Conservation*. Smithsonian Institution Press, Washington DC. pp. 155–163.

Beck, B. B., Castro, M. I., Stoinski, T. S., and Ballou, J. (2002). The effects of pre-release environments on survivorship in reintroduced golden lion tamarins. In: Kleiman, D. G., and Rylands, A. B. (eds.), *The Lion Tamarins of Brazil: Twenty-Five Years of Research and Conservation*. Smithsonian Institution Press, Washington DC. pp. 283–300.

Blackmore, S. (1999). *The Meme Machine*. Oxford University Press, Oxford.

Boesch, C. (1996). Three approaches for assessing chimpanzee culture. In: Russon, A. E., Bard, K. A., and Parker, S. (eds.), *Reaching into Thought: The Minds of the Great Apes*. Cambridge University Press, Cambridge. pp. 404–429.

Boesch, C., Marchesi, P., Marchesi, N., Fruth, B., and Joulian, F. (1994). Is nut cracking in wild chimpanzees a cultural behavior? *J. Hum. Evol.* 26:325–338.

Boesch, C., and Tomasello, M. (1998). Chimpanzee and human cultures. *Curr. Anthropol.* 39:591–614.

Bonner, J. T. (1980). *The Evolution of Culture in Animals*. Princeton University Press, Princeton, NJ.

Box, H. (1991). Training for life after release: simian primates as examples. In: Gipps, J. H. W. (ed.), *Beyond Captive Breeding: Reintroducing Endangered Mammals to the Wild. Symposium of the Zoological Society of London*. Clarendon Press, Oxford. pp. 111–123.

Boyd, R., and Richerson, P. J. (1985). *Culture and the Evolutionary Process*. University of Chicago Press, Chicago.

Boyd, R. and Richerson, P. J. (1996). Why culture is common, but cultural evolution is rare. *Proc. Br. Acad.* 88:77–93.

Bugnyar, T., and Huber, L. (1997). Push or pull: an experimental study on imitation in marmosets. *Anim. Behav.* 54:817–831.

Byrne, R. W. (2002). Imitation of novel complex actions: what does the evidence from animals mean? *Adv. Study Behav.* 31:77–105.

Byrne, R. W., and Russon, A. E. (1998). Learning by imitation: a hierarchical approach. *Behav. Brain Sci.* 21:667–721.

Caldwell, C. A., and Whiten, A. (2002). Evolutionary perspectives on imitation: is a comparative psychology of social learning possible? *Anim. Cogn.* 5:193–208.

Caldwell, C. A., and Whiten, A. (2003). Scrounging facilitates social learning in common marmosets, *Callithrix jacchus*. *Anim. Behav.* 65:1085–1092.

Caldwell, C. A., and Whiten, A. (2004). Testing for social learning and imitation in common marmosets, *Callithrix jacchus*, using an "artificial fruit." *Anim. Cogn.* 7:77–85.

Cavalli-Sforza, L. L., and Feldman, M. W. (1981). *Cultural Transmission and Evolution: A Quantitative Approach.* Princeton University Press, Princeton, NJ.

Coussi-Korbel, S., and Fragaszy, D. M. (1995). On the relation between social dynamics and social learning. *Anim. Behav.* 50:1441–1453.

Curio, E., Ernst, U., and Vieth, W. (1978). The adaptive significance of avian mobbing. II. Cultural transmission of enemy recognition in blackbirds: effectiveness and some constraints. *Z. Tierpsychol.* 48:184–202.

Custance, D. M., Whiten, A., and Fredman, T. (1999). Social learning of artificial fruit processing in enculturated capuchin monkeys. *J. Comp. Psychol.* 113:13–23.

Custance, D. M., Whiten, A., and Fredman, T. (2002). Social learning and primate reintroduction. *Int. J. Primatol.* 23:479–499.

Custance, D. M., Whiten, A., Sambrook, T., and Galdikas, B. (2001). Testing for social learning in the "artificial fruit" processing of wildborn orangutans (*Pongo pygmaeus*), Tanjung Puting, Indonesia. *Anim. Cogn.* 4:305–313.

Dawkins, R. (1976). *The Selfish Gene.* Oxford University Press, Oxford.

Day, R. L., Kendal, J., and Laland, K. N. (2001). Validating cultural transmission in cetaceans. *Behav. Brain Sci.* 24:330–331.

Dewar, G. (2003). The cue reliability approach to social transmission: designing tests for adaptive traditions. In: Fragaszy, D. M., and Perry, S. (eds.), *The Biology of Traditions: Models and Evidence.* Cambridge University Press, Cambridge. pp. 127–158.

Fragaszy, D. M., and Perry, S. (2003). Towards a biology of traditions. In: Fragaszy, D. M., and Perry, S. (eds.), *The Biology of Traditions: Models and Evidence.* Cambridge University Press, Cambridge. pp. 1–32.

Fragaszy, D. M., and Visalberghi, E. (1990). Social processes affecting the appearance of innovative behaviours in capuchin monkeys. *Folia Primatol.* 54:155–165.

Fragaszy, D. M., and Visalberghi, E. (2004). Socially-biased learning in monkeys. *Learn. Behav.*

Galef, B. G., Jr. (1988). Imitation in animals: history, definition, and interpretation of data from the psychological laboratory. In: Zentall, T. R., and Galef, B. G., Jr. (eds.), *Social Learning: Psychological and Biological Perspectives.* Lawrence Erlbaum, Hillsdale, NJ. pp. 3–28.

Galef, B. G., Jr. (1990). Tradition in animals: field observations and laboratory analyses. In: Beckoff, M., and Jamieson, D. (eds.), *Interpretations and Explanations in the Study of Behaviour: Comparative Perspectives.* Westview Press, Boulder. pp. 74–95.

Galef, B. G., Jr. (1992). The question of animal culture. *Hum. Nat.* 3:157–178.

Galef, B. G., Jr. (1995). Why behavior patterns that animals learn socially are locally adaptive. *Anim. Behav.* 49:1325–1334.

Galef, B. G., Jr. (2003). "Traditional" foraging behaviours of brown and black rats (*Rattus norvegicus* and *Rattus rattus*). In: Fragaszy, D. M., and Perry, S. (eds.), *The Biology of Traditions: Models and Evidence.* Cambridge University Press, Cambridge. pp. 159–186.

Galef, B. G., Jr., and Allen, C. (1995). A new model system for studying behavioural traditions in animals. *Anim. Behav.* 50:705–717.

Galef, B. G., Jr., Manzig, L. A., and Field, R. M. (1986). Imitation learning in budgerigars: Dawson and Foss 1965 revisited. *Behav. Proc.* 13:191–202.

Goodall, J. (1973). Cultural elements in a chimpanzee community. In: Menzel, E. W. J. (ed.), *Precultural Primate Behaviour.* Karger, Basel. pp. 144–184.

Haggerty, M. E. (1909). Imitation in monkeys. *J. Comp. Neurol. Psychol.* 19:337–455.

Hall, K. R. L. (1963). Observational learning in monkeys and apes. *Br. J. Psychol.* 54:201–226.

Helfman, G. S., and Schultz, E. T. (1984). Social traditions in a coral reef fish. *Anim. Behav.* 32:379–384.

Heyes, C. M. (1993). Imitation, culture and cognition. *Anim. Behav.* 46:999–1010.

Horowitz, A. (2003). Do humans ape? Or do apes human? Imitation and intention in humans (*Homo sapiens*) and other animals. *J. Comp. Psychol.* 117:325–336.

Itani, J., and Nishimura, A. (1973). The study of infrahuman culture in Japan. In: Menzel, E. W. J. (ed.), *Precultural Primate Behaviour.* Karger, Basel. pp. 26–50.

Kawai, M. (1965). Newly-acquired pre-cultural behavior of the natural troop of Japanese monkeys on Koshima Islet. *Primates* 6:1–30.

Kawamura, S. (1959). The process of sub-culture propagation among Japanese macaques. *Primates* 2:43–60.

Kroeber, A. L., and Kluckhorn, C. (1963). *Culture: A Critical Review of Concepts and Definitions.* Random House, New York.

Laland, K. N., and Hoppitt, W. (2003). Do animals have culture? *Evol. Anthropol.* 12:150–159.

Laland, K. N., and Kendal, J. (2003). What the models say about social learning. In: Fragaszy, D. M., and Perry, S. (eds.), *The Biology of Traditions: Models and Evidence.* Cambridge University Press, Cambridge. pp. 33–55.

Laland, K. N., and Plotkin, H. C. (1990). Social learning and social transmission of foraging information in Norway rats (*Rattus norvegicus*). *Anim. Learn. Behav.* 18:246–251.

Laland, K. N., Richerson, P. J., and Boyd, R. (1996). Developing a theory of animal social learning. In: Heyes, C. M., and Galef, B. G., Jr. (eds.), *Social Learning in Animals: The Roots of Culture.* Academic Press, London. pp. 129–154.

Laland, K. N., and Williams, K. (1997). Shoaling generates social learning of foraging information in guppies. *Anim. Behav.* 53:1161–1169.

Laland, K. N., and Williams, K. (1998). Social transmission of maladaptive information in the guppy. *Behav. Ecol.* 9:493–499.

Lefebvre, L. (1995a). Culturally-transmitted feeding behavior in primates: evidence for accelerating learning rates. *Primates* 36:227–239.

Lefebvre, L. (1995b). The opening of milk bottles by birds: evidence for accelerating learning rates, but against the wave-of-advance model of cultural transmission. *Behav. Proc.* 34:43–53.

Lefebvre, L., and Palameta, B. (1988). Mechanisms, ecology, and population diffusion of socially learned, food-finding behaviour in feral pigeons. In: Zentall, T. R., and Galef, B. G., Jr. (eds.), *Social Learning: Psychological and Biological Perspectives.* Lawrence Erlbaum, Hillsdale, NJ. pp. 141–163.

McGrew, W. C. (1992). *Chimpanzee Material Culture.* Cambridge University Press, Cambridge.

McGrew, W. C. (1998). Culture in nonhuman primates. *Annu. Rev. Anthropol.* 27:301–328.

Mesoudi, A., and Whiten, A. (2004). The hierarchical transformation of event knowledge in human cultural transmission. *J. Cogn. Cult.*

Mesoudi, A., Whiten, A., and Laland, K. N. (2004). Is human cultural evolution Darwinian? Evidence reviewed from the perspective of *The Origin of Species. Evolution* 58:1–11.

Panger, M. A., Perry, S., Rose, L., Gros-Louis, J., Vogel, E., MacKinnon, K. C., and Baker, M. (2002). Cross-site differences in foraging behaviour of white-faced capuchins (*Cebus capucinus*). *Am. J. Phys. Anthropol.* 119:52–66.

Perry, S., Baker, M., Fedigan, L., Gros-Louis, J., Jack, K., MacKinnon, K. C., Manson, J. H., Panger, M. A., Pyle, K., and Rose, L. (2003). Social conventions in wild white-faced capuchin monkeys: evidence for traditions in a neotropical primate. *Curr. Anthropol.* 44:241–268.

Reader, S. M., and Laland, K. N. (2000). Diffusion of foraging innovations in the guppy. *Anim. Behav.* 60:175–180.

Rendell, L., and Whitehead, H. (2001). Culture in whales and dolphins. *Behav. Brain Sci.* 24:309–324.

Richerson, P. J., and Boyd, R. (1998). The evolution of human ultra-sociality. In: Eibl-Eibisfeldt, I., and Salter, F. (eds.), *Ideology, Warfare, and Indoctrinability.* Berghahn Books, New York. pp. 71–95.

Roper, T. J. (1986). Cultural evolution of feeding behaviour in animals. *Sci. Prog.* 70:751–783.

Russell, C., and Russell, W. M. S. (1990). Cultural evolution of behaviour. *Neth. J. Zool.* 40:745–762.

Russon, A. E., and Galdikas, B. M. F. (1993). Imitation in free-ranging rehabilitant orangutans (*Pongo pygmaeus*). *J. Comp. Psychol.* 107:147–161.

Russon, A. E., and Galdikas, B. M. F. (1995). Constraints on great apes' imitation: model and action selectivity in rehabilitant orangutan (*Pongo pygmaeus*) imitation. *J. Comp. Psychol.* 109:5–17.

Spence, K. W. (1937). Experimental studies of learning and higher mental processes in infrahuman primates. *Psychol. Bull.* 34:806–850.

Stoinski, T. S., Beck, B., Bloomsmith, M. A., and Maple, T. L. (2003). A behavioural comparison of captive-born, reintroduced golden lion tamarins and their wild-born offspring. *Behaviour* 140:137–160.

Stoinski, T. S., and Whiten, A. (2003). Social learning by orangutans (*Pongo abelii* and *Pongo pygmaeus*) in a simulated food-processing task. *J. Comp. Psychol.* 117:272–282.

Stoinski, T. S., Wrate, J. L., Ure, N., and Whiten, A. (2001). Imitative learning by captive western lowland gorillas (*Gorilla gorilla gorilla*) in a simulated food-processing task. *J. Comp. Psychol.* 115:272–281.

Terkel, J. (1995). Cultural transmission in the black rat: pine cone feeding. *Adv. Study Behav.* 24:119–154.

Terkel, J. (1996). Cultural transmission of feeding behaviour in the black rat (*Rattus rattus*). In: Heyes, C. M., and Galef, B. G., Jr. (eds.), *Social Learning in Animals: The Roots of Culture.* Academic Press, San Diego. pp. 17–47.

Thorndike, E. L. (1898). Animal intelligence: an experimental study of the associative process in animals. *Psychol. Rev. Monogr.* 2.

Tomasello, M. (1990). Cultural transmission in tool use and communicatory signalling of chimpanzees. In: Parker, S., and Gibson, K. (eds.), *"Language" and Intelligence in Monkeys and Apes: Comparative Developmental Perspectives.* Cambridge University Press, Cambridge, pp. 274–311.

Tomasello, M. (1994). The question of chimpanzee culture. In: Wrangham, R. W., McGrew, W. C., de Waal, F. M. B., and Heltne, P. G. (eds.), *Chimpanzee Cultures.* Harvard University Press, Cambridge, MA. pp. 301–317.

Tomasello, M. (2000). *The Cultural Origins of Human Cognition.* Harvard University Press, Cambridge, MA.

Tomasello, M., and Call, J. (1997). *Primate Cognition.* Oxford University Press, New York.

Tomasello, M., Davis-Dasilva, M., Camak, L., and Bard, K. A. (1987). Observational learning of tool use by young chimpanzees. *Hum. Evol.* 2:175–183.

Tomasello, M., Kruger, A. C., and Ratner, H. H. (1993). Cultural learning. *Behav. Brain Sci.* 16:495–552.

van Schaik, C. P. (2003). Local traditions in orangutans and chimpanzees: social learning and social tolerance. In: Fragaszy, D. M., and Perry, S. (eds.), *The Biology of Traditions: Models and Evidence.* Cambridge University Press, Cambridge. pp. 297–328.

van Schaik, C. P., Ancrenaz, M., Borgen, G., Galdikas, B., Knott, C. D., Singleton, I., Suzuki, A., Utami, S. S., and Merrill, M. (2003). Orangutan cultures and the evolution of material culture. *Science* 299:102–105.

Visalberghi, E., and Fragaszy, D. M. (1990). Do monkeys ape? In: Parker, S., and Gibson, K. (eds.), *"Language" and Intelligence in Monkeys and Apes: Comparative Developmental Perspectives.* Cambridge University Press, Cambridge. pp. 247–273.

Visalberghi, E., and Fragaszy, D. M. (2002). Do monkeys ape? Ten years after. In: Dautenhahn, K., and Nehaniv, C. L. (eds.), *Imitation in Animals and Artifacts.* MIT Press, Cambridge, MA. pp. 471–499.

Voelkl, B., and Huber, L. (2000). True imitation in marmosets. *Anim. Behav.* 60:195–202.

Warden, C., and Jackson, T. (1935). Imitative behaviour in the rhesus monkey. *J. Genet. Psychol.* 46:103–125.

Whiten, A. (1998). Imitation of the sequential structure of actions by chimpanzees (*Pan troglodytes*). *J. Comp. Psychol.* 112:270–281.

Whiten, A., and Custance, D. M. (1996). Studies of imitation in chimpanzees and children. In: Heyes, C. M., and Galef, B. G., Jr. (eds.), *Social Learning in Animals: The Roots of Culture.* Academic Press, New York. pp. 291–318.

Whiten, A., Custance, D. M., Gomez, J. C., Texidor, P., and Bard, K. A. (1996). Imitative learning of artificial fruit processing in children (*Homo sapiens*) and chimpanzees (*Pan troglodytes*). *J. Comp. Psychol.* 110:3–14.

Whiten, A., Horner, V., and de Waal, F. B. M. (2005). Conformity to cultural norms of tool use in chimpanzees. *Nature* 437:737–740.

Whiten, A., Goodall, J., Mcgrew, W. C., Nishida, T., Reynolds, V., Sugiyama, Y., Tutin, C., Wrangham, R., and Boesch, C. (1999). Cultures in chimpanzees. *Nature* 399:682–685.

Whiten, A., Goodall, J., McGrew, W. C., Nishida, T., Reynolds, V., Sugiyama, Y., Tutin, C. E. G., Wrangham, R. W., and Boesch, C. (2001). Charting cultural variation in chimpanzees. *Behaviour* 138:1481–1516.

Whiten, A., and Ham, R. (1992). On the nature and evolution of imitation in the animal kingdom: reappraisal of a century of research. *Adv. Study Behav.* 21:239–283.

Whiten, A., Horner, V., Litchfield, C., and Marshall-Pescini, S. (2004). How do apes ape? *Learn. Behav.* 32:36–52.

Whiten, A., Horner, V., and Marshall-Pescini, S. (2003). Cultural panthropology. *Evol. Anthropol.* 12:92–105.

41

Tool Use and Cognition in Primates

Melissa Panger

INTRODUCTION

Tool use has been of interest to primatologists for decades because of what it might tell us about the evolution of tool use and cognition in primates, including our early hominin ancestors (*hominin* refers to members of the human clade or, in other words, components of the lineage that separated from the last common ancestor of *Homo* and *Pan* spp. 8–5 million years ago [mya] and which persist as *Homo sapiens* [Wood and Richmond 2000]). The most widely used definition of *tool use* is Beck's (1980:10):

> the external employment of an unattached environmental object to alter more efficiently the form, position, or condition of another object, another organism, or the user itself when the user holds or carries the tool during or just prior to use and is responsible for the proper and effective orientation of the tool.

Therefore, tool use involves the manipulation of both the *object of change* (the object being altered) and the *agent of change* (the tool) (Parker and Gibson 1977). Although tool use is exhibited by a variety of animals, it has received particular attention in primates because of their phylogenetic proximity to humans. Additionally, Parker and Gibson (1977) make a distinction between what they call "intelligent" tool use and tool use that is context-specific (see also Hall 1963, Weir et al. 2002). The type of tool use found among most animals is context-specific and represented by stereotyped behaviors (therefore, the applications, contexts, and types of tool used vary little among individuals and populations of the same species). An example of this type of tool use is the dropping of stones by Egyptian vultures (*Neophron pernopterus*) to crack open eggs. Stone dropping is the only type of tool use exhibited by Egyptian vultures, and the behavior varies little across individuals and populations. "Intelligent" tool use refers to a more flexible repertoire and is the type found among, although not limited to, some primates. For example, chimpanzees (*Pan troglodytes*) use sticks as probing tools; however, their tool use is not limited to this one type of tool (see below). Additionally, chimpanzee probing (and other) tools vary across individuals and populations.

Our understanding of primate tool use has expanded greatly in the last few decades, but it remains unclear how tool use relates to other cognitive abilities. The evolution of tool use in primates is also currently poorly understood.

In this chapter, I will provide an overview of primate tool use and will briefly investigate how tool use in primates is related to other cognitive abilities. I will then conclude by exploring the evolution of primate tool use.

PRIMATES THAT USE TOOLS

Many primates are highly manipulative creatures and regularly examine and handle any object with which they come into contact, thanks in part to their relatively large brains and opposable or semiopposable thumbs and toes (and, in some cases, highly manipulative lips and prehensile tails). Most object manipulation observed in primates does not involve tool use as defined above, where both the object of change and the agent of change are manipulated by the user. Object manipulation more often involves investigating through touch, shaking objects (as threats), carrying objects, licking objects, handling objects during feeding, and performing a variety of object-use behaviors. Object use is similar to tool use in that an individual alters an object using another object. The difference is that with object use only the object of change is manipulated by the user (i.e., the agent of change is not manipulated) (see Panger 1998, Parker and Gibson 1977).

Various forms of object manipulation have been studied extensively in captive primates, largely under a Piagetian framework of sensorimotor development (e.g., Fobes and King 1982a,b; Gibson et al. 2001; Matsuzawa 2001; Parker and McKinney 1999; Torigoe 1985; see Tomasello and Call 1997 for a review). A Piagetian framework is based on Piaget's theory of sensorimotor development, which states that certain cognitive skills can be measured by gauging object manipulation skills. Unfortunately, the full range of object manipulation in free-ranging primates is not clear at this time because, with notable exceptions (e.g., Boinski et al. 2000, Byrne and Byrne 1993, Fuentes 1992, Huffman and Hirata 2003, Panger 1998, Parker and Gibson 1977), research attention has been focused on tool-using behaviors and other types of object manipulation have most likely been underreported. Reports of object use from the wild include, but are not limited to, capuchin monkeys (*Cebus* spp.) pounding and/or rubbing objects against fixed substrates (e.g., Boinski et al. 2000, Panger 1998) and macaques (*Macaca* spp.) rubbing objects in their hands and against fixed substrates

(Fuentes 1992). Therefore, primates perform a variety of complex manipulative behaviors that do not fit traditional definitions of tool use. Although object manipulation is very common among primates, tool use is not.

To identify the primate "tool users," I surveyed the primate literature using the following sources: published bibliographies on nonhuman primate tool use from 1940 to 1992 (Williams 1978, 1984, 1992), PrimateLit (http://primatelit.library.wisc.edu, keyword *tool*) search from 1992 to 2002 (excluding abstracts, book reviews, and titles not dealing with tool use [e.g., "Rehabilitation: A Tool for Conservation"]), and books on animal cognition and tool use (i.e., Beck 1980, McGrew 1992, Tomasello and Call 1997). Care was taken not to use duplicate references or those focusing on negative data, that is, those reporting an absence of tool use. The references do not take into account variable research effort per genus and include data from captive and free-ranging individuals.

Using these search criteria, tool use has been reported in the following primate genera (listed in order from greatest to fewest reports): *Pan, Pongo, Cebus, Macaca, Gorilla, Papio, Hylobates, Colobus, Cercocebus, Cercopithecus, Ateles, Erythrocebus, Presbytis, Saimiri, Alouatta, Pithecia, Lagothrix, Leontopithecus, Saguinus,* and *Daubentonia.* Therefore, tool use has been reported in a variety of primate genera; however, repeated reports (over 25) of tool use are limited to *Pan, Pongo, Cebus, Macaca, Gorilla,* and *Papio* (Fig. 41.1) (see Tomasello and Call 1997 for a similar conclusion). These six genera are considered the "tool users" in this chapter.

Captive vs. Free-Ranging Data

Tool use in primates has been extensively studied under captive conditions, and many primate species have been trained or induced to use tools in captivity. The majority of captive tool-use studies, however, have involved the great apes and capuchin monkeys and, to a lesser degree, baboons and macaques. The range of tool use reported from primate laboratory studies is wide; in fact, it is too extensive to fully review here (for reviews see Beck 1980, McGrew 1992, Tomasello and Call 1997). Examples, however, include using clubs (*Cebus* spp. [Cooper and Harlow 1961 as reported in Visalberghi 1990], *Pan troglodytes* [Kohler 1927]), using probing tools (*Cebus* spp. [e.g., Westergaard and Fragaszy 1987, Westergaard et al. 1997], *Gorilla gorilla* [Parker et al. 1999; personal observation], *Macaca* [e.g., Anderson 1985, Tokida et al. 1994], *P. troglodytes* [e.g., Nash 1982, Yerkes 1943], *P. paniscus* [Jordan 1982], *Papio* [Westergaard 1992], *Pongo pygmaeus* [e.g., Chevalier-Skolnikoff 1983, Galdikas 1982]), and using sponging tools (*Cebus* spp. [Westergaard and Fragaszy 1987, Westergaard et al. 1995], *G. gorilla* [Fontaine et al. 1995], *P. troglodytes* [Kitahara-Frisch and Norikoshi 1982], *P. paniscus* [Jordan 1982]). Additionally, several captive primates have been taught to make stone tools that are similar, but not identical, to some of the earliest tools found in the archaeological record (*Cebus apella* [e.g.,

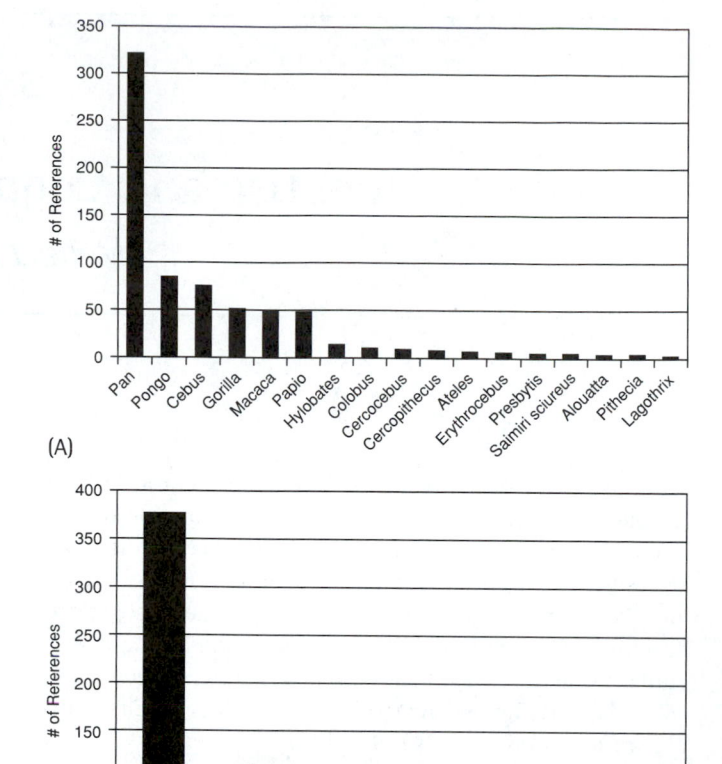

(A)

(B)

Figure 41.1 Number of tool-use references per primate genus. Numbers include both captive and free-ranging studies. (A) Initial survey results using literature from 1940–1992 based on Williams (1978, 1984, 1992). (B) Additional survey results using literature from 1940–2002 based on Beck (1980), McGrew (1992), Tomasello and Call (1997), Williams (1978, 1984, 1992), and PrimateLit (http://primatelit.library.wisc.edu, keyword 'tool'). Only references for the top six tool-using genera are reported.

Westergaard and Suomi 1996], *P. paniscus* [Schick et al. 1999, Toth et al. 1993], *P. pygmaeus* [Wright 1972]). These captive studies are important because they provide insight into the range of primate capabilities, but they do not necessarily reflect behaviors found among free-ranging counterparts.

Although many primates can be induced to use tools in captivity, spontaneous tool use and tool use under free-ranging conditions are relatively rare. The only primates that are known to spontaneously and habitually manufacture and use tools in all known free-ranging populations are chimpanzees (*Pan trolodytes*). The term *habitual* refers to tool-use events that are repeated by several individuals in a population over time (McGrew and Marchant 1997). The most common types of tool use observed in free-ranging chimpanzees are the manufacture and use of probing tools (made of stick, bark, grass, or other vegetation; e.g., Goodall 1964, McGrew 1974, McGrew and Marchant 1992), the manufacture and use of leaf sponges (e.g., Boesch and Boesch 1990), and the use of wood and stone as hammers and anvils (e.g., Boesch and Boesch 1981, Sakura and Matsuzawa 1991, Yamakoshi

1998; for review see McGrew 1992; see also Whiten et al. 1999). The last of these, hammer and anvil use, leaves a record that is recoverable through archaeological excavation (Mercader et al. 2002; see below). Chimpanzees are also known to use sticks and leaves for grooming purposes in addition to sticks as clubs and levers (Goodall 1986, McGrew 1992, Tomasello and Call 1997). The use of a variety of tool composites (sometimes referred to as "tool sets"), in which two or more tools are used to achieve a single goal (e.g., using a probing tool to remove nut kernels after the nuts have been cracked with a hammer), has also been reported in free-ranging chimpanzees (e.g., Boesch and Boesch 1990, Brewer and McGrew 1990, Sugiyama 1997). Nest building, a behavior found among all living great apes, is usually excluded from studies of tool use because it does not fit traditional definitions of tool use (e.g., Boesch and Boesch 1990, McGrew 1992). For similar reasons, other common primate behaviors, such as the breaking or dragging of branches during agonistic displays, are also often excluded from tool-use studies (e.g., Panger 1998).

Some free-ranging orangutan (*Pongo pygmaeus*) populations, like chimpanzees, also habitually and spontaneously manufacture and use tools made of sticks (e.g., Fox et al. 1999, van Schaik et al. 1996). The hammers, probes, and scrapers are used to access insects (adults and eggs/larvae of ants, stingless bees, and termites), honey, and fruits. The tools vary in size depending on the task (Fox et al. 1999). Bonobos (*Pan paniscus*) have been observed using sticks for grooming and large leaves as shields against the rain, although habitual tool use in free-ranging bonobos has never been reported (Ingmanson 1996). Gorillas (*Gorilla gorilla*) have yet to be seen manufacturing or using tools in the wild. It is clear that all of the great apes, however, will readily use tools in captivity (e.g., for orangutans, see Russon and Galdikas 1992; for gorillas, see Parker et al. 1999; for chimpanzees, see Kohler 1927; and for bonobos, see Jordan 1982) as discussed above.

Although capuchin tool use has been studied extensively under captive conditions, what their capabilities under free-ranging conditions are less well known. Several anecdotal reports suggest that tool use is part of the normal behavioral repertoire of many free-ranging capuchin populations (e.g., the use of a club by *Cebus capucinus* [Boinski 1988a, Chapman 1986], the use of leaves to protect their hands from caterpillars before being rubbed [Panger et al. 2002b], and the use of a hammer and anvil by *Cebus apella* [Boinski et al. 2000, Fernandes 1991] have been reported). Furthermore, in a 5-month study of free-ranging *C. capucinus*, Chevalier-Skolnikoff (1990) observed several bouts of tool use, and Phillips (1998) reported the use of leaf containers by more than one individual in her *Cebus albifrons* study group. Reports of tool use in free-ranging macaques and baboons are less common and currently limited to the use of probing tools (*Macaca* spp. [Fitch-Snyder and Carter 1993], *Papio* spp. [Oyen 1979]) and leaf napkins (*Macaca* spp. [Sinha 1997], *Papio* spp. [van Lawick-Goodall et al. 1973]). Among free-ranging primates, tool manufacture (i.e.,

modifying an object to improve its use as a tool) has been seen only in chimpanzees and orangutans.

Intraspecific Differences

Interindividual Differences in Tool Use

In primate species that are known tool users, not all members use tools equally. Among free-ranging chimpanzees and orangutans, individuals younger than 3 years do not normally exhibit tool use and proficient tool use does not develop until later in life (Boesch and Boesch 1981, Fox et al. 1999, Inoue-Nakamura and Matsuzawa 1997). This is most likely due to the time required for individuals to acquire the skill necessary to successfully perform complex manipulative behaviors. For capuchin monkeys, most data regarding age/sex class differences in tool use come from captive studies (see McGrew and Marchant 1997 and Visalberghi 1990 for reviews). In captivity, capuchin tool use is seen more often, although not exclusively, in adults/subadults (Anderson 1990, Visalberghi 1987). This is supported by anecdotal reports of tool use in wild capuchins, which primarily involve adults (Boinski 1988a, Boinski et al. 2000, Fernandes 1991; but see Phillips 1998).

Sex differences in tool use have also been reported in free-ranging primates, specifically in chimpanzees, with females showing significantly higher rates of common forms of tool use compared to males (e.g., termite fishing and nut cracking) (Boesch and Boesch 1981; Boesch and Boesch-Achermann 2000; McGrew 1979, 1992). Differences in the rates of tool use across sex classes in adult capuchins, however, are less obvious as no definitive variation in the tool use of adult males and females has been found in captivity. Tool-use data from other species are currently too sparse to make conclusions regarding sex differences in behavior.

Potential explanations of the variation in primate tool use among different age/sex classes are many. Most focus on differences in foraging behavior and not differences in cognitive abilities. It is important to note that most tool use observed in free-ranging primates involves some type of food processing, and within-population variation in foraging behavior has been documented in a wide range of animals, including invertebrates (Persons and Uetz 1999, Rusterholz and Erhardt 2000), birds (Dit Durell et al. 1993, Enoksson 1988, Whitehead and Tschirner 1992), reptiles (Camelleri and Shine 1990, Lovern 2000), fish (Magurran and Macias Garcia 2000, Sano 1993), and mammals, including primates (Boinski 1988b, Clutton-Brock 1977, Harrison 1983, Le Boeuf et al. 2000, Rose 1994), with most of the differences occurring across age and sex classes. Several hypotheses have been offered to explain why individuals belonging to the same species but different demographic categories might vary in their foraging behavior. The primary explanations include, but are not limited to, factors involving different metabolic demands, differences in body size and strength, variation in social status, differences in foraging skills and experience, and avoidance of feeding competition (Clutton-Brock 1977, Fragaszy and Boinski 1995, Rose 1994). Other differences that might also specifically influence the rates of tool use among different age/sex classes include the amount of time spent on the ground versus in the trees (since the ground is normally a more stable substrate than a tree branch, it is assumed that many types of tool are easier to use on the ground [McGrew 1992, McGrew and Marchant 1997]), opportunities to learn from others (van Schaik et al. 2003), and variations in social patterns which might influence the horizontal transfer of complex behaviors (see Chapter 40).

Differences Across Populations

In all primates that are known to use tools under free-ranging conditions, tool use varies across populations intraspecifically. Some of these differences involve the absence of a tool-use behavior at one site and its presence at another, which cannot be easily explained by ecological differences. For example, probing tools are used by orangutans at Suaq Balimbing but not by orangutans at other known sites (e.g., Fox et al. 1998). In chimpanzees, hammers are used in some West African populations to crack open hard-shelled nuts but are not used in other parts of Africa where the same nut species occur (e.g., Whiten et al. 1999). Among capuchins, tool-use behaviors such as leaf wrapping are found at some sites but are absent at others where ecological differences cannot easily account for the variation (e.g., Panger et al. 2002b).

In other cases, the same type of tool use can be found across sites but the behavior is used to process different food species. For example, chimpanzees in some populations use probing tools to access honey, while in other populations (where probing tools are used to process other food species) tools are not used for honey extraction (e.g., McGrew 1992, Whiten et al. 1999). Additionally, among some chimpanzee populations, there are variations in the tools themselves. For example, termite fishing is a common behavior in many chimpanzee populations, and although the behavior is broadly similar across populations, the probes used to termite fish vary at the interpopulation level. The variation includes differences across sites in the average size of the tool and type of raw material used (McGrew 1992). Most cross-population differences in primate tool use have been widely, although not wholly, accepted as cultural variation (see Chapter 40 volume).

HOW DOES TOOL USE CORRELATE WITH OTHER MEASURES OF COGNITION?

Tool use, especially among primates, has long been used as a proxy for overall cognitive ability, with *cognition* being defined as the ability to understand how and why one set of events or behaviors is related to another set of events or behaviors (Visalberghi and Tomasello 1998). In other words, tool-using primates have often been thought of as the "smartest" primates. However, there is a growing body of evidence suggesting that, although different primates may use similar tools, the cognitive route to using the tools varies

across genera (e.g., Adams-Curtis 1990, Garber and Brown in press, Jalles-Filho et al. 2001, Visalberghi and Fragaszy 1990, Visalberghi and Trinca 1989). Great apes seem to be capable of solving a variety of tool-use problems by initially working through the solution in their heads before physically solving the problem, while monkeys tend to poke, prod, bang, and push objects until an effective solution is found. These represent contrasts in insightful versus trial-and-error learning and indicate differences in understanding *how* something works versus *that* something works (Garber and Brown 2002).

These apparent differences in the approaches used to solve tool-use problems among primates bring into question the traditional view that tool use can be considered a proxy for overall cognitive abilities. This is an important point to clarify because a fuller understanding of how tool use is associated with other cognitive abilities relates to several notable topics in psychology and anthropology. For example, is the mind modular in its cognitive abilities (e.g., is there such thing as overall "intelligence," or can cognition be divided into analytically distinct and independent subunits)? Does tool use necessarily lead to higher overall cognitive abilities? Are social or ecological factors responsible for driving the evolution of primate, including hominin, cognition? Although I will not be able to answer these questions, in this section I will attempt to answer the narrower question: Does tool use correlate with other cognitive abilities? Answering this question is an important stepping-stone to gaining insight into these broader issues.

Common Measures of Cognition in Primates and How Each Relates to Tool Use

Although it is not the goal of this chapter to evaluate different cognitive tests, I will briefly review the literature on brain size and capabilities regarding discrimination learning, mirror self-recognition, tactical deception, and social learning processes. These five potential measures of cognition were chosen because they are some of the most commonly used measures found in the primate comparative cognition literature and data are available from a wide variety of primate species.

It is important to note that there are problems with attempts to compare cognition across species. For example, a negative performance on a test does not necessarily imply a lack of ability (Heyes 1998), the results of different tests can vary intraspecifically (Meador et al. 1987), animal cognition is often measured against a human standard (Zentall 2000), and there is inequality in research effort across species. To avoid getting mired in complexity in the limited space available, I review the primate comparative cognition literature under the following assumptions: (*1*) some measure of cognition can be obtained using brain size, tactical deception, the ability to imitate, and success on discrimination learning and mirror self-recognition tests and (*2*) such cognition can be adequately compared across species.

Brain Size

Because behavioral measures of cognition are often contentious, several researchers have turned to morphological criteria, specifically those related to brain size, to evaluate cognitive ability. The general idea is the bigger the brain, the greater the intelligence (Beran et al. 1999, Gibson et al. 2001, Jerison 1973, 2000). Although there is agreement that brain size is most likely related to cognitive ability, how to best measure brain size for cross-species comparisons is highly debatable. Brain size scales with negative allometry relative to body mass, meaning that small-bodied primates have relatively large brains compared to their large-bodied counterparts (Jerison 1973, Martin 1982, Radinsky 1982). Therefore, researchers interested in comparing brain size across taxa normally scale their data in some way. A common technique has been to calculate the ratio of whole brain size to body size or some biological variable, after scaling for body size (Jerison 1973, Milton 1981; see also Harvey and Krebs 1990). Others have investigated the relationship of specific structures in the brain to total brain size (Bauchot 1982, Dunbar 1995, Passingham 1975) or to other biological factors (e.g., diet [see Harvey and Krebs 1990], life history parameters [Allman et al. 1993], group size [Dunbar 1995]).

As stated above, which measure of brain size best corresponds to levels of cognition is questionable; and to date, there is no evidence to suggest that one method is better than another (e.g., Deaner et al. 2000, Gibson et al. 2001, Harvey and Krebs 1990). This makes the relationship between brain size and cognition speculative. For illustrative purposes and for the objectives of this chapter, I will consider two common techniques that have included data from a variety of primates to evaluate brain size: *encephalization quotient* (EQ, the ratio of an animal's brain weight to the brain weight of a "typical" animal of the same body mass) and neocortex size relative to whole brain size (the neocortex is the outer part of the brain that surrounds the cerebellum and brain stem).

Using Jerison's EQ (Jerison 1973), *Homo*, *Cebus*, and *Pan* have higher scores than do other primates but the other tool users (*Macaca*, *Papio*, *Pongo*, and *Gorilla*) do not separate out from other primates (see Fig. 41.2a). Looking at the size of the neocortex relative to the rest of the brain, again *Homo* and *Cebus*, along with *Pan*, do have higher scores relative to other primates (see Fig. 41.2b). With this measure, *Macaca* moves closer to *Homo* but *Gorilla* still falls in the middle with other primates (data for this measure were not available for *Pongo* and *Papio*).

Although not all of the primate tool users have the largest brains (using EQs and neocortex ratios), there does seem to be an overall relationship between tool use and brain size in primates (i.e., primates that use tools tend to have the largest brains; see Reader and Laland 2002).

Discrimination Learning Tests

One area of comparative cognition research that has received considerable attention is performance on discrimination

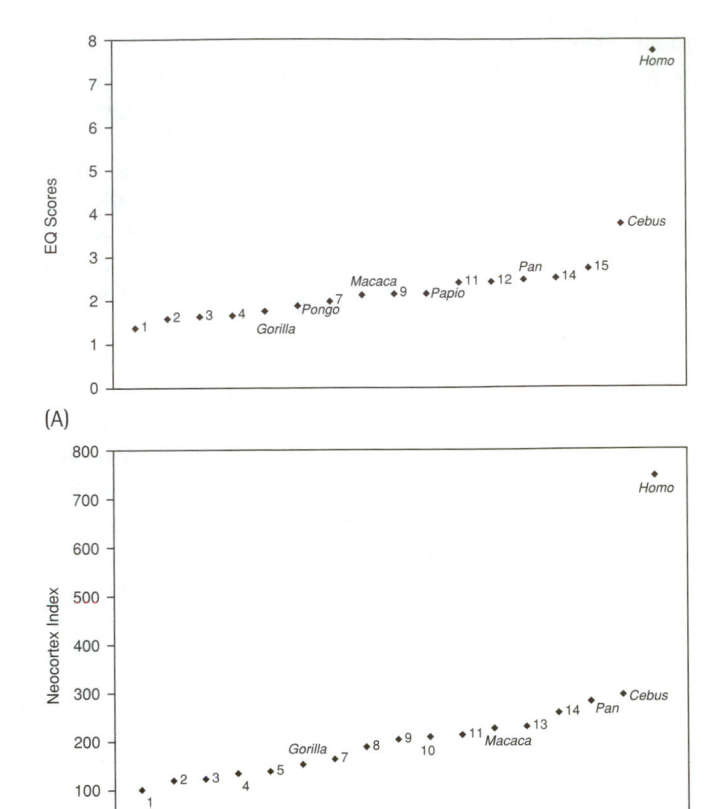

(A)

(B)

Figure 41.2 (A) Encephalization Quotients (EQ's) for a variety of primate genera. The points are rank ordered. Numbers correspond to: 1, *Presbytis*; 2, *Mandrillus*; 3, *Alouatta*; 4, *Saguinus*; 7, *Callicebus*; 9, *Cercopithecus*; 11, *Cercocebus*; 12, *Hylobates*; 13, *Pan*; 14, *Ateles*; 15, *Saimiri*. Data from Jerison (1973:392–393). (B) Neocortex indices for a variety of primate genera. The points are rank ordered. Numbers correspond to: 1, *Alouatta*; 2, *Aotus*; 3, *Callithrix*; 4, *Saguinus*; 5, *Callicebus*; 7, *Colous*; 8, *Pithecia*; 9, *Cercopithecus*; 10, *Ateles*; 11, *Saimiri*; 13, Cercopithecus; 14, Lagothrix. Data from Bauchot (1982:164–65).

learning tests in laboratory settings. In normal discrimination learning tests (e.g., having an animal learn to discriminate stimuli based on their position, color, shape, etc.), there is no clear evidence that apes do any better than monkeys or indeed that primates do any better than nonprimates (Meador et al. 1987, Tomasello and Call 1997). With more complicated discrimination learning tests, such as those requiring an individual to identify objects on the basis of conceptual categories (e.g., learning sets [tests that measure "learning to learn," applying a learned generalized concept to a new situation] and delayed response [tests that measure the response of individuals, after some delay, to a test in which they have previously been successful]), however, some variation across primate genera is reported (for reviews, see Fobes and King 1982a,b; Meador et al. 1987; Tomasello and Call 1997).

A summary of the available data on object learning set formation provided by Tomasello and Call (1997: Fig. 4.1) shows that primate tool-using species do not necessarily perform better than non-tool-using primates on these types of test. For example, in studies that exposed individuals to more than 200 trials, primates are ranked in the following order based on their performance on tests of learning set formation (the numbers in parentheses represent the percent of successful trials [second-trial performance]): *Pongo pygmaeus* (~63%), *Macaca* spp. (70%–90%), *Pan troglodytes* (~73%), *Callithrix jacchus* (~75%), *Cebus albifrons* (~75%), *Gorilla gorilla* (~75%), *Saimiri sciureus* (~80%), *Ateles geoffroyi* (~80%), *Lemur macaco* (~83%), and *Pan paniscus* (~90%) (data for *Papio* spp. were not available).

In reversal learning tests, in which the selection of one type of object is initially rewarded but the object being rewarded is reversed once an individual reaches a certain level of success, the great apes (*Gorilla*, *Pan*, and *Pongo*) and macaques (*Macaca*) do perform better than other primates but *Cebus* performs equally with *Hylobates* and worse than *Saimiri* and *Cercopithecus* (data for *Papio* are not available) (see Tomasello and Call 1997 for review). In tests that measure performance in delayed responses, the results again are mixed. After a 30 sec delay, the percentages of successful trials for different genera are as follows: *Cebus* ~45%, *Lagothrix* ~45%, *Lemur* ~58%, *Pan* ~61%, *Cercocebus* ~62%, *Cercopithecus* ~72%, *Gorilla* ~72%, *Mandrillus* ~76%, *Macaca* ~77%, *Hylobates* 80%, *Papio* ~91%, and *Pongo* ~95% (data from Tomasello and Call 1997: Fig. 4.3). Therefore, there is no clear link with performance on these discrimination learning tests and tool use in primates.

Mirror Self-Recognition

Gallup (1998) states that organisms aware of themselves are in a unique position to use their experience as a means of modeling the experience of others. Therefore, the concept of self, as it relates to the "theory of mind," or understanding the intentionality and the mental states of others, has been argued to be a cognitive ability that allows for social complexity and complex forms of social learning (see below). One way that the "theory of mind" has been traditionally tested in nonhuman animals is to examine an individual's response to its own reflection after extensive exposure to mirrors (i.e., mirror self-recognition tests). In some cases, the determination of self-recognition is based on the amount of self-directed behavior (e.g., grooming an area of the body that cannot usually be seen) versus social behavior (e.g., treating the reflection as if it were another individual). Self-recognition can also be determined by anesthetizing an individual and marking its face (or some other visually inaccessible body part) with a marker that cannot be smelled or felt but can be seen by the individual in a mirror (the "mark test"). Self-directed mirror-guided behavior is argued to be evidence of self-awareness (Gallup 1970, Parker and Mitchell 1994; but see Bekoff 2002, Heyes 1998, Tomasello and Call 1997). Mirror self-recognition has been reported in non-primate animals (e.g., bottlenose dolphins [Reiss and Marino 2001]), but most of this research has been conducted on nonhuman primates and infant humans.

Monkeys do not show consistent evidence for mirror self-recognition. For example, evidence for mirror self-

recognition was reported in *Saguinus oedipus* (Hauser et al. 1995), but the results were not replicated in a later study with the same species (Hauser et al, 2001). There have been some conflicting results with *Macaca nemestrina* also (see Tomasello and Call 1997). Overall, however, there appears to be little difference among tool-using monkeys and non-tool-using monkeys (and gibbons) in their response to mirrors (e.g., Inoue-Nakamura 1997). There does, however, seem to be a difference in the way that monkeys and great apes react to mirrors.

Chimpanzees, bonobos, and orangutans consistently, although not in every study (e.g., de Veer and van den Bos 1999, Povinelli et al. 1993), show strong evidence for mirror self-recognition (Gallup 1970, Hyatt and Hopkins 1994, Inoue-Nakamura 1997, Miles 1994, Savage-Rumbaugh 1986, Suarez and Gallup 1981). Gorillas, although successful in some studies (Inoue-Nakamura 1997, Patterson and Cohn 1994, Suarez and Gallup 1981), seem to fail more often than *Pan* and *Pongo* in showing self-directed behaviors in response to their reflections.

The research to date demonstrates that monkeys, including the tool-using monkeys *Cebus*, *Macaca*, and *Papio*, regularly "fail" mirror self-recognition tests, while chimpanzees and orangutans regularly "pass" them. Gorillas are less consistent in their responses and fail more often than the other great apes; however, overall they perform better than monkeys and gibbons. Therefore, the correlation among mirror self-recognition in primates seems more closely related to major taxonomic criteria than to tool use per se (Gallup 1970, Gibson et al. 2001), but more data are needed before firm conclusions can be made (see de Veer and van den Bos 1999).

Deception

Another measure recently used in comparative primate cognition studies is evidence of tactical deception. Byrne and Whiten (1992:611) define *tactical deception* as "acts from the normal repertoire of the agent, deployed such that another individual is likely to misinterpret what the acts signify, to the advantage of the agent" and suggest that deception can be used to measure one aspect of social expertise and intelligence (i.e., Machiavellian intelligence), in addition to providing evidence for the theory of mind (Byrne 1993; but see Heyes 1998, Tomasello and Call 1997). The types of deception described can be broadly placed into three categories: *active concealment* (e.g., chimpanzees being more quiet than normal when they are patrolling the border of their territory and the territory of another chimpanzee community), *active misleading* (e.g., a baboon making a warning bark toward a nonexistent predator to distract an aggressive conspecific), and *counter-deception* (e.g., a chimpanzee walking past a hidden choice food when dominant group members, which could "steal" the food, are in proximity).

Although Byrne and Whiten's well-known 1992 study, which surveyed evidence of tactical deception in primates using solicited anecdotal reports from researchers, suffers from some potential methodological problems (e.g., see

Tomasello and Call 1997), it stands as the most complete investigation of tactical deception in primates. In their study, Byrne and Whiten found evidence of deception in New World monkeys (*Saguinus*, *Cebus*, *Alouatta*, and *Ateles*), Old World monkeys (*Macaca*, *Cercocebus*, *Papio*, *Theropithecus*, *Cercopithecus*, and *Erythrocebus*), and apes (*Hylobates*, *Pongo*, *Gorilla*, and *Pan*) but not in any prosimian (Byrne and Whiten 1992).

Although tactical deception was found across the anthropoids, *Pan* and *Papio* are the only two genera that showed statistically higher than expected occurrences after controlling for research effort. Data from more controlled studies of tactical deception are limited to a few species and do not allow for comparisons across primates at this time (e.g., Anderson et al. 2001, Ducoing and Thierry 1998, Hirata and Matsuzawa 2001, Mitchell and Anderson 1997). Therefore, although some of the tool users have the highest rates of reported deception (i.e., *Papio* and *Pan*) based on available data, there is no clear overall link with tool use.

Social Learning Processes

Another measure often used in studies of primate comparative cognition is the ability to learn (or replicate) a behavior by watching another individual, that is, *social learning*. Psychologists define, although variously, several different ways in which an individual can learn by watching another's behavior. These social learning processes are often discussed in a spectrum-like order, from those perceived to be the least complex to those requiring greater cognitive abilities. For example, *stimulus enhancement* (when an individual is drawn to a conspecific [either its location or the stimuli with which it is interacting] and then learns something it would not have otherwise learned) and *social facilitation* (when the frequency of certain behaviors is increased in a social situation) are often viewed as simpler forms of social learning, while imitation is viewed as the most complex. *Imitation* generally refers to an individual attaining a goal by reproducing a model's behavior when the individual understands the "means" and the "ends" of the model's behavior (Byrne 1994, Heyes 1998, Tomasello and Call 1997, Whiten and Ham 1992). Of the many different types of social learning process, the ability to imitate has received the most research attention in primate cognition studies.

Evidence for social learning has been found in a variety of animal species, including dolphins and whales (Rendell and Whitehead 2001), birds (Lefebvre and Palameta 1988, Zeltner et al. 2000), rats (Galef 2003), and several primates (e.g., *Cebus capucinus* [Panger et al. 2002b], *Macaca fuscata* [Itani 1958], and *Pan troglodytes* [McGrew and Marchant 1992; see McGrew 1998 for review]). Regarding imitation specifically, however, the over-riding view, although debatable, in the literature is that monkeys, including the tool users—capuchins, macaques, and baboons—cannot truly imitate (Byrne 1994, Fragaszy and Visalberghi 1996, Galef 1992, Visalberghi and Fragaszy 1990; but see Custance et al. 1999) while great apes can (Byrne 1996, Heyes 1998, Inoue-Nakamura and Matsuzawa 1997, Russon and Galdikas 1992, Whiten and Ham 1992). Therefore, as with mirror-self recognition, which some would argue should align closely with imitation (e.g., Parker and Mitchell 1994), the correlation seems to fall along major taxonomic lines (see also Gibson et al. 2001) and not with tool use in general.

Overview

Therefore, the only cognitive measure correlated with tool use across all tool-using genera is brain size (see Table 41.1). For chimpanzees (*Pan troglodytes*), tool use is associated with large brain size, success in mirror self-recognition tests, and an ability to imitate and to socially deceive others but not necessarily with success in discrimination learning tests. Orangutans (*Pongo pygmaeus*) have relatively large brains, seem able to imitate, and are consistently successful in mirror self-recognition studies; but they do not necessarily exhibit more tactical deception or perform better on discrimination learning tests than non-tool-using primates. With the other tool-using primates, the link between tool use and other cognitive abilities is even less clear.

Gorillas (*Gorilla gorilla*) seem to be able to imitate and, in some cases, to recognize their reflection in mirrors; but they do not have larger relative brain sizes, exhibit more tactical deception, or show greater success with discrimination learning tests than many non-tool-using primates. Baboons (*Papio* spp.) show some of the highest rates of deception but do not necessarily do better than non-tool-using primates in the other measures of cognition. Capuchins (*Cebus* spp.) have large brain sizes that associate with their tool-using abilities,

Table 41.1 Overview of Results

COGNITIVE MEASURE	PAN	PONGO	CEBUS	GORILLA	MACACA	PAPIO
Tool use	■	■	■	▣	▣	▣
Large brain size	■	■	■	▣	▣	▣
Discrimination learning	▣	▣	□	▣	▣	–
Mirror self-recognition	■	■	□	▣	□	□
Deception	■	▣	▣	▣	▣	■
Imitation	■	■	□	■	□	□

■, genus consistently rates higher than most non-tool-using primates; ▣, genus consistently rates higher than *many* non-tool-using primates, or inconsistently rates higher than *most* non-tool-using primates; □, genus does not rate higher than many non-tool-using primates; –, no data available.

but they do not perform better than non-tool-using primates in other cognitive measures. Macaques (*Macaca* spp.) do not stand out from other non-tool-using primates in any of the cognitive measures reviewed here. Therefore, when the results are compiled, there is no evidence to suggest that a species' ability to solve problems using tools necessarily corresponds with many commonly used measures of cognition. Thus, tool use may not, by itself, imply higher levels of overall cognitive ability in primates. That is not to say that tool use may not have led to higher cognition in some primates, including hominoids, but instead that tool use by itself does not *necessarily* lead to higher overall cognitive ability.

THE EVOLUTION OF TOOL USE IN PRIMATES

Ever since Louis and Mary Leakey discovered crude stone tools in the 1930s and attributed them to early hominins (Leakey 1935), there has been intense research interest focused on trying to understand the evolution of tool use in our early hominin ancestors. However, little research effort has been put forth to understand the evolution of tool use more broadly among all primates. A first step toward this goal is to look at tool use among primates phylogenetically or, in other words, to consider which taxa use tools and how these taxa are related evolutionarily.

Examining the distribution of extant primate tool users and their evolutionary relationship to each other, parsimony suggests that tool use evolved at least three times during primate evolution: once in the line leading to capuchins, once in the line leading to macaques/baboons, and once in the line leading to hominoids (the great apes and humans). The fact that the different major taxonomic groups (monkeys versus apes) seem to solve tool-using problems differently (trial-and-error versus insightfully) supports the idea that tool use likely evolved independently in at least monkeys and hominoids. It is important to note that if tool use did evolve in the line leading to hominoids, then it is a retained trait in hominins and the earliest hominins living 5–8 mya would have been tool users (see Panger et al. 2002a).

Pinpointing exactly when and how many times tool use evolved in primates will be extremely difficult because direct evidence of tool use in extinct non-hominin primates is lacking. Archaeological techniques are used to study tool use in early hominins, but these techniques are severely limited to certain types of tool use, namely tool use that produces durable products or by-products that are recoverable in the archaeological record, most notably stone tools which first appear in the archaeological record about 2.5 mya (e.g., Harris 1983, Kibunjia et al. 1992, Semaw et al. 1997, Roche et al. 1999). Tools made of less durable material such as sticks, leaves, and grass are not likely to be preserved or recognized as tools in the archaeological or fossil records, especially those dating to prior than a few million years ago. Therefore, the types of tool most likely manufactured and/or used by non-human primates (extant and extinct) are likely

lost through time and not recoverable. The one possible exception is stone tool use similar to that seen in some chimpanzee populations.

Some chimpanzees in West Africa crack a variety of nut species using stone and wooden hammers and anvils. Chimpanzee nut-cracking behavior has been extensively studied at Taï National Park in Côte d'Ivoire (Boesch and Boesch-Achermann 2000). The chimpanzees at Taï crack five different nut species, the hardest of which is *Panda oleosa*. The chimpanzees almost always use stone hammers to open *Panda* nuts. For anvils, the chimpanzees normally use the exposed root systems of hardwood trees. While stone hammers are often moved between *Panda* trees, the anvil systems are immobile. During the nut-cracking season, adult chimpanzees spend up to 3 hr a day cracking nuts, and the same anvil systems are visited year after year (again see Boesch and Boesch-Achermann 2000). Therefore, large mounds of debris form around the anvils. Although the chimpanzees have never been seen intentionally modifying their stone tools, the stone hammers sometimes break during use.

Recent research by myself and colleagues has demonstrated that chimpanzee nut-cracking behavior leaves a clear signal in the archaeological record in the form of buried nut debris and broken pieces of stone hammers (Mercader et al. 2002). During the chimpanzee archaeological project at Taï, we recovered about 40 kg of buried nutshells and 479 pieces of buried stone at six anvil sites associated with one *Panda* nut tree. All of the buried shells and stones were limited to areas directly associated with the six anvil areas. The stone pieces were buried as deep as 20 cm below the surface and dated to a few hundred years ago. The majority of the stone pieces recovered were small in size (about half of them 1–10 mm), and all of the buried stone had been brought in from other locations. Most of the recovered stone pieces showed evidence that they had broken off of larger stones.

Although the data from the Taï chimpanzee archaeology project are preliminary, further excavation of chimpanzee tool sites could provide information important for studying past primate tool use. Just as traditional archaeology tells us about past humans, chimpanzee archaeology could add a time dimension to the study of chimpanzee tool use. For example, it could shed light on the age of chimpanzee nut-cracking behavior in West Africa (i.e., is it a recent or an ancient behavior?). Furthermore, chimpanzee archaeology might provide a model for identifying tool use in the early hominin record prior to 2.5 mya when intentionally modified stone tools first appear, thus providing important information for understanding the evolution of early hominin tool use (see Panger et al. 2002a). However, it is not likely to shed light on the broader question regarding the timing of the origin(s) of tool use in nonhuman primates.

As for why primate tool use evolved, because direct evidence is lacking, we again are limited to indirect data. Since phylogeny cannot explain the distribution of tool use among primates, the behavior likely evolved independently at least three times (as discussed above). Therefore, there has been a behavioral convergence among capuchins, macaques/baboons, and hominoids. Because most behavioral convergences are the result of a shared adaptive function, determining the function that tool use serves among free-ranging primates can help shed light on why the behavior might have evolved.

Primates use tools in a variety of contexts, but most of the tool use reported to date among free-ranging individuals involves the processing of foods, especially those difficult to access because of physical or chemical defenses (e.g., Boinski et al. 2000; Fox et al. 1999; McGrew 1992, 1993; Parker and Gibson 1977; van Schaik et al. 1996). Therefore, tool use may provide an adaptive advantage by allowing access to resources, some of which are especially important during periods of food scarcity, that would otherwise be inaccessible (Parker and Gibson 1977, Yamakoshi 1998). This has led researchers to argue that access to difficult-to-process food resources has played a large role in the evolution of primate, including hominin, tool use (Parker and Gibson 1977, van Schaik et al. 1999). It should also be noted that some of the earliest-known tools in the hominin archaeological record were most likely used to process foods (Bunn and Kroll 1986, Schick and Toth 1993, Toth 1997). However, more research focused on tool use among free-ranging primates and how their tool use specifically relates to different ecological factors is needed before this can be fully addressed.

SUMMARY

In summary, although object manipulation is common among primates, regular reports of tool use, including captive data, are limited to a few genera: *Pan*, *Pongo*, *Cebus*, and to a lesser extent *Gorilla*, *Macaca*, and *Papio*. Tool use has been observed under free-ranging conditions in *Pan*, *Pongo*, *Cebus*, and rarely in *Macaca* and *Papio* but never in *Gorilla*. *Pan*, specifically chimpanzees (*Pan troglodytes*), is the only primate genus that is known to habitually use tools in every known free-ranging population, and tool manufacture in the wild is currently limited to *Pan* and *Pongo*. In all of the primate genera known to use tools in the wild, their tool use varies at the population level and across age/sex classes. The great apes tend to solve tool-using problems insightfully, whereas monkeys tend to rely on trial and error to solve problems using tools. Therefore, although tool use can be found among both monkeys and apes, monkeys and apes seem to differ widely in their approach to tool use.

Tool use is often used as a proxy for overall cognitive ability; however, there is little evidence to support this. Primate tool users do not necessarily perform better than other primates using other common measures of cognition, such as performance on discrimination learning tests, mirror self-recognition, tactical deception, and social learning processes. The only common measure of cognition that shows some relationship to tool use is brain size, although the relationship is far from perfect. Therefore, there is little evidence to support the idea that a species' ability to solve problems using tools necessarily corresponds with greater overall cognitive ability.

Regarding the evolution of tool use in primates, most research focus has been limited to studying the evolution of tool use in our early hominin ancestors, and broader questions relating to the evolution of tool use in nonhuman primates have largely been ignored. Although direct evidence is lacking, indirect data using primarily a phylogenetic approach suggest that tool use evolved at least three times during primate evolution (once in the line leading to *Cebus*, once in the line leading to *Macaca/Papio*, and once in the line leading to the great apes and humans) and most likely evolved as a foraging adaptation.

REFERENCES

Adams-Curtis, L. E. (1990). Conceptual learning in capuchin monkeys. *Folia Primatol.* 54:129–137.

Allman, J. M., McLaughlin, T., and Hakeem, A. (1993). Brain structures and life-span in primate species. *Proc. Natl. Acad. Sci. USA* 90:3559–3563.

Anderson, J. R. (1985). Development of tool-use to obtain food in a captive group of *Macaca tonkeana*. *J. Hum. Evol.* 14:637–645.

Anderson, J. R. (1990). Use of objects as hammers to open nuts by capuchin monkeys (*Cebus apella*). *Folia Primatol.* 54:138–145.

Anderson, J. R., Kuroshima, H., Kuwahata, H., Fujita, K., and Vick, S. J. (2001). Training squirrel monkeys (*Saimiri sciureus*) to deceive: acquisition and analysis of behavior toward cooperative and competitive trainers. *J. Comp. Psychol.* 115:282–293.

Bauchot, R. (1982). Brain organization and taxonomic relationships in Insectivora and Primates. In: Armstrong, E. (ed.), *Primate Brain Evolution*. Plenum Press, New York. pp. 163–174.

Beck, B. (1980). *Animal Tool Behavior: The Use and Manufacture of Tools by Animals*. Garland Press, New York.

Bekoff, M. (2002). Animal reflections. *Nature* 419:255.

Beran, M., Gibson, K., and Rumbaugh, D. (1999). Predicting hominid intelligence from brain size. In: Corballis, M. C., and Lea, S. E. G. (eds.), *The Descent of Mind: Psychological Perspectives on Hominid Evolution*. Oxford University Press, New York. pp. 88–97.

Boesch, C., and Boesch, H. (1981). Sex differences in the use of natural hammers by wild chimpanzees: a preliminary report. *J. Hum. Evol.* 10:585–593.

Boesch, C., and Boesch, H. (1990). Tool use and tool making in wild chimpanzees. *Folia Primatol.* 54:86–99.

Boesch, C., and Boesch-Achermann, H. (2000). *The Chimpanzees of the Taï Forest*. Oxford University Press, Oxford.

Boinski, S. (1988a). Use of a club by a wild white-faced capuchin (*Cebus capucinus*) to attack a venomous snake (*Bothrops asper*). *Am. J. Primatol.* 14:177–179.

Boinski, S. (1988b). Sex differences in the foraging behavior of squirrel monkeys in a seasonal habitat. *Behav. Ecol. Sociobiol.* 23:177–186.

Boinski, S., Quatrone, R., and Swats, H. (2000). Substrate and tool use by brown capuchins in Suriname: ecological contexts and cognitive bases. *Am. Anthropol.* 102:741–761.

Brewer, S. M., and McGrew, W. C. (1990). Chimpanzee use of a tool-set to get set honey. *Folia Primatol.* 54:100–104.

Bunn, H. T., and Kroll, E. M. (1986). Systematic butchery by Plio-Pleistocene hominids at Olduvai Gorge, Tanzania. *Curr. Anthropol.* 27:431–452.

Byrne, J. M. (1993). Evolving a theory of mind: the nature of non-verbal mentalism in other primates. In: Baron-Cohen, S., Tager Flusberg, H., and Cohen, D. J. (eds.), *Understanding Other Minds*. Oxford University Press, Oxford. pp. 367–396.

Byrne, J. M. (1994). The evolution of intelligence. In: Slater, P. J. B., and Halliday, T. R. (eds.), *Behaviour and Evolution*. Cambridge University Press, Cambridge. pp. 223–265.

Byrne, J. M. (1996). The misunderstood ape: cognitive skills of the gorilla. In: Russon, A. E., Bard, K. A., and Parker, S. T. (eds.), *Reaching into Thought: The Minds of the Great Apes*. Cambridge University Press, Cambridge. pp. 111–130.

Byrne, J. M., and Whiten, A. (1992). Cognitive evolution in primates: evidence from tactical deception. *Man* 27:609–627.

Byrne, R. W., and Byrne, J. M. E. (1993). Complex leaf-gathering skills of mountain gorillas (*Gorilla g. beringei*): variability and standardization. *Am. J. Primatol.* 31:241–261.

Camelleri, C., and Shine, R. (1990). Sexual dimorphism and dietary divergences: differences in trophic morphology between male and female snakes. *Copeia* 1990:649–658.

Chapman, C. A. (1986). Boa constrictor predation and group response in white-faced *Cebus* monkeys. *Biotropica* 18:171–172.

Chevalier-Skolnikoff, S. (1983). Sensorimotor development in orang-utans and other primates. *J. Hum. Evol.* 12:545–561.

Chevalier-Skolnikoff, S. (1990). Tool use by wild *Cebus* monkeys at Santa Rosa National Park, Costa Rica. *Primates* 31:375–383.

Clutton-Brock, T. H. (1977). Some aspects of intraspecific variation in feeding and ranging behavior in primates. In: Clutton-Brock, T. H. (ed.), *Primate Ecology*. Academic Press, New York. pp. 539–556.

Cooper, L. R., and Harlow, H. F. (1961). Note on a cebus monkey' use of a stick as a weapon. *Psychol. Rep.* 8:418.

Custance, D., Whiten, A., and Fredman, T. (1999). Social learning of an artificial fruit task in capuchin monkeys (*Cebus apella*). *J. Comp. Psychol.* 113:13–23.

Deaner, R. O., Nunn, C. L., and van Schaik, C. P. (2000). Comparative tests of primate cognition: different scaling methods produce different results. *Brain Behav. Evol.* 55:44–52.

de Veer, M. W., and van den Bos, R. (1999). A critical review of methodology and interpretation of mirror self-recognition research in non-human primates. *Anim. Behav.* 58:459–468.

Dit Durell, S. E. A., Goss-Custard, J. D., and Caldow, R. W. G. (1993). Sex-related differences in diet and feeding method in the oystercatcher *Haematopus ostralegus*. *J. Anim. Ecol.* 62:205–215.

Ducoing, A. M., and Thierry, B. (1998). An experimental study of deception in semi-free-ranging Tonkean macaques. *Am. J. Primatol.* 45:177–178.

Dunbar, R. M. (1995). Neocortex size and group size in primates: a test of the hypothesis. *J. Hum. Evol.* 28:287–296.

Enoksson, B. (1988). Age- and sex-related differences in dominance and foraging behaviour of nuthatches *Sitta europaea*. *Anim. Behav.* 36:231–238.

Fernandes, M. E. B. (1991). Tool use and predation of oysters (*Crassostrea rhizophorea*) by the tufted capuchin, *Cebus apella apella*, in brackish water mangrove swamp. *Primates* 32:529–531.

Fitch-Snyder, H., and Carter, J. (1993). Tool use to acquire drinking water by free-ranging lion-tailed macaques (*Macaca silenus*). *Lab. Primate Newslett.* 32:1–2.

Fobes, J. L., and King, J. E. (1982a). Complex learning by primates. In: Fobes, J. L., and King, J. E. (eds.), *Primate Behavior*. Academic Press, New York. pp. 327–360.

Fobes, J. L., and King, J. E. (1982b). Measuring primate learning abilities. In: Fobes, J. L., and King, J. E. (eds.), *Primate Behavior*. Academic Press, New York. pp. 289–326.

Fontaine, B., Moisson, P. Y., and Wickings, E. J. (1995). Observations of spontaneous tool making and tool use in a captive group of western lowland gorillas (*Gorilla gorilla gorilla*). *Folia Primatol.* 65:219–223.

Fox, E. A., Sitompul, A. F., and van Schaik, C. P. (1998). Intelligent tool use in wild Sumatran orangutans. In: Parker, S. T., Mitchell, R. W., and Miles, H. L. (eds.), *The Mentality of Gorillas and Orangutans: Comparative Perspectives*. Cambridge University Press, Cambridge. pp. 99–116.

Fox, E. A., Sitompul, A. F., and van Schaik, C. P. (1999). Intelligent tool use in wild Sumatran orangutans. In: Parker, S. T., Mitchell, R. W., and Miles, H. L. (eds.), *The Mentalities of Gorillas and Orangutans: Comparative Perspectives*. Cambridge University Press, Cambridge. pp. 99–116.

Fragaszy, D. M., and Boinski, S. (1995). Patterns of individual diet choice and efficiency of foraging in wedge-capped capuchin monkeys (*Cebus olivaceus*). *J. Comp. Psychol.* 109:339–348.

Fragaszy, D., and Visalberghi, E. (1996). Social learning in monkeys: primate "primacy" reconsidered. In: Heyes, C. M., and Galef, B. G. (eds.), *Social Learning in Animals: The Roots of Culture*. Academic Press, San Diego. pp. 65–84.

Fuentes, A. (1992). Object rubbing in Balinese macaques (*Macaca fascicularis*). *Lab. Primate Newslett.* 31:14–15.

Galdikas, B. M. F. (1982). Orang-utan tool-use at Tanjung Puting Reserve, central Indonesian Borneo. *J. Hum. Evol.* 10:19–33.

Galef, B. G. (1992). The question of animal culture. *Hum. Nat.* 3:157–178.

Galef, B. G. (2003). "Traditional" foraging behaviors of Norway and black rats. In: Fragaszy, D., and Perry, S. (eds.), *The Biology of Traditions: Models and Evidence*. Cambridge University Press, Cambridge. pp. 159–186.

Gallup, G. G. (1970). Chimpanzee: self-recognition. *Science* 167:86–87.

Gallup, G. G. (1998). Can animals empathize? *Sci. Am.* 9:66–71.

Garber, P. A., and Brown, E. (2002). Experimental field study of tool use in wild capuchins (*Cebus capucinus*): learning by association or learning by insight? *Am. J. Primatol.*

Gibson, K., Rumbaugh, D., and Beran, M. (2001). Bigger is better: primate brain size in relationship to cognition. In: Falk, D., and Gibson, K. (eds.), *Evolutionary Anatomy of the Primate Cerebral Cortex*. Cambridge University Press, Cambridge. pp. 79–93.

Goodall, J. (1964). Tool-using and aimed throwing in a community of free-living chimpanzees. *Nature* 201:1264–1266.

Goodall, J. (1986). *The Chimpanzees of Gombe: Patterns of Behavior*. Harvard University Press, Cambridge, MA.

Hall, K. R. L. (1963). Tool-using performances as indicators of behavioral adaptability. *Curr. Anthropol.* 4:479–487.

Harris, J. W. K. (1983). Cultural beginnings: Plio-Pleistocene archaeological occurrences from the Afar, Ethiopia. *Afr. Archaeol. Rev.* 1:3–31.

Harrison, M. J. (1983). Age and sex differences in the diet and feeding strategies of the green monkey, *Cercopithecus sabaeus*. *Anim. Behav.* 31:969–977.

Harvey, P. H., and Krebs, J. R. (1990). Comparing brains. *Science* 249:140–146.

Hauser, M. D., Kralik, J., Botto-Mahan, C., Garrett, M., and Oser, J. (1995). Self-recognition in primates: phylogeny and the salience of species-typical features. *Proc. Natl. Acad. Sci. USA* 92:10811–10814.

Hauser, M. D., Miller, C. T., Liu, K., and Gupta, R. (2001). Cotton-top tamarins (*Saguinus oedipus*) fail to show mirror-guided self-exploration. *Am. J. Primatol.* 53:131–137.

Heyes, C. M. (1998). Theory of mind in nonhuman primates. *Behav. Brain Sci.* 21:101–148.

Hirata, S., and Matsuzawa, T. (2001). Tactics to obtain a hidden food item in chimpanzee pairs (*Pan troglodytes*). *Anim. Cogn.* 4:285–295.

Huffman, M. A., and Hirata, S. (2003). Ecological and biological foundations of primate behavioral traditions. In: Fragaszy, D. M., and Perry, S. (eds.), *The Biology of Traditions: Models and Evidence*. Cambridge University Press, Cambridge. pp. 267–296.

Hyatt, C. W., and Hopkins, W. H. (1994). Self-awareness in bonobos and chimpanzees: a comparative perspective. In: Parker, S. T., Mitchell, R. W., and Boccia, M. L. (eds.), *Self-Awareness in Animals and Humans: Developmental Perspectives*. Cambridge University Press, Cambridge. pp. 248–253.

Ingmanson, E. J. (1996). Tool-using behavior in wild *Pan paniscus*: social and ecological considerations. In: Russon, A. E., Bard, K. A., and Parker, S. T. (eds.), *Reaching into Thought: The Minds of the Great Apes*. Cambridge University Press, Cambridge. pp. 190–210.

Inoue-Nakamura, N. (1997). Mirror self-recognition in nonhuman primates: a phylogenetic approach. *Jpn. Psychol. Res.* 39:266–275.

Inoue-Nakamura, N., and Matsuzawa, T. (1997). Development of stone tool use by wild chimpanzees (*Pan troglodytes*). *J. Comp. Psychol.* 111:159–173.

Itani, J. (1958). On the acquisition and propagation of new food habits in the troop of Japanese monkeys at Takasakiyama. *Primates* 1:84–98.

Jalles-Filho, E., Grassetto, R., de Cunha, T., and Salm, R. A. (2001). Transport of tools and mental representation: is capuchin monkey tool behaviour a useful model of Plio-Pleistocene hominid technology? *J. Hum. Evol.* 40:365–377.

Jerison, H. J. (1973). *Evolution of the Brain and Intelligence*. Academic Press, San Diego.

Jerison, H. J. (2000). The evolution of intelligence. In: Sternberg, R. J. (ed.), *Handbook of Intelligence*. Cambridge University Press, Cambridge. pp. 216–244.

Jordan, C. (1982). Object manipulation and tool-use in captive pygmy chimpanzees (*Pan paniscus*). *J. Hum. Evol.* 11:35–39.

Kibunjia, M., Roche, H., Brown, F., and Leakey, R. E. F. (1992). Pliocene and Pleistocene archaeological sites west of Lake Turkana. *J. Hum. Evol.* 23:431–438.

Kitahara-Frisch, J., and Norikoshi, K. (1982). Spontaneous sponge-making in captive chimpanzees. *J. Hum. Evol.* 11:41–47.

Kohler, W. (1927). *The Mentality of Apes*. Kegan Paul, Trench, Trubner, London.

Leakey, L. S. B. (1935). *The Stone Age Races of Kenya*. Oxford University Press, London.

Le Boeuf, B. J., Crocker, D. E., Costa, D. P., Blackwell, S. B., Webb, P. M., and Houser, D. S. (2000). Foraging ecology of northern elephant seals. *Ecol. Monogr.* 70:353–382.

Lefebvre, L., and Palameta, B. (1988). Mechanisms, ecology, and population difussion of socially learned, food-finding behavior in feral pigeons. In: Zentall, T. R., and Galef, B. G. (eds.), *Social Learning: Psychological and Biological Perspectives*. Erlbaum, Hillsdale, NJ. pp. 141–164.

Lovern, M. B. (2000). Behavioral ontogeny in free-ranging juvenile male and female green anoles, *Anolis carolinensis*, in relation to sexual selection. *J. Herpetol.* 34:274–281.

Magurran, A. E., and Macias Garcia, C. (2000). Sex differences in behaviour as an indirect consequence of mating system. *J. Fish Biol.* 57:839–857.

Martin, R. D. (1982). Allometric approaches to the evolution of the primate nervous system. In: Armstrong, E., and Falk, D. (eds.), *Primate Brain Evolution: Methods and Concepts*. Plenum Press, New York. pp. 39–56.

Matsuzawa, T. (2001). Primate foundations of human intelligence: a view of tool use in nonhuman primates and fossil homininds. In: Matsuzawa, T. (ed.), *Primate Origins of Human Cognition and Behavior*. Springer-Verlag, Hong Kong. pp. 3–25.

McGrew, W. C. (1974). Tool use by wild chimpanzees in feeding upon driver ants. *J. Hum. Evol.* 3:501–508.

McGrew, W. C. (1979). Evolutionary implications of sex differences in chimpanzee predation and tool use. In: Hamburg, D. A., and McCown, E. R. (eds.), *The Great Apes*. Benjamin/Cummings, Menlo Park, CA. pp. 440–463.

McGrew, W. C. (1992). *Chimpanzee Material Culture: Implications for Human Evolution*. Cambridge University Press, Cambridge.

McGrew, W. C. (1993). The intelligent use of tools: twenty propositions. In: Gibson, K., and Ingold, T. (eds.), *Tools, Language and Cognition in Human Evolution*. Cambridge University Press, Cambridge. pp. 151–170.

McGrew, W. C. (1998). Culture in nonhuman primates? *Annu. Rev. Anthropol.* 27:301–328.

McGrew, W. C., and Marchant, L. F. (1992). Chimpanzees, tools, and termites: hand preference or handedness? *Curr. Anthropol.* 33:114–119.

McGrew, W. C., and Marchant, L. F. (1997). Using the tools at hand: manual laterality and elementary technology in *Cebus* spp. and *Pan* spp. *Int. J. Primatol.* 18:787–810.

Meador, D. M., Rumbaugh, D., Pate, J. L., and Bard, K. A. (1987). Learning, problem solving, cognition, and intelligence. In: Mitchell, G., and Erwin, J. (eds.), *Comparative Primate Biology*. Alan R. Liss, New York. pp. 17–83.

Mercader, J., Panger, M., and Boesch, C. (2002). Excavation of a chimpanzee stone tool site in the African rainforest. *Science* 296:1452–1455.

Miles, H. L. (1994). ME CHANTEK: the development of self-awareness in a signing orangutan. In: Parker, S. T., Mitchell, R. W., and Boccia, M. L. (eds.), *Self-Awareness in Animals and Humans: Developmental Perspectives*. Cambridge University Press, Cambridge. pp. 254–272.

Milton, K. (1981). Distribution patterns of tropical plant foods as an evolutionary stimulus to primate mental development. *Am. Anthropol.* 83:534–548.

Mitchell, R. W., and Anderson, J. R. (1997). Pointing, withholding information, and deception in capuchin monkeys (*Cebus apella*). *J. Comp. Psychol.* 111:351–361.

Nash, V. J. (1982). Tool use by captive chimpanzees at an artificial termite mound. *Zoo Biol.* 1:211–221.

Oyen, O. J. (1979). Tool-use in free-ranging baboons of Nairobi National Park. *Primates* 20:595–597.

Panger, M. A. (1998). Object-use in free-ranging white-faced capuchins (*Cebus capucinus*) in Costa Rica. *Am. J. Phys. Anthropol.* 106:311–321.

Panger, M. A., Brooks, A., Richmond, B. G., and Wood, B. (2002a). Older than the Oldowan? Rethinking the emergence of hominin tool use. *Evol. Anthropol.* 11:235–245.

Panger, M. A., Perry, S., Rose, L., Gros-Louis, J., Vogel, E., MacKinnon, K. L., and Baker, M. (2002b). Cross-site differences in foraging behavior of white-faced capuchins (*Cebus capucinus*). *Am. J. Phys. Anthropol.* 119:52–66.

Parker, S. T., and Gibson, K. (1977). Object manipulation, tool use and sensorimotor intelligence as feeding adaptations in *Cebus* monkeys and great apes. *J. Hum. Evol.* 6:623–641.

Parker, S. T., Kerr, M., Markowitz, H., and Gould, J. (1999). A survey of tool use in zoo gorillas. In: Parker, S. T., Mitchell, R. W., and Miles, H. L. (eds.), *The Mentalities of Gorillas and Orangutans*. Cambridge University Press, Cambridge. pp. 188–193.

Parker, S. T., and McKinney, M. L. (1999). *Origins of Intelligence: The Evolution of Cognitive Development in Monkeys, Apes, and Humans*. Johns Hopkins University Press, Baltimore.

Parker, S. T., and Mitchell, R. W. (1994). Evolving self-awareness. In: Parker, S. T., Mitchell, R. W., and Boccia, M. L. (eds.), *Self-Awareness in Animals and Humans: Developmental Perspectives*. Cambridge University Press, Cambridge. pp. 413–428.

Passingham, R. E. (1975). The brain and intelligence. *Brain Behav. Evol.* 11:1–15.

Patterson, F. G., and Cohn, R. H. (1994). Self-recognition and self-awareness in lowland gorillas. In: Parker, S. T., Mitchell, R. W., and Boccia, M. L. (eds.), *Self-Awareness in Animals and Humans: Developmental Perspectives*. Cambridge University Press, Cambridge. pp. 273–290.

Persons, M. H., and Uetz, G. W. (1999). Age and sex-based differences in the use of prey sensory cues in wolf spiders (Araneae: Lycosidae). *J. Insect Behav.* 12:723–736.

Phillips, K. (1998). Tool use in wild capuchin monkeys (*Cebus albifrons trinitatis*). *Am. J. Primatol.* 46:259–261.

Povinelli, D. J., Rulf, A. B., Landau, K. R., and Bierschwale, D. T. (1993). Self-recognition in chimpanzees (*Pan troglodytes*): distribution, ontogeny, and patterns of emergence. *J. Comp. Psychol.* 107:347–372.

Radinsky, L. (1982). Some cautionary notes on making inferences about relative brain size. In: Armstrong, E., and Falk, D. (eds.), *Primate Brain Evolution: Methods and Concepts*. Plenum Press, New York. pp. 29–37.

Reader, S. M., and Laland, K. N. (2002). Social intelligence, innovation, and enhanced brain size in primates. *Proc. Natl. Acad. Sci. USA* 99:4436–4441.

Reiss, D., and Marino, L. (2001). Mirror self-recognition in the bottlenose dolphin: a case of cognitive convergence. *Proc. Natl. Acad. Sci. USA* 98:5937–5942.

Rendell, L., and Whitehead, H. (2001). Culture in whales and dolphins. *Behav. Brain Sci.* 24:309–324.

Roche, H., Delagnes, A., Brugal, J. P., Feibel, C., Kibunjia, M., Mourrel, V., and Texier, P. J. (1999). Early hominid stone tool production and technical skill 2.34 myr ago in West Turkana, Kenya. *Nature* 399:57–60.

Rose, L. M. (1994). Sex differences in diet and foraging behavior in white-faced capuchins, *Cebus capucinus*. *Int. J. Primatol.* 15:63–82.

Russon, A. E., and Galdikas, B. M. F. (1992). Imitation in ex-captive orangutans (*Pongo pygmaeus*). *J. Comp. Psychol.* 107:147–161.

Rusterholz, H., and Erhardt, A. (2000). Can nectar properties explain sex-specific preferences in the Adonis Blue butterfly *Lysandra bellargus*? *Ecol. Entomol.* 25:81–90.

Sakura, O., and Matsuzawa, T. (1991). Flexibility of wild chimpanzee nut-cracking behavior using stone hammers an anvils: an experimental analysis. *Ethology* 87:237–248.

Sano, M. (1993). Foraging activities and diets of males and females in a haremic sandperch (Pisces: Pinguipedidae). *Mar. Ecol. Prog. Ser.* 98:55–59.

Savage-Rumbaugh, E. S. (1986). *Ape Language: From Conditioned Response to Symbol*. Columbia University Press, New York.

Schick, K. D., and Toth, N. (1993). *Making Silent Stones Speak*. Simon and Schuster, New York.

Schick, K. D., Toth, N., Garufi, G., Savage-Rumbaugh, E. S., Rumbaugh, D., and Sevcik, R. (1999). Continuing investigations into the stone tool-making and tool-using capabilities of a bonobo (*Pan paniscus*). *J. Archaeol. Sci.* 26:821–832.

Semaw, S., Renne, P., Harris, J. W. K., Feibel, C. S., Bernor, R. L., Fesseha, N., and Mowbray, K. (1997). 2.5-million-year-old stone tools from Gona, Ethiopia. *Nature* 385:333–336.

Sinha, A. (1997). Complex tool manufacture by a wild bonnet macaque, *Macaca radiata*. *Folia Primatol.* 68:23–25.

Suarez, S. D., and Gallup, G. G. (1981). Self-recognition in chimpanzees and orangutans, but not gorillas. *J. Hum. Evol.* 10:175–188.

Sugiyama, Y. (1997). Social tradition and the use of tool-composites by wild chimpanzees. *Evol. Anthropol.* 6:23–27.

Tokida, E., Tanaka, I., Takefushi, H., and Hagiwara, T. (1994). Tool-using Japanese macaques: use of stones to obtain fruit from a pipe. *Anim. Behav.* 47:1023–1030.

Tomasello, M., and Call, J. (1997). *Primate Cognition*. Oxford University Press, New York.

Torigoe, T. (1985). Comparison of object manipulation among 74 species of nonhuman primates. *Primates* 26:182–194.

Toth, N. (1997). The artefact assemblages in the light of experimental studies. In: Isaac, G. L. (ed.), *Koobi Fora Research Project. Plio-Pleistocene Archaeology*, vol. 5. Claredon Press, Oxford. pp. 361–388.

Toth, N., Schick, K. D., Savage-Rumbaugh, E. S., Sevcik, R., and Rumbaugh, D. (1993). *Pan* the tool-maker: investigations into the stone tool-making and tool-using capabilities of a bonobo (*Pan paniscus*). *J. Archaeol. Sci.* 20:81–91.

van Lawick-Goodall, J., van Lawick, H., and Packer, H. (1973). Tool-use in free living babbons in the Gombe National Park. *Nature* 241:212–213.

van Schaik, C. P., Ancrenaz, M., Borgen, G., Galdikas, B., Knott, C. D., Singleton, I., Suzuki, A., Utami, S. S., and Merrill, M. (2003). Orangutan cultures and the evolution of material culture. *Science* 299:102–105.

van Schaik, C. P., Deaner, R. O., and Merrill, M. Y. (1999). The conditions for tool use in primates: implications for the evolution of material culture. *J. Hum. Evol.* 36:719–741.

van Schaik, C. P., Fox, E. A., and Sitompul, A. F. (1996). Manufacture and use of tools in wild Sumatran orangutans: implications for human evolution. *Naturwissenschaften* 83:186–188.

Visalberghi, E. (1987). Acquisition of nut-cracking behaviour by 2 capuchin monkeys (*Cebus apella*). *Folia Primatol.* 49:168–181.

Visalberghi, E. (1990). Tool use in *Cebus. Folia Primatol.* 54:146–154.

Visalberghi, E., and Fragaszy, D. (1990). Do monkeys ape? In: Parker, S. T., and Gibson, K. (eds.), *"Language" and Intelligence in Monkeys and Apes*. Cambridge University Press, Cambridge. pp. 247–273.

Visalberghi, E., and Tomasello, M. (1998). Primate causal understanding in the physical and psychological domains. *Behav. Proc.* 42:189–203.

Visalberghi, E., and Trinca, L. (1989). Tool use in capuchin monkeys: distinguishing between performing and understanding. *Primates* 30:511–521.

Weir, A. A. S., Chappell, J., and Kacelnik, A. (2002). Shaping of hooks in New Caldonian crows. *Nature* 297:981.

Westergaard, G. C. (1992). Object manipulation and the use of tools by infant baboons (*Papio cynocephalus anubis*). *J. Comp. Psychol.* 106:398–403.

Westergaard, G. C., and Fragaszy, D. M. (1987). The manufacture and use of tools by capuchin monkeys (*Cebus apella*). *J. Comp. Psychol.* 101:159–168.

Westergaard, G. C., Greene, J. A., Babitz, M. A., and Suomi, S. J. (1995). Pestle use and modification by tufted capuchins (*Cebus apella*). *Int. J. Primatol.* 16:643–651.

Westergaard, G. C., Lundquist, A. L., Kuhn, H. E., and Suomi, S. J. (1997). Ant-gathering with tools by captive tufted capuchins (*Cebus apella*). *Int. J. Primatol.* 18:95–103.

Westergaard, G. C., and Suomi, S. J. (1996). Hand preference for stone artefact production and tool-use by monkeys: possible implications for the evolution of right-handedness in hominids. *J. Hum. Evol.* 30:291–298.

Whitehead, P. J., and Tschirner, K. (1992). Sex and age related variation in foraging strategies of magpie geese *Anseranas semipalmata. Emu* 92:28–32.

Whiten, A., Goodall, J., McGrew, W. C., Nishida, T., Reynolds, V., Sugiyama, Y., Tutin, C. E. G., Wrangham, R. W., and Boesch, C. (1999). Cultures in chimpanzees. *Nature* 399:682–685.

Whiten, A., and Ham, R. (1992). On the nature and evolution of imitation in the animal kingdom: reappraisal of a century of research. In: Slater, P. J. B., Rosenblatt, J. S., Beer, C., and Milinski, M. (eds.), *Advances in the Study of Behavior.* Academic Press, San Diego. pp. 239–283.

Williams, J. B. (1978). *Tool Use in Nonhuman Primates*. Primate Information Center, Seattle.

Williams, J. B. (1984). *Tool Use by Nonhuman Primates: A Bibliography, 1975–1983*. Primate Information Center, Seattle.

Williams, J. B. (1992). *Tool Use by Nonhuman Primates: A Bibliography, 1983–1992*. Primate Information Center, Seattle.

Wood, B., and Richmond, B. G. (2000). Human evolution: taxonomy and paleobiology. *J. Anat.* 197:19–60.

Wright, R. V. S. (1972). Imitative learning of a flaked stone technology—the case of an orangutan. *Mankind* 8:296–306.

Yamakoshi, G. (1998). Dietary responses to fruit scarcity of wild chimpanzees at Bossou, Guinea: possible implications for ecological importance of tool use. *Am. J. Phys. Anthropol.* 106:283–295.

Yerkes, R. M. (1943). *Chimpanzees: A Laboratory Colony*. Yale University Press, New Haven.

Zeltner, E., Klein, T., and Huber-Eicher, B. (2000). Is there social transmission of feather pecking in groups of laying hen chicks? *Anim. Behav.* 60:211–216.

Zentall, T. R. (2000). Animal intelligence. In: Sternberg, R. J. (ed.), *Handbook of Intelligence*. Cambridge University Press, Cambridge. pp. 197–215.

42

Primate Self-Medication

Michael A. Huffman

INTRODUCTION

Whether prosimian or great ape, a primate's existence is closely intertwined with that of parasites and pathogens. There is a great diversity of host–parasite relationships across the animal kingdom (see Clayton and Moore 1997, Moore 2002). Parasites and pathogens have a strong influence on their hosts, as is attested by the variety of effects they have on the overall behavior and reproductive fitness of individuals infected by them (e.g., Beckage 1997, Hart 1990, Holmes and Zohar 1990). While parasites are considered to be of little problem to the host (e.g., Ashford et al. 2000), when not kept in check the consequences can be great (cf. Allison 1982, Brack 1987, Huffman et al. 1996b, Toft et al. 1991). Hence, when the well-being of an individual is compromised to the point that daily activities no longer become possible or when reproduction and infant rearing are affected, it is unquestionably in that individual's interest to respond.

Growing awareness and interest in the way that hosts are able to suppress or prevent the deleterious effects of parasitism and other causes of illness via behavioral means has led to increased attention on the evolution and operation of such a system (e.g., Hart 1988; Huffman 1997, 2001; Lozano 1998; Engel 2002). This field of study is referred to as *zoopharmacognosy* (Rodriguez and Wrangham 1993, Berry et al. 1995) or *self-medication* (e.g., Huffman 1997, 2001; Lozano 1998). The original definition given by Rodriguez and Wrangham (1993) included only the use of medicinal plants and their pharmacological properties, hence the root word *pharmacognosy*. Advancement of this field in recent years has shed new light on known and new behaviors incorporating nonplant materials and indeed nonpharmacological means of parasite control using plants, making this term's general applicability obsolete. In this chapter, as in my previous works, the term *self-medication* is used, defined as those behavioral strategies by which animals avoid or suppress disease transmission and treat or control disease and/or its symptoms, thereby directly or indirectly enhancing their health and reproductive fitness. The study of self-medication is devoted to understanding how animals respond to potential threats to their health and how these behaviors are maintained within a population.

Behavioral strategies for health maintenance typically form what Hart (1990) calls "the front line defense" against invasion by agents of disease such as nematodes, microparasites (protozoa, bacteria, viruses, fungi), ectoparasites and biting insects. I have argued that such behavioral strategies come into play when physiological adaptations are insufficient and behavioral avoidance, limited contact with, or a direct response to illness is warranted (Huffman and Caton 2001). From either perspective, the maintenance of behavioral strategies that insure basic survival and thus indirectly or directly enhance reproductive fitness is universally adaptive to all animals.

Self-medication may be operated by innate mechanisms, socially transmitted from generation to generation, or indeed a combination of the two (Huffman and Hirata 2003). At the ultimate level, behavioral propensities to perform such actions or basic dietary preferences may be selected for their direct benefits to health maintenance and increased reproductive fitness. Lozano (1998) divided self-medication into prophylactic and therapeutic modes and predicted that the level of occurrence and modes of acquisition should be different. The use of the term *prophylactic* implies intentionality and an understanding of the cause of illness and how to prevent it. Such capabilities have not yet been demonstrated in nonhuman animals, and as Lozano himself admits, the use of this classification is premature as future research may well show that some presumptive prophylactic behaviors are in fact examples of therapeutic self-medication.

At our current level of understanding, health-maintenance and self-medicative behaviors can be classified into five levels: *(1) sick behaviors*, such as lethargy, depression, anorexia, reduction in grooming, behavioral fever (self-induced rise in body temperature, e.g., sun basking), basking behavior (Kluger et al. 1975, Hart 1988); *(2) optimal avoidance or reduction of the possibility for disease transmission*, such as avoidance of feces contaminated food, water, substrates; *(3) dietary selection of items with a preventative or health-maintenance effect*, such as items eaten routinely in small amounts or on a limited basis; *(4) ingestion of a substance for the curative treatment of a disease or the symptoms thereof*, such as use of toxic or otherwise biologically active items at low frequency or in small amounts, having little or no nutritional value; and *(5) external application of a substance to the body for the treatment or control of a disease or condition*.

Some of the most extensively documented examples of animal self-medication come from primates. This chapter reviews the current evidence for self-medication and other possible forms of health maintenance in primates. Behavioral categories to be discussed in this chapter fall within levels 2–5.

BEHAVIORAL AVOIDANCE OF DISEASE TRANSMISSION

Perhaps the most fundamental behavioral strategy for maintaining health is to avoid or decrease the likelihood of disease transmission. This can occur in various ways. Waterborne diseases pose a threat to primates directly dependent on streams and ponds as their main water source. *Papio hamadryas* living near the city of Taif, Saudia Arabia, are known to dig drinking holes in the sand directly adjacent to the murky algae-tainted watering sites of livestock. They patiently wait for the filtered water to seep through the sand (A. Bough and A. Mori, personal communications). This appears to be a common behavior of this species at many locations (e.g., Nelson 1960, Kummer 1995). *Theropithecus gelada* living in the Semien Highlands of Ethiopia often drink from the cliff springs where they spend their nights, instead of down in the river beds. In addition to the obvious benefits of filtering out unwanted slime, dirt, and debris, Mori and Bough believe this is likely an efficient way for baboons to avoid fecal contamination and water-transmitted parasites such as the blood fluke *Schistosoma mansoni* found in stagnant ponds and slow running streams in these areas.

Hausfater and Meade (1982) describe the routine changing of sleeping grove sites by *Papio cynocephalus* in Kenya's Amboseli National Park as a strategy for avoiding parasite infection via restricted contact with infective ova and larvae of nematodes (*Oesophagostomum*, *Strongyloides*, and *Trichostrongylus*). *Cercocebus albigena* defecate throughout the day and appear not to be able to avoid, or even try to avoid, contaminating vegetation that they later use for feeding, traveling, or sleeping (Freeland 1980). Freeland (1980) proposed an alternative way that these monkeys may avoid infection from intestinal protozoa. He showed that *C. albigena* ranged more widely during dry periods than rainy periods, when parasite-contaminated fecal material was likely to stay longer on the leaves at Ngogo in the Kibale Forest, Uganda. In this study, neither food density nor an inhibition to travel in the rain could explain this difference (Freeland 1980). Interestingly, 10 km away in Kanyawara, Olupot et al. (1997) found that this species traveled significantly longer distances during rainy than dry periods and that fruit density did influence travel patterns (Olupot et al. 1997). Differences in food distribution and availability between the two sites are cited as a likely reason for this difference, suggesting that either food takes priority over risk of parasite infection or risk of infection may vary between even neighboring habitats. A study by Gillespie (2004) in the same region showed that risk of parasite infection differs significantly for some species between neighboring logged and unlogged habitats.

DIETARY SELECTION AND HEALTH MAINTENANCE

Medicinal Foods

Selecting a proper diet incorporating non-nutritional bioactive elements, so-called medicinal foods (Huffman 1997, Huffman et al. 1998), should be an important part of disease control and prevention. Johns (1990) argues that the herbal medicines and modern pharmaceuticals used by humans today replaced the nonnutritive secondary compounds typically present in the diets of animals in the wild.

In spite of the potential ill effects of secondary plant compounds (Howe and Westley 1988, Wink et al. 1993), primates commonly ingest them (e.g., Carrai et al. 2003; Chapman and Chapman 2002; Glander 1975, 1982; Janzen 1978; McKey 1978; Milton 1979; Oates et al. 1977, 1980; Wrangham et al. 1998; Wrangham and Waterman 1983). Janzen (1978) first suggested the possibility that their incidental ingestion by non-human primates might help to combat parasite infection. Lozano (1991) proposed that parasites have a significant impact on the foraging decisions of animals, comparable to that of basic nutritional requirements.

A preliminary look at the diet of several primate species suggests that there may be a continuum from mildly bioactive and highly nutritious to minimally nutritious and highly bioactive items. Glander (1994) suggested that *Alouatta palliata* might benefit from the antiparasitic properties of various fig species in their diet. Freeland (1983) also noted figs as a possible mediating dietary factor in primate host–parasite interactions. Clinical trials show that *Ficus* spp. are effective against helminths in both humans (Hansson et al. 1986) and nonhuman animals (de Amorin et al. 1999). The susceptible parasites include *Ascaris* and *Trichuris* (Hansson et al. 1986), both of which have been noted in wild *A. palliata* (Stuart et al. 1998). Ficin is a proteolytic enzyme present in all figs (*Ficus* spp.) (Harborne et al. 1999), potentially active against intestinal parasites. Hansson et al. (1986) reported that concentrations as low as 0.05% of *F. glabrata* latex destroyed the cuticle of *Ascaris* and caused other lethal alterations in the parasites' morphology.

Pan and *Gorilla* across Africa frequently eat the pith and fruit of *Aframomum* (Zingiberaceae). In an extensive literature survey of the pharmacological properties of *Aframomum* eaten by *Gorilla*, a wide range of significant pharmacological properties, including potent bactericidal activities, were found (Cousins and Huffman 2002). Broad fungicidal and pathogen growth-inhibitory activities have also been reported (Adegoke and Skura 1994, Oloke et al. 1988).

Parasites were recognized to have a significant seasonal impact on the health and behavior of *Pan troglodytes schweinfurthii* at Mahale (Huffman et al. 1993, 1996b, 1997). This prompted myself and colleagues to form a multidisciplinary research consortium, The C.H.I.M.P.P. Group (Chemoethology of Hominoid Interactions with Medicinal Plants and Parasites; see Huffman 1994, Jacobsen and Hamel 1996), based at Kyoto University with affiliates worldwide, to conduct research on the pharmacologically based antiparasitic activity of plant foods in the diet of chimpanzees (e.g., Huffman et al. 1998; Koshimizu et al. 1993, Ohigashi et al. 1991b, 1994; Ohigashi 1995). Out of 172 native plant species eaten by chimpanzees at Mahale (Nishida and Uehara 1983), we found that 43 are used to treat parasitic or gastrointestinal illnesses traditionally by humans in Africa (Huffman et al. 1998). In 16 of these species, the same plant part(s) was utilized by both humans and chimpanzees, and 33% (20/63 plant parts) of the items ingested from them by chimpanzees corresponded to the parts utilized in ethnomedicine for the treatment of intestinal nematodes, dysentery, malaria, colic, and diarrhea and/or as an antiseptic (Huffman et al. 1998). Chimpanzees of M group eat these items only occasionally and typically in small amounts but significantly more frequently during rainy season months. This coincides with the identified period of reinfection by *Oesophagostomum stephanostomum*, the nodular worm associated with two therapeutic forms of self-medicative behavior at Mahale (see below). Of interest, in vitro and in vivo investigations of the activity of such "medicinal food" items detected significant anti-malarial and antischistosomal activity from extracts of *Combretum molle* (Combretaceae), *Erythrina abyssinica* (Papilionaceae), and *Stephania abyssinica* (Menispermaceae) (Huffman et al. 1998).

Krief and colleagues conducted a similar study on the health and pharmacological properties of the diet of Kanyawara group chimpanzees in the Kibale Forest, Uganda (Krief 2003, Krief et al. 2005). Significantly high activity was noted in compounds isolated from plant parts of three species only rarely eaten at Kanyawara, *Diospyros abyssinica* (Ebenaceae), *Uvariopsis congensis* (Annonaceae), and *Trichilia rubescens* (Meliaceae). From *T. rubescens* leaves, two novel compounds possessing significant antimalarial activities were discovered (Krief et al. 2004).

Phillips-Conroy (1996) described the ingestion of *Balanites aegyptiaca* (Balanitaceae) fruit by *Papio hamadryas* and *P. anubis* × *P. hamadryas* hybrid troops living in areas of acute susceptibility to infection by the blood fluke *Schistosoma mansoni* along the Awash River, Ethiopia. While the trees were equally abundant all along the course of the river in their study area, the fruits were eaten only by those troops in areas with the most likely risk of contact with the infective-stage cercaria that persist in warmer, slow-moving water sources where the secondary host snail is found. This led Phillips-Conroy to hypothesize a medicinal function for this food item. Cautious of not overinterpreting results, she left open both possibilities that this is either a dietary strategy with prophylactic effect or a curative treatment.

Subsequent laboratory experiments were conducted by Phillips-Conroy and Knopf (1986), feeding schistosome-infected mice with the active principle of the fruit, diosgenin (a plant hormone). Attempts were made to simulate both prophylactic ingestion (access to diosgenin in the diet before the infection continuing up to the end of the study) and curative ingestion (access only subsequent to infection up to the end of the study). The experiment was inconclusive in demonstrating an antischistosomal effect in either mode of ingestion. In fact, it was shown that for diosgenin-treated mice, liver weight (a pathology of the disease) and female schistosome egg output were greater than for control mice (Phillips-Conroy and Knopf 1986). The reader is cautioned, however, that these results do not disprove the original hypothesis that *Papio* benefit from the consumption of these fruits. As the investigators note, the interactions between parasite, hormone, and host are likely to be unique for each species. One important contribution this study has provided is a useful model for future testing.

The tendency for some, but not all, food items to contain active secondary compounds of potential health-maintenance value suggests that a category of "medicinal foods" exists. It is possible that some of the examples described above (e.g., *Trichilia rubescens*, *Diospyros abyssinica*) have remained obscure for lack of detailed behavioral and health-related data.

Geophagy

A recent detailed review of geophagy in primates has been presented by Krishnamani and Mahaney (2000); thus, the subject need only be briefly reviewed here. Clay contains none of the essential components of nutrition: protein, carbohydrates, lipids, and vitamins. Clay soils are fine-grained mineral deposits consisting mainly of the hydrous silicates of aluminum magnesium and/or iron. The chief groups of clay minerals are kaolinite, halloysite, illite, montmorillonite, and vermiculite, whose outstanding property is the capacity for holding water.

Geophagy is practiced by many nonhuman animal species and humans worldwide (e.g., Aufreiter et al. 1997, Bateson and LeBray 1978, Halstead 1968, Johns 1990, Klaus et al. 1998, Kreulen and Jager 1984, Wiley and Katz 1998). From the ethnographic literature, clay is consumed for its effectiveness as an antidiarrheal and is useful for the elimination of bitter taste and the prevention of stomachaches or vomiting after the consumption of large quantities of "famine foods" high in secondary compounds (Johns 1990). Currently, there

are at least 38 primate species for which geophagy has been reported (Krishnamani and Mahaney 2000). Similar to humans, in nonhuman primates it has been described as functioning to absorb tannins and other toxins acquired from a diet rich in secondary compounds, to relieve indigestion from a diet of restricted variation, and possibly to mediate endoparasitic infection (Davies and Baillie 1988, Knezevich 1998, Müller et al. 1997, Oates 1978, Wakibara et al. 2001). Overconsumption of clay, however, can lead to serious nutritional deficiencies and illness (e.g., Halstead 1968, Kreulen 1985).

William Mahaney and colleagues formed the Geophagy Research Project at York University (Toronto, Ontario) and have begun extensive studies in Africa (Rwanda, Uganda, Tanzania, Guinea), Asia (Sumatra, Borneo, Japan), and the Americas (Puerto Rico) to analyze soils ingested by primates. Thus far, studies of the mountain gorilla (*Gorilla gorilla beringei*) and rhesus monkey (*Macaca mulatta*) suggest that they ingest soil for rare minerals of possible nutritional value or to relieve diarrhea caused by dietary changes or parasitic infection (Mahaney et al. 1990, 1995, 1996, 1997). All clay ingested by humans and nonhuman primates analyzed thus far has been shown to have components closely resembling those of Kaopectate® (Pharmacia/Upjohn), a pharmaceutical commercially sold to treat gastrointestinal upset (Mahaney et al. 1990, 1995, 1996, 1997). Analogous in function to geophagy is charcoal eating, a behavior described for *Piliocolobus kirkii* on the island of Zanzibar, Tanzania, by Cooney and Struhsaker (1997) but also known to occur in other primates and, indeed, other mammals (Engel 2002).

THERAPEUTIC SELF-MEDICATIVE BEHAVIOR

At present, the African great apes offer the most detailed work on possible therapeutic self-medication in nonhuman primates. The hypothesis being developed from these investigations is that certain behaviors aid in the control of intestinal parasites and/or provide relief from related gastrointestinal upset. The two most clearly described behaviors are bitter pith chewing and leaf swallowing. Among the African great apes, one or both of these two proposed therapeutic behaviors have been documented from 15 sites and 24 communities at locations spanning their entire geographical distribution (Huffman 2001).

The majority of details discussed below for these two behaviors come from three study areas, Mahale (M group, now extinct K group), Gombe (Kasakela, now extinct Kahama), and Kibale (Kanywara, Ngogo), where research into chimpanzee self-medication began.

Parasites, Infection, and General Health at the Time of Self-Medication

A first step toward demonstrating therapeutic self-medication is identifying the disease being treated. In a longitudinal investigation of the intestinal parasite fauna of *Pan troglodytes schweinfurthii* conducted on the Mahale M group (Huffman et al. 1997), we showed that they are infected by three parasite species from three genera of nematode, *Strongyloides fuelleborni* (thread worm), *Trichuris trichiura* (whip worm), and *Oesophagostomum stephanostomum* (nodular worm); one genus of trematode, *Dicrocoelium lanceatum* (lancet fluke); and four genera of protozoa, *Entamoeba coli* (amoeba), *Endolimax nana* (amoeba), *Iodamoeba buetschlii* (amoeba), and *Giardia lamblia* (amoeba which causes giardiasis). All are parasites common to *Pan paniscus* and the other *Pan* subspecies in the wild (e.g., Ashford et al. 2000, File et al. 1976, Landsoud-Soukate et al. 1995, Murray et al. 2000).

Individuals of M group were monitored over time to detect weekly, monthly, and yearly changes in infection levels. Among all group members monitored in 1991–1992 and 1993–1994, a statistically significant seasonal difference was recognized only for the number of individuals infected by nodular worms (Fisher's exact test $p = 0.0001$ and $p = 0.01$, respectively; Huffman et al. 1997). Among all three nematode species detected, nodular worm infections were associated significantly more frequently with bitter pith chewing and leaf swallowing during the same two study periods than either whip worms or thread worms (Fisher's exact test for nodular and whip worms $p = 0.0142$ and for nodular and thread worms $p = 0.005$; Huffman et al. 1997).

Oesophagostomum stephanostomum is the most hazardous species of nodular worm found in the great apes (Brack 1987). Repeated infection occurs in the wild and causes significant complications including secondary bacterial infection, diarrhea, severe abdominal pain, weight loss, and weakness, resulting in high mortality (Brack 1987). Whip worm and thread worm infections usually are not serious and may go unnoticed in mild to moderate cases (Brack 1987). The serious effects of nodular worms on the host, however, suggest that it may be an important stimulus for bitter pith chewing and other anti-parasite behaviors.

Bitter Pith Chewing

Behavioral Ecology

Another important step toward demonstrating therapeutic self-medication is distinguishing its use from that of everyday food items and providing evidence for recovery from symptoms identified with the illness associated with its use. Bitter pith chewing is proposed to aid in the control of intestinal nematode infection via pharmacological action and relief from gastrointestinal upset (see Huffman 1997 for detailed review). The hypothesis that bitter pith chewing has medicinal value for chimpanzees was first proposed from detailed behavioral observations and parasitological and phytochemical analyses of patently ill individuals ingesting *Vernonia amygdalina* (Compositae) at Mahale (Huffman and Seifu 1989, Huffman et al. 1993).

Within the home range of M group, *Vernonia amygdalina* is neither abundant nor evenly distributed and usually occurs singly along or near streams. When ingesting the pith from young shoots, the outer bark and leaves are meticulously removed to chew on the exposed pith, from which they extract the bitter juice. The amount of pith ingested in a single bout is relatively small, ranging from portions of 5 to 120 cm × 1 cm. The entire process, depending on the amount ingested, takes anywhere from less than 1 to 8 min (in Huffman 1997). Often, mature conspecifics in proximity to an individual chewing *Vernonia* bitter pith show no interest in ingesting the pith (Huffman and Seifu 1989, Huffman et al. 1997). Infants, however, have been observed on occasion to taste small amounts of the pith discarded by their ill mothers. The ingestion of leaves or bark, the most abundant parts of the plant available for consumption by chimpanzees, is extremely rare.

At Mahale, use of *Vernonia amygdalina* has been recorded in all months except June and October (late dry season), demonstrating its year-round availability (Nishida and Uehara 1983). Despite this, plant use is highly seasonal and rare (Huffman 1997). General observations on the state of health at the time bitter pith was chewed were reported for four cases recorded between 1987 and 1996 in M group. Ill health was evidenced in all cases by the presence of diarrhea, malaise, and nematode infection (Huffman et al. 1997). Recovery within 20–24 hr from a lack of appetite, malaise, and constipation or diarrhea has also been reported in detail twice (Huffman and Seifu 1989, Huffman et al. 1993).

In one case, the intensity of the infection could be measured and was found to have dropped from an egg per gram (EPG) count of 130 to 15 within 20 hr. This was quite unusual when compared to seven other individuals with nodular worm infections monitored over the same period. Their nodular worm EPG levels conversely increased over time. The average increase was 69.9 EPG (standard deviation = 84, range 5–236) and reflects the rainy season trend for increased reinfection by nodular worms of chimpanzees at Mahale (Huffman et al. 1997).

Ethnomedicinal Evidence

Along with direct measurable evidence for a change in health conditions following a proposed self-medicative behavior, ethnomedicinal reports of plant activity and direct pharmacological analysis of compounds extracted from the specific plants used by the sick animals are essential steps toward demonstrating therapeutic self-medication. For numerous African ethnic groups, a concoction made from *Vernonia amygdalina* is prescribed treatment for malarial fever, schistosomiasis, amoebic dysentery, several other intestinal parasites, and stomachaches (Burkill 1985, Dalziel 1937, Huffman et al. 1996a, Watt and Breyer-Brandwinjk 1962).

The noted recovery time of 20–24 hr after bitter pith chewing in two individuals of M group is comparable to that of local human inhabitants, the Tongwe, who use cold concoctions of this plant as a treatment for malaria, intestinal parasites, diarrhea and stomach upset. These observations encouraged us to further investigate the possible contribution of plant secondary compounds in *Vernonia amygdalina* against parasitic infection. Phytochemical analysis of *V. amygdalina* samples collected at Mahale in 1989 and 1991 from plants used by sick chimpanzees revealed the presence of two major classes of bioactive compounds. From this work, a total of four known sesquiterpene lactones (vernodalin, vernolide, hydroxyvernolide, vernodalol), seven new stigmastane-type steroid glucosides (vernonioside A1–A4, B1–B3), and two freely occurring aglycones of these glucosides (vernoniol A1, B1) were isolated (Ohigashi et al. 1991a; Jisaka et al. 1992a,b, 1993a,b).

Phytochemical Evidence

The sesquiterpene lactones present in *Vernonia amygdalina*, are well known in many species of *Vernonia* for their antihelmintic, antiamoebic, antitumor, and antibiotic properties (Asaka et al. 1977; Gasquet et al. 1985; Jisaka et al. 1992a, 1993b; Kupchan et al. 1969; Toubiana and Gaudemer 1967). Koshimizu et al. (1993) also found inhibition of tumor promotion and immunosuppressive activities.

In vitro tests on the antischistosomal activity of the pith's most abundant steroid glucoside (vernonioside B1) and sesquiterpene lactone (vernodalin) showed significant inhibition of egg-laying capacity consistent with the observed decline in nodular worm EPG level 20 hr after a sick adult female in M group ingested *Vernonia amygdalina* pith (Jisaka et al. 1992b, Huffman et al. 1993).

In vitro tests on the antiparasitic activity of both the sesquiterpene lactones and the steroid glucosides using *Leishmania infantum* (produces visceral leishmaniasis), *Entamoeba histolytica* (produces amoebic dysentery), and a K1 multidrug-resistant strain of *Plasmodium falciparum* (produces falciparum malaria) have also been conducted and showed significant plasmodicidal activity (Ohigashi et al. 1994). In total, the evidence from parasitological, pharmacological, and ethnomedicinal observations is substantial and lends support to the hypothesis that bitter pith chewing is a therapeutic form of self-medication stimulated by, and controlling, nodule worm infection.

Leaf Swallowing

Behavioral Ecology

Leaf swallowing is currently proposed to control nodular worm infection (Huffman et al. 1996b) and relieve potential pain caused by tapeworm infection (Wrangham 1995). The substantiated mode of parasite control is a physical mechanism for the expulsion of parasites via the self-induced increase in gut motility that acts as a purge (Huffman and Caton 2001).

Leaf-swallowing behavior in the African great apes was first recorded for chimpanzees at Gombe and Mahale (Wrangham 1977, Wrangham and Nishida 1983). It came to the attention of Wrangham and Nishida at both sites that leaf swallowing was unlikely to provide any nutritional value after they found the folded, undigested leaves of *Aspilia*

mossambicensis (Compositae), *A. pluriseta*, and *A. rudis* in the dung. The interest generated in this behavior by the claim that *Aspilia* spp. leaves contained the powerful antibiotic compound thiarubrine A (Rodriguez et al. 1985; but see Page et al. 1997) stimulated many field researchers to look for leaf swallowing at their ape study sites (e.g., Boesch 1995, Dupain et al. 2002, Matsuzawa and Yamakoshi 1996, Pruetz and Johnson-Fulton 2003, Takasaki and Hunt 1987).

Leaf-swallowing behavior has since been observed in numerous populations of *Pan troglodytes schweinfurthii*, *P. t. troglodytes*, *P. t. verus*, *P. paniscus*, and *Gorilla gorilla graueri* across Africa (Huffman 2001). The number continues to increase yearly as fieldworkers venture into new study sites or look for the presence of folded, undigested leaves in the dung.

The behavior itself is quit distinct (see Huffman and Hirata 2004, online http://dx.doi.org/10.1007/s10329-003-0065-5 for mpg clip of this behavior). Leaves are taken from the distal end one at a time and are carefully folded by the tongue and palate as it is slowly pulled into the mouth, and then swallowed whole. The leaves' roughness makes them difficult to swallow, so folding them with the tongue and palate before swallowing is a necessary part of ingestion. An individual may swallow anywhere from one to 100 leaves in one bout, at a median rate of five leaves per minute (range 2.4–15.4 min, $n = 7$ bouts; data from Wrangham and Goodall 1989). An individual may do so more than once in a day and over several days (Huffman et al. 1996b, Wrangham and Nishida 1983). Leaves are typically swallowed within the first few hours after leaving the sleeping nest before the first meal (Boesch 1995, Wrangham and Goodall 1989, Wrangham and Nishida 1983, Huffman et al. 1996b) and, thus, on a relatively empty stomach. At Mahale, Gombe, Kibale, Lomako (Congo), and Fongoli (Senegal), where longitudinal data exist, leaf swallowing is an extremely rare behavior (Dupain et al. 2002, Pruetz and Johnson-Fulton 2003, Wrangham and Nishida 1983, Huffman 1997). Average rates of once every 69.0 hr ($n = 18$ bouts during 1,242 hr) and once every 81.9 hr ($n = 11$ bouts during 901.37 hr) for *Aspilia mossambicensis* at Mahale and once every 102.6 hr ($n = 10$ bouts during 1,026 hr) for *Aspilia* spp. at Gombe have been reported (Wrangham and Nishida 1983, Huffman 1997).

All 12 species whose leaves are swallowed whole at both Gombe and Mahale are available year-round. Based on botanical descriptions, this is expected to be the case for the more than 40 different species that were confirmed to be swallowed at sites across Africa. Nonetheless, interannual variation in the frequency of leaf swallowing at Mahale and Gombe is substantial. At Mahale, use is most common after the beginning of the rainy season (November–May), with peak frequencies in January and February being 10–12 times higher than those in other months (Wrangham and Nishida 1983).

At Kibale, direct observations were made only three times but systematic dung collection from December 1987 to May 1994 provided evidence ($n = 44$) for leaf swallowing by *Pan troglodyter schweinfurthii* in both Kanywara and Ngogo (Wrangham 1995). Unlike the distinct dry (June–October) and rainy (November–May) season pattern of Gombe and Mahale, rain falls more evenly throughout the year at Kanyawara, with two "wet" periods (approx. March–May and August–November) and two drier "transitional" periods (approx. December–February and June–July) but no extended dry season (Struhsaker 1997). Based on these criteria, no significant difference was detected in the frequency of leaf swallowing between the wet and transitional periods at Kanyawara (estimated by the author from Fig. 1 of Wrangham 1995).

Parasite Infection and General Health

Information on parasite infection and health at the time of leaf swallowing is so far available only from observations at Mahale and Kibale. This information varies in a number of important aspects, so these two sites will be discussed separately below.

Mahale. Behavioral profiles were collected concurrently with all cases of leaf swallowing observed during a 4-month study from December 1993 to February 1994 at Mahale (Huffman et al. 1996b). As part of this procedure, dung samples were systematically inspected macroscopically during focal-animal and ad libitum observations to check for the presence of whole leaves and parasites in the dung. General health was also monitored by stool type (diarrheal or normal) and collection of dung samples for microscopic parasitological examination. Nematode infection was demonstrated in 83% of all cases of leaf swallowing documented ($n = 10/12$ cases, 11 individuals, five from dung samples and seven from direct observation; Huffman et al. 1996b). Multiple-species infections were common, but nodular worm infections (78%) were most commonly associated with leaf-swallowing behavior, followed by those of thread worms (56%) and whip worms (33%). Infection by other intestinal parasite species was rare and inconsequential during this study period. As noted above, during the rainy season there is a significant trend for an increase in the number of individuals infected with nodular worms at Mahale. Thread worm infections remained low year-round, and whip worm infections were inconsistent from year to year in their rainy season–dry season prevalence of infection (Huffman et al. 1997). The presence of these two nematode species may have intensified symptoms of illness, but alone their impact on chimpanzees was not found to be substantial.

Symptoms which may be caused by infections from these nematodes (diarrhea, malaise, abdominal pain; Brack 1987, Anderson 1992) were verified from direct observation in seven of the eight chimpanzees at the time that leaves were swallowed (Huffman et al. 1996b, 1997). Nodular worms (body ca. 2.5 cm) were found in 3.7% ($n = 9/254$) of the dung collected from focal and nonfocal individuals. The occurrence of worms in the dung was rare and limited to individuals that displayed symptoms of malaise and diarrhea around the time leaf swallowing was observed.

Kibale. The most frequently occurring parasites in Kibale, Gombe, Mahale, and Budongo chimpanzees have been reported to be a nodular worm species, followed by thread worms and whip worms (Barrows 1996, File et al. 1976, Huffman et al. 1997, Rodriguez and Wrangham 1993). During a 78-month monitoring period of chimpanzee dung between 1987 and 1994 at Kanyawara, tapeworm proglottid fragments (*Bertiella studeri*) were found in 1% (16/1,696) of the total chimpanzee dungs inspected (Wrangham 1995). This occurrence of tapeworm fragments in the dung was restricted to a 6-month period (transitional–wet–transitional) referred to by Wrangham as the "period of tapeworm infection." Note, however, that the absence of tapeworm fragments in the dung does not rule out the possibility of tapeworm infection. Infection can be preceded by the presence of tapeworms for as long as 2 years before proglottids appear in the feces of humans (Beaver et al. 1984).

Direct observation of leaf swallowing at Kibale was observed only three times by Wrangham (1995). Parasite infection and a general evaluation of health around the time of leaf swallowing could only be inferred from macroscopic dung analysis. Of 12 dungs containing tapeworm fragments collected during this 6-month period, three were diarrheal and another three were scored as intermediate between diarrheal and normal. From this it was suggested that some individuals egesting tapeworm fragments may experience digestive disturbances, but it was not mentioned if these dungs contained leaves or not (Wrangham 1995).

Parasites and the Possible Functions and Mechanisms of Leaf Swallowing

The Mahale nodular worm hypothesis. The nematode control hypothesis derived from the Mahale leaf-swallowing data complements that of bitter pith chewing in function and target but differs in its proposed mechanism. At Mahale, both behaviors are sometimes displayed by the same individual on the same day and together may have a synergetic affect on nodular worm infection. In 1993–1994, six of the nine dungs ($n = 254$) found to contain worms also contained whole undigested leaves of *Aspilia mossambicensis*, *Trema orientalis*, or *Aneilema aequinoctiale*. The remaining three dungs containing worms but no leaves were collected from three of the individuals whose dung contained both leaves and worms earlier that day (Huffman et al. 1996b). Worms may continue to be expelled for a short while after leaves have been egested. In only one case were worms not found in the dung when leaves were present during the same study period. The relationship between the presence of both leaves and nodular worms in the dung was highly significant (Fisher's exact test, two-sided, $p = 0.0001$; Huffman et al. 1996b). From this it was concluded that a strong relationship between leaf swallowing and the expulsion of nodular worms exists.

The proposed mechanism for nodular worm expulsion is based solely on data collected from Mahale (Huffman and Caton 2001). Leaf swallowing appears to affect only nodular worms. Nodular worms inhabit the large intestine, where they attach by the buccal capsule. This attachment is not permanent, and worms move around the intestine in search of food and mates (Anderson 1992). Adult thread worms (2 mm) and whip worms (30–40 mm), on the other hand, are smaller and burrow into the mucosa of the small intestine and cecum, respectively, where they embed themselves firmly (Anderson 1992) and thus are not considered to be susceptible to mechanical removal by the leaves. The "control of nematode infection hypothesis" predicts that since nodular worm infections are typically self-limiting (Beaver et al. 1984), the total infection may be controllable if a chimpanzee periodically swallows leaves during the most likely period of reinfection. In general, for all nodular worm species, the peak period of reinfection is the rainy season (Krepel et al. 1995). At Mahale, the peak period of reinfection by nodular worms was found to occur 1–2 months after the onset of the rainy season around December or January (Huffman et al. 1997). This is also when chimpanzees tend to swallow leaves (and chew bitter pith) most frequently at Mahale and Gombe (Huffman et al. 1996b, 1997; Wrangham and Goodall 1989; Wrangham and Nishida 1983). The continued expulsion of adult worms over time from successive bouts of leaf swallowing may be partially responsible for the observed decline in total nodular worm egg burdens (directly correlated to the number of egg-laying adults) later in the rainy season (Huffman et al. 1997).

Nodular worms produce abdominal pain along with bowel irritation and diarrhea (Brack 1987), symptoms that could provide ample stimulus for leaf-swallowing behavior in chimpanzees infected with this nematode (Huffman and Seifu 1989; Huffman et al. 1993, 1996b).

The Kibale tapeworm hypothesis.   In 1993, significantly more leaves (8.5% of 271 dungs) were found in dungs during the 6-month period that tapeworm fragments were also found than in the dungs outside of this period (1.3% of 1,425 dungs collected between 1987–1992) (Mann-Whitney $z = 4.32$, $p < 0.001$; Wrangham 1995). Also, the proportion of dungs with both tapeworm fragments and whole leaves (21.7%, $n = 23$) was significantly higher than the proportion of dungs with whole leaves and no tapeworm fragments (3.6%, $n = 248$; binomial test $z = 4.46$, $p < 0.001$; Wrangham 1995). From this, Wrangham concluded that whole leaves in the gut increase the probability of tapeworm fragments being shed.

Whole leaves were found in dung during 21 months of the survey, while leaves and tapeworms were found together in only 33% of these months. It is possible that some tapeworm fragments went undetected. The greater overall occurrence of leaves in dung without tapeworms, however, suggests that the expulsion of proglottid fragments may not be the only function of leaf swallowing. Wrangham (1995) proposes that the stimuli for leaf swallowing as observed at Kibale could be abdominal pain caused by tapeworm infections and, thus, may function to alleviate pain.

Therapeutic self-medication by definition assumes the relief of discomfort as a driving motive behind its practice. In this light, further observational data from Kibale will help to expand upon our understanding of the antiparasitic function of leaf-swallowing behavior with regard to tapeworms.

Phytochemical Evidence Reevaluated

Thiarubrine A, a powerful antibiotic, antifungal, and anthelmintic secondary plant compound (Towers et al. 1985), was reported by Rodriguez et al. (1985) to be in the leaves of *Aspilia mossambicensis* and *A. pluriseta*. Rodriguez and Wrangham (1993) later proposed that thiarubrine A could act as a nematocide. Two alternative hypotheses were proposed to explain the function of swallowing *Aspilia* leaves based on the role of thiarubrine A: *(1)* buccal absorption (Newton and Nishida 1990) and *(2)* unchewed leaves as thiarubrine A–protective capsules (Rodriguez and Wrangham 1993). Both were proposed ad hoc to resolve the dilemma of how thiarubrine A would make it to the site of treatment without being destroyed by gastric acids.

However, after numerous attempts to replicate these phytochemical findings, Page and colleagues (1992, 1997) could detect no trace of thiarubrine A, B, D, E or their corresponding thiophenes in the leaves of 27 samples of three relevant *Aspilia* species collected from Mahale and Gombe; Nanyuki, Kenya; the Kampala District in Uganda; and greenhouse-grown specimens from Vancouver, Canada. Furthermore, analysis revealed no detectable antifungal activity against *Candida albicans* (Page et al. 1997), a human pathogenic yeast sensitive to thiarubrines (Towers et al. 1985). Messner and Wrangham (1996) found no nematocidal activity against infective L3-stage *Strongyloides* nematode larvae in the leaf extracts of *Rubia cordifolia* (Compositae). Neither of these phytochemical hypotheses provided an adequate universal explanation for leaf swallowing, and they have been abandoned (Huffman et al. 1996b, Page et al. 1997).

EXTERNAL APPLICATION OF SUBSTANCES TO THE BODY

Fur Rubbing

The health-maintenance and therapeutic self-medicative behaviors discussed to this point all have in common the ingestion of substances for internal action. Fur rubbing differs from these behaviors in that substance use is limited to external application. Both medicative and social functions of the behavior have been proposed in primates. Considerations of possible nutritional gain are irrelevant as the items used in these behaviors are applied to the body and not eaten. The wide taxonomic and geographic distribution of primates that are reported to display fur rubbing suggests that it could be even more widespread than currently reported among New World primates and prosimians (Table 42.1).

Table 42.1   Fur-Rubbing and Its Proposed Functions in Primates

SPECIES	SUBSTANCE USED	PARTS/CHEMICALS USED	LOCATION (SETTINGS)	PROPOSED FUNCTION
White-faced capuchins	*Citrus* sp.	Fruit	Nicoya Peninsula	Insect repellent
Cebus capucinus	*Soloanea teniflora*	Leaves, stem	Costa Rica (wild)[1]	
	Clematis dioica	Leaves, stem		
	Piper marginatum	Leaves, stem		
	Camponatus sericeiventris	Ant	Corcovado National Park Costa Rica (wild)[2]	Skin stimulation
	Gardening peat		Strasbourg, France (captive)[3]	Skin/pelage maintenance
Wedge-capped capuchins	Millipede	Whole insect	Ilanos, Venezuela (wild)[4]	Insect repellent
Cebus olivaceus	*Orthoporus dorsovittatus*	2 methyl–1,4-benzoquinone		
		2-methoxy–3 methyi–1		
		Benzoquinone		
Black-handed spider monkey	*Citrus aurantifolia*	Leaves	Barro Colorado Is.	Scent marking
Ateles geoffroyi	*Zanthozylum procerum*	Leaves	Panama. (wild)[5,6]	Insect repellent
	Zanthozylum belizense	Leaves		
Owl monkeys	*Piper marginatum*	Plant extract	DuMond Conservancy,	Insect repellent
Aotus boliviensis	Milliped	Whole insect	Florida (captive)[7]	
A. lemurinus griseimembra	*Anadenobolus monilicornis*			
A. nancymaae	Millipede extract	Benzoquinone, toluquinone, 2-methyl–1,4 – benzoquinine, hydroquinone extracts placed on filter paper		
	Onions, garlic and chives			
Black lemur	Millipede	Whole insect	Lokobe Forest—Nosy Be,	Control of ectoparasites
Eulemur macaco	*Charactopygus* sp.	Aldehydes, quinones	Madagascar (wild) [8]	
		Phenols, chlorine		

Source:
[1] Baker (1996); [2] Longino (1984); [3] Ludes and Anderson (1995); [4] Valderrama et al. (2000); [5] Campbell (2000); [6] Richard (1970); [7] Zito et al. (2002); [8] Birkinshaw (1999).

Behavioral Ecology

Fur rubbing has been observed both in the wild and under captive conditions (Table 42.1). To date, it has been reported for *Cebus capucinus*, *C. olivaceus*, *C. apella* (e.g., Baker 1996, Buckley 1983, Hill 1960, Ludes and Anderson 1995, Oppenheimer 1968, Valderrama et al. 2000), *Ateles geoffroyi*, *A. belzebuth* (Campbell 2000, Richard 1970, Klein 1972), *Aotus boliviensis*, *A. lemurinus griseimembra*, *A. nancymaae* (Zito et al. 2003), and *Eulemur macaco* (Birkinshaw 1999).

Plant material or insects can be applied directly or chewed and mixed with saliva, then rubbed frenziedly into the fur all over the body in a group and/or solitarily for some species. Baker (1996) studied fur rubbing in capuchins on the southeastern edge of the Nicoya Peninsula of Costa Rica. Fifty-three bouts were observed during a total of 1,140 hr (i.e., 1 bout/19.36 hr). Duration of time spent fur rubbing ranged from 10 sec to 41 min (mean 6.3 min). Solitary and group rubbing were observed. Capuchins used at least five different

plant genera including the fruits of four introduced *Citrus* spp. (Rutaceae) and the leaves and stems of *Sloanea terniflora* (Elaeocarpaceae), *Clematis dioica* L. (Ranunculaceae), and *Piper marginatum* (Piperaceae). *Citrus* (lemon, lime, orange) was the most frequently used genus, but use was often combined with other plant types. Citrus fruits were used exclusively during the dry season; however, fur rubbing occurred significantly more often in the rainy season than the dry season (chi squared = 15.8, $p < 0.05$; Baker 1996). The trend for an increase in fur rubbing during the rainy season was suggested to reflect the increase in risk of bacterial and fungal infection due to a rise in humidity. This suggests, but alone does not demonstrate, that the behavior is performed for the relief of symptoms caused by these ailments.

Aotus (Zito et al. 2003), *Cebus* (Longino 1984, Valderrama et al. 2000), and *Eulemur* (Birkinshaw 1999) are unique among primates for their use of millipede species (*Anadenobolus monilicornis*, *Charactopygus* sp., *Orthoporus dorsovittatus*) and ants as material for fur rubbing. Millipedes secrete caustic benzoquinones when threatened or attacked by predators. Valderrama et al. (2000) investigated fur rubbing in *C. olivaceus* at Fundo Pecuario Masaguaral in the llanos of central Venezuela and analyzed the chemistry and biological activities of the benzoquinone secretions in the laboratory. As many as four individuals will share one millipede, rolling on it, putting it gently in the mouth, and drooling with eyes apparently glazed over. Body rubbing with the apparent intent of picking up millipede secretions from conspecifics is also observed. The behavior can last from 5 sec to over 2 min. Individuals may anoint several times a day. There is not an agesex class bias in individuals performing this behavior. The behavior occurred most consistently during periods of high rainfall. The two benzoquinones chemically isolated in the laboratory from the millipede secretions are known to be effective at repelling insects and to have antimicrobial properties.

Any benefits gained from applying these substances to the skin are presumed to be immediate. As was found for Baker's study site, the wet, humid conditions of the rainy season may present stressful environmental conditions for *Cebus*. Valderrama et al. (2000) proposed that the capuchins were anointing themselves as protection possibly from insects. Positive social effects of this behavior in *Cebus apella* have also been described (Gilbert et al. 1998).

Ateles geoffroyi in the BCI group on Barro Colorado Island perform fur rubbing solitarily, applying plant material only to the sternal and axillary regions of the body (Campbell 2000). Differences in the behavior between *Cebus* and this species led Campbell (2000) to reject a medicinal hypothesis. Citing as evidence against the medicinal hypothesis were three main factors: *(1)* males tend to perform the behavior more frequently than females, *(2)* only a limited part of the body is anointed and therefore it is unlikely to be effective against insects, and *(3)* a lack of seasonal variation in the occurrence of the behavior in spite of strong seasonality in the prevalence of mosquitoes and ticks (Campbell 2000). Campbell notes that fur rubbing has been documented only in *A. geoffroyi* living on Barro Colorado Island and in captivity but not elsewhere in the wild where the species has also been extensively studied in the past (Klein 1972).

Phytochemical and Ethnobotanical Evidence

A number of the plant species identified by Baker (1996) as being used by capuchins for fur rubbing are known to contain secondary compounds with insecticidal, antiseptic, fungistatic, anti-inflammatory, anesthetic, and general dermatological activities. Richard (1970) suggested that the use of citrus fruits by *Ateles* might repel insects or act as an astringent. All plants share the quality of being pungent. Fur rubbing by capuchins and lemurs includes the use of a variety of pungent items, such as tobacco, onions, garlic (in captivity), millipedes, and formic acid (in the wild) (cf. Baker 1996, Valderrama et al. 2000).

It seems likely that this behavior has multiple functions based on differing environmental and pest pressures. Because the behavior can be induced readily under captive conditions, it seems that the species exhibiting this behavior have a behavioral predisposition to rub scented objects into the fur. Some reports suggest an almost hedonic pleasure derived from its practice. This ability to study the behavior easily under controlled conditions could provide important opportunities for future investigations into aspects of social learning and ontogenetic development as well as clinical evaluation of the effects of the behavior on health conditions that are not possible in the wild. Further direct evidence for a medicinal or insect-repellent effect of this behavior for more species at different sites is needed in order to clarify its role in health maintenance. Further research into the taxonomic distribution of this behavior in primates is also needed.

SUMMARY AND FUTURE DIRECTIONS

The evidence described in this chapter suggests that a variety of behavioral adaptations exist for the control of internal and possibly external parasites. Much of the detailed evidence in support of the developing hypotheses on self-medication in the African great apes comes from their response to changing levels of parasite infection. The applicability of this model to other primate species needs to be confirmed. Three of the greatest constraints on field investigations of self-medicative behavior are *(1)* the unpredictability of the behaviors' occurrence, *(2)* the unreliability of being able to consistently follow and observe sick individuals over time, and *(3)* restraints on subject manipulation. These constraints must somehow be overcome to test relevant hypotheses.

Systematic monitoring of individuals throughout the year for fluctuating levels of parasite infection is one effective means of identifying the predominant parasites and their likely period of increased impact on the host (e.g., Huffman et al. 1997). Relevant pathogens other than parasites also need to be investigated, but traditionally this has required capturing animals for the collection of blood and other tissues. The development of noninvasive techniques with wide applicability will allow for a better assessment of the range and impact of self-medicative behaviors and provide a wide range of useful tools for conservation biology.

Close long-term monitoring of the overall health of group members using detailed analysis of activity budgets and systematic inspection of general health is necessary to identify illness when it occurs and to understand the proximate effects of that illness on the individual as well as the function and effectiveness of any proposed self-medicative behavior (e.g., Huffman et al. 1997, Kaur and Huffman 2004, Krief et al. 2005). These procedures are straightforward and compatible with most primate field study designs. Experimental behavioral studies on captive primates can be useful to answer some questions that are not possible or are extremely difficult to answer in the wild, for example, individual acquisition and transmission of behaviors within the group (e.g., Huffman and Hirata 2004, Sgaravatti et al. 2004).

This chapter has shown than self-medication can take many forms and occurs widely among primates. An important task that lies ahead is to identify and evaluate not only the immediate outcome of such behaviors but also the long-term impact on individual fitness and reproductive success. While the current direct evidence for primate self-medication comes largely from work on the great apes, we should avoid the temptation to conclude that advanced cognitive ability is a prerequisite for exhibiting such behavior. Indeed, it would be unwise not to expect it across the entire animal kingdom (Engel 2002). Further efforts to better understand the diversity, function, and mechanisms of self-medicative behavior in all primates will be rewarded with a richer appreciation for our primate ancestors and no doubt provide important lessons for the future of our own species.

REFERENCES

Adegoke, G. O., and Skura, B. J. (1994). Nutritional profile and antimicrobial spectrum of the spice *Aframomum danielli*. *Plant Foods Hum. Nutr.* 45:175–182.

Allison, A. C. (1982). Co-evolution between hosts and infectious disease agents and its effect on virulence. In: Anderson, R. M., and May, R. M. (eds.), *Population Biology of Infectious Diseases*. Springer-Verlag, Berlin. pp. 245–267.

Anderson, R. C. (1992). *Nematode Parasites of Vertebrates. Their Development and Transmission*. C.A.B. International, Wallingford.

Asaka, Y., Kubota, T., and Kulkarni, A. B. (1977). Studies on a bitter principle from *Vernonia anthelmintica*. *Phytochemistry* 16:1838–1839.

Ashford, R. W., Reid, G. D. F., and Wrangham, R. W. (2000). Intestinal parasites of the chimpanzee *Pan troglodytes* in Kibale Forest, Uganda. *Ann. Trop. Med. Parasitol.* 84:337–340.

Aufreiter, S., Hancock, R., Mahaney, W. C., Stambolic-Robb, A., and Sanmugadas, K. (1997). Geochemistry and mineralogy of soils eaten by humans. *Int. J. Food Sci. Nutr.* 48:292–305.

Baker, M. (1996). Fur rubbing: use of medicinal plants by capuchin monkeys (*Cebus capucinus*). *Am. J. Primatol.* 38:263–270.

Barrows, M. (1996). A survey of the intestinal parasites of the primate in Budongo Forest, Uganda. Report to the Department of Veterinary Medicine, University of Glasgow.

Bateson, E. M., and LeBray, J. (1978). Clay eating by aboriginals of the Northern Territory. *Med. J. Aust.* 1(suppl.):60–61.

Beaver, P. C., Jung, R. C., and Cupp, E. W. (1984). *Clinical Parasitology*, 9th ed. Lea & Febiger, Philadelphia.

Beckage, N. (1997). *Parasites and Pathogens—Effects on Host Hormones and Behavior*. International Thomson Publishing, New York.

Berry, J. P., McFarren, M. A., and Rodriguez, E. (1995). Zoopharmacognosy: a "biorational" strategy for phytochemical prospecting. In: Gustine, D. L., and Flores, H. E. (eds.), *Phytochemicals and Health*. American Society of Plant Physiologists, Rockville, MD. pp. 165–178.

Birkinshaw, C. R. (1999). Use of millipedes by black lemurs to anoint their bodies. *Folia Primatol.* 70:170–171.

Boesch, C. (1995). Innovation in wild chimpanzees (*Pan troglodytes*). *Int. J. Primatol.* 16:1–16.

Brack, M. (1987). *Agents Transmissible from Simians to Man*. Springer-Verlag, Berlin.

Buckley, J. S. (1983). The feeding behavior, social behavior, and ecology of the white-faced monkey *Cebus capucinus* at Trujillo, northern Honduras, Central America [PhD thesis]. University of Illinois, Urbana.

Burkill, H. M. (1985). *The Useful Plants of West Tropical Africa*, 2nd ed., vol. 1. Royal Botanical Gardens, Kew.

Campbell, C. J. (2000). Fur rubbing behavior in free-ranging black-handed spider monkeys (*Ateles geoffroyi*) in Panama. *Am. J. Primatol.* 51:205–208 .

Carrai, V., Silvana, M., Borgognini-Tarli, S., Huffman, M. A., and Massimo, B. (2003). Increase in tannin consumption by sifaka (*Propithecus verreauxi verreauxi*) females during the birth season: a case for self medication in prosimians? *Primates* 44:61–66.

Chapman, C. A., and Chapman, L. J. (2002). Foraging challenges of red colobus monkeys: influence of nutrients and secondary compounds. *Comp. Biochem. Physiol.* 133:861–875.

Clayton, D. H., and Moore, J. (1997). *Host–Parasite Evolution. General Principles and Avian Models*. Oxford University Press, Oxford.

Cooney, D. O., and Struhsaker, T. T. (1997). Adsorptive capacity of charcoals eaten by Zanzibar red colobus monkeys: implications for reducing dietary toxins. *Int. J. Primatol.* 18:235–246.

Cousins, D., and Huffman, M. A. (2002). Medicinal properties in the diet of gorillas—an ethnopharmacological evaluation. *Afr. Stud. Monogr.* 23:65–89.

Dalziel, J. M. (1937). The useful plants of west tropical Africa. In: Hutchinson, J., and Dalziel, J. M. (eds.), *Appendix to Flora of West Tropical Africa*. Whitefriars Press, London.

Davies, N. D., and Baillie, I. C. (1988). Soil-eating in red leaf monkeys (*Presbytis rubicunda*) in Saba, northern Borneo. *Biotropica* 20:252–258.

de Amorin, A., Borba, H. R., Carauta, J. P. P., Lopes, D., and Kaplan, M. A. (1999). Antihelmintic activity of the latex of *Ficus* species. *J. Ethnopharm.* 64:255–258.

Dupain, J., Van Elsacker, L., Nell, C., Garcia, P., Ponce, F., and Huffman, M. A. (2002). New evidence for leaf swallowing and *Oesophagostomum* infection in bonobos (*Pan paniscus*). *Int. J. Primatol.* 23:1053–1062.

Engel, C. (2002). *Wild Health*. Houghton Mifflin, Boston.

File, S. K., McGrew, W. C., and Tutin, C. E. G. (1976). The intestinal parasites of a community of feral chimpanzees, *Pan troglodytes schweinfurthii*. *J. Parasitol.* 62:259–261.

Freeland, W. J. (1980). Mangabey (*Cercocebus albigena*) movement patterns in relation to food availability and fecal contamination. *Ecology* 61:1297–1303.

Freeland, W. J. (1983). Parasites and the coexistence of animal host species. *Am. Nat.* 121:223–236.

Gasquet, M., Bamba, D., Babadjamian, A., Balansard, G., Timon-David, P., and Metzger, J. (1985). Action amoebicide et anthelminthique du vernolide et de l'hydroxyvernolide isoles des feuilles de *Vernonia colorata* (Willd.) Drake. *Eur. J. Med. Chem. Theory* 2:111–115.

Gilbert, T. M., Brown, D. A., and Boysen, S. T. (1998). Social effects on behavior in capuchins (*Cebus apella*) [abstract]. *Am. J. Primatol.* 45:182.

Gillespie, T. (2004). Effects of human disturbance on primate–parasite dynamics [PhD thesis]. University of Florida, Gainsville.

Glander, K. E. (1975). Habitat description and resource utilization: a preliminary report on mantled howler monkey ecology. In: Tuttle, R. H. (ed.), *Socioecology and Psychology of Primates.* Mouton Press, The Hague. pp. 37–57.

Glander, K. E. (1982). The impact of plant secondary compounds on primate feeding behavior. *Ybk. Phys. Anthropol.* 25:1–18.

Glander, K. E. (1994). Nonhuman primate self-medication with wild plant foods. In: Etkin, N. L. (ed.), *Eating on the Wild Side: The Pharmacologic, Ecologic, and Social Implications of Using Noncultigens.* University of Arizona Press, Tuscon. pp. 239–256.

Halstead, J. A. (1968). Geophagia in man: its nature and nutritional effects. *Am. J. Clin. Nutr.* 21:1384–1393.

Hansson, A., Veliz, G., Naquira, C., Amren, M., Arroyo, M., and Arevalo, G. (1986). Preclinical and clinical studies with latex from *Ficus glabrata* Hbk, a traditional intestinal antehelminthic in the Amazon area. *J. Ethnopharm.* 17:105–138.

Harborne, J. B., Moss, G. P., and Baxter, H. (1999). *Phytochemical Dictionary.* Taylor & Francis, London.

Hart, B. L. (1988). Biological basis of the behavior of sick animals. *Neurosci. Biobehav. Rev.* 12:123–137.

Hart, B. L. (1990). Behavioral adaptations to pathogens and parasites: five strategies. *Neurosci. Biobehav. Rev.* 14:273–294.

Hausfater, G., and Meade, B. J. (1982). Alternation of sleeping groves by yellow baboons (*Papio cynocephalus*) as a strategy for parasite avoidance. *Primates* 23:287–297.

Hill, W. C. O. (1960). *Primates: Comparative Anatomy and Taxonomy, Cebidae,* vol. IV, pt. A. Edinburgh University Press, Edinburgh.

Holmes, J. C., and Zohar, S. (1990). Pathology and host behavior. In: Barnard, C. J., and Behnke, J. M. (eds.), *Parasitism and Host Behavior,* Taylor and Frances, London. pp. 34–63.

Howe, H. F., and Westley, L. C. (1988). *Ecological Relationships of Plants and Animals.* Oxford University Press, New York.

Huffman, M. A. (1994). The C.H.I.M.P.P. Group: a mutlidisciplinary investigation into the use of medicinal plants by chimpanzees. *Pan Afr. News* 1:3–5.

Huffman, M. A. (1997). Current evidence for self-medication in primates: a multidisciplinary perspective. *Ybk. Phys. Anthropol.* 40:171–200.

Huffman, M. A. (2001). Self-medicative behavior in the African great apes—an evolutionary perspective into the origins of human traditional medicine. *Bioscience* 51:651–661.

Huffman, M. A., and Caton, J. M. (2001). Self-induced increase of gut motility and the control of parasitic infections in wild chimpanzees. *Int. J. Primatol.* 22:329–346.

Huffman, M. A., Gotoh, S., Izutsu, D., Koshimizu, K., and Kalunde, M. S. (1993). Further observations on the use of *Vernonia amygdalina* by a wild chimpanzee, its possible effect on parasite load, and its phytochemistry. *Afr. Stud. Monogr.* 14:227–240.

Huffman, M. A., Gotoh, S., Turner, L. A., Hamai, M., and Yoshida, K. (1997). Seasonal trends in intestinal nematode infection and medicinal plant use among chimpanzees in the Mahale Mountains, Tanzania. *Primates* 38:111–125.

Huffman, M. A., and Hirata, S. (2003). Biological and ecological foundations of primate behavioral traditions. In: Fragaszy, D. M., and Perry, S. (eds.), *The Biology of Traditions.* Cambridge University Press, Cambridge. pp. 267–296.

Huffman, M. A., and Hirata, S. (2004). An experimental study of leaf swallowing in captive chimpanzees—insights into the origin of a self-medicative behavior and the role of social learning. *Primates* 45:113–118.

Huffman, M. A., Koshimizu, K., and Ohigashi, H. (1996a). Ethnobotany and zoopharmacognosy of *Vernonia amygdalina,* a medicinal plant used by humans and chimpanzees. In: Caligari, P. D. S., and Hind, D. J. N. (eds.), *Compositae: Biology and Utilization,* vol. 2. Royal Botanical Gardens Press, Kew. pp. 351–360.

Huffman, M. A., Page, J. E., Sukhdeo, M. V. K., Gotoh, S., Kalunde, M. S., Chandrasiri, T., and Towers, G. H. N. (1996b). Leaf-swallowing by chimpanzees, a behavioral adaptation for the control of strongyle nematode infections. *Int. J. Primatol.* 72:475–503.

Huffman, M. A., and Seifu, M. (1989). Observations on the illness and consumption of a possibly medicinal plant *Vernonia amygdalina* (Del.), by a wild chimpanzee in the Mahale Mountains National Park, Tanzania. *Primates* 30:51–63.

Huffman, M. A., Ohigashi, H., Kawanaka, M., Page, J. E., Kirby, G. C., Gasquet, M., Murakami, A., and Koshimizu, K. (1998). African great ape self-medication: a new paradigm for treating parasite disease with natural medicines? In: Ebizuka, Y. (ed.), *Towards Natural Medicine Research in the 21st Century.* Elsevier Science, Amsterdam. pp. 113–123.

Jacobsen, L., and Hamel, R. (1996). *International Directory of Primatology.* Wisconsin Regional Primate Research Center, Madison.

Janzen, D. H. (1978). Complications in interpreting the chemical defenses of trees against tropical arboreal plant-eating vertebrates. In: Montgomery, G. G. (ed.), *The Ecology of Arboreal Folivores.* Smithsonian Institution Press, Washington DC. pp. 73–84.

Jisaka, M., Kawanaka, M., Sugiyama, H., Takegawa, K., Huffman, M. A., Ohigashi, H., and Koshimizu, K. (1992a). Antischistosomal activities of sesquiterpene lactones and steroid glucosides from *Vernonia amygdalina,* possibly used by wild chimpanzees against parasite-related diseses. *Biosci. Biotech. Biochem.* 56:845–846.

Jisaka, M., Ohigashi, H., Takagaki, T., Nozaki, H., Tada, T., Hirota, M., Irie, R., Huffman, M. A., Nishida, T., Kaji, M., and Koshimizu, K. (1992b). Bitter steroid glucosides, vernoniosides A1, A2, and A3 and related B1 from a possible medicinal plant *Vernonia amygdalina,* used by wild chimpanzees. *Tetrahedron* 48:625–632.

Jisaka, M., Ohigashi, H., Takegawa, K., Hirota, M., Irie, R., Huffman, M. A., and Koshimizu, K. (1993a). Steroid glucosides from *Vernonia amygdalina,* a possible chimpanzee medicinal plant. *Phytochemistry* 34:409–413.

Jisaka, M., Ohigashi, H., Takegawa, K., Huffman, M. A., and Koshimizu, K. (1993b). Antitumor and antimicrobial activities of bitter sesquiterpene lactones of *Vernonia amygdalina,* a possible medicinal plant used by wild chimpanzees. *Biosci. Biotech. Biochem.* 57:833–834.

Johns, T. (1990). *With Bitter Herbs They Shall Eat It.* University of Arizona Press, Tucson.

Kaur, T., and Huffman, M.A. (2004). Descriptive urological record of chimpanzees (*Pan troglodytes schweinfurthii*) in the wild and limitations associated with using multi-reagent dipstick test strips. *J. Med. Primatol.* 33:187–196.

Klaus, G., Klaus-Hugi, C., and Schmid, B. (1998). Geophagy by large mammals at natural licks in the rain forest of the Dzanga National Park, Central African Republic. *J. Trop. Ecol.* 14:207–227.

Klein, L. L. (1972). The ecology and social organization of the spider monkey, *Ateles belzebuth* [PhD thesis]. University of California, Berkeley.

Kluger, M. J., Ringler, D. H., and Anver, M. R. (1975). Fever and survival. *Science* 188:166–168.

Knezevich, M. (1998). Geophagy as therapeutic mediator of endoparasitism in a free-ranging group of rhesus macaques (*Macaca mulatta*). *Am. J. Primatol.* 44:71–82.

Koshimizu, K., Ohigashi, H., Huffman, M. A., Nisihda, T., and Takasaki, H. (1993). Physiological activities and the active constituents of potentially medicinal plants used by wild chimpanzees of the Mahale Mountains, Tanzania. *Int. J. Primatol.* 14:345–356.

Krepel, H. P., Baeta, S., Kootstra, C. J., and Polderman, A. M. (1995). Reinfection patterns of *Oesophagostomum bifurcum* and hookworm after anthelmintic treatment. *Trop. Geogr. Med.* 47:160–163.

Kreulen, D. A. (1985). Lick use by large herbivores: a review of benefits and banes of soil consumption. *Mammal Rev.* 15:107–123.

Kreulen, D. A., and Jager, T. (1984). The significance of soil ingestion in the utilization of arid rangelands by large herbivores with special reference to natural licks on the Kalahari pan. In: Gilchrist, F. M. C., and Mackie, R. I. (eds.), *Herbivore Nutrition in the Subtropics and Tropics*. Science Press, Johannesburg. pp. 204–221.

Krief, S. (2003). Métabolites secondaires des plantes et comportement animal: surveillance sanitaire et observations de l'alimentation de chimpanzés (*Pan troglodytes schweinfurthii*) en Ouganda. Activités biologiques et étude chimique de plantes consommées [diss.]. Museum National d'Histoire Naturelle, Paris.

Krief, S., Huffman, M.A., Sévenet, T., Guillot, J., Bories, C., Hladik, C. M., and Wrangham, R. W. (2005). Non-invasive monitoring of the health of *Pan troglodytes schweinfurthii* in the Kibale National Park, Uganda. *Int. J. Primatol.* 26:467–490.

Krief, S., Martin, M.-T., Grellier, P., Kasenene, J., and Sévenet, T. (2004). Novel antimalarial compounds isolated in a survey of self-medicative behavior of wild chimpanzees in Uganda. *Antimicrob. Agents Chemother.* 48:3196–3199.

Krishnamani, R., and Mahaney, W. C. (2000). Geophagy among primates: adaptive significance and ecological consequences. *Anim. Behav.* 59:899–915.

Kummer, H. (1995). *In Quest of the Sacred Baboon—A Scientist's Journey*. Princeton University Press, Princeton, NJ.

Kupchan, S. M., Hemingway, R. J., Karim, A., and Werner, D. (1969). Tumor inhibitors XLVII. Vernodalin and vernomygdin, two new cytotoxic sesquiterpene lactones from *Vernonia amygdalina* Del. *J. Organ. Chem.* 34:3908–3911.

Landsoud-Soukate, J., Tutin, C. E. G., and Fernandez, M. (1995). Intestinal parasites of sympatric gorillas and chimpanzees in the Lope Reserve, Gabon. *Ann. Trop. Med. Parasitol.* 89:73–79.

Longino, J. T. (1984). True anting by the capuchin, *Cebus capucinus*. *Primates* 25:243–245.

Lozano, G. A. (1991). Optimal foraging theory: a possible role for parasites. *Oikos* 60:391–395.

Lozano, G. A. (1998). Parasitic stress and self-medication in wild animals. *Adv. Study Behav.* 27:291–317.

Ludes, E., and Anderson, J. R. (1995). "Peat-bathing" by captive white-faced capuchin monkeys (*Cebus capucinus*). *Folia Primatol.* 65:38–42.

Mahaney, W. C., Hancock, R. G. V., and Aufreiter, S. (1995). Mountain gorilla geophagy: a possible strategy for dealing with intestinal problems. *Int. J. Primatol.* 16:475–487.

Mahaney, W. C., Hancock, R. G. V., Aufreiter, S., and Huffman, M. A. (1996). Geochemistry and clay minerology of termite mound soil and the role of geophagy in chimpanzees of the Mahale Mountains, Tanzania. *Primates* 37:121–134.

Mahaney, W. C., Milner, M. W., Sanmugadas, K., Hancock, R. G. V., Aufreiter, S., Wrangham, R. W., and Pier, H. W. (1997). Analysis of geophagy soils in Kibale Forest, Uganda. *Primates* 38:159–176.

Mahaney, W. C., Watts, D. P., and Hancock, R. G. V. (1990). Geophagia by mountain gorillas (*Gorilla gorilla beringei*) in the Virunga Mountains, Rwanda. *Primates* 31:113–120.

Matsuzawa, T., and Yamakoshi, G. (1996). Comparison of chimpanzee material culture between Bossou and Nimba, West Africa. In: Russon, A. E., Bard, K. A., and Taylor Parker, S. (eds.), *Reaching into Thought. The Minds of the Great Apes*. University of Cambridge Press, Cambridge. pp. 211–232.

McKey, D. (1978). Soil, vegetation and seed-eating by black colobus monkeys. In: Montgomery, G. G. (ed.), *The Ecology of Arboreal Folivores*. Smithsonian Institution Press, Washington DC. pp. 423–437.

Messner, E. J., and Wrangham R. W. (1996). In vitro testing of the biological activity of *Rubia cordifolia* leaves on primate *Strongyloide* species. *Primates* 37:105–108.

Milton, K. (1979). Factors influencing leaf choice by howler monkeys: a test for some hypotheses of food selection by generalist herbivores. *Am. Nat.* 114:362–378.

Moore, J. (2002). *Parasites and the Behavior of Animals*. Oxford University Press, Oxford.

Müller, K.-H., Ahl, C., and Hartman, G. (1997). Geophagy in masked titi monkeys (*Callicebus personatus melanochir*). *Primates* 38:69–77.

Murray, S., Stem, C., Boudreau, B., and Goodall, J. (2000). Intestinal parasites of baboons (*Papio cynocephalus anubis*) and chimpanzees (*Pan troglodytes*) in Gombe National Park. *J. Zoo Wildl. Med.* 31:176–178.

Nelson, G. S. (1960). Schistosome infections as zoonoses in Africa. *Transcrip. R. Soc. Trop. Med. Hyg.* 54:301–314.

Newton, P. N., and Nishida, T. (1990). Possible buccal administration of herbal drugs by wild chimpanzees, *Pan troglodytes*. *Anim. Behav.* 39:798–801.

Nishida, T., and Uehara, S. (1983). Natural diet of chimpanzees (*Pan troglodytes schweinfurthii*): long term record from the Mahale Mountains, Tanzania. *Afr. Stud. Monogr.* 3:109–130.

Oates, J. F. (1978). Water–plant and soil consumption by guereza monkeys (*Colobus guereza*): a relationship between minerals and toxins in the diet? *Biotropica* 10:241–253.

Oates, J. F., Swain, T., and Zantovska, J. (1977). Secondary compounds and food selection by colobus monkeys. *Biochem. Syst. Ecol.* 5:317–321.

Oates, J. F., Waterman, P. G., and Choo, G. M. (1980). Food selection by the south Indian leaf-monkey *Presbytis johnii*, in relation to leaf chemistry. *Oecologia* 45:45–56.

Ohigashi, H. (1995). Plants used medicinally by primates in the wild and their physiologically active constituents. Report to the Ministry of Science, Education and Culture for 1994 Grant-in-Aid for Scientific Research (06303012).

Ohigashi, H., Huffman, M. A., Izutsu, D., Koshimizu, K., Kawanaka, M., Sugiyama, H., Kirby, G. C., Warhurst, D. C.,

Allen, D., Wright, C. W., Phillipson, J. D., Timmon-David, P., Delnas, F., Elias, R., and Balansard, G. (1994). Toward the chemical ecology of medicinal plant-use in chimpanzees: the case of *Vernonia amygdalina* Del. A plant used by wild chimpanzees possibly for parasite-related diseases. *J. Chem. Ecol.* 20:541–553.

Ohigashi, H., Jisaka, M., Takagaki, T., Nozaki, H., Tada, T., Huffman, M. A., Nishida, T., Kaji, M., and Koshimizu, K. (1991a). Bitter principle and a related steroid glucoside from *Vernonia amygdalina*, a possible medicinal plant for wild chimpanzees. *Agricult. Biol. Chem.* 55:1201–1203.

Ohigashi, H., Takagaki, T., Koshimizu, K., Nishida, T., Huffman, M. A., Takasaki, H., Jato, J., and Muanza, D. N. (1991b). Biological activities of plant extracts from tropical Africa. *Afr. Stud. Monogr.* 12:201–210.

Oloke, J. K., Kolawole, D. O., and Erhun, W. O. (1988). The antibacterial and antifungal activities of certain components of *Afromomum melegueta* fruits. *Fitoterapia* LIX:384–388.

Olupot, W., Chapman, C. A., Waser, P. M., and Isabirye-Basuta, G. (1997). Mangabey (*Cercocebus albigena*) ranging patterns in relation to fruit availability and the risk of parasite infection in Kibale National Park Uganda. *Am. J. Primatol.* 43:65–78.

Oppenheimer, J. R. (1968). Behavior and ecology of the white-faced monkey *Cebus capucinus*, on Barro Colorado Island, CZ [PhD thesis]. University of Illinois, Urbana.

Page, J. E., Balza, F. F., Nishida, T., and Towers, G. H. N. (1992). Biologically active diterpenes from *Aspilia mossambicensis*, a chimpanzee medicinal plant. *Phytochemistry* 31:3437–3439.

Page, J. E., Huffman, M. A., Smith, V., and Towers, G. H. N. (1997). Chemical basis for medicinal consumption of *Aspilia* leaves by chimpanzees: a re-analysis. *J. Chem. Ecol.* 23:2211–2225.

Phillips-Conroy, J. E. (1996). Baboons, diet, and disease: food plant selection and schistosomiaisis. In: Taub, D. M., and King, F. A. (eds.), *Current Perspectives in Primate Social Dynamics.* Van Nostrand Reinhold, New York. pp. 287–304.

Phillips-Conroy, J. E., and Knopf, P. M. (1986). The effects of ingesting plant hormones on schistosomiaisis in mice: an experimental study. *Biochem. Syst. Ecol.* 14:637–645.

Pruetz, J. D., and Johnson-Fulton, S. (2003). Evidence for leaf-swallowing behavior by savanna chimpanzees in Senegal—a new site record. *Pan Afr. News* 10:14–16.

Richard, A. (1970). A comparative study of the activity patterns and behavior of *Alouatta villosa* and *Ateles geoffroyi*. *Folia Primatol.* 12:241–263.

Rodriguez, E., Aregullin, M., Nishida, T., Uehara, S., Wrangham, R. W., Abramowski, Z., Finlayson, A., and Towers, G. H. N. (1985). Thiarubrin A, a bioactive constituent of *Aspilia* (Asteraceae) consumed by wild chimpanzees. *Experientia* 41:419–420.

Rodriguez, E., and Wrangham, R. W. (1993). Zoopharmacognosy: the use of medicinal plants by animals. In: Downum, K. R., Romeo, J. T., and Stafford, H. (eds.), *Recent Advances in Phytochemistry. Phytochemical Potential of Tropic Plants*, vol. 27. Plenum Press, New York. pp. 89–105.

Sgaravatti, A., Spiezio, C., Grassi, D., and Huffman, M. A. (2004). How do chimpanzees learn leaf-swallowing behavior? *Folia Primatol.* 75(suppl. 1):333.

Struhsaker. T. T. (1997). *Ecology of an African Rain Forest.* University Press of Florida, Gainesville.

Stuart, M., Pendergast, V., Rumfelt, S., Peirberg, S., Greenspan, L., Glander, K., and Clarke, M. (1998). Parasites of wild howlers (*Alouatta* spp.). *Int. J. Primatol.* 19:493–512.

Takasaki, H., and Hunt, K. (1987). Further medicinal plant consumption in wild chimpanzees? *Afr. Stud. Monogr.* 8:125–128.

Toft, C. A., Aeschlimann, A., and Bolis, L. (1991). *Parasite–Host Associations; Coexistence or Conflict?* Oxford University Press, Oxford.

Toubiana, R., and Gaudemer, A. (1967). Structure du vernolide, nouvel ester sesquiterpique isole de *Vernonia colorata*. *Tetrahedron Lett.* 14:1333–1336.

Towers, G. H. N., Abramowski, Z., Finlayson, A. J., and Zucconi, A. (1985). Antibiotic properties of thiarubrine-A, a naturally occurring dithiacyclohedadiene polyine. *Planta Med.* 3:225–229.

Valderrama, X., Robinson, J. G., Attygalle, A. B., and Eisner, T. (2000). Seasonal anointment with millipedes in a wild primate: a chemical defense against insects. *J. Chem. Ecol.* 26: 2781–2790.

Wakibara, J. V., Huffman, M. A., Wink, M., Reich, S., Aufreiter, S., Hancock, R. G. V., Sodhi, R., Mahaney, W. C., and Russell, S. (2001). Adaptive significance of geophagy for Japanese macaques (*Macaca fuscata*) at Arashiyama, Japan. *Int. J. Primatol.* 22:495–520.

Watt, J. M., and Breyer-Brandwinjk, M. G. (1962). *The Medicinal and Poisonous Plants of Southern and East Africa*. E. and S. Livingstone, Edinburgh.

Wiley, A. S., and Katz, S. H. (1998). Geophagy in pregnancy: a test of a hypothesis. *Curr. Anthropol.* 39:532–544.

Wink, M., Hofer, A., Bilfinger, M., Englert, E., Martin, M., and Schneider, D. (1993). Geese and dietary allelochemicals—food palatability and geophagy. *Chemoecology* 4:93–107.

Wrangham, R. W. (1977). Feeding behavior of chimpanzees in Gombe National Park, Tanzania. In: Clutton-Brock, T. H. (ed.), *Primate Ecology*. Academic Press, London. pp. 504–538.

Wrangham, R. W. (1995). Relationship of chimpanzee leaf-swallowing to a tapeworm infection. *Am. J. Primatol.* 37:297–303.

Wrangham, R. W., Conklin-Brittain, N.-I., and Hunt, K. D. (1998). Dietary reposnse of chimpanzees and cercopithecinae to seasonal variation in fruit abundance. I. Antifeedants. *Int. J. Primatol.* 19:946–970.

Wrangham, R. W., and Goodall, J. (1989). Chimpanzee use of medicinal leaves. In: Heltne, P. G., and Marquardt, L. A. (eds.), *Understanding Chimpanzees*. Harvard University Press, Cambridge, MA. pp. 22–37.

Wrangham, R. W., and Nishida, T. (1983). *Aspilia* spp. leaves: a puzzle in the feeding behavior of wild chimpanzees. *Primates* 24:276–282.

Wrangham, R. W., and Waterman, P. G. (1983). Condensed tannins in fruits eaten by chimpanzees. *Biotropica* 15:217–222.

Zito, M., Evans, S., and Weldon, P. J. (2003). Owl monkeys (*Aotus* spp.) self-anoint with plants and millipedes. *Folia Primatol.* 74:159–161.

43

Ethnoprimatology
Contextualizing Human and Nonhuman Primate Interactions
Linda D. Wolfe and Agustín Fuentes

Every group of non-human primates is increasingly dependent on its nearest human neighbours. What the humans think of these animals, the manner in which they are willing to live with them, whether or not they exploit them, ignore, them, or provide for them determines the ultimate well-being and survival of those non-human primates. Until recently, ecological questions regarding non-human primates have been treated as if the ecosystem in which they lived, was self-perpetuating, and separate from the human ecosystem. Clearly this is not case. Humans and non-human primates are engaged in a complex set of interactions. (Burton and Carroll 2005)

INTRODUCTION

The subject of this chapter is *ethnoprimatology*, the multifaceted interactions of human and nonhuman primates (Sponsel 1997, Sponsel et al. 2002). Ethnoprimatology should not be confused with *cultural primatology*, the study of primate cultural traditions (McGrew 1992), or *folkbiology*, the way ordinary people understand and categorize plants and animals (Medin and Atran 1999). The field of ethnoprimatology is important in so far as it provides a holistic view of human and nonhuman primates and, hopefully, will facilitate the survival of nonhuman primates, many of which are facing extinction (see Chapter 30). This holistic approach to the ethnoprimatology of humans and nonhuman primates includes, but is not exclusive to, hunting of nonhuman primates for food, keeping nonhuman primates as pets, bidirectional pathogen exchange, the impacts of habitat alteration/destruction and crop raiding, indigenous knowledge of nonhuman primate behavior, and the incorporation of nonhuman primates into myths, folklore, and other narratives.

Space does not permit an examination of all of the subjects named above, but there are two recent edited books with extensive bibliographies and chapters devoted to various topics concerning the interaction between human and nonhuman primates (Fuentes and Wolfe 2002, Paterson and Wellis 2005). Hunting, habitat destruction, crop raiding, and conservation are among the older concerns of primatologists and conservationists (see Chapter 30). The bidirectional

spread of infectious diseases and the consequences of ecotourism and primate pet ownership are more recent concerns. Janette Wallis and Lisa Jones-Engel have been instrumental in bringing these important issues to our attention (Wallis and Lee 1999, Jones-Engel et al. 2005).

In this chapter, we will provide first a broad overview of basic patterns of interconnections between human and nonhuman primates, including a few in-depth examples. The second section is a brief review of some of the ways in which nonhuman primates appear in the myth, folklore, and behavioral attitudes of peoples in different regions of the world (see App. 43.1). These two sections are meant as suggestive templates from which the reader may draw possible insight into perspectives on the human–nonhuman primate interface. While we intend to cover the wide area wherein human and nonhuman primates interact, many of our specific examples will come from Asia due to our areas of specialty and the, currently, larger body of ethnoprimatological literature focused on that region.

Human and Nonhuman Primate Interconnections

While human and nonhuman primates share a number of interconnections, at the most basic level there is a distinction introduced by geography. When envisioning the globe, there are relatively clear demarcations between zones of *sympatry* (geographic overlap) and zones of *allopatry* (lack of geographic overlap) (Fig. 43.1). The zones of sympatry fall primarily in Africa; southern, eastern, and southeastern Asia; and South/Central America and reflect the distribution of nonhuman primates for at least the last 10 millennia. The zones of allopatry are primarily above ~30° north of the equator (above the tropic of Cancer) except in East Asia, where two species of macaque monkeys (*Macaca mulatta* and *M. fuscata*) range relatively far north in China and Japan, and in North America, where nonhuman primate populations reach the tropic of Cancer (23.5° north).

Understanding the relationships between human and nonhuman primates requires a temporal context as well as a geographic one. Long-term sympatry, especially sympatry

Figure 43.1 Zones of Sympatry (■) and allopatry (□).

that involves common usage of habitat, can result in a form of coecology. That is, ecological pressures impact mammals in particular ways, and mammals that share many morphological and physiological facets, such as the anthropoid primates (monkeys, apes, and humans), may also share similar adaptations at a variety of levels. This is important in understanding the interconnections between human and nonhuman primates because long-term overlap and similarity in behavior/other modes of adaptation (even if slight) can impact human conceptualizations of "nature" and act to facilitate distinct patterns of integration/engagement between humans and other primates. Alternatively, areas with less time overlap, especially in zones of allopatry where overlap is mitigated by captivity or other forms of limited/selective exposure, specific patterns of strong association and incorporation are expected to be uncommon. However, as with all things human, this dichotomy is far from complete. With rapid urbanization in areas of sympatry and cultural shifts in the context and conceptualizations of nonhuman animals (especially apes) in zones of allopatry, these distinctions can become blurred or even reversed.

VARIANTS IN HUMAN–MONKEY INTERCONNECTIONS

Throughout much of southern and southeastern Asia, at least three species of macaques appear to coexist well alongside human populations. Two of these species, the rhesus macaque (*Macaca mulatta*) and the long-tailed macaque (*M. fascicularis*), may in fact be increasing their numbers in

areas of direct habitat overlap with humans relative to forested or other areas considered "natural" habitat for them. These increases are tied to both human-directed changes to the local ecologies which provide favorable habitat/ ecological features for the macaques and specific aspects of the macaques' "fit" within human cultural practice such that humans may also derive specific cultural benefits from the macaques (Fuentes et al. 2005). In some areas of Amazonia, local peoples differentially treat specific species of monkeys such that they are brought into human cultural practice and offered relative protection from predation (Cormier 2002, 2003; Lizarralde 2002; Shepard 2002). These behavioral and physical overlaps and interactions also create an environment where more than space is shared; there may also be commingling of pathogens. While it is generally assumed that human–nonhuman primate pathogen "sharing" has deleterious repercussions, our understanding of the patterns and contexts of these copathogen environments and their evolution remains rather incomplete (Engel et al. 2002, Jones-Engel et al. 2005). Regardless, if human and nonhuman primates overlap, gain "benefits" from (physiological, ecological, behavioral, cultural, or otherwise), exert "costs" on (physiological, ecological, behavioral, cultural, or otherwise), and impact one another over time via their sympatry, then one can envision each of the species having effects on the other, especially under conditions where the temporal variable is deep. There are four specific examples of interconnections that deserve attention in this context: temple/ urban monkeys, primates as prey, primates as pets, and the in-depth social and economic inclusion of primates into human lives.

Temple/Urban Monkeys

Temple/urban monkeys exhibit extensive sympatry with human populations and tend to reside around specific temple structures either in or around areas of human habitation. These situations can be in human cities, small local shrines and temples in villages and towns, or specific temple sites that double as sacred places and tourist locations. This pattern is primarily found in southern and southeastern Asia and is strongly associated with cultures that practice Hinduism and/or Buddhism and their associated rituals and temple construction patterns (Aggimarangee 1992, Fuentes et al. 2005, Wolfe 2002). However, in the case of purely, or primarily, tourist locations, these sites can also include nontemple sites where monkeys are residents. Such tourist sites can be found in Gibraltar in Europe (macaques); many places in Kenya, South Africa, and Tanzania (baboons and vervet monkeys); in Japan (macaques); and even in the U.S. state of Florida (macaques).

While construction of Hindu and Buddhist temples reflects a time depth of no more than one to four millennia (depending on location), human and macaque overlap in these areas has a much greater temporal depth. Evidence for habitat overlap between orangutans, macaques, and humans (and possible predation by humans) in southeastern Asia is clear from remains at Niah Cave in Borneo over 25,000 years ago (see Chapter 17; Sponsel et al. 2002). Given the prominent role of monkeys in Hindu and Buddhist myth (see below) and the high degree of security and nutritional assistance that humans provide to extant temple macaque populations, one can argue that there are human cultural benefits to this arrangement (Fuentes et al. 2005). This situation is strongly exemplified by the pattern of interactions and coexistence between *Macaca fascicularis* and humans on the island of Bali, Indonesia.

The macaques on Bali are found throughout the island, except in the extreme urban areas in the regency of Badung (such as the capital city, Denpasar). There appear to be at least 63 sites where macaques reside on the island (Fuentes et al. 2005). Each of these sites has one to three groups of macaques that either range fully within the site or use the site as part of their total home range. Each site has between 15 and over 300 monkeys. Macaque densities at these sites range from one to over 20 individuals/km², while human densities average over 500 individuals/km² across the island. Most importantly, over 68% of these sites are associated with a temple or shrine. These religious complexes can be as small as a simple shrine consisting of a few stones and an altar to elaborate temple complexes that are heavily used by Balinese and, in some cases, foreigners (Fuentes et al. 2005). Many of these macaque groups receive some provisioning at sites that contain a shrine or temple complex and have temple ceremonies several times per year. A component of Balinese Hinduism is the regular placement of offerings at shrines and temples. In most cases, these offerings consist of at least 30% edible elements. This suggests that a large

percentage of the macaque groups on Bali are food-enhanced (see Fa and Southwick 1988, Wheatley 1999), especially if one includes crop raiding as food enhancement. Presumably tied to this food availability, temple site populations tend to have larger group sizes than nontemple site populations. Importantly, it appears that specific land-use patterns and wet-rice agriculture combined with the complex temple and irrigation systems of the Balinese (Lansing 1991) have resulted in a mosaic of riparian forest corridors and small forest islands throughout much of Bali (Fuentes et al. 2005). This type of landscape fits remarkably well with the macaques' patterns of using riparian habitats and small forest clusters for residence, foraging, and dispersal. This landscape has been formed over at least the last few millennia, and the pattern of distribution of macaque populations across the island suggests that the macaques are exploiting it (Fuentes et al. 2005). In this scenario, the human niche construction, or creation of Balinese *place* (socially impacted space, such as the temple complexes), creates an ecology and habitat that is beneficial to the macaques via protection, nutrition, and ecology. Simultaneously, the humans gain specific cultural and economic benefits from the presence of the macaques in and around their temple sites (Fuentes et al. 2005, Wheatley 1999). This sympatry impacts the behavior and possibly other physiological facets of both primate species. Finally, there is also growing evidence of cross-species pathogen exchange and possibly coevolution to similar pathogen environments between humans and macaques facilitated via extensive sympatry resulting in mutual habitat use and behavioral interactions (Engel et al. 2002).

Primates as Prey

Primates are prey items for a diverse array of human cultures in zones of sympatry and are also captured by people from sympatric and allopatric zones for various human uses ranging from ingredients in traditional medicines to subjects of biomedical research. As relatively large mammals, especially in the neotropics, primates are a common choice for hunters (Alvard et al. 1997, Amman et al. 2000, Lizarralde 2002, Shepard 2002). However, there is evidence that humans do not take nonhuman primates randomly or solely in respect to optimal prey return models (Shepard 2002). In fact, it is common for human hunters to selectively hunt specific primates over others for very cultural reasons (Cormier 2002, 2003; Fuentes 2002; Lizarralde 2002; Shepard 2002). For example, the Matsigenka, who live in the Manu region of Peru, preferentially hunt larger monkeys, so much so that in areas where shotgun use is practiced, the larger species (spider [genus *Ateles*] and woolly [genus *Lagothrix*] monkeys) have become locally extinct (Alvard and Kaplan 1991). However, the howler monkeys (*Alouatta seniculus*) in Matsigenka territory, similar in size to the spider and woolly monkeys, remain at viable population levels and are taken about half as often as the capuchin monkeys, which are one-half their size (Shepard 2002). The

Matsigeneka have two reasons for this differential predation: the howlers are less "tasty" and are considered slow and lazy, characteristics that it is believed are passed on to children who eat the meat (Shepard 2002). However, this aversion to howlers is not common to all Amazonian peoples; in fact, some preferentially consume them (also for practical and cultural reasons) (Cormier 2002, 2003). These examples and others (see Social and/or Economic Inclusion of Nonhuman Primates, below.) suggest that humans selectively prey on primates, such that some species see a greater or lesser impact than others. Given these cases, one would assume various models proposing that selective predation by humans impacts the prey populations of nonhuman primates. In fact, recent overviews have demonstrated substantial behavioral and possibly morphological changes in primates due to extant or even past predation risks caused by nonhuman predators (Miller 2002). However, aside from general impacts on prey densities and local extinction events, there are few studies that have attempted to model the impact of human predation on nonhuman primate behavior, physiology, and morphology in a manner similar to the modeling of other animals' predation on primates (Alvard and Kaplan 1991). Unfortunately, this suggests that long-term, sustained predation by humans at a level below extinction-causing may impact the behavior and morphology of nonhuman primates, but we currently do not have sufficient data to reach a conclusion. It is also possible that selective hunting for body parts or other features that are used in traditional medicines may impact primate populations.

Primates as Pets

While *pets* is a substantial category, here we can narrowly define it as animals kept for companionship, enjoyment, or status rather than as working "domesticates" that contribute to the income, nutritional intake, or other functional facets of the humans who own them. Today, it is largely in areas of sympatry that we see primates as pets. This is due to the laws regarding the import/export of animals across international boundaries and the impact of the World Conservation Union (IUCN, see Chapter 30) classifications regarding the endangered/threatened status of a variety of primate species. However, the present situation is largely recent as nonhuman primates, especially small monkeys, were widely available in Europe during the early centuries following colonial expansion and both monkeys and apes were kept as pets in private menageries and homes during that time. Due to the costs of transport and the high levels of mortality during transport, it is unlikely that the zones of allopatry ever had the levels of primate pet ownership that we see in zones of sympatry. Today, pet ownership of select nonhuman primate species is widespread in southeastern Asia and Amazonia and occurs in southern Asia and other parts of South America and Africa as well. In Asia, the majority of primates owned as pets belong to the genera *Macaca* and *Nycticebus* (the slow loris) (Malone et al. 2002); in South America, nearly all of

the species found in Amazonia are kept as pets by at least some indigenous groups; and in sub-Saharan Africa, most pets are the medium–small monkeys such as vervets (genus *Chlorocebus*), guenons (genus *Cercopithecus*), and galagos (genera *Galago* and *Galagoides*). In North Africa, the sole African macaque, *Macaca sylvanus*, is kept as a pet; and this species has also had a small resurgence as a pet in Western Europe as well (although it remains illegal). Apes are held as status pets in Asia and Africa but have fairly low rates of survivorship in captivity as pets and are quite expensive and energy-intensive to keep relative to other nonhuman primates.

It is not clear that the keeping of pets impacts the "wild" populations of monkeys/apes in a way distinct from predation as pets are frequently by-products of hunting and their capture functions similarly to extraction via predation. Because the pet tradition is quite old in southeastern Asia (and probably South America), it is also quite likely that, as with the temple monkey overlap or even more so, there is substantial bidirectional pathogen transmission between humans and their primate pets (Jones-Engel et al. 2005). It is possible that there are physiological changes, even adaptations, in populations of human and nonhuman primates that overlap extensively, such as pets, humans, and the local populations from which the pet primates come, as a result of these intensive interactions.

Although not analogous to pet keeping, some nonhuman primates are currently raised in captivity and selectively bred for human biomedical experimentation. These colonies (primarily of macaque monkeys and a few other species) are selectively bred to create specific pathogen-free breeding colonies to provide test subjects for a variety of human disorders ranging from obesity to human immunodeficiency virus/acquired immunodeficiency syndrome. Many of these colonies are in zones of allopatry (primarily the United States and Europe), but increasing numbers are being established and run in zones of sympatry (e.g., Indonesia, Nepal, and Japan).

Social and/or Economic Inclusion of Nonhuman Primates

The term *social inclusion* refers to the integral inclusion of the nonhuman primate as a participant in facets of human society. While there are a number of human societies where this pattern takes place to a greater or lesser extent, the Guaja of Brazil are among the strongest examples (Cormier 2002, 2003). The Guaja, a Tupi-Guarani-speaking group in Amazonia, display a remarkable and intense inclusion of nonhuman primates, of many species, into their social fabric. Young monkeys are kept as pets, trained by children, and cared for by women and represent a significant culturally valued image of fertility (Cormier 2002, 2003). Young monkeys whose parents are killed for food will be brought to the village and "adopted." Women will bathe, breast-feed, and "wear" the monkeys; and throughout much of their development, the monkeys are carried and cared for by members of the Guaja village (Cormier 2002, 2003). The Guaja have a complex cosmology that involves ritual cannibalism (of howler monkeys), an emphasis on fertility and lactation, and the "wearing" of monkeys as signs of physical and social prowess for women. While patterns of carrying and raising monkeys have a considerable cost in terms of energy, movement, and foodstuffs, the returns go beyond the purely social. Young girls (and to a lesser extent boys) frequently practice mothering not only with their siblings but also with infant monkeys, with some girls as young as 5 years being the primary caretakers of infant monkeys. Boys gain important hunting experience through exposure to monkey movement and behavioral patterns including vocalizations (Cormier 2002, 2003). However, the mortality rate for the captive monkeys is fairly high at the infant level; and if they live into adulthood, they are no longer as tightly wound into the fabric of the Guaja's social lives and generally begin to display an array of aberrant and aggressive behaviors (although some do reproduce in the Guaja villages with others of their own species) (Cormier 2002, 2003).

Nonhuman primates are used as economic tools in southern, southeastern, and northeastern Asia. While there is reference to crop-picking macaques across much of southern Asia, the best current studies come from Thailand, where males of one species of macaque (*Macaca nemestrina*) are kept, raised, and trained to pick coconuts (Sponsel et al. 2002). In this case, humans in southern Thailand capture macaques as young individuals and then invest substantial time and energy to train them to be effective coconut pickers. This system can be highly efficient for the humans as a well-trained macaque can harvest 500–1,000 coconuts per day and do so with a lower cost and higher return than a human could (the other harvesters are young human boys). Sponsel et al. (2002) present this situation as an example of a possible adaptive shift from a predator–prey relationship to a cooperative human–monkey economic relationship. They argue that cultural and agricultural systems have favored a move from conflict between humans and macaques over crops to a synergistic relationship wherein the humans capture, train, and maintain the macaques as they serve their economic role (job). Obviously, the impact on the macaque population may be negative in the sense of individuals taken out (similar to predation), but the overlap between the humans and the trained macaques, in this case, can be seen almost as a direct functional domestication, or at least the use of a "wild" animal in a domestic capacity.

Monkey performance is a blend of the economic and social inclusion facets of the monkey–human relationship. Monkey performances include a variety of theatrical performances by trained macaques, where the monkeys mimic human cultural behavior via a series of staged interactions with their trainer, the audience, and in some cases other monkeys. Generally, the audience provides monetary contributions at the conclusion of the performance. These funds are the income of the trainer and are used to maintain both the monkey(s) and the trainer's family. While these

performances are widespread across Asia, they also occur in northern Africa and historically in Europe as well (Janson 1952). The most in-depth description of these practices comes from Emiko Ohnuki-Tierney's work on Japanese monkey performances. Ohnuki-Tierney (1987, 1995) presents the notion that in Japan the monkey performances, and the macaque monkeys themselves, act as a mirror for humanity, "playing a powerful role in their [humans'] deliberations of who they are as humans vis-à-vis animals and as a people vis-à-vis other peoples" (Ohnuki-Tierney 1995:297). As monkeys are the only animal addressed as *san*, the adult human address, and referred to as "humans minus three pieces of hair," there is an aspect of social inclusion, at least in the context of a special role for the monkeys, in Japanese culture that is distinct from the roles of other animals (Ohnuki-Tierney 1987, 1995). The integration of economic and cultural roles for macaques in Japan results in a distinct and complex cultural relationship between humans and monkeys that exists at the same time that increasing conflict over land and crops emerges as a predominant pattern of interaction between humans and "wild" macaques (Sprague 2002).

PRIMATES IN HUMAN MYTH AND FOLKLORE

Shepard (2002:114) suggests the following:

> Myths concerning the origins of biodiversity are a source of both entertainment and philosophical speculation. Animal species are anthropomorphized in folktales, spiritual beliefs, and personal anecdotes. Metaphors drawn from ecological processes are used to understand illness, death, and other aspects of the human condition, while human intentions and emotions are projected into the ecological sphere.

The symbolic worldviews of people who are sympatric with nonhuman primates make up an important aspect of the study of the primates themselves. Myths, epic poems, folklore, and so forth are important aspects of culture and impact the lives of the primates that exist in and around the humans. It is possible that aspects of human cultures might be used to create projects and programs to ensure the future survival of nonhuman primates. However, it is not always clear how, and if, primates' presence in these cultural narratives influences human attitudes and treatment of nonhuman primates or how such views might be used to promote conservation. Part of the problem is that these facets of human culture (myths and the like) are often analyzed independently of observations of the behavior and attitudes of people toward nonhuman primates. Because of the way academia has constructed its disciplines, moreover, conservationists and primatologists can be educated separately from folklorists and cultural anthropologists (and vice versa), and therefore, one discipline possesses very little of the knowledge held by another discipline. The hope of ethnoprimatology is that a more holistic view of human and nonhuman primates will suggest new opportunities to develop better programs to avert the extinction of nonhuman primates.

The remainder of this chapter presents a very brief view, by region, of some myths, epic poems, folklore, and human perspectives concerning nonhuman primates. Each region is differentially represented; and given space limitations, rather than attempt a broad survey in each zone, we selected specific points that we hope are fodder for possible incorporation into, or stimulation of, research, management, or conservation projects and perspectives. A more holistic view of the interconnections between human and nonhuman primates must include the human folkloric perspective and potentially stimulate the reader to attempt integration of these disparate components into analyses of human and nonhuman primate relations.

African Folktales and Some Current Perspectives

Africa is a large continent with many different cultures, and ethnographers of the nineteenth and twentieth centuries recorded some African folktales and myths related to nonhuman primates. Like those across the globe, the folktales of Africa often explain the world as it is and may have a moral or set of instructions for appropriate human behavior. Unfortunately, the many ethnographic accounts of African folklore do not tell us much about how folktales may have influenced the attitudes toward or treatment of nonhuman primates.

For example, the cosmology of ancient Egypt featured several *therianthropes* (animal–humans hybrids), including human forms with heads of birds, jackals, baboons, and so forth. Hapy, a son of *Horus*, is the baboon-headed therianthrope present on the canopic jars that held the lungs of a mummy. Animals, including baboon pets, were also mummified when the owner died (Hamilton-Paterson and Andrews 1978); but it is unclear how these historical contexts might impact any current behavior by humans in the same region.

Many folktales invoke the biobehavioral proximity between other primates and humans. Herskovits and Herskovits (1958) recorded many folktales of the Dahomean (a West African people). One tale explains why monkeys are similar to humans but still animals. Mawu, the creator god, after creating all the animals, told Monkey to work some clay and promised if he did he would become a man. Monkey became excited and forgot to work the clay. Consequently, Monkey had to stay an animal. Another story is about a monkey that tried to fool a diviner and was killed by a leopard as a consequence. There are also tales of why monkeys (perhaps baboons) have red buttocks, such as the folktale that states that a monkey received red hindquarters by sitting on a hot hoe to keep a woman from coming out after he swallowed her. The Nuer of Sudan have a story that starts with the premise that God made the monkey to be like a person. The monkeys are described as acting like humans by cooking their potatoes, watching their offspring play, and talking among themselves. When a crocodile killed a monkey, the other monkeys grieved (Huffman 1970).

Honey (1969) records a variety of South African tales of monkeys doing very human-like things. One tells of a monkey that received a magic bow and arrow and a fiddle. A wolf accused the monkey of stealing his bow and arrow, and a jury agreed. The wolf took the bow and arrow from the monkey. The monkey started to play his magic fiddle, and everyone started to dance. Exhausted, everyone agreed to give the monkey back his bow and arrow if he would just stop playing his fiddle. In another folktale, a jackal was caught in a trap set by a farmer who wanted to stop the killing of his lamb. The jackal convinced the monkey to let him down and try hanging in the trap. The jackal ran away, and the farmer shot the monkey. In yet another tale, a tailor is upset that Mouse had torn his clothes. Baboon took care of the problem and the tailor thanked him. Subsequently, Baboon changed his name from Jan to Baboon and engaged in foolish behavior. The tale ends "Since that time Baboon walks on all fours, having probably lost the privilege of walking erect through his foolish judgment" (Honey 1969:120).

Other folktales reflect interconnections related to foods. For example, a Kikuyu (Kenya) folktale, as recorded by Njururi (1966), begins with baboons crop raiding and the men of the village hunting them down. Peace was eventually made between the people and the baboons. The baboons, however, broke the peace agreement, which caused the men to start hunting them once again. The men chased the baboons away from the crops and up into the trees, where baboons and monkeys remained forever. In this tale, the baboons could talk and enter into an agreement with the people.

None of the above examples can truly inform us of how humans might behave toward nonhuman primates based on the folktales. However, Courlander (1975) reports on a Yoruba (Nigeria) legend relating monkeys and twins. The belief is that twins are sent by monkeys. It is important that the twins are well treated; otherwise, misfortune could come to the parents. On the other hand, "the arrival of twins is always welcomed because of their power to better the lot of the parents" (Courlander 1975:233). Because of the connection between monkeys and twins, twins and their parents are prohibited from eating the flesh of monkeys. Also, Morris (1998) reports that in Malawi baboons and monkeys are in an ambiguous category but recognized as being like humans. They are eaten but perhaps not as frequently as other animals. Women, however, "are less inclined [than men] to eat baboon meat" (Morris 1998:200). Some recent studies of the interaction between human and nonhuman primates include discussions of how African worldviews influence the treatment of sympatric nonhuman primates. For example, Saj et al. (2005) published evidence that local folklore can affect the treatment of nonhuman primates. They report that prior to the 1970s ursine colobus (*Colobus vellerosus*) were protected from hunting by the beliefs of villagers in Ghana of an association between their gods and the local monkeys. During this time, the monkeys were plentiful. In the early 1970s, a Christian sect known as the Saviour Church of Ghana encouraged converts to hunt monkeys, and the number of monkeys drastically declined. In 1974, village elders asked the Wildlife Division of Ghana for help in protecting the monkeys. Wildlife officers stepped in, and the monkeys received protection.

Like parts of Asia, Africa also has diverse ways in which humans view other primates. In Madagascar, Jolly (1980) reports visiting Lavalohalika Island to see the black lemurs (*Lemur macaco*) that guard an old temple. The people of Lavalohalika Island feed the lemurs bananas as part of their religious ceremonies, and killing the lemurs is forbidden. However, the aye-ayes (genus *Daubentonia*) of Madagascar are a dramatically distinct-looking primate with their "bat ears, beaver teeth, two skeleton's fingers, and a black and silver tail like an overgrown ostrich feather" (Jolly 1980:60). Perhaps because of their strange appearance, the Malagasy fear the aye-aye and believe that they cause bad luck or death (Colquhoun 2004). If an aye-aye comes to a village, it must be killed and the village burned and relocated. In this case, cultural perceptions result in two very different treatments of two nonhuman primate species.

In the chapter "Reflections on the Concept of Nature and Gorillas in Rwanda: Implications for Conservation," Sicotte and Uwengeli (2002) report on interviews with people who live near the Parc National des Volcans, the home of Rwanda's gorillas. The authors were interested in the "cultural construction of nature" in Rwanda, including the animals that inhabit "wild spaces." Sicotte and Uwengeli (2002:174) note that gorillas are absent from the folklore of the Rwandans despite "the fact that the Virungas and the spirits that inhabit them are the objects of stories." Rwandans do, however, recognize the similarity of gorillas and humans and are aware that many foreigners are very interested in gorillas. The people of the Virungas compare their care of cows with that given to gorillas by white people and their Rwandan employees. As a consequence, the gorillas may now have a "symbolic presence" they lacked in the past (Sciotte and Uwengeli 2002:177). It is this pattern of approach and inclusion of human perspective, especially in the context of folklore, which may result in a more engaged understanding of the human and nonhuman primate interconnections in any given locale.

Amazonian Views and a Central American Potential for Modern Myth

As in parts of Africa, many indigenous people of Central and South America lacked a system of writing; therefore, there are few written records of their folklore and myths. Ethnographers have, however, provided through their studies a record of some of the folklore concerning nonhuman primates. As we covered a few examples of Amazonian human and nonhuman primate interconnections earlier, we will give a brief review of those same people's myths regarding the sympatric monkeys.

According to Lizarralde (2002:85), until recently the Barí of the Parque Nacional Sierra de Perijá, Venezuela, "have eaten monkeys, kept them as pets, and used their teeth for decoration." Lizarralde (2002) reports that they believe that their creator god turned two Barí into monkeys as punishment for their misdeeds. Because the Barí want to acquire the characteristics of spider monkeys (genus *Ateles*) and capuchins (genus *Cebus*), they prefer to eat these monkeys and use their teeth for decoration. Howler monkeys (genus *Alouatta*), because they are considered slow and of low intelligence, are eaten only when no other monkeys can be found, and their teeth are not used for decoration. Infant spider monkeys are kept as pets and given family names.

As reviewed earlier, Loretta Cormier (2002) presents the relationship between the Guaja and the monkeys they hunt and eat as complex and related to both the enhanced image of the fertility of women and symbolic endocannibalism. Cormier (2002:80) suggests the following:

> Guajá identification with the physical and behavioral similarities between themselves and the monkeys is prerequisite for infant monkeys to fill their social role as substitute human beings and in their symbolic role as a version of the self to be cannibalized. Howler monkeys are the preferred game because they are considered to be most like the Guajá.

The Matsigenka of the Manu National Park (Peru), while slash-and-burn agriculturalists, also hunt and eat monkeys (Shepard 2002). The Matsigenka prefer the woolly (genus *Lagothrix*) and spider monkeys over the smaller capuchins and howler monkeys. Matsigenka mythology posits that all animals were originally humans and were "devolved" by the first shaman. Monkeys, however, are believed to be less devolved and more like humans than other animals.

In a different approach, Burton and Carroll (2005) provide an example of what they refer to as "by-product mutualism" in Honduras, where they describe a complex relationship between figs, howler monkeys (*Alouatta palliata*), and humans and their domesticants (pigs, bovids). The monkeys drop figs as they forage in the fig trees (*Ficus*), and the domesticants eat the figs that have fallen to the ground. Thus, the domesticants have access to a nutritious food source that would otherwise be beyond their reach. Everyone benefits in that the seeds of the fig trees are dispersed, the monkeys and the domesticants have fruit to eat, and the humans have less need to provide food to their domesticants. Burton and Carroll do not explicitly discuss the local people's perspectives on the foraging habits of their domesticants. They suggest that this complex interaction between monkeys, figs, and domesticants might be used to create an awareness, a "modern folktale" possibly, that might convince local people of the importance of conserving monkeys and fig trees.

Asian Buddhist and Hindu Influences

Asia has long had a system of writing, and there are many old documents on folklore, epic poems, and myths of various people's origins and history. Many of these documents have been translated and are available to speakers of modern languages.

The island of Bali (Indonesia) is a Hindu society with substantial differences in beliefs, artistic styles, and ceremonies from Hindu societies in southern Asia (such as India). However, as in India, Hanuman, the demigod monkey character from the epic poem *Ramayana*, is revered as a hero and loyal devotee of Lord Rama. As mentioned previously, the macaques of Bali are generally tolerated and provided with space and food at such tourist sites as the Wanara Wana Monkey Forest and Hindu temple complex located in the Padangtegal village of Ubud (Fuentes et al. 2005, Stephenson et al. 2002). Wheatley (1999) and Fuentes et al. (2005) suggest that in the temple areas human religious beliefs and experiential perspectives result in a high tolerance and interaction with the macaques.

At many Buddhist temples in China, macaques are traditional inhabitants and are often fed by the monks. Buddhists believe that Buddha had at least one incarnation as a monkey. There is, however, more than just the occurrence of an incarnation of Buddha as a monkey; there are the moral lessons that are connected to the accounts of the incarnations. The folktales of the incarnations of Buddha as different animals, including as a monkey, are known as *jatakas* (Khoroche 1989:xiv):

> adaptations of popular tales that are used to teach Buddhist morals. The ethical and moral superiority of animals over men is a typical theme. There are over five hundred *jatakas*. . . . And some if not all of these were already in existence in the third/second centuries BC.

A *jataka* entitled *The King of the Monkeys* is typical of the folktales involving the rebirth of the Buddha. In this folktale, the king of the monkeys (i.e., the Buddha) sacrifices himself so that the arrows of the men of the local king would not kill the monkeys. The king asked the king of the monkeys why he did not run away and save himself instead of injuring himself beyond healing to save the monkeys. The king of the monkeys responded that he was paying his debt for the blessing he had received from his subjects. Upon telling this to the king, the Buddha died and went to heaven. The folktale ends "So, then those who make a practice of good behavior can win over the hearts even of their enemies. With this in mind, he who wants to win people over completely should follow the ways of the good" (Khoroche 1989:192).

As reported by Chinese primatologist Zhao (1991), the macaques at the Buddhist temples can be dangerous to tourists who do not know how to act around them. Rather than eliminating the monkeys, however, it is important to find ways to protect the tourists and use the presence of the monkeys to educate people as to the value of biodiversity (Wolfe 1991, Zhao 2005). Moreover, because the *jatakas* are used to teach Buddhist morality, it is important that care is taken not to disparage the folktales. Understanding their contexts may enable researchers to work with the local people, governments, religious leaders, etc. to protect the monkeys and other wildlife that occupy areas around Buddhist temples while at the same time providing for the safety of temple visitors.

Burton (2002) recently explored the Chinese epic *Journey to the West* in which the Monkey King, or Sun WuKong (in Mandarin), is a central character. *Journey to the West* is based in part on the travels of Hsuan-tsang, a Buddhist priest who traveled from China to India in the seventh century to bring back documents of the teachings of Buddha (Dudbridge 1970). The round trip to and from India took 16 years (A.D. 629–645). The priest traveled alone and published a "geographical and ethnical account of the lands to the west" (Dudbridge 1970:11). Drawing on the travels of Hsuan-tsang and folklore presented in song and on stage, *Journey to the West*, or the *Hsi-yu Chi*, was written by Wu Ch'Eng-En (ca 1505–1580) in the sixteenth century (Wu Ch'Eng-En 1943). The first seven chapters tell of the early life and mischievousness of Monkey. The subsequent chapters concern the trials and tribulations of the trip from China to India and the heroism of Monkey. The *Hsi-yu Chi* combines elements of the travels of Hsuan-tsang, Buddhism, the *jatakas*, Daoism, Confucianism, and the monkey character Sun WuKong. Dudbridge (1970) examines the origin of the idea of Sun WuKong and suggests that Wu Ch'Eng-En based Sun WuKong on the demigod Hanuman of the Hindu epic poem the *Ramayana*. The similarities between Sun WuKong and Hanuman include immortality and the ability to fly and change shape. Because the story combines many aspects of the religions found in China, it is widely known and used to teach values to children. The image of Sun WuKong, moreover, is used in advertising in China, and Burton (2002) suggests that the image of the monkey king could also be used to promote conservation and assist in the survival of the monkeys of China.

In India, the epic poem the *Ramayana* is central in the Hindu belief system. In the *Ramayana*, the monkeys are among the main heroes and Hanuman, the loyal servant of Rama and possessor of supernatural abilities, is the most important of the monkey heroes (for more details see Wolfe 2002). Generally, monkeys in India are not killed outright; rather, their presence is tolerated, and people bring food to the monkeys that inhabit areas in and around temples (Wolfe 2002, personal observation). The monkeys are not, however, otherwise provisioned. While people do bring the monkeys food, it is not consistent throughout the week. On Tuesdays and Saturdays (the two days associated with the good deeds of Hanuman) and on religious occasions, the monkeys of the temple are brought more than they can eat, and on the other days little food is carried to them. For the most part, the monkeys are given food so that Hanuman and Lord Rama might look favorably upon their human providers.

In the 1980s, one of us (Linda Wolfe) studied the rhesus monkeys that live in and around the city of Jaipur, which is located in the state of Rajasthan in north-central India south of New Delhi (Wolfe 1992). Early in the twentieth century, as Jaipur began to grow outside of the walls of the city, the forest that surrounded the walled city was cut down. Unfortunately, the forest is where the monkeys had lived. By cutting down the forest, the people forced the monkeys to become more dependent on them for food. Nothing was done to preserve any of the forest or its wildlife. Moreover, there was no systematic attempt to monitor the health or demographics of the monkeys of Jaipur.

There was a dominance hierarchy among troops, moreover, and not all troops had access to the temple where the most food was deposited. As a consequence, there were monkeys that appeared to be in poor health and troops in which no offspring were born during the 1985 birth season. Our point is that here is a belief system in which the monkeys are valued because of their association with a popular legend but that association does not necessarily mean that they will be well treated. The legend of Hanuman is helpful to use as a rationale to start the process of the care of the monkeys of India, but that care needs to be given more systemically as they need both veterinary attention and a consistent food source.

While monkey performance has at times been an important facet of Japanese culture (see Social and/or Economic Inclusion of Nonhuman Primates, above), Japan seems to lack major epic narratives involving monkeys. While devoted Buddhists may know of the *jatakas* and the legend of the "journey to the west," the *jatakas* of Buddhism do not seem to be part of the cultural consciousness of the people of Japan. As discussed earlier, Ohnuki-Tierney (1987) argues that the meaning of the entertainment provided by the monkey dances and their trainers in modern Japan is culturally complex and can vary from individual to individual and city to city.

Wolfe and Gray (1982) report on an observation that occurred during a field study of Japanese macaques (*Macaca fuscata*) at the tourism site of Arashiyama, Japan. Despite articles in the popular literature that described the important role of females in the social system of Japanese macaques, the Japanese public was inordinately interested in the *zaru bosu* (or monkey boss). The authors argue that the interest in the monkey boss was twofold: first, Japan is a patriarchal society, and the interest in the monkey boss reflected that reality; second, Japanese society functions by consensus, and the perception of the individual decision-making power of the monkey boss contradicts that of society and, therefore, separates the natural animal world from the cultural life of humans.

In the literature on the creation myths of the Tibetan people, there are several variations on the account of their creation. Snellgrove (1987) has suggested that an old creation myth was later modified when Buddhism came to the Tibetan highlands sometime around A.D. 600. The old creation myth begins with the meeting of a wise and compassionate monkey and a rock demoness. At first, the monkey, who wanted to maintain his chastity, rejected the advances of the rock demoness. Eventually, however, the monkey succumbed to the requests of the rock demoness, and the results were the six original ancestors of the people we know today as the Tibetans (Boord 1993). With the arrival of Buddhists, the saintly monkey became associated with Avalokiteshvara (a bodhisattva) and the rock demoness with the Buddhist goddess Tara. Avalokiteshvara is today manifested as the Dalai Lama. The folktale of the monkey and the rock demoness persists today, although in the form altered by the more recent conversion of Tibetans to Buddhism.

It is abundantly evident that human cultural narratives play a very important role in establishing the context for human and nonhuman primate interconnections in Asia. However, these patterns are neither uniform nor ubiquitous across the broad expanse of the region.

A Modern North American Case

A final view of humans' cultural perceptions can be found in the United States of America. In 1938, the late Colonel Tooey released several rhesus monkeys (*Macaca mulatta*) along the Silver River near Ocala, Florida. Silver Springs, the headwater of the Silver River, is a popular recreation park. Tooey ran a boat ride and released the monkeys to enhance the excitement of the boat ride and increase his revenues (Wolfe 2002). The size of the monkey population increased, and between about 1982 and 1995 the Florida Game and Fresh Water Fish Commission (FGFWFC) began to oversee the trapping and disposal of the monkeys. Most of the local people opposed the trapping of the monkeys. The people of Silver Springs liked observing the monkeys and particularly enjoyed watching the mothers care for their infants. Although the FGFWFC did remove perhaps as many as 500 monkeys during the 1980s, there are still rhesus monkeys living in the area of the Silver River. One of us (wolfe 2002) attributed the survival of the rhesus macaques of the Silver River in large part to the objections of the local people to the trapping of the monkeys.

SOME CONCLUDING THOUGHTS

Human and nonhuman primates living in sympatry affect each other's lifeways. People who share a habitat with nonhuman primates tend to include those primates in their myths and folklore. It is the premise of ethnoprimatology that folklore, myth, and a broader understanding of the complex array of interconnections between human and nonhuman primates can be used to encourage diverse groups of people to cooperate with government officials, conservation, and management groups to ensure the survival of nonhuman primates. At the very least, understanding and contextualizing myths and folklore combined with detailed assessments of the actual patterns of interactions can give people from places of sympatry and the researchers involved in primatology access to possible reasons why they might want to see that the animals around them survive into the future. These reasons themselves may also be key components in effective management and conservation strategies.

The ethnoprimatological approach might help to overcome the cultural and perceptive isolation of conservationists and primatologists from those who work with local peoples in areas of sympatry. Holistic projects based on knowledge of the interaction between human and nonhuman primates, plans that include both the enhancement of the lives of local peoples and the needs of wildlife, are required if nonhuman primates are to survive in the face of the continuing growth of the human population.

REFERENCES

Aggimarangee, N. (1992). Survey for semi-tame colonies of macaques in Thailand. *Nat. Hist. Bull. Siam Soc.* 40:103–166.

Alvard, J., and Kaplan, H. (1991). Procurement technology and prey mortality among indigenous neotropical hunters. In: Stines, M. C. (ed.), *Human Predators and Prey Mortality*. Westview Press, Boulder. pp. 335–387.

Alvard, M. J., Roninson, J. G., Redford, K. H., and Kaplan, H. (1997). The sustainability of subsistence hunting in the neotropics. *Conserv. Biol.* 11:977–982.

Amman, K., Pearce, J., and Williams, J. (2000). *Bushmeat: Africa's Conservation Crisis*. World Society for the Protection of Animals, London.

Boord, M. (1993). Myths of origin. In: Willis, R. (ed.), *World Mythology*. Henry Holt and Company, New York. pp. 102–109.

Brunvand, J. H. (1996). Folklore. In: Brunvand, J. H. (ed.), *American Folklore: An Encylopedia*. Garland Publishing, New York. pp. 285–287.

Burton, F., and Carroll, A. (2005). By-product mutualism: conservation implications amongst monkeys, figs, humans, and their domesticants in Honduras. In: Paterson, J. (ed.), *Commensalism and Conflict: The Human–Primate Interface*. American Society of Primatology, Norman, OK. pp. 24–39.

Burton, F. D. (2002). Monkey king in China: basis for a conservation policy. In: Fuentes, A., and Wolfe, L. D. (eds.), *Primates Face to Face*. Cambridge University Press, Cambridge.

Colquhoun, I. C. (2004). Primates in the forest: Sakalava ethnoprimatology and synecological relations with black lemurs at Ambato Massif, Madagascar. In: Paterson, J. (ed.), *Commensalism and Conflict: The Human–Primate Interface* American Society of Primatologists, Norman, OK.

Cormier, L. A. (2002). Monkey as food, monkey as child: Guaja symbolic cannibalism. In: Fuentes, A., and Wolfe, L. D. (eds.), *Primates Face to Face*. Cambridge University Press, Cambridge. pp. 63–84.

Cormier, L. A. (2003). *Kinship with Monkeys: The Guaja Foragers of Eastern Amazonia*. Columbia University Press, New York.

Courlander, H. (1975). *A Treasury of African Folklore*. Crown Publishers, New York.

de Waal, F. B. M. (1997). Foreword. In: Mitchell, R. W., Thompson, N. S., and Miles, H. L. (eds.), *Anthropomorphism, Anecdotes, and Animals*. State University of New York Press, New York.

Dudbridge, G. (1970). *A Study of Antecedents to the Sixteenth-Century Chinese Novel*. Cambridge Univeristy Press, Cambridge.

Engel, G. A., Jones-Engel, L., Suaryana, K. G., Arta Putra, I. G. A., Schilliaci, M. A., Fuentes, A., and Henkel, R. (2002). Human exposures to Herpes B seropositive macaques in Bali, Indonesia. *Emerg. Infect. Dis.* 8:789–795.

Fa, J. E. and Southwick, C. H. (ed.) (1988). *Ecology and Behavior of Food-Enhanced Groups*. Alan R. Liss, NY.

Farrer, C. R. (1997). Myth. In: Green, T. A. (ed.), *Folklore: An Encyclopedia of Beliefs, Customs, Tales, Music and Art*, vol. II. ACB-Clio, Santa Barbara.

Fuentes, A. (2002). Monkeys, humans and politics in the Mentawai Islands: no simple solutions in a complex world. In: Fuentes, A., and Wolfe, L. D. (eds.), *Primates Face to Face: The Conservation Implications of Human–Non-Human Primate Interconnections*. Cambridge University Press, Cambridge. pp. 187–207.

Fuentes, A., Southern, M., and Suaryana, K. G. (2005). Monkey forests and human landscapes: is extensive sympatry sustainable for *Homo sapiens* and *Macaca fascicularis* in Bali? In: Paterson, J., and Wallis, J. (eds.), *Commensalism and Conflict: The Human–Primate Interface*. American Society of Primatology, Norman, OK. pp. 168–195.

Fuentes, A., and Wolfe, L. D. (2002). *Primates Face to Face: The Conservation Implications of Human–Non-Human Primate Interconnections*. Cambridge University Press, Cambridge.

Hamilton-Paterson, J., and Andrews, C. (1978). *Mummies*. Penguin Books, New York.

Herskovits, M. J., and Herskovits, S. (1958). *Dahomean Narrative: A Cross-Cultural Analysis*. Northwestern University Press, Evanston, IL.

Honey, J. A. (1969). *South-African Folk-Tales*. Negro Universities Press, New York.

Huffman, R. (1970). *Nuer Customs & Folk-Lore*. Frank Cass and Company London.

Janson, H. W. (1952). *Apes and Ape Lore in the Middle Ages and the Renaissance*. Studies of the Warburg Institute, vol. 20. University of London, London.

Jolly, A. (1980). *A World Like Our Own*. Yale University Press, New Haven.

Jones-Engel, L., Schillaci, M., Engel, G., Paputungan, U., and Froehlich, J. (2005). Characterizing primate pet ownership in Sulawesi: implications for disease transmission. In: Paterson, J. (ed.), *Commensalism and Conflict: The Human–Primate Interface*. American Society of Primatologists, Norman, OK.

Khoroche, P. (1989). *Once the Buddha Was a Monkey: Arya Sura's Jatakamala*. University of Chicago Press, Chicago.

Lansing, S. J. (1991). *Priests and Programmers: Technologies and Power in the Engineered Landscape of Bali*. Princeton University Press, Princeton, NJ.

Lizarralde, M. (2002). Ethnoecology of monkeys among the Bari of Venezuela: perception, use and conservation. In: Fuentes, A., and Wolfe, L. D. (eds.), *Primates Face to Face*. Cambridge University Press, Cambridge. pp. 85–100.

Malone, N., Purnama, A. R., Wedana, M., and Fuentes, A. (2002). Assessment of the sale of primates at Indonesian bird markets. *Asian Primates* 8:7–11.

McGrew, M. C. (1992). *Chimpanzee Material Culture*. Cambridge University Press, Cambridge.

Medin, D. L., and Atran, S. (eds.) (1999). *Folkbiology*. MIT Press, Cambridge, MA.

Miller, L. E. (2002). *Eat or Be Eaten: Predator Sensitive Foraging Among Primates*. Cambridge University Press, Cambridge.

Mitchell, R. W., Thompson, N. S., and Miles, H. L. (eds.) (1997). *Anthropomorphism, Anecdotes, and Animals*. State University of New York Press, New York.

Morris, B. (1998). *The Power of Animals*. Berg, Oxford.

Njururi, N. (1966). *Agikuyu*. Oxford University Press, London.

Ohnuki-Tierney, E. (1987). *The Monkeys as Mirror*. Princeton University Press, Princeton, NJ.

Ohnuki-Tierney, E. (1995). Representations of the monkey (Saru) in Japanese culture. In: Corby, R., and Theunissen, B. (eds.), *Ape, Man, Apeman: Changing Views Since 1600. Evaluative Proceedings of the Symposium Ape, Man, Apeman: Changing Views Since 1600, Leiden, the Netherlands, 28 June–1 July, 1993*. Department of Prehistory, Leiden University, Leiden. pp. 297–308.

Paterson, J. D., and Wallis, J. (eds.) (2005). *Commensalism and Conflict: The Human–Primate Interface*. American Society of Primatologists, Norman, OK.

Pentikainen, J. (1997). Epic. In: Green, T. A. (ed.), *Folklore: An Encyclopedia of Beliefs, Customs, Tales, Music and Art* vol. I. ACB-Clio, Santa Barbara. pp. 222–225.

Saj, T. L., Teichroeb, J. A., and Sicotte, P. (2005). The population status of the ursine colobus (*Colobus vellerosus*) at Boabeng-Fiema, Ghana. In: Paterson, J. (ed.), *Commensalism and Conflict: The Human–Primate Interface*, American Society of Primatologists, Norman, OK. pp. 350–374.

Shepard, G. H. (2002). Primates in Matsigenka: subsistence and world view. In: Fuentes, A., and Wolfe, L. D. (eds.), *Primates Face to Face*. Cambridge University Press, Cambridge. pp. 101–136.

Sicotte, P., and Uwengeli, P. (2002). Reflections on the concept of nature and gorillas in Rwanda: implications for conservation. In: Fuentes, A., and Wolfe, L. D. (eds.), *Primates Face to Face*. Cambridge University Press, Cambridge. pp. 163–182.

Snellgrove, D. (1987). *Indo-Tibetan Buddhism*. Shambhala, Boston.

Sponsel, L. E. (1997). The human niche: exploration in ethnoprimatology. In: Kinzey, W. G. (ed.), *New World Primates*. Walter de Gruyter, New York. pp. 143–165.

Sponsel, L. E., Ruttanadakul, N., and Natadecha-Sponsel, P. (2002). Monkey business? The conservation implications of macaque ethnoprimatology in southern Thailand. In: Fuentes, A., and Wolfe, L. D. (eds.), *Primates Face to Face*. Cambridge University Press, Cambridge. pp. 28–309.

Sprague, D. S. (2002). Monkey in the backyard: encroaching wildlife in rural communities in Japan In: Fuentes, A., and Wolfe, L. D. (eds.), *Primates Face to Face: The Conservation Implications of Human–Non-Human Primate Interconnections*. Cambridge University Press, Cambridge. pp. 254–272.

Stephenson, R. A., Kurashina, H., Iverson, T. J., and Chiang, L. N. (2002). *Visitor's perception of cultural improprieties in Bali, Indonesia. J. Nat. Park* 12:156–169.

Stone, K. F. (1996). Folktales. In: Brunvand, J. H. (ed.), *American Folklore: An Encylopedia*. Garland Publishing, New York. pp. 294–295.

Wallis, J., and Lee, D. R. (1999). Primate conservation: the prevention of disease transmission. *Int. J. Primatol.* 20:803–826.

Wheatley, B. P. (1999). *The Sacred Monkeys of Bali*. Waveland Press, Prospect Heights, IL.

Wolfe, L. D. (1991). Macaques, pilgrims, and tourists revisited. *Nat. Geogr. Res. Expl.* 7:241.

Wolfe, L. D. (1992). Feeding habits of the rhesus monkeys (*Macaca mulatta*) of Jaipur and Galta, India. *Hum. Evol.* 7:43–54.

Wolfe, L. D. (2002). Rhesus macaques: a comparative study of two sites, Jaipur, India and Silver Springs, Florida. In: Fuentes, A., and Wolfe, L. D. (eds.), *Primates Face to Face*. Cambridge University Press, Cambridge. pp. 310–330.

Wolfe, L. D., and Gray, J. P. (1982). Japanese monkeys and popular culture. *J. Pop. Cult.* 16:97–105.

Wu Ch'Eng-En (1943). *Monkey*. Waley, A. (trans.). Grove Press, New York.

Zhao, Q. K. (1991). Macaques and tourists at Mt. Emei, China. *Nat. Geogr. Res. Expl.* 7:115–116.

Zhao, Q. K. (2005). Management strategies and effects. In: Paterson, J., and Wallis, J. (eds.), *Commensalism and Conflict: The Human–Primate Interface*. American Society of Primatologists, Norman, OK. pp. 376–399.

APPENDIX 43.1: GLOSSARY

Anthropomorphism: There are many definitions, but one we prefer is the description of animal behavior "in human hence intentionalistic terms" (de Waal 1997:xv). (For a discussion of the various definitions and ramifications of anthropomorphisms or the lack thereof, see Mitchell et al. 1997.)

Epic: According to Pentikainen (1997:223), "In folkloristics, epic is a narrative couched in poetic language and subject to special rules of texture, structure, style and form. Usually, epics contain hundreds or thousands of verses and present a complex narrative full of wonders and heroism, often centered around the exploits of a main personage."

Ethnoprimatology: The study of the multifarious interaction of human and nonhuman primates, including, but not exclusive to, the image of primates in folklore, legends, and myths; influence of cultural beliefs on the hunting of primates; conservation ecology and the management of primate populations in their natural habitats; and so forth.

Folkbiology: A cross-cultural and interdisciplinary study of how ordinary people understand, perceive, and categorize the natural world (e.g., see Medin and Atran 1999). Folkbiologists are also concerned with the relationship between local worldviews, globalization, and conservation.

Folklore: A word coined in 1846 by English scholar William J. Thoms. There are several common features that comprise folklore: *(1)* usually the tales are transmitted orally, *(2)* the elements are traditional in form, *(3)* different versions or variants exist, and *(4)* the tales are anonymous as to origin (Brunvand 1996).

Folkloristics: The study of folklore (Brunvand 1996).

Folktale: "A fictional narrative varied in length and rich in symbolic and metaphorical meaning, oral in origin but now found more often in printed collections. . . . Folktales are filled with fantastic creatures, events and objects. . . . Folktales arise in an oral tradition; thus, one cannot identify original or authoritative texts. There are as many texts of any particular folktale as there are people who have told it over many generations" (Stone 1996:294).

Jatakas: Folktales "that are used to teach Buddhist morals" (Khoroche 1989:xiv). Literally, it means "stories of the re-birth of Buddha" (Khoroche 1989:xiv).

Legend: A localized and historicized story or narrative usually related orally about a "specific place and time with named characters" (Tangherlini 1996:437).

Myth: According to Farrer (1997:577–578) a myth is different from other forms of narrative in that it is a "sacred narrative that is most often believed to be true. Because myth is of such a sacred nature, many people object to the very word *myth* when it is attached to their own sacred narratives. For those in whose culture the myth is embedded, the truth of the myth may be literal or figurative. Literalists believe fully in the absolute truth of myth, whereas figurativists believe in myth as metaphor."

44

Where We Have Been, Where We Are, and Where We Are Going
The Future of Primatological Research

Christina J. Campbell, Agustín Fuentes, Katherine C. MacKinnon, Melissa Panger, and Simon K. Bearder

---

THE PLACE OF PRIMATOLOGY: EMBRACING INTERACTIONS WITH OTHER DISCIPLINES

Primatology is, by its very nature, interdisciplinary. Field studies and captive work are undertaken from intellectual foundations as varied as anthropology, psychology, behavioral ecology, zoology, anatomy, ethnology, genetics, epidemiology, evolutionary biology, conservation biology, and veterinary medicine. Indeed, it is the object of study—the primates themselves—that best define *primatology*, as opposed to a restrictive academic philosophy. Thus, questions addressing topics as diverse as the biomechanics of locomotion, behavioral variation in maternal care patterns, bidirectional pathogen transmission between primate species, and the selective pressures acting on primate social systems, to name a few, can all be effectively addressed from under the encompassing umbrella of our field.

Historically, this interdisciplinary nature has been of primary importance to the field; that is, the relevance of nonhuman primates to inquiries about human evolution and behavior is one of the foundations of primate studies. Pioneers Robert Yerkes (a psychologist), Clarence Ray Carpenter (an ethologist), Solly Zuckerman (an anatomist), and Kinji Imanishi (an entomologist, ecologist, and anthropologist) implemented the first studies of nonhuman primate behavior, thereby helping to shape primatology into a multidisciplinary field. Importantly, Sherwood Washburn's call for a comparative primatology (which supplemented studies of anatomy and human behavioral evolution with comparative data from primates living in the wild) stimulated several generations of field primatologists. Additionally, the paleoanthropologist Louis Leakey promoted the long-term ape research projects of Birute Galdikas, Diane Fossey, and Jane Goodall, which provided the basis for the growth in ape research. The last half of the twentieth century was indeed a fruitful time in our field, and as the number of primate species studied grew, the data increasingly revealed the massive variability within the primate order. Would this productive trajectory of investigations into the lives of nonhuman primates have occurred if the field did not have such an interdisciplinary background and character? Probably not, for the simple reason that many of the research questions asked could not have sprung forth from a more insular origin.

Modern evolutionary theory, specifically current behavioral ecological approaches, and more than a half-century of primatological data have placed us, at the beginning of the twenty-first century, in a position to make greater strides into the complexities of primate biology than ever before. Primatology continues to be wonderfully positioned to examine a wide range of topics and, in doing so, can draw from a wealth of information from varied data sets. This philosophical position provides not only insight but also a complex prism through which we can view the interwoven facets of primate behavior, ecology, and physiology.

We have seen in a number of chapters in this volume how a multidisciplinary focus is the modern standard in primate work. Whether researchers' home bases lie in anthropology, conservation biology, epidemiology, behavioral ecology, or genetics, a "cross-pollination" approach to research can only continue to strengthen primatology. Exciting results have come about from multidisciplinary collaborations, and we urge primatologists-to-be to keep this in mind as they begin their studies. In addition, we implore more seasoned veterans to incorporate varied lines of evidence when tackling their research questions. Being interdisciplinary is primatology's legacy, and further communication between and among the various disciplines involved is our continued charge.

NATURAL HISTORY AND THE SCIENTIFIC METHOD

Natural history, the study and description of organisms and natural objects (especially their origins, evolution, and interrelationships), has played a pivotal role in the biological sciences including primatology. Many of the earliest primatologists were natural historians who set out to broadly describe the behavior and ecology of primates living under natural conditions. Such studies are vital for any scientific field of study since the collection of general descriptive data represents the first step in the scientific method, which involves observing phenomena, detecting patterns from those observations, and developing testable hypotheses to try to explain the patterns observed.

Natural history data were the focus of primatology until the 1970s and 1980s, when there was a major shift toward hypothesis-driven research (the collection of a limited set of data used to test specific hypotheses). During this time, the two forms of research, natural history and hypothesis-driven research, were no longer considered simply steps in the scientific process but instead dichotomized, with greater value placed on hypothesis-driven research. This switch in focus occurred in many fields of study and was partly the result of what has been called "physics envy." As a result of this shift, natural history data are currently undervalued, and it can be challenging in primatology (and other biological fields) to

obtain funding specifically for the collection of broad behavioral and ecological data and to get natural history information published in leading primatological journals.

Natural history data, however, are important for a variety of reasons. For example, as stated above, collecting general descriptive data is an important first step in the scientific method and the information retrieved from broad-based studies drives hypothesis-driven research. The chapters in this book clearly illustrate that while we currently know a great deal about many primate species, we have only really scratched the surface of what there is to know about many of our primate cousins. Related to this, broad behavioral and ecological research is also important because as we learn more about prosimians, monkeys, and apes, exciting and novel hypotheses will be developed that we cannot currently fathom. Natural history data could provide important information for later studies as new "hot topics" and research goals emerge in the future. For example, recent studies of culture in a variety of primates have relied heavily on natural history data (see Chapter 40). The long-term data used in these studies were not collected to test specific hypotheses related to culture but as part of larger, broad-based studies, yet the data have been successfully used to support an argument for culture in nonhuman primates. Finally, we feel that natural history data are important because many primate populations are rapidly disappearing and, thus, there is an urgent need to collect as much information as possible from the declining populations before they go extinct. This is akin to the philosophy of "salvage ethnography" that played an important role in the collection of ethnographic information from a variety of cultures in the mid-part of the last century before those cultures were lost due to assimilation into an increasingly modernized world.

We do not wish for the primate research pendulum to swing completely away from hypothesis-driven research, but we believe the value of both types of research should be recognized. Ideally, funding opportunities to conduct natural history research on understudied taxa should be increased. However, at the very least, primatologists and primatologists-to-be would gain much by recognizing the importance of broad-based data and regularly collecting them while conducting hypothesis-driven research. In short, natural history has played an important role in the history of primatology, and we think it has an important role to play in the future of the field.

INTERACTIONS BETWEEN RESEARCHERS

Although researchers from diverse fields of study are carrying out ongoing primatological inquiry, their interactions with each other are often not as productive as we would hope. An interesting and informative parallel between the behavior of the animals that we study and our own reaction to some of our colleagues is competition. How can we arrange our research programs so as to minimize conflict, aggression, and territoriality and maximize collaboration? The answer is linked to the theory of competitive exclusion: if we occupy different niches, then we are less likely to compete! Indeed, it may be possible to arrange research themes that are complementary and, thus, benefit in much the same way as species that live together symbiotically. An effective way to achieve this comes from our ability to communicate over great distances and to keep each other informed about what we are doing and why. Now that the global network makes communication between researchers easier, such cooperation should be even more prominent. There are many present-day collegial situations among researchers. A fine example involves primatologists from a number of fields who work on capuchin monkeys (*Cebus* spp.). The "capuchin people" are famously cohesive, often overlap in terms of species and study topics, and share data (anecdotal and otherwise) back and forth. Multi-authored papers have been written as a result of fruitful collaborations, and a special satellite conference of the 2004 International Primatological Society meetings was organized, entitled "Capuchins: The State of the Art"; this conference brought together capuchin researchers from all over the world, contacts were made and reestablished between new and old friends, and as a result, future research ideas were stimulated.

The primatological field will surely be strengthened if we appreciate the myriad research niches that are available and the range of strengths and abilities of the researchers who may queue to fill them. Even people working on the same species and topic can work in different study areas or on different captive populations, and they are likely to produce synergistic results. Those interested in a particular taxonomic group can diversify by concentrating on different species or subspecies. Ironically, this becomes ever easier as the number of primate taxa increases year by year as a result of our more detailed understanding of cryptic species and more thorough exploration of wild places. An exciting shift in taxonomic concentration is currently under way as more and more people enter our exciting field. Gone are the days of everyone wanting to study the "famous" primate species (e.g., gorillas, chimpanzees, baboons). We are now witnessing an influx of those who are taking on the difficult task of studying species that have received less attention in the past, for example, many nocturnal species. A less species-centric approach is enormously rewarding and opens up previously unexplored research niches, enabling the discovery of new species and, for example, a dramatic new appreciation of phylogenetic relationships through new techniques such as molecular analysis. For those who are setting out in this fascinating field of study, we suggest keeping an open mind about where to go, what to do, and what to study. Try to avoid areas of work that are already heavily populated, but think instead of how to add extra pieces to the fascinating puzzle. Work on topics that complement the work of others and thereby benefit from the give and take of true collaboration. Also, of course, publish or publicize as much as possible to ensure that the results reach the widest public.

TECHNOLOGY

In addition to the increasing ability of researchers to collaborate due to faster and more efficient means of communication, one of the most exciting advances in recent primatological inquiry is the employment of new technologies. No longer is it sufficient to infer paternity from behavioral observations of copulations or to infer female reproductive state from behavior alone. Today, field researchers commonly employ techniques providing information on paternity and genetic relatedness through the analysis of genetic data. In addition, hormonal information indicating female reproductive state, ovarian cycle stage, male testosterone levels, and stress levels as indicated by cortisol levels is obtainable by those carrying out field studies. Both sources of information have been available for a while, but not until they could be obtained from sources collected noninvasively (e.g., from fecal samples) were these methods to be widely employed by field researchers. With increasing prevalence in primatological studies, such technologies will undoubtedly provide data that will challenge current theoretical ideologies. Already, genetic data are challenging the ubiquity of the "priority of access" model of male reproductive success. In addition, hormonal data are allowing us to reassess the uniqueness of human female sexuality and often suggest that female monkeys and apes do not show estrous cycles as they are repeatedly reported to in the literature (see Chapter 25). The expense of such technologies is the main reason they are not employed in all studies, but in the future it is reasonable to expect that the costs of performing such studies will decrease as the technologies become more common and more refined.

Technology is also playing a part in easing one of the more difficult components of primatological fieldwork: maintaining contact with the animals on a day-to-day basis. Those who study species such as howler monkeys (*Alouatta* spp.) benefit from being able to rely on the animals making a great deal of noise every morning, thus making them easy to locate on a regular basis. In most cases, however, there is no such daily announcement of the study group's location. In the past, this has meant that, in order to maintain contact, field primatologists have had to perform what has become the ritual dawn-to-dusk follow: waking up and being in the field prior to dawn to locate the animals where they were left the night before and following them all day until they settle down to sleep, only to repeat the process the next day. Clearly, continuing this for any great length of time leads to extreme fatigue, both mental and physical. Today and in the future, primatologists can benefit greatly from the increased use and availability of radiotelemetry. Radio collars for primates have become smaller and less obtrusive, batteries have a longer life span, and drugs used when anesthetizing wild primates have become safer. Many still question the danger associated with darting wild animals, especially those that are high in the forest canopy; and we are by no means implying that researchers should go out darting their animals without careful consideration of the many risks involved. The benefits of radiotracking extend beyond merely locating the study group however. Contact can be maintained with multiple groups by one researcher, allowing for studies on intergroup contacts and interactions. Following cryptic nocturnal species in many cases would be near to impossible if the animals were not fitted with radiotracking devices. Future advances in the field of radiotracking are potentially very exciting to primatologists. Further decreases in collar and battery size and the potential for subcutaneous devices will hopefully decrease the immediate impact that wearing a collar has on the daily life of a primate.

We are not suggesting that future primatological research will be conducted purely through technological means and that behavioral observations will not be important. Nor are we implying that long hours in the field should be dropped from the primatologist's repertoire. It is our collective experience that one only begins to see the intricate details of social interactions after hundreds, if not thousands, of hours in the field. We believe that technologies such as those described above can greatly benefit field primatologists, allowing us to test hypotheses that in the past we could only dream of testing without making assumptions unsupported by any empirical data. It is indeed an exciting time to be entering the primatological field!

CONSERVATION AND THE FUTURE OF THE PRIMATES THEMSELVES

Human and nonhuman primates share intertwined destinies. Nonhuman primates are our closest evolutionary relatives and are integrated into our mythologies, diets, and economies. There is no longer the possibility of studying a group or population of nonhuman primates without some element of human interaction. Nearly 90% of the world's primates are found in tropical forests, and these forests are being altered by human use faster and more dramatically than any other habitats on earth. Fifty percent of all primate species are a conservation concern to the Species Survival Commission and the World Conservation Union, and 20% are considered endangered or critically endangered. Human and nonhuman primates have been interacting for millions of years, and it is now up to humans to decide how much longer the other primates will be around to continue this pattern.

Accordingly, practically every practicing primatologist will agree that conservation is among the most important aspects of the field. Karen Strier eloquently lays out the current case for primate conservation, its practice and patterns, and the urgent need to act regarding endangered and threatened primate populations in Chapter 30. However, we stress two additional aspects related to primate conservation in this concluding chapter: the role of disease and the role of source-country primatologists.

Changing human social, economic, and political conditions and associated deforestation result in nonhuman primates coming into increasingly regular contact with humans. Hunting, logging, farming, the pet trade, ecotourism, and research all provide contact opportunities in which diseases can jump the species barrier. While research has demonstrated the devastating effect that human diseases (such as tuberculosis, polio, and respiratory viruses) can have on nonhuman primates in captivity, only recently has research begun to focus on the effects of these, and other, diseases on populations of wild nonhuman primates. It is most likely that one result of increased interactions with humans is that many wild nonhuman primate populations will be negatively impacted by human disease. This can be especially dangerous for those populations or species that are currently endangered or threatened with extinction. In the other direction, it has become apparent that many human diseases may have origins in nonhuman primates. For example, recent research suggests that the chimpanzee simian immunodeficiency virus (SIVcpz) made the jump to humans and mutated into the human immunodeficiency virus (HIV-1) at some point in the last century.

Source countries are those where nonhuman primate populations currently exist. The last few decades have seen some increase in the amount of trained and practicing primatologists from countries where nonhuman primates are endemic. It is our contention that this trend must continue and, in fact, needs to intensify if conservation actions are to become truly sustainable at the local level. Given the diverse and complex cultural, social, and political boundaries and hierarchies that characterize the modern world, effective and long-lived conservational activates are most likely to succeed if local populations and source country citizens are, in general, deeply invested and involved in their creation and implementation. It is our hope that the future of primatology continues to enhance the infrastructure and professional opportunities for primatologists from source countries as a major component of our conservation paradigm.

CONCLUSION

We hope that readers of this volume have become excited about the potential for future primatological study. It was our intent when putting these chapters together to provide a volume broad in theoretical and topical issues, and we are confident that we have achieved this goal. We thank all those who have contributed to this volume and indeed all those who have contributed to the myriad primatological studies discussed in each chapter. To those contemplating entering the primatological community, we wish you well in your studies and hope to see results from your future research incorporated into updated volumes of this kind.